THE UNIVERSITY OF ARIZONA SPACE SCIENCE SERIES

Tom Gehrels, General Editor

Planets, Stars and Nebulae, Studied with Photopolarimetry
Tom Gehrels, editor, 1974, 1133 pages

Jupiter
Tom Gehrels, editor, 1976, 1254 pages

Planetary Satellites
Joseph A. Burns, editor, 1977, 598 pages

Protostars and Planets
Tom Gehrels, editor, 1978, 756 pages

Asteroids
Tom Gehrels, editor, 1979, 1181 pages

Comets
Laurel L. Wilkening, editor, 1982, 766 pages

Satellites of Jupiter
David Morrison, editor, 1982, 972 pages

Venus
D. M. Hunten, L. Colin, T. M. Donahue and V. I. Moroz, editors, 1983, 1143 pages

Saturn
Tom Gehrels and Mildred S. Matthews, editors, 1984, 968 pages

Planetary Rings
Richard Greenberg and André Brahic, editors, 1984, 784 pages

Protostars and Planets II
David C. Black and Mildred S. Matthews, editors, 1985, 1293 pages

Satellites
Joseph A. Burns and Mildred S. Matthews, editors, 1986, 1021 pages

The Galaxy and the Solar System
Roman Smoluchowski, John N. Bahcall and Mildred S. Matthews, editors, 1986, 485 pages

Meteorites and the Early Solar System
John F. Kerridge and Mildred S. Matthews, editors, 1988, 1269 pages

Mercury
Faith Vilas, Clark R. Chapman and Mildred S. Matthews, editors, 1988, 794 pages

Origin and Evolution of Planetary and Satellite Atmospheres
S. K. Atreya, J. B. Pollack and M. S. Matthews, editors, 1989, 881 pages

Asteroids II
Richard P. Binzel, Tom Gehrels and Mildred S. Matthews, editors, 1989, 1258 pages

Uranus
Jay T. Bergstralh, Ellis D. Miner and Mildred S. Matthews, editors, 1991, 1076 pages

The Sun in Time
C. P. Sonett, M. S. Giampapa and M. S. Matthews, editors, 1991, 996 pages

Solar Interior and Atmosphere
A. N. Cox, W. C. Livingston and M. S. Matthews, editors, 1991, 1414 pages

Mars
H. H. Kieffer, B. M. Jakosky, C. W. Snyder and M. S. Matthews, editors, 1992, 1536 pages

Protostars and Planets III
E. H. Levy and J. I. Lunine, editors, 1993, 1596 pages

Resources of Near-Earth Space
J. S. Lewis, M. S. Matthews and M. L. Guerrieri, editors, 1993, 977 pages

Hazards Due to Comets and Asteroids
T. Gehrels, editor, 1994, 1300 pages

Neptune and Triton
D. P. Cruikshank, editor, 1995, 1247 pages

Cosmic Winds and the Heliosphere
J. R. Jokipii, C. P. Sonett and M. S. Giampapa, editors, 1997, 1052 pages

Venus II—Geology, Geophysics, Atmosphere, and Solar Wind Environment
S. W. Bougher, D. M. Hunten and R. J. Phillips, editors, 1997, 1376 pages

Pluto and Charon
S. Alan Stern and David J. Tholen, editors, 1997, 756 pages

Protostars and Planets IV
V. Mannings, A. P. Boss and S. S. Russell, editors, 2000, 1440 pages

PROTOSTARS AND PLANETS IV

PROTOSTARS AND PLANETS IV

Vincent Mannings
Alan P. Boss
Sara S. Russell

Editors

With 167 collaborating authors

THE UNIVERSITY OF ARIZONA PRESS
Tucson

About the front cover:

The main image shows the center of the Trapezium cluster in the Orion Nebula, with the four massive stars responsible for the ionization of the nebula and low-mass stars surrounded by shock fronts, ionized envelopes, and tails. The envelopes are produced by photoevaporation of remnant circumstellar proto-planetary disks. This false color *Hubble Space Telescope* (HST) Wide Field Planetary Camera2 (WFPC2) image shows Hα in red, [O III] in green, and [O I] in blue. (Bally, J., Sutherland, R. S., Devine, D., and Johnstone, D. 1998. *Astron. J.* 116:293)

About the back cover:

The foreground upper picture on the back cover is a near-infrared image of IRAS 04302+2247, a young stellar object in the Taurus L1536 cloud at a distance of 140 pc. This very young star is hidden from direct view and is seen only as circumstellar nebulosity. Bisecting the nebula is a 900-AU-long dust lane that is believed to be a large optically thick disk seen precisely edge-on. The composite NICMOS image shows reddening at the edges of the absorption lane, clearly indicating increased density toward the midplane. Dark clouds and bright wisps within the bright nebulosity suggest the presence of an infalling envelope surrounding the presumed disk. This figure is a deconvolved color composite of HST/NICMOS 1.1-μm, 1.6-μm, and 2.05-μm broadband images with an effective resolution of 15 AU. (Padgett, D. L., Brandner, W., Stapelfeldt, K. R., Strom, S. E., Terebey, S., and Koerner, D. 1999. *Astron. J.* 117:1490)

The foreground lower image on the back cover shows part of a compound chondrule from the CV3 meteorite Allende 5492-1. The round object in the center (appears pink in this image) is a barred olivine chondrule that is enclosed within a porphyritic olivine chondrule. Field of view: 2 millimeters. (Credit: National Museum of Natural History, Smithsonian Institution; image supplied by H. C. Connolly, Jr.)

The University of Arizona Press

First printing

∞This book is printed on acid-free, archival-quality paper.
Manufactured in the United States of America
05 04 03 02 01 00 6 5 4 3 2 1

Library of Congress Cataloging-in-Publication Data
Protostars and Planets IV/Vincent Mannings, Alan Paul Boss,
Sara Samantha Russell, editors.
p. cm.—(Space science series)
Refereed review chapters presented in early form at a conference held at the University of California, Santa Barbara in July 1998.
Includes bibliographical references and index.
ISBN 0-8165-2059-3 (cloth : acid-free paper)
1. Protostars 2. Planetology 3. Molecular clouds 4. Stars–Formation
5. Disks (Astrophysics) I. Title: Protostars and planets 4. II. Title: Protostars and planets four. III. Mannings, Vincent, 1962– . IV. Boss, Alan, 1951– .
V. Russell, Sara Samantha, 1966– . VI. Series.
QB806 .P78 2000
523.8′8–dc21 99-050922

British Cataloguing-in-Publication Data
A catalogue record for this book is available from the British Library

DEDICATION

We dedicate this volume to V. S. Safronov for his pioneering work on planet formation theory.

CONTENTS

PART V—Early Solar System and Planet Formation

PART VI—Comets and the Kuiper Belt

COLLABORATING AUTHORS

PREFACE

In the preface of a *Protostars and Planets* volume, it has naturally become traditional to marvel at the advances in star- and planet-formation research since publication of the previous book. We are happy, and fortunate, to follow this custom. The advances are truly legion. The great pace of progress since *Protostars and Planets III* (PPIII; 1993) is best illustrated by the fact that the current volume is the first to include chapters describing the discoveries of extrasolar planets, brown dwarfs, and Edgeworth-Kuiper Belt objects. It is also the first to include high-resolution optical and near-infrared images of protoplanetary disks. All fields, however, have progressed tremendously since PPIII, many of them assisted by great improvements in observational techniques and by the availability of new facilities such as the *Infrared Space Observatory*, the refurbished *Hubble Space Telescope*, and the 10-m Keck telescopes. Simultaneously, ever-higher speeds and decreasing costs of computers, coupled with increasingly sophisticated hydrodynamical and magnetohydrodynamical modeling, have benefited theoretical studies of molecular clouds, star formation, jets, and disks. Closer to home, petrologic and geochemical laboratory analyses of meteorites continue to yield fresh and important insights into conditions and processes within our Sun's early protoplanetary disk.

Protostars and Planets IV (PPIV) is intended for established researchers and, crucially, for young scientists entering the field: the imagination and hard work of the latter will help to make the discoveries to be listed by the editors of *Protostars and Planets V.* PPIV is both a textbook and a report card for every facet of present-day research in the formation of stars and planets. Context and progress are described for observational and theoretical studies of the origin, structure, and evolution of molecular clouds; the collapse of cores and the formation of protostars; the formation and properties of young binary stars; the properties of winds, jets, and molecular outflows from young stellar objects; the evolution of circumstellar envelopes and disks; grain growth in disks and the formation of planets; the properties of the early solar nebula.

In July of 1998 we organized a PPIV meeting at the University of California, Santa Barbara. The conference was funded in part by the Lunar and Planetary Institute. Some 500 scientists attended from 27 countries. The broad and lively meeting comprised 49 session talks and 350 poster presentations. We thank S. Vito and S. Peale for helping to organize the conference, and we thank P. Ho and G. Wasserburg for opening and closing the meeting, respectively. We also thank all the conference session chairs: F. Adams, M. Barlow, N. Evans, M. Holman, D. Jewitt, R. Larson, J. Lissauer, S. Lizano, M. Mayor, L. Rodríguez, A. Sargent, D. Saumon, M. Simon, S. Strom, J. Wasson, and D. Weintraub.

A set of PPIV research papers has been published as a special issue of *Icarus*. The present volume consists of 49 refereed review chapters by 167 authors. We take this opportunity to thank colleagues who helped with the organization of the book: F. Adams, I. Bonnell, P. Cassen, A. Ghez, D. Jewitt, R. Jones, T. Lee, L. Mundy, Y. Nakagawa, E. Ostriker, K. Stapelfeldt, J. Wasson, and S. Weidenschilling. We warmly thank all the PPIV authors for their hard work, for helping to pay for this book to be published, and for both tolerating and meeting our stringent deadlines. We are also very grateful to the referees for their constructive reports. Of course, the deepest gratitude to the many scientists who contributed to the PPIV book will be implicit in the use of this work by researchers during the five to ten years before PPV. Those years will see the advent of SIRTF, SOFIA, FIRST, and the Keck IR Interferometer, to name just a few of the upcoming facilities. Some of the advances in protostars and planets research will be routine and expected, while others are barely glimpsed right now. If we are lucky, a few will be complete surprises. We believe this book will help to advance the field.

Vincent Mannings

Alan P. Boss

Sara S. Russell

March 1999

PART I
Molecular Clouds and Star Formation

COMPRESSIBLE MHD TURBULENCE: IMPLICATIONS FOR MOLECULAR CLOUD AND STAR FORMATION

ENRIQUE VÁZQUEZ-SEMADENI
Universidad Nacional Autónoma de México

EVE C. OSTRIKER
University of Maryland

THIERRY PASSOT
Observatoire de Nice

CHARLES F. GAMMIE
Harvard-Smithsonian Center for Astrophysics

and

JAMES M. STONE
University of Maryland

We review recent results from numerical simulations and related models of MHD turbulence in the interstellar medium (ISM) and in molecular clouds. We discuss the implications of turbulence for the processes of cloud formation and evolution and the determination of clouds' physical properties. Numerical simulations of the turbulent ISM to date have included magnetic fields, self-gravity, parameterized heating and cooling, modeled star formation, and other turbulent inputs. The structures that form reproduce well the observed velocity-size scaling properties while predicting the nonexistence of a general density-size scaling law. Criteria for the formation of gravitationally bound structures by turbulent compression are summarized. For flows with equations of state $P \propto \rho^{\gamma}$, the statistics of the density field depend on the exponent γ. Numerical simulations of both forced and decaying MHD compressible turbulence have shown that the decay rate is comparable to the nonmagnetic case. For virialized clouds, the turbulent decay time is shorter than the gravitational free fall time, so wholesale cloud collapse is prevented only by ongoing turbulent inputs, a strong mean magnetic field, or both. Finally, perspectives for future work in this field are briefly discussed.

I. INTRODUCTION

A. Observational Motivation

Present-day star formation in our Galaxy is observed to take place in cold molecular clouds, which appear to be in a state of highly compressible magnetohydrodynamic (MHD) turbulence. Furthermore, the atomic gas within which the molecular clouds form is also turbulent on larger scales, from hundreds to thousands of parsecs (e.g., Braun 1999). In this chapter we review recent results in the areas of cloud formation, structure, and evolution as well as their implications for observed physical and statistical cloud properties; these results are obtained mainly from numerical simulations of compressible MHD turbulence and from related analytical models. An introductory review, including a more detailed discussion of turbulence basics and discussions (current up to 1997) on fractality and phenomenological models, has recently been given by Vázquez-Semadeni (1999). The review of Ostriker (1997) focuses on the role of MHD turbulence in the internal dynamical evolution of molecular clouds. The reviews of Heiles et al. (1993) and McKee et al. (1993) provide extensive discussions of the observations and general theory, respectively, of magnetic fields in star-forming regions.

The formation of molecular clouds probably cannot be considered separately from the formation of larger diffuse H I structures, because the former are often observed to have H I "envelopes" (e.g., Peters and Bash 1987; Shaya and Federman 1987; Wannier et al. 1991; Blitz 1993; Williams et al. 1995). This suggests that molecular clouds and clumps may be regarded as the "tips of the icebergs" in the general continuum interstellar density field of galaxies. The process of cloud formation quite possibly involves more than a single mechanism, including the passage of spiral density waves and the effects of combined large-scale instabilities (e.g., Elmegreen 1993*a*, 1995; Gammie 1996), operating preferentially in the formation of the largest high-density structures, and the production of smaller density condensations by either swept-up shells (Elmegreen and Elmegreen 1978; Vishniac 1983, 1994; Hunter et al. 1986), or by a generally turbulent medium (Hunter 1979; Hunter and Fleck 1982; Tohline et al. 1987; Elmegreen 1993*b*). In Section II of this chapter we discuss recent results on the generation of density fluctuations in a turbulent medium, obtained from numerical simulations and semiphenomenological models. Other mechanisms of cloud formation have been reviewed extensively by Elmegreen (1993*a*, 1995). Note, however, that discrete cloud coagulation mechanisms discussed there may not be directly applicable in the context of a dynamically evolving continuum, such as that considered here.

Structurally, molecular clouds are very complex, with volume-averaged H_2 densities $n(H_2)$ ranging from $\lesssim 50$ cm^{-3}, for giant molecular clouds (GMCs) of sizes several tens of parsecs (e.g., Blitz 1993), to $n(H_2) \gtrsim 10^5$ cm^{-3} for dense cores of sizes 0.03–0.1 pc (e.g., Wilson and Walmsley 1989; Plume et al. 1997; Pratap et al. 1997; Lada et al. 1997; Myers 1995). Kinetic temperatures are relatively constant, typically $T \sim 10$ K $\pm$ a few degrees (e.g., Pratap et al. 1997), although temperatures larger by factors of a few are found in the vicinity of star-forming regions (e.g., Torrelles et al. 1983; Solomon and Sanders 1985). Their internal velocity dispersions are generally supersonic, with Mach numbers up to $\gtrsim 10$, except for the smallest cores (size $R \lesssim 0.1$ pc) (e.g., Larson 1981; Blitz 1993 and references therein; Goodman et al. 1998). When available (and significant; see, e.g., Crutcher et al. 1993, 1996; Padoan and Nordlund 1999), Zeeman measurements give typical values of the magnetic field intensity of a few to a few tens of microgauss (e.g., Heiles et al. 1993; Crutcher et al. 1993; Troland et al. 1996; Crutcher 1999), consistent with near equipartition between the kinetic and magnetic energies.

Additionally, interstellar clouds seem to follow power law scaling laws between their average density, velocity dispersion, and size (Larson 1981; see also Blitz 1993 and references therein). Together with the near equipartition between kinetic and magnetic energy, these are normally interpreted as evidence for virialized magnetic support for the clouds (e.g., Shu et al. 1987; Mouschovias 1987; Myers and Goodman 1988*a*,*b*; see also Whitworth 1996), although they provide only arguments of self-consistency of the virial equilibrium hypothesis rather than conclusive proofs (Heiles et al. 1993). In addition to the clouds that are self-consistent with virialization, however, examples also exist of objects that appear to be highly disturbed (e.g., Carr 1987; Loren 1989; Plume et al. 1997), or simply regions away from map intensity maxima (Falgarone et al. 1992, 1998) that do not satisfy one or more of the scalings. Another example of such scalings is the mass spectrum of the clouds and clumps, which seems also to be a power law (e.g., Blitz 1993; Williams et al. 1995), although present galactic cloud identification surveys remain incomplete.

Molecular clouds also exhibit signatures of self-similarity or, more generally, multifractality. Scalo (1990) and Falgarone et al. (1991) have shown that cloud boundaries have projected fractal dimensions $\sim$1.4. Also, clouds exhibit hierarchical structure; that is, the densest structures are nested within larger, less dense ones in a self-similar fashion (e.g., Scalo 1985), at least at large scales. The self-similarity appears either to break down or, at least to change similarity exponents at some small scale, reported by Larson (1995) to be close to 0.05 pc in Taurus. That scale corresponds to the local Jeans length (see section III.B of the chapter by Meyer et al., this volume), although the latter point is controversial (e.g., Simon 1997; Bate et al. 1998). A similar break has also been reported at scales 0.25–0.5 pc (Blitz and Williams 1997), showing that this question is still largely open.

Such complexity strongly suggests itself as a manifestation of the turbulent regime that permeates molecular clouds. Thus, an understanding of the basics of compressible MHD turbulence and its incarnation in the interstellar case is necessary for explaining molecular cloud structure and evolution and for diagnosing the consequences of turbulent dynamics for the process of star formation. The disordered, nonlinear nature of interstellar MHD turbulence makes direct numerical simulation the most useful tool for developing this understanding. In section III of this chapter, we summarize recent results on cloud structure and discuss questions such as whether the physical conditions and scaling laws observed in clouds arise naturally in simulations of the turbulent ISM, and what simulations predict for energy balance in the clouds, density fluctuation statistics, and the general characterization of structure in these flows.

Evolutionary aspects of turbulence and the mechanisms of cloud support under highly nonlinear, strongly self-gravitating conditions have remained open questions since the original identification of turbulent molecular clouds in the interstellar medium (ISM). Maintenance of supersonic turbulence faces the well-known problem of an expected excessive dissipation in shocks (Goldreich and Kwan 1974). Arons and Max (1975) proposed that the "turbulent" motions in molecular clouds may actually be moderate-amplitude (sub-Alfvénic) MHD waves, arguing that shock formation and the associated dissipation would then be significantly diminished. Additionally, it has been suggested that even strong dissipation may be compensated by sufficient energy injection from embedded stars and other sources (e.g., Norman and Silk 1980; Scalo 1987; McKee 1989; Kornreich and Scalo 1999). Section IV in this chapter discusses recent numerical MHD results on calculating dissipation rates for turbulence with parameters appropriate for the cold ISM and the implications of these results for questions of cloud support against self-gravitating contraction. In section V we conclude with a summary and discussion of outstanding goals and challenges for future research.

B. MHD Compressible Turbulence Basics

Before proceeding to the next sections, a few words on the nature and parameters of molecular cloud turbulence are in order. Its properties are very different from laboratory incompressible turbulence. Besides the viscous and magnetic Reynolds numbers, which measure the magnitude of the nonlinear advection term $\mathbf{v} \cdot \nabla \mathbf{v}$ in the momentum equation as compared to the viscous and ohmic dissipation terms, additional nondimensional parameters are necessary to characterize the flow. Two of these parameters are the sonic (thermal) and Alfvénic Mach numbers $M_s \equiv u/c$ and $M_A \equiv u/v_A$, where u is a characteristic velocity of the flow, c is the isothermal sound speed, and $v_A = B/(4\pi\rho)^{1/2}$ is the Alfvén speed, with B the magnetic field strength and ρ the fluid density. The plasma beta, $\beta \equiv c^2\rho/(B^2/8\pi) = 2(M_A/M_s)^2$ (i.e., the ratio of the thermal to the magnetic pressure), is also frequently used to characterize the importance of magnetic fields. The last basic parameter, which characterizes the importance of self-gravity, is the Jeans number $n_J \equiv L/L_J = L/\sqrt{\pi c^2/G\rho}$; this compares the cloud linear scale L with the minimum Jeans-unstable wavelength L_J.

Other peculiarities of ISM turbulence are related to the driving mechanisms. In standard ("Kolmogorov") incompressible turbulence theory, energy is assumed to be injected at large scales; from there it cascades down to the small scales where it is dissipated. In the ISM, on the other hand, forcing occurs at all scales (e.g., Scalo 1987; Norman and Ferrara 1996). Some mechanisms operate at large scales ($\gtrsim$1 kpc), such as the galactic shear (Fleck 1981, although this is believed to be a very inefficient source;

see, e.g., Shu et al. 1987) and supershells. Others operate at intermediate scales ($\gtrsim$100 pc), such as expanding H II regions and supernova explosions. Yet others act at small scales (a few tenths of a parsec to a few parsecs), such as stellar winds or bipolar outflows. All these mechanisms are important sources of kinetic energy.

Small-scale dissipation mechanisms in the cold ISM are of several types. First, there is the usual viscous dissipation, which operates mostly in shocks and which remains finite even in the limit of vanishing viscosity. A significant amount of dissipation also results from ambipolar diffusion due to ion-neutral friction at high densities in the presence of a strong enough magnetic field (Kulsrud and Pearce 1969; see McKee et al. 1993; Myers and Khersonsky 1995). Cooling processes are also very efficient and may radiate much of the dissipated energy, but they are balanced by heating processes and may result in near-isothermal or quasipolytropic behavior (i.e., $P \propto \rho^{\gamma}$) (section II).

Even aside from all these astrophysical properties, compressible turbulence has many distinctive features related to the transfer of energy from large scales to the dissipation (thermalization) scale. The new ingredient, as compared to incompressible turbulence, is the existence of *potential* (or compressible) modes in addition to the solenoidal (vortical) ones. Because compressions and rarefactions can exchange energy between bulk kinetic and microscopic thermal parts, compressible flows do not in general conserve bulk kinetic energy. Strongly radiative shocks, which yield an irreversible energy loss, are the most important aspect of this feature. The presence of additional compressive and thermal degrees of freedom therefore excludes the use of dimensional analysis to determine the slope of the velocity spectrum, as is done (by assumption of an energy-conservative cascade through scales) in the so-called K41 theory (Kolmogorov 1941; Obukhov 1941; see, e.g., Landau and Lifshitz 1987), which produces the well-known "universal" spectrum of the form $d(u^2)/dk \equiv E(k) \propto k^{-5/3}$ for incompressible, nonmagnetic turbulence, where $k = 2\pi/\lambda$ is the wavenumber corresponding to wavelength λ. For magnetized, incompressible, strong turbulence, recent theoretical work suggests that the mean magnetic field leads to strong anisotropy in the cascade but the same averaged energy spectrum (Goldreich and Sridhar 1995).

Moreover, the K41 theory is based on the assumption of locality in Fourier space of the nonlinear cascade (i.e., transfer between Fourier modes of similar wavelengths), a hypothesis probably not valid in the compressible case, because coupling among very different scales occurs in shocks. In these highly intermittent (i.e., inhomogeneous in space and time) structures, all Fourier modes decay at the same rate (Kadomtsev and Petviashvili 1973; Landau and Lifshitz 1987), thus invalidating the notion of inviscid cascade along the "inertial range," defined as the range in Fourier space where the energy flux is constant. Note, however, that cascade processes in the presence of shocks have been discussed by

Kraichnan (1968) and Kornreich and Scalo (1999). The spectrum of the compressible modes appears to approach a Burgers (1974) spectrum of the form k^{-2}, which arises simply from the Fourier structure of the shocks. The corresponding configuration-space scaling for Burgers turbulence is $u_l \propto l^{1/2}$ (section III.A). This scaling has been observed in numerical simulations of strongly compressible turbulence even in the presence of the magnetic field (Passot et al. 1988, 1995; Gammie and Ostriker 1996; Balsara et al. 1997; Stone 1998).

II. TURBULENT CLOUD FORMATION

An important question is whether structures formed by either turbulent compressions or passages of single shock waves can become gravitationally unstable and collapse (Ögelman and Maran 1976; Elmegreen and Lada 1977; Elmegreen and Elmegreen 1978; Hunter 1979; Hunter and Fleck 1982; Hunter et al. 1986; Tohline et al. 1987; Stevens et al. 1992; Elmegreen 1993*b*; Klein and Woods 1998). As pointed out by Hunter et al. (1986), when both heating and cooling are present, the isothermal approximation, often used to describe radiative flows, is just one out of a continuum of possibilities. In cases where the heating and cooling rates are faster than the dynamical rates and are reasonably approximated by power law functions of the density and temperature, the gas can be described as a barotropic fluid with power law "equation of state" $P \propto \rho^{\gamma}$ (e.g., de Jong et al. 1980; Maloney 1988; Elmegreen 1991; Vázquez-Semadeni et al. 1996); we use the term "effective polytropic exponent" to refer to γ.[a] In general, γ is different from 1 for the global ISM (Myers 1978) and possibly even for molecular clouds (de Jong et al. 1980; Scalo et al. 1998). Note, however, that, because cooling laws are often approximated by piecewise power laws, the effective polytropic exponent is also approximately piecewise constant. Numerical simulations that explicitly include heating and cooling appropriate for atomic gas indeed show that the corresponding rates are faster than the dynamical rates by factors of at least 50 (Passot et al. 1995; Vázquez-Semadeni et al. 1996), confirming earlier estimations (Spitzer and Savedoff 1950; Elmegreen 1993*b*). However, those simulations depart from pure power law equations of state near star formation sites. For optically thin gas with rapid heating and cooling, an effective polytropic equation of state with parameterized γ may therefore be considered as the next level of refinement over isothermal or adiabatic laws; from the aforementioned simulations, γ is found to take values between

[a] We remark that this effective polytropic exponent, which describes the thermal behavior of gases with local heating and cooling in equilibrium, is different from the adiabatic exponent c_p/c_v (describing thermal behavior with no heating or cooling) and that its usage does not imply in any form that the system is in hydrostatic equilibrium. Also note that, strictly speaking, the polytropic equation is not properly an equation of state, because the ideal-gas equation of state is also satisfied at all times.

Mouschovias 1987), fixes the other relation. However, molecular clouds may not always be gravitationally bound (Blitz 1994); indeed, except at the largest scales, observed internal clumps within clouds are *not* gravitationally confined (Bertoldi and McKee 1992). Being highly dynamic entities, even bound clouds may generally not be in *static* virial equilibrium (Ballesteros-Paredes and Vázquez-Semadeni 1997).

Alternative explanations under turbulent conditions have been proposed as well. Kolmogorov-like arguments based on, for example, cascades of angular momentum (Henriksen and Turner 1984) or kinetic energy density (Ferrini et al. 1983; Fleck 1996), are reviewed in Vázquez-Semadeni (1999). Here we briefly discuss results from numerical simulations.

As mentioned in Section I.B, for highly compressible regimes the turbulent energy spectrum is expected to approach the form $E(k) \propto k^{-2}$ of shock-dominated Burgers turbulence. If we assume that the observed linewidths measure the root-mean-square turbulent velocity u_l over scales smaller than l, we can write $u_l^2 = 2\int_{2\pi/l}^{\infty} E(k)\,dk \propto \int_{2\pi/l}^{\infty} k^{-2}\,dk \propto l$, so that the observed velocity dispersion–size relation appears to emerge naturally. However, the identification of $\Delta v(l)$ with u_l is not trivial. While $\Delta v(l)$ is the linewidth within a beam of width l integrated over the whole line of sight, u_l is an average over an idealized ensemble (or, in practice, over all space) for volumes of size l^3. Thus, the identification of u_l with $\Delta v(l)$ may depend on each beam being dominated by a single component of scale $\sim l^3$; hierarchical density fluctuations could potentially produce the required structure. The data of Issa et al. (1990) and Falgarone et al. (1992), which include positions in the sky away from brightness maxima and still exhibit a similar relation (slope ~ 0.4), seem to support the turbulent origin of the Δv-R relationship (Issa et al. also considered random positions in their maps). Peng et al. (1998) have recently studied a sample of CCS clumps, most of which are reported to be gravitationally unbound yet seem to follow Larson's linewidth-size relation as well. However, Bertoldi and McKee (1992) found that the smaller, non-self-gravitating clumps in their large study sample did not follow Larson's linewidth-size relation.

Vázquez-Semadeni et al. (1997) have surveyed the clouds that appear in the 2D numerical simulations of Passot et al. (1995), finding a velocity dispersion-to-size relation with a logarithmic slope ~ 0.4 as well (albeit with a large scatter). The energy spectrum of those simulations is indeed of the form k^{-2}. To study the development of the energy spectrum, simulations of MHD turbulence in which driving is localized in wavenumber space at scales smaller than the box have been performed in slab geometry ("$1\frac{2}{3}$D") (Gammie and Ostriker 1996) and in fully 3D geometry (Stone 1999; Stone et al. 1998). For both cases, extended spectra develop in k-space above and below the range of driving frequencies, indicating the development of both "direct" and "inverse" cascades. The $1\frac{2}{3}$D models, which were evolved over very long times, show spectral slopes -2 or slightly steeper both above and below the forcing scale.

The 3D models, although more limited in dynamic range, show combined $u_k^2 + B_k^2$ slopes between -2 and $-\frac{5}{3}$ for k larger than the forcing scale, with the more negative values occurring in weaker magnetic field models. The slope of u_k^2 alone is -2 (stronger fields) or slightly steeper (weaker fields).

Overall, turbulent MHD simulations tend to evolve toward global energy equipartition. For example, Ostriker et al. (1998) and Stone et al. (1998) find perturbed magnetic energies between 30 and 60% of kinetic energies over a wide range of β. Kinetic and magnetic spectra are within factors of a few from each other (Passot et al. 1995), implying equipartition at all scales.

Vázquez-Semadeni et al. (1997) also find cloud mass spectra of the form $dN(M)/dM \propto M^{-1.44\pm0.1}$, consistent with the low end of observational estimates (e.g., Blitz 1993), suggesting that the simulations reproduce well a number of observational cloud properties, even though they are two-dimensional. Additionally, they find that Larson's density-size relation is not satisfied in general, but rather seems to be the upper envelope in a $\log\rho$-$\log R$ diagram of the clouds' locus. That is, it is satisfied only by the densest clouds at a given size (implying the largest column densities). Low-column-density clouds would naturally escape observational surveys that utilize limited amounts of integration time (Larson 1981; Kegel 1989; Scalo 1990), suggesting that this relation may be an observational artifact. The off-peak data of Falgarone et al. (1992) are consistent with this suggestion. In summary, these results can be interpreted as implying that a Δv-R relation comparable to the observed scalings may be established globally as a consequence of the development of compressible MHD turbulence, while the density-size relation may occur only for gravitationally bound clouds within this turbulent field. Further work is necessary to confirm this possibility.

B. Turbulent Pressure

Vázquez-Semadeni et al. (1998) have performed 2D and 3D numerical simulations of isothermal gravitational collapse in initially turbulent clouds, following the evolution of the velocity dispersion as the mean density increases during the collapse. They found power law behavior of the form $P_t \sim \rho^{\gamma_t}$ for the "turbulent pressure." In particular, for slowly collapsing magnetic simulations, $\gamma_t \sim 3/2$, consistent with the result of McKee and Zweibel (1995) for the adiabatic exponent of Alfvén waves upon slow compression. However, nonmagnetic and rapidly collapsing (shorter Jeans length) magnetic simulations have $\gamma_t \sim 2$. Gammie and Ostriker (1996) also verified the McKee and Zweibel scalings $P_{\rm wave} \propto \rho^{3/2}$ and $P_{\rm wave} \propto \rho^{1/2}$ for Alfvén wave pressure of, respectively, an adiabatically contracting medium and propagation along a density gradient. For freely evolving decay simulations, however, they found only a weak, variable correlation between perturbed magnetic pressure and density. These results are incompatible with the so-called "logatropic" equation $P \sim \ln\rho$, warning against its usage in dynamical situations.

C. The Density Field

The turbulent density field has a number of relevant statistical and physical properties. Besides the density-size scaling relation (Section III.A above), which may or may not be a true property of clouds, these exhibit hierarchical nesting (e.g., Scalo 1985; Houlahan and Scalo 1990, 1992; Vázquez-Semadeni 1994) and evidence for fractal (e.g., Falgarone et al. 1991) or even multifractal structure (Chappell and Scalo 1998). The spatial and statistical distribution of the density field is crucial in the study of star formation and the understanding of the stellar initial mass function (IMF).

To construct a real theory of the IMF (or at least of the cloud mass spectrum, if the actual masses of stars turn out to be rather independent of their parent clumps), it is necessary to have a complete knowledge of the density statistics. A simple theory based exclusively on the probability density function (PDF, also commonly referred to as the density distribution function) of the fluid density (Padoan 1995; Padoan et al. 1997*b*) has been criticized by Scalo et al. (1998), who point out that, in addition to the probability of occurrence of high-density sites, it is also necessary to know how much mass is contained in these fluctuations; this requires information on multipoint statistics.

As a first step toward the understanding of density fluctuations in the ISM, the PDF of the density in polytropic gas dynamics ($P \propto \rho^{\gamma}$) has been investigated as a function of the rms Mach number, γ, and the mean magnetic field strength (Vázquez-Semadeni 1994; Scalo et al. 1998; Passot and Vázquez-Semadeni 1998; Nordlund and Padoan 1999; Ostriker et al. 1999; Stone et al. 1998). In the isothermal case ($\gamma = 1$) the density PDF is close to a lognormal distribution for every value of the Mach number. For polytropic cases with $\gamma < 1$ or $\gamma > 1$, a power law develops at densities larger than the mean or smaller than the mean, respectively, the effect being enhanced as the Mach number increases. This behavior is a consequence of the dependence of the local Mach number on the density in each case (Passot and Vázquez-Semadeni 1998; Nordlund and Padoan 1998). It has been verified in one-dimensional numerical simulations of a forced polytropic gas, but it also appears to be supported in several dimensions and in the presence of thermal heating and cooling, which yield "effective" polytropic behavior (Scalo et al. 1998). In the limit of high Mach numbers or vanishing effective polytropic exponent, the behavior does not coincide with that of the Burgers equation (Passot and Vázquez-Semadeni 1998). One implication of these results is that it should be possible in principle to determine the actual effective γ of the medium from observational determinations of its PDF, provided the problem of deconvolving the projected PDF is solved and γ does not vary much in the observed region. Also, for example, the observation by Scalo et al. (1998) that full simulations of the ISM that include the turbulent magnetic field have PDF consistent with $\gamma < 1$ suggests that the MHD waves do not give appreciably large values of γ (as is also suggested by the $P \propto \rho^{1/2}$ scaling of propagating linear-amplitude waves; cf. McKee and Zweibel 1995).

The mean mass-averaged value of $\log(\rho/\overline{\rho})$ increases with the Mach number M_s; for isothermal models in 2.5D and 3D respectively, Ostriker et al. (1999) and Padoan et al. (1997*a*) both find a logarithmic dependence on M_s^2. In large-Mach-number, high-resolution 1D isothermal models by Passot and Vázquez-Semadeni (1998), a linear dependence is found between the variance of the density logarithm and M_s^2, leading to a mass-averaged value of $\log \rho/\overline{\rho}$ that varies like M_s^2; this difference with higher-dimensional simulations (linear vs. logarithmic scaling with M_s^2) likely arises because the 1D simulations have a purely compressive velocity field and the higher-dimension simulations do not. Stone et al. (1998) and Ostriker et al. (1999) show that the largest mean contrasts in the density logarithm occur in models with the strongest mean magnetic fields (see also Pouquet et al. 1990).

The spectrum of the density field has recently come under investigation. Padoan et al. (1997*a*) have reported a steep logarithmic slope, $\sim -2.6 \pm 0.5$ for low-resolution 3D simulations of isothermal turbulence. Scalo et al. (1998) have reported two regimes, one with a slope ~ -0.9 at low k and another with slope ~ -2.4 at high k for high-resolution 2D simulations of the ISM with heating and cooling. The steeper slopes may be due to inadequate resolution. On the other hand, Lazarian (1995) has produced an algorithm for deconvolving projected H I interferometric spectra, favoring a slope ~ -1. Equivalent work for molecular clouds and high-resolution simulations in the corresponding regimes are necessary to resolve this issue.

An interesting application has been given by Padoan et al. (1997*a*), who have shown that a turbulent density field with a power law spectrum (though steep) and a lognormal PDF produces simulated plots of extinction dispersion vs. mean extinction that compare well with the analogous observational diagram for an extinction map of a dark cloud. Padoan and Nordlund (1999) also suggest that models with weaker mean magnetic fields (such that $M_A \sim 10$) have extinction dispersion vs. mean extinction plots that show better agreement with observations than do models with stronger mean magnetic fields (such that $M_A \sim 1$); however, these models do not include self-gravity, which would affect the distribution of column densities (i.e., extinctions).

D. Correlations among Variables

Numerical simulations are especially useful in the investigation of the spatial correlation among physical variables, a crucial ingredient in the formation of stars. Of particular interest is the correlation of magnetic field strength with density, as well as the correlation between field direction and the topology of density features.

The topology of the clouds is extremely clumpy and filamentary; for an example, see Color Plate 1. In general, the magnetic field exhibits a morphology indicative of significant distortion by the turbulent motions, with greater magnetic field tangling in cases with weak mean magnetic fields (Ostriker et al. 1999; Stone et al. 1998). Magnetic fields may be either aligned with or perpendicular to density features (cf. Color Plate 1; Passot et al. 1995; Vázquez-Semadeni and Passot 1999; Ostriker et al. 1999; Stone et al. 1998), with the former trend most visible at the boundaries of supershells (Gazol and Passot 1998*a*), and the latter most prominent for strong-field simulations in which field kinks and density maxima coincide (Gammie and Ostriker 1996; Ostriker et al. 1999).

Globally, no clear trend between density and magnetic field intensity is found in the simulations, but very weak correlations are observed (Passot et al. 1995; Gammie and Ostriker 1996). For example, in the large-scale 2D ISM models of Passot et al. (1995), the field strength B varies from $\sim 10^{-2}$ μG in the low-density intercloud medium to ~ 25 μG in the densest clouds ($n \sim 50$ cm^{-3}, size $\sim$ several tens of parsecs), although vanishing field strengths are also found in those regions. In 3D simulations with weak mean magnetic fields ($B_0 = 1$ μG), Padoan and Nordlund (1999) found a large dispersion in the values of B as a function of density n, but a power law correlation $B \propto n^{0.4}$ in the upper envelope of this distribution. In stronger-mean-field simulations ($B_0 = 30$ μG), they found little variation of B with n. Padoan and Nordlund argue that the B-n upper-envelope correlation found in the former case supports the notion that the mean fields in molecular clouds are weak, because a similar B-n scaling has been observed for measured Zeeman field strengths (e.g., Crutcher 1998). However, because the range of densities over which the envelope correlation is found in simulations is much smaller than the density regime in which a power law B-n relation is observed in real clouds, and because the simulations do not include gravity, the conclusion remains controversial.

To explore how line spectra may vary spatially in simulated clouds, Padoan et al. (1998) have produced synthetic non-LTE (not in local thermodynamic equilibrium) spectra of various molecular transitions from models with weak and strong mean magnetic fields. They find that their super-Alfvénic model reproduces the observational trend of linewidth vs. integrated temperature found by Heyer et al. (1996), whereas the equipartition model gives a weaker trend. However, these models do not include self-gravity, which would affect the density distributions and density-velocity correlations. Thus, it remains uncertain whether the weak-mean-field model of molecular clouds advanced by Padoan and Nordlund (1999) truly provides a better fit to observations.

IV. CLOUD EVOLUTION

A. Dissipation Rates and Turbulence Maintenance

As already mentioned, the first observations of supersonic internal cloud velocities immediately led to the questions, which have persisted up to the present, of what creates these large-amplitude motions and how they are maintained. Since the large-scale ISM is itself turbulent, turbulent motions may be incorporated into cold clouds from their formation stages, being part of the same continuum. In addition, there may be ongoing inputs that tap energy from larger scales in the Galaxy or from smaller scales within the cloud, particularly associated with various aspects of star formation (see, e.g., Scalo 1987; Miesch and Bally 1994; Kornreich and Scalo 1999). One of the first steps in understanding cloud evolution and in relating this evolution to the initiation of star formation and dynamical feedback is to assess the rate of turbulent decay under the range of conditions representative of dark clouds and GMCs.

From dimensional analysis, the decay rate of kinetic energy per unit mass must scale as $\dot{E} \sim v^3/R$, where v is some characteristic speed (or weighted product of two or more speeds), and R is some characteristic scale. In incompressible turbulence (cf. section I.B), the only characteristic scales are that of the box (L), that of the flow velocity difference u_L on the largest scale, and the value of the small-scale viscosity ν. The velocity dispersion over the whole box $\sigma_v \sim u_L$ provided the spectral slope is -1 or steeper. The first two quantities determine the decay rate, σ_v^3/L, while the last sets the spatial dissipation scale to $l_{\rm min}/L \sim (\nu/L\sigma_v)^{3/4} \equiv {\rm Re}^{-3/4}$. For a nearly pressureless (i.e., highly compressible) fluid, again possessing the same three characteristic scales, the same dissipation rate would apply, except that energy would be transferred by a shock directly from the largest scale to the dissipation scale within a flow-crossing time. For a magnetized flow of finite compressibility, on the other hand, other velocity scales in addition to σ_v (namely, those associated with thermal pressure and magnetic stress, c and v_A) enter the problem and may potentially influence the scaling of the dissipation rate.

In molecular clouds the thermal pressure is very low, and therefore, strong dissipation in shocks is expected. However, from early observations until quite recently, it has widely been considered likely that the "cushioning" effect of magnetic fields would significantly reduce kinetic energy dissipation for motions transverse to the mean field, provided that turbulent velocities remain sub-Alfvénic. Low-dimension ($1\frac{2}{3}$D) simulations (Gammie and Ostriker 1996) provided some support for this idea in that they found a scaling of dissipation rate with β in quasisteady state as $\dot{E} \propto \beta^{1/4}$, such that magnetic fields of a few tens of microgauss could potentially provide a factor $\sim$3 reduction in dissipation compared to that in weak-field ($\beta = 1$) cases. However, very recent higher-dimension numerical simulations of both forced and decaying turbulence have shown that although some differences remain between weak-field and strong-field cases, dissipation rates are never substantially lower than the predictions of unmagnetized turbulence. We next describe specific results.

Recent 3D MHD decay simulations (Mac Low et al. 1998; Stone et al. 1998; Padoan and Nordlund 1999) have followed the evolution of Mach-5 turbulence with a variety of initial velocity spectra and β ranging from

0.02 to 2, and have also been compared with the results of unmagnetized models. These models have uniform initial **B**. Stone et al. (1998) also included simulations of decay from fully saturated turbulence and from saturated turbulence with initial density fluctuations suppressed. These experiments all show kinetic energy decay times (defined as the time for kinetic energy reduction by 50%) in the range $\sim$0.4 to $1 t_f$, where t_f is the flow-crossing time l/u_l on the main energy-containing scale. In Stone et al. (1998), the difference in decay time between the strongest-field case (β = 0.02, corresponding to 44 μG for $n_{H_2} = 10^3$ cm^{-3} and T = 10 K) and the unmagnetized run is less than a factor of two. In decay models, Mac Low et al. (1998) and Stone et al. (1998) find late-time power law dependence of the turbulent energy as $E \propto t^{-\eta}$, with η ranging between 0.8 and 1.

To assess turbulent decay rates for a quasisteady state, Stone et al. (1998) performed simulations in which a fixed mechanical power $\dot{E}_K$ is input to the flow in the form of random, uncorrelated velocity perturbations. A saturated state with energy level E_K is reached after a time $\sim t_f$, with only relatively small differences in the saturation energy between magnetized and unmagnetized models; the turbulence dissipation times $E_K/\dot{E}_K$ range between 0.5 and 0.7 t_f. Mac Low (1999) has found similar results, and suggests that the power law time dependence of decay models arises from a secular increase in the smallest scale of turbulence. The implications of the short turbulence dissipation time for potential cloud support are discussed below in section IV.B.

Kornreich and Scalo (1999) have recently considered the problem of turbulence maintenance in molecular clouds, comparing the average time between external shock wave passages through a cloud with the energy decay time, finding that they are comparable, and implying that this "shock pump" is capable of sustaining the cloud turbulence. An additional result is the proposal of a cascadelike mechanism for the compressible case, in which vorticity is generated behind shocks, which in turn rapidly produces new smaller-scale shocks, and so on. This is due to an interesting asymmetry of the evolution equations for the vorticity and divergence of the velocity field (for recent discussions, see Vázquez-Semadeni et al. 1996; Kornreich and Scalo 1999): the nonlinear transfer will produce compressible modes out of purely rotational ones, but the converse is not true. This can also be understood in terms of the well-known Kelvin theorem of conservation of circulation.

The mechanisms of vorticity production have also been investigated numerically. Simulations of compressible turbulence with purely potential forcing indicate that a negligible amount of kinetic energy is transferred from compressible to solenoidal modes (Kida and Orszag 1990; Vázquez-Semadeni et al. 1996). Vorticity generation behind curved shocks or shock intersections, as well as the vortex stretching term, do not seem to be efficient processes to maintain a nonnegligible level of vorticity in the flow. However, in the presence of the Coriolis force, thermal heating (through the baroclinic term), or magnetic field, near equipartition between solenoidal and compressible modes is easily obtained (Vázquez-Semadeni et al. 1996).

B. Turbulence and Cloud Support against Gravity

An issue that is often discussed in tandem with turbulent dissipation is the question of cloud support against self-gravity. Cold, dark clouds and GMCs have typical Jeans lengths ~2 pc, such that the whole cloud entities exceed the Jeans mass $M_J = \rho L_J^3$ by factors of a thousand or much more. This is just another way of stating that thermal pressure gradients would be powerless to prevent self-gravitating runaway. If not for the intervention of other dynamical processes, wholesale cloud collapse would be the rule.

"Turbulent pressure" is often invoked as a means to counter gravity. For a weakly compressible medium, the ρv^2 Reynolds stresses are naturally associated with this turbulent pressure. The effect of turbulence on the gravitational instability in the absence of a magnetic field was first investigated by Chandrasekhar (1951), who suggested an increase of the Jeans length, and later by Bonazzola et al. (1987, 1992) and Vázquez-Semadeni and Gazol (1995), who predicted a reversal of the Jeans criterion when certain conditions on the energy spectrum are met.

In a strongly compressible and radiative medium, the collisions of supersonically converging streams of gas are nearly inelastic, so the Reynolds stress does not itself act as an effective pressure. Turbulent motions can, however, generate turbulent magnetic fields (and vice versa); these fluctuating magnetic fields exert pressure and tension forces on the medium. Several authors (Shu et al. 1987; Fatuzzo and Adams 1993; McKee and Zweibel 1995) have suggested that, in particular, the time-dependent magnetic field perturbations associated with Alfvén waves could potentially be important in providing "wave pressure" support against gravity along the mean magnetic field direction in a cloud. This mean-field axis is the most susceptible to collapse; in the orthogonal directions, magnetic pressure suppresses gravitational instability in a homogeneous cloud as long as $n_J \equiv L/L_J < (\beta/2)^{-1/2}$ (Chandrasekhar and Fermi 1953) and suppresses instability in a cloud pancake of surface density Σ provided $\Sigma/B < 1/(2\pi\sqrt{G})$ (cf. Mouschovias and Spitzer 1976; Tomisaka et al. 1988).

An exact derivation of the Jeans criterion is hardly possible in the MHD turbulent case. An attempt in this direction has been made by considering the linear stability of a self-gravitating medium, permeated by a uniform magnetic field B_0 along which a finite-amplitude, circularly polarized Alfvén wave propagates. For perturbations along B_0 it is found that the Alfvén wave increases the critical Jeans length (Lou 1997). In the case of perturbations perpendicular to the mean field, Gazol and Passot (1999*b*) have shown that the medium is less stable in the presence of a moderate-amplitude Alfvén wave. For large-amplitude waves, however, McKee and Zweibel (1995) show that the waves have an isotropic stabilizing effect.

Gammie and Ostriker (1996) verified, using simulations in $1\frac{2}{3}$D, that Alfvén waves of sufficient amplitude (such that $n_J \lesssim \delta v_A/2c_s$) can indeed prevent collapse of slab clouds along the mean field direction. These simulations included both cases with decaying and cases with quasisteady forced turbulence. For 1D decay models, the turbulent dissipation rate is low enough that clouds with initial turbulent energy above the limit remain uncollapsed for times up to $t_g \equiv L_J/c$ ($\sim$10 Myr for typical conditions). However, more recent simulations performed in higher dimensions have shown that the greater dissipation rates quench turbulence too rapidly for magnetic fluctuations to prevent mean-field collapse. Because 3D dissipation times for both magnetized and unmagnetized flows are shorter than the flow-crossing time t_f (cf. section IV.A), they will also be less than the gravitational collapse times $\approx 0.3t_g$ for clouds that are virialized (e.g., observed cloud scalings yield $t_f \sim 0.5t_g$); thus, cloud support cannot be expected for unregenerated turbulence. Self-gravitating, magnetized $2\frac{1}{2}$D simulations of Ostriker et al. (1999) have already demonstrated this result directly and concluded that without ongoing energy inputs, only the strength of the mean magnetic field is important in determining whether or not clouds collapse in times <10 Myr. Numerical simulations have also been performed that allow for ongoing turbulent excitation. In unmagnetized 2D models, Léorat et al. (1990) showed that large-scale gravitational collapse can be prevented indefinitely provided a high enough Mach number is maintained by forcing, and the energy is injected at small enough scales. For magnetized models, Gammie and Ostriker (1996) found similar results for $1\frac{2}{3}$D simulations (as well as for unpublished 2D simulations).

Ballesteros-Paredes and Vázquez-Semadeni (1997) have measured the overall virial balance of clouds in a "survey" of the 2D high-resolution ISM simulations of Passot et al. (1995). It was found that the gravitational term appearing in the virial theorem is comparable to the sum of the other virial terms for the largest clouds but progressively loses importance on the average as smaller clouds are considered. Nevertheless, the scatter about this average trend is large, and a small fraction of small clouds have very large gravitational terms that overwhelm the others and induce collapse. In this scenario, the low efficiency of star formation is understood as a consequence of the intermittency of the turbulence.

C. Evolution of the Star Formation Rate

Numerical simulations on the large scales provide information on the star formation (SF) history as well (Vázquez-Semadeni et al. 1995; Gazol and Passot 1998*a*). In these models, stars form whenever a certain density threshold is exceeded, rather than being treated as a separate fluid (e.g., Chiang and Prendergast 1985; Rosen et al. 1993). A nontrivial result is the development of a self-sustained cycle of SF in which the turbulence, aided by self-gravity, contains enough power to produce star-forming clouds, while the energy injected by the stars is sufficient to regenerate the turbulence. Due to the strong nonlinearity of the SF scheme, this cycle is highly chaotic and particularly intermittent in the presence of supernovae. Self-propagating SF in supershells is very efficient, increasing the SF rate in active periods, but the destructive power of superbubbles is so large that the system requires a longer time to create new SF sites once the shells have dispersed. A similar result was obtained by Vázquez and Scalo (1989) in a simple model for gas-accreting galaxies. This behavior is consistent with recent observations suggesting that the SF history in galaxies is highly irregular (e.g., Grebel 1998). Another interesting point concerns the influence of the strength of the uniform component of the magnetic field B_0 (Gazol and Passot 1998*a*). The star formation rate is found to increase with B_0 (with $B_0 = 5\ \mu$G it is larger by a factor ~3 compared with the case $B_0 = 0$), as long as B_0 is not too large. For very large B_0, the SF rate decreases due to the rigidification of the medium.

V. CONCLUSIONS

In this chapter we have reviewed a vast body of results that address the problem of MHD compressible (MHDC) turbulence and its implications for problems of cloud and star formation. Most of these results are new, appearing after the last *Protostars and Planets* conference (PP III). Interstellar turbulence is inherently a multiscale and nonequilibrium phenomenon and thus appears to play a fundamental role in the formation, evolution, and determination of cloud structural properties. Most of the studies reviewed here have relied on direct numerical simulations of MHDC turbulence in a variety of regimes, ranging from isothermal to polytropic to fully thermodynamic, the latter including parameterized heating and cooling and modeled star formation. High-resolution 2D and 3D simulations of the ISM at "box" scales from 1 kpc down to a few pc have shown that the clouds formed in them reproduce several observational cloud properties well (such as clumpy structure and linewidth-size scalings), while suggesting that some others [e.g., Larson's (1981) density-size relation] may either arise from selection effects or require other special conditions. Simulations to date have shown that the old paradigm that MHDC turbulence should dissipate much more slowly than nonmagnetic turbulence may be incorrect, bringing back the necessity of strong large-scale fields for magnetic cloud support and of continued energy injection in order to sustain the turbulence. The idea for self-regulation of star formation by turbulent feedback remains promising, and it is also possible that physical processes not yet incorporated in simulations may reduce the turbulent dissipation rate.

A large number of questions remain unanswered, however. At the larger scales it is necessary to understand the interplay between turbulence, large-scale instabilities, and spiral waves in the formation of molecular clouds and complexes, as well as to investigate the processes that lead to cloud destruction. Quantitative assessments must be developed for the efficiency of turbulent excitation from internal and external sources. A theory of the IMF, or at least of the cloud mass spectrum, requires the knowledge and understanding of multipoint density statistics. Better understanding of the parameter dependence of these and other questions in cloud evolution and structure will be required to address the problem of "deriving" the star formation rate and efficiency, in answer to the oft-posed challenge of extragalactic astronomers. Comparison with observations is crucial to discriminate among models, but it will always face the degeneracy limitations associated with projection effects.

Future research in these areas is likely to include more physical processes (e.g., ambipolar diffusion, radiative transfer, and global environment effects, such as the galactic spiral potential and variations in external radiation with galactic position), to perform direct statistical comparison with observations of molecular lines, and polarization and IR measurements, and to work at higher resolutions in 3D, if the multiscale nature of the problem is to be captured adequately. With so much still untried, there is considerable ground to cover before reaching the long-range goal of synthesizing the results of disparate numerical simulations into a coherent theory of the turbulent ISM. Nevertheless, the important advances since PP III show the success of numerical methods in answering many long-standing questions in the theory of interstellar MHD turbulence and the opportunity for great progress before the next *Protostars and Planets*.

Acknowledgments We are grateful to C. F. McKee for helpful comments on this manuscript and to J. Scalo for enlightening discussions. This work has been supported in part by grants UNAM/DGAPA IN105295, CRAY/UNAM SC008397, NAG5380 from NASA, and by the CNRS "PCMI" National Program.

REFERENCES

Arons, J., and Max, C. E. 1975. Hydromagnetic waves in molecular clouds. *Astrophys. J. Lett.* 196:77–81

Ballesteros-Paredes, J., and Vázquez-Semadeni, E. 1997. Virial balance and scaling relations in turbulent MHD two-dimensional numerical simulations of the ISM. In *Star Formation Near and Far,* ed. S. Holt and L. Mundy (Woodbury, NY: AIP Press), pp. 81–84.

Ballesteros-Paredes, J., Vázquez-Semadeni, E., and Scalo, J. 1999. Clouds as turbulent density fluctuations. Implications for pressure confinement and spectral line data interpretation. *Astrophys. J.* 515:286–303.

Balsara, D. S., Crutcher, R. M., and Pouquet, A. 1997. Numerical simulations of MHD turbulence and gravitational collapse. In *Star Formation, Near and Far*, ed. S. Holt and L. Mundy (Woodbury, NY: AIP Press), pp. 89–92.

Bash, F. N., Green, E., and Peters, W. L., III. 1977. The galactic density wave, molecular clouds, and star formation. *Astrophys. J.* 217:464–472.

Bate, M. R., Clarke, C. J., and McCaughrean, M. J. 1998. Interpreting the mean surface density of companions in star-forming regions. *Mon. Not. Roy. Astron. Soc.* 297:1163–1181.

Bertoldi, F., and McKee, C. 1992. Pressure-confined clumps in magnetized molecular clouds. *Astrophys. J.* 395:140–157.
Blitz, L. 1993. Giant molecular clouds. In *Protostars and Planets III*, ed. E. H. Levy and J. I. Lunine (Tucson: University of Arizona Press), pp. 125–161.
Blitz, L. 1994. Simple things we don't know about molecular clouds. In *The Cold Universe*, ed. T. Montmerle, C. J. Lada, I. F. Mirabel, and J. Trân Thanh Vân (Gif-sur-Yvette: Editions Frontières), pp. 99–106.
Blitz, L., and Shu, F. H. 1980. The origin and lifetime of giant molecular cloud complexes. *Astrophys. J.* 238:148–157.
Blitz, L., and Williams, J. P. 1997. Molecular clouds are not fractal: A characteristic size scale in Taurus. *Astrophys. J. Lett.* 488:L145–L148.
Bonazzola, S., Falgarone, E., Heyvaerts, J., Pérault, M., and Puget, J. L. 1987. Jeans collapse in a turbulent medium. *Astron. Astrophys.* 172:293–298.
Bonazzola, S., Pérault, M., Puget, J. L., Heyvaerts, J., Falgarone, E., and Panis, J. F. 1992. Jeans collapse of turbulent gas clouds. A tentative theory. *J. Fluid Mech.* 245:1–28.
Boulares, A., and Cox, D. P. 1990. Galactic hydrostatic equilibrium with magnetic tension and cosmic-ray diffusion. *Astrophys. J.* 365:544–558.
Braun, R. 1999. Properties of atomic gas in spiral galaxies. In *Interstellar Turbulence, Proceedings of the 2nd Guillermo Haro Conference*, ed. J. Franco and A. Carramiñana (Cambridge: Cambridge University Press), pp. 12–19.
Burgers, J. M. 1974. *The Nonlinear Diffusion Equation* (Dordrecht: Reidel).
Carr, J. S. 1987. A study of clumping in the Cepheus OB 3 molecular cloud. *Astrophys. J.* 323:170–178.
Chandrasekhar, S. 1951. The gravitational instability in an infinite homogeneous turbulent medium. *Proc. Royal Soc. London* 210:26–29.
Chandrasekhar, S. 1961. *Hydrodynamic and Hydromagnetic Stability* (Oxford: Clarendon Press).
Chandrasekhar, S., and Fermi, E. 1953. Problems of gravitational instability in the presence of a magnetic field. *Astrophys. J.* 118:116–141.
Chappell, D., and Scalo, J. 1998. Multifractal scaling, geometrical diversity, and hierarchical structure in the cool interstellar medium. *Astrophys. J.*, in press.
Chiang, W.-H., and Prendergast, K. H. 1985. Numerical study of a two-fluid hydrodynamic model of the interstellar medium and Population I stars. *Astrophys. J.* 297:507–530.
Crutcher, R. M. 1999. Observations of magnetic fields in dense clouds: Implications for MHD turbulence and cloud evolution. In *Interstellar Turbulence, Proceedings of the 2nd Guillermo Haro Conference*, ed. J. Franco and A. Carramiñana (Cambridge: Cambridge University Press), pp. 213–217.
Crutcher, R. M., Troland, T. H., Goodman, A. A., Heiles, C., Kazès, I., and Myers, P. C. 1993. OH Zeeman observations of dark clouds. *Astrophys. J.* 407:175–184.
Crutcher, R. M., Troland, T. H., Lazareff, B., Kazès, I. 1996. CN Zeeman observations of molecular cloud cores. *Astrophys. J.* 456:217–224.
de Jong, T., Dalgarno, A., and Boland, W. 1980. Hydrostatic models of molecular clouds. I. Steady state models. *Astron. Astrophys.* 91:68–84.
Draine, B. T. 1978. Photoelectric heating of interstellar gas. *Astrophys. J. Suppl.* 36:595–619.
Elmegreen, B. G. 1991. Cloud formation by combined instabilities in galactic gas layers. Evidence for a Q threshold in the fragmentation of shearing wavelets. *Astrophys. J.* 378:139–156.
Elmegreen, B. G. 1993*a*. Formation of interstellar clouds and structure. In *Protostars and Planets III*, ed. E. H. Levy and J. I. Lunine (Tucson: University of Arizona Press), pp. 97–124.

Elmegreen, B. G. 1993*b*. Star formation at compressed interfaces in turbulent self-gravitating clouds. *Astrophys. J. Lett.* 419:L29–L32.

Elmegreen, B. G. 1995. Star formation on a large scale. In *Molecular Clouds and Star Formation*, ed. C. Yuan and Junhan You (Singapore: World Scientific), pp. 149–205.

Elmegreen, B. G., and Lada, C. J. 1977. Sequential formation of subgroups in OB associations. *Astrophys. J.* 214:725–741.

Elmegreen, B. G., and Elmegreen, D. M. 1978. Star formation in shock-compressed layers. *Astrophys. J.* 220:1051–1062.

Falgarone, E., Phillips, T. G., and Walker, C. K. 1991. The edges of molecular clouds: Fractal boundaries and density structure. *Astrophys. J.* 378:186–201.

Falgarone, E., Puget, J.-L., and Pérault, M. 1992. The small-scale density and velocity structure of molecular clouds. *Astron. Astrophys.* 257:715–730.

Falgarone, E., Panis, J.-F., Heithhausen, A., Pérault, M., Stutzki, J., Puget, J.-L., and Bensch, F. 1998. The IRAM key-project: Small-scale structure of pre-star-forming regions. I. Observational results. *Astron. Astrophys.* 331:669–696.

Fatuzzo, M., and Adams, F. C. 1993. Magnetohydrodynamic wave propagation in one-dimensional nonhomogeneous, self-gravitating clouds. *Astrophys. J.* 412:146–159.

Ferrini, F., Marchesoni, F., and Vulpiani, A. 1983. A hierarchical model for gravitational compressible turbulence. *Astrophys. Space Sci.* 96:83–93.

Field, G. B., Goldsmith, D. W., and Habing, H. J. 1969. Cosmic ray heating of the interstellar gas. *Astrophys. J. Lett.* 155:L149–L154.

Fleck, R. C. 1981. On the generation and maintenance of turbulence in the interstellar medium. *Astrophys. J. Lett.* 246:L151–L154.

Fleck, R. C. 1996. Scaling relations for the turbulent, non-self-gravitating, neutral component of the interstellar medium. *Astrophys. J.* 458:739–741.

Gammie, C. F. 1996. Linear theory of magnetized, viscous, self-gravitating gas disks. *Astrophys. J.* 462:725-731.

Gammie, C. F., and Ostriker, E. C. 1996. Can nonlinear hydromagnetic waves support a self-gravitating cloud? *Astrophys. J.* 466:814–830.

Gazol, A., and Passot, T. 1999*a*. A turbulent model for the interstellar medium. III. Stratification and supernova explosions. *Astrophys. J.* 518:748–759.

Gazol, A., and Passot, T. 1999*b*. Influence of Alfvén waves on the gravitational instability. *Astron. Astrophys.*, submitted.

Goldreich, P., and Kwan, J. 1974. Molecular clouds. *Astrophys. J.* 189:441–453.

Goldreich, P., and Sridhar, S. 1995. Toward a theory of interstellar turbulence. II: Strong Alfvénic turbulence. *Astrophys. J.* 438:763–775.

Goodman, A. A., Barranco, J. A., Wilner, D. J., and Heyer, M. H. 1998. Coherence in dense cores. II. The transition to coherence. *Astrophys. J.* 504:223–246.

Grebel, E. 1998. Star formation histories of Local Group dwarf galaxies. In *Dwarf Galaxies and Cosmology*, ed. T. X. Thuan, C. Balkowski, V. Cayatte, and J. Trân Thanh Vân (Gif-sur-Yvette: Editions Frontières), in press.

Heiles, C., Goodman, A. A., McKee, C. F., and Zweibel, E. G. 1993. Magnetic fields in star-forming regions: Observations. In *Protostars and Planets III*, ed. E. H. Levy and J. I. Lunine (Tucson: University of Arizona Press), pp. 279–326.

Henriksen, R. N., and Turner, B. E. 1984. Star cloud turbulence. *Astrophys. J.* 287:200–297.

Heyer, M. H., Carpenter, J. M., and Ladd, E. F. 1996. Giant molecular cloud complexes with optical H II regions: ^{12}CO and ^{13}CO observations and global cloud properties. *Astrophys. J.* 463:630–641.

Houlahan, P., and Scalo, J. 1990. Recognition and characterization of hierarchical interstellar structure. I. Correlation function. *Astrophys. J. Suppl.* 72:133–152.

Houlahan, P., and Scalo, J. 1992. Recognition and characterization of hierarchical interstellar structure. II. Structure tree statistics. *Astrophys. J.* 393:172–187.
Hunter, J. H., Jr. 1979. The influence of initial velocity fields upon star formation. *Astrophys. J.* 233:946–949.
Hunter, J. H., Jr., and Fleck, R. C. 1982. Star formation: The influence of velocity fields and turbulence. *Astrophys. J.* 256:505–513.
Hunter, J. H., Jr., Sandford, M. T., II, Whitaker, R. W., and Klein, R. I. 1986. Star formation in colliding gas flows. *Astrophys. J.* 305:309–332.
Issa, M., MacLaren, I., and Wolfendale, W. 1990. The size–line width relation and the mass of molecular hydrogen. *Astrophys. J.* 352:132–138.
Kadomtsev, B. B., and Petviashvili, V. I. 1973. Acoustic turbulence. *Sov. Phys. Dokl.* 18:115–116.
Kegel, W. H. 1989. The interpretation of correlation between observed parameters of molecular clouds. *Astron. Astrophys.* 225:517–520.
Kida, S., and Orszag, S. A. 1990. Energy and spectral dynamics in forced compressible turbulence. *J. Sci. Comput.* 5:85–125.
Klein, R. I., and Woods, D. T. 1998. Bending mode instabilities and fragmentation in interstellar cloud collisions: A mechanism for complex structure. *Astrophys. J.* 497:777–799.
Kolmogorov, A. N. 1941. The local structure of turbulence in an incompressible viscous fluid for very large Reynolds numbers. *Dokl. Akad. Nauk* 30:301–305.
Kornreich, P., and Scalo, J. 1999. The galactic shock pump. A source of supersonic internal motions in the cool interstellar medium. *Astrophys. J.*, in press.
Kraichnan, R. H. 1968. Lagrangian-history statistical theory for Burgers' equation. *Phys. Fluids* 11:265–277.
Kulsrud, R. M., and Pearce, W. P. 1969. The effect of wave-particle interactions on the propagation of cosmic rays. *Astrophys. J.* 163:567–576.
Lada, E. A., Evans, N. J., II, and Falgarone, E. 1997. Physical properties of molecular cloud cores in L1630 and implications for star formation. *Astrophys. J.* 488:286–306.
Landau, L. D., and Lifshitz, E. M. 1987. *Fluid Mechanics*. 2nd ed. (Oxford: Pergamon Press).
Larson, R. B. 1981. Turbulence and star formation in molecular clouds. *Mon. Not. Roy. Astron. Soc.* 194:809–826.
Larson, R. B. 1995. Star formation in groups. *Mon. Not. Roy. Astron. Soc.* 272:213–220.
Lazarian, A. 1995. Study of turbulence in H I using radiointerferometers. *Astron. Astrophys.* 293:507–520.
Léorat, J., Passot, T., and Pouquet, A. 1990. Influence of supersonic turbulence on self-gravitating flows. *Mon. Not. Roy. Astron. Soc.* 243:293–311.
Loren, R. B. 1989. The cobwebs of Ophiuchus. I. Strands of (C-13)O: The mass distribution. *Astrophys. J.* 338:902–924.
Lou, Y.-Q. 1997. Gravitational collapse in the presence of a finite-amplitude circularly polarized Alfvén wave. *Mon. Not. Roy. Astron. Soc.* 279:L67–L71.
Mac Low, M.-M. 1999. The energy dissipation rate of supersonic, magnetohydrodynamic turbulence in molecular clouds. *Astrophys. J.*, submitted.
Mac Low, M.-M., Klessen, R. S. Burkert, and Smith, M. D. 1998. Kinetic energy decay of supersonic and super-Alfvénic turbulence in star-forming clouds. *Phys. Rev. Lett.* 80:2754–2757.
Maloney, P. 1988. The turbulent interstellar medium and pressure-bounded molecular clouds. *Astrophys. J.* 334:761–770.
McKee, C. F. 1989. Photoionization-regulated star formation and the structure of molecular clouds. *Astrophys. J.* 345:782–801.

McKee, C. F., and Zweibel, E. G. 1992. On the virial theorem for turbulent molecular clouds. *Astrophys. J.* 399:551–562.
McKee, C. F., and Zweibel, E. G. 1995. Alfven waves in interstellar gasdynamics. *Astrophys. J.* 440:686–696.
McKee, C. F., Zweibel, E. G., Goodman, A. A., and Heiles, C. 1993. Magnetic fields in star-forming regions. In *Protostars and Planets III*, ed. E. H. Levy and J. I. Lunine (Tucson: University of Arizona Press), pp. 327–366.
Miesch, M. S., and Bally, J. 1994. Statistical analysis of turbulence in molecular clouds. *Astrophys. J.* 429:645–671.
Mouschovias, T. 1987. Star formation in magnetic interstellar clouds. I. Interplay between theory and observation. In *Physical Processes in Interstellar Clouds,* ed. G. E. Morfill and M. Scholer (Dordrecht: Reidel), pp. 453–489.
Mouschovias, T., and Spitzer, L. 1976. Note on the collapse of magnetic interstellar clouds. *Astrophys. J.* 210:326-327.
Myers, P. C. 1978. A compilation of interstellar gas properties. *Astrophys. J.* 225:380–389.
Myers, P. C. 1995. Star forming molecular clouds. In *Molecular Clouds and Star Formation*, ed. C. Yuan and Junhan You (Singapore: World Scientific), pp. 47–96.
Myers, P. C., and Goodman, A. A. 1988*a*. Evidence for magnetic and virial equilibrium in molecular clouds. *Astrophys. J. Lett.* 326:27–30.
Myers, P. C., and Goodman, A. A. 1988*b*. Magnetic molecular clouds: Indirect evidence for magnetic support and ambipolar diffusion. *Astrophys. J.* 329:392–405.
Myers, P. C., and Khersonsky, V. K. 1995. On magnetic turbulence in interstellar clouds. *Astrophys. J.* 442:186–196.
Nordlund, Å., and Padoan, P. 1999. The density PDFs of supersonic random flows. In *Interstellar Turbulence, Proceedings of the 2nd Guillermo Haro Conference*, ed. J. Franco and A. Carramiñana (Cambridge: Cambridge University Press), pp. 218–222.
Norman, C., and Ferrara, A. 1996. The turbulent interstellar medium: Generalizing to a scale-dependent phase continuum. *Astrophys. J.* 467:280–291.
Norman, C., and Silk, J. 1980. Clumpy molecular clouds: A dynamic model self-consistently regulated by T Tauri star formation. *Astrophys. J.* 238:158–174.
Obukhov, A. M. 1941. Of the distribution of energy in the spectrum of turbulent flow. *Comptes Rend. Acad. Sci. URSS* 32:19.
Ögelman, H. B., and Maran, S. P. 1976. The origin of OB associations and extended regions of high-energy activity in the Galaxy through supernova cascade processes. *Astrophys. J.* 209:124–129.
Ostriker, E. C. 1997. Turbulence and magnetic fields in star formation. In *Star Formation, Near and Far*, ed. S. Holt and L. Mundy (Woodbury, NY: AIP Press), pp 51–63.
Ostriker, E. C., Gammie, C. F., and Stone, J. M. 1999. Kinetic and structural evolution of self-gravitating, magnetized clouds: 2.5-dimensional simulations of decaying turbulence. *Astrophys. J.* 513:259–274.
Padoan, P. 1995. Supersonic turbulent flows and the fragmentation of a cold medium, *Mon. Not. Roy. Astron. Soc.* 277:377–388.
Padoan, P., and Nordlund, Å. 1999. A super-Alfvénic model for dark clouds. *Astrophys. J.*, in press.
Padoan, P., Jones, B. J. T., Nordlund, Å. 1997*a*. Supersonic turbulence in the interstellar medium: Stellar extinction determinations as probes of the structure and dynamics of dark clouds. *Astrophys. J.* 474:730–734.
Padoan, P., Nordlund, Å., and Jones, B. J. T. 1997*b*. The universality of the stellar initial mass function. *Mon. Not. Roy. Astron. Soc.* 288:145–152.

Padoan, P., Juvela, M., Bally, J., and Nordlund, Å. 1998. Synthetic molecular clouds from supersonic MHD and non-LTE radiative transfer calculations. *Astrophys. J.* 504:300–313.

Passot, T., and Vázquez-Semadeni, E. 1998. Density probability distribution in one-dimensional gas dynamics. *Phys. Rev. E* 58:4501–4510.

Passot, T., Pouquet, A., and Woodward, P. 1988. The plausibility of Kolmogorov-type spectra in molecular clouds. *Astron. Astrophys.* 197:228–234.

Passot, T., Vázquez-Semadeni, E., and Pouquet, A. 1995. A turbulent model for the interstellar medium. II. Magnetic fields and rotation. *Astrophys. J.* 455:536–555.

Peng, R., Langer, W. D., Velusamy, T., Kuiper, T. B. H., and Levin, S. 1998. Low-mass clumps in TMC-1: Scaling laws in the small-scale regime. *Astrophys. J.* 197:842–849.

Peters, W. L., and Bash, F. N. 1987. The correlation of spiral arm molecular clouds with atomic hydrogen. *Astrophys. J.* 317:646–652.

Plume, R., Jaffe, D. T., Evans, N. J., II, Martín-Pintado, J., and Gómez-González, J. 1997. Dense gas and star formation: Characteristics of cloud cores associated with water masers. *Astrophys. J.* 476:730–749.

Pouquet, A., Passot, T., and Léorat, J. 1990. Numerical simulations of compressible flows. Proceedings of the IAU symposium 147. In *Fragmentation of Molecular Clouds and Star Formation*, ed. E. Falgarone, F. Boulanger, and G. Duvert (Dordrecht: Kluwer), pp. 101–118.

Pratap, P., Dickens, J. E., Snell, R. L., Miralles, M. P., Bergin, E. A., Irvine, W. M., and Schloerb, F. P. 1997. A study of the physics and chemistry of TMC-1. *Astrophys. J.* 486:862–885.

Rosen, A., Bregman, J. N., and Norman, M. L. 1993. Hydrodynamical simulations of star-gas interactions in the interstellar medium with an external gravitational potential. *Astrophys. J.* 413:137–149.

Scalo, J. 1985. Fragmentation and hierarchical structure in the interstellar medium. In *Protostars and Planets II*, ed. D. C. Black and M. S. Matthews (Tucson: University of Arizona Press), pp. 201–296.

Scalo, J. 1987. Theoretical approaches to interstellar turbulence. In *Interstellar Processes*, ed. D. J. Hollenbach and H. A. Thronson, Jr. (Dordrecht: Reidel), pp. 349–392.

Scalo, J. 1990. Perception of interstellar structure. Facing complexity. In *Physical Processes in Fragmentation and Star Formation*, ed. R. Capuzzo-Dolcetta, C. Chiosi, and A. di Fazio (Dordrecht: Kluwer), pp. 151–177.

Scalo, J, Vázquez-Semadeni, E., Chappell, D., and Passot, T. 1998. On the probability density function of galactic gas. I. Numerical simulations and the significance of the polytropic index. *Astrophys. J.* 504:835–853.

Shaya, E. J., and Federman, S. R. 1987. The H I distribution in clouds within galaxies. *Astrophys. J.* 319:76–83.

Shu, F. H., Adams, F. C., and Lizano, S. 1987. Star formation in molecular clouds: Observation and theory. *Ann. Rev. Astron. Astrophys.* 25:23–81.

Simon, M. 1997. Clustering of young stars in Taurus, Ophiuchus, and the Orion Trapezium. *Astrophys. J. Lett.* 482:L81–L84.

Solomon, P. M., and Sanders, D. B. 1985. Star formation in a galactic context: The location and properties of molecular clouds. In *Protostars and Planets II*, ed. D. C. Black and M. S. Matthews (Tucson: University of Arizona Press), pp. 59–80.

Spitzer, L., Jr., and Savedoff, M. P. 1950. The temperature of interstellar matter. III. *Astrophys. J.* 111:593–608.

Stevens, I. R., Blondin, J. M., and Pollock, A. M. T. 1992. Colliding winds from early-type stars in binary systems. *Astrophys. J.* 386:265–287.

Stone, J. M. 1999. Direct numerical simulations of compressible MHD turbulence. In *Interstellar Turbulence, Proceedings of the 2nd Guillermo Haro Conference*, ed. J. Franco and A. Carramiñana (Cambridge: Cambridge University Press), pp. 267–271.

Stone, J. M., Ostriker, E. C., and Gammie, C. F. 1998. Dissipation in compressible MHD turbulence. *Astrophys. J. Lett.* 508:L99–L102.

Tohline, J. E, Bodenheimer, P. H., and Christodoulou, D. M. 1987. The crucial role of cooling in the making of molecular clouds and stars. *Astrophys. J.* 322:787–794.

Tomisaka, K., Ikeuchi, S., and Nakamura T. 1988. Equilibria and evolutions of magnetized, rotating, isothermal clouds. II. The extreme case: Nonrotating clouds. *Astrophys. J.* 335:239–262.

Torrelles, J. M., Rodríguez, L. F., Cantó, J., Marcaide, J., and Gyulbudaghian, A. L. 1983. A search for molecular outflows associated with peculiar nebulosities and regions of star formation. *Rev. Mex. Astron. Astrofis.* 8:147–154.

Troland, T. H., Crutcher, R. M., Goodman, A. A., Heiles, C., Kazes, I., and Myers, P. C. 1996. The magnetic fields in the Ophiuchus and Taurus molecular clouds. *Astrophys. J.* 471:302–307.

Vázquez, E. C., and Scalo, J. 1989. Evolution of the star formation rate in galaxies with increasing densities. *Astrophys. J.* 343:644–658.

Vázquez-Semadeni, E. 1994. Hierarchical structure in nearly pressureless flows as a consequence of self-similar statistics. *Astrophys. J.* 423:681–692.

Vázquez-Semadeni, E. 1999. Turbulence in molecular clouds. In *Millimeter and Submillimeter Astronomy: Chemistry and Physics in Molecular Clouds*, ed. W. F. Wall, A. Carramiñana, L. Carrasco, and P. F. Goldsmith (Dordrecht: Kluwer), pp. 161–186.

Vázquez-Semadeni, E., and Gazol, A. 1995. Gravitational instability in turbulent, non-uniform media. *Astron. Astrophys.* 303:204–210.

Vázquez-Semadeni, E., and Passot, T. 1999. Turbulence as an organizing agent in the ISM. In *Interstellar Turbulence, Proceedings of the 2nd Guillermo Haro Conference*, ed. J. Franco and A. Carramiñana (Cambridge: Cambridge University Press), pp. 223–231.

Vázquez-Semadeni, E., Passot, T., and Pouquet, A. 1995. A turbulent model for the interstellar medium. I. Threshold star formation and self-gravity. *Astrophys. J.* 441:702–725.

Vázquez-Semadeni, E., Passot, T., and Pouquet, A. 1996. Influence of cooling-induced compressibility on the structure of turbulent flows and gravitational collapse. *Astrophys. J.* 473:881–893.

Vázquez-Semadeni, E., Ballesteros-Paredes, J., and Rodríguez, L. F. 1997. A search for Larson-type relations in numerical simulations of the ISM. Evidence for non-constant column densities. *Astrophys. J.* 474:292–307.

Vázquez-Semadeni, E., Cantó, J., and Lizano, S. 1998. Does turbulent pressure behave as a logatrope? *Astrophys. J.* 492:596–602.

Vishniac, E. T. 1983. The dynamic and gravitational instabilities of spherical shocks. *Astrophys. J.* 274:152–167.

Vishniac, E. T. 1994. Nonlinear instabilities in shock-bounded slabs. *Astrophys. J.* 428:186–208.

Wannier, P. G., Lichten, S. M., Andersson, B.-G., and Morris, M. 1991. Warm neutral halos around molecular clouds. II. H I and CO (J = 1–0) observations. *Astrophys. J. Suppl.* 75:987–998.

Whitworth, A. 1996. The structure of the neutral interstellar medium: A theory of interstellar turbulence. In *Unsolved Problems of the Milky Way*, ed. L. Blitz and P. Teuben (Dordrecht:Kluwer) pp. 591–596.

Williams, J. P., Blitz, L., and Stark, A. A. 1995. The density structure in the Rosette molecular cloud: Signposts of evolution. *Astrophys. J.* 451: 252–274.
Wilson, T. L., and Walmsley, C. M. 1989. Small-scale clumping in molecular clouds. *Astron. Astrophys. Rev.* 1:141-176.
Wolfire, M. G., Hollenbach, D., McKee, C. F., Tielens, A. G. G. M., and Bakes, E. L. O. 1995. The neutral atomic phases of the interstellar medium. *Astrophys. J.*, 443:152–168.

CHEMICAL EVOLUTION OF PROTOSTELLAR MATTER

WILLIAM D. LANGER
Jet Propulsion Laboratory

EWINE F. VAN DISHOECK
Leiden Observatory

EDWIN A. BERGIN
Smithsonian Astrophysical Observatory

GEOFFREY A. BLAKE
California Institute of Technology

ALEXANDER G. G. M. TIELENS
Kapteyn Astronomical Institute

THANGASAMY VELUSAMY
Jet Propulsion Laboratory

and

DOUGLAS C. B. WHITTET
Rensselaer Polytechnic Institute

We review the chemical processes that are important in the evolution from a molecular cloud core to a protostellar disk. These cover both gas-phase and gas-grain interactions. The current observational and theoretical status of this field are discussed.

I. INTRODUCTION

The study of chemical evolution from interstellar gas and dust to planetary systems is a key to understanding the pathways and the processes leading to solar origins. The formation of stars and planetary systems begins with the collapse of a dense interstellar cloud core, a reservoir of gas and dust from which a protostar and circumstellar disk are assembled. Throughout these evolutionary stages, simple and complex molecules that are formed deplete onto grains, and a portion of these ices is ultimately incorporated into stars and planetary bodies. There is now substantial observational evidence of chemical evolution throughout the formation and evolution of protostars and protostellar disks. Clues to the degree of chemical processing also come from studies of comets in our own solar system (see the chapter by Irvine et al.). Much of the interest is driven by the inventory of carbon-bearing species that probably formed a steppingstone toward the organic inventory of comets, meteorites, and planetary bodies in the solar system and, by inference, in other planetary systems as well.

The wealth of new observations at various wavelengths show the need for chemical evolutionary models that cover the history of the formation of the protosolar nebula, from the forming dense core, to gas infalling onto and evolving in an accretion disk, onward to the planetary disk. Over the last decade, the following scenario has emerged. In the cold molecular cores prior to star formation, much of the chemistry is dominated by low-temperature gas-phase reactions leading to the formation of small radicals and unsaturated molecules. Small amounts of long carbon chains form at early times if the material is initially atomic-carbon rich. Gas-grain interactions are also important, because observations show that ices form in cold molecular clouds prior to star formation. Surface reactions play an important role in the formation of the ices. However, it is during the collapse phase that the density becomes so high that most molecules accrete onto the grains and are incorporated into an icy mantle. After the new star has formed, its radiation heats up the surrounding envelope, so that the molecules evaporate back into the gas phase, probably in a sequence according to their sublimation temperatures. In addition, the outflows from the young star penetrate the envelope, creating high-temperature shocks and lower-temperature turbulent regions in which both volatile and refractory material can be returned. These freshly evaporated molecules then drive a rich chemistry in the "hot cores" for a period of 10^5 yr. Some of this gas-phase and icy material can be incorporated through accretion into the circumstellar disks surrounding the young star. Finally, the envelope is dispersed by winds, ultraviolet photons, or both, leading to the appearance of photon-dominated regions in the case of high-mass young stars.

In this chapter the chemistry of the prestellar cores, collapse envelopes, and circumstellar disks is briefly discussed. The emphasis lies on a discussion of those species that are most characteristic of a particular evolutionary phase but are not necessarily the dominant species in terms of elemental abundances. More detailed reviews of various aspects include Tielens and Charnley (1997), van Dishoeck and Blake (1998), Hartquist et al. (1998), van Dishoeck (1998*b*), and van Dishoeck and Hogerheijde (1999).

II. OBSERVATIONS OF CORES AND ENVELOPES

A. Gas-phase Species

Nearly 120 different molecules have been detected in the interstellar and circumstellar gas, not counting the different isotopic varieties (see Ohishi 1997 for an overview). Most have been observed through their rotational transitions at (sub)millimeter wavelengths, but species like CH_4 and CO_2 can be measured only through infrared absorption lines. One of the most important detections of the last decade is the H_3^+ ion, a key player in the ion-molecule chemistry networks (Geballe and Oka 1996; McCall et al. 1998). New species continue to be discovered, either through systematic line surveys or through dedicated searches based on laboratory frequencies. For example, the frequency measurements of more than thirty new carbon chains in the laboratory (McCarthy et al. 1997) has stimulated deep searches for them in interstellar clouds (Langer et al. 1997).

A large number of different carbon chain molecules (including, e.g., ring chain carbenes $C_{2n+1}H_2$, cumulene carbenes H_2C_n, cyanopolyynes $HC_{2n+1}N$, methylpolyynes $H_2C_{2n+1}N$, methylcyanopolyynes $H_2C_{2n}N$; free radicals C_nH) have been detected in dark cloud cores with typical abundances of 10^{-9} with respect to hydrogen. Laboratory spectra of long carbon chains may reproduce observed features of some Diffuse Interstellar Bands (DIBs) (Kirkwood et al. 1998), and, hence, carbon chains may be ubiquitously present in the diffuse interstellar medium as well. The infrared emission spectrum of interstellar clouds is dominated by the vibrational modes of large polycyclic aromatic hydrocarbon molecules (PAHs), and as a class these species lock up a few percent of the elemental carbon (i.e., an abundance of $\sim 10^{-7}$). Both these classes of species are likely to be pervasive and stable components of the chemistry of protosolar systems, and, indeed, PAHs are known to be abundant in carbonaceous meteorites (Henning and Salama 1998).

1. Prestellar Cores. The cold, dark cloud cores are important sites with which to probe the interstellar chemistry that prevails before the onset of collapse. More than 100 dark cores have been identified through optical extinction and molecular line studies (cf. Benson et al. 1998 and references therein; Turner et al. 1998). The best-studied dark cores in terms of their chemical composition are TMC-1 and L134N. About half of the known interstellar molecules have been detected in TMC-1 (Ohishi et al. 1992; Ohishi and Kaifu 1998). Detailed maps in a number of species have most recently been provided by Pratap et al. (1997). The southern part of TMC-1 is particularly rich in unsaturated, long carbon chains, whereas the northern part is chemically similar to the high-latitude cloud L134N, consisting mostly of simple radicals and ions. In both clouds, significant chemical gradients have been found across the cores, which have been attributed either to age effects (see section III.A) or differential depletion of elements such as O, C, and S.

Figure 1 illustrates the richness of complex species detected in TMC-1, including the cumulene carbene H_2C_6 ($H_2C{=}(C{=})_4C$) (Langer et al. 1997), the carbon chain radical C_8H (Velusamy and Langer 1998), the shorter cumulene carbene chains H_2C_3, H_2C_4, and several other complex molecules. Although the relative abundances of the cumulene carbenes are distributed similarly to those of other complex carbon compounds such as

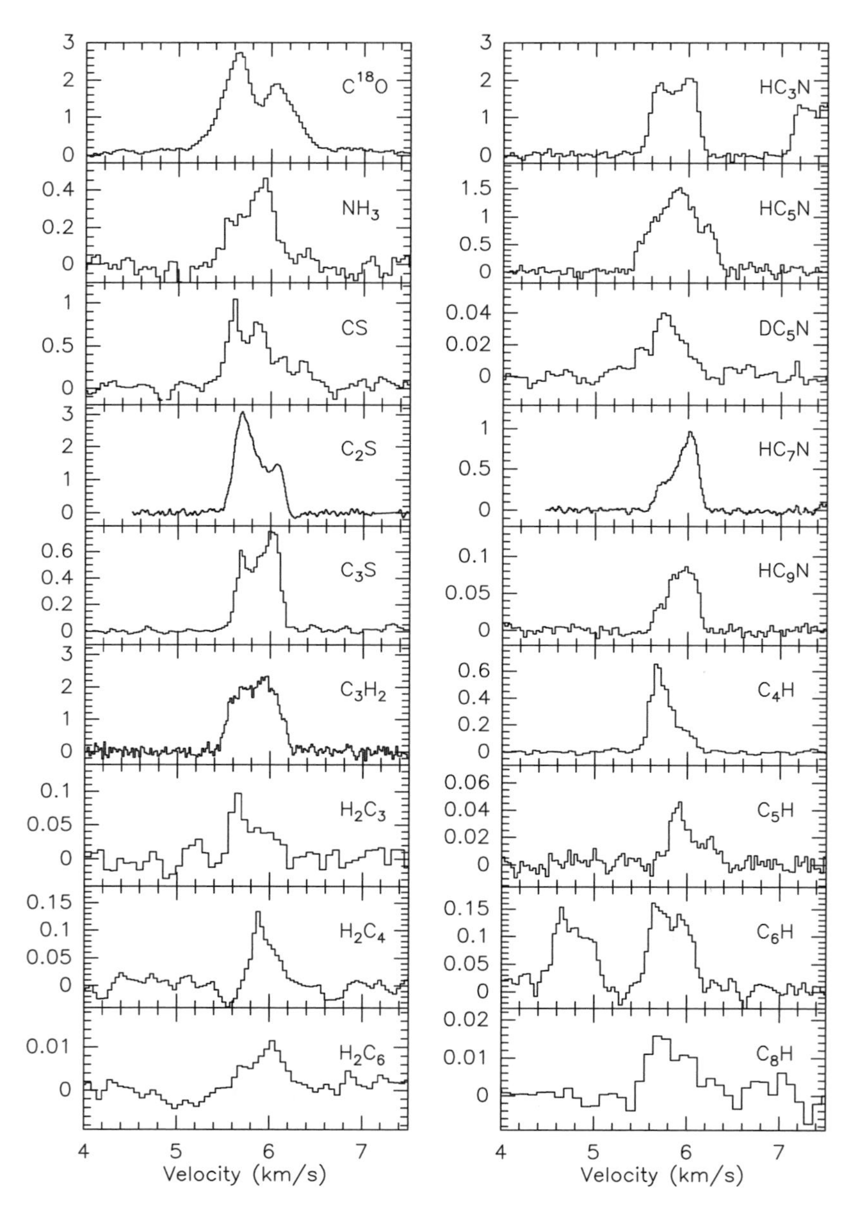

Figure 1. Examples of spectra (antenna temperature, T_A (K), versus frequency, as measured in velocity offset from $V_{\rm lsr}$) of complex molecules taken at the CCS peak emission (Langer et al. 1997; Velusamy and Langer 1998). Note the differences in the intensity ratios among these molecules at different velocities.

$HC_{2n+1}N$ and C_nH, the spectral shapes in Fig. 1 show that there are at least three prestellar fragments along this line of sight with different intensity ratios among these molecules. Indeed very high-spectral-resolution, well-sampled CCS maps confirm that the emission comes from a highly clumped gas, loosely associated with preprotostellar structures (Peng et al. 1998). Thus, chemical differentiation among the velocity components is large, indicating chemical evolution on small spatial scales (10,000 AU) owing, perhaps, to the density and grain-dependent chemical processing. It seems unlikely that fragments formed so close together in relative isolation should have such intrinsically different chemical composition without some chemical evolution, in which some of the fragments have evolved more rapidly because of initially higher density or dynamical factors.

Systematic surveys of characteristic molecules in a larger number of dark cores have been performed by Suzuki et al. (1992) using C_2S, HC_3N, and NH_3, and by Benson et al. (1998) in C_2S, c-C_3H_2, and N_2H^+. In both studies, the carbon chain molecules are found to correlate well with each other, but not with NH_3 nor N_2H^+. The C_2S/NH_3 or C_2S/N_2H^+ abundance ratios have been proposed as indicators of the amount of time that has passed since the gas was atomic-carbon rich, as illustrated in Fig. 2

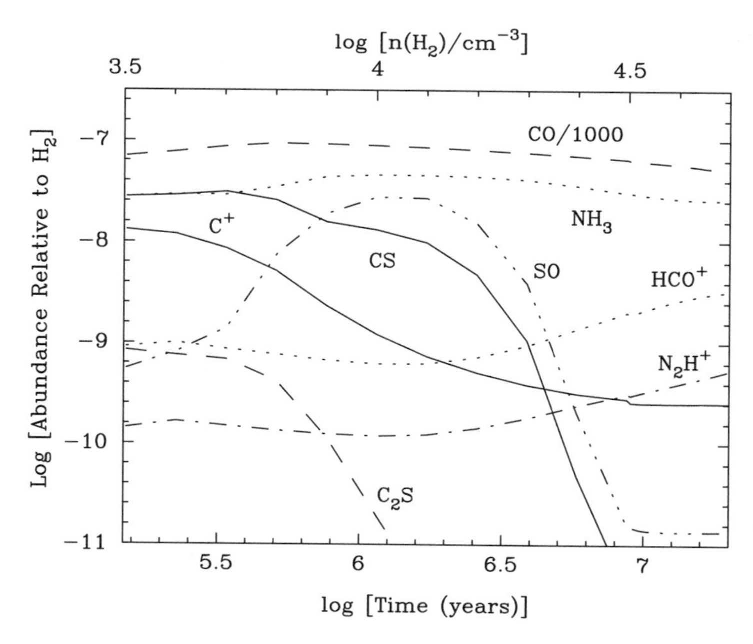

Figure 2. Time-dependent chemistry for the evolution in a collapsing dense core including depletion (based on Bergin and Langer 1997). This shows the "late time" chemistry, in which species such as CS and C_2S deplete onto grains while the nitrogen species such as N_2H^+ and NH_3, as well as HCO^+, survive.

(Bergin and Langer 1997). In this scenario TMC-1 is found to be one of the youngest clouds yet studied.

A long-standing question concerns the amount of depletion in these dark cores. An illustrative example is provided by the quiescent core L1498, studied by Kuiper et al. (1996) (see Color Plate 2). The C_2S, CS, and NH_3 maps show a chemically differentiated structure, with the carbon-rich molecules more abundant in the diffuse outer part and NH_3 dominant in the denser cloud center. Observationally, depletion is very difficult to prove, because every cloud has a "skin" in which the abundances are close to normal (Mundy and McMullin 1997). Even if the skin contains only a few percent of the total column density, the higher abundances can effectively mask any depletions deep inside. The case of L1498 provides some evidence for depletion of CO at the core center, since offsets between the $C^{18}O$ emission (Lemme et al. 1995) and the far-infrared continuum have been found (Willacy et al. 1998*a*). Another interesting example is provided by the dark cloud IC 5146, for which the comparison of the extinction A_V derived from infrared star counts with the $C^{18}O$ emission suggests depletion at $A_V > 20$ magnitudes (Kramer et al. 1998). Direct observations of ices in dark clouds indicate that the column density of solid CO may be comparable to that of gas-phase CO, and that up to 40% of the heavy elements may be condensed onto the grains (Chiar et al. 1998; Schutte 1999).

2. *Cold Collapse Envelopes.* The increasing densities and decreasing temperatures during collapse result in enhanced depletion of gas-phase molecules in the envelopes. The dust obscuration is too high at this stage for direct observations of ices with current instrumentation. Careful modeling of the line and dust emission on various scales appears the only way to probe the abundances in this phase. One of the best-studied cases is that of the deeply embedded young stellar object (YSO) NGC 1333 IRAS 4, where depletions of more than a factor of 10 have been inferred (Blake et al. 1995). Studies of other Class 0 YSOs suggest that this phase of high depletion is short lived, however (Hogerheijde et al. 1999).

Chemical differentiation such as seen toward L1498 is still observed in the outer, less dense parts of the collapsing envelopes. An example is provided by the deeply embedded object B335 (Velusamy et al. 1995), where CCS and other radicals are located predominantly in the outer infall envelope. CS is abundant farther in, while NH_3 is more abundant close to the circumstellar disk. Thus, we can trace the sequence by which these simple species are chemically transformed and freeze out onto dust to form icy grain mantles that populate planet-forming disks.

Systematic observations of molecules other than CO have been performed for only a few deeply embedded YSOs. For NGC 1333 IRAS 4A, only weak emission from other species is found, and most of it is associated with the outflow (Blake et al. 1995). In contrast, a wealth of molecular lines has been detected in the circumbinary envelope of IRAS 16293–2422

(Blake et al. 1994; van Dishoeck et al. 1995), including organics such as CH_3OH and CH_3CN in the warm, inner part of the envelope on scales of a few hundred AU that result from the interaction with the outflow. The optically thin lines of $H^{13}CO^+$ and $H^{13}CN$ trace the dense envelope on scales up to a few thousand AU, whereas the optically thick HCO^+ and HCN outline the walls of the outflow cavity. Apparently, a larger fraction of the envelope has been affected for IRAS 16293–2422 than for NGC 1333 IRAS 4A. A similar trend is found for three deeply embedded YSOs in Serpens (SMM1, SMM3, and SMM4) studied by Hogerheijde et al. (1999).

A systematic high-resolution study of the HCO^+ and CS lines in the envelopes of a set of more evolved, but still embedded low-mass objects in Taurus has been performed by Ohashi et al. (1991, 1996) and Hogerheijde et al. (1997, 1998) using a combination of single-dish and interferometer data. As for the deeply embedded objects, HCO^+ is found to be an excellent tracer of the inner envelope structure and mass, whereas N_2H^+ appears to trace preferentially the quiescent outer envelope (Bachiller and Peréz-Gutierrez 1997; Benson et al. 1998). N_2H^+ may be destroyed by proton transfer to CO and H_2O in regions where these molecules are not significantly condensed onto grains (Bergin et al. 1998). In B5 IRS1, HCN, but not HCO^+, also appears to avoid the inner part of the envelope (Langer et al. 1996), as does CS (Langer and Velusamy 1999). This distribution is consistent with the result of Blake et al. (1992), who found the CS/CO ratio in the inner envelope of HL Tau to be a factor of 25 to 50 less than that of the surrounding dense core.

Detailed chemical studies of the envelopes of high-mass YSOs have been performed for several objects, including W 3 IRS5 (Helmich and van Dishoeck 1997), NGC 2264 IRS1 (Schreyer et al. 1997) and AFGL 2591 (Carr et al. 1995; van der Tak et al. 1999). Most of them show a simple chemistry, with little or no evidence for high abundances of complex organic molecules such as those found in hot cores (Blake et al. 1987; Sutton et al. 1995). These studies are interesting because they can be combined with direct observations of solid-state species for the same line of sight (see section II.B below).

B. Ices

Interstellar ices have been detected in absorption by their vibrational bands in the infrared, either toward background field stars or, more often, embedded young stars (see Whittet 1993; Tielens and Whittet 1997 for overviews). The YSOs heat their surrounding dust to a few hundred K, providing a bright continuum against which the ices in the colder envelope can be seen in absorption. The availability of a grating-resolution spectrometer in space, the Short Wavelength Spectrometer (SWS) carried by the *Infrared Space Observatory* (ISO), represents an advance in our observational capability at least as significant as when such instruments

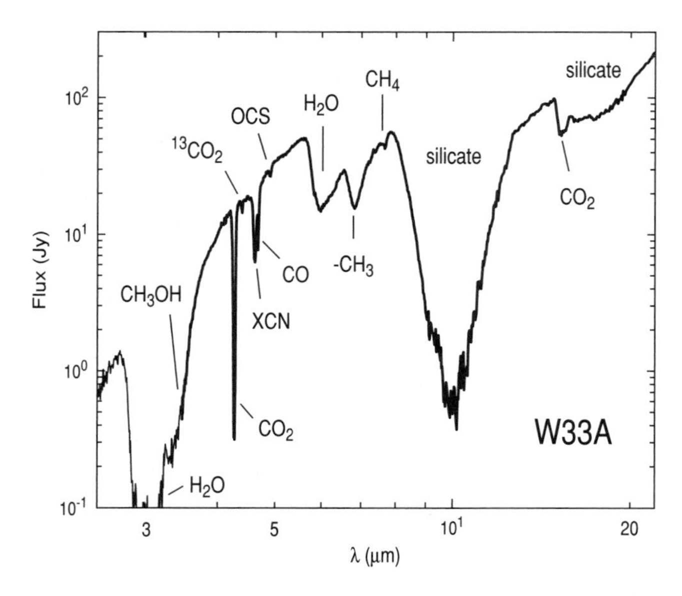

Figure 3. *Infrared Space Observatory* Short Wavelength Spectrometer (ISO-SWS) spectrum of the deeply embedded source W 33A in the wavelength range 2.4–20 μm. The mean spectral resolving power is approximately 1000. Various absorption features arising in silicate dust and icy mantles along the line of sight are labeled.

were first used on ground-based telescopes more than two decades ago, and allows an almost complete inventory of the major constituents of ices and organics.

1. Inventory of Ices. A sample spectrum taken with the ISO-SWS covering the wavelength range 2.4–20 μm at a mean resolving power $\lambda/\Delta\lambda \approx 1000$ toward the well-known embedded massive young star W 33A is shown in Fig. 3. The spectrum is dominated by strong, broad features centered near 3 and 10 μm, attributed to H_2O ice and silicates, the primary constituents of the mantles and cores of the grains, respectively. H_2O ice is represented by its bending mode at 6.0 μm as well as the stretching mode at 3 μm. The spectral signatures of various carbon-bearing molecules are also present in the spectrum. Prominent amongst these are the CO stretching mode at 4.67 μm and the stretching and bending modes of $^{12}CO_2$ at 4.27 and 15 μm, as well as the stretching mode of $^{13}CO_2$ at 4.39 μm. It is notable that signatures of CH-bonded carbon are less in evidence: only CH_3OH and CH_4 are securely identified.

Absorption at 4.62 μm is associated with CN bonds in an unidentified molecule or ion ("XCN"), and a weak feature at 4.9 μm is most likely identified with OCS. Unidentified features include a shallow absorption at 3.47 μm, buried in the "wing" of the deep ice feature, and the more prominent 6.85 μm band; these seem likely to be caused by CH stretching and deformation modes in hydrocarbons (Allamandola et al. 1992; Schutte et al. 1996), but no convincing fits with laboratory analogs have yet been reported. Solid NH_3 is difficult to detect, because the N–H feature at 2.95 μm is effectively blocked in objects such as W 33A, with deep 3.0 μm H_2O ice bands. A tentative detection of the 9.6 μm band has been made by Lacy et al. (1998) toward NGC 7538 IRS9, indicating an abundance of ~10% with respect to H_2O ice. Observations of other lines of sight suggest that the NH_3 abundance is no more than a few percent (Smith et al. 1989; Whittet et al. 1996). From theoretical considerations, O_2 and N_2 are also expected to be abundant, but because these species have no intrinsic dipole moment, they are very hard to detect in the infrared (Ehrenfreund and van Dishoeck 1998; Vandenbussche et al. 1999).

As SWS data reduction techniques continue to be refined, detailed analysis becomes feasible, encompassing also the weaker features of less abundant species in ice-rich sources such as W 33A and NGC 7538 IRS9. Of greatest interest are organic molecules of potential exobiological significance. Tentative evidence for an organic acid (R–COOH, most likely formic acid, HCOOH) has recently been presented (Schutte et al. 1996, 1999), and preliminary analysis indicates that it may be ubiquitous (Keane et al. 1999). Finally, detailed studies of the CH stretch region have placed upper limits on the abundances of ethane (C_2H_6) and ethanol (CH_3CH_2OH) in the mantles (Boudin et al. 1998).

2. Heating and Evaporation of Ices. The sensitivity of the mantle material to environment is studied by comparing observations for different sources. A problem is that both processed and unprocessed material may be present at different locations along the line of sight toward YSOs, so that it is also important to observe background field stars to give a measure of unprocessed mantle properties. Table I compares the ice abundances for five interstellar lines of sight: one quiescent dark cloud and four YSOs. All abundances are normalized to $N(H_2O) = 100$. Elias 16 is a field star behind the Taurus Dark Cloud in a line of sight close to TMC-1, where the kinetic gas temperature is $T_{gas} \approx 10$ K (Pratap et al. 1997). Each of the four YSOs has both cold and warm/hot gas and dust components in the line of sight (Mitchell et al. 1990). Sources are ordered from left to right in Table I in a sequence of declining solid CO abundance. Because CO is the most volatile of the primary constituents of the ices (in pure form it sublimes at ~16 K), its abundance varies inversely with the degree of thermal processing. Estimates of T_{gas} (Table I) are consistent with this conclusion: Solid CO is most abundant in the quiescent cloud and is undetected toward GL2591, where $T_{gas} > 30$ K.

Modeling the solid state features of molecules such as CO and CO_2 allows not only their abundances but also their molecular environment within the mantles to be examined. Distinct "polar" and "nonpolar" ice phases, dominated by H_2O and CO, respectively, have been identified

TABLE I

Abundances of C-bearing Molecules in Interstellar Ices[a]

Species	Dark Cloud	Embedded YSOs				Comets
	Elias 16[b]	NGC7538[b]	W 33A[b]	GL 2136[b]	GL 2591[b]	
CH_3OH	<3	4	22	6	4	0.3–5
CH_4	—	2	2	—	—	0.2–1.2
H_2CO	—	—	—	7	—	0–5
CO (total)	25	16	9	2	<1	5–7
(polar)	3	2	7	2	—	—
(nonpolar)	22	14	2	—	—	—
CO_2 (total)	18	24	14	13	12	3–20
(polar)	18	16	13	10	8	—
(nonpolar)	—	8	1	3	4	—
XCN[c]	<2	2	10	<0.5	—	0.02–0.1
HCOOH	—	3	—	—	—	0.05–0.1
OCS	—	—	0.3	—	—	—
C_2H_6	—	<0.5	—	—	—	0–2
CH_3CH_2OH	—	<1	—	—	—	—
T_{gas} (cold) K	10	26	23	17	38	—
(hot) K	—	180	120	580	1010	—
References[d]	1, 2	1, 3, 4	1, 4	1, 4	1, 4	5, 6, 7, 8

[a] Compared with the typical range seen in comets. All values are expressed as a percentage relative to H_2O ice. Dashes indicate a current lack of information.

[b] Elias 16 is a background star that samples a quiescent dark-cloud environment, whereas NGC 7538 IRS9, W 33A, GL 2136, and GL 2591 are embedded young stellar objects; all five objects are listed in a sequence from left to right of increasing thermal processing (as determined by decreasing solid CO abundance).

[c] XCN is a CN-bonded molecule of uncertain identity; the cometary value for XCN refers to HCN.

[d] References: [1] Whittet (1997) and refs. therein. [2] Whittet et al. (1998). [3] Schutte et al. (1996). [4] Gerakines et al. (1999). [5] Mumma et al. (1993, 1996). [6] Crovisier et al. (1996). [7] Wink et al. (1998). [8] Chapter by Irvine et al., this volume.

(Tielens et al. 1991; Chiar et al. 1995, 1998; de Graauw et al. 1996; Boogert et al. 1999; Gerakines et al. 1999). These ices evidently form under distinctly different conditions. Polar, H_2O-rich mantles accumulate in regions where atomic H has appreciable abundance in the gas, so that accreting CNO-group species typically become hydrogenated; nonpolar mantles accumulate in regions where essentially all the H is molecular and include species such as CO, O_2, and N_2, which may freeze directly out of the gas as well as form by surface catalysis (see section III.B).

The bending mode of $^{12}CO_2$ near 15 μm is an especially sensitive diagnostic of the ice environment and the degree of thermal processing (Ehrenfreund et al. 1996, 1998; Gerakines et al. 1999). SWS observa-

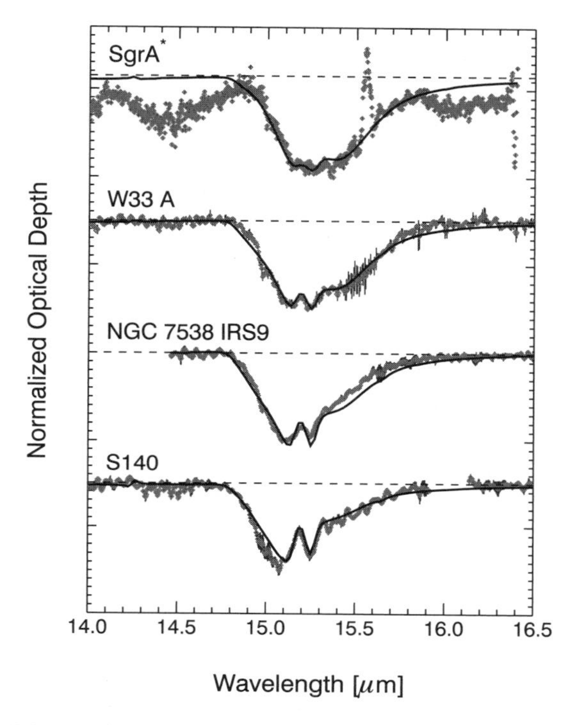

Figure 4. ISO-SWS spectra of the CO_2 bending mode near 15 μm, illustrating thermal evolution of interstellar ices. The sources are displayed in order of increasing gas temperature (T_{gas} = 16 K, 23 K, 26 K, and 28 K for SgrA*, W33A, NGC7538 IRS9, and S140, respectively). Solid lines are fits based on laboratory data for an ice mixture (H_2O:CH_3OH:CO_2 = 1:1:1) at various temperatures (Gerakines et al. 1999). The strengths of the narrow features near 15.12 and 15.25 μm increase systematically with temperature.

tions of four representative sources, ordered in sequence of increasing T_{gas}, are illustrated in Fig. 4; also shown are fits based on laboratory data. In polar ice mixtures, the 15-μm feature is broad [full width at half maximum (FWHM) = 10–25 cm^{-1}] with no substructure; however, sharp peaks (FWHM = 2–4 cm^{-1}) near 15.27 μm and 15.15 μm appear in nonpolar molecules containing CO_2 or in pure annealed CO_2. The systematic evolution of the profile with temperature evident in Fig. 4 provides a strong indication that thermal processing is an important process (perhaps the dominant process) in the environments of embedded YSOs. The presence of a nonpolar CO_2 component only in YSOs suggests that some segregation of mantle constituents (especially H_2O, CH_3OH, and CO_2) might be taking place as the grains are warmed. Further evidence for thermal processing arises from detailed study of the $^{13}CO_2$ stretch feature (Boogert et al. 1999), which shows structure consistent with segregation between CO_2 and polar molecules.

The overall abundances of CO_2 are surprisingly similar (Table I) in contrasting environments. Because CO_2 forms easily in the laboratory by UV photolysis of ice mixtures containing CO, it might be assumed that the CO_2 abundance should measure the degree of radiative processing. The detection of CO_2 ice in the dark cloud environment (Whittet et al. 1998) was thus something of a surprise: CO_2 formation can evidently occur in clouds remote from local embedded sources of radiation. Because CO_2 is located almost exclusively in the polar (H_2O-dominated) component of the ices in dark clouds, it must form simultaneously with H_2O. If surface catalysis is the primary route, a problem for chemical models (section III.B) is to explain why CO_2 is generally much more abundant than CH_3OH or CH_4, given that it must form in a hydrogenating environment. If photolysis is the primary route to CO_2, it must be driven by ambient photons from the interstellar radiation field or by cosmic rays.

An important motivation for the study of interstellar ices is to allow detailed comparison with comets (see Table I). Information on the composition of comets comes both from observations with ground-based and space-based telescopes and from *in situ* studies (see the chapter by Irvine et al. for an in-depth review). Note that the cometary abundances in Table I represent typical values/ranges and do not refer to any individual comet. Although preliminary, these results do suggest that cometary and interstellar abundances are broadly similar. Interstellar and nebular models for cometary origins predict systematically different abundances for many species. The generally low NH_3 and CH_4 abundances measured in comets, together with high CO/CH_4 and CH_3OH/CH_4 ratios and deuterium enrichment (see the chapter by Irvine et al.), are suggestive of a substantial contribution from interstellar ices.

III. CHEMICAL MODELS

A. Gas-phase Chemistry

The basic gas-phase molecular processes and chemical networks developed to explain the observed abundances have been described extensively in the literature (see Herbst 1995; van Dishoeck 1998*a*,*b* for recent overviews). At the low densities of interstellar clouds, only two-body reactions are important. Processes such as radiative association, photodissociation, dissociative recombination, and ion-molecule reactions can form, destroy, and rearrange molecular bonds.

Chemical models of star-forming regions require the specification of the initial chemical and physical conditions, namely density, temperature, radiation field, initial elemental abundances, and even geometry (e.g., Millar et al. 1997; Lee et al. 1996*a*). Of these, the density and temperature can be constrained through observations, while the radiation field is typically parameterized in terms of a "standard" interstellar radiation field (e.g., Habing 1968) attenuated according to the depth (e.g., visual extinction). The initial elemental abundances are based on observations of diffuse clouds such as ζ Oph (see Meyer 1997), but often additional depletion is assumed for dark clouds.

One of the principal aims of the modeling efforts is to develop sets of tracers that can be used to discriminate between various stages of dense core evolution. For pure gas-phase chemical models the main time-dependent aspects depend on how the various elemental pools (C, O, N, S, etc.) interact, both internally and with each other. The main driver of the chemistry starts with cosmic-ray ionization of the dominant molecular species, H_2, with an estimated rate of $\zeta_{H_2} \approx (1\text{–}5) \times 10^{-17}$ s^{-1} (Lepp 1992; van Dishoeck and Black 1986; Plume et al. 1998). Ionization of H_2 produces H_2^+, which reacts rapidly with H_2, forming H_3^+ and H. The H_3^+ ion then reacts with numerous atoms and molecules, leading to a rich chemistry.

The most important transition in the chemistry is the eventual production of CO from neutral and ionized atomic carbon. If dark cores are assumed to evolve from diffuse gas, the carbon is initially present as C^+. Because the $C^+ + H_2 \rightarrow CH^+ + H$ reaction is endothermic, C^+ can recombine with electrons to form neutral carbon within a few hundred years. C then reacts with H_3^+, OH, and O_2 to produce CO within 10^5–10^6 years at typical cloud densities of 10^4–10^5 cm^{-3}. This multistage process is quite important, because the creation of more complex organic species and carbon chains (e.g., CCS, HC_3N) requires large abundances of C and C^+ for carbon insertion reactions to be effective. Thus, the concentrations of complex molecules tend to peak at early times ($\sim 10^5$ years) and decline as equilibrium is approached (see Fig. 2). C^+ also reacts slowly via a radiative association reaction, $C^+ + H_2 \rightarrow CH_2^+ + h\nu$, providing another molecular formation pathway during the neutralization phase. It competes with carbon recombination only after the electron fractional abundance has decreased by one to two orders of magnitude.

The main time-dependent aspects of the other elemental pools, O, N, and S, depend on how their chemistry is linked to that of carbon. The oxygen chemistry begins with $H_3^+ + O \rightarrow OH^+ \rightarrow H_2O^+ \rightarrow H_3O^+$, where the latter dissociatively recombines to form OH, H_2O, and O with branching ratios that have only recently been measured in the laboratory (Williams et al. 1996; Vejby-Christensen et al. 1997). O_2 is produced through reaction of O and OH. The abundances of these simple oxygen-bearing molecules are low at early times because OH, O_2, and H_2O react with C and C^+. The primary N-bearing molecules (N_2, NH_3) form via simple ion-molecule and neutral-neutral reactions and have slow, steady buildup in concentration until equilibrium is reached. The most abundant sulfur-bearing molecules, CS and SO, exhibit slight evolutionary differences, because SO is produced mainly through a reaction of S with OH and O_2 and is destroyed by atomic carbon to produce CS. Thus, the chemistry of CS is linked to the carbon network, which leads to slightly higher abundances at earlier times compared to steady state, whereas SO has large concentrations in equilibrium. The reaction of SO with OH is expected to produce SO_2. However, the observed SO_2 abundance is generally low in dark clouds, indicating an incomplete understanding of the sulfur chemistry and the depletion of sulfur (Palumbo et al. 1997).

The above scenario leads to a dichotomy in the chemistry:

1. Species linked to the carbon chemistry, such as CS, CN, HCN, complex carbon chains, and organics, have larger abundances at earlier evolutionary stages.
2. Molecules that are independent or destructively linked to the carbon chemistry such as N_2, NH_3, N_2H^+, and SO exhibit higher concentrations near equilibrium.

It is these aspects of the chemistry that have been used to interpret observations of molecular clouds in terms of evolution in sources such as TMC-1 or L134N (Lee et al. 1996*a,b*; Millar et al. 1997; Pratap et al. 1997; Taylor et al. 1998). In the case of L1498 the chemical structure (Color Plate 2) reflects this dichotomy and is very well explained by the time-dependent models of Bergin and Langer (1997). Note, however, that this "chemical age" does not necessarily reflect the age of the core since its formation from the diffuse gas; it basically measures the time since a dynamical event destroyed molecules and effectively reset the chemical clock (e.g., Langer et al. 1995, Bergin et al. 1997). Other dynamical processes, such as turbulence (Xie et al. 1995) or outflows (Norman and Silk 1989) can also bring fresh atomic carbon inside the cores and lead to a "young" appearance.

B. Grain Chemistry

Four processes contribute to the composition and evolution of the observed interstellar ices in molecular clouds: surface chemistry of species accreted on interstellar grains, thermal processing of ice mantles driven by nearby newly formed stars, energetic processing of ices by far ultraviolet (FUV) photons or particle bombardment, and depletion of molecules produced in the gas phase and in shocks. Evidence for the importance of each of these has been claimed in the observational characteristics of interstellar ice mantles (see, e.g., Tielens and Whittet 1997 and Schutte 1999 for reviews).

1. Grain Surface Chemistry. Grain surface chemistry is dominated by hydrogenation and oxidation reactions of simple species accreted from the gas phase. While many of these reactions have activation barriers, they may still proceed on grain surfaces in view of the long timescale ($\sim$1 day) available for reaction between two reactive species on a grain surface before another reactive species is accreted from the gas phase (Tielens and Hagen 1982). In the gas phase, CO is the dominant C-bearing species. Other C-bearing compounds have observed (or calculated) abundances $\lesssim 10^{-4}$ relative to CO, except perhaps for atomic C, which may be as abundant as 0.01 (Bergin et al. 1997). Hence, the chemistry of CO is particularly relevant for the organic reservoir of interstellar ices.

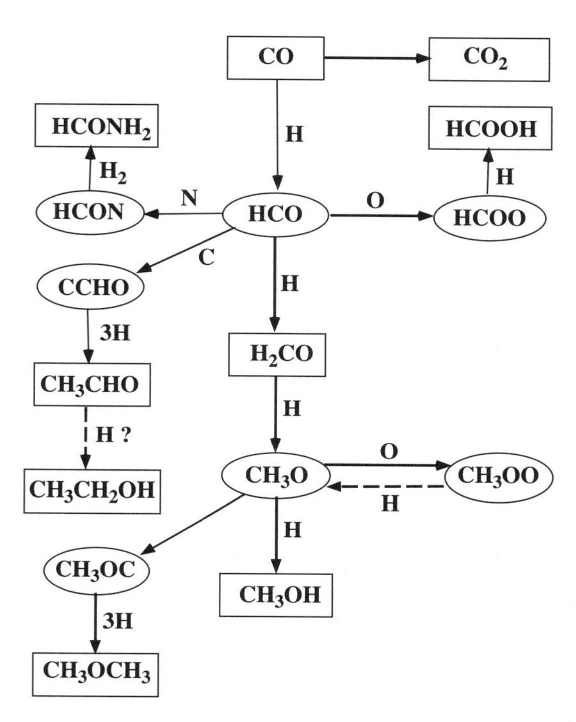

Figure 5. Grain surface chemistry routes involving CO (adapted from Tielens and Hagen 1982). Hydrogenation of CO mainly leads to the formation of H_2CO and CH_3OH. The accreted C, N, and O can also lead to the formation of a variety of complex species whose abundances depend only on the relative accretion rates of these species.

The key step in the surface chemistry of CO is hydrogenation to HCO (Fig. 5), which has a measured activation barrier of 1000 K (Tielens and Hagen 1982), which is known to occur in low-temperature matrices (van IJzendoorn et al. 1982). The resulting radical HCO will react on a grain with H to form H_2CO. Formaldehyde may be unstable to further hydrogenation to form methanol (Tielens and Allamandola 1987), and there is some experimental support for this (Hiraoka et al. 1994, 1998). Because of the high H flux in these experiments, most of the H formed H_2, and the measured efficiency of methanol formation was low. In contrast, CH_3OH is calculated to be the preferred endpoint of the hydrogenation of CO under typical dark cloud conditions (Charnley et al. 1997).

The intermediate HCO (formyl) and CH_3O (methoxy) radicals in this route can also react with various other accreted C, N, and O atoms, leading to a variety of complex species (Fig. 5; Tielens and Hagen 1982), including CH_3CHO (acetaldehyde) and CH_3OCH_3 (dimethyl ether). Like H_2CO, the former may be unstable to hydrogenation, leading to CH_3CH_2OH (ethanol) (Charnley 1998). Reaction of HCO with O results in the formation of HCOOH (formic acid). The reaction of N with HCO will lead to NCHO, which might be further hydrogenated to $HCONH_2$ (formamide; Tielens and Hagen 1982), or isomerize to its more stable form, HNCO (Charnley 1998). Observations of hot cores near massive stars, where the ices have evaporated off the grains, can form significant tests of these models.

The long timescale between the accretion of reactive radicals on grain surfaces may also allow oxidation of CO to occur. Laboratory studies by Tielens and Hagen (unpublished) and Grim and d'Hendecourt (1986) are somewhat contradictory, but the reaction of O with CO may well be the predominant grain surface chemistry route toward CO_2 in dark clouds (Tielens and Hagen 1982).

If atomic O indeed reacts readily with CO on a grain (or if the reaction of O with O_2 dominates), there will always be a reaction partner for atomic O, and the surface reaction of O and H to form H_2O is then blocked. Hydrogenation of O_2 (through H_2O_2) and its oxidation product, O_3 (through OH), are then the dominant routes toward H_2O in theoretical calculations of grain surface chemistry (Tielens and Hagen 1982). In the O_3 route, O_2 plays a catalytic role.

2. Thermal Processing. Thermal processing of interstellar ice mantles is important in star-forming regions, and the experimental data have recently been reviewed by Tielens and Charnley (1997). When an ice mantle is heated to a temperature near its sublimation point, outgassing will occur. Sublimation temperatures of a variety of mixed molecular ices have been measured in the laboratory (Sandford and Allamandola 1990, 1993; Kouchi 1990). In view of the difference in timescales, these sublimation temperatures have to be scaled down somewhat for the interstellar case, and representative values are summarized in Table II. In a mixed molecular ice where two (or more) components have similar concentrations but different sublimation temperatures, the evaporation of each component is largely regulated by its own sublimation behavior (Kouchi 1990; Sandford and Allamandola 1990). Thus, sublimation experiments on $CO/H_2O = 1/2$ and $CO_2/H_2O = 1/1$ ices show early release of CO (around 25 K) and CO_2 (around 85 K). A small fraction of the CO or CO_2 stays behind and is (partly) released during phase transformations in which the whole H_2O ice-crystal structure rearranges. This phase transformation may be associated with the amorphous-to-cubic (or, more properly, clathrate formation) transition of H_2O ice (around 120–130 K in laboratory experiments).

CH_3OH has a sublimation temperature ($\simeq$140 K in the laboratory) above the clathrate formation temperature. Hence, H_2O/CH_3OH mixtures, when heated to $\simeq$130 K, will segregate into rather pure CH_3OH and H_2O-clathrate domains (Blake et al. 1995) whose evaporation behavior

TABLE II
Sublimation Temperatures of Pure Ices[a]

Compound	Laboratory T_{sub} (K)	Dark Cloud Core T_{sub} (K)
N_2	22	13
O_2	22	13
CO	25	16
CH_4	30	18
CO_2	83	50
NH_3	95	55
CH_3OH	140	80
H_2O	150	90

[a] From Nakagawa (1980) and Sandford and Allamandola (1993).

(necessarily) follows that of the pure substance: CH_3OH and H_2O evaporation at 140 and 150 K, respectively, in the laboratory or 80 and 90 K in space. Recent experiments have shown that CH_3OH/CO_2 and $H_2O/CH_3OH/CO_2$ mixtures show similar segregation behavior (Ehrenfreund et al. 1998). The profiles of the interstellar CO_2 bands show evidence for this process around luminous protostars (Boogert et al. 1999; Gerakines et al. 1999, see Fig. 4).

3. Processing by Ultraviolet Photons. UV photolysis of ices containing H_2O, CO, NH_3, and CH_4 produces small radicals such as H, O, OH, N, NH, NH_2, C, CH, CH_2, and CH_3. These react with each other and the parent species, resulting in molecules such as H_2O_2, N_2H_2, N_2H_4, and CO_2 and radicals such as HCO, HO_2, C_2H_3, which are available for further reactions (e.g., Hagen et al. 1980; Gerakines et al. 1996; Bernstein et al. 1995). In mixtures with CO, the HCO channel leads to various interesting complex molecules such as H_2CO, HCOOH, $HCONH_2$, and $HCOCH_3$ (Allamandola 1987). This chemistry is very much akin to the surface hydrogenation scheme for CO, which also runs through HCO and hence gives rise to the same species. For the surface chemistry scheme, the relative abundances of the various end products reflect the relative accretion rates of the migrating radicals H, C, N, and O (Tielens and Hagen 1982). In the photolysis scheme, this is convolved with the UV destruction rate of the parent species, which can vary by an order of magnitude (Hagen et al. 1980; Gerakines et al. 1996). No quantitative comparison between these different formation routes has yet been made.

Mixtures containing CH_3OH (and H_2CO) are of particular interest for molecular complexity. UV photolysis of methanol leads to formaldehyde, which can thermally polymerize, particularly in the presence of NH_3, to polyoxymethylene (POM; see above; Schutte et al. 1993; Bernstein et al. 1995). In this process, many of the H atoms may be replaced by other functional groups. Furthermore, reactions of simple radicals such as CH_3 with the double-bonded O in formaldehyde can lead to products such as CH_3OCH_3. Similar reactions involving CO_2, another important (second-order) product of methanol photolysis, will lead to species such as CH_3OCHO. The presence of NH_3 in these mixtures has important consequences. Besides facilitating thermal polymerization of H_2CO (Schutte et al. 1993), photolysis will lead to efficient carbon-nitrogen bond formation (Bernstein et al. 1995). Analysis of residues shows that hexamethylenetetramine (HMT) is the dominant product of the photolysis of mixtures containing CH_3OH and NH_3, locking up as much as 40% of the initial nitrogen. For comparison, POM-like species lock up less than 1% of the carbon initially in methanol. Other compounds in the residue include ethers, ketones, and amides, which lock up about 5% of the C.

C. Gas-Grain Models

Both the gas-phase and grain-surface processes have been incorporated into chemical models that take the exchange between the two phases into account. For gas densities of more than 10^4 cm^{-3}, all the gas-phase molecules (except H_2, H_3^+, CO, N_2, and a few other weakly polar molecules) should accrete onto grain surfaces in 10^5 yr, a timescale short compared to the ages of molecular clouds but comparable to the timescales for protostar formation. At the densities characteristic of disk material, $>10^6$ cm^{-3}, the depletion rate is even faster, because it scales inversely with density. Chemical history may be even more complicated, because molecules that freeze onto grain surfaces can be reinjected into the gas phase by the desorption of grain mantles (e.g., Willacy and Williams 1993; Shalabiea and Greenberg 1994; Bergin and Langer 1997). Each of these scenarios makes specific predictions regarding the chemistry in the gas phase and frozen icy grain mantles in the disk accretion and early evolutionary phases, which eventually determines the makeup of solar system objects.

Two different regimes have been considered in the gas-grain models (Tielens and Charnley 1997). In the "accretion-limited" regime (Tielens and Hagen 1982), reactions are limited by the rate at which species are "delivered" to the grain surface, where they rapidly migrate and react. In the "reaction-limited" regime (e.g., Hasegawa and Herbst 1993*a*,*b*), the opposite holds: Many reactive species are present on a grain surface, and reaction is controlled by surface concentrations (as well as kinetic parameters). Most of the chemical models that include gas-grain interactions have been formulated in the "reaction-limited" regime using rate equations that mirror those used for gas-phase chemistry (e.g., Hasegawa and Herbst 1993 *a*,*b*; Shalabiea and Greenberg 1995). This approach is accurate only when a large number of reactive species exist on a single grain surface, because only average abundances are calculated. That condition is usually not met in interstellar clouds, because the accretion times are long, grains are small, and reactions are fast, so at most one reactive species is present on a grain at any time. The grain surface chemistry is therefore in the accretion-limited regime and can be properly treated only by a Monte Carlo method that determines the likelihood of two such species arriving from the gas in succession onto a particular grain in a steady-state model. Recently, an *ad hoc* reformulation of the reaction-limited approach has been proposed by Caselli et al. (1998) to remedy its shortcomings, and the consequences for the models have been discussed by Shalabiea et al. (1998). So far, no Monte Carlo method has been developed for use in time-dependent codes, because of computational restrictions.

In addition to the thermal desorption discussed above, other nonthermal desorption processes involving cosmic-ray spot heating, heating by the energy liberated by molecule formation, or photodesorption by ultraviolet or infrared radiation. Grain-grain collisions may play a role if stored radicals, produced for example by ultraviolet photolysis, are present in the ice (see Schutte 1996 for overview). These processes are effective only for volatile molecules, not for H_2O- and CH_3OH-rich ices, which contain strong hydrogen bonds. The outcome of the models then depends strongly on the adopted desorption mechanisms. For example, if the nonthermal mechanisms are assumed to be ineffective, N_2 is one of the few heavy species that is not significantly condensed onto grains in cold cores with $T \approx 10$ K, leading to significant N_2H^+ and NH_3 (see Fig. 2 and Color Plate 2).

Models appropriate for the cold outer envelopes, which include collapse dynamics, have been presented by Rawlings et al. (1992), Willacy et al. (1994), Shalabiea and Greenberg (1995), and Bergin and Langer (1997). The ions HCO^+ and N_2H^+ are good tracers of the envelopes, because their abundances remain high owing to the increase in the H_3^+ abundance when its main removal partners (CO, O, . . .) are depleted. Because CO and H_2O are also the main removal partners of N_2H^+, this ion avoids the warmer inner envelopes, where these molecules have been returned to the gas phase. In contrast, HCO^+ is removed primarily by electrons, making it a better tracer of the inner envelope, consistent with observations (Hogerheijde et al. 1997). Studies of the temperature structure and chemistry in the warmer, inner envelopes have been presented by Ceccarelli et al. (1996) and Doty and Neufeld (1997).

Gas-grain interactions and the formation of ices can also be significant in regions that have recently been subjected to shocks due to outflows from young stars. Bergin et al. (1998) predict that large quantities of water vapor are produced in shocked gas, as supported by ISO observations of strong emission from gaseous water toward Orion BN-KL (Harwit et al. 1998). In the cold postshock gas the abundant H_2O vapor will condense onto the dust grain surface, producing a water-dominated mantle. This mechanism, which requires some shielding from photodissociating photons, may also be able to reproduce the observed deuterium fractionation of water in comets. Further studies of species such as CH_3OH are needed to test the importance of this mechanism compared with other processes.

IV. CIRCUMSTELLAR DISKS

As shown in the previous sections, substantial variations in chemical abundances occur as the circumstellar envelope evolves. What happens to these species when they are incorporated into the dense circumstellar disks? With the increased sensitivity and resolution of telescopes, it is now also possible to probe the chemistry in the disks themselves on scales of a couple of hundred AU.

A. Observations

The best objects for study are YSOs that have already dispersed their envelopes and cloud cores, so that no confusion with the surrounding material is possible. Excellent examples are provided by DM Tau and GG Tau (Dutrey et al. 1997) and TW Hya (Kastner et al. 1997), where molecules such as CN, HCN, HNC, CS, HCO^+, C_2H, and H_2CO have been detected with single-dish telescopes. These molecules appear underabundant with respect to the standard prestellar cores, probably because of depletion of gas-phase molecules onto the surfaces of cold dust grains. An issue here is whether the molecules are passively stored on the grains during this period or whether grain surface chemistry modifies their composition.

The available single-dish data are suggestive of a scenario in which the distance at which depletion occurs is species-dependent, with the most volatile species remaining in the gas phase farthest from the central star. Further characterization of the chemistry in circumstellar disks would benefit from a combination of high spectral- and spatial-resolution observations. Current millimeter-wave interferometric arrays are now capable of imaging the more abundant molecules in circumstellar disks, as is illustrated by the observations of the ^{13}CO, CN, HCO^+, and HCN $1 \rightarrow 0$ transitions toward the T Tauri star LkCa 15, presented in Fig. 6 (Qi et al. 1999). Even at 2–3$''$ resolution (or a disk diameter of ~300 AU) there appear to be interesting morphological differences between species expected to follow different (photo)chemical paths, such as ^{13}CO and CN. The HCO^+ emission provides an important lower bound to the gas fractional ionization of a few $\times 10^{-10}$, while the ratio of the HCN, HCO^+, and ^{13}CO $1 \rightarrow 0$ emission to that of higher-J transitions detected with submillimeter telescopes indicates low kinetic temperatures of less than ~30 K. Future observations of disks around T Tauri and Herbig Ae stars at even higher spatial resolution will be able to measure directly any gas-phase chemical gradients that may exist in the icy planetesimal formation zone of circumstellar disks. In favorable cases, the millimeter-wave images may also be combined with infrared spectroscopy to constrain the icy budget.

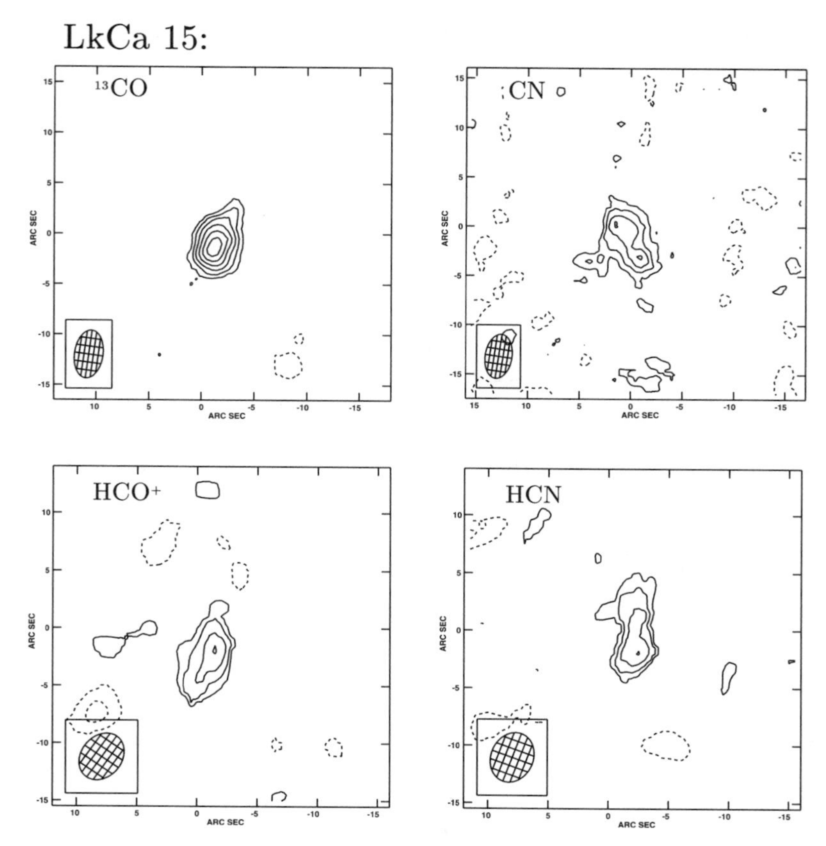

Figure 6. Owens Valley Radio Observatory (OVRO) Millimeter Array observations of the ^{13}CO, CN, HCO$^+$, and HCN 1–0 emission from LkCa 15 (Qi et al. 1999). The synthesized beam is depicted in the lower left-hand corner of each panel and is approximately $4.5'' \times 3''$ ($\sim$700 $\times$ 500 AU). Contour levels are spaced by roughly 0.6, 0.6, 0.35, and 0.4 K km s^{-1}, respectively, and begin at twice this level. In each case, the emission is centered on, or very close to, the position of the star, as measured by the thermal continuum emission from dust. The velocity ranges measured for the observed emission are similar to or less than those of CO (2–1) and arise from gas bound to the star.

B. Chemistry of Disks

Densities in the outer disk are typically greater than 10^6 cm^{-3} (they rise to 10^9 cm^{-3} at 100 AU), and temperatures are around 10 K, so depletion takes place in a short time, $<10^4$ years. The timescale for desorption by cosmic ray spot heating is about 10^6 years, so the chemistry is not able to reach equilibrium. Thus, it is necessary to consider the dynamical chemical evolution as material is transported inward and to consider a variety of desorption mechanisms (cf. Aikawa et al. 1997, 1998). The transport needs to be calculated along with the chemistry because of the changing physical conditions and timescales as material moves inward. Several groups have considered the chemical composition of a parcel of gas transported toward the center of the disk (cf. Bauer et al. 1997; Duschl et al. 1996) from 1000 AU inward. However, their major focus was on the very central regions, where the dust is destroyed (<10 AU), and a very limited set of initial conditions and gas-grain chemistry were considered.

Aikawa et al. (1997, 1998) and Willacy et al. (1998*b*) have calculated more comprehensive chemical and gas-dust models. For example, in the models of Aikawa et al., cosmic ray ionization drives the chemistry beyond 10 AU, because the surface density is too small to attenuate the cosmic rays there. These produce the H_3^+ and He^+ ions that initiate the chemistry and convert CO and N_2, initially the most abundant species, into CO_2, CH_4, NH_3, and HCN. In the cold regions the reaction products are depleted onto grains as part of the ice mantle. As matter accretes toward the inner, warmer regions, the ice mantle evaporates. Therefore, in regions beyond 20 AU, species such as CH_4 are in the form of ice. Inside 20 AU the methane is desorbed and converted into larger hydrocarbons. Clearly, the details of the transport and heating are critical for understanding the chemical distribution in the early stages of disk formation and evolution.

At small disk radii (<100 AU), irradiation by the central star will be significant, and possible sources of grain heating include disk surface heating by stellar photons and stellar X-rays. The impact of heating by stellar visible and UV photons on the vertical temperature structure has yet to be considered in the context of desorption. Chiang and Goldreich (1997) show that even at 100 AU distances, grains at disk surfaces that are irradiated by an (unattenuated) T Tauri star will be heated to temperatures in excess of the disk effective temperature because of the mismatch in the optical absorption and far-infrared emission efficiencies of grains. The high grain temperature (80 K for 1000 Å grains) will likely result in significant CO desorption. Photons absorbed in the surface layer will be reradiated down toward lower layers, producing a potentially significant column density of warm dust. Disk chemistry models are in their infancy, and much work remains to produce realistic 2D disk chemistry models.

V. SUMMARY

A great deal of progress in understanding interstellar chemistry has been made in the last several years since the *Protostars and Planets III* reviews. In large measure this progress has been due to the increased instrumental capabilities of interferometers, the thrust into space from ISO, new laboratory results for grain chemistry, and better theoretical models that couple dynamics and chemistry. We expect further improvements in the coming years with the advent of new larger millimeter arrays, space-based missions such as the *Far Infrared Submillimeter Telescope* (FIRST) and the *Next Generation Space Telescope,* and the advent of more sensitive ground-based near-IR instruments.

REFERENCES

Aikawa, Y., Umebayashi, T., Nakano, T., and Miyama, S. 1997. Evolution of molecular abundances in protoplanetary disks. *Astrophys. J. Lett.* 486:L51–L54.

Aikawa, Y., Umebayashi, T., Nakano, T., and Miyama, S. 1998. Molecular evolution in planet-forming circumstellar disks. *Faraday Disc.* 109:281–301.

Allamandola, L. J. 1987. Interstellar problems and matrix solutions. *J. Molec. Struct.* 157:255–273.

Allamandola, L. J., Sandford, S. A., Tielens, A. G. G. M., and Herbst, T. 1992. Infrared spectroscopy of dense clouds in the C–H stretch region: Methanol and "diamonds." *Astrophys. J.* 399:134–146.

Bachiller, R., and Peréz-Gutierrez, M. 1997. Shock chemistry in the young bipolar outflow L1157. *Astrophys. J.* 487:L93–L96.

Bauer, I., Finocchi, F., Duschl, W. J., Gail, H.-P., and Schloeder, J. P. 1997. Simulation of chemical reactions and dust destruction in protoplanetary accretion disks. *Astron. Astrophys.* 317:273–289.

Benson, P., Caselli, P., and Myers, P. C. 1998. Dense cores in dark clouds. XI. A survey for N_2H^+, C_3H_2, and CCS. *Astrophys. J.* 506:743–757.

Bergin, E. A., and Langer, W. D. 1997. Chemical evolution in preprotostellar and protostellar cores. *Astrophys. J.* 486:316–328.

Bergin, E. A., Goldsmith, P. F., Snell, R. L., and Langer, W. D. 1997. The chemical composition and evolution of giant molecular cloud cores: A comparison of observation and theory. *Astrophys. J.* 482:285–297.

Bergin, E. A., Neufeld, D. A., and Melnick, G. J. 1998. The postshock chemical lifetimes of outflow tracers and a possible new mechanism to produce water ice mantles. *Astrophys. J.* 499:777–792.

Bernstein, M. P., Sandford, S. A., Allamandola, L. J., Chang, S., and Scharberg, M. A. 1995. Organic compounds produced by photolysis of realistic interstellar and cometary ice analogs containing methanol. *Astrophys. J.* 454:327–344.

Blake, G. A., Sutton, E. C., Masson, C. R., and Phillips, T. G. 1987. Molecular abundances in OMC-1. The chemical composition of interstellar molecular clouds and the influence of massive star formation. *Astrophys. J.* 315:621–645.

Blake, G. A., van Dishoeck, E. F., and Sargent, A. 1992. Chemistry in circumstellar disks—CS toward HL Tauri. *Astrophys. J.* 391:L99–L103.

Blake, G. A., van Dishoeck, E. F., Jansen, D. J., Groesbeck, T., and Mundy, L. G. 1994. Molecular abundances and low-mass star formation. 1: Si- and S-bearing species toward IRAS 16293-2422. *Astrophys. J.* 428:680–692.

Blake, G. A., Sandell, G., van Dishoeck, E. F., Groesbeck, T., Mundy, L. G., and Aspin, C. 1995. A molecular line study of NGC 1333/IRAS 4. *Astrophys. J.* 441:689–701.

Boogert, A. C. A., Ehrenfreund, P., Gerakines, P. A., Tielens, A. G. G. M., Whittet, D. C. B., Schutte, W. A., van Dishoeck, E. F., de Graauw, T., Decin, L., and Prusti, T. 1999. ISO-SWS observations of interstellar solid $^{13}CO_2$: Heated ice and the galactic $^{12}C/^{13}C$ abundance ratio. *Astron. Astrophys.,* in press.

Boudin, N., Schutte, W. A., and Greenberg, J. M. 1998. Constraints on the abundances of various molecules in interstellar ice: Laboratory studies and astrophysical implications. *Astron. Astrophys.* 331:749–753.

Carr, J. S., Evans, N. J., Lacy, J. L., and Zhou, S. 1995. Observation of infrared and radio lines of molecules toward GL 2591 and comparison to physical and chemical models. *Astrophys. J.* 450:667–690.

Caselli, P., Hasegawa, T. I., and Herbst, E. 1998. A proposed modification of the rate equations for reactions on grain surfaces. *Astrophys. J.* 495:309–316.

Ceccarelli, C., Hollenbach, D. J., and Tielens, A. G. G. M. 1996. Far-infrared line emission from collapsing protostellar envelopes. *Astrophys. J.* 471:400–426.

Charnley, S. 1998. The organic chemistry of TMC-1 revisited. In *Formation and Evolution of Solids in Space,* ed. J. M. Greenberg and A. Li (Dordrecht: Kluwer), pp. 131–137.

Charnley, S. B., Tielens, A. G. G. M., and Rodgers, S. D. 1997. Deuterated methanol in the Orion compact ridge. *Astrophys. J. Lett.* 482:L203–L206.

Chiang, E. I., and Goldreich, P. 1997. Spectral energy distributions of T Tauri stars with passive circumstellar disks. *Astrophys. J.* 490:368–376.

Chiar, J. E., Adamson, A. J., Kerr, T. H., and Whittet, D. C. B. 1995. High resolution studies of solid CO in the Taurus dark cloud: Characterizing the ices in quiescent clouds, *Astrophys. J.* 455:234–243.

Chiar, J. E., Gerakines, P. A., Whittet, D. C. B., Pendleton, Y. J., Tielens, A. G. G. M., and Adamson, A. J. 1998. Processing of icy mantles in protostellar environments. *Astrophys. J.* 498:716–727.

Crovisier, J., Brooke, T. Y., Hanner, M. S., Keller, H. U., Lamy, P. L., Altieri, B., Bockelee-Morvan, D., Jorda, L., Leech, K., and Lellouch, E. 1996. The infrared spectrum of comet C/1995 O1 (Hale-Bopp) at 4.6 AU from the Sun. *Astron. Astrophys. Lett.* 315:L385–L388.

de Graauw, T., Whittet, D. C. B., Gerakines, P. A., et al. (32 authors). 1996. SWS observations of solid CO_2 in molecular clouds. *Astron. Astrophys. Lett.* 315:L345–L348.

Doty, S. D., and Neufeld, D. A. 1997. Models for dense molecular cloud cores. *Astrophys. J.* 489:122–142.

Duschl, W. J., Gail, H.-P., and Tscharnuter, W. M. 1996. Destruction processes for dust in protoplanetary accretion disks. *Astron. Astrophys.* 324:624–642.

Dutrey, A., Guilloteau, S., and Guelin, M. 1997. Chemistry of protosolar-like nebulae: The molecular content of the DM Tau and GG Tau disks. *Astron. Astrophys. Lett.* 317:L55–L58.

Ehrenfreund, P., and van Dishoeck, E. F. 1998. The search for solid O_2 and N_2 with ISO. *Adv. Space Res.* 21:15–20.

Ehrenfreund, P., Boogert, A. C. A., Gerakines, P. A., Jansen, D. J., Schutte, W. A., Tielens, A. G. G. M., and van Dishoeck, E. F. 1996. A laboratory database of solid CO and CO_2 for ISO. *Astron. Astrophys. Lett.* 315:L341–L344.

Ehrenfreund, P., Dartois, E., Demyk, K., and d'Hendecourt, L. 1998. Ice segregation toward massive protostars. *Astron. Astrophys. Lett.* 339:L17–L20.

Elias, J. H. 1978. A study of the Taurus dark cloud complex. *Astrophys. J.* 224:857–872.

Geballe, T. R., and Oka, T. 1996. Detection of H_3^+ in interstellar space. *Nature* 384:334–335.

Gerakines, P. A., Schutte, W. A., and Ehrenfreund, P. 1996. Ultraviolet processing of interstellar ice analogs. I. Pure ices. *Astron. Astrophys.* 312:289–305.

Gerakines, P. A., Whittet, D. C. B., Ehrenfreund, P., Boogert, A. C. A., Tielens, A. G. G. M., Schutte, W. A., Chiar, J. E., van Dishoeck, E. F., Prusti, T., Helmich, F. P., and de Graauw, T. 1999. ISO-SWS observations of solid carbon dioxide in molecular clouds. *Astrophys. J.,* in press.

Grim, R. J. A., and d'Hendecourt, L. B. 1986. Time-dependent chemistry in dense molecular clouds. IV. Interstellar grain surface reactions inferred from a matrix isolation study. *Astron. Astrophys.* 167:161–165.

Habing, H. J. 1968. The interstellar radiation density between 912 Å and 2400 Å. *Bull. Astron. Inst. Netherlands* 19:421–431.

Hagen, W., Allamandola, L. J., and Greenberg, J. M. 1980. Infrared absorption lines by molecules in grain mantles. *Astron. Astrophys.* 86:L1–L2.

Hartquist, T. W., Caselli, P., Rawlings, J. M. C., Ruffle, D. P., and Williams, D. A. 1998. The chemistry of star-forming regions. In *The Molecular Astrophysics of Stars and Galaxies,* ed. T. W. Hartquist and D. A. Williams (Oxford: Oxford University), pp. 101–137.

Harwit, M., Neufeld, D. A., Melnick, G. J., and Kaufman, M. J. 1998. Thermal water vapor emission from shocked regions in Orion. *Astrophys. J. Lett.* 497:L105–L108.

Hasegawa, T. I., and Herbst, E. 1993*a*. New gas-grain chemical models of quiescent dense interstellar clouds: The effects of H_2 tunneling reactions and cosmic ray induced desorption. *Mon. Not. Roy. Astron. Soc.* 261:83–102.

Hasegawa, T. I., and Herbst, E. 1993*b*. Three-phase chemical models of dense interstellar clouds: Gas dust particle mantles and dust particle surfaces. *Mon. Not. Roy. Astron. Soc.* 263:589–606.

Helmich, F. P., and van Dishoeck, E. F. 1997. Physical and chemical variations within the W3 star-forming region. II. The 345 GHz spectral line survey. *Astron. Astrophys. Suppl.* 124:205–253.

Henning, Th., and Salama, F. 1998. Carbon in the universe. *Science* 282:2204–2210.

Herbst, E. 1995. Chemistry in the interstellar medium. *Ann. Rev. Phys. Chem.* 46:27–53.

Hiraoka, K., Ohashi, N., Kihara, Y., et al. 1994. Formation of formaldehyde and methanol from the reactions of H atoms with solid CO at 10–20 K. *Chem. Phys. Lett.* 229:408–414.

Hiraoka, K., Miyagoshi, T., Takayama, T., Yamamoto, K., and Kihara, Y. 1998. Gas-grain processes for the formation of CH_4 and H_2O: Reactions of H–atoms with C, O, and CO in the solid phase at 12 K. *Astrophys. J.* 498:710–715.

Hogerheijde, M. R., van Dishoeck, E. F., Blake, G.A., and van Langevelde, H. J. 1997. Tracing the envelopes around embedded low-mass young stellar objects with HCO^+ and millimeter-continuum observations. *Astrophys. J.* 489:293–313.

Hogerheijde, M. R., van Dishoeck, E. F., Blake, G. A., and van Langevelde, H. J. 1998. Envelope structure on 700 AU scales and the molecular outflows of low-mass young stellar objects. *Astrophys. J.* 502:315–336.

Hogerheijde, M. R., van Dishoeck, E. F., Salverda, J., and Blake, G. A. 1999. Envelope structure of deeply embedded young stellar objects in the Serpens molecular cloud. *Astrophys. J.,* in press.

Kastner, J. H., Zuckerman, B., Weintraub, D. A., and Forveille, T. 1997. X-ray and molecular emission from the nearest region of recent star formation. *Science* 277:67–71.

Keane, J., Tielens, A. G. G. M., Boogert, A. C. A., and Whittet, D. C. B. 1999. *Astron. Astrophys.,* in preparation.

Kirkwood, D. A., Linnartz, H., Grutter, M. et al. 1998. Electronic spectroscopy of carbon chains and relevance to astrophysics. *Faraday Disc.,* 109:109–199.

Kouchi, A. 1990. Evaporation of H_2O-CO ice and its astrophysical implications. *J. Crystal Growth* 99:1220.
Kramer, C., Alves, J., Lada, C., Lada, E., Sievers, A., Ungerechts, H., and Walmsley, M. 1998. The millimeter wavelength emissivity in IC5146. *Astron. Astrophys. Lett.* 329:L33–L36.
Kuiper, T. B. H., Langer, W. D., and Velusamy, T. 1996. Evolutionary status of the pre-protostellar core L1498. *Astrophys. J.* 468:761–773.
Lacy, J. H., Faraji, H., Sandford, S. A., and Allamandola, L. J. 1998. Unraveling the 10 micron "silicate" feature of protostars: The detection of frozen interstellar ammonia. *Astrophys. J. Lett.* 501:L105–L109.
Langer, W. D., and Velusamy, T. 1999. CS Observations of B5 IRS1. In preparation.
Langer, W. D., Velusamy, T., Kuiper, T. B. H., Levin, S., Olsen, E., and Migenes, V. 1995. Study of structure and small-scale fragmentation in TMC-1. *Astrophys. J.* 453:293–307.
Langer, W. D., Velusamy, T., and Xie, T. 1996. The IRS 1 circumstellar disk, and the origin of the jet and CO outflow in B5. *Astrophys. J. Lett.* 468:L41–L44.
Langer, W. D., Velusamy, T., Kuiper, T. B. H., Peng, R., McCarthy, M. C., Travers, M. J., Kovacs, A., Gottlieb, C. A., and Thaddeus, P. 1997. First astronomical detection of the cumulene carbon chain molecule H_2C_6 in TMC-1. *Astrophys. J. Lett.,* 480: L63–L66.
Lee H.-H., Bettens, R. P. A., and Herbst, E. 1996*a*. Fractional abundances of molecules in dense interstellar clouds: A compendium of recent model results. *Astron. Astrophys. Suppl.* 119:111–114.
Lee H.-H., Herbst, E., Pineau des Forêts, G., Roueff, E., and Le Bourlot, J. 1996*b*. Photodissociation of H_2 and CO and time dependent chemistry in inhomogeneous interstellar clouds. *Astron. Astrophys.* 311:690–707.
Lemme, C., Walmsley, C. M., Wilson, T. L., and Muders, D. 1995. A detailed study of an extremely quiescent core: L 1498. *Astron. Astrophys.* 302:509–520.
Lepp, S.. 1992. The cosmic-ray ionization rate. In *Astrochemistry of Cosmic Phenomena,* ed. P. D. Singh (Dordrecht: Kluwer), pp. 471–475.
McCall, B. J., Geballe, T. R., Hinkle, K. H., and Oka, T. 1998. Detection of H_3^+ in the diffuse interstellar medium toward Cygnus OB2 No. 12. *Science* 279:1910–1913.
McCarthy, M. C., Travers, M. J., Kovacs, A., Gottlieb, C. A., and Thaddeus, P. 1997. Eight new carbon chain molecules. *Astrophys. J. Supp.* 113:105–120.
Meyer, D. M. 1997. Optical and ultraviolet observations of diffuse interstellar clouds. In *Molecules in Astrophysics: Probes and Processes,* IAU Symp. 178, ed. E. F. van Dishoeck (Dordrecht: Kluwer), pp. 407–419.
Millar, T. J., Farquhar, P. R. A., and Willacy, K. 1997. The UMIST database for astrochemistry 1995. *Astron. Astrophys. Suppl.* 121:139–185.
Mitchell, G. F., Maillard, J.-P., Allen, M., Beer, R., and Belcourt, K. 1990. Hot and cold gas toward young stellar objects. *Astrophys. J.* 363:554–573.
Mumma, M. J., Weissman, P. R., and Stern, S. A. 1993. Comets and the origin of the solar system: Reading the Rosetta stone. In *Protostars and Planets III,* ed. E. H. Levy and J. Lunine (Tucson: University Arizona Press), pp. 1177–1252.
Mumma, M. J., DiSanti, M. A., Dello Russo, N., Fomenkova, M., Magee-Sauer, K., Kaminski, C. D., and Xie, D. X. 1996. Detection of abundant ethane and methane, along with carbon monoxide and water, in comet C/1996 B2 Hyakutake: Evidence for interstellar origin. *Science* 272:1310–1314.
Mundy, L. G., and McMullin, J. P. 1997. Molecular depletions in cloud cores. In *Molecules in Astrophysics: Probes and Processes,* IAU Symp. 178, ed. E. F. van Dishoeck (Dordrecht: Kluwer), pp. 183–191.

Nakagawa, N. 1980. Interstellar molecules on dust mantles. In *Interstellar Molecules,* ed. B. H. Andrew (Dordrecht: Reidel), pp. 365–366.

Norman, C., and Silk, J. 1989. Clumpy molecular clouds. A dynamic model self-consistently regulated by T-Tauri star formation. *Astrophys. J.* 238:158–174.

Ohashi, N., Kawabe, R., Ishiguro, M., and Hayashi, M. 1991. Observations of 11 protostellar sources in Taurus with Nobeyama millimeter array: Growth of circumstellar disks. *Astron. J.* 102:2054–2065.

Ohashi, N., Hayashi, M., Kawabe, R., and Ishiguro, M. 1996. The Nobeyama millimeter array survey of young stellar objects associated with the Taurus molecular cloud. *Astrophys. J.* 466:317–337.

Ohishi, M. 1997. Observations of hot cores. In *Molecules in Astrophysics: Probes and Processes,* IAU Symp. 178, ed. E. F. van Dishoeck (Dordrecht: Kluwer), pp. 61–74.

Ohishi, M., and Kaifu, N. 1998. Chemical and physical evolution of dark clouds. Molecular spectral line survey toward TMC–1. *Faraday Disc.* 109:205–216.

Ohishi, M., Irvine, W. M., and Kaifu, N. 1992. Molecular abundance variations among and within cold dark molecular clouds. In *Astrochemistry of Cosmic Phenomena,* IAU Symp. 150, ed. P. D. Singh (Dordrecht: Kluwer), pp. 171–177.

Palumbo, M. E., Geballe, T. R., and Tielens, A. G. G. M. 1997. Solid carbonyl sulfide (OCS) in dense molecular clouds. *Astrophys. J.* 479:839–844.

Peng, R., Langer, W. D., Velusamy, T., Kuiper, T. B. H., and Levin, S. 1998. Low-mass clumps in TMC-1: Scaling laws in the small-scale regime. *Astrophys. J.* 497:842–849.

Plume, R., Bergin, E. A., Williams, J. P., and Myers, P. C. 1998. Electron abundances in dense cloud cores. Implications for star formation. *Faraday Disc.* 109: 47–60.

Pratap, P., Dickens, J. E., Snell, R. L., Miralles, M. P., Bergin, E. A., Irvine, W. I., and Schloerb, F. P. 1997. A study of the physics and chemistry of TMC-1. *Astrophys. J.* 486:862–885.

Rawlings, J. M. C., Hartquist, T. W., Menten, K. M., and Williams, D. A. 1992. Direct diagnosis of infall in collapsing protostars. I. The theoretical identification of molecular species with broad velocity distributions. *Mon. Not. Roy. Astron. Soc.* 255:471–485.

Ruffle, D. P., Hartquist, T. W., Taylor, S. D., and Williams, D. A. 1997. Cyanopolyynes as indicators of late-time chemistry and depletion in star-forming regions. *Mon. Not. Roy. Astron.* 291:235–240.

Qi, C., Sargent, A. I., and Blake, G. A. 1999. In preparation.

Sandford, S. A., and Allamandola, L. J. 1990. The physical and infrared spectral properties of CO_2 in astrophysical ice analogs. *Astrophys. J.* 355:357–372.

Sandford, S. A., and Allamandola, L. J. 1993. Condensation and vaporization studies of CH_3OH and NH_3 ices: Major implications for astrochemistry. *Astrophys. J.* 417:815–825.

Schreyer, K., Helmich, F. P., van Dishoeck, E. F., and Henning, T. 1997. A molecular line and infrared study of NGC 2264 IRS 1. *Astron. Astrophys.* 326:347–365.

Schutte, W. A. 1996. Formation and evolution of interstellar icy grain mantles. In *Cosmic Dust Connection,* ed. J. M. Greenberg (Dordrecht: Kluwer), pp. 1–42.

Schutte, W. A. 1999. Laboratory simulation of processes in interstellar ices. In *Formation and Evolution of Solids in Space,* ed. J. M. Greenberg and Al. Li (Dordrecht: Kluwer), pp. 177–201.

Schutte, W. A., Allamandola, L. J., and Sandford, S. A. 1993. Formaldehyde and organic molecule production in astrophysical ices at cryogenic temperatures. *Science* 104:118–137.

Schutte, W. A., Tielens, A. G. G. M., Whittet, D. C. B., Boogert, A. C. A., Ehrenfreund, P., de Graauw, T., Prusti, T., van Dishoeck, E. F., and Wesselius, P. R. 1996. The 6.0 and 6.8 μm absorption features in the spectrum of NGC 7538 IRS9. *Astron. Astrophys. Lett.* 315:L333–L336.

Schutte, W. A., Boogert, A. C. A., Tielens, A. G. G. M., et al. 1999. Weak ice absorption features at 7.24 and 7.41 μm in the spectrum of the obscured young stellar objects W 33A. *Astron. Astrophys.* 343:966–976.

Shalabiea, O. M., and Greenberg, J. M. 1994. Two key processes in dust/gas chemical modelling: Photoprocessing of grain mantles and explosive desorption. *Astron. Astrophys.* 290:266–278.

Shalabiea, O. M., and Greenberg, J. M. 1995. Chemical evolution of free-fall collapsing interstellar clouds: Pseudo and real time dependent models. *Astron. Astrophys.* 303:233–241.

Shalabiea, O. M., Caselli, P., and Herbst, E. 1998. Grain surface chemistry: Modified models. *Astrophys. J.* 502:652–660.

Smith, R. G., Sellgren, K., and Tokunaga, A. T. 1989. Absorption features in the 3 micron spectra of protostars. *Astrophys. J.* 344:413–426.

Suzuki, H., Yamamoto, S., Ohishi, M., Kaifu, N., Ishikawa, S., Hirahara, Y., and Takano, S. 1992. A survey of CCS, HC_3N, HC_5N, and NH_3 toward dark cloud cores and their production chemistry. *Astrophys. J.* 392:551–570.

Sutton, E. C., Peng, R., Danchi, W. C., Jaminet, P. A., Sandell, G., and Russell, A. P. G. 1995. The distribution of molecules in the core of OMC-1. *Astrophys. J. Supp.* 97:455–496.

Taylor, S. D., Morata, O., and Williams, D. A. 1998. The distribution of molecules in star-forming regions. *Astron. Astrophys.* 336:309–314.

Tielens, A. G. G. M., and Allamandola, L. 1987. Composition, structure and chemistry of interstellar dust. In *Interstellar Processes,* ed. D. J. Hollenbach and H. A. Thronson (Dordrecht: Kluwer), pp. 397–469.

Tielens, A. G. G. M., and Charnley, S. B. 1997. Circumstellar and interstellar synthesis of organic molecules. *Origins Life Evol. Biosphere* 27:23–51.

Tielens, A. G. G. M., and Hagen, W. 1982. Model calculations of the molecular composition of interstellar grain mantles. *Astron. Astrophys.* 114:245–260.

Tielens, A. G. G. M., and Whittet, D. C. B. 1997. Ices in star forming regions. In *Molecular Astrophysics: Probes and Processes,* IAU Symp. 178, ed. E. F. van Dishoeck (Dordrecht: Kluwer), pp. 45–60.

Tielens, A. G. G. M., Tokunaga, A. T., Geballe, T. R., and Baas, F. 1991. Interstellar solid CO: Polar and nonpolar interstellar ices. *Astrophys. J.* 381:181–199.

Turner, B. E., Lee, H. H., and Herbst, E. 1998. The physics and chemistry of small translucent molecular clouds. IX. Acetylenic chemistry, *Astrophys. J. Suppl.* 115:91–118.

Vandenbussche, B., Ehrenfreund, P., Boogert, A. C. A., et al. 1999. Constraints on the abundance of solid O_2 in dense clouds from ISO-SWS and ground-based observations. *Astron. Astrophys.,* submitted.

van der Tak, F. F. S., van Dishoeck, E. F., Evans, N. J., Bakker, E., and Blake, G. A. 1999. The impact of the massive young star GL 2591 on its surroundings. *Astrophys. J.,* submitted.

van Dishoeck, E. F. 1998*a*. The chemistry of diffuse and dark interstellar clouds. In *The Molecular Astrophysics of Stars and Galaxies,* ed. T. W. Hartquist and D. A. Williams (Oxford: Oxford University), pp. 53–99.

van Dishoeck, E. F. 1998*b*. What can ISO tell us about gas-grain chemistry? *Faraday Disc.* 109:31–46.

van Dishoeck, E. F., and Black, J. H. 1986. Comprehensive models of diffuse interstellar clouds: Physical conditions and molecular abundances. *Astrophys. J. Suppl.* 66:109–145.
van Dishoeck, E. F., and Blake, G. A. 1998. Chemical evolution of star-forming regions. *Ann. Rev. Astron. Astrophys.* 36:317–368.
van Dishoeck, E. F., and Hogerheijde, M. R. 1999. Models and observations of the chemistry near young stellar objects. In *Physics of Star Formation and Early Stellar Evolution,* ed. C. J. Lada and N. Kylafis (Dordrecht: Kluwer), in press.
van Dishoeck, E. F., Blake, G. A., Jansen, D. J., and Groesbeck, T. 1995. Molecular abundances and low-mass star formation. II. Organic and deuterated species toward IRAS 16293-2422. *Astrophys. J.* 447:760–782.
van IJzendoorn, L. J., Allamandola, L. J., Baas, F., and Greenberg, J. M. 1983. Visible spectroscopy of matrix isolated HCO: The $^2A''\Pi \leftarrow X^2A'$ transition. *J. Chem. Phys.* 78:7019–7028.
Vejby-Christensen, L., Andersen, L. H., Heber, O., Kella, D., Pedersen, H. B., Schmidt, H. T., and Zajfman, D. 1997. Complete branching ratios for the dissociative recombination of H_2O^+, H_3O^+, and CH_3^+. *Astrophys. J.* 483:531–540.
Velusamy, T., and Langer, W. D. 1998. Detection and evolution of complex hydrocarbons in TMC. *Bull. Am. Astron. Soc.* 193: 71.07 (abstract).
Velusamy, T., Kuiper, T. B. H., and Langer, W. D. 1995. CCS observations of the protostellar envelope of B335. *Astrophys. J. Lett.* 451:L75–78.
Whittet, D. C. B. 1993. Observations of molecular ices. In *Dust and Chemistry in Astronomy,* ed. T. J. Millar and D. A. Williams (Cambridge: Cambridge University Press), pp. 9–35.
Whittet, D. C. B., Smith, R. G., Adamson, A. J., Aitken, D. K., Chiar, J. E., Kerr, T. H., Roche, P. F., Smith, C. H., and Wright, C. M. 1996. Interstellar dust absorption features in the infrared spectrum of HH100–IR: Searching for the nitrogen component of the ices. *Astrophys. J.* 458:363–370.
Whittet, D. C. B., Gerakines, P. A., Tielens, A. G. G. M., Adamson, A. J., Boogert, A. C. A., Chiar, J. E., de Graauw, Th., Ehrenfreund, P., Prusti, T., Schutte, W. A., Vandenbussche, B., and van Dishoeck, E. F. 1998. Detection of abundant CO_2 ice in the quiescent dark cloud medium toward Elias 16. *Astrophys. J. Lett.* 498:L159–L163.
Willacy, K., and Williams, D. A. 1993. Desorption processes in molecular clouds: Quasi-steady-state chemistry. *Mon. Not. Roy. Astron. Soc.* 260:635–642.
Willacy, K., Rawlings, J. M. C., and Williams, D. A. 1994. Molecular desorption from dust in star-forming regions. *Mon. Not. Roy. Astron. Soc.* 269:921–927.
Willacy, K., Langer, W. D., and Velusamy, T. 1998*a*. Dust emission and molecular depletion in L1498. *Astrophys. J. Lett.* 507:L171–L175.
Willacy, K., Klahr, H. H., Millar, T. J., and Henning, Th. 1998*b*. Gas and grain chemistry in a protoplanetary disk. *Astron. Astrophys.* 338:995–1005.
Williams, T. L., Adams, N. G., Babcock, L., Herd, C. R., and Geoghegan, M. 1996. Production and loss of the water-related species H_3O^+, H_2O, and OH in dense interstellar clouds. *Mon. Not. Roy. Astron. Soc.* 282:413–420.
Wink, J. E., Bockelee-Morvan, N., Biver, N., Colom, P., Crovisier, J., Gerard, E., Rauer, H., Despois, D., Moreno, R., Paubert, G., Davies, J. K., and Dent, W. R. F. 1998. Detection of formic acid in comet C/1995 O1 Hale-Bopp. *IAU Circ.* No. 6599.
Xie, T. L., Allen, M., and Langer, W. D. 1995. Turbulent diffusion and its effects on the chemistry of molecular clouds. *Astrophys. J.* 440:674–685.

FROM PRESTELLAR CORES TO PROTOSTARS: THE INITIAL CONDITIONS OF STAR FORMATION

PHILIPPE ANDRÉ
CEA/DSM/DAPNIA, Service d'Astrophysique, Saclay, France

DEREK WARD-THOMPSON
University of Wales at Cardiff

and

MARY BARSONY
University of California at Riverside

The last decade has witnessed significant advances in our observational understanding of the earliest stages of low-mass star formation. The advent of sensitive receivers on large radio telescopes such as the James Clerk Maxwell Telescope (JCMT) and Institut de Radio Astronomie Millimétrique (IRAM) 30m telescope has led to the identification of young protostars at the beginning of the main accretion phase ("Class 0" objects) and has made it possible for the first time to probe the inner density structure of precollapse cores. Class 0 objects are characterized by strong, centrally condensed dust continuum emission at submillimeter wavelengths, very little emission shortward of $\sim$10 μm, and powerful jetlike outflows. Direct evidence for gravitational infall has been observed in several of them. They are interpreted as accreting protostars that have not yet accumulated the majority of their final stellar mass. In contrast to protostars, prestellar cores have flat inner density profiles, suggesting that the initial conditions for fast protostellar collapse depart sometimes significantly from a singular isothermal sphere. In the case of nonsingular initial conditions, the beginning of protostellar evolution is expected to feature a brief phase of vigorous accretion and ejection, which may coincide with Class 0 objects. In addition, submillimeter continuum imaging surveys of regions of multiple star formation, such as Ophiuchus and Serpens, suggest a picture according to which each star in an embedded cluster is built from a finite reservoir of mass and the associated initial mass function (IMF) is primarily determined at the prestellar stage of evolution.

I. INTRODUCTION

The formation of low-mass stars is believed to involve a series of conceptually different stages (e.g., Larson 1969; Shu et al. 1987). The first stage corresponds to the fragmentation of a molecular cloud into a number of gravitationally bound cores, which are initially supported against gravity by a combination of thermal, magnetic, and turbulent pressures (e.g., Mouschovias 1991; Shu et al. 1987). These prestellar condensations or fragments form and evolve as a result of a still poorly understood mechanism that may involve ambipolar diffusion (e.g., Mouschovias 1991), the dissipation of turbulence (e.g., Nakano 1998), and an outside impulse (e.g., Bonnell et al. 1997). Once such a condensation becomes gravitationally unstable and collapses, the main *theoretical* features of the ensuing dynamical evolution have been known since the pioneering work of Larson (1969). During a probably brief initial phase, the released gravitational energy is freely radiated away, and the collapsing fragment stays roughly isothermal. This "runaway" isothermal collapse phase tends to produce a strong central concentration of matter with a radial density gradient approaching $\rho \propto r^{-2}$ at small radii essentially independently of initial conditions (e.g., Whitworth and Summers 1985; Blottiau et al. 1988; Foster and Chevalier 1993). It ends with the formation of an opaque, hydrostatic protostellar object in the center (e.g., Larson 1969; Boss and Yorke 1995; Bate 1998). Numerical simulations in fact predict the successive formations of two hydrostatic objects, one before and one after the dissociation of molecular hydrogen (see Boss and Yorke 1995), but we will not distinguish between them here. The system then enters the main accretion phase, during which the central object builds up its mass ($M_\star$) from a surrounding infalling envelope (of mass $M_{\rm env}$) and accretion disk, while progressively warming up. In this chapter, we will refer to the system consisting of the central object plus envelope and disk as an accreting protostar. The youngest accreting protostars have $M_{\rm env} \gg M_\star$ and radiate the accretion luminosity $L_{\rm acc} \approx GM_\star \dot{M}_{\rm acc}/R_\star$. In the "standard" theory of isolated star formation (Shu et al. 1987, 1993), the collapse initial conditions are taken to be (static) singular isothermal spheroids ($\rho \sim (a^2/2\pi G)\, r^{-2}$, cf. Li and Shu 1996, 1997), there is no runaway collapse phase, and the accretion rate $\dot{M}_{\rm acc}$ is constant in time $\sim a^3/G$, where a is the effective isothermal sound speed. With other collapse initial conditions, the accretion rate is generally time dependent (see section III.D below).

Observations have shown that the main accretion phase is always accompanied by a powerful ejection of a small fraction of the accreted material in the form of prominent bipolar jets or outflows (e.g., Bachiller 1996). These outflows are believed to carry away the excess angular momentum of the infalling matter (e.g., the chapter by Königl and Pudritz, this volume). When the central object has accumulated most ($\gtrsim$90%) of its final, main-sequence mass, it becomes a pre-main-sequence (PMS) star, which evolves approximately at fixed mass on the Kelvin-Helmholtz contraction timescale (e.g., Stahler and Walter 1993). (Note that, during the protostellar accretion phase, stars more massive than a few 0.1 $M_\odot$ start burning deuterium whereas stars with masses in excess of $\sim$8 $M_\odot$ begin to burn hydrogen; see Palla and Stahler 1991.)

The details of the earliest stages outlined above are still poorly known. Improving our understanding of these early stages is of prime importance since to some extent they must govern the origin of the stellar initial mass function (IMF).

Observationally, it is by comparing the structure of starless dense cores with that of the envelopes surrounding the youngest stellar objects that one may hope to estimate the initial conditions for protostellar collapse. The purpose of this chapter is to review several major advances made in this field over the last decade, thanks mostly to ground-based (sub)millimeter continuum observations. We discuss results obtained on prestellar cores and young accreting protostars in sections II and III, respectively. We then combine these two sets of results and conclude in section IV.

II. PRESTELLAR CORES

A. Definition and Identification

The prestellar stage of star formation may be defined as the phase in which a gravitationally bound core has formed in a molecular cloud and evolves toward higher degrees of central condensation, but no central hydrostatic protostellar object exists yet within the core.

A pioneering survey of isolated dense cores in dark clouds was carried out in transitions of NH_3 by Myers and coworkers (see Myers and Benson 1983; Benson and Myers 1989, and references therein), who cataloged about 90 cores. These were separated into starless cores and cores with stars (Beichman et al. 1986), on the basis of the presence or absence of an embedded source detected by the *Infrared Astronomy Satellite* (IRAS). The starless NH_3 cores were identified by Beichman et al. as the potential sites of future isolated low-mass star formation. Other dense core surveys have been carried out by Clemens and Barvainis (1988), Wood et al. (1994), Bourke et al. (1995*a*,*b*), Lee and Myers (1999), and Jessop and Ward-Thompson (1999).

Using the 15-m James Clerk Maxwell Telescope (JCMT), Ward-Thompson et al. (1994) observed the 800-μm dust continuum emission from about 20 starless NH_3 cores from the Benson and Myers list, mapping four of the cores, and showed that they have larger full-width at half maximum (FWHM) sizes than, but comparable masses to, the envelopes of the youngest protostars (Class 0 sources; see section III below). This is consistent with starless NH_3 cores being prestellar in nature and the precursors of protostars (see also Mizuno et al. 1994). Ward-Thompson et al. also demonstrated that prestellar cores do not have density profiles which can be modeled by a single scale-free power law; instead they have flat inner radial density profiles, suggestive of magnetically supported cores contracting by ambipolar diffusion (see Mouschovias 1991, 1995, and references therein). Recent molecular line spectroscopy of several prestellar cores (e.g., Tafalla et al. 1998 and the chapter by Myers et al., this volume) appears to support the argument that they are contracting, but more slowly than the infall seen toward Class 0 protostars (e.g., Mardones et al. 1997; see section III.C).

The 800-μm study by Ward-Thompson et al. (1994) also suggests that wide-field submillimeter continuum imaging may be a powerful tool for searching for new prestellar cores in the future (cf. Ristorcelli et al. 1998).

B. Spectral Energy Distributions and Temperatures

The advent of the *Infrared Space Observatory* (ISO) of the European Space Agency (ESA) has allowed prestellar cores to be studied in the

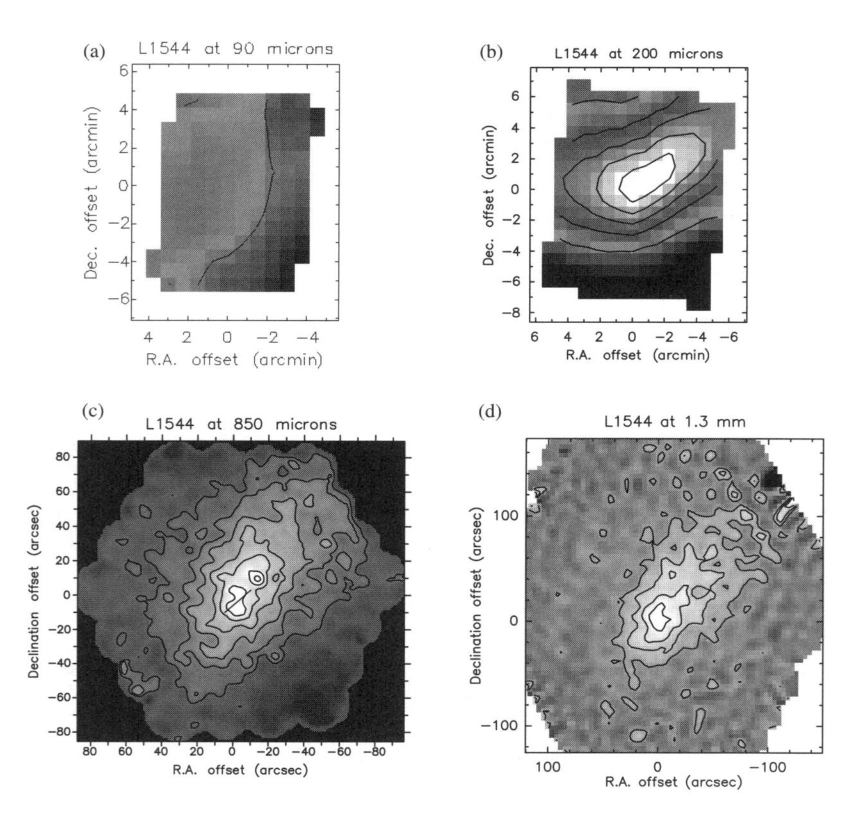

Figure 1. Dust continuum images of L1544 at 90 μm (a) and 200 μm (b) from ISOPHOT, at 850 μm (c) from SCUBA, and at 1.3 mm (d) from the IRAM 30 m. A polarization **E**-vector, perpendicular to the **B** field, is overlaid on the 850 μm image (c), as measured with the SCUBA polarimeter. The observed morphology is consistent with a magnetically supported core that has flattened along the direction of the mean magnetic field.

far-infrared (FIR) for the first time (Ward-Thompson and André, in prep.), since these cores were not detected by IRAS. Likewise, the Submillimetre Common User Bolometer Array (SCUBA) camera on the JCMT has allowed prestellar cores to be observed in the submillimeter region with a greater signal-to-noise ratio than ever before (Ward-Thompson et al., in prep.). Prestellar cores emit almost all of their radiation in the FIR and submillimeter regimes, so the combination of these two instruments provides a unique opportunity to study them.

As an illustration, Fig. 1 shows a series of images of the L1544 prestellar core at 90 and 200 μm (from ISO), 850 μm (from SCUBA), and 1.3 mm (from IRAM). The core is clearly detected at 200–1300 μm but is almost undetected at 90 μm. This shows that the core is very cold, and its dust temperature can be obtained from fitting a modified black body to the observed emission. The spectral energy distribution (SED) of L1544 in the FIR and submillimeter wavelength regimes is shown in Fig. 2. The solid line is a gray-body curve of the form

$$S_\nu = B_\nu(T_{\rm dust})\,[1 - \exp(-\tau_\nu)]\,\Delta\Omega$$

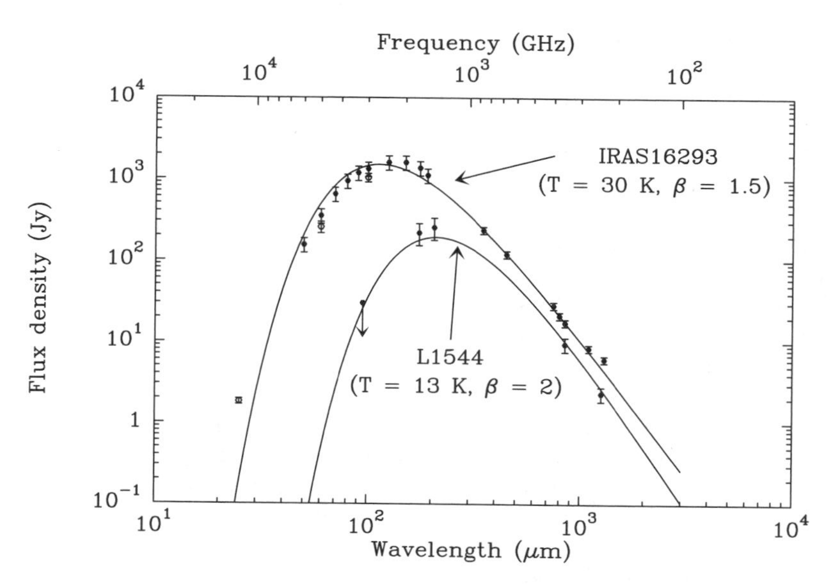

Figure 2. Spectral energy distributions of the prestellar core L1544 and the Class 0 protostar IRAS 16293, along with gray-body fits (see text). The L1544 SED is based on ISOPHOT, JCMT, and IRAM data from Ward-Thompson et al. (in prep.) The IRAS 16293 SED is based on IRAS, ISO-LWS (Correia et al., in prep.; see also Ceccarelli et al. 1998), and JCMT data (Sandell 1994). The fits suggest that both sources are opaque ($\tau_\nu > 1$) at $\lambda < 30$–50 μm. Note that a simple gray-body model cannot account for the 25-μm emission of IRAS 16293 and that a two-component model is required (Correia, in prep.).

where $B_\nu(T_{\rm dust})$ is the Planck function at frequency ν for a dust temperature $T_{\rm dust}$, $\tau_\nu = \kappa_\nu \Sigma$ is the dust optical depth through a mass column density Σ (see below), and $\Delta\Omega$ is the source solid angle. In this simple modeling, the dust opacity per unit (gas + dust) mass column density, κ_ν, is assumed to scale as ν^β, with $\beta = 1.5$–2, as usually appropriate in the submillimeter range (e.g., Hildebrand 1983). For L1544, a good fit to the SED is obtained with $T_{\rm dust} = 13$ K, $\beta = 2$, $\tau_\nu = 0.12$ at $\lambda = 100\ \mu$m, and $\Delta\Omega = 2.3 \times 10^{-7}$ sr. Similar results are obtained in other starless cores. This confirms the lack of any warm dust in such cores and consequently the lack of any embedded protostellar object. The (sub)millimeter data show a morphology similar to the ISO images at much higher resolution, indicating that the same dust is being observed at all wavelengths. Consequently the temperature derived from the SED is representative of all the emitting dust and can be used to convert submillimeter fluxes into estimates of dust masses and, hence, to gas masses.

C. Mass and Density Structure

Dust emission is generally optically thin at (sub)millimeter wavelengths and hence is a direct tracer of the mass content of molecular cloud cores. For an isothermal dust source, the total (gas + dust) mass $M(r < R)$ contained within a radius R from the center is related to the submillimeter flux density $S_\nu(\theta)$ integrated over a circle of projected angular radius $\theta = R/d$ by

$$M(r < R) \equiv \pi R^2 \langle \Sigma \rangle_R = [S_\nu(\theta)\, d^2]/[\kappa_\nu\, B_\nu(T_{\rm dust})],$$

where $\langle \Sigma \rangle_R$ is the average mass column density. The dust opacity κ_ν is somewhat uncertain, but the uncertainties are much reduced when appropriate dust models are used (see Henning et al. 1995 for a review). For prestellar cores of intermediate densities ($n_{\rm H_2} \lesssim 10^5$ cm^{-3}), κ_ν is believed to be close to $\kappa_{1.3} = 0.005$ cm^2 g^{-1} at 1.3 mm (e.g., Hildebrand 1983; Preibisch et al. 1993). In denser cloud cores and protostellar envelopes, grain coagulation and the formation of ice mantles make κ_ν a factor of $\sim$2 larger, i.e., $\kappa_{1.3} = 0.01$ cm^2 g^{-1} assuming a gas-to-dust mass ratio of 100 (e.g., Ossenkopf and Henning 1994). A still higher value, $\kappa_{1.3} = 0.02$ cm^2 g^{-1}, is recommended in protoplanetary disks (Beckwith et al. 1990; Pollack et al. 1994).

Following this method, FWHM masses ranging from $\sim$0.5 M$_\odot$ to $\sim$35 M$_\odot$ are derived for the nine isolated cores mapped by Ward-Thompson et al. (1999).

Although observed prestellar cores are generally not circularly or elliptically symmetric (see, e.g., Fig. 1), one can still usefully constrain their radial density profiles by averaging the (sub)millimeter emission in circular or elliptical annuli.

Figure 3a shows the azimuthally averaged radial intensity profile of the prestellar core L1689B at 1.3 mm, compared to the profile of a spherical

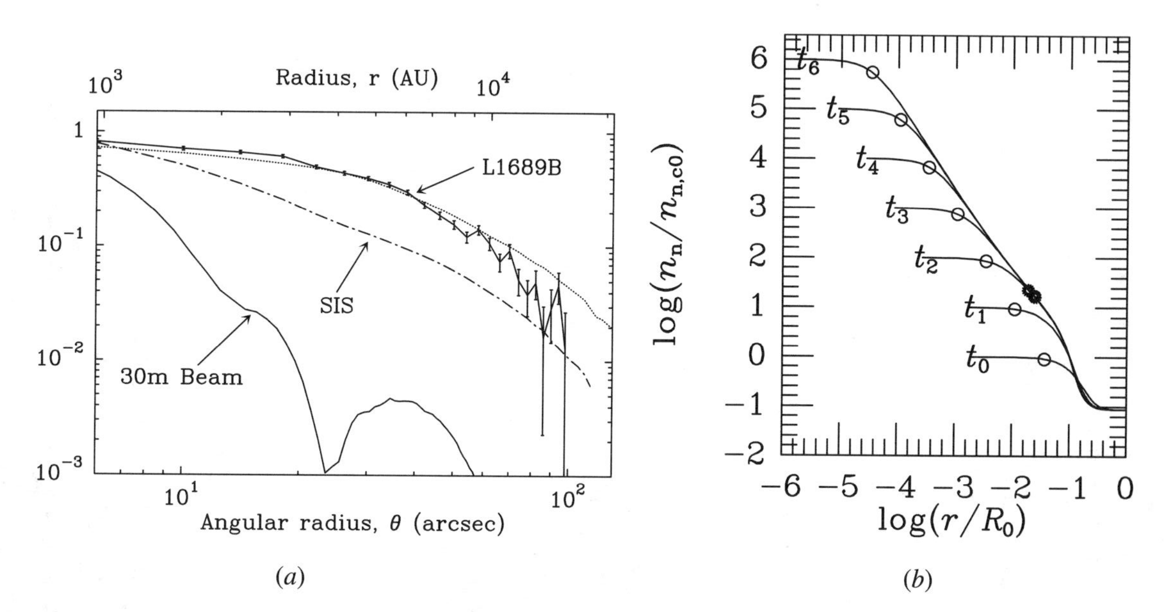

Figure 3. (a) (left) Radial intensity profile of L1689B at 1.3 mm, illustrating that prestellar cores have flat inner density profiles (from André et al. 1996). For comparison, the dotted curve shows a spherical isothermal model with $\rho(r) \propto r^{-1.2}$ for $r < 4000$ AU and $\rho(r) \propto r^{-2}$ for $r \geq 4000$ AU. The dash-dotted curve shows a model with $\rho(r) \propto r^{-2}$, such as a singular isothermal sphere (SIS). (b) (right) Typical density profiles expected for a magnetically supported core undergoing ambipolar diffusion at various times increasing from t_0 to t_6 (from Ciolek and Mouschovias 1994). The normalization values are $n_{n,c0} = 2.6 \times 10^3$ cm^{-3} and $R_0 = 4.29$ pc. Open circles mark the instantaneous radius of the uniform-density central region; starred circles mark the radius of the magnetically supercritical region (present only for $t \geq t_2$).

isothermal core model with $\rho(r) \propto r^{-2}$. The model intensity profile results from a complete simulation that takes into account both the observing technique (dual-beam mapping) and the reduction method (cf. André et al. 1996). We see that L1689B exhibits the familiar radial profile of prestellar cores, with a flat inner region, steepening toward the edges (Ward-Thompson et al. 1994; André et al. 1996; Ward-Thompson et al. 1999). In this representative example, the radial density profile inferred under the assumption of a constant dust temperature is as flat as $\rho(r) \propto r^{-0.4}$ (if the 3-D core shape is disklike) or $\rho(r) \propto r^{-1.2}$ (if the core shape is spheroidal) at radii less than $R_{\mathrm{flat}} \sim 4000$ AU and approaches $\rho(r) \propto r^{-2}$ only between $\sim$4000 AU and $\sim$15,000 AU.

More recently, it has been possible to constrain the outer density gradient of starless cores through absorption studies in the mid-infrared with the camera aboard ISO (ISOCAM) (e.g., Abergel et al. 1996, 1998; Bacmann et al. 1998). It appears that isolated prestellar cores are often characterized by sharp edges, steeper than $\rho \propto r^{-3}$ or $\rho \propto r^{-4}$, at radii $R \gtrsim 15{,}000$ AU.

These features of prestellar density structure (i.e., inner flattening and sharp outer edge) are qualitatively consistent with models of magnetically supported cores evolving through ambipolar diffusion prior to protostar formation (see Fig. 3b; e.g., Ciolek and Mouschovias 1994; Basu and Mouschovias 1995), although the models generally require fairly strong magnetic fields ($\sim$100 μG). Alternatively, the observed structure may also be explained by models of thermally supported self-gravitating cores interacting with an external UV radiation field (e.g., Falgarone and Puget 1985; Chièze and Pineau des Forêts 1987).

D. Lifetimes

Beichman et al. (1986) used the ratio of numbers of starless cores to numbers of cores with embedded IRAS sources to estimate their relative timescales. They found roughly equal numbers of cores with and without IRAS sources. They estimated the lifetime of the stage of cores with stars, based on T Tauri lifetimes and pre-main-sequence HR diagram tracks (e.g., Stahler 1988). Based on this, they estimated the lifetime of the prestellar core phase to be a few times 10^6 yr.

However, the lifetime of a prestellar core depends on its central density. Figure 4 (taken from Jessop and Ward-Thompson 1999) shows the estimated lifetime of starless cores for each of six dark cloud surveys mentioned in section II.A above, versus the mean volume density of cores in each sample. The lifespan of cores without stars in each sample was estimated from the fraction of cores with IRAS sources, using the same method as Beichman et al. (1986). An anticorrelation between lifetime and density is clearly apparent in Fig. 4. The solid line has the form $t \propto \rho^{-0.75}$, while the dashed line is of the form $t \propto \rho^{-0.5}$. These two forms are expected for cores evolving on the ambipolar diffusion timescale $t_{\mathrm{AD}} \propto x_e$ (where x_e is the ionization fraction, e.g., Nakano 1984), if the dominant ionization mechanism is UV ionization or cosmic ray ionization respec-

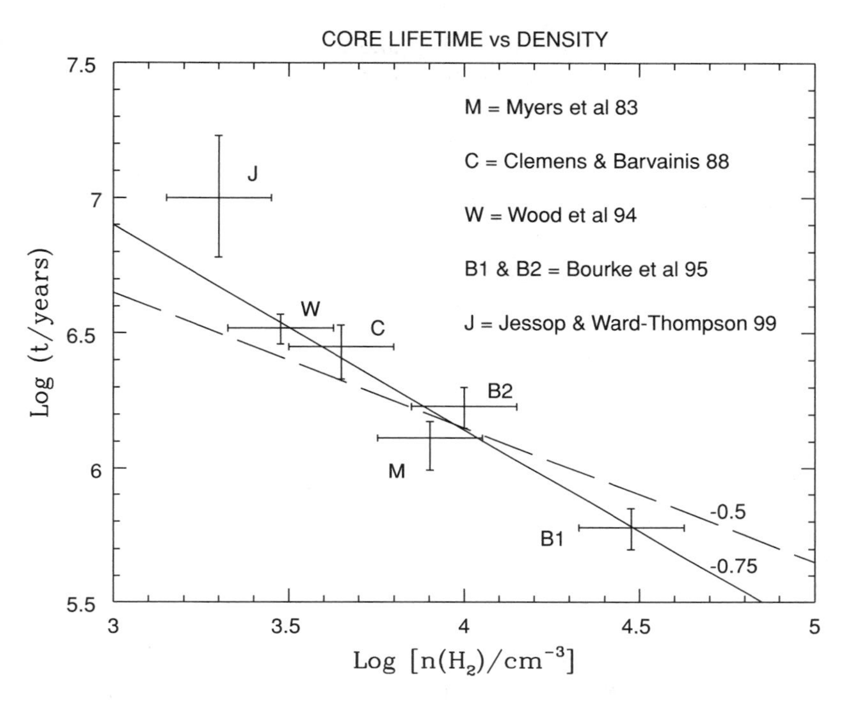

Figure 4. Plot of lifetime vs. mean density for six core samples compared with $t \propto \rho^{-0.75}$ and $t \propto \rho^{-0.5}$ as predicted by ambipolar diffusion models with different ionization mechanisms (Jessop and Ward-Thompson 1999).

tively (e.g., McKee 1989). Although the shape of the t vs ρ correlation is roughly consistent with expectations, Lee and Myers (1999) note that the lifetimes of cores at $n_{H_2} \sim 10^4$ cm^{-3} tend to be shorter than the predictions of ambipolar diffusion models by a factor $\gtrsim$3.

More details about the evolution of prestellar cores at higher densities can be inferred from the results of the submillimeter continuum SCUBA survey of Ward-Thompson et al. (in prep.). In this survey, 17 of the 38 NH_3 cores without IRAS sources from Benson and Myers (1989) were detected by SCUBA at 850 μm. Ward-Thompson et al. estimate that the 17 SCUBA detections all have central densities between $\sim 10^5$ and $\sim 10^6$ cm^{-3}, whereas the 21 nondetections must have lower central densities, typically between $\sim 10^4$ and $\sim 10^5$ cm^{-3} (see Benson and Myers 1989; Butner et al. 1995, and references therein). Consequently, they deduce that the lifetime of these two phases, that of central density increasing from $\sim 10^4$ to $\sim 10^5$ cm^{-3} and that of central density of $\sim 10^5$ cm^{-3} until the formation of a protostellar object at the center, must be roughly equal. This result can be compared with the predictions of ambipolar diffusion models.

Figure 3b is taken from Ciolek and Mouschovias (1994) and shows the radial density profile predicted by an ambipolar diffusion model at different evolutionary stages (t_0 to t_6). The stage at which the central density is $\sim 10^4$ cm^{-3} corresponds to time t_1, and the stage at which the central density is $\sim 10^5$ cm^{-3} corresponds to time t_2. The time at which a protostellar object forms is effectively t_6. In this model the time taken to go from t_1 to t_2 ($\sim 2 \times 10^6$ yr) is six times longer than the time taken to go from t_2 to t_6. Some discrepancy could perhaps be accounted for by the statistical errors associated with our source number counting technique, but the ratio between the two timescales should be fairly robust. The model predicts that SCUBA should only have detected one-seventh of the cores, whereas it detected half of the sample.

We are left with the conclusion that cores at central densities $\sim 10^5$ cm^{-3} evolve more slowly than ambipolar diffusion models predict, a trend opposite to that noted at low densities. These departures from models involving only a static magnetic field are perhaps due to turbulence, which generates nonstatic fields and modifies core support (e.g., Gammie and Ostriker 1996; Nakano 1998; Balsara et al. 1998).

E. Prestellar Condensations in Star-Forming Clusters

In regions of multiple star formation, submillimeter dust continuum mapping has revealed a wealth of small-scale cloud fragments, sometimes organized along filaments [e.g., Mezger et al. 1992; André et al. 1993; Casali et al. 1993; Launhardt et al. 1996; Chini et al. 1997*b*; Johnstone and Bally 1999]. Such fragmentation along filaments has not been observed in Taurus, but examples do exist in young embedded clusters forming primarily low-mass stars like ρ Ophiuchi (Motte et al. 1998). The individual fragments, which are denser ($\langle n \rangle \gtrsim 10^6$–$10^7$ cm^{-3}) and more compact (a few thousand AU in size) than the isolated prestellar cores discussed above, often remained totally undetected (in emission) by IRAS or ISO in the mid- to far-IR. Since molecules tend to freeze out onto dust grains at low temperatures and high densities, (sub)millimeter dust emission may be the most effective tracer of such condensations (e.g., Mauersberger et al. 1992).

The most centrally condensed of these starless fragments have been claimed to be isothermal protostars: collapsing condensations with no central hydrostatic object (see section I) (e.g., Mezger et al. 1992; Launhardt et al. 1996; Motte et al. 1998). Good examples are FIR 3 and FIR 4 in NGC 2024, OphA-SM1 and OphE-MM3 in ρ Oph, LBS 17-SM in Orion B, and MMS1 and MMS4 in OMC-3. This isothermal protostar interpretation remains to be confirmed, however, by observations of appropriate spectral line signatures (cf. the chapter by Myers et al. this volume).

Furthermore, the advent of large-format bolometer arrays now makes possible systematic studies of the genetic link between prestellar cloud fragments and young stars. Color Plate 3 is a 1.3-mm continuum wide-field mosaic of the ρ Oph cloud (Motte et al. 1998), showing a total of 100 structures with characteristic angular scales of $\sim 15''$–$30''$ (i.e., ~ 2000–4000 AU), which are associated with 59 starless condensations (undetected by ISO in the mid-IR) and 41 embedded young stellar objects (YSOs) (detected at IR or radio continuum wavelengths).

Comparison of the masses derived from the 1.3 mm continuum (from $\sim$0.05 $M_\odot$ to $\sim$3 $M_\odot$) with Jeans masses suggests that most of the 59 starless fragments are close to gravitational virial equilibrium, with $M/M_{\rm vir} \gtrsim$ 0.3–0.5, and will form stars in the near future. These prestellar condensations generally have flat inner density profiles, like isolated prestellar cores, but are distinguished by compact, finite sizes of a few thousand AU. The typical fragmentation lengthscale derived from the average projected separation between condensations is $\sim$6000 AU in ρ Oph. This is $\sim$5 times smaller than the radial extent of isolated dense cores in the Taurus cloud (see Gómez et al. 1993).

Figure 5 shows the mass distribution of the 59 ρ Oph prestellar fragments. It follows approximately $\Delta N/\Delta M \propto M^{-1.5}$ below $\sim$0.5 $M_\odot$, which is similar to the clump mass spectrum found by large-scale molecular line studies (e.g., Blitz 1993 and the chapter by Williams et al., this volume). The novel feature, however, is that the fragment mass spectrum found at 1.3 mm in ρ Oph appears to steepen to $\Delta N/\Delta M \propto M^{-2.5}$ above $\sim$0.5 $M_\odot$. A similarly steep mass spectrum above $\sim$0.5 $M_\odot$ was obtained by Testi and Sargent (1998) for compact 3-mm starless condensations in the Serpens core. These prestellar mass spectra resemble the shape of the *stellar* initial mass function (IMF), which is known to approach $\Delta N/\Delta M_\star \propto M_\star^{-2.7}$ for 1 $M_\odot \lesssim M_\star \lesssim$ 10 $M_\odot$ and $\Delta N/\Delta M_\star \propto M_\star^{-1.2}$ for 0.1 $M_\odot \lesssim M_\star \lesssim$ 1 $M_\odot$ (e.g., Kroupa et al. 1993; Tinney 1993, 1995; Scalo 1998; see also the chapter by Meyer et al., this volume). Given the factor of $\sim$2 uncertainty on the measured prestellar masses, such a resemblance is remarkable and

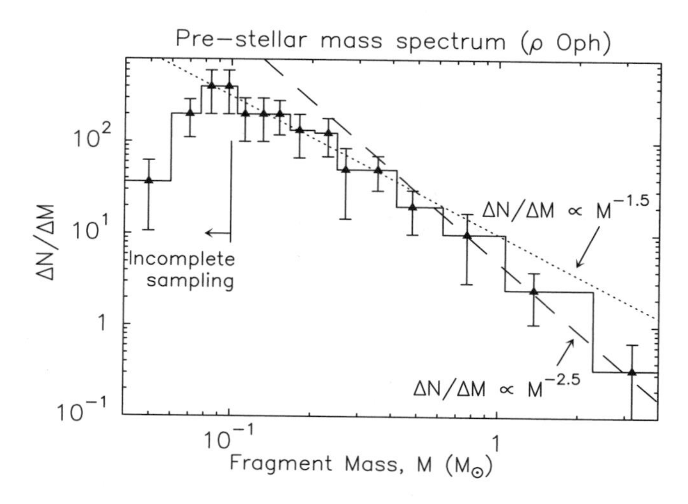

Figure 5. Mass spectrum of the 59 prestellar fragments extracted from the ρ Oph 1.3 mm continuum mosaic shown in Color Plate 3 (from Motte et al. 1998). For comparison, dotted and long-dashed lines show power laws of the form $\Delta N/\Delta M \propto M^{-1.5}$ and $\Delta N/\Delta M \propto M^{-2.5}$, respectively. This prestellar mass spectrum is remarkable in that it resembles the shape of the IMF.

suggests that the IMF of embedded clusters is primarily determined at the prestellar stage of star formation (see also section IV below).

III. THE YOUNGEST PROTOSTARS

A. Class 0 Protostars and Other YSO Stages

1. Infrared YSO Classes. In the near/mid-infrared, three broad classes of YSOs can be distinguished based on the slopes of their SEDs between 2.2 μm and 10–25 μm, $\alpha_{IR} = d \log(\lambda F_\lambda)/d \log(\lambda)$, which are interpreted in terms of an evolutionary sequence (Lada and Wilking 1984; Lada 1987). Going backward in time, Class III ($\alpha_{IR} < -1.5$) and Class II ($-1.5 < \alpha_{IR} < 0$) sources correspond to pre-main-sequence (PMS) stars ("weak" and "classical" T Tauri stars, respectively), surrounded by a circumstellar disk (optically thin in Class III and optically thick in Class II at $\lambda \lesssim$ 10 μm), but lacking a dense circumstellar envelope (see André and Montmerle 1994). The youngest YSOs detected at 2 μm are the Class I sources, which are characterized by $\alpha_{IR} > 0$ (e.g., Wilking et al. 1989) and the close association with dense molecular gas (e.g., Myers et al. 1987). Class I objects are now interpreted as relatively evolved protostars with typical ages $\sim$1–2 $\times$ 10^5 yr (e.g., Barsony and Kenyon 1992; Greene et al. 1994; Kenyon and Hartmann 1995), surrounded by both a disk and a diffuse circumstellar envelope of substellar ($\lesssim$0.1–0.3 $M_\odot$) mass (Whitney and Hartmann 1993; Kenyon et al. 1993*b*; André and Montmerle 1994; Lucas and Roche 1997). Their SEDs are successfully modeled in the framework of the "standard" theory of isolated protostars (e.g., Adams et al. 1987; Kenyon et al. 1993*a*), in agreement with the idea that they derive a substantial fraction of their luminosity from accretion (see also Greene and Lada 1996 and Kenyon et al. 1998).

2. Class 0 Protostars. Several condensations detected in submillimeter dust continuum maps of molecular clouds (such as those described in section II.E) appear to be associated with formed, hydrostatic YSOs and have been designated "Class 0" protostars (André et al. 1993). Specifically, Class 0 objects are defined by the following observational properties (André et al. 1993):

(i) Indirect evidence for a central YSO, as indicated by, e.g., the detection of a compact centimeter radio continuum source, a collimated CO outflow, or an internal heating source.
(ii) Centrally peaked but extended submillimeter continuum emission tracing the presence of a spheroidal circumstellar dust envelope (as opposed to just a disk).
(iii) High ratio of submillimeter to bolometric luminosity, suggesting that the envelope mass exceeds the central stellar mass: $L_{smm}/L_{bol} >$ 0.5%, where L_{smm} is measured longward of 350 μm. In practice, this often means an SED resembling a single-temperature blackbody at $T \sim$ 15–30 K (see Fig. 2).

Property (i) distinguishes Class 0 objects from the prestellar cores and condensations discussed in section II. In particular, deep VLA observations reveal no compact radio continuum sources in the centers of prestellar cores (Bontemps 1996; Yun et al. 1996). Properties (ii) and (iii) distinguish Class 0 objects from more evolved (Class I and Class II) YSOs. As shown by André et al. (1993), the L_{smm}/L_{bol} ratio should roughly track the ratio $M_{env}/M_{\star}$ of envelope to stellar mass and may be used as an evolutionary indicator (decreasing with time) for low-luminosity ($L_{bol} \lesssim 50\,L_{\odot}$) embedded YSOs. Criterion (iii) approximately selects objects that have $M_{env}/M_{\star} > 1$, assuming plausible relations between L_{bol} and $M_{\star}$ on the one hand and between L_{smm} and M_{env} on the other (see André et al. 1993; André and Montmerle 1994). [A roughly equivalent criterion is $M_{env}/L_{bol} > 0.1$ $M_{\odot}/L_{\odot}$.] Class 0 objects are therefore excellent candidates for being very young accreting protostars in which a hydrostatic core has formed but not yet accumulated the majority of its final mass. In practice, most of the confirmed Class 0 objects listed in Table I have $L_{smm}/L_{bol} \gg 0.5\%$ and are likely to be at the beginning of the main accretion phase with $M_{env} \gg M_{\star}$ (see Fig. 6b).

3. Evolutionary Diagrams for Embedded YSOs. Combining infrared and submillimeter data, it is therefore possible to define a complete, empirical evolutionary sequence (Class 0 → Class I → Class II → Class III) for low-mass YSOs, which likely correspond to conceptually different stages of evolution: (early) main accretion phase, late accretion phase, PMS stars with protoplanetary disks, PMS stars with debris disks (see André and Montmerle 1994). This sequence is quasicontinuous and may be parameterized by the "bolometric temperature," T_{bol}, defined by Myers and Ladd (1993) as the temperature of a blackbody having the same mean frequency as the observed YSO SED. Myers and Ladd proposed to use the L_{bol}–T_{bol} diagram for embedded YSOs as a direct analog to the HR diagram for optically visible stars. As shown by Chen et al. (1995, 1997), YSOs with known classes have distinct ranges of T_{bol}: <70 K for Class 0, 70–650 K for Class I, 650–2880 K for Class II, and >2880 K for Class III (e.g., Fig. 6a). The evolution of T_{bol} and L_{bol} from the Class 0 stage to the zero-age main sequence (ZAMS) has been modeled in the context of various envelope dissipation scenarios by Myers et al. (1998).

A perhaps more direct approach to tracking the circumstellar evolution of YSOs is to use the circumstellar mass $M_{c\star}$ derived from (sub)-millimeter continuum measurements of optically thin dust emission. Such measurements show that $M_{c\star}$ ($= M_{env} + M_{disk}$) is generally dominated by M_{env} in Class 0/Class I sources (e.g., Terebey et al. 1993) and decreases by a factor ~5–10 on average from one YSO class to the next (André and Montmerle 1994). In the spirit of the L_{smm}/L_{bol} evolutionary indicator of André et al. (1993), Saraceno et al. (1996*a*) proposed the L_{smm}–L_{bol} (or equivalently M_{env}–L_{bol}) diagram as an alternative evolutionary diagram for self-embedded YSOs. While L_{smm} and M_{env} are well correlated with L_{bol} for the majority of embedded YSOs (e.g., Reipurth et al. 1993),

TABLE I

Properties of Confirmed Class 0 Protostars

Object	α(2000)	δ(2000)	Dist. (pc)	L_{bol} ($L_\odot$)	M_{env} ($M_\odot$)	L_{smm}/L_{bol} (%)[a]	T_{bol} (K)	Outflow Manifestations	Infall	Structure[b]	References[c]
W3OH-TW[d]	02:27:04.7	+61:52:24	2200	10^3–10^4	~20	~1	≳80?	H_2O, radio	—	—	1
L1448-IRS2	03:25:22.4	+30:45:12	300	6	0.9	3	70?	CO	—	—	2
L1448-N	03:25:36.3	+30:45:15	300	11	2.3	3	70	CO, radio	Y?	D, B	3,4,5,6,7,8,9,10
L1448-C	03:25:38.8	+30:44:05	300	9	1.4	2	60	CO, radio, H_2	N?	D	3,4,5,11,12,8,10,13
NGC1333-IRAS2	03:28:55.4	+31:14:35	350	40	1.7	≲1	50	CO, H_2	Y?	—	14,15,9,16,17,13
SVS13B	03:29:03.1	+31:15:52	350	~7	2.7	~5	~30	SiO, H_2	—	—	18,19,20,21,22
NGC1333-IRAS4A	03:29:10.3	+31:13:31	350	14	7.	5	34	CO, radio	Y	D, B	23,24,15,16,25,26,27,28,10,13
NGC1333-IRAS4B	03:29:12.0	+31:13:09	350	14	2.7	3	36	CO, radio	Y	D, B	23,24,15,16,25,28,10,13
IRAS 03282	03:31:20.8	+30:45:31	300	1.5	0.6	5	35	CO, H_2	—	—	29,30,8
HH211-MM	03:43:56.8	+32:00:50	300	5	1.5	~4	~30	H_2	—	—	31,32
IRAM 04191	04:21:56.9	+15:29:46	140	0.15	0.5	12	18	CO, radio	Y	N?	33
L1527[e]	04:39:53.9	+26:03:10	140	2	0.4	0.7	60	CO, H_2	Y	D	34,35,36,37,38,6,10,13
HH114MMS	05:18:15.2	+07:12:03	450	≲25	2.8	≳1	~40	Radio, HH	—	—	19
RNO43-MM	05:32:19.4	+12:49:42	400	5	0.6	~5	36	CO, radio, H_2	N?	—	39,19,40,41,10
OMC3-MM6	05:35:23.5	−05:01:32	450	<60	12	≳2	~30	CO, H_2	—	—	42,43
L1641-VLA1[e]	05:36:22.8	−06:46:07	450	50	6.5	~3	70?	Radio, HH	—	—	44,19,45
NGC2023-MM1	05:41:24.8	−02:18:09	450	~8	1.8–4.6	~3–10	~30	CO	—	—	46,47
NGC2024-FIR5	05:41:44.5	−01:55:43	450	≳10	15.	≲30	20	CO	N?	D, B?	48,49,50,10
NGC2024-FIR6	05:41:45.2	−01:56:05	450	≳15	6.	≲10	20	CO	N?	—	48,49,50,10

TABLE I (CONT'D)

Object	$\alpha(2000)$	$\delta(2000)$	Dist. (pc)	L_{bol} ($L_\odot$)	M_{env} ($M_\odot$)	L_{smm}/L_{bol} (%)[a]	T_{bol} (K)	Outflow Manifestations	Infall	Structure[b]	References[c]
HH212-MM	05:43:51.5	−01:02:52	400	14	1.2	~2	70?	CO, H_2	—	—	39,19,51
HH24MMS	05:46:08.3	−00:10:42	450	5	4	10	20?	Radio, H_2	—	D	52,53,54,55,9,56
HH25MMS[e]	05:46:07.5	−00:13:36	450	6	0.5	5	34	CO, radio, H_2	Y?	—	54,57,10
NGC2264G-VLA2	06:41:11.1	+09:55:59	800	12	2	2	25	CO, radio	—	—	58,59,60
IRAS 08076[e]	08:09:32.8	−36:05:00	400	17	2.3	≳1	74	CO, H_2	—	—	61,16,62
BHR71-MM	12:01:36.3	−65:08:44	200	10	2.4	~3	56	CO, HH	—	—	63,62
IRAS 13036[e]	13:07:36.1	−77:00:05	200	1.7	0.3	≲2	60	CO	Y	—	62,64,13
VLA 1623	16:26:26.4	−24:24:30	160	1	0.7	10	<35	CO, radio, H_2	Y?	D, B?	65,66,67,68,69,70,9,10,13,71,22
IRAS 16293	16:32:22.7	−24:28:32	160	23	2.3	2	43	CO, radio	Y	D, B	72,73,74,75,26,27,76,77,10,13
Trifid-TC3[d]	18:02:07.0	−23:05:11	1680	~10^3	~60	~1	~30	SiO	—	—	78
L483-MM[e]	18:17:29.8	−04:39:38	200	9	0.3	~0.7	50	CO, radio	Y?	—	79,37,10,13
Serp-S68N	18:29:47.9	+01:16:46	310	6	1.0	3	40?	CO, CS	Y?	—	80,81,82,83,84,13
Serp-FIRS1[e]	18:29:49.9	+01:15:20	310	46	3	~1	51	CO, radio, H_2O	N?	N?	85,80,81,82,86,87,84,10,13,88
Serp-SMM4	18:29:57.1	+01:13:15	310	9	3	~3	35	CO, radio	Y	N?	85,81,82,84,89,10,13,88
Serp-SMM3[e]	18:29:59.7	+01:14:00	310	8	0.9	~1	40	CO, H_2	N?	N?	85,81,82,84,90,10,88
G34.24+0.13MM[d]	18:53:21.5	+01:13:45	3700	4000	100	~0.5	50	CH_3OH	—	—	91
L723-MM	19:17:53.7	+19:12:20	300	3	0.6	~4	50	CO, radio	N?	—	92,93,32,19,94,95,10
B335	19:37:00.8	+07:34:11	250	3	0.8	6	37	CO, radio	Y	D?	92,96,97,40,98,10,13,99

TABLE I (CONT'D)

Object	$\alpha(2000)$	$\delta(2000)$	Dist. (pc)	L_{bol} ($L_\odot$)	M_{env} ($M_\odot$)	L_{smm}/L_{bol} (%)[a]	T_{bol} (K)	Outflow Manifestations	Infall	Structure[b]	References[c]
S106-SMM	20:27:25.3	+37:22:46	600	$\gtrsim$24	$\lesssim$10	$\lesssim$8	$\gtrsim$20	Bip. H II, H_2O	—	—	100,70
L1157-MM	20:39:06.2	+68:02:22	440	11	0.5	~5	~60	CO, H_2	Y	D?	101,32,102,69,10,13
GF9–2	20:51:30.1	+60:18:39	200	0.3	~0.5	~10	$\lesssim$20	—	Y	—	103,104,105
CepE-MM	23:03:13.1	+61:42:26	730	75	7	~2	~60	CO, H_2	—	—	106,107
IRAS 23385[d]	23:40:54.5	+61:10:28	4900	16000	370	~0.7	$\gtrsim$40	SiO	—	—	108

[a] L_{smm} is the luminosity radiated longward of 350 μm; Class 0s are defined by $L_{smm}/L_{bol} > 0.5\%$ (see section III.A).

[b] Small-scale structure: D = presence of disk-like component; B = binary or multiple system; N = single object with no disk.

[c] References: (1) Wilner et al. 1995; (2) O'Linger et al. 1999; (3) Bachiller et al. 1990; (4) Curiel et al. 1990; (5) Bachiller et al. 1995; (6) Terebey et al. 1993; (7) Terebey and Padgett 1997; (8) Barsony et al. 1998; (9) Greaves et al. 1997; (10) Gregersen et al. 1997; (11) Bachiller et al. 1991*a*; (12) Bally et al. 1993; (13) Mardones et al. 1997; (14) Sandell et al. 1994; (15) Langer et al. 1996; (16) Hodapp and Ladd 1995; (17) Ward-Thompson et al. 1996; (18) Grossman et al. 1987; (19) Chini et al. 1997*a*; (20) Bachiller et al. 1998; (21) Lefloch et al. 1998; (22) Looney et al. 1999; (23) Sandell et al. 1991; (24) Blake et al. 1995; (25) Mundy et al. 1993; (26) Akeson and Carlstrom 1997; (27) Greaves and Holland 1998; (28) Lay et al. 1995; (29) Bachiller et al. 1991*b*; (30) Bachiller et al. 1994; (31) McCaughrean et al. 1994; (32) Motte 1998; (33) André et al. 1999; (34) Ladd et al. 1991; (35) Eiroa et al. 1994; (36) Bontemps et al. 1996*a*; (37) Myers et al. 1995; (38) Ohashi et al. 1997; (39) Zinnecker et al. 1992; (40) Anglada et al. 1992; (41) Bence et al. 1996; (42) Chini et al. 1997*b*; (43) Yu et al. 1997; (44) Pravdo et al. 1985; (45) Zavagno et al. 1997; (46) Sandell et al. 1999; (47) Launhardt et al. 1996; (48) Mezger et al. 1992; (49) Richer et al. 1992; (50) Wiesemeyer et al. 1997; (51) Zinnecker et al. 1998; (52) Chini et al. 1993; (53) Ward-Thompson et al. 1995*a*; (54) Bontemps et al. 1995; (55) Bontemps et al. 1996*b*; (56) Chandler et al. 1995; (57) Gibb and Davis 1998; (58) Gómez et al. 1994; (59) Ward-Thompson et al. 1995*b*; (60) Lada and Fich 1996; (61) Persi et al. 1994; (62) Henning and Launhardt 1998; (63) Bourke et al. 1997; (64) Lehtinen 1997; (65) André et al. 1990; (66) André et al. 1993; (67) Leous et al. 1991; (68) Dent et al. 1995; (69) Davis and Eislöffel 1995; (70) Holland et al. 1996; (71) Pudritz et al. 1996; (72) Walker et al. 1986; (73) Wootten 1989; (74) Mizuno et al. 1990; (75) Tamura et al. 1993; (76) Mundy et al. 1992; (77) Walker et al. 1994; (78) Cernicharo et al. 1998; (79) Fuller et al. 1995*a*; (80) McMullin et al. 1994; (81) Hurt and Barsony 1996; (82) White et al. 1995; (83) Wolf-Chase et al. 1998; (84) Hurt et al. 1996; (85) Casali et al. 1993; (86) Curiel et al. 1993; (87) Jenness et al. 1995; (88) Hogerheijde et al. 1999; (89) Bontemps 1996; (90) Herbst et al. 1997; (91) Hunter et al. 1998; (92) Davidson 1987; (93) Cabrit and André 1991; (94) Anglada et al. 1991; (95) Avery et al. 1990; (96) Chandler et al. 1990; (97) Cabrit et al. 1988; (98) Zhou et al. 1993; (99) Hirano et al. 1992; (100) Richer et al. 1993; (101) Umemoto et al. 1992; (102) Gueth et al. 1997; (103) Güsten 1994; (104) Wiesemeyer 1997; (105) Wiesemeyer et al. 1998; (106) Lefloch et al. 1996; (107) Ladd and Hodapp 1997; (108) Molinari et al. 1998.

[d] Candidate massive Class 0 object.

[e] Borderline Class 0.

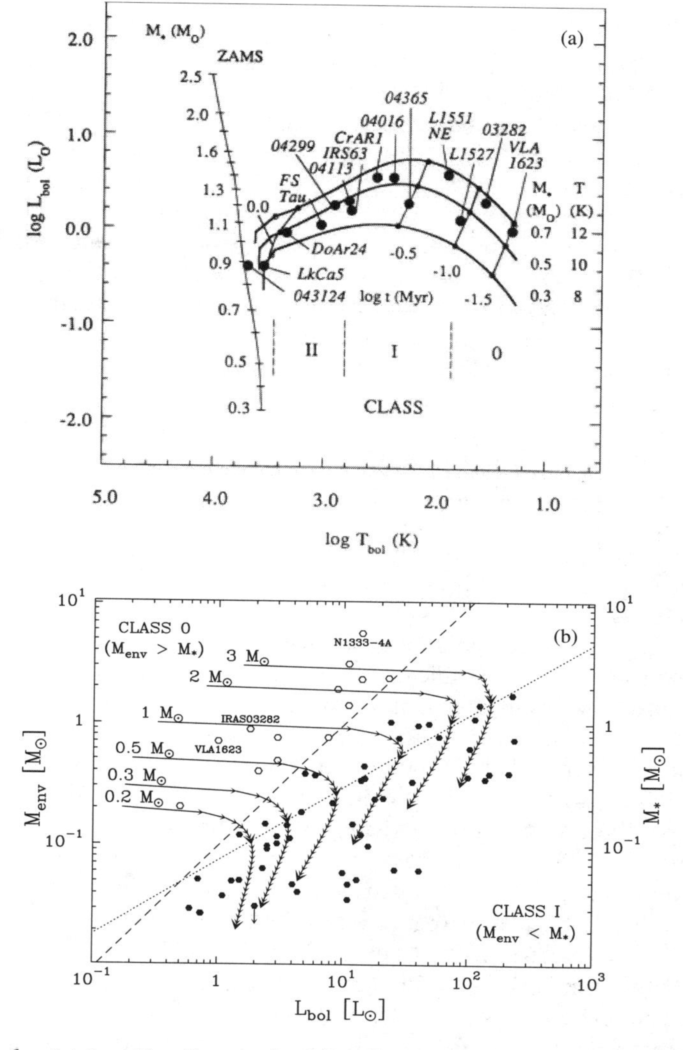

Figure 6. (a) $L_{\rm bol}$–$T_{\rm bol}$ diagram for 14 well-documented YSOs along with three model evolutionary tracks for various (final) stellar masses and cloud temperatures (from Myers et al. 1998). Four times, t(Myr), since the start of infall are indicated, at $\log t = -1.5, -1.0, -0.5$, and 0.0. (b) $M_{\rm env}$–$L_{\rm bol}$ diagram for a sample of Class I (filled circles) and confirmed Class 0 sources (open circles) mainly in Ophiuchus, Perseus, and Orion (adapted from André and Montmerle 1994 and Saraceno et al. 1996*a*). Evolutionary tracks shown are computed assuming that protostars form from bounded condensations of finite initial masses and have $L_{\rm bol} = GM_\star \dot{M}_{\rm acc}/R_\star + L_\star$, where $L_\star$ is the PMS birthline luminosity (e.g., Stahler 1988). $M_{\rm env}$ and $\dot{M}_{\rm acc} = M_{\rm env}/\tau$ (where $\tau = 10^5$ yr) have been assumed to decline exponentially with time (see Bontemps et al. 1996*a*). Small arrows are plotted on the tracks every 10^4 yr, big arrows when 50% and 90% of the initial condensation has been accreted. The dashed and dotted lines are two $M_\star$–$L_{\rm bol}$ relations marking the conceptual border zone between the Class 0 ($M_{\rm env} > M_\star$) and Class I ($M_{\rm env} < M_\star$) stages; the dashed line has $M_\star \propto L_{\rm bol}$ (cf. André et al. 1993; André and Montmerle 1994), while the dotted line has $M_\star \propto L_{\rm bol}^{0.6}$, as suggested by the accretion scenario adopted in the tracks.

Class 0 objects clearly stand out from Class I sources in this diagram as objects with excess (sub)millimeter emission, that is, excess circumstellar material (see Fig. 6b). Moreover, one may compare the locations of observed embedded sources in the $M_{\rm env}$–$L_{\rm bol}$ diagram with simple protostellar evolutionary tracks (Saraceno et al. 1996*a,b*). Qualitatively at least, scenarios in which the mass accretion rate decreases with time for a given protostar (Bontemps et al. 1996a; Myers et al. 1998; see also section III.D below) and increases with the mass of the initial precollapse fragment (e.g., Myers and Fuller 1993; Reipurth et al. 1993; Saraceno et al. 1996*b*) yield tracks in better agreement with observations than the constant-rate scenario discussed by Saraceno et al. (1996a). In particular, the peak accretion luminosity is reduced by a factor ~2–4 compared to the constant-rate scenario (cf. Bontemps et al. 1996*a*; Myers et al. 1998; Fig. 6b), which agrees better with observed luminosities (e.g., Kenyon and Hartmann 1995).

Inclination effects may *a priori* affect the positions of individual protostellar objects in these evolutionary diagrams. In particular, it has been claimed (e.g., Yorke et al. 1995; Sonnhalter et al. 1995; Men'shchikov and Henning 1997) that some Class I sources may potentially look like Class 0 objects when observed at high inclination angles to the line of sight. However, the fact that Class 0 objects are associated with outflows an order of magnitude more powerful than in Class I sources (see section III.D below) confirms that these two types of YSOs differ qualitatively from each other. Furthermore, some Class 0 sources are known to have small inclination angles (e.g., Cabrit and Bertout 1992; Greaves et al. 1997; Wolf-Chase et al. 1998). We also stress that existing (sub)millimeter maps of dust continuum and $C^{18}O$ emission provide direct evidence that both Class I and Class 0 objects are *self-embedded* in substantial amounts of circumstellar material distributed in spatially resolved, spheroidal envelopes (e.g., André and Montmerle 1994; Chen et al. 1997; Ladd et al. 1998; Dent et al. 1998; see also section III.B below). This material has the ability to absorb the optical and near IR emission from the underlying star-disk system and to reradiate it quasi-isotropically at longer far-IR and (sub)millimeter wavelengths. In such a self-embedded configuration, viewing angle effects are minimized, as confirmed by radiative transfer calculations (e.g., Efstathiou and Rowan-Robinson 1991; Yorke et al. 1995). Physically, this is because the bulk of the luminosity emerges at long wavelengths, where the envelope is effectively optically thin. However, the short-wavelength emission from the inner star-disk remains very dependent on viewing angle, implying that $T_{\rm bol}$ estimates should be somewhat more sensitive to orientation effects than $L_{\rm bol}$ and $L_{\rm smm}/L_{\rm bol}$.

4. Protostar Surveys and Lifetime Estimates. Based on the key attributes of Class 0 protostars (see section III.A.2 above), various strategies can be used to search for and discover new candidates: (sub)millimeter continuum mapping (e.g., Mezger et al. 1992; Casali et al. 1993; Sandell and Weintraub 1994; Reipurth et al. 1996; Launhardt et al. 1996; Chini et al. 1997*a*, *b*; Motte et al. 1998; André et al. 1999), high-resolution (HIRES) processing of the IRAS data (Hurt and Barsony 1996; O'Linger et al. 1999), deep radio continuum VLA surveys (e.g., Leous et al. 1991; Bontemps et al. 1995; Yun et al. 1996; Moreira et al. 1997; Gibb 1999), CO mapping (e.g., Bachiller et al. 1990; André et al. 1990; Bourke et al. 1997), and large-scale near IR/optical imaging of shocked H_2 and [S II] emission (e.g., Hodapp and Ladd 1995; Yu et al. 1997; Wilking et al. 1997; Gómez et al. 1998; Stanke et al. 1998; Phelps and Barsony 2000).

Class 0 objects appear to be short-lived compared to both prestellar cores/fragments (see section II above) and Class I near-IR sources. For example, in the ρ Oph main cloud, where the mapping study of Motte et al. (1998) with the Institut de Radio Astronomie Millimétrique (IRAM) 30-m telescope provides a reasonably complete census of both pre- and protostellar condensations down to $\lesssim 0.1$ $M_\odot$, there are only two good Class 0 candidates (including the prototypical object VLA 1623; see André et al. 1993), whereas there are 15 to 30 near-IR Class I sources (e.g., Wilking et al. 1989; André and Montmerle 1994; Greene et al. 1994; Barsony and Ressler 2000; Bontemps et al. 1999).

Under the assumption that ρ Oph is representative and forming stars at a roughly constant rate, the lifetime of Class 0 objects should thus be approximately an order of magnitude shorter than the lifetime of Class I sources (see Section III.A.1 above), i.e., typically ~ 1–3×10^4 yr. The jetlike morphology and short dynamical timescales of Class 0 outflows are consistent with this estimate (e.g., Barsony et al. 1998; see section III.D below). A lifetime as short as a few 10^4 yr supports the interpretation of Class 0 objects as very young accreting protostars (see, e.g., Fletcher and Stahler 1994 and Barsony 1994).

Regions such as the Serpens, Orion, and Perseus/NGC 1333 complexes seem to be particularly rich in Class 0 objects, and this is indicative of fairly high levels of ongoing, probably induced star formation activity (e.g., Hurt and Barsony 1996; Chini et al. 1997*b*; Yu et al. 1997; Barsony et al. 1998). For instance, we estimate the current star formation rate in Orion OMC-3 to be $\sim 2 \times 10^{-3}$ $M_\odot$ yr^{-1}, which is 1–2 orders of magnitude larger than the star formation rates characterizing the Trapezium, NGC 1333, and IC 348 near IR clusters when averaged over $\gtrsim 10^6$ yr periods (see Lada et al. 1996).

B. Density Structure of the Protostellar Environment

1. Envelope. In contrast to prestellar cores, the envelopes of low-mass Class 0 and Class I *protostars* are always found to be strongly centrally condensed and do not exhibit any inner flattening in their (sub)millimeter continuum radial intensity profiles. In practice, this means that, when protostars are mapped with the resolution of the largest single-dish telescopes, the measured peak flux density is typically a fraction $\gtrsim$20% of the flux integrated over five beam widths. For comparison, the same fraction is

$\lesssim$10% for prestellar cores. (Sub)millimeter continuum maps indicate that protostellar envelopes in regions of isolated star formation such as Taurus have radial density gradients generally consistent with $\rho(r) \propto r^{-p}$, with $p \sim 1.5$–2, over more than ~10,000–15,000 AU in radius (e.g., Walker et al. 1990; Ladd et al. 1991; Motte 1998; Motte and André 1999; see also the chapter by Mundy et al., this volume). The estimated density gradient thus agrees with most collapse models, which predict a value of p between 1.5 and 2 during the protostellar accretion phase (e.g., Whitworth and Summers 1985). Some studies have, however, inferred shallower density gradients ($p \sim 0.5$–1; e.g., Barsony and Chandler 1993; Chandler et al. 1998). In any case, the densities and masses measured for the envelopes around the bona fide Class I objects of Taurus appear to be consistent within a factor of ~4 with the predictions of the "standard" inside-out collapse theory (e.g., Shu et al. 1993) for ~10^5-yr-old, isolated protostars (Motte 1998; Motte and André 1999).

The situation is markedly different in star-forming clusters. In ρ Ophiuchi in particular, the circumstellar envelopes of Class I and Class 0 protostars are observed to be very compact: they merge with dense cores, other envelopes, or the diffuse ambient cloud at a finite radius $R_{\rm out} \lesssim 5000$ AU (Motte et al. 1998). This is $\gtrsim$3 times smaller than the collapse expansion wavefront at a "Class I age" of $\sim 2 \times 10^5$ yr in the standard theory of isolated protostars, emphasizing the fact that each YSO has a finite "sphere of influence" in ρ Oph. Similar results were obtained in the case of the Perseus Class 0 sources NGC 1333-4A, NGC 1333-2, L1448-C, and L1448-N (e.g., Motte 1998). Moreover, the envelopes of these Perseus protostars are 3 to 10 times denser than the singular isothermal sphere for a sound speed $a = 0.2$ km s^{-1}. This suggests that, prior to collapse, the main support against gravity was turbulent or magnetic in origin rather than purely thermal (see also Mardones et al. 1997).

2. *Disks and Multiplicity.* Many Class 0 protostars are seen to be multiple systems when viewed at subarcsecond resolution, sharing a common envelope and sometimes a circumbinary disk (e.g., Looney et al. 1999; see also col. 11 of Table I and the chapter by Mundy et al., this volume). These protobinaries probably formed by dynamical fragmentation during (or at the end of) the isothermal collapse phase (e.g., Chapman et al. 1992; Bonnell 1994; Boss and Myhill 1995). Interestingly enough, only cores with inner density profiles as flat as $\rho \propto r^{-1}$ or flatter (like observed prestellar cores; see section II), can apparently fragment during collapse (Myhill and Kaula 1992; Boss 1995; Burkert et al. 1997).

Despite the difficulty of discriminating between the disk and envelope components, existing (sub)millimeter continuum interferometric measurements suggest that the "disks" of Class 0 objects are a factor of $\gtrsim$10 less massive than their surrounding circumstellar envelopes (e.g., Chandler et al. 1995; Pudritz et al. 1996; Looney et al. 1999; Hogerheijde 1998; Motte 1998; and the chapter by Wilner and Lay, this volume).

C. Direct Evidence for Infall

Rather convincing spectroscopic signatures of gravitational infall have been reported for several Class 0 objects, confirming their protostellar nature (e.g., Walker et al. 1986; Zhou et al. 1993; Gregersen et al. 1997; Mardones et al. 1997; see also col. 10 of Table I). Inward motions can be traced by optically thick molecular lines, which should (locally) exhibit asymmetric self-absorbed profiles skewed to the blue (see the chapter by Myers et al., this volume). The interpretation is often complicated by the simultaneous presence of rotation, outflow, or both (e.g., Menten et al. 1987; Walker et al. 1994; Cabrit et al. 1996).

A comprehensive survey of a sample of 47 embedded YSOs in H_2CO (2_{12}–1_{11}) and CS (2–1) suggests that infall is more prominent in Class 0 than in Class I sources (Mardones et al. 1997). In these transitions, infall asymmetries are detected in 40–50% of Class 0 objects but less than 10% of Class I sources. This is qualitatively consistent with a decline of infall/accretion rate with evolutionary stage (see section III.D below). However, a more recent survey by Gregersen et al. (1999) using HCO^+ (3–2) finds no difference in the fraction of sources with "blue profiles" between Class 0 and Class I sources. The HCO^+ (3–2) line is more optically thick than H_2CO (2_{12}–1_{11}) and CS (2–1), and is therefore a better tracer of infall at advanced stages. This result shows that some infall is still present at the Class I stage, but it remains consistent with a decline of the net accretion rate with time. The outflow is so broad in Class I sources that there often appears to be little transfer of mass to the inner ~2000 AU radius region around these objects (e.g., Fuller et al. 1995*b*; Cabrit et al. 1996; Brown and Chandler 1999).

D. Decline of Outflow and Inflow with Time

1. Evolution from Class 0 to Class I. Most, if not all, Class 0 protostars drive powerful, "jetlike" CO molecular outflows (see, e.g., Bachiller 1996 and the chapter by Richer et al., this volume, for reviews). The mechanical luminosities of these outflows are often of the same order as the bolometric luminosities of the central sources (e.g., Curiel et al. 1990; André et al. 1993; Barsony et al. 1998). In contrast, while there is good evidence that some outflow activity exists throughout the accretion phase (e.g., Terebey et al. 1989; Parker et al. 1991; Bontemps et al. 1996*a*), the CO outflows from Class I sources tend to be much less powerful and less collimated than those from Class 0 objects.

In an effort to quantify the evolution of mass loss during the protostellar phase, Bontemps et al. (1996*a*) obtained and analyzed a homogeneous set of CO (2–1) outflow data around a large sample of low-luminosity ($L_{bol} < 50\ L_\odot$), nearby ($d < 450$ pc) self-embedded YSOs. Their results show that Class 0 objects lie an order of magnitude above the well-known (e.g., Cabrit and Bertout 1992) correlation between outflow momentum flux (F_{CO}) and bolometric luminosity (L_{bol}) holding for Class I sources (see the F_{CO}–L_{bol} diagram shown in Fig. 5 of Bontemps et al. 1996*a*). Furthermore, they found that F_{CO} was well correlated with M_{env} in their *entire* sample (including both Class I and Class 0 sources). The same correlation was noted independently on other source samples by Moriarty-Schieven et al. (1994), Hogerheijde et al. (1998), and Henning and Launhardt (1998). As argued by Bontemps et al. (1996*a*), this new correlation is independent of the F_{CO}–L_{bol} correlation and most likely results from a progressive decrease of outflow power with time during the accretion phase. This is illustrated in the normalized $F_{CO}c/L_{bol}$ versus $M_{env}/L_{bol}^{0.6}$ diagram of Fig. 7, which should be essentially free of any luminosity effect.

Since magnetocentrifugal accretion/ejection models of bipolar outflows (e.g., Shu et al. 1994; Ferreira and Pelletier 1995; Fiege and Henriksen 1996; Ouyed and Pudritz 1997) predict a direct proportionality

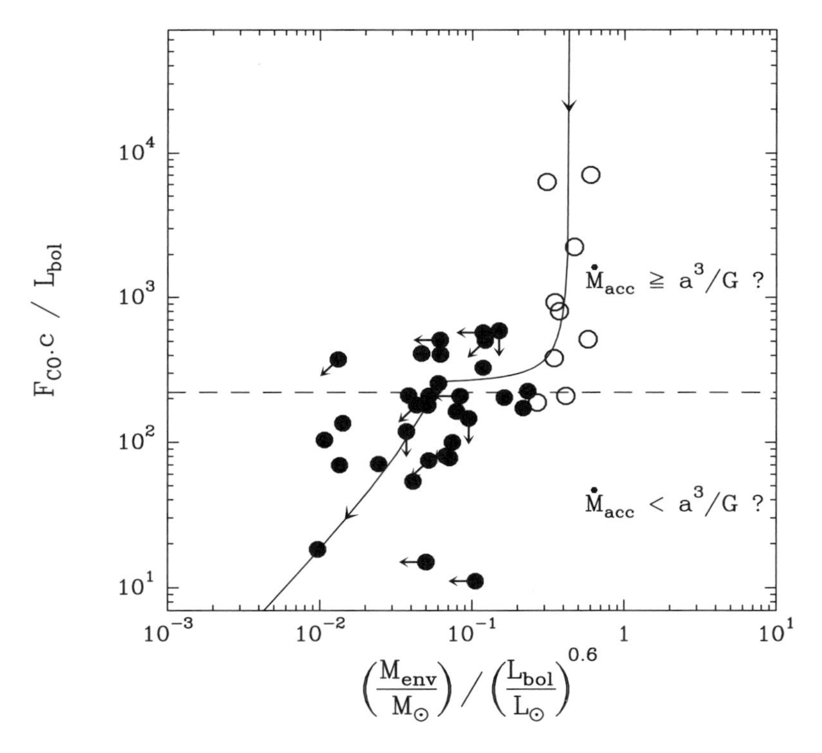

Figure 7. Normalized outflow momentum flux, $F_{CO}\ c/L_{bol}$, versus normalized envelope mass, $M_{env}/L_{bol}^{0.6}$, for a sample of Class 0 (open circles) and Class I (filled circles) objects (from Bontemps et al. 1996a). $F_{CO}\ c/L_{bol}$ can be taken as an empirical tracer of the accretion rate $\dot{M}_{acc}$, while $M_{env}/L_{bol}^{0.6}$ is an evolutionary indicator that decreases with time. This diagram should therefore mainly reflect the evolution of $\dot{M}_{acc}$ during the protostellar phase. The solid curve shows the accretion rate history predicted by the simplified collapse model of Henriksen et al. (1997).

between accretion and ejection, Bontemps et al. (1996*a*) proposed that the decline of outflow power with evolutionary stage seen in Fig. 7 reflects a corresponding decrease in the mass accretion rate. The results of Bontemps et al. (1996*a*) indicate that $\dot{M}_{\rm jet}$ declines from $\sim 10^{-6}$ M$_\odot$ yr^{-1} for the youngest Class 0 protostars to $\sim 2 \times 10^{-8}$ M$_\odot$ yr^{-1} for the most evolved Class I sources, suggesting a decrease in $\dot{M}_{\rm acc}$ from $\sim 10^{-5}$ M$_\odot$ yr^{-1} to $\sim 2 \times 10^{-7}$ M$_\odot$ yr^{-1} if realistic jet model parameters are adopted ($\dot{M}_{\rm jet}/\dot{M}_{\rm acc} \sim 0.1$–$0.3$, $V_{\rm jet} \sim 100$ km s^{-1}). These indirect estimates of $\dot{M}_{\rm acc}$ for Class 0 and Class I protostars should only be taken as indicative of the true evolutionary trend. Nevertheless, it is interesting to note that they agree well with independent estimates of the rates of envelope dissipation based on circumstellar mass versus age arguments (Ward-Thompson 1996; Ladd et al. 1998).

As illustrated by the evolutionary tracks of Fig. 6b, a decline of $\dot{M}_{\rm acc}$ with time does not imply a higher accretion luminosity for Class 0 objects compared to Class I sources, because the central stellar mass is smaller at the Class 0 stage and the stellar radius is likely to be larger (see Henriksen et al. 1997).

2. *Link with the Collapse Initial Conditions.* The apparent decay of $\dot{M}_{\rm acc}$ from the Class 0 to the Class I stage may be linked with the density structure observed for prestellar cores/condensations (Sect. II). (Magneto)hydrodynamic collapse models predict a time-dependent accretion history when the radial density profile at the onset of fast protostellar collapse differs from $\rho \propto r^{-2}$ (e.g., Foster and Chevalier 1993; Tomisaka 1996; McLaughlin and Pudritz 1997; Henriksen et al. 1997; Basu 1997; Safier et al. 1997; Li 1998; and Ciolek and Königl 1998). In particular, when starting from Bonnor-Ebert-like initial conditions resembling observed prestellar cores, these studies find that supersonic infall velocities develop prior to the formation of the central hydrostatic protostellar object at $t = 0$ (see, e.g., Foster and Chevalier 1993). Observationally, this early collapse phase should correspond to "isothermal protostars" (see section II.E for possible examples). During the protostellar accretion phase ($t > 0$), because of the significant infall velocities achieved at $t < 0$, $\dot{M}_{\rm acc}$ is initially higher than the standard $\sim a^3/G$ value obtained for the inside-out collapse of a static singular isothermal sphere (Shu 1977; see also section I). The accretion rate then converges toward the standard value of the Shu solution, and declines again below a^3/G at late times if the reservoir of mass is finite. By comparison with the rough estimates of $\dot{M}_{\rm acc}$ given above, it is tempting to identify the short period of energetic accretion ($\dot{M}_{\rm acc} \sim 10 \times a^3/G$) predicted by the models just after point-mass formation with the observationally defined Class 0 stage (Henriksen et al. 1997; see Fig. 7). In this view, the more evolved Class I objects would correspond to the longer period of moderate accretion/ejection, when the accretion rate approaches the standard value ($\dot{M}_{\rm acc} \lesssim a^3/G \sim 2 \times 10^{-6}$ M$_\odot$ yr^{-1} for a cloud temperature ~ 10 K).

Using simple pressure-free analytical calculations (justified, since the inflow becomes supersonic early on), Henriksen et al. (1997) could indeed find a good overall fit to the empirical accretion history inferred by Bontemps et al. (1996*a*) on the basis of CO outflow observations (see Fig. 7).

In the absence of magnetic fields, Foster and Chevalier (1993) showed that the timescale for convergence to the standard accretion rate of Shu (1977) depends on the radius of the flat inner core ($R_{\rm flat}$) relative to the outer radius of the initial precollapse condensation ($R_{\rm out}$). They found that a phase of constant $\sim a^3/G$ accretion rate is achieved only when $R_{\rm out}/R_{\rm flat} \gtrsim 20$, typically after ~ 10 free-fall times of the flat inner region, for a period lasting ~ 15 free-fall times when $R_{\rm out}/R_{\rm flat} = 20$ and progressively longer as $R_{\rm out}/R_{\rm flat}$ increases. Since observations suggest $R_{\rm out}/R_{\rm flat} \lesssim 3$ in Ophiuchus and $R_{\rm out}/R_{\rm flat} \gtrsim 15$ in Taurus (see sections II.C and III.B, and Motte et al. 1998), one may expect a marked time dependence of $\dot{M}_{\rm acc}$ in Ophiuchus but a reasonable agreement with the constant accretion rate of the self-similar theory of Shu et al. (1987, 1993) in Taurus. Indeed, Henriksen et al. (1997) note that there is a much better continuity between Class 0 and Class I protostars in Taurus than in Ophiuchus (see also André 1997).

Finally, we stress that the absolute values of $\dot{M}_{\rm acc}$ in the Class 0 and Class I stages are presently quite uncertain. An alternative interpretation of the evolution seen in Fig. 7 is that Class 0 protostars accrete at a rate roughly consistent with $\sim a^3/G$, while most Class I sources are in a terminal accretion phase with $\dot{M}_{\rm acc} \lesssim 0.1 \times a^3/G$, resulting from the finite effective reservoir of mass available to each object in clusters (e.g., Motte et al. 1998 and section III.B) or from the effects of outflows dispersing the envelope (e.g., Myers et al. 1998; Ladd et al. 1998). Distinguishing between this possibility and that advocated in the preceding paragraph will require direct measurements of the mass accretion rates.

IV. CONCLUSIONS AND IMPLICATIONS

The observational studies discussed in section II demonstrate that prestellar cores and fragments are characterized by flat inner radial density gradients. This, in turn, suggests that the initial conditions for fast protostellar collapse are nonsingular; that is, the density profile at the onset of collapse is not infinitely centrally condensed. (Sub)millimeter observations also set strong constraints on protostellar evolution (section III). The fact that young (Class 0) protostars drive more powerful outflows than evolved (Class I) protostars suggests that the mass accretion rate $\dot{M}_{\rm acc}$ decreases by typically a factor of $\sim$5–10 from the Class 0 to the Class I stage (section III.D). Such a decline in $\dot{M}_{\rm acc}$ during the protostellar accretion phase may be the direct result of a flattened initial density profile (see section III.D). Based on these observational constraints, we suggest that most protostars form in a dynamical rather than quasistatic fashion.

The results summarized in this chapter also have broader implications concerning, e.g., the origin of the IMF. As pointed out in section II.E, the prestellar condensations observed in regions of multiple star formation such as ρ Ophiuchi are finite-size structures, typically a few thousand AU in radius, which are clearly not scale-free. This favors a picture of star formation in clusters in which individual protostellar collapse is initiated in compact dense clumps, resulting from fragmentation and resembling finite-size Bonnor-Ebert cloudlets more than singular isothermal spheres. Such condensations may correspond to dense, low-ionization pockets decoupling themselves from the parent molecular cloud as a result of ambipolar diffusion or the dissipation of turbulence (e.g., Mouschovias 1991; Myers 1998; Nakano 1998). They would thus be free to undergo Jeans-like gravitational instabilities and collapse under the influence of external disturbances, of which there are many types in regions of multiple star formation (e.g., Pringle 1989; Whitworth et al. 1996; Motte et al. 1998; Barsony et al. 1998). By contrast, the low-density regions of the ambient cloud, being more ionized, would remain supported against collapse by static and turbulent magnetic fields. The typical separation between individual condensations should be of the order of the Jeans length in the parent cloud or core, in rough agreement with observations (e.g., Motte et al. 1998).

In this observationally driven scenario of fragmentation and collapse, stars are built from bounded fragments that represent finite reservoirs of mass. The star formation efficiency within these fragments is high: most of their "initial" masses at the onset of collapse end up in stars. If this is true, it implies that the physical mechanisms responsible for the formation of prestellar cores and condensations in molecular clouds, such as turbulent fragmentation (e.g., Padoan et al. 1997), play a key role in determining the IMF of embedded clusters. Such a picture, which we favor for regions of multiple star formation, is consistent with some theoretical scenarios of protocluster formation (e.g., Larson 1985; Klessen et al. 1998). It need not be universal, however, and it is in fact unlikely to apply to regions of isolated star formation such as the Taurus cloud. In these regions, protostars may accrete from larger reservoirs of mass, and feedback processes such as stellar winds may be more important in limiting accretion and defining stellar masses (e.g., Shu et al. 1987; Silk 1995; Adams and Fatuzzo 1996; Velusamy and Langer 1998).

With the advent of major new facilities at far-IR and (sub)millimeter wavelengths, the next decade promises to be at least as rich in observational discoveries as the past ten years. By combining the capabilities of a space telescope such as FIRST with those of a large ground-based interferometric array such as ALMA, it will be possible to study the detailed physics of complete samples of young protostars and precollapse fragments, in a variety of star-forming clouds, and down to the brown-dwarf mass regime. This should tremendously improve our global understanding of the initial stages of star formation in the Galaxy.

Acknowledgments We thank C. Correia for providing unpublished data shown in Fig. 2; G. Ciolek for Fig. 3b; P. Myers for Fig. 6a; and N. Grosso for assistance in preparing some of the figures. We are also grateful to F. Motte and the referee, N. Evans, for useful comments and suggestions.

REFERENCES

Abergel, A., Bernard, J. P., Boulanger, F., et al. 1996. ISOCAM mapping of the ρ Ophiuchi main cloud. *Astron. Astrophys.* 315:L329–L332.

Abergel, A., Bernard, J. P., Boulanger, F., et al. 1998. The dense core Oph D seen in extinction by ISOCAM. In *Star Formation with ISO,* ASP Conf. Ser. 132, ed. J. L. Yun and R. Liseau (San Francisco: Astronomical Society of the Pacific), pp. 220–229.

Adams, F. C., and Fatuzzo, M. 1996. A theory of the initial mass function for star formation in molecular clouds. *Astrophys. J.* 464:256–271.

Adams, F. C., Lada, C. J., and Shu, F. H. 1987. Spectral evolution of young stellar objects. *Astrophys. J.* 312:788–806.

Akeson, R. L., and Carlstrom, J. E. 1997. Magnetic field structure in protostellar envelopes. *Astrophys. J.* 491:254–266.

André, P. 1997. The evolution of flows and protostars. In *Herbig-Haro Flows and the Birth of Low Mass Stars,* IAU Symp. 182, ed. B. Reipurth and C. Bertout (Dordrecht: Kluwer), pp. 483–494.

André, P., and Montmerle, T. 1994. From T Tauri stars to protostars: Circumstellar material and young stellar objects in the ρ Ophiuchi cloud. *Astrophys. J.* 420:837–862.

André, P., Martín-Pintado, J., Despois, D., and Montmerle, T. 1990. Discovery of a remarkable bipolar flow and exciting source in the ρ Ophiuchi cloud core. *Astron. Astrophys.* 236:180–192.

André, P., Ward-Thompson, D., and Barsony, M. 1993. Submillimeter continuum observations of ρ Ophiuchi A: The candidate protostar VLA 1623 and prestellar clumps. *Astrophys. J.* 406:122–141.

André, P., Ward-Thompson, D., and Motte, F. 1996. Probing the initial conditions of star formation: The structure of the prestellar core L1689B. *Astron. Astrophys.* 314:625–635.

André, P., Motte, F., and Bacmann, A. 1999. Discovery of an extremely young accreting protostar in Taurus. *Astrophys. J. Lett.,* 513:L57–L60.

Anglada, G., Estalella, R., Rodríguez, L. F., Torrelles, J. M., López, R., and Cantó, J. 1991. A double radio source at the center of the outflow in L723. *Astrophys. J.* 376:615–617.

Anglada, G., Rodríguez, L. F., Cantó, J., Estalella, R., and Torrelles, J. M. 1992. Radio continuum from the powering sources of the RNO 43, HARO 4-255 FIR, B335, and PV Cephei outflows and from the Herbig-Haro object 32A. *Astrophys. J.* 395:494–500.

Avery, L. W., Hayashi, S. S., and White, G. J. 1990. The unusual morphology of the high-velocity gas in L723: One outflow or two? *Astrophys. J.* 357:524–530.

Bachiller, R. 1996. Bipolar molecular outflows from young stars and protostars. *Ann. Rev. Astron. Astrophys.* 34:111–154.

Bachiller, R., Martín-Pintado, J., Tafalla, M., Cernicharo, J., and Lazareff, B. 1990. High-velocity molecular bullets in a fast bipolar outflow near L1448/IRS3. *Astron. Astrophys.* 231:174–186.

Bachiller, R., André, P., and Cabrit, S. 1991*a*. Detection of the exciting source of the spectacular molecular outflow L1448 at $\lambda\lambda$ 1–3 mm. *Astron. Astrophys.* 241:L43–L46.

Bachiller, R., Martín-Pintado, J., and Planesas, P. 1991*b*. High-velocity molecular jets and bullets from IRAS 03282+3035. *Astron. Astrophys.* 251:639–648.

Bachiller, R., Terebey, S., Jarrett, T., Martín-Pintado, J., Beichman, C. A., and van Buren, D. 1994. Shocked molecular gas around the extremely young source IRAS 03282+3035. *Astrophys. J.* 437:296–304.

Bachiller, R., Guilloteau, S., Dutrey, A., Planesas, P., and Martín-Pintado, J. 1995. The jet-driven molecular outflow in L 1448. CO and continuum synthesis images. *Astron. Astrophys.* 299:857–868.

Bachiller, R., Guilloteau, S., Gueth, F., Tafalla, M., Dutrey, A., Codella, C., and Catsets, A. 1998. A molecular jet from SVS 13B near HH 7-11. *Astron. Astrophys.* 339:L49–L52.

Bacmann, A., André, P., Abergel, A., et al. 1998. An ISOCAM absorption study of dense cloud cores. In *Star Formation with ISO,* ASP Conf. Ser. 132, ed. J. L. Yun and R. Liseau (San Francisco: Astronomical Society of the Pacific), pp. 307–313.

Bally, J., Lada, E. A., and Lane, A. P. 1993. The L1448 molecular jet. *Astrophys. J.* 418:322–327.

Balsara, D., Ward-Thompson, D., Pouquet, A., and Crutcher, R. M. 1998. An MHD model of the interstellar medium and a new method of accretion onto dense star-forming cores. In *Interstellar Turbulence,* ed. J. Franco and A. Carramiñana (Cambridge: Cambridge University Press), pp. 261–266.

Barsony, M. 1994. Class 0 protostars. In *Clouds, Cores, and Low-mass Stars,* ASP Conf. Ser. 65, ed. D. P. Clemens and R. Barvainis (San Francisco: Astronomical Society of the Pacific), pp. 197–206.

Barsony, M., and Chandler, C. J. 1993. The circumstellar density distribution of L1551NE. *Astrophys. J.* 406:L71–L74.

Barsony, M., and Kenyon, S. J. 1992. On the origin of submillimeter emission from young stars in Taurus-Auriga. *Astrophys. J.* 384:L53–L57.

Barsony, M., and Ressler, M. 2000. The initial luminosity function for L1688: New mid-IR imaging photometry. *Astrophys. J.,* in preparation.

Barsony, M., Ward-Thompson, D., André, P., and O'Linger, J. 1998. Protostars in Perseus: Outflow induced fragmentation. *Astrophys. J.* 509:733–748.

Basu, S. 1997. A semianalytic model for supercritical core collapse: Self-similar evolution and the approach to protostar formation. *Astrophys. J.* 485:240–253.

Basu, S., and Mouschovias, T. C. 1995. Magnetic braking, ambipolar diffusion, and the formation of cloud cores and protostars. III. Effect of the initial mass-to-flux ratio. *Astrophys. J.* 453:271–283.

Bate, M. R. 1998. Collapse of a molecular cloud core to stellar densities: The first three-dimensional calculations. *Astrophys. J. Lett.* 508:L95–L98.

Beckwith, S. V. W., Sargent A. I, Chini, R. S., and Guesten, R. 1990. A survey for circumstellar disks around young stellar objects. *Astron. J.* 99:924–945.

Beichman, C. A., Myers, P. C., Emerson, J. P., et al. 1986. Candidate solar-type protostars in nearby molecular cloud cores. *Astrophys. J.* 307:337–349.

Bence, S. J., Richer, J. S., and Padman, R. 1996. RNO 43: A jet-driven super-outflow. *Mon. Not. Roy. Astron. Soc.* 279:866–883.

Benson, P. J., and Myers, P. C. 1989. A survey for dense cores in dark clouds. *Astrophys. J. Sup.* 71:89–108.

Blake, G. A., Sandell, G., van Dishoeck, E. W., Groesbeck, T. D., Mundy, L. G., and Aspin, C. 1995. A molecular line study of NGC 1333/IRAS 4. *Astrophys. J.* 441:689–701.

Blitz, L. 1993. Giant molecular clouds. In *Protostars and Planets III,* ed. E. H. Levy and J. I. Lunine (Tucson: University of Arizona Press), pp. 125–161.

Blottiau, P., Chièze, J. P., and Bouquet, S. 1988. An asymptotic self-similar solution for the gravitational collapse. *Astron. Astrophys.* 207:24–36.

Bonnell, I. A. 1994. Fragmentation and the formation of binary and multiple systems. In *Clouds, Cores, and Low-mass Stars,* ASP Conf. Ser. 65, ed. D. P. Clemens and R. Barvainis (San Francisco: Astronomical Society of the Pacific), pp. 115–124.

Bonnell, I. A., Bate, M. R., Clarke, C. J., and Pringle, J. E. 1997. Accretion and the stellar mass spectrum in small clusters. *Mon. Not. Roy. Astron. Soc.* 285:201–208.

Bontemps, S. 1996. Evolution de l'éjection de matière des proto-étoiles. Ph.D. Dissertation, University of Paris, XI.

Bontemps, S., André, P., and Ward-Thompson, D. 1995. Deep VLA search for the youngest protostars: A Class 0 source in the HH24-26 region. *Astron. Astrophys.* 297:98–102.

Bontemps, S., André, P., Terebey, S., and Cabrit, S. 1996*a*. Evolution of outflow activity around low-mass embedded young stellar objects. *Astron. Astrophys.* 311:858–872.

Bontemps, S., Ward-Thompson, D., and André, P. 1996*b*. Discovery of a jet emanating from the protostar HH 24 MMS. *Astron. Astrophys.* 314:477–483.

Bontemps, S., Nordh, L., Olofsson, G., André, P., Huldtgren, M., Kaas, A. A., Abergel, A., Blommaert, J., Boulanger, F., Burgdorf, M., Cenarsky, C. J., Cenarsky, D., Copet, E., Davies, J., Falgarone, E., Lagouche, G., Montmerle, T., Pérault, M., Persi, P., Prunti, T., Puget, J. L., and Sibille, F. 1999. Constraints on the initial mass function from ISOCAM observations of the ρ Ophiuchi embedded cluster. *Astron. Astrophys.*, submitted.

Boss, A. P. 1995. Gravitational collapse and binary protostars. *Rev. Mex. Astron. Astrofis. Ser. Conf.* 1:165–177.

Boss, A. P., and Myhill, E. A. 1995. Collapse and fragmentation of molecular cloud cores. III. Initial differential rotation. *Astrophys. J.* 451:218–224.

Boss, A. P., and Yorke, H. W. 1995. Spectral energy of first protostellar cores: Detecting "class $-$I" protostars with ISO and SIRTF. *Astrophys. J.* 439:L55–L58.

Bourke, T. L., Hyland, A. R., and Robinson, G. 1995*a*. Studies of star formation in isolated small dark clouds. I. A catalogue of southern Bok globules: Optical and IRAS properties. *Mon. Not. Roy. Astron. Soc.* 276:1052–1066.

Bourke, T. L., Hyland, A. R., Robinson, G., James, S. D., and Wright, C. M. 1995*b*. Studies of star formation in isolated small dark clouds. II. A southern ammonia survey. *Mon. Not. Roy. Astron. Soc.* 276:1067–1084.

Bourke, T. L., Garay, G., Lehtinen, K. K., et al. 1997. Discovery of a highly collimated molecular outflow in the southern Bok globule BHR 71. *Astrophys. J.* 476:781–800.

Brown, D. W., and Chandler, C. J. 1999. Circumstellar kinematics and the measurement of stellar mass for the protostars TMC1 and TMC1A. *Mon. Not. Roy. Astron. Soc.,* 303:855–863.

Burkert, A., Bate, M. R., and Bodenheimer, P. 1997. Protostellar fragmentation in a power-law density distribution. *Mon. Not. Roy. Astron. Soc.* 289:497–504.

Butner, H. M., Lada, E. A., and Loren, R. B. 1995. Physical properties of dense cores: DCO + observations. *Astrophys. J.* 448:207–225.

Cabrit, S., and André, P. 1991. An observational connection between circumstellar disk mass and molecular outflows. *Astrophys. J. Lett.* 379:L25–L28.

Cabrit, S., and Bertout, C. 1992. CO line formation in bipolar flows. III. The energetics of molecular flows and ionized winds. *Astron. Astrophys.* 261:274–284.

Cabrit, S., Goldsmith, P. F., and Snell, R. L. 1988. Identification of RNO 43 and B335 as two highly collimated bipolar flows oriented nearly in the plane of the sky. *Astrophys. J.* 334:196–208.

Cabrit, S., Guilloteau, S., André, P., Bertout, C., Montmerle, T., and Schuster, K. 1996. Plateau de Bure observations of HL Tauri: Outflow motions in a remnant circumstellar envelope. *Astron. Astrophys.* 305:527–540.

Casali, M. M., Eiroa, C., and Duncan, W. D. 1993. A second phase of star formation in the Serpens core. *Astron. Astrophys.* 275:195–200.

Ceccarelli, C., Caux, E., White, G. J., Molinari, S., Furniss, I., Liseau, R., and Nisini, B. 1998. The far infrared line spectrum of the protostar IRAS 16293-2422. *Astron. Astrophys.* 331:372–382.

Cernicharo, J., Lefloch, B., Cox, P., Cenarsky, D., Yusef-Zadeh, F., Mendez, D. I., Acosta-Pulido, J., López, R. J. G., and Heran, A. 1998. Induced massive star formation in the Trifid nebula? *Science* 282:462–465.

Chandler, C. J., Gear, W. K., Sandell, G., Hayashi, S., Duncan, W. D., and Griffin, M. J. 1990. B335: Protostar or embedded pre-main-sequence star? *Mon. Not. Roy. Astron. Soc.* 243:330–335.

Chandler, C. J., Koerner, D. W., Sargent, A. I., and Wood, D. O. S. 1995. Dust emission from protostars: The disk and envelope of HH 24 MMS. *Astrophys. J.* 449:L139–L142.

Chandler, C. J., Barsony, M., and Moore, T. J. T. 1998. The circumstellar envelopes around three protostars in Taurus. *Mon. Not. Roy. Astron. Soc.* 299:789–798.

Chapman, S. J., Davies, J. R., Disney, M. J., Nelson, A. H., Pongracic, H., Turner, J. A., Whitworth, A. P. 1992. The formation of binary and multiple star systems. *Nature* 359:207–210.

Chen, H., Myers, P. C., Ladd, E. F., and Wood, D. O. S. 1995. Bolometric temperature and young stars in the Taurus and Ophiuchus complexes. *Astrophys. J.* 445:377–392.

Chen, H., Grenfell, T. G., Myers, P. C., and Hughes, J. D. 1997. Comparison of star formation in five nearby molecular clouds. *Astrophys. J.* 478:295–312.

Chièze, J.-P., and Pineau des Forêts, G. 1987. The fragmentation of molecular clouds. II. Gravitational stability of low-mass molecular cloud cores. *Astron. Astrophys.* 183:98–108.

Chini, R., Krügel, E., Haslam, C. G. T., Kreysa, E., Lemke, R., Reipurth, B., Sievers, A., and Ward-Thompson, D. 1993. Discovery of a cold and gravitationally unstable cloud fragment. *Astron. Astrophys.* 272:L5–L8.

Chini, R., Reipurth, B., Sievers, A., Ward-Thompson, D., Haslam, C. G. T., Kreysa, E., and Lemke, R. 1997*a*. Cold dust around Herbig-Haro energy sources: Morphology and new protostellar candidates. *Astron. Astrophys.* 325:542–550.

Chini, R., Reipurth, B., Ward-Thompson, D., Bally, J., Nyman, L. A., Sievers, A., and Billawala, Y. 1997*b*. Dust filaments and star formation in OMC-2 and OMC-3. *Astrophys. J.* 474:L135–L138.

Ciolek, G. E., and Königl, A. 1998. Dynamical collapse of nonrotating magnetic molecular cloud cores: Evolution through point-mass formation. *Astrophys. J.* 504:257–279.

Ciolek, G. E., and Mouschovias, T. C. 1994. Ambipolar diffusion, interstellar dust, and the formation of cloud cores and protostars. III. Typical axisymmetric solutions. *Astrophys. J.* 425:142–160.

Clemens, D. P., and Barvainis, R. 1988. A catalog of small, optically selected molecular clouds: Optical, infrared, and millimeter properties. *Astrophys. J. Sup.* 68:257–286.

Curiel, S., Raymond, J. C., Rodríguez, L. F., Cantó, J., and Moran, J. M. 1990. The exciting source of the bipolar outflow in L1448. *Astrophys. J. Lett.* 365:L85–L88.

Curiel, S., Rodríguez, L. F., Moran, J. M., and Cantó, J. 1993. The triple radio continuum source in Serpens: The birth of a Herbig-Haro system? *Astrophys. J.* 415:191–203.

Davidson, J. A. 1987. Low-luminosity embedded sources and their environs. *Astrophys. J.* 315:602–620.

Davis, C. J., and Eislöffel, J. 1995. Near-infrared imaging in H_2 of molecular (CO) outflows from young stars. *Astron. Astrophys.* 300:851–869.

Dent, W. R. F., Matthews, H. E., and Walther, D. M. 1995. CO and shocked H_2 in the highly collimated outflow from VLA 1623. *Mon. Not. Roy. Astron. Soc.* 277:193–209.

Dent, W. R. F., Matthews, H. E., and Ward-Thompson, D. 1998. The submillimetre colour of young stellar objects. *Mon. Not. Roy. Astron. Soc.* 301:1049–1063.

Efstathiou, A., and Rowan-Robinson, M. 1991. Radiative transfer in axisymmetric dust clouds. II. Models of rotating protostars. *Mon. Not. Roy. Astron. Soc.* 252:528–534.

Eiroa, C., Miranda, L. F., Anglada, G., Estalella, R., and Torrelles, J. M. 1994. Herbig-Haro objects associated with extremely young sources in L1527 and L1448. *Astron. Astrophys.* 283:973–977.

Falgarone, E., and Puget, J.-L. 1985. A model of clumped molecular clouds. I. Hydrostatic structure of dense cores. *Astron. Astrophys.* 142:157–170.

Ferreira J., and Pelletier G. 1995. Magnetized accretion-ejection structures. III. Stellar and extragalactic jets as weakly dissipative disk outflows. *Astron. Astrophys.* 295:807–832.

Fiege, J. D., and Henriksen, R. N. 1996. A global model of protostellar bipolar outflow. I. *Mon. Not. Roy. Astron. Soc.* 281:1038–1054.

Fletcher, A. B., and Stahler, S. W. 1994. The luminosity functions of embedded stellar clusters. I: Method of solution and analytic results. *Astrophys. J.* 435:313–328.

Foster, P. N., and Chevalier, R. A. 1993. Gravitational collapse of an isothermal sphere. *Astrophys. J.* 416:303–311.

Fuller, G. A., Lada, E. A., Masson, C. R., and Myers, P. C. 1995*a*. The infrared nebula and outflow in Lynds 483. *Astrophys. J.* 453:754–760.

Fuller, G. A., Lada, E. A., Masson, C. R., and Myers, P. C. 1995*b*. The circumstellar molecular core around L1551 IRS 5. *Astrophys. J.* 454:862–871.

Gammie, C. F., and Ostriker, E. C. 1996. Can nonlinear hydromagnetic waves support a self-gravitating cloud? *Astrophys. J.* 466:814–830.

Gibb, A. G. 1999. A VLA search for embedded young stellar objects and protostellar candidates in L1630. *Mon. Not. Roy. Astron. Soc.,* 304:1–7.

Gibb, A. G., and Davis, C. J. 1998. The outflow from the class 0 protostar HH25MMS: Methanol enhancement in a well-collimated outflow. *Mon. Not. Roy. Astron. Soc.* 298:644–656.

Gómez, J. F., Curiel, S., Torrelles, J. M., Rodríguez, L. F., Anglada, G., and Girart, J. M. 1994. The molecular core and the powering source of the bipolar molecular outflow in NGC 2264G. *Astrophys. J.* 436:749–753.

Gómez, M., Hartmann, L., Kenyon, S. J., and Hewett, R. 1993. On the spatial distribution of pre-main-sequence stars in Taurus. *Astron. J.* 105:1927–1937.

Gómez, M., Whitney, B. A., and Wood, K. 1998. A survey of optical jets and Herbig-Haro objects in the ρ Ophiuchi cloud core. *Astron. J.* 115:2018–2027.

Greaves, J. S., and Holland, W. S. 1998. Twisted magnetic field lines around protostars. *Astron. Astrophys.* 333:L23–L26.

Greaves, J. S., Holland, W. S., and Ward-Thompson, D. 1997. Submillimeter polarimetry of Class 0 protostars: Constraints on magnetized outflow models. *Astrophys. J.* 480:255–261.

Greene, T. P., and Lada, C. J. 1996. Near-infrared spectra and the evolutionary status of young stellar objects: Results of a 1.1–2.4 μm survey. *Astron. J.* 112:2184–2221.

Greene, T. P., Wilking, B. A., André, P., Young, E. T., and Lada, C. J. 1994. Further mid-infrared study of the ρ Ophiuchi cloud young stellar population: Luminosities and masses of pre-main-sequence stars. *Astrophys. J.* 434:614–626.

Gregersen, E. M., Evans, N. J., II, Zhou, S., and Choi, M. 1997. New protostellar collapse candidates: An HCO+ survey of the Class 0 sources. *Astrophys. J.* 484:256–276.

Gregersen, E. M., Evans, N. J., II, Mardones, D., and Myers, P. C. 1999. Does infall end before the Class I stage? *Astrophys. J.,* in preparation.

Grossman, E. N., Masson, C. R., Sargent, A. I., Scoville, N. Z., Scott, S., and Woody, D. P. 1987. A possible protostar near HH 7-11. *Astrophys. J.* 320:356–363.

Gueth, F., Guilloteau, S., Dutrey, A., and Bachiller, R. 1997. Structure and kinematics of a protostar: mm-Interferometry of L 1157. *Astron. Astrophys.* 323:943–952.

Güsten, R. 1994. Protostellar condensations. In *The Cold Universe,* ed. T. Montmerle, C. J. Lada, I. F. Mirabel, and J. Trân Thanh Vân (Gif-sur-Yvette: Editions Frontières), pp. 169–177.

Henning, T., and Launhardt, R. 1998. Millimetre study of star formation in southern globules. *Astron. Astrophys.* 338:223–242.

Henning, T., Michel, B., and Stognienko, R. 1995. Dust opacities in dense regions. *Planet. Space Sci.* 43:1333–1343.

Henriksen, R. N., André, P., and Bontemps, S. 1997. Time-dependent accretion and ejection implied by prestellar density profiles. *Astron. Astrophys.* 323:549–565.

Herbst, T. M., Beckwith, S, and Robberto, M. 1997. A new molecular hydrogen outflow in Serpens. *Astrophys. J. Lett.* 486:L59–L62.

Hildebrand, R. H. 1983. The determination of cloud masses and dust characteristics from submillimetre thermal emission. *Quart. J. Roy. Astron. Soc.* 24:267–282.

Hirano, N., Kameya, O., Kasuga, T., and Umemoto, T. 1992. Bipolar outflow in B335. The small-scale structure. *Astrophys. J. Lett.* 390:L85–L88.

Hodapp, K.-W., and Ladd, E. F. 1995. Bipolar jets from extremely young stars observed in molecular hydrogen emission. *Astrophys. J.* 453:715–720.

Hogerheijde, M. R. 1998. The molecular environment of low-mass protostars. Ph.D. Dissertation, University of Leiden (Amsterdam: Thesis Publishers).

Hogerheijde, M. R., van Dishoeck, E. F., Blake, G. A., and van Langevelde, H. J. 1998. Envelope structure on 700 AU scales and the molecular outflows of low-mass young stellar objects. *Astrophys. J.* 502:315–336.

Hogerheijde, M. R., van Dishoeck, E. F., Salverda, J. M., and Blake, G. A. 1999. Envelope structure of deeply embedded young stellar objects in the Serpens molecular cloud. *Astrophys. J.,* 513:350–369.

Holland, W. S., Greaves, J. S., Ward-Thompson, D., and André, P. 1996. The magnetic field structure around protostars. Submillimetre polarimetry of VLA 1623 and S106-IR/FIR. *Astron. Astrophys.* 309:267–274.

Hunter, T. R., Neugebauer, G., Benford, D. J., Matthews, K., Lis, D. C., Serabyn, E., and Phillips, T. G. 1998. G34.24+0.13MM: A deeply embedded proto-B-star. *Astrophys. J. Lett.* 493:L97–L100.

Hurt, R. L., and Barsony, M. 1996. A cluster of Class 0 protostars in Serpens: An IRAS HIRES study. *Astrophys. J. Lett.* 460:L45–L48.

Hurt, R. L., Barsony, M., and Wootten, A. 1996. Potential protostars in cloud cores: H_2CO observations of Serpens. *Astrophys. J.* 456:686–695.

Jenness, T., Scott, P. F., and Padman, R. 1995. Studies of embedded far-infrared sources in the vicinity of H_2O masers-I. Observations. *Mon. Not. Roy. Astron. Soc.* 276:1024–1040.

Jessop, N., and Ward-Thompson, D. 1999. A far-infrared survey of molecular cloud cores. *Mon. Not. Roy. Astron. Soc.,* in press.

Johnstone, D., and Bally, J. 1999. JCMT/SCUBA sub-millimeter wavelength imaging of the integral-shaped filament in Orion. *Astrophys. J. Lett.* 510:L49–L53.

Kenyon, S. J., and Hartmann, L. 1995. Pre-main-sequence evolution in the Taurus-Auriga molecular cloud. *Astrophys. J. Suppl.* 101:117–171.

Kenyon, S. J., Calvet, N., and Hartmann, L. 1993*a*. The embedded young stars in the Taurus-Auriga molecular cloud. I. Models for spectral energy distributions. *Astrophys. J.* 414:676–694.

Kenyon, S. J., Whitney, B. A., Gomez, M., and Hartmann, L. 1993*b*. The embedded young stars in the Taurus-Auriga molecular cloud. II. Models for scattered light images. *Astrophys. J.* 414:773–792.

Kenyon, S. J., Brown, D. I., Tout, C. A., and Berlind, P. 1998. Optical spectroscopy of embedded young stars in the Taurus-Auriga molecular cloud. *Astron. J.* 115:2491–2591.

Klessen, R. S., Burkert, A., and Bate, M. R. 1998. Fragmentation of molecular clouds: The initial phase of a stellar cluster. *Astrophys. J. Lett.* 501:L205–L208.

Kroupa, P., Tout, C. A., and Gilmore, G. 1993. The distribution of low-mass stars in the Galactic disc. *Mon. Not. Roy. Astron. Soc.* 262:545–587.

Lada, C. J. 1987. Star formation: From OB associations to protostars. In *Star Forming Regions,* IAU Symp. 115, ed. M. Peimbert and J. Jugaku (Dordrecht: Reidel), pp. 1–18.

Lada, C. J., and Fich, M. 1996. The structure and energetics of a highly collimated bipolar outflow: NGC 2264G. *Astrophys. J.* 459:638–652.

Lada, C. J., and Wilking, B. 1984. The nature of the embedded population in the Rho Ophiuchi dark cloud. Mid-infrared observations. *Astrophys. J.* 287:610–621.

Lada, C. J., Alves, J., and Lada, E. A. 1996. Near-infrared imaging of embedded clusters: NGC 1333. *Astron. J.* 111:1964–1976.

Ladd, E. F., and Hodapp, K.-W. 1997. A double outflow from a deeply embedded source in Cepheus. *Astrophys. J.* 474:749–759.

Ladd, E. F., Adams, F. C., Casey, S., Davidson, J. A., Fuller, G. A., Harper, D. A., Myers, P. C., and Padman, R. 1991. Far-infrared and submillimeter wavelength observations of star-forming dense cores. II. Images. *Astrophys. J.* 382:555–569.

Ladd, E. F., Fuller, G. A., and Deane, J. R. 1998. $C^{18}O$ and $C^{17}O$ observations of embedded young stars in the Taurus molecular cloud. I. Integrated intensities and column densities. *Astrophys. J.* 495:871–890.

Langer, W. D., Castets, A., and Lefloch, B. 1996. The IRAS 2 and IRAS 4 outflows and star formation in NGC 1333. *Astrophys. J. Lett.* 471:L111–L114.

Larson, R. B. 1969. Numerical calculations of the dynamics of a collapsing protostar. *Mon. Not. Roy. Astron. Soc.* 145:271–295.

Larson, R. B. 1985. Cloud fragmentation and stellar masses. *Mon. Not. Roy. Astron. Soc.* 214:379–398.

Launhardt, R., Mezger, P. G., Haslam, C. G. T., Kreysa, E., Lemke, R., Sievers, A., and Zylka, R. 1996. Dust emission from star-forming regions. IV. Dense cores in the Orion B molecular cloud. *Astron. Astrophys.* 312:569–584.

Lay, O. P., Carlstrom, J. E., and Hills, R. E. 1995. NGC 1333 IRAS 4: Further multiplicity revealed with the CSO-JCMT interferometer. *Astrophys. J. Lett.* 452:L73–L76.

Lee, C. W, and Myers, P. C. 1999. A catalogue of optically selected cores. *Astrophys. J. Suppl.* 123:333.
Lefloch, B., Eislöffel, J., Lazareff, B. 1996. The remarkable Class 0 source Cep E. *Astron. Astrophys.* 313:L17–L20.
Lefloch, B., Castets, A., Cernicharo, J., and Loinard, L. 1998. Widespread SiO emission in NGC1333. *Astrophys. J. Lett.* 504:L109–L112.
Lehtinen, K. 1997. Spectroscopic evidence of mass infall towards an embedded infrared source in the globule DC 303.8-14.2. *Astron. Astrophys.* 317:L5–L8.
Leous, J. A., Feigelson, E. D., André, P., and Montmerle, T. 1991. A rich cluster of radio stars in the Rho Ophiuchi cloud cores. *Astrophys. J.* 379:683–688.
Li, Z.-Y. 1998. Formation and collapse of magnetized spherical molecular cloud cores. *Astrophys. J.* 493:230–246.
Li, Z.-Y., and Shu, F. H. 1996. Magnetized singular isothermal toroids. *Astrophys. J.* 472:211–224.
Li, Z.-Y., and Shu, F. H. 1997. Self-similar collapse of an isopedic isothermal disk. *Astrophys. J.* 475:237–250.
Looney, L. W., Mundy, L. G., and Welch, W. J. 1999. Unveiling the envelope and disk: A sub-arcsecond survey of circumstellar structures. *Astrophys. J.*, in press.
Lucas, P. W., and Roche, P. F. 1997. Butterfly star in Taurus: Structures of young stellar objects. *Mon. Not. Roy. Astron. Soc.* 286:895–919.
Mardones, D., Myers, P. C., Tafalla, M., Wilner, D. J., Bachiller, R., Garay, G. 1997. A search for infall motions toward nearby young stellar objects. *Astrophys. J.* 489:719–733.
Mauersberger, R., Wilson, T. L., Mezger, P. G., Gaume, R., Johnston, K. J. 1992. The internal structure of molecular clouds. III. Evidence for molecular depletion in the NGC 2024 condensations. *Astron. Astrophys.* 256:640–651.
McCaughrean, M. J., Rayner, J. T., and Zinnecker, H. 1994. Discovery of a molecular hydrogen jet near IC 348. *Astrophys. J.* 436:L189–L192.
McKee, C. F. 1989. Photoionization-regulated star formation and the structure of molecular clouds. *Astrophys. J.* 345:782–801.
McLaughlin, D. E., and Pudritz, R. E. 1997. Gravitational collapse and star formation in logotropic and nonisothermal spheres *Astrophys. J.* 476:750–765.
McMullin, J. P., Mundy, L. G., Wilking, B. A., Hezel, T., and Blake, G. A. 1994. Structure and chemistry in the northwestern condensation of the Serpens molecular cloud core. *Astrophys. J.* 424:222–236.
Men'shchikov, A. B., and Henning, T. 1997. Radiation transfer in circumstellar disks. *Astron. Astrophys.* 318:879–907.
Menten, K. M., Serabyn, E., Güsten, R., and Wilson, T. L. 1987. Physical conditions in the IRAS 16293-2422 parent cloud. *Astron. Astrophys.* 177:L57–L60.
Mezger, P. G., Sievers, A. W., Haslam, C. G. T., Kreysa, E., Lemke, R., Mauersberger, R., and Wilson, T. L. 1992. Dust emission from star forming regions. II. The NGC 2024 cloud core: Revisited. *Astron. Astrophys.* 256:631–639.
Mizuno, A., Fukui, Y., Iwata, T., Nozawa, S., Takana, T. 1990. A remarkable multilobe molecular outflow: Rho Ophiuchi East, associated with IRAS 16293-2422. *Astrophys. J.* 356:184–194.
Mizuno, A., Onishi, T., Hayashi, M. Ohashi, N., Sunada, K., Hasegawa, T., and Fukui, Y. 1994. Molecular cloud condensation as a tracer of low-mass star formation. *Nature* 368:719–721.
Molinari, S., Testi, L., Brand, J., Cesaroni, R., and Palla, F. 1998. IRAS 23385 +6053: A prototype massive Class 0 object. *Astrophys. J. Lett.* 505:L39–L42.
Moreira, M. C., Yun, J. L., Vázquez, R., and Torrelles, J. M. 1997. Thermal radio sources in Bok globules. *Astron. J.* 113:1371–1374.
Moriarty-Schieven, G. H., Wannier, P. G., Keene, J., Tamura, M. 1994. Circumprotostellar environments. 2: Envelopes, activity, and evolution. *Astrophys. J.* 436:800–806.

Motte, F. 1998. Structure des coeurs denses proto-stellaires: Étude en continuum millimétrique. Ph.D. Dissertation, University of Paris, XI.

Motte, F., André, P., and Neri, R. 1998. The initial conditions of star formation in the ρ Ophiuchi main cloud: Wide-field millimeter continuum mapping. *Astron. Astrophys.* 336:150–172.

Motte, F. and André, P. 1999. Density structure of isolated protostellar envelopes: A millimeter continuum survey of Taurus infrared protostars. *Astron. Astrophys.,* in preparation.

Mouschovias, T. Ch. 1991. Single-stage fragmentation and a modern theory of star formation. In *The Physics of Star Formation and Early Stellar Evolution,* ed. C. J. Lada and N. D. Kylafis (Dordrecht: Kluwer), pp. 449–468.

Mouschovias, T. Ch. 1995. Role of magnetic fields in the early stages of star formation. In *The Physics of the Interstellar Medium and Intergalactic Medium,* ed. A. Ferrara, C. F. McKee, C. Heiles, and P. R. Shapiro (San Francisco: Astronomical Society of the Pacific), 80:184–217.

Mundy, L. G., Wootten, H. A., Wilking, B. A., Blake, G. A., and Sargent, A. I. 1992. IRAS 16293-2422: A very young binary system? *Astrophys. J.* 385:306–313.

Mundy, L. G., McMullin, J. P., and Grossman, A. W. 1993. Observations of circumstellar disks at centimeter wavelengths. *Icarus* 106:11–19.

Myers, P. C. 1998. Cluster-forming molecular cloud cores. *Astrophys. J. Lett.* 496:L109–L112.

Myers, P. C., and Benson, P. J. 1983. Dense cores in dark clouds. II. NH_3 observations and star formation. *Astrophys. J.* 266:309–320.

Myers, P. C., and Fuller, G. A. 1993. Gravitational formation times and stellar mass distributions for stars of mass 0.3–30 $M_{\odot}$. *Astrophys. J.* 402:635–642.

Myers, P. C., and Ladd, E. F. 1993. Bolometric temperatures of young stellar objects. *Astrophys. J. Lett.* 413:L47–L50.

Myers, P. C., Fuller, G. A., Mathieu, R. D., Beichman, C. A., Benson, P. J., Schild, R. E., and Emerson, J. P. 1987. Near-infrared and optical observations of IRAS sources in and near dense cores. *Astrophys. J.* 319:340–357.

Myers, P. C., Bachiller, R., Caselli, P., Fuller, G. A., Mardones, D., Tafalla, M., and Wilner, D. J. 1995. Gravitational infall in the dense cores L1527 and L483. *Astrophys. J. Lett.* 449:L65–L68.

Myers, P. C., Adams, F. C., Chen, H., and Schaff, E. 1998. Evolution of the bolometric temperature and luminosity of young stellar objects. *Astrophys. J.* 492:703–726.

Myhill, E. A., and Kaula, W. M. 1992. Numerical models for the collapse and fragmentation of centrally condensed molecular cloud cores. *Astrophys. J.* 386:578–586.

Nakano, T. 1984. Contraction of magnetic interstellar clouds. *Fund. Cosm. Phys.* 9:139–231.

Nakano, T. 1998. Star formation in magnetic clouds. *Astrophys. J.* 494:587–604.

Ohashi, N., Hayashi, M., Ho, P. T. P., and Momose, M. 1997. Interferometric imaging of IRAS 04368+2557 in the L1527 molecular cloud core: A dynamically infalling envelope with rotation. *Astrophys. J.* 475:211–223.

O'Linger, J., Wolf-Chase, G. A., Barsony, M., and Ward-Thompson, D. 1999. L1448 IRS2: A HIRES-Identified Class 0 protostar. *Astrophys. J.* 515:696–705.

Ossenkopf, V., and Henning, T. 1994. Dust opacities for protostellar cores. *Astron. Astrophys.* 291:943–959.

Ouyed, R., and Pudritz, R. E. 1997. Numerical simulations of astrophysical jets from Keplerian disks. I. Stationary models. *Astrophys. J.* 482:712–732.

Padoan, P., Nordlund, A., and Jones, B. J. T. 1997. The universality of the stellar initial mass function. *Mon. Not. Roy. Astron. Soc.* 288:145–152.

Palla, F., and Stahler, S. W. 1991. The evolution of intermediate-mass protostars. I. Basic results. *Astrophys. J.* 375:288–299.

Parker, N. D., Padman, R., and Scott, P. F. 1991. Outflows in dark clouds. Their role in protostellar evolution. *Mon. Not. Roy. Astron. Soc.* 252:442–461.

Persi, P., Ferrari-Toniolo, M., Marenzi, A. R., Anglada, G., Chini, R., Krügel, E., and Sepulveda, I. 1994. Infrared images, 1.3 mm continuum and ammonia line observations of IRAS 08076-3556. *Astron. Astrophys.* 282:233–239.

Phelps, R., and Barsony, M. 2000. Herbig-Haro objects in Serpens and Ophiuchus. *Astron. J.,* in preparation.

Pollack, J. B., Hollenbach, D., Beckwith, S., Simonelli, D. P., Roush, T., and Fong, W. 1994. Composition and radiative properties of grains in molecular clouds and accretion disks. *Astrophys. J.* 421:615–639.

Pravdo, S. H., Rodríguez, L. F., Curiel, S., Cantó, J., Torrelles, J. M., Becker, R. H., and Sellgren, K. 1985. Detection of radio continuum emission from Herbig-Haro objects 1 and 2 and from their central exciting source. *Astrophys. J. Lett.* 293:L35–L38.

Preibisch, Th., Ossenkopf, V., Yorke, H. W., and Henning, Th. 1993. The influence of ice-coated grains on protostellar spectra. *Astron. Astrophys.* 279:577–588.

Pringle, J. E. 1989. On the formation of binary stars. *Mon. Not. Roy. Astron. Soc.* 239:361–370.

Pudritz, R. E., Wilson, C. D., Carlstrom, J. E., Lay, O. P., Hills, R. E., and Ward-Thompson, D. 1996. Accretions disks around Class 0 protostars: The case of VLA 1623. *Astrophys. J. Lett.* 470:L123–L126.

Reipurth, B., Chini, R., Krügel, E., Kreysa, E., and Sievers, A. 1993. Cold dust around Herbig-Haro energy sources: A 1300 μm survey. *Astron. Astrophys.* 273:221–238.

Reipurth, B., Nyman, L.-A., and Chini, R. 1996. Protostellar candidates in southern molecular clouds. *Astron. Astrophys.* 314:258–264.

Richer, J. S., Hills, R. E., and Padman, R. 1992. A fast CO jet in Orion B. *Mon. Not. Roy. Astron. Soc.* 254:525–538.

Richer, J. S., Padman, R., Ward-Thompson, D., Hills, R. E., and Harris, A. I. 1993. The molecular environment of S106 IR. *Mon. Not. Roy. Astron. Soc.* 262:839–854.

Ristorcelli, I., Serra, G., Lamarre, J. M., et al. 1998. Discovery of a cold extended condensation in the Orion A complex. *Astrophys. J.* 496:267–273.

Safier, P. N., McKee, C. F., and Stahler, S. W. 1997. Star formation in cold, spherical, magnetized molecular clouds. *Astrophys. J.* 485:660–679.

Sandell, G. 1994. Secondary calibrators at submillimeter wavelengths. *Mon. Not. Roy. Astron. Soc.* 262:839–854.

Sandell, G., and Weintraub, D. A. 1994. A submillimeter protostar near LkH-alpha 198. *Astron. Astrophys.* 292:L1–L4.

Sandell, G., Aspin, C., Duncan, W. D., Russell, A. P. G., and Robson, E. I. 1991. NGC 1333 IRAS 4. A very young, low-luminosity binary system. *Astrophys. J.* 376:L17–L20.

Sandell, G., Knee, L. B. G., Aspin, C., Robson, I. E., and Russell, A. P. G. 1994. A molecular jet and bow shock in the low mass protostellar binary NGC 1333-IRAS2. *Astron. Astrophys.* 285:L1–L4.

Sandell, G., Avery, L. W., Baas, F., Coulson, I., Dent, W. R. F., Friberg, P., Gear, W. P. K., Greaves, J., Holland, W., Jenness, T., Jewell, P., Lightfoot, J., Matthews, H. E., Moriarty-Schieven, G., Prestage, R., Robson, E. I., Stevens, J., Tilanus, R. P. J., and Watt, G. D. 1999. A jet-driven, extreme high-velocity outflow powered by a cold, low-luminosity protostar near NGC 2023. *Astrophys. J.,* 519:236–243.

Saraceno, P., André, P., Ceccarelli, C., Griffin, M., and Molinari, S. 1996*a*. An evolutionary diagram for young stellar objects. *Astron. Astrophys.* 309:827–839.

Saraceno, P., D'Antona, F., Palla, F., Griffin, M., and Tommasi, E. 1996*b*. The luminosity-mm flux correlation of Class I sources exciting outflows. In *The Role of Dust in the Formation of Stars,* ed. H. U. Käufl and R. Siebenmorgen (Berlin: Springer), pp. 59–62.

Scalo, J. 1998. The IMF revisited: A case for variations. In *The Stellar Initial Mass Function,* ASP Conf. Ser. 142, ed. G. Gilmore and D. Howell (San Francisco: Astronomical Society of the Pacific), pp. 201–236.

Shu, F. H. 1977. Self-similar collapse of isothermal spheres and star formation. *Astrophys. J.* 214:488–497.

Shu, F. H., Adams, F. C., and Lizano, S. 1987. Star formation in molecular clouds. Observation and theory. *Ann. Rev. Astron. Astrophys.* 25:23–81.

Shu, F., Najita, J., Galli, D., Ostriker, E., and Lizano S. 1993. The collapse of clouds and the formation and evolution of stars and disks. In *Protostars and Planets III,* ed. E. H. Levy and J. I. Lunine (Tucson: University of Arizona Press), pp. 3–45.

Shu, F. H., Najita, J., Ostriker, E., Wilkin, F., Ruden, S., and Lizano, S. 1994. Magnetocentrifugally driven flows from young stars and disks. I. A generalized model. *Astrophys. J.* 429:781–796.

Silk, J. 1995. A theory for the initial mass function. *Astrophys. J. Lett.* 438:L41–L44.

Sonnhalter, C., Preibisch, T., and Yorke, H. W. 1995. Frequency dependent radiation transfer in protostellar disks. *Astron. Astrophys.* 299:545–556.

Stahler S. W. 1988. Deuterium and the stellar birthline. *Astrophys. J.* 332:804–825.

Stahler, S. W., and Walter, F. M. 1993. Pre-main-sequence evolution and the birth population. In *Protostars and Planets III,* ed. E. H. Levy and J. I. Lunine (Tucson: University of Arizona Press), pp. 405–428.

Stanke, T., McCaughrean, M., and Zinnecker, H. 1998. First results of an unbiased H_2 survey for protostellar jets in Orion A. *Astron. Astrophys.* 332:307–313.

Tafalla, M., Mardones, D., Myers, P. C., Caselli, P., Bachiller, R., and Benson, P. J. 1998. L1544: A starless dense core with extended inward motions. *Astrophys. J.* 504:900–914.

Tamura, M., Hayashi, S. S., Yamashita, T., Duncan, W. D., and Hough, J. H. 1993. Magnetic field in a low-mass protostar disk. Millimeter polarimetry of IRAS 16293-2422. *Astrophys. J. Lett.* 404:L21–L24.

Terebey, S., and Padgett, D. L. 1997. Millimeter interferometry of Class 0 sources: Rotation and infall towards L1448N. In *Herbig-Haro Flows and the Birth of Low Mass Stars,* IAU Symp. 182, ed. B. Reipurth and C. Bertout (Dordrecht: Kluwer), pp. 507–514.

Terebey, S., Vogel S. N., Myers P. C. 1989. High-resolution CO observations of young low-mass stars. *Astrophys. J.* 340:472–478.

Terebey, S., Chandler, C. J., and André, P. 1993. The contribution of disks and envelopes to the millimeter continuum emission from very young low-mass stars. *Astrophys. J.* 414:759–772.

Testi, L., and Sargent, A. I. 1998. Star formation in clusters: A survey of compact millimeter-wave sources in the Serpens core. *Astrophys. J. Lett.* 508:L91–L94.

Tinney, C. G. 1993. The faintest stars: The luminosity and mass functions at the bottom of the main sequence. *Astrophys. J.* 414:279–301.

Tinney, C. G. 1995. The faintest stars: The luminosity and mass functions at the bottom of the main sequence: Erratum. *Astrophys. J.* 445:1017.

Tomisaka, K. 1996. Accretion in gravitationally contracting clouds. *Publ. Astron. Soc. Japan* 48:L97–L101.

Umemoto, T., Iwata, T., Fukui, Y., Mikami, H., Yamamoto, S., Kameyama, O., and Hirano, N. 1992. The outflow in the L1157 dark cloud: Evidence for shock heating of the interacting gas. *Astrophys. J. Lett.* 392:L83–L86.

Velusamy, T., and Langer, W. D. 1998. Outflow-infall interactions as a mechanism for terminating accretion in protostars. *Nature* 392:685–687.

Walker, C. K., Adams, F. C., and Lada, C. J. 1990. 1.3 millimeter continuum observations of cold molecular cloud cores. *Astrophys. J.* 349:515–528.

Walker, C. K., Lada, C. J., Young, E. T., Maloney, P. R., and Wilking, B. A. 1986. Spectroscopic evidence for infall around an extraordinary IRAS source in Ophiuchus. *Astrophys. J. Lett.* 309:L47–L51.

Walker, C. K., Narayanan, G., and Boss, A. P. 1994. Spectroscopic signatures of infall in young protostellar systems. *Astrophys. J.* 431:767–782.

Ward-Thompson, D. 1996. The formation and evolution of low mass protostars. *Astrophys. Space Sci.* 239: 151–170.

Ward-Thompson, D., Scott, P. F., Hills, R. E., and André, P. 1994. A submillimetre continuum survey of pre-protostellar cores. *Mon. Not. Roy. Astron. Soc.* 268:276–290.

Ward-Thompson, D., Chini, R., Krugel, E., André, P., and Bontemps, S. 1995*a*. A submillimetre study of the Class 0 protostar HH24MMS. *Mon. Not. Roy. Astron. Soc.* 274:1219–1224.

Ward-Thompson, D., Eiroa, C., and Casali, M. M. 1995*b*. Confirmation of the driving source of the NGC 2264G bipolar outflow: A Class 0 protostar. *Mon. Not. Roy. Astron. Soc.* 273:L25–L28.

Ward-Thompson, D., Buckley, H. D., Greaves, J. S., Holland, W. S., and André, P. 1996. Evidence for protostellar infall in NGC 1333-IRAS2. *Mon. Not. Roy. Astron. Soc.* 281:L53–L56.

Ward-Thompson, D., Motte, F., and André, P. 1999. The initial conditions of isolated star formation. III: Millimetre continuum mapping of prestellar cores. *Mon. Not. Roy. Astron. Soc.* 305:143–150.

White, G. J., Casali, M. M., and Eiroa, C. 1995. High resolution molecular line observations of the Serpens Nebula. *Astron. Astrophys.* 298:594–605.

Whitney, B. A., and Hartmann, L. 1993. Model scattering envelopes of young stellar objects. II: Infalling envelopes. *Astrophys. J.* 402:605–622.

Whitworth, A., and Summers, D. 1985. Self-similar condensation of spherically symmetric self-gravitating isothermal gas clouds. *Mon. Not. Roy. Astron. Soc.* 214:1–25.

Whitworth, A. P., Bhattal, A. S., Francis, N., and Watkins, S. J. 1996. Star formation and the singular isothermal sphere. *Mon. Not. Roy. Astron. Soc.* 283:1061–1070.

Wiesemeyer, H. 1997. The spectral signature of accretion in low-mass protostars. Ph. D. Dissertation, University of Bonn.

Wiesemeyer, H., Güsten, R., Wink, J. E., and Yorke, H. W. 1997. High resolution studies of protostellar condensations in NGC 2024. *Astron. Astrophys.* 320:287–299.

Wiesemeyer, H., Güsten, R., Cox, P., Zylka, R., and Wright, M. C. H. 1998. The pivotal onset of protostellar collapse: ISO's view and complementary observations. In *Star Formation with ISO,* ASP Conf. Ser. 132, ed. J. L. Yun and R. Liseau (San Francisco: Astronomical Society of the Pacific), pp. 189–194.

Wilking, B. A., Lada, C. J., and Young, E. T. 1989. IRAS observations of the Rho Ophiuchi infrared cluster: Spectral energy distributions and luminosity function. *Astrophys. J.* 340:823–852.

Wilking, B. A., Schwartz, R. D., Fanetti, T. M., and Friel, E. D. 1997. Herbig-Haro objects in the ρ Ophiuchi cloud. *Publ. Astron. Soc. Pacific* 109:549–553.

Wilner, D. J., Welch, W. J., and Forster, J. R. 1995. Sub-arcsecond Imaging of W3(OH) at 87.7 GHz. *Astrophys. J. Lett.* 449:L73–L76.

Wolf-Chase, G. A., Barsony, M., Wootten, H. A., Ward-Thompson, D., Lowrance, P. J., Kastner, J. H., and McMullin, J. P. 1998. The Protostellar Origin of a CS Outflow in S68N. *Astrophys. J. Lett.* 501:L193–L198.

Wood, D. O. S., Myers, P. C., and Daugherty, D. A. 1994. IRAS images of nearby dark clouds. *Astrophys. J. Sup.* 95:457–501.

Wootten, A. 1989. The duplicity of IRAS 16293-2422: A protobinary star? *Astrophys. J.* 337:858–864.

Yorke, H. W., Bodenheimer, P., and Laughlin, G. 1995. The formation of protostellar disks. II: Disks around intermediate-mass stars *Astrophys. J.* 443:199–208.

Yu, K. C., Bally, J., and Devine, D. 1997. Shock-excited H_2 flows in OMC-2 and OMC-3. *Astrophys. J. Lett.* 485:L45–L48.

Yun, J. L., Moreira, M. C., Torrelles, J. M., Afonso, J. M., and Santos, N. C. 1996. A search for radio continuum emission from young stellar objects in Bok globules. *Astron. J.* 111:841–845.

Zavagno, A., Molinari, S., Tommasi, E., Saraceno, P., and Griffin, M. 1997. Young stellar objects in Lynds 1641: A submillimetre continuum study. *Astron. Astrophys.* 325:685–692.

Zhou, S., Evans, N. J., II, Kömpe, C., and Walmsley, C. M. 1993. Evidence for protostellar collapse in B335. *Astrophys. J.* 404:232–246.

Zinnecker, H., Bastien, P., Arcoragi, J. P., and Yorke, H. W. 1992. Submillimeter dust continuum observations of three low luminosity protostellar IRAS sources. *Astron. Astrophys.* 265:726–732.

Zinnecker, H., McCaughrean, M. J., and Rayner, J. T. 1998. A symmetrically pulsed jet of gas from an invisible protostar in Orion. *Nature* 394:862–865.

THE STRUCTURE AND EVOLUTION OF MOLECULAR CLOUDS: FROM CLUMPS TO CORES TO THE IMF

JONATHAN P. WILLIAMS
National Radio Astronomy Observatory

LEO BLITZ AND CHRISTOPHER F. McKEE
University of California at Berkeley

We review the progress that has been made in observing and analyzing molecular cloud structure in recent years. Structures are self-similar over a wide range of scales, with similar power law indexes, independent of the star-forming nature of a cloud. Comparison of structures at parsec-scale resolution in a star-forming and non-star-forming cloud show that the average densities in the former are higher but the structural characteristics in each cloud are much the same. In gravitationally bound regions of a cloud, however, and at higher densities and resolution, the self-similar scaling relationships break down, and it is possible to observe the first steps toward star formation. High-resolution observations of the dense individual star-forming cores within the clumps hold the key to an empirical understanding of the origins of the stellar initial mass function.

I. INTRODUCTION

Molecular clouds are generally self-gravitating, magnetized, turbulent, compressible fluids. The puzzle of how stars form from molecular clouds begins with understanding the physics of such objects and how individual, gravitationally unstable cores condense within them.

In this review we describe advances in understanding that have been made since the last *Protostars and Planets* meeting (see Blitz 1993 in particular, but also Lada et al. 1993; Elmegreen 1993*a*; McKee et al. 1993; and Heiles et al. 1993). The predominant point of view at that time was that the inhomogeneous structure that had been observed even in the earliest complete maps of molecular clouds (e.g., Kutner et al. 1977; Blitz and Thaddeus 1980) could be described as a set of discrete clumps (Blitz and Stark 1986). These clumps themselves contain dense cores, which are the localized sites of star formation within the cloud (Myers and Benson 1983).

In the intervening years a growing number of papers have described an alternative view that clouds are scale-free and that their structure is best described as fractal (e.g., Bazell and Désert 1988; Scalo 1990; Falgarone et al. 1991). In this picture, the hierarchy of cores within clumps within clouds is simply an observational categorization of this self-similar structure. Here we contrast, but also try to reconcile, these two descriptions with a focus on the global questions that link star formation to the interstellar medium (ISM): What controls the efficiency and rate of star formation, and what determines the shape of the stellar initial mass function (IMF)?

We begin by describing the large-scale view of the molecular ISM, and move to progressively smaller scales. The study of molecular clouds is a broad topic, and to stay close to the title of this volume we concentrate on the structure and evolution of the star-forming regions in these clouds.

II. THE LARGE-SCALE VIEW

Advances in millimeter-wave receiver technology have made it possible to map molecular clouds rapidly at high sensitivity. The noise in an observation is directly proportional to the system temperature, and as receiver temperatures have decreased, it has become feasible to map entire complexes, degrees in angular size, at sub-arcminute resolution. Focal plane arrays (arrays of receivers observing neighboring points of the sky simultaneously) at the Five College Radio Astronomy Observatory (FCRAO) 14-m and National Radio Astronomy Observatory (NRAO) 12-m single-dish telescopes have increased the mapping speed by an order of magnitude. Up to four receivers operating at different frequencies can be used simultaneously at the Institut de Radio Astronomie Millimétrique (IRAM) 30-m telescope to observe the same position on the sky. The expansion of the millimeter-wave interferometers IRAM, the Owens Valley Radio Observatory (OVRO), and the Berkeley-Illinois-Maryland Array (BIMA) has dramatically increased the imaging quality and sensitivity of these instruments and has made high-resolution observations ($<10''$) of structure in molecular clouds considerably quicker and easier.

Most of the mass of the molecular ISM is in the form of giant molecular clouds (GMCs), with masses $\sim 10^{5-6}$ $M_\odot$, diameters $\sim$50 pc, and average densities $\langle n_{H_2} \rangle \sim 10^2$ cm^{-3} (e.g., Blitz 1993). The sharp cutoff at the upper end of the cloud mass distribution at $\sim 6 \times 10^6$ $M_\odot$ (Williams and McKee 1997) indicates that cloud masses are limited by some physical process, such as the tidal field of the Galaxy or the disrupting effect of massive stars within them.

The FCRAO telescope has recently completed a sensitive, high-resolution, unbiased study of the CO emission in the outer Galaxy (Heyer et al. 1998). There is no distance ambiguity in the outer Galaxy and much less blending of emission; therefore, this survey allows a more detailed investigation of the large-scale structure of the ISM (see Fig. 1) than earlier surveys of the inner Galaxy by Dame et al. (1987) and Sanders et al. (1985*a*). Heyer et al. (1998) confirm that there are large regions with little or no CO emission (Dame et al. 1986) and concur with earlier results that these regions have been cleared of molecular gas by the intense radiation

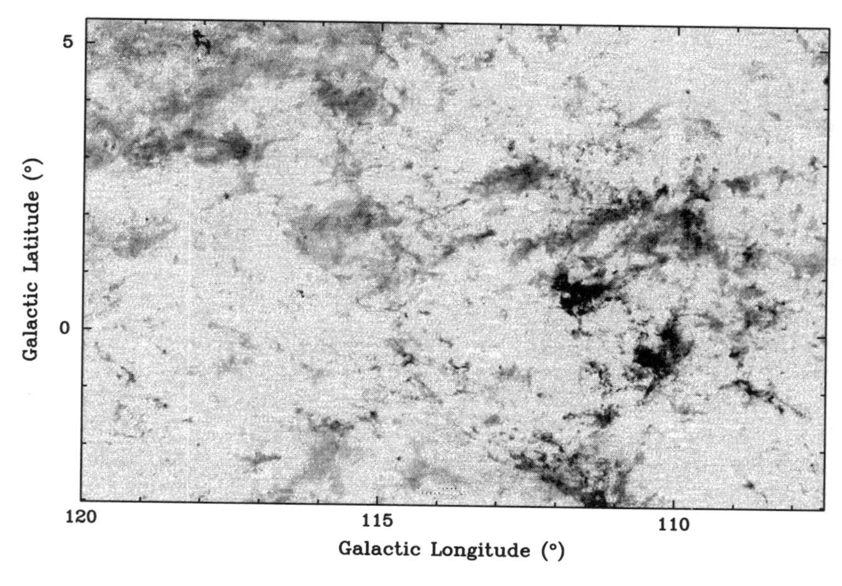

Figure 1. Velocity-integrated CO emission in the outer Galaxy (from Heyer et al. 1998). This map, made with the 15-beam receiver on the FCRAO telescope, contains over half a million spectra and shows a little less than one-third of their outer Galaxy survey. The gray scale ranges from 4 to 40 K km s^{-1}. The high spatial dynamic range of this survey shows the large-scale distribution of molecular gas in the ISM in exceptional detail.

fields and stellar winds from massive stars. Further recent observations of the structural imprint of massive star formation in individual star-forming regions are discussed by Patel et al. (1995), Carpenter et al. (1995*a*), and Heyer et al. (1996). Carpenter et al. (1995*b*) suggest that massive stars can act to compress gas and create dense cores that give rise to the next generation of star formation as originally proposed by Elmegreen and Lada (1977).

Heyer and Terebey (1998) and Digel et al. (1996) show that CO emission in the outer Galaxy is confined almost exclusively to the spiral arms, confirming the earlier results of Cohen et al. (1980). The former, most recent work shows that the ratio of emission in the arm to that in interarm regions is greater than 28:1. The absence of molecular gas in the interarm regions implies that molecular clouds form from a compressed atomic medium and have lifetimes that are less than an arm crossing time $\sim 10^7$ yr. These conclusions may not apply in the inner Galaxy, where it has been more difficult to isolate arm emission and calculate an arm-to-interarm ratio. Solomon and Rivolo (1989) have argued that about half the CO in the inner Galaxy is in clouds that are not actively forming stars and that this gas is not concentrated in spiral arms. However, Loinard et al. (1996) find an arm-to-interarm ratio $\sim$10:1 in M31 in the southwest "ring" $\sim$9 kpc from the galaxy center, which suggests that similar conclusions about cloud formation and lifetimes can be drawn in that galaxy too.

There have been several recent studies of H I halos around molecular clouds (e.g., Kuchar and Bania 1993; Williams and Maddalena 1996; Moriarty-Schieven et al. 1997), and a comprehensive survey is underway at the Dominion Radio Astronomy Observatory (DRAO) telescope (Normandeau et al. 1997). In order to form a GMC of mass 10^6 $M_\odot$ out of an atomic ISM of density $\langle n_H \rangle \sim 1$ cm^{-3}, gas must be accumulated from a volume $\sim$0.4 kpc in diameter. Since such a large region encompasses many atomic clouds, the density inhomogeneities in molecular clouds may simply reflect the initial nonuniform conditions of their formation rather than the first step in the fragmentation and condensation processes that lead to the creation of stars (Elmegreen 1985). To distinguish between the remnants of the formation of a cloud and the first steps toward star formation, it is necessary to analyze and compare structures in a number of different clouds.

III. CLOUD STRUCTURE AND SELF-SIMILARITY

A. A Categorization of Molecular Cloud Structure

Before we discuss the analysis of cloud structure, we first define an operational categorization into clouds, clumps, and cores. This categorization is not inconsistent with the fractal models for cloud structure that are discussed in Section III.E, although we argue that gravity introduces scales that limit the range of validity of the fractal description.

Molecular *clouds* are regions in which the gas is primarily molecular. Almost all known molecular clouds in the Galaxy are detectable in CO. Giant molecular clouds have masses $\gtrsim 10^4$ $M_\odot$, are generally gravitationally bound, and may contain several sites of star formation. However, there also exist small molecular clouds with masses $\lesssim 10^2$ $M_\odot$, such as the unbound high-latitude clouds discovered by Blitz et al. (1984), and the small, gravitationally bound molecular clouds in the Galactic plane cataloged by Clemens and Barvainis (1988). A small number of low-mass stars are observed to form in some of these clouds, but the contribution to their total star formation rate in the Galaxy is negligible (Magnani et al. 1995).

Clumps are coherent regions in *lbv* space (galactic longitude, galactic latitude, and radial velocity), generally identified from spectral line maps of molecular emission. Star-forming clumps are the massive clumps out of which stellar clusters form. Although most clusters are unbound, the gas out of which they form is bound (Williams et al. 1995). *Cores* are regions out of which single stars (or multiple systems such as binaries) form and are necessarily gravitationally bound. Not all material that goes into forming a star must come from the core; some may be accreted from the surrounding clump or cloud as the protostar moves through it (Bonnell et al. 1997).

We have discussed this categorization to describe observations in a uniform manner and to provide a clear link to the processes of star formation. In the following sections we adopt this classification and discuss the relationship between the structure of molecular clouds and their evolution toward star formation.

B. The Virial Theorem for Molecular Clouds

The condition for a molecular cloud or clump within it to be gravitationally bound can be inferred from the virial theorem, which may be written

$$\frac{1}{2}\ddot{I} = 2(\mathcal{T} - \mathcal{T}_0) + \mathcal{M} + \mathcal{W} \tag{1}$$

where I is the moment of inertia, $\mathcal{T}$ is the total kinetic energy (including thermal), $\mathcal{M}$ is the net magnetic energy, and $\mathcal{W}$ is the gravitational energy (see McKee et al. 1993 for more details). The moment of inertia term is usually neglected, but it may be significant for a turbulent cloud. In contrast to the terms on the right-hand side of the equation, it can be of either sign, and as a result its effects can be averaged out either by applying the virial theorem to an ensemble of clouds or clumps or by averaging over a time long compared with the dynamical time (assuming the cloud lives that long.) The kinetic energy term

$$\mathcal{T} = \int_{V_{\rm cl}} \left(\frac{3}{2}P_{\rm th} + \frac{1}{2}\rho v^2\right) dV \equiv \frac{3}{2}\overline{P}\,V_{\rm cl} \tag{2}$$

includes both the thermal and nonthermal pressure inside the cloud. The surface term can be expressed as $\mathcal{T}_0 = \frac{3}{2}P_0 V_{\rm cl}$, where P_0 is about equal to the total thermal plus nonthermal pressure in the ambient medium (McKee and Zweibel 1992). Finally, the gravitational term can be written as (Bertoldi and McKee 1992)

$$\mathcal{W} \equiv -3P_G V_{\rm cl} \tag{3}$$

where the "gravitational pressure" P_G is the mean weight of the material in the cloud. With these results, the steady-state, or time-averaged, virial theorem becomes

$$\overline{P} = P_0 + P_G\left(1 - \frac{\mathcal{M}}{|\mathcal{W}|}\right) \tag{4}$$

In this form the virial theorem has an immediate intuitive meaning: The mean pressure inside the cloud is the surface pressure plus the weight of the material inside the cloud, reduced by the magnetic stresses.

In the absence of an external gravitational field, $\mathcal{W}$ is the gravitational energy of the cloud,

$$\mathcal{W} = -\frac{3}{5}a\left(\frac{GM^2}{R}\right) \tag{5}$$

where a is a numerical factor, of order unity, which has been evaluated by Bertoldi and McKee (1992). The gravitational pressure is then

$$P_G = \left(\frac{3\pi a}{20}\right) G\Sigma^2 \rightarrow 5540\overline{A}_V^2 \text{ K cm}^{-3} \tag{6}$$

where Σ is the mean surface density of the cloud and $\overline{A}_V$ is the corresponding visual extinction. The numerical evaluation is for a spherical cloud with a $1/r$ density profile.

Magnetic fields play a crucial role in the structure and evolution of molecular clouds. For poloidal fields, the relative importance of gravity and magnetic field is determined by the ratio of the mass to the *magnetic critical mass*,

$$M_\Phi = \frac{c_\Phi \Phi}{G^{1/2}} \tag{7}$$

The value of the numerical factor c_Φ depends on the distribution of mass to flux. If the mass-to-flux ratio is constant, then $c_\Phi = 1/2\pi$ (Nakano and Nakamura 1978). In that case the ratio of the magnetic force to the gravitational force is (in our notation) $(M_\Phi/M)^2$ (Shu and Li 1997), which is invariant so long as the magnetic flux is frozen to the matter. For $M < M_\Phi$, the cloud is said to be *magnetically subcritical*. Such a cloud can never undergo gravitational collapse (so long as flux freezing holds), because the magnetic force always exceeds the gravitational force. Conversely, if $M > M_\Phi$, the cloud is *magnetically supercritical*, and magnetic fields cannot prevent gravitational collapse. Shu and Li (1997) have presented a general analysis of the forces in magnetized disks with a constant mass-to-flux ratio (which they term "isopedic" disks); their analysis applies even if the disks are nonaxisymmetric and time dependent. The distinction between magnetically subcritical and supercritical disks holds for nonisopedic disks as well, although the value of c_Φ may differ. For example, for a mass-to-flux distribution corresponding to a uniform field threading a uniform, spherical cloud, Tomisaka et al. (1988) find $c_\Phi \simeq 0.12$.

Toroidal fields can provide a confining force, thereby reducing the magnetic critical mass (Tomisaka 1991; Fiege and Pudritz 1998). However, the ratio of the toroidal field to the poloidal field cannot become too large without engendering instabilities (e.g., Jackson 1975).

If the cloud is supported by thermal and nonthermal pressure in addition to the magnetic stresses, then the maximum stable mass is the critical mass $M_{\rm cr}$, and the cloud is subcritical for $M < M_{\rm cr}$ and supercritical for $M > M_{\rm cr}$. For an isothermal cloud, the critical mass is given by $M_{\rm cr} \simeq M_\Phi + M_{\rm BE}$, where the Bonnor-Ebert mass $M_{\rm BE} = 1.18c_s^4/(G^3P_0)^{1/2}$ is the largest gravitationally stable mass at an exterior pressure P_0 for a nonmagnetic sphere (McKee 1989).

What do observations say about the importance of magnetic fields in clouds? The most detailed study is that of the cloud B1, for which Crutcher et al. (1994) found that the inner envelope was marginally magnetically subcritical, whereas the densest region was somewhat supercritical. The observational results were shown to be in good agreement with a numerical model that, however, did not include the observed nonthermal motions. Crutcher (1999) has summarized the observations of magnetic fields in several clouds and finds that they are magnetically supercritical; in this sample, which tends to focus on the central regions of clouds, there is no clear case in which a cloud is magnetically subcritical. If the magnetic field makes an angle θ with respect to the line of sight, then the observed field is smaller than the true value by a factor $\cos\theta$; on average, this is a factor $\frac{1}{2}$. After allowing for this, Crutcher (1999) finds that on average $\langle M/M_\Phi \rangle \simeq 2.5$. If the clouds are flattened along the field lines, then the observed area is smaller than the true area by factor $\cos\theta$ as well, so $M/M_\Phi \propto \cos^2\theta$; on average, this is a factor $\frac{1}{3}$. However, since clouds are observed to have substantial motions, they are unlikely to be highly flattened along field lines, so Crutcher concludes that $\langle M/M_\Phi \rangle \simeq 2$. The idea that GMCs are magnetically supercritical with $M/M_\Phi \simeq 2$ was suggested theoretically by McKee (1989); Bertoldi and McKee (1992) extended this argument to star-forming clumps, and Nakano (1998) argued that cores are magnetically supercritical. Crutcher (1999) also finds that the Alfvén Mach number of the motions, $m_A = \sigma\sqrt{3}/v_A$, is about unity, as inferred previously by others on the basis of less complete data (e.g., Myers and Goodman 1988).

With these results in hand, we can now address the issue of whether molecular clouds and their constituents are gravitationally bound. The virial theorem (eq. 4) enables us to write the total energy $E = \mathcal{T} + \mathcal{M} + \mathcal{W}$ as

$$E = \frac{3}{2}\left[P_0 - P_G\left(1 - \frac{\mathcal{M}}{|\mathcal{W}|}\right)\right]V_{\rm cl} \tag{8}$$

In the absence of a magnetic field, the condition that the cloud be bound (i.e., $E < 0$) is simply $P_G > P_0$. We shall use this criterion even for magnetized clouds, bearing in mind that using the total ambient gas pressure for P_0 is an overestimate and that our analysis is approximate because we have used the time-averaged virial theorem.

For GMCs the surface pressure is that of the ambient interstellar medium. In the solar vicinity, the total interstellar pressure, which balances the weight of the ISM, is about 2.8×10^4 K cm^{-3} (Boulares and Cox 1990). Of this, about 0.7×10^4 K cm^{-3} is due to cosmic rays; since they pervade both the ISM and a molecular cloud, they do not contribute to the support of a cloud and may be neglected. The magnetic pressure is about 0.3×10^4 K cm^{-3} (Heiles 1996), leaving $P_0 \simeq 1.8 \times 10^4$ K cm^{-3} as the total ambient gas pressure.

What is the minimum value of P_G for a molecular cloud? According to van Dishoeck and Black (1988), molecular clouds exposed to the local interstellar radiation field have a layer of C^+ and C^0 corresponding to a visual extinction of 0.7 mag. If we require at least $\frac{1}{3}$ of the carbon along a line of sight through a cloud to be in the form of CO in order for it to be considered "molecular," then the total visual extinction must be $\bar{A}_V > 2$ mag (allowing for a shielding layer on both sides). According to equation (6), this gives $P_G \gtrsim 2{\times}10^4$ K cm^{-3} $\sim P_0$, verifying that molecular clouds as observed in CO are at least marginally bound (e.g., Larson 1981). Note that if we defined molecular clouds as having a significant fraction of H_2 rather than CO, the minimum column density required would be substantially less and the clouds might not be bound. Furthermore, the conclusion that CO clouds are bound depends both on the interstellar pressure and on the strength of the far UV radiation field, and CO clouds may not be bound everywhere in the Galaxy or in other galaxies (Elmegreen 1993*b*).

GMCs in the solar neighborhood typically have mean extinctions significantly greater than 2 mag, and as a result P_G is generally significantly greater than P_0. For GMCs in the solar neighborhood, $P_G \sim 2 \times 10^5$ K cm^{-3}, an order of magnitude greater than P_0 (Blitz 1991; Bertoldi and McKee 1992; Williams et al. 1995). Thus, if GMCs are dynamically stable entities (the crossing time for a GMC is about 10^7 yr, smaller than the expected lifetime; see Blitz and Shu 1980), then GMCs must be self-gravitating. In the inner galaxy, where P_0 is expected to be greater, the typical GMC linewidths also appear to be somewhat greater than those found locally (Sanders et al. 1985*b*), and thus P_G is still comfortably greater than P_0.

For clumps within GMCs, the surface pressure is just the mean pressure inside the GMC, $P_G(\mathrm{GMC}) \propto \Sigma^2(\mathrm{GMC})$, so the virial theorem becomes $\bar{P}$ (clump) $\propto \Sigma^2(\mathrm{GMC}) + \Sigma^2(\mathrm{clump})$. Most clumps observed in ^{13}CO have $\Sigma(\mathrm{clump}) \ll \Sigma(\mathrm{GMC})$ and therefore are confined by pressure rather than gravity; on the other hand, much of the mass is in massive star-forming clumps that have $\Sigma(\mathrm{clump}) \gtrsim \Sigma(\mathrm{GMC})$ and are therefore gravitationally bound (Bertoldi and McKee 1992).

C. Structure Analysis Techniques

Molecular cloud structure can be mapped via radio spectroscopy of molecular lines (e.g., Bally et al. 1987), continuum emission from dust (e.g., Wood et al. 1994), or stellar absorption by dust (Lada et al. 1994). The first gives kinematical as well as spatial information and results in a three-dimensional cube of data, whereas the latter two result in two-dimensional datasets. Many different techniques have been developed to analyze these data; we discuss them briefly here.

Stutzki and Güsten (1990) and Williams et al. (1994) use the most direct approach and decompose the data into a set of discrete clumps, the former based on recursive triaxial Gaussian fits, and the latter by identifying peaks of emission and then tracing contours to lower levels. The resulting clumps can be considered to be the "building blocks" of the cloud and may be analyzed in any number of ways to determine a size-linewidth relation, mass spectrum, and variations in cloud conditions as a function of position (Williams et al. 1995). There are caveats associated with each method of clump deconvolution, however. Since the structures in a spectral line map of a molecular cloud are not, in general, Gaussian, the recursive fitting method of Stutzki and Güsten (1990) will tend to find and subsequently fit residuals around each clump, which results in a mass spectrum that is steeper than the true distribution. On the other hand, the contour tracing method of Williams et al. (1994) has a tendency to blend small features with larger structures and results in a mass spectrum that is flatter than the true distribution.

Heyer and Schloerb (1997) use principal component analysis to identify differences in line profiles over a map. A series of eigenvectors and eigenimages are created, which identify ever smaller velocity fluctuations and their spatial distribution, resulting in the determination of a size-linewidth relation. Langer et al. (1993) use Laplacian pyramid transforms (a generalization of the Fourier transform) to measure the power on different size scales in a map; as an application, they determine the mass spectrum in the B5 molecular cloud. Recently, Stutzki et al. (1998) have described a closely related Allan variance technique to characterize the fractal structure of two-dimensional maps. Houlahan and Scalo (1992) define an algorithm that constructs a structure tree for a map; this retains the spatial relation of the individual components within the map but loses information regarding their shapes and sizes. It is most useful for displaying and ordering the hierarchical nature of the structures in a cloud.

Adams (1992) discusses a topological approach to quantify the difference between maps. Various "output functions" (e.g., distribution of density, volume, and number of components as a function of column density; see Wiseman and Adams 1994) are calculated for each cloud dataset, and a suitably defined metric is used to determine the distance between these functions and therefore to quantify how similar clouds are or to rank a set of clouds.

A completely different technique was pioneered by Lada et al. (1994). They determined a dust column density in the dark cloud IC 5146 by star counts in the near infrared and mapped cloud structure over a much greater dynamic range ($A_V = 0$–32 mag) than a single spectral line map.

The most striking result of applying these various analysis tools to molecular cloud datasets is the identification of self-similar structures characterized by power law relationships between, most famously, the size and linewidth of features (Larson 1981) and the number of objects of a given mass (e.g., Loren 1989). Indeed, mass spectra are observed to follow a power law with nearly the same exponent, $x = 0.6$–0.8, where $dN/d \ln M \propto M^{-x}$, from clouds with masses up to 10^5 $M_\odot$ in the outer Galaxy to features in nearby high-latitude clouds with masses as small as 10^{-4} $M_\odot$ (Heyer and Terebey 1998; Kramer et al. 1998a; Heithausen et al. 1998). Since a power law does not have a characteristic scale, the implication is that clouds and their internal structure are scale-free. This is a powerful motivation for a fractal description of the molecular ISM (Falgarone et al. 1991, Elmegreen 1997*a*). On the other hand, molecular cloud maps do have clearly identifiable features, especially in spectral line maps when a velocity axis can be used to separate kinematically distinct features along a line of sight (Blitz 1993). These features are commonly called clumps, but there are also filaments (e.g., Nagahama et al. 1998), rings, cavities, and shells (e.g., Carpenter et al. 1995*a*).

D. Clumps

Clump decomposition methods such as those described above by Stutzki and Güsten (1990) and Williams et al. (1994) can be readily visualized and have an appealing simplicity. In addition, as for all automated techniques, these algorithms offer an unbiased way to analyze datasets and are still a valid and useful tool for cloud comparisons even if one does not subscribe to the notion of clumps within clouds as a physical reality (Scalo 1990).

In a comparative study of two clouds, Williams et al. (1994) searched for differences in cloud structure between a star-forming and a non-star-forming GMC. The datasets they analyzed were maps of ^{13}CO (1–0) emission with similar spatial (0.7 pc) and velocity resolution (0.68 km s^{-1}), but the two clouds, although of similar mass $\sim 10^5$ $M_\odot$, have very different levels of star formation activity. The first, the Rosette molecular cloud (RMC), is associated with an H II region powered by a cluster of 17 O and B stars and also contains several bright infrared sources from ongoing star formation deeper within the cloud (Cox et al. 1990). The second cloud, G216$-$2.5, originally discovered by Maddalena and Thaddeus (1985), contains no IRAS sources from sites of embedded star formation and has an exceptionally low far infrared luminosity-to-mass ratio (Blitz 1990), $L_{\rm IR}/M_{\rm cloud} < 0.07$ $L_\odot/M_\odot$, compared to more typical values of order unity (see Williams and Blitz 1998).

Almost 100 clumps were cataloged in each cloud, and sizes, linewidths, and masses were calculated for each. These basic quantities were found to be related by power laws with the same index for the two clouds, but with different offsets (Fig. 2) in the sense that for a given mass, clumps in the non-star-forming cloud are larger and have greater linewidths than in the star-forming cloud. The similarity of the power law indices suggests that, on these scales ($\sim$few pc) and at the low average densities ($\langle n_{\rm H_2}\rangle \sim 300$ cm^{-3}) of the observed clumps, the principal difference between the star-forming and non-star-forming clouds is the change of scale rather than the collective nature of the structures in each cloud.

Figure 2 shows that the kinetic energy of each clump in G216–2.5 exceeds its gravitational potential energy, and therefore no clump in the cloud is bound (although the cloud as a whole is bound). On the other

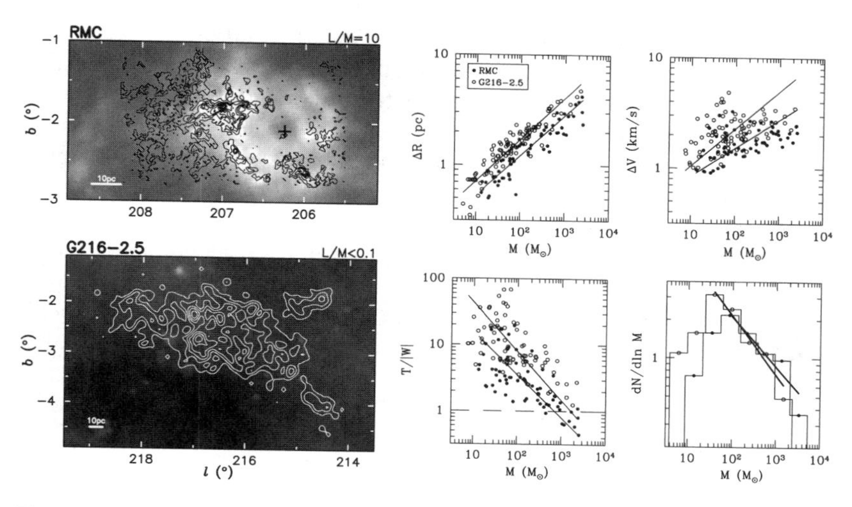

Figure 2. Structure in the Rosette (RMC) and G216−2.5 molecular clouds. The two left panels show contours of velocity-integrated CO emission (the starting level and increment are 15 K km s^{-1} for the RMC, 1.8 K km s^{-1} for G216−2.5), overlaid on a grayscale image of the IRAS 100-μm intensity (1.1 to 2.5 MJy sr^{-1}, same for both clouds). The Rosette cloud is infrared bright, indicative of its high star formation rate, but G216−2.5 has a very low infrared luminosity because of a lack of star formation within it. The four rightmost panels show power law relations between clump mass and size, linewidth, energy balance [i.e. the ratio of kinetic energy, $\mathcal{T} = 3M(\Delta v/2.355)^2/2$, to gravitational potential energy, approximated as $\mathcal{W} = -3GM^2/5R$], and number (i.e., the clump mass spectrum) for the two clouds. The solid circles represent clumps in the RMC, and open circles represent clumps in G216−2.5. Each relationship has been fitted by a power law: Note that the power law exponent is approximately the same for the clumps in each cloud despite the large difference in star formation activity.

hand, Williams et al. (1995) show that, for the Rosette molecular cloud, star formation occurs only in the gravitationally bound clumps in the cloud. Therefore, the lack of bound clumps in G216−2.5 may explain why no star formation is currently taking place within it.

Even in the Rosette cloud, most clumps are not gravitationally bound. These unbound clumps have similar density profiles, $n(r) \propto 1/r^2$, to the bound clumps (Williams et al. 1995), but contain relatively little dense gas as traced by CO (3–2) or CS (2–1) (Williams and Blitz 1998). The unbound clumps are "pressure confined" in that their internal kinetic pressure, which is primarily turbulent, is comparable to the mean pressure of the ambient GMC (Blitz 1991; Bertoldi and McKee 1992). Simulations suggest that these clumps are transient structures (see the chapter by Vázquez-Semadeni et al., this volume).

The nature of the interclump medium remains unclear. Blitz and Stark (1986) postulated that it is low-density molecular gas, but that suggestion has been questioned by Schneider et al. (1996). Blitz (1990, 1993) showed that atomic envelopes around molecular clouds are quite common and that the atomic gas plausibly pervades the interclump medium. Williams et al. (1995) show that the H I associated with the molecular gas in the Rosette has about the same turbulent pressure, and they suggest that it could serve as the confining medium.

Cloud, clump, and core density profiles are reflections of the physics that shape their evolution, but the density profiles of clouds and clumps have received scant attention. For clouds, which often are quite amorphous without a clear central peak, the density profile is often difficult to define observationally. For clumps, Williams et al. (1995) showed that surface density profiles of pressure-bound, gravitationally bound, and star-forming clumps all have similar power law indices, close to 1. Formally, the fits range from -0.8 to -1.2, but these differences do not appear to be significant. For a spherical cloud of infinite extent, $\Sigma(r) \propto r^{-1}$ implies $\rho(r) \propto r^{-2}$, suggesting that the (turbulent) pressure support is spatially constant. However, McLaughlin and Pudritz (1996) argued that for finite spheres, the volume density distribution can be considerably flatter than that inferred for infinite clumps. Density distributions inferred from observations also require consideration of beam-convolution effects. It is nevertheless astonishing that both strongly self-gravitating clumps and those bound by external pressure have such similar, perhaps identical density distributions. Why this should be so is unclear.

E. Fractal Structures

An alternate description of the ISM is based on fractals. High-spatial-dynamic-range observations of molecular clouds, whether by millimeter spectroscopy (e.g., Falgarone et al. 1998), the *Infrared Astronomy Satellite* (IRAS) (Bazell and Désert 1988), or using the *Hubble Space Telescope* (e.g., O'Dell and Wong 1996; Hester et al. 1996) show exceedingly complex patterns that appear to defy a simple description in terms of clouds, clumps, and cores; Scalo (1990) has argued that such loaded names arose from lower-dynamic-range observations and a general human tendency to categorize continuous forms into discrete units.

As we have discussed above, it seems that however one analyzes a molecular cloud dataset, one finds self-similar structures. Moreover, the highly supersonic linewidths that are observed in molecular clouds probably imply turbulent motions (see discussion in Falgarone and Phillips 1990), for which one would naturally expect a fractal structure (Mandelbrot 1982).

The fractal dimension D of a cloud boundary can be determined from the perimeter-area relation of a map, $P \propto A^{D/2}$. Many studies of the molecular ISM find a similar dimension, $D \simeq 1.4$ (Falgarone et al. 1991 and references therein). In the absence of noise, $D > 1$ demonstrates that cloud boundaries are fractal. That D is invariant from cloud to cloud (star-forming or quiescent, gravitationally bound or not) is perhaps related to the similarity in the mass spectrum index in many different molecular clouds (Kramer et al. 1998*a*; Heithausen et al. 1998). Fractal models have been used to explain both the observed mass spectrum of structures (Elmegreen and Falgarone 1996) and the stellar IMF (Elmegreen 1997*b*).

Probability density functions (PDFs) may be used to describe the distribution of physical quantities (such as density and velocity) in a region of space without resorting to concepts of discrete objects such as clouds, clumps, and cores. The density PDF is readily determined from simulations, and can be modeled semianalytically (Vázquez-Semadeni 1998), but it is very difficult to determine from observations, because projection and excitation effects result in maps of column density integrated over a limited range of volume density. We discuss column density PDFs in section III.F.

Velocities, albeit also projected, can be much more easily measured. Falgarone and Phillips (1990) show that the PDF of the velocity field can be determined from high signal-to-noise observations of a single line profile. The low-level, broad line wings that are observed in non-star-forming regions show that the probabilities of rare, high-velocity motions in the gas are greater than those predicted by a normal (Gaussian) probability distribution. This *intermittent* behavior is expected in a turbulent medium, and their detailed analysis shows that the deviations from the predictions for Kolmogorov turbulence are small, even though the basic assumptions of Kolmogorov turbulence, such as incompressibility and $B = 0$, are not satisfied. Miesch and Bally (1995) calculate velocity centroid PDFs from ^{13}CO observations of star-forming regions and also report non-Gaussian behavior. Lis et al. (1996) compare their results with a similar analysis applied to simulations of compressible turbulence; such work may be a promising avenue for exploring the role of turbulence in molecular clouds.

A key project at the IRAM 30-m telescope has been the investigation of the small-scale structure in pre-star-forming regions (Falgarone et al. 1998). Their study of many high-resolution, high-spatial-dynamic-range maps in a number of different tracers and transitions shows that the line core emission has a rather smooth and extended spatial distribution, but the line wing emission has a more filamentary distribution. These filaments also show very large local velocity gradients, $\gtrsim$10 km s^{-1} pc^{-1}, and have the greatest amount of small-scale structure.

Falgarone et al. (1998) also address a long-standing problem concerning the extreme smoothness of observed line profiles. Isotopic ratios indicate that CO lines, especially at line center, are very optically thick, yet line profiles are generally neither flat-topped nor self-reversed. Moreover, line profiles do not break up into separate components at high angular and spectral resolution. This has traditionally presented a problem for the clump-based picture of molecular clouds (Martin et al. 1984). Falgarone et al. interpret their results in terms of a macroturbulent model in which emission arises from a large number of cells with size $\lesssim$200 AU and densities ranging from $n_{H_2} \sim 10^3$ cm^{-3} for line-wing emission to $n_{H_2} \sim 10^5$ cm^{-3} for line-core emission. They find an anticorrelation between linewidth and intensity and speculate that as the turbulent motions dissipate, there is increased radiative coupling between cells. Macroturbulent models have also been explored by Wolfire et al. (1993), who include the effects of photodissociation, and Tauber (1996), who includes an interclump medium but finds no evidence for small-scale structures ($\lesssim$0.5 pc) in Orion A.

F. Departures from Self-Similarity

The universal self-similarity that is observed in all types of cloud, over a wide range in mass and star-forming activity, is remarkable, but a consequence of this universality is that it does not differentiate between clouds with different rates of star formation (or those that are not forming stars at all), and therefore it cannot be expected to explain the detailed processes by which a star forms. Star formation must be preceded by a departure from structural self-similarity.

An upper limit to the range over which self-similar scaling laws apply is set by the generalization of the Bonnor-Ebert mass to nonthermal linewidths. From an analysis of the clumps in Ophiuchus, Orion B, the Rosette, and Cep OB 3, Bertoldi and McKee (1992) found that $M_{\rm BE} = 1.18\sigma^4/(G^3P)^{1/2}$ is about constant for all the pressure-confined clouds in each GMC. Here, σ is the total (thermal + nonthermal) velocity dispersion. For all but Cep OB 3, for which the data are of low resolution, star formation was confined to clumps with $M \gtrsim M_{\rm BE}$, and essentially all the clumps with $M > M_{\rm BE}$ are forming stars. The maximum mass of a clump in these clouds was in the range (1–10) $M_{\rm BE}$ (for Cep OB 3, this statement applies only to the limited part of the cloud for which data were available).

On the small scale, there have long been suggestions that the thermal Bonnor-Ebert mass gives a scale that determines the characteristic mass of stars (Larson 1985). In order to determine whether such a scale is important in molecular clouds, Blitz and Williams (1997) examined how the structural properties of a large-scale, high-resolution ^{13}CO map of the Taurus molecular cloud obtained by Mizuno et al. (1995) varied as the resolution was degraded by an order of magnitude. In their work, they use the temperature histogram of the dataset to compare the cloud properties as a function of resolution. This is the most basic statistic and requires minimal interpretation of the data. In Fig. 3 we show the column density PDF, which is proportional to the integrated intensity histogram for optically thin emission, for this Taurus dataset at two resolutions and four other ^{13}CO maps of molecular clouds.

To compare the different cloud PDFs, Fig. 3 shows column densities that have been normalized by the peak, $N_{\rm peak}$, of each map. Each PDF has also been truncated at $N/N_{\rm peak} \simeq 0.15$–$0.25$ to show only those points with high signal-to-noise ratio. Within the Poisson errors (not shown for clarity),

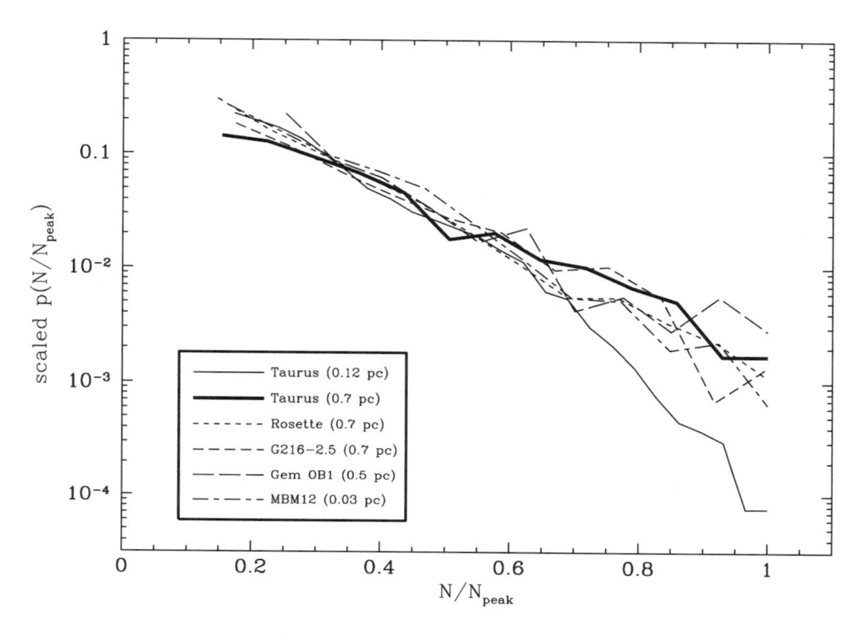

Figure 3. Column density PDF of ^{13}CO emission in five molecular clouds. Column densities have been normalized by the peak value in each dataset and then binned. The fraction of data points in each bin is plotted as a function of normalized column density. The ordinate has been arbitrarily scaled so as to align the different curves at low column densities. Note the similarity between the different datasets, except for the high-resolution Taurus map which has a lower fraction of data points with column densities within 70% of the peak.

the PDFs of the different clouds are all the same except for the higher-resolution Taurus PDF, for which there is a lower relative probability of having lines of sight with $N/N_{\rm peak} \gtrsim 0.7$. The low intensities of the ^{13}CO emission imply that the optical depth is small along all lines of sight in the map and is not responsible for this effect. Rather, from examination of the integrated intensity maps, Blitz and Williams (1997) show that the effect is due to a steepening of the column density profiles at small size scales (see also Abergel et al. 1994).

There are two immediate implications from Fig. 3. First, the common exponential shape for the column density PDF is another manifestation of the self-similar nature of cloud structure. It is a simple quantity to calculate and may provide a quick and useful test of the fidelity of cloud simulations.

Second, because the behavior of the Taurus dataset changes as it is smoothed, it cannot be described by a single fractal dimension over all scales represented in the map (see Williams 1998 for a demonstration that the intensity histogram of a fractal cloud is resolution invariant, as one would expect). There is other evidence for departures from self-similarity at similar size scales. Goodman et al. (1998) examine in detail the nature of the size-linewidth relation in dense cores as linewidths approach a constant, slightly greater than thermal, value in a central "coherent" region ~ 0.1 pc diameter (Myers 1983). Also, Larson (1995) finds that the two-point angular correlation function of T Tauri stars in Taurus departs from a single power law at a size scale of 0.04 pc (see also Simon 1997).

For gas of density $n_{H_2} \sim 10^3$ cm^{-3}, these size scales correspond to masses of order ~ 1 M$_\odot$, close to the thermal Bonnor-Ebert mass at a temperature $T = 10$ K. It must be emphasized that the above evidence for characteristic scales comes from studies of gravitationally bound, star-forming regions; self-similarity in unbound clouds continues to much smaller scales. Figure 3 shows that the column density PDF of the unbound, high-latitude cloud MBM12 is identical, at a resolution of 0.03 pc, to the other lower-resolution PDFs of star-forming GMCs. Similarly, the mass spectra of other high-latitude clouds follow power laws, $dN/d \ln M \propto M^{-x}$ with $x \simeq 0.6$–0.8, down to extremely low masses, $M \simeq 10^{-4}$ M$_\odot$ (Kramer et al. 1998*a*; Heithausen et al. 1998). It appears to be the action of gravity that creates the observed departures from self-similarity.

IV. THE CONNECTION BETWEEN CLOUD STRUCTURE AND STAR FORMATION

A. Star-Forming Clumps

Star-forming clumps are bound regions in a molecular cloud that form stellar clusters. Since most stars form in clusters, questions of star formation efficiency and rate are tied into the efficiency and rate of formation of star-forming clumps, and the IMF is related to the fragmentation of such clumps into individual star-forming cores.

The median column density of molecular gas in the outer Galaxy CO survey by Heyer et al. (1998) is only $N(H_2) \simeq 2 \times 10^{21}$ cm^{-2}, and most of the mass of a molecular cloud is contained within the low-column-density lines of sight, $A_V \lesssim 2$ (Carpenter et al. 1995*a*; Heyer et al. 1996). Such gas is ionized predominantly by the interstellar far ultraviolet radiation field ($h\nu < 13.6$ eV). McKee (1989) showed that this is true throughout the Galaxy: Most molecular gas is photoionized and therefore has a higher level of ionization than that due to cosmic rays alone. Since the rate of low-mass star formation may be governed by ambipolar diffusion, which in turn is determined by the ionization (e.g., Shu et al. 1987), low-mass star formation is "photoionization regulated": Most stars form in regions shielded from photoionization by column densities of dust corresponding to $A_V \gtrsim 3$–4. This idea has been applied to the formation of star clusters by Bertoldi and McKee (1996). It naturally accounts for the low average rate of star formation in the Galaxy, since only about 10% of the mass of a typical GMC is sufficiently shielded to have active star formation, and the ambipolar diffusion timescale is about 10 times the free fall time.

Observations of the ISM in the low-metallicity Magellanic Clouds have verified the prediction that molecular clouds there have about the same dust column densities, and therefore higher gas column densities, as Galactic molecular clouds (Pak et al. 1998). Li et al. (1997) have tested photoionization-regulated star formation by searching for evidence of recent star formation in low-extinction regions of L1630; they found no evidence for such star formation. On the other hand, Strom et al. (1993) did find evidence for distributed star formation in L1641; possible reasons for the discrepancy are discussed by Li et al. (1997). In addition, Nakano (1998) has questioned whether ambipolar diffusion plays any important role in low-mass star formation, correctly pointing out that cores are magnetically supercritical, so ambipolar diffusion is unimportant there. However, he did not address the issue of whether ambipolar diffusion was important prior to the formation of the cores. An observational determination of the role of ambipolar diffusion in low-mass star formation remains a challenge for the future.

B. Cores

The core that forms an individual star (or multiple stellar system) is the final stage of cloud fragmentation. Cores have typical average densities $n_{H_2} \sim 10^5$ cm^{-3} and can be observed in high-excitation lines or transitions of molecules with large dipole moments (Benson and Myers 1989) or via dust continuum emission at millimeter and submillimeter wavelengths (Kramer et al. 1998*b*).

Because of their high densities, the surface-filling fraction of cores is low, even in cluster-forming environments. Therefore, searches for cores have generally followed signs of star formation activity (such as IRAS emission, outflows, etc.), and there have been few unbiased searches (e.g., Myers and Benson 1983). However, increases in instrument speed have now made it possible to survey millimeter continuum emission over relatively large areas of the sky. There have been two very recent results in this regard, the first by Motte et al. (1998) at 1.3 mm using the array bolometer on the IRAM 30-m telescope, and the second by Testi and Sargent (1998) at 3 mm using the OVRO interferometer.

Motte et al. (1998) mapped the ρ Ophiuchi cloud, the closest rich cluster-forming region (see the chapter by André et al., this volume), and Testi and Sargent (1998) mapped the Serpens molecular cloud, a more distant, but more massive and somewhat richer, star-forming region (Fig. 4). In each case, the large-scale, high-resolution observations reveal a large number of embedded young protostars and also starless, dense condensations. Both studies find that the mass spectrum of the cores is significantly steeper, $x > 1.1$ (where $dN/d\ln M \propto M^{-x}$) than clump mass spectra, $x \simeq 0.6$–0.8. The core mass spectra resemble the slope, $x = 1.35$, of the stellar (Salpeter) IMF, which suggests a very direct link between cloud structure and star formation. However, it has not yet been

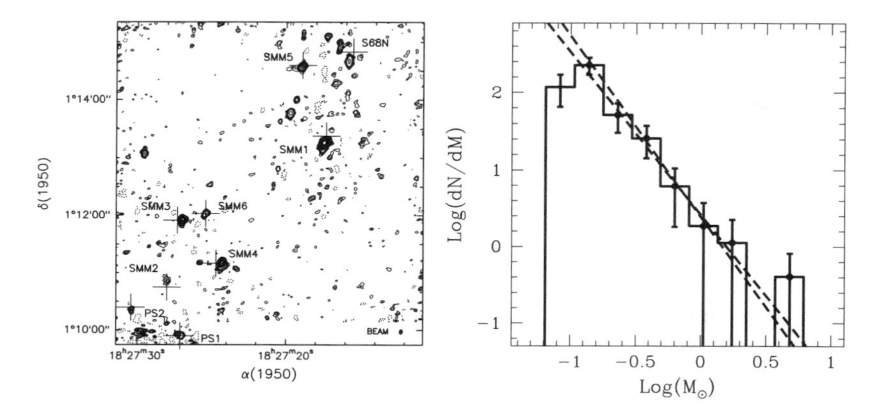

Figure 4. 3-m continuum emission in the Serpens molecular cloud (from Testi and Sargent 1998). This map is the result of a mosaic of 50 fields using the OVRO interferometer. The synthesized beam is $5.7'' \times 4.3''$ (FWHM) and is indicated by the filled ellipse in the lower right corner. The noise level is σ = 0.9 mJy beam^{-1} and contours begin and are in steps of 3σ. A large number of sources, each a dense dust condensation, are visible. The previously known far infrared and submillimeter sources are marked with crosses and labeled. The core mass spectrum has a slope $dN/dM \propto M^{-2.1}$ and more closely resembles the Salpeter IMF than the clump mass spectrum in molecular clouds.

shown that these starless cores are self-gravitating, which is an important step in establishing the link.

C. The Origin of the IMF

The spectrum, lifetime, and end state of a star are primarily determined by its mass. Consequently, the problem of understanding how the mass of a star is determined during its formation, and the origin of the IMF, has a very wide application in many fields, from galaxy evolution to the habitability of extrasolar planets. The form of the IMF is typically assumed to be invariant, but since it is directly measurable only locally, knowing how it comes about can help us predict how it might vary under different astrophysical environments.

Observations and theories of the IMF are discussed in detail in the chapter by Meyer et al., but here we briefly note the connection between the work described in this chapter and the IMF. Many explanations for the form of the IMF use as their starting point the mass spectrum of clouds and clumps as revealed by molecular line emission, $dN/d \ln M \propto M^{-x}$ with $x \simeq 0.5$. Most such structures, however, are not forming stars; the majority of stars form in clusters in a few of the most massive clumps in a cloud. An understanding of the origin of the IMF can only come about with a more complete picture of the formation of star-forming clumps and the fragmentation of these clumps down to individual star-forming cores. The unbiased continuum surveys by Motte et al. (1998) and Testi and Sargent

(1998) are an important step in this direction. As high-resolution studies of individual cores in cluster environments become more commonplace, the relationship between stellar mass and core mass can be determined. If the core mass spectrum is indeed similar to the stellar IMF, then the fraction of the mass of a core that goes into a star (the star formation efficiency of the core) is approximately independent of mass, and the stellar IMF is determined principally by the cloud fragmentation processes. By measuring the core mass spectrum in different clusters in a variety of different molecular clouds, the influence of the large-scale structure and environment on the IMF can be quantified.

V. SUMMARY

The study of the structure of molecular clouds is inextricably linked to the formation of stars. In the outer Galaxy the molecular gas is confined to spiral arms. The inner Galaxy also shows confinement of the molecular gas to the spiral arms, but there is some evidence for interarm molecular gas as well. The observed star formation rate and efficiency in the Galaxy may be the result of only about 10% of the molecular ISM being shielded from the interstellar UV radiation field to an extent that matter can drift through the magnetic field lines and condense into star-forming cores on timescales $\sim$10 Myr.

Molecular clouds, in the outer Galaxy at least, probably result from the compression of atomic gas entering a spiral arm. Thus, the density inhomogeneities in clouds may simply reflect the initial non-uniform conditions of the atomic ISM and need not be related to star formation. At moderate densities, $n_{H_2} \lesssim 10^3$ cm^{-3}, cloud structures are self-similar up to a scale set by self-gravity, and clump mass spectra have similar power law indexes, independent of the star-forming nature of the cloud.

As linewidths approach their thermal value, structures depart from the same self-similar description. This departure may mark the boundary between cloud evolution and star formation. Clusters of individual star-forming cores, with a mass spectrum that approaches the Salpeter IMF, are observed in the ρ Ophiuchi and Serpens clouds. In coming years we can expect there to be increased observational and theoretical effort to understand the structure, dynamics, and distribution of these cores in a variety of star-forming environments, which should lead to a better understanding of the relationship between the structure and evolution of molecular clouds and the initial mass function of stars.

Acknowledgments We enjoyed many informative discussions with both Edith Falgarone and Dick Crutcher, and we thank them also for a thorough reading of the manuscript. We are grateful to Mark Heyer for providing Fig. 1 and for a series of interesting conversations over the last several years. Finally, thanks to Leonardo Testi for Fig. 4 in advance of publication.

REFERENCES

Abergel, A., Boulanger, F., Mizuno, A., and Fukui, Y. 1994. Comparative analysis of the far-infrared and ^{13}CO (J = 1–0) emissions of the Taurus complex. *Astrophys. J. Lett.* 423:L59–L62.

Adams, F. C. 1992. A topological/geometrical approach to the study of astrophysical maps. *Astrophys. J.* 387:572–590.

Bally, J., Stark, A. A., Wilson, R. W., and Langer, W. D. 1987. Filamentary structure in the Orion molecular cloud. *Astrophys. J. Lett.* 312:L45–L49.

Bazell, D., and Désert, F. X. 1988. Fractal structure of interstellar cirrus. *Astrophys. J.* 333:353–358.

Benson, P. J., and Myers, P. C. 1989. A survey for dense cores in dark clouds. *Astrophys. J. Suppl.* 71:89–108.

Bertoldi, F., and McKee, C. F. 1992. Pressure-confined clumps in magnetized molecular clouds. *Astrophys. J.* 395:140–157.

Bertoldi, F., and McKee, C. F. 1996. Self-regulated star formation in molecular clouds. In *Amazing Light, A Volume Dedicated to Charles Hard Townes on His 80th Birthday,* ed. R. Y. Chiao (New York: Springer), pp. 41–44.

Blitz, L. 1990. The evolution of galactic giant molecular clouds. In *The Evolution of the Interstellar Medium,* ed. L. Blitz (San Francisco: Astronomical Society of the Pacific), pp. 273–289.

Blitz, L. 1991. Star forming giant molecular clouds. In *The Physics of Star Formation and Early Stellar Evolution,* ed. C. J. Lada and N. D. Kylafis (Dordrecht: Kluwer), pp. 3-33.

Blitz, L., 1993. Giant molecular clouds. In *Protostars and Planets III,* eds. E. H. Levy and J. I. Lunine (Tucson: University of Arizona Press), pp. 125–161.

Blitz, L., and Shu, F. H. 1980. The origin and lifetime of giant molecular cloud complexes. *Astrophys. J.* 238:148–157.

Blitz, L., and Stark, A. A. 1986. Detection of clump and interclump gas in the Rosette molecular cloud complex. *Astrophys. J. Lett.* 300:L89–L93.

Blitz, L., and Thaddeus, P. 1980. Giant molecular complexes and OB associations. I: The Rosette Molecular Complex. *Astrophys. J.* 241:676–696.

Blitz, L., and Williams, J. P. 1997. Molecular clouds are not fractal: A characteristic size scale in Taurus. *Astrophys. J. Lett.* 488:L145–L149.

Blitz, L., Magnani, L., and Mundy, L. 1984. High-latitude molecular clouds. *Astrophys. J. Lett.* 282:L9–L12.

Bonnell, I. A., Bate, M. R., Clarke, C. J., and Pringle, J. E. 1997. Accretion and the stellar mass spectrum in small clusters. *Mon. Not. Roy. Astron. Soc.* 285:201–208.

Boulares, A., and Cox, D. P. 1990. Galactic hydrostatic equilibrium with magnetic tension and cosmic-ray diffusion. *Astrophys. J.* 365:544–558.

Carpenter, J. M., Snell, R. L., and Schloerb, F. P. 1995*a*. Anatomy of the Gemini OB1 molecular cloud complex. *Astrophys. J.* 445:246–268.

Carpenter, J. M., Snell, R. L., and Schloerb, F. P. 1995*b*. Star formation in the Gemini OB1 molecular cloud complex. *Astrophys. J.* 450:201–216.

Clemens, D. P., and Barvainis, R. 1988. A catalog of small, optically selected molecular clouds: Optical, infrared, and millimeter properties. *Astrophys. J. Suppl.* 68:257–286.

Cohen, R. S., Cong, H.-I., Dame, T. M., and Thaddeus, P. 1980. Molecular clouds and galactic spiral structure. *Astrophys. J. Lett.* 239:L53–L56.

Cox, P., Deharveng, L., and Leene, A. 1990. IRAS observations of the Rosette nebula complex. *Astron. Astrophys.* 230:181–192.

Crutcher, R. M. 1999. Magnetic fields in molecular clouds: Observations confront theory. *Astrophys. J.,* submitted.

Crutcher, R. M., Mouschovias, T. M., Troland, T. H., and Ciolek, G. E. 1994. Structure and evolution of magnetically supported molecular clouds: Evidence for ambipolar diffusion in the Barnard 1 cloud. *Astrophys. J.* 427:839–847.

Dame, T. M., Ungerechts, H., Cohen, R. S., De Geus, E. J., Grenier, I. A., May, J., Murphy, D. C., Nyman, L.-Å., and Thaddeus, P. 1987. A composite CO survey of the entire Milky Way. *Astrophys. J.* 322:706–720.

Dame, T. M., Elmegreen, B. G., Cohen, R. S., and Thaddeus, P. 1986. The largest molecular cloud complexes in the first galactic quadrant. *Astrophys. J.* 305:892–908.

Digel, S. W., Lyder, D. A., Philbrick, A. J., Puche, D., and Thaddeus, P. 1996. A large-scale CO survey toward W3, W4, and W5. *Astrophys. J.* 458:561–575.

Elmegreen, B. G. 1985. Molecular clouds and star formation: An overview. In *Protostars and Planets II,* ed. D. C. Black and M. S. Matthews (Tucson: University of Arizona Press), pp. 97–161.

Elmegreen, B. G. 1993*a*. Formation of interstellar clouds and structure. In *Protostars and Planets III,* ed. E. H. Levy and J. I. Lunine (Tucson: University of Arizona Press), pp. 97–161.

Elmegreen, B. G. 1993*b*. The H to H_2 transition in galaxies: Totally molecular galaxies. *Astrophys. J.* 411:170–177.

Elmegreen, B. G. 1997*a*. Intercloud structure in a turbulent fractal interstellar medium. *Astrophys. J.* 477:196–203.

Elmegreen, B. G. 1997*b*. The initial stellar mass function from random sampling in a turbulent fractal cloud. *Astrophys. J.* 486:944–954.

Elmegreen, B. G., and Falgarone, E. 1996. A fractal origin for the mass spectrum of interstellar clouds. *Astrophys. J.* 471:816–821.

Elmegreen, B. G., and Lada, C. J. 1977. Sequential formation of subgroups in OB associations. *Astrophys. J.* 214:725–741.

Falgarone, E., and Phillips, T. G. 1990. A signature of the intermittency of interstellar turbulence: The wings of molecular line profiles. *Astrophys. J.* 359:344–354.

Falgarone, E., Phillips, T. G., and Walker, C. K. 1991. The edges of molecular clouds: Fractal boundaries and density structure. *Astrophys. J.* 378:186–201.

Falgarone, E., Panis, J.-F., Heithausen, A., Pérault, M., Stutzki, J., Puget, J.-L., and Bensch, F. 1998. The IRAM key-project: Small-scale structure of pre-star-forming regions. I. Observational results. *Astron. Astrophys.* 331:669–696.

Fiege, J. D., and Pudritz, R. E. 1998. Helical fields and filamentary molecular clouds. *Mon. Not. Roy. Astron. Soc.,* submitted.

Goodman, A. A., Barranco, J. A., Wilner, D. J., and Heyer, M. H. 1998. Coherence in dense cores. II. The transition to coherence. *Astrophys. J.* 504:223–246.

Heiles, C. 1996. A comprehensive view of the galactic magnetic field, especially near the sun. In *Polarimetry of the Interstellar Medium,* ed. W. G. Roberge and D. C. B. Whittet (San Francisco: Astronomical Society of the Pacific), pp. 457–468.

Heiles, C., Goodman, A. A., McKee, C. F., and Zweibel, E. G. 1993. Magnetic fields in star-forming regions: Observations. In *Protostars and Planets III,* ed. E. H. Levy and J. I. Lunine (Tucson: University of Arizona Press), pp. 279–326.

Heithausen, A., Bensch, F., Stutzki, J., Falgarone, E., and Panis, J.-F. 1998. The IRAM key project: Small-scale structure of pre-star forming regions. Combined mass spectra and scaling laws. *Astron. Astrophys. Lett.* 331:L65–L68.

Hester, J. J., et al. 1996. Hubble Space Telescope WFPC2 imaging of M16: Photoevaporation and emerging young stellar objects. *Astron. J.* 111:2349–2360.

Heyer, M. H., Carpenter, J. M., and Ladd, E. F. 1996. Giant molecular cloud complexes with optical H II regions: ^{12}CO and ^{13}CO observations and global cloud properties. *Astrophys. J.* 463:630–641.
Heyer, M. H., Brunt, C., Snell, R. L., Howe, J. E., Schloerb, F. P., and Carpenter, J. M. 1998. The Five College Radio Astronomy Observatory CO survey of the outer Galaxy. *Astrophys. J. Suppl.* 115:241–258.
Heyer, M. H., and Schloerb, F. P. 1997. Application of principal component analysis to large-scale spectral line imaging studies of the interstellar medium. *Astrophys. J.* 475:173–187.
Heyer, M. H., and Terebey, S. 1998. The anatomy of the Perseus spiral arm: ^{12}CO and IRAS imaging observations of the W3–W4–W5 cloud complex. *Astrophys. J.* 502:265–277.
Houlahan, P., and Scalo, J. 1992. Recognition and characterization of hierarchical interstellar structure. II: Structure tree statistics. *Astrophys. J.* 393:172–187.
Jackson, J. D. 1975. *Classical Electrodynamics* (New York:Wiley).
Kramer, C., Stutzki, J., Röhrig, R., and Corneliussen, U. 1998*a*. Clump mass spectra of molecular clouds. *Astron. Astrophys.* 329:249–264.
Kramer, C., Alves, J., Lada, C., Lada, E., Sievers, A., Ungerechts, H., and Walmsley, M. 1998*b*. The millimeter wavelength emissivity in IC5146. *Astron. Astrophys. Lett.* 329:L33–L36.
Kuchar, T. A., and Bania, T. M. 1993. A high-resolution H I survey of the Rosette nebula. *Astrophys. J.* 414:664–671.
Kutner, M. L., Tucker, K. D., Chin, G., and Thaddeus, P. 1977. The molecular complexes in Orion. *Astrophys. J.* 215:521–528.
Lada, C. J., Lada, E. A., Clemens, D. P., and Bally, J. 1994. Dust extinction and molecular gas in the dark cloud IC5146. *Astrophys. J.* 429:694–709.
Lada, E. A. Strom, K. M., and Myers, P. C. 1993. Environments of star formation: Relationship between molecular clouds, dense cores and young stars. In *Protostars and Planets III,* ed. E. H. Levy and J. I. Lunine (Tucson: University of Arizona Press), pp. 247–277.
Langer, W. D., Wilson, R. W., and Anderson, C. H. 1993. Hierarchical structure analysis of interstellar clouds using nonorthogonal wavelets. *Astrophys. J. Lett.* 408:L45–L48.
Larson, R. B. 1981. Turbulence and star formation in molecular clouds. *Mon. Not. Roy. Astron. Soc.* 194:809–826.
Larson, R. B. 1985. Cloud fragmentation and stellar masses. *Mon. Not. Roy. Astron. Soc.* 214:379–398.
Larson, R. B. 1995. Star formation in groups. *Mon. Not. Roy. Astron. Soc.* 272:213–220.
Li, W., Evans, N. J., and Lada, E. A. 1997. Looking for distributed star formation in L1630: A near-infrared (J, H, K) survey. *Astrophys. J.* 488:277–285.
Lis, D. C., Pety, J., Phillips, T. G., and Falgarone, E. 1996. Statistical properties of line centroid velocities and centroid velocity increments in compressible turbulence. *Astrophys. J.* 463:623–629.
Loinard, L., Dame, T. M., Koper, E., Lequeux, J., Thaddeus, P., and Young, J. S. 1996. Molecular spiral arms in M31. *Astrophys. J. Lett.* 469:L101–L104.
Loren, R. B. 1989. The cobwebs of Ophiuchus. I: Strands of ^{13}CO—The mass distribution. *Astrophys. J.* 338:902–924.
Maddalena, R., and Thaddeus, P. 1985. A large, cold, and unusual molecular cloud in Monoceros. *Astrophys. J.* 294:231–237.
Magnani, L., Caillault, J.-P., Buchalter, A., and Beichman, C. A. 1995. A search for T Tauri stars in high-latitude molecular clouds. II: The IRAS Faint Source Survey catalog. *Astrophys. J. Suppl.* 96:159–173.
Mandelbrot, B. B. 1982. *The Fractal Geometry of Nature* (San Francisco: Freeman).

Martin, H. M., Sanders, D. B., and Hills, R. E. 1984. CO emission from fragmentary molecular clouds: A model applied to observations of M17 SW. *Mon. Not. Roy. Astron. Soc.* 208:35–55.

McKee, C. F. 1989. Photoionization-regulated star formation and the structure of molecular clouds. *Astrophys. J.* 345:782–801.

McKee, C. F., and Zweibel, E. G. 1992. On the virial theorem for turbulent molecular clouds. *Astrophys. J.* 399:551–562.

McKee, C. F., Zweibel, E. G., Goodman, A. A., and Heiles, C. 1993. Magnetic fields in star-forming regions: Theory. In *Protostars and Planets III,* ed. E. H. Levy and J. I. Lunine (Tucson: University of Arizona Press), pp. 327–366.

McLaughlin, D. E., and Pudritz, R. E. 1996. A model for the internal structure of molecular cloud cores. *Astrophys. J.* 469:194–208.

Miesch, M. S., and Bally, J. 1994. Statistical analysis of turbulence in molecular clouds. *Astrophys. J.* 429:645–671.

Mizuno, A., Onishi, T., Yonekura, Y., Nagahama, T., Ogawa, H., and Fukui, Y. 1995. Overall distribution of dense molecular gas and star formation in the Taurus cloud complex. *Astrophys. J. Lett.* 445:L161–L165.

Moriarty-Schieven, G. H., Andersson, B.-G., and Wannier, P. G. 1997. The L1457 molecular/atomic cloud complex: H I and CO maps. *Astrophys. J.* 475:642–660.

Motte, F., André, P., and Neri, R. 1998. The initial conditions of star formation in the ρ Ophiuchi main cloud: Wide-field millimeter continuum mapping. *Astron. Astrophys.* 336:150–172.

Myers, P. C. 1983. Dense cores in dark clouds: III. Subsonic turbulence. *Astrophys. J.* 270:105–118.

Myers, P. C., and Benson, P. J. 1983. Dense cores in dark clouds. II: NH_3 observations and star formation. *Astrophys. J.* 266:309–320.

Myers, P. C., and Goodman, A. A. 1988. Evidence for magnetic and virial equilibrium in molecular clouds. *Astrophys. J.* 326:L27–L130.

Nagahama, T., Mizuno, A., Ogawa, H., and Fukui, Y. 1998. A spatially complete ^{13}CO $J = 1$–0 survey of the Orion A cloud. *Astron. J.* 116:336–348.

Nakano, T. 1998. Star formation in magnetic clouds. *Astrophys. J.* 494:587–604.

Nakano, T., and Nakamura, T. 1978. Gravitational instability of magnetized gaseous disks. *Publ. Astron. Soc. Japan* 30:679–681.

Normandeau, M., Taylor, A. R., and Dewdney, P. E. 1997. The Dominion Astrophysical Observatory galactic plane survey pilot project: The W3/W4/W5/HB 3 region. *Astrophys. J. Suppl.* 108:279–299.

O'Dell, C. R., and Wong, S. K. 1996. Hubble Space Telescope mapping of the Orion nebula. I. A survey of stars and compact objects. *Astron. J.* 111:846–855.

Pak, S., Jaffe, D. T., van Dishoeck, E. F., Johansson, L. E. B., and Booth, R. S. 1998. Molecular cloud structure in the Magellanic Clouds: Effect of metallicity. *Astrophys. J.* 498:735–756.

Patel, N., Goldsmith, P. F., Snell, R. L., Hezel, T., and Taoling, X. 1995. The large-scale structure, kinematics, and evolution of IC 1396. *Astrophys. J.* 447:721–741.

Sanders, D. B., Clemens, D. P., Scoville, N. Z., and Solomon, P. M. 1985*a*. Massachusetts—Stony Brook Galactic plane CO survey. I: (b,V) maps of the first Galactic quadrant. *Astrophys. J. Suppl.* 60:1–303.

Sanders, D. B., Scoville, N. Z., and Solomon, P. M. 1985*b*. Giant molecular clouds in the Galaxy II: Characteristics of discrete features. *Astrophys. J.* 289:373–387.

Scalo, J. 1990. Perception of interstellar structure: Facing complexity. In *Physical Processes in Fragmentation and Star Formation,* ed. R. Capuzzo-Dolcetta et al. (Dordrecht: Kluwer), pp. 151–171.

Schneider, N., Stutzki, J., Winnewisser, G., and Blitz, L. 1996. The nature of the molecular line wing emission in the Rosette molecular complex. *Astrophys. J. Lett.* 468:L119–L122.
Shu, F. H., and Li, Z.-Y. 1997. Magnetic forces in an isopedic disk. *Astrophys. J.* 475:251–259.
Shu, F. H., Adams, F. C., and Lizano, S., 1987. Star formation in molecular clouds: Observation and theory. *Ann. Rev. Astron. Astrophys.* 25:23–81.
Simon, M. 1997. Clustering of young stars in Taurus, Ophiuchus, and the Orion Trapezium. *Astrophys. J. Lett.* 482:L81–L84.
Solomon, P. M., and Rivolo, A. R. 1989. A face-on view of the first galactic quadrant in molecular clouds. *Astrophys. J.* 339:919–925.
Strom, K. M., Strom, S. E., and Merrill, K. M. 1993. Infrared luminosity functions for the young stellar population associated with the L1641 molecular cloud. *Astrophys. J.* 412:233–253.
Stutzki, J., and Güsten, R., 1990. High spatial resolution isotopic CO and CS observations of M17 SW: The clumpy structure of the molecular cloud core. *Astrophys. J.* 356:513–533.
Stutzki, J., Bensch, F., Heithausen, A., Ossenkopf, V., and Zielinsky, M. 1998. On the fractal structure of molecular clouds. *Astron. Astrophys.* 336:697–720.
Tauber, J. A. 1996. The smoothness of line profiles: A useful diagnostic of clump properties. *Astron. Astrophys.* 315:591–602.
Testi, L., and Sargent, A. I. 1998. The OVRO 3 mm continuum survey for compact sources in the Serpens core. *Astrophys. J. Lett.* 508:L91–L94.
Tomisaka, K. 1991. The equilibria and evolutions of magnetized, rotating, isothermal clouds. V: The effect of the toroidal field. *Astrophys. J.* 376:190–198.
Tomisaka, K., Ikeuchi, S., and Nakamura, T. 1988. The equilibria and evolutions of magnetized, rotating, isothermal clouds. I: Basic equations and numerical methods. *Astrophys. J.* 326:208–222.
van Dishoeck, E. F., and Black, J. H. 1988. The photodissociation and chemistry of interstellar CO. *Astrophys. J.* 334:771–802.
Vázquez-Semadeni, E. 1998. Turbulence as an organizing agent in the ISM. In *Interstellar Turbulence, Proceedings of the 2nd Guillermo Haro Conference,* ed. J. Franco and A. Carramiñana (Cambridge: Cambridge University Press), in press.
Williams, J. P. 1998. The structure of molecular clouds: Are they fractal? In *Interstellar Turbulence, Proceedings of the 2nd Guillermo Haro Conference,* ed. J. Franco and A. Carramiñana (Cambridge: Cambridge University Press), in press.
Williams, J. P., and Blitz, L. 1998. A multitransition CO and CS(2–1) comparison of a star-forming and a non-star-forming giant molecular cloud. *Astrophys. J.* 494:657–673.
Williams, J. P., de Geus, E. J., and Blitz, L. 1994. Determining structure in molecular clouds. *Astrophys. J.* 428:693–712.
Williams, J. P., Blitz, L., and Stark, A. A. 1995. The density structure in the Rosette molecular cloud: Signposts of evolution. *Astrophys. J.* 451:252–274.
Williams, J. P., and Maddalena, R. J. 1996. A large photodissociation region around the cold, unusual cloud G216-2.5. *Astrophys. J.* 464:247–255.
Williams, J. P., and McKee, C. F. 1997. The Galactic distribution of OB associations in molecular clouds. *Astrophys. J.* 476:166–183.
Wiseman, J. J., and Adams, F. C. 1994. A quantitative analysis of IRAS maps of molecular clouds. *Astrophys. J.* 435:708–721.
Wolfire, M. G., Hollenbach, D., and Tielens, A. G. G. M. 1993. CO(J = 1–0) line emission from giant molecular clouds. *Astrophys. J.* 402:195–215.
Wood, D. O. S., Myers, P. C., and Daugherty, D. A. 1994. IRAS images of nearby dark clouds. *Astrophys. J. Suppl.* 95:457–501.

THE STELLAR INITIAL MASS FUNCTION: CONSTRAINTS FROM YOUNG CLUSTERS, AND THEORETICAL PERSPECTIVES

MICHAEL R. MEYER
University of Arizona, Steward Observatory

FRED C. ADAMS
University of Michigan

LYNNE A. HILLENBRAND, JOHN M. CARPENTER
California Institute of Technology

and

RICHARD B. LARSON
Yale University

We summarize recent observational and theoretical progress aimed at understanding the origin of the stellar initial mass function (IMF), with specific focus on galactic star-forming regions. We synthesize data from various efforts to determine the IMF in very young, partially embedded stellar clusters and find that (1) no significant variations in the low–mass IMF have been observed between different star-forming regions, and (2) the mass distributions of young stars just emerging from molecular clouds are consistent with having been drawn from the IMF derived from field stars in the solar neighborhood. These results apply only to gross characterizations of the IMF (e.g., the ratio of high- to low-mass stars); present observations do not rule out more subtle regional differences. Further studies are required in order to assess whether or not there is evidence for a universal turnover near the hydrogen-burning limit. We also provide a general framework for discussing theories of the IMF and summarize recent work on several physical mechanisms that could play a role in determining the form of the stellar initial mass function.

I. THE STELLAR INITIAL MASS FUNCTION

A fundamental question in star formation is the origin of stellar masses. Considerable progress has been made in recent years in understanding the formation of single stars, and we now have a working paradigm of the process. Yet quantitative understanding of the distribution of stellar masses formed within molecular clouds remains elusive.

One estimate of the initial distribution of stellar masses comes from studies of volume-limited samples of stars in the solar neighborhood.

Combining a variety of parallactic, spectroscopic, and photometric techniques, a luminosity function is derived for main sequence stars. A main sequence mass-luminosity relationship is then applied in order to derive the *present-day mass function,* or PDMF, from the luminosity function. Next, the effects of stellar evolution are taken into account in order to derive the *initial mass function,* or IMF. In constructing the "solar neighborhood IMF," the higher-mass stars are typically drawn from young associations out to distances of a few kiloparsecs, whereas the lower mass stars are drawn from well-mixed disk populations at distances out to tens of parsecs. Furthermore, because of their short main sequence lifetimes, the high-mass stars used in constructing the IMF are quite young (0.01–1×10^8 yr), whereas lower-mass stars found in the solar neighborhood are systematically older (1–100×10^8 yr). For these reasons, even a consistent definition of the IMF requires the assumption that it does not vary in time and space within the disk of the galaxy. This implies that the molecular clouds that produced the stars currently found in the solar neighborhood each formed stars with the same IMF as the regions producing stars today. It is precisely this assumption we wish to test with the work reviewed below.

The first comprehensive determination of the IMF over the full range of stellar masses was given by Miller and Scalo (1979), subsequently updated by Scalo (1986). Significant revisions have been made more recently at the low-mass end (e.g., Kroupa 1998; Reid 1998) and at the high-mass end (Garmany et al. 1982; Massey 1998). The reader is referred to Scalo (1986) for a detailed discussion of the ingredients that go into deriving the IMF and to Scalo (1998) for a general review of the field. Here we refer to the IMF as the number of stars per unit logarithmic mass interval, and we will use the power law notation $\Gamma = d \log F(\log m_\star)/d \log m_\star$ to characterize the IMF over a fixed mass range. In this notation, the slope of Salpeter (1955) is $\Gamma = -1.35$.

The IMF deduced from studies of OB associations and field stars in the solar neighborhood exhibits two main features on which there is general agreement. First, for masses greater than about 5 $M_\odot$, the IMF has a nearly power-law form. Massey et al. (1995*a*; 1995*b*) find a slope $\Gamma = -1.3 \pm 0.3$ for massive stars in clusters and a steeper power law for high-mass stars in the field. Secondly, the mass function becomes flatter for masses < 1 $M_\odot$ (Kroupa et al. 1993). Deriving the field star IMF near 1 $M_\odot$ is severely complicated by corrections for stellar evolution, which require detailed knowledge of the star formation history of the galaxy. Young open clusters may be the best place to measure the IMF between 1 and 15 $M_\odot$ (e.g., Phelps and Janes 1993). At the lowest masses, there is considerable debate whether the IMF continues to rise, is flat, or turns over between 0.1 and 0.5 $M_\odot$ (Reid and Gizis 1997; Mera et al. 1996; Kroupa 1995; Gould et al. 1997). That the IMF seems to change from a pure power law to a more complex distribution between 1 and 5 $M_\odot$ provides an important constraint on theories for the origin of stellar masses. Of even greater importance would be the clear demonstration of a peak in the IMF at the low-mass end. Considerable observational effort has been focused on establishing whether or not such a peak exists and, if so, characterizing its location and width (see Fig. 1).

In addition, it is extremely important to know whether or not the time- and space-averaged distribution of masses characterizing the solar neighborhood is universal. Do all star-forming events give rise to the same distribution of stellar masses? If star formation is essentially a self-regulating process, then one might expect the IMF to be strictly universal. Alternatively, if stellar masses are determined only by the physical structure of the interstellar medium (e.g., fragmentation), then one might expect differences in the IMF that depend on local conditions, such as

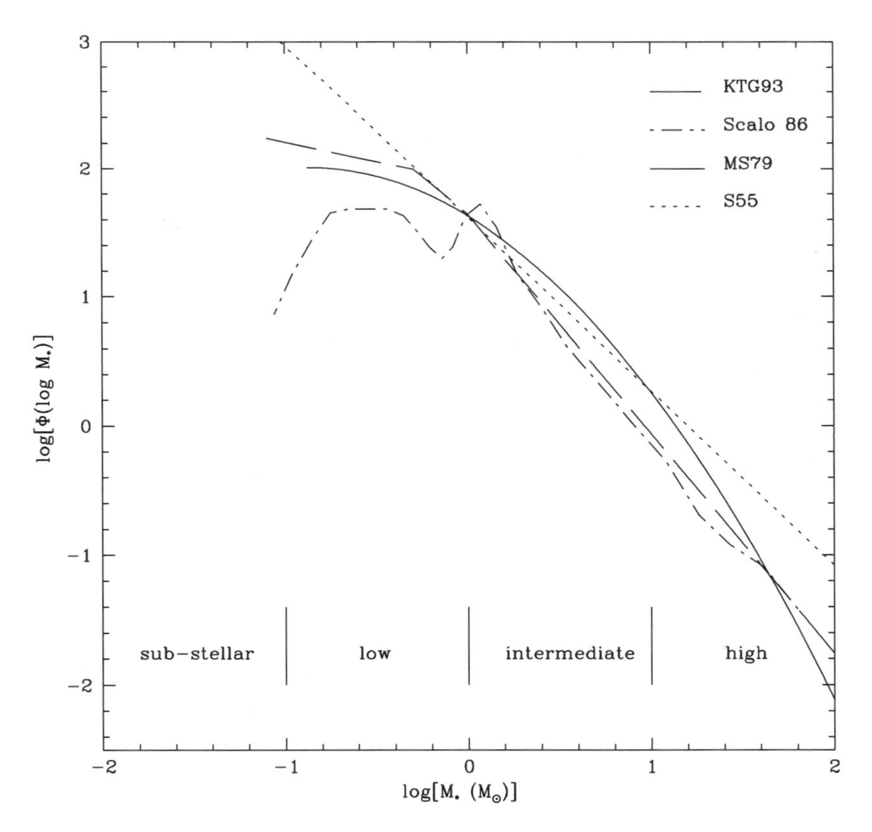

Figure 1. Initial mass function for field stars in the solar neighborhood taken from a variety of recent studies. These results have been normalized at 1 $M_{\odot}$. For the IMFs of Miller and Scalo (1979) (MS79) and of Scalo (1986), we have adopted 15 Gyr as the age of the Milky Way. Current work suggests that the upper end of the IMF (>5 $M_{\odot}$) is best represented by a power law similar to that of Salpeter (1955) (curve S55), whereas the low-mass end (<1 $M_{\odot}$) is flatter (Kroupa et al. 1993) (curve KTG93). The shape of the IMF fr om 1 to 5 $M_{\odot}$ is highly uncertain. See the references listed for details.

cloud temperature. Certainly the details of either process depend on the physical conditions of the clouds of gas and dust from which the stars form. The unanswered question is, how sensitively does the distribution of stellar masses depend on the initial conditions in the natal environment?

In this chapter, we outline the observational and theoretical progress that has been made toward understanding the origin of the IMF in star-forming regions. In section II, we review advances in deriving mass distributions for very young clusters and attempt to summarize the ensemble of results. In section III, we review various theoretical constructs that have been put forward to explain the expected shape of the IMF and its dependence on initial conditions. In section IV, we summarize our conclusions.

II. STELLAR MASS DISTRIBUTIONS IN STAR-FORMING REGIONS

Instead of providing a complete survey of the literature, we focus instead on a summary of approaches used and a comparison of results. We begin by describing why young clusters are helpful in understanding the origin of the IMF. We then outline techniques employed to derive stellar mass distributions in very young clusters, highlighting the advantages and disadvantages of each method. Next, we describe several direct comparisons of mass distributions assembled from the literature on star-forming regions. We then discuss a statistic which provides a gross characterization of the IMF, the ratio of high- to low-mass stars, and use data from the literature to constrain its variation. Finally, we compare work on star-forming regions with IMF studies of resolved stellar populations in the Milky Way and other galaxies.

A. The Utility of Young Clusters

Astronomers have long used studies of galactic clusters to answer questions concerning the formation and early evolution of stars (see the chapters by Clarke et al. and Elmegreen et al., this volume). Bound open clusters provide a useful starting point, but because they are rare and long-lived, it is thought that they do not contribute significantly to the field star population of the galactic disk (cf. Roberts 1957). In contrast, most young embedded clusters are thought to evolve into unbound associations, which comprise the majority of stars that populate the galactic disk (Lada and Lada 1991). Embedded clusters are useful for studying the form of the IMF for several practical reasons. First, because of their youth, evolutionary corrections needed to translate the present-day distribution of stellar masses into an IMF are minimized. Second, observations of such clusters are more sensitive to low-mass objects, because the mass-luminosity relationship for stars in the pre-main-sequence (PMS) phase is not as steep a function of mass as for stars on the main sequence. Third, because of their compactness, they occupy small projected areas on the sky, reducing the contamination by foreground stars that plague studies of larger, optically visible associations. Finally, the molecular cloud cores that contain embedded clusters provide natural screens against background stars, which would otherwise contaminate the sample. Thus, we can attempt to derive distributions of stellar masses for particular star-forming events associated with individual molecular cloud cores. We can then determine whether or not the IMF varies as a function of cloud conditions, providing insight into its origin (cf. the chapter by Williams et al., this volume).

B. Approaches Used to Study Emergent Mass Distributions of Young Clusters

Of course, partially embedded young clusters also present difficulties to astronomers interested in deriving their stellar mass distributions. First of all, ongoing star formation is typically observed in these clusters. As such, the observed mass distribution is only a "snapshot" of the IMF for the star-forming event and may not represent the integrated final product of the cloud core. In order to keep this distinction clear between the initial mass function (as defined from the solar neighborhood sample) and the observed mass distributions of embedded young clusters, we will refer to these "snapshots" of the IMF as *emergent mass distributions* (EMDs). Another complication in studying embedded clusters is the time-dependent nature of the mass-luminosity (*M-L*) relationship (required for translating an observed luminosity function into a mass distribution) for PMS stars. Third, large and variable obscuration makes it necessary to correct for extinction on an individual, star-by-star basis in many embedded clusters. Uncertainties in extinction-corrected absolute magnitudes can be reduced by observing embedded clusters at longer wavelengths. For example, interstellar extinction in the *K* band (2.2 μm) is $\times$ 10 smaller than in the *V* band (0.55 μm) and $\times$ 3 smaller than in the *J* band (1.25 μm). However, at wavelengths beyond 2.0 μm the observed flux from PMS cluster members is frequently contaminated by excess emission. The near-IR excess (associated with hot circumstellar dust located in the inner disk region) complicates the interpretation of monochromatic near-IR luminosity functions as *stellar* bolometric luminosity functions (e.g., Meyer et al. 1997). Most recent attempts to derive emergent mass distributions from observations of embedded young clusters are based on (1) modeling of monochromatic near infrared luminosity functions; (2) analysis of multicolor photometric data; (3) spectroscopic survey samples; or (4) some combination of these techniques. We discuss each of these methods (along with their pros and cons) below.

A comprehensive review of embedded cluster work published before *Protostars and Planets III* is given by Zinnecker et al. (1993). Many of these studies compared the observed distribution of *K* band magnitudes with that expected from a standard IMF convolved with a main-sequence *M-L* relationship (e.g., Lada et al. 1993; Greene and Young 1992). In contrast, Zinnecker et al. (1993) constructed models for the *evolution* of embedded cluster luminosity functions by transforming theoretical PMS evolutionary tracks into an observational plane. They generated synthetic *K* band luminosity functions (KLFs) for clusters with ages 0.3–2 Myr from an assumed input IMF. Ali and DePoy (1995; see also Megeath 1996) made allowances for spatially variable extinction and excess emission due to circumstellar disks in their analyses of KLFs. Lada and Lada (1995; see also Giovannetti et al. 1998) extended these approaches by considering continuous as opposed to discretized star formation.

Use of multicolor photometry increases the amount of information and thus decreases the number of assumptions needed to derive an EMD. Strom et al. (1993; see also Aspin et al. 1994) attempted to deredden individual embedded sources by assuming the same intrinsic color for each star; they also considered dereddened *J* band luminosity functions (as opposed to *K* band) to minimize the effect of infrared excess emission. Many studies of embedded young clusters assume an input mass distribution, which is combined with analysis of an observed luminosity function to derive the cluster age or age distribution. In principle, extensions of these techniques could place constraints on the stellar mass distribution. By making the shape of the input mass spectrum a variable, certain distributions could be ruled out if found inconsistent with the observed luminosity function for any reasonable input age distribution (e.g., Lada et al. 1998). In practice, treating both mass and age distributions as variables makes it difficult to find a unique best-fit solution.

Comeron et al. (1993) developed a novel technique for analyzing multicolor photometric data of embedded clusters. By modeling spectral energy distributions (SEDs) of embedded sources, one can estimate the extinction, intrinsic luminosities, and the spectral slopes of the SEDs. The spectral slopes are taken as an indicator of evolutionary state, and volume corrections are applied to the sample as a function of intrinsic luminosity. This latter step takes into account the fact that more luminous objects probe larger volumes in flux-limited samples. By adopting *M-L* relationships deemed appropriate for the evolutionary state of the objects in question, the luminosity function is then transformed into an emergent mass distribution. Comeron et al. (1996) expanded on this approach by adopting appropriate *M-L* relationships in a Monte Carlo fashion from an assumed age distribution. Meyer (1996) developed an alternate method that uses multicolor photometry to deredden individual embedded sources, explicitly taking into account the possibility of near infrared excess emission. Then by adopting an age (or age distribution) for a cluster, dereddened absolute *J* band magnitudes are used to estimate stellar masses. In this case, an extinction-limited sample is used to derive an emergent mass distribution that is not biased toward higher-mass stars seen more deeply into the cloud.

Ultimately, the construction of reliable IMFs for a large number of young clusters, forming under a wide range of conditions, requires a combination of deep photometric survey work and follow-up spectroscopic analysis. Spectra allow one to derive the photospheric temperatures for the embedded objects and place them in the HR diagram. Comparing the positions of sources in the HR diagram with PMS evolutionary models provides estimates of the mass and age distributions of the sample. Ideally, we would like to have *complete* spectroscopic samples for each cluster. In practice, often the best we can do is obtain spectra for a subset of the photometric sample (either representative or flux-limited), providing an estimate of the cluster age. With this constraint on the age-dependent *M-L* relationship, one can characterize the dereddened luminosity function in terms of the emergent mass distribution for the cluster and compare it to the solar neighborhood IMF. The spectroscopic techniques allow a more unambiguous accounting for the effects of extinction, infrared excess, and a stellar age distribution, yet the photometric techniques have the distinct advantage that they can be applied to fainter stellar populations at much greater distances. It is important to remember that in all of these methods the masses derived for individual stars depend sensitively on the adopted PMS evolutionary tracks. As a result, estimates of mass scales at which inflections are observed in the detailed distribution functions are necessarily uncertain until we have a better calibration of PMS evolutionary tracks.

C. Three Techniques for Direct Comparison of EMDs

Direct comparisons between observations of two independent star-forming regions provide the best means to uncover differences in the emergent mass distributions of clusters. Statistical tests, such as the Kolmogorov-Smirnov (KS) test (Press et al. 1993), between individual distributions are more informative than comparisons to uncertain analytic functions such as those derived for the field star IMF. Provided that studies of young clusters are performed in a uniform way, we can overcome uncertainties in the various techniques by making relative rather than absolute comparisons. However, care must be taken in applying such statistical tests, since any systematic differences in the observations between two studies can easily produce a significant signal in the KS test. Here we restrict ourselves to three datasets, each of which has been assembled using the same analysis, so that meaningful comparisons for different clusters can be made. We begin with the KLF analyses of IC 348 and NGC 1333. Next we review the multicolor photometric surveys of Ophiuchus and NGC 2024, informed by follow-up spectroscopy. Finally, we compare the extensive spectroscopic surveys for the Orion Nebula cluster and IC 348.

1. K Band Luminosity Functions: IC 348 vs. NGC 1333. IC 348 and NGC 1333 are both young clusters in the Perseus cloud, located at a distance of 320 pc (Herbig 1998).

Lada and Lada (1995) present near-IR imaging of IC 348; 380 sources are identified in excess of the measured field star population. Because of low extinction and the low fraction of association members exhibiting strong near-infrared excess emission ($<25\%$), the observed KLF gives a reasonable estimate of the stellar luminosity function, albeit convolved with an age distribution.

Lada et al. (1996) performed a similar near-IR survey of NGC 1333, identifying a "double cluster" of 94 sources. However, in this case, differential extinction seriously affects the observed KLF, and in addition, $\sim$50% of the association members displayed strong infrared excess emission. To correct for extinction, the cluster populations were dereddened using $(H - K)$ colors. This dereddened KLF was compared with that observed for IC 348 and with that published for the Trapezium cluster by Zinnecker et al. (1993), corrected for the difference in distance. The KLFs for NGC 1333 and the Trapezium are consistent with having been drawn from the same parent population.

Both clusters are thought to be the same age and to have similar IR excess frequencies, so this result suggests that the underlying mass functions are similar. Comparison of the KLFs for NGC 1333 and IC 348 yields a different result: There is only a 20% chance that they were drawn from the same distribution. Lada et al. (1996) point out that the observed KLFs of all three clusters could be derived from the same underlying mass functions but convolved with different age distributions. Without independent estimates of the cluster ages based on spectroscopic observations, it is difficult to draw robust conclusions.

2. Combining Spectroscopy and Photometry: Ophiuchus vs. NGC 2024. Comeron et al. (1993) conducted a K band imaging survey of the Ophiuchus cloud core ($d \sim 150$ pc) within the $A_V > 50$ mag molecular contour and obtained follow-up four-color photometry for all sources $K < 14.5$ mag. Strom et al. (1995) published a comparable multicolor imaging survey, focusing on the "aggregates" associated with dense cores located within the $A_V > 50$ mag contour. Greene and Meyer (1995) combined infrared spectroscopy of 19 embedded sources with photometry from Strom et al. (1995) in order to place them on the HR diagram and to estimate masses and ages directly. Williams et al. (1995*b*) obtained spectra for three candidate low-mass objects from the sample of Comeron et al. (1993) and compared the IMF results of Comeron et al. (1993) and Strom et al. (1995) and found them to be in broad agreement. Combining results from both studies of Ophiuchus yielded an emergent mass distribution from 0.1 to 5.0 $M_\odot$ with $\Gamma \sim -0.1$.

Comeron et al. (1996) performed a deep near-infrared survey of several fields toward the embedded cluster associated with NGC 2024 ($d \sim$ 470 pc). Assuming an age distribution for the cluster, they derive a mass function between 0.08 and 2.0 $M_\odot$ that is nearly flat with no evidence for a turnover at the low-mass end ($\Gamma = -0.2 \pm 0.1$), consistent with their results (1993) for Ophiuchus. Meyer et al. (1999) present a multicolor near-IR survey of the innermost 0.5 pc of NGC 2024, sampling 0.1 $M_\odot$ stars viewed through $A_V < 19.0$ mag. By combining this photometric survey with near-IR spectra for two dozen sources in the region, they constructed an HR diagram. Taking the age derived from the HR diagram as characteristic of the entire embedded population, they estimate the extinction toward each star and construct a stellar luminosity function complete down to 0.1 $M_\odot$.

Given that existing near infrared surveys of the embedded clusters associated with NGC 2024 and the Ophiuchus molecular cloud are of comparable sensitivity ($M_K < 4.5$ and $A_V < 19$ mag) and physical resolution ($d \sim 400$ AU), one can directly compare the derived luminosity functions. The KS test reveals that there is a small chance ($P = 0.04$) that the stellar luminosity distributions were drawn from the same parent population. Meyer et al. (1999), in making this comparison, investigate three factors that can affect the shape of the stellar luminosity function for an embedded cluster: age distribution, accretion properties, and emergent mass distribution. Comparison of the HR diagrams shows that the evolutionary states of both clusters are similar. Comparison of the *JHK* color-color diagrams for complete samples in each cluster reveals that they are consistent with having been drawn from the same parent population, suggesting that the accretion properties of NGC 2024 and Ophiuchus are also similar. Although the KS test suggests differences in the luminosity functions that could be attributed to differences in the mass functions, the uncertainties make it difficult to argue that the emergent mass distributions are significantly different.

3. Complete Spectroscopic Samples: IC 348 vs. the Orion Nebula Cluster. Herbig (1998) conducted an extensive photometric and spectroscopic survey of IC 348, placing ~80 optically visible stars on the HR diagram. The derived age distribution was used to study the mass distribution of a larger photometric sample in the $(V - I)$ vs. V color-magnitude diagram. Based on a sample of 125 stars fitted to the PMS evolutionary models of D'Antona and Mazzitelli (1994) in the $(V - I)$ vs. V color-magnitude diagram, the derived mass function is similar to the Scalo (1986) IMF down to 0.3 $M_\odot$. Luhman et al. (1998) independently conducted an infrared and optical spectroscopic study of the IC 348 region. Spectra were obtained for 75 sources from the photometric survey of Lada and Lada (1995). Luhman et al. claim completeness of the spectroscopic sample down to 0.1 $M_\odot$. Masses are estimated for the sources that lack spectra by adopting the *M-L* relationship suggested by the extant spectroscopic sample, and a corrected emergent mass distribution is constructed for the cluster.

The Orion Nebula Cluster (ONC) is the richest young cluster within 1 kpc and has been the target of several photometric studies in the last decade (Herbig and Terndrup 1986; McCaughrean and Stauffer, 1994; Ali and DePoy 1995). Hillenbrand (1997) presents results from an optical spectroscopic survey of nearly 1000 stars located within ~2 pc of the Trapezium stars. From the resulting distribution of stellar masses derived from the HR diagram (down to 0.1 $M_\odot$), Hillenbrand concludes that the cluster mass function turns over at ~0.2 $M_\odot$.

IC 348 and the ONC have the most complete EMDs derived to date among the very young clusters studied spectroscopically. As a result, it is desirable to compare them directly (Fig. 2). We use masses derived from the D'Antona and Mazzitelli (1994) tracks in this exercise, and note that even this direct comparison using identical observational techniques is "model independent" only to the extent that the clusters exhibit similar age distributions so that we are using comparable theory to translate observables into stellar masses. Comparison of the EMD derived globally for the ONC cluster (605 stars $A_V < 2.0$ mag) with that presented for IC 348 (73 stars $A_V < 5.0$ mag) over the mass range 0.1–2.5 $M_\odot$ indicates that the two mass distributions were not drawn from the same parent population ($P = 5.92 \times 10^{-5}$). However, when a restricted sample from the inner ONC is taken [133 stars within $r < 0.5$ pc, comparable to the physical size of the Luhman et al. (1998) region in IC 348], we cannot rule out that they were drawn from the same distribution ($P = 0.06$), despite

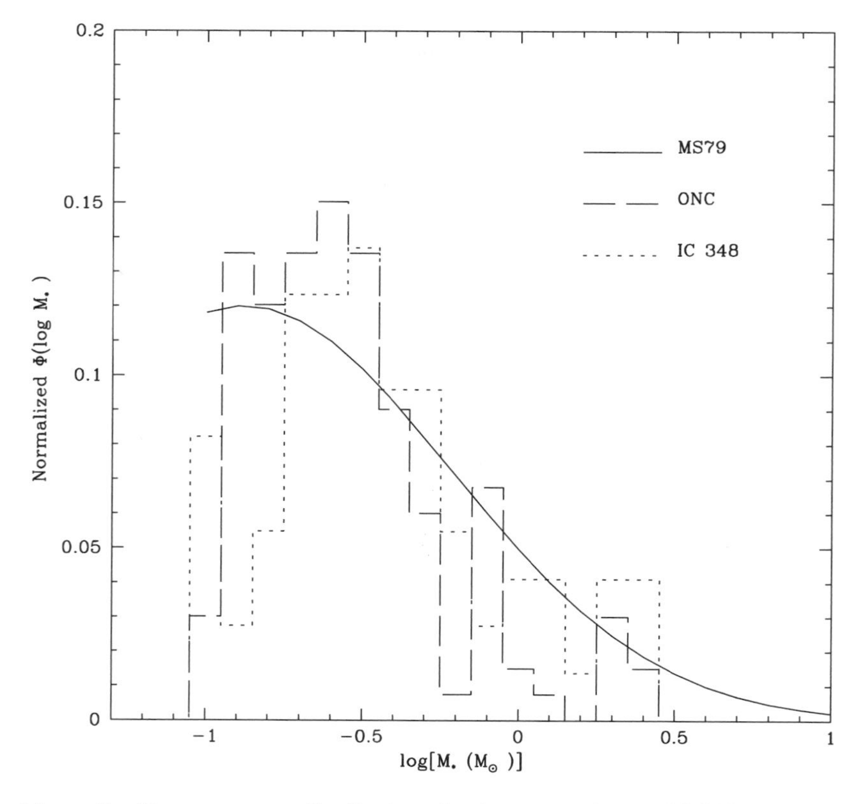

Figure 2. Emergent mass distributions for the young clusters IC 348 (Luhman et al. 1998; Herbig 1998) and the Orion Nebula Cluster (ONC) (Hillenbrand 1997). Also shown is the distribution of stellar masses derived from the lognormal form of the Miller and Scalo (1979) IMF (MS79). Given the uncertainties in the PMS tracks, we cannot conclude that the observed mass distributions are different. Both are broadly consistent with the field star IMF and suggest that the IMF below 0.3 $M_\odot$ is falling in logarithmic units.

$\times 100$ difference in central stellar density! This latter result could be due in part to the smaller sample considered; the smaller the differences between the EMDs of different regions, the larger the sample size needed to discern them.

D. Synthesis of Results for an Ensemble of Clusters

To study the IMF over the full range of stellar masses, we require photometric observations that are sensitive below the hydrogen-burning limit for the distance and age of the cluster. We also wish to sample the stellar population through some well-determined and significant value of extinction. Finally, we need follow-up spectroscopy in order to inform cluster age estimates, which are crucial for adopting an appropriate M-L relationship. Why have so few quantitative results emerged from the study of EMDs in very young clusters?

Since *Protostars and Planets III,* comprehensive spectroscopic studies of young stellar populations have been conducted toward a variety of star-forming regions (e.g., Alcala et al. 1997; Allen 1996; Hughes et al. 1994; Lawson et al. 1996; Walter et al. 1994; 1997). These studies, while providing crucial tests of many aspects of PMS evolution, are not ideally suited for studying emergent mass distributions. For example, Hα, X-ray, and variable star samples impose activity-related selection effects that make it difficult to assess completeness. Even in the well-studied Taurus dark cloud (Kenyon and Hartmann 1995), only now are complete samples down to the hydrogen-burning limit becoming available (Briceño et al. 1998).

Armed with deep photometric surveys, often we are still faced with a statistical problem. The clusters and aggregates found in typical molecular clouds contain only tens to hundreds of stars; populous regions like the ONC are rare. As demonstrated above, detailed comparisons of mass or luminosity functions based on sample sizes $\ll 1000$ are inconclusive unless gross differences exist in the underlying distributions. In this section we compare results from a variety of recent studies and ask whether the ensemble of results can tell us something about the shape of the IMF.

One approach is to use mass bins that are significantly wider than the errors in the assigned stellar masses from the methods described above. For example, in the study of nearby star-forming regions a particularly useful diagnostic is the ratio $\mathcal{R}$ of intermediate- to low-mass stars:

$$\mathcal{R} = N(1-10\ \mathrm{M}_\odot)/N(0.1-1\ \mathrm{M}_\odot)$$

Each of the techniques outlined above for characterizing the emergent mass distribution can be collapsed into such a ratio. In Table I, we present the $\mathcal{R}$ values for various regions for which the ratio of intermediate- to low-mass stars can be constructed, based on data assembled from the literature. We restrict this analysis to regions located within 1 kpc of the Sun, for which a flux-limited survey down to a well-defined completeness limit

TABLE I

Ratios of High- to Low-Mass Stars for Young Embedded Clusters

Name	$N_\star$	Age (yr)	A_V limit (mag)	$R_{int/low}$	D (pc)	$\rho_\star(pc^{-3})^a$	T^b_{gas}(K)
NGC2024 [c]	72	3×10^5	18.9	0.24 ± 0.14	0.5	2000–5000	50
Ophiuchus [d]	32	3×10^5	19.3	0.1 ± 0.04	0.5	500–1500	20
L1495 [e]	27	3×10^5	5.0	0.13 ± 0.05	1.0	15–45	10
OMC–2 [f]	107	1×10^6	5.0	0.07 ± 0.04	1.0	200–1000	50
BD+40°4124 [g]	32	3×10^5	20.0	0.45 ± 0.15	0.34	800–3000	30
R Cr A [h]	45	1×10^6	40.0	0.25 ± 0.07	0.5	500–1500	15
ONC [i]	133	3×10^5	2.0	0.07 ± 0.02	1.0	10,000	50
Mon R2 [j]	115	1×10^6	11.0	0.14 ± 0.10	0.8	1000–3000	35
IC 348 [k]	73	2×10^6	5.0	0.18 ± 0.06	1.0	100–700	12

[a] Averaged over a region of $r \sim 0.3$ pc.
[b] Inferred from NH_3 measurements when available.
[c] Meyer et al. (1999).
[d] Strom et al. (1995); Greene & Meyer (1995).
[e] Strom & Strom (1994); Luhman & Rieke (1998).
[f] Jones et al. (1994); Hillenbrand et al. (1997).
[g] Hillenbrand et al., 1995.
[h] Wilking et al. (1997); Wilking & Meyer (in prep).
[i] Hillenbrand (1997).
[j] Carpenter et al. (1997).
[k] Luhman et al. (1998); Herbig (1998).

exists, along with complementary spectroscopy to constrain the age distribution of the association members. In Fig. 3, we compare the ratio of intermediate- to low-mass stars derived for nine such regions to the same ratio predicted from various analytic forms of the IMF.

The range and errors in the values measured for the ratio $\mathcal{R}$ of intermediate- to low-mass stars suggest a conservative conclusion: Most extremely young, compact star-forming regions exhibit EMDs consistent with having been drawn from the field star IMF, within our ability to distinguish any differences. The small sizes of existing observational samples, and the coarse nature of the tests we have been able to perform, allow us to detect only gross differences between the observed mass distributions and various forms of the field star IMF. For example, it is clear

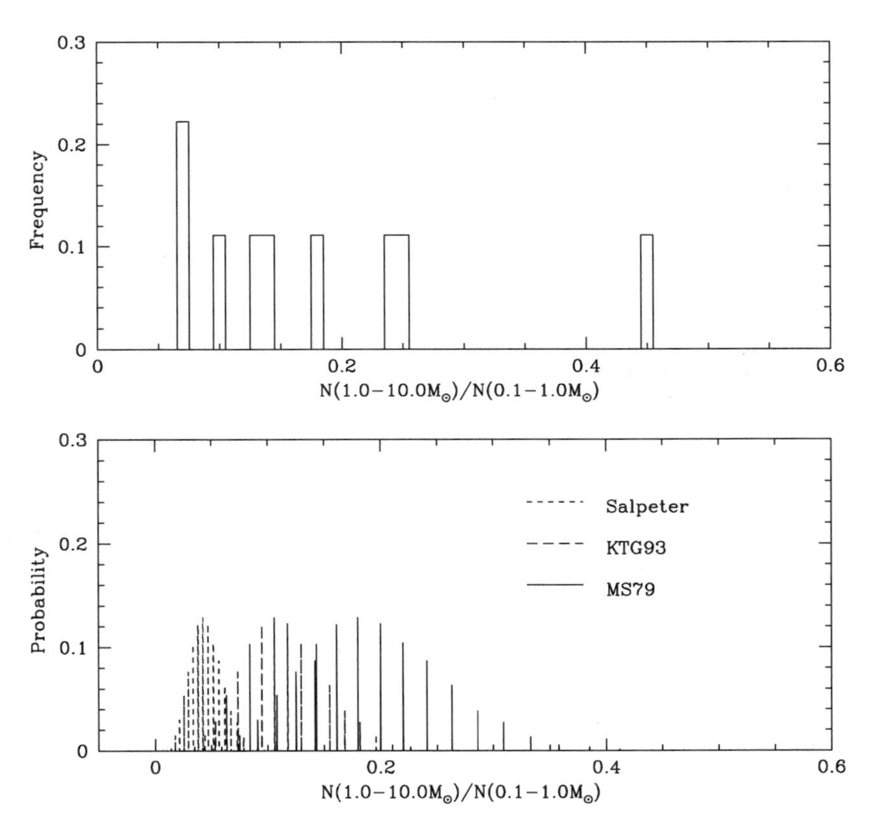

Figure 3. Ratio of intermediate ($1-10$ $M_\odot$) to low ($0.1-1$ $M_\odot$) mass stars for star-forming regions as listed in Table 1. Also shown are the expected distributions if the measurements were drawn from the Salpeter (1955), Miller and Scalo (1979) (MS79), or Kroupa et al. (1993) (KTG93) mass distributions. Based on results from the Kolmogorov-Smirnov (KS) test, the probabilities that the observed distributions were drawn from these parent populations are 3.28×10^{-8}, 0.121, and 0.053, respectively. We conclude that the observations are inconsistent with the Salpeter IMF extending over the range $0.1-10$ $M_\odot$.

that the Salpeter (1955) IMF does not hold below 1.0 $M_\odot$, because it predicts a ratio of intermediate- to low-mass stars of $\mathcal{R} = 0.04$, compared to $\mathcal{R} = 0.17$ for the Miller and Scalo (1979) IMF. Furthermore, the IMF does not have a sharp truncation at masses well above 0.1 $M_\odot$. If the Miller and Scalo (1979) IMF were truncated at 0.4 $M_\odot$, the expected ratio of intermediate- to low-mass stars would be 0.4, which is also excluded by the data. Finally, based on the information in Table I, we can can rule out dramatic dependencies of the IMF on environmental characteristics such as cloud temperature (as measured from molecular line observations) or mean stellar volume density (averaged over a region $r \sim 0.3$ pc).

E. Comparisons to Other Estimates of the Galactic IMF

Because differences in observational technique could introduce important systematic errors, it is dangerous (though interesting!) to compare results derived for young clusters in star-forming regions with other constraints on the IMF. Much recent work has focused on nearby open clusters with ages <1 Gyr that share some of the attributes that make star-forming regions excellent places to search for variations in the IMF. Bouvier et al. (1998; see also Williams et al. 1996; Zapatero-Osorio et al. 1997) have derived the IMF for the Pleiades open cluster well into the brown dwarf regime. Although the ratios $\mathcal{R}$ for the Pleiades (120 Myr) and the much younger Trapezium are both consistent with having been drawn from the field star IMF, the detailed mass distributions are much different [($P(d >$ obs$) = 1.9 \times 10^{-38}$!)]. This is probably due largely to differences in the techniques used and the adopted PMS tracks, although it could reflect true differences in the IMF. Hambly et al. (1995) studied the IMF below 1.0 $M_\odot$ in the 900-Myr-old, metal-rich cluster Praesepe, finding a slope $\Gamma = -0.5$ between 0.1–1.0 $M_\odot$, though Williams et al. (1995*a*) find a somewhat shallower $\Gamma = -0.34 \pm 0.25$. Pinfield et al. (1997) have extended this work down into the brown dwarf regime and find a steeper rise in the mass function. There has been additional work identifying low-mass stars and brown dwarf candidates in other clusters, including α Per (Zapatero-Osorio et al. 1996) and the twin clusters IC 2602/2391 (Stauffer et al. 1997), though detailed investigations of the IMF are not yet available.

Phelps and Janes (1993) derive the IMF between 1.4 and 9.0 $M_\odot$ for a dozen older open clusters observed in the disk of the Milky Way. Although most regions are consistent with a power law slope of $\Gamma = -1.4 \pm 0.13$, two regions exhibit significantly different IMFs (NGC 581 and NGC 663). Similar results are reported by Sagar and Griffiths (1998), though intriguing differences between clusters are found over small mass ranges. Further support for a roughly universal IMF comes from von Hippel et al. (1996), who demonstrate that the turnovers in luminosity functions observed in a variety of cluster environments are correlated with metallicity (as expected, given the dependence of the mass-luminosity relationship on metallicity). Studies of the low-metallicity Population II component of the Milky Way are in surprisingly good agreement concerning the shape of the low-mass IMF. Results from globular clusters (cf. Chabrier and Mera 1997; Cool 1998 and references therein), spheroid populations (Gould et al. 1998), and the bulge (Holtzman et al. 1998) indicate that $\Gamma \sim 0.0 \pm 0.5$, consistent with the field star IMF and our results for young clusters.

Massey and collaborators (1998 and references therein) also find that the high-mass end of the IMF does not change with metallicity; their results suggest that the IMF is well fitted by a similar power law in the Magellanic Clouds and the Milky Way. These results are generally confirmed by other studies of resolved galactic and extragalactic OB associations (e.g. Brown 1998; Bresolin et al. 1998). However, the mass distribution of "field" OB stars does seem to differ from that for stars found in clusters. In studies of the low-mass component of giant H II regions in the Milky Way (NGC 3603, Eisenhauer et al. 1998; W3, Megeath et al. 1996) and the Large Magellanic Cloud (LMC) (R136 in 30 Dor; Hunter et al. 1996; Brandl et al. 1996), the distributions of stellar masses are consistent with the field star IMF, at least down to 1.0 $M_{\odot}$. Deeper observations obtained at higher spatial resolution will be required to sample the IMF down to the hydrogen-burning limit.

Taken as a whole, the preponderance of the evidence suggests that crude characterizations of the IMF (such as the ratio of high- to low-mass stars) do not vary strongly as a function of metallicity, star-forming environment, or cosmic time. That said, there are indications of possible variations in the IMF: the ONC vs. IC 348; results for NGC 581 and 663 compared to other open clusters; and the massive star IMF derived from the field sample vs. results from associations. These differences may provide clues to help unravel the mystery surrounding the origins of stellar masses. Finally, we note that all the results discussed above are for the "system IMF," not accounting for unresolved binary stars. In order to quantify this effect, we need not only the distribution of masses for the composite systems but also the distribution of companion mass ratios (see the chapter by Mathieu et al., this volume). Knowing both the IMF and the distribution of companion mass ratios is crucial to understanding the process of star formation. However in what follows, we concentrate on theories that might explain the origin of the system IMF.

III. SOME THEORETICAL IDEAS ON THE ORIGIN OF THE IMF

Although the current theory of star formation remains incomplete, we can begin to consider approaches to the IMF problem. In this section we discuss briefly some current theoretical ideas that may be relevant to the origin of the stellar IMF. Given the limited space available, we can only discuss some of the many ideas that have been proposed. We also emphasize that we still have only rather speculative ideas to discuss. Additional theoretical reviews can be found elsewhere (e.g., Clarke 1998; Larson 2000; Elmegreen 1999).

A. Toward a General Formulation

Within the context of the current theory of star formation, it is often useful to conceptually divide the process that determines the IMF into two subprocesses:

1. The spectrum of initial conditions produced by molecular clouds or other star forming environment
2. The transformation between a given set of initial conditions and the properties of the final (formed) star

In this section, we outline one particular approach to constructing theories of the initial mass function. In an IMF theory, the final mass of a forming star must be determined through some physical mechanism. The identification of that mechanism, which could include many different physical processes, lies at the heart of determining a theory of the IMF. In practical terms, we need to specify the transformation between the initial conditions and the final stellar properties (Adams and Fatuzzo 1996; Khersonsky 1997). For example, the mass of the star $M_\star$ could be given by a "semiempirical mass formula" (SEMF) or "transfer function" of the form $M_\star = M_\star(\alpha_1, \alpha_2, \alpha_3, \ldots)$, where the α_j are the physical variables that determine the mass of the star. In the limit in which a large number of physical variables is required to determine stellar masses, we obtain a particularly simple result, suggesting a statistical approach to the calculation of the IMF.

We would like to find a relationship between the distributions of the initial variables and the resulting distribution of stellar masses (the IMF). For many (but certainly not all) cases of interest, the transformation can be written as a product of variables, i.e.,

$$M_\star = \prod_{j=1}^{N} \alpha_j \tag{1}$$

where the α_j represent the relevant physical variables (which could be the sound speed a, the rotation rate Ω, etc., taken to the appropriate powers). Each of these variables has a distribution $f_j(\alpha_j)$ with a mean value $\ln \overline{\alpha}_j \equiv \langle \ln \alpha_j \rangle$ and a corresponding variance σ_j^2.

In the limit of a large number N of variables, the composite distribution (the IMF) approaches a lognormal form. This behavior is a direct consequence of the central limit theorem (Richtmyer 1978) and has been invoked by many authors (Larson 1973; Elmegreen and Mathieu 1983; Zinnecker 1984; Adams and Fatuzzo 1996). Whenever a *large* number of independent physical variables are involved in the star formation process, the resulting IMF can be *approximately* described by a lognormal form. The departure of the IMF from a purely lognormal form depends on the shapes of the individual distributions f_j and on the number of relevant variables. However, in the limit that the IMF can be described to leading order by a lognormal form, there are simple relationships between the distributions of the initial variables and the shape parameters m_C and $\langle\sigma\rangle$ that determine the IMF. The mass scale m_C is determined by the mean values of the original variables α_j, and the dimensionless shape parameter $\langle\sigma\rangle$ is determined by the widths σ_j of the initial distributions, i.e., $m_C \equiv \prod_j \exp[\langle \ln \alpha_j \rangle]$ and $\langle\sigma\rangle^2 = \sum_j \sigma_j^2$.

In the limit that star formation involves a large number of statistically independent variables, we would obtain a "pure" lognormal distribution. In this limit the only relevant parameters are the total width of the distribution $\langle\sigma\rangle$ and the mass scale of the distribution m_C, and these parameters are constrained by observations of the IMF; the quantities $\langle\sigma\rangle$ and m_C are determined by the distributions of the physical variables in the problem. In a complete theory, we could calculate these initial distributions from *a priori* considerations. In the absence of a complete theory, however, we can use observations of the physical variables to estimate their distributions and hence determine $\langle\sigma\rangle$ and m_C.

Although the number of physical variables involved in the star formation process may be large, it is certainly not infinite. An important challenge of the future will be to identify the relevant physical variables in the problem unambiguously. In any case, the IMF will never *completely* converge to a lognormal form. Instead, the distribution will retain *tails,* departures from a lognormal form, at both the high-mass and low-mass ends. Even though the composite distribution does not obtain a purely lognormal form, however, the theoretical predictions for the mass scale m_C and total variance $\langle\sigma\rangle$ must be consistent with the constraints from the observed IMF.

B. Gravitational Instability and the Jeans Criterion

In order for a clump of gas to collapse and form stars, its self-gravity must overcome the effects that tend to prevent collapse. On large scales, turbulence and magnetic fields provide the dominant form of support against gravity, while in the densest core regions of molecular clouds, thermal pressure is the dominant supporting force. A minimal requirement for collapse to occur is thus the classical Jeans criterion that the self-gravity of a dense core must overcome its thermal pressure; for a given temperature and density, this implies a minimum size and mass, called the Jeans length and Jeans mass. For density fluctuations in an infinite uniform medium with density ρ and isothermal sound speed a, the minimum unstable mass is $\pi^{3/2}a^3/G^{3/2}\rho^{1/2}$ (Jeans 1929; Spitzer 1978). Although this result is not self-consistent, in that it neglects the overall collapse of the medium, dimensionally equivalent results are obtained from analyses of the stability of equilibrium configurations. For a sheet, disk, or filament with surface density μ, the minimum unstable mass is a few times $a^4/G\mu$ (Larson 1985). For an isothermal sphere confined by an external pressure P, the minimum unstable mass is $1.18\,a^4/G^{3/2}P^{1/2}$ (Ebert 1955; Bonnor 1956; Spitzer 1968). These expressions are all dimensionally equivalent, because $P = \rho a^2$ and because $P \sim G\mu^2$ in a self-gravitating configuration; thus they are basically different expressions for the same quantity, for which the term "Jeans mass" is a convenient name. If star-forming cores are created by turbulence in molecular clouds (Larson 1981), and if they are confined by a characteristic nonthermal pressure arising from the cloud formation process (Larson 1996; Myers 1998), then the best estimate of the Jeans mass may be the mass of a critically stable "Bonnor-Ebert sphere" bounded by this pressure. For a typical molecular cloud temperature of 10 K and a typical nonthermal cloud pressure of 3×10^5 cm^{-3} K, this quantity is about 0.7 M$_\odot$, similar to the observed typical stellar mass (Larson 2000).

Direct evidence for the existence of gravitationally bound clumps having about the Jeans size and mass in a star-forming cloud has been found in recent millimeter continuum mapping of the ρ Ophiuchi cloud (Motte et al. 1998). The observed clumps have masses between 0.05 and 3 M$_\odot$ and a mass spectrum that closely resembles the stellar IMF, becoming flatter below 0.5 M$_\odot$; thus, they may be the "direct progenitors of individual stars or systems" (Motte et al. 1998). The separations between these clumps are comparable to the predicted fragmentation scale of about 0.03 pc in the ρ Oph cloud, and this suggests that they have been formed by gravitational fragmentation of the cloud.

Since a forming star grows in mass by accretion from a surrounding envelope, the final mass that it attains depends on how much matter is accreted. Within the thermally supported inner region, the standard picture of nearly radial infall from a nearly spherical envelope may apply, but outside this region turbulence and magnetic fields dominate the dynamics, and further accretion may be inhibited by these effects. A rigorous treatment of the problem is possible in the case where the envelope is supported by a static magnetic field, as in the standard model in which collapse is initiated by slow ambipolar diffusion in a magnetically supported configuration. According to several recent studies, even in this case a central region having approximately the thermal Jeans mass begins to collapse dynamically at an early stage, producing a high initial accretion rate but leaving behind a magnetically supported envelope that is accreted more slowly (e.g., Basu 1997; Safier et al. 1997; Ciolek and Königl 1998). Most of the final stellar mass is probably acquired during the early phase of rapid accretion, but infall may not stop completely after a Jeans mass has been accreted. The final termination of accretion may be caused by the onset of a stellar wind, as discussed below.

C. The Role of Outflows

In this section, we explore the possibility that stars themselves determine their masses through the action of stellar winds and outflows. Protostellar outflows are sufficiently energetic to affect the overall support of a molecular cloud (Norman and Silk 1980) and have been conjectured to halt the inward accretion flow of a forming star (Shu 1985). This idea has been used as the basis for constructing various theoretical models of the IMF (e.g., Silk 1995; Nakano et al. 1995; Adams and Fatuzzo 1996).

Molecular clouds provide the initial conditions for the star-forming process. These clouds are supported against gravity both by turbulent motions and by magnetic fields. As the fields gradually diffuse outward, the clouds produce centrally condensed structures (molecular cloud cores), which represent the initial conditions for protostellar collapse. In the simplest picture, these cores can be characterized by two physical variables: the effective sound speed a and the rotation rate Ω. The effective sound speed generally contains contributions from both magnetic fields and turbulence, as well as the thermal contribution.

When molecular cloud cores undergo dynamic collapse, the central regions fall in first, and successive outer layers follow as pressure support is lost from below (Shu 1977). Because the initial state contains angular momentum, some of the infalling material collects into a circumstellar disk surrounding the forming star. The collapse flow is characterized by a mass infall rate $\dot{M} \approx a^3/G$, the rate at which the central star-disk system gains mass from the collapsing core. The total amount of mass available to a forming star is generally much larger than the final mass of the star.

In this rotating accretion flow, the ram pressure of the infall is weakest at the rotational poles of the object. The central star-disk system gains mass until it is able to generate a powerful stellar wind, which breaks through the infall at the rotational poles and thereby leads to a bipolar outflow configuration. Although the mechanism that generates these winds remains under study (see, e.g., the chapters by Shu et al. and by Königl and Pudritz, this volume), the characteristics of outflow sources have been well determined observationally (Padman et al. 1997 and the chapter by Richer et al., this volume). The basic working hypotheses of the "outflow conjecture" is that these outflows help separate nearly formed stars from the infalling envelope and thereby determine, in part, the final masses of the stars. In this scenario, the transformation between initial conditions and stellar masses is accomplished through the action of stellar winds and outflows. The central star-disk system gains mass at a well-defined mass infall rate. As the nascent star gains mass over time, it becomes more luminous and produces an increasingly more powerful stellar outflow. When the strength of this outflow becomes larger than the ram pressure of the infalling material, the star separates itself from the surrounding molecular environment and thereby determines its final mass.

We can use this idea to calculate a transformation between the initial conditions in a molecular cloud core and the final mass of the star produced by its collapse. Using the idea that the stellar mass is determined when the outflow strength exceeds the infall strength, we can write this transformation (the SEMF) in the form

$$L_\star M_\star^2 = 8 m_0 \gamma^3 \delta \frac{\beta}{\alpha\epsilon} \frac{a^{11}}{G^3 \Omega^2} = \Lambda \frac{a^{11}}{G^3 \Omega^2} \tag{2}$$

where the parameters α, β, γ, δ, and ϵ are efficiency factors (Adams and Fatuzzo 1996; Shu et al. 1987). This formula specifies a transformation between initial conditions (the sound speed a and the rotation rate Ω) and the final properties of the star (the luminosity $L_\star$ and the mass $M_\star$). Since the protostellar luminosity $L_\star$ as a function of mass is known, we can find the final stellar mass in terms of the initial conditions. In general, all of the quantities on the righthand side of equation (2) will have a distribution of values. These individual distributions ultimately determine the composite distribution of stellar masses $M_\star$.

In principle, the physical variables appearing in equation (2) can be measured observationally. If we use the observed distributions of these variables to estimate the shape parameters appearing in the IMF, we obtain $m_C \approx 0.25$ and $\langle\sigma\rangle \approx 1.8$ (Adams and Fatuzzo 1996). These values are in reasonably close agreement with those required to fit the Miller and Scalo (1979) IMF, namely $m_C = 0.1$ and $\langle\sigma\rangle = 1.57$. Although a quantitative comparison with observations is premature, this approach to the IMF contains some predictive power and is roughly consistent with observations.

The main weakness of this outflow approach is that the interaction between stellar outflows and the inward accretion flow has not yet been calculated. Highly collimated outflows can reverse the infall only along the poles of the system. The outflows must therefore widen with time and must be able to suppress accretion over most of the solid angle centered on the star. Although some observational evidence suggests that outflows can successfully reverse the inward accretion flow (Velusamy and Langer 1998), this issue remains open, on both the theoretical and observational fronts.

D. Hierarchical Fragmentation

Since the Jeans mass decreases with increasing density, it is possible that a collapsing cloud can fragment into successively smaller pieces as its density increases. This hypothesis of hierarchical fragmentation (Hoyle 1953) has formed the basis of many theories of the IMF. In principle, a wide spectrum of stellar masses can be produced in this way, depending on how many clumps stop subdividing and collapse directly into individual stars at each stage of the overall collapse. For example, if the probability of further subdivision of a cloud fragment is the same for each unit logarithmic increase in density, a lognormal IMF is produced (Larson 1973).

Numerical simulations of fragmentation in collapsing clouds suggest that hierarchical fragmentation is of limited importance. For example, a rotating cloud does not fragment significantly until it has collapsed to a disk, since the development of subcondensations is inhibited by pressure gradients and does not get ahead of the overall collapse (Tohline 1980; Monaghan and Lattanzio 1991). This result reflects the inconsistency of the original Jeans analysis, which neglected the overall collapse, and suggests that significant fragmentation does not occur until large-scale collapse has stopped and a near-equilibrium configuration has formed. Such a configuration may then fragment as expected from linear stability theory (Larson 1985). The formation of an equilibrium disk is an idealized case, but transient near-equilibrium structures such as sheets or filaments may often be created by turbulence in molecular clouds. Much of the observed structure of these clouds is in fact filamentary, and many observed star-forming cores may have formed by the fragmentation of filaments (Schneider and Elmegreen 1979). Thus, star-forming cores may form directly by a single stage of fragmentation, rather than through a series of stages of hierarchical fragmentation.

Hierarchical fragmentation may still be relevant to the formation of low-mass stars in the binary and small multiple systems that are the typical outcome of the collapse of Jeans-mass cloud cores (Larson 1995). Numerical simulations show that multiple systems containing several accreting and interacting "protostars" are often formed in this case (Burkert et al. 1997), and these objects are often surrounded by disks or spiral filaments that can fragment into yet smaller objects, so some amount of hierarchical fragmentation is possible. However, coalescence can also occur and reduce the number of fragments, and more work is needed to determine the final outcome (see the chapter by Bodenheimer et al., this volume). Such small-scale fragmentation processes may be responsible for the formation of the least massive stars and may help to determine the form of the lower IMF. At present, however, both observations and simulations suggest that only a small fraction of the mass participating in such processes goes into the smallest objects or "proto-brown dwarfs."

We can place this hierarchical fragmentation scenario into the general picture of section III.A. As a cloud core fragments into successively smaller pieces, we can follow one particular chain of fragmentation events to the final fragment mass. After one round of fragmentation, the piece that we are following has mass $M_1 = f_1 M_0$, where f_1 is a fraction of the original core mass M_0. After N iterations of this hierarchy, the fragment mass will be M_N. Although one could identify the final fragment mass M_N with the mass of the star $M_\star$ formed therein, in practice, the final fragment is not yet a star. Instead, it provides the initial conditions for the star formation process. To account for the additional physical processes that occur as the fragment collapses into a star, we can write $M_\star = f_E M_N = M_0 f_E f_1 f_2 \ldots f_N$, where f_E is the star formation efficiency factor of the final fragment. We thus obtain another SEMF, and section III.A applies. For example, the total variance of the distribution is given by $\langle \sigma \rangle^2 = \sigma_E^2 + \sum \sigma_j^2$ where the σ_j are the variances of the distributions of the fragmentation fractions f_j.

E. Accretion and Agglomeration Processes

Star formation could also involve continuing accretion or agglomeration processes, especially for the most massive stars, which typically form in dense clusters containing many less massive stars. Jeans fragmentation seems unlikely to be relevant to the formation of these stars, and some kind of accumulation process may instead be required (Larson 1982). Radial accretion of gas is inhibited for very massive stars because of the effects of radiation pressure (Wolfire et al. 1985), but rotation may allow infalling material to collect into a disk, which can then accrete onto the star (Jijina and Adams 1996). Alternatively, the material that builds massive stars may already be in the form of dense clumps, which may then accumulate into larger objects. An extreme case of this scenario would be the merging of already formed stars (see the chapter by Stahler et al., this volume).

No convincing prediction of the form of the upper IMF has yet emerged from any of these ideas, but because accretion or agglomeration processes have no preferred scale, they could in principle proceed in a scale-free manner and build up the observed power law upper IMF. As a simple example, if each star accretes matter from a diffuse medium at a rate proportional to the square of its mass, the upper IMF becomes a power law with a slope not very different from the original Salpeter form (Zinnecker 1982). Models involving the agglomeration of randomly moving clumps have also been studied, and these models can yield approximate power law mass functions (e.g., Nakano 1966; Silk and Takahashi 1979; Pumphrey and Scalo 1983; Murray and Lin 1996), but their predictions are sensitive to the assumed properties of the clumps and their evolution between collisions. Fractal concepts have been invoked to relate the form of the upper IMF to the possible self-similar structure of molecular clouds (Larson 1992; Elmegreen 1997), but in these theories strong assumptions must be made about how matter accumulates into stars at each level of the fractal hierarchy. Finally, it has been suggested that already formed stars might sometimes merge to form more massive stars, a mechanism that would readily overcome the problems posed by radiation pressure and winds (Bonnell et al. 1998, and the chapter by Stahler et al., this volume). The dynamics of systems dense enough for frequent mergers to occur is likely to be chaotic, but because no new mass scale is introduced, such processes might proceed in a scale-free way and build up a power law upper IMF.

In summary, accretion and agglomeration processes almost certainly play an important role in the formation of the most massive stars, and they could plausibly proceed in a scale-free way and build up a power law upper IMF. However, a quantitative understanding of these processes does not yet exist, and much more work will be needed before a reliable prediction of the slope of the upper IMF is possible.

IV. SUMMARY OF CONCLUSIONS AND FUTURE DIRECTIONS

In this chapter we have reviewed the techniques used to analyze recent observations of mass distributions in very young clusters and attempted to summarize results. We have also endeavored to review current theoretical arguments that have been put forth to explain these observations. While the observational results are still not definitive, several clear trends are emerging:

1. Detailed comparisons of emergent mass distributions for the best studied young clusters suggest that the IMF does not vary wildly from region to region, though more subtle differences could still be uncovered.
2. The ensemble of results that characterize the mass distributions of embedded clusters, such as the ratio of intermediate- to low-mass stars, is consistent with having been drawn from the field star IMF and *rule out* a single power law Salpeter IMF that extends from 0.1 to 10 $M_{\odot}$.
3. Although the evidence remains preliminary, the emergent mass distributions derived for the two best-studied young clusters (IC 348 and the ONC) hint at turnovers between 0.1 and 0.5 $M_{\odot}$.

Currently, several limitations prevent us from drawing more robust conclusions. Pre-main-sequence evolutionary tracks remain uncertain, so we cannot determine stellar masses with the requisite accuracy. We also need to measure the distribution of companion mass ratios to account properly for unresolved binaries, especially for the low-mass end of the IMF. Next, we must search for variations in the substellar mass function, as well as extreme environments (starburst analogs, etc.) in order to test whether the IMF is truly universal. Improved statistical methodology is also needed to help push our data to the natural limits imposed by information theory. Finally, it will soon become possible to combine detailed IMF studies in young clusters with determinations from other resolved stellar populations in the Milky Way and other galaxies in order to search for variations over cosmic time.

Theoretical progress is being made in understanding the origin of the IMF, but the problem is formidable. Stellar masses are determined by a complex interplay between the initial conditions in the natal cloud environment and the stars that are formed therein. In this review, we have suggested a general framework for discussing theories of the IMF, and we have outlined briefly some of the physical mechanisms that come into play, including gravitational instability and the Jeans scale, the effect of outflows, hierarchical fragmentation, and accretion or agglomeration processes. While it is still too premature to make meaningful comparisons between theories and observations of the IMF, it is our expectation that in the coming years such a comparison can be effected.

Acknowledgments We thank C. Lada and G. Rieke for reading an earlier version of this work, and an anonymous referee for a critical review of the manuscript. M. R. M., L. A. H., and J. M. C. would like to express their gratitude to S. Strom for encouraging them along the road, paved with good intentions, that led to our interest in the IMF.

REFERENCES

Adams, F. C., and Fatuzzo, M. 1996. A theory of the initial mass function for star formation in molecular clouds. *Astrophys. J.* 464:256–271.

Alcala, J. M., Krautter, J., Covino, E., Neuhauser, R., Schmitt, J., and Wichmann, R. 1997. A study of the Chamaeleon star-forming region from the ROSAT All–Sky Survey. II. The pre-main-sequence population. *Astron. Astrophys.* 319:184–200.

Ali, B., and DePoy, D. 1995. A 2.2 micron imaging survey of the Orion A Molecular Cloud. *Astron J.* 109:709–720.

Allen, L. E. 1996. Star formation in Lynds 1641. Ph.D. Thesis, University of Massachusetts.

Aspin, C., Sandell, G., and Russell, A. P. G. 1994. Near-IR imaging photometry of NGC 1333: I. The embedded PMS stellar population. *Astron. Astrophys. Suppl.* 106:165–198.

Basu, S. 1997. A semianalytic model for supercritical core collapse: Self-similar evolution and the approach to protostar formation. *Astrophys. J.* 485:240–253.

Bonnor, W. B. 1956. Boyle's law and gravitational instability. *Mon. Not. Roy. Astron. Soc.* 116:351–359.

Bonnell, I. A., Bate, M. R, and Zinnecker, H. 1998. On the formation of massive stars. *Mon. Not. Roy. Astron. Soc.* 298:93–102.

Bouvier, J., Stauffer, J. R., Martin, E. L., Barrado, Y., Navascues, D., Wallace, B., and Bejar, V. J. S. 1998. Brown dwarfs and very low-mass stars in the Pleiades cluster: A deep wide-field imaging survey. *Astron. Astrophys.* 336:490–502.

Brandl, B., Sams, B. J., Bertoldi, F., Eckart, A., Genzel, R., Drapatz, S., Hofmann, R., Loewe, M., and Quirrenbach, A. 1996. Adaptive optics near-infrared imaging of R136 in 30 Doradus: The stellar population of a nearby starburst. *Astrophys. J.* 466:254–273.

Bresolin, F., Kennicutt, R. C., Ferrarese, L., Gibson, B. K., Graham, J. A., Macri, L. M., Phelps, R. L., Rawson, D. M., Sakai, S., Silbermann, N. A., Stetson, P. B., Turner, A. M. 1998. A Hubble Space Telescope study of extragalactic OB associations. *Astron. J.* 116:119–130.

Briceño, C., Hartmann, L., Stauffer, J., and Martin, E. 1998. A search for very low mass pre-main sequence stars in Taurus. *Astron. J.* 115:2074–2091.

Brown, A. 1998. The initial mass function in nearby OB associations. In *The Stellar Initial Mass Function,* ed. G. Gilmore and D. Howell (San Francisco: *Astronomical Society of the Pacific*), pp. 45–59.

Burkert, A., Bate, M. R., and Bodenheimer, P. 1997. Protostellar fragmentation in a power-law density distribution. *Mon. Not. Roy. Astron. Soc.* 289:497–504.

Carpenter, J. M., Meyer, M. R., Dougados, C., Strom, S. E., and Hillenbrand, L. A. 1997. Properties of the Monoceros R2 stellar cluster. *Astron. J.* 114:198–221.

Chabrier, G., Mera, D. 1997. Determination of the globular cluster and halo stellar mass functions and stellar and brown dwarf densities. *Astron. Astrophys.* 328:83–94.

Ciolek, G. E., and Königl, A. 1998. Dynamical collapse of nonrotating magnetic molecular cloud cores: Evolution through point-mass formation. *Astrophys. J.* 504:257–279.

Clarke, C. 1998. Star formation theories and the IMF. In *The Stellar Initial Mass Function,* ed. G. Gilmore and D. Howell (San Francisco: Astronomical Society of the Pacific), pp. 189–200.

Comeron, F., Rieke, G. H., Burrows, A., and Rieke, M. J. 1993. The stellar population in the ρ Ophiuchi cluster. *Astrophys. J.* 416:185–203.

Comeron, F., Rieke, G. H., and Rieke, M. J. 1996. Properties of low-mass objects in NGC 2024. *Astrophys. J.* 473:294–303.

Comeron, F., Rieke, G. H., Claes, P., Torra, J., and Laureijs, R. J. 1998. ISO observations of candidate young brown dwarfs. *Astron. Astrophys.* 335: 522–532.

Cool, A. M. 1998. Measuring globular cluster mass functions with HST. In *The Stellar Initial Mass Function,* ed. G. Gilmore and D. Howell (San Francisco: Astronomical Society of the Pacific), pp. 139–156.

D'Antona, F., and Mazzitelli, I. 1994. New pre-main-sequence tracks for $M_* <$ 2.5 $M_\odot$ as tests of opacities and convection model. *Astrophys. J. Suppl.* 90:467.

Ebert, R. 1955. Temperatur des interstellaren Gases bei grossen Dichten. *Z. Astroph.* 37:217–232.

Eisenhauer, F., Quirrenbach, A., Zinnecker, H., and Genzel, R. 1998. Stellar content of the galactic starburst template NGC 3603 from adaptive optics observations. *Astrophys. J.* 498:278–292.

Elmegreen, B. G. 1997. The initial stellar mass function from random sampling in a turbulent fractal cloud. *Astrophys. J.* 486:944–954.

Elmegreen, B. G. 1999. Observations and theory of the stellar initial mass function. In *Unsolved Problems in Stellar Evolution,* ed. M. Livio (Cambridge: Cambridge University Press), pp. 59–85.

Elmegreen, B. G., and Mathieu, R. D. 1983. Monte Carlo simulations of the initial stellar mass function. *Mon. Not. Roy. Astron. Soc.* 203:305–315.

Garmany, C. D., Conti, P. S., and Chiosi, C. 1982. The initial mass function for massive stars. *Astrophys. J.* 263:777–790.

Giovannetti, P., Caux, E., Nadeau, D., and Monin, J. 1998. Deep optical and near-infrared imaging photometry of the Serpens cloud core. *Astron. Astrophys.* 330:990–998.

Gould, A., Bahcall, J. N., and Flynn, C. 1997. M dwarfs from Hubble Space Telescope star counts: III. The Groth Strip. *Astrophys. J.* 482:913–918.

Gould, A., Flynn, C., and Bahcall, J. N. 1998. Spheroid luminosity and mass functions from Hubble Space Telescope star counts. *Astrophys. J.* 503:798–808.

Greene, T. P., and Meyer, M. R. 1995. An infrared spectroscopic survey of the ρ Ophiuchi young stellar cluster: Masses and ages from the H–R diagram. *Astrophys. J.* 450:233–244.

Greene, T. P., and Young, E. T. 1992. Near-infrared observations of young stellar objects in the ρ Ophiuchi dark cloud. *Astrophys. J.* 395:516–528.

Hambly, N. C., Steele, I. A., Hawkins, M. R. S., and Jameson, R. F. 1995. The very low-mass main sequence in the galactic cluster Praesepe. *Mon. Not. Roy. Astron. Soc.* 273:505–512.

Herbig, G. 1998. The young cluster IC 348. *Astrophys. J.* 497:736–758.

Herbig, G. H., and Terndrup, D. M. 1986. The Trapezium cluster of the Orion Nebula. *Astrophys. J.* 307:609–618.

Hillenbrand, L. A. 1997. On the stellar population and star-forming history of the Orion Nebula cluster. *Astron. J.* 113:1733–1768.

Hillenbrand, L. A., Meyer, M. R., Strom, S. E, and Skrutskie, M. F. 1995. Isolated star-forming regions containing Herbig Ae/Be stars: I. The young stellar aggregate associated with BD+40 4124. *Astron. J.* 109:280–297.

Holtzman, J. A., Watson, A. M., Baum, W. A., Grillmair, C. J., Groth, E. J., Light, R. M., Lynds, R., and O'Neil, E. J., Jr., 1998. The luminosity function and initial mass function in the galactic bulge. *Astron. J.* 115:1946–1957.

Hoyle, F. 1953. On the fragmentation of gas clouds into galaxies and stars. *Astrophys. J.* 118:513–528.

Hughes, J., Hartigan, P., Krautter, J., and Keleman, J. 1994. The stellar population of the lupus clouds. *Astron. J.* 108:1071–1090.

Hunter, D. A., Vacca, W. D., Massey, P., Lynds, R., and O'Neil, E. J. 1996. Ultraviolet photometry of stars in the compact cluster R136 in the Large Magellanic Cloud. *Astron. J.* 113:1691–1699.

Jeans, J. H. 1929. *Astronomy and Cosmogony* (Cambridge: Cambridge University Press).
Jijina, J., and Adams, F. C. 1996. Infall collapse solutions in the inner limit: Radiation pressure and its effects on star formation. *Astrophys. J.* 462:874–887.
Jones, T. J., Mergen, J., Odewahn, S., Gehrz, R. D., Gatley, I., Merrill, K. M., Probst, R., and Woodward, C. E. 1994. A near-infrared survey of the OMC2 region. *Astron. J.* 107:2120–2130.
Kenyon, S. J., and Hartmann, L. 1995. Pre-main-sequence evolution in the Taurus-Auriga molecular cloud. *Astrophys. J. Supp.* 101:117–171.
Khersonsky, V. K. 1997. The connection between the interstellar cloud mass spectrum and the stellar mass spectrum in star-forming regions. *Astrophys. J.* 475:594–603.
Kroupa, P. 1995. Unification of the nearby and photometric stellar luminosity functions. *Astrophys. J.* 453:358–368.
Kroupa, P. 1998. The stellar mass function. In *Brown Dwarfs and Extra-solar Planets*, ed. R. Rebolo, E. Martin, and M. Zapatero-Osorio (San Francisco: Astronomical Society of the Pacific), pp. 483–494.
Kroupa, P., Tout, C. A., and Gilmore, G. 1993. The distribution of low-mass stars in the galactic disc. *Mon. Not. Roy. Astron. Soc.* 262:545–587.
Lada, C. J., and Lada, E. A. 1991. The nature, origin, and evolution of embedded star clusters. In *The Formation and Evolution of Star Clusters*, ed. K. Janes (San Francisco: Astronomical Society of the Pacific), pp. 3–22.
Lada, E. A., and Lada, C. J. 1995. Near-infrared images of IC 348 and the luminosity functions of young embedded star clusters. *Astron. J.* 109:1682–1692.
Lada, C. J., Young, E. T., and Greene, T. P. 1993. Infrared images of the young cluster NGC 2264. *Astrophys. J.* 408:471–483.
Lada, C. J., Alves, J., and Lada, E. A. 1996. Near-infrared imaging of embedded clusters: NGC 1333. *Astron. J.* 111:1964–1976.
Lada, E., Lada, C. J., and Muench, A. 1998. Infrared luminosity functions of embedded clusters. In *The Initial Mass Function*, ed. G. Gilmore and D. Howell (San Francisco: Astronomical Society of the Pacific), pp. 107–120.
Larson, R. B. 1973. A simple probabilistic theory of fragmentation. *Mon. Not. Roy. Astron. Soc.* 161:133–143.
Larson, R. B. 1981. Turbulence and star formation in molecular clouds. *Mon. Not. Roy. Astron. Soc.* 194:809–826.
Larson, R. B. 1982. Mass spectra of young stars. *Mon. Not. Roy. Astron. Soc.* 200:159–174.
Larson, R. B., 1985. Cloud fragmentation and stellar masses. *Mon. Not. Roy. Astron. Soc.* 214:379–398.
Larson, R. B. 1992. Towards understanding the stellar initial mass function. *Mon. Not. Roy. Astron. Soc.* 256:641–646.
Larson, R. B. 1995. Star formation in groups. *Mon. Not. Roy. Astron. Soc.* 272:213–220.
Larson, R. B. 1996. Star formation and galactic evolution. In *The Interplay between Massive Star Formation, the ISM, and Galaxy Evolution,* ed. D. Kunth, B. Guiderdoni, M. Heydari-Malayeri, and T. X. Thuan (Gif sur Yvette: Editions Frontieres), pp. 3–16.
Larson, R. B., 2000. Theoretical aspects of star formation. In *The Orion Complex Revisited,* ed. M. J. McCaughrean and A. Burkert (San Francisco: Astronomical Society of the Pacific), in press.
Lawson, W. A., Feigelson, E. D., and Huenemoerder, D. P. 1996. An improved H-R diagram for Chamaeleon I pre-main sequence stars. *Mon. Not. Roy. Astron. Soc.* 280:1071–1088.
Luhman, K., and Rieke, G. H., 1998. The low-mass initial mass function in young clusters: L1495E. *Astrophys. J.* 497:354–369.

Luhman, K., Rieke, G. H., Lada, C. J., and Lada, E. A. 1998. Low-mass star formation and the initial mass function in IC 348. *Astrophys. J.* 507:347–369.

Massey, P. 1998. The initial mass function of massive stars in the Local Group. In *The Stellar Initial Mass Function,* ed. G. Gilmore and D. Howell (San Francisco: Astronomical Society of the Pacific), pp. 17–44.

Massey, P., Lang, C., DeGioia-Eastwood, K., and Garmany, C. D. 1995*a*. Massive stars in the field and associations of the Magellanic Clouds: The upper mass limit, the initial mass function, and a critical test of main-sequence stellar evolutionary theory. *Astrophys. J.* 438:188–217.

Massey, P., Johnson, K. E., and DeGioia-Eastwood, K., 1995*b*. The initial mass function and massive star evolution in the OB associations of the northern Milky Way. *Astrophys. J.* 454:151–171.

McCaughrean, M. J., and Stauffer, J. R. 1994. High resolution near-infrared imaging of the Trapezium: A stellar census. *Astron. J.* 108:1382–1397.

Megeath, S. T. 1996. The effect of extinction on the K-band luminosity functions of embedded stellar clusters. *Astron. Astrophys.* 311:135–144.

Megeath, S. T., Herter, T., Beichman, C., Gautier, N., Hester, J. J., Rayner, J., and Shupe, D. 1996. A dense stellar cluster surrounding W3 IRS 5. *Astron. Astrophys.* 307:775–790.

Mera, D., Chabrier, G., and Baraffe, I. 1996. Determination of the low-mass star mass function in the galactic disk. *Astrophys. J.* 459:L87–L90.

Meyer, M. R., 1996. Stellar populations of deeply embedded young clusters: Near-infrared spectroscopy and emergent mass distributions. Ph.D. Thesis, University of Massachusetts.

Meyer, M. R., Calvet, N., and Hillenbrand, L. A. 1997. Intrinsic near-infrared excesses of T Tauri stars: Understanding the classical T Tauri locus. *Astron. J.* 114:288–300.

Meyer, M. R., Carpenter, J. M., Strom, S. E., and Hillenbrand, L. A. 1999. The embedded cluster associated with NGC 2024. *Astron. J.*, submitted.

Miller, G. E., and Scalo, J. M. 1979. The initial mass function and stellar birthrate in the solar neighborhood. *Astrophys. J. Supp.* 41:513–547.

Monaghan, J. J., and Lattanzio, J. C. 1991. A simulation of the collapse and fragmentation of cooling molecular clouds. *Astrophys. J.* 375:177–189.

Motte, F., André, P., and Neri, R. 1998. The initial conditions of star formation in the ρ Ophiuchi main cloud: Wide-field millimeter continuum mapping. *Astron. Astrophys.* 336:150–172.

Murray, S. D., and Lin, D. N. C. 1996. Coalescence, star formation, and the cluster initial mass function. *Astrophys. J.* 467:728–748.

Myers, P. C. 1998. Cluster-forming molecular cloud cores. *Astrophys. J. Lett.* 496:L109–L112.

Nakano, T. 1966. Fragmentation of a cloud and the mass function of stars in galactic clusters. *Prog. Theoret. Phys.* 36:515–529.

Nakano, T., Hasegawa, T., and Norman, C. 1995. The mass of a star formed in a cloud core: Theory and its application to the Orion A cloud. *Astrophys. J.* 450:183–195.

Norman, C. A., and Silk, J. 1980. Clumpy molecular clouds: A dynamic model self-consistently regulated by T Tauri star-formation. *Astrophys. J.* 238:158–174.

Padman, R., Bence, S. J., and Richer, J. S. 1997. Observational properties of molecular outflows. In *Herbig-Haro Flows and the Birth of Low Mass Stars,* IAU Symp. 182, ed. B. Reipurth and C. Bertout (Dordrecht: Kluwer), pp. 123–140.

Phelps, R. L., and Janes, K. 1993. Young open clusters as probes of the star-formation process: 2. Mass and luminosity functions of young open clusters. *Astron. J.* 106:1870–1884.

Pinfield, D. J., Hodgkin, S. T., Jameson, R. F., Cossburn, M. R., and von Hippel, T. 1997. Brown dwarf candidates in Praesepe. *Mon. Not. Roy. Astron. Soc.* 287:180–188.

Press, W., Teukolsky, S. A., Vetterling, W. T., and Flannery, B. P. 1993. *Numerical Recipes in C* (Cambridge: Cambridge University Press).

Pumphrey, W. A., and Scalo, J. M. 1983. Simulation models for the evolution of cloud systems: I. Introduction and preliminary simulations. *Astrophys. J.* 269:531–559.

Richtmyer, R. D. 1978. *Principles of Advanced Mathematical Physics* (New York: Springer).

Reid, N. 1998. All things under the sun: The lower main-sequence mass function from an empiricist's perspective. In *The Stellar Initial Mass Function,* ed. G. Gilmore and D. Howell (San Francisco: Astronomical Society of the Pacific), pp. 121–138.

Reid, N., and Gizis, J. 1997. Low-mass binaries and the stellar luminosity function. *Astron. J.* 113:2246–2269.

Roberts, M. S. 1957. The numbers of early-type stars in the galaxy and their relation to galactic clusters and associations. *Pub. Astron. Soc. Pacific* 69:59–64.

Safier, P. N., McKee, C. F., and Stahler, S. W. 1997. Star formation in cold, spherical, magnetized molecular clouds. *Astrophys. J.* 485:660–679.

Sagar, R., and Griffiths, W. K. 1998. Mass functions of five distant northern open star clusters. *Mon. Not. Roy. Astron. Soc.* 299:777–788.

Salpeter, E. E. 1955. The luminosity function and stellar evolution. *Astrophys. J.* 121:161–167.

Scalo, J. M. 1986. The stellar initial mass function. *Fund. Cosm. Phys.* 11:1–278.

Scalo, J. M. 1998. The IMF revisited: A case for variations. In *The Stellar Initial Mass Function,* ed. G. Gilmore and D. Howell (San Francisco: Astronomical Society of the Pacific), pp. 201–236.

Schneider, S., and Elmegreen, B. G. 1979. A catalog of dark globular filaments. *Astrophys. J. Supp.* 41:87–95.

Shu, F. 1977. Self-similar collapse of isothermal spheres and star formation. *Astrophys. J.* 214:488–497.

Shu, F. H. 1985. Star formation in molecular clouds. In *The Milky Way,* IAU Symp. No. 106, ed. H. van Woerden, W. B. Burton, and R. J. Allen (Dordrecht: Reidel), pp. 561–566.

Shu, F., Adams, F. C., and Lizano, S. 1987. Star formation in molecular clouds: Observations and theory. *Ann. Rev. Astron. Astrophys.* 25:23–81.

Silk, J. 1995. A theory for the initial mass function. *Astrophys. J.* 438:L41–L44.

Silk, J., and Takahashi 1979. A statistical model for the stellar initial mass function. *Astrophys. J.* 229:242–256.

Spitzer, L. 1968. Dynamics of interstellar matter and the formation of stars. In *Nebulae and Interstellar Matter* (*Stars and Stellar Systems,* Vol. 7), ed. B. M. Middlehurst and L. H. Aller (Chicago: University of Chicago Press), pp. 1–63.

Spitzer, L. 1978. *Physical Processes in the Interstellar Medium* (New York: Wiley-Interscience Press).

Stauffer, J. R., Hartmann, L. W., Prosser, C. F., Randich, S., Balachandran, S., Patten, B. M., Simon, T., and Giampapa, M. 1997. Rotational velocities and chromospheric/coronal activity of low-mass stars in the young open clusters IC 2391 and IC 2602. *Astrophys. J.* 479:776–791.

Strom, K. M., and Strom, S. E. 1994. A multiwavelength study of star formation in the L1495E cloud in Taurus. *Astrophys. J.* 424:237–256.

Strom, K. M., Strom, S. E., and Merrill, M. 1993. Infrared luminosity functions for the young stellar population associated with the L1641 molecular cloud. *Astron. J.* 412:233–253.

Strom, K. M., Kepner, J., and Strom, S. E. 1995. The evolutionary status of the stellar population in the ρ Ophiuchi cloud core. *Astrophys. J.* 438:813–829.
Tohline, J. E., 1980. The gravitational fragmentation of primordial gas clouds. *Astrophys. J.* 239:417–427.
Velusamy, T., and Langer, W. D. 1998. Outflow-infall interactions as a mechanism for terminating accretion in protostars. *Nature* 392:685–687.
von Hippel, T., Gilmore, G., Tanvir, N., Robinson, D., and Jones, D. H. 1996. The metallicity dependence of the stellar luminosity and initial mass function: HST observations of open and globular clusters. *Astron. J.* 112:192–200.
Walter, F. M., Vrba, R. J., Mathieu, R. D., Brown, A., and Myers, P. C. 1994. X-ray sources in regions of star formation: 5. The low mass stars of the upper Scorpius association. *Astron. J.* 107:692–719.
Walter, F. M., Vrba, R. J., Wolk, S. J., Mathieu, R. D., and Neuhauser, R. 1997. X-ray sources in regions of star formation: VI. The R CrA Association as viewed by Einstein. *Astron. J.* 114:1544–1554.
Wilking, B. A., McCaughrean, M. J., Burton, M. G., Giblin, T., Rayner, J., and Zinnecker, H. 1997. Deep infrared imaging of the R Coronae Australis cloud core. *Astron. J.* 114:2029–2042.
Williams, D. M., Rieke, G. H., and Stauffer, J. R. 1995*a*. The stellar mass function of Praesepe. *Astrophys. J.* 445:359–366.
Williams, D. M., Comeron, F., Rieke, G. H., and Rieke, M. J. 1995*b*. The Low-mass IMF in the ρ Ophiuchi cluster. *Astrophys. J.* 454:144–150.
Williams, D. M., Boyle, D. J., Morgan, W. T., Rieke, G. H., Stauffer, J. R., and Rieke, M. J. 1996. Very low mass stars and substellar objects in the Pleiades. *Astrophys. J.* 464:238–246.
Wolfire, M. G., and Cassenelli, J. P. 1986. The temperature structure in accretion flows onto massive protostars. *Astrophys. J.* 310:207–221.
Zapatero-Osorio, M. R., Rebolo, R., Martin, E. L., and Garcia Lopez, R. J. 1996. Stars approaching the substellar limit in the Alpha Persei open cluster. *Astron. Astrophys.* 305:519–526.
Zapatero-Osorio, M. R., Rebolo, R., Martin, E. L., Basri, G., Magazzu, A., Hodgkin, S. T., Jameson, R. F., and Cossburn, M. R. 1997. New brown dwarfs in the Pleiades cluster. *Astrophys. J.* 491:L81–L84.
Zinnecker, H. 1982. Prediction of the protostellar mass spectrum in the Orion near-infrared cluster. *Ann. N. Y. Acad. Sci.*) 395:226–235.
Zinnecker, H. 1984. Star formation from hierarchical cloud fragmentation: A statistical theory of the log-normal initial mass function. *Mon. Not. Roy. Astron. Soc.* 210:43–56.
Zinnecker, H., McCaughrean, M. J., and Wilking, B. A. 1993. The initial stellar population. In *Protostars and Planets III,* ed. E. Levy and J. Lunine (Tucson: University of Arizona Press), p. 429–495.

THE FORMATION OF STELLAR CLUSTERS

CATHIE J. CLARKE, IAN A. BONNELL
Institute of Astronomy at Cambridge

and

LYNNE A. HILLENBRAND
California Institute of Technology

We review recent work that investigates the formation of stellar clusters, ranging in scale from globular clusters through open clusters to the small-scale aggregates of stars observed in T associations. In all cases, recent advances in understanding have been achieved through the use of state-of-the-art stellar dynamical and gas dynamical calculations, combined with the possibility of intercomparison with an increasingly large dataset on young clusters. Among the subjects that are highlighted are the frequency of cluster-mode star formation, the possible relationship between cluster density and the highest stellar mass, subclustering, and the dynamical interactions that occur in compact aggregates, such as binary star formation. We also consider how the spectrum of stellar masses may be shaped by the process of competitive accretion in dense environments and how cluster properties, such as mass segregation and cluster morphology, can be used in conjunction with numerical simulations to investigate the initial conditions for cluster formation. Lastly, we contrast bottom-up and top-down scenarios for cluster formation and discuss their applicability to the formation of clusters on a range of scales.

I. INTRODUCTION

Observations indicate that stars frequently form in clustered environments: in rich clusters of many hundreds to many thousands of stars, or in smaller groups and aggregates containing up to a few tens of stars. It is only recently, however, that the properties of young clusters have begun to be well characterized. Cluster formation is important, therefore, insofar as it is a fundamental unit of star formation. Given the high stellar densities measured in young clusters and therefore the possible role of encounters, it is also increasingly clear that whether a star forms in a cluster or in isolation may be important in determining its fundamental properties, such as its mass, binarity, or possession of planets.

This chapter concentrates on the issue of how observed young clusters can be used to deduce the conditions in clusters at birth. In particular, it stresses the interplay between observations and numerical simulations, which allows one to address numerous questions regarding the initial shapes, mass distributions, and dynamical states of clusters, and to explore how likely it is for clusters to survive as bound structures. Significant observational and computational advances in recent years make this exercise particularly timely. On the observational front, deep wide-field imaging at infrared wavelengths and multifiber spectroscopy have brought a wealth of data concerning the states of clusters at increasingly young ages. Numerical simulations have also advanced considerably through the development of hydrodynamics software that can deal with the highly inhomogeneous conditions in star-forming gas. Of particular significance is the recent advent of special-purpose hardware, GRAPE, for calculation of gravitational forces (Okumura et al. 1993). This innovation has heralded a new era in N-body calculations; it is now straightforward to perform simulations (over tens of dynamical times) in which the number of particles matches the number of stars, even in the case of populous clusters containing many tens of thousands of stars.

The reason it is desirable to derive the basic characteristics of clusters at birth is because of the light such information sheds on how clusters form. Observational constraints on the age spread in clusters, the time sequence of star formation as a function of stellar mass, and the degree of subclustering are all important constraints on theoretical models. We defer a fuller discussion of current theoretical ideas until section VII, but here indicate some of the issues in order to motivate the intervening sections of the chapter.

Historically, cluster formation theories considered the monolithic top-down collapse of Jeans-unstable gas, and the main issue therefore concerned the number of fragments ("stars") formed during collapse (Hoyle 1953; Larson 1978). Such studies envisaged rather smooth initial conditions, and interest therefore focused on the amplification of initially linear density perturbations and on the efficiency of cooling during collapse. Two facts about the state of star-forming gas in molecular clouds, however, render this picture obsolete. First, the thermal energy content of the gas is negligible compared with the energy density in turbulence [assumed to be magnetohydrodynamic (MHD)]. Hence, the question of how pieces of the cloud collapse to form stars does not hinge primarily on cooling but instead on their ability to decouple from the magnetic field. Secondly, molecular clouds are extremely inhomogeneous (see, e.g. the chapter by Vázquez-Semadeni et al., this volume), consisting of a flocculent ensemble of structures within structures (for a hierarchical description of star-forming clouds, see the chapter by Elmegreen et al., this volume).

This inhomogeneity of the parent gas has several implications for cluster formation. For one thing, it renders trivial the question of why stars are clustered at birth, because at some level this reflects the structure of the star-forming gas, albeit modified by dynamical effects (see Klessen et al. 1998 for a first attempt to model cluster formation from highly inhomogeneous initial conditions). It should be noted in passing that the fractal dimension that characterizes the distribution of young stars is *not* equal to that of the gas, implying either that the star formation process engenders tighter clustering or else that stars form from the most tightly clustered component of the molecular gas (Larson 1995).

The complex density structure of star-forming clouds also raises questions as to the degree of coordination that is required to form a cluster. It is well known (e.g., Lada et al. 1984; Goodwin 1997*b*; see section V) that the formation of a *bound* cluster requires that a high fraction (30–50%) of the gas must be turned into stars before destructive feedback mechanisms from massive stars come into play; in practice, this means a high conversion efficiency within a few cluster dynamical times. Such locally coordinated star formation is a natural expectation in top-down scenarios (i.e., where the structure develops as a result of gravitational instabilities during collapse). It is not, however, the obvious outcome if star formation is taking place in an already highly structured environment, unless some external agent can synchronize the onset of star formation in a set of discrete, mutually independent clumps. Such triggered star formation is therefore an attractive possibility theoretically, and there are clear examples [such as in IC 1396 (Patel et al. 1998), the Rosette Molecular Cloud (White et al. 1997), IC 1805 (Heyer et al. 1996), Gemini OB1 (Carpenter et al. 1995), and in more isolated "bright rim" regions (Sugitani et al. 1991,1994)] where the location of young stellar objects (and clusters) in the dense gas swept up by expanding H II regions lends credence to this scenario (Elmegreen and Lada 1977). In other cases, however, the locations of young clusters give no hint of external triggering (e.g., Taurus, NGC 2264). Thus, a key question (whether cluster formation is induced or spontaneous) remains unanswered at the present time. Clearly, the derivation of cluster parameters at birth (particularly the age spread of stars within a cluster and the initial degree of subclustering) can shed considerable light on this question.

II. OBSERVATIONS OF YOUNG CLUSTERS

Clusters are useful laboratories for star formation studies, because they provide stellar samples of constant metallicity at approximately uniform distance. The task of identifying and characterizing clusters so young that they are still embedded in the molecular material from which they formed has been considerably aided within the past decade by near infrared imaging capabilities. Near infrared surveys penetrate through an order of magnitude more column density than does visual imaging and allow us to see clusters closer to the epoch of their formation. For example, Fig. 1 shows the infrared *H*-band image of Monoceros R2 (Carpenter et al 1997; see also Color Plate 4).

In what follows we consider only young stellar populations located within 1 kpc of the Sun and focus on infrared surveys, as summarized in

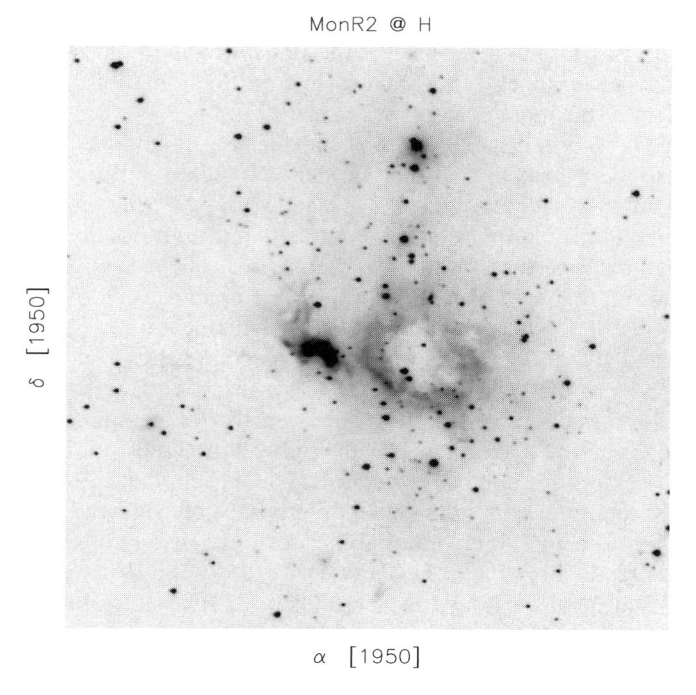

Figure 1. The Monoceros R2 cluster imaged in the near-infrared *H* band (from Carpenter et al. 1997). The field of view is $\sim 3.2 \times 3.2$ arcmin^2, corresponding to $\sim 0.8 \times 0.8$ pc^2 for a distance of 830 pc. The central cluster is completely embedded and contains >300 stars within a 0.4-pc diameter.

Table I. We distinguish between biased surveys or deep imaging of some interesting class of object (e.g., molecular outflows, IRAS point sources, Bok globules, OB or Herbig Ae/Be stars) and the often shallower, unbiased surveys of large regions containing molecular material. We highlight two issues: the relative importance of isolated vs. cluster mode star formation, and the apparent association of high-mass star formation with the formation of clusters. First, however, we briefly touch on some of the problems involved in the identification and characterization of clusters.

Clusters are usually identified via enhanced surface density relative to the background. An obvious disadvantage of this method is that since a given cluster subtends a smaller angle at larger distances, distant clusters are more readily identified, although this effect is partially offset by diminishing survey sensitivity at larger distances. Further problems are the correct subtraction of foreground and background sources, and the tendency for patchy absorption to act as a source of spurious clustering. As cluster surveys are extended to regions that are increasingly embedded (i.e., closer to the $t = 0$ of star formation), it becomes increasingly necessary to interpret clustering statistics in conjunction with molecular extinction

TABLE I
Summary of Clustering Parameters

Cluster	FWHM (pc)	Aspect Ratio	Central Density (stars pc^{-3})	N_{cl}	N^{*}/cl	References
			Targeted Surveys			
ONC/OMC-1	0.8	2:1	20,000	1	3500–5000	Hillenbrand and Hartmann 1998
Mon R2	0.4	2:1	9,000	1	>500	Carpenter et al. 1997
	$R_{eff} = \sqrt{A/\pi}$ or actual R_{cl} (pc)	Shape	Average Surface Density (stars pc^{-2})			
LkHa101	0.1	Irregular	~1500	1	50	Aspin and Barsony 1994
Herbig Ae/Be	0.1–0.7	Round	40–400	>30	5–50	Hillenbrand 1995; Testi et al. 1997, 1998
R CrA	0.3	Elongated	~200	1	40	Wilking et al. 1997
OMC-2	0.1	Round	230	1	33	Jones et al. 1994; Ali and DePoy 1995
ρ Oph A	<0.4	Roundish	>200	1	90	Comeron et al. 1993; Strom et al. 1995
S 106	0.3	Elongated	550	1	160	Hodapp and Rayner 1991
Serpens	0.2	Irregular	400	1	55	Giovannetti et al. 1998; Eiroa and Casali 1992
Luminous IRAS	0.2–0.6	Elongated	10–200	19	15–100	Carpenter et al. 1993
σ Ori	3.0	Round?	15	1	>350	Walter et al. 1998
			"All-Cloud" Surveys			
Taurus	0.5–1.1	Half round and half elongated	30–120	6	~15	Gomez et al. 1993; Luhman and Rieke 1998
L1630	0.3–0.9	Elongated round, irregular	70–450	4	20–300	Lada et al. 1991; Comeron et al 1996; Meyer et al. 1999
L1641—south	0.7	Round?	100	1	150	Strom et al. 1993
L1641—others	0.1–0.5	Round?	70–140	12	5–50	Chen and Tokunaga 1994 Hodapp and Deane 1993 Strom et al. 1993
NGC 2264	0.9	Elongated	40	2	~100	Piche 1993; Lada et al. 1993;
IC 348—main	0.5	Elongated?	200	1	160	Lada and Lada 1995
IC 348—others	0.1–0.2	Round?	70–270	8	5–20	Lada and Lada 1995
NGC1333	0.2	Elongated	360	2	~45	Lada 1996; Aspin and Sandell 1997; Aspin et al. 1994
Rosette	0.3–0.5	Roundish	40–100	7	10–30	Phelps and Lada 1997

maps. These not only allow one to distinguish between true clustering and the apparent clustering of sources in windows of low extinction but also allow more accurate subtraction of foreground and background sources. It should be stressed that in what follows the term "cluster" is used to describe apparent groupings of stars in projection; since kinematic data are not usually available, it is not possible to make the conventional distinction between clusters and associations on the basis of whether or not they are gravitationally bound. It should also be noted that the detection of clustering in molecular clouds is strongly affected by the age of the system. With velocity dispersions of 1–2 km s^{-1}, smaller and less dense clusters can disperse quickly, possibly causing us to have overestimated "typical" cluster membership numbers and projected densities.

The first large-scale near-infrared imaging survey of a molecular cloud is the oft-quoted work of Lada et al. (1991*b*), which covered over 50 pc^2 of the Orion B molecular cloud (see also Li et al. 1997). Subsequently, similar unbiased surveys have been conducted in several other star-forming regions: the Orion A cloud (Strom et al. 1993; Jones et al. 1994; Ali and DePoy 1995), NGC 2264 (Piche 1993; Lada et al. 1993; Strom et al. 1999), IC 348 (Lada and Lada 1995), NGC 1333 (Aspin et al. 1994; Aspin and Sandell 1997; Lada et al. 1996), the Rosette molecular cloud (Phelps and Lada 1997), R Coronae Australis (Wilking et al. 1997), Taurus (Itoh et al. 1996), and the most thoroughly studied region, Ophiuchus (Rieke et al. 1989; Barsony et al. 1989; Greene and Young 1992; Comeron et al. 1993; Strom et al. 1995; Barsony et al. 1997). Clusters are found in all cases, and generally there is an accompanying distributed population of young stars as well.

It is obviously of interest to assess what fraction of stars form in clusters. The strong clustering of massive stars has been evident for a long time (e.g., Blaauw 1964), but it is the advent of near-infrared imaging (as reviewed by Zinnecker et al. 1993) that has revealed that low-mass stars form abundantly in the vicinity of high-mass stars and thus share in the cluster environment at birth. The results of unbiased surveys of star-forming regions suggest that the fraction of star formation taking place in clusters varies quite strongly from place to place. This is particularly striking in the case of the Orion giant molecular cloud, where marked differences are found between the A and B clouds (see Meyer and Lada 1999 for a fuller discussion). In the Orion B cloud, almost all (96%) of associated infrared sources are thought to be in clusters. In the Orion A cloud, by contrast, there is a significant distributed population, with only 50–80% of the stellar population formed in clusters [the range depending on whether one does not or does, respectively, count the Orion Nebula Cluster (ONC)]. The Orion A result is more consistent with what has been found in other surveys of molecular clouds, where the fraction of stars located in projected density enhancements ("clusters") is 50–70% (Taurus, Gomez et al. 1993; NGC 2264, Piche et al. 1993; NGC 1333, Lada et al. 1996;

IC 348, Lada and Lada 1995). We note, however, that while large fractions of the most dense and "active" areas of many clouds have been surveyed in the near-infrared, in no case has the entirety of any giant molecular cloud been mapped. Thus, the fraction of stars observed to have formed in and out of clusters and aggregates is still uncertain. Significant progress on characterizing the cluster-forming properties of different regions is likely to come from analysis of data on star-forming regions contained in the near infrared all-sky surveys (2MASS, DENIS).

Although it is not clear why the fraction of star formation taking place in dense clusters should vary from cloud to cloud, all of the regions surveyed thus far seem to support a basic picture in which *the majority of star formation at all masses takes place in clusters*.

A similar picture, in which clustering is a common, but not ubiquitous, accompaniment to star formation, emerges from the biased surveys of localized regions associated with some indicator of very recent star formation. In L 1641, 25% of the young IRAS sources surveyed by Chen and Tokunaga (1994) were found to have near-infrared clusters. From the same survey, 63% of the outflow regions contain clusters, while in a broader survey Hodapp (1994) found 33% of molecular outflow sources to have clusters. Of 44 bright-rimmed clouds (regions thought to be examples of triggered star formation) containing IRAS sources surveyed by Sugitani et al. (1995), "most" are claimed to harbor small clusters. On the other hand, Carballo and Sahu (1994) found no evidence for clustering around the IRAS sources in their survey, and a similar null result was obtained from deep imaging of Bok globules (Yun and Clemens 1994).

An even higher incidence of clustering appears in the surveys of regions containing massive stars. For example, the unbiased surveys of Orion B reveal that clusters are associated only with the bright stars exciting the conspicuous nebulae in the region; stated in reverse, each of the high-mass stars in Orion B is accompanied by a cluster. A similar connection is suggested from the biased surveys. The highest incidence of clustering (19 of 20 cases) occurred in the survey of outer Galaxy IRAS sources radio-selected to contain OB stars (Carpenter et al. 1993). Similarly, the near infrared surveys of Herbig Ae/Be stars by Hillenbrand (1995) and Testi et al. (1997, 1998) (see also Aspin and Barsony 1994; Wilking et al. 1997), indicate that clusters are present around those Ae/Be stars with masses in excess of 3–5 $M_{\odot}$, with little evidence of clustering around less massive objects (see also the chapter by Stahler et al., this volume).

One possible correlation in the data is that between stellar density and the mass of the most massive cluster member (Hillenbrand 1995; Testi et al. 1999; see Fig. 2). Since clusters exhibit a rather small range of projected radii (see Table I and also Fig. 1 in Testi et al. 1999), this also translates into a correlation between the cluster membership number N and most massive star. It is at present unclear whether this correlation represents a

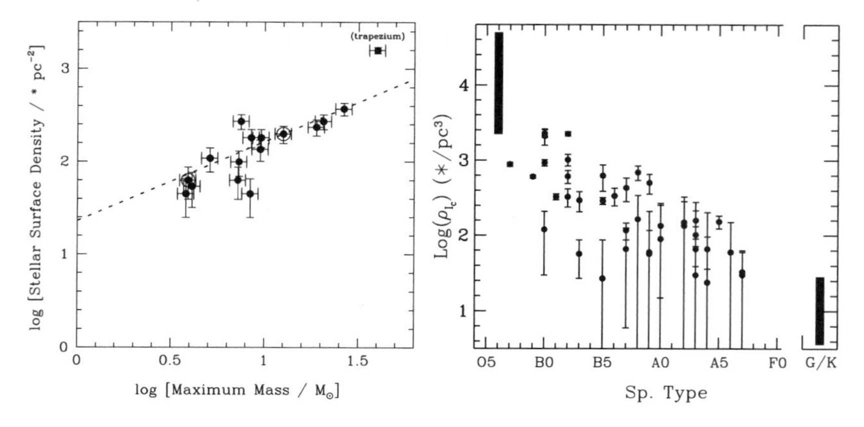

Figure 2. Quantification of clustering around Herbig Ae/Be stars. In the left panel, stellar surface density (pc^{-2}) is plotted against mass ($M_\odot$) of the most massive star (from Hillenbrand 1995); random errors of 10% in mass and $\sqrt{N}$ in star counts at K band are shown, along with a least-squares fit. In the right panel, stellar volume density (pc^{-3}) is plotted against spectral type of the Ae/Be star (from Testi et al. 1998); counting statistics in the I band source counts are shown. For scaling reference only, the innermost region of the Orion Nebula Cluster is plotted in the upper right of panel (a) and the upper left of panel (b). Regions containing more than one Ae/Be star do not occupy any preferred location in these diagrams.

genuine physical requirement of high density or N for massive star formation (see, e.g., Bonnell et al. 1998), or whether it is merely a consequence of random drawing from an initial mass function (IMF), which would imply that a given cluster is more likely to contain a massive star if it has high N. Distinguishing between these two possibilities will require a significant number of small-N clusters to compare with the mass distributions in large-N clusters.

A strong association is found between the location of clusters and of dense, massive molecular cloud cores. For example, all the clusters in Orion B are associated with CS cores (Lada et al. 1991*a*), as are those in Orion A (Strom et al. 1993). Moreover, those CS cores in L1630 that are associated with clusters contain a higher fraction of very dense gas ($\gtrsim 10^5$ cm^{-3}) than the clusterless cores (Lada et al. 1997). Likewise, in the Rosette molecular cloud, the seven embedded clusters discovered by Phelps and Lada (1997) are all associated with moderately massive molecular cores (as traced by ^{13}CO), although the majority of massive cores do not harbor clusters. These results may suggest that gas density, as opposed to mass, may be the critical factor in promoting cluster formation, although follow-up studies in a density-sensitive tracer such as CS or NH_3 are required in the Rosette region to confirm the hypothesis.

Cluster parameters as summarized in Table I are not directly comparable between the various regions, because of inconsistencies in the analyses. In particular, we emphasize that the values given for the number of stars, and hence the number density, are in all cases likely to be lower limits. To effect a rigorous comparison of the stellar populations emerging from molecular clouds, we ideally need surveys to uniform completeness in mass (to $<0.1\ M_\odot$) over a known range in age ($\sim$3 Myr) and through some given value of the extinction (10-20 mag, say). However, if one assumes from current databases that cluster sizes are good to a factor of 2 and cluster densities are good to a factor of 3–5, intercomparisons can be made.

The sizes of young (ages less than a few 10^6 yr) clusters appear fairly uniform [in the range 0.2–0.8 pc full width at half maximum (FWHM)] and, notably, are a factor of 5–10 smaller than the typical sizes of Galactic open clusters (with ages a few 10^7 to 10^9 yr; Phelps and Janes 1994; Janes 1988). Several young clusters within a kiloparsec of the Sun are sufficiently populous to rank as candidate proto-open clusters, although it is uncertain that they will remain bound once their component gas is removed (see section V). Note that we exclude OB associations from Table I and stress that these are considerably bigger (a few tens of pc). Cluster densities have a spread that is larger than the errors, spanning a few 10^2 to a few 10^4 stars pc^{-3}, the latter value corresponding to the core of the ONC. Such densities correspond to volume-averaged values in the range several 10^3 to several 10^5 molecules cm^{-3}, consistent with the strong correlation between clusters and concentrations of *dense* molecular gas.

Finally, we turn from a description of the gross parameters of young clusters to a brief mention of recent attempts to characterize their stellar content in detail. This exercise involves combined spectroscopy and photometry in order that stars can be individually dereddened and placed in theoretical HR diagrams, where their location can in principle (i.e., given well-determined theoretical tracks) be used to determine stellar masses and ages (cf. Hillenbrand 1997; Herbig 1998; Strom et al. 1999). Recently, this traditionally optical technique has been successfully applied in the near infrared to study deeply embedded populations: those obscured by 10–50 magnitudes of interstellar and circumstellar extinction (cf. Hodapp and Deane 1993; Greene and Meyer 1995; Carpenter et al. 1997; Hanson et al. 1997; Luhman and Rieke 1998; Meyer et al. 1999). The observationally intensive nature of this exercise means that few clusters have been studied in detail as yet. Clearly, the information yielded on the mass distribution (cf. the chapter by Meyer et al., this volume) and age spread of stars in clusters can be expected to have a major impact on cluster formation theories in the next few years. It is notable, for instance, that the relatively old cluster IC 348 appears to show evidence for ongoing star formation over a considerable period (up to 10 Myr; Lada and Lada 1995; Preibisch et al. 1996; Herbig 1998), whereas a high fraction of mass in the ONC would seem to have been converted into stars in less than a million years (Hillenbrand 1997).

III. CLUSTERS WITHIN CLUSTERS?

Images of young clusters often contain substructure that is readily identifiable by eye. Examples occur on a wide range of size scales. At one extreme, the "Super Star Clusters" (SSCs) observed in interacting galaxies such as the Antennae (Whitmore et al. 1998) comprise ensembles of tens to hundreds of clusters within a couple of hundred pc, while the SSCs themselves appear to be clustered in groups of a few. Nearby star-forming regions also contain a wealth of substructure (see, for example, Gomez et al. 1993; and the chapter by Elmegreen et al., this volume).

The issue of subclustering of stars at birth is a fundamental one, because it defines the local potential in which stars form and determines whether or not interactions between adjacent protostars (and associated gas and disks) play an important role in the star formation process. Compact clusters with few members are, however, short-lived against dynamical dissolution (see section IV), so, by the age at which clusters are observed in a relatively unobscured state (generally a million years or so), much of the original substructure may have been erased, although traces may remain in positional and velocity data. The link between the structure of observed clusters and the structure they had at birth therefore needs to be mapped out via numerical simulations (see, for example, Goodwin 1997*a* in the context of the LMC globular clusters).

Apart from these questions of how observed structure relates to structure at birth (which can be addressed by simulations), there is the equally important issue of how the statistical significance of apparent substructure is to be assessed. The eye is notoriously adept at picking out apparent groupings in randomly generated distributions of points. A commonly used statistic is the mean surface density of companions (MSDC), first applied to a star-forming region (in this case Taurus) by Larson (1995) and subsequently to several other star-forming regions (Simon 1997; Bate et al. 1998; Nakajima et al. 1998; Gladwin et al. 1999), although incompleteness in some cases limits the utility of this approach. The MSDC is related to the two-point correlation function (Peebles 1980) but has the advantage that it is not sensitive to the choice of average density in the surveyed region. It is simply computed, as a function of angular separation, by averaging the surface density of stars in annuli of appropriate radius placed in turn on each of the stars in the sample.

In Taurus, the MSDC can be fitted as a power law (of slope –0.6) for stellar separations in excess of around 0.04 pc (Larson 1995). A uniform stellar distribution gives rise to a flat MSDC (equal surface densities on all scales), so this result is immediate evidence for an inhomogeneous stellar distribution. A power law MSDC over a large dynamic range is, moreover, evidence for fractal clustering (an interpretation favored by Larson), although the observed MSDC over the limited dynamic range available in

Taurus is also consistent with clustering on a single scale (Bate et al. 1998). The conclusion that Taurus is indeed highly inhomogeneous is readily confirmed by visual inspection of the stellar distribution, which clearly shows the existence of discrete groupings containing around 15 stars in regions of typical size 0.5–1.1 pc (Herbig 1977; Gomez et al. 1993). Given the velocity dispersion measured in Taurus (e.g., Hartmann et al. 1986; Frink et al. 1997), these groups are not bound; this velocity dispersion is, however, consistent with these groups having expanded from very compact configurations over their assumed lifetimes. Thus, the existence of the Gomez groups is consistent with (but does not prove) an origin of stars in compact miniclusters.

Interpretation of the MSDC in clusters, as opposed to the more diffuse and irregular environment of Taurus, is complicated by the global decline of surface density with radius in this case. It turns out, however, that if the surface density declines with distance from the cluster center as R^{-1} or less steeply, then for clusters with no substructure the MSDC should still be approximately flat apart from possible edge effects (Bate et al. 1998). This convenient property means that in the ONC, for example, where the surface density declines with radius approximately as R^{-1} outside the core, the flatness or otherwise of the MSDC can still be used as a diagnostic of clustering.

The result for the ONC is that although the eye arguably can pick out apparent stellar groupings, the MSDC is essentially flat: i.e., the stellar distribution is statistically consistent with a smoothly declining density law with no subclustering. This is not to say, however, that subclustering is necessarily absent. Through generation of synthetic clusters, Bate et al. showed that over a limited region of parameter space (i.e., for miniclusters of a few times 10^4 AU in size), it was possible to hide a substantial fraction of the stars in miniclusters and yet produce an MSDC consistent with that observed. The range of size scales that can be hidden in this way shrinks with the membership number of the cluster, so, unless the cluster sizes are very finely tuned, the number of stars contained in each needs to be quite small (a few tens at most).

In summary then, there is no evidence for subclustering within the ONC, although there are patterns of subclustering that would not be ruled out by the observed MSDC. (Note, however, that the *massive* stars do appear to be clustered in the central regions: see section VI). Rough estimates suggest that this lack of subclustering may not necessarily rule out subclustering at birth; although Orion is generally believed to be younger than Taurus (Kenyon and Hartmann 1995), the higher stellar surface density means that subclusters would merge and lose their identity more rapidly during the dissolution process. Further modeling, using all the available kinematic and spatial data for the cluster, is required to rule out the possibility that the ONC was composed of an ensemble of subclusters at birth.

IV. DYNAMICAL INTERACTIONS IN COMPACT CLUSTERS

Miniclusters comprising N members dissolve due to point mass gravitational interactions on a timescale that is a strong positive function of N (van Albada 1968; Heggie 1974). Thus, point mass gravitational effects are the main agent of dissolution for small-N systems, where a central binary can interact and eject the majority of stars, whereas gas expulsion may predominate in larger-N systems (see section V). Compact, small-N clusters, therefore, are short-lived even if gas expulsion is neglected; for example, a cluster of 10 stars in a volume of radius 0.1 pc dissolves in less than a million years. This fact underlines the difficulty of assessing the level of subclustering at birth in star-forming regions, inasmuch as information on the smallest scales is rapidly erased, sometimes before the cluster becomes optically visible.

Cluster dissolution by point mass dynamics results from the formation of a central binary, which absorbs the potential energy of the cluster, thereby unbinding the other members. There is an overwhelming tendency for the two most massive stars to constitute the binary (van Albada 1968). Thus, *if* binaries form from small-N, nonhierarchical ensembles, their pairing statistics are well defined (McDonald and Clarke 1993): the binary fraction is a strongly increasing function of primary mass, and, unless the membership number of the minicluster is very small (3 or so), there is a strong tendency for stars to pair with companions of almost equal mass. McDonald and Clarke showed that a hallmark of binaries formed dynamically in such small clusters is that the mass distribution of secondaries does not depend on the primary's mass. This property can be tested for in binary samples with primaries of various masses. It is clear, however, (because most solar-type stars are binary primaries, whereas most OB binaries have high-mass secondaries) that this process cannot simultaneously account for both low- and high-mass binary statistics, unless the IMF is spatially variable.

In reality, of course, one would not expect interactions in such miniclusters to result purely from point mass gravity. For few-body clusters, the expected radii of circumstellar disks are a significant fraction of the mean interstellar separation (Pringle 1989; Clarke and Pringle 1991*b*), so hydrodynamic encounters with disk gas are to be expected at closest approach (Larson 1990; Heller 1993; Hall et al. 1996). Whereas the higher velocity dispersion in large-N clusters renders most such encounters disk destroying (rather than binary producing; Clarke and Pringle 1991*a*), the relatively slow encounters within small-N miniclusters can lead to a substantial binary fraction through star-disk capture (McDonald and Clarke 1995). If star-disk capture is the dominant binary production route, the dependence of binary fraction on primary mass is somewhat reduced, while the companion mass distribution reflects almost random pairing from the IMF.

In addition to the possible production of binaries, close encounters in miniclusters can have two further effects. The first is the destructive effect of star-disk encounters. It has been argued, for example by Mottmann (1977), that the Sun may have originated in a cluster, so that episodes of intense meteoritic bombardment, as evidenced by the cratering record of the terrestrial planets, would have followed perturbations to the Oort Cloud by stellar encounters. Simulations of star-disk encounters indicate that disks are truncated at about one-third of the stars' closest approach (Clarke and Pringle 1993), the pruned remnant being left with an exponential radial density profile (Hall 1997) similar to those observed in the "silhouette disks" in Orion (McCaughrean and O'Dell 1996). Such pruning would not only reduce the strength of disk emission (by reducing the mass and surface area of the disk), but would also shorten the disk lifetime (mainly due to the reduction in the disk's radial extent). It has been noted (e.g., Bouvier et al. 1997; Armitage 1996) that a wide range of disk lifetimes is necessary both to explain the coexistence of classical and weak-line T Tauri stars in the same region of the HR diagram and to explain the spread in rotation rates of stars on the zero-age main sequence (ZAMS).

The velocities acquired by stars during the dissolution of small clusters is of the order of the velocity at pericenter during a three-body encounter. Thus, while the majority of stars drift apart with a velocity that exceeds the cluster escape velocity by a factor of order unity, stars can be ejected from particularly close encounters with considerably larger velocities. Sterzik and Durisen (1995) have applied this model to the production of the dispersed population of X-ray sources detected by ROSAT in the vicinity of star-forming regions (Alcalá et al. 1996; Neuhauser 1997), arguing that these sources are weak-line T Tauri stars that were formed in the smaller volume currently occupied by the emission line (classical T Tauri) stars but were ejected by dynamical encounters in small clusters (see Feigelson 1996 for an alternative view). The combination of the apparent distances of these stars from their putative birthplaces and their ages derived from the HR diagram implies ejection velocities greater than $\sim$3 km/s, which requires miniclusters that comprise a few (i.e., 5–10) stars within a radius of 500–1000 AU. The close encounters (pericenter of about 0.5 AU) that are required to generate such velocities shave the disks to such small radii that the disk depletion timescale is considerably reduced. In the case of disks that are magnetically disrupted in their innermost regions, such tidal pruning in close encounters can lead to the system appearing as a weak-line T Tauri star even at the young age ($\sim 10^6$ years) inferred for the dispersed population of X-ray sources (Armitage and Clarke 1997). It should be noted, however, that many of the dispersed X-ray sources may be somewhat older foreground stars (see discussion by Briceno et al. 1997; Wichmann et al. 1997) and that proper motion data support the ejection hypothesis only in some cases (Neuhauser et al. 1998; Frink et al. 1997). Clearly the controversial question of what proportion of the ROSAT sources are indeed runaway T Tauri stars needs to be settled before one can assess the required ejection rates of T Tauri stars from star-forming regions and hence the number of compact miniclusters that are needed to generate this ejection rate.

In summary, then, several physical processes occurring in very compact miniclusters can profoundly affect the properties of the stars and their associated disks. These physical processes rely on small interstellar separations and relatively low velocity dispersions, and their role is thus negligible if estimated using the densities and velocity dispersions of large-scale star-forming regions (such as, for example, the Orion Nebula Cluster or the central regions of Taurus). If the stars in these regions were not considerably subclustered at birth, then close encounters would have played an insignificant role, and stars would have evolved essentially independently. On the other hand, if stars were tightly clustered at birth, then such clustering may provide solutions to a number of problems (e.g., those of binary formation, of the apparently large dispersion in disk lifetimes, and of the generation of runaway T Tauri stars).

V. THE ROLE OF GAS IN CLUSTERS

As discussed in section II, young stellar clusters are commonly associated with massive cores of molecular gas (e.g., Lada 1992; Lada et al. 1997). This gas constitutes the majority of the cluster mass in the youngest systems (typically 50–90% of their total mass; Lada 1991) but appears to be absent in older systems (e.g., IC 348 at $\approx 5 \times 10^6$ years; Lada and Lada 1995).

In addition to being a major contributor to the gravitational potential (and hence, by its removal, providing an obvious way to unbind the cluster), the gas can also interact with and be accreted by the stars. As pointed out by several authors (e.g., Zinnecker 1982; Larson 1992), accretion in a clustered environment may play an important role in shaping the observed spectrum, and segregation, of stellar masses. Bonnell et al. (1997) used smoothed particle hydrodynamics (SPH)/accretion particle simulations to study the evolution of clusters initially comprising a few (point mass) stars plus a distributed gas component. The stars excite gravitational wakes in the surrounding gas (cf. Gorti and Bhatt 1996) and gain mass by accretion (in these calculations no gas expulsion is included, so all the gas ultimately ends up on the stars).

The competitive accretion of gas by the various stars leads to an IMF in which the dynamic range of final stellar masses is large, even when the masses of the initial protostellar seeds are all set to the same value. The chief determinant of ultimate stellar mass is in this case the initial position of the protostellar seed in the cluster potential: seeds initially deep in the potential well acquire accreted mass rapidly from the start and then become hard to nudge from their central position because of their large masses. Seeds initially at large radii, conversely, accrete mass slowly; being low-mass objects, they are more likely to be flung out of the cluster by interactions with more massive stars and thus stop accreting altogether. Thus the interplay of hydrodynamic accretion and point mass gravitational interactions is such as to enhance the initial "advantage" of seeds located near the cluster core and generates a large dynamic range of stellar masses from arbitrary initial conditions. It is notable, in the context of the mass segregation observed in clusters (see section VI), that competitive accretion provides a natural way of producing the most massive stars in the cluster core and requires no gradients in the initial conditions.

It has also been argued (Bonnell et al. 1998) that massive stars *must* form in the centers of dense clusters. These authors consider systems that become extremely dense (up to 10^8 pc^{-3}) as they shrink because of the effects of continuing accretion of gas. In such high-density environments, massive stars can form via collisional build-up of protostellar fragments. An episode of vigorous mass loss is then invoked to clear the cluster of gas and cause it to reexpand (because these effects occur on timescales of $\approx 10^4$ years, these clusters are unlikely to be directly observable in their high-density phase). The formation of massive stars through collisions is an attractive scenario inasmuch as it avoids the classic problem of forming them by accretion (namely, that for stars more massive than around 10 $M_\odot$, accretion is halted by the action of radiation pressure on dust grains).

In reality, however, gas is lost from clusters in a variety of ways. Massive stars ($\gtrsim 8$ $M_\odot$) eject gas by the action of supernovae, photoionization, and stellar winds (Whitworth 1979; Tenorio-Tagle et. al. 1986; Franco et al. 1994). It is also becoming increasingly apparent that low-mass stars can provide effective feedback of energy into the surrounding medium through the action of energetic molecular outflows (see the chapter by Eislöffel et al., this volume). Note that whereas it is difficult to sustain the case that an *isolated* star can cut off its own accretion supply through the action of outflows (because the outflows are somewhat collimated, whereas accretion occurs over a large solid angle, and preferentially equatorially at small radii), obviously a set of randomly oriented outflow sources in a small cluster can inflict significant damage on the residual gas.

The fate of a particular cluster in response to gas loss depends on the initial gas fraction, the removal timescale, and the stellar velocity dispersion when the gas is dispersed (Lada et al. 1984; Pinto 1987; Verschueren and David 1989; Goodwin 1997*b*; see also the chapter by Elmegreen et al., this volume). If the gas comprises a significant fraction of the total mass ($\gtrsim 50\%$) and is removed quickly compared to the cluster crossing time, then the dramatic reduction in the binding energy, without affecting the stellar kinetic energy, results in an unbound cluster. Alternatively, if the gas is removed over several crossing times, then the cluster can adapt to the new potential and can survive with a significant fraction of its initial stars. For example, clusters with gas fractions as high as 80% can survive with approximately half of the stars if the gas removal occurs over four or more crossing times (Lada et. al. 1984).

The number and age distribution of Galactic clusters suggest that only a few percent of all Galactic field stars can have originated in bound clusters (Wielen 1971). However, the frequency of cluster-mode star formation (see section II) and the properties of Galactic field binaries (Kroupa 1995) indicate that most stars may form in clusters. The implication is that the lifetime of most young clusters is short, $\lesssim 10^7$ yr (Battinelli and Capuzzo-Dolcetta 1991), which is a natural consequence of rapid gas expulsion and low local star formation efficiency.

VI. THE INITIAL MASS DISTRIBUTIONS AND SHAPES OF CLUSTERS

A. Mass Segregation

A common observational finding in clusters is that the most massive stars tend to be concentrated in the central core (e.g., Mon R2: Carpenter et al. 1997; ONC: Hillenbrand and Hartmann 1998; NGC 6231: Raboud and Mermilliod 1998; NGC 2157 in the LMC: Fischer et al. 1998; SL666 and NGC 2098 in the LMC: Kontizas et al. 1998). This mass segregation is present even in the youngest clusters, suggesting that it represents the initial conditions of the cluster and is not due to its subsequent evolution. Order-of-magnitude arguments support this view: The timescale for mass segregation from two-body interactions (which drive the stellar kinetic energies toward equipartition and thus allow the massive stars to sink to the center) is approximately the relaxation time (Binney and Tremaine 1987; Bonnell and Davies 1998), which is typically very long (many crossing times) compared to the age of the cluster. However, the segregation timescale is inversely proportional to the stellar mass, so the most massive stars will segregate significantly faster than this. It is not therefore clear *a priori* whether the presence of massive subsystems in the cores of clusters (such as the Trapezium of OB stars in the ONC) is attributable to dynamical effects or segregation at birth.

Bonnell and Davies (1998) investigated this issue through N-body simulations of stellar clusters, exploring the timescale for massive stars to sink to the cluster center as a function of their initial location. Comparing these results with recent observations of the ONC (Hillenbrand 1997; Hillenbrand and Hartmann 1998) shows that the location of the massive stars (and, in particular, the existence of the Trapezium) *cannot* be accounted for by dynamical mass segregation but must reflect the initial conditions. The clearest indication of this result comes from repeated simulations based on different random realizations of the initial conditions. It was found that Trapezium-like systems were generated with significant frequency *only* if the massive stars were initially rather centrally condensed (i.e., within the innermost 10% of the stars for a 70% probability of Trapezium formation, or within 20% for a 10% probability). Note that these simulations did not include gas; it is possible in principle for the observed ONC to have expanded due to previous gas loss, in which case the shorter dynamical timescales in its initially denser configuration may have permitted more effective mass segregation.

As initially discussed by Zinnecker et al. (1993; see also Bonnell et al. 1998), simple Jeans-type arguments do not lead to the expectation that the most massive stars should form in the center of dense clusters. Since these regions have high densities, the associated Jeans mass is *low* unless the local temperature is anomalously high. Evolutionary effects, involving accretion and protostellar collisions, are probably required to build up massive stars in cluster cores (see section V).

B. Cluster Morphology

Simulations of cluster dynamics are often undertaken in spherical geometry, motivated in part by the shapes of globular clusters in the Galaxy. It is, however, well known that some clusters are significantly flattened, the best studied examples being the globular clusters in the LMC. Since some of these systems are both young (with ages, at less than 20 Myr, of order 10 crossing times) and significantly elliptical (projected axis ratio on the sky $\gtrsim 0.7$), it would seem likely that they would have originated from aspherical initial conditions. Analysis of the projected axis ratio distribution in these clusters suggests that their intrinsic shapes are triaxial (Han and Ryden 1994), indicating that velocity anisotropy, rather than rotational flattening, is responsible. Further examples of flattened young clusters are found among the SSCs ("super star clusters") that are conspicuous in images of some interacting galaxies (O'Connell et al. 1994). Here the strongly disturbed gas flows that are to be expected in galactic encounters make cluster formation from cloud-cloud collisions an attractive possibility (Murray and Lin 1992; Kimura and Tosa 1996), so that flattened clusters are a natural expectation in these environments. In the Galaxy, obvious examples of flattened young clusters are the ONC, Mon R2, and NGC 2024, where isophotal fitting of the outer regions yields a projected axis ratio of about 2:1 (Hillenbrand and Hartmann 1998; Carpenter et al. 1997; Lada et al. 1991).

The interest in examining the shapes of young clusters derives from the clues that those shapes might give as to the mechanism for cluster formation. Indeed, it is hard to think of an external trigger for cluster formation, whether cloud-cloud collisions or the sweeping up of gas by supernova blast waves or powerful stellar winds, that does not induce star formation in sheet/slab-like geometry. At first sight, it might appear most obvious to examine the shapes of the youngest embedded clusters in nearby star-forming regions, which are still associated with molecular material (see Table 1). This exercise is, however, complicated by the problem of patchy extinction, plus the difficulty of isophotal fitting in clusters that comprise relatively few stars. Therefore, the young globular clusters in the LMC are the best laboratories for studying this problem, since they are relatively populous and devoid of gas.

Clusters in which star formation is externally triggered, as by a shock wave, are unlikely to form in virial equilibrium, however, so that even the youngest of the LMC globular clusters would already have undergone a phase of violent relaxation. Numerical simulations are therefore required in order to relate the morphologies of observed clusters to the initial (i.e., pre-violent relaxation) configuration of the star-forming gas. This exercise (Boily et al. 1999; see also Aarseth and Binney 1978; Goodwin 1997*a*) yields the answer that apart from the thinnest initial configurations (i.e., sheets of scale height less than the mean interstellar separation, which are subject to two-body scattering on a dynamical timescale), the system retains a strong memory of its initial geometry during the violent relaxation process. The relation between "initial" and "final" (i.e., relaxed) morphologies is set by the principle of adiabatic invariance and yields the prediction that the initial geometry is substantially more flattened than that of the relaxed cluster. When applied to the LMC globulars, initial conditions that are flattened in the ratio of about 1:5 are required.

Although the degree of flattening that is required is quite substantial, it can be generated by gas swept up in shocks of relatively low Mach number. Since the density contrast induced in strong shocks is of the order of the square of the Mach number, one sees that far flatter configurations (axis ratio of order 10^{-4}) would be produced, for example, by colliding cold, thermally supported *homogeneous* clouds at relative velocities typical of the LMC. In the case of collisions between inhomogeneous clouds, density peaks carry momentum across the net symmetry plane and generate a buckled, and thus effectively, thicker, geometry. The initial morphologies deduced for the LMC globulars may thus be compatible with externally triggered cluster formation in clouds containing substantial preexisting density structure.

VII. THEORETICAL CONSIDERATIONS

In this section we lay out a very idealized conceptual framework for cluster formation and indicate where recent theoretical work can be installed into this framework.

In order to keep an open mind as to whether cluster formation is primarily a bottom-up or top-down process, we set up a general scenario in which the cluster progenitor gas, mass $M_{\rm clus}$, consists prior to cluster formation of an ensemble of dense lumps, mass M_J. Since molecular clouds are hierarchically structured, we define the mass scale M_J as being the mass of *thermally* supported lumps that are marginally Jeans stable. Substructure within such lumps is not gravitationally bound, whereas larger-scale structures are supported by superthermal random motions. We now suppose that some external trigger overruns the protocluster region, destabilizing lumps of mass M_J. Each lump then collapses to form a member of the eventual cluster (e.g., Klessen et al. 1998). If this destabilization promotes subfragmentation of the lumps, down to a mass scale $M_\star$, then the initial state of the cluster is one of an ensemble of miniclusters (mass M_J).

Stated in this general manner, one can consider cluster formation as occupying some position on a spectrum of possibilities. The extreme positions are top-down fragmentation (as envisaged, for example, in many models for globular cluster formation, e.g., Fall and Rees 1977; Murray and Lin 1989), in which case $M_J = M_{\rm clus}$, and bottom-up scenarios, in which case $M_J = M_\star$. We note that top-down fragmentation engenders structures that are coeval (to within a crossing time), whereas the age spread in bottom-up scenarios depends on the timescale on which discrete lumps are destabilized and is affected, for example, by the speed with which an external trigger overruns the region.

Before proceeding further, we here introduce some numbers that will motivate the following discussion. Hierarchical structures in molecular clouds obey a mass-radius ("Larson") relation of the form $M \propto R^2$, which corresponds to a hierarchy of self-gravitating structures that share the same kinetic pressure. As one descends such a hierarchy, structures of increasing density are characterized by a decreasing velocity dispersion, until eventually the scale is reached at which this velocity dispersion becomes subthermal. This scale represents the minimum mass of a self-gravitating structure within a cloud of given kinetic pressure (or, equivalently, M/R^2 for the parent GMC) and temperature, and is thus equal to M_J in the above nomenclature. Employing canonical values for the temperature and mass-radius relation in GMCs (respectively $T = 10$ K and $(M/{\rm M}_\odot) \sim (R/0.1 \text{ pc})^2$; Chieze 1987), one finds that M_J is of the order of a solar mass.

Thermally supported, self-gravitating clumps of around solar mass are indeed observed, in nearby star-forming clouds such as Taurus, as the dense cores traced by NH_3 (Benson and Myers 1989). The low masses of these cores implies that one would expect top-down fragmentation to be operative only in the generation of miniclusters (i.e., those comprising a small number of stars). More populous clusters must result from a bottom-up process: that is, the coordinated collapse of many such units. Cores that are currently forming clusters (such as those in Orion) have superthermal line widths and are thus presumed to be supported by Alfvénic turbulence (Harju et al. 1993). Myers (1998) has however suggested that these cores should contain pockets of thermally supported gas from which Alfvénic turbulence is excluded, arguing that regions can decouple from the turbulence on size scales less than the minimum turbulent wavelength (this being set by the requirement that the inverse frequency equals the ion-neutral collision time). We will return below to the issue of how such thermally supported pockets might be destabilized.

If one considers instead the environment in which the galactic globular clusters would have formed, with kinetic pressures characteristic of the proto-Galaxy and temperatures of 10^4 K (this marking the steep decline of the cooling function for primordial gas), one obtains a mass scale M_J of around 10^5–10^6 $M_\odot$ (Fall and Rees 1977). This mass is comparable to that of globular clusters, suggesting that star formation in globular clusters may well have been a top-down process.

The issue of hierarchical fragmentation in the top-down collapse of Jeans unstable gas has, however, a controversial history (see, for example, Hoyle 1953, Hunter 1962, and Layzer 1963 for early analytical arguments for and against opacity-limited fragmentation). Larson (1978) studied the problem numerically using a crude Lagrangian hydrodynamic code and concluded that fragmentation does not proceed down to the opacity limit, but instead reflects the number of Jeans masses in the gas at the initiation of collapse.

The production of clusters by top-down fragmentation thus requires that a clump initially containing one Jeans mass makes a rapid (i.e., less than dynamical timescale) transition so that it contains a large number of Jeans masses as it enters its collapse. This reduction in Jeans mass may be achieved via either cooling or compression, if the system remains spherically symmetric. Most plausible compression mechanisms, however, result in the system becoming approximately planar. The 2D Jeans mass then depends only on the temperature and column density, so the fragmentation of clouds that are swept up in shocks, for example, demands that such shocks cool to *less than* the original temperature (Lubow and Pringle 1993; Whitworth et al. 1994). In the context of globular cluster formation at primordial epochs, it has been suggested (e.g., Palla and Zinnecker 1987; Murray and Lin 1989) that protogalactic shocks activate nonequilibrium cooling (i.e., cooling by molecular hydrogen whose formation is catalyzed by a nonequilibrium concentration of electrons in rapidly cooling gas) and that this can effectively cool protoglobular clouds from 10^4 K to 100 K.

In the context of current star-forming clouds, no such dramatic cooling is required, since M_J is already in the stellar regime and thus subfragmentation, if it occurs, will involve only a small number of pieces. Whitworth and Clarke (1997) considered the response of Jeans-stable clumps to the mildly supersonic shocks induced by clump-clump collisions and concluded that cooling by dust in the dense gas behind the shock imposes close thermal coupling between the gas and dust. Whether or not this represents a "better than isothermal" shock (as required to promote subfragmentation) depends, of course, on the relation between the dust and gas temperatures in the unshocked clump, which is uncertain.

Bottom-up cluster formation places less stringent requirements on the interaction between clumps and external trigger (since the trigger only has to destabilize the clumps rather than initiate subfragmentation). Whitworth et al. (1998) have shown that the densities and temperatures of thermally supported clumps in molecular clouds place them close to, but somewhat outside, a regime in which dust cooling can dispose of the compressional heating generated by collapse in a free-fall time. It is interesting to note that if the mass-radius relation for molecular clouds were somewhat different, so that thermally supported clumps lay within this regime, then gas would not "hang up" at this scale but would instead collapse to a star in a free-fall time. If, conversely, thermally supported clumps lay far from this regime, then they would be extremely hard to destabilize, and the star formation rate would be correspondingly low. The proximity of observed dense cores to the dust cooling regime instead allows a situation where such cores are stable but may be destabilized by fairly modest perturbations. [See, for example, the suggestion of Clarke and Pringle (1997) that cores may be destabilized by external stirring, which widens the bandpass for cooling in optically thick lines.] Clearly, a situation where Jeans mass clumps are fairly stable (and hence may accumulate in a given region) but are then fairly easy to destabilize is an optimum one for producing clusters. Considerably more work is required, however, before the feasibility of such ideas can be established.

REFERENCES

Aarseth, S. J., and Binney, J. 1978. On the relaxation of galaxies and clusters from aspherical initial conditions. *Mon. Not. Roy. Astron. Soc.* 185:227–240.

Alcalá, J. M., Terranegra, L., Wichmann, R., et al. 1996. New weak-line T Tauri stars in Orion from the ROSAT all-sky survey. *Astron. Astrophys. Suppl.* 119:7–42.

Ali, B., and DePoy, D. L. 1995. A 2.2 micrometer imaging survey of the Orion A molecular cloud. *Astron. J.* 109:709–723.

Armitage, P. J. 1996. Accretion discs in T Tauri stars and interesting binaries. Ph.D. Thesis, University of Cambridge.

Armitage, P. J., and Clarke, C. J. 1997. The ejection of T Tauri stars from molecular clouds and the fate of circumstellar discs. *Mon. Not. Roy. Astron. Soc.* 285:540–545.

Aspin, C., and Sandell, G. 1997. Near-IR imaging photometry of NGC 1333: A 3-μm imaging survey. *Mon. Not. Roy. Astron. Soc.* 289:1–28.

Aspin, C., and Barsony, M. 1994. Near-IR imaging photometry of the $J - K >$ 4 sources in the Lk H_α 101 infrared cluster. *Astron. Astrophys.* 288:849–870.

Aspin, C., and Sandell, G., and Russell, A. P. G. 1994. Near-IR imaging photometry of NGC 1333. I. The embedded PMS stellar population. *Astron. Astrophys. Suppl.* 106:165–180.

Barsony, M., Kenyon, S. J., Lada, E. A., and Teuben, P. J. 1997. A near-infrared imaging survey of the ρ Ophiuchi cloud core. *Astrophys. J. Suppl.* 112:109–131.

Barsony, M., Carlstrom, J. E., Burton, M. G., Russell, A. P. G., and Garden, R. 1989. Discovery of new 2 micron sources in Rho Ophiuchi. *Astrophys. J. Lett.* 346:L93–96.

Bate, M. R., Clarke, C. J., and McCaughrean, M. 1998. Interpreting the mean surface density of companions in star-forming regions. *Mon. Not. Roy. Astron. Soc.* 297:1163–1174.

Battinelli, P., and Capuzzo-Dolcetta, R. 1991. Formation and evolutionary properties of the Galactic open cluster system. *Mon. Not. Roy. Astron. Soc.* 249:76–87.

Benson, P., and Myers, P. 1989. A survey for dense cores in dark clouds. *Astrophys. J. Suppl.* 71:89–104.

Binney, J., and Tremaine, S. 1987. *Galactic Dynamics* (Princeton: Princeton University Press).

Blaauw, A. The O associations in the solar neighborhood. 1964. *Ann. Rev. Astron. Astrophys.* 2:213–230.

Boily, C. M., Clarke, C. J., and Murray, S. D. 1999. Collapse and evolution of flattened stellar systems. *Mon. Not. Roy. Astron. Soc.* 302:399–414.

Bonnell, I. A., Bate, M. R., Clarke, C. J., and Pringle, J. E. 1997. Accretion and the stellar mass spectrum in small clusters. *Mon. Not. Roy. Astron. Soc.* 285:201–215.

Bonnell, I. A., Bate, M. R., Zinnecker, H. 1998. On the formation of massive stars. *Mon. Not. Roy. Astron. Soc.* 298:93–102.

Bonnell, I. A., Davies, M. B. 1998. Mass segregation in young stellar clusters. *Mon. Not. Roy. Astron. Soc.* 295:691–702.

Bouvier, J., Forestini, M., and Allain, S. 1997. The angular momentum evolution of low-mass stars. *Astron. Astrophys.* 326:1023–1035.

Briceno, C., Hartmann, L. W., Stauffer, J. R., Gagné, M., and Stern, R. A. 1997. X-ray surveys and the post-T Tauri problem. *Astron. J.* 113:840–856.

Carballo, R., and Sahu, M. 1994. Near-infrared observations of new young stellar objects from the IRAS Point Source Catalog. *Astron. Astrophys.* 289:131–144.

Carpenter, J. M., Snell, R. L., Schloerb, F. P., and Skrutskie, M. F. 1993. Embedded star clusters associated with luminous IRAS point sources. *Astrophys. J.* 407:657–671.

Carpenter, J. M., Snell, R. L., and Schloerb, F. P. 1995. Star formation in the Gemini OB1 molecular cloud complex. *Astrophys. J.* 450:201–220.

Carpenter, J. M., Meyer, M. R., Dougados, C., Strom, S. E., and Hillenbrand, L. A. 1997. Properties of the Monoceros R2 stellar cluster. *Astron. J.* 114: 198–216.

Chen, H., and Tokunaga, A. T. 1994. Stellar density enhancements associated with IRAS sources in L1641. *Astrophys. J. Suppl.* 90:149–164.

Chieze, J.-P. 1987. The fragmentation of molecular clouds. I. The mass-radius-velocity dispersion relations. *Astron. Astrophys.* 171:225–236.

Clarke, C. J., and Pringle, J. E. 1991*a*. Star-disc interactions and binary star formation. *Mon. Not. Roy. Astron. Soc.* 249:584–587.

Clarke, C. J., and Pringle, J. E. 1991*b*. The role of discs in the formation of binary and multiple star systems. *Mon. Not. Roy. Astron. Soc.* 249:588–594.

Clarke, C. J., and Pringle, J. E. 1993. Accretion disc response to a stellar fly-by. *Mon. Not. Roy. Astron. Soc.* 261:192–204.

Clarke, C. J., and Pringle, J. E. 1997. Thermal and dynamical balance in dense molecular cloud cores. *Mon. Not. Roy. Astron. Soc.* 288: 674–685.

Comeron, F., Rieke, G. H., Burrows, A., and Rieke, M. J. 1993. The stellar population in the ρ Ophiuchi cluster. *Astrophys. J.* 416:185–201.

Comeron, F., Rieke, G. H., and Rieke, M. J. 1996. Properties of low-mass objects in NGC 2024. *Astrophys. J.* 473:294–306.

Eiroa, C., and Casali, M. M. 1992. Near-infrared images of the Serpens cloud core: The stellar cluster. *Astron. Astrophys.* 262:468–481.

Elmegreen, B., and Lada, C. 1977. Sequential formation of subgroups in OB associations. *Astrophys. J.* 214:725–739.

Fall, S. M., and Ross, M. J. 1977. Survival and disruption of galactic substructure. *Mon. Not. Roy. Astron. Soc.* 181:37–50.

Feigelson, E. 1996. Dispersed T Tauri stars and galactic star formation. *Astrophys. J.* 468:306–314.

Fischer, P., Pryor, C., Murray, S., Mateo, M., and Richtler, T. 1998. Mass segregation in young Large Magellanic Cloud clusters. I. NGC 2157. *Astron. J.* 115:592–612.

Franco, J., Shore, S., and Tenorio-Tagle, G. 1994. On the massive star-forming capacity of molecular clouds. *Astrophys. J.* 436:795–811.

Frink, S., Roser, S., Neuhauser, R., and Sterzik, M. F. 1997. New proper motions of pre-main sequence stars in Taurus-Auriga. *Astron. Astrophys.* 325:613–624.

Giovannetti, P., Caux, E., Nadeau, D., and Monin, J.-L. 1998. Deep optical and near infrared imaging photometry of the Serpens cloud core. *Astron. Astrophys.* 330:990–1009.

Gladwin, P. P., Kitsionas, S., Boffin, H. M. J., and Whitworth, A. P. 1999. The structure of young star clusters. *Mon. Not. Roy. Astron. Soc.* 302:205–217.

Gomez, M., Hartmann, L., Kenyon, S., and Hewett, R. 1993. On the spatial distribution of pre-main-sequence stars in Taurus. *Astron. J.* 105:1927–1941.

Goodwin, S. P. 1997*a*. The initial conditions of young globular clusters in the Large Magellanic Cloud. *Mon. Not. Roy. Astron. Soc.* 286:669–682.

Goodwin, S. P. 1997*b*. Residual gas expulsion from young globular clusters. *Mon. Not. Roy. Astron. Soc.* 284:785–797.

Gorti, U., and Bhatt,, H. C. 1996. Dynamics of embedded protostar clusters in clouds. *Mon. Not. Roy. Astron. Soc.* 278:611–619.

Greene, T. P., and Meyer, M. R. 1995. An infrared spectroscopic survey of the ρ Ophiuchi young stellar cluster: Masses and ages from the H-R diagram. *Astrophys. J.* 450:233–248.

Greene, T. P., and Young, E. T. 1992. Near-infrared observations of young stellar objects in the Rho Ophiuchi dark cloud. *Astrophys. J.* 395: 516–529.

Hall, S. M. 1997. Circumstellar disc density profiles: A dynamic approach. *Mon. Not. Roy. Astron. Soc.* 287:148–155.

Hall, S. M., Clarke, C. J., and Pringle, J. E. 1996. Energetics of star-disc encounters in the non-linear regime. *Mon. Not. Roy. Astron. Soc.* 278:303–315.

Han, C., and Ryden, B. S. 1994. A comparison of the intrinsic shapes of globular clusters in four different galaxies. *Astrophys. J.* 433:80–91.

Hanson, M. M., Howarth, I. D., and Conti, P. S. 1997. The young massive stellar objects of M17. *Astrophys. J.* 489:698–711.

Harju, J., Walmsley, C. M., and Wouterloot, J. G. 1993. Ammonia clumps in the Orion and Cepheus clouds. *Astron. Astrophys. Suppl.* 98:51–61.

Hartmann, L., Hewett, R., Stahler, S., and Mathieu, R. D. 1986. Rotational and radial velocities of T Tauri stars. *Astrophys. J.* 309:275–291.

Heggie, D. C. 1974. The role of binaries in cluster dynamics. In *The Stability of the Solar System and Small Stellar Systems*, IAU Symp. 62, ed. Y. Kozai (Dordrecht: Kluwer), p. 225.

Heller, C. 1993. Encounters with protostellar disks. I. Disk tilt and the nonzero solar obliquity. *Astrophys. J.* 408:337–351.

Herbig, G. H. 1977. Radial velocities and spectral types of T Tauri stars. *Astrophys. J.* 214:747–758.

Herbig, G. H. 1998. The young cluster IC 348. *Astrophys. J.* 497:736–750.

Heyer, M. H., Brunt, C., Snell, R. L., et al. 1996. A massive cometary cloud associated with IC 1805. *Astrophys. J. Lett.* 464:L175–L177.

Hillenbrand, L. A. 1995. Herbig Ae/Be stars: An investigation of molecular environments and associated stellar populations. Ph.D. Thesis, University of Massachusetts.

Hillenbrand, L. A. 1997. On the stellar population and star-forming history of the Orion Nebula Cluster. *Astron. J.* 113:1733–1756.

Hillenbrand, L. A., and Hartmann, L. W. 1998. A preliminary study of the Orion Nebula Cluster structure and dynamics. *Astrophys. J.* 492:540–554.
Hodapp, K.-W. 1994. A K' imaging survey of molecular outflow sources. *Astrophys. J. Suppl.* 94:615–627.
Hodapp, K.-W., and Deane, J. 1993. Star formation in the L1641 North cluster. *Astrophys. J. Suppl.* 88:119–133.
Hodapp, K.-W., and Rayner, J. 1991. The S106 star-forming region. *Astron. J.* 102:1108–1129.
Hoyle, F. 1953. On the fragmentation of gas clouds into galaxies and stars. *Astrophys. J.* 118:513–523.
Hunter, C. 1962. The instability of the collapse of a self-gravitating gas cloud. *Astrophys. J.* 136:594–601.
Itoh, Y., Tamura, M., and Gatley, I. 1996. A near-infrared survey of the Taurus molecular cloud: Near-infrared luminosity function. *Astrophys. J.* 465: L129–132.
Janes, K. A., Tilley, C., and Lynga, G. 1988. Properties of the open cluster system. *Astron. J.* 95:771–784.
Jones, T. J., Mergen, J., Odewahn, S., Gehrz, R. D., Gatley, I., Merrill, K. M., Probst, R., and Woodward, C. E. 1994. A near-infrared survey of the OMC2 region. *Astron. J.* 107:2120–2132.
Kenyon, S., and Hartmann, L. 1995. Pre-main-sequence evolution in the Taurus-Auriga molecular cloud. *Astrophys. J. Suppl.* 101:117–134.
Kimura, T., and Tosa, M. 1996. Collision of clumpy molecular clouds. *Astron. Astrophys.* 308:979–989.
Klessen, R., Burkert, A., and Bate, M. 1998. Fragmentation of molecular clouds: The initial phase of a stellar cluster. *Astrophys. J. Lett.* 501: L205–208.
Kontizas, M., Hatzidimitriou, D., Bellas-Velidis, I., Gouliermis, D., Kontizas, E., and Cannon, R. D. 1998. Mass segregation in two young clusters in the Large Magellanic Cloud: SL 666 and NGC 2098. *Astron. Astrophys.* 336: 503–515.
Kroupa, P. 1995. Inverse dynamical population synthesis and star formation. *Mon. Not. Roy. Astron. Soc.* 277:1491–1511.
Lada, C. J. 1991. The formation of low mass stars: Observations. In *The Physics of Star Formation and Early Stellar Evolution*, ed. C. J. Lada and N. D. Kylafis (Dordrecht: Kluwer), p. 329–346.
Lada, C. J., Margulis, M., and Dearborn, D. 1984. The formation and early dynamical evolution of bound stellar systems. *Astrophys. J.* 285:141–155.
Lada, C. J., Young, E. T., and Greene, T. P. 1993. Infrared images of the young cluster NGC 2264. *Astrophys. J.* 408:471–483.
Lada, C. J., Alves J., and Lada, E. A. 1996. Near-infrared imaging of embedded clusters: NGC 1333. *Astron. J.* 111:1964–1973.
Lada, E. A. 1992. Global star formation in the L1630 molecular cloud. *Astrophys. J. Lett.* 393:L25–L28.
Lada, E. A., and Lada, C. J. 1995. Near-infrared images of IC 348 and the luminosity functions of young embedded star clusters. *Astron. J.* 109:1682–1699.
Lada, E. A., Bally, J., and Stark, A. A. 1991*a*. An unbiased survey for dense cores in the Lynds 1630 molecular cloud. *Astrophys. J.* 368:432–444.
Lada, E. A., DePoy, D. L., Evans, N. J., and Gatley, I. 1991*b*. A 2.2 micron survey in the L1630 molecular cloud. *Astrophys. J.* 371:171–184.
Lada, E. A., Evans, N. J. II, and Falgarone, E. 1997. Physical properties of molecular cloud cores in L1630 and implications for star formation. *Astrophys. J.* 488:286–299.
Larson, R. B. 1978. Calculations of three-dimensional collapse and fragmentation. *Mon. Not. Roy. Astron. Soc.* 184:69–80.

Larson, R. B. 1990. Formation of star clusters. In *Physical Processes in Fragmentation and Star Formation*, eds. R. Capuzzo-Dolcetta, C. Chiosi, and A. DiFazio (Dordrecht: Kluwer), p. 389–399.

Larson, R. B. 1992. Towards understanding the stellar initial mass function. *Mon. Not. Roy. Astron. Soc.* 256:641–653.

Larson, R. B. 1995. Star formation in groups. *Mon. Not. Roy. Astron. Soc.* 272:213–222.

Layzer, D. 1963. On the fragmentation of self-gravitating gas clouds. *Astrophys. J.* 137:351–357.

Li, W., Evans, N. J. II, and Lada, E. A. 1997. Looking for distributed star formation in L1630: A near-infrared (J, H, K) survey. *Astrophys. J.* 488:277–282.

Lubow, S. H., and Pringle, J. E. 1993. The gravitational stability of a compressed slab of gas. *Mon. Not. Roy. Astron. Soc.* 263:701–709.

Luhman, K., and Rieke, G. H. 1998. The low-mass initial mass function in young clusters: L1495E. *Astrophys. J.* 497:354–367.

McCaughrean, M., and O'Dell, R. 1996. Direct imaging of circumstellar disks in the Orion Nebula. *Astron. J.* 111:1977–1992.

McDonald, J. M., and Clarke, C. J. 1993. Dynamical biasing in binary star formation: Implications for brown dwarfs in binaries. *Mon. Not. Roy. Astron. Soc.* 262:800–811.

McDonald, J. M., and Clarke, C. J. 1995. The effect of star-disc interactions on the binary mass-ratio distribution. *Mon. Not. Roy. Astron. Soc.* 275:671–683.

Meyer, M. R., Carpenter, J. M., Hillenbrand, L. A., and Strom, S. E. 1999. The embedded cluster associated with NGC 2024: Near-IR spectroscopy and emergent mass distribution. *Astron. J.*, submitted.

Meyer, M. R., and Lada, E. A. 1999. The stellar populations in the L1630 (Orion B) Cloud. In *The Orion Complex Revisited*, ASP Conf. Ser., ed. M. McCaughrean, in press.

Mottmann, J. 1977. The origin of the late heavy bombard of Moon, Mars and Mercury. *Icarus* 31:412–425.

Murray, S. D., and Lin, D. N. C. 1989. The fragmentation of proto-globular clusters. I. Thermal instabilities. *Astrophys. J.* 339:933–950.

Murray, S. D., and Lin, D. N. C. 1992. Globular cluster formation. The fossil record. *Astrophys. J.* 400:265–279.

Myers, P. 1998. Cluster-forming molecular cloud cores. *Astrophys. J.* 496: L109–L111.

Nakajima, Y., Tachihara, K., Hanawa, T., and Nakano, M. 1998. Clustering of pre-main-sequence stars in the Orion, Ophiuchus, Chamaeleon, Vela, and Lupus star-forming regions. *Astrophys. J.* 497:721–735.

Neuhauser, R. 1997. Low-mass pre-main sequence stars and their X-ray emission. *Science* 276:1363–1369.

Neuhauser, R., Wolk, S. J., Torres, G., et al. 1998. Optical and X-ray monitoring, Doppler imaging, and space motion of the young star Par 1724 in Orion. *Astron. Astrophys.* 334:873–890.

O'Connell, R.W., Gallagher, J. S., and Hunter, D. A. 1994. Hubble Space Telescope imaging of super-star clusters in NGC 1569 and NGC 1705. *Astrophys. J.* 433:65–80.

Okumura, S. K., Makino, J., Ebisuzaki, T., et al. 1993. Highly parallelized special-purpose computer, GRAPE-3. *Pub. Astron. Soc. Japan* 45: 329–342.

Palla F., and Zinnecker, H. 1987. Non-equilibrium cooling of a hot primordial gas cloud. In *Starbursts and Galaxy Evolution*, eds. T. X. Thuan, T. Montmerle, J. Tran Thanh Van (Gif sur Yvette: Editions Frontieres) pp. 533–546.

Patel, N. A., Goldsmith, P. F., Heyer, M. H., Snell, R. L., and Pratap, P. 1998. Origin and evolution of the Cepheus bubble. *Astrophys. J.* 507: 241–253.

Peebles, P. J. E. 1980. *The Large Scale Structure of the Universe* (Princeton: Princeton University Press), pp. 138–256.
Phelps, R. L., and Janes, K. 1994. Young open clusters as probes of the star formation process. 1: An atlas of open cluster photometry. *Astrophys. J. Suppl.* 90:31–49.
Phelps, R. L., and Lada, E. A. 1997. Spatial distribution of embedded clusters in the Rosette molecular cloud: Implications for cluster formation. *Astrophys. J.* 477:176–188.
Piche, F. 1993. A near-infrared survey of the star forming region NGC 2264. *Pub. Astron. Soc. Pacific* 105:324–341.
Pinto, F. 1987. Bound star clusters from gas clouds with low star formation efficiency. *Pub. Astron. Soc. Pacific* 99:1161–1174.
Preibisch, T., Zinnecker, H., and Herbig, G. H. 1996. ROSAT X-ray observations of the young cluster IC348. *Astron. Astrophys.* 310:456–470.
Pringle, J. E. 1989. On the formation of binary stars. *Mon. Not. Roy. Astron. Soc.* 239:631–640.
Raboud, D., and Mermilliod, J. C. 1998. Evolution of mass segregation in open clusters: Some observational evidences. *Astron. Astrophys.* 333: 897–912.
Rieke, G. H., Ashok, N. M., and Boyle, R. P. 1989. The initial mass function in the Rho Ophiuchi cluster. *Astrophys. J. Lett.* 339: L71–L74.
Simon, M. 1997. Clustering of young stars in Taurus, Ophiuchus, and the Orion Trapezium. *Astrophys. J. Lett.* 482: L81–L84.
Sterzik, M. F., and Durisen, R. H. 1995. Escape of T Tauri stars from young stellar systems. *Astron. Astrophys.* 304:L9–L12.
Strom, K. M., Strom, S. E., and Merrill, M. 1993. Infrared luminosity functions for the young stellar population associated with the L1641 molecular cloud. *Astrophys. J.* 412:233–247.
Strom, K. M., Kepner, J., and Strom, S. E. 1995. The evolutionary status of the stellar population in the ρ Ophiuchi cloud core. *Astrophys. J.* 438:813–828.
Sugitani, K., and Ogura, K. 1994. A catalog of bright-rimmed clouds with IRAS point sources: Candidates for star formation by radiation-driven implosion. 2: The southern hemisphere. *Astrophys. J. Suppl.* 92: 163–185.
Sugitani, K., Fukui, Y., and Ogura, K. 1991. A catalog of bright-rimmed clouds with IRAS point sources: Candidates for star formation by radiation-driven implosion. I—The northern hemisphere. *Astrophys. J. Suppl.* 77: 59–72.
Sugitani, K., Tamura, M., and Ogura, K. 1995. Young star clusters in bright-rimmed clouds: Small-scale sequential star formation? *Astrophys. J. Lett.* 455: L39–L42.
Tenorio-Tagle, G., Bodenheimer, P., Lin, D., and Noriega-Crespo, A. 1986. On star formation in stellar systems. I. Photoionization effects in protoglobular clusters. *Mon. Not. Roy. Astron. Soc.* 221:635–650.
Testi, L., Palla, F., Prusti, T., Natta, A., and Maltagliati, S. 1997. A search for clustering around Herbig Ae/Be stars. *Astron. Astrophys.* 320:159–169.
Testi, L., Palla, F., and Natta, A. 1998. A search for clustering around Herbig Ae/Be stars. II. Atlas of the observed sources. *Astron. Astrophys. Suppl.* 133:81–100.
Testi, L., Palla, F., and Natta, A. 1999. The onset of cluster formation around Herbig Ae/Be stars. *Astron. Astrophys.* 342:515–522.
van Albada, T. S. 1968. The evolution of small stellar systems and the implications for double star formation. *Bull. Astron. Inst. Neth.* 19:479–490.
Verschueren, W., and David, M. 1989. The effect of gas removal on the dynamical evolution of young stellar clusters. *Astron. Astrophys.* 219:105–119.

Walter, F. M., Wolk, S. J., and Sherry, W. 1998. The σ Orionis cluster. In *Cool Stars, Stellar Systems, and the Sun X*, ed. R. Donahue and J. Bookbinder (San Francisco: Astronomical Society of the Pacific), CD-1793.
White, G. J., Lefloch, B., Fridlund, C. V. M., et al. 1997. An observational study of cometary globules near the Rosette nebula. *Astron. Astrophys.* 323:931–949.
Whitmore, B. C., Zhang, Q., Leitherer, C., Fall, M., Schweizer, F., and Miller, B. W. 1998. The luminosity function of young star clusters in the "Antennae" galaxies (NGC 4038/4039). American Astronomical Society Meeting 194, No. 12.05.
Whitworth, A. P., and Clarke, C. J. 1997. Cooling behind mildly supersonic shocks in molecular clouds. *Mon. Not. Roy. Astron. Soc.* 291:578–589.
Whitworth, A. P. 1979. The erosion and dispersal of massive molecular clouds by young stars. *Mon. Not. Roy. Astron. Soc.* 186:59–71.
Whitworth, A. P., Bhattal, A. S., Chapman, S. J., Disney, M. J., and Turner, J. A. 1994. Fragmentation of shocked interstellar gas layers. *Astron. Astrophys.* 290:421–435.
Whitworth, A. P., Boffin, H. M. J., and Francis, N. 1998. Gas cooling by dust during dynamical fragmentation. *Mon. Not. Roy. Astron. Soc.* 299:554–560.
Wichmann, R., Krautter, J., Covino, E., et al. 1997. The T Tauri star population in the Lupus star forming region. *Astron. Astrophys.* 320:185–198.
Wielen, R. 1971. The age distribution and total lifetimes of galactic clusters. *Astron. Astrophys.* 13:309–321.
Wilking, B. A., McCaughrean, M. J., Burton, M. G., Giblin, T., Rayner, J. T., and Zinnecker, H. 1997. Deep infrared imaging of the R Coronae Australis cloud core. *Astron. J.* 114:2029–2042.
Yun, J. L., and Clemens, D. P. 1994. Near-infrared imaging survey of young stellar objects in Bok globules. *Astron. J.* 108:612–624.
Zinnecker, H. 1982. Prediction of the protostellar mass spectrum in the Orion near-infrared cluster. In *Symposium on the Orion Nebula to Honour Henry Draper*, ed. A. E. Glassgold et al. (New York: New York Academy of Sciences), pp. 226–240.
Zinnecker, H., McCaughrean, M. J., and Wilking, B. A. 1993. The initial stellar population. In *Protostars and Planets III*, ed. E. H. Levy and J. I. Lunine (Tucson: University of Arizona Press), pp. 429–441.

OBSERVATIONS AND THEORY OF STAR CLUSTER FORMATION

BRUCE G. ELMEGREEN
IBM Watson Research Center

YURI EFREMOV
University of Moscow, Sternberg Astronomical Institute

RALPH E. PUDRITZ
McMaster University

and

HANS ZINNECKER
Astrophysikalisches Institut Potsdam

Young stars form on a wide range of scales, producing aggregates and clusters with various degrees of gravitational self-binding. The loose aggregates have a hierarchical structure in both space and time that resembles interstellar turbulence, suggesting that these stars form in only a few turbulent crossing times with positions that map out the previous gas distribution. Dense clusters, on the other hand, are often well mixed, as if self-gravitational motion has erased the initial fine structure. Nevertheless, some of the youngest dense clusters also show subclumping, so it may be that all stellar clustering is related to turbulence. Some of the densest clusters may also be triggered. The evidence for mass segregation of the stars inside clusters is reviewed, along with various explanations for this effect. Other aspects of the theory of cluster formation are reviewed as well, including many specific proposals for cluster formation mechanisms. The conditions for the formation of bound clusters are discussed. Critical star formation efficiencies can be as low as 10% if the gas removal process is slow and if the stars are born at subvirial speeds. Environmental conditions, particularly pressure, may affect the fraction and masses of clusters that end up bound. Globular clusters may form like normal open clusters but in conditions that prevailed during the formation of the halo and bulge, or in interacting and starburst galaxies today. Various theories for the formation of globular clusters are summarized.

I. INTRODUCTION

The advent of large array cameras at visible to infrared wavelengths, and the growing capability to conduct deep surveys with semiautomated searching and analysis techniques, have led to a resurgence in the study of stellar clusters and groupings in the disk and halo of our Galaxy, in nearby galaxies, and in distant galaxies. The complementary aspect of the cluster formation problem, namely the structure of molecular clouds and complexes, is also being realized by submillimeter continuum mapping and comprehensive millimeter wave surveys. Here we review various theories about the origin of star clusters and the implications of young stellar clustering in general, and we discuss the requirements for gravitational self-binding in open and globular clusters.

Previous reviews of cluster formation were in Wilking and Lada (1985), Larson (1990), Lada (1993), E. Lada et al. (1993), and Zinnecker et al. (1993).

II. STRUCTURE OF YOUNG STELLAR GROUPS

Stellar groupings basically come in two types: bound and unbound. Some of the unbound groups could be loose aggregates of stars that formed in the dense cores of weakly bound or unbound cloud structures, such as the Taurus star-forming complex. Other unbound groups could be dispersed remnants of inefficient star formation in strongly self-gravitating clouds or cloud cores.

This section discusses loose aggregates of young stars first, considering hierarchical structure and a size-time correlation, and then more compact groups of stars: cluster formation in dense cloud cores, along with the associated process of stellar mass segregation, and the effects (or lack of effects) of high stellar densities on disks, binaries, and the stellar initial mass function.

A. Hierarchical Structure in Clouds and Stellar Groups

1. Gas Structure. Interstellar gas is structured into clouds and clumps on a wide range of scales, from substellar masses (10^{-4} $M_\odot$) that are hundredths of a parsec in size to giant cloud complexes (10^7 $M_\odot$) that are as large as the thickness of a galaxy. This structure is often characterized as consisting of discrete clumps or clouds with a mass spectrum that is approximately a power law, $n(M)dM \propto M^{-\alpha}dM$, with α in the range from $\sim$1.5 to $\sim$1.9 (for the smaller scales, see Heithausen et al. 1998 and Kramer et al. 1998, with a review in Blitz 1993; for the larger scales, see Solomon et al. 1987, Dickey and Garwood 1989, and the review in Elmegreen 1993).

Geometrical properties of the gas structure can also be measured from power spectra of emission line intensity maps (Stutzki et al. 1998). The power law nature of the result implies that the emission intensity is self-similar over a wide range of scales. Self-similar structure has also been found on cloud perimeters, where a fractal dimension of $\sim 1.3 \pm 0.3$ was determined (Beech 1987; Bazell and Désert 1988; Scalo 1990; Dickman et al. 1990; Falgarone et al. 1991; Zimmermann and Stutzki 1992, 1993; Vogelaar and Wakker 1994). This power law structure includes both gas that is self-gravitating and gas that is non-self-gravitating, so the origin is not purely gravitational fragmentation. The most likely source is some combination of turbulence (see review in Falgarone and Phillips 1991), agglomeration with fragmentation (Carlberg and Pudritz 1990; McLaughlin and Pudritz 1996), and self-gravity (de Vega et al. 1996).

Interstellar gas is not always fractal. Shells, filaments, and dense cores are not fractal in their overall shapes; they are regular and have characteristic scales. Generally, the structures that form as a result of specific high-pressure events, such as stellar or galactic pressures, have an overall shape that is defined by that event, whereas the structures that are formed as a result of turbulence are hierarchical and fractal.

2. *Stellar Clustering.* Stellar clustering occurs on a wide range of scales, like most gas structures, with a mass distribution function that is about the same as the clump mass distribution function in the gas. For open clusters, it is a power law with α in the range from $\sim$1.5 (van den Bergh and Lafontaine 1984; Elson and Fall 1985; Bhatt et al. 1991) to $\sim$2 (Battinelli et al. 1994; Elmegreen and Efremov 1997), and for OB associations, it is a power law with $\alpha \sim 1.7$–2, as determined from the luminosity distribution function of H II regions in galaxies (Kennicutt et al. 1989; Comeron and Torra 1996; Feinstein 1997; Oey and Clarke 1998).

It is important to recognize that the gas and stars are not just clumped, as in the old plum pudding model (Clark 1965); they are *hierarchically* clumped, meaning that small pieces of gas and small clusters are usually contained inside larger pieces of gas and larger clusters through a wide range of scales. Scalo (1985) reviewed these observations for the gas, and Elmegreen and Efremov (1998) reviewed the corresponding observations for the stars. Figure 1 shows a section of the Large Magellanic Cloud (LMC) southwest of 30 Doradus. Stellar clusterings are present on many scales, and many of the large clusters contain smaller clusters inside of them, down to the limit of resolution.

Stellar clumping on a kiloparsec scale was first studied by Efremov (1978), who identified "complexes" of Cepheid variables, supergiants, and open clusters. Such complexes trace out the recent $\sim$30–50 million years of star formation in the largest cloud structures, which are the H I "superclouds" (Elmegreen and Elmegreen 1983, 1987; Elmegreen 1995*b*) and "giant molecular associations" (GMAs) (Rand and Kulkarni 1990) that are commonly found in spiral arms. The sizes of star complexes and the sizes of superclouds or GMAs are about the same in each galaxy, increasing regularly with galaxy size from $\sim$300 pc or less in small galaxies like the LMC to $\sim$600 pc or more in large galaxies like the Milky Way (Elmegreen et al. 1996). Each star complex typically contains several OB associations in addition to the Cepheids, because star formation continues in each supercloud, first in one center and then in another, for the whole period of

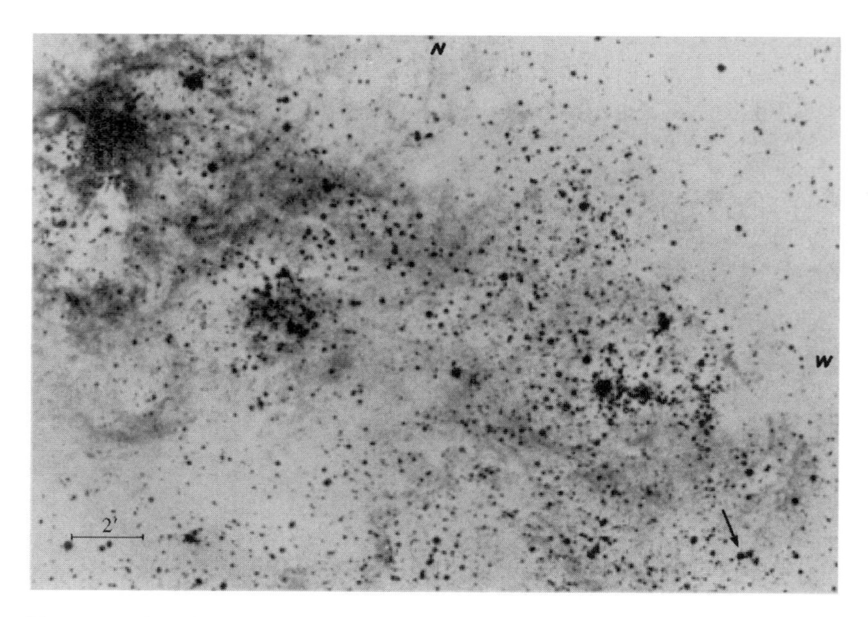

Figure 1. Star field southwest of 30 Dor in the Large Magellanic Cloud, showing hierarchical structure in the stellar groupings. Image from Efremov (1989).

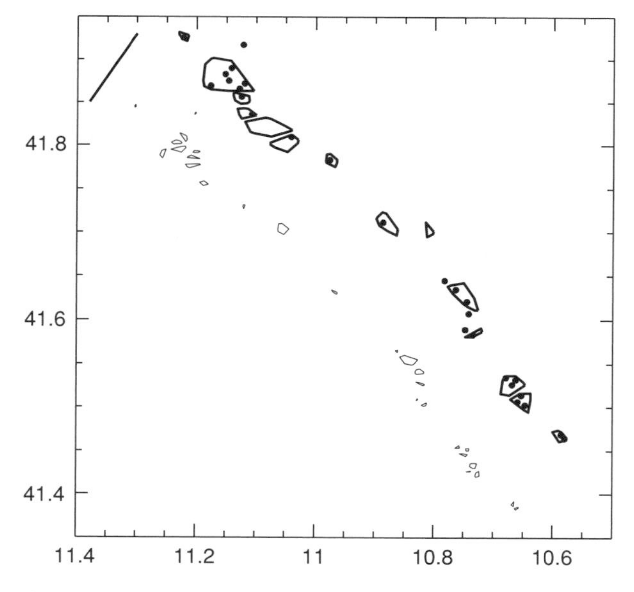

Figure 2. OB associations and star complexes along the western spiral arm of M31. The black dots show the positions of the OB associations inside the heavy outlines of the star complexes. The faint outlines to the lower left of this show the actual OB associations in relation to each other, shifted to the southeast by the length of the diagonal line for clarity (from Battinelli et al. 1996).

~30–50 Myr. This overall timescale is smaller in smaller galaxies because the complexes are smaller, as Battinelli and Efremov (1999) confirmed for the LMC.

Star complexes are well studied in the Milky Way and local galaxies (see reviews in Efremov 1989, 1995). In our Galaxy, they were examined most recently by Berdnikov and Efremov (1989, 1993) and by Efremov and Sitnik (1988). The latter showed that 90% of the young (~10 Myr) clusters and associations in the Milky Way are united into the same star complexes that are traced by Cepheids and supergiants. A similarly high fraction of hierarchical clustering has been found in M31 (Efremov et al. 1987; Battinelli 1992; Magnier et al. 1993; Battinelli et al. 1996), M81 (Ivanov 1992), the LMC (Feitzinger and Braunsfurth 1984), and many other galaxies (Feitzinger and Galinski 1987; Battinelli 1991; Elmegreen and Salzer 1999).

A map showing two levels in the hierarchy of stellar structures along the southwest portion of the western spiral arm in M31 is shown in Fig. 2 (from Battinelli et al. 1996). The smaller groups, which are Orion-type OB associations, are shown with faint outlines displaced 0.1° to the south of the larger groups for clarity; the smaller groups are also shown as dots with their proper positions inside the larger groups. Evidently, most of the OB associations are within the larger groups.

The oldest star complexes are sheared by differential galactic rotation and appear as flocculent spiral arms if there are no strong density waves (Elmegreen and Efremov 1996). When there are density waves, the complexes form in the arm crests, somewhat equally spaced, as a result of gravitational instabilities (Elmegreen and Elmegreen 1983; Elmegreen 1994; Rand 1995; Efremov 1998).

3. Hierarchical Clustering of Stars on Smaller Scales. Hierarchical clustering of stars continues from star complexes to OB associations, down to very small scales. Infrared observations reveal embedded clusters of various sizes and densities in star-forming regions. Many of these, as discussed in the next section, are extremely dense and deeply embedded in self-gravitating molecular cores. Others are more open and clumpy, as if they were simply following the hierarchical gas distribution around them. A good example of the latter is in the Lynds 1641 cloud in the Orion association, which has several aggregates that each comprise 10–50 young stars, plus a dispersed population throughout the cloud (Strom et al. 1993; Hodapp and Deane 1993; Chen and Tokunaga 1994; Allen 1995).

The distribution of young, X-ray active stars around the sky (Guillout et al. 1998) is also irregular and clumpy on a range of scales. The low-mass young stars seen in these X-ray surveys are no longer confined to dense cores. Sterzik et al. (1995), Feigelson (1996), Covino et al. (1997), Neuhäuser (1997), and Frink et al. (1997, 1998) found that the low-mass membership in small star-forming regions extends far beyond the previously accepted boundaries. This result is consistent with the hierarchical clustering model.

4. Two Examples of Hierarchical Stellar Structure: Orion and W3. There are many observations of individual clusters that are part of a hierarchy on larger scales. The Orion region overall contains at least five levels of hierarchical structure. On the largest scale (first level), there is the so-called local arm or Orion-Cygnus spur, which has only young stars (Efremov 1997) and is therefore a sheared star formation feature, not a spiral density wave (Elmegreen and Efremov 1996); in contrast, the Sagittarius-Carina arm also has old stars (Efremov 1997). The largest local condensation (second level) in the Orion-Cygnus spur is Gould's Belt, of which Orion OB1 is one of several similar condensations (third level) (Pöppel 1997). Inside Orion OB1 are four subgroups (fourth level) (Blaauw 1964). The youngest of them, including the Trapezium cluster, contains substructure as well (fifth level): one region is the BN/KL region, perhaps triggered by θ^1C, and another is near OMC-1S (Zinnecker et al. 1993). The main Trapezium cluster may have no substructure, though (Bate et al. 1998).

A similar hierarchy with five levels surrounds W3. On the largest scale is the Perseus spiral arm (first level), which contains several giant star formation regions separated by 1–2 kpc; the W3 complex is in one of them, and the NGC 7538 complex is in another. The kiloparsec-scale condensation surrounding W3 (second level) contains associations Cassiopeia OB6 and Perseus OB1 (which is below the galactic plane and includes the double cluster h and χ Per), and these two associations form a stellar complex. The association Cas OB8, which includes a compact group of five clusters (Efremov 1989, Fig. 16 and Table 7 on p. 77) may also be a member of this complex, as suggested by the distances and radial velocities. Cas OB6 is the third level for W3. Cas OB6 consists of the two main star-forming regions W4 (fourth level) and W5, and W4 has three condensations at the edge of the expanded H II region, in the associated molecular cloud (Lada et al. 1978). W3 is one of these three condensations and therefore represents the fifth level in the hierarchy. The hierarchy may continue further too, because W3 contains two apparently separate sites of star formation, W3A and W3B (Wynn-Williams et al. 1972; Normandeau et al. 1997).

Most young, embedded clusters resemble Orion and W3 in this respect. They have some level of current star formation activity, with an age possibly less than 10^5 years, and are also part of an older OB association or other extended star formation up to galactic scales, with other clusters forming in the dense parts here and there for a relatively long time.

5. Cluster Pairs and Other Small-Scale Structure. Another way this hierarchy appears is in cluster pairs. Many clusters in both the LMC (Bhatia and Hatzidimitriou 1988; Kontizas et al. 1989; Dieball and Grebel 1998; Vallenari et al. 1998) and Small Magellanic Cloud (SMC) (Hatzidimitriou and Bhatia 1990) occur in distinct pairs with about the same age. Most of these binary clusters are inside larger groups of clusters and stellar complexes. However, the clumps of clusters and the clumps of Cepheids in the LMC do not usually coincide (Efremov, 1989, p. 205; Battinelli and Efremov, 1999).

Some embedded clusters also have more structure inside of them. For example, star formation in the cloud G35.20–1.74 has occurred in several different and independent episodes (Persi et al. 1997), and there is also evidence for noncoeval star formation in NGC 3603, the most massive visible young cluster in the Galaxy (Eisenhauer et al. 1998). W33 contains three separate centers or subclusters of star formation inside of it that have not yet merged into a single cluster (Beck et al. 1998). The same is true in 30 Dor and the associated cluster NGC 2070 (Seleznev 1995), which appears to have triggered a second generation of star formation in the adjacent molecular clouds (Hyland et al. 1992; Walborn and Blades 1997; Rubio et al. 1998; Walborn et al. 1999). Similarly, NGC 3603 has substructure with an age difference of $\sim$10 Myr, presumably from triggering too (Brandner et al. 1997). Lada and Lada (1995) found eight small subclusters with 10 to 20 stars each in the outer parts of IC 348. Piche (1993) found two levels of hierarchical structure in NGC 2264: two main clusters with two subclusters in one and three in the other. The old stellar cluster M67 still apparently contains clumpy outer structure (Chupina and Vereshchagin 1998). Some subclusters can even have slightly different ages: Strobel (1992) found age substructure in 14 young clusters, and Elson (1991) found spatial substructure in 18 rich clusters in the LMC.

Evidence that subclustering did not occur in dense globular clusters was recently given by Goodwin (1998), who noted from numerical simulations that initial substructure in globular clusters would not be completely erased during the short lifetimes of some of the youngest in the LMC. Because these young populous clusters appear very smooth, their initial conditions had to be somewhat smooth and spherical too.

The similarity between the loose clustering properties of many young stellar regions and the clumpy structure of weakly self-gravitating gas appears to be the result of star formation following the gas in hierarchical clouds that are organized by supersonic turbulence. Turbulence also implies motion and, therefore, a size-dependent crossing time for the gas. We shall see in the next section that this size-dependent timescale might also apply to the duration of star formation in a region.

B. Star Formation Timescales

The duration of star formation tends to vary with the size S of the region as something like the crossing time for turbulent motions; that is, it increases about as $S^{0.5}$. This means that star formation in larger structures takes longer than star formation in subregions. A schematic diagram of this time-size pattern is shown in Fig. 3. The largest scale is taken to be that of a flocculent spiral arm, which is typically $\sim$100 Myr old, as determined from the pitch angle (Efremov and Elmegreen 1998).

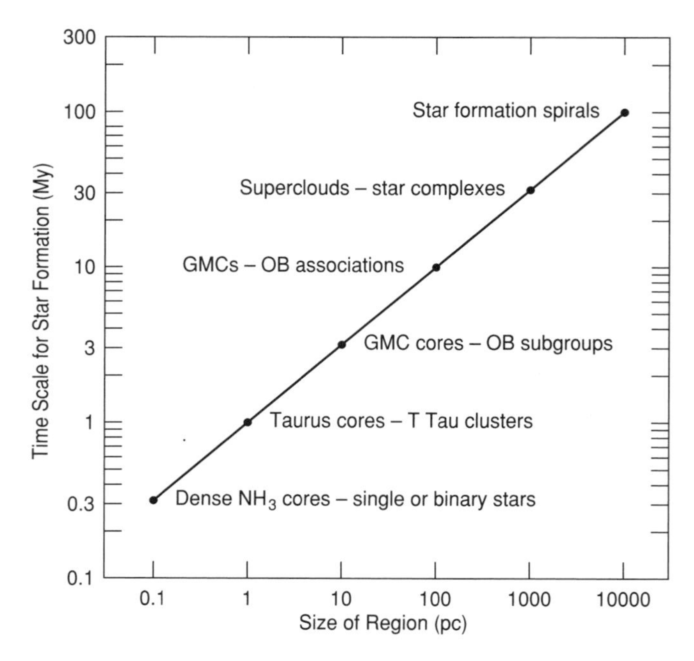

Figure 3. Schematic diagram of the relationship between the duration of star formation and the region size (from Efremov and Elmegreen 1998). Larger regions of star formation form stars for a longer total time.

This relationship between the duration of star formation and the region size implies that clusters that form together in small regions will usually have about the same age, within perhaps a factor of three of the turbulent crossing time of the small scale, whereas clusters that form together in larger regions will have a wider range of ages, proportional to the crossing time on the larger scale. Figure 4 shows this relationship for 590 clusters in the LMC (Efremov and Elmegreen 1998). Plotted on the ordinate is the average difference in age between all pairs of clusters whose deprojected spatial separations equal the values on the abscissa. The average age difference between clusters increases with their spatial separation. In the figure, the correlation ranges between 0.02° and 1° in the LMC, which corresponds to a spatial scale of 15 to 780 pc. The correlation disappears above 1°, perhaps because the largest scale for star formation equals the Jeans length or the disk thickness. A similar duration-size relation is also observed within the clumps of clusters in the LMC. Larger clumps of clusters have larger age dispersions (Battinelli and Efremov 1999).

The correlations between cluster birth times and spatial scale are reminiscent of the correlation between internal crossing time and size in molecular clouds. The crossing time in a molecular cloud or cloud clump is about the ratio of the radius (half width at half maximum size) to the Gaussian velocity dispersion. The data for several molecular cloud surveys are

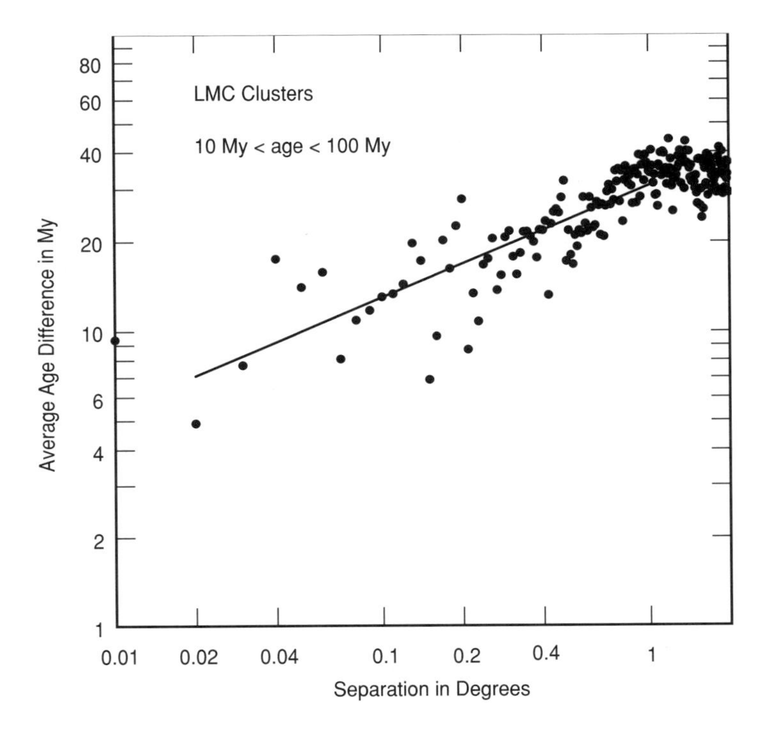

Figure 4. The average age differences between clusters in the LMC are plotted vs. their deprojected angular separations, for clusters in the age interval from 10 to 100 Myr (from Efremov and Elmegreen 1998). Clusters that are close together in space have similar ages. The line is a fit to the data given by $\log \Delta t\,(\mathrm{yr}) = 7.49 + 0.38 \log S\,(\mathrm{deg})$.

shown in Fig. 5, with different symbols for each survey. On the top is a plot of the Gaussian linewidth versus size, and on the bottom is the crossing time versus size. Smaller clouds and clumps have smaller crossing times, approximately in proportion to size $S^{0.5}$. Overlaid on this plot, as large crosses, are the age difference versus separation points for LMC clusters from Fig. 4. Evidently, the cluster correlation fits in nicely at the top part of the molecular cloud crossing time-size relation.

These correlations underscore our perception that both cloud structure and at least some stellar clusterings come from interstellar gas turbulence. The cluster age differences also suggest that star formation is mostly finished in a cloud within only ~2 to 3 turbulent crossing times, which is very fast. In fact, this time is much faster than the magnetic diffusion time through the bulk of the cloud, which is ~10 crossing times in a uniform medium with cosmic ray ionization (Shu et al. 1987) and even longer if UV light can get in (Myers and Khersonsky 1995) and if the clouds are clumpy (Elmegreen and Combes 1992). Thus, magnetic diffusion does not regulate the formation of stellar groups. It may regulate only the formation

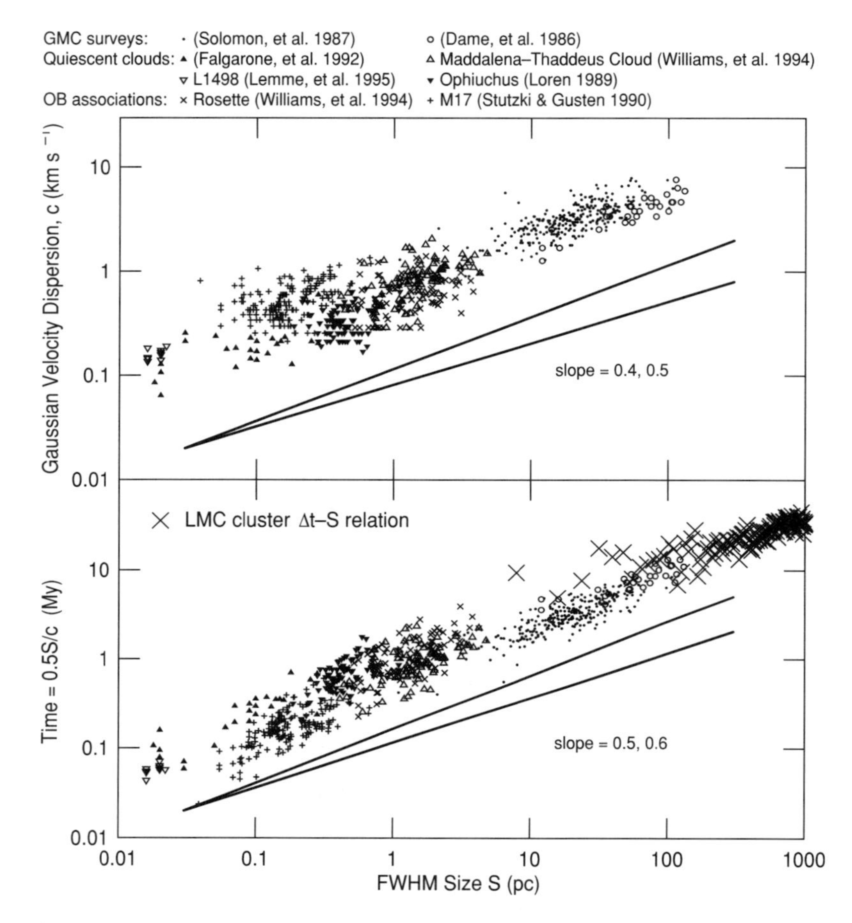

Figure 5. The molecular cloud size-linewidth relation for the Milky Way is shown at the top, considering many different surveys, as indicated by the symbol types, and the ratio of half of the size to the Gaussian linewidth is shown at the bottom. This latter ratio is the crossing time in the cloud; it scales about as the square root of the cloud size. Superposed on this crossing time-size relation is the age difference vs. size relation shown in Fig. 4, using clusters from the LMC. If the size-linewidth relations for the two galaxies are comparable to within a factor of 2, this diagram suggests that the duration of star formation in a region is approximately equal to the turbulent crossing time, at least on the large scales considered here (from Efremov and Elmegreen 1998).

of individual stars, which occurs on much smaller scales (Shu et al. 1987; but see Nakano 1998).

Star formation in a cluster may begin when the turbulent energy of the cloud dissipates. This is apparently a rapid process, as indicated by recent numerical simulations of supersonic magnetohydrodynamic (MHD) turbulence, which show a dissipation time of only 1–2 internal crossing times (Mac Low et al. 1998; Stone et al. 1998). Most giant molecular clouds have similar turbulent and magnetic energies (Myers and Goodman 1988), and they would be unstable without the turbulent energy (McKee et al. 1993), so the rapid dissipation of turbulence should lead to a similarly rapid onset of star formation (e.g., McLaughlin and Pudritz 1996). The turbulence has to be replenished for the cloud to last more than several crossing times.

The observed age-size correlation is significantly different from what one might expect from simple crossing time arguments in the absence of turbulence. If the velocity dispersion were independent of scale, as for an isothermal fluid without correlated turbulent motions, the slope of the age-size correlation would be 1.0, not $\sim$0.35. The correlation is also not due to stochastic self-propagating star formation, which would imply a diffusion process for the size of a star formation patch and give a spatial scale that increased as the square root of time. In that case the slope on Fig. 4 would be 2.

The duration-size relation for stellar groupings implies that OB associations and 10^5 $M_\odot$ GMCs are not physically significant scales for star formation—only regions that are large enough to have statistically sampled the high-mass end of the IMF and young enough to have these OB stars still present. Regions with such an age tend to have a certain size, $\sim$100 pc, from the size-time relation, but the cloud and star formation processes need not be physically distinct.

The time scale vs. size correlations for star formation should not have the same coefficients in front of the power laws for all regions of all galaxies. This coefficient should scale with the total turbulent ISM pressure to the inverse $\frac{1}{4}$ power, based on the relations $P \sim GM^2/R^4$ and $\Delta v^2 \sim GM/(5R)$ for self-gravitating gas (Chièze 1987; Elmegreen 1989). Thus, regions with pressures higher than the local value by a factor of 10^2–10^4 should have durations of star formation shorter than the local regions by a factor of 3–10, for the same spatial scale. This result corresponds to the observation for starburst galaxies that the formation time of very dense clusters, containing the mass equivalent of a whole OB association, is extraordinarily fast, on the order of 1–3 Myr, whereas in our Galaxy it takes $\sim$10 Myr to form an aggregate of this mass. Similarly, high-pressure cores in GMCs (section II.C) should form stars faster than low-pressure regions with a similar mass or size.

There are many observations of the duration of star formation in various regions, both active and inactive. In the Orion Trapezium cluster, the age spread for 80% of the stars is very short, less than 1 Myr (Prosser et al. 1994), as it is in L1641 (Hodapp and Deane 1993). It might be even shorter for a large number (but not necessarily a large fraction) of stars in NGC 1333, because of the large number of jets and Herbig-Haro objects that are present today (Bally et al. 1996). In NGC 6531 as well, the age spread is immeasurably small (Forbes 1996). Other clusters have larger age spreads. Hillenbrand et al. (1993) found that, whereas the most massive stars (80 $M_\odot$) in NGC 6611 (=M16) have a 1-Myr age spread around a mean age of $\sim$2 Myr, there are also pre-main-sequence stars and a star of 30 $M_\odot$ with an age of 6 Myr. The cluster NGC 1850 in the LMC has an age spread of 2–10 Myr (Caloi and Cassatella 1998), and in NGC 2004 there are evolved low-mass stars in the midst of less evolved high-mass stars (Caloi and Cassatella 1995). In NGC 4755 the age spread is 6–7 Myr, based on the simultaneous presence of both high- and low-mass star formation (Sagar and Cannon 1995). One of the best-studied clusters for an age spread is the Pleiades, where features in the luminosity function (Belikov et al. 1998) and synthetic HR diagrams (Siess et al. 1997) suggest continuous star formation over ~30 Myr when it formed (~100 Myr ago). This is much longer than the other age spreads for small clusters; it may have another explanation, including the possibility that the Pleiades primordial cloud captured some stars from a neighboring, slightly older, star-forming region (e.g., Bhatt 1989). Recall that the age spreads are much larger than several Myr for whole OB associations and star complexes, as discussed above.

C. Clusters in Dense Molecular Cores

1. Cluster Densities. Infrared, X-ray, and radio continuum maps reveal dense clusters of young stars in many nearby GMC cores. Reviews of embedded infrared clusters, including three-color *JHK* images, have been published by E. Lada et al. (1993) and Zinnecker et al. (1993).

Most observations of embedded young clusters have been made with *JHK* imagery. A list of some of the regions studied is in Table I. These clusters typically have radii of ~0.1 pc to several tenths of a parsec, and contain several hundred cataloged stars, making the stellar densities on the order of several times 10^3 pc^{-3} or larger. For example, in the Trapezium cluster the stellar density is ~5000 stars pc^{-3} (Prosser et al. 1994) or higher (McCaughrean and Stauffer 1994), and in Mon R2 it is ~9000 stars pc^{-3} (Carpenter et al. 1997). Perhaps the more distant clusters in this list are slightly larger, as a result of selection effects.

Some clusters, such as W3, NGC 6334, Mon R2, M17, CMa OB1, S106, and the maser clusters, contain massive, even O-type stars in the pre-UCH II phase or with H II regions. Others, such as ρ Ophiuchi, contain primarily low-mass stars. Although the mass functions vary a little from region to region, there is no reason to think at this time that the spatially averaged IMFs in these clusters are significantly different from the Salpeter (1955), Scalo (1986), or Kroupa et al. (1993) functions. Thus, the clusters with high-mass stars also tend to have low-mass stars (Zinnecker et al. 1993), although not all of the low-mass stars are seen yet, and clusters with primarily low-mass stars are not populous enough to contain a relatively rare massive star (see review of the IMF in Elmegreen 1998*a*).

Embedded X-ray clusters have been found in NGC 2024 (Freyberg and Schmitt 1995), IC 348 (Preibisch et al. 1996), IC 1396 (Schulz et al. 1997), and the Mon R2 and Rosette molecular clouds (Gregorio-Hetem et

TABLE I
Embedded Young Clusters

Region	Reference
ρ Ophiuchi	Wilking and Lada 1983; Wilking et al. 1985, 1989; Greene and Young 1992; Comeron et al. 1993; Barsony et al. 1997
R Coronae Austrinae	Wilking et al. 1985; Wilking et al. 1997
Serpens	White et al. 1995; Hurt and Barsony 1996; Giovannetti et al. 1998
M17	C. Lada et al. 1991; Hanson et al. 1997; Chini and Wargau 1998
L1630	E. Lada et al. 1991; Lada 1992; Li et al. 1997
Trapezium OMC2	Ali and DePoy 1995
Monoceros R2	Carpenter et al. 1997
Rosette	Phelps and Lada 1997
NGC 281	Henning et al. 1994; Megeath and Wilson 1997
NGC 1333	Aspin et al. 1994; Lada et al. 1996
NGC 2264	C. Lada et al. 1993; Piche 1993
NGC 2282	Horner et al. 1997
NGC 3576	Persi et al. 1994
NGC 6334	Tapia et al. 1996
IC 348	Lada and Lada 1995
W3 IRS5	Megeath et al. 1996
S106	Hodapp and Rayner 1991
S255	Howard et al. 1997
S269	Eiroa and Casali 1995
BD 40° 4124	Hillenbrand et al. 1995
LkHα101	Aspin and Barsony 1994
G35.20-1.74	Persi et al. 1997
H_2O and OH maser sources	Testi et al. 1994
19 IRAS sources	Carpenter et al. 1993

al. 1998). These show X-ray point sources that are probably T Tauri stars, some of which are seen optically. The presence of strong X-rays in dense regions of star formation increases the ionization fraction over previous estimates that were based only on cosmic ray fluxes. At higher ionization fractions, magnetic diffusion takes longer, and this may slow the star formation process. For this reason, Casanova et al. (1995) and Preibisch et al. (1996) suggested that X-rays from T Tauri stars lead to self-regulation of the star formation rate in dense clusters. On the other hand, Nakano (1998) suggests that star formation occurs quickly, by direct collapse, without any delay from magnetic diffusion. X-rays can also affect the final accretion phase from the disk. The X-ray irradiation of protostellar disks can lead to better coupling between the gas and the magnetic fields and more efficient angular momentum losses through hydromagnetic winds (cf. the chapter by Königl and Pudritz, this volume). Such a process might increase the efficiency of star formation. The full implications of X-ray radiation in the cluster environment are not understood yet.

A stellar density of 10^3 $M_\odot$ pc^{-3} corresponds to an H_2 density of $\sim 10^4$ cm^{-3}. Molecular cores with densities of 10^5 cm^{-3} or higher (e.g., Lada 1992) can easily make clusters this dense. Measured star formation efficiencies are typically 10–40% (e.g., Greene and Young 1992; Megeath et al. 1996; Tapia et al. 1996). Gas densities of $\sim 10^5$ cm^{-3} also imply extinctions of $A_V \sim 40$ mag on scales of ~ 0.2 pc, which are commonly seen in these regions, and they imply masses of ~ 200 $M_\odot$ and virial velocities of ~ 1 km s^{-1}, which is the typical order of magnitude of the gas velocity dispersion of cold star-forming clouds in the solar neighborhood. There should be larger and smaller dense clusters too, of course, rather than a *characteristic* cluster size that is simply the average value seen locally, because unbiased surveys, as in the LMC (Bica et al. 1996), show a wide range of cluster masses with power law mass functions, i.e., no characteristic scale (cf. section II.A).

2. *Cluster Effects on Binary Stars and Disks.* The protostellar binary fraction is lower in the Trapezium cluster than the Taurus-Auriga region by a factor of ~ 3 (Petr et al. 1998), and lower in the Pleiades cluster than in Taurus-Auriga as well (Bouvier et al. 1997). Yet the binary frequencies in the Trapezium and Pleiades clusters are comparable to that in the field (Prosser et al. 1994). This observation suggests that most stars form in dense clusters and that these clusters reduce an initially high binary fraction at starbirth (e.g., Kroupa 1995*a*; Bouvier et al. 1997).

The cluster environment should indeed affect binaries. The density of $n_{star} = 10^3$ stars pc^{-3} in a cloud core of size $R_{core} \sim 0.2$ pc implies that objects with this density will collide with each other in one crossing time if their cross section is $\sigma \sim (n_{star} R_{core})^{-1} \sim 0.005$ pc^2, which corresponds to a physical size of 10^3–$10^4 \left(R_{core}(\mathrm{pc}) n_{star}/10^3\right)^{-1/2}$ AU. This is the scale for long-period binary stars.

Another indication that a cluster environment affects binary stars is that the peak in the separation distribution for binaries is smaller (90 AU) in the part of the Scorpius-Centaurus association that contains early-type stars than it is (215 AU) in the part of the Scorpius-Centaurus association that contains no early-type stars (Brandner and Köhler 1998). This observation suggests that dissipative interactions leading to tighter binaries, or perhaps interactions leading to the destruction of loose binaries, are more important where massive stars form.

Computer simulations of protostellar interactions in dense cluster environments reproduce some of these observations. Kroupa (1995*a*) got the observed period and mass-ratio distributions for field binaries by following the interactions between 200 binaries in a cluster with an initial radius of 0.8 pc. Kroupa (1995*b*) also got the observed correlations between eccentricity, mass ratio, and period for field binaries using the same initial conditions. Kroupa (1995*c*) predicted further that interactions will cause stars to be ejected from clusters, and the binary fraction among these ejected stars will be lower than in the remaining cluster stars (see also Kroupa 1998). These simulations assume that all stars begin as binary members and interactions destroy some of these binaries over time.

Another point of view is that the protostars begin as single objects and capture each other to form binaries. In this scenario, McDonald and Clarke (1995) found that disks around stars aid with the capture process, and they reproduced the field binary fraction in model clusters with 4–10 stars (see review by Clarke 1996). According to this simulation, the cluster environment should affect disks too. There are indeed observations of this nature. Mundy et al. (1995) suggested that massive disks are relatively rare in the Trapezium cluster, and Nürnberger et al. (1997) found that protostellar disk mass decreases with stellar age in the Lupus young cluster, but not in the Taurus-Auriga region, which is less dense. When massive stars are present, as in the Trapezium cluster, UV radiation can photoionize the neighboring disks, and this is a type of interaction as well (Johnstone et al. 1998).

3. Cluster Effects on the IMF? The best examples of cluster environmental effects on star formation have been limited, so far, to binaries and disks. Nevertheless, there are similar suggestions that the cluster environment can affect the stellar mass as well, and, in doing so, affect the initial stellar mass function (e.g., Zinnecker 1986). For example, computer simulations have long been able to reproduce the IMF using clump (Silk and Takahashi 1979; Murray and Lin 1996) or protostellar (Price and Podsiadlowski 1995; Bonnell et al. 1997) interaction models of various types.

There is no direct evidence for IMF variations with cluster density, however (e.g., Massey and Hunter 1998; Luhman and Rieke 1998). Even in extremely dense globular clusters, the IMF seems normal at low mass (Cool 1998). This may not be surprising, because protostellar condensations are very small compared to the interstar separations, even in globular clusters (Aarseth et al. 1988), but the suggestion that massive stars are made by coalescence of smaller protostellar clumps continues to surface (see Zinnecker et al. 1993, and the chapter by Stahler et al., this volume).

Another indication that cluster interactions do not affect the stellar mass comes from the observation by Bouvier et al. (1997) that the rotation rates of stars in the Pleiades cluster are independent of the presence of a binary companion. These authors suggest that the rotation rate is the result of accretion from a disk, so the observation implies that disk accretion is not significantly affected by companions. Presumably this accretion would be even less affected by other cluster members, which are more distant than the binary companions. Along these lines, Heller (1995) found in computer simulations that interactions do not destroy protostellar disks, although they may remove half of their mass.

There is a way that could have gone unnoticed in which the cluster environment may affect the IMF. This is in the reduction of the thermal Jeans mass at the high pressure of a cluster-forming core. A lower Jeans mass might shift the turnover mass in the IMF to a lower value in dense clusters than in loose groups (Elmegreen 1997, 1999).

D. Mass Segregation in Clusters

One of the more perplexing observations of dense star clusters is the generally centralized location of the most massive stars. This has been observed for a long time and is usually obvious to the eye. For young clusters, it cannot be the result of "thermalization," because the timescale for that process is longer than the age of the cluster (e.g., Bonnell and Davies 1998). Thus it is an indication of some peculiar feature of starbirth.

The observation has been quantified using color gradients in 12 clusters (Sagar and Bhatt 1989), and by the steepening of the IMF with radius in several clusters (Pandey et al. 1992), including Tr 14 (Vazquez et al. 1996), the Trapezium in Orion (Jones and Walker 1988; Hillenbrand 1997; Hillenbrand and Hartmann 1998), and, in the LMC, NGC 2157 (Fischer et al. 1998), SL 666, and NGC 2098 (Kontizas et al. 1998). On the other hand, Carpenter et al. (1997) found no evidence from the IMF for mass segregation in Mon R2 at $M < 2\ M_{\odot}$, but noted that the most massive star ($10\ M_{\odot}$) is near the center nevertheless. Raboud and Mermilliod (1998) found a segregation of the binary stars and single stars in the Pleiades, with the binaries closer to the center, presumably because of their greater mass. A related observation is that intermediate-mass stars always seem to have clusters of low-mass stars around them (Testi et al. 1998), as if they needed these low-mass stars to form by coalescence, as suggested by these authors.

There are many possible explanations for these effects. The stars near the center could accrete gas at a higher rate and end up more massive (Larson 1978, 1982; Zinnecker 1982; Bonnell et al. 1997); they (or their predecessor clumps) could coalesce more (Larson 1990, Zinnecker et al. 1993, Bonnell et al. 1998, and the chapter by Stahler et al., this volume), or the most massive stars and clumps forming anywhere could migrate to the center faster because of a greater gas drag (Larson 1990, 1991; Gorti and Bhatt 1995, 1996; Saiyadpour et al. 1997). A central location for the most massive pieces is also expected in a hierarchical cloud (Elmegreen 1999). The centralized location of binaries could be the result of something different: the preferential ejection of single stars that have interacted with other cluster stars (Kroupa 1995*c*). The presence of low-mass stars around high-mass stars could have a different explanation too: High-mass stars are rare, so low-mass stars are likely to form before a high-mass star appears, whatever the origin of the IMF.

III. CLUSTER FORMATION MODELS

A. Bound Clusters as Examples of Triggered Star Formation?

Section II.A considered loose stellar groupings as a possible reflection of hierarchical cloud structure, possibly derived from turbulent motions, and it considered dense cluster formation in cloud cores separately, as if this process were different. In fact, the two types of clusters and the processes that lead to them could be related. Even the bound clusters, which presumably formed in dense cloud cores, have a power law mass distribution, and it is very much like the power law for the associations that make H II regions, so perhaps both loose and dense clusters get their mass from cloud hierarchical structure. The difference might be simply that dense clusters form in cloud pieces that get compressed by an external agent.

There are many young clusters embedded in cores at the compressed interfaces between molecular clouds and expanded H II regions, including many of those listed in Table I here, as reviewed in Elmegreen (1998*b*). For example, Megeath and Wilson (1997) recently proposed that the embedded cluster in NGC 281 was triggered by the H II region from an adjacent, older, Trapezium-like cluster, and Sugitani et al. (1995) found embedded clusters inside bright-rimmed clouds. Compressive triggering of a cluster can also occur at the interface between colliding clouds, as shown by Usami et al. (1995). A case in point is the S255 embedded cluster (Zinnecker et al. 1993; Howard et al. 1997; Whitworth and Clarke 1997).

Outside compression aids the formation of clusters in several ways. It brings the gas together, so that the stars end up in a dense cluster, and it also speeds up the star formation processes by increasing the density. These processes can be independent of the compression and may be the same as in other dense regions that were not rapidly compressed; the only point is that they operate more rapidly in compressed gas than in lower-density gas. The external pressure may also prevent or delay cloud disruption by newborn stars, allowing a large fraction of the gas to be converted into stars and thereby improving the chances that the cluster will end up self-bound (cf. section IV; Elmegreen and Efremov 1997; Lefloch et al. 1997).

Cloud cores should also be able to achieve high densities on their own, without direct compression. This might take longer, but the usual processes of energy dissipation and gravitational contraction can lead to the same overall core structure as the high pressure from an external H II region. Heyer et al. (1997) discussed the morphology of dense molecular cores and cluster formation in the outer Galaxy, showing that new star clusters tend to form primarily in the self-gravitating, high-pressure knots that occur here and there amid the more loosely connected network of lower-pressure material. Many of these knots presumably reached their high densities spontaneously.

B. Spontaneous Models and Large-Scale Triggering

The most recent development in cluster formation models is the direct computer simulation of interacting protostars and clumps, leading to clump and stellar mass spectra (Klessen et al. 1998). Earlier versions of this type of problem covered protostellar envelope stripping by clump collisions (Price and Podsiadlowski 1995), general stirring and cloud support by moving protostars with their winds (Tenorio-Tagle et al. 1993), and gas removal from protoclusters (Theuns 1990).

The core collapse problem was also considered by Boss (1996), who simulated the collapse of an oblate cloud, forming a cluster with $\sim$10 stars. A detailed model of thermal instabilities in a cloud core, followed by a collapse of the dense fragments into the core center and their subsequent coalescence, was given by Murray and Lin (1996). Patel and Pudritz (1994) considered core instability with stars and gas treated as separate fluids, showing that the colder stellar fluid destabilized the gaseous fluid.

Myers (1998) considered magnetic processes in dense cores and showed that stellar-mass kernels could exist at about the right spacing for stars in a cluster and not be severely disrupted by magnetic waves. Whitworth et al. (1998) discussed a similar characteristic core size at the threshold between strong gravitational heating and grain cooling on smaller scales, and turbulence heating and molecular line cooling on larger scales.

Some cluster formation models proposed that molecular clouds are made when high-velocity clouds impact the Galactic disk (Tenorio-Tagle 1981). Edvardsson et al. (1995) based this result on abundance anomalies in the ζ Sculptoris cluster. Lepine and Duvert (1994) considered the collision model because of the distribution of gas and star formation in local clusters and OB associations, whereas Phelps (1993) referred to the spatial distribution, ages, velocities, and proper motions of 23 clusters in the Perseus arm. Comeron et al. (1992) considered the same origin for stars in Gould's Belt based on local stellar kinematics. For other studies of Gould's Belt kinematics, see Lindblad et al. (1997) and De Zeeuw et al. (1999).

Other origins for stellar clustering on a large scale include triggering by spiral density waves, which is reviewed in Elmegreen (1994, 1995*a*). According to this model, Gould's Belt was a self-gravitating condensation in the Sagittarius-Carina spiral arm when it passed us $\sim$60 Myr ago, and is now in the process of large-scale dispersal as it enters the interarm region, even though there is continuing star formation in the Lindblad ring and other disturbed gas from this condensation (see Elmegreen 1993).

The evolution of a dense molecular core during the formation of its embedded cluster is unknown. The core could collapse dynamically while the cluster stars form, giving it a total lifetime comparable to the core crossing time, or it could be somewhat stable as the stars form on smaller scales inside of it. Indeed there is direct evidence for gas collapse onto individual stars in cloud cores (Mardones et al. 1997; Motte et al. 1998), but not much evidence for the collapse of whole cores (except perhaps in W49: see Welch et al. 1987; De Pree et al. 1997).

IV. CONDITIONS FOR THE FORMATION OF BOUND CLUSTERS

A. Critical Efficiencies

The final state of an embedded cluster of young stars depends on the efficiency, ϵ, of star formation in that region: that is, on the ratio of the final stellar mass to the total mass (stars + gas) in that part of the cloud. When this ratio is high, the stars have enough mass to remain gravitationally bound when the residual gas leaves, forming a bound cluster. When this ratio is low, random stellar motions from the time of birth disperse the cluster in a few crossing times, following the expulsion of residual gas. The threshold for self-binding occurs at a local efficiency of about 50% (von Hoerner 1968). This result is most easily seen from the virial theorem $2T + \Omega = 0$ and total energy $E = T + \Omega$ for stellar kinetic and potential energies, T and Ω. Before the gas expulsion, $E = \Omega_{\rm before}/2 < 0$ from these two equations. In the instant after *rapid* gas expulsion, the kinetic energy and radius of the cluster are approximately unchanged, because the stellar motions are at first unaffected, but the potential energy changes because of the sudden loss of mass (rapid gas expulsion occurs when the outflowing gas moves significantly faster than the virial speed of the cloud). To remain bound thereafter, E must remain less than zero, which means that during the expulsion, the potential energy can increase by no more than the addition of $|\Omega_{\rm before}|/2$. Thus, immediately after the expulsion of gas, the potential energy of the cluster, $\Omega_{\rm after}$, has to be less than half the potential energy before, $\Omega_{\rm before}/2$. Writing $\Omega_{\rm before} = -\alpha G M_{\rm stars} M_{\rm total}/R$ and $\Omega_{\rm after} = -\alpha G M^2_{\rm stars}/R$ for the same α and R, we see that this constraint requires $M_{\rm stars} > M_{\rm total}/2$ for self-binding. Thus the efficiency for star formation, $M_{\rm stars}/M_{\rm total}$, has to exceed about $\frac{1}{2}$ for a cluster to be self-bound (see also Mathieu 1983; Elmegreen 1983).

Another way to write this is in terms of the expansion factor for radius, $R_{\rm final}/R_{\rm before}$, where $R_{\rm final}$ is the cluster radius after the gas-free cluster readjusts its virial equilibrium. A cluster is bound if $R_{\rm final}$ does not become infinite. Hills (1980) derived $R_{\rm final}/R_{\rm initial} = \epsilon/(2\epsilon - 1)$, from which we again obtain $\epsilon > 0.5$ for final self-binding with efficiency ϵ. Danilov (1987) derived a critical efficiency in terms of the ratio of cluster radius to cloud radius; this ratio has to be <0.2 for a bound cluster to form.

There can be many modifications to this result, depending on the specific model of star formation. One important change is to consider initial stellar motions that are less than their virial speeds in the potential of the cloud because the cloud is supported by both magnetic and kinematic energies, whereas the star cluster is supported only by kinematic energy. This modification was considered by Lada et al. (1984), Elmegreen and

Clemens (1985), Pinto (1987), and Verschueren (1990), who derived a critical efficiency for isothermal clouds that may be approximated by the expression

$$2(1-\epsilon)\ln\left(\frac{\epsilon}{1-\epsilon}\right)+1+\epsilon = 1.5t^2$$

where $t = a_s/a_{VT} < 1$ is the ratio of the stellar velocity dispersion to the virial. This expression gives ϵ between 0.29 at $t = 0$ and 0.5 at $t = 1$. Other cloud structures gave a similar range for ϵ. This result is the critical star formation efficiency for the whole cloud; it assumes that the stars fall to the center after birth, and have a critical efficiency for binding in the center equal to the standard value of 0.5.

A related issue is the question of purely gravitational effects that arise in a cluster-forming core once the stars comprise more than ~30% of the gas. In this situation, the stars may be regarded as a separate (collisionless) "fluid" from the gas. The stability of such two-fluid systems was considered by Jog and Solomon (1984) and by Fridman and Polyachenko (1984). The Jeans length for a two-component fluid is smaller than that for either fluid separately. Dense stellar clusters might therefore fragment into subgroups, perhaps accounting for some of the substructure that is observed in young embedded star clusters (Patel and Pudritz 1994).

Lada et al. (1984) also considered the implications of slow gas removal on cluster self-binding. They found that gas removal on timescales of several cloud crossing times lowers the required efficiency by about a factor of 2 and, when combined with the effect of slow starbirth velocities, lowers the efficiency by a combined factor of ~4. For clouds in which stars are born at about half the virial speed, and in which gas removal takes ~4 crossing times, the critical efficiency for the formation of a bound cluster may be only ~10%.

Another way to lower the critical efficiency is to consider gas drag on the stars that form. Gas drag removes stellar kinetic energy and causes the stars to sink to the bottom of the cloud potential well, just like a low birth velocity. Saiyadpour et al. (1997) found that the critical efficiency can be only 0.1 in this case. Gas accretion also slows down protostars and causes them to sink to the center (Bonnell and Davies 1998).

It follows from these examples that the critical efficiency for self-binding can be between ~0.1 and 0.5, depending on the details of the star formation process.

B. Bound Clusters vs. Unbound OB Associations

The onset of massive star formation should mark the beginning of rapid cloud dispersal, because ionizing radiation is much more destructive per unit stellar mass than short-lived winds from low-mass stars (e.g., see Whitworth 1979). According to Vacca et al. (1996), the ionizing photon luminosity scales with stellar mass approximately as M^4. In that case, the total Lyman continuum luminosity from stars with luminosities in the range $\log L$ to $\log L + d \log L$ increases approximately as $L^{0.66}$ for a Salpeter IMF (in a Salpeter IMF, the number of stars in a logarithmic interval, $n\,[M_{\rm star}]\, d \log M_{\rm star}$, is proportional to $M_{\rm star}^{-1.35} d \log M_{\rm star}$). Thus, the total ionizing luminosity increases with cluster mass more rapidly than the total cloud mass, and cloud destruction by ionization follows the onset of massive star formation.

If massive stars effectively destroy clouds, then the overall efficiency is likely to be low wherever a lot of massive stars form [unless they form *preferentially* late, as suggested by Herbig (1962), and not just randomly late]. Thus, we can explain both the low efficiency and the unboundedness of an OB association: The destructive nature of O-star ionization causes both. We can also explain why all open clusters in normal galaxy disks have small masses, generally less than several times 10^3 $M_\odot$ in the catalog of Lynga (1987; e.g., see Battinelli et al. 1994): Low-mass star-forming regions are statistically unlikely to produce massive stars. Discussions of this point are in Elmegreen (1983), Henning and Stecklum (1986), and Pandey et al. (1990).

The idea that massive stars form late in the development of a cluster goes back to Herbig (1962) and Iben and Talbot (1966), with more recent work by Herbst and Miller (1982) and Adams et al. (1983). However, Stahler (1985) suggested that the observations have a different explanation, and the rare massive stars should form later than the more common low-mass stars anyway, for statistical reasons (Schroeder and Comins 1988; Elmegreen 1999).

The efficiency of star formation has been estimated for several embedded clusters, giving values such as 25% for NGC 6334 (Tapia et al. 1996), 6–18% for W3 IRS5 (Megeath et al. 1996), 2.5% for Serpens (White et al. 1995), 19% for NGC 3576 (Persi et al. 1994), and 23% for ρ Ophiuchi (Greene and Young 1992), to name a few.

C. Variation in Efficiency with Ambient Pressure

Variation in the efficiency from region to region could have important consequences, because it might affect the fraction of star formation going into bound clusters (in addition to the overall star formation rate per unit gas mass). One consideration is that the efficiency may increase in regions of high pressure (Elmegreen et al. 1993; Elmegreen and Efremov 1997). This is because the virial velocity of a gravitationally bound cloud increases with pressure and mass as $V_{\rm VT} \sim (PM^2)^{1/8}$, as may be determined from the relationships $V_{\rm VT}^2 \sim GM/(5R)$ and $P \sim GM^2/R^4$ for radius R. If the pressure increases and the virial velocity follows, then clouds of a given mass are harder to destroy with H II regions, which push on material with a fixed velocity of about 10 km s^{-1}. In fact, a high fraction of star formation in starburst galaxies, which generally have a high pressure, could be in the form of bound clusters (Meurer et al. 1995).

The lack of expansion of H II regions in virialized clouds with high velocity dispersions also means that the massive stars will not ionize much. They will ionize only the relatively small mass of high-density gas initially around them.

We can determine the average pressures in today's globular clusters from their masses and sizes using the relationship $P \sim GM^2/R^4$. This gives $P \sim 10^6$–$10^8 k_B$ (Harris and Pudritz 1994; Elmegreen and Efremov 1997), which is 10^2–10^4 times the local total ISM pressure. If the pressures of star-forming regions in the Galactic halo were this high when the globular clusters formed, and the globular cluster cloud masses were higher than those near OB associations by a factor of ~ 10 (to account for the higher globular cluster masses), then the velocity dispersions in globular cluster cores had to be larger than the velocity dispersion in a local GMC by a factor $\left(M^2 P\right)^{1/8} = \left(10^2 \times 10^4\right)^{1/8} \sim 5.6$. This puts the dispersion close to 10 km s^{-1}, making the globular cluster clouds difficult to disrupt by H II regions.

V. GLOBULAR CLUSTER FORMATION

Globular clusters in the halos of galaxies are denser, smoother, and more massive than open clusters in galactic disks, and the globulars are also much older, but they have about the same power law mass distribution function as open clusters at the high-mass end, and of course both are gravitationally bound systems. We are therefore faced with the challenging question of whether the similarities between these two types of clusters are more important than their differences. If so, then they may have nearly the same formation mechanisms, modified in the case of the globulars by the peculiar conditions in the early universe. If the differences are too great for a unified model, then we need a unique formation theory for globular clusters.

The history of the theory on this topic is almost entirely weighted toward the latter point of view, because the full mass distribution function for globular clusters is essentially Gaussian (when plotted as linear in number vs. logarithm in mass or luminosity; e.g., Harris and Racine 1979; Abraham and van den Bergh 1995), with a *characteristic mass* of several hundred thousand $M_\odot$. Nearly all of the early models have attempted to explain this mass. For example, Peebles and Dicke (1968), Peebles (1984), Rosenblatt et al. (1988), and Padoan et al. (1997) regarded globular clusters as primordial objects produced by density fluctuations in the expanding universe. Peebles and Dicke (1968) thought the characteristic mass was a Jeans mass. Other models viewed globulars as secondary objects, formed by thermal instabilities in cooling halo gas (Fall and Rees 1985; Murray and Lin 1992; Vietri and Pesce 1995), gravitational instabilities in giant bubbles (Brown et al. 1995), or the shocked layers between colliding clouds (Zinnecker and Palla 1987; Shapiro et al. 1992; Kumai et al. 1993;

Murray and Lin 1992). Schweizer (1987) and Ashman and Zepf (1992) suggested that many globulars formed during galaxy mergers. This could explain the high specific frequency of globular clusters (number per unit galaxy luminosity; Harris and van den Bergh 1981) in ellipticals compared to spirals if the ellipticals formed in mergers. However, Forbes et al. (1997) found that galaxies with high specific frequencies of globular clusters have lower cluster metallicities, whereas the opposite might be expected in the merger model. Also, McLaughlin (1999) has suggested that the specific frequency of globular clusters is the same everywhere when X-ray halo gas and stellar evolution are included.

There is another point of view if the globular cluster mass function is not primordial but evolved from an initial power law. This is a reasonable hypothesis, because low-mass globulars evaporate and get dispersed first, depressing an initial power law at low mass to resemble a Gaussian after a Hubble time (Surdin 1979; Okazaki and Tosa 1995; Elmegreen and Efremov 1997). Observations of young globular clusters, forming in starburst regions, also show a power law luminosity function with a mixture of ages (Holtzman et al. 1992; Whitmore and Schweizer 1995; Meurer et al. 1995; Maoz et al. 1996; Carlson et al. 1998), and the high-mass end of the old globular systems is nearly a power law too (Harris and Pudritz 1994; McLaughlin and Pudritz 1996; Durrell et al. 1996).

In that case, there is a good possibility that old globular clusters formed in much the same way as young open clusters: that is, in dense cores that are part of a large-scale hierarchical gas structure derived from cloud collisions (Harris and Pudritz 1994; McLaughlin and Pudritz 1996) or turbulent motions (Elmegreen and Efremov 1997). Direct observations of globular cluster luminosity functions at cosmological distances should be able to tell the difference between formation models with a characteristic mass and those that are scale free.

Another model for globular cluster formation suggests that they are the cores of former dwarf galaxies (Zinnecker et al. 1988; Freeman 1993), "eaten" by the large galaxy during dissipative collisions. The globulars NGC 6715, Terzan 7, Terzan 8, and Arp 2, which are comoving with the Sagittarius dwarf galaxy, are possible examples (Ibata et al. 1995; Da Costa and Armandroff 1995). Other dwarf galaxies have globular cluster systems too (Durrell et al. 1996), so the globulars around large galaxies may not come from the cores of the dwarfs but from the dwarf globulars themselves. It remains to be seen whether this formation mechanism can account for the globular cluster luminosity function.

VI. CONCLUSIONS

1. Loose hierarchical clusters form when the associated gas is only weakly self-gravitating and clumped in this fashion before the star formation begins. Dense clusters come from strongly self-gravitating gas, which

may be triggered, and which also may be gravitationally unstable to bulk collapse.

2. Cluster formation is often quite rapid, requiring only a few internal crossing times to make most of the stars. This conclusion follows from the relatively small age differences between nearby clusters and from the hierarchical structure of embedded and young stellar groups. Such structure would presumably get destroyed by orbital mixing if the region were much older than a crossing time.

3. Dense cluster environments seem to affect the formation or destruction of protostellar disks and binary stars, but not the stellar initial mass function.

4. Bound clusters require a relatively high star formation efficiency. This is not a problem for typically low-mass open clusters, but it requires something special, such as high pressure, for a massive globular cluster.

Acknowledgments We would like to thank the conference organizers for the opportunity to write this review. Helpful comments on the manuscript were provided by W. Brandner, D. McLaughlin, and P. Kroupa. Y. E. appreciates partial support from the Russian Foundation for Basic Research and the Council for Support of Scientific Schools. The research of R. P. is supported by the National Science and Engineering Research Council of Canada (NSERC). Travel for H. Z. was supported by the Deutsche Forschungsgemeinschaft (DFG).

REFERENCES

Aarseth, S. J., Lin, D. N. C., and Papaloizou, J. C. B. 1988. On the collapse and violent relaxation of protoglobular clusters. *Astrophys. J.* 324:288–310.

Abraham, R. G., and van den Bergh, S. 1995. A Gauss-Hermite expansion of the galactic globular cluster luminosity function. *Astrophys. J.* 438:212–222.

Adams, M. T., Strom, K. M., and Strom, S. E. 1983. The star-forming history of the young cluster NGC 2264. *Astrophys. J. Suppl.* 53:893–936.

Ali, B., and DePoy, D. L. 1995. A 2.2 micrometer imaging survey of the Orion A molecular cloud. *Astron. J.* 109:709–720.

Allen, L. E. 1995. Star formation in Lynds 1641. Ph.D. Dissertation, University of Massachusetts at Amherst.

Ashman, K. M., and Zepf, S. E. 1992. The formation of globular clusters in merging and interacting galaxies. *Astrophys. J.* 384:50–61.

Aspin, C., and Barsony, M. 1994. Near-IR imaging photometry of the $J - K > 4$ sources in the Lk Hα 101 infrared cluster. *Astron. Astrophys.* 288:849–859.

Aspin, C., Sandell, G., and Russell, A. P. G. 1994. Near-IR imaging photometry of NGC 1333. I. The embedded PMS stellar population. *Astron. Astrophys.* 106:165–198.

Bally, J., Devine, D., and Reipurth, B. 1996. A burst of Herbig-Haro flows in NGC 1333. *Astrophys. J. Lett.* 473:L49–L53.

Barsony, M., Kenyon, S. J., Lada, E. A., and Teuben, P. J. 1997. A near-infrared imaging survey of the ρ Ophiuchi cloud core. *Astrophys. J. Suppl.* 112:109–191.

Bate, M. R., Clarke, C. J., and McCaughrean, M. J. 1998. Interpreting the mean surface density of companions in star-forming regions. *Mon. Not. Roy. Astron. Soc.* 297:1163–1181.

Battinelli, P. 1991. A new identification technique for OB associations: OB associations in the Small Magellanic Cloud. *Astron. Astrophys.* 244:69–74.

Battinelli, P. 1992. OB associations in four stellar fields of M 31. *Astron. Astrophys.* 258:269–271.

Battinelli, P., and Efremov, Y. N. 1999. Comparison of distribution of Cepheids and clusters in the LMC. *Astron. Astrophys.* 346:778–784.

Battinelli, P., Brandimarti, A., and Capuzzo-Dolcetta, R. 1994. Integrated photometric properties of open clusters. *Astron. Astrophys. Suppl.* 104:379–390.

Battinelli, P., Efremov, Y. N., and Magnier, E. A. 1996. An objective determination of blue star groupings in the Andromeda Galaxy. *Astron. Astrophys.* 314:51–58.

Bazell, D., and Désert, F. X. 1988. Fractal structure of interstellar cirrus. *Astrophys. J.* 333:353–358.

Beck, S. C., Kelly, D. M., and Lacy, J. H. 1998. Infrared ionic line emission in W33. *Astron. J.* 115:2504–2508.

Beech, M. 1987. Are Lynds dark clouds fractals? *Astrophys. Space Sci.* 133:193–195.

Belikov, A. N., Hirte, S., Meusinger, H., Piskunov, A. E., and Schilbach, E. 1998. The fine structure of the Pleiades luminosity function and pre-main sequence evolution. *Astron. Astrophys.* 332:575–585.

Berdnikov, L. N., and Efremov, Y. N. 1989. Groupings of Cepheids in the Galaxy. *Sov. Astron.* 33:274–279.

Berdnikov, L. N., and Efremov, Y. N. 1993. Contours of constant density and z height for Cepheids. *Sov. Astron. Lett.* 19:389–394.

Bhatia, R.K., and Hatzidimitriou, D. 1988. Binary star clusters in the Large Magellanic Cloud. *Mon. Not. Roy. Astron. Soc.* 230:215–221.

Bhatt, B. C., Pandey, A. K., and Mahra, H. S. 1991. Integrated luminosity distribution of galactic open clusters. *J. Astrophys. Astron.* 12:179–185.

Bhatt, H. C. 1989. Capture of field stars by molecular clouds. *Astron. Astrophys.* 213:299–302.

Bica, E., Claria, J. J., Dottori, H., Santos, J. F. C., Jr., and Piatti, A. E. 1996. Integrated UBV photometry of 624 star clusters and associations in the Large Magellanic Cloud. *Astrophys. J. Suppl.* 102:57–73.

Blaauw, A. 1964. The O associations in the solar neighborhood. *Ann. Rev. Astron. Astrophys.* 2:213–247.

Blitz, L. 1993. Giant molecular clouds. In *Protostars and Planets III,* ed. E. H. Levy and J. I. Lunine (Tucson: University of Arizona Press), pp. 125–161.

Bonnell, I. A., and Davies, M. B. 1998. Mass segregation in young stellar clusters. *Mon. Not. Roy. Astron. Soc.* 295:691–698.

Bonnell, I. A., Bate, M. R., Clarke, C. J., and Pringle, J. E. 1997. Accretion and the stellar mass spectrum in small clusters. *Mon. Not. Roy. Astron. Soc.* 285:201–208.

Bonnell, I. A., Bate, M. R., and Zinnecker, H. 1998. On the formation of massive stars. *Mon. Not. Roy. Astron. Soc.* 298:93–102.

Boss, A. P. 1996. Collapse and fragmentation of molecular cloud cores. IV. Oblate clouds and small cluster formation. *Astrophys. J.* 468:231–240.

Bouvier, J., Rigaut, F., and Nadeau, D. 1997. Pleiades low-mass binaries: Do companions affect the evolution of protoplanetary disks? *Astron. Astrophys.* 323:139–150.

Brandner, W., and Köhler, R. 1998. Star formation environments and the distribution of binary separations. *Astrophys. J. Lett.* 499:L79–L82.

Brandner, W., Grebel, E. K., Chu, Y. H., and Weis, K. 1997. Ring nebula and bipolar outflows associated with the B1.5 supergiant Sher 25 in NGC 3603. *Astrophys. J. Lett.* 475:L45–L48.

Brown, J. H., Burkert, A., and Truran, J. W. 1995. On the formation of globular clusters. II. Early dynamical evolution. *Astrophys. J.* 440:666–673.

Caloi, V., and Cassatella, A. 1995. On the evolutionary status of the LMC cluster NGC 2004. *Astron. Astrophys.* 295:63–74.

Caloi, V., and Cassatella, A. 1998. Evolutionary status and age spread in the young LMC cluster NGC 1850A. *Astron. Astrophys.* 330:492–504.

Carlberg, R. G., and Pudritz, R., 1990. Magnetic support and fragmentation of molecular clouds. *Mon. Not. Roy. Astron. Soc.* 247:353–366.

Carlson, M. N., Holtzman, J. A., Watson, A. M., Grillmair, C. J., Mould, J. R., Ballester, G. E., Burrows, C. J., Clarke, J. T., Crisp, D., Evans, R. W., Gallagher, J. S., III, Griffiths, R. E., Hester, J. J., Hoessel, J. G., Scowen, P. A., Stapelfeldt, K. R., Trauger, J. T., and Westphal, J. A., 1998. Deep Hubble Space Telescope observations of star clusters in NGC 1275. *Astron. J.* 115:1778–1790.

Carpenter, J. M., Snell, R. L., Schloerb, F. P., and Skrutskie, M. F. 1993. Embedded star clusters associated with luminous IRAS point sources. *Astrophys. J.* 407:657-679.

Carpenter, J. M., Meyer, M. R., Dougados, C., Strom, S. E., Hillenbrand, L. A. 1997. Properties of the Monoceros R2 stellar cluster. *Astron. J.* 114:198–221.

Casanova, S., Montmerle, T., Feigelson, E. D., and Andre, P. 1995. ROSAT X-ray sources embedded in the ρ Ophiuchi cloud core. *Astrophys. J.* 439:752–770.

Chen, H., and Tokunaga, A. T. 1994. Stellar density enhancements associated with IRAS sources in L1641. *Astrophys. J. Suppl.* 90:149–172.

Chièze, J.P. 1987. The fragmentation of molecular clouds. I. The mass-radius-velocity dispersion relations. *Astron. Astrophys.* 171:225–232.

Chini, R., and Wargau, W. F. 1998. Young stellar objects and abnormal extinction within M17. *Astron. Astrophys.* 329:161–168.

Chupina, N. V., and Vereshchagin, S. V. 1998. Stellar clumps within the corona in the open cluster M 67. *Astron. Astrophys.* 334:552–557.

Clark, B. G. 1965. An interferometric investigation of the 21-cm hydrogen line absorption. *Astrophys. J.* 142:1398–1422.

Clarke, C. J. 1996. The formation of binaries in small N clusters. In *The Origins, Evolution, and Destinies of Binary Stars in Clusters,* ASP Conf. Ser. 90, ed. E. F. Milone and J.-C. Mermilliod (San Francisco: Astronomical Society of the Pacific), pp. 242–251.

Comeron, F., and Torra, J. 1996. The galactic distribution and luminosity function of ultracompact H II regions. *Astron. Astrophys.* 314:776–784.

Comeron, F., Torra, J., and Gomez, A. E. 1992. The characteristics and origin of the Gould's Belt. *Astrophys. Space Sci.* 187:187–195.

Comeron, F., Rieke, G. H., Burrows, A., and Rieke, M. J. 1993. The stellar population in the ρ Ophiuchi cluster. *Astrophys. J.* 416:185–203.

Cool, A. M. 1998. Measuring globular cluster mass functions with HST. In *The Stellar Initial Mass Function,* ed. G. Gilmore, I. Parry, and S. Ryan (Cambridge: Cambridge University Press), pp. 139–156.

Covino, E., Alcala, J. M., Allain, S., Bouvier, J., Terranegra, L., and Krautter, J. 1997. A study of the Chamaeleon star-forming region from the ROSAT all-sky survey. III. High resolution spectroscopic study. *Astron. Astrophys.* 328:187–202.

Da Costa, G. S., and Armandroff, T. E. 1995. Abundances and kinematics of the globular cluster systems of the Galaxy and of the Sagittarius Dwarf. *Astron. J.* 109:2533–2552.

Dame, T., Elmegreen, B. G., Cohen, R., and Thaddeus, P. 1986. The largest molecular cloud complexes in the first galactic quadrant. *Astrophys. J.* 305:892–908.

Danilov, V. M. 1987. The dynamics of forming open stellar clusters. *Astron. Zh.* 64:656–659.

De Pree, C. G., Mehringer, D. M., and Goss, W. M. 1997. Multifrequency, high-resolution radio recombination line observations of the massive star-forming region W49A. *Astrophys. J.* 482:307–333.

de Vega, H., Sánchez, N., and Combes, F. 1996. Self gravity as an explanation of the fractal structure of the interstellar medium. *Nature* 383:56–58.

De Zeeuw, P. T., Hoogerwerf, R., de Bruijne, J. H. J., Brown, A. G. A., and Blaauw, A. 1999. A Hipparcos census of nearby OB associations. *Astrophys. J.* 117:354–399.

Dickey, J. M., and Garwood, R. W. 1989. The mass spectrum of interstellar clouds. *Astrophys. J.* 341:201–207.

Dickman, R. L., Horvath, M. A., and Margulis, M. 1990. A search for scale-dependent morphology in five molecular cloud complexes. *Astrophys. J.* 365:586–601.

Dieball, A., and Grebel, E. 1998. Binary star clusters in the Large Magellanic Cloud. In *New Views of the Magellanic Clouds,* IAU Symp. 190, ed. Y.-H. Chu, D. Bohlender, J. Hesser, and N. Suntzeff (San Francisco: Astronomical Society of the Pacific), in press.

Durrell, P. R., McLaughlin, D. E., Harris, W. E., and Hanes, D. A. 1996. Globular cluster systems in dwarf elliptical galaxies. I. The dE, N galaxy NGC 3115 DW1. *Astrophys. J.* 463:543–554.

Edvardsson, B., Pettersson, B., Kharrazi, M., and Westerlund, B. 1995. Abundance analysis and origin of the ζ Sculptoris open cluster. *Astron. Astrophys.* 293:75–86.

Efremov, Y. N. 1978. Star complexes. *Sov. Astron. Lett.* 4:66.

Efremov, Y. N. 1989. *Star Formation Origins in Galaxies* (Moscow: Nauka).

Efremov, Y. N. 1995. Star complexes and associations: fundamental and elementary cells of star formation. *Astron. J.* 110:2757–2770.

Efremov, Y. N. 1997. Concentration of Cepheids and open clusters in the Spiral arms of the Galaxy. *Astron. Lett.* 23:579–584.

Efremov, Y. N. 1998. The Car-Sgr arm as outlined by superclouds and the grand design of the Galaxy. *Astron. Astrophys. Trans.* 15:3–17.

Efremov, Y. N., and Elmegreen, B. G. 1998. Hierarchical star formation from the time-space distribution of star clusters in the Large Magellanic Cloud. *Mon. Not. Roy. Astron. Soc.* 299:588–594.

Efremov, Y. N., and Sitnik, T. G. 1988. Young stellar-gas complexes in the Galaxy. *Sov. Astron. Lett.* 14:347–352.

Efremov, Y. N., Ivanov, G. R., and Nikolov, N. S. 1987. Star complexes and associations in the Andromeda galaxy. *Astrophys. Space Sci.* 135:119–130.

Eiroa, C., and Casali, M. M. 1995. The S 269 stellar cluster. *Astron. Astrophys.* 303:87–94.

Eisenhauer, F., Quirrenbach, A., Zinnecker, H., and Genzel, R. 1998. Stellar content of the Galactic starburst template NGC 3603 from adaptive optics observations. *Astrophys. J.* 498:278–292.

Elmegreen, B. G. 1983. Quiescent formation of bound galactic clusters. *Mon. Not. Roy. Astron. Soc.* 203:1011–1020.

Elmegreen, B. G. 1989. A pressure and metallicity dependence for molecular cloud correlations and the calibration of mass. *Astrophys. J.* 338:178–196.

Elmegreen, B. G. 1993. Formation of interstellar clouds and structure. In *Protostars and Planets III,* ed. E. H. Levy and J. I. Lunine (Tucson: University of Arizona Press), pp. 97–124.

Elmegreen, B. G. 1994. Supercloud formation by gravitational collapse of magnetic gas in the crest of a spiral density wave. *Astrophys. J.* 433:39–47.
Elmegreen, B. G. 1995*a*. Density waves and star formation: Is there triggering? In *The Formation of the Milky Way,* ed. E. J. Alfaro and A. J. Delgado (Cambridge: Cambridge University Press) pp. 28–38.
Elmegreen, B. G. 1995*b*. Large scale star formation. In *The 7th Guo Shoujing Summer School on Astrophysics: Molecular Clouds and Star Formation,* ed. C. Yuan and Hunhan You (Singapore: World Press), pp. 149–205.
Elmegreen, B. G. 1997. The initial stellar mass function from random sampling in a turbulent fractal cloud. *Astrophys. J.* 486:944–954.
Elmegreen, B. G. 1998*a*. Observations and theory of the initial stellar mass function. In *Unsolved Problems in Stellar Evolution,* ed. M. Livio (Cambridge: Cambridge University Press), in press.
Elmegreen, B. G. 1998*b*. Observations and theory of dynamical triggers for star formation. In *Origins of Galaxies, Stars, Planets and Life,* ASP Conf. Ser. 148, ed. C. E. Woodward, H. A. Thronson, and M. Shull (San Francisco: Astronomical Society of the Pacific), pp. 150–183.
Elmegreen, B. G. 1999. The initial stellar mass function from random sampling in hierarchical clouds II: Statistical fluctuations and a mass dependence for starbirth positions and times. *Astrophys. J.* 515:323–336.
Elmegreen, B. G., and Clemens, C. 1985. On the formation rate of galactic clusters in clouds of various masses. *Astrophys. J.* 294:523–532.
Elmegreen, B. G., and Combes, F. 1992. Magnetic diffusion in clumpy molecular clouds. *Astron. Astrophys.* 259:232–240.
Elmegreen, B. G., and Efremov, Y. N. 1996. An extension of hierarchical star formation to galactic scales. *Astrophys. J.* 466:802–807.
Elmegreen, B. G., and Efremov, Y. N. 1997. A universal formation mechanism for open and globular clusters in turbulent gas. *Astrophys. J.* 480:235–245.
Elmegreen, B. G., and Efremov, Y. N. 1998. Hierarchy of interstellar and stellar structures and the case of the Orion star-forming region. In *The Orion Complex Revisited,* ed. M. J. McCaughrean and A. Burkert (San Francisco: Astronomical Socity of the Pacific), in press.
Elmegreen, B. G., and Elmegreen, D. M. 1983. Regular strings of H II regions and superclouds in spiral galaxies: Clues to the origin of cloudy structure. *Mon. Not. Roy. Astron. Soc.* 203:31–45.
Elmegreen, B. G., and Elmegreen, D. M. 1987. H I superclouds in the inner Galaxy. *Astrophys. J.* 320:182–198.
Elmegreen, B. G., Kaufman, M., and Thomasson, M. 1993. An interaction model for the formation of dwarf galaxies and $10^8 M_\odot$ clouds in spiral disks. *Astrophys. J.* 412:90–98.
Elmegreen, B. G., Elmegreen, D. M., Salzer, J., and Mann, H. 1996. On the size and formation mechanism of star complexes in Sm, Im, and BCD Galaxies. *Astrophys. J.* 467:579–588.
Elmegreen, D. M., and Salzer, J. J., 1999. Star-forming complexes in a sample of spiral and irregular galaxies. *Astron. J.* 117: 764–777.
Elson, R. A. W. 1991. The structure and evolution of rich star clusters in the Large Magellanic Cloud. *Astrophys. J. Suppl.* 76:185–214.
Elson, R. A., and Fall, S. M. 1985. A luminosity function for star clusters in the Large Magellanic Cloud. *Pub. Astron. Soc. Pacific* 97:692–696.
Falgarone, E., and Phillips, T. G. 1991. Signatures of turbulence in the dense interstellar medium. In *Fragmentation of Molecular Clouds and Star Formation,* ed. E. Falgarone, F. Boulanger, and G. Duvert (Dordrecht: Kluwer), pp. 119–138.
Falgarone, E., Phillips, T., and Walker, C. K. 1991. The edges of molecular clouds: Fractal boundaries and density structure. *Astrophys. J.* 378:186–201.

Falgarone, E., Puget, J. L., and Pérault, M. 1992. The small-scale density and velocity structure of quiescent molecular clouds. *Astron. Astrophys.,* 257:715–730.

Fall, S. M., and Rees, M. J. 1985. A theory for the origin of globular clusters. *Astrophys. J.* 298:18–26.

Feigelson, E. 1996. Dispersed T Tauri stars and galactic star formation. *Astrophys. J.* 468:306–322.

Feinstein, C. 1997. H II regions in southern spiral galaxies: The H α luminosity function. *Astrophys. J. Suppl.* 112:29–47.

Feitzinger, J. V., and Braunsfurth, E. 1984. The spatial distribution of young objects in the Large Magellanic Cloud: A problem of pattern recognition. *Astron. Astrophys.* 139:104–114.

Feitzinger, J. V., and Galinski, T. 1987. The fractal dimension of star-forming sites in galaxies. *Astron. Astrophys.* 179:249–254.

Fischer, P., Pryor, C., Murray, S., Mateo, M., and Richtler, T. 1998. Mass segregation in young Large Magellanic Cloud clusters. I. NGC 2157. *Astron. J.* 115:592–604.

Forbes, D. 1996. Star formation in NGC 6531: Evidence from the age spread and initial mass function. *Astron. J.* 112:1073–1084.

Forbes, D. A., Brodie, J. P., and Grillmair, C. J. 1997. On the origin of globular clusters in elliptical and cD galaxies. *Astron. J.* 113:1652–1665.

Freeman, K. H. 1993. Globular clusters and nucleated dwarf ellipticals. In *The Globular Cluster Galaxy Connection,* ASP Conf. Ser. 48, ed. G. H. Smith and J. P. Brodie (San Francisco: Astronomical Society of the Pacific), pp.608–614.

Freyberg, M. J., and Schmitt, J. H. M. M. 1995. ROSAT X-ray observations of the stellar clusters in NGC 2023 and NGC 2024. *Astron. Astrophys.* 296:L21–L24.

Fridman, A. M., and Polyachenko, V. L. 1984. In *Physics of Gravitating Systems I. Equilibrium of Stability* (Berlin: Springer Verlag).

Frink, S., Roeser, S., Neuhäuser, R., and Sterzik, M. F. 1997. New proper motions of pre-main sequence stars in Taurus-Auriga. *Astron. Astrophys.* 325:613–622.

Frink, S., Roeser, S., Alcala, J. M., Covino, E., and Brandner, W. 1998. Kinematics of T Tauri stars in Chamaeleon. *Astron. Astrophys.* 338:442–451.

Giovannetti, P., Caux, E., Nadeau, D., and Monin, J.-L. 1998. Deep optical and near infrared imaging photometry of the Serpens cloud core. *Astron. Astrophys.* 330:990–998.

Goodwin, S. P. 1998. Constraints on the initial conditions of globular clusters. *Mon. Not. Roy. Astron. Soc.* 294:47–60.

Gorti, U., and Bhatt, H. C. 1995. Effect of gas drag on the dynamics of protostellar clumps in molecular clouds. *Mon. Not. Roy. Astron. Soc.* 272:61–70.

Gorti, U., and Bhatt, H. C. 1996. Dynamics of embedded protostar clusters in clouds. *Mon. Not. Roy. Astron. Soc.* 278:611–616.

Gregorio-Hetem, J., Montmerle, T., Casanova, S., and Feigelson, E. D. 1998. X-rays and star formation: ROSAT observations of the Monoceros and Rosette molecular clouds. *Astron. Astrophys.* 331:193–210.

Greene, T. P., and Young, E. T. 1992. Near-infrared observations of young stellar objects in the Rho Ophiuchi dark cloud. *Astrophys. J.* 395:516–528.

Guillout, P., Sterzik, M. F., Schmitt, J. H. M. M, Motch, C., Egret, D., Voges, W., and Neuhäuser, R. 1998. The large-scale distribution of X-ray active stars. *Astron. Astrophys.* 334:540–544.

Hanson, M. M., Howarth, I. D., and Conti, P. S. 1997. The young massive stellar objects of M17. *Astrophys. J.* 489:698–718.

Harris, W. E., and Racine, R. 1979. Globular clusters in galaxies. *Ann. Rev. Astron. Astrophys.* 17:241–274.

Harris, W. E., and van den Bergh, S. 1981. Globular clusters in galaxies beyond the local group. I. New cluster systems in selected northern ellipticals. *Astron. J.* 86:1627–1642.
Harris, W. E., and Pudritz, R. E. 1994. Supergiant molecular clouds and the formation of globular cluster systems. *Astrophys. J.* 429:177–191.
Hatzidimitriou, D., and Bhatia, R. K. 1990. Cluster pairs in the Small Magellanic Cloud. *Astron. Astrophys.* 230:11–15.
Heithausen, A., Bensch, F., Stutzki, J., Falgarone, E., and Panis, J. F. 1998. The IRAM Key Project: Small scale structure of pre-star forming cores. Combined mass spectra and scaling laws. *Astron. Astrophys.* 331:65–68.
Heller, C. H. 1995. Encounters with protostellar disks. II. Disruption and binary formation. *Astrophys. J.* 455:252–259.
Henning, T., and Stecklum, B. 1986. Self-regulated star formation and the evolution of stellar systems. *Astrophys. Space Sci.* 128:237–251.
Henning, T., Martin, K., Reimann, H.-G., Launhardt, R., Leisawitz, D., and Zinnecker, H. 1994. Multi-wavelength study of NGC 281 A. *Astron. Astrophys.* 288:282–292.
Herbig, G. H. 1962. Spectral classification of faint members of the Hyades and the dating problem in galactic clusters. *Astrophys. J.* 135:736–747.
Herbst, W., and Miller, D. P. 1982. The age spread and initial mass function of NGC 3293: Implications for the formation of clusters. *Astron. J.* 87:1478–1490.
Heyer, M., Snell, R., and Carpenter, J. 1997. Barometrically challenged molecular clouds. *Bull. Am. Astron. Soc.* 191:121.03.
Hillenbrand, L. A. 1997. On the stellar population and star-forming history of the Orion Nebula Cluster. *Astron. J.* 113:1733–1768.
Hillenbrand, L. A., and Hartmann, L. 1998. A preliminary study of the Orion Nebula Cluster structure and dynamics. *Astrophys. J.* 492:540–553.
Hillenbrand, L. A., Massey, P., Strom, S. E., and Merrill, K. M. 1993. NGC 6611: A cluster caught in the act. *Astron. J.* 106:1906–1946.
Hillenbrand, L. A., Meyer, M. R., Strom, S. E., and Skrutskie, M. F. 1995. Isolated star-forming regions containing Herbig Ae/Be stars. 1: The young stellar aggregate associated with BD +40° 4124. *Astron. J.* 109:280–297.
Hills, J. G. 1980. The effect of mass loss on the dynamical evolution of a stellar system: Analytic approximations. *Astrophys. J.* 235:986–991.
Hodapp, K.-W., and Deane, J. 1993. Star formation in the L1641 North cluster. *Astrophys. J. Suppl.* 88:119–135.
Hodapp, K.-W. R., and Rayner, J. 1991. The S106 star-forming region. *Astron. J.* 102:1108–1117.
Holtzman, J. A., Faber, S. M., Shaya, E. J., Lauer, T. R., Grothe, J., Hunter, D. A., Baum, W. A., Ewald, S. P., Hester, J. F., Light, R. M., Lynds, C. R., O'Neil, E. J., Jr., and Westphal, J. A. 1992. Planetary Camera observations of NGC 1275: Discovery of a central population of compact massive blue star clusters. *Astron. J.* 103:691–702.
Horner, D. J., Lada, E. A., and Lada, C. J. 1997. A near-infrared imaging survey of NGC 2282. *Astron. J.* 113:1788–1798.
Howard, E. M., Pipher, J. L., and Forrest, W. J. 1997. S255–2: The formation of a stellar cluster. *Astrophys. J.* 481:327–342.
Hurt, R. L., and Barsony, M. 1996. A cluster of Class 0 protostars in Serpens: An IRAS HIRES study. *Astrophys. J. Lett.* 460:L45–L48.
Hyland, A. R., Straw, S., Jones, T. J., and Gatley, I. 1992. Star formation in the Magellanic Clouds. IV: Protostars in the vicinity of 30 Doradus. *Mon. Not. Roy. Astron. Soc.* 257:391–403.
Ibata, R. A., Gilmore, G., and Irwin, M. J. 1995. Sagittarius: The nearest dwarf galaxy. *Mon. Not. Roy. Astron. Soc.* 277:781–800.

Iben, I., Jr., and Talbot, R. J. 1966. Stellar formation rates in young clusters. *Astrophys. J.* 144:968–977.

Ivanov, G. R. 1992. Stellar associations in M81. *Mon. Not. Roy. Astron. Soc.* 257:119–124.

Jog, C. J., and Solomon, P. M. 1984. Two-fluid gravitational instabilities in a galactic disk. *Astrophys. J.* 276:114–126.

Johnstone, D., Hollenbach, D., and Bally, J. 1998. Photoevaporation of disks and clumps by nearby massive stars: Application to disk destruction in the Orion Nebula. *Astrophys. J.* 499:758–776.

Jones, B. F., and Walker, M. F. 1988. Proper motions and variabilities of stars near the Orion Nebula. *Astron. J.* 95:1755–1782.

Kennicutt, R. C., Edgar, B. K., and Hodge, P. W. 1989. Properties of H II region populations in galaxies. II: The H II region luminosity function. *Astrophys. J.* 337:761–781.

Klessen, R. S., Burkert, A., and Bate, M. R. 1998. Fragmentation of molecular clouds: The initial phase of a stellar cluster. *Astrophys. J. Lett.* 501:L205–L208.

Kontizas, E., Xiradaki, E., and Kontizas, M. 1989. The stellar content of binary star clusters in the LMC. *Astrophys. Space Sci.* 156:81–84.

Kontizas, M., Hatzidimitriou, D., Bellas-Velidis, I., Gouliermis, D., Kontizas, E., and Cannon, R. D. 1998. Mass segregation in two young clusters in the Large Magellanic Cloud: SL 666 and NGC 2098. *Astron. Astrophys.* 336:503–517.

Kramer, C., Stutzki, J., Röhrig, R., and Corneliussen, U. 1998. Clump mass spectra of molecular clouds. *Astron. Astrophys.* 329:249–264.

Kroupa, P. 1995*a*. Inverse dynamical population synthesis and star formation. *Mon. Not. Roy. Astron. Soc.* 277:1491–1506.

Kroupa, P. 1995*b*. The dynamical properties of stellar systems in the Galactic disc. *Mon. Not. Roy. Astron. Soc.* 277:1507–1521.

Kroupa, P. 1995*c*. Star cluster evolution, dynamical age estimation and the kinematical signature of star formation. *Mon. Not. Roy. Astron. Soc.* 277:1522–1540.

Kroupa, P. 1998. On the binary properties and the spatial and kinematical distribution of young stars. *Mon. Not. Roy. Astron. Soc.* 298:231–242.

Kroupa, P., Tout, C. A., and Gilmore, G. 1993. The distribution of low-mass stars in the Galactic disc. *Mon. Not. Roy. Astron. Soc.* 262:545–587.

Kumai, Y., Basu, B., and Fujimoto, M. 1993. Formation of globular clusters from gas in large-scale unorganized motion in galaxies. *Astrophys. J.* 404:144–161.

Lada, C. J. 1993. The formation of low mass stars: Observations. In *The Physics of Star Formation and Early Stellar Evolution,* ed. C. J. Lada and N. D. Kylafis (Dordrecht: Kluwer), p. 329.

Lada, C. J., Elmegreen, B. G., Cong, H., and Thaddeus, P. 1978. Molecular clouds in the vicinity of W3, W4, and W5. *Astrophys. J. Lett.* 226:L39–L42.

Lada, C. J., Margulis, M., and Dearborn, D. 1984. The formation and early dynamical evolution of bound stellar systems. *Astrophys. J.* 285:141–152.

Lada, C. J., Depoy, D. L., Merrill, K. M., and Gatley, I. 1991. Infrared images of M17. *Astrophys. J.* 374:533–539.

Lada, C. J., Young, E. T., and Greene, T. P. 1993. Infrared images of the young cluster NGC 2264. *Astrophys. J.* 408:471–483.

Lada, C. J., Alves, J., and Lada, E. A. 1996. Near-infrared imaging of embedded clusters: NGC 1333. *Astron. J.* 111:1964–1976.

Lada, E. A. 1992. Global star formation in the L1630 molecular cloud. *Astrophys. J. Lett.* 393:L25–L28.

Lada, E. A., and Lada, C. J. 1995. Near-infrared images of IC 348 and the luminosity functions of young embedded star clusters. *Astron. J.* 109:1682–1696.

Lada, E. A., Evans, N. J., II, Depoy, D. L., and Gatley, I. 1991. A 2.2 micron survey in the L1630 molecular cloud. *Astrophys. J.* 371:171–182.
Lada, E. A., Strom, K. M., and Myers, P. C. 1993. Environments of star formation: Relationship between molecular clouds, dense cores and young stars. In *Protostars and Planets III,* ed. E. H. Levy and J. I. Lunine (Tucson: University of Arizona Press), pp. 245–277.
Larson, R. B. 1978. Calculations of three-dimensional collapse and fragmentation. *Mon. Not. Roy. Astron. Soc.* 184:69–85.
Larson, R. B. 1982. Mass spectra of young stars. *Mon. Not. Roy. Astron. Soc.* 200:159–174.
Larson, R. B. 1990. Formation of Star Clusters. In *Physical Processes in Fragmentation and Star Formation,* ed. R. Capuzzo-Dolcetta, C. Chiosi, and A. Di Fazio (Dordrecht: Kluwer), pp. 389–399.
Larson, R. B. 1991. Some processes influencing the stellar initial mass function. In *Fragmentation of Molecular Clouds and Star Formation,* ed. E. Falgarone, F. Boulanger, and G. Duvert (Dordrecht: Kluwer), p. 261.
Lefloch, B., Lazarell, B., and Castets, A. 1997. Cometary globules. III. Triggered star formation in IC 1848. *Astron. Astrophys.* 324:249–262.
Lemme, C., Walmsley, C. M., Wilson, T. L., and Muders, D. 1995. A detailed study of an extremely quiescent core: L 1498. *Astron. Astrophys.* 302:509–520.
Lepine, J. R. D., and Duvert, G. 1994. Star formation by infall of high velocity clouds on the galactic disk. *Astron. Astrophys.* 286:60–71.
Li, W., Evans, N. J., II, and Lada, E. A. 1997. Looking for distributed star formation in L1630: A near-infrared (*J, H, K*) Survey. *Astrophys. J.* 488:277–285.
Lindblad, P. O., Palous, J., Loden, K., and Lindegren, L. 1997. The kinematics and nature of Gould's belt: A 30 Myr old star forming region. In *Hipparcos,* Proceedings of the ESA Symposium ESA SP-402, ed. B. Battrick, scientific coord.: M. A. C. Perryman and P. L. Bernacca (Noordwijk: European Space Agency), pp. 507–511.
Loren, B. R. 1989. The cobwebs of Ophiuchus. I. Strands of (C-13)O: The mass distribution. *Astrophys. J.* 338:902–924.
Luhman, K. L., and Rieke, G. H. 1998. The low-mass initial mass function in young clusters: L1495E. *Astrophys. J.* 497:354–369.
Lynga, G. 1987. *Catalogue of Open Cluster Data,* 5th edition (Strasbourg: CDS).
Mac Low, M.-M., Klessen, R. S., Burkert, A., and Smith, M. D. 1998. Kinetic energy decay rates of supersonic and super-Alfvenic turbulence in star-forming clouds. *Phys. Rev. Lett.* 80:2754–2757.
Magnier, E. A., Battinelli P., Lewin, W. H. G., Haiman, Z., van Paradijs, J., Hasinger, G., Pietsch, W., Supper, R., and Truemper, J. 1993. Automated identification of OB associations in M31. *Astron. Astrophys.* 278:36–42.
Maoz, D., Barth, A. J., Sternberg, A., Filippenko, A. V., Ho, L. C., Macchetto, F. D., Rix, H. W., and Schneider, D. P. 1996. Hubble Space Telescope ultraviolet images of five circumnuclear star-forming rings. *Astron. J.* 111:2248–2264.
Mardones, D., Myers, P. C., Tafalla, M., Wilner, D. J., Bachiller, R., and Garay, G. 1997. A search for infall motions toward nearby young stellar objects. *Astrophys. J.* 489:719–733.
Massey, P., and Hunter, D. A. 1998. Star formation in R136: A cluster of O3 stars revealed by Hubble Space Telescope spectroscopy. *Astrophys. J.* 493:180–194.
Mathieu, R. D. 1983. Dynamical constraints on star formation efficiency. *Astrophys. J. Lett.* 267:L97–101.
McCaughrean, M. J., and Stauffer, J. R. 1994. High resolution near-infrared imaging of the Trapezium: A stellar census. *Astron. J.* 108:1382–1397.

McDonald, J. M., and Clarke, C. J. 1995. The effect of star-disc interactions on the binary mass-ratio distribution. *Mon. Not. Roy. Astron. Soc.* 275:671–684.
McKee, C. F., Zweibel, E. G., Goodman, A. A., and Heiles, C. 1993. Magnetic fields in star-forming regions: Theory. In *Protostars and Planets III,* ed. E. H. Levy and J. I. Lunine (Tucson: University of Arizona Press), pp. 327–366.
McLaughlin, D. E. 1999. The efficiency of globular cluster formation. *Astron. J.* 117:2398–2427.
McLaughlin, D. E., and Pudritz, R. E. 1996. The formation of globular cluster systems. I. The luminosity function. *Astrophys. J.* 457:578–597.
Megeath, S. T., Herter, T., Beichman, C., Gautier, N., Hester, J. J., Rayner, J., and Shupe, D. 1996. A dense stellar cluster surrounding W3 IRS 5. *Astron. Astrophys.* 307:775–790.
Megeath, S. T., and Wilson, T. L. 1997. The NGC 281 west cluster. I. Star formation in photoevaporating clumps. *Astron. J.* 114:1106–1120.
Meurer, G. R., Heckman, T. M., Leitherer, C., Kinney, A., Robert, C., and Garnett, D. R. 1995. Starbursts and star clusters in the ultraviolet. *Astron. J.* 110:2665–2691.
Motte, F., Andre, P., and Neri, R. 1998. The initial conditions of star formation in the ρ Ophiuchi main cloud: Wide-field millimeter continuum mapping. *Astron. Astrophys.* 336:150–172.
Mundy, L. G., Looney, L. W., and Lada, E. A. 1995. Constraints on circumstellar disk masses in the Trapezium cluster. *Astrophys. J. Lett.* 452:L137–L140.
Murray, S. D., and Lin, D. N. C. 1992. Globular cluster formation: The fossil record. *Astrophys. J.* 400:265–272.
Murray, S. D., and Lin, D. N. C. 1996. Coalescence, star formation, and the cluster initial mass function. *Astrophys. J.* 467:728–748.
Myers, P. C. 1998. Cluster-forming molecular cloud cores. *Astrophys. J. Lett.* 496:L109–L112.
Myers, P. C., and Goodman, A. A. 1988. Evidence for magnetic and virial equilibrium in molecular clouds. *Astrophys. J. Lett.* 326:L27–L30.
Myers, P. C., and Khersonsky, V. K. 1995. On magnetic turbulence in interstellar clouds. *Astrophys. J.* 442:186–196.
Nakano, T. 1998. Star formation in magnetic clouds. *Astrophys. J.* 494:587–604.
Neuhäuser, R. 1997. Low mass pre main sequence stars and their X-ray emission. *Science* 276:1363–1370.
Normandeau, M., Taylor, A. R., and Dewdney, P. E. 1997. The Dominion Radio Astrophysical Observatory Galactic Plane Survey Pilot Project: The W3/W4/W5/HB 3 region. *Astrophys. J. Suppl.* 108:279–299.
Nürnberger, D., Chini, R., and Zinnecker, H. 1997. A 1.3mm dust continuum survey of Hα selected T Tauri stars in Lupus. *Astron. Astrophys.* 324:1036–1045.
Oey, M. S., and Clarke, C. J. 1998. On the form of the H II region luminosity function. *Astron. J.* 115:1543–1553.
Okazaki, T., and Tosa, M. 1995. The evolution of the luminosity function of globular cluster systems. *Mon. Not. Roy. Astron. Soc.* 274:48–60.
Padoan, P., Jimenez, R., and Jones, B. 1997. On star formation in primordial protoglobular clouds. *Mon. Not. Roy. Astron. Soc.* 285:711–717.
Pandey, A. K., Paliwal, D. C., and Mahra, H. S. 1990. Star formation efficiency in clouds of various masses. *Astrophys. J.* 362:165–167.
Pandey, A. K., Mahra, H. S., and Sagar, R. 1992. Effect of mass segregation on mass function of young open clusters. *Astr. Soc. India* 20:287–295.
Patel, K., and Pudritz, R. E. 1994. The formation of stellar groups and clusters in molecular cloud cores. *Astrophys. J.* 424:688–713.
Peebles, P. J. E. 1984. Dark matter and the origin of galaxies and globular star clusters. *Astrophys. J.* 277:470–477.

Peebles, P. J. E., and Dicke, R. H. 1968. Origin of the globular star clusters. *Astrophys. J.* 154:891–908.
Persi, P., Roth, M., Tapia, M., Ferrari-Toniolo, M., and Marenzi, A. R. 1994. The young stellar population associated with the H II region NGC 3576. *Astron. Astrophys.* 232:474–484.
Persi, P., Felli, M., Lagage, P. O., Roth, M., and Testi, L. 1997. Sub-arcsec resolution infrared images of the star forming region G 35.20–1.74. *Astron. Astrophys.* 327:299–308.
Petr, M. G., Coude du Foresto, V. T., Beckwith, S. V. W., Richichi, A., and McCaughrean, M. J. 1998. Binary stars in the Orion Trapezium cluster core *Astrophys. J.* 500:825–837.
Phelps, R. L. 1993. Young open clusters as probes of the star formation process. Ph.D. Dissertation, Boston University.
Phelps, R. L., and Lada, E. A. 1997. Spatial distribution of embedded clusters in the Rosette Molecular Cloud: Implications for cluster formation. *Astrophys. J.* 477:176–182.
Piche, F. 1993. A near-infrared survey of the star forming region NGC 2264. *Pub. Astron. Soc. Pacific* 105:324–324.
Pinto, F. 1987. Bound star clusters from gas clouds with low star formation efficiency. *Pub. Astron. Soc. Pacific* 99:1161–1166.
Pöppel, W. 1997. The Gould Belt system and the local interstellar medium. *Fund. Cosmic Phys.,* 18:1–272.
Preibisch, T., Zinnecker, H., and Herbig, G. H. 1996. ROSAT X-ray observations of the young cluster IC 348. *Astron. Astrophys.* 310:456–473.
Price, N. M., and Podsiadlowski, P. 1995. Dynamical interactions between young stellar objects and a collisional model for the origin of the stellar mass spectrum. *Mon. Not. Roy. Astron. Soc.* 273:1041–1068.
Prosser, C. F., Stauffer, J. R., Hartmann, L., Soderblom, D. R., Jones, B. F., Werner, M. W., and McCaughrean, M. J. 1994. HST photometry of the Trapezium cluster. *Astrophys. J.* 421:517–541.
Raboud, D., and Mermilliod, J. C. 1998. Investigation of the Pleiades cluster. IV. The radial structure. *Astron. Astrophys.* 329:101–114.
Rand, R. J. 1995. Berkeley-Illinois-Maryland Array observations of molecular spiral structure in M100 (NGC 4321). *Astron. J.* 109:2444–2458.
Rand, R. J., and Kulkarni, S. 1990. M51: Molecular spiral arms, giant molecular associations, and superclouds. *Astrophys. J. Lett.* 349:L43–L46.
Rosenblatt, E. I., Faber, S. M., and Blumenthal, G. R. 1988. Pregalactic formation of globular clusters in cold dark matter. *Astrophys. J.* 330:191–200.
Rubio, M., Barba, R. H., Walborn, N. R., Probst, R. G., Garcia, J., and Roth, M. R. 1998. Infrared observations of ongoing star formation in the 30 Doradus nebula and a comparison with Hubble Space Telescope WFPC 2 images. *Astron. J.* 116:1708–1718.
Sagar, R., and Bhatt, H. C. 1989. Radial distribution of the integrated light and photometric colors in open star clusters. *J. Astrophys. Astron.* 10:173–182.
Sagar, R., and Cannon, R. D. 1995. A deep UBVRI CCD photometric study of the moderately young southern open star cluster NGC 4755 = κ Crucis. *Astron. Astrophys.* 111:75–84.
Saiyadpour, A., Deiss, B. M., and Kegel, W. H. 1997. The effect of dynamical friction on a young stellar cluster prior to the gas removal. *Astron. Astrophys.* 322:756–763.
Salpeter, E. E. 1955. The luminosity function and stellar evolution. *Astrophys. J.* 121:161–167.
Scalo, J.M. 1985. Fragmentation and hierarchical structure in the interstellar medium. In *Protostars and Planets II,* ed. D. C. Black and M. S. Matthews (Tucson: University of Arizona Press), pp. 201–296.

Scalo, J. M. 1986. The stellar initial mass function. *Fundam. Cosmic Phys.* 11:1–278.

Scalo, J. 1990. Perception of interstellar structure: Facing complexity. In *Physical Processes in Fragmentation and Star Formation,* ed. R. Capuzzo-Dolcetta, C. Chiosi, and A. Di Fazio (Dordrecht: Kluwer), pp. 151–178.

Schroeder, M. C., and Comins, N. F. 1988. Star formation in very young galactic clusters. *Astrophys. J.* 326:756–760.

Schulz, N. S., Berghöfer, T. W., and Zinnecker, H. 1997. The X-ray view of the central part of IC 1396. *Astron. Astrophys.* 325:1001–1012.

Schweizer, F. 1987. Star formation in colliding and merging galaxies. In *Nearly Normal Galaxies,* ed. S. M. Faber (New York: Springer-Verlag), pp. 18–25.

Seleznev, A. F. 1995. The structure of the halo of the star cluster NGC 2070. *Astr. Lett.* 21:663–669.

Shapiro, P. R., Clocchiatti, A., and Kang, H. 1992. Magnetic fields and radiative shocks in protogalaxies and the origin of globular clusters. *Astrophys. J.* 389:269–285.

Shu, F. H., Adams, F. C., and Lizano, S. 1987. Star formation in molecular clouds: Observation and theory. *Ann. Rev. Astron. Astrophys.* 25:23–81.

Siess, L., Forestini, M., and Dougados, C. 1997. Synthetic Hertzsprung-Russell diagrams of open clusters. *Astron. Astrophys.* 324:556–565.

Silk, J., and Takahashi, T. 1979. A statistical model for the initial stellar mass function. *Astrophys. J.* 229:242–256.

Solomon, P. M., Rivolo, A. R., Barrett, J., and Yahil, A. 1987. Mass, luminosity, and line width relations of Galactic molecular clouds. *Astrophys. J.* 319:730–741.

Stahler, S. W. 1985. The star formation history of very young clusters. *Astrophys. J.* 293:207–215.

Sterzik, M. F., Alcala, J. M., Neuhäuser, R., and Schmitt, J. H. M. M. 1995. The spatial distribution of X-ray selected T Tauri stars. I. Orion. *Astron. Astrophys.* 297:418–426.

Stone, J. M., Ostriker, E., and Gammie, C. F. 1998. Dissipation in compressible magnetohydrodynamic turbulence. *Astrophys. J. Lett.* 508:L99–L102.

Strobel, A. 1992. Age subgroups in open clusters. *Astron. Astrophys.* 253:374–378.

Strom, K. M., Strom, S. E., and Merrill, K. M. 1993. Infrared luminosity functions for the young stellar population associated with the L1641 molecular cloud. *Astrophys. J.* 412:233–253.

Stutzki, J., and Güsten, R. 1990. High spatial resolution isotopic CO and CS observations of M17 SW: The clumpy structure of the molecular cloud core. *Astrophys. J.* 356:513–533.

Stutzki, J., Bensch, F., Heithausen, A., Ossenkopf, V., and Zielinsky, M. 1998. On the fractal structure of molecular clouds. *Astron. Astrophys.* 336:697–720.

Sugitani, K., Tamura, M., and Ogura, K. 1995. Young star clusters in bright-rimmed clouds: Small-scale sequential star formation? *Astrophys. J. Lett.* 455:L39–L41.

Surdin, V. G. 1979. Tidal destruction of globular clusters in the Galaxy. *Sov. Astron.* 23:648–653.

Tapia, M., Persi, P., and Roth, M. 1996. The embedded stellar population in northern NGC 6334. *Astron. Astrophys.* 316:102–110.

Tenorio-Tagle, G. 1981. The collision of clouds with the galactic disk. *Astron. Astrophys.* 94:338–344.

Tenorio-Tagle, G., Munoz-Tunon, C., and Cox, D. P. 1993. On the formation of spheroidal stellar systems and the nature of supersonic turbulence in star-forming regions. *Astrophys. J.* 418:767–773.

Testi, L., Felli, M., Persi, P., and Roth, M. 1994. Near-infrared images of galactic masers I. Association between infrared sources and masers. *Astron. Astrophys.* 288:634–646.

Testi, L., Palla, F., and Natta, A. 1998. A search for clustering around Herbig Ae/Be stars. II. Atlas of the observed sources. *Astron. Astrophys. Suppl.* 133:81–121.

Theuns, T. 1990. A combination of SPH and N-body2 for gas dynamics in star clusters. *Astrophys. Space Sci.* 170:221–224.

Usami, M., Hanawa, T., and Fujimoto, M. 1995. High-velocity oblique cloud collisions and gravitational instability of a shock-compressed slab with rotation and velocity shear. *Pub. Astron. Soc. Japan* 47:271–285.

Vacca, W. D., Garmany, C.D., and Shull, J.M. 1996. The Lyman-continuum fluxes and stellar parameters of O and early B-type stars. *Astrophys. J.* 460:914–931.

Vallenari, A., Bettoni, D., and Chiosi, C. 1998. Clusters in the west side of the bar of the Large Magellanic Cloud: Interacting pairs? *Astron. Astrophys.* 331:506–518.

van den Bergh, S., and Lafontaine, A. 1984. Luminosity function of the integrated magnitudes of open clusters. *Astron. J.* 89:1822–1824.

Vazquez, R. A., Baume, G., Feinstein, A., and Prado, P. 1996. Investigation on the region of the open cluster Tr 14. *Astron. Astrophys. Suppl.* 116:75–94.

Verschueren, W. 1990. Collapse of young stellar clusters before gas removal. *Astron. Astrophys.* 234:156–163.

Vietri, M., and Pesce, E. 1995. Yet another theory for the origin of halo globular clusters and spheroid stars. *Astrophys. J.* 442:618–627.

Vogelaar, M. G. R., and Wakker, B. P. 1994. Measuring the fractal structure of interstellar clouds. *Astron. Astrophys.* 291:557–568.

von Hoerner, S. 1968. The formation of stars. In *Interstellar Ionized Hydrogen,* ed. Y. Terzian (New York: Benjamin), pp. 101–170.

Walborn, N. R., and Blades, J. C. 1997. Spectral classification of the 30 Doradus stellar populations. *Astrophys. J. Suppl.* 112:457–485.

Walborn, N. R., Barba, R. H., Brandner, W., Rubio, M., Grebel, E. K., and Probst, R. G. 1999. Some characteristics of current star formation in the 30 Doradus Nebula revealed by HST/NICMOS. *Astron. J.* 117:225–237.

Welch, W. J., Dreher, J. W., Jackson, S. M., Terebey, S., and Vogel, S. N. 1987. Star formation in W49A: Gravitational collapse of a molecular cloud core toward a ring of massive stars. *Science* 238:1550–1555.

White, G. J., Casali, M. M., and Eiroa, C. 1995. High resolution molecular line observations of the Serpens Nebula. *Astron. Astrophys.* 298:594–605.

Whitmore, B. C., and Schweizer, F. 1995. Hubble space telescope observations of young star clusters in NGC-4038/4039, "the antennae" galaxies. *Astron. J.* 109:960–980.

Whitworth, A. P. 1979. The erosion and dispersal of massive molecular clouds by young stars. *Mon. Not. Roy. Astron. Soc.* 186:59–67.

Whitworth, A. P., and Clarke, C. J. 1997. Cooling behind mildly supersonic shocks in molecular clouds. *Mon. Not. Roy. Astron. Soc.* 291:578–584.

Whitworth, A. P., Boffin, H. M. J., and Francic, N. 1998. The thermodynamics of dense cores. In *Star Formation with the Infrared Space Observatory,* ASP Conf. Ser. 132, ed. J. Yun and R. Liseau (San Francisco: Astronomical Society of the Pacific), pp. 183–188.

Wilking, B. A., and Lada, C. J. 1983. The discovery of new embedded sources in the centrally condensed core of the Rho Ophiuchi dark cloud: The formation of a bound cluster. *Astrophys. J.* 274:698–716.

Wilking, B. A., and Lada, C. J. 1985. The formation of bound stellar clusters. In *Protostars and Planets II,* ed. D. C. Black and M. S. Matthews (Tucson: University of Arizona), pp. 297–319.

Wilking, B. A., Harvey, P. M., Joy, M., Hyland, A. R., and Jones, T. J. 1985. Far-infrared observations of young clusters embedded in the R Coronae Australis and Rho Ophiuchi dark clouds. *Astrophys. J.* 293:165–177.

Wilking, B. A., Lada, C. J., and Young, E. T. 1989. IRAS observations of the Rho Ophiuchi infrared cluster: Spectral energy distributions and luminosity function. *Astrophys. J.* 340:823–852.

Wilking, B. A., McCaughrean, M. J., Burton, M. G., Giblin, T., Rayner, J. T., and Zinnecker, H. 1997. Deep infrared imaging of the R Coronae Australis cloud core. *Astron. J.* 114:2029–2042.

Williams, J. P., de Geus, E. J., and Blitz, L. 1994. Determining structure in molecular clouds. *Astrophys. J.* 428:693–712.

Wynn-Williams, C. G., Becklin, E. E., and Neugebauer, G. 1972. Infra-red sources in the H II region W 3. *Mon. Not. Roy. Astron. Soc.* 160:1–14.

Zimmermann, T., and Stutzki, J. 1992. The fractal appearance of interstellar clouds. *Physica A* 191:79–84.

Zimmermann, T., and Stutzki, J. 1993. Fractal aspects of interstellar clouds. *Fractals* 1:930–938.

Zinnecker, H. 1982. Prediction of the protostellar mass spectrum in the Orion near-infrared cluster. In *Symposium on the Orion Nebula to Honor Henry Draper,* ed. A. E. Glassgold, P. J. Huggins, and E. L. Schucking (New York: New York Academy of Science), pp. 226–235.

Zinnecker, H. 1986. IMF in starburst regions. In *Light on Dark Matter, Astrophysics and Space Science Library,* Vol. 124, ed. F. P. Israel (Dordrecht: Reidel), pp. 277–278.

Zinnecker, H., Keable, C. J., Dunlop, J. S., Cannon, R. D., and Griffiths, W. K. 1988. The nuclei of nucleated dwarf elliptical galaxies: Are they globular clusters? In *The Harlow Shapley Symposium on Globular Cluster Systems in Galaxies,* IAU Symp. 126, ed. J. E. Grindley and A. G. Davis Philip (Dordrecht: Kluwer), pp. 603–604.

Zinnecker, H., McCaughrean, M. J., and Wilking, B. A. 1993. The initial stellar population. In *Protostars and Planets III,* ed. E. H. Levy and J. I. Lunine (Tucson: University of Arizona), pp. 429–495.

Zinnecker, H., and Palla, F. 1987. Star formation in proto-globular cluster clouds. In *ESO Workshop on Stellar Evolution and Dynamics in the Outer Halo of the Galaxy,* ed. M. Azzopardi and F. Matteucci (Garching: ESO), pp. 355–361.

OBSERVATIONS OF INFALL IN STAR-FORMING REGIONS

PHILIP C. MYERS
Harvard-Smithsonian Center for Astrophysics

NEAL J. EVANS II
University of Texas at Austin

and

NAGAYOSHI OHASHI
Academia Sinica Institute of Astronomy and Astrophysics

Evidence of inward gas motions in star-forming regions is now abundant, due to improvements in the sensitivity and resolution of millimeter-wavelength telescopes and spectrometers. "Infall asymmetry," a characteristic signature of certain line profiles predicted for contracting clouds, has been detected in numerous dense cores, some with no embedded stars and some with deeply embedded Class 0 and Class I objects. Infall asymmetry is fairly common in cores with highly embedded sources; further studies of less embedded sources are needed to trace the evolution of infall. In several starless cores, infall asymmetry is more extended than expected for "inside-out" gravitational collapse and may indicate speeds greater than expected from most models of ambipolar diffusion. Interferometric observations of flattened envelopes, using optically thin lines, reveal infall and rotation as velocity gradients along the minor and major axes, on size scales $<$1000 AU. Observed line profiles toward B335 match detailed radiative transfer models of emission from cores with embedded protostars undergoing inside-out collapse. Models that include modest rotation can match the observations of L1527, and IRAS 16293–2422 can be modeled with substantial rotation. Models of infall in starless cores, cluster-forming cores, and cores dominated by magnetic fields and turbulent motions are being developed, but they remain to be tested against observations.

I. INTRODUCTION

The origin of stars is one of the oldest problems in astrophysics. In the last three decades this problem has begun to yield to improvements in a wide variety of observational techniques and to improvements in theoretical understanding of the gravitational, magnetic, and turbulent interactions in interstellar clouds. Star-forming regions of dense gas have been identified through observations of spectral lines of CO (Wilson et al. 1970; Kutner et al. 1977), NH_3 (Cheung et al. 1968; Ho et al. 1979; Myers and

Benson 1983), and over 100 other species. Candidate protostars have been identified through observations of far infrared and submillimeter emission (Keene et al. 1980; Lada and Wilking 1984; Beichman et al. 1986; André et al. 1993). Theoretical models detail how protostars form from dense gas by gravitational infall (Larson 1969; Penston 1969; Mouschovias 1976; Shu 1977; Boss 1982; Terebey et al. 1984; Hartmann 1998).

Despite this progress, the link is weak between the "initial conditions," or gas properties of dense cores, and the "protostars" that they form. The key information needed to account for star formation and to distinguish among competing models consists of the spatial distribution of inward gas motions and the evolution of the distribution over time. It is now becoming possible to study this process of gravitational infall through Doppler spectroscopy of molecular spectral lines that trace sufficiently dense gas, as observed with fine spectral and spatial resolution. In this chapter we describe recent progress in the study of such infall, primarily from an observational viewpoint. Closely related discussions are given by Myers (1997*a*,*b*) and Evans (1999).

The plan of this chapter is as follows. We describe the main observational techniques used to study infall (section II) and surveys of embedded sources and starless cores to identify kinematic infall candidates (section III). We then describe maps of these candidates (section IV), interferometric observations of selected regions (section V), and comparison of observations and models (section VI).

II. INFERRING INWARD MOTIONS

Spectroscopic methods of inferring motion along the line of sight rely on distinguishing velocities of absorption and emission in a line profile, to obtain the sign and speed of such motions. A spectral signature of inward motion, "infall asymmetry," is observable if the foreground infalling gas has lower excitation temperature than the background gas and if the foreground gas has sufficient optical depth (e.g., Lucas 1976; Leung and Liszt 1976; Leung and Brown 1977). If these conditions are met, the line will be skewed to the blue or double-peaked, with a stronger blue peak. The detailed line shape depends on the velocity field.

In the simplest model, two uniform layers approach each other (Myers et al. 1996). If the approach speed is less than the velocity dispersion, the absorption appears as a dip between the brighter blue peak and the fainter red peak. As the approach speed increases, more of the red peak is absorbed by the foreground layer, and the blue peak becomes brighter while the red peak disappears into a wing or shoulder (Fig. 1). If the approach speed is much greater than the velocity dispersion, as may occur for the Larson-Penston type collapse, two separated peaks of nearly equal strength will appear (Zhou 1992), and optically thin lines will also be double-peaked. If a static envelope surrounds the infall zone, as in inside-

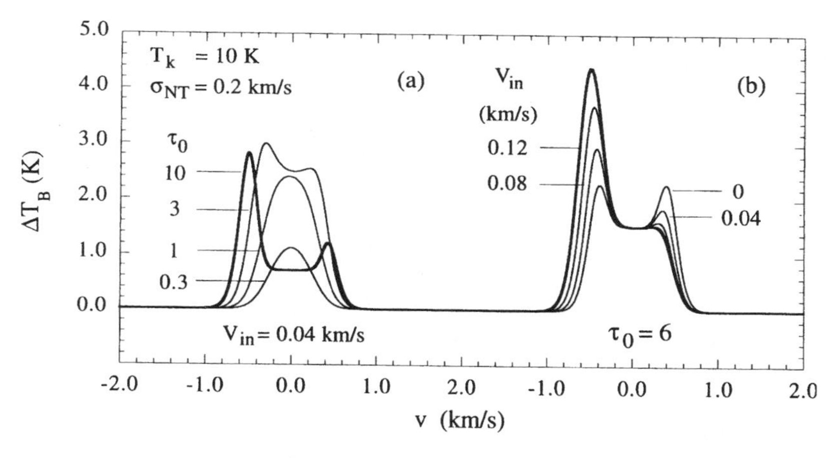

Figure 1. Variation of infall asymmetry (a) with peak optical depth τ_0 and (b) with infall speed V_{in}. In a radiative transfer model of two uniform layers with velocity dispersion σ and approach speed V_{in}, the line profile is symmetric for $\tau_0 < 1$, but its peak skews to the blue as τ_0 increases beyond 1. For $V_{in} < \sigma$ and increasing τ_0, the profile has two peaks, with increasing ratio of blue to red peak intensity. As V_{in} increases for fixed $\tau_0 > 1$, the blue-red intensity ratio increases until the red peak disappears into a red shoulder (Myers et al. 1996).

out collapse (Shu 1977), the dip velocity is independent of infall speed, and the line profile evolves as the infall zone expands (Zhou 1992; Gregersen 1998). If two unrelated clouds with similar velocities lie along the line of sight, their spectrum in an optically thick line can have two peaks, giving spurious infall asymmetry. However, observation of an optically thin line will then reveal two components, not one component as in true infall asymmetry.

If the observed system is known to have a flattened geometry, then further simplification is possible, using maps of optically thin lines. Then rotation and infall motions separate cleanly, revealed as velocity gradients along the projected major and minor axes (section V).

III. SURVEYS FOR INFALL ASYMMETRY

The earliest reports of infall asymmetry came from observations of emission in the $J = 1$–0 lines of ^{12}CO and ^{13}CO from gas associated with luminous infrared sources and H II regions (Snell and Loren 1977; Loren et al. 1981; Phillips et al. 1981). However, these observations have not been generally accepted as indicating infall, because of confusion from outflows, because of numerous CO spectra with reverse infall asymmetry toward similar sources, and because the low-density gas traced by these lines appears too extended for star-forming infall. More recently, observations of the $J = 5$–4 and 2–1 lines of CS toward IRAS 16293 − 2426 (Walker et al.

1986) were reported as consistent with the "inside-out" collapse model of Shu (1977). Menten et al. (1987) challenged this claim, based on maps of CS line profiles that showed a reversal of the blue-red asymmetry. They interpreted their data by a combination of rotation and absorption by a foreground cloud. Further complications arose as this source was shown to have a quadrupolar outflow (Mizuno et al. 1990). See section VI for further discussion.

More recent observations of the deeply embedded submillimeter source in B335 (Keene et al. 1983) in lines of CS and H_2CO (Zhou et al. 1993, 1994; Choi et al. 1995) indicate a detailed match of spectral profiles with predictions of the inside-out collapse model (Shu 1977). Similar spectra of the low-luminosity source 04368+2557 in the core L1527 are reasonably consistent with radiative transfer models based on inside-out collapse (Myers et al. 1995; Zhou 1995). More detailed discussions of B335 and L1527 are given in section VI. Further claims of infall based on spectral profiles toward individual sources have been presented by Moriarty-Schieven et al. (1995), Hurt et al. (1996), Ward-Thompson et al. (1996), Wolf-Chase and Gregersen (1997), Lehtinen (1997), Tafalla et al. (1998), Wiesemeyer (1997), Williams et al. (1999), and Williams and Myers (1999).

On the other hand, observations of infall asymmetry in a few selected objects may arise from peculiarities of source structure or from the confusing effects of outflows. Further, a few objects are not necessarily representative. Thus, it is necessary to survey objects selected by objective criteria, to search for a statistically significant excess of sources with infall asymmetry. The excess can be defined as $(N_b - N_r)/N$, where N_b is the number of sources with a blue-skewed profile, N_r is the number with a red-skewed profile, and N is the total number in the sample (Mardones et al. 1997). If the blue-skewed profiles were caused by rotation or outflows rather than collapse, then a sufficiently large sample, having a random distribution of angles between the line of sight and their rotation and outflow axes, should have no significant excess. Two recent surveys, conducted with this motivation, show a substantial excess of sources with infall asymmetry, supporting the collapse interpretation.

A survey of 23 Class 0 objects (André et al. 1993) was made in the $J = 4$–3 and 3–2 lines of HCO^+, and 19 of these were observed in the optically thin $J = 3$–2 line of $H^{13}CO^+$ (Gregersen et al. 1997). The spectral energy distributions of all of the sources satisfied within their uncertainties $T_{bol} < 70$ K, a value found by Chen et al. (1995) to correspond to known Class 0 sources. T_{bol} is the effective temperature of a blackbody with the same mean frequency as the observed spectrum (Myers and Ladd 1993). Gregersen et al. (1997) found nine sources with infall asymmetry but three with the opposite asymmetry, a "blue excess" of about 0.26. This result supports the interpretation of blue-skewed profiles as indicators of collapse.

A similar survey was carried out by Mardones et al. (1997). The sources were restricted to lie within 400 pc of the Sun and to have $T_{\rm bol} <$ 200 K. The sample of Mardones et al. (1997) included many of the Gregersen et al. (1997) Class 0 sources, but it also included 24 Class I sources (Lada and Wilking 1984), with $70 < T_{\rm bol} < 200$ K. The survey used the 1–0 line of N_2H^+ as the optically thin tracer (Caselli et al. 1995) and two independent lines as thick tracers: the 2_{12}–1_{11} line of H_2CO (47 sources) and the 2–1 line of CS (37 sources).

Mardones et al. (1997) found that each of the two optically thick tracer lines shows a statistically significant excess (0.39 for H_2CO; 0.53 for CS), when the sample is restricted to Class 0 sources. This result corroborates the findings of Gregersen et al. (1997), and indeed the three spectral line surveys agree as to the presence or absence of infall asymmetry, for most sources observed in common. No significant excess was seen among Class I sources, suggesting that infall asymmetry is much more prevalent in Class 0 than in Class I sources.

Because the infall signature should become more difficult to detect at later stages, when opacities drop, infall in Class I sources might be harder to detect in lines of modest opacity. In addition, the orientation of a flattened envelope can affect the spectral appearance. An edge-on envelope will produce deeper self-absorption than a pole-on envelope. Edge-on envelopes are indeed suggested, because the outflow lies primarily in the plane of the sky, in the two best infall candidates: B335 (Hirano et al. 1988) and L1527 (Zhou et al. 1996, Tamura et al. 1996).

To test these possibilities, Gregersen et al. (1998) used the more opaque HCO^+ 3–2 line to survey 20 of the Class I sources observed by Mardones et al. (1997). Using the same criteria as Mardones et al. (1997), and combining with results from Gregersen et al. (1997), Gregersen et al. (1998) find that among 23 Class I sources, the excess in the HCO^+ line is 0.32, essentially the same as the value (0.33) found for the full sample of Class 0 sources. While further work is necessary, it is clear that infall cannot be ruled out because of the absence of an infall signature in any particular line. Important goals of future work are to understand how lines of different optical depth can give profiles with significantly different asymmetry, to study the time evolution of infall, and to test the evolutionary interpretation of the classes described by Adams et al. (1987).

The surveys of Mardones et al. (1997) and Gregersen et al. (1997) have identified or confirmed most of the known kinematic infall candidates. The following 13 cores have infall asymmetry in at least two lines, and no reverse infall asymmetry, among the lines of CS 2–1, H_2CO 2_{12}–1_{11}, HCO^+ 3–2, and HCO^+ 4–3: NGC 1333 IRAS 4A; NGC 1333 IRAS 4B; L1527; I13036–7644; VLA 1623; WL22; I16293–2422; L483; Serpens SMM4; Ser SMM5; B335; L1157; and L1251B. Three of these cores (L1527, B335, L1157) have lines that are "thermally dominated" and are associated with single sources, whereas the rest are dominated by turbulent motions and are generally associated with clusters or small groups of young stars.

A survey of 19 regions at distances <1 kpc, known to be sites of multiple star formation, and distinct from the sources studied by Mardones et al. (1997), was carried out in lines of CS 2–1 and N_2H^+ 1–0 (Williams and Myers 1999*a*). No statistically significant excess of sources with blue asymmetry was found, perhaps due in part to the relatively low optical depth of the CS 2–1 line in these sources: Very few of the CS spectra showed significant self-absorption of either blue or red sign. However, one remarkable source, Cepheus A, shows infall asymmetry extended over $\sim$0.2 pc, substantially greater than expected from simple models of gravitational infall.

The foregoing surveys are based on observations of infrared point sources and are therefore sensitive to the motions associated with already formed, or forming, stars. Another useful approach is to survey starless cores, or regions that show no evidence of an associated star but have unusually high density and column density. These cores are good candidates to form stars in the future, and observations in infall tracer lines may be expected to show infall asymmetry. Starless cores do not have embedded point sources to mark survey target positions, but their infall signatures, if detected, will not suffer confusion from stellar winds and outflows. A systematic survey of 224 starless cores, drawn from the catalog of Lee and Myers (1999), has been carried out in lines of CS and N_2H^+, indicating infall asymmetry in numerous cores (Lee et al. 1999). A survey of a smaller sample of 17 cores, using HCO^+ $J = 3$–2 (Gregersen 1998), found a significant excess of sources with infall asymmetry. All the sources with infall asymmetry were detected in the submillimeter continuum by Ward-Thompson et al. (1994), who called them "pre-protostellar."

IV. MAPS OF INFALL ASYMMETRY

Detailed tests of consistency between observations of infall asymmetry and models of collapse require maps in both optically thick and thin spectral lines. The maps reveal the center, shape, and extent of the zone of infall asymmetry and allow comparison to models. Mapping observations are also needed to discriminate infall from rotation and bipolar outflows, each of which can create the appearance of infall asymmetry in line profiles at particular positions. For example, a differentially rotating cloud with rotation axis in the plane of the sky can have line profiles with blue asymmetry on one side of the axis and red asymmetry on the other, while a nonrotating, contracting cloud will have blue asymmetry in all directions from the center of infall. Similarly, emission by low-velocity gas in the blue lobe of a bipolar outflow region can be absorbed by quiescent foreground gas, yielding spectra with infall asymmetry. The spectra arising from these various motions may be identical in a particular map position, but the motions can be readily distinguished by comparing the maps (Adelson and Leung 1988; Cabrit and Bertout 1986; Walker et al. 1994; Myers et al. 1995; Gregersen et al. 1997).

An important consideration is the sensitivity of infall asymmetry to line optical depth and excitation temperature. The spatial extent of the infall asymmetry in a cloud can vary greatly from tracer to tracer, as the optical depth of the tracer increases. For example, the extent of the infall asymmetry in the starless core L1544 increases from ~0.01 pc in the 1–0 line of N_2H^+ to ~0.1 pc in the 2–1 line of CS, probably because of differences in the spatial extent of the optically thick emission of the two species (Tafalla et al. 1998; Williams et al. 1998). Therefore, maps in more than one optically thick tracer are needed to distinguish spatial variations in optical depth from spatial variations in the velocity field.

In the nearest star-forming complexes, the 10–30$''$ angular resolution of large filled-aperture antennas is 0.007–0.02 pc at a distance of 140 pc, suitable for probing the inward motions expected for gravitational infall. For example, the diameter of a low-mass NH_3 core is ~0.1 pc, which is well resolved in one of the nearest star-forming regions. On the other hand, resolution of circumstellar envelope gas in more distant star-forming regions, or of the motions associated with circumstellar disks in nearby regions, requires finer resolution than filled-aperture telescopes can provide. For these observations it is necessary to use interferometers. With angular resolution generally 1–5$''$, these instruments can probe much finer scales than can the filled-aperture telescopes. In a few cases, single-antenna and interferometer maps have been combined (e.g., Zhou et al. 1996), yielding maps sensitive to a range of spatial scales spanning a factor of ~30 (e.g., Gueth and Guilloteau 1999). Such interferometric observations are discussed in section V.

We illustrate several features of infall asymmetry with maps of three distinctly different sources. L1544 is a starless dense core with no known IRAS point source or embedded T Tauri star. The nonthermal motions in its spectral lines that trace dense gas ($n > 10^4$ cm^{-3}) are smaller than the thermal motions of the molecule of mean mass, so it is "thermally dominated" (Myers et al. 1991). Its likely stellar product, if any, is a single star or binary, as opposed to a stellar group or cluster, based on the star-forming character of its neighboring cores in the Taurus complex. L1527 is another thermally dominated core in Taurus, but with a single embedded Class 0 IRAS source 04368+2557. L1251B is a core whose lines are dominated by turbulent motions and that has a Class I IRAS source and an associated group of ~5 additional young stellar objects (Hodapp 1994).

Figure 2 illustrates the variety of spectral profiles taken at the position of peak emission in L1544, ranging from extremely optically thick (HCO^+ 1–0) with strong infall asymmetry, to optically thin ($C^{34}S$ 2–1) with no infall asymmetry. This sequence indicates that the optical depth of the spectral line is critically important to the degree of infall asymmetry.

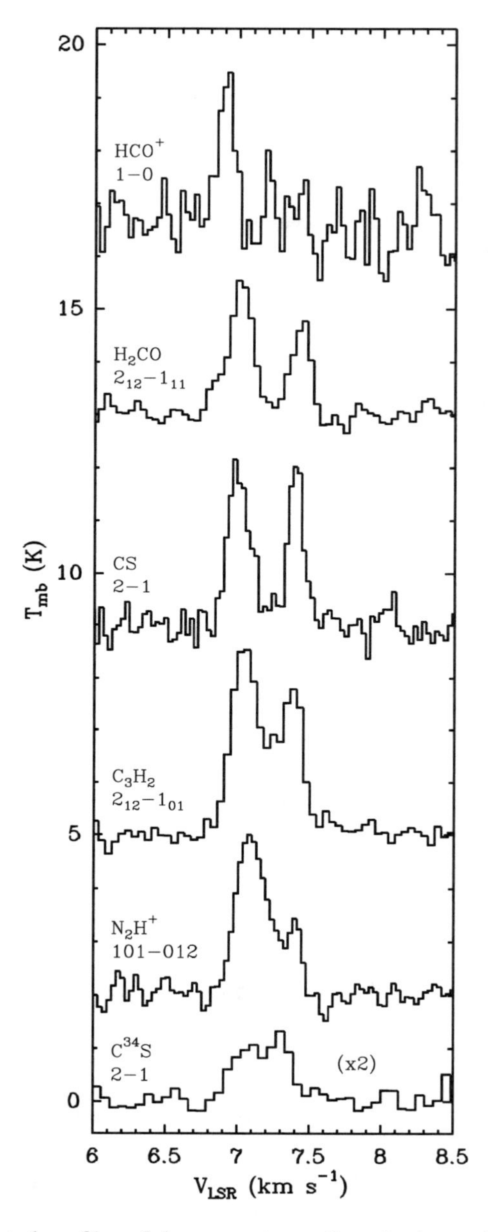

Figure 2. Spectral profiles of dense gas tracer lines in the starless core L1544. These lines vary from HCO^+ 1–0, with high optical depth and strong infall asymmetry, to $C^{34}S$ 2–1, with low optical depth and no infall asymmetry (Tafalla et al. 1998).

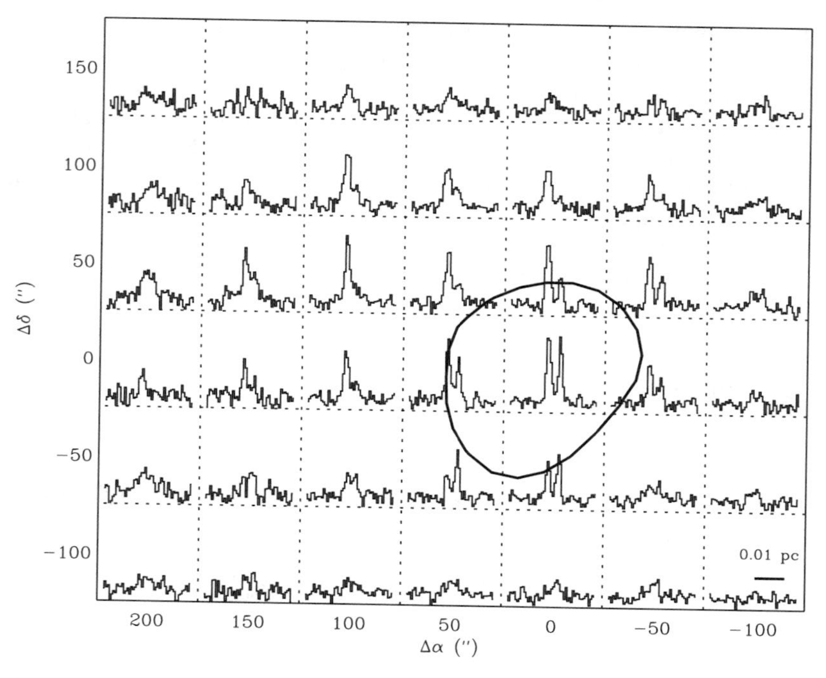

Figure 3. Spectral profiles of CS 2–1 in L1544, in relation to the half-maximum integrated intensity contour of the optically thin N_2H^+ 1–0 line, taken as a tracer of the dense core gas. Each spectral window has velocity range 5.7–8.7 km s^{-1} and brightness temperature range −0.5–4.0 K. The (0,0) position, epoch 1950, is $(\alpha, \delta) = (05^h01^m12^s.5, 25°06'40'')$. Profiles with infall asymmetry are extended over ~0.15 pc, substantially greater than expected from "inside-out collapse" and substantially greater than the extent of the dense core (Tafalla et al. 1998).

Figure 3 shows a grid of spectra illustrating the large extent of infall asymmetry in the CS 2–1 line, ~0.15 pc, or well beyond the extent of the dense core, ~0.04 pc, as indicated by the half-maximum (HM) integrated intensity contour of the optically thin N_2H^+ 1–0 emission. This core definition is arbitrary but useful, because it allows comparison with N_2H^+ maps of other cores. The CS map has no well-defined center of line asymmetry, and the position of maximum asymmetry lies outside the HM of the dense core emission (Tafalla et al. 1998). In contrast to this behavior, interferometric observations of the N_2H^+ emission shows that infall asymmetry and line intensity peak at the same position. Remarkably, the N_2H^+ line indicates an inward speed similar to that of the much more extended CS infall asymmetry, ~0.1 km s^{-1} (Williams et al. 1999).

It is difficult to account for the large extent of the infall asymmetry in L1544 with the simplest models of gravitational infall or of gravitational motion limited by ambipolar diffusion (Tafalla et al. 1998). If the extent of the L1544 infall asymmetry were comparable to the diameter of the rarefaction wave associated with inside-out collapse (Shu 1977), then a protostellar point source with luminosity of several $L_\odot$ should have already formed, whereas none is known. Similarly, models of ambipolar diffusion, such as the detailed numerical model of Ciolek and Mouschovias (1995), predict line-of-sight inward speeds <0.02 km s^{-1} at the largest observed radius $\sim$0.08 pc, where the observed speed is as much as 0.1 km s^{-1}. It may be possible for ambipolar diffusion to account for these motions under

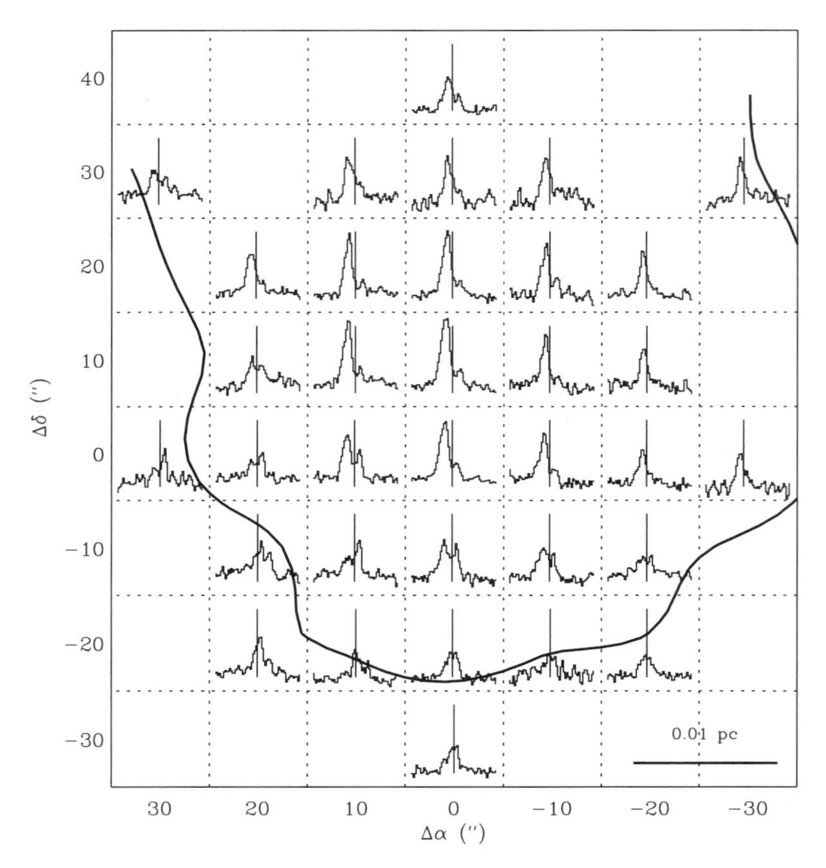

Figure 4. Spectral profiles of H_2CO 2_{12}–1_{11} in the "thermal" core L1527, in relation to the half-maximum integrated intensity contour of the optically thin N_2H^+ 1–0 line. Each spectral window has velocity range 4.5–7.5 km s^{-1} and brightness temperature range −0.8–5.0 K. The vertical line marks the velocity of optically thin N_2H^+ emission. The (0,0) position, epoch 1950, is $(\alpha, \delta) = (04^h36^m49^s.3, 25°57'16'')$. Profiles with infall asymmetry have extent $\sim$0.04 pc (less than that of the dense core, $\sim$0.06 pc) and are well centered on the single embedded Class 0 source, at position offset (3.9″, 4.6″) (Mardones et al. 1999).

different initial conditions (Ciolek and Basu 1999). On the other hand, lines of the ions HCO^+, DCO^+, and N_2H^+ all show infall asymmetry in L1544, thus ruling out ambipolar diffusion in its most extreme form, in which the ions are stationary while the neutrals flow inward.

Alternatively, such extended, subsonic, inward motion may start as a pressure-driven flow, arising from differential dissipation of turbulence. Starless cores may gradually dissipate the nonthermal motions in their linewidths, due, e.g., to ion-neutral friction or to nonlinear interactions among magnetohydrodynamic (MHD) waves (Nakano 1998). If turbulence dissipates at a greater rate in the core than in its surroundings, the resulting inward pressure gradient should lead to an inward flow. If the resulting increase in core density further increases the turbulent dissipation, then the flow should be self-sustaining (Myers and Khersonsky 1995, Myers and Lazarian 1998). The predicted flow speeds are similar to the sound speed and larger than those expected for ambipolar diffusion.

Figure 4 shows spectra of the 2_{12}–1_{11} line of H_2CO in L1527, in relation to the Class 0 IRAS source 04368+2557 and the half-maximum contour of its N_2H^+ integrated intensity (Mardones et al. 1999). The L1527 and L1544 cores are similar in their thermally dominated linewidths and in their dense core sizes ($\sim$0.05 pc) but different in their extents of infall asymmetry (0.04 and 0.15 pc). In L1527, the infall asymmetry is much more localized than in L1544, as expected if the inward motions in L1527 are primarily gravitational in origin. The infall asymmetry is more pronounced in the north–south direction, whereas the outflow is concentrated in the east–west direction, as is evident in the H_2CO spectral line wings. Thus, confusion by the outflow cannot account for most of the spectra with infall asymmetry. As discussed in section VI, L1527 resembles B335 in the relatively good agreement of its spectral profiles, spatial extent of infall asymmetry, and stellar luminosity with the inside-out collapse model. On the other hand, many H_2CO spectra in L1527 have ratios of blue and red peak line intensities significantly greater than the model predicts (see section VI).

Figure 5 shows spectra of the 2_{12}-1_{11} line of H_2CO in L1251B, in relation to the Class I IRAS source 22377+7455 and the half-maximum contour of its N_2H^+ integrated intensity (Mardones et al. 1999). The infall asymmetry is about as extended as the dense core, thus resembling L1527 but with a wider line, suggestive of higher turbulence, than in L1527. Furthermore, a second zone of infall asymmetry is seen about one core diameter to the east, where no infrared source is known and where a map of N_2H^+ emission indicates very little dense gas (Caselli et al. 1999). In this respect, L1251B shows infall asymmetry beyond the zone of dense core gas and so resembles L1544. Extended infall asymmetry, beyond the dense core half-maximum boundary, is also seen in a zone about $1'$ NW of NGC 1333 IRS4A (Mardones et al. 1999).

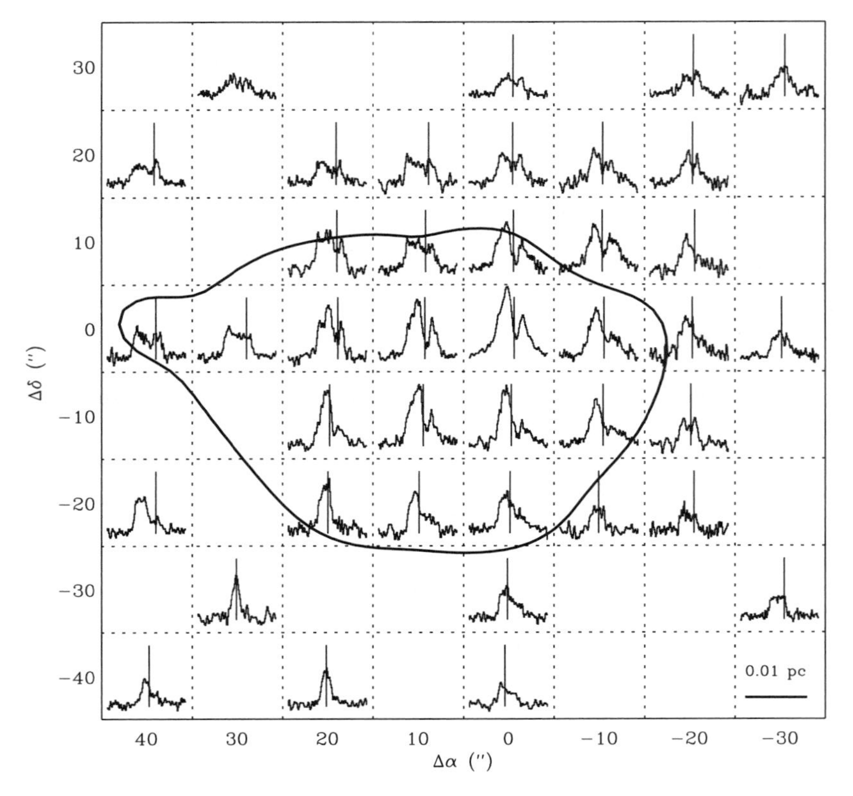

Figure 5. Spectral profiles of H_2CO 2_{12}–1_{11} in the "turbulent" core L1251B, in relation to the half-maximum integrated intensity contour of the optically thin N_2H^+ 1–0 line. Each spectral window has velocity range −7.0 to −1.0 km s^{-1} and brightness temperature range −0.8 to 4.0 K. The vertical line marks the velocity of optically thin N_2H^+ emission. The (0,0) position, epoch 1950, is $(\alpha, \delta) = (22^h37^m40^s.8, 74°55'50'')$. Profiles with infall asymmetry have extent similar to that of the dense core (~0.06 pc) and are well centered on the embedded IRAS source and small stellar group at position offset (0,0) (Mardones et al. 1999).

V. INTERFEROMETRIC IMAGING OF INFALL CANDIDATES

A. Overview

In addition to single-dish telescopes, interferometers are powerful tools to observe infall. The finer angular resolution provided by interferometers enables us to distinguish infall from rotation or outflow, motions sometimes confused in filled-aperture maps.

As with filled-aperture observations, moderately optically thick molecular lines have been observed with interferometers to detect the infall asymmetry. Two good examples are B335 (^{13}CO 1–0, Chandler and Sargent 1993; HCO^+ 1–0 and HCN 1–0, Choi et al. 1998) and L1527

(^{13}CO1–0, Zhou et al. 1996; H_2CO 3_{12}–2_{11}, Wilner et al. 1997). These observations show clear infall asymmetries, which are consistent with the prediction from collapse models. Some sources, however, show only blueshifted peaks in interferometric observations, even though they show clear infall asymmetries in single-dish observations. This could be because optical depths of their redshifted emissions are too high (Choi et al. 1998) or because the redshifted emissions are more spatially extended than the corresponding blueshifted emissions, so the redshifted emissions are resolved out by interferometers (Ohashi et al. 1997*a*). It is also important to note that central dips of infall asymmetries observed with interferometers can be spurious, because the interferometer's spatial filtering resolves out the extended ambient gas (e.g., Gueth and Guilloteau 1999).

The ability of interferometers to image with fine angular resolution is valuable in observing infall. Interferometers can resolve the kinematic structure of gaseous envelopes. In fact, interferometric observations have shown that gaseous envelopes often have flattened structures, oriented perpendicular to the associated outflow axis (Sargent and Beckwith 1987, 1991; Sargent et al. 1988; Mundy et al. 1992; Ohashi et al. 1991, 1996*b*). This flattened envelope is substantially larger than the rotationally supported circumstellar disk, of radius $\sim$100 AU, which it may contain. If infalling motions exist in such a flattened envelope, blueshifted emission due to infall will be observed in the far side of the envelope and redshifted emission will be observed in the near side. Thus, infall can be identified as a velocity gradient along the projected minor axis of the flattened envelope. Rotation is well distinguished from infall, because rotation contributes to a velocity gradient along the projected major axis of the envelope.

The strategy just outlined allows us to image infalling motions directly. Thus, interferometric observations of HL Tauri indicate a flattened envelope, perpendicular to the associated optical jet, and a velocity gradient along the minor axis of the envelope. This pattern was interpreted as indicating infall (Hayashi et al. 1993). It is possible for outflow to contaminate infall, because outflow shows the same velocity gradient as infall (Cabrit et al. 1996). Nevertheless, the fraction of emission from outflowing gas may be small, judging from velocity-channel maps, because the *entire* flattened envelope shows a simple kinematic pattern suggestive of infall, with emission basically residing in the flattened envelope at *all* velocities. After the observations of HL Tauri, more sources, including infall candidates identified through single-dish observations, have been observed with interferometers, demonstrating that infalling envelopes represent distinctive kinematic features in position-velocity diagrams (Ohashi et al. 1996*a*, 1997*a*,*b*; Ohashi 1997; Saito et al. 1996; Momose et al. 1998). In addition to the velocity gradient along the minor axis, these observations show that linewidths of the observed line emission get wider toward the inner radii of the envelopes, consistent with acceleration toward the central gravitational source. Note that these observations have been made in relatively optically thin lines, such as $C^{18}O$ or $H^{13}CO^+$, because optically thin lines allow us to examine the whole velocity structure of the envelope without self-absorption. Because some sources show kinematic structure due to infall in their interferometric maps even though no infall asymmetries have been observed with single-dish telescopes, systematic searches for infall candidates with interferometers are very important.

B. Physical Properties of Envelopes around Infall Candidates: The Case of L1551 IRS5

Interferometric imaging of envelopes around infall candidates provides information on the detailed physical properties of infalling envelopes. Infalling envelopes tend to be characterized by elongated, flattened structures with a scale of thousands of AU, while their kinematics are explained by dynamical infall with slow rotation.

Figure 6 shows an excellent example of interferometric images of an infalling envelope, that shows the above characteristics, obtained in $C^{18}O$ 1–0 from L1551 IRS5 with the Nobeyama Millimeter Array (Momose et al. 1998). The envelope around L1551 IRS5 has a clear flattened structure, 2400 AU $\times$ 1100 AU in size, which is elongated in the direction perpendicular to the axis of the associated outflow (Fig. 6a). The southern half of the envelope is blueshifted while its northern half is redshifted, as shown in the intensity-weighted mean velocity map (Fig. 6b); the overall velocity gradient of the envelope is in the north-south direction, which is different from the direction of either the minor or the major axis. This suggests that both infall and rotation exist in the envelope, because pure infall yields a velocity gradient along the minor axis, but pure rotation yields a gradient along the major axis.

Detailed velocity structures of infall and rotation can be examined by using position-velocity (PV) diagrams. The PV diagram along the major axis of the envelope (Fig. 6c) shows a velocity shift due to rotation. A remarkable feature is that the amount of velocity shift increases as the position approaches the central star, indicative of rotation velocity getting higher toward inner radii. On the other hand, the PV diagram along the cut offset by $1.65''$ southwest (the far side of the envelope) from the major axis (Fig. 6d) shows mainly blueshifted emission, getting much bluer as the position approaches the center. This feature can be explained by infalling motions, which accelerate toward the center. In addition, the most blueshifted emission is not located at the center but is slightly shifted to the southeast ($\Delta < 0''$ in Fig. 6d). This deviation is naturally explained by rotation.

A simple analysis on the assumption of a geometrically thin envelope shows that the rotation velocity $V_{\rm rot} \propto r^{-1}$, which is the case of angular momentum conservation, while the radial dependence of the infall velocity is consistent with dynamical infall ($V_{\rm inf} \propto r^{-0.5}$) onto a 0.1–0.5-$M_\odot$ central star. The infall and rotation velocities at $r = 700$ AU are 0.5 km s^{-1} and 0.24 km s^{-1}, respectively. The infall rate was estimated to be 6.4×10^{-6} $M_\odot$ yr^{-1}.

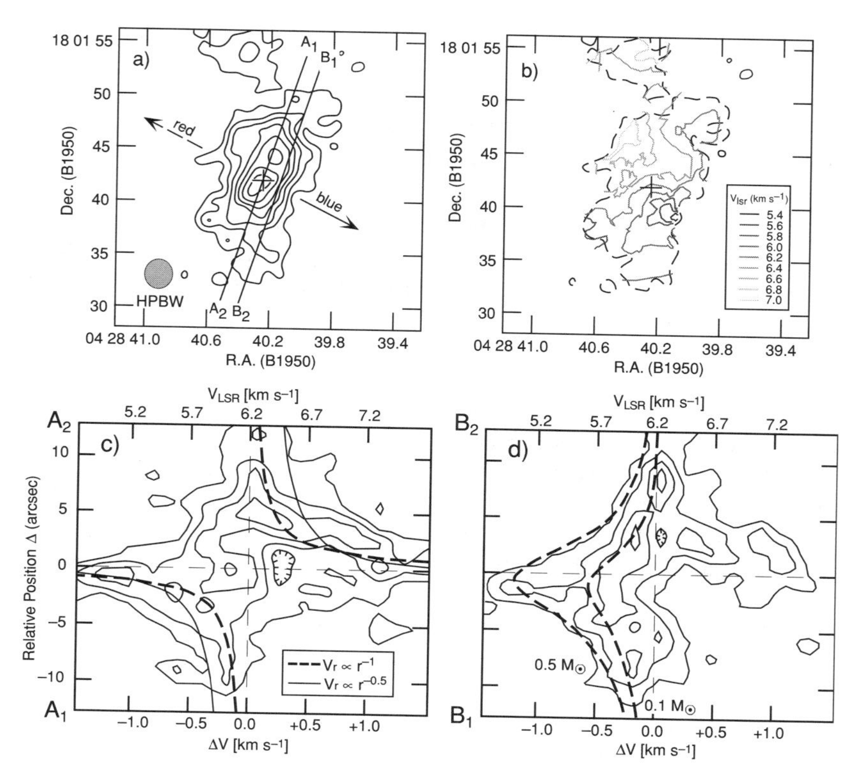

Figure 6. Interferometric $C^{18}O$ 1–0 maps of L1551 IRS5 (Momose et al. 1998). The angular resolution is $2.8'' \times 2.5''$, corresponding to ~370 AU, while the velocity resolution is ~0.2 km s^{-1}. The contour spacing is 1.5σ, starting from 1.5σ in (a), (c), and (d). (a) The total intensity map. The direction of the outflow measured by optical observations (Stocke et al. 1988) is indicated by arrows. The cross indicates the position of IRS5. (b) The intensity-weighted mean velocity map. Bluer velocities are drawn in darker grey contours. The systemic velocity of IRS5 is 6.2 km s^{-1}. The 1.5σ contour of the total intensity is drawn in the dashed contour. (c) The position-velocity diagram along the major axis of the $C^{18}O$ envelope [A_1A_2 in (a)]. The dashed curves indicate the distribution of the rotation velocity proportional to r^{-1}, while the thin solid curves show the Keplerian rotation velocity yielded by a 0.15-$M_\odot$ central star. (d) The position-velocity diagram along the cut B_1B_2 in (a) offset by $1.65''$ southwest from the major axis. The two dashed curves indicate distributions of dynamical infall velocities; the inner curve is the velocity yielded by a 0.1-$M_\odot$ central star, and the outer one is that yielded by a 0.5-$M_\odot$ star.

Note that there is an additional redshifted emission in Fig. 6d. This emission originates from infalling gas in the near side of the envelope, which is observable when the envelope is geometrically thick and nearly edge-on with respect to observers. In fact, model infalling envelopes with vertical structures can reproduce both blueshifted and redshifted emission in the diagram (see Fig. 8 in Momose et al. 1998). Such a geometrically thick, nearly edge-on, flattened, infalling structure was directly imaged toward L1527 (Ohashi et al. 1997*a*; Wilner et al. 1997; Choi et al. 1998).

C. Specific Angular Momenta of Infalling Envelopes: The Typical Size Scale for Dynamical Collapse

In a dynamically infalling envelope with slow rotation, the specific angular momentum of each gas element is considered to be conserved, as was shown in the case of L1551 IRS5, until the infall motion shifts to the centrifugally supported motion around the radius at which the rotation velocity is comparable to the infall velocity. On the other hand, star forming dense cores with larger scales often show consistency with solid-body rotation, which seems not to conserve the specific angular momentum of each element. On what size scale does specific angular momentum start being conserved in a dense core?

Ohashi et al. (1997*b*) examined the size dependence of local specific angular momentum in dense cores over a wide range of sizes (200 AU to 80,000 AU in radius). In Fig. 7, local specific angular momenta, $j_{\rm local} = V_{\rm rot}R_{\rm rot}$, where $V_{\rm rot}$ is the rotation velocity at a radius of $R_{\rm rot}$, are plotted for three kinds of objects in Taurus (i.e., infalling envelopes, rotationally supported disks, and NH_3 dense cores) as a function of the rotation radius.

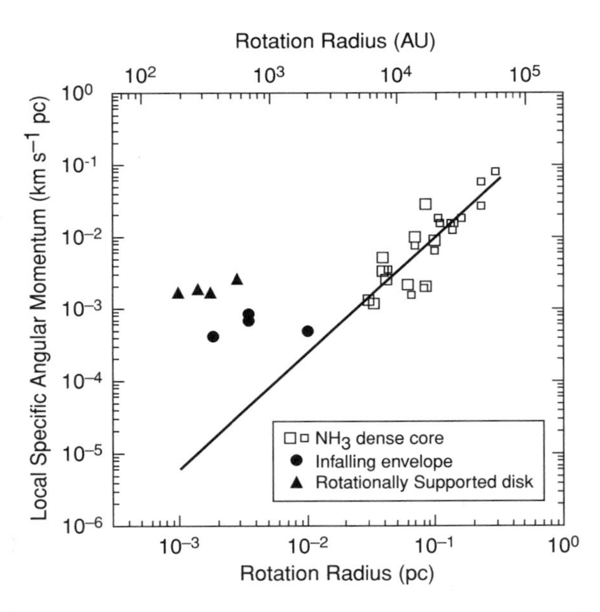

Figure 7. Local specific angular momentum plotted as a function of the rotation radius. Open squares, filled circles, and filled triangles indicate data for NH_3 dense cores, infalling envelopes, and rotationally supported disks, respectively. (NH_3 cores not associated with the Taurus molecular cloud are indicated by small squares.) The solid line shows a power law relation with an index of 1.6 for all the NH_3 cores (see Goodman et al. 1993).

Figure 7 shows a clear difference in the size dependence of the local specific angular momentum between the large scales of the dense cores and the smaller scales of envelopes and disks. The local specific angular momentum for the dense cores systematically varies with radius as a power law with an index of $\sim$1.6 (Goodman et al. 1993), but it is almost constant at $\sim 10^{-3}$ km s^{-1} pc for infalling envelopes and rotationally supported disks. This result suggests that a star-forming dense core can be divided into two zones by the dependence of the local specific angular momentum: an inner region, where gas has relatively constant angular momentum, and an outer region, where the power law relation applies. Figure 7 suggests that the two zones are divided at $\sim$0.03 pc (6000 AU) in radius.

Why do the infalling envelopes and rotationally supported disks have similar specific angular momenta in Fig. 7? Suppose the angular momentum distribution of pre-star-forming dense cores, from which the envelopes and disks formed, is consistent with the power law relation. If a dynamically collapsing region of a dense core is limited to only within 0.03 pc, then the specific angular momentum of a gas element in the infalling region will be set to $\sim 10^{-3}$ km s^{-1} pc, resulting in the similar specific angular momenta of the envelopes and disks in Fig. 7 (see Ohashi et al. 1997*b* for more details). This radius of 0.03 pc encloses a typical stellar mass in Taurus; namely, a Bonnor-Ebert condensation at $T = 10$ K with a radius of 0.03 pc contains a mass of $\sim$0.6 $M_{\odot}$.

VI. COMPARISON OF OBSERVATIONS AND MODELS

A. Expected Collapse Signatures

Even if we restrict ourselves to the spherical collapse of an isolated, nonfragmenting, isothermal cloud core, there are many different models for how the collapse proceeds. Here we consider the question of whether any of these predict the observed properties of star-forming regions. The main uncertainties in the modeling are in the initial conditions and in what processes are included. In pioneering work on protostellar collapse, Larson (1969) and, independently, Penston (1969) developed similarity solutions for the collapse of a uniform-density cloud with a fixed outer boundary. The solutions indicated that such a cloud would develop large infall velocities ($3.3a$, where a is the sound speed) at all radii and a power law density gradient [$n(r) \propto r^{-\alpha}$] with $\alpha = 2$. In contrast, Shu (1977) began with an equilibrium configuration for a singular isothermal sphere, in which the density gradient has reached $\alpha = 2$ by a quasistatic process, presumably ambipolar diffusion (Lizano and Shu 1989). If collapse begins in such a centrally condensed configuration, the collapse begins at the center, forming an opaque core, and propagates outward (the "inside-out" collapse), leading to an infalling inner region, with α approaching 1.5 asymptotically, surrounded by a static envelope with the original $\alpha = 2$. The Larson-Penston and Shu solutions represent the extremes of a whole class of solutions (Hunter 1977; Whitworth and Summers 1985). The main observable differences between the solutions of Larson-Penston and Shu are the nature of "precollapse" cores (uniform vs. centrally condensed) and the velocities of collapse (high and extended over the core vs. low and confined to the inner regions at early stages).

In a seminal paper, Zhou (1992) used the Larson-Penston and Shu solutions to predict line profiles of CS transitions as a function of time, identifying the particular features indicative of collapse. The higher velocities in the Larson-Penston solution make the peaks more separated than in the Shu solution, and the resulting linewidths from the Larson-Penston solution are inconsistent with observations of low-mass star-forming cores (Zhou 1992). Thus, most recent comparisons of observations to models have focused on the Shu model. However, regions forming higher-mass stars do have wider lines, and Zhou's conclusion should not be overgeneralized. In addition, Hunter (1977) noted that numerical calculations tended to give results between the two extreme similarity solutions. Further, calculations with nonsingular isothermal spheres tend toward the Larson-Penston solution, but only at the center, and $\dot{M}$ decreases with time after core formation unless the outer radius is more than 20 times the core radius (Foster and Chevalier 1993).

In another approach, Walker et al. (1994) took a numerical hydrodynamic calculation of collapse with rotation as input to a line simulation program to predict line profiles of CS lines. They were limited to local thermodynamic equilibrium (LTE) excitation, but they were able to consider the effects of viewing angle in aspherical geometries. They noted that maps of the line centroid velocity displayed a distinctive "blue bulge" when infall dominated over rotation.

B. The Theoretician's Dream: B335

B335 is a relatively isolated, spherical globule with a Class 0 infrared source (Keene et al. 1983) of low luminosity, $L \sim 3\ L_{\odot}$ (Chandler et al. 1990). The narrow lines imply a turbulent velocity much less than the sound speed, and rotation is very slow, with angular frequency $\Omega \sim 1.4 \times 10^{-14} s^{-1}$ (Frerking et al. 1987). In other words, B335 looks like the perfect test case.

H_2CO absorption in a $\Delta J = 0$ transition at 6-cm wavelength has provided evidence that the density increases toward the center, and detailed models of the spatial variation of absorption have indicated that power laws with $\alpha = 1.5$ match the observations best (Zhou et al. 1990). This value is consistent with the inner part of an inside-out collapse, so the full solution, including the static envelope, was tested. It provided an even better fit to the data. The models predicted that the $\Delta J = 1$ transitions of H_2CO at millimeter wavelengths should appear in emission. Observations of these lines verified the predictions and showed a double-peaked profile with a stronger blue peak. These observations were modeled with the full solution (Shu 1977), producing model line profiles consistent with both H_2CO and CS (Zhou et al. 1993, 1994). The luminosity predicted from the infall rate implied by the data matched that of the source, leading Zhou et al. to claim B335 as the first clear-cut case of a collapsing protostar.

Subsequent observational work has generally confirmed the existence of collapse in B335 (Chandler and Sargent 1993; Velusamy et al. 1995), with the caveat (Velusamy et al. 1995; Wilner et al. 1999) that the cloud may have significant clumpy structure superposed on the overall density gradient. Meanwhile, the modeling was improved by including thermal and turbulent line broadening in a self-consistent way, using Monte Carlo techniques (Choi et al. 1995). By considering a grid of models, Choi et al. established the best values, within the context of the Shu model, for the infall age (1.3×10^5 yr), mass of the protostellar core (0.37 $M_\odot$), and abundances of CS and H_2CO. The fact that the physical conditions are so well established leads to the idea that B335 can be used as a test bed for interstellar chemistry (Rawlings et al. 1994; Hartstein and Liseau 1998), and that other collapse models (section VI.D) can also be tested against the observations of B335.

C. Rotating Collapse: L1527 and IRAS 16293–2422

As discussed in section IV, L1527 is probably the second best case for being a collapsing protostar (Zhou et al. 1994; Myers et al. 1995). Myers et al. used an inside-out collapse model to predict line profiles of several lines toward the central object. The model is generally consistent with the observations, except that some lines have a larger blue/red asymmetry than predicted. Gregersen et al. (1997) used a 1D Monte Carlo code to model the line profiles of HCO^+, including rare isotopomers. They found a reasonable fit to the inner parts of the line, including the deep self-absorption, but the blue/red ratio of the observed line was again higher than predicted. In addition, contamination by the outflow was clearly a problem, because the observed lines of the main isotope had extensive wings not produced by the models.

The rotation rate in L1527, $\Omega \sim 1.1 \times 10^{-13}$ s^{-1} (Goodman et al. 1993), is about eight times greater than in B335, prompting Zhou et al. (1996) to use a generalization of the inside-out collapse model that includes rotation (Terebey et al. 1984) to model interferometric observations of ^{13}CO and $C^{18}O$. The model that fitted the single-dish and interferometer data best gave an infall age (10^5 yr) and stellar mass (0.36 $M_\odot$) similar to those of B335. The Terebey et al. (1984) model predicts that a disk should appear at a distance from the forming star approximated by the centrifugal radius (R_c), where infall equals rotation. For the best-fitting model of Zhou et al. (1996), R_c = 34 AU.

Ohashi et al. (1997*a*) provided more clear evidence for infall and rotation in L1527, finding evidence in their interferometer maps of $C^{18}O$ for a flattened envelope about 2000 AU in radius (see section V). Similar results were found by Choi et al. (1998) in interferometer maps of HCO^+ and HCN. By comparing position-velocity diagrams to models with rotation and infall, Ohashi et al. (1997*a*) were able to separate the effects of the two motions. At 2000 AU, infall (0.3 km s^{-1}) dominated rotation (0.05 km s^{-1}), suggesting that the flattening was not caused by rotation. Moreover, the sense of rotation is opposite to what is seen on larger scales. The age is similar to that found by Zhou et al. (1996), but the infall rate is slower and the central mass is smaller (0.1 $M_\odot$).

How much does rotation affect the models of B335? Zhou (1995) used the Terebey et al. (1984) models to compare to a spatial grid of CS spectra of B335 and found that some changes of the blue/red ratio at off-center positions could be explained. The parameters such as age and mass were not much affected, and the predicted $R_c = 3$ AU, too small to be detected.

Zhou (1995) also applied the Terebey et al. (1984) models to IRAS 16293–2422, finding that rotating infall provided an excellent fit to the spatial array of CS spectra (Menten et al. 1987), with rotation $\Omega \sim 3 \times 10^{-13}$ s^{-1}, still more rapid than in L1527. The combination of rotation and infall produces the reversal of the blue/red asymmetry used by Menten et al. (1987) to argue against collapse. Thus, Zhou's (1995) modeling supported the original claim of Walker et al. (1986) that this source is collapsing. Recently, Narayanan et al. (1998) used the blue-bulge technique on IRAS 16293–2422, also supporting the original claim of collapse. They modeled a large grid of spectral lines with Terebey et al. (1984) models and an LTE, aspherical code, assuming $\Omega = 5 \times 10^{-13} s^{-1}$. The best fit age is 6×10^4 yr, and the predicted $R_c = 300$ AU, about half the binary separation. Zhou (1995) found numbers within a factor of 2 of these, suggesting that Terebey et al. (1984) models provide a reasonable first approximation even for such rapid rotation. However, the fast rotation and evident formation of a binary indicates that models for single-star formation cannot be applied literally to this object.

D. Other Collapse Models

Recent interest in this subject has called forth a new group of collapse models, distinguished by different initial conditions or by inclusion of different processes. Stimulated by observational evidence for flatter density profiles (Ward-Thompson et al. 1994) and sharp outer edges (Abergel et al. 1996) in precollapse cores, Henriksen et al. (1997) considered collapse from a broken power law with three regimes: $\alpha = 0$ for small r, $\alpha = 2$ at intermediate r, and $\alpha > 2$ at large r. They found an initial phase of rapid $\dot{M}$ (coinciding with the Class 0 phase), relaxing to the constant $\dot{M}$ of the Shu models at later times (matching the Class I sources in Taurus). They also argued that sources in the ρ Ophiuchi cluster have a different accretion history from sources in Taurus. André (1997) has summarized these developments.

McLaughlin and Pudritz (1996) argued that the properties of precollapse cores are better described as a logatropic sphere, in which the total pressure, including nonthermal pressure, follows $P/P_c = 1 + A \ln(\rho/\rho_c)$ where A is a constant. The density distributions in such objects are well approximated by power laws with $\alpha = 1$. McLaughlin and Pudritz (1997) considered inside-out collapse from such a configuration and found $n(r) \propto r^{-1.5}$ and $v(r) \propto r^{-0.5}$ at small radii, inside the infall radius. Although the form is the same as in the Shu solution, the density increases in absolute terms inside the infall radius rather than decreasing, and the velocities are lower than in the Shu solution, providing a possible test of which model better describes the observations. Also, the accretion rate is not constant; instead, $\dot{M} \propto t^3$, implying very slow growth at early times. With this model, the infall age of B335 would increase to 1.2×10^6 yr. Since Henriksen et al. (1997) favor ages around 10^4 yr for Class 0 sources such as B335, there are now estimates spanning two orders of magnitude, but we consider the largest age estimates unlikely.

Magnetic fields will no doubt modify the nature of the precollapse core (e.g., Mouschovias 1995; Li and Shu 1996) as well as the collapse (Li and McKee 1996; Li 1998; Ciolek and Königl 1998). Safier et al. (1997) and Li (1998) considered magnetic effects in a spherical cloud and found that a long, quasistatic phase precedes a short dynamical phase, in which the inner envelope collapses rapidly from the outside in to form the protostar. This phase is followed by an inside-out collapse of the rest of the envelope. A magnetic accretion shock (Li and McKee 1996) is driven outward, causing a pileup of material on a scale of about 10^3 AU. The density profile in the precollapse phase approaches a power law that is slightly steeper than that of an isothermal sphere ($\alpha \sim 2.4$).

Although spherical clouds provide some insight, the precollapse cores of magnetic clouds are likely to be flattened (Fiedler and Mouschovias 1993; Li and Shu 1996), necessitating aspherical radiative transfer models. In addition, collapse from an initially spherical, magnetized cloud will produce flattened structures on scales of $\sim 10^3$ AU (e.g., Galli and Shu 1993*a*,*b*). These latter structures, commonly called "pseudodisks," may account for the flattened mass concentrations seen in some collapsing regions (e.g., L1527; see section V). In addition, Hartmann et al. (1996) have considered the collapse from sheetlike structures more generally.

Wiesemeyer (1997) has included some of these aspherical effects, using an accelerated Λ-iteration method to solve the radiative transfer in cylindrical geometry. He produced a grid of simulated line profiles for both sheet collapse (Hartmann et al. 1996) and a model with a magnetic accretion shock (Li and McKee 1996). He compared these simulations to observations of two condensations within the L1082 (GF9) globular filament, favoring the sheet collapse picture in one and the magnetic accretion shock picture in the other.

It is clear that theoretical models incorporating magnetic fields, rotation, or both must be compared with observational data via detailed line profile modeling using aspherical radiative transfer codes. Such codes are, at present, time consuming, and the larger number of free parameters, including both theoretical parameters and issues such as orientation relative to the observer, make comparison with observations problematic. Very complete data sets, with tracers sensitive to different aspects of the region, will be necessary to constrain the full range of theoretical imagination.

VII. SUMMARY

The quality and quantity of evidence for inward motions in star-forming regions is much greater than in years past, and their study with observations and models is now a serious undertaking, with activity by numerous groups. Our main conclusions follow.

1. Surveys of candidate protostars yield significant numbers of sources with infall asymmetry: a skewing of an optically thick line profile toward the blue when the line-forming gas has sufficient inward motions, optical depth, and gradient in excitation temperature. At least 13 infall candidates have infall asymmetry in two or more lines among rotational lines of CS, H_2CO, and HCO^+. Most of these candidates have substantial turbulent motions, unlike the best-known infall source B335. Infall asymmetry is fairly common in the most embedded sources; further studies of less embedded sources are needed to trace the evolution of infall.

2. Infall asymmetry is also seen in several starless cores. As in cores with embedded sources the incidence depends on the spectral line used. The best-studied starless core, L1544, shows infall asymmetry on scales from 0.01 to 0.1 pc in numerous lines. The largest extent is too large for consistency with inside-out collapse. The infall speed in a two-layer model is 0.05–0.1 km s^{-1}, which appears too fast for consistency with ambipolar diffusion on the largest scales.

3. Single-dish maps of kinematic infall candidates generally show a zone of extent 0.03–0.1 pc where infall asymmetry predominates. In L1251B and NGC 1333 IRS 4, the zone of infall asymmetry is more extended than the corresponding map of optically thin dense gas, suggesting large-scale inward motions that are larger than expected from simple gravitational infall. In L1527 and B335, infall asymmetry is less extended than the dense core map and is well centered on the embedded young stellar object, suggesting consistency with gravitational infall.

4. Interferometric observations achieve higher resolution than single-dish maps can ($<$1000 AU in nearby regions), and they filter out large-scale structures. Many interferometric studies interpret optically thin line emission in the context of a flattened structure, identifying infall from a velocity gradient along the projected minor axis and rotation from a velocity gradient along the projected major axis. Such observations of L1551 IRS5

suggest a combination of infall and rotation ($\sim$0.5 km s^{-1} and 0.2 km s^{-1} respectively at radius 700 AU), with each speed increasing inward.

5. Detailed radiative transfer modeling of line profiles has been attempted for only a few cores with embedded sources, including B335, L1527, and IRAS 16293–2422. The observed line profiles of B335 agree reasonably well with models of "inside-out" gravitational collapse. L1527 and IRAS 16293–2422 require models with rotation. None of the models currently include outflows, turbulence, or magnetic fields. Models assuming different initial density profiles have yet to be thoroughly tested against observations. Theoretical models will be needed for comparison to many cores, including starless cores and cores which form multiple sources.

In the coming years we may expect more observational studies of infall, extending our knowledge in space and time, for both isolated and clustered regions of star formation. Interferometers will reveal details on size scales of circumstellar disks (cf. the chapter by Wilner and Lay, this volume) and of the interaction between infall and outflow. Observations of starless cores and of Class I sources will reveal the earliest and latest indications of infall, and will help to understand infall evolution. Observations of higher rotational lines at submillimeter wavelengths will probe infall in gas with higher densities. Observations of submillimeter continuum emission will provide important information on the density distribution, without the uncertainties that depletion and opacity cause in interpreting molecular line data. High-resolution observations of cluster-forming regions may allow us to resolve simultaneous inward motions onto multiple centers. We expect that we will also achieve a fuller understanding of the role of turbulence and magnetic fields on the motions and structure of infalling gas in star-forming regions.

Acknowledgments We thank E. Gregersen, M. Hayashi, C. W. Lee, D. Mardones, M. Momose, P. T. P. Ho, M. Tafalla, J. P. Williams, and D. Wilner for helpful discussions and for sharing unpublished data. This work was supported in part by NASA grants NAG5-3401 and NAG5-7203 and NSC grant NSC89-2112-M-001-021.

REFERENCES

Abergel, A., Benard, J. P., Boulanger, F., Cesarsky, C., Desert, F. X., Falgarone, E., Lagache, G., Perault, M., Puget, J.-L., Reach, W. T., Nordh, L., Olofsson, G., Huldtgren, M., Kaas, A. A., Andre, P., Bontemps, S., Burgdorf, M., Copet, E., Davies, J., Montmerle, T., Persi, P., and Sibile, F. 1996. ISOCAM mapping of the ρ Ophiuchi main cloud. *Astron. Astrophys.* 315:L329–L332.

Adelson, L. M., and Leung, C. M. 1988. On the effects of rotation on interstellar molecular line profiles. *Mon. Not. Roy. Astron. Soc.* 235:349–364.

Adams, F. C., Lada, C. J., and Shu, F. H. 1987. Spectral evolution of young stellar objects. *Astrophys. J.* 312:788–806.

André, P. 1997. The evolution of flows and protostars. In *Herbig-Haro Flows and the Birth of Low-Mass Stars,* IAU Symp. 182, ed. B. Reipurth and C. Bertout (Dordrecht: Kluwer), pp. 483–494.

André, P., Ward-Thompson, D., and Barsony, M. 1993. Submillimeter continuum observations of Rho Ophiuchi A. The candidate protostar VLA 1623 and prestellar clumps. *Astrophys. J.* 406:122–141.

Beichman, C. A., Myers, P. C., Emerson, J. P., Harris, S., Mathieu, R., Benson, P. J., and Jennings, R. E. 1986. Candidate solar-type protostars in nearby molecular cloud cores. *Astrophys. J.* 307:337–349.

Boss, A. P. 1982. Hydrodynamical models of presolar nebula formation. *Icarus* 51:623–632.

Cabrit, S., and Bertout, C. 1986. CO line formation in bipolar flows. I. Accelerated outflows. *Astrophys. J.* 307:313–323.

Cabrit, S., Guilloteau, S., André, P., Bertout, C., Montmerle, T., and Schuster, K. 1996. Plateau de Bure observations of HL Tau: Outflow motions in a remnant circumstellar envelope. *Astron. Astrophys.* 305:527–540.

Caselli, P., Myers, P. C., and Thaddeus, P. 1995. Radio-astronomical spectroscopy of the hyperfine structure of N_2H^+. *Astrophys. J. Lett.* 455:L77–L80.

Caselli, P., Benson, P. J., Myers, P. C. and Tafalla, M. 1999. Dense cores in dark clouds. XIII. N_2H^+ maps. In preparation.

Chandler, C. J., and Sargent, A. I. 1993. The small-scale structure and kinematics of B335. *Astrophys. J. Lett.* 414:L29–L32.

Chandler, C. J. , Gear, W. K., Sandell, G., Hayashi, S., Duncan, W. D., Griffin, M. J., and Hazell, A. S. 1990. B335: Protostar or embedded pre-main-sequence star? *Mon. Not. Roy. Astron. Soc.* 243:330–335.

Chen, H., Myers, P. C., Ladd, E. F., and Wood, D. O. 1995. Bolometric temperature and young stars in the Taurus and Ophiuchus complexes. *Astrophys. J.* 445:377–392.

Cheung, A. C., Rank, D. M., Townes, C. H., Thornton, D. D., and Welch, W. J. 1968. Discovery of interstellar ammonia. *Phys. Rev. Lett.* 21:1701.

Choi, M., Evans, N. J., Gregersen, E., and Wang, Y. 1995. Modeling line profiles of protostellar collapse in B335 with the Monte Carlo method. *Astrophys. J.* 448:742–747.

Choi, M., Panis, J-F., and Evans, N. J., II. 1998. BIMA survey of protostellar collapse candidates in HCO^+ and HCN lines. *Astrophys. J.,* in press.

Ciolek, G. E., and Basu, S. 1999. Consistency of ambipolar diffusion models with infall in the L1544 protostellar core. *Astrophys. J.,* submitted.

Ciolek, G. E., and Königl, A. 1998. Dynamical collapse of nonrotating magnetic molecular cloud cores: Evolution through point mass formation. *Astrophys. J.* 504:257–279.

Ciolek, G. E., and Mouschovias, T. C. 1995. Ambipolar diffusion, interstellar dust, and the formation of cloud cores and protostars. IV. Effect of ultraviolet ionization and magnetically controlled infall rate. *Astrophys. J.* 454:194–216.

Evans, N. J., II. 1999. Physical conditions in regions of star formation. *Ann. Rev. Astron. Astrophys.* 37, in press.

Fiedler, R. A., and Mouschovias, T. C. 1993. Ambipolar diffusion and star formation: Formation and contraction of axisymmetric cloud cores. II. Results. *Astrophys. J.* 415:680–700.

Foster, P. N., and Chevalier, R. A. 1993. Gravitational collapse of an isothermal sphere. *Astrophys. J.* 416:303–311.

Frerking, M. A., Langer, W. D., and Wilson, R. W. 1987. The structure and dynamics of Bok globule B335. *Astrophys. J.* 313:320–345.

Galli, D., and Shu, F. H. 1993*a*. Collapse of magnetized molecular cloud cores. I. Semianalytical solution. *Astrophys. J.* 417:220–258.

Galli, D., and Shu, F. H. 1993*b*. Collapse of magnetized molecular cloud cores. II. Numerical results. *Astrophys. J.* 417:243–258.

Goodman, A. A., Benson, P. J., Fuller, G. A., and Myers, P. C. 1993. Dense cores in dark clouds. VIII. Velocity gradients. *Astrophys. J.* 406:528–547.

Gregersen, E. M. 1998. Collapse and beyond: An investigation of the star formation process. Ph.D. Dissertation, University of Texas at Austin.

Gregersen, E. M., Evans, N. J., Zhou, S., and Choi, M. 1997. New protostellar collapse candidates: An HCO^+ survey of the Class 0 sources. *Astrophys. J.* 484:256–276.

Gregersen, E. M., Evans, N. J., Mardones, D. M., and Myers, P. C. 1998. Does infall end at the Class I stage? *Astrophys. J.,* submitted.

Gueth, F., and Guilloteau, S. 1999. The jet-driven molecular outflow of HH 211. *Astron. Astrophys.* 343:571.

Hartmann, L. W. 1998. *Accretion Processes in Astrophysics* (Cambridge: Cambridge University Press).

Hartmann, L., Calvet, N., and Boss, A. 1996. Sheet models of protostellar collapse. *Astrophys. J.* 464:387–403.

Hartstein, D., and Liseau, R. 1998. Monte Carlo simulations at very high optical depth: Non-LTE transfer in H_2O in the protostellar object B335. *Astron. Astrophys.* 332:703–713.

Hayashi, M., Ohashi, N., and Miyama, S. M. 1993. A dynamically accreting gas disk around HL Tauri. *Astrophys. J. Lett.* 418:L71–L74.

Henriksen, R., André, P., and Bontemps, S. 1997. Time-dependent accretion and ejection implied by prestellar density profiles. *Astron. Astrophys.* 323:549–565.

Ho, P. T., Barrett, A. H., Myers, P. C., Matsakis, D. N., Chui, M. F., Townes, C. H., Cheung, A. C., and Yngvesson, K. S. 1979. Ammonia observations of the Orion molecular cloud. *Astrophys. J.* 234:912–921.

Hodapp, K.-W. 1994. A KU imaging survey of molecular outflow sources. *Astrophys. J. Suppl.* 94:615–649.

Hunter, C., 1977. The collapse of unstable isothermal spheres. *Astrophys. J.* 218:834–845.

Hurt, R. L., Barsony, A. W., and Wootten, H. A. 1996. Potential protostars in cloud cores: H_2CO observations. *Astrophys. J.* 456:686–695.

Keene, J., Harper, D. A., Hildebrand, R. H., and Whitcomb, S. E. 1980. Far-infrared observations of the globule B335. *Astrophys. J. Lett.* 240:L43–L46.

Keene, J., Davidson, J. A., Harper, D. A., Hildebrand, R. H., Jaffe, D. T., Loewenstein, R. F., Low, F. J., and Pernic, R. 1983. Far-infrared detection of low-luminosity star formation in the Bok globule B335. *Astrophys. J. Lett.* 274:L43–L47.

Kutner, M. L., Tucker, K. D., Chin, G., and Thaddeus, P. 1977. The molecular complexes in Orion. *Astrophys. J.* 215:521–528.

Lada, C. J., and Wilking, B. A. 1984. The nature of the embedded population in the Rho Ophiuchi dark cloud. Mid-infrared observations. *Astrophys. J.* 287:610–621.

Larson, R. B. 1969. Numerical calculations of the dynamics of a collapsing protostar. *Mon. Not. Roy. Astron. Soc.* 145:271–295.

Lee, C. W., and Myers, P. C. 1999. A catalog of optically selected cores. *Astrophys. J. Suppl.,* in press.

Lee, C. W., Myers, P. C., and Tafalla, M. 1999. A survey for infall motions toward starless cores. I. CS (2–1) and N_2H^+ (1–0) observations. *Astrophys. J.,* in press.

Lehtinen, K. 1997. Spectroscopic evidence of mass infall towards DC 303.8-14.2. *Astron. Astrophys.* 317:L5–L8.

Leung, C. M., and Brown, R. B. 1977. On the interpretation of carbon monoxide self-absorption profiles seen toward embedded stars in dense interstellar clouds. *Astrophys. J. Lett.* 214:L73–L78.

Leung, C. M., and Liszt, H. 1976. Radiation transport and non-LTE analysis of interstellar molecular lines. I. Carbon monoxide. *Astrophys. J.* 208:732–746.

Li, Z.-Y. 1998. Formation and collapse of magnetized spherical molecular cores. *Astrophys. J.* 493:230–246.

Li, Z.-Y., and McKee, C. F. 1996. Hydromagnetic accretion shocks around low-mass protostars. *Astrophys. J.* 464:373–386.

Li, Z.-Y., and Shu, F. H. 1996. Magnetized singular isothermal toroids. *Astrophys. J.* 472:211–224.

Lizano, S., and Shu, F. 1989. Molecular cloud cores and bimodal star formation. *Astrophys. J.* 342:834–854.

Loren, R. B., Plambeck, R. L., Davis, J. H., and Snell, R. L. 1981. High resolution $J = 2–1$ and $J = 1–0$ carbon monoxide self-reversed line profiles toward molecular clouds. *Astrophys. J.* 245:495–511.

Lucas, R. 1976. CO line formation in turbulent interstellar clouds where a small velocity gradient is present. *Astron. Astrophys.* 46:473–475.

Mardones, D., Myers, P. C., Tafalla, M., Wilner, D. J., Bachiller, R., and Garay, G. 1997. A search for infall motions toward nearby young stellar objects. *Astrophys. J.* 489:719–733.

Mardones, D., Myers, P. C., Tafalla, M., Wilner, D. J., Bachiller, R., and Garay, G. 1999. *Astrophys. J.,* in preparation.

McLaughlin, D. E., and Pudritz, R. E. 1996. A model for the internal structure of molecular cloud cores. *Astrophys. J.* 469:194–208.

McLaughlin, D. E., and Pudritz, R. E. 1997. Gravitational collapse and star formation in logotropic and nonisothermal spheres. *Astrophys. J.* 476:750–765.

Menten, K. M., Serabyn, E., Güsten, R., and Wilson, T. L. 1987. Physical conditions in the IRAS 16293–2422 parent cloud. *Astron. Astrophys.* 177:L57–L60.

Mizuno, A., Fukui, Y., Iwata, T., Nozawa, S., and Takano, T. 1990. A remarkable multilobe molecular outflow: Rho Ophiuchi East, associated with IRAS 16293–2422, *Astrophys. J.* 356:184–194.

Momose, M., Ohashi, N., Kawabe, R., Nakano, T., and Hayashi, M. 1998. Aperture synthesis $C^{18}O$ ($J = 1-0$) observations of L1551 IRS5: Detailed structure of the infalling envelope. *Astrophys. J.* 504:314–333.

Moriarty-Schieven, G. H., Wannier, P. G., Mangum, J. G., Tamura, M., and Olmsted, V. K. 1995. Circum-protostellar environments. III. Gas densities and kinetic temperatures. *Astrophys. J.* 455:190–201.

Mouschovias, T. C. 1976. Nonhomologous contraction and equilibria of self-gravitating, magnetic interstellar clouds embedded in an intercloud medium: Star formation. II. Results. *Astrophys. J.* 207:141–158.

Mouschovias, T. C. 1995. Role of magnetic fields in the early stages of star formation. In *Physics of the Interstellar Medium and the Intergalactic Medium,* ASP Conf. Ser. 80, ed. A. Ferrara, C. Heiles, C. F. McKee, and P. Shapiro (San Francisco: Astronomical Society of the Pacific), pp. 184–217.

Mundy, L. G., Wootten, A., Wilking, B. A., Blake, G. A., and Sargent, A. I. 1992. *Astrophys. J.* 385:306–313.

Myers, P. C. 1997*a*. Early star formation. In *Star Formation Near and Far,* ed. S. S. Holt and L. G. Mundy (New York: AIP Press), pp. 41–50.

Myers, P. C. 1997*b*. Turbulence and collapse in star-forming molecular clouds. In *CO: Twenty-five Years of Millimeter-wave Spectroscopy,* ed. W. B. Latter, S. E. Radford, P. R. Jewell, J. G. Mangum, and J. Bally (Dordrecht: Kluwer), pp. 137–147.

Myers, P. C., and Benson, P. J. 1983. Dense cores in dark clouds. II. NH_3 observations and star formation. *Astrophys. J.* 266:309–320.
Myers, P. C., and Khersonsky, V. K. 1995. On magnetic turbulence in interstellar clouds. *Astrophys. J.* 442:186–196.
Myers, P. C., and Ladd, E. F. 1993. Bolometric temperatures of young stellar objects. *Astrophys. J. Lett.* 413:L47–L50.
Myers, P. C., Ladd, E. F., and Fuller, G. A. 1991. Thermal and nonthermal motions in dense cores. *Astrophys. J. Lett.* 372:L95–L98.
Myers, P. C., Bachiller, R., Caselli, P., Fuller, G. A., Mardones, D., Tafalla, M., and Wilner, D. J. 1995. Gravitational infall in the dense cores L1527 and L483. *Astrophys. J. Lett.* 449:L65–L68.
Myers, P. C., and Lazarian, A. 1998. Turbulent cooling flows in molecular clouds. *Astrophys. J. Lett.* 507:L157–L160.
Myers, P. C., Mardones, D., Tafalla, M., Williams, J. P., and Wilner, D. J. 1996. A simple model of spectral-line profiles from contracting clouds. *Astrophys. J. Lett.* 465:133–136.
Narayanan, G., Walker, C. K., and Buckley, H. D. 1998. The "blue-bulge" infall signature towards IRAS 16293–2422. *Astrophys. J.* 496:292–310.
Nakano, T. 1998. Star formation in magnetic clouds. *Astrophys. J.* 494:587–604.
Ohashi, N. 1997. Dynamically infalling envelopes around low-mass protostar candidates. In *Star Formation Near and Far,* ed. S. S. Holt and L. G. Mundy (New York: AIP Press), pp. 93–96.
Ohashi, N., Kawabe, R., Hayashi, M., and Ishiguro, M. 1991. Observations of 11 protostellar sources in Taurus with Nobeyama millimeter array: Growth of circumstellar disks. *Astron. J.* 102:2054–2065.
Ohashi, N. Hayashi, M., Ho, P. T. P., Momose, M., and Hirano, N. 1996*a*. Possible infall in the gas disk around L1551-IRS5. *Astrophys. J.* 466:957–963.
Ohashi, N., Hayashi, M., Kawabe, R., and Ishiguro, M. 1996*b*. The Nobeyama millimeter array survey of young stellar objects associated with the Taurus molecular cloud. *Astrophys. J.* 466:317–337.
Ohashi, N. Hayashi, M., Ho, P. T. P., and Momose, M. 1997*a*. Interferometric imaging of IRAS 04368+2557 in the L1527 molecular cloud core: A dynamically infalling envelope with rotation. *Astrophys. J.* 475:211–223.
Ohashi, N. Hayashi, M., Ho, P. T. P., Momose, M., Tamura, M., Hirano, N., and Sargent, A. I. 1997*b*. Rotation in the protostellar envelopes around IRAS 04169+2702 and IRAS 04365+2535: The size scale for dynamical collapse. *Astrophys. J.* 488:317–329.
Penston, M. V. 1969. Dynamics of self-gravitating gaseous spheres. III. *Mon. Not. Roy. Astron. Soc.* 144:425–448.
Phillips, T. G., Knapp, G. R., Wannier, P. G., Huggins, P. J., Werner, M. W., Neugebauer, G., and Ennis, D. 1981. Temperatures of galactic molecular clouds showing CO self-absorption. *Astrophys. J.* 245:512–528.
Rawlings, J. M. C., Evans, N. J., II, and Zhou, S., 1994. Gas-grain interactions in the low mass star-forming region B335. *Astrophys. Space Sci.* 216:155–157.
Safier, P. N., McKee, C. F., and Stahler, S. W. 1997. Star formation in cold, spherical, magnetized molecular clouds. *Astrophys. J.* 485:660–679.
Saito, M., Kawabe, R., Kitamura, Y., and Sunada, K. 1996. Imaging of an infalling disklike envelope around L1551 IRS 5. *Astrophys. J.* 473:464–469.
Sargent, A. I., and Beckwith, S. V. W. 1987. Kinematics of the circumstellar gas of HL Tauri and R Monocerotis. *Astrophys. J.* 323:294–305.
Sargent, A. I., and Beckwith, S. V. W. 1991. The molecular structure around HL Tauri. *Astrophys. J. Lett.* 382:L31–L35.
Sargent, A. I., Beckwith, S., Keene, J., and Masson, C. 1988. Small-scale structure of the circumstellar gas around L1551 IRS 5. *Astrophys. J.* 333:936–942.

Shu, F. H. 1977. Self-similar collapse of isothermal spheres and star formation. *Astrophys. J.* 214:488–497.

Snell, R. L., and Loren, R. B. 1977. Self-reversed CO profiles in collapsing molecular clouds. *Astrophys. J.* 211:122–127.

Stocke, J. T., Hartigan, P. M., Strom, S. E., Strom, K. M., Anderson, E. R., Hartmann, L. W., and Kenyon, S. J. 1998. A detailed study of the Lynds 1551 star formation region. *Astrophys. J. Suppl.* 68:229–255.

Tafalla, M., Mardones, D., Myers, P. C., Caselli, P., Bachiller, R., and Benson, P. J. 1998. L1544: A starless dense core with extended inward motions. *Astrophys. J.* 504:900–914.

Tamura, M., Ohashi, N., Hirano, N., Itoh, Y., and Moriarty-Schieven, G. H. 1996. Interferometric observations of outflows from low-mass protostars in Taurus. *Astron. J.* 112:2076–2085.

Terebey, S., Shu, F. H., and Cassen, F. 1984. The collapse of the cores of slowly rotating isothermal clouds. *Astrophys. J.* 286:529–551.

Velusamy, T., Kuiper, T. B. H., Langer, W. D. 1995. CCS observations of the protostellar envelope of B335. *Astrophys. J. Lett.* 451:L75–L78.

Walker, C. K., Lada, C. J., Young, E. T., Maloney, P. R., and Wilking, B. A. 1986. Spectroscopic evidence for infall around an extraordinary IRAS source in Ophiuchus. *Astrophys. J. Lett.* 309:L47–L51.

Walker, C. K., Narayanan, G., and Boss, A. P. 1994. Spectroscopic signatures of infall in young protostellar systems. *Astrophys. J.* 431:767–782.

Ward-Thompson, D., Scott, P. F., Hills, R. E., and André, P. 1994. A submillimetre continuum survey of pre-protostellar cores. *Mon. Not. Roy. Astron. Soc.* 268:276–290.

Ward-Thompson, D., Buckley, H. D., Greaves, J. S., Holland, W. S., and André, P. 1996. Evidence for protostellar infall in NGC 1333-IRAS2. *Mon. Not. Roy. Astron. Soc.* 281:L53–L56.

Whitworth, A., and Summers, D. 1985. Self-similar condensation of spherically symmetric self-gravitating isothermal gas clouds. *Mon. Not. Roy. Astron. Soc.,* 214:1–25.

Wiesemeyer, H. 1997. The spectral signature of accretion in low-mass protostars. Ph.D. Dissertation, University of Bonn.

Williams, J. P., and Myers, P. C. 1999*a*. A survey for inward motions in the dense gas around young stellar clusters. *Astrophys. J.* 511:208–217.

Williams, J. P., and Myers, P. C. 1999*b*. A contracting, turbulent, starless core in the Serpens cluster. *Astrophys. J. Lett.* 518:L37–L40.

Williams, J. P., Myers, P. C., Wilner, D. J., Di Francesco, J., and Tafalla, M. 1999. A high resolution study of the slowly contracting starless core L1544. *Astrophys. J. Lett.,* 513:L61–L64.

Wilner, D. J., Mardones, D., and Myers, P. C. 1997. Interferometric imaging of dense gas tracers in the protostellar collapse candidate L1527. In *Star Formation Near and Far,* ed. S. S. Holt and L. G. Mundy (New York: AIP Press), pp. 109–112.

Wilner, D. J., Myers, P. C., Mardones, D., and Tafalla, M. 1999. *Astrophys. J.,* in preparation.

Wilson, R. W., Jefferts, K. B., and Penzias, A. A. 1970. Carbon monoxide in the Orion nebula. *Astrophys. J. Lett.* 161:L43–L44.

Wolf-Chase, G. A., and Gregersen, E. 1997. Possible infall around the intermediate-mass young stellar object NGC2264 IRS. *Astrophys. J. Lett.* 479:L67–L70.

Zhou, S. 1992. In search of evidence for protostellar collapse: A systematic study of line formation in low-mass dense cores. *Astrophys. J.* 394:204–216.

Zhou, S. 1995. Line formation in collapsing cloud cores with rotation and applications to B335 and IRAS 16293–2422. *Astrophys. J.* 442:685–693.

Zhou, S., Evans, N. J., II, Butner, H. M., Kutner, M. L., Leung, C. M., and Mundy, L. G. 1990. Testing star formation theories: VLA observations of H_2CO in the Bok globule B335. *Astrophys. J.* 363:168–179.

Zhou, S., Evans, N. J., II, Kömpe, C., and Walmsley, C. M. 1993. Evidence for protostellar collapse in B335. *Astrophys. J.* 404:232–246.

Zhou, S., Evans, N. J., II, Kömpe, C., and Walmsley, C. M. 1994. Erratum re: Evidence for protostellar collapse in B335. *Astrophys. J.* 421:854–855.

Zhou, S., Evans, N. J., II, Wang, Y., Peng, R., and Lo, K. Y. 1994. A $C^{18}O$ survey of dense cores in the Taurus molecular cloud: Signatures of evolution and protostellar collapse. *Astrophys. J.* 433:131–148.

Zhou, S., Evans, N. J., II, and Wang, Y. 1996. Small-scale structure of candidates for protostellar collapse. I. BIMA observations of L1527 and CB 54. *Astrophys. J.* 466:296–308.

POLARIZED LIGHT FROM STAR-FORMING REGIONS

DAVID A. WEINTRAUB
Vanderbilt University

ALYSSA A. GOODMAN
Harvard University

and

RACHEL L. AKESON
University of California at Berkeley

Studies of polarized radiation from molecular clouds and the environments of protostars are providing information about the orientations and strengths of magnetic fields and the sizes and compositions of dust grains in these environments. This relationship between dust grains, magnetic fields, and polarized light exists because dust grains aligned by magnetic fields will polarize background starlight, which otherwise would be unpolarized, and will emit intrinsically polarized thermal radiation; also, even unaligned dust grains will scatter photons, so the dusty environments of young stars often produce reflection nebulae. Importantly, some of the observational results are providing the first ever measurements of magnetic field alignments and strengths on scales from 1000 AU to tens of parsecs in dense cores and molecular clouds. Such measurements are particularly valuable for understanding the mechanisms of star formation and the importance of magnetic fields and outflows. In addition, polarization studies are driving ideas that the size distributions and compositions of populations of dust grains around young stellar objects may be quite different from those assumed for astrophysical dust in the interstellar medium. This astrophysically important information is unavailable, at present, from any other type of measurements.

I. INTRODUCTION

Polarization studies have progressed enormously since *Protostars and Planets III*, including a huge expansion in the measurement of polarized thermal emission from dust grains at far infrared through millimeter-wavelengths. Linear and circular polarization observations, and the modeling of those observations in terms of the relative contributions of scattering and dichroic extinction and emission, contribute to our understanding of the relationships among magnetic fields, circumstellar disks, and young stellar outflows. This work also contributes to our understanding of the properties of dust grains in the interstellar medium (ISM), in dark clouds, and in protostellar nebulae. In addition, thermal emission polarimetry now informs us about the large- and small-scale magnetic field structure in star-forming regions. In the 1990s we have witnessed enormous improvements in computational modeling and the development of multiple techniques for obtaining polarimetric data on many telescopes and at many wavelengths.

In the first part of this chapter we discuss how observations of polarized background starlight, at both optical and near infrared wavelengths, and polarized thermal emission from dust, at far infrared and millimeter wavelengths, can be used to chart magnetic fields in star-forming regions. In the second part we describe how nebulae around young stellar objects (YSOs) can be studied by measuring the polarized, scattered light that emerges from these dusty environments at optical and near infrared wavelengths.

II. MAPPING MAGNETIC FIELDS

Magnetic fields introduce asymmetries into physical systems that often can be detected via polarimetric observations. In the ISM and in circumstellar environments, magnetic fields impose these asymmetries by aligning dust particles more or less with the direction of local field lines. As will be shown below, magnetic fields draw dust grains into line more successfully in some regions than in others. With any degree of success though, the grains become partially aligned along some preferred (magnetic field) direction. Consequently, background starlight passing near them will become partially polarized, and light emitted by these dust grains will be partially polarized.

From the standpoint of polarimetry as it was done in 1998, there are two kinds of star-forming regions. "Dark cloud complexes" harbor embedded protostars found alone or in small groups. The densest parts of these regions are apparent as "dark" clouds ($0.5 < A_V < 10$ mag) on optical photographs and as moderate-level ($\sim$10–100 MJy Sr^{-1} at 100 μm) extended emission in the *Infrared Astronomy Satellite* (IRAS) survey. To date, polarimetry of background starlight has been the principal tool used in mapping magnetic fields in dark clouds. "Massive star-forming regions," on the other hand, are regions where most embedded protostars are found in large clusters ($\gg$10 stars forming together). Optically, these regions often show up as high-extinction black patches interspersed with bright reflection or emission nebulae. The peak amounts of thermal dust emission from these regions (typically $>10^4$ MJy Sr^{-1} at 100 μm) far exceed those in dark clouds. At present, magnetic fields in massive star-forming regions can be mapped with thermal emission polarimetry at parsec and subparsec scales. However, in all but the brightest, warmest peaks of dark clouds, thermal emission polarimetry is still not feasible. This situation should change over the next few years, when satellite-based far infrared polarimetry becomes a reality and as ground-based millimeter and submillimeter instruments become more sensitive.

A. Dark Cloud Complexes

In order to understand what polarization maps can tell us about dark clouds, it is necessary to fully understand how and why the dust in dark clouds polarizes background starlight. If every dust grain along the line of sight to a background star were the same size, shape, and composition and were aligned to the same degree by a magnetic field whose direction with respect to the line of sight never changed, polarization maps would be easy to interpret. In that case the ensemble of photons passing through a sea of similar grains would become systematically more highly polarized after each encounter. Because grains tend to align with their shortest axis parallel to the field (see Fig. 1; see also http://cfa-www.harvard.edu/~agoodman/ppiv/, which contains additional figures not shown in this chapter), the polarization direction observed would give the direction of the magnetic field as projected onto the plane of the sky. In this simple case, the percentage of polarization observed (p) would rise linearly with the number of grains, which in turn is proportional to the extinction, A_V.

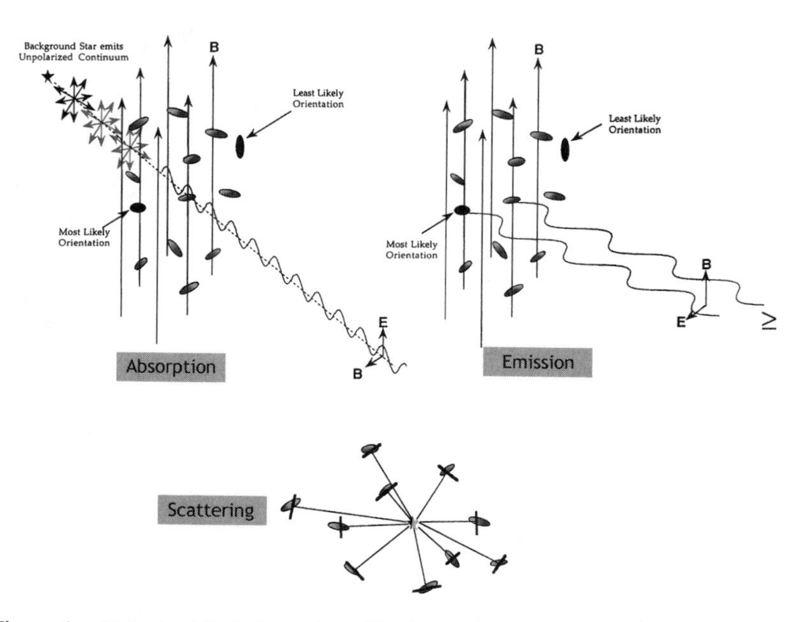

Figure 1. Polarized light is produced by interstellar dust grains in three different ways: dichroic absorption by aligned dust grains, thermal radiation by aligned dust grains, and scattering of light by dust grains. The first two mechanisms rely on the systematic alignment of dust grains, usually by a magnetic field. In most magnetic alignment scenarios, the grain's short axis prefers to be parallel to the field (see Lazarian et al. 1997). Usually, scattered light is polarized, in the plane of the sky, along the direction perpendicular to the ray from the illuminating source to the scatterer.

Reality, however, is not so simple. Grains are not all similar, they are not all equally susceptible to alignment, field lines are not straight, and mechanisms other than magnetic fields can also align grains. Therefore, we need models of the grain distribution, the alignment distribution, and the field structure before we can learn about magnetic fields from polarization maps.

Dark clouds and star-forming regions represent a special challenge in the interpretation of background starlight polarization maps, because grains in those regions seem to have properties unlike those in the general ISM (see Whittet 1992 and references therein). In the larger-scale ISM, the interpretation of polarization maps (see, e.g., the Mathewson and Ford 1970 map of the local Galactic region), is easier, because grain properties "average out" over long lines of sight (see Heiles 1996).

Figure 2 shows an optical polarization map of Taurus, superimposed on the IRAS Sky Survey Atlas (ISSA) 100-μm map. The most striking feature of the Taurus polarization map is the smoothness of the overall pat-

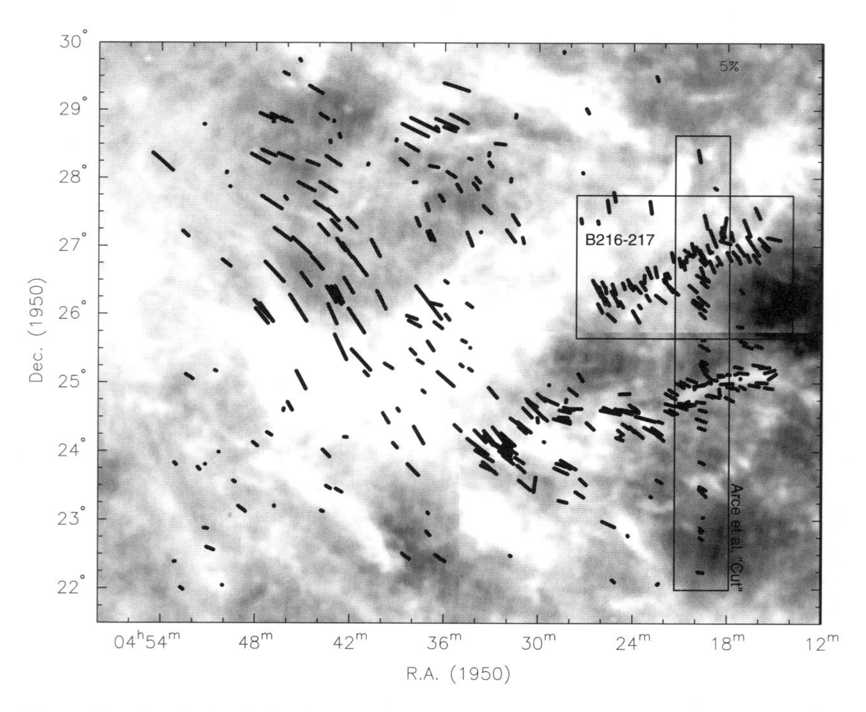

Figure 2. Optical polarization map of Taurus superimposed on an ISSA 100-μm map. The horizontal black rectangle shows the location of the B216-217 dark cloud complex studied by Heyer et al. (1987) and Goodman et al. (1992); see Fig. 3. The vertical black rectangle shows the "cut" through Taurus studied by Arce et al. (1998); see Fig. 5. The polarization vectors shown are taken from Moneti et al. 1984; Heyer et al. 1987; Goodman et al. 1990; and Arce et al. 1998. (Figure courtesy of Héctor Arce.)

tern. There are no obvious topological changes in the field caused by the presence of the dark clouds (Goodman et al. 1990). Recent magnetohydrodynamic (MHD) modeling shows that this lack of structural change can be expected in the case where the field and kinetic energies are roughly equal (Ostriker et al. 1999). Nonetheless, one cannot help wondering whether the field *inside* dark clouds might be in a different direction or have a different dispersion than that in clouds' surroundings. Since background starlight polarimetry cannot be used where background stars are not visible, optical observations are useful only up to about $A_V \sim 2$ mag. Near infrared polarimetry of background stars, though, can see into as much as $A_V \sim 40$ mag (see, e.g., Wilking et al. 1979).

In cold dark clouds, near-infrared polarization maps look exactly the same as expected by extrapolating from optical polarimetry obtained around the clouds' peripheries. Figure 3 shows a comparison of optical and near-IR polarization position angles for the dark cloud B216-217 (region highlighted by the horizontal rectangle in Fig. 2). The near-IR map, which covers interior portions of the cloud too opaque for optical polarimetry, looks simply like a shrunken version of the optical map (see Goodman et

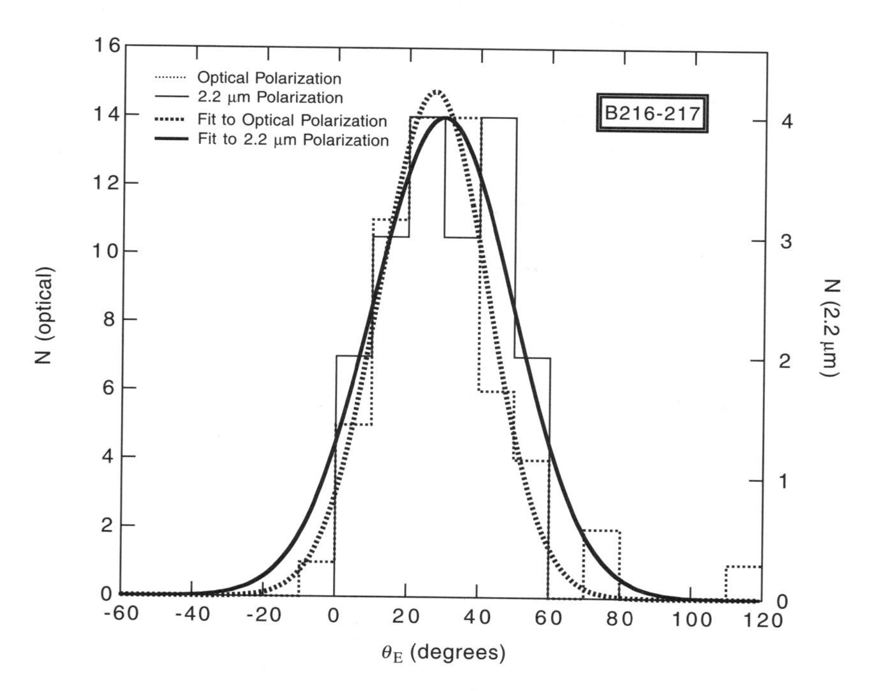

Figure 3. Distribution of polarization position angle in the dark cloud B216-217. Fits of the Myers and Goodman (1991) model for polarization dispersion give statistically indistinguishable results for the optical data (Heyer et al. 1987) and the near-infrared data (Goodman et al. 1992). (Figure reproduced from Goodman et al. 1992.)

al. 1992). Fits to the polarization position angle distributions are identical within the errors (Fig. 3). So either the field in the dense part of B216-217 is topologically exactly the same as in the lower-density cloud environment, or the polarization observations are somehow insensitive to the dust in the dark cloud.

Polarization-extinction relations show, definitively, that polarimetry measurements of background starlight are insensitive to magnetic fields in dark clouds (Goodman et al. 1995; Gerakines et al. 1995; Arce et al. 1998). As is shown in Fig. 4, polarization percentage does not rise with extinction in cold dark clouds. This means that while cold dark clouds contribute a great deal of extinction, they contribute no polarization, so

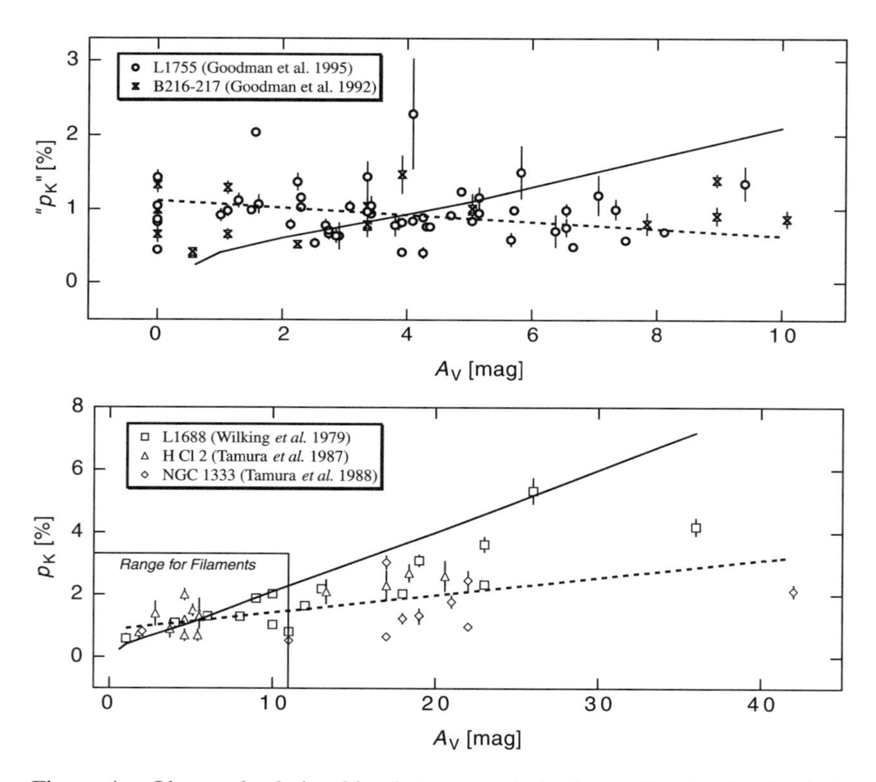

Figure 4. Observed relationships between polarization and extinction in dark clouds. Top panel: Data for filamentary dark clouds, L1755 and B216-217. Lower panel: Data for L1688, HCl 2, and NGC 1333, dark clouds which are closer to round, in projection, than L1755 and B216-217 and produce higher extinction. The box in the lower panel shows how small the range of polarization-extinction space covered by the "filamentary" clouds in the top panel is in comparison with the range for "round" clouds. Both panels show polarization near 2.2 microns on the y axis. Dashed lines in each panel show error-weighted least-squares linear fits to the data points. Solid lines show the predictions of the Jones et al. (1992) model, assuming equal nonuniform and uniform magnetic field energy. (Figure based on Goodman et al. 1995.)

polarization maps are insensitive to the field there. Why? It is likely that grain alignment mechanisms are far more efficient in the warmer, lower-density ISM than they are in dark clouds (see Lazarian et al. 1997). Thus, the polarization along a line of sight through a dark cloud is strongly dominated by dust interactions that do not take place in the dense interiors of dark clouds. It is as if the dark cloud is invisible from the standpoint of polarizing background starlight (see Goodman et al. 1995).

The polarization-extinction plots in Fig. 5 represent an attempt to quantify the conditions where interstellar material goes from being "good" at polarizing background starlight to being "bad." In the "good" regions it is safe to use background starlight polarimetry to map fields; in the "bad" regions it is not. The plots show that a polarization-extinction relation constructed for the "cut" through Taurus shown in Fig. 2 exhibits two clear trends. The stars background to the "general low-density ISM" show a linear rise in polarization with extinction, while those behind the cold dark clouds show a flat p–A_V relation like those shown using near-IR data in Fig. 4. These two trends cross at $A_V = 1.3 \pm 0.2$ mag, and Arce et al. (1998) identify this point as the "breakpoint" between good and bad regimes. This result implies that almost all stars form in regions of the ISM where background starlight polarimetry cannot effectively map the magnetic field, and other techniques need to be explored.

The "other technique" best suited for studying both dark clouds and warmer dense gas is thermal emission polarimetry. As mentioned above, this technique is currently most feasible in the warmer regions (which produce more thermal emission). The astute reader may have noticed that the lower panel of Fig. 4 does show a small rise in polarization with extinction for the warmer clouds, and may wonder whether that means that background starlight polarimetry is all right to use there. Unfortunately, the answer is still no. In those regions, there is tremendous confusion of the field in the region of interest, which typically lies at >1 kpc distance with fields all along the line of sight to that region. An example of this problem can be seen by comparing the optical polarization map of M17 in Schulz et al. (1981), which shows a messy, bimodal distribution of polarization vectors, with the far-infrared polarization map from the Kuiper Airborne Observatory (KAO) (Dotson 1996) and the 800-μm polarization map from the James Clerk Maxwell Telescope (JCMT) (Vallée and Bastien 1996), which show very smooth distributions of vectors for the same region (see Figure 4 in Goodman 1995).

B. Magnetic Fields on Large Scales in Molecular Clouds

Because observations of selective absorption rely on sufficient flux from background sources, these measurements become difficult in dense regions with high extinction. However, the same aligned dust grains that produce selective extinction in the optical and near-infrared emit polarized radiation in the far infrared to millimeter regimes (Fig. 1). The cross sections for absorption and scattering at far-infrared and longer wavelengths

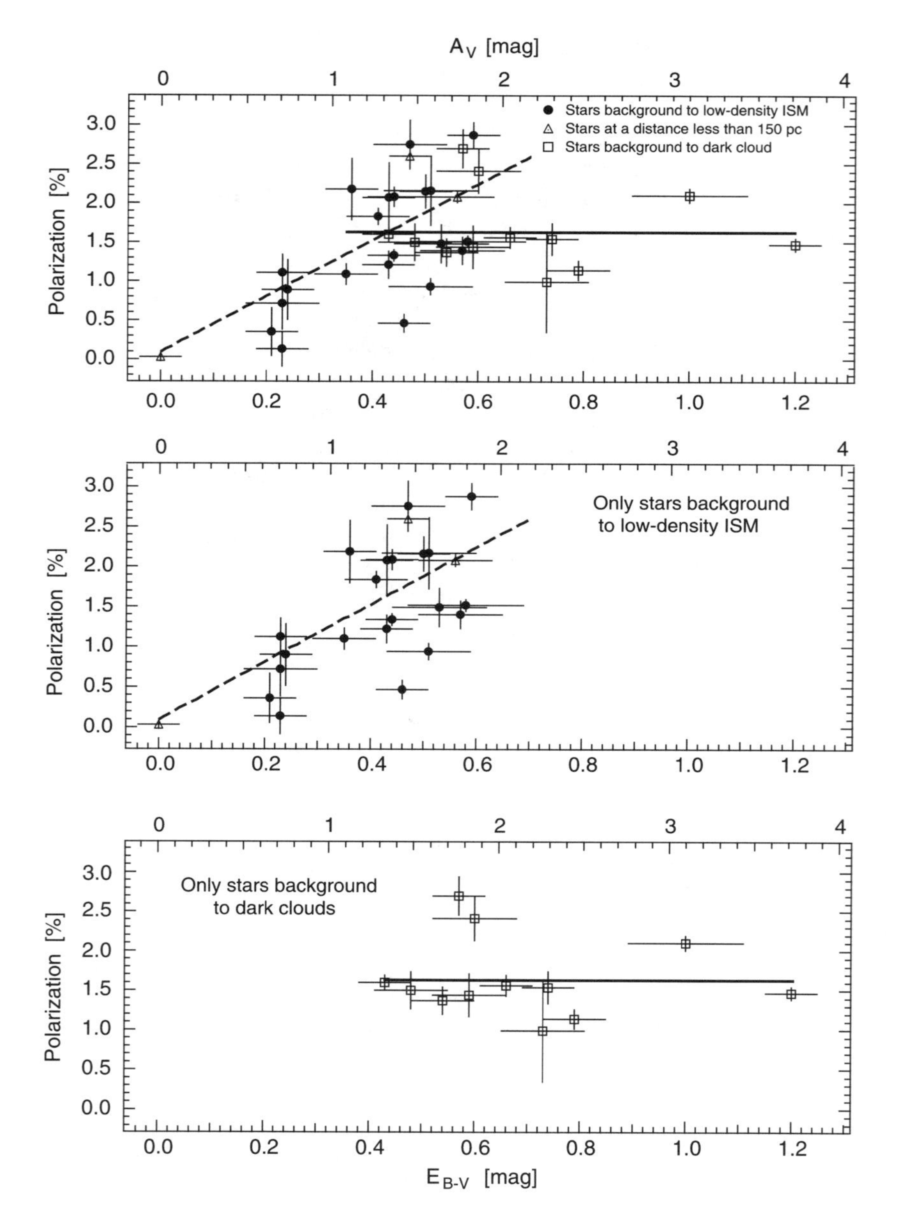

Figure 5. Observed relation between polarization and extinction, where $A_V = 3.1E_{B-V}$. The lines are least-squares linear fits, weighted by the uncertainty in p. The dashed line is the fit to points representing the stars background to the low-density ISM, which gives $p = (0.09 \pm 0.06) + (3.58 \pm 0.13)E_{B-V}$, with a correlation coefficient of 0.68. The solid line is the fit to the points representing the stars background to dark clouds, which gives $p = (1.61 \pm 0.13) + (0.03 \pm 0.15)E_{B-V}$, with a correlation coefficient of 0.79. (Figure reproduced from Arce et al. 1998.)

are very small, ensuring in almost all cases that the polarization is due to emission from aligned grains. The significant column densities necessary to detect far-infrared to millimeter-wavelength polarized emission generally restricts these observations to warm molecular clouds and areas of active star formation. For example, a column density $>5 \times 10^{22}$ cm^{-2} ($A_V \sim 25$) is necessary with current instrumentation for detection of polarized emission at 350 μm (D. A. Schleuning, personal correspondence). Thus, one advantage of these measurements is that the polarized emission is not affected by low-column-density material. In this section, we will discuss observations of polarized thermal emission from dust made over large regions of molecular clouds, and will discuss the Orion molecular cloud in detail.

Observations in the far-infrared are sensitive mainly to warm ($T \sim 50$ K) dust, whereas (sub)millimeter observations probe the entire dust column. Given the resolution currently available at far-infrared and submillimeter wavelengths, these observations trace the field structure on scales from 0.05 pc in Orion to 1 pc in Sagittarius B2.

One of the most remarkable results of far-infrared polarimetry is that polarized emission is detected from the vast majority of locations observed (Hildebrand 1996). This implies that aligned grains are the rule rather than the exception in warm molecular clouds. Based on over 500 independent measurements, Hildebrand (1996) found that the maximum polarization observed is 9%, whereas the median value of p is near 2%. The observed magnetic field direction is generally orderly over the entire map region (Fig. 6), which covers from 0.3 pc in DR21 to 5 pc in Sagittarius B2 (see, e.g., Dotson 1996 and Vallée and Bastien 1996 for M17; Dowell 1997 and Novak et al. 1997 for Sgr B2; Minchin and Murray 1994 and Glenn et al. 1998 for DR 21; Greaves et al. 1999).

Another attribute of the far-infrared and (sub)millimeter observations with several pointings within a cloud is that the polarization percentage (p) decreases toward the regions of highest optical depth. Several pieces of information are known about this so-called "polarization hole" effect:

1. The decrease in p is more than expected from opacity effects.
2. In contrast to optical observations, the polarized flux increases in the regions of highest total flux up to extinctions of at least $A_V = 300$ (Schleuning et al. 1996), although p decreases.
3. The polarization percentage is not correlated with dust temperature (Dotson 1996).

Various proposed explanations for polarization holes include small-scale (compared to the resolution) field structure, decreased grain alignment with higher density, and spherical grain growth.

As a specific example of polarized emission observations of a massive star-forming region, we discuss measurements of the Orion molecular cloud (OMC), including the observed polarization hole. Many groups have observed selected objects or limited regions within OMC, but the

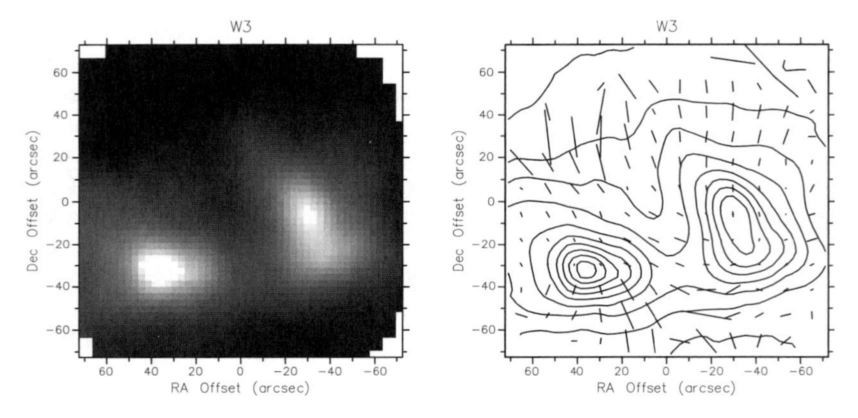

Figure 6. W3 IRS 4 and 5: a) Gray scale image of total flux at 850 μm. b) Magnetic field vectors overlaid on contours of total flux at 850 μm; vector length is proportional to percentage polarization, and direction traces **B**. The field shows large-scale structure connecting the two cores, and polarization minima at the star-forming sites suggest field tangling. At a distance of 2.3 kpc, the largely uniform field covers ~1.4 pc. This map was obtained from 2 hrs of integration with the Submillimetre Common User Bolometer Array (SCUBA) Polarimeter at the JCMT. (Figure courtesy of Antonio Chrysostomou and Jane Greaves.)

large-scale (1 pc) field structure was mapped only recently by Schleuning (1998). The magnetic field structure is extremely orderly on this scale although there are symmetric deviations of 25° from the average field direction (Fig. 7). Schleuning interprets this curvature as field pinching with a radius of 0.5 pc. The Kleinmann-Low (KL) nebula, which includes the energetic object IRc2, has been observed by many groups in polarized emission at wavelengths from 100 μm to 3 mm. Single-dish observations over several positions near KL invariably find that p decreases toward IRc2 (Gonatas et al. 1990; Leach et al. 1991; Schleuning et al. 1996; Aitken et al. 1997). Recent interferometric observations at millimeter wavelengths by Rao et al. (1998) at higher resolution reveal significant structure in the polarized emission near IRc2 (Fig. 7). Thus, the polarization hole at Orion KL appears to be due to averaging of varying polarization directions within the single-dish beams; however, more high-resolution observations are necessary to determine whether the decrease in p is always due to unresolved field structure.

C. Observations of Individual Embedded Protostellar Objects

Observing the magnetic field structure toward individual YSOs requires high resolution and sensitivity to relatively cool (<50 K) dust. Current telescope sensitivity levels limit detections of polarized emission to regions with high dust columns (>10^{24} cm^{-2} at 1 mm). Although only a handful of YSOs have been observed in polarized, millimeter wavelength emission to date, the number is increasing as telescope sensitivities in-

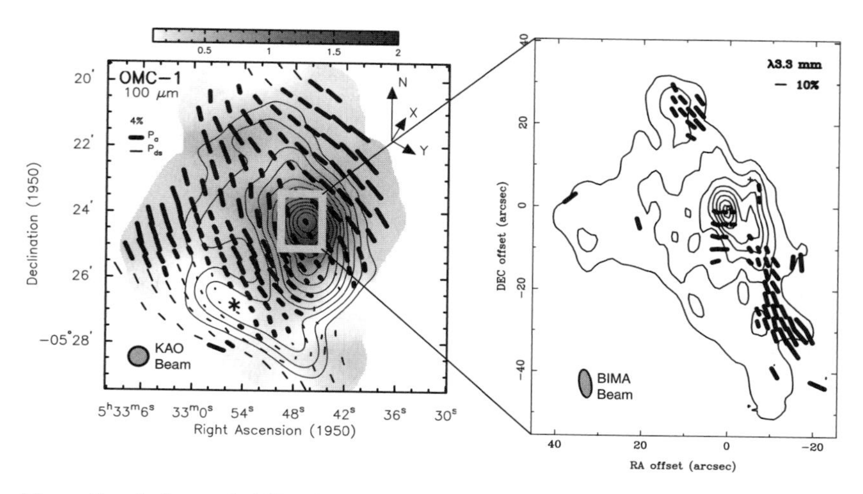

Figure 7. *Left panel:* 100-micron polarization as observed with the KAO (Schleuning 1998). The bar at the top shows total polarized intensity. *Right panel:* 3-mm polarization as observed with the Berkeley-Illinois-Maryland Array (BIMA) interferometer (Rao et al. 1998). The intensity peak of both maps is located near the position of IRc2. Note that at high spatial resolution, the apparent polarization hole near IRc2 shows significant nonuniform structure in the orientation of polarization vectors.

crease. Sources with a range of stellar mass have been observed, although the observations tend to concentrate on younger, embedded sources such as the Class 0 YSOs. These observations are the best method for measuring the field structure in protostellar envelopes and disks directly.

One aspect of star formation involving magnetic field structure is the modeling of bipolar outflows, which, it is generally assumed, require the presence of magnetic fields. Comparison of the measured magnetic field and outflow directions may provide important constraints on the theories. In some sources, such as NGC 1333 IRAS 4A (see, e.g., Akeson and Carlstrom 1997; Minchin et al. 1995) and NGC 7538 IRS 11 (Minchin and Murray 1994), the field direction and the outflow appear to be very well aligned. However, in other sources, such as VLA 1623, which drives a well-collimated molecular outflow, the magnetic field direction and the outflow are nearly orthogonal (Holland et al. 1996). Some authors (Greaves et al. 1997, Minchin et al. 1996) suggest that the observed angle between the magnetic field and the outflow may be a function of the viewing angle. We caution, however, that outflow inclination angles often are poorly determined and that some of the sources are binaries with multiple outflows, but the polarization has been measured only for the entire system. We also note that polarization observations measure the field direction in the plane of the sky, and any observed correlations on the sky do not guarantee correlation in three dimensions.

Most observations of magnetic fields toward YSOs are unable to resolve structure within the protostellar disk and envelope. At millimeter wavelengths, a typical single-dish resolution of 10–15″ yields a beam of 1500–2000 AU in Taurus, yet low-mass protostellar disks are generally $\leq$100 AU in diameter (see the chapter by Wilner and Lay, this volume) and envelopes are thousands of AU. These observations generally include emission from the protostellar envelope and disk, and even contributions from the surrounding core for measurements at higher frequencies. Thus, although fields can be characterized as poloidal or toroidal depending on the relative angles of the outflows or disks, the measured fields are weighted averages over all field components present in a beam. High-resolution interferometric observations have been made for NGC 1333 IRAS 4A (Fig. 8). These 3-mm observations are sensitive only to emission from the densest parts of the protostellar disk and envelope, with a lower limit to the H_2 density of 10^8 cm^{-3} in the observed material. These observations, in which structure was observed in the polarized emission on scales of 1000 AU, are evidence against a decrease in grain alignment at higher densities. Akeson and Carlstrom (1997) modeled the data using an hourglass field morphology within the envelope, where the field was aligned parallel to the outflow. Evidence for an hourglass field on a much larger (0.1 pc) scale has been seen for the high-mass source W3 IRS5 (Greaves et al. 1994) (Fig. 6).

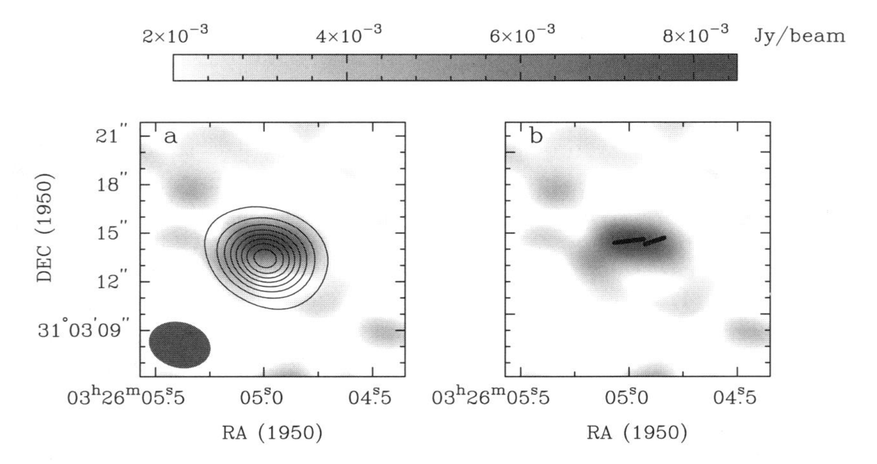

Figure 8. NGC 1333 IRAS 4A. (a) The linearly polarized emission (gray scale) plotted with the total intensity (contours) for NGC 1333/IRAS 4A (from Akeson and Carlstrom 1997). The contour levels are 25 mJy beam^{-1}. The full width at half maximum (FWHM) of the beam (3.7″ by 2.8″ at a position angle of 74°) is shown in the lower left. (b) The linearly polarized emission (gray scale) and two vectors of the polarization angle represent the range of angles seen in the data. The vectors shown are at angles of 97° and 109°(from Akeson and Carlstrom 1997).

Although (sub)millimeter polarimetry of individual sources (e.g., Tamura et al. 1995) has begun to reveal the role of the magnetic field, high-resolution observations using interferometers, as well as observations of many more sources (perhaps utilizing the new submillimeter bolometer arrays, such as SCUBA on the JCMT), are needed to constrain the role of the magnetic field.

III. MAPPING CIRCUMSTELLAR NEBULAE

Reflection nebulae, which are intrinsically polarized because we are observing scattered photons, are commonly found to surround YSOs when these objects are observed at optical and near infrared wavelengths (see Fig. 1). In this section, we concentrate on observations of the spatially resolved reflection nebulae, not on polarimetric observations of the central stars of these nebulae. Polarization observations of the stars themselves, taken through diaphragms centered on the stars, will invariably also include some light from the circumstellar nebula. In fact, aperture polarimetric observations that show unusually high polarization values provide a strong rationale for seeking polarimetric maps of such objects (although it also true that objects with low polarization values in aperture measurements may show significant polarization in maps). Aperture polarimetry work has been extensively reviewed in recent years for Herbig Ae/Be stars (Grinin 1994); for T Tauri stars (Bastien 1988); and for YSOs (Bastien 1996). In addition, a compilation of all published polarimetric maps prior to 1990 is given by Bastien and Ménard (1990).

In general, reflection nebulae can be observed at a given wavelength if enough dust grains are present in the circumstellar environment, if the dust grains are large enough in comparison to the wavelength, and if the YSO is an abundant source of photons at the selected wavelength. Under these conditions, some fraction of the light emerging from the protostellar photosphere scatters one or more times in the dust zone before emerging into interstellar space as polarized light. From these simple considerations, it is clear that the polarization signature of a reflection nebula, or the lack thereof, provides information about the density and size distribution of circumstellar dust and about the position of the central light source.

A. Scattering Models

For most of the 1990s, the working model for interpreting polarization maps obtained by observers has been that developed by Whitney and Hartmann (1992, 1993). In these and similar models (e.g., Wood et al. 1998), a Monte Carlo code is used to calculate polarization maps in dusty, circumstellar environments, using Mie scattering theory and assuming spherical dust particles. The work of Whitney and Hartmann built on the original work of Ménard (1990), who was the first to compute such Monte Carlo models, and the recognition by Bastien and Ménard (1988) that multiple scattering is essential in interpreting polarization maps. Additional work has been done by Fischer et al. (1994, 1996) and by Lucas and Roche (1997, 1998). The most important achievement of these models is in developing maps that produce qualitatively good matches to observed polarization maps of reflection nebulae around YSOs (Fig. 9).

The ability of a population of dust grains to scatter (and thereby polarize) starlight depends on the sizes, shapes, and compositions of dust grains present in an environment. The maximum degree of linear polarization ($p_{\max}$) as well as the total amount of scattered light at a given wavelength, depends strongly on the size of the dust grains relative to the wavelength of the incident photons. If the size parameter x is small, where $x = 2\pi a/\lambda$ and a is the radius of a spherical dust grain, the particles are small compared to the wavelength and thus are in the Rayleigh scattering regime, wherein the scattering efficiency decreases as λ^4. This is unlikely to be the situation for dust grains and scattering in the optical or infrared.

In many cases, YSOs appear as bright reflection nebulae in the optical but as point sources in the near infrared. In these cases, the dust grains are

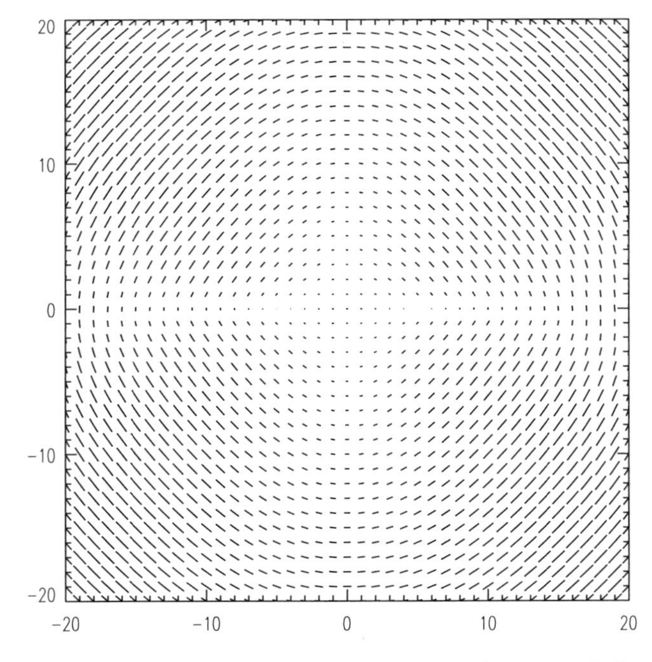

Figure 9. A simulated reflection nebula with a source located at (0,0) and a disk aligned horizontally at $y = 0$. Most of the light from the central source singly scatters out of the nebula, producing the centrosymmetric pattern. Some of the light multiply scatters into and out of the central disk plane, producing a linear region of low-amplitude vectors along $y = 0$; this region is referred to as the "polarization disk." The small vectors along the disk plane are the result of the convolution of the true polarization signature of the nebula with the point spread function of the telescope. (Courtesy of Philip Lucas.)

extremely effective at scattering 0.5-μm photons but equally ineffective at scattering photons at $\lambda = 1$–$2\ \mu$m. For such nebulae, dust distributions designed to match the ISM, such as that of Mathis et al. (1977) or that of Fischer et al. (1994), which have maximum grain sizes $a \leq 0.25\ \mu$m and steep power laws describing the number of dust grains of a given size (e.g., $N(a) \propto a^{-3.5}$), and thus have almost no large dust grains, clearly are not appropriate. In these cases, strong evidence exists for a population of larger dust grains with different albedo and polarization properties. Alternative "large-grain" models that have been proposed include that of Pendleton et al. (1990), with a maximum grain size of 0.8 μm, the grains of Kim et al. (1994) with a maximum grain size of 10 μm, and the very large grains of Huard et al. (1997*a*).

Most dust models include dust grains that are fairly reflective in the optical but become increasingly dark in the near infrared. These models also appear to contradict what is observed. For example, Whitney et al. (1997) estimate a dust albedo $w \simeq 0.3$–0.4 at 2.2 μm from their polarization models, whereas the Mathis et al. (1979) distribution is barely half this value at the same wavelength. Similarly, Leinert et al. (1991) report a value of $w \geq 0.32$. Lucas and Roche (1998) argue that icy mantles on dust grains, which are common in molecular clouds and around YSOs, would tend to increase the dust albedo above that predicted for bare grains.

Thus, the next generation of models will need to include scattering properties of large grains (Fig. 10 and 11). These will serve as an excellent complement to small-grain models and will thereby allow observers to better determine the size and material properties of circumstellar dust grains around YSOs.

B. Identifying YSOs, Disk Orientations, and Outflow Sources

One important way in which the models have contributed thus far is in showing that polarization "disks" mark the optically thick regions (i.e., the equatorial disk planes) of reflection nebulae (Fig. 9). These polarization disks are regions close to the central stars, where the polarization patterns clearly depart from centrosymmetry. Usually, the vectors in these regions are characterized by low (~10%) amplitudes and are roughly perpendicular to the symmetry axis of the bipolar nebula.

Bastien and Ménard (1988, 1990) demonstrated that polarization disks can be explained as the result of multiple scattering. In environments where stars have optically thick but spatially thin disks, photons that originate at the stellar photospheres cannot propagate radially outward within the disk plane; instead, they travel through the optically thin polar regions (in many cases evacuated by outflows) and scatter into the disk from above and below. Photons that scatter into the disk at sufficiently large radial distances from the star (where the disk itself has become optically thin) can scatter out of the nebula into our line of sight. This scenario produces maps with regions of low polarization that define the disk plane. In addition, the maps show centrosymmetry in which the original sources

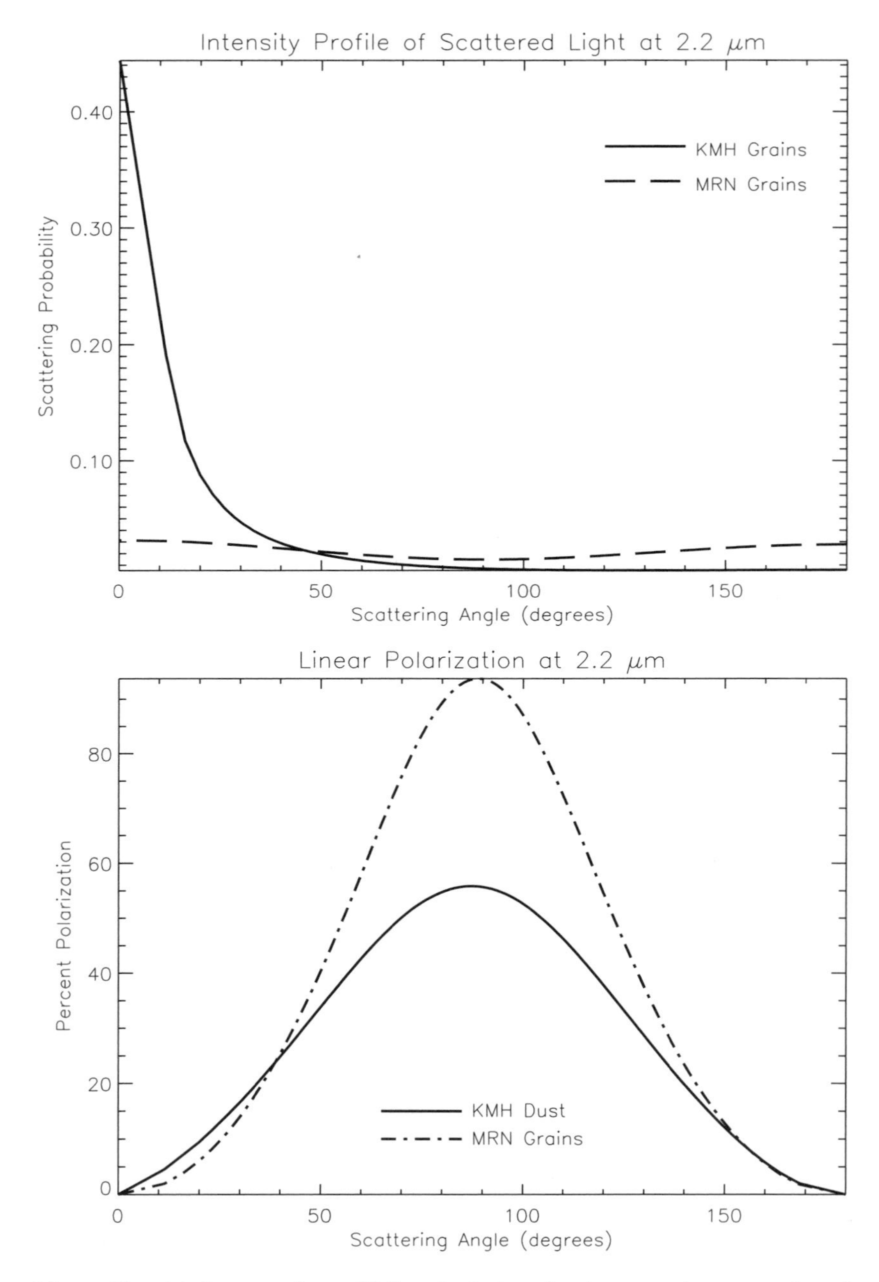

Figure 10. (a) A comparison of Mie calculations for the scattering properties of small (MRN; Mathis et al. 1979) and large (KHM; Kim et al. 1994) grains at 2.2 μm. The large grains are much more efficient at forward-scattering these long wavelength photons. The small grains are better at back-scattering (scattering angle >90°), but their scattering efficiency is low at all angles at 2.2 μm. (b) The MRN grains produce more highly polarized photons than the KMH grains at almost all scattering angles (T. C. Huard, 1998, unpublished).

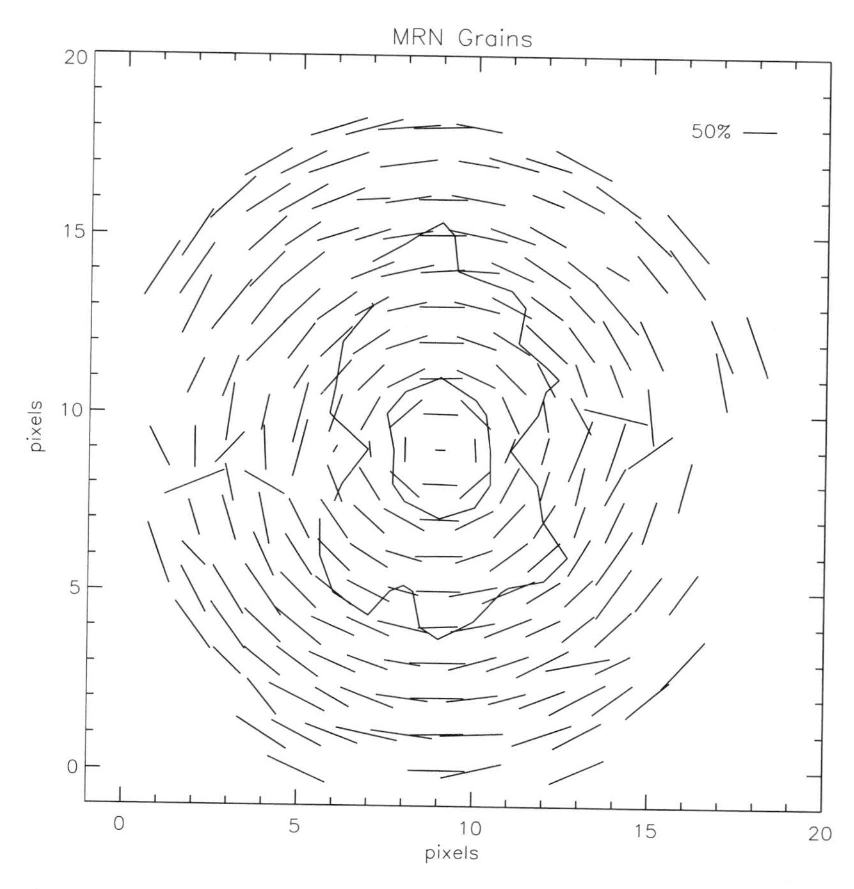

Figure 11. Simulated reflection nebulae at 2.2 μm, using (a) small (MRN; Mathis et al. 1979) and (b) large (KMH; Kim et al. 1994) grain models (part b is on page 264). Each model follows 10^7 photons as they scatter through a disk of 100 AU, seen edge-on, and an envelope of 10,000 AU radii. Contours for the scattered intensity are at the 10- and 100-photon levels. Note how the model with KMH grains shows a much brighter nebula with far more photons back-scattered toward the disk plane. This occurs because the KMH grains scatter so many more photons, even though the MRN grains are intrinsically better at back-scattering (T. C. Huard, 1998; unpublished).

of photons (i.e., the protostars) lie at the origins (the polarization centroids) of the centrosymmetric maps. Thus, maps such as these are well suited for determining whether an intensity source is a (proto)star or a reflection nebula and whether an intensity source is actually the source of scattered photons in the nebula.

If circumstellar disks are the fairly small (radii ≤ 100 AU) accretion disks expected in protostellar collapse (e.g., Terebey et al. 1984) and spectral energy distribution models (e.g., Adams et al. 1987), then in most

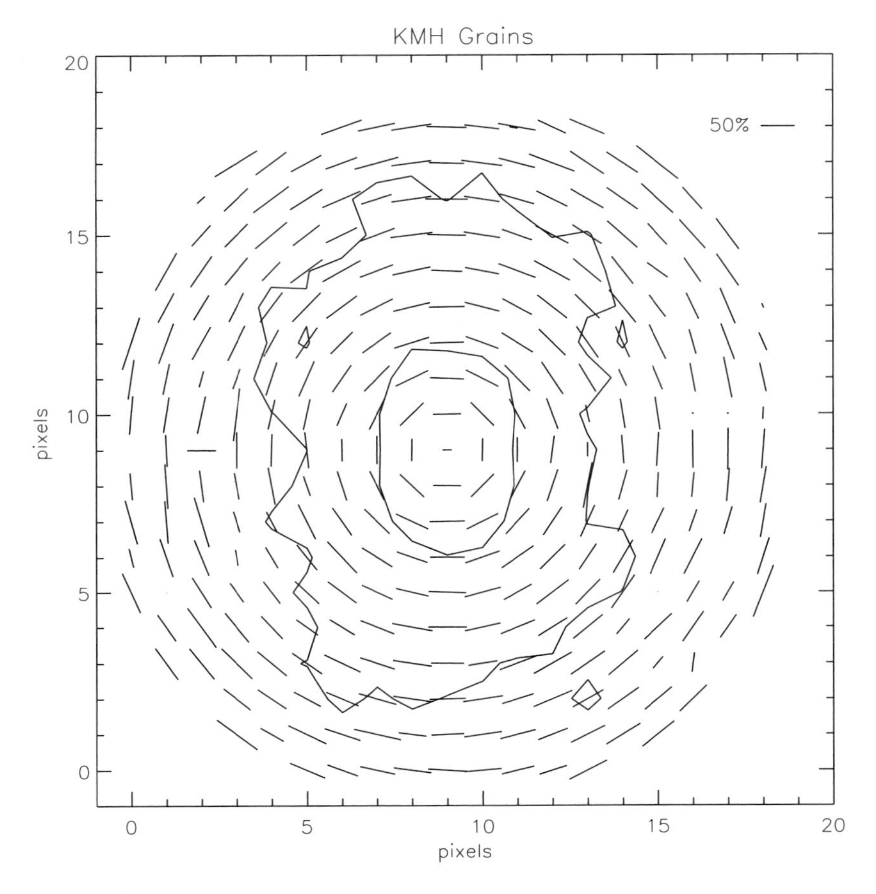

Figure 11. continued

cases, the observed light from polarization disks probably does not represent scattering from these disks themselves. Rather, because of the limited spatial resolution of most observations, a single pixel in an observational map almost certainly includes photons that scatter from material immediately above and below such disks (Whitney et al. 1997; Lucas and Roche 1998). However, it is possible that circumstellar disks are much larger, with radii of 10^3–10^4 AU, comparable to the pseudodisks modeled by Galli and Shu (1993) and to observations obtained by many workers in the submillimeter and millimeter regimes (see Weintraub et al. 1999). Many of the published polarization maps could, in fact, be resolving these large disks. In addition to identifying the presence and projected orientation of circumstellar disks, polarization maps also have been used to determine the inclination angles of the disks, through comparisons of the bipolar reflection nebulae with the Monte Carlo disk models.

The technique of using the polarization centroid to locate an embedded protostar has been applied successfully to polarization studies of the dark cloud L1287 (Weintraub and Kastner 1993), who discovered a bipolar reflection nebula centered on the position of IRAS 00338+6312. Notably, the IRAS source position is distinctly offset from the positions of the FU Orionis stars RNO 1B/1C, which previously had been identified as sources of the observed outflow and as identical with IRAS 00338+6312. Minchin et al. (1991*d*) and Kastner et al. (1992) demonstrated that only one of the three intensity peaks in the Juggler nebula (AFGL 2136) was a source (Fig. 12). Using the polarization map, they identified a bipolar reflection

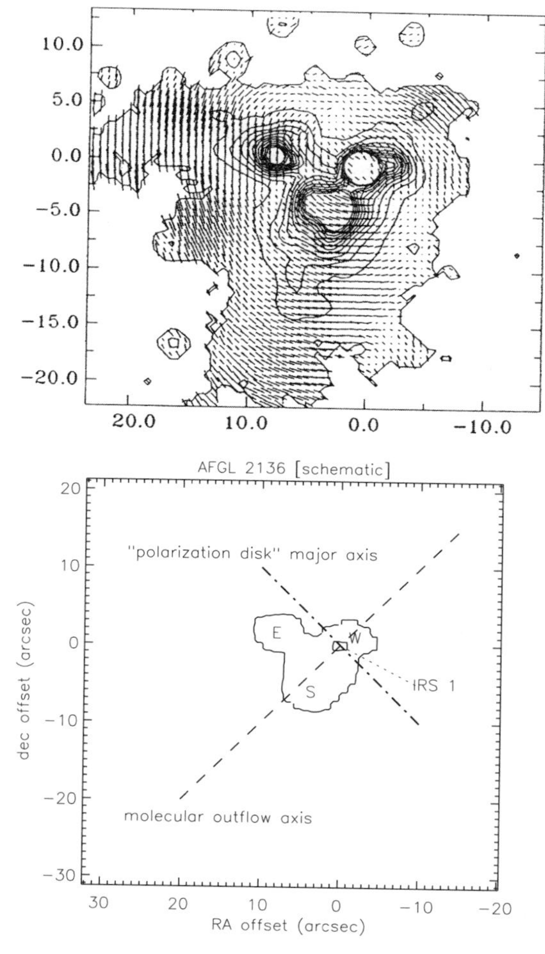

Figure 12. (a) Reflection nebula surrounding AFGL 2136 IRS 1. The base level contour is at 0.2 mJy arcsec^{-2}, and the contour step is 0.3 mJy arcsec^{-2}. Offsets are in arcsec from AFGL 2136 IRS 1. Note that two of the three bright intensity peaks are pure scattered light; these peaks mark limb-brightened walls of an outflow cone oriented at position angle 135°. Note the "polarization disk" at position angle 45° (from Minchin et al. 1991*d*). (b) schematic of AFGL 2136 (from Kastner and Weintraub 1996).

nebula likely produced by a molecular outflow and successfully predicted the outflow orientation from that of the polarization disk (Kastner et al. 1994).

In other cases, where the presence of a molecular outflow was established prior to the polarization mapping (for example, HL Tauri, IRc2), the polarization disks always lie perpendicular to the known orientations of the outflows. Similar work in identifying likely outflow sources in bipolar reflection nebulae has been carried out for SVS 2 and SVS 20 in Serpens (Huard et al. 1997*b*), WK 34 in AFGL 437 (Weintraub and Kastner 1996), NGC 2024 FIR 4 (Moore and Yamashita 1995), GSS 30 (Weintraub et al. 1993; Tamura et al. 1991), R Mon (Minchin et al. 1991*a*), AFGL 2591 (Minchin et al. 1991*b*), and OMC-1 (Minchin et al. 1991*c*). It is also generally true that in cases where only aperture polarimetry exists, the position angle of the polarization of the unresolved star plus nebula will indicate the orientation of an outflow.

C. Polarization of Line Emission

The reflection nebula in OMC-1 shows a centrosymmetric pattern centered approximately on the position of IRc2 (Minchin et al. 1991*c*). This reflection nebula has been detected in both continuum and in molecular line emission from shocked H_2 in the $v = 1-0$ S (1) line. Hough et al. (1986) (see also Burton et al. 1991; Chrysostomou et al. 1994) discovered that the inner portion of the reflection nebula, when observed in the H_2 line, is dichroically polarized. This is due to the presence of an extended region of material emitting S (1) photons that lies behind a medium of aligned grains. For now, this region of Orion is the only region known to show polarized line emission. Such regions may prove valuable for determining structure (i.e., alignment of grains) and magnetic field strengths, especially if additional regions of polarized line emission are discovered.

D. Circular Polarization

The measurement of circular polarization (CP) in the near infrared has been pushed forward in the last few years through improved instrumentation. The Chamaeleon infrared nebula (Gledhill et al. 1996) and GSS 30 in ρ Ophiuchi (Chrysostomou et al. 1997) show evidence for CP in the near infrared at levels of 1–2%, whereas CP levels in OMC-1 (Bailey et al. 1998) reach an astounding 17%; most YSOs, however, show CP levels below 0.1% (Bastien et al. 1989). Several mechanisms have been suggested for producing CP, all of which involve polarization by scattering; however, the discovery of large ($>$1%) CP is quite significant and will provide a challenge to the models. These methods for producing CP include multiple scattering off absorptive grains, single scattering off large, (magnetically) aligned grains, scattering of unpolarized incident photons off nonspherical, (magnetically) aligned grains, and scattering of linearly polarized incident photons off spherical grains.

IV. CONCLUSIONS

In the 1990s, polarization studies have moved into the mainstream as an observational tool for astrophysics. Specifically, polarization studies provide information about dust grains. Measurements of intrinsically unpolarized starlight that becomes polarized as the result of passage through a dark cloud informs us directly that layers of dust grains exist in the cloud and that these dust grains have been aligned, probably by a magnetic field inside the cloud, producing a dichroic filter that polarizes background starlight. These same dust grains produce thermal radiation, which, because of the preferred orientation imparted by the magnetic field, is intrinsically polarized. Finally, in the vicinity of YSOs, where dust grains grow in the shielded environments of cloud cores and dense, circumstellar disks, randomly oriented and sized dust grains will scatter photons from YSOs. These scattered light nebulae provide information about the sizes and types of dust grains, about the locations of YSOs that are detectable only through reflected light from the nebulae, and about the orientations of circumstellar disks and thereby the position angles of molecular outflows, where such outflows exist.

Acknowledgments The research of D. A. W. on polarization work is supported by a NASA Origins of Solar Systems grant. A. A. G. is grateful to the NASA-ISO program and the NSF Young Investigator program for supporting this work. R. L. A. gratefully acknowledges support from the Miller Institute for Basic Research in Science. We thank Héctor Arce for his assistance with the Taurus data used to prepare Fig. 2.

REFERENCES

Adams, F. C., Lada, C. J., and Shu, F. H. 1987. Spectral evolution of young stellar objects. *Astrophys. J.* 312:788–806.

Aitken, D. K., Smith, C. H., Morre, T. J. T., Roche, P. F., Fujiyoshi, T., and Wright, C. M. 1997. Mid- and far-infrared polarimetric studies of the core of OMC-1: The inner field configuration. *Mon. Not. Roy. Astron. Soc.* 286:85–96.

Akeson, R. L., and Carlstrom, J. E. 1997. Magnetic field structure in protostellar envelopes. *Astrophys. J.* 491:254–266.

Arce, H. G., Goodman, A. A., Bastien, P., Manset, N., and Sumner, M. 1998. The polarizing power of the interstellar medium in Taurus. *Astrophys. J. Lett.* 499:L93–L96.

Bailey, J., Chrysostomou, A., Hough, J. H., Gledhill, T. M., McCall, A., Clark, S., Ménard, F., and Tamura, M. 1998. Circular polarization in star-formation regions: Implications for biomolecular homochirality. *Science* 281:672–675.

Bastien, P. 1988. Polarization properties of T Tauri stars and other pre-main sequence objects. In *Polarized Radiation of Circumstellar Origin*, ed. G. V. Coyne (Tucson: Vatican Observatory/University of Arizona Press), pp. 541–582.

Bastien, P. 1996. Polarization of young stellar objects. In *Polarimetry of the Interstellar Medium*, ed. W. G. Roberge and D. C. B. Whittet (San Francisco: Astronomical Society of the Pacific), pp. 297–314.
Bastien, P., and Ménard, F. 1988. On the interpretation of polarization maps of young stellar objects. *Astrophys. J.* 326:334–338.
Bastien, P., and Ménard, F. 1990. Parameters of disks around young stellar objects from polarization observations. *Astrophys. J.* 364:232–241.
Bastien, P., Robert, C., and Nadeau, R. 1989. Circular polarization in T Tauri stars. II. New observations and evidence for multiple scattering. *Astrophys. J.* 339:1089–1092.
Burton, M. G., Minchin, N. R., Hough, J. H., Aspin, C., Axon, D. J., and Bailey, J. A. 1991. Molecular hydrogen polarization images of OMC-1. *Astrophys. J.* 375:611–617.
Chrysostomou, A., Hough, J. H., Burton, M. G., and Tamura, M. 1994. Twisting magnetic fields in the core region of OMC-1. *Mon. Not. Roy. Astron. Soc.* 268:325–334.
Chrysostomou, A., Ménard, F., Gledhill, T. M., Clark, S., Hough, J. H., McCall, A., and Tamura, M. 1997. Polarimetry of young stellar objects. II. Circular polarization of GSS 30. *Mon. Not. Roy. Astron. Soc.* 285:750–758.
Dotson, J. L. 1996. Polarization of the far-infrared emission from M17. *Astrophys. J.* 470:566–576.
Dowell, C. D. 1997. Far-infrared polarization by absorption in the molecular cloud Sagittarius B2. *Astrophys. J.* 487:237–247.
Fischer, O., Henning, T., and Yorke, H. W. 1994. Simulation of polarization maps. I. Protostellar envelopes. *Astron. Astrophys.* 284:187–209.
Fischer, O., Henning, T., and Yorke, H. W., 1996. Simulation of polarization maps. II. The circumstellar environment of pre-main sequence objects. *Astron. Astrophys.* 308:863–885.
Galli, D., and Shu, F. H. 1993. Collapse of magnetized molecular cloud cores. I. Semianalytical solution. *Astrophys. J.* 417:220–242.
Gerakines, P. A., Whittet, D. C. B., and Lazarian, A. 1995. Grain alignment in the Taurus dark cloud. *Astrophys. J. Let.* 455:171–175.
Gledhill, T., Chrysostomou, A., and Hough, J. H. 1996. Linear and circular imaging polarimetry of the Chamaeleon infrared nebula. *Mon. Not. Roy. Astron. Soc.* 282:1418–1436.
Glenn, J., Walker, C. K., and Young, E. T. 1998. Magnetic fields in star formation regions: 1.3 mm continuum polarimetry. *Bull. Amer. Astron. Soc.* 192:1017 (abstract).
Gonatas, D. P., Engargiola, G. A., Hildebrand, R. H., Platt, S. R., Wu, X. D., Davidson, J. A., Novak, G., Aitken, A. K., and Smith, C. 1990. The far-infrared polarization of the Orion Nebula. *Astrophys. J.* 357:132–137.
Goodman, A. A. 1995. The future of magnetic field mapping in the interstellar medium. In *Airborne Astronomy Symposium on the Galactic Ecosystem: From Gas to Stars to Dust,* ed. M. R. Haas, J. A. Davidson, and E. F. Erickson (San Francisco: Astronomical Society of the Pacific), p. 45.
Goodman, A. A., Bastien, P., Myers, P. C., and Ménard, F. 1990. Optical polarization maps of star-forming regions in Perseus, Taurus and Ophiuchus. *Astrophys. J.* 359:363–377.
Goodman, A. A., Jones, T. J., Lada, E. A., and Myers, P. C. 1992. The structure of magnetic fields in dark clouds: Infrared polarimetry in B216-217. *Astrophys. J.* 399:108–113.
Goodman, A. A., Jones, T. J., Lada, E. A., and Myers, P. C. 1995. Does near-infrared polarimetry reveal the magnetic field in cold dark clouds? *Astrophys. J.* 448:748–765.

Greaves, J. S., Holland, W. S., and Ward-Thompson, D. 1997. Submillimeter polarimetry of class 0 protostars: Constraints on magnetized outflow models. *Astrophys. J.* 480:255–261.

Greaves, J. S., Murray, A. G., and Holland, W. S. 1994. Investigating the magnetic field structure around star formation cores. *Astron. Astrophys.* 284:L19–L22.

Greaves, J. S., Holland, W. S., Minchin, N. R., Murray, A. G., and Stevens, J. A. 1999. Submillimetre polarization constraints on dust grain alignment. *Astron. Astrophys.* 344:668–674.

Grinin, V. P. 1994. Polarimetric activity of Herbig Ae/Be stars. In *The Nature and Evolutionary Status of Herbig Ae/Be Stars*, ed. P. S. The, M. R. Perez, and P. J. van den Heuvel (San Francisco: Astronomical Society of the Pacific), p. 63.

Heiles, C. 1996. The local direction and curvature of the galactic magnetic field derived from starlight polarization. *Astrophys. J.* 462:316–325.

Heyer, M. H., Vrba, F. J., Snell, R. L., Schloerb, F. P., Strom, S. E., Goldsmith, P. F., and Strom, K. M. 1987. The magnetic evolution of the Taurus molecular clouds. I. Large-scale properties. *Astrophys. J.* 321:855–876.

Hildebrand, R. H. 1996. Problems in far-infrared polarimetry. In *Polarimetry of the Interstellar Medium*, ed. W. G. Roberge and D. C. B. Whittet (San Francisco: Astronomical Society of the Pacific), pp. 254–268.

Holland, W. S., Greaves, J. S., Ward-Thompson, D., and André, P. 1996. The magnetic field structure around protostars: Submillimetre polarimetry of VLA 1623 and S106-IR/FIR. *Astron. Astrophys.* 309:267–274.

Hough, J. H., Axon, D. J., Burton, M. G., Gatley, I., Sato, S., Bailey, J., McCaughrean, M. J., McLean, I. S., Nagata, M., Allen, D., Garden, R. P., Hasegawa, T., Hayashi, M., Kaifu, N., Morimoto, M., and Walther, D. 1986. Infrared polarization in OMC-1. Discovery of a molecular hydrogen reflection nebula. *Mon. Not. Roy. Astron. Soc.* 222:629–644.

Huard, T. C., Weintraub, D. A., and Kastner, J. H. 1997*a*. Polarization modelling of protostellar environments with large dust grain distributions. *Bull. Amer. Astron. Soc.* 190:4106 (abstract).

Huard, T. C., Weintraub, D. A., and Kastner, J. H. 1997*b*. Bipolar outflow sources in the Serpens core: SVS 2 and SVS 20. *Mon. Not. Roy. Astron. Soc.* 290:598–606.

Jones, T. J., Klebe, D., and Dickey, J. M. 1992. Infrared polarimetry and the galactic magnetic field II: Improved models. *Astrophys. J.* 389:602–615.

Kastner, J. H., and Weintraub, D.A. 1996. Water ice in the disk around the protostar AFGL 2136 IRS 1. *Astrophys. J. Lett.* 466:L103–L106.

Kastner, J. H., Weintraub, D. A., and Aspin, C. A. 1992. The Juggler: A three-lobed near-IR reflection nebula toward CRL 2136 = OH17.6+0.2. *Astrophys. J.* 389:357–368.

Kastner, J. H., Weintraub, D. A., Snell, R. L., Sandell, G., Aspin, C., Hughes, D., and Baas, F. 1994. The massive molecular outflow from CRL 2136 IRS 1. *Astrophys. J.* 425:695–706.

Kim, S., Martin, P. G., and Hendry, P. D. 1994. The size distribution of interstellar dust particles as determined from extinction. *Astrophys. J.* 422:164–175.

Lazarian, A., Goodman, A. A., and Myers, P. C. 1997. On the efficiency of grain alignment in dark clouds. *Astrophys. J.* 490:273–280.

Leach, R. W., Clemens, D. P., Kane, B. D., and Barvainis, R. 1991. Polarimetric mapping of Orion using Millipol: Magnetic activity in BN/KL. *Astrophys. J.* 370:257–262.

Leinert, C. H., Haas, M., and Lenzen, R. 1991. LkHα 198 and V376 Cassiopeiae. Speckle interferometric and polarimetric observations of circumstellar dust. *Astron. Astrophys.* 246:180–194.

Lucas, P. W., and Roche, P. F. 1997. Butterfly star in Taurus: Structures of young stellar objects. *Mon. Not. Roy. Astron. Soc.* 286:895–919.

Lucas, P. W., and Roche, P. F. 1998. Imaging polarimetry of Class I young stellar objects. *Mon. Not. Roy. Astron. Soc.* 299:699–722.

Mathewson, D. S., and Ford, V. L. 1970. Polarization observations of 1800 stars. *Mem. Roy. Astron. Soc.* 74:139.

Mathis, J. S., Rumpl, W., and Nordsieck, K. H., 1977. The size distribution of interstellar grains. *Astrophys. J.* 217:425–433.

Ménard, F. 1990. Étude de la polarisation causée par des grains dans les enveloppes circumstellaires denses. Ph.D. Thesis, Université de Montréal.

Minchin, N. R., Bonifacio, V. H. R., and Murray, A. G. 1996. Submillimetre polarimetric observations of S140 and GL2591: Investigating the role of viewing angle on observed polarization position angles. *Astron. Astrophys.* 315:L5–L8.

Minchin, N. R., Hough, J. H., McCall, A., Aspin, C., Yamashita, T., and Burton, M. G. 1991*a*. Near-infrared imaging polarimetry of bipolar nebulae. III—R Mon/NGC 2261. *Mon. Not. Roy. Astron. Soc.* 249:707–715.

Minchin, N. R., Hough, J. H., McCall, A., Aspin, C., Hayashi, S. S., Yamashita, T., and Burton, M. G. 1991*b*. Near-infrared imaging polarimetry of bipolar nebulae. II—GL 2591. *Mon. Not. Roy. Astron. Soc.* 251:508–521.

Minchin, N. R., Hough, J. H., McCall, A., Burton, M. G., McCaughrean, M. J., Aspin, C., Bailey, J. A., Axon, D. J., and Sato, S. 1991*c*. Near-infrared imaging polarimetry of bipolar nebulae. I—The BN-KL region of OMC-1. *Mon. Not. Roy. Astron. Soc.* 248:715–729.

Minchin, N. R., Hough, J. H., Burton, M. G., and Yamashita, T. 1991*d*. Near-infrared imaging polarimetry of bipolar nebulae. IV—GL 490, GL 2789 and GL 2136. *Mon. Not. Roy. Astron. Soc.* 251:522–528.

Minchin, N. R., and Murray, A. G. 1994. Submillimetre polarimetric mapping of DR21 and NGC 7358-IRS 11: Tracing the circumstellar magnetic field. *Astron. Astrophys.* 286:579–587.

Minchin, N. R., Sandell, G., and Murray, A. G. 1995. Submillimetre polarimetric observations of NGC 1333 IRAS 4A and 4B: Tracing the circumstellar magnetic field. *Astron. Astrophys.* 293:L61–L64.

Moneti, A., Pipher, J. L., Helfer, H. L., McMillan, R. S., and Perry, M. L. 1984. Magnetic field structure in the Taurus dark cloud. *Astrophys. J.* 282:508–515.

Moore, T. J. T., and Yamashita, T. 1995. A near-infrared reflection nebula associated with NGC 2024 FIR 4. *Astrophys. J.* 440:722–727.

Myers, P. C., and Goodman, A. A. 1991. On the dispersion in direction of interstellar polarization. *Astrophys. J.* 373:509–524.

Novak, G., Dotson, J. L., Dowell, C. D., Goldsmith, P. F., Hildebrand, R. H., and Schleuning, D. A. 1997. Polarized far-infrared emission from the core and envelope of the Sagittarius B2 molecular cloud. *Astrophys. J.* 487:320–327.

Ostriker, E., Gammie, C., and Stone, J. 1999. Kinetic and structural evolution of self-gravitating, magnetized clouds: 2.5-dimensional simulations of decaying turbulence. *Astrophys. J.* 513:259–274.

Pendleton, Y. J., Tielens, A. G. G. M., and Werner, M. W. 1990. Studies of dust grain properties in infrared reflection nebulae. *Astrophys. J.* 349:107–119.

Rao, R., Crutcher, R. M., Plambeck, R. L., and Wright, M. C. H. 1998. High resolution millimeter-wave mapping of linearly polarized dust emission: Magnetic field structure in Orion. *Astrophys. J. Lett.* 502:L75–L78.

Schleuning, D. A. 1998. Far-infrared and submillimeter polarization of OMC-1: Evidence for magnetically regulated star formation. *Astrophys. J.* 493:811–825.

Schleuning, D. A., Dowell, C. D., and Platt, S. R. 1996. Array polarimetry of the Orion Nebula from the Caltech Submillimeter Observatory. In *Polarimetry of the Interstellar Medium*, ed. W. G. Roberge and D. C. B. Whittet (San Francisco: Astronomical Society of the Pacific), pp. 254–268.

Schulz, A., Lenzen, R., Schmidt, T., and Proetel, K. 1981. Polarization of starlight in M17. *Astron. Astrophys.* 95:94–99.

Tamura, M., Gatley, I., Joyce, R. R., Ueno, M., Suto, H., and Sekiguchi, M. 1991. Infrared polarization images of star-forming regions. I. The ubiquity of bipolar structure. *Astrophys. J.* 378:611–627.

Tamura, M., Hough, J. H., and Hayashi, S. S. 1995. 1 millimeter polarimetry of young stellar objects: Low-mass protostars and T Tauri stars. *Astrophys. J.* 448:346–355.

Terebey, S., Shu, F. H., and Cassen, P. 1984. The collapse of the cores of slowly rotating isothermal clouds. *Astrophys. J.* 286:529–551.

Vallée, J. P., and Bastien, P. 1996. Extreme-infrared (800 μm) polarimetry of the M17-SW molecular cloud with the JCMT. *Astron. Astrophys.* 313:255–268.

Weintraub, D. A., and Kastner, J. H. 1993. The exciting young stellar object for the molecular outflow at the core of L1287. *Astrophys. J.* 411:767–772.

Weintraub, D. A., and Kastner, J. H. 1996. The deeply embedded source that drives the protostellar outflow in AFGL 437: Evidence from near-infrared polarimetric imaging. *Astrophys. J.* 458:670–679.

Weintraub, D. A., Kastner, J. H., Griffith, L. L., and Campins, H. C. 1993. Near-infrared, polarimetric imaging of the bipolar lobes of GSS 30: Protostellar infall or outflow? *Astron. J.* 105:271–283.

Weintraub, D. A., Sandell, G., Huard, T. L., Kastner, J. H., van den Ancker, M. E., and Waters, R. 1999. Submillimeter imaging of T Tauri's circumbinary disk and the discovery of a protostar in Hind's Nebula. *Astrophys. J.*, in press.

Whitney, B., and Hartmann, L. 1993. Model scattering envelopes of young stellar objects. II. Infalling envelopes. *Astrophys. J.* 402:605–622.

Whitney, B., and Hartmann, L. 1992. Model scattering envelopes of young stellar objects. I. Method and application to circumstellar disks. *Astrophys. J.* 395:529–539.

Whitney, B., Kenyon, S. J., and Gomez, M. 1997. Near-infrared imaging polarimetry of embedded young stars in the Taurus-Auriga molecular cloud. *Astrophys. J.* 485:703–734.

Whittet, D. C. B. 1992. *Dust in the Galactic Environment* (Bristol: IOP Publishing).

Wilking, B. A., Lebofsky, M. J., Kemp, J. C., and Rieke, G. H. 1979. Infrared polarimetry in the Rho Ophiuchi dark cloud. *Astron. J.* 84:199–203.

Wood, K., Kenyon, S. J., Whitney, B., and Turnbull, M. 1998. Optical and near-infrared model images of the circumstellar environments of classical T Tauri stars. *Astrophys. J.* 497:404–418.

THE LOW-MASS STELLAR POPULATION OF THE ORION OB1 ASSOCIATION, AND IMPLICATIONS FOR THE FORMATION OF LOW-MASS STARS

FREDERICK M. WALTER
State University of New York at Stony Brook

JUAN M. ALCALÁ
Osservatorio Astronomico di Capodimonte

RALPH NEUHÄUSER
Max-Planck-Institut für extraterrestrische Physik

MICHAEL STERZIK
European Southern Observatory

and

SCOTT J. WOLK
Harvard-Smithsonian Center for Astrophysics

The OB associations, which are fossil star formation regions, retain the end-products of the star formation process in an open, unobscured environment. Observations of the low-mass stars in OB associations provide a far clearer picture of the results of the star formation process than do observations of embedded regions of ongoing star formation. We review the X-ray and optical surveys of the fossil star formation regions in the Orion OB1 association. Low-mass pre-main-sequence stars not only abound in the known regions of recent star formation but are also distributed over a much larger volume. The pre-main-sequence stars have a narrow age spread, and the mass function extends down to substellar masses. The clustering of pre-main-sequence stars around σ Ori may represent an evolved version of the Orion Nebula Cluster. We speculate about the effect of the OB environment on the initial mass function and the formation of planetary systems like the solar system.

I. STAR FORMATION IN T AND OB ASSOCIATIONS

A long time ago[a], in a part of the Galaxy far, far away, our Sun was born. The isotopic abundances in the chondrules and calcium-aluminum-rich

[a] About 4.6 Gyr.

inclusions (CAIs) (see the chapter by Goswami and Vanhala, this volume), in particular the existence of very short-lived parent radionuclides, provide clues about the environment in which our Sun and its planets were born. Among the plausible sources for these short-lived nuclides are an asymptotic giant branch (AGB) star, a supernova, a Wolf-Rayet (W-R) star, or a combination thereof (Cameron 1993; see also Lee et al. 1998 and the chapter by Glassgold et al., this volume). If the Sun formed in close proximity to a supernova or a W-R star, then the Sun most likely formed in an OB association, not in the quieter confines of a T association.

The T associations[b] are young unbound groups of low-mass stars with ages of a few million years (Myr), characterized by dust clouds and T Tauri stars. Most of what we know about the formation of low-mass stars is based on observations of T associations, because they are closer than OB associations, and the stars are easier to study. That only a few T Tauri stars were known to be associated with the nearby OB associations was taken to mean that low-mass stars did not form in great numbers in the OB associations. However, we now know that only a small percentage of the low-mass stars form in T associations: Most form in the OB associations, which are striking concentrations of short-lived, bright, high-mass stars.

The very different environments of OB and T associations will influence the formation of the low-mass stars and subsequent evolution of their protoplanetary disks. To infer the general properties of low-mass pre-main-sequence (PMS) stars and their protoplanetary disks from observations of T associations alone may lead to biased conclusions. Our purpose in this review is to draw attention to the low-mass star formation in OB associations.

We concentrate here on the low-mass stars of the Orion OB1 association (Brown et al. 1998 review the high-mass stars). Our focus is on the more exposed parts of the complex, the fossil star formation regions (SFRs) of the Ori OB1a and Ori OB1b subassociations. We will not discuss Ori OB1c, which surrounds the sword, or Ori OB1d, the Orion Nebula Cluster (ONC), a dynamically young, partially embedded, active SFR associated with the Orion A cloud (Hillenbrand 1997).

We begin with a brief review of low-mass PMS stars. In section II we review the techniques used to identify low-mass PMS stars. We discuss the Orion OB1 association in sections III and IV and speculate about the implications for star and planet formation in section V.

A. The Low-Mass PMS Stars in T Associations

T Tauri and other stars of its class are characterized by strong Balmer emission lines, ultraviolet and infrared continuum excesses, and erratic variability. They are generally found near dark clouds. Joy (1945) first

---

[b] Under T associations we include all star-forming regions except those dominated by OB stars. This encompasses such regions as Taurus-Auriga, Chamaeleon I, and ρ Ophiuchi, which include B and Ae stars.

identified these stars as a distinct class of objects. Ambartsumian (1947) concluded that these T Tauri stars were recently formed stars, and introduced the term "T association" for their groupings. Herbig (1962) arrived at similar conclusions. Cohen and Kuhi (1979) created a spectral atlas of the then-known classical T Tauri stars (cTTSs). Herbig and Bell (1988) have produced the most current catalog of low-mass PMS stars. Recent reviews of the cTTSs and their evolution include those by Basri and Bertout (1993) and Stahler and Walter (1993).

In the 20 years since the launch of the *Einstein* Observatory, X-ray imaging observations provided a new view of SFRs. Not only were some cTTSs found to be strong soft X-ray sources, but many other X-ray sources present in these SFRs were identified optically as low-mass PMS stars. Most of these newly identified PMS stars lacked the strong line emission and the UV and near IR continuum excesses of the cTTSs and were thought to be post-T Tauri stars (Herbig 1978): stars that have lost their circumstellar material and were evolving towards the zero-age main sequence (ZAMS). Placement on the Hertzsprung-Russell (HR) diagram showed that many of these stars had ages of a few Myr, comparable to those of the cTTSs. Walter (1986) called these the naked T Tauri stars (nTTSs), stars that had become unveiled, but they are more commonly referred to as the weak T Tauri stars (wTTSs).[c]

Spurred in part by the realization that X-ray surveys could reveal a more complete population of low-mass PMS stars, much effort went into studying the nearby T associations, such as Taurus (Walter et al. 1988; Neuhäuser et al. 1995*b*); Chamaeleon (Feigelson et al. 1993; Alcalá et al. 1995); Corona Australis (Walter et al. 1997); ρ Ophiuchi (Montmerle et al. 1983; Casanova et al. 1995); Lupus (Wichmann et al. 1997; Krautter et al. 1997). These studies have shown, among other things, that

- The low-mass population is dominated, at least at ages of more than about 1 Myr, by stars that do not have circumstellar disks (Walter et al. 1988; Strom et al. 1989).
- Low-mass stars can form either in small clumps (cluster mode) or individually (distributed mode) (Strom et al. 1990).
- Most stars form in multiple systems. The binary fraction in T associations seems to be twice that of the field (Leinert et al. 1993; Ghez et al. 1993; Simon et al. 1993; Kohler and Leinert 1998).
- Low-mass stars may be found far from their natal clouds or may not be associated with any recognizable cloud (Walter et al. 1988; Alcalá et al. 1995; Neuhäuser et al. 1995*c*; Kastner et al. 1997).
- The distribution of rotational periods is bimodal, with both slow rotators, whose periods are set by disk-breaking, and a rapidly rotating

[c] The definition of the wTTSs is that the Hα equivalent width is less than 10 Å, while the nTTSs lack evidence for circumstellar material. All nTTSs are wTTSs; the converse is not always true.

population, representing stars that have spun up as they contract along their Hayashi tracks (Edwards et al. 1993; Bouvier et al. 1997).
- There is a large spread of stellar ages in T associations.

The large age spread suggests that the star formation lifetime of a molecular cloud is related to the sound-crossing timescale of the cloud (Herbig 1978) or to the ambipolar diffusion timescales. Given the typical 1 to 2 km s^{-1} velocity dispersion within associations (Herbig 1977; Hartmann et al. 1986; Frink et al. 1997), one would expect the older population to have dispersed far from the cloud. The lack of these older stars gives rise to the post-T Tauri problem. Palla and Galli (1997), however, noted that star formation is not a steady process but runs slowly and less efficiently in the first few million years of the cloud lifetime. Hence, there should be only few post-TTSs, but many coeval cTTSs and nTTSs.

B. The Low-Mass PMS Stars in OB Associations

While the T associations produce perhaps a few thousand stars over their 10–30 Myr lifetimes, the OB associations are far more productive, though for a shorter interval. OB associations are loose, easily identifiable concentrations of bright, high-mass stars (see Humphries 1978; Blaauw 1964). Ambartsumian (1947) showed that their typical mass densities of $<0.1\ M_\odot\ pc^{-3}$ are unstable to galactic tidal forces, and therefore they must be young. This conclusion is supported by the ages derived from the HR diagrams for these associations.

Blaauw (1991) reviewed the nearby OB associations and their relation to local star formation. The older OB associations retain a fossil record of star formation processes in a giant molecular cloud. The very process of formation of the massive stars disperses the giant molecular cloud and thereby disrupts further star formation.

One can estimate the importance of OB associations for low-mass star formation. Within 500 pc of the Sun lie three OB associations (Orion OB1, Scorpius-Centaurus-Lupus, and Perseus OB2) and several T associations (Taurus-Auriga, Chamaeleon, Corona Australis, Lupus, TW Hydrae). Assuming a Miller-Scalo mass function in the OB associations, and counting stars in the T associations, one can show that over 90% of the low-mass stars with ages less than about 10 Myr are likely to be members of OB associations.

It had long been thought that star formation was bimodal, in the sense that high-mass stars did not form in T associations and that low-mass stars did not form in great numbers in OB associations (e.g., Larson 1986; Shu and Lizano 1988). We now know that the latter is not true. Hα surveys suggest that Hα-emitting stars not only are found in the actively star-forming parts of giant molecular clouds (e.g., Haro 1953; Duerr et al. 1982; Strom et al. 1990) but also abound in the fossil OB associations (e.g., Kogure et al. 1989; Nakano et al. 1995).

Walter et al. (1994) investigated the low-mass population of the Upper Scorpius association (de Geus et al. 1989), a 5-Myr-old association at a distance of ~140 pc. Starting with *Einstein* X-ray observations, they found a low-mass PMS population, whose properties appear significantly different from those found in T associations:

- The space density of PMS stars is higher than that in Taurus by about a factor of 3.
- The low-mass stars seem to be coeval, at an age of 1–2 Myr.
- The low-mass PMS population is largely devoid of circumstellar material and near-IR excesses, even at this age.
- The distribution of rotational periods is not peaked, suggesting that the association is observed during an epoch when all the stars are spinning up (Adams et al. 1998).
- Between 10 and 0.3 $M_{\odot}$, the mass function $d \log N/d \log M$ is consistent with the field star initial mass function (Miller and Scalo 1979).

Sciortino et al. (1998) reached similar conclusions about the Upper Scorpius association based on *Röntgen Satellite* (ROSAT) observations. Brandner and Köhler (1998) suggested that the binary fraction of PMS stars in OB associations is about half that observed in the T associations and is comparable to that of the field. They note that the binary fraction may be smaller yet near the OB stars.

The nearest giant molecular cloud complex and site of active star formation is in Orion: the Orion A and B clouds and the Orion OB1 association. All stages of the star formation process can be found here, from deeply embedded protoclusters to fully exposed OB associations. The different modes of star formation occurring in these clouds (clustered, distributed, isolated) allow us to learn more about the influence of the environment on the star formation process.

We study the Ori OB1a and OB1b fossil SFRs, because in these regions star formation is complete, all the stars are visible (few embedded sources remain), and the stars are at their final masses. Yet these fossil SFRs are sufficiently young (2–10 Myr) that the full population remains cospatial and that spatial substructure in the star formation process may still be detectable.

II. HOW TO FIND LOW-MASS PMS STARS IN ASSOCIATIONS

To study the global processes of low-mass star formation, one must identify and sample all the populations of low-mass PMS stars. The embedded sources, the cTTSs, and the nTTSs may represent different populations, with different spatial and age distributions. All the populations, or their evolutionary descendents, are present (and none are hidden) in fossil SFRs. The cTTSs tend to be readily identifiable either photometrically, because of their variability or their prominent near-IR continuum excesses, or spectroscopically, through their strong emission lines. However, the vast majority of the low-mass PMS stars are not so easily discovered. Their coronal X-ray and chromospheric emission line fluxes are little stronger than those of active ZAMS stars; they have no continuum excesses; and they are no more variable than heavily spotted, active late-type stars. We review methods of searching for the nTTSs and wTTSs and of identifying the complete PMS population of an association.

A. X-Ray Surveys

Low-mass PMS stars have X-ray luminosities between 10^{29} and 10^{31} erg s^{-1}, reflecting the strong magnetic activity of these stars. The X-ray surface flux scales with the stellar mass (Walter 1996), which gives an apparent rotation-activity correlation, because stellar rotation rates increase with increasing mass. There is no clear evidence that either the X-ray surface flux or the ratio of the X-ray flux to the bolometric flux correlates with stellar rotation.[d] The surface flux of the PMS stars is somewhat lower than the "saturated" value seen in the rapidly rotating dwarfs in the Pleiades (Stauffer et al. 1994*a*).

The high X-ray luminosities, at a characteristic temperature of about 1 keV, are easily detectable in short pointed X-ray imaging observations, or in flux-limited surveys like the ROSAT All-Sky Survey (RASS) (e.g., Voges et al. 1996). Although most cTTSs have been detected as X-ray emitters, they tend to be fairly heavily absorbed, so their mean observed fluxes in soft X-rays are lower than those of the nTTSs. Neuhäuser et al. (1995*a*) argue that the cTTSs are also intrinsically less X-ray luminous than the nTTSs; the detection rate for cTTSs in the RASS is only 15% (Neuhäuser et al. 1995*a*). The detection efficiency of unveiled stars is unknown, but deep pointings seem to recover 70–80% of the photometrically identifiable PMS stars (Walter et al. 1998). These X-ray surveys complement optical Hα emission line surveys, which yield mainly cTTSs.

The complete sky coverage of the RASS to a flux limit of $\approx$1–2 $\times 10^{-13}$ erg cm^{-2} s^{-1} has permitted unbiased analyses of the spatial distribution of X-ray active stars, including low-mass PMS stars. However, the complete identification and classification of all detected sources requires optical spectroscopy of each of the possible counterparts of an X-ray source. Neuhäuser et al. (1995*b*) and Sterzik et al. (1995) established a discrimination criterion, based on X-ray spectral appearance and f_x/f_v of known TTSs, for selecting PMS candidates. Sterzik et al. (1995) applied it to map the large-scale spatial distribution of young stars in a 700-deg^2 field centered on the Orion SFR. The spatial distribution of the stellar X-ray sources selected in this way gives a qualitative idea of the

[d] This is not surprising. Among the Pleiades G stars (Stauffer et al. 1994*a*), a rotation-activity relation is seen only for $v \sin i < 15$ km s^{-1}. Nearly all PMS stars rotate more rapidly.

general morphology of the SFR, by tracing the X-ray active young stellar population.[e]

The $\sim 10^{-14}$ erg cm^{-2} s^{-1} limiting fluxes of existing X-ray observations, together with the $f_X/f_v \sim 10^{-3}$ typically seen in the PMS stars, means that the X-ray pointings can select optical counterparts down to about $V = 15$. At the distance and age of Ori OB1, this corresponds to ≈ 0.5 M$_\odot$.

B. Spectroscopy

The most certain way to identify low-mass PMS stars is by their optical spectra.

The principal spectroscopic characteristic of low-mass PMS stars is the 6707 Å line of Li I. This line is generally considered to be an indicator of youth, because the depletion of Li in the outer layers of convective low-mass stars is very rapid. Lithium is easily destroyed by convective mixing in the stellar interiors when the temperature at the bottom of the convective layer reaches about 2.5×10^6 K (Bodenheimer 1965; D'Antona and Mazzitelli 1994). The surface Li abundance is indeed anticorrelated with the stellar age in convective stars (Duncan 1981). Among the G and K stars, there is no significant Li destruction on the convective track, so lithium can be used as a criterion to identify low-mass PMS stars of G and K spectral types. Many PMS stars do exhibit abundances ($\log[n_{\rm Li}] = 3$) consistent with undepleted cosmic material (e.g., Basri et al. 1991).

PMS lithium burning is expected among the M stars (e.g., Pinsonneault et al. 1990). Walter et al. (1997) observed this among the lowest-mass PMS stars of the Corona Australis association, but the depletion was greater than expected by the models. They noted that this could be accounted for by the fact that the presence of an active chromosphere/transition region will overionize Li relative to local thermodynamic equilibrium (LTE), giving an apparent underabundance (Houdebine and Doyle 1995). In the CrA association, the M stars with the largest Hα emission equivalent widths do exhibit the smallest apparent Li abundances.

The nTT stars exhibit strong chromospheric emission at Hα, Ca II K and H, and Mg II k and h. The transition region emission lines are at saturated levels. The saturated chromospheric Hα emission surface flux of order 10^7 erg cm^{-2} s^{-1} (averaged over the star) is sufficient to cause an emission line above the photosphere only among the late K and M stars.

[e] The main source of contamination of X-ray surveys is by active ZAMS stars, which have X-ray colors and activity levels similar to those of the PMS stars, and which are expected to dominate among field star X-ray emitters (ages up to about 10^8 years; Briceño et al. 1997). Such stars are still young, but not necessarily PMS; their ages must still be determined from observations of the space motions and parallaxes. A second source of contamination is active binary stars and RS Canum Venaticorum systems, for which high levels of X-ray activity are maintained by tidal coupling of the close binary system.

Although such activity appears commonplace among the PMS stars, it is not unique and cannot be used to identify a star as PMS. Walter and Barry (1991) discuss the timescales for the decay of the chromospheric and coronal emissions.

The cTTSs show far stronger levels of chromospheric and transition region emission, including the hydrogen Balmer lines, the Ca II K and H resonance lines, the Fe II λ 4924 Å, and the fluorescence lines of Fe I $\lambda\lambda$ 4063 and 4132 Å in emission (Herbig 1962). We now know that much of this flux arises not in a compact, solarlike chromosphere but in an extended atmosphere and accretion flows (e.g., Hartmann et al. 1994; Muzerolle et al. 1998). Bastian et al. (1983) defined a phenomenological working definition for cTTSs as having an Hα emission equivalent width >5 Å to discriminate against dMe stars, and in general a 10-Å Hα emission equivalent width is used to discriminate between the cTTSs and the nTT or wTT stars. The emission line spectra of cTTSs also often have forbidden lines of [S II] and [O I] that, together with a broad Hα emission line profile and P Cygni line profiles, are diagnostics of strong winds (e.g., Mundt 1984; Hartigan et al. 1995).

Radial velocities are powerful tools for establishing membership in a particular cluster or association, because the dispersion in radial velocities of an association is 1–2 km s^{-1} (e.g., Hartmann et al. 1986). Proper motions (e.g., Jones and Herbig 1979; Frink et al. 1997) can also be used to establish likely membership.

Star formation regions are big, so efficient searches must survey large solid angles. Traditionally, objective prism surveys with photographic plates have been used to survey for Hα emission line objects. Most known cTTSs have been discovered in such surveys. Herbig et al. (1986) used objective prism techniques at Ca II K and H. They could identify emission from M stars, but the bright $\lambda\lambda$ 3900–4000 Å continua of the K stars greatly reduce sensitivity for hotter stars. Because of the generally low spectral resolution (>10 Å), objective techniques are not suitable for measuring relatively weak absorption lines, such as the important Li I line.

Once limited to single objects, high-resolution spectroscopy is now commonly undertaken in survey mode, due to the maturation of optical fiber technology and multiobject spectroscopy. One can obtain high-resolution spectra of hundreds of objects within a field as large as a square degree simultaneously. At these higher resolutions, weak absorption lines such as Li I λ 6707 Å can be resolved from the Ca I λ 6717 Å line, and radial velocities can be measured for all the objects in the field. Examples of such data are given by Walter et al. (1998).

C. Wide-Field Imaging Photometry

In an area with little interstellar and circumstellar reddening, optical color-magnitude diagrams (CMDs) are a powerful tool for selecting PMS stars.

Large CCD arrays coupled with small (1-meter class) telescopes can map out large areas of the sky fairly efficiently. Star identification algorithms such as DAOPHOT are robust in rejecting objects with nonstellar point spread functions and provide very reliable photometric results at the mean stellar densities observed in the Orion OB1 association (about 2 stars brighter than $V = 18$ per arcmin^2).

In practice, we use the known PMS stars to define the PMS locus in the V vs. $V - R$ or V vs. $V - I$ CMD. As the reddening vector is nearly parallel to the PMS locus in these CMDs, it is not important to know the reddening in order to identify the PMS candidates. Although the X-ray sources calibrate the PMS locus to about $V = 15$, the optical photometry extends to $V \sim 20$ even in short exposures.

For example, we have obtained CCD photometry of a 2200-arcmin^2 region in Orion's belt (Ori OB1b), near σ Ori. Completeness of these images, as determined by number counts, was $V = 22$, with some stars as faint as magnitude $V = 23$ detected in at least three colors. The data (Fig. 1) bifurcate into two distinct groups: a group of stars, including most of the X-ray sources, which lies in a diagonal band across the diagram, and a set of background stars below the band. These data will be discussed in section IV.

III. LOW-MASS STARS IN ORI OB1: LARGE SCALES

The Orion complex covers about 25° in declination (including the λ Ori region) and one hour in right ascension. Surveys for low-mass stars over this entire area are important for studies of the subassociation boundaries, the history of star formation, and the kinematics of the association and its interactions with the molecular gas.

A. Hα Surveys

The Kiso objective prism survey in Orion (Wiramihardja et al. 1989; 1991; 1993; Kogure et al. 1989) revealed 1157 Hα emission line objects over an area of 150 deg^2 from 5.15^h to 5.85^h in right ascension and from $-13.0°$ to $+2.8°$ in declination, to a limiting magnitude of $V = 17.5$. The magnitude distribution, peaked around $V = 15$, supports a classical T Tauri nature for these objects. Kogure et al. (1992) followed up with low-dispersion spectroscopic observations of 34 emission line stars in Ori OB1b and concluded that they were indeed T Tauri stars based on H Balmer and Ca II K emission lines. Nakano and McGregor (1995) obtained near IR photometry for a number of these stars and also concluded that they were mostly T Tauri stars.

The spatial distribution of the Kiso Hα emission line objects is shown in the left panel of Fig. 2. The concentrations of emission line objects coincide with the general location of the NGC 2023 and NGC 2024 clusters,

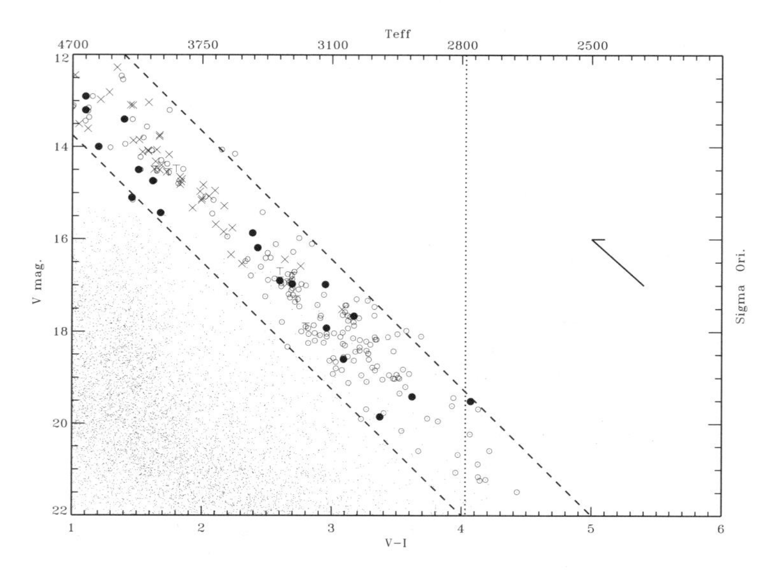

Figure 1. The color-magnitude diagram for all stars (dots) in the 2200-square-arcminute region centered on σ Ori. Effective temperatures are derived from Bessell and Stringfellow (1995). The dashed lines trace the expected pre-main-sequence locus. The vertical dotted line indicates the brown dwarf cutoff derived in Baraffe et al. (1998) with corrections for "missing opacity" and mean reddening. The crosses indicate X-ray sources confirmed to be PMS by their spectra. Filled circles are photometrically selected and spectroscopically confirmed PMS candidates. "T's" indicate classical T Tauri stars among these confirmed candidates. Open circles are photometric candidate PMS stars whose $V - R$ and $R - I$ colors are consistent with PMS nature (uncircled dots within the PMS locus have discrepant $R - I$ colors), but for which we have no spectra. The $A_V = 1$ reddening vector is indicated. The clear separation of the PMS stars from the general distribution of background stars demonstrates the efficacy of using wide-field optical photometry to identify low-mass PMS stars in OB associations.

and with the OB1c, OB1b, and OB1a associations. Note that emission line objects are also found far outside the limits of the OB1 associations.

B. ROSAT Survey Results

The surface density distribution of X-ray-selected PMS star candidates in a 700-deg^2 field around the Orion molecular clouds (Sterzik et al. 1995) shows peaks associated with subgroup associations (OB1a, OB1b, OB1c, and λ Ori). The spatial extent of the density peaks is consistent with dispersal times between 2 and 10 Myr, the ages of the stellar components in these regions. Surprisingly, the largest fraction of the PMS star candidates in this sample is not in this "clustered" population but is distributed widely over an area many times greater than that of the molecular gas or the OB stars. A large number of sources are seemingly unrelated to any molecular

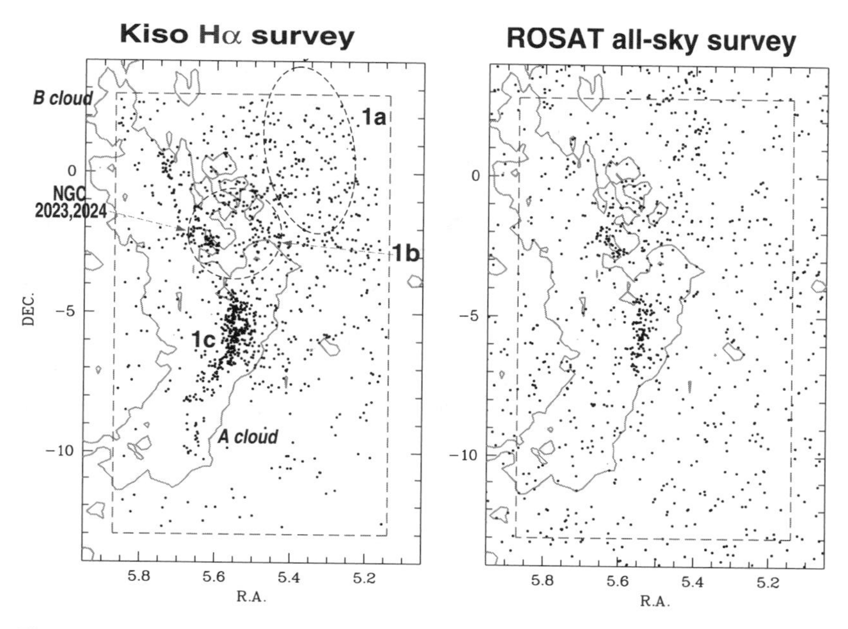

Figure 2. Spatial distribution of the Hα emission line objects found in the Kiso survey (left panel) and of the ROSAT all-sky survey X-ray sources in Orion (right panel). The position of the OB1 associations as well as the NGC 2023 and NGC 2024 clusters (Zinnecker et al. 1993) are indicated in the left panel. The outlines of the CO survey by Maddalena et al. (1986) are also shown. The dashed square in each panel indicates the extent of the Kiso survey. While the strongest density enhancements in both the Hα and X-ray source densities are associated with the Orion A cloud, significant numbers of sources extend out beyond the molecular cloud.

clouds or fossil SFR. This distributed population has also been found around other nearby SFRs such as Taurus-Auriga (Wichmann et al. 1996), Lupus (Krautter et al. 1997), and Chamaeleon (Alcalá et al. 1995).

To explore the nature of these PMS candidates, Sterzik et al. (1997) extended their earlier analysis to a larger area around Orion. They interpreted the spatial distribution of PMS candidates (Color Plate 5) in the framework of a stellar population model of the galaxy. The 6482 RASS sources in the $\sim$5000 deg^2 field include a subsample of 1467 PMS candidates. The densest regions (up to 10 stars/deg^2) coincide with centers of active star formation, such as in the Orion nebula region, near λ Ori, and in well-known star-forming sites in the Taurus clouds. In all these cases, the expected sensitivity for selecting young stars is verified. In addition, other clusters (e.g., around NGC 1788; (see section IV.C) that have not previously been recognized as prominent SFRs are also detected with this method and are likely to harbor a high fraction of PMS stars.

Color Plate 5 shows that the Orion and Taurus SFRs are projected against an approximately 10° wide strip of apparently young stars (density <1 star/deg^2 above the RASS detection limit). Although Orion and Taurus are at different distances from the Sun, they seem to be connected by a broad lane that extends further southeastward. This contiguous structure is not symmetric about the galactic plane but rather follows the mean location of the Gould Belt, as defined by early-type stars in this direction (Blaauw 1991). The surface density of young star candidates drops down to a background value of about 0.1 candidate stars/deg^2 near $b_{\rm II} = 0°$, and below that value at high galactic latitudes.

Based on the morphology and surface density distribution of X-ray-selected young star candidates, and detailed comparisons with the predictions of a galactic X-ray population model (Guillout et al. 1998), Sterzik et al. (1997) show that the X-ray population consists of a mixture of three distinct populations:

1. The *clustered population* comprises the dense regions associated with sites of active or recent star formation (e.g., OB1a, OB1b, OB1c, λ Ori, NGC 1788).
2. The strip connecting Orion and Taurus coincides with Gould's Belt.
3. The *background population*, having a density ~0.1 stars/deg^2 near the galactic plane, is likely dominated by ZAMS stars.

C. Optical Followups to the ROSAT Survey

The spatial distribution of 671 PMS candidates in the RASS sample is shown in the right panel of Fig. 2. There is an apparent spatial coincidence between those regions of high X-ray source density and those of high densities of Hα emission objects from the Kiso survey. An immediate conclusion would be that many of the emission line objects are detected in X-rays, but this is not the case. Fewer than 5% of the Kiso emission line stars are coincident with RASS PMS candidate X-ray sources. Although the Kiso survey goes deeper than the expected optical magnitude of PMS stars detectable in the RASS, over 90% of the RASS X-ray sources are not coincident with an emission line star. We note that only a handful of emission line objects in the λ Ori region (Duerr et al. 1982) coincide with RASS sources.

The lack of coincidences between the RASS sources and the Hα emitters is a consequence of the fact that cTTSs are more difficult to detect in soft X-rays than nTTSs, perhaps because X-rays are efficiently absorbed in the dense circumstellar envelopes of cTTSs (Walter and Kuhi 1981) or because nTTSs rotate faster than cTTSs (Bouvier et al. 1993, 1995). The important implication is that most of the PMS stars are not emission line objects.

Alcalá et al. (1996) observed a spatially unbiased sample of 181 RASS sources, using intermediate-resolution long-slit spectroscopy and photoelectric photometry. They identified 112 stars that showed Li absorption and a late-type spectrum. These low-mass PMS star candidates have a spatial distribution indistinguishable from that shown in the right panel of Fig. 2.

More recently, Alcalá et al. (1998) placed a representative subsample of these stars in the HR diagram assuming a distance of 460 pc and found that they fall well above the main sequence with typical T Tauri masses and ages ($0.8 < M_\star/\mathrm{M}_\odot < 3.4$; $0.2 < \tau_{\rm age}(\mathrm{Myr}) < 7$). They found that the stars with stronger Li I (λ 6707 Å) lines tend to concentrate toward the Orion molecular clouds, but they did not find any other correlations between the spatial locations, ages, or other stellar parameter. The lack of stars with masses less than 0.8 $\mathrm{M}_\odot$ in this sample is simply due to the flux limit of the RASS and the approximately constant f_x/f_v of PMS stars.

To verify the Li measurements (which are blended at moderate resolution) and to obtain radial velocities, which can be used to distinguish association members from nonmembers, Alcalá et al. (1999) undertook high-resolution ($R = 25{,}000$) spectroscopic observations of many of the 112 stars in the Alcalá et al. (1996) sample.

The distribution of radial velocities of the RASS stars in Orion that have strong Li and appear to be PMS is shown in Fig. 3. The radial velocity distribution is broad with a velocity dispersion of about 9 km s^{-1}

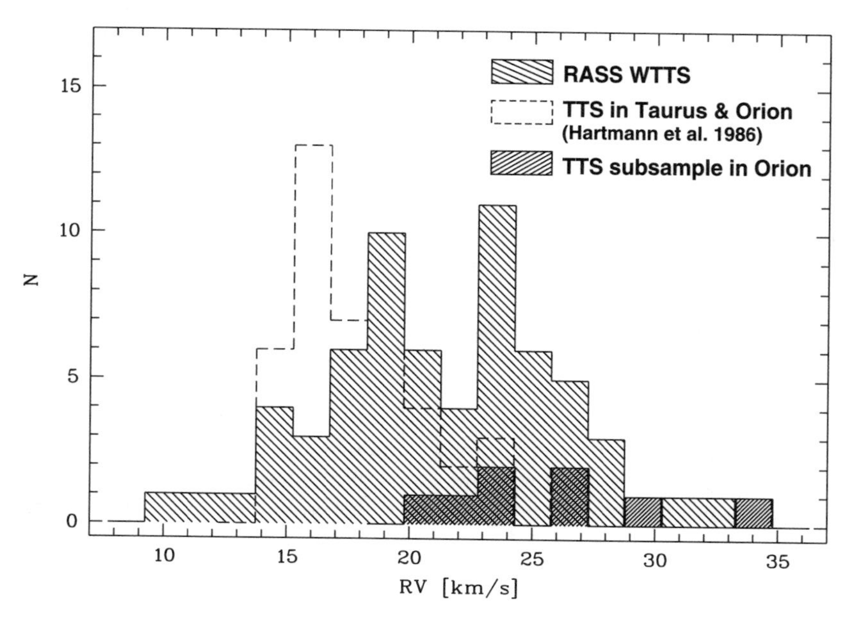

Figure 3. Radial velocity distribution of the RASS PMS candidates observed with high resolution. The radial velocity of the Orion association is about 25 km s^{-1}; the peak near 18 km s^{-1} is consistent with the radial velocity of the nearby Taurus SFR (Hartmann et al. 1986). This suggests that the RASS survey has detected two discrete populations of low-mass PMS stars.

and apparently shows a double peak, one at $\approx$25 km s^{-1} and the other at $\approx$18 km s^{-1}. The former coincides with the mean radial velocity for Orion, while the latter appears to coincide with the radial velocity of the Taurus clouds. This suggests that the RASS PMS sample in Orion is a juxtaposition of two distinct groups of stars, one associated with the Orion SFR. The relation of the other group to the Taurus SFR is unclear, as these stars would lie a minimum of 50–80 pc from the currently active star formation in the Taurus clouds. This group might be related to the Gould's Belt population (section III.B).

There is no statistical difference in the lithium strength between the two radial velocity groups; most of the stars have lithium abundances comparable to those of low-mass PMS stars. We conclude that most of these stars in the RASS sample are indeed PMS, and that at least some are demonstrably members of the Orion association.

IV. LOW-MASS STARS IN ORI OB1: SMALL SCALES

On scales of a few degrees or less, the Orion OB1 association breaks up into individual subassociations. Orion is sufficiently young that the subassociations retain their individual identities. One can investigate the timescales of the star formation process and study the initial mass function and mass segregation. Aside from the obvious differences in age and total mass, just how similar are these subassociations?

A. The Belt

A series of ROSAT Position-Sensitive Proportional Counter (PSPC) pointings (Walter 1994) in the belt of Orion revealed hundreds of X-ray sources, of which many are now confirmed low-mass PMS stars, based on spectroscopic and photometric followups. We have concentrated our efforts on the region surrounding σ Ori (Walter et al. 1998). The ROSAT PSPC and High-Resolution Imager (HRI) observations reveal over 100 X-ray point sources within 1° of σ Ori, a member of Ori OB1b and a Trapezium-like system.

Walter et al. (1998) obtained spectra of most of the optical counterparts of the X-ray sources near σ Ori and of a randomly selected sample of nearby stars. Among these ~300 stars, they identified 104 likely PMS stars within 30 arcmin of σ Ori. Primary identification was made on the basis of a strong Li I λ 6707 Å absorption line. The Hα strengths range from an emission equivalent width of 77 Å in a K1 star to normal photospheric absorption. The distribution of radial velocities is strongly peaked at the 25 km s^{-1} velocity of the OB association (Fig. 4).

The color-magnitude diagram (Fig. 1) for 0.6 deg^2 surrounding σ Ori shows a clear PMS locus, well separated from the background galactic stars. The narrowness of the PMS locus suggests coevality, but an age less than the 1.7-Myr age of the OB association.

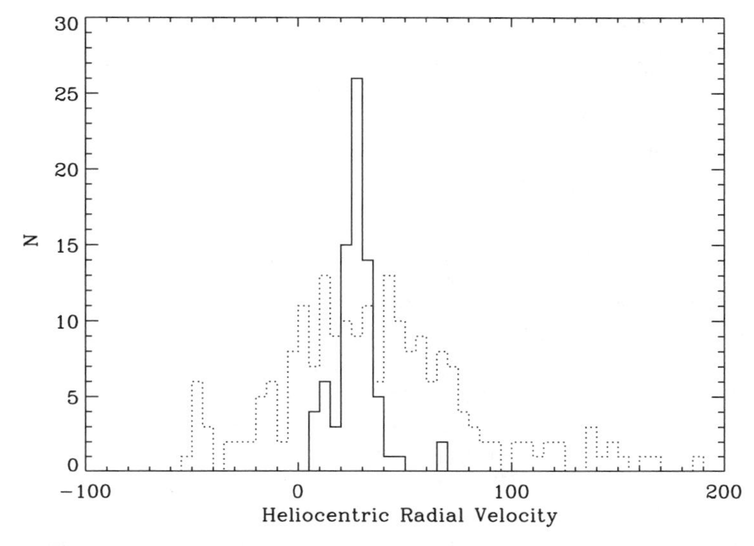

Figure 4. The radial velocity distribution of stars within 30 arcmin of σ Ori. Radial velocities were determined by cross-correlating the spectra with those of the dusk or dawn sky. Uncertainties are about ±5 km s^{-1} for spectra with high S/N. The spectroscopically identified PMS stars (solid histogram) are well fitted as a Gaussian distribution of mean 25 km s^{-1} with σ = 5 km s^{-1}. The secondary peak at 12 km s^{-1} is due to a systematic shift of M star velocities and may be an artifact of using a sky spectrum as a velocity template. The radial velocities of the other stars in the sample (dotted histogram) have a mean of 31 km s^{-1} with σ = 37 km s^{-1}. It is clear from the radial velocities that the PMS stars in this field are all members of the same association.

The spectroscopic sample, which is spatially uniform and statistically complete to about $V = 15$, has a surface density of 120 PMS stars/deg^2 ($10 < V < 15$) and shows evidence for clustering. The centroid of the PMS star distribution coincides with the position of σ Ori. Summation of the stars into radial bins centered on σ Ori shows that the distribution is flat for the non-PMS stars, but that the radial distribution of the PMS stars is peaked at σ Ori. The inferred cluster radius (where the density of PMS stars reaches zero) is about 0.5 deg (3.3 pc). If all the PMS stars are distributed in this way, then the space density, in the magnitude range $12 < V < 19$, is about 400 stars/deg^2 (or more, if many stars are multiple). The total inferred mass of this group of stars is about half that of the ONC. The σ Ori cluster is the second youngest cluster now known after the ONC and may be an evolved analog of the ONC.

Wolk (1996) showed that about 30% of these stars are slow rotators and 70% are rapid rotators. Similarly, about 30% of the spectroscopically identified PMS stars have strong Hα emission and appear to be cTTSs. This suggests that, at an age of about 2 Myr, 30% of the low-mass stars near σ Ori retain their circumstellar disks. This is a higher fraction than seen in Upper Scorpius (Walter et al. 1994) at a similar age.

The existence of the σ Ori cluster implies that there is substructure in Ori OB1b. As there is no evidence for evolutionary differences between the early-type stars in Ori OB1b (Brown et al. 1994), it may be that Ori OB1b formed through the merging of several Trapezium-like clusters that formed at more or less the same time. Observations are under way to test this hypothesis.

B. Ori OB1a

Observations of five other regions in Ori OB1 reveal similar concentrations of PMS stars, with space densities ($V < 15$) from about 40 to 150 stars/deg^2, comparable to that near σ Ori. We find the highest space density of PMS stars within the Ori OB1a association, near $5^{\rm h}24^{\rm m} + 1°$. This sample of stars includes no slow rotators (rotation periods longer than 4 days; Wolk 1996) and only one classical T Tauri star, suggesting that essentially all circumstellar disks have dissipated by the ~10-Myr age of this association. The CMD is similar to that for the σ Ori region, except that the PMS stars appear older. For a 330-pc distance (Brown et al. 1998), the PMS stars lie above the 10-Myr locus expected from the age of the B stars. In general, the low-mass stars in OB associations appear younger than the high-mass stars. This effect is evident in σ Ori, in the λ Ori region (Dolan 1998), and in the Upper Scorpius association (Walter et al. 1994).

C. NGC 1788

A significant density enhancement of RASS PMS candidates is present near the reflection nebula NGC 1788, which coincides with the CO-clump #13 of Maddalena et al. (1986) at $\alpha \approx 76°$ and $\delta \approx -3.5°$. This reflection nebula is centered on a cluster of stars and illuminated by the B9 V star HD 293815 (Witt and Schild 1986). The cluster seems heavily obscured by foreground material, consistent with gas column densities $>5 \times 10^{22} N_{\rm H_2}$ derived from ^{12}CO/^{13}CO line ratio observations (Knapp et al. 1977). This high column density is likely to hide most X-ray sources in the cloud itself. One cTTS, LkHα 333, is known in its vicinity. A surface density of $\approx$4 sources/deg^2 in an elongated structure of 3° length and 1.5° width is detected in the RASS. A preliminary analysis of the recent HRI pointing resulted in the discovery of more than 50 additional X-ray sources, probably embedded in the molecular cloud. The relatively small width of the density enhancement would indicate a diffusion age of less than 5 Myr, if these stars have formed in a central cluster with a dispersion velocity of 2 km s^{-1}. More support for the existence of a young cluster near NGC 1788 follows from recent near-infrared imaging by Dougados et al. (in preparation).

Unlike the fossil SFRs, NGC 1788 appears to be a region of ongoing star formation far from the Orion A and B clouds. Its relation to the Orion complex is not known.

D. Runaway T Tauri Stars and the Distributed Population

Not all PMS stars are found in or near molecular clouds. RASS observations have shown that a distributed population of PMS stars appears to be a common characteristic of SFRs. Although such a distributed population may have formed locally (e.g., Feigelson 1996), Sterzik et al. (1995) suggested that the PMS stars in the distributed population were ejected from their birth clouds with high velocities. They called such stars "runaway TTSs" (raTTSs). Few-body encounters can happen early in the lifetime of a multiple protostellar system, so they are also of relevance in establishing the fraction of binary (and triple) PMS stars. Many on-cloud PMS stars are multiples, so the multiplicity must be established very early in the PMS phase (e.g., Leinert et al. 1993; Ghez et al. 1993; Mathieu 1994; and the chapter by Mathieu et al., this volume). In star-star or star-cloud encounters such as those studied by Sterzik and Durisen (1995), ejected raTTSs are either single stars or close binaries and should be on average less massive than average TTSs. Gorti and Bhatt (1996) modeled the ejection of protostars in encounters of protostars with clouds and found that some protostars can be ejected in such a way. Kroupa (1995) showed that several percent of the members of a cluster as rich as the Trapezium can be ejected by close encounters with velocities exceeding 5 km s^{-1}.

The characterization of any individual star as an raTTS requires detailed knowledge of their space motions and ages. There are a few well-studied stars in Orion whose space motions, locations, and ages indicate that they may well be raTTSs. These include Par 1540 (Marschall and Mathieu 1988), which may have been ejected from the ONC $\sim 10^5$ years ago; Par 1724 (V1321 Ori; Neuhäuser et al. 1998), which is moving north at about 10 km s^{-1} and may also have been ejected from the ONC $\approx 10^5$ yr ago; and RXJ0511.2+1031 (Magazzù et al. 1997; Neuhäuser et al. 1997), which may have originated in the λ Ori region.

These examples show that there are indeed very young stars, far away from star-forming clouds, whose space motions point back to those clouds. Runaway TTSs certainly do exist, and if they are numerous, the raTTSs may be the bridge between the star formation, occurring on small scales, and the observed large-scale distribution of PMS stars observed after a few million years.

V. APPLICATIONS AND IMPLICATIONS

A. The Low End of the Mass Function

Very low-mass (VLM) objects (the lowest-mass stars as well as substellar-mass objects) are quite bright when young. A 2-Myr-old star at the hydrogen-burning limit (0.08 $M_\odot$) is only about 10 times less luminous than the Sun (Burrows et al. 1997). At the distance of Orion, such stars will have $V \approx 19$, which is easily observable with small telescopes. By the age of the Pleiades, VLM stars have faded by about 3 magnitudes at V and are some 200–400 K cooler. So, whereas IR studies and large telescopes (e.g., Stauffer et al. 1994*b*; Zapatero-Osorio et al. 1996) are required to detect VLM objects in clusters like α Persei and the Pleiades, a more modest approach will yield similar results in nearby SFRs.

At the 2-Myr age of Ori OB1, a star near the H-burning limit will have $V - R \sim 1.2$ (Baraffe et al. 1998). Our deep photometry of the σ Ori field (Fig. 1) shows that the empirical PMS locus seems to continue into the brown dwarf regime. To determine whether the red colors are intrinsic or are due to extreme reddening, Wolk (in preparation) has obtained low-resolution spectra of several of these objects. The filled circle in Fig. 1 redward of the theoretical brown dwarf limit is of spectral type M6 with minimal reddening. M6 is the spectral type of the Pleiades brown dwarfs PPl15 and Teide 1, confirmed by the "lithium test" (Basri et al. 1996). Because a younger object of the same spectral type must be less massive, it is highly likely that our object is of substellar mass. It is unlikely that all the other very red objects are subject to extreme reddening. The 2200 arcmin^2 area included in Fig. 1 contains eight additional brown dwarf candidates both fainter and redder than the candidate discussed above.

Color-magnitude diagrams of 2200 arcmin^2 around δ Ori and of 1100 arcmin^2 around ϵ Ori reveal nine photometric brown dwarf candidates. This is a density of about 12 brown dwarfs/deg^2, and suggests that there may be an abundance of substellar-mass objects in OB associations.

B. The Initial Mass Function

The easiest time to determine the initial mass function (IMF), the relative numbers of stars as a function of mass, is early in the life of an association or a cluster, before high-mass stars burn out and before dynamical friction segregates the masses and ejects the lower-mass systems. Studies of unobscured fossil SFRs, the OB associations, afford all of the advantages of studying the IMF in embedded clusters (see the chapter by Meyer et al., this volume) and none of the disadvantages encountered in highly obscured regions, where the stars may not yet have reached their final masses. At ages of a few Myr, one can readily see to, and below, the hydrogen-burning limit. The ability to identify PMS stars independent of their activity levels gives the opportunity to identify a (statistically) complete population of PMS stars. In OB associations we can directly observe and measure the IMF from O stars through substellar-mass objects.

The published data suggest that there is a universal slope to the mass function in associations and that it is similar to the field star mass function (Miller and Scalo 1979). Walter and Boyd (1981) found that the mass function in Taurus approximated the field star mass function from a few solar masses to about 0.3 $M_{\odot}$; Walter et al. (1994) found the same in Upper Scorpius. Hillenbrand (1997) found the mass function in the ONC to agree overall with the Miller and Scalo IMF, though with some differences in detail. This universality should not be surprising, because most stars are born in associations, which disperse to populate the field. However, the question of whether there are small differences in the mass function between associations remains open.

One place where differences may exist is at the low-mass end of the mass function. Brown dwarfs appear to be rare in T associations (e.g., Stauffer et al. 1991), although recently Luhman et al. (1997) and Neuhäuser and Comerón (1998) reported finding brown dwarfs in the ρ Ophiuchi and Chamaeleon I SFRs, respectively. On the other hand, there appears to be a high density of VLM objects in Ori OB1b. This may be a simple consequence of the differences between the T and OB association environments. A low-mass protostar in a T association is able to accrete mass for a time set by the accretion process and the local environment. In an OB association, however, the mass accretion can be terminated abruptly as the local cloud is disrupted by nearby massive stars (see Walter et al. 1994) through either winds or supernovae. So, whereas in a T association a protostar may accumulate a significant fraction of the mass within its original Jeans radius, a low-mass protostar in an OB association may never accrete that last fraction of a solar mass, and VLM stars may end up with substellar masses. OB associations may be the place to search for brown dwarfs.

There is some evidence that this may indeed be the case. In the Taurus T association, the typical PMS star is spectral type K7–M0, suggesting a mass function peaking near 0.3–0.4 $M_\odot$. Walter et al. (1994) found the typical low-mass PMS star in Upper Scorpius to be early M and found that the mass function extended to 0.2 $M_\odot$. Hillenbrand (1997) finds a mass function peaking at 0.2 $M_\odot$ in the ONC. If the mass functions are different in OB and T associations, it will have profound consequences for where one looks for brown dwarfs, or perhaps for large planets.

C. Disk Survival Times and Implications for Planets

We cannot yet directly detect planets orbiting PMS stars in Orion or in any SFR. We cannot yet detect terrestrial-size planets anywhere. We can detect circumstellar disks, which are likely to be a necessary ingredient for planet formation, but we do not know how the disk turns into planets. Models that qualitatively explain the distribution of planets in the solar system fail to predict the existence of giant planets close to their stars (but see the chapter by Lin et al., this volume). An observer of stars and disks can only speculate about how planets might fare in an OB environment, but can describe those conditions under which planet formation must proceed.

Elsewhere in this volume, Hollenbach et al. discuss the evaporating disks seen in *Hubble Space Telescope* (HST) images of the Orion Nebula. In a simple picture, longer-lived disks provide more time for planets to form, and disks in T associations may survive longer than disks in OB associations. About a million years are needed to form a Jupiter (see review by Wuchterl et al., this volume). In the ONC, with a nominal age of about 0.5 Myr, about half the low-mass PMS stars seem to have disks (Hillenbrand 1997). In the older σ Ori cluster (nominal age 1.7 Myr), about 30% seem to retain disks, based on using rotation periods and Hα emission as disk proxies, while by the age of Ori OB1a there is little evidence for disk survival. The disk survival times do appear to be short in OB associations.

Short disk survival times may not impede the formation of terrestrial planets: small bodies can form very quickly and need not accumulate in the presence of gas (Weidenschilling and Cuzzi 1993; Lissauer and Stewart 1993). Indeed, short disk survival times may be advantageous for terrestrial-sized planets. Lin et al. (1996) suggest that giant planets tend to migrate inward in the presence of a disk, sweeping ahead of them all planetesimals that may have formed in the inner planetary system (see also reviews by Lin et al. and by Ward and Hahn, this volume). Short disk survival times may prevent such orbital migration and protect small inner planets. Thus, one could make a case that planetary systems like our own may be most likely to form in environments like OB associations.

VI. SUMMARY

There is a wealth of knowledge to be gained about the global processes of star formation from studying the low-mass population of OB associations. The IMF is best determined in fossil star formation regions, because all the stars are readily countable and have reached their final masses. The ages of the low-mass stars can be estimated more accurately (subject to systematic uncertainties in the evolutionary models) than can the ages of massive stars already on the main sequence, permitting studies of the coevality of star formation across the association. The apparent difference in ages between the high- and low-mass stars may provide information about the triggers and timescales of low-mass star formation. The radial velocities of the low-mass stars can often be measured more precisely than can those of the rapidly rotating O and B stars and can provide more definitive measures of the kinematics of OB associations. The spatial distribution of the numerous low-mass stars can yield insights into the substructuring of the associations.

Low-mass stars, revealed by X-ray and Hα surveys, abound in the fossil SFRs of Orion OB1. The PMS stars not only concentrate in the known subassociations but also are distributed over a much larger volume than are the OB stars. Optical photometry shows that the PMS stars have a narrow age spread and extend down to substellar masses. The high density of brown dwarf candidates in Ori OB1b may be a consequence of star formation in the OB environment. The ONC is not the only young cluster in Orion; σ Ori has its own cluster. Subclustering may be common: The belt of Orion may be the amalgamation of ONC-like systems that formed at about the same time.

Most low-mass stars in our galaxy likely formed in large OB associations like the Orion OB1 association. There is circumstantial evidence that our Sun may have formed in an OB association 4.6 Gyr ago. If so, then star formation in the OB environment does not preclude the formation of planetary systems, and we can be optimistic that planetary systems like our own are common in the Galaxy.

REFERENCES

Adams, N. R., Walter, F. M., and Wolk, S. J. 1998. Rotation periods of low-mass stars of the Upper Scorpius OB association. *Astron. J.* 116:237–244.

Alcalá, J. M., Krautter, J., Schmitt, J. H. M. M., Covino, E., Wichmann, R., and Mundt, R. 1995. A study of the Chamaeleon star-forming region from the ROSAT all-sky survey. I: X-ray observations and optical identifications. *Astron. Astrophys. Suppl.* 114:109–134.

Alcalá, J. M., Terranegra, L., Wichmann, R., Chavarria, C., Krautter, J., Schmitt, J. H. M. M., Moreno-Corral, M. A., de Lara, E., and Wagnmer, R. M. 1996. New weak-line T Tauri stars in Orion from the ROSAT all-sky survey. *Astron. Astrophys. Suppl.* 119:7–24.

Alcalá, J. M., Chavarría-K C., and Terranegra, L. 1998. On the nature of the ROSAT X-ray selected weak-line T Tauri stars in Orion. *Astron. Astrophys.* 330:1017–1028.

Alcalá, J. M., Covino, E., Torres, G., Sterzik, M. F., Pfeiffer, M., and Neuhäuser, R. 1999. High-resolution spectroscopy of ROSAT weak-line T Tauri stars in Orion. *Astron. Astrophys.,* submitted.

Ambartsumian, V. A. 1947. *Stellar Evolution and Astrophysics* (Yerevan: Academy of Sciences of the Armenian S.S.R.).

Baraffe, I., Chabrier, G., Allard, F., and Hauschildt, P. H. 1998. Evolutionary models for solar metallicity low-mass stars: Mass-magnitude relationships and color-magnitude diagrams. *Astron. Astrophys.* 337:403–412.

Basri, G., and Bertout, C. 1993. T Tauri stars and their accretion disks. In *Protostars and Planets III,* ed. E. H. Levy and J. I. Lunine (Tucson: University of Arizona Press), pp. 543–566.

Basri, G., Martin, E., and Bertout, C. 1991. The lithium resonance line in T Tauri stars. *Astron. Astrophys.* 252:625–638.

Basri, G. Marcy, G. W., and Graham, J. R. 1996. Lithium in brown dwarf candidates: The mass and age of the faintest Pleiades stars. *Astrophys. J.* 458:600–609.

Bastian, U., Finkenzeller, U., Jascheck, C., and Jascheck, M. 1983. The definition of T Tauri and Herbig Ae/Be stars. *Astron. Astrophys.* 126:438–439.

Bessell, M. S., and Stringfellow, G. S. 1993. The faint end of the stellar luminosity function. *Ann. Rev. Astron. Astrophys.* 31:433–471.

Blaauw, A. 1964. The O associations in the solar neighborhood. *Ann. Rev. Astron. Astrophys.* 2:213–247.

Blaauw, A. 1991. OB associations and the fossil record of star formation. In *The Physics of Star Formation and Early Stellar Evolution,* ed. C. J. Lada and N. D. Kylafis (Dordrecht: Kluwer), pp. 125–154.

Bodenheimer, P. 1965. Studies in stellar evolution. II. Lithium depletion during the pre-main-sequence contraction. *Astrophys. J.* 142:451–461.

Bouvier, J., Cabrit, S., Fernández, M., Martín, E. L., Matthews, J. M. 1993. COYOTES I: Multisite UBVRI photometry of 24 pre-main sequence stars of the Taurus-Auriga cloud. *Astron. Astrophys. Suppl.* 101:485–498.

Bouvier, J., Covino, E., Kovo, O., Martín, E. L., Matthews, J. M., Terranegra, L., and Beck S. C. 1995. COYOTES II: Spot properties and the origin of photometric period variations in T Tauri stars. *Astron. Astrophys.* 229:89–107.

Bouvier, J., Forestini, M., and Allain, S. 1997. The angular momentum evolution of low-mass stars. *Astron. Astrophys.* 326:1023–1043.

Brandner, W., and Köhler, R. 1998. Star formation environments and the distribution of binary separations. *Astrophys. J. Lett.* 499: L79–L82.

Briceño, C., Hartmann, L. W., Stauffer, J. R., Gagné, M., Stern, R. A., and Caillault, J. -P. 1997. X-ray surveys and the post-T Tauri problem. *Astron. J.* 113:740–751.

Brown, A. G. A., de Geus, E. J., and de Zeeuw, P. T. 1994. The Orion OB1 association. I: Stellar content. *Astron. Astrophys.* 289:101–120.

Brown, A. G. A., Walter, F. M., and Blaauw, A. 1998. The large-scale distribution and motions of older stars in Orion. In *The Orion Complex Revisited,* ed. A. Burkert and M. McCaughrean, in press.

Burrows, A., Marley, M., Hubbard, W. B., Lunine, J. I., Guillot, T., Saumon, D., Freedman, R., Sudarsky, D., and Sharp, C. 1997. A nongray theory of extrasolar giant planets and brown dwarfs. *Astrophys. J.* 491:856–875.

Cameron, A. G. W. 1993. Nucleosynthesis and star formation. In *Protostars and Planets III,* ed. E. H. Levy and J. I. Lunine (Tucson: University of Arizona Press), pp. 47–73.

Casanova, S., Montmerle, T., Feigelson, E. D., and André, P. 1995. ROSAT X-ray sources embedded in the ρ Ophiuchi cloud core. *Astrophys. J.* 439:752–770.

Cohen, M., and Kuhi, L. V. 1979. Observational studies of pre-main-sequence evolution. *Astrophys. J. Suppl.* 41:743–843.

D'Antona, F., and Mazzitelli, I. 1994. New pre-main-sequence tracks for $M \leq 2.5 M_{\odot}$ as tests of opacities and convection model. *Astrophys. J. Suppl.* 90:467–500.

Dolan, C. 1998. Poster presented at *Protostars and Planets IV* meeting.

Duerr, R., Imhoff, C. L., and Lada, C. J. 1982. Star formation in the λ Orionis region. I: The distribution of young objects. *Astrophys. J.* 261:135–150.

Duncan, D. K. 1981. Lithium abundances, K line emission, and ages of nearby solar type stars. *Astrophys. J.* 248:651–669.

Edwards, S., Strom, S. E., Hartigan, P., Strom, K. M., Hillenbrand, L. A., Herbst, W., Attridge, J., Merrill, K. M., Probst, R., and Gatley, I. 1993. Angular momentum regulation in low-mass young stars surrounded by accretion disks. *Astron. J.* 106:372–382.

Feigelson, E. D. 1996. Dispersed T Tauri stars and galactic star formation. *Astrophys. J.* 468:306–322.

Feigelson, E. D., Casanova, S., Montmerle, T., and Guibert, J. 1993. ROSAT X–ray study of the Chamaeleon I dark cloud. I: The stellar population. *Astrophys. J.* 416:623–646.

Frink, S., Röser, S., Neuhäuser, R., and Sterzik, M. F. 1997. New proper motions of pre-main-sequence stars in Taurus-Auriga. *Astron. Astrophys.* 325:613–622.

de Geus, E. J., de Zeeuw, P. T., and Lub, J. 1989. Physical parameters of stars in the Scorpio-Centaurus OB association. *Astron. Astrophys.* 216:44–61.

Ghez, A. M., Neugebauer, G., and Matthews, K. 1993. The multiplicity of T Tauri stars in the star forming regions Taurus-Auriga and Ophiuchus-Scorpius: A 2.2 micron speckle imaging survey. *Astron. J.* 106:2005–2023.

Gorti, U., and Bhatt, H. C. 1996. Dynamics of embedded protostar clusters in clouds. *Mon. Not. Roy. Astron. Soc.* 278:611–616.

Guillout, P., Sterzik, M. F., Schmitt, J. H. M. M., Motch, C., Egret, D., Voges, W., and Neuhäuser, R. 1998. The large-scale distribution of X-ray active stars. *Astron. Astrophys.* 334:540–544.

Haro, G. 1953. Hα emission stars and peculiar objects in the region of the Orion Nebula. *Astrophys. J.* 117:73–83.

Hartigan, P., Edwards, S., and Ghandour, L. 1995. Disk accretion and mass loss from young stars. *Astrophys. J.* 452:736–768.

Hartmann, L., Hewett, R., Stahler, S., and Mathieu, R. D. 1986. Rotational and radial velocities of T Tauri stars. *Astrophys. J.* 309:275–293.

Hartmann, L., Hewett, R., and Calvet, N. 1994. Magnetospheric accretion models for T Tauri stars. I: Balmer line profiles without rotation. *Astrophys. J.* 426:669–687.

Herbig, G. H. 1962. The properties and problems of T Tauri stars and related objects. *Adv. Astron. Astrophys.* 1:47–103.

Herbig, G. H. 1977. Radial velocities and spectral types of T Tauri stars. *Astrophys. J.* 214:747–758.

Herbig, G. H. 1978. The post-T Tauri stars. In *Problems of Physics and Evolution of the Universe,* ed. L. Myrzoyan (Yerevan: Academy of Sciences of the Armenian S.S.R.), pp. 171–183.

Herbig, G. H., and Bell, K. R. 1988. Third catalog of emission line stars of the Orion population. *Lick Obs. Bull.* 1111.

Herbig, G. H., Vrba, F. J., and Rydgren, A. E. 1986. A spectroscopic survey of the Taurus-Auriga dark clouds for pre-main-sequence stars having Ca II H,K emission. *Astron. J.* 91:575–582.

Hillenbrand, L. A. 1997. On the stellar population and star-forming history of the Orion Nebula Cluster. *Astron. J.* 113:1733–1768.

Houdebine, E. R., and Doyle, J. G. 1995. Observation and modelling of main sequence star chromospheres. IV: The chromospheric contribution to Li I lines in active dwarfs. *Astron. Astrophys.* 302:861–869.

Humphries, R. M. 1978. Studies of luminous stars in nearby galaxies. I: Supergiants and O stars in the Milky Way. *Astrophys. J. Suppl.* 38:309–350.

Jones, B. F., and Herbig, G. H. 1979. Proper motions of T Tauri variables and other stars associated with the Taurus-Auriga dark clouds. *Astron. J.* 84:1872–1889.

Joy, A. H. 1945. T Tauri variables. *Astrophys. J.* 102:168–195.

Kastner, J. H., Zuckerman, B., Weintraub, D. A., and Forveille, T. 1997. X-ray and molecular emission from the nearest region of recent star formation. *Science* 277:67–71.

Knapp, G. R., Kuiper, T. B. H., Knapp, S. L., and Brown, R. L. 1977. CO observations of galactic reflection nebulae. *Astrophys. J.* 214:78–85.

Kohler, R., and Leinert, C. 1998. Multiplicity of T Tauri stars in Taurus after ROSAT. *Astron. Astrophys.* 331:977–988.

Kogure, T., Yoshida S., Wiramihardja, S. D., Nakano, M., Iwata, T., and Ogura, K. 1989. Survey observations of emission-line stars in the Orion region. II: The Kiso area A-0903. *Pub. Astron. Soc. Japan* 41:1195–1213.

Kogure, T., Ogura, K., Nakano, M., and Yoshida, S. 1992. Spectroscopic observations of emission-line stars in the Orion. I: Ori OB 1b region. *Pub. Astron. Soc. Japan* 44:91–99.

Krautter J., Wichmann, R., Schmitt, J. H. M. M., Alcalá, J. M., Neuhäuser, R., and Terranegra, L. 1997. New weak-line T Tauri stars in Lupus. *Astron. Astrophys. Suppl.* 123:329–352.

Kroupa, P. 1995. Star cluster evolution, dynamical age estimation, and the kinematical signature of star formation. *Mon. Not. Roy. Astron. Soc.* 277:1522–1540.

Larson, R. B. 1986. Bimodal star formation and remnant-dominated galactic models. *Mon. Not. Roy. Astron. Soc.* 218:409–428.
Lee, T., Shu, F. H., Shang, H., Glassgold, A. E., and Rehm, K. E. 1998. Protostellar cosmic rays and extinct radioactivities in meteorites. *Astrophys. J.* 506:898–912.
Leinert, C., Zinnecker, H., Weitzel, N., Christou, J., Ridgway, S. T., Jameson, R., Haas, M., and Lenzen, R. 1993. A systematic approach for young binaries in Taurus. *Astron. Astrophys.* 278:129–149.
Lin, D. N. C., Bodenheimer, P., Richardson, D. C. 1996. Orbital migration of the planetary companion of 51 Pegasi to its present location. *Nature* 380:606–607.
Lissauer, J. J., and Stewart, G. R. 1993. Growth of planets from planetesimals. In *Protostars and Planets III,* ed. E. H. Levy and J. I. Lunine (Tucson: University of Arizona Press), pp. 1061–1088.
Luhman, K. L., Liebert, J., and Rieke, G. H. 1997. Spectroscopy of a young brown dwarf in the ρ Ophiuchi Cluster 1. *Astrophys. J. Lett.* 489: L165–L168.
Maddalena, R. J., Morris, M., Moscowitz, J., and Thaddeus, P. 1986. The large system of molecular clouds in Orion and Monoceros. *Astrophys. J.* 303:375–391.
Magazzù, A., Martin, E. L., Sterzik, M. F., Neuhäuser, R., Covino, E., and Alcalá, J. M. 1997. Search for young low-mass stars in a ROSAT selected sample south of the Taurus-Auriga molecular clouds. *Astron. Astrophys.* 124:449–467.
Marschall, L. A., and Mathieu, R. D. 1988. Parenago 1540: A pre-main-sequence double-lined spectroscopic binary near the Orion Trapezium. *Astron. J.* 96:1956–1964.
Mathieu, R. D. 1994. Pre-main-sequence binary stars. *Ann. Rev. Astron. Astrophys.* 32:465–530.
Miller, G. E., and Scalo, J. M. 1979. The initial mass function and stellar birthrate in the solar neighborhood. *Astrophys. J. Suppl.* 41:513–547.
Montmerle, T., Koch-Miramond, L., Falgarone, E., and Grindlay, J. E. 1983. Einstein observations of the ρ Ophiuchi dark cloud: An X-ray Christmas tree. *Astrophys. J.* 269:182–201.
Mundt, R. 1984. Mass loss in T Tauri stars: Observational studies of the cool parts of their stellar winds and expanding shells. *Astrophys. J.* 280:749–770.
Muzerolle, J., Hartmann, L., and Calvet, N. 1998. Emission-line diagnostics of T Tauri magnetospheric accretion. I: Line profile observations. *Astron. J.* 116:455–468.
Nakano, M., and McGregor, P. J. 1995. In *Future Utilisation of Schmidt Telescopes,* ASP Conf. Ser. 84, ed. J. Chapman, R. Cannon, S. Harrison, and B. Hidayat, (San Francisco: Astronomical Society of the Pacific). pp. 376–378.
Nakano, M., Wiramihardja, S. D., and Kogure, T. 1995. Survey observations of emission-line stars in the Orion Region. V: The outer regions. *Pub. Astron. Soc. Japan* 47:889–896.
Neuhäuser, R., and Comerón, F. 1998. ROSAT X-ray detection of a young brown dwarf in the Chamaeleon I dark cloud. *Science* 282:83–85.
Neuhäuser, R., Sterzik, M. F., Schmitt, J. H. M. M., Wichmann, R., and Krautter, J. 1995*a*. ROSAT survey observation of T Tauri stars in Taurus. *Astron. Astrophys.* 297:391–417.
Neuhäuser, R., Sterzik, M. F., Schmitt, J. H. M. M., Wichmann, R., and Krautter, J. 1995*b*. Discovering new weak-line T Tauri stars in Taurus-Auriga with the ROSAT All-Sky Survey. *Astron. Astrophys.* 295:L5–L8.
Neuhäuser, R., Sterzik, M. F., Torres, G., and Martin, E. L. 1995*c*. Weak-line T Tauri stars south of Taurus. *Astron. Astrophys.* 299:L13–L16.

Neuhäuser, R., Torres, G., Sterzik, M. F., and Randich, S. 1997. Optical high-resolution spectroscopy of ROSAT-detected late-type stars south of the Taurus molecular clouds. *Astron. Astrophys.* 325:647–663.

Neuhäuser, R., Wolk, S. J., Torres, G., Preibisch, Th., Stout-Batalha, N. M., Hatzes, A., Frink, S., Wichmann, R., Covino, E., Alcalá, J. M., Brandner, W., Walter, F. M., and Sterzik, M. F. 1998. Optical and X-ray monitoring, Doppler imaging, and space motion of the young star Par 1724 in Orion. *Astron. Astrophys.* 334:873–894.

Palla, F., and Galli, F. 1997. Post-T Tauri stars: A false problem. *Astrophys. J. Lett.* 476:L35–L38.

Pinsonneault, M. H., Kawaler, S. D., and Demarque, P. 1990. Rotation of low-mass stars: A new probe of stellar evolution. *Astrophys. J. Suppl.* 74:501–550.

Sciortino, S., Damiani, F., Favata, F., and Micela, G. 1998. An X-ray study of the PMS population of the Upper Sco-Cen association. *Astron. Astrophys.* 332:825–841.

Shu, F. H., and Lizano, S. 1988. The evolution of molecular clouds. In *Interstellar Matter: Proceedings of the Second Haystack Observatory Meeting,* ed. J. M. Moran and P. T. P. Ho (New York: Gordoin and Breach), pp. 65–74.

Simon, M., Ghez, A. M., and Leinert, C. 1993. Multiplicity and the ages of the stars in the Taurus star-forming region. *Astrophys. J. Lett.* 408:L33–L36.

Stauffer, J. R., Herter, T., Hamilton, D., Rieke, G. H., Rieke, M. J., Probst, R. G., and Forrest, W. 1991. Spectroscopy of Taurus cloud brown dwarf candidates. *Astrophys. J. Lett.* 367:L23–L26.

Stauffer, J. R., Caillault, J.-P., Gagne, M., Prosser, C. F., and Hartmann, L. W. 1994*a*. A deep-imaging survey of the Pleiades with ROSAT. *Astrophys. J. Suppl.* 91:625–657.

Stauffer, J. R., Hamilton, D., and Probst, R. G. 1994*b*. A CCD-based search for very low mass members of the Pleiades cluster. *Astron. J.* 108:155–159.

Stahler, S. W., and Walter, F. M. 1993. Pre-main-sequence evolution and the birth population. In *Protostars and Planets III,* ed. E. H. Levy and J. I. Lunine (Tucson: University of Arizona Press), pp. 405–428.

Sterzik, M. F., and Durisen, R. H. 1995. Escape of T Tauri stars from young stellar systems. *Astron. Astrophys.* 304:L9–L12.

Sterzik, M. F., Alcalá, J. M., Neuhäuser, R., and Schmitt, J. H. M. M. 1995. The spatial distribution of X-ray-selected T Tauri stars. I. Orion. *Astron. Astrophys.* 297:418–426.

Sterzik, M. F., Alcalá, J. M., Neuhäuser, R., and Durisen R. H. 1997. The large-scale distribution of X-ray sources in Orion. *The Orion Complex Revisited,* ed. A. Burkert and M. McCaughrean, in press.

Strom, K. M., Wilkin, F. P., Strom, S. E., and Seaman, R. L. 1989. Lithium abundances among solar-type pre-main-sequence stars. *Astron. J.* 98:1444–1450.

Strom, K. M., Strom, S. E., Wilkin, F. P., Carrasco, L., Cruz-Gonzalez, I., Recillas, E., Serrano, A., Seaman, R. L., Stauffer, J. R., Dai, D., and Sottile, J. 1990. A study of the stellar population in the Lynds 1641 dark cloud. IV: The Einstein X-ray sources. *Astrophys. J.* 362:168–190.

Voges, W., Boller, T., Dennerl, K., Englhauser, J., Gruber, R., Haberl, F., Paul, J., Pietsch, W., Trümper, J., and Zimmermann, H. U. 1996. Identification of the ROSAT all-sky survey sources. In *Röntgenstrahlung from the Universe,* MPE Report 263, ed. H. U. Zimmermann, J. Trümper, and H. Yorke (Garching: Max-Planck-Institut für extraterrestrische Physik), pp. 637–640.

Walter, F. M. 1986. X-ray sources in regions of star formation I: The naked T Tauri stars. *Astrophys. J.* 306:573–586.

Walter, F. M. 1994. Star formation in Orion (the constellation). In *The Soft X-Ray Cosmos,* ed. E. M. Schlegel and R. Petre (New York: API Conf. Proc. 313), pp. 282–284.

Walter, F. M. 1996. Coronal magnetic activity in low-mass pre-main-sequence stars. In *Magnetodynamic Phenomena in the Solar Atmosphere: Prototypes of Stellar Magnetic Activity,* ed. Y. Uchida, T. Kosugi, and H. S. Hudson (Dordrecht: Kluwer), pp. 395–396.

Walter, F. M., and Barry, D. C. 1991. Pre- and main-sequence evolution of solar activity. In *The Sun in Time,* ed. C. P. Sonnett, M. S. Giampapa, and M. S. Matthews (Tucson: University of Arizona Press), pp. 633–657.

Walter, F. M., and Boyd, W. T. 1981. Star formation in Taurus-Auriga: The high-mass stars. *Astrophys. J.* 370:318–323.

Walter, F. M., and Kuhi, L. 1981. The smothered coronae of T Tauri stars. *Astrophys. J.* 250:254–261.

Walter, F. M., Brown, A., Mathieu, R. D., Myers, P. C., and Vrba, F. J. 1988. X-ray sources in regions of star formation. III: Naked T Tauri stars associated with the Taurus-Auriga complex. *Astron. J.* 96:297–325.

Walter, F. M., Vrba, F. J., Mathieu, R. D., Brown, A., and Myers, P. C. 1994. X-ray sources in regions of star formation. III: The low-mass stars of the Upper Scorpius Association. *Astron. J.* 107:692–719.

Walter, F. M., Vrba, F. J., Wolk, S. J., Mathieu, R. D., and Neuhäuser, R. 1997. X-ray sources in regions of star formation. VI: The R CrA association as viewed by EINSTEIN. *Astron. J.* 114:1544–1554.

Walter, F. M., Wolk, S. J., and Sherry, W. 1998. The σ Orionis cluster. In *Cool Stars, Stellar Systems, and the Sun X,* ASP Conf. Ser., ed. R. Donahue and J. Bookbinder (San Francisco: Astronomical Society of the Pacific), CD-1793.

Weidenschilling, S. J., and Cuzzi, J. N. 1993. Formation of planetesimals in the solar nebula. In *Protostars and Planets III,* ed. E. H. Levy and J. I. Lunine (Tucson: University of Arizona Press), pp. 1031–1060.

Wichmann, R., Krautter, J., Schmitt, J. H. M. M., Neuhaeuser, R., Alcala, J. M., Zinnecker, H., Wagner, R. M., Mundt, R., and Sterzik, M. F. 1996. New weak-line T Tauri stars in Taurus-Auriga. *Astron. Astrophys.* 312:439–454.

Wichmann, R., Krautter, J., Covino, E., Alcalá, J. M., Neuhëuser, R., and Schmitt, J. H. M. M. 1997. The T Tauri star population in the Lupus star-forming region. *Astron. Astrophys.* 320:185–195.

Wiramihardja, S. D., Kogure, T., Yoshida, S., Ogura, K., and Nakano, M. 1989. Survey observations of emission-line stars in the Orion region. I: The Kiso area A-0904. *Pub. Astron. Soc. Japan* 41:155–174.

Wiramihardja, S. D., Kogure, T., Yoshida, S., Nakano, M., Ogura, K., and Iwata, T. 1991. Survey observations of emission-line stars in the Orion region. III: The Kiso areas A-0975 and A-0976. *Pub. Astron. Soc. Japan* 43:27–73.

Wiramihardja, S. D., Kogure, T., Yoshida, S., Ogura, K., and Nakano, M. 1993. Survey observations of emission-line stars in the Orion region. IV. The Kiso areas A-1047 and A-1048. *Pub. Astron. Soc. Japan* 45:643–653.

Witt, A. N., and Schild, R. E. 1986. CCD surface photometry of bright reflection nebulae. *Astrophys. J. Suppl.* 62:839–852.

Wolk, S. J. 1996. Watching the stars go 'round and 'round. Ph.D. Thesis, State University of New York at Stony Brook, unpublished.

Zapatero-Osorio, M. R., Rebolo, R., and Martin, E. L. 1996. Brown dwarfs in the Pleiades cluster: A CCD-based R, I survey. *Astron. Astrophys.* 317:164–170.

Zinnecker, H., McCaughrean, M. J., and Wilking, B. A. 1993. The initial stellar population. In *Protostars and Planets III,* ed. E. H. Levy and J. I. Lunine (Tucson: University of Arizona Press), pp. 429–496.

HOT MOLECULAR CORES AND THE EARLIEST PHASES OF HIGH-MASS STAR FORMATION

STAN KURTZ
Universidad Nacional Autónoma de México

RICCARDO CESARONI
Osservatorio Astrofisico di Arcetri

ED CHURCHWELL
University of Wisconsin at Madison, Washburn Observatory

PETER HOFNER
University of Puerto Rico and Arecibo Observatory, NAIC

and

C. MALCOLM WALMSLEY
Osservatorio Astrofisico di Arcetri

We review recent observational results in the field of high-mass star formation. Special attention is given to the earliest evolutionary stages of massive stars through the study of hot, dense molecular cores and ultracompact H II regions. The characteristics of these two classes of objects are illustrated in some detail, emphasizing the role of hot cores as sites of massive star formation. Finally, a possible scenario for an evolutionary sequence from hot cores to ultracompact H II regions is proposed.

I. INTRODUCTION

High-mass stars ($\gtrsim$10 $M_\odot$) have gained steadily increasing attention in star formation research. As our understanding of low-mass star formation improves, and as observational power increases, we are better equipped to approach the study of massive star formation. The formation of high-mass stars—more distant, more heavily obscured, more rapid in their evolution, and more strongly interacting with their environment—is often considered to present a greater challenge than low-mass star formation. Yet the topic demands our attention. The profound effects that massive stars have on the interstellar medium, and their contribution to the galactic environment and evolution, attest to their important place in astronomical studies.

In this review we present an overview of the results obtained with radio continuum and line observations of OB star-forming regions. Our discussion is confined to the dense, hot environment in the inner parts of molecular clouds, where massive star formation is proceeding. Historically, ultracompact (UC) H II regions have been the primary means to investigate the earliest phases of OB star evolution. Recent studies suggest that "hot cores" represent an even earlier stage in the star formation process. These two classes of objects form the heart of massive star formation. We place special emphasis on them in this review and ultimately propose an evolutionary scenario leading from one to the other. We neglect many important topics such as theory, bimodal star formation, comparison with low-mass star formation, and the important contributions of recent infrared observations, to name only a few. The reader is directed to a number of very relevant chapters elsewhere in this volume, which include the theory of intermediate- and high-mass star formation (Stahler et al.) and various observational aspects of low-mass star formation (André et al.; Myers et al.; Mundy et al.). We first discuss some of the strategies used to locate massive star-forming regions (section II); we then discuss some of the results and implications of these searches, first for the ionized gas component (i.e., UC H II regions; section III) and then for the molecular gas component (i.e., hot cores; section IV). We examine the evolutionary link between them in section V.

II. THE SEARCH FOR HIGH-MASS (PROTO)STARS

The evolution of massive stars is characterized by shorter timescales than those of low-mass objects. The precise duration of each evolutionary phase is unclear, but a lower limit to the timescale for the formation of a massive protostar through collapse is the free-fall time. This is $\gtrsim 10^4$ yr, which implies a nonnegligible probability of finding a protostar in a typical high-mass star-forming cloud. Observational investigation is hence feasible.

Typically, the role of observations is twofold. Surveys are important for statistical purposes and to determine the time spent in each phase, while detailed physical and chemical studies of individual objects are needed to clarify the mechanisms of star formation. In view of these goals, a proper observational approach to the study of young massive stars must rely on two basic choices: a suitable tracer and a sensible sample of targets.

A massive star forms from the contraction of at least 10 $M_\odot$ concentrated in a region comparable to the mean star separation observed in clusters, i.e., $\lesssim 0.05$ pc (McCaughrean and Stauffer 1994; Hillenbrand 1997; Testi et al. 1997). The consequent release of gravitational energy in such a small region must increase the gas and dust temperature. As a consequence, observations of molecular transitions excited at high densities and temperatures, $\gtrsim 10^7$ cm^{-3} and $\gtrsim 100$ K (Blake et al. 1996; Cesaroni et al. 1991; Churchwell et al. 1990; Hofner et al. 1999*b*; Wyrowski 1997) or of the (sub)millimeter continuum (Jenness et al. 1995; Olmi et al. 1996*b*; Wyrowski et al. 1997; Shepherd et al. 1997; Carral et al. 1997; Hunter et al. 1998) are appropriate for investigating the dusty molecular envelopes around forming stars.

While the choice of the tracer is relatively easy, the selection of a sample of young and luminous (i.e., massive) targets is nontrivial. It is commonly accepted that phenomena such as maser emission, IRAS point sources, UC H II regions, and molecular outflows are signposts of high-mass star formation (see e.g., Wright et al. 1996 and references therein for Orion). However, apart from UC H II regions, a physical connection between such phenomena and the birth of massive stars is far from established. Nevertheless, these signposts are useful criteria for selecting candidate massive star formation regions, and several such searches have been made. We classify them in three distinct types depending on the selection criteria, and discuss them below.

A. Searches for Protostars toward UC H II Regions

The tendency of OB stars to form in clusters and associations is well known. Moreover, the presence of a range of stellar ages and even continuing star formation within a pre-existing association has been established (Blaauw 1991 and references therein). Although debate persists over their lifetimes, UC H II regions probably represent a relatively young stage of stellar development. Clustering is evident even at this early stage, with DR21, W3, W49, W51, NGC 6334, and NGC 7538 being some of the better known examples. Given the relatively short timescales characteristic of massive star formation, and the observational evidence cited above, it seems clear that some (nearly) coeval massive star formation must occur. As such, the sites of young massive stars, as traced by UC H II regions, are likely candidates in the search for massive (proto)stars in earlier evolutionary states. Indeed, many of the known hot cores were first identified by Churchwell et al. (1990) using a sample of UC H II regions as their targets.

B. Searches for Protostars toward IRAS Point Sources

Since the advent of the IRAS (*Infrared Astronomy Satellite*) era, the *IRAS Point Source Catalog* has been used to identify samples of objects on the basis of their far-infrared (FIR) colors. The most successful applications of this method to massive star formation have probably been the searches for H_2O masers (Braz and Epchtein 1987; Scalise et al. 1989; Palla et al. 1991, 1993) and UC H II regions (Wood and Churchwell 1989*a*; Kurtz et al. 1994; Miralles et al. 1994; Kurtz and Churchwell 2000). In both cases, the high detection rates achieved demonstrate the efficacy of the method. The method is clearly empirical, however, and the physical connection between the FIR emission and the radio objects remains uncertain. This is reflected in the variety of color and flux density criteria that have been proposed. This caveat notwithstanding, selection criteria based on the FIR emission are plausible, because very young massive stars must be better seen at $\sim$100 μm, where their continuum emission is expected to peak. Some contamination is likely, because a variety of objects may present an infrared appearance similar to that of massive-star formation regions. Nevertheless, one can expect that a worthwhile fraction of IRAS selected objects represent bona fide massive (proto)stars, particularly if the IRAS criteria are combined with other massive-star formation indicators.

C. Searches for Protostars toward H_2O Masers

Various evidence indicates that most water masers are associated with the formation of massive stars. Extensive surveys in the 22-GHz masing line of H_2O (Comoretto et al. 1990; Brand et al. 1994; Matthews et al. 1985) indicate that nearly all ($\gtrsim$90%) H_2O masers in star-forming regions have IRAS counterparts with luminosities above 10^4 $L_\odot$ (see Fig. 10b of Palagi et al. 1993), indicating zero-age main-sequence (ZAMS) stars earlier than B1. Water masers are also associated with molecular outflows (Felli et al. 1992; Xiang and Turner 1995) and possibly with disks (Torrelles et al. 1996, 1997, 1998; Fiebig et al. 1996). The high temperatures (several hundred K) and densities (10^8–10^{10} cm^{-3}) required to excite the maser line suggest that H_2O masers arise in the hot, dense surroundings of an embedded young star. These characteristics make H_2O masers excellent targets in the search for luminous embedded young stellar objects (YSOs).

III. THE IONIZED COMPONENT: UC H II REGIONS

In this section we discuss some of the results obtained from searches for young massive stars, focusing in particular on the ionized component: UC H II regions. These nebulae represent one of the earliest manifestations of OB stars.

A. UC H II Regions and the Lifetime Problem

UC H II regions have received increased observational and theoretical attention in recent years, particularly stemming from the seminal works of Wood and Churchwell (1989*a,b*). The interest in these objects is partly because they are unambiguous indicators of young early-type stars. The "compactness" (i.e., the small diameter) of the H II region is *presumed* to be proof of its youthfulness. However, any plausible definition of "UC H II region" must also require large electron densities or, equivalently, emission measures, to distinguish H II regions ionized by O or early B stars from those caused by other processes (e.g., shocks associated with mass loss). In the following we somewhat arbitrarily define UC H II regions to be ionized by stars of spectral type B0 or earlier and characterized by diameters $\lesssim$0.1 pc and electron densities $\gtrsim 10^4$ cm^{-3}. From an observational point of view, such a definition implies that at a wavelength of 1.3 cm and a distance of 5 kpc, an optically thin UC H II region would present a continuum flux density ~100 mJy and a diameter of ~4″. Such a region would be detectable (at a 1-mJy limit) anywhere in the Galaxy (see Fig. 3 of Churchwell 1991). The spectral type B0 is chosen because the Lyman continuum flux begins to fall rapidly with decreasing effective temperature at this point (Panagia 1973). We caution, however, that if the cutoff were placed somewhat lower (at B2 or B3), the number of UC H II regions might increase significantly.

Much of the research on UC H II regions has touched in some way on the "lifetime problem" (Wood and Churchwell 1989a). One expects the H II region surrounding an O star to be overpressurized relative to its surroundings due to its high temperature (~10^4 K). It should thus expand at roughly the sound speed of the ionized material (c_{II} ~10 km s^{-1}), giving a "lifetime" of order r/c_{II} for an H II region of radius r. This lifetime is of order 10^4 yr. Many more UC H II regions are seen than this naive view predicts; hence their lifetimes must be longer than anticipated. It is worthwhile to assess the state of this issue, to see whether it still constitutes a "problem."

B. Radio Continuum Surveys of UC H II Regions

The Wood and Churchwell (1989*a,b*) studies proposed color criteria that identified over 1600 IRAS-selected UC H II candidates. Many followup studies searched for radio continuum counterparts to these IRAS sources and, in some cases, suggested that the color criteria had significantly overestimated the number of UC H II regions and hence of young, early-type stars. These surveys were made in both "biased" (toward selected targets) and "unbiased" modes. Among the former are the Kurtz et al. (1994) and Kurtz and Churchwell (2000) surveys; examples of the latter are the galactic plane Very Large Array (VLA) surveys of Zoonematkermani et al. (1990) and Helfand et al. (1992) at 1.4 GHz and by Becker et al. (1994) at 5 GHz. Although low frequencies are not optimal for detecting UC H II regions, the unbiased nature of the latter surveys allows a valuable check of the color criteria.

The main result of the galactic plane surveys is the identification of ~470 probable UC H II regions between 350° and 40° longitude. The galactic latitude distribution of these regions is much narrower [full width at half maximum (FWHM) ≃ 0.25° or 35 pc at a distance of 8 kpc] than that of sources defined solely by the IRAS color criteria (1°). White et al. (1991) interpret this as meaning that these radio surveys detect sources with larger Lyman continuum luminosities (i.e., O stars), while about one-third of the Wood and Churchwell (1989*b*) sample would contain nearby late B stars. Ramesh and Sridharan (1997) correlate single-dish radio continuum surveys with the IRAS UC H II candidates and conclude that only about one-fourth of the Wood and Churchwell sample are UC H II regions.

A weakness of the continuum surveys is that they lack distance estimates for the sources. This problem has been partly resolved by Codella et al. (1994), and Bronfman et al. (1996), who report spectral line velocities. Codella et al. (1994) studied a sample of IRAS sources detected in radio recombination lines, most of which satisfy the UC H II color criteria. They develop a 60-μm flux density cutoff (of 100 Jy) below which the probability of association with an H II region drops considerably. Forty-four percent of the Wood and Churchwell candidates fall below this cutoff. Bronfman et al. (1996) searched for CS emission toward $\sim$1430 of the Wood and Churchwell sources (90% of the total) and had detections in about 60% of these, suggesting that at least 40% of the candidates are not massive embedded stars.

From these studies, one can reasonably conclude that from 25 to 75% of the IRAS sources satisfying the Wood and Churchwell (1989*b*) criteria are not associated with early-type stars. A correction by a factor $\sim$4 of the number of embedded massive stars in the Galaxy, and hence on the lifetime of the UC phase of H II regions, is not negligible. However, several further considerations are in order.

First, multiple UC sources are frequently found in a single IRAS field. The number of embedded stars may be nearly double the number of IRAS-identified candidates for this reason. Also, the expected number of UC H II regions can vary by factors of two or three depending on the assumptions made for the number of ionizing photons available as well as the details of the expansion of the H II region.

Another consideration is that several surveys suggest that IRAS criteria actually *under*estimate the number of galactic UC H II regions. Ellingsen et al. (1996; see also Caswell 1996) find that only half of the roughly 50 CH_3OH masers they detected can be identified with IRAS sources with UC H II region colors. They suggest that the Wood and Churchwell (1989*b*) criteria underestimate the number of UC H II regions by a factor of two. The galactic plane surveys mentioned above suggest that about one-third of the Wood and Churchwell sample is contaminated by lower-mass stars. Yet these same surveys detect a significant number of UC H II regions without any IRAS counterpart: Becker et al. (1994) conclude that about one-third of existing UC H II regions are missing from the IRAS catalog. Part of the problem may be incompleteness of the *Point Source Catalog* in areas of high confusion; it would be useful to study MSX and IRAS HIRES data.

In conclusion, it seems likely that the actual number of galactic UC H II regions differs by a factor of 2 or 3 from the estimate of Wood and Churchwell (1989*b*). It seems unlikely, however, that they overcounted by the factor of 10–20 that is required to resolve the lifetime problem. Therefore the disparity seems real, and at least six proposals have been made to resolve it. They fall into two general categories: those that question the underlying assumptions of the argument, and those that propose physical mechanisms to extend the life of these regions beyond that predicted by classical expansion. We consider these in turn, beginning with the latter category.

C. Solutions to the Lifetime Problem

1. Mechanisms to Extend the Lifetime. Wood and Churchwell (1989*a*) noted that the ram pressure of infalling matter, bow shocks, or both might extend the time in the UC phase. Since then, other theories have been developed, including photoevaporating circumstellar disks (Hollenbach et al. 1994; Yorke and Welz 1996; Richling and Yorke 1997; the chapter by Hollenbach et al., this volume) and mass-loaded flows (Dyson et al. 1995; Lizano et al. 1996). The photoevaporating disk model makes the reasonable assumption that a recently formed massive star will have a remnant accretion disk. The combined effect of a stellar wind and the ionizing photon flux causes photoevaporation of the disk. In this model, the ionized gas is not *confined* on timescales of 10^5 yr but rather is *replenished* by the evaporating disk material. Similarly, mass-loaded flows act to replenish, rather than confine, the ionized gas. In this case, clumps of neutral gas, embedded within the H II region and reasonably supposed to exist within a clumpy cloud, are the source of the replenishing gas. Through a combination of photoionization and hydrodynamic ablation, this clump material is injected into the H II component.

Observational tests of these models will be challenging. For disk photoevaporation, the so-called weak wind case generally applies to unresolved sources, and strong observational confirmation is unlikely. The strong wind case, applicable to resolved sources, does make specific predictions for the spectral index of the emission. High-quality, high-angular-resolution images will be required to test these predictions, and, practically speaking, the results are likely to be somewhat ambiguous. The detection of the so-called partially ionized globules (PIGs) that may give rise to mass-loaded flows is perhaps even more challenging. These neutral globules might be seen in various molecular tracers. However, to detect such emission in the midst of a UC H II region and possibly confused with the surrounding molecular material would be extremely challenging. The externally ionized nature of the PIGs might be confirmed by measurement of their spectral index (Pastor et al. 1991), but again, such observations would be difficult.

Much of the theoretical work on UC H II regions has avoided the issue of the observed morphologies. Notable exceptions are the work of Dyson et al. (1995) and also that of Kessel et al. (1998). In the former case, kinematic information would be useful to test their model for the cometary morphology, but it is not clear whether the expected velocity gradients can be uniquely distinguished from those predicted by other models.

2. Reassessing the Lifetime Problem Assumptions. The underlying assumptions of the lifetime problem have been questioned by De Pree et al. (1995), who point out that recent molecular line data suggest that more realistic values for the density and temperature of the circumstellar molecular gas are 10^7 cm^{-3} and 100 K, increasing the ambient pressure by a factor of 400 over that assumed by Wood and Churchwell (1989*a*) and resulting in smaller radii, both of the initial Strömgren sphere and of the final expanded region in pressure equilibrium with the ISM. In fact, H II regions would still be classified as UC even when they had reached pressure equilibrium with such a high-pressure environment. Akeson and Carlstrom (1996) develop this argument further, both underscoring the denser, warmer cores that we now find and also refining the expansion model beyond that of the strong shock approximation. An even more detailed examination of the expansion phase is given by García-Segura and Franco (1996).

Xie et al. (1996) note that the parameters suggested by De Pree et al. would result in regions with emission measures of order 10^{11} pc cm^{-6}. To date, the highest emission measures found are $\lesssim 10^{10}$ pc cm^{-6}. Such regions would remain optically thick into the millimeter range, however, and existing surveys would probably not detect them. Xie et al. note that turbulent pressure (whether hydrodynamic or hydromagnetic in nature) can provide the high ambient pressures needed to confine the H II region, and, because the pressure does not arise from very high densities, the prediction of large emission measures is avoided.

IV. THE MOLECULAR ENVIRONMENT: CLUMPS AND CORES

Probably the most significant finding in recent studies of high-mass star-forming regions is the detection of dense gas around high-mass star formation signposts such as those discussed in section II. The most salient features are the extent of these gaseous clumps (almost 1 pc in size) and the existence of embedded denser, hotter molecular cores. The latter are very often positionally coincident with H_2O (and sometimes OH) masers but often do not seem to be physically connected with the nearby UC H II region (if any). Figure 1 shows an example of this: The UC H II region G29.96–0.02 (seen in the 1.3-cm continuum) is embedded in a molecular clump traced by the $C^{34}S$ (5–4) line and lies close to a compact core mapped in CH_3CN (6–5).

A single nomenclature to describe the molecular environment in star-forming regions has not yet emerged in the literature. We adopt the term "molecular clump" to signify the large ($\sim$1-pc) regions of molecular gas, and we use "hot core" to signify the high-density, hot condensations within the molecular clumps. In this section we discuss the molecular gas in high-mass star formation regions, with special attention to hot cores and to their role as massive star formation sites. A more extensive review is given by Evans (1999).

A. Molecular Clumps

There is compelling evidence that high-mass star formation is related to the formation of clusters and associations inside molecular clouds. O stars do not form "as single spies" but "in battalions"; hence, understanding

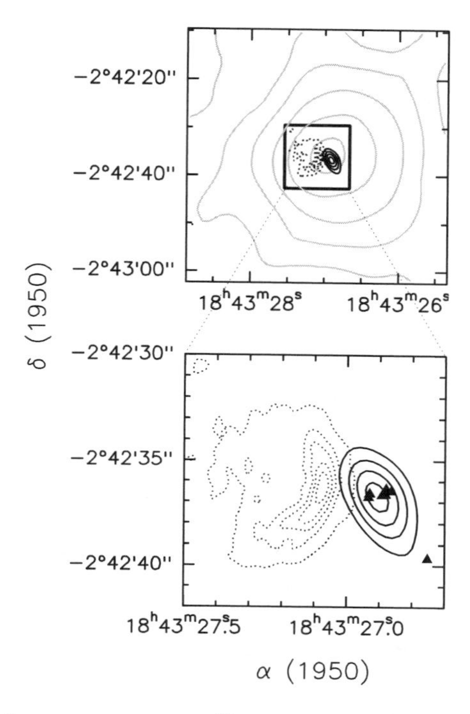

Figure 1. Top: Comparison of the $C^{34}S$ (5–4) (gray contours), CH_3CN (6–5) (black contours), and 1.3-cm continuum (dotted contours) emission toward the UC H II region G29.96–0.02. Bottom: Enlargement of the region where the hot core and the UC H II region are located; the triangles indicate the position of the H_2O masers.

O-star formation involves understanding cluster formation (e.g., Blaauw 1991; Zinnecker et al. 1993). It is useful to compare the mass distribution in young clusters with that observed in the associated molecular clouds and to determine the gravitational effect of the stars upon the gas and *vice versa*.

For the Orion Nebula/Trapezium cluster, Hillenbrand and Hartmann (1998) find evidence that the cluster is elongated north-south, as is the background molecular cloud, suggesting a relation between the two. Within the central 2 pc of the cluster they estimate 1800 $M_\odot$ in the stellar component. The mass of the background cloud has been inferred from molecular emission maps (Goldsmith et al. 1997; Bally et al. 1987; Castets et al. 1990) and is consistent with a surface density of between 1000 and 1600 $M_\odot$ pc^{-2}, which is within a factor of $\sim$2 of the stellar surface density mentioned above. One concludes that gas and stellar masses are similar and that a dynamical effect of one upon the other is probable. It is worth noting that the stellar mass within 0.16 pc of the Trapezium cluster corresponds to a volume density of $n_{H_2} \simeq 4 \times 10^5$ cm^{-3}, similar to the n_{H_2} of molecular clumps in regions of high-mass star formation.

Molecular aggregates similar to OMC1 can be observed throughout most of the Galaxy, in contrast to the stellar component of the Orion Nebula cluster. Thus, using H_2O masers, UC H II regions, or IRAS sources as tracers of high-mass star formation, several studies have been made to search for high-mass ($>$1000 $M_\odot$) compact ($\lesssim$1 pc) clumps and to estimate their parameters. Examples are the works of Churchwell et al. (1992), Cesaroni et al. (1991), Plume et al. (1992, 1997), Hauschildt et al. (1993), Zinchenko et al. (1995), Molinari et al. (1996), and Juvela (1996). Of these, the most extensive studies are those of Plume and collaborators, who observed the CS (2–1), (3–2), (5–4), and (7–6) transitions toward 150 galactic water masers. The critical densities of these transitions range between 4×10^5 cm^{-3} and 2×10^7 cm^{-3}, and thus they are well suited for density estimates of molecular clumps.

Plume et al. (1997) find a mean density of 8×10^5 cm^{-3} in the clumps, similar to the estimates cited above for Orion. They also derived clump diameters and hence masses for a few sources. For the mass estimates, they used three approaches: the estimated number density and size (M_n); the estimated CS column density and size (M_N) with an assumed CS abundance; and the virial theorem (M_v). They find that on average M_n and M_N agree, though with considerable dispersion. On the other hand, M_v is typically an order of magnitude smaller than M_n, perhaps due to low volume filling factors.

Although it is debatable, we would argue that due to the filling factor problem, more reliable masses for dense molecular clumps can be derived either from measurements of optically thin CO isotopomers or based on millimeter dust emission. Both of these approaches have difficulties, mainly stemming from the use of a surrogate for molecular hydrogen. For six well-studied objects, in Fig. 2 we compare masses derived from submillimeter continuum maps (Hunter 1997) with masses based on $C^{17}O$ observations (Wyrowski 1997). Although there is considerable dispersion, the various methods generally agree within an order of magnitude. These differences are at least in part due to temperature gradients on angular scales smaller than the single-dish resolutions used for the studies of Fig. 2.

The studies cited above demonstrate that molecular clumps of size ~1 pc and mass ~10^4 $M_\odot$ are common in high-mass star-forming regions. Their masses are of the same order as the virial mass; hence, they are likely to be gravitationally bound though not necessarily stable. How long will such gas clumps survive? Plume et al. (1997) estimate a period of 10^7 yrs for the conversion of the molecular gas into stars. We take this to be a maximum value because ablation by H II regions, supernovae, etc., is not included (cf. Franco et al. 1994). A minimum lifetime is given by the free-fall time of ~10^4 yrs.

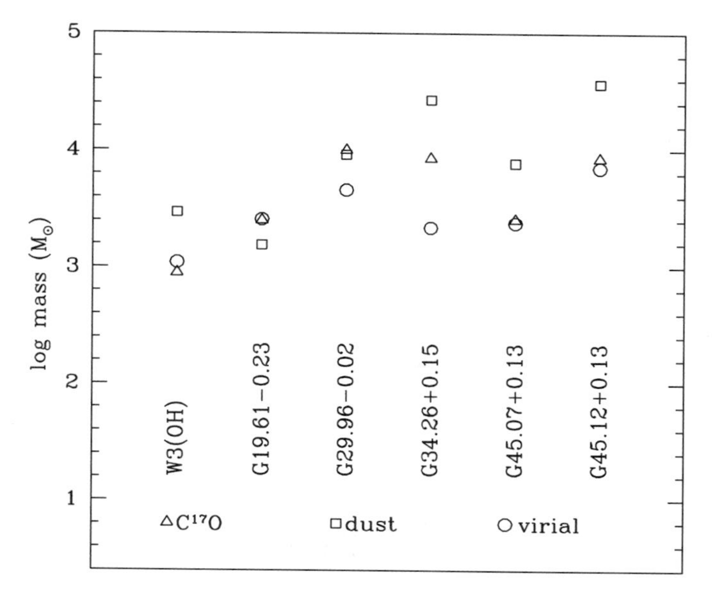

Figure 2. Hot core masses as derived by different methods. $M_{C^{17}O}$ and M_v are based upon $C^{17}O$ measurements reported by Wyrowski (1997). M_{dust} is based upon submillimeter continuum measurements by Hunter (1997). The methods generally agree to within one order of magnitude.

B. Hot Cores

The molecular gas associated with UC H II regions and H_2O masers often shows strong emission in lines excited at temperatures >100 K and densities $>10^6$ cm^{-3}, even when the observations are carried out with low angular resolution ($\gtrsim 10''$). This suggests that such lines arise from compact, optically thick hot cores, the existence of which is confirmed by high-angular-resolution ($\lesssim 1''$) observations.

Hot cores (HCs) may be defined physically as having diameters $\lesssim$0.1 pc, densities $\gtrsim 10^7$ cm^{-3}, and temperatures $\gtrsim$100 K. Observationally, they are characterized by large molecular line optical depths and high line brightness temperatures (when resolved). HCs are also rich in rare molecular species.

In Table I we list all HCs known to date with their main physical parameters. These parameters have been determined by a variety of means and must be taken with caution; for a given object the temperature estimate can easily vary by a factor of 2 and the mass by an order of magnitude, depending on the method used. Furthermore, different sources have often been observed in different tracers, each of which might probe distinct regions of the core, as clearly demonstrated for the Orion HC by Wright et al. (1996; see their Fig. 2). The core luminosity is also a difficult parameter to determine. In many cases it is derived from the FIR flux densities of the

TABLE I
Known Hot Cores and Their Physical Parameters

Source Name	Distance (kpc)	T_{kin} (K)	Diameter (pc)	Mass[a] ($M_\odot$)	L_{IR}[a] ($L_\odot$)	Ref.[b] Code
Orion-KL	0.5	300	0.05	10	$1.0\ 10^5$	1
SgrB2N[c]	8.5	200	0.1	2000	$6.5\ 10^6$	2 3
SgrB2M[c]	8.5	200	0.1	2000	"	2
G5.89–0.39[c]	4.0	90	0.18	2800	$7.1\ 10^5$	4 5
G9.62+0.19	5.7	>100	0.059	55–160	$4.4\ 10^5$	6 7
G10.47+0.03[c]	5.8	150	0.078	2200	$5.0\ 10^5$	6 8 9 10
G10.62–0.38[c]	6.0	144	0.05	1100[d]	$1.1\ 10^6$	11 12
G19.62–0.23	3.5	230	0.032	450	$1.6\ 10^5$	13
G29.96–0.02	7.4	100	0.052	460	$1.4\ 10^6$	6 8 14
G31.41+0.31	7.9	110	0.080	2500	$2.6\ 10^5$	6 8 9
G34.26+0.15	3.8	250	0.066	1400	$6.3\ 10^5$	4 15 16 17 18
G45.07+0.13[c]	8.3	140	0.027	<7800	$1.1\ 10^6$	19
G45.12+0.13[c]	8.3	120	0.056	<37000	$1.3\ 10^6$	19 20
G45.47+0.05[c]	8.3	90	0.056–0.3	250	$1.1\ 10^6$	20
IRAS 20126+4104	1.7	200	0.012	10	$1.3\ 10^4$	21
DR 21(OH) MM1	3.0	>80	0.05	350	$5.0\ 10^4$	22 23
W3(H_2O)	2.2	220	0.014	10	$1.0\ 10^5$	24
W51 e2[c]	8.0	140	0.093	200–400	$1.5\ 10^6$	25 26
W51 e8[c]	8.0	130	0.096	<200	"	25 26
W51 N-Dust	8.0	200	0.1	400	"	25 26

[a] Mass and luminosity estimates are derived respectively from (sub)millimeter continuum emission and from infrared flux densities.
[b] References: (1) Wright et al. 1996. (2) Wilson et al. 1996. (3) Peng et al. 1993. (4) Akeson and Carlstrom 1996. (5) Gómez et al. 1991. (6) Cesaroni et al. 1994*a*. (7) Hofner et al. 1996. (8) Cesaroni et al. 1998. (9) Olmi et al. 1996*b*. (10) Garay et al. 1993. (11) Ho et al. 1994. (12) Keto et al. 1988. (13) Cesaroni et al. 1999. (14) Hofner et al., in preparation. (15) Millar et al. 1997. (16) Hunter et al. 1998. (17) Garay and Rodríguez 1990. (18) Heaton et al. 1989. (19) Hunter et al. 1997. (20) Hofner et al. 1998. (21) Cesaroni et al. 1997. (22) Mangum et al. 1991. (23) Mangum et al. 1992. (24) Wyrowski et al. 1997. (25) Zhang and Ho 1997. (26) Zhang et al. 1998*a*.
[c] With embedded UC H II region.
[d] Derived from $C^{18}O$.

associated IRAS source; this very likely overestimates the actual contribution of the core, because the IRAS beam also includes any adjacent UC H II region, diffuse galactic plane IR emission, and possibly other discrete sources unrelated to the region. These caveats notwithstanding, one sees from Table I that the core masses and luminosities span more than two orders of magnitude, while diameters and temperatures vary by at most a factor of three. We plot the mass vs. luminosity of the known hot cores in Fig. 3; the trend toward higher luminosities for heavier cores is clearly evident.

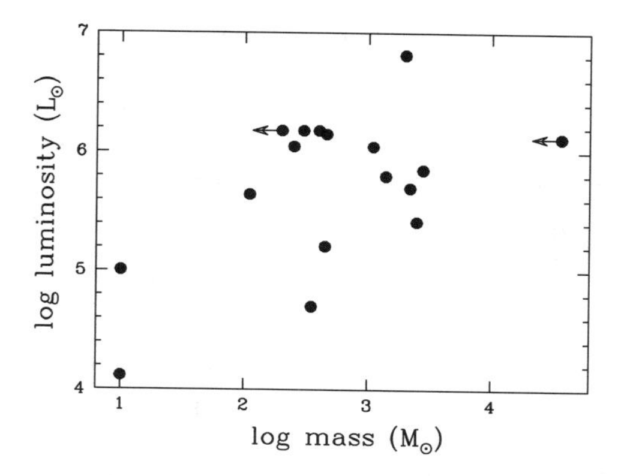

Figure 3. The mass-luminosity distribution of hot cores. The data plotted are taken from Table I. As discussed in the text, there is considerable uncertainty, particularly in the core masses. Nevertheless, a clear trend toward higher luminosities for heavier cores is seen. The two "light cores" (Orion-KL and IRAS 20126) have conspicuously lower masses than the heavier cores.

Although single-dish spectral-line surveys furnish important information on the physical status and chemical composition of HCs (Mauersberger et al. 1988; Churchwell et al. 1992; Cesaroni et al. 1992; Olmi et al. 1993; Helmich 1996; Olmi et al. 1996*a*; Wyrowski 1997; Hunter 1997; Hatchell et al. 1998), the high angular resolution of interferometric observations makes them a more powerful tool for investigating the detailed nature of HCs. The tracers used are low-abundance (and therefore low-optical-depth) molecules or high-excitation transitions of more common species, and millimeter continuum emission. Given the small number of HCs with detailed analysis of interferometric data, we will describe the results for the best-studied objects rather than discussing general properties.

The location of the energy source heating the HCs is a question of significant interest. It is premature to make a general statement, but for at least some sources it appears that the heat source is internal. In the G10.47+0.03 complex, for example, three UC H II regions are embedded in a core seen in ammonia and methyl cyanide (Cesaroni et al. 1994*a*, 1998; Olmi et al. 1996*b*). Subarcsecond maps in the NH_3 (4,4) line show that the kinetic temperature increases toward the center of the core, where the two most compact H II regions are located. This suggests that the embedded high-mass stars are heating the molecular gas. Two other HCs (G29.96–0.02 and G31.41+0.31) were mapped by Cesaroni et al. (1998) at $\sim 0.4''$ angular resolution, and in both cases the temperature increases toward the center. These observations argue in favor of *internal* core heating. Existing observations of the W51 region (Ho and Young 1996; Zhang and Ho 1997; Zhang et al. 1998*a*) lack the angular resolution to detect a temperature gradient. They do find that the gas *density* increases toward the center, as expected if star formation is occurring close to the core centers. Theoretical models of Kaufman et al. (1998) examine both internal and external heating and, for the Orion HC, support the former. Finally, we urge caution in the use of "internal" versus "external" to describe HC heating. In particular, an "external" heat source and the clump it heats may have arisen from collapse within a *single* core. In this sense, "external" would not mean "unrelated genetically."

Another question is the form of the initial mass function (IMF) in these objects. High-mass stars very likely exist inside the cores, as proved by the core luminosity and, in some cases (see Table I), by embedded UC H II regions. If the stellar content of the cores were described by a Miller and Scalo (1979) IMF, the presence of massive stars would imply a *total* stellar mass at least an order of magnitude greater than the mass of the core itself. If this were true, then the hot core virial mass would be substantially greater than its mass determined by dust observations. Inspection of Fig. 2 shows that this is not the case. Therefore, the number of low-mass stars within HCs is much smaller than predicted by the standard IMF. The idea that HCs form "mainly" high-mass stars thus seems reasonable. We caution, however, that lower-mass stars belonging to an OB association would not necessarily be contained within the hot core and hence would not necessarily contribute to the warm dust emission. That the virial masses are in rough agreement with the masses derived from dust observations is suggestive of a truncated IMF, but it is not a proof of it.

Hot cores are likely sites of star formation. The main evidence for this claim is that

1. They are hot and internally heated, as shown by the observed temperature gradients.
2. They are centrally dense, as expected for collapse.
3. Embedded stellar sources are seen from their FIR or radio continuum emission.
4. Maser emission (mostly H_2O) is detected in the core, possibly tracing an outflow or a disk.

C. Kinematics of Hot Cores

The kinematics of HCs should provide important information regarding the physical processes occurring during the star formation process, including both the necessary collapse of molecular material and the possibility of outflows.

Strong observational evidence indicates that powerful molecular outflows exist in high-mass star-forming regions (e.g., Shepherd and Churchwell 1996*a,b*) and may have properties similar to those from low-mass stars, the main difference being the mass involved, which is an order of magnitude larger (see Table 1 of Churchwell 1997*b*). A discussion of the nature of outflows associated with massive star formation may be found in Churchwell (1997*a,b*); see also the chapter by Richer et al., this volume.

It is clear that, regardless of the details of contraction, conservation of angular momentum will force the gas to form rotating structures around the center of collapse. The detection of such "disks" in HCs would support the existence of embedded newly born stars. In recent years, increasing evidence for rotation, collapse, or both in massive objects has been found, so it is not unreasonable to believe that circumstellar disks may form around higher-mass stars as well as their lower-mass counterparts (see the chapter by Natta et al., this volume). Although the resolution presently attainable with interferometers is not sufficient to resolve these disks, observations in high-density and -temperature tracers such as NH_3 (4,4) and CH_3CN have shown that the inner part of HCs is slightly elongated; moreover, a systematic velocity shift of the line peak is found across the major axis of these structures. Evidence for Keplerian rotation is seen in IRAS 20126+4104 (Zhang et al. 1998*b*), strongly suggesting the presence of a disk. In cases where a molecular outflow or jet is present, the axis of the flow is perpendicular to the direction of the velocity gradient in the core, lending strong support to the disk hypothesis (see Fig. 4). Examples of these studies are Ho et al. (1994), Cesaroni et al. (1994*b*; 1997), Olmi et al. (1996*b*), Akeson and Carlstrom (1996), Wyrowski et al. (1997), and Zhang and Ho (1997).

There is also evidence for collapse in HCs, mostly from inverse P Cygni profiles in single-dish spectra. Interferometric observations show red-shifted absorption against the continuum of an embedded UC H II region. This has been found in at least four objects (see Keto et al. 1988 for G10.62–0.38; Wink et al. 1994 for W3 (OH); Zhang and Ho 1997, Zhang et al. 1998*a* for W51; and Hofner et al. 1999*b* for G45.47). The interpretation is not always clear-cut. In the case of W3 (OH), for example, the collapse suggested by absorption profiles is not supported by maser proper motions (Bloemhof et al. 1992).

In conclusion, rotation and collapse seem to be quite common in HCs, supporting the idea of star formation at the center while we are seeing the remnants of collapse.

D. Chemistry in Hot Cores

It was recognized some two decades ago (e.g., Sweitzer 1978; Genzel et al. 1982) that the Orion HC had unusually high abundances of ammonia and other saturated species. This led to the suggestion that one was observing (albeit in some processed form) the contents of grain ice mantles that had recently ($\sim 10^4$ yrs) been evaporated and released into the gas phase. Reviews on the gas- and solid-phase composition of HCs have been given by van Dishoeck et al. (1993), Millar (1997), Ohishi (1997), van Dishoeck and Blake (1998), and the chapter by Langer et al., this volume.

In this section we highlight some new developments. The most important of these is probably the information on the composition, in both

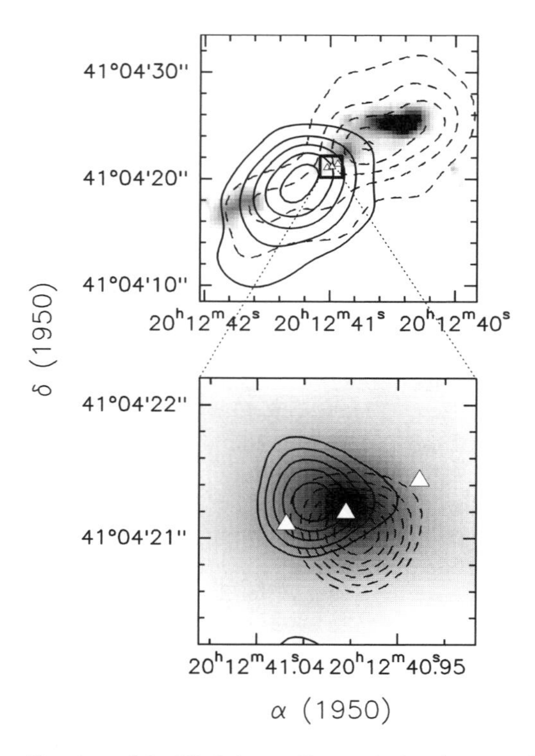

Figure 4. Top: Overlay of the H_2 2.1-μm line emission (gray scale) and HCO^+ (1–0) emission integrated under the blue (full contours) and red (dashed contours) line wings in IRAS 20126+4104 (Cesaroni et al. 1997). Bottom: Enlargement of the region where the hot core is located; the gray scale represents the 1.3-mm continuum emission, and the full and dashed contours correspond to the CH_3CN (12–11) emission integrated, respectively, under the blue and red wing of the $K = 8$ line (Cesaroni et al. 1999). The triangles indicate the position of the H_2O masers. Note the perpendicularity between the H_2-jet/HCO^+-outflow and the rotating disk seen in CH_3CN.

solid and gas phases, obtained from the *Infrared Space Observatory* (ISO). This has led to an improved knowledge of the main species in interstellar ices (Whittet et al. 1996) and to the detection of gas-phase water at high (>200 K) temperatures in absorption toward young massive embedded stars (Helmich et al. 1996; van Dishoeck and Helmich 1996). This suggests a gas-phase abundance ratio $[H_2O]/[H_2] \simeq 3 \times 10^{-5}$, a factor of 2 smaller than that estimated for solid water (van Dishoeck et al. 1993; cf. Gensheimer et al. 1996), but we doubt that this difference is significant. Thus, the thesis that HC material has recently been liberated from grain surfaces seems confirmed. Nevertheless, it is clear that some of the species observed are "daughter species" produced by gas-phase processes subsequent to evaporation. Distinguishing between "daughters" and "parents" could be worthwhile, because suitable models (e.g., Charnley 1997) relate the daughter/parent abundance ratios to the age of the hot molecular gas. We note, however, that care must be taken with age arguments of hot cores based on chemical reactions, because rates may have been significantly enhanced by the greater thermal energies.

Making use of such models requires both a vast spectral coverage and high angular resolution. For Orion there is a set of "classical" single-dish studies, which, when completed, will cover the entire submillimeter range available to ground-based telescopes (Ziurys and McGonagle 1993; Serabyn and Weisstein 1995; Schilke et al. 1997, 1998). Other sources for which less detailed studies are available are Sagittarius B2 (Nummelin 1996), W3 (Helmich and van Dishoeck 1997), G34.26+0.15 (Macdonald et al. 1996) and a selection of 14 UC H II regions (Hatchell et al. 1998). Interferometrically, Blake et al. (1996) observed Orion-KL, covering 4 GHz with roughly 2″ and 1 MHz resolution. They found excitation gradients on scales of 1000 AU and suggest that source I in the IRc2 complex is the main energy source. High angular resolution is of greater importance for HCs more distant than Orion, such as W3 (H_2O), which in many ways is similar to the Orion HC (Helmich and van Dishoeck 1997; Helmich et al. 1994). Wyrowski et al. (1997) found that W3 (H_2O) contains both an "oxygen-rich" peak similar to the Orion "compact ridge" (where methanol, methyl formate, and dimethyl ether are abundant) and a "nitrogen-rich" peak (with species such as ethyl and vinyl cyanide) similar to the Orion HC (see Caselli et al. 1993 for a possible explanation).

Are there differences between the abundances found in HCs of differing physical characteristics? A first cautious answer to this question appears to be "no." The molecular species and abundances in G10.47+0.03 and G31.41+0.31 (see Wyrowski 1997) are approximately the same as those in Orion, Sgr B2-N, and G34.26+0.15 (see also Mehringer and Snyder 1996; Miao and Snyder 1997). Moreover, the abundances do not seem to have a strong dependence on temperature. Light HCs such as Orion do not have chemical properties that distinguish them greatly from heavy HCs such as Sgr B2-N and G10.47+0.03.

Our tentative conclusion is that the abundance pattern is strikingly similar in objects whose mass, luminosity, and location in the Galaxy vary widely. We suggest that this means that dust ice mantle composition does not vary enormously between the inner Galaxy and the solar neighborhood. Moreover, we suspect that we are mainly observing parent rather than daughter species. This implies that most of the sources are younger than 10^5 yr, since the ion-molecule chemistry (driven by cosmic rays) requires a timescale of this order to break down the parent species.

E. Masers, UC H IIs, and Hot Cores

A fruitful line of research in recent years has been the study of molecular masers found in the environs of UC H II regions and HCs. Especially important are studies of H_2O, OH, and CH_3OH masers. Large single-dish databases of H_2O masers have been compiled (Cesaroni et al. 1988; Comoretto et al. 1990; Brand et al. 1994) that are useful in statistical studies of the correlation of masers with UC H II regions and HCs. Ultimately, however, interferometric observations are essential to study the detailed physical relationship between the masers and the star-forming core. Relatively few workers have begun to examine the detailed physical connection between masers and other star formation signposts; Codella et al. (1997) are one exception. Their ammonia observations toward water masers suggest that the maser emission arises from HCs rather than UC H II regions. The possibility that masers arise in circumstellar disks has long been discussed; recent observations of NGC 7538 (Minier et al. 1998), among others, support this view.

Methanol masers were discovered much more recently than H_2O and OH masers, and as a result their place(s) in the star formation process is not as well understood. Southern Hemisphere observations are extensive, but lack the supporting observations that are more plentiful in the north. Conversely, many well-observed northern regions have received less attention in CH_3OH maser studies; high-resolution (interferometric) data are particularly lacking. Among the more extensive surveys are those of Bachiller et al. (1990), Menten (1991), Kalenskii et al. (1993), Norris et al. (1993), Slysh et al. (1994), Caswell et al. (1995), van der Walt et al. (1996), Caswell (1996), and Ellingsen et al. (1996). The latter two are noteworthy because they are sensitivity limited surveys of galactic plane regions. Further work is needed to clarify the nature of the masers they detected, but it seems clear that CH_3OH masers are a significant feature of massive star-forming regions and will be an important observational probe in future studies of UC H II regions and HCs.

V. FROM HOT CORES TO UC H II REGIONS

If HCs represent the natal environment of massive stars, then an interesting question is the evolutionary sequence connecting them with UC H II regions. The latter should follow the former in direct succession, but the precise details of the transition, and regions observed to be undergoing such a transition are lacking, although a few candidates have been found.

A region presenting massive YSOs in different evolutionary stages is G9.62+0.19. This region, shown in Fig. 5, has a chain of UC H II regions of different sizes (and possibly of different ages) associated with a hot core traced by the NH_3 (4,4) emission, probably corresponding to an earlier phase, and possibly near the point of becoming a UC H II region.

Another candidate for a hot core on the verge of developing a UC H II region is IRAS 20126+4104 (Cesaroni et al. 1997; see Fig. 4), which is associated with a molecular outflow and probably a rotating disk. The bottom panel of Fig. 4 shows a 1.3-mm image ($0.7''$ resolution) in the $J = 12-11$, $K = 8$ transition of CH_3CN (526 K above ground). An offset of

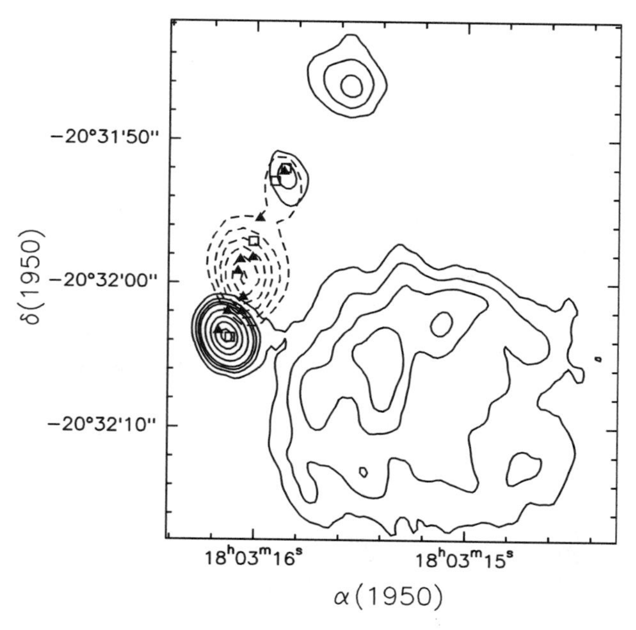

Figure 5. Maps of the NH_3 (4,4) line (dashed contours) and 1.3-cm continuum emission (solid contours) toward G9.62+0.19 (Cesaroni et al. 1994*a*). The positions of the H_2O (Hofner and Churchwell 1996) and OH (Forster 1993) masers are indicated respectively by the filled triangles and empty squares.

0.4″ (0.0033 pc) is seen between the blue- and redshifted CH_3CN emission, perpendicular to the HCO^+ (1–0) outflow (upper panel). This is consistent with the outflow emanating from the rotating disk seen in CH_3CN. Recent NH_3 observations by Zhang et al. (1998*b*) support the interpretation of rotation and show that the "disk" is much larger than seen in CH_3CN. The source is weak at centimeter wavelengths, where it appears as a double source, possibly in the form of thermal jets (Hofner et al. 1999*a*).

G9.62 and IRAS 20126 represent the search for HCs on the verge of becoming UC H II regions. Another approach is to search for the youngest UC H II regions, which have just made the transition from HCs, to see what artifacts of their earlier state may remain. De Pree et al. (1998) searched the Sgr B2 Main region in the 7-mm continuum, and detected some 20 extremely small ($\sim 10^{-3}$ pc), very high-emission-measure ($\sim 10^9$ pc cm^{-6}) H II regions. Such high-emission-measure objects may be recently born UC H II regions. Other promising sites to search are those UC H II regions still embedded in HCs listed in Table I, particularly those unresolved even by subarcsecond beams, such as G10.47+0.03 or W51 e8. We note that some HCs, e.g., W3 (OH/H_2O), may be powered by B stars of sufficiently late spectral type that they have a negligible Lyman continuum flux. Other HCs (see Table I) are so luminous that they probably are powered by a (proto) O star. In these latter cases, one might hope to see a transition state from a hot core to a UC H II region. With enough such regions detected, either evolved HCs or nascent UC H II regions, we can hope to begin to understand the transition between the two.

In fact, the transition phase from HC to UC H II region may be intimately related to their lifetimes, and hence to the number of such objects present at any one time. Similar to the estimates for UC H II regions, one may estimate HC lifetimes based on an assumed massive star formation rate (10^{-2} yr^{-1}) and the number of HCs thought to be present at any one time. Using the number of detected HCs (19; see Table I) as an extreme lower limit, and the number of candidate high-mass Class 0 objects ($\sim$575; see Ramesh and Sridharan 1997) as an upper limit, one estimates the HC lifetime between 1.9×10^3 and 5.7×10^4 yr.

What is the physical mechanism responsible for the transition from HC to UC H II region? Mass accretion rates of 10^{-4}–10^{-5} $M_\odot$ yr^{-1} can effectively quench the development of an H II region (Walmsley 1995). As discussed in section IV.C, there is considerable evidence to suggest that such infall is a significant feature of HCs. Assuming free-fall collapse of the molecular gas onto the embedded massive star, one finds that the Strömgren radius depends dramatically on the mass accretion rate (see Yorke 1979). One thus speculates that the transition from HC to UC H II region occurs because of some modification in the accretion mechanism, such as a gap in the influx of matter or departure from spherical symmetry (on which the previous result relies). Formation of disklike structures on scales $\gtrsim$100 AU might suffice.

VI. CONCLUSIONS

Recent high-frequency, high-angular-resolution studies have allowed us to probe hot, dense molecular cloud cores on scales $\lesssim$1000 AU, where OB stars are forming. We have reexamined the lifetime problem of UC H II regions and conclude that, though the problem is real, probable solutions have been proposed. We have reviewed the most recent observational findings concerning the formation and the earliest evolutionary stages of high-mass stars. The most salient result is the detection of hot, compact, dense cores, rich in rare molecular species and positionally coincident with various masers (especially H_2O). These HCs are often (but not always) associated with nearby or (in some cases) embedded UC H II regions, which suggests a connection between the two. Indeed, there is strong support for the idea that HCs represent the natal environment of high-mass stars. It is tempting to speculate that HCs will eventually evolve into UC H II regions, but a definite answer to this question will come only from more sensitive and higher-resolution observations capable of revealing the temperature, density, and velocity structure of the HCs. This goal will be achieved with the forthcoming generation of instruments, such as ALMA and FIRST.

Acknowledgments C. M. W. acknowledges travel support from ASI grants ARS-96-66 and ARS-98-116. S. E. K. acknowledges partial support from CONACyT grant 25470-E and DGAPA projects IN109696 and IN102395.

REFERENCES

Akeson, R. L., and Carlstrom, J. E. 1996. Lifetimes of ultracompact H II regions: High-resolution methyl cyanide observations. *Astrophys. J.* 470:528–536.

Bachiller, R., Menten, K. M., Gómez-González, J., and Barcia, A. 1990. The 44 GHz methanol masers: Results of an extensive survey in the $7_0 \rightarrow 6_1A^+$ line. *Astron. Astrophys.* 240:116–122.

Bally, J., Stark, A. A., Wilson, R. W., and Langer, W. D. 1987. Filamentary structure in the Orion molecular cloud. *Astrophys. J. Lett.* 312:L45–L49.

Becker, R. H., White, R. L., Helfand, D. J., and Zoonematkermani, S. 1994. A 5 GHz VLA survey of the galactic plane. *Astrophys. J. Suppl.* 91:347–387.

Blaauw, A. 1991. OB associations and the fossil record of star formation. In *The Physics of Star Formation and Early Stellar Evolution,* ed. C. J. Lada and N. D. Kylafis (Dordrecht: Kluwer), pp. 125–154.

Blake, G. A., Mundy, L. G., Carlstrom, J. E., Padin, S., Scott, S. L., Scoville, N. Z., and Woody, D. P. 1996. A λ = 1.3 mm aperture synthesis molecular line survey of Orion Kleinmann-Low. *Astrophys. J. Lett.* 472:L49–L52.

Bloemhof, E. E., Reid, M. J., and Moran, J. M. 1992. Kinematics of W3(OH): First proper motions of OH masers from VLBI measurements. *Astrophys. J.* 397:500–519.

Brand, J., Cesaroni, R., Caselli, P., Catarzi, M., Codella, C., Comoretto, G., Curioni, G. P., Curioni, P., Di Franco, S., Felli, M., Giovanardi, C., Olmi, L., Palagi, F., Palla, F., Panella, D., Pareschi, G., Rossi, E., Speroni, N., and Tofani, G. 1994. The Arcetri catalogue of H_2O maser sources update. *Astron. Astrophys. Suppl.* 103:541–572.

Braz, M. A., and Epchtein, N. 1987. New detections of probable massive pre-main sequence stars in the southern galactic plane. *Astron. Astrophys.* 176:245–254.

Bronfman, L., Nyman, L.-Å., and May, J. 1996. A CS(2–1) survey of IRAS point sources with color characteristics of ultra-compact H II regions. *Astron. Astrophys. Suppl.* 115:81–95.

Carral, P., Kurtz, S. E., Rodríguez, L. F., De Pree, C., and Hofner, P. 1997. Detection of 7 millimeter sources near cometary H II regions. *Astrophys. J. Lett.* 486:L103–L106.

Caselli, P., Hasegawa, T., and Herbst, E. 1993. Chemical differentiation between star-forming regions. The Orion hot core and compact ridge. *Astrophys. J.* 408:548–558.

Castets, A., Duvert, G., Dutrey, A., Bally, J., Langer, W. D., and Wilson, R. W. 1990. A multi-transition study of carbon monoxide in the Orion A molecular cloud. *Astron. Astrophys.* 234:469–486.

Caswell, J. L. 1996. A new survey for 6.6 GHz methanol masers. *Mon. Not. Roy. Astron. Soc.* 279:79–87.

Caswell, J. L., Vaile, R. A., Ellingsen, S. P., Whiteoak, J. B., and Norris, R. P. 1995. Galactic methanol masers at 6.6 GHz. *Mon. Not. Roy. Astron. Soc.* 272:96–138.

Cesaroni, R., Palagi, F., Felli, M., Catarzi, M., Comoretto, G., Di Franco, S., Giovanardi, C., and Palla, F. 1988. A catalogue of H_2O maser sources north of $\delta = -30°$. *Astron. Astrophys. Suppl.* 76:445–458.

Cesaroni, R., Walmsley, C. M., Kömpe, C., and Churchwell, E. 1991. Molecular clumps associated with ultracompact H II regions. *Astron. Astrophys.* 252:278–290.

Cesaroni, R., Walmsley, C. M., and Churchwell, E. 1992. Hot ammonia toward ultracompact H II regions. *Astron. Astrophys.* 256:618–630.

Cesaroni, R., Churchwell, E., Hofner, P., Walmsley, C. M., and Kurtz, S. 1994*a*. Hot ammonia towards compact H II regions. *Astron. Astrophys.* 288:903–920.

Cesaroni, R., Olmi, L., Walmsley, C. M., Hofner, P., and Churchwell, E. 1994*b*. A massive young embedded object associated with the UC H II region G31.41+0.31. *Astrophys. J. Lett.* 435:L137–L140.

Cesaroni, R., Felli, M., Testi, L., Walmsley, C. M., and Olmi, L. 1997. The disk-outflow system around the high-mass (proto)star IRAS 20126+4104. *Astron. Astrophys.* 325:725–744.

Cesaroni, R., Hofner, P., Walmsley, C. M., and Churchwell, E. 1998. Subarcsecond structure of hot cores in the NH_3 (4,4) line. *Astron. Astrophys.* 331:709–725.

Cesaroni, R., Felli, M. Jenness, T., Neri, R., Olmi, L., Robberto, M., Testi, L., and Walmsley, C. M. 1999. Unveiling the disk-jet system in the massive (proto)star IRAS 20126+4104. *Astron. Astrophys.* 345:949–964.

Charnley, S. B. 1997. Sulfuretted molecules in hot cores. *Astrophys. J.* 481:396–405.

Churchwell, E. 1991. Newly formed massive stars. In *The Physics of Star Formation and Early Stellar Evolution,* ed. C. J. Lada and N. D. Kylafis (Dordrecht: Kluwer), pp. 221–268.

Churchwell, E. 1997*a*. Origin of the mass in massive star outflows. *Astrophys. J. Lett.* 479:L59–L61.

Churchwell, E. 1997*b*. Models of hot molecular cores. In *Herbig-Haro Flows and the Birth of Low Mass Stars,* IAU Symp. 182, ed. B. Reipurth and C. Bertout (Dordrecht: Kluwer), pp. 525–536.

Churchwell, E., Walmsley, C. M., and Cesaroni, R., 1990. A survey of ammonia and water vapor emission from ultracompact H II regions. *Astron. Astrophys. Suppl.* 83:119–144.

Churchwell, E., Walmsley, C. M., and Wood, D. O. S. 1992. Hot, dense molecular gas associated with ultracompact H II regions. *Astron. Astrophys.* 253:541–556.

Codella, C., Felli, M., and Natale, V. 1994. H II regions and IRAS PSC sources: The reliability of the association. *Astron. Astrophys.* 284:233–240.

Codella, C., Testi, L., and Cesaroni, R. 1997. The molecular environment of H_2O masers: VLA ammonia observations. *Astron. Astrophys.* 325:282–294.

Comoretto, G., Palagi, F., Cesaroni, R., Felli, M., Bettarini, A., Catarzi, M., Curioni, G. P., Curioni, P., Di Franco, S., Giovanardi, C., Massi, M., Palla, F., Panella, D., Rossi, E., Speroni, N., and Tofani, G. 1990. The Arcetri atlas of H_2O maser sources. *Astron. Astrophys. Suppl.* 84:179–225.

De Pree, C. G., Rodríguez, L. F., and Goss, W. M. 1995. Ultracompact H II regions: Are their lifetimes extended by dense, warm environments? *Rev. Mex. Astron. Astrof.* 31: 39–44.

De Pree, C. G., Goss, W. M., and Gaume, R. A. 1998. Pressure confined ultracompact H II regions in Sgr B2 Main. *Astrophys. J.* 500:847–852.

Dyson, J. E., Williams, R. J. R., and Redman, M. P. 1995. Clumpy ultra compact H II regions. I. Fully supersonic wind-blown model. *Mon. Not. Roy. Astron. Soc.* 277: 700–704.

Ellingsen, S. P., von Bibra, M. L., McCulloch, P. M., Norris, R. P., Deshpande, A. A., and Phillips, C. J. 1996. A survey of the galactic plane for 6.7 GHz methanol masers. I. $l = 325° - 335° b = -0.°53 - +0.°53$. *Mon. Not. Roy. Astron. Soc.* 280:378–396.

Evans, N. J. 1999. Physical conditions in regions of star formation. *Ann. Rev. Astron. Astrophys.*, in press.

Felli, M., Palagi, F., and Tofani, G. 1992. Molecular outflows and H_2O masers: What type of connection? *Astron. Astrophys.* 255:293–322.

Fiebig, D., Duschl, W. J., Menten, K. M., and Tscharnuter, W. M. 1996. The masing environment of star forming object IRAS00338+6312: Disk, outflow, or both? *Astron. Astrophys.* 310:199–210.

Forster, J. R. 1993. Some interesting OH/H_2O masers. In *Astrophysical Masers,* ed. A. W. Clegg and G. E. Nedoluha (Berlin: Springer-Verlag), pp. 108–111.

Franco, J., Shore, S. N., and Tenorio-Tagle, G. 1994. On the massive star-forming capacity of molecular clouds. *Astrophys. J.* 436:795–799.

Garay, G., and Rodríguez, L. F. 1990. The compact molecular core toward G34.3+0.2: VLA observations in the (2,2) and (3,3) lines of ammonia. *Astrophys. J.* 362:191–201.

Garay, G., Moran, J. M., and Rodríguez, L. F. 1993. Compact ammonia sources toward the G10.5+0.0 H II region complex. *Astrophys. J.* 413: 582–592.

García-Segura, G., and Franco., J. 1996. From ultracompact to extended H II regions. *Astrophys. J.* 469: 171–188.

Gensheimer, P. D., Mauersberger, R., and Wilson, T. L. 1996. Water in galactic hot cores. *Astron. Astrophys.* 314:281–294.

Genzel, R., Downes, D., Ho, P. T. P., and Bieging, J. H. 1982. NH_3 in Orion-KL: A new interpretation. *Astrophys. J. Lett.* 259:L103–L107.

Goldsmith, P. F., Bergin, E. A., and Lis, D. C. 1997. Carbon monoxide and dust column densities: The dust-to-gas ratio and structure of three giant molecular cloud cores. *Astrophys. J.* 491:615–637.

Gómez, Y., Rodríguez, L. F., Garay, G., and Moran, J. M. 1991. The dense molecular envelope around the compact H II region G5.89–0.39 (W28 A2). *Astrophys. J.* 377:519–525.

Hatchell, J., Thompson, M. A., Millar, T. J., and Macdonald, G. H. 1998. A survey of molecular line emission towards ultracompact H II regions. *Astron. Astrophys. Suppl.* 133:29–49.

Hauschildt, H., Güsten, R., Phillips, T. G., Schilke, P., Serabyn, E., and Walker, C. K. 1993. First detection of CS(10–9) in galactic star forming cores. *Astron. Astrophys.* 273:L23–L26.

Heaton, B. D., Little, L. T., and Bishop, I. S. 1989. The "ultracompact hot core" of G34.3+0.15: Arcsecond resolution ammonia observations. *Astron. Astrophys.* 213:148–154.

Helfand, D. J., Zoonematkermani, S., Becker, R. H., and White, R. L. 1992. Compact radio sources near the galactic plane. *Astrophys. J. Suppl.* 80:211–255.

Helmich, F. P. 1996. Dense molecular gas around massive young stars. Ph.D. Thesis, University of Leiden.

Helmich, F. P., and van Dishoeck, E. F. 1997. Variations within the W3 star-forming region. II. The 345 GHz line survey. *Astron. Astrophys. Suppl.* 124:205–253.

Helmich, F. P., Jansen, D. J., de Grauuw, T., Groesbeck, T. D., and van Dishoeck, E. F. 1994. Physical and chemical variations within the W3 star-forming region. 1: SO_2, CH_3OH, and H_2CO. *Astron. Astrophys.* 283:626–634.

Helmich, F. P., van Dishoeck, E. F., Black, J. H., de Graauw, T., Beintema, D. A., Heras, A. M., Lahuis, F., Morris, P. W., and Valentijn, E. A. 1996. Detection of hot abundant water toward AFGL 2591. *Astron. Astrophys.* 315:L173–L176.

Hillenbrand, L. A. 1997. On the stellar population and star-forming history of the Orion Nebula cluster. *Astron. J.* 113:1733–1768.

Hillenbrand, L. A., and Hartmann, L. W. 1998. A preliminary study of the Orion nebula cluster structure and dynamics. *Astrophys. J.* 492:540–553.

Ho, P. T. P., and Young, L. M. 1996. The contracting molecular cores e1 and e2 in W51. *Astrophys. J.* 472:742–754.

Ho, P. T. P., Terebey, S., and Turner, J. L. 1994. The rotating molecular core in G10.6–0.4: Synthesis maps in $^{12}C^{18}O$. *Astrophys. J.* 423:320–325.

Hofner, P., and Churchwell, E. 1996. A survey of water maser emission toward ultracompact H II regions. *Astron. Astrophys. Suppl.* 120:283–299.

Hofner, P., Kurtz, S., Churchwell, E., Walmsley, C. M., and Cesaroni, R. 1996. Massive star formation in the hot, dense cloud core of G9.62+0.19. *Astrophys. J.* 460:359–371.

Hofner, P., Cesaroni, R., Rodríguez, L. F., and Martí, J. 1999*a*. A double system of ionized jets in IRAS 20126+4104. *Astron. Astrophys.* 345:L43–L46.

Hofner, P., Peterson, S., and Cesaroni, R. 1999*b*. Ammonia absorption toward the ultracompact H II regions G45.12+0.13 and G45.47+0.05. *Astrophys. J.* 514:899–908.

Hollenbach, D., Johnstone, D., Lizano, S., and Shu, F. 1994. Photoevaporation of disks around massive stars and application to ultracompact H II regions. *Astrophys. J.* 428:654–669.

Hunter, T. R. 1997. A submillimeter imaging survey of ultracompact H II regions. Ph.D. Thesis (California Institute of Technology, Pasadena).

Hunter, T. R., Phillips, T. G., and Menten, K. M. 1997. Active star formation toward the ultracompact H II regions G45.12+0.13 and G45.07+0.13. *Astrophys. J.* 478:283–294.

Hunter, T. R., Neugebauer, G., Benford, D. J., Matthews, K., Lis, D. C., Serabyn, E., and Phillips, T. G. 1998. G34.24+0.13MM: A deeply embedded proto-B-star. *Astrophys. J. Lett.* 493:L97–L100.

Jenness, T., Scott, P. F., and Padman, R. 1995. Studies of embedded far-infrared sources in the vicinity of H_2O masers. I. Observations. *Mon. Not. Roy. Astron. Soc.* 276:1024–1040.

Juvela, M. 1996. Studies of dense molecular cores in regions of massive star formation. IV. Multitransition CS-study towards southern H_2O masers in the longitude range $l = 308° - 360°$. *Astron. Astrophys. Suppl.* 118:191–226.

Kalenskii, S. V., Bachiller, R., Berulis, I. I., Val'tts, I. E., Gomez-Gonzalez, J., Martin-Pintado, J., Rodriguez-Franco, A., and Slysh, V. I. 1993. Search for methanol masers at 44 GHz. *Sov. Astron.* 36:517–523.

Kaufman, M. J., Hollenbach, D. J., and Tielens, A. G. G. M. 1998. High-temperature molecular cores near massive stars and the application to the Orion hot core. *Astrophys. J.* 497:276–287.

Kessel, O., Yorke, H., and Richling, S. 1998. Photoevaporation of protostellar disks. III. The appearance of photoevaporating disks around young intermediate mass stars. *Astron. Astrophys.* 337: 832–846.

Keto, E. R., Ho, P. T. P., and Haschick, A. D. 1988. The observed structure of the accretion flow around G10.6–0.4. *Astrophys. J.* 324:920–930.

Kurtz, S., and Churchwell, E. 2000. Ultracompact H II regions. III: The outer galaxy. *Astrophys. J. Suppl.*, in preparation.

Kurtz, S., Churchwell, E., and Wood, D. O. S. 1994. Ultracompact H II regions. II: New high-resolution radio images. *Astrophys. J. Suppl.* 91:659–712.

Lizano, S., Cantó, J., Garay, G., and Hollenbach, D. 1996. Photoevaporated flows from H II regions. *Astrophys. J.* 468:739–748.

Macdonald, G. H., Gibb, A. G., Habing, R. J., and Millar, T. J. 1996. A 330–360 GHz spectral survey of G34.3+0.15. I. Data and physical analysis. *Astron. Astrophys. Suppl.* 119:333–367.

Mangum, J. G., Wootten, A., and Mundy, L. G. 1991. Synthesis imaging of the DR 21(OH) cluster. I. Dust continuum and $C^{18}O$ emission. *Astrophys. J.* 378:576–585.

Mangum, J. G., Wootten, A., and Mundy, L. G. 1992. Synthesis imaging of the DR 21(OH) cluster. II. Thermal ammonia and water maser emission. *Astrophys. J.* 388:467–488.

Matthews, H. E., Olnon, F. M., Winnberg, A., and Baud, B. 1985. H_2O sources near the galactic plane: A pilot systematic survey. *Astron. Astrophys.* 149: 227–238.

Mauersberger, R., Wilson, T. L., and Henkel, C. 1988. A multilevel study of ammonia in star-forming regions. IV. Emission and absorption toward W3(OH). *Astron. Astrophys.* 201:123–130.

McCaughrean, M. J. and Stauffer, J. R. 1994. High resolution near-infrared imaging of the Trapezium: A stellar census. *Astron. J.* 108:1382–1397.

Mehringer, D., and Snyder, L. E. 1996. The location of complex molecules in G34.3+0.2: Further evidence for grain-surface chemistry. *Astrophys. J.* 471:897–902.

Menten, K. M. 1991. The discovery of a new, very strong and widespread interstellar methanol maser line. *Astrophys. J. Lett.* 380:L75–L78.

Miao, Y., and Snyder, L. E. 1997. Full synthesis observations of CH_3CH_2CN in Sagittarius B2: Further evidence for grain chemistry. *Astrophys. J. Lett.* 480:L67–L70.

Millar, T. J. 1997. Models of hot molecular cores. In *Molecules in Astrophysics: Probes and Processes,* IAU Symp. 178, ed. E. F. van Dishoeck (Dordrecht: Kluwer), pp. 75–88.

Millar, T. J., Macdonald, G. H., and Gibb, A. G. 1997. A 330–360 GHz spectral survey of G34.3+0.15 II. Chemical modelling. *Astron. Astrophys.* 325:1163–1173.

Miller, G. E., and Scalo, J. M. 1979. The initial mass function and stellar birthrate in the solar neighborhood. *Astrophys. J. Suppl.* 41: 513–547.

Minier, V., Booth, R. S., and Conway, J. E. 1998. Observations of methanol masers in NGC7538: Probable detection of a circumstellar disk. *Astron. Astrophys.* 336:L5–L8.

Miralles, M. P., Rodríguez, L. F., and Scalise, E. 1994. Radio continuum, ammonia and water maser observations of bright, unassociated IRAS point sources. *Astrophys. J. Supp.* 92:173–188.

Molinari, S., Brand, J., Cesaroni, R., and Palla, F. 1996. A search for precursors of ultracompact H II regions in a sample of luminous IRAS sources. I. Association with ammonia cores. *Astron. Astrophys.* 308:573–587.

Norris, R. P., Whiteoak, J. B., Caswell, J. L., Wieringa, M. H., and Gough, R. G. 1993. Synthesis imaging of 6.7 GHz methanol masers. *Astrophys. J.* 412:222–232.

Nummelin, A. 1996. Millimeter-wave observations of the SgrB2 Molecular Cloud, Technical Report 256L (Göteborg: Chalmers University).

Ohishi, M. 1997. Observations of hot cores. In *Molecules in Astrophysics: Probes and Processes,* IAU Symp. 178, ed. E. F. van Dishoeck (Dordrecht: Kluwer), pp. 61–74.

Olmi, L., Cesaroni, R., and Walmsley, C. M. 1993. Ammonia and methyl cyanide in hot cores. *Astron. Astrophys.* 276:489–506.

Olmi, L., Cesaroni, R., and Walmsley, C. M. 1996*a*. CH_3CN towards G10.47+0.03 and G31.41+0.31. *Astron. Astrophys.* 307:599–608.

Olmi, L., Cesaroni, R., Neri, R., and Walmsley, C. M. 1996*b*. High resolution CH_3CN observations towards hot cores. *Astron. Astrophys.* 315:565–577.
Palagi, F., Cesaroni, R., Comoretto, G., Felli, M., and Natale, V. 1993. Classification and statistical properties of galactic H_2O masers. *Astron. Astrophys. Suppl.* 101:153–193.
Palla, F., Brand, J., Cesaroni, R., Comoretto, G., and Felli, M. 1991. Water masers associated with dense molecular clouds and ultracompact H II regions. *Astron. Astrophys.* 246:249–263.
Palla, F., Cesaroni, R., Brand, J., Caselli, P., Comoretto, G., and Felli, M. 1993. H_2O masers associated with dense molecular clouds and ultracompact H II regions. II: The extended sample. *Astron. Astrophys.* 280:599–608.
Panagia, N. 1973. Some physical parameters of early-type stars. *Astron. J.* 78:929–934.
Pastor, J., Cantó, J., and Rodríguez, L. F., 1991. The radio continuum spectrum of partially ionized globules. *Astron. Astrophys.* 246:551–558.
Peng, Y., Vogel, S. N., and Carlstrom, J. E. 1993. Deuterated ammonia in Sagittarius B2. *Astrophys. J.* 418:255–262.
Plume, R., Jaffe, D. T., and Evans, N. J., II., 1992. A survey of CS $J = 7 \rightarrow 6$ in regions of massive star formation. *Astrophys. J. Suppl.* 78:505–515.
Plume, R., Jaffe, D. T., Evans, N. J., II, Martín-Pintado, J., and Gómez-González, J. 1997. Dense gas and star formation: Characteristics of cloud cores associated with water masers. *Astrophys. J.* 476:730–749.
Ramesh, B., and Sridharan, T. K. 1997. Reliable galaxy-wide identification of ultracompact H II regions. *Mon. Not. Roy. Astron. Soc.* 284: 1001–1006.
Richling, S., and Yorke, H. 1997. Photoevaporation of protostellar disks II. The importance of UV dust properties and ionizing flux. *Astron. Astrophys.* 327:317–324.
Scalise, E., Jr., Rodríguez, L. F., and Mendoza-Torres, E. 1989. Water-vapor maser emission from bright, unassociated IRAS point sources. *Astron. Astrophys.* 221:105–109.
Schilke, P., Groesbeck, T. D., Blake, G. A., and Phillips, T. G. 1997. A line survey of Orion-KL from 325 to 360 GHz. *Astrophys. J. Suppl.* 108:301–337.
Schilke, P., Benford, D. J., Hunter, T. R., Li, D. C., and Phillips, T. G. 1998. A line survey of Orion-KL from 607 to 725 GHz. *Astrophys. J. Suppl.*, submitted.
Serabyn, E., and Weisstein, E. W. 1995. Fourier transform spectroscopy of the Orion molecular cloud core. *Astrophys. J.* 451:238–251.
Shepherd, D. S., and Churchwell, E. 1996*a*. High velocity molecular gas from high-mass star formation regions. *Astrophys. J.* 457:267–276.
Shepherd, D. S., and Churchwell, E. 1996*b*. Bipolar molecular outflows in massive star formation regions. *Astrophys. J.* 472:225–239.
Shepherd, D. S., Churchwell, E., and Wilner, D. J. 1997. A high spatial resolution study of the ON2 massive star-forming region. *Astrophys. J.* 482:355–371.
Slysh, V. I., Kalenskii, S. V., Val'tts, I. E., and Otrupcek, R. 1994. The Parkes survey of methanol masers at 44.07 GHz. *Mon. Not. Roy. Astron. Soc.* 268:464–474.
Sweitzer, J. S. 1978. On the excitation of interstellar ammonia in the Kleinmann-Low nebula. *Astrophys. J.* 225:116–129.
Testi, L., Palla, F., Prusti, T., Natta, A., and Maltagliati, S. 1997. A search for clustering around Herbig Ae/Be stars. *Astron. Astrophys.* 320:159–166.
Torrelles, J. M., Gómez, J. F., Rodríguez, L. F., Curiel, S., Ho, P. T. P., and Garay, G. 1996. The thermal radio jet of Cepheus A HW2 and the water maser distribution at $0\rlap{.}''08$ scale (60 AU). *Astrophys. J. Lett.* 457:L107–L111.
Torrelles, J. M., Gómez, J. F., Rodríguez, L. F., Ho, P. T. P., Curiel, S., and Vázquez, R. 1997. A radio jet–H_2O maser system in W75N(B) at a 200 AU

scale: Exploring the evolutionary stages of young stellar objects. *Astrophys. J.* 489:744–752.

Torrelles, J. M., Gómez, J. F., Rodríguez, L. F., Curiel, S., Anglada, G., and Ho, P. T. P. 1998. Radio continuum H_2O maser systems in NGC 2071: H_2O masers tracing a jet (IRS1) and a rotating protoplanetary disk of radius 20 AU (IRS3). *Astrophys J.* 505:756–765.

van der Walt, D. J., Retief, S. J. P., Gaylard, M. J., and Macleod, G. C. 1996. A search for 5_1–6_0 A^+-methanol masers towards faint IRAS sources. *Mon. Not. Roy. Astron. Soc.* 282:1085–1095.

van Dishoeck, E. F., and Blake, G. A. 1998. Chemical evolution of star forming regions. *Ann. Rev. Astron. Astrophys.,* in press.

van Dishoeck, E. F., and Helmich, F. P. 1996. Infrared absorption of H_2O towards massive young stars. *Astron. Astrophys.* 315:L177–L180.

van Dishoeck, E. F., Blake, G. A., Draine, B. T., and Lunine, J. I. 1993. The chemical evolution of protostellar and protoplanetary matter. In *Protostars and Planets III,* ed. E. H. Levy and J. I. Lunine (Tucson: University of Arizona Press), pp. 163–241.

Walmsley, C. M. 1995. Dense cores in molecular clouds. *Rev. Mex. Astron. Astrof. Ser. de Conf.* 1:137–148.

White, R. L., Becker, R. H., and Helfand, D. J. 1991. The infrared properties of compact galactic radio sources: The young and the restless. *Astrophys. J.* 371:148–162.

Whittet, D. C. B., Schutte, W. A., Tielens, A. G. G. M., Boogert, A. C. A., de Grauuw, T., Ehrenfreund, P., Gerakines, P. A., Helmich, F. P., Prusti, T., and van Dishoeck, E. F. 1996. An ISO SWS view of interstellar ices: First results. *Astron. Astrophys.* 315:L357–L360.

Wilson, T. L., Snyder, L. E., Comoretto, G., Jewell, P. R., and Henkel, C. 1996. A new molecular core in Sgr B2. *Astron. Astrophys.* 314:909–916.

Wink, J. E., Duvert, G., Guilloteau, S., Güsten, R., Walmsley, C. M. and Wilson, T. L. 1994. The molecular surroundings of W3(OH). *Astron. Astrophys.* 281:505–516.

Wood, D. O. S., and Churchwell, E. 1989*a*. The morphologies and physical properties of ultracompact H II regions. *Astrophys. J. Suppl.* 69:831–895.

Wood, D. O. S., and Churchwell, E. 1989*b*. Massive stars embedded in molecular clouds: Their population and distribution in the galaxy. *Astrophys. J.* 340:265–272.

Wright, M. C. H., Plambeck, R. L., and Wilner, D. J. 1996. A multi-line aperture synthesis study of Orion-KL. *Astrophys. J.* 469:216–237.

Wyrowski, F. 1997. Radio observations of high mass star forming regions: hot cores and photon dominated regions. Ph.D. Thesis, University of Cologne.

Wyrowski, F., Hofner, P., Schilke, P., Walmsley, C. M., Wilner, D. J., and Wink, J. E. 1997. Millimeter interferometry towards the ultracompact H II region W3(OH). *Astron. Astrophys.* 320:L17–L20.

Xiang, D., and Turner, B. E. 1995. Newly discovered galactic H_2O masers associated with outflows. *Astrophys. J. Suppl.* 99:121–133.

Xie, T., Mundy, L. G., Vogel, S. N., and Hofner, P. 1996. On turbulent pressure confinement of ultracompact H II regions. *Astrophys. J. Lett.* 473:L131–L134.

Yorke, H. W. 1979. The evolution of protostellar envelopes of masses 3 and 10 solar masses. I. Structure and hydrodynamic evolution. *Astron. Astrophys.* 80: 308–316.

Yorke, H. W., and Welz, A. 1996. Photoevaporation of protostellar disks I. The evolution of disks around early B stars. *Astron. Astrophys.* 315:555–564.

Zhang, Q., and Ho, P. T. P. 1997. Dynamical collapse in W51 massive cores: NH_3 observations. *Astrophys. J.* 488:241–257.
Zhang, Q., Ho, P. T. P., and Ohashi, N. 1998*a*. Dynamical collapse in W51 massive cores: CS(3–2) and CH_3CN observations. *Astrophys. J.* 494:636–656.
Zhang, Q., Hunter, T. R., and Sridharan, T. K. 1998*b*. A rotating disk around a high-mass young star. *Astrophys. J. Lett.* 505:L151–L154.
Zinchenko, I., Mattila, K., and Toriseva, M. 1995. Studies of dense molecular cores in regions of massive star formation. II. CS $J = 2 \rightarrow 1$ survey of southern H_2O masers in the longitude range $l = 260°–310°$. *Astron. Astrophys. Suppl.* 111:95–114.
Zinnecker, H., McCaughrean, M. J., and Wilking, B. A. 1993. The initial stellar population. In *Protostars and Planets III,* ed. E. H. Levy and J. I. Lunine (Tucson: University of Arizona Press), pp. 429–495.
Ziurys, L. M., and McGonagle, D. 1993. The spectrum of Orion-KL at 2 millimeters (150–160 GHz). *Astrophys. J. Suppl.* 89:155–187.
Zoonematkermani, S., Helfand, D. J., Becker, R. H., White, R. L., and Perley, R. A. 1990. A catalog of small-diameter radio sources in the galactic plane. *Astrophys. J. Suppl.* 74:181–224.

THE FORMATION OF MASSIVE STARS

S. W. STAHLER
University of California at Berkeley

F. PALLA
Osservatorio Astrofisico di Arcetri

and

P. T. P. HO
Harvard-Smithsonian Center for Astrophysics

We review observational and theoretical evidence bearing on the origin of high-mass stars. The conventional theory of protostellar infall successfully accounts for objects between about 2 and 10 $M_\odot$. After formation, these stars undergo quasistatic contraction with radiative interiors. In this stage, they correspond to Herbig Ae and Be stars. More massive objects have no pre-main-sequence phase and disrupt their infalling envelopes through strong winds and radiation pressure. Nevertheless, there is spectroscopic evidence for ongoing collapse in a number of systems associated with compact H II regions. The striking concentration of massive stars toward the centers of young clusters provides an important clue. The indication is that stars well in excess of 10 $M_\odot$ arise from the coalescence of already existing cluster members of lower mass. Coalescence may occur after the merger of dense molecular cores that contain the lower-mass stars.

I. OBSERVATIONAL BACKGROUND

Our picture of stellar birth is most securely established for objects of solar and subsolar mass. Pre-main-sequence stars in this regime, the well-studied T Tauri class, exist in abundance within a few hundred parsecs. Conversely, one must cast one's observational net out to the ends of the Galaxy to sample a comparable number of O and B-type stars. It is true that the high luminosity of a massive star renders it conspicuous even at such distances and beyond. On the other hand, the object's prodigious radiative output, coupled with the momentum of its stellar wind, quickly and effectively disperses natal cloud matter. This circumstance has impeded progress in understanding the formation mechanism.

All theoretical considerations must begin with the facts at hand. In this limited review, we will focus exclusively on our own Galaxy. Hence, we will be unable to cover such fascinating topics as the starburst phenomenon, whose proper understanding may well shed light on massive stars generally. In any case, it has long been known that most Galactic O and B stars reside within loose associations ranging from tens to hundreds of parsecs in diameter (Blaauw 1991). Their large internal velocities, typically about 4 km s^{-1}, doom these systems to expansion and eventual dispersal. The very nomenclature of OB associations reflects the historical tendency to focus on their brightest and most massive components. On the other hand, the infrared array surveys undertaken over the past decade have revealed a multitude of low-mass members (e.g., McCaughrean and Stauffer 1994). It is now clear that the very luminous stars in a typical association represent the tail of a much broader distribution. Where sufficient data are available, the latter resembles the field-star initial mass function (Brown et al. 1994; Massey et al. 1995*a*). This agreement bolsters the long-held view that most stars in the Galaxy originate from OB associations (Roberts 1957; Miller and Scalo 1978).

Discerning the properties of such a stellar group requires first a measure of its distance. The bulk of association members lie on the main sequence, so one may utilize the venerable technique of spectroscopic parallax. Establishing a distance then allows one to assess the masses of the visible stars. Here, both spectroscopic and dynamic methods are available. In the former, one fits model atmospheres to the observed narrowband spectrum. The free parameters in such models are the effective temperature $T_{\rm eff}$ and the surface gravity $g \equiv GM_\star/R_\star^2$. In principle, one could also obtain $T_{\rm eff}$ directly from a broadband color. For massive stars, however, such color indexes as $B - V$ change very little with spectral type and are unreliable as temperature indicators (Massey et al. 1995*b*). The model atmosphere calculation also gives the bolometric correction, so that the M_V value now yields an estimate for $L_{\rm bol}$. The star's mass then follows from its position within the main sequence in the $L_{\rm bol} - T_{\rm eff}$ plane. For post-main-sequence objects, one finds the evolutionary track that passes through the star's position in the diagram (see, e.g., Schaller et al. 1992). Alternatively, one may determine $R_\star$ through the blackbody relation between $L_{\rm bol}$ and $T_{\rm eff}$. The mass then follows from knowledge of the surface gravity: $M_\star = gR_\star^2/G$ (Kudritzki and Hummer 1990; Maeder and Conti 1994).

For O stars that are members of binaries, one may also obtain the mass through orbital Doppler shifts. Burkholder et al. (1997) have recently studied a sample of early-type spectroscopic binaries for this purpose. They find good agreement between the dynamical mass determination and that obtained spectroscopically. Unfortunately, their sample only extends to stars up to about 15 $M_\odot$. For objects of higher mass, the surface gravity technique generally yields smaller $M_\star$ values than does placement on evolutionary tracks. The origin of this discrepancy, which can be up to a factor of 1.5, is not yet clear. With the possible exception of the Pistol Star in the Galactic center (Figer et al. 1995), the current upper mass limit is held by stars in the clusters Trumpler 14 and 16. Kudritzki et al. (1992) used the spectroscopic method to derive masses of 100 $M_\odot$ for several objects classified as O3.

The techniques just outlined have allowed detailed examination of the few dozen OB associations within the nearest few kpc, and rapidly improving surveys of more distant systems (Garmany 1994). It is hardly surprising that the best-studied group is also the closest. This is the complex of associations strung out along the Orion molecular cloud, at a distance of 450 pc. The nucleus of each association contains a tight grouping of several O stars (Brown et al. 1994). Within the Orion B cloud, the grouping also coincides with a dense agglomeration of molecular gas, as evidenced through CS emission (Lada et al. 1997). This gas contains numerous infrared sources that represent low-mass stars in the act of formation. The most famous collection of high-mass stars is, of course, the Trapezium that powers the Orion Nebula. Here, the brightest object is θ^1 C, which has a spectral type of O6 and a mass of 33 $M_\odot$. The glowing Nebula itself is energized by the ultraviolet emission from this object and its companions. Even more massive stars are located in the OB1b subgroup, with ζ Ori A topping the list at 49 $M_\odot$.

The relative proximity of the Orion region has also allowed the most complete census of the low-mass complement to the brightest stars. The Trapezium itself sits near the center of a heavily populated and dense cluster, whose full extent was first revealed in the near-infrared (McCaughrean et al. 1991). About half the cluster members are optically visible, apparently because the O stars have already cleared away much of the ambient gas. Hillenbrand (1997) has succeeded in placing over 900 of these objects in the HR diagram. The stellar density near the cluster's center is extremely high, some 10^4 pc^{-3}, but is probably not the highest in the Galaxy. One rival is NGC 3603, an H II region located in the Carina spiral arm, at a distance of 7 kpc. Here one finds six stars with mass about 50 $M_\odot$ in a volume less than 0.03 pc^3, and a total cluster luminosity of order 10^7 $L_\odot$ (Eisenhauer et al. 1998). By comparison, the luminosity of the Orion Nebula cluster is less by two orders of magnitude.

The infrared surveys have uncovered other, less spectacular, aggregates with a central concentration of massive stars. The heavy obscuration in these regions is a hallmark of extreme youth. With the progressive loss of gas through stellar winds and radiation, such systems may expand to become the more familiar OB associations. However, it is important to bear in mind that not all massive stars are found inside associations or clusters. About 25 percent of the population are true field stars with large galactic latitudes. The radial velocities of these objects exhibit significant dispersion about the local value expected from galactic rotation alone. Statistical analysis of the velocities reveals that most objects are leaving the plane rather than entering it from above or below (Gies 1987). Thus, the stars were likely born in associations, but with speeds that were higher than average.

An important subset of the field objects are the runaway OB stars. These have extremely high spatial velocities, typically from 50 to 150 km s^{-1}. They are usually situated far from the galactic plane. The proper motions of these stars can usually be traced back to known OB associations. The real problem is their velocities, which indicate that the objects were once subjected to strong forces. Thus, each runaway could have been a member of a close binary pair, propelled to high speed when its companion detonated as a supernova. Alternatively, the stars might have been ejected from the central cluster of their parent associations as a result of close encounters with neighbors. We shall later return to consider both possibilities in more detail.

II. RESULTS FROM PROTOSTAR THEORY

A. Evolution at Intermediate Masses

This sketch of the observational landscape suggests that O and B stars *may* form in a cloud medium similar to that yielding low-mass objects, but somewhat enhanced in density. Unfortunately, the degree of enhancement is difficult to discern using the standard molecular line techniques. When a new cluster of O stars has been revealed, as in the case of the Trapezium, there is little gas left in the immediate vicinity. On the other hand, infrared sources of high luminosity have poorly defined stellar properties and often prove under closer examination to be collections of lower-mass objects. One promising strategy is to focus on the brightest, pointlike drivers of outflows and maser emission. Using this approach, Molinari et al. (1998) recently discovered IRAS 23385+6053, a deeply embedded object with a strong molecular outflow and an $L_{\rm bol}$ of order 10^4 $L_\odot$. Although no measure of volume density is yet available, the spectral energy distribution indicates an A_V of 2000 mag.

Millimeter continuum studies also suggest an especially dense environment for the most luminous young stars (e.g., Mooney et al. 1995). Let us, therefore, assume provisionally that such objects originate from more massive and compact analogs to the cloud cores known to be responsible for T Tauri stars (Myers and Benson 1983). Suppose we speculate further that the ensuing collapse process is essentially the same. Then the properties of the growing protostar follow almost entirely from $\dot{M}$, the rate of mass infall near the core's center. This rate, in turn, is given by

$$\dot{M} \approx \frac{a^3}{G} \tag{1}$$

where a is the sound speed, and where the equality holds to within an order of magnitude (see, e.g., Foster and Chevalier 1993).

A remarkable feature of equation (1) is that the predicted infall rate is *independent* of the parent cloud density. Even if our putative cores have a density three or four orders of magnitude larger than those normally encountered, $\dot{M}$ depends solely on a and therefore on the ambient temperature. This simplification results from our tacit assumption that the cloud configuration was in near force balance between gravity and thermal pressure prior to collapse. The infall process then spreads outward from the center at the sound speed a. A cloud of fixed mass M with a higher density ρ has a shorter free-fall time (proportional to $\rho^{-1/2}$) but brings in correspondingly less mass to the center during this interval. Magnetic support in the precollapse configuration alters this result in detail, but not quantitatively (Li 1998).

Within our adopted framework, then, the description of protostar evolution rests on a proper assessment of cloud temperature. Observers routinely measure kinetic temperatures of order 100 K in such environments as Orion BN-KL or W3, an H II region 2.4 kpc distant (Cesaroni et al. 1994; Hofner et al. 1996). The most reliable values of T follow from a level-population analysis designed to reproduce observed line strengths. Similar temperatures are needed to drive chemical reaction networks (e.g., Caselli et al. 1993). All such environments, however, are critically affected by the presence of an existing massive star or stellar group. We are rather interested in the temperature just *before* star formation. Until our "denser cores" are discovered and analyzed in detail, this information is unavailable.

The simplest and most conservative assumption is that the precollapse temperature is independent of core density and mass and is similar to well-observed values in clouds producing low-mass stars. These values range from 10 to 20 K in nearby complexes and yield, through equation (1), $\dot{M}$ values from about 10^{-6} to 10^{-5} $M_{\odot}$ yr^{-1}. A feasible and instructive theoretical program is to use such rates to build up protostars through intermediate and higher masses. This was the approach initiated by Palla and Stahler (1991, 1992). We may briefly summarize their main results, which have since been corroborated and extended by other workers (Beech and Mitalas 1994; Bernasconi and Maeder 1996).

The structure of a protostar reflects both the addition of successive layers of infalling matter and the heat injection from nuclear processes. The gas either impacts the star directly (after passing through an accretion shock) or first lands on a circumstellar disk and only later spirals to the surface. In either case, the deepening gravitational potential well ensures that the specific entropy of the newly added gas rises inexorably with time. This circumstance would normally force the protostar to be stable against thermal convection. Beginning at subsolar masses, however, deuterium that has been accreted along with the interstellar hydrogen ignites in the deep interior. The heat input from deuterium fusion renders low-mass protostars fully convective (Stahler et al. 1980).

Convection persists as long as fresh deuterium can be dragged down toward the hot center. When the star's mass reaches a value between 2 and 3 $M_{\odot}$, its interior reverts to a state of radiative stability. The change occurs because of the decline in opacity at higher temperatures. In more detail, convection disappears first in a shell roughly midway to the surface. This *radiative barrier* blocks the transport of deuterium to the center via turbulent diffusion. Nuclear fusion therefore shuts down. At this point, the protostar consists of a thermally inert central core exhausted of deuterium, along with a mantle of deuterium-rich gas extending to the surface.

Continued addition of mass drives the interior temperature ever higher. When the base of the mantle reaches a temperature of about 10^6 K, deuterium again ignites. The burning, now confined to a thick shell, drives convection out to the subphotospheric surface layers. This energy input to the mantle dramatically swells the radius of the protostar. However, the rising interior temperature and falling opacity guarantee the eventual end of convection. Both the burning shell and the convective mantle gradually retreat to the surface, vanishing by the time the total mass has reached 4 $M_\odot$.

Thus far the protostar's major source of luminosity has been the kinetic energy released at the accretion shock. Note again that this shock also covers the disk surfaces if the infall has significant angular momentum (Cassen and Moosman 1981; Stahler et al. 1994). At about the same time that the burning shell disappears, the star begins to contract under the influence of self-gravity. Contraction releases additional energy. The protostar's total luminosity is now

$$L_\star = \frac{GM_\star \dot{M}}{R_\star} + L_{\rm int} \tag{2}$$

Here, the first term is the accretion luminosity contributed by infalling matter, while the second is the component generated by contraction. In a radiatively stable protostar, $L_{\rm int}$ varies as $M_\star^{11/2} R_\star^{-1/2}$ (Cox and Giuli 1968). Hence the second term in equation (2) rapidly overwhelms the first as $M_\star$ increases. The contracting object is beginning to resemble a pre-main-sequence star.

The rise of $L_{\rm int}$ means that the star's thermal energy is escaping at an accelerated rate, so the contraction process itself speeds up. The protostar's central temperature also increases sharply and soon exceeds 10^7 K. At this point, ordinary hydrogen ignites via the CN cycle. The heat from this source stops further contraction, and the star effectively joins the zero-age main sequence (ZAMS). The precise mass for this event depends weakly on the assumed rate of infall. For $\dot{M} = 1 \times 10^{-5}$ $M_\odot$ yr^{-1}, the mass is 8 $M_\odot$; the mass would be 14 $M_\odot$ if $\dot{M}$ were larger by an order of magnitude. The calculations also show that the evolutionary sequence is essentially unaltered if the infalling mass traverses a disk instead of landing directly on the surface.

We have been imagining a situation in which the mass buildup persists in a steady manner. For any given star, however, $\dot{M}$ must vanish at some point, allowing the object to contract quasistatically. The manner by which infall ceases is still problematic for any stellar mass. It may be that the powerful jets and outflows observed to emanate from embedded stars are responsible for this reversal (see below). Magnetic support of the parent cloud could also provide the ultimate barrier (Mouschovias 1976). Whatever the process, subsequent pre-main-sequence evolution is largely unaffected if $\dot{M}$ dies off in a period brief compared to the time scale for contraction. Under this assumption, we may select various models from the sequence of protostars and follow their evolution at fixed mass (Palla and Stahler 1993).

Each protostar model, then, represents the *initial state* of a pre-main-sequence star of the same mass. Divested of its dusty infalling envelope, the star is now optically visible and occupies a well-defined position in the HR diagram. The entire sequence of protostars yields the *birthline* in the diagram, i.e., the locus from which all pre-main-sequence contraction begins (Stahler 1983; Palla and Stahler 1990). We have just seen that protostars with more than about 10 $M_{\odot}$ are burning hydrogen. In the HR diagram, the upper end of the birthline intersects the main sequence at just this point (see Fig. 1). More massive stars have no contracting pre-main-sequence phase. This theoretical prediction seems to accord well

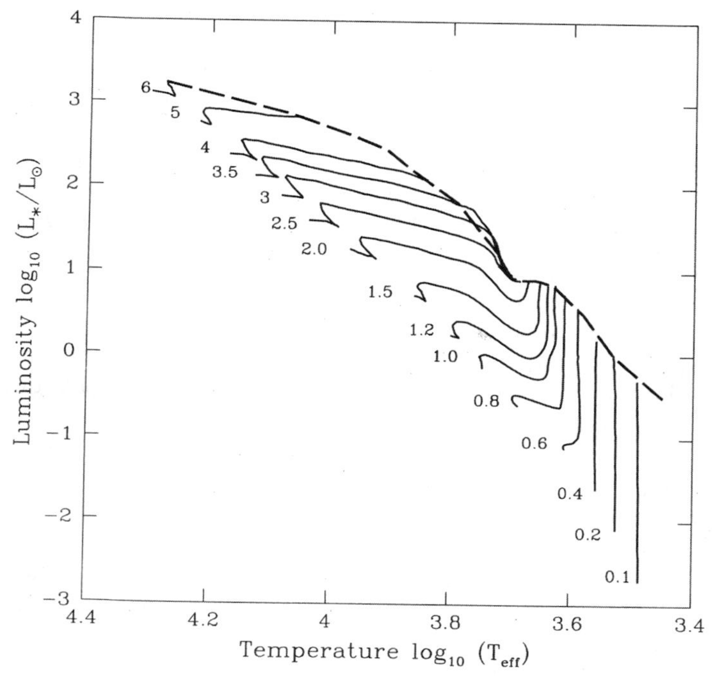

Figure 1. Theoretical pre-main-sequence tracks in the HR diagram (Palla and Stahler 1999). Each track is labeled by the corresponding stellar mass, in units of $M_{\odot}$. All the tracks start at the birthline, indicated by the dashed curve, and end at the zero-age main sequence (ZAMS), which has been omitted for clarity.

with observations of Herbig Ae and Be stars, which indeed have a mass limit of about the right value (see, e.g, Hillenbrand et al. 1993).

Pre-main-sequence stars from 2 to 10 $M_\odot$ begin with radiative interiors, apart from their outer mantles of deuterium-rich gas. These mantles quickly disappear as the contraction proceeds, and they do not affect subsequent evolution. However, the contraction itself is quite different from that envisioned in the original studies of the pre-main-sequence phase (Ezer and Cameron 1967). For masses up to about 4 $M_\odot$, the star first appears *below* its classical track in the HR diagram. It then moves upward and to the left before joining the more horizontal path that leads to the ZAMS. This *luminosity jump* is the outward manifestation of an internal readjustment in entropy (Stahler 1989). That is, the structure built up through accretion is incompatible with uniform, homologous contraction. The star expels a thermal pulse as the pattern of contraction shifts.

Corroboration of the evolutionary calculations with the data on Herbig Ae and Be stars has only just begun. One interesting prediction is that the ages of these objects should be much less than previously thought. Here, we start the clock at the birthline, so that the total contraction age shrinks to zero for masses near 10 $M_\odot$. Now the record of star formation in a cluster may be read from the age distribution of its members. Application of the new evolutionary tracks to well-observed clusters may substantially alter our conception of their history and of cluster formation in general.

B. Disruption of Infall

We have seen that the rise of L_{int} in equation (2) causes protostars of sufficient mass to contract even while they are still accreting. That is, the Kelvin-Helmholtz time characterizing the contraction, $t_{KH} \equiv GM_\star^2/R_\star L_\star$, becomes shorter than the accretion timescale, $t_{acc} \equiv M_\star/\dot{M}$. The fact that this transition occurs at a fairly well-defined $M_\star$ value simply reflects the sensitivity of $L_\star$ to mass. Heavier protostars are burning hydrogen and have the structure of main-sequence objects. However, they are still cloaked in dusty envelopes. Once infall ends, these stars appear directly on the ZAMS in the HR diagram.

Our theoretical picture has thus far ignored the structure and dynamics of the diffuse envelope. In particular, we must consider the mechanical effect of the protostar's own radiation on the infalling gas. The accretion shock at the stellar and disk surfaces produces photons at optical and ultraviolet wavelengths. These are absorbed by the envelope's dust grains and reradiated into the infrared. Trapping of this infrared radiation creates thermal pressure that retards infall. The stellar luminosity also raises the envelope's temperature until the grains sublimate at the dust destruction front (Stahler et al. 1980). The direct impact of short-wavelength photons on grains at this front provides an additional retarding force.

Radiation pressure begins to become significant in protostars of intermediate mass. Infall cannot be stopped, however, as long as the accretion component dominates the total luminosity. Any retardation of the flow diminishes $\dot{M}$, which in turn lowers $L_\star$. Once the luminosity falls, the flow picks up again. Numerical simulations of collapsing clouds with central stars display this effect. For stars in the proper mass range, $\dot{M}$ oscillates strongly without actually reversing (Yorke and Krügel 1977).

The feedback between radiation pressure and infall is cut once $L_{\rm int}$ dominates $L_\star$. Again, the transition occurs just where $t_{\rm KH}$ falls below $t_{\rm acc}$, i.e., for protostars that have joined the main sequence. The internal luminosity is now supplied by hydrogen fusion rather than contraction, but it still scales as $M_\star^{11/2} R_\star^{-1/2}$. Hence the retarding force on the envelope increases rapidly with mass. Within the context of spherical collapse, the result is inevitable. Wolfire and Cassinelli (1987) analyzed the dynamics of envelopes falling onto stars with $M_\star$ exceeding 60 $M_\odot$. They found that infall was reversed unless $\dot{M}$ were arbitrarily increased to at least 10^{-3} $M_\odot$ yr^{-1} *and* grains were depleted in abundance by an order of magnitude. These extreme conditions are unlikely to occur, so we are faced with a basic dilemma. (For earlier discussion of this issue, see Larson and Starrfield 1971, and Kahn 1974.)

There is little doubt that a massive star produces more than enough energy to expel its envelope. In fact, the inequality $t_{\rm KH} < t_{\rm acc}$ implies that the radiated luminosity can easily unbind the parent cloud before the star has gained appreciable mass. The real issue is the *efficiency* of energy transfer. Here, any departure of the infall from spherical symmetry is likely to play a major role (Nakano 1989).

Consider, for example, the collapse of a rotating cloud. As we have mentioned, fluid elements far from the central axis depart from radial trajectories and strike the equatorial disk. Let us suppose that the cloud's rotation rate is sufficiently large that the disk edge lies beyond the dust destruction front. Then the outer portion of the flow never attains the temperature that would create high radiation pressure. Once an element has entered the disk, the relatively large optical thickness could shield it from further heating as it spirals inward. Note, however, that O stars with no protostellar envelopes, such as those in the Trapezium, destroy both their own disks and those of neighboring stars through photoevaporation (Hollenbach et al 1994; Johnstone et al. 1998). Hence, this accretion scenario can work only if the radiation color temperature is significantly lowered before the photons strike the disk.

That portion of the infall closer to the rotation axis must still bear the full brunt of the radiation pressure. In spherical symmetry, this gas brakes and reverses its velocity. The fluid within a two-dimensional flow can respond by veering off laterally. Calculating this effect self-consistently is a difficult (and still unsolved) problem in radiation hydrodynamics. Jijina and Adams (1996) have presented a simplified, but instructive solution in which they retain an isotropic radiation field. They find that incoming elements can penetrate only to a "turnaround" surface that grows with time.

The density in their steady-state flow formally diverges at this locus. More realistically, the strong deceleration would create a standing shock front. This would effectively replace the accretion shock in lower-mass protostars, but it would be spatially removed from the star and disk. We may picture incoming gas entering the shock obliquely, then turning toward the equator within a condensed shell of postshock matter. It remains to be seen whether such a flow pattern is stable or whether the shell tends to break up.

Stellar winds act as a further impediment to infall. Those from visible O and B stars are radiatively driven and have the highest velocities along the main sequence (Ignace et al. 1998). There is no apparent reason why massive protostars should not also produce such flows while still accreting. Theoretical work in this area has focused almost exclusively on the mechanical interaction of winds with static envelopes. Here we may broadly distinguish two cases. Wind material of relatively low speed (below about 300 km s^{-1}) quickly cools after impact to produce thin, dense shells. Traditionally, there has been much interest in these curved surfaces in relation to low-mass stellar jets (Cantó 1980). The faster winds from O and B stars cannot radiate as efficiently after the shock, because of the ionization of metallic coolants. In the extreme case of strict energy conservation, these winds inflate hot bubbles that expand into the external medium (Königl 1982).

How a wind of *any* speed impacts a *collapsing* envelope is still poorly understood. In rotating collapse, the centrifugal repulsion lowers both the infall density and ram pressure within an expanding central column. The diminished density alone would cause an initially isotropic wind to lengthen in the polar direction (Mellema and Frank 1997). Wilkin and Stahler (1998) have recently calculated the steady-state interaction of a spherical wind and a rotating, collapsing flow. They assumed the existence of a narrow, shocked shell, so the calculation would need to be modified for massive stellar winds. Nevertheless, their general results are of relevance.

The two most important characteristics of rotating infall are that it is anisotropic and that it is time dependent. The second feature stems from the fact that the specific angular momentum in the parent cloud increases away from the central axis. Because the collapse process in a marginally unstable configuration spreads outward at the sound speed a, the angular momentum being transported inward steadily rises. Concurrently, the region of lowered density and ram pressure also expands.

The infall is nearly spherical relatively early in the collapse. At this stage, the wind material cannot progress far before it falls back to the star and disk. In the calculation of Wilkin and Stahler, this fallback occurs through motion within the narrow shell. As the impinging ram pressure falls, the shell elongates, especially in the axial direction. Eventually, the tips of the configuration erupt in a bipolar fashion. The wind now flows out unimpeded within a solid angle that continually widens.

Returning to massive stars, the collapse must already be in a relatively advanced stage by the time a large $M_\star$ has accumulated. Thus, we expect centrifugal distortion of the infall to be severe, and for the wind to have little difficulty in breaking out within the polar region. Calculations assuming a hot bubble rather than a thin shell arrive at a similar result, at least for static envelopes (Frank and Noriega-Crespo 1994). In reality, of course, both stellar winds and radiation pressure operate simultaneously. Our discussion indicates that the only portion of the infall with a chance to survive is that which impacts the outer disk. It remains to be seen whether this material can indeed spiral inward and reach the stellar surface.

III. EVIDENCE FOR INFALL

A. Observational Strategy

The current difficulties in understanding how infall proceeds would be greatly alleviated by direct, empirical measurements of cloud collapse. Unfortunately, the problem is inherently difficult for *any* stellar mass. First, true dynamical infall occurs only at a radius r that is relatively close to the central object. Second, the magnitude of the infall speed scales as $(M_\star/r)^{1/2}$. For solar-type protostars, the envelope gas reaches a speed of 1 km s^{-1} only at $r = 0.004$ pc, or about 9″ at a distance of 100 pc. Direct imaging of the region is therefore infeasible, and one turns to spectroscopic techniques. However, the amount of matter actually participating in collapse is typically small compared with surrounding material in the telescope beam. Absorption by cooler foreground gas and confusion with outflows, rotation, or turbulence cast doubt on putative detections.

The search for infall around low-mass protostars has recently concentrated on the detection of asymmetric line profiles. This technique, resurrected by Zhou et al. (1993), utilizes the fact that any substantial radial motion separates in velocity the foreground and background halves of the envelope. As the line becomes optically thick, it emanates chiefly from the nearer parts of both halves. Suppose, in addition, that the protostar creates a substantial gradient in gas temperature. Then the detected portion of the foreground half will be relatively cool, as it lies farther from the illuminating source. Conversely, the background half will appear hotter. The resulting asymmetric line profile is a promising diagnostic for infall. Several groups have carried out surveys; these include observations at multiple lines designed to test the predicted radiative transfer effect (e.g., Mardones et al. 1997; see also the chapter by Myers et al., this volume). Interferometric studies are also addressing the problem of spatial resolution.

Observations of high-mass protostars have several distinct advantages. The greater depth of the gravitational potential implies that the magnitude of infall speed increases. We also expect a greater amount of mass involved in contraction and collapse. The observational benefit is an enhanced optical depth in any molecular line. In addition, more species and higher-excitation transitions are amenable to study. The greater luminosity of the central star elevates the temperature of the ambient gas. In some cases, the relevant lines are seen in absorption against an H II region surrounding the source itself. The foreground gas is then separated kinematically from background material and can be examined more readily.

Unfortunately, there are also serious problems that are peculiar to the high-mass case. Because massive stars often form in clusters, the amount of diffuse gas being observed may be very large, from 10^2 to 10^4 $M_\odot$. The inevitable result is further confusion regarding multiple exciting sources, as well as a potentially complex morphology. There is also the fundamental constraint that massive stars are intrinsically rare, so that the distance to the star of interest may be several kpc. Thus, even when the collapsing volume is physically large, it subtends a small angle. Researchers have again turned to interferometers for detailed examination of infall around high-mass stars.

B. Infall at Large and Small Scales

While many molecular clouds associated with massive stars have now been observed (see the chapter by Kurtz et al., this volume), few show any indication for rapid, inward motion. Two exceptions are the complexes known as W49 and W51. Both contain a large number of ultracompact H II regions. The small sizes of such regions imply relatively short lifetimes. Thus, the simultaneous presence of many regions calls for some kind of global synchronization of the massive-star formation process. We shall return later to consider this issue.

Consider the case of W51. The first evidence for large-scale motion came from Berkeley-Illinois-Maryland Array (BIMA) observations of HCO^+ (Rudolph et al. 1990). Red shifted absorption features were seen; these appear to grow stronger at higher angular resolution. The indication was that the entire molecular cloud complex, containing some 4×10^4 $M_\odot$, could be undergoing global collapse. Here, the maximum infall speed actually observed was 20 km s^{-1}, over a sizescale of 1 pc.

This original interpretation suffered from a number of difficulties. First, the infall velocity did not appear to vary either with radial or transverse distance. Second, the amount of material presumed to be undergoing collapse seemed far too high, at least in comparison with the much smaller total mass of the existing stellar population. New interferometric observations, using a variety of molecular lines, have now shown that the dynamical effects are more localized than previously thought, and also more complicated. The molecules employed have been NH_3 (Zhang and Ho 1995; Ho and Young 1996), along with CS and CH_3CN (Zhang et al. 1998).

The more recent observations show that infall, if present, is confined to a few distinct, compact cores. At these sites, one finds broadening of the

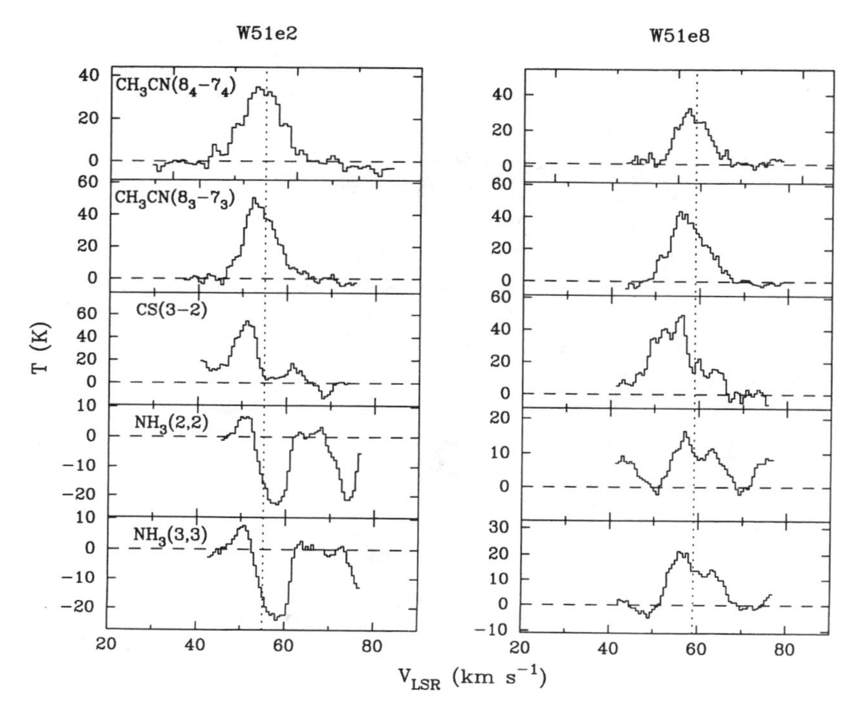

Figure 2. Spectral line profiles toward W51e2 (left) and W51e8 (right). The various molecular transitions are indicated on the left and become more optically thick from top to bottom. Notice how the red wing goes into absorption in W51e2, where there is an embedded, bright H II region.

spectral lines, with this broadening always centered on the exciting star. When resolved with sufficient angular resolution ($< 1''$), position-velocity diagrams show the classic "C" or "O" shapes consistent with radial motion projected along the line of sight. Furthermore, the redshifted portion of the line is seen in absorption when there is an intervening H II region (see Fig. 2). Away from the H II region, both the blue- and redshifted portions are seen in emission and at equal intensity. This behavior suggests that we are observing the front and rear halves of a contracting or collapsing surface. Note finally that those lines exhibiting the broadening are more optically thick and elevated in temperature. This trend is also consistent with infall arising from the warmer and denser cloud interior.

From the resolved radial motion, the total gravitating mass at each site is of order 200 $M_\odot$, within a region of size 0.1 pc. Thus, it appears that both diffuse and stellar matter are important dynamically. Note that the original idea of global collapse was largely based on an absorption feature in HCO^+, redshifted by 8 km s^{-1}. This absorption apparently stems from a foreground screen that is both cool and tenuous. The more recent observations suggest instead that most of the complex is *not* undergoing collapse. The localized motion appears instead to be infall onto individual stars, continuing after the initial formation epoch. Note that the associated ram pressure is large enough that it could, in principle, suppress expansion of each star's H II region. However, a quantitative model of the stellar environment is still lacking.

Other isolated regions within W51 appear to be contracting without any central star. These regions could be at an earlier evolutionary phase. Similar line broadening, both with and without stars, is seen in the OB cluster G10.6–0.64 (Ho and Haschick 1986; Keto et al. 1987, 1988; Ho et al. 1994). Here again, we find redshifted absorption and blueshifted emission, all associated with denser and hotter gas. Further study of clouds at high angular resolution should reveal other examples in the near future.

In the case of low-mass stars, there are also signs of contraction over length scales as small as 100 AU. Studies of ^{13}CO emission indicate both rotation and infall within flattened structures that may represent circumstellar disks (see the chapter by Myers et al., this volume). The difficulty in applying molecular-line techniques to massive stars stems from the angular resolution. A length of 100 AU at a distance of 1 kpc subtends only 0.1″. At this resolution, the minimum brightness detectable in a line is a few times 10^3 K. Probing the kinematics requires that we turn to nonthermal emission, such as masers.

Several recent experiments have succeeded in using Doppler shifts in the 22-GHz maser line of H_2O to probe motion within 100 AU of massive stars. Torrelles et al. (1996, 1997, 1998*a,b*) mapped the masers in W75N, NGC 2071, and Cepheus A, utilizing the VLA in its largest configuration. Here, an angular resolution of 0.08″ was achieved. Each of these systems contains multiple O and B stars, radio continuum jets, and extensive molecular outflows. In each case, Torrelles et al. found a disklike distribution of masers surrounding a B star. The configurations are oriented perpendicular to the axes of the radio jets and centered on the jet sources. Detected motions of the masers indicate some combination of rotation and contraction. These results, while extremely promising, represent but a first step. Within the next few years, very long-baseline interferometry (VLBI) studies will be able to supply proper motions for the maser sources. Combined observations of transverse and radial velocities should greatly improve our understanding of the motion around massive stars.

IV. FORMATION WITHIN CLUSTERS

A. Mass Segregation

Our quantitative, theoretical efforts have thus far been a straightforward extension of those successfully applied to lower-mass objects. The disruptive effect of O and B stars is so great, however, that such an approach may be fundamentally flawed. It may simply be impossible for a star well in excess of 10 $M_\odot$ to continue building up in a steady manner. Could it be that the final mass accumulates much more rapidly, so as to avoid the problems associated with winds and radiation pressure? One can begin to address this issue by looking once more at the environments where O and B stars actually form.

We have emphasized from the start that most massive objects are not born in isolation but within clusters and associations. We also noted that the latter are unbound, freely expanding entities. Careful measurement of the proper motions sometimes allows us to reconstruct the original stellar distribution (Blaauw 1991). For example, the Upper Scorpius association was a football-shaped swarm, some 45 pc in length, about 4.5×10^6 yr ago. This dimension is in accord with those of present-day molecular cloud complexes. Such findings reinforce the belief that the majority of O and B stars originate in such complexes. Moreover, the birth sites plausibly consist of relatively tight, deeply embedded clusters like those of Orion.

Most clusters contain only low-mass stars. Those few that remain gravitationally bound following dispersal of their gas may be examined in detail through the optical spectra of their stars. All such studies find a strong tendency toward *mass segregation* (e.g., Raboud and Mermilliod 1998); that is, the average stellar mass increases toward the system's center, which is also the densest region. This trend is evident, for example, in the Pleiades. Although the entire cluster of some 800 stars has a radius of 10 pc, the brightest members all reside within a much smaller core. In this case, the system age is about 10^8 yr, long enough for dynamical relaxation to have played a significant role. Gravitational scattering gradually leads to energy equipartition, so that heavier, slower-moving stars sink toward the deepest region of the potential well. Meanwhile, they impart higher velocities, and wider orbits, to stars of lesser mass.

The clusters spawning O and B stars are much younger than the Pleiades. Their embedded nature makes a complete stellar census impossible. Nevertheless, there is growing evidence that mass segregation still occurs. Consider, for example, the Monoceros Molecular Cloud. Within this complex lies the populous young cluster NGC 2264. Since the gas here has been partially cleared away, a large fraction of the cluster members are optically visible. Careful mapping of the stellar density reveals *two* concentrations: one surrounding the brightest O star, S Mon, and another associated with HD 47887, again the most massive object in its vicinity.

Several recent studies of Herbig Ae and Be stars have revealed significant trends in clustering and mass segregation. Hillenbrand (1995) first noted a correlation between the mass of such a star and the number of nearby, lower-mass objects. Her finding has been confirmed and extended by Testi et al. (1997, 1998*a,b*), who examined 44 fields around stars ranging in spectral type from A7 to O7. With increasing mass of the central star, the volume density of surrounding objects also rises. Thus, A-type stars are never accompanied by conspicuous groups. Conversely, rich clusters with densities exceeding 10^3 pc^{-3} only appear around stars earlier than B5, corresponding to a mass near 6 $M_\odot$.

Could this trend represent an evolutionary effect? The typical Ae star in these surveys is at least several million years old, so the absence of clustering might conceivably reflect prior dispersal through dynamical relaxation. The recent studies, however, find no correlation between the age of a star and the presence of a cluster; only the mass seems to count. In addition, the clusters surrounding early-type stars all have similar sizes, regardless of their total membership. The typical cluster radius, about 0.2 pc, is reminiscent of molecular cloud cores.

Returning to the case of Orion, we note again that optical spectra are available for about half the stars surrounding the Trapezium. The partial clearing of gas within this Orion Nebula Cluster must be due to the O stars themselves. Placement of the visible members in the HR diagram indicates a cluster age of about 1×10^6 yr (Hillenbrand 1997). In a recent series of N-body simulations of the system, Bonnell and Davies (1998) find that dynamical relaxation is insignificant over this time. Yet mass segregation is undeniably present, as can be seen by the central location of the Trapezium stars themselves. More generally, Figure 6 of Hillenbrand and Hartmann (1998) shows how stars in excess of 5 $M_\odot$ crowd toward the cluster's geometrical center.

The primary question, then, is not how massive stars find their way to the densest, inner region of a cluster. The problem is rather why they preferentially *form* in that location (Zinnecker et al. 1993). Addressing this issue theoretically will require broadening our focus from the mutual interactions of the member stars to the structure and dynamics of their parent cloud medium. Again, the Orion Nebula Cluster illustrates this important point. Here the distribution of stars is not spherically symmetric, as one would expect in a relaxed swarm of point masses. Instead, the contours of projected stellar surface density form concentric ellipses, with an aspect ratio of about 2:1 (Hillenbrand and Hartmann 1998, Fig. 3). The orientation of these ellipses coincides with the filamentary structure evident in CO maps of the nearby molecular gas (Bergin et al. 1996).

One should hardly be surprised, at least in hindsight, that the stellar distribution in a system this young retains a memory of the diffuse medium from which it arose. After all, a similar match holds in much looser systems of low-mass objects such as that of Taurus-Auriga (Gomez et al. 1993). It seems profitable, in any case, to view the birth of massive stars as an event associated with especially dense and heavily populated clusters. Molecular line studies of clouds harboring O and B stars have demonstrated vividly how coherent structures are shredded and dispersed within a few times 10^6 yr (Leisawitz et al. 1989). On the other hand, low-mass stars form over an interval longer by an order of magnitude (Strom et al. 1993). It appears, then, that both the parent cloud and its emerging cluster slowly condense throughout the longer timescale. Only when the central density crosses some threshold does the rapid formation of massive stars begin.

B. Binarity and Stellar Coalescence

If most high-mass stars indeed form in the middle of dense clusters, then such crowded regions might also be expected to produce an elevated number of binary and higher-order multiple systems. Now the origin of binaries in general is still a matter of uncertainty, so the reasoning here cannot be ironclad. Nevertheless, there are several intriguing lines of evidence bearing on this issue.

A major advance in our understanding of early stellar evolution has been the discovery that binarity is the rule, rather than the exception. This fact has long been known with regard to main-sequence objects. According to Duquennoy and Mayor (1991), 57 percent of G dwarfs in the solar neighborhood have at least one companion. The corresponding figure is estimated to be 42 percent for M stars (Fischer and Marcy 1992). Optical and infrared studies of nearby cloud complexes find a similar, or even higher, incidence of multiplicity among T Tauri stars (for a review, see Mathieu 1994). Hartigan et al. (1994) and Brandner and Zinnecker (1997) have examined the ages of several dozen wide T Tauri binaries by placing both components in the theoretical HR diagram. The estimated ages are sufficiently close that the typical binary must form *in situ* rather than through the accidental binding of two stars initially dispersed within the parent cluster.

There is no reason why the same, local formation mechanism should not apply to binaries containing O and B stars. What are the properties of such systems? Let us begin with q, the ratio of secondary to primary mass. In the better studied G-dwarf binaries, the typical q value is well under unity. Indeed, the distribution of secondary masses is consistent with the field star initial mass function (IMF) (Duquennoy and Mayor 1991). For M stars the distribution is much flatter (Fischer and Marcy 1992). Recall that the primary in the latter case has a mass no greater than 0.5 $M_\odot$, so the secondaries represent an even lower-mass population. The flattening of the q distribution could thus reflect the similar feature displayed in the IMF for stars of a few times 0.1 $M_\odot$ (Scalo 1986). In summary, the evidence is that both components of present-day binaries have masses drawn statistically from the same distribution.

Complete surveys of binarity among O stars are unavailable. Nevertheless, it is clear that the secondary masses are not distributed according to the IMF. If they were, there would be only a tiny probability for the companion to be an O- or even a B-type star, yet such doubly massive pairs are not hard to find. Eggleton et al. (1989) have confirmed and quantified the statistical anomaly of this situation. Close examination shows that, in many of the tighter pairs amenable to spectroscopic study, the components are exchanging mass. A notable exception is S Mon, the O7 star in the NGC 2264 cluster. Radial velocity measurements and interferometry reveal that a detached companion is orbiting with a period of 24 yr. The system is both a spectroscopic and astrometric binary, so that one may obtain both masses, which are 35 and 24 $M_\odot$ (Gies et al. 1997).

Within associations, the fraction of O and B stars that are visible binaries appears similar to that of lower-mass systems (McAlister et al. 1993; Mason et al. 1998). Our initial expectation of a higher multiplicity may therefore be mistaken. However, we must bear in mind the severe difficulty of finding solar-type companions to massive objects, considering both the luminosity contrast with the primary and the greater distances involved. There is also a severe lack of data for systems that are too wide for spectroscopic detection, yet too close for the components to be spatially resolved.

It is interesting that the proportion of binaries declines markedly for the field O stars located well outside associations (Mason et al. 1998). Among runaway O and B stars, the fraction is essentially zero. We mentioned earlier the theory for these objects that posits an original close binary. One star leaves the system at high speed after its even more massive companion detonates as a supernova. Detailed evolutionary calculations show that the initially more massive star (the primary) transfers a great deal of mass to the secondary before the primary explodes (for a review, see van den Heuvel 1993). Usually the supernova does not disrupt the binary. Instead, the system's center of mass accelerates from the recoil accompanying ejection of the supernova shell. The binary thereafter consists of the now-massive secondary along with a neutron star or black hole. Such, at least, is the prediction of theory. In fact, no such system has ever been detected.

We are left with the alternative hypothesis: that the runaways originate from clusters as a result of close, dynamical encounters. Such high-speed ejections are actually a well-known effect in stellar dynamics (Chapter 8 of Binney and Tremaine 1987). Most common are three-body encounters, in which a single star approaches a pre-existing binary pair. After a complicated interaction, either the incident star or one of the pair leaves at a speed comparable to the internal orbital velocities in the original binary. Whatever its composition, the pair that remains must tighten as a result of the encounter, as a result of energy conservation.

Leonard and Duncan (1990) have followed these dynamical ejections in N-body simulations of massive clusters. They found that binary-binary encounters play a significant role in this case. More intriguing is the fact that the ejected objects are generally the *least* massive of those involved in the encounter. This rule is also obeyed in clusters of ordinary stars, whenever a broad mass spectrum is present (Van Albada 1968). The implication for runaway O and B stars is that the relevant central region of the parent cluster is severely deficient in low-mass members (Clarke and Pringle 1992). In addition, the original binaries must have been biased toward unit mass ratio, just as the observations indicate.

The emerging picture is that the centers of evolving, dense clusters become efficient factories for the production of both single and multiple massive stars. If we are to reject the traditional notion that such objects form through the collapse of individual cloud cores, then their precursors must be stars already crowded into the region. Could it be that the highest-mass stars form through the coalescence of low- and intermediate-mass cluster members?

One potential difficulty here is the time required for stellar encounters. The radii of stars are so tiny that direct collisions are highly improbable, even in the densest cluster environments. In this regard, Larson (1990) has emphasized that gravitational focusing increases the effective cross section, as does the presence of circumstellar disks (see also Clarke and Pringle 1993). Bonnell et al. (1998) have recently advocated a picture in which the stars competitively sweep up ambient, diffuse gas. These studies indicate that in a region like the Orion Nebula Cluster, with a stellar density of order 10^4 pc^{-3}, one would expect roughly one collision every 10^6 yr. Thus, the timescale *per se* may not be an insurmountable problem.

In the absence of a detailed physical model, the actual outcome of such close encounters is unclear. It is plausible that the two stars first form a binary, whose orbit then decays. However, the N-body studies demonstrate that even a few tight binaries forming at a cluster's center quickly and efficiently disperse the remaining stars (Terlevich 1987). Clarke and Pringle (1993) have pointed out that collisions between stars and their disks would alleviate this problem, since the excess energy would go into disruption of the disks rather than ejection of stars. For the disks to soak up enough energy, they must have masses comparable to those of their parent stars (McDonald and Clarke 1995). Such massive structures are not presently observed, but theory does posit their existence at a stage when the star is still forming out of its molecular cloud core (Stahler et al. 1994).

Our discussion thus indicates a link between the formation of massive stars and the general problem of binary formation. More importantly, these theoretical considerations stress once more the critical role of the cloud environment. If massive objects indeed form through coalescence, then the interacting units are unlikely to be bare stars, or even stars with disks. The relevant clusters are so young that most of the lower-mass objects near their centers are still deeply embedded, whether or not they are accreting protostars. The colliding units may therefore be dense cores containing single stars or binaries. Since the cross section for core interaction is much greater, the collision rate would be interestingly high even in the moderately crowded clusters forming Be stars. Price and Podsiadlowski (1995) have considered the encounters of cloud fragments containing protostars, and hypothesize that one or both stars will be stripped entirely of its envelope. However, gauging the true outcome of a core-core interaction will require more detailed calculations that include the magnetic support of these structures.

Assuming that the embedded stars remain within the merged cores, theory must also confront the matter of how these objects lose enough energy and angular momentum to finally coalesce. An interesting early effort in this regard is that of Boss (1984), who followed the orbital evolution of a binary embedded within an idealized, uniform-density medium. Here, the point masses are subject to gravitational torquing from the nonaxisymmetric background gas. After taking into account the varying phase angle between the binary and its medium, Boss found interesting, quasiperiodic behavior in the orbital motion but no secular loss of angular momentum. The implication is that the phase angle does not vary; that is, the binary itself generates the disturbance that brakes its motion. Pursuing this line of reasoning, Gorti and Bhatt (1996) have invoked dynamical friction with an ambient cloud core. Both these calculations were directed toward the issue of close binary formation, but they also represent promising first steps for the problem at hand.

It is evident that a number of difficult theoretical problems must be confronted before coalescence becomes a fully quantitative model. The important point for now is that understanding massive stars requires a departure from traditional ideas. It is no longer profitable to think of these objects as arising from the same basic mechanism that produces their low-mass counterparts. Massive star formation appears instead to be a critical event in the life of a populous cluster, an event touched off when the central density, in both gas and stars, exceeds some threshold value. This new picture should help provide a conceptual framework as we continue to probe the origin of the still enigmatic stars of highest mass.

REFERENCES

Beech, M., and Mitalas, R. 1994. Formation and evolution of massive stars. *Astrophys. J. Suppl.* 95:517–534.

Bergin, E. A., Snell, R. L., and Goldsmith, P. F. 1996. Density structure in giant molecular cloud cores. *Astrophys. J.* 460:343–358.

Bernasconi, P., and Maeder, A. 1996. About the absence of a proper zero-age main sequence for massive stars. *Astron. Astrophys.* 307:829–839.

Binney, J., and Tremaine, S. 1987. *Galactic Dynamics* (Princeton: Princeton University Press).

Blaauw, A. 1991. OB associations and the fossil record of star formation. In *The Physics of Star Formation and Early Stellar Evolution*, ed. C. J. Lada and N. D. Kylafis (Dordrecht: Kluwer Academic Publishers), p. 125.

Bonnell, I. A., and Davies, M. B. 1998. Mass segregation in young stellar clusters. *Mon. Not. Roy. Astron. Soc.* 295:691–698.

Bonnell, I. A., Bate, M. R., and Zinnecker, H. 1998. The formation of massive stars. *Mon. Not. Roy. Astron. Soc.* 298:93–102.

Boss, A. P. 1984. Angular momentum transfer by gravitational torques and the evolution of binary protostars. *Mon. Not. Roy. Astron. Soc.* 209:543–567.

Brandner, W., and Zinnecker, H. 1997. Physical properties of 90 AU to 250 AU pre-main-sequence binaries. *Astron. Astrophys.* 321:220–228.

Brown, A. G. A., de Geus, E. J., and de Zeeuw, P. T. 1994. The Orion OB1 association. I: Stellar content. *Astron. Astrophys.* 289:101–120.

Burkholder, V., Massey, P., and Morrell, N. 1997. The "mass discrepancy" for massive stars: Tests of models using spectroscopic binaries. *Astrophys. J.* 490:328–342.

Cantó, J. 1980. A stellar wind model for Herbig-Haro objects. *Astron. Astrophys.* 86:327–338.

Caselli, P., Hasegawa, T. I., and Herbst, E. 1993. Chemical differentiation between star-forming regions: The Orion hot core and compact ridge. *Astrophys. J.* 408:548–558.

Cassen, P., and Moosman, A. 1981. On the formation of protostellar disks. *Icarus* 48:353–376.

Cesaroni, R., Churchwell, E., Hofner, P., Walmsley, C. M., and Kurtz, S. 1994. Hot ammonia toward compact H II regions. *Astron. Astrophys.* 288:903–920.

Clarke, C. J., and Pringle, J. E. 1992. The implications of runaway OB stars for high-mass star formation. *Mon. Not. Roy. Astron. Soc.* 255:423–430.

Clarke, C. J., and Pringle, J. E. 1993. Accretion disc response to a stellar fly-by. *Mon. Not. Roy. Astron. Soc.* 261:190–202.

Cox, J. P., and Giuli, R. T. 1968. *Principles of Stellar Evolution and Nucleosynthesis*, Vol. 2 (New York: Gordon and Breach).

Duquennoy, A., and Mayor, M. 1991. Multiplicity among solar-type stars in the solar neighborhood. II: Distribution of the orbital elements in an unbiased sample. *Astron. Astrophys.* 248:485–524.

Eggleton, P. P., Tout, C. A., and Fitchett, M. J. 1989. The distribution of visual binaries with two bright components. *Astrophys. J.* 347:998–1011.

Eisenhauer, F., Quirrenbach, A., Zinnecker, H., and Genzel, R. 1998. Stellar content of the galactic starburst template NGC 3603 from adaptive optics observations. *Astrophys. J.* 498:278–292.

Ezer, D., and Cameron, A. G. W. 1967. A study of solar evolution. *Canadian J. Phys.* 43:1497–1517.

Fischer, D.A., and Marcy, G. W. 1992. Multiplicity among M dwarfs. *Astrophys. J.* 396:178–194.

Figer, D. F., McLean, I. S., and Morris, M. 1995. Two new Wolf-Rayet stars and a luminous blue variable star in the Quintuplet (AFGL 2004) near the Galactic Center. *Astrophys. J. Lett.* 447:L29–L32.

Foster, P. N., and Chevalier, R. A. 1993. Gravitational collapse of an isothermal sphere. *Astrophys. J.* 416:303–311.

Frank, A., and Noriega-Crespo, A. 1994. The collimation of jets and bipolar outflows in young stellar objects: Inertial confinement. *Astron. Astrophys.* 290:643–648.

Garmany, C. D. 1994. OB associations: Massive stars in context. *Publ. Astron. Soc. Pacific* 106:25–37.

Gies, D. R. 1987. The kinematical and binary properties of association and field O stars. *Astrophys. J. Suppl.* 64:545–563.

Gies, R. G., Mason, B. D., Bagnuolo, W. G., Hahula, M. E., Hartkopf, W. I., McAlister, H. A., Thaller, M. L., McKibben, W. P., and Penny, L. R. 1997. The O-type binary 15 Monocerotis nears periastron. *Astrophys. J. Lett.* 475:L49–L52.

Gomez, M., Hartmann, L., Kenyon, S. J., and Hewett, R. 1993. On the spatial distribution of pre-main-sequence stars in Taurus. *Astrophys. J.* 105:1927–1937.

Gorti, U., and Bhatt, H. C. 1996. Orbital decay of protostellar binaries in molecular clouds. *Mon. Not. Roy. Astron. Soc.* 283:566–576.

Hartigan, P., Strom, K. M., and Strom, S. E. 1994. Are wide pre-main-sequence binaries coeval? *Astrophys. J.* 427:961–977.
Hillenbrand, L. A. 1995. Herbig Ae/Be stars: An investigation of molecular environment and associated stellar populations. Ph.D. Thesis, University of Massachusetts.
Hillenbrand, L. A. 1997. On the stellar population and star-forming history of the Orion Nebula cluster. *Astron. J.* 113:1733–1768.
Hillenbrand, L. A., and Hartmann, L. 1998. A preliminary study of the Orion Nebula Cluster structure and dynamics. *Astrophys. J.* 492:540–565.
Hillenbrand, L. A., Massey, P., Strom, S. E., and Merrill, K. M. 1993. NGC6611: A cluster caught in the act. *Astron. J.* 106:1906–1946.
Ho, P. T. P., and Haschick, A. D. 1986. Molecular clouds associated with compact H II regions. III: Spin-up and collapse in the core of G10.6–0.4. *Astrophys. J.* 304:501–514.
Ho, P. T. P., and Young, L. M. 1996. The contracting molecular cores e1 and e2 in W51. *Astrophys. J.* 472:742–754.
Ho, P. T. P., Terebey, S., and Turner, J. L. 1994. The rotating molecular core in G10.6–0.4: Synthesis maps in $^{12}C^{18}O$. *Astrophys. J.* 423:320–325.
Hofner, P., Kurtz, S., Churchwell, E., Walmsley, C. M., and Cesaroni, R. 1996. Massive star formation in the hot, dense cloud core of G9.62+0.19. *Astrophys. J.* 460:359–371.
Hollenbach, D., Johnstone, D., Lizano, S., and Shu, F. 1994. Photoevaporation of disks around massive stars and application to ultracompact H II regions. *Astrophys. J.* 428:654–669.
Ignace, R., Brown, J. C., Richardson, L. L., and Cassinelli, J. P. 1998. Inference of steady stellar wind $v(r)$ laws from optically thin emission lines. II: Occultation effects and the determination of intrinsic stellar properties. *Astron. Astrophys.* 330:253–264.
Jijina, J., and Adams, F. 1996. Infall collapse solutions in the inner limit: Radiation pressure and its effects on star formation. *Astrophys. J.* 462:874–887.
Johnstone, D., Hollenbach, D., and Bally, J. 1998. Photoevaporation of disks and clumps by nearby massive stars: Application to disk destruction in the Orion Nebula. *Astrophys. J.* 499:758–776.
Kahn, F. D. 1974. Cocoons around early-type stars. *Astron. Astrophys.* 37:149–162.
Keto, E. R., Ho, P. T. P., and Haschick, A. D. 1987. Temperature and density structure of the collapsing core of G10.6–0.4. *Astrophys. J.* 318:712–728.
Keto, E. R., Ho, P. T. P., and Haschick, A. D. 1988. The observed structure of the accretion flow around G10.6–0.4. *Astrophys. J.* 324:920–930.
Königl, A. 1982. On the nature of bipolar sources in dense molecular clouds. *Astrophys. J.* 261:115–134.
Kudritzki, R. P., and Hummer, D. G. 1990. Quantitative spectroscopy of hot stars. *Ann. Rev. Astron. Astrophys.* 28:303–345.
Kudritzki, R. P., Hummer, D. G., Pauldrich, A. W. A., Puls, J., Najarro, F., and Imhoff, C. 1992. Radiation-driven winds of hot luminous stars. X: The determination of stellar masses, radii, and distances from terminal velocities and mass-loss rates. *Astron. Astrophys.* 257:655–662.
Lada, E. A., Evans, N. J., and Falgarone, E. 1997. Physical properties of molecular cloud cores in L1630 and implications for star formation. *Astrophys. J.* 488:286–306.
Larson, R. B. 1990. Formation of star clusters. In *Physical Processes in Fragmentation and Star Formation*, ed. R. Capuzzo-Dolcetta et al. (Dordrecht: Kluwer), pp. 389–399.
Larson, R. B., and Starrfield, S. 1971. On the formation of massive stars and the upper limit of stellar masses. *Astron. Astrophys.* 13:190–197.

Leisawitz, D., Bash, F. N., and Thaddeus, P. 1989. A CO survey of regions around 34 open clusters. *Astrophys. J. Suppl. Ser.* 70:731–812.

Leonard, P. J. T., and Duncan, M. J. 1990. Runaway stars from young star clusters containing initial binaries. II: A mass spectrum and a binary energy spectrum. *Astron. J.* 99:608–616.

Li, Z. Y. 1998. Self-similar collapse of magnetized molecular cloud cores with ambipolar diffusion and the "magnetic flux problem" in star formation. *Astrophys. J.* 497:850–858.

Maeder, A., and Conti, P. S. 1994. Massive star populations in nearby galaxies. *Ann. Rev. Astron. Astrophys.* 32:227–276.

Mardones, D., Myers, P. C., Tafalla, M., Wilner, D. J., Bachiller, R., and Garay, G. 1997. A search for infall motions toward nearby young stellar objects. *Astrophys. J.* 489:719–733.

Mason, B. D., Gies, D. R., Hartkopf, W. I., Bagnuolo, W. G., Ten Brummeelaar, T., and McAlister, H. A. 1998. ICCD speckle observations of binary stars. XIX: An astrometric/spectroscopic survey of O stars. *Astron. J.* 115:821–847.

Massey, P. Johnson, K. E., and DeGioia-Eastwood, K. 1995*a*. The initial mass function and massive star evolution in the OB associations of the northern Milky Way. *Astrophys. J.* 454:151–171.

Massey, P., Lang, C. C., DeGioia-Eastwood, K., and Garmany, C. D. 1995*b*. Massive stars in the field and associations of the Magellanic Clouds: The upper mass limit, the initial mass function, and a critical test of main-sequence stellar evolutionary theory. *Astrophys. J.* 438:188–217.

Mathieu, R. D. 1994. Pre-main-sequence binary stars. *Ann. Rev. Astron. Astrophys.* 32:465–530.

McAlister, H. A., Mason, B. D., Hartkopf, W. I., and Shara, M. M. 1993. ICCD speckle observations of binary stars. X: A further survey for duplicity among the bright stars. *Astron. J.* 106:1639–1655.

McCaughrean, M. J., and Stauffer, J. R. 1994. High-resolution near-infrared imaging of the Trapezium: A stellar census. *Astron. J.* 108:1382–1397.

McCaughrean, M. J., Zinnecker, H., Aspin, C., and McLean, I. 1991. Low mass pre-main-sequence clusters in regions of massive star formation. In *Astrophysics with Infrared Arrays*, ed. R. Elston (San Francisco: Astronomical Society of the Pacific), pp. 238–240.

McDonald, J. M., and Clarke, C. J. 1995. The effect of star-disc interactions on the binary mass-ratio distribution. *Mon. Not. Roy. Astron. Soc.* 275:671-684.

Mellema, G., and Frank, A. 1997. Outflow collimation in young stellar objects. *Mon. Not. Roy. Astron. Soc.* 292:795–807.

Miller, G. E., and Scalo, J. M. 1978. On the birthplaces of stars. *Publ. Astron. Soc. Pacific* 90:506–513.

Molinari, S., Testi, L., Brand, J., Cesaroni, R., and Palla, F. 1998. IRAS 23385 +6053: A prototype massive Class 0 protostar. *Astrophys. J.*, in press.

Mooney, T., Sievers, A., Mezger, P. G., Solomon, P. M., Kreysa, E., Haslam, C. G. T., and Lemke, R. 1995. Dust emission from star-forming regions. IV: A study of IR-strong cloud cores. *Astron. Astrophys.* 299:869–884.

Mouschovias, T. C. 1976. Nonhomologous contraction and equilibrium of self-gravitating, magnetic interstellar clouds embedded in an intercloud medium. Star formation. II. *Astrophys. J.* 207:141–158.

Myers, P. C., and Benson, P. J. 1983. Dense cores in dark clouds. II: NH_3 observations and star formation. *Astrophys. J.* 266:309–320.

Nakano, T. 1989. Conditions for the formation of massive stars through nonspherical accretion. *Astrophys. J.* 345:464–471.

Palla, F., and Stahler, S. W. 1990. The birthline for intermediate-mass stars. *Astrophys. J.* 360:47–50.

Palla, F., and Stahler, S. W. 1991. The evolution of intermediate-mass protostars. I: Basic results. *Astrophys. J.* 375:288–299.
Palla, F., and Stahler, S. W. 1992. The evolution of intermediate-mass protostars. II: Influence of the accretion flow. *Astrophys. J.* 392:667–677.
Palla, F., and Stahler, S. W. 1993. The pre-main-sequence evolution of intermediate-mass stars. *Astrophys. J.* 418:414–425.
Palla, F., and Stahler, S. W. 1999. Star formation in the Orion Nebula Cluster. *Astrophys. J.*, preprint.
Price, N. M., and Podsiadlowski, P. 1995. Dynamical interactions between young stellar objects and a collisional model for the origin of the stellar mass spectrum. *Mon. Not. Roy. Astron. Soc.* 273:1041–1068.
Raboud, D., and Mermilliod, J. C. 1998. Evolution of mass segregation in open clusters: Some observational evidences. *Astron. Astrophys.* 333:897–909.
Roberts, M. S. 1957. The number of early-type stars in the galaxy and their relation to galactic clusters and associations. *Pub. Astron. Soc. Pacific* 69:59–64.
Rudolph, A., Welch, W. J., Palmer, P., and Dubrulle, B. 1990. Dynamical collapse of the W51 star-forming region. *Astrophys. J.* 363:528–546.
Scalo, J. 1986. The stellar initial mass function. *Fund. Cosm. Phys.* 11:1–278.
Schaller, G., Schaerer, D., Meynet, G., and Maeder, A. 1992. New grids of stellar models from 0.8 to 120 solar masses at $Z = 0.020$ and $Z = 0.001$. *Astron. Astrophys. Suppl.* 96:269–331.
Stahler, S. W. 1983. The birthline for low-mass stars. *Astrophys. J.* 274:822–829.
Stahler, S. W. 1989. Luminosity jumps in pre-main-sequence stars. *Astrophys. J.* 347:950–958.
Stahler, S. W., Korycansky, D. G., Brothers, M. J., and Touma, J. 1994. The early evolution of protostellar disks. *Astrophys. J.* 431:341–358.
Stahler, S. W., Shu, F. H., and Taam, R. E. 1980. The evolution of protostars. I: Global formulation and results. *Astrophys. J.* 241:637–654.
Strom, K. M, Strom, S. E., and Merrill, K. M. 1993. Infrared luminosity functions for the young stellar population associated with the L1641 molecular cloud. *Astrophys. J.* 412:233–253.
Terlevich, E. 1987. The evolution of n-body open clusters. *Mon. Not. Roy. Astron. Soc.* 224:193–225.
Testi, L., Palla, F., Prusti, T., Natta, A., and Maltagliati, S. 1997. A search for clustering around Herbig Ae/Be stars. *Astron. Astrophys.* 320:159–166.
Testi, L., Palla, F., and Natta, A. 1998*a*. A search for clustering around Herbig Ae/Be stars. II: Atlas of the observed sources. *Astron. Astrophys.*, in press.
Testi, L., Palla, F., and Natta, A. 1998*b*. The onset of cluster formation around Herbig Ae/Be stars. *Astron. Astrophys.*, submitted.
Torrelles, J. M., Gomez, J., Rodriguez, L. F., Curiel, S., Ho, P. T. P., and Garay, G. 1996. The thermal radio jet of Cepheus A HW2 and the water maser distribution at 0.08″ scale (60 AU). *Astrophys. J. Lett.* 457:L107–L111.
Torrelles, J. M., Gomez, J., Rodriguez, L. F., Ho, P. T. P., Curiel, S., and Vazquez, R. 1997. A radio jet-H_2O maser system in W75N(B) at a 200 AU scale: Exploring the evolutionary stages of young stellar objects. *Astrophys. J.* 489:744–752.
Torrelles, J. M., Gomez, J., Rodriguez, L. F., Curiel, S., Anglada, G., and Ho, P. T. P. 1998*a*. Radio continuum-H_2O maser systems in NGC 2071: H_2O masers tracing a jet (IRS 1) and a rotating protoplanetary disk of radius 20 AU (IRS 3). *Astrophys. J.*, in press.
Torrelles, J. M., Gomez, J., Garay, G., Rodriguez, L. F., Curiel, S., Cohen, R. J., and Ho, P. T. P. 1998*b*. Systems with H_2O maser and 1.3 cm continuum in Cepheus A. *Astrophys. J.*, in press.
van Albada, T. S. 1968. Statistical properties of early-type double and multiple stars. *Bull. Astron. Inst. Netherlands* 20:47–88.

van den Heuvel, E. P. J. 1993. Massive close binaries: Observational characteristics. *Sp. Sci. Reviews* 66:309–322.

Wilkin, F. P., and Stahler, S. W. 1998. The confinement and breakout of protostellar winds: Quasi-steady solution. *Astrophys. J.* 502:661–675.

Wolfire, M. G., and Cassinelli, J. P. 1987. Conditions for the formation of massive stars. *Astrophys. J.* 319:850–867.

Yorke, H. W., and Krügel, E. 1977. The dynamical evolution of massive protostellar clouds. *Astron. Astrophys.* 54:183–194.

Zhang, Q., and Ho, P. T. P. 1995. Ammonia maser in a molecular outflow toward W51. *Astrophys. J. Lett.* 450:L63–L66.

Zhang, Q., Ho, P. T. P., and Ohashi, N. 1998. Dynamical collapse in W51 massive cores: CS(3–2) and CH_3CN observations. *Astrophys. J.* 494:636–656.

Zhou, S., Evans, N. J. II, Kompe, C., and Walmsley, C. M. 1993. Evidence for protostellar collapse in B335. *Astrophys. J.* 404:232–246.

Zinnecker, H., McCaughrean, M. J., and Wilking, B. A. 1993. The initial stellar population. In *Protostars and Planets III*, ed. E. H. Levy and J. I. Lunine (Tucson: University of Arizona Press), pp. 429–495.

PART II
Circumstellar Envelopes and Disks

THE STRUCTURE AND EVOLUTION OF ENVELOPES AND DISKS IN YOUNG STELLAR SYSTEMS

LEE G. MUNDY AND LESLIE W. LOONEY
University of Maryland at College Park

and

WILLIAM J. WELCH
University of California at Berkeley

This chapter presents an observational overview of the characteristics and evolution of young stellar systems that are forming singly or in loose clusters and groups. Starting with the nature of fragmentation and multiplicity in these environments, we look at the structure of envelopes and disks around the youngest stars and explore the evolution of these structures from the earliest stage of stellar core formation to the emergence of an optical star. Evidence for the kinematics of disks and large-scale flattened structures is discussed and placed in an overall context.

I. INTRODUCTION

The early evolution of young stellar systems is characterized by dramatic growth and change that take the system from a collapsing prestellar core to a revealed star with 90% or more of its final mass. During this period, the molecular core, which originally extended from several thousand to 10,000 AU, falls inward, feeding a central star and a circumstellar disk. In many cases the core forms two or more stars rather than a single star, and both circumstellar and circumbinary structures may form (see also the chapters by Bodenheimer et al. and Mathieu et al. in this volume).

This chapter focuses on observational studies of the environments of young, forming stars in isolated and loose group environments. By "isolated" we refer to both individual stars and multiple stellar systems that are forming at separations of 15,000 AU or more from other systems. This separation is large enough that neighboring systems do not compete gravitationally for material or interact tidally during formation, unlike the situation that occurs in cluster-mode star formation (Bonnell et al. 1997; see also the chapter by Clarke et al., this volume). This definition of "isolated" is broader than in many previous works, where the term refers only to studies of single-star formation (see, e.g., Shu et al. 1993). The reality is that multiplicity and grouping are common among young stars (Simon et al. 1992; Ghez et al. 1997; Patience et al. 1998). In fact, the median separation of main-sequence binaries ($\sim$30 AU; Duquennoy and Mayor 1991) corresponds to less than an arcsecond in the nearest star-forming clouds; hence, it is often difficult to know whether a given system is single or multiple. A good example of this is the L1551 IRS5 system, which was initially considered a prototypical single-star system but now is argued to be a binary with a projected separation of 45 AU (Looney et al. 1997; Rodriguez et al. 1998).

The earliest stages of star formation, which take place before the formation of a stellar core, are covered in the chapter by André et al. in this volume. This chapter will concentrate on the environments of stellar systems which have formed a protostellar core. We will look at systems in stages of evolution from a deeply embedded protostar to a nearly uncovered young star.

II. FRAGMENTATION AND MULTIPLICITY

Molecular clouds are inhomogeneous on a wide range of scales. Facets of their structure can be observed utilizing molecular line and dust emission (cf. the chapter by Williams et al., this volume), but a complete understanding has proven elusive. The connection between structures within a cloud and the eventual formation of stars in that material is of particular interest here. Detailed studies indicate that the largest structures, containing hundreds to thousands of solar masses, are gravitationally bound as measured by the virial theorem, but most structures with ten solar masses or less are either not gravitationally bound or are bound only in the presence of a high external pressure (Bertoldi and McKee 1992; Williams and Blitz 1998). In the first case, the large structures must fragment into lower-mass cores to form stars; in the second case, it is unclear whether the cores are on the verge of gravitational collapse or whether they are transient unbound structures. Without knowledge of which pieces become stars, it is difficult to predict the distribution and multiplicity of the resulting stellar population.

An alternative approach for studying the star-forming capacity of clouds is to look at the distribution of the youngest embedded stars. These systems, which are still embedded in their natal cores, trace the number and distribution of bound cores that have formed stars over the last 10^5 to 10^6 years. The relative positions and extents of the cores provide information about the fragmentation process and the spatial scales of stellar clustering during formation (Motte et al. 1998; Testi and Sargent 1998). This information is different from that derived in cloud structure studies, where the star-forming capacity of the structures is uncertain, and from the results of studies of the distribution of young optical and near-infrared stars (Gomez et al. 1993; Simon 1997; Nakajima et al. 1998), where one sees the star formation history integrated over more than 10^6 years.

Recent wide-field continuum maps of a portion of the Perseus cloud NGC 1333 (Sandell and Knee 1998) and of the main ρ Ophiuchi cloud (Motte et al. 1998) provide good examples of the structure in active star formation regions. Figure 1 shows the $\lambda = 850\ \mu$m map of the NGC 1333

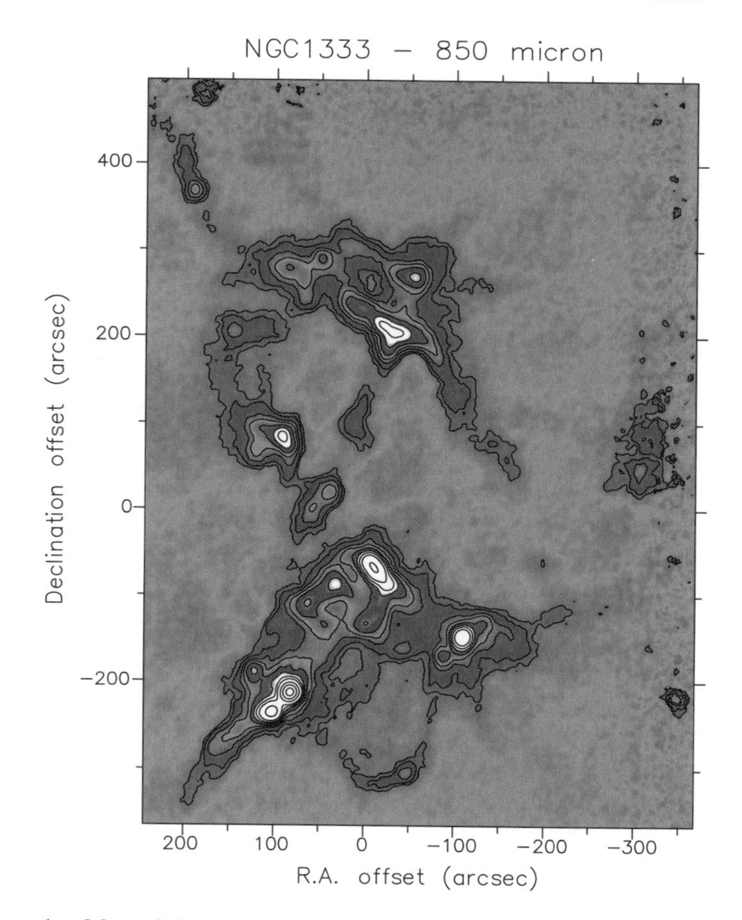

Figure 1. Map of the $\lambda = 850\ \mu$m continuum emission from the NGC 1333 SVS13 region (G. Sandell and L. B. G. Knee, personal communication, 1998). The brightest emission appears as white in the grayscale image. The data were acquired with the Submillimetre Common User Bolometer Array (SCUBA) on the James Clerk Maxwell Telescope (JCMT).

region. The emission at $\lambda = 850\ \mu$m arises primarily from dust. The amount of emission increases with increasing temperature and column density; hence, when young stars are present, bright peaks in the distribution closely trace the locations of young embedded stars. In Fig. 1, there are roughly 18 discrete peaks over an area of roughly 65 square arcminutes. The strongest peaks can be identified with known embedded systems: SVS 13, HH 1–2, NGC 1333 IRAS 2, NGC 1333 IRAS 4, and NGC 1333 IRAS 7. In ρ Ophiuchi, roughly 100 small-scale structures were identified.

Higher-resolution continuum images of many of these systems uncover substructure. For example, Chini et al. (1997) find three separate sources within the HH 1–2 system and the SVS 13 system (see also

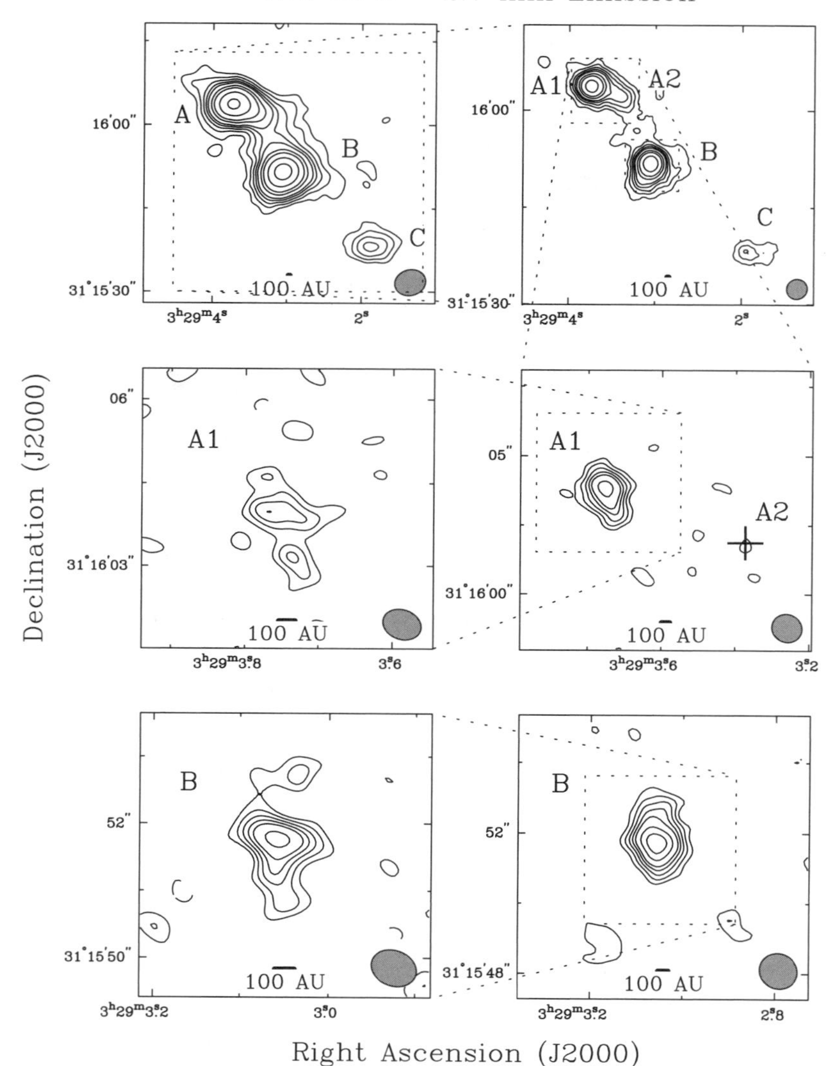

Figure 2. Images of the $\lambda = 2.7$ mm continuum emission from the NGC 1333 SVS13 region from Looney et al. (1998). The four upper panels show images of the system with increasing resolution: 5″ in the upper left, 3.1″ in the upper right, 1.1″ in the middle right, and 0.6″ in the middle left. The bars at the bottom of each panel show 100 AU linear distance. The lower two panels show 1.1″ (right) and 0.6″ (left) images of the SVS13B source. The source designations, A, B, and C, are indicated in the top two panels.

Bachiller et al. 1998); the NGC 1333 IRAS 4 system divides into four sources at arcsecond resolution (Sandell et al. 1991; Lay et al. 1995; Lefloch et al. 1998). Subarcsecond resolution images of several sources in this region confirm this multiplicity. Figures 2 and 3 show interferometric images of the SVS 13 and NGC 1333 IRAS 4 systems; the six panels

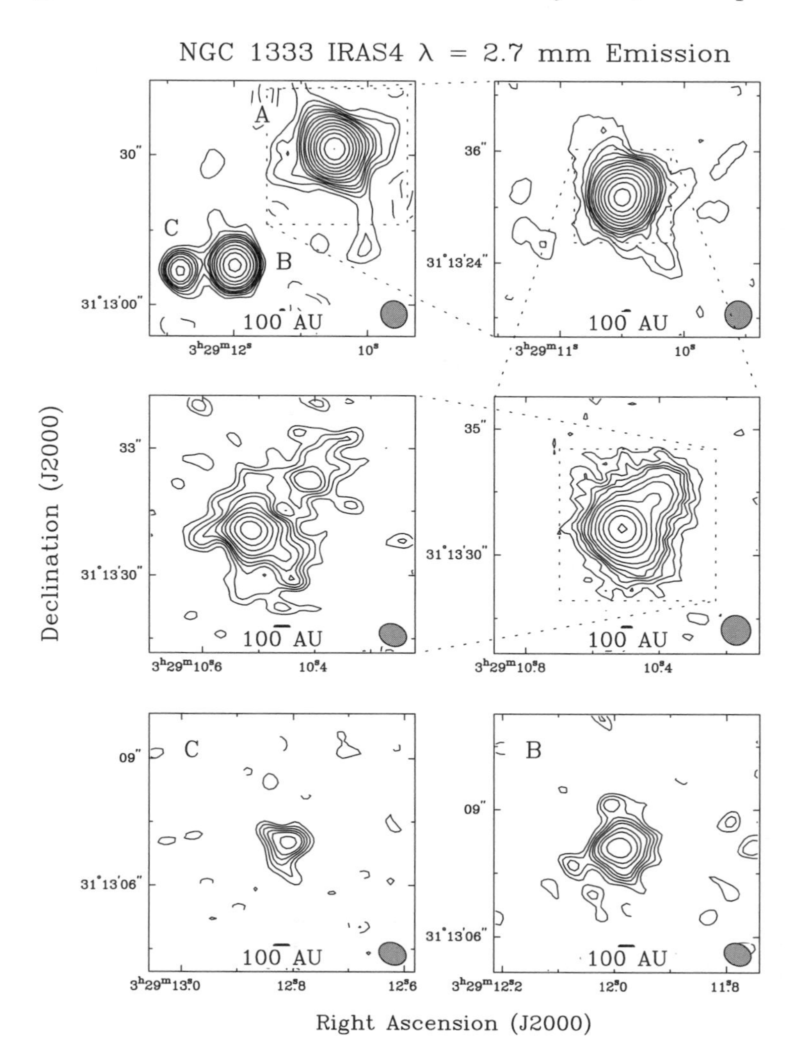

Figure 3. Images of the λ = 2.7 mm continuum emission from the NGC 1333 IRAS 4 region from Looney et al. (1998). The four upper panels show images of the system with increasing resolution: 5.2″ in the upper left, 3.0″ in the upper right, 1.1″ in the middle right, and 0.6″ in the middle left. The bars at the bottom of each panel show 100 AU linear distance. The upper panels are centered on IRAS 4A; the lower two panels show 0.6″-resolution images of IRAS 4B and 4C. The source designations, A, B, and C, are indicated in the top left panel.

within each figure range in resolution from roughly 5″ to 0.6″ (Looney et al. 1998). Multiplicity on a range of scales is clearly present. These images are part of a 0.6″-resolution millimeter-wavelength survey of six deeply embedded systems in Ophiuchus and Perseus (Looney et al. 1998) that found that all systems were multiples on a $\leq$10,000 AU scale, with a total of seventeen sources identified in the six systems. Several systems were binary on 100–700 AU scales. Due to the resolution limit of the survey, little is known about multiplicity on scales $<$100 AU.

Morphologically, three types of multiple systems can be identified in the abovementioned works: independent envelope, common envelope, and common disk systems. The characteristics of the systems are defined by the distribution of the circumstellar material. Independent envelope systems exhibit clearly distinct centers of gravitational concentration with separations of $\geq$6000 AU; the components are within a larger surrounding core of low-density material. Common envelope systems have one primary core of gravitational concentration, which breaks into multiple objects at separations of 150–3000 AU. Common disk systems have separations of $\leq$100 AU and typically have circumbinary disklike distributions of material. Table I summarizes these characteristics.

There are several clear connections between these morphological distinctions and other works. The study of the separation distribution of optical binaries by Larson (1995) found a knee in the distribution at 0.04 pc (8250 AU), which was posited to correlate with the Jeans size. Larson suggested that systems on that scale and larger were formed by fragmentation and separate collapse, exactly the structure found in the independent envelope systems. This scenario of prompt initial fragmentation is not new (see, e.g., Larson 1978; Pringle 1989; Bonnell et al. 1991); it was discussed recently by Bonnell et al. (1997) in the context of small cluster formation. The critical issue is that the collapse is initiated in a system that contains multiple Jeans masses in a weakly condensed configuration; one example of such a system would be a prolate Gaussian distribution with several Jeans masses along the long axis and one Jeans mass across the short axis. This mass concentration before fragmentation is needed to get sufficient mass on the appropriate size scale for systems with ~6000 AU separations. At separations much larger than 6000 AU, independent inside-out or quasistatic collapse from the original cloud (Shu 1977; Fiedler and Mouschovias 1993; Galli and Shu 1993) is viable.

TABLE I
Characteristic Scale of Multiplicity

Property	Scale
Independent envelope	$\geq$6000 AU
Common envelope	150–3000 AU
Common disk	$\leq$100 AU

The common envelope systems can be linked with models for the fragmentation of moderately centrally condensed spherical systems (Burkert and Bodenheimer 1993; Boss 1995, 1997; the chapter by Bodenheimer et al., this volume). In this case, the models find fragmentation in the dense central region within an overall single core. The primary requirement for fragmentation is that the central region have a fairly flat distribution; evidence of this flat region is erased soon after the collapse occurs. Thus, the forming multiple system is embedded within a single centrally condensed core. Finally, the common disk systems are similar to models of high-angular-momentum systems (Artymowicz and Lubow 1994; Bate and Bonnell 1997). The close stellar systems represent the fragmentation of early disks. The distribution of material between circumstellar and circumbinary structures depends on the angular momentum of the infalling material.

III. THE STRUCTURE OF ENVELOPES

The primary parameters that define the characteristics of a circumstellar envelope are size, density structure, temperature structure, mass, and kinematics. These characteristics can, in principle, be observationally determined from the molecular emission associated with the gas and the thermal continuum emission from the dust. There are, however, significant difficulties and uncertainties associated with the various methods available for deriving these parameters.

A. Molecular Emission

Molecular emission arises from a variety of species present in the gas. Differences in molecular structure and abundance make selected species particularly useful in determining specific physical parameters. For example, ^{13}CO or $C^{18}O$ are generally good H_2 column density tracers because their abundance is reasonably constant, they are easily thermalized, and the transitions are generally optically thin. Common molecules with large dipole moments, such as CS, H_2CO, HCN, and HCO^+, are good probes of the gas density and can be useful probes of the column density in some situations. Symmetric or slightly asymmetric top molecules, such as NH_3, H_2CO, and CH_3CN, provide measures of the gas temperature. The molecular line shapes yield information about the kinematics of the gas; the kinematics of the envelope is the focus of the chapter by Myers et al., this volume. In principle, by mapping the emission from several transitions in representative species from each of these key groups, one can derive a complete picture of the envelope structure (e.g., Fuller et al. 1995; Ladd et al. 1998).

Two examples of what can be done utilizing multiple molecular probes are provided by detailed work on B335 (Zhou et al. 1993; Choi et al. 1995) and IRAS 16293−2422 (Zhou 1995; Narayanan et al. 1998). B335 is an isolated globule forming a low-mass star, which is argued to be currently undergoing collapse. The authors fitted multiple transitions of CS and H_2CO to the inside-out collapse model for the envelope (Shu 1977; Shu et al. 1993); they found that the $r^{-1.5}$ density distribution within the infall radius, and r^{-2} density profile beyond that radius, produced good fits to their line profiles and maps for an infall radius of 0.03 pc. IRAS 16293−2422 is a deeply embedded young binary system with a separation of about 750 AU. Zhou (1995) fits the inside-out collapse model to this system, and Narayanan et al. (1998) fit the rotating collapse solution of Terebey et al. (1984) to transitions of CS and HCO^+. The latter find that the data are best fitted for an infall radius of 0.03 pc.

It is important to note that such model fitting shows that the data are consistent with the model, but it does not prove that the model is unique. In a survey of embedded low-mass young stars, Hogerheijde et al. (1997) found that the HCO^+ emission from the envelopes was well described with either the inside-out collapse or by a simple power law density distribution with a power law index of 1 to 3.

B. Continuum Emission

Observations of the continuum from millimeter through mid-infrared wavelengths provide information about the dust distribution and temperature. The millimeter and submillimeter wavelengths are particularly good for determining the column density, because the emission is linearly dependent on dust temperature. Because of the exponential temperature dependence in the Planck function, the dust temperature is best determined from continuum observations from 450 μm to the mid-infrared wavelengths. Unfortunately, measurements of the column density and temperature structure are not decoupled from each other, so maps at several wavelengths spaced across the millimeter to mid-infrared regime and flux measurements covering the broad spectral energy distribution are essential. Physical density must be inferred for an assumed envelope geometry.

The simplest approach for modeling the dust emission is to assume a power law envelope (e.g., Adams et al. 1987; Walker et al. 1990; Adams 1991; Yun and Clemens 1991). In a power law envelope, the density and temperature are assumed to follow a power law dependence on radius: $\rho \propto r^{-p}$, and $T \propto r^{-q}$, respectively; the dust opacity is assumed to follow a power law in frequency: $\kappa_\nu \propto \nu^\beta$. At long wavelengths, where the Rayleigh-Jeans limit applies, the dust emission is then proportional to $\nu^{2+\beta} r^{-(p+q)}$. Modifications to the density power law have also been used, such as assuming the density distribution from Terebey et al. (1984). Such simplifying assumptions are very useful in situations where limited data are available or little detail is known about the source structure.

When a large body of data is available on an object, a full radiative transfer treatment is more appropriate and yields more realistic models. In such treatments, grain temperatures for a distribution of grain sizes are calculated assuming a central heating source (Rowan-Robinson 1980; Wolfire and Cassinelli 1986; Butner et al. 1994). In these models, the temperature profile in the inner regions of the envelope rises more steeply than $r^{-0.4}$ and is not well represented by a single power law. In the outer parts of the envelope, the temperature profile is reasonably approximated by a power law, $r^{-0.4}$, until external heating from the ambient radiation field dominates. For low-luminosity sources, external heating can be important over a significant portion of the core (Motte 1998). In addition to solving for dust temperatures, the radiative transfer models allow consideration of more specific source geometry. For example, Men'shchikov and Henning (1997) model the L1551 IRS5 system utilizing a Gaussian core and power law density envelope with outflow cavities. With this multicomponent model they can fit observations including far-infrared maps, the broad spectral energy distribution, interferometer visibility data, scattered-light images, and dust polarization measurements.

An approach that measures the spatial structure more directly is to fit the visibilities measured by interferometers (Keene and Masson 1990; Lay et al. 1994; Looney et al. 1998). Interferometers intrinsically measure the Fourier transform of the sky brightness. The Fourier transform of the emission from an optically thin power law envelope, $F(r) \propto r^{-(p+q)}$, is the visibility amplitude, $V(d) \propto d^{(p+q-3)}$, where d is the (u,v) distance corresponding to the projected interferometer baseline length (Welch et al. 1999). This expression is valid for $1.5 < p + q < 3$, which is a reasonable range for expected parameters (Foster and Chevalier 1993; Wilner et al. 1995). Thus, the visibility is a power law in (u,v) distance. Different (u,v) distances trace different spatial scales in an object. Figure 4 shows

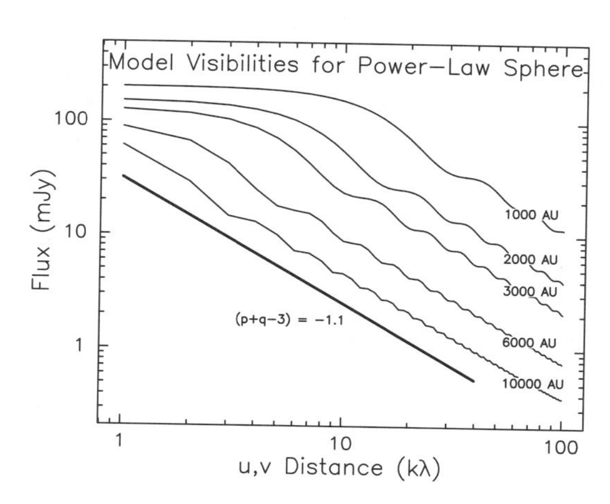

Figure 4. Plots of model visibilities for power law envelopes with different outer cutoff radii. The curves are labeled with the outer radius in AU. The lower thick line is the analytic request for an infinite power law distribution, as explained in the text. The envelope is optically thin.

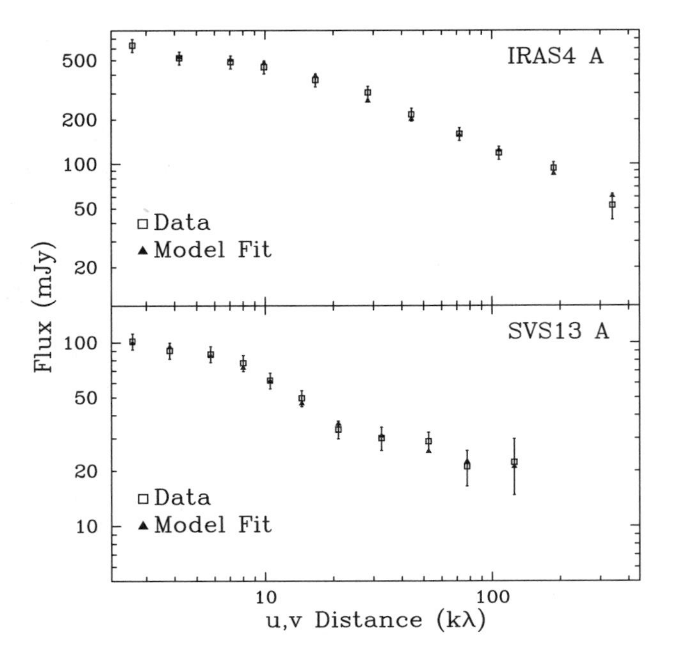

Figure 5. Plots of annular averaged (u, v) data and fits for the NGC 1333 IRAS 4A and NGC 1333 SVS 13A systems from Looney et al. (1998). The model parameters for the fit to IRAS 4A are density power law index = 1.8, outer radius = 2000 AU, point source flux = 22mJy, and envelope mass = 4.2 $M_{\odot}$. The model parameters for the fit to SVS 13A are density power law index = 1.6, outer radius = 4000 AU, point source flux = 18 mJy, and envelope mass = 0.8 $M_{\odot}$.

an example of model visibility amplitudes for an envelope with different outer radius cutoffs. The outer radius cutoffs cause the curves to flatten at smaller (u,v) distances, corresponding to bigger spatial scales, and the abrupt outer edge creates the oscillations with (u,v) distance. Because of these effects, more detailed modeling is valuable in deriving $p + q$.

Figure 5 shows the data and fits for two young embedded systems (Looney et al. 1998). In these fits to the real data, a central compact source is included, to represent a circumstellar disk embedded within the envelope, and the dust temperature is solved self-consistently. The best-fit density power law indexes for these two systems are in the range 1.2–2.0. The masses in the envelopes are 4.2 and 0.8 $M_{\odot}$ for IRAS 4A and SVS13A, respectively. If the fitted point sources are assumed to be HL Tauri type disks at the distance of Perseus, they correspond to masses of 0.08 to 0.15 $M_{\odot}$.

C. Difficulties and Broad Results

Methods for determining the structure of envelopes with embedded young stars are not as simple as they first appear. Often significant cross-correlations between parameters in the models limit the ability to derive quantities. For example, density and temperature have a significant cross-correlation in molecular excitation calculations, and radial density law and outer radius cutoff are intertwined in dust continuum modeling. In addition, observations have limitations due to optical depth in molecular lines and the presence of foreground or background structures. Continuum observations with both single dishes and interferometers are insensitive to large-scale structures due to beam switching in the single-dish case and lack of very short spacings in the interferometer data.

More serious difficulties arise, however, because sources are not as simple as the model assumptions. Source geometry is a significant issue. Nearly all young embedded stars have outflows, which create cavities in the envelope and walls at the envelope-outflow boundary; one or more circumstellar disks may be present within the envelope; circumbinary structures may surround systems. These geometrical considerations affect any method for deriving broad envelope structure and require extensive auxiliary information to constrain models. Molecular line tracers are subject to abundance variations driven by outflow shocks, gas heating, grain mantle evaporation, and depletions of gas-phase molecules onto grains. These abundance variations can be factors of 10 or greater, causing selected shocked gas to light up or regions of molecular depletion to disappear completely (van Dishoeck and Blake, 1998). The use of dust emission to probe envelope structure may suffer related difficulties, because grain properties can change as a function of position in the envelope due to grain mantle accumulation and grain-grain coagulation.

All these complications limit the derivation of definitive envelope properties. To first order, $\rho \propto r^{-1.5}$ to r^{-2} often provides a reasonable fit to most data, but it is difficult to distinguish between $r^{-1.5}$ to r^{-2} density profiles or to establish the radius at which $r^{-1.5}$ infall envelopes switch over to r^{-2} as expected in most dynamical models (Foster and Chevalier 1993; Shu et al. 1993). There are indications that some systems may have r^{-1} density gradients in their envelopes (Chandler et al. 1998), but more detailed studies of these systems are needed. Temperature gradients derived from self-consistent radiative transfer models yield reasonable matches to the observed spectral energy distribution. Continued detailed observations and modeling are needed to improve our understanding of envelope structure further.

IV. EARLY DISK AND ENVELOPE EVOLUTION

The evolution of the mass and size of the envelope and disk is broadly understood but only modestly constrained in detail by observations to date. At the simplest level, the envelope starts as the dominant mass component, and its mass must decrease with time in order to produce eventually an optically visible star. Dynamical models of envelope collapse (Larson 1969;

Penston 1969; Shu 1977; Hunter 1977; Foster and Chevalier 1993; Shu et al. 1993) all suggest that the envelope starts with $\rho \propto r^{-2}$ over most (or all) of its extent. As dynamical collapse of the central region progresses, a free-fall density profile, $\rho \propto r^{-1.5}$, develops and moves out into the envelope. For the inside-out model collapse of Shu and collaborators, the density at fixed radius within the infall region evolves as $\rho(\text{fixed } r) \propto (t)^{-1/2}$, where t is time. From an observational viewpoint, this means that the mass within a given beam evolves roughly as $(\text{time})^{-1/2}$, once the free-fall region encompasses the beam. The mass evolution in other models, though not so simple, follows a similar trend. Of course, outflows also clear out envelope material, but their importance to early envelope evolution is dependent on the collimation of the stellar outflow.

The expected evolution of a disk is less well understood than for an envelope. The centrifugal radius (the radius at which centrifugal support from rotation in the parent core becomes important) sets a characteristic disk size. However, the size and mass of a disk should be initially small and increase with time, as progressively higher-angular-momentum material falls into the center of the system and as angular momentum from earlier infall material is redistributed (Terebey et al. 1984; Ruden and Lin 1986; Lin and Pringle 1990). The predicted sizes for disks are in the range of tens to a few hundred AU, but the detailed surface density distribution and size depend on the angular momentum content of the infall material, the viscosity of the material within the disk, and the nature of the processes that control angular momentum redistribution within the disk (Cassen and Moosman 1981; Stahler et al. 1994). None of these are well enough understood to make definitive disk models.

The data to date are sufficient to illustrate broad trends in the evolution. Dividing sources according to the class designation (Adams et al. 1987; André et al. 1993), there is a clear trend of decreasing envelope mass with increasing class numeral (André and Montmerle 1994; Ohashi et al. 1997*a,b*; Hogerheijde et al. 1997; Ladd et al. 1998). This can be roughly represented as

Class 0: $M_{\rm env} \sim 0.2$–$3\ {\rm M}_\odot$
Class I: $M_{\rm env} \sim 0.02$–$0.3\ {\rm M}_\odot$
Class II: $M_{\rm env} \leq 0.03\ {\rm M}_\odot$

Although the class system is a rough evolutionary sequence, it does not provide a pure comparison between envelope mass and age, because the definitions of the classes refer in part to the amount of emission from the envelope. It is, however, the best that can be done until independent age estimates can be made for embedded systems.

Typical disk masses can likewise be estimated for different classes. The masses are best known for Class II sources, where the typical value is $\sim 0.02\ {\rm M}_\odot$ (Beckwith et al. 1990; André and Montmerle 1994; Osterloh and Beckwith 1995). Less data exist for Class I and Class I/II sources, but typical disk masses appear to be similar to those of Class II sources (Terebey et al. 1993; Osterloh and Beckwith 1995). Disk masses are poorly determined for Class 0 sources because of the difficulty of separating the disk and envelope components, but it appears that their disks are not significantly more massive than for Class I and II sources (Looney et al. 1998). The disks in Class 0 sources are typically $\leq 10\%$ as massive as their envelopes. Based on the Class II sources, which have the most measurements, there is a large range in disk mass within a class. This range may be associated with age, but it more likely depends on several additional factors, including the angular momentum content of the parent core and multiplicity of the system.

These broad estimates for envelope and disk masses support the expected trend: The circumstellar mass in Class 0 and I systems is dominated by the envelope; Class I to II transition objects have comparably massive envelopes and disks; and the disk dominates the circumstellar mass in Class II systems. This trend is also supported by detailed studies of individual systems (Terebey et al. 1993; Looney et al. 1998). Unfortunately, the time evolution of the disk mass is not well constrained by the observations. There does not appear to be an embedded phase during which low-mass stars have a massive disk (or at least that phase must be very short-lived), but it is not yet clear whether disk mass builds up monotonically throughout the lifetime of the envelope. Scenarios in which the disk mass varies nonmonotonically with time because of accretion events are not ruled out (Bell and Lin 1994; Hartmann and Kenyon 1996).

Measurements of disk and envelope sizes are problematic due to the nature of power law distributions. Observations are now beginning to resolve circumstellar disks in nearby young systems (Lay et al. 1994, 1997; Mundy et al. 1996; Wilner et al. 1996; chapter by Wilner and Lay, this volume); however, the estimated outer radius of the disk depends on the steepness of the power law surface density distribution. Typical values are from 70 to 160 AU radius, with the outer radius poorly constrained for steeper surface density profiles. The definition of the outer radius is further complicated by molecular observations, which typically find disk sizes of 500 to 800 AU (see section V). Envelope outer radii are also poorly defined because of the power law nature of the density profile, but some systems may show discrete outer boundaries (Motte et al. 1998). A systematic approach to quantifying disk and envelope sizes is essential to make progress on understanding the evolution of size.

More high-resolution surveys are needed to establish better the properties of disks and envelopes, and how they evolve. Differences in the properties of single vs. multiple systems are largely unknown for Class 0 and Class I sources because it has been difficult to identify which systems are multiple. Disk properties of Class II sources can be affected by the multiplicity of a system (Beckwith et al. 1990; Jensen et al. 1994, 1996; Osterloh and Beckwith 1995; chapter by Mathieu et al., this volume), but it is unclear whether disk, and perhaps envelope, properties of multiple systems are different during earlier evolutionary stages.

V. DISK KINEMATICS

Observational studies of disk kinematics have advanced significantly in the past five years, primarily due to increases in the resolution and sensitivity of the millimeter wavelength arrays. The first solid observations of Keplerian rotation in a circumstellar disk were made for the HL Tauri system (Sargent and Beckwith 1987, 1991). This result confirmed the simple theoretical expectation that stable circumstellar disks, unless significant in mass compared to the star, should exhibit nearly Keplerian rotation. Deviations from Keplerian motion are expected because of the transport of material inward onto the star, but within a rotationally supported disk those velocities should be a small fraction of the basic orbital speed unless the disk is dynamically unstable.

There are, however, additional velocity fields present in the inner few hundred AU of a stellar system. The HL Tauri system illustrates these complications. Subsequent observations of the HL Tauri system by Hayashi et al. (1993) found that the velocities were predominantly inflow motions rather than rotation. They fitted a rotation speed of 0.2 km s^{-1} at 700 AU and an infall motion of 1 km s^{-1}, consistent with the outer disk dynamically accreting onto the inner disk and star. The key distinction between the velocity fields arising from rotation and infall is the orientation of the velocity gradient: For rotation the strongest velocity gradient is along the major axis, whereas for infall it is along the minor axis of the projected disk. Clear discussions of the kinematics of rotation and infall are given by Beckwith and Sargent (1993) and Ohashi et al. (1997*b*). The stellar outflow is another source of kinematics that must be considered. Cabrit et al. (1996) argue that the velocity field seen in their images of the HL Tauri system is likely a combination of infall and outflow motion, where the outflow motion arises from material that has been entrained in the stellar jet. Unfortunately, the outflow motions of entrained material are not necessarily as systematic as rotational motions and hence are more difficult to disentangle. Thus, in general, it is necessary to consider rotational, infall, and outflow motions as possibilities. Surveys of the kinematics of T Tauri systems indicate that rotation generally dominates, but a wide variety of velocity fields are seen (Koerner and Sargent 1995; D. W. Koerner and A. I. Sargent, personal communication, 1998).

The best Keplerian velocity curves have been found for optical T Tauri star (Class II) systems, where the lack of a significant envelope minimizes the material involved in infall or outflow activity. Three good examples of Keplerian rotation are provided by the recent observations of the DM Tau, GM Aur, and LkCa 15 systems. The DM Tau system is well fitted by pure Keplerian rotation ($V \propto r^{-0.5\pm0.1}$), with an outer disk radius of 525 AU and a stellar mass of 0.6–0.85 $M_\odot$ (Dutrey et al. 1998). The DM Tau ve-

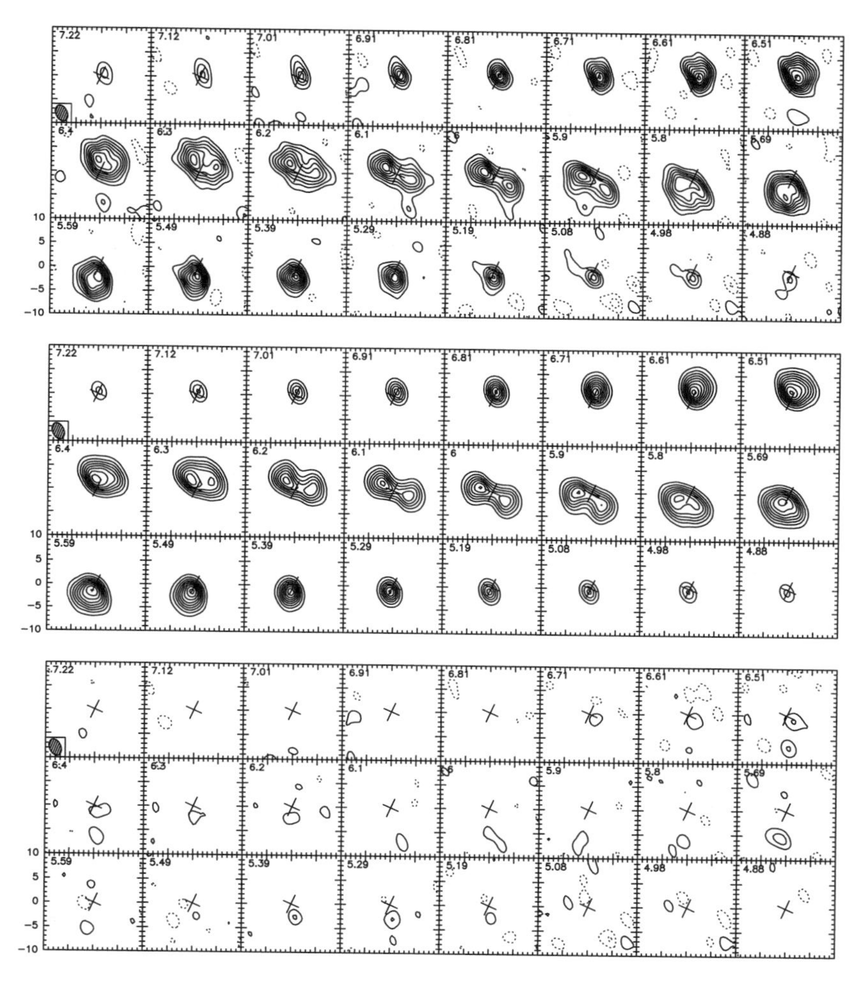

Figure 6. Images of the CO $J = 2$–1 emission, model, and residuals for the DM Tauri system from Guilloteau and Dutrey (1998). The upper set of panels shows the observed emission, with each panel labeled with velocity. The middle panels display the emission from the best-fitted model. The lower panels display the differences between the observations and the model. The angular resolution of the observations is $3.5 \times 2.4''$.

locity field, shown in Fig. 6, is well represented by Keplerian rotation with an outer disk radius of 850 AU and a central stellar mass of 0.4–0.6 $M_\odot$ (Guilloteau and Dutrey 1998). The LkCa 15 velocity field, shown in Fig. 7, is consistent with Keplerian rotation with an outer radius of 220 AU and a central stellar mass of 0.7 $M_\odot$ (A. I. Sargent and D. W. Koerner, personal communication, 1998). As displayed in Figs. 6 and 7, these papers fit velocity channel images, so they are able to set limits on other non-Keplerian motions. For DM Tau, Guilloteau and Dutrey (1998) find that the residual turbulent component of the velocity field is ~0.1 km sec^{-1}, or

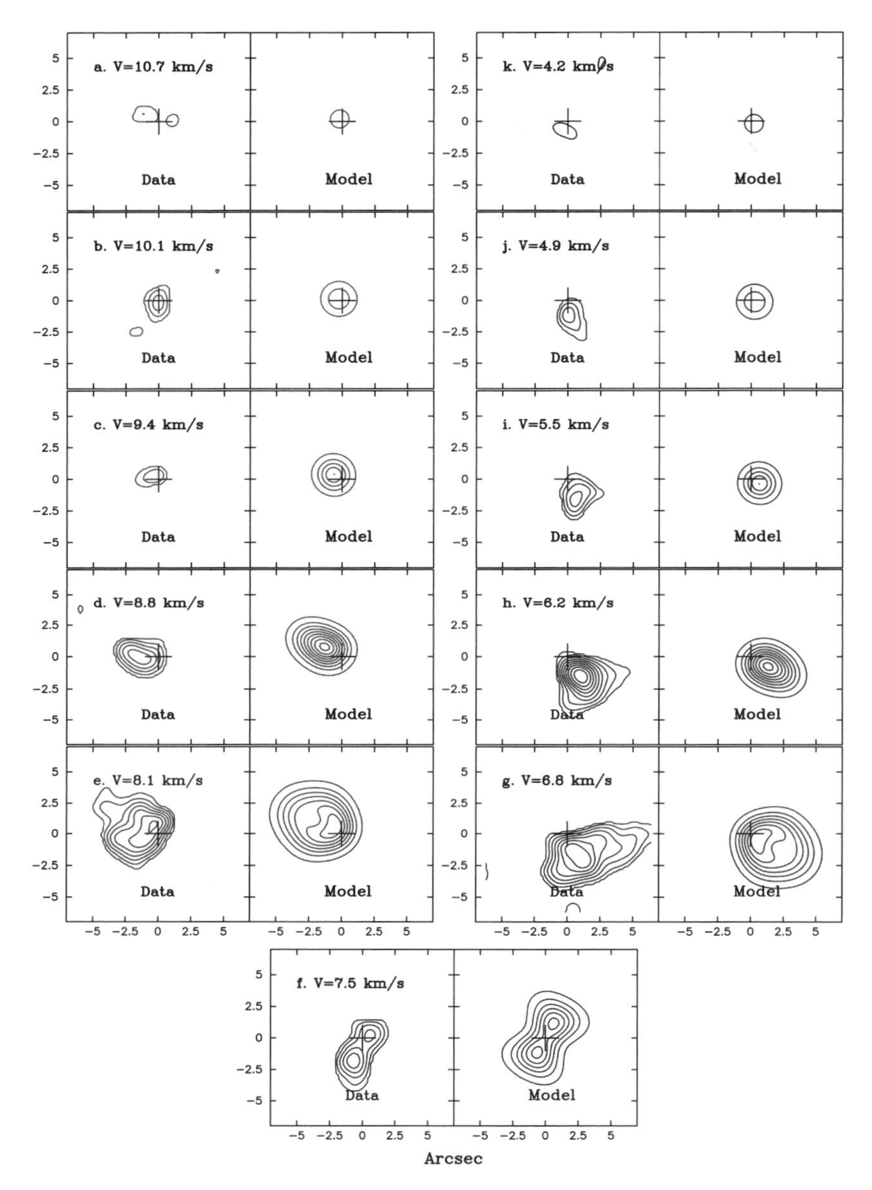

Figure 7. Images of the CO emission and models for the LkCa 15 system from Sargent and Koerner (in preparation). For each pair of panels, the observations are on the left and the model on the right. The panels are labeled with velocity.

0.2–0.3 times the expected thermal width. All these disks are rotationally supported over most, if not all, of their extent.

Systems with even modest envelopes show much more complicated velocity structure and morphology. Observations of three IRAS sources embedded in the Taurus cloud (Ohashi et al. 1997*a,b*) show thick disk-like structures with sizes (diameters) of 1400 to 2200 AU that are oriented perpendicular to the outflow axes. In two of the sources, the velocity field is argued to be a combination of rotation and infall with infall dominant; in the third case, the motion is consistent with rotation. One of these sources, L1527 (IRAS 04368+2557), shows a central "X" structure, which is likely associated with the cavity walls of the outflow. The L1551 IRS5 system also has a central elongated structure 2400 AU in size that is dominated by infall motions, with lesser rotational motion (Momose et al. 1998); it is argued that the flattened envelope is infalling at a diluted free-fall speed (due to pressure or magnetic support) into a rotationally supported disk inside the centrifugal radius. A more deeply embedded (Class 0) system studied by Gueth et al. (1997), L1157, shows weak evidence of rotation in the inner-500-AU flattened core and evidence of infall and outflow interaction in the surrounding envelope. The coherent progression of structures seen in these systems allows a simple picture of infall and rotation. On the 700- to 1500-AU radial scale, the envelope falls into a flattened, thick disk. Material within this thick disk is primarily infalling but has a modest rotational component. The molecular disk on scales from 200 to 800 AU is composed of infalling material with high specific angular momentum and material that has acquired high angular momentum from transport mechanisms within the disk. Most of the infalling material falls onto the disk at radii less than 150 AU and moves inward within the disk, corresponding to the compact disk observed in the continuum. This picture qualitatively matches the expectations of collapse models with significant rotational or magnetic support (Terebey et al. 1984; Li and Shu 1997; Galli and Shu 1993; Fiedler and Mouschovias 1993). The different scales of structure are linked: the large-scale molecular disk traces the small fraction of material that has high angular momentum; the small-scale continuum disk traces the bulk of the material. These two sizescales are divided observationally, because continuum observations have difficulty imaging the large-scale disk due to the low surface density, whereas the CO and isotopic CO observations have difficulty seeing the inner 80–100 AU of the disk due to the lines becoming optically thick. The quantitative exploration of the connection between the disks traced by CO and continuum emission is a project for the future.

VI. SUMMARY AND CONCLUSIONS

1. Evidence of core fragmentation is seen on a range of scales. The broad morphology of cores fits with current ideas about prompt fragmentation, fragmentation during collapse, and high-angular-momentum scenarios. We suggest that there are natural scales of multiplicity in young embedded systems: separate envelope systems on scales $\geq$6000 AU, common envelope systems on scales from 150 to 3000 AU, and common disk systems on scales $\leq$100 AU.

2. It is difficult to determine the density structure of envelopes with embedded sources accurately. Radial density dependences of $r^{-1.5}$ to r^{-2} are consistent with the data for most young systems, but other power law indexes can also fit. Better observational measurements of the radial density are needed to test current theoretical models, but this will not be easy. The presence of disks, outflows, and multiple structures on different scales, as well as the possibility of multiplicity, complicate the source geometry. The likelihood of strong molecular abundance variations, and the possibility of significant alterations of grain properties within circumstellar material also affect the integrity of the standard molecular and continuum probes.

3. It is clear that envelope mass decreases in going from Class 0 to Class I to Class II sources, but it is unclear how uniquely and precisely that trend is related to age. Current evidence suggests that disk masses in Class I and classical T Tauri systems are not very different. Disk masses in embedded systems are problematic because of the strong envelope emission; the current best estimate is that their disks are not significantly more massive than those in Class I or Class II sources. The time evolution of envelope and disk masses is still an uncertain area that needs more work.

4. Kinematic studies of molecular disks are revealing rotational and infall motions. In some systems, the infall motion is quite significant; in others, pure Keplerian rotation is seen. Pure Keplerian disks with outer radii of 200–800 AU have been measured in several T Tauri systems. The observations support a simple picture in which these large-scale Keplerian disks seen in molecules are high-angular-momentum extensions of the smaller-scale, more massive disks seen in continuum emission. The flattened structures 1000 to 2000 AU in size in embedded systems appear to be partially supported envelopes that are feeding material onto the interior disk structures. More observational studies are needed to solidify these connections and quantify the properties and evolution of the structures.

Acknowledgments We thank A. Sargent, A. Dutrey, D. Koerner, S. Stahler, and C. Chandler for comments and conversations. We give a special thanks to P. André for his critical reading of the text. We thank A. Sargent, D. Koerner, A. Dutrey, and G. Sandell for providing figures from their recent works. We acknowledge support for this work from NSF grant 96-13716 and NASA Origins grant NAG-54429.

REFERENCES

Adams, F. C. 1991. Asymptotic theory for the spatial distribution of protostellar emission. *Astrophys. J.* 382:544–554.

Adams, F. C., Lada, C. J., and Shu, F. C. 1987. Spectral evolution of young stellar objects. *Astrophys. J.* 312:788–806.

André, P., and Montmerle, T. 1994. From T Tauri stars to protostars: Circumstellar material and young stellar objects in the ρ Ophiuchi cloud. *Astrophys. J.* 420:837–862.

André, P., Ward-Thompson, D., and Barsony, M. 1993. Submillimeter continuum observations of ρ Ophiuchi A: The candidate protostar VLA 1623 and prestellar clumps. *Astrophys. J.* 406:122–141.

Artymowicz, P., and Lubow, S. H. 1994. Dynamics of binary-disk interaction. I: Resonances and disk gap sizes. *Astrophys. J.* 421:651–667.

Bachiller, R., Guilloteau, S., Gueth, F., Tafalla, M., Dutrey, A., Codella, C., and Castets, A. 1998. A molecular jet from SVS 13B near HH 7–11. *Astron. Astrophys.* 339:L49–L52.

Bate, M. R., and Bonnell, I. A. 1997. Accretion during binary star formation. II: Gaseous accretion and disc formation. *Mon. Not. Roy. Astron. Soc.* 285:33–48.

Beckwith, S. V. W., and Sargent, A. I. 1993. Molecular line emission from circumstellar disks. *Astrophys. J.* 402:280–291.

Beckwith, S. V. W., Sargent, A. I., Chini, R. S., and Guesten, R. 1990. A survey of circumstellar disks around young stellar objects. *Astron. J.* 99:924–945.

Bell, K. R., and Lin, D. N. C. 1994. Using FU Orionis outbursts to constrain self-regulated protostellar disk models. *Astrophys. J.* 427:987–1004.

Bertoldi, F., and McKee, C. F. 1992. Pressure-confined clumps in magnetized molecular clouds. *Astrophys. J.* 395:140–157.

Bonnell, I., Martel, H., Bastien, P., Arcoragi, J-P., and Benz, W. 1991. Fragmentation of elongated cylindrical clouds. III: Formation of binary and multiple systems. *Astrophys. J.* 377:553–558.

Bonnell, I. A., Bate, M. R., Clarke, C. J., and Pringle, J. E. 1997. Accretion and the stellar mass spectrum in small clusters. *Mon. Not. Roy. Astron. Soc.* 285:201–208.

Boss, A. 1995. Gravitational collapse and binary protostars. *Rev. Mex. Astron. Astrofis.* 1:165–177.

Boss, A. P. 1997. Collapse and fragmentation of molecular cloud cores. V: Loss of magnetic field support. *Astrophys. J.* 483:309–319.

Burkert, A., and Bodenheimer, P. 1993. Multiple fragmentation in collapsing protostars. *Mon. Not. Roy. Astron. Soc.* 264:798–806.

Butner, H. M., Natta, A., and Evans II, N. J. 1994. Spherical disks: Moving toward a unified source model for L1551. *Astrophys. J.* 420:326–335.

Cabrit, S., Guilloteau, S., André, P., Bertout, C., Montmerle, T., and Schuster, K. 1996. Plateau de Bure observations of HL Tauri: Outflow motions in a remnant circumstellar envelope. *Astron. Astrophys.* 305:527–540.

Cassen, P., and Moosman, A. 1981. On the formation of protostellar disks. *Icarus* 48:353–376.

Chandler, C. J., Barsony, M., and Moore, T. J. T. 1998. The circumstellar envelopes around three protostars in Taurus. *Mon. Not. Roy. Astron. Soc.* 299:789–798.

Chini, R., Reipurth, B., Sievers, A., Ward-Thompson, D., Haslam, C. G. T., Kreysa, E., and Lemke, R. 1997. Cold dust around Herbig-Haro energy sources: Morphology and new protostellar candidates. *Astron. Astrophys.* 325:542–550.

Choi, M., Evans II, N. J., Gregersen, E. M., and Wang, Y. 1995. Modeling line profiles of protostellar collapse in B335 with the Monte Carlo method. *Astrophys. J.* 448:742–747.

Duquennoy, A., and Mayor, M. 1991. Multiplicity among solar type stars in the solar neighborhood. II: Distribution of the orbital elements in an unbiased sample. *Astron. Astrophys.* 248:485–524.

Dutrey, A., Guilloteau, S., Prato, L., Simon, M., Duvert, G., Schuster, K., and Menard, F. 1998. CO study of the GM Aurigae Keplerian disk. *Astron. Astrophys.* 338:L63–L66.
Fiedler, R. A., and Mouschovias, T. C. 1993. Ambipolar diffusion and star formation: Formation and contraction of axisymmetric cloud cores. II: Results. *Astrophys. J.* 415:680–700.
Foster, P. N., and Chevalier, R. A. 1993. Gravitational collapse of an isothermal sphere. *Astrophys. J.* 416:303–311.
Fuller, G. A., Ladd, E. F., Padman, R., Myers, P. C., and Adams, F. C. 1995. The circumstellar molecular core around L1551 IRS 5. *Astrophys. J.* 454:862–871.
Galli, D., and Shu, F. H. 1993. Collapse of magnetized molecular cloud cores. II: Numerical results. *Astrophys. J.* 417:243–258.
Ghez, A., McCarthy, D. W., Patience, J. L., and Beck, T. L. 1997. The multiplicity of pre-main-sequence stars in southern star-forming regions. *Astrophys. J.* 481:378–385.
Gomez, M., Hartmann, L., Kenyon, S. J., and Hewett, R. 1993. On the spatial distribution of pre-main-sequence stars in Taurus. *Astron. J.* 105:1927–1937.
Gueth, F., Guilloteau, S., Dutrey, A., and Bachiller, R. 1997. Structure and kinematics of a protostar: mm-interferometry of L1157. *Astron. Astrophys.* 323:943–952.
Guilloteau, S., and Dutrey, A. 1998. Physical parameters of the Keplerian protoplanetary disk of DM Tauri. *Astron. Astrophys.* 339:467–476.
Hartmann, L., and Kenyon, S. J. 1996. The FU Orionis phenomenon. *Ann. Rev. Astron. Astrophys.* 34:207–240.
Hayashi, M., Ohashi, N., and Miyama, S. M. 1993. A dynamically accreting gas disk around HL Tau. *Astrophys. J. Lett.* 418:L71–L74.
Hogerheijde, M. R., van Dishoeck, E. F., Blake, G. A., and van Langevelde, H. J. 1997. Tracing the envelope around embedded low-mass young stellar objects with HCO^+ and millimeter continuum observations. *Astrophys. J.* 489:293–313.
Hunter, C. 1977. The collapse of unstable isothermal spheres. *Astrophys. J.* 218:834–845.
Jensen, E. L. N., Mathieu, R. D., and Fuller, G. A. 1994. A connection between submillimeter continuum flux and separation in young binaries. *Astrophys. J. Lett.* 429:L29–L32.
Jensen, E. L. N., Mathieu, R. D., and Fuller, G. A. 1996. The connection between submillimeter continuum flux and binary separation in young binaries: Evidence for interaction between stars and disks. *Astrophys. J.* 458:312–326.
Keene, J., and Masson, C. R. 1990. Detection of a 45 AU radius source around L1551-IRS5: A possible accretion disk. *Astrophys. J.* 355:635–644.
Koerner, D. W., and Sargent, A. I. 1995. Imaging the small-scale circumstellar gas around T Tauri stars. *Astron. J.* 109:2138–2145.
Ladd, E. F., Fuller, G. A., and Deane, J. R. 1998. $C^{18}O$ and $C^{17}O$ observations of embedded young stars in the Taurus molecular cloud. I: Integrated intensities and column densities. *Astrophys. J.* 495:871–890.
Larson, R. B. 1969. Numerical calculations of the dynamics of a collapsing protostar. *Mon. Not. Roy. Astron. Soc.* 145:271–295.
Larson, R. B. 1978. Calculations of three-dimensional collapse and fragmentation. *Mon. Not. Roy. Astron. Soc.* 184:69–85.
Larson, R. B. 1995. Star formation in groups. *Mon. Not. Roy. Astron. Soc.* 272:213–220.
Lay, O. P., Carlstrom, J. E., Hills, R. E., and Phillips, T. G. 1994. Protostellar accretion disks resolved with the CSO-JCMT interferometer. *Astrophys. J. Lett.* 434:L75–L78.

Lay, O. P., Carlstrom, J. E., and Hills, R. E. 1995. NGC1333 IRAS 4: Further multiplicity revealed with the CSO-JCMT interferometer. *Astrophys. J. Lett.* 452:L73–L76.

Lay, O. P., Carlstrom, J. E., and Hills, R. E. 1997. Constraints on the HL Tauri protostellar disk from millimeter and submillimeter wave interferometry. *Astrophys. J.* 489:917–927.

Lefloch, B., Castets, A., Cernicharo, J., Langer, W. D., and Zylka, R. 1998. Cores and cavities in NGC 1333. *Astron. Astrophys.* 334:269–279.

Li, Z. Y., and Shu, F. H. 1997. Self-similar collapse of an isopedic isothermal disk. *Astrophys. J.* 475:237–250.

Lin, D. N. C., and Pringle, J. E. 1990. The formation and initial evolution of protostellar disks. *Astrophys. J.* 358:515–524.

Looney, L. W., Mundy, L. G., and Welch, W. J. 1997. High-resolution $\lambda = 2.7$ millimeter observations of L1551 IRS 5. *Astrophys. J. Lett.* 484:L157–L160.

Looney, L. W., Mundy, L. G., and Welch, W. J. 1998. Unveiling the envelope and disk: A sub-arcsecond survey. *Astrophys. J.*, submitted.

Men'shchikov, A. B., and Henning, T. 1997. Radiative transfer in circumstellar disks. *Astron. Astrophys.* 318:879–907.

Momose, M., Ohashi, N., Kawabe, R., Nakano, T., and Hayashi, M. 1998. Aperture synthesis $C^{18}O$ $J = 1$–0 observations of L1551 IRS5: Detailed structure of the infalling envelope. *Astrophys. J.* 504:314–333.

Motte, F. 1998. Structure des coeurs denses proto-stellaires: Etude en continuum millimetrique. Ph.D. thesis. *J. Astron. Francais* 57:82.

Motte, F., André, P., and Neri, R. 1998. The initial conditions of star formation in the ρ Ophiuchi main cloud: Wide-field millimeter continuum mapping. *Astron. Astrophys.* 336:150–172.

Mundy, L. G., Looney, L. W., Erickson, W., Grossman, A., Welch, W. J., Forster, J. R., Wright, M. C. H., Plambeck, R. L., Lugten, J., and Thornton, D. D. 1996. Imaging the HL Tauri disk at $\lambda = 2.7$ millimeters with the BIMA array. *Astrophys. J. Lett.* 464:L169–L172.

Nakajima, Y., Tachihara, K., Hanawa, T., and Nakano, M. 1998. Clustering of pre-main-sequence stars in the Orion, Ophiuchus, Chamaeleon, Vela, and Lupus star-forming regions. *Astrophys. J.* 497:721–735.

Narayanan, G., Walker, C. K., and Buckley, H. D. 1998. *Astrophys. J.* 496:292–310.

Ohashi, N., Masahiko, H., Ho, P. T. P., Momose, M., Tamura, M., Hirano, N., and Sargent, A. I. 1997*a*. Rotation in the protostellar envelopes around IRAS 04169+2702 and IRAS 04365+2535: The size scale for dynamical collapse. *Astrophys. J.* 488:317–329.

Ohashi, N., Hayashi, M., Ho, P. T. P., and Momose, M. 1997*b*. Interferometric imaging of IRAS 04368+2557 in the L1527 molecular cloud core: A dynamically infalling envelope with rotation. *Astrophys. J.* 475:211–223.

Osterloh, M., and Beckwith, S. V. W. 1995. Millimeter-wave continuum measurements of young stars. *Astrophys. J.* 439:288–302.

Patience, J., Ghez, A. M., Reid, I. N., Weinberger, A. J., and Matthews, K. 1998. The multiplicity of the Hyades and its implications for binary star formation and evolution. *Astron. J.* 115:1972–1988.

Penston, M. V. 1969. Dynamics of self-gravitating gaseous spheres. III. *Mon. Not. Roy. Astron. Soc.* 144:425–448.

Pringle, J. E. 1989. On the formation of binary stars. *Mon. Not. Roy. Astron. Soc.* 239:361–370.

Rodriguez, L. F., D'Alessio, P., Wilner, D. J., Ho, P. T. P., Torrelles, J. M., Curiel, S., Gomez, Y., Lizano, S., Pedlar, A., Canto, J., and Raga, A. C. 1998. Compact protoplanetary disks in a binary system in L1551. *Nature* 395:355–357.

Rowan-Robinson, M. 1980. Radiative transfer in dust clouds. I: Hot-centered clouds associated with regions of massive star formation. *Astrophys. J. Suppl.* 44:403–426.

Ruden, S. P., and Lin, D. N. C. 1996. The global evolution of the primordial solar nebula. *Astrophys. J.* 308:883–901.

Sandell, G., Aspin, C., Duncan, W. D., Russell, A. P. G., and Robson, E. I. 1991. NGC1333 IRAS4: A very young, low-luminosity binary system. *Astrophys. J. Lett.* 376:L17–L20.

Sargent, A. I., and Beckwith, S. V. W. 1987. Kinematics of the circumstellar gas of HL Tauri and R Monocerotis. *Astrophys. J.* 323:294–305.

Sargent, A. I., and Beckwith, S. V. W. 1991. The molecular structure around HL Tauri. *Astrophys. J. Lett.* 382:L31–L35.

Shu, F. H. 1977. Self-similar collapse of isothermal spheres and star formation. *Astrophys. J.* 214:488–497.

Shu, F.H., Najita, J., Galli, D., Ostriker, E., and Lizano, S. 1993. The collapse of clouds and the formation and evolution of stars and disks. In *Protostars and Planets III,* ed. E. H. Levy and J. I. Lunine (Tucson: University of Arizona Press), pp. 3–45.

Simon, M. 1997. Clustering of young stars in Taurus, Ophiuchus, and the Orion Trapezium. *Astrophys. J. Lett.* 482:L81–L84.

Simon, M., Chen, W. P., Howell, R. R., Denson, J. A., and Slowik, D. 1992. Multiplicity among the young stars in Taurus. *Astrophys. J.* 384:212–219.

Stahler, S. W., Korycansky, D. G., Brothers, M. J., and Touma, J. 1994. The early evolution of protostellar disks. *Astrophys. J.* 431:341–358.

Terebey, S., Shu, F. H., and Cassen, P. 1984. The collapse of slowly rotating isothermal clouds. *Astrophys. J.* 286:529–551.

Terebey, S., Chandler, C. J., and André, P. 1993. The contribution of disks and envelopes to the millimeter continuum emission from very young low-mass stars. *Astrophys. J.* 414:759–772.

Testi, L., and Sargent, A. I. 1998. Star formation in clusters: A survey of compact millimeter-wave sources in the Serpens core. *Astrophys. J. Lett.* 508:L91–L94.

van Dishoeck, E. F., and Blake, G. A. 1998. Chemical evolution of star-forming regions. *Ann. Rev. Astron. Astrophys.* 36:317–368.

Walker, C. K., Adams, F. C., and Lada, C. J. 1990. 1.3-millimeter continuum observations of cold molecular cloud cores. *Astrophys. J.* 349:515–528.

Welch, W. J., Mundy, L. G., and Looney, L. W. 1999. *Astrophys. J.,* in preparation.

Williams, J. P., and Blitz, L. 1998. A multitransition CO and CS (2–1) comparison of a star-forming and a non-star-forming giant molecular cloud. *Astrophys. J.* 491:657–673.

Wilner, D. J., Welch, W. J., and Forster, J. R. 1995. Sub-arcsecond imaging of W3(OH) at 87.7 GHz. *Astrophys. J. Lett.* 449:L73–L76.

Wilner, D. J., Ho, P. T. P., and Rodriguez, L. F. 1996. Subarcsecond VLA observations of HL Tauri: Imaging the circumstellar disk. *Astrophys. J. Lett.* 470:L117–L121.

Wolfire, M. G., and Cassinelli, J. P. 1986. The temperature structure in accretion flows into massive protostars. *Astrophys. J.* 310:207–221.

Yun, J. L., and Clemens, D. P. 1991. Radial dust density profiles in small molecular clouds. *Astrophys. J.* 381:474–483.

Zhou, S., Evans, N. J., II, Kompe, C., and Walmsley, C. M. 1993. Evidence for protostellar collapse in B335. *Astrophys. J.* 404:232–246.

Zhou, S. 1995. Line formation in collapsing cloud cores with rotation and applications to B335 and IRAS 16293−2422. *Astrophys. J.* 442:685–693.

EVOLUTION OF DISK ACCRETION

NURIA CALVET
Harvard-Smithsonian Center for Astrophysics and Centro de Investigaciones de Astronomía

LEE HARTMANN
Harvard-Smithsonian Center for Astrophysics

and

STEPHEN E. STROM
National Optical Astronomy Observatories

We review the present knowledge of disk accretion in young low-mass stars and, in particular, the mass accretion rate $\dot{M}$ and its evolution with time. The methods used to obtain mass accretion rates from ultraviolet excesses and emission lines are described, and the current best estimates of $\dot{M}$ for classical T Tauri stars and for objects still surrounded by infalling envelopes are given. We argue that the low mass accretion rates of the latter objects require episodes of high mass accretion rate to build the bulk of the star. Similarity solutions for viscous disk evolution suggest that the inner disk mass accretion rates can be self-consistently understood in terms of the disk mass and size if the viscosity parameter $\alpha \sim 10^{-2}$. Close companion stars may accelerate the disk accretion process, resulting in accretion onto the central star in ≤ 1 Myr; this may help explain the number of very young stars that are not currently surrounded by accretion disks (the weak-emission T Tauri stars).

I. INTRODUCTION

The initial angular momenta of star-forming molecular cloud cores must be responsible for the ultimate production of binary (and multiple) stellar systems and circumstellar disks. Unless the protostellar cloud core is very slowly rotating, much or most of the stellar mass is likely to land initially on a disk and must subsequently be accreted from the disk onto the protostellar core. In the early phases of this accretion, the circumstellar disk may be relatively massive in comparison with the central protostar, so gravitational instabilities may be important in angular momentum transport and consequent disk evolution. At later phases, disk evolution is likely to be driven by viscous processes, perhaps limited by condensation of bodies. Stellar magnetic fields may ultimately halt the accretion of material onto the central star. There are substantial uncertainties in our theoretical understanding of these processes, and we must rely on observations for guidance.

In this chapter we review the present knowledge of the disk accretion process around low-mass stars, in particular the rate at which mass is transferred onto the star, $\dot{M}$, and its evolution with time. Knowledge of $\dot{M}$ will help put lower limits on the disk mass at a given age, independent of uncertainties in dust opacities. The value of $\dot{M}$ puts constraints on disk physics, in particular on temperatures and surface densities at a given age, and thus on conditions that obtain during the time when solid bodies agglomerate. We begin with a description of the different methods of determining disk accretion rates and the values of $\dot{M}$ obtained, applying these methods to stars in different groups and environments. We then discuss the implications of the observational results for disk accretion physics and evolution; in particular we investigate whether the data are consistent with simple models of viscous disk evolution.

II. DETERMINATION OF $\dot{M}$

A. Infrared Excesses

When the infrared excess emission in classical T Tauri stars (cTTSs) was recognized as emission from dust at low temperatures distributed in a circumstellar disk, it was thought that the excess energy would be a direct measurement of the accretion luminosity ($L_{\rm acc} = G\dot{M}M_\star/R_\star$, where $M_\star$ and $R_\star$ are the stellar mass and radius) (Rucinski 1985). However, it soon became apparent that their spectral energy distributions (SEDs) did not follow the law $\lambda F_\lambda \propto \lambda^{-4/3}$ expected for standard accretion disks but were much flatter (Rydgren and Zak 1987; Kenyon and Hartmann 1987). It has now been realized that the major agent heating the disks of many cTTSs is irradiation by the central star (Adams and Shu 1986; Kenyon and Hartmann 1987; Calvet et al. 1991, 1992; Chiang and Goldreich 1997; D'Alessio et al. 1998). The optically thick disks of cTTSs are probably "flared"; that is, they have "photospheres" that curve away from the disk midplane. This flaring makes the disk more efficient in absorbing light from the central star; this extra heating in turn helps to increase the disk scale height and thus increases the flaring. Self-consistent calculations (in various approximations) indicate that the flaring is especially important at large radii (Kenyon and Hartmann 1987; Chiang and Goldreich 1997; D'Alessio et al. 1998). This makes it extremely difficult, if not impossible, to say anything about accretion energy release, and thus accretion rates, in outer disk emission.

For many cTTSs it is very difficult to extract quantitative estimates of mass accretion rates even from emission from the inner, flat parts of the disk. The basic reason is that the accretion luminosities of many cTTSs are less than the luminosity the optically thick inner disk produces as a result of absorbing light from the central star.

The effective temperature T of the disk is determined by internal viscous dissipation and external irradiation by the central star; at a given radius r,

$$T^4(r) \sim T_v{}^4 + T_i{}^4 \tag{1}$$

where $T_v = [(3GM_\star \dot{M}/8\pi\sigma r^3)f]^{1/4}$ is the effective temperature that would result from accretion without irradiation, with $f = 1 - \sqrt{(R_\star/r)}$, and $T_i^4 = (2/3\pi)T_\star^4(R_\star/r)^3$ is the effective temperature for the case of irradiation only (the flat disk approximation). In these expressions, we assume that the mass accretion rate through the inner disk is constant and equal to the rate at which mass is transferred onto the star. The mass accretion rate at which $T_i \sim T_v$ is

$$\dot{M}_c \sim 2 \times 10^{-8}\ \mathrm{M_\odot\ yr^{-1}} (T_\star/4000\ \mathrm{K})^4 (R_*/2\ \mathrm{R_\odot})^3 (M_\star/0.5\ \mathrm{M_\odot})^{-1}$$

calculated for typical Taurus cTTS parameters (spectral types K7–M0, age $\sim$ 1 Myr), and $r \sim 3R_\star$. The median value of $\dot{M}$ in cTTSs is $\sim 10^{-8}\ \mathrm{M_\odot\ yr^{-1}}$ (see below), so this implies that irradiation dominates the inner disk heating for a large number of the stars and determines the amount of flux excess. Only for objects with significantly high $\dot{M}$ will the (near-infrared) disk emission depend on the mass accretion rate.

Other considerations affect a disk's near-infrared emission. It was originally thought that cTTS disks extended all the way into the star and that disk material joined the star through a "hot boundary layer," where half of the accretion energy was dissipated, while the other half was emitted in the disk (cf. Bertout 1989). It has now become apparent that for typical values of the stellar magnetic field ($<$1 kG, Basri et al. 1992) and of the mass accretion rate, the inner disk of a cTTS can be disrupted by the magnetic field (Königl 1991 and the chapter by Najita et al., this volume); material falls onto the star along the magnetic field lines, forming a magnetosphere, and merges with the star through an accretion shock at the stellar surface. In support of this model, fluxes and line profiles of the broad (permitted) emission lines in cTTSs are well explained by the magnetospheric flow (Calvet and Hartmann 1992; Hartmann et al. 1994; Edwards et al. 1994; Muzerolle et al. 1998*a*,*b*,*c*; and the chapter by Najita et al., this volume), while the ultraviolet and optical excess fluxes are well accounted for by the accretion shock emission (Calvet and Gullbring 1998).

For truncation radii $R_i \sim$ 3–5 $R_\star$, the maximum temperature in a disk with typical parameters will be ~1200–850 K, so that disk emission drops sharply below ~2–3.5 μm. The calculation of the disk luminosity from the observed excess also depends on the cosine of the inclination angle, which is generally unknown. As an illustration of these points, Meyer et al. (1997) have shown that the so-called "cTTS locus" in *JHK* and *HKL* diagrams (that is, the region populated by cTTSs outside that corresponding to reddened main-sequence stars; see the chapter by Meyer et al., this volume) could be explained in terms of emission by irradiated accretion disks with inner holes of different sizes (but inside corotation radii) and a random distribution of inclinations. However, no one-to-one correlation could be found between the excess and $\dot{M}$. In a sample of stars in Taurus with known reddening and $\dot{M}$, the largest excesses in the locus were produced by stars with the highest $\dot{M}$, but the converse was not true because of the effects of holes and inclination.

Finally, in objects surrounded by infalling dusty envelopes, emission from the dust destruction radius, which peaks at $\sim 2\ \mu$m, may contribute significantly to the *K* band and longward (Calvet et al. 1997), hiding the intrinsic disk emission.

The difficulty of deriving reliable measures of $\dot{M}$ from excess infrared emission requires the use of alternative methods of estimating $\dot{M}$ values. We will consider such methods in the next two sections.

B. Veiling Luminosities

The best evidence for cTTS disk accretion generally comes from the interpretation of the ultraviolet and optical spectra. The photospheric absorption lines are "veiled"; that is, they are less deep than those in standard stars of the same spectral type. This veiling is produced by (mostly) continuum emission from a region hotter than the stellar photosphere. The luminosity of the veiling or excess continuum in cTTSs is typically $\sim$10% of the total stellar luminosity. It is difficult to account for this extra energy with a stellar origin, and impossible in the case of most extreme cTTSs, where the excess continuum luminosity is several times the stellar luminosity. This conclusion is reinforced by the existence of the weak-emission T Tauri stars (wTTSs). The wTTSs have similar masses and ages as the cTTSs in many regions but do not exhibit the UV and optical veiling continuum of cTTSs, showing that the excess emission is not an intrinsic property of young stars. The veiling emission is observed only when there is excess near-infrared emission (Hartigan et al. 1995, and the chapter by Najita et al., this volume), strongly supporting the idea that accretion from a disk is occurring, producing both the infrared excess and the hot continuum, as originally envisaged by Lynden-Bell and Pringle (1974).

In the magnetospheric model, the disk is truncated in the inner regions, which is precisely where an accretion disk would emit most of its energy. The total disk accretion luminosity is expected to be about

$$L_{\rm acc}({\rm disk}) \sim GM_\star \dot{M}/(2R_i) + L_{\rm diss}$$

where $L_{\rm diss}$ is any energy that might be dissipated in the disk by the stellar magnetic field lines (Kenyon et al. 1996). If we assume that the effect of the stellar magnetic field is to reduce the angular momentum of the disk material substantially, so that it starts out nearly at rest at R_i before falling onto the star along the magnetic field lines, then the infalling material should dissipate its energy at the stellar surface at a rate $\sim(GM_\star \dot{M}/R_\star)(1 - R_\star/R_i)$. For disk truncation radii $\sim$3–5 $R_\star$, most of the accretion energy is released at the stellar surface in the hot accretion shock, whose radiation is observed as the veiling continuum (Königl 1991; Calvet and Gullbring 1998).

Accretion luminosities can be obtained from measurements of the veiling of the absorption lines by the following procedure, in which excess and intrinsic photospheric emission components are separated. The observed fluxes are fitted by the scaled, dereddened fluxes of a standard star, assumed to be of the same spectral type as the object star, plus a continuum flux, which produces the observed veiling at each absorption line wavelength. This fit yields the spectrum of the excess continuum and the reddening toward the star. The total excess luminosity is calculated from the measured luminosity using a model to extrapolate the emission to unobserved wavelengths. Finally, the mass accretion rate can be calculated from stellar mass and radius estimated from the location of the star in the HR diagram and comparison with evolutionary tracks.

Accretion luminosities for cTTSs in Taurus have been determined from veiling measurements by a number of authors, including Hartigan et al. (1991), Valenti et al. (1993), and Hartigan et al. (1995). The mass accretion rates derived from these studies range from $\sim10^{-7}$ M$_\odot$ yr^{-1} for Hartigan et al. (1995) to $\sim10^{-8}$ M$_\odot$ yr^{-1} for Valenti et al. (1993). More recently, Gullbring et al. (1998) remeasured accretion rates for a sample of Taurus cTTSs using spectrophotometry covering the blue and Balmer jump region of the spectrum down to the atmospheric cutoff, and found their results to agree with the previous lower estimates. In a sample of 17 cTTSs in Taurus, they found a median value of $\dot{M} \sim 10^{-8}$ M$_\odot$ yr^{-1}, with a factor of $\sim$3 in estimated error.

These differences reflect the cumulative effects of several differences in working assumptions rather than a fundamental difference in approach. Among them we can list the following:

1. Adopted evolutionary tracks.
2. Adopted model of accretion; the magnetosphere model predicts more luminosity per unit mass accretion rate than the old boundary layer model does.
3. Adopted physics; some treatments assumed that a substantial fraction of the emitted accretion energy was absorbed by the star, which seems unlikely (Hartmann et al. 1997).
4. Method used to determine extinction corrections; the photospheres of T Tauri stars show color anomalies relative to standards, which render the determination of reddening uncertain, and in particular, Gullbring et al. (1998) found large color anomalies (probably due to spots on the stellar surface or to unresolved, cooler companion stars) in some

nonaccreting wTTSs often used as templates, which would cause the extinction to be overestimated in all objects.

To provide a powerful tool for determining mass accretion rates for large samples of stars for which short-wavelength spectrophotometry is difficult to obtain, Gullbring et al. (1998) determined the relationship between the accretion luminosity and the excess luminosity in the U photometric band, L_U, for the stars in their sample. Figure 1a shows the excellent relationship between log $L_{\rm acc}$ and log L_U. A least-squares fit to the line yields

$$\log (L_{\rm acc}/{\rm L}_\odot) = 1.04^{+0.04}_{-0.18} \log (L_U/{\rm L}_\odot) + 0.98^{+0.02}_{-0.07} \qquad (2)$$

The spectrophotometric sample from which equation (2) was derived is formed by stars in the K5–M2 range, with most of the stars being K7–M0. The application of this calibration to a wider spectral type/mass range remains to be confirmed, but theoretical models of accretion shock emission from stars of differing mass and radii indicate that for the characteristic energy fluxes found in the accretion columns of cTTSs, the spectrum of the excess emission does not depend on the underlying star, and the proportion of the total excess luminosity in the U band ($\sim$10%) displayed by equation (2) holds (Calvet and Gullbring 1998).

Mass accretion rates have been determined for a larger sample of TTSs in Taurus using equation (2) by Hartmann et al. (1998). The median mass accretion rate is $\sim 10^{-8}$ M$_\odot$ yr^{-1}, similar to that of the spectrophotometric sample, but the errors are larger, given the lack of simultaneity in the photometric measurements and the high degree of variability of cTTSs. Hartmann et al. (1998) also used spectral types and photometry from the compilation of Gauvin and Strom (1992) to estimate mass accretion rates for K5–M3 stars in the Chamaeleon I association. The median mass accretion rate in Chamaeleon I is $\sim 4 \times 10^{-9}$ M$_\odot$ yr^{-1}. A histogram of the mass

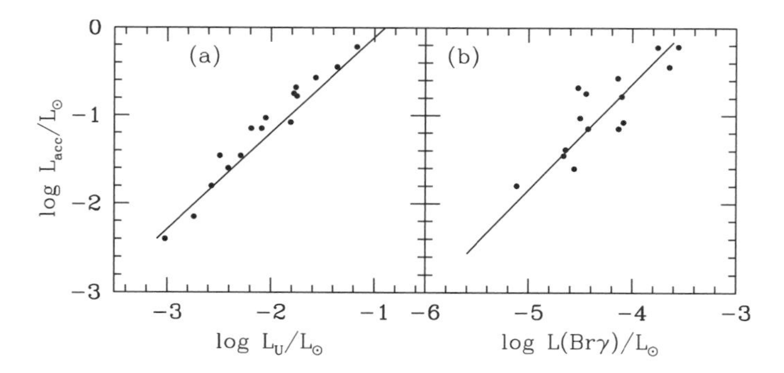

Figure 1. Relationship between accretion luminosity and (a) the excess luminosity in the U band, and (b) the luminosity in Brγ for a sample of cTTSs in Taurus. Data from Gullbring et al. (1998) and Muzerolle et al. (1998*c*).

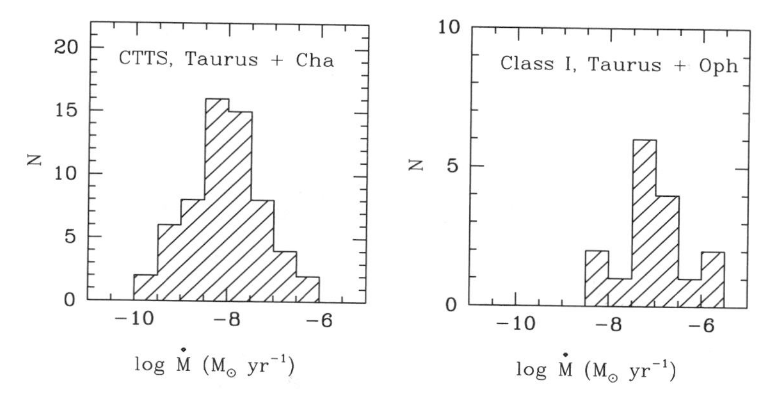

Figure 2. Histogram showing the distribution of mass accretion rates: (a) for T Tauri stars in Taurus and Chamaeleon I, with $\dot{M}$ determined from blue spectra or U magnitudes, and (b) for Class I sources in Taurus and ρ Ophiuchi, determined from Brγ measurements.

accretion rate for TTSs in Taurus and Chamaeleon I determined from the ultraviolet excess is presented in Fig. 2a.

C. Magnetospheric Emission Lines

The measurement of mass accretion rates from the optical and ultraviolet excess fluxes, either spectrophotometrically or from broad-band photometry, is sensitive to reddening corrections. In heavily extincted stars, the UV fluxes are either very uncertain or unobservable; thus, in early stages of evolution such as the infall/protostellar phase, or in very young, dense regions of star formation, such methods cannot be used. For these objects, it is necessary to devise methods to measure $\dot{M}$ in spectral regions that are less affected by intervening dust.

Most emission lines present in the spectra of young objects are thought to be produced in the magnetospheric flow (see the chapter by Najita et al., this volume). Since the material flows through the magnetosphere with a rate similar to $\dot{M}$ in the inner disk, the emission fluxes of the lines formed in this flow are expected to depend on the mass accretion rate. Theoretical models show this to be the case, but other parameters, such as the unknown temperature structure and the characteristic size, play a role too, as well as the optical depth of the line (Muzerolle et al. 1998*a*). For these reasons, empirical correlations between line luminosities and accretion luminosities have been investigated, leaving the interpretation to future theoretical work.

Muzerolle et al. (1998*b*,*c*) have undertaken spectroscopic studies in the red and infrared of the sample of stars with accretion rates determined from spectrophotometric measurements and have found remarkably good correlations between the luminosity of the Ca II triplet 8542, Paβ, and Brγ

lines with accretion luminosities. Figure 1b shows the correlation between L_{acc} and the luminosity in Brγ, as an example. Least-squares fits to the data give

$$\log(L_{acc}/L_\odot) = (0.85 \pm 0.12)\log(L_{Ca\,II\,8542}/L_\odot) + (2.46 \pm 0.46) \quad (3)$$

$$\log(L_{acc}/L_\odot) = (1.03 \pm 0.16)\log(L_{Pa\beta}/L_\odot) + (2.80 \pm 0.58) \quad (4)$$

and

$$\log(L_{acc}/L_\odot) = (1.20 \pm 0.21)\log(L_{Br\gamma}/L_\odot) + (4.16 \pm 0.86) \quad (5)$$

These subsidiary calibrators of the accretion rate provide, for the first time, the means to determine accretion rates in heavily extincted objects.

Muzerolle et al. (1998*c*) used luminosities of the Brγ line for extincted Class II sources in ρ Oph from Greene and Lada (1996) to make the first determination of accretion luminosities for these sources and found that the distribution of accretion luminosities is similar to that in Taurus. Using approximate spectral types from Greene and Meyer (1995), the estimated median mass accretion rate is $\sim 1.5 \times 10^{-8}$ $M_\odot yr^{-1}$, although the span of spectral types covered is wider than in the Taurus sample.

Even more far reaching, Muzerolle et al. (1998*c*) determined accretion luminosities for the deeply embedded Class I objects (sources still surrounded by infalling envelopes) for the first time, using Brγ luminosities. Because typically $A_V \sim 20$–30 for Class I objects, significant extinction is expected at Brγ ($A_K \sim 2$–3), which introduces large uncertainties in the determination of the line luminosities. The light from the central star plus disk and from the envelope itself is both absorbed and scattered by the infalling envelope, in proportions that depend on uncertain parameters such as the geometry of the inner envelope (Calvet et al. 1997). Following Kenyon et al. (1993*b*), Muzerolle et al. (1998*c*) calculate a correction factor $K - K_0$ by assuming that the central object has $J - K$ colors similar to cTTSs and estimating the intrinsic J magnitude from the bolometric luminosity.

Figure 3 shows accretion luminosities estimated by Muzerolle et al. (1998*c*) for a small number of Class I sources in Taurus and in ρ Oph that have data on Brγ (Greene and Lada 1996) plotted against the total luminosities. The accretion luminosities of Class I sources are significantly lower than the system bolometric luminosity; in fact, the mean accretion luminosity is ~10–20% of the mean bolometric luminosity of the sample. This fact implies that the luminosity of Class I is dominated by the stellar component, and it gives a natural explanation to the similarity of the distributions of luminosities of Class I and optically visible T Tauri stars in Taurus (Kenyon et al. 1990).

Determining the mass accretion rate from L_{acc} requires knowledge of $M_\star/R_\star$. We can obtain this ratio assuming that the stars are still on the

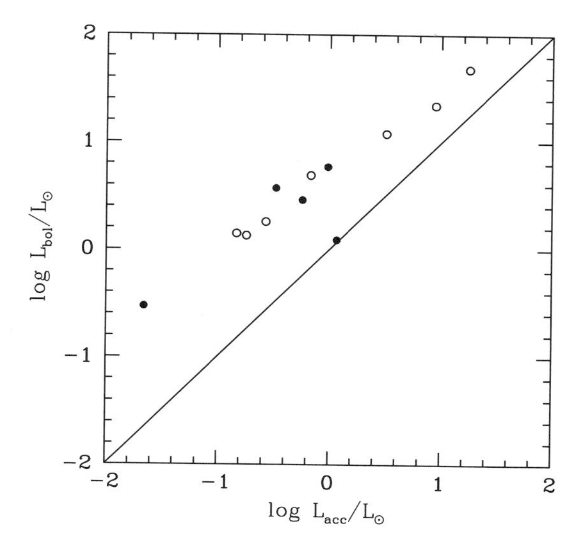

Figure 3. Relationship between accretion luminosity and bolometric luminosity for Class I sources in Taurus (filled circles) and in ρ Oph (open circles) (from Muzerolle et al. 1998*c*).

birthline (Stahler 1988), which seems justified, because they are still accumulating mass from the envelope and disk. We used the observed bolometric luminosity to locate the object along the birthline (calculated for an infall rate of 2×10^{-6} $M_{\odot}yr^{-1}$). Figure 2b shows the distribution of $\dot{M}$ thus obtained for Class I sources in Taurus and ρ Oph.

The mean mass infall rate in the envelope, $\dot{M}_i$, in Taurus has been estimated, from fitting the spectral energy distributions and scattered light images of Class I sources, to be $\sim 4 \times 10^{-6}$ $M_{\odot}yr^{-1}$ (Kenyon et al. 1993*a*), a value that is consistent with theoretical expectations (Shu et al. 1987). The present determination of disk accretion rate shown in Fig. 2b indicates that many Class I objects are slowly accreting from their disks, despite the fact that mass is being deposited in the outer disk at a higher rate. This discrepancy was first recognized by Kenyon et al. (1990) as the so-called luminosity problem of Class I sources. If infall were spherical, the luminosity of the system would be given by an accretion luminosity $\sim GM_{\star}\dot{M}_i/R_{\star}$. In this case, however, the predicted accretion luminosity should be of order 10 $L_{\odot}$, but the mean luminosity of Taurus Class I sources is ~1–2 $L_{\odot}$ (Kenyon et al. 1990, 1994). Kenyon et al. (1990) pointed out that the mass accretion rate in the disk onto the star does not necessarily equal the mass infall rate from the envelope onto the disk, since they are regulated by different physical processes. The imbalance between the infall and accretion rates in this picture leads to an accumulation of mass in the disk. These disks could eventually become gravitationally unstable (Larson 1984), with consequent rapid accretion until sufficient mass has been emptied out of the disk. Kenyon et al. (1990) suggested that these episodes could be related to the FU Orionis outbursts.

D. The FU Orionis Disks

The FU Orionis outbursts are now recognized as a transient phase of high mass accretion in the disk around a forming low-mass star (see review by Hartmann and Kenyon 1996 and references therein). Although FU Ori objects have been mostly studied as isolated phenomena so far, it is becoming increasingly clear that episodes of high mass accretion rate may be a crucial, if not dominant, process in the formation of stars. We briefly review here the determination of $\dot{M}$ for these objects.

The canonical FU Ori objects were discovered from their increase in brightness by several magnitudes over timescales of months to years (Herbig 1977), during which the luminosity increased from values typical of cTTSs to a few hundred $L_\odot$. Because the emission is dominated by the accretion disk in FU Ori objects, the accretion luminosity can be readily determined from the observed SED. Additional information is required to obtain $M_\star/R_\star$ and $\dot{M}$ independently, which can be estimated from the surface temperature in the inner disk ($R \sim$ 2–3 $R_\star$), $T_{\rm max} \sim 7000$ K$(L_{\rm acc}/100\ {\rm L}_\odot)^{1/4}(R_\star/2\ {\rm R}_\odot)^{1/2}$ (from T_v in equation [1]) and measurements of the rotational velocity. Using this method, $\dot{M} \sim 10^{-4}\ {\rm M}_\odot\ {\rm yr}^{-1}$ has been inferred for the canonical objects for which outbursts have been observed, consistent with their high luminosities, $>100\ {\rm L}_\odot$.

A significant number of objects have been identified as FU Ori objects in recent years, mostly at IR wavelengths. Photometric variability indicative of outbursts has been detected in only a few objects; their classification as FU Ori systems has been based mostly on the presence of the near-infrared first-overtone bands of CO in deep absorption, comparable to late-type giants and supergiants and to the canonical FU Ori objects. Most of these objects are in very early stages of evolution, as is their prototype L1551 IRS 5, embedded in infalling envelopes and associated with Herbig-Haro objects, jets, and/or molecular outflows (Elias 1978; Graham and Frogel 1985; Reipurth 1985; Staude and Neckel 1991; Kenyon et al. 1993*c*; Hanson and Conti 1995; Hodapp et al. 1996; Reipurth and Aspin 1997; Sandell and Aspin 1998). The luminosities of the objects identified so far as members of the FU Ori class range from $\sim$10 $L_\odot$ to 800 $L_\odot$. Since $\dot{M} \sim 2 \times 10^{-6}\ {\rm M}_\odot {\rm yr}^{-1}(L_{\rm acc}/10\ {\rm L}_\odot) \times [(M_\star/R_\star)/0.18\ {\rm M}_\odot/{\rm R}_\odot]^{-1}$, this range implies $\dot{M} \geq$ few $\times 10^{-6}\ {\rm M}_\odot {\rm yr}^{-1}$ (assuming they are on the birthline, with typical $M_\star/R_\star \sim 0.18\ {\rm M}_\odot/{\rm R}_\odot$). This value is consistent with the presence of CO in absorption, because disk atmosphere calculations indicate that the temperature of the continuum-forming region is higher than the surface temperature (neglecting any wind contribution) for $\dot{M} > 10^{-6}\ {\rm M}_\odot {\rm yr}^{-1}$ (Calvet et al. 1992).

The nature of the low-luminosity FU Ori objects and their relationship to the canonical objects remains to be elucidated. One possibility is that disks undergo instabilities that drive outbursts of different magnitude. Alternatively, these objects could be in the phase of decay from a canonical FU Ori outburst. An argument in favor of the second possibility comes from comparison of the mass loss rate required to drive the molecular outflow and the present mass accretion rate. For L1551 IRS5 and PP13S (L = 10 $L_\odot$ and 30 $L_\odot$, respectively), the momentum flux of the molecular flow is a few $\times 10^{-4}$ $M_\odot yr^{-1}$ km s^{-1} (Moriarty-Schieven and Snell 1988; Sandell and Aspin 1998). For typical velocities of the jet of a few hundred km s^{-1}, and assuming momentum conservation, the momentum flux implies mass loss rates of the same order as the inferred mass accretion rates, while both theoretically and observationally, the ratio between the mass loss rate and mass accretion rate is close to ~0.1 (Calvet 1998). This discrepancy may imply that the mass accretion rate of the disk was much higher when the material giving rise to the molecular outflow was ejected.

III. THE EVOLUTION OF ACCRETION

A. Observed $\dot{M}$ vs. Age

Figure 4 shows mass accretion rate vs. age for cTTSs in Taurus, ρ Ophiuchi, and Chamaeleon I, for which significant infall from an envelope has stopped. We also show the range of $\dot{M}$ covered by Class I sources in Taurus, assuming a median age of 0.1 Myr (estimated from the ratio of the number of Class I sources to T Tauri stars in Taurus, and adopting a mean age of 1 Myr for the latter; Kenyon et al. 1990). The data indicate a clear trend of accretion decaying with time (Hartmann et al. 1998; see also Hartigan et al. 1995), even in the relatively short age spread covered by the observations. There is also a very large spread in mass accretion rate at a given age, which makes it very difficult to quantify the rate of decay. Hartmann et al. (1998) show that the slope of a least-squares fit to the data is highly dependent on the errors of $\dot{M}$ and age. If errors are assumed larger in $\dot{M}$ than in age (the most likely situation, cf. Hartmann et al. 1998), then the slope is ≈ -1.5 (with large uncertainty).

B. Disk Masses

We can use the observed mean decay of $\dot{M}$ with time to obtain an estimate for the mass of the disk for comparison with masses estimated from dust emission. The amount of mass accreted by a cTTS from the present time to infinity constitutes a lower limit to the mass remaining in the disk. If $\dot{M}(t) \sim \dot{M}(t_o)(t/t_o)^{-\eta}$, then this limit to the disk mass is $M_{\rm acc} = \dot{M}(t_o)t_o/(\eta - 1) \sim 2\dot{M}(t_o)t_o$, with $\eta \sim 1.5$. This gas mass can

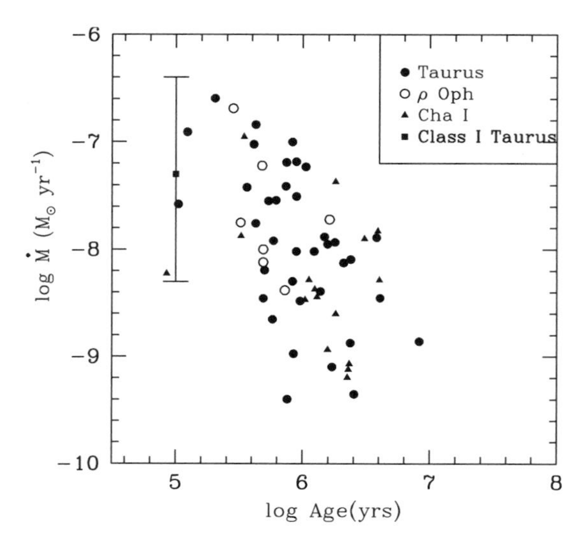

Figure 4. Observed mass accretion rate vs. age for cTTSs in Taurus, Chameleon I, and ρ Ophiuchi. Mass accretion rates have been obtained by the methods described in section II. Ages for the cTTSs have been estimated from the position in the HR diagram and comparison with the theoretical tracks from D'Antona and Mazzitelli (1994, CMA case). Luminosities and spectral types were taken from Kenyon and Hartmann (1995) for Taurus, from Gauvin and Strom (1992) for Cha I, and from Greene and Meyer (1995) for ρ Oph. The mean and dispersion of the estimated mass accretion rates for Class I sources is also shown for comparison (see section II.C). The mean age for the Class I sources is assumed to be 0.1 Myr.

be compared to the disk mass estimated from dust emission in the submillimeter and millimeter range, M_{mm}, which is very dependent on the assumed opacities (Beckwith et al. 1990; Osterloh and Beckwith 1995, masses corrected by a factor of 2.5; Hartmann et al. 1998). The comparison (for single stars) yields $\log(2\dot{M} \times \mathrm{age}/M_{\mathrm{mm}}) = -0.07 \pm 0.21$, indicating that masses inferred from the current mass accretion rates and age estimates, which probe the gas, are consistent with disk masses estimated independently from millimeter-wave dust opacities, which in turn suggests that the dust opacities used in the latter estimates are appropriate.

C. Early Stages

Figure 4 indicates that the mean mass accreted onto the star during the cTTS phase is $\sim 10^{-2}$ $M_{\odot}$. This suggests that by the time stars reach the optically visible cTTS stage, the remaining disk mass is relatively low, and little mass is added to the forming star. This picture implies that the bulk of stellar accretion must occur during the (highly extincted) infall phase, when the disk is being continually replenished by the collapsing envelope. However, for the infall phase, our results suggest that disk accretion rates during quiescent phases are only slightly higher than those in the cTTS stage, precluding addition of more than ~0.01 $M_\odot$ during these phases. Significant additions to the mass of the growing protostar must therefore come via material accreted during transient episodes of high accretion.

André et al. (1993) suggest that the Class 0 objects are in the earliest phases of envelope collapse, when central accretion rates are expected to be the highest; they argue that the Class 0 sources are the true protostars. These very heavily extincted objects have somewhat higher luminosities than the mean Class I luminosity in Taurus and so may have higher accretion rates onto the central star. However, Class 0 sources tend to be less frequent than Class I sources, especially in Taurus; thus, the lifetime of the Class 0 phase may be too short to account for most of the stellar accretion.

The high-accretion-rate episodes required to explain the accretion of the stellar mass can be attributed to instabilities in quiescent, low-$\dot{M}$ disks, which result in outbursts and transient periods of high mass accretion rates. Several models have been presented to trigger those outbursts. The most accepted model is that of thermal instabilities in the inner disk (Lin and Papaloizou 1985; Clarke et al. 1990; Kawazoe and Mineshige 1993; Bell and Lin 1994; Bell et al. 1995). This model can naturally explain the occurrence of outbursts during the infall phase, because it requires a high background mass accretion rate from the outer disk, of the order of a few $\times 10^{-6}$ $M_\odot yr^{-1}$, to match the observations. Moreover, according to this model, outbursts will be triggered in the disk as long as mass is deposited in the outer disk at this rate, ensuring that mass will be transferred from the initial cloud into the star.

The number of currently known FU Ori objects is insufficient to explain the formation of typical stars mostly through outbursts (Hartmann and Kenyon 1996). However, the known sample of FU Ori disks is very incomplete, because the identifying characteristic of high-mass-accretion-rate disks is the near-infrared bands of CO in absorption and many objects may be too heavily extincted to be detected at ~2.2 μm. More and more sensitive observations of embedded objects in the near-infrared are necessary to test this hypothesis.

D. Viscous Evolution in TTS Disks

Once the main infall phase is over and the envelope is no longer feeding mass and angular momentum into the disk, we expect disk evolution to be driven mostly by viscous processes, namely, those in which the angular momentum transport is provided by a turbulent viscosity. In this and the next section, we attempt to interpret the observed properties of cTTS disks in terms of viscous evolution, with the aim of understanding the main physical processes at play and the role of initial and boundary conditions. Similarly, physical models for disk evolution help us relate properties and evolution of the inner disk, as measured by $\dot{M}(t)$, to those of the outer disk, such as radius and mass, R_d and M_d.

The disk angular momentum is

$$J_d = \int_0^{M_d} dM\, \Omega R^2 \propto M_d R_d^{1/2} \tag{6}$$

where Ω is the Keplerian angular velocity. If we neglect the (small amount of) angular momentum being added to the star, or the possible angular momentum loss to an inner disk wind (see Shu et al. 1994; see the chapter by Shu et al., this volume), then the disk angular momentum is approximately constant. In turn, this requires $R_d \propto M_d{}^{-2}$, so angular momentum conservation implies that the disk expands as the mass of the disk is accreted to the star. Evolution occurs on the viscous time scale, $t_{\rm visc} \sim R^2/\nu$, where ν is the viscosity (Pringle 1981). If $\nu \propto R^\gamma$, then $dR_d/dt \sim R_d/t_{\rm visc} \propto R_d{}^{\gamma-1}$, so $R_d \propto t^{1/(2-\gamma)}$, $M_d \propto t^{-1/2(2-\gamma)}$, and $\dot{M} \propto t^{-(5/2-\gamma)/(2-\gamma)}$. Therefore and in principle, from the observed decay of $\dot{M}$ with time, $\dot{M}(t) \propto t^{-\eta}$, we can obtain $\gamma = (2\eta - 5/2)/(\eta - 1)$, and the evolution of radius and mass of the disk can be predicted.

As Lynden-Bell and Pringle (1974) showed, analytic similarity solutions describing the evolution of disk properties exist for the case of power law viscosity (see also Lin and Bodenheimer 1982). These analytical similarity solutions have been applied by Hartmann et al. (1998) as a first approximation to the evolution of T Tauri disks. More complex models have been used by Stepinski (1998) to consider similar issues; whether the observational constraints justify approaches with more assumed parameters is not clear. Hartmann et al. (1998) argue that the use of a power law viscosity can be justified on approximate grounds. Using the α prescription for the viscosity (Shakura and Sunyaev 1973), $\nu = \alpha c_s H \propto c_s{}^2/\Omega(R) \propto T(R) R^{3/2}$, where c_s is the sound speed and H the scale height. If $T(R) \propto R^{-q}$, then $\nu \propto R^{3/2-q}$. With $q \sim 1/2$, corresponding to irradiated disks at large distances from the star (Kenyon and Hartmann 1987; D'Alessio et al. 1998), and also found by empirical fitting to apply to most disks in cTTSs (Beckwith et al. 1990), then $\gamma \sim 1$, which is roughly consistent with the observed slope of the $\dot{M}$ vs. age data, $\eta \approx 1.5$ (section III.A).

Hartmann et al. (1998) calculated similarity solutions for viscous evolution for a range of initial conditions applicable to cTTS disks. Figures 5a and 5b show the evolution of mass accretion rate and disk mass for a subset of models in this study. Initial disk masses have been taken as $M_d(0) = 0.1\, \mathrm{M}_\odot$, consistent with small disk masses remaining after disk-draining episodes of high $\dot{M}$ (section III.C). Values for other values of $M_d(0)$ can be obtained by simple scaling. Model results are shown for three values of the initial radius (1, 10, and 100 AU), which cover an order of magnitude in angular momentum, consistent with the spread of angular momentum between half of the binaries in the solar neighborhood (Duquennoy and Mayor 1991). The calculations assume a temperature of

10 K at 100 AU, as suggested by irradiated disk calculations (D'Alessio et al. 1998), and a central stellar mass of 0.5 $M_\odot$. The viscosity parameter α has been taken as 0.01, except when stated otherwise. The observed values of disk masses and mass accretion rates, shown in Figs. 5a and 5b, lie within the region bounded by the assumed range of initial conditions. Disks with larger initial radii take longer to start evolving, because the viscous timescale is $t_{\rm visc} \sim R^2/\nu \propto R$.

The surface density of the similarity solutions behaves with radius as

$$\Sigma \propto \frac{e^{-(R/R_1)/(t/t_s)}}{(R/R_1)} \tag{7}$$

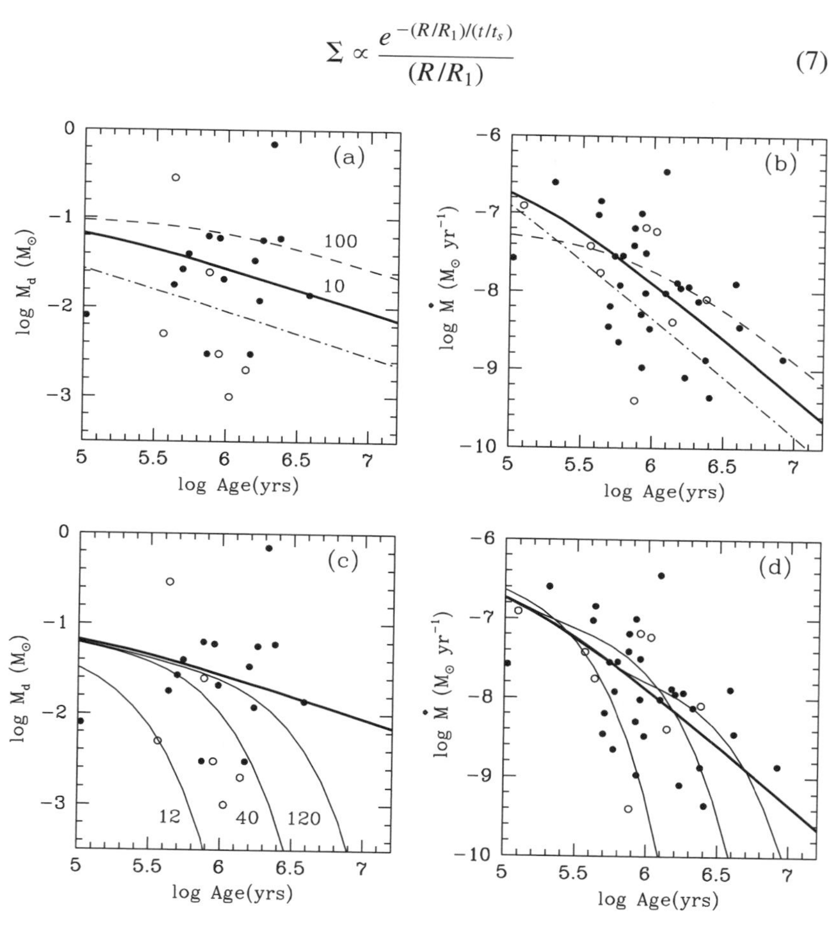

Figure 5. Similarity solution for disk evolution, with $\nu \propto R$, compared to observations. The upper panels show evolution for isolated disks: (a) disk mass vs. time; (b) $\dot{M}$ vs. time. Models shown have initial disk mass of 0.1 $M_\odot$ and initial disk radii (marked) 1, 10, and 100 AU. The lower panels show evolution with binary companions: (c) disk mass vs. time; (d) $\dot{M}$ vs. time. Models are shown for initial disk mass and radius of 0.1 $M_\odot$ and 10 AU and for three binary separations: 30 AU, 100 AU, and 300 AU (corresponding to truncation radii of 12, 40, and 120 AU, marked). The corresponding evolution for the isolated disk is shown for comparison (heavy line). Binaries are indicated by open circles.

where R_1 and t_s are characteristic radius and time (see Hartmann et al. 1998 for details); it varies as $\propto R^{-1}$ at small radii and falls sharply at large distances. This last property determines important differences in the disk "sizes" measured at different wavelengths and naturally explains the observed disparity between the optical and millimeter-wave sizes. Disk radii measured at millimeter wavelengths are of the order of a few hundred AU (Dutrey et al. 1996), whereas the radii of the disks seen in silhouettes in the Orion Nebula cluster at 0.6 μm are much larger, ~500–1000 AU (McCaughrean and O'Dell 1996).

Figure 6 shows the predicted radii of models in Fig. 5 at 2.7 mm and 0.6 μm, compared to the observations. Circles indicate the Dutrey et al. (1996) observations, while the error bar indicates the range of sizes of the Orion silhouettes. To calculate the millimeter sizes, the two-dimensional brightness distribution of the disk model has been convolved with a Gaussian with the appropriate beam size. Radii at other wavelengths correspond to the radii where the optical depth is ~1 at the given wavelength (Hart-

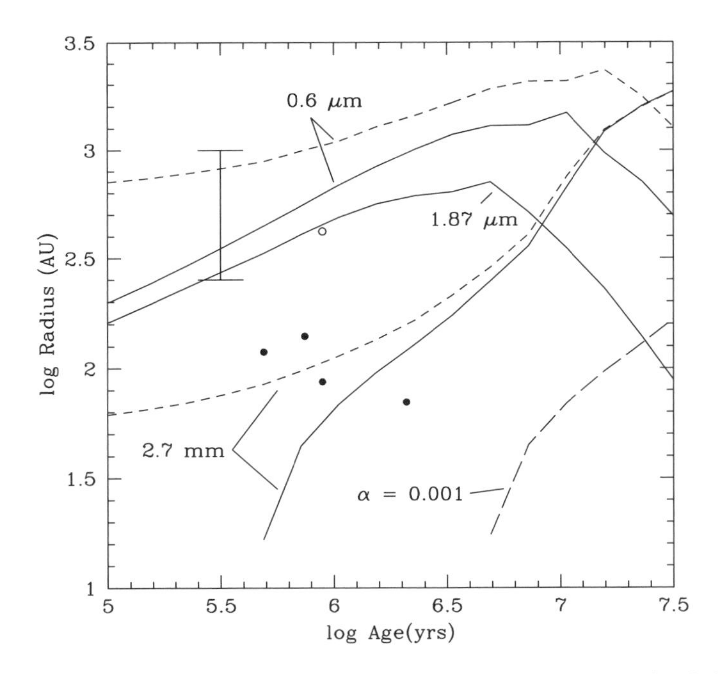

Figure 6. Characteristic disk sizes for viscous evolution as observed at 0.6 μm, 1.87 μm, and 2.7 mm. Models are shown for an initial mass of 0.1 $M_\odot$ and initial radii 10 AU (solid) and 100 AU (short dashes). A model with initial radii 10 AU but $\alpha = 0.001$ is shown for comparison (long dashes). Data from Dutrey et al. (1996) are shown as circles, and the error bar indicates the range of sizes measured in the disks seen in silhouettes in the Orion Nebula cluster (McCaughrean and O'Dell 1996). Binaries are indicated by open circles. See text.

mann et al. 1998). The theoretical predictions compare well with the observations. Because dust opacity increases rapidly towards the optical, the outer tenuous regions of the disk can effectively absorb background light and produce a large apparent size. In contrast, in the millimeter range these outer cooler, low-density regions contribute little to the surface brightness, and the observed sizes are consequently smaller (Hartmann et al. 1998). The larger millimeter-wave size of the disk in the binary (open circle) may be indicative of a circumbinary disk, for which the present calculations do not apply.

Figure 6 shows the predicted size of one of the models at 1.87 μm. At an age of $\sim$0.5 Myr, typical of the Orion Nebula cluster, the infrared sizes are $\sim$20% smaller than the optical sizes, in good agreement with observations (McCaughrean et al. 1998). The sharp decline of surface density with radius predicted by viscously evolving models naturally explains the observed sizes, without the need for heavily truncated edges.

Figure 6 also shows with long dashes the predictions for the observed sizes in the millimeter-wave range of a disk model with $\alpha = 0.001$. Since $t_{\rm visc} \sim R^2/\nu \propto 1/\alpha$, disks with small α take much longer to start evolving and growing, resulting in sizes much smaller than observed at the typical ages of the young population. Thus, measurements of disk sizes as a function of time will place important constraints on the characteristic value of α in cTTS disks.

Disk evolution could be considerably different if angular momentum is lost from outer disk regions through a wind (e.g., Pudritz and Norman 1983, Königl 1989). We have argued elsewhere (Hartmann 1995) that this is not the case. Similarly, coagulation of disk material into bodies that sweep the gas clear will significantly modify this simple picture of disk evolution. Nevertheless, it is encouraging that the observations can be explained with a viscosity ($\alpha \sim 10^{-2}$) comparable to that estimated in simulations of the Balbus-Hawley magnetorotational instability (Stone et al. 1996; Brandenburg et al. 1996).

E. Effects of Companion Stars

There is a large spread in the mass accretion rates and disk masses as a function of age. Some of this range is probably due to uncertainties in age determinations, errors in accretion rates (for example, due to ignoring inclination effects), time variability, and a range in initial conditions. However, the potential effects of companion stars cannot be ignored; given that approximately two-thirds of all systems are binaries (Duquennoy and Mayor 1991), it is important to consider the effects of a binary companion on the evolution of disk accretion when comparing with observations.

A companion star may prevent the formation of a disk in its immediate vicinity, and it will tend to open gaps on either side of its orbit (cf. Artymowicz and Lubow 1994; and the chapter by Lubow and Artymowicz, this volume). These "initial" effects of a binary companion can strongly limit disk structure and accretion, but it is important to realize that there may be secondary effects as well. Specifically, even with a relatively distant companion, the inner structure of a viscously evolving disk will eventually be affected by the companion, even though the tidal forces are negligible in this region. The reason is that the isolated viscously evolving disk can accrete only if its outer regions expand to take up the necessary angular momentum. In the similarity solution described above, the mass accretion rate decreases as a power law in time, because as the disk empties out, it also expands and thus has an increasingly long viscous time. In contrast, if a binary companion limits the expansion of the disk, then once the disk reaches its maximum size, the viscous time remains constant, so the (inner) regions empty out exponentially with time. (Note that these considerations are relevant only to circumstellar disks, not to circumbinary disks, which can expand.)

Figures 5c and 5d show a very simple calculation of this type of effect, using the same power law viscosity used in the standard model, but now not allowing the disk to expand beyond a certain outer radius, using the boundary conditions discussed by Pringle (1991). One observes that when the disk expands to the limiting radius, the accretion rate first increases slightly, and then drops precipitously as the disk empties out rapidly.

The significance of this estimate can be seen by noting that the median binary separation is roughly 30 AU (Duquennoy and Mayor 1991). With the reference disk model used, and estimating a truncation radius about one-third of the binary separation (Artymowicz and Lubow 1994), this would mean that even if all binaries originally had circumstellar disks, half of those binaries would have their disks empty out by an age of 1 Myr. These estimates are roughly consistent with the percentage of $\sim$1-Myr wTTSs in Taurus, $\sim$45% (Kenyon and Hartmann 1995); the predicted fraction of wTTSs could be even higher if, indeed, the fraction of binaries in Taurus is higher than in the solar neighborhood (Simon et al. 1995).

These estimates of binary effects on disk evolution are rough, and the model ignores the effects of the (significant) eccentricities of binary orbits (Artymowicz and Lubow 1994), but they serve to illustrate the importance of identifying stellar companions to understand disk evolution in individual systems. Figures 5c and d do not show a very strong correlation of mass accretion rates with binarity, though there is a significant effect on disk masses (cf. Jensen et al. 1994; Mathieu et al. 1995; Osterloh and Beckwith 1995). In any case, many of the systems shown have not been studied carefully for potential companions, and in general much work remains to be done in this area.

IV. SUMMARY AND IMPLICATIONS

Figure 7 summarizes the ideas presented so far in a sketch of disk evolution with time for a single star. Perhaps after a short initial period of

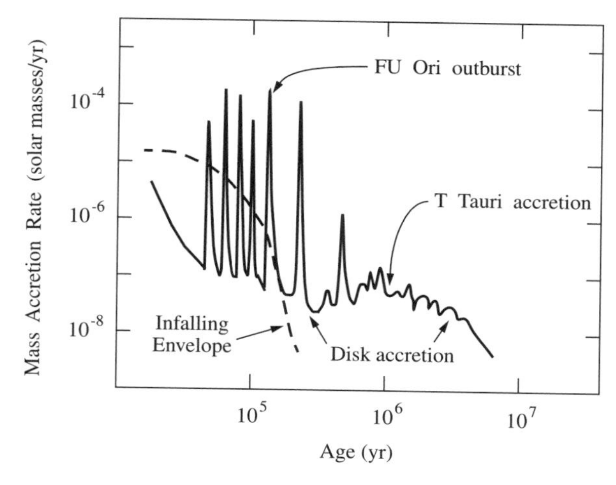

Figure 7. Sketch of disk evolution with time, summarizing the ideas presented in this chapter. The disk remains most of the time in a quiescent state, punctuated by episodes of high $\dot{M}$ as long as the envelope feeds mass to the disk. When infall ceases, the disk evolves viscously, and $\dot{M}$ slowly decreases with time.

high mass accretion rate, the disk remains most of the time in a quiescent state. Episodes of high mass accretion rate are triggered mostly during the phase in which the disk is still immersed in the infalling envelope, in which we expect most of the star to be built. After the infall ceases and the star emerges as a T Tauri star, the disk evolves viscously, $\dot{M}$ slowly decreases with time, at the same time that the disk expands and its mass decreases.

There are several implications of these results for star and planet formation. First, disk accretion during the protostar phase appears to be highly variable. This may call into question theories of the birthline, or the initial position of stars in the HR diagram (Stahler 1988; Hartmann et al. 1997), which assume steady accretion at the rates of infall of the protostellar envelope. It is conceivable that planetesimals or other bodies form in the disk during this phase but are swept into the star as the disk accretes, perhaps partly accounting for some of the accretion variability; our observations really probe only energy release in the inner disk, near the star. Second, there appears to be a wide variety of disk masses and accretion rates (say, a range of an order of magnitude) during the T Tauri phase, produced in part by differences in initial angular momenta. This may mean that any consequent planetary systems that form could have quite different properties. Third, the ability of viscous accretion disk models to explain the observations so far suggests that substantial migration of material occurs during the T Tauri phase; this migration, as disks actively accrete, may be important in explaining some of the extrasolar planets that lie close to their stars. Fourth, the presence of binary companions obviously does not always prevent disk formation, but they may accelerate (circumstellar) disk accretion. Improved millimeter and submillimeter interferometry, as well as infrared speckle searches for companion stars and improved radial velocity studies to search for close, low-mass stellar companions, will lead to a greatly improved understanding of disk evolution.

Acknowledgments We thank several people for useful discussions and valuable insight, including James Muzerolle, Erik Gullbring, Paola D'Alessio, Cesar Briceño, Ray Jayawardhana, Suzan Edwards, Lynne Hillenbrand, Michael Meyer, Bo Reipurth, and David Wilner. This work was supported in part by NASA grant NAG5-4282.

REFERENCES

Adams, F. C., and Shu, F. H. 1986. Infrared spectra of rotating protostars. *Astrophys. J.* 308:836–853.

André, P., Ward-Thompson, D., and Barsony, M. 1993. Submillimeter continuum observations of ρ Ophiuchi A: The candidate protostar VLA 1623 and prestellar clumps. *Astrophys. J.* 406:122–141.

Artymowicz, P., and Lubow, S. H. 1994. Dynamics of binary-disk interaction. I: Resonances and disk gap size. *Astrophys. J.* 421:651–667.

Basri, G., Marcy, G.W., and Valenti, J. A. 1992. Limits on the magnetic flux of pre-main-sequence stars. *Astrophys. J.* 390:622–633.

Beckwith, S. V. W., Sargent, A. I., Chini, R. S., and Guesten, R. 1990. A survey for circumstellar disks around young stellar objects. *Astron. J.* 99:924–945.

Bell, K. R., and Lin, D. N. C. 1994. Using FU Orionis outbursts to constrain self-regulated protostellar disk models. *Astrophys. J.* 427:987–1004.

Bell, K. R., Lin, D. N. C., Hartmann, L., and Kenyon, S. J. 1995. The FU Orionis outburst as a thermal accretion event: Observational constraints for protostellar disk models. *Astrophys. J.* 444:376–395.

Bertout, C. 1989. T Tauri stars: Wild as dust. *Ann. Rev. Astron. Astrophys.* 27:351–395.

Brandenburg, A., Nordlund, A., Stein, R.F., and Torkelsson, U. 1996. The disk accretion rate for dynamo-generated turbulence. *Astrophys. J. Lett.* 458:L45–L48.

Calvet, N. 1997. Properties of the winds of T Tauri stars. In *Herbig-Haro Flows and the Birth of Low Mass Stars,* IAU Symp. 182, ed. B. Reipurth and C. Bertout (Dordrecht: Kluwer Academic Publishers), pp. 417–432.

Calvet, N. 1998. Disk accretion in pre-main-sequence stars. In *Some Like It Hot!*, AIP Conf. 431, ed. S. S. Holt and T. R. Kallman (Woodbury, NY: AIP Press), pp. 495–504.

Calvet. N., and Gullbring, E. 1998. The structure and emission of the accretion shock in T Tauri stars. *Astrophys. J.* 509:802–818.

Calvet, N., and Hartmann, L. 1992. Balmer line profiles for infalling T Tauri envelopes. *Astrophys. J.* 386:239–247.

Calvet, N., Patiño, A., Magris C. G., and D'Alessio, P. 1991. Irradiation of accretion disks around young objects. I: Near-infrared CO bands. *Astrophys. J.* 380:617–630.

Calvet, N., Magris C. G., Patiño, A., and D'Alessio, P. 1992. Irradiation of accretion disks around young objects. II: Continuum energy distribution. *Rev. Mex. Astron. Astrofis.* 24:27–42.

Calvet, N., Hartmann, L., and Strom, S. E. 1997. Near-infrared emission of protostars. *Astrophys. J. Lett.* 481:L912–L917.

Chiang, E. I., and Goldreich, P. 1997. Spectral energy distributions of T Tauri stars with passive circumstellar disks. *Astrophys. J.* 490:368–376.

Clarke, C. J., Lin, D. N. C., and Pringle, J. E. 1990. Pre-conditions for disc-generated FU Orionis outbursts. *Mon. Not. Roy. Astron. Soc.* 242:439–446.

D'Alessio, P., Cantó, J., Calvet, N., and Lizano, S. 1998. Accretion disks around young objects. I: The detailed vertical structure. *Astrophys. J.* 500:411–427.

D'Antona, F., and Mazzitelli, I. 1994. New pre-main-sequence tracks for M less than or equal to 2.5 solar mass as tests of opacities and convection model. *Astrophys. J. Suppl.* 90:467–500.

Duquennoy, A., and Mayor, M. 1991. Multiplicity among solar-type stars in the solar neighbourhood. II: Distribution of the orbital elements in an unbiased sample. *Astron. Astrophys.* 248:485–524.

Dutrey, A., Guilloteau, S., Duvert, G., Prato, L., Simon, M., Schuster, K., and Menard, F. 1996. Dust and gas distribution around T Tauri stars in Taurus-Auriga. I: Interferometric 2.7mm continuum and CO_{13} $J = 1$–0 observations. *Astron. Astrophys.* 309:493–504.

Edwards, S., Hartigan, P., Ghandour, L., and Andrulis, C. 1995. Spectroscopic evidence for magnetospheric accretion in classical T Tauri stars. *Astron. J.* 108:1056–1070.

Elias, J. H. 1978. A study of the IC 5146 dark cloud complex. *Astrophys. J.* 223:859–875.

Gauvin, L. S., and Strom, K. M. 1992. A study of the stellar population in the Chamaeleon dark clouds. *Astrophys. J.* 385:217–231.

Graham, J. A., and Frogel, J. A. 1985. An FU Orionis star associated with Herbig-Haro object 57. *Astrophys. J.* 289:331–341.

Greene, T. P., and Lada, C. J. 1996. Near-infrared spectra and the evolutionary status of young stellar objects: Results of a 1.1–2.4 micron survey. *Astron. J.* 112:2184–2221.

Greene, T. P., and Meyer, M. R. 1995. An infrared spectroscopic survey of the ρ Ophiuchi young stellar cluster: Masses and ages from the H-R diagram. *Astrophys. J.* 450:233–244.

Gullbring, E., Hartmann, L., Briceño, C., and Calvet, N. 1998. Disk accretion rates for T Tauri stars. *Astrophys. J.* 492:323–341.

Hanson, M. M., and Conti, P. S. 1995. Identification of ionizing sources and young stellar objects in M17. *Astrophys. J. Lett.* 448:L45–L48.

Hartigan, P., Strom, S. E., Edwards, S., Kenyon, S. J., Hartmann, L., Stauffer, J., and Welty, A. D. 1991. Optical excess emission in T Tauri stars. *Astrophys. J.* 382:617–635.

Hartigan, P., Edwards, S., and Ghandour, L. 1995. Disk accretion and mass loss from young stars. *Astrophys. J.* 452:736–768.

Hartmann, L. 1995. Observational constraints on disk winds. In *Circumstellar Disks, Outflows and Star Formation, Rev. Mex. Astron. Astrofis. Ser. Conf.* 1:285–291.

Hartmann, L., and Kenyon, S. J. 1996. The FU Orionis phenomenon. *Ann. Rev. Astron. Astrophys.* 34:207–240.

Hartmann, L., Hewett, R., and Calvet, N. 1994. Magnetospheric accretion models for T Tauri stars. I: Balmer line profiles without rotation. *Astrophys. J.* 426:669–687.

Hartmann, L., Cassen, P., and Kenyon, S. J. 1997. Disk accretion and the stellar birthline. *Astrophys. J.* 475:770–785.
Hartmann, L., Calvet, N., Gullbring, E., and D'Alessio, P. 1998. Accretion and the evolution of T Tauri disks. *Astrophys. J.* 495:385–400.
Herbig, G. H. 1977. Eruptive phenomena in early stellar evolution. *Astrophys. J.* 217:693–715.
Hodapp, K-W., Hora, J. L., Rayner, J. T., Pickles, A. J., and Ladd, E. F. 1996. An outburst of a deeply embedded star in Serpens. *Astrophys. J.* 468:861–870.
Jensen, E. L. N., Mathieu, R. D., and Fuller, G. A. 1994. A connection between submillimeter continuum flux and separation in young binaries. *Astrophys. J. Lett.* 429:L29–L32.
Kawazoe, E., and Mineshige, S. 1993. Unstable accretion disks in FU Orionis stars. *Pub. Astron. Soc. Pacific* 45:715–725.
Kenyon, S. J., and Hartmann, L. 1987. Spectral energy distributions of T Tauri stars: Disk flaring and limits on accretion. *Astrophys. J.* 323:714–733.
Kenyon, S. J., and Hartmann, L. 1995. Pre-main-sequence evolution in the Taurus-Auriga molecular cloud. *Astrophys. J. Suppl.* 101:117–171.
Kenyon, S. J., Hartmann, L., Strom, K. M., and Strom, S. E. 1990. An IRAS survey of the Taurus-Auriga molecular cloud. *Astron. J.* 99:869–887.
Kenyon, S. J., Calvet, N., and Hartmann, L. 1993*a*. The embedded young stars in the Taurus-Auriga molecular cloud. I: Models for the spectral energy distribution. *Astrophys. J.* 414:676–694.
Kenyon, S. J., Whitney, B. A., Gomez, M., and Hartmann, L. 1993*b*. The embedded young stars in the Taurus-Auriga molecular cloud. II: Models for scattered light images. *Astrophys. J.* 414:773–792.
Kenyon, S. J., Hartmann, L., Gomez, M., Carr, J. S., and Tokunaga, A. 1993*c*. RNO 1B/1C: A double FU Orionis system. *Astron. J.* 105:1505–1510.
Kenyon, S., Gomez, M., Marzke, R. O., and Hartmann, L. 1994. New pre-main-sequence stars in the Taurus-Auriga molecular cloud. *Astron. J.* 108:251–261.
Kenyon, S. J., Yi, I., and Hartmann, L. 1996. A magnetic accretion disk model for the infrared excesses of T Tauri stars. *Astrophys. J.* 462:439–455.
Königl, A. 1989. Self-similar models of magnetized accretion disks. *Astrophys. J.* 342:208–223.
Königl, A. 1991. Disk accretion onto magnetic T Tauri stars. *Astrophys. J. Lett.* 370:L39–L43.
Larson, R. B. 1984. Gravitational torques and star formation. *Mon. Not. Roy. Astron. Soc.* 206:97–207.
Lin, D. N. C., and Bodenheimer, P. 1982. On the evolution of convective accretion disk models of the primordial solar nebula. *Astrophys. J.* 262:768–779.
Lin, D. N. C., and Papaloizou, J. C. B. 1985. On the dynamical origin of the solar system. In *Protostars and Planets II*, ed. D. C. Black and M. S. Matthews (Tucson: University of Arizona Press), pp. 981–1072.
Lynden-Bell, D., and Pringle, J. E. 1974. The evolution of viscous discs and the origin of the nebular variables. *Mon. Not. Roy. Astron. Soc.* 168:603–638.
Mathieu, R., Adams, F. C., Fuller, G. A., Jensen, E. L. N., Koerner, D. W., and Sargent, A. 1995. Submillimeter continuum observations of the T Tauri spectroscopic binary GW Orionis. *Astron. J.* 109:2655–2669.
McCaughrean, M. J., and O'Dell, C. R. 1996. Direct imaging of circumstellar disks in the Orion Nebula. *Astron. J.* 111:1977–1986.
McCaughrean, M. J., Chen, H., Bally, J., Erickson, E., Thompson, R., Rieke, M., Schneider, G., Stolovy, S., and Young, E. 1998. High-resolution near-infrared imaging of the Orion 114-426 silhouette disk. *Astrophys. J. Lett.* 492:L157–L161.

Meyer, M. R., Calvet, N., and Hillenbrand, L. A. 1997. Intrinsic near-infrared excesses of T Tauri stars: Understanding the classical T Tauri star locus. *Astrophys. J.* 114:288–300.
Muzerolle, J., Calvet, N., and Hartmann, L. 1998*a*. Magnetospheric accretion models for the hydrogen emission lines of T Tauri stars. *Astrophys. J.* 492:743–753.
Muzerolle, J., Hartmann, L., and Calvet, N. 1998*b*. Emission-line diagnostics of T Tauri magnetospheric accretion. I: Line profile observations. *Astron. J.* 116:455–468.
Muzerolle, J., Hartmann, L., and Calvet, N. 1998*c*. A Brγ probe of disk accretion in T Tauri stars and embedded young stellar objects. *Astrophys. J.*, in press.
Moriarty-Schieven, G.H., and Snell, R. 1988. High-resolution images of the L1551 molecular outflow. II: Structure and kinematics. *Astrophys. J.* 332:364–378.
Osterloh, M., and Beckwith, S. V. W. 1995. Millimeter-wave continuum measurements of young stars. *Astrophys. J.* 439:288–302.
Pringle, J. E. 1981. Accretion discs in astrophysics. *Ann. Rev. Astron. Astrophys.* 19:137–162.
Pringle, J. E. 1991. The properties of external accretion discs. *Mon. Not. Roy. Astron. Soc.* 248: 754–759.
Pudritz, R. E., and Norman, C. A. 1983. Centrifugally driven winds from contracting molecular disks. *Astrophys. J.* 274:677–697.
Reipurth, B. 1985. Herbig-Haro objects and FU Orionis eruptions: The case of HH 57. *Astron. Astrophys.* 143:435–442.
Reipurth, B., and Aspin, C. 1997. Infrared spectroscopy of Herbig-Haro energy sources. *Astron. J.* 114:2700–2707.
Rucinski, S. M. 1985. IRAS observations of T Tauri and post-T Tauri stars. *Astron. J.* 90:2321–2330.
Rydgren, A. E., and Zak, D. S. 1987. On the spectral form of the infrared excess component in T Tauri systems. *Pub. Astron. Soc. Pacific* 99:141–145.
Sandell, G., and Aspin, C. 1998. PP13S, a young, low-mass FU Orionis-type pre-main-sequence star. *Astron. Astrophys.* 333:1016–1025.
Shakura, N. I., and Sunyaev, R. A. 1973. Black holes in binary systems: Observational appearance. *Astron. Astrophys.* 24:337–355.
Shu, F. H., Adams, F. C., and Lizano, S. 1987. Star formation in molecular clouds: Observation and theory. *Ann. Rev. Astron. Astrophys.* 25:23–81.
Shu, F. H., Najita, J., Ostriker, E., Wilkin, F., Ruden, S. P., and Lizano, S. 1994. Magnetocentrifugally driven flows from young stars and disks. I: A generalized model. *Astrophys. J.* 429:781–796.
Simon, M., Ghez, A. M., Leinert, C. H., Cassar, L., Chen, W. P., Howell, R. R., Jameson, R. F., Matthews, K., Neugebauer, G., and Richichi, A. 1995. A lunar occultation and direct imaging survey of multiplicity in the Ophiuchus and Taurus star-forming regions. *Astrophys. J.* 443:625–637.
Stahler, S. W. 1988. Deuterium and the stellar birthline. *Astrophys. J.* 332:804–825.
Staude, H. J., and Neckel, T. 1991. RNO 1B: A new FUor in Cassiopeia. *Astron. Astrophys.* 244:L13–L16.
Stepinski, T. F. 1998. Diagnosing properties of protoplanetary disks from their evolution. *Astrophys. J.*, in press.
Stone, J. M., Hawley, J. F., Gammie, C. F., and Balbus, S. A. 1996. *Astrophys. J.* 463:656.
Valenti, J. A., Basri, G., and Johns, C. M. 1993. T Tauri stars in blue. *Astron. J.* 106:2024–2050.

DISK DISPERSAL AROUND YOUNG STARS

DAVID J. HOLLENBACH
NASA-Ames Research Center

HAROLD W. YORKE
Jet Propulsion Laboratory

and

DOUG JOHNSTONE
Canadian Institute for Theoretical Astrophysics

We review the evidence pertaining to the lifetimes of planet-forming disks and discuss possible disk dispersal mechanisms: (1) viscous accretion of material onto the central source, (2) close stellar encounters, (3) stellar winds, and (4) photoevaporation by ultraviolet radiation. We focus on (3) and (4) and describe the quasi-steady-state appearance and the overall evolution of disks under the influence of winds and radiation from the central star and of radiation from external OB stars. Viscous accretion likely dominates disk dispersal in the inner disk ($r \lesssim 10$ AU), whereas photoevaporation is the principal process of disk dispersal outside of $r \gtrsim 10$ AU. Disk dispersal timescales are compared and discussed in relation to theoretical estimates for planet formation timescales. Photoevaporation may explain the large differences in the hydrogen content of the giant planets in the solar system. The commonly held belief that our early sun's stellar wind dispersed the solar nebula is called into question.

I. INTRODUCTION

The question of planet formation is intimately related to the formation and evolution of disks around condensing protostars. When considering the evolution of dust grains through the planetesimal stage up to the creation of (proto)planets, it is important to realize that the disks themselves are continually evolving. Indeed, very short-lived disks may not have sufficient time to produce planets.

In the standard theory of low-mass star formation and disk evolution, a rotating, collapsing molecular core accretes at a rate $\dot{M}_c \sim 10^{-6}$–10^{-5} $M_\odot$ yr^{-1}, initially directly onto the protostar but soon mainly onto an orbiting accretion disk around the protostar (Terebey et al. 1984). This rapid accretion quickly builds up the disk mass M_d until gravitational instabilities set in, when the disk mass is roughly $\sim 0.3\ M_\star$, where $M_\star$ is the mass of the protostar (Laughlin and Bodenheimer 1994; Yorke et al. 1995). The gravitational instabilities produce spiral density waves, which allow angular momentum to be transported outwards. The disk material can then be transferred onto the protostar at nearly the same rate as the disk accretes material from the molecular core; although this process may be quite episodic, the disk might roughly "hover" at the critical gravitationally unstable limit, $M_d \sim 0.3\ M_\star$, as the star grows. The result at the end of the accretion phase could be, for example, a 1-$M_\odot$ star with a $\sim$0.3-$M_\odot$ disk.

Once the cloud stops accreting onto the disk, the epoch of disk and stellar growth ceases. As the disk mass falls below the level required for gravitational instability, the transfer of angular momentum drops considerably, and other, less efficient processes for transport take over, possibly turbulent or magnetic (Balbus and Hawley 1991). This phase might be identified with the Class II, or T Tauri, phase of disk evolution, and observations of the excess blue emission from the accretion of the disk onto the star suggest median mass accretion rates of order $10^{-8}\ M_\odot\ yr^{-1}$ with an order of magnitude scatter (Hartmann et al. 1998). As the disk loses mass, this rate should decrease. Whether the lifetime of the disk is thereafter determined by the evolution of the viscosity in the disk, which controls the accretion rate onto the star, or whether other processes such as winds and ultraviolet radiation eventually dominate disk dispersal, is part of the subject of this chapter.

Observational estimates of disk masses around young pre-main-sequence stars typically give values that are less than 10% the mass of the central star (Adams et al. 1990; Beckwith et al. 1990); often a value closer to 1% can be inferred. An example of an observationally well-determined and relatively high-mass disk is that of GM Aurigae (Koerner et al. 1993), a star of 0.72 $M_\odot$ with a disk of 0.09 $M_\odot$. In spite of some uncertainty in the mass estimates, these values for the later stages of disk evolution are in contrast to the theoretical values $\sim$0.3 $M_\star$ discussed above for disks still accreting material from the parent cloud. This might imply nonnegligible disk evolution after accretion stops.

The dust in the inner disks around low-mass stars apparently disappears in about 10^7 years, whereas disks around intermediate-mass stars appear to be depleted in somewhat shorter timescales (Strom 1995). Using pre-main-sequence evolutionary models to calculate the ages of star-forming clusters, and counting the numbers of stars found with and without observable inner disks, Strom showed that for very young ($\sim 10^6$ yr old) clusters such as Ophiuchus, almost all pre-main-sequence stars have disks, whereas for older regions the fraction of stars with disks drops as the stellar mass rises. A small fraction, perhaps 10%, of low-mass stars still have disks beyond 10 Myr. Therefore, the (inner) disk dispersal timescale for low-mass ($M_\star \lesssim 3\ M_\odot$) stars may be of order $\tau_d \sim 10^7$ years.

These constraints on the evolution of the disk can be compared to models of the formation of giant planets in the solar system. Current theoretical thinking envisages a typical giant planet in the outer solar system to form by (1) the accumulation of solid (cometlike) planetesimals to ever larger bodies (Safronov 1969; Wetherill 1980; Nakagawa et al. 1983; timescale $\lesssim 10^5$ years, Weidenschilling 1984), until (2) a runaway growth occurs for the largest planetary embryos (Greenberg et al. 1978, 1984), with the runaway halted only after the depletion of the planetesimals in a given embryo's "feeding zone" and the embryos become gravitationally isolated from one another (at a mass $\sim 1\ M_\oplus$ in the giant planet region for a "minimum" solar nebula); followed by (3) a much slower agglomeration of individual embryos into the core of each of the current planets (Wetherill and Stewart 1986), until (4) the core reaches a critical mass ($\sim$15–20 $M_\oplus$) and begins to gather in gas much faster than it gathers solids (Mizuno 1980; Bodenheimer and Pollack 1986; Pollack et al. 1996), followed yet later by (5) the cessation of gas and solid accretion, when tidal forces inside the Hill sphere suffice to open up a gap around the planet (Papaloizou and Lin 1984; Lin and Papaloizou 1985, 1986*a*,*b*). Protostellar disks can therefore be partially destroyed by the transformation of the gas and dust into planets.

However, it is clear that considerably more mass was dispersed by other mechanisms than was congealed into planets. If the solar nebula had only the "minimum mass" needed to form the known planets ($\sim 0.013\ M_\odot$ according to Hayashi et al. 1985), the timescale needed to assemble the giant planets ranges from $\sim 10^8$ yr for Jupiter to $\sim 10^{11}$ yr for Neptune (Nakagawa et al. 1983), much too long to explain many observed features of the solar system (such as the existence of the asteroid belt and of Neptune). Lissauer (1987, 1993) has proposed that the timescale problem can be alleviated by postulating a solar nebula that contained much more mass than the "minimum" value. Increases by a factor of 5–10 in the giant-planet zones would push the embryo isolation mass to or beyond the critical core mass, which would eliminate the slow process (3) of embryo accumulation, allowing Jupiter, Saturn, Uranus, and Neptune to form on timescales of $\sim 10^6$–10^8 yr. The core growth occurs then on timescales of $\sim 3 \times 10^6 (r/10\ \mathrm{AU})^{3/2}$ yr and the gas accretion onto the cores at times $\gtrsim 10^6$–10^7 yr.

Even if the solar nebula originated with only the minimum mass, there is clear evidence for disk dispersal by mechanisms other than planet formation. In accounting for the elemental abundances in the solar planets, one finds that the hydrogen and helium are deficient by a large factor compared to the original cosmic abundances that formed the planets. Clearly, the early solar nebula incorporated many of the heavier elements in the terrestrial planets and rocky cores, but substantial hydrogen- and helium-rich gas was dispersed by other mechanisms. Therefore, we conclude that planet formation is a minor disk dispersal mechanism. This chapter focuses instead on the dispersal timescales of other mechanisms in order to understand the time available for the above processes of planet formation to operate.

A number of mechanisms, both internal and external, disperse the gas and dust disks around young stars or protostars. These processes can be so efficient that the disks may not have sufficient time to form planets, or the planets formed may be dispersed with the nebular material. For example, viscosity in the disk (see section II.A) coupled with the tidal force of a relatively massive disk on embedded giant planets may result in inwardly migrating planets that can be accreted, along with gas and dust, onto the central star (Lin et al. 1996). Close encounters with binary or cluster companions can strip and truncate the outer disks and can liberate the outer planets (section II.B). Stellar winds have often been postulated as a likely mechanism to entrain and push the gas and dust in the disk and to drive it into the interstellar medium (section II.C). The ultraviolet radiation from the central star or an external star can heat the outer disks and cause them to evaporate on relatively short timescales (section II.D). Obviously, understanding disk dispersal timescales is crucial to predicting the final products of planetary formation.

II. DISK DISPERSAL MECHANISMS

In our discussion of disk dispersal mechanisms, we shall focus on the dispersal of disk material at distances $\gtrsim$1 AU from the central star, where terrestrial and giant planets may form and where the bulk of the mass of the young disk presumably resides, assuming a surface density distribution $\Sigma \propto r^{-\beta}$ with $\beta < 2$. We also focus on the early stages of disk evolution and the dispersal of the bulk of the original disk, rather than the later stages of dispersal of remnant disks, such as the one around β Pic. There are four basic processes for disk dispersal other than planet formation: (A) accretion of disk material onto the central source, (B) stripping due to close stellar encounters, (C) the removal of disk material due to the effects of stellar or disk winds, and (D) photoevaporation due to the ultraviolet radiation from the central source and/or close companions. All of these interactions operate concurrently, and we will discuss how winds can enhance photoevaporation. Considerable progress has been made since the last *Protostars and Planets* meeting in understanding the last of these mechanisms, so we devote a large proportion of our review to photoevaporation.

A. Disk Accretion

Radial mass flow in disks is a well-studied, although still not completely understood, phenomenon. The transport of angular momentum outward allows disk material to flow radially inward. The angular momentum can be transported, for example, by turbulence or magnetic fields, and often the physics of the evolution is subsumed in a parameter α that fixes the effective viscosity, which drives radial accretion through the disk.

Parameterizing the viscosity by $\nu = \alpha c_s H$ (Pringle and Rees 1972; Shakura and Sunyaev 1973), where scale height H and sound speed c_s are given by $H = rc_s/(GM_\star/r)^{1/2}$ and $c_s = (kT/m_H)^{1/2}$, we find $\nu \propto r$, if the temperature in the disk follows a $T \propto r^{-1/2}$ power law. Given that the viscous timescale $t_\nu \simeq r^2/\nu$, it follows that $t_\nu \propto r$:

$$t_\nu \simeq 10^5\,\mathrm{yr}\left(\frac{\alpha}{0.01}\right)^{-1}\left(\frac{r}{10\,\mathrm{AU}}\right)$$

The viscous dispersal timescale is plotted in Fig. 1 for two values of α (10^{-3} and 10^{-2}) thought to best represent the effective viscosity in protoplanetary disks (see, e.g., Hartmann et al. 1998). Assuming that disk angular momentum is constant and that the disk mass resides mostly at the outer disk radius r_d, it follows that $r_d(t) \propto M_d(t)^{-2}$. Therefore, as the disk accretes matter onto the central star, it preserves its angular momentum by expanding to greater size. For times greater than the initial timescale $t_{\nu 0}$,

$$r_d \simeq r_{d0}\left(\frac{t}{t_{\nu 0}}\right) \qquad \text{and} \qquad M_d(t) \simeq M_{d0}\left(\frac{t}{t_{\nu 0}}\right)^{-1/2}$$

where r_{d0} and M_{d0} are the initial radius and mass of the disk at $t_{\nu 0}$. As the disk grows, the viscous timescale in the outer regions, where the mass resides, gets longer. Therefore, viscous accretion may account for the

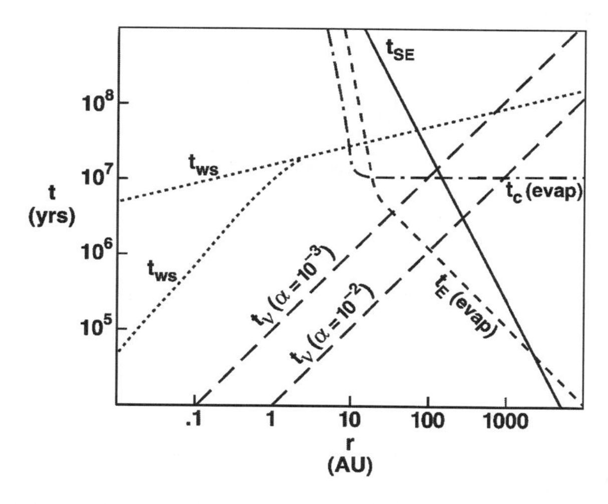

Figure 1. Timescales for disk dispersal: t_ν is the viscous timescale for $\alpha = 10^{-2}$ and 10^{-3}; t_{SE} is the stellar encounter (tidal stripping) timescale for Trapezium cluster conditions (see sections II.B and III); t_{ws} and t'_{ws} are stellar wind stripping timescales for wind and disk parameters summarized in section III; t_c(evap) is the photoevaporation timescale by the central star (strong wind case), and t_E(evap) is the photoevaporation timescale for an external star (Trapezium conditions) for the conditions summarized in sections II.D and III.

dispersal of the inner parts of disks, but it becomes insignificant for the outer disks and cannot explain the complete dispersal of the disk.

For $M_{d0} = 0.1\ \mathrm{M}_\odot$ the mass accretion rate onto the central star, or the accretion mass loss rate from the disk, is given for $t > t_{\nu 0}$

$$\dot{M}_\nu \simeq 5 \times 10^{-7}\ \mathrm{M}_\odot\ \mathrm{yr}^{-1} \left(\frac{\alpha}{0.01}\right)\left(\frac{r_{d0}}{10\ \mathrm{AU}}\right)^{-1}\left(\frac{t}{t_{\nu 0}}\right)^{-3/2}$$

B. Stellar Encounters

The effects of close stellar encounters on protostellar disks have been discussed by Clarke and Pringle (1993), Heller (1995), Hall et al. (1996), Larwood (1997), and Bonnell and Kroupa (1998). Typically, the disk (and any outer planets) is stripped to about $\frac{1}{3}$ the impact parameter r_p in a single close encounter. From this we can estimate the timescale t_{SE} to truncate a disk to the radius r_d: $t_{\mathrm{SE}} \simeq 1/n_\star \sigma v$, where $n_\star$ is the density of stars, $\sigma \simeq \pi(3r_d)^2$ is the collision cross section, and v is the velocity dispersion of stars:

$$t_{\mathrm{SE}} \simeq 2 \times 10^7\ \mathrm{yr} \left(\frac{n_\star}{10^4\ \mathrm{pc}^{-3}}\right)^{-1}\left(\frac{v}{1\ \mathrm{km\ s}^{-1}}\right)^{-1}\left(\frac{r_d}{100\ \mathrm{AU}}\right)^{-2}$$

This timescale is also shown in Fig. 1.

C. Effects of Winds on Disks

Although protostellar winds and outflows were a great surprise when first observed some two decades ago, we are now beginning to understand the origin of the winds and the wind environment in the vicinity of circumstellar disks. During and after the cloud collapse phase, which builds up the disk mass, the disk actively accretes onto the protostar ($t \lesssim 10^7$ yr) and creates a strong wind. The interaction of the rotating magnetic field with the inner disk produces this wind, whose mass loss rate ($\dot{M}_w \sim 10^{-7}\ \mathrm{M}_\odot\ \mathrm{yr}^{-1}$ [for $t \sim 10^6$ yr] to $10^{-8}\ \mathrm{M}_\odot\ \mathrm{yr}^{-1}$ [for $t \sim 10^7$ yr]) is about 0.3–0.5 times the accretion rate onto the protostar (Shu et al. 1988; Königl 1995; Ouyed and Pudritz 1997*a*,*b*). The terminal wind speeds are of order 100–200 km s^{-1}. The mass lost in these strong winds is disk material ejected via coupling between the (weakly) ionized disk gas and the outward-bending magnetic field. These disk winds are not spherically symmetric but are magnetically collimated away from the disk plane toward the poles. Later, as the accretion subsides, the magnetic activity in the chromosphere of the young, low-mass star drives a (spherically symmetric) stellar wind, which is likely much more active in the early ($\sim 10^7$ yr) epoch than the current solar wind ($\dot{M}_w \sim 10^{-14}\ \mathrm{M}_\odot\ \mathrm{yr}^{-1}$).

As an upper limit to the dispersing power of disk winds, we will ignore their collimation and consider the momentum transferred by the collision of spherically symmetric winds with the disk (Fig. 2). The scale height of cold [$c_s \sim 0.1\,(GM_\star/r)^{1/2}$] protostellar disks is of order $H \sim 0.1r$, which

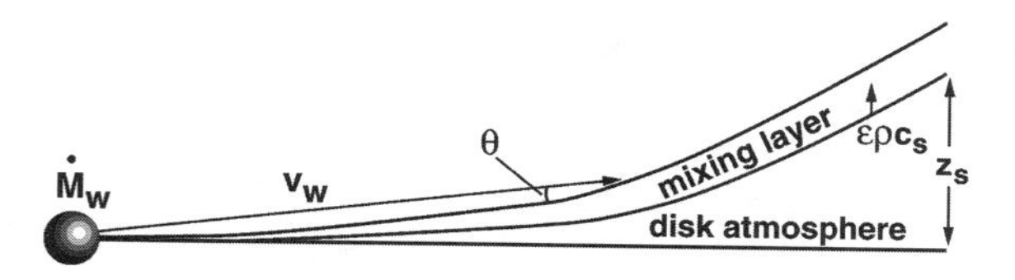

Figure 2. Schematic representations of the stripping of disk material due to the shear caused by the impact of a stellar wind.

implies that the fraction of solid angle subtended by cold disks is of order 0.1. Assuming that stripping occurs in a thin mixing layer at a height $z_s(r)$ above the disk and that disk gas flows into this mixing region at a velocity of ϵc_s, where ϵ must be less than unity and has been estimated to be $\simeq$0.01–0.1 by Canto and Raga (1991), the mass loss rate can be approximated by $\dot{M}_{\rm ws} = 2\int_{r_\star}^{r_d} \epsilon\rho(r, z_s)\, c_s(r)\, 2\pi\, r\, dr$. In the thin mixing layer at $z_s(r)$, the normal component of the wind ram pressure $(\sin\theta)^2\,(\dot{M}_w v_w/4\pi r^2)$ equals the disk gas pressure $\rho(r, z_s)\, c_s^2(r)$, where θ is the angle between the incident wind and the wind shock, which traces the top of the mixing layer. Substituting this relation into the mass loss equation and assuming that $c_s(r)$ declines with r, we conclude that the maximum mass loss rate $\dot{M}_{\rm ws}$ and the minimum dispersal timescale $t_{\rm ws}$ from wind stripping are given by

$$\dot{M}_{\rm ws} \simeq 4\times10^{-2}\,\dot{M}_w \left(\frac{v_w}{100\ {\rm km\ s^{-1}}}\right)\left(\frac{\epsilon\overline{(\sin\theta)^2}}{10^{-4}}\right)\left(\frac{T(r_d)}{100\ {\rm K}}\right)^{-1/2}$$

and

$$t_{\rm ws} \simeq 10^7\ {\rm yr}\left(\frac{r}{1\ {\rm AU}}\right)^{1/4}\left(\frac{\epsilon\,\overline{(\sin\theta)^2}}{10^{-4}}\right)^{-1}\left(\frac{M_d}{M_{\rm min}}\right)v_{w7}^{-1}\dot{M}_{w-8}^{-1}$$

where $v_{w7} = v_w/100\ {\rm km\ s^{-1}}$, $\dot{M}_{w-8} = \dot{M}_w/10^{-8}\ {\rm M_\odot\ yr^{-1}}$, $\overline{(\sin\theta)^2}$ is the integral-averaged value of $(\sin\theta)^2$, and $M_{\rm min} \simeq 0.01\ {\rm M_\odot}$ is the minimum mass solar nebula. We have numerically computed $\overline{(\sin\theta)^2} \sim 10^{-3}$ for a reference model with $\dot{M}_{w-8} = 1$, $v_{w7} = 1$, and $M_d = M_{\rm min}$.

Stellar wind dispersal has been postulated as a mechanism for the dispersal of stellar disks (Cameron 1973; Horedt 1978, 1982; Elmegreen 1979). However, as can be seen in Fig. 1, where various disk dispersal timescales are compared to one another, wind stripping should not be an effective disk dispersal mechanism for a reasonably flat (cold) disk, assuming $\epsilon(\sin\theta)^2 \lesssim 10^{-4}$.

Three separate arguments support this conclusion. First, Richling and Yorke (1997) have done numerical simulations that include the ultraviolet-induced photoevaporation of the disk (see section II.D below) acting simultaneously with (isotropic) wind stripping. The ultraviolet-heated disk atmosphere and photoevaporative flow deflect and collimate the wind and reduce the wind stripping significantly. A low degree of mixing between the winds and the disk outflows is found. Although no numerical simulations exist for the particular configuration of interacting T Tauri disks and winds, one can infer from the results of Richling and Yorke (for disks around high-mass stars) that the value for $\epsilon(\sin\theta)^2$ is indeed low. Secondly, we note that the strong protostellar (disk) winds collimate quickly, probably by magnetic forces, and for this reason alone are unlikely to aid in the removal of material from the outer disk; see, for example, recent Hubble Space Telescope (HST) images of DG Tau (Kepner et al. 1993) and HH30 (Stapelfeldt et al. 1994; Burrows et al. 1996). Finally, if the powerful winds around young stars are powered by disk accretion, it is unlikely that they can remove the entire disk.

Further work should be done to test for instabilities that could significantly raise ϵ and to study the parameter space of disk properties exhaustively to test the variations in $(\sin\theta)^2$. In the absence of these results, the wind dispersal timescales plotted on Fig. 1 [$t_{\rm ws} \equiv \Sigma(r)/\dot{\Sigma}(r)$, where $\Sigma(r)$ is the surface density at r] are only improved versions of previous work, still uncertain by perhaps an order of magnitude. Nevertheless, we suggest that wind stripping is probably not as significant as viscous accretion and photoevaporation.

D. Photoevaporation

Disk photoevaporation was first proposed for high-mass stars with disks, because these stars have extremely powerful Lyman continuum fluxes. One of the main motivations was to explain the relatively long lifetimes of small ($r \lesssim 10^{17}$ cm) and dense ($n_e \gtrsim 10^4$–10^5 cm^{-3}) ultracompact H II (UC H II) regions. The high densities and temperatures of these regions compared to molecular clouds as a whole suggest that UC H II regions are overpressurized and should expand at velocities in excess of $c_s \sim$ 10 km s^{-1}. Without replenishment of this gas, the dynamical timescale of $\sim 10^{3.5}$ yr predicts that these regions should quickly expand and become more diffuse. However, Wood and Churchwell (1989) performed a radio survey of UC H II regions and found that 10–20% of all O stars are in the UC H II phase. Given that O stars live approximately 10^{6-7} yr, this implies that dense UC H II regions last for $10^{5.5}$ yr, two orders of magnitude longer than the dynamical timescale.

Whereas the idea of using the photoevaporation process as a means for eroding disks around young massive stars had been mentioned occasionally in the literature (e.g., Bally and Scoville 1982), Hollenbach et al. (1993, 1994) first applied the idea rigorously as a model for ultracompact H II regions. The first numerical studies of this process (Yorke and Welz 1993, 1994) have been supplemented by more realistic simulations including the effects of UV dust scattering (Richling and Yorke 1997) and photoevaporating the disks of close companions in a multiple system

(Richling and Yorke 1998, 1999). Hollenbach et al. (1994) proposed that UC H II regions live for $10^{5.5}$ yr because they are constantly being replenished by a dense circumstellar reservoir: the orbiting disk. Although the high-pressure gas expands away from the disk at speeds of 10–50 km s^{-1}, the disk reservoir maintains a quasi-steady-state density and intensity profile. The photoevaporating disk model matches well with the morphology, emission measure, and spectrum of numerous UC H II regions. As we shall demonstrate below in section II.D.1, the mass loss rate $\dot{M}_{ph}$ from the disk around an O star with a Lyman continuum photon luminosity $\Phi_i \sim 10^{49}$ s^{-1} is of order a few times 10^{-5} $M_\odot$ yr^{-1}. Therefore, if massive O stars are born with disks with mass ~0.1–0.3 $M_\star$, the timescale to photoevaporate the disk is $10^{5.5}$ years. Although this timescale is long enough to explain UC H II regions, it is sufficiently short to have significant impact on the formation of planets around high-mass stars.

Surface disk material evaporates efficiently when it is heated to temperatures such that the average thermal speeds of the atoms or ions are comparable to the escape speed from the gravitational potential. Photoionization of hydrogen by Lyman continuum (extreme ultraviolet or EUV, $h\nu > 13.6$ eV) photons heats the gas at the surface of the disk to temperatures $T \sim 10^4$ K, forming an ionized atmosphere or corona above the disk. Beyond a critical distance from the central star, $r_g \simeq GM_\star/c_s^2$ ($\sim 10^{14}$ $[M_\star/M_\odot]$ cm), the heated gas becomes unbound and produces a thermal disk wind. Massive stars radiate copious EUV photons, which rapidly photoevaporate their outer disks. Young low-mass stars also have sufficient Lyman continuum flux to photoionize their outer disk material and create an appreciable photoevaporative flow. Alternatively, nearby massive stars in star-forming regions produce large fluxes of both EUV and FUV (for ultraviolet, 6 eV $< h\nu <$ 13.6 eV) photons, which can heat and evaporate the circumstellar disks around young low-mass stars.

1. Photoevaporation by the Central Star: Physical Processes and Semianalytical Models. The ultraviolet flux from the central star can be divided into the EUV, capable of ionizing hydrogen, and FUV, capable of dissociating H_2 and CO and ionizing C. The EUV flux ionizes and heats the surface of the disk to 10^4 K. The FUV flux penetrates the H II region created by the EUV flux and heats a primarily neutral layer of H and H_2 to temperatures of order 100–3000 K, depending on the magnitude of the flux and the density of the gas. These regions in the interstellar medium are often called photodissociation regions, or PDRs (e.g., Hollenbach and Tielens 1997).

Previous work has treated the effect of EUV photons, and we shall focus on that case here. However, the FUV photoevaporation by the central star is nearly completely analogous, and we will remark on the small differences for this case as they arise. We will also normalize our equations for the case of the photoevaporation by a low-mass ($M_\star = 1$ $M_\odot$) star, with an estimated EUV photon luminosity of 10^{41} s^{-1}. However, the equations are equally applicable to high-mass stars (indeed, as noted above, they were originally derived for this case).

Weak stellar wind case. If the central star has a "weak" stellar wind (we shall quantitatively describe the criterion for a weak wind below), then photoionization of the neutral hydrogen on the surface of the inner region of the disk should result in the formation of a bound ionized atmosphere with $T \sim 10^4$ K (c.f. Fig. 3a). Inside $r_g \simeq 10$ AU $M_\star/\mathrm{M}_\odot$, where the sound velocity is less than the escape velocity, the atmosphere can be approximated as hydrostatic and isothermal. The density of ionized hydrogen $n(r, z)$ in the inner region depends on the density at the base of the H II atmosphere $n_0(r)$ and the height z above the disk according to $n(r, z) = n_0(r)\exp(-z^2/H^2)$, where the scale height is given by: $H(r) = r^{3/2}/r_g^{1/2}$ (Hollenbach et al. 1994). Note that $H(r_g) = r_g$.

Outside of r_g the evaporated material flows from the disk at approximately the sound speed $v \sim c_s \simeq 10$ km s^{-1}. Here, $n(r, z) \simeq n_0(r)$ yields a reliable estimate near the disk ($z < r$), where the flow has little opportunity to expand. The mass loss rate from the disk $\dot{M}_{\rm ph}$ is given by $\dot{M}_{\rm ph} = 2m_{\rm H}c_s \int_{r_g}^{\infty} 2\pi n_0(r) r\, dr$, where the first factor of 2 accounts for both faces of the disk and $m_{\rm H}$ is the mass per hydrogen nucleus. A similar re-

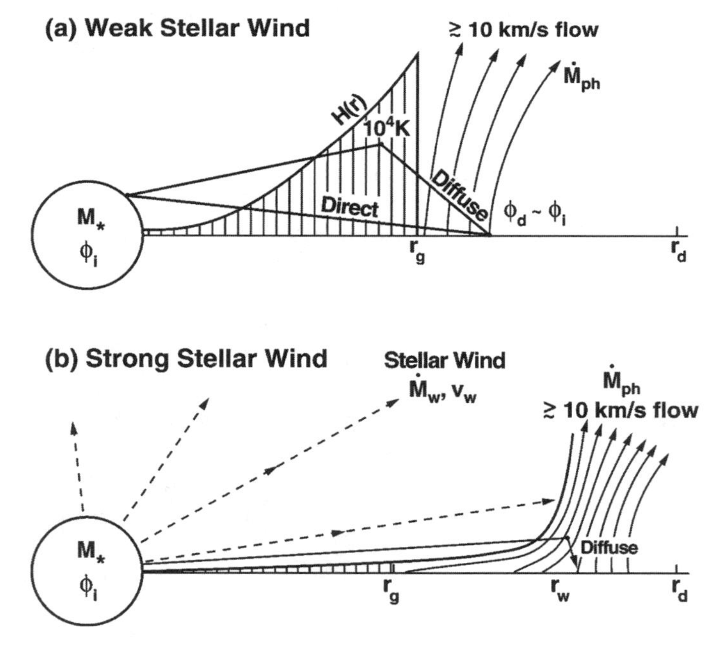

Figure 3. Schematic representation of photoevaporation by the central star. On the top (a) is the case with insignificant stellar wind. On the bottom (b) is the strong wind case. The scale height $H(r)$ of the disk is shown in (a), along with typical paths of direct and diffuse EUV photons.

lationship holds for flows controlled by FUV, where the sound speed is of order 1 to 3 km s^{-1} and r_g is correspondingly increased.

Estimating the mass loss rate and dispersal timescale of a disk by photoevaporation therefore reduces to determining $n_0(r)$, constrained by the above equations for $n(r, z)$ and $H(r)$, such that the UV photons are just absorbed at the base of the H II atmosphere or flow. For EUV photons, the recombination of electrons and protons in the ionized H II surface layers leads temporarily to H atoms, which absorb EUV photons and become reionized. Therefore, both dust and recombinations provide a source of opacity, which attenuates the incident EUV flux. A self-regulating feedback mechanism is established. The base density $n_0(r)$ adjusts itself so that EUV photons are nearly completely absorbed by the time they reach the base (i.e., there is an ionization front at the base). If $n_0(r)$ were lower than this value, the EUV photons would penetrate deeper into the dense neutral disk and would reach a higher $n_0(r)$. On the other hand, if the density were higher than this value, then the recombinations or dust in the overlying gas would provide such high opacity that the EUV photons would not penetrate to the base, which is contradictory.

In the case of EUV photons, both the recombinations in the atmosphere and flow as well as the scattering from dust in these regions provide a source of *diffuse* EUV photons. Approximately one-third of the recombinations is to the ground state of H, which produces isotropically radiated EUV photons. Hollenbach et al. (1994) show that the atmosphere absorbs a significant fraction of the EUV photon luminosity, Φ_i, of the central star and that the diffuse field dominates the direct EUV flux from the star in determining $n_0(r)$. In the case of FUV photons, dust scattering alone provides a diffuse source of photons, but this case has not yet been quantitatively analyzed. The dust grains that dominate the scattering and absorption at EUV and FUV wavelengths have typical radii ~100 Å. These small dust particles tend to remain in the upper atmospheres of the disks for substantial periods before settling and coagulating (Weidenschilling 1984, see also section II.D.4). However, if they do clump together and rain out from the upper layers of the disk in the absence of turbulence (e.g., Cameron 1995), the scattering could be reduced significantly.

Hollenbach et al. (1994) find an analytic solution for the case of EUV photons in the bound ($r < r_g$) region. They find that $n_0 \propto r^{-3/2}$ in the static region, whereas beyond the static region the power law steepens to $n(r) \propto r^{-\gamma}$, where $\gamma \gtrsim \frac{5}{2}$. Most of the mass loss occurs, then, in the region just outside of r_g. A complete analysis finds

$$\dot{M}_{\rm ph} = 4 \times 10^{-10}\ {\rm M_\odot\ yr^{-1}} \left(\frac{\Phi_i}{10^{41}\ {\rm s^{-1}}}\right)^{1/2} \left(\frac{M_\star}{\rm M_\odot}\right)^{1/2}$$

where $\Phi_i \sim 10^{41}$ s^{-1} is a crude estimate of the EUV luminosity of a young (less than 10^7 yr old) solar-type star (see section II.D.3 below). Both within and beyond the static region the ionized hydrogen is maintained primarily from diffuse photons produced in the atmosphere. The dominance of the diffuse field means that the exact shape of the thin disk does not enter (shadowing is not important). In addition, the radiative transfer ensures that $n_0(r)$ and $\dot{M}_{\rm ph}$ do not depend on the surface density distribution of the disk, as long as it is sufficient to maintain a dense neutral disk under the H II surface.

Numerical simulations of disks photoionized by a central EUV source with a weak stellar wind have been performed by Yorke and Welz (1994, 1996) and by Richling and Yorke (1997). Richling and Yorke (1997) include the effects of dust scattering and find a similar power law dependence of $\dot{M}_{\rm ph}$ on Φ_i (exponent ≈ 0.58) and comparable mass loss rates for high-mass stars ($\Phi_i \sim 10^{49}$ s^{-1}) in the range $10^{-7} \lesssim \dot{M}_{\rm ph}/\,{\rm M}_\odot\,{\rm yr}^{-1} \lesssim 10^{-5.5}$. However, the numerical simulations employed finite disks with central holes, whereas the disks discussed here are considered to extend from the stellar surface until $r \gg r_g$.

Strong stellar wind case. In the case of a strong wind from the central star, the ram pressure of the stellar wind, $\rho_w v_w^2$, will be too high for a static atmosphere to achieve a full scale height above the disk. Instead, the stellar wind will suppress the height of the ionized layer out to a radius $r_w > r_g$, where the thermal pressure of the ionized hydrogen flowing from the disk balances $\rho_w v_w^2$ (see Fig. 3b). The photoevaporated material can freely flow vertically off the disk at radii beyond $r_w \simeq 10$ AU $\dot{M}_{w-10}^2 v_{w7}^2 \Phi_{41}^{-1}$, where $\dot{M}_{w-10} = \dot{M}_w/10^{-10}$ M$_\odot$ yr^{-1}, $v_{w7} = v_w/100$ km s^{-1}, and $\Phi_{41} = \Phi_i/10^{41}$ s^{-1}. The criterion for a strong wind, $r_w > r_g$, can be written that the stellar wind mass loss rate $\dot{M}_w$ exceeds a critical value $\dot{M}_{\rm cr} = 4 \times 10^{-11} \Phi_{41}^{1/2} M_0^{1/2} v_{w7}^{-1}$ M$_\odot$ yr^{-1}, where $M_0 = M_\star/\,$M$_\odot$ (Hollenbach et al. 1994). Although there are large uncertainties in both $\dot{M}_w$ and Φ_i for young low-mass stars, it appears that the strong wind condition may often be met. With fewer uncertainties, it also is generally met for O and B stars (van Buren 1985).

The overall effect of the stellar wind is to allow EUV flux to penetrate more readily to the disk surface in the flow ($r > r_g$) region, which enhances the photoevaporative mass loss rate. Hollenbach et al. (1994) find, for the strong wind conditions, $\dot{M}_w > \dot{M}_{\rm cr}$,

$$\dot{M}_{\rm ph} \simeq 4 \times 10^{-10}\ {\rm M}_\odot\ {\rm yr}^{-1} \left(\frac{\dot{M}_w}{10^{-10}\ {\rm M}_\odot\ {\rm yr}^{-1}}\right)\left(\frac{v_w}{100\ {\rm km\ s}^{-1}}\right)$$

The photoevaporative mass loss rate from the outer disk now becomes slightly more than the mass loss rate in the assumed isotropic stellar wind. We emphasize that these approximate analytical solutions hold only when the disk extends to $r_d > r_w$.

The photoevaporation timescale t_c(evap) caused by the central star in a strong wind case is also plotted in Fig. 1. Here we have assumed that $r_w > r_d$, so $n_0 \propto r^{-3/2}$ from r_g to r_d (Hollenbach et al. 1994), and that $\beta = 3/2$, $\Phi_i = 10^{41}$ s^{-1}, and a minimum mass nebula. Since $\Sigma \propto r^{-3/2}$

is assumed, the photoevaporative timescale becomes independent of r for $r > r_g$:

$$t_c(\mathrm{evap}) \simeq 10^7 \ \mathrm{yr} \left(\frac{\Phi_i}{10^{41} \ \mathrm{s}^{-1}} \right)^{-1/2} \left(\frac{\Sigma_0}{\Sigma_0(\mathrm{min})} \right)$$

where Σ_0 is the disk surface density at a fiducial radius and $\Sigma_0(\mathrm{min})$ is the corresponding value for the minimum solar nebula.

A strong wind also affects the radiative transfer in two important ways. First, if it is partially neutral, it is a source of opacity for the EUV photons trying to travel from the star to r_g. Therefore, it could significantly lower the ratio of EUV to FUV photons arriving at r_g from the star. Secondly, the ionized material in the wind will recombine and thereby produce a "diffuse" source of EUV and FUV photons above the disk. Many of the EUV photons will be absorbed "on the spot" by the neutral component of the wind. These effects should be addressed quantitatively in the future.

Although no analytic solution has been proposed for a central star that radiates only FUV photons, the outline of the solution is clear. Whereas both hydrogen atoms and dust are opacity and scattering sources for EUV photons, dust alone dominates the FUV opacity. Dust also scatters the FUV (the albedo of interstellar dust is typically $\sim$0.3–0.6 in the FUV), and the isotropic component of the scattering produces a diffuse FUV field. The critical gravitational radius is now further out, because $r_g \propto c_s^{-2}$ and the PDR surface gas is cooler than 10^4 K. Again, the EUV and FUV flux from the wind should be evaluated.

The above analytic solutions (Hollenbach et al. 1994) made simple analytic approximations for the dynamics of an evaporating disk (e.g., $v_z = 0$ for $r < r_g$ and $v_z = c_s$ for $r > r_g$) and for the radiative transfer solution for the diffuse EUV field. In order to consider various effects such as disk evolution, non-steady-state heating/cooling, and ionization/recombination, as well as more detailed hydrodynamic interactions and radiative transfer, numerical simulations are required.

2. Photoevaporation by the Central Star: Numerical Simulations. The first numerical simulations of the evolution of circumstellar disks subjected to the influence of EUV photons from the central source were discussed by Yorke and Welz (1993) as a model for UC H II regions. Although the effects of disk rotation, self-gravity, IR heating of the dust, time-dependent heating and cooling, and hydrogen ionization and recombination were included in the 2D axially symmetric hydrodynamic code, these early models considered the direct stellar EUV flux only; sharp shadows were cast by the disk. Including the effects of the diffuse EUV field produced by recombinations directly into the ground state (Yorke and Welz 1994, 1996) allowed the ionization front to wrap around and envelop the disk, but these "soft" EUV photons were able to heat the ionized material in the disk's shadow regions only to $\approx$2000 K, much lower than the value ≈8000 K of the ionized material directly illuminated by the central star. Richling and Yorke (1997) found that if the effects of UV dust scattering were included, the "shadow" regions quickly disappeared and the ionized gas attained temperatures ≈8000 K everywhere. The mass loss rate was enhanced by a factor of about 2 for UV dust opacities $\kappa_{\rm uv} = 200$ cm^2 g^{-1}. The top two panels of Fig. 4 show results from their numerical simulations of stars with no stellar winds. This scattering result is for the case of high-mass stars, which produce relatively high ($\sim$1) dust optical depths in the disk atmosphere. Solar type stars will produce much lower dust optical depths, and will show much less effect from dust scattering.

For the cases with strong stellar winds, Yorke and Welz (1996) and Richling and Yorke (1997) find a much weaker dependence of the disk mass loss rate with wind velocity (a power law exponent 0.1 rather than 1) when compared to the analytic models discussed here and by Hollenbach et al. (1994). Because, however, the numerical simulations consider cases of very massive disks $M_d \gtrsim 0.2\ M_\star$ with a central cavity and a finite outer radius r_d of order r_w, they are not readily comparable to the analytic results, which assume $r_d \gg r_w$ and no central cavity. In the numerical simulations the originally isotropic stellar wind interacts with the outflowing photoevaporation flow from the disk to produce shocks in both the wind material and the photoevaporation flow, which are separated by a contact discontinuity (see bottom panel of Fig. 4, from Kessel et al. 1998). The shock-heated wind material seeks the easiest path of escape, which is along the rotational axis. A hydrodynamically collimated bipolar outflow results, the opening angle of which increases slowly with increasing wind velocity. For warm ($T \approx 10^4$ K), low-velocity ($v_w \approx 50$ km s^{-1}) winds, the mass loss rate due to photoevaporation $\dot{M}_{\rm ph}$ actually *decreases* with increasing stellar wind mass loss $\dot{M}_w$, because the wind itself absorbs EUV photons.

3. Photoevaporation of the Solar Nebula by the Sun and the Formation of the Giant Planets. The introduction to this chapter presents the current picture of giant planet formation and concludes that (1) this process is a minor disk dispersal mechanism and (2) $M_d \approx$ several $M_{\rm min}$. As discussed by Shu et al. (1993), several serious issues remain in this picture. What dispersed all the residual gas in this enhanced-mass nebula? Removing $\sim$0.1 M$_\odot$ of gas stored in a thin disk by a T Tauri wind (mass loss rate $\sim 10^{-8}$ M$_\odot$ yr^{-1} for 10^7 yr) is a questionable proposition, whereas transporting this much material inward to the center by viscous accretion would push any planet (like Jupiter) that has cleared a gap around itself also into the sun (for the basic dynamical mechanisms, see Goldreich and Tremaine 1979, 1980; Lin and Papaloizou 1979, 1985; Hourigan and Ward 1984; Ward 1986).

Even if we ignore the timescale problem for accumulating critical cores for the giant planets and assume that the giant planets somehow formed under conditions that resembled the minimum solar nebula, we

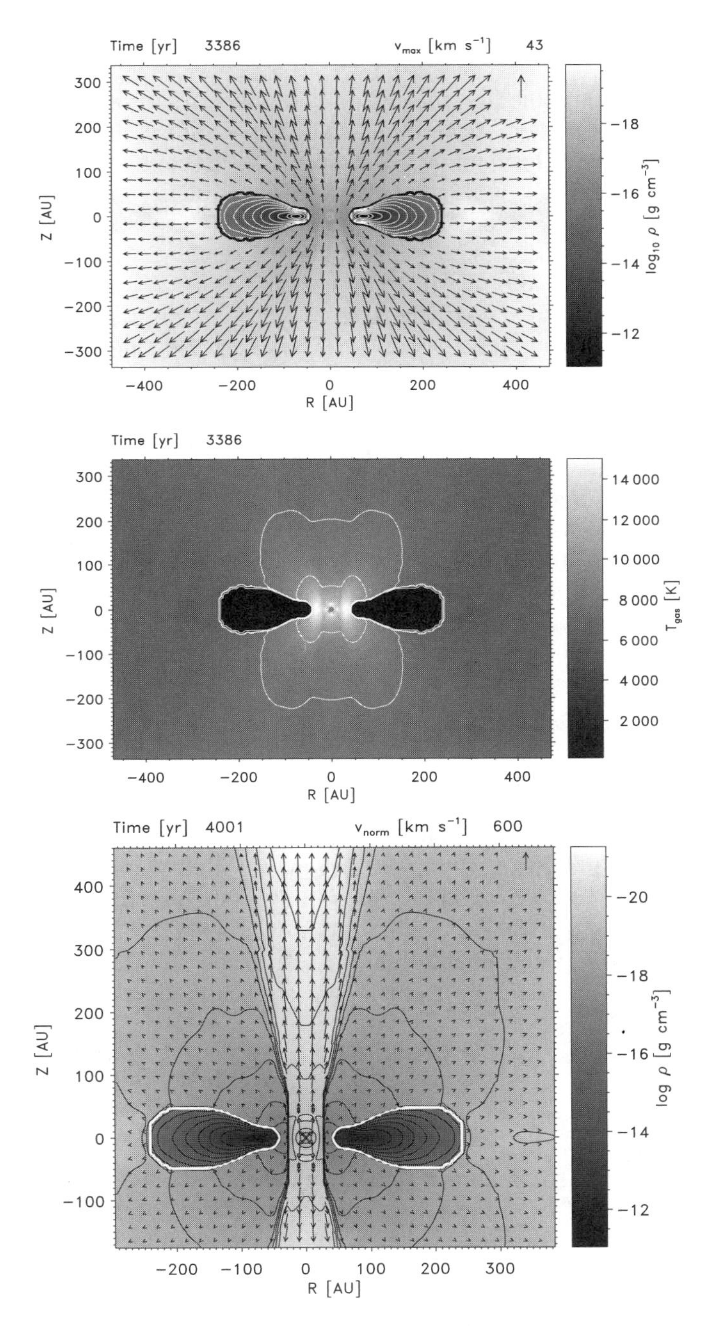

Figure 4. Numerical simulations of the photoionization and photoevaporation of circumstellar disks by a central EUV source (Richling and Yorke 1997). Top: Density (gray scale + contours), ionization front (thick line) and velocity structure (defined at tips of arrows; the velocity scale is given at the upper right of each frame) are shown long after a quasi-steady-state flow has been achieved. Middle: The corresponding temperature structure is displayed. Bottom: The resulting quasi-steady-state structure of a simulation that included the effects of a central stellar wind (Kessel et al. 1998).

run into a second difficulty: the delicate timing between core growth and nebula dispersal that must seemingly be achieved to yield the differential results that we observe for the masses of the gaseous envelopes of the four giant planets. The rock-ice cores of the giant planets are all about 15 Earth masses ($M_\oplus$), but the hydrogen/helium envelopes for Jupiter, Saturn, Uranus, and Neptune contain $\sim$300, $\sim$75, $\sim$2, and $\sim$2 $M_\oplus$ respectively (Shu et al. 1993). Note that there is a sharp cutoff in hydrogen/helium mass between Saturn and Uranus, even though the core masses are similar and are presumably large enough in Uranus and Neptune to initiate the rapid accumulation of hydrogen and helium gas. The problem arises because most theories of disk dispersal by thermal evaporation or solar wind erosion (e.g., Cameron 1978; Elmegreen 1978; Horedt 1978, 1980, 1982; Sekiya et al. 1980; Pechernikova and Vitjazev 1981; Ohtsuki and Nakagawa 1988) predict gas loss rates that vary smoothly with time and radial position in the disk, whereas the usual core instability picture demonstrates that the process of runaway gas gathering proceeds very quickly once it gets going. Adjusting the parameters appropriately to explain Jupiter and Neptune will typically leave Uranus and Saturn complete mysteries. The problem then becomes to identify the gas removal mechanism that would satisfactorily explain the sharp differences in envelope masses between the gas-rich giants, Jupiter and Saturn, and the gas-poor giants, Uranus and Neptune.

Shu et al. (1993) point out an interesting coincidence between $r_g \approx$ 10 AU for the solar system and the radial position that separates the gas-rich giants from the gas-poor ones. The transition between photoionized gas being gravitationally trapped in the solar nebula and flowing off in an evaporative wind occurs roughly at the orbital distance of Saturn. If the photoevaporative rate of mass loss $\dot{M}_{\rm ph}$ beyond r_g is sufficiently large, no hydrogen and helium may be left in the nebula when Uranus and Neptune acquired large enough core masses to enter the rapid gas-gathering phase.

As we have shown in section II.D.1 above, the EUV photoevaporative mass loss rate caused by the central star varies as $\Phi^{1/2}$. If the photoionization rate $\Phi_{\rm i}$ of the primitive sun equals 10^{41} s^{-1} for $\sim$$10^7$ yr, an amount of mass comparable to a minimum solar nebula could have been removed by such a process from the outer disk beyond $r_g \approx 10$ AU in this period. The rate $\Phi_{\rm i} = 10^{41}$ s^{-1} exceeds the ultraviolet output of the present Sun by two orders of magnitude, but it is compatible with the observed output of classical T Tauri stars (cTTSs) (see, e.g., Gahm et al. 1979; Bertout 1989). To explain the loss of gases from the Earth's primitive atmosphere by hydrodynamic "blow-off," other workers have postulated EUV rates $\Phi_{\rm i} \sim 10^{41}$ s^{-1} from the early sun lasting for a few hundred million years (Sekiya et al. 1980; Hunten 1985, 1993; Zahnle and Kasting 1986), so assuming this rate for 10^7 yr is conservative in comparison.

Mass accretion from the inner disk onto the stellar surface (Bertout et al. 1988), possibly aided by magnetic fields (Königl 1991), may provide the simplest explanation for the ultraviolet excesses of cTTSs. According to the empirical evidence presented by Strom et al. (1989, 1993), the cTTS phase lasts about 10^7 yr. If the cTTSs have outer disks that do not contain much more gas than a minimum solar nebula (see, e.g., Beckwith et al. 1990; Terebey et al. 1993), then photoevaporation may well cause the giant planets that form beyond r_g to be gas-poor giants of the ilk of Uranus and Neptune.

We discussed in section II.D.1 how the early sun may have had a strong, partially ionized wind that may have absorbed EUV photons from the sun but may itself have produced an EUV and FUV diffuse field. This effect has not been studied, and may well affect the mass loss rate and, to a much lesser extent, r_g. However, the basic point that significant mass loss occurs at a radius outside Saturn's orbit but inside that of Uranus remains valid.

Shu et al. (1993) speculated that the solar nebula actually formed with a mass much greater than the minimum mass. Giant planets that formed early, when the outer disk was still massive, were pushed by the viscous evolution of the disk into the protosun. Eventually, the combination of viscous accretion and photoevaporation lowered the gas mass exterior to Jupiter's orbit to a value less than a few times Jupiter's mass. At this point, the gas could no longer push the giant planets into the protosun, and the remaining gases were dispersed or accreted onto the giant planets. By this way of thinking, the "minimum solar nebula" does not define the mass of Jupiter; rather the mass of Jupiter defines the conditions of the "protoplanetary" nebula. When the role of photoevaporation is added, this line of thought has the additional noteworthy implication that the resultant disk properties may also automatically produce transitional-case planets like Saturn and gas-poor planets like Uranus and Neptune.

4. Photoevaporation of Circumstellar Disks by an External Star. For low-mass stars born in clusters, a significant fraction of the UV photons incident on the protoplanetary disks may be produced by nearby massive stars and may significantly enhance the photoevaporation process. The "proplyds" (externally ionized *pro*to*pl*anetar*y* *d*isk*s*) in the Orion Nebula are the best example of this scenario and have been extensively studied (Bally et al. 1998; O'Dell 1998 and references therein). The Trapezium cluster, whose most massive member, θ^1Ori C, resides at the cluster center and produces the H II region, contains over 700 stars with a peak density of 5×10^4 pc^{-3} in the central 0.1-pc-diameter core (McCaughrean and Stauffer 1994). Within the H II region over 150 proplyds are visible as small ($r \lesssim 100$ AU) ionized envelopes surrounding low-mass stars. Churchwell et al. (1987), using radio continuum observations, first proposed that evaporating circumstellar disks produced the envelopes. More recently, the HST has imaged numerous disks as silhouettes, either against the nebula background (McCaughrean and O'Dell 1996) or against the background ionized envelope (Bally et al. 1998). The disks glow in [O I] 6300 Å emission as well (Bally et al. 1998).

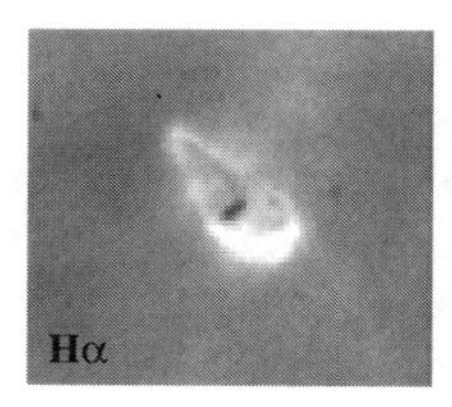

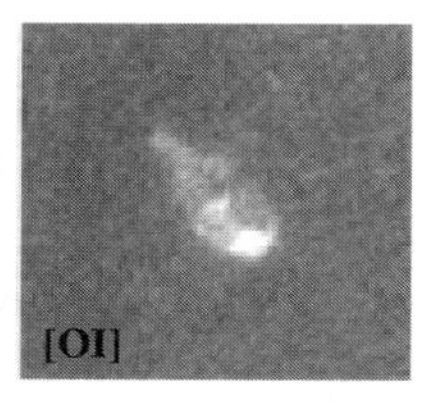

Figure 5. Hubble Space Telescope images of HST 182-413 (HST 10) in Hα (left) and [O I] 6300 Å (right). The disk appears in absorption for most optical lines except [O I]. The ionization front, seen clearly in Hα, is offset from the disk surface (Johnstone et al. 1998).

In Fig. 5 we show two narrowband images of proplyd HST 182-413 (HST 10). The head-tail appearance is characteristic of all proplyds, with the head pointing toward θ^1Ori C (although not always exactly). HST 182-413 contains a disk ($r_d \simeq 100$ AU) visible as an edge-on silhouette in all filters except [O I] and H_2 (Chen et al. 1998), where the disk emits. Surrounding the disk, yet well offset, is an envelope of ionized gas (the teardrop shape bright emission). As we shall discuss below, neutral material is removed from the disk surface at a rate higher than the incident EUV flux can keep entirely ionized, producing the offset ionization front.

In this section we discuss the photoevaporation of circumstellar disks by external stars using the Orion Nebula observations as a guide. The first assumption made in models of externally illuminated disks is that the disk radius $r_d > r_g$, so that evaporation occurs. For externally illuminated disks, this means that r_g effectively drops out of the problem. The incident EUV or FUV flux sets up a flow of constant mass flux from the disk surface for $r > r_g$. In contrast to photoevaporative flows caused by the central star, where the mass loss was dominated by the *inner* $r \sim r_g$ or r_w region close to the source, the mass loss for external illumination is dominated by the *outer* $r \sim r_d$ region, where most of the surface area lies.

Since most of the mass loss is from $r \sim r_d$, the evaporation from the disk can be approximated by the evaporation from a sphere of radius r_d. Pressure gradients in the flow will cause the streamlines to diverge rapidly beyond r_d and approximate spherical outflow. The "surface" of the disk (sphere), for the purposes of photoevaporation, can be defined as the sonic surface, where the disk flow changes from subsonic to supersonic flow. As discussed in section II.D.1, it lies at the point where the EUV or FUV flux is attenuated to the point of just heating the gas to a temperature such that the sound speed equals the escape speed. If EUV dominates at the sonic point, either dust or recombinations can attenuate the flux. If FUV dominates, then dust is the prime opacity source. Let n_0 be the gas density at the disk surface or sonic point. Then the photoevaporative mass loss from the sphere is given by $\dot{M}_{\rm ph} \simeq m_{\rm H} n_0 c_s 4\pi r_d^2$, where c_s is the sound speed in the heated (10^4 K ionized or $\sim 10^3$ K neutral) flow. Therefore, to calculate $\dot{M}_{\rm ph}$, one need only estimate n_0. The base density n_0 is given by the condition that the EUV or FUV be (just) absorbed in the flow.

If dust dominates the EUV or FUV opacity, then the dust optical depth $\tau_{\rm UV}$ in the flow is proportional to the gas column density N_D in the flow. Because the density falls off rapidly due to spherical divergence, $N_D \simeq n_0 r_d$. For interstellar dust, $N_D \simeq 10^{21}$ cm^{-2} gives $\tau_{\rm UV} \simeq 1$. Given that N_D is fixed to give $\tau_{\rm UV} \sim$ few, it follows that $n_0 \propto r_d^{-1}$; the base density is smaller for larger disks. In the case $n_0 r_d \sim$ constant, $\dot{M}_{\rm ph}$ is quite independent of the incident UV flux (as long as it is sufficient to heat the gas so that $c_s \sim$ constant), and $\dot{M}_{\rm ph} \propto r_d$.

If recombinations dominate the EUV opacity, then the recombinations in a radial column ($\sim\frac{1}{3}\alpha_r n_0^2 r_d$, where α_r is the recombination coefficient for 10^4 K ionized gas) balance the incident EUV flux ($F_i = \Phi_i/(4\pi d^2)$, where d is the distance to the EUV source). In this case, $\dot{M}_{\rm ph}$ depends on the incident EUV flux ($\dot{M}_{\rm ph} \propto F_i^{1/2}$) and $\dot{M}_{\rm ph} \propto \Phi_i^{1/2} d^{-1} r_d^{3/2}$.

Johnstone et al. (1998) and Störzer and Hollenbach (1999) have showed that the photoevaporation rates of disks around low-mass stars illuminated by nearby massive stars may either be dominated by the EUV or the FUV photon flux from the high-mass star. In the case of EUV-dominated flow, a thin PDR region ($\Delta r \lesssim r_d$) is produced at the disk surface by the FUV flux, but the neutral gas moves subsonically through the PDR to the ionization front (IF), where the flow is ionized, heated to 10^4 K, and accelerated to supersonic speeds, $v_f \gtrsim 10$ km s^{-1}. The mass loss rate, if recombinations dominate the opacity, is given from the above

$$\dot{M}_{\rm ph}^{\rm EUV} \simeq 7 \times 10^{-9}\ {\rm M_\odot\ yr^{-1}} \left(\frac{\Phi_i}{10^{49}\ {\rm s^{-1}}}\right)^{1/2} \left(\frac{r_d}{10\ {\rm AU}}\right)^{3/2} \left(\frac{10^{17}\ {\rm cm}}{d}\right)$$

where $\Phi_i \sim 10^{49}$ s^{-1} for an early O star as nearby neighbor.

On the other hand, in FUV-dominated flows, the FUV produces sufficient PDR pressure that a supersonic neutral flow is launched from the disk surface. The ionizing EUV flux cannot penetrate this opaque flow until its density has dropped significantly. The ionized H II flow commences in an IF at a standoff distance $r_{\rm IF} \gtrsim 2r_d$. Moving outwards from the disk surface, the supersonic neutral wind first passes through a shock, decelerates to subsonic speeds, forms a thick subsonic layer, and then passes through a stationary IF at $r_{\rm IF}$ where the flow is heated to 10^4 K and reaccelerated to supersonic ($v_f \gtrsim 10$ km s^{-1}) speeds. The mass loss rate in this case can be approximated

$$\dot{M}_{\rm ph}^{\rm FUV} \simeq 2 \times 10^{-8}\ {\rm M_\odot\ yr^{-1}} \left(\frac{N_D}{5 \times 10^{21}\ {\rm cm^{-2}}}\right)\left(\frac{r_d}{10\ {\rm AU}}\right)$$

where N_D ($\sim 5 \times 10^{21}$ cm^{-2} in the numerical studies of Störzer and Hollenbach 1999) is the column density from the ionization front to the sonic point (disk surface).

Given the UV flux incident on the disk and the disk size r_d, the photoevaporation model predicts $r_{\rm IF}$, with only one "free" parameter, the UV opacity of the dust grains in the flow. Approximately 10 proplyds have been observed in Orion, where both r_d and $r_{\rm IF}$ are directly measured with some accuracy (Johnstone et al. 1998). Störzer and Hollenbach (1999) used these sources to derive an average UV dust opacity in the evaporating flows and found a dust cross section per hydrogen nucleus of $\sigma_{\rm UV} \simeq 8 \times 10^{-22}$ cm^2. This value is only a factor of 2 or 3 less than diffuse interstellar values, and compatible with Orion Nebula or dense molecular cloud values (Draine and Bertoldi 1996). Therefore, consistent with the expectations of settling and coagulation of dust particles in disks (Weidenschilling 1984; Weidenschilling and Cuzzi 1993), the abundance of small ($\sim$100 Å) dust particles, which dominate the UV opacity, is relatively unchanged in the upper atmospheres of these young ($t \lesssim 10^6$ yr) disks at 10–100 AU.

Using this value for $\sigma_{\rm UV}$, Störzer and Hollenbach (1999) derive the parameter space for EUV- and FUV-dominated flows for the proplyds in Orion. FUV-dominated flows occur roughly when 10 AU $\lesssim r_d \lesssim$ 100 AU and when 0.01 pc $\lesssim d \lesssim$ 0.3 pc, where d is the distance to the dominant UV source θ^1 Ori C. For large distances or small disks, the FUV cannot heat the PDR to sufficient temperature to drive evaporative flow (i.e., $c_s < v_{\rm escape}$). For small distances or large disks, the EUV penetrates close to the disk surface and determines the mass loss.

Johnstone et al. (1998) and Störzer and Hollenbach (1999) analyzed the observational data from approximately 40 proplyds and showed that many of the flows, including HST 182-413 (Fig. 5), are likely to be FUV-dominated. Störzer and Hollenbach (1998, 1999) further showed that the FUV-dominated flow models match perfectly with the observed intensities of the optical emission near the IF as well as the neutral [O I] 6300 Å and H_2 2-μm intensities coming from the disk surface. Henney and Arthur (1998) show that a photoevaporation model fits the observed falloff in optical emission with distance from the proplyd. Johnstone et al. (1998) proposed that the observed tails of the comet-shaped ionization fronts could be caused by the diffuse UV radiation driving an FUV-dominated flow from the shadowed side of the proplyd. Figure 6 shows a numerical confirmation of this proposal from Richling and Yorke (1999). The numerical study used such a large disk ($r_d \sim$ 300 AU) that the front side is EUV-dominated. Nevertheless, on the back side an FUV-dominated flow with an IF shaped like the observed tails is seen in the simulation. All these results give added confidence in the validity of the FUV-dominated photoevaporation models.

Assuming that $\sigma_{\rm UV}$ is approximately the same for all the observed proplyds, the theoretical models are able to predict the disk sizes from the observed sizes of the ionization fronts. Johnstone et al. (1998) derived the disk radii for approximately 30 proplyds in which $r_{\rm IF}$ was measured but where the disk could not be seen. The disk radii ranged from 10 to 100 AU.

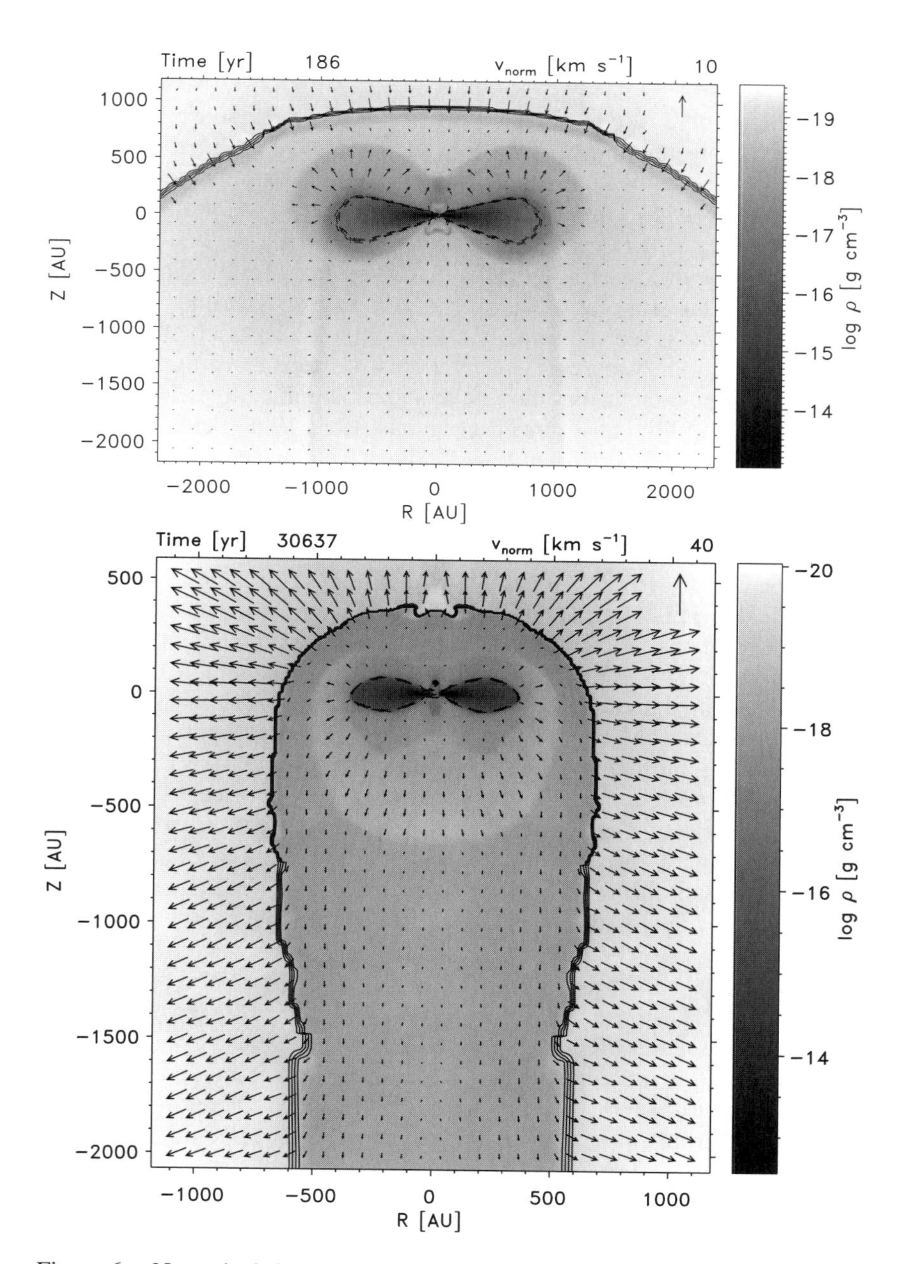

Figure 6. Numerical simulations of the photoionization and photoevaporation of circumstellar disks by an external EUV and FUV source (Richling and Yorke 1998*b*). The top frame depicts an early stage after the sudden turn-on of the source located above the disk, with fluxes $f_{\rm EUV} = 6.3 \times 10^{12}$ cm^{-2} s^{-1} and $f_{\rm FUV} = 1.6 \times 10^{13}$ cm^{-2} s^{-1}. The appearance at this time of a second simulation with a higher FUV flux $f_{\rm FUV} = 1.3 \times 10^{14}$ cm^{-2} s^{-1} is nearly identical. The bottom frame shows the quasi-steady-state result of this simulation. Note the significant reduction of disk size due to removal of the loosely bound outer disk material. The FUV-dominated flow on the back side creates a cometlike tail.

If the proplyds are in circular orbits around the UV source θ^1 Ori C, so that d is constant in time, Johnstone et al. (1998) and Störzer and Hollenbach (1999) showed how the current disk mass M_d could be estimated from the current $\dot{M}_{\rm ph}$ and the disk illumination time t_i. Assuming that the surface density in the disk scales as $\Sigma \propto r^{-\beta}$, where $1 < \beta < 2$, most of the mass in the disk is at $r \sim r_d$, the disk evaporates from outside in, and the timescale for evaporation $t_{\rm evap}$ increases as the disk shrinks. (Note that $t_{\rm evap} = M_d/\dot{M}_{\rm ph}$ is not the timescale for complete dispersal, but only for the mass and radius to decrease by factors of order 2). If the disks initially are large enough that they have shrunk significantly after time t_i, $M_d \simeq \dot{M}_{\rm ph} t_i$. Millimeter-wave observations (Lada 1998) detect two disks with $M_d \sim 0.01$ $M_\odot$ and limit several others to smaller masses. Observed and theoretical estimates (Johnstone et al. 1998; Störzer and Hollenbach 1999) of $\dot{M}_{\rm ph} \gtrsim 10^{-7}$ $M_\odot$ yr^{-1} then limit $t_i < 10^5$ yr. In the case of circular orbits, it appears that θ^1 Ori C must be less than 10^5 years old, surprisingly small compared to the $\sim 10^6$-yr age estimated for the low-mass stars in the cluster (Hillenbrand 1997).

On the other hand, the proplyds are likely to be on much more eccentric orbits around θ^1 Ori C. (Note that the cluster potential is dominated by the distributed mass for $d \gtrsim 0.05$ pc, so the orbits are not Keplerian in any case.) Störzer and Hollenbach (1999) considered the opposite extreme of radial orbits and showed that, in this case, the disks may shrink and lose mass as they approach θ^1 Ori C. However, inside a critical distance $d_{\rm cr}$, their dynamical time is shorter than $t_{\rm evap}$, so they move quickly through the central region "frozen" in mass and size. For disks that have experienced significant shrinkage and which lie inside of $d_{\rm cr}$ (as most of the ~ 40 well-studied proplyds do), the current mass of the disk is approximately $M_d \simeq \dot{M}_{\rm ph}(d_{\rm cr}/v)$, where $\dot{M}_{\rm ph}$ is evaluated at $d_{\rm cr}$ and where $d_{\rm cr}/v$ is the crossing time t_c for a proplyd moving at v through the central region. Hillenbrand and Hartmann (1998) estimate $v \simeq 2.3$ km s^{-1}, or $t_c \simeq 10^5$ yr. Störzer and Hollenbach (1999) derive theoretically that $d_{\rm cr} \simeq 0.2$ pc and that $\dot{M}_{\rm ph}$ at $d_{\rm cr}$ is roughly fitted by the analytic approximation to numerical models $\dot{M}_{\rm ph} \simeq 10^{-7}(r_d/100\ {\rm AU})^{1-1.5}$ $M_\odot$ yr^{-1}. Therefore, the disk masses for shrunken proplyds passing near ($d < d_{\rm cr}$) θ^1Ori C are given by $M_d \simeq 10^{-2}(r_d/100\ {\rm AU})^{1-1.5}$ $M_\odot$; the disk masses are roughly proportional to their sizes. Since most of the 40 well-studied proplyds have $r_d \lesssim 100$ AU, the models predict $M_d \lesssim 10^{-2}$ $M_\odot$, in agreement with observation. In this case of radial orbits, there is no need for θ^1Ori C to be exceptionally young. The Trapezium cluster extends to $d \sim 2$ pc, has a half-mass radius of $d \sim 0.5$ pc, and a typical crossing time across the whole cluster of $\sim 10^6$ yr (Hillenbrand and Hartmann 1998). If θ^1Ori C is $\sim 10^6$ years old and the entire cluster is on radial orbits, then each of the ~ 1000 proplyds in the cluster has passed close to θ^1Ori C roughly one time and has photoevaporated to $r_d \sim 10$–100 AU. We presently observe mostly the nearby bright objects inside of $d_{\rm cr}$. Those foreground proplyds that have not passed through the central regions and have stayed outside the H II region ($d \gtrsim 0.6$–1 pc), perhaps because their orbits are more circular, remain large ($r_d \sim 100$–500 AU) and are observed as silhouettes (McCaughrean and O'Dell 1996). In the case of the Trapezium cluster, one O star has significantly affected the disk evolution of perhaps hundreds of low-mass stars like the Sun through the process of photoevaporation.

Although it appears that most stars in the Galaxy are formed in clusters (Lada et al. 1991; Lada 1992; Bonnell and Kroupa 1998) and that these clusters have sizes 0.1–1 pc, it is not clear how many stars form in a typical cluster. If that number is greater than about 100, then it is very likely that an O or B star will form among its members, and that photoevaporation by an an external star will affect the fate of most planet-forming disks in the Galaxy. In Fig. 1 we have also plotted the photoevaporation timescale t_E(evap) as a function of r for disks around low-mass stars in clusters like the Trapezium cluster.

III. CONCLUSIONS

Figure 1 shows the timescales for various gas (and small dust) dispersal mechanisms in planet-forming disks as a function of the position r in the disk. The stellar encounter lifetime $t_{\rm SE}$ is plotted for Trapezium-like conditions ($n_\star = 10^4$ pc^{-3} and $v = 1$ km s^{-1}) and clearly is only important in the outer regions, $r \gtrsim 100$ AU, of planet-forming disks in dense clusters. The viscous timescale t_ν is plotted for two representative values of α and are seen to be dominant in the inner, $r < 10$ AU, parts of disks. The wind timescales, t_{ws} and t'_{ws}, are plotted assuming a minimum-mass ($M_d \simeq 0.01$ M$_\odot$) disk with $\beta = 3/2$ and with $\epsilon(\sin\theta)^2 = 10^{-4}$. Wind mass loss rates of $\dot{M}_w = 10^{-8}$ M$_\odot$ yr^{-1} are assumed for t_{ws}; in the t'_{ws} case, we have assumed that the wind mass loss rate declines with time as $\dot{M}_w = 10^{-6}(10^5\ {\rm yr}/t)$ M$_\odot$ yr^{-1} for $t < 10^7$ yr. The winds are assumed to be spherically symmetric. Although there is considerable uncertainty in $\epsilon(\sin\theta)^2$, it appears that stellar wind stripping is not particularly effective, in contrast to the commonly held view of the dispersal of the solar nebula. The photoevaporation timescale caused by the central star, t_c(evap), is plotted assuming $\Phi_i = 10^{41}$ s^{-1}, the "strong wind" case, and a minimum-mass nebula with $\beta = 3/2$. The photoevaporation timescale caused by an external star, t_E(evap), is plotted for Trapezium conditions, assuming radial orbits, a cluster of size 1 pc, and a minimum-mass nebula with $\beta = 3/2$. The photoevaporation timescales become very long at $r < 10$ AU because the stellar gravity holds the disk material and prevents evaporation. Photoevaporation is generally dominant for $r > 10$ AU. Photoevaporation acting together with viscous spreading of material will remove the entire disk in timescales of order 10^7 years for the nominal parameters assumed in Fig. 1.

The rapid cutoff between Saturn and Uranus in the hydrogen and helium gas content of the giant planets may be explained by the photoevaporation of the outer, $r > 10$ AU, solar nebula early in its history before Uranus and Neptune could accumulate the gas. Shu et al. (1993) proposed that the early sun provided the UV photons. However, it is also quite possible that the solar nebula could have spent its early life in a stellar cluster, where a nearby O or B star may have photoevaporated the outer nebula in timescales $t \lesssim 10^7$ years.

Acknowledgments This work was conducted at the authors' three affiliated institutes and has been supported in part through the NASA Astrophysics Theory Program, which supports a joint Center for Star Formation Studies at NASA/Ames Research Center, UC Berkeley, and UC Santa Cruz; in part through NASA sponsorship at Jet Propulsion Laboratory, California Institute of Technology, within the "Origins" Program; and in part through funding by the Natural Sciences and Engineering Research Council of Canada.

REFERENCES

Adams, F. C., Emerson, J. P., and Fuller, G. A. 1990. Submillimeter photometry and disk masses of T Tauri disk systems. *Astrophys. J.* 357:606–620.

Balbus, S. A., and Hawley, J. F. 1991. A powerful local shear instability in weakly magnetized disks. I: Linear analysis. II: Nonlinear evolution. *Astrophys. J.* 376:214–233.

Bally, J., and Scoville, N. Z. 1982. Structure and evolution of molecular clouds near H II regions. II: The disk constrained H II region, S106. *Astrophys. J.* 255:497–509.

Bally, J., Sutherland, R. S., Devine, D., and Johnstone, D. 1998. Externally illuminated young stellar environments in the Orion Nebula: Hubble Space Telescope planetary camera and ultraviolet observations. *Astron J.* 116:293–321.

Beckwith, S. V. W., Sargent, A. I., Chini, R. S., and Guesten, R. 1990. A survey for circumstellar disks around young stellar objects. *Astron J.* 99:924–945.

Bertout, C. 1989. T Tauri stars: Wild as dust. *Ann. Rev. Astron. Astrophys.* 27:351–395.

Bertout, C., Basri, G., and Bouvier, J. 1988. Accretion disks around T Tauri stars. *Astrophys. J.* 330:350–373.

Bodenheimer, P., and Pollack, J. B. 1986. Calculations of the accretion and evolution of giant planets: The effects of solid cores. *Icarus* 67:391–408.

Bonnell, I., and Kroupa, P. 1998. Dynamical interactions in dense stellar clusters. In *The Orion Complex Revisited,* Astronomical Society of the Pacific Conf. Series, ed. M. J. McCaughrean and A. Burkert, in press.

Burrows, C. J., Stapelfeldt, K. R., Watson, A. M., Krist, J. E., Ballester, G. E., Clarke, J. T., Crisp, D., Gallagher, J. S., Griffiths, R. E., Hester, J. J., Hoessel, J. G., Holtzman, J. A., Mould, J. R., Scowen, P. A., Trauger, J. T., and

Westphal, J. A. 1996. Hubble Space Telescope observations of the disk and jet of HH 30. *Astrophys. J.* 473:437–451.
Cameron, A. G. W. 1973. Accumulation processes in the primitive solar nebula. *Icarus* 18:407–450.
Cameron, A. G. W. 1978. Physics of the primitive solar nebula and of giant gaseous protoplanets. In *Protostars and Planets,* ed. T. Gehrels and M. S. Matthews (Tucson: University of Arizona Press), pp. 453–487.
Cameron, A. G. W. 1995. The first ten million years in the solar nebula. *Meteoritics* 30:133–161.
Canto, J., and Raga, A. 1991. Mixing layers in stellar outflows. *Astrophys. J.* 372:646–658.
Chen, H., Bally, J., O'Dell, C. R. McCaughrean, M. J., Thompson, R. L., Rieke, M., Schneider, G., and Young, E. T. 1998. 2.12 micron molecular hydrogen emission from circumstellar disks embedded in the Orion Nebula. *Astrophys. J. Lett.* 492:L173–L176.
Churchwell, E., Felli, M., Wood, D. O. S., and Massi, M. 1987. Solar system sized condensations in the Orion Nebula. *Astrophys. J.* 321:516–529.
Clarke, C. J., and Pringle, J. E. 1993. Accretion disc response to a stellar fly-by. *Mon. Not. Roy. Astron. Soc.* 261:190–202.
Draine, B. T., and Bertoldi, F. 1996. Structure of stationary photodissociation fronts. *Astrophys. J.* 468:269–289.
Elmegreen, B. G. 1978. On the interaction between a strong stellar wind and a surrounding disk nebula. *Moon Plan.* 19:261–267.
Elmegreen, B. G. 1979. On the disruption of a protoplanetary disk nebula by a T Tauri like solar wind. *Astron. Astrophys.* 80:77–78.
Gahm, G. F., Fredga, K., Liseau, R., and Dravins, D. 1979. The far-UV spectrum of the T Tauri star RU Lupi. *Astron. Astrophys.* 73:L4–L6.
Goldreich, P., and Tremaine, S. 1979. The excitation of density waves at the Lindblad and corotation resonances by an external potential. *Astrophys. J.* 233:857–871.
Goldreich, P., and Tremaine S. 1980. Disk–satellite interactions. *Astrophys. J.* 241:425–441.
Greenberg, R., Hartmann, W. K., Chapman, C. R., and Wacker, J. F. 1978. Planetesimals to planets: Numerical simulation of collisional evolution. *Icarus* 35:1–26.
Greenberg, R., Weidenschilling, S., Chapman, C. R., and Davis, D. R. 1984. From icy planetesimals to outer planets and comets. *Icarus* 59:87–113.
Hall, S. M., Clarke, C. J., and Pringle, J. E. 1996. Energetics of star-disc encounters in the nonlinear regime. *Mon. Not. Roy. Astron. Soc.* 278:303–320.
Hartmann, L., Calvet, N., Gullbring, E., and D'Alessio, P. 1998. Accretion and the evolution of T Tauri disks. *Astrophys. J.* 495:385–400.
Hayashi, C., Nakazawa, K., and Nakagawa, Y. 1985. Formation of the solar system. In *Protostars and Planets II,* ed. D. C. Black and M. S. Matthews (Tucson: University of Arizona Press), pp. 1100–1153.
Heller, C. H. 1995. Encounters with protostellar disks. II: Disruption and binary formation. *Astrophys. J.* 455:252–259.
Henney, W. J., and Arthur, S. J. 1998. Modeling the brightness profiles of the Orion proplyds. *Astron J.* 116:322–335.
Hillenbrand, L. 1997. On the stellar population and star-forming history of the Orion Nebula cluster. *Astron J.* 113:1733–1768.
Hillenbrand, L., and Hartmann, L. 1998. A preliminary study of the Orion Nebula Cluster structure and dynamics. *Astrophys. J.* 492:540–553.
Hollenbach, D., and Tielens, A. G. G. M. 1997. Dense photodissociation regions (PDRs). *Ann. Rev. Astron. Astrophys.* 35:179–216.

Hollenbach, D., Johnstone, D., and Shu, F. 1993. Photoevaporation of disks around massive stars and ultracompact H II regions. In *Massive Stars: Their Lives in the Interstellar Medium,* ed. J. P. Cassinelli and E. B. Churchwell, Astronomical Society of the Pacific Conf. Series 35, pp. 26–34.

Hollenbach, D., Johnstone, D., Lizano, S., and Shu, F. 1994. Photoevaporation of disks around massive stars and application to ultracompact H II regions. *Astrophys. J.* 428:654–669.

Horedt, G. P. 1978. Blowoff of the protoplanetary cloud by a T Tauri-like solar wind. *Astron. Astrophys.* 64:173–178.

Horedt, G. P. 1980. Mass loss from planetary protoatmospheres and from the protoplanetary nebula. *Astron. Astrophys.* 92:267–272.

Horedt, G. P. 1982. Mass loss from the protoplanetary nebula. *Astron. Astrophys.* 110:209–214.

Hourigan, K., and Ward, W. R. 1984. Radial migration of preplanetary material: Implications for the accretion time scale problem. *Icarus* 60:29–39.

Hunten, D. M. 1985. Blowoff of an atmosphere and possible application to Io. *Geophys. Res. Lett.* 12:271–273.

Hunten, D. M. 1993. Atmospheric evolution of the terrestrial planets. *Science* 259:915–920.

Johnstone, D., Hollenbach, D., and Bally, J. 1998. Photoevaporation of disks and clumps by nearby massive stars: Application to disk destruction in the Orion Nebula. *Astrophys. J.* 499:758–776.

Kepner, J., Hartigan, P., Yang, C., and Strom, S. 1993. Hubble Space Telescope images of the subarcsecond jet in DG Tauri. *Astrophys. J. Lett.* 415:L119–L121.

Kessel, O., Richling, S., Yorke, H. W. 1998. Photoevaporation of protostellar disks. III: The appearance of photoevaporating disks around young intermediate stars. *Astron. Astrophys.* 337:832–846.

Koerner, D. W., Sargent, A. I., and Beckwith, S. V. W. 1993. A rotating gaseous disk around the T Tauri star GM Aurigae. *Icarus* 106:2–10.

Königl, A. 1991. Disk accretion onto magnetic T Tauri stars. *Astrophys. J. Lett.* 370:39–43.

Königl, A. 1995. Disk-driven hydromagnetic winds in young stellar objects. *Rev. Mex. Astron. Astrofis. Ser. Conf.* 1:275–283.

Lada, E. A. 1992. Global star formation in the L1630 molecular cloud. *Astrophys. J. Lett.* 393:L25–L28.

Lada, E. A. 1998. Observations of star formation: The role of embedded clusters. In *Origins,* ed. C. E. Woodward, J. M. Shull, and H. A. Thronson (San Francisco: ASP Publishing), pp. 198–202.

Lada, E. A., Evans, N. J., De Poy, D. L., and Gatley, I. 1991. A 2.2-micron survey in the L1630 molecular cloud. *Astrophys. J.* 371:171–182.

Larwood, J. D. 1997. The tidal disruption of protoplanetary accretion discs. *Mon. Not. Roy. Astron. Soc.* 290:490–504.

Laughlin, G., and Bodenheimer P. 1994. Nonaxisymmetric evolution in protostellar disks. *Astrophys. J.* 436:335–354.

Lin, D. N. C., and Papaloizou, J. 1979. Tidal torques on accretion discs in binary systems with extreme mass ratios. *Mon. Not. Roy. Astron. Soc.* 186:799–812.

Lin, D. N. C., and Papaloizou, J. 1985. On the dynamical origin of the solar system. *Protostars and Planets II,* ed. D. C. Black and M. S. Matthews (Tucson: University of Arizona Press), 981–1071.

Lin, D. N. C., and Papaloizou, J. 1986*a*. On the tidal interaction between protoplanets and the primordial solar nebula. II: Self-consistent nonlinear interaction. *Astrophys. J.* 307:395–409.

Lin, D. N. C., and Papaloizou, J. 1986*b*. On the tidal interaction between protoplanets and the protoplanetary disk. III: Orbital migration of protoplanets. *Astrophys. J.* 309:846–857.

Lin, D. N. C., Bodenheimer P., and Richardson, D. C. 1996. Orbital migration of the planetary companion of 51 Pegasi to its present location. *Nature* 380:606–607.

Lissauer, J. J. 1987. Timescales for planetary accretion and the structure of the protoplanetary disk. *Icarus* 69:249–265.

Lissauer, J. J. 1993. Planet formation. *Ann. Rev. Astron. Astrophys.* 31:129–174.

McCaughrean, M. J., and O'Dell, C. R. 1996. Direct imaging of circumstellar disks in the Orion Nebula. *Astron J.* 111:1977–1986.

McCaughrean, M. J., and Stauffer, J. R. 1994. High-resolution near-infrared imaging of the Trapezium: A stellar census. *Astron J.* 106:1382–1397.

Mizuno, H. 1980. Formation of the giant planets. *Prog. Theor. Phys.* 64:544–557.

Nakagawa, Y., Hayashi, C., and Nakazawa, K. 1983. Accumulation of planetesimals in the solar nebula. *Icarus* 54:361–376.

O'Dell, C. R. 1998. Observational properties of the Orion Nebula proplyds. *Astron. J.* 115:263–273.

Ohtsuki, K., and Nakagawa, Y. 1988. Accumulation process of planetesimals to the planets. *Prog. Theor. Phys. Suppl.* 96:239–255.

Ouyed, R., and Pudritz, R. 1997*a*. Numerical simulations of astrophysical disks. I: Stationary models. *Astrophys. J.* 482:712–732.

Ouyed, R., and Pudritz, R. 1997*b*. Numerical simulations of astrophysical disks. II: Episodic outflows. *Astrophys. J.* 484:794–809.

Papaloizou, J., and Lin, D. N. C. 1984. On the tidal interaction between protoplanets and the primordial solar nebula. I: Linear calculation of the role of angular momentum exchange. *Astrophys. J.* 285:818–834.

Pechernikova, G. V., and Vitjazev, A. V. 1981. Thermal dissipation of gas from the protoplanetary cloud. *Adv. Space Res.* 1:55–60.

Pollack, J. B., Hubickyj, O., Bodenheimer P., Lissauer, J. J., Podolak, M., and Greenzweig, Y. 1996. Formation of the giant planets by concurrent accretion of solids and gas. *Icarus* 124:62–85.

Pringle, J. E., and Rees, M. J. 1972. Accretion disc models for compact X-ray sources. *Astron. Astrophys.* 21:1–9.

Richling, S., and Yorke, H. W. 1997. Photoevaporation of protostellar disks. II: The importance of UV dust properties and ionizing flux. *Astron. Astrophys.* 327:317–324.

Richling, S., and Yorke, H. W. 1998. Photoevaporation of protostellar disks. IV. Externally illuminated disks. *Astron. Astrophys.* 340:508–520.

Richling, S., and Yorke, H. W. 1999. Photoevaporation of protostellar disks. IV: Circumstellar disks under the influence of both EUV and FUV. *Astrophys. J.*, submitted.

Safronov, V. S. 1969. *Evolution of the protoplanetary cloud and formation of Earth and the Planets* (Moscow: Nauka).

Sekiya, M., Nakazawa, K., and Hayashi, C. 1980. Dissipation of the rare gases contained in the primordial Earth's atmosphere. *Earth Plan. Sci. Lett.* 50:197–201.

Shakura, N. I., and Sunyaev, R. A. 1973. Black holes in binary systems: Observational appearance. *Astron. Astrophys.* 24:337–355.

Shu, F. H., Lizano, S., Ruden, S. P., and Najita, J. 1988. Mass loss from rapidly rotating magnetic protostars. *Astrophys. J.* 328:L19–L23.

Shu, F. H., Johnstone, D., and Hollenbach, D. 1993. Photoevaporation of the solar nebula and the formation of the giant planets. *Icarus* 106:92–101.

Stapelfeldt, K. R., Burrows, C. J., Krist, J., Watson, A. M., Trauger, J. T., Ballester, G. E., Casertano, S., Clarke, J. T., Crisp, D., Evans, R. W., Gallagher, J. S., Griffiths, R. E., Hester, J. J., Hoessel, J. G., Holtzman, J., Mould, J. R., Scowen, D. A., and Westphal, J. A. 1994. WFPC2 observations of the HH 30 disk and jet. *Bull. Amer. Astron. Soc.* 185:4802.

Störzer, H., and Hollenbach, D. 1998. On the [O I] lambda 6300-line emission from the photoevaporating circumstellar disks in the Orion Nebula. *Astrophys. J. Lett.* 502:L71–L74.

Störzer, H., and Hollenbach, D. 1999. PDR models of photoevaporating circumstellar disks and application to the proplyds in Orion. *Astrophys. J.*515:669–684.

Strom, K. M., Strom, S. E., Edwards, S., Cabrit, S., and Skrutskie, M. F. 1989. Circumstellar material associated with solar-type pre-main-sequence stars: A possible constraint on the timescale for planet building. *Astron J.* 97:1451–1470.

Strom, S. E., 1995. Initial frequency, lifetime and evolution of YSO disks. *Rev. Mex. Astron. Astrofis.* 1:317–328.

Strom, S. E., Edwards, S., and Skrutskie, M. F. 1993. Evolutionary time scales for circumstellar disks associated with intermediate- and solar-type stars. In *Protostars and Planets III,* ed. E. H. Levy and J. I. Lunine (Tucson: University of Arizona Press), pp. 837–866.

Terebey, S., Shu, F. H., and Cassen, P. 1984. The collapse of the cores of slowly rotating isothermal clouds. *Astrophys. J.* 286:529–551.

Terebey, S., Chandler, C. J., and André, P. 1993. The contribution of disks and envelopes to the millimeter continuum emission from very young low-mass stars. *Astrophys. J.* 414:759–772.

van Buren, D. 1985. The initial mass function and global rates of mass, momentum, and energy input to the interstellar medium via stellar winds. *Astrophys. J.* 294:567–577.

Ward, W. R. 1986. Density waves in the solar nebula: Differential Lindblad torque. *Icarus* 67:164–180.

Weidenschilling, S. J. 1984. Evolution of grains in a turbulent solar nebula. *Icarus* 60:553–567.

Weidenschilling, S. J., and Cuzzi, J. N. 1993. Formation of planetesimals in the solar nebula. In *Protostars and Planets III,* pp. 1031–1060.

Wetherill, G. W. 1980. Formation of the terrestrial planets. *Ann. Rev. Astron. Astrophys.* 18:77–113.

Wetherill, G. W., and Stewart, G. R. 1986. The early stages of planetesimal accumulation. *Lun. Plan. Sci.* 17:939.

Wood, D. O. S., and Churchwell, E. 1989. The morphologies and physical properties of ultracompact H II regions. *Astrophys. J. Suppl.* 69:831–895.

Yorke, H. W., and Welz, A. 1993. In *Star Formation, Galaxies and the Interstellar Medium,* ed. J. Franco, F. Ferrini, and G. Tenorio-Tagle (Cambridge University Press), pp. 239–244.

Yorke, H. W., and Welz, A. 1994. In *Numerical Simulations in Astrophysics.* ed. J. Franco, S. Lizano, L. Aguilar, and E. Daltabuit (Cambridge University Press), pp. 318–326.

Yorke, H. W., and Welz, A. 1996. Photoevaporation of protostellar disks. I: The evolution of disks around early B stars. *Astron. Astrophys.* 315:555–564.

Yorke, H. W., Bodenheimer P., and Laughlin, G. 1995. The formation of protostellar disks. II: Disks around intermediate-mass stars. *Astrophys. J.* 443:199–208.

Zahnle, K .J., and Kasting, J. F. 1986. Mass fractionation during transonic escape and implications for loss of water from Mars and Venus. *Icarus* 68:462–480.

EFFECTS OF ENERGETIC RADIATION IN YOUNG STELLAR OBJECTS

ALFRED E. GLASSGOLD
New York University

ERIC D. FEIGELSON
Pennsylvania State University

and

THIERRY MONTMERLE
Centre d'Études de Saclay

Astronomical observations provide compelling evidence for enhanced magnetic activity in T Tauri stars and some protostars in the form of strong X-ray and nonthermal radio emission. A typical value for the X-ray luminosity of a low-mass young stellar object (YSO) is in the range $L_X \sim 10^{28}$–10^{30} erg s^{-1} or $L_X/L_{\rm bol} \sim 10^{-4}$, but L_X can increase by orders of magnitude during a flare. Irradiation at such power levels can have important effects on the physical and chemical properties of surrounding material. We discuss in detail the X-ray ionization of circumstellar accretion disks and cloud cores, where X-rays can dominate other ionization sources out to $\gtrsim$0.01 pc = 2000 AU. The X-rays can also induce a wide variety of changes in the chemical and physical properties of the environment and may promote magnetohydrodynamic (MHD) turbulence and accretion in disks. Radio emission from MeV electrons implies that energetic particles are also produced by YSOs, and a variety of meteoritic measurements support their presence.

I. INTRODUCTION

The formation and early evolutionary stages of low-mass (Sunlike) stars and their disks are usually viewed from the perspective of low-temperature astronomy. Observational evidence for gravitational collapse of molecular gas is found through millimeter lines; dust in the circumstellar disks of pre-main-sequence stars is studied through continuum emission at infrared through millimeter wavelengths; and jets and bipolar outflows are seen in optical through millimeter emission lines and thermal radio continuum emission. The stages of young stellar object (YSO) evolution are frequently classified by their infrared spectra: Class 0 protostars are in their youngest infalling phase; Class I protostars and Class II classical T

Tauri stars are dominated by star-disk interactions; and Class III weak-lined T Tauri stars have little or no disk material (Lada 1991; André and Montmerle 1994).

There is extensive observational evidence that high-energy processes coexist with cold material in young stellar objects. Variable keV X-ray emission with $L_X \sim 10^{-4}$–$10^{-3}\,L_{\rm bol}$ is nearly ubiquitous in Class II and Class III YSOs (T Tauri stars) and is detected at similar or higher levels in several Class I protostars. This emission requires the presence of a large volume of high-density plasma at a temperature of about 10^7 K. The plasma must be magnetically confined, probably in large loops on a scale $>R_\star$. Nonthermal centimeter radio continuum emission is seen in several dozen YSOs at levels 3 to 6 orders of magnitude higher than in the contemporary Sun. A comprehensive review of over 200 optical, X-ray, and radio studies of magnetic activity and associated high-energy processes is given by Feigelson and Montmerle (1999).

The key to the puzzle of the coexistence of keV-MeV radiation with much colder matter (≤ 1 eV particles) is the presence of dynamic magnetic fields that lead to violent reconnection phenomena. The physical context is analogous to the outer layers of the Sun, where magnetic fields generated in the solar interior are stirred by subphotospheric motions until reconnection produces flares, which heat trapped gas to X-ray temperatures and accelerate particles to MeV or GeV energies via shocks and plasma processes. The YSO situation, however, is considerably more complex (and uncertain) than the contemporary Sun. YSO magnetic fields may be produced by a dynamo or may be inherited from the parent molecular cloud. Magnetic field footprints may lodge in the star, connect the star to the disk, or link portions of the disk.

Our goal in this chapter is to discuss some of the consequences of high-energy processes for the gas and dust in the circumstellar envelopes and disks of mainly low-mass YSOs. X-rays will ionize the gas and promote gas-field coupling, alter the solids in dust particles, and induce chemical reactions. Energetic flare particles will initiate nuclear reactions with the nuclei of disk atoms, and flare shocks will affect disk material in complex ways. Many of these effects have yet to be examined in depth. After summarizing the astronomical observations of YSO X-rays, we will concentrate on the ionization effects of X-rays on the gaseous parts of the circumstellar environment, the subject of recent theoretical studies. We then discuss X-ray-induced chemistry, irradiation of dust, and links to meteoritics.

II. ASTRONOMICAL OBSERVATIONS

The first evidence for extremely hot plasma in YSOs was detected in the Orion Trapezium as early as 1972 with the *UHURU* satellite, although the understanding that the source consisted of hundreds of individual low-mass stars only emerged with the *Einstein, Roentgen Satellite* (ROSAT), and *Advanced Satellite for Cosmology and Astrophysics* (ASCA) imaging X-ray telescopes of the 1980s and 1990s. All major nearby sites of star formation have now been studied in the X-ray band, including the Taurus-Auriga, Ophiuchus, Chamaeleon, Corona Australis, and Perseus clouds, portions of the Scorpius-Centaurus and Orion associations, as well as the more distant Rosette and Monoceros molecular clouds (e.g., Feigelson et al. 1993; Strom and Strom 1994; Gagné et al. 1995; Casanova et al. 1995; Preibisch et al. 1996; Preibisch 1997; Neuhäuser 1997; Gregorio-Hetem et al. 1998).

Figure 1 is an example of the central portion of a cluster of low-mass YSOs observed in X-rays by ROSAT. It is typical of X-ray images of star-forming molecular clouds, which show dozens of faint variable sources associated with an optical or infrared YSO. Many of the X-ray sources are previously unnoticed, optically visible weak-lined T Tauri stars (wTT stars, e.g., Walter et al. 1988), also known as Class III YSOs from their infrared properties (Lada 1991). Optically, they are spectral type G–M stars lying on Hayashi tracks above the main sequence in the Hertzsprung-Russell diagram. They are characterized by little or no

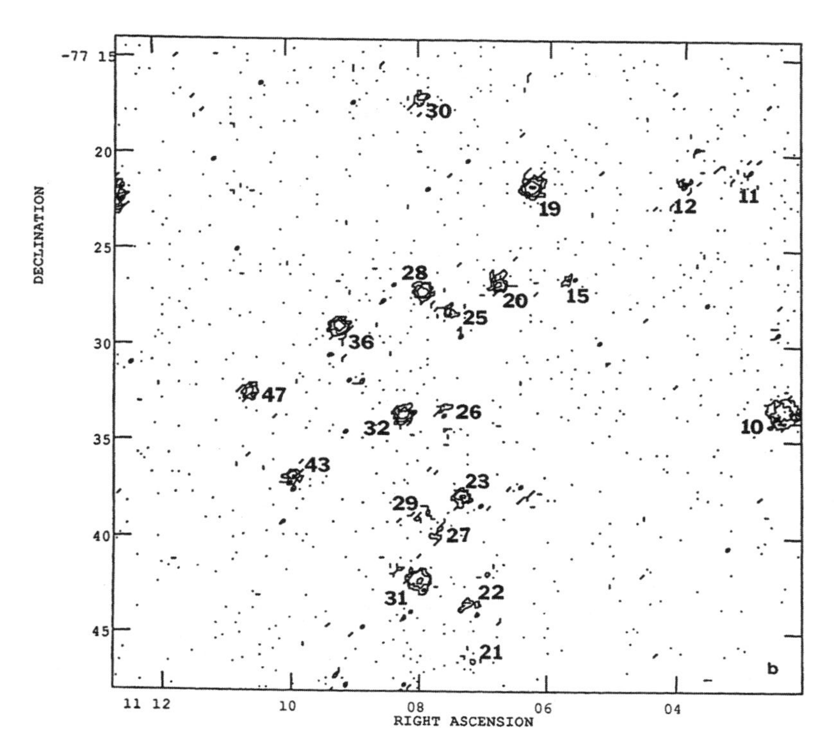

Figure 1. ROSAT image of the southern portion of the Chamaeleon I cloud, showing dozens of soft X-ray sources associated with T Tauri stars (Feigelson et al. 1993).

infrared excess, strong Li λ 6707 absorption lines, large cool starspots, strong chromospheric absorption, and weak Hα emission lines. Their soft X-ray luminosities (energies $\sim$ 0.2–2 keV) are typically in the range 10^{28}–10^{30} erg s^{-1}, or factors of 10^2–10^3 above contemporary solar levels.

Virtually all of these X-ray sources vary on timescales of hours to days. Some show slow aperiodic variation with amplitude variations of a factor ≤ 2, but at any moment, several percent of the sources exhibit high-amplitude rapid flares with peak $L_X \simeq 10^{30}$–10^{32} erg s^{-1}. One YSO is known with quiescent emission of $L_X \simeq 2 \times 10^{32}$ erg s^{-1} (Preibisch et al. 1998). A typical flare lasts one to several hours, with a fast rise and a slow decay. The particular flare illustrated in Fig. 2 had a somewhat shorter decay time than usual and is reminiscent of the common impulsive flares seen in the Sun. Modeling the X-ray spectrum requires a multitemperature plasma, with temperatures around 2–20 MK during low states to roughly 60 MK for stronger sources and flares (e.g., Skinner et al. 1997; Preibisch et al. 1998).

A large fraction of classical T Tauri stars (cTT stars, or Class II YSOs) are detected with X-ray characteristics that are very similar to those of Class III stars. This may indicate that the presence (in Class II) or absence (in Class III) of a disk is not critical to X-ray production or absorption

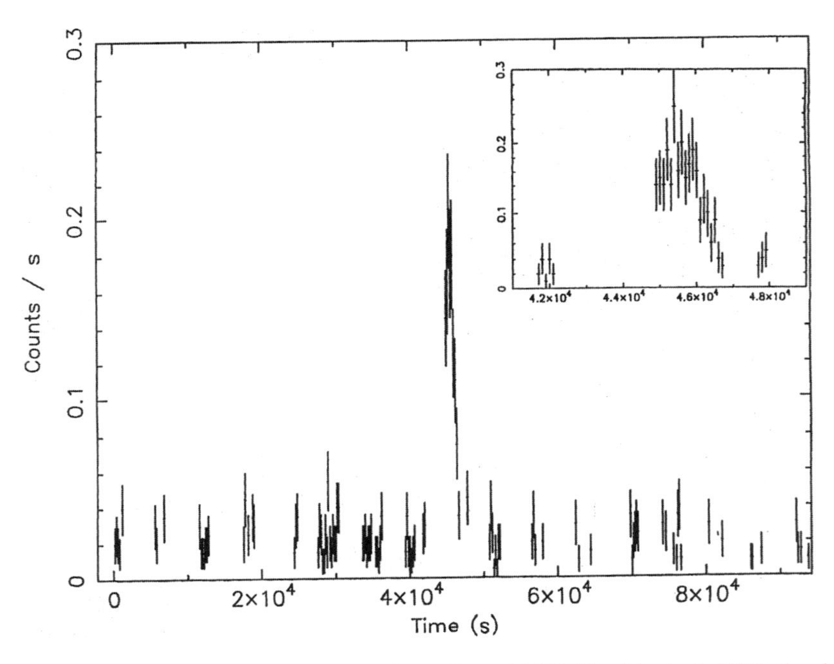

Figure 2. ASCA light curve of flare in wTT star V 826 Tauri in the L 1551 cloud (Carkner et al. 1996).

and that the magnetic footprints lie on the stellar surface (Feigelson et al. 1991). However, simple astrophysical modeling of the radiative cooling of the 10^{30}–10^{31} erg s^{-1} X-ray flares in Class II stars consistently indicates loop sizes of order 10^{11} cm, much larger than the 10^9–10^{10}-cm loops seen on the Sun and sometimes larger than the X-ray-emitting star itself. Field lines linking the star and disk at the corotation radius may explain the strongest of these powerful flares, if not X-ray flares in general (Hayashi et al. 1996; Shu et al. 1997).

Recent observations have detected X-ray emission from several Class I protostars with estimated ages $\approx 10^5$ yr (Casanova et al. 1995; Koyama et al. 1996; Grosso et al. 1997). The detected Class I YSOs tend to have harder spectra than T Tauri stars, and their flare luminosities can be considerably higher, in the range $L_X \simeq 10^{30}$ to $>10^{33}$ erg s^{-1}. However, the detected protostars probably represent the top of the protostellar luminosity function, because many other Class I sources have not been detected at these levels (Carkner et al. 1998).

Another important observational indicator of high-energy magnetic activity in YSO systems is the detection of nonthermal radio continuum emission in several dozen Class III stars (see the review by André 1996). Surveys with the National Radio Astronomy Observatory (NRAO) Very Large Array (VLA) indicate that roughly 20% of weak-lined T Tauri stars have radio luminosity densities $L_r(8\,\mathrm{GHz}) \geq 10^{15.5}$ erg s^{-1} Hz^{-1} (Chiang et al. 1996). This is ~2 orders of magnitude above the most active late-type main sequence stars and is many orders of magnitude above contemporary solar levels. The radio emission is often variable and, in a few cases, circularly polarized (e.g., White et al. 1992). This shows that the emission is gyrosynchrotron radiation from mildly relativistic electrons spiraling in well-ordered magnetic fields. The same mechanism is present in other magnetically active late-type stars, such as dMe flare stars, RS Canum Venaticorum binaries, and the Sun.

Nonthermal radio emission has not yet been detected in Class II YSOs, arguably because emission from magnetic loops near the star suffers free-free absorption by the partially ionized outflows of cTT stars (André 1996). Class I stars frequently show extended free-free emission from the base of their bipolar outflows (see the review by Rodríguez 1997). Two protostars do exhibit circularly polarized, variable emission denoting gyrosynchrotron emission: T Tauri South is resolved into two lobes of opposite helicity, suggesting particle acceleration in the outflow (Ray et al. 1997), while CrA IRS 5 exhibits rapid polarized flaring, indicative of magnetic reconnection events near the star (Feigelson et al. 1998).

Optical studies have recently confirmed the presence of strong magnetic fields on the surfaces of Class III and some Class II YSOs. Zeeman effect measurements show surface fields in several stars with $Bf \simeq 1$–3 kG, where f is the surface filling factor (Johns-Krull et al. 1998; Guenther et al. 1998). Periodic photometric variations and Doppler imaging studies reveal rotationally modulated, cool starspots covering up to half of the stellar surface (e.g., Bouvier et al. 1995; Johns-Krull et al. 1999).

It appears likely that *all* low-mass YSOs possess strong magnetic fields near the stellar surface that extend out to the circumstellar disk. These fields may be the sites of powerful magnetic reconnection events that produce ionizing X-rays and accelerate electrons to MeV energies. Reconnection of large field structures is a natural consequence of theories of low-mass star formation, where stellar field lines are linked to the disk and collimate both inflows and outflows (see the chapters by Shu et al. and by Königl and Pudritz, this volume). In view of the numerous qualitative similarities between X-ray and radio emission mechanisms in YSOs and in the Sun, it is also reasonable to infer that YSO flares produce MeV protons and heavier ions, analogous to solar flare cosmic rays.

III. X-RAY INTERACTION PHYSICS

The principal effects of X-rays on the cold interstellar medium (ISM) are ionization, excitation, and heating. Ionization can help couple the gas and ambient magnetic fields. It may also induce a characteristic X-ray photochemistry and alter the properties of dust. Early research on these issues dates back to the beginning of X-ray astronomy (see the review by Dalgarno and McCray 1972). Soft X-rays and low-energy cosmic rays were then considered the most promising ionization sources for interstellar clouds shielded from the interstellar far UV starlight. While the underlying ionization mechanism is due to electronic collisions in both cases, X-rays are absorbed by much smaller columns of matter than cosmic rays are. Thus, galactic cosmic rays are able to provide a more global, low-level ionization of the ISM, whereas X-rays can produce a high level of localized ionization. A stellar X-ray source in a molecular environment is surrounded first by ionized and then by atomic and molecular material (e.g., Halpern and Grindlay 1980; Lepp and McCray 1983). Lepp et al. (1985) showed that X-irradiated interstellar gas can, under certain conditions, occur in three stable phases: coronal gas ($T > 10^5$ K), warm gas ($T \approx 10^4$ K), and cold gas ($T < 10^2$ K), which can become molecular. The molecular phase has not yet been fully studied, and X-ray-induced chemistry is a subject of renewed interest (e.g., Maloney et al. 1996; Lepp and Dalgarno 1996; Yan and Dalgarno 1997), as discussed in section VI.

An intriguing early suggestion was that X-rays play a role in regulating the rate of star formation through the inhibition of gravitational collapse by the suppression of ambipolar diffusion in an X-ray-ionized medium (Silk and Norman 1983). However, we will see below that stellar X-ray ionization is generally not effective for the bulk of a molecular cloud core (≥ 0.1 pc). The present consensus is that galactic far UV and cosmic rays are more important than X-rays for coupling magnetic fields and interstellar cloud material on this spatial scale (McKee 1989).

However, X-rays can influence other steps in early stellar evolution on scales <0.1 pc, such as infall, accretion, and outflow.

The photoelectric effect is the most important interaction for X-rays in the 1–10 keV energy range relevant for YSOs. Most of the X-rays ($>80\%$ above 1 keV) are absorbed by low-abundance elements heavier than H and He. If the irradiated matter is only weakly ionized, the primary photoelectron ejected from a heavy element, which typically has an energy $\sim$1 keV, is able to collisionally ionize and excite many H and He atoms (and H_2 molecules in a molecular region), generating lower-energy secondary electrons, which produce further ionization. The role of secondary electrons in X-ray ionization and heating was first considered for atomic regions by Silk and Werner (1969).

X-rays and cosmic rays ionize interstellar and circumstellar gas in different ways. Cosmic rays primarily eject outer-shell electrons and thus affect light and heavy atoms in much the same fashion. Most cosmic-ray ionization comes from abundant H and He atoms. In contrast, X-rays are absorbed more strongly by heavy atoms, from which they eject K- and L-shell electrons. This is accompanied by the Auger effect, in which the de-excitation of an atom with $Z > 3$ is accompanied by the ejection of several electrons after an internal cascade. The mean number of Auger electrons ($\mathscr{A}$) increases with Z, from $\mathscr{A} = 1$ for $Z \leq 10$ (Ne) to $\mathscr{A} = 4.75$ for Fe (Kaastra and Mewe 1993). The primary fast X-ray photoelectron also produces additional ionization of atoms and molecules. The mean energy to make an ion pair in weakly ionized, cosmic-abundance gas is $\Delta\epsilon \approx 35$ eV. About 40% of this energy goes into ionization, 45% into excitation, and 15% into heat for an electron fraction $x_e = 10^{-4}$ in an atomic gas (Shull and Van Steenberg 1985). Thus 1 keV of incident X-ray energy produces a cascade of $\mathscr{S} \approx 1000/35 \approx 30$ secondary electrons, and the number of secondary electrons $\mathscr{S}$ is generally much greater than $\mathscr{A}$, the number of Auger electrons. The energy of the Auger electrons may be high ($\sim$1 keV) if they come from the lower atomic levels, but more often they come from higher levels and emerge with comparatively low energy. In practice, their effects have usually been ignored.

A valuable source of X-ray absorption cross sections is the tabulation of Henke et al. (1993), which is updated and available on line at www-cxro.lbl.gov/optical_constants/asf.html. It is only slightly changed from the earlier edition used by Morrison and McCammon (1983). The absorption cross section is usually expressed as the total photoelectric cross section per H nucleus, $\sigma(E) = \sum_k x_k \, \sigma_k(E)$, which is a sum over the elemental cross sections $\sigma_k(E)$ weighted by the abundances x_k of each atomic species. A typical photoelectric cross section has a K-shell ionization threshold, then decreases approximately as E^{-3}, so that the sum displays a discontinuity when a new elemental ionization threshold energy is reached. $\sigma(E)$ is fairly independent of the depletion of many heavy elements onto grains, as long as the grains are not too large (Fireman 1974; Morrison and McCammon 1983).

The total photoelectric cross section decreases rapidly with X-ray energy E and, after smoothing, can be fitted by a power law

$$\sigma(E) = \tilde{\sigma}\,(E/\mathrm{keV})^{-n} \tag{1}$$

For cosmic abundances, normal gas-to-dust ratio, and X-ray energies 1–20 keV, the fit coefficients are $\tilde{\sigma} = 2.27 \times 10^{-22}$ cm^2 and $n = 2.485$ (Glassgold et al. 1997). The corresponding optical depth for absorption of an X-ray is $\tau(E) = \left(N_{\mathrm{H}}/10^{22}\mathrm{cm}^{-2}\right) \tilde{\sigma}\,(E/\mathrm{keV})^{-n}$. For $E = 1$ keV and solar abundances, this becomes $\tau(1\,\mathrm{keV}) = N_{\mathrm{H}}/4.41 \times 10^{21}\mathrm{cm}^{-2}$. If the usual conversion between N_{H} and A_V for diffuse interstellar clouds is used, then $\tau(1\,\mathrm{keV}) = 1$ occurs when $A_V \simeq 2$.

Two additional effects can be important. First, X-rays above $E \sim 2$ keV can also ionize when they scatter through a large angle (e.g., Halpern and Grindlay 1980). For the energy range of interest, the process is essentially Thomson scattering, although it is conventionally referred to as Compton ionization. The Compton ionized electron acquires only a small amount of energy, $\sim E^2/(m_{\mathrm{e}}c^2)$, not enough to make any secondaries until $E \sim 100$ keV. The energy transfer is much smaller than in the photoelectric effect, especially for the heavy atoms responsible for most of the absorption cross section, and Compton energy losses do not become competitive until $E \sim 20$ keV. The second effect is that some of the energy absorbed by high-Z atoms goes into fluorescence instead of the primary and Auger electrons. The fluorescent yield increases with Z and reaches about 35% for Fe. However, much of the fluorescent energy still goes into the production of secondary electrons, because the fluorescence occurs mainly at X-ray energies. A reasonably accurate calculation of the ionization and heating of cool interstellar or circumstellar matter by YSO X-rays can be made by assuming that all of the X-ray energy goes into secondary electron production, with a small correction for heating. However, important X-ray diagnostics are contained in the fluorescent radiation at longer wavelengths (e.g., Shapiro and Bahcall 1981; Lepp and McCray 1983).

IV. X-RAY IONIZATION OF THE ENVIRONMENT

YSO X-rays can partially ionize nearby circumstellar and interstellar matter through the processes just described. We can estimate the distance from the YSO within which X-ray ionization will dominate that produced by galactic cosmic rays. Following Krolik and Kallman (1983), we write the secondary electron contribution to the ionization rate at a distance r from a stellar X-ray source of luminosity L_{X}, temperature kT_{X}, and low-energy cutoff E_0, as

$$\zeta_{\mathrm{X}} = [(L_{\mathrm{X}}/kT_{\mathrm{X}} 4\pi r^2)\,\sigma(kT_{\mathrm{X}})](kT_{\mathrm{X}}/\Delta\epsilon)\,J(\tau, x_0) \tag{2}$$

In this equation, $\sigma(kT_{\mathrm{X}})$ is the total photoelectric absorption cross section in equation (1) evaluated at $E = kT_{\mathrm{X}}$; $\Delta\epsilon$ is the energy to make an ion pair; and $J(\tau, x_0)$ is an attenuation factor (called J_h by Krolik and Kallman 1983) that depends on $\tau = \tau(kT_X)$ and $x_0 = E_0/kT_X$:

$$J(\tau, x_0) = \int_{x_o}^{\infty} dx\, x^{-n}\, e^{-(x+\tau x^{-n})} \tag{3}$$

The x^{-n} factors come from the cross section fit in equation (1), and the other exponential comes from the thermal bremsstrahlung spectrum.

The factors in equation (2) can be understood as follows. The first (between square brackets) is the primary ionization rate, assuming that all of the photons contribute the same as the one with the mean energy, $E = kT_X$. The next factor is the mean number of secondary electrons produced by each primary photoelectron, $(kT_X/\Delta\epsilon)$. Note that the last factor, the attenuation factor J, is not unity at zero optical depth; that is, $J(0, x_0) \equiv J_0 \neq 1$. When $x_0 \ll 1$, $J_0 \approx 1/(n-1)x_0^{n-1} \gg 1$, but J_0 may be <1 in the likely situation that x_0 is a finite fraction of kT_X.

To estimate the maximum range of X-ray interactions, we need the ionization rate at zero optical depth at a distance r from a source of X-ray luminosity $L_{29} = L_X/(10^{29}\text{erg s}^{-1})$:

$$\zeta_X = 1.4 \times 10^{-10}\,\text{s}^{-1}\,[\sigma(kT_X)/\sigma(1\,\text{keV})]\,L_{29}\,J_0\,(r/\text{AU})^{-2} \tag{4}$$

The total ionization rate due to external sources is the sum of contributions from X-rays, cosmic rays, and UV radiation. Each of these is attenuated by characteristic processes that depend on the properties of the medium; we have already discussed the physics of X-ray absorption. Low-energy cosmic rays may be partially excluded from dense regions by magnetic scattering (Skilling and Strong 1976; Cesarsky and Völk 1978), especially if stellar winds are present (Parker 1960). Many attempts have been made to determine the effective cosmic ray ionization rate in UV-shielded clouds, e.g., by chemical modeling of molecular line observations (see the review by Lepp 1992), but this important parameter is still not known to within better than an order of magnitude. This difficulty is exemplified by two recent analyses of ionization in molecular cloud cores. One group concludes that the ionization rates may differ between cores by 1–2 orders of magnitude (Caselli et al. 1998), while another group concludes that the ionization rates are essentially the same (Williams et al. 1998).

We express the cosmic ray ionization rate in UV-shielded cores as $\zeta_{CR} = \zeta_{-17} \times 10^{-17}\,\text{s}^{-1}$, and we use equation (4) to define the distance at which X-ray and cosmic-ray ionization become equal,

$$r_{\max} \approx 0.02\,\text{pc}\,[(\sigma(kT_X)/\sigma(1\,\text{keV}))\,L_{29}\,J_0/\zeta_{-17}]^{1/2} \tag{5}$$

T Tauri stars have typical temperatures $kT_X = 1$ keV and X-ray luminosities in the range $L_{29} = 0.1$–100 (see section II). Taking into account variations in YSO X-ray properties and the uncertainty in ζ_{-17}, the overall range in $r_{\max}$ is about a factor of 100. For a typical YSO, X-ray ionization dominates cosmic ray ionization out to $\simeq$1000 AU. For the more luminous

X-ray sources, it may dominate the ionization over much of the cloud core shielded from interstellar UV radiation, especially if they are grouped in clusters with interstar distances $d_\star \sim$ a few $\times$ 0.01 pc. In the dense ρ Ophiuchi cloud Core F region, where $d_\star$ is small, Carkner (1998) finds that $\zeta_X > \zeta_{-17}$ over several percent of the cloud volume.

For considerations of ionization, the effective L_X for a given YSO may be closer to the peak luminosity of occasional powerful flares rather than to the time-averaged luminosity. The recombination coefficient for a molecular gas is $\beta \approx 10^{-7} cm^3 s^{-1}$, so that regions with $n_H = 10^4$ cm^{-3} and $x_e < 10^{-6}$ will recombine on a timescale >30 yr. This timescale is larger than the duty cycle of flares, which, although poorly known, is measured in days to months (Montmerle et al. 1983). A YSO may thus have an X-ray-dominated ionization region that is considerably larger than 0.01 pc = 2000 AU, which would encompass nearby protostellar cores.

Similar considerations should apply to the environs of the most X-ray-luminous low-mass YSOs (Preibisch 1998) and of intermediate-mass YSOs such as Herbig Ae/Be stars and zero-age main sequence (ZAMS) B stars, for which $L_X \simeq 10^{30}-10^{32}$ erg s^{-1} (Caillault and Zoonematkermani 1989; Zinnecker and Preibisch 1994). Although B stars are more luminous in UV than in X-rays, the UV is absorbed much closer to the star. Hard X-rays from W3 have recently been observed by ASCA (Hofner & Churchwell 1997) and may produce substantial environmental effects in this region. The variable X-ray emission from θ^1 Orionis has recently been shown by Gagné et al. (1997) to be periodic on the same timescale as other spectral lines (15 days). Babel and Montmerle (1997*a,b*) have been able to account for the observations within the framework of a magnetically confined wind.

Although r_{max} gives a rough estimate of the maximum extent of X-ray ionization, the detailed effects of YSO X-rays depend on the attenuation factor J in equation (2). Glassgold et al. (1997) show that, on ignoring the effects of the low-energy cutoff, J is well approximated for all relevant values of τ (even $\tau \ll 1$) by an asymptotic expansion of the integral in equation (3),

$$J = A\tau^{-a} e^{-B\tau^b} \tag{6}$$

where $a, A, B,$ and b are fit parameters that depend on abundances and the energy band under consideration. For solar abundances and energies around 1 keV ($n = 2.485$), $A = 0.800$, $a = 0.570$, $B = 1.821$, and $b = 0.287$.

Returning to equation (3), we see that (because $n > 1$) J becomes singular at $\tau = 0$ for $E_0 = 0$, which is unphysical because the photoelectric effect requires a minimum energy, such as 13.6 eV to ionize H. More importantly, soft X-rays are likely to be absorbed near the source. Thus the low-energy cutoff E_0 included in equations (2) and (3) must be at least as large as 0.0136 keV and surely much larger for directions that intersect major mass components of the YSO. We are assured by the detection of X-rays from most YSOs that E_0 is not so high that all X-rays are absorbed very close to the star. When the line of sight absorption to a YSO can be measured with optical-infrared photometry, the result usually agrees with E_0 determined from X-ray spectra. Nonetheless, it is likely that E_0 varies with direction for a given YSO because of the asymmetries associated with the accretion disk and with the inflows and outflows.

Figure 3 shows how the ionization rate depends on the optical depth, $\tau(1\ \text{keV})$. All the curves are calculated numerically for $E_0 = 1$ keV, except the one with solid triangles, which is based on equation (3) without a low-energy cutoff. The dotted line is an exact calculation (with no cutoff) to show the accuracy of the asymptotic calculation. The solid curves for photons with $kT_X = 1$ and 5 keV become flat below $\tau = 1$ due to the

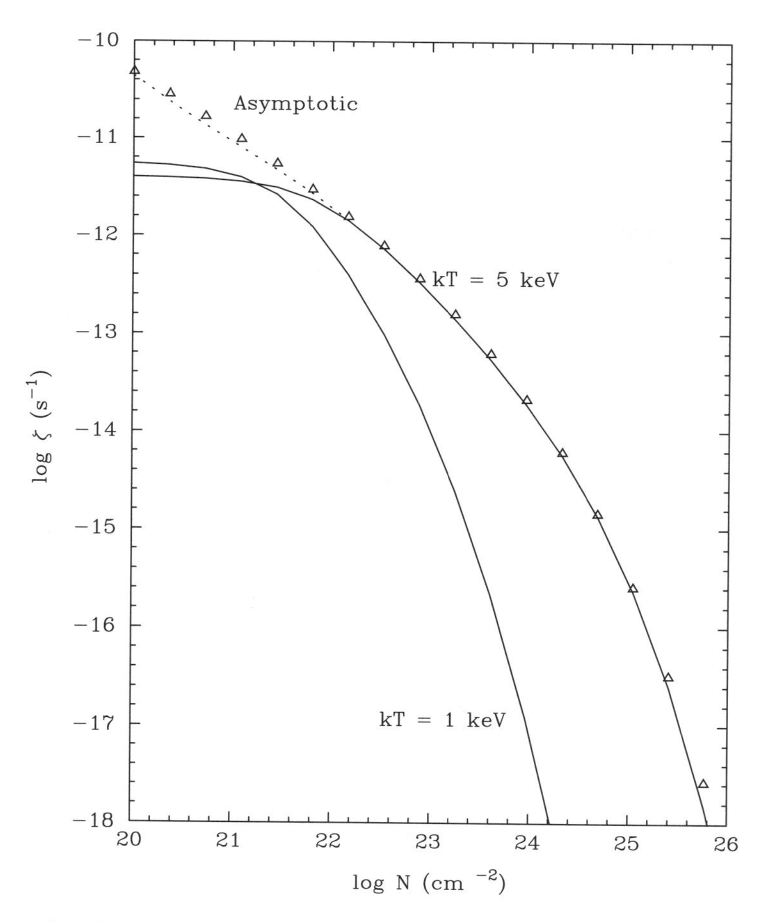

Figure 3. X-ray ionization rate vs. optical depth τ (adapted and corrected from Fig. 1 of Glassgold et al. 1997).

cut-off. The curve for $kT_X = 1$ keV is only about 40% higher in this region than the curve for $kT_X = 5$ keV, despite the fact that the cross section is 55 times larger than at the higher energy, because the 1-keV cutoff applied to a $kT_X = 1$ keV spectrum removes most of the photons. Figure 3 also shows the greater penetrability of the high-energy photons at large optical depths, leading to ionization rates for $kT_X = 5$ keV that are more than a factor of 100 larger than for $kT_X = 1$ keV.

V. X-RAY EFFECTS ON CIRCUMSTELLAR DISKS

Complex patterns of infall, accretion, and outflow are present in YSOs at distances where X-ray ionization may play a role. We concentrate on the accretion disk, where the results have been most fully developed. This problem has potential importance for understanding the currently unknown origin of YSO accretion disk viscosity. The viscosity may have a magnetohydrodynamic (MHD) origin of the type advocated by Balbus and Hawley (1991), who proposed that differential rotation drives the perpendicular magnetic field of an accretion disk unstable. If well-coupled to the largely neutral disk material, this "magnetorotational" instability then makes the disk turbulent. X-ray irradiation of the disk is important because the X-rays can supply the ionization needed to couple the magnetic field to the neutral disk material.

Gammie (1996) addressed the possible role of the Balbus-Hawley instability for YSO disk accretion, assuming that galactic cosmic rays are the dominant source of ionization. Takano and his collaborators (e.g., Umebayashi and Takano 1990) had established that cosmic rays can ionize those parts of disks with mass surface column densities $\Sigma \leq 200$ g cm^{-2}. Since both theory and observations suggest that the inner regions of YSO accretion disks are much thicker, $\Sigma \sim 2000$ g cm^{-2}, Gammie concluded that only the outer layers of disks would be well coupled to magnetic fields. A potential problem for this attractive idea is that low-energy cosmic rays are likely to be excluded from the disk by the strong magnetized winds of YSOs (Parker 1960).

Glassgold et. al. (1997) made the first calculations of X-ray ionization of a YSO accretion disk using equations (2) and (3). They adopted the steady, minimum-solar-nebula disk model described by Hayashi et al. (1985), with mass density and temperature given by

$$\rho(r, z) = \rho_0 (r/\mathrm{AU})^{-q} e^{-z^2/2H^2} \qquad T = T_0 (r/\mathrm{AU})^{-p} \tag{7}$$

where r and z are cylindrical coordinates. Hayashi et al. (1985) chose the power-law indexes as $q = 2.75$ and $p = 0.5$, so that the midplane density and temperature at 1 AU are $\rho_0 = 1.4 \times 10^{-9}$ g cm^{-3} and $T_0 = 280$ K. The vertical scale height parameter H and surface mass density Σ are then expressed by the power laws

$$\Sigma = \Sigma_0 (r/\mathrm{AU})^{-1.5} \qquad H = H_0 (r/\mathrm{AU})^{1.25} \tag{8}$$

with $\Sigma_0 = 1700\ \mathrm{g\ cm^{-2}}$ and $H_0 = 5.0 \times 10^{11}$ cm if all of the hydrogen is molecular. When integrated from $r_1 = 0.35$ AU to $r_2 = 35$ AU, the mass of the disk is $M_D = 0.013\ M_\odot$. The midplane volume density of hydrogen nuclei at 1 AU is $n_H = 5.8 \times 10^{14}\ \mathrm{cm^{-3}}$ (using $\rho/n_H = 1.425\, m_H$). The values of p, T_0, and M_D are consistent with millimeter-wave measurements of YSO disks (e.g., Beckwith et al. 1990; André and Montmerle 1994).

Glassgold et al. (1997) assumed that the disk was geometrically thin ($H \ll r$) and planar (not flared) and that the X-ray source is elevated a height $z_S = 12\ R_\odot$ above the disk midplane, as suggested by the x-wind model (Shu et al. 1994) and roughly consistent with the large loop sizes inferred from X-ray and radio flares (section II). The X-rays then encounter the disk at a small angle $\sim z_S/r$. Glassgold et al. (1997) measured vertical distances with the vertical column density,

$$N_\perp(r, z) = \int_z^\infty n_H(r, z')\, dz' \qquad (9)$$

They assumed that the dust particles had settled to the midplane of the disk, following the calculations of Weidenschilling and Cuzzi (1993), and ignored the z dependence of temperature, which may affect the local density structure (e.g., Bell et al. 1997) and temperature-dependent reaction rates. Their main result was that hard X-rays penetrate to large vertical column densities towards the disk midplane, typically at 1 AU to $N_\perp \sim 10^{24}$ and $10^{25}\ \mathrm{cm^{-2}}$ for $kT_X = 1$ and 5 keV, respectively. This ionization level is sufficient to produce an electron fraction in excess of the estimated critical electron fraction needed to couple magnetic fields and disk matter (Blaes and Balbus 1994, Gammie 1996). Because only an outer layer of the disk is ionized, Glassgold et al. (1997) recovered Gammie's picture of layered MHD accretion, where an interior dead zone develops at small radial distances around $r \leq 5$ AU.

A full 3D calculation of X-ray transport and ionization in axially symmetric disks has been developed by Igea and Glassgold (1999) using a Monte Carlo method. It is capable of treating flared disks and includes Compton scattering as well as photoelectric absorption. Figure 4 gives a typical result for the ionization rate at $r = 1$ AU in the minimum-solar-nebula disk described above, irradiated by X-rays with $L_X = 10^{29}$ erg $\mathrm{s^{-1}}$. The three curves plot the ionization rate vs. $N_\perp$ for thermal spectra with $kT_X = 3$, 5, and 8 keV and for a fixed low-energy cutoff of $E_0 = 1$ keV. The X-ray-emitting region is modeled as a ring at height $z_S = 10\ R_\odot$ and radius $r_S = 10\ R_\odot$. All three curves show three components: a flat portion at small $N_\perp < 10^{20}\ \mathrm{cm^{-2}}$ (or vertical heights $z > 5H_0$), where absorption is negligible; a smooth decline due to absorption for $10^{21} < N_\perp < 2 \times 10^{23}\ \mathrm{cm^{-2}}$ ($2 < z < 4.5H_0$); and a "scattering shoulder" for $N_\perp > 2 \times 10^{23}\ \mathrm{cm^{-2}}$, where the ionization rate declines less rapidly due to Compton scattering. The remarkable similarity of the curves at all optical depths depends critically on the use of the vertical column depth $N_\perp$ as the

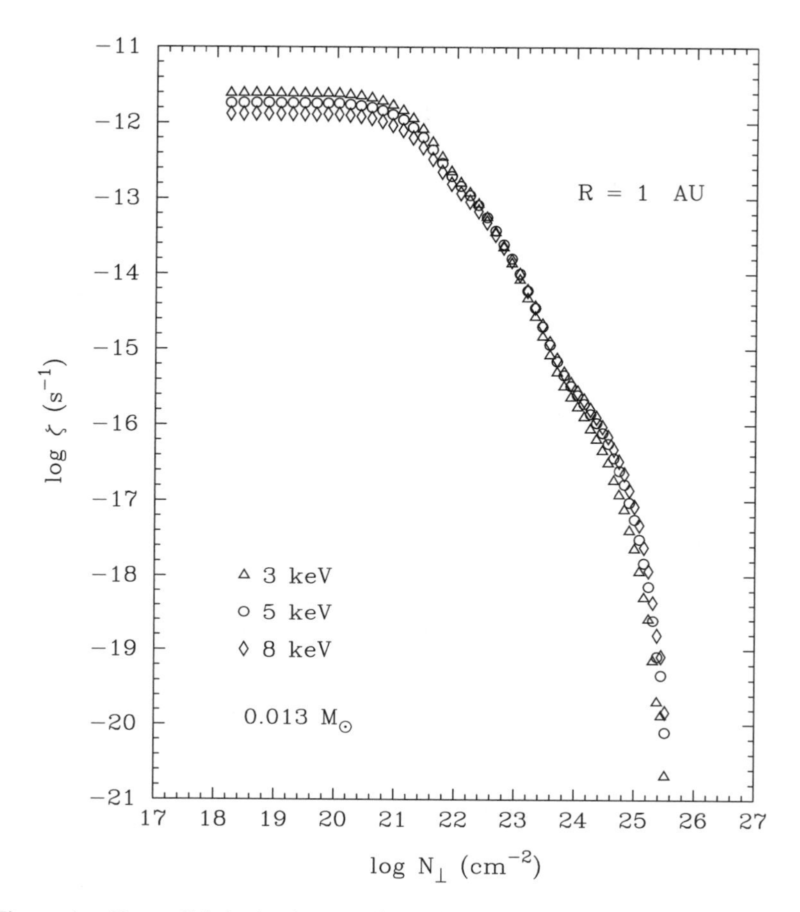

Figure 4. X-ray disk ionization rate in a minimum-solar-nebula disk (Igea and Glassgold 1999).

independent variable. The scaling displayed in Fig. 4 is also found at other disk radii and is insensitive to changes in disk parameters such as disk mass M_D, surface density Σ_0, and slope q. The absolute level of the ionization scales linearly with L_X and inversely with the radial distance r, i.e., as the incident flux.

These results show that, over a large range of column densities ($N_\perp < 2 \times 10^{24}$ cm^{-2}), the ionization rate is larger than the nominal value due to galactic cosmic rays, $\zeta_{CR} = 10^{-17}$s^{-1}. However, as remarked earlier, cosmic rays may not penetrate into the inner disk due to wind modulation. Dead zones are also expected to be present in the interiors of X-ray-ionized disks. For example, Fig. 4 shows that the ionization rate falls precipitously for $N_\perp > 10^{25}$ cm^{-2}, whereas the half-disk thickness in this case corresponds to $N_\perp = 10^{26}$ cm^{-2}. Thus, at 1 AU, no more than the outer 10% of the disk mass can be coupled to the magnetic field. Because Σ decreases with r, the dead zone disappears at large radii, and the entire outer disk will be ionized by X-rays. Full penetration of X-rays to the midplane first occurs for the model in Fig. 4 at $r = 5$ AU. The occurrence of a dead zone can be expected to have important implications for the dynamics of YSO disk accretion (Gammie 1996; chapter by Stone et al., this volume). It would therefore be of great interest to include the effects of X-ray ionization in numerical MHD simulations of disk dynamics.

VI. X-RAY-INDUCED CHEMISTRY

When photons from an X-ray source penetrate into a cloud, they produce a sequence of chemical zones with a wide range of physical and chemical properties that range from hot, highly ionized to cool, lightly ionized gas (section III). This progression can be understood in terms of the variation in the local ionization parameter, defined here as $\xi \equiv \zeta / n$. A familiar example of the use of an ionization parameter arises in calculating the electron fraction in molecular regions dominated by long-lived molecular ions such as HCO^+. In this case, the electron fraction is given approximately as $x_e = (\xi/\beta)^{1/2}$, where β is the effective dissociative recombination coefficient. The ionization parameter changes with distance into a cloud due to geometric dilution and absorption of the incident radiation by the intervening medium, which may have considerable density variation. The combined effects of the energy dependence of the absorption cross section, equation (1), and the broad spectrum of the source tend to produce gradual rather than sharp phase transitions that extend over a large range of column densities. This situation may be contrasted with a conventional photodissociation region (produced by far UV radiation longward of the Lyman limit), which has a narrow transition of the order of several magnitudes of visual extinction and no fully ionized zone.

The X-ray chemistry of YSOs is a particularly challenging but largely undeveloped area of astrochemistry, primarily because account has to be taken of the special geometry and the time dependence and complex history of the flows important in star formation. However, in the absence of definitive studies of YSO X-ray chemistry, considerable insight can be gained from steady-state chemical models of interstellar clouds irradiated by X-rays (e.g., Maloney et al. 1996; Lepp and Dalgarno 1996; Sternberg, Yan, and Dalgarno 1995; Yan and Dalgarno 1997). This research has been largely motivated by circumnuclear clouds in active galactic nuclei (AGNs). Despite the much larger X-ray luminosities, $L_X \sim 10^{44}$ erg s^{-1}, the large distances of the clouds from the AGN (hundreds of pc) bring the ionization parameter into the range relevant for galactic clouds near strong X-ray sources and for molecular cloud cores irradiated by YSO X-rays. For example, Maloney et al. (1996) consider the range $\xi = 10^{-19}$–10^{-24} cm^3 s^{-1} for the case $n = 10^5$ cm^{-3}, which brackets the standard cosmic-ray ionization parameter $\xi_{CR} \sim 10^{-22}$ cm^3 s^{-1} for this density.

According to these studies, molecule formation occurs over a large range of ξ (or, at constant density, X-ray dilution and absorption). Its pace quickens in a warm, partially ionized region ($T \sim 5000$ K, $x_e \sim 0.01$) that cools and becomes more neutral with decreasing ξ, and thus also more molecular. In the warmer regions, the chemistry makes use of neutral reactions promoted by X-ray heating of the gas, whereas in cooler regions the X-rays induce molecular synthesis by ion-molecule reactions.

Each of the chemical zones of a cloud exposed to an X-ray source provides a variety of spectral-line diagnostics determined by the local physical conditions and escape probabilities. Certain fine-structure transitions provide some of the strongest lines in all regions, notably [O I] 63 μm and to a lesser extent [S II] 35 μm and [Fe II] 26 μm. The [C II] 158-μm line is an important diagnostic for the cooler regions ($T \leq 100$ K). The rovibrational transitions of H_2 are also emitted strongly by the warm parts of the X-ray transition region (Gredel and Dalgarno 1995; Tiné et al. 1997). Of special interest are molecular ions produced by secondary electrons, such as HeH^+ (Roberge and Dalgarno 1982) and H_3^+ (Draine and Woods 1991). Maloney et al. (1996) show that the 149-μm rotational transition of HeH^+ should be emitted strongly from the warm parts of an X-ray-irradiated cloud and that the 4.35-μm line of H_3^+ is quite weak. Searches for some of these infrared lines are underway using the Short Wavelength Spectrometer (SWS) instrument on the *Infrared Space Observatory* (ISO) around YSOs (T. Montmerle, unpublished) and around strong X-ray sources near the center of the Milky Way (Maloney et al. 1997).

The millimeter rotational transitions of molecules and radicals provide additional chemical diagnostics in regions near YSO X-ray sources. Phillips and Lazio (1995) observed that the HCO^+ emission near the galactic center X-ray source 1E 1740.7–2942 is peaked away from the source, and suggested that this is due to enhanced destruction by X-rays. Lepp and Dalgarno (1996) and Yan and Dalgarno (1997) calculate that the primary destruction mechanism for HCO^+ is dissociative recombination with electrons. They also indicate that radicals such as OH, CH, CN, and NO may provide additional evidence of X-ray-induced chemistry, as suggested earlier by Krolik and Kallman (1983).

Very Long Baseline Array measurements of water masers at 22 GHz have become a powerful tool for understanding the activity at the centers of external galaxies. A beautiful example is NGC 4258, where the observations provide compelling evidence for a subparsec Keplerian disk and a central black hole (e.g., Miyoshi et al. 1995; Greenhill et al. 1995). The masers are believed to be located in the transition region between the atomic and molecular phases in a circumnuclear disk heated by X-rays from the central AGN (Neufeld et al. 1994; Neufeld and Maloney 1995).

One difficulty in demonstrating the presence of X-ray-induced chemistry is that the suggested diagnostics are often ambiguous. The most obvious source of confusion is cosmic rays, which produce chemically active secondary electrons in much the same way as X-rays do. Similarly, any external energy source that heats the gas to several thousand K will initiate a neutral chemistry that is characteristic of that temperature and not of any specific heating agency. One sure solution to this problem is spectroscopic observations at high spatial resolution, because cosmic rays, X-rays, UV, and shocks all operate on different spatial scales. The HCO^+ observations of 1E 1740.7–2942 (Phillips and Lazio 1995) provide a good illustration of this point and also call for higher-spatial-resolution mapping of this source.

It has long been recognized that observations of DCO^+ as well as the isotopes of HCO^+ can be used to estimate electron fractions in molecular clouds (e.g., Guélin et al. 1982; Wootten et al. 1982; Caselli et al. 1998; Williams et al. 1998). It would be of great interest to search for spatial gradients of these species in the vicinity of YSO X-ray sources. The spatial scale of $\sim$0.02 pc in equation (5) corresponds to an angular resolution $\approx 30''$ for nearby star-forming regions, so that existing millimeter arrays can probe regions where X-rays contribute to the ionization of cloud cores. Observational study of disk chemistry has only just begun (e.g., Dutrey et al. 1997). Highly sensitive planned arrays, such as the NRAO Millimeter Array and European Large Southern Array, should be effective for such studies.

Observational evidence for X-ray-induced disk chemistry has been discussed by Kastner et al. (1997), who detected millimeter emission from CO, HCN, CN, and HCO^+ around the nearby ($d \sim 50$ pc) T Tauri star TW Hydrae. They attribute the large CN/HCN ratio and the high HCO^+ abundance to the large X-ray luminosity of the star, noting that similar results are found in planetary nebulae illuminated by very hot central white dwarfs. The conclusion is not definitive, however, because comparable molecular abundances are seen in the non-X-ray-emitting T Tauri stars DM Tauri and GG Tauri (Dutrey et al. 1997). J. Kastner (personal communication) is developing a clearer picture based on detailed chemical modeling.

VII. X-RAY INTERACTIONS WITH DUST

The interaction of X-rays with astrophysical solids is not yet well understood, especially for YSOs, where the dust is at various stages of evolution and may experience complicated thermal and chemical processing. However, valuable insights may be acquired from the many laboratory studies of "radiation damage" of solids by X-rays (e.g., Cazaux 1997). When an X-ray interacts with a heavy atom in a solid, the first step is the same as in a gas (cf. section III): A photoelectron is ejected, and the excited ion decays by the Auger process and, with lower probability, by fluorescence. In a solid the resulting photoelectron and several Auger electrons propagate in a dense medium of heavy atoms, in contrast to the dilute interstellar gas, dominated by light atoms. The range of an electron in a solid can be much smaller than the size of a grain; for example, it is 200–300 Å at 1 keV (Dwek and Smith 1996). Thus, a substantial fraction of the absorbed X-ray energy may be deposited in the grain, converted into phonons, and eventually reradiated in the infrared. The transient heating of the grain can alter its spectroscopic features and also lead to the evaporation of the grain mantle. Electrons can escape from small grains, increase the grain charge, and thus affect the sticking probability of gas-phase species.

The electron transport process is critical for understanding the physical and chemical properties of dust irradiated by X-rays. The problem is considerably more complicated than it is for gases, and it is exacerbated by the uncertainty in our knowledge of the structure of dust in the different astrophysical situations relevant for YSOs. The most complete work to date on the energy deposition of X-rays in interstellar dust is by Dwek and Smith (1996). They considered a wide range of photon energies, from 10^{-2} to 10^3 keV, and spherical grains of radius a = 50–2000 Å with silicate and carbon composition. For a given radius, essentially all of the X-ray energy is absorbed for photons below a certain energy. For example, a small grain with a = 50 Å absorbs all photons with energy $E < 0.1$ keV, whereas a large grain with a = 2000 Å absorbs all photons with $E < 2$ keV. In related work, Laor and Draine (1993) examined the question of how X-irradiation affects the strength of the 10-μm silicate features in AGNs.

Voit (1991) analyzed the thermal and infrared spectral properties of X-irradiated spherical silicate and carbon grains, taking into account evaporative as well as radiative cooling. Very small grains are the carriers of the ubiquitous "unidentified infrared" (UIR) bands in the 3–13 μm diffuse emission of interstellar gas, and are widely believed to be large carbonaceous molecular compounds, either polycyclic aromatic hydrocarbons (PAHs), containing 100–1000 C atoms (Boulanger et al. 1998), or kerogens (i.e., coals, composites of PAHs and aliphatic hydrocarbons with impurities such as O, N, or S; Papoular et al. 1996). Voit (1992) concluded that grains with $a <$ 10 Å evaporate completely, which may explain the rarity of the 3–13-μm UIR features in the spectra of active galactic nuclei. Vaporization of very small grains may also occur in the vicinity of X-ray-emitting YSOs, and the expected weakening of the UIR features is currently being searched for in the ρ Oph cloud and in the circumstellar material of Herbig Ae/Be stars with ISO (Corcoran and Montmerle 1998).

Pursuing a promising new direction of research, Gougeon (1998) has exposed a wide variety of materials to the >5-keV beam of the European Synchrotron Radiation Facility in Grenoble, including graphite, PAHs, coal, SiC, SiO_2, and olivine. The experimental radiation dose corresponds roughly to that obtained at 0.1 pc from a TTS having $L_X \sim 10^{30}$ erg s^{-1} for a period of 10^8 yr. A wide variety of effects are found, which still have to understood in astrophysical terms. The results for PAHs are particularly interesting, showing dehydrogenation and breakage of aromatic rings with consequent changes in the infrared spectra. Other materials, such as silicates, are relatively unaffected by X-irradiation. Thus, specific signatures of X-ray-irradiated grains may exist and be sought with planned observatories such as the Shuttle Infrared Telescope Facility.

The interaction of X-rays with dust has many chemical implications, which have yet to be investigated in detail. A central issue is grain charging, which may affect gas-dust interactions and the coupling of dust to the magnetic field. The evaporation of grain mantles by X-ray irradiated grains is a potentially important desorption mechanism. Léger et al. (1985) invoked heavy cosmic rays as a way of solving the long-standing problem of the rapid condensation of molecules onto cold interstellar grain surfaces. E. Bergin et al. (personal communication) have recently generalized this theory to treat the molecule depletion problem in protoplanetary disks exposed to YSO X-rays. In order to be effective at the level suggested by observations of CO in the outer parts of accretion disks, they invoke "spot heating" to achieve grain temperatures high enough to remove CO molecules efficiently. Because disk grains are likely to be fluffy coagulates, localized regions of the grain may be heated sufficiently to cause efficient desorption.

VIII. IRRADIATION OF THE EARLY SOLAR SYSTEM

The observations summarized in section II show that YSOs, starting with Class I objects and extending in evolution all the way through the ZAMS, have high levels of magnetic activity. This conclusion is not restricted to a small subsample of YSOs but appears to be a universal property of stars in the process of formation. Independently of evolutionary stage, YSOs undergo huge flares that are much more powerful than those in the contemporary Sun. The detection of extremely high levels of gyrosynchrotron emission is particularly important in this connection, because it directly reveals *in situ* particle acceleration up to MeV energies in flares. It is thus reasonable to suppose that the solar nebula was subject to intense fluxes of keV X-rays, MeV particles, and shocks produced by violent magnetic reconnection events. Possible consequences include chondrule melting (Feigelson 1982; Levy and Araki 1989; Cameron 1995), correlated particle track and spallogenic ^{21}Ne excesses in some meteoritic grains (Woolum and Hohenberg 1993; Rao et al. 1997), and the production of short-lived isotopic anomalies seen in meteoritic inclusions.

The evidence for ^{41}Ca, ^{26}Al, ^{53}Mn, and other short-lived nuclides in components of primitive meteorites requires high-energy processes in or near the solar nebula. Nucleosynthesis by solar energetic particles was first discussed by Fowler et al. (1962) and subsequently pursued by many others (Clayton et al. 1977; Lee 1978; Dwek 1978; Feigelson 1982; Wasserburg 1985; Clayton and Jin 1995). This "local irradiation scenario" fell out of favor partly because estimates of the particle fluence required to produce ^{26}Al by nuclear reactions in the asteroid belt were too small by several orders of magnitude. External seeding scenarios, which posit the injection of nucleosynthetic material by a nearby supernova, asymptotic giant branch star, or Wolf-Rayet star, then gained ascendency (e.g., Cameron and Truran 1977; Wasserburg 1985; Cameron 1985; Cameron 1993; Wasserburg et al. 1994; Harper 1996).

The local irradiation scenario has recently been revived in the context of the convincing astronomical evidence discussed in section II for enhanced flaring in YSOs. Shu et al. (1997) and Lee et al. (1998) use the x-wind model (e.g., Shu et al. 1994; chapter by Shu et al., this volume) to locate the source of accelerated particles just inside the inner radius of the accretion disk, where magnetic reconnection is likely to occur. The reconnection events that produce the MeV electrons seen in radio continuum emission will also produce MeV particles that bombard nebular solids in the inner disk. Radioactive nuclei can then be blown out to larger radii by the x-wind (Shu et al. 1996). This model nicely solves the fluence problem by having the stellar energetic particles produce the radioactivity at several stellar radii instead of several AU. The flares may also affect the thermal properties of the ejected solids.

Lee et al. (1998) have analyzed the production of short-lived isotopes with this local irradiation model based on the x-wind model. With reasonable scalings of X-ray and particle fluxes from the current Sun to YSOs, they find that proton reactions can produce ^{41}Ca, ^{53}Mn, and ^{138}La in amounts that are consistent with measurements of meteorites. However, they are unable to obtain the high measured abundances of ^{26}Al with proton or α reactions. Instead, they invoke flare-generated ^{3}He nuclei using reactions such as ^{24}Mg(^{3}He,p)^{26}Al. The present, most simple form of the model cannot account for all observed isotopic anomalies, and many basic issues still need to be addressed. The chapters by Goswami and Vanhala, this volume, and by Jones et al., this volume, provide a broader discussion of these issues for both the local irradiation and external seeding scenarios.

IX. CONCLUDING REMARKS

The astronomical observations summarized in section II show that low-mass YSOs, from embedded Class I protostars to Class III stars approaching the main-sequence, exhibit levels of magnetic activity orders of magnitude above that seen in main-sequence stars like the Sun. YSOs undergo powerful magnetic reconnection flares that probably require organized magnetic structures larger than the star itself. Radio gyrosynchrotron emission is even more elevated than X-ray emission and directly reveals *in situ* acceleration of MeV particles within the YSO. We can thus reiterate with confidence the conclusion reached after the *Einstein Observatory* era: "Invoking high levels of energetic particle or radiation fluxes associated with magnetic activity in the early Sun is no longer *ad hoc*" (Feigelson et al. 1991).

The most exciting observational results since the *Protostars and Planets III* review (Montmerle et al. 1993) are the detections of powerful X-ray flares in several Class I protostars and gyrosynchrotron flares in one Class I system. Further advances in this area are expected from the *Chandra X-ray Observatory* and the *X-ray Multi-Mirror Mission* (XMM). Uncertainty remains regarding the exact location and nature of flaring within the complex magnetic fields likely to be present in YSOs. On one hand, most of the Class II and Class III phenomenology can be explained by enhanced solar-type magnetic activity: magnetic reconnection events in multipolar fields rooted in the stellar photosphere. However, a persuasive case emerges that associates flaring events, particularly in protostars, with star-disk magnetic fields in the neighborhood of the disk corotation radius. Reconnection may also occur within the disk corona and would have different implications for the effects of X-irradiation and particle bombardment of disk material.

Substantial theoretical progress has also been made since *Protostars and Planets III*. The recent calculations presented in sections III–V show that disk ionization due to YSO X-rays greatly exceeds that of galactic cosmic rays. The X-rays penetrate surprisingly deep into the disk, due in part to Compton scattering. Within a few AU, the outer layers of the disk are partially ionized, and even the disk midplane is irradiated at larger radii. Thus, X-ray ionization will influence the magnetohydrodynamic properties of protoplanetary disks. There are also strong indications that YSO X-rays will alter both the molecular chemistry and the solid state characteristics of cold material in the environs of YSOs (sections VI–VII). In addition to the significant theoretical challenge of X-ray chemistry for YSOs, there are important opportunities for millimeter and infrared telescopes with both spectroscopic capability and high spatial resolution. Observations are needed to detect definitively the unusual molecular ions, molecular radicals, and altered grains that are expected to exist in the immediate vicinity of X-ray-luminous YSOs.

Finally, it is likely that effects of magnetic flaring events that occurred when the Sun was a YSO have already been detected in ancient meteorites (section VIII). It is important to recall again that the detection of radio gyrosynchrotron emission directly demonstrates that MeV particle acceleration occurs in YSOs. Although the link between magnetic reconnection particles and meteoritics has been discussed for decades, the quantitative study of this fascinating subject has only just begun.

Acknowledgments The research of A. E. G. is supported in part by the National Science Foundation, and E. D. F. is supported by the National Aeronautics and Space Administration. The authors would like to thank the reviewer, Francesco Palla, for helpful comments.

REFERENCES

André, P. 1996. Radio emission as a probe of large-scale magnetic structures around young stellar objects. In *Radio Emission from the Stars and the Sun*, ed. A. R. Taylor and J. M. Paredes (San Francisco: Astronomical Society of the Pacific), pp. 273–284.

André, P., and Montmerle, T. 1994. From T Tauri stars to protostars: Circumstellar material and young stellar objects in the ρ Ophiuchi cloud. *Astrophys. J.* 420:837–862.

Babel, J., and Montmerle, T. 1997*a*. X-ray emission from Ap-Bp stars: A magnetically confined wind-shock model for IQ Aur. *Astron. Astrophys.* 323:121–138.

Babel, J., and Montmerle, T. 1997*b*. On the periodic X-ray emission from the O7 V star θ^1 Orionis C. *Astrophys. J. Lett.* 485:L29–L32.

Balbus, S. A., and Hawley, J. F. 1991. A powerful local shear instability in weakly magnetized disks. I: Nonlinear evolution. *Astrophys. J.* 376:223–233.

Beckwith, S. V., Sargent, A. I., Chini, R. S., and Güsten, R. 1990. A survey for circumstellar disks around young stellar objects. *Astron. J.* 99:924–945.

Bell, K. R., Cassen, P. M., Klahr, H. H., and Henning, T. 1997. The structure and appearance of protostellar accretion disks: Limits on disk flaring. *Astrophys. J.* 486:372–387.

Blaes, O. M., and Balbus, S. A. 1994. Local shear instabilities in weakly ionized, weakly magnetized disks. *Astrophys. J.* 421:163–177.

Boulanger, F., Abergal, A., Bernard, J. P., Cesarsky, D., Puget, J. L., Reach, W. T., Ryler, C., Cesarsky, C. J., Sauvage, M., Tran, D., Vigreux, L., Falgarone, E., Lequeux, J., Pevault, M., and Rouan, D. 1998. The nature of small interstellar dust particles. In *Star Formation with the Infrared Space Observatory*, ed. J. L. Yun and R. Liseau (San Francisco: Astronomical Society of the Pacific, pp. 15–23.

Bouvier, J., Covino, E., Kovo, O., Martín, E. L., Matthews, J. M., Terranegra, J., and Beck, S. C. 1995. COYOTES II: Spot properties and the origin of photometric period variations in T Tauri stars. *Astron. Astrophys.* 299:89–107.

Caillault, J. P., and Zoonematkermani, S. 1989. Detection of a dozen X-ray-emitting main-sequence B6–A3 stars in Orion. *Astrophys. J. Lett.* 338:L57–L60.

Cameron, A. G. W., and Truran, J. W. 1977. The supernova trigger for formation of the solar system. *Icarus* 30:447–461.

Cameron, A. G. W. 1985. Formation and evolution of the primitive solar nebula. In *Protostars and Planets II*, ed. D. C. Black and M. S. Matthews (Tucson: University of Arizona Press), pp. 1073–1099.

Cameron, A. G. W. 1993. Nucleosynthesis and star formation. In *Protostars and Planets III*, ed. E. H. Levy and J. I. Lunine (Tucson: University of Arizona Press), pp. 47–73.

Cameron, A. G. W. 1995. The first ten million years in the solar nebula. *Meteoritics* 30:133–161.

Carkner, L., Feigelson, E. D., Koyama, K., Montmerle, T., and Reid, N. 1996. X-ray emitting T Tauri stars in the L1551 cloud. *Astrophys. J.* 464:286–305.

Carkner, L., Kozak, J., and Feigelson, E. D. 1998. X-ray survey of very young stellar objects. *Astron. J.* 116:1933–1939.

Carkner, L. 1998. Pre-main-sequence stellar magnetic activity: Evolution and effects. Ph.D. Dissertation, Pennsylvania State University.

Casanova, S., Montmerle, T., Feigelson, E. D., and André, P. 1995. ROSAT X-ray sources embedded in the ρ Ophiuchi cloud core. *Astrophys. J.* 439:752–770.

Caselli, P., Walmsley, C. M., Terzieva, R., and Herbst, E. 1998. The ionization fraction in dense cloud cores. *Astrophys. J.* 499:234–249.
Cazaux, J. 1997. A physical approach to the radiation damage mechanisms induced by X-rays in X-ray microscopy and related techniques. *J. Microscopy* 188,2:106–124.
Cesarsky, C. J., and Volk, H. J. 1978. Cosmic ray penetration into molecular clouds. *Astron. Astrophys.* 70:367–377.
Chiang, E., Phillips, R. B., and Lonsdale, C. J. 1996. A λ3.6-cm radio survey of low-mass, weak-line T Tauri stars in Taurus-Auriga. *Astron. J.* 111:355–364.
Clayton, D. D., Dwek, E. and Woosley, S. E. 1977. Isotopic anomalies and proton irradiation in the early solar system. *Astrophys. J.* 214:300–315.
Clayton, D. D., and Jin, L. 1995. Interpretation of ^{26}Al in meteoritic inclusions. *Astrophys. J. Lett.* 451:L87–L91.
Corcoran, M., and Montmerle, T. 1998. ISOCAM observations of 6.2 microns and 7.7 microns dust emission from four X-ray emitting Herbig Ae/Be stars. In *Star Formation with the Infrared Space Observatory*, ed. J. Yun and R. Liseau (San Francisco: Astronomical Society of the Pacific), pp. 338–341.
Dalgarno, A., and McCray, R. A. 1972. Heating and ionization of H I regions. *Ann. Rev. Astron. Astrophys.* 10:375–426.
Draine, B. T., and Woods, D. T. 1991. Supernova remnants in dense clouds. I: Blast-wave dynamics and X-ray irradiation. *Astrophys. J.* 383:621–638.
Dutrey, A., Guilloteau, S., and Guélin, M. 1997. Chemistry of proto-solar-like nebulae: The molecular content of the DM Tau and GG Tau disks. *Astron. Astrophys. Lett.* 317:L55–L58.
Dwek, E. 1978. Proton-associated alpha-irradiation in the early solar system: A possible ^{41}K anomaly. *Astrophys. J.* 221:1026–1031.
Dwek, E., and Smith, R. K. 1996. Energy deposition and photoelectric emission from the interaction of 10 eV to 1 MeV photons with interstellar dust particles. *Astrophys. J.* 459:686–700.
Feigelson, E. D. 1982. X-ray emission from young stars and implications for the early solar system. *Icarus* 51:155–163.
Feigelson, E. D., Giampapa, M. S., and Vrba, F. J. 1991. Magnetic activity in pre-main-sequence stars. In *The Sun in Time*, ed. C. P. Sonett, M. S. Giampapa, and M. S. Matthews (Tucson: University of Arizona Press), pp. 658–681.
Feigelson, E. D., Casanova, S., Montmerle, T., and Guibert, J. 1993. ROSAT X-ray study of the Chamaeleon I dark cloud. I: The stellar population. *Astrophys. J.* 416:623–646.
Feigelson, E. D., Carkner, L., and Wilking, B. A. 1998. Circularly polarized radio emission from an X-ray protostar. *Astrophys. J. Lett.* 494:L215–L218.
Feigelson, E. D., and Montmerle, T. 1999. High-energy processes in young stellar objects. *Ann. Rev. Astron. Astrophys.*, in press.
Fireman, E. L. 1974. Interstellar absorption of X-rays. *Astrophys. J.* 187:57–60.
Fowler, W., Greenstein, J., and Hoyle, F. 1962. Nucleosynthesis during the early history of the solar system. *Geophys. J. Roy. Astron. Soc.* 6:148–220.
Gagné, M., Caillault, J. P., and Stauffer, J. R. 1995. Deep ROSAT HRI observations of the Orion nebula region. *Astrophys. J.* 445:280–313.
Gagné, M., Caillault, J. P., Stauffer, J. R., and Linsky, J. L. 1997. Periodic X-ray emission from the O7 V star θ^1 Orionis C. *Astrophys. J. Lett.* 478:L87–91.
Gammie, C. F. 1996. Layered accretion in T Tauri disks *Astrophys. J.* 457:355–362.
Glassgold, A. E., Najita, J., and Igea, J. 1997. X-ray ionization of protoplanetary disks (and erratum). *Astrophys. J.* 480:344–350 (and 485:902).
Gougeon, S. 1998. Contributions expérimentales à l'astrophysique en rayons X. Ph. D. Thesis, University of Paris VII.

Gredel, R., and Dalgarno, A. 1995. Infrared response of H_2 to X-rays. *Astrophys. J.* 446:852–859.
Greenhill, L. J., Jiang, D. R., Moran, J. M., Reid, M. J., Lo, K. Y., Claussen, N. J. 1995. Detection of a subparsec diameter disk in the nucleus of NGC 4258. *Astrophys. J.* 440:619–627.
Gregorio-Hetem, J., Montmerle, T., Casanova, S., and Feigelson, E. D. 1998. X-rays and star formation: ROSAT observations of the Monoceros and Rosette molecular clouds. *Astron. Astrophys.* 331:193–210.
Grosso, N., Montmerle, T., Feigelson, E. D., André, P., Casanova, S., and Gregorio-Hetem, J. 1997. An X-ray superflare from an infrared protostar. *Nature* 387:56–58.
Guélin, M., Langer, W. D., and Wilson, R. W. 1982. The state of ionization in dense molecular clouds. *Astron. Astrophys.* 107:107–127.
Guenther, E. W., Lehmann, H., Emerson, J. P., and Staude, J. 1998. Measurements of magnetic field strength on T Tauri stars. *Astron. Astrophys.* 341:768–783.
Halpern, J. P., and Grindlay, J. E. 1980. X-ray photoionized nebulae. *Astrophys. J.* 242:1041–1055.
Harper, C. L., Jr. 1996. Astrophysical site of the origin of the solar system inferred from extinct radionuclide abundances. *Astrophys. J.* 426:1026–1038.
Hayashi, C., Nakazawa, K., and Nakagawa, Y. 1985. Formation of the solar system. In *Protostars and Planets II*, ed. D. C. Black and M. S. Mathews (Tucson: University of Arizona Press), pp. 1100–1153.
Hayashi, M. R., Shibata, K., and Matsumoto, R. 1996. X-ray flares and mass outflows driven by magnetic interaction between a protostar and its surrounding disk. *Astrophys. J. Lett.* 468:L37–L40.
Henke, B. L., Gullikson, E. M., and Davis, J. C. 1993. X-ray interactions: Photoabsorption, scattering, transmission, and reflection at $E = 50$–30,000 eV, $Z = 1$–92. *At. Data Nuc. Data Tables* 54:181–342.
Hofner, P., and Churchwell, E. 1997. Hard X-ray emission from the W3 core. *Astrophys. J. Lett.*, 486:L39–L42.
Igea, J., and Glassgold, A. E. 1999. X-ray ionization of the disks of young stellar objects. *Astrophys. J.*, 518:848–858.
Johns-Krull, C. M., Hatzes, A. P. 1997. The classical T Tauri star Sz 64: Doppler imaging and evidence for magnetospheric accretion. *Astrophys. J.* 487:896–915.
Johns-Krull, C. M., Valenti, J. A., and Koresko, C. 1999. Measuring the magnetic field on the classical T Tauri star BP Tauri. *Astrophys. J.* 516: 900–915.
Kaastra, J. S., and Mewe, R. 1993. X-ray emission from thin plasmas. I: Multiple Auger ionization and fluorescence processes from Be to Zn. *Astron. Astrophys. Suppl.* 97:443–482.
Kastner, J. H., Zuckerman, B., Weintraub, D. A., and Forveille, T. 1997. X-ray and molecular emission from the nearest region of recent star formation. *Science* 277:67–71.
Koyama, K., Ueno, S., Kobayashi, N., and Feigelson, E. 1996. Detection of hard X-rays from a cluster of protostars. *Pub. Astron. Soc. Japan* 48:L87–L92.
Krolik, J. H., and Kallman, T. R. 1983. X-ray ionization and the Orion molecular cloud. *Astrophys. J.* 267:610–624.
Laor, A., and Draine, B. T. 1993. Spectroscopic constraints on the properties of dust in active galactic nuclei. *Astrophys. J.* 401:441–468.
Lada, C.J. 1991. Formation of low-mass stars: Observations. In *The Physics of Star Formation and Early Stellar Evolution*, NATO ASI Series Vol. 342, ed. C. J. Lada and N. Kylafis (Dordrecht: Kluwer), pp. 329–363.
Lee, T. 1978. A local proton irradiation model for isotopic anomalies in the solar system. *Astrophys. J. Lett.* 224:L217–L226.

Lee, T., Shu, F. H., Shang, H., Glassgold, A. E., and Rehm, K. E. 1998. Protostellar cosmic rays and extinct radioactivities in meteorites. *Astrophys. J.*, in press.

Léger, A., Jura, M., and Omont, A. 1985. Desorption from interstellar grains. *Astron. Astrophys.* 144:147–160.

Lepp, S., and McCray, R. 1983. X-ray sources in molecular clouds. *Astrophys. J.* 269:560–567.

Lepp, S., McCray, R., Shull, J. M., Woods, D. T., and Kallman, T. 1985. Thermal phases of interstellar and quasar gas. *Astrophys. J.* 288:58–64.

Lepp, S. 1992. The cosmic-ray ionization rate. In *Astrochemistry of Cosmic Phenomena*, ed. P. Singh (Dordrecht: Reidel), pp. 471–475.

Lepp, S., and Dalgarno, A. 1996. X-ray-induced chemistry of interstellar clouds. *Astron. Astrophys.* 306:L21–L24.

Levy, E. H., and Araki, S. 1989. Magnetic reconnection flares in the protoplanetary nebula and the possible origin of meteorite chondrules. *Icarus* 81:74–91.

Maloney, P. R., Hollenbach, D. R., and Tielens, A. G. 1996. X-ray irradiated molecular gas. I: Physical processes and general results. *Astrophys. J.* 466:561–584.

Maloney, P. R., Colgan, S. W., and Hollenbach, D. J. 1997. Probing galactic center black hole candidates with far-infrared spectroscopy. *Astrophys. J. Lett.* 482:L41–L44.

McKee, C. F. 1989. Photoionization-regulated star formation and the structure of molecular clouds. *Astrophys. J.* 345:782–801.

Miyoshi, M., Moran, J., Herrnstein, J., Greenhill, L., Nakai, N., Diamond, P., and Inoue, M. 1995. *Nature* 373:127–130.

Montmerle, T., Koch-Miramond, L., Falgarone, E., and Grindlay, J. E. 1983. Einstein observations of the ρ Ophiuchi dark cloud: An X-ray Christmas tree. *Astrophys. J.* 269:182–201.

Montmerle, T., Feigelson, E. D., Bouvier, J., and André, P. 1993. Magnetic fields, activity and circumstellar material around young stellar objects. *Protostars and Planets III*, ed. E. H. Levy and J. I. Lunine (Tucson: University Arizona Press), pp. 689–717.

Morrison, R., and McCammon, D. 1983. Interstellar photoelectric absorption cross sections, .03–10 keV. *Astrophys. J.* 270:119–122.

Neufeld, D. A., Maloney, P. R., and Conger, S. 1994. Water maser emission from X-ray-heated circumnuclear gas in active galaxies. *Astrophys. J. Lett.* 436:L127–L130.

Neufeld, D. A., and Maloney, P. R. 1995. The mass accretion rate through the masing molecular disk in the active galaxy NGC 4258. *Astrophys. J. Lett.* 447:L17–L20.

Neuhäuser, R. 1997. Low-mass pre-main stars and their X-ray emission. *Science* 276:1363–1369.

Papoular, R., Conard, J., Guillois, O., Nenner, I., Reynaud, C., and Rouzaud, J.-N. 1996. A comparison of solid-state carbonaceous models of cosmic dust. *Astron. Astrophys.* 315:222–236.

Parker, E. N. 1960. The hydrodynamic theory of solar corpuscular radiation and stellar winds. *Astrophys. J.* 132:821–866.

Phillips, J. A., and Lazio, T. J. W. 1995. Images of $HCO^+(1-0)$ emission in a molecular cloud near 1E 1740.7–2942. *Astrophys. J. Lett.* 442:L37–L39.

Preibisch, T., Zinnecker, H., and Herbig, E. 1996. ROSAT X-ray observations of the young cluster IC 348. *Astron. Astrophys.* 310:456–473.

Preibisch, T. 1997. ROSAT coronal temperatures of young late-type stars. *Astron. Astrophys.* 320:525–539.

Preibisch, T., Neuhäuser, R., and Stanke, T. 1998. SVS 16: The most X-ray-luminous young stellar object. *Astron. Astrophys.* 338:923–932.

Rao, M. N., Garrison, D. H., Palma, R. L., and Bogard, D. D. 1997. Energetic proton irradiation history of the HED parent body regolith and implications for ancient solar activity. *Meteoritics* 32:531–543.

Ray, T. P., Muxlow, T. W., Axon, D. J., Brown, A., Corcoran, D., Dyson, J., and Mundt, R. 1997. Large-scale magnetic fields in the outflow from the young stellar object T Tauri S. *Nature* 384:415–417.

Roberge, W., and Dalgarno, A. 1982. The formation and destruction of HeH^+ in astrophysical plasmas. *Astrophys. J.* 255:489–496.

Rodríguez, L. F. 1997. Thermal radio jets. In *Herbig-Haro Flows and the Birth of Low Mass Stars*, ed. B. Reipurth and C. Bertout (Dordrecht: Kluwer), pp. 83–92.

Shapiro, P. R., and Bahcall, J. N. 1981. X-ray absorption and the post-Auger decay spectrum of multielectron atoms. *Astrophys. J.* 245:335–349.

Shu, F., Najita, J., Ostriker, E., Wilkin, F., Ruden, S., and Lizano, S. 1994. Magnetocentrifugally driven flows from young stars and disks. I: A generalized model. *Astrophys. J.* 429:781–796.

Shu, F. H., Shang, H., and Lee, T. 1996. Toward an astrophysical theory of chondrites. *Science* 271:1545–1552.

Shu, F. H., Shang, H., Glassgold, A. E., and Lee, T. 1997. X-rays and fluctuating X-winds from protostars. *Science* 277:1475–1479.

Shull, J. M., and Van Steenberg, M. E. 1985. X-ray secondary heating and ionization in quasar emission-line clouds. *Astrophys. J.* 298:268–274.

Silk, J., and Norman, C. 1983. X-ray emission from pre-main-sequence stars, molecular clouds, and star formation. *Astrophys. J. Lett.* 272:L49–L53.

Silk, J., and Werner, M. 1969. Heating of H I regions by soft X-rays. *Astrophys. J.* 156:186–191.

Skilling, J., and Strong, A. W. 1976. On cosmic ray diffusion and anisotropy. *Astron. Astrophys.* 53:253–258.

Skinner, S. L., Güdel, M., Koyama, K., and Yamauchi, S. 1997. ASCA observations of the Barnard 209 dark cloud and an intense X-ray flare on V773 Tauri. *Astrophys. J.* 486:886–902.

Sternberg, A., and Dalgarno, A. 1995. Chemistry in dense photon-dominated regions. *Astrophys. J. Suppl.* 99:565–607.

Strom, K. M., and Strom, S. E. 1994. A multiwavelength study of star formation in the L1495E cloud in Taurus. *Astrophys. J.* 424:237–256.

Tiné, S., Lepp, S., Gredel, R., and Dalgarno, A. 1997. Infrared response of H_2 to X-rays in dense clouds. *Astrophys. J.* 481:282–295.

Umebayashi, T., and Nakano, T. 1990. Magnetic flux loss from interstellar clouds. *Mon. Not. Roy. Astron. Soc.* 243:103–113.

Voit, G. M. 1991. Energy deposition by X-ray photoelectrons into interstellar molecular clouds. *Astrophys. J.* 377:158–170.

Voit, G. M. 1992. Destruction and survival of polycyclic aromatic hydrocarbons in active galaxies. *Mon. Not. Roy. Astron. Soc.* 258:841–848.

Walter, F. M., Brown, A., Mathieu, R. D., Myers, P. C., and Vrba, F. J. 1988. X-ray sources in regions of star formation. III: Naked T Tauri stars associated with the Tauris-Auriga complex. *Astron. J.* 96:297–325.

Wasserburg, G. J. 1985. Short-lived nuclei in the early solar system. *Protostars and Planets II*, ed. D. C. Black and M. S. Matthews (Tucson: University of Arizona Press), pp. 703–737.

Wasserburg, G. J., Busso, M., Gallino, R., and Raiteri, C. M. 1994. Asymptotic giant branch stars as a source of short-lived radioactive nuclei in the solar nebula. *Astrophys. J.* 424:412–428.

Weidenschilling, S. J., and Cuzzi, J. N. 1993. Formation of planetesimals in the solar nebula. In *Protostars and Planets III*, ed. E. H. Levy and J. I. Lunine (Tucson: University of Arizona Press), pp. 1031–1060.

White, S. M., Pallavicini, R., and Kundu, M. R. 1992. Radio flares and magnetic fields on weak-line T Tauri stars. *Astron. Astrophys.* 257:557–566.
Williams, J. P., Bergin, E. A., Caselli, P., Myers, P. C., and Plume, R. 1998. The ionization fraction in dense molecular gas. I. Low-mass cores. *Astrophys. J.* 503:689–699.
Woolum, D. S., and Hohenberg, C. 1993. Energetic particle environment in the early solar system: Extremely long pre-compaction meteoritic ages or an enhanced early particle flux. In *Protostars and Planets III*, ed. E. H. Levy and J. I. Lunine (Tucson: University of Arizona Press), pp. 903–919.
Wooten, A., Loren, R. B., and Snell, R. L. 1982. A study of DCO^+ emission regions in interstellar clouds. *Astrophys. J.* 255:160–175.
Yan, M., and Dalgarno, A. 1997. The molecular cloud near the hard X-ray source 1E 1740.7-2942. *Astrophys. J.* 481:296–301.
Zinnecker, H., and Preibisch, T. 1994. X-ray emission from Herbig Ae/Be stars: A ROSAT survey. *Astron. Astrophys.* 292:152–164.

SPECTROSCOPY OF INNER PROTOPLANETARY DISKS AND THE STAR-DISK INTERFACE

JOAN R. NAJITA
National Optical Astronomy Observatories

SUZAN EDWARDS
Smith College

GIBOR BASRI
University of California at Berkeley

and

JOHN CARR
Naval Research Laboratory

We review the recent contribution of high-resolution spectroscopy to advances in our understanding of the region within a few AU of young low-mass stars. In particular, we describe the role played by strong stellar magnetic fields in the resolution of the angular momentum problem for star formation, and our ability to probe the properties of planet formation environments.

I. INTRODUCTION

The region within a few AU of accreting young stars is of great interest for star and planet formation. Within this region, stars accrete from and interact with their surrounding disks, energetic winds emerge, and planet formation may already be under way. The study of the dynamics and detailed structure of this region is currently restricted to the realm of high-resolution spectroscopy, given the small angular scale that it subtends at the distance of the nearest star-forming regions. In this chapter we review the contribution of high-resolution spectroscopy to advances in our understanding of the star-disk interface and the properties of disks at planet formation distances.

One of the most striking features of accreting young stars is the simultaneous presence of disk accretion and energetic outflows among stars of all masses, a result that was recognized at the time of *Protostars and Planets III* (Edwards et al. 1993*b*). Since that time, results from high-resolution spectroscopy have altered and enhanced our understanding of the nature of the interaction between stars and their accretion disks as well as the consequences of that interaction for the mass and angular momentum evolution of young stars. In the case of T Tauri stars (TTSs), which are low-mass stars in the final stages of disk accretion, the earlier picture, in which accreting disk matter joined the star through an equatorial boundary layer (Lynden-Bell and Pringle 1974), has given way to a new picture with broader explanatory power. A wide array of observational evidence now supports a picture in which a strong, organized stellar magnetic field truncates the inner disk, and disk material from the truncation region accretes onto the star along closed stellar field lines in a funnel flow. The stellar rotation rate is slowed and possibly regulated by the magnetic interaction with the disk so that the star maintains its slow rotation rate even while actively accreting. In at least one version of this picture, mass loss is integral to the whole process: an energetic wind also emerges from the truncation region along open field lines, contributing significantly to angular momentum loss from the system. The observational evidence from high-resolution spectroscopy that supports this paradigm shift for T Tauri stars is outlined in section II.

Also since *Protostars and Planets III* the advent of sensitive infrared spectrographs with large-format arrays has opened up new opportunities to probe the properties of protoplanetary disks within a few AU of the star. The ability to study planet formation environments is of even greater interest today given the discovery of extrasolar planets (see the chapter by Marcy et al., this volume). The unexpected presence of giant planets at small orbital radii and the diversity in planetary masses and eccentricities have challenged traditional theories of planet formation and underscored the possibility of significant dynamical evolution in young planetary systems. These results emphasize the need for observational studies of young disk systems in order to better understand the physical and dynamical conditions under which planets form. As we describe in section III, work to date using infrared spectroscopy demonstrates our current capability to study disk dynamics and physical properties within a few AU, the same range of radii currently probed by precision radial velocity searches for extrasolar planets.

II. THE STAR-DISK INTERFACE

The classical T Tauri stars (cTTSs) offer an excellent opportunity to probe the star-disk interface. These young, low-mass stars undergoing disk accretion are optically revealed, allowing their disk accretion rates to be assessed from the magnitude of their continuum veiling (Gullbring et al. 1998) and the dynamics of the interface region to be probed by spectroscopic study. The first suggestion that disk matter might accrete onto cTTSs along stellar magnetic field lines was made by Bertout et al. (1988), who interpreted the observation of rotationally modulated hot "spots" as evidence for nonaxisymmetric funnel flows. However, it was the subsequent confrontation of theory with the well-known fact that cTTSs rotate much below breakup ($v \sin i \sim 15$ km s^{-1}; e.g., Hartmann and Stauffer 1989), despite ongoing disk accretion, that eventually led to an understanding of the importance of star-disk magnetic coupling for stellar mass and angular momentum evolution. A key observational constraint (Edwards et al. 1993*a*; Bouvier et al. 1993; Choi and Herbst 1996) was the discovery that cTTSs also rotate more slowly than stars of comparable spectral type and age that show no evidence for disk accretion (weak T Tauri stars; wTTSs).

As discussed initially by Königl (1991), and expanded and refined by others (Shu et al. 1994; Cameron and Campbell 1996; Armitage and Clarke 1996), a likely explanation for the slow rotation of accreting young stars is that a strong, organized stellar magnetic field truncates and couples to the inner disk at a few stellar radii, thereby regulating the rotation of the star. Although the proposed models differ in their treatment of the radial extent of the coupling region and the angular momentum redistribution that allows continued slow stellar rotation, all assume the existence of strong (kG), organized fields and the accretion of disk matter from the truncation region in funnel flows that terminate in shocks at the stellar surface. We review in the next section the spectroscopic evidence for strong fields in young low-mass stars.

A. Stellar Magnetic Fields

T Tauri stars have long been thought to be magnetically active, based on their location on pre-main-sequence convective tracks coupled with numerous surrogate diagnostics of magnetic dynamos (X-ray and nonthermal radio emission, photometric evidence for large starspots; see the chapter by Glassgold et al., this volume). Direct measurements of TTS field strengths are complicated by rotational broadening, which dominates over Zeeman broadening at optical wavelengths. As a result, the first direct indication of strong fields (Basri et al. 1992) ignored line profiles and used a more indirect measure, the correlation of equivalent width enhancement with Zeeman sensitivity (i.e., the number and distribution of Zeeman components for a given line and their Landé g factors), a technique that is sensitive to the average field strength over the stellar surface. To date, field strengths in the 1–2.5 kG range have been measured for approximately three TTSs using this technique (Basri et al. 1992; Guenther et al. 1998). The large uncertainties associated with these values ($\sim$1 kG) stem from the dominance of rotational broadening and the resulting sensitivity to uncertainties in the stellar model.

The detectability of kG fields is much enhanced by going to long wavelengths where Zeeman broadening ($\propto \lambda^2$) dominates over rotational broadening ($\propto \lambda$). In their study of the Zeeman-sensitive Ti I λ 2.2 μm line in the cTTS BP Tauri, Johns-Krull et al. (1999*a*) find a line profile with a clear signature of Zeeman broadening and derive an averaged field strength of $Bf = 3.3$ kG, in excellent agreement with values required by theories of magnetospheric accretion. Because Zeeman broadening dominates at 2.2 μm, the surface distribution of field strengths and filling factors can be extracted from the line profile. Analysis of the Ti I line implies a distribution of filling factors and field strengths up to 5–10 kG.

Insight into the large-scale organization of the stellar field is provided in part by Doppler imaging from time-resolved high-resolution spectroscopy. Doppler imaging of two wTTSs (Hatzes 1995; Joncour et al. 1994) and one cTTS (Johns-Krull and Hatzes 1997) suggests the presence of large cool spots concentrated near the rotational poles, as would be expected from a strong dipole component, although other interpretations are also possible.

Spectropolarimetry contributes additional diagnostic power to the measurement of field strength and geometry. Because the Zeeman σ components in magnetized regions are oppositely polarized, a net field polarity on the stellar surface can be identified by, and field strengths measured from, the wavelength shift between the right and left circularly polarized components of the line profile. Since the measurement is differential in nature, modest line splittings can be detected even in the presence of rapid rotation. Although the detection of circular polarization is typically compromised if regions of opposite polarity are mixed on the visible hemisphere (as in the case of solar-type activity), rapid rotation provides a significant advantage in this case. The detection of net polarization in a given velocity interval of the profile is more likely because the interval probes a restricted region on the stellar surface. In the TTS photospheric spectrum, the polarization is typically weak, reflecting a lack of organization in the photospheric field as a whole (Brown and Landstreet 1981; Johnstone and Penston 1987; Donati et al. 1997). For example, circular polarization has been reported in the optical spectrum of two wTTSs (Donati et al. 1997), where rapid rotation and broad spectral coverage were used to detect polarization at 0.2% of the continuum.

In contrast to the situation for the photospheric field, work by Johns-Krull et al. (1999*b*) on the cTTS BP Tau provides clear evidence for a strong, organized field component that is associated with accretion onto the stellar surface. Although in this star circular polarization is not detected in Zeeman-sensitive photospheric lines, the He I λ 5876 Å emission line, which likely forms in accretion shocks at the base of funnel flows (see section II.B), *is* strongly polarized (at ~10% of the continuum), and the wavelength shift implies a field strength $\geq$2.4 kG. The strong polarization of the He I line argues that fields participating in accretion are globally organized (e.g., in a dipole-like configuration), as would be necessary if they are to couple to an inner disk several stellar radii away. The existence of such large-scale fields (magnetic loops extending to several $R_\star$) is also supported by evidence from radio interferometry of wTTSs (Phillips 1992; Phillips et al. 1996). The contrasting lack of polarization in the photospheric lines indicates that the organized fields cover a small fraction of the stellar surface. In the rest of this section, we discuss the evidence that the strong, organized field component controls accretion onto the stellar surface and plays an important role in the origin of winds.

B. Kinematic Clues from Permitted Atomic Emission Lines

Spectroscopically, cTTSs are characterized by a rich permitted-emission-line spectrum, with strong lines of H I and numerous neutral and singly ionized metals in the optical and near infrared and a wide range of ionization states in the ultraviolet. The densities and temperatures required to excite these lines indicate formation within $\sim 10R_{\star}$ of the stellar surface. The utility of permitted-emission-line profiles as probes of the dynamics of the star-disk interface depends critically on our ability to disentangle different kinematic components from profiles of composite origin. The composite nature of the permitted lines is indicated by the variety of shapes found among lines of different optical depth, excitation, and ionization. In Fig. 1 we illustrate a sequence of near-simultaneous profiles from eight lines in the high-accretion-rate cTTS DF Tau. The profiles consist of multiple emission and absorption components and display an astonishing range in breadth, symmetry, and morphology. Kinematic features indicating outflow within a few stellar radii are seen in lines of Hα, Hβ, and Mg II, and the presence of simultaneous outflow and infall is seen in the blueshifted and redshifted absorptions at Na D.

Whereas the profile sequence shown in Fig. 1 for DF Tau is typical for a high-accretion-rate cTTS, the velocities and depths of the blueshifted and redshifted absorption components are particularly time variable among the high-accretion-rate stars. There are, however, some systematic trends in profile morphology displayed by cTTSs as a function of accretion rate (Edwards et al. 1994). In Fig. 2 we illustrate the range of profiles among some of the strongest cTTS emission lines: Hα, Hδ, and He I λ 5876 Å. In the remainder of this section, we use the profiles shown in Figs. 1 and 2 to demonstrate that funnel flows, accretion shocks, and winds contribute to the composite permitted-line profiles in cTTSs. We also note that the composite permitted-emission-line profiles in cTTSs almost certainly have additional kinematic components whose origin is as yet unidentified.

1. Funnel Flows. The most compelling spectroscopic evidence for funnel flows is redshifted "inverse P Cygni" (IPC) absorption features (Walker 1972; Edwards et al. 1994). These are especially prominent in lines of the upper Balmer, Paschen, and Brackett series but are also frequent at Na D and O I λ 7773 Å (e.g., Hamann and Persson 1992; Najita et al. 1996*a*). The observed infall velocities of several hundred km s^{-1} are readily explained as magnetic accretion from a disk truncated at several stellar radii and are inconsistent with accretion through an equatorial boundary layer. This interpretation of the line profiles is strengthened by similarities between observed profiles and those predicted by radiative transfer models of idealized funnel flows (Muzerolle et al. 1998*a,b*; Hartmann et al. 1994). The predicted profiles are broad and centrally peaked

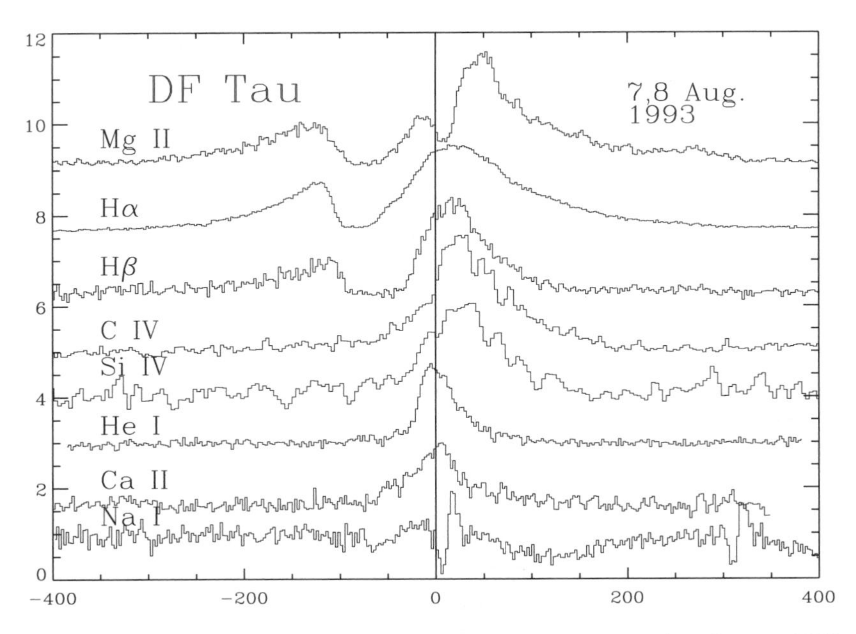

Figure 1. A series of line profiles taken at virtually the same time for the cTTS DF Tau. The Mg II, C IV, and Si IV profiles were obtained by the Goddard High Resolution Spectrograph (GHRS) on the *Hubble Space Telescope* (HST), and the others were obtained at Lick Observatory with the Hamilton echelle. Note the blueshifted wind component and broad emission wings in the top three most opaque diagnostics, the redshifted peak for the next two (high-temperature) diagnostics, and the nearly central peak for the next two less opaque optical diagnostics. The narrow central absorption in Mg II and emission in Na I are probably not associated with the TTS. There is a high-velocity redshifted absorption in Na I.

with a distinct blueward asymmetry. Formation of the redshifted IPC absorption component signifying supersonic infall is found to be sensitive to a variety of factors, including inclination and thermalization effects, and is predicted to be rare in optically thick lines such as Hα. These general characteristics are found to be representative of Balmer and Na D profiles in a snapshot survey of cTTSs covering the full range of disk accretion rates (Edwards et al. 1994). Specifically, Edwards et al. (1994) find (1) IPC absorption minima in upper Balmer lines and Na D in about 60% of cTTSs, independent of accretion rate; (2) blueward asymmetry in upper Balmer lines in about 90% of cTTSs; (3) IPC and blueward asymmetry only at Hα among cTTSs of low disk accretion rates. These similarities between observed and model profiles suggest that funnel flows are ubiquitous among cTTSs of all disk accretion rates.

The models of Hartmann and collaborators are also able to reproduce the observed luminosities of the Balmer and some metal lines with reasonable estimates for truncation radii and mass accretion rates. This

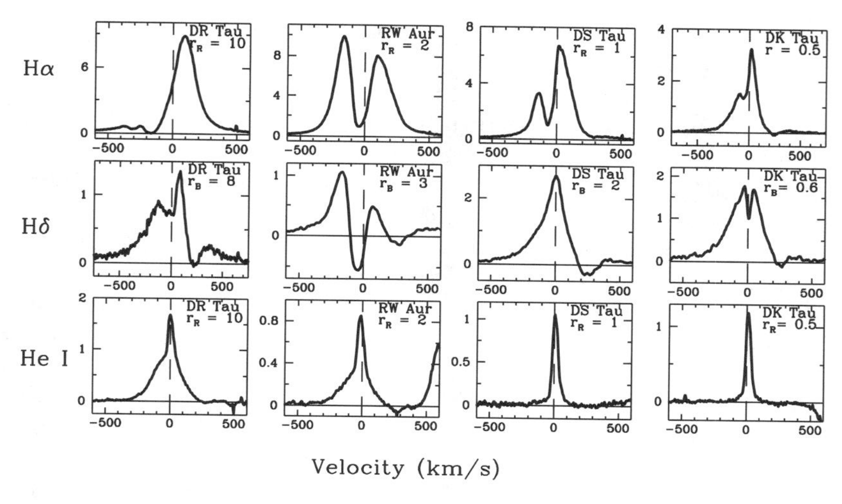

Figure 2. Trends as a function of continuum veiling are shown for representative permitted emission profiles in four cTTSs. Veiling levels vary from 10 times to 0.5 times the photospheric flux, corresponding roughly to variations in mass accretion rate over 1.5 orders of magnitude. At Hα, blueshifted absorption attributed to formation in an inner wind is seen most prominently in stars with high veiling, whereas redshifted absorption attributed to magnetic funnel flows is only seen at Hα in cTTSs with low veiling levels. Redshifted absorption is common in the the higher Balmer lines such as Hδ and is found at most all veiling levels. Higher Balmer lines also often show additional absorption components near the stellar velocity that appear to bear no relation to veiling. Strong metallic features such as He I λ 5876 Å are narrower than the Balmer lines and display a two-component emission profile. There is a tendency for stars with high veiling to have a prominent broad component, whereas stars with low veiling have a profile dominated by a narrow component.

implies that the bulk of the emission in the hydrogen lines arises in accretion flows, not in winds as was previously thought (Hartmann et al. 1994). However, it will be necessary to develop radiative transfer models that include a rigorous treatment of the rotation and thermal structure of funnel flows before we can accurately determine the fraction of the permitted-line luminosity that arises in these flows. A first step in this direction has been taken by Martin (1996), who computed the temperature and ionization structure in funnel flows and found that adiabatic compression is a significant heat source. The predicted range of temperatures within the funnel flow (3000–7000 K) is considerably lower than what Hartmann et al. (1994) find is necessary to account for the hydrogen line luminosity, if it arises entirely in a funnel flow. Further study of additional heating processes [e.g., magnetohydrodynamic (MHD) wave heating] and the radiative properties of funnel flows is required before definitive conclusions can be made. At present the similarities between theoretical and observed profiles are heartening, but physically self-consistent models are needed in order to establish the uniqueness of the funnel flow interpretation for these permitted-line diagnostics.

2. Accretion Shocks. Although magnetic accretion models successfully describe emission profiles in some permitted lines, the majority of the metallic lines are characterized by a morphology with a symmetric core and a distinctive two-component structure, quite different from the profiles predicted for funnel flows (Basri 1987; Hamann and Persson 1992). Figure 2 illustrates the two-component morphology for representative He I λ 5876 Å profiles, which are well characterized as the sum of a narrow component (NC) and a broad component (BC). The kinematic properties of the NC are fairly consistent among large samples of cTTSs, characterized by half-widths ≤ 50 km s^{-1}, symmetric structure, and velocities close to the stellar rest velocity. In contrast, there is considerable variety in the structure of the BC, characterized by half-widths from one to several hundred km s^{-1}, often with blueward asymmetries (Batalha et al. 1996; Edwards 1997). There is some indication that BC emission is more prominent in stars with high accretion rates (Muzerolle et al. 1998*b*), although variations in veiling and line profiles in individual stars show no correlation between the instantaneous accretion rate and the relative proportion of BC to NC emission (Edwards 1997).

A likely interpretation of the two-component structure, based largely on the observed kinematics, is that the NC arises in postshock gas at the base of magnetic accretion columns, whereas the BC arises in infalling gas in the funnel flow. Evidence supporting this interpretation includes the following: (1) An analysis of Fe I and Fe II lines in the cTTS DR Tau suggests that the NC and BC arise in regions with differing opacities and temperatures (Beristain et al. 1998). (2) Idealized funnel flow models can account for the strength and morphology of the BC in lines of Na D and the Ca IR triplet (Muzerolle et al. 1998*b*). (3) The NCs in the highest ionization states seen in the UV from *Hubble Space Telescope* (HST) Goddard High Resolution Spectrograph (GHRS) spectra, such as C IV and Si IV (see Fig. 1), are redshifted compared to lower ionization states, as would be expected in immediate postshock gas, and their line fluxes can be accounted for by emission in the expected postshock conditions (Calvet et al. 1996). The postshock gas is also the likely source of the excess continuum veiling of TTSs, which was formerly attributed to emission from an equatorial boundary layer. In contrast to funnel flows, very little theoretical work has been done to date on magnetic accretion shocks in cTTSs. The exploration of the basic shock structure and prediction of the resulting continuum radiation by Calvet and Gullbring (1998) constitute a promising first step, and we expect additional significant advances in the near future.

The combination of emission features believed to arise in either the funnel flow or accretion shock also provides important insight into the topology of the magnetic field. In a typical cTTS the spectrum *always* shows evidence for an NC (Batalha et al. 1996; Beristain et al. 1998), infall (Edwards et al. 1994), and continuum veiling. The continuous presence of all three phenomena in most cTTSs suggests a large-scale uniformity in the distribution of accretion footpoints. While this could arise from either a dipole geometry or a complex magnetic topology that covers much of the stellar surface, the detection of circular polarization in He I λ 5876 Å in BP Tau (Johns-Krull et al. 1999*b*; section II.A), a line dominated by NC emission from the base of the accretion shock, argues strongly for the former interpretation.

3. Inner Winds. Spectroscopic evidence for winds emerging from near-stellar regions comes from blueshifted absorption components in profiles of Hα, Na I D, Ca II H and K, and Mg II h and k (Kuhi 1964; Giampapa et al. 1981). Evidence for winds on more extended scales is provided by the blueshifted forbidden lines, which trace flows that have been collimated into jets at distances of tens of AU from the star and in some instances are spatially resolved (Hartigan et al. 1995; Hirth et al. 1997). Both the inner and extended wind diagnostics correlate with disk accretion rates in cTTSs, and neither are found in nonaccreting wTTSs. For the inner winds, this trend is demonstrated by a progression in the permitted-line absorption features with disk accretion rate, as illustrated in Figs. 1 and 2 for Hα. Classic P Cygni profiles are found among stars with high accretion rates, progressing to less deep and less blueshifted absorption features in stars with average accretion rates, and to weak or absent wind signatures among cTTSs with the lowest accretion rates. This progression is qualitatively consistent with the correlation found between disk accretion rates and mass loss rates inferred from the forbidden-line luminosities (Hartigan et al. 1995), suggesting that both inner winds and extended winds traced by forbidden lines are powered by accretion.

It is of considerable interest to compare mass loss rates estimated by inner and extended wind diagnostics to ascertain whether they trace the same flow (Calvet 1997) and to compare the mass loss rates of each to disk accretion rates. Of these quantities, the cTTS disk accretion rates are probably the best constrained. The most recent disk accretion rates evaluated from optical continuum veiling average 10^{-8} $M_\odot$ yr^{-1}, but span a wide range of values from 10^{-7}–10^{-9} $M_\odot$ yr^{-1} (see the chapter by Calvet et al., this volume). Mass loss rates from forbidden-line luminosities are more uncertain, although average values are estimated to be about 10^{-9} $M_\odot$ yr^{-1} (Hartigan et al. 1995). Mass loss rates from inner winds are the least well defined.

Inner wind properties were traditionally estimated from permitted-line luminosities and profiles under the assumption that the lines form solely in the wind. For example, Giovanardi et al. (1991) used the hydrogen and Na I D lines to estimate wind mass loss rates, with wind temperature primarily constrained by hydrogen line intensity and column density by Na I absorption strength. They found that inner winds are cool (5000–7000 K) and neutral with mass loss rates 10^{-8}–10^{-7} $M_\odot$ yr^{-1} in the most extreme cTTSs, comparable to that of associated molecular outflows. The recent realization that the bulk of the hydrogen emission may not form in the wind has loosened the constraints on inner wind properties. The lack of significant emission from the wind, coupled with the presence of blueshifted wind features in the Hα and Na I D lines and their absence in the higher Balmer or Brackett lines, implies that inner wind temperatures are lower than previous estimates. In the likely event that photoionization is dominant over collisional ionization (i.e., for $T_w < 10^4$ K), the lower temperatures imply larger mass loss rates than previously estimated for a given Na I optical depth (e.g., Najita et al. 1996*a*).

Thus, despite the reinterpretation of hydrogen line luminosities, the indication is that inner winds are dynamically significant components of accreting systems, with mass loss rates that are appreciable in comparison to disk accretion rates. Nevertheless, definitive mass loss rates from the inner wind have not yet been established. An improved determination of inner wind mass loss rates will probably require thermal and excitation analyses for specific wind dynamical models. With these, we will be able to make more robust quantitative comparisons to mass loss rates from forbidden lines and disk accretion rates, and thereby potentially distinguish between the various models proposed to explain the existence of jets and winds from accreting young stars.

C. Synoptic Studies

Deeper insight into the composite origin of the emission regions at the star-disk interface can be acquired through variability studies. The spectral features of cTTSs can vary on a wide range of timescales, ranging from minutes to years, and thus temporal data can convey information that individual spectra cannot. In particular, synoptic studies provide insight on the size of the region that is changing, on the extent of the changes that can take place in it, and on the timescale of those changes. Ideally, one can even follow events from one region as they have an impact on neighboring regions after some transmittal time. Synoptic observations over many stellar rotation periods can uncover periodic phenomena, thereby providing an additional tool with which to probe the phase relationships between the dynamical components discussed in the previous section. Moreover, synoptic observations of photospheric features can also be used to construct Doppler images of the stellar surface in order to study the properties of hot and cool spots (see section II.A).

An outstanding example of periodic variations in emission line features is demonstrated by the TTS SU Aurigae. The depths of both a blueshifted absorption component (in Hα and Hβ) and a redshifted absorption component (in Hβ) are observed to vary at the three-day rotation period of the star but are, remarkably, 180° out of phase (Giampapa et al.

1993; Johns and Basri 1995*a*). In addition, the velocity of the redshifted component varies at the stellar rotation period and is at a maximum when the feature is strongest (Petrov et al. 1996). This behavior can be explained by a stellar magnetosphere that is tilted $\sim 15°$ from the rotation axis of the star and that couples to a disk at an x-point (Shu et al. 1994) from which both a funnel flow and a magnetocentrifugal wind emerge. Because the tilt favors inflow in one region of the magnetosphere and outflow 180° away in azimuth, the rotation of the magnetosphere at the rotation period of the star can account for the observed period and phase difference (Johns and Basri 1995*a*).

In a second example, the cTTS Sz 68, an infall feature at Hα is found to vary at the stellar rotation period of 4 days, but wind features are not detected (Johns-Krull and Hatzes 1997). As a comment on these studies, it is noteworthy that the large-scale symmetry of TTS magnetospheres (as indicated, e.g., by the rough stability of hydrogen line luminosity with rotational phase) implies that the rotationally modulated features detected in the above examples must represent fairly small perturbations on top of the large-scale structure. Such small perturbations are consistent with the existence of "hot spots," which represent a small fraction ($\sim 1\%$) of the accretion luminosity onto the star (Bertout et al. 1996).

A broader synoptic study with thorough time coverage of seven late-type cTTSs did not find further examples of periodic modulation, but it did determine that moderately blueshifted absorption components are the least variable part of the line profile, suggesting that winds are more quiescent than the funnel flow region and form well beyond it (Johns and Basri 1995*b*). This is generally consistent with most wind theories, which place the origin of the wind beyond the magnetosphere.

The many spectroscopic monitoring studies of cTTSs have opened our eyes to the very time-dependent and complex nature of the star-disk interface. For example, a recent study of ultraviolet spectral features in the cTTS BP Tau reveals that recombination lines vary systematically, possibly with the stellar rotation period, whereas lines dominated by collisional excitation vary irregularly (Gómez de Castro and Franqueira 1997). These distinctions might arise if the recombination lines form in accretion spots that show rotational modulation, while the strong permitted lines might have significant contributions from flares. (See also Johns-Krull and Basri 1997; Hessman and Guenther 1997; Gullbring et al. 1996.) We anticipate that, in the future, synoptic monitoring programs will play increasingly important roles in unravelling the complex, time-dependent nature of the star-disk interaction region. Synoptic monitoring programs with comprehensive time coverage that include lines with a wide range of optical depths would be of considerable interest.

Recent work on DQ Tau provides a final example of the exciting use of synoptic observations, in this case as a probe of the star-disk interaction and accretion dynamics in close binary systems (Mathieu et al. 1997; Basri et al. 1997). With the tight orbit implied by the 16-day period ($8R_\star$ at periastron), there is little room for disks around each star, and the circumbinary disk is classically expected to be held off by a gap of about 3 times the separation of the two stars. In this situation, accretion is perhaps not expected, and the fact that DQ Tau is a cTTS system may appear puzzling. However, in the case of DQ Tau, both the broadband flux and emission equivalent widths vary with the orbital period of the system, with the flux and equivalent width reaching maxima near periastron passage. These properties are in excellent agreement with recent dynamical models that suggest that accreting material from the circumbinary disk can cross the gap in streams, thereby perpetuating stellar accretion beyond the stage of gap formation (Artymowicz and Lubow 1996). Spectroscopic evidence for the presence of gas in gaps in close binary systems (including DQ Tau) is discussed in section III.C. These results illustrate the kind of dynamical issues that may prove critical in understanding binary star formation and the role that high-resolution spectroscopy may play in resolving these issues.

D. Summary

There is now convincing evidence that, during the T Tauri phase, accreting disk matter joins the star by infall along stellar magnetic field lines. This picture is based both on direct evidence that young stars possess stellar fields of the requisite strength and geometry to control accretion onto the star, and on the commonly observed dynamical signatures of gas infalling onto stellar surfaces. In addition, there is spectroscopic evidence for outflowing gas in close proximity to the disk truncation region. Moreover, the correlation between inner wind activity and disk accretion rate argues that inner winds are an integral part of the mass accretion process. In the coming decade we anticipate that the study of the star-disk interface in cTTSs will progress beyond the cataloging of the empirical phenomena that diagnose the presence of funnel flows, accretion shocks, and inner winds, and will provide a quantitative assessment of the physical conditions and mass flow rates in the region where stellar magnetic fields couple to disks and inner winds are generated.

Although considerable progress has been made in recognizing the role of magnetic accretion in the angular momentum evolution of young stars, consensus has yet to be achieved on the role of winds in this process (see, e.g., the chapters by Shu et al. and by Königl and Pudritz, this volume; Camenzind 1990; Ferreira et al. 1997). Much theoretical effort has recently been devoted to the origin of collimated flows (bipolar flows and jets) at large distances, but the fact that the same systems also possess dynamically significant inner winds is sometimes overlooked. In the x-wind theory, which addresses the origin of both inner winds and spatially extended collimated flows, disk matter reaching the truncation region is loaded onto stellar field lines that diverge into funnel flows and magnetocentrifugal inner winds. Through the generation of these flows, the stellar field mediates the angular momentum redistribution that allows the star to both spin slowly and grow in mass. Thus, in the context of this theory, the strong stellar fields, funnel flows, and inner winds discussed in this section all play a fundamental role in the mass and angular momentum evolution of young stars.

While much of the work to date has focused on the role of stellar fields in the mass and angular momentum evolution of single, low-mass stars, recent work has begun to expand beyond these boundaries. For example, strong stellar magnetic fields may also be important in various aspects of planet formation. Lin et al. (1996) have suggested that the truncation of disks by a strong stellar field may play a role in halting inward planetary migration. An important issue for the future is the degree to which the current picture for the inner few $R_\star$ of cTTSs applies to stars at earlier evolutionary phases, of higher masses, or possessing close companions. Remarkably, the spectroscopic diagnostics of the star-disk interface appear to be indistinguishable between stars with and without close companions (see the chapter by Mathieu et al., this volume), implying that the isolation of stellar components from the circumbinary disk through gap formation is not an impediment to the emergence of winds and funnel flows. Basri et al. (1997) have hypothesized that accretion streams traverse the gap and are channeled onto the stars via magnetic funneling when they near the star (see section II.C), but more work is needed to understand how funnel flows and winds can exist in the absence of stable accretion disks.

III. THE INNER DISK

While there is now abundant evidence for the existence of circumstellar disks around young low-mass stars (see the chapters by Wilner and Lay and by McCaughrean et al., this volume), our understanding of the detailed properties of disks, especially at planet formation distances $\lesssim 5$ AU, is still in its infancy. Early spectroscopic studies of this region in very young low-mass stars undergoing an outburst of mass accretion (i.e., FU Ori objects) focused on demonstrating the existence of circumstellar disks. As discussed at *Protostars and Planets III,* compelling spectroscopic evidence for rotating disks extending close to stellar surfaces had been obtained from high-spectral-resolution studies of these objects in the optical and near-infrared (Hartmann et al. 1993; see also Hartmann and Kenyon 1996 for a more recent update).

Since *Protostars and Planets III,* it has become apparent that it is possible to carry out spectroscopic studies of circumstellar disks in a broader class of young stars, including the more quiescent T Tauri stars, and over a larger range of disk temperatures and radii. This is good news, given the considerable motivation for detailed studies of the dynamics and physical and chemical structure of disks. Theoretical studies of a wide range of problems involving disk physics (e.g., the timescale and physical processes governing planet formation; see the chapter by Wuchterl et al., this volume) typically rely on assumed disk properties (e.g., the minimum-mass solar nebula). The ability to measure disk properties directly and examine the range of variation among systems would allow us, among other things, to test and distinguish between theories of planet formation in order to better address fundamental questions, such as the existence of solar systems like our own.

For example, the formation of stellar or planetary companions out of disk material is predicted to alter the physical structure of the disk, creating gaps through tidal torques and the excitation of spiral density waves (see the chapters by Lubow and Artymowicz and by Lin et al., this volume). Developing probes of disk physical structure in order to obtain observational evidence in support of this picture is an important step toward confirming this particular aspect of planet formation theories. Measurements of disk molecular abundances are of interest as well, because they can provide unique constraints on disk physical conditions and can constrain the extent of chemical processing that occurs in disks. Because measurements of disk chemical abundances would provide an observational context in which to interpret cometary abundances, which carry the fossil record of the conditions in the protosolar nebula, they may also provide important clues to the origin of our own solar system (see the chapters by Langer et al. and by Irvine et al., this volume).

Due to the small angular sizes involved, the study of the detailed properties and physical structure of disks at $\lesssim$5 AU is currently restricted to the realm of high-resolution spectroscopy, and spectroscopy of molecules in the infrared is particularly well suited to this task. At the high densities and warm temperatures (2000–150 K) characteristic of disks at $\sim$0.1–5 AU around low- to intermediate-mass stars, molecules are expected to be abundant in the gas phase and sufficiently excited to produce a rich vibrational/rotational spectrum at near- and mid-infrared wavelengths. The ability of high-spectral-resolution studies to resolve individual lines improves the detectability of weak spectral features and provides the kinematic information by which the emitting region can be located in the disk. By measuring multiple resolved profiles, excitation temperatures and column densities can be determined as a function of disk radius.

We are only now beginning to address questions such as those described above with the advent of sensitive, high-resolution infrared spectrographs in the last decade [e.g., the Cryogenic Echelle Spectrograph (CSHELL) at the NASA Infrared Telescope Facility (IRTF), CGS4 at the United Kingdom Infrared Telescope (UKIRT)]. In addition to the technological challenge of developing spectrographs with the necessary sensitivity, the study of planet formation environments is intrinsically challenging given the likelihood that protoplanetary disks are optically thick in the continuum at the expected vertical column densities at AU distances (e.g., $N_{\rm H}$ = 1500 g cm^{-2} at 1 AU for the minimum-mass solar nebula). As a result, it is necessary to target for study optically thick regions with significant vertical temperature structure (a disk atmosphere) in order to detect measurable line emission or absorption. An alternative approach is to focus on regions of low column density, created, for example, as part of the companion formation process, and search for line emission.

High-resolution work to date on inner disks has focused on H_2O and CO, molecules that are likely to be among the most abundant in disks. In the following, we give examples of studies that use these diagnostics to probe both disk atmospheres and low-column-density regions. While we focus on emission diagnostics in this review, strong absorption in disk atmospheres is also straightforward to measure in disk systems with very high mass accretion rates (e.g., FU Ori objects and stars with associated Herbig-Haro objects; Hartmann and Kenyon 1996; Reipurth and Aspin 1997). The results demonstrate the viability of high-resolution spectroscopy for the study of inner disk properties and illustrate the kind of diagnostic information that can currently be extracted from the data.

A. CO Overtone Emission

Due to its high dissociation energy, the CO molecule is expected to be abundant even at fairly high temperatures ($<$5000 K). In addition, the excitation temperatures and critical densities of the first overtone bands ($\Delta v = 2$; $\lambda \sim 2.3$ μm) make these transitions excellent probes of the warm, high-density ($n_H = 10^{10-15}$ cm^{-3}) conditions that characterize inner disks ($<$0.1 AU).

Following upon the heels of the first detection of CO overtone emission from young stars (Scoville et al. 1983), early low-spectral-resolution surveys for overtone emission (e.g., Geballe and Persson 1987; Carr 1989) targeted young stars with energetic outflows, detecting a large fraction of sources ($\sim$20%) in emission over a range of stellar masses (1–10 $M_\odot$). The unusual combination of circumstances—*molecular* emission from gas at warm temperatures and high density—restricted possible explanations for the emission and eventually led to proposals for an origin in circumstellar disks (Scoville et al. 1983; Carr 1989; Calvet et al. 1991) or outflowing winds (e.g., Carr 1989). Since excitation studies based on the low-resolution data were unable to distinguish between the two scenarios, higher-resolution observations capable of resolving line profiles have been used to obtain a more definitive diagnosis. Such studies are now straightforward with current high-resolution ($R \geq 20{,}000$) infrared spectrographs.

High-spectral-resolution studies generally find a spectral shape that strongly suggests a disk origin (Chandler et al. 1993, Carr et al. 1993, Najita et al. 1996*b*), particularly in the case of the sources studied in detail (e.g., WL16 and 1548c27). These observations, in fact, currently provide some of the best evidence for the existence of rotating disks around young stars. Excitation and spectral synthesis modeling of the overtone emission in objects covering a range of stellar masses shows that an origin in a vertical temperature inversion region in a Keplerian disk provides an excellent fit to the strength and shape of the emission (Fig. 3; Carr et al. 1993, Najita et al. 1996*b*, 1999).

The model fits also place useful constraints on stellar properties (e.g., masses) and the thermal and physical structure of the inner disk. For low- and intermediate-mass stars, we typically find that the temperature distribution of the emitting gas in the inversion region is $T_d \sim r^{-q}$, where $q = 0.4$–0.8, the column density of the emitting gas typically decreases with radius over the range $\sigma_H \sim 0.1$–1000 g cm^{-2}, with the emission originating within $R \simeq 0.4$ AU. The origin of the inversion is presently unclear. Irradiation by the central star has been investigated by Calvet et al. (1991) and D'Alessio et al. (1998), but the importance of irradiation by the

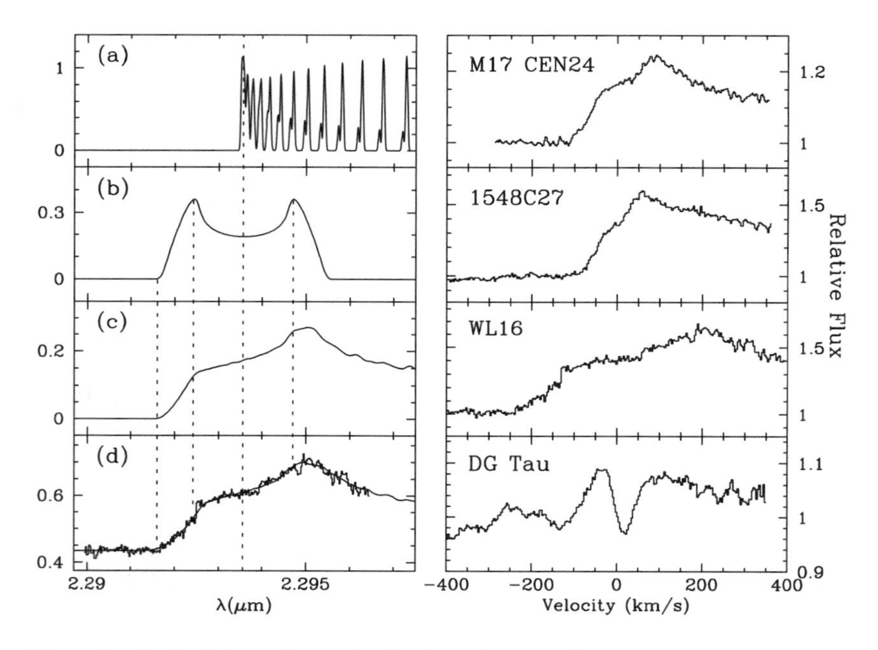

Figure 3. Left: Schematic synthesis of bandhead emission from a rotating disk. The rest distribution of CO lines near the $v = 2$–0 bandhead (a) is convolved with the double-horned profile of a single isolated line from an inclined Keplerian disk (b) to produce the characteristic profile of bandhead emission from a rotating disk (c). Actual excitation and spectral synthesis modeling, when compared with spectra for objects such as WL16 (d), demonstrates that an origin in a temperature inversion region in a Keplerian disk provides an excellent fit to the data. Right: The signature of bandhead emission from a rotating disk (c) is commonly observed in young stars over a range of masses: DG Tau ($\lesssim$1 $M_\odot$), WL16 ($\sim$2 $M_\odot$), 1548c27 ($\sim$4 $M_\odot$), M17 CEN 24 ($\sim$10 $M_\odot$). In the DG Tau spectrum, the spectral signature of the disk is complicated by strong stellar CO bandhead absorption.

magnetospheric accretion shock has yet to be explored in detail. As noted by Carr et al. (1993), the inversion might also be due to hydromagnetic or hydrodynamic wave heating or to turbulent shear and subsequent heating generated by a stellar wind flowing over the disk.

Other proposed explanations for the origin of the emission encounter various difficulties. For example, the scenario in which the emission arises in an optically thin disk (e.g., Carr 1989; Chandler et al. 1995) is less likely, since the absence of continuum emission over the range of disk temperatures from which the overtone emission arises cannot explain the strong near-IR excesses that accompany overtone emission. One of the primary difficulties facing the proposal that overtone emission arises in a wind is the likely inability of winds to produce the fairly symmetric emission line profiles that are observed, rather than P Cygni-like profiles with blueshifted absorption components (due to absorption by the wind of the hotter stellar and inner disk continuum), or a global line asymmetry due to the occultation of the receding flow by an optically thick inner disk (see also Najita et al. 1996*b*). In fact, the inability to detect winds via the overtone bands is somewhat of a puzzle, given previous theoretical predictions of abundant CO in warm ($\sim$2000 K) stellar winds (Glassgold et al. 1991) and the indirect confirmation of low wind temperatures implied by the revised understanding that hydrogen line emission originates in funnel flows rather than winds (see section II and the discussion in Najita et al. 1996*a*).

The more recent suggestion that overtone emission arises from gas accreting along stellar magnetic field lines (Martin 1997) is interesting, because funnel flows, like winds, should originate from a cool region of the disk that is likely to have abundant CO, at least in the case of low-mass T Tauri stars. Although funnel flows may certainly produce some emission, the claim that they are solely responsible for the emission, and their relevance to the detection of emission from high-mass objects, both require further theoretical study as well as detailed spectral synthesis and comparison to observed line profiles. Unlike the hydrogen emission lines, which are produced near the stellar surface, overtone emission is more likely to arise from the cooler outer region of the funnel flow. Since the extent of the CO-emitting region will be restricted by thermal and photodissociation processes, thermal models of funnel flows that consider additional heating sources (e.g., MHD wave heating; see section II.B), and photodissociation models that include the UV flux of the accretion shock at the stellar surface will be needed to calculate accurate line intensities and profiles. More realistic dynamical models for funnel flows (including rotation) and radiative transfer that treats the background continuum of the star and disk are also required.

Variability studies can also constrain the origin of the overtone emission. Dramatic variations are known to be possible from the comparison of observations that are typically widely separated in time. For example, at different epochs DG Tau has been observed to be strongly in emission (Carr 1989) and to have nearly zero emission equivalent width (Greene and Lada 1996). However, systematic variability studies are limited. The only synoptic monitoring study to date is the low-resolution study of Biscaya et al. (1997), who studied known overtone emission sources of low and high mass. Based on low-resolution spectra, these authors find significant variability of the emission equivalent width on timescales as short as a few days. If the variability timescale represents rotational modulation due to azimuthal structure, then this indicates that the overtone emission arises within several stellar radii.

The recent popularity of unbiased, moderate-resolution spectroscopic surveys designed primarily for the spectral classification of young stars (e.g., Greene and Lada 1996; Meyer 1996; Luhman and Rieke 1998) has yielded, as a by-product, improved statistics on the frequency of CO overtone emission from young stars. Studies of low-mass systems find that, consistent with the early results, CO overtone emission typically arises only in the most energetic low-mass systems, which constitute only a few percent of the sources in a given cluster. The infrequency of the emission is probably due to the high column density needed to produce detectable overtone emission. Since absorption in a late-type stellar photosphere could reduce the unresolved emission equivalent width in some fraction of sources, Carr and Najita carried out a small high-resolution survey of sources in Taurus to search for additional overtone emission sources. Only one of ~15 sources was detected strongly in emission, but weaker emission cannot be ruled out without further analysis. The incidence rate of emission among young high-mass stars is less clear given the small number of systems studied, but the current indication from studies of rich young clusters is that overtone emission may be more common among young high-mass stars than low-mass stars (e.g., ~10% in M17; Hanson et al. 1997).

B. Hot Water Emission

Water vapor is also expected to be abundant in disks, and an excellent probe of the high-density conditions between the thermal dissociation radius ($<$0.1 AU; ~2500 K) and the ice condensation radius ($\gtrsim$5 AU; ~150 K). In this region, H_2O is a strong molecular coolant, as well as the dominant source of infrared atmospheric opacity in the event of grain settling out of the upper disk atmosphere. Given its high abundance and the large number of radiative transitions capable of sampling a wide range in temperature, H_2O is an excellent diagnostic of the properties of circumstellar disks over a range of disk radii.

Telluric absorption typically presents a significant challenge to ground-based studies of H_2O. However, because the higher excitation states of H_2O have transitions far from the vibrational band centers (e.g., at 1.4 and 1.9 μm), ground-based observations of water that is very much *hotter* than the Earth's atmosphere can be made in the wings of the H_2O bands. Water emission is detectable at low spectral resolution if the emission is very strong. For example, the shape of the 1–2.5-μm spectrum of SVS-13, which shows water emission in broad local maxima around the 1.4- and 1.9-μm band centers, is a useful probe of the physical conditions of the gas (Carr et al. 1999). Preliminary modeling of the low-resolution data confirms that the emission arises from hot (~2000 K) gas, i.e., at temperatures characteristic of disk radii <0.3 AU.

Much weaker water emission can be detected at high spectral resolution by studying individual, resolved lines located far from the vibrational band centers. Using this approach, hot H_2O emission has been detected in a small number of young stellar objects known to be strong CO overtone emission sources (Carr et al. 1999; Fig. 4). High-resolution spectroscopy in the 2-μm window reveals that the water lines are consistently narrower than the CO lines, in agreement with a common origin for the CO and water lines in a differentially rotating disk with an outwardly decreasing temperature gradient. In such a situation, molecules with lower dissociation temperatures (e.g., H_2O) will tend to have narrower widths than those

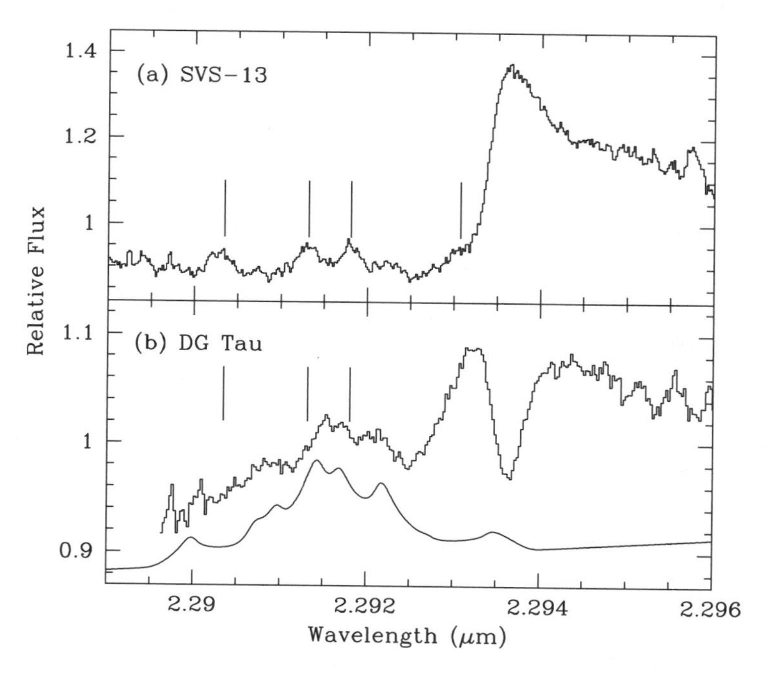

Figure 4. Hot water emission from young stars. Individual emission lines (positions marked by vertical lines) are detected near the CO bandhead in (a) SVS-13 and (b) DG Tau. The water lines are narrower than the CO lines, in agreement with a common origin for the CO and water lines in a differentially rotating disk with an outwardly decreasing temperature gradient. In DG Tau, the water spectrum (histogram) is dominated by the strongest line (at λ 2.2918 μm), which is broad and double-peaked. The rotational broadening of all the lines produces a blend of features that are well reproduced in the synthetic spectrum (solid line).

with higher dissociation temperatures (e.g., CO). In two sources, the expected double-peaked disk line profiles are observed (e.g., DG Tau; Fig. 4).

Because the water and CO overtone transitions probe similar physical conditions, the water emission is presumed to also originate in a temperature inversion in the disk atmosphere. More detailed modeling is needed to confirm this. At the moment, the primary obstacle to improved spectral synthesis modeling is the lack of complete water line lists in the near-infrared. Ongoing efforts are likely to produce rapid improvement in this area (e.g., Viti et al. 1997). Nevertheless, the ability to study multiple diagnostics (CO, H_2O) that originate from a common region in the disk atmosphere indicates the exciting future possibility of carrying out disk chemical abundances studies.

C. CO Fundamental Emission

The low-J CO fundamental ($v = 1$–0) lines at 4.6 μm are good tracers of disk structure at $0.2-2$ AU because of their sensitivity to low column densities of gas ($\ll 1$ g cm^{-2}) at temperatures characteristic of this range of disk radii (1500–300 K). Due to the difficulties of working in the M band (high thermal background, variable atmospheric transparency), work to date on the fundamental transitions has been limited to only the brightest few low-mass young stars. Work by Carr, Najita, and Mathieu has targeted predominantly known spectroscopic binary systems in order to search for low-column-density regions (gaps) created by the dynamical interaction of companions with the disk.

The existence of gaps in disks around some young stars has been inferred from their spectral energy distributions (SEDs), both in known spectroscopic binaries (e.g., Mathieu et al. 1991; Jensen and Mathieu 1997) and in systems without known stellar companions (e.g., Marsh and Mahoney 1992). The SEDs in these systems show deficits of continuum emission over a limited range of wavelengths, consistent with a much reduced continuum optical depth over a range of disk radii. In spectroscopic binary systems the implied gap sizes are generally consistent with those expected for dynamical clearing by the binary. For systems without known stellar companions, the altered disk structure could be due to unseen, perhaps planetary, companions.

Since structure in SEDs can arise for reasons other than disk gaps, it is difficult to use SEDs alone to diagnose the existence and properties of gaps. Less ambiguous studies of disk radial structure are possible with spectroscopy of residual gas in the gap, which will appear in emission against the optically thin or absent dust continuum. Because the system parameters of spectroscopic binaries are typically well known (e.g., stellar masses, sometimes inclination), spectroscopy of disk gas offers the considerable advantage that the radial location and extent of the emitting gas can be pinpointed using the kinematic information in velocity-resolved profiles. The viability of this approach is demonstrated by the detection of CO fundamental emission from a number of close binaries, the first detections in low-mass pre-main-sequence stars. The resulting line profiles and relative line strengths are generally consistent with emission from residual gas in a gap created by the binary.

The results for the double-lined spectroscopic binary DQ Tau are representative of the current state of affairs. Figure 5 shows spectra of the $v = 1–0$ $R(3)$ and P(18) and $v = 2–1$ $R(10)$ and $R(11)$ lines from DQ Tau, along with model disk emission spectra. The line profiles locate the gas at ~0.03–0.4 AU. Remarkably, this is exactly the radial extent expected for a dynamically cleared gap in this system (Mathieu et al. 1997). The average excitation temperature is 1150 K with a typical disk column density of 5×10^{-4} g cm^{-2}. Thus, the emission is produced by an extremely small amount of material, a total gas mass of just 10^{-5} $M_\oplus$. Interestingly, DQ Tau is one close binary system in which the expected dip in the SED is not apparent. Mathieu et al. (1997) postulate that a small amount of optically thin dust (corresponding to a total gas mass of

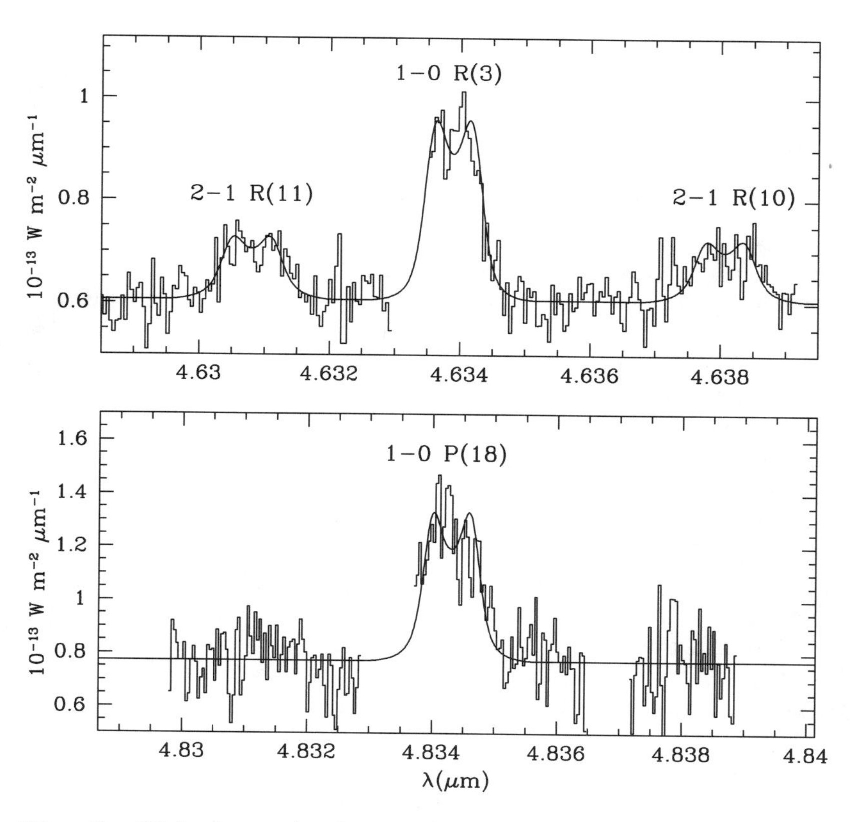

Figure 5. CO fundamental emission from the double-lined spectroscopic binary DQ Tau. Spectra of the v=1–0 $R(3)$ and v=2–1 $R(10)$ and $R(11)$ lines (top) and v=1–0 $P(18)$ lines (bottom) are shown, along with synthetic spectra of emission from an optically thin disk.

1.6×10^{-4} $M_{\oplus}$ at 1000 K) resides in the cleared region. This is remarkably close to the mass required to produce the CO emission. The lower total mass implied by the fundamental observations could be due in part to the assumption of thermal level populations, subthermal excitation being more likely.

Of possible relevance to the origin of the emission in DQ Tau are the observed modulation of the optical continuum veiling and emission line strengths at the orbital period. These favor a scenario in which accretion streams from the circumbinary disk cross the gap and accrete onto the stars via magnetic funneling (Basri et al. 1997; see section II.C). Perhaps some of the gas that produces the CO fundamental emission also originates in the accretion streams.

Overall, the detection rate for the CO fundamental lines is very high; five of the eight systems studied have been detected. The high detection rate, compared to that of CO overtone emission, probably reflects the lower column densities needed to produce measurable CO fundamental emission. Most of the detected systems are known spectroscopic binaries, but one is not known to possess stellar companions. These results indicate the potential of applying this technique to search for disk gaps induced by giant protoplanetary companions.

IV. FUTURE PROSPECTS

The results described above illustrate the contribution of high-resolution optical and infrared spectroscopy to the exciting developments in our understanding of inner accretion disks. Many of the outstanding issues can be further explored, and potentially resolved, with the next generation of spectrographs on large telescopes. For example, the light-gathering power of large telescopes, when coupled with the large wavelength coverage of cross-dispersed echelle spectrographs, is a powerful combination for studies of inner accretion disks. Tremendous progress in our understanding of stellar magnetic field strength and geometry is expected from infrared Zeeman measurements that will be made with instruments such as the Keck NIRSPEC. The ability to measure multiple Zeeman-sensitive lines simultaneously will allow the unambiguous detection of Zeeman splitting and a robust determination of the surface distribution of magnetic field strengths and filling factors. Through Doppler imaging of accretion shock diagnostics (e.g., He I λ 5876 Å) in polarized light, it will be possible to map out the spatial distribution of accretion footpoints and examine their long-term stability.

Effective, long-term campaigns for spectroscopic monitoring of the star-disk interface will be possible with queue-scheduled telescopes such as the Hobby-Eberly Telescope. The ability to simultaneously monitor large numbers of lines that span a range in optical depth will aid in identifying rotationally modulated phenomena with which to study the dynamical relationship between stellar rotation and inflows and outflows. The same data may also reveal true time-variable phenomena that will probe the time stability of the star-disk interface. Possibly fruitful comparisons may be made with new dynamical models that specifically address the time-dependent nature of accretion and outflow (e.g., Hayashi et al. 1996; Goodson et al. 1997; Miller and Stone 1997).

Dramatic developments are also expected in the study of planet formation environments. The impact of 8-m class ground-based telescopes on this field can be illustrated by considering the possibility of detecting forming protoplanets via their dynamical impact on the parent disk, i.e., through the detection of line emission from a disk gap created by the protoplanet. If the CO fundamental work described in section III is extended to lower companion masses, the sensitivity of large ground-based telescopes translates into the ability to detect protoplanets with masses as low as a Jupiter mass at an orbital radius of 1 AU around T Tauri stars 140 pc away. Until the development of milliarcsecond imaging capability in the thermal infrared, this technique offers possibly the best opportunity for detecting forming protoplanets.

Spectroscopic capability in the thermal infrared (>4 μm) will be critical to the study of planet formation environments at large radial distances. Not only does the Planck function for disk material at AU distances peak in the mid-infrared, but this spectral region contains important suites of molecules with transitions that are well excited at the lower densities and temperatures of disks at several AU (e.g., rotational H_2O lines). Despite its great potential for studies of planetary origins, the mid-infrared has remained largely unexplored because of the severe limitations imposed by large thermal backgrounds and strong telluric absorption. Thus, a large-aperture, cooled telescope in space with spectroscopic capability has the potential for significant breakthroughs. Studies of disk structure and physical properties comparable to those possible from the ground could be carried out in star-forming regions as distant as Orion, enabling studies of thousands of potentially planet-forming systems (Carr and Najita 1997). In addition, the ability to observe above the Earth's atmosphere would allow the study of cool, molecular transitions that are inaccessible from the ground (e.g., H_2O, CH_4), but which are excellent, often unique, probes of the physical and chemical properties of planet formation environments. The large number of 8-m class telescopes to be available at the start of the next millennium offers the exciting opportunity to address these issues, as well as new issues of fundamental importance that have yet to emerge.

REFERENCES

Artymowicz, P. and Lubow, S. H. 1996. Mass flow through gaps in circumbinary disks. *Astrophys. J. Lett.* 467:L77–L80.

Armitage, P. J., and Clarke, C. J. 1996. Magnetic braking of T Tauri stars. *Mon. Not. Roy. Astron. Soc.* 280:458–468.

Basri, G. 1987. The T Tauri stars. In *Cool Stars, Stellar Systems, and the Sun V,* ed. J. Linsky and W. Stencil (Berlin: Springer-Verlag), pp. 411–420.
Basri, G., Marcy, G. W., and Valenti, J. A. 1992. Limits on the magnetic flux of pre-main sequence stars. *Astrophys. J.* 390:622–633.
Basri, G., Johns-Krull, C. M., and Mathieu, R. D. 1997. The classical T Tauri spectroscopic binary DQ Tau. II. Emission line variations with orbital phase. *Astron. J.* 114:781–792.
Batalha, C. C., Stout-Batalha, N. M., Basri, G., and Terra, M. 1996. The narrow emission lines of T Tauri stars. *Astrophys. J. Suppl.* 103:211–233.
Beristain, G., Edwards, S., and Kwan, J. 1998. Permitted iron emission lines in the classical T Tauri star DR Tauri. *Astrophys. J.* 499:828–852.
Bertout, C., Basri, G., and Bouvier, J. 1988. Accretion disks around T Tauri stars. *Astrophys. J.* 330:350–373.
Bertout, C., Harder, S., Malbet, F., and Mennessier, C. 1996. Photometric observations of YY Orionis: New insight into the accretion process. *Astron. J.* 112:2159–2167.
Biscaya, A. M., Rieke, G. H., Narayanan, G., Luhman, K. L., and Young, E. T. 1997. First-overtone CO variability in young stellar objects. *Astrophys. J.* 491:359–365.
Bouvier, J., Cabrit, S., Fernandez, M., Martin, E. L., and Matthews, J. M. 1993. COYOTES—I. The photometric variability and rotational evolution of T-Tauri stars. *Astron. Astrophys.* 272:176–206.
Brown, D. N., and Landstreet, J. D. 1981. A search for weak longitudinal magnetic fields on late-type stars. *Astrophys. J.* 246:899–904.
Calvet, N. 1997. Properties of the winds of T Tauri stars. In *Herbig-Haro Flows and the Birth of Low Mass Stars,* IAU Symp. 182, eds. B. Reipurth and C. Bertout (Dordrecht: Reidel), pp. 417–432.
Calvet, N., and Gullbring, E. 1998. The structure and emission of the accretion shock in T Tauri stars. *Astrophys. J.,* in press.
Calvet, N., Patino, A., Magris, G., and D'Alessio, P. 1991. Irradiation of accretion disks around young objects. I: Near-infrared CO bands. *Astrophys. J.* 380:617–630.
Calvet, N., Hartmann, L., Hewett, R., Valenti, J., Basri, G., and Walter, F. 1996. C IV in classical T Tauri stars. In *Cool Stars, Stellar Systems and the Sun,* ASP Conf. Proc. 109, ed. R. Pallavicini and A. Dupree (San Francisco: Astronomical Society of the Pacific), pp. 419–420.
Cameron, A. C., and Campbell, C. G. 1996. Rotational evolution of magnetic T Tauri stars with accretion disks. *Astron. Astrophys.* 274:309–318.
Camenzind, M. 1990. Magnetized disk winds and the origin of bipolar outflows. In *Reviews in Modern Astronomy, Vol. 3,* ed. G. Klare (Berlin: Springer-Verlag), pp. 234–265.
Carr, J. S. 1989. Near-infrared CO emission in young stellar objects. *Astrophys. J.* 345:522–535.
Carr, J. S., and Najita, J. 1997. Studying the origins of planetary systems with NGST. In *Science with the NGST,* ASP Conf. Ser. 133, ed. E. Smith and A. Kortakar (San Francisco: Astronomical Society of the Pacific), pp. 163–171.
Carr, J. S., Tokunaga, A. T., Najita, J., Shu, F. H., Glassgold, A. E. 1993. The inner-disk and stellar properties of the young stellar object WL16. *Astrophys. J. Lett.* 411:L37–L40.
Carr, J. S., Najita, J., and Tokunaga, A. T. 1999. Discovery of hot water emission from young stars. In preparation.
Chandler, C. J., Carlstrom, J. E., Scoville, N. Z., Dent, W. R. F., and Geballe, T. R. 1993. Infrared CO emission from young stars: High resolution spectroscopy. *Astrophys. J. Lett.* 412:L71–L74.

Chandler, C. J., Carlstrom, J. E., and Scoville, N. Z. 1995. Infrared CO emission from young stars: Accretion disks and neutral winds. *Astrophys. J.* 446:793–800.

Choi, P. I., and Herbst. W. 1996. Rotation periods of stars in the Orion Nebula cluster: The bimodal distribution. *Astron. J.* 111:283–298.

D'Alessio, P., Canto, J., Calvet, N. and Lizano, S. 1998. Accretion disks around young objects. I. The detailed vertical structure. *Astrophys. J.* 500:411–427.

Donati, J.-F., Semel, M., Carter, B. D., Rees, D. E., and Cameron, A. C. 1997. Spectropolarimetric observations of active stars. *Mon. Not. Roy. Astron. Soc.* 291:658–682.

Edwards, S. 1997. Magnetospherically mediated accretion in classical T Tauri stars: Paradigm for low mass stars undergoing disk accretion? In *Herbig-Haro Flows and the Birth of Low Mass Stars,* IAU Symp. 182, ed. B. Reipurth and C. Bertout (Dordrecht: Reidel), pp. 433–442.

Edwards, S., Strom, S., Hartigan, P., Strom, K., Hillenbrand, L., Herbst, W., Attridge, J., Merrill, M., Probst, R., and Gatley, I. 1993*a*. Angular momentum regulation in low mass young stars surrounded by accretion disks. *Astron. J.* 106:372–382.

Edwards, S., Ray, T., and Mundt, R. 1993*b*. Energetic mass outflows from young stars. In *Protostars and Planets III,* ed. E. H. Levy and J. I. Lunine (Tucson: University of Arizona Press), pp. 567–602.

Edwards, S., Hartigan, P., Ghandour, L., and Andrulis, C. 1994. Spectroscopic evidence for magnetospheric accretion in classical T Tauri stars. *Astron. J.* 104:1056–1070.

Ferreira, J., Pelletier, G., and Appl, S. 1997. Braking of a protostar driving magnetic X-winds. In *Herbig-Haro Objects and the Birth of Low Mass Stars,* IAU Symp. 182, ed. F. Malbet and A. Castets (Dordrecht: Reidel), pp. 112–114.

Geballe, T. R., and Persson, S. E. 1987. Emission from CO band heads in young stellar objects. *Astrophys. J.* 312:297–302.

Giampapa, M., Calvet, N., Imhoff, C. L., and Kuhi, L. V. 1981. IUE observations of pre-main-sequence stars. I. Mg II and Ca II resonance linefluxes for T Tauri stars. *Astrophys. J.* 251:113–125.

Giampapa, M., Basri, G. Johns, K., and Imhoff, C. 1993. A synoptic of H-alpha line profile in the T Tauri star SU Aurigae. *Astrophys. J. Suppl.* 89:321–344.

Giovanardi, C., Gennari, S., Natta, A., and Stanga, R. 1991. Infrared spectroscopy of T Tauri stars. *Astrophys. J.* 367:173–181.

Glassgold, A. E., Mamon, G. A., and Huggins, P. J. 1991. The formation of molecules in protostellar winds. *Astrophys. J.* 373:254–265.

Gómez de Castro, A. and Franquiera, M. 1997. Accretion and UV variability in BP tauri. *Astrophys. J.* 482:465–469.

Goodson, A. P., Winglee, R. M., and Boehm, K.-H. 1997. Time-dependent accretion by magnetic young stellar objects as a launching mechanism for stellar jets. *Astrophys. J.* 489:199–209.

Greene, T. P., and Lada, C. J. 1996. Near-infrared spectra and the evolutionary status of young stellar objects: Results of a 1.1–2.4 μm survey. *Astron. J.* 112:2184–2221.

Guenther, E., Lehman, H., Emerson, J., and Staude, J. 1998. The magnetic accretion scenario in young stellar objects. *Astron. Astrophys.,* in press.

Gullbring, E., Petrov, P., Ilyin, I., Tuominen, I., Gahm, G., and Loden, K. 1996. Line profile variations of the classical T Tauri star BP Tauri. *Astron. Astrophys.* 314:835–845.

Gullbring, E., Hartmann, L., Briceno, C., and Calvet, N. 1998. Disk accretion rates for T Tauri stars. *Astrophys. J.* 492:323–341.

Hamann, F., and Persson, S. E. 1992. Emission-line studies of young stars. I. The T Tauri stars. *Astrophys. J. Suppl.* 82:247–283.

Hanson, M. M., Howarth, I. D., and Conti, P. S. 1997. The young massive stellar objects of M17. *Astrophys. J.* 489:698–718.
Hartigan, P., Edwards, S., and Ghandour, L. 1995. Disk accretion and mass loss from young stars. *Astrophys. J.* 452:736–768.
Hartmann, L., and Kenyon, S. 1996. The FU Orionis phenomenon. *Ann. Rev. Astron. Astrophys.* 34:207–240.
Hartmann, L., and Stauffer, J. R. 1989. Additional measurements of pre-main sequence stellar rotation. *Astron. J.* 97:873–880.
Hartmann, L., Kenyon, S., and Hartigan, P. 1993. Young stars: episodic phenomena, activity, and variability. In *Protostars and Planets III,* ed. E. H. Levy and J. I. Lunine (Tucson: University of Arizona Press), pp. 497–518.
Hartmann, L., Hewett, R., and Calvet, N. 1994. Magnetospheric accretion models for T Tauri stars: Balmer line profiles without rotation. *Astrophys. J.* 426:669–687.
Hatzes, A. P., 1995. Doppler imaging of the cool spot distribution on the weak T Tauri star V410 Tauri. *Astrophys. J.* 451:784–794.
Hayashi, M. R., Shibata, K., and Matsumoto, R. 1996. X-ray flares and mass outflows driven by magnetic interaction between a protostar and its surrounding disk. *Astrophys. J. Lett.* 468:L37–L40.
Hessman, F. V., and Guenther, E. W. 1997. The highly veiled T Tauri stars DG Tau, DR Tau, and DI Cep. *Astron. Astrophys.* 321:497–512.
Hirth, G. A., Mundt, R., and Solf, J. 1997. Spatial and kinematic properties of the forbidden emission line region of T Tauri stars. *Astron. Astrophys. Suppl.* 126:437–469.
Jensen, E. L. N., and Mathieu, R. D. 1997. Evidence for cleared regions in the disks around pre-main-sequence spectroscopic binaries. *Astron. J.* 114:301–316.
Johns, C., and Basri, G. 1995*a*. The line profile variability of SU Aurigae. *Astrophys. J.* 449:341–364.
Johns, C., and Basri, G. 1995*b*. Hamilton echelle spectra of young stars. II. Time series analysis of H(alpha) variations. *Astron. J.* 109:2800–2816.
Johns-Krull, C. M., and Hatzes, A. P. 1997. The classical T Tauri star Sz68: Doppler imaging and evidence for magnetospheric accretion. *Astrophys. J.* 487:896–915.
Johns-Krull, C. M., Valenti, J. A., and Koresko, C. 1999*a*. Measuring the magnetic field on the classical T Tauri star BP Tauri. *Astrophys. J.*, in press.
Johns-Krull, C. M., Valenti, J. A., Hatzes, A. P., and Kanaan, A. 1999*b*. Spectropolarimetry of magnetospheric accretion on the classical T Tauri star BP Tauri. *Astrophys. J. Lett.,* in press.
Johnstone, R. M., and Penston, M. V. 1987. Follow-up Zeeman observations of the T Tauri star RU Lup. *Mon. Not. Roy. Astron. Soc.* 227:797–800.
Joncour, I., Bertout, C., and Bouvier, J. 1994. Doppler imaging of the T Tauri star HDE 283572. *Astron. Astrophys.* 291:L19–L22.
Königl, A. 1991. Disk accretion onto magnetic T Tauri stars. *Astrophys. J. Lett.* 370:L39–L43.
Kuhi, L. 1964. Mass loss from T Tauri stars. *Astrophys. J.* 140:1409–1433.
Lin, D. N. C., Bodenheimer, P., and Richardson, D. C. 1996. Orbital migration of the planetary companion of 51 Pegasi to its present location. *Nature* 380:606–607.
Luhman, K. L., and Rieke, G. H. 1998. The low-mass initial mass function in young clusters: L1495E. *Astrophys. J.* 197:354–369.
Lynden-Bell, D., and Pringle, J. E. 1974. The evolution of viscous discs and the origin of the nebular variables. *Mon. Not. Roy. Astron. Soc.* 168:603–638.
Marsh, K. A., and Mahoney, M. J. 1992. Evidence for unseen companions around T Tauri stars. *Astrophys. J. Lett.* 395:L115–L118.
Martin, S. C. 1996. The thermal structure of magnetic accretion funnels in young stellar objects. *Astrophys. J.* 470:537–550.

Martin, S. C. 1997. A funnel flow origin for CO bandhead emission in young stellar objects. *Astrophys. J. Lett.* 478:L33–L36.

Mathieu, R. D., Adams, F. C., and Latham, D. W. 1991. The T Tauri spectroscopic binary GW Orionis. *Astrophys. J.* 101:2184–2198.

Mathieu, R. D., Stassun, K., Basri, G., Jensen, E. L. N., Johns-Krull, C. M., Valenti, J. A., and Hartmann, L. W. 1997. The classical T Tauri spectroscopic binary DQ Tau. I. Orbital elements and light curves. *Astron. J.* 113:1841–1854.

Meyer, M. 1996. The stellar populations of deeply embedded young clusters: Near-infrared spectroscopy and emergent mass distributions. Ph.D. Thesis, University of Massachusetts, Amherst.

Miller, K. A., and Stone, J. M. 1997. Magnetohydrodynamic simulations of stellar magnetosphere–accretion disk interaction. *Astrophys. J.* 489:890–902.

Muzerolle, J., Calvet, N., and Hartmann, L. 1998*a*. Magnetospheric accretion models for the hydrogen emission lines of T Tauri stars. *Astrophys. J.* 492:743–753.

Muzerolle, J., Hartmann, L., and Calvet, N. 1998*b*. Emission-line diagnostics of T Tauri magnetospheric accretion. I. Line profile observations. *Astron. J.* 116:455–468.

Najita, J., Carr, J. S., Tokunaga, A. T. 1996*a*. High-resolution spectroscopy of Brγ emission in young stellar objects. *Astrophys. J.* 456:292–299.

Najita, J., Carr, J. S., Glassgold, A. E., Shu, F. H., and Tokunaga, A. T. 1996*b*. Kinematic diagnostics of disks around young stars: CO overtone emission from WL16 and 1548C27. *Astrophys. J.* 462:919–936.

Najita, J., Greene, T. P., and Lada, C. J. 1999. High resolution spectroscopy of CO overtone emission from massive young stars. In preparation.

Petrov, P. P., Gullbring, E., Ilyin, I., Gahm, G., Tuominen, I., Hackman, T., and Loden, K. 1996. The gas flows of SU Aurigae. *Astron. Astrophys.* 314:821–834.

Phillips, R. B. 1992. VLBI of single stars: Characteristics of their magnetically dominated radio emission. In *Cool Stars, Stellar Systems, and the Sun VII,* ASP Conf. Ser. 26, ed. M. Giampapa (San Francisco: Astronomical Society of the Pacific), pp. 309–318.

Phillips, R. B., Lonsdale, C. J., Feigelson, E. D. and Deeney, B. D. 1996. Polarized radio emission from the multiple T Tauri system HD 283447. *Astron. J.* 111:918–929.

Reipurth, B., and Aspin, C. 1997. Infrared spectroscopy of Herbig-Haro energy sources. *Astron. J.* 114:2700–2707.

Scoville, N., Kleinmann, S. G., Hall, D. N. B., and Ridgway, S. T. 1983. The circumstellar and nebular environment of the Becklin-Neugebauer Object: λ = 2.5 micron spectroscopy. *Astrophys. J.* 275:201–224.

Shu, F. H., Najita, J., Ostriker, E., Wilkin, F., Ruden, S., and Lizano, S. 1994. Magnetocentrifugally driven flows from young stars and disks. I. A generalized model. *Astrophys. J.* 429:781–796.

Viti, S., Tennyson, J., and Polyansky, O. L. 1997. A spectroscopic linelist for hot water. *Mon. Not. Roy. Astron. Soc.* 287:79–86.

Walker, M. 1972. Studies of extremly young clusters: Spectroscopic observations of the ultraviolet-excess stars in the Orion nebula cluster and NGC 2264. *Astrophys. J.* 175:89–116.

HIGH-RESOLUTION OPTICAL AND NEAR-INFRARED IMAGING OF YOUNG CIRCUMSTELLAR DISKS

MARK J. McCAUGHREAN
Astrophysikalisches Institut Potsdam

KARL R. STAPELFELDT
Jet Propulsion Laboratory

and

LAIRD M. CLOSE
University of Hawaii

In the past five years, observations at optical and near-infrared wavelengths obtained with the *Hubble Space Telescope* and ground-based adaptive optics have provided the first well-resolved images of young circumstellar disks, which may form planetary systems. We review these two observational techniques and highlight their results by presenting prototype examples of disks imaged in the Taurus-Auriga and Orion star-forming regions. As appropriate, we discuss the disk parameters that may be typically derived from the observations as well as the implications that the observations may have on our understanding of, for example, the role of the ambient environment in shaping the disk evolution. We end with a brief summary of the prospects for future improvements in space- and ground-based optical/IR imaging techniques, and what impact they may have on disk studies.

I. DIRECT IMAGING OF CIRCUMSTELLAR DISKS

The Copernican demotion of humankind away from the center of our local planetary system also provided the shift in perspective required to understand its cosmogony. Once it was apparent that the solar system comprised a number of planets in essentially circular and coplanar orbits around the Sun, theories for its formation were developed involving condensation from a rotating disk-shaped primordial nebula (the Urnebel). The so-called "Kant-Laplace nebular hypothesis" eventually held sway in the latter half of the twentieth century after lengthy competition with rival "catastrophic" theories (see Koerner 1997 for a review) and was subsequently vindicated by the discovery of analogs to the Urnebel around young stars elsewhere in the galaxy.

These circumstellar disks were first detected indirectly via infrared excess emission in spectral energy distributions (SEDs), polarization mapping, kinematic asymmetries in stellar winds, UV boundary layer emission, and the presence of bipolar collimated jets. These diverse observations could be reconciled with each other only by invoking a flattened disk of dust and gas around the young star, a model tied together with early evidence for molecular gas in Keplerian rotation around the young source HL Tauri (see Beckwith and Sargent 1993, 1996 for reviews).

However, to reinforce the paradigm more viscerally, direct images of circumstellar disks were sought. The coronagraphic detection of an edge-on disk around the main-sequence star β Pictoris (Smith and Terrile 1984) raised early hopes, but unfortunately, surveys of larger samples of nearby main-sequence stars yielded no other disk images (Smith et al. 1992; Kalas 1996). This result is now understood to be a consequence of the especially large optical depth of the β Pic disk compared to other debris disk sources. Models show that, in general, disks are not expected to be detected around main-sequence stars unless an order-of-magnitude improvement can be made in suppressing the stellar light in coronagraphic imaging (Kalas and Jewitt 1996). More recently, studies of the somewhat younger main-sequence star HR 4796A have shown that debris disks with high enough optical depths can be imaged via direct emission from warm dust particles at thermal-IR wavelengths (Jayawardhana et al. 1998; Jura et al. 1998; Koerner et al. 1998), as well as in scattered light at near-IR wavelengths (Schneider et al. 1999).

However, there are important caveats concerning the β Pic and HR 4796A disks. They are relatively evolved, optically thin, effectively gas-free, consisting of only small amounts (a few lunar masses) of reprocessed dust, possibly replenished by collisions between much larger asteroids or planetesimals. Thus, while interesting in the context of later disk evolution after the main planet-building phase, they provide little insight into the nature of young, primordial disks. Also, at $\sim$16 and 70 pc respectively, they are considerably closer than the nearest star-forming regions, making their structure relatively easy to resolve.

For more recently formed disks, there is a clear observational challenge: At the $\sim$150-pc distance of the nearby low-mass star-forming regions such as Taurus-Auriga, our own 60-AU-diameter solar system would subtend only 0.4 arcsecond, and it would subtend only 0.1 arcsecond at the 500-pc distance to the Orion giant molecular clouds, the nearest site of ongoing high- and low-mass star formation. The knowledge that the young solar system was probably significantly larger, also encompassing a then optically thick proto-Kuiper Belt, alleviates matters somewhat, but nevertheless, resolving even a several-hundred-AU-diameter circumstellar disk at a distance of several hundred parsecs still requires subarcsecond spatial resolution.

The Very Large Array (VLA) provides this resolution at centimeter wavelengths, but emission from gas in thermal equilibrium in a disk at 10–50 K is impossible to detect until the millimeter/submillimeter regime is reached. Millimeter interferometer arrays including the Berkeley-Illinois-Maryland Array (BIMA), the Owens Valley Radio Observatory (OVRO), Plateau de Bure, and Nobeyama now provide ~arcsec resolution, giving indications of the density and velocity structure in disks in Taurus-Auriga (e.g., Mundy et al. 1996; Koerner and Sargent 1995; Guilloteau et al. 1997; Kitamura et al. 1997; and the chapters by Mundy et al. and by Wilner and Lay, this volume), while planned larger interferometer arrays will deliver 0.1-arcsecond resolution and better.

The same fiducial 0.1-arcsecond resolution is the diffraction limit of a 2.4-m telescope at 1-μm wavelength, which can now be achieved via space-based imaging with the *Hubble Space Telescope* (HST) and ground-based adaptive optics (AO). Although the bulk of the disk gas cannot be detected directly in emission at optical and near-IR wavelengths, the scattering and absorbing properties of the associated dust can be used to obtain images of the disk. This chapter reviews the following: how young disks can be seen at optical and near-IR wavelengths; the relative merits and demerits of the various imaging techniques; example objects, including edge-on disks where the central star is completely obscured, disks seen in silhouette against bright emission lines in H II regions, and disks where the central star or binary is clearly visible; the inferences that can be made as to the structure, mass, and evolutionary status of the disks; the influence of environment on the disks; and the impact of future instrumentation developments on the field.

II. DETECTABILITY AND TECHNIQUES

A. The Detectability of Disks at Optical and Near-IR Wavelengths

Millimeter and thermal-IR observations image the gas and warm dust in a disk directly in emission, while optical and near-IR observations rely on either dust scattering or absorption. At these wavelengths, the scattering cross section and opacity of typical dust grains is high, and by virtue of their high masses and densities, young circumstellar disks should be optically thick. Thus, the full geometrical area of a young disk is available to scatter light from the central star(s) (from the top surface of the disk or from more tenuous dust in an associated envelope) or to be seen in absorption against a bright background.

Observations of scattered or absorbed starlight are much more sensitive to small amounts of dust than are observations of emission at longer wavelengths. For typical interstellar medium (ISM) dust grains, a uniform surface density disk with radius 100 AU, containing as little as a few Earth masses of gas and dust, would still have an optical depth at visible wavelengths greater than unity. Thus, in principle, nebulosity can be seen around young stellar objects to a limiting mass hundreds of times smaller than currently achievable in millimeter continuum surveys. A caveat is that the central star is also bright at optical and near-IR wavelengths, generally presenting a severe contrast problem, as seen for the debris disks around main-sequence stars. However, under favorable circumstances, the disk may be seen near edge-on, obscuring the star completely and leaving the disk clearly visible.

B. Imaging with the *Hubble Space Telescope*

Free from Earth's atmosphere, the HST has a diffraction-limited point spread function (PSF) over an order of magnitude in wavelength ($\sim$2000 Å–2 μm). This inherently stable, well-characterized PSF, combined with a clean optical train, high throughput, and low background, yields very deep, very high-contrast images anywhere on the sky. However, the resolution of the 2.4-m diameter HST is ultimately limited compared with that of the current generation of 8–10-m ground-based telescopes.

C. Imaging with Ground-Based Adaptive Optics

On large ground-based telescopes the spatial resolution is degraded by atmospheric seeing effects to the arcsecond level, significantly worse than the intrinsic optical and near-IR diffraction limit. Speckle techniques have long been used to retrieve diffraction-limited resolution for bright objects, employing post-detection processing of large numbers of very short exposure frames. Images of a small sample of very bright circumstellar disks around both young and evolved stars have been made in this way (e.g., Beckwith et al. 1984, 1989 for HL Tauri; Falcke et al. 1996 for η Carinae; Osterbart et al. 1997 for the Red Rectangle).

Substantial improvement in dynamic range and sensitivity can be made using AO techniques. A relatively bright star (either the object of interest itself or a nearby companion) is monitored at high speed, allowing seeing-induced motions and distortions to be compensated for using a flexible mirror, yielding near diffraction-limited images in real time, which can be further improved through post-detection deconvolution. AO techniques can be applied to the largest available ground-based telescopes, implying resolutions approaching 0.025 arcsecond at 1 μm for 10-m class telescopes. Currently 0.1-arcsecond resolution is routinely achieved at 2 μm on 4-m class telescopes (see, for example, Graves et al. 1998).

Since a suitably bright nearby star must be available for wavefront sensing, sky coverage is limited, although in the specific case of young stellar objects (YSOs) within a few hundred parsecs the source itself is usually bright enough to be used, as long as the disk is not seen edge-on. A more fundamental issue is the instability in the low-level halo surrounding the diffraction-limited core due to variable and imperfect correction. This is particularly problematic for studies of faint nebular emission around bright stars.

III. A MENAGERIE OF DISKS

In this section, we present a detailed discussion of the observations of several prototype disk systems with both the HST and ground-based AO. These are arranged according to how well the central star is suppressed, starting with edge-on systems, where the star is unseen; moving to non-edge-on silhouette systems, where the contrast of the disk can be enhanced using a narrowband filter; and finishing with systems where the stars are bright enough to interfere with the interpretation of the disk structure.

A. Edge-on Systems

Disks seen edge-on offer the best opportunity to study their structure, because the bright central star is not visible directly at optical or near-IR wavelengths. For a typical optically thick flared disk with a scale height on the order of 10% of the radius, the central star will be occulted in about 10% of all possible viewing angles. Although several such sources are now known, their frequency of occurrence appears to be less than expected from these simple geometrical arguments, probably because current catalogs of young sources (e.g., optically visible T Tauri stars) are biased against edge-on disk systems by definition. Also, since these systems, by definition, lack a bright central star, they are generally very difficult to study using speckle or AO techniques.

A defining characteristic of sources with edge-on disks is a bipolar reflection nebula structure, with two lobes of scattered light separated by a dark lane. These nebulae have sharp brightness gradients perpendicular to the lane and show their greatest elongation in a direction parallel to it. In addition, edge-on disk sources are all relatively faint in the near-IR; they have significant optical/near-IR polarization (>2%); and any associated jets lie close to the plane of the sky and thus have low radial velocities. We now describe a number of template systems in order of evolutionary status.

1. IRAS 04302+2247: A Disk Forming Within an Extended Envelope. IRAS 04302+2247 is a very young star embedded in the Taurus L 1536 molecular cloud. Lucas and Roche (1997) showed it to consist of a central dust lane and a bipolar reflection nebula, modeling the latter as a circumstellar envelope. More detailed near-IR images made using the Near Infrared Camera and Multi-Object Spectrometer (NICMOS) on the HST (Color Plate 6, top left; Padgett et al. 1999*a*) show the central dust lane to be sharply defined, with reddening at its edges clearly indicating increased density toward the midplane. The observed length of the dust lane suggests that a disk has formed whose outer radius could be as much as 450 AU. Irregular bright and dark streamers within the two nebula lobes possibly indicate material infalling onto the disk. A rotating molecular gas structure has been resolved within the dust lane by millimeter interferometry (Padgett et al. 1999*b*), while millimeter continuum measurements imply a circumstellar mass $\sim 10^{-2}$ $M_{\odot}$. Models including this disk mass yield a vertical extent consistent with the observed thickness of the dust lane. Thus, the dust lane of IRAS 04302+2247 appears to trace a large, young disk, although a surrounding envelope structure remains essential to account for the infrared SED and the full extent of scattered light. Therefore, this source is likely to be at an early evolutionary stage.

2. Edge-on Silhouette Disks: Orion 182-413 and 114-426. The silhouette disks in the Orion Nebula are compact structures seen in absorption against the bright nebulosity of the H II region, and they provide an important group of coeval disks for comparative studies. In most cases, the disk is not oriented edge-on, and a young Trapezium Cluster star is seen at the center; these are discussed further in section III.B. Two sources are seen edge-on, however. Orion 182-413 (also known as HST 10; O'Dell and Wen 1994; Bally et al. 1998*a*; see also Fig. 5 in the chapter by Hollenbach et al., this volume) contains a dark absorption bar some 200 AU across seen within a larger bright globule, the latter being ionized by the OB stars at the center of the cluster. The presumed central star remains undetected even at 4 μm, and although no reflected continuum nebular lobes are seen, these would be difficult to detect against the bright ionized globule. Interestingly, the dark absorption bar is seen in emission in the v = 1–0 S(1) line of H_2 at 2.122 μm (Chen et al. 1998), directly confirming the presence of molecular gas in the disk.

A clearer example is found in the much larger Orion 114-426. Optical HST images (McCaughrean and O'Dell 1996) show a dark, thick silhouette 1000 AU across, with faint continuum reflection nebulosities either side of the midplane. In the near-IR, the central star again remains undetected at least to 4 μm, and NICMOS images at 1–2 μm show that the dust lane appears smaller and thinner, with the bipolar reflection nebulosities much brighter and more symmetric (Color Plate 6, top right; McCaughrean et al. 1998). The long axis of the silhouette shrinks by 20% from 0.6 to 1.9 μm, too small a change to be due to plausible intrinsic radial density gradients in the original disk structure; rather, abrupt truncation at some outer radius is required. The nondetection of the central star can be combined with disk models to yield a lower-limit disk mass of 5×10^{-4} $M_{\odot}$ (McCaughrean et al. 1998), and a comparison of the dust lane thickness with scattered-light models for edge-on disks suggests a mass on the order of 0.006 $M_{\odot}$. Thus far the disk has not been unambiguously detected in the millimeter continuum, which would allow the mass to be established directly (Bally et al. 1998*b*). Finally, relatively little envelope material is seen at high latitudes above the 114-426 disk. This result could be interpreted as implying that Orion 114-426 represents the stage of YSO circumstellar material evolution immediately after IRAS 04302+2247, but it may alternatively be more related to environmental effects.

3. HH 30: A Prototype Disk-Jet System. Disks and jets associated with young stars are expected to be oriented perpendicularly, and recent high-resolution images have confirmed this expectation at the subarcsecond level. For example, HST images of Haro 6-5B and DG Tauri B show compact bipolar nebula structures and dust lanes several hundred AU in diameter (Krist et al. 1998; Padgett et al. 1999*a*). The sharpest and most symmetric system is HH 30 (see Color Plate 6, bottom left; Burrows et al. 1996; Ray et al. 1996), the source of a spectacular jet in the Taurus L 1551 molecular cloud (Mundt et al. 1990).

HH 30 was the first YSO disk in which the vertical structure was clearly resolved. The disk is seen to flare (become thicker) with increasing radial distance, confirming longstanding predictions from disk structure theory and SED fitting (Lynden-Bell and Pringle 1974; Kenyon and Hartmann 1987; Bell et al. 1997). Modeling of the isophotes (Burrows et al. 1996; Wood et al. 1998) indicates that the disk is vertically hydrostatic, with a scale height $H = 15$ AU at $r = 100$ AU; furthermore, Burrows et al. (1996) found that its scale height may flare with radius more rapidly than expected for a steady-state accretion disk. The disk outer radius is 225 AU, and has inclination angle of $\sim 7°$ with respect to the line of sight.

The disk molecular gas component and its kinematics have been resolved via millimeter interferometry by Stapelfeldt and Padgett (1999); millimeter continuum measurements have yielded a surprisingly small total disk mass of $\sim 10^{-3}$ $M_{\odot}$ assuming nominal dust properties, similar to the mass derived independently from the HST images. This mass is much less than typically inferred for Herbig-Haro jet sources (Reipurth et al. 1993). Implications are that the HH 30 disk must be largely optically thin to its own emergent thermal-IR radiation, and also that steady accretion would consume the system in less than 0.1 Myr. Apparently, therefore, HH 30 represents a young star nearing the end of its disk accretion phase.

4. HK Tau/c: A Circumstellar Disk in a Binary System. High-resolution imaging of the faint companion star to HK Tauri shows it to be a nebulous edge-on disk (Color Plate 6, bottom right; Koresko 1998; Stapelfeldt et al. 1998), making it the first such object seen in a young binary system. HK Tau/c is very small, with a 1.5-arcsecond size, corresponding to a disk radius of 105 AU, and an extremely narrow 0.2-arcsecond dust lane. Objects like HK Tau/c would be difficult to recognize as edge-on disks beyond the nearest star-forming clouds.

Modeling of the isophotes (Stapelfeldt et al. 1998) indicates a disk inclination of $\sim 5°$ and a scale height $H = 3.8$ AU at $r = 50$ AU, making the disk significantly flatter than that of HH 30. For its estimated age of 0.5 Myr, the disk mass appears to be extremely small: Stapelfeldt et al. (1998) find it to be 10^{-4} $M_{\odot}$ (0.1 $M_{\rm Jup}$) based on HST optical wavelength observations, while Koresko (1998) derives 10^{-3} $M_{\odot}$ from ground-based near-IR speckle holography. With its small apparent mass (1–10% that of the minimum-mass solar nebula) and little evidence for accretion, HK Tau/c may represent a disk that has already been partially dissipated.

The mass and radius of the HK Tau/c disk have likely been reduced by tidal effects of the nearby primary star. However, it will be difficult to evaluate the importance of these perturbations quantitatively until the orbital parameters of the binary are established. Statistically, the physical separation of the two stars is likely to be about twice the projected separation, or 700 AU. This would make the observed disk radius about 15% of the distance between the components and would require the disk plane to be inclined to the stellar orbit plane. The HK Tauri system therefore provides an interesting case for the application of binary/disk interaction theory (see the chapter by Lubow and Artymowicz, this volume).

B. Silhouette Disks around Directly Visible Stars

As discussed above, silhouette disks are seen in projection against a bright background screen. Even when not edge-on, they can still be seen at reasonably high contrast as long as the background illumination is provided by a bright emission line (e.g., from an H II region), because a narrowband filter can be used to accept all the light from the emission line, while simultaneously suppressing the bulk of the continuum light from the central star. In addition to the edge-on disks seen in the Orion Nebula, others are seen at inclinations that allow their central star to be detected directly (Fig. 1; McCaughrean and O'Dell 1996).

A combination of optical, near-, and mid-IR photometry for the central stars shows that they are typically young (1–2 Myr) and low-mass (0.3–1.5 $M_\odot$), all with excess thermal-IR emission, indicating hot dust near the inner edge of the disk (McCaughrean and O'Dell 1996; Hayward and McCaughrean 1997). In this sense, they appear to be counterparts to classical T Tauri stars, although, unlike most T Tauri stars, the disks are seen directly and their outer radii accurately determined from the silhouette, with the non-edge-on Orion Nebula silhouette disks ranging in diameter from 50 to 500 AU.

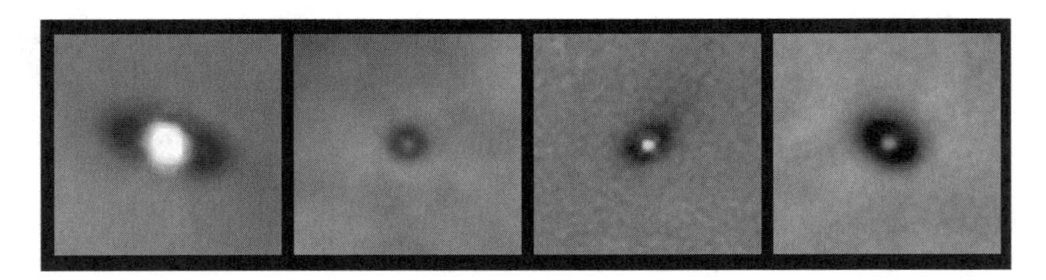

Figure 1. HST images of four silhouette disks in the Orion Nebula. From left to right: Orion 218-354, 167-231, 121-1925, and 183-405. All are seen against the nebular Hα background with either the Wide Field Camera or Planetary Camera in these relatively shallow survey images. Each panel is 2.0 arcsecond or 900 AU square, cf. the 60-AU diameter of the solar system. From McCaughrean and O'Dell (1996).

Because these disks are seen partly face-on, it has been possible to examine their surface density profiles, at least near the edges, and McCaughrean and O'Dell (1996) showed that the disks are best fitted with an opaque inner section and relatively sharp exponential edge. Since the inner section is highly optically thick, standard radial surface density laws with $\Sigma(r) \propto r^p$ (with p in the range -0.75 to -1.5; Shakura and Sunyaev 1973; Adams et al. 1988) cannot be ruled out there, but the disks do appear to have been truncated at some outer radius. Possible causes are discussed in the following paragraphs.

1. Internal Disk Evolution. Simple gravity-dominated models show that as circumstellar disks form via inside-out collapse, they comprise two distinct parts: an inner disk, rotating in quasi-equilibrium, and an outer component, contracting dynamically (Saigo and Hanawa 1998). The presence of a sharp transition zone between the two parts is predicted, and although it is possible that this corresponds to the edges seen in the silhouette images, there appears to be little evidence for more extended infalling envelopes around the Orion disks; such envelopes may have been removed by external effects. Also, such a simple model may be significantly modified if magnetic fields play an important role in the collapse.

2. OB Star Ionizing Flux and Stellar Wind. Near the center of the Orion Nebula, low-mass stars are often accompanied by compact nebulae, externally ionized by the Trapezium OB stars. These so-called "proplyds" (protoplanetary disks) are thought to have disks at their centers (O'Dell et al. 1993; O'Dell and Wong 1996; Bally et al. 1998*a*; Henney et al. 1996; Johnstone et al. 1998; see also the chapter by Hollenbach et al., this volume). The silhouette disks are simply thought to be far enough from the OB stars to escape ionization, but it is nevertheless plausible that smaller-scale effects are at work. For example, Orion 121-1925 has a faint tail that points away from the brightest OB star θ^1Ori C, perhaps being driven off by the stellar wind. Such an effect should truncate the disk, although modeling is required to determine the likely profile.

3. Star/Disk Interactions. The Trapezium Cluster contains many low-mass stars, and tidal interactions between a star/disk system and interloper stars should lead to distortion, tidal tails, and truncation of the disk (Heller 1995; Hall et al. 1996; Larwood et al. 1996). Models by Hall (1997) show that repeated encounters with stars in a cluster can convert an original power law radial profile into one with a sharper exponential edge, as observed for the silhouette disks.

A more global form of tidal stripping may truncate a disk at the radius where its Keplerian velocity is of the same order as the cluster velocity dispersion, inasmuch that the latter reflects the general gravitational potential of the cluster. The Keplerian velocities at the outer edges of the silhouettes, calculated from their radii and central stellar masses, are typically $\sim$1.5–2.5 km s^{-1}. By comparison, the 3D cluster velocity dispersion is $\sim$4 km s^{-1} (Jones and Walker 1988), and a proper analysis would be required to determine whether or not the field corresponding to that velocity would indeed truncate the disks at the observed radii.

4. Pressure Balance. An edge may also be expected when the cold gas in the disk comes into pressure equilibrium with the hot ionized gas of the H II region. Assuming a temperature and density for the ionized gas of 10^4 K and 10^4 cm^{-3} respectively, the corresponding disk density would be 10^6–10^7 cm^{-3} for reasonable disk temperatures of 100–10 K. McCaughrean and O'Dell (1996) show that such a volume density converts to a column density roughly equivalent to an extinction of $A_V \sim 1$ (i.e., as observed at the disk edge), for plausible disk thicknesses.

5. Ram Pressure Stripping. Finally, a rotating disk translating through the H II region at a few km s^{-1} should be compressed on the side where the rotation and translation velocities are in opposition and stripped on the side where they combine, leading to truncation and a steady-state asymmetry in the disk. Similar ram pressure stripping has been modeled and observed for spiral galaxies (e.g., Kritsuk 1983; Phookun and Mundy 1995).

Most of these effects are external, and it is clear that the environment in which stars, disks, and planetary systems form might play a crucial role in defining their parameters (see also the chapter by Walter et al., this volume). Planets are thought to begin forming within the first 1 Myr or so, and thus it is important to know whether the effects of environment at that early stage are important in promoting or retarding their formation. The panoply of destructive effects apparently present in the Orion Nebula might suggest that disks would have a hard time surviving long enough to form planets. However, as Hall (1997) has shown, although repeated star-star encounters in a cluster will sharply truncate the outer edge of a disk, cutting it off from any larger-scale reservoir of material, a significant fraction of the disk material is actually moved *inwards*, piling up within a few tens of AU, where planets are thought to form. Also, spiral instability waves excited by these close encounters might cause material to accumulate preferentially, thus seeding the formation of planets.

To date, the Orion Nebula is the only site where silhouette disks have been seen; surveys in other H II regions and reflection nebulae have so far been unsuccessful (Stapelfeldt et al. 1997*b*). There are two reasons why this might the case. On the one hand, emission lines from fainter H II regions or continuum light from reflection nebulae might simply provide insufficient contrast to see a silhouette; on the other, the Orion Nebula might be seen at a special time, when disks around the low-mass stars have only just been uncovered by the OB stars but have not yet been destroyed by them.

C. Disks Seen in the Presence of Bright Starlight

Most disk systems will be neither edge-on nor located in bright H II regions. These systems will typically be viewed at inclinations that allow the relatively bright central star to be seen directly along the line of sight, thus swamping the faint scattered light from the disk. Furthermore, if the disk is not flared, the inner disk may cast the outer regions in shadow.

The detection of a disk in such systems is difficult but not impossible, with a stable, largely symmetric, and well-calibrated PSF the basic requirement. The visibility of the faint disk nebulosity increases rapidly with distance from the central star, and estimates of the detectable disk mass and size completeness curves can be made for a given inclination, using empirical PSFs (Close et al. 1998*b*). For example, assuming an ISM-like distribution of dust grains (Mathis and Whiffen 1989), Mie scattering calculations show that a $\sim$0.01-$M_\odot$ disk at an inclination of 45° is detectable at radii $\gtrsim$1 arcsecond from the central star, provided 0.1-arcsecond resolution can be achieved. For a fully sampled pixel scale (e.g., $\sim$0.05 arcsecond), such a disk has a typical per-pixel surface brightness 10^5 times fainter than the central star. At the 150-pc distance to the nearest star-forming regions, only the largest disks ($\gtrsim$100 AU radius) have been studied so far. Imaging of smaller disks is challenging.

As discussed in the introduction, millimeter interferometry provides an important tool for studying the outer parts of large circumstellar disks, with the advantages that emission from the central star is negligible and that the gas kinematics can be studied directly. For example, the millimeter emission from GG Tauri (Dutrey et al. 1994), GM Aurigae (Dutrey et al. 1998), DM Tauri (Saito et al. 1995; Guilloteau and Dutrey 1998), and UY Aurigae (Duvert et al. 1998) have all been resolved and found to be consistent with Keplerian disks. HL Tau appears to have signs of rotation as well (Sargent and Beckwith 1991), although infall (Hayashi et al. 1993) and perhaps outflow (Cabrit et al. 1996) have also been observed, underlining how complex the kinematics can be. Since all of these systems are known to have disks in Keplerian rotation with resolved diameters of a few arcseconds, they are natural candidates for direct imaging in light scattered by associated dust. All have been imaged by the HST in the optical and by AO in the near-IR, with HL Tau, GM Aur, GG Tau, and UY Aur all detected. Surprisingly, the DL Tau and DM Tau disks have yet to be seen; it is plausible that the dust in these systems has already condensed to form grains too large ($>$10 μm) to scatter efficiently.

1. HL Tau: A Young Embedded Source. HL Tau is likely to be the youngest such system. Although the central star is obscured by $\sim$24 magnitudes of visual extinction (Stapelfeldt et al. 1995; Beckwith and Birk 1995; Close et al. 1997*a*), it is not an edge-on system, with inclination estimates varying between 20 and 40° (cf. Mundy et al. 1996). The large visual extinction is apparently caused by the dusty upper part of the disk/envelope. There is significant infall in the envelope (Hayashi et al. 1993), and a strong 300 km s^{-1} outflow (Mundt et al. 1990); these features, combined with strong emission lines and continuum veiling (Cohen and Kuhi 1979; Basri and Batalha 1990) all attest to the youth of HL Tau.

Dereddened near-IR colors suggest a very young (~0.1 Myr) low-mass (0.7 $M_\odot$) pre-main-sequence (PMS) star (Close et al. 1997*a*).

HL Tau is unique in that its disk is dense (total mass ~0.1 $M_\odot$) and large (~150 AU radius) enough to be directly resolved in submillimeter and millimeter continuum emission from dust (Lay et al. 1994; Mundy et al. 1996; the chapter by Wilner and Lay, this volume). AO imaging at 1.2 μm with 0.2-arcsecond spatial resolution shows a similar extent to that seen in the millimeter continuum (radius ~150 AU; position angle ~125°) with an inclination of ~20°. Both HST and AO observations detect cavities above and below the disk (Stapelfeldt et al. 1995; Close et al. 1997*a*), most likely produced by the fast (300 km s^{-1}) jet (see Color Plate 7, top left). As these cavities expand, the PMS star should become directly visible in the optical.

2. GM Aur: An Older Circumstellar Disk. The GM Aur disk has been detected by the HST (Stapelfeldt et al. 1997*a*) in the optical and via AO in the near-IR (Close et al. 1998*b*). Compared to HL Tau, GM Aur is a rather evolved T Tauri star with little envelope material visible. Indeed, weaker extinction toward the source and a lack of strong veiling or outflow argue for a more moderate ~1 Myr age for the system. The AO images (see Color Plate 7, top right) suggest that GM Aur has a ~300 AU radius disk, enhanced on its nearside due to forward scattering, while the HST images show a larger radius, ~450 AU, and an inclination of 20–30° from preliminary modeling.

3. GG Tau: A Circumbinary Disk. Perhaps one of the most interesting surprises resulting from direct imaging of disks was their detection around binary T Tauri stars (see the chapter by Mathieu et al., this volume). The best known example is GG Tau, where millimeter interferometry revealed a large circumbinary disk in Keplerian rotation, albeit only marginally resolved (Dutrey et al. 1994). AO images at 0.1-arcsecond resolution (Color Plate 7, bottom left; Roddier et al. 1996) show a ring inclined at 55° and an inner hole with a radius of ~190 AU. Scattered-light models of the system have been presented by Wood et al. (1999). The inner hole size is some 2.7 times larger than the binary semimajor axis, consistent with models that show that the lower-mass companion in a binary system should eject disk material via tidal effects, clearing such a hole (Artymowicz and Lubow 1995; and the chapter by Lubow and Artymowicz, this volume). The inner hole in a circumbinary disk moves the peak nebular brightness off the central star, suggesting that, in general, it may be relatively easier to detect disks by direct imaging in scattered light in binary systems than around single stars.

Once the inclination of a circumbinary disk is known and its orientation and velocity profile modeled from the millimeter maps, the total mass of the binary can be estimated: for GG Tau, Dutrey et al. (1994) estimated a total mass of 1.2 $M_\odot$. This result can be independently checked by monitoring the relative motion of the two binary stars (separation

0.25 arcsecond) via speckle and AO observations (Leinert et al. 1993; Ghez et al. 1993, 1995; Roddier et al. 1996). The presently measured orbital velocity of $\sim$1.45 AU yr^{-1} (under the assumption that the stars are coplanar with the 35° inclination disk) suggests that the binary is close to minimum separation and contains a total mass of $\sim$1.5 $M_{\odot}$. Distance and velocity uncertainties yield a possible 40% error in this mass estimate, making it consistent with that determined from the millimeter kinematics. Interestingly, both estimates are considerably higher than the masses derived from the stellar fluxes using current PMS evolution models (D'Antona and Mazzitelli 1994; Roddier et al. 1996).

4. UY Aur: A Large Circumbinary Disk. Millimeter interferometry has also revealed a circumbinary disk around UY Aur (Dutrey et al. 1996; Duvert et al. 1998), as also seen via AO imaging (Color Plate 7, bottom right; Close et al. 1998*a*). Compared to GG Tau, the UY Aur disk is larger, with an inner hole radius $\sim$420 AU. This is unsurprising, however, because the UY Aur binary has a larger projected separation of 0.88 arcsecond, corresponding to a semimajor axis of $\sim$190 AU, and thus the observations confirm the prediction that wider binaries should have larger gaps in their disks.

The millimeter interferometry suggests that the disk gas is in Keplerian rotation around a 1.2-$M_{\odot}$ binary (Duvert et al. 1998), while the binary orbital motion yields the slightly higher mass of 1.6$\pm$0.5 $M_{\odot}$. As with GG Tau, the kinematic masses are higher than those derived from stellar fluxes combined with PMS tracks. In both GG Tau and UY Aur, in addition to the large (300–500 AU) circumbinary disks, much smaller (5–10 AU) circumstellar disks must be present around each binary component in order to account for near-IR excesses seen in their SEDs (Close et al. 1998*a*). Short estimated lifetimes for these inner disks makes it likely that they are replenished via accretion from the outer large circumbinary disks, in line with theoretical expectations (cf. Duvert et al. 1998; Close et al. 1998*a*; the chapter by Lubow and Artymowicz, this volume).

Finally, the large size of the UY Aur disk makes it possible to obtain imaging polarimetry, a difficult proposition requiring a stable and sharp PSF (Close et al. 1997*b*). AO imaging polarimetry with 0.09-arcsecond resolution indicates that the light from the UY Aur disk is strongly polarized ($\sim$80%) because of single scattering off mainly small ($<$0.1 μm) dust grains (Potter et al. 1998). The observed polarization pattern around UY Aur was found to be in good agreement with the dust grains distributed in a disk. Other distributions, such as a spherical envelope, can be rejected by comparison with suitable models (Potter et al. 1998).

IV. PHYSICAL PROPERTIES OF DISKS

Models of the distribution of light scattered by circumstellar dust grains for a variety of density distributions have been presented by Whitney and

Hartmann (1992, 1993), Fischer et al. (1996), and Wood et al. (1998). For circumstellar disks, power law formulations are conventionally adopted for the radial dependence of surface density, $\Sigma(r) = \Sigma_0(r/r_0)^p$, and scale height, $H(r) = H_0(r/r_0)^\beta$. A vertically isothermal, pressure-supported disk in a gravitational potential dominated by its central star then follows a density law

$$\rho(r, z) \propto \frac{M_d}{H_0 R_o^2}\left(\frac{r}{r_0}\right)^\alpha \exp(-z^2/2H(r)^2) \tag{1}$$

where r and z are cylindrical coordinates, M_d is the disk mass, R_o is the disk outer radius, $\alpha \equiv p - \beta$, and r_0 is a fiducial radius. Combined with assumptions about the dust opacity for scattering/absorption, the dust albedo, and a scattering phase function, the modeling task is largely reduced to calculation of column density integrals and projection angles, although proper treatment of polarization can add significant complexity. In this section, we discuss the extent to which such modeling of the optical/near-IR observations of disks has succeeded in determining some of their more important physical parameters (see also the chapter by Wilner and Lay, this volume, for constraints on disk parameters from modeling of millimeter and submillimeter images and visibilities).

A. Outer Radius

Observations in scattered light do not necessarily reveal the full radius of the disk, as demonstrated by Orion 114-426, where the polar scattering lobes seen in the near-IR are somewhat smaller than the disk outer radius as seen in silhouette (McCaughrean et al. 1998). Also, the radial extent of a disk traced in molecular line maps made in CO isotopes is generally larger than seen in scattered light (Stapelfeldt and Padgett 1999; Padgett et al. 1999*b*). In some cases, observations may be insufficiently sensitive to detect faint nebulosity to the full radial extent of the system. Alternatively, outer disk regions can be shadowed by the inner disk if the flaring function turns over (i.e., dH/dr becomes negative). Conversely, however, the average size of disks seen via scattering or absorption is significantly larger than inferred from modeling of the infrared SEDs (Beckwith et al. 1990).

B. Inclination

Theoretically, an edge-on disk should have bipolar reflection nebulae of equal brightness, while at increasing inclinations the farside of the disk will appear progressively smaller and fainter than the nearside. The inclination of an observed disk can thus be derived by comparison with a grid of models, although this simple trend can break down if significant scattered light from an envelope is present. The inclination derived in this way can be independently checked using measurements of the kinematics in an associated jet, assuming perpendicularity.

C. Mass

In principle, the brightness of the scattered light can be used to estimate the disk mass, but, as found for the corresponding absorption in the silhouette disks, very little material is required to account for the observations, because of the large optical depths at optical/IR wavelengths. Thus, only lower limits are derived, typically just $\sim 10^{-5}$ $M_{\odot}$ for the scattered light. However, in an edge-on system, the nebular structure is strongly affected by the total mass in the disk, with the thickness of the central dust lane increasing monotonically with disk mass, as shown in Fig. 2. Thus, high-resolution images of an edge-on disk can be compared with a grid of models to determine its mass. For the two edge-on systems where millimeter continuum observations are currently available, IRAS 04302+2247 and HH 30, the millimeter-derived masses are in reasonable agreement with those obtained from scattered-light models, giving some confidence in the applicability of a disk density law to these systems and in the current knowledge of dust opacities at optical and millimeter wavelengths.

D. Radial Density Profile

The radial density distribution is not well determined by observations at wavelengths where disks are optically thick. It is only in the very outermost sections of the silhouette disks, for example, that their structure can be traced, revealing their strongly truncated outer edges (see section III.B). In HH 30 and HK Tau/c, models using only the nebular light distribution as a constraint suggest a surface density weakly *increasing* with radius, although the same models produce too little extinction to obscure the central star. To reproduce the minimum necessary extinction, a radially

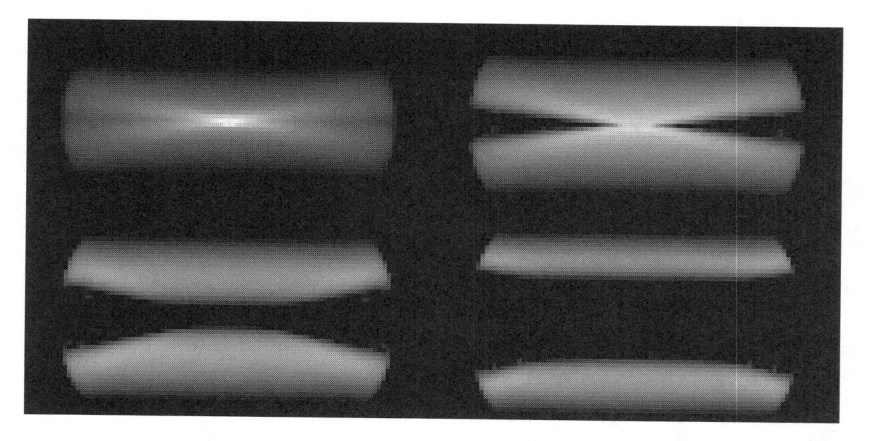

Figure 2. Scattering models at $\lambda = 0.8$ μm for optically thick, nearly edge-on circumstellar disks. The increase in apparent dust lane thickness as a function of disk mass is illustrated by models with 10^{-6} $M_{\odot}$ (top left), 3×10^{-5} $M_{\odot}$ (top right), 10^{-3} $M_{\odot}$ (lower left), and 3×10^{-2} $M_{\odot}$ (lower right) of gas and dust.

decreasing surface density is required, but since p is seen to be strongly degenerate with the scale height index β, only a weak conclusion can be drawn, namely that $p < -0.3$ in both systems.

E. Vertical Structure and Its Radial Dependence

In edge-on systems, gradients in the nebular brightness adjacent to the dark lane allow the scale height near the disk outer radius to be derived. This parameter is interesting because it can be directly related to the local temperature in a disk that is vertically hydrostatic; its radial variation is diagnostic of the radial temperature structure. In addition, changes in the vertical distribution of scattering dust grains may accompany the initial stages of particle growth, which lead to planetesimal formation.

Burrows et al. (1996) and Wood et al. (1998) independently derived identical values for the reference scale height H_0 in the HH 30 disk, and while a weak coupling of H_0 with other model parameters is seen, the systematic uncertainty introduced is less than 20%. Unfortunately, these modeling efforts have also shown that the radial dependence of scale height (the exponent β) cannot be uniquely determined from a single image, because the outer disk, where the scale height is largest, obscures both the central star and the inner disk from direct view. Thus, the vertical profile is not observed over a large enough range of radii to allow β to be uniquely solved for, although in the future it may be possible to solve for β uniquely via combined modeling of multiwavelength image data and the disk SED.

F. Circumstellar Dust Properties

When comparing the models of the scattered-light nebulae associated with circumstellar disks with the observations, the assumed dust grain properties can also be tested. For example, it is widely expected that grain growth will occur in dense circumstellar regions, and this growth might be enough to affect the nebula models. As a starting point, most authors use the properties of normal interstellar grains, which are well characterized both observationally and theoretically.

In general, any distribution of circumstellar dust grains will have a mix of small and large grains. If all these grains were homogeneous spheres with radii in the range 0.01–1 μm, exact Mie scattering theory (e.g., Bohren and Huffman 1983) could be used to calculate their scattering properties in in the optical/near-IR wavelength regime. However, it is likely that real circumstellar grains have more complex, possibly fractal geometries and consist of an aggregate of different materials (Ossenkopf 1991; Kozasa et al. 1993), making them hard to model. Fortunately, Rouleau (1996) has recently shown that the convoluted surface of such a grain can be well approximated by a spherical grain with the same total surface area. This presumably accounts for the success of Mathis and Whiffen (1989) in reproducing the observed ISM extinction curve from UV to far-IR wavelengths (cf. Mathis 1990) using a simple power law distribution of spherical grains, combined with a mix of different dielectric constants of various minerals and vacuum, produce a single effective dielectric constant.

Close et al. (1998*a*) created a similar recipe for circumstellar grains, using a composition of amorphous carbon, graphite, and silicates, and 80% vacuum by volume, similar to model A of Mathis and Whiffen (1989). A power law distribution of radii was assumed, with $n(a) = Ca^{-\eta}$ and $n(a)da$ equal to the number density of grains per hydrogen atom with radii between a and da. The size distribution has upper and lower limits at $a_{\max}$ and $a_{\min}$ respectively. Close et al. (1998*a*) used this grain prescription in their models for the scattered light seen in the circumbinary disks around GG Tau and UY Aur systems, where the inclination was well established from the millimeter kinematics. They found the best results were obtained using a power law index of $\eta = 4.7$, and radii ranging from 0.03 μm up to 0.5–0.6 μm, i.e., similar to ISM grains. A significant population of grains with radii >0.6 μm was found to be unlikely, because such large grains would increase the contrast between the near and far sides of the disk to levels higher than observed. Burrows et al. (1996) and Stapelfeldt et al. (1998) derived similar results for their models of the HH 30 and HK Tau/c edge-on disks, although the presence of a population of somewhat larger ($\sim$1 μm) grains was implied based on the apparently enhanced forward scattering at 0.8-μm wavelength. Nevertheless, the larger millimeter-sized grains inferred by millimeter observations apparently do not play a major role in the scattering process: it is possible that they may have sunk to the central (optically thick) plane of the disks, remaining hidden from view in scattered light, but nevertheless still dominating the millimeter continuum emission.

V. FUTURE DEVELOPMENTS

Almost all of the results presented in this review have been obtained since the *Protostars and Planets III* meeting in 1990. Clearly, before the next meeting in the series, substantial progress will be made in studying the most important phases in the formation and evolution of young circumstellar disks and protoplanetary systems.

Foremost is the need to find more examples of disks around young stellar objects at all evolutionary stages, by continuing surveys of nearby dark clouds and H II regions. Existing techniques will be used to the full, including AO on the ground and the HST in space. The future HST servicing mission will see the installation of the wide-field fully sampled Advanced Camera for Surveys at optical wavelengths and the possible return to operation of NICMOS in the near-IR.

Second, improvements in angular resolution are needed to enable more detailed studies of nearby sources and to extend the surveys to encompass the more distant star-forming regions with reasonable linear resolution. In the near term, this will be achieved on the ground by equipping the 8–10-m class telescopes with AO systems; in the longer term, significant gains will be made by the passively cooled IR-optimized Next Generation Space Telescope (NGST) and through multitelescope optical/IR interferometry, using, for example, the Keck, the European Southern Observatory's Very Large Telescope Interferometer (VLTI), and the Large Binocular Telescope (LBT) on the ground and the Space Interferometry Mission (SIM) in space. Malbet et al. (1998) have very recently shown the way forward in this regard, using the Palomar Testbed Interferometer to resolve thermal emission from a disk around FU Ori at near-IR wavelengths at spatial scales around 4 milliarcsecond, or just 2 AU at 450 pc.

Finally, there is a strong case for increased-contrast imaging in order to study low-surface-brightness scattered light in the presence of a bright central star. There are several known examples of YSOs with disks that have been resolved via millimeter interferometry, and yet show just bare PSFs in HST and AO images at optical/near-IR wavelengths. These systems, including DL Tau, DM Tau, CY Tau, V 892 Tau, MWC 480, LkCa 15, and AS 209, clearly have disks, but current optical/near-IR instrumentation is simply not up to the contrast challenge. Future ameliorating developments will include improved forms of coronagraphy and optical/IR nulling interferometry on the ground and in space.

Acknowledgments The research described in this paper was carried out under grants from the Deutsche Forschungsgemeinschaft to the Astrophysikalisches Institut Potsdam, from the National Aeronautics and Space Administration to the Jet Propulsion Laboratory, California Institute of Technology, and from the National Science Foundation to the University of Hawaii.

REFERENCES

Adams, F. C., Shu, F. H., and Lada, C. J. 1988. The disks of T Tauri stars with flat infrared spectra. *Astrophys. J.* 326:865–883.

Artymowicz, P., and Lubow, S. H. 1996. Mass flow through gaps in circumbinary disks. *Astrophys. J. Lett.* 467:L77–L80.

Artymowicz, P., and Lubow, S. H. 1995. Interaction of young binaries with protostellar disks. In *Disks and Outflows around Young Stars*, ed. S. V. W. Beckwith, J. Staude, A. Quetz, and A. Natta (Heidelberg: Springer), pp. 115–131.

Bally, J., Sutherland, R. S., Devine, D., and Johnstone, D. 1998*a*. Externally illuminated young stellar environments in the Orion Nebula: Hubble Space Telescope Planetary Camera and ultraviolet observations. *Astron. J.* 116:293–321.

Bally, J., Testi, L., Sargent, A. I., and Carlstrom, J. 1998*b*, Disk mass limits and lifetimes of externally irradiated young stellar objects embedded in the Orion Nebula. *Astron. J.* 116:854–859.

Basri, G., and Batalha, C. 1990. Hamilton echelle spectra of young stars. I: Optical veiling. *Astrophys. J.* 363:654–669.

Beckwith, S. V. W., and Birk, C. C. 1995. Vertical structure in HL Tauri. *Astrophys. J. Lett.* 449:L59–L63.

Beckwith, S. V. W., and Sargent, A. I. 1993. The occurrence and properties of disks around young stars. In *Protostars and Planets III*, ed. E. H. Levy and J. I. Lunine (Tucson: University of Arizona Press), pp. 521–541.

Beckwith, S. V. W. and Sargent, A. I. 1996. Circumstellar disks and the search for neighbouring planetary systems. *Nature* 383:139–144.

Beckwith, S. V. W., Skrutskie, M. F., Zuckerman, B., and Dyck, H. M. 1984. Discovery of solar system-size halos around young stars. *Astrophys. J.* 287:793–800.

Beckwith, S. V. W., Koresko, C. D., Sargent, A. I., and Weintraub, D. A. 1989. Tomographic imaging of HL Tauri. *Astrophys. J.* 343:393–399.

Beckwith, S. V. W., Sargent, A. I., Güsten, R., and Chini, R. 1990. A survey for circumstellar disks around young stellar objects. *Astron. J.* 99:924– 945.

Bell, K. R., Cassen, P. M., Klahr, H. H., and Henning, T. 1997. The structure and appearance of protostellar accretion disks: Limits on disk flaring. *Astrophys. J.* 486:372–387.

Bohren, C. F., and Huffman, D. R. 1983. *Absorption and Scattering of Light by Small Particles* (New York: John Wiley & Sons).

Burrows, C. J., Stapelfeldt, K. R., Watson, A. M., Krist, J. E., Ballester, G. E., Clarke, J. T., Crisp, D., Gallagher, J. S., Griffiths, R. E., III, Hester, J. J., Hoessel, J. G., Holtzman, J. A., Mould, J. R., Scowen, P. A., Trauger, J. T., and Westphal, J. A. 1996. Hubble Space Telescope observations of the disk and jet of HH 30. *Astrophys. J.* 473:437–451.

Cabrit, S., Guilloteau, S., André, P., Bertout, C., Montmerle, T., and Schuster, K. 1996. Plateau de Bure observations of HL Tauri: Outflow motions in a remnant circumstellar envelope. *Astron. Astrophys.* 305:527–540.

Chen, H., Bally, J., O'Dell, C. R., McCaughrean, M. J., Thompson, R. I., Rieke, M., Schneider, G., and Young, E. T. 1998. 2.12 μm molecular hydrogen emission from circumstellar disks embedded in the Orion Nebula. *Astrophys. J. Lett.* 492:L173–L176.

Close, L., Roddier, F., Hora, J. L., Graves, J. E., Northcott, M., Roddier, C., Hoffman, W. F., Dayal, A., Fazio, G. G., and Deutsch, L. K. 1997*a*. Adaptive optics infrared imaging polarimetry and optical HST imaging of Hubble's variable nebula (R Monocerotis/NGC 2261): A close look at a very young active Herbig Ae/Be star. *Astrophys. J.* 489:210–221.

Close, L., Roddier, F., Northcott, M. J., Roddier, C., and Graves, J. E. 1997*b*. Adaptive optics 0.2 arcsec resolution infrared images of HL Tau: Direct images of an active accretion disk around a protostar. *Astrophys. J.* 478:766–777.

Close, L., Dutrey, A., Roddier, F., Guilloteau, S., Roddier, C., Northcott, M., Ménard, F., Duvert, G., Graves, J. E., and Potter, D. 1998*a*. Adaptive optics imaging of the circumbinary disk around the T Tauri binary UY Aur: Estimates of the binary mass and circumbinary dust grain size distribution. *Astrophys. J.* 499:883–888.

Close, L. M., Roddier, F. J., Roddier, C. A., Graves, J. E., Northcott, M. J., Northcott, M. J., and Potter, D. 1998*b*. Scientific results from the University of Hawaii Adaptive Optics Program. In *Adaptive Optical Systems Technologies*, Proc. SPIE 3353, ed. D. Bonaccini and R. K. Tyson (Bellingham, WA: SPIE—The International Society for Optical Engineering), pp. 406–416.

Cohen, M., and Kuhi, L. V. 1979. Observational studies of pre-main-sequence evolution. *Astrophys. J. Suppl.* 41:743–843.

D'Antona, F., and Mazzitelli, I. 1994. New pre-main-sequence tracks for $M \leq 2.5$ $M_\odot$ as tests of opacities and convection model. *Astrophys. J. Suppl.* 90:467–500.

Dutrey, A., Guilloteau, S., and Simon, M. 1994. Images of the GG Tauri rotating ring. *Astron. Astrophys.* 286:149–159.

Dutrey, A., Guilloteau, S., Duvert, G., Prato, L., Simon, M., Schuster, K., and Ménard, F. 1996. Dust and gas distribution around T Tauri stars in Taurus-Auriga. I. Interferometric 2.7 mm continuum and ^{13}CO $J = 1$–0 observations. *Astron. Astrophys.* 309:493–504.

Dutrey, A., Guilloteau, S., Prato, L., Simon, M., Duvert, G., Schuster, K., and Ménard, F. 1998. CO study of the GM Aurigae Keplerian disk. *Astron. Astrophys.* 338:L63–L66.

Duvert, G., Dutrey, A., Guilloteau, S., Ménard, F., Schuster, K., Prato, L., and Simon, M. 1998. Disks in the UY Aurigae binary. *Astron. Astrophys.* 322:867–874.

Falcke, H., Davidson, K., Hofmann, K.-H., and Weigelt, G. 1996. Speckle-masking imaging polarimetry of η Carinae: Evidence for an equatorial disk. *Astron. Astrophys.* 306:L17–L20.

Fischer, O., Henning, T., and Yorke, H. W. 1996. Simulation of polarization maps. II. The circumstellar environment of pre-main sequence objects. *Astron. Astrophys.* 308:863–885.

Ghez, A. M., Neugebauer, G., and Matthews, K. 1993. The multiplicity of T Tauri stars in the star forming regions Taurus-Auriga and Ophiuchus-Scorpius: A 2.2μm speckle imaging survey. *Astron. J.* 106:2005–2023.

Ghez, A. M., Weinberger, A. J., Neugebauer, G., Matthews, K., and McCarthy, D. W., Jr. 1995. Speckle imaging measurements of the relative tangential velocities of the components of T Tauri binary stars. *Astron. J.* 110:753–765.

Graves, J. E., Northcott, M. J., Roddier, F. J., Roddier, C. A., and Close, L. M. 1998. First Light for Hokupa'a: 36-element Curvature AO system at the University of Hawaii. In *Adaptive Optical Systems Technologies*, Proc. SPIE 3353, ed. D. Bonaccini and R. K. Tyson (Bellingham, WA: SPIE—The International Society for Optical Engineering), pp. 34–44.

Guilloteau, S., Dutrey, A., and Gueth, F. 1997. Disks and outflows as seen from the IRAM interferometer. In *Herbig-Haro Flows and the Birth of Stars*, IAU Symp. 182, eds. B. Reipurth and C. Bertout (Dordrecht: Kluwer), pp. 365–380.

Guilloteau, S., and Dutrey, A. 1998. Physical parameters of the Keplerian protoplanetary disk of DM Tauri. *Astron. Astrophys.* 339:467–476.

Hall, S. M. 1997. The energetics of star-disc encounters and the disc density profiles. Ph.D. Thesis, University of Cambridge.

Hall, S. M., Clarke, C. J., and Pringle, J. E. 1996. Energetics of star-disc encounters in the non-linear regime. *Mon. Not. Roy. Astron. Soc.* 278:303–320.

Hayashi, M., Ohashi, N., and Miyama, S. M. 1993. A dynamically accreting gas disk around HL Tauri. *Astrophys. J. Lett.* 418:L71–L74.

Hayward, T. L., and McCaughrean, M. J. 1997. A search for thermal infrared emission from three silhouette disks in Orion. *Astron. J.* 113:346–353.

Heller, C. H. 1995. Encounters with protostellar disks. II. Disruption and binary formation. *Astrophys. J.* 455:252–259.

Henney, W. J., Raga, A. C., Lizano, S., and Curiel, S. 1996. A two-wind interaction model for proplyds. *Astrophys. J.* 465:216–230.

Jayawardhana, R., Fisher, S., Hartmann, L., Telesco, C., Piña, R., and Fazio, G. 1998. A dust disk surrounding the young star HR 4796A. *Astrophys. J. Lett.* 503:L79–L82.

Johnstone, D., Hollenbach, D., and Bally, J. 1998. Photoevaporation of disks and clumps by nearby massive stars: Application to disk destruction in the Orion Nebula. *Astrophys. J.* 499:758–776.

Jones, B. F., and Walker, M. F. 1998. Proper motions and variabilities of stars near the Orion Nebula. *Astron. J.* 95:1755–1782.

Jura, M., Malkan, M., White, R., Telesco, C., Piña, R., and Fisher, R. S. 1998. A protocometary cloud around HR 4796A? *Astrophys. J.* 505:897–902.

Kalas, P. 1996. A coronagraphic survey for circumstellar disks around main sequence and pre-main sequence stars. Ph.D. Thesis, University of Hawaii.

Kalas, P., and Jewitt, D. 1996. The detectability of β Pic-like circumstellar disks around nearby main sequence stars. *Astron. J.* 111:1347–1355.

Kenyon, S. J., and Hartmann, L. 1987. Spectral energy distributions of T Tauri stars: Disk flaring and limits on accretion. *Astrophys. J.* 323:714–733.

Kitamura, Y., Saito, M., Kawabe, R., and Sunada, K. 1997. NMA imaging of envelopes and disks around low mass protostars and T Tauri stars. In *Herbig-Haro Flows and the Birth of Low Mass Stars*, IAU Symp. 182, ed. B. Reipurth and C. Bertout (Dordrecht: Kluwer), pp. 381–390.

Koerner, D. W. 1997. Analogs of the early solar system. In *Planetary and Interstellar Processes Relevant to the Origins of Life*, ed. D. C. B. Whittet (Dordrecht: Kluwer), pp. 157–184.

Koerner, D. W., and Sargent, A. I. 1995. Imaging the small-scale circumstellar gas around T Tauri stars. *Astron. J.* 109:2138–2145.

Koerner, D. W., Ressler, M. E., Werner, M. W., and Backman, D. E. 1998. Mid-infrared imaging of a circumstellar disk around HR 4796: Mapping the debris of planetary formation. *Astrophys. J. Lett.* 503:L83–L87.

Koresko, C. D. 1998. A circumstellar disk in a pre-main sequence binary star. *Astrophys. J. Lett.* 507:145–148.

Kozasa, T., Blum, J., Okamoto, H., and Mukai, T. 1993. Optical properties of dust aggregates. II. Angular dependence of scattered light. *Astron. Astrophys.* 276:278–288.

Krist, J. E., Stapelfeldt, K. R., Burrows, C. J., et al. 1998. Hubble Space Telescope WFPC2 imaging of FS Tauri and Haro 6-5B. *Astrophys. J.* 501:841–852.

Kritsuk, A. G. 1983. Dynamics of the sweeping of interstellar clouds from a rotating galaxy as it moves in the intergalactic medium. *Astrophysics* 19:263–270.

Larwood, J., Nelson, R. P., Papaloizou, J. C. B., and Terquem, C. 1996. The tidally induced warping, precession and truncation of accretion discs in binary systems: Three-dimensional simulations. *Mon. Not. Roy. Astron. Soc.* 282:597–613.

Lay, O. P., Carlstrom, J. E., Hills, R. E., and Phillips, T. G. 1994. Protostellar accretion disks resolved with the JCMT–CSO interferometer. *Astrophys. J. Lett.* 434:L75–L78.

Leinert, C., Zinnecker, H., Weitzel, N., Christou, J., Ridgway, S. T., Jameson, R., Haas, M., and Lenzen, R. 1993. A systematic approach for young binaries in Taurus. *Astron. Astrophys.* 278:129–149.

Lucas, P. W., and Roche, P. F. 1997. Butterfly star in Taurus: Structures of young stellar objects. *Mon. Not. Roy. Astron. Soc.* 286:895–919.

Lynden-Bell, D., and Pringle, J. E. 1974. The evolution of viscous discs and the origin of nebular variables. *Mon. Not. Roy. Astron. Soc.* 168:603–638.

Malbet, F., Berger, J.-P., Colavita, M. M., Koresko, C. D., Beichman, C., Boden, A. F., Kulkarni, S. R., Lane, B. F., Mobley, D. W., Pan, X. P., Shao, M., Van Belle, G. T., and Wallace, J. K. 1998. FU Orionis resolved by infrared long-baseline interferometry at a 2 AU scale. *Astrophys. J. Lett.* 507:L149–L152.

Mathis, J. S. 1990. Interstellar dust and extinction. *Ann. Rev. Astron. Astrophys.* 28:37–70.

Mathis, J. S., and Whiffen, G. 1989. Composite interstellar grains. *Astrophys. J.* 341:808–822.

McCaughrean, M. J., and O'Dell, C. R. 1996. Direct imaging of circumstellar disks in the Orion Nebula. *Astron. J.* 111:1977–1986.

McCaughrean, M. J., Rayner, J. T., Zinnecker, H., and Stauffer, J. R. 1996. Circumstellar disks in the Trapezium Cluster. In *Disks and Outflows around Young Stars*, ed. S. V. W. Beckwith, J. Staude, A. Quetz, and A. Natta (Heidelberg: Springer), pp. 33–43.

McCaughrean, M. J., Chen, H., Bally, J., Erickson, E., Thompson, R. I., Rieke, M., Schneider, G., Stolovy, S., and Young, E. T. 1998. High-resolution near-infrared imaging of the Orion 114-426 silhouette disk. *Astrophys. J. Lett.* 492:L157–L161.

Mundt, R., Bührke, T., Solf, J., Ray, T. P., and Raga, A.C. 1990. Optical jets and outflows in the HL Tauri region. *Astron. Astrophys.* 232:37–61.

Mundy, L. G., Looney, L. W., Erickson, W., Grossman, A., Welch, W. J., Forster, J. R., Wright, M. C. H., Plambeck, R. L., Lugten, J., and Thornton, D. D. 1996. Imaging the HL Tauri disk at $\lambda = 2.7$ millimeters with the BIMA array. *Astrophys. J. Lett.* 464:L169–L173.

O'Dell, C. R., and Wen, Z. 1994. Postrefurbishment mission Hubble Space Telescope images of the core of the Orion Nebula: Proplyds, Herbig-Haro objects, and measurements of a circumstellar disk. *Astrophys. J.* 436:194–202.

O'Dell, C. R., Wen, Z., and Hu, X. 1993. Discovery of new objects in the Orion Nebula on HST images: Shocks, compact sources, and protoplanetary disks. *Astrophys. J.* 410:696–700.

O'Dell, C. R., and Wong, S.-K. 1996. Hubble Space Telescope mapping of the Orion Nebula. I. A survey of stars and compact objects. *Astron. J.* 111:846–855.

Ossenkopf, V. 1991. Effective-medium theories for cosmic dust grains. *Astron. Astrophys.* 251:210–219.

Osterbart, R., Langer, N., and Weigelt, G. 1997. High-resolution imaging of the bipolar nebula Red Rectangle: Evidence for unstable mass transfer in a close binary system. *Astron. Astrophys.* 325:609–612.

Padgett, D. L., Brandner, W., Stapelfeldt, K. R., Strom, S. E., Terebey, S., and Koerner, D. 1999*a*. HST/NICMOS imaging of disks and envelopes around very young stars. *Astron. J.*. 117:1490–1504.

Padgett, D. L., Brandner, W., Stapelfeldt, K. R., and Koerner, D. 1999*b*. Anatomy of a butterfly: Disk structure models for IRAS 04302+2247. In preparation.

Phookun, B., and Mundy, L. G. 1995. NGC 4654: A Virgo cluster spiral interacting with the intracluster medium. *Astrophys. J.* 453:154–161.

Potter, D. E., Close, L. M., Roddier, F., Roddier, C., Graves, J. E., and Northcott, M. 1998. The first resolved polarimetry map of a circumbinary disk. *Astrophys. J. Lett.*, submitted.

Ray, T. P., Mundt, R., Dyson, J. E., Falle, S. A. E. G., and Raga, A. 1996. HST observations of jets from young stars. *Astrophys. J. Lett.* 468:L103–L106.

Reipurth, B., Chini, R., Krugel, E., Kreysa, E., and Seivers, A. 1993. Cold dust around Herbig-Haro energy sources: A 1300 micron survey. *Astron. Astrophys.* 273:221–238.

Roddier, C., Roddier, F., Northcott, M. J., Graves, J. E., and Jim, K. 1996. Adaptive optics imaging of GG Tauri: Optical detection of the circumbinary ring. *Astrophys. J.* 463:326–335.

Rouleau, F. 1996. Electromagnetic scattering by compact clusters of spheres. *Astron. Astrophys.* 310:686–698.

Saigo, K., and Hanawa, T. 1998. Similarity solution for formation of a circumstellar disk through the collapse of a flattened rotating cloud. *Astrophys. J.* 493:342–350.

Saito, M., Kawabe, R., Ishiguro, M., Miyama, S. M., Hayashi, M., Handa, T., Kitamura, Y., and Omodaka, T. 1995. Aperture synthesis ^{12}CO and ^{13}CO observations of DM Tauri: 350 AU radius circumstellar gas disk. *Astrophys. J.* 453:384–392.

Sargent, A. I., and Beckwith, S. V. W. 1991. The molecular structure around HL Tauri. *Astrophys. J. Lett.* 382:L31–L35.

Schneider, G., Smith, B. A., Becklin, E. E., Koerner, D. W., Meier, R., Hines, D. C., Lowrance, P. J., Terrile, R. J., Thompson, R. I., and Rieke, M. 1999. NICMOS imaging of the HR 4796A circumstellar disk. *Astrophys. J. Lett.* 513:L127–L130.

Shakura, N. I., and Sunyaev, R. A. 1973. Black holes in binary systems. Observational appearance. *Astron. Astrophys.* 24:337–355.

Smith, B. A., and Terrile, R. J. 1984. A circumstellar disk around Beta Pictoris. *Science* 226:1421–1424.

Smith, B. A., Fountain, J. W., and Terrile, R. J. 1992. An optical search for Beta Pictoris-like disks around nearby stars. *Astron. Astrophys.* 261:499–502.

Stapelfeldt, K. R., and Padgett, D. L. 1999. OVRO millimeter array observations of the HH 30 circumstellar disk. In preparation.

Stapelfeldt, K. R., Burrows, C. J., Krist, J. E., Trauger, J. T., Hester, J. J., Holtzmann, J. A., Ballester, G. E., Casertano, S., Clarke, J. T., Crisp, D., Evans, R. W., Gallagher, J. S., Griffiths, R. E., III, Hoessel, J. G., Mould, J. R., Scowen, P. A., Watson, A. M., and Westphal, J. A. 1995. WFPC2 imaging of the circumstellar nebulosity of HL Tauri. *Astrophys. J.* 449:888–893.

Stapelfeldt, K. R., Burrows, C. J., Krist, J. E., and the WFPC2 Science Team. 1997*a*. Hubble Space Telescope imaging of the disks and jets of Taurus young stellar objects. In *Herbig-Haro Flows and the Birth of Low Mass Stars*, IAU Symp. 182, ed. B. Reipurth and C. Bertout (Dordrecht: Kluwer), pp. 355–364.

Stapelfeldt, K. R., Sahai, R., Werner, M., and Trauger, J. 1997*b*. An HST imaging search for circumstellar matter in young nebulous clusters. In *Planets beyond the Solar System and the Next Generation of Space Missions*, ASP Conf. Ser. 119, ed. D. R. Soderblom (San Francisco: Astronomical Society of the Pacific), pp. 131–134.

Stapelfeldt, K. R., Krist, J. E., Ménard, F., Bouvier, J., Padgett, D. L., and Burrows, C. J. 1998. An edge-on circumstellar disk in the young binary system HK Tauri. *Astrophys. J. Lett.* 502:L65–L68.

Whitney, B. A., and Hartmann, L. 1992. Model scattering envelopes of young stellar objects. I: Method and application to circumstellar disks. *Astrophys. J.* 395:529–539.

Whitney, B. A., and Hartmann, L. 1993. Model scattering envelopes of young stellar objects. II: Infalling envelopes. *Astrophys. J.* 402:605–622.

Wood, K., Crosas, M., and Ghez, A. M. 1999. GG Tauri's circumbinary disk: Models for near-infrared scattered light images and molecular line profiles. *Astrophys. J.*, 516:335–341.

Wood, K., Kenyon, S. J., Whitney, B., and Turnbull, M. 1998. Optical and near-infrared model images of circumstellar environments of classical T Tauri stars. *Astrophys. J.* 497:404–418.

SUBARCSECOND MILLIMETER AND SUBMILLIMETER OBSERVATIONS OF CIRCUMSTELLAR DISKS

DAVID J. WILNER
Harvard-Smithsonian Center for Astrophysics

and

OLIVER P. LAY
University of California at Berkeley

Millimeter and submillimeter observations provide a valuable tool for probing the cool, dusty material in the disks around young stars. With ongoing instrumental upgrades, the technique of interferometry now obtains subarcsecond resolution at these wavelengths, capable of spatially resolving the size scales associated with planet formation in nearby star-forming dark clouds. We describe in detail the subarcsecond millimeter and submillimeter observations of the well-known low-mass young stellar objects HL Tau and L1551 IRS5 to illustrate what may be learned from such observations. High-resolution multifrequency continuum observations yield estimates of surface mass density distributions and optical depths, providing important constraints on accretion mechanisms as well as starting points for modeling the evolution of the primitive solar nebula. Drawing on observations of several low-mass pre-main-sequence systems, we examine power law disk models, departures from simple power law structure, and the influence of companion bodies. We also discuss observations of maser emission as subarcsecond probes of gas kinematics in some sources.

I. INTRODUCTION

A large fraction of young stars exhibits emission from small particles thought to be distributed in disks with properties similar to the early solar system. These disks provide a direct link between the birth of stars and the formation of planetary systems. By the time of the *Protostars and Planets III* conference, spectral energy distributions (SEDs) of young stellar objects (YSOs), spanning wavelengths from optical to millimeter, were commonly modeled as arising from combinations of stellar photospheres, extended envelopes, and circumstellar disks of solar system dimensions. Yet there existed little, if any, direct support for the disk paradigm in the form of spatially resolved observations, largely because of the high angular resolution required. In nearby star-forming dark clouds such as Taurus-Auriga and Ophiuchus (distance 140 pc), a circumstellar disk with extent comparable to the planetary system around our Sun subtends less than 0.5″ (70 AU). While progress continues to be made in characterizing the physical properties of the circumstellar environment through increasingly complex modeling of spatially unresolved measurements, observations with high angular resolution are ultimately essential to validate the underlying constructs of these models.

The intervening years have seen remarkable instrumental advances that allow observations with subarcsecond resolution at optical, infrared, and millimeter wavelengths. With these new capabilities, it has become possible to observe directly the disks that had been long invoked to explain the photometric and spectroscopic peculiarities of young stars. At optical wavelengths, where the *Hubble Space Telescope* (HST) provides the required resolution, spectacular observations in the Orion Nebula resolve a few circumstellar disks as dark ovals in dramatic silhouette against the nebular background (McCaughrean and O'Dell 1996). In Taurus, sources such as HH 30 (Burrows et al. 1996) and HK Tau/c (Stapelfeldt et al. 1998) are viewed nearly equator-on, obscuring the stellar photosphere and allowing scattered light from the central star to provide exceptionally clear pictures of the disk geometry. For these sources, it is important to keep in mind that the extended emission arises entirely from scattering. In rare cases, the scattered light shows a correspondence with extended cold gas imaged in low-lying transitions of CO (e.g., GM Aurigae, Stapelfeldt et al. 1995*b*; Dutrey et al. 1998). However, no information may be extracted from these tracers on the inner disk regions of high column density that remain opaque.

The disk material beyond a few stellar radii is at low temperatures, less than a few hundred K, and emits most of its energy at far-infrared through millimeter wavelengths. In this chapter we review the development and impact of subarcsecond-resolution millimeter and submillimeter observations of circumstellar material. We refer the reader to the chapter by McCaughrean et al., this volume, for a discussion of optical and infrared observations of disks, and to the chapter by Mundy et al., this volume, for a review of millimeter emission from protostellar envelopes and the relationship of disks to these larger-scale structures.

A. Millimeter/Submillimeter Emission

At millimeter and submillimeter wavelengths, emission from heated dust in the circumstellar environment of a low-mass star dominates any emission from the stellar photosphere by many orders of magnitude. Surveys with single-dish telescopes detect millimeter emission from a substantial fraction of pre-main-sequence systems, and the inferred disk properties are thought to correspond to analogs of the protosolar nebula before the onset of planet formation (Beckwith et al. 1990). However, these low-resolution

($>7''$) measurements cannot separate the disk flux component from the extended envelope, which is largely optically thin at these wavelengths and often dominates the disk contribution in young systems. Interferometers act as spatial filters, so observations using long baselines are insensitive to the extended envelopes. The combination of wavelength and resolution therefore make millimeter and submillimeter interferometry an especially powerful probe of the compact disks that surround young stellar objects.

The push toward subarcsecond resolution at millimeter and submillimeter wavelengths capable of resolving the structure of the circumstellar disks has been made possible through advances in instrumentation, including low-noise receivers, wider bandwidths, and longer interferometer baselines. Table I summarizes the subarcsecond capabilities of existing and future arrays. An important point is that as the synthesized beam becomes smaller, the sensitivity to surface brightness goes down; there is simply less flux from the smaller patch of sky. When observing a source with brightness temperature T_B at a wavelength λ with a Gaussian beam of angular diameter θ, the flux density in the beam (in the Rayleigh-Jeans regime) is given by

$$\left(\frac{S}{\mathrm{mJy}}\right) = 5.7\left(\frac{\lambda}{3\ \mathrm{mm}}\right)^{-2}\left(\frac{T_B}{100\ \mathrm{K}}\right)\left(\frac{\theta}{0.1''}\right)^{2}$$

High-resolution observations of low-brightness emission therefore require very high sensitivity. Much of the recent progress comes from observations of dust continuum emission, where full advantage may be taken of the maximum available bandwidths. Spectral line emission, by contrast, is generally limited in bandwidth by the velocities of Doppler motions, and currently only very high brightness temperatures may be detected from these narrow-bandwidth signals at subarcsecond resolution. Maser action provides one natural mechanism for establishing very high-brightness spectral line emission, and we describe observations of masers in the circumstellar environment in section IV.

The effect of atmospheric fluctuations has not been included in Table I. These fluctuations disrupt the wavefronts and impose a "seeing" limit on the angular resolution that may be achieved. It is only recently that the calibration techniques of fast switching (between the target and a reference source) and water vapor radiometry have been introduced to overcome the seeing limit. Advances in this technology will continue to improve instrument performance and greatly increase the fraction of time available for observations with subarcsecond resolution. The next generation of millimeter arrays planned for deployment in Chile offers spectacular increases in sensitivity, approximately two orders of magnitude for dust continuum emission (a factor of 10^4 in observing speed), and will revolutionize the field.

TABLE I

High-resolution Capabilities of Current and Future Millimeter and Submillimeter Arrays

Array[a]	Antennas No. × diam.	Maximum Baseline	λ^b /mm	Highest Resolution[c]	Sensitivity[d]		
					Flux/mJy	Mass[e] /$M_\odot$	Brightness[f] /K
BIMA	10 × 6 m	2 km	3.0	0.3″	1.0	3.5×10^{-4}	1.9
			1.3	0.13″	3.0	8.6×10^{-5}	5.7
IRAM	5 × 15 m	400 m	3.0	1.5″	0.4	1.4×10^{-4}	0.03
			1.3	0.7″	1.4	4.0×10^{-5}	0.06
NMA	6 × 10 m	600 m	3.0	1.0″	1.0	3.5×10^{-4}	0.17
			2.0	0.7″	2.0	2.1×10^{-4}	0.14
OVRO	6 × 10 m	400 m	3.0	1.5″	0.5	1.8×10^{-4}	0.04
			1.3	0.7″	1.2	3.4×10^{-5}	0.09
CSO–JCMT	10 m, 15 m	160 m	0.85	1.0″	4	3.2×10^{-5}	0.05
			0.65	0.8″	10	3.6×10^{-5}	0.13
VLA (Q)	13 × 25 m	35 km	7.0	0.04″	0.06	2.7×10^{-4}	35
	27 × 25 m	7 km	7.0	0.02″	0.03	1.3×10^{-4}	70
SMA	8 × 6 m	500 m	1.3	0.5″	0.4	1.1×10^{-5}	0.05
			0.85	0.3″	1.0	8.0×10^{-6}	0.11
			0.37	0.15″	17	1.1×10^{-5}	2.0
MMA/	50 × 12 m[g]	3 km	1.3	0.1″	0.002	5.5×10^{-8}	0.008
LSA/			0.85	0.06″	0.004	2.8×10^{-8}	0.013
LMSA			0.37	0.025″	0.02	1.3×10^{-8}	0.09

[a] BIMA: Berkeley–Illinois–Maryland Association; IRAM: Institut de Radio Astronomie Millimétrique; NMA: Nobeyama Millimeter Array; OVRO: Owens Valley Radio Observatory; CTS–JCMT: Caltech Submillimeter Observatory–James Clerk Maxwell Telescope Interferometer (part-time); VLA: Very Large Array (part-time; upgrade to 27 Q band antennas and baseline expansion underway); SMA: Smithsonian Submillimeter Array (future); MMA/LSA/LMSA: Millimeter Array, Large Southern Array, and Large Millimeter and Submillimeter Array.

[b] Only selected wavebands shown for CSO–JCMT, VLA, SMA, and MMA/LSA/LMSA.

[c] Defined as λ/ max. baseline.

[d] 1 σ sensitivity for 8 hours of integration time, based on observatory descriptions at time of writing.

[e] Assumes unresolved optically thin emission with an average physical temperature of 100 K at a distance of 140 pc, $\kappa_\nu = 0.1(250\ \mu\text{m}/\lambda)\ \text{cm}^2\ \text{g}^{-1}$.

[f] Minimum detectable brightness temperature within highest resolution beam, using formula in main text.

[g] A nominal size for one array.

B. Models of Dust Emission from Disks

Compact millimeter continuum emission from low-mass stars is commonly attributed to dusty circumstellar disks, and a model is required to interpret the observations. Following the work of Adams et al. (1988) and Beckwith et al. (1990), it has been the common practice to assume that the disks are flat, thin, and circularly symmetric, with dust emission originating from an inner radius, $r_{\rm in}$, at the edge of the dust destruction zone, to an outer cutoff radius, $r_{\rm out}$. The temperature T and surface density Σ are assumed to follow power law dependencies with the radius:

$$\Sigma(r) = \Sigma_{50}\left(\frac{r}{50\ \mathrm{AU}}\right)^{-p} \qquad T(r) = T_{50}\left(\frac{r}{50\ \mathrm{AU}}\right)^{-q}$$

We adopt the value of 50 AU as a representative radius in the disk appropriate for the normalization. The flux (units: W Hz^{-1} ster^{-1}) emitted by a disk element of area dA is

$$dS = \frac{1}{D^2} B_\nu(T)(1 - e^{-\tau}) \cos i \ dA$$

where D is the distance to the star, $B_\nu(T)$ is the Planck function, i is the inclination of the disk ($i = 0$ is face-on), and the optical depth $\tau = \Sigma(r)\kappa_\nu / \cos i$. The dust emissivity κ_ν (units: $\mathrm{cm}^2\ \mathrm{g}^{-1}$) is parameterized as a power law in frequency: $\kappa_\nu = \kappa_0(\nu/\nu_0)^\beta$, with the index β depending on the nature (size, structure and composition) of the emitting grains (see the chapter by Beckwith et al., this volume). For small interstellar grains, simple physical considerations suggest that $\kappa_\nu \propto \nu^2$ (e.g. Draine and Lee 1984), whereas detailed calculations of the dust mixtures expected in the disk environment suggest a less steep frequency dependence may be appropriate (e.g., Pollack et al. 1994). Note that grains large compared to the observing wavelength have emissivity that depends only on physical cross section; i.e., $\beta = 0$. The brightness distribution of the model disk as seen by an observer is generated by mapping each area element to its appropriate sky coordinates, a transformation that depends on the position angle γ and inclination i of the disk.

The power law indexes for disks are expected to lie in the ranges $0 < p < 2, 0.5 < q < 0.75, 0 < \beta < 2$. Although a temperature distribution of the form $\mathrm{T} \propto r^{-3/4}$ is thought to characterize both viscous accretion disks and simple reprocessing disks (Adams et al. 1988), the spectral energy distributions often imply significantly less steep radial temperature gradients; in particular, values of q as low as 0.5 are inferred for "flat spectrum" sources. Several explanations for these nonstandard temperature distributions have been proposed, including variations of geometric flaring whereby the disk intercepts more stellar energy (Kenyon and Hartmann 1987; Natta 1993; Calvet et al. 1994; Chiang and Goldreich 1997) as well as more exotic ideas such as wave heating through spiral instabilities (Shu et al. 1990).

Regardless of the nature of the heating mechanism, in the limit of very low frequency and very low optical depth, $\tau \ll 1$, the dust continuum emission is proportional to a product of the column density and temperature: $dS \propto r^{-(p+q)}$. For $p + q < 2$ the outer radii will be the dominant source of emission. For higher frequencies, this effect is accentuated as the disk becomes increasingly optically thick towards the axis ($dS \propto r^{-q}$ for optically thick emission). For low-mass stars, the inner cutoff radius is of order 0.05 AU, where heating destroys grains. This inner radius has no significant effect on the millimeter and submillimeter brightness distribution, because the emission is predicted to be optically thick and the solid angle subtended by this region is very small.

The flat, thin, circularly symmetric model is of course an extreme simplification. For example, the dust properties probably change as a function of radius, and it is unlikely that there is a sharp cutoff at the outer radius. It is also likely that geometry plays an important role through flaring and shadowing, in which case the simple model will break down at high inclinations, and a three-dimensional model becomes necessary. Although detailed and complex disk models are becoming available, the simple power law model described above lends itself more readily to a comparison with current data. There are eight free parameters: γ, i, r_{out}, p, q, β, $(\Sigma_{50}\kappa_0)$, and T_{50}. Note that it is not possible to separate the contributions from the surface density and the dust emissivity normalizations. In addition, a minimum disk temperature, T_{min}, is usually specified, to account for external heating (Dutrey et al. 1996) or back-warming from an envelope (Natta 1993). Robust estimates for the disk parameters can be obtained by fitting the simple model to the data. The disk mass is given by $\int 2\pi r \Sigma \, dr$. The dust emissivity index β is a possible diagnostic for grain growth in the disk (Beckwith and Sargent 1991), and the value of p relates the observations to models of the accretion process and planet formation.

A simple steady disk model illustrates the connection between disk structure, as quantified by the value of p, and predictions of accretion theory. For a thin disk with material fed in at large radii at a constant rate $\dot{M}$, conservation of mass and angular momentum combine to relate the disk surface density Σ to $\dot{M}$, r, and the kinematic viscosity ν, by

$$\Sigma = \left[\frac{\dot{M}}{3\pi\nu}\right]\left[1 - \left(\frac{R_\star}{r}\right)^{1/2}\right]$$

where $R_\star$ is the stellar radius. The kinematic viscosity is often parameterized in terms of a velocity and a scale length of the angular momentum transport process, $\nu = \alpha c_s H$, where c_s is the local sound speed, H is the scale height, and α (<1) is a dimensionless number. For $r \gg R_\star$ and α = constant, the surface density may be represented by a power law, $\Sigma \propto (r^{3/2} T_m)^{-1} \sim r^{-p}$, where T_m is the local midplane temperature, whose value is governed by the balance of heating, largely through viscous processes, and cooling, through radiative losses. For an isothermal disk, $\Sigma \propto r^{-3/2}$. On the other hand, an optically thick disk with opacities appropriate to a mix of low-temperature ices and grains will develop $T_m \propto r^{-3/2}$ and $\Sigma \approx$ constant (Lin and Papaloizou 1985). Of course, if the mass flux through the disk is not constant, or if any other model assumptions are strongly violated, then the viscosity will manifest itself less directly in the surface density distribution.

II. HL TAU AND L1551 IRS 5: LOW-MASS PROTOTYPES

HL Tauri and L1551 IRS 5 are two of the best-studied low-mass embedded protostars, and we adopt them as examples to illustrate what can be learned from millimeter and submillimeter data with subarcsecond resolution. In section III we discuss subarcsecond-resolution observations of other low-mass pre-main-sequence stars, building on the information developed here.

A. Background and Early Work

Both HL Tau and L1551 IRS 5 are located in the Taurus-Auriga molecular cloud complex at a distance of $\sim$140 pc (Elias 1978), so that $1''$ corresponds to 140 AU. Both have well-developed outflows (Mundt et al. 1987; Snell et al. 1980) and large envelopes of molecular gas. Early observations with the Owens Valley Radio Observatory (OVRO) array provided the first estimates of the masses of the disks around these objects. For HL Tau, the 2.7-mm flux measured by Beckwith et al. (1986) and Sargent and Beckwith (1987, 1991) was about 0.11 Jy, with associated mass of 0.14 $M_\odot$. For L1551 IRS 5, the 2.7-mm flux measured by Keene and Masson (1990) in an unresolved component was 0.13 Jy, with associated mass of 0.16 $M_\odot$. Note that conversion of an unresolved flux into a mass implicitly assumes specific profiles of surface density and temperature within the disk, as well as an estimate of the dust emissivity, and the results must be viewed with appropriate caution.

The spectral energy distribution of HL Tau is well modeled by either a thin accretion disk (Beckwith et al. 1990) or a disk and residual extended envelope (Calvet et al. 1994). The nature of the envelope is still unclear. Early OVRO observations of the ^{13}CO $J = 1$–0 transition revealed a flattened circumstellar structure of size $\sim 30''$ (4000 AU), which was interpreted as a circumstellar disk with a Keplerian rotation curve (Beckwith et al. 1986; Sargent and Beckwith 1987, 1991). More recent ^{13}CO studies indicate that this structure is more likely a magnetically produced "pseudodisk" with kinematics dominated by infall (Hayashi et al. 1993) or by entrained outflow material (Cabrit et al. 1996). All these studies showed the presence of a continuum component, unresolved by beams as small as $2.7''$ (Woody et al. 1989; Ohashi et al. 1991), indicating that most of the circumstellar material was concentrated in a much more compact structure. Although there is an envelope of molecular gas, HL Tau is not a deeply embedded object; indeed, the protostar was originally thought to be optically visible, and only through recent HST observations has it been shown that this is scattered light, and the protostar is hidden (Stapelfeldt et al. 1995*a*).

Keene and Masson (1990) were the first to show that L1551 IRS 5 consisted of an extended envelope and a bright, compact core. The latter appeared pointlike in the 2.6″ synthesized beam of the three-element OVRO array, but they were able to estimate a radius of 45 ± 20 AU, based on the measured flux and the expected dust temperature as a function of distance from the central source. Butner et al. (1994) modeled the spectral energy distribution of L1551 IRS 5 but were unable to separate the contributions of the disk and envelope. The spectral energy distribution suggests that this source is younger and more embedded than HL Tau.

B. CSO-JCMT: Visibility Curves

The first millimeter or submillimeter instrument capable of subarcsecond resolution was the Caltech Submillimeter Observatory–James Clerk Maxwell Telescope (CSO-JCMT) Interferometer. With only a single fixed baseline, the interferometer does not sample sufficient Fourier components to synthesize an image of the brightness distribution. Instead, models must be fitted directly to the measured Fourier components. Although the physical baseline is fixed, the baseline vector projected along the line of sight determines the Fourier component to which the interferometer is sensitive. As the target rises above the horizon, transits, and sets, this projected baseline vector traces out an ellipse (Fig. 1a). Also shown in the figure are schematic contours for the Fourier transform of a disklike object (remem-

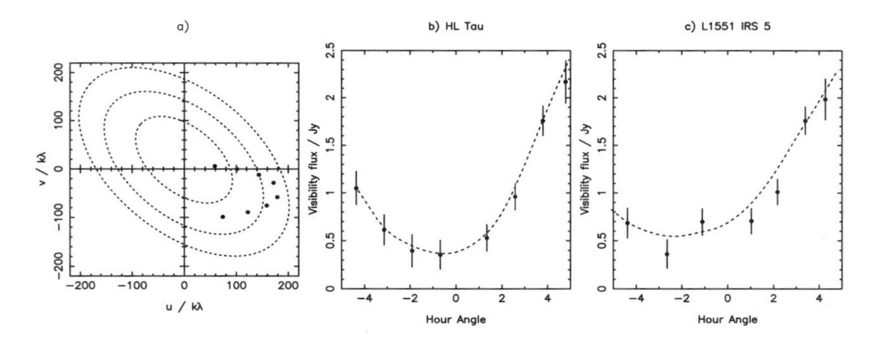

Figure 1. CSO-JCMT Interferometer data for HL Tau and L1551 IRS 5 at 0.87 mm, from Lay et al. (1994). (a) The $u - v$ plane showing the sampled Fourier components (black squares) and contours of the Fourier transform of a schematic disk. The measured visibility amplitudes are plotted for HL Tau (b) and L1551 IRS 5 (c), each with best-fitting elliptical Gaussian model (dashed line).

ber that the major axis in the Fourier transform plane corresponds to the minor axis on the sky, and vice versa). Figures 1b and 1c show the Fourier amplitudes for an observing wavelength of 0.87 mm measured as a function of hour angle for HL Tau and L1551 IRS 5 (Lay et al. 1994). An unresolved pointlike object would have a constant Fourier amplitude equal to its total flux, so both objects have been clearly resolved. The dashed lines show the best-fitting elliptical Gaussian models. This simple model gives a flux, major and minor radii, and a position angle.

The spatial filtering of the interferometer means that the fluxes are almost totally insensitive to the extended envelope emission. It was found that the disk is the source of almost all of the 0.87-mm continuum emission from HL Tau, but only 25% of that from L1551 IRS 5, indicating that the latter is much more embedded, and therefore probably younger. The major radii to half-maximum brightness were determined to be 60^{+10}_{-8} AU (HL Tau) and 80^{+20}_{-10} AU (L1551 IRS 5), and an upper limit of 52 AU was set on the minor radius in each case (i.e., the inclination is not well constrained). The position angles of 126° (HL Tau) and 162° (L1551 IRS 5) are almost perpendicular to the outflow directions from each source, consistent with the accretion disk interpretation of the data, although the profiles of surface density and temperature were not well constrained. Data have since been obtained at 460 GHz with the CSO-JCMT Interferometer (Lay et al. 1997), and the implications of these additional observations for disk structure are discussed in section II.E.

C. Subarcsecond Images with Millimeter Arrays

The first subarcsecond image of the HL Tau dust disk was obtained by Mundy et al. (1996) at 2.7 mm using the six-element Berkeley-Illinois-Maryland Association (BIMA) array with two outrigger stations linked by optical fiber. The upper panel of Fig. 2 shows contours of 2.7-mm emission superposed on the optical nebulosity observed by HST. The image shows a southeast-northwest extension, consistent with the CSO-JCMT result. The optical emission traces the cavity evacuated by the bipolar flow from the central star, and the millimeter emission coincides nicely with the sharp edge of dust extinction.

Figure 3 shows the 2.7-mm emission from L1551 IRS5 imaged at a variety of resolutions with the ten-element BIMA array (Looney et al. 1997). On the largest scales the emission is dominated by the inner regions of the envelope and has close to circular symmetry. At an angular resolution of about 1″, the source appears elongated at a position angle of 157°, very similar to the CSO-JCMT measurement. At the highest resolution of 0.3″ these larger structures are almost entirely resolved out, and the emission is consistent with two point sources. These are coincident with peaks in free-free emission that originate in ionized outflowing gas, most likely marking the two stars of a protobinary system (Bieging and Cohen 1985; Rodriguez et al. 1986). Thus the millimeter continuum features, in order of descending size scale, may be identified with an extended

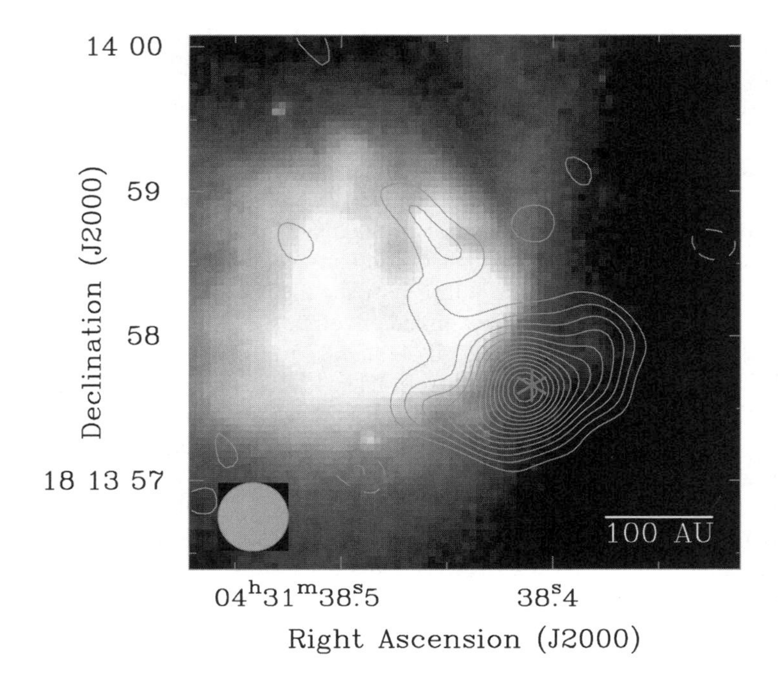

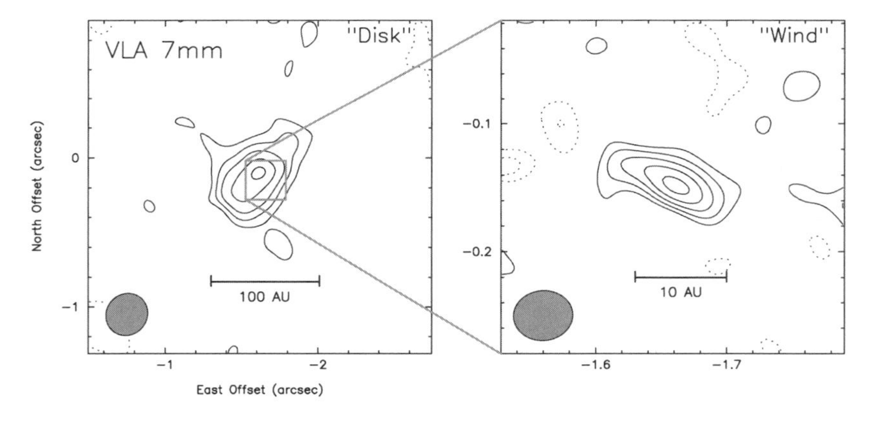

Figure 2. Subarcsecond views of HL Tau. Upper: Contours of 2.7-mm dust continuum emission from HL Tau imaged with the BIMA array (Looney et al. 1999) superposed on the *V* band nebulosity imaged with the HST (Stapelfeldt et al. 1995*a*). The millimeter emission traces the high-column-density extincting material that hides the direct starlight. The asterisk marks the position of the protostar, defined by the centimeter radio emission. Lower: Two images of the 7-mm emission from HL Tau obtained with the VLA (Wilner et al. 1999). The image on the right shows the map from data limited to baselines $> 1000\mathrm{k}\lambda$, which highlights the compact ionized wind component. The image on the left shows a lower-resolution map obtained after subtracting the wind component, which isolates dust emission from the disk.

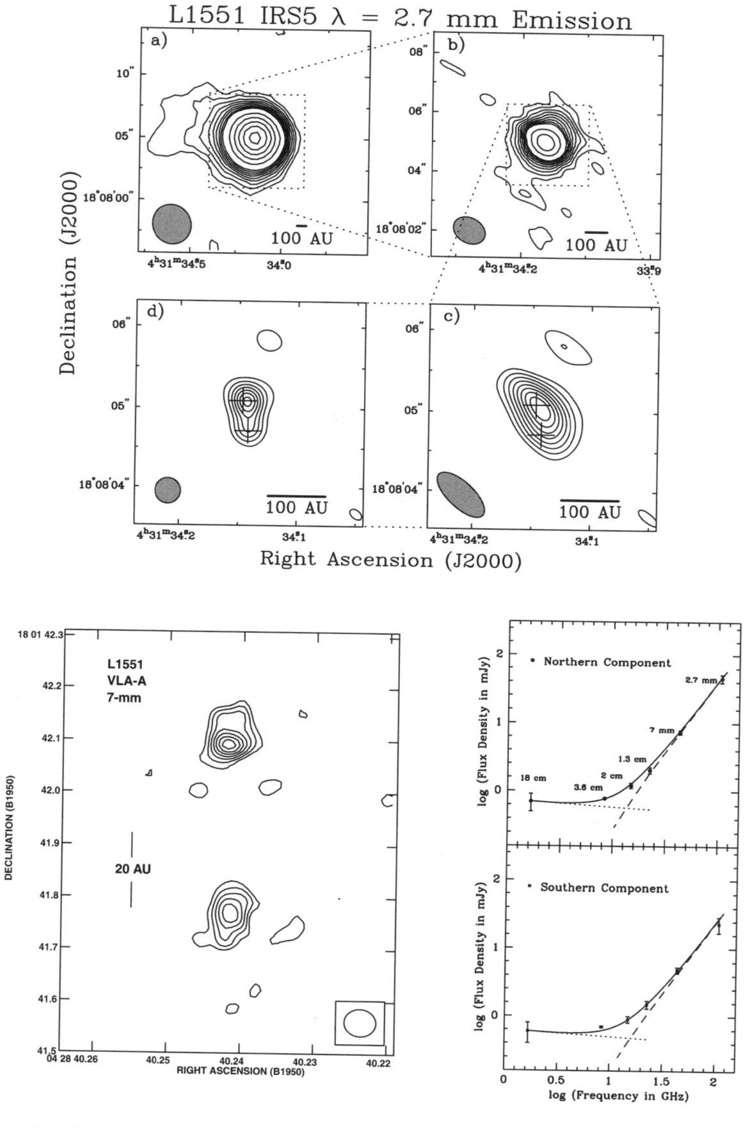

Figure 3. Dust emission from L1551 IRS5. Upper: Maps of the 2.7-mm emission from L1551 IRS5 observed with increasing angular resolution using the BIMA array (Looney et al. 1997). The images show that the L1551 system is composed of two circumstellar disks located within a flattened circumbinary structure embedded in an approximately spherical large-scale envelope. Lower: The VLA 7-mm image with 50-milliarcsecond (7 AU) resolution clearly resolves the individual dust disks surrounding the members of the protobinary system (Rodriguez et al. 1998). The spectral decomposition demonstrates that the long centimeter wavelengths are dominated by optically thin free-free emission, while emission at wavelengths shorter than about 7 mm originates in heated dust in the circumstellar disks.

envelope, a circumbinary structure, and the individual circumstellar disks surrounding each member of the binary.

D. Very Large Array

The Very Large Array (VLA) of the National Radio Astronomy Observatory has the potential to advance our understanding of the protostellar environment through observations at an angular resolution unmatched by other currently operating facilities. In the most extended antenna configuration currently available, the synthesized beam at the shortest operating wavelength of 7 mm is 40 milliarcseconds, comparable to the orbital radius of Jupiter at 140 pc distance. An upgrade to include the Pie Town VLBA antenna is under way and will soon halve the synthesized beam size. Since dust emissivity drops steeply toward long wavelengths, the thermal emission from disks is not strong. Nonetheless, the sensitivity available is sufficient to detect planetary masses of interstellar dust in reasonable integration times, and the low dust optical depth allows penetration of high-surface-density regions. However, care must be taken to separate dust and free-free emission at these wavelengths.

The high-resolution VLA data provide stringent tests of the inferences from lower resolution. The lower panels of Fig. 2 show VLA 7-mm images of HL Tau at two resolutions (Wilner et al. 1999). At this long wavelength, emission from ionized plasma is comparable to dust emission from the circumstellar disk. The right panel shows the image obtained from baselines >1000 kλ, which highlights the small-scale high-surface-brightness structure identified with the ionized protostellar wind; this component accounts for $\sim 30\%$ of the 7-mm flux. The left panel shows a lower-resolution image obtained after subtracting the wind component, which isolates emission from the dusty disk. The structure and orientation of the disk emission are similar to that observed at 2.7 mm at lower resolution by Mundy et al. (1996). Remarkably, the disk emission is almost entirely absent in the highest-resolution VLA image; all that can be seen is emission from the relatively bright wind component. The flux density and angular size of this structure vary with frequency approximately as predicted by the biconical wind theory (see the chapter by Eislöffel et al., this volume, for details).

The VLA 7-mm maps of HL Tau in Fig. 2 demonstrate the dramatic effect of surface brightness sensitivity on the appearance of a disk at high angular resolution. A disk with a steep surface density profile would have substantial emission at radii close to the central star, due to the confluence of high surface density and temperature. A disk with a flat surface density profile, on the other hand, exhibits much weaker emission at small radii. For HL Tau, the 7-mm data agree best with models having surface density falling off less steeply than r^{-1}. These models match the extended emission visible at $0.25''$ resolution and the absence of detectable disk emission at 40-milliarcsecond resolution. The long-wavelength data support a shallow falloff in disk emission from large radii well into the giant planet zone.

The VLA 7-mm map of L1551 IRS5 shown in Fig. 3 has angular resolution of 50 milliarcseconds, about an order of magnitude higher than the previous millimeter observations described in section II.C. This image clearly resolves two individual millimeter sources with projected separation 45 AU. Two arguments support the contention that the 7-mm emission traces a disk around each of the two stars. First, the 7-mm sources are elongated approximately in the north-south direction, but centimeter emission and other outflow tracers extend east-west, consistent with expectations for disks and jets. Second, the 7-mm flux densities are too large to be accounted for by ionized plasma and are more likely due to dust, as illustrated by the resolved spectra of each source, shown in Fig. 3. The average brightness temperatures are high, about 400 K, which suggests these compact disks are relatively hot and have substantial optical depth even with the low dust emissivity expected at this long wavelength. The 7-mm observations, together with the constraints of the bolometric luminosity and the radio continuum flux densities, can be modeled as viscous disks with radii $\sim$10 AU, total masses of $\sim$0.06 $M_\odot$ (north component) and $\sim$0.03 $M_\odot$ (south component). These viscous disks are hotter in the interior than at the surfaces, and the millimeter observations penetrate deeper in the disk and "see" hotter material than infrared observations that do not penetrate as deeply (D'Alessio et al. 1998). It is interesting that both disks have similar dimensions and masses, properties that may reflect the way mass is accreted from circumbinary material. The relatively small size of these disks could inhibit the formation of outer, icy planets at distances similar to those of Uranus and Neptune.

E. Multifrequency Analysis

It is possible to fit a range of different disk models to any one of the existing data sets, because the distribution of emission depends on a combination of dust emissivity, temperature, and surface density. To break this degeneracy, it is clearly advantageous to fit models simultaneously in both the spatial and the spectral domain. This approach will set tighter constraints than fitting the spectral energy distribution and spatially resolved data sets separately. The accretion disk model of section I.B is unlikely to remain a good representation of the disk at small radii (see, e.g., Bell et al. 1997), so it may not be appropriate to use infrared data to draw conclusions about the larger-scale properties of the disk that are probed by millimeter and submillimeter data. Similarly, centimeter-wave data are contaminated by free-free emission in some cases and may be included only if a reliable spectral decomposition has been performed.

The first attempt was made by Mundy et al. (1996), who made a joint analysis of the 3-mm BIMA data and the 0.87-mm CSO-JCMT data for HL Tau. The position angle was fixed, and both the temperature distribution and dust opacity index were first estimated from the spectral energy distribution. This leaves the different combinations of outer radius, inclination, and surface density distribution as free parameters. Each accretion disk model was first convolved with a 0.5″ circular Gaussian, representing the synthesized beam. An elliptical Gaussian model was then fitted to this model image and a value of χ^2 obtained by comparing these fitted parameters to those obtained from the real data. The model disks were also Fourier-transformed to compare them with the CSO-JCMT data, and the χ^2 from this fit was simply averaged with that from the 3-mm data. The outer radius (90–180 AU) and inclination (20–55°) were not tightly constrained, but it was found that shallow surface density profiles ($p < 1$) were favored.

Lay et al. (1997) performed the most extensive multifrequency analysis of HL Tau. They fitted accretion disk models directly to the measured visibilities at 220 GHz (λ = 1.4 mm OVRO data), 345 GHz (λ = 0.87 mm CSO-JCMT data) and 460 GHz (λ = 0.65 mm CSO-JCMT data), and a disk flux measured at 110 GHz (λ = 2.7 mm BIMA data). A model for each combination of the nine free parameters was Fourier-transformed and compared to the data, and a relative likelihood was calculated. This method avoids any intermediate Gaussian fitting and accounts for uncertainties in the data, such as the flux calibration at each frequency, but is computationally intensive; over 5 million different models were evaluated at each frequency. The likelihood distributions obtained for each parameter are shown in Fig. 4. Surprisingly, this analysis favors relatively high values of the power law index p, corresponding to centrally concentrated density distributions. Such models predict that the apparent disk size (e.g., obtained from a Gaussian fit) will decrease at longer wavelengths as the emission becomes optically thin. This is not consistent

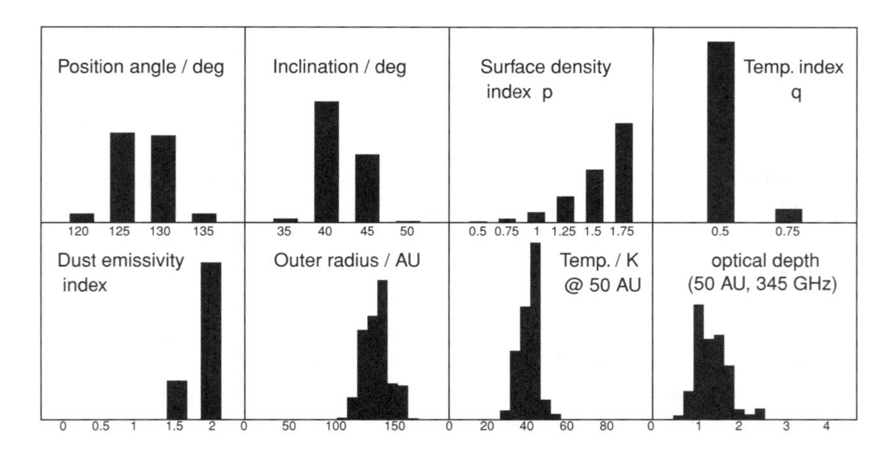

Figure 4. Histograms showing the relative likelihood of different disk model parameter values for HL Tau, based on fits to data at 110, 220, 345, and 460 GHz, from Lay et al. (1997).

with the sizes determined from the VLA and BIMA data, which appear very similar to the submillimeter values. This disagreement points to the possible inadequacy of the disk model described in section I.B. More sophisticated models that incorporate multiple dust components and allow for flaring need to be explored to see whether they can reconcile the long-wavelength data (7 and 3 mm) with the shorter wavelengths (1.1, 0.87, and 0.65 mm).

III. OTHER EXAMPLES

While HL Tau and L1551 IRS5 are the brightest millimeter disk sources in Taurus, millimeter interferometers routinely detect compact emission from other low-mass pre-main-sequence stars, and these observations are increasingly providing information at subarcsecond scales. Data from a large sample are crucial for identifying general properties, free from the idiosyncrasies of the "prototype" objects described in section II, and for addressing the complications introduced by evolution and environment. We refer the reader to the chapter by Natta et al., this volume, for results from millimeter imaging of Herbig Ae/Be stars, a class of more massive pre-main-sequence objects.

A. Low-Mass Pre-Main-Sequence Stars

The most extensive millimeter interferometer observations of pre-main-sequence disks to date was performed by Dutrey et al. (1996) at 2.7 mm using the Institut de Radio Astronomie Millimétrique (IRAM) Plateau de Bure interferometer. From a survey of 33 T Tauri stars at 2.7 mm, angular sizes typically $1.5''$ were inferred from fitting visibilities for those sources with sufficient signal to noise. In the context of power law disk models, these results suggest relatively flat surface density distributions and outer radii of 100 to 200 AU, similar to the properties derived for the HL Tau disk. Guilloteau et al. (1997) describe followup data at 1.3 mm obtained for a subset of the IRAM survey sources. Because the angular size at a particular wavelength can be produced by various combinations of radial temperature and opacity distributions, as discussed in section II.E, spatially resolved observations at multiple wavelengths are needed. The 1.3-mm data apparently show angular sizes comparable to those found at longer wavelengths, despite the large difference in dust opacity. This result is most easily understood if the brightness distributions are similar at the two wavelengths, consistent with shallow radial emissivity profiles, though this is not the only possibility.

The GM Aur system, which is located in a region of low extinction near the edge of the Taurus complex, provides an especially nice example. Dutrey et al. (1998) present high-resolution images of GM Aur from the IRAM Plateau de Bure interferometer, taking advantage of the recent baseline extensions and capabilities for operation in the 1.3-millimeter

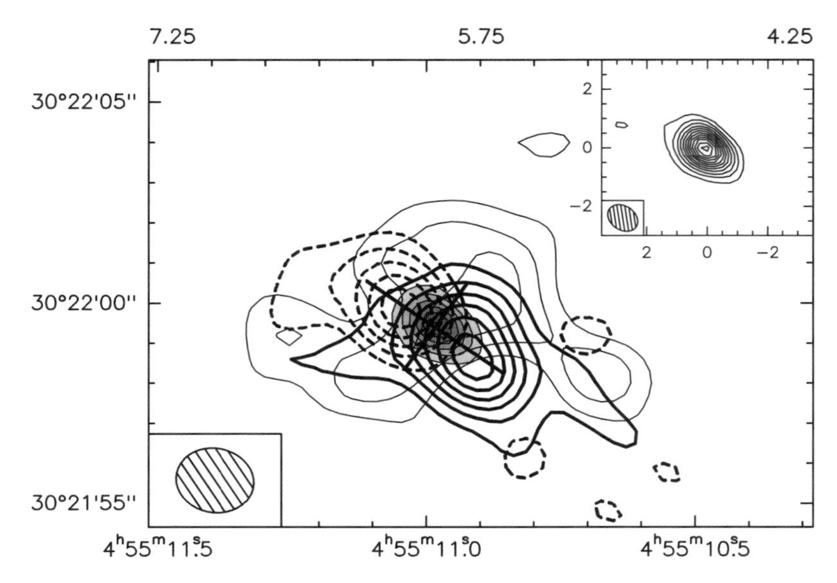

Figure 5. IRAM Plateau de Bure images of GM Aur. The 1.3-mm dust continuum emission is fully resolved and shown in grayscale; the cross marks the position, orientation, and aspect ratio of the continuum peak. Contours of ^{12}CO J = 2–1 line emission from redshifted (dashed), systemic (light), and blueshifted (dark) velocities show the rotation of the circumstellar disk along the axis of elongation. The inset (upper right) shows the 1.3-mm continuum emission in contour form together with the $0.7'' \times 0.6''$ synthesized beam. (Courtesy of A. Dutrey.)

atmospheric window to resolve the disk material fully (Fig. 5) For the continuum emission, these observations indicate angular sizes of $\sim 1.1''$ at both 2.7 mm and 1.3 mm. According to simple power law models, the observed sizes imply a minimum outer radius of ~200 AU and a surface density distribution that cannot be too steep, though a value for the power law index p as high as 1.5 is not excluded. The high-sensitivity images of the ^{12}CO J = 2–1 line build on the early results from Koerner et al. (1993) using the OVRO array to define the kinematics, inclination, and orientation of the extended cold gas in Keplerian rotation around the central star in this system.

For DO Tau, a system where the CO observations of Koerner and Sargent (1995) suggest a centrifugally supported disk within a radius of 350 AU, an early attempt using the VLA at 7 mm provided subarcsecond-resolution continuum observations that bear on the nature of dust particles (Koerner et al. 1995). Although plagued by atmospheric phase fluctuations, dust emission at 7 mm was clearly detected, unresolved in the $0.6''$ synthesized beam. These data, coupled with additional high-resolution observations at millimeter and centimeter wavelengths from OVRO and the VLA, allow limits to be set on any contribution of ionized plasma, and they densely sample the long-wavelength spectral energy distribution. The emission is well into the Rayleigh-Jeans regime, and the bulk of the dust is very likely optically thin at wavelengths longer than 3 mm, so the spectral slope directly constrains β, the power law index of the dust opacity. The result $\beta \approx 0.5$, with small fractional error, provides evidence for grains with sizes of order 1 millimeter or larger, perhaps the first stage of planetesimal formation. A goal of future millimeter observations will be to elucidate changes in dust properties in large samples of circumstellar disks, eventually taking account of spatially resolved radial variations.

B. Multiple Systems: UZ Tau, T Tau, and GG Tau

Multiplicity clearly plays an important role in shaping the circumstellar environment and may affect the development of disk structure and subsequent evolution. The UZ Tau system was among the first multiple pre-main-sequence disk systems to be spatially resolved through millimeter interferometry (Jensen et al. 1996a; Mathieu et al., this volume). This hierarchical quadruple system consists of a spectroscopic binary (UZ Tau E) and a 50-AU binary (UZ Tau W) separated by a projected distance of 500 AU. Images with the OVRO array at 1.3 mm and 3 mm show strong dust emission around the compact binary, typical of single stars, and weaker dust emission around the wider separation pair. Disk-binary interactions may explain the relative weakness of dust continuum emission in systems with separations in the range 30 to 300 AU (Jensen et al. 1996*b*). Wide binaries may be able to support individual circumstellar disks with radii of order 100 AU, whereas closer binaries may be able to maintain circumbinary structures with radii larger than about 100 AU.

The young stellar object T Tauri, the prototype of the class, is also a binary. The system consists of an optically visible star (T Tau N) and an infrared companion (T Tau S) with a projected separation of 100 AU. The combination of strong millimeter flux and disparity between components has prompted considerable observational attention, with reports of subarcsecond observations at 2.7 mm with the BIMA array (Akeson et al. 1998) and at 1.3 mm with the OVRO array and at 0.87 mm with the CSO-JCMT Interferometer (Hogerheijde et al. 1997). Surprisingly, these high-resolution observations show that the compact dust emission in the T Tau system is confined to the surroundings of the visible star, T Tau N. Multifrequency analysis using the technique described in section II.E suggests properties mostly similar to those derived for disks around single pre-main-sequence stars. However, the disk is quite small, with outer radius no more than 70 AU, possibly because the T Tau N disk is truncated by the presence of T Tau S. If so, it is possible that the variable infrared nature of T Tau S may be explained by obscuration by the tenuous outer edges of the T Tau N disk. The reason for the relative absence of dust continuum emission associated with T Tau S is not known.

Perhaps the most interesting direction for subarcsecond millimeter imaging of circumstellar disks is to uncover major departures from simple power law structure. One important prediction is for large gaps or inner holes in disks due to density waves driven by companion bodies. The hierarchical quadruple system GG Tau provides a nice illustration of this idea. Dutrey et al. (1994) discovered a huge hole, extending 160 to 180 AU in radius, in the disk surrounding the northern close binary pair in the GG Tau system (see Fig. 5 in the chapter by Mathieu et al., this volume). A beautiful correspondence exists between the central hole in the circumbinary disk and near-infrared scattered light imaged with adaptive optics (Roddier et al. 1996). The parameters of this circumbinary structure are well explained by resonances in the disk-binary interaction (see Artymowicz et al. 1991 and the chapter by Lubow and Artymowicz, this volume).

The circumbinary holes observed in systems like GG Tau are the large-scale analogs of the effects of giant planets in formation. Calculations by Takeuchi et al. (1996), among others, show that the formation of a Jupiter-mass object at a radius of 5 AU will quickly clear out the inner disk and leave a largely empty hole. The large area of the hole makes it much more readily detectable than the object responsible for clearing it. Other global distortions are also potentially observable with sufficient resolution and sensitivity. For example, one idea put forward to explain the flat infrared spectra of some T Tauri stars is an $m = 1$ spiral perturbation that allows for higher temperatures at large radii (Shu et al. 1990). Such global distortions may be detectable with existing capabilities.

IV. MASERS

A. Maser Properties

Maser emission can provide a tracer of protostellar disk kinematics with very high spatial resolution. When the physical conditions are right, a population inversion can be created between two states of an atom or molecule, and the process of stimulated emission provides coherent amplification of a light source (see Elitzur 1992 for more details). The radiation is strongly beamed, producing small spots of emission; these are the projection of long columns of gas along which the line-of-sight velocity dispersion is very small. The spots are typically milliarcseconds across and have brightness temperatures in excess of 1000 K. An interferometer can measure the relative position of the maser emission centroid at each velocity to very high accuracy; for example, the centroid for a $1''$ synthesized beam can be determined to within 10 milliarcsec for a signal-to-noise ratio of 100. Two examples are shown in Fig. 6, corresponding to SiO maser emission at 86 GHz from Orion IRc2 (Baudry et al. 1998) and hydrogen recombination line emission from MWC 349 (White et al. 1998).

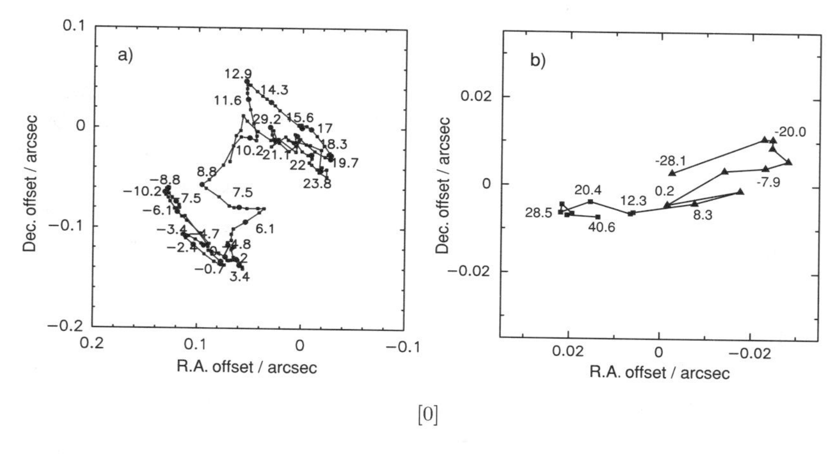

[0]

Figure 6. (a) Positions of ^{28}SiO $v = 1$, $J = 2$–1 maser centroids for Orion IRc2 (Baudry et al. 1998). The labels correspond to the local standard of rest (LSR) velocity, and adjacent velocity channels have been connected. The units are in arcseconds. (b) Positions of the H30α recombination line centroids for MWC 349 (White et al. 1999).

In addition to the kinematic information, masers can potentially provide information about physical conditions (e.g., temperature, density, and infrared flux) in the environments where they are found. This requires a detailed model of the maser mechanism, and there is still debate about specific examples (e.g., Elitzur 1992; Bujarrabal 1994; Thum et al. 1994; Baudry et al. 1998). It is clear that the subarcsecond spatial resolution provided by interferometers has greatly helped in the interpretation of masing regions, for which previously only a spectral profile had been available.

B. SiO

The Orion IRc2 region is the best studied example of high-mass star formation. A large bipolar outflow is oriented in the NW-SE direction and is believed to originate from the object observed in the radio continuum known as Source I. The ^{28}SiO $v = 1, J = 2$–1 maser emission has a flux in excess of 1000 Jy at 86 GHz and was first mapped by Plambeck et al. (1990), who demonstrated that the data are consistent with an expanding and rotating disk with a radius of 80 AU, oriented perpendicular to the outflow. Subsequent observations (Wright et al. 1995; Guilloteau et al. 1992; Baudry et al. 1998) show that the broad pattern of emission is preserved over time, although the detailed positions have changed. These changes are probably due to variations in the relative amplitude of the spots that contribute to each centroid. The current scenario is that the maser emission arises at the interface between the rotating protostellar disk and the ionized wind blowing out from Source I. Wright et al. (1995) also mapped the more weakly masing $v = 0, J = 2$–1 ground state transition with 0.5″

resolution, which appears to trace a larger (~1000 AU diameter) flared disk. However, the combination of insufficient angular resolution and non-thermal emission suggests care in interpretation. Recent observations of the SiO maser emission made with the VLBA directly resolve the spatial distribution of maser spots for the first time, and they appear to outline a conical outflow cavity (Doeleman et al. 1999; Greenhill et al. 1998). The kinematics do not appear consistent with the disk inferred from the maser centroid locations, and the physical nature of the structure centered on Source I remains unclear.

C. Hydrogen

The young, high-mass star MWC 349 is similar to the Orion maser source in many ways. It also has an ionized bipolar outflow and a characteristic double-peaked spectrum. In this case, population inversion arises from the recombination of electrons with hydrogen ions, which generates a whole series of masing transitions as the electron cascades down to the ground state. Planesas et al. (1992) were the first to map the H30α line emission at 232 GHz, and argued that it arose from the ionized surface of a disk seen close to edge-on, oriented roughly E–W. The map shown in Fig. 6b is a more recent version, clearly showing red- and blueshifted clumps separated by 60 milliarcsec (70 AU for a distance of 1.2 kpc). A similar map was made for the H26α transition at 353 GHz by the CSO-JCMT Interferometer. The data are consistent with Keplerian rotation about a central mass of 26 $M_\odot$.

D. H_2O

Unlike the extremely rare SiO and hydrogen masers described above, H_2O masers are commonly found to be associated with young stellar objects. The 22-GHz H_2O 6_{16}–5_{23} transition masers require temperatures of 200 to 1000 K and densities from 10^8 to 10^{10} cm^{-3}, well matched to the circumstellar environment. The individual maser spots may be relatively easily localized to very high accuracy using the VLA, and in some cases proper motions of the maser spots can be measured using very long-baseline techniques. However, the kinematics of the H_2O masers are often complex and difficult to interpret. This difficulty is often exacerbated by the variability of maser features, many of which persist no longer than a few weeks. In some sources, bound rotational motions have been claimed from the maser spot kinematics (e.g., NGC 2071 IRS3; Torrelles et al. 1998), while in other cases the maser spots show clear evidence for bipolar outflow motions (e.g., L1448C; Chernin 1995). From VLA observations of several H_2O maser clusters in the W75(N) and NGC 2071 regions, Torrelles et al. (1997) have suggested that H_2O masers may trace rotating disks in younger objects, while the maser spots in older objects preferentially trace outflowing jets. A conclusive test of this proposed evolutionary scheme will require observations of H_2O maser distributions around many more young stars.

V. FUTURE PROSPECTS

The new instrumentation at millimeter and submillimeter wavelengths makes possible the investigation of circumstellar structure at subarcsecond resolution in many sources. At this resolution, the most interesting structures are the disks capable of forming solar systems like our own. The last decade has witnessed the basic confirmation of the disk paradigm through direct imaging observations, and the studies reviewed in this chapter represent just the first glimpses at the complexities that become apparent at this physical scale. There is room for substantial progress in many areas.

It has not yet been possible to address evolutionary trends with the limited data currently available. The mechanism for viscous accretion remains unknown, and physical models need to be further developed that take into account the spatial constraints now becoming obtainable. The next generation of millimeter and submillimeter arrays will bring spectacular gains in sensitivity, making subarcsecond imaging routine for protostellar systems over the full planetary formation timescale, and it will be feasible to search for gaps in disks indicative of clearing by large bodies.

The understanding of dust properties has advanced considerably over the last decade, but there is a long way to go before the nature and distribution of dust within disks are fully characterized. This will only be possible through high-resolution studies spanning infrared to millimeter wavelengths, from which the spectral properties of different dust components can be separated. In addition, nonmasing spectral line emission will become detectable at subarcsecond resolution, providing a wealth of further information on kinematics, density, temperature, and chemistry within the protostellar environment.

REFERENCES

Adams, F. C., Shu, F. H., and Lada, C. J. 1988. The disks of T Tauri stars with flat infrared spectra. *Astrophys. J.* 326:865–883.

Akeson, R. L., Koerner, D. W., and Jensen, E. L. N. 1998. A circumstellar dust disk around T Tau N: Sub-arcsecond imaging at $\lambda = 3$mm. *Astrophys. J.,* in press.

Artymowicz, P., Clarke, C. J., Lubow, S. H., and Pringle, J. E. 1991. The effect of an external disk on the orbital elements of a central binary. *Astrophys. J. Lett.* 370:L35–L88.

Baudry, A., Herpin, F., and Lucas, R. 1998. ^{29}SiO ($v = 0$) and ^{28}SiO ($v = 1$) $J = 2$–1 maser emission from Orion IRc2. *Astron. Astrophys.* 335:654–660.

Beckwith, S. V. W., and Sargent, A. I. 1991. Particle emissivity in circumstellar disks. *Astrophys. J.* 381:250–258.

Beckwith, S., Sargent, A. I., Scoville, N. Z., Masson, C. R., Zuckerman, B., and Phillips, T. G. 1986. Small-scale structure of the circumstellar gas of HL Tauri and R Monocerotis. *Astrophys. J.* 309:755–761.

Beckwith, S. V. W., Sargent, A. I., Chini, R. S., and Guesten, R. 1990. A survey for circumstellar disks around young stellar objects. *Astron. J.* 99:924–945.

Bell, K. R., Cassen, P. M., Klahr, H. H., and Henning, T. 1997. The structure and appearance of protostellar accretion disks: Limits on disk flaring. *Astrophys. J.* 486:372–387.

Bieging, J. H., and Cohen, M. 1985. Multifrequency radio images of L1551 IRS 5. *Astrophys. J. Lett.* 289:L5–L8.

Bujarrabal, V. 1994. Numerical calculations of SiO maser emission. I. Intensity and variability. *Astron. Astrophys.* 285:953–970.

Burrows, C. J., Stapelfeldt, K. R., Watson, A. M., Krist, J. E., Ballester, G. E., Clarke, J. T., Crisp, D., Gallagher, J. S., III, Griffiths, R. E., Hester, J. J., Hoessel, J. G., Holtzman, J. A., Mould, J. R., Scowen, P. A., Trauger, J. T., and Westphal, J. A. 1996. Hubble Space Telescope observations of the disk and jet of HH 30. *Astrophys. J.* 473:437–451.

Butner, H. M., Natta, A., and Evans, N. J. II. 1994. "Spherical" disks: Moving toward a unified source model for L1551. *Astrophys. J.* 420:326–335.

Cabrit, S., Guilloteau, S., Andre, P., Bertout, C., Montmerle, T., and Schuster, K. 1996. Plateau de Bure observations of HL Tauri: Outflow motions in a remnant circumstellar envelope. *Astron. Astrophys.* 305:527–540.

Calvet, N., Hartmann, L., Kenyon, S. J., and Whitney, B. A. 1994. Flat spectrum T Tauri stars: The case for infall. *Astrophys. J.* 434:330–340.

Chernin, L. M. 1995. Water masers in the L1448C outflow. *Astrophys. J. Lett.* 440:L97–L99.

Chiang, E. I., and Goldreich, P. 1997. Spectral energy distributions of T Tauri stars with passive circumstellar disks. *Astrophys. J.* 490:368–376.

D'Alessio, P., Canto, J., Calvet, N., and Lizano, S. 1998. Accretion disks around young objects. I. The detailed vertical structure. *Astrophys. J.* 500:411–427.

Doeleman, S. S., Lonsdale, C. J., and Pelkey, S. 1999. A molecular outflow traced by SiO masers in Orion KL. *Astrophys. J. Lett.* 510:L55–L58.

Draine, B., and Lee, H. M. 1984. Optical properties of interstellar graphite and silicate grains. *Astrophys. J.* 285:89–108.

Dutrey, A., Guilloteau, S., and Simon, M. 1994. Images of the GG Tauri rotating ring. *Astron. Astrophys.* 286:149–159.

Dutrey, A., Guilloteau, S., Duvert, G., Prato, L., Simon, M., Schuster, K., and Menard, F. 1996. Dust and gas distribution around T Tauri stars in Taurus-Auriga. I. Interferometric 2.7 mm continuum and ^{13}CO $J = 1$–0 observations. *Astron. Astrophys.* 309:493–504.

Dutrey, A., Guilloteau, S., Prato, L., Simon, M., Duvert, G., Schuster, K., and Menard, F. 1998. CO study of the GM Aurigae Keplerian disk. *Astron. Astrophys.* 338:L63–L66.

Elias, J. H. 1978. A study of the Taurus dark cloud complex. *Astrophys. J.* 224:857–872.

Elitzur, M. 1992. *Astronomical Masers* (Dordrecht: Kluwer).

Greenhill, L. J., Gwinn, C. R., Schwartz, C., Moran, J. M., and Diamond, P. J. 1998. Coexisting conical bipolar and equatorial outflows from a high mass protostar. *Nature* 396:650–653.

Guilloteau, S., Delannoy, J., Downes, D., Greve, A., Guelin, M., Morris, D., Radford, S. J. E., Wink, J., Cernicharo, J., Forveille, T., Garcia-Burillo, S., Neri, R., Blondel, J., Perrigourad, A., Plathner, D., and Torres, M. 1992. The IRAM interferometer on Plateau de Bure. *Astron. Astrophys.* 262:624–633.

Guilloteau, S., Dutrey, A., and Gueth, F. 1997. Disks and outflows as seen from the IRAM Interferometer. In *Herbig-Haro Flows and the Birth of Low-Mass Stars*, ed. B. Reipurth and C. Bertout (Dordrecht: Kluwer), pp. 365–380.

Hayashi, M., Ohashi, N., and Miyama, S. M. 1993. A dynamically accreting gas disk around HL Tauri. *Astrophys. J. Lett.* 418:L71–L74.

Hogerheijde, M. R., van Langevelde, H. J., Mundy, L. G., Blake, G. A., and van Dishoeck, E. F. 1997. Subarcsecond imaging at 267 GHz of a young binary system: Detection of a dust disk of radius less than 70 AU around T Tauri N. *Astrophys. J. Lett.* 490:L99–L102.

Jensen, E. L. N., Mathieu, R. D., and Fuller, G. A. 1996*a*. The connection between submillimeter continuum flux and binary separation in young binaries: Evidence of interaction between stars and disks. *Astrophys. J.* 458:312–326.

Jensen, E. L. N., Koerner, D. W., and Mathieu, R. D. 1996*b*. High-resolution imaging of circumstellar gas and dust in UZ Tauri: Comparing binary and single-star disk properties. *Astron. J.* 111:2431–2438.

Keene, J. and Masson, C. R. 1990. Detection of a 45 AU radius source around L1551-IRS5: A possible accretion disk. *Astrophys. J.* 355:635–644.

Kenyon, S. J., and Hartmann, L. 1987. Spectral energy distributions of T Tauri stars: Disk flaring and limits on accretion. *Astrophys. J.* 323:714–733.

Koerner, D. W., and Sargent, A. I. 1995. Imaging the small-scale circumstellar gas around T Tauri stars. *Astron. J.* 109:2138–2145.

Koerner, D. W., Sargent, A. I., and Beckwith, S. V. W. 1993. A rotating gaseous disk around the T Tauri star GM Aurigae. *Icarus* 106:2–10.

Koerner, D. W., Chandler, C. J., and Sargent, A. I. 1995. Aperture synthesis imaging of the circumstellar dust disk around DO Tauri. *Astrophys. J. Lett.* 452:L69–L72.

Lay, O. P., Carlstrom, J. E., Hills, R. E., and Phillips, T. G. 1994. Protostellar accretion disks resolved with the JCMT-CSO interferometer. *Astrophys. J. Lett.* 434:L75–L78.

Lay, O. P., Carlstrom, J. E., and Hills, R. E. 1997. Constraints on the HL Tauri protostellar disk from millimeter- and submillimeter-wave interferometry. *Astrophys. J.* 489:917–927.

Lin, D. N. C., and Papaloizou, J. 1985. On the dynamical origin of the solar system. In *Protostars and Planets II*, ed. D. C. Black and M. S. Matthews (Tucson: University of Arizona Press), pp. 981–1072.

Looney, L. W., Mundy, L. G., and Welch, W. J. 1997. High-resolution $\lambda = 2.7$ millimeter observations of L1551 IRS 5. *Astrophys. J. Lett.* 484:L157–L160.

Looney, L., Mundy, L., and Welch, W. 1999. Unveiling the circumstellar envelope and disk: A sub-arcsecond survey of circumstellar structures. *Astrophys. J.*, in press.

McCaughrean, M. J. and O'Dell, C. R. 1996. Direct imaging of circumstellar disks in the Orion Nebula. *Astron. J.* 111:1977–1986.

Mundt, R., Brugel, E. W., and Buhrke, T. 1987. Jets from young stars: CCD imaging, long-slit spectroscopy, and interpretation of existing data. *Astrophys. J.* 319:275–303.

Mundy, L. G., Looney, L. W., Erickson, W. Grossman, A., Welch, W. J., Forster, J. R., Wright, M. C. H., Plambeck, R. L., Lugten, J., and Thornton, D. D. 1996. Imaging the HL Tauri disk at $\lambda = 2.7$ millimeters with the BIMA array. *Astrophys. J. Lett.* 464:L169–L173.

Natta, A. 1993. The temperature profile of T Tauri disks. *Astrophys. J.* 412:761–770.

Ohashi, N., Kawabe, R., Ishiguro, M., and Hayashi, M. 1991. Observations of 11 protostars in Taurus with the Nobeyama millimeter array: Growth of circumstellar disks. *Astron. J.* 102:2054–2065.

Plambeck, R. L., Wright, M. C. H., and Carlstrom, J. E. 1990. Velocity structure of the Orion-IRc2 SiO maser: Evidence for an 80 AU diameter circumstellar disk. *Astrophys. J. Lett.* 348:L65–L68.

Planesas, P., Martin-Pintado, J., and Serbayn, E. 1992. Positions of the radio recombination line masers in MWC 349. *Astrophys. J. Lett.* 386:L23–L26.

Pollack, J. B., Hollenbach, D., Beckwith, S., Simonelli, D. P., Roush, T., and Welsey, F. Composition and radiative properties of grains in molecular clouds and accretion disks. *Astrophys. J.* 421:615–639.

Roddier, C., Roddier, F., Northcott, M. J., Graves, J. E., and Jim, K. 1996. Adaptive optics imaging of GG Tauri: Optical detection of the circumbinary ring. *Astrophys. J.* 463:326–335.

Rodriguez, L. F., Canto, J., Torrelles, J. M., and Ho, P. T. P. 1986. The double radio source associated with L1551 IRS 5: Binary system or ionized circumstellar torus? *Astrophys. J. Lett.* 301:L25–L28.

Rodriguez, L. F., D'Alessio, P., Wilner, D. J., Ho, P. T. P., Torrelles, J. M., Curiel, S., Gomez, Y., Lizano, S., Pedlar, A., Canto, J., and Raga, A. C. 1998. Compact protoplanetary disks in a binary system in L1551. *Nature,* in press.

Sargent, A. I., and Beckwith, S. 1987. Kinematics of the circumstellar gas of HL Tauri and R Monocerotis. *Astrophys. J.* 323:294–305.

Sargent, A. I., and Beckwith, S. 1991. The molecular structure around HL Tauri. *Astrophys. J. Lett.* 382:L31–L35.

Shu, F. H., Tremaine, S., Adams, F. C., and Ruden, S. P. 1990. Sling amplification and eccentric gravitational instabilities in gaseous disks. *Astrophys. J.* 358:495–514.

Snell, R. L., Loren, R. B., and Plambeck, R. L. 1980. Observations of CO in L1551: Evidence for stellar wind driven shocks. *Astrophys. J. Lett.* 239:L17–L22.

Stapelfeldt, K. R., Burrows, C. J., Krist, J. E., Trauger, J. T., Hester, J. J., Holtzmann, J. A., Ballester, G. E., Casertano, S., Clarke, J. T., Crisp, D., Evans, R. W., Gallagher, J. S., III, Griffiths, R. E., Hoessel, J. G., Mould, J. R., Scowen, P. A., Watson, A. M., and Westphal, J. A. 1995*a*. WFPC2 imaging of the circumstellar nebulosity of HL Tauri. *Astrophys. J.* 449:888–893.

Stapelfeldt, K. R., Burrows, C. J., Koerner, D., Krist, J., Watson, A. M., Trauger, J. T., and the WFPC2 Team Investigation Definition Team. 1995*b*. WFPC2 imaging of GM Aurigae: A circumstellar disk seen in scattered light. American Astronomical Society Meeting, 187, #113.04.

Stapelfeldt, K. R., Krist, J. E., Menard, F., Bouvier, J., Padgett, D. L., and Burrows, C. J. 1998. An edge-on circumstellar disk in the young binary system HK Tauri. *Astrophys. J. Lett.* 502:L65–L69.

Takeuchi, T., Miyama, S. M., and Lin, D. N. C. 1996. Gap formation in protoplanetary disks. *Astrophys. J.* 460:832–847.

Thum, C., Matthews, H. E., Harris, A. I., Tacconi, L. J., Schuster, K. F., and Martín-Pintado, J. 1994. Detection of H21α maser emission at 662 GHz in MWC349. *Astron. Astrophys.* 288:L25–L28.

Torrelles, J. M., Gomez, J. F., Rodriguez, L. F., Ho, P. T. P., Curiel, S., and Vazquez, R. 1997. A Radio jet–H_2O maser system in W75 N(B) at a 200 AU scale: Exploring the evolutionary stages of young stellar objects. *Astrophys. J.* 489:744–752.

Torrelles, J. M., Gomez, J. F., Rodriguez, L. F., Curiel, S., Anglada, G., and Ho, P. T. P. 1998. Radio continuum–H_2O maser systems in NGC 2071: H_2O masers tracing a jet (IRS 1) and a rotating protoplanetary disk of radius 20 AU (IRS 3). *Astrophys. J.,* in press.

White, S. M., Welch, W. J., Vogel, S. N., and Lim, J. 1999, in preparation.

Wilner, D. J., Ho, P. T. P., and Rodriguez, L. F. 1999. 7 millimeter imaging of HL Tauri at 5 AU resolution. *Astrophys. J.*, in preparation.

Woody, D. P., Scott, S. L., Scoville, N. Z., Mundy, L. G., Sargent, A. I., Padin, S., Tinney, C. G., and Wilson, C. D. 1989. Interferometric observations of 1.4 mm continuum sources. *Astrophys. J. Lett.* 337:L41–L44.

Wright, M. C. H., Plambeck, R. L., Mundy, L. G., and Looney, L. W. 1995. SiO emission from a 1000 AU disk in Orion KL. *Astrophys. J. Lett.* 455:L185–L188.

DUST PROPERTIES AND ASSEMBLY OF LARGE PARTICLES IN PROTOPLANETARY DISKS

STEVEN V. W. BECKWITH
Max-Planck-Institut für Astronomie and Space Telescope Science Institute

THOMAS HENNING
Astrophysical Institute and University Observatory at Jena

and

YOSHITSUGU NAKAGAWA
Kobe University

Recent research on the buildup of rocks from small dust grains has reaffirmed that grain growth in protoplanetary disks should occur quickly. Calculation of growth rates have been made for a variety of growth processes and generally predict high probabilities of sticking in low-velocity collisions, which may be brought about in a number of ways in protoplanetary disks. Laboratory experiments have measured sticking coefficients for some materials, largely confirming the calculations. Although the detailed velocity fields of disks are not well understood, many of the important processes leading to particle collisions and grain growth have been studied theoretically and demonstrate likely paths by which dust is assembled into planets. Calculations of the radiative properties of particles with various size distributions show that large particles should produce observable changes in the spectral energy distributions of disks. Changes of the sort predicted are, in fact, observed, but their interpretation is ambiguous; there are other ways to produce the observed changes that do not require grain growth, so the evidence is currently inconclusive. The major uncertainties can be overcome with the next generation of millimeter-wave interferometers, and it seems likely that a firm case for grain growth could be established within a decade.

I. INTRODUCTION

Planets are believed to have been built from tiny interstellar dust grains by the steady accumulation of the smaller particles, at first through sticking, but later aided by gravity between larger planetesimals and finally by the accretion of gas onto the rocky cores. Such growth should occur naturally in the circumstellar disks created when young stars are born. The disks provide protected "wombs" for the development of the large bodies during the course of a few million years or more. Although disks are not the only places where planets might have grown, they are theoretically well suited for planet formation and observed to be ubiquitous, so they are the clear favorite for planetary birth sites at the time of this review. A demonstration of the growth of dust into large bodies is the one open issue that remains to demonstrate that circumstellar disks, hundreds of examples of which are routinely studied, do spawn planetary systems. Such a demonstration would also imply that planetary systems are common features of nearby stars, just as disks are commonly found around young stellar objects.

In the earliest stages, particles are built up by coagulation and sticking as they bump into one another in the disks. In about 10^4 yr, particles can grow from micron size to meter size through a variety of collisional processes: Brownian motion, settling, turbulence, and radial migration. The dominant mode depends on physical conditions within the disk. After another 10^4 years or so, the small bodies can attain sizes of order kilometers. Beyond 1 km, gravitational attraction causes these planetesimals to grow by pairwise collisions and later by runaway growth until some planets will be present after 1 Myr (Weidenschilling 1988; Weidenschilling and Cuzzi 1993; Lissauer 1993). Accretion of gas should create gaps in the disks, which may present a natural limit to the growth of planets over very long times. Figure 1 shows the three stages schematically. The timescales are uncertain but still well enough constrained to be shorter than the observed lifetimes of disks around young stars (e.g., Strom et al. 1993). If the theory is even approximately correct, there should be large bodies within currently observed circumstellar disks.

The buildup of a small, rocky planet is longer than the lifetime of any observer by several orders of magnitude, so it is not possible to watch this evolution. Instead, we must be content to compare the particle properties of disks of different ages and infer the growth of large bodies from trends in the size distribution with age. Even the indirect detection of planetary-mass bodies in circumstellar disks is difficult when the disk mass is greater

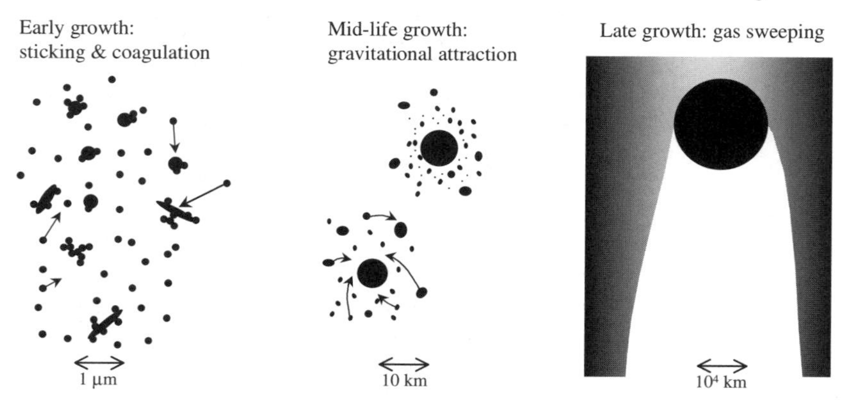

Figure 1. Sequence of events leading to planet formation: grain growth through particle-particle interactions, growth of planetesimals by two-body interactions aided by gravity, and the accretion of gas via gravitational attraction.

than the planetary mass, but there is hope that the indirect effects of larger bodies will be observable in a decade or sooner.

The theory of grain coagulation delineates the timescales over which to study disks and provides some limits on what is plausibly observed. The theory is checked by laboratory experiments for small-scale growth, where major bottlenecks might occur. It is possible to create tiny grains similar to interstellar dust in the laboratory under physical regimes like those thought to occur in disks. The growth rates of these particles can be checked against the theory to ensure that the physics of grain growth is well understood. Experimentally measured growth rates for small particles largely verify the calculations and lead us to believe that the growth of the smallest particles is reasonably well understood. It is difficult to verify the growth of larger bodies. It will, therefore, be important to detect some large bodies near other stars to check the calculations of early planet evolution.

There are enough known disks around nearby young stars to provide good samples for detection of grain growth. There are now methods for discerning the growth of dust grains to pebbles and rocks (Miyake and Nakagawa 1993) via the submillimeter opacities. These methods have had limited application to date but give interesting hints of grain growth.

This review discusses several aspects of this problem, including the physics of the growth process, the observable effects on radiation from collections of the particles, and the extant observations designed to see evidence of grain growth. There has been progress on all these fronts since the last *Protostars and Planets* meeting in 1990, spurred by an increasing interest in extrasolar planet formation that is driven in turn by detections of extrasolar planets. Most of this progress has been in laboratory work and new calculations of the properties of small particles as they pertain to coagulation and growth. Already at the time of *Protostars and Planets III* there were tantalizing hints that particle growth had been detected in circumstellar disks; a few new observations attempt to clarify these hints. We will concentrate on progress made since that conference.

II. GRAIN EVOLUTION IN DISKS

A. Structure of Protoplanetary Disks

The idealized infinitely thin, continuous disk assumed in early theories (e.g., Lynden-Bell and Pringle 1974; Pringle 1981; Adams and Shu 1986) has given way to a disk with vertical and radial structure and with gaps or holes needed to account for various observations (Bodenheimer 1995; Papaloizou and Lin 1995). Stellar radiation provides at least minimal heating, which is often enhanced by accretion energy as material from the disk falls onto the star. Disks heated mainly by accretion are referred to as *active,* whereas those heated entirely by the stars are called *passive.* The spectral energy distributions (SEDs) of most simple disks can be fitted reasonably well, assuming they are flared (Kenyon and Hartmann 1987), are centrally heated from radiation near the star (Chiang and Goldreich 1997), and have radially decreasing temperatures (Adams et al. 1988). There are striking exceptions to this picture: HL Tau, for example, and almost all of the "flat-spectrum" sources. These presumably have a variety of components (such as dense, disk-like regions; outflow cavities; and the surrounding molecular cloud cores) that contribute to the SEDs and complicate the analysis (Calvet et al. 1994).

Disks usually have outer radii between a few AU and a few hundred AU (Appenzeller et al. 1984; Edwards et al. 1987; Lay et al. 1997), often with sharp outer boundaries (O'Dell and Wenn 1994; McCaughrean and O'Dell 1996). Young disks are thought to be optically thick at all wavelengths shortward of 100 μm. Typical disk masses are between 0.001 and 0.1 $M_\odot$, although there is some uncertainty about this range (Adams et al. 1990; Beckwith et al. 1990; Henning et al. 1993; André and Montmerle 1994; Osterloh and Beckwith 1995). These properties accord well with those assumed for the disk in the early solar system from which our planets were born (Safronov 1969; Hayashi 1981; Cameron 1988).

Disk structure is important to the study of particle growth in two ways. First, the presence of convection, turbulence eddies, and other velocity gradients can strongly affect the growth of particles, because the growth rates depend on the relative particle velocities. Second, observations of disks must either take disk structure into account for their interpretation or else be designed to be relatively unaffected by the structure. These effects are discussed in the sections to follow. There is enough freedom in models of disks to produce many kinds of local structure, and important effects such as turbulence eddies, sedimentation, and radial migration of matter are currently beyond observation. They are also a matter of debate among theorists. While several instability mechanisms might in principle maintain nebular turbulence, the presence of turbulence and the dominance of any energy source has yet to be demonstrated definitively (Cabot et al. 1987; Balbus and Hawley 1991*a*,*b*; Dubrulle 1992, 1993; Tscharnuter and Boss 1993; Nakamoto and Nakagawa 1994, 1995; Kley et al. 1993; Stone and Balbus 1996; Klahr et al. 1998). As a result, there is considerable uncertainty about the actual evolution of particles in disk theories, and a wide range of physical conditions needs to be considered in theories and laboratory studies of grain growth relevant to planet formation.

B. Physical Processes Leading to Grain Growth

In the outer regions of an accretion disk, most of the interstellar dust remains unaltered during passage through the weak shock front when the disk is first created. Only volatile ice mantles around the refractory grain cores partly sublimate, but water ice efficiently forms when the water vapor cools behind the shock (Lunine et al. 1991; Neufeld and Hollenbach 1994). Pollack et al. (1994) summarize the expected grain components, their optical properties, and sublimation temperatures immediately after disk formation. The main species are (1) olivines ($Mg_xFe_{2-x}SiO_4$ with

$x = 1.4$), (2) orthopyroxenes ($Mg_xFe_{1-x}SiO_3$ with $x = 0.7$), (3) quartz (SiO_2), (4) metallic iron (Fe), (5) troilite (FeS), (6) refractory and volatile organics, and (7) water ice. Henning and Stognienko (1996) update these calculations by taking into account more recent determinations of optical constants and a possible fluffy structure of the grains. They also found that the grain opacities depend on the fraction of iron that is in iron grains or silicates.

If the grains are transported inward during the viscous evolution, the higher temperatures will anneal and eventually destroy them (see, e.g., Finocchi et al. 1997; Gail 1998). Gail (1998) assumes chemical equilibrium and concludes that crystalline silicates, iron, and aluminum compounds are the most important grain components in the warmer parts of the disk. Such equilibrium condensation calculations have been extensively performed (see, e.g., Grossman 1972, Cameron and Fegley 1982, Yoneda and Grossman 1995). The results are uncertain, however, because a good understanding of the chemical processes during grain condensation, growth, and destruction requires a kinetic approach that cannot yet be applied, owing to unknown reaction rates for oxygen-rich chemistry (see also Prinn 1993).

The condensation of ice onto refractory cores in the outer regions of the disk is important for the growth of submicron-sized grains but is irrelevant to growth from micron-sized grains to meter- and kilometer-sized planetesimals (e.g., Preibisch et al. 1993). The surface structure of the dust grains may be considerably modified by repeated condensation and sublimation caused by transport processes or heating events. Such "flares" were also discussed as a possible mechanism for the production of chondrules found in meteorites (see Hewins et al. 1996).

For much of the last 30 years, planetesimal formation was thought to be the result of a gravitational instability in a dense dust layer, leading to fragmentation into kilometer-sized bodies (Safronov 1969; Goldreich and Ward 1973). Weidenschilling (1980) was the first to note the difficulty with the particle layer instability. Cuzzi et al. (1993) and Weidenschilling (1995) demonstrated that such an instability is unlikely to occur. Particles must grow to at least meter size before they can decouple from shear-induced turbulence and slow to the low velocities required for the instability to occur. The radial velocity dispersion induced by drag will further delay the onset of instability until the mean size is in the range of 10–100 m. Thus, the formation of planetesimals has to be explained by the collisional aggregation of particles.

It is difficult to calculate this process from first principles, because there are complex feedback mechanisms in the two-fluid system whose relative importance is not well known: aerodynamical drag, thermal structure coupled to changing dust opacities, charge state of the medium, and coupling to magnetic fields, for example. It is imperative to understand particle growth from collisions between particles, however, if we want to estimate the likelihood of creating planetesimals and planets.

Particles collide because they are moving relative to one another. The sources of relative velocities in laminar nebulae are Brownian motion, $v_{\rm th}$, sedimentation, v_s, and radial drift, v_r. In turbulent systems, turbulence-driven motion, $v_{\rm tur}$, can additionally be of importance for grains/bodies of certain sizes. The equations for the relative velocities are (Weidenschilling 1977; Völk et al. 1980; Mizuno et al. 1988)

$$\Delta v_{\rm th} = \sqrt{8kT/\pi\mu} \quad \text{with} \quad \mu = (m_1 m_2)/(m_1 + m_2) \tag{1}$$

$$\Delta v_s = \Omega^2 z |\tau_{f1} - \tau_{f2}| \tag{2}$$

$$\Delta v_r = 2\Omega \Delta V |\tau_{f1} - \tau_{f2}| \tag{3}$$

$$\Delta v_{\rm tur} \simeq u_s/t_s |\tau_{f1} - \tau_{f2}| \quad \text{for} \quad \tau_{f1}, \tau_{f2} < t_s \tag{4}$$

$$\langle \Delta v_{\rm tur} \rangle = 1.33 u_l \sqrt{\tau_f / t_l} \quad \text{for} \quad t_l > \tau_{f1}, \tau_{f2} > t_s \tag{5}$$

where Ω is the Keplerian angular frequency; T the temperature; m_1 and m_2 the particle masses; τ_{f1} and τ_{f2} the friction times (see section III.A for a more detailed discussion), $\tau_f = \max(\tau_{f1}, \tau_{f2})$; z the vertical distance from the midplane; ΔV the velocity difference between Keplerian velocity and rotational gas velocity; u_s and t_s the velocity and turnover time of the smallest turbulent eddy; u_l and t_l the velocity and turnover time of the largest turbulent eddy. In the case of turbulence-driven motion, we have distinguished between two cases of the coupling of the grains to the gas ($\langle \Delta v_{\rm tur} \rangle$ is the rms value of the relative velocities).

A comparison of equations (1) through (5) shows that the particles with equal friction times must couple to the turbulence [second case, equation (5)] to obtain substantial relative velocities; otherwise, only the Brownian motion is important. The collisions caused by Brownian motion are relevant only for small (about micron-sized) grains because of the inverse mass dependence. On the other hand, an efficient coupling to the turbulence in a disk with $\alpha = 0.01$ [where α is the ratio of viscosity to the product of sound speed and scale height (the standard α-disk model; see the chapter by Stone et al., this volume)], is possible for millimeter-sized grains to meter-sized bodies, calculated for the disk model of Bell et al. (1997) at 10 AU radius and considering the friction time for compact particles. Grains with very small friction times essentially comove with the gas. Friction times larger than the turnover time of the smallest eddies are needed for the start of an efficient growth by turbulence. In a laminar nebula, a dispersion in friction times is always required for an efficient growth.

C. Homogeneous Growth Regime

Most models for collisional grain growth in protoplanetary accretion disks assume a locally homogeneous distribution of particles. There are classically two types of models: radial one-dimensional (1D) models that average the density distribution vertically through the disk (see, e.g., Schmitt et al. 1997), and vertical 1D models for the coagulation of particles by settling to the midplane at a given radius (see, e.g., Weidenschilling and Cuzzi 1993).

Several authors have studied the 1D radial case, each incorporating more detail into the calculations (see, e.g., Morfill 1988; Mizuno et al. 1988; Mizuno 1989). Ruden and Pollack (1991) treated the dynamical evolution of a protoplanetary disk under the hypothesis that the only source for the turbulent viscosity is thermal convection. They discussed grain settling and coagulation in the outer solar system when turbulence ceases. Schmitt et al. (1997) coupled the disk and dust evolution and distinguished three phases during the first 100 years of disk evolution: an early phase, in which the smallest particles disappeared quickly due to Brownian motion-driven coagulation and produced a relatively narrow mass distribution; an intermediate phase of self-similar growth due to turbulence, where the size distribution can be described by scaling laws; and a late phase, where the most massive particles decouple from the gas and drift motions become an important source of relative velocities. Stepinski and Valageas (1996) consider the global evolution of single-sized noncoagulating particles in radial models with α between 0.01 and 0.001. They conclude that particles larger than 0.1 cm are not really entrained in the gas. They also studied the formation and radial distribution of kilometer-sized icy planetesimals, assuming that the size distribution is always narrowly peaked around a mean value for a given radial location and time (Stepinski and Valageas 1997).

A shortcoming of radial models is the inability to treat the formation of a dust subdisk by settling. Weidenschilling (1980), Nakagawa et al. (1981, 1986), and Weidenschilling and Cuzzi (1993) developed numerical 1D models for settling with coagulation. The main steps of the evolution are as follows: (1) Micron-sized grains uniformly suspended in the gas grow by Brownian motion. (2) Grain aggregates begin to settle and sweep up smaller ones in a kind of runaway growth. (3) The aggregates grow to centimeter size and create a dense dust layer (Weidenschilling 1997). Weidenschilling (1997) treated the growth from microscopic grains to cometesimals at 30 AU in detail. The dependence of drag-induced velocities on particle size produced a stage in which most of the mass was concentrated in a narrow size range between 10 and 100 m. Dubrulle et al. (1995) found that turbulence does not influence the timescale of sedimentation but merely determines the equilibrium scale height of the dust subdisk. The smallest grains that can experience some kind of sedimentation in the presence of turbulence have sizes of the order of 150α cm, resulting in a size limit of about 1 cm for $\alpha = 0.01$. Sekiya (1998) showed that for millimeter-sized aggregates even very weak turbulence stirs up the dust aggregates. This calculation considered only locally produced turbulence, corresponding to the calculations by Cuzzi et al. (1993) and Champney et al. (1995), not an α-disk.

Radial transport becomes important over the timescale of settling. The results of the 1D investigations show that consistent 2D/3D models or at least 1+1D models are necessary for an adequate description of the dust dynamics.

D. Inhomogeneous Growth Regime

In the classical models, turbulence is treated only statistically, as a source of random velocities important for diffusion and the calculation of turbulence-induced relative velocities. These models assume that particles remain uniformly distributed in a turbulent velocity field. This prediction is valid only if the typical length scales are larger than the eddy scales.

Squires and Eaton (1991) were the first to show that turbulence can lead to particle concentration. Cuzzi et al. (1996) calculated this particle concentration for a protoplanetary disk with weak Kolmogorov-type turbulence. Concentrations of 10^5 to 10^7 occur for particles with radii between 0.1 and 1 mm in convergence zones between eddies. The particle-trapping mechanism may have dramatic consequences for the formation of chondrules, because of its size-selective nature (see Cuzzi et al. 1996).

Klahr and Henning (1997) showed that in certain vortical flows, particles are trapped and concentrated in the gas flow. Slowly rotating eddies and a gradient of the vertical component of the gravitational force produce this trapping. The gradient causes particles in the upper half of an eddy to fall faster than in the lower half. If this concentration is stronger than the dispersion of grains due to centrifugal forces, the particles are concentrated in the interior of the eddies. The strongest effect was observed for millimeter-sized particles, which can be concentrated by a factor of 100 within 100 years. Three-dimensional calculations by Klahr et al. (1998) indicate that the particle concentration mechanism is not just the result of 2D approximations. Figure 2 shows the results of a calculation in which an initially homogeneous distribution becomes inhomogeneous after 160

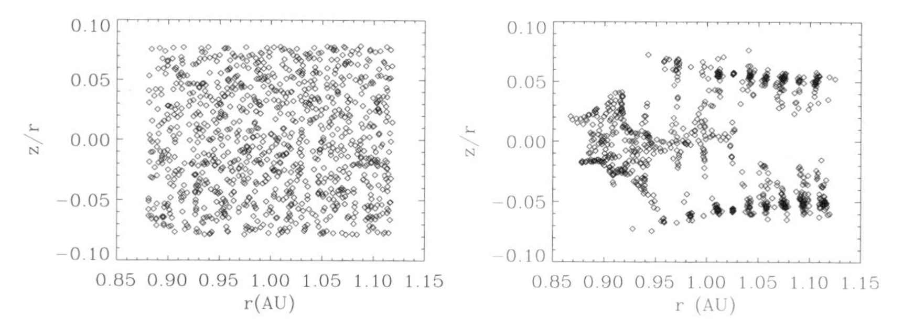

Figure 2. Left: An initially homogeneous distribution of grains. Right: Distribution that has become quite inhomogeneous after 160 yr because of trapping in eddies (Klahr et al. 1998).

years. However, we should note that the possibility to concentrate or to disperse small dust particles depends on the structure of the flow pattern (3D disk turbulence vs. slow and long-lived vortical circulation) and the stability of the vortices (Adams and Watkins 1995; Klahr and Henning 1997; Klahr et al. 1998; Brandenburg and Hodgson 1998).

Tanga et al. (1996; see also the earlier related study by Barge and Sommeria 1995) considered one more possibility of particle concentration via vorticity. They introduced an *ad hoc* system of regular vortices driven by differential rotation of the disk with rotational axes parallel to its angular momentum. In this case, the concentration mechanism is based on Coriolis forces, and much larger particles would be concentrated compared with the cases discussed before.

III. HOW WELL IS PARTICLE GROWTH UNDERSTOOD?

A. Aerodynamic Behavior of Particles in a Dilute Gas

The interaction of dust particles and small bodies with sizes below 1 km is determined by their aerodynamic coupling to the dilute gas of the protoplanetary accretion disk and not by gravitational forces. This coupling can be characterized by a friction (or response) time τ_f, which, for dust particles and aggregates, is much smaller than the orbital period; that is, the particles are strongly coupled to the gas. For the conditions in protoplanetary disks, the coupling of submicron- to centimeter-sized particles to the gas is dominated by collisions of grains with individual molecules. In this regime, the molecular mean free path, $l_{\rm mfp}$, is larger than the characteristic particle size, a. The friction time of a spherical dust particle of mass m and geometrical cross section σ_g is given by

$$\tau_f = \frac{m}{\frac{4}{3}(\delta P)\sigma_g \rho_g v_g} \tag{6}$$

Here δP is the momentum transfer efficiency factor, and the quantities ρ_g and v_g denote the gas density and the mean thermal velocity of the molecules, respectively. The material parameter δP is always close to 1. Experiments by Hutchins et al. (1995) and Blum et al. (1996) verified the validity of (6) and showed that a single value $\delta P = 1.11 \pm 0.17$ describes both spherical SiO_2 grains and small aggregates. These experiments confirmed the results of numerical Monte Carlo simulations on the aerodynamic properties of fractal aggregates (Meakin et al. 1989).

The strength of gas-grain coupling depends on the ratio of mass to cross section, so τ_f depends on whether the particles are compact (spheroids, say) or fluffy, open structures (Ossenkopf 1993). If the aggregates are fractals, we can introduce two fractal dimensions D_m and D_σ

$$m \propto R^{D_m} \quad \text{and} \quad \sigma_g \propto R^{D_\sigma} \tag{7}$$

and show that $\tau_f \propto R^{(D_m - D_\sigma)}$. In three-dimensional systems, $1 \leq D_\sigma \leq D_m \leq 3$. Note that $D_m = 3$ and $D_\sigma = 2$ for compact particles, so that $\tau_f \propto R$.

B. Aggregation Process

Ballistic cluster-cluster aggregation (BCCA) is the process in which clusters that collide stick together without restructuring and with sticking probability 1. This process leads to the production of open, fluffy aggregates. Numerical simulations of BCCA yield D_m values between 1.81 and 2.00, depending on the numerical algorithm and the selection of collision partners (Meakin 1984). Changing the sticking probability does not dramatically change D_m. Collisions of polydisperse clusters always result in somewhat higher fractal dimensions than collisions between monodisperse clusters (Brown and Ball 1985). Ballistic particle-cluster aggregation (BPCA), in which clusters grow by the addition of individual particles, give values of D_m close to 3.0. The dimension D_σ must always be smaller than D_m and is often assumed to be close to 2. This means that BCCA clusters have friction times that are nearly independent of radius. For such aggregates and typical conditions in the solar nebula at 1 AU, the relative velocities are between 10^{-3} m s^{-1} (Brownian motion) and 10^{-2} m s^{-1} (sedimentation). For turbulence-driven BPCA growth, the mean mass of the mass distribution scales with the evolutionary time t as t^6, whereas the scaling for BCCA particles is exponential with time (Mizuno et al. 1988).

Numerical simulations of the growth process driven by Brownian motion under conditions appropriate for protoplanetary accretion disks (Kempf et al. 1999) show only a small dispersion of the friction times, leading to small relative velocities and slow growth. The numerical calculations also demonstrated that the friction time increases rather slowly with radius: $\tau \propto R^{0.2}$; $D_m \sim 1.8$, $D_\sigma \sim 1.6$. No such simulations exist for turbulence-driven growth or coagulation driven by drift velocities. Experiments on the aggregation of micron-sized dust grains in a turbulent gas (collisional velocities in the experiment ~20 cm s^{-1}) and during sedimentation (collisional velocities ~1 cm s^{-1}) confirm the formation of fractal dust grains with D_m between 1.9 and 1.7 in the two regimes (Blum et al. 1998 and references therein). The slow increase of τ_f means that grain growth is also slow.

The friction time, τ_f, must increase markedly ($D_m > 2$) to produce efficient, turbulence-driven coagulation ($\tau_f > t_s$); see equation (5). Rapid growth will also occur when there is a reservoir of small particles (e.g., particles accreted from the molecular cloud core), leading to a BPCA-type growth process, or during "runaway" growth, where very large particles (decoupled from the mass distribution) sweep up a large amount of small grains (Safronov 1972; Weidenschilling and Cuzzi 1993; Schmitt et al. 1997; Weidenschilling 1997). In this case, the statistical coagulation equation cannot be used any longer (see, e.g., Tanaka and Nakazawa 1994).

Experimental results suggest the possibility of enhanced coagulation rates of small iron particles by the presence of a magnetic field (Nuth et al. 1994; Nuth and Wilkinson 1995). Whenever a magnetic field of about 100 gauss was present, the iron particles coagulated very rapidly and formed complicated aggregates. The relevance of this process for protoplanetary accretion disks depends on the possibility of the magnetization of small iron grains.

C. Critical Velocities for Grain Sticking

The growth process is governed by the sticking probability, S, and the strength of the adhesion forces holding the aggregates together. It is generally assumed that van der Waals forces are responsible for the sticking of small particles. The sticking probability for two colliding particles depends on the collision velocity, masses, shapes, and material properties. There may also be a critical velocity for sticking, v_{cr}. The existence of a critical velocity implies that there is a steep transition from the velocity range where sticking occurs ($v < v_{cr}$, $S \sim 1$) to the velocity range where the particles bounce ($v > v_{cr}$, $S \sim 0$). Even for the simplest case of spherical particles, the calculated and measured values of v_{cr} disagree. There are many complicating factors to be understood in these collisions, so this disagreement is not surprising.

Chokshi et al. (1993; see also Dominik and Tielens 1997) modeled the dynamics of sticking collisions between dust particles. Two colliding particles form a contact "neck," and the released energy due to the decrease of surface energy accelerates the particles towards each other. Elastic compression decelerates the particles, reverses the motion, and finally leads to acceleration in the opposite direction. When the grains again reach the point of first contact, they are still bound by a contact neck. After further separation, the binding neck becomes unstable and the contact can rupture. Whether or not rupture occurs depends on the amount of kinetic energy dissipated during the collision. In the model developed by Chokshi et al. (1993), energy dissipation is assumed to be dominated by elastic surface waves. The critical velocity they derived for central collisions between two spherical grains is

$$v_{cr} = 1.07 \frac{\gamma^{5/6}}{E^{1/3} r^{5/6} \rho^{1/2}} \tag{8}$$

where γ is the surface energy per unit area, r is the reduced radius [$r = r_1 r_2/(r_1 + r_2)$, where r_1 and r_2 are the radii of the colliding particles], and ρ is the mass density of the spheres. E is a material quantity that depends on the Poisson ratios and Young's moduli and reflects the elastic properties.[a]

---

[a] The factor in equation (8) has to be 1.07 instead of the 3.86 erroneously given by Chokshi et al. (1993), and the functional dependence has to be $E^{1/3}$ instead of E as misprinted by Dominik and Tielens (1997).

A compilation of material properties for astrophysically relevant systems is given by Dominik and Tielens (1997). For micron-sized quartz spheres, the critical velocity is calculated to be $\sim$5 cm s^{-1}.

Poppe et al. (1997, 1998) imaged the trajectories of micron- and submicron-sized particles colliding with a target to determine the sticking characteristics for different materials. For spherical silica particles with smooth surfaces and with $r = 0.6\ \mu$m, the critical velocity is 1.2 m s^{-1}, which is an order of magnitude higher than predicted by equation (8). The shape of the particles is an important property for sticking. The sticking probability increases significantly for irregularly shaped dust grains, especially for submicron grains. In some experiments, no critical velocity is seen, and a significant fraction of irregular particles stick even at impact velocities exceeding 20 m s^{-1}.

Measurements with an atomic force microscope demonstrate that the pull-off forces are in agreement with the theoretical predictions. The discrepancy between experimental and theoretical values for $v_{\rm cr}$ in the case of the spherical particles is probably caused by an underestimate of the actual energy dissipation rate in the theoretical calculations. The general conclusion from the experiments is that micron- and submicron-sized grains stick with high probability as long as the relative velocities are below a few m s^{-1}, a limit that should not be reached by such particles in protoplanetary disks with α values below 10^{-1}.

An interesting discovery of the experimental work is that collisional charging occurs. The charge transfer in a collision between micron-sized silicon particles and a flat quartz target surface results in averaged values of separated elementary charges per impact energy of 10^{-5} to 10^{-4} C J^{-1}. The charging is approximately proportional to the collision energy. Therefore, electrostatic forces between charged grains may be important for the growth process, but the sign and magnitude of the charges are difficult to predict *a priori*.

Bridges et al. (1996) and Supulver et al. (1997) addressed the further growth from centimeter-sized particles to meter-sized bodies. They found that several types of water frost-coated surfaces stick together when brought into contact at low enough sticking velocity, $<$0.4 cm s^{-1}. Their experiments show that the structure of frost is important for sticking. They concluded that the formation of surface layers of frost is a necessary step for providing a sticking mechanism for the growth of larger particles in low-temperature regions of protoplanetary accretion disks. It might be interesting to note that the formation of frosty surfaces plausibly results from sweepup of tiny grains onto the surfaces of the larger particles in relative motion. The layer would be regenerated following compacting bounces.

If larger particles (centimeter- to meter-sized) reach relative velocities of the order 10 m s^{-1} or more, the collisions may cause compaction and disruption. High velocities for the larger particles would lead to fragmentation and the production of a reservoir of small particles.

D. Restructuring, Compaction, and Disruption in Aggregate-Aggregate Collisions

Aggregates with open, BCCA-type structures cannot grow indefinitely through collisions. Collisions with larger grains or other aggregates should lead to restructuring and compaction at a certain mass and velocity threshold. Compaction is necessary to increase the fractal dimension ($D_m > 2$), which increases the frictional coupling and leads to a more efficient growth regime. This was the motivation for Weidenschilling (see, e.g., Weidenschilling and Cuzzi 1993) to start with a fractal dimension of 2.11 in the size range $1\,\mu$m–0.1 mm and then to introduce compaction starting with grain sizes of 1 cm. Although such a compaction is likely, the values of the size limits are unknown, and the change from $D_m = 2.11$ to 3 is arbitrary. Furthermore, compaction may not occur only in collisions; recurrent heating events could also create compact particles.

Theoretical calculations by Dominik and Tielens (1997) of compaction by collisions between aggregates predict a sequence of events that depends on the ratio between impact energies and critical energies for the different processes: (1) sticking without restructuring, (2) losing monomers, (3) maximum compression, and (4) catastrophic disruption. Their evaluation of the different processes showed that rolling is the most efficient mechanism for restructuring.

For velocities below 1 cm s^{-1}, the sticking probabilities in collisions of spherical SiO_2 aggregates are close to unity (Wurm and Blum 1998). There is no restructuring in turbulence-driven growth below collisional velocities of 20 cm s^{-1}. Measurements of the rolling friction between SiO_2 aggregates and a massive target by Wurm (1997) give a lower limit of 5.0×10^{-10} N, about a factor of 5 larger than the value used by Dominik and Tielens (1997). Bouncing and fragmentation dominated the results of collisions down to $\sim$1 m s^{-1}. Between 0.1 and 1 m s^{-1} (and especially close to the critical velocity), restructuring occurred.

The sequence found in the experiments is in qualitative agreement with the theoretical predictions of Dominik and Tielens (1997), but the threshold velocities for the different processes are higher by at least a factor of 2 than the calculated values, mainly because the calculations underestimate the frictional forces. These results imply that at least in the beginning of the sedimentation-driven growth, no restructuring will occur, and the aggregates have to reach centimeter sizes before compaction occurs.

IV. OBSERVATIONAL CONSEQUENCES OF PARTICLE SIZE

A. Theoretical Changes of Particle Opacity

As dust particles grow by collisions, the particle mass opacity function κ_ν (cm^2 g^{-1}) changes. Changes in opacity can in principle be observed in circumstellar disks, thus providing the signature needed to demonstrate that disks breed large particles and planets. The opacity at any wavelength λ will not change until a significant fraction of the particle mass is in particles with sizes of order λ or greater.

The flux density, F_ν, from a disk viewed face-on at distance D is

$$F_\nu = \frac{1}{D^2} \int_{r_0}^{R_D} B_\nu(T(r))(1 - e^{-\tau_\nu(r)}) 2\pi r \, dr \tag{9}$$

where the interior and exterior disk radii are r_0 and R_D, B_ν is the Planck function, T is the temperature, τ_ν is the optical depth, and r is the radial variable (cf Beckwith et al. 1990). The optical depth can be written in terms of the surface density, Σ, and mass opacity coefficient, κ_ν, as $\tau_\nu(r) = \kappa_\nu \Sigma(r)$. If the particles are small and the wavelengths are much larger than the size of the largest particles, the disk should be transparent, $\tau_\nu < 1$, and the emission is in the Rayleigh-Jeans regime, $B_\nu \approx 2kT\nu^2$. In this limit,

$$F_\nu \approx \kappa_\nu \nu^2 \frac{4\pi k}{D^2} \int_{r_0}^{R_D} T(r)\Sigma(r) r \, dr \tag{10}$$

Regardless of the disk structure, the flux density is directly proportional to the particle mass opacity, κ_ν. Although the absolute value of κ_ν is difficult to determine without a knowledge of $T(r)$, $\Sigma(r)$, and R_D (r_0 is usually not relevant in this limit), the frequency dependence can be observed directly.

Particle emission in most disks becomes optically thin at millimeter wavelengths. The emission emerges from the outer parts of disks, where T is a few tens of kelvins, and Σ should be small, so the approximations made in equation (10) are valid (Beckwith et al. 1990). Observations of disks over a range of wavelengths (between 0.6 and 2.7 mm, say) should provide a direct measure of the emissivity exponent, β. If a mixture of particle types and sizes is present, the observed value is an average over the different constituents.

At long wavelengths, $\lambda > 0.1$ mm, κ_ν is expected to scale as a power of the frequency: $\kappa_\nu \propto \nu^\beta$. For compact spherical particles smaller than the observing wavelength, $\beta = 2$ for metals and insulators under a wide range of conditions (Bohren and Huffman 1983; Emerson 1988). The power law exponent, β, can be as small as 1 for certain types of materials (amorphous carbonaceous material, for example) and even smaller over limited ranges, but it must go to 2 at long enough wavelengths to fulfill causality relations, and it is expected that $\beta \approx 2$ at millimeter wavelengths for interstellar dust particles. Careful attempts to measure both κ_ν and β in interstellar clouds have, indeed, yielded $\beta \approx 2$ and values of $\kappa_\nu = (0.002\text{–}0.004)(\lambda/1.3\text{ mm})^{-2}$ cm^2 g^{-1} (Hildebrand 1983; Draine and Lee 1984).

Rocks, asteroids, and planets are opaque to radiation at $\lambda \sim 1$ mm, in which case $\beta = 0$. Pebbles with sizes of order 1 mm should have exponents with intermediate values, $0 < \beta < 2$. If dust grains grow large

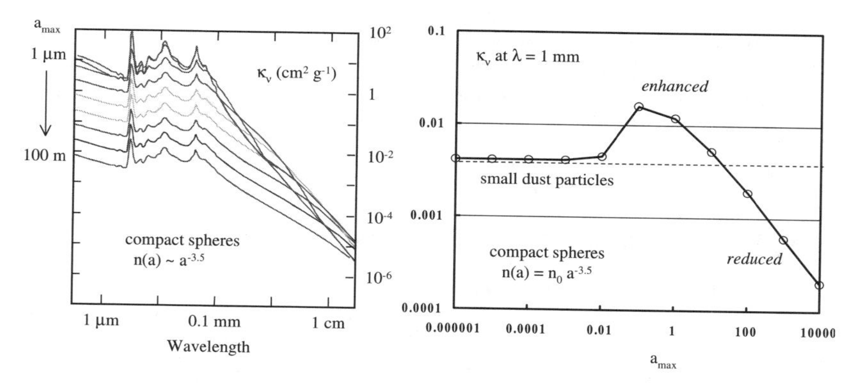

Figure 3. Left: Variation of the mass opacity coefficient, κ_ν, as a function of the maximum radius, $a_{\max}$, for a collection of spherical particles whose number density, n, is a power law in particle radius, a, with a maximum radius $a_{\max}$. The change in slope at long wavelengths is equivalent to a change in β. Right: The absolute value of κ_ν at a wavelength of 1 mm as the maximum particle radius changes from dust (0.01 μm) to boulders (100 m). (After Figs. 5 and 6 of Miyake and Nakagawa 1993.)

enough to put most of the mass in bodies larger than pebbles, they should have an observable signature at these wavelengths. Because the radiation does not penetrate far below the surface of a large body (i.e., most of the material never interacts with the radiation), the absolute value of the opacity per unit mass must decrease as the particles grow.

Calculations of β and κ_ν for spherical particles show that they decrease as the particles grow (Miyake and Nakagawa 1993; Krügel and Siebenmorgen 1994; Henning et al. 1995*b*). Figure 3 illustrates the mass opacity coefficients, $\kappa_\nu(a)$, for spheres of radii distributed as a power law with different maximum sizes, $a_{\max}$. In addition to the resonant absorption by silicates at 10 and 20 μm, that by ice at 3 μm, and possibly other solid-state features, the opacity is a power law with $\beta = 2$ for $\lambda > 100$ μm. For $\lambda \gg a$ (Rayleigh scattering), $\kappa_\nu(a)$ is independent of size a, whereas for $\lambda \ll a$ (i.e., in the geometrical cross section regime), $\kappa_\nu(a)$ is independent of ν but has $1/a$ dependence. For $\lambda \simeq a$, $\kappa_\nu(a)$ is enhanced in comparison with an extrapolation from the Rayleigh scattering regime, and $\kappa_\nu(a)$ deviates from a power law. Including the contribution of large particles with $a \simeq 1$ mm or larger, Miyake and Nakagawa (1993) showed that $\beta \leq 1$ at millimeter wavelengths. The absolute value of κ_ν can also be enhanced by particles with sizes similar to the observing wavelength.

Pollack et al. (1985, 1994) proposed a model of dust particles that includes troilite (FeS) and some organics as major species in addition to silicates, water ice, and metallic iron. They found variations in the behavior of κ_ν in the millimeter range, depending on the contributions of different constituents. The opacity is generally dominated by silicates, but troilite and organic components can be more important than the silicates over restricted wavelength ranges. Opacity enhancements can also occur in specific materials as the result of lattice resonances. Because both changes in the particle chemistry and changes in particle size can produce observable changes in the bulk opacities, it is not easy to demonstrate particle growth simply on the basis of changes in the opacities (Ossenkopf and Henning 1994; Henning et al. 1995*b*).

B. Laboratory Measurements of Particle Opacity

The interpretation of observations of circumstellar disks depends on a knowledge of the mass absorption coefficients of particle materials at the appropriate wavelengths, generally submillimeter to centimeter. It is only recently that laboratory groups have measured optical data for cosmic dust analogs at the temperatures and wavelengths appropriate for circumstellar disks. Agladze et al. (1995, 1996) measured absorption spectra of crystalline enstatite and forsterite grains and amorphous silicate grains with $a \simeq 0.1$–1 μm at $T = 1.2$–30 K for $\lambda = 0.7$–2.9 mm and found that the mass opacity coefficients, κ_ν, for amorphous silicates were up to about 10 times the values that are usually adopted for interstellar dust particles (Draine and Lee 1984). The power law behavior also varies with T: β varies between 1.5 and 2.5 for the amorphous $2MgO{\cdot}SiO_2$ and $MgO{\cdot}SiO_2$ but has a nearly constant value of 1.2, independent of temperature, for amorphous $MgO{\cdot}2SiO_2$. This behavior comes from resonant absorption by low-lying two-level lattice states whose excitation depends on the temperature; such states are overlooked in the classic theory of solids that predicts $\beta = 2$. Thus, in addition to the other complicating factors determining mass opacity (size, shape, clustering effects, the presence of mantles, etc.), the material structure is enough to alter the observable properties. This result will complicate the interpretation of observations. Data on amorphous and crystalline silicates (Jäger et al. 1994; Dorschner et al. 1995; Mutschke et al. 1998; Jäger et al. 1998), oxides (Henning et al. 1995*a*), sulfides (Begemann et al. 1994), and ices (Hudgins et al. 1993; Preibisch et al. 1993; Ehrenfreund et al. 1997) are of special importance for protoplanetary accretion disks (see Henning and Mutschke 1997 and Mennella et al. 1998 for a discussion of the temperature effects). Electronic databases exist for ices and refractory solids and can be accessed by http://www.strw.leidenuniv.nl (ices) and http://www.astro.uni-jena.de (refractory solids).

V. DO WE SEE PARTICLE GROWTH IN DISKS?

Circumstellar disks should contain large particles and planetesimals. There is little doubt that many observed disks are a few Myr old, and it should take only a few percent of this time to grow rocks from tiny dust grains. By a few Myr, much of the solid matter could be tied up in rocks and larger bodies, and we have seen that a shift in the mean particle sizes

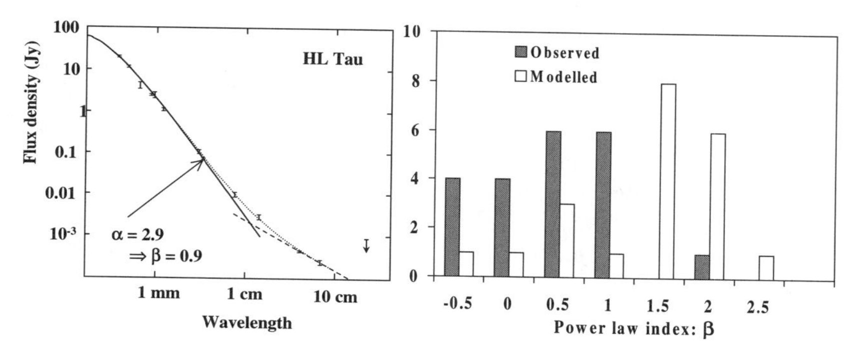

Figure 4. Left: Spectral energy distribution of HL Tau from about 350 μm to 6 cm wavelength. The solid line fits the short wavelengths (thermal emission by dust) and the dashed line fits the long wavelengths (free-free emission from ionized gas). The long-wavelength spectral index of the dust is 2.9, meaning that $\beta = 0.9$ with no corrections for optically thick emission (Wilner et al. 1996). Right: Distribution of β's derived in two ways for a larger sample of disks. The "observed" β's are derived directly from the spectral index as described in the text: $\beta_{\rm obs} = \alpha - 2$. The "model" β's are derived by assuming a surface density distribution as discussed in the text (Beckwith and Sargent 1991).

will have observable consequences for the spectral energy distributions. Do we see these effects?

Changes in the SEDs of disks relative to those typical of interstellar clouds *are* observed (Beckwith and Sargent 1991; Mannings and Emerson 1994; Koerner et al. 1995). The flux densities, F_ν, of disks tend to fall more slowly at wavelengths longer than about 400 μm than do those of interstellar clouds, with the spectral index, α (where $F_\nu \propto \nu^\alpha$), typically between 2 and 3, whereas $\alpha \sim 4$ for the interstellar medium. Assuming optically thin emission in the Rayleigh-Jeans limit, the exponent, β, is often between 0 and 1 for disks compared to the theoretical value of 2 for small particles in the long-wavelength limit. The righthand part of Fig. 4 shows the distribution of emissivity exponents, β, determined from the observed spectral slope of disk emission near λ 1 mm. Many of the disks in this sample have $\beta \leq 1$. It is tempting to interpret this change as the result of particle growth.

Unfortunately, there are many adjustable parameters in the disk models, and several different parameter combinations usually produce acceptable fits to the same SEDs (Thamm et al. 1994). A model typically has the following parameters: r_0, R_D, κ_ν, θ (inclination to line of sight), T_0 and q [$T(r) = T_0(r/r_0)^{-q}$], Σ_0 and p [$\Sigma(r) = \Sigma_0(r/r_0)^{-p}$]. There are rarely more than ten data points in an SED to be fitted. The SEDs are degenerate with respect to the models and almost certainly inadequate by themselves to demonstrate grain growth in disks.

Fortunately, some of these parameters affect the millimeter-wavelength emission weakly or can be determined reasonably well from other wavelengths. The inner radius of the disk, r_0, is unimportant for wavelengths beyond 20 μm in almost all cases. Both T_0 and q are well constrained by the optically thick emission between 10 and 100 μm. The disk inclination, θ, is unimportant for the optically thin (at millimeter wavelengths) emission but important for the optically thick ($\lambda < 100$ μm) part of the SED. In some cases, θ can be estimated directly from images.

The remaining parameters, Σ_0, p, R_D, and κ_ν, play a role in the long-wavelength emission and hence the derivation of the emissivity exponent, β, from the observed spectral slope, α. In the simplest case of $p = 0$ (constant surface density); $\tau_\nu < 1$ for all wavelengths longer than 100 μm, say; and $T(r)$ large enough for all r to make the Rayleigh-Jeans approximation valid for millimeter wavelengths, $\beta = \alpha - 2$ [see equation (10)], the spectral index gives the emissivity exponent directly. But if the density decreases at least as fast as $r^{-\frac{3}{2}}$ ($p = \frac{3}{2}$), there is substantial contribution to the long-wavelength emission from the inner parts of the disk. If $p > \frac{3}{2}$, the inner parts of the disk dominate the emission, making R_D irrelevant but implying that most of the emission at any wavelength comes from radii near the radius at which the disk becomes optically thick (see Beckwith et al. 1990 for a detailed discussion of these effects). In this case, the spectral index, α, depends on β *and* on both p and q. Specifically, Beckwith et al. (1990) derived β in terms of the model parameters as

$$\beta \approx (\alpha - 2)\left(1 + \frac{p}{(2-q)\ln\left[\dfrac{2}{(2-p)\overline{\tau}_\nu}\right]}\right) \tag{11}$$

where $\overline{\tau}_\nu = \kappa_\nu M_D / \cos\theta\pi R_D^2$ is the average optical depth at frequency ν, and M_D is the mass of the disk. [Note that Beckwith et al. (1990) derive the spectral index from νF_ν, not F_ν as used here.] Equation (11) depends strongly on p, which cannot be determined from SEDs alone. For this reason, the emissivity exponent derived from an SED is model dependent.

The righthand part of Fig. 4 includes the distribution of β's derived from model fits, assuming that the surface density is proportional to $r^{-3/2}$. Including the surface density distribution tends to increase the derived value of the emissivity exponent relative to the observed spectral slope. For example, for 1.3-mm observation of a disk with $M_D = 0.03$ $M_\odot$, $R_D =$ 100 AU, and $\kappa_\nu = 0.02$ cm^2 g^{-1}, the average optical depth $\overline{\tau}_\nu = 0.24$, and $\beta \approx 1.36(\alpha - 2)$. Any conclusions about particle growth depend on knowledge of how much optically thick parts of the disks contribute to the spectral energy distribution at long wavelengths. It is probably for this reason that the samples of disks with measured α's at long wavelengths have not been enlarged much since the work of Beckwith and Sargent (1991) and Mannings and Emerson (1994).

To measure the optical depth, it is essential that the disk emission be resolved at wavelengths where it is relatively transparent so that the surface density distribution may be determined directly from the distribution of optical depth. Resolving the disks is time consuming with the present suite of millimeter facilities. Several groups have now managed to resolve the emission from one star, HL Tauri, by carrying out novel experiments with existing millimeter-wave telescopes. Lay et al. (1994, 1997) combined the Caltech Submillimeter Observatory (CSO) and the James Clerk Maxwell Telescope (JCMT) into a single-baseline interferometer to observe at 0.65 and 0.87 mm and combined it with data from the Owens Valley Radio Observatory (OVRO) millimeter interferometer at 1.3 mm. Mundy et al. (1996) extended the Berkeley-Illinois-Maryland Association (BIMA) array to measure a size of the disk at 2.7 mm wavelength, and Wilner et al. (1996) used the Very Large Array (VLA) to get subarcsecond resolution of the disk at 7 mm. Each group resolved the emission but with beam sizes of the same order as the disk size, and the submillimeter observations had u-v plane coverage too poor to map the brightness distribution. All groups measured the average brightness temperature and overall disk orientation and constrained subsequent model fitting much more than is possible from the SEDs alone.

Unfortunately, the combined data are not consistent with simple power law models for disk optical depth. The apparent size of the HL Tau disk is approximately constant from 0.65 mm to 7 mm. Because the optical depths should drop by factors of 10 (for $\beta = 1$) to 100 ($\beta = 2$) over this wavelength range, any centrally concentrated density distribution, $p \geq 1$, would immediately imply smaller sizes at longer wavelengths. Lay et al. (1997) discuss the modifications in the models that might reconcile the data. They conclude that a mixture of particle sizes, so that there is no unique value of β, may explain the data. We note that there would have to be substantial contributions to the emission from particles of order 1 cm for this conclusion to hold. Although such particles are not yet the size of planetesimals, they are more than four orders of magnitude larger than typical interstellar grains and would be present only if there were substantial particle growth in the disk. HL Tau is one of the youngest disks (Cohen 1983; Beckwith et al. 1990), so demonstration of particle growth in this system would imply that it occurs quickly as theory suggests.

HL Tau is also one of the more complicated star/disk systems to interpret, owing to the large amount of surrounding material and the likelihood that high optical depths are contributing to the millimeter-wave emission (cf. Stapelfeldt et al. 1995; Beckwith and Birk 1995; Weintraub et al. 1995; Cabrit et al. 1996; Menshchikov et al. 1999). Mundy et al. (1996) observe a brightness temperature of 29 K from the HL Tau disk at 2.7 mm with an outer radius of order 150 AU. This temperature is very close to the physical temperature of the outer regions of the disk: Beckwith et al. (1990) derive $T = 32$ K at a disk radius of 100 AU. Since the brightness temperature, T_B, is related to the physical temperature by $T_B = T(1 - e^{-\tau})$, there must be a substantial contribution from optically thick dust that is not easily taken into account in the model fits.

Because HL Tau is much younger, more massive, and considerably more complex than most of the disks in Fig. 4, the derived β's may still be valid in many cases. However, because the first direct measurement of the optical depth distribution shows that the model assumptions are invalid, we cannot conclude yet that the long-wavelength spectral indexes tell us much about the particle properties.

The evidence for grain growth is, therefore, inconclusive. There are quite a few disks with spectral indexes that are inconsistent with optically thin emission from very small grains, but the spectral indexes are not unambiguous indicators of large grains. Contributions from high-optical-depth emission within the disks can flatten the SEDs in similar ways as changes in the grain size distribution. Furthermore, $\beta = 1$ is not sufficient to demonstrate grain growth; certain kinds of materials can produce an index of 1 even for small grains. There are quite a few disks for which the spectral index implies $\beta < 1$, but none have been resolved at millimeter wavelengths.

These observations are difficult, because the required spatial resolution is much higher than that routinely available with existing millimeter-wave interferometers. This situation will change during the next decade. The proposed construction of a large millimeter-wave interferometer in the southern hemisphere (the American MMA and European LSA projects) will yield a combination of sensitivity, resolution, and wavelength coverage allowing all the disks in the dark clouds within 300 pc of Earth to be easily resolved. Mapping the disk emission should provide a direct measure of the optical depth, assuming that the physical temperature can be observed independently (from molecular lines, say). From the optical depths at different wavelengths, it should be possible to demonstrate that a population of large grains or rocks does contribute to the emission, if such a population exists. This avenue is the most obvious hope for a demonstration of grain growth in disks.

In the context of planet formation, it is gratifying that the evidence favors grain growth in some disks, even if this evidence is insufficient to establish the presence of large particles unambiguously. The situation is similar to that for the disks themselves in the last decade, when many observations were consistent with the existence of circumstellar disks with the right properties for planet formation, but the evidence was insufficient for proof beyond reasonable doubt. Now, there is no doubt that disks exist with the general properties then inferred, and a separate line of research has uncovered (albeit indirectly) planets themselves orbiting other stars. It is natural to suppose that disks are the necessary precursors to planets. The next generation of millimeter-wave observatories may well prove the evolutionary link by demonstrating particle growth, too.

REFERENCES

Adams, F. C., and Shu, F. H. 1986. Infrared spectra of rotating protostars. *Astrophys. J.* 308:856–833.

Adams, F. C., and Watkins, R. 1995. Vortices in circumstellar disks. *Astrophys. J.* 451:314–327.

Adams, F. C., Lada, C. J., and Shu, F. H. 1988. The disks of T Tauri stars with flat infrared spectra. *Astrophys. J.* 326:865–883.

Adams, F. C., Emerson, J. P., and Fuller, G. A. 1990. Submillimeter photometry and disk masses of T Tauri disk systems. *Astrophys. J.* 357:606–620.

Agladze, N. I., Sievers, A. J., Jones, S. A., Burlitch, J. M., and Beckwith, S. V. W. 1995. Reassessment of millimetre-wave absorption coefficients in interstellar silicate grains. *Nature* 372:243–245.

Agladze, N. I., Sievers, A. J., Jones, S. A., Burlitch, J. M., and Beckwith, S. V. W. 1996. Laboratory results on millimeter-wave absorption in silicate grain materials at cryogenic temperatures. *Astrophys. J.* 462:1026–1040.

André, P., and Montmerle, T. 1994. From T Tauri stars to protostars: Circumstellar material and young stellar objects in the ρ Ophiuchi cloud. *Astrophys. J.* 420:837–862.

Appenzeller, I., Jankovics, I., and Östreicher, R. 1984. Forbidden-line profiles of T Tauri stars. *Astron. Astrophys.* 141:108–115.

Balbus, S. A., and Hawley, J. F. 1991*a*. A powerful local shear instability in weakly magnetized disks. I. Linear analysis. *Astrophys. J.* 376:214–222.

Balbus, S. A., and Hawley, J. F. 1991*b*. A powerful local shear instability in weakly magnetized disks. II. Nonlinear evolution. *Astrophys. J.* 376:223–233.

Barge, P., and Sommeria, J. 1995. Did planet formation begin inside persistent gaseous vortices? *Astron. Astrophys.* 295:L1–L4.

Beckwith, S. V. W., and Birk, C. 1995. Vertical disk structure in HL Tauri. *Astrophys. J.* 449:L59–L63.

Beckwith, S. V. W., and Sargent, A. I. 1991. Particle emissivity in circumstellar disks. *Astrophys. J.* 381:250–258.

Beckwith, S. V. W., Sargent, A. I., Chini, R. S., and Güsten, R. 1990. A survey for circumstellar disks around young stellar objects. *Astron. J.* 99:924–945.

Begemann, B., Dorschner, J., Henning, Th., Mutschke, H., and Thamm, E. 1994. A laboratory approach to the interstellar sulfide dust problem. *Astrophys. J.* 423:L71–L74.

Bell, K. R., Cassen, P. M., Klahr, H. H., and Henning, Th. 1997. The structure and appearance of protostellar accretion disks: Limits on disk flaring. *Astrophys. J.* 486:372–387.

Blum, J., Wurm, G., Poppe, T., and Heim, L.-O. 1999. Aspects of laboratory dust aggregation with relevance to the formation of planetesimals. In *Laboratory Astrophysics and Space Research,* ed. P. Ehrenfreund, and H. Kochan (Dordrecht: Kluwer), pp. 399–424.

Blum, J., Wurm, G., Kempf, S., and Henning, Th. 1996. The Brownian motion of dust in the solar nebula: an experimental approach to the problem of preplanetary dust aggregation. *Icarus* 124:441–451.

Bodenheimer, P. 1995. Angular momentum evolution of young stars and disks. *Ann. Rev. Astron. Astrophys.* 33:199–238.

Bohren, C. F., and Huffman, D. R. 1983. *Absorption and Scattering of Light by Small Particles* (New York: Wiley).

Brandenburg, A., and Hodgson, L. S. 1998. Turbulence effects in planetesimal formation. *Astron. Astrophys.* 330:1169–1174.

Bridges, F. G., Supulver, K. D., Lin, D. N. C., Knight, R., and Zafra, M. 1996. Energy loss and sticking mechanisms in particle aggregation in planetesimal formation. *Icarus* 123:422–435.

Brown, W., and Ball, R. 1985. Computer simulation of chemically limited aggregation. *J. Phys. A* 18:L517–L521.

Cabot, W., Canuto, V. M., Hubickyj, O., and Pollack, J. B. 1987. The role of turbulent convection in the primitive solar nebula. II. Results. *Icarus* 69:387–422.

Cabrit, S., Guilloteau, S., André, P., Bertout, C., Montmerle, T., and Schuster, K. 1996. Plateau de Bure observations of HL Tauri: Outflow motions in a remnant circumstellar envelope. *Astron Astrophys.* 305:527–540.

Calvet, N., Hartmann, L., Kenyon, S. J., and Whitney, B. A. 1994. Flat spectrum T Tauri stars: The case for infall. *Astrophys. J.* 434:330–340.

Cameron, A. G. W. 1988. Origin of the solar system. *Ann. Rev. Astron. Astrophys.* 26:441–472.

Cameron, A. G. W., and Fegley, M. B. 1982. Nucleation and condensation in the primitive solar nebula. *Icarus* 52:1–13.

Champney, J. M., Dobrovolskis, A. R., and Cuzzi, J. N. 1995. A numerical turbulence model for multiphase flows in the protoplanetary nebula. *Phys. Fluids* 7:1703–1711.

Chiang, E., and Goldreich, P. 1997. Spectral energy distributions of T Tauri stars with passive circumstellar disks. *Astrophys. J.* 490:368–376.

Chokshi, A., Tielens, A. G. G. M., and Hollenbach, D. 1993. Dust coagulation. *Astrophys. J.* 407:806–819.

Cohen, M. 1983. HL Tauri and its circumstellar disk. *Astrophys. J. Lett.* 270:L69–L71.

Cuzzi, J. N., Dobrovolskis, A. R., and Champney, J. M. 1993. Particle-gas dynamics in the midplane of a protoplanetary nebula. *Icarus* 106:102–134.

Cuzzi, J. N., Dobrovolskis, A. R., and Hogan, R. C. 1996. Turbulence, chondrules, and planetesimals. In *Chondrules and the Protoplanetary Disk,* ed. R. Hewins, R. Jones, and E. R. D. Scott (Cambridge: Cambridge University Press), pp. 35–43.

Dominik, C., and Tielens, A. G. G. M. 1997. The physics of dust coagulation and the structure of dust aggregates in space. *Astrophys. J.* 480:647–673.

Dorschner, J., Begemann, B., Henning, Th., Jäger, C., and Mutschke, H. 1995. Steps towards interstellar silicate mineralogy. II. Study of Mg-Fe silicate glasses of variable composition. *Astron. Astrophys.* 300:503–520.

Draine, B. T., and Lee, H. M. 1984. Optical properties of interstellar graphite and silicate grains. *Astrophys. J.* 285:89–108.

Dubrulle, B. 1992. A turbulent closure model for thin accretion disks. *Astron. Astrophys.* 266:592–604.

Dubrulle, B. 1993. Differential rotation as a source of angular momentum transfer in the solar nebula. *Icarus* 106:59–76.

Dubrulle, B., Morfill, G., and Sterzik, M. 1995. The dust subdisk in the protoplanetary nebula. *Icarus* 114:237–246.

Edwards, S., Cabrit, S., Strom, S., Heyer, I., Strom, K., and Anderson, E. 1987. Forbidden line and Hα profiles in T Tauri spectra: A probe of anisotropic mass outflows and circumstellar disks. *Astrophys. J.* 321:473–495.

Ehrenfreund, P., Boogert, A. C. A., Gerakines, P. A., Tielens, A. G. G. M., and van Dishoeck, E. F. 1997. Infrared spectroscopy of interstellar apolar ice analogs. *Astron. Astrophys.* 328:649–671.

Emerson, J. P. 1988. Infrared emission processes. In *Formation and Evolution of Low Mass Stars,* ed. A. K. Dupree and M. T. V. Lago (Dordrecht: Kluwer), pp. 21–44.

Finocchi, F., Gail, H.-P., and Duschl, W. 1997. Chemical reactions in protoplanetary disks. II. Carbon dust oxidation. *Astron. Astrophys.* 325:1264–1279.

Gail, H.-P. 1998. Chemical reactions in protoplanetary accretion disks. IV. Multicomponent dust mixture. *Astron. Astrophys.* 332:1099–1122.

Goldreich, P., and Ward, W. R. 1973. The formation of planetesimals. *Astrophys. J.* 183:1051–1061.

Grossman, L. 1972. Condensation in the primitive solar nebula. *Geochim. Cosmochim. Acta* 38:47–64.

Hayashi, C. 1981. Structure of the solar nebula, growth and decay of magnetic fields and effects of magnetic and turbulent viscosities on the nebula. *Prog. Theor. Phys. Suppl.* 70:35–53.

Henning, Th., Begemann, B., Mutschke, H., and Dorschner, J. 1995*a*. Optical properties of oxide dust grains. *Astron. Astrophys. Suppl. Ser.* 112:143–149.

Henning, Th., Michel, B., and Stognienko, R. 1995*b*. Dust opacities in dense regions. *Plan. Space Sci,* 43:1333–1343.

Henning, Th., and Mutschke, H. 1997. Low-temperature infrared properties of cosmic dust analogues. *Astron. Astrophys.* 327:743–754.

Henning, Th., Pfau, W., Zinnecker, H., and Prusti, T. 1993. A 1.3 mm survey for circumstellar dust around young Chamaeleon objects. *Astron. Astrophys.* 276:126–138.

Henning, Th., and Stognienko, R. 1996. Dust opacities for protoplanetary accretion disks: Influence of dust aggregates. *Astron. Astrophys.* 311:291–303.

Hewins, R. H., Jones, R. H., and Scott, E. R. D. 1996. *Chondrules and the Protoplanetary Disk* (Cambridge: Cambridge University Press).

Hildebrand, R. H. 1983. The determination of cloud masses and dust characteristics from submillimetre thermal emission. *Quart. J. Roy. Astron. Soc.* 24:267–282.

Hudgins, D. M., Sandford, S. A., Allamandola, L. J., and Tielens, A. G. G. M. 1993. Mid- and far-infrared spectroscopy of ices: Optical constants and integrated absorbances. *Astrophys. J. Suppl. Ser.* 86:713–870.

Hutchins, D. K., Harper, M. H., and Felder, R. L. 1995. Slip correction measurements for solid spherical particles by modulated dynamic light scattering. *Aerosol. Sci. Technol.* 22:202–218.

Jäger, C., Mutschke, H., Begemann, B., Dorschner, J., and Henning, Th. 1994. Steps towards interstellar silicate mineralogy. I. Laboratory results of a silicate glass of mean cosmic composition. *Astron. Astrophys.* 292:641–655.

Jäger, C., Molster, F. J., Dorschner, J., Henning, Th., Mutschke, H., and Waters, L. B. F. M. 1998. Steps towards interstellar silicate mineralogy. IV. The crystalline revolution. *Astron. Astrophys.* 339:904–916.

Kempf, S., Pfalzner, S., and Henning, Th. 1999. N-particle simulations of dust growth: I. Growth driven by Brownian motion. *Icarus,* in press.

Kenyon, S. J., and Hartmann, L. 1987. Spectral energy distributions of T Tauri stars: Disk flaring and limits on accretion. *Astrophys. J.* 323:714–733.

Klahr, H. H., and Henning, Th. 1997. Particle-trapping eddies in protoplanetary accretion disks. *Icarus* 128:213–229.

Klahr, H. H., Henning, Th., and Kley, W. 1999. On the azimuthal structure of thermal convection in circumstellar disks. *Astrophys. J.,* 514:325–343.

Kley, W., Papaloizou, J. C. B., and Lin, D. N. C. 1993. On the momentum transport associated with convective eddies in accretion disks. *Astrophys. J.* 416:679–689.

Koerner, D. W., Chandler, C. J., and Sargent, A. I. 1995. Aperture synthesis imaging of the circumstellar disk around DO Tauri. *Astrophys. J. Lett.* 452:L69–L72.

Krügel, E., and Siebenmorgen, R. 1994. Dust in protostellar cores and stellar disks. *Astron. Astrophys.* 288:929–941.

Lay, O. P., Carlstrom, J. E., Hills, R. E., and Phillips, T. G. 1994. Protostellar accretion disks resolved with the JCMT-CSO interferometer. *Astrophys. J. Lett.* 434:L75–L78.

Lay, O. P., Carlstrom, J. E., and Hills, R. E. 1997. Constraints on the HL Tauri protostellar disk from millimeter- and submillimeter-wave interferometry. *Astrophys. J.* 489:917–927.

Lissauer, J. 1993. Planet formation. *Ann. Rev. Astron. Astrophys.* 31:129–174.

Lunine, J. I., Engel, S., Rizk, B., and Horanyi, M. 1991. Sublimation and reformation of icy grains in the primitive solar nebula. *Icarus* 94:333–344.

Lynden-Bell, D., and Pringle, J. E. 1974. The evolution of viscous discs and the origin of the nebular variables. *Mon. Not. Roy. Astron. Soc.* 168:603–637.

Mannings, V., and Emerson, J. P. 1994. Dust in discs around T Tauri stars: Grain growth? *Mon. Not. Roy. Astron. Soc.* 267:361–378.

McCaughrean, M. J., and O'Dell, C. R. 1996. Direct imaging of circumstellar disks in the Orion Nebula. *Astron. J.* 111:1977–1986.

Meakin, P. 1984. Effects of cluster trajectories on cluster-cluster aggregation: A comparison of linear and Brownian trajectories in two- and three-dimensional simulations. *Phys. Rev. A* 29:997–999.

Meakin, P., Donn, B., and Mulholland, G. 1989. Collisions between point masses and fractal aggregates. *Langmuir* 5:510–518.

Mennella, V., Brucato, J. R., Colangeli, L., Palumbo, P., Rotundi, A., and Bussoletti, E. 1998. Temperature dependence of the absorption coefficient of cosmic analog grains in the wavelength range 20 microns to 2 millimeters. *Astrophys. J.* 496:1058–1066.

Menshchikov, A. B., Henning, Th., and Fischer, O. 1999. Self-consistent model of the dusty torus around HL Tau. *Astrophys. J.,* in press.

Miyake, K., and Nakagawa, Y. K. 1993. Effects of particle size distribution on opacity curves of protoplanetary disks around T Tauri stars. *Icarus* 106:20–41.

Mizuno, H. 1989. Grain growth in the turbulent accretion disk solar nebula. *Icarus* 80:189–201.

Mizuno, H., Markiewicz, W. J., and Völk, H. J. 1988. Grain growth in turbulent protoplanetary accretion disks. *Astron. Astrophys.* 195:183–192.

Morfill, G. E. 1988. Protoplanetary accretion disks with coagulation and evaporation. *Icarus* 75:371–379.

Mundy, L. G., Looney, L. W., Erickson, W., Grossman, A., Welch, W. J., Forster, J. R., Wright, M. C. H., Plambeck, R. L., Lugten, J., and Thornton, D. D. 1996. Imaging the HL Tauri disk at $\lambda = 2.7$ millimeters with the BIMA array. *Astrophys. J. Lett.* 464:L169–L173.

Mutschke, H., Begemann, B., Dorschner, J., Gürtler, J., Gustafson, B., Henning, Th., and Stognienko, R. 1998. Steps towards interstellar silicate mineralogy. III. The role of aluminium in circumstellar amorphous silicates. *Astron. Astrophys.* 333:188–198.

Nakagawa, Y. K., Nakazawa, K., and Hayashi, C. 1981. Growth and sedimentation of dust grains in the primordial solar nebula. *Icarus* 45:517–528.

Nakagawa, Y., Sekiya, M., and Hayashi, C. 1986. Settling and growth of dust particles in a laminar phase of a low-mass solar nebula. *Icarus* 67:355–390.

Nakamoto, T., and Nakagawa, Y. 1994. Formation, early evolution, and gravitational stability of protoplanetary disks. *Astrophys. J.* 421:640–651.

Nakamoto, T., and Nakagawa, Y. 1995. Growth of protoplanetary disks around young stellar objects. *Astrophys. J.* 445:330–336.

Neufeld, D. A., and Hollenbach, D. J. 1994. Dense molecular shocks and accretion onto protostellar disks. *Astrophys. J.* 428:170–185.

Nuth, J. A., III, and Wilkinson, G. M. 1995. Magnetically enhanced coagulation of very small iron grains: A correction of the enhancement factor due to dipole-dipole interactions. *Icarus* 117:431–434.

Nuth, J. A., III, Berg, O., Faris, J., and Wasilewski, P. 1994. Magnetically enhanced coagulation of very small iron grains. *Icarus* 107:155–163.

O'Dell, C. R., and Wen, Z. 1994. Postrefurbishment mission Hubble Space Telescope images of the core of the Orion Nebula: Proplyds, Herbig-Haro objects, and measurements of a circumstellar disk. *Astrophys. J.* 436:194–202.

Ossenkopf, V. 1993. Dust coagulation in dense molecular clouds: The formation of fluffy aggregates. *Astron. Astrophys.* 280:617–646.

Ossenkopf, V., and Henning, Th. 1994. Dust opacities for protostellar cores. *Astron. Astrophys.* 291:943–959.

Osterloh, M., and Beckwith, S. V. W. 1995. Millimeter-wave continuum measurements of young stars. *Astrophys. J.,* 439:288–302.

Papaloizou, J. C. B., and Lin, D. N. C. 1995. Theory of accretion disks. I: Angular momentum transport processes. *Ann. Rev. Astron. Astrophys.* 33:505–540.

Pollack, J. P., McKay, C. P., and Christofferson, B. M. 1985. A calculation of the Rosseland mean opacity of dust grains in primordial solar system nebulae. *Icarus* 64:471–492.

Pollack, J. B., Hollenbach, D., Beckwith, S., Simonelli, D.P., Roush, T., and Fong, W. 1994. Composition and radiative properties of grains in molecular clouds and accretion disks. *Astrophys. J.* 421:615–639.

Poppe, T., Blum, J., and Henning, Th. 1997. Generating a jet of de-agglomerated small particles in vacuum. *Rev. Sci. Instrum.* 68:2529–2533.

Poppe, T., Blum, J., and Henning, Th. 1998. Analogous experiments on the stickiness of micron-sized preplanetary dust. *Astrophys. J.,* submitted.

Preibisch, T., Ossenkopf, V., Yorke, H. W., and Henning, Th. 1993. The influence of ice-coated grains on protostellar spectra. *Astron. Astrophys.* 279:577–588.

Pringle, J. E. 1981. Accretion discs in astrophysics. *Ann. Rev. Astron. Astrophys.* 19:137–162.

Prinn, R. G. 1993. Chemistry and evolution of gaseous circumstellar disks. In *Protostars and Planets III,* ed. E. H. Levy and J. I. Lunine (Tucson: University of Arizona Press), pp. 1005–1028.

Ruden, S. P., and Pollack, J. B. 1991. The dynamical evolution of the protosolar nebula. *Astrophys. J.* 375:740–760.

Safronov, V. 1969. *Evolution of the Protoplanetary Cloud and the Formation of the Earth and Planets,* NASA TTF-667 (translation from Russian).

Safronov, V. S. 1972. *Evolution of the Protoplanetary Cloud and Formation of the Earth and Planets* (Moscow: Nauka Press).

Schmitt, W., Henning, Th., and Mucha, R. 1997. Dust evolution in protoplanetary accretion disks. *Astron. Astrophys.* 325:569–584.

Sekiya, M. 1998. Quasi-equilibrium density distributions of small dust aggregations in the solar nebula. *Icarus* 133:298–309.

Squires, K. D., and Eaton, J. K. 1991. Preferential concentration of particles by turbulence. *Phys. Fluids A* 3:1169–1178.

Stapelfeldt, K. R., Burrows, C. J., Krist, J. E., Trauger, J. T., Hester, J. J., Holtzmann, J. A., Ballester, G. E., Casertano, S., Clarke, J. T., Crisp, D., Evans, R. W., Gallagher, J. S., III, Griffiths, R. E., Hoessel, J. G., Mould, J. R., Scowen, P. A., Watson, A. M., and Westphal, J. A. 1995. WFPC2 imaging of the circumstellar nebulosity of HL Tauri. *Astrophys. J.* 449:888–893.

Stepinski, T. F., and Valageas, P. 1996. Global evolution of solid matter in turbulent protoplanetary disks. I. Aerodynamics of solid particles. *Astron. Astrophys.* 309:301–312.

Stepinski, T. F., and Valageas, P. 1997. Global evolution of solid matter in turbulent protoplanetary disks. II. Development of icy planetesimals. *Astron. Astrophys.* 319:1007–1019.

Stone, J.M., and Balbus, S.A. 1996. Angular momentum transport in accretion disks via convection. *Astrophys. J.* 464:364–372.

Strom, S. E., Edwards, W., and Skrutskie, M. F. 1993. Evolutionary time scales for circumstellar disks associated with intermediate- and solar-type stars. In *Protostars and Planets III,* ed. E. H. Levy and J. I. Lunine (Tucson: University of Arizona Press), pp. 837–866.

Supulver, K. D., Bridges, F. G., Tiscareno, S., Lievore, J., and Lin, D. N. C. 1997. The sticking properties of water frost produced under various ambient conditions. *Icarus* 129:539–554.

Tanga, P., Babiano, A., Dubrulle, B., and Provenzale, A. 1996. Forming planetesimals in vortices. *Icarus* 121:158–170.

Tanaka, H., and Nakazawa, K. 1994. Validity of the statistical coagulation equation and runaway growth of protoplanets. *Icarus* 107:404–412.

Thamm, E., Steinacker, J., and Henning, Th. 1994. Ambiguities of parametrized dust disk models for young stellar objects. *Astron. Astrophys.* 287:493–502.

Tscharnuter, W. M., and Boss, A. P. 1993. Formation of the protosolar nebula. In *Protostars and Planets III,* ed. E. H. Levy and J. I. Lunine (Tucson: University of Arizona Press), pp. 921–938.

Völk, H. J., Jones, F. C., Morfill, G. E., and Röser, S. 1980. Collisions between grains in a turbulent gas. *Astron. Astrophys.* 85:316–325.

Weidenschilling, S. J. 1977. Aerodynamics of solid bodies in the solar nebula. *Mon. Not. Roy. Astron. Soc.* 180:57–70.

Weidenschilling, S. J. 1980. Dust to planetesimals: Settling and coagulation in the solar nebula. *Icarus* 44:172–189.

Weidenschilling, S. J. 1988. Formation processes and time scales for meteorite parent bodies. In *Meteorites and the Early Solar System* (Tucson: University of Arizona Press), pp. 348–371.

Weidenschilling, S. J. 1995. Can gravitational instability form planetesimals? *Icarus* 116:433–435.

Weidenschilling, S. J. 1997. The origin of comets in the solar nebula. *Icarus* 127:290–306.

Weidenschilling, S. J., and Cuzzi, J. N. 1993. Formation of planetesimals in the solar nebula. In *Protostars and Planets III,* ed. E. H. Levy and J. I. Lunine (Tucson: University of Arizona Press), pp. 1031–1060.

Weintraub, D. A., Kastner, J. H., and Whitney, B. A. 1995. In search of HL Tauri. *Astrophys. J. Lett.* 452:L141–L145.

Wilner, D. J., Ho, P. T. P., and Rodriguez, L. F. 1996. Subarcsecond VLA observations of HL Tauri: Imaging the circumstellar disk. *Astrophys. J. Lett.,* 470:L117–L121.

Wurm, G. 1997. Experimentelle Untersuchungen zu Bewegung und Agglomerationsverhalten mikrometergrosser Teilchen in protoplanetaren Scheiben. Ph.D. Thesis, University of Jena.

Wurm, G., and Blum, J. 1998. Experiments on preplanetary dust aggregation. *Icarus* 132:125–136.

Yoneda, S., and Grossman, L. 1995. Condensation of CaO-MgO-Al_2O_3-SiO_2 liquids from cosmic gas. *Geochim. Cosmochim. Acta* 59:3413–3444.

PROPERTIES AND EVOLUTION OF DISKS AROUND PRE-MAIN-SEQUENCE STARS OF INTERMEDIATE MASS

ANTONELLA NATTA
Osservatorio Astrofisico di Arcetri

VLADIMIR P. GRININ
Crimean Astrophysical Observatory and St. Petersburg University

and

VINCENT MANNINGS
Jet Propulsion Laboratory

This chapter discusses the properties of the immediate circumstellar environment of pre-main-sequence stars of intermediate mass, or Herbig Ae/Be (HAeBe) stars, with particular emphasis on the properties and evolution of the circumstellar disks. HAeBe stars cover a large range of spectral types and luminosities; this variety has implications for their environments, which, by the time a star becomes optically visible, are very different in early B stars (HBe; $M_\star \gtrsim 5$ $M_\odot$) and in stars of later spectral types (HAe; $M_\star \lesssim 5$ $M_\odot$). A variety of recent infrared and millimeter observations are reviewed. They indicate that HBe stars generally lack clear evidence of disks and are often found inside large cavities, depleted of dust and gas. We interpret these observations as evidence of a rapid evolution of the circumstellar environment, possibly caused by the strong stellar radiation fields. In contrast, circumstellar disks appear to be associated with a large number of cataloged HAe stars. We discuss the properties and evolution of these disks within the context of possible grain growth and planet formation.

I. INTRODUCTION

Disks of gas and dust are associated with the majority of classical T Tauri systems (TTSs) in nearby star-forming regions (see the chapters by Mundy et al. and by Wilner and Lay, this volume). The TTSs have spectral types M and K and corresponding stellar masses in the range $0.25 \lesssim M_\star / \mathrm{M}_\odot \lesssim 1$. Herbig (1960) was first to identify pre-main-sequence (PMS) stars of earlier spectral types and greater masses, and he compiled a list of 26 candidate young objects in the spectral range F0 to B0. These stars are now invariably known as "Herbig Ae/Be stars" (or HAeBe stars), and the number of candidates has expanded to about 300 stars (Thé et al. 1994*a*; de Winter

1996). We will adopt here an "evolutionary" definition of HAeBe; namely, all stars of spectral type approximately in the range A–B that lie on PMS evolutionary tracks in the HR diagram. This criterion is applied very loosely to stars earlier than about B2, which remain on the zero-age main sequence (ZAMS) only for a very short time. The current sample includes stars of mass ~2–20 $M_\odot$, with luminosities of a few $L_\odot$ to $\sim 10^4$ $L_\odot$ and effective temperatures from 8000 to 30,000 K. The ages of cataloged HAeBe stars range from $\lesssim 10^5$ to about 10^7 yr.

HAeBe stars comprise an inhomogeneous class of objects not only because of their masses, luminosities, and temperatures but also because of their evolutionary histories. Like the TTSs, Herbig stars of spectral types A and late B (HAe, for simplicity) become optically visible long before arriving at the main sequence, and their pre-main-sequence evolution can be studied in detail. Conversely, stars with spectral types earlier than about B5 (henceforth HBe) never emerge from enveloping circumstellar material throughout their brief PMS phases (Palla and Stahler 1993; see also the chapter by Stahler et al. in this volume). Also, as for O stars, their high luminosities and hard radiation fields have a disruptive effect on their environments. A common property of all HAeBe stars, and one that differentiates them from stars of lower mass, is that their entire PMS evolution occurs along radiative tracks.

Much progress in our knowledge of HAeBe stars has been made since the time of the *Protostars and Planets III* conference. Many results are summarized in the proceedings of the conference on HAeBe stars held in Amsterdam in 1993 (Thé et al. 1994*b*), and recent work has been covered in a number of reviews such as those by Pérez and Grady (1997) and by Waters and Waelkens (1998). We refer the reader to these reviews for complete summaries of the observational properties of HAeBe stars.

In this work, we concentrate on just one aspect of intermediate-mass stars, namely the existence, properties, and evolution of circumstellar disks during the PMS phase. Understanding disks is of crucial importance in any study of star and planet formation. In recent years, disks around HAe stars have attracted increasing attention, because these stars have masses similar to those of the main-sequence disk sources β Pictoris (A5), α Piscis Austrini (A3), and α Lyrae (A0; Vega). These latter stars, which presumably passed through a Herbig Ae phase en route to the main sequence, appear to be surrounded by disks of debris produced by ongoing collisions and disruptions of large solid bodies (cf. Backman and Paresce 1993). The debris disks may be visible signatures of stellar environments in which planets have already been created (see the chapter by Lagrange et al., this volume, and references therein), raising the possibility that disks around their HAe progenitors could be sites of planet formation.

Strom et al. (1993) discussed the properties and evolutionary timescales for disks around intermediate-mass as well as solar-mass stars. They concluded that the inference of disks around stars of mass >3 $M_\odot$ was very uncertain. In fact, in the years immediately following, hard evidence for disks around HAeBe stars was very elusive, and they were the subject of intense debate, as we will discuss in section II. The situation has been drastically changed by recent observations in different areas, among them single-dish and interferometric observations at millimeter wavelengths, *Infrared Space Observatory* (ISO) data, new forbidden-line surveys, and long-term photometric and polarimetric monitoring.

II. THE DISK DEBATE

Although it is clear that HAeBe stars are accompanied by circumstellar dust and gas, there has been much uncertainty on the geometry of the circumstellar environment and, more specifically, on the association of intermediate-mass stars with disks. We will give in this section a brief summary of the "disk debate" (see also Waters and Waelkens 1998). If HAeBe stars do not have (and never had) circumstellar disks, we are forced to conclude that the mechanism of formation of massive stars differs from that of solar-mass stars, which is generally agreed to require the presence of an accretion disk (Shu et al. 1987). Coalescence of lower-mass stars in dense clusters, for example, has been proposed as an alternative formation mechanism by Bonnell et al. (1998) (see the chapter by Stahler et al., this volume).

In the following, we define "disks" as geometrically thin accretion and/or reprocessing structures (including the debris disks of Vega-like stars). The definition of "envelope" is much less precise. We will use it for any distribution of dust (and gas) that subtends a substantial solid angle at the star and has low optical depth at visible wavelengths. Often, envelopes are identified with infalling envelopes (Terebey et al. 1984). When this is the case, we will explicitly use the expression "infalling envelope."

A. Spectral Energy Distribution

In TTSs, strong evidence for circumstellar disks comes from the shape of the spectral energy distribution (SED) at infrared and millimeter wavelengths (see Beckwith and Sargent 1993*b*). HAeBe stars have SEDs similar to those observed for TTSs. Hillenbrand et al. (1992) proposed a classification of HAeBe stars in three groups. Group I corresponds to objects with SEDs well fitted by the emission expected from flat circumstellar accretion disks. Group II has SEDs that can be best understood in terms of a star+disk system surrounded by a roughly spherical envelope of dust and gas. Group III stars have very small infrared excesses, often consistent with free-free emission from stellar winds. These three groups may correspond to an evolutionary sequence (from Group II to I to III), in which the amount of circumstellar matter decreases with time.

The Hillenbrand et al. (1992) interpretation was challenged by numerous authors on various grounds. Hartmann et al. (1993) pointed out that the high accretion rates derived for Group I stars by Hillenbrand et al. (1992) imply an emission in the near-infrared and visual much larger than observed. The possibility that a dominant contribution to the mid-infrared emission of HAeBe stars could come from transiently heated particles, as observed in reflection nebulae, was suggested (Hartmann et al. 1993; Prusti et al. 1993; Natta and Krügel 1995).

Several teams of authors found that all the SEDs of HAeBe stars can instead be interpreted as the emission of spherical envelopes of various optical depths and density profiles (Berrilli et al. 1992; Miroshnichenko et al. 1997; Pezzuto et al. 1997). Far-infrared, high-spatial-resolution observations proved that the 50- and 100-μm emission of many stars belonging to Group I is in fact extended over scales ranging from $\sim$0.03 to $\sim$0.3 pc, inconsistent with the disk interpretation (Natta et al. 1993; Di Francesco et al. 1994). Most sources, especially the HBe objects, are also found to be extended in the 1.3-mm continuum (Henning et al. 1998). Detailed radiative transfer models of some of those stars (Natta et al. 1993), however, showed that the combined spatial and spectral constraints could not be satisfied by spherical envelopes alone, but that both disks and envelopes had to contribute to the observed SEDs: the disk emission in the near- and mid-infrared, the envelope emission in the far-infrared. Observations at different wavelengths may therefore be sensitive to different source components with greatly different spatial scales.

B. The Disk-Wind Connection

A strong, albeit indirect, line of evidence for disks in TTSs is provided by the correlation of wind indicators, such as the intensity of forbidden lines, with disk indicators, such as near-IR excess emission (Cabrit et al. 1990). This correlation has found an explanation in the theory of accretion-driven mass loss (see, for example, Edwards et al. 1993), where disk accretion, coupled with the stellar magnetic fields, provides the energy source for the outflow. The correlation of forbidden-line intensity with excess infrared luminosity observed in TTSs extends to HAeBe stars, suggesting that disk accretion is also driving the mass loss in these stars (Corcoran and Ray 1998).

Forbidden lines in TTSs tend to have blueshifted profiles. The most widely accepted explanation is that an optically thick disk occults the redshifted emission of a stellar wind or other outflow close to the star (Appenzeller et al. 1983, 1984; Edwards et al. 1987). The situation in HAeBe stars is much more controversial. The line profiles tend to be much more symmetric than in TTSs, with strongly blueshifted profiles confined to the most embedded objects (Böhm and Catala 1994; Corcoran and Ray 1997). However, there is a slight tendency to blueshifted profiles. Corcoran and Ray (1997) claim that their results are consistent with the prediction of the two-component forbidden-line emission model of TTSs (Hamann 1994; Hirth et al. 1994; Hartigan et al. 1995; Kwan and Tademaru 1995), where the high-velocity (jet) component is present only in the more embedded, younger objects and disappears first, leaving only the low-velocity component, emitted by a poorly collimated disk wind. In HAeBe stars this low-velocity emission is broader and more symmetric than in TTSs because of the faster rotation of the disk.

III. DISKS AS SEEN WITH MILLIMETER INTERFEROMETRY

A fundamental step forward in proving the presence of circumstellar disks in HAeBe stars has come from interferometric observations at $\lambda \sim 1$ mm and $\lambda \sim 3$ mm (Di Francesco et al. 1997; Mannings and Sargent 1997; Mannings et al. 1997). The detection of compact millimeter emission on scales of 1–2 arcsec is considered reliable evidence that the emitting dust is in a disk, rather than in a spherical envelope around the star, since the optical depth of such an envelope would have to be far larger than observed in order to account for the measured flux (see Beckwith et al. 1990; Mannings and Sargent 1997). Table I provides a summary of the millimeter interferometric observations of HAeBe stars obtained up to now. For each star, we give spectral type, distance, effective temperature, luminosity, age, and mass. The last two quantities were derived by comparing the position of the stars on the HR diagram with the predictions of PMS (Palla and Stahler 1993) and main-sequence (Schaller et al. 1992) evolutionary tracks (see discussions in van den Ancker et al. 1998 and Fuente et al. 1998). The wavelength of the observations is given in column 10, and the measured flux in column 11.

A. Disk Masses

The simplest way to estimate disk masses from the observed millimeter continuum fluxes is to assume that the emission comes from optically thin, isothermal dust at temperature T_D:

$$M_D = d^2 \frac{F_{\rm mm}}{\kappa_{\rm mm} B_{\rm mm}(T_D)} \tag{1}$$

where d is the distance of the source, $F_{\rm mm}$ the observed flux, $B_{\rm mm}$ the Planck function at T_D, and $\kappa_{\rm mm}$ the dust opacity per gram of gas at the observed wavelength. We have compared equation (1) to model-predicted millimeter fluxes for disks of different M_D. The disk structure is described in terms of power law temperature and surface density profiles (Beckwith et al. 1990; Natta 1993), and we have varied the model parameters (central star, disk mass, outer radius, temperature, and surface density profile) over a large range of possible values. We found that equation (1) recovers the correct value of M_D (within a factor of $2-3$) for disks with $M_D \lesssim 0.3$ $M_\odot$, provided that one allows T_D to vary as a function of the spectral type of the star as indicated in Table II. The disk masses M_D derived with this prescription are given in Table I, column 13. We have adopted here (and in the following) $\kappa_{1.3\ \rm mm} = 1$ cm^{-2} g^{-1} of dust (as suggested by Ossenkopf and Henning 1994), and a wavelength dependence $\kappa \propto \lambda^{-\beta}$ with $\beta = 1$, unless otherwise specified. We also assume a dust-to-gas mass ratio of

TABLE I
Disk Masses from Interferometer Measurements

(1) Source	(2) Spectral Type	(3) Ref.	(4) Distance (pc)	(5) Ref.	(6) Log T_* (K)	(7) Log L_* ($L_\odot$)	(8) Age (10^6 yr)	(9) M_* ($M_\odot$)	(10) λ_{obs} (mm)	(11) F_ν (mJy)	(12) Ref.	(13) M_D^a ($M_\odot$)
CQ Tau	F2	(R98)	100	(E97)	3.83	0.54	>30	1.3	1.3	143±8	(M99)	0.03
LkHα259	A9:	(C76)	850	(M68)	3.87	2.03	0.1	3.5	2.6	<6.0	(M99)	<0.6
V1318 Cyg	A8:	(H95b)	1000	(S91)	–	–	<0.1	–	2.7	19±6	(D97)	1.1
Elias 3−1	A6:	(Z94)	140	(E78)	3.91	1.11	>10	2	2.7	42±10	(D97)	0.1
Mac CH12	A5:	(HBC)	845	(M68)	3.91	1.05	>10	2	2.6	10.0±1.4	(M99)	0.8
MWC 758 (HD36112)	A5	(H95a)	200	(E97)	3.91	1.35	5	2	1.3	82.5±5.5	(M99)	0.05
PV Cep	A5:	(C81)	440	(W81)	3.91	2.14	0.3	3.5	2.7	35.6±0.7	(M98)	0.8
MWC 480 (HD 31648)	A3	(J91)	130	(E97)	3.94	1.51	4	2	1.3	279±7	(M97*b*)	0.07
UX Ori	A3	(H72)	430	(W78)	3.94	1.47	3	2.3	1.3	16.4±1.6	(N99)	0.04
HD 245185	A2	(V98)	430	(W78)	3.95	1.39	7	2	2.6	6.5±1.2	(M97)	0.1
MWC 863	A1	(H88)	120	(E97)	3.97	1.48	5	2.3	2.6	13.7±2.2	(M97)	0.02
AB Aur	A0	(B93)	144	(E97)	3.98	1.68	2.5	2.2	2.7	10.6±0.4	(M97)	0.02
HD 34282	A0	(C49)	160	(E97)	3.98	1.42	>7	2.2	2.6	24±3	(M99)	0.05
HD 163296	A0	(H88)	120	(E97)	3.98	1.48	>6	2.3	1.3	441±12	(M97)	0.08

TABLE I (continued)
Disk Masses from Interferometer Measurements

T Ori	B9	(H92)	460	(W78)	4.02	2.15	0.7	3.5	1.3	<6	(N98)	<0.01
MWC 614 (HD 179218)	B9	(S66)	240	(E97)	4.02	2.50	0.1	4.3	1.3	71±7	(M99)	0.04
LkHα198	B8:	(C85)	600	(R68)	4.08	2.32	0.5	3.5	2.7	<4.8	(D97)	<0.1
V594 Cas (BD+61 154)	B8	(F85)	650	(H70)	4.08	2.59	0.1	4	2.7	11±2	(M99)	0.3
V376 Cas	B5	(C79)	600	(R68)	4.19	3.05	0.1	5	2.7	<4.8	(D97)	<0.07
V1686 Cyg	B5	(H95b)	1000	(S91)	4.19	3.26	<0.1	6	2.7	<4.8	(D97)	<0.2
BD+31 643	B5	(E97)	330	(E97)	4.19	2.87	0.1	5	2.6	<2.7	(M98)	<0.01
BD+40 4124	B2	(H95b)	1000	(S91)	4.34	3.95	<0.1	9	2.7	<4.8	(D97)	<0.1
MWC 137	B0:	(S81)	1100	(C92)	4.48	4.24	0.1	14	2.7	<11	(M98)	<0.15
R Mon	B0	(C79)	800	(H82)	4.48	4.16	0.1	14	2.7	13.0±1.3	(M98)	0.1
MWC 1080	B0	(C79)	2200	(L88)	4.48	5.21	0.05–6	20	2.7	<5.4	(M98)	<0.2

[a] Mass of both gas and dust, assuming a gas-to-dust ratio of 100, by mass. Upper limits are 3σ limits.

References to Table I: (B93) Böhm and Catala 1993; (C49) Cannon and Mayall 1949; (Cs76) Cohen and Kuhi 1976; (C79) Cohen and Kuhi 1979; (C81) Cohen et al. 1981; (C85) Chavarria-K. 1985; (C92) Cahn et al. 1992; (D97) Di Francesco et al. 1997; (E78) Elias 1978; (E97) Hipparcos Catalogue; (F85) Finkenzeller 1985; (HBC) Herbig and Bell 1988; (H70) Hagen 1970; (H72) Herbig and Rao 1972; (H82) Herbst et al. 1982; (H88) Houk and Smith-Moore 1988; (H92) Hillenbrand et al. 1992; (H95*a*) N. Houk, personal communication, 1995, with B. Zuckerman; (H95*b*) Hillenbrand et al. 1995; (J91) Jaschek et al. 1991; (L88) Levreault 1988; (M68) MacConnell 1968; (M97) Mannings and Sargent 1997; (M97*b*) Mannings, et al. 1997; (M98) V. Mannings and A. I. Sargent, personal communication; (M99) Mannings and Sargent 1999; (N98) A. Natta, unpublished; (N99) Natta et al. 1999*b*; (R68) Racine 1968; (R98) Rostopchina 1998; (S66) Slettebak 1966; (S81) Sabbadin and Hamzaoglu 1981; (S91) Shevchenko et al. 1991; (V98) M. E. van den Ancker, personal communication; (W78) Warren and Hesser 1978; (W81) Whitcomb et al. 1981; (Z94) Zinnecker and Preibisch 1994.

TABLE II
Outer Disk Temperature[a]

(1)	(2)	(3)	(4)	(5)	(6)	(7)	(8)	(9)
	B0	B2	B5	B8	A0	A3	F0	M5
T_D(K)	215	101	56	38	28	23	16	15

[a] Value of T_D to use in equation (1) to compute M_D from $F_{\rm mm}$ for stars of different spectral types.

0.01. The $\kappa_{1.3\ \rm mm}$ value is consistent, within the uncertainties, with that computed by Pollack et al. (1994), who proposed a dust model for protoplanetary disks based on the composition of the primitive solar nebula. For a discussion of the uncertainty in κ at millimeter wavelengths, see, for example, Mannings and Emerson (1994) and Pollack et al. (1994).

Virtually all stars in our sample with spectral type later than B8 (12 out of 16) are detected by interferometers, in contrast with the detection of only one star (R Monocerotis) out of 7 earlier than B8 (see Fig. 1). This, in itself, could simply reflect a selection effect, arising from the sensitivity of the millimeter continuum measurements, because on average HBe stars are farther away than HAe stars. In fact, the derived disk mass of HAe stars has an average value of $M_D \sim 0.06$ M$_\odot$, with a spread of more than one order of magnitude. The smallest values of M_D are $\sim$ 0.02 M$_\odot$. The upper limits on M_D for HBe stars are ~0.1–0.2 M$_\odot$, with the single exception of BD+31 346, for which $M_D \lesssim 0.01$ M$_\odot$. However, if we consider the ratio of the disk mass to stellar mass $M_D/M_\star$ (see Fig. 2, panel b), we find that Ae stars tend to have higher values than the upper limits set for HBe stars.

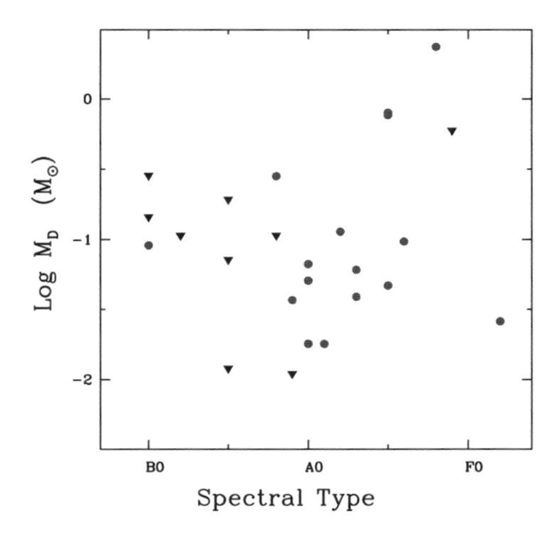

Figure 1. Disk mass (dust+gas) as a function of the spectral type of the star for HAeBe stars with millimeter interferometric measurements. Detections are shown by circles, 3σ upper limits by triangles.

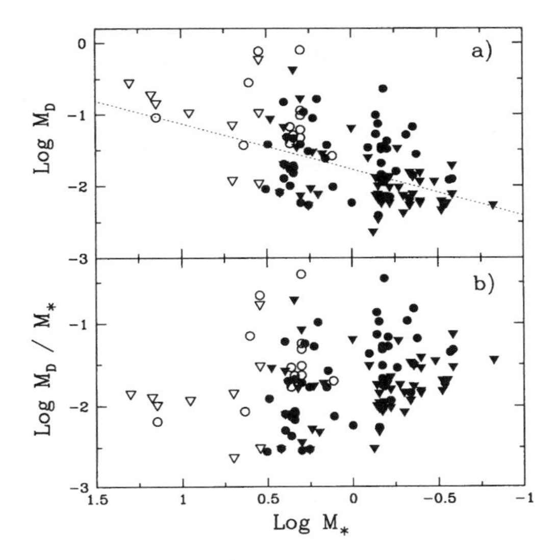

Figure 2. (a) $M_{\rm disk}$, the disk mass (dust + gas) is shown as a function of the mass of the star for HAeBe and T Tauri stars in Taurus. The stellar masses of HAeBe stars have been computed as described in the text; TTS masses are from Beckwith et al. (1990). (b) Ratio of the disk mass over the stellar mass as function of $M_\star$. In both panels detections are shown by circles, 3σ upper limits by triangles. Filled symbols refer to stars with single-dish observations, open symbols to interferometric measurements. The dashed line in (a) is the best-fitting curve, which has slope 0.64 ± 0.14. Note that the two samples (TTSs and HAeBe stars) do not provide a complete coverage of the mass spectrum, because there is a lack of stars in the interval $\sim$0.8–1.3 $M_\odot$.

R Mon, the only early Be star detected with the interferometers, has a ratio $M_D/M_\star$ ($\sim$0.007) lower than any of the HAe stars in our sample.

The highest values of M_D and of $M_D/M_\star$ are found in the most embedded stars (such as PV Cephei, MacC H12, and V1318 Cygni). This may be a true effect (more embedded objects may have more massive disks), but it is also possible that M_D is overestimated because the temperature of the outer disk is higher than we have assumed, due to heating by surrounding dust ("backwarming"; see the discussion of this effect in Butner et al. 1994).

B. Frequency of Occurrence of HAeBe Disks

There is a clear difference between the frequency of occurrence of disks around HBe stars (where it may be close to zero) and that around HAe stars. In order to estimate the latter, we cannot rely on the ratio of detections to nondetections in interferometric observations, because the sample of observed stars included only objects known to have strong millimeter flux from single-dish observations. However, for most HAe stars the interferometers recover, within the uncertainties, all the flux measured by single-dish telescopes. Natta et al. (1997) collected single-dish mm observations of a sample of 30 HAe stars in the age range 10^5–10^7 yr. The sample was put together to study disk masses in HAe stars with and without strong photometric variability. Their sample includes only stars for which millimetric observations (either detections or nondetections) were found in the literature. Insofar as this sample could be considered representative, it indicates that the frequency of occurrence of disks is very large, between 75 and 100%.

C. The Gas Disk

Observations in the emission lines of ^{13}CO (1–0), CO (1–0), and CO (2–1) with the Owens Valley Radio Observatory (OVRO) interferometer show that in 5 out of 11 cases a gaseous disk, as well as a dusty disk, is associated with HAe stars (Mannings and Sargent 1997, 1999; Mannings et al. 1997). For AB Aurigae, HD 163296, MWC 480, and MWC 758 the CO lines display the typical velocity pattern expected in a Keplerian rotating disk: an ordered velocity gradient directed along the major axis of the extended gas structure (cf. model predictions by Beckwith and Sargent 1993*a*). For the A2e system MWC 480, this interpretation is confirmed by detailed kinematic modeling (Mannings et al. 1997).

The CO emission is usually more extended than the continuum emission, and it has been possible to derive sizes and aspect ratios (Mannings and Sargent 1997, 1999; Mannings et al. 1997) from interferometric CO observations of AB Aur, HD 163296, MWC 480, and MWC 758. The CO disks have radii ranging from $<$85 AU in CQ Tau to 700 AU in MWC 480.

Masses derived from CO lines, assuming a ratio CO/H_2 as in molecular clouds, are significantly smaller than those derived from continuum emission. The discrepancy could be due to molecule depletion onto grains at low temperatures or to optical depth effects in the CO lines (cf. Dutrey et al. 1996). It is also possible, in principle, that gas is globally depleted with respect to dust. Sensitive followup observations using optically thin lines of CO isotopomers and molecules less easily depleted onto grains than CO are necessary to improve estimates of gaseous disk masses. In the meantime, estimates of dust and gas masses derived using continuum fluxes appear to be more reliable.

D. Comparison with T Tauri Disks

It is interesting to compare the masses of disks found around HAe stars with those detected in TTSs. We have recomputed disk masses in TTSs (a total of 87 stars) from the observed millimeter fluxes (Beckwith et al. 1990; Osterloh and Beckwith 1995) using equation (1) with T_D = 15 K, as appropriate for stars of spectral type later than F0 (see Table II), and $\kappa_{1.3\ \mathrm{mm}} = 1\ \mathrm{cm}^{-2}\ \mathrm{g}^{-1}$ of dust. To the HAeBe stars reported in Table I we have also added 15 HAe stars with single-dish millimeter measurements from Natta et al. (1997). In spite of the large dispersion of values for any given spectral type, there is a statistically significant trend (significant at the $4-5\sigma$ level according to three different statistical tests: the Cox proportional hazard model, the generalized Kendall's tau, and Spearman's rho test) of decreasing disk mass in stars of later spectral type: M_D decreases by about an order of magnitude between A0 and M7. Not surprisingly, because spectral type and mass are not independent, we find a correlation with similar statistical significance if we plot $M_{\rm disk}$ as a function of the stellar mass $M_\star$ (Fig. 2). While $M_{\rm disk}$ seems to decrease with $M_\star$, the ratio of disk mass to stellar mass is roughly constant for stars in the range A0–M7 (or, alternatively, in the mass range ~4–0.3 $M_\odot$), with an average value of ~0.04 and a large dispersion (Fig. 2). Note again the low values of $M_D/M_\star$ for early-B stars (i.e. for masses $\gtrsim 5$ $M_\odot$). We do not know whether this constancy has implications of any relevance for the process of disk formation and evolution. We just note that this value is much smaller than the stability threshold against self-gravitational perturbations ($M_D/M_\star \sim 0.24$; Shu et al. 1990). It seems that PMS stars of all masses have relatively low-mass disks.

IV. THE MID-IR SPECTRUM OF HAeBe

The range of wavelengths between ~5 and ~15 μm (the mid-IR) contains important information on the circumstellar environment of HAeBe stars. In particular, we find in the mid-IR a prominent silicate band with a peak at about 10 μm, together with the series of so-called unidentified IR bands (UIB) at 6.2, 7.8, 8.6, 11.3, and 12.5 μm, often attributed to polycyclic aromatic hydrocarbons (PAHs). Although this wavelength interval is only partially accessible from the ground, ISO obtained high-quality spectra that cover the whole mid-IR for a large number of HAeBe stars. We have collected a sample of 30 HAeBe stars with known mid-IR spectra (Wooden 1994; Prusti et al. 1999; Siebenmorgen et al. 1998, 1999; M. E. van den Ancker, personal communication). This sample is not complete and will grow in number and statistical significance as more ISO spectra are published. However, it is sufficient to illustrate some of the HAeBe star properties relevant to our discussion.

A. Unidentified Infrared Bands

UIBs are seen in practically all the early-type stars observed so far by ISO [personal communication by the SWS (Short Wavelength Spectrometer) team]; they are weaker in stars of later spectral type, where they can be detected only in high-quality, high-resolution spectra. UIB emission arises in regions of low optical depth exposed to UV radiation; it is often extended and not centered on the exciting star (see Siebenmorgen et al. 1999).

In some stars, the UIBs and, more generally, the emission of transiently heated species dominate the emission in the mid-IR. However, this seems to be true only for a few, relatively old, early-type stars. In most

HAe stars, the mid-IR flux is due to a strong continuum, probably emitted by grains in thermal equilibrium with the local radiation field.

B. The Silicate Feature

The silicate feature is present in virtually all (75%) sources of spectral type A0 or later in our sample, but only in 4 of the 14 with spectral type earlier than A0. Wooden (1994) was first to point out that silicate emission is common among later-type HAeBe stars, while it is generally absent in early-type sources whose spectra are dominated by UIBs.

Silicate emission in the 10-μm region is related to the existence of hot ($T_D \gtrsim 500$ K) grains with size less than a few μm and optical depths at the feature peak $\tau_{\rm sil} \sim 0.5$–1. This corresponds to $A_V \sim 2.5$–5 for standard interstellar grains (Cohen and Witteborn 1985). A spherical envelope of dust around the star can reproduce the HAe observed silicate emission feature if its inner radius is controlled by dust sublimation (which for silicates occurs at about 1500 K; Berrilli et al. 1992; Miroshnichenko et al. 1997; Pezzuto et al. 1997).

Optically thick disks cannot account for the silicate emission. In fact, if we consider a three-component star+disk+envelope system, the continuum mid-IR emission due to the disk dilutes the silicate emission of the envelope until it disappears (Natta 1993). More realistic disk models, which include the effects on the emitted spectrum of a disk atmosphere, have been proposed for TTSs by Calvet et al. (1991) and, more recently, by Chiang and Goldreich (1997). The disk atmosphere, heated by the stellar radiation to temperatures higher than the effective temperature of the underlying disk (analogous to a stellar chromosphere), produces a strong silicate emission feature on top of the continuum emission of the thick disk itself. For TTSs, these models account well for several observed properties, and they link the presence of silicate emission directly with the existence of a circumstellar disk. Similar models could presumably account also for the silicate emission observed in HAeBe stars.

The absence of silicate emission among HBe stars indicates that these stars lack the hot and optically thin dust component that emits the feature. If HBe stars are surrounded by dusty envelopes, the envelopes must have large inner holes, unless grains have grown to sizes greater than a few μm (but see section VII.C). In fact, spherical envelope models, which extend to the sublimation radius and can account for the observed SEDs of HBe stars, all predict strong silicate emission (Berrilli et al. 1992; Miroshnichenko et al. 1997; Pezzuto et al. 1997). In order to suppress the silicate emission in a B0 star (which has a typical dust sublimation radius $\sim 8 \times 10^{13}$ cm), the inner radius of the envelope must be about 0.02 pc, or 700 times larger than the dust sublimation radius. Such large cavities around several HBe stars have been inferred from far-infrared (Natta et al. 1993) and millimeter (Fuente et al. 1998) observations. If HBe stars have disks, the lack of silicate emission implies that they cannot sustain extended chromospheres, which would likely give rise to strong features.

V. THE ENVIRONMENT OF UXORs

Additional detailed information on the circumstellar environment of HAeBe stars is provided by a large subgroup of highly variable stars, named UXORs after the first such identified star, UX Orionis. UXORs are defined by their especially large photometric variability, with deep irregular minima in which a star can fade by as much as 2–3 magnitudes. At present, this subclass includes a few tens of objects, mostly HAe stars. Van den Ancker et al. (1998) estimate that about one-third of the HAe stars may be UXORs. To our knowledge, no HBe star is a UXOR. UXORs show spectroscopic evidence of sporadic, high-redshift absorption events (see the chapter by Grady et al., this volume).

A. An Inhomogeneous Environment

The most striking result in studies of UXORs is that the circumstellar dust seems to be distributed in an extremely inhomogeneous fashion. The most accepted interpretation of their variability (but see also Thé 1994; Eaton and Herbst 1995) is based on the variable circumstellar extinction model suggested by Wenzel (1969) and modified by Grinin (1988). According to this model, the brightness minima are caused by optically thick dusty clumps in orbit around the star, which sporadically intersect the stellar radiation along the line of sight to the star. The "bluing effect" (i.e., the fact that the radiation first becomes redder and then bluer as the star fades) and the behavior of the polarization (which is maximum when the light intensity reaches a minimum) can be explained by scattering of the stellar light by a flattened and rather optically thin torus of dust around the star, seen edge-on by the observer, within which the postulated clumps are embedded. The contribution of this scattered component to the total radiation observed at any given time increases when the direct (unpolarized) radiation of the star is absorbed by the intervening dust clump.

The dust mass of an individual clump, which produces a deep brightness minimum when intersecting the line of sight, is estimated to be $M_c \sim 10^{20}$–10^{21} g (Voshchinnikov and Grinin 1991; Meeus et al. 1998), comparable with the masses of the largest comets in the solar system (Festou et al. 1993). We can roughly estimate the total mass of dust in clumps from the expression $M_D \sim \phi(Rz/r_c^2)M_c$, where ϕ is the probability that one clump occults the star, r_c and M_c are the radius and mass of an individual clump, and R and z are the radius and scale height of the volume occupied by the clumps, respectively. Assuming a clump radius of the order of the stellar radius, and taking ϕ to be $\sim$0.1 and z to be $\sim$0.5R (see below), we obtain a dust mass $\lesssim 10^{-6}$ M$_\odot$ within a distance of 100 AU from the star. The short timescales ($\lesssim$1 day) observed in several episodes of variability can be accounted for only if clumps exist within a few AU of the star (Grinin 1994; Hutchinson et al. 1994; Sitko et al. 1994). The relation of the UXORs phenomena to the presence of dust very near to the star may account for the lack of UXOR among HBe stars (van den Ancker et al. 1998).

The estimates of the mass of dust in the optically thin torus that scatters the stellar radiation range from $\sim 10^{-8}$ (Voshchinnikov 1998) to $\sim 10^{-6}$ $M_{\odot}$ (Friedemann et al. 1994).

B. The Origin of UXORs

UXORs have many properties in common with nonvariable HAe stars. The SEDs at infrared and millimeter wavelengths have very similar shapes, and the ratio of IR to bolometric luminosity is comparable (Hillenbrand et al. 1992). The mass of dust estimated from millimeter observations does not depend on the level of the UXOR activity of the star (Natta et al. 1997). Moreover, of the four highly variable stars for which millimeter interferometer observations exist, three (UX Ori, HD 34282, and CQ Tau) have compact emission and disk masses similar to the other HAe stars. In fact, the clumps and the optically thin tori together contain only a very small fraction of the millimeter-observed amount of dust ($\sim 10^{-6}$ vs. $\gtrsim 10^{-4}$ $M_{\odot}$; see Table I), most of which must reside in an optically thick circumstellar disk.

This suggests that the complex and highly inhomogeneous environments of UXORs are in fact common to all HAe stars and that only those seen through a clumpy torus are UXORs. If so, the fact that about one-third of HAe stars are UXORs (van den Ancker et al. 1998) is consistent with a torus opening angle of order 35°.

Recently, Grinin et al. (1998) have suggested that binarity can be an important aspect of the UXOR phenomenon. They have found a cyclic variability with periods of a few years in some UXORs and interpret it as due to the large-scale perturbations produced in a postulated circumbinary disk by the presence of a companion star (or, possibly, a giant planet) at distances of 3–10 AU from the star.

It is, on the other hand, possible that UXORs represent a later evolutionary phase rather than a purely geometrical effect. This hypothesis is based on the interpretation that sporadic, high-velocity redshifted absorption events measured in a wide array of metal lines (and in Hα) are due to the infall and evaporation of planetesimals or protocomets in star-grazing orbits (Pérez et al. 1993; Grady et al. 1996*a*; Grinin et al. 1994; Sorelli et al. 1996; and the chapter by Grady et al., this volume). High-velocity infall in metal lines characterizes evolved systems such as β Pic (Lagrange et al., this volume), where it has been interpreted as evidence of evaporation of large bodies in star-grazing orbits, and in several A-shell main-sequence stars (Grady et al. 1996*b*). Further support for this interpretation comes from the detection of crystalline silicates, similar to those detected in comets (Waelkens et al. 1996; Malfait et al. 1998*b*; see also the chapter by Grady et al., this volume). However, Natta et al. (1997) found that UXORs do not generally appear to be older than non-UXOR HAe stars.

VI. THE CIRCUMSTELLAR ENVIRONMENT OF HBe STARS: RAPID EVOLUTION?

As discussed in the previous sections, HBe stars seem to lack in general the compact continuum emission expected from a circumstellar disk; additionally, very few HBe stars show silicate emission, whereas strong UIBs are always present in their spectra; no HBe star shows UXOR activity. Maps at far infrared and millimeter wavelengths, both in the continuum and in molecular lines, show that the stars are often located in the centers of large empty cavities; either the emission comes from the matter at the edges of the cavities, or, in some cases, it is clearly not associated with the optical star (Fuente et al. 1998; Henning et al. 1998; Di Francesco et al. 1998). On the other hand, at a very early stage of their evolution, both dense envelopes and circumstellar disks exist around early B stars [Kurtz et al., this volume; see also the interferometric mm detection of R Mon by Mannings and Sargent (personal communication) reported in Table I]. It is therefore tempting to ascribe the different environments of optically visible HAeBe stars of different spectral types to a much more rapid environmental evolution in the most massive and luminous stars. Fuente et al. (1998), in their millimeter study of a small group of HBe stars, find a timescale of less than 10^6 yr for complete dispersal of surrounding dense gas and dust.

Radiation fields can both halt the accretion, as soon as the infall rate decreases below a critical value, and affect the circumstellar disk itself. The effect of radiation pressure on the infalling envelope during the protostellar phase has been discussed by various authors (see Stahler et al., this volume). Radiation pressure reverses the infalling motion as soon as the mass infall rate decreases below a threshold value $\dot{M}_{\rm lim}^{\rm in}$. We follow Jijina and Adams (1996) and estimate $\dot{M}_{\rm lim}^{\rm in} \sim L_\star/(v_{\rm in}\, c)$, where $v_{\rm in}$ is the infall velocity at the dust sublimation radius, $L_\star$ is the stellar luminosity, and c is the speed of light. Figure 3 shows $\dot{M}_{\rm lim}^{\rm in}$ for stars of various masses as they evolve on the HR diagram from the birthline to the ZAMS. The survival of the infalling envelope against radiation pressure for a star of given mass depends on how quickly $\dot{M}_{\rm in}$ decreases during the PMS evolution. For example, a star of 5 $M_\odot$ disperses the infalling envelope as soon as $\dot{M}_{\rm in}$ decreases below $\sim 3 \times 10^{-7}$ $M_\odot$ yr^{-1}. A star of 2 $M_\odot$ can preserve an infalling envelope until the infall rate decreases well below 10^{-8} $M_\odot$ yr^{-1}. Radiation pressure can easily account for the large cavities, almost devoid of matter, in which many HBe stars lie. The further evolution of these stars may involve the formation of shells of enhanced density where the pressure of the surrounding matter equals the radiation pressure; these regions will emit strong PAH features (see, for example, the HD 97300 ring of PAHs in Siebenmorgen et al. 1998).

The dispersal of the infalling envelope may trigger a rapid evolution of the circumstellar disk, which will accrete onto the star on a viscous timescale (typically, $\sim 3 \times 10^5$ yr for the outer disk of a B0 star). The

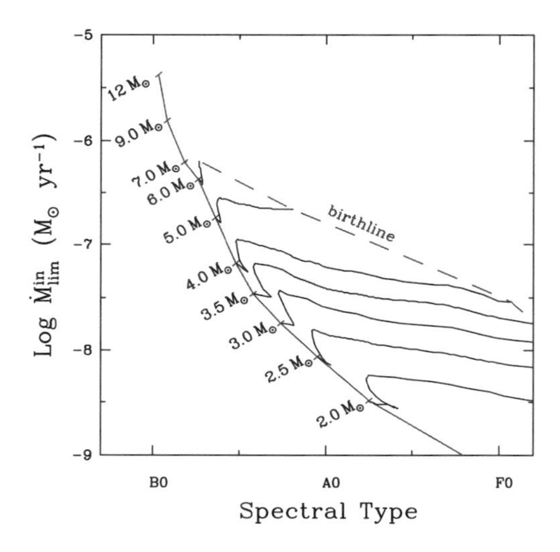

Figure 3. Limit on the infall mass rate as a function of the spectral type of the star for stars of different mass. Each solid line corresponds to the evolutionary track on the HR diagram of a star of given mass, as labeled. The birthline for an infall rate of 10^{-5} $M_\odot$ yr^{-1} is shown as a dashed line, the ZAMS by the thick solid line. Infalling envelopes with mass infall rates smaller than $\dot{M}^{in}_{lim}$ are dispersed by radiation pressure.

direct effect of radiation pressure on the circumstellar disk morphology can also be important. Recent work by Drew et al. (1998) and Proga et al. (1998) describes how, if the star has an accretion disk in Keplerian rotation, radiation pressure will drive a low-velocity, high-density equatorial disk wind. It is not clear how this will affect an underlying disk chromosphere or whether the wind itself will contain hot dust. UV photons from the star can also affect the disk structure (Hollenbach et al. 1994; see the chapter by Hollenbach et al., this volume). Interferometric observations of the dust continuum emission do not rule out that residual "light" disks (i.e., with a ratio of disk to stellar mass much smaller than the typical value ~0.04 observed in lower-mass stars) may exist around HBe stars. Such disks could emit the large near- and mid-IR fluxes that characterize many HAeBe stars (Hillenbrand et al. 1992). If so, it is likely that the disks are vertically compressed and eroded in their outer parts by the effect of radiation pressure and UV radiation.

The location of HBe stars in clusters may also explain the rapid evolution of the HBe environment and the absence of massive circumstellar disks. Hillenbrand (1995) and Testi et al. (1997, 1999) have shown that stars earlier than about B7 tend to be found in clusters, while stars of later spectral type tend to form in small groups. At the moment, however, it is not clear whether the density of stars in these clusters is large enough to have significant gravitational influence on the evolution of the immediate surroundings of the HBe stars in the very short timescales we derive, and we favor the idea that the fast evolution is caused by the stellar radiation field.

VII. DISK EVOLUTION IN HAe SYSTEMS

The circumstellar environment of HAe stars differs significantly from that of HBe stars. For many stars, there is convincing evidence of circumstellar disks whose evolution is very likely controlled not by the stellar radiation field but by physical processes occurring within the disk itself. As for HBe stars, the disk matter can accrete onto the star in a viscous timescale or can be blown away in a stellar wind. Gravitational perturbations by companion stars may dissipate the disk, entirely or in part, depending on the separation between the stars (see the chapter by Lubow and Artymowicz, this volume). Alternatively, and of more interest to us, the grain component may grow and accumulate to bodies of very large sizes (1–10 km, planetesimals). These, in turn, might coalesce to form terrestrial planets and the cores of giant planets (Lissauer 1993). The gas originally in the disk will either accrete onto and form the giant planets or be dispersed from the system. This scenario, used to explain the formation of the solar system, leads to the formation of a secondary or debris disk, where grains are continuously formed by fragmentation of larger bodies; secondary disks have little dust mass (most of the original disk dust is locked into planets) and no gas. This type of secondary disk is observed in Vega-like stars, the best known example being β Pic (Lagrange et al., this volume). Hence, disk evolution can be studied by examining the dependence on time of the mass of small grains and gas in the disk and the growth of grains from sizes typical of ISM grains to much larger sizes. Since different parts of the disk may evolve on different timescales, we will consider separately the existing evidence for evolution of the outer disk and of the inner disk.

A. The Inner Disk

The inner regions (less than about a few AU) of a typical circumstellar disk are optically thick at all wavelengths; their emission peaks at near- and mid-infrared wavelengths. The dissipation of the inner disk, due either to an actual decrease of the disk mass or to the coagulation of the grains into larger bodies, will cause the region to become optically thin at these wavelengths, with a consequent decrease of the emitted flux.

The evolution of the inner disk is best seen if the timescale of disk dissipation differs for the conditions of the inner disk and of the outer disk. The inner regions of the disk may become optically thin (and therefore "disappear") before the outer parts of the disk if the timescale of the agglomeration of grains into large bodies is shorter in the relatively denser material. Pollack et al. (1996) discuss the effect of the formation of a Jupiter-size planet in the inner disk around a solar-mass star. The planet will clear a gap in the disk; material outside the gap can no longer enter the inner disk (but see also the chapter by Lubow and Artymowicz, this volume), while the small grains and gas in the gap accrete onto the planet with timescales that can be quite short ($\sim 5 \times 10^5$ yr for Jupiter conditions; see Boss 1996). The signature of such processes is a spectral energy distribution characterized by a "deficit" of emission (a gap) in the mid-infrared with respect to longer wavelengths.

There are no objects with a clear mid-infrared gap among the HAeBe stars studied by Hillenbrand et al. (1992). Waelkens et al. (1994) and Malfait et al. (1998*a*) find a large number of stars with IR excesses characterized by a dip at 10 μm among isolated (in general ZAMS) HAe stars. They interpret their results in terms of an evolutionary sequence similar to that just outlined. This very attractive idea needs to be pursued further.

B. The Outer Disk

The evolution of the outer disk is best studied by means of its millimeter emission. If disks evolve into planetary systems, we expect to see a significant decrease of the millimeter flux. If grains coagulate into much larger bodies (size $\gtrsim$ a few cm), the mass of dust in the disk does not change, but $\kappa_{1.3\ \mathrm{mm}}$ decreases (Miyake and Nakagawa 1993; Pollack et al. 1994). Values of M_D derived from the measured millimeter flux via equation (1) with invariant κ_{mm} should decrease with time as the outer disk evolves.

We show in Fig. 4 disk masses (derived as in section III.A) as a function of the age of the star for many stars of spectral type about A belonging to different groups. Note that in this figure we have plotted *dust* masses only rather than dust+gas masses, since the ratio of dust to gas is likely to vary widely from group to group.

The youngest stars in Fig. 4 are the HAe stars (circles) from Table I and Natta et al. (1997). A second group of stars is formed by Vega-like stars of spectral type A (squares) studied by Sylvester et al. (1996). These stars lie in the same region of the diagram occupied by HAe stars; only in one case (HD 218396) do we derive an upper limit ($M_{\mathrm{dust}} \lesssim 1.2 \times 10^{-5}$ M$_\odot$) that is significantly lower than in HAe stars. We have added to Fig. 4 dust mass estimates for the three best-studied Vega-like stars, β Pic, α PsA, and α Lyr (Holland et al. 1998), and for HR 4796 A (Greaves et al. 1999). The arrows at an age of a few hundred Myr show the upper limits recently derived (Natta et al. 1999*a*) for a group of main-sequence A shell stars, which show evidence of infalling gas in their spectra (Grady et al. 1996*b*). Finally, we have added upper limits from Zuckerman and Becklin (1993) for stars in the Pleiades and UMa clusters.

Figure 4 indicates that there is no evidence that M_{dust} changes over the time interval between 10^5 and 10^7 yr (i.e., over the PMS evolution of the stars). If we divide the stars (both HAe and Vega-like) into two groups, one with ages below 3 Myr (20 stars) and one with ages above 3 Myr (29 stars), we find that the two groups have similar distributions of M_{dust}. The fraction of nondetections is not smaller in the first (younger) group ($\frac{9}{20}$) than in the second (older) one ($\frac{7}{29}$). Note that a similar result (i.e., the lack of a trend of disk mass vs. age up to 10^7 yr) was found

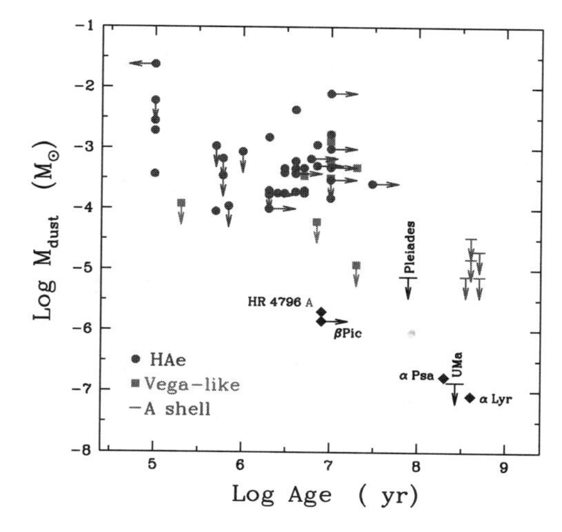

Figure 4. Dust mass as function of the age of the star for stars of spectral type about A. HAe stars are shown by circles, Vega-like stars from Sylvester et al. (1996) by squares. The arrows at age $\sim 3\times10^8$ yr plot upper limits to $M_{\rm dust}$ for a small group of main-sequence A-shell stars. Diamonds show the dust content of the best-known Vega-like stars β Pic (age from Crifo et al. 1997), α PsA (age from Barrado y Navascués et al. 1997), α Lyr (age from Backman and Paresce 1993) and HR 4796 A (age from Stauffer et al. 1995). The upper limits derived by Zuckerman and Becklin (1993) for the Pleiades and Ursa Major are also shown. Limits are shown by arrows. Ages for PMS stars (both HAe and Vega-like) have been determined by comparing their positions on the HR diagram to the theoretical evolutionary tracks of Palla and Stahler (1993). The observed HR location has been determined uniformly by us from available visual photometry, using Hipparcos distances whenever available.

by Beckwith et al. (1990) for TTSs in Taurus-Auriga. The transition from relatively massive disks, which characterize the PMS evolution of HAe stars, to debris disks seems to occur at about 10^7 yr and to be very fast. A fast transition time was found for naked TTSs by Wolk and Walter (1996). At later times, the debris disks show a slow decline of mass with time, as noted by Zuckerman and Becklin (1993).

C. Grain Growth; Formation of Planetesimals and Planets

Some evidence that growth to millimeter-sized grains may happen earlier in the evolution is provided by the observed shape of the SEDs at submillimeter and millimeter wavelengths [cf. the similar work for TTS disks by Beckwith and Sargent (1991) and Mannings and Emerson (1994)]. The values of the spectral index α, determined from single-dish observations in the range 350–1300 μm (see Table III), are $\alpha \gtrsim 2.8$ in all the early B stars in which disks have not been detected. On the contrary, stars of spectral types later than A0 with detected disks have $\alpha \lesssim 2.3$. The relation between α and the dust opacity law (which is usually written as $\kappa \propto \lambda^{-\beta}$ in the submillimeter and millimeter range) is not uniquely defined, since

TABLE III

Millimeter/Submillimeter Spectral Indexes

(1) Source	(2) Spectral Type	(3) Ref.	(4) Distance (pc)	(5) Ref.	(6) Log $T_\star$ (K)	(7) Log $L_\star$ ($L_\odot$)	(8) Age (10^6yr)	(9) M_* ($M_\odot$)	(10) λRange (mm)	(11) α ($F_\nu \propto \nu^\alpha$)	(12) Ref.
CQ Tau	F2	(R98)	100	(E97)	3.83	0.54	>30	1.3	0.45–1.3	1.98±0.20	(M98)
V1318 Cyg	A8:	(H95b)	1000	(S91)	–	–	<0.1	–	0.35–1.3	2.88±0.28	(S93)
Elias 3−1	A6:	(Z94)	140	(E78)	3.91	1.11	>10	2	0.35–1.3	2.18±0.10	(M94)
PV Cep	A5:	(C81)	440	(W81)	3.91	2.14	0.3	3.5	0.35–1.3	2.55±0.04	(M98)
MWC 758 (HD 36112)	A5	(H95a)	200	(E97)	3.91	1.35	5	2	0.45–1.3	2.80±0.21	(M98)
MWC 480 (HD 31648)	A3	(J91)	130	(E97)	3.94	1.51	4	2	0.35–1.3	2.20±0.07	(M98)
UX Ori	A3	(H72)	430	(W78)	3.94	1.47	3	2.3	1.7–2.3	2.1±0.2	(N99)
MWC 863	A1	(H88)	120	(E97)	3.97	1.48	5	2.3	0.35–1.3	2.33±0.15	(M98)
AB Aur	A0	(B93)	144	(E97)	3.98	1.68	2.5	2.2	0.45–1.3	2.44±0.08	(S93)
HD 163296	A0	(H88)	120	(E97)	3.98	1.48	>6	2.3	0.35–1.3	1.94±0.04	(M94)
LkHα198	B8:	(C85)	600	(R68)	4.08	2.32	0.5	3.5	0.45–1.3	3.30±0.21	(M94)
VY Mon	B8:	(HBC)	800	(H82)	4.05	2.88	0.1	5	0.80–1.3	2.25±0.14	(S93)
LkHα 234	B6	(S72)	1000	(F84)	4.11	2.74	0.1	4.5	0.45–1.3	2.75±0.10	(S93)
V376 Cas	B5	(C79)	600	(R68)	4.19	3.05	0.1	5	0.35–1.3	2.92±0.10	(M98)
BD+40 4124	B2	(H95b)	1000	(S91)	4.34	3.95	<0.1	9	0.35–1.3	3.12±0.34	(S93)
MWC 137	B0:	(S81)	1100	(C92)	4.48	4.24	0.1	14	0.35–1.3	2.94±0.12	(M98)
R Mon	B0	(C79)	800	(H82)	4.48	4.16	0.1	14	0.35–1.3	2.72±0.15	(M94)
CoD−42°11721	B0	(D90)	400	(D90)	4.48	4.45	6	15	0.45–1.3	3.28±0.27	(S93)
MWC 1080	B0	(C79)	2200	(L88)	4.48	5.21	0.01–6	22	0.35–1.3	2.81±0.38	(S93)

References to Table III: (B93) Böhm and Catala 1993; (C79) Cohen and Kuhi 1979; (C81) Cohen et al. 1981; (C85) Chevarria-K. 1985; (C92) Cahn et al. 1992; (D90) de Winter and Thé 1990; (E97) Hipparcos Catalogue; (F84) Finkenzeller and Mundt 1984; (H72) Herbig and Rao 1972; (H82) Herbst et al. 1982; (H88) Houk and Smith-Moore 1988; (H95a) N. Houk 1995, personal communication with B. Zuckerman; (H95b) Hillenbrand et al. 1995; (J91) Jaschek et al. 1991; (L88) Levreault 1988; (M94) Mannings 1994; (M98) V. Mannings et al., personal communication; (N99) Natta et al. 1999*b*; (R68) Racine 1968; (R98) Rostopchina 1998; (S72) Strom et al. 1972; (S91) Shevchenko et al. 1991; (S93) A. Sandell, personal communication; (W78) Warren and Hesser 1978; (W81) Whitcomb et al. 1981; (Z94) Zinnecker and Preibisch 1994.

for a given value of β, α depends upon the temperature and surface density profiles as well as the disk mass and outer radius (see the discussion in the chapter by Beckwith et al., this volume). Mannings (1994) showed that in the case of the A0 system HD 163296, the observed low value of $\alpha = 1.94$ could be accounted for only if $\beta \lesssim 1$, which, in turn, requires grain sizes in the range ~1–3 mm (cf. Miyake and Nakagawa 1993; Pollack et al. 1994). HD 163296 has been detected with the OVRO interferometer and has a disk mass of ~0.08 $M_\odot$ (see Table I). Low values of α occur for sources of all ages in our HAe sample (which, however, includes only one star, VY Mon, younger than 10^6 yr).

These results need to be confirmed by detailed analysis of more objects, because they provide a very interesting clue to the evolutionary status of HAe disks. Namely, theoretical calculations predict that in the disk midplane grains grow very rapidly from micron-size to kilometer-size bodies (see the chapter by Beckwith et al., this volume, and references therein; also Lissauer 1993). The characteristic timescale for this process depends on the initial grain size distribution, the density of the circumstellar matter, its turbulent velocity, and several other parameters (again, see the chapter by Beckwith et al., this volume and references therein). For the conditions of the primitive solar nebula, grain growth from sizes typical of grains in the interstellar medium to meter-size (and larger) occurs in less than 10^5 yr (the age of our youngest stars) at distances of 30 AU from the Sun, and it is faster at shorter distances (Weidenschilling 1997; Schmitt et al. 1997). Based on these results, it is not easy to understand how disk grains can have typical millimeter sizes over a long period of time. Some help may come from observations of UXORs. As discussed in section V, the UXOR phenomena, which may be in fact common to all HAe stars, seem to require the simultaneous presence of large dust clumps (with properties typical of planetesimals) in orbit around the star and of a much more massive disk of grains of a few millimeters in size at most.

It is tempting to speculate that most visible HAe stars have disks where grain growth has already occurred, reaching a sort of "bimodal" size distribution: Although most of the original disk dust mass is in grains with typical sizes of a few millimeters, a small fraction of it resides in large bodies (planetesimals). There is little dust mass in grains with size intermediate between these two limits. This situation should be very stable, lasting for the majority of the PMS life of the star. Kenyon and Luu (1998, 1999) discuss how, at the distance of the Kuiper Belt, the growth from about 1 to 100-km size is accompanied by the formation of a very large amount of relatively small grains, whose size spectrum is, however, very uncertain. Once the largest objects grow to more than ~1000 km, then the system loses the small grains very quickly. It is possible then that the transition from "normal" HAe disks to β Pic-like structures is linked to the formation of very large planetesimals. Then, the timescale for this to happen around A-type stars must be of order 10^7 yr. It is not clear whether a similar scenario could apply also to the inner disk.

VIII. SUMMARY

We have discussed in this chapter several topics concerning disks around HAeBe stars. There is today convincing evidence that many young intermediate-mass emission line stars, like their lower-mass counterparts, the classical TTSs, are surrounded by disks for most of their pre-main-sequence evolution.

The most massive and luminous stars among the HAeBe class, however, may form an exception, because they do not show obvious evidence of disks. The environment of these systems evolves relatively quickly, compared with Ae systems, possibly under the action of the strong stellar radiation fields.

A large part of this chapter has been dedicated to HAeBe stars of lower mass and luminosity. These stars are found very often to be surrounded by disks, which account for most of the observed millimeter fluxes. These disks are roughly similar to those of TTSs, with masses $\sim$0.06 $M_{\odot}$, radii in the range 100–300 AU, and ratios of disk to stellar mass of $\sim$0.04. In a large fraction of HAe systems, there is evidence that a small fraction of the circumstellar dust is in clumps of $\sim 10^{20}$–10^{21} g orbiting around the star. However, most of the dust mass is in small grains (less than about a few millimeters) distributed throughout the circumstellar disk.

There is no evidence that the circumstellar environment of HAe stars "evolves" significantly during the PMS phase (between $\sim 10^5$ and $\sim 10^7$ yr). At 10^7 yr, we find a large number of HAe stars with "normally" massive disks of dust and gas, as well as a few Vega-like stars (including β Pic), which retain a secondary debris disk after the original gas and dust disk has disappeared. If the growth of grains from submicron to planetesimal size occurs very rapidly, as most calculations indicate, then the conditions to initiate such growth in stars of intermediate mass may be present only after a period of about 10^7 yr. Two different kinds of observations, however, indicate that this might not be the case. Disk emission in HAe stars is characterized by a shallow spectral index in the submillimeter and millimeter, which is interpreted as an indication of growth of grains to millimeter sizes early in the PMS phase. The large-amplitude optical variability and sporadic infalling phenomena characteristic of UXORs are observed among HAe stars of any age. We are tempted to speculate that some kind of evolution has already occurred in all HAe stars by the time they become optically visible. A "bimodal" grain size distribution may have been reached where, although most of the original disk dust mass is in grains with typical sizes of a few millimeters, a small fraction of it resides in very large bodies (planetesimals). This situation is very stable, lasting for most of the PMS life of the star.

Clearly, much work is needed to understand the evolution of the environments of both HBe and HAe stars. The body of observational data is growing and is beginning to supply crucial information. In contrast, theoretical models of many of the processes involved are still very preliminary. Greater effort in this direction is certainly required.

Acknowledgments We are indebted to Mario van den Ancker for information on the stellar parameters of the HAeBe stars quoted in this paper and for having shared with us preliminary information on the ISO spectra of many of them. Many colleagues have kindly provided unpublished data to us. We thank Steven Beckwith, Thomas Henning, Scott Kenyon, Anneila Sargent, and Malcolm Walmsley for interesting discussions at various stages of this project. A. N. and V. P. G. were partly supported by ASI grant ARS-96-66 and CNR grant 97.00018.CT02 to the Osservatorio di Arcetri. V. M. is supported by the NASA *Origins* program.

REFERENCES

Appenzeller, I., Krautter, J., and Jankovics, I. 1983. Spectroscopy and infrared photometry of Southern T Tauri Stars. *Astron. Astrophys. Suppl.* 53:291–309.

Appenzeller, I., Jankovics, I., and Östreicher, R. 1984. Forbidden profiles of T Tauri stars. *Astron. Astrophys.* 141:108–115.

Backman, D. E., and Paresce, F. 1993. Main-sequence stars with circumstellar solid material: The Vega phenomenon. In *Protostars and Planets III*, ed. E. H. Levy and J. I. Lunine (Tucson: University of Arizona Press), pp. 1253–1304.

Barrado y Navascués, D., Stauffer, J. R., Hartmann, L., and Balachandran, S. C. 1997. The age of Gliese 879 and Fomalhaut. *Astrophys. J.* 475: 313–321.

Beckwith, S. V. W., and Sargent, A. I. 1991. Particle emissivity in circumstellar disks, *Astrophys. J.* 381:250–258.

Beckwith, S. V. W., and Sargent, A. I. 1993*a*. Molecular line emission from circumstellar disks. *Astrophys. J.* 402:280–291.

Beckwith, S. V. W., and Sargent, A. I. 1993*b*. The occurrence and properties of disks around young stars. In *Protostars and Planets III*, ed. E. H. Levy and J. I. Lunine (Tucson: University of Arizona Press), pp. 521–542.

Beckwith, S. V. W., Sargent, A. I., Chini, R. S., and Güsten, R. 1990. A survey for circumstellar disks around young stellar objects. *Astron. J.* 99:924–945.

Berrilli, F., Corciulo, G., Ingrosso, G., Lorenzetti, D., Nisini, B., and Strafella, F. 1992. Infrared emission from dust structures surrounding Herbig Ae/Be stars. *Astrophys. J.* 398:254–272.

Böhm, T., and Catala, C. 1993. A spectral atlas of the Herbig Ae star AB Aurigae. The visible domain from 391 nm to 874 nm. *Astron. Astrophys. Suppl.* 101:629–713.

Böhm, T., and Catala, C. 1994. Forbidden lines in Herbig Ae/Be stars: The [O I] 6300.31 and 6363.79 Å lines. I. Observations and qualitative analysis. *Astron. Astrophys.* 290:167–175.

Bonnell, I. A., Bate, M. R., and Zinnecker, H. 1998. On the formation of massive stars. *Mon. Not. Roy. Astron. Soc.* 298:93–102.

Boss, A. P. 1996. Extra-solar planets. *Physics Today* 49:32–38.

Butner, H. M., Natta, A., and Evans, N. J. II. 1994. "Spherical" disks: Moving toward a unified source model for L1551. *Astrophys. J.* 420:326–335.

Cabrit, S., Edwards, S., Strom, S. E., and Strom, K. M. 1990. Forbidden line emission and infrared excess in T Tauri stars: Evidence for accretion-driven mass loss? *Astrophys. J.* 352:687–700.

Cahn, J. H., Kaler, J. B., and Stanghellini, L. 1992. A catalogue of absolute fluxes and distances of planetary nebulae. *Astron. Astrophys. Suppl.* 94:399–452.

Calvet, N., Patino, A., Magris, G. C., and D'Alessio, P. 1991. Irradiation of accretion disks around young objects. I. Near-infrared CO bands. *Astrophys. J.* 380:617–630.

Cannon, A. J., and Mayall, M. W. 1949. The Henry Draper extension. *Harvard Ann.* 112.

Chavarria-K., C. 1985. Herbig's Ae and Be stars LkHα 198, a flare star candidate. *Astron. Astrophys.* 148:317–322.

Chiang, E., and Goldreich, P. 1997. Spectral energy distributions of T Tauri stars with passive circumstellar disks. *Astrophys. J.* 490:368–376.

Cohen, M., and Kuhi, L. V. 1976. Spectrophotometric studies of young stars. I. The Cepheus IV association. *Astrophys. J.* 210:365–376.

Cohen, M., and Kuhi, L. V. 1979. Observational studies of pre-main-sequence evolution. *Astrophys. J. Suppl.* 41:743–843.

Cohen, M., and Witteborn, F. C. 1985. Spectrophotometry at 10 microns of T Tauri stars. *Astrophys. J.* 294:345–356.

Cohen, M., Kuhi, L. V., Spinrad, H., and Harlan, E. A. 1981. Continuing changes in the peculiar nebulosity object PV Cephei. *Astrophys. J.* 245:920–926.

Corcoran, M., and Ray, T. P. 1997. Forbidden emission lines in Herbig Ae/Be stars. *Astron. Astrophys.* 321:189–201.

Corcoran, M., and Ray, T. P. 1998. Wind diagnostics and correlations with the near-infrared excess in Herbig Ae/Be stars. *Astron. Astrophys.* 331:147–161.

Crifo, F., Vidal-Madjar, A., Lallement, R., Ferlet, R., and Gerbaldi, M. 1997. β Pictoris revisited by Hipparcos. *Astron. Astrophys.* 320: L29–L32.

de Winter, D. 1996. Observational aspects of Herbig Ae/Be stars and of candidate young A and B stars. Ph.D. Thesis, University of Amsterdam.

de Winter, D., and Thé, P. S. 1990. The physical properties of the Hα emission line stars CD-42 11721, KK Oph and XX Oph. *Astrophys. Space Sci.* 166:99–121.

Di Francesco, J., Evans, N. J., II, Harvey, P. M., Mundy, L. G., and Butner, H. M. 1994. Constraining circumstellar environments: High-resolution far-infrared observations of Herbig Ae/Be stars. *Astrophys. J.* 432:710–719.

Di Francesco, J., Evans, N. J., II, Harvey, P. M., Mundy, L. G., Guilloteau, S., and Chandler, C. J. 1997. Millimeter and radio interferometry of Herbig Ae/Be stars. *Astrophys. J.* 482:433–441.

Di Francesco, J., Evans, N. J., II, Harvey, P. M., Mundy, L. G., and Butner, H. M. 1998. High-resolution far-infrared studies of intermediate mass pre-main-sequence objects. *Astrophys. J.,* in press (Dec. 10).

Drew, J. E., Proga, D., and Stone, J. 1998. A radiation-driven disc wind model for massive young stellar objects. *Mon. Not. Roy. Astron. Soc.* 296:L6–L10.

Dutrey, A., Guilloteau, S., Duvert, G., Prato, L., Simon, M., Schuster, K, and Menard, F. 1996. Dust and gas distribution around T Tauri stars in Taurus-Auriga. I. Interferometric 2.7mm continuum and ^{13}CO J = 1–0 observations. *Astron. Astrophys.* 309:493–504.

Eaton, N. L. and Herbst, W. 1995. An ultraviolet and optical study of accreting pre-main-sequence stars: UXORs. *Astron. J.* 110: 2369–2377.

Edwards, S. E., Cabrit, S., Strom, S. E., Heyer, I., Strom, K. M., and Anderson, E. 1987. Forbidden lines and Hα profiles in T Tauri spectra: A probe of anisotropic mass outflows and circumstellar disks. *Astrophys. J.* 321:473–495.

Edwards, S. E., Ray, T. P., and Mundt, R. 1993. Energetic mass outflows from young stars. In *Protostars and Planets III*, ed. E. H. Levy and J. I. Lunine (Tucson: University of Arizona Press), pp. 567–602.
Elias, J. H. 1978. A study of the Taurus dark cloud complex. *Astrophys. J.* 224:857–872.
Festou, M. C., Rickman, H., and West, R. M. 1993. Comets. I. Concepts and observations. *Astron. Astrophys. Rev.* 4:363–449.
Finkenzeller, U. 1985. Rotational velocities, spectral types, and forbidden lines of Herbig Ae/Be stars. *Astron. Astrophys.* 151:340–348.
Finkenzeller, U., and Mundt, R. 1984. The Herbig Ae/Be stars associated with nebulosity. *Astron. Astrophys. Suppl.* 55:109–141.
Friedemann, C., Reimann, H. G., and Guertler, J. 1994. Cloudy circumstellar dust shells around young variable stars. *Astrophys. Space Sci.* 212:221–229.
Fuente, A., Martín-Pintado, J., Bachiller, R., Neri, R., and Palla, F. 1998. Progressive dispersal of the dense gas in the environment of early-type and late-type Herbig Ae-Be stars. *Astron. Astrophys.* 334:252–263.
Grady, C., Pérez, M., Talavera, A., Bjorkman, K. S., de Winter, D., Thé, P. S., Molster, F. J., van den Ancker, M. E., Sitko, M. L., Morrison, N. D., Beaver, M. L., McCollum, B., and Castelaz, M. W. 1996*a*. The β Pictoris phenomenon among Herbig Ae/Be stars. *Astron. Astrophys. Suppl.* 120:157–177.
Grady, C., Pérez, M., Talavera, A., McCollum, B., Rawley, L. A., England, M. N., and Schlegel, M. 1996*b*. The β Pictoris phenomenon in A-shell stars: Detection of accreting gas. *Astrophys. J. Lett.* 471:L49–L52.
Greaves, J. S., Mannings, V., and Holland, W. S. 1999. The dust and gas content of a disk around the young star HR4796A. *Icarus* PPIV special issue, submitted.
Grinin, V. P. 1988. On the nature of the blue emission visible in deep minima of the young stars. *Pis'ma v Astron. Zn.* 14:65–69.
Grinin, V. P. 1994. Polarimetric activity of Herbig Ae/Be stars. In *The Nature and Evolutionary Status of Herbig Ae/Be Stars*, ASP Conf. Ser. 62, ed. P. S. Thé, M. R. Pérez, and E. P. J. van den Heuvel (San Francisco: Astronomical Society of the Pacific), pp. 63–70.
Grinin, V. P., Rostopchina, A. N., and Shakhovskoy, D. N. 1998. On the nature of cyclic variability of the UX Ori type stars. *Astron J. Lett.*, 24, 802–807.
Grinin, V. P., Thé, P. S., de Winter, D., Giampapa, M., Rostopchina, A. N., Tambovtseva, L. V., and van den Ancker, M. 1994. The β Pictoris phenomenon among young stars: The case of the Herbig Ae star UX Orionis. *Astron. Astrophys.* 292:165–174.
Hagen, G. 1970. An atlas of open cluster colour-magnitude diagrams. *Publ. David Dunlop Obs.* 4:1–22.
Hamann, F. 1994. Emission-line studies of young stars. 4. The optical forbidden lines. *Astrophys. J. Suppl.* 93:485–518.
Hartigan, P., Edwards, S. E., and Ghandour, L. 1995. Disk accretion and mass loss from young stars. *Astrophys. J.* 452:736–768.
Hartmann, L., Kenyon, S. J., and Calvet, N. 1993. The excess infrared emission of Herbig Ae/Be stars—Disks or envelopes? *Astrophys. J.* 407:219–231.
Henning, T., Burkert, A., Launhardt, R., Leinert, C., and Stecklum, B. 1998. Infrared imaging and millimetre continuum mapping of Herbig Ae/Be and FU Orionis stars. *Astron. Astrophys.* 336:565–585.
Herbig, G. H. 1960. The spectra of Be and Ae-type stars associated with nebulosity. *Astrophys. J. Suppl.* 4:337–368.
Herbig, G. H., and Bell, K. R. 1988. Third catalog of emission-line stars of Orion population. *Lick Obs. Bull.* 1111:1–90.
Herbig, G. H., and Rao, K. N. 1972. Second catalog of emission-line stars of Orion population. *Astrophys. J.*, 174:401–423.

Herbst, W., Holtzman, J. A., and Phelps, B. E. 1982. Optical monitoring of Orion population stars. I. Results for some T Tauri and Herbig Ae/Be stars. *Astron. J.* 87:1710–1729.
Hillenbrand, L. A. 1995. Herbig Ae/Be stars: An investigation of molecular environments and associated stellar populations. Ph.D. Thesis, University of Massachusetts.
Hillenbrand, L. A., Strom, S. E., Vrba, F. J., and Keene, J. 1992. Herbig Ae/Be stars—Intermediate-mass stars surrounded by massive circumstellar accretion disks. *Astrophys. J.* 397:613–643.
Hillenbrand, L. A., Meyer, M. R., Strom, S. E., and Skrutskie, M. F. 1995. Isolated star-forming regions containing Herbig Ae/Be stars. 1: The young stellar aggregate associated with BD+40 4124. *Astron. J.* 109:280–297.
Hirth, G. A., Mundt, R., and Solf, J. 1994. Jet flows and disk winds from T Tauri stars: The case of CV Tau. *Astron. Astrophys.* 285:929–942.
Holland, W. S., Greaves, J. S., Zuckerman, B., Webb, R. A., McCarthy, C., Coulson, I. M., Walther, D. M., Dent, W. R. F., Gear, W. K., and Robson, I. 1998. Submillimetre images of dusty debris around nearby stars. *Nature* 392:788–790.
Hollenbach, D., Johnstone, D., Lizano, S., and Shu, F. 1994. Photoevaporation of disks around massive stars and application to ultracompact HII regions. *Astrophys. J.* 428:654–669.
Houk, N., and Smith-Moore, A. 1988. *Michigan Spectral Survey,* Vol. 4. (Ann Arbor: University of Michigan Press).
Hutchinson, M. G., Albinson, J. S., Barrett, P., Davies, J. K., Evans, A., Goldsmith, M. J., and Maddison, M. J. 1994. Photometry and polarimetry of pre-main-sequence stars. *Astron. Astrophys.* 285:883–896.
Jaschek, C., Jaschek, M., Egret, D., and Andrillat, Y. 1991. Anomalous infrared emitters among A-type stars. *Astron. Astrophys.* 252:229–236.
Jijina, J., and Adams, F. C. 1996. Infall collapse solutions in the inner limit: Radiation pressure and its effects on star formation. *Astrophys. J.* 462:874–887.
Kenyon, S. J., and Luu, J. X. 1998. Accretion in the early Kuiper Belt. I. Coagulation and velocity evolution. *Astron. J.* 115:2136–2160.
Kenyon, S. J., and Luu, J. X. 1999. Accretion in the early Kuiper Belt. II. *Astron. J.,* submitted.
Kwan, J., and Tademaru, E. 1995. Disk winds from T Tauri stars. *Astrophys. J.* 454:382–393.
Levreault, R. M. 1988. A search for molecular outflows toward the pre-main-sequence objects. *Astrophys. J. Suppl.* 67:283–371.
Lissauer, J. J. 1993. Planet formation. *Ann. Rev. Astron. Astrophys.* 31:129–174.
MacConnell, D. J. 1968. A study of the Cepheus IV association. *Astrophys. J. Suppl.* 16:275–298.
Malfait, K., Bogaert, E., and Waelkens, C. 1998*a*. An ultraviolet, optical and infrared study of Herbig Ae/Be stars. *Astron. Astrophys.* 331:211–223.
Malfait, K., Waelkens, C., Waters, L. B. F. M., Vandenbussche, B., Huygen, E., and de Graauw, M. S. 1998*b*. The spectrum of the young star HD 100546 observed with the Infrared Space Observatory. *Astron. Astrophys.* 332:L25–L28.
Mannings, V. 1994. Submm observations of Herbig Ae/Be systems. *Mon. Not. Roy. Astron. Soc.* 271:587–600.
Mannings, V., and Emerson, J. P. 1994. Dust in disks around T Tauri stars: Grain growth? *Mon. Not. Roy. Astron. Soc.* 267:361–378.
Mannings, V., and Sargent, A. I. 1997. A high-resolution study of gas and dust around young intermediate-mass stars: Evidence for circumstellar disks in Herbig Ae systems. *Astrophys. J.* 490:792–802.

Mannings, V., and Sargent, A. I. 1999. High-resolution studies of gas and dust around young intermediate-mass stars: Paper II. *Astrophys. J.*, in press.
Mannings, V., Koerner, D. W., and Sargent, A. I. 1997. A rotating disk of gas and dust around a young counterpart to β Pictoris. *Nature* 388:555–557.
Meeus, G., Waelkens, C., and Malfait, K. 1998. HD 139614, HD 142666 and HD 144432: Evidence for circumstellar disks. *Astron. Astrophys.* 329:131–136.
Miroshnichenko, A., Ivezić, Z., and Elitzur, M. 1997. On protostellar disks in Herbig Ae/Be stars. *Astrophys. J. Lett.* 475:L41–L44.
Miyake, K., and Nakagawa Y. 1993. Effects of particle size distribution on opacity curves of protoplanetary disks around T Tauri stars. *Icarus* 106:20–41.
Natta, A. 1993. The temperature profile of T Tauri disks. *Astrophys. J.* 412:761–770.
Natta, A., and Krügel, E. 1995. PAH emission from Herbig Ae/Be and T Tauri stars. *Astron. Astrophys.* 302:849–860.
Natta, A., Palla, F., Butner, H. M., Evans, N. J., II, and Harvey, P. M. 1993. IR studies of circumstellar matter around Herbig Ae/Be and related stars. *Astrophys. J.* 406:674–691.
Natta, A., Grinin, V. P., Mannings, V., and Ungerechts, H. 1997. The evolutionary status of UX Orionis-type stars. *Astrophys. J.* 491: 885–890
Natta, A., Grinin, V. P., Mannings, V., and Ungerechts, H. 1999*a*. Cold dust in A-shell stars. In preparation.
Natta, A., Prusti, T. Neri, R. Thi, W. F., Grinin, V. P., and Mannings, V. 1999*b*. The circumstellar environment of UX Ori. *Astron. Astrophys.*, submitted.
Ossenkopf, V., and Henning, T. 1994. Dust opacities for protostellar cores. *Astron. Astrophys.* 291:943–959.
Osterloh, M., and Beckwith, S. V. W. 1995. Millimeter-wave continuum measurements of young stars. *Astrophys. J.* 439:288–302.
Palla, F., and Stahler, S. W. 1993. The pre-main-sequence evolution of intermediate-mass stars. *Astrophys. J.* 418:414–425.
Peŕez, M. R., and Grady, C. A. 1997. Observational overview of young intermediate-mass objects: Herbig Ae/Be stars. *Space Sci. Rev.* 82:407–450.
Pérez, M. R., Grady, C. A., and Thé, P. S. 1993. Ultraviolet spectral variability in the Herbig Ae Star HD 5999. Part II. The accretion interpretation. *Astron. Astrophys.* 274:381–390.
Pezzuto, S., Strafella, F., and Lorenzetti, D. 1997. On the circumstellar matter distribution around Herbig Ae/Be stars. *Astrophys. J.* 485:290–307.
Pollack, J. B., Hollenbach, D., Beckwith, S. V. W., Simonelli, D. P., Roush, E., and Fong, W. 1994. Composition and radiative properties of grains in molecular clouds and accretion disks. *Astrophys. J.* 421:615–639.
Pollack, J. B., Hubickyj, O., Bodenheimer, P., Lissauer, J. J., Podolak, M., and Greenzweig, Y. 1996. Formation of the giant planets by concurrent accretion of solids and gas. *Icarus* 124:62–85.
Proga, D., Stone, J. M., and Drew, J. E. 1998. Radiation-driven winds from luminous accretion disks. *Mon. Not. Roy. Astron. Soc.* 295: 595–617.
Prusti, T., Natta, A., and Krügel, E. 1993. Very small dust grains in the circumstellar environment of Herbig Ae/Be stars. *Astron. Astrophys.* 275:527–533.
Prusti, T., Grinin, V. P., and Natta, A. 1999. Variable infrared emission from the circumstellar environment of young, intermediate mass stars. In preparation.
Racine, R. 1968. Stars in reflection nebulae. *Astron. J.* 73:233–245.
Rostopchina, A. N. 1999. The location of the UX Ori-type stars in the Herzsprung-Russell diagram. *Astron. Rep.*, 43:113–118.
Sabbadin, F., and Hamzaoglu, E. 1981. Photographic and spectroscopic observations of planetary nebulae. *Astron. Astrophys.* 94:25–28.

Schaller, G., Schaerer, D., Meyenet, G., and Maeder, A. 1992. New grids of stellar models from 0.8 to 120 solar masses at $Z = 0.020$ and $Z = 0.001$. *Astron. Astrophys. Suppl.* 96:269–331.
Schmitt, W., Henning, T., and Mucha, R. 1997. Dust evolution in protoplanetary accretion disks. *Astron. Astrophys.* 325:569–584.
Shevchenko, V. S., Ibragimov, M. A., and Chernysheva, T. L., 1991. RSF 2 Cyg—a star-forming region associated with the extremely young cluster NGC 6910 and the Ae/Be Herbig stars BD+40 4124 and BD+41 3731. *Astron. Zn.* 68:466–479.
Shu, F. H., Tremaine, S., Adams, F. C., and Ruden, S. P. 1990. Sling amplification and eccentric gravitational instabilities in gaseous disks. *Astrophys. J.* 358:495–514.
Siebenmorgen, R., Natta, A., Krügel, E., and Prusti, T. 1998. A ring of organic molecules around HD 97300. *Astron. Astrophys.* 339:134–140.
Siebenmorgen, R., Natta, A., Krügel, E., and Prusti, T. 1999. Herbig Ae/Be stars in the mid-infrared: Spectrophotometry with ISOCAM. In preparation.
Sitko, M. L., Halbedel, E. M., Lawrence, G. F., Smith, J. A., and Yanow, K. 1994. *Astrophys. J.* 432:753–762.
Slettebak, A. 1966. Axial rotation in later B-type emission-line stars. *Astrophys. J.* 145:121–125.
Sorelli, C., Grinin, V. P., and Natta, A. 1996. Infall in Herbig Ae/Be stars: What Na D lines tell us. *Astron. Astrophys.* 309:155–162.
Stauffer, J. R., Hartmann, L. W., and Barrado y Navascués, D. 1995. An age estimate for the β Pictoris analog HR 4796A. *Astrophys. J.* 454:910–916.
Strom, S. E., Strom, K. M., Yost, J., Carrasco, L., and Grasdalen, G. 1972. The nature of the Herbig Ae and Be-type stars associated with nebulosity. *Astrophys. J.* 173:353–366.
Strom, S. E., Edwards, S., and Skrutskie, M. F. 1993. Evolutionary time scales for circumstellar disks associated with intermediate- and solar-type stars. In *Protostars and Planets III*, ed. E. H. Levy and J. I. Lunine (Tucson: University of Arizona Press), pp. 837–866.
Sylvester, R. J., Skinner, C. J., Barlow, M. J., and Mannings, V. 1996. Optical, infrared and millimetre-wave properties of Vega-like systems. *Mon. Not. Roy. Astron. Soc.* 279:915–939.
Terebey, S., Shu, F. H., and Cassen, P. 1984. The collapse of the cores of slowly rotating isothermal clouds. *Astrophys. J.* 286:529–551.
Testi, L., Palla, F., Prusti, T., Natta, A., and Maltagliati, S. 1997. A search for clustering around Herbig Ae/Be stars. *Astron. Astrophys.* 320:159–166.
Testi, L., Palla, F., and Natta, A. 1999. The onset of cluster formation around Herbig Ae/Be stars. *Astron. Astrophys.*, in press.
Thé, P. S. 1994. The photometric behavior of Herbig Ae/Be stars and its interpretation. In *The Nature and Evolutionary Status of Herbig Ae/Be Stars*, ASP Conf. Ser. 62, ed. P. S. Thé, M. R. Pérez, and E. P. J. van den Heuvel (San Francisco: Astronomical Society of the Pacific), pp. 23–30.
Thé, P. S., de Winter, D., and Pérez, M. 1994*a*. A new catalogue of members and candidate members of the Herbig Ae/Be (HAEBE) stellar group. *Astron. Astrophys. Suppl.* 104:315–339.
Thé, P. S., Pérez, M., and van den Heuvel, E. P. J. 1994*b*. *The Nature and Evolutionary Status of Herbig Ae/Be Stars,* ASP Conf. Ser. 62 (San Francisco: Astronomical Society of the Pacific).
van den Ancker, M. E., de Winter, D., and Tjin A Djie, H. R. E. 1998. Hipparcos photometry of Herbig Ae/Be stars. *Astron. Astrophys.* 330:145–154.
Voshchinnikov, N. V. 1998. On the properties of dust shells around Herbig Ae/Be stars. *Astron. Rep.* 42:54–60.

Voshchinnikov, N. V., and Grinin, V. P. 1991. Dust around young stars: Model of envelope of the Ae Herbig star WW Vulpeculae. *Astrofizika* 34:181–187.
Waelkens, C., Bogaert, E., and Waters, L. B. F. M. 1994. Spectral evolution of Herbig Ae/Be stars. In *The Nature and Evolutionary Status of Herbig Ae/Be Stars*, ASP Conf. Ser. 62, ed. P. S. Thé, M. R. Pérez, and E. P. J. van den Heuvel. (San Francisco: Astronomical Society of the Pacific), pp. 405–408.
Waelkens, C., Waters, L. B. F. M., de Graauw, M. S., Huygen, E., Malfait, K., Plets, H., Vandenbussche, B., Beintema, D. A., Boxhoorn, D. R., Habing, H. J., Heras, A. M., Kester, D. J. M., Lahuis, F., Morris, P. W., Roelfsema, P. R., Salama, A., Siebenmorgen, R., Trams, N. R., van der Bliek, N. R., Valentijn, E. A., and Wesselius, P. R. 1996. SWS observations of young main-sequence stars with dusty circumstellar disks. *Astron. Astrophys.* 315:L245–L248.
Warren, W. H., Jr., and Hesser, J. E. 1978. A photometric study of the Orion OB1 association. III. Subgroup analyses. *Astrophys. J. Suppl.* 36:497–672.
Waters, L. B. F. M., and Waelkens, C. 1998. Herbig Ae/Be stars. *Ann. Rev. Astron. Astrophys.* 36:233–267.
Weidenschilling, S. J. 1997. The origin of comets in the solar nebula: a unified model. *Icarus* 127:290–306.
Wenzel, W. 1969. Extremely young stars. In *Non-Periodic Phenomena in Variable Stars*, IAU Coll., ed. L. Detre (Budapest: Acad. Press), pp. 61–73.
Whitcomb, S. E., Gatley, I., Hildebrand, R. H., Keene, J., Sellgren, K., and Werner, M. W. 1981. Far-infrared properties of dust in the reflection nebula NGC 7023. *Astrophys. J.* 246:416–425.
Wolk, S. J., and Walter, F. M. 1996. A search for protoplanetary disks around naked T Tauri stars. *Astron. J.* 111:2066–2076.
Wooden, D. 1994. 8μm–13μm spectrophotometry of Herbig Ae/Be stars. In *The Nature and Evolutionary Status of Herbig Ae/Be Stars*, ASP Conf. Ser. 62, ed. P. S. Thé, M. R. Pérez, and E. P. J. van den Heuvel. (San Francisco: Astronomical Society of the Pacific), pp. 138–139.
Zinnecker, H., and Preibisch, T. 1994. X-ray emission from Herbig Ae/Be stars: A ROSAT survey. *Astron. Astrophys.* 292:152–164.
Zuckerman, B., and Becklin, E. E. 1993. Submillimeter studies of main-sequence stars. *Astrophys. J.* 414:793–802.

TRANSPORT PROCESSES IN PROTOSTELLAR DISKS

JAMES M. STONE
University of Maryland at College Park

CHARLES F. GAMMIE
University of Illinois at Urbana

and

STEVEN A. BALBUS AND JOHN F. HAWLEY
University of Virginia at Charlottesville, Virginia Institute of Theoretical Astronomy

The current understanding of angular momentum transport processes internal to protostellar accretion disks is reviewed. Recent numerical simulations of nonlinear accretion disk hydrodynamics have revealed that vertical convection, and hydrodynamic turbulence in general, are ineffective sources of angular momentum transport. However, the existence of a local, linear instability in weakly magnetized disks has shifted the focus to the role of magnetic fields and the presence of magnetohydrodynamical (MHD) turbulence. Numerical simulations demonstrate that the instability saturates as MHD turbulence with vigorous angular momentum transport. The existence and strength of the turbulence depend on the ionization fraction within the disk. The instability does not require perfect coupling between the magnetic field and gas, and it may operate on scales up to ~ 1 AU in protostellar disks with ionization fractions $f \equiv n_e/n_{\rm H}$ as low as $f \sim 10^{-13}$. Beyond this radius, however, the existence of MHD turbulence is increasingly problematic. Gravitational torques and global hydrodynamic waves remain as viable transport mechanisms.

I. INTRODUCTION

One of the primary evolutionary stages of star formation is accretion through a disk. Disks are ubiquitous because it is difficult to remove angular momentum from the gas during the infall stage, whereas it is easy to remove thermal energy via radiative cooling. Thus, the end product of infall will often be a rotationally supported thin disk. Disks are important for a variety of reasons: Their presence modifies the spectrum of young stellar objects; they regulate the mass accretion rate onto the protostar; they are thought to be the source of powerful winds and outflows; and, ultimately, they provide the environment in which planetary systems form. Moreover, accretion through a disk seems to be a long-lived stage of star formation. Hydrodynamical studies of the infall stage show that disks form very early in the evolution of a collapsing core (within one free-fall time, e.g., Bodenheimer et al., this volume), and observationally, disks remain associated with objects near the end of the star-forming process (weak-line T Tauri stars, e.g., Strom et al. 1993).

An understanding of the structure and dynamics of protostellar accretion disks is therefore an essential component of a complete theory of star formation. Most importantly, the mechanism for angular momentum transport in disks must be identified, for without it accretion cannot occur. Moreover, the transport mechanism is intimately linked with basic physical properties of the disk: its vertical structure, magnetization, temperature, and turbulent velocity dispersion.

Understanding angular momentum transport in accretion disks from first principles has proven challenging. Nevertheless, there has been considerable progress in the past few years, driven primarily by two advances. The first is the use of high-resolution numerical simulations to study the three-dimensional hydrodynamics and magnetohydrodynamics (MHD) of accretion disks for many dynamical times. As we discuss in this chapter, these simulations have taught us much about the nonlinear dynamics of disks. The second advance is the recognition of the fundamental role of magnetic fields in differentially rotating systems, beginning with the (re)discovery of a powerful, local, linear instability in weakly magnetized accretion disks (Balbus and Hawley 1991, 1992*b*). Although this magnetorotational instability (MRI) had been studied much earlier in its global form within the context of magnetized Couette flows (Velikhov 1959; Chandrasekhar 1960), its importance for accretion disks went unappreciated for decades. Three-dimensional numerical simulations have shown that the instability leads to fully developed MHD turbulence and significant outward angular momentum transport. The implications of this instability for the structure and evolution of protostellar accretion disks are very much a subject of active inquiry.

Most of this review will focus on discoveries made concerning the local fluid dynamical processes that govern *internal* transport. We will discuss the progress made in understanding the roles that convection and nonlinear hydrodynamical instabilities play (or, more appropriately, do *not* play) in contributing to transport. We will describe our current understanding of the MRI in ionized and partially ionized disks.

We will confine our discussion to the dynamics of Keplerian protostellar disks, which are much less massive than the central, isolated protostar. Of course, if the disk is massive or if tidal interactions are important, gravitational torques can dominate the transport. Although much of the stellar material may be processed this way, the majority of a disk's lifetime is probably spent in the low-mass phase. Gravitational mechanisms are briefly discussed in section II.C, and are reviewed more extensively by Bodenheimer et al. (this volume) and Adams and Lin (1993).

Internal torques are not the only possible transport mechanism. Significant angular momentum loss can also occur through winds. Königl and

Pudritz (this volume) review the substantial and important progress made in studies of protostellar disk winds and jets. Our current understanding, however, of the relationship between internal disk dynamics and outflow processes remains incomplete.

There are several recent reviews that provide further details on the dynamics of protostellar disks. Adams and Lin (1993) reviewed disks in *Protostars and Planets III*. Papaloizou and Lin (1995*b*) discussed transport processes in disks. Lin and Papaloizou (1996) focus on observational constraints. Balbus and Hawley (1998) review accretion disk turbulence and self-consistent disk dynamo theory.

II. BASIC TRANSPORT MECHANISMS

Because of the ubiquity of disk accretion in systems ranging from protostars to active galactic nuclei (AGNs), it has long been assumed that outward angular momentum transport must be an unavoidable consequence of the internal dynamics of disks. Recognizing that molecular viscosity was many orders of magnitude too small to account for observed accretion rates, Shakura and Sunyaev (1973) introduced the ansatz, based on dimensional arguments, that there must exist an "anomalous" stress tensor component $T_{R\phi}$ whose magnitude scales with the gas pressure P according to

$$T_{R\phi} = \alpha P \tag{1}$$

where α is a dimensionless constant and R and ϕ are the radial and angular coordinates. A similar approach was taken by Lynden-Bell and Pringle (1974), who assumed an anomalous Stokes viscosity

$$\nu_t = \alpha c_s H \tag{2}$$

where c_s is the sound speed in the disk and H is the vertical scale height. The subscript t emphasizes that from the beginning, the anomalous viscosity was assumed to arise from turbulence in the disk. This parameterization made possible the construction of vertically averaged "α-disk" models (see review of Pringle 1981), which have proved extremely useful. Nonetheless, a fundamental question went unanswered in these early models: What is the source of this anomalous (turbulent) viscosity? Three categories seem natural: (1) hydrodynamic turbulence driven either by vertical convection or nonlinear instabilities, (2) MHD turbulence, and (3) nonlocal mechanisms (e.g., gravitational torques in self-gravitating disks; nonaxisymmetric waves). We consider each in turn.

A. Hydrodynamical Turbulence

There are two primary difficulties with the suggestion that purely hydrodynamic turbulence drives angular momentum transport in protostellar accretion disks. The first is to decide what drives the turbulence. This is a nontrivial issue, because hydrodynamic disks are linearly stable by the Rayleigh criterion $\partial L/\partial R > 0$, where L is the specific angular momentum. The second is to ensure that turbulence actually transports angular momentum outwards. Again, this is not immediately obvious, since outward transport requires more than the presence of radial and azimuthal velocity fluctuations. It requires that they be well correlated (Prinn 1990; Balbus and Hawley 1998).

In protostellar disks there is a simple mechanism for hydrodynamic turbulence, namely vertical convection driven by efficient radiative cooling of the surface layers of the disk. Adams and Lin (1993) suggested that nonlinear dissipation of modes identified through a linear analysis (Ruden et al. 1988) might give rise to $\alpha \sim 10^{-2}$ to 10^{-3}. However, a linear analysis of nonaxisymmetric disturbances in the shearing-sheet approximation (Ryu and Goodman 1992) showed that these modes actually transport angular momentum *inward*. Two-dimensional axisymmetric simulations of the nonlinear evolution of convective modes by Kley et al. (1993) produced inward transport, but the authors were careful to note that nonaxisymmetric structure might lead to something very different.

With the increase in available computational power, direct hydrodynamical simulations of three-dimensional convective disk flows became possible, and several studies were carried out (Cabot and Pollack 1992; Cabot 1996; Stone and Balbus 1996). These simulations begin with a three-dimensional patch of a shearing disk that uses Cartesian geometry while retaining gravitational and Coriolis forces. The disk is initially convectively unstable. Stone and Balbus (1996) found that at late time the convective cells in the R-Z plane look much the same as do those in a nonrotating fluid. However, in the ϕ-Z plane the cells are stretched by shear into long sheets. Figure 1 shows the angular momentum transport rate (characterized by the α parameter) over 40 orbits in a disk in which vertical convective motions are sustained by heat input at the equator. Two results are immediately obvious. The first is that the magnitude of the transport rate is very small, generally $|\alpha| < 10^{-4}$. The second is that the time-averaged value of α is negative, implying net inward angular momentum transport. These results show that even when vigorous convective motions are driven in a protostellar disk, the turbulence does not produce significant angular momentum transport.

Convective instabilities are one way to drive hydrodynamic turbulence in an accretion disk; nonlinear shear instabilities have been proposed as another (Dubrulle 1993). The idea is that the nonlinear instability observed in certain types of high-Reynolds-number Couette flows is a generic breakdown; Keplerian disks should be similarly unstable. Recently, it has become possible to test this conjecture directly with numerical hydrodynamical experiments. Balbus et al. (1996) computed a series of local simulations of both differentially rotating and simple shear

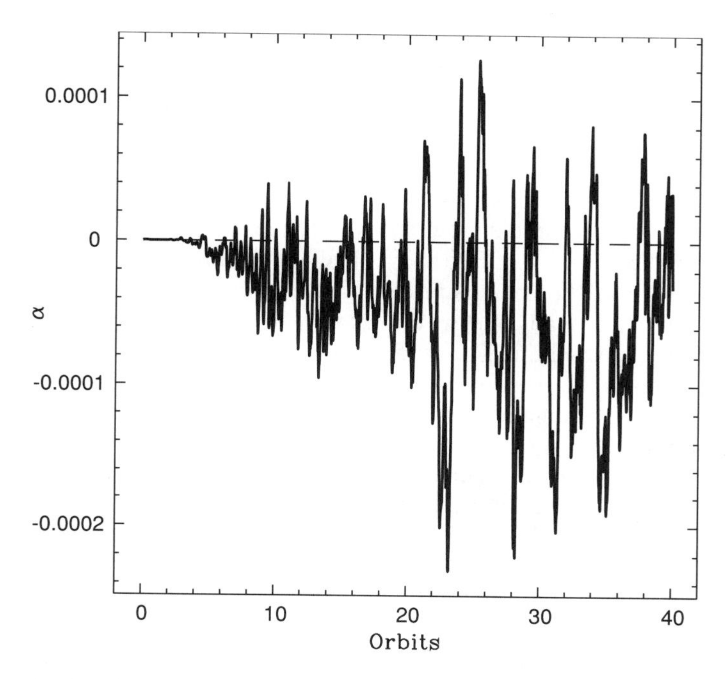

Figure 1. Evolution of the Reynolds stress normalized by the pressure at the disk midplane (equivalent to the α parameter) in a disk in which vertical convection is driven. Note that α is both very small and, on average, negative.

flows. In these Keplerian flow simulations, despite substantial initial perturbations, turbulence inevitably decayed.

Nonlinear shear instabilities are known to have a critical Reynolds number above which instability sets in. In fact, the *linear* Rayleigh instability itself has a large critical Reynolds number in Couette experiments (Drazin and Reid 1981). Thus, one might be concerned that numerical simulations have insufficient resolution (or equivalently, too low an effective Reynolds number) to capture the development of the nonlinear instability. However, such concerns can be addressed. First, relatively low-resolution simulations have no difficulty reproducing high-Reynolds-number nonlinear instability where it is known to exist from laboratory experiments, namely in a simple Cartesian shear flow. Since the Keplerian system differs only by the addition of orbital dynamical forces, the absence of nonlinear instabilities in differentially rotating flows cannot be due entirely to the presence of numerical viscosity. Second, experiments with a wide range of numerical resolutions (as many as 256^3 grid zones) and with both the ZEUS (Stone and Norman 1992) and the PPM (Colella and Woodward 1984) hydrodynamics algorithms (which have very different numerical dissipation properties) all give the same result: Finite-amplitude initial perturbations quickly die out in Keplerian shear flows, and they die out in remarkably detailed numerical agreement, regardless of which code is used (Hawley et al. 1999).

Balbus et al. (1996) show that the simulation results can be understood from a simple analysis of the dynamical equations themselves. The reader may refer to this paper for technical details, but the gist of the argument is as follows. In both simple shear and differentially rotating flows the shear itself provides a source of free energy to drive turbulence, but energy is not the whole story. The critical question is whether or not the turbulent stress tensor can dynamically sustain the velocity fluctuations required to transport angular momentum. The crucial difference between simple shear and disk systems is the streamwise momentum conservation equation. In differential rotation, angular momentum fluctuations feed off the background angular momentum gradient, which has the opposite sign from the angular *velocity* gradient energy source. Outward transport of angular momentum is directed "uphill" against the angular momentum gradient, and this reduces the amplitude of the angular velocity fluctuations needed for the turbulent transport itself. In contrast, a simple shear flow has only a velocity gradient; there is no significant distinction between linear momentum and linear velocity. In simple shear the dynamical fluctuations that produce transport are sustained, since tapping into the cross-stream velocity gradient is always a source for streamwise fluctuations.

The stabilizing influence of differential rotation can be demonstrated by evolving local disk models that vary only in the choice of background angular velocity gradient, characterized by $\Omega \propto R^{-q}$ (Balbus et al. 1996). Keplerian flows ($q = 1.5$) are both linearly and nonlinearly stable. In models in which the background angular velocity gradient is small ($q \rightarrow 2$), the linear restoring forces are weaker, and the turbulence decays more slowly. Finally, when $q = 2$, angular momentum fluctuations can be sustained: The system becomes nonlinearly unstable. Indeed, the moment equations of constant-angular-momentum flow are formally equivalent to those of simple Cartesian shear. For $q > 2$, both the angular velocity and the angular momentum decrease with radius, and the system is Rayleigh unstable. Thus, the nonlinear instability manifests itself only very near the boundary between linear instability and linear stability. It is only in this domain, where linear restoring forces are vanishingly small, that nonlinear effects hold sway.

The implications of these results go beyond the assertion that Keplerian flows are not susceptible to the same sort of instability that afflicts shear flows. They indicate that inward convective transport and nonlinear stability of Keplerian flows are different sides of the same coin. The same orbital dynamics that make the disk stable also prevent significant transport of angular momentum even when fluctuations are maintained from sources other than differential rotation. An explicit demonstration is provided by a simulation in which turbulence is driven by direct random forcing (Hawley and Balbus 1997). In this simulation, there is *no* net angular momentum transport, despite the presence of substantial velocity perturbations. In conclusion: (1) Hydrodynamical turbulence does not develop spontaneously within a Keplerian disk, and (2) if such turbulence is provided by some means (e.g., convection or random "stirring"), significant outward transport of angular momentum is neither guaranteed, nor, it would appear, likely.

B. MHD Turbulence

Unlike purely hydrodynamic disks, weakly magnetized accretion disks are subject to a powerful local linear instability (Balbus and Hawley 1991, 1992*b*). The fastest-growing modes are amplified at a rate of order the orbital frequency, so that the perturbation energy is amplified by a factor of 10^4 *per orbit*. This enormous growth rate [it is likely that no linear instability driven by shear in the disk can grow faster (Balbus and Hawley 1992*a*)] implies that the nonlinear stage of the MRI must have important consequences for disks.

Again, numerical simulations are required to study the nonlinear evolution. The first results from two-dimensional simulations showed that in axisymmetry, disks with a mean vertical field evolve into the channel solution, consisting of two oppositely directed streams (Hawley and Balbus 1991, 1992). Goodman and Xu (1994) showed that this channel solution is an exact nonlinear solution to the shearing-sheet MHD equations. More importantly, they showed that in three dimensions, the channel solution is subject to disruptive "parasitic" instabilities.

Because of the importance of nonaxisymmetric modes to the nonlinear outcome of the MRI, fully three-dimensional simulations are essential. The first such results (Hawley et al. 1995) borrowed a technique from Couette flow studies (Lees and Edwards 1972) to treat a local patch of the disk, incorporating the shear of the radial boundaries. We refer to this system as the "shearing box." Simulations of homogeneous boxes with a variety of field strengths and configurations confirm the results of the linear analyses and find that the nonlinear outcome of the MRI is MHD turbulence, which amplifies and sustains an initially weak field. It also provides for vigorous angular momentum transport. For example, Fig. 2 plots $\alpha = T_{R\phi}/P$ in a typical example studied by Hawley et al. (1995). In contrast to the convective turbulence shown in Fig. 1, strong outward transport of angular momentum is sustained by the MRI.

Brandenburg et al. (1995) and Hawley et al. (1996) addressed the question of dynamo activity. They carried out simulations of initially self-contained magnetic fields in shearing boxes. (The two studies used different boundary conditions.) Both studies found that the MRI leads to dynamo activity and sustains magnetic fields against dissipation for many resistive decay times. Significantly, the dynamo action cannot be described by a kinematic theory. Lorentz forces can never be ignored in determining the

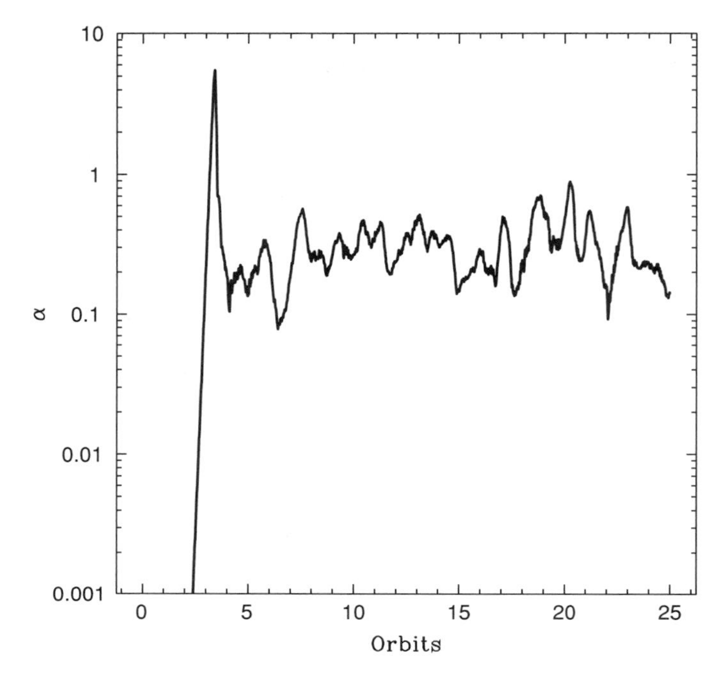

Figure 2. Evolution of the total (Reynolds plus Maxwell) stress normalized by the pressure at the disk midplane (equivalent to α) in a three-dimensional local simulation of a weakly magnetized disk. The MRI is able to sustain strong angular momentum transport apparently indefinitely.

velocity field, even when the field is weak. The point is that there is nothing with respect to which the field can be "weak"; the strength of the magnetic field simply sets the length scale at which the Lorentz force is significant.

Vertically stratified shearing boxes were the subject of further three-dimensional simulations (Brandenburg et al. 1995; Stone et al. 1996). Stratification does not change most of the properties of the instability reported for homogeneous boxes (e.g., MHD turbulence still results). Most importantly, buoyancy is *not* a significant saturation mechanism; instead, local dissipation dominates.

Going beyond the local shearing box representation is difficult because of the large dynamic range between the scale height H and the outer radius in a thin disk. Moreover, the very low densities encountered in the envelope of the disk result in very high Alfvén speeds. Miller and Stone (1999), using special numerical techniques that limit the Alfvén speed, have studied a local patch of a disk extending five scale heights above and below the midplane. Although most of the dissipation still occurs within the disk, buoyancy leads to a strongly magnetized, low-density corona. This may have important dynamical consequences for magnetically driven winds from the disk.

Fully global simulations are now being carried out though in their infancy. Matsumoto and Shibata (1997) modeled a thick disk embedded in a vertical field and found that rapid infall of the disk led to strong outflows near the axis. More recently, Armitage (1998) and Hawley and Balbus (1999) have followed the development of turbulence and angular momentum transport in global simulations. No calculation to date has been carried through to the point where a classical global α-disk has emerged.

To conclude this brief summary, nonlinear MHD simulations all indicate that the MRI produces outward angular momentum transport in disks. Values of α found in the current simulations range from 5×10^{-3} to ~ 0.5 (the range in these values is related to whether or not there is a net flux of magnetic field).

Since nearly all of the studies described above use the assumption of ideal MHD (i.e., that the magnetic field is frozen to the fluid), one may ask whether the results are applicable to protostellar disks, which are only weakly ionized. There are two processes capable of suppressing the MRI in circumstellar disks: ambipolar diffusion (Blaes and Balbus 1994) and resistivity (Jin 1996; Balbus and Hawley 1998). Ambipolar diffusion tends to be more important at higher field strength and lower density. For parameters typical of circumstellar disk models, both can be important (Hayashi 1981).

The resistivity of partially ionized plasma is (e.g., Blaes and Balbus 1994)

$$\eta = 230 \left(\frac{n_n}{n_e} \right) T^{1/2} \text{ cm}^2 \text{ s}^{-1} \tag{3}$$

where n_n is the neutral number density, n_e is the electron number density, and T is the temperature. A measure of its relative importance is given by the magnetic Reynolds number

$$\text{Re}_M = \frac{H c_s}{\eta} = \frac{c_s^2}{\Omega \eta} \tag{4}$$

Now, on dimensional grounds, we expect that the dissipation rate associated with a harmonic disturbance of wavenumber k to be ηk^2. The resistive stabilization of linear disturbances occurs when

$$\eta k^2 \simeq \Omega \tag{5}$$

a result supported by more detailed calculations (Papaloizou and Terquem 1997; Balbus and Hawley 1998). The most rapidly growing wavenumber of the MRI is approximately Ω / v_A; therefore, the linear stability will be significantly affected by resistivity when

$$\eta \simeq \frac{v_A^2}{\Omega} \tag{6}$$

corresponding to

$$\mathrm{Re}_M \simeq \frac{c_s^2}{v_A^2} \sim \beta \tag{7}$$

Thus, the linear growth of the instability should be strongly affected when the magnetic Reynolds number is less than the plasma β parameter.

The instability can be completely quenched if

$$\eta \left(\frac{2\pi}{H}\right)^2 \simeq \Omega \tag{8}$$

because dissipation rates would then exceed growth rates for all vertical wavenumbers. This leads to

$$\mathrm{Re}_M \simeq 4\pi^2 \tag{9}$$

as the threshold for damping at all wavelengths smaller than the disk scale height. On scales ~ 1 AU in the solar nebula, this requires an ionization fraction $f \lesssim 10^{-13}$. Only a very small ionization fraction is enough to trigger linear instability. In protostellar disks, there are many regions where f is small but larger than the value required for completely suppressing the MRI. The question then becomes whether the resulting turbulence can be sustained. Thus, studies of the nonlinear evolution of the MRI in resistive disks are important for protostellar systems.

Several groups have begun to investigate the nonlinear structure of the MRI in highly resistive disks. Two-dimensional simulations have recently been presented by Sano et al. (1998), who studied the resistive range $0.3 \leq \mathrm{Re}'_M \leq 3$ (where $\mathrm{Re}'_M \equiv v_A^2/\eta\Omega$, rather than the definition given in equation [4]) using purely vertical fields and a variety of field strengths. As expected from a linear analysis (Jin 1996), the instability does not grow for sufficiently small Re_M. At Reynolds numbers above this linear limit, Sano et al. (1998) discovered that the two-dimensional channel solution can saturate via reconnection across the radial streams. Fleming et al. (1999) have studied the three-dimensional evolution of the instability in resistive disks for a variety of initial field strengths and topologies. When the disk contains a mean vertical field supported by currents external to the computational domain, ohmic dissipation can never completely destroy the background field. In this case, MHD turbulence is sustained above a Reynolds number of around 10^3. Between this value and $\mathrm{Re}_M \sim 4\pi^2$, the instability is present and produces periodic behavior characterized by growth of the channel solution, saturation, and then decay of the fluctuations until the field again becomes weak. At this point, the channel solution emerges again, and the cycle repeats. At low Re_M, reconnection across the streams in the channel solution is an effective saturation mechanism, as was seen in the two-dimensional simulations of Sano et al. (1998).

For simulations with no mean field, however, the evolution can be affected by much larger values of the Reynolds number. Turbulence initially generated by the instability dies away, due to field dissipation, below a magnetic Reynolds number between 10^4 and 10^3. This suggests that in the absence of a mean field, MHD turbulence in protostellar disks can be inhibited by ohmic dissipation at much larger magnetic Reynolds numbers than expected from the linear analysis.

In low-density regions of the disk, where the neutral-ion collision time is longer than a gyroperiod, the dynamics of a weakly ionized disk are described by ambipolar diffusion. To study the MRI in this regime, MacLow et al. (1995) assume that the ion inertia can be ignored and that the ion density is a power law of the neutral density; they treat the effect of ambipolar diffusion as a nonlinear diffusion term to the induction equation (e.g., Shu 1992). Their simulations show suppression of the linear MRI when the ion-neutral diffusion rate is sufficiently large. Similar results with similar assumptions were obtained by Brandenburg et al. (1995) in three-dimensional simulations. More recently, Hawley and Stone (1998) have reported comprehensive three-dimensional simulations of two-fluid disks in which the ions and neutrals are evolved as separate fluids coupled through a drag term. They find that significant turbulence in the neutral fluid is produced only when the ion-neutral collision frequency is $\gtrsim$100 times the orbital frequency. The criterion for the linear modes to grow is only that these frequencies be comparable (Blaes and Balbus 1994); again, the nonlinear criterion for significant transport is more stringent.

C. Other Mechanisms

Gravitational instability can also create nonaxisymmetric structures that transfer angular momentum by gravitational and Reynolds stresses. The key parameter that determines whether or not gravitationally driven transport is possible is Toomre's Q parameter, where $Q \leq 1$ implies local gravitational instability:

$$\begin{aligned} Q \equiv \frac{c_s \Omega}{\pi G \Sigma} &= 56 \left(\frac{M_\star}{\mathrm{M}_\odot}\right)^{\frac{1}{2}} \left(\frac{R}{\mathrm{AU}}\right)^{-\frac{3}{2}} \left(\frac{T}{100\ \mathrm{K}}\right)^{\frac{1}{2}} \left(\frac{\Sigma}{10^3 \mathrm{g\ cm}^{-2}}\right) \\ &\simeq \left(\frac{H}{R}\right)\left(\frac{M_\star}{M_{\mathrm{disk}}}\right) \end{aligned} \tag{10}$$

where for the last equality we have taken $M_{\mathrm{disk}} \simeq \pi R^2 \Sigma$. Local or global gravitational instability usually requires Q close to 1 somewhere in the disk (i.e., a cold, massive disk).

The nonlinear outcome of gravitational instability in circumstellar disks is not yet understood. Several ideas have been suggested, including fragmentation (e.g., Nelson et al. 1998; Bodenheimer et al., this volume), rapid rearrangement of the disk material to a more stable configuration (e.g., Laughlin and Bodenheimer 1994), and persistent instability due to cooling (e.g., Gammie 1996*a*). There is little question that fully developed nonaxisymmetric gravitational instabilities can be a potent source of angular momentum transport. In contrast to shear-driven turbulence, it is not clear how gravitational disturbances approach a dynamically steady state.

Nonaxisymmetric hydrodynamic waves provide another possible mechanism to transport both energy and angular momentum throughout the body of the disk. This is an extensive subject; here we offer only a brief summary of key points. A comprehensive review of tidally excited waves in disks is given in Lin and Papaloizou (1993).

The hydrodynamic turbulent stress tensor is $T_{R\phi} \equiv \langle \rho u_R u_\phi \rangle$, where ρ is the mass density and $\mathbf{u}$ is the noncircular component of the velocity $\mathbf{v}$; i.e., $\mathbf{v} = R\Omega\hat{\boldsymbol{\phi}} + \mathbf{u}$. The angle brackets represent an average over ϕ and suitable radial and vertical scales. In shear turbulence (although not in unmagnetized disks!) the turbulent stress tensor is a significant source of angular momentum transport because the radial and azimuthal velocity fluctuations are well correlated. Wave transport depends on precisely the same correlation tensor. For axisymmetric waves, the velocity components are uncorrelated (more accurately, they are destructively out of phase); for nonaxisymmetric waves, the stress is typically linear in the azimuthal to radial wavenumber:

$$\langle u_R u_\phi \rangle \sim (m/kR)u^2 \tag{11}$$

The equivalent α parameter (pertaining to transport but not to energy dissipation) would be

$$\alpha \sim (m/kR)\mathcal{M}^2 \tag{12}$$

where $\mathcal{M}$ is the Mach number of the velocity fluctuation. Trailing spiral waves transport angular momentum outward. Even in the extreme case of velocity amplitudes that approach c_s, the α parameter of WKB waves will only be of order $1/kR$. When the wave velocity amplitudes are highly subsonic, the resulting α is even smaller yet.

Other potential difficulties faced by any wave transport mechanism are that waves must be present throughout the bulk of the disk (either because they can propagate or because they are directly excited *in situ*) and that such disturbances need to be dissipated to exchange angular momentum with the disk. Because waves tend to refract, it is not a simple matter for them to propagate throughout the interior of the disk if they are excited near the outer edge by a companion. Longer-radial-wavelength disturbances fare better in this regard, as they are less prone to refraction.

Different types of waves display different propagation behavior. Near the midplane of a disk, the local dispersion relation for hydrodynamical WKB waves is

$$(\omega - m\Omega)^4 - (\omega - m\Omega)^2(k_Z^2a^2 + k_R^2a^2 + \kappa^2) + k_Z^2a^2\kappa^2 = 0 \tag{13}$$

where k_R and k_Z are the radial and vertical wavenumbers, ω the wave frequency, Ω the local rotation frequency, a the adiabatic sound speed, and κ

is the epicyclic frequency. At large wavenumbers, two distinct branches of solutions emerge from the dispersion relation: density waves (rotationally modified sound waves),

$$(\omega - m\Omega)^2 - (k_Z^2 a^2 + k_R^2 a^2 + \kappa^2) = 0 \tag{14}$$

and inertial waves (Vishniac and Diamond 1989),

$$(\omega - m\Omega)^2 = \left(\frac{k_Z^2}{k_Z^2 + k_R^2}\right)\kappa^2 \tag{15}$$

Density waves can steepen and form shocks (Sawada et al. 1986; Różyczka and Spruit 1993); the idea that shocks might be a possible source of angular momentum transport in disk galaxies is an old one (Roberts 1969; Shu et al. 1973). Two-dimensional simulations suppress the effects of refraction and show relatively strong spiral structure (Różyczka and Spruit 1993). More recent results of tidally driven three-dimensional simulations (Yukawa et al. 1997) are less dramatic. The status of spiral shock accretion remains open, pending more refined numerical studies.

Inertial waves were first considered in the context of accretion disks by Vishniac and Diamond (1989), and they can be resonantly excited (Goodman 1993; Lubow and Ogilvie 1998). However, these waves also propagate slowly and tend to be refracted upwards. A strong case for substantial angular momentum transport by inertial waves has yet to be made.

Long-wavelength bending waves of the disk, which will be excited when the orbital plane of the companion differs from the disk plane (Papaloizou and Lin 1995*a*; Papaloizou and Terquem 1995; Terquem 1998), represent yet another possibility. These waves are not internally propagating WKB modes, so they are not represented by equation (13). It is a complicated matter simply to write down a general dispersion relation for bending waves ("warps"), and we shall not present such a formula here (see, e.g., Papaloizou and Lin 1995*a*).

Terquem (1998) has recently analyzed the properties of bending waves in a viscous Keplerian disk. Because such waves need to be damped to interact with the disk, some form of dissipation needs to be present. For linear waves, some viscosity is presumably associated with the turbulent transport in the disk, but this means that the waves themselves cannot be the entire basis of disk transport. They can alter the effective α, however, as a matter of principle, by a factor of order unity. If the waves reach nonlinear amplitude, they may be self-dissipating, leading to a scenario in some respects similar to spiral shocks.

In summary, wave propagation by itself is less efficient than fully developed MHD turbulence as angular momentum transport mechanism, but the excitation process is strongest in the outer regions of protostellar disks, where MHD turbulence may be ineffective. The interplay between waves and MHD turbulence is an underexplored (indeed, largely unexplored) process that may be quite important for the evolution of binary disks.

III. TRANSPORT PROCESSES AND THE RADIAL STRUCTURE OF DISKS

From the results discussed in section II, it is clear that the nature and efficacy of the angular momentum transport mechanism strongly depend on physical conditions within the protostellar disk, which will vary rapidly with both radius and height. Thus, transport across the entire disk, from protostar to outer edge, will not be mediated by a single mechanism. In the following subsections, we divide the disk into three regions in which the dynamics of angular momentum transport may be dominated by significantly different processes. The divisions are somewhat arbitrary, and the actual structure is almost certainly more complex than we have assumed. Figure 3 is a sketch of the radial structure described in the following sections.

A. The Circumstellar Region: Star-Disk Interaction

Within the innermost few stellar radii, the dynamics of the disk will be strongly influenced by interaction with the central star. There is substantial evidence that protostars have strong stellar magnetic fields (e.g., Montmerle et al. 1993). Thus, the star-disk interaction will be a complex interplay driven by the large inertia of the accretion flow and the strong magnetic forces within the protostellar magnetosphere.

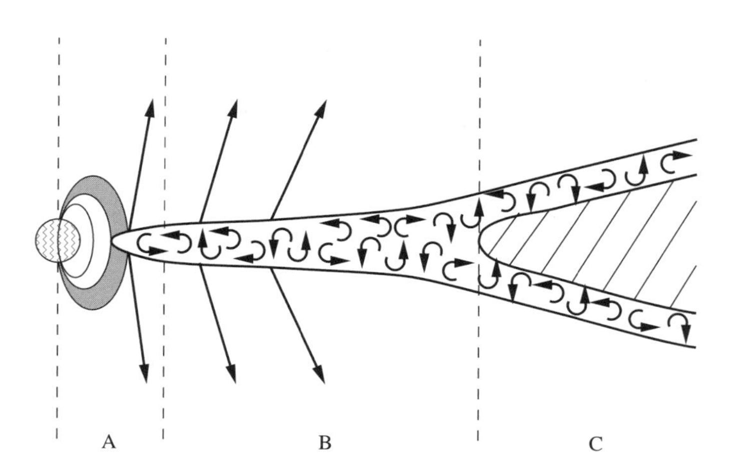

Figure 3. Possible radial structure of a protostellar accretion disk, as discussed in the text. Region A is dominated by the interaction of the stellar magnetosphere and the disk. In region B the ionization fraction is large enough that the disk is dominated by MHD turbulence throughout. In region C the central regions of the disk are decoupled from the magnetic field and therefore quiescent, resulting in layered accretion. Beyond region C, the disk merges with the ambient cloud core and can no longer be considered as an isolated structure. Outflows are probably associated with the inner parts of the disk (regions A and B). The locations of the transitions between regions are highly uncertain.

The theory of the interaction of a stellar magnetosphere and an accretion disk is still being developed (Pringle and Rees 1972; Ghosh and Lamb 1979; Lovelace et al. 1995; Shu et al. 1994). It seems likely, both on theoretical and observational grounds, that within the Alfvén radius R_A (at which point the Alfvén speed in the disk is comparable to the orbital velocity), the disk will be disrupted by the stellar field. The accretion flow will no longer be in the equatorial plane but instead will flow along stellar field lines to the magnetic poles. Precisely how plasma is loaded onto the stellar field lines is poorly understood; it may require a large effective (perhaps turbulent) resistivity. Near R_A, the disk will be threaded by stellar field lines, and a strong magnetically driven outflow may be produced (see the chapter by Shu et al., this volume). Magnetic coupling between the star and disk can control the rotation rate of the central star (Königl 1991; Lovelace et al. 1995).

Angular momentum transport in this star-disk interface can be regulated by several processes. The most likely are (1) MHD turbulence, (2) torques exerted by the rigid magnetospheric field, and (3) magnetically driven winds and outflows. The large anomalous resistivity required to load the disk plasma onto stellar field lines suggests that the internal dynamics in this region are turbulent and the magnetic field is highly tangled, implying that MHD turbulence must be present to some degree.

The dynamics of the circumstellar region may be the most complex in the entire system. For this reason, numerical simulations have become an increasingly important investigative tool. Hayashi et al. (1996) have used simulations to study the evolution of a Keplerian disk embedded in a pure dipole field anchored in the central star. Shearing of the field lines produces expansion of the magnetosphere (Lovelace et al. 1995; Lynden-Bell and Boily 1994), and reconnection ejects isolated plasmoids. As evidence for such reconnection events, these authors point to recently observed X-ray flares in T Tauri stars. Goodson et al. (1997) use a nested mesh code to follow the evolution of a similar problem over large spatial scales. As a result, they find not only plasmoid ejection but also global collimation of these objects into jetlike structures. Finally, Miller and Stone (1997) follow the axisymmetric evolution of a Keplerian disk in the magnetosphere of a central star. Rapid radial collapse of the disk driven by the MRI or magnetic braking is always observed, and polar cap accretion flows occur in some circumstances. Although substantial progress has been made, important questions (such as identifying the *dominant* transport mechanism in this region) remain for future studies.

B. The Inner Disk: MHD Turbulence

Because the transport properties of the disk depend crucially on the degree of magnetization, protostellar disks must have complex radial structure. The key parameter is the ionization fraction, f. If the resistivity of the gas is small enough, and the neutral-ion collision rate rapid enough, the

MRI is triggered. As discussed in section II.B, an ionization fraction as low as 10^{-13} is sufficient to trigger the linear instability on AU scales in standard solar nebula models. However, linear instability is not necessarily a guarantee that MHD turbulence will be maintained at significant levels. Nonlinear MHD simulations of resistive plasmas show that in the absence of a net vertical field, turbulence triggered by the instability ultimately dies away for Reynolds numbers below $\mathrm{Re}_M \lesssim 10^4$.

From these results it seems that the extent over which MHD turbulence will be important in a typical protostellar disk is likely to be rather small, somewhere between 0.1 and 1 AU. To see this, write the magnetic Reynolds number in the form

$$\mathrm{Re}_M \simeq T^{1/2} R_{\mathrm{AU}}^{3/2} \left(\frac{n_e}{n_n}\right)_{-12} \left(\mathrm{M}_\odot / M\right)^{1/2} \tag{16}$$

where R_{AU} is the disk radius in AU units, $M/\mathrm{M}_\odot$ is the central mass in units of solar mass, and n_e/n_n is the ionization fraction in units of 10^{-12}. At solar abundances, thermal ionization of the disk at the onset of magnetic coupling will be controlled almost entirely by potassium, with an ionization potential $\Phi = 4.341$ eV (Allen 1973). The solution to the Saha equation may be simply expressed in this regime as

$$\frac{n_e}{n_n} = a^{1/2} T^{3/4} \left(\frac{2.4 \times 10^{15}}{n_n}\right)^{1/2} \exp(-50{,}370/2T) \tag{17}$$

where a is the abundance of potassium ($\sim 10^{-7}$ if roughly solar), and n_n is the density of the dominant neutral species, generally molecular hydrogen for $T < 2000$ K. The Boltzmann factor makes the ionization exquisitely sensitive to temperature, acting as an "on-off" switch for $T \sim 1500$.

Take a simple disk model described by a surface temperature T_s, surface density Σ, midplane temperature T, and midplane optical depth τ,

$$T_s^4 = \frac{3}{4}\tau T^4 = \frac{3}{4}\tau \frac{3GM\dot{M}}{8\pi R^3 \sigma} \qquad \dot{M} = 3\pi\Sigma\alpha s H \tag{18}$$

This leads to

$$T = 1100 \left(\frac{\alpha}{10^{-2}}\right)^{1/3} \left(\frac{\Sigma\tau}{10^6}\right)^{1/3} R_{\mathrm{AU}}^{-1/2} \tag{19}$$

The principal uncertainty is the combination $\Sigma\tau$; fortunately, T depends rather weakly on the disk parameters. A midplane temperature as high as 1700 K would give rise to $\mathrm{Re}_M \simeq 4 \times 10^6$ at 1 AU; such a disk would be easily magnetized within this radius. If the temperature drops below 1400, however, Re_M approaches 10^4, the regime of suppressed turbulence.

This result is appropriate to disks with no net vertical field. If the disk contains a net field from an external source (e.g., a dipole field from the central protostar, or the remnant interstellar field), then, as discussed in section II.B, significant transport can persist at a much lower Re_M (Fleming et al. 1999).

In summary, the heating of the disk by its own dynamical dissipation permits MHD turbulence on scales up to 1 AU or so. In principle, this region could be as small as 0.1 AU if turbulence cannot sustain thermal ionization levels; at this radius, stellar heating alone will maintain requisite ionization levels.

C. The Outer Disk: Layered Accretion

While the thermally ionized inner portions of the disk can be understood with some confidence, the ionization fraction is much less certain in those regions of the disk where nonthermal ionization dominates. Cosmic rays may be an important source of ionization, if they are not excluded from the surface of the disk by surrounding magnetic structures. Once in the disk, cosmic rays are exponentially attenuated below a column of $\simeq 10^2$ g cm^{-2}, thus suggesting the possibility of "layered accretion" (Gammie 1996*b*), in which the surface layers are turbulent but the center of the disk is a quiescent "dead zone." X-rays are another possible source of ionization (e.g., Glassgold et al. 1997), and they are likewise strongly attenuated toward the midplane. Decay of ^{26}Al and ^{40}K is another possible source of ionization (Umebayashi 1983), although estimates suggest that it is not quite sufficient to provide good coupling of disk to magnetic field. Of course, the mean abundance of these elements is uncertain, and they may be concentrated by dust settling into particular regions in the disk (although not self-consistently, if the disk is turbulent). Uncertainties also remain on the recombination side. In particular, small dust grains are capable of soaking up most of the free charges in the gas and affecting the ion chemistry. In sum, the ionization state of the bulk of circumstellar disks is uncertain and an interesting area for future research; it may govern the evolution of circumstellar disks.

In any event, it seems likely that beyond $R \gtrsim 0.1$–1 AU, substantial portions of the centers of circumstellar disks are not well coupled to the magnetic field and are therefore not heated by MHD turbulence. Because these parts of the disk can suddenly *become* well coupled once the temperature rises above 10^3 K, there is the possibility of unstable accretion. An excursion in temperature above 10^3 K will be rewarded by turbulence and further heating, causing surrounding regions also to rise in temperature. This nonlinear switch is a robust feature of any model that incorporates angular momentum transport by MHD turbulence. Its consequences have been explored by Gammie (1996*b*), who finds that, combined with gravitational instability (which provides the "match" to ignite the disk), it leads to unsteady accretion similar to that observed in FU Orionis objects.

Many aspects of the outer disk region remain to be studied. The vertical structure produced by layered accretion in the disk has yet to be explored. As shown by two-fluid studies of the MRI in weakly ionized disks (Hawley and Stone 1998) or in the presence of large ohmic dissipation (Sano et al. 1998; Fleming et al. 1999), the nonlinear saturation may not always result in fully developed MHD turbulence. Moreover, transport by hydrodynamic waves is always a possibility, especially if a companion object is present. Finally, in the outer regions of the disk, infall may be occurring. The distinction between the disk and its environment becomes blurred, and it is no longer tenable to view the disk as an isolated structure.

IV. PROBLEMS FOR THE FUTURE

A detailed understanding of the internal dynamics of protostellar accretion disks is fundamental and essential to a theory of star formation. There are several areas in which progress is needed before this can be achieved. The following are several of the most important areas:

Better treatment of the thermodynamics: Because the ionization fraction of the disk is so sensitive to temperature, it is important to improve the treatment of the thermodynamics of the disk in MHD studies. The problem is difficult, because it requires proper modeling of the microphysical dissipation (which supplies the heat input) as well as heat transport via both turbulence in the optically thick core of the disk and radiation in the optically thin corona (making the problem in many ways similar to stellar convection).

Modeling the ionization fraction: Whenever the MRI can operate, it should dominate over all other internal transport processes. Thus, an important question is, in what regions of the disk is the ionization fraction too low to support the MRI? This question requires coupling the ionization balance equations with the dynamics (which will necessarily have to include a precise treatment of the thermodynamics, given the sensitivity of the ionization rate to temperature). Moreover, the outcome will depend on uncertain quantities such as cosmic ray fluxes at the disk surface, irradiance by high-energy photons, and ionization and recombination processes in a dusty gas. Nonetheless, modeling the ionization fraction will be an important extension of current MHD studies.

Global dynamical models: Almost all the dynamical models to date focus on local patches of the disk. Fully global models of thin disks have yet to be constructed, although global cylindrical flows have just begun to be studied (Armitage 1998). This is a promising area for rapid progress in the next few years.

Relative importance of internal and external torques: Once fully global models incorporating both the internal dynamics and external envelope are possible, investigating the relative transport contributions of magnetically driven outflows (i.e., external torques) and MHD turbulence (i.e., internal torques) becomes feasible. Can the MHD dynamo in the disk generate an external field that drives an outflow? Or is an externally supplied field, perhaps the remnant from the molecular cloud, required? How does the vertical structure of the disk affect the launching of the wind? Given the importance of protostellar outflows to star-forming regions, this is one of the major outstanding questions of the field.

Disk formation and evolution: Time-explicit MHD simulations can follow the evolution of disks only for relatively short times, perhaps hundreds of orbits, simply because of the limits imposed by current computer performance. Several key issues, however, can be addressed only through much longer time evolution. It may soon be possible to improve on the standard α formalism and provide a better parameterization of angular momentum transport in terms of the physical quantities in the disk. Time-implicit methods could then follow disk evolution on much longer time scales (much in the same way that stellar evolution is studied using a parameterization for convective heat transport).

While many challenges remain, our understanding of local transport processes in circumstellar disks has improved considerably since *Protostars and Planets III.* We know the mechanism by which accretion disks become turbulent. It is possible, in principle, to construct global dynamical models in which angular momentum transport arises spontaneously. We foresee a time when predictive protostellar disk studies can be pursued ab initio. This would truly be a milestone.

Acknowledgments J. S. acknowledges support from NASA grant NAG-54278 and NSF grant AST-9528299. J. H. and S. B. acknowledge support from NASA grants NAG-53058, NAGW-4431, and NSF grant AST-9423187. Computational resources for some of this work were provided under an NSF metacenter grant from NPACI and NCSA.

REFERENCES

Adams, F. C., and Lin, D. 1993. Transport processes and the evolution of disks. In *Protostars and Planets III*, ed. E. Levy and J. Lunine (Tucson: University of Arizona Press), pp. 721–748.

Allen, C. W. 1973. *Astrophysical Quantities* (London: Athlone Press).

Armitage, P. J. 1998. Turbulence and angular momentum transport in global accretion disk simulations. *Astrophys. J. Lett.* 501:L189–L192.

Balbus, S. A., and Hawley, J. F. 1991. A powerful local shear instability in weakly magnetized disks. I. Linear analysis. *Astrophys. J.* 376:214–233.

Balbus, S. A., and Hawley, J. F. 1992*a*. Is the Oort A-value a universal growth rate limit for accretion disk shear instabilities? *Astrophys. J.* 392:662–666.

Balbus, S. A., and Hawley, J. F. 1992*b*. A powerful local shear instability in weakly magnetized disks. IV. Nonaxisymmetric perturbations. *Astrophys. J.* 400:610–621.

Balbus, S. A., and Hawley, J. F. 1998. Instability, turbulence, and enhanced transport in accretion disks. *Rev. Mod. Phys.* 70:1–53.
Balbus, S. A., Hawley, J. F., and Stone, J. M. 1996. Nonlinear stability, hydrodynamical turbulence, and transport in disks. *Astrophys. J.* 467:76–86.
Blaes, O. M., and Balbus, S. A. 1994. Local shear instabilities in weakly ionized, weakly magnetized disks. *Astrophys. J.* 421:163–177.
Brandenburg, A., Nordlund, A., Stein, R., and Torkelsson, U. 1995. Dynamo-generated turbulence and large-scale magnetic fields in a Keplerian shear flow. *Astrophys. J.* 446:741–754.
Cabot, W. 1996. Numerical simulations of circumstellar disk convection. *Astrophys. J.* 465:874–886.
Cabot, W., and Pollack, J. 1992. *Geophys. Astrophys. Fluid Dyn.* 64:97. 97
Chandrasekhar, S. 1960. *Proc. Nat. Acad. Sci.* (USA) 46:253.
Colella, P., and Woodward, P. R. 1984. The piecewise parabolic method (PPM) for gas-dynamical simulations. *J. Comput. Phys.* 54:174.
Drazin, P. G., and Reid, W. H. 1981. *Hydrodynamic Stability* (Cambridge: Cambridge University Press).
Dubrulle, B. 1993. Differential rotation as a source of angular momentum transfer in the solar nebula. *Icarus* 106:59.
Fleming, T., Stone, J. M., and Hawley, J. F. 1999. The effect of resistivity on the nonlinear stage of the magnetorotational instability in accretion disks. *Astrophys. J.*, in press.
Gammie, C. F. 1996*a*. Nonlinear outcome of gravitational instability in optically thick disks. In *Accretion Phenomena and Related Outflows,* ed. D. T. Wickramasinghe, L. Ferrario, and G. Bicknell (San Francisco: Astronomical Society of the Pacific), p. 704.
Gammie, C. F. 1996*b*. Layered accretion in T Tauri disks, *Astrophys. J.* 457:355–362.
Ghosh, P., and Lamb, F. K. 1979. Accretion by rotating magnetic neutron stars. II. Radial and vertical structure of the transition zone in disk accretion. *Astrophys. J.* 232:259–276.
Glassgold, A. E., Najita, J., and Igea, J. 1997. X-ray ionization of protoplanetary disks. *Astrophys. J.* 480:344–350.
Goodman, J. 1993. A local instability of tidally distorted accretion disks. *Astrophys. J.* 460:596–613.
Goodman, J., and Xu, G. 1994. Parasitic instabilities in magnetized, differentially rotating disks. *Astrophys. J.* 432:213–223.
Goodson, A. P., Winglee, R. M., and Boehm, K.-H. 1997. Time-dependent accretion by magnetic young stellar objects as a launching mechanism for stellar jets. *Astrophys. J.* 489:199–209.
Hawley, J. F., and Balbus, S. A. 1991. A powerful local shear instability in weakly magnetized disks. II. Nonlinear evolution. *Astrophys. J.* 376:223–233.
Hawley, J. F., and Balbus, S. A. 1992. A powerful local shear instability in weakly magnetized disks. III. Long term evolution in a shearing sheet. *Astrophys. J.* 400:595–621.
Hawley, J. F., and Balbus, S. A. 1997. Three dimensional simulations of accretion disks. In *Accretion Phenomena and Related Outflows,* ed. D. Wickramasinghe, L. Ferrario, and G. Bicknell (San Francisco: Astronomical Society of the Pacific), pp. 179–189.
Hawley, J. F., and Balbus, S. A. 1999. Angular momentum transport, local and global simulations. In *Astrophysical Discs,* ed. J. Sellwood and J. Goodman (Cambridge: Cambridge University Press), p. 108.
Hawley, J. F., and Stone, J. M. 1998. Nonlinear evolution of the magnetorotational instability in ion-neutral disks. *Astrophys. J.* 501:758–771.

Hawley, J. F., Gammie, C. F., and Balbus, S. A. 1995. Local three-dimensional magnetohydrodynamic simulations of accretion disks. *Astrophys. J.* 440:742–763.

Hawley, J. F., Gammie, C. F., and Balbus, S. A. 1996. Local three-dimensional simulations of an accretion disk hydromagnetic dynamo. *Astrophys. J.* 464:690–703.

Hawley, J. F., Balbus, S. A., and Winters, W. F. 1999. *Astrophys. J.*, in press.

Hayashi, C. 1981. Structure of the solar nebula growth and decay of magnetic fields and effects of magnetic and turbulent viscosities on the nebula. *Prog. Theor. Phys.* 70:35.

Hayashi, M. R., Shibata, K., and Matsumoto, R. 1996. X-ray flares and mass outflows driven by magnetic interactions between a protostar and its surrounding disk. *Astrophys. J. Lett.* 468:L37–L40.

Jin, L. 1996. Damping of the shear instability in magnetized disks by ohmic diffusion. *Astrophys. J.* 457:798–804.

Kley, W., Papaloizou, J. C. B., and Lin, D. N. C. 1993. On the angular momentum transport associated with convective eddies in accretion disks. *Astrophys. J.* 416:679–688.

Königl, A. 1991. Disk accretion onto magnetic T Tauri stars. *Astrophys. J. Lett.* 370:L39–L42.

Laughlin, G., and Bodenheimer, P. 1994. Nonaxisymmetric evolution in protostellar disks. *Astrophys. J.* 436:335–354.

Lees, A. W., and Edwards, S. F. 1972. The computer study of transport processes under extreme conditions. *J. Phys. C.* 5:1921.

Lin, D. N. C., and Papaloizou, J. C. B. 1993. On the tidal interaction between protostellar disks and companions. In *Protostars and Planets III,* ed. E. H. Levy and J. I. Lunine (Tucson: University of Arizona Press), pp. 749–835.

Lin, D. N. C., and Papaloizou, J. C. B. 1996. Theory of accretion disks. II: Application to observed systems. *Ann. Rev. Astron. Astrophys.* 34:703–748.

Lovelace, R. V. E., Romanova, M. M., and Bisnovatyi-Kogan, G. S. 1995. Spin-up/spin-down of magnetized stars with accretion discs and outflows. *Mon. Not. Roy. Astron. Soc.* 275:244–254.

Lubow, S. H., and Ogilvie, G. I. 1998. Three-dimensional waves generated at Lindblad resonances in thermally stratified disks. *Astrophys. J.* 504:983–995.

Lynden-Bell, D., and Boily, C. 1994. Self-similar solutions up to flashpoint in highly wound magnetostatics. *Mon. Not. Roy. Astron. Soc.* 267:146–152.

Lynden-Bell, D., and Pringle, J. E. 1974. The evolution of viscous discs and the origin of the nebular variables. *Mon. Not. Roy. Astron. Soc.* 168:603–637.

MacLow, M.-M., Norman, M. L., Königl, A., and Wardle, M. 1995. Incorporation of ambipolar diffusion into the Zeus magnetohydrodynamic code. *Astrophys. J.* 442:726–735.

Matsumoto, R., and Shibata, K. 1997. Three dimensional global MHD simulations of magnetized disks and jets. In *Accretion Phenomena and Related Outflows,* ed. D. Wickramasinghe, L. Ferrario, and G. Bicknell, pp. 443–447.

Miller, K. A., and Stone, J. M. 1997. Magnetohydrodynamic simulations of stellar magnetosphere–accretion disk interaction. *Astrophys. J.* 489:890–902.

Miller, K. A., and Stone, J. M. 1999. The formation and structure of a strongly magnetized corona above a weakly magnetized accretion disk. *Astrophys J.*, in press.

Montmerle, T., Feigelson, E., Bouvier, J., and André, P. 1993. Magnetic fields, activity and circumstellar material around young stellar objects. In *Protostars and Planets III,* ed. E. H. Levy and J. I. Lunine (Tucson: University of Arizona Press), pp. 689–717.

Nelson, A. F., Benz, W., Adams, F. C., and Arnett, D. 1998. Dynamics of circumstellar disks. *Astrophys. J.* 502:342–371.
Papaloizou, J. C. B., and Lin, D. N. C. 1995*a*. On the dynamics of warped accretion disks. *Astrophys. J.* 438:841–851.
Papaloizou, J. C. B., and Lin, D. N. C. 1995*b*. Theory of accretion disks. I: Angular momentum transport processes. *Ann. Rev. Astron. Astrophys.* 33:505–540.
Papaloizou, J. C. B., and Terquem, C. 1995. On the dynamics of tilted disks around young stars. *Mon. Not. Roy. Astron. Soc.* 274:987–1001.
Papaloizou, J. C. B., and Terquem, C. 1997. On the stability of an accretion disk containing a toroidal magnetic field: The effect of resistivity. *Mon. Not. Roy. Astron. Soc.* 287:771–789.
Pringle, J. E. 1981. Accretion disks in astrophysics. *Ann. Rev. Astron. Astrophys.* 19:137–162.
Pringle, J. E., and Rees, M. J. 1972. Accretion disk models for compact X-ray sources. *Astron. Astrophys.* 21:1–9.
Prinn, R. G. 1990. On neglect of nonlinear momentum terms in solar nebula accretion disk models. *Astrophys. J.* 348:725–729.
Roberts, W. W. 1969. Large scale shock formation in spiral galaxies and its implications on star formation. *Astrophys. J.* 158:123–143.
Różyczka, M., and Spruit, H. C. 1993. Numerical simulations of shock driven accretion. *Astrophys. J.* 417:677–686.
Ruden, S. P., Papaloizou, J. C. B., and Lin, D. N. C. 1988. Axisymmetric perturbations of thin gaseous disks. I. Unstable convective modes and their consequences for the solar nebula. *Astrophys. J.* 329:739–763.
Ryu, D., and Goodman, J. 1992. Convective instability in differentially rotating disks. *Astrophys. J.* 388:438–450.
Sano, T., Inutsuka, S., and Miyama, S. 1998. A saturation mechanism of magnetorotational instability due to ohmic dissipation. *Astrophys. J. Lett.* 506:L57–L60.
Sawada, K., Matsuda, T., and Huchisa, I. 1986. Spiral shocks on a Roche lobe overflow in a semi-detached binary system. *Mon. Not. Roy. Astron. Soc.* 219:75–88.
Shakura, N. J., and Sunyaev, R. A. 1973. Black holes in binary systems. Observational appearances. *Astron. Astrophys.* 24:337–355.
Shu, F. H. 1992. *The Physics of Astrophysics, Vol. 2: Gas Dynamics* (Mill Valley: University Science Books).
Shu, F. H., Milione, V., and Roberts, W. W. 1973. Nonlinear gaseous density waves and galactic shocks. *Astrophys. J.* 183:819–842.
Shu, F., Najita, J., Ruden, S., and Lizano, S. 1994. Magnetocentrifugally driven flows from young stars and disks. II. Formulation of the dynamical problem. *Astrophys. J.* 429:797–807.
Stone, J. M., and Balbus, S. A. 1996. Angular momentum transport in accretion disks via convection. *Astrophys. J.* 464:364–372.
Stone, J. M., and Norman, M. L. 1992. Zeus-2D: A radiation magnetohydrodynamics code for astrophysical flows in two space dimensions. I. The hydrodynamic algorithms and tests. *Astrophys. J. Suppl.* 80:753–790.
Stone, J. M., Hawley, J., Gammie, C., and Balbus, S. 1996. Three-dimensional magnetohydrodynamical simulations of vertically stratified accretion disks. *Astrophys. J.* 463:656–673.
Strom, S. E., Edwards, S., and Skrutskie, M. F. 1993. Evolutionary timescales for circumstellar disks associated with intermediate- and solar-type stars. In *Protostars and Planets III,* ed. E. H. Levy and J. I. Lunine (Tucson: University of Arizona Press), pp. 837–866.

Terquem, C. 1998. The response of accretion disks to bending waves: Angular momentum transport and resonances. *Astrophys. J.* 509:819–835.
Umebayashi, T. 1983. The densities of charged particles in very dense interstellar clouds. *Prog. Theor. Phys.* 69:480.
Velikhov, E. P. 1959. Stability of an ideally conducting liquid flowing between cylinders rotating in a magnetic field. *Sov. Phys. JETP* 36:995.
Vishniac, E. T., and Diamond, P. 1989. A self-consistent model of mass and angular momentum transport in accretion disks. *Astrophys. J.* 347:435–447.
Yukawa, H., Boffin, H., and Matsuda, T. 1997. Spiral shocks in three-dimensional accretion discs. *Mon. Not. Roy. Astron. Soc.* 292:321–330.

INFALLING PLANETESIMALS IN PRE-MAIN-SEQUENCE STELLAR SYSTEMS

CAROL A. GRADY
Eureka Scientific and National Optical Astronomy Observatories/Space Telescope Imaging Spectrograph

MICHAEL L. SITKO
University of Cincinnati

RAY W. RUSSELL AND DAVID K. LYNCH
The Aerospace Corporation

MARTHA S. HANNER
Jet Propulsion Laboratory

MARIO R. PÉREZ
Space Applications Corporation

KAREN S. BJORKMAN
University of Toledo

and

DOLF de WINTER
Centro de Astrofísica de Canarias, España

Recent ultraviolet and optical observations have identified a large class of late-stage pre-main-sequence (PMS) and zero-age main sequence (ZAMS) stars having circumstellar material with a disklike geometry. These structures possess many of the characteristics expected for objects that will eventually evolve into planetary systems, including high-velocity infalling gas features exhibiting prominent departures from "cosmic" abundances. A complementary view of the near-stellar disk material can be obtained by probing the mineralogy of the small grains via infrared spectroscopy. We review the ground-based and *Infrared Space Observatory* (ISO) data on the small-grain components of the inner disk regions. The available models for infall activity in these stars are compared with the spectroscopic data on the gas and dust and the mineralogy of interstellar dust particles.

I. INTRODUCTION

These are exciting times in the study of planetary system formation, with a steadily expanding inventory of exoplanet detections and imaging of dust disks around nearby young and main-sequence stars. Although these discoveries imply that our solar system is far from unique, linking the data for the protoplanetary and debris disks to mature planetary systems requires a demonstration that disk evolution proceeds via planetesimal production and growth to the formation of planets (Beckwith and Sargent 1996; see chapter by Beckwith et al., this volume). Theoretical studies of planet formation indicate that planetesimals grow, via runaway accretion, to lunar-sized ($\approx$2000 km) embryos in 10^5 years (see, e.g., Boss 1996). Recent gas giant planet formation studies have suggested that most of the action in planet formation occurs over 1–16 Myr, with formation of planets similar to Jupiter in $t \leq 10$ Myr (Pollack et al. 1996), within the time interval during which infrared (IR) and optical emission line studies have demonstrated that circumstellar material remains detectable around both solar-mass and intermediate-mass stars (Strom et al. 1989; van den Ancker et al. 1997). Direct imaging of exoplanetesimals is not feasible with current or foreseeable technology, because such bodies have substantially less surface area than micron-sized grains distributed in a disk and thus are inefficient IR emitters. However, such bodies may be indirectly detectable.

A. β Pictoris: The Prototypical Disk System

The nearby [$d = 19.28$ pc, $t \geq 8$ Myr (Crifo et al. 1997)] β Pic system provided the first indication that exoplanetary systems are among field stars near the solar system (see the chapter by Lagrange et al., this volume). As a result of 15 years of intensive study following its detection as an IR excess source by the *Infrared Astronomy Satellite* (IRAS) and subsequent imaging of the disk (Smith and Terrile 1984), the β Pic system is currently the best-studied external disk system, with strong indications that the disk already harbors one or more giant planets (Heap et al. 2000; Artymowicz 1997). The star also possesses IR silicate emission features resembling cometary materials (Knacke et al. 1993) and optical polarization (Gledhill et al. 1991) similar to the zodiacal light in our solar system (Leinert et al. 1981). β Pic also has spectroscopic absorption features that are due to circumstellar gas at rest with respect to the star, as well as redshifted components in UV and optical spectra (Ferlet et al. 1987; Hobbs 1987). These components have been interpreted as absorption by the comae of swarms of cometlike bodies. Multiyear fluctuations in the frequency of exocometary activity may point to perturbations produced by gas giant planets (Deleuil et al. 1995; Mouillet et al. 1997). Spectroscopic signatures for the presence of planets are of particular interest, because such markers for the presence of planets can sample a larger volume of our galaxy than can be surveyed by imaging techniques.

B. The Other Intermediate-mass Stars

Although its exact age remains controversial, there is general agreement that β Pic is a main-sequence (MS) star (Crifo et al. 1997; Lecavelier des Etangs et al. 1997; Jura et al. 1993; Backman et al. 1997). If the infall activity in the β Pic disk is a remnant of earlier disk evolutionary phases, similar activity must be present around other young A and B stars and should also be detectable in association with young solar analogs: the T Tauri stars (TTSs).

There is a population of pre-main-sequence (PMS) to zero-age main-sequence (ZAMS) A and B stars that are young β Pic analogs: the Herbig Ae/Be (HAeBe) stars (Waters and Waelkens 1998; Pérez and Grady 1997; Thé et al. 1994). The nearest HAeBe stars are 100–140 pc from the solar system (van den Ancker et al. 1997, 1998). These stars are surrounded by circumstellar dust and gas disks with typical radii of several hundred AU (see the chapter by Natta et al., this volume, for a discussion of disk imaging, masses, and the gas content; Mannings et al. 1997; Mannings and Sargent 1997; Henning et al. 1998), comparable in diameter to the lower-mass dust disks around MS systems, with typical radii of 100 to several hundred AU (Holland et al. 1998; Jayawardhana et al. 1998; Koerner et al. 1998; Greaves et al. 1998). Compared to the TTSs, the HAeBe stars offer a number of advantages for line-of-sight absorption studies similar to the β Pic studies. First, the HAeBe stars have substantially higher T_{eff} and $L_{\star}$ than young Suns. Coupled with the minimal foreground interstellar extinction and low circumstellar optical depths toward the nearest HAeBe stars, the luminosity of these objects permits us to carry out line-of-sight studies from the IR through the far UV. The UV-bright HAeBe stars are also quiescent, with little indication of stellar activity, continuum veiling, or any of the other complications of T Tauri spectra. HAeBe stars rotate more rapidly than young Suns, resulting in washing out of the photospheric absorption features and enhancing of the contrast of circumstellar features against the stellar background. To date, the bulk of the UV absorption line studies of circumstellar gas around young stars, which provide a wider range of elemental diagnostics than are accessible in the optical, have been carried out for HAeBe stars and a few nearby, ostensibly main-sequence A and B stars with circumstellar gas.

II. THE SPECTROSCOPIC DATA

The first detections of redshifted absorption features in the spectra of HAeBe stars date to 1992 (Welty et al. 1992; Graham 1992; Pérez et al. 1993; Grady et al. 1993, 1996*a,b*; Grinin et al. 1994, 1996). The inventory of stars with infall detections continues to expand. At the time of writing, 13 of 33 HAeBe stars with *International Ultraviolet Explorer* (IUE) data have infall detections, as well as similar activity in three nonemission B stars (Grady et al. 1996*a*) and optical detections toward an additional

12 stars (Grady et al. 1996*a*; Grinin et al. 1994, 1996; Graham 1992). The optical detections are preferentially associated with objects with a history of large amplitude ($\Delta V > 2$ mag.) photometric and correlated polarimetric variability (Bibo and Thé 1991; Herbst et al. 1994, who termed such systems UXORs after the class prototype UX Orionis; see van den Ancker et al. 1999 and the chapter by Natta et al., this volume). At this time, redshifted features have been observed in *all* of the better-monitored UXORs and have subsequently been detected in one star, AB Aurigae, not currently exhibiting such variability but with a history of large-amplitude light variations in the 1920s (Grady et al. 1999*a*).

The close association of the UXOR phenomenon and infall activity suggests that the phenomenon is associated with the circumstellar disk. Both the IUE data (35% of sources) and the frequency of large-amplitude variability among the Hipparcos *(High Precision Parallax Collecting Satellite)* HAeBe stars (35%, van den Ancker et al. 1998) are consistent with detection of disk material up to 15° above the disk midplane, if we model the disks as having constant opening angles. More flattened distributions are possible if the disks are flared, as has been observed in HST observations of T Tauri stars (McCaughrean et al., this volume). These limits are compatible with the inclination data provided by the *polarimetrically* identified systems with infall, and suggest that infall activity is restricted to the disk plane among the HAeBe stars (Grinin and Rostopchina 1996; Grady et al. 1996*a,* 1997, 1999*a,b*). The true inclination dependence of infall activity in these stars is likely to be further constrained as data on disk inclinations, based on imaging of the disks, are obtained.

A. Characteristics of the Infalling Gas

Infalling circumstellar gas is observed in a wide range of atomic ions in the spectra of HAeBe stars and related MS stars (Figs. 1–3). Redshifts as large as 200–400 km s^{-1} are typical, with velocities as high as 500 km s^{-1} in some cases. For the HAeBe stars, redshifted absorption is usually observed in tandem with absorption at the stellar radial velocity and blueshifted absorption features due to a stellar wind. The strength of the absorption, velocity range over which infall can be detected, and ions exhibiting infall are typically reduced in the MS stars with "shell" spectra, relative to the PMS stars. In several systems (HD 100546, AB Aur, HD 37806), infalling gas is observed in transitions of higher-excitation gas than the outflowing material, suggesting that the infalling gas has substantially higher density than the high-velocity wind material. In some cases the neutral atomic gas is restricted to the infalling material (Fig. 2).

Multiple ionization stages, including neutrals through triply ionized species, of a given element (e.g., C I, C II, C IV; see Fig. 2) are usually seen *in the same velocity range* and frequently with similar or only slightly different profiles. Triply ionized species (C IV, Si IV) can be produced only

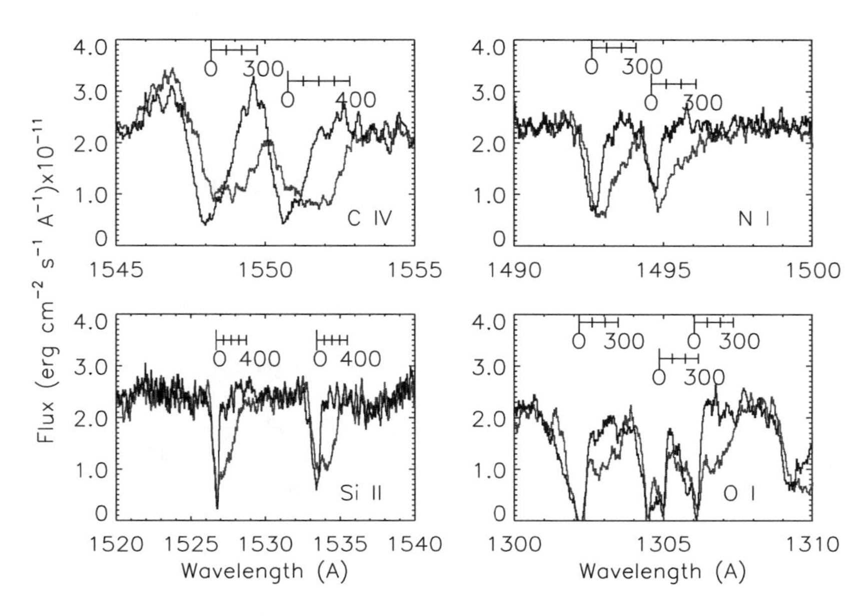

Figure 1. Infalling circumstellar gas profiles for HD 100546 at two epochs: March 7, 1995 (high-infall state, thick line) and May 25, 1995 (thin line). Infalling gas is seen over a wide range of ionization stages, ranging from neutral atomic gas species such as O I and N I, through intermediate ionization stages such as Si II, and including species that are produced by collisional ionization in the radiation field of the B9 star, such as C IV.

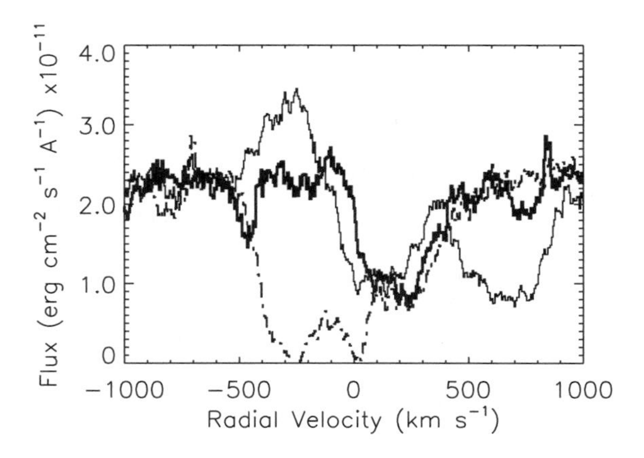

Figure 2. Infalling gas is routinely observed in multiple ionization stages of the same element. In the case of HD 100546 and other HAeBe stars, neutral atomic gas may be restricted to the infalling material. Collisionally ionized species, such as C IV, may also be preferentially observed in the infalling gas. Here C I (4) (thick line) and C IV (thin line) in HD 100546 from March 7, 1995, are shown with C II (dot-dash) in the same spectrum. The zero point of the radial velocity scale corresponds to the laboratory wavelength for the shorter-wavelength component of the C IV resonance doublet, the longer-wavelength member of the C II doublet, and the transition to the 0-eV level of the ground configuration in C I (4).

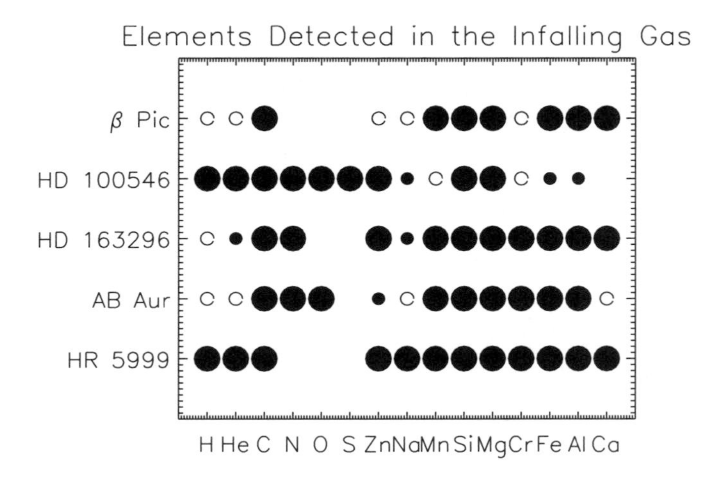

Figure 3. Schematic display of the elements detected *in the infalling gas,* ordered by gas-phase sublimation temperature (after Jenkins 1987), toward some of the nearest HAeBe stars with the most complete wavelength coverage and β Pic. Filled symbols indicate firmly detected elements, with the small symbols indicating the elements that have anomalously weak covering factors. Open symbols indicate elemental nondetections. Gaps indicate elements lacking suitable data coverage. Several of the stars (e.g., HD 163296 and AB Aur) have emission in H I, He I, and Na I (AB Aur). With the exception of HR 5999, the displayed stars lack nearby stellar companions. The stars are ordered roughly from bottom to top in terms of increasing system age.

by collisional ionization in the spectrum of a late B or A star (Bruhweiler et al. 1989). While there is a tendency for the neutral atomic gas features to be observed at lower velocities (typically ≤ 150 km s^{-1}), and higher ionization stages can range up to the maximum observed velocity, we do not see a trend of progressive ionization with increasing infall velocity.

When normalized either by a photospheric comparison spectrum or by a spectrum of the same star in a low-infall state, the infalling gas profiles have uniform absorption depth for a given ion, largely independent of transition excitation and oscillator strength. The infalling gas features usually show flat absorption troughs which are substantially above the instrumental scattered light levels. These features are similar to those seen toward β Pic (Vidal-Madjar et al. 1994; Vidal-Madjar and Ferlet 1995) and have been interpreted as detection of optically thick, saturated absorption from gas clouds that do not fully obscure the star. The depth of the saturated absorption measures the percentage obscuration of the stellar disk, or *covering factor,* and can be used to probe the spatial extent of the gas clouds in different elements.

B. Composition Constraints on the Infalling Material

We have begun to probe the composition of the infalling material. Although the high optical depths and saturation of the infalling gas profiles preclude conventional column density estimates, departures from expected abundances can be noted where absorption in strong transitions of elements is missing or anomalously weak. This approach to constraining the composition of the infalling material was first exploited by Grinin et al. (1994) and Sorelli et al. (1996). They noted that the detection of infalling Na I with high covering factors in the absence of heavy continuum line blanketing and prominent infall signatures in transitions of H I and He I indicated that the infalling material showed significant departures from cosmic abundances. Subsequent analysis of high-infall-state data for HD 100546 showed that the infalling material in that system was underabundant in Fe, Cr, Mn, and Al relative to Si, Mg, and Zn (Grady et al. 1997), with Cr underabundant relative to Zn by at least a factor of 12. The UV data for this system implied that any silicates in the disk should be strongly Mg-rich, a result borne out by *Infrared Space Observatory* (ISO) Long Wavelength Spectrometer (LWS) data (Malfait et al. 1998*b*).

Similar analyses have been carried out for the younger HR 5999 (0.6 Myr), AB Aur (2–4 Myr, Grady et al. 1999*a*), HD 163296 ($4^{+6}_{-2.5}$ Myr, Grady et al. 1999*b*), and HD 100546 ($t \geq 10$ Myr) systems (Fig. 3). These systems exhibit prominent UV line blanketing due to Fe II transitions and show no indication of underabundances in the siderophile and super-refractory elements. Covering factors for the magnesium or silicon ions tend to be comparable, and typically larger than for the iron ions, suggesting that the silicates in these systems may also be Mg-rich. The effect is not, however, as dramatic as for HD 100546. A larger sample of objects is needed to determine whether the trend of more prominent departures from cosmic abundances is an evolutionary effect for all young A and B star systems or is peculiar to HD 100546.

C. Variability and Synoptic Observations

At present, only a few HAeBe stars have been observed on a daily or higher-frequency basis, as opposed to more sporadic observations separated by months to years. The dense time series indicate that infall events are comparatively frequent in young HAeBe stars and become more sporadic with increasing system age (Table I). Episodes of enhanced infall typically last tens of hours (>35 hours, AB Aur, Fig. 4; >48 hours, BF Ori, de Winter et al. 1999; 24–48 hours, HR 5999, HD 142666; and 15–$\geq$ 45 hours, HD 163296, Grady et al. 1999*b*). The UV data provide some suggestion that episodes of enhanced infall are correlated with suppression of the high-velocity ($v \geq 200$ km s^{-1}) portion of the stellar wind, although lower-velocity wind material is apparently unaffected.

III. MODELING THE INFALL ACTIVITY

A number of models for accretion activity have been proposed for PMS stars. In this section we review their applicability to the data.

TABLE I

Comparison of Star-grazing Comets with Falling Evaporating Bodies (FEB) around HAeBe Stars and Young MS Stars

Characteristic	Sun-grazers[a]	Falling Evaporating Bodies	
		HAeBe Stars	MS/Zodiacal Disks
Maximum radial velocity	≈ 600 km/s	0–+500 km/s	0–200 km/s(A-shell stars); 0–400 km/s (β Pic)
Detection method	Coronagraphic imaging	Line-of-sight (LOS) absorption	LOS absorption
Post/preperiastron detection ratio	0 out of >60 comets (through mid-1998)	Infall detected	Infall predominates
		Outflow blends with stellar/disk wind	Outflow events sporadic
Species observed	Neutral atomic gas	Neutrals to triply ionized species	Neutrals to triply ionized species
Masses	Rubble (meter-sized)	Gas mass ≈ cometary in single observation	
Variability	Orbital time	Hours (high velocity) Events last 20–60 hrs	Hours (high velocity) Events last 20–60 hrs
Frequency	2/month	1/24 hours ($t < 3$ Myr) 1/5 days (AB Aur) 1–1.5/5 days (HD 163296) Sporadic (HD 100546)	High but uncertain (A-shell stars) Episodic (β Pic)

[a] Based on two years of observation by the *Solar and Heliospheric Observatory* (SOHO) Large Angle Spectrometric Coronagraph (D. A. Biesecker, private communication, 1998).

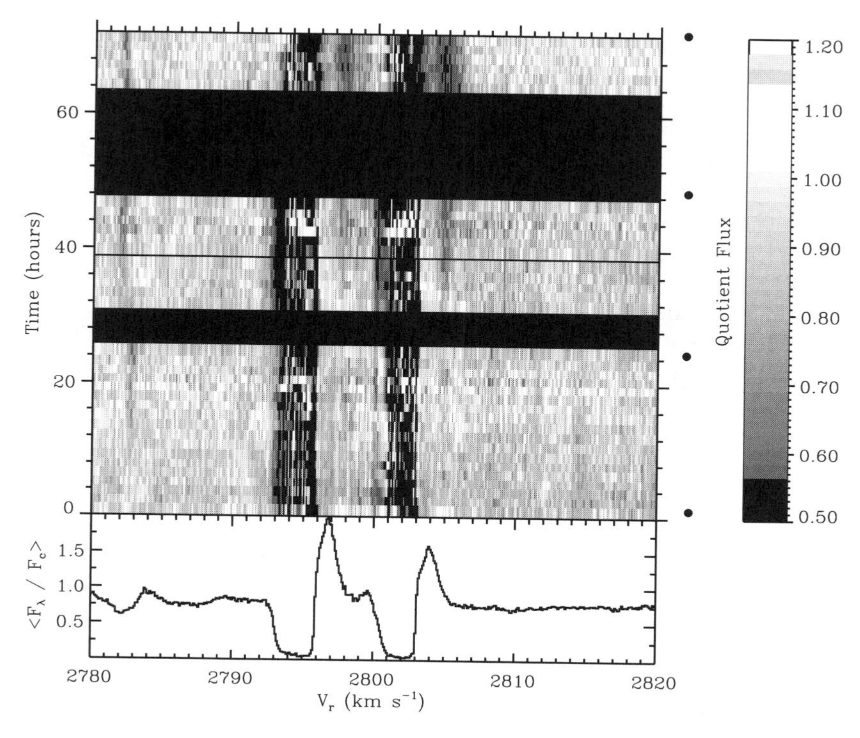

Figure 4. Dynamic spectral image of the Mg II resonance doublet in the Herbig Ae star AB Aurigae from a dense time series obtained by IUE in 1990 (from Pérez and Grady 1997). Time increases upward, and wavelength increases to the right. The spectral data have been normalized by the mean spectrum (shown at the bottom of the figure). Gaps in data coverage are denoted by black horizontal bands. The dark absorption in both members of the Mg II resonance doublet are due to saturated absorption in the stellar wind, which is typically observed in spectra of this star. During the time series, enhanced absorption redward of line center developed to the right of each of the Mg II emission components, beginning halfway through the series and continuing until the end of the observations.

A. Accretion Models

Star formation studies predict that the star grows in mass by accretion of material from a circumstellar disk, which can be modeled as a classical accretion disk. This disk may be embedded, especially if the system is still comparatively young, in a larger, and possibly spherical, envelope, which may continue to rain material onto the disk, but with velocities below 1 km s^{-1} (Bodenheimer 1997; see the chapter by Mundy et al., this volume). Higher infall velocities and ionization stages are expected if the infall is accompanied by classical accretion disk activity. Early on, the bulk composition of the disk should resemble the molecular cloud from which it formed. Accretion of material onto the star at this point should, therefore, be dominated by those elements that are numerically most abundant in the disk, and the infalling material should closely follow the stellar abundance pattern. This implies that the derived abundances should be dominated by volatiles such as hydrogen and helium.

The magnetospherically channeled accretion models (see Edwards et al. 1994) further predict that the bulk of the accreted material should have remained in the gas phase since disk formation and that accretion should be modulated by the stellar rotation period, with the bulk of the accretion onto localized hot spots on the stellar photosphere. An important prediction of this model is that accretion activity should be observed at high latitude and not be restricted to the disk midplane. This latitude independence is observed in classical TTSs (Edwards et al. 1994), where there is also the expectation that the outer convective region in the stellar atmosphere should produce a magnetosphere.

B. Star-grazing Comets

The star-grazing cometesimal hypothesis was initially developed for β Pic (Beust et al. 1990, 1996) to account for the presence of episodic high-velocity infall in a wide range of transitions of refractory elements (see the chapter by Lagrange et al., this volume). This model predicts that a large velocity extent should be seen in any ionic species for which the ratio of the radiation pressure force to that of gravity, $F_{\rm rad}/F_{\rm grav} \geq 1$. This model also predicts that infall events should be preferentially detected in the disk plane, unless the system is sufficiently old that both an Oort cloud has been populated and there has been enough time for bodies resident in an Oort cloud to be perturbed into cometary orbits ($t \approx 10$ Myr, with full isotropization in $\approx$100 Myr; see Duncan et al. 1987). The model also predicts that any species that can be produced by outgassing of primary volatiles or the dissociation and subsequent vaporization of grain materials should be represented in the infalling material, and that the abundance pattern should resemble cometary materials rather than the full stellar abundance pattern (see Greenberg 1998). For older systems, resonances introduced into the disk by the presence of massive planets are expected to cause episodic infall events and perturbation of swarms of bodies (Roques et al. 1994; Lecavelier des Etangs et al. 1996; Mouillet et al. 1997). The swarms should produce absorption features at characteristic velocities that are a function of the orbital characteristics of the perturbing bodies.

C. Comparison with the Data

Infalling material with velocities up to 60% of the free-fall velocity from infinity, in both the HAeBe stars and in the older A-shell star systems, is preferentially observed close to the stellar equatorial plane (Grady et al. 1996*a*,*b*, 1997, 1999*a*,*b*). Infall activity in the HAeBe stars does not exhibit the latitude independence seen in young solar analogs (Edwards et al. 1994). The infall events do not seem to show any sustained, periodic modulation on the stellar rotation period (Grady et al. 1999*a*,*b*). Detection of multiple ionization stages over the same velocity range (Fig. 2) is consistent with the behavior of bow shocks. The presence of infall activity associated with stars having extremely inhomogeneous disks (the UXORs; see discussion by Natta et al., this volume), as well as the similarity of the infall activity in ionization, velocity, and profile shapes to that observed toward MS stars with debris disks, such as β Pic, suggests that the infalling gas is not associated with accretion disk activity.

The presence of neutral atomic gas species at velocities of order 100 km s^{-1} (e.g., deep in the star's gravitational well), where such species should have extremely short photoionization lifetimes, of the order of seconds to minutes (Sorrelli et al., 1996), provides one of the most important clues to the source of the infalling material. The presence of such species is consistent with *in situ* production of the gas. Transport to the observed location of the neutral atomic material in the form of diatomic gases (such as H_2 and CO) is excluded, because the photodissociation lifetimes for these species are even shorter than the photoionization lifetimes of their daughter products. An origin in grains is implied both by the prominence of refractory elements in the infalling material and the small covering factors for primary volatiles (e.g., He I) and by the expectation that such elements will be preferentially depleted onto grains under the conditions of a protoplanetary disk (Owen 1997) around HAeBe stars. Thus, the infalling material is likely to be preferentially associated with grains and/or larger solid bodies.

The covering factor data also provide additional information on the nature of the solid material. There is a trend for refractory species (Mg II, Si II, Ca II, Al II, Al III) to have similar covering factors to the carbon ions. Both sets of ions can have large covering factors (60–80%). Semirefractory species, such as Zn II and S II (when detected), exhibit smaller covering factors, roughly comparable to the iron ion covering factors. N I and He I have the smallest covering factors, with the He I data consistent with obscuration of at most 10% of the stellar disk. If we assume that the bulk of the iron is associated with semirefractory grains such as FeS and FeO, the Zn with ZnS, and the carbon with refractory organic materials, the covering factor data suggest a good correlation with the sublimation temperatures of simple minerals, under the assumption that the parent bodies have not been substantially heated above 350–400 K over the evolutionary lifetimes of the stars ($\approx$1–10 Myr). If the N I and other volatile (see section V below) gas species (O I, H I, and C I) originate in low-temperature icy condensates, their presence implies that heating has not been systematically $>$120 K. They may also be produced by the dissociation of CHON particles (see section V.B.4). Regardless of the origin of the N I, the presence of He I (Grinin et al. 1994, 1996) in several HAeBe star spectra is consistent with release of helium trapped in icy clathrate structures. The suggestion that the silicates in the parent bodies are preferentially Mg-rich, together with the presence of large amounts of carbon originating in a refractory parent, is also consistent with predictions for the composition of comets (Greenberg 1998) and with data on solar system comets such as P/Halley and C/Hale-Bopp. We conclude, therefore, that the infall signatures reflect the presence of high levels of star-grazing cometary activity.

IV. DYNAMICS OF THE INFALLING MATERIAL

The infalling material is routinely observed to have velocities nearing the free-fall velocity (hundreds of km s^{-1}). In contrast, other grain acceleration mechanisms produce velocities orders of magnitude smaller, such as Poynting-Robertson drag (tens of m s^{-1}), the behavior of small grains in the presence of gas drag, or infall from a larger, extended envelope ($v \leq$ 1 km s^{-1}).

As a consequence of the strong radiation fields of these stars, grains that are smaller than several microns are subject to radiation pressure-driven outflow (Sitko et al. 1994) rather than infall. The presence of grains that are smaller than the blowout radius would therefore imply a replenishment mechanism, because under equilibrium conditions such grains are expected to be removed from the immediate vicinity of the star and thus not contribute appreciably to the polarization or the IR emission spectrum of the inner disk. Submicron grains in the disks of HAeBe stars are manifested via their optical (Grinin et al. 1991) and UV (Bjorkman et al. 1995) polarization characteristics and via the presence in some systems of mild UV excesses consistent with dust-scattered light. The most dramatic signatures of such grains are the IR emission features due to small grains (see section V below). Particularly for the UXORs, where the inner disks are apparently quite inhomogeneous, such grains are likely to be a by-product of the cometary activity producing the infall signatures.

The velocity data imply that the solid bodies associated with the gas features are sufficiently large to maintain free-fall velocities until they disintegrate (Grinin et al. 1994; Sorelli et al. 1996), providing a lower bound to the size of the parent bodies. An upper bound to the size of the parent bodies is provided by the composition of the infalling material. The mixture of low-temperature condensate dissociation products, super-refractory grains, silicates, and intermediate-temperature materials suggests the presence of parent bodies that, while sufficiently large to achieve substantial fractions of the free-fall velocity, either are not large enough or have not had enough time for density differentiation to occur. The observed nonequilibrium mixture of materials is consistent with cometary composition and meets our expectations for planetesimals.

V. THE SMALL-GRAIN COMPONENT OF THE DISKS

UV–IR spectra of HAeBe star disks provide ample evidence for the presence of small grains in the immediate vicinity of the star. Some of the nearest HAeBe stars lack associated reflection nebulosity or provide no indication of extensive envelopes (Henning et al. 1998). ISO data have greatly augmented the sample of stars with polycyclic aromatic hydrocarbon (PAH) and "unidentified IR" (UIR) band detections, but the data indicate that for these stars the band features do not provide the bulk of the 2–8 μm IR emission (Waelkens et al. 1996; Waters and Waelkens 1998; van den Ancker et al. 1999). Moreover, not all HAeBe stars exhibit PAH features (van den Ancker et al. 1999).

The more extensive silicate data can also provide additional constraints on the location of some of the small grains. Silicate emission bands are routinely detected in the spectra of many HAeBe stars and have been spectacularly revealed by the ISO Short Wave Spectrograph (SWS) (Waelkens et al. 1996; van den Ancker et al. 1999), where in some cases the optically thin emission dominates the IR excess of the star shortward of 60 μm. The SWS data, however, lack spatial resolution on a scale comparable to the size of the disks inferred from the millimeter data. Ground-based IR measurements of the 9.5-μm Si–O band show that such features are routinely detected in small apertures [e.g., the 3.4-arcsec diameter hole used by the BASS spectrometer at NASA Infrared Telescope Facility (IRTF)], corresponding to a radial distance $r \leq 200$ AU for stars, such as HD 163296, at distances of 120 pc. Widening the aperture (to 11″, Sitko 1981) does not increase the detected flux or change the silicate features, implying that the bulk of the silicate emission is located within 200 AU of the star. In addition, the continuum fluxes in the disks resolved with millimeter-wavelength interferometry are the same as those derived from single-dish measurements (see the chapter by Natta et al., this volume), indicating that the dust emission is not coming from extended material around these stars.

Model fits to the ISO IR emission spectra of several of the nearest HAeBe stars are consistent with optically thin, warm ($T = 210$ K) silicate grains (Malfait et al. 1998*b*; van den Ancker et al. 1999). Such warm grains must be comparatively near the star, and thus provide a probe of the solid state material in the *innermost portions of the disk.* In particular, grains that are smaller than the radiation-pressure blowout radius require replenishment on timescales that are short compared to the PMS evolution of stars. These transient grains hold the key to understanding the mineralogical composition of the dust, which gives rise to the infalling gas signatures that are observed.

A. Size Distribution

Extinction by dust in the HAeBe stars is unlike that of the general interstellar medium. This difference is because the mean grain size is considerably larger in these disks than in interstellar dust (Sitko et al. 1981, 1984, 1994). This difference in grain size affects the strength and shape of the 10-μm silicate band. Small, optically thin grains produce a strong emission feature, whereas large, optically thick (in the continuum) grains do not. Interstellar grains (with a large component of small particles) are observed to produce an emission band with a band/continuum ratio of about 3, as in the Trapezium. The same feature seen in solar system comets has a smaller ratio, usually less than 2 (Hanner et al. 1994), whereas that observed in the zodiacal light of the solar system has a ratio of about 1.15 (Reach et al. 1996). Such a trend toward a smaller ratio is expected as the spectrum becomes dominated by larger and larger particles. Smaller grains are removed by radiation pressure from the star, whereas larger ones are not.

The majority of HAeBes observed so far have band/continuum ratios intermediate between those of Trapezium dust and solar system zodiacal dust and are similar to those of comet comae. Because of the large ratio of luminosity to mass in the HAeBe stars, small grains (less than a few μm in size) are easily removed by radiation pressure on timescales of centuries. The fact that we see a large silicate emission band while grain blowout is so efficient indicates that the small grains must be replenished on similar timescales by the erosion of larger bodies that were presumably built at an earlier epoch. This is consistent with the grains coming from cometary objects in the HAeBe stars.

B. Mineralogical Composition

The IR spectra of the stars showing cometary activity contain a wealth of information concerning the mineralogical composition of the refractory and volatile solids that are present and are presumably the source of the high-velocity gas. The solids detected fall into three broad classes: refractory inorganics, refractory organics, and ice.

1. Mineralogy of the Silicates. These categories are also representative of the material in solar system comets and (except for ice) are present in interplanetary dust particles (IDPs) of (presumed) cometary origin. The most ubiquitous spectral feature seen in these stars is the 10-μm emission band, due primarily to silicate materials (Waelkens et al. 1996; van den Ancker et al. 1999; Butner et al. 1998; Malfait et al. 1998*b*). The same feature is also present in solar system comets (Fig. 5). By comparing the spectral structure observed in these objects with those of materials prepared in the laboratory (i.e., Stephens and Russell 1979; Stephens et al. 1995; Jager et al. 1994; Dorschner et al. 1995; Henning et al. 1995; Hallenbeck et al. 1998), and with anhydrous chondritic interplanetary dust particles (IDPs) of known mineralogy (Sanford and Walker 1985; Bradley et al. 1992), it is possible to identify the features seen in the stellar spectra with general mineralogical classes. This emission band is consistent with a mixture of silicate material composed primarily of pyroxenes, $(Mg,Fe,Ca)SiO_3$, and olivine, $(Mg,Fe)_2SiO_4$. The long-wavelength wing of this band, with a local maximum near 11.2 μm, is due to crystalline olivine (Fig. 5). When spectral information longward of 14 μm is included, the identification can be made even more specific. In HD 100546 (the Rosetta stone of HAeBe IR spectra), the data indicate that the crystalline olivine is forsterite, Mg_2SiO_4 (Sitko et al. 1996; Waelkens and Waters 1997; Malfait et al. 1998*b*; Sitko et al. 1999).

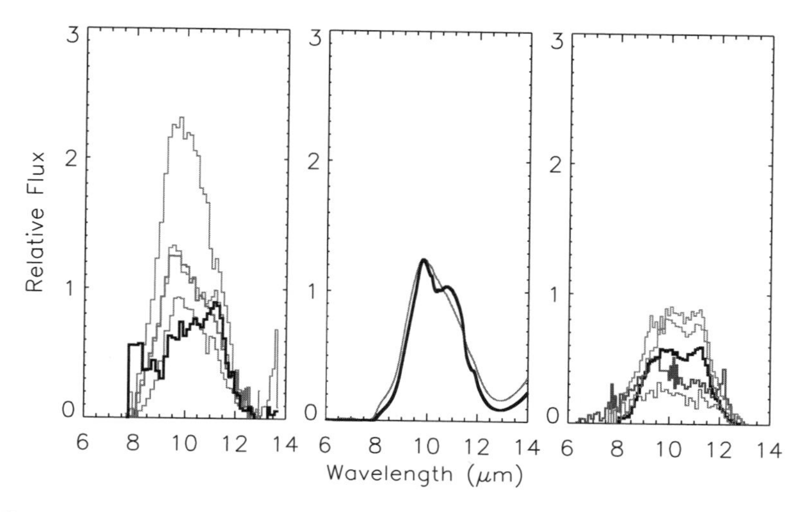

Figure 5. Continuum-normalized and continuum-subtracted 10-μm silicate profiles of a range of HAeBe stars (left), laboratory silicates (center), and comets (right). The astronomical data were normalized to a local continuum fit as a graybody at 8.5 and 12.3 μm. The most prominent emission, peaking at 9.5 μm, is seen in young ($t \approx 1$ Myr) systems such as UX Ori. Intermediate emission levels are seen in BF Ori (2–3 Myr), AB Aur (4 Myr), and HD 163296 ($4^{+6}_{-2.5}$ Myr), with the most prominent feature at 11.2 μm in HD 100546 (>10 Myr, dark, bold). The central panel shows laboratory Mg-silicate smokes that have been annealed 2 and 7 hours at 1030 K (S. L. Hallenbeck and J. A. Nuth III, private communication, 1998). The right panel shows silicate emission features from solar system comets Wilson, Borrelly, Levy, Bradfield, and Halley (bold, dark).

Even stronger than the olivine feature in the majority of the 10-μm bands in HAeBes is another feature that peaks near 9.5 μm (Fig. 6). In laboratory-produced silicates, IDPs, and in C/Hale-Bopp (Wooden et al. 1998) this feature is due to pyroxenes, with enstatite, $MgSiO_3$, being a likely candidate. The time-of-flight impact ionization mass spectrometers on the *Giotto* (PIA) and *Vega* (PUMA) Comet Halley probes found that the silicates in Halley were Mg-rich and Fe-poor (Jessberger and Kissell 1991; Schulze et al. 1997 and sources therein). Also, detailed chemical analysis of IDPs with cometlike (HAeBe-like) spectral features indicate that most of the Mg is tied up in silicates (Bradley et al. 1992). Thus, the silicates in both HAeBes and comets are Mg-rich, and that is where most of the Mg resides.

2. Crystallinity of the Silicates. The crystallinity of the olivine in the spectra of β Pic, HAeBe stars, and solar system comets is perplexing, because it is not seen in the interstellar medium grains from which these objects formed. Where is it made? It appears that amorphous presolar grains must be annealed in the inner protostellar nebula (or condensed there directly from the gas), transported to the region where comet formation is occurring, and then mixed with ice grains that probably condensed only at

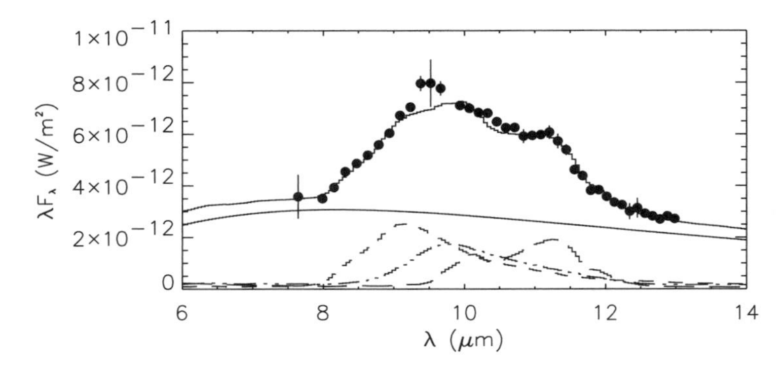

Figure 6. The spectrum of HD 163296 compared to a model composed of glassy enstatite (small dash), glassy olivine (dot-dot-dash), crystalline olivine (large dash), and an underlying blackbody continuum component (solid line), all at $T = 440$ K from Sitko et al. (1999).

temperatures ≤ 30 K (Crovisier et al. 1997). The mechanism for producing such mixing is not well understood, but a similar mechanism is also required to explain the degree of deuterated water in Comet Hyakutake (Bockelée-Morvan et al. 1998). This difficulty is analogous to that of getting melted silicates (i.e., chondrules) into the volatile-rich material that eventually gave us the carbonaceous chondrites. Many models have been suggested; none are without difficulties (Hewins 1997).

3. Iron. The IR spectra of HAeBes, chemical analysis of the "cometary" IDPs, and mass spectrometer data on Halley indicate that Fe is not exclusively tied up in silicates but occurs in significant abundance in a variety of other materials. In the case of both Halley dust (Schulze et al. 1997) and the "cometary" IDPs, Fe is primarily present as metal (kamacite), oxides (magnetite), and sulfides (see Bradley et al. 1992). Fe-rich oxides (but not silicates) are also consistent with the IR spectra of HD 100546 (Malfait et al. 1998*b*), HD 163296, and AB Aur (van den Ancker et al. 1999) as well as with the UV data.

4. Refractory Organics. The spectra of some HAeBes have emission features located at 3.28, 6.2, 7.7, 8.6, and 11.3 μm, which are indicative of organic materials. Sometimes additional bands at 3.4 and 3.5 μm are also present. The most popular explanation of these features are that they arise from vibrational transitions in PAHs. Other related organics, such as hydrogenated amorphous carbon (HAC) and quenched carbonaceous composites (QCC), have also been suggested. Among those HAeBes with infalling gas, HD 100546 stands out with very strong organic features. In another HAeBe, HD 97048 (not known to have infalling gas at the present time), speckle observations of the 3.5-μm band indicate that the emitters are located within 8 AU of the star (Roche et al. 1986).

Comets contain a significant component of refractory organic material. In Halley dust, a significant fraction of the refractory material is organic, composed of a mixture of C, H, O, and N ("CHON"). Cluster analysis of the PUMA-1 and PUMA-2 data indicates the presence of pure carbon, hydrocarbons and polymers of cyanopolyynes, and multicarbon monoxides (Fomenkova et al. 1994). In the IDPs, the refractory organic material is present as metal (FeNi) carbide and graphitic carbon as well as a disordered carbonaceous matrix (Bradley 1994). This carbonaceous matrix also contains N with an N/C ratio of about 0.1, i.e., chondritic (Keller et al. 1995). The CHON material may thus be one of the most important sources of N seen in the infalling gas. This CHON material is probably the source of most of the atomic and ionic C and N observed in the UV spectra of the accreting gas in these systems, and may be related to the organic emission features seen in the IR of the HAeBe stars. It is important to note that the ubiquitous IR bands seen in the HAeBe stars and the interstellar medium are weak (or absent) in solar system comets. The 6.2-, 7.7-, 8.6-, and 11.3-μm bands that are so apparent in the ISO spectrum of HD 100546 are not visible in the spectrum of Comet Hale-Bopp (see Fig. 5 of Malfait et al. 1998*b*). Nor is the 3.3-μm band seen in the ISO spectrum of Hale-Bopp (Crovisier et al. 1997). Although the 3.2–3.6-μm region of comets does possess an emission band, it seems to be dominated by methanol and the 3.4-μm emitter (whatever its nature), with the 3.3-μm band being the weakest, if present at all (Fig. 3 of Bockelée-Morvan et al. 1995). Joblin et al. (1997) have estimated the lifetimes of gas-phase neutral PAHs at 1 AU from the sun to be on the order of only 15 seconds. PAHs are, however, found in organic-rich meteorites, along with fullerenes and fulleranes (Becker and Bunch 1997). The chemical processing of organics in both the protosolar nebula and larger condensed bodies is a ripe field for further investigation.

5. *Ices.* Finally, we know that $\approx$50% of the mass of comets consists of water ice (which does not survive in captured IDPs), and ought to be present in the dust in accreting-gas HAeBes. Although the high temperatures and small optical depths make observing the 3.1-μm ice absorption band difficult in these stars, the HD 100546 ISO data for $50 \leq \lambda \leq 80$ μm are consistent with emission by cold ice (Malfait et al. 1998*b*). The ice may also be a reservoir for trapped volatile gases in clathrate hydrates.

The evaporation of cometary materials from parent bodies on highly eccentric orbits around HAeBe stars will naturally give rise to the observed redshifted absorption features, partial covering factors, and wide range of observed atomic ions.

VI. IMPLICATIONS

Both the infalling circumstellar gas and the small dust grains that can be remotely sensed around young, intermediate-mass stars bear a close resemblance to cometary materials in our own solar system (see Table II). Dynamically, the closest analog to the infalling bodies are the Kreutz

TABLE II
Signatures of Cometary Activity in HAeBe Disks

Observation	Interpretation
Redshifted absorption features:	**Infalling material:**
Neutral species with short lifetimes	Local production via grain dissociation
Multiple Ionization Stages with similar profiles	Production of collisionally ionized species in bow shocks
Ballistic velocities	Solid parent bodies not subject to drag
Mixture of elements in redshifted gas	Parent bodies not differentiated
Duration of episodes ≥ 20 hours	Swarms/families detected, not individual bodies
Transition to episodic infall at ≈ 6 Myr	Bombardment rate drops with clearing of the disk
IR solid state features:	**Grains with $a \leq$ few μm:**
Broad 9.5- and 18-μm features	Amorphous olivine
Sharp 11.2–33-μm features	Partially crystalline olivine
69-μm emission (HD 100546)	Forsterite (e.g., crystalline Mg-rich olivine)
Bulge in SED near 21 μm	FeO
9.5–10 μm and longer-λ features	Pyroxene
60–80-μm features	Crystalline water ice
3.28, 6.2, 7.7, 8.6, and 11.3 μm; 3.4, 3.5 μm	Polycyclic aromatic hydrocarbons (PAHs)

family of Sun-grazers, which not only have their eccentricities pumped up by encounters with Jupiter but also exhibit rich emission spectra from refractory elements (see Table I).

Infalling circumstellar gas, similar to β Pic's, is routinely detected in PMS (HAeBe) and young main-sequence (shell star) systems viewed within $\approx 14°$ of their disk midplanes. This limit is consistent with recent determinations, from imaging of the disk, of the system inclinations for HR 4796A (Koerner et al. 1998) and Fomalhaut (Holland et al. 1998). Infall detections are routine among the 1–4-Myr-old systems, but they show signs of transition to more episodic infall between 6 (HD 163296) and 10 Myr (HD 100546), similar to that reported for β Pic (Deleuil et al. 1995) (Fig. 7). There are only more sporadic detections of infalling material in the lines of sight to older, dated systems, with one detection toward the primary of a Lindroos (1986) double at $t = 79$ Myr, and infalling material present in the UV spectrum of Pleione ($t \approx 100$ Myr). There are some suggestions that close binaries make the transition to main-sequence

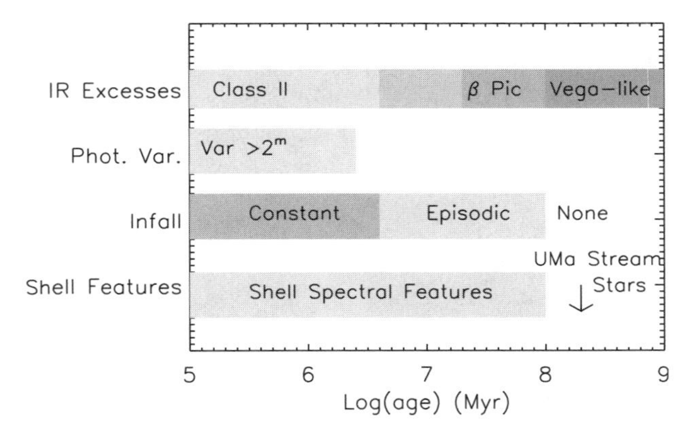

Figure 7. The evolutionary interval over which high-covering-factor infall is detectable in UV/optical spectra of young, intermediate-mass stars spans the age range, from the point at which the stars become directly observable at ≈ 0.1 Myr with Lada (1987) Class II IR excesses to the point at which the IR excess resembles that of Vega. The frequency of infall detections ramps down rapidly from 3–4 Myr to the age of β Pic. Infalling gas and material at the system velocity have not been reported for stars older than 200 Myr (Lecavelier des Etangs et al. 1997). Similarly, dust detections become increasingly scarce for stars in this age range (Abraham et al. 1998).

star levels of cometary activity earlier than isolated systems do, with low levels of infall present in 51 Oph (t = 0.3 Myr, van den Ancker et al. 1997), and HD 176386 (t = 2 Myr, Wilking et al. 1985; Grady et al. 1993).

Models for the β Pic system suggest that both long-term fluctuations in the infall rate and the presence of preferred velocities for the star-grazing cometesimals are signatures of perturbations on the disk by one or more Jupiter-mass planets (Deleuil et al. 1995; Beust et al. 1996). The HAe star data suggest that episodic infall, together with the development of prominent mid-IR "gaps" in the IR spectral energy distributions (Malfait et al. 1998*a*) become macroscopically detectable by 10 Myr (HD 100546). If multiyear fluctuations in the frequency of optical minima in younger HAe stars with UXOR light variations (Shevchenko et al. 1993*a,b*; Shevchenko and Herbst 1998; Grinin et al. 1998) represent similar activity, the effects of large bodies on the disk become detectable by 1–4 Myr in systems with no known close *stellar* companions. These observational constraints are broadly consistent with theoretical estimates of the time needed to produce a Jupiter (Pollack et al. 1996) and are consistent with identification of TMR-1C as a planet (Terebey et al. 1998).

Studies of the circumstellar material around intermediate-mass stars have greatly benefited from access to data from the FUV through the far IR. We anticipate that more detailed studies of many of these objects will continue with current and new instrumentation and will begin to give us a full view of the dynamics of these systems as forming planetary systems. However, infall activity and the related signatures of larger bodies within circumstellar disks do not appear limited to intermediate-mass stars. Several weak-line T Tauri stars have been identified with large-amplitude polarimetric and photometric variability similar to that observed toward suitably oriented young HAeBe stars (Gahm et al. 1993; Alcalá et al. 1993; Guenther and Hessman 1993; Grinin et al. 1995). These stars have evolutionary ages spanning $3 \leq t \leq 10$ Myr (Gahm et al. 1993; van den Ancker et al. 1998), making them contemporaries of the HAeBe systems. Comparatively weak optical emission features, and IR excesses indicative of central clearing of the disks (Alcalá et al. 1993), argue against gas-phase accretion from a classical accretion disk in these systems. However, these objects exhibit prominent infall signatures in transitions such as Hα and Na I D. The available data for these stars suggests that we are seeing β Pic-like cometary activity in the disks of these stars. The presence of cometary activity in one system with an age of $\geq$10 Myr suggests that high rates of cometary bombardment continue for longer times around these isolated young sunlike stars than for the HAeBe stars. The presence of such activity has important implications for the potential delivery of volatiles to terrestrial planet cores in such a disk, and thus for the potential development of life-bearing worlds. We anticipate that detailed studies of these objects will be as pivotal to studies of planetary system evolution in the next decade as β Pictoris has proven to be in studies of intermediate-mass stars. Exciting times lie ahead.

Acknowledgments Support for this study for C. A. G. was provided by NASA Contract NASW 4756 and PO Z78611Z to Eureka Scientific, and HST STIS GTO funding through support to AURA/NOAO in response to NASA A/O OSSA-4-84 through the Hubble Space Telescope Project at Goddard Space Flight Center. M. L. S. was supported by the University Research Council and Physics Department of the University of Cincinnati. D. K. L. and R. W. R. were supported by the Aerospace Corporation's Independent Research and Development Program. The work of M. S. H. was carried out at the Jet Propulsion Laboratory, California Institute of Technology, under contract with NASA. This study is based on observations made with the *International Ultraviolet Explorer,* The European Southern Observatory, the NASA Infrared Telescope Facility, the *Infrared Space Observatory,* and the Cerro Tololo Interamerican Observatory. We wish to thank S. Hallenbeck and J. Nuth for contributing some of their prepublication laboratory measurements of silicate emission, D. Biesecker for a valuable conversation regarding Sun-grazing comets, and J. Nuth and D. Devine for critical readings of the manuscript. We wish to thank the anonymous referee for his or her helpful comments. This study has made use of SIMBAD and the Astrophysics Data System.

REFERENCES

Abraham, P., Leinert, C., Burkert, A., Lembke, D., and Henning, T. 1998. Search for cool circumstellar matter in the Ursae Majoris group with ISO. *Astron. Astrophys.* 338:91–96.

Alcalá, J. M., Covino, B., Franchini, M., Kreeter, Terranegra, L., Wichmann, R. 1993. T-Chamaeleontis—a weak-line YY Orionis star? *Astron. Astrophys.* 272:225–234.

Artymowicz, P. 1997. β Pictoris: An early solar system? *Ann. Rev. Earth Planet. Sci.* 25:175.

Backman, D. E., Werner, M. W., Rieke, G. H., and Van Cleve, J. E. 1997. Exploring planetary debris around solar-type stars. In *From Stardust to Planetesimals,* ASP Conf. Ser. 122, ed. Y. J. Pendleton and A. G. G. M. Tielens (San Francisco: Astronomical Society of the Pacific), pp. 49–66.

Becker, L., and Bunch, T. E. 1997. Fullerenes, fulleranes, and polycyclic aromatic hydrocarbons in the Allende meteorite. *Meteor. Planet. Sci.* 32:479–487.

Beckwith, S. V. W., and Sargent, A. I. 1996. Circumstellar disks and the search for neighbouring planetary systems. *Nature* 383:139–144.

Beust, H., Vidal-Madjar, A., Ferlet, F., Lagrange-Henri, A. M. 1990. The β Pictoris circumstellar disk. X. numerical simulations of infalling evaporating bodies. *Astron. Astrophys.* 236:202–216.

Beust, H., Lagrange, A. M., Plazy, I., and Mouillet, D. 1996. The β Pictoris circumstellar disk. XXII. Investigating the model of multiple cometary infalls. *Astron. Astrophys.* 310:181–198.

Bibo, E. A., and Thé, P. S. 1991. The type of variability of Herbig Ae/Be stars. *Astron. Astrophys. Suppl.* 89:319–334.

Bjorkman, K. S., Meade, M. R., Babler, B. L., Anderson, C. M., Code, A. D., Fox, G. K., Johnson, J. J., Sanders, W. T., Weitenbeck, A. J., Zellner, N. E. B., Nordsieck, K. H., Pozdell, S., Hanson, J., Ager, W., Putman, M., Lupie, O. L., Edgar, R. J., Johansen, K. A., and Kobalnicky, H. A. 1995. Diagnosing the circumstellar environments of HD 163296. *Bull. Am. Astron. Soc.* 187:10604.

Bockelée-Morvan, D., Brooke, T. Y., and Crovisier, J. 1995. On the origin of the 3.2- to 3.6-μm emission feature in comets. *Icarus* 116:18–39.

Bockelée-Morvan, D., Gautire, D., Lis, D. C., Young, K., Keene, J., Phillips, T., Owen, T., Crovisier, J., Goldsmith, P. F., Bergin, E. A., Despois, D., and Wootten, A. 1998. Deuterated water in Comet C/1996 B2 (Hyakutake) and its implications for the origins of comets. *Icarus* 133:147–162.

Bodenheimer, P. 1997. The role of dust in star and planet formation: Theory. In *From Stardust to Planetesimals*, ASP Conf. Ser. 122, ed. Y. J. Pendleton and A. G. G. M. Tielens (San Francisco: Astronomical Society of the Pacific), 122:37–48.

Boss, A. P. 1996. Extra-solar planets. *Phys. Today* 49:32–38.

Bradley, J. P. 1994. Nanometer-scale mineralogy and petrography of fine-grain aggregates in anhydrous interplanetary dust particles. *Geochim. Cosmochim. Acta* 58:2123–2134.

Bradley, J. P., Humecki, H. J., and Germani, M. S. 1992. Combined infrared and analytical electron microscope studies of interplanetary dust particles. *Astrophys. J.* 394:643–651.

Bruhweiler, F. C., Grady, C. A., Chiu, W. A. 1989. Highly ionized species and circumstellar shells in B8-A1 stars. *Astrophys. J.* 340:1038–1048.

Butner, H. M., Walker, H. J., and Wooden, D. H. The dust around Vega-like stars: A bridge from YSOs to comets. In *Formation and Evolution of Solids in Space,* ed. T. Mayo Greenberg and A. Li (Dordrecht: Kluwer), 1999 p. 505.

Crifo, F., Vidal-Madjar, A., Lallement, R., Ferlet, R., Gerbaldi, M. 1997. β Pictoris revisited by Hipparcos. Star properties. *Astron. Astrophys.* 320:L29–L32.

Crovisier, J., Leech, K., Bockelée, D., Brooke, T. Y., Hanner, M. S., Altieri, B., Keller, H. U., and Lellouch, E. 1997. The spectrum of Comet Hale-Bopp (C/1995 O1) observed with the Infrared Space Observatory at 2.9 astronomical units from the sun. *Science* 275:1904–1907.

Deleuil, M., de Winter, D., Grady, C. A., van den Ancker, M. E., Pérez, M. R., and Eiroa, C. 1995. 10 years of UV observations of the β Pictoris. In *Circumstellar Disk in Circumstellar Dust Disks and Planet Formation*, ed. A. Vidal-Madjar and R. Ferlet (Paris: Editions Frontières), pp. 29–34.

de Winter, D., et al. 1999. Episodic accretion around the Herbig Ae star BF Ori: Evidence for the presence of extra-solar comets. *Astron. Astrophys.,* in press.

Dorschner, J., Begemann, B., Henning, T., Jager, C., and Mutschke, H. 1995. Steps toward interstellar silicate mineralogy. II. Study of Mg-Fe-silicate glasses with variable composition. *Astron. Astrophys.* 300:503–520.

Duncan, M., Quinn, T., and Tremaine, S. 1987. The formation and extent of the solar system comet cloud. *Astron. J.* 94:1330–1338.

Edwards, S., Hartigan, P., Ghandour, L., and Andrulis, C. 1994. Spectroscopic evidence for magnetospheric accretion in classical T Tauri stars. *Astron. J.* 108:1056–1070.

Ferlet, R., Vidal-Madjar, A., Hobbs, L. M. 1987. The β Pictoris circumstellar disk. V. Time variations of the Ca II-K line. *Astron. Astrophys.* 185:267–270.

Fomenkova, M. N., Chang, S., and Muhkin, L. M. 1994. Carbonaceous components in the Comet Halley dust. *Geochim. Cosmochim. Acta* 58:4503–4512.

Gahm, G. F., Liseau, R., Gullbring, and E., Hartstein, D. 1993. The circumstellar gleam from the T Tauri star RY Lup. *Astron. Astrophys.* 279:477–484.

Gledhill, T. M., Scarrott, S. M., and Wolstencroft, R. D. 1991. Optical polarization in the disc around β Pictoris. *Mon. Not. Roy. Astron. Soc.* 252: P50–P54.

Grady, C. A., Pérez, M. R., and Thé, P. S. 1993. The accreting circumstellar gas envelope of HD 176386: A young star in the R Coronae Austrinae star formation region. *Astron. Astrophys.* 274:847–850.

Grady, C. A., Pérez, M. R., Talavera, A., Bjorkman, K. S., de Winter, D., Thé, P. S., Molsler, F. J., van den Ancker, M. E., Sitko, M. L., Morrison, N. D., Beaver, M. L., McCollum, B., and Castelaz, M. W. 1996*a*. The β Pictoris phenomenon among Herbig Ae/Be stars: UV and optical high dispersion spectra. *Astron. Astrophys. Suppl.* 120:157–177.

Grady, C. A., Pérez, M. R., Talavera, A., McCollum, B., Rawley, L. A., England, M. N., and Schlegl, M. 1996*b*. The β Pictoris phenomenon in A-shell stars: Detection of accreting gas. *Astrophys. J. Lett.* 471:L49–L52.

Grady, C. A., Sitko, M. L., Bjorkman, K. S., Pérez, M. R., Lynch, D. K., Russell, R. W., and Hanner, M. S. 1997. The star-grazing extrasolar comets in the HD 100546 system. *Astrophys. J.* 483:449–456.

Grady, C. A., Pérez, M. R., Bjorkman, K. S., and Massa, D. 1999*a*. Transient infall events in the AB Aur disk: The β Pictoris phenomenon at 2–4 Myr. *Astrophys. J.,* 511:925–931.

Grady, C. A., et al. 1999*b*. The β Pic phenomenon at 4 Myr: HD 163296. *Astrophys. J.,* submitted.

Graham, J. A. 1992. Clumpy accretion onto pre-main-sequence stars. *Pub. Astron. Soc. Pacific* 104:479–488.

Greaves, J., Holland, W., Moriarty-Schieven, G., Dent, W., Zuckerman, B., McCarthy, C., Webb, R., Butner, H. M., Gear, W., and Walker, W. 1998. A dust ring around ϵ Eridani: Analogue to the young solar system. *Astrophys. J. Lett.*, submitted.

Greenberg, J. M. 1998. Making a comet nucleus. *Astron. Astrophys.* 330:375–380.
Grinin, V. P., and Rostopchina, A. N. 1996. Orientation of circumstellar disks and the statistics of Hα profiles of Ae/Be Herbig stars. *Astron. Rep.* 40:171–178.
Grinin, V. P., Kiselev, N. N., Chernova, G. P., Minikulov, N. K., and Voshchinnikov, N. V. 1991. The investigations of "zodiacal light" of isolated A-Herbig stars with nonperiodic Algol-type minima. *Astrophys. Space Sci.* 186:283–298.
Grinin, V. P., Thé, P. S., de Winter, D., Giampapa, M., Rostopchina, A. N., Tamboutseva, L. U., and van den Ancker, M. E. 1994. The β Pictoris phenomenon among young stars. 1: The case of the Herbig Ae star UX Orionis. *Astron. Astrophys.* 292:165–174.
Grinin, V. P., Kolotilov, E. A., and Rostopchina, A. N. 1995. Dust around young stars. Photopolarimetric observations of the T Tauri star BM Andromedae. *Astron. Astrophys. Suppl.* 112:457–473.
Grinin, V. P., Kozlova, O. V., Thé, P. S., and Rostopchina, A. N. 1996. The β Pictoris phenomenon among young stars. III. The Herbig Ae stars WW Vulpeculae, RR Tauri, and BF Orionis. *Astron. Astrophys.* 309:474–480.
Grinin, V. P., Rostopchina, A. N., and Shakhovskoy, D. N. 1998. On the nature of cyclic variability of the UX Ori type stars. *Astron. Lett.* 24:802.
Guenther, E., and Hessman, F. V. 1993. Variable redshifted He I absorption lines in BM Andromedae. *Astron. Astrophys.* 276:L25–L28.
Hallenbeck, S. L., Nuth, J. A., III, and Daukantas, P. L. 1998. Mid-infrared spectral evolution of amorphous magnesium smokes annealed in vacuum: Comparison to cometary spectra. *Icarus* 131:198–209.
Hanner, M. S., Lynch, D. K., and Russell, R. W. 1994. The 8–13 micron spectra of comets and the composition of silicate grains. *Astrophys.J.* 425:274–285.
Heap, S. R., Lindler, D. J., Lanz, Th., Woodgate, B., Cornett, R., Hubeny, I., Maran, S. P. 2000. Hubble/STIS coronagraphic imagery of β Pictoris. *Astrophys. J.,* submitted.
Henning, T., Begemann, B., Mutschke, H., and Dorschner, J. 1995. Optical properties of oxide dust grains. *Astron. Astrophys. Suppl.* 112:143–149.
Henning, T., Burkert, A., Launhardt, R., Leinert, C., and Stecklum, B. 1998. Infrared imaging and millimetre continuum mapping of Herbig Ae/Be and FU Orionis stars. *Astron. Astrophys.* 336:565–586.
Herbst, W., Herbst, D. K., and Grossman, E. J. 1994. Catalogue of UBVRI photometry of T Tauri stars and analysis of the causes of their variability. *Astron. J.* 108:1906–1923.
Hewins, R. H. 1997. Chondrules. *Ann. Rev. Earth Planet. Sci.* 25:61–83.
Hobbs, L. M. 1987. Observations of gaseous circumstellar disks III. *Astrophys. J.* 308:854–858.
Holland, W. S., Greaves, J. S., Zuckerman, B., Webb, R. A., McCarthy, C., Coulson, I. M., Walther, D. M., Dent, W. R. F., Gear, W. K., and Robson, I. 1998. Submillimetre images of dusty debris around nearby stars. *Nature* 392:788–790.
Jager, C., Mutschke, H., Begemann, B., Dorschner, J., and Henning, T. 1994. Steps toward interstellar silicate mineralogy. I. Laboratory results of a silicate glass of mean cosmic composition. *Astron. Astrophys.* 292:641–655.
Jayawardhana, R., Fisher, S., Hartmann, L., Telesco, C., Pina, R., and Fazio, G. 1998. A dust disk surrounding the young A star HR 4796A. *Astrophys. J. Lett.* 503:L79–L82.
Jenkins, E. B. 1987. Element abundances in the interstellar medium. *Astrophys. Space Sci.* 134:533–560.
Jessberger, E. K., and Kissell, J. 1991. Chemical properties of cometary dust and a note on carbon isotopes. In *Comets in the Post-Halley Era*, ed. R. L. Newburn, Jr., M. Neugebauer, and J. Rahe (Dordrecht: Kluwer), pp.1075– 1092.

Joblin, C., Boissel, P., and de Parseval, P. 1997. Polycyclic aromatic hydrocarbons lifetime in cometary environments. *Planet. Space Sci.* 45:1539–1542.
Jura, M., Zuckerman, B., Becklin, E. E., and Smith, R. C. 1993. Constraints on the evolution of remnant protostellar dust debris around HR 4796. *Astrophys. J. Lett.* 418:L37–L40.
Keller, L. P., Thomas, K. L., Bradley, J. P., and McKay, D. S. 1995. Nitrogen in interplanetary dust particles. *Meteoritics* 30:526–527 (abstract).
Knacke, R. F., Fajardo-Acosta, S. B., Telesco, C. M., Hackwell, J. A., Lynch, D. K., Russell, R. W. 1993. The silicates in the disk of β Pictoris. *Astrophys. J.* 418:440–450.
Koerner, D. W., Ressler, M., Werner, M., and Backman, D. E. 1998. Mid-infrared imaging of a circumstellar disk around HR 4796: Mapping the debris of planetary formation. *Astrophys. J.* 503:L83–86.
Lada, C. J. 1987. Star Formation: From OB associations to protostars. In *Star Forming Regions*, ed. M. Peimbert and J. Jugaka (Dordrecht: Kluwer), p. 1.
Lecavelier des Etangs, A., Scholl, H., Roques, F., Sicardy, B., and Vidal-Madjar, A. 1996. Perturbations of a planet on the β Pictoris circumstellar dust disk. *Icarus* 123:168–179.
Lecavelier des Etangs, A., Ferlet, R., and Vidal-Madjar, A. 1997. A search for β Pictoris-like Ca II circumstellar gas around Ursa Major stream stars. *Astron. Astrophys.* 328:602–605.
Leinert, C., Richter, I., Pitz, E., and Planck, B. 1981. The zodiacal light from 1.0 to 0.3 AU as observed by the Helios space probes. *Astron. Astrophys.* 103:177–188.
Lindroos, K. P. 1986. A study of visual double stars with early-type primaries. V. Post-T Tauri secondaries. *Astron. Astrophys.* 156:223–233.
Malfait, K., Bogaert, E., Waelkens, C. 1998*a*. An ultraviolet, optical and infrared study of Herbig Ae/Be stars. *Astron. Astrophys.* 331:211–223.
Malfait, K., Waelkens, C., Waters, L. B. F. M., Vandenbussche, B., Huygens, E., and de Graauw, M. S. 1998*b*. The spectrum of the young star HD 100546 observed with the Infrared Space Observatory. *Astron. Astrophys.*, 332:L25–L28.
Mannings, V., and Sargent, A. I. 1997. A high-resolution study of gas and dust around young intermediate-mass stars: Evidence for circumstellar disks in Herbig Ae systems. *Astrophys. J.* 490:792–802.
Mannings, V., Koerner, D. W., and Sargent, A. I. 1997. A rotating disk of gas and dust around a young counterpart to β Pictoris. *Nature* 388:555–557.
Mouillet, D., Larwood, J. D., Papaloizou, J. C. B., and Lagrange, A. M. 1997. A planet on an inclined orbit as an explanation of the warp in the β Pictoris disc. *Mon. Not. Roy. Astron. Soc.*, 292:896–904.
Owen, T. C. 1997. From planetesimals to planets: Contributions of icy planetesimals to planetary atmospheres. In *From Stardust to Planetesimals*, ASP Conf. Ser. 122, ed. Y. J. Pendleton and A. G. G. M. Tielens (San Francisco: Astronomical Society of the Pacific), pp. 435–452.
Pérez, M. R., and Grady, C. A. 1997. Observational overview of young intermediate-mass objects: Herbig Ae/Be stars. *Space Sci. Rev.* 82:407–450.
Pérez, M. R., Grady, C. A., and Thé, P. S. 1993. Ultraviolet spectral variability in the Herbig Ae star HR 5999. Part XI. The accretion interpretation. *Astron. Astrophys.* 274:381–390.
Pollack, J. B., Hubickyj, O., Bodenheimer, P., Lissauer, J. J., Podolak, M. and Greenzweig, Y. 1996. Formation of the giant planets by concurrent accretion of solids and gas. *Icarus* 124:62–85.
Reach, W. T., Abergel, A., Boulanger, F., Dèsert, F.-X., Perault, M., Bernard, J.-P., Blommaert, J., Cesarsky, C., Cesarsky, D., Metcalfe, L., Puget, J.-L., Sibille,

F., and Vigroux, L. 1996. Mid-infrared spectrum of the zodiacal light. *Astron. Astrophys. Lett.* 315:L381–L384.

Roche, P. F., Allen, D. A., and Bailey, J. A. 1986. The spatial extent and nature of the 3-μm emission features in HD 97048 and CPD $-56^{\circ}8032$. *Mon. Not. Roy. Astron. Soc.* 220:7P–11P.

Roques, F., Scholl, H., Sicardy, B., and Smith, B. A. 1994. Is there a planet around β Pictoris? Perturbations of a planet on a circumstellar dust disk. 1: The numerical model. *Icarus* 108:37–58.

Sanford, S. A., and Walker, R. M. 1985. Laboratory infrared transmission spectra of individual interplanetary dust particles from 2.5 to 25 microns. *Astrophys. J.* 291:838–851.

Schulze, H., Kissel, J., and Jessberger, E. K. 1997. Chemistry and mineralogy of Comet Halley's dust. In *From Stardust to Planetesimals*, ASP Conf. Ser. 122, ed. Y. J. Pendleton and A. G. G. M. Tielens (San Francisco: Astronomical Society of the Pacific), pp. 397–414.

Shevchenko, V. S. and Herbst, W. 1998. Photometric variations of T Tauri and Herbig Ae/Be stars: Classification and interpretation. *Astron. J.,* in preparation.

Shevchenko, V. S., Grankin, K. N., Ibragimov, M. A., Melnikov, S. Y., and Yakubov, S. D. 1993*a*. Periodic phenomena in Ae/Be Herbig stars lightcurves. Part 1. Light curves classification and digital analysis methods. *Astrophys. Space Sci.* 202:121–136.

Shevchenko, V. S., Grankin, K. N., Ibragimov, M. A., Melnikov, S. Y., and Yakubov, S. D. 1993*b*. Periodic phenomena in Ae/Be Herbig stars lightcurves. Part 2. Results and probable interpretation for selected stars. *Astrophys. Space Sci.* 202:137.

Sitko, M. L. 1981. Spectral energy distributions of hot stars with circumstellar dust. *Astrophys. J.* 247:1024–1038.

Sitko, M. L., Savage, B. D., and Meade, M. R. 1981. Ultraviolet observations of hot stars with circumstellar dust shells. *Astrophys. J.* 246:161–183.

Sitko, M. L., Simon, T., and Meade, M. R. 1984. Ultraviolet spectroscopy of hot stars with infrared excesses: NGC 2264-W46, W90, and W100. *Publ. Astron. Soc. Pacific* 96:54–61.

Sitko, M. L., Halbedel, E. M., Lawrence, G. F., Smith, J. A., and Yanow, K. 1994. Variable extinction in HD 45677 and the evolution of dust grains in pre–main sequence disks. *Astrophys. J.* 432: 753–762.

Sitko, M. L., Lynch, D. K., Russell, R. W., Hanner, M. S., and Grady, C. A. 1996. Partially crystalline silicate dust in protostellar disks. In *From Stardust to Planetesimals: Contributed Papers,* ed. M. E. Kress, A. G. G. M. Tielens, and Y. J. Pendleton, NASA CP 3343, (Moffett Field, California: NASA/Ames), pp.19–22.

Sitko, M. L., Grady, C. A., Lynch, D. K., Russell, R. W., and Hanner, M. S. 1999. Cometary dust in the debris disks of HD 31648 and HD 163296: Two "baby" β Pics. *Astrophys. J.,* in press.

Smith, B. A., and Terrile, R. J. 1984. A circumstellar disk around β Pictoris. *Science* 226:1421–1424.

Sorelli, C., Grinin, V. P., and Natta, A. 1996. Infall in Herbig Ae/Be stars: What Na D lines tell us. *Astron. Astrophys.* 309:155–162.

Stephens, J. R., and Russell, R. W. 1979. Emission and extinction of ground and vapor-condensed silicates from 4 to 14 microns and the 10 micron silicate feature. *Astrophys. J.* 228:780–786.

Stephens, J. R., Blanco, A., Bussoletti, E., Colangeli, L., Fonti, S., Mennella, V., and Orofino, V. 1995. Effect of composition on IR spectra of synthetic amorphous silicate cosmic dust analogues. *Planet. Space Sci.* 43:1241–1246.

Strom, K. M., Strom, S. E., Edwards, S., Cabrit, S., Skrutskie, M. F. 1989. Circumstellar material associated with solar-type pre-main-sequence stars—A possible constraint on the timescale for planet building. *Astron. J.* 97:1451–1470.

Terebey, S., van Buren, D., Padgett, D. L., Hancock, T., and Brundage, M. 1998. A candidate protoplanet in the Taurus star-forming region. *Astrophys. J. Lett.* 507:L71.

Thé, P. S., de Winter, D., and Pérez, M. R. 1994. A new catalogue of members and candidate members of the Herbig Ae/Be (HAEBE) stellar group. *Astron. Astrophys. Suppl.* 104:315–339.

van den Ancker, M. E., Thé, P. S., Tjin A Djie, H. R. E., Catala, C., de Winter, D., Blondel, P. F. C., and Waters, L. B. F. M. 1997. Hipparcos data on Herbig Ae/Be stars: An evolutionary scenario. *Astron. Astrophys.* 324:L33–L36.

van den Ancker, M. E., de Winter, D., and Tjin A Djie, H. R. E., 1998. Hipparcos photometry of Herbig Ae/Be stars. *Astron. Astrophys.* 330:145–154.

van den Ancker, M. E., Bouwman, J., Wesselius, P. R., Waters, L. B. F. M., Dougherty, S. M., and van Dishoeck, E. E. 1999. ISO-SWS observations of circumstellar dust in Herbig Ae/Be stars. *Astron. Astrophys.,* in press.

Vidal-Madjar, A., and Ferlet, R. 1995. β Pictoris: The gaseous circumstellar disk. In *Circumstellar Disks and Planet Formation*, ed. A. Vidal-Madjar and R. Ferlet (Paris: Editions Frontières), pp. 7–18.

Vidal-Madjar, A., Lagrange-Henri, A. M., Feldman, P. D., Beust, H., Lissauer, J. J., Deleuil, M., Ferlet, R., Gry, C., Hobbs, L. M., and McGrath, M. A. 1994. HST-GHRS observations of β Pictoris: Additional evidence for infalling comets. *Astron. Astrophys.* 290:245–258.

Waelkens, C., and Waters, L. B. F. M. 1997. First ISO-SWS results: From stardust to planetesimals. In *From Stardust to Planetesimals,* ASP Conf.Ser. 122, ed. Y. J. Pendleton and A. G. G. M. Tielens (San Francisco: Astronomical Society of the Pacific), pp. 67–74.

Waelkens, C., Waters, L. B. F. M., de Graauw, M. S., Huygen, E., Malfait, K., Plets, H., Vandenbussche, B., Beintma, D. A., Boxhoorn, D. R., Habing, H. J., Heras, A. M., Kester, D. J. M., Lahuis, F., Morris, P. W., Roelfsma, D. R., Salama, A., Siebenmorgen, R., Trams, N. R., van der Bliek, N. R., Valentijn, B. A., and Wesselius, P. R. 1996. SWS observations of young main-sequence stars with dusty circumstellar disks. *Astron. Astrophys.* 315:L245–L248.

Waters, L. B. F. M., and Waelkens, C. 1998. The Herbig Ae/Be stars. *Ann. Rev. Astron. Astrophys.* 36:233.

Welty, A. D., Barden, S. C., Huenemoerder, D. P., and Ramsey, L. W. 1992. BF Orionis—Evidence for an infalling circumstellar envelope. *Astron. J.* 103:1673–1678.

Wilking, B. A., Harvey, P. M., Joy, M., Hyland, A. R., and Jones, T. J. 1985. Far-infrared observations of young clusters embedded in the R Coronae Australis and ρ Ophiuchi dark clouds. *Astrophys. J.* 293:165–177.

Wooden, D. H., Harker, D. E., Woodward, C. E., Batner, H. M., Kuike, C., Witteborn, F. C., and McMurtry, C. W. 1999. Silicate mineralogy of the dust in the inner coma of Comet C/1995 01 (Hale-Bopp) pre- and postperihelion. *Astrophys. J.* 517:1034–1058.

PLANETARY MATERIAL AROUND MAIN-SEQUENCE STARS

ANNE-MARIE LAGRANGE
Laboratoire d'Astrophysique de Grenoble

DANA E. BACKMAN
Franklin and Marshall College

and

PAWEL ARTYMOWICZ
Stockholm Observatory

Main-sequence circumstellar dust systems resembling the IRAS-discovered prototypes Vega and β Pictoris have been found to be common, occurring around at least 15% of nearby field stars of types A–K. Defining characteristics of these objects include low dust luminosity and optical depth; small gas/dust mass ratio, such that dust dynamics are approximately Keplerian; and short dust lifetimes relative to star ages. The dust is clearly "second generation"; that is, not primordial but released from larger parent bodies such as asteroids or comets. Recent images of some of these objects show them to be disks with central gaps about the size of the planetary region of our solar system, as had been inferred previously. High-resolution imaging has revealed structure in some of the disks, implying the influence of planet masses.

A few systems such as β Pic have circumstellar gas that also must be "second generation"; however, a general connection between circumstellar dust and gas in main-sequence systems is yet to be established. Observations and elaboration of models of β Pic's transient absorption line features have confirmed that they can be explained in terms of infalling evaporating planetesimals. Planetesimals on star-grazing trajectories almost certainly imply the existence of planet-mass perturbers. It is reasonable to expect the same phenomenon to occur around other stars, especially younger ones, but interpretation of observations so far in these terms is not straightforward.

Stellar age estimates indicate that many of these disks are a few $\times 10^7$ to a few $\times 10^8$ yr old, corresponding to the hypothesized timespan for construction and heavy bombardment of planets in our solar system. The amount of dust appears generally to decrease with system age, and theoretical calculations imply that internal processes completely dominate disk evolution relative to effects such as erosion by the interstellar medium (ISM). Models of a postulated original massive Kuiper Belt and its collision debris production correspond in size and dust optical depth to the Vega /β Pic disks.

I. INTRODUCTION

We shall focus on observations and models of planetary systems in a late evolutionary stage, when original optically thick protoplanetary disks have disappeared, planet-mass objects have formed, and remnant planetesimals are sources of "second-generation" circumstellar (CS) dust and gas. Previous reviews regarding these objects are by Backman and Paresce (1993), Artymowicz (1997), Plets (1997), Waelkens and Waters (1998), Lagrange (1998), and Vidal-Madjar et al. (1998).

We suggest the following defining "Vega /β Pic-like" system characteristics, which will be discussed immediately below:

1. Bolometric $L_{\rm dust}/L_\star \ll 1$
2. M(dust + gas) $\ll 0.01$ M$_\odot$ (the minimum-mass solar nebula)
3. Dust dynamics that are not controlled by gas; $M_{\rm gas} \ll 10 M_{\rm dust}$
4. Grain destruction times that are less than the age of the star

Such cool, thin, nearly gasless systems are generally harder to detect than the protoplanetary disks found in and near star-forming regions and thus are presently less well known and less well studied than their younger counterparts.

Criteria 1, 2, and 3 ensure that a massive protostellar disk is not present; the existing CS material is optically thin; most of the gas has condensed, has dispersed, or is segregated from the dust; and solid particles are in Keplerian motion modified by radiation pressure (RP). Herbig Ae/Be (HAeBe) stars and classical T Tauri (TT) systems should be excluded by these criteria. In addition, the gas/dust ratio in criterion 3 is meant to eliminate systems in which grains are condensing at cosmic abundances in winds and jets, as well as "shell" stars with CS gas envelopes but no measurable dust.

Criterion 4 is crucial, requiring that observed grains be continuously replenished from reservoirs of larger objects and thus be second-generation and not primordial. Inference of second-generation *gas* around β Pic will be covered in section IV.B.2. It is important to point out that second-generation material around a star indicates the existence not only of planetesimals but also of larger masses capable of sending small bodies into fragmenting collisions and star-grazing orbits ("Falling Evaporating Bodies," or FEBs) to release dust and gas. Our solar system (SS) is an example of an old main-sequence (MS) system of second-generation dust and gas. An external observer detecting the SS dust and active comet comae could thereby infer the existence of remnant planetesimals and their planetary perturbers.

Note that, despite the phrase "main sequence" in our chapter title, we prefer a definition of Vega/β Pic debris disk systems based on CS physical processes rather than the exact evolutionary state of the primary star. A star may or may not have reached MS status by the time its circumstellar disk is cleared of gas. Observations of pre-main-sequence (PMS) objects show that substantial protoplanetary disks generally disappear in less than

10 Myr (e.g., Strom et al. 1993). In comparison, zero-age main sequence (ZAMS) status is reached after 5 or 15 Myr for high- and low-mass A stars, respectively, and 30+ Myr for solar-mass stars. An age of 10 Myr might serve as a rough statistical dividing line between systems with dense gaseous protoplanetary disks and the more evolved, optically thin planet-forming disks with which we are primarily concerned.

A modest number of CS dust systems have been found in the field at relatively large distances, e.g. HD 98800 (50 pc), HR 4796A (70 pc), HD 141569 (100 pc), and 51 Ophiuchi (130 pc), guaranteed by selection effects to contain much more dust than in typical systems nearer to us. These objects have ages less than a few tens of Myr; for lack of better terms we shall call these "old pre-MS" (OPMS) or "young MS" (YMS) stars. They are interesting because they represent the transition between PMS stars in star-forming regions and MS stars in the solar neighborhood. Such systems will be discussed in section V.B.

II. OBSERVATIONS

A. β Pictoris

A crucial bit of new information regarding this system is a revised distance from Hipparcos (*High-Precision Parallax-Collecting Satellite*) data of 19.28 ± 0.19 pc and hence an absolute V magnitude of 2.42 ± 0.03 (Crifo et al. 1997). This yields a bolometric luminosity for β Pic of 8.5 $L_\odot$ (cf. previous value of about 6 $L_\odot$), assuming an effective temperature of 8200 K, with little or no inferred CS dust extinction.

Analysis of several photospheric lines (Ca II, Cr II, Fe II) seems to show that the star has solar composition in metallic elements (Lanz et al. 1995; Holweger et al. 1997) in contrast with previous claims. The implications of this will be discussed in the context of star ages and disk evolution (section III.A.2).

1. β Pictoris—Circumstellar Dust. High-signal/noise optical images of β Pic's outer disk show five asymmetries at the few-percent level between the NE and SW wings (Kalas and Jewitt 1995): radial extent, surface brightness at a given radius, thickness at a given radius, and wing-tilt (position angle of the midplane), plus the so-called "butterfly" asymmetry in which disk thickness perpendicular to the midplane varies among the quadrants. The wing-tilt asymmetry can be explained by a combination of disk orientation and nonisotropic scattering phase function. The other surface brightness (dust density) asymmetries possibly reveal the dynamical influence of large masses in or near the disk.

High-spatial-resolution optical images ($\simeq$0.1 arcsec) obtained with adaptive optics (AO) techniques (Mouillet et al. 1997*a*) and *Hubble Space Telescope* (HST) (Burrows et al. 1995) trace the disk to within about 1 arcsec ($\simeq$20 AU) from the star. New HST Space Telescope Imaging Spectrograph (STIS) images (Fig. 1; Heap et al. 1999) reach in to $r \simeq 15$ AU and constrain the vertical distribution of dust in the disk to have a sharp

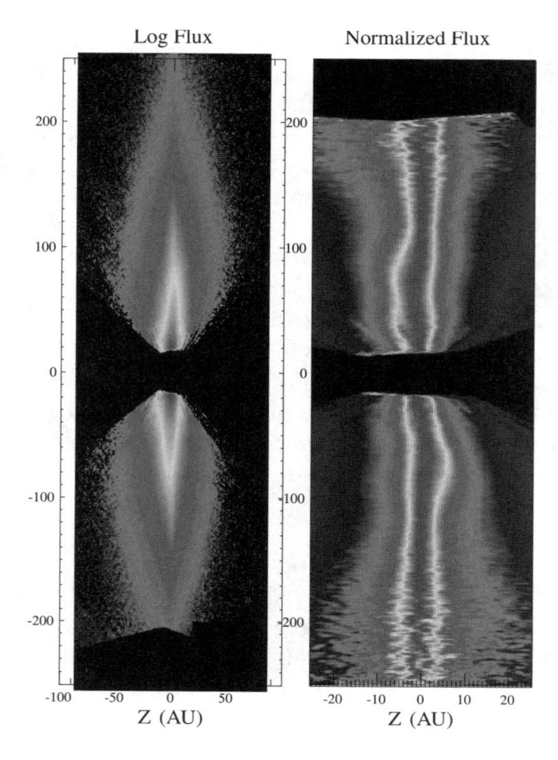

Figure 1. HST/STIS coronagraphic visible-light image of the β Pictoris disk (Heap et al. 1999), traced to within 15 AU of the star (middle of panel). To show the asymmetric bulge out to about 100 AU more clearly, the image was stretched 4× in the vertical direction and renormalized to the maximum flux in each charge-coupled device (CCD) column.

maximum at the disk midplane. The high-spatial-resolution images also show an inner 3° warp (or asymmetric bulge) in the midplane relative to that of the outer disk. Both the disk thickness and the warp are most easily explained by the influence of planetary or substellar companions.

Lagage and Pantin (1994*a*) mapped the thermal emission at 10 μm from the disk in the region $r < 100$ AU for the first time and directly showed the central low-density "gap" or void. The disk position angle in those images is the same as the warped disk noticed later via scattered light at optical wavelengths. They also detected a large asymmetry in thermal-IR surface brightness between the inner disk wings (discussed further in section IV.A.1). Harvey et al. (1996) resolved the outer disk at 50 μm in Kuiper Airborne Observatory (KAO) observations, obtaining results consistent with the *Infrared Astronomy Satellite* (IRAS) scale size (Aumann 1991). Submillimeter (sub-mm) photometry and assumption of a plausible mass absorption coefficient led to an estimated total grain mass for the disk of 7×10^{23} kg ($\simeq$0.1 $M_{\oplus}$) (Zuckerman and Becklin 1993*b*).

The Submillimetre Common User Bolometer Array (SCUBA) camera at the James Clerk Maxwell Telescope (JCMT) produced sub-mm images of the β Pic disk. A prominent "blob" of emission is seen in those images at $r \simeq 30$ arcsec (600 AU) (Fig. 2; Holland et al. 1998). The dust in the "blob" would need a net cross section area comparable to all the dust in the rest of the disk if it is heated by the star; no other local heat source at that position is detected in the near-IR to a deep limit of $K \simeq +19.5$ (Kalas et al., in preparation). In particular, it cannot be due to a brown dwarf or super-Jupiter.

IR aperture photometry at 2–13 μm plus spectrophotometric observations in the silicate bands (Knacke et al. 1993; Aitken et al. 1993) confirmed the previously detected silicate emission and inferred the presence of about 10^{-6} $M_\oplus$ of small ($\simeq$2 μm) grains within 30 AU of the star. The observed spectral feature has a significantly different shape from that of silicates in the interstellar medium (ISM) or around asymptotic giant branch (AGB) and PMS stars but resembles that of solar system comets, such as P/Halley, containing partially crystalline silicates (Knacke et al. 1993).

2. β Pictoris—Circumstellar Gas. Spectroscopic monitoring of β Pic's gas (summarized in Lagrange 1995, 1998) has continued to investigate the stable gas ring/disk lying within 1 AU of the star and to characterize absorption-line variability.

HST observations (Vidal-Madjar et al. 1994; Lagrange et al. 1996) detected variability in lines of various ions. Recently, variability was also found in one neutral atomic species, C I (Jolly et al. 1999). Long campaigns of medium- and high-resolution ($R = 3 \times 10^4$ to 10^5) spectroscopy from the ground and from space with the HST, as well as ultrahigh-resolution ($R = 10^6$) observations from the ground, revealed the presence of typical groupings of infall velocities of the Ca II lines. Very low-velocity (VLV) features have velocities relative to the star of less than 10 km s^{-1}; low-velocity (LV) features 10–30 km s^{-1}; and high-velocity (HV) features $\geq$80 km s^{-1} (Lagrange et al. 1996; Beust et al. 1998; Petterson and Tobin 1999). LV and HV features are also seen in ultraviolet lines. These velocity groupings are associated with different variability timescales; the higher the velocity, the shorter the timescale. The small number ($\leq$10%) of blueshifted events with respect to redshifted ones has been confirmed, and the overall rate of events was estimated to be as much as a few hundred per year (Ferlet et al. 1993). The sizes of infalling clouds depend on the specific ion considered. The conditions in a typical infalling Ca II cloud are $n_e \geq 10^6$ cm^{-3} and $T_e \geq 10^4$ K.

About 15 ions and neutral species have been identified in the stable gas with HST (Lagrange et al. 1998). Radio observations yield an upper limit to the H I content of $\leq$15 $M_\oplus$ and a ring gas/dust mass ratio estimated to be $\leq$10 (Freudling et al. 1995). Cold ($\simeq$20 K) molecular CO was also detected with HST (Vidal-Madjar et al. 1994; Jolly et al. 1998), although all attempts failed at radio wavelengths (Savoldini and Galletta 1994; Dent et al. 1995; Liseau and Artymowicz 1998). Most Ca II ions are probably located 0.3–1 AU from the star, but the CO is located much farther away ($\geq$10 AU; Jolly et al. 1998), possibly within the main dust disk.

3. Photometry. Small light variations ($\leq$0.06 mag) that occurred in 1981 were reported by Lecavelier des Etangs et al. (1995). The β Pic light curve, carefully monitored during several years for photometric calibration purposes by the Swiss ESO telescope, showed a brightening of 0.06 mag lasting about 10 days, with a central drop a few hours long, and finally a return to the long-term average value. These variations were tentatively attributed to a planet or a giant comet cloud (Lecavelier des Etangs et al. 1997*c*; Lamers et al. 1997).

B. General Surveys for CS Dust and Follow-Up Observations

Surveys of volume-limited stellar samples show that at least 15% of A–K MS stars have some far-IR dust excess with fractional dust luminosity ($f_d \equiv L_d/L_\star$) greater than or equal to α Lyr's value of $\sim 2 \times 10^{-5}$ (Backman and Paresce 1993; Plets 1997 [IRAS]; Dominik et al. 1998*a* [*Infrared Space Observatory* (ISO)]; Fajardo-Acosta et al. 1999 [ISO]). A few stars with f_d up to a few $\times 10^{-3}$, comparable to β Pic, were found. Even larger values ($f_d \simeq 0.1$ or more) were measured for some B–A stars that are also emission-line stars and probably quite young. They evidently should be classified as PMS or OPMS stars (e.g., Sylvester et al. 1996; Dunkin et al. 1997*b*; Malfait et al. 1998), confirmed in some cases by ages derived from Hipparcos data (van den Ancker et al. 1998).

The selection of real MS stars is thus critical to the interpretation of surveys. Mannings and Barlow (1998) explicitly chose MS stars over about half the sky at $|b| \geq 10°$ from the Michigan Spectral Catalog of HD stars and found 70 new candidates with IR excess in the IRAS Faint Source Catalog. M-type MS stars could have dust disks as commonly as do stars of earlier spectral type, but their low luminosity makes detection of dust impossible at present around all but the closest systems. IRAS and ISO observations have revealed a few examples (Backman et al. 1998 and references therein; Fajardo-Acosta et al. 1999).

Photometric observations across a wide range of wavelengths constrained system spectral energy distributions (SEDs) and hence the disk properties (e.g., Sylvester et al. 1996; Sylvester and Skinner 1996; Sylvester et al. 1997; Fajardo-Acosta et al. 1998*b*). Complementary observations were made to define the characteristics of the central stars (Dunkin et al. 1997*a*,*b*). Waelkens et al. (1996) found spectroscopic signatures of dust composition in ISO SWS observations.

Dominik et al. (1998*a*), Becklin et al. (1998), Abraham et al. (1998), and Gaidos (1999) found evidence in IRAS and ISO data of evolution toward lower CS dust density with timescales of a few $\times 10^8$ yr or less; these results will be discussed further in section III.B.2. Lynch and Russell (in preparation) searched for but found no warm (3.5–9 μm) excesses toward stars 51 Pegasi, 47 Ursae Majoris, υ Andromedae, and τ Bootis, which have known or suspected planetary companions.

A coronagraphic optical imaging survey was performed on more than 50 MS stars and did not detect any disks (Kalas and Jewitt 1995) except around BD+31°643 (age $\leq 10^7$ yr; Kalas and Jewitt 1997). These negative results are comparable to those of Smith et al. (1992). The general nondetection of optical disks, including even the prototype Vega and α Piscis Austrini systems, could be attributed to a small scattering cross section of CS dust (Kalas and Jewitt 1996), consistent with the IR SEDs, and does not require that the β Pic grains be substantially different from grains around other stars. Recently, disks were imaged at near-IR wavelengths around ρ^1 Cancri (a planet-bearing star resembling 51 Peg) and the OPMS star HD 141569 (sections II.D.4 and V.B.2, respectively).

C. General Surveys for CS Gas and Follow-Up Observations

Stars found to have CS absorption lines similar to β Pic's were reobserved with the HST. HR 10, HR 2174, and 51 Oph exhibit narrow UV absorptions arising from excited levels indicating the presence of CS gas at distances greater than 15 $R_\star$ (Dunkin et al. 1997*b*; Lecavelier des Etangs et al. 1997*a*). 51 Oph exhibits other spectroscopic similarities to β Pic that will be discussed in section V.B.4. Spectral monitoring of HR 10 (Welsh et al. 1998) confirmed variable redshifted absorptions (Lagrange-Henri et al. 1990) and, moreover, detected some blueshifted transients. The variability is interpreted as the result of comet evaporation, although no detailed models have been calculated as of this writing.

Other attempts to detect CS gas and/or spectroscopic variability were made in samples selected via IRAS (Cheng et al. 1992, 1995) and *International Ultraviolet Explorer* (IUE) criteria (Grady et al. 1996*a*). A few stars were observed to exhibit shell features, with 2 Andromedae also showing spectral variability in HST data (Cheng et al. 1997).

A survey for CO emission (Zuckerman et al. 1995*a*) around younger stars including some we label OPMS (section V.B) yielded upper limits (e.g., HR 4796A, HD 98800) or positive detections (e.g., HD 141569), leading the authors to conclude that gas in these systems dissipates on timescales shorter than about 10^7 years. This is consistent with the negative results of Lecavelier des Etangs et al. (1997*d*) for CS gas from a spectroscopic survey of a large number of stars in the Ursa Major Stream (age $\simeq$ 300 Myr).

D. Observations of Individual Stars

1. Vega (α Lyrae). Reanalysis of IRAS slow-scan observations produced a revised scale for the 60-μm dust emission of 35 arcsec diameter vs. the previous 23 arcsec (van der Bliek et al. 1994). This indicates that the dominant grains have characteristic sizes of order 10 μm rather than 100 μm and thus shorter lifetimes than previously calculated. Zuckerman and

Becklin (1993*b*) estimated the minimum dust mass to be only 4×10^{22} kg from sub-mm observations. SCUBA sub-mm images (Fig. 2; Holland et al. 1998) show an emission peak that is not centered on the star position, but the offset is not much more than 1σ and is possibly an artifact. Atmosphere models and the very low $v \sin i$ value indicate that the star is viewed pole-on (Gray and Garrisson 1987), consistent with the nearly circular aspect of the 60-μm dust emission.

Dent et al. (1995) failed to detect CO down to a column density limit of 1.6×10^{14} cm^{-2}. An upper limit of 10^{12} cm^{-2} had been placed previously on Ca II absorption (Hobbs 1986).

2. Fomalhaut (α Piscis Austrini). This disk was resolved by Harvey et al. (1996) at 50 and 90 μm. Fajardo-Acosta et al. (1997, 1998*a*) mapped the disk with ISO, showing 60-μm emission extending about 30–80 arcsec (200–500 AU) from the star, attributed to approximately 10-μm-size grains radiating at $T \geq 50$ K. The disk was mapped at sub-mm wavelengths (Fig. 2; Holland et al. 1998), clearly revealing a hollow central cavity. The position angle of the main axis of the ellipsoid is consistent among the SCUBA sub-mm images, ISO 60-μm maps, and slow-scan profile information from IRAS (Gillett 1986).

No CS gas has been detected at optical, UV, or radio wavelengths. In particular, the reported detection of CS gas based on IUE data by Cheng et al. (1994) was not confirmed in HST spectra (Ferlet et al. 1995). Dent et al. (1995) set a CO upper limit of 2.4×10^{14} cm^{-2}.

3. ϵ Eridani. A SCUBA sub-mm image (Fig. 2; Greaves et al. 1998) of the dust disk around this very nearby cool MS star (3 pc, K2 V) shows a definite ring and central clearing with a radius of about 30 AU. The size of the ring agrees with the scale derived from IRAS slow-scan observations (Aumann 1991). The sub-mm image shows significant azimuthal variations in the ring's surface brightness. Such asymmetries should be erased within a dust collision timescale of order 10^6 years. If real, they may provide evidence of either recent episodes of dust release or shepherding by larger bodies.

4. ρ^1 Cancri. This system, estimated to be $\simeq$5 Gyr old, is particularly interesting because it has both CS dust with modest luminosity ($f_d \sim 5 \times 10^{-5}$) and at least one planet ($a = 0.11$ AU, $M \sin i = 0.84$ M_{Jup}). ISO observations detected the disk at 60–90 μm, modeled as due to $\simeq 4 \times 10^{-5}$ $M_\oplus$ of 10-μm dust grains lying between 35 and 60 AU from the star with a definite central cavity (Dominik et al. 1998*b*). Recently Trilling and Brown (1998) imaged the disk in scattered light at 2.3 μm (Color Plate 8c).

III. EVOLUTIONARY ISSUES

A. Stellar Evolution

1. Ages of Stars: General. Reliable ages for field stars are still difficult to obtain, yet without ages for the MS CS dust and gas systems their place in

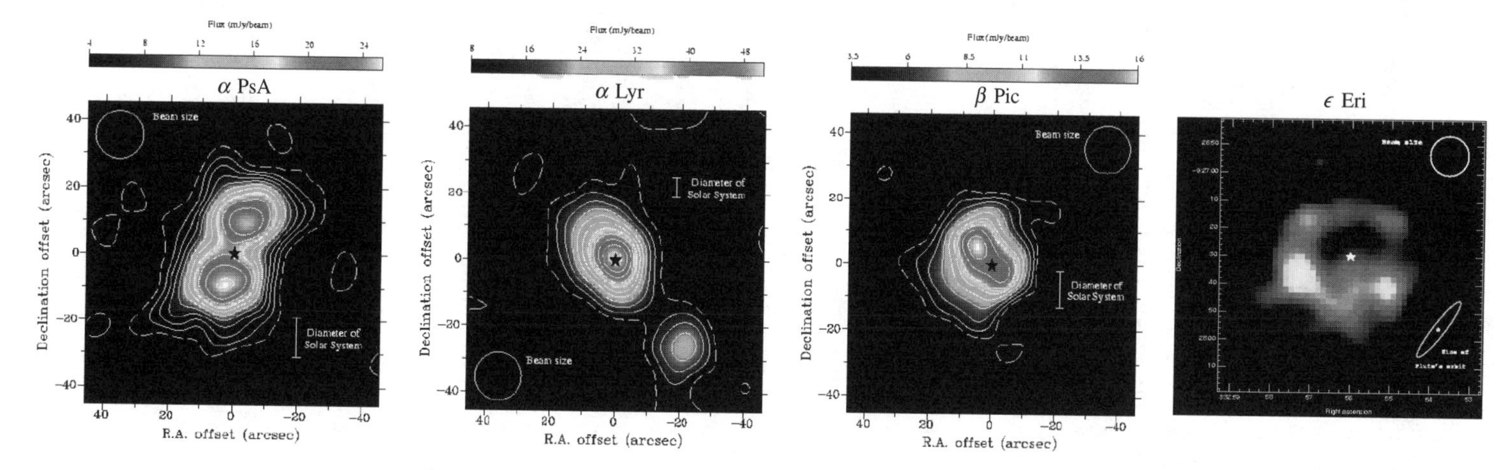

Figure 2. SCUBA 850-μm images of the disks around α PsA, α Lyr, β Pic (Holland et al. 1998; copyright *Nature,* reprinted with permission), and ϵ Eri (Greaves et al. 1998; copyright *Astrophysical Journal Letters,* reprinted with permission).

stellar and planetary evolution cannot be ascertained. Stars in open clusters have supposedly well-determined ages, but no post-ZAMS cluster is close enough for easy detection of CS dust by either IRAS or ISO (e.g., Backman et al. 1998). Hipparcos parallaxes should allow distance determinations that will challenge the quality of the stellar interior models upon which isochrones are based. One problem with isochrone ages is that they are bivalued, meaning that a star occupies a position in the HR diagram above the ZAMS both before and after the ZAMS. Although the distinction between MS and giant stars is usually straightforward, the distinction between MS and PMS stars sometimes is not, and this determination is more difficult for earlier spectral types.

It is statistically unlikely that any of the stars within 25 pc ($2 \times$ the volume containing β Pic) are much younger than a few $\times$ 10 Myr. This result applies generally for all spectral classes and can be derived by dividing the lifetime of a typical member of each class by the number of stars in that class within the volume considered. Discovery of stars apparently younger than 10 Myr in the TW Hydrae association (Webb et al. 1998) supports the above argument, because this group lies at $d \simeq 55$ pc and thus is in a $10\times$ larger sample volume. Surveys of age indicators such as Li abundance and Ca II activity (e.g., Henry et al. 1996) show that the age histogram for nearby solar-type stars is approximately flat, with few objects younger than 10^8 yr.

2. Ages of Stars: Individual. Vega has a luminosity roughly $3 \times L_{\rm ZAMS}$ for its temperature. Comparison of its properties with evolutionary tracks for $M_\star = 2.5$ M$_\odot$ (Palla and Stahler 1993; Bressan et al. 1993) indicate that it could be either a PMS object with an age of about 2.5 Myr or a post-ZAMS object with an age of about 350 Myr. Holweger and Rentzsch-Holm (1995) have claimed that Vega's surface abundances indicate that it is still accreting and therefore in a PMS stage. We prefer the post-ZAMS age for Vega, based on (1) the previous argument about minimum plausible ages for nearby stars, (2) the certain absence of G, K, and M stars as young as 2.5 Myr within 50 pc, and (3) comparison of the detection likelihood implied by the two timescales.

The question of the age of β Pic produces the same type of controversy as for α Lyr. β Pic's luminosity was interpreted as either representing pre-ZAMS status with substantial extinction (Lanz et al. 1995) or post-ZAMS with different stellar parameters (Paresce 1991). Recent determinations (section II.A) that the star's luminosity is close to ZAMS and that its metallicity is approximately solar are still consistent with a wide range of possible ages from about 10^7 to a few $\times 10^8$ yr.

An age estimate of 200 Myr for α PsA is obtained from the more easily determined age of its K4 V proper-motion companion Gl 879 (Barrado y Navascués et al. 1997). An age of 800 Myr for ϵ Eri is found from recently recalibrated Ca II H + K activity indices (Henry et al. 1996).

B. Evolution of Dust and Dust Parent Bodies

Creation and destruction processes affecting MS CS solid material have been studied in some depth (Backman and Paresce 1993; Artymowicz 1994, 1997; Artymowicz and Clampin 1997). The timescale for loss of solids from each system, usually based on Poynting-Robertson (P-R) drag and collisions as the two most readily calculable effects, is much shorter than the stellar age for all the prototype Vega /β Pic systems.

1. Radiation Pressure, Radial Motions, and Dust Avalanches. Radiation pressure will force some grains into eccentric orbits after release from parent bodies. The collision timescale, including consideration of dust planar motions (eccentricities) as well as vertical motions (inclinations), is approximately $t_{\rm coll} \simeq P_{\rm orb}/(12\tau_{\perp})$, where $P_{\rm orb}$ is the local orbital period and $\tau_{\perp}$ is the local geometric optical depth perpendicular to the disk. This shortens $t_{\rm coll}$ relative to the value derived from vertical motions alone and strengthens the argument that observed grains must be replenished and are not primordial.

Erosion of disk particles releases fine dust subject to radiative acceleration and ejection. Orbits will be hyperbolic for grains with an RP index $\beta \equiv F_{\rm rad}/F_{\rm grav} > \frac{1}{2}$ if released from large parent bodies in circular orbits. These grains can impact other disk particles at high relative velocity, causing efficient cratering and breakup. Such a collision debris swarm should grow exponentially as the dust avalanche traverses the disk radially. Overall erosion rates much faster than implied by "normal" collisions occur if the dust disk exceeds a critical midplane radial optical depth, roughly estimated to be about that of the β Pic disk (Artymowicz 1996). In denser disks the destruction rate grows exponentially with f_d.

2. Dust Quantity and Disk Evolution. The fact that at least 15% of nearby A, F, G, and K stars have far-IR dust excesses indicates that this phenomenon must extend in some cases over Gyr timescales. Nevertheless, a general decrease of f_d with age for PMS and MS objects is observed, discussed by Zuckerman and Becklin (1993*b*) and Holland et al. (1998). The similar A-type stars HR 4796A, β Pic, Fomalhaut, and Vega form a sequence of f_d decreasing monotonically from ages of 10 to 350 Myr (Fig. 3). Dominik et al. (1998*a*) and Becklin et al. (1998) both found evidence in ISO surveys of nearby field stars that the fraction with detectable far-IR excesses drops from about 50% to about 15% after approximately 500 Myr. In apparent contradiction to these results, Abraham et al. (1998) claimed that only 1 of 9 stars in the Ursa Major Stream (age $\simeq$ 300 Myr) showed dust emission, but four of their ISO targets (all without far-IR excess) may not actually be members of the Stream (cf. Soderblom and Mayor 1993). Gaidos (1999) examined IRAS data on solar-type stars selected to be younger than 800 Myr based primarily on X-ray flux. He found detections and limits consistent with the amount of terrestrial-temperature dust predicted by a simple model of the SS heavy bombardment era.

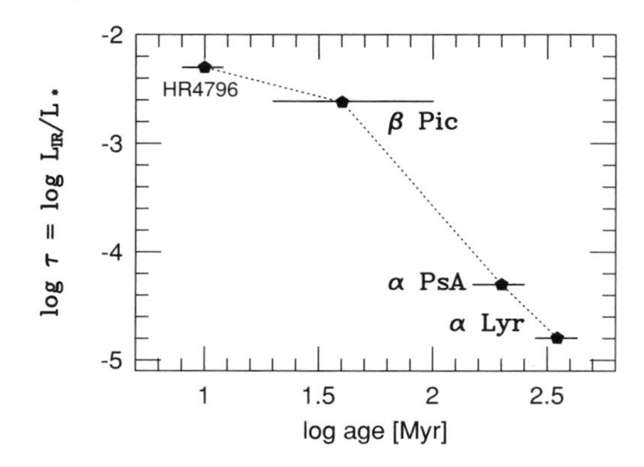

Figure 3. f_d versus age for four stars of similar spectral type (see text).

Artymowicz (1994, 1996, 1997) argued that the dustiness of Vega-type (low-f_d) systems is self-limited by the dust avalanche process and proposed that gas-poor disks, in which gas drag is unimportant for dust kinematics, cannot be much dustier than β Pic (if geometrically similar). Systems with much larger f_d could be disks that have gas-controlled dynamics. The critical lower limit to the gas/dust mass ratio is calculated to be of order 10, and this amount of gas would be currently undetectable in most MS debris disks. That hypothesis may be strengthened by indications that the dustiness of IR-excess systems is bimodal (data collected in Artymowicz 1996). Only a few systems (e.g., β Pic, HR 4796A, HD 141569) are in the range $f_d = 10^{-3}$–10^{-2}, whereas the neighboring logarithmic bins contain many more stars with lower (Vega-like) or higher values of f_d (stars generally known to be young PMS objects; cf. Dunkin et al. 1997*b*).

3. Interaction with the ISM: Nature vs. Nurture. Hypotheses that interaction with ISM grains might be important in the evolution of Vega-like disks is supported by the discovery of streams of large ISM grains in the vicinity of Jupiter (Grün et al. 1994). Artymowicz and Clampin (1997), however, calculated that internal erosion should generally dominate in MS disks over the influence of ISM grains because of (1) the strong repulsive RP force exerted by luminous stars on absorbing sub-μm ISM grains and (2) high dust density in disks compared to the ISM. Only at the very outskirts of the prototype disks (radii exceeding $\simeq$400 AU) can the ISM bombardment be important, possibly capable of causing slight disk asymmetries.

IV. DISK MODELS AND INTERPRETATIONS

A. β Pictoris: Circumstellar Dust Models

1. Disk Morphology. If one assumes a single grain population (Artymowicz 1997), then current models do not easily fit both the scattered light images from HST/STIS (Heap et al. 1999) and the thermal-IR images of Lagage and Pantin (1994*a*) and Pantin et al. (1997). The thermal-IR data indicate that dust optical thickness peaks or reaches a plateau at 30–60 AU. Those IR images also show an asymmetry larger than a factor of 2 between the NE and SW wings' 12-μm thermal emission at those distances.

The visible STIS data, in contrast, match a larger gap with peak dust density near $r \simeq 120$ AU and a much more symmetric distribution. The disk optical depth in a model of the scattered light by Artymowicz et al. (1999) equals $\tau_\perp(r) = 2\ \tau_m / \sqrt{(r/r_m)^{-2p} + (r/r_m)^{2q}}$, where $r_m = 120$ AU is a characteristic radius approximately coinciding with maximum $\tau_\perp(r) \simeq 0.01$. The STIS data support $p = 2$ (i.e., $\tau_\perp \sim r^2$ near the star) and $q = 3$ (i.e., a steep falloff, $\tau_\perp \sim r^{-3}$, at $r \geq r_m$). The significantly different sizes of the central gap deduced from scattered optical light versus thermal-IR data may imply two separate grain populations or one with radially varying physical and optical properties.

2. *Dust Composition and Origin.* The gray color in the visible and near-IR plus the silicate emission feature (section II.A.1) suggest a similarity of β Pic dust to cometary and interplanetary (zodiacal) dust in our system. Artymowicz (1997) showed that zodiacal light particles distributed as in the β Pic disk could also reproduce its observed linear polarization properties, implying a similar porosity or roughness of grain surfaces. On the other hand, the albedo of typical β Pic grains seems significantly higher than that of typical SS particles, most of which scatter 10% or less of visible sunlight. Based on models combining properties of the thermal and scattered radiation, the main β Pic disk has an albedo of $A \sim 0.3$–0.4 (Backman et al. 1992; Burrows et al. 1995) to $A > 0.5$ (Artymowicz et al. 1989). These observations are best explained by common mineral mixtures such as slightly darkened ices or bright silicates (Fe- and C-poor olivines and/or pyroxenes, e.g., Mg-rich pyroxenes with mass ratio Fe/(Mg+Fe) $\lesssim$0.2). Although water ice is an important constituent of planetesimals, its vulnerability to potent UV photo-sputtering by β Pic and its mechanical fragility make it an unlikely material for the observed grains (Artymowicz 1994, 1997).

Collisions involving bodies up to the size of planetesimals, erosive grain impacts, and thermal evaporation of cometlike bodies are probably all effective in generating fresh dust in debris disks. The estimated overall dust grinding rate at the location of maximum dust density in β Pic's disk corresponds to a system destruction timescale of roughly 50 Myr, about 10^4 times longer than the disk replenishment time. This fits within the wide range of estimated ages for the star and is independent evidence that β Pic and its disk are not much older than 100 Myr. The total mass of the original β Pic planetesimal disk required for continuous dust replenishment is $\sim$100 $M_\oplus$ (Artymowicz 1997), of order that originally present in the SS (section VI.A).

Lecavelier des Etangs et al. (1996), Lecavelier des Etangs (1998), and Li and Greenberg (1998) proposed disk models in which most of the dust is released by comet evaporation at only a few $\times 10$ AU from the star but can be driven by RP to produce a wedge disk out to $r > 1000$ AU. The grains would spend most of their time at their respective apoapses, so the outer disk would be relatively prominent. The evaporation of comets could also provide a source of the observed molecular CO column density.

B. β Pictoris: Circumstellar Gas Models

Comets (FEBs) as the origin of both the stable and the variable gas are the only mechanism that has been successfully modeled.

1. Falling Evaporating Body Models for Transient Absorption Features. The general idea of FEBs is that the observed variable gas is due to evaporation of star-grazing comets (reviewed in Beust 1994). Dust released from active comets evaporates and produces the observed metallic gas, which is subjected to RP, gravity, and collisions with the mostly neutral gas (H I, O I) around the comet. RP causes the size of the parabolic absorbing cloud (coma) to differ from one ion to another, ranging from $\leq 10\%$ of the stellar diameter to larger than the star (Vidal-Madjar et al. 1994; for up-to-date values of the RP index β for various species see Lagrange et al. 1998).

Simulations of individual events reproduce the variable features in terms of shape, velocities, and depths (Fig. 4b; Beust et al. 1998 and references therein). Typical distances of evaporation are respectively 30, 10–20, and $\leq 10\ R_\star$ for the VLV, LV, and HV features, and evaporation rates are of order 3×10^7 kg/s, consistent with scaled values obtained from observations of small SS Kreutz-family (sungrazer) comets.

The FEB scenario explains the presence of highly ionized species as due to high pressure and temperature on the parabolic front (e.g., Beust and Tagger 1993; Mouillet and Lagrange 1995), although no full simulation including all the ionization processes has yet been completed. It may also explain, via details of the comet dynamics (Beust and Lissauer 1994; see also section IV.C), the puzzling observations that Ca II H is deeper than Ca II K (Lagrange et al. 1996) and that blueshifted variable features are rare (e.g., Crawford et al. 1998).

The simulations have difficulty reproducing the observed LV feature timescales via the evaporation of a single body. The lines should be detectable only over a few hours, whereas actual events last significantly longer. One possible explanation is that we observe a "family" of bodies crossing the line of sight on close orbits, perhaps resembling Comet Shoemaker-Levy 9's fragment train (Beust et al. 1996).

Beust et al. (1998) found that the mean-motion resonance mechanism (section IV.C) proposed for propelling comets into the star implies that the comets' periastra undergo a gradual decrease until the comets start to evaporate and produce the observed variable absorptions. Comets of

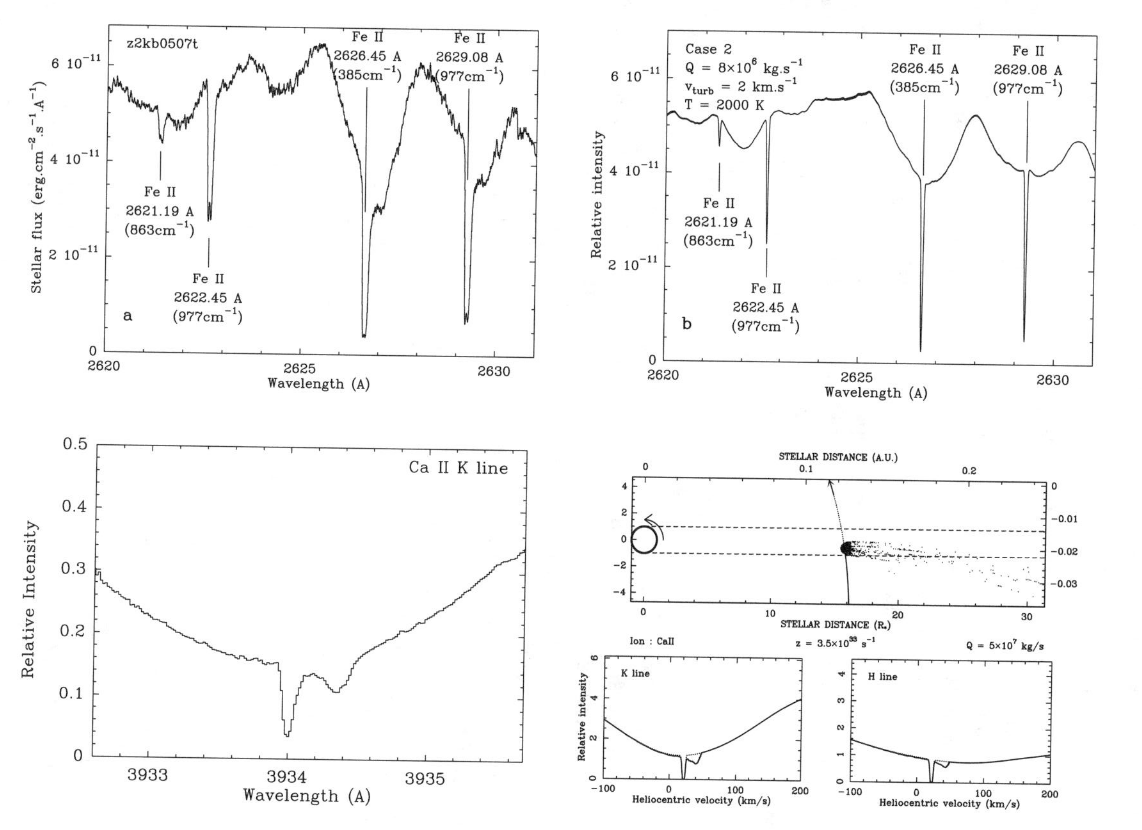

Figure 4. (a, top) Observed and simulated stable lines in the β Pic system; data from HST/GHRS. Left: observed Fe II lines arising from various low energy levels. Right: simulated spectra (see text). Note that in the observed data, redshifted absorptions are present in addition to the stable components; the former are not simulated (from Lagrange et al. 1998). (b, bottom) Observed and simulated variable lines in the β Pic system. Left: observed Ca II K line, data from ESO/CAT: a low-velocity absorption is seen in addition to the strong stable central absorption. Right: simulated Ca II K and H lines. The geometrical configuration of the comet is also shown (see text) (from Beust 1994).

1-km size can survive enough passages to produce the LV features at about 0.5 AU, but some comets must be ≃10 km in size to survive many more close passages before arriving at a few stellar radii, where the HV features are produced.

2. FEB Models versus Shell Models for the Stable Gas. Observed species that suffer strong RP and do not have saturated lines are very unstable and must be continuously replaced. Similarly, the lifetime of CO to photodissociation is small. Thus, it is clear that gas must be continuously produced to resupply the observed stable component, near the star for the atomic gas and further away for CO.

Simulations of dynamics of gas released within 0.5 AU of the star and subjected to RP, gravity, and interaction with a ring of neutral gas at ≃0.5 AU reproduce the observed line depths and chemical abundances (Fig. 4a; Lagrange et al. 1998). The gas production rate needed to sustain the lines is $\simeq 3 \times 10^7$ kg/s. This neutral gas rate is compatible with the observed upper limits, in particular for H I, and is coincidentally similar to the estimated evaporation rate in an average FEB event (previous subsection).

Comet evaporation is a natural source of the atomic gas. The required evaporation rate is similar to that needed to explain the variable lines, and the number of comets necessary (≃1000/yr in all directions from the star) is plausible for a young system (Lagrange et al. 1998). Also, this would explain the presence of CO through slow evaporation of comets at larger distances from the star (Lecavelier des Etangs et al. 1996). The gas/dust ratio deduced from the CO abundance and lifetimes of the dust and gas appear to be consistent with both gas and dust being released by the same process (e.g., Jolly et al. 1998).

An alternative explanation of the stable gas as originating in a wind cannot be excluded but is unlikely. The gas production rate would correspond to a mass loss rate of about 10^{-16} $M_\odot$/yr, compatible with an A-type stellar wind (Babel 1995). However, this model would imply that all stars, and in particular A-type stars having higher mass loss rates (e.g. Sirius, 10^{-12} $M_\odot$/yr; Bertin et al. 1995), should also be surrounded by stable gas, but this is not the case. Also, the inferred presence in the stable ring of species such as H I and O I that do not suffer much RP is inconsistent with a wind model.

Another explanation of the stable gas as a "shell structure" around the star seems inadequate. Maintaining gas in a hydrostatic shell suggests rotational support to balance the star's gravity. Although they are stable on dynamical timescales, such shells would be subject to viscosity effects, causing both accretion onto the star and outward spreading. The lifetime of shells with radii less than 1 AU is expected to be shorter than 10^4 yr based on the Balbus-Hawley magnetic instability assuming viscosity parameter $\alpha \sim 0.01$.

C. Model Results: Indirect Evidence for Planets in Disks

The possibility that the Vega/β Pic disks are locations of ongoing or completed planet formation is one of the most exciting aspects of these systems. There are several indirect lines of evidence regarding the presence of massive bodies in or near some of these disks.

The SEDs of the prototype CS dust systems all imply transitions from outer high-density zones to inner low-density zones, but at a wide range of temperatures such that ice sublimation is an unlikely explanation for all of them. A real possibility is that planets redirect grains (cf. Roques et al. 1994 for the case of β Pic) or consume grains as they drift via P-R drag toward central stars (cf. Liou et al. 1996 for the SS). α Lyr is an especially important case, because its grain collision timescale is similar to or longer than the P-R drag timescale (e.g., Backman and Paresce 1993). The observed lack of hot grains around α Lyr implies that something must be eliminating the grains as they drift inward, or the inner zone would be filled in relatively rapidly.

Nothing short of planetary gravitational perturbation seems adequate to explain the FEB triggering mechanism in the β Pic system. Possible mechanisms and their virtues and difficulties include (1) close encounter with a single planet (Beust et al. 1991), requiring a planet on a highly eccentric orbit (but note that the 16 Cyg B and 70 Vir systems contain such objects); (2) the Kozai mechanism (Bailey et al. 1992; Thomas and Morbidelli 1996), as in the origin of the Kreutz-family comets in the SS (but this would produce symmetric infall, which is not observed); (3) secular resonance perturbation by a planet (Levison et al. 1995), assuming a planetary configuration similar to that in the SS (the mechanism is much less efficient for other possible planetary arrangements); (4) mean-motion resonance, especially the 4:1 resonance, which seems to be an efficient and generic mechanism and one that can be an asymmetric source of star-grazing comets (Beust and Morbidelli 1996, 1998). Beust and Morbidelli (1998) also propose an additional Earth-mass planet to account for the rarely observed blueshifted events.

The structure of the β Pic disk is the best understood of the prototypes because of high-resolution imaging. Burrows et al. (1995) have proposed that gravitational perturbation of the disk by a massive companion (planet or brown dwarf) on an inclined orbit could explain a warp like that observed, reminiscent of previous work by Whitmire et al. (1988) regarding the outer disk. The simulations made by Mouillet et al. (1997*b*) showed that a physical warp at radii $\simeq$ 50 AU can be explained by a companion located between 1 and 20 AU from the star with a mass, respectively, between 10^{-2} and $10^{-5} M_\star$ (6000 and 6 $M_\oplus$) for a system age of 200 Myr. Finally, the finite thickness of the β Pic disk (Artymowicz et al. 1989; Kalas and Jewitt 1995) implies that at least 1000-km objects must be embedded in the disk to maintain the requisite dynamical heating.

V. COMPARISONS WITH PRE-MS, MS, AND POST-MS SYSTEMS

A. PMS Stars and Possible FEBs

Young PMS stars (ages $\leq$1–10 Myr, depending on mass) exhibit strong IR excesses and spectral features that indicate many characteristic differences from the Vega/β Pic disks: (1) The disk masses may be as high as 0.01 $M_\odot$ for HAeBe stars (Mannings and Sargent 1997; and the chapter by Natta et al., this volume) and Class II TT stars (Andre and Montmerle 1994). (2) Accretion rates deduced from the SEDs are high: 10^{-8}–10^{-6} $M_\odot$/yr for HAeBe stars and 10^{-9}–10^{-7} $M_\odot$/yr for TT stars, many orders of magnitude larger than the gas accretion rate in β Pic. (3) Winds and other gas ejection evident in PMS systems from blueshifted lines and P Cygni profiles are not observed in Vega/β Pic systems. (4) Photometric variability with amplitude as high as several magnitudes is observed in UX Ori-type objects (see chapter by Natta et al., this volume), substantially more than the possible sporadic variability of β Pic. For detailed descriptions of these PMS characteristics see Herbst et al. (1994) and Waelkens and Waters (1998) regarding TT stars and HAeBe stars, respectively.

A predictable evolutionary step in the dissipation of PMS disks is formation of kilometer-scale planetesimals (e.g., Lissauer 1993). Calculations indicate that resonant processes could have caused asteroid as well as comet infall to the sun (Farinella et al. 1994). It is therefore reasonable to expect asteroids and comets to be present around evolved PMS stars, creating at least some CS dust through collisions or evaporation and producing effects analogous to the β Pic FEB phenomenon. Whether FEB-generated spectral variability would be detectable in specific systems depends on (1) the efficiency of the process in producing strong enough absorption lines (2) the overall stellar activity. HAeBe stars usually exhibit strong, probably star-related variable spectral features, such as P Cygni profiles, that could mask FEB activity.

A number of HAeBe stars have been reported by Grady et al. (this volume) as exhibiting infall signatures superficially similar to those of β Pic. Those authors attribute this variability to FEB events at typical distances of $10R_\star$, but this comparison should be regarded as tentative for examples based on a limited number of low-resolution, low-signal/noise IUE spectra. Also, in many cases it is difficult to untangle stellar activity from genuine FEB signatures. It is very important to note that for some of the PMS FEB candidate stars (e.g., AB Aur), variable spectral features have instead been attributed to differentially rotating and chromospherically driven winds or magnetically channeled accretion (Catala et al. 1993; Boehm and Catala 1995). There is also some evidence that the variability timescale is related to stellar rotation, supporting alternatives to FEB models. Therefore, the application of the FEB hypothesis to a large number of HAeBe stars (30%; Grady et al. 1996*b*) and the proposed relation between infall event frequency and stellar age should not be considered proven (but see also the chapter by Grady et al., this volume.)

A few stars (UX Ori, Grinin et al. 1994; HD 100546, Grady et al. 1997) show more convincing β Pic-like features, specifically variable redshifted absorptions in metallic lines separable from features classically attributed to stellar activity. The variable features of ionized elements are similar to β Pic's in terms of shapes and velocity ranges, the infalling clouds have covering factors smaller than the stellar surface, and they contain overionized species (C IV, Al III). There are also some important differences: (1) Variable lines are usually stronger, implying much more gas in the variable component than for β Pic; an estimate for the case of UX Ori gives about 10^{12} 3-meter bodies evaporating in less than 3 days at about $10R_\star$ (Grinin et al. 1996), more than 100$\times$ the mass in a typical β Pic event. (2) Event rates are generally higher. (3) Variability timescales seem to be longer, although data on short timescales are sparse. (4) Strong variable lines are observed in neutral species (e.g. Na I, Hα, C I), whereas β Pic, even though cooler, does not show detectable amounts of variable neutrals. Sorelli et al. (1996) showed for UX Ori-type stars that Na I absorption events can be reproduced with magnetic accretion as well as by comet evaporation provided the magnetic field is $\simeq$600 gauss (below current detection limits).

A first attempt to apply the FEB simulations to a PMS FEB candidate, HD 100546, shows that it is difficult for an FEB model to reproduce the observed depths of the variable lines successfully, because the high stellar luminosity means high RP on the ions. Also, these stars usually have robust stellar winds that will strongly interact with the evaporated material. Comet evaporation rates much higher than in β Pic are needed to reproduce this star's variable Mg II features observed with IUE (Lagrange and Beust 1999).

A further critical question is the relative frequency of blueshifted and redshifted features. One should expect to detect approximately as many blueshifted as redshifted events in a system without significant dynamical asymmetry. Grady et al. (this volume) attribute the general predominance of PMS redshifted events to complete evaporation of FEBs before periapsis. However, even around a late B star a comet as large as P/Halley can be expected to survive periapsis passage easily at distances corresponding to velocities below about 100 km s^{-1}. Also, preperiapsis destruction raises dynamical difficulties because the bodies are more likely to become stargrazers in a gradual way rather than at first passage (as mentioned in section IV.B.1). Finally, the marked dominance of redshifted over blueshifted transients in β Pic requires that we must be viewing the system with a special orientation. If an asymmetry toward redshifted transients is seen at low velocities in many systems, then FEBs cannot be the general explanation for the spectroscopic phenomena.

In conclusion, it is probable that FEB events are occurring around some PMS stars, but their detection should not be as straightforward as in the case of β Pic. The interpretation of some HAeBe stars' variability as due to FEBs is questionable, but a few show variable features convincingly similar to those of β Pic. The latter stars obviously deserve detailed monitoring and modeling for comparison with β Pic's spectroscopic variability.

B. Borderline Cases: OPMS and YMS Stars

These are stars evidently evolved beyond the TT/HAeBe stage but with denser dust disks than the Vega/β Pic prototypes, possibly including a significant gas component. These were designated "old PMS" (OPMS) and "young MS" (YMS) systems in section I.

1. HR 4796A. This A0 V star at $d = 67$ pc has the highest IR disk fractional luminosity (5×10^{-3}) among MS A-type stars (Jura 1991; Jura et al. 1993). Spectroscopy of the nearer M-type companion (HR 4796B, $r \simeq 500$ AU) gives an age estimate of 8–10 $\times$ 10^6 yr for HR 4796A, assuming the stars are physically linked (Stauffer et al. 1995; Jura et al. 1998).

Koerner et al. (1998) and Jayawardhana et al. (1998) imaged a radially narrow disk at 10 and 20 μm oriented in the direction of component B, clearly visible in more recent near-IR HST Near Infrared Camera and Multi-Object Spectrometer (NICMOS) images (Color Plate 8a; Schneider et al. 1999). The IR data sets are modeled by a ring of material around a central gap about 55 AU in radius, confirming Jura et al. (1995). Sparse dust in the gap at a temperature of 200–400 K is resolved. The outer disk seems surprisingly narrow, quite unlike the similarly dense β Pic disk, which has been traced beyond $r = 1000$ AU. Dynamical model calculations are needed to see whether the influence of the stellar companion is sufficient to truncate the disk at its outer edge or whether nearer shepherding objects are required.

An upper limit of about 7 $M_\oplus$ of gas has been obtained from observations in molecular emission lines (Zuckerman et al. 1995*a*). A gas mass exceeding the dust mass by less than an order of magnitude (presently undetectable; Zuckerman et al. 1995*a*) could lengthen the dust destruction timescale enough to make it possible that the dust observed in this system is actually primordial. If gas dynamics are truly important there, the HR 4796A disk may be a very promising laboratory where distinctive spiral density waves and resonantly truncated disk edges could reveal the location(s) of planet(s).

2. HD 141569. This A0 star, located at about 100 pc, exhibits a fractional disk luminosity of $f_d = 8 \times 10^{-3}$. Van den Ancker et al. (1998) determined its age to be $\gtrsim 10$ Myr, but it may be younger than that, based on the strong Hα emission line (Dunkin et al. 1997*b*). It is usually classified as an HAeBe star. Recently, a disk (Color Plate 8b) was found around this object at near-IR wavelengths with HST/NICMOS (Weinberger et al. 1999; Augereau et al. 1999). As in HR 4796A, there is a ring of material around an apparent central gap, in this case with sizes $\Delta r \simeq 60$ AU and $r \simeq 260$ AU, respectively. Also, like HR 4796A, this star is a member of a double system, with a separation of about 700 AU.

3. HD 98800. This remarkable quadruple K-star system now has a good age determination of 10 Myr (5–20 Myr) based on its Hipparcos distance of 47 pc, indicating that these stars are in a PMS but post-TT evolutionary stage (Soderblom et al. 1998).

The HD 98800 dust has a fractional luminosity of $f_d \simeq 0.1$ (Zuckerman and Becklin 1993*a*; Sylvester et al. 1996), approaching the maximum possible value for a flat or flared disk passively reprocessing stellar radiation. The stars are arranged as two spectroscopic binary systems separated by 0.8 arcsec (40 AU projected) near periapsis of a very eccentric ($e = 0.993$) orbit (Torres et al. 1995; Soderblom et al. 1996). Koerner et al. (in preparation) obtained 10-μm images showing that the dust is primarily around the B component of this system. Some of the dust is known to be at least millimeter-size from sub-mm and mm detections of the system (Sylvester and Skinner 1996).

The explanation for the high dust luminosity and, therefore, density could be any of the following or a combination: (1) Enough of the original protoplanetary gas disk remains to preserve the dust from erosional avalanches; (2) an avalanche is in progress as we watch; (3) the close stellar encounter allows better heating of the outer parts of the disk or disks, substantially increasing the dust detectability; (4) the stellar encounter (possibly the first since the system formed) is disrupting a Kuiper Belt/Oort Cloud structure, causing massive collision rates, vaporization of planetesimals, and sudden release of grains. The latter possibility calls to mind the "Nemesis" hypothesis, which explained supposed periodic mass extinctions on Earth by proposing that the Sun has a low-mass stellar companion producing comet showers at perihelion passages in a long-period eccentric orbit.

4. 51 Ophiuchi. Photometric and spectrophotometric data at 5–22 μm (Fajardo-Acosta et al. 1993) showed the presence of silicate emission attributed to grains smaller than 5 μm and with a total mass of silicates of 8×10^{-5} $M_\oplus$. The emission strength suggests that the typical grains around 51 Oph are larger and hotter than the ones around β Pic. The emission-feature mineralogy resembles that in β Pic and in SS comets. Lagage and Pantin (1994*b*) did not resolve the 10-μm emission [full width at half-maximum (FWHM) = 0.9 arcsec], confirming that the dominant dust population is mostly closer to the star than in the β Pic system.

CS gas absorptions in a complex Ca II feature were confirmed through ultrahigh-resolution spectra (Crawford et al. 1997). The presence of highly ionized species (Grady and Silvis 1993; Lecavelier des Etangs et al. 1997*b*) may be attributed to shock excitation in FEB comae. 51 Oph also shows cold C I lines with a column density of 5×10^{13} cm^{-2} (Lecavelier des Etangs et al. 1997*b*). The short lifetime of C I indicates that species could be produced from photodissociation of CO released from comets.

C. λ Bootis, B-A Shell, and Post-MS Stars

The low metallicity proposed for β Pic (Paresce 1991) led King and Patten (1992) to associate β Pic with the (surface) metal-deficient A-type λ Bootis stars, tentatively attributed to accretion of gas after gas-dust separation (Venn and Lambert 1990). Dunkin et al. (1997*a*) looked for λ Boo depletion patterns in a sample of A-type MS stars with high f_d values. Only two showed subsolar metallicity, and they were not as strongly depleted as classical λ Boo stars. Comparison of a new list of λ Boo stars (Paunzen 1997) with IR excess stars from the catalogs of Mannings and Barlow (1998) and Oudmaijer et al. (1992) yields only a few stars in common. Thus, there seems to be no strong connection between either spectroscopic characteristics of λ Boo and β Pic-like stars or between λ Boo status and far-IR dust excesses.

The old classification of β Pic as a shell star led to consideration of a possible link between B-A stellar shells and IR excesses. Among the 17 *main-sequence* A-shell stars presently known and detected in the IRAS Faint Source Catalog, 4 have 12-μm excess: β Pic, HR 9043, HR 3310, and HR 3989; the latter two stars are binaries. Cheng et al. (1991) found an IR excess for HR 10 using IRAS one-dimensional co-added data, and Fajardo-Acosta et al. (1998*b*) recently detected a 20-μm excess around HR 2174. This yields 6 stars out of 17 with detected IR excesses, a fraction not much larger than that for ordinary field A stars, although these small-number statistics do not allow clear conclusions. More sensitive IR photometric observations of a larger sample of A-type MS shell stars are needed.

Individual giant stars have been discovered with dust envelopes arguably connected to disks of planetary debris (Judge et al. 1987; Skinner et al. 1995; Fekel et al. 1996). Searches were conducted in IRAS data for giant-luminosity descendants of MS stars to check whether far-IR excesses occur around them as often as around MS stars (Jura 1990; Zuckerman et al. 1995*b*; Plets et al. 1997). Their results were that far-IR dust excesses around ordinary red giants are significantly rarer than around their MS progenitors, indicating either that the disks decay substantially during the MS lifetimes of ordinary stars or that the rise in luminosity as stars leave the MS quickly destroys dust and dust parent bodies.

VI. COMPARISONS WITH OUR SOLAR SYSTEM

A. The Kuiper Belt

A hypothetical small-grain component of the Kuiper Belt (KB) could be the SS analog of the Vega/β Pic MS disks, especially if that structure originally contained more mass than it does at present. Although its existence has been inferred for many years, direct observations have only recently verified these speculations and begun to determine KB properties (see reviews by Weissman 1995; Weissman and Levison 1998; chapter by Farinella et al., this volume). More than 90 large (100–300 km diameter) objects have been found so far in the KB at $r = 30$–50 AU, and one object of this size class was discovered with a semimajor axis of 85 AU (Luu and Jewitt 1998; chapter by Jewitt and Luu, this volume). KB bodies with diameters in the few $\times$ 10 km size range are conjectured to be the source population for short-period comets. Objects of this size at about 40 AU have been reported in deep HST images (Cochran et al. 1995), but they await confirmation.

Model KBs consistent with these data and with IRAS and *Cosmic Background Explorer* (COBE) limits on a cold component of the zodiacal emission (Backman et al. 1995; Stern 1996*b*; Teplitz et al. 1999; chapter by Farinella et al., this volume) have grain populations orbiting at $r = 30$–100 AU in equilibrium between various removal processes and replenishment by collisions of a few $\times$ 0.1 $M_\oplus$ of comet nuclei. These models predict a fractional luminosity f_d in the range 10^{-7} to 10^{-6} for SS KB dust and show that dust optical depth $\tau_\perp$ just outside Neptune's orbit could be as much as an order of magnitude higher than at Earth's orbit without violating observational constraints due to warm foreground zodiacal dust. It is possible that HR 4796A, HD 141569, the known MS CS disks, and the SS have similar morphologies: outer disks or rings surrounding central regions of lower density corresponding to planetary zones.

The amount of material observed or inferred in the known ("inner") KB falls far below the type of extrapolation (e.g., Tremaine 1989) that led Kuiper to hypothesize the existence of the Belt in the first place. In fact, Stern (1996*a*), Stern and Colwell (1997), and Kenyon and Luu (1998) have calculated that known large KB objects and Pluto-Charon could not have formed quickly enough before Neptune grew large and overstirred the KB velocity dispersion, unless the primordial inner KB had a mass of at least 10 $M_\oplus$. The density of collisionally produced dust in such a disk would have resembled those in the prototype Vega and β Pic disks, because the dust radiating area scales approximately with the square of the total mass of colliding parent bodies (falling short of that at higher densities because of grain-grain collisions). Such a massive original "inner" KB could evolve into a system like the present KB in the age of the SS (Davis and Farinella 1997; chapter by Farinella et al., this volume).

B. Zodiacal Dust

Artymowicz (1997) discussed a number of similarities and differences between β Pic and SS zodiacal dust. Important contrasts include the facts that the SS dust disk (1) appears geometrically several times thicker in the planetary region than the β Pic disk and may be more symmetric; (2) seems to contain more Fe and C in grains; (3) has a slight negative radial gradient of mean albedo whereas β Pic may have a positive gradient;

(4) has a characteristic grain size of a few $\times 10$ μm versus a few $\times 1$ μm for β Pic. These differences contrast with significant similarities (section IV.A.2) and may simply reflect the earlier evolutionary stage of β Pic as well as some real compositional differences.

Of the four prototype MS CS dust systems, only β Pic so far has detectable dust within the central gap, with model temperatures up to at least 350 K (Fajardo-Acosta et al. 1993). Upper limits on the amount of warm dust around other stars (e.g., Vega, Aumann et al. 1984) are generally high due to lack of spatial resolution and low contrast of dust flux to stellar photospheres at short IR wavelengths. The best IRAS photometric sensitivity at 12 and 25 μm to terrestrial-temperature dust around the nearest stars is several hundred times the optical depth in the SS zodiacal cloud. A few cases of warm CS MS dust have been found and studied recently by Fajardo-Acosta et al. (1998*b*) via mid-IR spectrophotometry. Mid-IR images of HR 4796A (Koerner et al. 1998) resolved a central warm dust population radiating at 200–400 K.

Detection of warm dust might be easiest in young systems during the 10^7–10^8-yr timescale expected for the formation of planets (Wetherill 1991; Lissauer 1993; Gaidos 1999) continuing into the long "end game" of heavy bombardment. Grains produced by planetesimals with as little as 1% of the mass of the SS terrestrial planets could be easily detected from distances of tens of parsecs in the mid- and far-IR (Witteborn et al. 1982). This corresponds to the amount of interplanetary material estimated to have been present in the terrestrial zone a few hundred Myr after the formation of the SS.

Asymmetries and planetary wakes in exozodiacal clouds can be used to infer masses and locations of embedded planets (Dermott et al. 1998). Note, however, that the brightness of a 0.3-AU diameter patch of the SS zodiacal cloud would approximately equal that of Earth at both visual and IR wavelengths; it is estimated that warm dust at more than about 10 $\times$ SS density would challenge detectability of Earthlike planets via planned space-based mid-IR interferometers such as NASA's *Terrestrial Planet Finder* (TPF; Beichman et al. 1999). It is critical to such efforts that nearby stars be surveyed for terrestrial-temperature dust at sensitivities an order of magnitude below present limits.

VII. SUMMARY AND FUTURE PROSPECTS

The most important new results concerning the Vega/β Pic stars may be summarized as follows:

1. Vega/β Pic disks have been confirmed to be common; the true frequency of occurrence of solid material in orbit around normal stars may be much higher than the current detection sensitivity-limited estimate of $\sim$15%.

2. There is evidence that the amount of dust, quantified by fractional dust luminosity f_d, decreases with age from PMS stars to MS stars.

3. The creation and evolution of Vega/β Pic disks are understood theoretically to be governed primarily by internal processes such as collisions, radiation pressure, evaporation, and influence of possible planets, rather than by external effects such as erosion by the ISM.

4. Disks are now resolved around a few stars mainly via thermal-IR and sub-mm imaging, whereas optical/near-IR images providing higher spatial resolution allow examination of the fine structure of detectable disks. Imaging over a wide range of wavelengths is necessary to constrain the disks' properties further.

5. Central gaps, regions mostly if not completely lacking dust, are observed to be normal. Structures at the edges of the gaps (sharp disk truncation, asymmetries, warps, waves) could be good indicators of the presence of planets.

6. Detailed models of the β Pic environment support the idea that planetesimals/comets and probably also larger bodies govern production of both the CS gas and dust. Dynamical models involving planets reproduce this disk's inner warp. Specific planetary configurations can explain the triggering of FEBs.

7. A few "borderline" cases have been identified that are objects evolutionarily between PMS systems and Vega/β Pic disks. In such systems gas may still play an important role. The study of these objects is important, because they trace a crucial transition in planetary system formation.

Newly available large (10-m) single-aperture telescopes equipped with adaptive optics and observing at optical or thermal-IR wavelengths (10–20 μm) plus space observatories such as the *Next Generation Space Telescope* (NGST) will soon lead to more disk detections and also resolution of finer structures on scales of a few $\times 10^{-2}$ to 10^{-1} arcsec. Gaps created by jovian planets around β Pic should be detectable in this regime. Long-baseline (≥ 100 m) optical-IR (ESO VLTI; Keck) or sub-mm (ALMA) interferometers will give milliarcsecond resolution, and their sensitivity is expected to be sufficient to study this type of disk. These observations will allow a better description of the individual systems and provide much tighter constraints on the orbits and masses of planets we infer to be present around β Pic, HR 4796A, and other similar systems. Access to a large number of disks will help constrain the evolutionary timescales involved in the formation and evolution of planetary systems.

Extensive spectroscopic searches for FEB events around stars with various ages will probably be performed in the next few years, because they give precious although indirect information regarding the dynamics of planetesimals and planets. This will help constrain in detail the processes involved in planetary system construction.

Finally, more direct detection of planets in these disks obviously would be of great importance. The radial velocity techniques are restricted for the moment to late-type stars. Other techniques such as photometric surveys will need to be used for the earlier-type stars. Some planned dedicated tools, such as dark-speckle (Labeyrie 1995) or very high-order adaptive optics (Angel 1994), might lead to planet imaging in the future.

Acknowledgments A.-M. L. thanks J. C. Augereau, H. Beust, J. Bouvier, D. Bockelée-Morvan, C. Catala, M. Gerbaldi, A. Lecavelier, F. Ménard, D. Mouillet, and J. Zorec for providing information and fruitful discussions. D. B. thanks S. Fajardo-Acosta, P. Kalas, D. Koerner, and M. Brown for discussions; L. Caroff and D. Goorvitch (NASA-Ames), C. Beichman (Caltech/IPAC), and R. Joseph (U. Hawaii/IRTF) for work space and hospitality during a research leave from F&M College; STScI GO grant 05885.02-94A for partial support; and E. Roberts for help editing this manuscript. P. A. thanks S. Heap and E. Pantin for joint discussions and the NFR (Swedish Nat. Sci. Res. Council) for generous research grants. Special thanks to A. Lecavelier, C. Dominik, P. Kalas, and S. Fajardo-Acosta for their reading of the draft manuscript, and to the referee M. Barlow for very helpful and constructive comments.

REFERENCES

Abraham, P., Leinert, C., Burkert, A., Lemke, P., and Henning, T. 1998. Search for cool circumstellar matter in the Ursae Majoris group with ISO. *Astron. Astrophys.* 338:91–96.

Aitken, D. K., Moore T. J. T., and Roche, P. F. 1993. Mid-infrared spectroscopy of β Pic: Constraints on the dust grain size. *Mon. Not. Roy. Astron. Soc.* 265:L41–L44.

André, P., and Montmerle, T. 1994. From T Tauri stars to protostars: Circumstellar material and young stellar objects in the ρ Ophiuchi cloud. *Astrophys. J.* 420:837–862.

Angel, R. 1994. Groundbased imaging of extrasolar planets using adaptive optics. *Nature* 368:203–207.

Artymowicz, P. 1994. Modeling and understanding the dust around β Pic. In *Circumstellar Dust Disks and Planet Formation,* ed. R. Ferlet and A. Vidal-Madjar (Gif sur Yvette Cedex: Editions Frontières), pp. 47–66.

Artymowicz, P. 1996. Vega-type systems. In *The Role of Dust in the Formation of Stars,* ed. H. U. Käufl and R. Siebenmorgen (Berlin: Springer), pp. 137–48.

Artymowicz, P. 1997. β Pic: An early solar system? *Ann. Rev. Earth Planet. Sci.* 25:175–219.

Artymowicz, P., and Clampin, M. 1997. Dust around main-sequence stars: Nature or nurture by the interstellar medium? *Astrophys. J.* 490:863–878.

Artymowicz, P., Burrows, C., and Paresce, F. 1989. The structure of the β Pic circumstellar disk from combined IRAS and coronagraphic observations. *Astrophys. J.* 337:494–513.

Artymowicz, P., Heap, S. L., and Pantin, E. 1999. Structure of the β Pictoris disk from STIS/HST imaging. *Astron. Astrophys.*, in preparation.

Augereau, J. C., Lagrange, A.-M., and Mouillet, D. 1999. HST observations of the HD 141569 disk. *Astron. Astrophys.*, in preparation.

Aumann, H. H. 1991. Circumstellar material in main-sequence stars. In *The Infrared Spectral Region of Stars,* ed. C. Jaschek and Y. Andrillat (Cambridge: Cambridge University Press), pp. 363–379.

Aumann, H. H., Gillett, F. C., Beichman, C. A., De Jong, T., Houck, J. R., Low, F. J., Neugebauer, G., Walker, R. G., and Wesselius, P. R. 1984. Discovery of a shell around α Lyrae. *Astrophys. J. Lett.* 278:L23–L27.

Babel, J. 1995. Multi-component radiatively driven winds from A and B stars. I. The metallic wind of a main sequence A star. *Astron. Astrophys.* 301:823–839.

Backman, D. E., and Paresce, F. 1993. Main sequence stars with CS solid material: The Vega phenomenon. In *Protostars and Planets III,* ed. E. H. Levy and J. I. Lunine (Tucson: University of Arizona Press), pp. 1253–1304.

Backman, D. E., Gillett, F. C., and Witteborn, F. C. 1992. Infrared observations and thermal models of the β Pic disk. *Astrophys. J.* 385:670–679.

Backman, D. E., Dasgupta, A., and Stencel, R. E. 1995. Model of a Kuiper Belt small grain population and resulting far-infrared emission. *Astrophys. J. Lett.* 450:L35–L38.

Backman, D. E., Fajardo-Acosta, S. B., Stencel, R. E., and Stauffer, J. R. 1998. Dust disks around main sequence stars. *Astrophys. Space Sci.* 255:91–101.

Bailey, M. E., Chambers, J. E., and Hahn, G. 1992. Origin of stargrazers: A frequent cometary end-state. *Astron. Astrophys.* 257:315–322.

Barrado y Navascués, D., Stauffer, J. R., Hartmann, L., and Balachandran, S. C. 1997. The age of Gliese 879 and Fomalhaut. *Astrophys. J.* 475:313–321.

Becklin, E. E., Silverstone, M., Chary, R., Hare, J., Zuckerman, B., Spangler, C., and Sargent, A. 1998. Dust around solar mass stars. *Astrophys. Space Sci.* 255:113–118.

Beichman, C. A., and the TPF Science Working Group. 1999. *The Terrestrial Planet Finder (TPF): A NASA Origins Program to Search for Habitable Planets,* JPL Publication 99-3 (Pasadena: Jet Propulsion Laboratory).

Bertin, P., Lamers, H. J. G. L. M., Vidal-Madjar, A., Ferlet, R., Lallement, R. 1995. HST-GHRS observations of Sirius A. III. Detection of a stellar wind from Sirius A. *Astron. Astrophys.* 302:899–906.

Beust, H. 1994. β Pic: The falling evaporating body model. In *Circumstellar Dust Disks and Planet Formation,* ed. R. Ferlet and A. Vidal-Madjar (Gif sur Yvette Cedex: Editions Frontières), pp. 35–46.

Beust, H., and Lissauer, J. 1994. The effects of stellar rotation on the absorption spectra of comets orbiting β Pic. *Astron. Astrophys.* 282:804–810.

Beust, H., and Morbidelli, A. 1996. Mean-motion resonances as a source for infalling comets toward β Pic. *Icarus* 120:358–370.

Beust, H., and Morbidelli, A. 1998. Falling evaporating bodies in the β Pic system. *Icarus,* submitted.

Beust, H., and Tagger, M. 1993. A hydrodynamical model for infalling evaporating bodies in the β Pic circumstellar disk. *Icarus* 106:42–58.

Beust, H., Vidal-Madjar, A., and Ferlet, R. 1991. The β Pic protoplanetary system. XII. Planetary perturbations in the disk and star-grazing bodies. *Astron. Astrophys.* 247:505–515.

Beust, H., Lagrange, A.-M., Plazy, F., and Mouillet, D. 1996. The β Pic circumstellar disk. XXII. Investigating the model of multiple cometary infalls. *Astron. Astrophys. Suppl.* 310:181–198.

Beust, H., Lagrange, A.-M., Crawford, I. A., Goudard, C., Spyromilio, J., and Vidal-Madjar, A. 1998. The β Pic circumstellar disk. XXV. The Ca II absorption lines and the falling evaporating bodies model revisited using UHRF observations. *Astron. Astrophys.* 338:1015–1030.

Boehm, T., and Catala, C. 1995. Rotation, winds and active phenomena in Herbig Ae/Be stars. *Astron. Astrophys.* 301:155–169.

Bressan A., Fagotto, F., Bertelli, G., and Chiosi, C. 1993. Evolutionary sequences of stellar models with new radiative opacities. II. $Z = 0.02$. *Astron. Astrophys. Suppl.* 100:647–664.

Burrows, C., Krist, J. E., Stapelfeldt, K. R., and the WFPC2 Investigation Definition Team. 1995. HST observations of the β Pic disk. *Bull. Am. Astron. Soc.* 187:32.05 (abstract).
Catala, C., Boehm, T., Donati, J.-F., and Semel, M. 1993. Circular polarization and variability in the spectra of Herbig Ae/Be stars. I: The Fe II 5018 Å and He I 5876 Å lines of AB Aurigae. *Astron. Astrophys.* 278:187–198.
Cheng, K.-P., Grady, C. A., and Bruhweiler, F. C. 1991. A search for circumstellar dust around HR 10, a proposed β Pic star. *Astrophys. J. Lett.* 366:L87–L90.
Cheng, K.-P., Bruhweiler, F. C., Kondo, Y., and Grady, C. A. 1992. Newly identified main-sequence A stars with circumstellar dust. *Astrophys. J. Lett.* 396:L83–L86.
Cheng, K.-P., Bruhweiler, F. C., and Kondo, Y. 1994. A search for ultraviolet circumstellar gas absorption features in α Piscis Austrinus (Fomalhaut), a possible β Pic system. *Astrophys. J. Lett.* 425:L33–L36.
Cheng, K.-P., Neff, J. E., and Bruhweiler, F. C. 1995. A search for planetary system candidates: Nearby A stars with dust disks and circumstellar gas. *Astrophys. Space Sci.* 223:143.
Cheng, K.-P., Bruhweiler, F. C., and Neff, J. E. 1997. Detection of β Pic-like gaseous infall in 2 Andromedae. *Astrophys. J.* 481:866–871.
Cochran, A., Levison, H. F., Stern, S. A., and Duncan, M. J. 1995. The discovery of Halley-sized Kuiper Belt objects using the Hubble Space Telescope. *Astrophys. J.* 455:342–346.
Crawford, I. A., Craig, N., and Welsh, B. Y. 1997. The velocity structure of the local interstellar medium probed by ultra-high-resolution spectroscopy. *Astron. Astrophys.* 317:889–897.
Crawford, I. A., Beust, H., and Lagrange, A.-M. 1998. Detection of a strong transient blueshifted absorption component in the β Pictoris disc. *Mon. Not. Roy. Astron. Soc.* 294:L31–L34.
Crifo, F., Vidal-Madjar, A., Lallement, R., Ferlet, R., and Gerbaldi, M. 1997. β Pic revisited by Hipparcos. Star properties. *Astron. Astrophys.* 320:L29–L32.
Davis, D. R., and Farinella, P. 1997. Collisional evolution of Edgeworth-Kuiper Belt objects. *Icarus* 125:50–60.
Dent, W. R. F., Greaves, J. S., Mannings, V., Coulson, I. M., and Walther, D. M. 1995. A search for molecular gas components in prototypal Vega-excess systems. *Mon. Not. Roy. Astron. Soc.* 277:L25–L29.
Dermott, S. F., Grogan, K., Holmes, E. K., and Wyatt, M. C. 1998. Signatures of planets. In *Exozodiacal Dust Workshop,* ed. D. E. Backman, L. J. Caroff, S. A. Sandford, and D. H. Wooden, NASA Conference Publication 1998-10155 (Moffett Field: NASA-Ames Research Center), pp. 59–83.
Dominik, C., and the HJHVEGA Consortium. 1998*a*. Vega-like stars: Grain removal, replenishment, and recent ISO observations. *Astrophys. Space Sci.* 255:103–111.
Dominik, C., Laureijs, R. J., Jourdain de Muizon, M., and Habing, H. J. 1998*b*. A Vega-like disk associated with the planetary system of ρ^1 Cnc. *Astron. Astrophys.* 329:L53–L56.
Dunkin, S. K., Barlow, M. J., and Ryan, S. G. 1997*a*. High-resolution spectroscopy of Vega-like stars. I. Effective temperatures, gravities and photospheric abundances. *Mon. Not. Roy. Astron. Soc.* 286:604–616.
Dunkin, S. K., Barlow, M. J., and Ryan, S. G. 1997*b*. High-resolution spectroscopy of Vega-like stars. II. Age indicators, activity and circumstellar gas. *Mon. Not. Roy. Astron. Soc.* 290:165–185.
Fajardo-Acosta, S. B., Telesco, C. M., and Knacke, R. F. 1993. Detection of silicates in the 51 Ophiuchi system. *Astrophys. J. Lett.* 417:L33–L36.

Fajardo-Acosta, S. B., Stencel, R. E., and Backman, D. E. 1997. Infrared Space Observatory mapping of 60 μm dust emission around Vega-type systems. *Astrophys. J. Lett.* 487:L151–L154.

Fajardo-Acosta, S. B., Stencel, R. E., and Backman, D. E. 1998*a*. Infrared Space Observatory mapping of 60 μm dust emission around Vega-type systems: Erratum. *Astrophys. J.* 503:193.

Fajardo-Acosta, S. B., Telesco, C. M., and Knacke, R. F. 1998*b*. Infrared photometry of β Pic type systems. *Astron. J.* 115:2101–2121.

Fajardo-Acosta, S. B., Thakur, N., Stencel, R. E., and Backman, D. E. 1999. Infrared Space Observatory photometric search of main sequence stars for Vega-type systems. *Astrophys. J.*, in press.

Farinella, P., Froeschlé, Ch., Froeschlé, C., Gonczi, R., Hahn, G., Morbidelli, A., and Valsecchi, G. B. 1994. Asteroids falling onto the Sun. *Nature* 371:315–317.

Fekel, F. C., Webb, R. A., White, R. J., and Zuckerman, B. 1996. HDE 233517: Lithium and excess infrared emission in giant stars. *Astrophys. J. Lett.* 462:L95–L98.

Ferlet, R., Lagrange-Henri, A.-M., and Beust, H. 1993. The β Pic protoplanetary system. XIV. Simultaneous observations of the Ca II lines; evidence for diffuse and broad absorption features. *Astron. Astrophys.* 267:137–144.

Ferlet, R., Lecavelier des Etangs, A., Vidal-Madjar, A., Bertin, P., Deleuil, M., Lagrange-Henri, A.-M., and Lallement, R. 1995. HST-GHRS observations of α Piscis Austrini. Evidence of no gas content in the circumstellar environment. *Astron. Astrophys.* 297:L5–L8.

Freudling, W., Lagrange, A.-M., Vidal-Madjar, A., Ferlet, R., and Forveille, T. 1995. Gas around β Pictoris: An upper limit on the H I content. *Astron. Astrophys.* 301:231–235.

Gaidos, E. J. 1999. Observational constraints on late heavy bombardment episodes around young solar analogs. *Astrophys. J. Lett.* 510:L131–L134.

Gillett, F. C. 1986. IRAS observations of cool excess around main sequence stars. In *Light on Dark Matter, Astrophysics and Space Science Library,* Vol. 124, ed. F. P. Israel (Dordrecht: Reidel), pp. 61–69.

Grady, C. A., and Silvis, J. M. S. 1993. The circumstellar gas surrounding 51 Ophiuchi: A candidate proto-planetary system similar to β Pic. *Astrophys. J. Lett.* 402:L61–L64.

Grady, C. A., Pérez, M. R., Talavera, A., McCollum, B., Rawley, L. A., England, M. N., and Schlegel, M. 1996*a*. The β Pic phenomenon in A-shell stars: Detection of accreting gas. *Astrophys. J. Lett.* 471:L49–L52.

Grady, C. A., Pérez, M. R., Talavera A., Bjorkman, K. S., De Winter, D., Thé, P. -S., Molster, F. J., Van den Ancker, M. E., Sitko, M. L., Morrison, N. D., Beaver, M. L., McCollum, B., and Castelaz, M. W. 1996*b*. The β Pic phenomenon among Herbig Ae/Be stars. UV and optical high dispersion spectra. *Astron. Astrophys. Suppl.* 120:157–177.

Grady, C. A., Sitko, M. L., Bjorkman, K. S., Perez, M. R., Lynch, D. K., Russell, R. W., and Hanner, M. S. 1997. The stargrazing extrasolar comets in the HD 100546 system. *Astrophys. J.* 483:449–456.

Gray, R. O., and Garrisson, R. F. 1987. The early A type stars—refined MK classification, confrontation with Strömgren photometry, and the effects of rotation. *Astrophys. J. Suppl.* 65:581–602.

Greaves, J. S., Holland, W. S., Moriarty-Schieven, G., Jenness, T., Dent, W. R. F., Zuckerman, B., McCarthy, C., Webb, R. A., Butner, H. M., Gear, W. K., and Walker, H. J. 1998. A dust ring around ϵ Eridani: Analog to the young solar system. *Astrophys. J. Lett.* 506:L133–L137.

Grinin, V. P., Thé, P. S., and de Winter, D. 1994. The β Pic phenomenon among young stars. I. The case of the Herbig Ae star UX Orionis. *Astron. Astrophys.* 292:165–174.

Grinin, V. P., Natta, A., and Tambovtseva, L. 1996. Evaporation of star-grazing bodies in the vicinity of UX Ori-type stars. *Astron. Astrophys.* 313:857–865.

Grün, E., Gustafson, B., Mann, I., Baguhl, M., Morfill, G. E., Staubach, P., Taylor, A., and Zook, H. A. 1994. Interstellar dust in the heliosphere. *Astron. Astrophys.* 286:915–924.

Harvey, P., Smith, B. J., and Difrancesco, J. 1996. Far-infrared constraints on dust shells around Vega-like stars. *Astrophys. J.* 471:973–978.

Heap, S. R., Linder, D. J., Lanz, T. M., Woodgate, B., Cornett, R., Hubeny, I., and Maran, S. P. 1999. STIS coronagraphic observations of β Pictoris. *Astrophys. J.*, in press.

Henry, T. J., Soderblom, D. R., Donahue, R. A., and Baliunas, S. L. 1996. A survey of Ca II H and K chromospheric emission in southern solar-type stars. *Astron. J.* 111:439–456.

Herbst, W., Herbst, D. K., and Grossman, E. 1994. Catalogue of UBVRI photometry of T Tauri stars and analysis of the causes of their variability. *Astron. J.* 108:1906–1923.

Hobbs, L. M. 1986. Observations of gaseous circumstellar disks. *Astrophys. J.* 308:854–858.

Holland, W. S., Greaves, J. S., Zuckerman, B., Webb, R. A., McCarthy, C., Coulson, I. M., Walther, D. M., Dent, W. R. F., Gear, W. K., and Robson, I. 1998. Submillimetre images of dusty debris around nearby stars. *Nature* 392:788–790.

Holweger, H., and Rentzsch-Holm, I. 1995. High-resolution spectroscopy of λ Bootis stars and "dusty" normal A stars: Circumstellar gas, rotation, and accretion. *Astron. Astrophys.* 303:819–832.

Holweger, H., Hemple, M., Van Thiel, T., and Kaufer, A. 1997. The surface composition of β Pic. *Astron. Astrophys.* 320:L49–L52.

Jayawardhana, R., Fisher, S., Hartmann, L., Telesco, C., Pina, R., and Fazio, G. 1998. A dust disk surrounding the young A star HR 4796A. *Astrophys. J. Lett.* 503:L79–L82.

Jolly, A., McPhate, J. B., Lecavelier, A., Lagrange, A. M., Lemaire, J. L., Feldman, P. D., Vidal-Madjar, A., Ferlet, R., Malmasson, D., and Rostas, F. 1998. HST-GHRS observations of CO and CI in the β Pic circumstellar disk. *Astron. Astrophys.* 329:1028–1034.

Jolly, A., et al. 1999. HST-STIS observations of variable CI in the β Pic circumstellar disk. *Astron. Astrophys.*, in preparation.

Judge, P. G., Jordan, C., and Rowan-Robinson, M. 1987. δ Andromedae (K3 III)—An IRAS source with an unusual ultraviolet spectrum. *Mon. Not. Roy. Astron. Soc.* 224:93–106.

Jura, M. 1990. The absence of circumstellar dust debris around G giants. *Astrophys. J.* 365:317–320.

Jura, M. 1991. The dust debris around HR 4796. *Astrophys. J. Lett.* 383:L79–L80.

Jura, M., Zuckerman, B., Becklin, E. E., and Smith, R. C. 1993. Constraints on the evolution of remnant protostellar dust debris around HR 4796. *Astrophys. J. Lett.* 418:L37–L40.

Jura, M., Ghez, A. M., White, R. J., McCarthy, D. W., Smith, R. C., and Martin, P. G. 1995. The fate of the solid matter orbiting HR 4796A. *Astrophys. J.* 445:451–456.

Jura, M., Malkan, M., White, R., Telesco, C., Pina, R., and Fisher, R. S. 1998. A protocometary cloud around HR4796A? *Astrophys. J.* 505:897–902.

Kalas, P., and Jewitt, D. 1995. Asymmetries in the β Pic dust disk. *Astron. J.* 110:794–804.

Kalas, P., and Jewitt, D. 1996. The detectability of β Pic-like CS disks around nearby MS stars. *Astron. J.* 111:1347–1427.

Kalas, P., and Jewitt, D. 1997. A candidate dust disk surrounding the binary stellar system BD +31°643. *Nature* 386:52–54.

Kenyon, S. J., and Luu, J. X. 1998. Accretion in the early Kuiper Belt. I. Coagulation and velocity evolution. *Astron. J.* 115:2136–2160.

King, J. R., and Patten, B. M. 1992. β Pic and the λ Bootis stars—Testing the accretion hypothesis. *Mon. Not. Roy. Astron. Soc.* 256:571–574.

Knacke, R. F., Fajardo-Acosta, S. B., Telesco, C. M., Hackwell, J. A., Lynch, D. K., and Russell, R. W. 1993. The silicates in the disk of β Pic. *Astrophys. J.* 418:440–450.

Koerner, D. W., Ressler, M. E., Werner, M. W., and Backman, D. E. 1998. Mid-infrared imaging of a circumstellar disk around HR 4796: Mapping the debris of planetary formation. *Astrophys. J. Lett.* 503:L83–L88.

Labeyrie, A. 1995. Images of exo-planets obtainable from dark speckles in adaptive telescopes. *Astron. Astrophys.* 298:544–548.

Lagage, P. O., and Pantin, E. 1994*a*. Dust depletion in the inner disk of β Pic as a possible indicator of planets. *Nature* 369:628–630.

Lagage, P. O., and Pantin, E. 1994*b*. Sub-arcsec 10 μm imaging of β Pic and other star disk candidates. *Experimental Astron.* 3:57–60.

Lagrange, A.-M. 1995. Observations of disks around main sequence stars (β Pic). *Astrophys. Space Sci.* 223:19–43.

Lagrange, A.-M. 1998. Star formation and β Pic. In *The Scientific Impact of the Goddard High Resolution Spectrograph,* ed. J. C. Brandt, T. B. Ake, III, and C. C. Petersen, ASP Conf. Ser. 143 (San Francisco: Astronomical Society of the Pacific), pp. 83–98.

Lagrange, A.-M., and Beust, H. 1999. Applying the FEB model to HD 100546. *Astron. Astrophys.*, in preparation.

Lagrange, A.-M., Plazy, F., Beust, H., Mouillet, D., Deleuil, M., Ferlet, R., Spyromilio, J., Vidal-Madjar, A., Tobin, W., Hearnshaw, J. B., Clark, M., and Thomas, K. W. 1996. The β Pic circumstellar disk. XXI. Results of the December 1992 spectroscopic campaign. *Astron. Astrophys.* 310:547–563.

Lagrange, A.-M., Beust, H., Mouillet, D., Deleuil, M., Feldman, P. D., Ferlet, R., Hobbs, L., Lecavelier des Etangs, A., Lissauer, J. J., McGrath, M. A., McPhate, J. B., Spyromilio, J., Tobin, W., and Vidal-Madjar, A. 1998. The β Pic circumstellar disk. XXIV. Clues to the origin of the stable gas. *Astron. Astrophys.* 329:1091–1108.

Lagrange-Henri, A.-M., Beust, H., Ferlet, R., Vidal-Madjar, A., and Hobbs, L. M. 1990. HR 10: A new β Pic-like star? *Astron. Astrophys.* 227:L13–L16.

Lamers, H. J. G. L. M., Lecavelier des Etangs, A., and Vidal-Madjar, A. 1997. β Pic light variations. II. Scattering by a dust cloud. *Astron. Astrophys.* 328:321–330.

Lanz, T., Heap, S. R., and Hubeny, I. 1995. HST/GHRS observations of the β Pic system: Basic parameters and the age of the system. *Astrophys. J. Lett.* 447:L41–L44.

Lecavelier des Etangs, A. 1998. Planetary migration and sources of dust in the β Pic disk. *Astron. Astrophys.* 337:501–511.

Lecavelier des Etangs, A., Deleuil, M., Vidal-Madjar, A., Ferlet, R., Nitschelm, C., Nicolet, B., and Lagrange-Henri, A.-M. 1995. β Pic: Evidence for light variations. *Astron. Astrophys.* 299:557–562.

Lecavelier des Etangs, A., Vidal-Madjar, A., and Ferlet, R. 1996. Dust distribution in disks supplied by small bodies: Is the β Pictoris disk a gigantic multi-cometary tail? *Astron. Astrophys.* 307:542–550.

Lecavelier des Etangs, A., Deleuil, M., Vidal-Madjar, A., Lagrange-Henri, A.-M., Backman, D., Lissauer, J. J., Ferlet, R., Beust, H., and Mouillet, D. 1997*a*.

HST-GHRS observations of candidate β Pic-like circumstellar gaseous disks. *Astron. Astrophys.* 325:228–236.

Lecavelier des Etangs, A., Vidal-Madjar, A., Backman, D. E., Deleuil, M., Lagrange, A.-M., Lissauer, J. J., Ferlet, R., Beust, H., and Mouillet, D. 1997*b*. Discovery of C I around 51 Ophiuchi. *Astron. Astrophys.* 324:L39–L42.

Lecavelier des Etangs, A., Vidal-Madjar, A., Burki, G., Lamers, H. J. G. L. M., Ferlet, R., Nitschelm, C., and Sevre, F. 1997*c*. β Pic light variations. I. The planetary hypothesis. *Astron. Astrophys.* 328:311–320.

Lecavelier des Etangs, A., Ferlet, R., and Vidal-Madjar, A. 1997*d*. A search for β Pic-like Ca II circumstellar gas around Ursa Major Stream stars. *Astron. Astrophys.* 328:602–605.

Levison, H. F., Duncan, M. J., and Wetherill, G. W. 1995. Secular resonances and cometary orbits in the β Pic system. *Nature* 372:441–444.

Li, A., and Greenberg, J. M. 1998. A comet dust model for the β Pictoris disk. *Astron. Astrophys.* 331:291–313.

Liou, J., Zook, H. A., and Dermott, S. F. 1996. Kuiper Belt dust grains as a source of interplanetary dust particles. *Icarus* 124:429–440.

Liseau, R., and Artymowicz, P. 1998. High sensitivity search for molecular gas in the β Pic disk. On the low gas-to-dust mass ratio of the circumstellar disk around β Pictoris. *Astron. Astrophys.* 334:935–942.

Lissauer, J. 1993. Planet formation. *Ann. Rev. Astron. Astrophys.* 31:129–174.

Luu, J. X., and Jewitt, D. C. 1998. Optical and infrared reflectance spectrum of Kuiper Belt object 1996 TL 66. *Astrophys. J. Lett.* 494:L117–L120.

Malfait, K., Bogaert, E., and Waelkens, C. 1998. An ultraviolet, optical and infrared study of Herbig Ae/Be stars. *Astron. Astrophys.* 331:211–223.

Mannings, V., and Barlow, M. J. 1998. Candidate main-sequence stars with debris disks: A new sample of Vega-like sources. *Astrophys. J.* 497:330–341.

Mannings, V., and Sargent, A. 1997. A high-resolution study of gas and dust around young intermediate-mass stars: Evidence for circumstellar disks in Herbig Ae systems. *Astrophys. J.* 490:792–802.

Mouillet, D., and Lagrange, A.-M. 1995. The β Pictoris circumstellar disk: Some physical parameters of the gaseous component. *Astron. Astrophys.* 297:175–182.

Mouillet, D., Lagrange, A.-M., Beuzit, J.-L., and Renaud, N. 1997*a*. A stellar coronograph for the COME-ON-PLUS adaptive optics system. II. First astronomical results. *Astron. Astrophys.* 324:1083–1090.

Mouillet D., Larwood, J. D., Papaloizou, J. C. B., and Lagrange, A.-M., 1997*b*. A planet on an inclined orbit as an explanation of the warp in the β Pic disc. *Mon. Not. Roy. Astron. Soc.* 292:896–904.

Oudmaijer, R. D., Van Der Veen, W. E. C. J., Waters, L. B. F. M., Trams, N. R., Waelkens, C., and Engelsman, E. 1992. SAO stars with infrared excess in the IRAS Point Source Catalog. *Astron. Astrophys. Suppl.* 96:625–643.

Palla, F., and Stahler, S. W. 1993. The pre-main-sequence evolution of intermediate-mass stars. *Astrophys. J.* 418:414–425.

Pantin, E., Lagage, P. O., and Artymowicz, P. 1997. Mid-infrared images and models of the β Pictoris dust disk. *Astron. Astrophys.* 327:1123–1136.

Paresce, F. 1991. On the evolutionary status of β Pictoris. *Astron. Astrophys.* 247:L25–L27.

Paunzen, E. 1997. On the evolutionary status of λ Bootis stars using Hipparcos data. *Astron. Astrophys.* 326:L29–L32.

Petterson, O. K. L., and Tobin, W. 1999. β Pictoris: The variable Ca II H & K absorptions 1994–1996. *Mon. Not. Roy. Astron. Soc.*, submitted.

Plets, H., 1997. A systematic study of the occurrence of circumstellar dust around main-sequence stars and giants. Ph.D. Thesis, Katholieke Universitet Leuven.

Plets, H., Waelkens, C., Oudmaijer, R. D., and Waters, L. B. F. M. 1997. Giants with infrared excess. *Astron. Astrophys.* 323:513–523.

Roques, F., Scholl, H., Sicardy, B., and Smith, B. 1994. Is there a planet around β Pic? Perturbations of a planet on a circumstellar disk. I. The numerical model. *Icarus* 108:37–58.

Savoldini, M., and Galletta, G. 1994. CO depletion in the protoplanetary disk of β Pictoris. *Astron. Astrophys.* 285:467–468.

Schneider, G., Smith, B. A., Becklin, E. E., Koerner, D. W., Meier, R., Hines, D. C., Lowrance, P. J., Terrile, R. J., Thompson, R. I., and Rieke, M. 1999. NICMOS imaging of the HR 4796 A circumstellar disk. *Astrophys. J. Lett.*, in press.

Skinner, C. J., Sylvester, D. J., Graham, J. R., Barlow, M. J., Meixner, M., Keto, E., Arens, J. F., and Jernigan, J. G. 1995. The dust disk around the Vega-excess star SAO 26804. *Astrophys. J.* 444:861–873.

Smith, B. A., Fountain, J. W., and Terrile, R. J. 1992. An optical search for β Pic-like disks around nearby stars. *Astrophys. J.* 261:499–502.

Soderblom, D. R., and Mayor, M. 1993. Stellar kinematic groups. I. The Ursa Major group. *Astron. J.* 105:226–249.

Soderblom, D. R., Henry, T. J., Shetrone, M. D., Jones, B. F., and Saar, S. H. 1996. The age-related properties of the HD 98800 system. *Astrophys. J.* 460:984–992.

Soderblom, D. R., King, J. R., Siess, L., Noll, K. S., Gilmore, D. M., Henry, T. J., Nelan, E., Burrows, C. J., Brown, R. A., Perryman, M. A. C., Benedict, G. F., McArthur, B. J., Franz, O. G., Wasserman, L. H., Jones, B. F., Latham, D. W., Torres, G., and Stefanik, R. P. 1998. HD 98800: A unique stellar system of post-T Tauri stars. *Astrophys. J.* 498:385–393.

Sorelli, C., Grinin, V. P., and Natta, A. 1996. Infall in Herbig Ae/Be stars: What Na D lines tell us. *Astron. Astrophys.* 309:155–162.

Stauffer, J. R., Hartmann, L. W., and Barrado y Navascués, D. 1995. An age estimate for the β Pictoris analog HR 4796A. *Astrophys. J.* 454:910–916.

Stern, S. A. 1996*a*. On the collisional environment, accretion time scales, and architecture of the primordial Kuiper belt. *Astron. J.* 112:1203–1211.

Stern, S. A. 1996*b*. Signatures of collisions in the Kuiper disk. *Astron. Astrophys.* 310:999–1010.

Stern, S. A., and Colwell, J. E. 1997. Accretion in the Edgeworth-Kuiper Belt: Forming 100–1000 km radius bodies and beyond. *Astron. J.* 114:841–849.

Strom, S., Edwards, S., and Skrutskie, M. F. 1993. Evolutionary time scales for circumstellar disks associated with intermediate- and solar-type stars. In *Protostars and Planets III,* ed. E. H. Levy and J. I. Lunine (Tucson: University of Arizona Press), pp. 837–866.

Sylvester, R. J., and Skinner, C. J. 1996. Optical, infrared and millimetre-wave properties of Vega-like stars. II. Radiative transfer modeling. *Mon. Not. Roy. Astron. Soc.* 283:457–470.

Sylvester, R. J., Skinner, C. J., Barlow, M. J., and Mannings, V. 1996. Optical, infrared and millimetre-wave properties of Vega-like stars. *Mon. Not. Roy. Astron. Soc.* 279:915–939.

Sylvester, R. J., Skinner, C. J., and Barlow, M. J. 1997. Optical, infrared and millimetre-wave properties of Vega-like stars. III. Models with thermally spiking grains. *Mon. Not. Roy. Astron. Soc.* 289:831–846.

Teplitz, V. L., Stern, S. A., Anderson, J. D., Rosenbaum, D., Scalise, R. J., and Wentzler, P. 1999. IR Kuiper Belt constraints. *Astrophys. J.*, in press.

Thomas, F., and Morbidelli, A. 1996. The Kozai resonance in the outer solar system and the dynamics of long-period comets. *Celestial Mechan. Dynam. Astron.* 64:209–229.

Torres, G., Stefanik, R. P., Latham, D. W., and Mazeh, T. 1995. Study of spectroscopic binaries with TODCOR. IV. The multiplicity of the young nearby star HD 98800. *Astrophys. J.* 452:870–878.

Tremaine, S. 1989. Dark matter in the solar system. In *Baryonic Dark Matter,* ed. D. Lynden-Bell and G. Gilmore (Dordrecht: Kluwer), pp. 37–65.

Trilling, D. E., and Brown, R. H. 1998. A circumstellar dust disk around a star with a known planetary companion. *Nature* 395:775–777.

van den Ancker, M. E., de Winter, D., and Tjin A Djie, H. R. E. 1998. Hipparcos photometry of Herbig Ae/Be stars. *Astron. Astrophys.* 330:145–154.

van der Bliek, N. S., Prusti, T., and Waters, L. B. F. M. 1994. Vega: Smaller dust grains in a larger shell. *Astron. Astrophys.* 285:229–232.

Venn, K. A., and Lambert, D. L. 1990. The chemical composition of three λ Bootis stars. *Astrophys. J.* 363:234–244.

Vidal-Madjar, A., Lagrange-Henri, A.-M., Feldman, P. D., Beust, H., Lissauer, J. J., Deleuil, M., Ferlet, R., Gry, C., Hobbs, L. M., McGrath, M. A., McPhate, J. B., and Moos, H. W. 1994. HST-GHRS observations of β Pic: Additional evidence for infalling comets. *Astron. Astrophys.* 290:245–258.

Vidal-Madjar, A., Lecavelier des Etangs, A., and Ferlet, R. 1998. β Pic, a young planetary system? *Planet. Space Sci. Rev.* 46:629–648.

Waelkens, C., and Waters, R. 1998. Herbig Ae/Be stars. *Ann. Rev. Astron. Astrophys.* 36:233–266.

Waelkens, C., Waters, L. B. F. M., De Graauw, M. S., Huygen, E., Malfait, K., Plets, H., Vandenbussche, B., Beintema, D. A., Boxhoorn, D. R., Habing, H. J., Heras, A. M., Kester, D. J. M., Lahuis, F., Morris, P. W., Roelfsema, P. R., Salama, A., Siebenmorgen, R., Trams, N. R., van der Bliek, N. R., Valentijn, E. A., and Wesselius, P. R. 1996. SWS observations of young main-sequence stars with dusty circumstellar disks. *Astron. Astrophys.* 315:L245–L248.

Webb, R. A., Zuckerman, B., Platais, I., Patience, J., Schwartz, M., and White, R. J. 1998. Discovery of seven nearby T Tauri stars in the TW Hydrae association. *Bull. Am. Astron. Soc.* 192:10.07 (abstract).

Weinberger, A. J., Becklin, E. E., Schneider, G., Smith, B. A., Lowrance, P. J., Silverstone, M. D., Zuckerman, B., and Terrile, R. J. 1999. The circumstellar disk of HD 141569 imaged with NICMOS. *Astrophys. J. Lett.*, in press.

Weissman, P. R. 1995. The Kuiper Belt. *Ann. Rev. Astron. Astrophys.* 33:327–358.

Weissman, P. R., and Levison, H. F. 1998. The population of the trans-Neptunian region. In *Pluto and Charon,* ed. S. A. Stern and D. J. Tholen (Tucson: University of Arizona Press), pp. 559–604.

Welsh, B. Y., Craig, N., Crawford, I. A., and Price, R. J. 1998. β Pic-like circumstellar disk gas surrounding HR 10 and HD 85905. *Astron. Astrophys.* 338:674–682.

Wetherill, G. W. 1991. Occurrence of Earth-like bodies in planetary systems. *Science* 253:535–538.

Whitmire, D. P., Matese, J. J., and Tomley, L. J. 1988. A brown dwarf companion as an explanation of the asymmetry in the β Pictoris disk. *Astron. Astrophys.* 203:L13.

Witteborn, F. C., Bregman, J. D., Lester, D. F., and Rank, D. M. 1982. A search for fragmentation debris near Ursa Major Stream stars. *Icarus* 50:63–71.

Zuckerman, B., and Becklin, E. E. 1993*a*. Infrared observations of the remarkable main-sequence star HD 98800. *Astrophys. J. Lett.* 406:L25–L28.

Zuckerman, B., and Becklin, E. E. 1993*b*. Submillimeter studies of main-sequence stars. *Astrophys. J.* 414:793–802.

Zuckerman, B., Forveille, T., and Kastner, J. H. 1995*a*. Inhibition of giant-planet formation by rapid gas depletion around young stars. *Nature* 373:494–496.

Zuckerman, B., Kim, S. S., and Liu, T. 1995*b*. Luminosity class III stars with excess far-infrared emission. *Astrophys. J. Lett.* 446:L79–L83.

PART III
Young Binaries

MULTIPLE FRAGMENTATION OF PROTOSTARS

PETER BODENHEIMER
Lick Observatory

ANDREAS BURKERT
Max-Planck-Institut für Astronomie

RICHARD I. KLEIN
University of California at Berkeley and
Lawrence Livermore National Laboratory

and

ALAN P. BOSS
Carnegie Institution of Washington

Most stars are members of binary or multiple-star systems, yet our understanding of multiple star formation is rudimentary at best. Multiple star formation must be understood before we can claim to understand fully the formation of single stars, or of planets in either single- or multiple-star systems. Theoretical mechanisms for the formation of binary and multiple star systems, as well as recent numerical models, are reviewed here. Comparisons of results obtained with different numerical codes are presented. Observations of main-sequence and pre-main-sequence stars, as well as binary protostars, imply that most binary stars form by fragmentation during the collapse of dense molecular cloud cores.

I. INTRODUCTION

Observations of the properties of binary and multiple-star systems provide the ultimate test of any theory of how these systems came into being. Accordingly, we begin with a brief summary of those properties that seem to be the most relevant for choosing between the various mechanisms proposed for binary and multiple-star formation.

A. Main-Sequence Binaries

Duquennoy and Mayor (1991) surveyed all known F7–G9 dwarf stars within about 22 pc of the Sun, providing a relatively unbiased sample of main-sequence (MS) stars in the Sun's neighborhood. Fischer and Marcy (1992) performed a similar survey of M dwarf stars within 20 pc.

1. Frequency. Duquennoy and Mayor (1991) found that their sample had a ratio of number of single:binary:triple:quadruple (= s:b:t:q) systems of 57:38:4:1, for companions with a mass ratio $q = M_2/M_1 > 0.1$. They also found hints for very low-mass companions ($q < 0.1$) in about 10% of the primaries, and concluded that each primary star had about 0.5 companions, on average. Fischer and Marcy (1992) found that the M dwarfs had a similar average number of companions (0.55) and a similar system ratio of s:b:t:q = 58:33:7:1. However, Leinert et al. (1997) found the average number of companions to be ~0.32 for their sample of M dwarfs within 5 pc of the Sun.

2. Distribution with Period. The orbital period (P) distribution of the Duquennoy and Mayor (1991) sample was Gaussian-like as a function of log P with a median $P \approx 180$ yr. The periods ranged from less than a day to over 10^6 yr. A similar distribution was found by Fischer and Marcy (1992) for the M dwarfs. Multiple systems were invariably hierarchical in structure, with both long- and short-period components.

3. Mass Ratio. The number of systems as a function of q in the Duquennoy and Mayor (1991) sample is Gaussian with a maximum near $q = 0.23$, similar to a distribution drawn from the field star initial mass function (Kroupa et al. 1990). Fischer and Marcy (1992) found that their sample of low-mass dwarfs was deficient in companions of even lower mass (below $q \sim 0.5$), suggesting that brown-dwarf companions are relatively rare. Mazeh et al. (1992) reanalyzed Duquennoy and Mayor's (1991) data for short-period (< 3000 days) binaries and found that the distribution was relatively uniform, or even rising with q, for $0.1 < q < 0.9$, implying a different distribution for long-period and short-period systems.

4. Orbital Eccentricity. Binaries with $P < 11$ days in the Duquennoy and Mayor (1991) sample had circular orbits, consistent with circularization produced by tidal dissipation during the lifetimes (10^9–10^{10} yr) of these stars. Longer-period systems had a range of eccentricities scattered between $e \approx 0.1$ and $e \approx 0.9$.

5. Age. Patience et al. (1998) searched 167 bright stars in the Hyades cluster (age $\approx 6 \times 10^8$ yr) for companions with a mass ratio as low as $q \approx 0.23$. They found that each primary had an average of at least 0.46 companions, close to the results for G and M dwarfs, which have a mean age of several Gyr. A sample of 144 primaries in the even younger (~10^8 yr) Pleiades cluster also showed a binary frequency similar to that of the solar neighborhood (Bouvier et al. 1997).

B. Pre-Main-Sequence Binaries

1. Frequency. The binary frequency for young stars in the Taurus, Ophiuchus, Lupus, Chamaeleon, and Corona Australis star-forming regions is about two times higher than that of the Duquennoy and Mayor (1991) sample for separations of 15–1800 AU (Ghez et al. 1997). Köhler and Leinert (1998) found the same enhancement factor for pre-main-sequence (PMS) stars in Taurus in about the same range of separations. However, the Orion Trapezium cluster has a binary frequency in good agreement with that of MS stars (Petr et al. 1998; Simon et al. 1999). Brandner and Köhler (1998) found statistically different distributions of binary separations in different regions of the Upper Scorpius OB association and suggested that the MS distribution could result from the superposition of distinct distributions produced in varied star-forming regions.

2. *Distribution with Period.* Mathieu (1994) showed that PMS stars have a distribution of periods similar to that of MS stars, given the limitations of the PMS sample at the time.

3. *Mass Ratio.* These have not yet been determined reliably for PMS stars, for a variety of reasons (see Mathieu 1994).

4. *Orbital Eccentricity.* Mathieu (1994) showed that PMS stars have a distribution of eccentricities as a function of P that is very similar to that of MS stars, except that circular orbits occur only for P less than ~ 5 days, consistent with the reduced amount of time for tidal circularization in these young stars.

5. *Age.* T Tauri stars have ages on the order of a few times 10^6 yr, and their high binary frequency implies that binary formation has already finished by this age. Brandner and Zinnecker (1997) found that of eight spatially resolved PMS binaries that could be compared to theoretical evolutionary tracks, all eight pairs were coeval to within theoretical and observational uncertainty (see also Hartigan et al. 1994 and chapter by Mathieu et al., this volume).

C. Protostellar Binaries

Observations of binary and multiple protostars are as yet too few to provide a statistical summary of their properties. Fuller et al. (1996) found an embedded binary protostar in Taurus with a separation of 2800 AU and an age of $\sim 5 \times 10^3$ yr, much less than its $P \sim 10^5$ yr. Lay et al. (1995) detected a hierarchical quadruple (or higher-order) protostar, while Wootten (1989; see also Mundy et al. 1992 and Walker et al. 1993), Looney et al. (1997), and Terebey et al. (1998) found protostellar binaries with separations of about 840 AU, 50 AU, and 42 AU, respectively.

II. INITIAL CONDITIONS

The observational evidence strongly suggests that binary formation is closely related to the star formation process itself. Although it is well established that star formation takes place in molecular cloud cores (Myers 1985, 1987), the details of the initial conditions are still not well understood. The typical core, which is close to virial equilibrium, has a scale of about 0.1 pc, a mass in the range of 1 to a few solar masses, and a composition near solar. The hydrogen is in the form of H_2, the total number density is 10^4–10^5 cm^{-3}, the temperature is about 10 K, and the core is slowly rotating. These conditions give a thermal Jeans mass also of order 1 $M_\odot$ (Larson 1985). The parameters for describing the initial conditions include α and β (which are, respectively, the ratios of thermal and of rotational energy to the absolute value of the gravitational potential energy), as well as the cloud shape, the density distribution $\rho(r)$, and the angular momentum distribution. High-resolution observations of starless precollapse cores in regions of isolated star formation show that all of them are centrally condensed with Gaussian or power law density profiles (Walker et al. 1990; Ladd et al. 1991; Fuller and Myers 1992; Ward-Thompson et al. 1994; Henriksen et al. 1997). For example, the isolated prestellar core L1689B (André et al. 1996) has a core ($r \leq 4000$ AU) with $\rho \propto r^{-0.4}$ or $\rho \propto r^{-1.2}$, depending on the assumed geometry, and a mass of order 0.3 $M_\odot$. Outside the core region, the density decreases as r^{-2}. In contrast, Motte et al. (1998) find that the clustered star formation region ρ Ophiuchi has 10 compact prestellar clumps with steep r^{-2} density profiles everywhere, although 19 other cores in the region have the less steep profiles in the central regions. These results indicate that the density profiles for prestellar cores could depend on the environment.

In general, prestellar cores appear to be elongated and irregular. Myers et al. (1991) deduced that many cloud cores are actually prolate with apparent aspect ratios of 2:1; some evidence for oblate cores also exists (Boss 1996). On scales of molecular cores, rotation has been detected in about 50% of all cases, with the values for the angular velocity ranging between $\Omega \approx 10^{-13}$ and 3×10^{-15} s^{-1} (Goodman et al. 1993; see review by Bodenheimer 1995). The dependence of angular velocity on the radius of a core is $\Omega \sim R^{-0.4}$, with a large scatter. The typical ratio $\beta \sim 0.02$ is too small to explain the observed axis ratios; also, the observed velocity gradients within individual cores are consistent with the assumption of uniform rotation.

The initial conditions for protostellar collapse could be strongly affected by magnetic fields, which can support molecular cloud regions against collapse and which control the rate at which they evolve into dense cores. As described by Shu et al. (1993), the material gradually drifts relative to the magnetic field due to ambipolar diffusion, the turbulence decays, the cloud undergoes a slow contraction on the diffusion timescale, and a quiet, dense protostellar core forms. This conclusion is in agreement with observations that dense prestellar cores ($n \approx 10^5$ cm^{-3}) are characterized by low linewidths, of order 0.2 to 0.5 km s^{-1} (Benson and Myers 1989; Myers and Goodman 1988; Lemme et al. 1995), in which the thermal contribution dominates. During the phase of magnetically supported contraction the core acquires an r^{-2} density distribution (Lizano and Shu 1989; Tomisaka et al. 1990). Starting from the center, magnetic effects gradually become unimportant, and the cloud experiences an inside-out collapse. If the initial conditions are produced in this manner, the portion of the cloud that begins to collapse will contain close to

1 thermal Jeans mass, and depending on the details of the initial structure, it may not fragment.

Shu et al. (1993) note that the theoretical r^{-2} density profile represents an asymptotic result and that, in reality, cores would become gravitationally unstable in their central regions before the singular density cusp develops. The results of detailed two-dimensional cloud contraction models (Ciolek and Mouschovias 1995; Hujeirat 1998; Ciolek and Königl 1998) indeed show that the density profile becomes quite close to $\rho \propto r^{-2}$ at least in the outer regions, but the central regions retain a relatively uniform density. As a result, magnetically regulated prestellar cores will enter the dynamical collapse phase with $\rho(r)$ flatter than r^{-2} in the central region, at least qualitatively in agreement with observations. Note also that magnetic braking induces solid body rotation during the equilibrium contraction phase, as long as the magnetic braking time is shorter than the ambipolar diffusion time (Basu and Mouschovias 1994).

III. PHYSICAL PROCESSES

We discuss now the important physical processes that can influence the occurrence of fragmentation during the collapse of a cloud starting from initial conditions similar to those just described.

A. Radiation Effects

Two different phases of protostellar collapse are generally considered in fragmentation simulations. The first, the isothermal phase, corresponds to low-density ($\rho = 10^{-19}$–10^{-13} g cm^{-3}) gas that is optically thin to infrared radiation. A balance between the rates of heating by gravitational compression and cooling through emission by grains is obtained at a temperature of about 10 K over this entire density range (Hayashi 1966; de Jong et al. 1980; Whitworth and Clarke 1997). This phase is favorable for fragmentation. The general criterion for a uniform-density cloud with uniform rotation to fragment in the isothermal phase is $\alpha\beta < 0.12$ (Hayashi et al. 1982; Miyama et al. 1984). Recent highly resolved numerical simulations (Tsuribe and Inutsuka 1999) show that α must be less than 0.5 for fragmentation to occur, with only a weak dependence on β. However, if the cloud is centrally condensed, the region of (α, β) space where fragmentation occurs is reduced (Boss 1993).

The second phase starts in the central regions of the protostar or fragment, when it becomes dense enough to be optically thick. The density is high enough by this point that the free-fall time rapidly becomes short compared with the diffusion time for the radiation, so the radiation becomes trapped in the infalling matter, and the collapse is essentially adiabatic. The cloud becomes less susceptible to fragmentation, because increased pressure effects reduce the perturbation amplitude during this phase. The general criterion for fragmentation in a cloud of initially uniform density becomes $\alpha < 0.09\beta^{0.2}$ for a gas with $\gamma = 1.4$ (Hachisu et al. 1987; Tohline 1981; Boss 1981; Miyama 1992). The dust grains are important both for cooling during the isothermal phase and for radiative opacity during the optically thick phases. Opacity calculations for the relevant temperature regime and approximately solar composition are provided by Pollack et al. (1985), Alexander and Ferguson (1994), and Preibisch et al. (1993).

The transition between isothermal and adiabatic phases is important with regard to the minimum mass of a fragment produced during the collapse. During the isothermal phase the Jeans mass decreases as the density increases; however, once the fragment becomes optically thick and heats, the Jeans mass begins to rise. Thus, opacity effects determine the minimum mass of a fragment, which is estimated to be $\approx 0.01\ M_\odot$ (Low and Lynden-Bell 1976; Rees 1976; Silk 1977; Boss 1988*a*). Although grain opacities in the range $10\ K < T < 100$ K are not well known, this result is not sensitive to the precise value.

B. Equation of State and Thermodynamics

The generally cool and low-density protostellar material obeys an ideal-gas equation of state during the phases when fragmentation occurs. At temperatures below about 2000 K, the hydrogen is in molecular form. With a chemical composition close to solar, the equation of state will be an ideal gas with $\gamma \approx 1.67$ in the range $10 < T < 100$ K, and $\gamma \approx 1.4$ in the range $100 < T < 2000$ K. During the process of H_2 dissociation, the value of γ falls to about 1.1, and the gas becomes dynamically unstable and undergoes a second collapse. When dissociation is nearly complete ($T \approx 10^4$ K), $\gamma \approx 1.67$ again, and the hydrostatic stellar core forms. At higher temperatures the hydrogen ionizes, but no further dynamical collapse results (Larson 1969).

C. Gravitational Effects

At the onset of collapse, the molecular cloud core is expected to have a nonaxisymmetric structure, with either ordered or random density and velocity fluctuations. As collapse proceeds in near free fall, the collapse timescale and the timescale for growth of the perturbations is the same, and the amplitudes do not grow relative to the background, unless the perturbations are very large. The central regions of the cloud collapse most rapidly, leading to a density distribution that is highly centrally peaked at later times. Angular momentum is approximately conserved during this phase for a given mass element, because the gravitational torques associated with the (generally small) nonaxisymmetric effects do not transfer much angular momentum.

However, once filamentary or flattened structures develop and the collapse is slowed or halted, fragmentation is likely (Larson 1985). The collapse can be stopped, for example, by pressure effects associated with entering the adiabatic phase or by rotational effects associated with the spinup of the gas combined with pressure effects in the direction parallel to the rotation axis. In the case that rotation becomes important before the adiabatic phase is reached, the central density peak of the cloud flattens into a disk, the overall collapse is halted, and nonaxisymmetric structures, such as spiral arms, grow and develop into fragments (Larson 1978). Once pressure effects have become important, if filaments or rings have formed, they are likely to fragment; otherwise, rotational effects are not likely to be large enough to induce fragmentation during the adiabatic phase, and the end result will be a central star in equilibrium surrounded by a disk.

However, this disk can also become gravitationally unstable. Although the indicator (Safronov 1960; Toomre 1964) known as $Q = \kappa c_s/(\pi G \sigma)$, where κ is the epicyclic frequency, c_s is the sound speed, and σ is the local surface density, is a local criterion for axisymmetric stability, the minimum Q value in a nearly Keplerian disk serves as a useful indicator for the outcome of nonaxisymmetric instability. Extensive analytical and numerical modeling of gravitationally unstable disks (Cassen et al. 1981; Larson 1984; Adams et al. 1989; Papaloizou and Savonije 1991; Tomley et al. 1991, 1994; Heemskerk et al. 1992; Laughlin and Bodenheimer 1994; Miyama et al. 1994; Laughlin and Różyczka 1996; Nelson et al. 1998) has shown that spiral waves develop and that effective angular momentum transfer takes place for values of $Q_{\rm min}$ between 1 and 3. However, the amplitude of the waves saturates (Laughlin et al. 1997, 1998); only if $Q_{\rm min}$ falls below $\approx$1 do the spiral waves grow sufficiently to result in fragmentation. For $1 < Q_{\rm min} < 2$, the disk evolution time is on the order of a few local dynamical times (t_d), and $Q_{\rm min}$ tends to increase. As the disk accretes mass from the infalling envelope, $Q_{\rm min}$ tends to decrease. Throughout most of the evolution of a protostar the mass accretion timescale is considerably longer than t_d, so it is unlikely that conditions for fragmentation of the disk can be reached. Exceptions could occur if the accretion is strongly episodic, or if the disk is in an early phase when there still is rapid infall. Then fragmentation can be driven by the interaction of a disk with an embedded bar or binary (Bonnell 1994; Bonnell and Bate 1994*a*).

D. Magnetic Effects

As the discussion in section II indicates, magnetic forces are not particularly important in supporting a cloud core against gravity; the fields are, however, important in shaping the density profile at the onset of collapse. During collapse, ambipolar diffusion and the decrease in conductivity result in a decrease in the importance of magnetic effects. Because of ohmic resistivity the timescale for magnetic flux loss becomes shorter than the free fall time at densities above $\sim 10^{11}$ cm^{-3} (Umebayashi and Nakano 1990), while fragmentation typically begins at densities somewhat higher than that value. Also during collapse, the magnetic braking time becomes longer than the free fall time, and the interior of the cloud spins up and develops differential rotation (Basu and Mouschovias 1994). Three-dimensional collapse calculations that take into account magnetic pressure effects but not the magnetic stress, and that include an approximate model of ambipolar diffusion (Boss 1997*a*), show that fragmentation is not prevented by magnetic effects in a rapidly rotating cloud with an initial Gaussian density distribution.

IV. FORMATION MECHANISMS

Although three-dimensional hydrodynamic calculations to examine the collapse and fragmentation of molecular clouds have been in evidence for over a decade, we are still at an early stage in pinpointing the dominant mechanism for binary and multiple-star formation. In this section we provide a summary of some of the suggested processes and then consider some of the results of detailed numerical simulations on fragmentation during the collapse phase.

A. Brief Survey

Binary formation may occur through the processes of (i) capture, (ii) fission, (iii) fragmentation during the protostar collapse, (iv) disk fragmentation, and (v) prompt initial fragmentation. Item (iii) will be discussed in more detail in section IV.B.

Binary formation by direct capture, either in the presence of a third body to absorb the excess energy or in the presence of tidal dissipation induced in a close encounter, has been shown to be an inefficient process (Tassoul 1978; Boss 1988*b*) in the galactic disk. Larson (1990) suggested that capture could occur in a very compact young cluster such as the Orion Trapezium, with the dissipation provided through direct interaction of the captured star with a circumstellar disk. Clarke and Pringle (1991) pointed out that the rate would be too small to account for all of the Trapezium binaries and that the period distribution would not be as observed. However, in small clusters of 4–10 fragments, star-disk interactions could produce a reasonable binary fraction (McDonald and Clarke 1995). Note that in this case the binary components would be essentially coeval, as observed, because the age difference would not be greater than the cluster dynamical time.

Binary formation by fission occurs when a star or newly formed stellar core in its quasistatic contraction phase undergoes a rotational instability and divides into two distinct masses. In such a process, spin angular momentum is converted into orbital angular momentum. Because such young equilibrium objects in the mass range of 1 $M_\odot$ are known, from protostellar theory and from observations, to have radii less than 10 $R_\odot$, the outcome would be a close binary. Numerous objections to this theory have arisen, which are summarized by Tassoul (1978) and Bodenheimer et al. (1993), and numerical simulations of the process do not show binary formation (Williams and Tohline 1988).

Disk fragmentation can occur in an equilibrium circumstellar disk if the minimum Q value approaches 1 (see section III.C, which discusses the important question of whether the required initial condition can be reached). Numerical examples have been provided by Adams and Benz (1992), Miyama et al. (1994), Laughlin and Różyczka (1996), and Boss (1997*b*). Typical fragment masses are in the range 0.01–0.1 $M_{\odot}$ (i.e, in the brown dwarf range), and orbital separations are 5–100 AU. The calculations have not been followed far enough for orbital properties to be determined. However, external perturbations can also induce fragmentation in a disk; in this case the Q value could be >1. Numerical simulations (Bonnell et al. 1992; Boffin et al. 1998; Watkins et al. 1998*a,b*) show that star-disk encounters, as well as disk-disk encounters, can result in the triggering of new condensations in a disk.

"Prompt initial fragmentation" is a triggered star formation process introduced by Pringle (1989) as a necessary alternative for binary formation in view of the fact that the $\rho \propto r^{-2}$ profiles generated by magnetic effects in contracting cloud cores are unlikely to fragment during the subsequent collapse. He envisions cloud-cloud collisions resulting in binary formation on wide elliptical orbits. Subsequent interactions of the components with each other's disks would reduce the orbital separation. Numerical simulations (Chapman et al. 1992; Turner et al. 1995) show that cooling by molecules and dust, which occurs in the density range 10^{-22}–10^{-20} g cm^{-3}, is an important element of this process. Off-center collisions of gravitationally stable clouds of ≈ 200 $M_{\odot}$ result in a shocked layer, which cools and becomes unstable to the formation of several fragments; the system later evolves into a binary protostar, each component with a surrounding disk. Klein and Woods (1998) have considered a modification of this scenario, involving fragmentation by bending-mode instabilities in cloud collisions, and leading to porous structure in clouds.

B. Fragmentation

Fragmentation during protostar collapse can produce binaries with a wide range of periods. The nature of the problem can be appreciated when one considers that the phase space of initial parameters for molecular clouds is vast, consisting of possible variations in α, β, the initial distributions of velocity, density, and angular momentum, the form of initial perturbations, and the size and shape of clouds. A wide variety of (often idealized) initial conditions has been employed, such as a sphere in uniform rotation with a density perturbation in the form of an $m = 2$ mode with 10% amplitude, a prolate spheroid with a random perturbation (Boss 1993), or a filamentary cloud that is rotating end over end (Zinnecker 1990). The phase space would be somewhat reduced if the initial conditions were chosen to match the observed properties of molecular cloud cores approximately (section II).

Calculations have been performed under several different physical approximations. Many calculations have treated the isothermal phase of collapse (Bonnell et al. 1991; Bonnell and Bastien 1992; Burkert and Bodenheimer 1993, 1996; Nelson and Papaloizou 1993; Monaghan 1994; Sigalotti and Klapp 1994, 1997; Boss 1996; Truelove et al. 1997). Radiation has been treated in simplified approximations, such as the diffusion and the Eddington approximations, applicable to optically thick regions (Myhill and Kaula 1992; Boss 1993). The transition to the optically thick regime has also been approximated (Bonnell 1994; Bate et al. 1995; Klein et al. 1998) by a change in the equation of state from isothermal to adiabatic at the appropriate density ($\rho_{\rm crit} \approx 10^{-13}$ g cm^{-3}). Cooling during collapse has been considered in smoothed particle hydrodynamics (SPH) simulations by Monaghan and Lattanzio (1991), Chapman et al. (1992), and Turner et al. (1995). The cooling that is employed is appropriate to the density regime 10^{-22}–10^{-20} g cm^{-3}, so such calculations apply to clouds that fall in the mass range 100–200 $M_\odot$ rather than around 1 $M_\odot$. The adiabatic phase has been considered by Miyama (1992).

The resulting fragmentation occurs in various modes. A common occurrence is the simple formation of a wide binary with separation $D \approx$ 100–1000 AU, on an eccentric orbit. However, other results have been found: Monaghan and Lattanzio (1991) find six fragments in a ringlike formation; Bonnell et al. (1991), Boss (1991), and Chapman et al. (1992) find hierarchical multiple systems; Bonnell et al. (1992) find fragmentation of a disk induced by the tidal perturbation from a close companion; Monaghan (1994) finds multiple fragmentation in long filaments; Burkert and Bodenheimer (1993) find that a bar connecting two binary components fragments into several pieces along its length; Boss (1996) and Klapp and Sigalotti (1998) find that a small cluster develops; Bonnell and Bate (1994*a*) and Burkert and Bodenheimer (1996) find that an interior binary can induce further fragmentation in a circumbinary disk. Problems concerning numerical resolution, which may affect many of these studies, are discussed in section V.

Two major theoretical problems have been discussed in connection with these simulations. First, strongly peaked, centrally condensed density distributions are relatively stable to fragmentation. In particular, cores with $\rho \propto r^{-2}$ and uniform rotation (Tsai and Bertschinger 1989), which have been observed in some star-forming regions (section II), appear to be stable against fragmentation. This conclusion is particularly perplexing in view of the fact that all star-forming regions exhibit a substantial fraction of binary systems (Mathieu 1994; and the chapter by Mathieu et al., this volume). The discrepancy has been alleviated to some extent by (1) observations that at least some prestellar cores have density distributions that become relatively flat near the center (André et al. 1996) and (2) numerical simulations (Burkert et al. 1997) in which an initial density profile $\rho \propto r^{-1}$ with uniform rotation does fragment. The other significant problem is that most simulations produce wide binaries, whereas close ($D \leq 1$ AU) binaries appear to be difficult to form. Several mechanisms have been suggested. A promising one is the orbital decay of a wide binary by gravitational torques exerted on it by surrounding spiral arms in the circumbinary disk (Artymowicz et al. 1991). Numerical calculations have demonstrated that fission is extremely difficult (section IV.A). Hierarchical fragmentation has not been studied with sufficient numerical resolution to determine whether it is a likely process. One mode of such fragmentation would involve subfragmentation in the adiabatic phase of fragments formed in the isothermal phase; however, the range of initial conditions for the subfragmentation is quite limited (section III.A). Fragmentation during the second collapse phase (section III.B) has been suggested (Larson 1972; Bonnell and Bate 1994*b*), but more recent 3D simulations by Bate (1998) indicate that fragmentation may not occur during this phase because of prior efficient transfer of angular momentum by gravitational torques. Further investigation is also needed on other proposed mechanisms, such as fragmentation in a cloud with low initial angular momentum, disk fragmentation induced by a companion, or gravitational interactions in a compact small cluster of fragments (see Bodenheimer 1995).

V. NUMERICAL CALCULATIONS

A. Introduction

The physics of gravitational collapse and fragmentation is essential for obtaining an understanding of the formation of stars and galaxies, yet gaseous flows undergoing fragmentation naturally develop a substantial variation in lengthscale, up to a factor of 10^4 or more. This enormous dynamic range presents a formidable obstacle to obtaining an accurate numerical solution, because the flow must remain well resolved throughout the evolution. Fixed-resolution methods are not efficient for such simulations; variable-resolution methods are required. This section discusses a minimum requirement on resolution, known as the Jeans condition, and then explores a series of highly resolved test cases, carried out to determine whether two or more independent codes reach the same result on the same problem.

Four basic approaches have been developed for employing variable resolution in the solution of 3D self-gravitational problems. The SPH technique is a gridless, Lagrangian technique in which gravitating particles of fixed mass carry fluid properties (see, e.g., Monaghan 1992). SPH provides a means of achieving a large dynamic range in scale that is much easier to implement than many of the high-order finite-difference approaches. It concentrates its resolution and computational effort in high-density regions. Also, it is easy to track the transfer of angular momentum for a particular mass element. However, for similar computational times, SPH provides poorer resolution of shocks and poorer resolution of low-density regions than do grid-based codes (Kang et al. 1994). Shocks are an integral feature of calculations of astrophysical hydrodynamics, and this restricts the utility of SPH for such problems (Klein and McKee 1994). In addition, although the spatial redistribution of particles at each step yields a variable spatial resolution, the mass resolution of SPH codes remains fixed. Recent SPH work in the study of star formation has been carried out by Bate and Burkert (1997). A second approach to achieve variable resolution is simply to use a grid in spherical coordinates, which naturally affords increased resolution in the form of decreased cell volume about the origin. When coupled with a nonuniform radial cell dimension and rezoning, this method is useful for studying collapse that produces structure near the origin (Boss 1996; Sigalotti and Klapp 1997). A third approach, a multiple-grid method, is utilized by Burkert and Bodenheimer (1993, 1996). In this scheme, a series of smaller grids of increasingly finer resolution is concentrically positioned about the center of the computational volume. These grids remain fixed throughout the calculation. In typical problems, this method overresolves the cloud center during the early stages of collapse and may underresolve structure outside the preset locations of the finer grids. The fourth approach, called adaptive mesh refinement (AMR), was introduced into astrophysical problems by Klein et al. (1990) and Klein and McKee (1994) and generalized to include self-gravity by Truelove et al. (1997, 1998). This scheme uses grids at multiple levels of resolution. Linear resolution varies by integral refinement factors (usually 4) between levels, and a given subgrid is always fully contained within one at the next coarser level. The method stems from the seminal work of Berger and Oliger (1984). The AMR method can employ multiple spatially unconnected grids at a given level of refinement. Most importantly, the AMR method dynamically resizes and repositions these grids and inserts new, finer ones within them according to adjustable refinement criteria. Fine grids are automatically removed as flow conditions require less resolution. Thus, AMR naturally follows developing structure and fronts, placing high resolution where it is needed most. This powerful approach, when combined with an underlying higher-order hydrodynamics method such as the Godunov scheme, provides a natural means of resolving structure accurately over an enormous dynamic range in scale and also has the ability to follow strong fronts with great economy.

B. Jeans Condition

Recently Truelove et al. (1997, 1998) introduced the Jeans condition as an important grid refinement criterion for self-gravitational hydrodynamics to avoid artificial fragmentation in numerical simulations. This condition arises because perturbations on scales above the Jeans length, $\lambda_J \equiv (\pi c_s^2/G\rho)^{1/2}$, are physically unstable (Jeans 1902, 1928). Discretization of the equations for self-gravitational hydrodynamics introduces perturbations on scales of the order of the grid resolution Δx. If $\Delta x > \lambda_J$, these perturbations might result in artificial fragmentation. It is therefore essential to keep λ_J as resolved as possible in order to diminish the initial amplitude of perturbations that exceed this scale. Resolution of λ_J also suppresses numerical effects that simulate a viscosity and its accompanying effect of artificially slowing the collapse.

Defining the Jeans number $J \equiv (\Delta x/\lambda_J)$, Truelove et al. (1997) found that keeping $J \leq 0.25$ avoided artificial fragmentation in the particular case of an isothermal collapse spanning seven decades of density, the approximate range separating typical molecular cloud cores from nonisothermal protostellar fragments. It is important to note, however, that the Jeans condition is a necessary but not, in general, sufficient condition to ensure convergence. It is necessary to perform an appropriate convergence study to determine the necessary and sufficient value for J required to avoid artificial fragmentation in a particular simulation.

For a uniform Cartesian grid, the condition $J \leq 0.25$ is equivalent to requiring that the mass of a cell should not exceed $\frac{1}{64}$ of the Jeans mass (note that the precise value required will depend on the problem being considered). Boss (1998) extended the Jeans condition to the case of a nonuniformly spaced, spherical coordinate grid, where there are three grid spacings, $\Delta x_r = \Delta r$, $\Delta x_\theta = r\Delta\theta$, and $\Delta x_\phi = r \sin\theta\Delta\phi$, and a fourth grid length that is related to the cell mass, $\Delta x = (\Delta x_r \Delta x_\theta \Delta x_\phi)^{1/3}$. Boss (1998) found that as long as $\Delta x < \lambda_J/4$, artificial fragmentation could be avoided, even if Δx_r, Δx_θ, or Δx_ϕ exceeded $\lambda_J/4$, implying the primacy of a Jeans-mass condition over a Jeans-length condition for a spherical grid; for a Cartesian grid they are equivalent. A closely related Jeans condition has been developed for SPH calculations (Bate and Burkert 1997). Here the minimum resolvable mass must be less than the Jeans mass. In practice, if $N_{\rm neigh}$ is the number of particles within two smoothing lengths (typically 50), then $2N_{\rm neigh}$ times the particle mass must be less than the Jeans mass.

C. Code Intercomparisons for Isothermal Collapse

The uniform-density cloud studied by Burkert and Bodenheimer (1993), Bate and Burkert (1997), and Truelove et al. (1998), and the Gaussian profile studied by Boss (1991, 1993, 1998), Klapp et al. (1993), Burkert and Bodenheimer (1996), and Truelove et al. (1997) have emerged as important benchmark test cases for code intercomparisons because of the diversity of the results obtained. The test involving the uniform cloud has the following parameters: $M = 1\,{\rm M}_\odot$, $R = 5 \times 10^{16}$ cm, $\rho_0 = 10^{-17.4}$ g cm^{-3}, $\alpha = 0.26$, $\beta = 0.16$, $c_s = 0.167$ km s^{-1}, and a rotation rate $\Omega = 7.2 \times 10^{-13}$ rad s^{-1}. The cloud is perturbed by $\rho \rightarrow \rho \times [1 + 0.1\cos(2\phi)]$, an $m = 2$ mode with 10% amplitude. This cloud begins with $\lambda_J = 1.17R$.

Burkert and Bodenheimer (1993) calculated the isothermal collapse over a density increase of more than six decades. The result was two local density maxima (a binary) separated by about 10^{16} cm. A bar subsequently formed between the components and fragmented into nine small condensations. Truelove et al. (1998) attacked this problem using an initial resolution of R_{32} (32 cells per initial cloud radius) and dynamically refined it so as to ensure $J_{\max} = 0.25$ throughout the simulation. They found a binary with characteristics similar to that of Burkert and Bodenheimer (1993). Each fragment (Color Plate 9) has a mass $M_f = 0.032\ M_\odot$, with a binary separation of 8.6×10^{15} cm.

Truelove et al. (1998) continued the calculation until ρ had increased by 8.1 decades. At this point, the code used resolution at the level of R_{131072}, representing six levels of refinement beyond the original R_{32}; the finest cells are only 5.4 solar radii in size. Each member of the binary continues to collapse into a thinner and thinner filamentary structure and does not subfragment. Thus, the solution to the purely isothermal, inviscid collapse is a pair of singular filaments.

As in Burkert and Bodenheimer (1993), the binary is connected by a thin bar (Color Plate 9). The bar is also expected to collapse to a linear singularity in the absence of "arresting" agents of heating and viscosity (both numerical and artificial). The high-resolution calculations show no evidence of subfragmentation in the bar, in contrast to the results of Burkert and Bodenheimer (1993), who used fixed finest resolution and thus eventually violated the Jeans condition in the highest-density zones. The resulting poor resolution most likely slowed the collapse of the bar onto itself and allowed numerical perturbations to grow. Thus, in the pure isothermal case the bar fragmentation is most likely a numerical artifact, although in calculations in which heating is included above $\rho_{\rm crit}$ and in which the Jeans condition is satisfied, the collapse of the bar is slowed and fragmentation does result (Bate and Burkert 1997).

Recent work on the same problem by Boss (Fig. 1), with sufficiently high resolution in his spherical code to satisfy the Jeans condition for most of the simulation, finds excellent agreement with Truelove et al. (1998), with a basic morphology of a thin bar with binary fragments. In a lower-resolution calculation in which the Jeans condition is not satisfied in the vicinity of the forming binary pair, the bar begins to develop a third (low-mass) clump at its center. Thus, it appears that in this particular test case all groups find a binary pair connected by a bar with similar morphology. Multiple fragmentation along the bar is not found in purely isothermal calculations that satisfy the Jeans condition.

The Gaussian cloud has initial conditions identical to those of the uniform cloud, except that the radial density dependence is $\rho(r) = \rho_c e^{-(r/R_1)^2}$, where $\rho_c = 10^{-16.8}$g cm^{-3} and $R_1 = 0.58R$. The sound speed $c_s = 0.19$ km s^{-1}. Boss (1991, 1993) found the Gaussian cloud to fragment into a binary, then into a hierarchical quadruple system. Klapp et al. (1993) and Burkert and Bodenheimer (1996) also obtained the quadruple system, but in the latter paper the quadruple then evolved into an inner triple and an outer binary. Truelove et al. (1997) found that

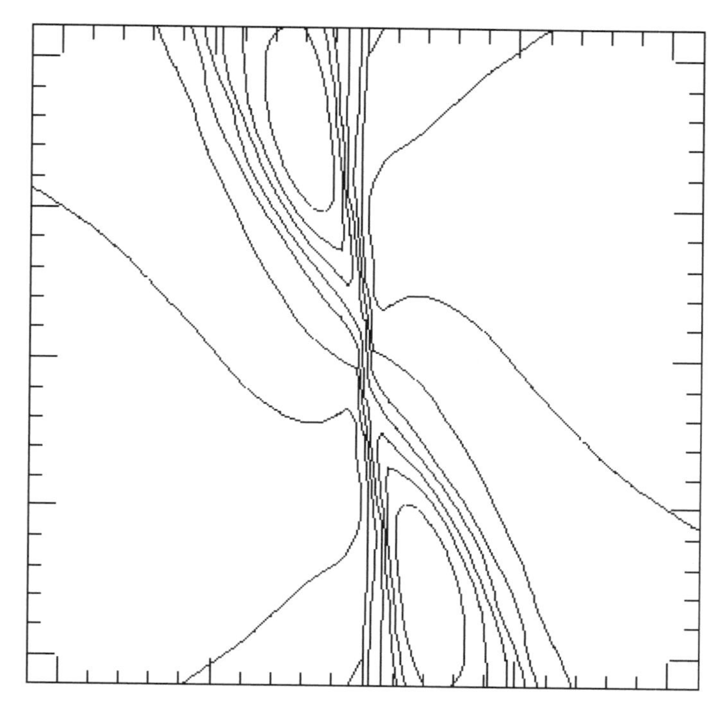

Figure 1. Density contours in the equatorial plane at time $t = 1.376 \times 10^{12}$ s for the same cloud that is shown in Color Plate 9, calculated with the Boss code. A binary protostar has formed. The maximum density $= 6.3 \times 10^{-14}$ g cm^{-3}, the box radius $= 5.8 \times 10^{15}$ cm, and the contours are separated by factors of 2.

with their highest spatial resolution, ensuring that $\Delta x < \lambda_J/4$, the Gaussian cloud collapsed to form a very thin filament without fragmentation (Color Plate 10). Boss (1998) obtained a binary system while maintaining $\Delta x < \lambda_J/4$.

Clearly, the earlier calculations did not have sufficient resolution to satisfy the Jeans condition. The main reason the Burkert and Bodenheimer (1996) results differ from those of Truelove et al. (1997) is that in the former simulation an artificial viscosity was specifically introduced to halt the collapse of the very densest regions, to allow the calculations to continue to later times. The effect is roughly similar to that of including heating. However, in the pure isothermal case the collapse of the filament cannot be stopped, and it continues, on ever shorter timescales, until a singularity is reached.

The isothermal collapse of the Gaussian cloud has been recalculated by Boss with considerably higher spatial resolution than he previously used. The grid is moved inward during the collapse, keeping all four spherical Jeans lengths less than $\lambda_J/4$. The result is shown in Fig. 2. At a

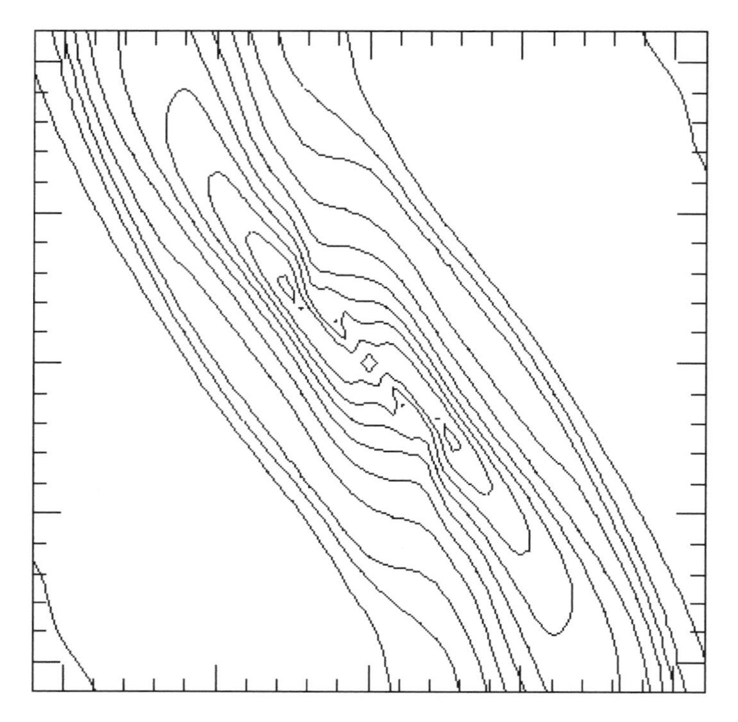

Figure 2. Density contours in the equatorial plane at $t = 7.050 \times 10^{11}$ s, for the same cloud that is shown in Color Plate 10, calculated with several times the spatial resolution of model I of Boss (1998). An isothermal filament has formed with maximum density $= 4.0 \times 10^{-11}$ g cm^{-3}. The box radius is 3.0×10^{14} cm, and the contours are separated by factors of 2.

maximum density comparable to that of the AMR calculation, a collapsing isothermal bar is seen with similar morphology.

Thus, it appears that the evolutionary outcome of the isothermal collapse of a Gaussian cloud is the singular isothermal filament. This result is consistent with Inutsuka and Miyama (1992, 1997), who show that perturbed, unstable, inviscid, isothermal clouds form filaments that increase in density faster than linear perturbations along them can grow to fragments. Neither set of calculations shows the complex multiple fragmentation seen, for example, by Burkert and Bodenheimer (1996). However, the latter case is not comparable, because the isothermal assumption was in fact broken by the introduction of artificial viscosity, which allowed the calculation to proceed to a time beyond the bar formation stage.

D. Uniform Clouds with Heating

Recent work using SPH (Bate and Burkert 1997) and high-resolution AMR (Fisher et al. 1998; Klein 1998; Klein et al. 1999) have followed the collapse of an initially rigidly rotating, uniform isothermal cloud. They use an equation of state that makes the transition from an isothermal equation in the optically thin regime to a polytrope in the optically thick regime in a smooth fashion. The initial conditions are identical to the isothermal uniform cloud previously discussed, including the $m = 2$ perturbation of 10% amplitude. The AMR calculations of Klein (1998) and Fisher et al. (1998) followed the subsequent evolution of the fragments over dynamical (orbital) timescales, while still adhering to the Jeans condition. The cloud initially collapses to an isothermal disk, a strong isothermal shock above the disk plane is established, and an elongated filamentary bar forms, with the first signs of fragmentation in it occurring by 1.41×10^{12} s. At about $t = 1.46 \times 10^{12}$ s, the isothermal bar becomes optically thick, and the accretion flow onto the bar is arrested because of pressure effects, resulting in the growth of nonaxisymmetric perturbations in the bar. Fragmentation in the bar results in the formation of binary, spherical cores. The core-bar system is embedded in an outer two-armed spiral, derivative of the initial $m = 2$ perturbation. The low angular momentum of shocked gas accreted in the bar allows the bar to be directly accreted onto the cores. The binary separation decreases as this mass is accreted, and the bar starts to dissipate.

Later, at $t = 1.51 \times 10^{12}$ s, protostellar disks have formed around the cores; the cores continue to grow by direct accretion from these surrounding disks. The disks are attached to the long outer spiral. The scale of the disks, of order 100 AU, is consistent with observations of gaseous disks surrounding single T Tauri stars and debris disks surrounding systems such as β Pictoris. The situation is shown at $t = 1.6 \times 10^{12}$ s (Color Plate 11). The cores, spiral arms, and disks have 20%, 27%, and 2%, respectively, of the total mass of the cloud at this time. The fact that the initial bar did not fragment further after binary formation is consistent with the results of Bate and Burkert (1997), who, using a slightly different equation of state, found that such fragmentation occurred only if $\rho_{crit} > 3 \times 10^{-14}$ g cm^{-3}.

E. Power Law Clouds with Heating

A further significant test case, involving the comparison of an SPH code with a grid code, was provided by Burkert et al. (1997). This work also considers the problem of fragmentation in clouds with significant density concentration toward the center. The initial core had mass 1.0 $M_\odot$, $R = 5 \times 10^{16}$ cm, and a density distribution $\rho \propto r^{-1}$. They showed that with an initial density perturbation of the form $m = 2$ and a 10% amplitude, and with $\alpha = 0.25$, $\beta = 0.23$, fragmentation into four objects occurs in the inner region of the cloud, on a scale of 100 AU. The calculation was done with an Eulerian grid code, with fixed nested grids such that the resolution on the innermost grid was $10^{-3}R$, where R is the total radius of the cloud. The same problem was calculated by use of an SPH code with 200,000 particles. The resolution element in such a calculation is the smoothing length, which varies through the volume; in this case it

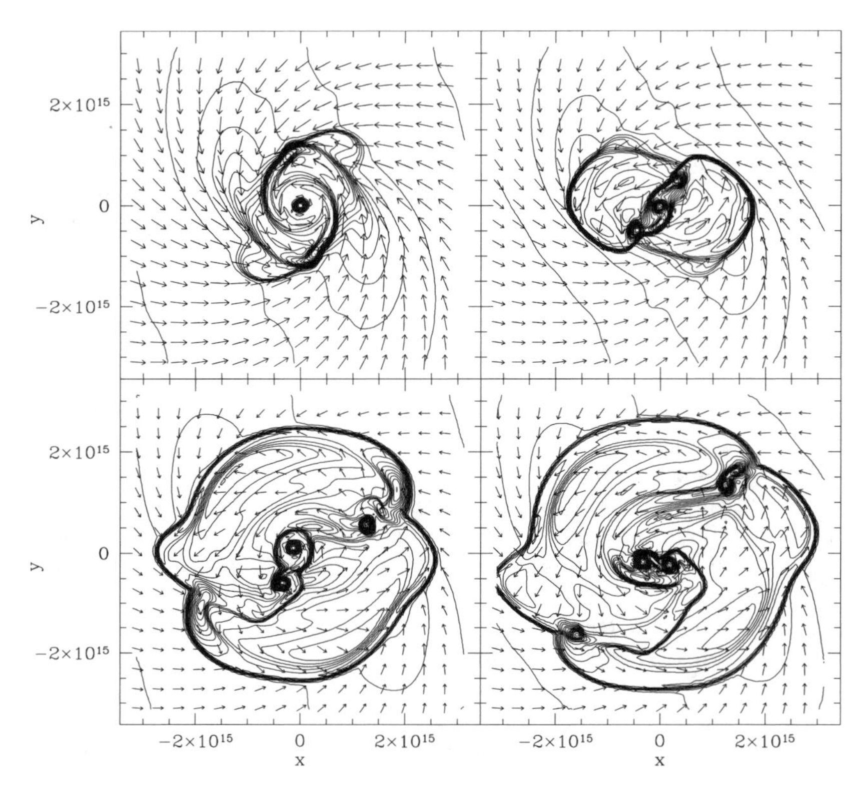

Figure 3(a). Contours of equal density in the equatorial (x, y) plane at four times during the fragmentation of a cloud with an initial $\rho \propto r^{-1}$, calculated with a nested grid code. Velocity vectors are shown with length proportional to speed. The times, maximum densities, contour intervals, and maximum velocities are as follows: (upper left) $t = 1.176 \times 10^{12}$ s; log $\rho_{\rm max} = -10.3$; Δ log $\rho = 0.166$; $V_{\rm max} = 1.95 \times 10^5$ cm s^{-1}. (upper right) $t = 1.190 \times 10^{12}$ s; log $\rho_{\rm max} = -10.2$; Δ log $\rho = 0.17$; $V_{\rm max} = 2.57 \times 10^5$ cm s^{-1}. (lower left) $t = 1.209 \times 10^{12}$ s; log $\rho_{\rm max} = -10.1$; Δ log $\rho = 0.173$; $V_{\rm max} = 3.17\times 10^5$ cm s^{-1}. (lower right) $t = 1.217 \times 10^{12}$ s; log $\rho_{\rm max} = -10.1$; Δ log $\rho = 0.173$; $V_{\rm max} = 3.67 \times 10^5$ cm s^{-1}. From Burkert et al. (1997).

was adjusted to be $10^{-3}R$ at a density of 10^{-12} g cm^{-3}, which is about the point where fragmentation starts. However, the spatial resolution of the two codes can be quite different at other densities. Figure 3(a) shows the onset of fragmentation in the grid code, in a region of the core that contains about 12% of the total mass. Interactions among the spiral arms result in the formation of an inner unstable triple, which ejects one component, which later merges with one of the binary components that has formed in the outer spiral arm. At the end of the simulation the inner binary has masses 0.055 and 0.025 M$_\odot$, and the outer binary has masses 0.017 and 0.011 M$_\odot$. Clearly there will be further orbital evolution and

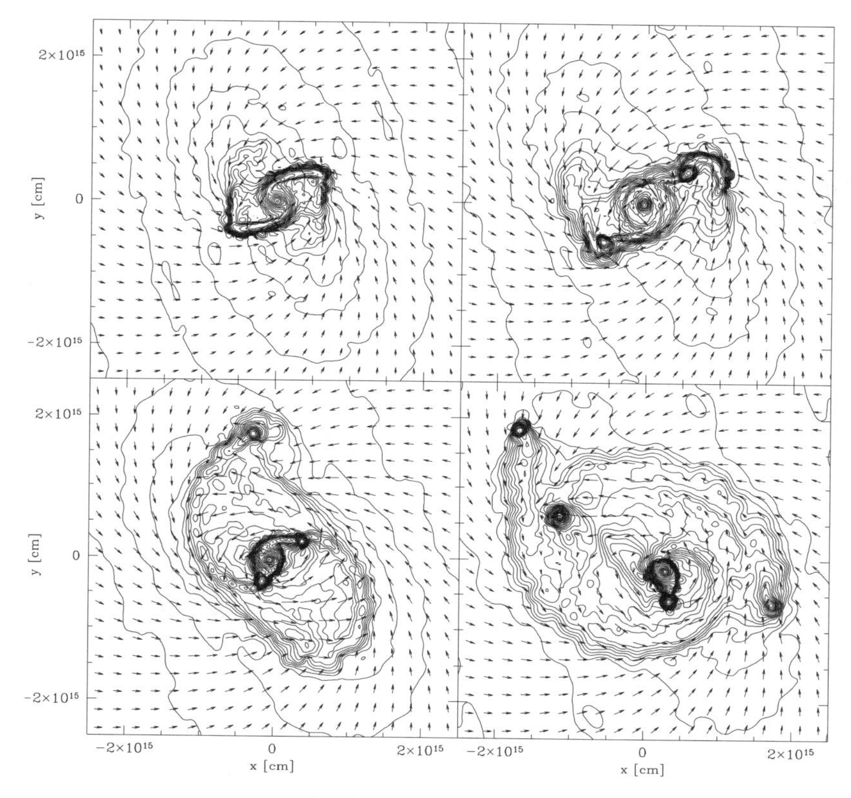

Figure 3(b). The evolution of the same cloud that is shown in part (a), calculated with an SPH code. Symbols and curves have the same meaning as in part (a). (upper left) $t = 1.135 \times 10^{12}$ s; log $\rho_{max} = -10.0$; Δ log $\rho = 0.25$; $V_{max} = 2.44 \times 10^5$ cm s^{-1}. (upper right) $t = 1.160 \times 10^{12}$ s; log $\rho_{max} = -9.7$; Δ log $\rho = 0.25$; $V_{max} = 2.88 \times 10^5$ cm s^{-1}. (lower left) $t = 1.175 \times 10^{12}$ s; log $\rho_{max} = -9.6$; Δ log $\rho = 0.25$; $V_{max} = 3.66 \times 10^5$ cm s^{-1}. (lower right) $t = 1.190 \times 10^{12}$ s; log $\rho_{max} = -9.5$; Δ log $\rho = 0.25$; $V_{max} = 3.38 \times 10^5$ cm s^{-1}. From Burkert et al. (1997).

accretion with possible mergers, further fragmentation, and ejections. Figure 3(b) shows the parallel evolution in the SPH code, which shows fragmentation on the same scale and by the same process but with some differences in detail. Both calculations satisfy the Jeans condition, although the form assumed for the heating above ρ_{crit} was somewhat different in the two calculations.

VI. SUMMARY

The goal of detailed fragmentation calculations is to explain the properties of the multiple systems that are described in section I. Although various plausible ideas have been investigated, an overall comparison between theory and observations has not yet been established. It is true that the relatively broad distribution of rotational velocities of molecular cloud cores could in part explain the broad period distribution of binary stars, but there is probably not a one-to-one correspondence between cloud spin angular momentum and resulting binary orbital angular momentum. Orbital interactions in a system of multiple fragments in a dense protostellar cluster could also play a role in the determination of the period distribution. Such interactions could also be responsible for the broad eccentricity distribution, although theoretical models have not progressed far enough to provide a statistical comparison. It is particularly difficult for theory to explain the very closest binary systems; possibilities include orbital decay of long-period systems, hierarchical fragmentation, or three-body captures in a dense system of protostellar fragments. The distribution of mass ratios likewise has not been well explained. Although numerical simulations do show that fragments with a mass difference of roughly a factor of 10 can form in a collapsing cloud, it is not clear how the mass ratio depends on initial conditions or how the mass ratio will evolve as a result of subsequent accretion (see the chapter by Lubow and Artymowicz, this volume). Very generally, the observational evidence is consistent with the idea that fragmentation during collapse of a rotating cloud is the basic process.

A further important issue is the binary frequency and the conditions under which single stars are formed. It has been suggested (chapter by Mathieu et al., this volume) that the density of the star-forming region has an effect on the binary fraction. However, in a given star formation region it is still not clear under what conditions single stars are formed. The possibilities include (1) ejection from a system of multiple fragments as a result of a close encounter, (2) a low-angular-momentum initial cloud that fails to fragment in the isothermal collapse phase and evolves to a single stellar core plus disk, (3) variations in the initial conditions of standard cloud cores, which on the average, observationally, have $\alpha = 0.4$ and $\beta = 0.01$. Boss (1993) calculated the collapse of initially prolate clouds with α and β in this range. Initial axis ratios of 1.5 and 2 led to an outcome of a single star and a binary, respectively. All these possibilities require further investigation.

What is the future for fragmentation calculations? Computing power is now becoming sufficient that, coupled with the numerical techniques developed over the past few years, the demanding three-dimensional collapse calculations can be performed with adequate spatial resolution and improved physics. Actual cloud cores are not in general represented by the highly idealized and symmetric initial conditions assumed in many of the test calculations done so far. Initial clouds with irregular shapes, with a residual spectrum of turbulent velocities, and with observed values of α and β should be considered. The earlier parameter surveys with radiation transport should be recalculated with improved spatial resolution, because the transition from the isothermal regime to the adiabatic regime is of particular significance with regard to the outcome of fragmentation. Of less importance are magnetic fields, which have possible influence on the earlier stages of collapse but probably not on the actual fragmentation phase. Finally, a much closer connection between the results of theoretical calculations and observational quantities must be established. Detailed line profiles for high-density tracers, such as CS and H_2CO, which are being used to establish the existence of collapse as well as rotational motions in protostars, should be calculated from the 3D models. Further diagnostics include synthetic infrared spectral energy distributions as well as continuum maps that correspond to the observed bands in the submillimeter and millimeter regions. Such detailed observational comparisons should provide further clues as to the origin of multiple systems.

REFERENCES

Adams, F. C., and Benz, W. 1992. Gravitational instabilities in circumstellar disks and the formation of binary companions. In *Complementary Approaches to Double and Multiple Star Research*, ed. H. McAllister and W. Hartkopf (San Francisco: Astronomical Society of the Pacific), pp. 185–194.

Adams, F. C., Ruden, S. P., and Shu, F. H. 1989. Eccentric gravitational instabilities in nearly Keplerian disks. *Astrophys. J.* 347:959–975.

Alexander, D. R., and Ferguson, J. W. 1994. Low-temperature Rosseland opacities. *Astrophys. J.* 437:879–891.

André, P., Ward-Thompson, D., and Motte, F. 1996. Probing the initial conditions of star formation: The structure of the prestellar core L1689B. *Astron. Astrophys.* 314:625–635.

Artymowicz, P., Clarke, C. J., Lubow, S. H., and Pringle, J. E. 1991. The effect of an external disk on the orbital elements of a central binary. *Astrophys. J. Lett.* 370:L35–L38.

Basu, S., and Mouschovias, T. C. 1994. Magnetic braking, ambipolar diffusion, and the formation of cloud cores and protostars. I. Axisymmetric solutions. *Astrophys. J.* 432:720–741.

Bate, M. R. 1998. Collapse of a molecular cloud core to stellar densities: The first three-dimensional calculations. *Astrophys. J. Lett.* 508:L95–L98.

Bate, M. R., and Burkert, A. 1997. Resolution requirements for smoothed particle hydrodynamics calculations with self-gravity. *Mon. Not. Roy. Astron. Soc.* 288:1060–1072.

Bate, M. R., Bonnell, I., and Price, N. 1995. Modelling accretion in protobinary systems. *Mon. Not. Roy. Astron. Soc.* 277:362–376.

Benson, P. J., and Myers, P. C. 1989. A survey for dense cores in dark clouds. *Astrophys. J. Suppl.* 71:89–108.

Berger, M. J., and Oliger, J. 1984. Adaptive mesh refinement for hyperbolic partial differential equations. *J. Comput. Phys.* 53:484–512.

Bodenheimer, P. 1995. Angular momentum evolution of young stars and disks. *Ann. Rev. Astron. Astrophys.* 33:199–238.

Bodenheimer, P., Ruzmaikina, T., and Mathieu, R. 1993. Stellar multiple systems: Constraints on the mechanism of origin. In *Protostars and Planets III*, ed. E. H. Levy and J. I. Lunine (Tucson: University of Arizona Press), pp. 367–404.

Boffin, H. M. J., Watkins, S. J., Bhattal, A. S., Francis, N., and Whitworth, A. P. 1998. Numerical simulations of protostellar encounters. I. Star–disc encounters. *Mon. Not. Roy. Astron. Soc.* 300:1189–1204.

Bonnell, I. 1994. A new binary formation mechanism. *Mon. Not. Roy. Astron. Soc.* 269:837–848.

Bonnell, I., and Bastien, P. 1992. Fragmentation of elongated cylindrical clouds. V. Dependence of mass ratios on initial conditions. *Astrophys. J.* 401:654–666.

Bonnell, I., and Bate, M. R. 1994*a*. Massive circumbinary disks and the formation of multiple systems. *Mon. Not. Roy. Astron. Soc.* 269:L45–L48.

Bonnell, I., and Bate, M. R. 1994*b*. The formation of close binary systems. *Mon. Not. Roy. Astron. Soc.* 271:999–1004.

Bonnell, I., Martel, H., Bastien, P., Arcoragi, J.-P., and Benz, W. 1991. Fragmentation of elongated cylindrical clouds. III. Formation of binary and multiple systems. *Astrophys. J.* 377:553–558.

Bonnell, I., Arcoragi, J.-P., Martel, H., and Bastien, P. 1992. Fragmentation of elongated cylindrical clouds. IV. Clouds with solid-body rotation about an arbitrary axis. *Astrophys. J.* 400:579–594.

Boss, A. P. 1981. Collapse and fragmentation of rotating adiabatic clouds. *Astrophys. J.* 250:636–644.

Boss, A. P. 1988*a*. Protostellar formation in rotating interstellar clouds. VII. Opacity and fragmentation. *Astrophys. J.* 331:370–376.

Boss, A. P. 1988*b*. Binary stars: Formation by fragmentation. *Comment. Astrophys.* 12:169–190.

Boss, A. P. 1991. Formation of hierarchical multiple protostellar cores. *Nature* 351:298–300.

Boss, A. P. 1993. Collapse and fragmentation of molecular cloud cores. I. Moderately centrally condensed cores. *Astrophys. J.* 410:157–167.

Boss, A. P. 1996. Collapse and fragmentation of molecular cloud cores. IV. Oblate clouds and small cluster formation. *Astrophys. J.* 468:231–240.

Boss, A. P. 1997*a*. Collapse and fragmentation of molecular cloud cores. V. Loss of magnetic field support. *Astrophys. J.* 483:309–319.

Boss, A. P. 1997*b*. Giant planet formation by gravitational instability. *Science* 276:1836–1839.

Boss, A. P. 1998. The Jeans mass constraint and the fragmentation of molecular cloud cores. *Astrophys. J. Lett.* 501:L77–L81.

Bouvier, J., Rigaut, F., and Nadeau, D. 1997. Pleiades low-mass binaries: Do companions affect the evolution of protoplanetary disks? *Astron. Astrophys.* 323:139–150.

Brandner, W., and Köhler, R. 1998. Star formation environments and the distribution of binary separations. *Astrophys. J. Lett.* 499:L79–L82.

Brandner, W., and Zinnecker, H. 1997. Physical properties of 90 AU to 250 AU pre-main-sequence binaries. *Astron. Astrophys.* 321:220–228.

Burkert, A., and Bodenheimer, P. 1993. Multiple fragmentation in collapsing protostars. *Mon. Not. Roy. Astron. Soc.* 264:798–806.

Burkert, A., and Bodenheimer, P. 1996. Fragmentation in a centrally condensed protostar. *Mon. Not. Roy. Astron. Soc.* 280:1190–1200.

Burkert, A., Bate, M. R., and Bodenheimer, P. 1997. Protostellar fragmentation in a power-law density distribution. *Mon. Not. Roy. Astron. Soc.* 289:497–504.

Cassen, P. M., Smith, B. F., Miller, R. H., and Reynolds, R. T. 1981. Numerical experiments on the stability of preplanetary disks. *Icarus* 48:377–392.

Chapman, S., Pongracic, H., Disney, M., Nelson, A., Turner, J., and Whitworth, A. 1992. The formation of binary and multiple star systems. *Nature* 359:207–210.

Ciolek, G., and Königl, A. 1998. Dynamical collapse of nonrotating magnetic molecular cloud cores: Evolution through point-mass formation. *Astrophys. J.* 504:257–279.

Ciolek, G., and Mouschovias, T. C. 1995. Ambipolar diffusion, interstellar dust, and the formation of cloud cores and protostars. IV. Effect of ultraviolet ionization and magnetically controlled infall rate. *Astrophys. J.* 454:194–216.

Clarke, C. J., and Pringle, J. E. 1991. Star-disk interactions and binary star formation. *Mon. Not. Roy. Astron. Soc.* 249:584–587.

de Jong, T., Dalgarno, A., and Boland, W. 1980. Hydrostatic models of molecular clouds. *Astron. Astrophys.* 91:68–84.

Duquennoy, A., and Mayor, M. 1991. Multiplicity among solar-type stars in the solar neighbourhood. II. Distribution of the orbital elements in an unbiased sample. *Astron. Astrophys.* 248:485–524.

Fischer, D. A., and Marcy, G. W. 1992. Multiplicity among M dwarfs. *Astrophys. J.* 396:178–194.

Fisher, R., Klein, R. I., and McKee, C. F. 1998. In preparation.

Fuller, G. A., and Myers, P. C. 1992. Dense cores in dark clouds. VII. Line width–size relation. *Astrophys. J.* 384:523–527.

Fuller, G. A., Ladd, E. F., and Hodapp, K.-W. 1996. Lynds 1527: An embedded protobinary system in Taurus. *Astrophys. J. Lett.* 463:L97–L100.

Ghez, A. M., McCarthy, D. W., Patience, J. L., and Beck, T. L. 1997. The multiplicity of pre-main-sequence stars in southern star-forming regions. *Astrophys. J.* 481:378–385.

Goodman, A., Benson, P., Fuller, G., and Myers, P. 1993. Dense cores in dark clouds. VIII. Velocity gradients. *Astrophys. J.* 406:528–547.

Hachisu, I., Tohline, J. E., and Eriguchi, Y. 1987. Fragmentation of rapidly rotating gas clouds. I. A universal criterion for fragmentation. *Astrophys. J.* 323:592–613.

Hartigan, P., Strom, K. M., and Strom, S. E. 1994. Are wide pre-main-sequence binaries coeval? *Astrophys. J.* 427:961–977.

Hayashi, C. 1966. Evolution of protostars. *Ann. Rev. Astron. Astrophys.* 4:171–192.

Hayashi, C., Narita, S., and Miyama, S. M. 1982. Analytic solutions for equilibrium of rotating isothermal clouds. *Prog. Theor. Phys.* 68:1949–1966.

Heemskerk, M. H. M., Papaloizou, J. C. B., and Savonije, G. J. 1992. Non-linear development of $m = 1$ instabilities in a self-gravitating gaseous disk. *Astron. Astrophys.* 260:161–174.

Henriksen, R., André, P., and Bontemps, S. 1997. Time-dependent accretion and ejection implied by pre-stellar density profiles. *Astron. Astrophys.* 323:549–565.

Hujeirat, A. 1998. Ambipolar diffusion in star-forming clouds. *Astron. Astrophys.* 334:742-745.

Inutsuka, S., and Miyama, S. M. 1992. Self-similar solutions and the stability of collapsing isothermal filaments. *Astrophys. J.* 388:392–399.

Inutsuka, S., and Miyama, S. M. 1997. A production mechanism for clusters of dense cores. *Astrophys. J.* 480:681–693.

Jeans, J. H. 1902. The stability of a spherical nebula. *Phil. Trans. A* 199:1–53.

Jeans, J. H. 1928. *Astronomy and Cosmogony* (London: Cambridge University Press).

Kang, H., Ostriker, J. P., Cen, R., Ryu, D., Hernquist, L., Evrard, A. E., Bryan, G. L., and Norman, M. L. 1994. A comparison of cosmological hydrodynamic codes. *Astrophys. J.* 430:83–100.

Klapp, J., and Sigalotti, L. DiG. 1998. Collapse and fragmentation models of oblate molecular cloud cores. III. Formation of small protostellar clusters. *Astrophys. J.* 504:158–169.

Klapp, J., Sigalotti, L. DiG., and DeFelice, F. 1993. Formation of multiple protostellar systems. *Astron. Astrophys.* 273:175–184.

Klein, R. I. 1998. Self-gravitational hydrodynamics with 3-D adaptive mesh refinement: The collapse and fragmentation of molecular clouds. *J. Comput. App. Math.*, in press.

Klein, R. I., and McKee, C. F. 1994. The hydrodynamics of cloud interactions. In *Numerical Simulations in Astrophysics,* ed. J. Franco, S. Lizano, L. Aguilar, and E. Daltabuit (Cambridge: Cambridge University Press), pp. 251–266.

Klein, R. I., and Woods, D. T. 1998. Bending mode instabilities and fragmentation in interstellar cloud collisions: A mechanism for complex structure. *Astrophys. J.* 497:777–799.

Klein, R. I., McKee, C. F., and Colella, P. 1990. Turbulent stripping of interstellar clouds by interaction with supernova remnants. In *The Evolution of the Interstellar Medium*, ASP Conf. Ser. 12, ed. L. Blitz (San Francisco: Astronomical Society of the Pacific), pp. 117–136.

Klein, R. I., Fisher, R., McKee, C. F., and Truelove, K. 1999. Gravitational collapse and fragmentation in molecular clouds with adaptive mesh refinement hydrodynamics. In *Numerical Astrophysics*, ed. S. Miyama, K. Tomisaka, and T. Hanawa (Dordrecht: Kluwer), pp. 131–140.

Köhler, R., and Leinert, C. H. 1998. Multiplicity of T Tauri stars in Taurus after ROSAT. *Astron. Astrophys.* 331:977–988.

Kroupa, P., Tout, C. A., and Gilmore, G. 1990. The low-luminosity stellar mass function. *Mon. Not. Roy. Astron. Soc.* 244:76–85.

Ladd, E. F., Adams, F. C., Casey, S., Davidson, J. A., Fuller, G. A., Harper, D. A., Myers, P. C., and Padman, R. 1991. Far-infrared and submillimeter wavelength observations of star-forming dense cores. II. Images. *Astrophys. J.* 382:555–569.

Larson, R. B. 1969. The dynamics of a collapsing proto-star. *Mon. Not. Roy. Astron. Soc.* 145:271–295.

Larson, R. B. 1972. The collapse of a rotating cloud. *Mon. Not. Roy. Astron. Soc.* 156:437–458.

Larson, R. B. 1978. Calculations of three-dimensional collapse and fragmentation. *Mon. Not. Roy. Astron. Soc.* 184:69–85.

Larson, R. B. 1984. Gravitational torques and star formation. *Mon. Not. Roy. Astron. Soc.* 206:197–207.

Larson, R. B. 1985. Cloud fragmentation and stellar masses. *Mon. Not. Roy. Astron. Soc.* 214:379–398.

Larson, R. B. 1990. Formation of star clusters. In *Physical Processes in Fragmentation and Star Formation*, ed. R. Capuzzo-Dolcetta, C. Chiosi, and A. DiFazio (Dordrecht: Kluwer), pp. 389–400.

Laughlin, G., and Bodenheimer, P. 1994. Nonaxisymmetric evolution in protostellar disks. *Astrophys. J.* 436:335–354.

Laughlin, G., and Różyczka, M. 1996. The effect of gravitational instabilities on protostellar disks. *Astrophys. J.* 456:279–291.

Laughlin, G., Korchagin, V., and Adams, F. C. 1997. Spiral mode saturation in self-gravitating disks. *Astrophys. J.* 477:410–423.

Laughlin, G., Korchagin, V., and Adams, F. C. 1998. The dynamics of heavy gaseous disks. *Astrophys. J.* 504:945–966.

Lay, O. P., Carlstrom, J. E., and Hills, R. E. 1995. NGC 1333 IRAS 4: Further multiplicity revealed with the CSO-JCMT interferometer. *Astrophys. J. Lett.* 452:L73–L76.

Leinert, C. H., Henry, T., Glindemann, A., and McCarthy, D. W., Jr. 1997. A search for companions to nearby southern M dwarfs with near-infrared speckle interferometry. *Astron. Astrophys.* 325:159–166.

Lemme, C., Walmsley, C. M., Wilson, T. L., and Muders, D. 1995. A detailed study of an extremely quiescent core: L1498. *Astron. Astrophys.* 302:509–520.

Lizano, S., and Shu, F. H. 1989. Molecular cloud cores and bimodal star formation. *Astrophys. J.* 342:834–854.

Looney, L. W., Mundy, L. G., and Welch, W. J. 1997. High-resolution $\lambda = 2.7$ millimeter observations of L1551 IRS 5: A protobinary system? *Astrophys. J. Lett.* 484:L157–L160.

Low, C., and Lynden-Bell, D. 1976. The minimum Jeans mass, or when fragmentation must stop. *Mon. Not. Roy. Astron. Soc.* 176:367–390.

Mathieu, R. 1994. Pre-main-sequence binary stars. *Ann. Rev. Astron. Astrophys.* 32:465–530.

Mazeh, T., Goldberg, D., Duquennoy, A., and Mayor, M. 1992. On the mass ratio distribution of spectroscopic binaries with solar-type primaries. *Astrophys. J.* 401:265–268.

McDonald, J. M., and Clarke, C. J. 1995. The effect of star-disc interactions on the binary mass-ratio distribution. *Mon. Not. Roy. Astron. Soc.* 275:671–684.

Miyama, S. M. 1992. Criteria for the collapse and fragmentation of rotating clouds. *Pub. Astron. Soc. Japan* 44:193–202.

Miyama, S. M., Hayashi, C., and Narita, S. 1984. Criteria for collapse and fragmentation of rotating isothermal clouds. *Astrophys. J.* 279:621–632.

Miyama, S. M., Nakamoto, T., Kikuchi, N., Inutsuka, S., Kobayashi, K., and Takeuchi, T. 1994. The stability of circumstellar disks. In *Numerical Simulations in Astrophysics*, ed. J. Franco, S. Lizano, L. Aguilar, and E. Daltabuit (Cambridge: Cambridge University Press), pp. 305–312.

Monaghan, J. J. 1992. Smoothed particle hydrodynamics. *Ann. Rev. Astron. Astrophys.* 30:543–574.

Monaghan, J. J. 1994. Vorticity, angular momentum, and cloud fragmentation. *Astrophys. J.* 420:692–704.

Monaghan, J. J., and Lattanzio, J. C. 1991. A simulation of the collapse and fragmentation of cooling molecular clouds. *Astrophys. J.* 375:177-189.

Motte, F., André, P., and Neri, R. 1998. The initial conditions of star formation in the ρ Ophiuchi main cloud: Wide-field millimeter continuum mapping. *Astron. Astrophys.* 336:150–172.

Mundy, L. G., Wootten, A., Wilking, B. A., Blake, G. A., and Sargent, A. I. 1992. IRAS 16293-2422: A very young binary system? *Astrophys. J.* 385:306–313.

Myers, P. C. 1985. Molecular cloud cores. In *Protostars and Planets II*, ed. D. C. Black and M. S. Matthews (Tucson: University of Arizona Press), pp. 81–103.

Myers, P. C. 1987. Dense cores and young stars in dark clouds. In *Star Forming Regions*, ed. M. Peimbert and J. Jugaku (Dordrecht: Reidel), pp. 33–44.

Myers, P. C., and Goodman, A. A. 1988. Evidence for magnetic and virial equilibrium in molecular clouds. *Astrophys. J. Lett.* 326:L27–L30.

Myers, P. C., Fuller, G. A., Goodman, A. A., and Benson, P. J. 1991. Dense cores in dark clouds. VI. Shapes. *Astrophys. J.* 376:561–572.

Myhill, E., and Kaula, W. M. 1992. Numerical models for the collapse and fragmentation of centrally condensed molecular cloud cores. *Astrophys. J.* 386:578–586.

Nelson, A. F., Benz, W., Adams, F. C., and Arnett, D. 1998. Dynamics of circumstellar disks. *Astrophys. J.* 502:342–371.

Nelson, R. P., and Papaloizou, J. C. B. 1993. Three-dimensional hydrodynamic simulations of collapsing prolate clouds. *Mon. Not. Roy. Astron. Soc.* 265:905–920.

Papaloizou, J. C. B., and Savonije, G. J. 1991. Instabilities in self-gravitating gaseous disks. *Mon. Not. Roy. Astron. Soc.* 248:353–369.

Patience, J., Ghez, A. M., Reid, I. N., Weinberger, A. J., and Matthews, K. 1998. The multiplicity of the Hyades and its implications for binary star formation and evolution. *Astron. J.* 115:1972–1988.

Petr, M. G., Du Foresto, V. C., Beckwith, S. V. W., Richichi, A., and McCaughrean, M. J. 1998. Binary stars in the Orion Trapezium cluster core. *Astrophys. J.* 500:825–837.

Pollack, J. B., McKay, C. P., and Christofferson, B. M. 1985. A calculation of the Rosseland mean opacity of dust grains in primordial solar system nebulae. *Icarus* 64:471–492.

Preibisch, T., Ossenkopf, V., Yorke, H. W., and Henning, T. 1993. The influence of ice-coated grains on protostellar spectra. *Astron. Astrophys.* 279:577–588.

Pringle, J. E. 1989. On the formation of binary stars. *Mon. Not. Roy. Astron. Soc.* 239:361–370.

Rees, M. J. 1976. Opacity-limited hierarchical fragmentation and the masses of protostars. *Mon. Not. Roy. Astron. Soc.* 176:483–486.

Safronov, V. S. 1960. On the gravitational instability in flattened systems with axial symmetry and non-uniform rotation. *Ann. Astrophys.* 23:901–904.

Shu, F. H., Najita, J., Galli, D., Ostriker, E., and Lizano, S. 1993. The collapse of cores and the formation and evolution of stars and disks. In *Protostars and Planets III*, ed. E. H. Levy and J. I. Lunine (Tucson: University of Arizona Press), pp. 3–45.

Sigalotti, L. DiG., and Klapp, J. 1994. Gravitational collapse and fragmentation of centrally condensed protostellar cores. *Mon. Not. Roy. Astron. Soc.* 268:625–640.

Sigalotti, L. DiG., and Klapp, J. 1997. Collapse and fragmentation models of prolate molecular cloud cores. I. Initial uniform rotation. *Astrophys. J.* 474:710–718.

Silk, J. 1977. On the fragmentation of cosmic gas clouds. II. Opacity-limited star formation. *Astrophys. J.* 214:152–160.

Simon, M., Close, L. M., and Beck, T. L. 1999. Adaptive optics imaging of the Orion Trapezium cluster. *Astron. J.*, 117:1375–1386.

Tassoul, J.-L. 1978. *Theory of Rotating Stars* (Princeton: Princeton University Press).

Terebey, S., Van Buren, D., Padgett, D. L., Hancock, T., and Brundage, M. 1998. A candidate protoplanet in the Taurus star-forming region. *Astrophys. J. Lett.* 507:L71–L74.

Tohline, J. E. 1981. The collapse to equilibrium of rotating adiabatic spheroids. I. Protostars. *Astrophys. J.* 248:717–726.

Tomisaka, K., Ikeuchi, S., and Nakamura, T. 1990. The equilibria and evolutions of magnetized, rotating isothermal clouds. IV. Quasi-static evolution. *Astrophys. J.* 362:202–214.

Tomley, L., Cassen, P., and Steiman-Cameron, T. 1991. On the evolution of gravitationally unstable protostellar disks. *Astrophys. J.* 382:530–543.

Tomley, L., Steiman-Cameron, T., and Cassen, P. 1994. Further studies of gravitationally unstable protostellar disks. *Astrophys. J.* 422:850–861.

Toomre, A. 1964. On the gravitational stability of a disk of stars. *Astrophys. J.* 139:1217–1238.

Truelove, J. K., Klein, R. I., McKee, C. F., Holliman, J. H., Howell, L. H., and Greenough, J. A. 1997. The Jeans condition: A new constraint on spatial resolution in simulations of isothermal self-gravitational hydrodynamics. *Astrophys. J. Lett.* 489:L179–L183.

Truelove, J. K., Klein, R. I., McKee, C. F., Holliman, J. H., Howell, L. H., Greenough, J. A., and Woods, D. T. 1998. Self-gravitational hydrodynamics with three-dimensional adaptive mesh refinement: Methodology and applications to molecular cloud collapse and fragmentation. *Astrophys. J.* 495:821–852.

Tsai, J., and Bertschinger, E. 1989. Stability of the expansion wave solution of collapsing isothermal spheres. *Bull. Am. Astron. Soc.* 21:1089 (abstract).

Tsuribe, T., and Inutsuka, S. 1999. Criteria for fragmentation of rotating isothermal clouds revisited. *Astrophys. J. Lett.*, in press.

Turner, J. A., Chapman, S. J., Bhattal, A. S., Disney, M. J., Pongracic, H., and Whitworth, A. P. 1995. Binary star formation: Gravitational fragmentation followed by capture. *Mon. Not. Roy. Astron. Soc.* 277:705–726.

Umebayashi, T., and Nakano, T. 1990. Magnetic flux loss from interstellar clouds. *Mon. Not. Roy. Astron. Soc.* 243:103–113.

Walker, C. K., Adams, F. C., and Lada, C. J. 1990. 1.3 millimeter continuum observations of cold molecular cloud cores. *Astrophys. J.* 349:515–528.

Walker, C. K., Carlstrom, J. E. and Bieging, J. H. 1993. The IRAS 16293–2422 cloud core: A study of a young binary system. *Astrophys. J.* 402:655–666.

Ward-Thompson, D., Scott, P. F., Hills, R. E., and André, P. 1994. A submillimetre continuum survey of pre-protostellar cores. *Mon. Not. Roy. Astron. Soc.* 268:276–290.

Watkins, S. J., Bhattal, A. S., Boffin, H. M. J., Francis, N., and Whitworth, A. P. 1998*a*. Numerical simulations of protostellar encounters. II. Coplanar disc-disc encounters. *Mon. Not. Roy. Astron. Soc.* 300:1205–1213.

Watkins, S. J., Bhattal, A. S., Boffin, H. M. J., Francis, N., and Whitworth, A. P. 1998*b*. Numerical simulations of protostellar encounters. III. Non-coplanar disc-disc encounters. *Mon. Not. Roy. Astron. Soc.* 300:1214–1224.

Whitworth, A. P., and Clarke, C. J. 1997. Cooling behind mildly supersonic shocks in molecular clouds. *Mon. Not. Roy. Astron. Soc.* 291:578–584.

Williams, H. A., and Tohline, J. E. 1988. Circumstellar ring formation in rapidly rotating protostars. *Astrophys. J.* 334:449–464.

Wootten, A. 1989. The duplicity of IRAS 16293–2422: A protobinary star? *Astrophys. J.* 337:858–864.

Zinnecker, H. 1990. Pre-main-sequence binaries. In *ESO Workshop on Low-Mass Star Formation and Pre-Main-Sequence Objects*, ed. B. Reipurth (Garching: European Southern Observatory), pp. 447–469.

YOUNG BINARY STARS AND ASSOCIATED DISKS

ROBERT D. MATHIEU
University of Wisconsin at Madison

ANDREA M. GHEZ
University of California at Los Angeles

ERIC L. N. JENSEN
Swarthmore College

and

MICHAL SIMON
State University of New York at Stony Brook

The typical product of the star formation process is a multiple-star system, most commonly a binary star. Binaries have provided the first dynamical measures of the masses of pre-main-sequence (PMS) stars. These measurements have established that T Tauri-like stars are indeed of solar mass or less, and they have provided preliminary support for the mass calibrations of theoretical PMS evolutionary tracks. Surprisingly, in some star-forming regions PMS binary frequencies have been found to be higher than among main-sequence solar-type stars. The binary frequency in the Taurus star-forming region is a factor of 2 in excess of the field, although other regions show no excess. Observations suggest that the difference between PMS and main-sequence binary frequencies is not an evolutionary effect; thus, recent attention has focused on correlations between binary frequency and initial conditions (e.g., stellar density or cloud temperatures). Accretion disks are common among young binary stars. Binaries with separations between 1 and 100 AU have substantially less submillimeter emission than closer or wider binaries, suggesting that such binaries have dynamically truncated their associated disks. Direct evidence of dynamical clearing has been seen in several binaries, most notably GG Tauri. Remarkably, PMS binaries of all separations show evidence of long-lived circumstellar disks and continued accretion at stellar surfaces. This strongly suggests that the circumstellar disks are replenished from circumbinary disks or envelopes, perhaps through recently hypothesized accretion streams across dynamically cleared gaps. The frequent presence of either circumstellar or circumbinary disks suggests that planet formation can occur in binary environments. That planets may form around stars in wide binaries is already established by their discovery. Circumbinary disk masses around very short-period binaries are ample to form planetary systems such as our own. The nature of planetary systems among the most frequent binaries, with separations between 10 and 100 AU, is less clear given the observed reduction in disk mass. However, even these systems have disks with masses adequate for the formation of terrestrial-like planets.

I. INTRODUCTION

During the last decade, we have gone from *suspecting* that most pre-main-sequence (PMS) stars are binary stars to *knowing* that most PMS stars are binary stars. Advances in high-angular-resolution infrared imaging technology have enabled large surveys for binaries in a variety of star-forming regions. The consistent result is that binary stars are abundant among young stars, indeed perhaps remarkably so.

The implications of this for star and planet formation research are enormous. From an observational point of view we ignore the presence of companion stars at our own peril. A companion star may substantially contaminate the light attributed to the primary, thereby leading to incorrect luminosity, temperature, and extinction measurements; erroneous age and mass estimates; incorrect disk models; and so on. From a theoretical point of view, we must recognize that binary star formation is the primary branch of the star formation process. While our presence in orbit around a single star will always drive interest in single-star formation, a general understanding of star formation must focus on multiple-star formation. Similarly, the typical environment for planet formation may be a binary star. The ubiquity of binary systems suggests that the question of planet formation in binary systems is critical for setting the overall frequency of planets.

In this contribution we will focus on those observations of PMS binaries as they relate to (1) probes of evolutionary models, (2) binary star populations, and (3) protoplanetary disks and the potential for planet formation. Typically, consideration will be restricted to binaries with primaries having masses less than 3 $M_\odot$ (with primaries of solar mass or less predominating) and ages of less than 10 million years. Observations of such PMS binaries have been comprehensively reviewed by Mathieu (1994; see also review papers in Milone and Mermilliod 1996), and thus we will concentrate our attention on observations made since then. Many theoretical issues of binary formation are addressed in the chapter by Bodenheimer et al., this volume.

II. PRE-MAIN-SEQUENCE STELLAR PROPERTIES

Typically, the masses and ages of young stars are inferred from comparisons with PMS evolutionary tracks. The mass and age calibrations of such tracks are largely untested against observations. Young binary stars provide powerful tools for assessing the validity of these models. In particular, their orbital motions yield *direct* measures of their stellar masses, while under assumptions of coeval formation the derived ages of binary components test relative age calibrations of the tracks.

High-angular-resolution observations from both speckle interferometry and the *Hubble Space Telescope* (HST) Fine Guidance Sensors have begun to contribute significantly to the determination of astrometric orbits (Ghez et al. 1995; Thiébaut et al. 1995; Simon et al. 1996). These studies clearly show relative motion of the two stars in many PMS binaries, and curvature appears in the relative motions of several. These motions are larger than the velocity dispersions in star-forming regions (SFRs) and thus generally argue for an orbital origin.

Ghez et al. (1995) treated the observed relative motions statistically to obtain a typical system mass among a set of astrometric binaries. Their results are shown in Fig. 1. The trend in the data of increasing relative velocity with decreasing separation also strongly suggests orbital motion. The comparison curves represent the expected values of relative velocity as a function of binary separation for a set of binary stars observed at random orientations. Although any given datum reflects only the state of a binary at a certain moment in its orbit, and thus cannot reliably provide binary mass, the median mass of 1.7 $M_\odot$ found from the ensemble should be a valid estimate of the typical binary total mass. This value empirically supports the theoretical conclusion that most PMS stars have solar masses or less.

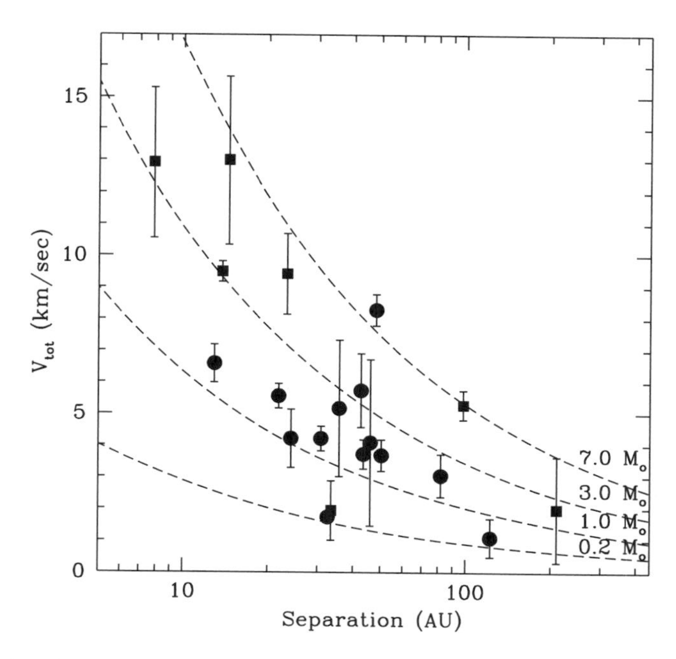

Figure 1. Relative velocities of binary stars' components as a function of their mean separation. The observed velocities are consistent with orbital motion, because they (1) decrease with increasing separation and (2) are generally greater for systems with higher-mass primary stars ($M_1 > 1$ $M_\odot$ and $M_1 \leq 1$ $M_\odot$ are plotted as squares and circles, respectively). The measurements are compared to those expected from a set of randomly oriented binary stars with total masses of 0.2, 1, 3, and 7 $M_\odot$. Although any individual total mass estimate is unreliable because of projection effects, the sample has an average total dynamical mass of ~1.7 $M_\odot$. (Taken from Ghez et al. 1995.)

Eclipsing binaries are particularly powerful routes to the measurement of *both* stellar masses and radii. Until recently only EK Cephei had been studied in detail (Popper 1987; Claret et al. 1995; Martín and Rebolo 1993). Unfortunately, the PMS secondary of this system is very near the zero-age main sequence (ZAMS) and thus places little constraint on evolutionary tracks. A second eclipsing binary, TY Coronae Australis, recently has been intensively studied by two groups (Casey et al. 1998; Corporon et al. 1996, and references therein). This system consists of a Herbig Be star of 3.16 $M_\odot$ on the main sequence and a cool (4900 K) 1.64 $M_\odot$ secondary star. The 2.1 $R_\odot$ radius of the secondary, along with its association with the R CrA dark cloud and lithium absorption, identify it as PMS. Comparison with evolutionary models places it at the base of its Hayashi track, with

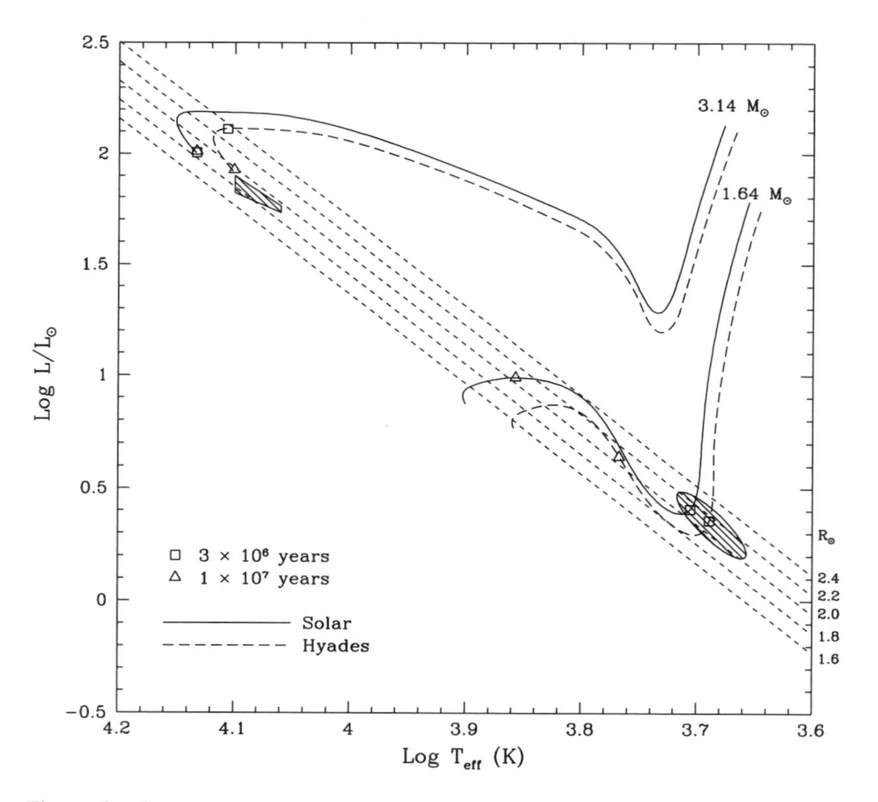

Figure 2. Location of TY CrA primary and secondary stars in the theoretical HR diagram. The hatched regions designate high-confidence domains for the primary and secondary based on light-curve analyses. Dotted lines are drawn at constant radii. Solid and dashed lines correspond to pre-main-sequence tracks of Swenson et al. (1994), calculated for the masses of the TY CrA components at solar (solid curve) and Hyades (dashed curve) compositions. Open boxes and triangles mark isochrone points at ages of 3×10^6 yr and 1×10^7 yr, respectively. (Taken from Casey et al. 1998.)

an age of 3 Myr (Fig. 2). Casey et al. (1998) tested three sets of evolutionary tracks [those of Swenson et al. (1994; Fig. 2), Claret (1995), and D'Antona and Mazzitelli (1994)] against the TY CrA secondary. In all three cases solar-composition 1.64 $M_\odot$ tracks are consistent with the observed physical parameters. Thus, the secondary star represents the first quantitative dynamical test of PMS evolutionary tracks, which they pass without contradiction. Unfortunately, the accuracies of the derived physical parameters were not adequate to distinguish critically *among* evolutionary tracks. Discovery and study of additional PMS eclipsing binaries are sorely needed.

Stellar masses can also be measured via orbital motions of disk gas. The circumstellar disks of the single stars DM Tauri and GM Aurigae, and the circumbinary disks of GG Tauri and UY Aurigae have been used in this way (Dutrey et al. 1994, 1998; Guilloteau and Dutrey 1998; Guilloteau et al. 1999; Duvert et al. 1998). In well-positioned systems, the precision of this technique can be less than 10%, so that uncertainty in the distance to a given system is the limiting factor in the determination of its mass.

The *relative* mass calibrations of evolutionary models can be tested with careful analyses of PMS double-lined binaries, which provide very accurate dynamical mass ratios. Adopting the observations and analysis procedures of Lee (1992), Figueiredo (1997) has used the PMS binary 162814–2427 to test sets of PMS evolutionary models with different input physics; Lee also studied four other double-lined systems. Both Lee and Figueiredo find that the relative mass calibrations of evolutionary tracks are consistent with the observed mass ratios of the binaries, presuming coeval formation of the component stars. However, Figueiredo notes that the observational uncertainties are considerably higher than the theoretical ones. *More generally, it should be appreciated that, even given accurate mass determinations, meaningful comparison with theory is severely limited by uncertainties in the effective temperatures and luminosities of the weighed stars.* (See Fig. 3 for an example of one aspect of the temperature difficulty.)

Recently, Prato (1998) has extended this observational test into the near infrared. Infrared observations permit the detection of cooler companions than do optical observations, offering the advantages of (1) converting single-lined systems into double-lined systems and thus expanding the sample and (2) providing double-lined systems with large mass ratios, which give more leverage in testing the models. Prato has obtained *H* band high-resolution spectroscopy of the previously single-lined system NTT 155913–2233, and detected an M5 companion to the K5 primary. The derived mass ratio is about 2, the largest mass ratio yet measured among PMS binaries. Interestingly, only the Swenson et al. (1994) models are roughly (within 1σ) consistent with the binary system. The components are neither coeval nor consistent with the mass limits using the D'Antona and Mazzitelli (1994) tracks.

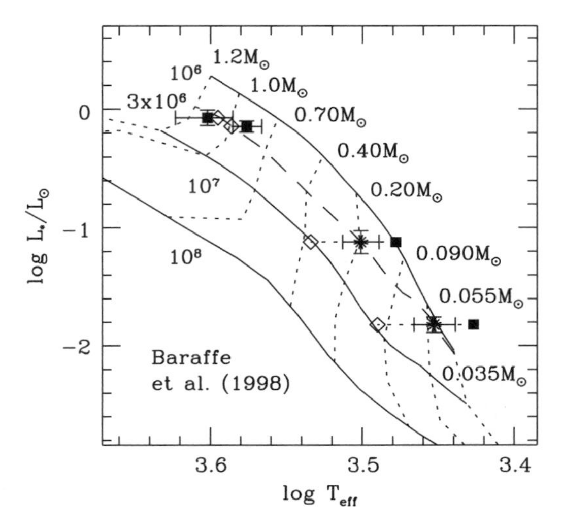

Figure 3. The stars of the GG Tau quadruple system compared with the theoretical evolutionary tracks of Baraffe et al. (1998; $\alpha = 1.9$). The range of plausible temperatures for each component are determined using a dwarf temperature scale (solid squares; Leggett et al. 1996) and a giant temperature scale (open diamonds; Perrin et al. 1998). Since the dwarf and giant temperature scales are nearly identical for the two hottest components, these two stars define an isochrone (dashed line) that can be used to test evolutionary models and the T Tauri temperature scale at lower masses. Of the models tested, the Baraffe et al. (1998) models yield the most consistent ages using a temperature scale intermediate between that of dwarfs and giants. These tracks and the implied coeval temperature scale (asterisks) yield a substellar mass of $0.044 \pm 0.006\ M_{\odot}$ for the lowest-mass component of GG Tau. (Taken from White et al. 1999.)

Similarly, White et al. (1999) conducted a test of evolutionary models by requiring that the components of the quadruple GG Tau be the same age. This system is particularly useful, because its components span a wide range in mass and extend across the stellar/substellar boundary, a region where both the evolutionary models and the PMS temperature scale are very uncertain. Of the evolutionary models tested, they find the Baraffe et al. (1998) models yield the most consistent ages using a temperature scale intermediate between that of giants and dwarfs for PMS M stars (Fig. 3). With this model, the coldest component of the GG Tau system, GG Tau Bb, is substellar with a mass of 50 M_{Jup}.

At present we have no empirical measure of the absolute ages of PMS stars, so age determination relies on the application of evolutionary models to the temperatures and luminosities of the stars. As stressed by Simon et al. (1993), applying this procedure to the combined light of binaries must lead to biased results, primarily toward younger ages because of enhanced luminosity. Typical errors are on the order of a factor of 2, although errors as large as an order of magnitude are possible (Brandner and Zinnecker 1997; Ghez et al. 1997*b*).

Assuming a given evolutionary model, the inferred relative ages of the components of binaries provide an empirical test of coevality, which is useful for distinguishing between different modes of binary star formation. Hartigan et al. (1994) determined the ages of components of wide binaries and found that in roughly one-third of the cases the secondary was significantly younger than the primary. Brandner and Zinnecker (1997) did not find any such age differences in a sample of somewhat closer binaries. Ghez et al. (1997*b*) noted that many of the binary components in the Hartigan et al. (1994) sample were themselves binaries, leading to biased age estimates. When they considered the subset of the Hartigan et al. (1994) sample not known to be triples, all of the binary components were coeval to within the uncertainties.

III. YOUNG BINARY POPULATIONS

The early multiplicity surveys of the Taurus and Ophiuchus dark cloud complexes revealed an apparent difference between the binary star fractions of PMS and main-sequence stars (Ghez et al. 1993; Leinert et al. 1993; Simon et al. 1995). This difference was found to be particularly pronounced within the separation range of 1–150 AU, where the young stars were found to be twice as likely to be members of binary systems as the older stars observed by Duquennoy and Mayor (1991). Interpretations of this unexpected finding have included (1) observational selection effects, especially differences in detection limits between the surveys and differences in sample populations [e.g., classical T Tauri stars (cTTSs) vs. weak-lined T Tauri stars (wTTSs)], (2) formation differences among SFRs (e.g., Reipurth and Zinnecker 1993), (3) varying period distributions (Leinert et al. 1993), or (4) the disruption of primordial multiple-star systems over time (Ghez et al. 1993). Alternatively, Mathieu (1996) questioned whether the excesses were statistically significant. The variety of interpretations led to many subsequent surveys in order to increase both the sample sizes and the numbers of SFRs studied (see Table I). These surveys have shed light on many of the questions raised.

A. Is the Excess of Young Companion Stars Real?

The question of whether or not the excess frequency of PMS binaries is real can be divided into two parts. First, is the high fraction of binary stars in Taurus and Ophiuchus statistically significant? Second, is this really different from the binary fraction in the solar neighborhood? The issue of the latter question is incompleteness, and it raises the related question of how incomplete surveys should be compared.

Addressing the first question requires observation of larger samples of stars. Köhler and Leinert (1998) roughly doubled the sample studied in

TABLE I
Recent PMS Binary Surveys

Region	N	Sep. Range (AU)	Bin. Freq.	Result/Field	Ref.
Taurus	178	19–1900	0.43 ± 0.05	1.93 ± 0.23	Köhler and Leinert (1998)
Ophiuchus	87	15–1800	0.20 ± 0.05	1.9 ± 0.5	Ghez et al. (1997*a*)
Upper Scorpius	118	19–86	0.36 ± 0.05	1.57 ± 0.34	Köhler et al. (1998)
Chamaeleon	195	120–1800	0.14 ± 0.03	0.97 ± 0.31	Brandner et al. (1996)
Cha/Lupus/CrA	104	15–1800	0.27 ± 0.06	2.2 ± 0.5	Ghez et al. (1997*a*)
IC 348	67	37–2560	0.19 ± 0.05	0.83 ± 0.3	Duchêne et al. (1999*a*)
NGC 2024, 2068, and 2071	99	138–1050	0.15 ± 0.04	1.3 ± 0.4	Padgett et al. (1997)
Orion Trapezium	50	138–828	0.14 ± 0.05	1.3 ± 0.4	Padgett et al. (1997)
Orion Trapezium	45	63–225	0.059 ± 0.04	0.7 ± 0.5	Petr et al. (1998)
Orion Trapezium	292	132–264	0.03 ± 0.01	1.0 ± 0.3	Simon et al. (1999)
α Per	102	26–595	0.05 ± 0.02	0.3 ± 0.4	Patience et al. (1998*a*)
Pleiades	144	11–910	0.28 ± 0.04	1.04 ± 0.14	Bouvier et al. (1997)
Hyades	167	5–50	0.30 ± 0.06	2.1 ± 0.6	Patience et al. (1998*b*)
Praesepe	50	26–595	0.16 ± 0.06	0.9 ± 0.4	Patience et al. (1998*a*)

the Taurus SFR by surveying 75 new PMS stars discovered by *Röntgen Satellite* (ROSAT), most of which are classified as wTTSs.[a] Their survey is comparable in sensitivity to the previous infrared speckle work done in

[a] It should be noted that distances, and hence luminosities and ages, of the ROSAT-selected stars are controversial; Briceño et al. (1997) argued that the majority are actually foreground ZAMS stars. Neuhäuser and Brandner (1998) studied the small sample of these stars that are sufficiently bright to have Hipparcos (*High Precision Parallax Collecting Satellite*) parallaxes and found them to be younger than 1.6×10^7 years, but in front of their presumed birth place, possibly due to ejection. Nonetheless, Köhler and Leinert (1998) point out that if their sample does include foreground stars with a binary frequency like that of the nearby solarlike stars, then the binary frequency of the wTTSs remaining in the Taurus SFR must be even higher than the observed value.

Taurus (flux ratio $\Delta K \sim 3$ mag) and results in a binary frequency (BF, number of multiples divided by number of systems, hereinafter and in Table I) consistent with the earlier surveys, reinforcing the conclusion that Taurus does indeed have a remarkably high BF.

The expanded Taurus sample has also addressed the concern that cTTSs and wTTSs might have differing binary frequencies. This would constitute a serious selection effect, because the early surveys comprised primarily cTTSs (which dominated the available catalogs such as Herbig and Bell 1988), whereas in fact the X-ray-selected wTTSs appear to outnumber the cTTSs (e.g., Walter et al. 1994; Neuhäuser 1997; Walter et al., this volume). However, Köhler and Leinert (1998) find no significant difference between the cTTSs and wTTSs in either their BF or in their distribution of separations.

Although the measured binary frequency for PMS stars in Taurus is a factor of 2 larger than that measured for stars in the solar neighborhood, this does not necessarily mean that the parent populations are different. The discrepancy between the two binary frequencies could be due to a difference in sensitivity to low-mass companions. In particular, it is possible that PMS star surveys are detecting very low-mass stars that are relatively more luminous when they are young (e.g., Burrows et al. 1993; Malkov et al. 1998). Surveys of main-sequence stars generally have well-defined mass ratio (q) sensitivity limits. In the case of the Duquennoy and Mayor (1991) survey of solar-mass stars (referred to here as DM91), the spectroscopic portion of the survey, which covers periods less than ~10,000 days or, equivalently, semimajor axes less than ~10 AU, is complete down to $q > 0.1$. DM91's sensitivity to longer periods or more widely separated systems, identified by direct imaging, is limited to $q > 0.3$. In contrast, the limits of PMS star surveys, which have generally been carried out at a single wavelength (typically 2.2 μm), are harder to characterize in terms of mass and thus are generally described in terms of limiting flux ratios or flux densities. Estimates of the secondary star masses from these single-wavelength measurements involve several assumptions. In particular, one has to assume that the two stars have the same age, the same line-of-sight extinction, and no infrared excess (e.g., Meyer and Beckwith 1998). Follow-up studies, which resolve the binary stars at multiple wavelengths, show that most systems have $q > 0.3$ (Hartigan et al. 1994; Brandner and Zinnecker 1997; Ghez et al. 1997*b*), suggesting that the high young-star BF is unlikely to arise from a multitude of binary star systems with $q < 0.3$. Still, the lack of understanding of the true mass limits of the young surveys and the limited depth of the DM91 main-sequence survey at comparable separations are major weaknesses in the discussion of the relative BFs of the PMS and main-sequence stars.

B. Does the Binary Population Evolve in Time?

If the PMS and main-sequence binary frequencies do differ, a possible explanation is the disruption of primordial multiple-star systems over time

(Ghez et al. 1993). This has led several groups to pursue observations of binary frequencies in open clusters with different ages. Four clusters have now been studied intensively by J. Bouvier and collaborators using adaptive optics and by J. Patience and collaborators using speckle imaging: α Per (~50–70 Myr), Pleiades (~80–120 Myr), Hyades (~600 Myr), and Praesepe (~600 Myr).

The *K* band speckle imaging surveys of the Hyades (Patience et al. 1998*b*), α Per, and Praesepe (Patience et al. 1998*a*) span a wide range of spectral types (A0–K5). These studies have a uniform mass ratio limit over their separation range, which fortuitously matches that of DM91; thus their uncorrected BF for F7–G9 stars is directly comparable to DM91's sample. They find the cluster binary frequencies both to be statistically consistent with the solar neighborhood population and to be significantly lower than the BF in Taurus. The near-infrared adaptive-optics studies of G and K stars in the Pleiades (Bouvier et al. 1997) and Praesepe (Bouvier et al., in preparation) have less uniform mass ratio limits, ranging from 0.6 to less than 0.1 over the separation range (0.08–6.9″) studied. Over the separation range reaching $q = 0.3$ (0.3–6.9″) and limiting the BF to $q > 0.3$ ratios results in a BF of 0.14 for the Pleiades, comparable to that reported by Duquennoy and Mayor (1991) in this range (0.17); Praesepe produces similar results. The lack of change in BFs within this age sequence of clusters indicates that evolution is not a strong effect, at least after ~50–70 Myr.

C. Does the Binary Formation Outcome Vary among the SFRs?

The apparent overabundance of young companions in Taurus with respect to the solar neighborhood can also be explained if Taurus-like SFRs are not the origin of most stars in our solar neighborhood. This would require other SFRs both to be less efficient at either forming or maintaining binary stars *and* to be the dominant contributors to the field population. Consistent with this line of reasoning, the majority of field stars have been suggested to originate in dense stellar clusters in giant molecular clouds and not in low-stellar-density regions such as the Taurus dark cloud complex (e.g., Lada et al. 1991). Furthermore, high-density SFRs might plausibly have lower binary fractions, either by inhibiting binary formation or by promoting their rapid destruction [e.g., through encounters (Kroupa 1995) or erosion of circumstellar disks (Hall 1997)].

Orion, the closest giant molecular cloud, is three times as distant as Taurus, so systems similar to the closest binaries resolved in Taurus are not currently resolved in Orion. Nonetheless there is still a large overlap in the separation/period range studied so far. Petr et al. (1998) and Simon et al. (1999) have studied the innermost region of the Orion Trapezium cluster using high-resolution imaging techniques at *K*. They find a BF similar to that of the nearby solarlike stars (Table I), although the uncertainty is large because of the small numbers in the samples. Padgett et al. (1997) used *V* and *I* band HST images of the Trapezium, and also of NGC 2024, 2068, and 2071, to investigate the BF. In both samples, they measured a BF somewhat higher than that of the solarlike stars, but only at a $\sim 1\sigma$ level. While these estimates are lower bounds because they do not correct for incompleteness, taken together with the results of Prosser et al. (1994) they suggest that the BF in the clusters of the Orion OB association is similar to that of the nearby solarlike stars, at least in the range of separations surveyed to date.

It is tempting to conclude that the high densities of the young clusters have led to lower binary frequencies than in Taurus, and more generally that binary frequency is anticorrelated with stellar density. Several other regions in Table I are consistent with this hypothesis [e.g., the dark cloud complexes Chamaeleon, Corona Australis, and Lupus (Ghez et al. 1997*a*); but see IC348 (Duchêne et al. 1999*a*)], but the precisions of the measured BFs are lower, as are the confidences of the consequent conclusions. In addition, the present samples do not clearly discriminate between stellar density and other environmental factors. Thus, Durisen and Sterzik (1994) have suggested that cloud temperature may play a key role; in the existing samples the higher stellar densities are found in OB associations, which might arguably have higher-temperature environments than dark cloud complexes. Isolating the critical factors in determining binary frequencies will be a challenging task.

Finally, none of these surveys covers the entire separation range of binary populations. As such, differences in measured binary frequencies could be the result of differing separation distributions, perhaps also linked to initial conditions. In this regard, Brandner and Köhler (1998) report that the binary separation distribution peaks at 90 AU within the youngest subgroup in Scorpius OB and at 215 AU toward an older subgroup in the same association. As always, small sample sizes are a concern, as is the possibility of contamination by field stars at the larger separations. The frequency of the shortest-period binaries (e.g., $P < 100$ days) has also received attention. At present, orbital solutions exist for nearly 30 spectroscopic binaries. In the most recent summary, Mathieu (1996) reports seven binaries with $P < 100$ days among 91 PMS stars, for a frequency of 8% ± 3%, the same as found for main-sequence solar-type stars. Even so, an excess comparable to those found in the high-angular-resolution surveys is not excluded by these data. Spectroscopic surveys of larger, carefully selected samples are needed. High-precision long-timescale surveys will be especially fruitful; their sensitivity to long-period systems will both increase the expected number of detections, improving statistical significance, and provide important cases for combined spectroscopic-astrometric orbital solutions.

IV. DISKS IN YOUNG BINARY SYSTEMS: STRUCTURE

The frequency of binary companions is a critical datum with respect to assessing the prospects for the formation of planets, in part because of the effect of companions on protoplanetary disks. The observational case for disks in young binaries is well established (see Mathieu 1994). Low-spatial-resolution observations of cTTS binaries reveal many of the classic signatures of disk material and accretion, such as excess emission at near infrared through millimeter wavelengths, spectral veiling, Balmer and forbidden emission lines, and polarization. Disk material may be located around individual stellar companions (circumstellar disks) as well as around entire binary star systems (circumbinary disks). Theoretical calculations of binary-disk interactions predict that companions will truncate both circumstellar and circumbinary disks; circumstellar disks will have outer radii of 0.2–0.5 times the binary semimajor axis a, and circumbinary disks will have inner radii of $2a$–$3a$, with the exact values depending on eccentricity, mass ratio, and disk viscosity (Artymowicz and Lubow 1994; see also the chapter by Lubow and Artymowicz, this volume). Recent observations have made significant progress toward delineating such structures of disks in binary environments and the potential for planet formation in these disks.

A. Disk Masses

Millimeter or submillimeter wavelength measurements of dust continuum emission allow a measurement of the total disk mass present in a system, because at least part of the disk is optically thin at these wavelengths. The first systematic survey of millimeter emission from a large number of young stars (Beckwith et al. 1990; Beckwith and Sargent 1993) suggested that millimeter fluxes from close binaries might be lower than those from wider binaries. The subsequent discovery of many more young binaries and further millimeter and submillimeter observations allowed more detailed investigation of this question. Jensen et al. (1994, 1996*b*), Osterloh and Beckwith (1995), and Nürnberger et al. (1998) found that millimeter fluxes (and by extension, disk masses) are significantly lower among binaries with separations of 1–100 AU than among wider binaries or single stars. Binaries wider than 100 AU have a distribution of millimeter fluxes indistinguishable from that of single stars. Finally, there is no evidence for a diminished disk mass around many PMS spectroscopic binaries (separations of less than 1 AU), including GW Ori, UZ Tau E, AK Sco, DQ Tau, and V4046 Sgr (Mathieu et al. 1995, 1997; Jensen et al. 1996*a*,*b*).

The amount of reduction in millimeter flux among the 1–100 AU binaries is consistent with their circumstellar disks being truncated at 0.2–0.5 times the binary separation, as predicted by theory. However, the surface densities of these circumstellar disks are poorly constrained. The low millimeter fluxes give typical disk mass upper limits of a few times 10^{-3} $M_\odot$, while the presence of IRAS 12-, 25-, and 60-μm emission from most of the binaries requires the presence of at least tenuous circumstellar disks with $M_{\rm disk} \gtrsim 10^{-5}$ $M_\odot$.[b]

---

[b] It should be noted that the absolute disk masses are highly uncertain due to the poorly determined dust opacities, gas-to-dust ratios, and disk surface density profiles (e.g., Beckwith and Sargent 1993).

These observations suggest the rather intuitive picture that binaries much wider or much closer than typical disk radii do not significantly alter disk structure, whereas those binaries whose separations are comparable to disk radii substantially modify the associated disks. The breakpoint of roughly 100 AU is similar to the sizes of disks seen in millimeter aperture synthesis images (e.g., Koerner and Sargent 1995) and optical/IR scattered-light images (McCaughrean et al., this volume).

Images of $\lambda = 1.3$ mm continuum emission from the young quadruple system UZ Tauri empirically confirm this general picture on all three scales (Fig. 4; Jensen et al. 1996*a*). UZ Tau E is a spectroscopic binary with

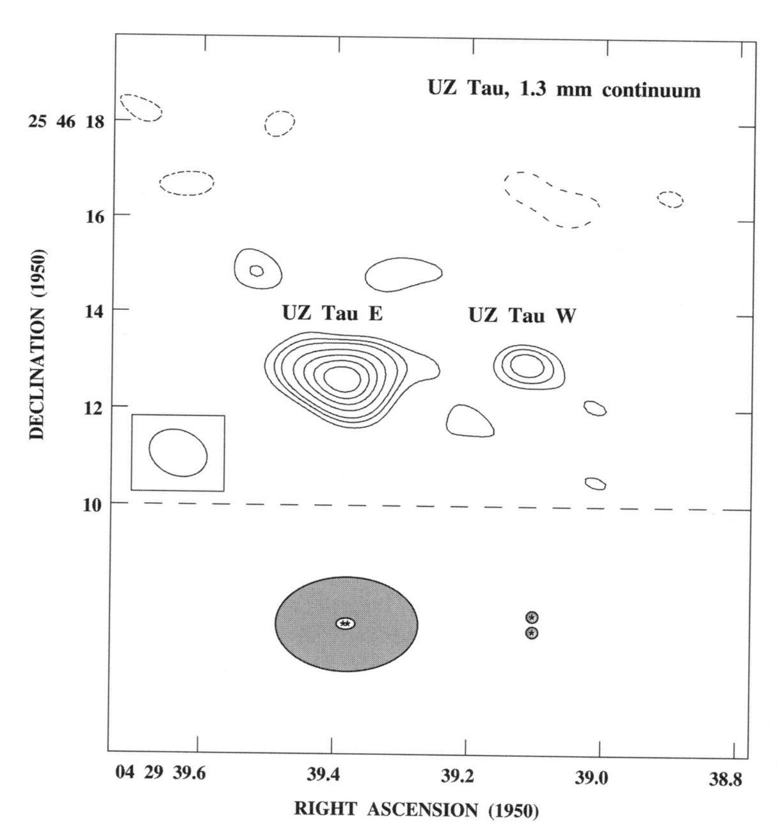

Figure 4. Owens Valley Radio Observatory (OVRO) $\lambda = 1.3$ mm continuum map (top) of UZ Tau, a young quadruple system, along with a schematic model for the system (bottom). UZ Tau W is a 50-AU binary, separated by 500 AU from the sub-AU binary UZ Tau E. UZ Tau E has a massive circumbinary disk, seemingly undisturbed by either the central close binary or by UZ Tau W located several disk radii away in projection. In contrast, UZ Tau W, a binary with a separation similar to typical disk radii, shows substantially less disk emission, though it retains some mass in one or two unresolved circumstellar disks. (Adapted from Jensen et al. 1996*a*.)

a projected semimajor axis of $a \sin i = 0.1$ AU (Mathieu et al. 1996), and UZ Tau W is a binary with a projected separation of 50 AU (Ghez et al. 1993). The two binaries are separated by roughly 500 AU. UZ Tau E has strong 1.3-mm emission that is resolved with a radius of ~170 AU and estimated mass of 0.06 $M_\odot$. This circumbinary disk has a size and mass similar to those seen around other young stars and, by extension, similar to the early solar nebula. It is not evidently affected either by the presence of an embedded binary with separation much smaller than its radius or by the presence of a companion (UZ Tau W) at a separation much larger than its radius. In marked contrast, UZ Tau W, with a separation comparable to a typical disk size, has millimeter emission that is greatly reduced both in flux and in spatial extent. The unresolved millimeter emission must arise from circumstellar disks around one or both of the stars in the 50-AU binary. It is noteworthy that, though the circumstellar disks are reduced in mass and size, they are still present. We note that unresolved observations of the quadruple system would see a "normal" millimeter flux (i.e., one comparable to that from single stars), whereas the distribution of disk material is in fact much more complex.

B. Circumbinary Disks and Disk Clearing

The millimeter surveys indicate that *massive* circumbinary disks are rare in binaries wider than a few tens of AU. As noted above, most binaries with separations of 1–100 AU are undetected at millimeter wavelengths, placing stringent limits on any circumbinary material. In addition, Dutrey et al. (1996) made $\lambda = 2.7$ mm interferometric observations of 18 binaries in Taurus-Auriga and detected circumbinary emission from only one, UY Aur. (Note that the circumbinary disk around GG Tau had been previously discovered and was not included in this study.) In the widest binaries, stable circumbinary material would lie at orbital distances of many hundreds of AU and might be difficult to detect because of its cold temperatures and low surface densities. Nonetheless, substantial circumbinary disks would have been detectable among binaries with separations of tens of AU.

The exceptions are notable, though. Both GG Tau (projected separation 40 AU) and UY Aur (projected separation 120 AU) have circumbinary material that is clearly seen in both millimeter interferometric maps (Dutrey et al. 1994; Duvert et al. 1998; Guilloteau et al. 1999) and near-infrared adaptive-optics images (Roddier et al. 1996; Close et al. 1998; see also the chapter by McCaughrean et al., this volume). The material in both of these circumbinary disks clearly shows Keplerian rotation. In GG Tau, the circumbinary disk is resolved in both CO and continuum emission (Color Plate 23). Detailed observations demonstrate that its circumbinary disk has two components: a narrow ring with sharp edges, including about 80% of the mass, and an extended disk reaching as far as 800 AU. The extended disk is cooler than the ring, consistent with heating by the star and circumbinary disk. The total circumbinary mass is 0.12 $M_\odot$, about 10% of the stellar mass. In contrast, the circumbinary disk in UY Aur is resolved only in CO; its continuum emission is compact and is therefore presumed to be associated with only one of the binary components.

CO observations show both circumbinary disks clearly separated from circumstellar disks by regions of very low surface density, indicative of disk clearing. Furthermore, the circumbinary disks have inner radii consistent with theory under reasonable assumptions about the binary orientation. The near-infrared scattered-light adaptive-optics images also suggest the presence of radial structures in these gaps that could arise from small amounts of infalling material; however, in the case of GG Tau these structures are not confirmed by optical HST images (J. Krist, personal communication). A third example of a directly detected circumbinary disk is found in an optical scattered-light image of the main-sequence B5 star BD+31°643, which has a projected binary star separation of ~200 AU (Kalas and Jewitt 1997).

As noted in Section IV.A, many PMS spectroscopic binaries have strong millimeter emission that, given their small semimajor axes, requires the presence of massive circumbinary disks. High-spatial-resolution interferometric maps of UZ Tau E (Fig. 4) resolve the millimeter emission, confirming that a massive disk surrounds the binary. Mid-infrared emission further reveals the presence of circumbinary material around yet more close binaries. The measured masses of the circumbinary disks of DQ Tau (0.02 $M_\odot$), UZ Tau E (0.06 $M_\odot$), and GW Ori (0.3 $M_\odot$) all exceed the minimum mass of the solar nebula. All have semimajor axes $\lesssim$1 AU, so these systems could have stable planetary orbits in their circumbinary disks at distances as small as $\approx$3 AU.

Such close binary systems are also excellent probes of disk clearing. The innermost regions of disks are, as yet, inaccessible to imaging at the distances of the nearest star-forming regions. However, these hot inner disks contribute essentially all of the near-infrared ($\lambda \simeq 2.2$–$5\ \mu$m) excess emission in the spectral energy distributions (SEDs) of young stellar objects. Thus, the lack of a near-infrared excess indicates a lack of hot disk material. Jensen and Mathieu (1997) studied the spectral energy distributions of all known young spectroscopic binaries with disks in order to search for disk clearing. Indeed, some of the binaries (V4046 Sgr, 162814−2427) have *no* near-infrared excess emission but substantial mid- and far-infrared excesses, the signature of cleared inner disks. The inferred sizes of these inner holes are consistent with the sizes expected given the binary orbits.[c]

Surprisingly, however, several binaries (UZ Tau E, DQ Tau, AK Sco) show relatively smooth power law SEDs, as expected from a continuous accretion disk. The near-infrared opacity of dust is very large, so relatively

[c] It is worth noting that care must be taken in interpreting small depressions in SEDs as evidence for cleared inner disks, particularly when no known companion is present, because these dips can also be caused by dust grain opacity and vertical temperature structure effects in a circumstellar disk (Boss and Yorke 1993, 1996).

little material is required to produce the observed near-infrared emission. This leaves open the possibility that these inner disks have been dynamically cleared but not with 100% efficiency. Such a situation could arise if material were to leak steadily from circumbinary disks into a cleared gap. Recent near-infrared CO observations of DQ Tau also reveal the presence of hot gas near the stars (Carr et al., in preparation).

C. Circumstellar Disks

The existence of circumstellar disks has always been implicit in the discovery of binaries among cTTSs, presuming that cTTS diagnostics are indeed indicative of accretion at stellar surfaces. The outstanding issue is whether circumstellar disks surround both the primary and the secondary stars, and how their relative accretion rates compare.

These questions can be answered best through observations capable of resolving the binary systems. As an example, recent observations of HK Tau by speckle imaging (Koresko 1998) and the HST Wide Field Planetary Camera 2 (WFPC2) (Stapelfeldt et al. 1998) provide direct evidence of a circumsecondary disk in a wide PMS binary star. (The HST image is shown in Color Plate 6.) In both observations the secondary is observed as two elongated reflection nebulosities separated by a dark lane, well matched to scattered-light models of an optically thick circumstellar disk seen close to edge-on. The disk has a radius of $\sim$100 AU, roughly one-third the projected separation of the binary. Statistical arguments for the true orbital elements suggest that dynamical truncation of the circumsecondary disk is a possibility (Stapelfeldt et al. 1998). Likewise, mid-infrared observations of the somewhat more evolved A0 star HR 4796A reveal a circumprimary disk, still present at an age of $\sim$8 Myr despite the presence of a companion star located 500 AU away (Koerner et al. 1998; Jayawardhana et al. 1998). If the companion has an eccentric orbit and is currently near apastron, it could influence the disk outer radius, but the observed confinement of the disk material in a narrow annulus roughly 60–80 AU from the primary is puzzling, perhaps suggesting the presence of one or more unseen companions (Schneider et al. 1999). An additional influence of a distant companion star can be to warp the disk (Terquem and Bertout 1993; Larwood et al. 1996). Telesco et al. (1999) note that their images of HR 4796 hint at a warp in the disk. Finally, the secondaries of both HR 4796 and HK Tau show indirect evidence of a circumstellar disk, and thus these systems appear to support both circumprimary and circumsecondary disks.

Circumstellar disks in binary stars have also been identified in observations that separate the emission for the primary and secondary stars but do not resolve the individual disks, using the same indirect measurements used to assess the presence of a disk in single stars, such as infrared and ultraviolet excesses and strong emission lines. For the widest binaries, separated by $\gtrsim$50 AU, Brandner and Zinnecker (1997), Prato and Simon (1997), Prato (1998), and Duchêne et al. (1999*b*) have investigated the occurrence of circumstellar disks through a spectroscopic study of the individual components' emission line (Hα and Brγ) characteristics. Of the combined 49 binaries whose combined light has been classified as cTTSs, 43 include two cTTS stars, suggesting that they harbor both circumprimary and circumsecondary disks.

Observations with the HST/WFPC2 and ground-based speckle imaging have permitted investigations of circumstellar disks among 31 binaries with separations of 10 AU to 100 AU (Ghez et al. 1997*b*; White and Ghez 1999) using the individual components' ultraviolet and infrared excesses as proxies for the presence of circumstellar disks. These studies have also shown that for the majority of pairs with accretion, both components show similar accretion signatures. Interestingly, among the remainder (29%) of the active pairs only the primary retains an accreting disk, suggesting that primaries may have somewhat longer-lived disks. Moreover, the excesses of the primaries are generally larger than or comparable to those of the secondaries, indicating that the primary stars are experiencing larger accretion rates. If primary disks do indeed have longer lifetimes and higher accretion rates than do secondary disks, then the primary disks must be more massive or preferentially replenished. However, near-infrared excesses do not provide meaningful measures of disks' masses, and so the specific masses of primary and secondary disks are unknown. Nonetheless, the presence of large accretion rates would suggest that massive circumstellar disks may survive in close binary systems.

High-spatial-resolution interferometric observations allow the millimeter emission to be localized even in binary stars with separations $\lesssim$100 AU. Three clear cases of circumstellar disks based on their compact millimeter emission are GG Tau (Guilloteau et al. 1999), UZ Tau W (Jensen et al. 1996*a*; see Fig. 4, section IV.A), and T Tau N (Akeson et al. 1998). In the case of GG Tau, Guilloteau et al. (1999) report that the emission is consistent with tidally truncated circumstellar disks of radius $R \sim$ 4–20 AU and total mass $\geq 1.5 \times 10^{-4}$ $M_\odot$.

Circumstellar disks may even exist in the closest binary star systems. As discussed in section V, PMS spectroscopic binaries can show infrared excesses, high-amplitude photometric variability, Hα equivalent widths in excess of 100 Å, heavy veiling, and large ultraviolet excesses. These observations are indicative of accretion and material very near the stars, which may be in the form of either accretion disks or accretion flows.

Finally, an important but relatively unknown property of circumstellar disks is their spatial orientation in the binary system. Disk alignment is one of the few observable properties that can distinguish among competing models of binary formation. It is also critically important for the long-term stability of planetary systems. However, the challenge of resolving circumstellar disks has limited our knowledge of disk alignment. A notable exception is the HK Tauri system (Stapelfeldt et al. 1998; Koresko 1998), where the resolved disk appears not to be coplanar with the binary orbital plane. Disk alignment can also be probed by polarimetry, even in systems where the disks themselves are unresolved. Scattering off a disk introduces a net polarization, so the polarization position angle (PA) of the combined light from disk and star indicates the PA of the disk on the sky (Koerner and Sargent 1995). Monin et al. (1998) compiled polarimetric measurements of 8–37″ binaries from the literature and measured the polarization of some closer systems. They found one system (GI/GK Tau) to have misaligned polarization vectors, while two other systems were consistent with being parallel. Jensen and Mathieu (in preparation) have made K band imaging polarimetric measurements of all $>1''$ young binaries with infrared excess in Taurus and Ophiuchus. Preliminary results show misaligned disks in HK Tau (with both components detected, indicating an unresolved disk around the primary) and in DK Tau. Currently this technique is limited to binaries wider than about 1″ because of the lack of sensitive, stable polarimeters combined with adaptive optics, so it cannot probe binaries with separations much less than typical disk radii to see whether the disks in closer binaries are aligned. A change in alignment properties with binary separation may shed light on whether there are different formation mechanisms for the closer and wider binaries.

V. DISKS IN YOUNG BINARY SYSTEMS: ACCRETION FLOWS

The observation of accretion diagnostics in cTTS binaries of *all* separations suggests that accretion continues at stellar surfaces, with the accreting material presumably flowing from circumstellar disks. However, the theoretical expectation has been that the balance of viscous and resonant forces at the inner edge of a circumbinary disk would prevent any flow of circumbinary material across the gap. The consequence of continued accretion at the stellar surfaces would thus ultimately be exhaustion of the circumstellar disks and the cessation of accretion. Thus one might expect the accretion timescale for binaries to differ from that of single stars and, indeed, to differ between close and wide binaries. In fact, there is little observational evidence for this; for example, Simon and Prato (1995) find no difference in the frequency of accretion diagnostics between single and binary stars. Such long-lived accretion from circumstellar disks is an outstanding puzzle, one that may lead to a much more dynamic view of disk evolution.

The widest binaries (>100 AU) whose circumstellar masses appear to be unaffected by their distant companions (section IV.A) are likely to have accretion histories very similar to those of single stars. Thus the existence of active accretion in wide binaries is not a surprise, where here we define “wide” operationally as a separation several times greater than typical disk radii. More surprising, perhaps, is an apparent correlation in the presence of accretion onto primary and secondary stars [see above discussion of Brandner and Zinnecker (1997); Prato and Simon (1997); Prato (1998); Duchêne et al. (1999*b*); White and Ghez (1999)]. These studies suggest that the components of binary systems typically retain their disks for similar lengths of time. Prato and Simon (1997) suggested that a common circumbinary envelope may replenish and maintain circumstellar disks around both stars, although evidence for such envelopes has not been found for many of the binaries in these studies.

In their HST study of four cTTSs and two wTTSs with separations of 10 AU to 50 AU, Ghez et al. (1997*b*) also found that both components of three close cTTS binaries (including UZ Tau W) show infrared excesses suggestive of circumstellar disks and that two stars have measurable ultraviolet excesses indicative of active accretion. The ultraviolet excesses from the two stars suggest that high accretion rates continue even in binary systems with separations much less than typical disk radii.

Perhaps most remarkably, the observational diagnostics for accretion are present among even the very closest binary stars. The star UZ Tauri E is arguably one of the most classic of cTTSs. It is a high-amplitude photometric variable [and indeed was noted as one of the most active T Tauri stars by Herbig (1977)], has an emission spectrum with Hα equivalent widths in excess of 100 Å, is often heavily veiled, and has a large ultraviolet excess. Analyses have suggested accretion rates as high as 2×10^{-6} $M_\odot$/yr (Hartigan et al. 1995). In addition, it is surrounded by a massive disk, which has been resolved at millimeter wavelengths, and it has a power law spectral energy distribution with excess emission at all infrared wavelengths. Despite these paradigmatic diagnostics for a disk accreting onto a single PMS star, UZ Tau E is a spectroscopic binary with a period of 19 days (Mathieu et al. 1996). Furthermore, with a maximum periastron separation of 0.1 AU, it is clear that the suggested accretion rate cannot be fed solely by an unreplenished circumstellar disk, although there is ample material in the circumbinary disk.

UZ Tau E is not a unique case. The cTTSs AK Scorpii and DQ Tauri have similar orbital periods and show diagnostics of accretion and material very near the stars. Indeed, the infrared spectral energy distribution of DQ Tau is one of the best examples among cTTSs of a power law (Mathieu et al. 1997), typically interpreted as a continuous disk. Like UZ Tau E, both of these binaries have massive circumbinary disks. The issue is whether the circumbinary material can be tapped to supply the material accreting onto the stellar surfaces. Photometric and spectroscopic monitoring of DQ Tau may have provided a clue to the tapping mechanism. Photometric monitoring has revealed periodic brightenings with a period of 15.8 days. This is precisely the same as the orbital period, with the brightenings occurring at periastron passage. During these brightenings the system becomes bluer, the veiling increases, and emission line strengths increase (Mathieu et al. 1997; Basri et al. 1997). Together, all these results point toward an increased mass accretion rate at periastron passage.

Basri, Mathieu, and collaborators have argued that these results are consistent with the presence of accretion streams from the circumbinary disk to at least one of the stellar surfaces. Recent theoretical work of Artymowicz and Lubow (1996) and Lubow and Artymowicz (this volume) has suggested that such streams may develop if the circumbinary disks are sufficiently viscous and warm. In this scenario, the increase in luminosity is due to the deposition of kinetic energy from the infalling streams and the consequent heating of a region near the stars. For a binary with elements similar to DQ Tau, their simulations argue that the streams will be pulsed, with maximum accretion rates at periastron passage, as observed. Clearly, the next observational forefront in the study of disks in binary systems is angular resolution on scales much smaller than companion separations, so that the dynamic environment within the binary orbit can be imaged.

VI. L1551 IRS5: OPENING THE WINDOW ON PROTOBINARIES

An exciting observational frontier lies in the direction of binaries in formation. Several very wide pairs have been found among embedded sources. The case of L1551 IRS5 shows well the prospects in the application of forefront technology to embedded sources.

L1551 IRS5 is arguably the canonical Class I system: a protostar still undergoing infall from an envelope. It has long been known to be double at centimeter wavelengths, but the interpretation as a binary system has not been secure. Recent millimeter observations demonstrate that the system is a protobinary with a projected separation of 45 AU (Looney et al. 1997; Rodriguez et al. 1998). The 7-mm Very Large Array (VLA) observations with 7-AU resolution are particularly impressive (Rodriguez et al. 1998). As shown in Fig. 5, not only is the binary clearly evident but the circumstellar emissions are marginally resolved as well. That these are in fact circumstellar disks is indicated by their flux levels being well above the extrapolation of the centimeter spectral energy distribution and by their orientation perpendicular to the centimeter outflow emission.

Combining the wealth of observations of L1551 IRS5, Looney et al. (1997) describe the system as having three main components: a large-scale envelope, a disk or extended structure with a size scale of order 150 AU, and the inner binary system with two circumstellar disks. They suggest an envelope mass and inner and outer radii of roughly 0.44 $M_{\odot}$, 30 AU, and 1300 AU, respectively. The nature of the circumbinary structure remains uncertain. First resolved by Lay et al. (1994), the emission is reasonably fitted with a Gaussian model with a very rough mass of 0.04 $M_{\odot}$. The VLA observations indicate circumstellar disk radii of 10 AU, with disk masses of 0.06 $M_{\odot}$ and 0.03 $M_{\odot}$. These disk masses are uncertain due to contamination by free-free emission, but Rodriguez et al. (1998) argue that this overestimate is no more than a factor of 4. As such, the disk masses remain comparable with the minimum mass required to form a planetary system like our own.

It is notable that such large circumstellar masses are found in a system with projected separation of 45 AU, given that at later evolutionary stages such binaries typically do not have detectable millimeter flux

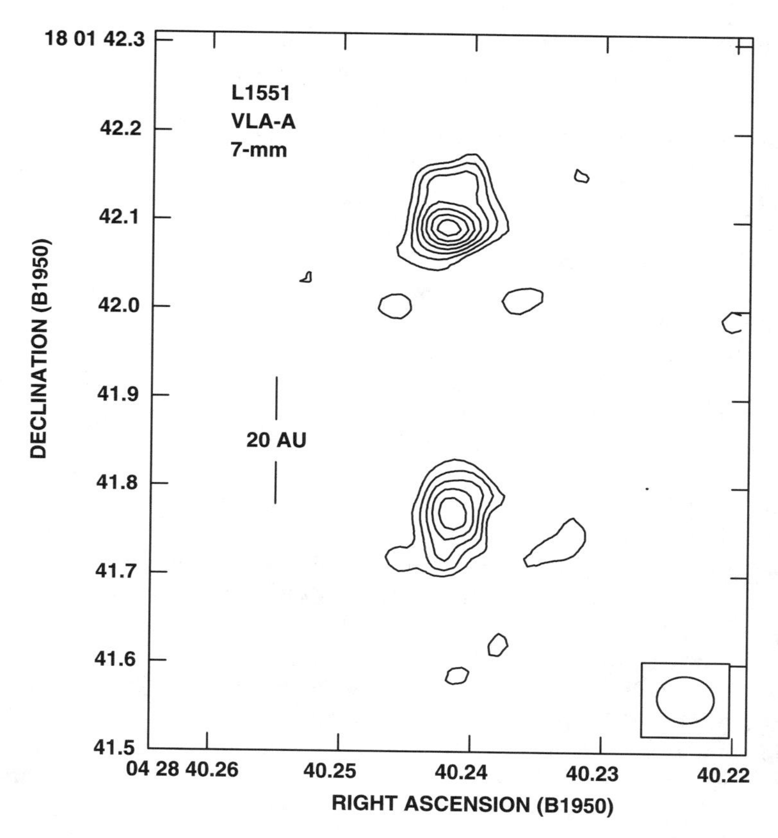

Figure 5. VLA image of the L1551 IRS5 region at 7 mm. The two compact sources are interpreted as protoplanetary disks in a gravitationally bound protobinary system. The masses of the individual disks are of order 0.05 $M_{\odot}$, adequate to form planetary systems like our own. However, the spatially resolved projected semimajor axes are only 10 AU, perhaps due to dynamical truncation by the stars. The half-power contour of the VLA beam was $\approx 0.05''$, opening a new forefront in high-resolution observations of young binary systems. (Taken from Rodriguez et al. 1998.)

(section IV.B). Indeed, given radii of only 10 AU, the inferred disk masses are quite large. Perhaps these substantial circumstellar disks are maintained by rapid accretion from the envelope, possibly via an accretion stream from a circumbinary disk.

VII. CLOSING THOUGHTS ON PLANET FORMATION

A wide variety of investigations in the last several years has made it clear that disks are common in binary systems of all separations. There is no significant difference in the frequency of near-infrared excess emission between binaries and single stars (Moneti and Zinnecker 1991; Simon and

Prato 1995; White and Ghez 1999), nor in the frequency of detected IRAS 12-, 25-, or 60-μm emission (Jensen et al. 1996*b*). This leaves open the very real possibility that, although disks in binary systems may be constrained in size by the presence of the companion, the remaining circumstellar disk material may be similar in temperature and surface density to that in disks around single stars. The clear implication is that, in a significant fraction of binary systems, planet formation may be able to proceed relatively undisturbed. Here we look more closely at exactly which binaries may be likely or unlikely to form planetary systems, based on our current knowledge.

In our own solar system, most of the planet formation has occurred in the range of 0.1–30 AU. If this holds true in other systems, what fraction of binaries could form similar planetary systems? A typical circumstellar disk radius is predicted by theory to be no more than 0.3 times the binary semimajor axis, allowing 30-AU disks in any binary wider than about 100 AU. Adopting the period distribution for main-sequence G stars (Duquennoy and Mayor 1991), this includes roughly 37% of all binary systems. Narrowing the disk radius to 5 AU (i.e., including only the terrestrial planets plus the asteroid belt) reduces the minimum binary separation to 17 AU and includes 57% of all binaries. These numbers assume that planet formation is a relatively local event, proceeding independently of conditions elsewhere in the system, which may not be the case. Nonetheless, it is clear that the possibility of planet formation in even relatively close binary systems is by no means ruled out by current data. These numbers also do not include the roughly 10% of very small-separation binaries ($a \lesssim 0.3$ AU) with circumbinary disks that begin at or inside 1 AU; such systems could form circumbinary planetary systems.

In wider (hundreds of AU) systems, the question has already been answered. Four of the eight extrasolar planetary systems known to date are in members of wide binary systems, with binary separations ranging from $\sim$200 to 1000 AU (see the chapter by Marcy et al., this volume).

Finally, we note that more subtle effects may be important for the long-term prospects of planets in binary systems. Little is currently known about the relative alignment of circumstellar disks and the binary orbital plane. Holman and Wiegert (1999; see also Wiegert and Holman 1997; Innanen et al. 1997) found that a planetary system around one member of a wide binary is stable for longer times if the planetary orbits and binary orbit are coplanar; in their calculations, planetary systems inclined to the binary orbit become unbound in 10^7–10^8 yr, while a coplanar planetary system survives for the full 10^9-yr integration. Investigations of disk alignment in binary systems will be important in determining the long-term stability of planets in binaries and therefore their prospects for evolving and supporting life.

Acknowledgments We would like to gratefully acknowledge assistance with figures provided by A. Dutrey, L. Rodriguez, and R. White.

R. D. M. was supported in part by NSF grant AST-941715. Funding for A. M. G.'s contribution was provided by NAG5-6975. E. L. N. J. and R. D. M. were supported by the NSF's Life in Extreme Environments program through grant AST 97-14246. M. S. was supported by NSF grant AST 94-17191.

REFERENCES

Akeson, R. L., Koerner, D. W., and Jensen, E. L. N. 1998. A circumstellar dust disk around T Tau N: Subarcsecond imaging at $\lambda = 3$ millimeters. *Astrophys. J.* 505:358–362.

Artymowicz, P., and Lubow, S. H. 1994. Dynamics of binary-disk interaction. 1: Resonances and disk gap sizes. *Astrophys. J.* 421:651–667.

Artymowicz, P., and Lubow, S. H. 1996. Mass flow through gaps in circumbinary disks. *Astrophys. J. Lett.* 467:L77–L80.

Baraffe, I., Chabrier, G., Allard, F., and Hauschildt, P. H. 1998. Evolutionary models for solar metallicity low-mass stars: Mass-magnitude relationships and color-magnitude diagrams. *Astron. Astrophys.* 337:403–412.

Basri, G., Johns-Krull, C. M., and Mathieu, R. D. 1997. The classical T Tauri spectroscopic binary DQ Tau. II. Emission line variations with orbital phase. *Astron. J.* 114:781–792.

Beckwith, S. V. W., and Sargent, A. I. 1993. The occurrence and properties of disks around young stars. In *Protostars and Planets III,* ed. E. H. Levy and J. I. Lunine (Tucson: University of Arizona Press), pp. 521–541.

Beckwith, S. V. W., Sargent, A. I., Chini, R. S., and Güsten, R. 1990. A survey for circumstellar disks around young stellar objects. *Astron. J.* 99:924–945.

Boss, A. P., and Yorke, H. W. 1993. An alternative to unseen companions to T Tauri stars. *Astrophys. J. Lett.* 411:L99–L102.

Boss, A. P., and Yorke, H. W. 1996. Protoplanetary disks, mid-infrared dips, and disk gaps. *Astrophys. J.* 469:366–372.

Bouvier, J., Rigaut, F., and Nadeau, D. 1997. Pleiades low-mass binaries: Do companions affect the evolution of protoplanetary disks? *Astron. Astrophys.* 323:139–150.

Brandner, W., and Köhler, R. 1998. Star formation environments and the distribution of binary separations. *Astrophys. J. Lett.* 499:L79–L82.

Brandner, W., and Zinnecker, H. 1997. Physical properties of 90 AU to 250 AU pre-main-sequence binaries. *Astron. Astrophys.* 321:220–228.

Brandner, W., Alcalá, J. M., Kunkel, M., Moneti, A., and Zinnecker, H. 1996. Multiplicity among T Tauri stars in OB and T associations. Implications for binary star formation. *Astron. Astrophys.* 307:121–136.

Briceño, C., Hartmann, L. W., Stauffer, J. R., Gagné, M., and Stern, R. A. 1997. X-rays surveys and the post-T Tauri problem. *Astron. J.* 113:740–752.

Burrows, A., Hubbard, W. B., Saumon, D., and Lunine, J. I. 1993. An expanded set of brown dwarf and very low mass star models. *Astrophys. J.* 406:158–171.

Casey, B. W., Mathieu, R. D., Vaz, L. P. R., Andersen, J., and Suntzeff, N. B. 1998. The pre-main-sequence eclipsing binary TY Coronae Australis: Precise stellar dimensions and tests of evolutionary models. *Astron. J.* 115:1617–1633.

Claret, A. 1995. Stellar models for a wide range of initial chemical compositions until helium burning. I. From $X = 0.60$ to $X = 0.80$ for $Z = 0.02$. *Astron. Astrophys. Suppl.* 109:441–446.

Claret, A., Gimenez, A., and Martin, E. L. 1995. A test case of stellar evolution: The eclipsing binary EK Cephei. A system with accurate dimensions, apsidal motion rate and lithium depletion level. *Astron. Astrophys.* 302:741–744.

Close, L. M., Dutrey, A., Roddier, F., Guilloteau, S., Roddier, C., Northcott, M., Menard, F., Duvert, G., Graves, J. E., and Potter, D. 1998. Adaptive optics imaging of the circumbinary disk around the T Tauri binary UY Aurigae: Estimates of the binary mass and circumbinary dust grain size distribution. *Astrophys. J.* 499:883–888.

Corporon, P., Lagrange, A. M., and Beust, H. 1996. Further characteristics of the young triple system TY Coronae Austrinae. *Astron. Astrophys.* 310:228–234.

D'Antona, F., and Mazzitelli, I. 1994. New pre-main-sequence tracks for $M \leq 2.5M_{\odot}$ as tests of opacities and convection model. *Astrophys. J. Suppl.* 90:467–500.

Duchêne, G., Bouvier, J., and Simon, T. 1999*a*. Low-mass binaries in the young cluster IC348: Implications for binary formation and evolution. *Astron. Astrophys.* 343:831–840.

Duchêne, G., Monin, J.-L., Bouvier, J., and Ménard, F. 1999*b*. Differential accretion in Taurus PMS binaries. *Astron. Astrophys.,* submitted.

Duquennoy, A., and Mayor, M. 1991. Multiplicity among solar-type stars in the solar neighbourhood. II. Distribution of the orbital elements in an unbiased sample. *Astron. Astrophys.* 248:485–524.

Durisen, R. H., and Sterzik, M. F. 1994. Do star forming regions have different binary fractions? *Astron. Astrophys.* 286:84–90.

Dutrey, A., Guilloteau, S., and Simon, M. 1994. Images of the GG Tauri rotating ring. *Astron. Astrophys.* 286:149–159.

Dutrey, A., Guilloteau, S., Duvert, G., Prato, L., Simon, M., Schuster, K., and Menard, F. 1996. Dust and gas distribution around T Tauri stars in Taurus-Auriga. I. Interferometric 2.7mm continuum and ^{13}CO $J = 1 \rightarrow 0$ observations. *Astron. Astrophys.* 309:493–504.

Dutrey, A., Guilloteau, S., Prato, L., Simon, M., Duvert, G., Schuster, K., and Menard, F. 1998. CO study of the GM Aurigae Keplerian disk. *Astron. Astrophys. Lett.* 338:L63–L66.

Duvert, G., Dutrey, A., Guilloteau, S., Menard, F., Schuster, K., Prato, L., and Simon, M. 1998. Disks in the UY Aurigae binary. *Astron. Astrophys.* 332:867–874.

Figueiredo, J. 1997. The pre–main-sequence spectroscopic binary NTTS 162814 −2427: Models versus observations. *Astron. Astrophys.* 318:783–790.

Ghez, A. M., McCarthy, D. W., Patience, J. L., and Beck, T. L. 1997*a*. The multiplicity of pre-main-sequence stars in southern star-forming regions. *Astrophys. J.* 481:378–385.

Ghez, A. M., Neugebauer, G., and Matthews, K. 1993. The multiplicity of T Tauri stars in the star forming regions Taurus-Auriga and Ophiuchus-Scorpius: A 2.2 micron speckle imaging survey. *Astron. J.* 106:2005–2023.

Ghez, A. M., White, R. J., and Simon, M. 1997*b*. High spatial resolution imaging of pre-main-sequence binary stars: Resolving the relationship between disks and close companions. *Astrophys. J.* 490:353–367.

Ghez, A. M., Weinberger, A. J., Neugebauer, G., Matthews, K., and McCarthy, D. W., Jr. 1995. Speckle imaging measurements of the relative tangential velocities of the components of T Tauri binary stars. *Astron. J.* 111:753–765.

Guilloteau, S., and Dutrey, A. 1998. Physical parameters of the Keplerian protoplanetary disk of DM Tauri. *Astron. Astrophys.* 339:467–476.

Guilloteau, S., Dutrey, A., and Simon, M. 1999. GG Tau: The ring world. *Astron. Astrophys.,* in press.

Hall, S. M. 1997. Circumstellar disc density profiles: A dynamic approach. *Mon. Not. Roy. Astron. Soc.* 287:148–154.

Hartigan, P., Strom, K. M., and Strom, S. E. 1994. Are wide pre-main-sequence binaries coeval? *Astrophys. J.* 427:961–977.

Hartigan, P., Edwards, S., and Ghandour, L. 1995. Disk accretion and mass loss from young stars. *Astrophys. J.* 452:736–768.

Herbig, G. H. 1977. Eruptive phenomena in early stellar evolution. *Astrophys. J.* 217:693–715.

Herbig, G. H., and Bell, K. R. 1988. Third catalog of emission-line stars of the Orion population. *Lick Obs. Bull.* 1111:1–90.

Holman, M., and Wiegert, P. 1999. Long-term stability of planets in binary systems. *Astrophys. J.* 117:621–628.

Innanen, K. A., Zheng, J. Q., Mikkola, S., and Valtonen, M. J. 1997. The Kozai mechanism and the stability of planetary orbits in binary star systems. *Astron. J.* 113:1915–1919.

Jayawardhana, R., Fisher, S., Hartmann, L., Telesco, C., Pina, R., and Fazio, G. 1998. A dust disk surrounding the young A star HR 4796A. *Astrophys. J. Lett.* 503:L79–L82.

Jensen, E. L. N., and Mathieu, R. D. 1997. Evidence for cleared regions in the disks around pre-main-sequence spectroscopic binaries. *Astron. J.* 114:301–316.

Jensen, E. L. N., Mathieu, R. D., and Fuller, G. A. 1994. A connection between submillimeter continuum flux and separation in young binaries. *Astrophys. J. Lett.* 429:L29–L32.

Jensen, E. L. N., Koerner, D. W., and Mathieu, R. D. 1996*a*. High-resolution imaging of circumstellar gas and dust in UZ Tauri: Comparing binary and single-star disk properties. *Astron. J.* 111:2431–2438.

Jensen, E. L. N., Mathieu, R. D., and Fuller, G. A. 1996*b*. The connection between submillimeter continuum flux and binary separation in young binaries: Evidence of interaction between stars and disks. *Astrophys. J.* 458:312–326.

Kalas, P., and Jewitt, D. 1997. A candidate dust disk surrounding the binary stellar system BD+31°643. *Nature* 386:52–54.

Koerner, D. W., and Sargent, A. I. 1995. Imaging the small-scale circumstellar gas around T Tauri stars. *Astron. J.* 109:2138–2145.

Koerner, D. W., Ressler, M. E., Werner, M. W., and Backman, D. E. 1998. Mid-infrared imaging of a circumstellar disk around HR 4796: Mapping the debris of planetary formation. *Astrophys. J. Lett.* 503:L83–L86.

Köhler, R., and Leinert, C. 1998. Multiplicity of T Tauri stars in Taurus after ROSAT. *Astron. Astrophys.* 331:977–988.

Köhler, R., Leinert, C., and Zinnecker, H. 1998. Multiplicity of T Tauri stars in different star-forming regions. *Astron. Gesellschaft Meeting Abstr.* 14:17.

Koresko, C. D. 1998. A circumstellar disk in a pre-main-sequence binary star. *Astrophys. J.* 507:145–148.

Kroupa, P. 1995. Star cluster evolution, dynamical age estimation and the kinematical signature of star formation. *Mon. Not. Roy. Astron. Soc.* 277:1522–1540.

Lada, E. A., Evans, N. J., DePoy, D. L., and Gatley, I. 1991. A 2.2 micron survey in the L1630 molecular cloud. *Astrophys. J.* 371:171–182.

Larwood, J. D., Nelson, R. P., Papaloizou, J. C. B., and Terquem, C. 1996. The tidally induced warping, precession and truncation of accretion discs in binary systems: Three-dimensional simulations. *Mon. Not. Roy. Astron. Soc.* 282:597–613.

Lay, O. P., Carlstrom, J. E., Hills, R. E., and Phillips, T. G. 1994. Protostellar accretion disks resolved with the JCMT-CSO interferometer. *Astrophys. J. Lett.* 434:L75–L78.

Lee, C.-W. 1992. Double-lined pre-main-sequence binaries: A test of pre-main-sequence evolutionary theory. Ph.D. Thesis, University of Wisconsin—Madison.

Leggett, S. K., Allard, F., Berriman, G., Dahn, C. C., Hauschildt, P. H. 1996. Infrared spectra of low-mass stars: Toward a temperature scale for red dwarfs. *Astrophys. J. Suppl.* 104:117–143.

Leinert, C., Zinnecker, H., Weitzel, N., Christou, J., Ridgway, S. T., Jameson, R., Haas, M., and Lenzen, R. 1993. A systematic search for young binaries in Taurus. *Astron. Astrophys.* 278:129–149.

Looney, L. W., Mundy, L. G., and Welch, W. J. 1997. High-resolution $\lambda = 2.7$ millimeter observations of L1551 IRS 5. *Astrophys. J. Lett.* 484:L157–L160.

Malkov, O., Piskunov, A., and Zinnecker, H. 1998. On the luminosity ratio of pre-main sequence binaries. *Astron. Astrophys.* 338:452–454.

Martín, E. L., and Rebolo, R. 1993. EK Cephei B: A test object for pre-ZAMS models of solar-type stars. *Astron. Astrophys.* 274:274–278.

Mathieu, R. D. 1994. Pre-main-sequence binary stars. *Ann. Rev. Astron. Astrophys.* 32:465–530.

Mathieu, R. D. 1996. Binary frequencies among pre-main-sequence stars. In *The Origins, Evolutions and Destinies of Binary Stars in Clusters,* ASP Conf. Ser. 90, ed. E. F. Milone and J.-C. Mermilliod (San Francisco: Astronomical Society of the Pacific), pp. 231–241.

Mathieu, R. D., Adams, F. C., Fuller, G. A., Jensen, E. L. N., Koerner, D. W., and Sargent, A. I. 1995. Submillimeter continuum observations of the T Tauri spectroscopic binary GW Orionis. *Astron. J.* 109:2655–2669.

Mathieu, R. D., Martin, E. L., and Magazzu, A. 1996. UZ Tau E: A new classical T Tauri spectroscopic binary. *Bull. Am. Astron. Soc.* 188:60.05.

Mathieu, R. D., Stassun, K., Basri, G., Jensen, E. L. N., Johns-Krull, C. M., Valenti, J. A., and Hartmann, L. W. 1997. The classical T Tauri spectroscopic binary DQ Tau. I. Orbital elements and light curves. *Astron. J.* 113:1841–1854.

Meyer, M. R., and Beckwith, S. V. W. 1998. The environments of pre-main-sequence stars: Brown dwarf companions and circumstellar disk evolution. In *Brown Dwarfs and Extrasolar Planets,* ASP Conf. Ser. 134, ed. R. Rebolo, E. L. Martin, Z. Osorio, and M. Rosa (San Francisco: Astronomical Society of the Pacific), p. 245.

Milone, E. F., and Mermilliod, J.-C. 1996. *The Origins, Evolution, and Destinies of Binary Stars in Clusters*, ASP Conf. Ser. 90 (San Francisco: Astronomical Society of the Pacific).

Moneti, A., and Zinnecker, H. 1991. Infrared imaging photometry of binary T Tauri stars. *Astron. Astrophys.* 242:428–432.

Monin, J.-L., Ménard, F., and Duchêne, G. 1998. Using polarimetry to check rotation alignment in PMS binary stars. Principles of the method and first results. *Astron. Astrophys.* 339:113–122.

Neuhäuser, R. 1997. Low-mass pre-main sequence stars and their X-ray emission. *Science* 276:1363–1370.

Neuhäuser, R., and Brandner, W. 1998. HIPPARCOS results for ROSAT-discovered young stars. *Astron. Astrophys. Lett.* 330:L29–L32.

Nürnberger, D., Brandner, W., Yorke, H. W., and Zinnecker, H. 1998. Millimeter continuum observations of X-ray selected T Tauri stars in Ophiuchus. *Astron. Astrophys.* 330:549–558.

Osterloh, M., and Beckwith, S. V. W. 1995. Millimeter-wave continuum measurements of young stars. *Astrophys. J.* 439:288–302.

Padgett, D. L., Strom, S. E., and Ghez, A. 1997. Hubble Space Telescope WFPC2 observations of the binary fraction among pre-main-sequence cluster stars in Orion. *Astrophys. J.* 477:705–710.

Patience, J., Ghez, A. M., Reid, I. N., and Matthews, K. 1998*a*. Multiplicity survey of α Persei: Studying the effects and evolution of companions. *Bull. Am. Astron. Soc.* 192:54.05.

Patience, J., Ghez, A. M., Reid, I. N., Weinberger, A. J., and Matthews, K. 1998*b*. The multiplicity of the Hyades and its implications for binary star formation and evolution. *Astron. J.* 115:1972–1988.

Perrin, G., Coude du Foresto, V., Ridway, S. T., Mariotti, J.-M., Traub, W. A., Carleton, N. P., and Lacasse, M. G. 1998. Extension of the effective temperature scale of giants to types later than M6. *Astron. Astrophys.* 331:619–626.

Petr, M. G., Coude du Foresto, V., Beckwith, S. V. W., Richichi, A., and McCaughrean, M. J. 1998. Binary stars in the Orion Trapezium cluster core. *Astrophys. J.* 500:825–837.

Popper, D. M. 1987. A pre-main sequence star in the detached binary EK Cephei. *Astrophys. J. Lett.* 313:L81–L83.

Prato, L. 1998. Pre-main-sequence binaries and evolution of their disks. Ph.D. Thesis, State University of New York, Stony Brook.

Prato, L., and Simon, M. 1997. Are both stars in a classic T Tauri binary classic T Tauri stars? *Astrophys. J.* 474:455–463.

Prosser, C. F., Stauffer, J. R., Hartmann, L., Soderblom, D. R., Jones, B. F., Werner, M. W., and McCaughrean, M. J. 1994. HST photometry of the Trapezium cluster. *Astrophys. J.* 421:517–541.

Reipurth, B., and Zinnecker, H. 1993. Visual binaries among pre–main sequence stars. *Astron. Astrophys.* 278:81–108.

Roddier, C., Roddier, F., Northcott, M. J., Graves, J. E., and Jim, K. 1996. Adaptive optics imaging of GG Tauri: Optical detection of the circumbinary ring. *Astrophys. J.* 463:326–335.

Rodriguez, L. F., D'Alessio, P. F., Wilner, D. J., Ho, P. T. P., Torrelles, J. M., Curiel, S., Gomez, Y., Lizano, S., Pedlar, A., Canto, J., and Raga, A. C. 1998. Compact protoplanetary disks in a binary system in L1551. *Nature* 395:355–357.

Schneider, G., Smith, B. A., Becklin, E. E., Koerner, D. W., Meier, R., Hines, D. C., Lowrance, P. J., Terrile, R. J., Thompson, R. I., and Rieke, M. 1999. NICMOS imaging of the HR 4796A circumstellar disk. *Astrophys. J. Lett.* 513:L127–L130.

Simon, M., and Prato, L. 1995. Disk dissipation in single and binary young star systems in Taurus. *Astrophys. J.* 450:824–829.

Simon, M., Ghez, A. M., and Leinert, C. 1993. Multiplicity and the ages of the stars in the Taurus star-forming region. *Astrophys. J. Lett.* 408:L33–L36.

Simon, M., Ghez, A. M., Leinert, C., Cassar, L., Chen, W. P., Howell, R. R., Jameson, R. F., Matthews, K., Neugebauer, G., and Richichi, A. 1995. A lunar occultation and direct imaging survey of multiplicity in the Ophiuchus and Taurus star-forming regions. *Astrophys. J.* 443:625–637.

Simon, M., Holfeltz, S. T., and Taff, L. G. 1996. Measurement of T Tauri binaries using the Hubble Space Telescope Fine Guidance Sensors. *Astrophys. J.* 469:890–897.

Simon, M., Close, L. M., and Beck, T. L. 1999. Adaptive optics imaging of the Orion Trapezium cluster. *Astron. J.* 117:1375–1386.

Stapelfeldt, K. R., Krist, J. E., Menard, F., Bouvier, J., Padgett, D. L., and Burrows, C. J. 1998. An edge-on circumstellar disk in the young binary system HK Tauri. *Astrophys. J. Lett.* 502:L65–L69.

Swenson, F. J., Faulkner, J., Rogers, F. J., and Iglesias, C. A. 1994. The Hyades lithium problem revisited. *Astrophys. J.* 425:286–302.

Telesco, C. M., Fisher, R. S., Pina, R. K., Knacke, R. F., Dermott, S. F., Wyatt, M. C., Crogan, K., Ghez, A. M., Prato, L., Hartmann, L. W., and Jayawardhana,

R. 1999. Mid-infrared imaging with Keck II of the HR4796A circumstellar disk. *Astrophys. J.,* in press.

Terquem, C., and Bertout, C. 1993. Tidally-induced warps in T-Tauri disks. Part one: First order perturbation theory. *Astron. Astrophys.* 274:291–303.

Thiébaut, E., Balega, Y., Balega, I., Belkine, I., Bouvier, J., Foy, R., Blazit, A., and Bonneau, D. 1995. Orbital motion of DF Tauri from speckle interferometry. *Astron. Astrophys. Lett.* 304:L17–L20.

Walter, F. M., Vrba, F. J., Mathieu, R. D., Brown, A., and Myers, P. C. 1994. X-ray sources in regions of star formation. 5: The low mass stars of the upper Scorpius association. *Astron. J.* 107:692–719.

White, R. J., and Ghez, A. M. 1999. A comparison of circumprimary and circumsecondary disks in young binary systems. *Bull. Am. Astron. Soc.* 193:73.11.

White, R. J., Ghez, A. M., Schultz, G., and Reid, I. N. 1998. Spatially resolved spectroscopy of the PMS quadruple GG Tau: Evidence for a substellar companion. *Astrophys. J.,* in press.

Wiegert, P. A., and Holman, M. J. 1997. The stability of planets in the α Centauri system. *Astron. J.* 113:1445–1450.

INTERACTIONS OF YOUNG BINARIES WITH DISKS

STEPHEN H. LUBOW
Space Telescope Science Institute

and

PAWEL ARTYMOWICZ
Stockholm Observatory

The environment of a binary star system may contain two circumstellar disks, one orbiting each of the stars, and a circumbinary disk orbiting about the entire binary. The disk structure and evolution are modified by the presence of the binary. Resonances emit waves and open disk gaps. The binary's total mass and mass ratio as well as orbital elements can be modified by the disks. Signatures of these interactions provide observational tests of the dynamical models. The interaction of young planets with protoplanetary disks circularizes the orbits of Jupiter-mass planets and may produce much more massive extrasolar planets on eccentric orbits.

I. INTRODUCTION

The study of the interaction of young binary stars with disks is motivated by two factors: the high frequency of binarity in young stars and the high frequency of disks around young stars. Most stars, including young stars, are found in binary star systems. Although the binary formation process is as yet not well understood (see Bodenheimer et al. 1993; the chapter by Bodenheimer et al., this volume), binarity appears to be established among the youngest observed stars (Ghez et al. 1993; Mathieu 1994).

General arguments suggest that disks are an inevitable consequence of the star formation process. The standard picture of single-star formation involves the collapse of a rotating cloud (Shu et al. 1987). A disk is a natural product of the collapse, because centrifugal forces prevent much of the cloud material from falling directly onto the central star. Instead, the material settles onto a centrifugally supported disk, from which accretion can occur onto the central star (Lynden-Bell and Pringle 1974).

Considerable observational evidence now exists for the presence of disks around young stars. Observationally determined disk frequencies in star-forming regions exceed 50% (see review by Sargent and Beckwith 1994). The images of disks taken with the *Hubble Space Telescope* (HST) provide overwhelming evidence for the existence of disks (e.g., McCaughrean and O'Dell 1996; Burrows et al. 1996). Resolved circumbinary disks are now directly observed (e.g., Dutrey et al. 1994; Roddier et al. 1996).

The scale of a typical binary star separation is about 30–50 AU for field stars (Duquennoy and Mayor 1991), which is less than the typical T Tauri disk size of a few hundred AU. Consequently, one expects that binary star systems will usually interact strongly with disks (Beckwith et al. 1990). Except for the shortest-period systems (periods less than about 8 days), binaries are eccentric with typical eccentricity of about 0.3 (see Mathieu 1994).

These interactions give rise to several phenomena important in evolutionary and observational contexts. In this review we emphasize the recent theoretical results obtained after the publication of a previous extensive review by Lin and Papaloizou (1993).

II. GENERAL DISK PROPERTIES

A. Disk Configuration

There are two types of disks in a binary environment: circumstellar (CS) disks, which surround only one star, and circumbinary (CB) disks, which surround the entire binary. Two CS disks, one around each star, and one CB disk can be present in a binary. The CS and CB disks are separated by a tidally produced gap (e.g., Lin and Papaloizou 1993; Artymowicz and Lubow 1994).

The exact size of the gap depends on several factors, including the binary eccentricity, disk turbulent viscosity, and sound speed, as will be described later. Subject to certain assumptions, the inner edge of the CB disk is typically $1.8a$ to $3a$, for binary semimajor axis a. The circumprimary and circumsecondary disk outer edges have typical radii of $0.35a$ to $0.5a$ and $0.2a$ to $0.3a$, respectively, for a binary with a mass ratio of 3:1.

Support for this picture comes from studies of millimeter (and submillimeter) emission of young binary systems, which is characteristic of material at several tens of AU from a central star. For binaries with (projected) separations of that scale, there is a systematic deficit of emission (Beckwith et al. 1990; Jensen and Mathieu 1997). On the other hand, near-IR emission, characteristic of regions close to each star, often does not show a corresponding deficit. This result is suggestive of gaps produced on the size scale of the binary.

The most direct evidence for gaps in disks comes from observations of binaries GG Tauri and UY Aurigae. In each case of these two systems, both millimeter interferometry and IR imaging clearly show the existence of a CB disk and a depletion of material near the central regions where the binary is located (Duvert et al. 1998; Close et al. 1998). For both binaries, there are observational signatures of circumstellar disks and accretion onto the stars (Hartigan et al. 1995; Hartmann et al. 1998).

An important related issue is whether these gaps are completely clear of material or whether material might flow through the gap and influence the evolution of the binary (section III.F)

In most cases, this review will assume that the disks are coplanar with the binary orbit. Although this assumption is a natural one, given the geometry of a simple-minded cloud collapse model, it need not actually be realized. Recent theoretical and observational results suggest that noncoplanarity sometimes occurs (cf. section III.E)

B. Evolutionary State

In the earliest phases of stellar evolution (Classes 0 and I of Lada and Shu 1990), a young protobinary might be expected to be embedded in its parent cloud. Cloud material accretes onto the binary and the disks. The infalling material affects the disk structure and accretion properties (e.g., Cassen and Moosman 1981).

Modeling binary evolution in this early phase has emphasized the evolution of the binary orbit and mass ratio, assuming the binary orbit remains circular at all times. The main result is that the binary mass ratio responds to the specific angular momentum (angular momentum per unit mass) of the accreting cloud material, relative to the specific angular momentum of the binary (Artymowicz 1983; Bate and Bonnell 1997). If the material falls in with low specific angular momentum (lower than approximately the binary specific angular momentum), then it is accreted preferentially onto the primary star. The reason is that the matter falls inward toward the deeper potential well. With high-angular-momentum accretion, the material preferentially accretes onto the secondary star. The high angular momentum of the material holds it at larger radii, where it can more easily encounter the secondary (for a physical argument see section III).

The semimajor axis evolution of a circular-orbit binary is dominated by gravitational torques of the accreting material and the advection of angular momentum carried by the accreting material. Gravitational torques cause the orbit to shrink. Generally, it is found that the advection of angular momentum dominates over gravitational torques. So the overall sense of the orbit evolution is that the orbit shrinks when material with low specific angular momentum is accreted, and the orbit expands when material with high specific angular momentum is accreted.

Figure 1 displays a set of smoothed particle hydrodynamics (SPH) results of the accretion flow for different values of specific angular momentum in the inflowing material. The embedding cloud that supplies material is likely to have its specific angular momentum increase with radius in order to be stable dynamically. Consequently, the specific angular momentum of material supplied to the binary will likely increase in time,

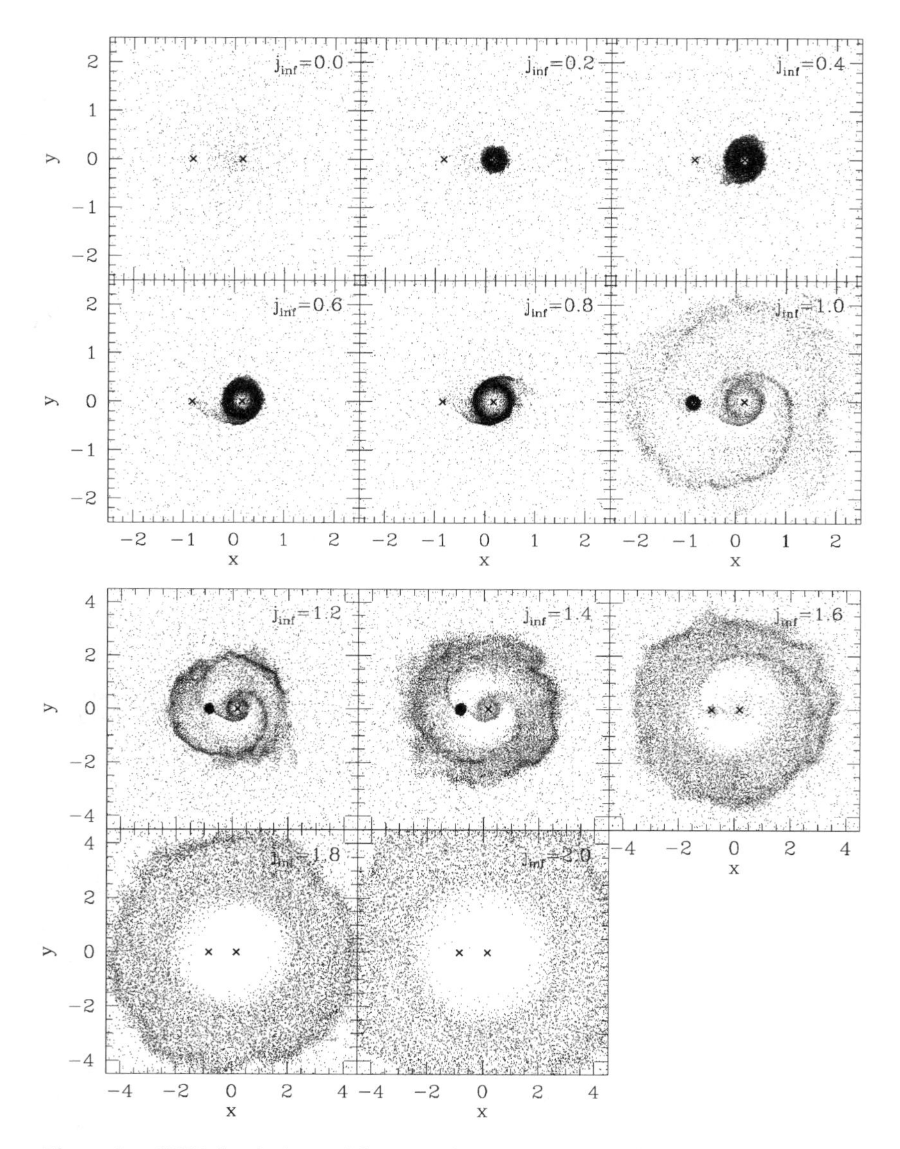

Figure 1. SPH simulations of the accretion onto the protobinary star of mass ratio 0.2 for different values of specific angular momentum in the inflowing material, $j_{\rm inf}$, determining the partitioning of the flow between the circumstellar disks (and stars). (Taken from Bate and Bonnell 1997.)

as material collapses onto the binary from greater and greater distances. The frames in Fig. 1 can then be considered to represent a sequence in time.

Starting with the lowest-angular-momentum inflow (earliest times), the sequence reveals that a Bondi-Hoyle (Bondi and Hoyle 1944) accretion column develops about each star, followed by the development of disks around first the primary star and then the secondary star. For inflow with somewhat higher specific angular momentum, a circumbinary disk forms, but the angular momentum is still low enough to drive an inflow into the binary. Ongoing observational work on embedded sources will provide more constraints on this early evolutionary phase (e.g., Looney et al. 1997).

For sufficiently high specific angular momentum, the disk settles into Keplerian orbits that surround the entire binary. Subsequent evolution occurs on the viscous timescale of the disk.

At later times, the binary consists of T Tauri (or Herbig Ae/Be) stars and is no longer surrounded by the cloud (Class II). Remnant disks are present in the system. The system may have a circumbinary disk, as seen in the last frame of Fig 1. In addition, remnant circumstellar disks may be present, which may have been formed during a phase of accretion of lower specific angular momentum, as seen in the middle frames of Fig. 1. To date, more attention has been given to the properties of binary systems at this state, because such systems are more easily observed. The remainder of this review will concentrate on such systems.

C. Accretion and Decretion Disks

There are two types of disks that are relevant to the binaries: accretion disks and decretion disks. The circumstellar disks are essentially standard accretion disks. Material flows inward as angular momentum flows outward by means of turbulent viscosity. The angular momentum is carried to the outer edge of the circumstellar disk, where it is removed through tidal torques on the binary.

Decretion disks are less well known. For such disks, the central torque provided by the binary prevents material from accreting. Instead, the binary loses angular momentum to the disk, and the disk then expands outward as the binary contracts (Lin and Papaloizou 1979; Pringle 1991; Lubow and Artymowicz 1996).

A circumbinary disk can in principle behave as a decretion disk. However, initially, during a transient stage, the circumbinary disk can behave as an accretion disk until the disk density profile adjusts to that appropriate for a decretion disk. Another effect is that material might penetrate the gap surrounding the binary and so accrete. Such an effect appears in two-dimensional simulations of disks (Artymowicz and Lubow 1996*b*). In practice, a circumbinary disk might then be intermediate between a pure accretion and a pure decretion disk.

D. Disk Viscosity Mechanisms

The models for disks described here are based on a simple effective viscosity prescription. The disk viscosity is considered to be an anomalous viscosity resulting from turbulence (see review by Pringle 1981). The disk turbulence could be due to gravitational instability (e.g., Lin and Pringle 1987). The Toomre criterion for gravitational instability in a Keplerian disk can approximately be expressed as $M_d > (H/r)M$, for a disk of mass M_d, thickness-to-radius ratio H/r, surrounding a binary or single star of mass M. Gravitational instabilities are almost certainly important during the earliest phases of star formation, because of the rapid mass buildup of the disk. However, for parameters of some observed systems such as GG Tauri (Dutrey et al. 1994), gravitational instability is also possible.

Another promising mechanism for producing this turbulence is a magnetic shearing instability (Balbus and Hawley 1991). Since this instability relies on the effects of a magnetic field, the exact nature of this instability is complex, and the use of a simple viscosity is a simplification. Dynamos may play a role, as well as various other magnetic instabilities (Tout and Pringle 1992; Hawley et al. 1996).

In application to protostellar disks, this instability depends on the disk being somewhat ionized. It appears possible that protostellar disks are sometimes not sufficiently ionized to be turbulent throughout. Instead, the turbulence may be restricted to the upper layers of the disk that are sufficiently ionized (Gammie 1996; Glassgold et al. 1997).

For the present purposes, these complications will be ignored, and a simple turbulent viscosity will be assumed. The level of disk turbulence is characterized by the Shakura-Sunyaev dimensionless parameter $\alpha = \nu/c_s H$, where ν is the kinematic turbulent viscosity, H is the disk thickness, and c_s is the disk sound speed (Shakura and Sunyaev 1973).

III. EFFECTS OF A BINARY ON DISKS

A. Gravitational Interactions

The tidal forces caused by the binary generally act to distort a disk from its circular form. Since most binaries are eccentric, they bring about time-dependent distortions that occur at the binary frequency and its harmonics. At special locations in the disk that are resonance points, strong interactions between the binary and disk often occur. It is these resonant interactions that usually dominate.

The theory of resonances for disks has been developed as a result of work by Goldreich and Tremaine (1980). From this linear theory, analytic expressions for the torques exerted at resonances can be obtained. To understand how resonances arise in the theory, consider a decomposition of the binary potential Φ into a sum of rigidly rotating Fourier components:

$$\Phi(r, \theta, t) = \sum_{m,l} \phi_{m,l}(r) \cos(m\theta - l\Omega_b t) \tag{1}$$

where cylindrical coordinates (r, θ) are centered on the binary center of mass in the inertial frame, and Ω_b is 2π divided by the binary orbital period.

For a circular-orbit binary, only diagonal (i.e., $m = l$) terms are nonzero. Setting $m = l$ in equation (1), we see that the potential is static in the frame of the binary, where $\theta - \Omega_b t$ is a constant. In this decomposition, nondiagonal (i.e., $l \neq m$) elements arise only to the extent that the binary is eccentric, which is typically the case. The magnitude of $\phi_{m,l}$ scales with eccentricity as $e^{|m-l|}$. At resonances, waves are launched, which exert torques on disk material as they damp.

Two types of resonances are relevant:

1. Lindblad resonances occur where $\Omega(r) = l\Omega_b/(m \pm 1)$. For $m = l$, their primary effect is to truncate disks. For $m \neq l$, they are called eccentric Lindblad resonances. They can truncate a disk and usually increase the binary eccentricity.
2. Corotational resonances occur where $\Omega(r) = l\Omega_b/m$. They generally damp binary eccentricity. (Technical note: The dynamical effects of corotational resonances depend on the radial derivative of disk vorticity divided by surface density, Σ. The sign of this quantity might seem ill determined, because the distribution of Σ is not known. However, in the case of a disk with a gap, the steep gradient in the density at the disk edge determines the sign of the effects of the corotational resonance. The edges generally produce the dominant corotational torques and act to damp eccentricity.)

B. Wave Propagation

At Lindblad resonances (LRs), waves are launched that carry energy and angular momentum from the binary. The disk experiences changes in its angular momentum in regions of space where the waves damp. Several mechanisms have been proposed that cause wave damping. The detailed conditions in the disk (which vary over time) determine which mechanism dominates.

Shocks provide one mechanism of wave damping (Spruit 1987; Yuan and Cassen 1994; Savonije et al. 1994). This form of damping may play a role, because a disk is often found in simulations to be truncated near resonances that produce strong waves. Waves that are mildly nonlinear near a resonance might steepen as they propagate away from the resonance and then shock. Such effects are likely to be more important for colder disks. Circumstellar disks that surround protostars may be warm enough that severe damping by this mechanism does not occur. Shocks could provide some accretion over a broad region of a protostellar disk.

Turbulent disk viscosity may act as a dissipation source. However, the level of damping it produces is highly uncertain, especially if magnetic stresses play an important role in providing the turbulence. It is unclear whether the damping of waves due to magnetohydrodynamic (MHD) turbulence can be profitably described through a viscosity. Another complication is that the eddy turnover timescale for the largest turbulent eddies may be longer than the wave period, resulting in a reduced effective disk turbulent viscosity (e.g., Lubow and Ogilvie 1998).

Sound waves can lose energy by radiative damping (Cassen and Woolum 1996). In the absence of radiative damping, the waves would induce adiabatic density fluctuations and propagate without energy loss. When effects of radiative losses are included in the cycle of wave compression and decompression, there is a net loss of energy. The wave radial damping length is roughly equal to the disk thickness times the ratio of the disk cooling time to the local orbit period in the disk. The range of plausible protostellar disk conditions permits rapid or slow decay to occur by this process.

Nearly all studies of waves in disks have regarded the disk as two-dimensional. That is, dynamical effects in the vertical direction are often ignored or simplified. In many circumstances, disks are likely to be optically thick in the vertical direction, and substantial vertical temperature variations could occur. For these conditions, conventional wisdom was that sound waves (p modes) would be launched at a Lindblad resonance, just as in the vertically isothermal case. The waves would then experience a propagation speed that varies with height. As a result, the waves would refract up into the disk atmosphere, where they would shock, over a radial distance scale of the order of the disk thickness (e.g., Lin et al. 1990).

However, recent semianalytic 3D results provide a different picture. For a disk having a midplane temperature that is large compared to the disk surface temperature, f-mode waves are launched by Lindblad resonances (Lubow and Ogilvie 1998). The waves are naturally confined vertically by the increasing vertical gravity with height from the disk plane and do not propagate vertically (the wave is vertically evanescent). As the f mode propagates radially away from the resonance, it behaves like a surface gravity wave. The wave energy becomes increasingly confined (channeled) close to the disk surface. The wave amplitude grows, and wave damping by shocks can sometimes occur (Ogilvie and Lubow 1999).

As a consequence of the damping mechanisms described above, it appears plausible that the waves launched from resonances are damped locally, somewhat near the resonance in protostellar circumbinary disks. Nonlinear simulations provide some support for local damping (e.g., Artymowicz and Lubow 1994). (However, numerical artifacts may be playing a role.) We will often assume local damping in the discussion below.

C. Gap Formation

Where wave damping occurs, the angular momentum of the disk material changes. This process leads to the opening of a gap in the disk (Lin and Papaloizou 1979; Goldreich and Tremaine 1980). Gap clearing is observed in astrophysical disks, among others in planetary rings perturbed by satellites (e.g., Lissauer et al. 1981). Viscous effects in the disk act to fill the gaps. The criterion for gap opening at any radius in a viscous disk is that the viscous torque (which acts to smooth density variations) must be less than the resonant torque. Both torques are typically evaluated using linear theory. In the case of an eccentric binary with moderate e, many resonances are present, but only one or two dominate. The gap opening criterion has been found to agree with the results of SPH simulations of eccentric binaries (Artymowicz and Lubow 1994).

If one assumes that the resonant torque is exerted locally near the resonance, the disk edge location is determined by the location of the weakest resonance for which the resonant torque equals or exceeds the viscous torque. If waves are damped nonlocally, then the viscosity-balancing resonance is still the location where the disk is effectively truncated, although the overall maximum density of the disk may be found farther away from the perturber (Takeuchi et al. 1996). In the extreme case of negligible viscosity and little other wave damping, there would only be a minimal gap (due to orbit crossing). In any case, consideration of the optical depth of the disk is needed to predict the position of the edge observed in a particular wavelength band. As a rule, the disks are sufficiently opaque to make the torque-balance definition of the gap size useful.

Figure 2 shows the expected locations of the inner edge of a typical CB disk for various values of binary eccentricity and disk turbulent viscosity. The dominant resonance responsible for disk truncation in a binary system having $e \sim 0.2$ and $\alpha \sim 0.01$ is typically the $m = 2$, $l = 1$ outer

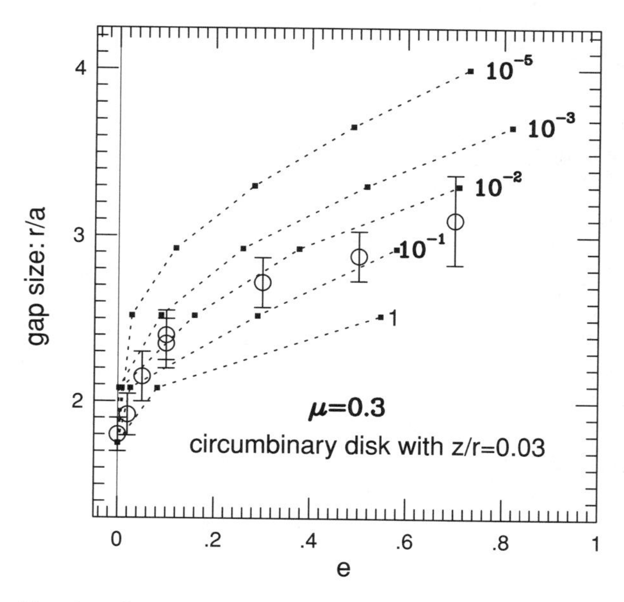

Figure 2. The size of a gap around a binary system with eccentricity e, in terms of binary semimajor axis a, for different α parameters for viscosity of the disk (numbers give $\log \alpha$).

Lindblad resonance (LR). All the barlike, $m = 2$ gravitational field harmonics decrease slowly with distance, in comparison with $m > 2$ harmonics. The noneccentric $(m, l) = (2, 2)$ LR is so strong that it easily clears the gas from its vicinity.

The importance of the next-strongest (2,1) LR is connected with the fact that binaries with $e = 0.2$–0.5 spend relatively much time near apastron, where their angular speed decreases to $\Omega_b/(1 + e)^2 \sim \Omega_b/2$. This gives rise to a strong $l = 1$ harmonic, which rotates uniformly at the speed $l\Omega_b/m = \Omega_b/2$. Whenever that resonance is too strong compared with viscosity, as happens at intermediate and high eccentricity e of the binary, the disk recedes further to the regions of higher-m LRs with $l = 1$.

Plots similar to Fig. 2, showing the expected disk sizes for circumprimary and circumsecondary disks (Artymowicz and Lubow 1994), demonstrate that as the eccentricity increases, the $m = 2$ inner eccentric LRs with increasing l number ($l > 2$) play a dominant role in CS disk truncation.

Jensen et al. (1996) and Jensen and Mathieu (1997) have successfully fit the infrared spectral energy distributions of pre-main-sequence (PMS) binaries with a simple model of disks with gaps. The standard protostellar disks are sharply truncated at radii corresponding to the torque balance. Exceptional cases, such as AK Scorpii, require, however, that the gap be partially filled with dust, probably embedded in gas flowing through the gap from a CB disk (see the chapter by Mathieu et al., this volume).

D. Generation of Disk Eccentricity and Features

A disk can become eccentric as a result of its interaction with the binary. The eccentricity can arise as an instability in a system with a binary in circular orbit or through direct driving by an eccentric binary.

One example of an eccentric instability is that which occurs in CS disks of superhump binaries (Whitehurst 1988). The instability occurs through a process of mode coupling and relies on the effects of the 3:1 resonance in the disk (Lubow 1991). For this instability to operate, the binary mass ratio must be fairly extreme, greater than 5:1. On the other hand, it cannot be too extreme, or the instability will be damped by viscosity. This or other mode-coupling instabilities in principle could also operate in a CB disk at the 1:3 resonance [see equation (35) of Lubow 1991, with $m = 1$].

An eccentric binary with unequal stellar masses drives eccentricity in an initially nearly axisymmetric CB disk via its one-armed bar potential with $(m, l) = (1, 0)$. The disk disturbance follows the slow apsidal motion of the binary (Artymowicz and Lubow 1996*a*). The one-sidedness of disk forcing has a twofold effect. Initially the eccentricity of the disk's edge, denoted as e_d, grows as

$$\dot{e}_d = -\frac{15}{16} e\mu(1-\mu)(1-2\mu)(a/a_d)^3(1-e_d^2)^{-2}\sin\varpi \quad (2)$$

where a and e are the binary's semimajor axis and eccentricity, a_d is the semimajor axis of the disk inner edge, and μ is the binary mass parameter (secondary mass divided by total mass). The longitude of periastron of disk orbits, with respect to the stellar periastron, is denoted as ϖ. Typically, disk eccentricity grows significantly in a few hundred orbital periods P of the binary, while $a_d \approx$ constant. During this phase, ϖ is locked at a stable value $\varpi = 3\pi/2$, corresponding to the perpendicular relative orientation of the two (disk and binary) apsidal lines. However, as e_d grows, the locking action of the lopsided potential weakens, and the standard prograde precession of the disk (due to the quadrupole moment of the double star) dominates. Both the analytical and SPH results show that the disk edge begins to precess around the binary when $e_d/e \sim 0.2$–0.7. Afterwards, eccentricity oscillates with a precessional period ($\sim 10^2$–$10^3 P$), driven up and down by the $\sin\varpi$ factor in equation (3). In numerical calculations, the CB disk typically attains $e_d/e = 0.5$–1. In addition to the direct driving described by equation (2), eccentric instabilities may play a role. Mode-coupling instabilities appear to saturate in the end state of the SPH simulations, which, apart from the periodic driving, resembles the free $m = 1$ mode studied by Hirose and Osaki (1993). In this mode, pressure gradients between adjacent rings synchronize their precession, counteracting the differential quadrupole-induced precession.

Figure 3 presents the eccentric disk evolution in snapshots from a very long SPH simulation of a CB disk around a binary with $e = 0.8$ and $\mu = 0.3$. The upper left panel of the figure illustrates the initial growth of e, while the remaining three panels show a quasistationary (oscillating) eccentricity of the dense, precessing, disk edge. The binary itself is always shown at periastron and, because of its very high eccentricity, is not well resolved in this figure.

Streamlike features emanating from the disk edge toward the otherwise empty circumbinary gap are nonresonant in origin. They are seen only on the receding part of the gas trajectories, following the periastron, when the perturbation by the binary is greatest. Since the binary is revolving much faster than the disk material, its main ($m = l$) harmonics create disturbances that are later enhanced by the kinematics of orbits crossing at caustics. These disturbances lead to local shocking of gas along the feathery features that are colliding with the disk rim. The shocks occur before the material starts descending toward the binary along the elliptic outline of the disk edge (hence the asymmetry between the two halves of the elliptic edge). The cuspy features at the disk rim are potentially observable. Their number contains information about the mean gap size, a_d/a, and the mean binary separation a (which may not be easy to establish otherwise from observations of wide binaries). They also act as potential sites for the streams of gas flowing onto the binary, described later.

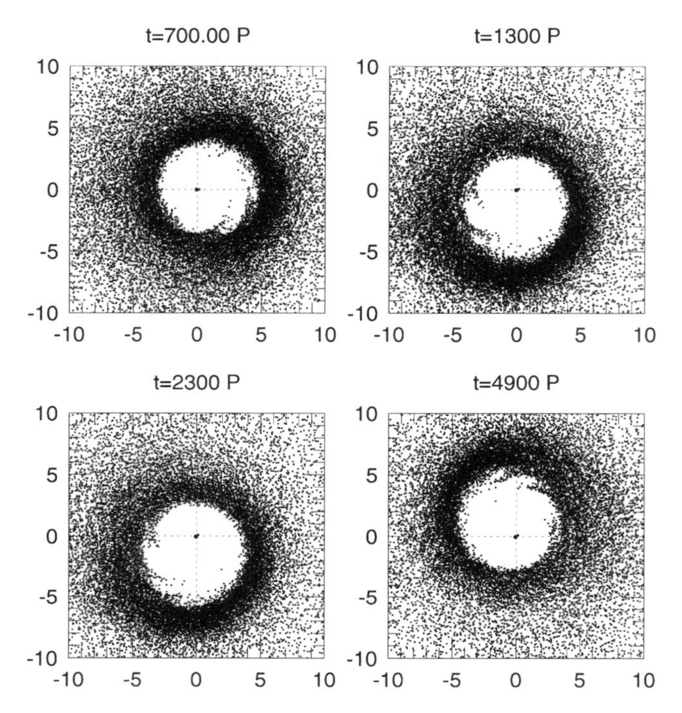

Figure 3. Top views of an SPH simulation of an eccentric binary ($e = 0.8$) with mass ratio 3:7. Axes are scaled in units of a, and time is given in binary orbital periods P. Slow precession and evolution of disk eccentricity are apparent, as well as the transient features generated at the disk edge by the binary.

E. Noncoplanar Disks

The binary tidal field might be expected to force a disk to coplanarity because of differential precession. Differential precession occurs because the disk nodal precession rate varies with radius. Consider a disk as a collection of ballistic orbits that initially all have the same tilt and line of nodes, so the disk is of a tilted planar form. Over time, different annuli in the disk would develop different nodal phasing, and the disk coherence would be lost. For a fluid disk, one might expect that the strong nonplanar shearing motions would give rise to strong dissipation (Papaloizou and Pringle 1983), which would lead to coplanarity with the orbit plane of the binary.

For a fluid disk, there is also the possibility that the disk could act as a coherently precessing body. Theory (Papaloizou and Terquem 1995) and simulations (Larwood et al. 1996) suggest that this is indeed possible, provided that the radial sound crossing time in the disk is shorter than the timescale for differential precession. For conditions in binaries, coherent precession typically requires that the ratio of the disk thickness to radius be greater than about 0.05. When the ratio lies in the range of 0.05 to 0.1, mildly warped structures develop. For much thinner disks, with thickness ratio of 0.02 or less, there is a strong disruption of the disk structure by differential precession.

The origin of the nonplanarity is unclear. It may be established by star formation in widely separated pairs. Observations of visual binaries suggest that nonalignment of the stellar spin axes occurs for separations greater than about 40 AU (Hale 1994). The stellar spin axis direction may reflect the orientation of a preexisting surrounding disk, which was subsequently accreted. Consequently, widely separated binaries might well have formed with misaligned disks.

A recent HST image of HK Tau provides evidence for a misaligned, tidally truncated disk in a binary system (Stapelfeldt et al. 1998). Although the disk's orientation is clear from the image (nearly edge-on), the binary's orientation is unknown. From statistical arguments concerning the binary orbit, the observed disk size suggests that the disk is tidally truncated. With moderate binary eccentricity ($e < 0.5$), the degree of misalignment is likely to be at least 20 degrees.

For this system, the disk thickness-to-radius ratio appears to be in the range for coherent precession with warping. Furthermore, the disk precession period is plausibly of the order of the age of the binary (several million years). The expected warping might then be related to the observed asymmetries.

For young stars of high luminosity (mass exceeding a few $M_\odot$, L greater than $\sim$10 $L_\odot$) the radiation from the central star can induce a warp instability in a CS disk (Armitage and Pringle 1997). The warp is strongest in the inner parts of the disk but may become significant in the outer parts.

F. Mass Flow through Gaps

Until the mid-1990s, the consensus, based in part on numerical case studies, was that there is essentially no flow of gas from the CB disk onto the central binary. Therefore the evolution of CS disks was thought to be independent of the CB disk. One of the natural consequences would be a faster depletion of the CS disks in binaries, because of their shorter viscous evolution time. However, the evidence from the accretion rates onto PMS spectroscopic binary stars (presumably from the CS disks) was in disagreement with naive expectations (Mathieu 1994).

It now appears that stars in single and binary systems may accrete similar amounts of mass from surrounding disks. Two-dimensional simulations of disks in binary-disk systems with moderately thin and viscous disks have revealed the presence of a flow through the gap from the CB disk, generally in the form of two gas streams (Artymowicz and Lubow 1996*a,b*). One reason the old paradigm was inadequate was its reliance on a one-dimensional approximation for the physical quantities. The mass density, resonant torque, and viscous torque were treated as functions of radius only, by considering their azimuthal averages. The two-dimensional studies showed that at certain locations in azimuth, the conditions may be such as to permit mass to penetrate the gap.

The flow process for moderately eccentric binaries can be understood in terms of an effective potential. The inner edge of the CB disk is maintained by the outer Lindblad resonance of the "simplified potential" or the combined $(m, l) = (0, 0)$ and $(2, 1)$ potential (see Artymowicz and Lubow 1996*a* for comparative simulations with the full and the simplified potentials). This bisymmetric potential is static in a frame that rotates at the rate $\Omega_b/2$, or one-half the mean angular speed of the binary. Its effective potential (gravitational potential plus centrifugal barrier) has two saddle points, which are unstable to small perturbations. They correspond closely to the collinear Lagrange points L2 and L3 in the circular restricted three-body problem of celestial mechanics. Free particles that have a small inward velocity would flow inward as a stream from the saddle points, similar to what occurs in a classical Roche overflow process. The finite enthalpy of the disk gas provides an expanding flow that produces the two streams, which are directed toward the stars. We have proposed that an "efficient" flow of gas from the CB disk can sometimes occur. (An efficient flow is one that produces an inward mass flow rate that is comparable to the rate that would be produced in the absence of the binary.)

The CB disk is typically truncated just outside the (2,1) LR by the resonant torques it produces. In this case, the gas passes through the vicinity of points at which penetration inward is easiest (i.e., the saddle points corotating with the simplified potential), located inside the LR. An efficient flow requires (and results from) the Bernoulli constant of the gas at the disk edge being sufficient to overcome the effective potential barrier between the LR and corotation point. This condition can be satisfied in disks which are sufficiently warm (larger enthalpy) or viscous (disk edge closer to the center and endowed with larger effective energy). We have found such flows in SPH simulations of disks with relative thickness ratio $H/r \approx c_s/\Omega r \geq 0.05$ and $\nu \sim 10^{-4}\Omega r^2$.

Figure 4 shows the snapshots from one SPH simulation, modeling a binary system with nearly equal masses (mass parameter $\mu = 0.44$) and eccentricity $e = 0.5$. The crosses denote the saddle points of the $(0, 0) + (2, 1)$ potential, and the added line segments indicate the preferred direction along which gas flows through these points. The direction is determined by taking into account the underlying potential and the Coriolis force, but neglecting gas enthalpy and velocity before the passage (cf. Lubow and Shu 1975). We see that the simplified binary potential indeed determines the number of streams, their rotation at one-half the mean binary rate, and the pitch angle of the streamlike features.

The mismatch of the streams' angular velocity with that of the binary causes a marked time dependence of the mass accretion rate onto the stars (or their CS disks). Although the binary pulls on the disk most strongly

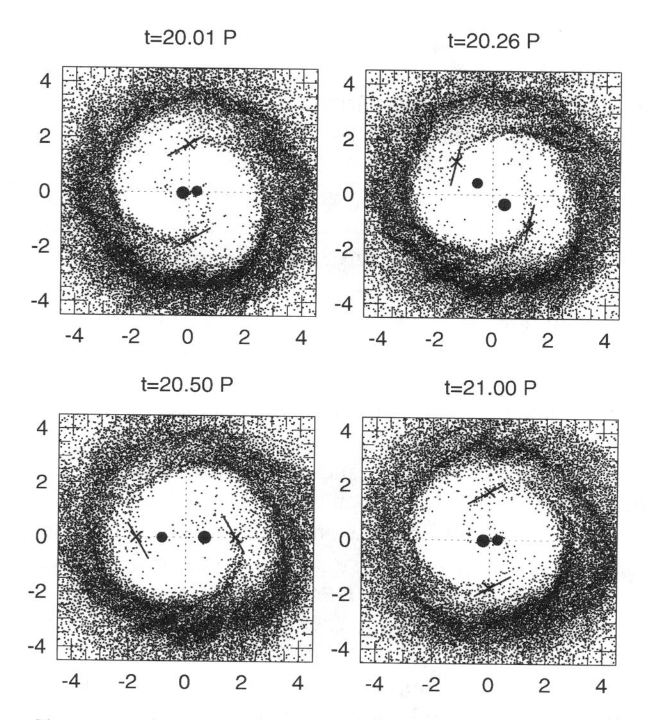

Figure 4. Circumbinary disk with half-thickness $H/r \approx c_s/\Omega r = 0.05$ transferring mass onto the central eccentric ($e = 0.5$) binary with nearly equal components (mass ratio 11:14).

at apastron, the material does not impact the binary in Fig. 4 until approximately periastron. The profiles of $\dot{M}(t)$ depend sensitively on the binary's orbital elements, making the variability a valuable tool in analyzing the properties of unresolved PMS systems (Artymowicz and Lubow 1996*b*).

Near the stars, in the case of moderate orbital eccentricity, the classical Roche effective potential (which corotates with the mean motion of the binary) exerts control over the flow. The flow is channeled by this potential in the vicinity of the classical outer Lagrange points L2 and L3. The L3 point (on the side of the less massive component) allows easier penetration because its effective potential barrier is lower. The L3 point therefore produces a higher accretion rate than the L2 point [as found by Artymowicz (1983) and by Bate and Bonnell (1997)]. Two major consequences of this flow are (i) accretion rate reversal, which may cause the less massive star to become more luminous (due to a higher accretional luminosity), and (ii) mass equalization. The importance of these effects for interpretation of observations is not yet clear (Clarke 1996). In principle, there are implications for the distribution of mass ratios and the observed binary frequency among all types of stars, because actively accreting stars with small companions may be easier to detect as binaries.

Several observations, summarized below, support the existence of a mass flow through gaps in CB disks.

Near-IR images of classical T Tauri binaries GG Tau and UY Aur (Roddier et al. 1996; Close et al. 1998) revealed the inner edges of the CB disks. The inner edge locations based on millimeter observations (Dutrey et al. 1994; Duvert et al. 1998) agree with the IR imaging and theoretical expectations, and the disk rotation is consistent with Keplerian. The IR observations provide weak evidence for the presence of material in the central gap. A high-velocity feature may be associated with gas streams in UY Aur.

A double-lined cTTS close binary, DQ Tau, does not have much space for CS disks, yet it shows all the signs of active accretion, presumably through the CB gap. Continuum and line fluxes undergo modulations on the period of the binary (Mathieu et al. 1997; Basri et al. 1997). The highest fluxes occur near binary periastron, a behavior consistent with the SPH simulations, if the enhanced emission is due to the gas stream impact onto the stars or very small CS disks. The dynamical model is consistent with the variability of spectral features, which provides an independent test for the model.

Spectral energy distributions of several PMS binaries (Jensen and Mathieu 1997) show no evidence for a binary-produced gap. This may indicate that the dust emission from the material flowing through the gap masks its presence.

Observations of 144 G and K dwarfs in Pleiades (Bouvier et al. 1997) showed no significant differences in the rotational velocity distributions for single and binary stars. This suggests that accretion onto both types of stars is similar and, in the case of binaries, likely proceeds from a CB disk.

Embedded IR companions to T Tauri stars (Koresko et al. 1997; Meyer et al. 1997) are plausibly interpreted in terms of the mass flow from CB disks onto the secondary companions. Cases in which the more massive star is apparently accreting more gas are common (Monin et al. 1998).

IV. EFFECTS OF DISKS ON A BINARY

The gravitational interaction of the bulk of the disk with the binary changes its orbital elements. Gas streams can also change the orbital elements, both by gravitational torque contributions and by direct impact (advection of mass, energy, and angular momentum). The potentially important effects of gas streams depend on the still unknown details of where and with what velocity the streams hit the CS disks or stars, in the highly dynamic environment of an eccentric binary. (However, in section V we mention preliminary work on protoplanets, treated as binaries with low eccentricities and extreme mass ratios.) In this section we describe the long-range effects of the disks, which do not include effects of gas streams.

A. Binary Separation and Eccentricity

In a binary with mass parameter $\mu = 0.3$ and eccentricity $e = 0.1$, studied by Artymowicz et al. (1991), by far most of the semimajor axis and eccentricity driving is due to an outer LR of the $(m, l) = (2, 1)$ potential component (which also keeps the disk edge in equilibrium). In such a case, the gravitational torque T_{ml} is equal to the axisymmetric viscous torque, $T_{21} = 3\pi\Sigma\nu\Omega r^2$, evaluated at radius r in the disk just outside of the edge region (maximum wave damping region). This relationship provides a means of estimating the rate of decrease of the binary semimajor axis a (cf. Lubow and Artymowicz 1996):

$$\frac{\dot{a}}{a} = -\frac{6l}{m}\frac{\alpha(H/r)^2}{\sqrt{1-e^2}\mu(1-\mu)}\frac{a}{r}q_d\Omega_b \tag{3}$$

where $q_d = \pi r^2\Sigma/M$ quantifies the disk-binary mass ratio, and M is the total mass of the binary. This robust formula is consistent with the numerical results reported by Artymowicz et al. (1991), giving $\dot{a}_b/a \sim -10^{-3}q_d\Omega_b$ for the binary with $\mu = 0.3$ and eccentricity $e = 0.1$.

The eccentricity evolution of a binary is dominated by the effects of the CB disk. In general, eccentricity grows at a rate

$$\dot{e} = \frac{1-e^2}{e}\left(\frac{l}{m} - \frac{1}{\sqrt{1-e^2}}\right)\frac{\dot{a}}{a} \tag{4}$$

This equation implies that a CB disk shepherded by the (2,1) potential in a low eccentricity binary has $e\dot{e} \simeq -\dot{a}/(2a)$, in good agreement with the SPH results of Artymowicz et al. (1991).

The numerical ratio between rates $\dot{e}/e$ and $\dot{a}/a$ was analyzed by Lubow and Artymowicz (1996), who concluded that the ratio of these rates increases linearly with e up to a maximum found at $e \sim 0.03$ (for their assumed disk parameters), then decreases as $1/e$. The maximum driving corresponds to a short timescale for doubling eccentricity, of the order of several hundred binary orbit periods.

At higher values of eccentricity, many resonances are excited, including resonances that damp eccentricity (eccentric inner LRs and corotational resonances). At an eccentricity of about 0.5 to 0.7, these resonances nearly cancel, and the mean rate of growth of eccentricity diminishes considerably (Lubow and Artymowicz 1993).

In addition, in systems with eccentric disks, the disk forces the binary eccentricity to oscillate on the relative precessional timescale. This occurs because of the disk's (1,0) one-sided forcing, which is similar in basic physics to the reverse influence of the eccentric binary on the disk, discussed previously. It is not uncommon to find in SPH simulations that this leads to temporary (nonsecular) eccentricity damping periods in many binary-disk systems.

B. Implications for Planet Formation

A paradigm in the planet formation theory was that a planet (specifically Jupiter) stops accreting gas when it opens a gap in the primordial nebula (e.g., Lin and Papaloizou 1993). Artymowicz and Lubow (1996*b*) suggested that this does not normally happen (cf. Artymowicz 1998 for a sample SPH calculation). Instead, simulations suggest that matter flows through an otherwise nearly evacuated gap, much as in the case of a binary star (although the dynamics may be different). This may provide a method of formation in standard solar nebulae for the "superplanets," or planets with masses $\sim$5 M_{Jup} (Jupiter masses) or greater. Newly discovered extrasolar planets around 14 Herculis, 70 Virginis, Gl 876, and HD 114762 may fall in this mass range. This somewhat arbitrary mass definition for superplanets assumes a solar-mass companion. In the theory, the mass ratio, rather than the mass, is relevant. Consequently, the dynamical properties we ascribe to superplanets may be appropriate to 70 Vir, since its minimum mass ratio is in the range for superplanets. Mutually compatible results supporting this general scenario have since been obtained with five modern hydrodynamical schemes in eight implementations (Artymowicz et al. 1999; Kley 1999; Bryden et al. 1999; Lubow et al. 1999; chapter by Lin et al., this volume). Results show that there is no fundamental impediment in the growth process of a giant protoplanet via disk accretion to become a superplanet or possibly even a brown dwarf-mass body (although not a brown dwarf according to formative and structural definition). The difficulties are more "practical": The mass of the gas in the disk or its longevity may be insufficient, or the disk may have too little viscosity or pressure to supply efficient flow. For instance, relatively thin and low-viscosity disks tend to prevent the growth of superplanets (see the chapter by Lin et al., this volume). Sufficiently high-mass perturbers may limit further accretion by tidal effects. For standard moderate-mass nebulae (with mass several times the minimum solar nebula, viscosity parameter $\alpha \sim 10^{-3}$–10^{-2}, and $H/r \approx 0.05$) one finds that Jupiter could easily double its mass within a few Myr. Present masses of the solar system giant planets may thus indicate that they formed late in the life of the solar nebula and could not capture the dwindling supply of mass. An apparent lack of their large-scale inward migration supports speaks of the same. That may not be a universal outcome, however.

Color Plate 12 demonstrates the flow, simulated with the PPM (piecewise parabolic method) for a disk with $\alpha = 0.004$ and a Jupiter-mass protoplanet in circular orbit. A top view of a fragment of the 2D disk with embedded Jupiter-mass planet is shown. Two wakes, which are shock waves, penetrate the gap. Disk gas hitting restricted sections of the wakes close to the Lagrange points delimiting the Roche lobe (white oval) is brought to a near-standstill in the frame corotating with the planet and falls toward the planet.

An intriguing correlation between the mass (ratio) of the planet and its orbital eccentricity exists among the currently confirmed radial-velocity planetary candidates (Mazeh et al. 1997; Marcy and Butler 1998). We reproduce in Fig. 5 the distribution of extrasolar planets and solar system planets, including the pulsar planets, in the m-e plane. This distribution, unless an artifact caused by small statistics, shows that superplanets avoid circular orbits and that the majority of Jupiter-class and smaller planets prefer circularity.

In the case of planets, the gaps are much smaller than in the case of binary stars, and the tidally disturbed material experiences many more resonances from the higher-m components of the planet's potential. Still, at some critical mass (called the crossover mass) the (2,1) potential harmonic dominates (the gap must extend to at least the 1:2 and 2:1 commensurabilities), and the same mechanism as described for stars may rapidly deform the initially nearly circular (super)planet orbits into ellipses of intermediate e ($e > 0.2$). On the other hand, planets unable to open gaps, typically less massive than Neptune, suffer strong and model-independent eccentricity damping by coorbital Lindblad resonances (Ward 1986; Artymowicz 1993). In addition, the Jupiter-class planets that open a modest gap are also circularized by corotational torques (Goldreich and Tremaine 1980). Thus, although the value of the crossover mass (which likely depends on disk temperature and viscosity) has yet to be determined in high-resolution calculations taking into account mass flows, its existence is well established. Based on early SPH calculations of only the gravitational interactions, the crossover mass in a standard solar nebula was determined to be $\sim$10 M_{Jup} (Artymowicz 1992). In more recent work, we found evidence that gas inflows such as seen in Color Plate 12 generally damp eccentricities. Since the flow is inefficient at masses exceeding $\sim$10 Jupiter masses (in a standard solar nebula), the crossover mass may indeed be close to that value. However, given the diversity of nebular properties, we do not expect a unique $e(m)$ functional dependence. Rather, a pure disk-planet interaction in an ensemble of systems would result in a switch from very low to moderately high eccentricities over a finite range of masses, probably centered near $m = 10$ M_{Jup}. Should such a mass-eccentricity pattern be confirmed by future, improved statistics, it would strongly argue for the importance of binary-disk interactions in shaping the orbits of exoplanets. Conversely, a large percentage of Jupiter-like or smaller planetary candidates found on high-e orbits around single stars would require a complementary mechanism for eccentricity generation: planet-planet interaction (e.g., Weidenschilling and Marzari 1996; Lin and Ida 1997; Levison et al. 1998).

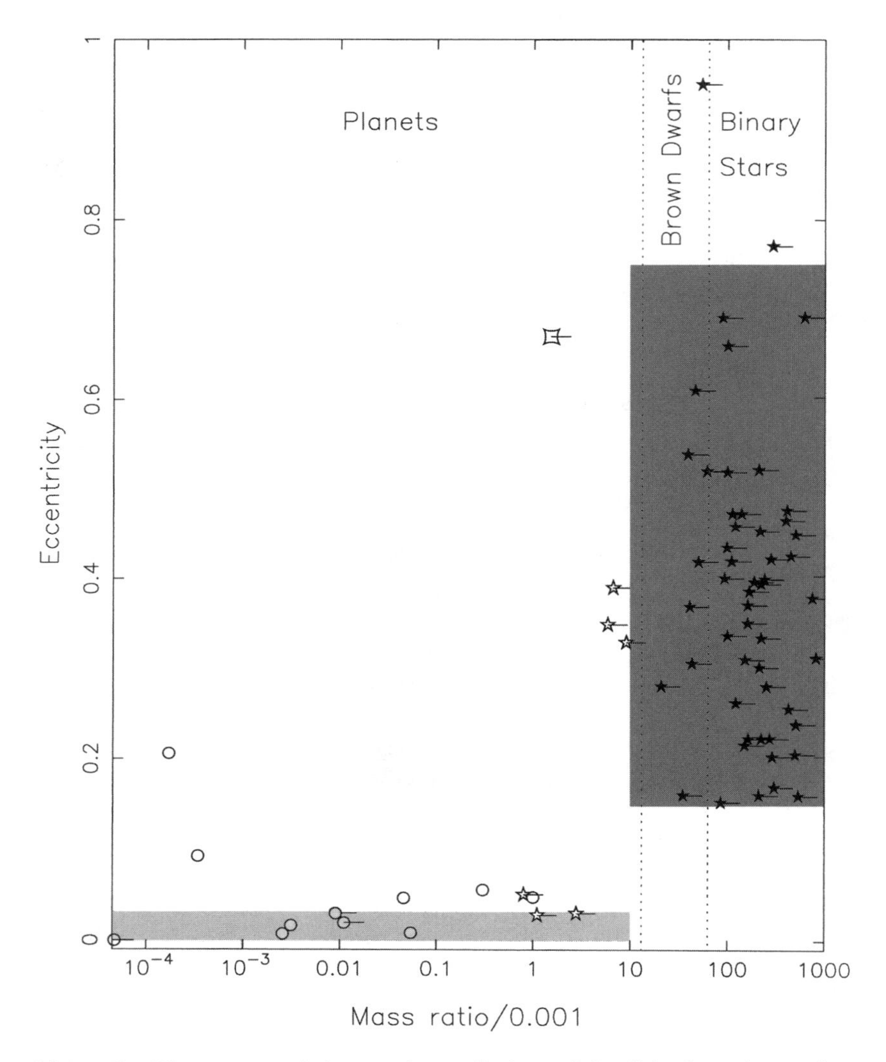

Figure 5. The *m*-*e* correlation vs. the predictions of the disk-planet interaction theory, in which the eccentricity is damped by protoplanetary disks at mass smaller than approximately 10 Jupiter masses and rapidly grows at larger masses. Potentially tidally circularized companions have been omitted (cf. Artymowicz 1998; Artymowicz et al. 1999).

Independent evidence of disk-planet interaction in extrasolar systems, resulting in migration, is provided by the short-period "hot" Jupiters (Lin et al. 1996; chapter by Lin et al., this volume).

V. SUMMARY

Disks play an important role in the evolution of young binaries. At the earliest phases of evolution, most of the material accreted by a sufficiently close binary (semimajor axis less than about 20 AU) passes through a circumbinary disk (see Fig. 1). The evolution occurs in the dynamical collapse timescale of about 10^5 years. During the later T Tauri phase, the evolution occurs on a viscous timescale of the disk. The binary undergoes eccentricity excitation because of its interactions with the disk.

Many aspects of disk-binary interactions are directly observable and provide a potentially powerful diagnostic tool. Resonances emit spiral waves and clear gaps (Fig. 2). Simulations and analytic theory indicate that circumbinary disks sometimes become eccentric as a result of their interactions with the central binaries (Fig. 3). Small-scale features at the disk edge result from interactions with an eccentric binary.

Mass sometimes flows as gas streams from the circumbinary disk, through the central gap, onto the binary (see Fig. 4). This flow can cause the secondary star to appear to be the more luminous (luminosity reversal) and can cause the secondary to accrete more mass than the primary (tending toward mass equalization). The flow has additional consequences on the orbital evolution of the binary that remain to be determined. There are several observational indications of the existence of these flows.

The analysis of binary-disk interactions can be extended to the case of planet-disk interactions. Mass flow onto a young Jupiter can occur despite the presence of a gap (Color Plate 12). It appears possible that some planets can grow to mass ratios with their stars comparable with or larger than the minimum mass ratios of the newly discovered extrasolar superplanets, such as those in 70 Vir or 14 Her. Unlike planet-planet interaction, planet-disk interactions may produce substantial eccentricity for high-mass planets and low eccentricity for low-mass planets. Current observations provide some support for this correlation, but the sample size is small (Fig. 5). The pattern of population of the mass-eccentricity plane by exoplanets should be indicative of which process is typically responsible for the shape of their orbits.

Acknowledgments This work was supported by NASA Grants NAGW-4156 and NAG5-4310. S. L. acknowledges support from The Isaac Newton Institute for Mathematical Sciences of Cambridge University. P. A. acknowledges the STScI visitor program and Swedish Natural Science Research Council research grants. We thank the referee, Matthew Bate, for comments.

REFERENCES

Armitage, P. J., and Pringle, J. E. 1997. Radiation-induced warping of protostellar accretions disks. *Astrophys. J. Lett.* 488:L47–L50.

Artymowicz, P. 1983. The role of accretion in binary star formation. *Acta Astron.* 33:223.

Artymowicz, P. 1992. Dynamics of binary and planetary system interaction with disks: Eccentricity evolution. *Pub. Astron Soc. Pacific* 104:769–774.

Artymowicz, P. 1993. Disk-satellite interaction via density waves and the eccentricity evolution of bodies embedded in disks. *Astrophys. J.* 419:166–180.

Artymowicz, P. 1998. On the formation of eccentric superplanets. In *Brown Dwarfs and Extrasolar Planets*, ASP Conf. Ser., 134, ed. R. Rebolo, E. L. Martin, and M. R. Zapatero-Osorio (San Francisco: Astronomical Society of the Pacific), pp. 152–161.

Artymowicz, P., and Lubow, S. H. 1994. Dynamics of binary-disk interaction: I. Resonances and disk gap sizes. *Astrophys. J.* 421:651–667.

Artymowicz, P., and Lubow, S. H. 1996*a*. Interaction of young binaries with protostellar disks. In *Disks and Outflows around Young Stars*, ed. S. Beckwith, J. Staude, A. Quetz, and A. Natta (Berlin: Springer), p. 115 (p. 242 in CD-ROM version).

Artymowicz, P., and Lubow, S. H. 1996*b*. Mass flow through gaps in circumbinary disks. *Astrophys. J. Lett.* 467:L77–L80.

Artymowicz, P., Clarke, C., Lubow, S. H., and Pringle, J. E. 1991. The effect of an external disk on the orbital elements of a central binary. *Astrophys. J. Lett.* 370:L35–L38.

Artymowicz, P., Lubow, S. L., and Kley, W. 1999. Planetary systems and their changing theories. In *Planetary Systems–The Long View*, ed. L. Celnikier et al. (Gif sur Yvette: Editions Frontières), in press.

Balbus, S. A., and Hawley, J. F. 1991. A powerful local shear instability in weakly magnetized disks. I. Linear analysis. II. Nonlinear evolution. *Astrophys. J.* 376:214–222.

Basri, G., Johns-Krull, C. M., and Mathieu, R. D. 1997. The classical T Tauri spectroscopic binary DQ Tau. II. Emission line variations with orbital phase. *Astron. J.* 114:781–792.

Bate, M., and Bonnell, I. 1997. Accretion during binary star formation. II. Gaseous accretion and disc formation. *Mon. Not. Roy. Astron. Soc.* 285:33–48.

Beckwith, S.V.W., Sargent, A. I., Chini, R. S., and Guesten, R. 1990. A survey for circumstellar disks around young stellar objects. *Astron. J.* 99:924–945.

Bodenheimer, P., Ruzmaikina, T., and Mathieu, R. D. 1993. Stellar multiple systems—constraints on the mechanism of origin. In *Protostars and Planets III*, ed. E. H. Levy and J. I. Lunine (Tucson: University of Arizona Press), p. 367–404.

Bondi, H., and Hoyle, F. 1944. On the mechanism of accretion by stars. *Mon. Not. Roy. Astron. Soc.* 104:273.

Bouvier, J., Rigaut, F., and Nadeau, D. 1997. Pleiades low-mass binaries: Do companions affect the evolution of protoplanetary disks? *Astron. Astrophys.* 323:139–150.

Burrows, C. J., Stapelfeldt, K. R., Watson, A., Krist, J. E., Ballester, G. E., Clarke, J. T., Crisp, D., Gallagher, J. S., Griffiths, R. E., Hester, J. J., Hoessel, J. G., Holtzman, J. A., Mould, J. R., Scowen, P. A., Trauger, J. T., and Westphal, J. A. 1996. Hubble Space Telescope observations of the disk and jet of HH 30. *Astrophys. J.* 473:437–451.

Bryden, G., Chen, X., Lin, D.N.C., Nelson, R. P., and Papaloizou, J.C.B. 1999. Tidally induced gap formation in protostellar disks: Gap clearing and suppression of protoplanetary growth. *Astrophys. J.,* in press.

Cassen, P., and Moosman. A. 1981. On the formation of protostellar disks. *Icarus* 48:353–376.

Cassen, P., and Woolum, D. S. 1996. Radiatively damped density waves in optically thick protostellar disks. *Astrophys. J.* 472:789-799.

Clarke, C. 1996. Dynamical processes in binary star formation. *NATO ASI 477,* ed. R. Wijers, M. Davies, and C. Tout (Dordrecht: Kluwer), p. 31.

Close, L. M., Dutrey, A., Roddier, F., Guilloteau, S., Roddier, C., Northcott, M., Menard, F., Duvert, G., Graves, J. E., and Potter, D. 1998. Adaptive optics imaging of the circumbinary disk around the T Tauri binary UY Aurigae: Estimates of the binary mass and circumbinary dust grain size distribution. *Astrophys. J.* 499:883–888.

Duquennoy, A., and Mayor, M. 1991. Multiplicity among solar-type stars in the solar neighbourhood. II. Distribution of the orbital elements in an unbiased sample. *Astron. Astrophys.* 248:485–524.

Dutrey, A., Guilloteau, S., and Simon, M. 1994. Images of the GG Tauri rotating ring. *Astron. Astrophys.* 286:149–159.

Duvert, G., Dutrey, A., Guilloteau, S., Menard, F., Schuster, K., Prato, L., and Simon, M. 1998. Disks in the UY Aurigae binary. *Astron. Astrophys.* 332:867–874.

Gammie, C. F. 1996. Layered accretion in T Tauri disks. *Astrophys. J.* 457:355–362.

Ghez, A. M., Neugebauer, G., and Matthews, K. 1993. The multiplicity of T Tauri stars in the star forming regions Taurus-Auriga and Ophiuchus-Scorpius: A 2.2 micron speckle imaging survey. *Astrophys. J.* 106:2005–2023.

Glassgold, A. E., Najita, J., and Igea, J. 1997. X-ray ionization of protoplanetary disks. *Astrophys. J.* 480:344–350.

Goldreich, P., and Tremaine, S. 1980. Disk-satellite interactions. *Astrophys. J.* 241:425–441.

Hale, A. 1994. Solar type binary systems. *Astron. J.* 107:306–332.

Hartigan, P., Edwards, S., and Ghandour, L. 1995. Disk accretion and mass loss from young star. *Astrophys. J.* 452:736–768.

Hartmann, L., Calvet, N., Gullbring, E., and D'Alessio, P. 1998. Accretion and the evolution of T Tauri disks. *Astrophys. J.* 495:385–400.

Hawley, J. F., Gammie, C. F., and Balbus, S. A. 1996. Local 3D simulations of an accretion disk hydromagnetic dynamo. *Astrophys. J.* 464:690–703.

Hirose, M., and Osaki, Y. 1993. Superhump periods in SU Ursae Majoris stars: Eigenfrequency of the eccentric mode of an accretion disk. *Pub. Astron. Soc. Pacific* 45:595–604.

Jensen, E. L. N., and Mathieu, R. D. 1997. Evidence for cleared regions in the disks around pre-main-sequence spectroscopic binaries. *Astron. J.* 114:301–316.

Jensen, E. L. N., Mathieu, R. D., and Fuller, G. A. 1996. The connection between submillimeter continuum flux and binary separation in young binaries: Evidence of interaction between stars and disks. *Astrophys. J.* 458:312–326.

Kley, W. 1999. Mass accretion onto protoplanets through gaps. *Mon. Not. Roy. Astron. Soc.*, in press.

Koresko, C. D., Harvey, P. M., Christou, J. C., Fugate, R. Q., and Li, W. 1997. The infrared companions of T Tauri stars. *Astrophys. J.* 480:741–753.

Lada, C. J., and Shu, F. H. 1990. The formation of sunlike stars. *Science* 248:564–572.

Larwood, J. D., Nelson, R. P., Papaloizou, J. C. B., and Terquem, C. 1996. The tidally induced warping, precession and truncation of accretion discs in binary systems: Three-dimensional simulations. *Mon. Not. Roy. Astron. Soc.* 282:597–613.

Levison, H. F., Lissauer, J. J., and Duncan, M. J. 1998. Modeling the diversity of outer planetary systems. *Astron. J.* 116:2067–2077.

Lin, D. N. C., and Ida, S. 1997. On the origin of massive eccentric planets. *Astrophys. J.* 477:781–791.

Lin, D. N. C., and Papaloizou, J. C. B. 1979. On the structure of circumbinary accretion disks and the tidal evolution of commensurable satellites. *Mon. Not. Roy. Astron. Soc.* 188:191–201.

Lin, D. N. C., and Papaloizou, J. C. B. 1993. On the tidal interaction between protostellar disks and companions. In *Protostars and Planets III,* ed. E. H. Levy and M. S. Matthews (Tucson: University of Arizona Press), pp. 749–836.

Lin, D. N. C., and Pringle, J. E. 1987. A viscosity prescription for a self-gravitating accretion disc. *Mon. Not. Roy. Astron. Soc.* 225:607–613.

Lin, D. N. C., Papaloizou, J. C. B., and Savonije, G. J. 1990. Propagation of tidal disturbance in gaseous accretion disks. *Mon. Not. Roy. Astron. Soc.* 365:748–756.

Lin, D. N. C., Bodenheimer, P., and Richardson, D. C. 1996. Orbital migration of the planetary companion of 51 Pegasi to its present location. *Nature* 380:606–607.

Lissauer, J. J., Shu, F. H., and Cuzzi, J. N. 1981. Moonlets in Saturn's rings. *Nature* 292:707–711.

Looney, L. G., Mundy, L. W., and Welch, W. J. 1997. High resolution 2.7 millimeter observations of L1551 IRS 5. *Astrophys. J. Lett.* 484:L157–L160.

Lubow, S. H. 1991 A model for tidally driven eccentric instabilities in fluid disks. *Astrophys. J.* 381:259–267.

Lubow, S. H., and Artymowicz, P. 1993. Eccentricity evolution of a binary embedded in a disk. In *Binaries as Tracers of Stellar Evolution*, ed. A. Duquennoy and M. Mayor (Cambridge: Cambridge University Press), p. 145.

Lubow, S. H., and Artymowicz, P. 1996. Young binary star/disk interactions. *NATO ASI 477,* ed. R. Wijers, M. Davies, and C. Tout (Dordrecht: Kluwer), p. 53.

Lubow, S. H., and Ogilvie, G. I. 1998. 3D waves generated at Lindblad resonances in thermally stratified disks. *Astrophys. J.* 504:983–995.

Lubow, S. H., Seibert, M., and Artymowicz, P. 1999. Disk accretion onto high-mass planets. *Astrophys. J.,* in press.

Lubow, S. H., and Shu, F. H. 1975. Gas dynamics of semi-detached binaries. *Astrophys. J.* 198:383–405.

Lynden-Bell, D., and Pringle, J. 1974. The evolution of viscous discs and the origin of the nebular variables. *Mon. Not. Roy. Astron. Soc.* 168:603–637.

Marcy, G. W., and Butler, P. R. 1998. Detection of extrasolar giant planets. *Ann. Rev. Astron. Astrophys.* 36:57–98.

Mathieu, R. D. 1994. Pre-main-sequence binary stars. *Ann. Rev. Astron. Astrophys.* 32:465–530.

Mathieu, R. D., Stassun, K., Basri, G., Jensen, E. L. N., Johns-Krull, C. M., Valenti, J. A., and Hartmann, L. W. 1997. The classical T Tauri spectroscopic binary DQ Tau. I. Orbital elements and light curves. *Astron. J.* 113:1841–1854.

Mazeh, T., Mayor, M., and Latham, D. 1997. Eccentricity versus mass for low-mass secondaries and planets. *Astrophys. J.* 478:367–370.

McCaughrean, M. J., and O'Dell, C. R. 1996. Direct imaging of circumstellar disks in the Orion Nebula. *Astron. J.* 111:1977–1986.

Meyer, M. R., Beckwith, S. V. W., Herbst, T. M., and Robberto, M. 1997. The transitional pre-main-sequence object DI Tauri: Evidence for a substellar companion and rapid disk evolution. *Astrophys. J.* 489:L173–L177.

Monin, J.-L., Duchene, G., and Geoffray, H. 1998. Circumstellar environment of PMS binaries. Preprint.

Ogilvie, G. I., and Lubow, S. H. 1999. The effect of an isothermal atmosphere on the propagation of three-dimensional waves in a thermally stratified disk. *Astrophys. J.* 515:767–775.

Papaloizou, J. C. B., and Pringle, J. 1983. The time-dependence of non-planar accretion disks. *Mon. Not. Roy. Astron. Soc.* 202:1181–1194.

Papaloizou, J. C. B., and Terquem, C. 1995. On the dynamics of tilted discs around young stars. *Mon. Not. Roy. Astron. Soc.* 274: 987–1001.

Pringle, J. E. 1981. Accretion discs in astrophysics. *Ann. Rev. Astron. Astrophys.* 19:137–162.

Pringle, J. E. 1991. The properties of external accretion discs. *Mon. Not. Roy. Astron. Soc.* 248:754–759.

Roddier, C., Roddier, F., Northcott, M. J., Graves, J. E., and Jim, K. 1996. Adaptive optics imaging of GG Tauri: Optical detection of the circumbinary ring. *Astrophys. J. Lett.* 463:326–335.

Sargent, A. I., and Beckwith, S. V. W. 1994. The detection and study of preplanetary disks. *Astrophys. Space Sci.* 212,1:181–189.

Savonije, G. J., Papaloizou, J. C. B., and Lin, D. N. C. 1994. On tidally induced shocks in accretion discs in close binary systems. *Mon. Not. Roy. Astron. Soc.* 268:13–28.

Shakura, N. I., and Sunyaev, R. A. 1973. Black holes in binary systems. Observational appearance. *Astron. Astrophys.* 24:337–355.

Shu, F. H., Adams, F., and Liszano, S. 1987. Star formation in molecular clouds—observation and theory. *Ann. Rev. Astron. Astrophys.* 25:23–81.

Spruit, H. 1987. Stationary shocks in accretion disks. *Astron. Astrophys.* 184:173–174.

Stapelfeldt, K. R., Krist, J. E., Menard, F., Bouvier, J., Padgett, D. L., and Burrows, C. J. 1998. An edge-on circumstellar disk in the young binary system HK Tauri. *Astrophys. J.* 502:L65–L68.

Takeuchi, T., Miyama, S. M., and Lin, D. N. C. 1996. Gap formation in protoplanetary disks. *Astrophys. J.* 460:832–847.

Tout, C. A., and Pringle, J. E. 1992. Accretion disc viscosity—a simple model for a magnetic dynamo. *Mon. Not. Roy. Astron. Soc.* 259:604–612.

Ward, W. R. 1986. Density waves in the solar nebula—differential Lindblad torque. *Icarus* 67:164–180.

Weidenschilling, S. J., and Marzari, F. 1996. Gravitational scattering as a possible origin for giant planets at small stellar distances. *Nature* 384:619–621.

Whitehurst, R. 1988. Numerical simulations of accretion disks. I. Superhumps—a tidal phenomenon of accretion disks. *Mon. Not. Roy. Astron. Soc.* 232:35–51.

Yuan, C., and Cassen, P. 1994. Resonantly driven nonlinear density waves in protostellar disks. *Astrophys. J.* 437:338–350.

PART IV
Jets and Outflows

DISK WINDS AND THE ACCRETION-OUTFLOW CONNECTION

ARIEH KÖNIGL
University of Chicago

and

RALPH E. PUDRITZ
McMaster University

We review recent observational and theoretical results on the relationship between circumstellar accretion disks and jets in young stellar objects. We then present a theoretical framework that interprets jets as accretion-powered, centrifugally driven winds from magnetized accretion disks. Recent progress in the numerical simulation of such outflows is described. We also discuss the structure of the underlying magnetized protostellar disks, emphasizing the role that large-scale, open magnetic fields can play in angular momentum transport.

I. INTRODUCTION

Two of the most remarkable aspects of star formation are the presence of disks and of energetic outflows already during the earliest phases of protostellar evolution. There is now strong evidence for an apparent correlation between the presence of outflows and of actively accreting disks, which suggests that there is a physical link between them. The prevalent interpretation is that the outflows are powered by accretion and that magnetic stresses mediate the inflow and outflow processes and eject some of the inflowing matter from the disk surfaces. If disks are threaded by open magnetic field lines, then the outflows can take the form of centrifugally driven winds. Such highly collimated winds carry angular momentum and may, in principle, play an important role in the angular momentum budget of disks and their central protostars.

This review concentrates on the developments in the study of outflows and their relationship to circumstellar disks that have occurred since the publication of *Protostars and Planets III*. The reader may consult Königl and Ruden (1993) in that volume and Pudritz et al. (1991) for reviews of earlier work. In section II we summarize the observational findings on outflows, disks, and magnetic fields, and their implications. Section III deals with the general theory of magnetized outflows, and section IV describes numerical simulations of disk-driven magnetohydrodynamic (MHD) winds. In section V we consider the theory of magnetized protostellar disks. Our conclusions are presented in section VI.

II. OBSERVATIONAL BACKGROUND

A. Bipolar Outflows and Jets

Bipolar molecular outflows and narrow atomic jets are ubiquitous phenomena in protostars. There are now more than 200 bipolar CO sources known (see the chapter by Richer et al., this volume); they typically appear as comparatively low-velocity ($\lesssim$25 km s^{-1}) and moderately collimated (length-to-width ratios $\sim$3–10) lobes, although several highly collimated CO outflows that exhibit high velocities ($>$40 km s^{-1}) near the flow axis have now been detected. The mass outflow rate exhibits a continuous increase with the bolometric luminosity of the driving source for $L_{\rm bol}$ in the range ~1–10^6 L$_\odot$. Molecular outflows are present through much of the embedded phase of protostars and, in fact, appear to be most powerful and best collimated during the earliest (Class 0) protostellar evolutionary phase (Bontemps et al. 1996).

The bipolar lobes are generally understood to represent ambient molecular material that has been swept up by the much faster, highly supersonic jets that emanate from the central star/disk system (see the chapters by Eislöffel et al. and by Hartigan et al., this volume). Jets associated with low-luminosity ($L_{\rm bol} < 10^3$ L$_\odot$) young stellar objects (YSOs) have velocities in the range ~150–400 km s^{-1}, large ($>$20) Mach numbers, and opening angles as small as $\sim$3–5° on scales of 10^3–10^4 AU. The inferred mass outflow rates are ~10^{-10}–10^{-8} M$_\odot$ yr^{-1}. A significant number of outflows has also been detected by optical observations of intermediate-mass ($2 \lesssim M_\star/{\rm M}_\odot \lesssim 10$) Herbig Ae/Be stars and other high-luminosity sources (Mundt and Ray 1994; Corcoran and Ray 1997). The jet speeds and mass outflow rates in these YSOs are, respectively, a factor $\sim$2–3 and ~10–100 higher than in low-$L_{\rm bol}$ objects. The total momentum delivered by the jets, taking into account both the density corrections implied by their partial ionization state and the long lifetimes indicated by the detection of parsec-scale outflows, appears to be consistent with that measured in the associated CO outflows (e.g., Hartigan et al. 1994; Eislöffel and Mundt 1997). A critical review of the physical mechanisms of coupling the jets and the surrounding gas is given in Cabrit et al. (1997).

The momentum discharge deduced from the bipolar outflow observations is typically a factor ~10^2 higher than the radiation pressure thrust $L_{\rm bol}/c$ produced by the central source (e.g., Lada 1985), which rules out radiative acceleration of the jets. Because the bolometric luminosity of protostars is by and large due to accretion, and because the ratio of jet kinetic luminosity to thrust is of the order of the outflow speed (~10^{-3}c), it follows that the jet kinetic luminosity is on average a fraction ~ 0.1 of the rate at which gravitational energy is liberated by accretion. This high ejection efficiency is most naturally understood if the jets are driven magnetically.

Magnetic fields have also been implicated in jet collimation. A particularly instructive case is provided by *Hubble Space Telescope* (HST) observations of the prototypical disk/jet system HH 30 (Burrows et al. 1996). The jet in this source can be traced to within $\lesssim$30 AU from the star and appears as a cone with an opening angle of 3° between 70 and 700 AU. The narrowness of the jet indicates some form of intrinsic collimation, because external density gradients would not act effectively on these small scales. Magnetic collimation is a likely candidate, made even more plausible by the fact that the jet appears to recollimate: its apparent opening angle decreases to 1.9° between 350 and 10^4 AU. Similar indications of recollimation have also been found in other jets. Given that any inertial confinement would be expected to diminish with distance from the source, this points to the likely role of intrinsic magnetic collimation.

B. Connection with Accretion Disks

The evidence for disks around YSOs and for their link to outflows has been strengthened by a variety of recent observations. These include systematic studies of the frequency of disks by means of infrared and millimeter surveys and further interferometric mappings (now comprising also the submillimeter range), which have resolved the structure and velocity field of disks down to scales of a few tens of AU (see the chapter by Wilner and Lay, this volume). High-resolution images of disks in several jet sources have also been obtained in the near infrared using adaptive optics and in the optical using the HST (see the chapter by McCaughrean et al., this volume).

Stellar jets are believed to be powered by the gravitational energy liberated in the accretion process and to be fed by disk material. This picture is supported by the strong apparent correlation that is found (e.g., Cabrit et al. 1990; Cabrit and André 1991; Hartigan et al. 1995) between the presence of outflow signatures (such as P Cygni line profiles, forbidden-line emission, thermal radio radiation, or well-developed molecular lobes) and accretion diagnostics (such as ultraviolet, infrared, and millimeter-wavelength emission excesses, or inverse P Cygni line profiles). Further support is provided by the apparent decline in outflow activity with stellar age, which follows the similar trend exhibited by the disk frequency (see the chapters by André et al. and by Mundy et al., this volume) and mass accretion rate (see the chapter by Calvet et al., this volume). Whereas virtually every Class 0 source has an associated outflow, a survey of optical and molecular outflows in the Taurus-Auriga cloud (Gomez et al. 1997) found an incidence rate of $\gtrsim$60 % among Class I objects but only ~10% among Class II ones (and none in Class III objects). The inference that jets are powered by accretion and originate in accretion disks is strengthened by the evidence for disks in the youngest YSOs in which outflows are detected. For example, submillimeter interferometric observations of VLA 1623, one of the youngest known Class 0 sources, imply the presence of a circumstellar disk of radius $<$175 AU and mass $\geq$0.03 $M_\odot$ (Pudritz et al. 1996).

Corcoran and Ray (1998) demonstrated that the correlation between [O I] λ 6300 Å line luminosity (an outflow signature) and excess infrared luminosity (an accretion diagnostic) originally found in Class II sources extends smoothly to YSOs with masses of up to ~10 $M_\odot$ and spans 5 orders of magnitude in luminosity. It is noteworthy that correlations of the type $\dot{M} \propto L_{\rm bol}^{0.6}$ that apply to both low-luminosity and high-luminosity YSOs have been established in several independent studies for the mass *accretion* rate (from IR continuum measurements; Hillenbrand et al. 1992; see, however, Hartmann et al. 1993; Bell 1994; Miroshnichenko et al. 1997; and Pezzuto et al. 1997 for alternative interpretations of the infrared emission in Herbig Ae/Be stars), the *ionized* mass *outflow* rate in the jets (from radio continuum observations; Skinner et al. 1993), and the bipolar *molecular outflow* rate (from CO line measurements; Levreault 1988). Taken together, these relationships suggest that a strong link between accretion and outflow exists also in high-mass YSOs and that the underlying physical mechanism is basically the same as in low-mass objects (see Königl 1999).

Strong evidence for a disk origin of jets is available for the energetic outflows associated with FU Orionis outbursts (see the chapters by Bell et al. and by Calvet et al., this volume). The outbursts have been inferred to arise in young YSOs that are still rapidly accreting, although it is possible that they last into the Class II phase. The duration of a typical outburst is $\sim 10^2$ yr, and during that time the mass accretion rate (as inferred from the bolometric luminosity) is $\sim 10^{-4}$ $M_\odot yr^{-1}$, with the deduced mass outflow rate $\dot{M}_{\rm wind}$ (at least in the most powerful sources like FU Ori and Z CMa) being a tenth as large. The ratio $\dot{M}_{\rm wind}/\dot{M}_{\rm acc} \approx 0.1$ is similar to that inferred in Class II YSOs and again points to a rather efficient outflow mechanism. Detailed spectral modeling demonstrates that virtually all the emission during an outburst is produced in a rotating disk. Furthermore, the correlation found in the prototype FU Ori between the strength and the velocity shift of various photospheric absorption lines can be naturally interpreted in terms of a wind accelerating from the disk surface (Calvet et al. 1993; Hartmann and Calvet 1995). Because of the comparatively low temperatures ($\sim$6000 K) in the wind acceleration zone, thermal pressure and radiative driving are unimportant; magnetic driving is thus strongly indicated. The recurrence time of outbursts has been estimated to lie in the range $\sim 10^3$–10^4 yr, and if these outbursts are associated with the large-scale bow shocks detected in parsec-scale jets (e.g., Reipurth 1991), then a value near the lower end of the range is implied. In that case most of the stellar mass would be accumulated through this process, and, correspondingly, most of the mass and momentum ejected over the lifetime of the YSO would originate in a disk-driven outflow during the outburst phases (Hartmann 1997).

C. Magnetic Fields in Outflow Sources

The commonly accepted scenario for the origin of low-mass protostars is that they are produced from the collapse of the inner regions of molecular clouds that are supported by large-scale magnetic fields (and likely also hydromagnetic waves). In this picture, a gravitationally unstable inner core forms as a result of mass redistribution by ambipolar diffusion and subsequently collapses dynamically (see review by McKee et al. 1993). The mass accretion rates predicted by this picture are consistent with the inferred evolution of young YSOs (e.g., Ciolek and Königl 1998). Basic support for this scenario is provided by far-infrared (e.g., Hildebrand et al. 1995) and submillimeter (e.g., Greaves et al. 1994, 1995; Schleuning 1998; Greaves and Holland 1998) polarization measurements, which reveal an ordered, hourglass-shaped field morphology on subparsec scales, consistent with the field lines being pulled in at the equatorial plane of the contracting core. Moreover, H I and OH Zeeman measurements (e.g., Crutcher et al. 1993, 1994, 1996) are consistent with the magnetic field having the strength to support the bulk of the cloud against gravitational collapse.

There now exist measurements of magnetic fields in the flows themselves at large distances from the origin (see the chapter by Eislöffel et al., this volume). In particular, the strong circular polarization detected in T Tau S in two oppositely directed nonthermal emission knots separated by 20 AU indicates a field strength of at least several gauss (Ray et al. 1997). This high value can be attributed to a magnetic field that is advected from the origin by the associated stellar outflow and that dominates the internal energy of the jet. This observation thus provides direct evidence for the essentially hydromagnetic character of jets.

III. MHD WINDS FROM ACCRETION DISKS

A. Basic MHD Wind Theory

The theory of centrifugally driven winds was first formulated in the context of rotating, magnetized stars (Schatzman 1962; Weber and Davis 1967; Mestel 1968). Using 1D, axisymmetric models, it was shown that such stars could lose angular momentum by driving winds of this type. This idea was applied to magnetized accretion disks in the seminal paper of Blandford and Payne (1982). Every annulus of a Keplerian disk may be regarded as rotating close to its "breakup" speed, so disks are ideal drivers of outflows when sufficiently well magnetized. The removal of disk angular momentum allows matter to move inward and produces an accretion flow. In a steady state, field lines must also slip radially out of the accreting gas and maintain their fixed position in space; this constraint necessitates diffusive processes. As we discuss in Section V.B, strong field diffusivity is a natural attribute of the partially ionized regions of protostellar disks, and it could counter both the advection of the field lines by the radial inflow and their winding-up by the differential rotation in the disk.

Consider, for the moment, the simplest possible description of a magnetized, rotating gas threaded by a large-scale, open field (characterized by an even symmetry about the midplane $z = 0$). The equations of stationary, axisymmetric, ideal MHD are the conservation of mass (continuity equation); the equation of motion with conducting gas of density ρ subject to forces associated with the pressure, p, the gravitational field (from the central object, whose gravitational potential is Φ), and the magnetic field $\mathbf{B}$; the induction equation for the evolution of the magnetic field in the moving fluid; and the solenoidal condition on $\mathbf{B}$:

$$\nabla \cdot (\rho \mathbf{V}) = 0 \tag{1}$$

$$\rho \mathbf{V} \cdot \nabla \mathbf{V} = -\nabla p - \rho \nabla \Phi + \frac{1}{4\pi}(\nabla \times \mathbf{B}) \times \mathbf{B} \tag{2}$$

$$\nabla \times (\mathbf{V} \times \mathbf{B}) = 0 \tag{3}$$

$$\nabla \cdot \mathbf{B} = 0 \tag{4}$$

Consider the **angular momentum equation** for axisymmetric flows. This is described by the ϕ component of equation (2). Ignoring stresses that would arise from turbulence, and noting that neither the pressure nor the gravitational term contributes, we find

$$\rho \mathbf{V}_{\mathrm{p}} \cdot \nabla (r V_{\phi}) = \frac{\mathbf{B}_{\mathrm{p}}}{4\pi} \cdot \nabla (r B_{\phi}) \tag{5}$$

where we have broken the magnetic and velocity fields into poloidal and toroidal components: $\mathbf{B} = \mathbf{B}_{\mathrm{p}} + B_{\phi}\hat{\mathbf{e}}_{\phi}$ and $\mathbf{V} = \mathbf{V}_{\mathrm{p}} + V_{\phi}\hat{\mathbf{e}}_{\phi}$.

Important links between the velocity field and the magnetic field are contained in the induction equation (3), whose solution is

$$\mathbf{V} \times \mathbf{B} = \nabla \chi \tag{6}$$

where χ is some scalar potential. This shows that the electric field due to the bulk motion of conducting gas in the magnetic field is derivable from an electrostatic potential. This has two important ramifications. The first is that, because of axisymmetry, the toroidal component of this equation must vanish ($\partial\chi/\partial\phi = 0$). This forces the poloidal velocity vector to be parallel to the poloidal component of the magnetic field, $\mathbf{V}_{\mathrm{p}} || \mathbf{B}_{\mathrm{p}}$. This, in turn, implies that there is a function k, the mass load of the wind, such that

$$\rho \mathbf{V}_{\mathrm{p}} = k \mathbf{B}_{\mathrm{p}} \tag{7}$$

Substitution of this result into the continuity equation (1), and then use of the solenoidal condition [equation (4)], reveals that k is a constant along a surface of constant magnetic flux; that is, it is conserved along field lines. This function can be more revealingly cast by noting that the wind mass loss rate passing through an annular section of the flow of area dA is $d\dot{M}_w = \rho V_p \, dA$, whereas the amount of poloidal magnetic flux through this same annulus is $d\Psi = B_p \, dA$. Thus, the mass load per unit time and per unit magnetic flux, which is preserved along each streamline emanating from the rotor (a disk in this case), is

$$k = \frac{\rho V_p}{B_p} = \frac{d\dot{M}_w}{d\Psi} \tag{8}$$

The mass load is determined by the physics of the underlying rotor, which is its source.

A second major consequence of the induction equation follows from the poloidal part of equation (6). Taking the dot product of it with $\mathbf{B}_p$ and using equation (7), it can be easily proved that the function $\omega = \Omega - (kB_\phi/\rho r)$, where $\Omega = V_\phi/r$, is also a constant along a magnetic flux surface. In order to evaluate ω, note that $B_\phi = 0$ at the disk midplane by the assumed even symmetry. Thus, ω equals Ω_0, the angular velocity of the disk at the midplane. We thus have a relation between the toroidal field in a rotating flow and the rotation of that flow,

$$B_\phi = \frac{\rho r}{k}(\Omega - \Omega_0) \tag{9}$$

Let us now examine the angular momentum equation. Returning to the full equation (5) and applying equation (7) and the constancy of k along a field line, we obtain

$$\mathbf{B}_p \cdot \nabla\left(rV_\phi - \frac{rB_\phi}{4\pi k}\right) = 0 \tag{10}$$

Hence the angular momentum per unit mass,

$$l = rV_\phi - \frac{rB_\phi}{4\pi k} \tag{11}$$

is constant along a streamline. This shows that the specific angular momentum of a magnetized flow is carried by both the rotating gas (first term) and the twisted field (second term). The value of l may be found by eliminating the toroidal field between equations (9) and (11) and solving for the rotation speed of the flow

$$rV_\phi = \frac{lm^2 - r^2\Omega_o}{m^2 - 1} \tag{12}$$

where the Alfvén Mach number m of the flow is defined as $m^2 = V_p^2/V_{Ap}^2$, with $V_{Ap} = B_p/(4\pi\rho)^{1/2}$ being the poloidal Alfvén speed of the flow. The Alfvén surface is the locus of the points $r = r_A$ on the outflow field lines where $m = 1$. The flow along any field line essentially corotates with the rotor until this point is reached. From the regularity condition at the Alfvén critical point [where the denominator of equation (12) vanishes], it follows that the conserved specific angular momentum satisfies

$$l = \Omega_0 r_A^2 \tag{13}$$

If we imagine following a field line from its footpoint at a radius r_0, the Alfvén radius is at a distance $r_A(r_0)$ from the rotation axis and constitutes the lever arm for the back torque that this flow exerts on the disk. The other critical points of the outflow are where the outflow speed V_p equals the speed of the slow and fast magnetosonic modes in the flow (at the so-called SM and FM surfaces).

Finally, a generalized version of Bernoulli's equation may be derived by taking the dot product of the equation of motion with $\mathbf{B}_p$. We then find that the specific energy

$$E = \frac{1}{2}(V_p^2 + \Omega^2 r^2) + \Phi + h + \Omega_0(\Omega_0 r_A^2 - \Omega r^2) \tag{14}$$

where h is the enthalpy per unit mass, is also a field-line constant.

The terminal speed $V_p = V_\infty$ corresponds to the region where the gravitational potential and the rotational energy of the flow are negligible. Since for cold flows the specific enthalpy may be ignored, we can infer from equation (14) that

$$V_\infty \simeq 2^{1/2}\Omega_0 r_A \tag{15}$$

a result first obtained by Michel (1969) for 1D flows. The important point regarding outflow speeds from disks is that $V_\infty/\Omega_0 r_0 \approx 2^{1/2} r_A/r_0$: The asymptotic speed is larger than the rotor speed by a factor that is approximately the ratio of the lever arm to the footpoint radius.

B. Connection with Underlying Accretion Disk

We now apply the angular momentum conservation relation (5) to calculate the torque exerted on a thin accretion disk by the external magnetic field. The vertical flow speed in the disk is negligible, so only the radial inflow speed V_r and the rotation speed V_ϕ (Keplerian for thin disks) contribute. On the righthand side, both the radial and vertical magnetic contributions come into play, so

$$\frac{\rho V_r}{r_0}\frac{\partial(r_0 V_\phi)}{\partial r_0} = \frac{B_r}{4\pi r_0}\frac{\partial(r_0 B_\phi)}{\partial r_0} + \frac{B_z}{4\pi}\frac{\partial B_\phi}{\partial z} \tag{16}$$

In other words, specific angular momentum is removed from the inward accretion flow by the action of two types of magnetic torque. The first term on the righthand side represents radial angular momentum associated with the radial shear of the toroidal field, whereas the second term is vertical transport due to the vertical shear of the toroidal field. In a thin disk, and for typical field inclinations, the second term will dominate. Note that the first term vanishes at the disk midplane, because $B_r = 0$ there. Now, following standard thin-disk theory, vertical integration of the resulting equation gives a relation between the disk accretion rate, $\dot{M}_{acc} = -2\pi\Sigma V_r r_0$, and the magnetic torques acting on its surfaces (subscript s):

$$\dot{M}_{acc} \frac{d(r_0 V_\phi)}{dr_0} = -r_0^2 B_{\phi,s} B_z \tag{17}$$

Angular momentum is thus extracted out of disks threaded by open magnetic fields. The angular momentum can be carried away either by torsional Alfvén waves or, when the magnetic field lines are inclined by more than 30° from the vertical, by a centrifugally driven wind. By rewriting equation (11) as $rB_\phi = 4\pi k(rV_\phi - l)$ and using the derived relations for k and l, the disk angular momentum equation can be cast into its most fundamental form,

$$\dot{M}_{acc} \frac{d(\Omega_0 r_0^2)}{dr_0} = \frac{d\dot{M}_{wind}}{dr_0} \Omega_0 r_A^2 \left[1 - \left(\frac{r_0}{r_A}\right)^2\right] \tag{18}$$

This equation shows that there is a crucial link between the mass outflow in the wind and the mass accretion rate through the disk:

$$\dot{M}_{acc} \simeq (r_A/r_0)^2 \dot{M}_{wind} \tag{19}$$

We have arrived at the profoundly useful expression of the idea that, if viscous torques in the disk are relatively unimportant, the rate at which the disk loses angular momentum ($\dot{J}_d = \dot{M}_{acc} \Omega_0 r_0^2$) is exactly the rate at which it is carried away by the wind ($\dot{J}_w = \dot{M}_{wind} \Omega_0 r_A^2$).

The value of the ratio r_A/r_0 is ~3 for typical parameters, so one finds $\dot{M}_{wind}/\dot{M}_{acc} \simeq 0.1$, which is in excellent agreement with the observations (section II). The explanation of this relationship is thus intimately linked to the disk's angular momentum loss to the wind.

C. Flow Initiation and Collimation

Stellar winds usually require hot coronae to get started, whereas winds from disks, which effectively rotate near "breakup," do not. In the case of a thin disk, even a cool atmosphere will suffice as long as the field lines emerging from the disk make an angle $\leq 60°$ to the surface. This follows from Bernoulli's equation by comparing the variations in the effective gravitational potential and the kinetic energy of a particle that moves along a field line near the disk surface (Blandford and Payne 1982; Königl and Ruden 1993; Spruit 1996; but see Ogilvie and Livio 1998).

The collimation of an outflow, as it accelerates away from the disk, arises to a large extent from the hoop stress of the toroidal field component. From equations (9) and (15) one sees that at the Alfvén surface $|B_\phi| \simeq B_\mathrm{p}$ and that in the far field (assuming that the flow opens up to radii $r \gg r_\mathrm{A}$) $|B_\phi/B_\mathrm{p}| \simeq r/r_\mathrm{A}$. Thus, the inertia of the gas, forced to corotate with the outflow out to r_A, eventually causes the jet to self-collimate through the $\mathbf{j}_z \times \mathbf{B}_\phi$ force (with $\mathbf{j}$ being the current density), which is known as the z-pinch in the plasma physics literature.

The detailed radial structure of the outflow is deduced by balancing all forces perpendicular to the field lines and is described by the so-called Grad-Shafranov equation. This is a complicated nonlinear equation for which no general solutions are available. Because of the mathematical difficulties (e.g., Heinemann and Olbert 1978), the analytic studies have been characterized by simplified approaches, including separation of variables (e.g., Tsinganos and Trussoni 1990; Sauty and Tsinganos 1994), self-similarity (e.g., Blandford and Payne, 1982; Bacciotti and Chiuderi 1992; Contopoulos and Lovelace 1994; Lynden-Bell and Boily 1994), and previously "guessed" magnetic configurations (e.g., Pudritz and Norman 1983; Lery et al. 1998), as well as examinations of various asymptotic limits to the theory (e.g., Heyvaerts and Norman 1989; Appl and Camenzind 1993; Ostriker 1997). Pelletier and Pudritz (1992) constructed non-self-similar models of disk winds, including cases where the wind emerges only from a finite portion of the disk.

Self-similarity imposes a specific structure on the underlying disk. In the Blandford and Payne (1982) model, all quantities scale as power laws of spherical radius along a given radial ray. This directly implies that the Alfvén surface is conical and, similarly, that the disk's scale height $H(r)$ scales linearly with r. Furthermore, inasmuch as the problem contains a characteristic speed, namely the Kepler speed in the disk, one infers that $V_\mathrm{A} \propto c_\mathrm{s} \propto V_\mathrm{r} \propto V_\infty \propto V_\mathrm{K} \propto r^{-1/2}$, i.e., that the Alfvén, sound, radial inflow, terminal outflow, and Kepler speeds, respectively, are all proportional to one another. The scaling $c_\mathrm{s} \propto V_\mathrm{K}$ implies that the disk temperature has the virial scaling $T \propto r^{-1}$. The scaling $V_\mathrm{r} \propto V_\mathrm{K}$ as well as the $H(r)$ relation imply that, for a constant mass accretion rate $\dot{M}_\mathrm{acc} = -2\pi(2H\rho)V_\mathrm{r}r$, the density $\rho \propto r^{-3/2}$. Next, the scaling of the disk Alfvén speed $V_\mathrm{A} \propto V_\mathrm{K}$ together with the density result imply that the disk poloidal field (and hence also B_ϕ) scales as $r^{-5/4}$. In turn, the mass load $k_0 \propto r_0^{-3/4}$ and the wind mass loss rate $\dot{M}_\mathrm{wind}(r_0) \propto \ln r_0$. More general self-similar models may be constructed by making the more realistic assumption that mass is lost from the disk so that the accretion rate is not constant; $\dot{M}_\mathrm{acc}(r) \propto r^{-\mu}$. Adopting the radial scaling of the disk magnetic field, $B_0(r) \propto r^{-\nu}$, one finds that self-similarity imposes the scaling $\mu = 2(\nu - 1.25)$ [see equation (17)].

The work of Heyvaerts and Norman (1989, 1997) and others has shown that in general two types of collimation are possible, depending on the asymptotic behavior of the electric current $I = (c/2)rB_\phi \propto r^2\rho\Omega_0$. If $I \rightarrow 0$ as $r \rightarrow \infty$, then the field lines are space-filling paraboloids, whereas if this limit for the current is finite, then the flow is collimated to cylinders. The character of the flow therefore depends on the boundary conditions at the disk.

Finally, we note that the stability of magnetized jets with toroidal magnetic fields is strongly assisted by the jet's poloidal magnetic field, which acts as a spinal column for the jet (e.g., Appl and Camenzind 1992). However, the question of whether such jets would develop a kink instability, and whether this would affect their collimation properties, is still being debated (e.g., Spruit et al. 1997).

IV. NUMERICAL SIMULATIONS OF DISK WINDS

The advent of numerical simulations has finally made it possible to study the rich, time-dependent behavior of MHD disk winds. This allows one to test the stationary theory presented above as well as to search for the conditions that give rise to episodic outflows. These simulations represent perhaps the main advance in the subject since *Protostars and Planets III*. The published simulations generally assume ideal MHD and may be grouped into two classes: (1) dynamic MHD disks, in which the structure and evolution of the magnetized disk is also part of the simulation, and (2) stationary MHD disks, in which the underlying accretion disk does not change and provides fixed boundary conditions for the outflow problem.

A. Dynamic Disks and Winds

The first numerical calculations of disk winds were published by Uchida and Shibata (1985) and Shibata and Uchida (1986). These simulations modeled a magnetized disk in sub-Keplerian rotation and showed that a rapid radial collapse develops in which the initially poloidal field threading the disk is wound up due to the differential rotation. The vertical Alfvén speed, being smaller than the free-fall speed, implies that a strong, vertical toroidal field pressure gradient $\partial B_\phi^2/\partial z$ must rapidly build up. This force results in the transient ejection of coronal material above and below the disk as the spring uncoils. The work of Uchida and Shibata (1985) was confirmed by Stone and Norman (1994) using their ZEUS-2D ideal MHD code (Stone and Norman 1992).

If the mean magnetic field energy density is less than the thermal pressure in the disk ($V_A < c_s$), then a strong magnetorotational (Balbus-Hawley, BH) instability will develop (e.g., Balbus and Hawley 1991). In 2D, this leads to a vigorous radial channel flow and rapid outward transport of angular momentum. Stone and Norman (1994) ran a series of ZEUS-2D simulations for a uniform magnetic field threading a wedge-shaped disk and a surrounding corona (models defined by four parameters). Their simulations investigated three cases (see also Bell and Lucek 1995 and Matsumoto et al. 1996): (a) sub-Keplerian rotation, (b) a Kepler disk with strong disk field, and (c) a Kepler disk with a weak field. In case (a), rapid collapse immediately ensues with an expanding, transient outflow appearing in 2.5 orbits (reproducing Uchida and Shibata 1985). In case (b), the disk is BH stable, but collapse occurs anyway, because of the very strong braking of the disk due to the external MHD torque. In case (c) one again sees rapid radial collapse of the disk, this time because of a strong BH instability.

The 2D channel flow does not, however, persist in 3D. In that case the BH instability develops into a fully turbulent flow, and the inflow rate is significantly reduced (see the chapter by Stone et al., this volume).

B. Stationary Disks and Winds

The outflow problem can be clarified by focusing on the more restricted question of how a wind is accelerated and collimated for a prescribed set of fixed boundary conditions on the disk. Keplerian disks in 3D are stable on many tens of orbital times, and this justifies the stationary disk approach: The launch and collimation of jets from their surfaces occurs in only a few inner-disk rotation times. Groups that have taken this route include Ustyugova et al. (1995), Ouyed et al. (1997), Ouyed and Pudritz (1997*a,b*), Romanova et al. (1997), and Meier et al. (1997). The published simulations differ in their assumed initial conditions, such as the magnetic field distribution on the disk, the plasma β ($\equiv P_{\rm gas}/P_{\rm mag}$) just above the disk surfaces, the state of the initial disk corona, and the handling of the gravity of the central star. Broadly speaking, all of the existing calculations show that winds from accretion disks can indeed be launched and accelerated, much along the lines suggested by the theory presented in section III. The results differ, however, in the degree to which flow collimation occurs.

Ustyugova et al. (1995) and Romanova et al. (1997) employed a magnetic configuration described by a monopole field centered beneath the disk surface. Nonequilibrium initial conditions, as well as a softened gravitational potential, were used. Relatively low resolution simulations of flows with $\beta \gg 1$ (Ustyugova et al. 1995) showed that collimated, nonstationary outflows develop. Similar simulations in the strong-field ($\beta < 1$) regime (Romanova et al. 1997) resulted in stationary, but uncollimated outflows on these scales.

Simulations by Ouyed et al. (1997) and Ouyed and Pudritz (1997*a,b*) (see Pudritz and Ouyed 1997 for a review) employed the ZEUS-2D code and studied two different magnetic configurations. The coronal gas was initially taken to be in hydrostatic equilibrium with the central object as well as in pressure balance with the top of the disk. Two initial magnetic configurations were adopted, each chosen so that no Lorentz force is exerted on the hydrostatic corona (i.e., such that $\mathbf{j} = 0$). These are the potential field configuration of Cao and Spruit (1994) and a uniform, vertical field that is everywhere parallel to the disk rotation axis. The gravity was unsoftened in these calculations, (500×200) spatial zones were used, and simulations were run up to $400t_{\rm i}$ (where $t_{\rm i}$ is the Kepler time for an orbit at the inner edge of the disk, $r_{\rm i}$). This model is described by five parameters set at $r_{\rm i}$: three to describe the initial corona (e.g., $\beta_{\rm i} = 1.0$), as well as two parameters to describe the disk physics (e.g., the injection speed $V_{\rm inj}$ of the material from the disk into the base of the corona). The initial conditions correspond to turning on the rotation of the underlying Keplerian disk at $t = 0$.

The first thing that happens is the launch and propagation of a brief, transient, torsional Alfvén wave front, which sweeps out from the disk surface but leaves the corona largely undisturbed. An outflow also begins almost immediately and develops into a stationary or an episodic jet. These jetlike outflows are highly collimated and terminate in a jet shock. A bow shock, driven by the jet, pushes through the corona. A noticeably empty cavity dominates most of the volume behind the bow shock: the cavity is filled with the toroidal magnetic field that is generated by the jet itself. This may provide an explanation of the extended bipolar cavities that surround highly collimated jets. For a fiducial injection speed of $10^{-3} V_{\rm K}$, the potential field configuration develops into a stationary outflow with many (but not all, since the flow is not self-similar) of the characteristics of the Blandford and Payne (1982) solution. The Alfvén surface is correctly predicted by the analysis in section III. The Alfvén and fast-magnetosonic Mach numbers reach 5 and 1.6, respectively, at $z = 10r_{\rm i}$, with the toroidal-to-poloidal field strength ratio being ~ 3 on this scale. Outflow takes place only on field lines that are inclined by less than 60° with respect to the disk surface, as predicted by Blandford and Payne (1982).

Figure 1 (adapted from Ouyed et al. 1997) shows that the poloidal field lines in the initial state are collimated toward the rotation axis by the hoop stress of the jet's toroidal field. The figure also shows the propagation of the jet-driven bow shock and the eventual creation of a cylindrically collimated jet with well-determined Alfvén and fast-magnetosonic critical surfaces in the acceleration region above the disk (the slow-magnetosonic surface is too close to the disk to be resolved in this figure).

For the same boundary and initial conditions, the initially vertical field configuration leads to the development of a jet that is episodic over the $400t_{\rm i}$ duration of the simulation (Ouyed et al. 1997; Ouyed and Pudritz 1997*b*). Even though the initial configuration is highly unfavorable to jet formation, jet production occurs. The reason is the effect of the toroidal field in the corona, which is concentrated toward the inner edge of the disk (where the Kepler rotation is the greatest). It therefore exerts a radial pressure force $\partial(B_\phi^2/8\pi)/\partial r$ that opens up field lines at larger radii. As long as $\beta \simeq 1$ in this region, field lines are pliable enough to move, and a jet is launched. Episodic knots are produced on a timescale $\tau_{\rm knot} \simeq r_{\rm jet}/V_{{\rm A},\phi}$, which is reminiscent of a type of kink instability. Regions of high toroidal field strength and low density separate and confine the knots, which in turn have high density and low toroidal field strength.

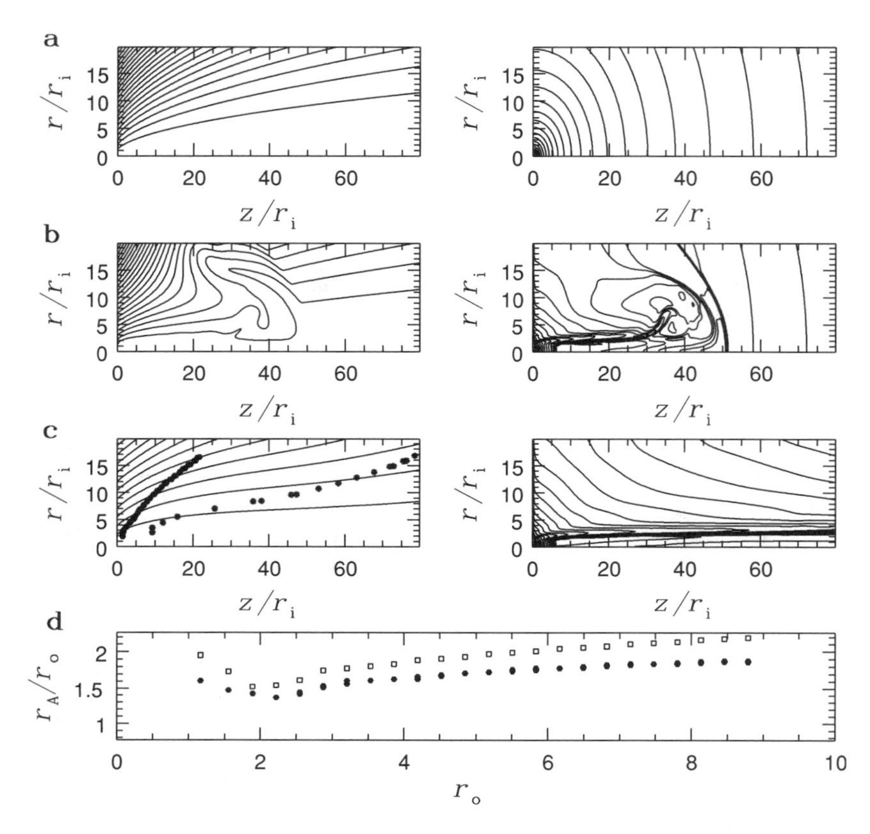

Figure 1. Numerical simulations of disk-driven MHD outflows (adapted from Ouyed et al. 1997). In frame (a), the left panel shows the initial magnetic configuration corresponding to the "potential field" solution, whereas the right panel displays the initial isodensity contours of the corona. The flow injection speed is $V_{\rm inj} = 10^{-3} V_{\rm K}$. Frames (b) and (c) show the evolution of the initial magnetic and density structures (the left and right panels, respectively) at 100 and 400 inner time units. Frame (c) also displays the locations of the Alfvén critical surface (filled hexagons) and the fast-magnetosonic surface (stars). In frame (d), the Alfvén lever arm ($r_{\rm A}/r_0$) found in the simulation is shown (filled hexagons) as a function of the location (r_0) of the footpoints of the field lines on the disk and is compared to the prediction from the steady-state theory of section III.A (squares).

Based on an extensive set of simulations, Ouyed and Pudritz (1999) concluded that the main factor that determines whether jets are stationary or episodic is not the initial field configuration, but rather the mass loading of the wind. Figure 2 shows a simulation with the same potential configuration parameters as in Fig. 1 except for a reduced injection speed (and hence mass load) of $V_{\rm inj} = 10^{-5} V_{\rm K}$. Clear episodic behavior is now seen in this flow.

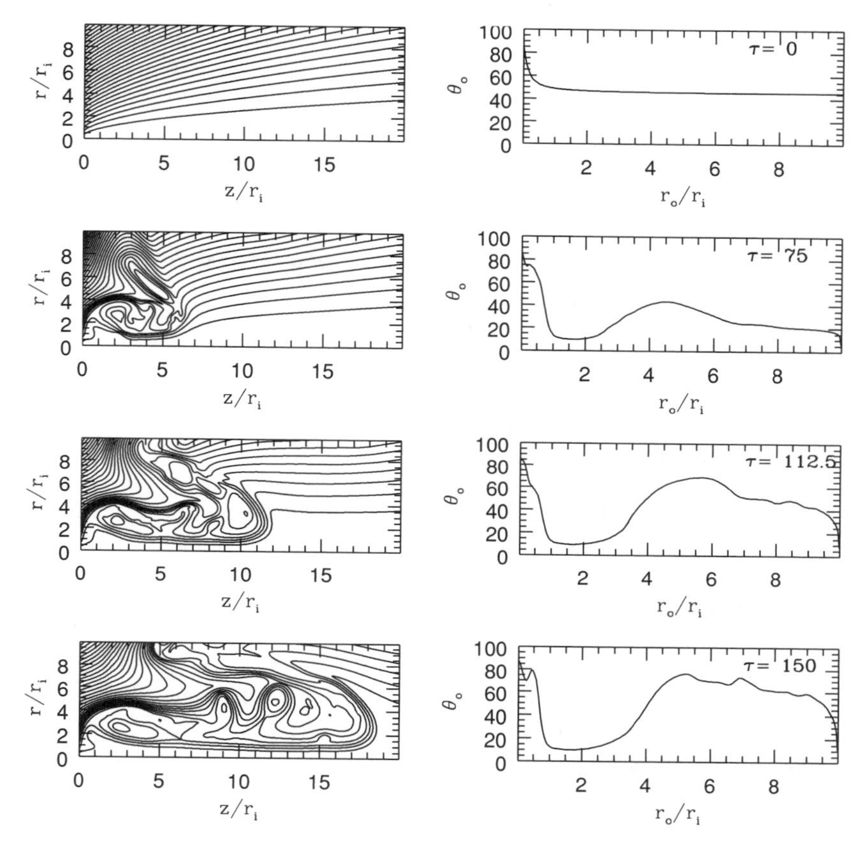

Figure 2. MHD outflow simulations with the same parameters and magnetic configuration as in Fig. 1, except $V_{\rm inj} = 10^{-5} V_{\rm K}$ (adapted from Ouyed and Pudritz 1999). The left panels show the magnetic field structure and knot formation at times 0.0, 75.0, 112.5, and 150.0 $t_{\rm i}$. The right panels show the angle θ_0 between the field lines and the disk surface at these 4 times. Only field lines that open up to an angle $\theta_0 \leq 60°$ are seen to drive an outflow.

Nonsteady outflows could conceivably also arise in highly conducting disk regions, where the winding-up of the field lines might eject gas through a strong $|B_\phi|$ pressure gradient (e.g., Contopoulos 1995).

C. Far-field Behavior of MHD Jets

Simulations of MHD jets propagating into a uniform medium show that, in certain respects, they differ dramatically from hydrodynamic jets. Clarke et al. (1986), Lind et al. (1989), and Kössel et al. (1990) showed that the strong toroidal field of a jet prevents matter from spraying sideways on encountering the jet shock. Rather, jet material decelerated by a Mach disk and a strong annular shock is focused mainly forward into a nose cone. This contrasts with the backflowing cocoon that characterizes purely hydrodynamic, low-density jets. Hydromagnetic jets are generally preceded by stronger, more oblique shocks and advance far more quickly into the surrounding medium than their hydrodynamic counterparts (see Kössel et al. 1990 for more detail). MHD jets may thus explain why strong transverse expansions have not been measured in bipolar molecular lobes (e.g., Masson and Chernin 1993; Cabrit et al. 1997).

V. MAGNETIZED PROTOSTELLAR ACCRETION DISKS

A. Magnetic Angular Momentum Transport

One of the primary reasons why open, ordered magnetic fields are important in accreting astrophysical systems is that they can mediate a vertical transport of angular momentum from accretion disks. In section III we discussed the angular momentum transport by centrifugally driven winds, but it is important to realize that angular momentum can be removed from the disk surfaces even if the field line inclination with respect to the vertical is not large enough for the wind-launching condition to be satisfied, so long as the field is attached to a "load" that exerts a back torque on the disk. In that case the disk can lose angular momentum through torsional Alfvén waves propagating away from the disk surfaces (the "magnetic braking" mechanism; see Mouschovias 1991 for a review). When the magnetic fields are well below equipartition with the gas pressure in a differentially rotating disk, the magnetorotational (BH) instability develops on a dynamical time scale and produces a magnetic stress-dominated turbulence that gives rise to radial transport of angular momentum characterized by an effective viscosity parameter α in the range $\sim$0.005–0.5 (see the chapter by Stone et al., this volume).

Numerical simulations (see section IV.A) have demonstrated that all of the above processes could contribute to the angular momentum transport in magnetized disks. One can estimate the relative roles of vertical wind transport and radial turbulent transport by taking the ratio of the external wind torque, computed in section III, and the viscous torque associated with magnetocentrifugal turbulence, which can be written as $(B_z^2/4\pi P_{\rm gas})(r_{\rm A}/\alpha H)$ (Pelletier and Pudritz 1992). Thus, even if the ordered field threading the disk were far below equipartition with the disk gas pressure, a wind torque could still dominate a turbulent torque simply because of its large lever arm (the Alfvén radius $r_{\rm A}$, which greatly exceeds the disk scale height H). We note in passing that a long effective lever arm also characterizes angular momentum transport by spiral waves in the disk.

The requirement that, in a steady state, the torque exerted by a large-scale magnetic field at the surface of a thin and nearly Keplerian accretion disk balances the inward angular momentum advection rate can be written as $\dot{M}_{\rm acc} = 2r^{5/2}|B_z B_{\phi,{\rm s}}|/(GM_\star)^{1/2}$ [see equation (17)]. Assuming a rough equality between the vertical and azimuthal surface field components, this relation implies that, at a distance of 1 AU from a solar-mass protostar, a 1-G field (the value indicated by meteoritic data for the protosolar nebula; Levy and Sonnett 1978) could induce accretion at a rate of $\sim 2 \times 10^{-6}$ $M_\odot$ yr^{-1}, which is compatible with the mean values inferred in embedded protostars. For comparison, the minimum-mass solar nebula model implies a thermal pressure at 1 AU that corresponds to an equipartition magnetic field of $\sim$18 G, which implies that a $\sim$1-G field could be readily anchored in the protosolar disk at that location.

B. Wind-Driving Protostellar Disk Models

The properties of protostellar accretion disks and of integrated disk/wind systems have, so far, been derived only under highly simplified assumptions. Among the papers that can be consulted on this topic are Königl (1989, 1997), Wardle and Königl (1993), Ferreira and Pelletier (1993*a,b,* 1995; also Ferreira 1997), Lubow et al. (1994), Li (1995, 1996), and Reyes-Ruiz and Stepinski (1996). A plausible origin for an open disk magnetic field is the interstellar field that had originally threaded the magnetically supported molecular cloud and that was subsequently carried in by the collapsing core. This picture is favored over disk-dynamo interpretations because the latter typically produce closed (quadrupolar) field configurations, although we note that scenarios for opening dynamo-generated field lines have been considered in the literature (e.g., Tout and Pringle 1996; Curry et al. 1994).

The theory and simulations of the previous sections assumed ideal MHD. This approximation is adequate for the surface layers of protostellar disks, but the disk interiors are typically weakly ionized, and their study entails the application of multifluid MHD. For typical parameters of disks around solar-mass YSOs, one can distinguish between the low-density regime (on scales $r \lesssim 100$ AU), in which the current in the disk is carried by metal ions and electrons (whose densities are determined from the balance between ionizations by cosmic rays and recombinations on grain surfaces), and the high-density regime (on scales $r \lesssim 1$–10 AU), where the current is carried by small charged grains or by ions and electrons that recombine without grains. We restrict our attention to the low-density regime and concentrate on the case in which the magnetic field $\mathbf{B}$ is "frozen" into the electrons and diffuses relative to the dominant neutral component as a result of an ion–neutral drift (ambipolar diffusion).

To derive a self-consistent steady-state disk/wind configuration, one combines the mass, momentum (radial, vertical, and angular), and energy conservation relations, together with Maxwell's equations and the generalized Ohm's law, and imposes the requirements that the outflow pass through the relevant critical points (see section IV). So far, only simple prescriptions for the disk thermal structure (isothermal or adiabatic) and conductivity (ambipolar diffusion in the density regime where the ion density is constant, or ohmic diffusivity parametrized using a "turbulence" prescription) have been considered. These are probably adequate for obtaining the basic structure of the disk, but more realistic calculations are needed to model the transition region between the disk and the wind correctly.

The vertical structure of a generic centrifugal wind-driving disk (or "active" surface layer; see section V.C) can be divided into three distinct zones (see Fig. 3): a quasihydrostatic region near the midplane of the disk, where the bulk of the matter is concentrated and most of the field-line bending takes place; a transition zone, where the inflow gradually diminishes with height; and an outflow region that corresponds to the base of the wind. The first two regions are characterized by a radial inflow and sub-Keplerian rotation, whereas the gas at the base of the wind flows out with $V_\phi > V_K$. The boundary conditions for the stationary disk simulations discussed in section IV.B arise from the physical properties of this latter region.

What determines the disk structure, and how does the magnetic field extract the angular momentum of the accreting gas? The quasihydrostatic region is matter dominated, with the ionized plasma and magnetic field being carried around by the neutral material. The ions are braked by a magnetic torque, which is transmitted to the neutral gas through the frictional (ambipolar diffusion) drag; therefore $V_{i\phi} < V_\phi$ in this region (with the subscript i denoting ions). The neutrals thus lose angular momentum to the field, and their back reaction leads to a buildup of the azimuthal field component $|B_\phi|$ away from the midplane. The loss of angular momentum enables the neutrals to drift toward the center, and in doing so they exert a radial drag on the field lines. This drag must be balanced by magnetic tension, so the field lines bend away from the rotation axis. This bending builds up the ratio B_r/B_z, which needs to exceed $1/\sqrt{3}$ at

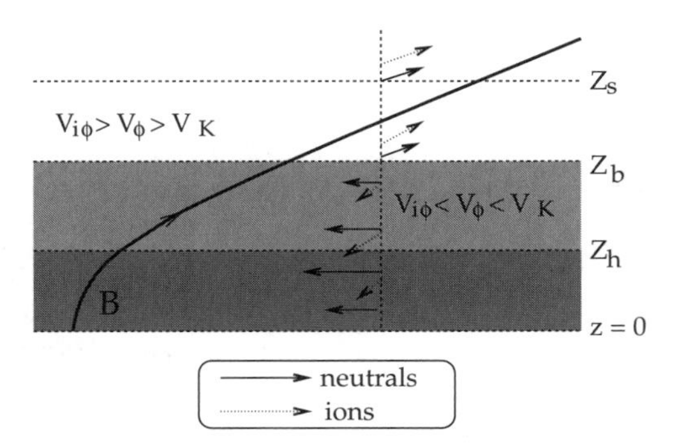

Figure 3. Schematic diagram of the vertical structure of an ambipolar diffusion-dominated disk, showing a representative field line and the poloidal velocities of the neutral (solid arrowheads) and the ionized (open arrowheads) fluid components. Note that the poloidal velocity of the ions vanishes at the midplane ($z = 0$) and is small for both fluids at the base of the wind ($z = Z_b$). The relationship between the azimuthal velocities is also indicated.

the disk surface to launch a centrifugally driven wind. The magnetic tension force, transmitted through ion-neutral collisions, contributes to the radial support of the neutral gas and causes it to rotate at sub-Keplerian speeds.

The growth of the radial and azimuthal field components on moving away from the midplane results in a magnetic pressure gradient that tends to compress the disk. The vertical compression by the combined magnetic and tidal stresses is, in turn, balanced by the thermal pressure gradient. The magnetic energy density becomes dominant as the gas density decreases, marking the beginning of the transition zone (at $z = Z_\mathrm{h}$). The field above this point is nearly force free $[(\nabla \times \mathbf{B})\times\mathbf{B} \approx 0]$, so the field lines, which vary only on a length scale $\sim r$, are locally straight. This is the basis for the adoption of force-free initial field configurations in the simulations reviewed in section IV.B.

The field angular velocity, given by $\omega = (V_{\mathrm{i}\phi} - V_{\mathrm{iz}}B_\phi/B_\mathrm{z})/r$, is a field-line constant (see section III.A). The ion angular velocity $V_{\mathrm{i}\phi}/r$ differs somewhat from ω but still changes only slightly along the field. Because the field lines bend away from the symmetry axis, the cylindrical radius r, and hence $V_{\mathrm{i}\phi}$, increase along any particular field line, whereas V_ϕ decreases because of the near-Keplerian rotation law. Eventually a point is reached where $(V_{\mathrm{i}\phi} - V_\phi)$ changes sign. At this point, the magnetic stresses on the neutral gas are small, and its angular velocity is almost exactly Keplerian. Above this point, the field lines overtake the neutrals and transfer angular momentum back to the matter, and the ions start to push the neutrals out in both the radial and the vertical directions. It is thus natural to identify the base Z_b of the wind with the location where the angular velocity of the field lines becomes equal to the Keplerian angular velocity. The mass outflow rate is fixed by the height Z_s of the effective sonic point of the wind.

As was first recognized by Wardle and Königl (1993), one can derive key constraints on a viable solution by neglecting the vertical component of the neutral velocity, which is generally a good approximation throughout much of the disk column. This simplifies the problem by transforming the radial and azimuthal components of the neutral momentum equation into algebraic relations. One important constraint that can be derived in this way expresses the intuitively obvious condition that the neutrals must be able to couple to the magnetic field on an orbital timescale if magnetic torques are to play a role in the removal of angular momentum from the disk. In the pure ambipolar diffusion regime, this condition is expressed by the requirement that the neutral-ion coupling time $1/\gamma\rho_\mathrm{i}$ (where $\gamma \approx 3.5 \times 10^{13}$ cm^3 g^{-1} s^{-1} is the collisional coupling coefficient) be shorter than the dynamical time. This is equivalent to requiring that the neutral-ion coupling parameter $\eta \equiv \gamma\rho_\mathrm{i}r/V_\mathrm{K}$ satisfy

$$\eta > 1 \tag{20}$$

This condition is quite general and is also relevant to disk models in which magnetic torques associated with small-scale magnetic fields transfer angular momentum radially through the disk [see the chapter by Stone et al., this volume; in the latter case equation (20) identifies the linearly unstable regime of the magnetorotational instability, although evidently a much higher minimum value of η is required for significant nonlinear growth].

Additional parameter constraints that can be derived by using the hydrostatic approximation for a wind-driving disk model in the pure ambipolar diffusion regime are given by

$$(2\eta)^{-1/2} \lesssim a \lesssim 2 \lesssim \epsilon\eta \lesssim V_K/2c_s \tag{21}$$

where $a \equiv B_0/(4\pi\rho_0)^{1/2}c_s$ is the ratio of the midplane (subscript 0) Alfvén speed to the isothermal sound speed c_s and $\epsilon \equiv -V_{r0}/c_s$ is the normalized midplane inflow speed. The first inequality corresponds to the requirement that the disk remain sub-Keplerian, the second to the wind-launching condition at the disk surface $B_r(Z_b) \gtrsim B_0/\sqrt{3}$, and the third to the requirement that the base of the wind lie well above a density scale height in the disk. The second and third inequalities together imply that the vertical magnetic stress dominates the gravitational tidal stress in confining the disk. This is a generic property of this class of disk solutions that does not depend on the nature of the magnetic diffusivity. The last inequality expresses the requirement that the ambipolar diffusion heating rate at the midplane not exceed the rate $\rho_0|V_{r0}|V_K/2r$ of gravitational potential energy release. In turn, c_s/V_K must be $\ll 1/(1+\epsilon)$ to guarantee that the disk is in near-Keplerian motion and geometrically thin. Upper limits can also be placed on the density at the sonic point to ensure that the bulk of the disk material is hydrostatic and that $\dot{M}_{\rm wind}$ does not exceed $\dot{M}_{\rm acc}$. Similar constraints are derived by applying this analysis to the other diffusivity regimes identified above. The solutions that satisfy these constraints tend to have $\epsilon \lesssim 1$ and $a \gtrsim 1$. Furthermore, the magnetic field in these models, whose magnitude is essentially determined from the condition that all the angular momentum liberated by the accreting matter is transported by a centrifugally driven wind, automatically lies in a "stability window," where it is strong enough not to be affected by the magnetorotational instability but not so strong as to be subject to the radial interchange instability (Königl and Wardle 1996).

As we have noted, the steady-state disk models are useful for guiding the choice of boundary conditions at the base of the outflow in numerical simulations of winds from quasistationary disks. In a complementary approach (which typically employs a highly simplified treatment of the disk outflow and therefore is not suitable for a detailed study of the disk wind), one can generalize the steady-state models and investigate the time evolution of wind-driving accretion disks (e.g., Lovelace et al. 1994; Königl 1997). The time-dependent models can be utilized to explore the feedback effect between the magnetic flux distribution, which affects the angular momentum transport in the disk, and the radial inflow induced by the magnetic removal of angular momentum, which can modify the flux distribution through field line advection.

C. Outflows from the Protostellar Vicinity

As one moves to within a few AU from the protostar, the disk hydrogen column density can become so large ($\gtrsim$192 g cm^{-2}) that cosmic rays are excluded from the disk interior. (For reference, the column density of the minimum-mass solar nebula model is 1.7×10^3 g cm^{-2} at 1 AU.) At larger columns the ionization rate is dominated by the decay of radioactive elements (notably ^{26}Al, if it is present, and ^{40}K; Stepinski 1992), but the charge density generally becomes too low for the magnetic field to be effectively coupled to the matter. Consequently, the interior of the disk becomes inert, and magnetically mediated accretion can proceed only through "active" surface layers that extend to a depth of $\sim$96 g cm^{-2} on each side of the disk. This scenario applies both to a wind-driving disk (Wardle 1997) and to the case where small-scale magnetic fields produce an effective viscosity within the disk (Gammie 1996). Galactic cosmic rays could reach the disk along the open magnetic field lines that thread it, but a super-Alfvénic outflow along these field lines would tend to exclude them. External ionization could, however, also be effected by stellar X-rays (see the chapter by Glassgold et al., this volume) as well as by fast particles accelerated in stellar flares. In addition, heating by stellar irradiation could contribute to the collisional ionization of the surface layers (e.g., D'Alessio et al. 1998).

The entire disk can recover an adequate coupling with the magnetic field once the interior temperature becomes high enough for collisional ionization to be effective. This first occurs when the temperature increases above $\sim 10^3$ K and potassium is rapidly ionized (e.g., Umebayashi and Nakano 1981). For disks in which magnetic fields dominate the angular momentum transport, it may be possible that Joule dissipation could maintain the requisite degree of ionization for efficient gas–field coupling [$\eta > 1$; see equation (20)]. Li (1996) first explored this possibility for the inner regions of wind-driving protostellar disks and concluded that, for a disk that is in the ambipolar diffusion regime on scales $\lesssim$1 AU, a self-consistent model can be constructed only if the accretion rate is very high ($\gtrsim 10^{-5}$ M$_\odot$ yr^{-1}).

A possible mechanism for recoupling the gas to the field in the vicinity of the YSO is the thermal ionization instability originally discussed in the context of dwarf novae and more recently invoked as a possible explanation of FU Orionis outbursts (e.g., Bell and Lin 1994). In this picture, accretion in the innermost disk proceeds in a nonsteady fashion, with a "gate" at $r \lesssim 0.25$ AU opening every $\sim 10^3$ yr or so after the accumulated column density has become large enough to trigger the instability. During the "high" phase of the instability (which lasts $\sim 10^2$ yr and is identified with an outburst), the gas is hot ($T \gtrsim 10^4$ K) and almost completely ionized, and mass rains in at a rate $\sim$(1–30)$\times 10^{-5}$ $M_\odot$ yr^{-1}, whereas during the "low" phase the temperature and degree of ionization decline sharply and the accretion rate drops to $\sim$(1–30)$\times 10^{-8}$ $M_\odot$ yr^{-1}. In the context of the magnetized disk model, the increase in the accretion rate as the gas becomes highly ionized can be attributed to the reestablishment of good coupling between the field and the matter, which allows the field to extract the angular momentum of the accreting gas. This could account both for the marked increase in $\dot{M}_{acc}$ and for the strong disk outflow that accompanies it (see section II.B). The magnetic recoupling idea may be relevant to these outbursts even if the wind is not the dominant angular momentum transport mechanism in the disk provided that the viscosity has a magnetic origin (see the chapter by Stone et al., this volume). We note, however, that the relatively large (~ 0.1) $\dot{M}_{wind}/\dot{M}_{acc}$ ratio indicated in some of the outburst sources is consistent with the bulk of the disk angular momentum being removed by the wind.

At distances of a few stellar radii, the stellar magnetic field could be strong enough to drive outflows from the disk. Various scenarios have been considered in this connection, including individual magnetic loops ejecting diamagnetic blobs through a magnetic surface drag force (King and Regev 1994); a steady-state configuration in which mass is transferred to the star near the corotation radius and an outflow is driven along adjacent (but disconnected from the star) field lines (Shu et al. 1994); and time-dependent ejection associated with the twisting, expansion, and subsequent reconnection of field lines that connect the star with the disk (Hayashi et al. 1996; Goodson et al. 1997). Numerical simulations are rapidly reaching a stage at which they could be utilized to identify the relevant mechanisms and settle many of the outstanding questions. There are already indications from some of the existing simulations that magnetically channeled accretion from the disk to the star is most likely to occur when the disk carries a strong axial magnetic field that reconnects with the stellar field at an equatorial x-point; the resulting field configuration appears to be quasisteady and gives rise to sturdy centrifugally driven outflows along open field lines (Hirose et al. 1997; Miller and Stone 1997).

Stellar field-driven outflows are the subject of the chapter by Shu et al., this volume. This class of models is based on the realization that stellar magnetic field lines can be inflated and opened up through an interaction with a surrounding disk, and that a centrifugal wind can be driven out along the opened field lines. In these scenarios, the outflow typically originates near the inner disk radius, where the disk is truncated by the stellar magnetic stresses. This contrasts with the scenarios considered in this chapter, wherein a disk-driven wind originates (and can contribute to the disk angular momentum transport) over a significant range of radii. It is worth noting in this connection that the massive inflows that characterize FU Orionis outbursts are expected to crush the respective stellar magnetospheres, so the strong outflows that are inferred to originate from the circumstellar disks during these outbursts are unlikely to be driven along stellar field lines (Hartmann and Kenyon 1996). Coupled with the apparent inadequacy of thermal pressure and radiative driving, this argument provides strong support for the relevance of disk-driven hydromagnetic winds that are not associated with a stellar magnetic field to the observed outflows from FU Orionis outburst sources. The ramifications of this argument become even stronger if a significant fraction of the mass accumulation in solar-type stars occurs through such outbursts during the early phases of their evolution (Hartmann 1997). To be sure, the "distributed" disk-wind models discussed in this chapter are meant to apply to YSOs in general and not just when they undergo an outburst, but only during an FU Orionis outburst does the disk become sufficiently luminous that an unambiguous observational signature of an extended disk wind can be obtained.

VI. CONCLUSIONS

Centrifugally driven winds from disks threaded by open magnetic field lines provide the most efficient way of tapping the gravitational potential energy liberated in the accretion process to power an outflow. The fact that such winds "automatically" carry away angular momentum and thus facilitate (and possibly even control) the accretion process makes them an attractive explanation for the ubiquity of jets in YSOs and in a variety of other accreting astronomical objects. One of the key findings of recent numerical simulations of MHD winds from disks is that such outflows are indeed easy to produce and maintain under a variety of surface boundary conditions. These simulations have also verified the ability of such outflows to self-collimate and give rise to narrow jets, as well as a variety of other characteristics that are consistent with YSO observations. In the case of protostellar disks, the presence of open field lines is a natural consequence of their formation from the collapse of magnetically supported molecular cloud cores. Although a stellar magnetic field that threads the disk could in principle also play a similar role, this possibility is unlikely to apply to the strong outflows associated with FU Orionis outbursts. (More generally, it is worth noting that, in contrast to disk field-driven outflows, a scenario that invokes a stellar field may not represent a universal mechanism, because many cosmic jet sources are associated with a black hole that does not provide an anchor for a central magnetic field.) Much progress has also been achieved in constructing global MHD disk/wind models, but the full elucidation of this picture and of its consequences for star formation remains a challenge for the future.

Acknowledgments We thank Eric Blackman, Vincent Mannings, Rachid Ouyed, and the anonymous referee for helpful comments on the manuscript. This work was supported in part by NASA grant NAG 5-3687

(A. K.) and by an operating grant from the Natural Science and Engineering Research Council of Canada (R. P.).

REFERENCES

Appl, S., and Camenzind, M. 1992. The stability of current carrying jets. *Astron. Astrophys.* 256:354–370.

Appl, S., and Camenzind, M. 1993. The structure of MHD jets: A solution to the non-linear Grad-Shafranov equation. *Astron. Astrophys.* 274:699–706.

Bacciotti, F., and Chiuderi, C. 1992. Axisymmetric magnetohydrodynamic equations: Exact solutions for stationary incompressible flows. *Phys. Fluids B* 4:35–43.

Balbus, S. A., and Hawley, J. F. 1991. A powerful local shear instability in weakly magnetized disks. I. Linear analysis. II. Nonlinear evolution. *Astrophys. J.* 376: 214–233.

Bell, A. R., and Lucek, S. G. 1995. Magnetohydrodynamic jet formation. *Mon. Not. Roy. Astron. Soc.* 277:1327–1340.

Bell, K. R. 1994. Reconciling accretion scenarios with inner holes: The thermal instability and the 2μm gap. In *The Nature and Evolution of Herbig Ae/Be Stars*, ASP Conf. Ser. 62, ed. P. S. Thé, M. R. Pérez, and E. P. J. van den Heuvel (San Francisco: Astronomical Society of the Pacific), pp. 215–218.

Bell, K. R., and Lin, D. N. C. 1994. Using FU Orionis outbursts to constrain self-regulated protostellar disk models. *Astrophys. J.* 427:987–1004.

Blandford, R. D., and Payne, D. G. 1982. Hydromagnetic flows from accretion discs and the production of radio jets. *Mon. Not. Roy. Astron. Soc.* 199:883–903.

Bontemps, S., André, P., Terebey, S., and Cabrit, S. 1996. Evolution of outflow activity around low-mass embedded young stellar objects. *Astron. Astrophys.* 311:858–872.

Burrows, C. J., Stapelfeldt, K. R., Watson, A. M., Krist, J. E., Ballester, G. E., Clarke, J. T., Crisp, D., Gallagher, J. S., III, Griffiths, R. E., Hester, J. J., Hoessel, J. G., Holtzman, J. A., Mould, J. R., Scowen, P. A., Trauger, J. T., and Westphal, J. A. 1996. Hubble Space Telescope observations of the disk and jet of HH 30. *Astrophys. J.* 473:437–451.

Cabrit, S., and André, P. 1991. An observational connection between circumstellar disk mass and molecular outflows. *Astrophys. J. Lett.* 379:L25–L28.

Cabrit, S., Edwards, S., Strom, S. E., and Strom, K. M. 1990. Forbidden line emission and infrared excesses in T Tauri stars—Evidence for accretion-driven mass loss? *Astrophys. J.* 354:687–700.

Cabrit, S., Raga, A. C., and Gueth, F. 1997. Models of bipolar molecular outflows. In *Herbig-Haro Flows and the Birth of Low Mass Stars*, IAU Symp. 182, ed. B. Reipurth and C. Bertout (Dordrecht: Kluwer), pp. 163–180.

Calvet, N., Hartmann, L., and Kenyon, S. J. 1993. Mass loss from pre-main-sequence accretion disks. I. The accelerating wind of FU Orionis. *Astrophys. J.* 402:623–634.

Cao, X., and Spruit, H. C. 1994. Magnetically driven wind from an accretion disk with low-inclination field lines. *Astron. Astrophys.* 287:80–86.

Ciolek, G. E., and Königl, A. 1998. Dynamical collapse of nonrotating magnetic molecular cloud cores: Evolution through point-mass formation. *Astrophys. J.* 504:257–279.

Clarke, D. A., Norman, M. L., and Burns, J. O. 1986. Numerical simulations of a magnetically confined jet. *Astrophys. J. Lett.* 311:L63–L67.

Contopoulos, J. 1995. A simple type of magnetically driven jets: An astrophysical plasma gun. *Astrophys. J.* 450:616–627.

Contopoulos, J., and Lovelace, R. V. E. 1994. Magnetically driven jets and winds: Exact solutions. *Astrophys. J.* 429:139–152.

Corcoran, M., and Ray, T. 1997. Forbidden emission lines in Herbig Ae/Be stars. *Astron. Astrophys.* 321:189–201.

Corcoran, M., and Ray, T. 1998. Wind diagnostics and correlations with the near-infrared excess in Herbig Ae/Be stars. *Astron. Astrophys.* 331:147–61.

Crutcher, R. M., Troland, T. H., Goodman, A. A., Heiles, C., Kazés, I., and Myers, P. C. 1993. OH Zeeman observations of dark clouds. *Astrophys. J.* 407:175–184.

Crutcher, R. M., Mouschovias, T. C., Troland, T. H., and Ciolek, G. E. 1994. Structure and evolution of magnetically supported molecular clouds: Evidence for ambipolar diffusion in the Barnard 1 cloud. *Astrophys. J.* 427:839–847.

Crutcher, R. M., Roberts, D. A., Mehringer, D. M., and Troland, T. H. 1996. H I Zeeman measurements of the magnetic field in Sagittarius B2. *Astrophys. J. Lett.* 462:L79–L82.

Curry, C., Pudritz, R. E., and Sutherland, P. G. 1994. On the global stability of magnetized accretion disks. I. Axisymmetric modes. *Astrophys. J.* 434:206–220.

D'Alessio, P., Cantó, J., Calvet, N., and Lizano, S. 1998. Accretion disks around young objects. I. The detailed vertical structure. *Astrophys. J.* 500:411–427.

Eislöffel, J, and Mundt, R. 1997. Parsec-scale jets from young stars. *Astron. J.* 114:280–287.

Ferreira, J. 1997. Magnetically-driven jets from Keplerian accretion discs. *Astron. Astrophys.* 319:340–359.

Ferreira, J., and Pelletier, G. 1993*a*. Magnetized accretion-ejection structures. I. General statements. *Astron. Astrophys.* 276:625–636.

Ferreira, J., and Pelletier, G. 1993*b*. Magnetized accretion-ejection structures. II. Magnetic channeling around compact objects. *Astron. Astrophys.* 276:637–647.

Ferreira, J., and Pelletier, G. 1995. Magnetized accretion-ejection structures. III. Stellar and extragalactic jets as weakly dissipative disk outflows. *Astron. Astrophys.* 295:807–832.

Gammie, C. F. 1996. Layered accretion in T Tauri disks. *Astrophys. J.* 457:355–362.

Gomez, M., Whitney, B. A., and Kenyon, S. J. 1997. A survey of optical and near-infrared jets in Taurus embedded sources. *Astron. J.* 114:1138–1153.

Goodson, A. P., Winglee, R. M., and Böhm, K.-H. 1997. Time-dependent accretion by magnetic young stellar objects as a launching mechanism for stellar jets. *Astrophys. J.* 489:199–209.

Greaves, J. S., and Holland, W. S. 1998. Twisted magnetic field lines around protostars. *Astron. Astrophys. Lett.* 333:L23–L26.

Greaves, J. S., Murray, A. G., and Holland, W. S. 1994. Investigating the magnetic field structure around star formation cores. *Mon. Not. Roy. Astron. Soc.* 284:L19–L22.

Greaves, J. S., Holland, W. S., and Murray, A. G. 1995. Magnetic field compression in the Mon R2 cloud core. *Mon. Not. Roy. Astron. Soc.* 297:L49–L52.

Hartigan, P., Morse, J., and Raymond, J. 1994. Mass-loss rates, ionization fractions, shock velocities, and magnetic fields of stellar jets. *Astrophys. J.* 136:124–143.

Hartigan, P., Edwards, S., and Ghandour, L. 1995. Disk accretion and mass loss from young stars. *Astrophys. J.* 452:736–768.

Hartmann, L. 1997. The observational evidence for accretion. In *Herbig-Haro Flows and the Birth of Low Mass Stars*, IAU Symp. 182, ed. B. Reipurth and C. Bertout (Dordrecht: Kluwer), pp. 391–405.
Hartmann, L., and Calvet, N. 1995. Observational constraints on FU Ori winds. *Astron. J.* 109:1846–1855.
Hartmann, L., and Kenyon, S. J. 1996. The FU Orionis phenomenon. *Ann. Rev. Astron. Astrophys.* 34:207–240.
Hartmann, L., Kenyon, S. J., and Calvet, N. 1993. The excess infrared emission of Herbig Ae/Be stars: Disks or envelopes? *Astrophys. J.* 407:219–231.
Hayashi, M. R., Shibata, K., and Matsumoto, R. 1996. X-ray flares and mass outflows driven by magnetic interaction between a protostar and its surrounding disk. *Astrophys. J. Lett.* 468:L37–L40.
Heinemann, M., and Olbert, S. 1978. Axisymmetric ideal MHD stellar wind flow. *J. Geophys. Res.* 83:2457–2460.
Heyvaerts, J., and Norman, C. A. 1989. The collimation of magnetized winds. *Astrophys. J.* 347:1055–1081.
Heyvaerts, J., and Norman, C. A. 1997. Asymptotic structure of rotating MHD winds and its relation to wind boundary conditions. In *Herbig-Haro Flows and the Birth of Low Mass Stars*, IAU Symp. 182, ed. B. Reipurth and C. Bertout (Dordrecht: Kluwer), pp. 275–290.
Hildebrand, R. H., Dotson, J. L., Dowell, C. D., Platt, S. R., Schleuning, D., Davidson, J. A., and Novak, G. 1995. Far-infrared polarimetry. In *Airborne Astronomy Symposium on the Galactic Ecosystem*, ASP Conf. Ser. 73, ed. M. R. Haas, J. A. Davidson, and E. F. Erickson (San Francisco: Astronomical Society of the Pacific), pp. 97–104.
Hillenbrand, L. A., Strom, S. E., Vrba, F. J., and Keene, J. 1992. Herbig Ae/Be stars—intermediate-mass stars surrounded by massive circumstellar accretion disks. *Astrophys. J.* 397:613–643.
Hirose, S., Uchida, Y., Shibata, K., and Matsumoto, R. 1997. Disk accretion onto a magnetized young star and associated jet formation. *Pub. Astron. Soc. Japan* 49:193–205.
King, A. R., and Regev, O. 1994. Spin rates and mass loss in accreting T Tauri stars. *Mon. Not. Roy. Astron. Soc.* 268:L69–L73.
Königl, A. 1989. Self-similar models of magnetized accretion disks. *Astrophys. J.* 342:208–223.
Königl, A. 1997. Magnetized accretion disks and the origin of bipolar outflows. In *Accretion Phenomena and Related Outflows*, ASP Conf. Ser. 121, ed. D. T. Wickramasinghe, G. V. Bicknell, and L. Ferrario (San Francisco: Astronomical Society of the Pacific), pp. 551–560.
Königl, A. 1999. Theory of bipolar outflows from high-mass young stellar objects. *New Astron. Rev.* 43:67–77.
Königl, A., and Ruden, S. P. 1993. Origin of outflows and winds. In *Protostars and Planets III*, ed. E. H. Levy and J. I. Lunine (Tucson: University of Arizona Press), pp. 641–688.
Königl, A., and Wardle, M. 1996. A comment on the stability of magnetic wind-driving accretion discs. *Mon. Not. Roy. Astron. Soc.* 279:L61–L64.
Kössel, D., Müller, E., and Hillebrandt, W. 1990. Numerical simulations of axially symmetric magnetized jets. *Astron. Astrophys.* 229:401–415.
Lada, C. J. 1985. Cold outflows, energetic winds, and enigmatic jets around young stellar objects. *Ann. Rev. Astron. Astrophys.* 23:267–317.
Lery, T., Heyvaerts, J., Appl, S., and Norman, C. A. 1998. Outflows from magnetic rotators. I. Inner structure. *Astron. Astrophys.* 337:603–624.
Levreault, R. M. 1988. Molecular outflows and mass loss in the pre-main-sequence stars. *Astrophys. J.* 330:897–910.

Levy, E. H., and Sonnett, C. P. 1978. Meteorite magnetism and early solar system magnetic fields. In *Protostars and Planets*, ed. T. Gehrels (Tucson: University of Arizona Press), pp. 516–532.

Li, Z.-Y. 1995. Magnetohydrodynamic disk-wind connection: Self-similar solutions. *Astrophys. J.* 444:848–860.

Li, Z.-Y. 1996. Magnetohydrodynamic disk-wind connection: Magnetocentrifugal winds from ambipolar diffusion-dominated accretion disks. *Astrophys. J.* 465:855–868.

Lind, K. R., Payne, D. G., Meier, D. L., and Blandford, R. D. 1989. Numerical simulations of magnetized jets. *Astrophys. J.* 344:89–103.

Lovelace, R. V. E., Romanova, M. M., and Newman, W. I. 1994. Implosive accretion and outbursts of active galactic nuclei. *Astrophys. J.* 437:136–143.

Lubow, S. H., Papaloizou, J. C. B., and Pringle, J. 1994. Magnetic field dragging in accretion discs. *Mon. Not. Roy. Astron. Soc.* 267:235–240.

Lynden-Bell, D., and Boily, C. 1994. Self-similar solutions up to flashpoint in highly wound magnetostatics. *Mon. Not. Roy. Astron. Soc.* 267:146–152.

Masson, C. R., and Chernin, L. M. 1993. Properties of jet-driven molecular outflows. *Astrophys. J.* 414:230–241.

Matsumoto, R., Uchida, Y., Hirose, S., Shibata, K., Hayashi, M. R., Ferrari, A., Bodo, G., and Norman, C. 1996. Radio jets and the formation of active galaxies: Accretion avalanches on the torus by the effect of a large-scale magnetic field. *Astrophys. J.* 461:115–126.

McKee, C. F., Zweibel, E. G., Goodman, A. A., and Heiles, C. 1993. Magnetic fields in star-forming regions: Theory. In *Protostars and Planets III*, ed. E. H. Levy and J. I. Lunine (Tucson: University of Arizona Press), pp. 327–366.

Meier, D., Edgington, S., Godon, P., Payne, D., and Lind, K. 1997. A magnetic switch that determines the speed of astrophysical jets. *Nature* 388:350–352.

Mestel, L. 1968. Magnetic braking by a stellar wind. *Mon. Not. Roy. Astron. Soc.* 138:359–391.

Michel, F. C. 1969. Relativistic stellar-wind torques. *Astrophys. J.* 158:727–738.

Miller, K. A., and Stone, J. M. 1997. Magnetohydrodynamic simulations of stellar magnetosphere–accretion disk interaction. *Astrophys. J.* 489:890–902.

Miroshnichenko, A., Ivezić, Ž., and Elitzur, M. 1997. On protostellar disks in Herbig Ae/Be stars. *Astrophys. J. Lett.* 475:L41–L44.

Mouschovias, T. C. 1991. Cosmic magnetism and the basic physics of the early stages of star formation. In *The Physics of Star Formation and Early Stellar Evolution*, ed. C. J. Lada and N. D. Kylafis (Dordrecht: Kluwer), pp. 61–122.

Mundt, R., and Ray, T. M. 1994. Optical outflows from Herbig Ae/Be stars and other high luminosity young stellar objects. In *The Nature and Evolution of Herbig Ae/Be Stars*, ASP Conf. Ser. 62, ed. P. S. Thé, M. R. Pérez, and E. P. J. van den Heuvel (San Francisco: Astronomical Society of the Pacific), pp. 237–252.

Ogilvie, G. I., and Livio, M. 1998. On the difficulty of launching an outflow from an accretion disk. *Astrophys. J.* 499:329–339.

Ostriker, E. 1997. Self-similar magnetocentrifugal disk winds with cylindrical asymptotics. *Astrophys. J.* 486:291–306.

Ouyed, R., and Pudritz, R. E. 1997*a*. Numerical simulations of astrophysical jets from Keplerian accretion disks. I. Stationary models. *Astrophys. J.* 482:712–732.

Ouyed, R., and Pudritz, R. E. 1997*b*. Numerical simulations of astrophysical jets from Keplerian accretion disks. II. Episodic outflows. *Astrophys. J.* 484:794–809.

Ouyed, R., and Pudritz, R. E. 1999. Numerical simulations of astrophysical jets from Keplerian accretion disks. III. The effects of mass loading. *Mon. Not. Roy. Astron. Soc.,* in press.

Ouyed, R., Pudritz, R. E., and Stone, J. M. 1997. Episodic jets from black holes and protostars. *Nature* 385:409–414.

Pelletier, G., and Pudritz, R. E. 1992. Hydromagnetic disk winds in young stellar objects and active galactic nuclei. *Astrophys. J.* 394:117–138.

Pezzuto, S., Strafella, F., and Lorenzetti, D. 1997. On the circumstellar matter distribution around Herbig Ae/Be stars. *Astrophys. J.* 485:290–307.

Pudritz, R. E., and Norman, C. A. 1983. Centrifugally driven winds from contracting molecular disks. *Astrophys. J.* 274:677–697.

Pudritz, R. E., and Ouyed, R. 1997. Numerical simulations of jets from accretion disks. In *Herbig-Haro Flows and the Birth of Low Mass Stars*, IAU Symposium 182, ed. B. Reipurth and C. Bertout (Dordrecht: Kluwer), pp. 259–274.

Pudritz, R. E., Pelletier, G., and Gomez de Castro, A. I. 1991. The physics of disk winds. In *The Physics of Star Formation and Early Stellar Evolution*, ed. C. J. Lada and N. D. Kylafis (Dordrecht: Kluwer), pp. 539–564.

Pudritz, R. E., Wilson, C. D., Carlstrom, J. E., Lay, O. P., Hills, R. E., and Ward-Thompson, D. 1996. Accretion disks around class 0 protostars: The case of VLA 1623. *Astrophys. J. Lett.* 470:L123–L126.

Ray, T., Muxlow, T. W. B., Axon, D. J., Brown, A., Corcoran, D., Dyson, J., and Mundt, R. 1997. Large-scale magnetic fields in the outflow from the young stellar object T Tauri S. *Nature* 385:415–417.

Reipurth, B. 1991. Observations of Herbig-Haro objects. In *Low Mass Star Formation and Pre–Main Sequence Objects*, ed. B. Reipurth (Munich: European Southern Observatory), pp. 247–279.

Reyes-Ruiz, M., and Stepinski, T. F. 1996. Axisymmetric two-dimensional computation of magnetic field dragging in accretion disks. *Astrophys. J.* 459:653–665.

Romanova, M. M., Ustyugova, G. V., Koldoba, A. V., Chechetkin, V. M., and Lovelace, R. V. E. 1997. Formation of stationary magnetohydrodynamic outflows from a disk by time-dependent simulations. *Astrophys. J.* 482:708–711.

Sauty, C., and Tsinganos, K. 1994. Nonradial and nonpolytropic astrophysical outflows. III. A criterion for the transition from jets to winds. *Astron. Astrophys.* 287:893–926.

Schatzman, E. 1962. A theory of the role of magnetic activity during star formation. *Ann. Astrophys.* 25:18–29.

Schleuning, D. A. 1998. Far-infrared and submillimeter polarization of OMC-1: Evidence for magnetically regulated star formation. *Astrophys. J.* 493:811–825.

Shibata, K., and Uchida, Y. 1986. A magnetohydrodynamical mechanism for the formation of astrophysical jets. II. Dynamical processes in the accretion of magnetized mass in rotation. *Pub. Astron. Soc. Japan* 38:631–660.

Shu, F. H., Najita, J., Ostriker, E., Wilkin, F., Ruden, S., and Lizano, S. 1994. Magnetocentrifugally driven flows from young stars and disks. I. A generalized model. *Astrophys. J.* 429:781–796.

Skinner, S. L., Brown, A., and Stewart, R. T. 1993. A high-sensitivity survey of radio continuum emission from Herbig Ae/Be stars. *Astrophys. J. Suppl.* 87:217–265.

Spruit, H. C. 1996. Magnetohydrodynamic jets and winds from accretion disks. In *Evolutionary Processes in Binary Stars,* NATO ASI Ser. C. 477, ed. R. A. M. J. Wijers, M. B. Davies, and C. A. Tout (Dordrecht: Kluwer), pp. 249–286.

Spruit, H. C., Foglizzo, T., and Stehle, R. 1997. Collimation of magnetically driven jets from accretion discs. *Mon. Not. Roy. Astron. Soc.* 288:333–342.

Stepinski, T. F. 1992. Generation of dynamo magnetic fields in the primordial solar nebula. *Icarus* 97:130–141.

Stone, J. M., and Norman, M. L. 1992. ZEUS-2D: A radiation magnetohydrodynamics code for astrophysical flows in two space dimensions. II. The magnetohydrodynamic algorithms and tests. *Astrophys. J. Suppl.* 80:791–818.

Stone, J. M., and Norman, M. L. 1994. Numerical simulations of magnetic accretion disks. *Astrophys. J.* 433:746–756.

Tout, C. A., and Pringle, J. E. 1996. Can a disc dynamo generate large-scale magnetic fields? *Mon. Not. Roy. Astron. Soc.* 281:219–225.

Tsinganos, K., and Trussoni, E. 1990. Analytic studies of collimated winds. I. Topologies of 2-D helicoidal hydrodynamic solutions. *Astron. Astrophys.* 231:270–276.

Uchida, Y., and Shibata, K. 1985. A magnetohydrodynamic mechanism for the formation of astrophysical jets. I. Dynamical effects of the relaxation of nonlinear magnetic twists. *Pub. Astron. Soc. Japan* 37:31–46.

Umebayashi, T., and Nakano, T. 1981. Fluxes of energetic particles and the ionization rate in very dense interstellar clouds. *Pub. Astron. Soc. Japan* 33:617–635.

Ustyugova, G. V., Koldoba, A. V., Romanova, M. M., Chechetkin, V. M., and Lovelace, R. V. E. 1995. Magnetohydrodynamic simulations of outflows from accretion disks. *Astrophys. J. Lett.* 439:L39–L42.

Wardle, M. 1997. Magnetically driven winds from protostellar disks. In *Accretion Phenomena and Related Outflows*, ASP Conf. Ser. 121, ed. D. T. Wickramasinghe, G. V. Bicknell, and L. Ferrario (San Francisco: Astronomical Society of the Pacific), pp. 561–565.

Wardle, M., and Königl, A. 1993. The structure of protostellar accretion disks and the origin of bipolar flows. *Astrophys. J.* 410:218–238.

Weber, E. J., and Davis, L. 1967. The angular momentum of the solar wind. *Astrophys. J.* 148:217–227.

X-WINDS: THEORY AND OBSERVATIONS

FRANK H. SHU
University of California at Berkeley

JOAN R. NAJITA
Space Telescope Science Institute

HSIEN SHANG
University of California at Berkeley

and

ZHI-YUN LI
University of Virginia at Charlottesville

We review the theory of x-winds in young stellar objects (YSOs), and we compare its predictions with a variety of astronomical observations. Such flows arise magnetocentrifugally from accretion disks when their inner edges interact with strongly magnetized central stars. X-winds collimate logarithmically slowly into jets, and their interactions with the surrounding molecular cloud cores of YSOs yield bipolar molecular outflows.

I. INTRODUCTION

Strong magnetic fields can considerably enhance the mass loss $\dot{M}_w$ of thermally driven winds from the surfaces of rapidly rotating stars (Mestel 1968). Even if the resultant flow is quite cold, Hartmann and MacGregor (1982) demonstrated that $\dot{M}_w$ could have almost arbitrarily large values, dependent only on the ratio of the azimuthal and radial field strengths at the position where the gas is injected onto open field lines near the equator of a protostar that rotates near breakup. Shu et al. (1988) assigned the cause for the protostar to spin at breakup to a circumstellar disk that abuts against the surface of the central object and accretes onto it at a high rate $\dot{M}_D$. They also replaced Hartmann and MacGregor's (1982) arbitrary angle of injection for gas velocity and magnetic field direction with the requirement that in steady state, the wind mass loss rate must be a definite fraction f of the disk accretion rate $\dot{M}_w = f\dot{M}_D$ (see below).

Blandford and Payne (1982) advanced an influential self-similar model of centrifugally driven winds from the surfaces of magnetized accretion disks. Pudritz and Norman (1983) applied these pure disk wind models to bipolar outflows, and Königl (1989) investigated how the wind might smoothly join a pattern of accretion flow inside the disk. Heyvaerts and Norman (1989) studied how the winds might collimate asymptotically into jets, while Uchida and Shibata (1985) and Lovelace et al. (1991) advocated alternative driving mechanisms in which magnetic pressure gradients play a bigger role in the acceleration of a disk-blown wind.

Motivated by the problem of binary X-ray sources, a parallel line of research developed concerning how magnetized stars accrete from surrounding disks. Ghosh and Lamb (1978) used order-of-magnitude arguments to show that a strongly magnetized star would truncate the surrounding accretion disk at a larger radius than the stellar radius $R_\star$ and divert the equatorial flow along closed field-line funnels toward the polar caps. Although Ghosh and Lamb thought that this inflow would spin the central object up faster than if the accretion disk had extended right up to the stellar surface, Königl (1991) made the surprising and insightful suggestion that the process might torque down the star and account for the relatively slow rate of spin of observed T Tauri stars. Observational support for magnetospheric accretion was subsequently marshaled by Edwards et al. (1994) and Hartmann et al. (1994).

Arons, McKee, and Pudritz (Arons 1986) proposed that a centrifugally driven outflow accompanies the funnel inflow (see also Camenzind 1990). Independently, Basri (unpublished) arrived observationally at the same suggestion by extending the synthesis work by Bertout et al. (1988) on ultraviolet excesses in T Tauri stars. Shu et al. (1994*a*) put these ideas together into a concrete proposal that generalized the earlier x-wind model of Shu et al. (1988). A related proposal, the so-called "magnetic propeller" (e.g., Li and Wickramasinghe 1997; Lovelace et al. 1999), has been invoked recently to explain the spindown of the cataclysmic variable AE Aquarii (Wynn et al. 1997) and the Rossi X-ray Timing Explorer (RXTE) observations of X-ray pulsars GX 1+4 and GRO J1744-28 (Cui 1997). A quick and somewhat oversimplified summary might be that x-wind theory adds the possibility of outflow to magnetospheric accretion models, while the magnetic propeller idea adds the possibility of time dependence.

II. GENERALIZED X-WIND MODEL

In the generalized x-wind model, if the star has mass $M_\star$ and magnetic dipole moment $\mu_\star$, the gas disk is truncated at an inner radius,

$$R_{\rm x} = \Phi_{\rm dx}^{-4/7}\left(\frac{\mu_\star^4}{GM_\star\dot{M}_D^2}\right)^{1/7} \quad (1)$$

A disk of solids may extend inward of $R_{\rm x}$ to the evaporation radius of calcium-aluminum silicates and oxides (see Shang et al. 1997 and Meyer et al. 1997). In equation (1), Φ_{dx} is a dimensionless number of order unity that measures the amount of magnetic dipole flux that has been pushed by the disk accretion flow to the inner edge of the disk. In the closed dead-zone model of Najita and Shu (1994) and Ostriker and Shu (1995), $\Phi_{dx} = 2\overline{\beta}_w f^{1/2}$, with $f = \dot{M}_w/\dot{M}_D$ and $\overline{\beta}_w$ defined as the streamline-averaged ratio of magnetic field to wind mass flux. This expression for the coefficient Φ_{dx} holds when magnetic flux is swept toward R_x from both larger and smaller radii (see below), and the magnetic flux trapped in a small neighborhood of R_x is $\frac{3}{2}$ times larger than the pure dipole value.

For a Keplerian disk, the inner edge of the disk rotates at angular speed

$$\Omega_x = \left(\frac{GM_\star}{R_x^3}\right)^{1/2} \tag{2}$$

To satisfy mass and angular momentum balance, the disk accretion divides at R_x into a wind fraction, $\dot{M}_w = f\dot{M}_D$, and a funnel-flow fraction, $\dot{M}_\star = (1-f)\dot{M}_D$, where

$$f = \frac{1 - \overline{J}_\star - \tau}{\overline{J}_w - \overline{J}_\star} \tag{3}$$

In equation (3) $\overline{J}_\star$ and $\overline{J}_w$ are, respectively, specific angular momenta nondimensionalized in units of $R_x^2\Omega_x$ and averaged over funnel and wind streamlines, and τ is the negative of the viscous torque of the disk, $-\mathcal{T}$, acting on its inner edge and measured in units of $\dot{M}_D R_x^2 \Omega_x$. In the approximation that the embedded stellar fields force the x-region to be only weakly differentially rotating (Shu et al. 1994*a,b*), τ is small compared to unity.

In steady state, the star is regulated to corotate with the inner disk edge,

$$\Omega_\star = \Omega_x \tag{4}$$

If corotation did not hold, the system would react to reduce the discrepancy between $\Omega_\star$ and Ω_x, where Ω_x is given by equation (2) with R_x determined by how much magnetic field there is to maintain a certain average standoff distance for the stellar magnetosphere [see equation (1)]. For example, suppose the star turns faster than the inner edge of the disk, $\Omega_\star > \Omega_x$. The field lines attached to both would then continuously wrap into ever tighter trailing spirals, with the field lines adjacent to the star tugging it backward in the sense of rotation. This tug decreases the star's angular rate of rotation $\Omega_\star$ to more nearly equal the rate Ω_x. Conversely, imagine that $\Omega_\star < \Omega_x$. With the star turning more slowly than the inner edge of the disk, the field lines attached to both would continuously wrap into ever tighter leading spirals, with the field lines adjacent to the star tugging it forward in the sense of rotation. This tug increases the star's angular rate of rotation $\Omega_\star$, again to more nearly equal the rate Ω_x. In true steady state, $\Omega_\star = \Omega_x$, and the funnel flow field lines acquire just enough of a trailing spiral pattern (but without continuously wrapping up) that the excess of material angular momentum brought toward the star by the inflowing gas is transferred outward by magnetic torques to the footpoints of the magnetic field in the disk (Shu et al. 1994*a,b*).

The x-wind gas gains angular momentum, and the funnel gas loses angular momentum, at the expense of the matter at the footpoint of the field in, respectively, the outer and inner parts of the x-region. As a consequence, this matter, and the field lines across which it is diffusing, pinch toward the middle of the x-region. In reality, Shu et al. (1997) point out that the idealized steady state of exact corotation probably cannot be maintained because of dissipative effects. Two surfaces of null poloidal field lines (labeled as "helmet streamer" and "reconnection ring" in Fig. 1) mediate the topological behavior of dipolelike field lines of the star, opened field lines of the x-wind, and trapped field lines of the funnel inflow emanating from the x-region. Across each of the null surfaces, which begin or end on "Y-points" (called "kink points" by Ostriker and Shu 1995), the poloidal magnetic field suffers a sharp reversal of direction. By Ampère's law, large electric currents must flow out of the plane of the figure along the null surfaces. Nonzero electrical resistivity would lead to the dissipation of these currents and to the reconnection of the oppositely directed field lines (see, e.g., Biskamp 1993). The resultant reduction of the trapped

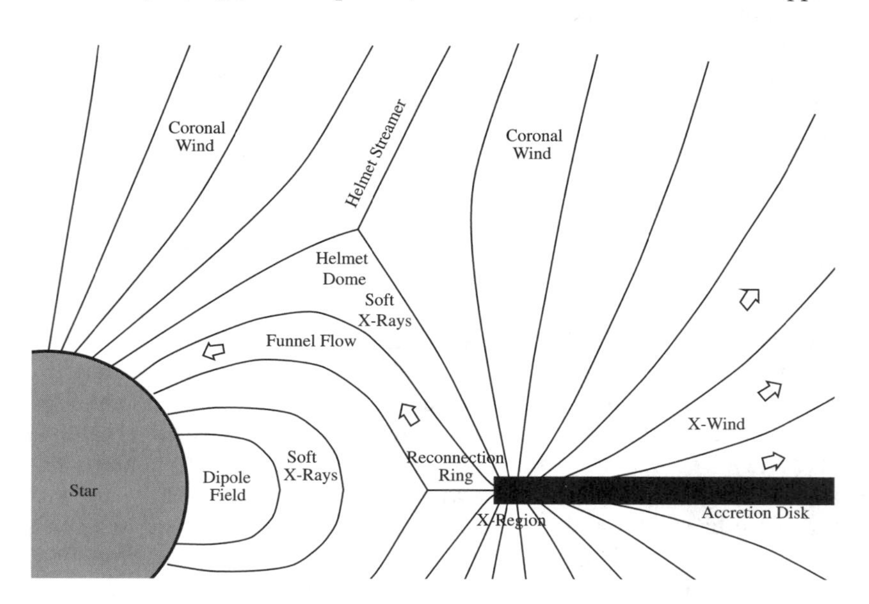

Figure 1. Schematic drawing of the x-wind model. Coronal winds from the star and the disk may help the x-wind to open field lines surrounding the helmet streamer, but this aspect of the configuration is not central to the model.

magnetic flux in the x-region as the fan of field lines presses into the evacuated region of the annihilated fields would change the numerical value of the coefficient Φ_{dx} in equation (1).

When R_x changes, the angular speed Ω_x of the footpoint of magnetic field lines in the x-region will vary according to equation (2). However, the considerable inertia of the star prevents its angular velocity $\Omega_\star$ from changing on the timescale of magnetic reconnection at the null surfaces of the magnetosphere. The resulting shear when $\Omega_\star \neq \Omega_x$ will stretch and amplify field lines attached to both the star and the disk. The poloidal field will bulge outward from the increased magnetic pressure, inserting more magnetic flux into the fan of field lines emanating from the x-region [see the simulations of Linker and Mikić (1995) and Hayashi et al. (1996)]. Dynamo action inside the star would presumably replace the upward-rising dipolelike poloidal fields. This dynamo action would be enhanced by the wrapping of field lines between the star and disk. Averaged over long times, we envisage a secular balance, with enough dynamo-generated poloidal field being inserted into the x-region to balance the rate of field dissipation at the null surfaces.

III. THE CENTRAL MACHINE: CHECKS WITH OBSERVATIONS

In lieu of spatially resolved interferometric images (which will become available within the next two decades), the picture just presented of the central machine that drives collimated outflows from young stellar objects (YSOs) is subject to many different kinds of spectroscopic checks. Here we comment only on four crucial observations: (a) the demonstration that rapidly rotating disks do extend almost to the surfaces of YSOs of sun-like masses, (b) the demonstration that the central stars are magnetized to the extent necessary to truncate the disks at an inner radius R_x compatible with locking stellar rotation rates to their observed low values, (c) the demonstration that the final accretion process onto the central stars is magnetically channeled, and (d) the demonstration that such funnel flows are accompanied by outflows that have the requisite geometry and phase relationship implied by x-wind theory.

A. Existence of Rapidly Rotating Inner Disks

The best evidence to date for the existence of inner disks comes from high-resolution spectroscopy of CO overtone emission in young stars (see the chapter by Najita et al., this volume). Through detailed modeling of CO overtone emission it is possible to constrain the outer and, in particular, inner radii for the emission. For example, in the case of the Class I source WL16, the overtone emission is found to arise from disk radii between 5 and 30 $R_\odot$. The outer cutoff of 30 $R_\odot$ probably arises from lack of excitation and has little other physical significance. The inner cutoff of 5 $R_\odot$

may arise either because of CO dissociation or because of a real truncation of the disk at an inner edge.

The latter interpretation is compatible with the emitting area required to explain the strength of the Brackett γ line at line center. The dramatic agreement between the Brγ profile for this source (Najita et al. 1996) and model Brγ spectra for accreting magnetospheres (Muzerolle et al. 1998; see also section III.C below) is suggestive evidence in favor of the claim that WL16 has a strong stellar magnetosphere that truncates its surrounding disk, and through which disk matter accretes onto the stellar surface (Fig. 1). From such evidence, we conclude that one-half of the two crucial ingredients needed for x-wind theory to hold, a rapidly rotating, gaseous, truncated inner disk, does exist in at least one well-studied source found in nature.

B. Strong Magnetization of the Central Stars

If one eliminates R_x from equations (1), (2), and (4), one gets the magnetic dipole moment of the star,

$$\mu_\star = \Phi_{dx}(GM_\star)^{5/6}\dot{M}_D^{1/2}\Omega_\star^{-7/6} \tag{5}$$

required to enforce, in steady state, a given stellar spin rate $\Omega_\star$ in a YSO of mass $M_\star$ and disk accretion rate $\dot{M}_D$. We should not expect equation (5) to hold when $\dot{M}_D$ drops to very low rates, because the remaining reservoir of mass and angular momentum coming through the disk may not then suffice to overcome the inertia of the central star and continue to regulate its spin period. Typical conditions for a classical T Tauri star might be marginal for the purpose at hand: $M_\star = 0.5\,M_\odot$, $\dot{M}_D = 3\times10^{-8}M_\odot\,yr^{-1}$, and $\Omega_\star = 2\pi/(8\,dy)$ (Bertout 1989; Edwards et al. 1993; Gullbring et al. 1998). With $\overline{\beta}_w \sim 1$ and $f \sim \frac{1}{3}$, so that $\Phi_{dx} \sim 2/\sqrt{3}$, and an equatorial field strength $B_\star$ given ideally by $B_\star = \mu_\star/R_\star^3$, where $R_\star = 2\,R_\odot$ is a typical stellar radius, equation (5) now implies $B_\star = 2.1$ kG, a measurable value.

Johns-Krull et al. (1999) have determined the Zeeman broadening of magnetically sensitive lines in the classical T Tauri star BP Tauri. [Several other sources show similarly large fields (C. M. Johns-Krull, personal communication, 1998); see also the early work of Basri et al. 1992.] Johns-Krull and coworkers (1999) have compared their results with the predicted relation (5) and find good agreement. These quantitative measures of the stellar magnetic field buttress the inference from X-ray flaring activity that T Tauri stars are strongly magnetized (Montmerle et al. 1993). One of the more remarkable recent developments in this field is the discovery that even Class I sources are strong X-ray emitters (see the chapter in this volume by Glassgold et al.), indicating that stellar magnetic activity begins early in young stars. We conclude that the other half of the two crucial ingredients needed for x-wind theory to hold, a strong stellar magnetic field, also does exist for several sunlike YSOs found in nature.

C. Magnetically Channeled Accretion onto the Central Star

Since the work of Ghosh and Lamb (1978), workers in the field of X-ray binaries have concurred that if the central object is sufficiently strongly magnetized, the last stages of mass accretion from a surrounding disk may be channeled magnetically onto the central (neutron) star rather than continue to the star's surface by viscous inspiraling. Only in the case of YSOs, however, do we have direct spectroscopic evidence of this process in the shape of predicted line profiles.

It now appears that the bulk of the hydrogen line emission in young stars arises not in outflowing gas, as was long believed, but in inflowing gas located within several stellar radii of the stellar surface. Outflow signatures are more often seen in Hα and Na I, and inflow signatures are commonly seen in the higher Balmer and Brackett γ lines (Edwards et al. 1994; Najita et al. 1996). The comparison of hydrogen line profiles with models of the kinematics and excitation conditions expected in magnetospheric accretion flows supports the idea that the emission originates in hot gas inflowing along stellar magnetic field lines (Muzerolle et al. 1998; see also the chapter in this volume by Najita et al.).

The reader should not conclude from this description, however, that the funnel flow in x-wind theory is identical to the magnetically channeled infall envisioned by Ghosh and Lamb (1978; see also the comments of Shu et al. 1994*a* and Ostriker and Shu 1995). In Ghosh and Lamb's (1978) theory, all stellar field lines are closed, and the braking torques exerted by field lines that thread distant radii in the disk balance, in steady state, the accelerating torques exerted by field lines that thread nearby radii in the disk. The "steady state" in this picture is intrinsically statistical and violent in nature (and therefore not computed in any detail), because it involves rapid field wrapping and magnetic reconnection. Because higher latitudes on the star are tugged more slowly by field lines that thread the disk more distantly than field lines from lower latitudes, such a picture would presumably result in a central object that has considerable differential rotation and no well-defined stellar period, contrary to many observations of the rotation properties of classical T Tauri stars (e.g., Vogel and Kuhi 1981; Rydgren and Vrba 1983; Walter et al. 1988; Bouvier et al. 1993; Johns-Krull 1996).

In x-wind theory, closed stellar field lines that thread a nearly uniformly rotating, small, x-region of the disk (funnel-flow field lines) transfer in steady state the excess angular momentum contained in the inflowing matter, not to the star but to the inner portions of the x-region. This excess angular momentum is then removed from the outer portions of the x-region by the opened field lines of the YSO outflow. [Angular-momentum transport from the inner portions to the outer portions of the x-region is probably accomplished by the magnetorotational instability (Balbus and Hawley 1991) and has not yet been examined in any detail.] In pure x-wind theory, the opened field lines came originally from the star (see Fig. 1), and therefore the magnetocentrifugally driven outflow bears definite geometric and phase relationships to the magnetocentrifugally driven inflow.

D. Geometric and Phase Relationships

As mentioned above, outflow and inflow absorption components are often observed simultaneously and are now believed to arise in the cooler wind and in accretion footpoints, respectively. Periodograms of the hydrogen lines in T Tauri stars can be used to probe the dynamical relationship between the wind, the funnel flow, and the rotation period of the star. Synoptic monitoring by Johns and Basri (1995*b*) of the T Tauri star SU Aurigae revealed inflow and outflow components to be variable at the photometric (rotation) period of the star but separated in phase by 180°. Johns-Krull and Hatzes (1997) have since found another system with similar properties. If the funnel flow and collimating wind emanate in these objects from the surrounding accretion disks, then the source regions for the flowing gas corotate with the central stars, in agreement with the prediction that $\Omega_\star = \Omega_x$. Moreover, the phase lag of 180° is consistent with an origin for both components in the funnel flow and x-wind induced by a tilted magnetic dipole, where field lines in one part of the magnetosphere naturally bend inward, promoting accretion, and field lines 180° away bend outward, promoting mass loss (see Fig. 15 of Johns and Basri 1995*b*). Unfortunately, the majority of objects studied by Johns and Basri (1995*a*) do not show such periodicity, indicating that even if $\Omega_\star$ does equal Ω_x as a long-term average, $\Omega_\star$ may not generally equal Ω_x instantaneously.

IV. OUTLINE OF MATHEMATICAL FORMULATION

A mathematical formulation of the steady-state problem when $\Omega_\star = \Omega_x$ was given in outline by Shu et al. (1988) and in detail by Shu et al. (1994*b*), who nondimensionalized the governing equations by introducing R_x, Ω_x, and $\dot{M}_w/4\pi R_x^3\Omega_x$, respectively, as the units of length, time, and density. Assuming axial symmetry and time independence in a frame that corotates with Ω_x, we may then introduce cylindrical coordinates (ϖ, φ, z) and a streamfunction $\psi(\varpi, z)$ that allows the satisfaction of the equation of continuity, $\nabla \cdot (\rho \mathbf{u}) = 0$, in the meridional plane:

$$\rho u_\varpi = \frac{1}{\varpi}\frac{\partial \psi}{\partial z}, \qquad \rho u_z = -\frac{1}{\varpi}\frac{\partial \psi}{\partial \varpi} \tag{6}$$

Field freezing in the corotating frame, $\mathbf{B} \times \mathbf{u} = 0$, implies that the magnetic field is proportional to the mass flux, $\mathbf{B} = \beta \rho \mathbf{u}$, where β is a scalar. The condition of no magnetic monopoles, $\nabla \cdot \mathbf{B} = 0$, now requires that β be conserved on streamlines, i.e.,

$$\beta = \beta(\psi) \tag{7}$$

Similarly, the conservation of total specific angular momentum requires that the amount carried by matter in the laboratory frame, $\varpi(\varpi\Omega_x + u_\varphi)$, where Ω_x is replaced by 1 in dimensionless equations, plus the amount carried by Maxwell torques, $-\varpi B_\varphi \mathbf{B}$, per unit mass flux, $\rho\mathbf{u}$, equals a function of ψ alone:

$$\varpi[(\varpi + u_\varphi) - \beta^2 \rho u_\varphi] = J(\psi) \tag{8}$$

Finally, conservation of specific "energy" in the corotating frame for a gas with dimensionless isothermal sound speed $\epsilon \equiv a_x/R_x\Omega_x$, results in Bernoulli's theorem:

$$\frac{1}{2}|\mathbf{u}|^2 + \mathcal{V}_{\text{eff}} + \epsilon^2 \ln \rho = H(\psi) \tag{9}$$

Up to an arbitrary constant, defined so that $\mathcal{V}_{\text{eff}} = 0$ at the x-point $\varpi = 1$ and $z = 0$, $\mathcal{V}_{\text{eff}}$ is the dimensionless gravitational potential plus centrifugal potential (measured in units of $GM_\star/R_x = \Omega_x^2 R_x^2$):

$$\mathcal{V}_{\text{eff}} = \frac{3}{2} - \frac{1}{(\varpi^2 + z^2)^{1/2}} - \frac{\varpi^2}{2} \tag{10}$$

Equations (7)–(9), with β, J, and H arbitrary functions of ψ, represent formal integrations of the governing set of equations. The remaining equation for the transfield momentum balance, the so-called Grad-Shafranov equation, cannot be integrated analytically and reads

$$\nabla \cdot (\mathcal{A}\nabla\psi) = \mathcal{Q} \tag{11}$$

where $\mathcal{A}$ is the Alfvén discriminant,

$$\mathcal{A} \equiv \frac{\beta^2\rho - 1}{\varpi^2\rho} \tag{12}$$

and $\mathcal{Q}$ is a source function for the internal collimation (or decollimation) of the flow:

$$\mathcal{Q} = \rho\left[\frac{u_\varphi}{\varpi}J'(\psi) + \rho|\mathbf{u}|^2\beta\beta'(\psi) - H'(\psi)\right] \tag{13}$$

with primes denoting differentiation with respect to the argument ψ.

The quantity $\mathcal{A} = 0$, i.e., $\beta^2\rho = 1$, when the square of the flow speed $|\mathbf{u}|^2$ equals the square of the Alfvén speed, $|\mathbf{B}|^2/\rho$. For sub-Alfvénic flow, $\mathcal{A} > 0$; for super-Alfvénic flow, $\mathcal{A} < 0$. If $\mathcal{A}$ were freely specifiable (which it is not), equation (11) would resemble the time-independent heat conduction equation. What is spreading in the meridional plane of our problem, however, is not heat, but streamlines.

V. FIXING THE FREE FUNCTIONS

When combined with Bernoulli's equation (9), the Grad-Shafranov equation (11) has three possible critical surfaces associated with it, corresponding to slow magnetohydrodynamic (MHD), Alfvén, and fast MHD crossings (Weber and Davis 1967). Thus, when applied to the problem of the x-wind, equation (11) is a second-order partial differential equation of elliptic type interior to the fast surface and of hyperbolic type exterior to this surface (see Heinemann and Olbert 1978 and Sakurai 1985). The unknown functions $\beta(\psi)$, $J(\psi)$, and $H(\psi)$ are to be determined self-consistently so that the three crossings of the critical surfaces are made smoothly. This does not fix all three functions uniquely, because the loci of the critical surfaces in (ϖ, z)-space are not known in advance. It turns out that we can choose one of the loci freely. Alternatively, we can freely specify one of the functions β, J, or H. The method of Najita and Shu (1994) fixes in advance the locus of the Alfvén surface and determines all other quantities self-consistently from this parameterization. Shang and Shu (1998) have invented a simplified procedure in which the function $\beta(\psi)$ is specified in advance; then $J(\psi)$, $H(\psi)$, and the loci of the slow, Alfvén, and fast surfaces are found as part of the overall solution (including force balance with the opened field lines of the star and dead zone). The mathematical procedure of choosing an appropriate $\beta(\psi)$ is a substitute for the physical problem of how to load matter onto field lines in the x-region where the ideal MHD approximation of field freezing breaks down (Shu et al. 1994*a,b*).

Apart from a trivial replacement of $\dot{M}_\star$ for $\dot{M}_\mathrm{w}$, the funnel flow behaves somewhat differently from the x-wind. The funnel flow has a slow MHD crossing but probably no Alfvén or fast MHD crossings. (In steady state the star and disk can communicate with each other along closed field lines by means of the latter two signals.) As a consequence, both $\beta(\psi)$ and $J(\psi)$ can be freely specified for the funnel flow, although $J(\psi)$ must ultimately be made self-consistent with the physical assumption that no spinup or spindown of the star occurs in steady state (that is, that $\Omega_\star$ remains equal to Ω_x as a long-term average). In these circumstances, when the star is small, Ostriker and Shu (1995) show that $J(\psi)$ is likely to be nearly zero on every funnel-flow streamline ψ = constant. In other words, $\bar{J}_\star \approx 0$, and any excess angular momentum brought to the star by the matter inflow is transferred back to the disk by the magnetic torques of a trailing spiral pattern of funnel field lines.

VI. THE COLD LIMIT

The overall problem is mathematically tractable, because the parameter $\epsilon \equiv a_\mathrm{x}/R_\mathrm{x}\Omega_\mathrm{x}$ is much smaller than unity (typically, $\epsilon \approx 0.03$). This leads to many simplifications; in particular, to the use of matched asymptotic expansions for solving and connecting different parts of the flow. For example, we may show that the slow MHD crossing must be made by matter-carrying streamlines within a fractional distance ϵ of R_x (unity in our nondimensionalization), and that to order ϵ^2, $H(\psi)$ may be approximated as zero. In the limit $\epsilon \to 0$, the gas that becomes the x-wind (or the funnel flow) emerges with linearly increasing velocities in a fan of streamlines from the x-region as if from a single point.

In this approximation, the function $\beta(\psi)$ cannot be chosen completely arbitrarily if the magnetic field, mass flux, and mass density do not diverge on the uppermost streamline $\psi \to 1$ as the x-wind leaves the x-region. For modeling purposes, Shang and Shu (1998) adopt the following distribution of magnetic field to mass flux:

$$\beta(\psi) = \beta_0(1 - \psi)^{-1/3} \tag{14}$$

where β_0 is a numerical constant related to the mean value of β averaged over streamlines:

$$\overline{\beta} \equiv \int_0^1 \beta(\psi)d\psi = \frac{3}{2}\beta_0 \tag{15}$$

The reader should not worry that equation (14) implies $\beta \to \infty$ as $\psi \to 1$. This singular behavior merely reflects the fact that the magnetic field $\mathbf{B} = \beta\rho\mathbf{u}$ is nonzero on the last x-wind streamline, where, by definition, ρ must become vanishingly small while $\mathbf{u}$ remains finite.

In what follows, we choose β_0 so that $f \approx \frac{1}{3}$ or $\frac{1}{4}$, with equation (3) implying that $\overline{J}_w \approx 3$ or 4 if $\overline{J}_\star \approx 0$ and $\tau \approx 0$. In other words, when the stellar magnetic field is strong enough to truncate the disk via a balanced x-wind and funnel flow, and when one-third to one-fourth of the matter drifting toward the inner edge of the disk becomes entangled in the one-third of the field lines in the x-region that have a proper outward orientation to launch an outflow, then the streamline-averaged location of the Alfvén surface is a factor $\sqrt{\overline{J}_w} = \sqrt{3}$, or 2 larger than the launch radius R_x. The conclusion that $\overline{J}_w \approx 3$ or 4 is important when we compare the predicted terminal velocities of stellar jets, $(2\overline{J}_w - 3)^{1/2}R_x\Omega_x$ averaged over all streamlines, with measured values (see below).

VII. ASYMPTOTIC COLLIMATION INTO JETS

A. Streamline Shape

Given the form $\beta(\psi)$ from equation (14), the loci of streamlines at large distances from the origin may be recovered in spherical polar coordinates $[r = (\varpi^2 + z^2)^{1/2},\ \theta = \arctan(\varpi/z)]$ from the asymptotic analysis of Shu et al. (1995):

$$r = \frac{2\overline{\beta}}{C}\cosh[F(C, 1)] \qquad \sin\theta = \mathrm{sech}[F(C, \psi)] \tag{16}$$

where different values of the dimensionless current C correspond to different locations on any streamline ψ = constant, and where $F(C, \psi)$ is the integral function

$$F(C, \psi) \equiv \frac{1}{C}\int_0^{\psi} \frac{\beta(\psi)d\psi}{[2J(\psi) - 3 - 2C\beta(\psi)]^{1/2}} \tag{17}$$

Notice that $r \rightarrow \infty$ when $C \rightarrow 0$ and *vice versa*.

The factor $2\overline{\beta}$ in equation (16) is replaced by $6\overline{\beta}$ [see the discussion of Shang and Shu (1998)] if the dead zone is completely opened, as in Fig. 1, rather than completely closed, as assumed in the calculations of Ostriker and Shu (1995) and Shu et al. (1995). In the former case, the hoop stresses of the toroidally wrapped fields on the uppermost streamlines of the x-wind are balanced by the magnetic pressure of a bundle of longitudinal field lines, coming from the star, that carries the same dimensionless magnetic flux, $2\pi\overline{\beta}$, as their opened counterparts in the x-wind. In the latter case, there are additional longitudinal field lines roughly parallel and antiparallel to the polar axis carrying oppositely directed flux $\pm 2\pi\overline{\beta}$ from the opened field lines of the dead zone. In reality, the system probably alternates between states in which outbursts, similar to coronal mass ejections, completely open the field lines of the dead zone and states in which finite resistivity helps to reconnect the oppositely directed open lines and produce a completely closed configuration for the dead zone.

B. Density and Velocity Fields

The density and velocity fields associated with equation (16) are obtained from

$$\rho = \frac{C}{\beta\varpi^2} \qquad v_w = (2J - 3 - 2C\beta)^{1/2} \tag{18}$$

where $\varpi \equiv r \sin\theta$ and v_w is the wind speed in an inertial frame. The solid and short dashed lines in Fig. 2 show, respectively, isodensity contours and flow streamlines for a case $\beta_0 = 1$ computed on four different scales by the simplified approximate procedure discussed by Shang and Shu (1998). For the model, $\overline{J}_w = 3.73$, obtainable from the average ϖ^2 of the location of the Alfvén surface marked by the inner set of long dashes in Fig. 2 [see equation (8) when $\beta^2\rho = 1$]. With $\overline{J}_\star$ and τ assumed to be zero, equations (3) and (15) imply $f = 1/\overline{J}_w = 0.268$ and $\overline{\beta}_w = \frac{3}{2}$. The associated value of $\Phi_{dx} = 2\overline{\beta}_w f^{1/2} = 1.55$.

The outermost density contour in the fourth panel is $\rho = 10^{-8}$ in units of $\dot{M}_w/4\pi\Omega_x R_x^3$. The Alfvén and fast surfaces formally asymptote to infinity as the upper streamline $\psi = 1$ is approached, where the density becomes vanishingly small. Because numerical computations are difficult in this limit, the actual uppermost streamline displayed is $\psi = 0.98$ rather

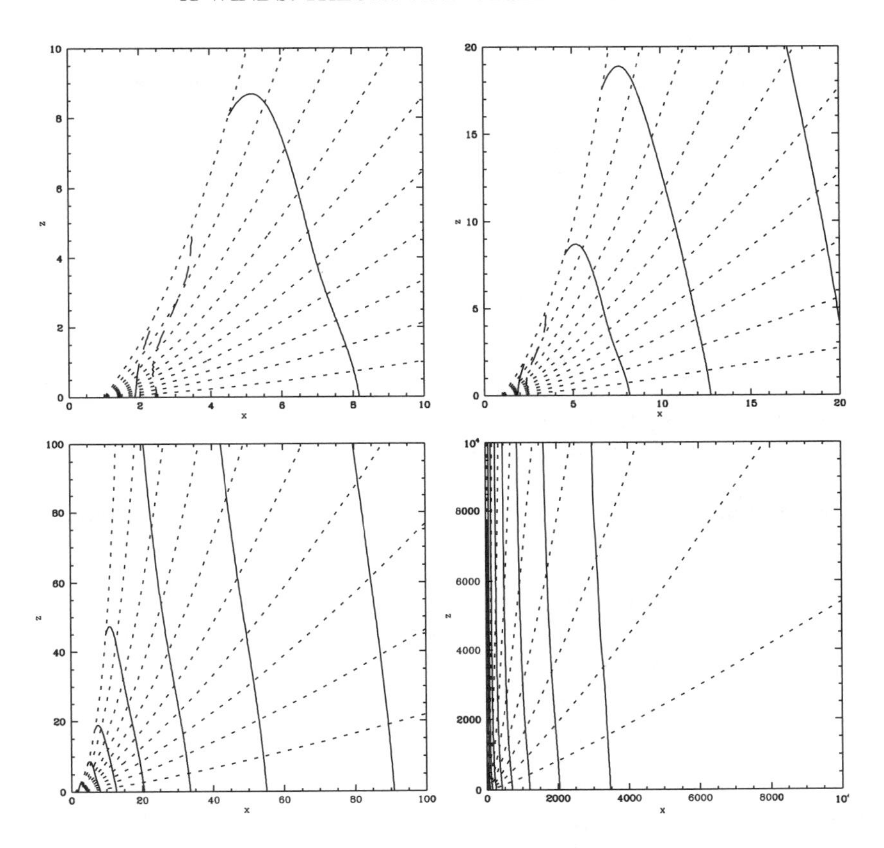

Figure 2. Isodensity contours (solid curves) and streamlines (dotted curves) for a cold x-wind with $\beta(\psi) = \beta_0(1-\psi)^{-1/3}$, where $\beta_0 = 1$. Isodensity contours are spaced logarithmically in intervals of $\Delta \log_{10} \rho = 0.5$, and streamlines are spaced so that successive dotted lines contain an additional 10% of the total mass loss in the upper hemisphere of the flow. The loci of the Alfvén and fast surfaces are marked by dashed lines. The empty space inside the uppermost streamline, $\varpi \le \varpi_1$, is filled with open field lines from the central star that asymptotically have the field strength, $B_z = 2\bar{\beta}/\varpi_1^2$.

than $\psi = 1$. Although streamlines collimate logarithmically slowly, isodensity contours become cylindrically stratified fairly quickly, roughly as $\rho \propto \varpi^{-2}$ [cf. equation (18)]. Since the radiative emission of forbidden lines is highly biased toward regions of moderately high density, the flow will appear more collimated than it actually is (see Shang et al. 1998).

The dimensional units of length and velocity, R_x and $\Omega_x R_x$, are $\sim$0.06 AU and $\sim$100 km s^{-1} in typical application; thus, at a distance of $\sim$600 AU, about 50% of the streamlines are flowing at speeds $\sim$200 km s^{-1}

within an angle $\theta \sim 4.5°$ of the polar axis, while another 50% exit in a wide-angle wind. In contrast, high-spatial-resolution imaging with the *Hubble Space Telescope* shows emission-line jets from young stars appearing to collimate perfectly within tens of AU of the star (e.g., Burrows et al. 1996). If our explanation is correct, the effect is partly an optical illusion arising from the strong cylindrical density stratification of the x-wind flow (see Fig. 2 of Shang et al. 1998).

The relative lack of streamline collimation at tens of AU scales, as opposed to density collimation, provides potentially a key discriminating test of the model. As seen in Fig. 3, the model predicts that forbidden lines arising in the x-wind should have larger velocity widths at the base of the flow, where a typical line of sight encounters flow velocities at a variety of angles, than farther up the length of the jet, where the flow vectors are better oriented along a single direction as the x-wind continues to collimate. Long-slit spectroscopy by Reipurth and Heathcote (1991) and Bacciotti et al. (1996) for several jets shows exactly this behavior. Unfortunately, the light seen near the base of the flow in embedded sources probably arrives by scattering from surrounding dust particles, so the observed effect appears to occur at a larger physical scale than predicted by the theory. Whether the basic kinematics has been affected by observing the phenomenon via "mirrors" remains to be ascertained, perhaps by spectropolarimetry. Better source candidates for a cleaner test of our prediction are the "naked jets" now found in evolved H II regions (B. Reipurth, personal communication, 1998).

The conventional interpretation of line profiles at the base of the flow being wider than in the jet proper invokes a mixing of fast and slow winds (e.g., Kwan and Tademaru 1988; Hirth et al. 1997). Such mixing of fast and slow material might occur because of turbulent entrainment as the lowermost x-wind streamlines interact with a flared accretion disk (Li and Shu 1996*a*). An x-wind could also interact with a slow wind driven by photoevaporation of a nebular disk (e.g., Shu et al. 1993). In all cases, however, as long as the source of the slower-moving material lies in a flattened distribution, appreciable mixing is easily understood only if the fast flow has a substantial equatorially directed component on the size scale of the inferred accretion disks.

VIII. BIPOLAR MOLECULAR OUTFLOWS

The phenomenon of bipolar molecular outflows (Snell et al. 1980) caught theorists by surprise, because infall rather than outflow had been expected to define the process of star formation. Since 1980, more than 200 molecular outflows have been observed (see, e.g., the reviews of Lada 1985; Fukui et al. 1993; Wu et al. 1996; Cabrit et al. 1997). By now, this ubiquitous phenomenon has become an integral part of the standard paradigm for star formation (Shu et al. 1987). Here, we limit our discussion to recent developments since the publication of *Protostars and Planets III*.

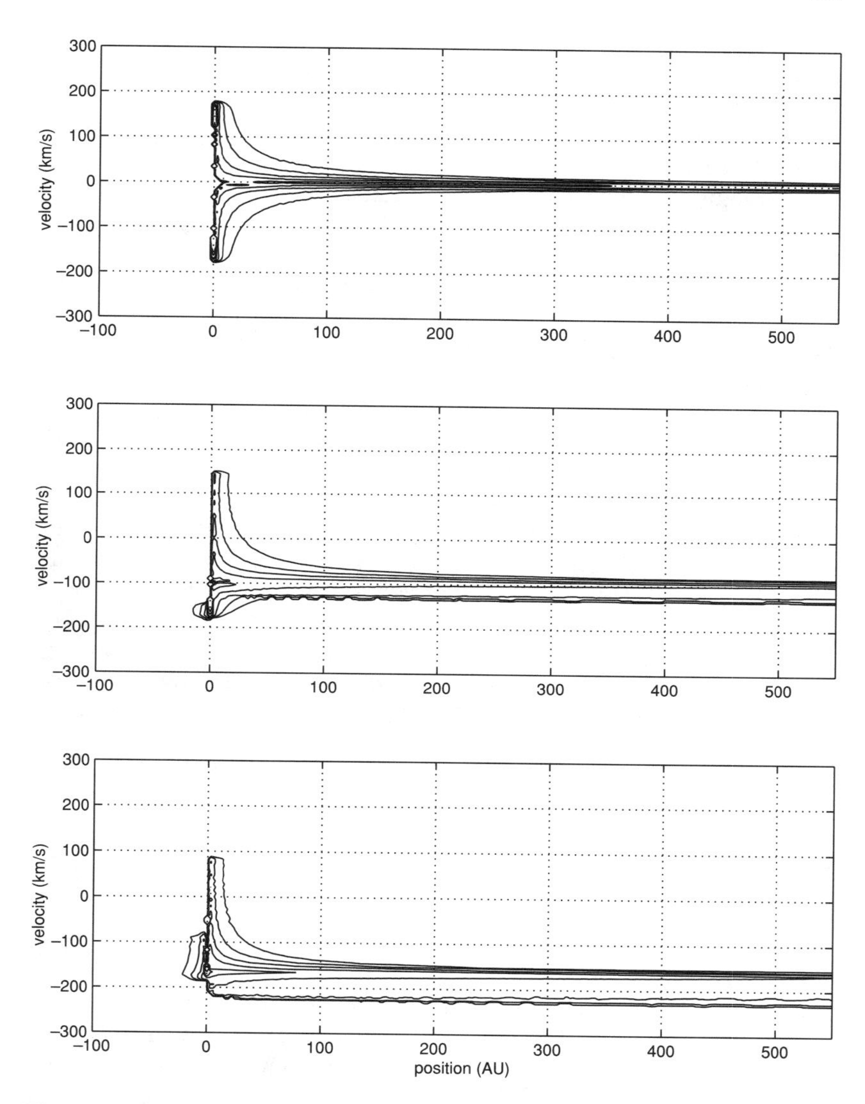

Figure 3. Position-velocity diagrams for [S II] λ 6731 emission when the synthetic spectrum is taken with a long slit placed along the length of the jet but displaced laterally by $1.5R_x$ with respect to its central axis. The different figures correspond to inclination angles $i = 90°$ (top), 60° (middle), and 30° (bottom). The range of projected terminal velocities seen in the models compares well with observed values. By the same token, the fact that measured terminal velocities in stellar jets have a limited range of values implies that the streamlines cannot be launched from a wide variety of disk radii of differing centrifugal speeds, as is implicit in many disk-wind models.

IX. THE AMBIENT MEDIUM INTO WHICH OUTFLOWS PROPAGATE

Low-mass stars form in the dense cores of molecular clouds (Myers 1995). These cores have a typical size ~0.1 pc, a number density $\sim 3 \times 10^4$ cm^{-3}, a mass ranging from a small fraction of a solar mass to about 10 $M_\odot$, and an axial ratio for flattening of typically 2:1. For isolated cores, the last fact implies that agents other than isotropic thermal or turbulent pressures help to support cores against their self-gravity, although it is not yet clear observationally whether the true shapes are oblate, prolate, or triaxial (P. C. Myers, personal communication, 1998). Observed cloud rotation rates are generally too small to account for the observed flattening (Goodman et al. 1993). This leaves magnetic fields, which are believed for other reasons to play a crucial role in contemporary star formation.

In one scenario, the weakening of magnetic support in the central part of a molecular cloud by ambipolar diffusion leads to the continued contraction of a cloud core with ever-growing central concentration (Nakano 1979; Lizano and Shu 1989; Basu and Mouschovias 1994). Li and Shu (1996*b*) referred to the cloud configuration when the central isothermal concentration first becomes formally infinite as the "pivotal" state. This state separates the nearly quasistatic phase of core evolution from the fully dynamic phase of protostellar accretion (see the first two stages depicted in Fig. 11 of Shu et al. 1987). The numerical simulations indicate that the pivotal states have several simplifying properties, which motivated Li and Shu (1996*b*) to approximate them as scale-free equilibria with power law radial dependences for the density and magnetic flux function:

$$\rho(r, \theta) = \frac{a^2}{2\pi G r^2} R(\theta) \qquad \Phi(r, \theta) = \frac{4\pi a^2 r}{G^{1/2}} \phi(\theta) \tag{19}$$

In equation (19), we have adopted a spherical polar coordinate system (r, θ, φ), and a is the isothermal sound speed of the cloud, while $R(\theta)$ and $\phi(\theta)$ are the dimensionless angular distribution functions for the density and magnetic flux, given by force balance along and across field lines. The resulting differential equations and boundary conditions yield a linear sequence of possible solutions, characterized by a single dimensionless free parameter, H_0, which represents the fractional overdensity supported by the magnetic field above that supported by thermal pressure. Comparison with the typical degree of elongation of observed cores suggests that the overdensity factor $H_0 \approx 0.5 - 1$. In Fig. 4, we plot isodensity contours and field lines for the case $H_0 = 0.5$. The presence of very low-density regions at the poles of the density toroid has profound implications for the shape and kinematics of bipolar molecular outflows. (See also Torrelles et al. 1983, 1994.)

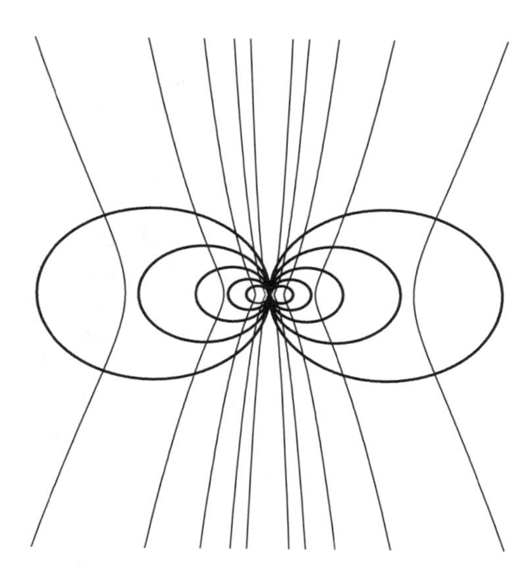

Figure 4. Isodensity contours (heavy curves) and magnetic field lines (light curves) in a meridional plane for a molecular cloud core in the pivotal state with an overdensity factor of $H_0 = 0.5$.

X. MECHANISM FOR MOMENTUM TRANSFER

Observations of one of the earliest types of young stellar objects, the so-called "Class 0" sources, indicate that a high-speed wind is turned on early in the main accretion phase of star formation (Bontemps et al. 1996). Theoretically, the wind should create two wind-blown bubbles, one on each side of the accretion disk, which continues to feed the central stellar object through the equatorial region. The bubbles are bound by the swept-up ambient core material and are expected to become elongated in the direction along the rotation/magnetic axis, where the wind momentum flux is expected to be largest and the ambient density lowest. To determine the shape and kinematics of the swept-up material (i.e., the molecular outflow), one needs to know how momentum is transferred from the wind to the ambient medium. Generally speaking, the ambient medium is swept up by a forward shock that runs ahead of the wind bubble. At the same time, the free wind runs into a reverse shock. Between the shocked ambient medium and the shocked wind material lies a contact discontinuity, across which pressure balance must be maintained.

Depending on the cooling timescale, the shocked material could be either radiative or adiabatic. For an ambient medium shocked with a relatively low shock speed of order 10 km s^{-1} and a relatively high preshock density of order 10^3 cm^{-3}, the cooling timescale is always much shorter than the time to sweep up a mass comparable to the wind mass (Koo and McKee 1992). The short cooling timescale leads to a radiative shocked ambient medium, which forms a thin shell around the wind bubble. For a wind speed of order 300 km s^{-1} or less, the shocked wind material will become radiative as well (Koo and McKee 1992). With these assumptions, the free protostellar wind is bound by two thin layers of shocked material. The two layers tend to slide relative to each other, exciting Kelvin-Helmholtz instabilities at the interface. Full development of fluid instabilities may lead to a well-mixed shell of two different types of shocked material, creating an essentially ballistic putty. Each segment of mixed radiative shell absorbs all of the wind momentum imparted to it. This local conservation of vector momentum forms the basis of a simple theory for molecular outflows (Shu et al. 1991; see also Wilkin 1997). A more realistic treatment, where the shell is not treated as spatially thin because of the "cushioning" effect of the embedded magnetic fields, will be more complicated but deserves investigation [see Li and Shu's (1996*a*) treatment of the related problem of the interaction of a wide-angle wind with a flared disk].

XI. A SIMPLE THIN-SHELL THEORY

In the simple theory developed by Shu et al. (1991), wind momentum flux per steradian of the form

$$K(r, \theta) = \frac{\dot{M}_w v_w}{4\pi} P(\theta) \tag{20}$$

propagates into a static molecular cloud core of density given by the first of equations (19). In equation (20), $\dot{M}_w$ and v_w are the mass loss rate and average speed of the wind, and $P(\theta)$ is a normalized angular distribution function. The model makes an implicit assumption that the directions of the spin vector $\mathbf{\Omega}_x$ of the accretion disk, which defines the axis of the bipolar outflow, and the ambient magnetic field $\mathbf{B}$, which defines the axis of the molecular toroid, are aligned. Presumably this alignment arises because magnetic braking in the prior epoch of molecular cloud core formation was more efficient at reducing components of precollapse $\mathbf{\Omega}$ perpendicular to $\mathbf{B}$ than at reducing the component parallel to $\mathbf{B}$ (see, e.g., Mouschovias and Paleologou 1980).

If we assume the same polar axis for the functions $P(\theta)$ and $R(\theta)$, momentum conservation in every direction θ yields the expansion velocity of the swept-up shell as

$$v_s = \left(\frac{\dot{M}_w}{2\dot{M}_a}\right)^{1/2} (a v_w)^{1/2} \left[\frac{P(\theta)}{R(\theta)}\right]^{1/2} \tag{21}$$

where $\dot{M}_a \equiv a^3/G$ is a measure of the mass accretion rate onto the central protostellar object (Shu 1977). Note that the shell expansion speed is independent of radius r, a special property of the r^{-2} density profile of the ambient medium. Aside from factors of order unity, equation (21) predicts a characteristic bipolar molecular outflow speed that is the geometric mean of the wind speed and the cloud sound speed. Since the wind speed v_w is typically a few hundred km s^{-1} (Edwards et al. 1994) and the sound speed a is of the order of a fraction of a km s^{-1}, we expect a characteristic molecular outflow speed of order 10 km s^{-1}, as observed (Fukui et al. 1993). The simple model also naturally accounts for an otherwise mysterious "Hubble" expansion, i.e., a flow velocity proportional to the distance from the central object that is often observed in molecular outflow lobes (e.g., Lada and Fich 1996).

At the time of their proposal, the functional forms of $P(\theta)$ and $R(\theta)$ were not known to Shu et al. (1991). Masson and Chernin (1992) therefore criticized their model on the grounds that "reasonable" choices for $P(\theta)$ and $R(\theta)$ yielded masses for the swept-up material that should increase with increasing line-of-sight velocities v rather than decline as a power law, $dm/dv \propto v^{\alpha}$, with the exponent α close to -1.8, as determined empirically for several well-observed outflows. Masson and Chernin (1992) argued that such a steep decrease of mass with velocity would be difficult to reproduce with any model of outflow cavities carved by wide-angle winds, in essence because the portion of the lobe that travels fastest also travels farthest and is therefore likely to sweep up the greatest amount of matter, not the least amount. This argument presupposes that the main θ dependence enters in $P(\theta)$ and not in $R(\theta)$, and that neither angular distribution is extreme.

The assumption that P is a function only of θ neglects the continuing collimation of the x-wind, which occurs logarithmically slowly according to the asymptotic analysis of section VII. Under the approximation that C can be treated as a constant, rather than a quantity that varies logarithmically with r, and that the dependences of $\beta(\psi)$ and $J(\psi)$ on ψ can be ignored for the streamlines (ψ not near 1) that interact most strongly with the ambient cloud, equation (18) implies that $P(\theta) \propto \sin^{-2}\theta$. This extreme behavior for $P(\theta)$, very large near the poles $\theta = 0$ and π, is paired with an equally extreme behavior for $R(\theta)$, vanishingly small near the poles. For the particular case of $H_0 = 0.5$ shown in Fig. 4, we plot the resulting lobe shape in Fig. 5a and the mass-velocity distribution in Fig. 5b. Both properties are in reasonably good agreement with observations, leading to a refutation of the primary argument that bipolar molecular outflows must be driven by highly collimated jets, despite the objection that such jets have difficulty accounting for the relatively poor collimation of most molecular outflow lobes (Bachiller 1996; Cabrit et al. 1997).

XII. STABILITY OF PARSEC-LONG JETS

The small amount of ambient matter swept to high velocities, which solves the original objection of Masson and Chernin (1992) to the bipolar outflow model of Shu et al. (1991), arises because the heaviest part of the x-wind encounters only the low-density material of the poles of the molecular toroid [where $R(\theta)$ ideally becomes vanishingly small]. As a consequence, it is easy for the x-wind to break out of the molecular cloud core in these directions (see Fig. 5) and, thus, to account for the parsec-long jets discovered by Bally and Devine (1997).

Because the x-wind is launched within a few stellar radii of the star, even logarithmically slow collimation yields highly directed jets at parsec distances ($\sim 3 \times 10^6\ R_x$). For $r = 3 \times 10^6$, equation (16) implies that an undeflected $\psi = 0.50$ streamline is collimated within an angle $\theta = 0.9°$ of the polar axis for the example illustrated in Fig. 3. In other words, at a distance ~1 pc from the central source, 50% of the undisturbed outflow is encased within a cylinder of radius ~0.015 pc. At very large distances from the central source the difference between x-winds and pure stellar jets narrows considerably.

It has often been asked how jets can maintain their integrity for such long paths. The glib answer provided by all proponents of magnetocentrifugal driving is that the flow is self-collimated. Self-collimation occurs

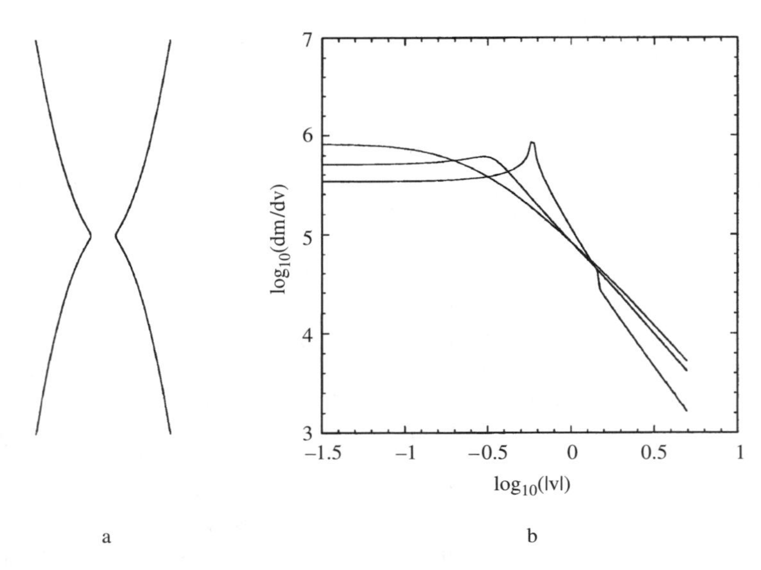

Figure 5. (a) Shape of swept-up lobes in the meridional plane for a wind power $\propto \sin^{-2}\theta$ and a core density distribution $\propto R(\theta)$ of a singular model with $H_0 = 0.5$. The dashed line indicates schematically the trajectory of the dense parts of the x-wind seen as optical jets that easily burst through the rarefied polar regions of the surrounding molecular toroid. (b) The logarithm of the swept-up mass at each logarithmic interval of the line-of-sight velocity for inclination angles $\vartheta = 80°, 45°$ (middle line), and $10°$ with respect to the axis of the bipolar flow. The case $\vartheta = 80°$ has a nearly discontinuous step due to the waist of the two lobes in (a).

by the hoop stresses of the toroidal magnetic fields that grow in strength relative to the poloidal components as the flow distance from the source increases. However, such magnetic confinement schemes have long been known to the fusion community to be unstable with respect to kinking and sausaging (see Fig. 10.6 in Jackson 1975; see also Eichler 1993). A solution to the problem has also long been known and is the principle behind working Tokamaks: the introduction of dynamically important levels of longitudinal magnetic fields along the length of the plasma column to stabilize the kinking and sausaging motions (see Figures 10.7 and 10.8 of Jackson 1975). Such dynamically important levels of longitudinal fields along the central flow axis are exactly what fill the hollow core of the jet in the x-wind model of Fig. 2.

Acknowledgments This research is funded by a grant from the National Science Foundation and by the NASA Astrophysics Theory Program, which supports a joint Center for Star Formation Studies at NASA/Ames Research Center, the University of California at Berkeley, and the University of California at Santa Cruz.

REFERENCES

Arons, J. 1986. Centrifugally driven winds from X-ray pulsars. In *Plasma Penetration into Magnetospheres*, ed. N. Kylafis, J. Papamastorakis, and J. Ventura (Iraklion: Crete University Press), pp. 115–124.

Bacciotti, F., Hirth, G. A., and Natta, A. 1996. The optical jet of RW Aurigae: Excitation temperature and ionization state from long-slit spectra. *Astron. Astrophys.* 310:309–314.

Bachiller, R. 1996. Bipolar molecular outflows from young stars and protostars. *Ann. Rev. Astron. Astrophys.* 34:111–154.

Balbus, S., and Hawley, J. 1991. A powerful local shear instability in weakly magnetized disks. I. Linear analysis. *Astrophys. J.* 376:214–233.

Bally, J., and Devine, D. 1997. Giant Herbig-Haro flows. In *Herbig-Haro Flows and the Birth of Low Mass Stars*, Proc. IAU Symp. 182, ed. B. Reipurth and C. Bertout (Dordrecht: Kluwer), pp. 29–38.

Basri, G., Marcy, G. W., and Valenti, J. A. 1992. Limits on the magnetic flux of pre-main-sequence stars. *Astrophys. J.* 390:622–633.

Basu, S., and Mouschovias, T. 1994. Magnetic braking, ambipolar diffusion, and the formation of cloud cores and protostars. I. Axisymmetric solutions. *Astrophys. J.* 432:720–741.

Bertout, C. 1989. T Tauri stars—Wild as dust. *Ann. Rev. Astron. Astrophys.* 27:351–395.

Bertout, C., Basri, G., and Bouvier, J. 1988. Accretion disks around T Tauri stars. *Astrophys. J.* 330:350–373.

Biskamp, D. 1993. *Nonlinear Magnetohydrodynamics* (Cambridge: Cambridge University Press).

Blandford, R. D., and Payne, D. G. 1982. Hydromagnetic flows from accretion discs and the production of radio jets. *Mon. Not. Roy. Astron. Soc.* 199:883–903.

Bontemps, S., Andre, P., Terebey, S., and Cabrit, S. 1996. Evolution of outflow activity around low-mass embedded young stellar objects. *Astron. Astrophys.* 311:858–872.

Bouvier, J., Cabrit, S., Fernandez, M., Martin, E. L., and Matthews, J. M. 1993. Coyotes-I—The photometric variability and rotational evolution of T Tauri stars. *Astron. Astrophys.* 272:176–206.

Burrows, C. J., Stapelfeldt, K. R., Watson, A. M., Krist, J. E., Ballester, G. E., Clarke, J. T., Crisp, D., Gallagher, J. S., III, Griffiths, R. E., Hester, J. J., Hoessel, J. G., Holtzman, J. A., Mould, J. R., Scowen, P. A., Trauger, J. T., and Westphal, J. A. 1996. Hubble Space Telescope observations of the disk and jet of HH 30. *Astrophys. J.* 473:437–451.

Cabrit, S., Raga, A. C., and Gueth, F. 1997. Models of bipolar molecular outflows. In *Herbig-Haro Flows and the Birth of Low Mass Stars*, Proc. IAU Symp. 182, ed. B. Reipurth and C. Bertout (Dordrecht: Kluwer), pp. 163–180.

Camenzind, M. 1990. Magnetized disk-winds and the origin of bipolar outflows. In *Reviews in Modern Astronomy*, Vol. 3, ed. G. Klare (Berlin: Springer), pp. 234–265.

Cui, W. 1997. Evidence for "propeller" effects in X-ray pulsars GX 1+4 and GRO J1744-28. *Astrophys. J. Lett.* 482:L163–L166.

Edwards, S., Strom, S. E., Hartigan, P., Strom, K. M., Hillenbrand, L. A., Herbst, W., Attridge, J., Merreil, K. M., Probst, R., and Gatley, I. 1993. Angular momentum regulation in low-mass young stars surrounded by accretion disks. *Astron. J.* 106:372–382.

Edwards, S., Hartigan, P., Ghandour, L., and Andrulis, C. 1994. Spectroscopic evidence of magnetic accretion in classical T Tauri stars. *Astrophys. J.* 108:1056–1070.

Eichler, D. 1993. Magnetic confinement of jets. *Astrophys. J.* 419:111–116.

Fukui, Y., Iwata, T., Mizuno, A., Bally, J., and Lane, A. 1993. Molecular outflows. In *Protostars and Planets III*, ed. E. H. Levy and J. I. Lunine (Tucson: University of Arizona Press), pp. 603–639.

Ghosh, P., and Lamb, F. K. 1978. Disk accretion by magnetic neutron stars. *Astrophys. J. Lett.* 223:L83–L87.

Goodman, A. A., Benson, P. J., Fuller, G. A., and Myers, P. C. 1993. Dense cores in dark clouds. VIII. Velocity gradients. *Astrophys. J.* 406:528–547.

Gullbring, E., Hartmann, L., Briceno, C., and Calvet, N. 1998. Disk accretion rates for T Tauri stars. *Astrophys. J.* 492:323–341.

Hartmann, L., and MacGregor, K. B. 1982. Protostellar mass and angular momentum loss. *Astrophys. J.* 259:180–192.

Hartmann, L., Hewitt, R., and Calvet, N. 1994. Magnetospheric accretion models for T Tauri stars. 1: Balmer line profiles without rotation. *Astrophys. J.* 426:669–687.

Hayashi, M. R., Shibata, K., and Matsumoto, R. 1996. X-ray flares and mass outflows driven by magnetic interaction between a protostar and its surrounding disk. *Astrophys. J. Lett.* 468:L37–L40.

Heinemann, M., and Olbert, S. 1978. Axisymmetric ideal MHD stellar wind flow. *J. Geophys. Res.* 83:2457–2460.

Heyvaerts, J., and Norman, C. 1989. The collimation of magnetized winds. *Astrophys. J.* 347:1055–1081.

Hirth, G. A., Mundt, R., and Solf, J. 1997. Spatial and kinematic properties of the forbidden emission line region of T Tauri stars. *Astron. Astrophys. Suppl.* 126:437–496.

Jackson, J. D. 1975. *Classical Electrodynamics* (New York: Wiley).

Johns, C. M., and Basri, G. 1995*a*. Hamilton echelle spectra of young stars. II. Time series analysis of H(alpha) variations. *Astron. J.* 109:2800–2816.

Johns, C. M., and Basri, G. 1995*b*. The line profile variability of SU Aurigae. *Astrophys. J.* 449:341–364.
Johns-Krull, C. M. 1996. Limits on the differential rotation of T Tauri stars. *Astrophys. J.* 306:803–810.
Johns-Krull, C. M., and Hatzes, A. P. 1997. The classical T Tauri star Sz 68: The Doppler imaging and evidence of magnetospheric accretion. *Astrophys. J.* 487:896–915.
Johns-Krull, C. M., Valenti, J. A., and Koresko, C. 1999. Measuring the magnetic field on the classical T Tauri star BP Tauri. *Astrophys. J.* 516:900–915.
Königl, A. 1989. Self-similar models of magnetized accretion disks. *Astrophys. J.* 342:208–223.
Königl, A. 1991. Disk accretion onto magnetic T Tauri stars. *Astrophys. J. Lett.* 370:L39–L43.
Koo, B.-C., and McKee, C. F. 1992. Dynamics of wind bubbles and superbubbles. I. Slow winds and fast winds. *Astrophys. J.* 388:93–126.
Kwan, J., and Tademaru, E. 1988. Jets from T Tauri stars. Spectroscopic evidence and collimation mechanism. *Astrophys. J. Lett.* 332:L41–L44.
Lada, C. J. 1985. Cold outflows, energetic winds, and enigmatic jets around young stellar objects. *Ann. Rev. Astron. Astrophys.* 23:267–317.
Lada, C. J., and Fich, M. 1996. The structure and energetics of a highly collimated bipolar outflow: NGC 2264G. *Astrophys. J.* 459:638–652.
Li, J., and Wickramasinghe, D. T. 1997. Disc accretion onto magnetic stars: Slow rotator and propeller. In *Accretion Phenomena and Related Outflows*, IAU Colloq. 163, ASP Conf. Ser. 121, ed. D. T. Wickramasinghe, L. Ferrario, and G. V. Bicknell (San Francisco: Astronomical Society of the Pacific), pp. 241–250.
Li, Z. Y., and Shu, F. H. 1996*a*. Interaction of wide-angle MHD winds with flared disks. *Astrophys. J.* 468:261–268.
Li, Z. Y., and Shu, F. H. 1996*b*. Magnetized singular isothermal toroids. *Astrophys. J.* 472:211–224.
Linker, J. A., and Mikić, Z. 1995. Disruption of a helmet streamer by photospheric shear. *Astrophys. J. Lett.* 438:L45–L48.
Lizano, S., and Shu, F. H. 1989. Molecular cloud cores and bimodal star formation. *Astrophys. J.* 342:834–854.
Lovelace, R. V. E., Berk, H. L., and Contopoulos, J. 1991. Magnetically driven jets and winds. *Astrophys. J.* 379:696–705.
Lovelace, R. V. E., Romanova, M. M., and Bisnovatyi-Kogan, G. S. 1999. Magnetic propeller outflows. *Astrophys. J.* 514:368–372.
Masson, C., and Chernin, L. 1992. Properties of swept-up molecular outflows. *Astrophys. J. Lett.* 387:L47–L50.
Mestel, L. 1968. Magnetic braking by a stellar wind—I. *Mon. Not. Roy. Astron. Soc.* 138:359–391.
Meyer, M. R., Calvet, N., and Hillenbrand, L. A. 1997. Intrinsic near-infrared excesses of T Tauri stars: Understanding the classical T Tauri star locus. *Astron. J.* 114:288–300.
Montmerle, T., Feigelson, E., Bouvier, J., and Andre, P. 1993. Magnetic fields, activity and circumstellar material around young stellar objects. *Protostars and Planets III*, ed. E. H. Levy and J. I. Lunine (Tucson: University of Arizona Press) pp. 689–717.
Mouschovias, T. C., and Paleologou, E. V. 1980. Magnetic braking of an aligned rotator during star formation. *Astrophys. J.* 237:877–899.
Muzerolle, J., Calvet, N., and Hartmann, L. 1998. Magnetospheric accretion models for the hydrogen emission lines of T Tauri stars. *Astrophys. J.* 492:743–753.

Myers, P. C. 1995. Star-forming molecular clouds. In *Molecular Clouds and Star Formation,* ed. C. Yuan and J. You (Singapore: World Scientific), pp. 47–96.

Najita, J. R., and Shu, F. H. 1994. Magnetocentrifugally driven flows from young stars and disks. III: Numerical solutions of the sub-Alfvenic region. *Astrophys. J.* 429:808–825.

Najita, J. R., Carr, J. S., and Tokunaga, A. T. 1996. High-resolution spectroscopy of Br gamma emission in young stellar objects. *Astrophys. J.* 456:292–299.

Nakano, T. 1979. Quasistatic contraction of magnetic protostars due to magnetic flux leakage. I. Formulation and an example. *Pub. Astron. Soc. Japan* 31:697–712.

Ostriker, E. C., and Shu, F. H. 1995. Magnetocentrifugally driven flows from young stars and disks. IV. The accretional funnel and dead zone. *Astrophys. J.* 447:813–828.

Pudritz, R. E., and Norman, C. A. 1983. Centrifugally driven winds from contracting molecular disks. *Astrophys. J.* 274:677–697.

Reipurth, B., and Heathcote, S. 1991. The jet and energy source of HH 46/47. *Astron. J.* 246:511–534.

Rydgren, A. E., and Vrba, F. J. 1983. Periodic light variations in four pre-main-sequence K stars. *Astrophys. J.* 267:191–198.

Sakurai, T. 1985. Magnetic stellar winds: A 2-D generalization of the Weber-Davis model. *Astron. Astrophys.* 152:121–129.

Shang, H., and Shu, F. H. 1998. X-wind made simple. *Astrophys. J.*, in preparation.

Shang, H., Shu, F., Lee, T., and Glassgold, A. E. 1997. Protostellar cosmic rays and extinct radioactivities. In *Low Mass Star Formation from Infall to Outflow,* Poster Proc. IAU Symp. No. 182, ed. F. Malbet and A. Castets (Grenoble: Laboratorie d'Astrophysique), pp. 312–315.

Shang, H., Shu, F. H., and Glassgold, A. E. 1998. Synthetic images and long-slit spectra of protostellar jets. *Astrophys. J. Lett.* 493:L91–L94.

Shu, F. H. 1977. Self-similar collapse of isothermal spheres and star formation. *Astrophys. J.* 214:488–497.

Shu, F. H., Adams, F. C., and Lizano, S. 1987. Star formation in molecular clouds: Observation and theory. *Ann. Rev. Astron. Astrophys.* 25:23–81.

Shu, F. H., Lizano, S., Ruden, S., and Najita, J. 1988. Mass loss from rapidly rotating magnetic protostars. *Astrophys. J. Lett.* 328:L19–L23.

Shu, F. H., Ruden, S. P., Lada, C. J., and Lizano, S. 1991. Star formation and the nature of bipolar outflows. *Astrophys. J. Lett.* 370:L31–L34.

Shu, F. H., Johnstone, D., and Hollenbach, D. 1993. Photoevaporation of the solar nebula and the formation of the giant planets. *Icarus* 106:92–101.

Shu, F., Najita, J., Ostriker, E., Wilkin, F., Ruden, S., and Lizano, S. 1994*a*. Magnetocentrifugally driven flows from young stars and disks. I. A generalized model. *Astrophys. J.* 429:781–796.

Shu, F. H., Najita, J., Ruden, S. P., and Lizano, S. 1994*b*. Magnetocentrifugally driven flows from young stars and disks. II. Formulation of the dynamical problem. *Astrophys. J.* 429:797–807.

Shu, F. H., Najita, J., Ostriker, E. C., and Shang, H. 1995. Magnetocentrifugally driven flows from young stars and disks. V. Asymptotic collimation into jets. *Astrophys. J. Lett.* 455:L155–L158.

Shu, F. H., Shang, H., Glassgold, A. E., and Lee, T. 1997. X-rays and fluctuating x-winds. *Science* 277:1475–1479.

Snell, R. L., Loren, R. B., and Plambeck, R. L. 1980. Observations of CO in L1551—evidence for stellar wind driven shocks. *Astrophys. J. Lett.* 239:L17–L22.

Torrelles, J. M., Rodriguez, L. F., Canto, J., Carral, P., Marcaide, J., Moran, J. M., and Ho, P. T. P. 1983. Are interstellar toroids the focusing agent of the bipolar molecular outflow? *Astrophys. J.* 217:214–230.

Torrelles, J. M., Gomez, J. F., Ho, P. T. P., Rodriguez, L. F., Anglada, G., and Canto, J. 1994. The puzzling distribution of the high-density molecular gas in HH 1-2: A contracting interstellar toroid? *Astrophys. J.* 435:290–312.

Uchida, Y., and Shibata, K. 1985. Magnetodynamical acceleration of CO and optical bipolar flows from the region of star formation. *Pub. Astron. Soc. Japan* 37:515–535.

Vogel, S. N., and Kuhi, L. V. 1981. Rotational velocities of pre–main sequence stars. *Astrophys. J.* 245:960–976.

Walter, F. M., Brown, A., Mathieu, R. D., Myers, P. C., and Vrba, F. J. 1988. X-ray sources in regions of star formation. III. Naked T Tauri stars associated with the Taurus-Auriga complex. *Astron. J.* 96:297–325.

Weber, E. J., and Davis, L. 1967. The angular momentum of the solar wind. *Astrophys. J.* 148:217–227.

Wilkin, F. P. 1997. Bow shocks in non-uniform medium. In *Low Mass Star Formation from Infall to Outflow*, Poster Proc. IAU Symp. 182, ed. F. Malbet and A. Castets (Grenoble: Laboratorie d'Astrophysique), p. 190.

Wu, Y., Huang, M., and He, J. 1996. A catalogue of high velocity molecular outflows. *Astron. Astrophys. Suppl.* 115:283–284.

Wynn, G. A., King, A. R., and Horne, K. 1997. A magnetic propeller in the cataclysmic variable AE Aquarii. *Mon. Not. Roy. Astron. Soc.* 286:436–446.

COLLIMATION AND PROPAGATION OF STELLAR JETS

JOCHEN EISLÖFFEL
Thüringer Landessternwarte Tautenburg

REINHARD MUNDT
Max-Planck-Institut für Astronomie

THOMAS P. RAY
Dublin Institute for Advanced Studies

and

LUIS FELIPE RODRÍGUEZ
Universidad Nacional Autónoma de México

Most studies of jets from young stars have concentrated on how they propagate and interact with their surroundings on scales of a few thousand AU to several parsecs (see, for example, the chapter by Hartigan et al., this volume). Ultimately, however, it is the region closest to the star (on angular scales $\leq 1.0''$ for the nearest star-forming regions) that we have to probe if we are to obtain clues as to the origin of the jets themselves. On the observational side, a large amount of progress has been made towards understanding the "central engine." We discuss various aspects of jets from young stars, notably measurements of their width and opening angle close to the source, the so-called "microjets" as well as their thermal and nonthermal radio emission. Finally, we conclude by reviewing the rapidly growing area of molecular hydrogen studies of outflows.

I. INTRODUCTION

Jets from young stars are among the most spectacular manifestations of star formation. They may also be an important ingredient in the process whereby stars form, because they can in principle remove excess angular momentum from accreted material and hence prevent young stellar objects (YSOs) from spinning up to breakup velocity.

In recent years, major progress has been made in understanding the collimation and propagation of YSO jets and outflows, both on large and small scales. On small scales (i.e., less than a few hundred AU), the parameters of jets close to their source are of considerable interest because they may help us to choose between different jet models (see the chapters by Königl and Pudritz and by Shu et al., this volume). Although their acceleration regions cannot yet be resolved, ground-based and *Hubble Space Telescope* (HST) observations allow us to measure jet diameters relatively close to their parent YSO, providing us with new constraints on jet collimation (see section II). In addition, spectroscopic observations in the close vicinity of apparently more evolved YSOs have recently led to the discovery of the so-called "small-scale jets," which are seen as the natural extensions of the forbidden-emission-line regions of classical T Tauri stars (cTTSs). Such jets have typical sizes of a few tenths of an arcsecond to a few arcseconds; they seem to be much more common, and persist for much longer, than their prominent large-scale counterparts (see section III). On similar lengthscales thermal, and in a few rare cases possibly nonthermal, radio jets have been found associated with young stars (see section IV). Moreover, because of their very high resolution, radio observations provide us with valuable information on jet collimation and help characterize the nature of the outflowing gas.

Since *Protostars and Planets III,* we now understand much more about the interaction of YSO jets with their environment. In particular, near-infrared observations of molecular hydrogen emission have given us a new window on where and how this interaction takes place (see section V). HST is allowing us to resolve knots within jets for the first time, and recent ground-based imaging with large charge-coupled device (CCD) arrays have found the not surprising (in retrospect!) result that many outflows are much larger than originally thought. These "parsec-scale jets" have kinematical ages of several times 10^4 years and are apparently capable of dumping so much momentum in their parental molecular cloud that they may be able to disrupt it and thus influence the star formation process on large scales (see, e.g., Eislöffel and Mundt 1997; the chapter by Hartigan et al., this volume).

In the following, we will concentrate on discussing recent developments in our understanding of the collimation and propagation of YSO jets close to their source (i.e., typically within a few hundred AU). Jets on larger scales are dealt with by Hartigan et al. (this volume). To complete the picture, an appreciation of the immediate environments of YSOs is also needed; however, that is beyond the scope of this chapter (see the chapter by Mundy et al., this volume).

II. COLLIMATION OF JETS

A. Principal Observational Problems

In order to get constraints on jet collimation from direct images, one has to measure the width of a jet as a function of distance from its source. Obviously, one would like to start measuring the width as close as possible to the YSO. From a theoretical point of view, "close to the source" should ideally be a few stellar radii (e.g., 10^{12} cm), which would require a spatial resolution of 0.5 milliarcsecond (for $d = 140$ pc). This is, however, far beyond current and near future technology. A width measurement close to the source is not only limited by the available spatial resolution of the telescope and instrument (typically $0.1''$–$1''$) but to a large degree by the source properties. Many YSOs are heavily embedded in gas and dust as well as being surrounded by a circumstellar disk. For these reasons, the jets often become optically observable only a few arcseconds from the YSO, where the circumstellar extinction is sufficiently small. Even when the source is optically visible, direct starlight or bright reflection nebulae may make it very difficult, because of contrast problems, to detect or resolve a jet in its vicinity. These contrast problems are not, in general, eliminated by the use of narrow-band filters, because large amounts of continuum light from the source, or nearby reflection nebula, are always registered as well as the line emission from the jet. Disentangling the contribution of the various components is not easy!

In essence, therefore, jet sources are surrounded by an unobservable region (a "black box") with an angular size ranging from 0.1 to several arcseconds (10^3–10^4 stellar radii for the nearest YSOs). The size of the "black box" depends critically on our spatial resolution and instrumentation as well as the source properties. In Fig. 1, however, we illustrate that despite this problem, one can still derive useful observational constraints on the collimation process from width vs. distance measurements. In this figure we distinguish two cases. In case A the jet is rather poorly collimated close to the source and so rapidly widens into a relatively broad jet. It then decreases its opening angle at distances from the source that are comparable to the size of the "black box." In case B the "final" opening angle and high degree of collimation are already achieved in a region

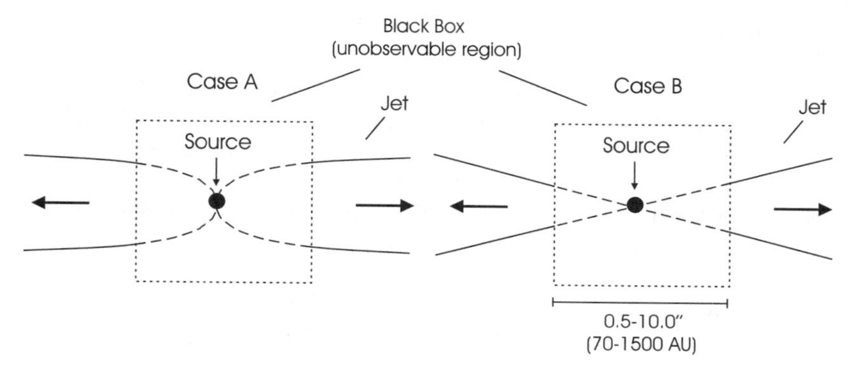

Figure 1. Two schemes for the possible collimation of YSO jets on small scales. The "black box" (dotted line) around the source outlines the region around the source within which the jet is unobservable due to the source properties (e.g., extinction) or due to limited spatial resolution.

that is much smaller than the size of the "black box." In the following we will show that there is strong evidence that case A applies for most known sources.

B. Ground-based Imaging

The first measurements of jet widths, as a function of distance from the source, were published by Mundt et al. (1991) and Raga et al. (1991). They used a one-dimensional deconvolution technique to measure widths in [S II] $\lambda\lambda$ 6716,6731 Å and Hα emission line frames. Recent HST studies by Ray et al. (1996) have given practically the same results for the three objects in common (see below). In total, 15 jets and counterjets were studied by these authors. They defined two opening angles to describe the collimation properties of the jets: (1) an "initial" opening angle α_i, which is the (average) opening angle of the jet within the "black box," and (2) an average opening α_a, which is the average opening angle of the jet outside the "black box." The typical values of α_a ranged from 0° to 3°. To determine α_i, a jet width measurement just outside the "black box" is needed, and one has to assume that the jet has zero width at the source.

The studies by Mundt et al. (1991), Raga et al. (1991), and Eislöffel et al. (1994*a*) showed that for most cases α_i is much larger than α_a (for 70% of the jets $\alpha_i \geq 3\alpha_a$). In fact, for 14 of the 16 jets in their sample, case A applied as opposed to case B. We note that, for some of the jets studied, the size of the "black box" was rather larger (as much as 15″), because no jet emission was observed close to the IR source, probably because of extinction. Interestingly, these authors found in some cases large asymmetries in α_i (and α_a) between the two sides of a bipolar outflow.

C. HST Observations

Ray et al. (1996) have published the only measurements of jet widths based on images taken with the restored HST using narrow-band emission line filters ([S II], Hα). These data confirm the findings of the ground-based studies and indicate that case A applies for the five jets and counterjets studied by them. Of the four flows they examined (HH 30, HL Tau, HH 1, and HH 34) the jet could be traced closest to the star for HH 30. In this case, width measurements were possible up to 1.5″ from the source for the jet and 1.0″ for the counterjet of this bipolar outflow (1.0″ = 140 AU for d = 140 pc).[a] Closer again ($\leq$ 0.25″) to the star, the emission from the bipolar jets is absorbed by the dust in the circumstellar disk (see also Burrows et al. 1996). In Color Plate 13 we illustrate the results of the width measurements for the bright jet and faint counter-

[a] We now regard the uncorrected full width at half maximum (FWHM) jet width measurements in Ray et al. (1996) with values $\leq$ 0.25″ to be unreliable in the case of the Wide Field CCD images. This is due to a software problem in coping with the poorly sampled point spread function (PSF).

jet of HH30. At large distances from the source, the jet only slowly increases its (average) width, and an opening angle $\alpha_a = 2°$ is derived. For the counterjet the corresponding α_a value is 5°. Because the jet and counterjet still have a very large full width at half maximum (FWHM) of 0.35″ as close as 2.0″ and 1.0″ from the source respectively, their initial opening angles are much larger ($\alpha_i \approx 10°$ and 20°). Moreover, because jet widths have been defined in terms of FWHM, the actual width of the jet at zero intensity is probably twice as large again and therefore α_i correspondingly larger. Furthermore α_i gives the average opening angle within the "black box" and therefore the jet opening angle could in reality be even larger closer to the source. Such large opening angles are expected according to various theoretical models.

The jets of HL Tau and HH 34 have, at the edge of their "black boxes," similar observed angular widths as the HH 30 jets (about 0.4″). The angular sizes of their "black boxes" are approximately the same, so α_i is correspondingly similar. It is important to determine whether the behavior of these jets close to their sources is representative of Herbig-Haro (HH) jets as a whole. To see whether this is the case, narrowband HST imaging of less extincted flows (like RW Aurigae; Mundt and Eislöffel 1998) is called for. In principle it is possible to measure jet widths, or at least get upper limits, as close as 0.1″ from T Tauri stars (TTSs) like RW Aur.

Theoretical models suggest that the actual jet channel could be narrower (if bow shocks in the flow splashed sideways) or wider than the observed jet width (if the jet density falls off rapidly away from the jet axis; see the chapter by Shu et al., this volume). If these models are correct, the jet opening angles could accordingly scale and be somewhat smaller or larger than those measured directly.

D. Forbidden-Emission-Line Profiles of T Tauri Stars

Information on the opening angles and collimation properties of YSO jets within 0.2–1″ (30–140 AU) from their source can also be derived from the forbidden-emission-line (FEL) profiles of TTSs. General details of what FEL profiles tell us about small-scale jets from TTSs will be discussed in section III (below). As will be shown there, the FEL profiles of many TTSs, in, for example, [O I] λ 6300 or [S II] $\lambda\lambda$ 6716 and 6731, show two velocity components: (1) a so-called high-velocity component (HVC), which is very probably formed in a jet, and (2) a low-velocity component (LVC), which might result from a disk wind (e.g., Kwan and Tademaru 1988). Here, we concentrate only on the properties of the HVC (i.e., the jet component). Various long-slit spectroscopic studies of TTSs in the nearby Taurus-Auriga dark cloud (e.g., Hirth et al. 1997) have shown that the HVC emission originates at different distances from the source for different emission lines. In their sample of 12 TTSs the centroid of the [O I] λ 6300 emission is located at an average distance of 0.2″, whereas that of the [S II] λ 6731 and [N II] λ 6584 lines are factors of 3 and 3.5

times further away, respectively. This is probably because the jet streamlines diverge with distance from the source, giving rise to a drop in density. In particular, because the [O I] line has a critical electron density roughly 100 times higher than the [S II] line, it will form much closer to the source. It follows that a comparison of the HVC width (velocity width) between the [O I] and [S II] lines may tell us whether the opening angle (collimation) changes on scales of 0.2–0.6″.

To derive the opening angle of a jet from the measured linewidth, we make the simplifying assumptions that a jet can be approximated by a homogeneously filled cone with the same velocity V_j along the various diverging streamlines and that it has an opening angle α. Let us suppose that the jet is at an angle γ to the line of sight; then, providing $\gamma > \alpha/2$, the linewidth (at FWHM) is given by

$$\Delta V(\text{FWHM}) \approx 2f_{1/2}V_j\,|\sin\gamma\,\sin(\alpha/2)| \tag{1}$$

where $f_{1/2}$ is a correction factor (≈ 0.9) to convert full width at zero intensity into FWHM (see Mundt et al. 1990 for details). The opening angles derived in this manner agree reasonably well with those derived from ground-based and space-based observations (Mundt and Hirth 1999). This comparison is, however, possible only for a few cases. Mundt and Hirth (1999) have compiled the FEL line profile data for a large sample of TTSs (about 60) from their own work (Hirth et al. 1997) and from the literature (e.g., Hartigan et al. 1995). The average linewidth of the HVC in the [O I], [S II] and [N II] lines is about 70 km s^{-1} (FWHM) in those 10–20 objects in which the HVC and LVC are sufficiently well resolved. According to equation (1), this velocity width corresponds to $\alpha \approx 25°$. These lines form at typical distances of 0.2–0.7″ from the source, and therefore the opening angle derived from their linewidth is somewhat larger than what is observed for HH 30 at similar distances. Unless the opening angles derived from the FEL widths are totally meaningless, this result implies that the small opening angle of the HH 30 jet at around 1″ from the source is not representative of TTSs in general, consistent with the rather small linewidth of its jet in [S II] λ 6731 of ≤ 10 km s^{-1} (Mundt et al. 1990).

A very important question is whether there are significant changes in linewidths with distance from the star, such as between the [O I] line formation region (at about 0.2″) and the [S II] line formation region (at about 0.6″). Mundt and Hirth (1999) found only six objects in the literature that are suitable to answer this question. The number of objects is limited, not only because many TTSs show [O I] emission alone (i.e., other forbidden lines are not detected) but also because the LVCs and HVCs are often insufficiently separated to measure their linewidths reliably. In the six TTSs found, the ratios of [O I] to [S II] linewidths are 0.9, 1.0, 1.25, 1.3, 1.35, and 1.8, with an average value of 1.27. The data would suggest then that the opening angles of TTS jets often decrease on scales of 0.2–0.6″. Again this provides strong support for case A of Fig. 1. Even more interesting in this respect is the rather large linewidth of 170 km s^{-1} (FWHM) of the [O I] λ 5577 line of DG Tau, which is nearly twice as large as that of the [N II] λ 6583 line (Hartigan et al. 1995). This [O I] line has a critical electron density of 10^8 cm^{-3}; that is, 10^4 times higher than that of the [N II] line. If the emission of both lines forms mainly at the critical electron density, and if that density drops in the jet with r^{-2}, then the [O I] λ 5577 line originates at about 10^{-2} arcsecond from the YSO. At that location the jet has a full opening angle of about 50° for V_j = 300 km s^{-1}.

III. SMALL-SCALE FLOWS CLOSE TO THE SOURCE

As already mentioned, most jet sources are highly embedded, so it is impossible, at least optically, to study their immediate environments. That said, there are several outflows from T Tauri and Herbig Ae/Be stars where one can trace the jet right back to its source. Often, presumably because their sources are more evolved than those of their embedded counterparts, such flows are not very extended (hence their nickname "microjets"; Solf 1997). Direct narrow-band imaging of microjets is difficult to carry out for the reasons mentioned above, i.e., the presence of bright sources and compact reflection nebulae. As demonstrated by Solf and Böhm (1993) and by Böhm and Solf (1994), however, long-slit spectroscopy is a very effective means of "imaging" such flows. With intermediate-dispersion spectrograms, the continuum light from the star can be subtracted with a high degree of accuracy; allowance can even be made for stellar absorption lines using a suitable template spectrum. The resultant spectra for individual lines are then effectively 1D position-velocity diagrams, and the various line ratios can give us information on physical conditions in the emitting gas.

We begin by reviewing what studies of the forbidden emission lines in classical T Tauri stars tell us about the outflow components close to the source before considering their more massive counterparts, the Herbig Ae/Be stars.

A. The Classical T Tauri Stars

As is the case in standard (i.e., more extended) HH objects or jets, small-scale flows close to the star are best seen in forbidden emission lines (FELs) such as the [O I] $\lambda\lambda$ 6300, 6363 and the [S II] $\lambda\lambda$ 6716, 6731 doublets. Often the FELs are found to be blueshifted in classical T Tauri stars (Hartigan et al. 1995). Moreover, double-peaked profiles are frequently seen, especially for those sources with relatively large near-infrared excesses. The high-velocity components (or HVCs) have radial velocities normally between −50 and −200 km s^{-1}, whereas the low-velocity components (LVCs) typically range from −5 to −20 km s^{-1}. Note that these velocities are with respect to the systemic velocities of the stars. The usual explanation for the blueshifted bias in the statistics of FELs is obscuration of the redshifted flow by a circumstellar disk (e.g., Edwards et al. 1993), as was realized by Appenzeller et al. (1984) many years ago.

It should be stressed that from an observational perspective, it is often problematic to distinguish the LVC from the HVC, since the width of both components can be comparable or larger than their radial velocity separations. Moreover, in those cases where the flow is oriented close to the plane of the sky, the HVC and LVC obviously tend to merge. Despite these difficulties, distinct differences can be seen in the properties of the HVC and the LVC. For example, Hamann (1994) has shown that the gas in the LVC is normally of higher density and lower excitation than that of the HVC. To illustrate some of the other characteristics of the FEL regions of classical T Tauri stars, we will use V536 Aquilae (see Hirth et al. 1997) as an example.

V536 Aql is a classical T Tauri star that is known to be double (Ageorges et al. 1994). The separation between the two components is 0.54″ at a position angle of 17°. A few faint HH knots have been seen in its vicinity (Mundt and Eislöffel 1998), indicating an outflow direction of approximately 110° (i.e., perpendicular to the binary axis). Figure 2 shows continuum subtracted position-velocity maps of V536 Aql in various FELs taken at a position angle of 90° (from Hirth et al. 1997). It can immediately be seen that this outflow is bipolar, although the redshifted flow is fainter, as one might expect. Both an HVC (at about -80 km s^{-1}) and an LVC (at about -7 km s^{-1}) are observed in the [O I] λ 6300 and [S II] λ 6731 lines, although only the HVC is seen in the [N II] λ 6583 line, as is commonly the case with microjets (Hirth et al. 1997).

In Fig. 3 the spatial offset (with respect to the stellar continuum), width, and integrated intensity of the FELs in Fig. 2 are plotted against radial velocity. It is evident that the emission centroid of the HVC has a larger spatial offset in [N II] than in [S II], which in turn has a bigger offset than that for the [O I] line. Moreover, these offsets are factors of 3 or more larger than the corresponding values for the LVC. We emphasize that these results are typical of other classical T Tauri stars.

In the past there were several attempts to model both the LVC and the HVC in terms of a single outflow, the appearance of two components then arising from geometric effects (e.g., Edwards et al. 1987; Hartmann and Raymond 1989; Ouyed and Pudritz 1993). However, it is now generally agreed that the HVC corresponds to a jet, as first proposed by Kwan and Tademaru (1988, 1995; but see also Hirth et al. 1994). This is particularly evident in those sources where an extended jet is present, because the radial velocity of the extended jet is seen to join smoothly to that of the HVC. The nature of the LVC, however, is less certain, although it has been suggested that it is a disk wind (Kwan and Tademaru 1995).

The fact that the gas in the LVC is normally of higher density and lower excitation than that in the HVC can explain why, for example, the LVC is not seen in the [N II] λ 6583 line. As pointed out by Hirth et al.

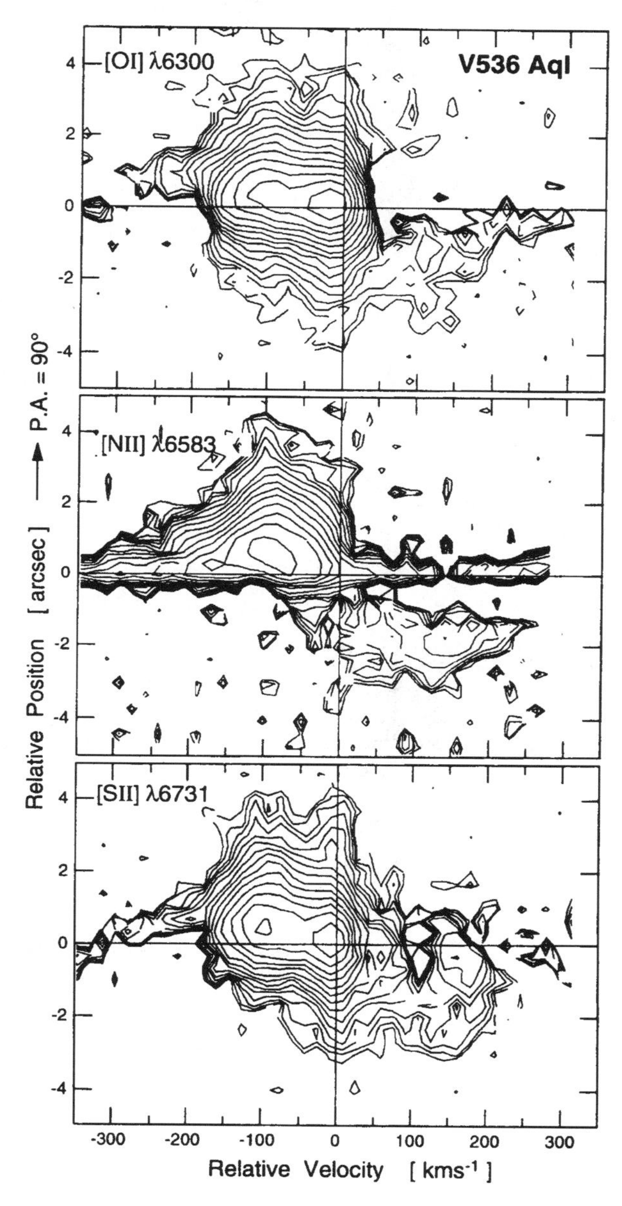

Figure 2. Position-velocity maps of the [O I] λ 6300, [N II] λ 6583, and [S II] λ 6731 lines for V536 Aql taken with a long-slit position angle of 90° (from Hirth et al. 1997). In each case the stellar continuum has been subtracted from the line emission, although, because of broad Hα wings, the subtraction is not perfect in the vicinity of the [N II] λ 6583 line. Contour lines are of relative intensity with a logarithmic spacing of $\sqrt{2}$. The radial velocity in this figure and Fig. 3 is with respect to the systemic velocity of the star.

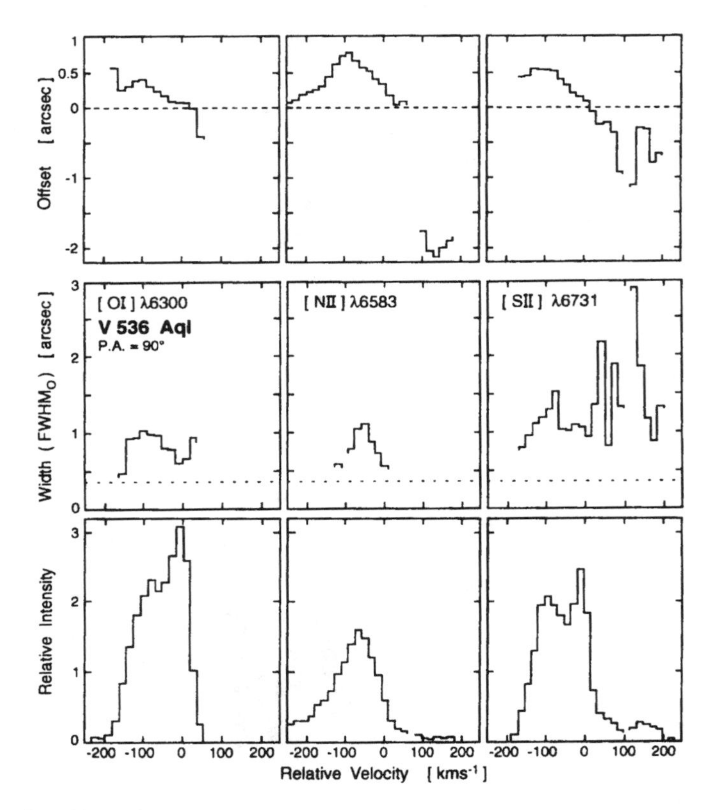

Figure 3. Plots of spatial width, offset, and relative intensity for the FEL regions of V536 Aql against radial velocity. The plots are extracted from the maps in Fig. 2.

(1997), we can also understand the increasing offset in the emission centroid in the HVC in going from [O I] to [S II] to [N II] if the electron density of the jet component is decreasing while at the same time its excitation is increasing with distance from the source. Note that the critical density for the [O I] line ($N_e^{crit} \approx 10^6$ cm^{-3}) is some 10^2 times greater than that for the [S II] doublet, which in turn has a higher critical electron density than the [N II] line. Strong emission from the [N II] line requires, however, quite high excitation, and though it is clear that in a jet with diverging streamlines the density will usually decrease with distance, it is not obvious why its excitation should increase for the first few hundred AU from the source.

V536 Aql is representative of many other classical T Tauri outflows in that one can detect an inner region of the HVC, extending perhaps up to a few arcseconds in length (the microjet), and several fainter HH condensations further out. Other examples are given in Hirth et al. (1997), Lavalley et al. (1997), and Solf (1997). Although such flows appear short in comparison to classical jets (see the chapter by Hartigan et al., this volume), no doubt their brightness distribution has a large role to play in their appearance, and in reality they are much more extensive. Note in particular that many of the so-called classical jets, such as HH 30 (Ray et al. 1996), decrease drastically in brightness with increasing distance from their source. Microjets may well behave in a similar manner.

B. The FEL Regions of Herbig Ae/Be Stars

In comparison to the classical T Tauri stars, we know much less about the FEL regions of Herbig Ae/Be stars. Several recent studies (e.g., Corcoran and Ray 1997, 1998*b*) have started to address this imbalance. The central question is, To what degree can close parallels be drawn between Herbig Ae/Be stars and their lower-mass counterparts? There has even been some debate in the literature as to whether the forbidden emission in Herbig Ae/Be stars arises in the same way as in the classical T Tauri stars. Some authors have argued that enhanced chromospheric activity may be responsible for their forbidden emission (e.g., Böhm and Catala 1995). Moreover, others have questioned whether disks are even necessary to explain the observational characteristics of these stars (e.g., Hartmann et al. 1993; Miroshnichenko et al. 1997).

As previously mentioned, the absence of high-velocity redshifted forbidden lines in classical T Tauri stars (Appenzeller et al. 1984; Edwards et al. 1987) has long been taken as evidence of opaque circumstellar disks. To test whether similar asymmetries are seen in intermediate-mass young stars, Corcoran and Ray (1997) undertook a spectroscopic study of 56 Herbig Ae/Be stars, half of which were observed to have detectable [O I] emission. It was found that those stars with [O I] emission could be divided into four groups, as determined by line profiles and radial velocities. Roughly 15% of the sample showed both distinct HVC and LVC blueshifted FELs; 50%, low-velocity blueshifted emission with centroid velocities in the range $-55 \text{ km s}^{-1} \leq v_c \leq -10 \text{ km s}^{-1}$; 25%, unshifted ($|v_c| \leq 5 \text{ km s}^{-1}$) symmetrical FELs; and finally, the remaining 10% showed low-velocity ($10 \text{ km s}^{-1} \leq v_c \leq 15 \text{ km s}^{-1}$) redshifted emission. No Herbig Ae/Be star was found to possess strongly redshifted FELs. Obviously, the clear tendency towards blueshifted velocities, and the velocities observed, implicitly suggest not only a similar origin for the FELs in Herbig Ae/Be stars as in classical T Tauri stars but also the presence of occluding disks. Corcoran and Ray (1997) also found an apparent link between the degree of embeddedness and the typical velocity shift in the sense that those stars that were more embedded tended to have higher blueshifted velocities.

Further evidence that the FELs arise in the same way in the Herbig Ae/Be stars as in the T Tauri stars comes from comparisons of their near-infrared colors with the relative strengths of their FELs (Corcoran and Ray

1998*b*). Such studies show, for example, that the equivalent width of the [O I] λ 6300 line in Herbig Ae/Be stars is related to near-infrared colors in the same fashion as in the T Tauri stars (for T Tauri stars see Edwards et al. 1993). As argued by Corcoran and Ray (1998*b*), this strongly supports the idea that the winds from Herbig Ae/Be stars arise in a similar way as those from T Tauri stars, i.e. through accretion-driven mass loss. Corcoran and Ray (1998*b*) also found that the [O I] λ 6300 line luminosity correlates better with excess infrared luminosity than with stellar luminosities, again supporting the idea that Herbig Ae/Be winds are accretion driven. If the Herbig Ae/Be stars and T Tauri stars with forbidden-line emission lines are grouped together, the correlation between mass loss rate (as indicated by the strength of the forbidden-line emission) and infrared excess spans nearly 5 orders of magnitude in luminosity and a range of masses from 0.5 $M_\odot$ to approximately 10 $M_\odot$ (Corcoran and Ray 1998*b*). An obvious implication is that the same outflow model must be applicable to the Herbig Ae/Be stars with forbidden-line emission and to the classical T Tauri stars. To illustrate the close parallels between the FEL regions in classical T Tauri stars and those in Herbig Ae/Be stars, we consider the case of LkHα 233 (Corcoran and Ray 1998*a*).

LkHα233 is an A5 Herbig Ae/Be star at a distance of 880 pc that is associated with an X-shaped bipolar nebula. Long-slit observations in the vicinity of its [S II] $\lambda\lambda$ 6716/6731 lines are shown in Fig. 4. An approximately $\pm 5''$ long high-velocity ($V_{\rm rad} \sim \pm$ 120 km s^{-1}) jet can be seen. It is on the same angular scale as the T Tauri microjets, and, as with V536 Aql, imaging shows that it is associated with a few faint HH knots. It is interesting to note that while the redshifted counterjet is observed to begin $0.7''$ from the center of the stellar continuum emission, the blueshifted jet can be traced right back to the continuum peak. This asymmetry is most naturally interpreted as being due to the occultation of the counterjet by a circumstellar disk with an upper limit to its projected radius of about 600 AU. The jet, at a position angle of 250°, is perpendicular to the inferred disk orientation based on polarization measurements and bisects the optical bipolar nebula associated with this star.

Close to the star itself ($\leq 2''$), the [S II] $\lambda\lambda$ 6716, 6731 emission is resolved into an HVC and an LVC with systemic radial velocities of about -125 and -25 km s^{-1}, respectively. The HVC can obviously be identified with the extended jet, whereas the broad, compact LVC is probably a disk wind, following the suggestion of Kwan and Tademaru (1995).

Finally, we should mention that HH outflows from intermediate-mass stars, such as LkHα 233 and the source of HH 80/81 (see section IV) have higher mass loss rates and velocities than their lower-mass counterparts (Mundt and Ray 1994). As one might expect, for example, their velocities tend to scale up with the escape velocity from the YSO. The reader is referred to Mundt and Ray (1994) for a more detailed account.

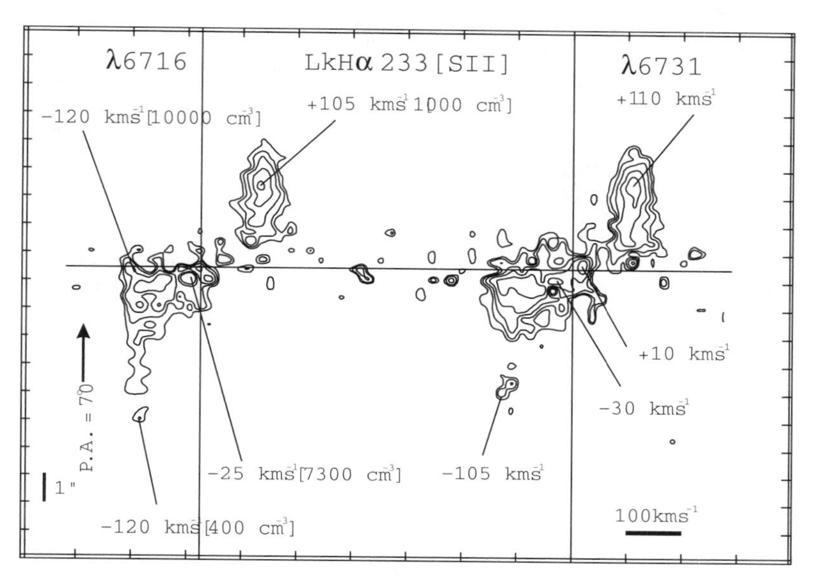

Figure 4. A position-velocity diagram of LkHα 233 in the region of the [S II] $\lambda\lambda$ 6716/6731 lines. The slit has been oriented along the outflow axis. The velocities are relative to the stellar rest velocity (vertical lines), electron densities are indicated in square brackets ([ ]), and the spacing of the continuum levels is logarithmic, scaling by factors of $\sqrt{2}$. The horizontal line marks the center of the continuum emission as determined from the fitting procedure.

IV. RADIO JETS

A. Introduction

The pioneering studies of Cohen et al. (1982), made with the Very Large Array (VLA), showed that at centimeter wavelengths the continuum emission from many T Tauri stars is dominated by free-free (thermal) emission, most probably originating in ionized, collimated outflows. These sources are referred to in the literature as thermal jets, to distinguish them from the nonthermal (synchrotron) jets emanating from the nuclei of active galaxies, quasars, and X-ray binaries. The thermal emission can be observed using sensitive radio interferometers with angular resolutions in the range of $0.1''$, and its study provides information on the nature of the outflowing gas very close to the star. Thermal jets are now known to be present in T Tauri stars as well as in less evolved, heavily obscured young stars of both low (Rodríguez et al. 1990) and high mass (Martí et al. 1993). The phenomenon seems to be characteristic of outflow sources: Rodríguez and Reipurth (1996, 1998) made relatively deep integrations (1 to 2 hr) with the VLA at 3.6 cm, detecting the possible exciting source in 11 out of 14 regions studied. Recent detailed reviews concerning thermal jets have been given by Anglada (1996) and Rodríguez (1997).

The high angular resolution and accurate positional information of the radio observations, together with the fact that they are practically unaffected by dust obscuration, enable this type of observation to discriminate, identify, and provide accurate positions for the exciting sources of outflows and to study the collimation very close to the star.

On the order of 100 radio sources detected at the centers of outflow regions are most likely thermal jets. However, only about 20 of them have been imaged with high angular resolution. In these sources (Anglada 1996) it is found that the major axis of the jet usually aligns well with the indicators of large-scale outflow (molecular bipolar outflows and HH objects).

B. Some Recent Results on Thermal Jets

To give the reader an idea of what information can be obtained from the study of thermal jets, we briefly discuss here recent results from a few selected sources.

HL Tauri is among the brightest millimeter continuum sources in the nearby Taurus complex (distance of 140 pc) and has been the subject of many studies to image its 100-AU protoplanetary disk (e.g., Wilner et al. 1996). Very high angular-resolution (0.04″) observations made by Wilner et al. (1999) at 7 mm with the VLA in its widest configuration resolve out the extended but relatively faint emission from the disk and reveal the presence of the compact, relatively bright thermal jet (see Fig. 5). This image implies that the collimation is present within a few AU from the star, and this result constitutes at present the most stringent direct upper limit for the collimation scale of jets. Note, however, that these limits are still consistent with the initial opening angles inferred from the optical observations near the source (see section II).

Various observations suggest that, in some regions of star formation, quadrupolar outflows (i.e., flows that appear to be close superpositions in the sky of two distinct bipolar outflows) are observed. One of the best examples of quadrupolar outflows is HH 111, with optical (Reipurth 1989), infrared (Davis et al. 1994), and CO outflows (Cernicharo and Reipurth 1996). Here the two distinct flows appear to be nearly perpendicular to each other. Recent VLA observations at 3.6 cm with angular resolution of 0.2″ (Rodríguez and Reipurth 1999) indicate that, in addition to the previously known elongation along the optical HH flow, the radio source shows weaker elongations approximately in the north-south direction, closely aligned with the axis of the second outflow. Thus, the HH 111 thermal jet is also quadrupolar (see Fig. 6). The centroids of the two jets appear to coincide in projection within 0.1″ (i.e., 50 AU at the distance of the source), again pointing to collimation very close to the exciting stars.

Finally, we mention the HH 80–81 region in Sagittarius, which is the site of a highly collimated bipolar thermal radio jet that emanates from a young massive star (Martí et al. 1993). This jet has been monitored for

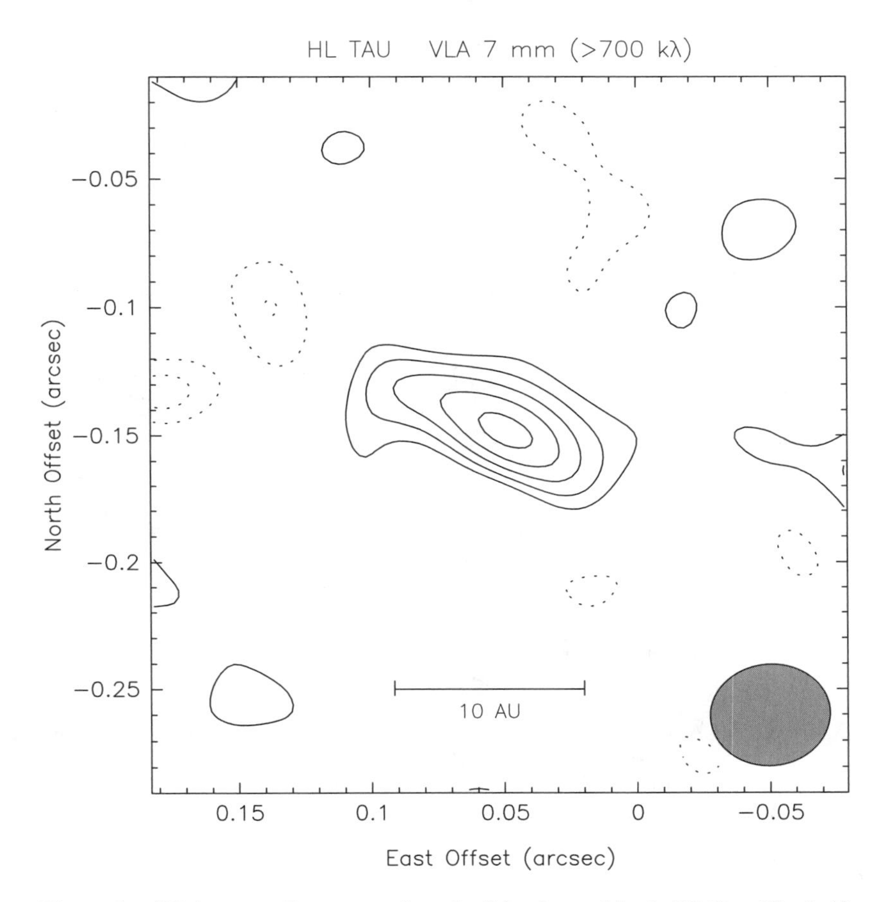

Figure 5. VLA map at 7 mm wavelength of the thermal jet in HL Tau. The half-power contour of the synthesized beam is shown in the bottom right corner. Contours are −3, −2, 2, 3, 4, 5, and 6 times 0.18 mJy beam^{-1}. This image implies that collimation is present within a few AU from the star.

four epochs (1990, 1994, 1995, and 1997) by Martí et al. (1998). These authors computed the difference maps of the 1994, 1995, and 1997 epochs with respect to 1990. From these difference maps (Fig. 7) they find a pair of condensations that travel away from the central source at a velocity of $\sim$500 km s^{-1}. These condensations appear to be ejected periodically, about every 10 years, with the next ejection of a pair of condensations expected for around the year 2000.

C. Mass Loss Rate Estimates

It is also possible to estimate the mass loss rate in the jet, following Reynolds (1986). For this, one assumes a pure hydrogen jet with constant opening angle, terminal velocity, and ionization fraction as well as constant electron temperature, taken to be equal to 10^4 K. One further

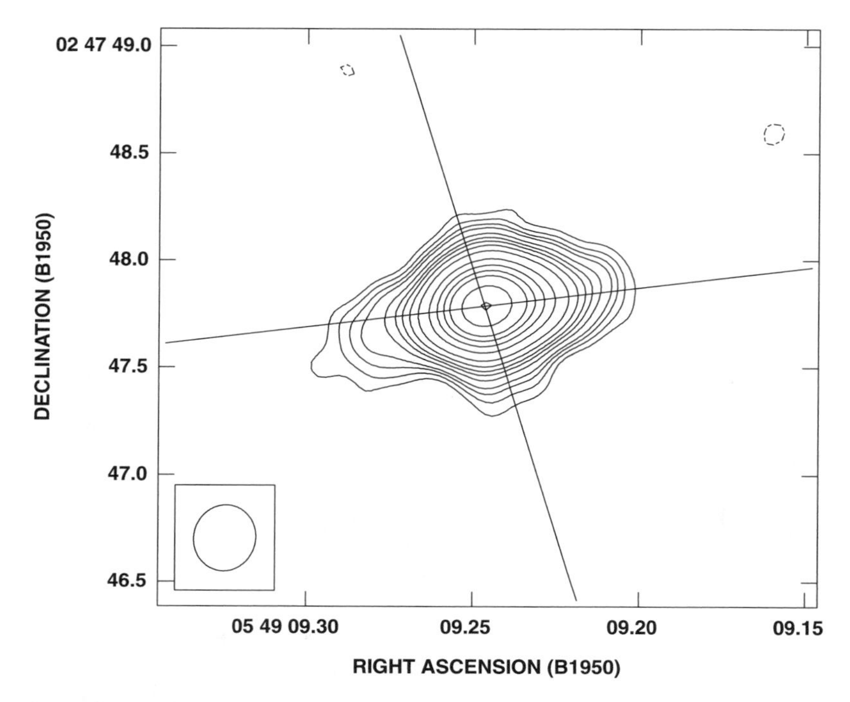

Figure 6. VLA map at 3.6 cm wavelength of the quadrupolar thermal jet at the core of the quadrupolar outflow in HH 111. Contours are −3, 3, 4, 5, 6, 8, 10, 12, 15, 20, 30, 40, 50, 60, 80, and 100 times the rms noise of 5.4 μJy beam^{-1}. The half-power contour of the synthesized beam is also shown. The straight lines give the position angles of the two large-scale flows (optical and infrared) in the region.

assumes that the jet axis is perpendicular to the line of sight (that is, with an inclination angle of $i = 90°$). As regards the last assumption it should be noted that variations in i from 45° to 90° change the mass loss estimate by less than 10%. Moreover, to be identified as a jet, the flow normally has an elongated appearance in the sky, so it is unlikely that i has values that are much smaller than 45°.

Under these assumptions, the mass loss rate in the jet is given by

$$\dot{M}_{-6} = 1.9 v_8 x_0^{-1} S_{\rm mJy}^{0.75} \nu_9^{-0.45} d_{\rm kpc}^{1.5} \theta_0^{0.75} \tag{2}$$

where $\dot{M}_{-6}$ is the mass loss rate in 10^{-6} $M_\odot$ yr^{-1}, v_8 is the terminal velocity of the jet in 10^3 km s^{-1}, x_0 is the ionization fraction, $S_{\rm mJy}$ is the observed flux density in mJy, ν_9 is the observed frequency in GHz, $d_{\rm kpc}$ is the distance in kpc, and θ_0 is the opening angle in radians.

The opening angle is estimated to be

$$\theta_0 = 2 \tan^{-1}(\theta_{\rm min}/\theta_{\rm maj}) \tag{3}$$

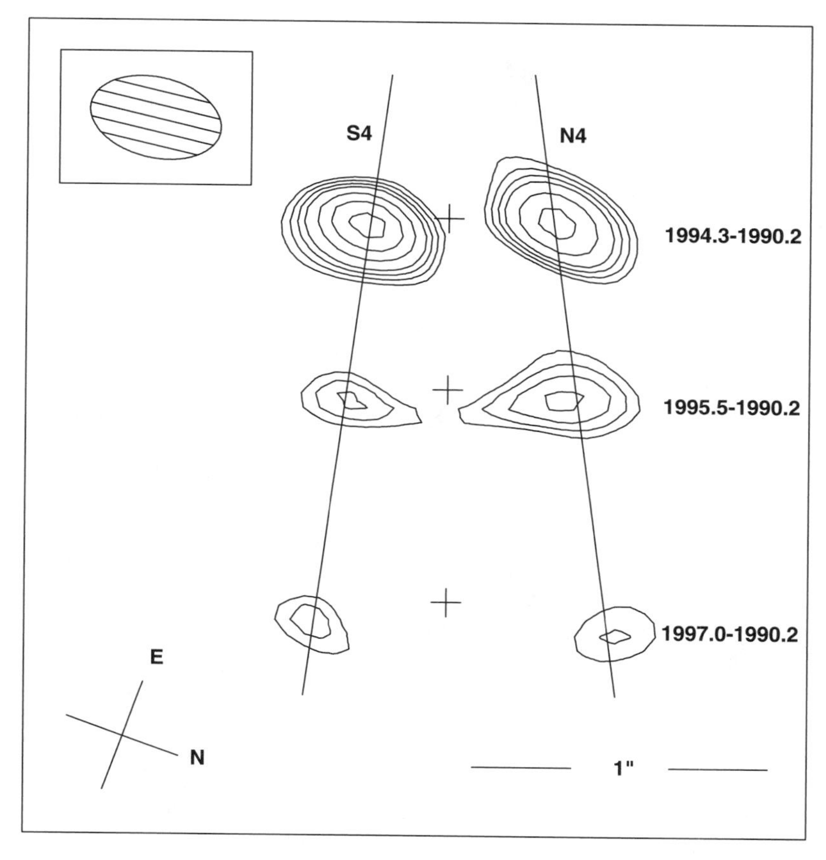

Figure 7. Difference maps of the thermal radio jet in the HH 80–81 complex for the epochs 1994.3–1990.2, 1995.5–1990.2, and 1997.0–1990.2. The crosses indicate the position of the central source, whose B1950.0 coordinates are $\alpha = 18^h16^m12\overset{s}{.}997$ and $\delta = -20°48'48\overset{''}{.}27$. Contour levels are 5, 6, 7, 8, 10, 12 and 14 times the rms noise of 26, 25 and 22 μJy beam^{-1} of each difference image, respectively. The half-power contour of the synthesized beam is also shown.

where θ_{min} and θ_{maj} are the deconvolved minor and major axes of the jet. The terminal velocity can be obtained from proper motions and radial velocity information obtained from studies of the jets at larger scales. If this information is not available, an educated guess has to be made. If the star is of low (that is, solar) mass, a value of $v_8 = 0.3$ is adopted. If the star is more massive, values of $v_8 = 0.5$–1 are usually assumed.

We should comment that the momentum rate in the ionized jet is normally found to be an order of magnitude smaller than that observed in the larger-scale molecular outflow. This result has been taken to indicate that jets are only partially ionized and that the total momentum rate is much larger than the momentum rate in ionized gas.

D. Nonthermal Emission

Although it has been known for some time (e.g., André 1996) that the radio emission from weak-line T Tauri stars (wTTSs) is nonthermal ($S_\nu \propto \nu^\alpha$, $\alpha \leq 0$), nonthermal emission from classical T Tauri stars (Anglada 1996) or Herbig Ae/Be stars (Skinner and Brown 1993) appears to be virtually unknown. For wTTSs, very long-baseline interferometry (VLBI) has shown that the emission comes from compact regions ($\leq 30R_\star$) with brightness temperatures in some cases exceeding 10^9 K. Moreover, circular polarization has been observed in a few wTTSs (e.g., Skinner 1993), implying the presence of ordered magnetic fields. In the case of the wTTSs, the radiation is almost certainly gyrosynchrotron emission from mildly relativistic electrons (Lorentz factors $\gamma \approx 2$–5), and explosive magnetic field reconnection seems to be the only viable means of accelerating particles to such energies.

Although there is some debate over the origin of the various outflows in the T Tauri region, it is generally agreed that T Tauri S, the southern embedded companion to the optical star, is an outflow source. It is thus interesting that T Tauri S appears, at least some of the time, to have a nonthermal radio spectrum and circularly polarized radio emission (Skinner and Brown 1994; Philips et al. 1993). Recently Ray et al. (1997) observed the circular polarization of T Tauri S to be on a much more extended scale than seen in wTTSs; they found evidence for ordered magnetic fields at distances of tens of AU (rather than tens of stellar radii) from the source. Assuming the emission is gyrosynchrotron radiation, the inferred field strengths at these distances are quite large (a few gauss). Moreover, extrapolating back a dipole field to the source would imply inordinately large field strengths near the stellar surface. So it would seem much more likely that the fields are convected as part of an outflow.

V. MOLECULAR HYDROGEN OUTFLOWS

A. Morphology of H_2 Outflows

Molecular hydrogen (H_2) is an important tracer of low-velocity shocks in outflows from young stars. It has been detected both in optically visible, highly collimated HH jets (see, e.g., Hartigan et al. 1989; Stapelfeldt et al. 1991; Davis et al. 1994; Eislöffel et al. 1994*b*; Noriega-Crespo et al. 1996) and in poorly collimated molecular (CO) outflows (see, e.g., Bally et al. 1993; Hodapp 1994; Davis et al. 1998). Not all HH flows show strong H_2 emission, although its absence is probably due to small amounts of molecular material being present. Deeply embedded outflows from Class 0 and Class I sources, on the other hand, are probably much younger and are still plowing their way through a dense molecular environment. In such flows we usually find bright and bow-shaped H_2 knots (see, e.g., Davis and Eislöffel 1995; Eislöffel et al. 1996; Davis et al. 1997*b*; Ladd and Hodapp

1997). These bright knots are interpreted as working surfaces in the flow. In addition, such flows often show fainter H_2 emission extending along both sides of the flow axis from the source to the bow shocks (Davis and Eislöffel 1995; Dent et al. 1995). This emission may arise in the interaction of the flow with the ambient medium in shear layers along the flow channel walls.

B. Excitation of H_2 in Outflows

The molecular hydrogen emission in outflows is almost certainly due to postshock, radiatively cooling gas. The shock waves may arise within the supersonic flows, they may be driven into the surrounding medium along the flow channel, or they could arise at the head of the flow where the jet plows through the ambient medium. Various types of shock structures may be present in these shock waves. Given a sufficiently high shock speed, a "J-shock" may develop, in which a jump in velocity, density, pressure, temperature, and entropy takes place. Such a shock front therefore appears as a discontinuity. In a "C-shock," by comparison, the shock energy is dissipated via ambipolar diffusion between ions and neutrals; hence, the state of physical quantities within the shock structure will change continuously (Draine and McKee 1993). Other suggested excitation mechanisms for the H_2 gas include shocks with magnetic precursors (e.g., Hartigan et al. 1989; Carr 1993) and H_2 fluorescence (Fernandes and Brand 1995). Only recently, systematic work has started to incorporate magnetic fields into J- and C-type bow shock models (MacLow and Smith 1997; Smith and MacLow 1997; Stone 1997). All of the above-mentioned models (and several more) have been invoked to explain H_2 spectroscopy of outflows and have led to widely differing results and interpretations (e.g., Moorhouse et al. 1990; Smith et al. 1991*a*; Gredel 1994; Eislöffel et al. 1996). Indeed, it seems that the excitation of H_2 in collimated outflows is still not well understood.

C. Kinematics of H_2 in Outflows

Only a few studies of the kinematics of H_2 in outflows have been made so far. In particular, high-resolution spectroscopy of the 1–0 S(1) line at 2.12 μm has been used to investigate radial velocities and line profiles. Early spectroscopy of OMC-1 (Smith et al. 1991*b*) and several knots in HH 1/2 and HH 46/47 (Zinnecker et al. 1989), lacked spatial resolution. This is in contrast with more recent studies, for example of L1448 (Davis and Smith 1996*a*), DR21 (Davis and Smith 1996*b*), and HH 32 (Davis et al. 1996), which have permitted detailed comparison of observation with theory.

Such studies provide important insights into the geometry of the shock structures. Wide, double-peaked profiles are expected in bow shocks, whereas wide, but single-peaked, Gaussian profiles are expected in turbulent boundary layers. On the other hand, narrower symmetric lines are expected in planar shocks (Smith and Brand 1990*a,b*). Shock speed and outflow orientations can be predicted from centroid velocities and observed asymmetries in the line profiles (see, e.g., Hartmann and Raymond 1984). Determination of the outflow orientation is especially important, because the outflow energy can be estimated only if the inclination angle of a flow is known. The observations of L1448 (Davis and Smith 1996*a*) indeed show wide, double-peaked profiles, at least in the bright knots,

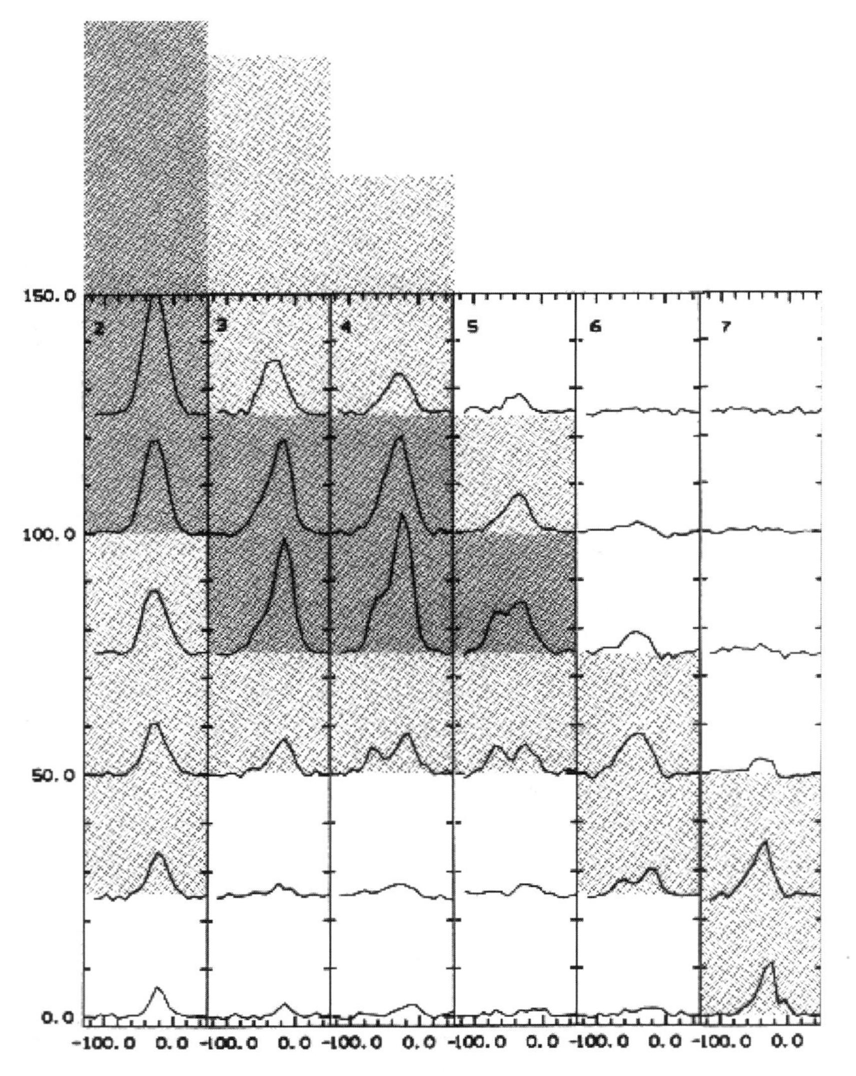

Figure 8. High-resolution spectra of knot A in L1448 in the 1–0 S(1) line of H_2 at 2.12 μm, superimposed onto a gray-scale image of knot A. Double-peaked line profiles, a typical signature of bow shocks, are clearly indicated (from Davis and Smith 1996*a*).

which resemble bow shocks (see Fig. 8). Also in HH 32 (Davis et al. 1996), the H_2 line profiles indicate a high-velocity bow shock pointed away from us, and are consistent with the result derived from optical high-resolution spectroscopy (Solf et al. 1986; Hartigan et al. 1987). On the other hand, single-peaked Gaussian line profiles have been seen in the spectra of DR21 (Davis and Smith 1996*b*) and also in parts of L1448. These suggest instead that shear layers between the collimated flow and its environment are responsible for the excitation of the H_2 gas.

Recently, the first proper motion measurements of H_2 knots in HH 1 (Noriega-Crespo et al. 1997) and HH 46/47 (Micono et al. 1998, see Fig. 9) have also been made. They show that the H_2 knots in these flows move at tangential velocities of 100 to 400 km s^{-1} (i.e., similar to the velocities measured for the optical knots). Because the observed velocities of the H_2 knots are clearly higher than the shock speeds necessary for the dissociation of the H_2 molecules, we are led to the conclusion that the H_2 gas is comoving with the outflow. Hence, we are seeing molecular H_2 gas in the jet, or entrained ambient gas that has already been accelerated to a major fraction of the jet velocity.

D. Comparison of H_2 and CO Outflows

A comparison of the H_2 image and the CO map of a flow may give us interesting information about its propagation. Because the cooling times of H_2 may be as short as a year, the H_2 image shows us where the interactions

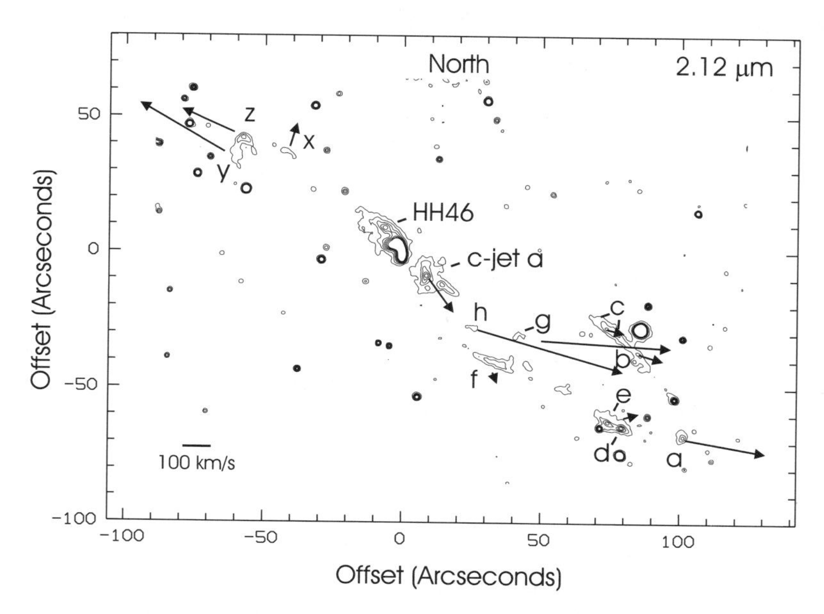

Figure 9. Contour plot of the HH 46/47 outflow in the 1–0 S(1) line of H_2 at 2.12 μm. Proper motion vectors indicate the velocities for several condensations in this outflow (from Micono et al. 1998).

between the flow and its surroundings are "currently" taking place. In contrast, CO can maintain an excited state easily for a long time, because of its low excitation potential of only 5 K, and therefore shows the coadded history of the outflow. Recent interferometric observations indicate that at least some CO outflows are also fairly well collimated (see Bachiller 1996; Richer et al., this volume, and references therein). By comparing such CO maps and H_2 images, an intimate spatial correlation between the two molecules is found: In almost all objects the CO peaks just upstream of the bright H_2 bows. H_2 knots along the flow between the source and these bows seem to line up along the sides of the CO flow lobes (Davis and Eislöffel 1995; Davis et al. 1997*a*). This suggests that the CO outflow mainly consists of ambient gas that has been swept up and entrained in the (leading) working surfaces of the highly collimated jet, while some (usually much smaller fraction) is ambient material accelerated in turbulent shear layers along the flow channel walls.

REFERENCES

Ageorges, N., Menard, F., Monin, J.-L., and Eckart, A. 1994. SHARP images of the pre–main sequence star V536 Aquilae: A highly polarized binary. *Astron. Astrophys. Lett.* 283:L5–L8.

André, P. 1996. Radio emission as a probe of large-scale magnetic structures around young stellar objects. In *Radio Emission from the Stars and the Sun*, ASP Conf. Ser. 93, ed. A. R. Taylor and J. M. Paredes (San Francisco: Astronomical Society of the Pacific), pp. 273–293.

Anglada, G. 1996. Radio jets in young stellar objects. In *Radio Emission from the Stars and the Sun*, ASP Conf. Ser. 93, ed. A. R. Taylor and J. M. Paredes (San Francisco: Astronomical Society of the Pacific), pp. 3–14.

Appenzeller, I., Jankovics, I., and Östreicher, R. 1984. Forbidden-line profiles of T Tauri stars. *Astron. Astrophys.* 141:108–115.

Bachiller, R. 1996. Bipolar molecular outflows from young stars and protostars. *Ann. Rev. Astron. Astrophys.* 34:111–154.

Bally, J., Lada, E. A., and Lane, A. P. 1993. The L1448 molecular jet. *Astrophys. J.* 418:322–327.

Böhm, K. H., and Solf, J. 1994. A sub-arcsecond-scale spectroscopic study of the complex mass outflows in the vicinity of T Tauri. *Astrophys. J.* 430:277–290.

Böhm, T., and Catala, C. 1995. Rotation, winds and active phenomena in Herbig Ae/Be stars. *Astron. Astrophys.* 301:155–169.

Burrows, C. J., Stapelfeldt, K. R., Watson, A. M., Krist, J. E., Ballester, G. E., Clarke, J. T., Crisp, D., Gallagher, J. S., III, Griffiths, R. E., Hester, J. J., Hoessel, J. G., Holtzman, J. A., Mould, J. R., Scowen, P. A., Trauger, J. T., and Westphal, J. A. 1996. Hubble Space Telescope observations of the disk and the jet of HH30. *Astrophys. J.* 453:437–451.

Carr, J. S. 1993. The H_2 velocity field in Herbig-Haro 7-11. *Astrophys. J.* 406:553–562.

Cernicharo, J., and Reipurth, B. 1996. Herbig-Haro jets, CO flows, and CO bullets: The case of HH 111. *Astrophys. J. Lett.* 460:L57–L60.

Cohen, M., Bieging, J. H., and Schwartz, P. R. 1982. VLA observations of mass loss from T Tauri stars. *Astrophys. J.* 253:707–715.

Corcoran, M., and Ray, T. P. 1997. Forbidden emission lines in Herbig Ae/Be stars. *Astron. Astrophys.* 321:189–201.

Corcoran, M., and Ray, T. P. 1998*a*. Spectroscopic discovery of a bipolar jet from the Herbig Ae/Be star LkHα233. *Astron. Astrophys.* 336:535–538.

Corcoran, M., and Ray, T. P. 1998*b*. Wind diagnostics and correlations with the near-infrared excess in Herbig Ae/Be stars. *Astron. Astrophys.* 331:147–161.

Davis, C. J., and Eislöffel, J. 1995. Near-infrared imaging in H_2 of molecular (CO) outflows from young stars. *Astron. Astrophys.* 300:851–869.

Davis, C. J., and Smith, M. D. 1996*a*. Echelle spectroscopy of shocked H_2 in the L1448 outflow. *Astron. Astrophys.* 309:929–938.

Davis, C. J., and Smith, M. D. 1996*b*. Near-IR imaging and spectroscopy of DR21: A case for supersonic turbulence. *Astron. Astrophys.* 310:961–969.

Davis, C. J., Mundt, R., and Eislöffel, J. 1994. Near-infrared imaging of the jets and flows associated with Herbig-Haro objects HH 91, HH 110, and HH 111. *Astrophys. J. Lett.* 437:L55–L58.

Davis, C. J., Eislöffel, J., and Smith, M. D. 1996. Near-infrared and optical observations of an obliquely viewed bow shock: AS353A/HH32. *Astrophys. J.* 463:246–253.

Davis, C. J., Eislöffel, J., Ray, T. P., and Jenness, T. 1997*a*. Prompt entrainment in the variable molecular jet from RNO 15-FIR. *Astron. Astrophys.* 324:1013–1019.

Davis, C. J., Ray, T. P., Eislöffel, J., and Corcoran, D. 1997*b*. Near-IR imaging of the molecular outflows in HH24-26, L1634 (HH240-241), L1660 (HH72) and RNO15FIR. *Astron. Astrophys.* 324:263–275.

Davis, C. J., Moriarty-Schieven, G., Eislöffel, J., Hoare, M. G., and Ray, T. P. 1998. Observations of shocked H_2 and entrained CO in outflows from luminous young stars. *Astron. J.* 115:1118–1134.

Dent, W. R. F., Matthews, H. E., and Walther, D. M. 1995. CO and shocked H_2 in the highly collimated outflow from VLA1623. *Mon. Not. Roy. Astron. Soc.* 277:193–209.

Draine, B. T., and McKee, C. F. 1993. Theory of interstellar shocks. *Ann. Rev. Astron. Astrophys.* 31:373–432.

Edwards, S., Cabrit, S., Strom, S., Heyer, I., Strom, K., and Anderson, E. 1987. Forbidden line and Hα profiles in T Tauri spectra: A probe of anisotropic mass outflows and circumstellar disks. *Astrophys. J.* 321:473–495.

Edwards, S., Ray, T. P., and Mundt, R. 1993. Energetic mass outflows from young stars. In *Protostars and Planets III*, ed. E. H. Levy and J. I. Lunine (Tucson: University of Arizona Press), pp. 567–602.

Eislöffel, J., Mundt, R., and Böhm, K.-H. 1994*a*. Structure and proper motions in Herbig-Haro objects 1 and 2. *Astron. J.* 108:1042–1055.

Eislöffel, J., Davis, C. J., Ray, T. P., and Mundt, R. 1994*b*. Near-infrared observations of the HH 46/47 system. *Astrophys. J. Lett.* 422:L91–L95.

Eislöffel, J., Smith, M. D., Davis, C. J., and Ray, T. P. 1996. Molecular hydrogen in the outflow from Cep E. *Astron. J.* 112:2086–2093.

Eislöffel, J., and Mundt, R. 1997. Parsec-scale jets from young stars. *Astron. J.* 114:280–287.

Fernandes, A. J. L., Brand, P. W. J. L. 1995. Shock diagnostics in Herbig-Haro 7: Evidence for H_2 fluorescence. *Mon. Not. Roy. Astron. Soc.* 274:639–656.

Gredel, R. 1994. Near-infrared spectroscopy and imaging of Herbig-Haro objects. *Astron. Astrophys.* 292:580–592.

Hamann, F. 1994. Emission-line studies of young stars. 4: The optical forbidden lines. *Astrophys. J. Suppl.* 93:485–518.

Hartigan, P., Raymond, J., and Hartmann, L. 1987. Radiative bow shock models of Herbig-Haro objects. *Astrophys. J.* 316:323–348.
Hartigan, P., Curiel, S., and Raymond, J. 1989. Molecular hydrogen and optical images of HH 7-11. *Astrophys. J. Lett.* 347:L31–L35.
Hartigan, P., Edwards, S., and Ghandour, L. 1995. Disk accretion and mass loss from young stars. *Astrophys. J.* 452:736–768.
Hartmann, L., and Raymond, J. C. 1984. A high-resolution study of Herbig-Haro objects 1 and 2. *Astrophys. J.* 276:560–571.
Hartmann, L., and Raymond, J. C. 1989. Wind-disk shocks around T Tauri stars. *Astrophys. J.* 337:903–916.
Hartmann, L., Kenyon, S. J., and Calvet, N. 1993. The excess infrared emission of Herbig Ae/Be stars—Disks or envelopes? *Astrophys. J.* 407:219–231.
Hirth, G. A., Mundt, R., and Solf, J. 1994. Jet flows and disk winds from T Tauri stars: the case of CW Tau. *Astron. Astrophys.* 285:929–942.
Hirth, G. A., Mundt, R., and Solf, J. 1997. Spatial and kinematic properties of the forbidden emission line region of T Tauri stars. *Astron. Astrophys. Suppl.* 126:437–469.
Hodapp, K.-W. 1994. A K′ imaging survey of molecular outflow sources. *Astrophys. J. Suppl.* 94:615–649.
Kwan, J., and Tademaru, E. 1988. Jets from T Tauri stars: Spectroscopic evidence and collimation mechanism. *Astrophys. J. Lett.* 332:L41–L44.
Kwan, J., and Tademaru, E. 1995. Disk winds from T Tauri stars. *Astrophys. J.* 454:382–393.
Ladd, E. F., and Hodapp, K.-W. 1997. A double outflow from a deeply embedded source in Cepheus. *Astrophys. J.* 474:749–759.
Lavalley, C., Cabrit, S., Dougados, C., Ferruit, P., and Bacon, R. 1997. Subarcsecond morphology and kinematics of the DG Tauri jet in the [O I] λ 6300 line. *Astron. Astrophys.* 327:671–680.
MacLow, M.-M., and Smith, M. D. 1997. Non-linear development and observational consequences of Wardle C-shock instabilities. *Astrophys. J.* 491:596–614.
Martí, J., Rodríguez, L. F., and Reipurth, B. 1993. HH 80-81: A highly collimated Herbig-Haro complex powered by a massive young star. *Astrophys. J.* 416:208–217.
Martí, J., Rodríguez, L. F., and Reipurth, B. 1998. Proper motions of the inner condensations in the HH 80-81 thermal radio jet. *Astrophys. J.* 502:337–341.
Micono, M., Davis, C. J., Ray, T. P., Eislöffel, J., and Shetrone, M. D. 1998. Proper motions and variability of the H_2 emission in the HH 46 system. *Astrophys. J. Lett.* 494:L227–L230.
Miroshnichenko, A., Ivezic, Z., and Elitzur, M. 1997. On protostellar disks in Herbig Ae/Be stars. *Astrophys. J. Lett.* 475:L41–L44.
Moorhouse, A., Brand, P. W. J. L., Geballe, T., and Burton, M. G. 1990. Velocity profiles of high-excitation molecular hydrogen lines. *Mon. Not. Roy. Astron. Soc.* 242:88–91.
Mundt, R., and Eislöffel, J. 1998. T Tauri stars associated with Herbig-Haro objects and jets. *Astron. J.*, 116:860–867.
Mundt, R., and Hirth, G. A. 1999. Properties of T Tauri star winds derived from forbidden emission line properties. In preparation.
Mundt, R., and Ray, T. P. 1994. Optical outflows from Herbig Ae/Be stars and other high luminosity young stellar objects. In *The Nature and Evolutionary Status of Herbig Ae/Be Stars*, ASP Conf. Ser. 62, ed. P. S. Thé, M. R. Perez, and E. P. J. van den Heuvel (San Francisco: Astronomical Society of the Pacific), pp. 237–257.

Mundt, R., Ray, T. P., Bührke, T., Raga, A. C., and Solf, J. 1990. Optical jets and outflows in the HL Tau region. *Astron. Astrophys.* 232:37–61.

Mundt, R., Ray, T. P., and Raga, A. C. 1991. Collimation of stellar jets—constraints from the observed spatial structure. II. Observational results. *Astron. Astrophys.* 252:740–761.

Noriega-Crespo, A., Garnavich, P. M., Raga, A. C., Cantó, J., and Böhm, K.-H. 1996. HH110 jet near-infrared imaging: The outflow mixing layer. *Astrophys. J.* 462:804–812.

Noriega-Crespo, A., Garnavich, P. M., Curiel, S., Raga, A. C., and Ayala, S. 1997. The proper motions of the warm molecular hydrogen gas in Herbig-Haro 1. *Astrophys. J. Lett.* 486:L57–L60.

Ouyed, R., and Pudritz, R. E. 1993. Forbidden line formation in hydromagnetic disk winds. I. Oblique shocks. *Astrophys. J.* 419: 255–267.

Philips, R. B., Lonsdale, C. J., and Feigelson, E. D. 1993. Magnetic non-thermal activity in the T Tauri system. *Astrophys. J. Lett.* 403:L43–L46.

Raga, A. C., Mundt, R., and Ray, T. P. 1991. Collimation of stellar jets—constraints from the observed spatial structure. I. Data analysis methods. *Astron. Astrophys.* 252:733–739.

Ray, T. P., Mundt, R., Dyson, J. E., Falle, S. A. E. G., and Raga, A. C. 1996. HST observations of jets from young stars. *Astrophys. J. Lett.* 468:L103–L106.

Ray, T. P., Muxlow, T. W. B., Axon, D. J., Brown, A., Corcoran, D., Dyson, J., and Mundt, R. 1997. Large-scale magnetic fields in the outflow from the young stellar object T Tauri S. *Nature* 385:415–417.

Reipurth, B. 1989. The HH 111 jet and multiple outflow episodes from young stars. *Nature* 340:42–45.

Reynolds, S. P. 1986. Continuum spectra of collimated, ionized stellar winds. *Astrophys. J.* 304:713–720.

Rodríguez, L. F. 1997. Thermal radio jets. In *Herbig-Haro Flows and the Birth of Low Mass Stars.* IAU Symp. 182, ed. B. Reipurth and C. Bertout (Dordrecht: Kluwer), pp. 83–92.

Rodríguez, L. F., and Reipurth, B. 1996. VLA detection of the exciting sources of HH 34, HH 114, and HH 199. *Rev. Mex. Astron. Astrofis.* 32:27–33.

Rodríguez, L. F., and Reipurth, B. 1998. VLA Detection of the exciting sources of HH 83, HH 117, HH 124, HH 192, HH 300, HH 366, and HH 375. *Rev. Mex. Astron. Astrofis.* 34:13–19.

Rodríguez, L. F., and Reipurth, B. 1999. A quadrupolar thermal jet in HH 111. In preparation.

Rodríguez, L. F., Ho, P. T. P., Torrelles, J. M., Curiel, S., and Cantó, J. 1990. VLA observations of the Herbig-Haro 1-2 system. *Astrophys. J.* 352:645–653.

Skinner, S. L. 1993. Circularly polarised radio emission from the T Tauri star Hubble 4. *Astrophys. J.* 408:660–667.

Skinner, S. L., and Brown, A. S. R. T. 1993. A high sensitivity survey of radio continuum emission from Herbig Ae/Be stars. *Astrophys. J. Suppl.* 87:217–265.

Skinner, S. L., and Brown, A. S. R. T. 1994. The enigmatic T Tauri radio source. *Astron. J.* 107:1461–1468.

Smith, M. D., and Brand, P. W. J. L. 1990*a*. Cool C-shocks and high-velocity flows in molecular clouds. *Mon. Not. Roy. Astron. Soc.* 242:495–504.

Smith, M. D., and Brand, P. W. J. L. 1990*b*. Signatures of C-shocks in molecular clouds. *Mon. Not. Roy. Astron. Soc.* 243:498–503.

Smith, M. D., and MacLow, M. -M. 1997. The formation of C-shocks: Structure and signatures. *Astron. Astrophys.* 326:801–810.

Smith, M. D., Brand, P. W. J. L., and Moorhouse, A. 1991*a*. Bow shocks in molecular clouds—H_2 line strengths. *Mon. Not. Roy. Astron. Soc.* 248:451–456.

Smith, M. D., Brand, P. W. J. L., and Moorhouse, A. 1991*b*. Shock absorbers in bipolar outflows. *Mon. Not. Roy. Astron. Soc.* 248:730–740.

Solf, J. 1997. Spectroscopic signatures of micro-jets. In *Herbig-Haro Flows and the Birth of Low-Mass Stars*, IAU Symposium No. 182, ed. B. Reipurth and C. Bertout (Dordrecht: Kluwer), pp. 63–72.

Solf, J., and Böhm, K.-H. 1993. High-resolution long-slit spectral imaging of the mass outflows in the immediate vicinity of DG Tauri. *Astrophys. J. Lett.* 410:L31–L34.

Solf, J., Böhm, K. -H., and Raga, A. C. 1986. Kinematical and hydrodynamical study of the HH32 complex. *Astrophys. J.* 305:795–804.

Stapelfeldt, K. R., Scoville, N. Z., Beichman, C. A., Hester, J. J., and Gautier, T. N. 1991. Near-infrared emission-line images of three Herbig-Haro objects. *Astrophys. J.* 371:226–236.

Stone, J. M. 1997. The Wardle instability in interstellar shocks. I. Non-linear dynamical evolution. *Astrophys. J.* 487:271–282.

Wilner, D. J., Ho, P. T. P., and Rodríguez, L. F. 1996. Sub-arcsecond observations of HL Tauri: Imaging the circumstellar disk. *Astrophys. J. Lett.* 470:L117–L121.

Wilner, D. J., Ho, P. T. P., and Rodríguez, L. F. 1999. Observations of the jet/disk system in HL Tau at 7mm. In preparation.

Zinnecker, H., Mundt, R., Geballe, T. R., and Zealey, W. J. 1989. High spectral resolution observations of the H_2 2.12μm line in Herbig-Haro objects. *Astrophys. J.* 342:337–344.

SHOCK STRUCTURES AND MOMENTUM TRANSFER IN HERBIG-HARO JETS

PATRICK HARTIGAN
Rice University

JOHN BALLY, BO REIPURTH, AND JON A. MORSE
University of Colorado

Herbig-Haro jets record the mass ejection and accretion history of young stars and provide important clues as to how stars form. The use of outflows to constrain the physics of star formation requires an understanding of how shocks within a jet transfer momentum to the ambient medium. Our understanding of how this momentum transfer occurs is improving at a rapid pace, driven by (1) spectacular high-spatial-resolution *Hubble Space Telescope* (HST) images, (2) large-format ground-based charge-coupled devices (CCDs) with wide fields of view, and (3) velocity-resolved images taken with Fabry-Perot spectrometers and image slicers that enable radial velocities to be measured over a large field of view. HST images of jets resolve the spatial structure of the cooling zones behind the shocks in jets clearly for the first time, and enable us to identify shock fronts and to follow proper motions of subarcsecond structures. Wide-field CCDs have shown that outflows from young stars can extend dozens of light-years from their sources, which are often multiple systems that drive multiple jets. Velocity and line excitation maps of jets probe the physical conditions within shocked gas and make possible quantitative comparisons with theoretical models of the flow dynamics. Studies of jets within H II regions are in their infancy, but such objects offer a unique opportunity to observe entire outflows as they are illuminated by ambient ultraviolet light.

I. INTRODUCTION

Mass loss in the form of winds or jets accompanies newly formed stars from the time a protostar first appears within a molecular cloud (e.g., Bachiller 1996). Millimeter-wavelength observations of CO frequently reveal poorly collimated bipolar molecular outflows of low velocity (3 to 30 km s^{-1}) in the vicinity of forming stars (e.g., Tamura et al. 1996; Moriarty-Schieven et al. 1995). Intermediate velocity ($\sim$50–100 km s^{-1}) CO "bullets" are also sometimes observed in bipolar outflows (Cernicharo and Reipurth 1996), and analogous shock-excited emission from infrared lines of H_2 are found toward many outflows (e.g., Davis and Eislöffel 1995). Along lines of sight that are relatively unobscured by dust, we can observe optical emission line nebulae known as Herbig-Haro (HH) objects. These objects mark the locations where the highest-velocity gas ($>$100 km s^{-1}) cools behind shock waves in the outflow, and they often trace a series of bow shocks in highly collimated jets (e.g., Reipurth et al. 1997*b*). Hence, HH objects, near-IR emission from H_2, and CO outflows all appear to be manifestations of the mass loss produced during the early stages of the life of a star.

There is growing evidence that strong winds or jets from young stars are powered by unsteady massive disk accretion events (Reipurth 1989; Hartmann et al. 1993). The resulting intermittent outflows produce a chain of internal shocks where faster flow components overtake slower ejecta. Mass outflow rates increase as accretion rates increase (Hartigan et al. 1995), so shocks in jets provide a fossil record of the time evolution of the accretion history of a young star. Where the jet interacts with the ambient medium, it creates external shocks at the ends or the sides of an outflow cavity and accelerates the surrounding gas to produce molecular outflows (e.g., Chernin and Masson 1995; Cernicharo and Reipurth 1996).

Outflows appear to play a fundamental role in the process of star formation. In order for a young star to accrete mass, either from spherical infall from the molecular cloud or through a disk, it must lose the angular momentum brought in by the accreting material. Paradoxically, young stars that actively accrete large amounts of material rotate more slowly than their counterparts that do not accrete (Edwards et al. 1993; Bouvier et al. 1993). Hence, accretion disks are able to discard any excess angular momentum very efficiently. As outflow rates are tied to accretion rates, it seems likely that the outflows play a role in removing angular momentum from protostars, without which stars might not form at all except under special initial conditions of almost no angular momentum in the parent molecular cloud core.

HH flows also provide intriguing clues to other unsolved issues in star formation. In several cases, jets have been found to emerge at nearly right angles from a pair of young stars that have projected spatial separations of $<$1500 AU (Gredel and Reipurth 1993; Reipurth et al. 1993). The presence of a jet around both components of a binary suggests that each star has its own accretion disk, and the lack of alignment means that the rotation axes of the two disks are probably not parallel (jets have been observed to emerge perpendicular to the disk plane; Burrows et al. 1996). An increasing number of Herbig-Haro energy sources appear to be double or multiple systems that reside in exceptionally overdense regions, where interactions between adjacent stars must play a role. Such interactions may be fundamental in determining the symmetries (or lack thereof) of outflows.

Jets may also regulate future star formation within a molecular cloud. The energy deposited into the molecular cloud by outflows from young stars affects the dynamics of the cloud and may increase turbulence enough to inhibit gravitational collapse of additional protostars. This feedback of kinetic energy deposited into the cloud by outflows is likely to be most important where jets are common, such as near clusters of young stars.

The recent literature contains a number of articles and even entire volumes devoted to reviews of Herbig-Haro objects (e.g., the Chamonix conference proceedings; Reipurth and Bertout 1997), accretion and outflow phenomena (Livio 1997), and shock waves (Draine and McKee 1993). In this chapter we will focus on the most recent developments that are changing the way we look at how young stellar outflows interact with their environment. The closely related topic of how accretion disks collimate infalling material into a supersonic jet is also of great interest to studies of star formation, and is reviewed in the chapters by Eislöffel et al. and by Shu et al., this volume.

Several major developments in the study of outflows in general, and of Herbig-Haro objects in particular, have taken place within the last few years:

1. The high angular resolution of the *Hubble Space Telescope* (HST) resolves the transverse extents of stellar jets and the structure of many shocks (Burrows et al. 1996; Heathcote et al. 1996; Ray et al. 1996; Reipurth et al. 1997*b*; Hester et al. 1998). A key to understanding these images has been to realize that collisionally excited Hα from the immediate postshock region marks the location of shocks in the flow, whereas forbidden-line emission follows behind the shock as the gas cools and recombines. Being able to observe where the shock fronts occur within a jet has eliminated much of the speculation that often accompanies interpretation of images. HST has allowed us to measure proper motions on timescales short compared to the cooling time in the postshock layer. Therefore, for the first time, we can reliably distinguish between true proper motion and photometric variations in the intensity of the emission from postshock gas.

2. Interactions of the gas within jets and with the surrounding medium are controlled by the flow dynamics, which are revealed to us via proper motions, radial velocities, and emission line ratios at each point in the flow. Measuring these quantities across an entire outflow and presenting the information in a comprehensible manner is a major challenge. One of the most useful ways to display kinematic data for a spatially extended object such as a jet or HH object is to construct an image of the region at each radial velocity. Within the last few years it has become possible to make velocity images of jets using either a Fabry-Perot spectrometer (Morse et al. 1994) or an image slicer (Lavalley et al. 1997). These data sets provide a powerful means to test any numerical model of collimated outflow quantitatively.

3. Because star formation is highly clustered, HH flows also tend to be clustered. For example, the NGC 1333 cloud core shows a "burst" of HH flows emerging from numerous embedded young stars (Bally et al. 1996*b*). When all of the substructure is taken into account, this single

2-pc diameter region contains several hundred individual shocks, produced by several dozen active outflows from about 100 low-mass young stars that have recently formed in this cloud core. A significant fraction of the surface area of the cloud is covered by visible shocks, demonstrating that such shocks, produced by outflows from low-mass young stars, must have a profound impact on the surrounding environment as they burrow cavities through the cloud core, entrain cloud material, and ultimately deposit their energy into the intercloud medium.

4. Within the past few years, dozens of *parsec-scale* HH flows driven by low-mass pre-main-sequence stars have been found in nearby star-forming regions (Bally and Devine 1994; Bally et al. 1996*a*; Reipurth et al. 1997*a*; Eislöffel and Mundt 1997). These flows are up to 10 times longer than previously recognized HH flows and sometimes extend over a degree on the plane of the sky. Giant flows frequently show S-shaped point symmetry, evidence that the orientation of the jets changes on timescales shorter than the time required to accrete the mass of the star (Gomez et al. 1997).

5. Finally, any high-velocity gas within an H II region will become visible as it is ionized by the ambient ultraviolet radiation. Within large H II regions such as the Orion Nebula, it is difficult to separate high-velocity material from the bright stationary nebular gas, but this separation can be accomplished with Fabry-Perot spectroscopy (O'Dell et al. 1997*a*). Jets may also become ionized locally in the immediate vicinity of the young O and B stars that drive them. These jets are frequently one-sided and are powered by sources that suffer relatively low obscuration. Such irradiated jets provide a new way to investigate the wind formation and jet collimation mechanisms, because with these systems we observe the entire outflow, not just the portion that radiates behind a shock (Reipurth et al. 1998).

In this chapter we shall consider each of the new developments listed above, particularly how they relate to jet propagation and momentum transfer. Emission lines from shocks provide the principal means by which we can study stellar jets; we consider the physics of these radiative shocks and measures of mass loss rates in section II. The momentum transfer within jets, as defined by morphologies, radial velocities, and proper motions of the shocks, is covered in section III, with a brief discussion of molecular emission in section IV. Large-scale HH flows are discussed in section V, multiple jets in section VI, and externally ionized jets in section VII.

II. EMISSION-LINE DIAGNOSTICS OF HH SHOCKS

Because it is possible to measure radial velocities and emission-line ratios at any position within an HH flow that radiates, we can determine the physical conditions throughout these flows directly. Such observations give powerful constraints on models of jets, but because HH objects often emit dozens of lines at optical wavelengths, each with a resolved velocity profile that varies spatially across the object, it is easy to become overwhelmed by the sheer volume of information. In this section we will consider how to interpret spectra of HH objects and explore what these data tell us about conditions in the flows.

A. Structure of Radiative Shocks

The high radial velocities and similarity of HH spectra to those of supernova remnants prompted Schwartz (1975) to propose that HH objects were radiative shocks. Astronomers call a shock "radiative" if it cools by emitting radiation on a timescale short compared with cooling by adiabatic expansion. Examples of *non*-radiative shocks include some supernova remnants, where the gas behind the shock is too hot and rarefied to radiate emission lines efficiently, so the shock is detectable only through radiation of X-ray and radio continuum or lines of H excited by collisional excitation immediately at the shock (Chevalier et al. 1980).

The basic structure of a radiative shock is shown in Fig. 1. Within the frame of reference of the shock, considered here as an infinitesimally thin interface, preshock material enters from the left with a density n_0 at a velocity V, known as the shock velocity. Because $V > c_s$, the sound speed in the preshock gas, the postshock material to the right of the shock is unable to communicate to the preshock gas via sound waves that the preshock gas is about to encounter a denser, slower portion of the flow. As a result, the preshock gas suddenly undergoes a jump in density and temperature, known as a J-shock. By balancing momentum, energy, and mass across the shock one can derive the Rankine-Hugoniot jump conditions (Zel'dovich and Razier 1966), which show that strong shocks (those

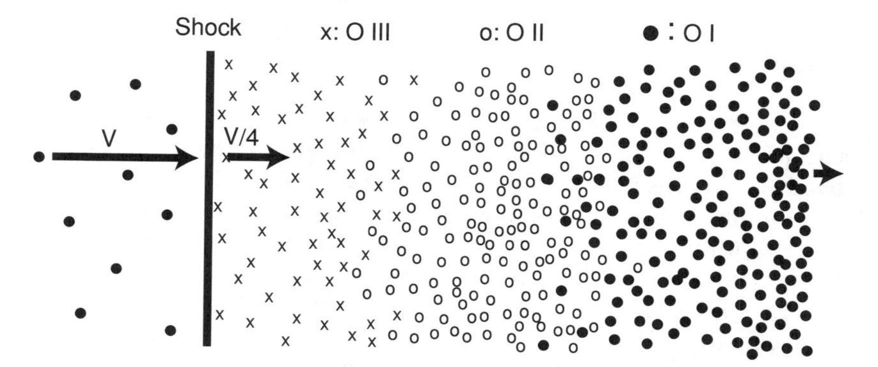

Figure 1. The structure of a radiative shock. Material enters the shock front at left with a velocity V and is slowed by a factor of 4 after crossing the shock. The density jumps by a factor of 4 at the shock and increases gradually as the gas cools and slows. As an example, cooling zones of oxygen are depicted by the crosses, circles, and dots. This diagram is drawn in the frame of reference of the shock; for HH objects viewed from the star, all components of this diagram would appear to move at a velocity $10V$–$20V$.

with high Mach numbers $M = V/c_s$) in atomic gas undergo an increase in density by a factor of 4 across the shock. Because the mass flux ρV is conserved across the shock, the postshock velocity drops to $V/4$ with respect to the shock front. Most of the kinetic energy of the preshock gas for high-Mach-number shocks goes into heat; the postshock temperature is proportional to V^2. Incident ions carry more kinetic energy than the electrons do, so at the shock the ion temperature initially exceeds the electron temperature, but the two temperatures later equilibrate through Coulomb collisions.

Several processes can modify the simple picture just described. Though the supersonic speed of the preshock gas prevents it from encountering any sound waves from the postshock gas, the postshock gas *can* affect the preshock gas through various processes, called precursors. One of the most important of these is a radiative precursor, where ultraviolet emission from the hot postshock gas emerges from the shock and ionizes the preshock gas (Sutherland et al. 1993; Dopita 1995; Morse et al. 1996). When ionized preshock gas enters the shock, energy that would have gone into ionizing the gas now goes into increasing the postshock temperature. Hence, emission from a preionized shock resembles that from a higher-velocity neutral shock (Cox and Raymond 1985). In the absence of an external illuminating source, the transition from neutral to preionized shocks occurs around 100 km s^{-1} (Shull and McKee 1979; Hartigan et al. 1987).

Magnetic precursors may be particularly important in molecular shocks. In this case, magnetic waves propagate ahead of the shock and excite any ions in the preshock gas, which then collide with and gradually heat the neutral molecular component before it encounters the shock. When the field is very strong, it may eliminate the J-shock altogether and leave a smooth increase of temperature and density known as a C-shock (Draine 1980; Draine and McKee 1993; Hollenbach 1997). Magnetic fields also influence the physics at the shock by redirecting the highest-energy ions back into the preshock gas. This process is thought to generate cosmic rays, and it may produce a precursor that heats the preshock gas, but this effect has not been explored in any detail for HH objects. Finally, some of the energy that would go into heating the gas at the shock compresses the magnetic field instead, so magnetic fields reduce the postshock temperature to levels expected from a lower-velocity nonmagnetic shock.

Thus far we have discussed only the physics at the shock front. Immediately behind the shock any neutral H suddenly encounters a very hot environment. For a wide range of energies there is a 10–20% chance that an H atom will become collisionally excited before it is ionized (Chevalier and Raymond 1978; Chevalier et al. 1980; Hartigan 1997). *This process is extremely important for HH shocks, because it implies that strong, sharp Hα features in emission line images denote the positions of shock fronts in the outflows.*

As the gas radiates emission lines, it gradually cools and recombines. Species such as O III are followed by O II and finally O I in well-defined cooling zones behind the shock (Raymond 1979; Fig. 1). These cooling zones are nowhere near ionization equilibrium, so, to predict emission line fluxes and ionization fractions accurately, one must construct a numerical code that follows the time dependence of the cooling in each line of interest. Forbidden lines such as [O I] λ 6300, [N II] λ 6583, and [S II] λ 6731 are excited by hot electrons in the postshock gas.

The gas in HH objects is largely transparent to optical photons, but radiative transfer can be important at ultraviolet wavelengths. When the postshock gas cools below $\sim 10^4$ K, it usually becomes opaque to Lyman continuum photons, and ultraviolet resonance lines are scattered in some cases (Hartigan et al. 1998). If enough ultraviolet light is present near the shock, this energy can propagate downstream and reionize a sizable fraction of the gas after it recombines. All numerical codes must account for these radiative processes, which affect the observed line fluxes. Fortunately, the physics of ionization and collisional excitation is understood well, and different numerical codes give about the same answers when applied to identical test problems (Ferland 1995). Abundances within HH objects appear to be close to solar, without any significant depletion of refractory elements onto grains (Hartigan et al. 1998; Beck-Winchatz et al. 1996).

B. Measures of Shock Velocities, Densities, and Magnetic Fields in HH Objects

The most important parameters that control the physics of the line emission from HH objects are the geometry and velocity of the shock, the preshock density, and possibly the magnetic field. In practice, observed fluxes, linewidths, and images are combined with emission line ratios to estimate the shock parameters. Line ratios alone indicate electron densities and temperatures within HH objects as they do in other emission nebulae, but it is not always straightforward to interpret these because the gas emits over a range of densities and temperatures behind a radiative shock.

The following describes the procedure that one typically follows to model an HH object. The best way to get an overview of the object is to investigate emission line images and attempt to identify the location and shape of the shocks present. As described above, sharp Balmer emission outlines the locations of shocks except for high shock velocities, where the preshock gas is ionized and it is easy to identify the shocks via high-excitation lines. Low-excitation lines such as those from O I and S II should follow in the cooling zone behind the shock, typically 100–1000 AU for most HH objects. If the shock is curved like a bow, the highest-excitation lines, such as [O III] λ 5007, should occur near the apex of the bow. In fact, because [O III] lines occur only when the shock velocity exceeds $\sim$90–100 km s^{-1}, the point where [O III] emission vanishes along the bow indicates that the velocity of material across the shock has decreased to $\sim$100 km s^{-1} at that point.

The above information, together with the observed shape of the bow shock, gives a measure of the shock velocity (e.g., Morse et al. 1992). Another measure of the shock velocity for a bow shock are the widths of the emission lines, which equal the shock velocity for any emission line that emits over the entire bow shock independent of the viewing angle and preshock density (Hartigan et al. 1987). Linewidths are good diagnostics of the shock velocity, provided that thermal or magnetic broadening is small compared with the shock velocity; that is the case for most emission lines, except perhaps for the collisional component of the Balmer lines, which emits at a high temperature immediately behind the shock and can broaden H emission line profiles (e.g., Morse et al. 1993).

Proper motions and radial velocities indicate the orientation angle of the flow, and this information is also contained within the emission line profiles (Heathcote and Reipurth 1992; Hartigan et al. 1987). The shock velocities implied by the emission linewidths and line ratios are significantly lower than the space velocities measured by radial velocities and proper motions. Hence, HH objects move into material that already moves away from the young star at hundreds of km s^{-1}. For this reason, shocks that are strong enough to generate [O III] are fairly rare in HH objects; they occur only in the brightest bow shocks that also have large linewidths.

One of the best diagnostics of the preshock density is the total observed flux of a bright line such as Hα. While the volume emission coefficient ϵ (erg cm^{-3} s^{-1}) typically scales as N^2, where N is the density, sizes of cooling zones behind shocks are proportional to N^{-1}, so the fluxes F of most lines tend to scale linearly with N. This dependence makes sense when we consider that a certain number of photons are emitted for each atom that crosses the shock, so doubling the preshock density tends to make most lines twice as bright. The dependence is not exact for each line, however, because the densities may exceed the critical densities for some lines and induce collisional quenching. For these lines, $\epsilon \sim N$, and $F \sim$ constant. Even though the preshock density may be in the low-density limit for a given transition, the postshock density may increase past the high-density limit as the gas cools and compresses.

Both line fluxes and line ratios need to be dereddened, and for HH objects this correction is not always easy because Hα/Hβ ratios in low-velocity shocks typically range as high as 5–6, much larger than their recombination value of about 3 (Osterbrock 1989). The intrinsic Balmer decrement can be assumed to be $\sim$3 only in high-excitation HH objects, where the Balmer emission from collisional excitation at the shock contributes negligibly to the total flux. When an H II region exists nearby, the Balmer decrement there gives a reddening. Other combinations of HH emission lines are sometimes used to estimate reddenings, but these are either inaccurate (in the case of comparing emission from different atoms) or difficult to measure (such as lines that originate from the same upper level, such as the blue [S II] doublet and the [S II] lines around 1 μm). If UV observations are available, one can use the shape of the two-photon continuum to estimate a reddening (Hartigan et al. 1998).

The magnetic field B is probably the most difficult parameter to estimate, because emission lines from a shock with a strong magnetic field resemble those from a lower-velocity nonmagnetic shock (Hartigan et al. 1994). However, if one has already measured the shock velocity and density by the methods described above, it is possible to estimate the component of the magnetic field along the shock by comparing the density in the [S II]-emitting region with that expected if no magnetic fields were present. Because B is proportional to N, the magnetic field in the cooling zone increases as the postshock gas cools and compresses, so the magnetic pressure $B^2/8\pi$ also increases. The additional magnetic pressure from even a weakly magnetized shock can reduce the density in the [S II]-emitting region by an order of magnitude compared with an identical nonmagnetic shock. For this reason any secondary shocks within the cooling zone should be strongly magnetized (Hartigan et al. 1998). When applied to bow shocks, which are typically 0.1 pc from their exciting stars, these methods have given preshock magnetic fields of 30–100 μG (Morse et al. 1992, 1993, 1994), similar to estimates of the magnetic field within dark clouds (e.g., Heiles et al. 1993).

Not all predictions of emission lines from shock models are equally reliable. For example, the Ca II infrared triplet at 8500 Å is difficult to model correctly because the fluxes in these lines depend upon the amount of resonant scattering in the H and K lines, which pump the 4p level. Similarly, the fluxes from C I are difficult to quantify because molecules may become important coolants at low temperatures.

C. Mass Loss Rates

We would like to be able to measure the mass loss rates within jets as accurately as possible because the ratio of mass accretion to mass outflow rates is one of the parameters predicted by theories of jet formation and collimation (e.g., Najita and Shu 1994). Unfortunately, while most methods applied to a group of jets can sort the mass loss rates from highest to lowest, no method can claim to be accurate to better than a factor of 3–10.

Because we can measure the cross-sectional area A of most jets, as well as the velocity v and density ρ, it is not hard to calculate $\rho v A$, which is the mass loss rate. The first difficulty with this method is that what is actually measured from the emission line ratios is the electron density N_e. To convert N_e to the total density N we must divide by the ionization fraction X of the gas where the emission line is measured. The ionization fraction in the [S II]-emitting region (where N_e is measured) is typically 3% or so in the best shock models (Hartigan et al. 1994), though some analytical approximations give values closer to 10% (Raga 1991).

The main difficulty now becomes how to interpret this mass loss rate. Within the frame of reference of the shock, the velocity is inversely proportional to the density, so it does not matter where we measure ρv. However, in a stellar jet the shock itself moves outward at a velocity close to that of the flow speed. Because the gas is compressed behind the shock, the mass loss rate appears much higher there than it does in front of the shock. This inherent clumpy nature of an outflow creates uncertainty in mass loss estimates. One procedure is to correct for the compression produced by the shock to estimate an "average" mass loss rate in the jet (Hartigan et al. 1994).

Other methods commonly used to estimate mass loss rates depend on the total luminosity in a particular emission line. One procedure assumes that the dominant coolant below about 5000 K for any atomic shock is the [O I] 63-μm line (Hollenbach 1985). Hence, one can relate the luminosity $L_{[\text{O I}]63\ \mu\text{m}}$ directly to the mass loss rate by $L_{63\ \mu\text{m}} = 3kT\dot{M}/(2\mu m_{\text{H}})$, where k is Boltzmann's constant, T is the temperature where the line emits, m_{H} is the mass of the hydrogen atom, μ is the mean molecular weight, and $\dot{M}$ is the mass loss rate. An analogous expression can be derived for other emission lines by introducing a constant f to the right side of the equation or by calculating the number of photons of a particular line that are emitted for each atom that crosses the shock (see Hartigan et al. 1995 for a discussion). The trouble with all these methods is that stellar jets typically have multiple shocks along the jet, and each of these shocks generates its own cooling zone and line emission. Hence, these methods tend to overestimate mass loss rates because each atom in the jet may pass through many shocks as it moves away from the star.

Another approach is simply to add up all the emission observed within a given aperture and convert this to a total mass, assuming standard abundance ratios. Measures of the velocity then give the mass loss rate. The drawback with this method is that not all of the gas within the flow radiates; we observe only the portion that has been heated recently by a shock. Moreover, no single emission line traces both the hottest and the coolest regions behind a radiative shock. In this regard, study of jets within H II regions may prove useful, because with sufficiently strong ambient ultraviolet radiation we can be assured of seeing all of the high-velocity gas (see section VII).

III. MOMENTUM TRANSFER WITHIN STELLAR JETS

A. Bow Shocks and Hα Arcs within Jets

The first indication that HH flows transfer momentum along their axes via bow shocks came from observations of large emission line widths of up to 200 km s^{-1} in many objects (Schwartz 1981). Bow shocks are a natural explanation for large linewidths in small objects because material is pushed away from the axis of the flow as it enters the bow. Triangular position-velocity diagrams and double-peaked emission line profiles characteristic of bow shocks were subsequently observed in a number of objects (Solf et al. 1986; Hartigan et al. 1987).

As more HH objects were imaged over larger fields of view, several clear examples of bows emerged, many lying along the axis of narrower collimated jets. It was initially thought that stellar jets may transfer momentum to their surroundings in an analogous manner to what was seen in numerical models of extragalactic jets (Norman et al. 1982; Cioffi and Blondin 1992), where large, hot, backflowing cocoons may excite internal shocks along the jet. However, no kinematic evidence exists for the backflowing cocoon in stellar jets or for an extended region of million-degree gas surrounding the jet. The most important differences between stellar and extragalactic jets seem to be that stellar jets are denser than extragalactic jets, and therefore cool radiatively and do not form hot cocoons. Also, stellar jets are much denser than their surroundings, and essentially plow through the interstellar medium as a series of bulletlike objects (e.g., Reipurth et al. 1997*b*).

When two fluids collide supersonically, a shock should propagate into each fluid. In the case of HH objects the "forward" shock, which accelerates the slower material, is the bow shock, and the "reverse" shock, which slows the faster gas closer to the star, is the Mach disk. The Mach disk is probably a time-variable structure and may be subject to a number of instabilities (Blondin et al. 1990; Stone and Norman 1993). Theoretically, both Mach disks and bow shocks should be radiative (Hartigan 1989), and both shocks have been successfully identified in a number of objects both via their distinctively different line ratios and by their differing kinematics (Morse et al. 1992; Reipurth and Heathcote 1992). In most cases the bow shock has a higher shock velocity than does the Mach disk, indicating that the jet is denser than the medium into which it propagates.

Though the emission line widths are larger and more high-excitation lines exist in spectra of bright bow shocks in HH flows than occur within jets, neither the linewidths nor the line excitations are as high as they should be if the bow shocks accelerate stationary gas. This fact, together with images of multiple bow shocks, led to the idea that HH flows consist of a series of nested bow shocks. Such systems have now been imaged with great clarity with HST (Burrows et al. 1996; Heathcote et al. 1996; Ray et al. 1996; Reipurth et al. 1997*b*; Hester et al. 1998). They also show the kinematics expected for dense flows moving into the wakes of previous mass ejections (Hartigan et al. 1990; Morse et al. 1994).

The transfer of momentum from the jet laterally is of great interest because this process may drive molecular outflows. Jets were found to have higher radial velocities along their axes than they have at their edges, which led to the idea that material is entrained in a turbulent manner along the edges of jets and along bow shocks (Solf 1987; Raga and Cabrit 1993; Raymond et al. 1994; see Hartigan et al. 1996 for a discussion). However, a series of bow shocks would also set up a velocity field with the highest velocities along the axis of the flow. To distinguish between the two models, one must look to the shock velocities, which should be highest along the apex for bow shocks but highest in a mixing layer at the edge of the jet for entrainment. When ground-based images showed that the highest Hα/[S II] line ratios occurred along the edges of the flow, it seemed as if entrainment produced much of the observable emission in HH objects, because this ratio increases with shock velocity (Hartigan et al. 1993).

However, new HST images of jets show that Hα is remarkably sharp spatially, not at all like a cooling zone behind a fast, turbulent shock. The morphologies of these Hα arcs (Fig. 2) appear exactly like the wings of bow shocks, though sometimes these arcs appear only along one side of the jet, as if the flow axis varied with time (see section V). Strong Hα occurs in the HST images because this emission is concentrated near the shock front, whereas the [S II] emission is more extended spatially within the cooling zone behind the shock. At the edges of the HH 111 jet, the Hα arcs suddenly become fainter at a distance of about 200 AU from the axis of the jet, as would occur if the preshock density of the medium that surrounds the jet is much lower than that within the jet (Fig. 2). Hence, this

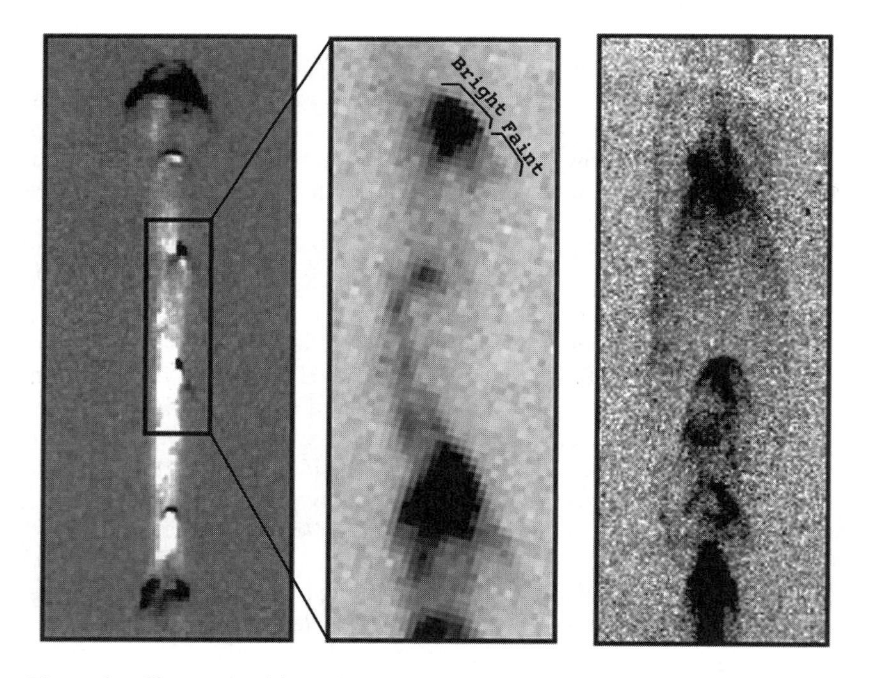

Figure 2. Hα arcs and bow shocks within the HH 111 jet (from Reipurth et al. 1997*b*). Left: The inner portion of the jet appears in this image formed from the difference of an Hα image and an [S II] image. White areas have stronger [S II] emission than Hα, and black areas are relatively stronger at Hα. Hα is especially strong in arcs that mark the location of bow shocks in the flow. Middle: Wings of bow shocks are visible to the side of the jet in this Hα image. These protruding wings are fainter than the main Hα knot in the jet and indicate that the density exterior to the main part of the jet is low (i.e., the surrounding material does not confine the flow). Right: A series of nested bow shocks makes up the middle portion of the HH 111 jet.

ambient medium plays no significant role in keeping the jet collimated, in agreement with images that show that jets are collimated within a few hundred AU from their driving stars (Burrows et al. 1996; Ray et al. 1996).

While "prompt" entrainment (where material is pushed ahead by a series of shock waves; DeYoung 1986) dominates the dynamics of most HH flows, at least one object exists in which turbulent entrainment appears to be important. The HH 110 outflow appears as a diffuse, curving outflow in emission line images (Reipurth and Olberg 1991; Raga and Canto 1995; Noriega-Crespo et al. 1996). Attempts to identify the exciting source for this flow failed until it was realized that the flow arises from a highly collimated jet that has undergone a complete change of direction after impacting the edge of a dense molecular cloud (Reipurth et al. 1996). The interface between the molecular cloud and the redirected jet appears quite irregular, and shocks appear both in the HH flow and within the molecular cloud. This region is probably the best example to date of a turbulent HH flow. Younger HH flows that are obscured optically and become visible through shocked H_2 emission are other places to look for turbulent entrainment (Davis and Smith 1996*b*; section IV).

B. Radial Velocity and Excitation Maps of HH Flows

Together with the proper motions discussed in the next section, radial velocities show how the gas moves within HH flows. In the case of a single bow shock, material is pushed ahead at the highest velocity at the apex of the bow. However, this position will be the highest radial velocity only if the bow moves directly toward or away from us. In the limit where the bow moves in the plane of the sky, the radial velocity at the apex will be zero, and the largest radial velocities, both positive and negative, will be displaced away from the apex along the axis of the bow. Many papers have explored how velocities should vary with position along a long slit for a bow shock (Raga and Böhm 1986; Solf et al. 1986; Hartigan et al. 1990). For most viewing angles, larger linewidths occur near the apex of a bow than appear in the wings.

As discussed above, shocks with markedly differing velocities, such as a bow shock and Mach disk, can often be identified with relative ease by comparing two emission line images such as Hα and [S II]. The spatial structure of a bow shock should also show excitation gradients, with the highest-ionization lines occurring at the apex of the bow. Such gradients are observed (e.g., HH 34, Morse et al. 1992; HH 1, Hester et al. 1998), where high-ionization lines such as [O III] λ 5007 emit only near the apexes of bow shocks.

Within the last few years, it has become possible to measure radial velocities over entire HH flows with Fabry-Perot spectrometers (e.g., Morse et al. 1992) and image slicers (Lavalley et al. 1997). A combination of radial velocities, proper motions, line profiles, and line excitations gives a large set of constraints that any theoretical model must satisfy. An example of the remarkable power of such observations is shown in Color Plate 14, where the velocity field of the L 1551 jet shows clearly that the bright knot at the end of the jet is a bow shock, with large linewidths near the apex and slower gas along the edges. In contrast, the kinematics around HH 29 in the same outflow is much more complex, showing several bright knots with large linewidths contained within a larger curved structure. In HH 29 it appears that the large, curved bow shock seen in Hα has wrapped around these slower, dense clumps in the flow seen in [S II] (see also Fridlund et al. 1998). This interpretation predicts that the knots marked in Color Plate 14 should have relatively slower proper motions, larger linewidths, and higher excitations as compared with adjacent bow shock gas, as is observed.

C. Proper Motions in Jets

Herbig-Haro objects represent one of the few classes of spatially extended astronomical objects in which both structural changes and proper motions can be measured with relative ease. The HH 29 region described above is a striking example of a shock whose morphology changes dramatically within a year or two in response to the transfer of momentum from a powerful outflow to its surroundings. Proper motions for the closest HH objects, such as HH 29, become apparent in ground-based images within a few years, but the interval is reduced to less than a year with the superb 0.05–0.1$''$ angular resolution of the HST. Though proper motions of HH objects have been measured since the early 1980s (e.g., Herbig and Jones 1981), it is only with the high angular resolution of HST that we can begin to distinguish true proper motions from photometric variations in the intensity of emitting gas within the postshock cooling zone.

The wide field of view of modern CCDs also provides the first opportunity to determine the proper motions of entire parsec-scale outflows (Devine et al. 1997; Reipurth et al. 1997*a*). When combined with three-dimensional spectroscopic mapping of the radial velocity fields of the associated shock systems of entire outflows, proper motions will enable us to diagnose the full spatial and velocity structure of outflows in five of the six phase space dimensions.

Several general conclusions can be drawn from recent results of proper motion studies:

1. Photometric variations are more apparent in Hα than they are in [S II]. This difference makes sense when we consider that most of the Hα emission from low-velocity shocks comes from collisional excitation in the immediate vicinity of the shock front. Hence, any sudden increase in the preshock density, such as would occur in a clumpy flow, would produce a corresponding increase in the Hα intensity. In contrast, warm gas behind HH shocks typically radiates over an interval from 10 to 100 years after it passes through the shock, so it is relatively unaffected by short-term variations in the preshock density.

2. Proper motions generally decline with increasing distance from the jet axis.

3. There is a correlation between the shock orientation and the observed proper motion. Bow shocks that curve back toward the source tend to have the greatest proper motions. Reverse bow shocks (those that curve away from the source as would be the case for a shock around a stationary or slowly moving obstacle) usually have low proper motions but can exhibit large radial velocities and velocity dispersions (e.g., HH 29; Color Plate 14).

4. Large differential motions exist within single bow shock systems. Such motions are often associated with multiple clumps that lie within larger structures that appear to have fragmented. The rapid development of instabilities within shocks and their cooling layers is one way to fragment bow shocks into clumps.

IV. MOLECULAR EMISSION FROM JETS

Observations of shock-excited H_2 emission in outflows provide a crucial link between high-velocity, optically visible Herbig-Haro objects and lower-velocity molecular outflows. Near-infrared lines, such as the 2.12 μm S(1) line of H_2, probe the physical conditions in shocks with velocities ranging from about 20 km s^{-1} in purely hydrodynamic shocks to over 100 km s^{-1} in magnetized C-shocks, where strong magnetic fields cushion the shock and enable H_2 molecules to survive higher velocities that would otherwise dissociate molecules (Hollenbach 1997). These molecular shocks are typically invisible optically, either because the optical emission is intrinsically fainter than the H_2 emission in these flows or because the extinction is an order of magnitude higher at optical wavelengths than it is at 2 μm.

Examples of obscured YSOs with extensive H_2 outflows include the Class 0 sources L1448C (Bally et al. 1993; Davis et al. 1994*a*; Davis and Smith 1996*a*), IC 348-IR (McCaughrean et al. 1994), L1634-IR (Eislöffel 1997), and IRAS 05413−0104, the source of the spectacular HH 212 outflow in Orion B (Zinnecker et al. 1998). Though portions of these flows are sometimes visible as Herbig-Haro objects, the full morphology appears only in deep H_2 images. The H_2 emission in these flows traces bow shocks that move into the downstream medium and shocks that propagate back into the jet. H_2 emission is also a potent tracer of outflows from high-luminosity YSOs (10^4–10^5 $L_\odot$), which tend to be located in more distant and relatively obscured clouds. Spectacular examples include the fingers of H_2 emission associated with the massive OMC1 outflow behind the Orion Nebula (Allen and Burton 1993) and the molecular bubble and jet that emanate from massive young stars in Cepheus A (Hartigan et al. 1996).

Recent important results from studies of shock-excited H_2 emission include the following.

1. Outflows accompany even the youngest protostars. The vast majority of Class 0 and extreme Class I protostars are associated with outflows

and shock systems (Yu et al. 1997; Bontemps et al. 1996). Owing to the large extinction toward these objects, such outflows are best traced with H_2 images.

2. The more highly embedded, younger outflows generally have lower velocities and higher densities than the older Herbig-Haro flows. This behavior may arise because outflows driven by accretion should have terminal velocities comparable to the escape velocities in their acceleration regions. Younger T Tauri stars have relatively larger radii and correspondingly lower escape velocities.

3. Outflows from low-mass protostars such as the IRAS 05413−0104, the exciting source of HH 212 (Zinnecker et al. 1998), tend to be better collimated and more jetlike than outflows from high-mass YSOs such as Cepheus A (Hartigan et al. 1996), OMC 1, and DR 21 (Garden et al. 1991).

4. Shock-excited H_2 emission can trace sites of momentum transfer between stellar jets and their associated CO outflows. In these cases CO emission surrounds the stellar jet in a sheathlike morphology (Reipurth and Cernicharo 1995; Nagar et al. 1997). Molecular emission also sometimes accompanies the wings of high-excitation bow shocks (HH 32, Davis et al. 1996; HH 1, Davis et al. 1994*b*).

5. Shock-excited H_2 also occurs ahead of some optically visible low-excitation HH bow shocks (e.g., HH 7, Carr 1993). This emission has kinematics and spatial distribution characteristic of a magnetic precursor or a C-shock.

V. PARSEC-SCALE HH FLOWS

Perhaps the most unexpected result in HH research in recent years has been the realization that jets are not the tiny flows once envisaged but can extend many parsecs. The first parsec-scale Herbig-Haro flow discovered was the HH 34 system located about 1.5° south of the Orion Nebula (Bally and Devine 1994). The outflow consists of a 20′-long chain of HH objects, which corresponds to a projected length of 2.7 pc at a distance of 460 pc. Radial velocity and proper motion measurements for several giant outflow complexes such as HH 34 (Devine et al. 1997) and HH 111 (Reipurth et al. 1997*a*; Fig. 3) show that all components move away from the central source and decline systematically in radial velocity and proper motion with increasing projected distance from the source. The dynamical ages of these giant flows are $\tau_{\rm dyn} = 10^4\ d_{\rm pc}/v_{100}$ years, where d is the distance from the source in parsecs and v_{100} is the apparent velocity in units of 100 km s^{-1}. These ages range from over 10^4 years to nearly 10^5 years, comparable to the accretion time for a typical low-mass star. Reipurth et al. (1997*a*) discuss over 20 examples of outflows traced by their HH objects with lengths ranging from 1 to over 7 parsecs. Hence, the morphology and distribution of the shocks in an HH flow complex trace the mass loss history from the YSO *over a timescale comparable to its accretion time.*

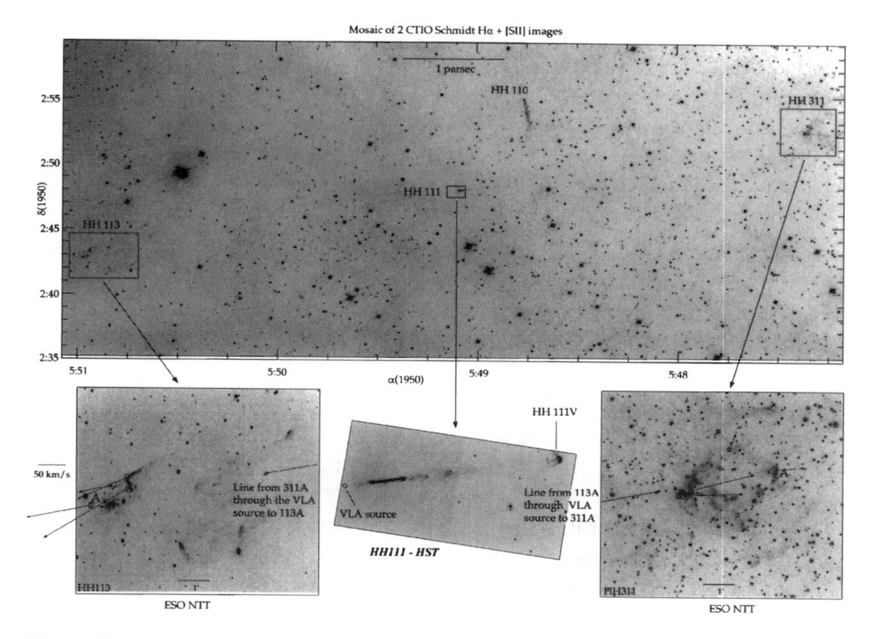

Figure 3. A composite Hα + [S II] image showing the 7-parsec-long outflow associated with the HH 111 jet. The upper (left) panel is a mosaic of two CTIO Curtis Schmidt CCD images showing the nearly 1° long Herbig-Haro flow. The middle frame (right) shows a *Hubble Space Telescope* image of the HH 111 jet. The lower left (bottom right) and lower right (top right) frames show CCD images of HH 113 and HH 311 at the far ends of the flow, obtained with the ESO NTT. Superimposed vectors show the proper motions of the brightest knots. See Reipurth et al. (1997*b*, 1998) for details.

Most parsec-scale flows consist of more or less regularly spaced HH objects, implying that major outbursts occur every several hundred to a thousand years, comparable to the expected intervals between massive accretion events known as FU Ori outbursts (Reipurth 1989; Hartmann et al. 1993). Though many show S-shaped point reflection symmetry about the central source (Gomez et al. 1997), a few show C-shaped symmetry (Bally et al. 1996*a*), similar to some extragalactic jets.

Systematic trends in the properties of HH flows depend on the projected separation of the HH objects from their driving sources. HH objects close to their sources often contain highly collimated, fast-moving, but low-excitation knots that trace the inner jet. Farther out, HH objects tend to be bow-shaped and exhibit higher-excitation line emission. The most distant shocks appear highly fragmented with amorphous morphologies and have lower velocities than HH objects closer to the source. The sizes of HH objects also tend to increase with increasing distance from the driving source, and a few HH bow shocks are nearly 1 parsec in extent (e.g., the terminal bow shock HH 401 of the HH 1/2 system; Ogura 1995).

The observed properties of parsec-scale HH flows require several kinds of mass-loss variability in the source. The periodicity, low excitation, and high proper motions of knots in the inner jets can be explained by a variable ejection velocity, whereas the S-shaped symmetry seen in the extended outflow complexes (cf. HH 34 and HH 315) requires a variable ejection angle. The mass loss rate and possibly the degree of collimation may also be time dependent.

Many parsec-scale HH flows are larger than their CO counterparts, which tend to have dimensions comparable to the host cloud cores (0.05 to 0.3 pc in CO). It is likely that close to their sources jets entrain CO-bearing gas and produce bipolar CO outflows, whereas farther from the sources jets entrain predominantly atomic gas.

In several parsec-scale HH flows, shock-excited optical emission is seen toward low-obscuration regions where galaxies and rich star fields are visible. These outflows have punched out of their parent cloud cores and are pumping energy and momentum into the interclump medium of giant molecular clouds or into the surrounding interstellar medium. It may be possible to constrain the nature of the intercloud medium of giant molecular clouds by analyzing the terminal working surfaces of giant HH flows. The low extinction toward the lobes of several of the parsec-scale outflows and the large velocities observed for some flow components imply that UV and X-ray techniques may be used to investigate some outflows from young stars. It may be possible to detect highly ionized species such as O VI, Si III, and C IV either in absorption against bright background stars or, in some cases, directly in emission.

The large number of known HH objects, their high surface area covering factor, and their large angular scales imply that outflows profoundly affect the molecular cloud environment in active star formation regions such as NGC 1333 in Perseus and in Orion (Bally et al. 1996*b*; Reipurth et al. 1997*a*). Terminal working surfaces penetrate the cloud volume, blow out of their host cores, shock, dissociate, and even ionize the material they encounter. Outflows may be important in the overall stability of clouds. Their shocks may dissociate molecules and be responsible for the large observed abundances of neutral or ionized carbon deduced from submillimeter and far infrared observations, and they may generate turbulence in molecular clouds, helping to support the clouds against gravitational collapse.

VI. MULTIPLE JETS AND MULTIPLE SOURCES

Most stars are found in binaries. Recent studies have documented that this is also true for young pre-main-sequence stars (e.g., Reipurth and Zinnecker 1993; Ghez et al. 1993). The peak of the separation distribution function for main-sequence stars is around 30 AU (Duquennoy and Mayor 1991), which for the nearest star-forming regions translates into subarcsecond separations. This is an important fact, because 30 AU is smaller than the typical disk sizes around young stars; consequently, disks are likely to play significant roles in the early evolution of young binaries. One would expect that there are two types of disks among binary systems: circumstellar disks, which surround individual binary components, and circumbinary disks, which encompass entire binaries. For certain separations it is likely that individual disks orbit individual components, while a circumbinary disk at the same time surrounds the whole binary, with a tidally induced gap between the inner and outer disks (Mathieu 1994; Artymowicz and Lubow 1994).

An increasing number of double jets are being discovered (Reipurth et al. 1993; Gredel and Reipurth 1994), documenting that very young jet sources can have two simultaneously active components. Other evidence for active young binaries come from the discovery of quadrupolar molecular outflows (Avery et al. 1990). The Near Infrared Camera and Multi-Object Spectrometer (NICMOS) on HST has recently been used to image jet sources, which are among the youngest stars known. The new images resolve several binaries, in some cases with binary jets emerging at large projected angles to one another (Reipurth et al., in preparation). If we assume that a jet is launched perpendicular to its accretion disk (a likely assumption, but proven in only a few cases thus far; e.g., Ohashi and Hayashi 1996), then these observations imply that the circumstellar accretion disks may not be coplanar. The episodic behavior of binary jets may provide clues about mutual disk interactions and the fueling of circumstellar disks from circumbinary disks.

Figure 4 shows a recent HST NICMOS observation at 1.6 μm of the HH 1 jet, which demonstrates that another flow emanates from a cavity near the embedded HH 1 source. A third flow source (located just outside the figure) is deeply embedded and is seen only at centimeter wavelengths; the three sources together are likely to form a nonhierarchical triple system, in which strong tidal forces must act and vary on orbital timescales. Such timescales are mostly short compared to the duration of outflow activity; therefore, jet axes should be affected by precession. Jets may therefore carry information on the orbital history of very young binaries. A recently discovered case in which binary motion may influence the ejection direction of one of the lobes of a binary jet is the outflow from the T Tauri star Haro 6-10 (Devine et al. 1999).

VII. JETS WITHIN H II REGIONS

One of the difficulties in studying stellar jets and HH objects is that we see only the portion of the flow that cools after passing through shocks. However, if the jet lies within an H II region, then all of the gas in the flow can become visible because, depending on its density, the jet may be fully photoionized. The advantage of being able to observe the entire outflow

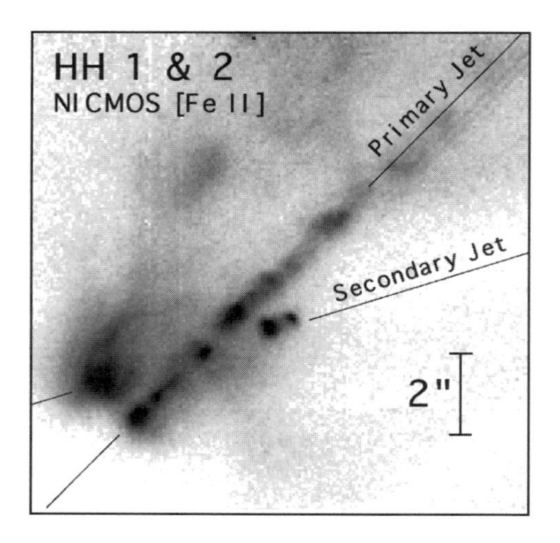

Figure 4. An infrared image obtained with the NICMOS camera aboard the *Hubble Space Telescope* at a wavelength of 1.6 μm showing the region around the energy source of the HH 1/2 jet. The primary jet is clearly visible in [Fe II] emission all the way to the location of the embedded VLA source. An arc-shaped reflection nebula at lower left opens up toward a secondary HH flow. Hence, a companion to the VLA source must be located near the apex of the nebula. The two flow axes are indicated. The scale bar of 2″ corresponds to ~920 AU at the distance to these jets. Adapted from Reipurth et al. (in preparation).

is reduced somewhat by the difficulty in distinguishing the jet from the ambient gas, though flows can be identified in this manner through Fabry-Perot spectroscopy (O'Dell et al. 1997*a*). Emission line ratios are more complex in these regions because the excitation is a mix of photoionization and shock heating, and the preshock ionization is affected by the ultraviolet light that ionizes the H II region. As for normal bow shocks (section III), a bow shock within an H II region should have its highest excitation lines at the apex. This was indeed found to be the case for most of the HH objects studied with HST within the Orion Nebula (O'Dell et al. 1997*b*). When the bow shock is not illuminated uniformly by ambient ultraviolet light, as occurs within HH 203/204 in Orion, the line excitation maps show a more complex morphology (Color Plate 15).

We are currently identifying a group of young stellar objects for which it will be possible to study the kinematics of jets close to the star, where the jets become collimated (Reipurth et al. 1998). These jets lie within H II regions, so much of the gas within them is heated by photoionization and can be distinguished from the rest of the nebular gas by its Doppler motion. Surprisingly, only one side of the jet has been found in many of these sources. Similar objects exist in the outskirts of the Orion Nebula and near the NGC 1333 reflection nebula, which is illuminated by a B star. All indications are that these irradiated jets are common, but they simply have escaped detection up to now because they are embedded in bright H II regions or reflection nebulae.

VIII. CONCLUDING REMARKS

The wealth of new observations of Herbig-Haro jets discussed in this chapter make it clear that these flows play a significant role in the star formation process, both close to the star, where the accretion disk loses angular momentum and accelerates the jet in sudden accretion events, and far from the star, as the jet transfers its energy to the surrounding molecular cloud. To a large extent, conditions within HH flows are determined by events that occurred close to the young star in the recent past; when read properly, HH flows provide a unique record of this past.

Several exciting developments should occur in the study of HH jets in the immediate future. As the number of jets observed and reobserved with HST increases, we can expect a series of spectacular movies of HH motion that can be compared directly to similar numerical simulations. Such data will undoubtedly revolutionize the field in unexpected ways. Additional spectroscopy and images over large fields of view will continue to clarify which dynamical processes are important in HH flows. Many of the basics of HH dynamics, particularly the roles of magnetic fields and molecular cooling in the flows, remain uncertain at this point. Finally, studies of binary jets and of flows within H II regions alluded to in this chapter have great promise in opening new avenues of research. We have just begun to appreciate the major effect stellar jets have on the dynamics and morphologies of molecular clouds, the intercloud medium, and future sites of star formation.

REFERENCES

Allen, D., and Burton, M. 1993. Explosive ejection of matter associated with star formation in the Orion Nebula. *Nature* 363:54–56.

Artymowicz, P., and Lubow, S. H. 1994. Dynamics of binary-disk interaction. 1: Resonances and disk gap sizes. *Astrophys. J.* 421:651–667.

Avery, L. W., Hayashi, S. S., and White, G. J. 1990. The unusual morphology of the high-velocity gas in L723—one outflow or two? *Astrophys. J.* 357:524–530.

Bachiller, R. 1996. Bipolar molecular outflows from young stars and protostars. *Ann. Rev. Astron. Astrophys.* 34:111–154.

Bally, J., and Devine, D. 1994. A parsec-scale "superjet" and quasi-periodic structure in the HH 34 outflow? *Astrophys. J. Lett.* 428:L65–L68.

Bally, J., Lada, E., and Lane, A. 1993. The L1448 molecular jet. *Astrophys. J.* 418:322–327.

Bally, J., Devine, D., and Alten, V. 1996*a*. A parsec-scale Herbig-Haro jet in Barnard 5. *Astrophys. J.* 473:921–928.

Bally, J., Devine, D., and Reipurth, B. 1996*b*. A burst of Herbig-Haro flows in NGC 1333. *Astrophys. J. Lett.* 473:L49–L52.

Beck-Winchatz, B., Böhm, K. H., and Noriega-Crespo, A. 1996. The gas-phase abundances of heavy elements and the destruction of dust grains in Herbig-Haro shock waves. *Astron. J.* 111:346–354.

Blondin, J., Königl, A., and Fryxell, B. 1990. The structure and evolution of radiatively cooling jets. *Astrophys. J.* 360:370–386.

Bontemps, S., Andre, P., Terebey, S., and Castets, S. 1996. Evolution of outflow activity around low-mass embedded young stellar objects. *Astron. Astrophys.* 311:858–872.

Bouvier, J., Cabrit, S., Fernandez, M., Martin, E., and Matthews, J. 1993. COYOTES. Part one. Multisite UBVRI photometry of 24 pre-main-sequence stars of the Taurus-Auriga cloud. *Astron. Astrophys. Suppl.* 101:485–498.

Burrows, C. J., Stapelfeldt, K. R., Watson, A. M., Krist, J. E., Ballester, G. E., Clarke, J. T., Crisp, D., Gallagher, J. S., III, Griffiths, R. E., Hester, J. J., Hoessel, J. G., Holtzman, J. A., Mould, J. R., Scowen, P. A., Trauger, J. T., and Westphal, J. A. 1996. Hubble Space Telescope observations of the disk and jet of HH 30. *Astrophys. J.* 473:437–451.

Carr, J. 1993. The H_2 velocity field in Herbig-Haro 7-11. *Astrophys. J.* 406:553–562.

Cernicharo, J., and Reipurth, B. 1996. Herbig-Haro jets, CO flows, and CO bullets: The case of HH 111. *Astrophys. J. Lett.* 460:L57–L60.

Chernin, L., and Masson, C. 1995. Momentum distribution in molecular outflows. *Astrophys. J.* 455:182–189.

Chevalier, R., and Raymond, J. 1978. Optical emission from a fast shock wave—the remnants of Tycho's supernova and SN 1006. *Astrophys. J. Lett.* 225:L27–L30.

Chevalier, R., Kirchner, R., and Raymond, J. 1980. The optical emission from a fast shock wave with application to supernova remnants. *Astrophys. J.* 235:186–195.

Cioffi, D. F., and Blondin, J. 1992. The evolution of cocoons surrounding light, extragalactic jets. *Astrophys. J.* 392:458–464.

Cox, D., and Raymond, J. 1985. Preionization-dependent families of radiative shock waves. *Astrophys. J.* 298:651–659.

Davis, C., and Eislöffel, J. 1995. Near-infrared imaging in H_2 of molecular (CO) outflows from young stars. *Astron. Astrophys.* 300:851–869.

Davis, C., and Smith, M. 1996*a*. Echelle spectroscopy of shocked H_2 in the L1448 outflow. *Astron. Astrophys.* 309:929–938.

Davis, C. and Smith, M. 1996*b*. Near-IR imaging and spectroscopy of DR21: A case for supersonic turbulence. *Astron. Astrophys.* 310:961–969.

Davis, C., Dent, H., Matthews, H., Aspin, C., and Lightfoot, J. 1994*a*. Submillimetre and near-infrared observations of L 1448—a curving H_2 jet with multiple bow shocks. *Mon. Not. Roy. Astron. Soc.* 266:933–944.

Davis, C., Eislöffel, J., and Ray, T. 1994*b*. Near-infrared imaging of HH 1/2 in shocked molecular hydrogen and (Fe II). *Astrophys. J. Lett.* 426:L93–L95.

Davis, C., Eislöffel, J., and Smith, M. 1996. Near-infrared and optical observations of an obliquely viewed bow shock: AS 353A/HH 32. *Astrophys. J.* 463:246–253.

Devine, D., Bally, J., Reipurth, B., and Heathcote, S. 1997. Kinematics and evolution of the giant HH34 complex. *Astron. J.* 114:2095–2111.

Devine, D., Reipurth, B., Bally, J., and Balonek, T. 1999, in preparation.

DeYoung, D. 1986. Mass entrainment in astrophysical jets. *Astrophys. J.* 307:62–72.

Dopita, G. 1995. Photoionizing shocks. In *The Analysis of Emission Lines*, ed. R. E. Williams and M. Livio (Cambridge: Cambridge University Press), pp. 65–82.

Draine, B. 1980. Interstellar shock waves with magnetic precursors. *Astrophys. J.* 241:1021–1038.

Draine, B., and McKee, C. 1993. Theory of interstellar shocks. *Ann. Rev. Astron. Astrophys.* 31:373–432.

Duquennoy, A., and Mayor, M. 1991. Multiplicity among solar-type stars in the solar neighbourhood. II. Distribution of the orbital elements in an unbiased sample. *Astron. Astrophys.* 248:485–524.

Edwards, S., Strom, S. E., Hartigan, P., Strom, K. M., Hillenbrand, L., Herbst, W., Attridge, J., Merrill, K., Probst, R., and Gatley, I. 1993. Angular momentum regulation in low-mass young stars surrounded by accretion disks. *Astron. J.* 106:372–382.

Eislöffel, J. 1997. Molecular hydrogen emission in embedded flows. In *Herbig-Haro Flows and the Birth of Low Mass Stars*, IAU Symp. 182, ed. B. Reipurth and C. Bertout (Dordrecht: Kluwer), pp. 93–102.

Eislöffel, J., and Mundt, R. 1997. Parsec-scale jets from young stars. *Astron. J.* 114:280–287.

Ferland, G. 1995. The Lexington benchmarks for numerical simulations of nebulae. In *The Analysis of Emission Lines,* ed. R. E. Williams and M. Livio (Cambridge: Cambridge University Press), pp. 83–96.

Fridlund, C. V. M., Liseau, R., and Gullbring, E. 1998. The physical structure of the Herbig Haro object HH 29. *Astron. Astrophys.* 330:327–335.

Garden, R., Geballe, T., Gatley, I., and Nadeau, D. 1991. A spectroscopic study of the DR 21 outflow source. II. The vibrational H_2 line emission. *Astrophys. J.* 366:474–483.

Ghez, A. M., Neugebauer, G., and Matthews, K. 1993. The multiplicity of T Tauri stars in the star forming regions Taurus-Auriga and Ophiuchus-Scorpius: A 2.2 micron speckle imaging survey. *Astron. J.* 106:2005–2023.

Gomez, M., Kenyon, S., and Whitney, B. 1997. The bipolar optical outflow associated with PV Cephei. *Astron. J.* 114:265–271.

Gredel, R., and Reipurth, B. 1993. Near-infrared observations of the HH 111 region. *Astrophys. J. Lett.* 407:L29–L32.

Gredel, R., and Reipurth, B. 1994. An infrared counter-flow in the HH 111 jet complex. *Astron. Astrophys. Lett* 289:L19–L22.

Hartigan, P. 1989. The visibility of the Mach disk and the bow shock of a stellar jet. *Astrophys. J.* 339:987–999.

Hartigan, P. 1997. HST observations of the HH 47 and HH 111 stellar jets. In *Accretion Phenomena and Related Outflows,* IAU Colloq. 163, ASP Conf. Ser. 121, ed. D. T. Wickramasinghe, G. V. Bicknell, and L. Ferrario (San Francisco: Astronomical Society of the Pacific), pp. 536–545.

Hartigan, P., Raymond, J., and Hartmann, L. 1987. Radiative bow shock models of Herbig-Haro objects. *Astrophys. J.* 316:323–348.

Hartigan, P., Raymond, J., and Meaburn, J. 1990. Observations and shock models of the jet and Herbig-Haro objects HH 46/47. *Astrophys. J.* 362:624–633.

Hartigan, P., Morse, J., Heathcote, S., Cecil, G., and Raymond, J. 1993. Observations of entrainment and time variability in the HH 47 jet. *Astrophys. J. Lett.* 414:L121–L124.

Hartigan, P., Morse, J., and Raymond, J. 1994. Mass-loss rates, ionization fractions, shock velocities, and magnetic fields of stellar jets. *Astrophys. J.* 436:125–143.

Hartigan, P., Edwards, S., and Ghandour, L. 1995. Accretion and outflow from young stars. *Astrophys. J.* 452:736–768.

Hartigan, P., Carpenter, J., Dougados, C., and Skrutskie, M. 1996. Jet bow shocks and clumpy shells of H_2 emission in the young stellar outflow Cepheus A. *Astron. J.* 111:1278–1285.

Hartigan, P., Morse, J., Tumlinson, J., Raymond, J., and Heathcote, S. 1998. FOS optical and ultraviolet spectroscopy of the bow shock HH 47A. *Astrophys. J.*, in press.

Hartmann, L., Kenyon, S., and Hartigan, P. 1993. Young stars: Episodic phenomena, activity and variability. In *Protostars and Planets III,* ed. E. H. Levy and J. I. Lunine (Tucson: University of Arizona Press), pp. 497–518.

Heathcote, S., and Reipurth, B. 1992. Kinematics and evolution of the HH 34 complex. *Astron. J.* 104:2193–2212.

Heathcote, S., Morse, J., Hartigan, P., Reipurth, B., Schwartz, R. D., Bally, J., and Stone, J. 1996. Hubble Space Telescope observations of the HH 47 jet: Narrowband images. *Astron. J.* 112:1141–1168.

Heiles, C., Goodman, A., McKee, C., and Zweibel, E. 1993. Magnetic fields in star-forming regions: Observations. In *Protostars and Planets III,* ed. E. H. Levy and J. I. Lunine (Tucson: University of Arizona Press), pp. 279–326.

Herbig, G., and Jones, B. 1981. Large proper motions of the Herbig-Haro objects HH 1 and HH 2. *Astron. J.* 86:1232–1244.

Hester, J., Stapelfeldt, K., and Scowen, P. 1998. Hubble Space Telescope wide field planetary camera 2 observations of HH 1-2. *Astron. J.* 116:372–395.

Hollenbach, D. 1985. Mass loss rates from protostars and OI (63 micron) shock luminosities. *Icarus* 61:36–39.

Hollenbach, D. 1997. The physics of molecular shocks in YSO outflows. In *Herbig-Haro Flows and the Birth of Low Mass Stars*, IAU Symp. 182, ed. B. Reipurth and C. Bertout (Dordrecht: Kluwer), pp. 181–198.

Lavalley, C., Cabrit, S., Dougados, C., Ferruit, P., and Bacon, R. 1997. Subarcsecond morphology and kinematics of the DG Tauri jet in the [O I]λ6300 line. *Astron. Astrophys.* 327:671–680.

Livio, M. 1997. The formation of astrophysical jets. In *Accretion Phenomena and Related Outflows,* IAU Colloq. 163, ASP Conf. Ser. 121, ed. D. T. Wickramasinghe, G. V. Bicknell, and L. Ferrario (San Francisco: Astronomical Society of the Pacific), pp. 845–866.

Mathieu, R. D. 1994. Pre-main-sequence binary stars. *Ann. Rev. Astron. Astrophys.* 32:465–530.

McCaughrean, M., Rayner, J., and Zinnecker, H. 1994. Discovery of a molecular hydrogen jet near IC 348. *Astrophys. J. Lett.* 436:L189–L192.

Moriarty-Schieven, G. H., Butner, H. M., and Wannier, P. G. 1995. The L1551NE molecular outflow. *Astrophys. J. Lett.* 445:L55–L58.

Morse, J., Hartigan, P., Cecil, J., Raymond, J., and Heathcote, S. 1992. The bow shock and Mach disk of HH 34. *Astrophys. J.* 399:231–245.

Morse, J., Heathcote, S., Cecil, G., Hartigan, P., and Raymond, J. 1993. The bow shock and Mach disk of HH 111V. *Astrophys. J.* 410:764–776.

Morse, J., Hartigan, P., Heathcote, S., Raymond, J., and Cecil, G. 1994. Fabry-Perot observations and new models of the HH 47A and HH 47D bow shocks. *Astrophys. J.* 425:738–754.

Morse, J., Raymond, J., and Wilson, A. 1996. On the viability of fast shocks as an ionization mechanism in active galaxies. *Pub. Astron. Soc. Pacific* 108:426–440.

Nagar, N. M., Vogel, S., Stone, J., and Ostriker, E. 1997. Kinematics of the molecular sheath of the HH 111 optical jet. *Astrophys. J. Lett.* 482:L195–L198.

Najita, J., and Shu, F. 1994. Magnetocentrifugally driven flows from young stars and disks. 3: Numerical solution of the sub-Alfvenic region. *Astrophys. J.* 429:808–825.

Noriega-Crespo, A., Garnavich, P., Raga, A., Cantó, J., and Böhm, K.-H. 1996. HH 110 jet near-infrared imaging: The outflow mixing layer. *Astrophys. J.* 462:804–812.
Norman, M. L., Smarr, L. L., Winkler, K.-H., and Smith, M. 1982. Structure and dynamics of supersonic jets. *Astron. Astrophys.* 113:285–302.
O'Dell, C. R., Hartigan, P., Bally, J., and Morse, J. 1997*a*. High velocity features in the Orion Nebula. *Astron. J.* 114:2016–2028.
O'Dell, C. R., Hartigan, P., Lane, W. M., Wong, S. K., Burton, M. G., Raymond, J., and Axon, D. J. 1997*b*. Herbig Haro objects in the Orion Nebula. *Astron. J.* 114:730–743.
Ogura, K. 1995. Giant bow shock pairs associated with Herbig-Haro jets. *Astrophys. J. Lett.* 450:L23–L26.
Ohashi, N., and Hayashi, M. 1996. High resolution observations of disks around protostellar sources with the Nobeyama Millimeter Array. In *Disks and Outflows around Young Stars*, ed. S. Beckwith et al. (Berlin: Springer-Verlag), pp. 44–57.
Osterbrock, D. 1989. *Astrophysics of Gaseous Nebulae and Active Galactic Nuclei*, (Mill Valley, CA: University Science Books), p. 73.
Raga, A. 1991. A new analysis of the momentum and mass-loss rates of stellar jets. *Astron. J.* 101:1472–1475.
Raga, A., and Böhm, K.-H. 1986. Predicted long-slit, high-resolution emission-line profiles from interstellar bow shocks. II. Arbitrary bow shock orientation. *Astrophys. J.* 308:829–835.
Raga, A., and Cabrit, S. 1993. Molecular outflows entrained by jet bowshocks. *Astron. Astrophys.* 278:267–278.
Raga, A., and Canto, J. 1995. The initial stages of an HH jet/cloud core collision. *Rev. Mex. Astron. Astrofis.* 31:51–61.
Ray, T., Mundt, R., Dyson, J. E., Falle, S. A. E. G., and Raga, A. 1996. HST observations of jets from young stars. *Astrophys. J. Lett.* 468:L103–L106.
Raymond, J. 1979. Shock waves in the interstellar medium. *Astrophys. J. Suppl.* 39:1–27.
Raymond, J., Morse, J., Hartigan, P., Curiel, S., and Heathcote, S. 1994. Entrainment by the jet in HH 47. *Astrophys. J.* 434:232–236.
Reipurth, B. 1989. The HH 111 jet and multiple outflow episodes from young stars. *Nature* 340:42–45.
Reipurth, B., and Bertout, C., eds. 1997. *Herbig-Haro Flows and the Birth of Low Mass Stars*, IAU Symp. 182, (Dordrecht: Kluwer).
Reipurth, B., and Cernicharo, J. 1995. Herbig-Haro jets at optical, infrared and millimeter wavelengths. *Rev. Mex. Astron. Astrofis. Ser. Conf.* 1:43–58.
Reipurth, B., and Heathcote, S. 1992. Multiple bow shocks in the HH 34 system. *Astron. Astrophys.* 257:693–700.
Reipurth, B., and Olberg, M. 1991. Herbig-Haro jets and molecular outflows in L1617. *Astron. Astrophys.* 246:535–550.
Reipurth, B., and Zinnecker, H. 1993. Visual binaries among pre-main sequence stars. *Astron. Astrophys.* 278:81–108.
Reipurth, B., Heathcote, S., Roth, M., Noriega-Crespo, A., and Raga, A. C. 1993. A new Herbig-Haro flow in the HH 1-2 complex. *Astrophys. J. Lett.* 408:L49–L52.
Reipurth, B., Raga, A., and Heathcote, S. 1996. HH 110: The grazing collision of a Herbig-Haro flow with a molecular cloud core. *Astron. Astrophys.* 311:989–996.
Reipurth, B., Bally, J., and Devine, D. 1997*a*. Giant Herbig-Haro flows. *Astron. J.* 114:2708–2735.

Reipurth, B., Hartigan, P., Heathcote, S., Morse, J. A., and Bally, J. 1997*b*. Hubble Space Telescope images of the HH 111 jet. *Astron. J.* 114:757–780.

Reipurth, B., Bally, J., Fesen, R., and Devine, D. 1998. Protostellar jets irradiated by massive stars. *Nature* 396:343–345.

Schwartz, R. D. 1975. T Tauri nebulae and Herbig-Haro nebulae—evidence for excitation by a strong stellar wind. *Astrophys. J.* 195:631–642.

Schwartz, R. D. 1981. High dispersion spectra of Herbig-Haro objects—evidence for shock wave dynamics. *Astrophys. J.* 243:197–203.

Shull, J., and McKee, C. 1979. Theoretical models of interstellar shocks. I. Radiative transfer and UV precursors. *Astrophys. J.* 227:131–149.

Solf, J. 1987. The kinematic structure of the HH 24 complex derived from high-resolution spectroscopy. *Astron. Astrophys.* 184:322–328.

Solf, J., Böhm, K.-H., and Raga, A. 1986. Kinematical and hydrodynamical study of the HH 32 complex. *Astrophys. J.* 305:795–804.

Stone, J., and Norman, M. 1993. Numerical simulations of protostellar jets with nonequilibrium cooling. II. Models of pulsed jets. *Astrophys. J.* 413:210–220.

Sutherland, R., Bicknell, G., and Dopita, M. 1993. Shock excitation of the emission-line filaments in Centaurus A. *Astrophys. J.* 414:510–526.

Tamura, M., Ohashi, N., Hirano, N., Itoh, Y., and Moriarty-Schieven, G. H. 1996. Interferometric observations of outflows from low-mass protostars in Taurus. *Astron. J.* 112:2076–2085.

Yu, K., Bally, J., and Devine, D. 1997. Shock-excited H_2 flows in OMC-2 and OMC-3. *Astrophys. J. Lett.* 485:L45–L48.

Zel'dovich, Y. B., and Razier, Y. P. 1966. *Physics of Shock Waves and High-Temperature Hydrodynamic Phenomena*, (New York: Academic Press), p. 45.

Zinnecker, H., McCaughrean, M., and Rayner, J. 1998. A symmetrically pulsed jet of gas from an invisible protostar in Orion. *Nature* 394:862–864.

MOLECULAR OUTFLOWS FROM YOUNG STELLAR OBJECTS

JOHN S. RICHER
Cavendish Laboratory, Cambridge

DEBRA S. SHEPHERD
California Institute of Technology

SYLVIE CABRIT
DEMIRM, Observatoire de Paris

RAFAEL BACHILLER
IGN Observatorio Astronómico Nacional

and

ED CHURCHWELL
University of Wisconsin at Madison

We review some aspects of the bipolar molecular outflow phenomenon. In particular, we compare the morphological properties, energetics, and velocity structures of outflows from high- and low-mass protostars and investigate to what extent a common source model can explain outflows from sources of very different luminosities. Many flow properties, in particular the CO spatial and velocity structure, are broadly similar across the entire luminosity range, although the evidence for jet entrainment is still less clear-cut in massive flows than in low-mass systems. We use the correlation of flow momentum deposition rate with source luminosity to estimate the ratio f of mass ejection to mass accretion rate. From this analysis it appears that a common driving mechanism could operate across the entire luminosity range. However, we stress that for the high-mass YSOs, the detailed physics of this mechanism and how the ejected wind/jet entrains ambient material remain to be addressed. We also briefly consider the alternative possibility that high-mass outflows can be explained by the recently proposed circulation models, and we discuss several shortcomings of those models. Finally, we survey the current evidence on the nature of the shocks driven by YSOs during their pre-main-sequence evolution.

I. INTRODUCTION

Stars of all masses undergo energetic, generally bipolar, mass loss during their formation. Diagnostics of mass loss using optical and near-infrared line emission provide the most dramatic images of the violent birth of stars, but molecular flows, predominantly traced by their CO emission lines, provide the strongest constraints on models of cloud collapse and star formation, because molecular flows are the massive and predominantly cool reservoir where most of the flow momentum is eventually deposited and so provide a fossil record of the mass loss history of the protostar. In contrast, the optical and near-infrared emission arises from hot shocked gas, which cools in a few years and hence traces only the currently active shocks in the flow. In addition, because there is negligible extinction at millimeter wavelengths, even the youngest, most deeply embedded objects can be studied.

Since the last review of this subject in the *Protostars and Planets III* conference proceedings (Fukui et al. 1993), many breakthroughs in understanding the nature of these objects have been forthcoming, primarily as a result of improvements at millimeter- and submillimeter-wavelength observatories; we refer the reader in particular to volume 1 of the *Revista Mexicana de Astronomía y Astrofísica Conference Series* (Pismis and Torres-Peimbert 1995) and the proceedings of IAU Symposium 182 (Reipurth and Bertout 1997). More sensitive receivers and focal-plane arrays have allowed wide-field imaging of outflows, showing that they often extend to many parsecs in size, sometimes even beyond their natal cloud boundaries (e.g., Padman et al. 1997). In addition, the availability of mosaicked interferometric images of many outflows at 1–2″ resolution has led to significant breakthroughs in understanding the small-scale structure of outflows (e.g., Bachiller et al. 1995; Cernicharo and Reipurth 1996; Gueth et al. 1996) and provide strong constraints on viable outflow-acceleration models, in particular on the role of jets and the nature of the entrainment mechanism (Cabrit et al. 1997; Shu and Shang 1997). Finally, multiline studies (Bachiller and Pérez Gutiérrez 1997) have shown that shocks driven by flows chemically process the interstellar medium (ISM) at a rapid rate, modifying its composition and having profound implications for chemical and dynamical models of molecular clouds.

In this chapter we review some of the recent results on outflows from both low- and high-mass protostars and focus on a comparison of their physical properties. In section II we summarize the observed flow morphologies and velocity structures and look for evidence that a common mechanism could reproduce the observations of all such flows. In section III the possible entrainment and ejection mechanisms for outflows are addressed, and the evidence for a common mechanism is discussed based on an analysis of the flow energetics. Finally, in section IV the evidence for, and nature of, shocks in outflows is presented.

II. OUTFLOW STRUCTURE

Molecular outflows come in a wide variety of shapes and sizes. This is hardly surprising, given the likely diversity in the mass and multiplicities of the driving protostars, the outflow ages, and molecular environments.

Although at least 200 outflows have been cataloged (Wu et al. 1996), there is still no large homogeneous sample that has been mapped with good resolution and sensitivity, so much of our current understanding is heavily influenced by a few well-studied examples. This is even more true for high-mass systems, where perhaps only 10 or so outflows have been studied in detail, although rapid progress is being made using millimeter-wave interferometers to study flows in the often complex environment of high-mass star formation (Shepherd et al. 1998). Consequently, our current estimates of "typical" outflow properties may be far from accurate.

A. Low-mass Systems

It is now clear that stellar jets, with speeds of 100–300 km s^{-1} and densities of order 10^3 cm^{-3}, are responsible for accelerating much of the molecular gas in many of the youngest low-mass outflow systems (e.g., Richer et al. 1992; Padman and Richer 1994; Bachiller et al. 1995). However, there is also good evidence for momentum being deposited into the flows by wind components with wider opening angles and that this component perhaps becomes relatively more powerful as the flows age (e.g., Bence et al. 1998).

Some of the first clear evidence for a jet-dominated outflow origin was identified in the flows from L1448C (Bachiller et al. 1990; Guilloteau et al. 1992; Dutrey et al. 1997) and NGC 2024-FIR5 (Richer et al. 1989, 1992). In L1448, a spectacular SiO jet is seen emanating from the driving source, aligned with the large-scale CO flow and unresolved across its width. In the NGC 2024-FIR5 flow, the fastest gas is seen to form an elongated jetlike feature on the axis of nested shells of lower-velocity gas. The collimation ratio q, defined as the width of the flow to the distance to the driving source, is as high as 30 for the high-velocity (30 km s^{-1}) gas in this source, but only 4 or so for the lower-velocity (5 km s^{-1}) envelope. Many other flows are now known that show this structure, with high-velocity elongated components tracing jetlike activity lying inside cavities of lower-velocity gas; examples include L1157 (Gueth et al. 1996), HH 111 (Cernicharo and Reipurth 1996; Nagar et al. 1997), and HH 211 (Gueth and Guilloteau 1999). All of these flows appear to be driven by very young, low-mass objects, based on their low luminosities and nondetection at even infrared wavelengths. The CO flow from HH 211 (Fig. 1) is perhaps the most striking image to date of this phenomenon, showing an unresolved CO jet with high-velocity gas ($v > 10$ km s^{-1} with no correction for inclination) lying within an ovoid cavity of slower-moving gas (<10 km s^{-1}). HH 211 is probably also one of the youngest low-mass outflows known, having a dynamical age $\tau_d = 0.07$ pc/10 km s^{-1} $=$ 7000 years; if the source is inclined close to the plane of the sky, as is expected given the clear separation of red and blue outflow lobes and the relatively modest projected flow speeds, the true dynamical age could well be a factor of 5 or so lower. It appears that HH 211 is at the very start of its main accretion phase, and this perhaps explains

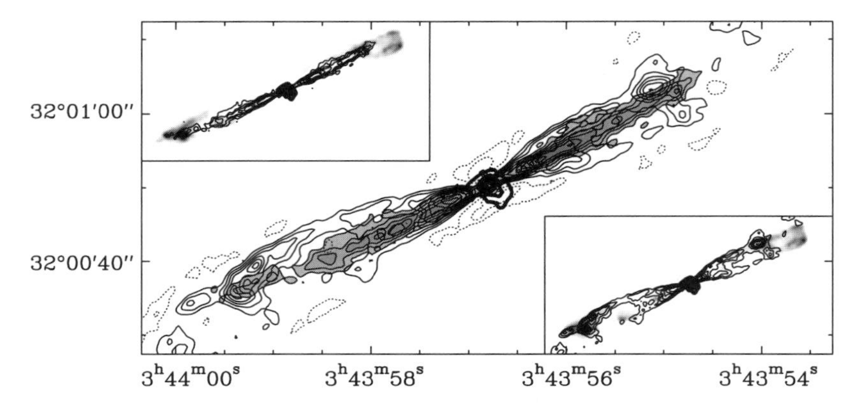

Figure 1. The HH 211 molecular jet mapped with the Plateau de Bure interferometer. The main panel shows the high-velocity CO jet in gray scale superposed on the lower-velocity outflowing gas, which forms a cavity around the jet. The thick contours show the 1.3-mm continuum emission. The panels top left and lower right show the fast and slow CO emission overlaid on the shocked H_2 line emission. Data taken from McCaughrean et al. (1994) and Gueth and Guilloteau (1999).

the relative simplicity of the outflow structure. Note that the full opening angle at the base of the flow is only 22°.

It is natural to identify these so-called "molecular jets" as the deeply embedded counterparts of the Herbig-Haro (HH) jets seen in less obscured systems and as the neutral counterparts of the ionized jets seen in the radio (see the chapters by Eislöffel et al. and by Hartigan et al., this volume). Important work by Raga (1991) and Hartigan et al. (1994) demonstrated that HH jets were denser and hence more powerful than initial estimates suggested (e.g., Mundt et al. 1987), so that the total momentum flux in HH jets integrated over the lifetime of a typical source is sufficient to drive the observed CO flows (e.g., Richer et al. 1992; Mitchell et al. 1994). This unified picture of jet-driven outflows is entirely consistent with the observed CO structures discussed above (Masson and Chernin 1993; Cabrit et al. 1997; Smith et al. 1997; Gueth and Guilloteau 1999). However, we caution that the actual composition of the driving jet, whether primarily atomic or molecular, is unknown. It is still unclear whether the CO jets seen in sources such as HH 211 and NGC 2024-FIR5 arise (1) from the body of a jet, where molecules have formed in the gas phase (Glassgold et al. 1989; Smith et al. 1997); (2) from molecules formed in the postshock region of shocks in the jet; or (3) from ambient molecular gas turbulently entrained along the jet's edge and at internal working surfaces (Raga et al. 1993; Taylor and Raga 1995).

RNO 43-FIR is another low-luminosity outflow source and is probably a flow in middle age (Bence et al. 1996; Cabrit et al. 1988): the driving source, with $L_\star = 6\ L_\odot$, is a heavily embedded Class 0 protostar invisible even at 2 μm, but the flow extends over a total size of 3 pc. The outflow axis is close to the plane of the sky, so the image of CO integrated intensity shown in Fig. 2 shows both blue- and redshifted sides of the flow. The dynamical age is $\tau_{\rm d} \sim 1.5\,{\rm pc}/10\,{\rm km\ s^{-1}} = 1.5 \times 10^5$ years, which is an upper limit, because of the small but unknown inclination angle. The CO flow is much more complex than in HH 211, and this most likely reflects the clumpy nature of the molecular gas through which the driving jet has passed; nonetheless, a very approximate S-shape

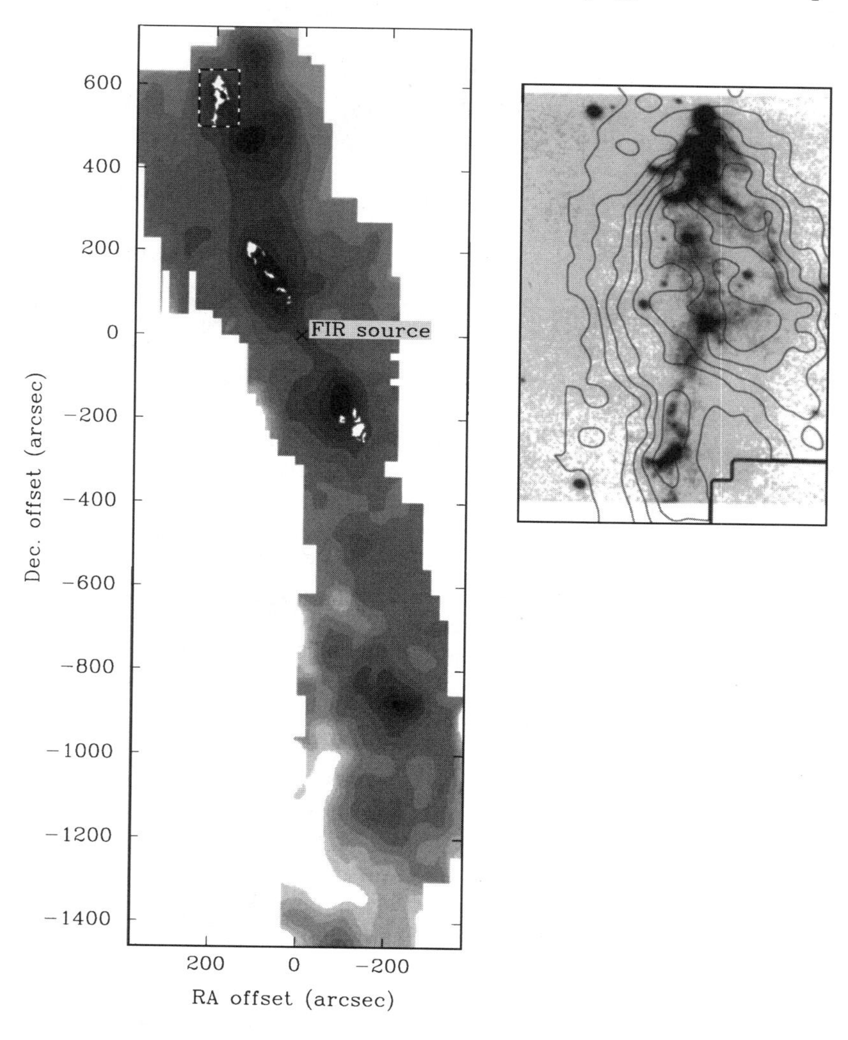

Figure 2. CO 2–1 image (in gray scale) of the RNO 43 molecular outflow (Bence et al. 1996); the driving far-IR (FIR) source is at the origin, marked with a cross. The solid white patches show the extent of the Hα emission. Note the large flow extent and correlation of the Hα emission with the CO hot spots. The panel to the right shows a detail from the northernmost Hα emission patch, showing the bow shock-shaped Hα structure in gray scale, and CO 2–1 contours overlaid.

symmetry can be seen about the driving source, both in the spatial and velocity data, suggesting that the jet direction has changed over time (Bence et al. 1996). A precessing-jet model, with a full cone angle of 27° and a precession timescale of order 3×10^4 years, provides a reasonable description of the overall flow properties (Bence et al. 1996); such precession could be driven by a binary companion in an inclined orbit.

The evidence that this flow is jet-driven, even at a distance of 1.5 pc from the driving source, is demonstrated by the Hα images of the object, which show strong emission coincident with the brightest CO points (see Fig. 2). In addition, at the northernmost CO feature in the flow, bright Hα coincident with a bow shock-shaped CO feature suggests that the whole of the RNO 43 outflow can be explained by the gradual sweeping up of a clumpy molecular cloud by a powerful stellar jet whose ejection axis varies slowly with time. It is also interesting to note that parsec-scale flows such as RNO 43 are the natural counterparts to the parsec-scale Herbig-Haro flows (Reipurth et al. 1997) seen in wide-field optical imaging: If there is molecular gas in the jet's path, then flows such as RNO 43 result, whereas jets in essentially empty space such as HH 34 (Devine et al. 1997) show only optical emission. This points strongly to the jets being primarily atomic in composition. However, in the HH 111 outflow system, CO "bullets" associated with the optical jet and having similar velocities to the optical gas have been detected far beyond the molecular cloud boundary (Cernicharo and Reipurth 1996). This suggests that, in some cases, the jets may have a molecular component.

Not all low-mass flows are jet-dominated. Many of the older flows show CO emission dominated by low-velocity cavities with little evidence for elongated high-velocity CO jet features. Examples include L43 (Bence et al. 1998), L1551 (Moriarty-Schieven and Snell 1988), and B5 (Velusamy and Langer 1998). These flows are typically 10^{5-6} years old and associated with nebulosities visible in the optical or at 2 μm, suggesting that they are Class I or II protostars, older than the Class 0 objects responsible for the HH 211, RNO 43, and NGC 2024-FIR5 outflows. The lack of high-velocity CO and obvious jetlike features in these flows (in contrast to their younger counterparts such as HH 211), as well as the presence of much wider cavities at the base of the flows, suggest that the jet power has declined over time and that a wider-opening-angle wind is now primarily responsible for driving the outflow. However, even in these older sources there is usually some evidence for weak jet activity. L1551 has an optical jet and extended Herbig-Haro emission, apparently, on one of its cavity walls (Davis et al. 1995; Fridlund and Liseau 1998), as well as fast CO emission suggestive of a jet origin (Bachiller et al. 1994), and B5 has optical jets stretching 2.2 pc away from the driving source. However, the L43 flow shows no signs of shocks, jets, or very fast CO, and this may be a true "coasting" flow, which is no longer being accelerated by a stellar wind or jet.

The recent interferometric images of the B5 outflow (Velusamy and Langer 1998) reveal a beautiful example of wide, hollow cavities at the base of the outflow, much like the reflection nebulosities seen in the near infrared in many of these systems; in order to reproduce the clearly separated red- and blueshifted emission lobes, the CO must be flowing along these cavity walls, presumably being accelerated by a poorly collimated radial wind from the star. The entire outflow driven by B5 has a dynamical age of 10^6 years and a collimation factor of about 5. The cavity opening angle at the protostar is extremely large, in the range 90–125°, and Velusamy and Langer (1998) suggest that if this angle further broadens with time, it may ultimately cut off the accretion flow. This idea is consistent with most of the available data: The young flows such as HH 211, RNO 43, and L1157 have opening angles less than 30° or so, whereas the older flows such as L1551, L43, and B5 are significantly greater than 90°. With maps at good resolution of a larger sample of outflows, it will be possible to test the hypothesis that flow opening angle is a measure of the source age.

B. High-mass Systems

Our understanding of massive flows is beginning to change, because we are starting to find a few isolated systems that can be studied in depth. However, we are still observationally biased toward older flows, which are easier to identify and study. This bias is likely to diminish in the future as more massive young flows are identified (e.g., Cesaroni et al. 1997; Molinari et al. 1998*a*; Zhang et al. 1998).

Most luminous YSOs have relatively wide-opening-angle outflows as defined by their CO morphology. Although the statistics are poor, because few massive flows have been studied with sufficient resolution to determine the morphology adequately, collimation factors q for seven well-mapped flows produced by YSOs with $L_{\rm bol} > 10^3\ L_\odot$ range from 1 to 1.8 (NGC 7538 IRS1: Kameya et al. 1989; HH 80–81: Yamashita et al. 1989; NGC 7538 IRS9: Mitchell et al. 1991; GL 490: Mitchell et al. 1995; Ori A: Chernin and Wright 1996; W75N: Davis et al. 1998; G192.16: Shepherd et al. 1998). The dynamical timescales for these outflows range from 750 years to $\sim 2 \times 10^5$ years, and there is no obvious dependence of flow collimation on age. In comparison, collimation factors in low-mass outflows range from ~ 1 to as high as 10, with a typical value being ~ 2 or 3 (Fukui et al. 1993 and references therein). It appears that more luminous YSOs do not in general produce very well-collimated CO outflows, and this result is independent of outflow age, unlike outflows produced by low-luminosity YSOs. This may be because outflows from luminous YSOs tend to break free of their molecular cloud core at a very early stage in the outflow process. Hence, massive molecular flows are frequently the truncated base of a much larger outflow that extends well beyond the cloud boundaries (e.g., HH 80–81: Yamashita et al. 1989; DR21: Russell et al. 1992; G192.16: Devine et al. 1999).

The outflow from Orion A is perhaps the best-known and best-studied high-mass outflow system. The driving source is believed to be an O star, and the estimated age of the flow is ~750 years, which makes this one of the youngest known outflows. The flow differs significantly from those produced by low-luminosity sources and represents a spectacular example of the outflow phenomenon. The CO outflow is poorly collimated at all velocities and spatial resolutions, and the H_2 emission is dispersed into a broad fan shape that is unlike any other known outflow. Its morphology is highly suggestive of an almost isotropic, explosive origin. McCaughrean and Mac Low (1997) model the H_2 bullets as a fragmented stellar wind bubble using the fragmentation model of Stone et al. (1995) and suggest that the bullets are caused by several young sources within the BN-KL cluster. Chernin and Wright (1996) argue that the flow is driven by a single massive YSO, source I. The estimated opening angle, corrected for inclination, is approximately 60° for the blue lobe and 120° for the red lobe.

The outflow from G192.16 is perhaps more typical of outflows from B stars. Figure 3 shows an interferometric CO $J = 1$–0 image of the outflow together with a closeup of the 3-mm continuum emission showing a flattened distribution of hot dust (Shepherd et al. 1998; Shepherd and Kurtz 1999). The 3-pc-long molecular flow represents the truncated base of a much larger flow, identified by Hα and [S II] emission, that extends almost 5 pc from the YSO (Devine et al. 1999). The CO flow is $\sim 10^5$ years old and appears to be driven by an early B star. Despite the very different masses and energies involved, the CO morphology looks very similar to

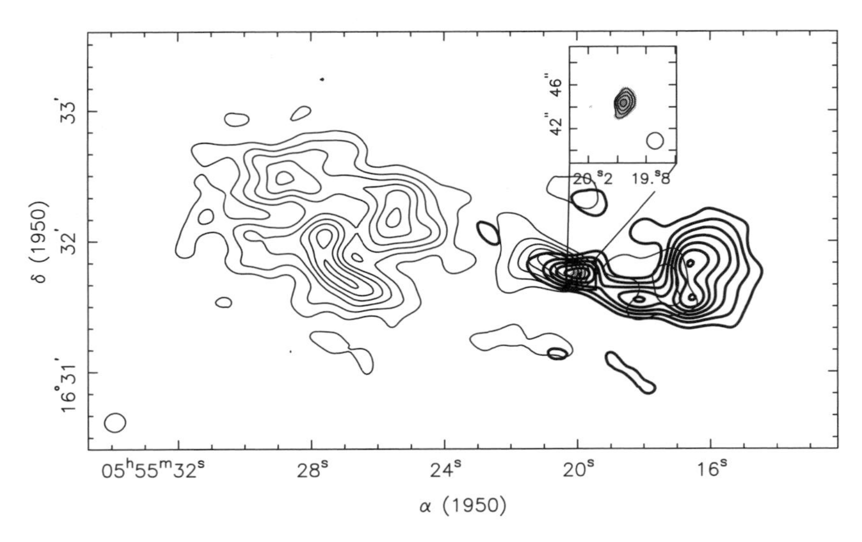

Figure 3. The G192.16 outflow mapped with the Owens Valley Radio Observatory (OVRO) interferometer. Contours of redshifted CO (thick lines) and blueshifted CO (thin lines) delineate the bipolar flow emanating from a dense core, traced by the central dust continuum emission represented in gray scale.

L1551-IRS5: the extent of the G192.16 CO outflow is 2.6×0.8 pc, approximately twice that of L1551, but the mass in the outflow (≈ 95 $M_\odot$) is much greater. The opening angle at the base of the flow is $\sim 90°$.

The present lack of well-collimated CO outflows with $q > 3$ from YSOs with $L_{bol} > 10^3$ $L_\odot$ does not mean that jets and well-collimated structures are not present in these massive sources. For example, the central source in HH 80–81 ($L_{bol} \sim 2 \times 10^4$ $L_\odot$) powers the largest known Herbig-Haro jet, with a total projected length of 5.3 pc, assuming a distance of 1.7 kpc (Martí et al. 1993, 1995 and references therein). However, the CO flow appears poorly collimated, with $q \sim 1$ when mapped at moderate resolution (Yamashita et al. 1989). Also, the biconical thermal radio jet from Cepheus A HW2 ($L_{bol} \sim 10^4$ $L_\odot$) appears to be responsible for at least part of the complicated molecular flow seen in CO and shock-enhanced species such as H_2, SiO, and SO (e.g., Doyon and Nadeau 1988; Martín-Pintado et al. 1992; Hughes 1993; Torrelles et al. 1993; Rodríguez et al. 1994; Rodríguez 1995; Garay et al. 1996; Hartigan et al. 1996; Narayanan and Walker 1996). The HH 80–81 and Cepheus A HW2 systems demonstrate that high-luminosity YSOs can produce well-collimated jets like those found in association with less luminous stars, even though the CO flow may appear chaotic or poorly collimated. Other examples of possible jets in massive outflows include IRAS 20126 and W75N IRS1. IRAS 20126 ($L_{bol} \sim 1.3 \times 10^4$ $L_\odot$) appears to drive a compact jet seen in SiO and H_2 (Cesaroni et al. 1997). W75N IRS1 ($L_{bol} \sim 1.4 \times 10^5$ $L_\odot$) shows H_2 2.12-μm shock-excited emission at the end and sides of the CO lobe, with a morphology and emission characteristics highly suggestive of a jet bow shock (Davis et al. 1998). However, this "bow shock" is 0.3 pc wide, i.e. 30 times larger than the H_2 2.12-μm bow shock at the end of the HH 211 flow (cf. Fig. 1). It is not fully clear whether such wide bows are created by protostellar jets or by a low-collimation wind component. Scaling up from current hydrodynamical jet simulations (e.g. Suttner et al. 1997), one would need a jet radius of ~ 0.03 pc at 1.3 pc from the star, hence a jet opening angle $\sim 2.6°$.

C. Outflow Velocity Structure

From the above discussion, we conclude that jet activity, bow shocks, and CO cavities are common to outflows from low- and high-mass systems. There is some evidence that high-mass systems are less well collimated than low-mass ones, but that may be due to selection effects: Young, high-mass systems with small opening angles may simply be missing from the small sample currently known (although the Orion outflow does appear to be very young). This conclusion suggests it is sensible to consider the possibility that a common driving mechanism is responsible for all outflows.

Several authors have noted that molecular flows seem to be characterized by a power law dependence of flow mass $M_{CO}(v)$ as a function of velocity. The power law exponent γ (where $M_{CO}(v) \propto v^\gamma$) is typically ~ -1.8 for most low-mass outflows, although the slope often steepens at

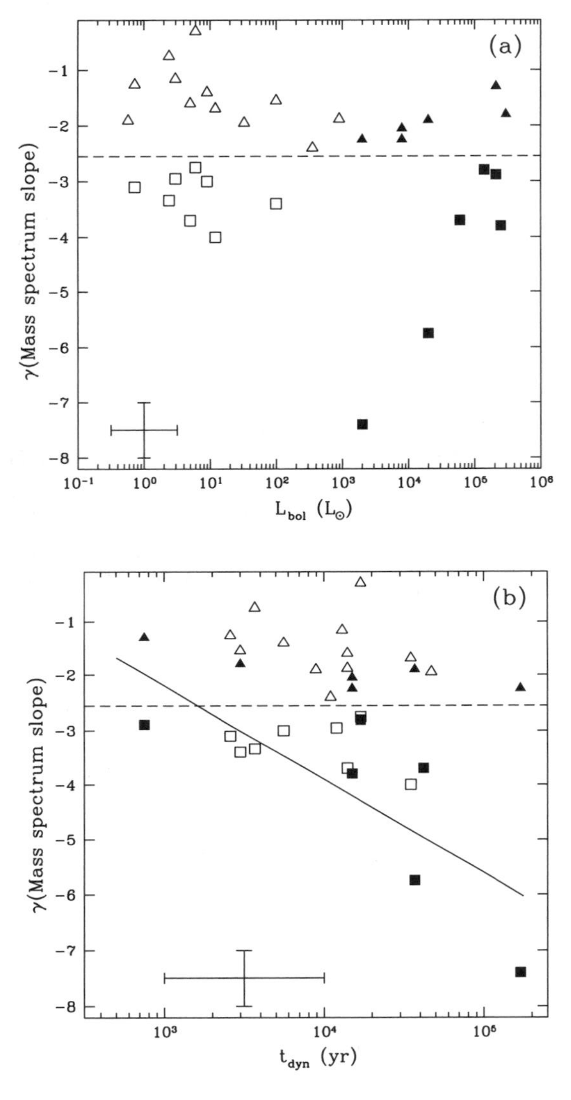

Figure 4. The slope γ of the mass spectrum $M(v)$ plotted as a function of (a) source bolometric luminosity and (b) flow dynamical age τ_d. Triangles represent γ for gas with projected speeds less than 10 km s^{-1} relative to the source, and squares are for gas moving at more than 10 km s^{-1}. The dashed, horizontal line separates the low- and high-velocity γ's for clarity. Sources with $L_{bol} < 10^3$ $L_\odot$ are plotted with open symbols, those with $L_{bol} > 10^3$ $L_\odot$ with filled symbols. Representative error bars are displayed in the lower left corner of each plot. The solid line in part (b) is a linear least-squares fit (slope -1.7 ± 0.6) to high-velocity γ's in luminous sources vs. log(τ_d). The sources plotted here are VLA 1623, IRAS 03282, L1448-C, L1551-IRS5, NGC 2071-IRS1, and Ori A IRC2 (Cabrit and Bertout 1992); TMC-1 and TMC-1A (Chandler et al. 1996); L379-IRS1-S (Kelly and MacDonald 1996); NGC 2264G (Lada and Fich 1996); G5.89-0.39 (Acord et al. 1997). Cep E (Smith et al. 1997) HH 251-254, NGC 7538-IRS9, W75N-IRS1, and NGC 7538-IRS1 (Davis et al. 1998); G192.16 (Shepherd et al. 1998); and HH 26IR, LBS 17-H, G35.2-0.74, HH 25MMS (A. Gibb, personal communication).

velocities greater than 10 km s^{-1} from v_{LSR} (e.g., Masson and Chernin 1992; Rodríguez et al. 1982; Stahler 1994; Chandler et al. 1996; Lada and Fich 1996; A. Gibb, personal communication).

Figure 4 plots (a) γ vs. L_{bol} and (b) γ vs. the dynamical timescale t_{dyn} for a new compilation of 22 sources with luminosities ranging from 0.58 $L_{\odot}$ to 3×10^5 $L_{\odot}$. Triangles represent slopes derived from gas moving less than 10 km s^{-1} relative to v_{LSR} while squares represent slopes derived from gas moving more than 10 km s^{-1} relative to v_{LSR}. Sources with $L_{bol} < 10^3$ $L_{\odot}$ are plotted with open symbols, while those with $L_{bol} > 10^3$ $L_{\odot}$ are plotted with filled symbols. Both well-collimated and poorly collimated outflows are represented in the sample.

The most striking result from Fig. 4a is that γ's for low-velocity gas are similar in sources of all luminosities. This suggests that a common gas acceleration mechanism may operate over nearly six decades in L_{bol}. In addition, there is a clear separation between γ's in high- and low-velocity gas, which supports the interpretation that there are often two distinct outflow velocity components, perhaps corresponding to a recently accelerated component and a slower, coasting component.

Hydrodynamic simulations of jet-driven outflows from low-luminosity YSOs predict such a change of slope at high velocity. They also predict that γ should steepen over time, possibly due to the collection of a reservoir of low-velocity gas (Smith et al. 1997). Figure 4b reveals marginal evidence that the mass spectrum in flows from luminous YSOs does become steeper with time, in both the low- and high-velocity ranges. The solid line in Fig. 4b is a linear least-squares fit (slope -1.7 ± 0.6) to high-velocity γ's in luminous sources versus $\log(\tau_d)$. There is no indication of time evolution of the mass spectrum slopes in outflows from low-luminosity sources.

The decrease of γ with time in more luminous sources may be due to a difference in the driving mechanism, or it may simply be more prominent because the mass outflow rate is several orders of magnitude greater than in outflows from low-luminosity YSOs (thus allowing more precise determination of γ) and because their flow ages cover a broader range, from 750 to 2×10^5 yrs.

III. TESTS OF PROTOSTELLAR WIND AND ACCRETION MODELS

A. Flow Energetics

It is well known that outflow energetics correlate reasonably well with L_{bol} over the entire observed luminosity range. In Fig. 5a, we show a recent compilation of the mean momentum deposition rate F_{CO} as a function of bolometric luminosity of the driving star. It must be remembered that F_{CO} is the time-averaged force required to drive the CO outflow: $F_{CO} = M_{CO} v_{CO}/\tau_d$, where we assume that τ_d is a good approximation

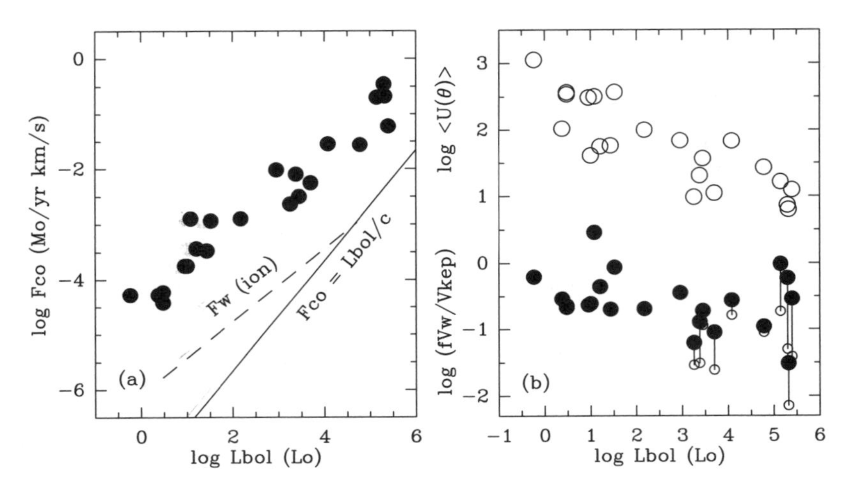

Figure 5. (a) Average momentum flux in the CO flows as a function of source luminosity. The solid line shows the force available in stellar photons (assuming single scattering), and the dashed line the force available from the ionized wind components (data from Panagia 1991). (b) The factor fv_w/v_K derived for the same set of sources is plotted as solid circles (small open circles assume that luminous sources are on the zero-age main sequence. Large open circles show the required $\langle U(\theta)\rangle$ factors for the same objects if the circulation models are applicable (see text for details).

to the flow age. If the CO-emitting material is accelerated by a separate stellar wind (or jet), the wind momentum flux F_w may be quite different from F_{CO}, depending on the nature of the wind-cloud interaction. There are clearly large uncertainties in the measured flow properties, primarily due to difficulties in estimating the inclination angle of the outflow, the optical depth and excitation of the CO, and the correctness of the assumption that the dynamical timescale is a good estimate of the flow age (Cabrit and Bertout 1992; Padman et al. 1997). However, as seen in Fig. 5a, the correlations are roughly consistent with a single power law (here of slope ~ 0.7) across the full range of source luminosities. It has been suggested that this correlation argues for a common entrainment or driving mechanism for molecular flows of all masses; but this is by no means a compelling argument, given that we would surely expect most physically reasonable outflow mechanisms to generate more powerful winds if the source mass and luminosity are increased.

Regardless of the details of how the stellar wind or jet entrains material, if the shock cooling times are short compared to flow dynamical timescales (as we expect for wind speeds less than 300 km s^{-1}; Dyson 1984, or if there is efficient mixing at the wind/molecular gas interface; Shu et al. 1991), then the wind and molecular flow momenta will be equal. This is often called the momentum-conserving limit. Then molecular outflows represent a good opportunity to test proposed protostellar ejection mechanisms. In particular, we show in this section that they can be used to estimate the ratio f of ejection to accretion rates that would be necessary if the wind is accretion-powered.

We first recall that optically thin, line-driven radiative winds (such as those present in main-sequence O and B stars) are insufficient to drive the flows if the flows are momentum-driven. The solid line in Fig. 5a shows the maximum momentum flux available in stellar photons in the single-scattering limit, L_{bol}/c. It falls short of the observed amount in molecular flows by one to three orders of magnitude. If flow lifetimes have been underestimated, the discrepancy is reduced, but not by a sufficient amount. In principle, higher momentum flux rates could be reached with multiple scattering. Wolf-Rayet stars ($L_{bol} \sim$ a few 10^5 $L_\odot$) have a wind force reaching 20–50L_{bol}/c, similar to the momentum rate in molecular flows from sources of comparable luminosity (cf. Fig. 5a). Such high values are attributed to multiple scattering of each photon by many lines closely spaced in frequency (e.g., Gayley et al. 1995 and references therein). However, excitation conditions in protostellar winds are very different from those in hot, ionized Wolf-Rayet winds. The dashed line in Fig. 5a shows the typical momentum flux in the ionized component of protostellar winds, inferred from recombination lines or radio continuum data (Panagia 1991): the values lie a factor of 10 below F_{CO}, implying that the driving winds must be 90% neutral. If dust grains instead provide the dominant opacity source in protostellar winds, comparison with winds from cool giants and supergiants might be more appropriate. Their wind velocities are $\simeq$5–30 km s^{-1}, and mass loss rates do not exceed 10^{-4} $M_\odot$ yr^{-1}, hence $F_w \leq 0.5 - 3L_{bol}/c$ for a typical $L_{bol} \sim 5 \times 10^4$ $L_\odot$. Calculations by Netzer and Elitzur (1993) show that 10 times larger mass loss rates could in principle be achieved in oxygen-rich stars, where silicates dominate the opacity curve. The wind force could then become comparable to the molecular outflow momentum rate for $L_{bol} \sim 5 \times 10^4$ $L_\odot$. Hence, it is just possible that luminous stars ($L_\star > 5 \times 10^4$ $L_\odot$) could drive their flows by radiative acceleration if dust opacity plays a significant role; further work is needed to investigate whether such models are viable. In lower-luminosity sources, however, the opacity required to lift material above escape speeds largely exceeds typical values for circumstellar dust. Therefore, molecular outflow sources with $L_\star < 5 \times 10^4$ $L_\odot$ must possess an efficient nonradiative outward momentum source.

It thus appears most likely that both low- and high-mass systems possess an efficient nonradiative wind generation mechanism in their embedded protostellar phase. If energetic bipolar winds are the chief means of angular momentum loss during the main accretion phase for stars of all mass, this is not surprising. However, the details of the wind ejection mechanism could differ; in particular, massive accreting stars are likely to have thinner convective layers and probably rotate faster, so that the magnetic field and accretion geometry close to the star may be very different from those in low-mass stars.

There exist quite a number of efficient accretion-powered wind mechanisms from the stellar surface, disk, or disk-magnetosphere boundary. The most efficient models use a strong magnetic field in the star or disk to drive the wind and to carry off angular momentum from the accreting gas (e.g., see reviews by Königl and Pudritz and by Shu et al., this volume). This wind is further collimated by magnetic or hydrodynamic processes (e.g., Mellema and Frank 1997), generating a high-Mach number wind or jet with a speed of order 200–800 km s^{-1}. A fraction $f < 1$ of the accretion flow $\dot{M}_a$ is ejected in the wind: $\dot{M}_w = f \dot{M}_a$.

A different scenario recently explored by Fiege and Henriksen (1996*a*, *b*) is that molecular flows are not predominantly swept up by an underlying wind but represent infalling gas that has been deflected into polar streams by magnetic forces. Only a small fraction of the infalling gas actually reaches the star to produce accretion luminosity. In that case, $f > 1$. This model has been invoked in particular to explain the large flow masses $M_{\rm CO} \sim 10 M_\star$ observed in flows from luminous sources. In the following we use molecular flow observations to set constraints on these two classes of proposed models.

B. The Ejection/Accretion Ratio in Protostellar Winds

Two simple, independent methods have been used in the literature to estimate f, assuming that the driving mechanism is steady over the source lifetime. First, in low-mass protostars where the luminosity is accretion-dominated, it is possible to use the observed correlation of flow force with $L_{\rm bol}$ (Fig. 5a) to derive the ejection fraction f. The source bolometric luminosity is $L_{\rm bol} = GM_\star \dot{M}_a / R_\star = \dot{M}_a v_K^2$, where v_K is the Keplerian speed at the stellar surface. The flow and wind force (assuming momentum conservation) is $F_{\rm CO} = f \dot{M}_a v_w$. Hence $F_{\rm CO}/L_{\rm bol} = f v_w / v_K^2$. A value $f \sim 0.1$ is inferred in both Class 0 and Class I low-luminosity objects (Bontemps et al. 1996). Alternatively, if one estimates $M_\star$ from $L_{\rm bol}$ via the zero-age main-sequence (ZAMS) relationship (which is probably valid only for sources with $L_{\rm bol} > 10^3$ L$_\odot$), one can use the accumulated flow momentum to infer f, using: $P_{\rm CO} = v_w M_w = f M_\star v_w$. Values of f ranging from 0.1 to 1 are inferred for a wind speed of 150 km s^{-1} (Masson and Chernin 1994; Shepherd and Churchwell 1996). However, we point out that wind speeds are unlikely to remain constant over the whole $L_{\rm bol}$ range. There is evidence for higher wind velocities in luminous flow sources: for example, proper motions up to 1400 km s^{-1} are seen in HH 80–81 (Martí et al. 1995), and Z Canis Majoris shows optical jet emission with speeds up to 650 km s^{-1} (Poetzel et al. 1989). These are significantly higher than the 100–200 km s^{-1} typically seen in low-mass sources.

To reexamine this issue in a homogeneous way over the whole luminosity range, we plot in Fig. 5b the values of $f v_w / v_K$ obtained using a new combination of the above two methods (Cabrit and Shepherd, in preparation). The solid circles are derived assuming the luminosity is accretion-dominated, whereas the open circles (for sources more luminous than 10^3 L$_\odot$) assume the ZAMS relationship given above. Typically, v_K ranges from 100 km s^{-1} in low-luminosity sources to 800 km s^{-1} in high-luminosity sources. A rather constant value $fv_w/v_K \sim 0.3$ seems to hold over the whole range of $L_{\rm bol}$. This value is in line with both popular magnetohydrodynamic (MHD) ejection models. In the x-wind model (Shu et al. 1994, and chapter in this volume), the wind is launched close to the stellar surface ($v_w \sim v_K$), and a large fraction of the accreting gas is ejected ($f \sim 0.3$). In the self-similar disk-wind models (Ferreira 1997; see the chapter by Königl and Pudritz, this volume), less material is ejected ($f \sim 0.03$), but the long magnetic lever arm accelerates it to many times the Keplerian speed ($v_w \sim 10v_K$). Thus we conclude that the energetics of the flows over the entire luminosity range are broadly consistent with a unified MHD ejection model for all flow luminosities, but we reiterate that given the very different physics involved around high- and low-mass protostars, the details of such a model for high-mass systems are still unclear.

We stress again that this plot assumes perfect momentum conservation in the wind/flow interaction. There could be strong deviations from this key assumption. First, we should keep in mind the possibility that massive flows enter the energy-driven regime: for high wind speeds, the gas cooling behind the shock will be slow, and the snowplow or momentum-conserving flow will turn into an energy-driven one. In that case, the shell momentum can exceed the momentum in the wind itself by a factor of order $v_w/v_{\rm CO}$ (Cabrit and Bertout 1992; Dyson 1984), so reducing markedly the momentum requirement of the driving wind. Of course, there are objections to energy-driven flows, in particular their inability to reproduce the bipolar velocity fields of many flows (Masson and Chernin 1992); but given the apparently poorer collimation of high-mass outflows, this issue should perhaps be reexamined. Second, if the flow is entrained in a jet bow shock, the efficiency of momentum transfer will depend on the ratio of ambient to jet density: it will be close to 1 only if the jet is less dense than the ambient medium. Highly overdense jets will pierce through the cloud without depositing much of their momentum (Chernin et al. 1994), and in that case the ratio fv_w/v_K plotted in Fig. 5b would have to be increased. We conclude that for current protostellar jet models to apply, we have the additional condition that most of the jet momentum must be in a component that is not significantly denser than the ambient molecular cloud on scales of 0.1–1 pc.

The foregoing analysis also provides important constraints on accretion rates in protostellar objects of various masses. If $fv_w/v_K \sim 0.3$, then the values of $F_{\rm CO}$ in Fig. 5a show that $\dot{M}_a = 3F_{\rm CO}/v_K$ must range from a few 10^{-6} M$_\odot$ yr^{-1} at $L_{\rm bol} \sim 1$ L$_\odot$ to a few 10^{-3} M$_\odot$ yr^{-1} at $L_{\rm bol} \sim 10^5$ L$_\odot$. Then protostellar sources would not be characterized by a single infall rate across the whole stellar mass range, and massive stars would form with much higher infall rates than low-mass stars (Cabrit and Shepherd, in preparation).

The wealth of data available on low-mass systems also allows one to break the samples down by estimated age, and so look for evolution of outflow properties with stellar age. Bontemps et al. (1996) made an important study of low-mass systems in Taurus and Ophiuchus and found evidence for a secular decline in outflow power with age, while f remained constant. Intriguingly, the best correlation of outflow power was with circumstellar mass (as measured by the millimeter continuum flux) rather than with source bolometric luminosity; Saraceno et al. (1996) also presented a similar correlation between millimeter continuum flux and outflow kinetic luminosity in systems with $L < 10^3\ L_\odot$. These results strongly suggest that the accretion rate and the outflow strength both decline in proportion to the disk and envelope mass.

C. Deflected Infall Models

Although the above analysis shows that f need not be necessarily higher in high-mass outflows, the very large masses involved, combined with the inefficiency of entrainment and momentum transfer by dense jets (especially once they escape their parent clouds), has led to some discussion of whether the sweeping up of ambient molecular gas by an accretion-driven wind is a viable mechanism for these objects (e.g., Churchwell 1997).

A recent class of outflow models, termed circulation models, can naturally generate outflow masses much greater than the stellar mass. In these, most of the infalling circumstellar material is diverted magnetically at large radii into a slow-moving outflow along the polar direction, while infall proceeds along the equatorial plane (Fiege and Henriksen 1996*a*,*b*). The main attraction of these models is that they generate large outflow masses for even small stellar masses and can generally explain the observed opening angles and velocity structures seen in high-mass systems. In particular, self-similar models predict that the velocity and density laws should take the form $V(r,\theta) = U(\theta)\sqrt{GM_\star/r}$ and $\rho(r,\theta) = \mu(\theta)M_\star r_0^{-3}(r/r_0)^{2\alpha-0.5}$, where $0.25 \geq \alpha > -0.5$, r_0 is an unspecified radial scale, and $U(\theta)$ and $\mu(\theta)$ are dimensionless functions. It is then straightforward to show that the force and mass flux in the outflow are related by $F_{CO}/\dot{M}_{CO} = \sqrt{GM_\star/R_{CO}}\langle U(\theta)\rangle$ where $\langle U(\theta)\rangle = \int U(\theta)^2\mu(\theta)d\omega / \int U(\theta)\mu(\theta)d\omega$ is the density-weighted average velocity over the outflow solid angle. The inferred $\langle U(\theta)\rangle$ (using the same $M_\star$ as for our estimates of $f\,v_w/v_K$) is plotted in the top part of Fig. 5b as open circles. It is clear that high values (between 1000 and 10) are required. Generalized circulation models that include Poynting flux driving yield values of $\langle U(\theta)\rangle$ between 5 and 200 (Lery et al., in preparation). In model cases in which radiation transport is important in setting up the flow, a power law slope of 0.8 is predicted between F_{CO} and L_{bol}, which is close to the observed slope ~ 0.7 (see also Henriksen 1994). Observations seem to indicate a systematic decline of $\langle U(\theta)\rangle$ with L_{bol}, which would also have to be explained.

There are several concerns about these circulation models. First, there are many unipolar CO outflows known, such as NGC 2024-FIR5 (Richer et al. 1992), and the almost-unipolar HH 46–47 system (Chernin and Masson 1991). These are naturally explained by swept-up wind models if the protostar is forming on the edge of a cloud or close to an H II region interface: the jet or wind propagating into the cloud will then sweep up a large CO flow, while in the opposite direction little evidence for a CO lobe will be seen. In circulation models, anisotropic solutions may also occur, but it is unclear why the weaker lobe would necessarily be on the side where the large-scale cloud density is low. Second, in some objects such as B5 (Velusamy and Langer 1998), there is an apparent lack of molecular material in the equatorial plane that can feed a circulation flow. Third, as discussed in section II, in some high-mass systems such as HH 80–81 there is direct evidence for fast jets and bow shock entrainment of molecular gas. Consequently, it seems more probable, given the evidence presented, that even high-mass outflows can be generated by the sweeping up of ambient cloud material by an accretion-driven stellar wind or jet. The details of the MHD driving mechanism in these cases, and of the momentum transfer between the wind and the jet, remain open issues.

IV. SHOCK CHEMISTRY AND ENERGETICS

The interaction between a supersonic protostellar wind and surrounding quiescent material is expected to drive strong shock fronts. Shocks can be of type C (continuous) or J (jump), depending on the shock velocity, the magnetic field, and the ionization fraction of the preshock gas (Draine and McKee 1993; Hollenbach 1997). In the last few years, spectacular gains in sensitivity in the millimeter and IR domains have allowed us for the first time to witness the chemical and thermal effects of these shocks in molecular outflows. These observations yield direct estimates of the flow age, energetics, and entrainment conditions, which represent an important new step toward a complete description of the outflow phenomenon.

A. Chemical Processing of ISM in Molecular Flows

1. Theoretical Expectations. By compressing and heating the gas, shock waves trigger new chemical processes, which lead to a specific "shock chemistry" (see the chapter by Langer et al., this volume). The most active molecular chemistry is expected to occur in C-shocks, because they increase the temperature to moderate values of about 2000 K in a thick layer, where molecules can survive and reactions that overcome energy barriers can proceed. In particular, the very reactive OH radical can be formed by $O + H_2 \rightarrow OH + H$ (which has an energy barrier of 3160 K) and will contribute to the formation of H_2O by further reaction with H_2: $OH + H_2 \rightarrow H_2O + H$ (energy barrier: 1660 K). In dissociative J-shocks, molecules are destroyed in the hot ($T \sim 10^5$ K) thin postshock layer and reform only over longer timescales, in a plateau of gas at ~400 K. Since some of these chemical processes are fast, and the cooling times are short, the chemical composition of the shocked regions is expected to be strongly time dependent.

Shocks also process dust grains. In the most violent J-shocks, destruction of grain cores and thermal sputtering inject refractory elements (such as Si and Fe) into the gas phase (e.g., Flower et al. 1996). In slower C-type shocks, nonthermal sputtering will inject refractory and volatile species mainly from the grain mantles into the gas phase (e.g., Flower and Pineau des Forêts 1994). The entrance of this fresh material, together with the high abundance of OH, will produce oxides such as SO and SiO (see Bachiller 1996; van Dishoeck and Blake 1998; and references therein). As the shocked gas cools, the dominant reactions will again be those of the usual low-temperature chemistry, and depletion onto dust grain surfaces will reduce the abundances of some of the newly formed molecules (e.g., H_2O, see Bergin et al. 1998). However, the chemical composition of both the gas and the solid phases will remain altered with respect to preshock ones.

2. *Observations.* The chemical effects of shocks have been observed in numerous outflows from low-mass Class 0 objects, which are particularly energetic and contain shocked regions well separated spatially from the quiescent protostellar envelope. Recent examples include IRAS 16293 (Blake et al. 1994; van Dishoeck et al. 1995), NGC 1333 IRAS4 (Blake et al. 1995), and NGC 1333 IRAS2 (Langer et al. 1996; Blake 1997; Bachiller et al. 1998).

A comprehensive study of many different species has recently been carried out on L1157 (Bachiller and Pérez Gutiérrez 1997). The abundances of many molecules (e.g., CH_3OH, H_2CO, HCO^+, NH_3, HCN, HNC, CN, CS, SO, and SO_2) are observed to be enhanced by factors ranging from a few to a few hundred. The extreme case is SiO, which is enhanced by a factor of $\sim 10^6$. There are significant differences in spatial distribution among the different species: some molecules, such as HCO^+ and CN, peak close to the central source, whereas SO and SO_2 have a maximum in the more distant shocks, with OCS having the most distant peak. Other molecules, such as SiO, CS, CH_3OH, and H_2CO, show an intermediate behavior. Such differences cannot be attributed solely to excitation conditions; an important gradient in chemical composition is observed along the outflow. It is very likely that this strong gradient is related to the time dependence of shock chemistry. As an example, consider the chemistry of SO, SO_2, and OCS, which has been recently modeled by Charnley (1997). It is believed that sulfur is released from grains in the form of H_2S and that it is then oxidized to SO and SO_2 in a few 10^3 yr. The formation of OCS needs a few 10^4 yr. This is in general agreement with observations, because the SO/H_2S and SO_2/H_2S ratios do increase with distance from the source (i.e., with time), and OCS emission is only observed in the most distant position (i.e., the oldest shock). Hence, chemical studies are of high potential to constrain the age and time evolution of molecular outflows.

B. Shock Cooling and Energetics

Millimeter observations of shock-enhanced molecules trace chemically processed gas that has already cooled down to 60–100 K, as indicated, e.g., by multiline NH_3 studies (Bachiller et al. 1993; Tafalla and Bachiller 1995). Emission from hotter postshock gas, on the other hand, is important to obtain information on the instantaneous energy input rate and preshock conditions in outflows.

1. Hot ($T \geq 1000$ K) Shocked Gas. Hollenbach (1985) suggested that the [O I] 63-μm line should offer a useful, extinction-insensitive measure of dissociative J-shocks in molecular flows. Because [O I] 63 μm is the main coolant below ~5000 K, its intensity is roughly proportional to the mass flux into the J-shock, $\dot{M}_{JS}$, through the relation $L_{[O\ I]}/L_{\odot} = 10^4 \times \dot{M}_{JS}/M_{\odot}yr^{-1}$, as long as the line remains optically thin (i.e., $n_0 V_{JS} < 10^7$ km s^{-1} cm^{-3}, where n_0 is the preshock density and V_{JS} is the J-shock velocity).

First detections of [O I] 63 μm in outflows were obtained with the Kuiper Airborne Observatory (KAO) toward three HH objects and five highly collimated Class 0 outflows (Cohen et al. 1988; Ceccarelli et al. 1997). Since the advent of the *Infrared Space Observatory* (ISO), the Long Wavelength Spectrometer (LWS) has revealed [O I] 63-μm emission in at least 10 more HH objects and molecular outflows (Liseau et al. 1997; Saraceno et al. 1998). These authors find a surprisingly good correlation between *current* values of $\dot{M}_{JS}$ derived from [O I] 63 μm and *time-averaged* $\dot{M}_{ave}$ values derived from millimeter observations of the outflow, assuming ram pressure equilibrium at the shock (i.e., $F_{CO} = \dot{M}_{ave} \times V_{JS}$). Both values agree for V_{JS} ~100 km s^{-1}. The dispersion in this correlation, roughly a factor of 3, is of the same order as the uncertainties in F_{CO} caused by opacity and projection effects (Cabrit and Bertout 1992). Hence, CO-derived momentum rates in outflows do not appear to suffer from large systematic errors.

With the development of large-format near-IR arrays in the early 1990s, it also became possible to map molecular outflows in the 2.12-μm $v = 1$–0 S(1) line of H_2, a tracer of hot (~2000 K) shocked molecular gas. In both low-luminosity and high-luminosity outflows, the H_2 emission delineates single or multiple bow-shaped features associated with the leading edge of the CO emission (Davis and Eislöffel 1995; Davis et al. 1998), as illustrated in Fig. 1 for HH 211; in a few cases, H_2 emission also traces collimated jets and cavity walls (e.g., Bally et al. 1993; Eislöffel et al. 1994). Observed surface brightnesses and rotational temperatures ~1500–2500 K indicate moderate-velocity shocks, either J-shocks of speed 10–25 km s^{-1} or C-shocks with v_s ~ 30 km s^{-1} and low filling factor (Smith 1994; Gredel 1994). In particular, the morphology, line profile shapes, intensity, and proper motions of H_2 2.12-μm bows are well reproduced by hydrodynamical simulations of jets propagating into the surrounding cloud, where H_2 2.12-μm emission arises mostly in the nondissociative wings of the bow shock (Raga et al. 1995; Micono et al. 1998; Suttner et al. 1997).

If these bow shocks are also where most of the slow molecular outflow is being accelerated, and if H_2 emission dominates the cooling (as expected, e.g., in 2000-K molecular gas at densities of 10^5–10^8 cm^{-3}), then $L(H_2)/L_{CO}$ should be of order unity (see, e.g., Hollenbach 1997). In the five flows studied by Davis and Eislöffel (1995), the observed ratio $L(H_2)/L_{CO}$ has a median value of ~0.4, but it covers a very broad range from 0.001 to 30. The discrepancies could be caused by uncertainties in 2-μm extinction (corrections typically amount to 10–100 for A_V = 20–50 mag); by the use of unreliable L_{CO} estimates; or, in the case of very low ratios, by a strong decrease in outflow power over time (W75N; see Davis et al. 1998). Hence the H_2 2.12-μm line alone is not a sufficient diagnostic of the outflow entrainment process.

2. *Warm ($T \sim$ 300–1000 K) molecular Gas.* The ISO mission has led to the detection of a new component of warm postshock gas at $T \sim$ 300–1000 K in several outflows. Figure 6 shows a map of the L1157 outflow in the $v = 0$–0 S(5) pure rotational line of H_2 at 6.9 μm, obtained with ISOCAM (Cabrit et al. 1998). A series of bright emission spots are seen along the outflow axis. They coincide spatially with hot shocked gas emitting in the H_2 2.12-μm line (Eislöffel and Davis 1995) and with the various peaks of shock-enhanced molecules identified by Bachiller and Pérez Gutiérrez (1997; see section IV.A). However, they trace an intermediate temperature regime of ~800 K, considerably lower than the 2000 K observed in rovibrational H_2 lines. Warm H_2 at 700–800 K was also found with the ISO Short Wavelength Spectrometer (SWS) in two outflows from very luminous sources, Cepheus A and DR 21 (Wright et al. 1996; Smith et al. 1998). Finally, warm CO at $T \sim$ 330–1600 K was detected in high-J lines (J_{up} = 14 to 28) with ISO-LWS toward five outflows of various luminosities, and H_2O and OH lines were detected in two cases (Nisini et al. 1996, 1998; Ceccarelli et al. 1997).

The observed emission fluxes and temperatures in H_2 and CO are well explained by nondissociative J-shocks with $v_s \sim 10$ km s^{-1} or slow C-shocks with $v_s \sim$ 10–25 km s^{-1}, and $n_0 \sim 10^4$–3×10^5 cm^{-3} (e.g., Wright et al. 1996; Nisini et al. 1998; Cabrit et al. 1998). One important constraint is the rather low $[H_2O]/[H_2]$ abundance ratio ~ 1–2×10^{-5} observed in HH54 and IRAS16293 (Liseau et al. 1997; Ceccarelli et al. 1997); steady-state C-shocks would predict complete conversion of O into water. The relatively high $[OH]/[H_2O]$ ratio $\sim \frac{1}{4}$–$\frac{1}{10}$ in these two flows suggests that the shock age is too short for conversion to be complete and points to the need for time-dependent C-shock models for proper interpretation of the data (e.g., Chièze et al. 1998).

Mid- and far-infrared emission from this warm molecular gas component appears more tightly correlated with L_{CO} than the 2-μm H_2 lines.

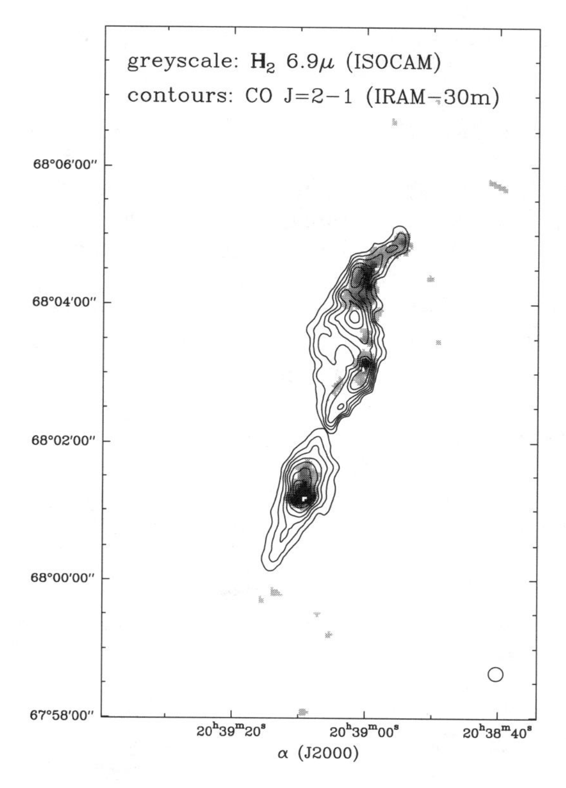

Figure 6. The L1157 outflow mapped in the 6.9-μm pure rotational line of H_2 (adapted from Cabrit et al. 1998) with CO(2–1) contours superimposed (from Bachiller and Pérez Gutiérrez 1997).

In four out of the five outflows studied by Nisini et al. (1998), the FIR CO luminosity represents 10–30% of the flow kinetic luminosity, in good agreement with C-shock calculations in the inferred density and velocity range (Kaufman and Neufeld 1996). The only large discrepancy is observed in IC 1396N, an object contaminated by photodissociation region (PDR) emission (Molinari et al. 1998*a*). In L1157, the H_2 luminosity of warm gas is also around 10% of L_{CO} (Cabrit et al. 1998). Hence, these slow shocks seem sufficient to drive the whole outflows. Detailed comparisons between H_2 and FIR-CO lines in the same objects are now under way to narrow the range of possible shock models further and perhaps allow us to discriminate between wide-angle wind and jet scenarios for the entrainment of outflows.

V. CONCLUSIONS

We have shown that the current data on the structure and energetics of molecular outflows suggest broad similarities across the entire luminosity range, from 1 to 10^5 $L_\odot$. If the flows are swept up by a stellar wind or jet, we find that $\dot{M}_w v_w / \dot{M}_a v_K$ has a value of about 0.3 for all flows, perhaps suggesting that flows have a common drive mechanism. However, it remains unclear whether the MHD disk and x-wind models that have been used to explain low-mass outflows are appropriate in the very different physical regime of high-mass YSOs. While observational data continue to improve these constraints, there remains an urgent need for a larger sample of molecular outflows, particularly from high-mass stars, to be fully mapped at high resolution; at the moment it is very possible that our estimates of the properties of high-mass systems are biased by strong selection effects. Single-dish data, especially from the new focal-plane arrays, plus interferometric images at millimeter wavelengths will continue to accumulate. However, only when the large millimeter interferometer (ALMA) is operational will it be possible to acquire high-resolution data quickly enough to study large samples of outflows in detail.

The nature of the shocks that drive outflows is slowly becoming clearer. We now have diagnostics of all the temperature components in the outflows, from the 2000-K gas seen in the 2-μm H_2 lines through the several-hundred-kelvin component recently detected by ISO, to the cool massive component seen in the millimeter waveband, where most of the momentum is eventually deposited. The relationship between these components is providing valuable tests of the outflow mechanism, although a fuller understanding will require observations at higher angular resolution than ISO provided. The SOFIA and especially the FIRST missions will provide valuable data in this area.

Acknowledgments D. S. S. would like to thank Andy Gibb and Chris Davis for providing unpublished outflow data for a number of sources and A. Gibb, A. Sargent, and L. Testi for useful discussions. J. S. R. acknowledges support from the Royal Society.

REFERENCES

Acord, J. M., Walmsley, C. M., and Churchwell, E. 1997. The extraordinary outflow toward G5.89 − 0.39. *Astrophys. J.* 475:693–704.

Bachiller, R. 1996. Bipolar molecular outflows from young stars and protostars. *Ann. Rev. Astron. Astrophys.* 34:111–154.

Bachiller, R., and Pérez Gutiérrez, M. 1997. Shock chemistry in the young bipolar outflow L1157. *Astrophys. J. Lett.* 487:93–97.

Bachiller, R., Cernicharo, J., Martín-Pintado, J., Tafalla, M., and Lazareff, B. 1990. High-velocity molecular bullets in a fast bipolar outflow near L1448/IRS3. *Astron. Astrophys.* 231:174–186.

Bachiller, R., Martín-Pintado, J., and Fuente, A. 1993. High-velocity hot ammonia in bipolar outflows. *Astrophys. J. Lett.* 417:45–48.

Bachiller, R., Tafalla, M., and Cernicharo, J. 1994. Successive ejection events in the L1551 molecular outflow. *Astrophys. J. Lett.* 425:93–96.

Bachiller, R., Guilloteau, S., Dutrey, A., Planesas, P., and Martín-Pintado, J. 1995. The jet-driven molecular outflow in L1448. CO and continuum synthesis images. *Astron. Astrophys.* 299:857–868.

Bachiller, R., Codella, C., Colomer, F., Liechti, S., and Walmsley, C. M. 1998. Methanol in protostellar outflows. Single-dish and interferometric maps of NGC 1333/IRAS 2. *Astrophys. J.* 335:266–276.

Bally, J., Devine, D., Hereld, M., and Rauscher, B. J. 1993. Molecular hydrogen in the IRAS 03282+3035 stellar jet. *Astrophys. J. Lett.* 418:75–78.

Bence, S. J., Richer, J. S., and Padman, R. 1996. RNO 43: A jet-driven super-outflow. *Mon. Not. Roy. Astron. Soc.* 279: 866–883.

Bence, S. J., Padman, R., Isaak, K. G., Wiedner, M. C., and Wright, G. S. 1998. L43: The late stages of a molecular outflow. *Mon. Not. Roy. Astron. Soc.* 299:965–976.

Bergin, E. A., Neufeld, D. A., and Melnick, G. J. 1998. The postshock chemical lifetimes of outflow tracers and a possible new mechanism to produce water ice mantles. *Astrophys. J.* 499:777–792.

Blake, G.A. 1997. High angular resolution observations of the gas phase composition of young stellar objects. In *Molecules in Astrophysics: Probes and Processes*, IAU Symp. 178 (Dordrecht: Kluwer), ed. E. van Dishoeck, pp. 31–44.

Blake, G. A., van Dishoeck, E. F., Jansen, D. J., Groesbeck, T. D., and Mundy, L. G. 1994. Molecular abundances and low-mass star formation. 1: Si- and S-bearing species toward IRAS 16293−2422. *Astrophys. J.* 428:680–692.

Blake, G. A., Sandell, G., van Dishoeck, E. F., Groesbeck, T. D., Mundy, L. G., and Aspin, C. 1995. A molecular line study of NGC 1333/IRAS 4. *Astrophys. J.* 441:689–701.

Bontemps, S., André, P., Terebey, S., and Cabrit, S. 1996. Evolution of outflow activity around low-mass young stellar objects. *Astron. Astrophys.* 311:858–872.

Cabrit, S., and Bertout, C. 1992. CO line formation in bipolar flows. 3. The energetics of molecular flows and ionized winds. *Astron. Astrophys.* 261:274–284.

Cabrit, S., Goldsmith, P. F., and Snell, R. L. 1988. Identification of RNO 43 and B335 as two highly collimated bipolar flows oriented nearly in the plane of the sky. *Astrophys. J.* 334:196–208.

Cabrit, S., Raga, A., and Gueth, F. 1997. Models of bipolar molecular outflows. In *Herbig-Haro Flows and the Birth of Low Mass Stars*, IAU Symp. 182, ed. B. Reipurth and C. Bertout (Dordrecht: Kluwer), pp. 163–180.

Cabrit, S., Couturier, P., Andre, P., Boulade, O., Cesarsky, C. J., and Lagage, P. O., Sauvage, M., Bontemps, S., Nordh, L., Olofsson, G., Boulanger, F., Sibille, F., and Siebenmorgen, R. 1998. Mid-infrared emission maps of bipolar outflows with ISOCAM: An in-depth study of the L1157 outflow. In *Star Formation with the Infrared Space Observatory,* ASP Conf. Ser. 132, ed. J. Yun and R. Liseau (San Francisco Astronomical Society of the Pacific), pp. 326–329.

Ceccarelli, C., Haas, M. R., Hollenbach, D. J., and Rudolph, A. L. 1997. OI 63 micron-determined mass-loss rates in young stellar objects. *Astrophys. J.* 476:771–780.

Cernicharo, J., and Reipurth, B. 1996. Herbig-Haro jets, CO flows, and CO bullets: The case of HH 111. *Astrophys. J. Lett.* 460:57–61.

Cesaroni, R., Felli, M., Testi, L., Walmsley, C. M., and Olmi, L. 1997. The disk-outflow system around the high-mass (proto)star IRAS 20126+4104. *Astron. Astrophys.* 325:725–744.

Chandler, C. J., Terebey, S., Barsony, M., Moore, T. J. T., and Gautier, T. N. 1996. Compact outflows associated with TMC-1 and TMC-1A. *Astrophys. J.* 471:308–320.
Charnley, S. B. 1997. Sulfuretted molecules in hot cores. *Astrophys. J.* 481:396–405.
Chernin, L. M., and Masson, C. R. 1991. A nearly unipolar CO outflow from the HH 46-47 system. *Astrophys. J. Lett.* 382:93–96.
Chernin, L. M., and Wright, M. C. H. 1996. High-resolution CO observations of the molecular outflow in the Orion IRc2 region. *Astrophys. J.* 467:676–683.
Chernin, L., Masson, C., Gouveia Dal Pino, E. M., and Benz, W. 1994. Momentum transfer by astrophysical jets. *Astrophys. J.* 426:204–214.
Chièze, J.-P., Pineau des Forêts, G., and Flower, D. R. 1998. Temporal evolution of MHD shocks in the interstellar medium. *Mon. Not. Roy. Astron. Soc.* 295:672–682.
Churchwell, E. 1997*b*. Origin of the mass in massive star outflows. *Astrophys. J. Lett.* 479:59–61.
Cohen, M., Hollenbach, D. J., Haas, M. R., and Erickson, E. F. 1988. Observations of the 63 micron forbidden O I line in Herbig-Haro objects. *Astrophys. J.* 329:863–873.
Davis, C. J., and Eislöffel, J. 1995. Near-infrared imaging in H_2 of molecular (CO) outflows from young stars. *Astron. Astrophys.* 300:851–869.
Davis, C. J., Mundt, R., Ray, T. P., and Eislöffel, J. 1995. Near-infrared and optical imaging of the L1551-IRS5 region—The importance of poorly collimated outflows from young stars. *Astron. J.* 110:766–775.
Davis, C. J., Moriarty-Schieven, G., Eislöffel, J., Hoare, M. G., and Ray, T. P. 1998. Observations of shocked H_2 and entrained CO in outflows from luminous young stars. *Astron. J.* 115:1118–1134.
Devine, D., Bally, J., Reipurth, B., and Heathcote, S. 1997. Kinematics and evolution of the giant HH34 complex. *Astron. J.* 114:2095–2111.
Devine, D., Bally, J., Reipurth, B., Shepherd, D. S., and Watson, A. M. 1999. A giant Herbig-Haro flow from a massive young star in G192.16−3.82. *Astron. J.*, in press.
Doyon, R., and Nadeau, D. 1988. The molecular hydrogen emission from the Cepheus A star-formation region. *Astrophys. J.* 334:883–890.
Draine, B. T., and McKee, C. F. 1993. Theory of interstellar shocks. *Ann. Rev. Astron. Astrophys.* 31:373–432.
Dutrey, A., Guilloteau, S., and Bachiller, R. 1997. Successive SiO shocks along the L1448 jet axis. *Astron. Astrophys.* 325:758–768.
Dyson, J. E. 1984. The interpretation of flows in molecular clouds. *Astrophys. Space Sci.* 106:181–197.
Eislöffel, J., and Davis, C. J., 1995. Near-infrared imaging in H_2 of molecular (CO) outflows from young stars. *Astrophys. Space Sci.* 233:59–62.
Eislöffel, J., Davis, C. J., Ray, T. P., and Mundt, R. 1994. Near-infrared observations of the HH 46/47 system. *Astrophys. J. Lett.* 422:91–93.
Ferreira, J. 1997. Magnetically-driven jets from Keplerian accretion discs. *Astron. Astrophys.* 319:340–359.
Fiege, J. D., and Henriksen, R. N. 1996*a*. A global model of protostellar bipolar outflow—I. *Mon. Not. Roy. Astron. Soc.* 281:1038.
Fiege, J. D., and Henriksen, R. N. 1996*b*. A global model of protostellar bipolar outflow—II. *Mon. Not. Roy. Astron. Soc.* 281:1055.
Flower, D. R., and Pineau des Forêts, G. 1994. Grain-mantle erosion in magnetohydrodynamic shocks. *Mon. Not. Roy. Astron. Soc.* 268:724.
Flower, D. R., Pineau des Forêts, G., Field, D., and May, P. W. 1996. The structure of MHD shocks in molecular outflows: Grain sputtering and SiO formation. *Mon. Not. Roy. Astron. Soc.* 280:447.

Fridlund, C. V. M., and Liseau, R. 1998. Two jets from the protostellar system L1551 IRS5. *Astrophys. J. Lett.* 599:75–77.
Fukui, Y., Iwata, T., Mizuno, A., Bally, J., and Lane, A. P. 1993. Molecular outflows. In *Protostars and Planets III*, ed. E. H. Levy and J. I. Lunine (Tucson: University of Arizona Press), pp. 603–639.
Garay, G., Ramirez, S., Rodríguez, L. F., Curiel, S., and Torrelles, J. M. 1996. The nature of the radio sources within the Cepheus A star-forming region. *Astrophys. J.* 459:193–208.
Gayley, K. G., Owocki, S. P., and Cranmer, S. R. 1995. Momentum deposition in Wolf-Rayet winds: Nonisotropic diffusion with effective gray opacity. *Astrophys. J.* 442:296–310.
Glassgold, A. E., Mamon, G. A., and Huggins, P. J. 1989. Molecule formation in fast neutral winds from protostars. *Astrophys. J. Lett.* 336:29–31.
Gredel, R. 1994. Near-infrared spectroscopy and imaging of Herbig-Haro objects. *Astron. Astrophys.* 292:580–592.
Gueth, F., and Guilloteau, S. 1999. The jet-driven molecular outflow of HH211. *Astron. Astrophys.* 343:571–584.
Gueth, F., Guilloteau, S., and Bachiller, R. 1996. A precessing jet in the L1157 molecular outflow. *Astron. Astrophys.* 307:891–897.
Guilloteau, S., Bachiller, R., Fuente, A., and Lucas, R. 1992. First observations of young bipolar outflows with the IRAM interferometer—2 arcsec resolution SiO images of the molecular jet in L1448. *Astron. Astrophys. Lett.* 265:49–52.
Hartigan, P., Morse, J., and Raymond, J. 1994. Mass loss rates, ionization fractions, shock velocities and magnetic fields of stellar jets. *Astrophys. J.* 436:125–143.
Hartigan, P., Carpenter, J. M., Dougados, C., and Skrutskie, M. F. 1996. Jet bow shocks and clumpy shells of H_2 emission in the young stellar outflow Cepheus A. *Astron. J.* 111:1278–1285.
Henriksen, R.N. 1994. Theory of bipolar outflows. In *The Cold Universe*, ed. T. Montmerle, C. J. Lada, I. F. Mirabel, and J. Tran Thanh Van (Gif-sur-Yvette: Editions Frontières), pp. 241–254.
Hollenbach, D. 1985. Mass loss rates from protostars and OI (63 micron) shock luminosities. *Icarus* 61:36–39.
Hollenbach, D. 1997. The physics of molecular shocks in YSO outflows. In *Herbig-Haro Flows and the Birth of Low-Mass Stars*, IAU Symp. 182, ed. B. Reipurth and C. Bertout (Dordrecht: Kluwer), pp. 181–198.
Hughes, S.M.G. 1993. Is Cepheus A East a Herbig-Haro object? *Astron. J.* 105:331–338.
Kameya, O., Hasegawa, T. I., Hirano, N., and Takakubo, K. 1989. High velocity flows in the NAC7538 molecular cloud. *Astrophys. J.* 339:222–230.
Kaufman, M. J., and Neufeld, D. A. 1996. Far-infrared water emission from magnetohydrodynamic shock waves. *Astrophys. J.* 456:611–630.
Kelly, M. L., and Macdonald, G. H. 1996. Two new young stellar objects with bipolar outflows in L379. *Mon. Not. Roy. Astron. Soc.* 282:401–412.
Lada, C. J., and Fich, M. 1996. The structure and energetics of a highly collimated bipolar outflow: NGC 2264G. *Astrophys. J.* 459:638–652.
Langer, W. D., Castets, A., and Lefloch, B. 1996. The IRAS 2 and IRAS 4 outflows and star formation in NGC 1333. *Astrophys. J. Lett.* 471:L111—L115.
Liseau, R., Giannini, T., Nisini, B., Saraceno, P., Spinoglio, L., Larsson, B., Lorenzetti, D., and Tommasi, E. 1997. Far-IR spectrophotometry of HH flows with the ISO long-wavelength spectrometer. In *Herbig-Haro Flows and the Birth of Low Mass Stars*, IAU Symp. 182, ed. B. Reipurth and C. Bertout (Dordrecht: Kluwer), pp. 111–120.

Martí, J., Rodríguez, L. F., and Reipurth, B. 1993. HH 80-81: A highly collimated Herbig-Haro complex powered by a massive young star. *Astrophys. J.* 416:208–217.

Martí, J., Rodríguez, L. F., and Reipurth, B. 1995. Large proper motions and ejection of new condensations in the HH 80-81 thermal radio jet. *Astrophys. J.* 449:184-187.

Martín-Pintado, J., Bachiller, R., and Fuente, A. 1992. SiO emission as a tracer of shocked gas in molecular outflows. *Astron. Astrophys.* 254:315–326.

Masson, C. R., and Chernin, L. M. 1992. Properties of swept-up molecular outflows. *Astrophys. J. Lett.* 387:L47–L50.

Masson, C. R., and Chernin, L. M. 1993. Properties of jet-driven molecular outflows. *Astrophys. J.* 414:230–241.

Masson, C. R., and Chernin, L. M. 1994. Observational constraints on outflow models. In *Clouds, Cores, and Low-Mass Stars*, ASP Conf. Ser. 65, ed. D. Clemens and R. Barvainis (San Francisco: Astronomical Society of the Pacific), pp. 350–359.

McCaughrean, M., and Mac Low, M.-M. 1997. The OMC-1 molecular hydrogen outflow as a fragmented stellar wind bubble. *Astron. J.* 113:391–400.

McCaughrean, M. J., Rayner, J. T., and Zinnecker, H. 1994. Discovery of a molecular hydrogen jet near IC 348. *Astrophys. J. Lett.* 436:L189–L192.

Mellema, G., and Frank, A. 1997. Outflow collimation in young stellar objects. *Mon. Not. Roy. Astron. Soc.* 292:795–807.

Micono, M., Davis, C. J., Ray, T. P., Eislöffel, J., Shetrone, M. D. 1998. Proper motions and variability of the H_2 emission in the HH 46/47 system. *Astrophys. J. Lett.* 494:L227–L230.

Mitchell, G. F., Maillard, J. P., and Hasegawa, T. I. 1991. Episodic outflows from high-mass protostars. *Astrophys. J.* 371:342–356.

Mitchell, G. F., Hasegawa, T. I., Dent, W.R.F., and Matthews, H. E. 1994. A molecular outflow driven by an optical jet. *Astrophys. J. Lett.* 436:177–180.

Mitchell, G. F., Lee, S. W., Maillard, J. P., Matthews, H. E., Hasegawa, T. I., and Harris, A. I. 1995. A multitransitional CO study of GL 490. *Astrophys. J.* 438:794–812.

Molinari, S., Saraceno, P., Nisini, B., Giannini, T., Ceccarelli, C., White, G. J., Caux, E., and Palla, F. 1998*a*. Shocks and PDRs in an intermediate mass star forming globule: The case of IC1396N. *Star Formation with the Infrared Space Observatory,* in ASP Conf. Ser. 132, ed. J. Yun and R. Liseau (San Francisco: Astronomical Society of the Pacific), pp. 390–394.

Molinari, S., Testi, L., Brand, J., Cesaroni, R., and Palla, F. 1998*b*. IRAS 23385 +6053: A prototype massive Class 0 object. *Astrophys. J. Lett.* 505:39–42.

Moriarty-Schieven, G. H., and Snell, R. L. 1988. High-resolution images of the L1551 molecular outflows. II. Structure and kinematics. *Astrophys. J.* 332:364–378.

Mundt, R., Brugel, E. W., and Bührke, T. 1987. Jets from young stars: CCD imaging, long-slit spectroscopy and interpretation of existing data. *Astrophys. J.* 319:275–303.

Nagar, N. M., Vogel, S. N., Stone, J. M., and Ostriker, E. C. 1997. Kinematics of the molecular sheath of the HH 111 optical jet. *Astrophys. J. Lett.* 482:195–198.

Narayanan, G., and Walker, C. K. 1996. Evidence for multiple outbursts from the Cepheus A molecular outflow. *Astrophys. J.* 466:844–865.

Netzer, N., and Elitzur, M. 1993. The dynamics of stellar outflows dominated by interaction of dust and radiation. *Astrophys. J.* 410:701–713.

Nisini, B., Lorenzetti, D., Cohen, M., Ceccarelli, C., Giannini, T., Liseau, R., Molinari, S., Radicchi, A., Saraceno, P., Spinoglio, L., Tommasi, E., Clegg, P. E., Ade, P.A.R., Armand, C., Barlow, M. J., Burgdorf, M., Caux, E., Cerulli,

P., Church, S. E., di Giorgio, A., Fischer, J., Furniss, I., Glencross, W. M., Griffin, M. J., Gry, C., King, K. J., Lim, T., Naylor, D. A., Texier, D., Orfei, R., Nguyen-Q-Rieu, Sidher, S., Smith, H. A., Swinyard, B. M., Trams, N., Unger, S. J., and White, G. J. 1996. LWS-spectroscopy of Herbig Haro objects and molecular outflows in the Cha II dark cloud. *Astron. Astrophys.* 315:L321–L324.

Nisini, B., Giannini, T., Molinari, S., Saraceno, P., Caux, E., Ceccarelli, C., Liseau, R., Lorenzetti, D., Tommasi, E. and White, G. J. 1998. High-J CO lines from YSOs driving molecular outflows. In *Star Formation with the Infrared Space Observatory*, ASP Conf. Ser. 132, ed. J. Yun and R. Liseau (San Francisco: Astronomical Society of the Pacific), pp. 256–264.

Padman, R., and Richer, J. S. 1994. Interactions between molecular outflows and optical jets. *Astrophys. Space Sci.* 216:129–134.

Padman, R., Bence, S., and Richer, J. 1997. Observational properties of molecular outflows. In *Herbig-Haro Flows and the Birth of Low Mass Stars,* IAU Symp. 182, ed. B. Reipurth and C. Bertout (Dordrecht: Kluwer), pp. 123–140.

Panagia, N. 1991. Ionized winds from young stellar objects. In *The Physics of Star Formation and Early Stellar Evolution*, ed. C. J. Lada and N. Kylafis (Dordrecht: Kluwer), pp. 565–593.

Pismis, P., and Torres-Peimbert, S. 1995. *Revista Mexicana de Astronomía y Astrofísica Serie de Conferencías* vol. 1.

Poetzel, R., Mundt, R., and Ray, T. P. 1989. Z CMa—a large-scale high velocity bipolar outflow traced by Herbig-Haro objects and a jet. *Astron. Astrophys. Lett.* 224:13–16.

Raga, A. 1991. A new analysis of the momentum and mass-loss rates of stellar jets. *Astron. J.* 101:1472–1475.

Raga, A. C., Cantó, J., Calvet, N., Rodríguez, L. F., and Torrelles, J. M. 1993. A unified stellar jet/molecular outflow model. *Astron. Astrophys.* 276:539–548.

Raga, A. C., Taylor, S. D., Cabrit, S., and Biro, S. 1995. A simulation of an HH jet in a molecular environment. *Astron. Astrophys.* 296:833–843.

Reipurth, B., and Bertout, C. 1997. *Herbig-Haro Flows and the Birth of Low-Mass Stars,* IAU Symp. 182 (Dordrecht: Kluwer).

Reipurth, B., Bally, J., and Devine, D. 1997. Giant Herbig-Haro flows. *Astron. J.* 114:2708–2735.

Richer, J. S., Hills, R. E., Padman, R., and Russell, A.P.G. 1989. High-resolution molecular line observations of the core and outflow in Orion B. *Mon. Not. Roy. Astron. Soc.* 241:231–246.

Richer, J. S., Hills, R. E., and Padman, R. 1992. A fast CO jet in Orion B. *Mon. Not. Roy. Astron. Soc.* 254:525–538.

Rodríguez, L. F. 1995. Subarcsecond observations of radio continuum from jets and disks. *Rev. Mex. Astron. Astrofís. Ser. Conf.* 1:1–10.

Rodríguez, L. F., Carral, P., Moran, J. M., and Ho, P. T. P. 1982. Anisotropic mass outflow in regions of star formation. *Astrophys. J.* 260:635–646.

Rodríguez, L. F., Garay, G., Curiel, S., Ramírez, S., Torrelles, J. M., Gómez, Y., and Velázquez, A. 1994. Cepheus A HW2: A powerful thermal radio jet. *Astrophys. J. Lett.* 430:L65–L68.

Russell, A. P. G., Bally, J., Padman, R., and Hills, R. E. 1992. Atomic and molecular outflow in DR 21. *Astrophys. J.* 387:219–228.

Saraceno, P., André, P., Ceccarelli, C., Griffin, M., and Molinari, S. 1996. An evolutionary diagram for young stellar objects. *Astron. Astrophys.* 309:827–839.

Saraceno, P., Nisini, B., Benedettini, M., Ceccarelli, C., di Giorgio, A. M., Giannini, T., Molinari, S., Spinogl, L., Clegg, P. E., Correia, J. C., Griffin, M. J., Leeks, S. J., White, G. J., Caux, E., Lorenzetti, D., Tommasi, E., Liseau, R., and Smith, H. A. 1998. LWS observations of pre–main sequence objects.

In *Star Formation with the Infrared Space Observatory*, ASP Conf. Ser. 132, ed. J. Yun and R. Liseau (San Francisco Astronomical Society of the Pacific), pp. 233–237.

Shepherd, D. S., and Churchwell, E. 1996. Bipolar molecular outflows in massive star-formation regions. *Astrophys. J.* 472:225–239.

Shepherd, D. S., and Kurtz, S. E. 1999. A 1000-AU rotating disk around the massive young stellar object a192.16. *Astrophys. J.* in press

Shepherd, D. S., Watson, A. M., Sargent, A. I., and Churchwell, E. 1998. Outflows and luminous YSOs: A new perspective on the G192.16 massive bipolar outflow. *Astrophys. J.* 507:861–873.

Shu, F. H., and Shang, H. 1997. Protostellar X-rays, jets, and bipolar outflows. In *Herbig-Haro Flows and the Birth of Low-Mass Stars,* IAU Symp. 182, ed. B. Reipurth and C. Bertout (Dordrecht: Kluwer), pp. 225–239.

Shu, F. H., Ruden, S. P., Lada, C. J., and Lizano, A. 1991. Star formation and the nature of bipolar outflows. *Astrophys. J. Lett.* 370:L31–L34.

Shu, F., Najita, J., Ostriker, E., Wilkin, F., Ruden, S., and Lizano, S. 1994. Magnetocentrifugally driven flows from young stars and disks. 1: A generalized model. *Astrophys. J. Lett.* 429:781–796.

Smith, M. D. 1994. Jump shocks in molecular clouds—speed limits and excitation levels. *Mon. Not. Roy. Astron. Soc.* 266:238–246.

Smith, M. D., Suttner, G., and Yorke, H. W. 1997. Numerical hydrodynamic simulation of jet-driven bipolar outflows. *Astron. Astrophys.* 323:223–230.

Smith, M. D., Eislöffel, J., and Davis, C. J. 1998. ISO observations of molecular hydrogen in the DR21 bipolar outflow. *Mon. Not. Roy. Astron. Soc.* 297:687–691.

Stahler, S. W. 1994. The kinematics of molecular outflows. *Astrophys. J.* 422:616–620.

Stone, J. M., Xu, J., and Mundy, L. G. 1995. Formation of bullets by hydrodynamical instabilities in stellar outflow. *Nature* 377:315–317.

Suttner, G., Smith, M. D., Yorke, H. W., and Zinnecker, H. 1997. Multi-dimensional numerical simulations of molecular jets. *Astron. Astrophys.* 318:595–607.

Tafalla, M., and Bachiller, R. 1995. Ammonia emission from bow shocks in the L1157 outflow. *Astrophys. J. Lett.* 443:37–40.

Taylor, S. D., and Raga, A. C. 1995. Molecular mixing layers in stellar outflows. *Astron. Astrophys.* 296:823–832.

Torrelles, J. M., Verdes-Montenegro, L., Ho, P. T. P., Rodríguez, L. F., and Jorge, C. 1993. From bipolar to quadrupolar—the collimation processes of the Cepheus A outflow. *Astrophys. J.* 410:202–217.

van Dishoeck, E. F., and Blake, G. A. 1998. Chemical evolution of star-forming regions. *Ann. Rev. Astron. Astrophys.,* 36:317–368.

van Dishoeck, E. F., Blake, G. A., Jansen, D. J., and Groesbeck, T. D. 1995. Molecular abundances and low-mass star formation. II. Organic and deuterated species toward IRAS 16293-2422. *Astrophys. J.* 447:760–787.

Velusamy, T., and Langer, W. D. 1998. Outflow-infall interactions as a mechanism for terminating accretion in protostars. *Nature* 392:685–687.

Wright, C. M., Drapatz, S., Timmermann, R., Van Der Werf, P. P., Katterloher, R., and De Graauw, T. 1996. Molecular hydrogen observations of Cepheus A West. *Astron. Astrophys. Lett.* 315:L301–L304.

Wu, Y., Huang, M., and He, J. 1996. A catalogue of high velocity molecular outflows. *Astron. Astrophys. Suppl.* 115:283–284.

Yamashita, T., Suzuki, H., Kaifu, N., Tamura, M., Mountain, C. M., and Moore, T. J. T. 1989. A new CO bipolar flow and dense disk system associated with the infrared reflection nebula GGD 27 IRS. *Astrophys. J.* 347:894–900.

Zhang, Q., Hunter, T. R., and Sridharan, T. K. 1998. A rotating disk around a high-mass young star. *Astrophys. J. Lett.* 505:L151–L154.

PART V

Early Solar System and Planet Formation

THE FU ORIONIS PHENOMENON AND SOLAR NEBULA MATERIAL

K. R. BELL, P. M. CASSEN
NASA Ames Research Center

J. T. WASSON
University of California at Los Angeles

and

D. S. WOOLUM
California State University at Fullerton

We summarize astronomical, meteoritic, and theoretical evidence relating to the FU Orionis phenomenon. This evidence suggests that at early times (the first few 10^5 yr), the solar nebula experienced a hot phase characterized by high accretion rates (the "FU Ori epoch"), punctuated by episodic outbursts of enhanced mass flow through the inner part of the disk ($\lesssim$0.3 AU). Throughout this epoch, disk midplane temperatures exceeded 1000 K at 1 AU. Diminishing infall from the cloud core led to decreasing mass flux throughout the disk. When mass flow decreased below the value critical for outburst (5×10^{-7} $M_\odot$ yr^{-1}, as suggested by thermal ionization instability models), outbursts ceased and the T Tauri epoch began. Outburst timescales are too long to explain calcium- and aluminum-rich inclusion (CAI) and chondrule formation. Volatility-dependent fractionation patterns seen in meteoritic materials suggest that solids formed beginning during a hot epoch when temperatures exceeded 1400 K, and the presence of volatiles in chondrites argues that this process continued until the nebula had cooled to below 400 K. The thermal ionization instability model for FU Ori outbursts is in quantitative agreement with astronomical observations. Its results imply that the terrestrial region of the nebula reached the hot end of this range only during a time when mass flow through the disk was high enough to trigger outbursts (i.e., the FU Ori epoch) and reached the cool end of this range only during the later T Tauri epoch. According to the models, heating of material in the terrestrial planet region during individual FU Ori outbursts would be limited to surface layers of the nebula, leaving midplane materials (which are at $\gtrsim$1000 K) largely unaffected. Alternative FU Ori models should be developed, particularly if compositional differences among chondrite clans are attributable to episodic heating.

I. INTRODUCTION

Beginning in 1936, the young star FU Orionis increased by four magnitudes in optical luminosity (Herbig 1966). Two additional young stellar objects (YSOs) with similar outburst histories were identified, and the class of FU Ori objects was defined by Herbig (1977). A recent review of the phenomenon can be found in Hartmann and Kenyon (1996). Time spent in outburst varies but is on the order of a hundred years. Spectral fitting of outburst sources and the development of numerical models have led to the view that the enhanced outburst luminosity is emitted primarily by the inner, circumstellar accretion disk ($r \lesssim 0.3$ AU) rather than by the central star. About a dozen FU Ori systems (or "Fuors") have so far been identified.

Herbig (1977) was the first to suggest that the outburst luminosity might have affected solid materials embedded in the solar nebula and left evidence in the meteoritic record. In this chapter, we review meteoritic and astrophysical evidence that bears on this subject. That many kinds of meteorites show elemental fractionations correlated with differences in volatility is evidence for the formation of solid material at a time when large regions of the solar nebula were hot ($T \gtrsim 1000$ K). On the other hand, astrophysical evidence indicates that T Tauri disks are considerably cooler than this at $r = 1$–3 AU. The existence of FU Orionis objects argues that T Tauri disks were subject to more extreme conditions earlier in their histories. Models and observations motivate the identification of an FU Ori epoch, preceding the T Tauri epoch, during which inner parts of the disks were hot enough to vaporize silicates.

Based on the models described in this chapter, we define *the FU Ori epoch* to be the time during which mass is accreted through the circumstellar disk at a rate high enough to induce periodic outbursts, $\gtrsim 10^{-6}\ M_\odot\ yr^{-1}$. Such a mass flow is expected to be sustained by infall from a molecular cloud core, so this epoch is associated with embedded protostars. During this time, disks spend roughly 90% of their time in the quiescent phase and 10% in outburst, with decades-long outbursts alternating with centuries-long periods of quiescence. This phase may be several hundred thousand years long.

We also define *the T Tauri epoch* as the period after which the placental cloud has dispersed and the remaining disk accretes onto the star at a declining rate. The disk may still be optically thick and actively accreting at a lower mass accretion rate ($\dot{M} \lesssim 10^{-7}\ M_\odot\ yr^{-1}$), especially during the early "classical" T Tauri stages. This phase may last several million years (Strom et al. 1993; Calvet, Hartmann, and Strom, this volume).

Dust, asteroid, and planetesimal formation could have been a continuous process through both the FU Ori and the T Tauri phases, with chemical, mineralogical, and morphological signatures determined by local conditions at the epoch and radial position of formation. Solar nebula conditions are best preserved in chondritic meteorites ("chondrites"). The spread in ages deduced from refractory calcium- and aluminum-rich inclusions ("CAIs") and chondrules suggests that the solar nebula was actively producing new meteoritic material, as well as agglomerating existing material, over a several-million-year period (Podosek and Cassen 1994;

MacPherson et al. 1995; Wadhwa and Russell, this volume). According to the models discussed below, widespread fractionations attributable to a hot nebula could have occurred only during the FU Ori epoch, whereas the making of chondrules and the final assembling of chondritic parent bodies continued through the T Tauri epoch.

Note that the actual formation of individual CAIs and chondrules apparently occurred on timescales of minutes to days (Hewins et al. 1988; Grossman et al. 1988; Wasson 1996; see also the chapter by Jones et al., this volume), much shorter than the years to decades for development and duration of FU Ori events. Thus, there is no obvious direct link between chondrule formation, which appears to involve local events, and FU Ori outbursts, whose effects are large-scale. It may be that chondrules were formed throughout both FU Ori and T Tauri epochs.

We begin in section II by presenting what is known about the outburst cycle and summarizing evidence that systems subject to outburst are younger than classical T Tauri systems. In section III we discuss meteoritic evidence regarding nebular temperatures, episodic heating events, and the preservation of primitive material in the solar nebula. In section IV we present astronomical constraints on T Tauri disk temperatures and argue that these disks are too cool to have produced the high-temperature fractionation suggested by the meteoritic evidence, thereby implicating earlier formation during the hotter FU Ori epoch. In section V we discuss what current models suggest about conditions in protostellar accretion disks at different epochs and what recent numerical simulations suggest about the effect of outbursts on material embedded in the solar nebula. Conclusions are summarized in section VI.

II. FU ORI OUTBURSTS

We begin with a discussion of the outburst cycle and evidence that Fuors are systematically younger than T Tauri systems. We then trace the development of the thermal disk instability model and summarize alternatives.

A. Observations: The Disk as Luminosity Source

During outburst, luminosities increase by a factor of roughly a hundred. Light curves of each of the three sources for which the outburst has been monitored are distinct: FU Ori rose to peak luminosity in 18 months and has gradually declined by about a factor of two since 1937. The rise time of V1057 Cyg was two years, followed by an exponential decay over decades. The light curve of V1515 Cyg is smooth with a longer rise timescale of decades. Sources known as Exors have lower-intensity (30 $\times$ increase) and shorter-lived (months to years) outbursts that may operate with a mechanism similar to Fuors (Herbig 1989; Lehman et al. 1995).

Given a birth rate of one low-mass star per kpc^3 per hundred years and a detection of roughly six outbursts in 60 years, each system is expected to experience about ten outbursts during its lifetime (Hartmann and Kenyon

1996). Time between FU Ori outbursts then depends on the assumed duration of the FU Ori epoch. Confining the outburst epoch to the first 10^5 yr suggests an interval of $\sim$10,000 yr between outbursts. Periodic spacing of Herbig-Haro objects (ejecta associated with FU Ori events) and recent numerical simulations, however, indicate 500–1000-yr intervals (Bally et al. 1995 and Bell and Lin 1994, respectively) suggesting that either the FU Ori epoch is a relatively short-lived phase ($<10^5$ yr) or that not all low-mass YSOs experience outbursts.

The luminosity of a system in outburst is inferred to originate from a disk because (1) optical and near-IR spectral energy distributions (SEDs) are broader than a single-temperature blackbody and have the shape predicted by a constant-mass flux accretion disk model with $\dot{M} \approx 10^{-4}$ $M_\odot$ yr^{-1} (Hartmann and Kenyon 1985; Kenyon et al. 1988; Calvet et al. 1991*a*; Turner et al. 1997); (2) the equivalent spectral type varies with wavelength, hotter photospheres being seen at shorter wavelengths, as expected for a disk that radiates over a range of temperatures (Herbig 1977); (3) absorption lines are rotationally split, with splitting greater at shorter wavelengths, consistent with absorption in the atmosphere of an optically thick, self-luminous, Keplerian disk in which temperature decreases with distance from the central star (Hartmann and Kenyon 1987*a*, *b*; Welty et al. 1990).

FU Orionis systems (of which about a dozen have been identified) are deduced to be younger than T Tauri systems. Observations of long-wavelength excess emission support the model that Fuors are surrounded by dusty envelopes with extinctions much higher than those associated with T Tauri systems (Weintraub et al. 1991; Kenyon and Hartmann 1991; Turner et al. 1997). They are also commonly embedded and associated with arcs of nebulosity (Herbig 1977; Goodrich 1987) as well as with Herbig-Haro objects (Bally et al. 1995), jets (Reipurth 1989), and massive outflows (Croswell et al. 1987; Welty et al. 1992; Calvet et al. 1993; Evans et al. 1994), all phenomena typical of particularly young systems. A recent survey of high-luminosity embedded Herbig-Haro sources supports this hypothesis by finding that a large fraction of the sources have spectral features typical of Fuors (Reipurth and Aspin 1997). Optically visible Fuors are deduced to be viewed pole-on (Kenyon et al. 1988), presumably through optically thin cavities cleared by outflows. The effect of inclination is evident in the FU Ori source L1551 IRS5, which has a large extinction due to its edge-on orientation.

Outbursts thus appear to occur in a phase of evolution that predates the classical T Tauri epoch. We associate this outburst phase with the embedded Class 0 or Class I protostars (as defined in Lada and Wilking 1984 and André and Montmerle 1994), for which mass fluxes through the disk are expected to be higher than for the older, Class II T Tauri objects. Because disk luminosity scales linearly with mass flux, these high-mass-flux protostellar sources should be on average ten times brighter. Nevertheless, surveys of embedded sources in Taurus-Auriga show that protostars have luminosities similar to revealed sources (e.g., Kenyon et al. 1990, 1993). This "luminosity problem" may be explained by the FU Orionis cycle, in which, 90 percent of the time, sources are in a quiescent phase with high rates of mass flux ($>5 \times 10^{-7}$ $M_\odot$ yr^{-1}) only beyond half an AU or so (see the next subsection). At smaller radii, mass is transported at a greatly reduced rate, so the output luminosity is much less than what would be expected from a steadily accreting disk. Thus the disk could be hot at the midplane of the terrestrial planet region but not so luminous as to conflict with observations.

B. Models: The Thermal Ionization Instability

In this section we principally discuss the thermal ionization instability model for FU Orionis outbursts. Most alternative explanations, discussed at the end of this section, are not developed enough to be critically tested by observation.

Detailed theoretical work has been carried out on the model of a disk-based thermal ionization instability, which was largely developed and refined since *Protostars and Planets III.* In this model, mass flows stably through the disk and onto the central star as long as hydrogen remains neutral. When mass being transported to the inner fraction of an AU exceeds a critical rate, however, hydrogen becomes ionized, and a thermally unstable regime is entered. The sensitive temperature dependence of opacity due to H^- absorption ($\kappa \sim T^{10}$), leads to increased optical depth and less efficient cooling. The gas heats progressively on a rapid thermal timescale until full ionization has occurred, and the opacity has entered a new, stable regime where H^- opacity no longer dominates. In the higher-temperature regime, mass flow is enhanced, because turbulent velocities and mixing lengths increase with temperature (viscosity is proportional to the local sound speed). Alternatively, magnetic viscosity may be enhanced as the gas becomes ionized. Mass flow onto the star during outburst exceeds the mass being supplied at the outer edge of the outburst region, and the inner disk drains. When the optical depth decreases to the point where the disk cools back into the thermally unstable regime, the reverse instability occurs (temperature drops, opacity drops, optical depth drops), the disk returns to a quiescent state, and material accretes onto the star at a rate below that being supplied. During the quiescent phase, the inner disk refills on the slow viscous timescale in preparation for another outburst.

The concept for the thermal ionization instability as applied to protoplanetary disks was developed by various researchers, including Lin and Papaloizou (1985), and Clarke et al. (1989, 1990). Repeating outbursts of proper magnitude and timescale are reproduced by a disk supplied with a constant mass flux above a critical value for outburst (Bell 1993; Kawazoe and Mineshige 1993). We use the results of the most detailed numerical study of the phenomenon (Bell and Lin 1994; Bell et al. 1995), which concludes that the critical mass flux is 5×10^{-7} $M_\odot$ yr^{-1}. This value is independent of the viscous efficiency, but somewhat dependent on the assumed stellar mass, 1 $M_\odot$, and radius, 3 $R_\odot$. Models that match observationally constrained timescales and duty cycle of time spent in outburst and quiescence (1/10) have $\alpha = 10^{-4}$, where hydrogen is neutral and 10^{-3} where ionized, with a smooth transition between states.[a]

Several objections to the disk outburst model (e.g., Simon and Joyce 1988; Herbig 1989) were answered by showing that an ionization front that moves radially reproduces observed evolution of colors and velocity line widths naturally (Bell et al. 1995). Self-regulated outbursts initiate spontaneously at the inner edge of the disk (as confirmed by the two-dimensional hydrodynamic simulations of Kley and Lin 1996, 1999) and propagate outward to a radius determined by the stellar mass and the steady input mass flux rate, generally $\lesssim 0.3$ AU (Bell and Lin 1994). Outside the thermally unstable region, mass flux continues steadily at the supplied rate during both outburst and quiescence. The light curve of V1515 Cyg can be explained by self-regulated outbursts; however, the rapid rise times of FU Ori and V1057 Cyg require initiation by perturbations of surface density at the outer edge of the unstable region (Bell et al. 1995) such as might be caused by interactions with orbiting bodies embedded in the disk (Clarke and Syer 1996).

During the FU Ori epoch, mass flux through the solar nebula is above the critical value, $\dot{M} > 5 \times 10^{-7}\ M_\odot\ \mathrm{yr}^{-1}$ (Bell and Lin 1994). The disk is replenished with infall from the placental molecular cloud (Kenyon and Hartmann 1991) which is expected to be in the range of $(1\text{–}10) \times 10^{-6}\ M_\odot\ \mathrm{yr}^{-1}$ (Shu et al. 1987). While the innermost disk alternates between quiescence and outburst, mass spirals in steadily at the supplied rate beyond about half an AU. Midplane temperatures during both outburst and quiescent phases may exceed 1000 K as far out as several AU (see section V.A). Once infall ceases, however, the mass of the disk declines, and midplane temperatures decrease. When mass flux onto the star drops below the critical value, outbursts cease, and the system enters the T Tauri epoch.

Alternate explanations for the outbursts include dynamical instabilities in rapidly rotating stars (Larson 1980) and thermal relaxation in young stars (Stahler 1989). The absorption line splitting observed in Fuor sources can be fit by an isolated, rotating star (Herbig 1989; Petrov and Herbig 1992), although disk wind models can also fit the line profiles (Calvet et al. 1993; Hartmann and Calvet 1995). Low-resolution smoothed-particle hydrodynamic (SPH) calculations suggest that enhanced disk mass flow might occur during the close passage of a binary star (Bonnell and Bastien 1992), but more detailed calculations suggest that passage sufficiently close to reproduce the observed rapid rise times would permanently disrupt the disk (Clarke and Pringle 1993). None of the above alternative models

[a] The efficiency factor α characterizes the viscosity, which is assumed to be the result of mixing due to turbulence of an unspecified nature (e.g., Shakura and Sunyaev 1973; Stone et al. this volume). We use the formulation $\nu = \alpha c_s^2 \Omega^{-1}$ where c_s is the thermal sound speed and Ω is the Keplerian frequency.

can account for multiple outbursts per star. It has also been suggested (J. T. Wasson, unpublished) that episodic outbursts may be caused by irregular accretion onto the disk through the addition of fresh interstellar material to even relatively mature T Tauri systems. Finally, Gammie (1996) has suggested a magnetically controlled mass transport model that shares some of the features of the mechanism discussed here (Stone et al., this volume). Gravitational instabilities in an early, massive disk also have the potential to provide episodic mass flow and would operate on the relatively short timescales required. None of these ideas has been quantitatively developed.

III. METEORITIC EVIDENCE

It has long been recognized that the components of the least altered chondritic meteorites preserve the record of their formation in the solar nebula. They might therefore provide evidence regarding conditions in the nebula and the possibility that it experienced an FU Ori epoch. In addition to containing chondrules, chondrites are broadly similar in that their elemental abundances are approximately the same as those of the sun for all but the most volatile elements. The systematic compositional deviations from solar abundances provide the basis for distinguishing different classes of chondrites. Each class possesses distinctive physical properties (such as chondrule abundance and size), as well as a particular isotopic composition of the abundant element oxygen. These various properties suggest that each chondrite group formed in a unique place or time, and they motivate the study of the groups' formation in the light of astrophysical models of the nebula.

Because nebular temperatures would be high throughout the FU Ori epoch, we consider the evidence of chondrite properties regarding nebular temperatures. Although important information is obtained by the study of individual grains, we focus our discussion on the bulk properties of chondrite groups, which are most likely to be representative of global phenomena. Much of the material summarized in this section is discussed in greater detail in previous volumes of the *Protostars and Planets* series (Boynton 1978, 1985; Wasson 1978; Palme and Boynton 1993), as well as in several chapters of the book *Meteorites and the Early Solar System* (Kerridge and Matthews 1988) and in Wasson (1985) and Taylor (1992).

In the following subsections, we distinguish two fundamentally different types of heating effects. In section A, we summarize evidence that chondrites formed in a nebula cooling from 1400 at least to 400 K. In section B, we comment on the potential impact of episodic heating events on meteorite parent bodies. Finally, in section C, we briefly discuss the issue of the preservation of meteoritic material throughout the nebular epoch.

A. Nebular Temperature Estimates from Meteoritic Data

The overall uniformity of isotopic compositions found in bulk samples of all types of planetary materials, despite noteworthy deviations in trace phases, indicates that homogenization occurred on a broad scale. This pervasive homogeneity is sometimes cited as evidence for a hot phase of nebular evolution with temperatures above about 1600 K,[b] at which temperature most material is vaporized in a nebula with $p(H_2) = 10^{-5}$ atm (Larimer and Anders 1967; Grossman 1972; Yoneda and Grossman 1995). Mixing in a completely gaseous nebula would have certainly obliterated the isotopic diversity that is expected to have accrued from the many nucleosynthetic sources of solar system material and isotopic fractionation processes that could occur in the interstellar medium. In fact, the deviations (usually at the microscopic level) from this pervasive uniformity allow the identification of presolar material. Unfortunately, it is difficult to determine to what degree the homogeneity itself predated the formation of the solar system. It seems likely that large-scale nebular evaporation was an early event limited to the inner solar system (i.e., the terrestrial planet region), while large regions of the outer solar system remained cool enough to preserve characteristics of a presolar existence (e.g., Bernatowicz and Zinner 1997; Mumma et al. 1993).

The mineralogy of chondritic meteorites having unequilibrated phases is generally consistent with that expected to result from condensation from a vapor of solar composition (Grossman and Larimer 1974). This supports the idea that these meteorites contain material largely derived from parts of the solar nebula that were vaporized at one time. Relative abundances of the most refractory[c] lithophile elements are identical within each chondrite group, but absolute abundances are variable among the different chondrite classes. Most carbonaceous chondrites are enhanced in refractory lithophiles relative to CI chondrites, while the ordinary and enstatite chondrites are depleted (e.g., Larimer and Wasson 1988*a*).

For moderately volatile elements (those with condensation temperatures ranging from 1200 K down to about 600 K), abundance patterns in several chondritic groups suggest mechanical isolation according to volatility. Within each group, these patterns are uniform (i.e., they are independent of the degree of metamorphic heating) and are therefore thought to reflect global nebular conditions as opposed to either local nebular conditions or parent body effects (Wasson and Chou 1974; Palme et al. 1988; Palme and Boynton 1993). These patterns are consistent with the progressive isolation of condensed material from vaporized material in a slowly cooling nebula (Wai and Wasson 1977; Cassen 1996). Although the three ordinary chondrite groups, H, L, and LL, and in some respects the enstatite

[b] A detailed calculation of various condensation temperatures, including their dependence on local gas pressures (typically ± 100 K for the range expected in nebular regimes), is presented in Yoneda and Grossman (1995).

[c] Refractories include Al, Ca, Ti, and the rare earths; their oxides make up about 7 wt% of the solids. They form stable oxide condensates at temperatures higher than do the common elements Mg, Si, and Fe, responsible for 85–90 wt% of the solids.

groups, EH and EL, have remarkably similar compositions and structures, they differ in their oxidation states and in their relative abundances of siderophile elements. Mechanical segregation of nebula components that formed at high temperatures is indicated (Larimer and Wasson 1988*b*). Grossman (1996) identifies and discusses six volatility-related fractionation patterns and presents evidence that these patterns were established prior to the formation of chondrules.

In addition to evidence of nebular temperatures from chondritic meteorites, there also exists evidence that some groups of iron meteorites were formed from material that was fractionated at high temperatures prior to the formation of their parent bodies. Iron meteorites are commonly deficient in moderately volatile siderophiles such as Ge and S (which, as FeS, should dissolve in a metallic melt) (Scott and Wasson 1975; Scott 1979; Palme et al. 1988); some groups are severely deficient (Groups IV A and IV B). That these fractionations were at least partially inherited from preplanetary material (and thus indicate nebular temperatures), rather than produced exclusively by planetary (i.e., igneous) processes, is supported by chemical, crystallization, and isotopic evidence (Scott et al. 1996; Chen and Wasserburg 1996). The correlation of siderophile abundance with volatility (Kelly and Larimer 1977; Palme et al. 1988) implies condensation from a hot (>1300 K) gas.

The above evidence has been interpreted in terms of widespread condensation from a globally hot nebula. The existence of exclusively refractory components (the CAIs) indicates that the isolation of high-temperature phases occurred. Distinctive patterns of rare earth elements in some (Group II) CAIs have been explained by condensation at about 1500 K in an environment in which some refractory oxides were already condensed (Boynton 1975, 1978). Although the individual inclusions must have been formed by local, brief processes, their precursor materials may well have been isolated in a high-temperature nebula.

Theoretical studies have identified conditions under which the most refractory elements (e.g., Ca, Al) would fractionate from other less refractory, common elements (e.g., Mg, Fe, Si) but not from each other, while the moderately volatile elements would be monotonically fractionated according to volatility as observed. Because fine dust is the major source of nebular opacity, regions of the nebula hot enough for the common rock-forming elements to be vaporized tend to be locally optically thin; if surrounded by opaque layers, these regions will be roughly isothermal. The temperature of such a region would be set by the value at its edge, namely the vaporization temperature of the opacity-producing dust: ≈1400 K (Morfill 1985; Boss 1990; Cassen 1994); this environment permits the condensation (or preservation without volatilization) and isolation in solid form of the more refractory elements. The models discussed in section II predict that the inner few AU of the nebular midplane might have experienced such temperatures continuously throughout the FU Ori epoch (also section V.A below).

At larger radii (or at later times), the nebula would be cool enough for the major rock-forming elements to be condensed in the form of fine dust. There the nebula would be optically thick, and substantial spatial temperature gradients would exist. Roughly speaking, the nebula may thus exist in one of two thermal states: a high-temperature, low-opacity regime, in which silicates and metals are vaporized and only the most refractory minerals, such as those constituting the CAIs, can exist as solids, and a lower-temperature, high-opacity regime in which the condensed solids are characterized by compositional variations according to volatility (Wasson 1985; Cassen 1996). Note that the separation of the most refractory minerals at an early nebular stage does not, by itself, explain the observed physical characteristics and distribution of CAIs. If made of material fractionated during an FU Ori epoch, these discrete objects must have been reheated and cooled rapidly in a lower-temperature environment (Stolper and Paque 1986) and eventually mixed with low-temperature phases.

The evidence cited above for the survival of meteoritic material from a hot phase of nebular history is not unequivocal. Even in the most primitive meteorites, there are few materials that are recognizable as primary condensates (Kerridge 1993). Some parent body alteration occurred (thermal or aqueous) in most meteorites (Zolensky et al. 1997), thereby obscuring primitive mineralogies. Also, the degree to which fractionation patterns might be produced by chondrule formation itself has been debated extensively in the literature (Larimer and Anders 1967; Wasson and Chou 1974; Grossman 1996). Although many arguments (summarized in Palme et al. 1988 and Cassen 1996) favor fractionation to be the result of nebular condensation, it would be imprudent to consider the issue closed, for the following reasons:

1. It is difficult to distinguish the consequences of nebular condensation from those of evaporation and recondensation in repetitive, localized heating events.
2. The variation in O-isotopic compositions among the different chondrite groups (Clayton 1993) may suggest that the solar nebula formed from compositionally different reservoirs of interstellar materials or that the composition of accreting presolar material shifted with time. The distinct chondrite classes might therefore reflect the relative contributions of these reservoirs, sampled to varying degrees over time (Choi and Wasson 1997).
3. Systematic fractionation of material may also occur due to grain-grain segregations based on physical (i.e., aerodynamic or magnetic) properties.
4. Finally, we mention that it is conceivable that chondrite properties derive not from condensation in some large region of the nebula but from condensation of the oxidizing vapor produced by collisions of large, primitive bodies at high relative velocity. Such ideas have been rejected in the past because (a) collisions are inefficient at vaporiza-

tion of the target and (b) the predicted unvaporized fragments have not been found in the meteorite record.

The evidence for low-temperature phases in primitive meteorites is as significant as the evidence for high temperatures. Calculations of volatile element condensation (e.g., Larimer 1973) show that agglomeration of solid material must have continued until nebular temperatures had declined to levels at least as low as 420 K to condense 90% of the most volatile metals (e.g., Cd, In, Bi), as observed in some primitive, relatively unaltered, chondrites. Evidence for minor condensation of water in primitive chondrites such as Semarkona or Renazzo (Hutchison et al. 1987) requires temperatures below about 350 K, at which point hydrated silicates condense if constituent elements are available as individual atoms, radicals, or fine amorphous solids. The incorporation of substantial fractions (tens of percent) of water in CM and CI meteorites, as has been inferred from fractionated O-isotopic compositions (Clayton and Mayeda 1984; Clayton et al. 1991; Leshin et al. 1997), requires even lower temperatures. Thus, it may be concluded that nebular agglomeration and parent body accumulation continued to well below 400 K.

B. The Possibility of Episodic Heating

It has long been a cosmochemical mystery as to why there are such big compositional gaps between the carbonaceous chondrite and ordinary chondrite clans and between these and the enstatite chondrite clan. An example of these gaps in bulk composition is shown in Fig. 1, a plot of $\Delta^{17}O$ vs. FeO/(FeO + MgO). The latter quantity is a measure of the degree of oxidation of the chondrite; complete oxidation of Fe leads to a ratio of $\approx$0.40. The quantity $\Delta^{17}O$ is a measure of the excess in the $^{17}O/^{16}O$ ratio compared to terrestrial samples. In alignment with the widespread view that the nebula cooled monotonically, it is commonly presumed that these differences are attributable to formation of the clans at different distances from the Sun or at different times in nebular history.

There exists, however, another possibility, with implications that we discuss in this section. Many young stars form in very dense, chaotic environments; in the Trapezium region, for example, the mean separation of stars is only 10^4 AU (Zinnecker et al. 1993). In such an environment, accretion onto each core may have been irregular, with order-of-magnitude variations in the rate of infall. If our Sun formed in such an environment, it may be that periods of high rates of infall alternated with periods of lower rates throughout the (1–3) $\times$ 10^6 yr history of the solar nebula. We have therefore the possibility of a series of FU Ori epochs (during each of which the nebula would experience several outbursts) rather than the single epoch that we deduce from the assumption of a monotonically declining infall rate. Given the possibility of episodic infall, it may be suggested that the differing abundances of the bulk compositions of the clans demonstrated in Fig.1 could have been acquired by variations in the composition

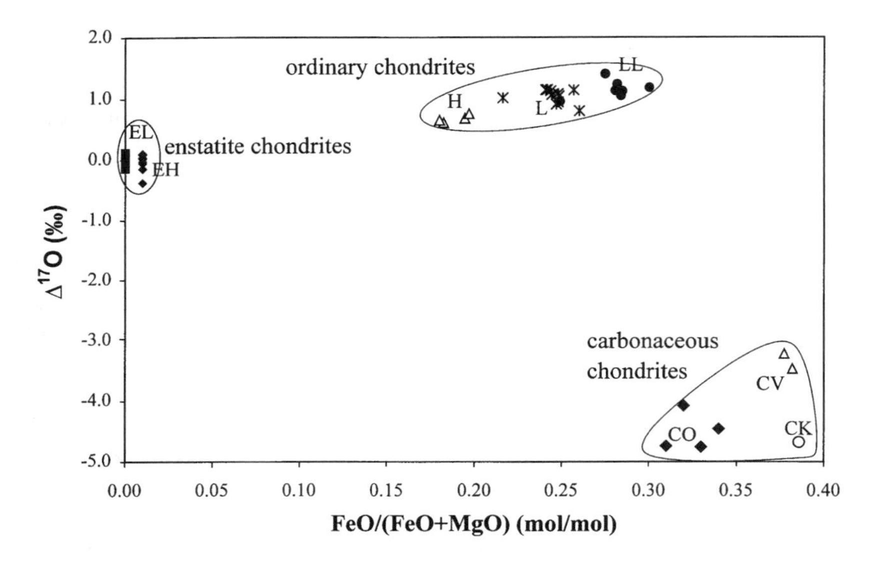

Figure 1. Bulk properties of the various chondrite clans are well separated on a plot of $\Delta^{17}O$ (a measure of the excess ^{17}O relative to a terrestrial standard; data mainly from Clayton et al. 1991 and the references in Clayton 1993) vs. the degree of oxidation as measured by the FeO/(FeO + MgO) ratio (data mainly from Jarosewich 1990). CI, CR, and CM classes are not plotted; they have experienced extensive aqueous alteration on their parent asteroids and thus may not preserve the composition of oxygen in the original nebular materials. EL chondrites typically contain unmeasurably low FeO contents and are arbitrarily plotted at zero.

of material accreted episodically over a several-million-year period (Wasson 1990).

Further, could episodic temperature excursions of the nebula as a whole be responsible for the melting of meteoritic parent bodies? Detailed calculation of the heating of asteroidal bodies by an external source have not been performed, but the heating depth can be estimated. Given typical thermal diffusivities of chondritic materials of about 10^{-2} cm^2 s^{-1}, the diffusion depth would be about 60 m in 100 years [$d \approx (D \cdot t)^{1/2}$]. However, the first kilometer-size chondritic objects formed in the solar nebula are likely to have been highly porous, with porosities similar to or even greater than the 50–60% observed in analogous deposits produced by major dust storms on Earth. As a result, if the surface of the object melted, the melt would be much more dense than the underlying materials and would tend to flow into the interior, thus permitting heat transport somewhat deeper than suggested by simple diffusion alone (Wasson 1992). A melt can be formed only if the vapor pressure is elevated above nebular values. This might occur through the formation of a refractory, confining crust created by the evaporation of the common elements. Occasional rupturing of this shell would permit the explosive escape of internal atmospheres that would carry off Na, K, H_2O, and other volatiles. Under these circumstances it might be possible to differentiate kilometer-size bodies during FU Ori outbursts. These bodies would have been much smaller than sizes previously considered as possible parent bodies of the igneously formed meteorites such as the eucrites, the magmatic irons (e.g., groups II AB and III AB), the pallasites, and so forth (J. T. Wasson, unpublished).

If this process occurred, unaltered chondrites might be preserved in larger or in more compact bodies, assuming that ^{26}Al contents are sufficiently low. The lower the porosity and the greater the strength of the body, the shallower the penetration depth of the melt. If only shallow layers are vaporized or penetrated by melt, the parts of the bodies deeper than about 100 m would not be appreciably heated during FU Ori events having durations of about 100 yr.

The ideas presented in this section are in contrast to the detailed outburst models discussed elsewhere in this chapter, which assume a monotonically declining accretion rate. A quantitative model for episodic infall remains to be developed. There is at present no observational evidence for infall onto mature T Tauri systems sufficient to induce FU Ori outburst.

C. The Preservation of Solar Nebula Material

If fractionation patterns created during early epochs are retained by material currently existing in the solar system, the material must have been preserved throughout a million years or more of nebular activity. The problem of the preservation of solids from the solar nebula has long been recognized in the literature (Weidenschilling 1977). Bodies smaller than a kilometer in diameter may be removed from the nebula through two mechanisms. The very smallest particles are coupled to the gas and participate in its steady inward spiraling movement. The orbits of large bodies decay because of the headwind they experience from the gas component, which rotates at less than Keplerian speed as a result of pressure support. This gas drag is greatest for meter-sized bodies; original estimates suggested that these bodies might spiral into the Sun from 1 AU in as short as 10^2 yr (Weidenschilling 1977). Bodies larger than 10 km circle the Sun essentially independently of the gas. Their orbits do not decay unless their masses become large enough to interact gravitationally with the gaseous component of the disk (Lin et al. 1996; Ward 1997).

It should be realized that loss of solids by gas drag is potentially important throughout the nebular lifetime, not just during the relatively short FU Ori epoch. The drag-induced radial velocity for a given particle size is proportional to the radial pressure gradient in the gas and to the presence or absence of turbulence but is not otherwise strongly dependent on nebular parameters. The most obvious way to preserve material in the solar nebula is by rapid accumulation through the meter-size range (during which drift is greatest) into bodies greater than 1 km in size.

Coagulation theory predicts that rapid growth will, in fact, occur in a laminar nebula. In a minimum-mass solar nebula, the sedimentation of material occurs on the timescale of

$$t_{\rm sed} \approx 2000(2\pi/\Omega) \approx 2000\,(r/1\ {\rm AU})^{3/2}\ {\rm yr}$$

(Hayashi et al. 1985; Weidenschilling and Cuzzi 1993). Large bodies settle faster than small ones, and if sticking is efficient, they sweep up material to become roughly centimeter-sized on reaching the midplane (Hayashi et al. 1985). Once in the particle-rich midplane, a body's radial drift in a laminar nebula is inhibited, because the collective action of the dense particle layer speeds up the orbiting gas, thereby reducing drag on individual particles (Nakagawa et al. 1986; Cuzzi et al. 1993). Although studies indicate that turbulence prevents the midplane dust layer from becoming thin enough for gravitational clumping to play a significant role in the growth of planetesimals (Weidenschilling 1980, 1988; Cuzzi et al. 1993; Dubrulle et al. 1995; Sekiya 1998), relative velocities between bodies of different sizes allow larger bodies rapidly to sweep up the smaller. This process of "drift-augmented accumulation" allows seed bodies to grow to kilometer sizes on timescales of a few thousand years at 1 AU (Cuzzi et al. 1993). During the period from condensation to incorporation into stable bodies, radial drift may be as small as $\sim 10^{-2}$ AU if the layer attains a dust-to-gas mass ratio of between 1 and 10. In less particle-rich midplanes, planetesimal growth will be slower (Supulver and Lin 1999).

Results of studies of particle evolution in turbulent nebulae are more complex but suggest less favorable growth rates (e.g., Weidenschilling and Cuzzi 1993). In the presence of turbulence, sedimentation rates will be much slower, and the enhancement of particle densities at the midplane will be lower. Also, radial drift will be much larger. In a turbulent nebula, the growth of bodies to greater than meter size may occur in turbulent eddies (Cuzzi et al. 1996; Klahr and Henning 1997).

There may be other factors that mitigate the effects of drag. For instance, systematic outward flow fields, which oppose drag-induced drift, might be established by the turbulent transport of angular momentum. Two examples have been identified:

1. Cuzzi et al. (1993) found that the layer of gas just above and below a dense layer of solids at the midplane would move outward through exchange of angular momentum with the particle layer. Small particles entrained in such a flow would move outward; the bulk of solids at the midplane would still drift inward, albeit at a slower rate than an isolated particle.
2. Global turbulent transport might even produce a net outward flow at the midplane. It has been shown that relatively efficient viscous transport ($\alpha \gtrsim 10^{-3}$) might produce a net outward flow at the midplane even in the absence of a dense particle layer, because the midplane layer gains angular momentum at the expense of higher layers in the nebula (Kley et al. 1993; Różyczka et al. 1994). The generality of this process in turbulent disks remains to be demonstrated.

Finally, it is possible that much solid material was lost by drift into the Sun but was continually replenished by fresh material from large radial distances (Stepinski and Valageas 1996, 1997; Supulver and Lin 1998) or by reprocessed material entrained in a wind from near the Sun (Liffman and Brown 1995; Shu et al. 1996). The former process must have occurred and would be expected to bring primarily primitive material from the colder parts of the nebula to the asteroid belt, but it cannot explain the preservation of high-temperature components. The consequences for the latter process for planetesimal accumulation have yet to be explored. For the present purposes, we adopt the point of view that chondritic precursor material was created continuously during the cooling phase of the nebula and was preserved against radial migration by prompt incorporation into kilometer-sized bodies.

IV. NEBULAR TEMPERATURE ESTIMATES FROM ASTRONOMICAL DATA

We have reviewed evidence from the meteoritic record that nebular midplane temperatures at one time exceeded 1400 K and that the components of chondrites continued to form until temperatures dropped below 400 K. In this section, we use observationally deduced properties of T Tauri systems to estimate disk midplane temperatures, T_{mid}. We find that estimated midplane temperatures of revealed T Tauri systems are concentrated at the cool end of this temperature range and thus T Tauri systems appear to be nearing the end of their chondrite formation period.

As deduced from spectral energy distributions, surface temperatures of disks around T Tauri systems lie primarily in the range $50 < T_{obs} < 300$ K at 1 AU (e.g., Beckwith et al. 1990). These are photospheric temperatures, however, not the internal temperatures that are most relevant to the thermal history of the solids in the disks. The internal temperatures of T Tauri disks can be estimated by models that incorporate both photometric observations over a broad range of wavelengths and high-resolution spectra in the visible and ultraviolet. Woolum and Cassen (1999; also Cassen and Woolum 1998) use available data and models (Beckwith et al. 1990; Valenti et al. 1993; Osterloh and Beckwith 1995; Gullbring et al. 1998; Hartmann et al. 1998) to calculate midplane temperature profiles for 27 T Tauri disks. Their method makes use of steady-state accretion disk theory (e.g., Lynden-Bell and Pringle 1974) and evaluates radiative transport by treating the disk as a plane parallel, gray atmosphere. The observed disk luminosity is taken to be the combined effect of accretion and externally produced reprocessed radiation. Deduced midplane temperatures at 1 AU for these sources are shown in Fig. 2. The differently hatched columns are used to distinguish two sets of available and overlapping evaluations of mass flux (gray for Valenti et al. 1993 vs. black for Hartmann et al. 1998). Most of the midplane temperatures cluster between $200 < T_{mid} < 700$ K,

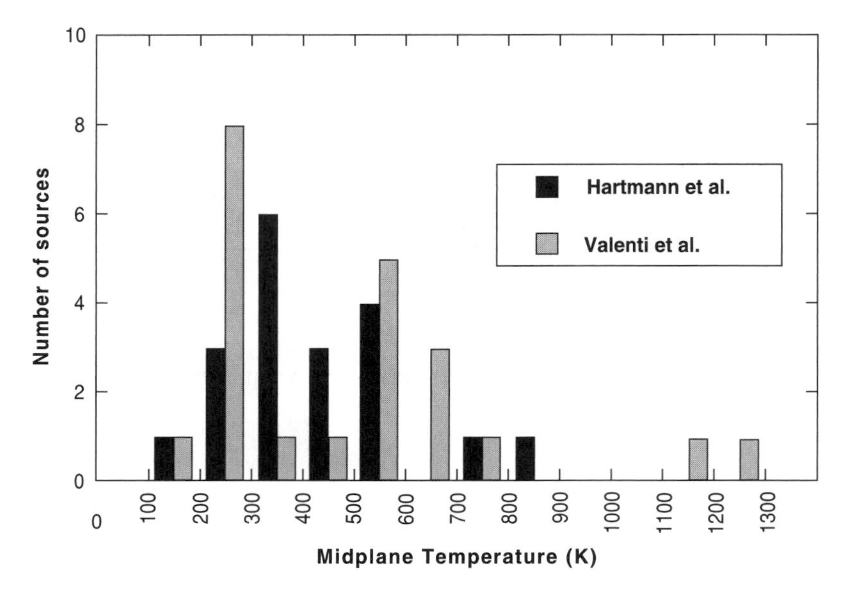

Figure 2. Histogram of midplane temperatures at 1 AU as derived for 27 classical T Tauri sources. Disk mass fluxes are taken from Hartmann et al. (1998) and Valenti et al. (1993). A given source appears twice if cited in both references.

above the effective surface temperatures noted above but well below silicate vaporization temperatures.

Although many assumptions and approximations are involved in these estimates, one gains some confidence that these are reasonable estimates of the internal thermal states of observed T Tauri disks, because they are derived from a large set of stars for which the data are internally consistent. There are effects that might cause midplane temperature to be systematically over- or underestimated. Nevertheless, such factors are not likely to alter the conclusion from Fig. 2 that these mature disks are not currently hot enough to vaporize silicates at the midplanes of their terrestrial planet regions.

It should be noted that it is difficult to determine the mass flux onto the star in systems with disk luminosities in excess of the central star; mass accretion rates above several times 10^{-7} $M_\odot$ yr^{-1} are difficult to quantify. Hotter disks have therefore been selectively excluded from the sample shown in Fig. 2, most of which have accretion rate a few times 10^{-8} $M_\odot$ yr^{-1} or less.[d] Also, in many cases, the disk masses of these systems are deduced from millimeter measurements to be less than the 10^{-2} $M_\odot$ estimated to accrete through the disk during a single FU Orionis outburst. Epochs earlier than the T Tauri phase (i.e., the FU Ori epoch)

[d] Recent critical analysis of the method for deriving stellar mass fluxes from veiling have led to the conclusion that, in contrast to earlier higher-mass flux results, these low mass fluxes are typical of T Tauri systems (Gullbring et al. 1998).

are expected to feature both greater accretion rates and higher disk masses than those of the sample analyzed here. During this earlier phase, internal disk temperatures would have been correspondingly higher.

The intensely observed star HL Tauri provides a good example of a younger, more active system. The disk and obscuring envelope have been modeled in detail (Beckwith et al. 1989; Hayashi et al. 1993; Calvet et al. 1994; Mundy et al. 1996; Close et al. 1997; Lay et al. 1997), resulting in estimates of infall of 3–5 $\times$ 10^{-6} $M_{\odot}$ yr^{-1} and a disk mass of 0.04 $M_{\odot}$. According to Beckwith et al. (1990), the total luminosity of the system is only 8 $L_{\odot}$, about a factor of 6 less than what would be expected if material was being accreted through the disk and onto the star at the envelope infall rate. For this reason, among others, Lin et al. (1994) suggested that HL Tau is an FU Orionis object in quiescence, in which case the stellar mass flux would be less than the mass flux through most of the disk. If most of the total luminosity of the system can be attributed to accretion (Calvet et al. 1994), midplane temperatures are sufficiently hot to vaporize silicates at 1 AU and are in the range of 800–900 K at 2 AU. Close et al. (1997) infer a much higher luminosity for HL Tau, because they estimate a very high near infrared extinction (7.7 magnitudes), which leads them to infer that stellar accretion is comparable to infall accretion. Taking their value of 45 $L_{\odot}$ for the system luminosity at face value, one would conclude that midplane temperatures reach silicate vaporization even out at 2 AU.

Whichever disk model is adopted, the technique outlined in Cassen and Woolum (1998), along with the high inferred infall rate, suggests that midplane temperatures in the HL Tau disk are high enough to vaporize silicates. The mass infall rates onto the disks of embedded or partially embedded protostars have been estimated to be comparable to that of HL Tau (Calvet et al. 1994;Whitney and Hartmann 1993; Whitney et al. 1997; Kenyon et al. 1990, 1993); thus, high midplane temperatures are likely to be a common occurrence for this class of objects. HL Tau is inferred to be only a few $\times$ 10^5 yr of age. It is more embedded than most T Tauri systems, and therefore probably represents an earlier phase of evolution. It may either be currently in the FU Ori epoch or in transition between FU Ori and T Tauri epochs.

These results suggest that, although mature T Tauri disks are too cool in the midplanes of their terrestrial planet regions to explain the meteoritic evidence for a hot solar nebula, they evolved from a hotter state. We suggest that this hotter state occurred during the FU Ori epoch, when the mass flux was larger.

V. NEBULAR MODELS

In this section we outline the evolution of physical conditions in the solar nebula during the FU Ori and T Tauri epochs as derived through viscous models. We also examine the impact of the outbursts themselves on midplane conditions. For the outbursts we adopt the thermal ionization instability model. These models have been shown to be in general agreement with astronomical constraints and are the only ones sufficiently developed to provide quantitative estimates of nebular thermal states.

A. Thermal Disk Structure: FU Ori and T Tauri Epochs

The structure of viscous disks has been investigated extensively. We make use of recent models that calculate temperature, density, and shape and consider various mass flow rates and viscous efficiencies (Bell et al. 1997; D'Alessio et al. 1998). We focus on the thermal structure of the terrestrial planet region as the most relevant for the formation of meteorite parent bodies and planets.

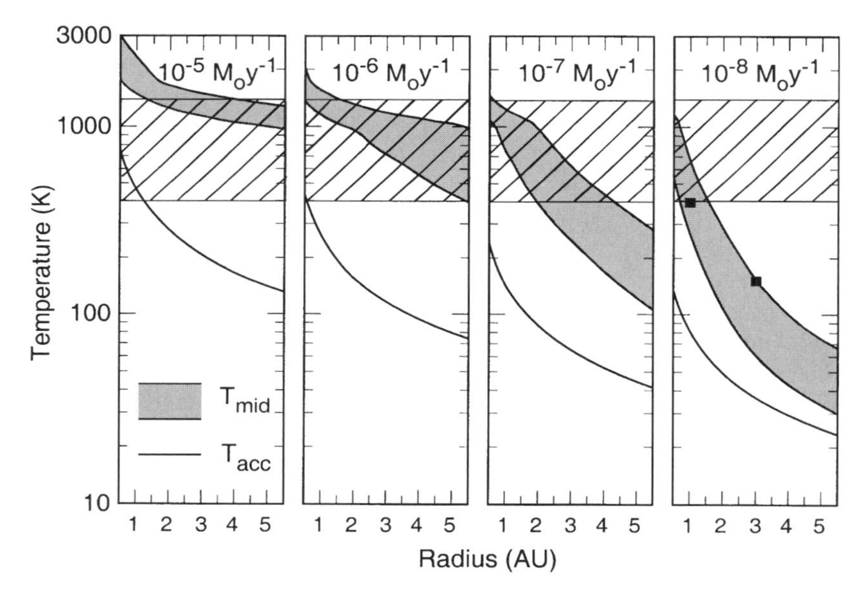

Figure 3. Locally generated surface and midplane temperature profiles for mass fluxes as marked (data from Bell et al. 1997). The first two panels correspond to the FU Ori epoch, the second two to the T Tauri epoch. The shaded areas show the range of midplane temperatures, $T_{\rm mid}$, given by models with α varying from 10^{-2} (lower edge) to 10^{-4} (upper edge). The lowest curved line marks the disk surface temperature, $T_{\rm acc}$, due to locally generated accretion energy (reprocessed energy is not included). The hatched areas correspond to the temperature range of the formation of meteoritic components; the upper edge (1400 K) marks the refractory fractionation temperature constraint, the lower edge (400 K) the agglomeration temperature. Meteorite components can form in the radial ranges where the shaded and hatched regions overlap. Solid squares in the last panel mark the centroid of midplane temperatures derived for T Tauri systems at 1 and 3 AU (400 K and 150 K, respectively). These squares coincide with midplane temperatures in the final panel, suggesting that revealed T Tauri systems are nearing the end of their chondrite formation phase.

Figure 3 shows temperature profiles for disks with four different radially constant mass fluxes. Nebular surface temperature due to the locally generated accretion flux, T_{acc}, is shown by the lowest curve. The shaded regions indicate midplane temperatures, T_{mid}, derived from viscous efficiency ranging from $\alpha = 10^{-2}$ (lower line) to 10^{-4} (upper). These may be taken as the range of midplane temperatures as suggested by theory for each mass flux. The hatched regions refer to the range of temperatures derived from meteoritic constraints. The upper limit, 1400 K, corresponds to the midplane temperature required to evaporate most rock-forming elements, and the lower limit, 400 K, corresponds to an inferred agglomeration temperature (section III.A).

Meteorite parent bodies could form at radii where hatched and shaded regions of Fig. 3 overlap. At the highest mass flux shown (corresponding to the earliest times), only the most refractory material could condense in the terrestrial planet region. The FU Ori epoch probably continued through mass fluxes down to $\lesssim 10^{-6}$ $M_{\odot}$ yr^{-1}, by which time chondrites could begin forming throughout the entire 1–5-AU region. Parent bodies could continue to form until midplane temperatures dropped to 400 K. The solid-forming region of the nebula thus moves slowly to smaller radii as mass flux through the solar nebula declines. Preservation of material condensed over a range of conditions from the early, hot nebula through later, cooler stages could provide conditions favorable for establishing observed meteoritic fractionation patterns, as summarized above in section III.A. Mixing of material formed at different times would produce the full range of thermal histories seen in any given meteorite.

The nebular temperatures shown in Fig. 3 include neither the effects of surface irradiation nor the effects of an FU Ori outburst. As discussed above, current models suggest that only the inner fraction of an AU of the nebula would have been directly subject to the thermal ionization instability. Most of the nebula would be affected only through surface illumination; this effect is examined in the next subsection. The temperatures in Fig. 3 therefore represent minimum temperatures experienced by the nebula during each epoch.

In the last panel of this figure, black squares are shown at 1 AU and 3 AU, at typical midplane temperatures derived for revealed T Tauri systems (section IV; Cassen and Woolum 1998). The model yields such low midplane temperatures only when the mass flux has dropped to 10^{-8} $M_{\odot}$ yr^{-1}. T Tauri nebulae are therefore inferred to be nearing the end of their active chondrite formation phases.

B. Effects of Outbursts on Material in the Terrestrial Planet Region

We now return to the FU Ori epoch and argue that material at the nebular midplane is not likely to be greatly heated by the outbursts themselves. The models presented in section II.B suggest that the thermally unstable region is confined to a distance of a fraction of an AU from the protosun.

Temperatures in the outburst region will for decades be much higher than the vaporization temperature of all solid material; very little would therefore be expected to survive in the inner one-third AU. In these models, most of the disk will be affected during the outburst only by an increase in surface illumination. Recent results suggest that, while this heating may increase the temperatures of surface layers and thus alter the visible chemistry of the disk, midplane temperatures will be largely unaffected (Bell 1999). In this section we discuss the models that led to this conclusion.

During outburst, radiation from the central star is negligible; the inner disk itself becomes the dominant source of radiation that illuminates the outer disk. The detailed shape of the disk is therefore critical in determining the redistribution of radiation in the system. The distance from midplane to Rosseland mean photosphere of an annulus with given thermal and density structure is determined principally by the local opacity. The radial trend of the thickness, and thus the disk's shape, is determined by the temperature dependence of that opacity (Bell et al. 1997). The innermost annuli of disks in outburst (where most of the luminosity is emitted) are fully ionized, so they have opacities that rise steeply with temperature ($\kappa \sim T^{10}$). The disk flares very strongly in this region; most of the outburst radiation is emitted from the inside of a cone-shaped surface and directed out along the poles of the system. Models suggest that the disk thickness reaches a peak at about three stellar radii (0.04 AU) and then decreases to a minimum at about 0.25 AU (detailed figures of the outburst region are given in Turner et al. 1997 and Bell 1999). For the remainder of the disk, the thickness increases only slowly with radius. Because of the funneling effect of the shape of the innermost annuli, the surface illumination of most of the disk is greatly reduced from what it would be if the radiation were emitted isotropically.

After the outburst fades, the disk's surface is illuminated both by the central star and by radiation emitted from the disk surface visible across the axis of the system (Bell 1999). The surface illumination during quiescence can be considerable and further diminishes the contrast between outburst and quiescent states. The reprocessing of stellar light has been shown to be important for disks with relatively low intrinsic luminosities, i.e., low mass fluxes (Adams and Shu 1986; Kenyon and Hartmann 1987; Ruden and Pollack 1991; Calvet et al. 1991*b*, 1992; D'Alessio et al. 1997; Chiang and Goldreich 1997). For higher mass fluxes, however, like those expected throughout the FU Ori epoch, the disk, even during quiescence, can contribute a luminosity equal to that of the star, which must be taken into account when calculating the surface illumination.

Figure 4 shows the difference in disk surface temperature between quiescent and outbursting phases (from Bell 1999). The data are snapshots taken from a time-dependent model sequence being fed at the rate of 10^{-6} $M_\odot$ yr^{-1}. The two curves indicate the disk surface temperature during quiescence (lower, thin line) and outburst (upper, thick line) including both locally generated (accretion) and externally provided (reprocessing)

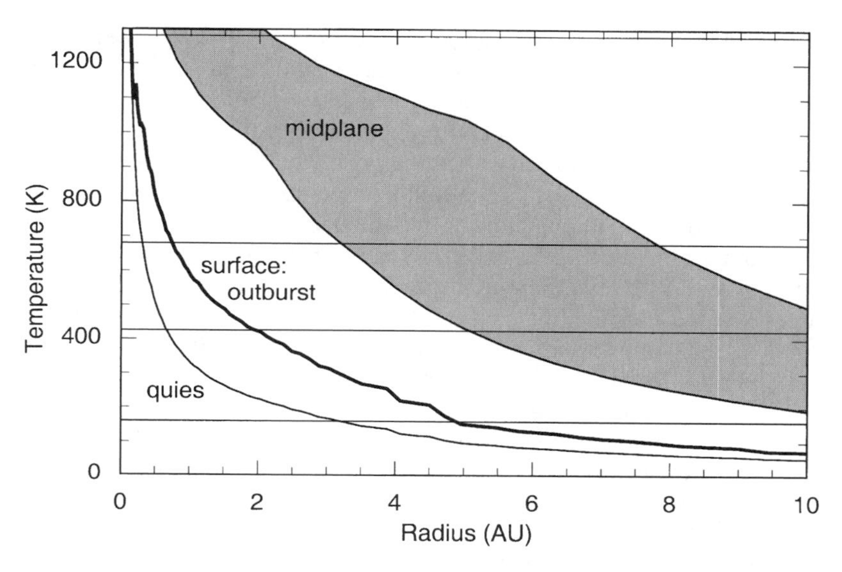

Figure 4. Disk temperatures during the FU Ori epoch (data from Bell 1999). Surface temperatures during outburst (heavy line) and quiescence (light line), including both locally generated accretion flux and surface irradiation from both disk and central star. The range of midplane temperatures expected during quiescence, as estimated by nebular models, is indicated by the shaded region. Horizontal lines indicate destruction temperatures of various solid species (top to bottom: olivine, troilite, refractory organics, and water ice).

components. During outburst, the surface flux increases by about a factor of 10 over that during quiescence. Horizontal lines indicate condensation/destruction temperatures for olivine, troilite, refractory organics, and H_2O ice.[e] Irradiation during outburst will move the evaporation fronts of these species at the surface layers to larger radii than during quiescence. For example, H_2O ice could exist at the surface in to almost 3 AU during quiescence but would be evaporated out to 5 AU during outburst. Luminosity absorbed and reradiated down onto the disk by the circumstellar envelope will be a source of heating that is not accounted for in these models. Recent calculations indicate that such illumination could raise surface temperatures substantially (Natta 1993; Butner et al. 1994; Chick et al. 1996; Chick and Cassen 1997; D'Alessio et al. 1997; Bell and Chick 1997).

The range of midplane temperatures (corresponding to α varying from 10^{-4} to 10^{-2}) for a disk accreting 10^{-6} $M_{\odot}$ yr^{-1} is shown in Fig. 4 as a shaded area; this range has been calculated, as in the previous figure, without considering the effect of surface illumination. In all cases, at each radius, the midplane temperature is considerably hotter than the

[e] Temperatures are 1280, 680, 425, and 160 K, respectively, and are taken from Pollack et al. (1994) for disks at densities of 10^{-10} g cm^{-3}.

corresponding surface temperature during outburst (heavy line). Particles (at regions where $T > 1000$ K, these would be largely magnesium silicates, Fe-Ni metal, and refractory phases) embedded at the nebular midplane are therefore not likely to experience dramatic heating during outburst. The effects of surface heating during outburst can be felt at the midplane only at radii where the midplane temperature does not greatly exceed the effective temperature of the illumination. Thus, volatile species in the outer regions of the nebula, beyond several AU, where the optical depth is relatively low, may be those most affected by an outburst. With larger surface illumination (e.g., with more intense outbursts or in the event of substantial reprocessing from the envelope), it would become more difficult to shield material at the midplane.

If illumination during outburst is, in fact, in excess of what is predicted by current models, the question of whether the duration of outburst is sufficient to heat through to the midplane becomes important. Because only changes in surface illumination greater than the locally generated accretion flux could appreciably affect the midplane temperatures, a representative timescale to produce midplane thermal response is given by an annulus's energy content divided by its accretion flux,

$$\tau_{\mathrm{therm}} \approx \frac{C_V \Sigma T_{\mathrm{mid}}}{\sigma T_{\mathrm{acc}}^4} \approx (\alpha\Omega)^{-1}$$

where Σ is the mass surface density and C_V is the specific heat at constant volume. The final relation results from the definition of α and steady-state accretion disk theory and therefore applies only to optically thick regions of the nebula. At 1 AU with $\alpha = 10^{-2}$, the thermal timescale is 15 yr; at 3 AU, it is 75 yr. Thus only material in the inner several AU of the disk have the potential to be heated to the midplane during a 100-year outburst.

VI. SUMMARY

Astronomical evidence suggests that circumstellar disks evolve from an early, hot state of high mass accretion rate to a cooler T Tauri state. The hot state, the FU Ori epoch, which probably lasts only a few 10^5 yr, is punctuated by outbursts resulting from episodes of enhanced mass flux through the inner disk. For the early solar system there is also evidence, largely derived from the volatility-dependent fractionations of bulk meteorites, that planetary material survived from a hot epoch ($T >$ 1400 K) and that the formation of solids continued to low temperatures ($T < 400$ K). It can therefore be argued that the formation of chondritic components began early in nebular evolution (the FU Ori epoch) and continued until a later, cooler period (the T Tauri epoch). Unless our solar nebula had a unique history, one can also conclude that T Tauri disks presently contain the analogs of meteorite parent bodies.

Outburst timescales are too long to be directly responsible for the formation of discrete CAIs and chondrules, but one could postulate that the characteristics of chondritic material reflect cooling from individual FU Ori outbursts, with different chondrite classes being produced after each outburst. Subsequent outbursts might then result in the differentiation of the most porous parent bodies. These hypotheses are not, however, in agreement with current astrophysical models, which assume monotonically decreasing accretion rates:

1. During the FU Ori epoch, midplane temperatures even in the quiescent phase are too hot to allow the existence of any but the most refractory materials.
2. Because models indicate that the active part of the disk during outburst is confined to less than an AU, the outburst's main effect for most of the nebula is produced by surface illumination. The large optical depths predicted by the models would severely mute the midplane effects of outbursts in the terrestrial planet region.

Perhaps the currently developed models are oversimplified, resting as they do on various assumptions (e.g., that infall from the placental cloud declines monotonically with time). Models based on other mechanisms (e.g., for the transport of angular momentum), with different consequences for thermal evolution, might prove more realistic. Despite the weight of current evidence, it is still possible that the observed chemical fractionations in chondrites are really the result of the cumulative effects of localized thermal fluctuations rather than evidence for global thermal evolution. Models that cool down between outbursts, for example, would offer a richer context for the establishment of meteoritic diversity.

Nevertheless, at present we discern no compelling reason to dismiss either the current astrophysical models based on gradually decreasing mass flux or the interpretation of the meteoritic fractionations in terms of global nebular thermal evolution. The nebular outburst models explain the astronomical data and are consistent with meteoritic evidence for a hot nebula. Moreover, they provide falsifiable predictions, e.g., low-mass-flux T Tauri systems without substantial envelopes should not go into outburst. Models suggest that a hot phase of nebular evolution, which we associate with the FU Ori epoch, precedes the cooler T Tauri epoch. Currently developed outburst simulations argue further that disks during the FU Ori epoch are sufficiently optically thick that the outbursts themselves would have little impact on material embedded at the nebular midplane.

It is clear that much remains unresolved in the reconciliation of astrophysical and meteoritic constraints. The current nebular models and, in particular, the ionization instability outburst models, should be expanded and refined to improve the reliability of their predictions for thermal evolution. Alternative outburst models should be developed to the point where they can be tested against observational data. To constrain astrophysical models of the impact of FU Ori outbursts on solar nebula material better, the effects of global nebular conditions, including possible episodic thermal excursions, need to be distinguished from those of local processes in the meteoritic record.

Acknowledgments This research was supported by NAG5-4610 (K. R. B.), RTOP 344-30-51-01 (P. M. C.), NAG5-4331 (J. T. W.), and NAG5-4945 (D. S. W.). We benefited from discussion with Jeff Cuzzi, Lee Hartmann, Sara Russell, and Diane Wooden. K. R. B. particularly thanks Wooden for preparation of Figs. 3 and 4.

REFERENCES

Adams, F. C., and Shu, F. H. 1986. Infrared spectra of rotating protostars. *Astrophys. J.* 308:836–853.

André, P., and Montmerle, T. 1994. From T Tauri stars to protostars: Circumstellar material and young stellar objects in the ρ Ophiuchi cloud. *Astrophys. J.* 420:837–862.

Bally, J., Devine, D., Fesen, R. A., and Lane, A. P. 1995. Twin Herbig-Haro jets and molecular outflows in L1228. *Astrophys. J.* 454:345–369.

Beckwith, S. V. W., Koresko, C. D., Sargent, A. I., and Weintraub, D. A. 1989. Tomographic imaging of HL Tauri. *Astrophys. J.* 343:393–399.

Beckwith, S. V. W., Sargent, A. I., Chini, R. S., and Güsten, R. 1990. A survey for circumstellar disks around young stellar objects. *Astron. J.* 99:924–945.

Bell, K. R. 1993. The dynamics of FU Orionis outbursts: Time-dependent accretion through protostellar disks. Ph.D. thesis, University of California, Santa Cruz.

Bell, K. R. 1999. Reprocessing in luminous disks. *Astrophys. J.,* in press.

Bell, K. R., and Chick, K. C. 1997. The radiative impact of FU Orionis outbursts on protostellar envelopes. In *Herbig-Haro Flows and the Birth of Low Mass Stars*, IAU Symp. 182, ed. B. Reipurth and C. Bertout (Dordrecht: Kluwer), pp. 407–416.

Bell, K. R., and Lin, D. N. C. 1994. Using FU Orionis outbursts to constrain self-regulated protostellar disk models. *Astrophys. J.* 427:987–1004.

Bell, K. R., Lin, D. N. C., Hartmann, L. W., and Kenyon, S. J. 1995. The FU Orionis outburst as a thermal accretion event: Observational constraints for protostellar disk models. *Astrophys. J.* 444:376–395.

Bell, K. R., Cassen, P. M., Klahr, H. H., and Henning, T. 1997. The structure and appearance of protostellar accretion disks: Limits on disk flaring. *Astrophys. J.* 486:372–387

Bernatowicz, T., and Zinner, E., eds. 1997. *Proceedings of the Conference on the Astrophysical Implications of the Laboratory Study of Presolar Materials* (Woodbury, NY: American Institute of Physics).

Bonnell, I., and Bastien, P. 1992. A binary origin for FU Orionis stars. *Astrophys. J. Lett.* 401:L31–L34.

Boss, A. P. 1990. 3D solar nebula models: Implications for earth origin. In *Origin of the Earth*, ed. H. E. Newsome and J. H. Jones (New York: Oxford University Press), pp. 3–15.

Boynton, W. V. 1975. Fractionation in the solar nebula: Condensation of yttrium and the rare earth elements. *Geochim. Cosmochim. Acta* 39:569–584.

Boynton, W. V. 1978. The chaotic solar nebula: Evidence for episodic condensation in several distinct zones. In *Protostars and Planets*, ed. T. Gehrels (Tucson: University of Arizona Press), pp. 427–438.
Boynton, W. V. 1985. Meteoritic evidence concerning conditions in the solar nebula. In *Protostars and Planets II*, ed. D. C. Black and M. S. Matthews (Tucson: University of Arizona Press), pp. 772–787.
Butner, H. M., Natta, A., and Evans, N. J. 1994. "Spherical" disks: Moving toward a unified source model for L1551. *Astrophys. J.* 420:326–335.
Calvet, N., Hartmann, L., and Kenyon, S. J. 1991*a*. On the near-infrared spectrum of FU Orionis. *Astrophys. J.* 383:752–756.
Calvet, N., Patiño, A., Magris, G. C., and D'Alessio, P. 1991*b*. Irradiation of accretion disks around young objects. I. Near-infrared CO bands. *Astrophys. J.* 380:617–630.
Calvet, N., Magris, C. G., Patiño, A., and D'Alessio, P. 1992. Irradiation of accretion disks around young objects. II. Continuum energy distribution. *Rev. Mex. Astron. Astrofís.* 24:27–43.
Calvet, N., Hartmann, L., and Kenyon, S. J. 1993. Mass loss from pre-main-sequence accretion disks. I. The accelerating wind of FU Orionis. *Astrophys. J.* 402:623–634.
Calvet, N., Hartmann, L., Kenyon, S. J., and Whitney, B. A. 1994. Flat spectrum T Tauri stars: The case for infall. *Astrophys. J.* 434:330–340.
Cassen, P. M. 1994. Utilitarian models of the solar nebula. *Icarus* 112:405–429.
Cassen, P. M. 1996. Models for the fractionation of moderately volatile elements in the solar nebula. *Meteoritics* 31:793–806.
Cassen, P. M., and Woolum, D. S. 1998. Internal temperatures of the solar nebula inferred from astronomical observations of circumstellar disks. *Meteorit. Planet. Sci.* 33:A28–A29 (abstract).
Chen, J. H., and Wasserburg, G. J. 1996. Live ^{107}Pd in the early solar system and implications for planetary evolution. In *Earth Processes: Reading the Isotopic Code,* Geophysical Monograph 95, ed. A. Basu and S. Hart (Washington: American Geophysical Union), pp. 1–20.
Chiang, E. I., and Goldreich, P. 1997. Spectral energy distributions of T Tauri stars with passive circumstellar disks. *Astrophys. J.* 490:368–376.
Chick, K. M., and Cassen, P. M. 1997. Thermal processing of interstellar dust grains in the primitive solar environment. *Astrophys. J.* 477:398–409.
Chick, K. M., Pollack, J. B., and Cassen, P. M. 1996. The transport of thermal radiation in a protostellar envelope. *Astrophys. J.* 461:956–971.
Choi B.-G., and Wasson, J. T. 1997. Episodic accretion of solar nebula: Meteoritic clues. *Meteorit. Planet. Sci.* 31:A28 (abstract).
Clarke, C. J., and Pringle, J. E. 1993. Accretion disc response to a stellar fly-by. *Mon. Not. Roy. Astron. Soc.* 261:190–202.
Clarke, C. J., and Syer, D. 1996. Low mass companions to T Tauri stars: A mechanism for rapid-rise FU Orionis outbursts. *Mon. Not. Roy. Astron. Soc. Lett.* 278:L23–L27.
Clarke, C. J., Lin, D. N. C., and Papaloizou, J. 1989. Accretion disc flows around FU Orionis stars. *Mon. Not. Roy. Astron. Soc.* 236:495–503.
Clarke, C. J., Lin, D. N. C., and Pringle, J. E. 1990. Pre-conditions for disc-generated FU Orionis outbursts. *Mon. Not. Roy. Astron. Soc.* 242:439–446.
Clayton, R. N., 1993. Oxygen isotopes in meteorites. *Ann. Rev. Earth Planet. Sci.* 21:115–149.
Clayton, R. N., and Mayeda, T. K. 1984. The oxygen isotope record in Murchison and other carbonaceous chondrites. *Earth Planet. Sci. Lett.* 67:151–161.
Clayton, R. N., Mayeda, T. K., Goswami, J. N., and Olsen, E. J. 1991. Oxygen isotope studies of ordinary chondrites. *Geochim. Cosmochim. Acta* 55:2317–2337.

Close, L. M., Roddier, F., Northcott, M. J., Roddier, C., and Graves, J. E. 1997. Adaptive optics $0''.2$ resolution infrared images of HL Tauri: Direct images of an active accretion disk around a protostar. *Astrophys. J.* 478:766–777.

Croswell, K., Hartmann, L., and Avrett, E. 1987. Mass loss from FU Orionis objects. *Astrophys. J.* 312:227–242.

Cuzzi, J. N., Dobrovolskis, A. R., and Champney, J. M. 1993. Particle-gas dynamics in the midplane of a protoplanetary nebula. *Icarus* 106:102–134.

Cuzzi, J. N., Dobrovolskis, A. R., and Hogan, R. C. 1996. Turbulence, chondrules, and planetesimals. In *Chondrules and the Protoplanetary Disk,* ed. R. H. Hewins, R. H. Jones, and E. R. D. Scott (Cambridge: Cambridge University Press), pp. 35–43.

D'Alessio, P., Calvet, N., and Hartmann, L. W. 1997. The structure and emission of accretion disks irradiated by infalling envelopes. *Astrophys. J.* 474:397–406.

D'Alessio, P., Canto, J., Calvet, N., and Lizano, S. 1998. Accretion disks around young objects. I. The detailed vertical structure. *Astrophys. J.* 500:411–427.

Dubrulle, B., Morfill, G., and Sterzik, M. 1995. The dust subdisk in the protoplanetary nebula. *Icarus* 114:237–246.

Evans, N. J., II, Balkum, S., Levreault, R. M., Hartmann, L., Kenyon, S. 1994. Molecular outflows from FU Orionis star. *Astrophys. J.* 424:793–799.

Gammie, C. F. 1996. Layered accretion in T Tauri disks. *Astrophys. J.* 457:355–362.

Goodrich, R. W. 1987. The ring-shaped nebulae around FU Orionis stars. *Pub. Astron. Soc. Pacific* 99:116–125.

Grossman, J. N. 1996. Chemical fractionations of chondrites: Signatures of events before chondrite formation. In *Chondrules and the Protoplanetary Disk,* ed. R. H. Hewins, R. H. Jones, and E. R. D. Scott (Cambridge: Cambridge University Press), pp. 243–253.

Grossman, J. N., Rubin, A. E., Nagahara, H., and King, E. A. 1988. Properties of chondrules. In *Meteorites and the Early Solar System,* ed. J. F. Kerridge and M. S. Matthews (Tucson: University of Arizona Press), pp. 619–659.

Grossman, L. 1972. Condensation in the primitive solar nebula. *Geochim. Cosmochim. Acta* 39:597–619.

Grossman, L., and Larimer, J. W. 1974. Early chemical history of the solar system. *Rev. Geophys. Space Phys.* 12:71–101.

Gullbring, E., Hartmann, L. W., Briceño, C., and Calvet, N. 1998. Disk accretion rates for T Tauri stars. *Astrophys. J.* 492:323–341.

Hartmann, L., and Calvet, N. 1995. Observational constraints on FU Ori winds. *Astron. J.* 109:1846–1855.

Hartmann, L., and Kenyon, S. J. 1985. On the nature of FU Orionis objects. *Astrophys. J.* 299:462-478.

Hartmann, L., and Kenyon, S. J. 1987*a*. Further evidence for disk accretion in FU Orionis objects. *Astrophys. J.* 312:243–253.

Hartmann, L., and Kenyon, S. J. 1987*b*. High spectral resolution infrared observations of V1057 Cygni. *Astrophys. J.* 322:393–398.

Hartmann, L., and Kenyon, S. J. 1996. The FU Orionis phenomenon. *Ann. Rev. Astron. Astrophys.* 34:207–240.

Hartmann, L., Calvet, N., Gullbring, E., and D'Alessio, P. 1998. Accretion and the evolution of T Tauri disks. *Astrophys. J.* 495:385–400.

Hayashi, C., Nakazawa, K., and Nakagawa, Y. 1985. Formation of the solar system. In *Protostars and Planets II*, ed. D. C. Black and M. S. Matthews (Tucson: University of Arizona Press), pp. 1100–1153.

Hayashi, M., Ohashi, N., and Miyama, S. M. 1993. A dynamically accreting gas disk around HL Tauri. *Astrophys. J. Lett.* 418:L71–L76.

Herbig, G. H. 1966. On the interpretation of FU Orionis. *Vistas Astron.* 8:109–125.

Herbig, G. H. 1977. Eruptive phenomena in early stellar evolution. *Astrophys. J.* 217:693–715.

Herbig, G. H. 1989. FU Orionis eruptions. In *ESO Workshop on Low Mass Star Formation and Pre–Main Sequence Objects*, No. 33, ed. B. Reipurth (Garching: European Southern Observatory), pp. 233–246.

Hewins, R. H., Jones, R., and Scott, E. R. D., eds. 1988. *Chondrules and the Protoplanetary Nebula* (Cambridge: Cambridge University Press).

Hutchison, R., Alexander, C. M. O., and Barber, D. J. 1987. The Semarkona meteorite—First recorded occurrence of smectite in an ordinary chondrite, and its implications. *Geochim. Cosmochim. Acta* 51:1875–1882.

Jarosewich, E. 1990. Chemical analyses of meteorites: A compilation of stony and iron meteorite analyses. *Meteoritics* 25:323–337.

Kawazoe, E., and Mineshige, S. 1993. Unstable accretion disks in FU Orionis stars. *Pub. Astron. Soc. Japan* 45:715–725.

Kelly, W. R., and Larimer, J. W. 1977. Chemical fractionations in meteorites. VIII. Iron meteorites and the cosmochemical history of the metal phase. *Geochim. Cosmochim. Acta* 41:93–111.

Kenyon, S. J., and Hartmann, L. W. 1987. Spectral energy distributions of T Tauri stars: Disk flaring and limits on accretion. *Astrophys. J.* 323:714–733.

Kenyon, S. J., and Hartmann, L. W. 1991. The dusty envelopes of FU Orionis variables. *Astrophys. J.* 383:664–673.

Kenyon, S. J., Hartmann, L., and Hewett, R. 1988. Accretion disk models for FU Orionis and V1057 Cygni—Detailed comparisons between observations and theory. *Astrophys. J.* 325:231–251.

Kenyon, S. J., Hartmann, L. W., Strom, K. M., and Strom, S. E. 1990. An IRAS survey of the Taurus-Auriga molecular cloud. *Astron. J.* 99:869–887.

Kenyon, S. J., Calvet, N., and Hartmann, L. 1993. The embedded young stars in the Taurus-Auriga molecular cloud. I. Models for spectral energy distributions. *Astrophys. J.* 414:676–694.

Kerridge, J. F. 1993. What can meteorites tell us about nebular conditions and processes during planetesimal accretion? *Icarus* 106:135–150.

Kerridge, J. F., and Matthews, M. S., ed. 1988. *Meteorites and the Early Solar System* (Tucson: University of Arizona Press).

Klahr, H. H., and Henning, T. 1997. Particle-trapping eddies in protoplanetary accretion disks. *Icarus* 128:213–229.

Kley, W., and Lin, D. N. C. 1996. The structure of the boundary layer in protostellar disks. *Astrophys. J.* 461:933–950.

Kley, W., and Lin, D. N. C. 1999. Evolution of an FU Orionis outburst in protostellar disks. *Astrophys. J.* 518:833–847.

Kley, W., Papaloizou, J. C. B., and Lin, D. N. C. 1993. On the angular momentum transport associated with convective eddies in accretion disks. *Astrophys. J.* 416:679–688.

Lada, C. J., and Wilking, B. A. 1984. The nature of the embedded population in the Rho Ophiuchi dark cloud—Mid-infrared observations. *Astrophys. J.* 287:610–621.

Larimer, J. W. 1973. Chemistry of the solar nebula. *Space Sci. Rev.* 15:103–119.

Larimer, J. W., and Anders, E. 1967. Chemical fractionations in meteorites. II. Abundance patterns and their interpretations. *Geochim. Cosmochim. Acta* 31:1239–1270.

Larimer, J. W., and Wasson, J. T. 1988*a*. Refractory lithophile elements. In *Meteorites and the Early Solar System*, ed. J. F. Kerridge and M. S. Matthews (Tucson: University of Arizona Press), pp. 394–415.

Larimer, J. W., and Wasson, J. T. 1988*b*. Siderophile element fractionation. In *Meteorites and the Early Solar System*, ed. J. F. Kerridge and M. S. Matthews (Tucson: University of Arizona Press), pp. 416–435.

Larson, R. B. 1980. The FU Orionis mechanism. *Mon. Not. Roy. Astron. Soc.* 190:321–335.

Lay, O. P., Carlstrom, J. E., and Hills, R. E. 1997. Constraints on the HL Tauri protostellar disk from millimeter- and submillimeter-wave interferometry. *Astrophys. J.* 489:917–927.

Lehman, T., Reipurth, B., and Brandner, W. 1995. The outburst of T Tauri star EX Lup in 1994. *Astron. Astrophys.* 300:L9–L12.

Leshin, L. A., Rubin, A. E., and McKeegan, K. D. 1997. The oxygen isotopic composition of olivine and pyroxene from CI chondrites. *Geochim. Cosmochim. Acta* 61:835–845.

Liffman, K., and Brown, M. 1995. The motion and size sorting of particles ejected from a protostellar accretion disk. *Icarus* 116:275–290.

Lin, D. N. C., and Papaloizou, J. 1985. On the dynamical origin of the solar system. In *Protostars and Planets II*, ed. D. C. Black and M. S. Matthews (Tucson: University of Arizona Press), pp. 981–1072.

Lin, D. N. C., Hayashi, M., Bell, K. R., and Ohashi, N. 1994. Is HL Tau an FU Orionis system in quiescence? *Astrophys. J.* 435:821–828.

Lin, D. N. C., Bodenheimer, P. H., and Richardson, D. C. 1996. Orbital migration of the planetary companion of 51 Pegasi to its present location. *Nature* 380:606–607.

Lynden-Bell, D., and Pringle, J. E. 1974. The evolution of viscous discs and the origin of the nebular variables. *Mon. Not. Roy. Astron. Soc.* 168:603–638.

MacPherson, G. J., Davis, A. M., and Zinner, E. K. 1995. The distribution of aluminum-26 in the early solar system—A reappraisal. *Meteoritics* 30:365–386.

Morfill, G. E., 1985. Physics and chemistry in the primitive solar nebula. In *Birth and Infancy of Stars*, ed. R. Lucas and A. Omont (Amsterdam: North Holland), pp. 693–794.

Mumma, M. J., Weissman, P. R., and Stern, S. A. 1993. Comets and the origin of the solar system: Reading the Rosetta stone. In *Protostars and Planets III,* ed. E. H. Levy and J. I. Lunine (Tucson: University of Arizona Press), pp. 1177–1252.

Mundy, L. G., Looney, L. W., Erickson, W., Grossman, A., Welch, W. J., Forster, J. R., Wright, M. C. H., Plambeck, R. L., Lugten, J., and Thornton, D. D. 1996. Imaging the HL Tauri disk at $\lambda = 2.7$ millimeters with the BIMA array. *Astrophys. J. Lett.* 464:L169–L173.

Nakagawa, Y., Sekiya, M., and Hayashi, C. 1986. Settling and growth of dust particles in a laminar phase of a low-mass solar nebula. *Icarus* 67:375–390.

Natta, A. 1993. The temperature profile of T Tauri disks. *Astrophys. J.* 412:761–770.

Osterloh, M., and Beckwith, S. V. W. 1995. Millimeter-wave continuum measurements of young stars. *Astrophys. J.* 439:288–302.

Palme, H., and Boynton, W. V. 1993. Meteoritic constraints on conditions in the solar nebula. In *Protostars and Planets III*, ed. E. H. Levy and J. I. Lunine (Tucson: University of Arizona Press), pp. 979–1004.

Palme, H., Larimer, J. W., and Lipschutz, M. E. 1988. Moderately volatile elements. In *Meteorites and the Early Solar System,* ed. J. F. Kerridge and M. S. Matthews (Tucson: University of Arizona Press), pp. 436–461.

Petrov, P. P., and Herbig, G. H. 1992. On the interpretation of the spectrum of FU Orionis. *Astrophys. J.* 392:209–217.

Podosek, F. A., and Cassen, P. 1994. Theoretical, observational, and isotopic estimates of the lifetime of the solar nebula. *Meteoritics* 29:6–25.

Pollack, J. B., Hollenbach, D., Beckwith, S., Simonelli, D. P., Roush, T., and Fong, W. 1994. Composition and radiative properties of grains in molecular clouds and accretion disks. *Astrophys. J.* 421:615–639.

Reipurth, B. 1989. The HH111 jet and multiple outflow episodes from young stars. *Nature* 340:42–44.
Reipurth, B., and Aspin, C. 1997. Infrared spectroscopy of Herbig-Haro energy sources. *Astron. J.* 114:2700–2707.
Różyczka, M. N., Bodenheimer, P. H., and Bell, K. R. 1994. A numerical study of viscous flows in axisymmetric α-accretion disks. *Astrophys. J.* 423:736–747.
Ruden, S. P., and Pollack, J. B. 1991. The dynamical evolution of the protosolar nebula. *Astrophys. J.* 375:740–760.
Scott, E. R. D. 1979. Origin of anomalous iron meteorites. *Mineral. Mag.* 43:415–421.
Scott, E. R. D., and Wasson, J. T. 1975. Classification and properties of iron meteorites. *Rev. Geophys. Space Phys.* 13:527–546.
Scott, E. R. D., Haack, H., and McCoy, T. J. 1996. Core crystallization and silicate-metal mixing in the parent body of the IVA iron and stony-iron meteorites. *Geochim. Cosmochim. Acta* 60:1615–1631.
Sekiya, M. 1998. Quasi-equilibrium density distributions of small dust aggregations in the solar nebula. *Icarus* 133:298–309.
Shakura, N. I., and Sunyaev, R. A. 1973. Black holes in binary systems. Observational appearance. *Astron. Astrophys.* 24:337–355.
Shu, F. H., Adams, F. C., and Lizano, S. 1987. Star formation in molecular clouds—Observation and theory. *Ann. Rev. Astron. Astrophys.* 25:23–81.
Shu, F. H., Shang, H., and Lee, T. 1996. Toward an astrophysical theory of chondrites. *Science* 271:1545–1552.
Simon, T., and Joyce, R. R. 1988. Infrared photometry of V1057 Cygni (1971–87). *Pub. Astron. Soc. Pacific* 100:1549–1554.
Stahler, S. W. 1989. Luminosity jumps in pre-main-sequence stars. *Astrophys. J.* 347:950–958.
Stepinski, T. F., and Valageas, P. 1996. Global evolution of solid matter in turbulent protoplanetary disks. I. Aerodynamics of solid particles. *Astron. Astrophys.* 309:301–312.
Stepinski, T. F., and Valageas, P. 1997. Global evolution of solid matter in turbulent protoplanetary disks. II. Development of icy planetesimals. *Astron. Astrophys.* 319:1007–1019.
Stolper, E., and Paque, J. M. 1986. Crystallization sequences of Ca-Al inclusions from Allende: The effects of cooling rates and maximum temperatures. *Geochim. Cosmochim. Acta* 50:1785–1806.
Strom, S. E., Edwards, S., and Skrutskie, M. F. 1993. Evolutionary timescales for circumstellar disks associated with intermediate- and solar-type stars. In *Protostars and Planets III*, ed. E. H. Levy and J. I. Lunine (Tucson: University of Arizona Press), pp. 837–866.
Supulver, K. D., and Lin, D. N. C. 1999. Formation of icy planetesimals in a turbulent solar nebula. *Icarus,* in press.
Taylor, S. R. 1992. *Solar System Evolution: A New Perspective* (Cambridge: Cambridge University Press).
Turner, N. J. J., Bodenheimer, P., and Bell, K. R. 1997. Models of the spectral energy distributions of FU Orionis stars. *Astrophys. J.* 480:754–766.
Valenti, J. A., Basri, G., and Johns, C. M. 1993. T Tauri stars in the blue. *Astron. J.* 106:2024–2050.
Wai, C. M., and Wasson, J. T. 1977. Nebular condensation of moderately volatile elements and their abundances in ordinary chondrites. *Earth Planet. Sci. Lett.* 36:1–28.
Ward, W. R. 1997. Protoplanet migration by nebula tides. *Icarus* 126:261–281.
Wasson, J. T. 1978. Maximum temperatures during the formation of the solar nebula. In *Protostars and Planets,* ed. T. Gehrels (Tucson: University of Arizona Press), pp. 488–501.

Wasson, J. T. 1985. *Meteorites: Their Record of Early Solar System History* (New York: Freeman).
Wasson, J. T. 1990. Episodic accretion and high-temperature events. *Meteoritics* 25:418–419 (abstract).
Wasson, J. T. 1992. Planetesimal heating by FU Orionis-type events. *Am. Geophys. Union Trans.* 73:336 (abstract).
Wasson, J. T. 1996. Chondrule formation energetics and length scales. In *Chondrules and the Protoplanetary Nebula*, ed. R. H. Hewins, R. H. Jones, and E. R. D. Scott (Cambridge: Cambridge University Press), pp. 45–51.
Wasson, J. T., and Chou, C.-L. 1974. Fractionation of moderately volatile elements in ordinary chondrites. *Meteoritics* 9:69–84.
Weidenschilling, S. J. 1977. Aerodynamics of solid bodies in the solar nebula. *Mon. Not. Roy. Astron. Soc.* 180:57–70.
Weidenschilling, S. J. 1980. Dust to planetesimals—Settling and coagulation in the solar nebula. *Icarus* 44:172–189.
Weidenschilling, S. J. 1988. Formation processes and timescales for meteorite parent bodies. In *Meteorites and the Early Solar System*, ed. J. F. Kerridge and M. S. Matthews (Tucson: University of Arizona Press), pp. 348–371.
Weidenschilling, S. J., and Cuzzi, J. N. 1993. Formation of planetesimals in the solar nebula. In *Protostars and Planets III*, ed. E. H. Levy and J. I. Lunine (Tucson: University of Arizona Press), pp. 1031–1060.
Weintraub, D. A., Sandell, G., and Duncan, W. D. 1991. Are FU Orionis stars younger than T Tauri stars? Submillimeter constraints on circumstellar disks. *Astrophys. J.* 382:270–289.
Welty, A. D., Strom, S. E., Strom, K. M., Hartmann, L. W., Kenyon, S. J., Grasdalen, G. L., and Stauffer, J. R. 1990. Further evidence for differential rotation in V1057 Cygni. *Astrophys. J.* 349:328–334.
Welty, A. D., Strom, S. E., Edwards, S., Kenyon, S. J., and Hartmann, L. W. 1992. Optical spectroscopy of Z Canis Majoris, V1057 Cygni, and FU Orionis: Accretion disks and signatures of disk winds. *Astrophys. J.* 397:260–276.
Whitney, B. A., and Hartmann, L. 1993. Model scattering envelopes of young stellar objects. II. Infalling envelopes. *Astrophys. J.* 402:605–622.
Whitney, B. A., Kenyon, S. J., and Gomez, M. 1997. Near-infrared imaging polarimetry of embedded young stars in the Taurus-Auriga molecular cloud. *Astrophys. J.* 485:703–734.
Woolum, D. S., and Cassen, P. M. 1999. Astronomical constraints on nebular temperatures: Implications for planetesimal formation. *Meteorit. Planet. Sci.*, in press.
Yoneda, S., and Grossman, L. 1995. Condensation of CaO–MgO–Al_2O_3–SiO_2 liquids from cosmic gases. *Geochim. Cosmochim. Acta* 59:3413–3444.
Zinnecker, H., McCaughrean, M. J., and Wilking, B. A. 1993. The initial stellar population. In *Protostars and Planets III*, ed. E. H. Levy and J. I. Lunine (Tucson: University of Arizona Press), pp. 429–495.
Zolensky, M. E., Krot, A. N., and Scott, E. R. D., eds. 1997. *Workshop on Parent-Body and Nebular Modification of Chondritic Materials*, LPI Tech. Rep. 97–02, Part 1 (Houston: Lunar and Planetary Institute).

FORMATION OF CHONDRULES AND CAIs: THEORY VS. OBSERVATION

RHIAN H. JONES
University of New Mexico at Albuquerque

TYPHOON LEE
Academia Sinica, Institute of Earth Science and Institute of Astronomy and Astrophysics

HAROLD C. CONNOLLY, JR.
California Institute of Technology

STANLEY G. LOVE
NASA Johnson Space Center

and

HSIEN SHANG
University of California at Berkeley

Chondrules and calcium- and aluminum-rich inclusions (CAIs) are small pieces of rock that probably formed in the solar nebula when fine-grained dustballs were melted and crystallized in short-lived heating and cooling events. Many different models have been proposed to explain the heat source for these events. Studies of the mineralogical, compositional, and isotopic properties of chondrules and CAIs provide important observational constraints that any viable model must explain. Several models may be dismissed, at least in their current formulations, because they are unable to meet fundamental constraints. We examine three currently popular models for chondrule and CAI formation in more detail: nebular lightning, the shock wave model, and the x-wind model. None of the models is entirely satisfactory, and the origin of chondrules and CAIs remains enigmatic.

I. INTRODUCTION

Chondrules are nearly round, submillimeter-sized objects consisting mainly of Mg-Fe silicates. They appear to have solidified by the quick cooling of melt droplets in the early solar system. They are the dominant component in the most common and primitive classes of meteorites, the chondrites, occupying as much as 80% of the total volume (Fig. 1). A second group of similar, but less abundant, objects is the Ca-Al-rich inclusions (CAIs), which consist mainly of Ca-Al silicates and oxides. Because chondrites are considered to be the building blocks of the terrestrial planets, the formation of chondrules (and CAIs) is an important episode in the formation of our solar system (Hewins et al. 1996). Most researchers agree that solid precursors of chondrules and CAIs were melted during rapid heating events that lasted for intervals of the order of minutes; then the molten droplets cooled rapidly, over intervals of hours to days. Some chondrules and CAIs appear to have gone through this cycle more than once. However, most questions concerning the origin of chondrules and CAIs still lack definitive answers. The most fundamental question is, Where and how did the heating take place? Most researchers favor an environment in which chondrules and CAIs were free-floating objects within the solar nebula, but some still prefer planetary surface environments. Wherever they formed, there is no agreement on the most viable heating mechanism.

In the last decade, substantial progress toward solving the riddle of chondrule and CAI origins has been made. The improved understanding of the formation history of sunlike stars and their accretion disks means that chondrule formation can be addressed in a more rigorous astrophysical framework. There have also been major advances in microbeam tech-

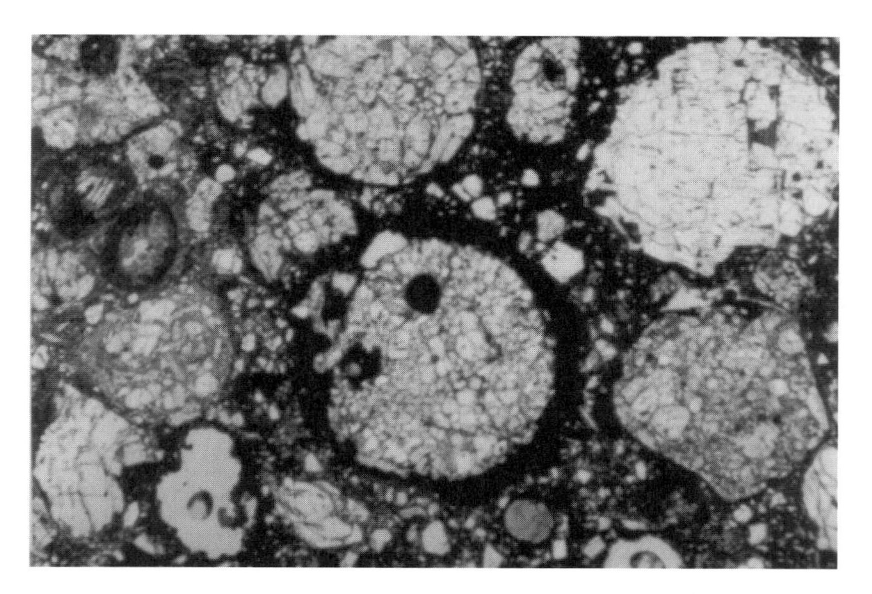

Figure 1. Texture of an unequilibrated ordinary chondrite (Ragland, LL3), showing chondrules (light) and fine-grained silicate-rich matrix which shows as black. In this transmitted-light image, metal and sulfide minerals are opaque and are also black. The chondrule in the center has a well-defined, fine-grained rim. Photo is 3 mm across.

niques that perform *in situ* analyses of the chemical and isotopic compositions of chondrule constituents on a scale of tens of micrometers or less. A third impetus has come from more sophisticated simulation experiments that have explored the effect of different parameters such as crystallization seeds, chemistry, and thermal history on the formation of chondrule textures, compositions, and phase relationships.

In this chapter we first distill a set of constraints for chondrule and CAI formation models from simulation experiments, petrographic examination, and *in situ* analyses. Then we briefly discuss why many proposed models, at least in their most naïve form, can be dismissed, because they violate some of the most critical observational constraints. Finally, we explore in more detail the three models we consider to be most viable: lightning, the nebula shock model, and the x-wind model. We suggest additional petrographic information that could potentially be used as discriminatory tests for these models.

II. OBSERVATIONAL CONSTRAINTS

In Table I, we summarize the constraints for the chondrule and CAI formation models described below and the observations that support them. The interpretation of some observations is more ambiguous than others, and we have indicated those that are not generally accepted.

Note that there are several different types of CAIs. Types B and C CAIs are coarse-grained and crystallized from melt droplets (Fig. 2). Other types of CAIs are fine-grained and have irregular shapes. In this chapter we discuss only the formation of molten (igneous) CAIs; this restriction is implied in most of the discussion where we use the term CAI.

A. Chondrites

Chondrule properties should be regarded in the context of the chondritic rocks in which they are observed. In brief, there are three main classes of chondrites: ordinary (O), carbonaceous (C), and enstatite (E). Within these classes there are 12 well-established groups. In addition, there are two less well-established groups, Rumuruti-like (R) and Kakangari-like (K) (see Table II). The groups are characterized by properties such as sizes and abundances of various components (chondrules, CAIs, matrix, etc.), bulk chemistry, and O isotopic composition (e.g., Brearley and Jones 1998), and the variation between the groups indicates that each one formed in a localized region. Each chondrite group may represent a sample as restricted as a single parent body from the asteroid belt.

B. Abundances of Chondrules and CAIs

Chondrules are the most abundant component of most types of chondrites (up to 80 vol%: Table II), providing evidence that chondrule formation affected a large proportion of the material that accreted at 2–4 AU. We infer

TABLE I
Observational Constraints for Chondrule and CAI Formation Models

Constraint	Supporting observations
Essential constraints for which a model must account:	
High efficiency	High chondrule abundances
Nebular timescales only	Isotopic dating
Low ambient temperature	Presence of volatiles
Short heating time (minutes)	Retention of volatiles; preservation of relict grains; experimental reproduction of textures
Peak temperatures ~2000 K	Experimental reproduction of textures
Short cooling time (hours)	Experimental reproduction of textures; zoning in minerals; presence of glass
Localized process	Limited size range in each chondrite group; differences in O isotopic composition
Multiple episodes; recycling	Relict grains; compound chondrules; igneous rims
Magnetic field	Remanent magnetization
Constraints not firmly accepted by meteoriticists:	
Chondrule formation ~2 Myr after CAI formation	Short-lived radioisotope data
Elevated gas pressure	Stability of molten chondrule
Variable nebular oxidation states	Variable oxidation states of chondrules
Well-defined observations for which a heating model does not necessarily need to account:	
Size sorting	Restricted size range in each chondrite group
Presence of dust that escapes heating	Fine-grained rims

that the chondrule-forming process was common and efficient, although the high abundance of chondrules is also consistent with an additional process that selectively concentrates millimeter-sized particles in chondritic parent bodies. In comparison, CAIs are far less abundant. Igneous CAIs occur almost exclusively in CV chondrites, in which Type B CAIs constitute ~3 vol% and Type C CAIs are very rare. Chondrule abundances are reported differently by different authors, because of some lack of agreement about defining what constitutes a chondrule. Chondrules are usually recognizable as having at least some properties derived from a droplet origin, such as partially rounded outlines, the presence of glass, and tex-

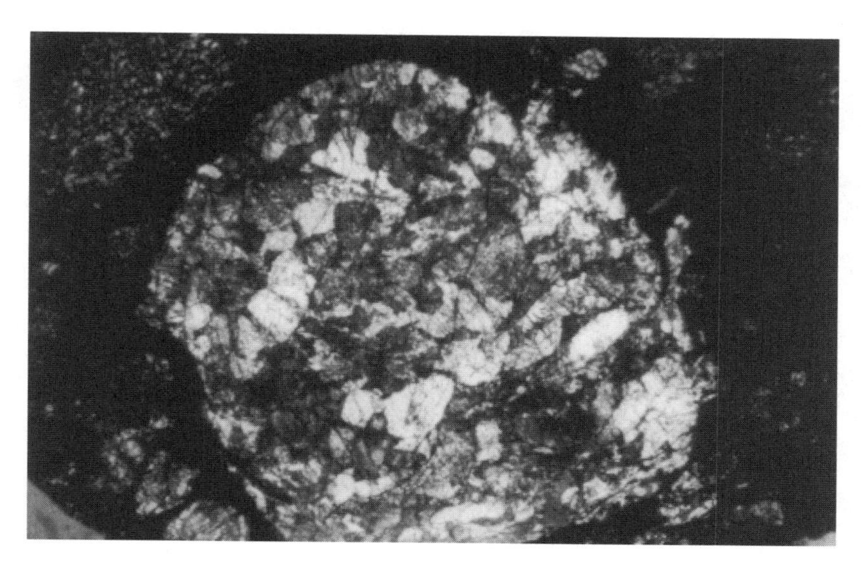

Figure 2. A Type B CAI from the CV3 meteorite Leoville (inclusion 3537–2). This object is approximately 8 mm across and is composed largely of melilite and pyroxene with minor amounts of spinel and anorthite. The chondrite matrix is black.

tures indicating crystallization from melts (igneous textures). In addition to chondrules, many chondrite groups also contain a volumetrically significant population of isolated grains and mineral fragments (usually olivine and pyroxene), which, in many cases, are inferred to be broken pieces of chondrules. Mineral fragments are included in the chondrule abundance data for the various chondrite groups in Table II.

C. Sizes of Chondrules and CAIs

Chondrules from all chondrite groups show a very restricted size range. Mean diameters range between 0.2 and 1.0 mm in all known chondrite groups, with the exception of CH chondrites (Table II). The highly reduced CH chondrites are an extraordinary chondrite group in many respects; there is some doubt that they are true chondrites formed in the solar nebula (Wasson and Kallemeyn 1990). The actual spread in chondrule sizes is wide, from "microchondrules" ($<40\ \mu$m) (Krot and Rubin 1996; Krot et al. 1997) to extraordinarily large ones, up to 5 cm across (Prinz et al. 1988; Weisberg et al. 1988), but such extreme sizes are rare. The restricted size range indicates either a heating mechanism that is specific to the submillimeter size range or the presence of a concentration mechanism for this size range that may be independent of the heating mechanism. Possible size selection mechanisms include turbulent eddies (Cuzzi et al. 1996; Paque and Cuzzi 1997) and aerodynamic coupling

TABLE II

Chondrule Abundances and Sizes in the Various Chondrite Groups

	Carbonaceous (C) chondrites							Ordinary (O) chondrites			Enstatite (E) chondrites		Other chondrite groups	
	CI	CM	CR	CO	CV	CK	CH[b]	H	L	LL	EH	EL	R	K
Chondrule abundance[a] (vol%)	$\ll$1	20	50–60	48	45	15	~70	60–80	60–80	60–80	60–80	60–80	>40	27
Chondrule mean diameter (mm)	—	0.3	0.7	0.15	1.0	0.7	0.02	0.3	0.7	0.9	0.2	0.6	0.4	0.6

[a] Chondrule abundance includes mineral fragments.

[b] Properties of the CH group may not reflect nebular processes.

The names of the O and E chondrite groups indicate the relative abundances of metallic iron present: high (H and EH), low (L and EL), and low total iron, low metal (LL). For the C class, the groups are named after a typical chondrite fall in the group, e.g., CI chondrites are similar to the Ivuna meteorite.

of gas and dust. Metallic particles also show restricted size ranges in individual chondrite groups, and silicate and metal particle size distributions may be consistent with aerodynamic sorting (e.g., Skinner and Leenhouts 1993). Chondrules in individual chondrites show lognormal size-frequency distributions, and the mean chondrule diameter is very well defined for each chondrite group. The mean chondrule size for each group must be observed empirically and is not related to any other chemical or physical properties. If a size-sorting mechanism was operative, it must have been localized and efficient. Sorting may have taken place before, during, or after chondrule formation.

For CAIs, size data have not been quantified very clearly, but the largest (up to 2.5 cm in size: Clarke et al. 1970) are found in the CV chondrites, which also have the largest chondrules; CO, CM, and ordinary chondrites contain smaller CAIs (<1 mm).

D. Chondrule and CAI Compositions

Most chondrules are ferromagnesian chondrules, dominated by Fe, Mg, Si, and O. There is considerable compositional variation, particularly in terms of FeO/MgO ratios and SiO_2 contents (e.g., Grossman et al. 1988). FeO-poor chondrules are reduced and contain metallic Fe as well as FeO-bearing silicate minerals, whereas FeO-rich chondrules are more oxidized and contain only minor metallic Fe. Within each of these types, higher SiO_2 contents lead to crystallization of a higher ratio of pyroxene relative to olivine. Most ferromagnesian chondrules also contain glass. There are arguments that all ferromagnesian chondrule compositions can be derived from FeO-rich material by a combination of evaporative loss and loss of metal beads during prolonged melting events (e.g., Sears et al. 1996). However, the siderophile ("metal-loving") element and other fractionations observed in chondrules cannot be explained entirely by such a process (Grossman et al. 1988; Grossman 1996). Also, this scenario is inconsistent with the very short heating times that chondrules experienced (see section II.G below).

Aluminum-rich (Al-rich) chondrules constitute <5% of all chondrules and occur in most chondrite types. Compositionally, they span the gap between ferromagnesian chondrules and CAIs, but their relationship to both these other types of objects is not well understood (Bischoff and Keil 1984; MacPherson and Russell 1997). They typically contain plagioclase feldspar, pyroxene, olivine, and glass.

Igneous CAIs are classified into two types, B and C, that have different proportions of the minerals melilite, spinel, pyroxene, and anorthite feldspar. Unlike chondrules, none of these CAIs contain any appreciable quantities of glass. The minerals are described as refractory, meaning that they are among the first minerals predicted to condense, under equilibrium conditions, from a cooling gas of solar composition, and among the last minerals to evaporate on heating. Although refractory during condensation, CAIs have melting temperatures lower than many chondrules. The differences between Type B and C CAIs probably reflect differences in precursor assemblages rather than a genetic relationship between the two types.

E. Oxygen Isotopic Compositions of CAIs and Chondrules

Oxygen isotopic variations are widespread in the solar system. In an oxygen three-isotope diagram (Fig. 3), different constituents of unequilibrated chondrites all have their own oxygen isotopic variation patterns. The anhydrous (water-free) minerals (AM) in carbonaceous chondrites (CC) populate a line with a slope of 1 (CCAM line). These data may be interpreted as a mixture of two or more distinct components with different oxygen isotopic compositions, such as a nucleosynthetic component from supernovae (^{16}O-rich) and the nebular gas (^{16}O-poor; Clayton et al. 1973). Alternatively, the data may also be explained as a mass-independent isotopic fractionation effect (Thiemens 1996).

Chondrules from different meteorite groups have limited ranges and plot in different parts of the O isotope diagram, suggesting different formation sites or times for chondrules in different chondrite groups (Clayton et al. 1983; Clayton 1993). Oxygen isotopic compositions of chondrules all lie relatively close to the terrestrial fractionation line (TF), although

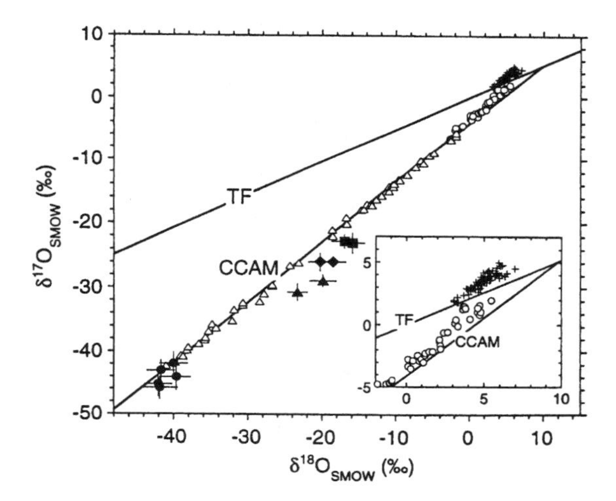

Figure 3. Oxygen isotopic compositions of chondrules and CAIs from ordinary and carbonaceous chondrites. Reprinted with permission from K. D. McKeegan et al. (1998), *Science* 280:414–418. Copyright (1998) American Association for the Advancement of Science. $\delta^{17}O_{SMOW}$ and $\delta^{18}O_{SMOW}$ refer to deviations per thousand, relative to standard mean ocean water (SMOW). Data for chondrules from ordinary chondrites (crosses) lie above the terrestrial fractionation line (TF), whereas data for chondrules from carbonaceous chondrites (circles) lie below TF (see inset). Data for whole CAIs from carbonaceous chondrites (open triangles) fall along a line of slope ~1, known as the carbonaceous chondrite anhydrous minerals (CCAM) line. Recent data for CAIs from ordinary chondrites (filled symbols) lie on the same line.

measurable compositional spreads are observed. In contrast, oxygen isotopic compositions of CAIs show a much wider spread along the CCAM line. McKeegan et al. (1998) have shown that oxygen isotopic compositions of the rare CAIs in ordinary chondrites plot near the same slope 1 line as the CAIs in carbonaceous chondrites. A possible interpretation of this result seems to be that all CAIs from different groups of meteorites share a common origin.

F. Chondrule and CAI Precursors

The immediate precursor material of chondrules probably consisted of individual dustballs, each one having a unique composition. Components of the dustball are thought to be fine-grained. In porphyritic chondrules, relict grains tens of microns in size are preserved (see subsection II.G), while the majority of chondrule precursor material melted. This suggests that most of the host chondrule precursor material was relatively fine-grained. Although there is no constraint on a lower limit for precursor grain size, porphyritic textures require preservation of crystal nuclei from the precursor dust, which probably requires initial grain sizes in the range of micrometers (see section II.G).

G. Thermal Histories of Chondrules and CAIs

Thermal histories of chondrules and CAIs are a function of four important features: ambient temperature, peak temperatures of melting, duration of melting, and cooling rates. An understanding of these variables has been achieved through a comparison of the textures, mineral chemistries (including elemental zonation within crystals), and bulk compositions of the natural objects with synthetic analogs produced in the laboratory. We use the expression "integrated T/t" when we describe the temperature and the duration of a heating process. This is an important parameter, because, for kinetically driven processes, the effect of heating to a high peak T for a short time can be equivalent to that of heating to a lower T for a longer time. The thermal histories we describe below are summarized in Fig. 4.

Earlier experiments designed to constrain thermal histories of chondrules emphasized the effect of cooling rate to produce variations in texture. However, it is now apparent that one key to interpreting chondrule and CAI textures is the number of crystal nuclei present in the melt as it begins to cool (e.g., Lofgren 1996). For example, radiating and barred textures are produced from melts with few nuclei remaining, whereas porphyritic textures are the product of a melt in which many nuclei survived (Fig. 5). One source of crystal nuclei is incomplete melting of precursor dust grains. When this is the case, all other parameters being equal, porphyritic textures represent a lesser degree of heating (lower integrated T/t) than barred textures. It is also possible, however, that chondrules collided with dust grains while they were molten and that these "seed" grains controlled the texture, especially if no nuclei derived from precursor dust survived melting (Connolly and Hewins 1995).

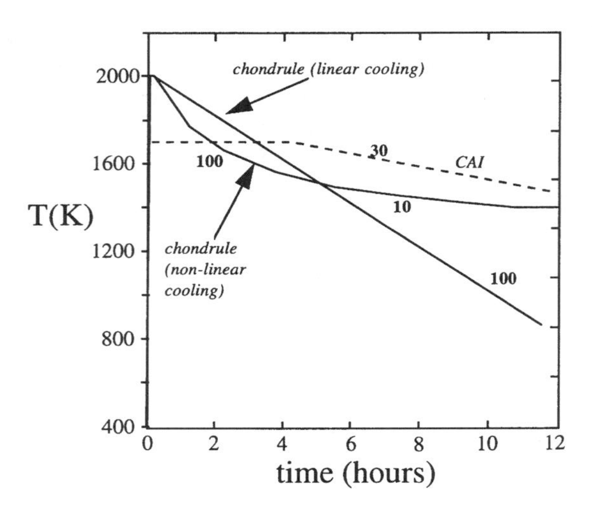

Figure 4. Schematic thermal histories for chondrules (solid lines) and igneous CAIs (dashed line). Cooling rates in kelvins per hour are indicated. Inferred thermal histories for CAIs have more extended times at peak temperatures and lower cooling rates than those for chondrules.

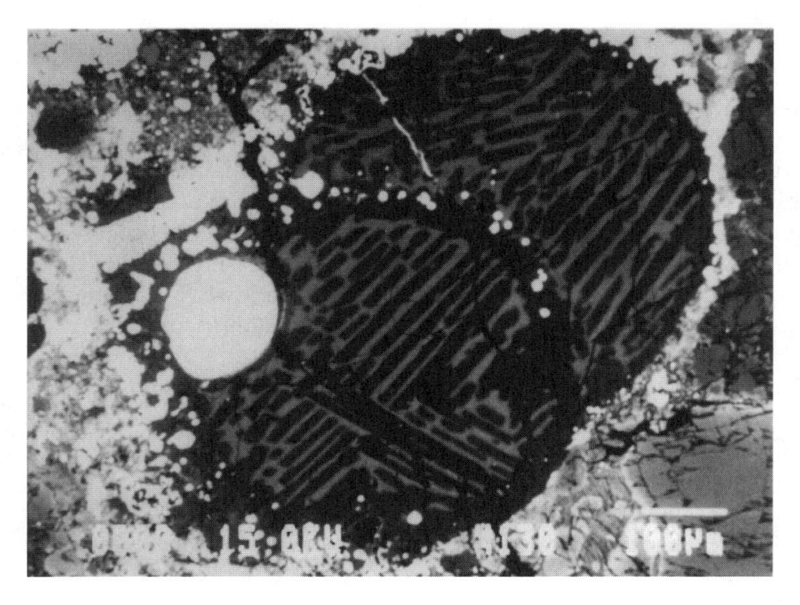

Figure 5. Textures of chondrules and experimental analogs. All the images are back-scattered electron images, in which phases with higher average atomic numbers appear brighter. (a) A compound chondrule from the Semarkona (LL3) ordinary chondrite, consisting of a pair of barred olivine chondrules. The dark gray grains are elongate olivine crystals, and the light gray interstitial material is glassy mesostasis. Both chondrules have abundant metal blebs (white) around their rims.

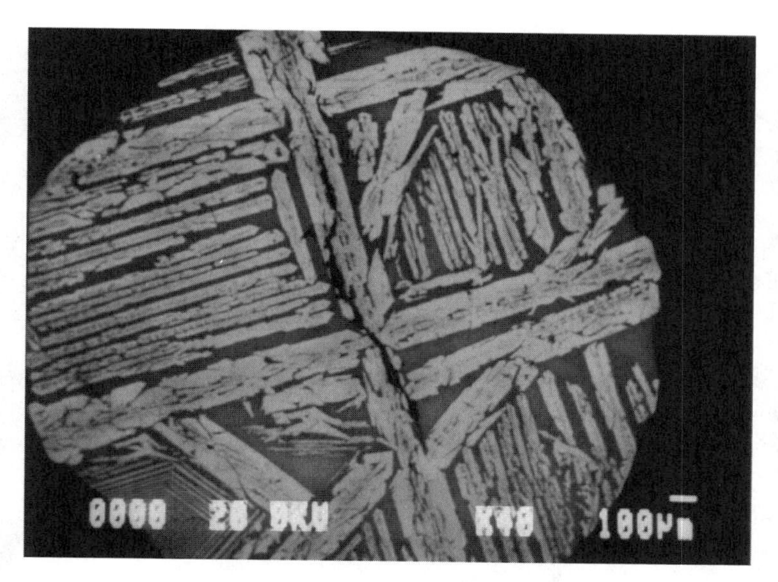

Figure 5. (b) Synthetic barred olivine chondrule showing a similar texture to (a). This texture can be produced in flash-melting experiments, in which the peak temperature is maintained for only seconds to minutes.

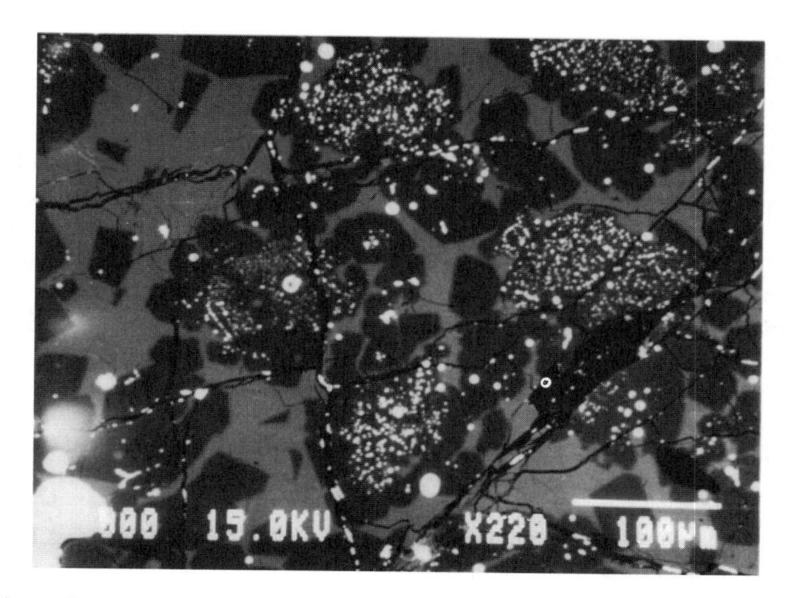

Figure 5. (c) FeO-poor (type IA) porphyritic olivine chondrule from the Semarkona (LL3) ordinary chondrite. Olivine grains are dark gray and show straight crystal faces; mesostasis glass is light gray. Metal (white) occurs in two associations: as larger blebs occurring throughout the olivine and mesostasis, and as hosts of small blebs in the interiors of "dusty" olivine grains. The small blebs were produced by reduction of more FeO-rich relict olivine grains that were incorporated into the host chondrule precursor assemblage and escaped melting.

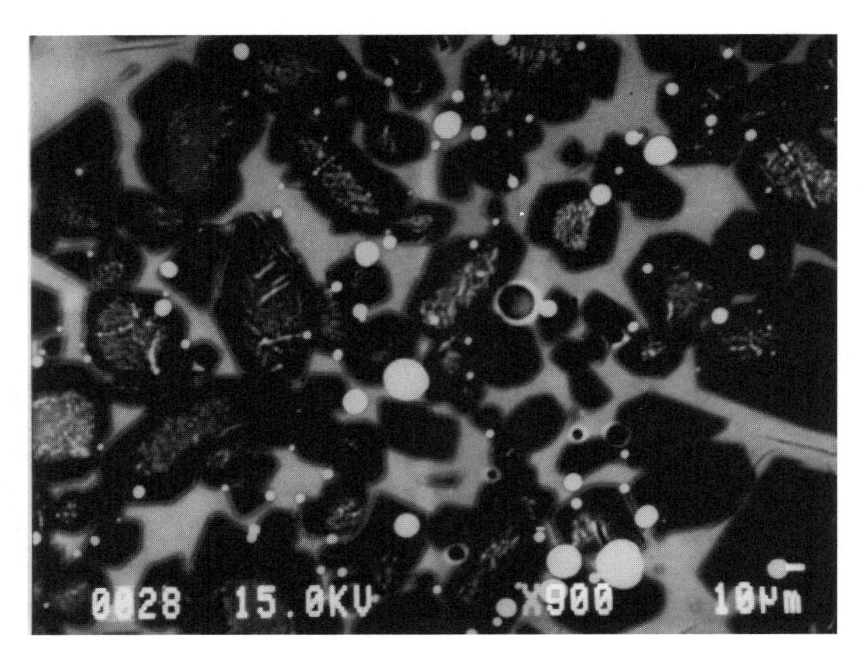

Figure 5. (d) Synthetic FeO-poor chondrule showing a similar texture to (c). This experiment contained carbon in its starting material, which, upon heating, reduced the FeO in the unmelted silicate grains to Fe metal. Scale bars in (a), (b), and (c) are 100 microns, in (d) is 10 microns.

1. Ambient Nebular Temperature. The presence of moderately volatile elements such as Na and S within chondrules, assuming that they were inherited from their precursors, provides a constraint on the premelting ambient nebular gas temperature within the region of chondrule formation (Grossman 1988; Wasson 1996; Zanda et al. 1996; Connolly and Love 1998; Rubin et al. 1999). This temperature must have been below that at which S (usually in the form of FeS) and Na evaporate: ~650 K and ~970 K, respectively (Wasson 1985).

CAIs contain almost no volatile elements. The ambient nebular temperature for CAIs is constrained to be around 1400–1500 K.

2. Duration of the Heating Event. The melting interval for chondrules was originally postulated to have lasted for several hours (Hewins 1989; Lofgren 1996) but more recent experiments have restricted this time to less than a few minutes in order to facilitate retention of moderately volatile elements such as Na and S. Experiments conducted at a range of pressures and oxygen fugacities have shown that if heating from low ambient nebular temperatures was slow, more than a few minutes, or if melting at peak T was longer than several minutes, chondrules would not retain these volatile elements (Radomsky and Hewins 1990; Yu et al. 1996; Hewins et al. 1997; Connolly et al. 1998; Yu and Hewins 1998).

Another independent constraint on the duration of chondrule melting is the presence of relict grains that were not in contact with melt for a long enough time to dissolve. Survival of relict grains in chondrules is a function of the integrated T/t close to peak temperatures, and it constrains the maximum time that appreciable melt was present to tens of seconds to several minutes (Connolly et al. 1994; Greenwood and Hess 1996; Hewins and Connolly 1996).

3. Peak Temperatures. Estimates of peak melting temperatures are obtained by comparisons between chondrule simulation experiments, the calculated equilibrium liquidus temperatures of chondrules (temperatures of complete melting) and chondrule textures (Hewins and Radomsky 1990; Lofgren 1996; Connolly et al. 1998). The total range of estimated peak temperatures of chondrule formation is from ~1800 to 2200 K. Pivotal to these estimates is the production of barred olivine textures (Hewins and Radomsky 1990), which, if no external seeds interacted during formation, require very precise melting and cooling conditions (Lofgren and Lanier 1990; Radomsky and Hewins 1990). The peak T for a given composition cannot be estimated precisely because of the importance of kinetic factors such as the duration of the heating interval and the precursor grain size.

For molten CAIs, T/t paths determined from experiments are significantly different from those determined for chondrules (Fig. 4). Based on growth characteristics of melilite, the dominant silicate mineral in Type B CAIs, the maximum melting temperature is inferred to be ~1700 K with a duration of several hours (Stolper 1982; Stolper and Paque 1986; Beckett et al. 1988; Paque and Lofgren 1993). However, initial melting conditions similar to those experienced by chondrules have not yet been fully tested on CAI compositions, and the possibility of similar heating cycles for melted CAIs and chondrules should not be eliminated (Paque 1995). Although Type B CAIs appear to contain several generations of spinel (Meeker 1995; Connolly and Burnett 1998), dissolution times have not been estimated.

Any heating mechanism that heats particles in good thermal contact with the nebular gas might be expected to produce a narrow range of peak temperatures. This is because, at nebular pressures, molecular hydrogen dissociates at temperatures comparable to chondrule and CAI peak temperatures, e.g. 1700–2000 K at 10^{-4} atm (Liepmann and Roshko 1957; Scott et al. 1996; Wasson 1996). The large energy per unit mass required to dissociate hydrogen makes it an excellent "thermostat" for both the gas and any solid materials in good thermal contact with it.

4. Cooling. Variations in cooling rate affect the textures and mineral chemistries of chondrules and CAIs, as well as bulk compositions, if evaporation occurred during this interval. In general, chondrules cooled much quicker than CAIs: Approximate linear cooling rates are 50–1000°/hr for chondrules and 2–50°/hr for CAIs (Lofgren and Russell 1986; Stolper and

Paque 1986; Radomsky and Hewins 1990; Jones and Lofgren 1993; Lofgren 1996; Hewins et al. 1997). These cooling rates are relevant to the temperature range in which a significant proportion of the object is molten (~1600–2000 K for chondrules). Preservation of chemically zoned mineral grains and the presence of glass not only are indicators of rapid cooling but also argue that cooling continued down to low ambient temperatures and that no significant reheating or annealing occurred. Small values of the integrated T/t for a reheating or annealing event (e.g., 1200 K for a few seconds or 800 K for a few hours) would be difficult to detect. However, longer durations or a higher temperature would result in homogenization of zoning by solid-state diffusion, and glass would crystallize (devitrify) (Connolly and Hewins 1992; Jones and Lofgren 1993; DeHart and Lofgren 1996). Wasson (1996) proposed that observed chondrule textures are produced by multiple low-temperature annealing events on largely glassy precursors, but chondrule textures have not been reproduced experimentally in this manner.

Although most chondrule and CAI analog experiments have determined linear cooling rates, it is more likely that cooling curves in any natural process would be nonlinear (Yu et al. 1996; Yu and Hewins 1998). Microstructures in plagioclase and pyroxene in some chondrules record cooling rates of tens of degrees per hour in the temperature range 1450–1600 K (Weinbruch and Müller 1995). Typical nonlinear cooling curves may be like the one illustrated in Fig. 4.

Cooling rates of chondrules and CAIs are slower than would be expected from simple radiative heat loss to the surrounding ambient nebular gas, implying that some type of heat-buffering effect accompanied their formation. This is commonly attributed to high dust/gas ratios relative to a gas of solar composition, which insulates the formation regions (Wood 1984; Grossman 1988; Sahagian and Hewins 1992). Alternatively, thermal buffering could also have been achieved through a hot, compressed gas (as in the shock wave model: see section III.B.2). The estimated cooling rates, as well as short times at peak temperatures, are inconsistent with models that require heating of large volumes of dusty clouds (e.g., >100 km in size), which would not allow sufficiently rapid cooling (Sahagian and Hewins 1992; Boss and Graham 1993; Wasson 1996). This concern further supports the interpretation that chondrule formation was a fairly localized process.

H. Oxygen Fugacity and Pressure of Ambient Nebular Gas

Chondrules and CAIs provide limited information about the surrounding gas during the formation interval. During the short heating times for chondrules, there would be only limited exchange of material between chondrules and gas, even at high temperatures, and the chondrules would not necessarily achieve equilibrium with the gas. Evidence for such limited interactions comes from the fact that O isotopic compositions of some chondrules that have undergone more intense heating and melting (e.g., barred olivine chondrules) indicate a higher degree of equilibration with the gas than those of porphyritic chondrules do (Clayton et al. 1983).

The oxidation states of chondrules vary greatly. If chondrule oxidation states were controlled by the surrounding gas, FeO-poor chondrules record conditions similar to that of a gas of solar composition, but FeO-rich chondrules record much more oxidizing conditions. Increases in oxidation states have been explained by the evaporation of fine dust in different concentrations. Alternatively, the degree of reduction of an individual chondrule may be controlled to a large extent by the abundance of reducing phases, particularly C, in the precursor assemblage (Connolly et al. 1994; Hewins 1997), and hence may record only limited information about the surrounding gas.

Little is known about the oxidation state of CAIs. Determinations of the oxidation state of Ti in hibonite suggest that they formed under reducing conditions and that they could have equilibrated with a gas of solar composition (Beckett et al. 1988).

Chondrules may also provide clues concerning the local gas pressure within their formation region, although this constraint is very uncertain. During condensation, liquids of chondrule compositions are stable against evaporation only at pressures greater than $\sim 10^{-3}$ atm (Wood and Hashimoto 1993; Ebel and Grossman 1997, 1998), significantly higher than canonical values of $\sim 10^{-6}$ atm. Synthetic chondrules lose most of their Na within 10 minutes of melting at pressures of 10^{-5} atm (Yu and Hewins 1998), so retention of Na in chondrules also implies an elevated pressure. Local pressures of 10^{-3} atm may be obtained by enhancing the dust/gas ratio to 100–1000 times that of a gas of solar composition. Alternatively, chondrules and CAIs could have formed in a region of the nebula that was at a higher total pressure.

I. Multiple Heating and Recycling

At least 25% of chondrules provide evidence that they have experienced multiple heating events (e.g., Rubin and Krot 1996). This evidence includes the presence of (1) coarse-grained rims, (2) compound chondrules, which consist of two or more chondrules joined together (Fig. 5a), and (3) relict grains (Fig. 5c). Coarse-grained, or igneous, rims have been interpreted as partially melted dusty mantles on primary chondrules (Rubin and Krot 1996). Although many compound chondrules may be produced by collisions of partially molten objects, certain types consist of a primary chondrule that is entirely entrained in a larger secondary chondrule. The secondary chondrule is considered to have formed by heating and melting of fine-grained dust that had accreted onto the surface of the primary (Wasson et al. 1995). Relict grains have chemistries, textures, and O isotopic compositions that indicate that they are recycled from a previous generation of chondrules (Jones 1996; Jones and Danielson 1997; Jones et al. 1998). Hence, disruptive collisions occurred commonly between heating events. An estimate of the number density of chondrules in the chondrule-forming region is of the order of 1×10^{-3} cm^{-3} (Wasson 1993).

Many Type B CAIs contain a thin, multilayered rim, commonly referred to as a Wark-Lovering rim, which probably formed when the CAIs were rapidly melted and quickly cooled after the object had crystallized (MacPherson et al. 1988; Davis and MacPherson 1996). Multiple heating events may also be necessary to reproduce spinel rims (Paque et al. 1998). There is still some question as to whether CAIs contain relict grains. Some studies suggest that relict grains occur (Meeker 1995; Paque 1990; Connolly and Burnett 1998).

J. Fine-grained Rims

Many chondrules have fine-grained rims, consisting of a complex mixture of silicate, metal, and sulfide material that is similar in composition to the chondrite matrix (the fine-grained material that occurs interstitially to chondrules, CAIs, etc.). This material is interpreted as dust that adhered to the chondrules between the intervals of chondrule formation and accretion and escaped the heating process that formed chondrules. There is considerable variability in rim thicknesses between different chondrite groups, although within individual chondrite groups chondrule rim thicknesses are more uniform. Thick rims (up to ~20% of the chondrule diameter) are observed on chondrules in CM and CV chondrites, whereas in ordinary and enstatite chondrites rims are very narrow or nonexistent. Rim thicknesses generally show a positive correlation with chondrule diameters (Metzler et al. 1992; Paque and Cuzzi 1997), and this correlation has been used to argue for chondrule formation at sites local to the parent bodies (Morfill et al. 1998). Some Type B CAIs also have fine-grained accretionary rims.

K. Magnetism

Chondrules preserve a record of magnetic fields of variable strength that were present during their formation (e.g., Sugiura et al. 1979; Sugiura and Strangway 1982; Nagata and Funaki 1981; Morden and Collinson 1992). The magnetic component, which is randomly oriented, was probably acquired when chondrules cooled through the Curie temperature of their magnetic minerals. The field strengths recorded are fairly strong, between 0.1–0.7 mT (~1–10 G). CAIs do not contain magnetic minerals, so the magnetic properties of the environment in which they formed is unknown.

L. Timing of Chondrule and CAI Formation

One of the most fundamental questions about chondrules and CAIs is the timing of the formation events. There are two questions to consider: when chondrule formation occurred relative to CAI formation, and the absolute age of chondrules and CAI relative to the history of the solar system.

The question of the relative ages of CAIs and chondrules does not have a straightforward answer, because the ideal chronometer does not exist. The ^{26}Al chronometer sheds some light, but there are problems with interpreting the data. Details are discussed by MacPherson et al. (1995) and the chapter by Wadhwa and Russell, this volume. One problem with the ^{26}Al data is that most of the chondrules in which abundances have been measured are the relatively rare Al-rich variety. A simple interpretation of the ^{26}Al data is that CAIs are the oldest processed material in the solar system and that chondrules were formed at least 2 Myr later than most CAIs. The problem of mixing material formed 2 Myr apart into the same chondrite parent body constitutes a severe challenge to many nebular models as well as chondrule/CAI formation models.

Other short-lived radioisotopes that have been measured in both chondrules and CAIs include ^{53}Mn and ^{129}I (see Swindle et al. 1996 and the chapter by Wadhwa and Russell, this volume). Instead of dating chondrule and CAI formation events themselves, both systems may record secondary events that postdate formation times and occurred either in the nebula or on parent bodies. The results are not very systematic but are generally consistent with the notion that most secondary activities were over within about 10 Myr of CAI formation.

The problem with absolute age dating of CAIs and chondrules is that most action happened within the first 0.01 Gyr (i.e. 0.2%) of the 4.5-Gyr history of the solar system. To establish a chronology of events in such a short interval so long ago taxes conventional radionuclide dating techniques to their limit. So far only the $^{235}U/^{207}Pb$ system has been applied to CAI and chondrule dating successfully with time resolution approaching 1 Myr. The Pb-Pb age of CAIs most often quoted is 4.566 Gyr (Allègre et al. 1995), which is slightly higher than the earlier results of Chen and Wasserburg (1981), 4.559 Gyr. No suitable minerals for U/Pb dating occur as crystallization products of chondrule melts. However, phosphate minerals that occur in chondrites as a secondary product of thermal metamorphism are suitable and can be used to define lower limits for chondrule formation ages. The oldest chondritic phosphate ages have been determined in H4 meteorites (Göpel et al. 1994) and are only 3–5 Myr younger than the Allende CAI age.

All chronometers point to chondrule and CAI formation very early, within about 10 Myr of initial CAI formation. This timescale is comparable to the lifetime of the nebula (Podosek and Cassen 1994). Because many planetary processes that have been suggested for chondrule formation would also be expected to persist later in solar system history, even up to the present day, this evidence suggests that chondrules and CAIs formed in the solar nebula.

M. Chondrule and CAI Formation: The Same Process?

One question that arises from the above discussion is whether molten CAIs and chondrules were formed by the same process. Despite the obvious differences in precursor mineral compositions, several other properties distinguish the two processes: CAI formation apparently occurred in a significantly earlier epoch than chondrule formation, and CAIs appear to have undergone slightly different thermal histories, including slower cooling rates, compared with chondrules. At present it is not possible to distinguish between two possibilities: (1) a single process that was operative over a period of several million years in the solar nebula and was sufficiently variable in nature to incorporate the constraints of both types of objects, or (2) different processes for CAI and chondrule formation. This question is left open in the following discussion.

III. THEORETICAL MODELS FOR CHONDRULE AND CAI FORMATION

A. Models We Dismiss

Many diverse heating mechanisms for chondrules have been proposed over the last century, but no single model is clearly preferable to all others. Most models have not addressed formation of CAIs specifically. The status of a variety of models was summarized by Boss (1996). In this section we briefly discuss the predictions of some of the models that we consider to be less viable. Each of these has one or more serious inconsistencies with the observational constraints summarized in Table I.

One model that has been investigated repeatedly is the impact model. In this scenario, chondrules are spheres of melt produced in impacts between planetesimals at relative velocities of 5 km s^{-1} or more, possibly analogous to crystalline spherules observed in the lunar regolith (e.g., Symes et al. 1998). Several difficulties with this mechanism have been discussed by Taylor et al. (1983). This model produces the observed restricted range of chondrule peak temperatures only by coincidence, and it is inconsistent with multiple episodes of chondrule recycling. Although recent models (Hood 1998; Weidenschilling et al. 1998) support planetesimal velocities sufficient for impact melting, impact melt production rates on asteroids are extremely small (Keil et al. 1997). The preponderance of chondrules in common meteorites argues for a more efficient process.

A related model is one in which chondrules are produced as droplets formed in the collision of molten kilometer-sized planetesimals (Sanders 1996). This model has many of the same difficulties as the impact melt hypothesis, with the additional challenges of preserving chondritic compositions and maintaining relict grains in melt for the very long timescales associated with kilometer-sized bodies.

Another potential chondrule formation model is ablation of molten droplets from meteors entering transient protoplanet atmospheres, from high-speed planetesimals traversing the nebula, or from planetesimals caught in the young Sun's bipolar outflow jets (Liffman 1992). Support for the requisite high velocities was cited above. A difficulty with this model is a droplet's predicted cooling time once free of its parent body. The cooling time should be similar to the drag heat pulse durations of the shock wave model: tens of seconds rather than minutes to hours. Chondrule recycling is also difficult to imagine in the context of this model.

Chondrules may also have been heated in a locally hot inner protoplanetary nebula. This model is difficult to reconcile with cold precursors, short heating and cooling times, variations among chondrules from different parts of the nebula, elevated pressures, and variable oxygen fugacities.

It has been suggested that chondrules were heated by periodic activity on or near the young Sun, analogous to FU Ori outbursts (e.g., Huss 1988; Bell et al. 1995, and the chapter by Bell et al., this volume). The observed outbursts, however, last very much longer than the allowed range of chondrule heating and cooling times: Outbursts have peak durations of months followed by decades-long declines.

Chondrules could also be interstellar precursors thermally processed in the accretion shock, the discontinuity where infalling molecular cloud material struck the protoplanetary disk (Wood 1984; Ruzmaikina and Ip 1996). This scenario has difficulty explaining locally variable chondrule compositions. It also is inconsistent with chondrule recycling, because each chondrule could only encounter the shock exactly once.

B. More Viable Models

We consider three models to be more viable for chondrule and CAI formation: lightning, the shock wave model, and the x-wind model. All of these models describe processes that only occur in the nebula, so they all automatically conform to the constraint of occurring only on nebular timescales.

1. Lightning. The popularity of lightning as a chondrule formation mechanism has waxed and waned for some time (e.g., Cameron 1966; Whipple 1966; Love et al. 1995; Horanyi et al. 1995; Horanyi and Robertson 1996; Pilipp et al. 1998; Gibbard et al. 1997; Desch and Cuzzi 1999). In a terrestrial thunderstorm, hailstones that have grown to millimeter size and begun falling under the influence of gravity collide with 100-μm ice crystals, which are too small to fall efficiently. In each collision, $\sim 10^5$ electrons are preferentially transferred to the ice crystal (e.g., Keith and Saunders 1989). The continuing rainout of positively charged hailstones leads to a growing large-scale separation of charge and an increasing vertical electric field. When the field reaches a critical threshold value ($\sim 10^5$ V m^{-1}), the electrical resistance of the air breaks down and current, in the form of a lightning bolt, flows to cancel the charge separation.

Lightning in the asteroidal region of the protoplanetary nebula is envisioned to operate analogously. Here, gas turbulence, gas convection, or vertical (i.e., perpendicular to the midplane) solar gravity produces size-segregated motions of chondritic-composition particles, which are assumed to transfer charge as do ice particles in earthly thunderclouds.

Large-scale charge separation builds until the nebular gas breaks down. Reports of lightning in terrestrial volcanic plumes and dust storms (with diverse particle properties) and in outer planet atmospheres (with compositions similar to the protoplanetary nebula) support the idea of nebular lightning. So do observations of "red sprites" and "blue jets" (e.g., Sentman et al. 1995), discharges observed in the clear air above active terrestrial thunderstorms, some of which are seen at altitudes where the pressure ($\sim 10^{-5}$ bar) is comparable to that of the nebula.

Lightning is attractive as a chondrule formation mechanism, because it is consistent with most of the observed characteristics of chondrules. Lightning works in cold environments. The formulation of Morfill et al. (1993) suggests that durations of nebular discharges should be ~ 100 s, much longer than those of terrestrial lightning bolts and consistent with chondrule peak heating times. Discharge temperatures could have been buffered by hydrogen dissociation. The spatial scale of nebular discharges is not well constrained and could have been large enough to produce the observed chondrule cooling rates. Lightning can work locally and repeatedly and can produce magnetic fields that might explain the remanent magnetism of chondrules. Fine dust caught in lightning bolts would have been more strongly heated than millimeter-sized chondrule precursors (because of the former's greater ratio of cross-sectional area to mass). The dust might have evaporated, recondensing immediately after the discharge along with volatiles boiled from neighboring chondrules. The gas pressure within the hot discharge channel would be higher than in the surrounding nebula.

Despite its good agreement with the observations, concerns about the model's physical feasibility have kept lightning from being broadly accepted as the source of chondrules. Some lightning models (e.g., Desch and Cuzzi 1999) suggest discharge durations of milliseconds (as in terrestrial lightning), which are difficult to reconcile with chondrule textures (e.g., Brownlee et al. 1983) and cooling times. The electrical conductivity of the nebula may have been high enough to "short-circuit" large-scale charge separation, effectively preventing nebular lightning (e.g., Gibbard et al. 1997). If lightning did occur, its energy flux may have been too small to melt silicates (e.g., Love et al. 1995). Some models of nebular turbulence provide too little energy to produce the inferred preponderance of chondrules (e.g., Weidenschilling 1996, 1997). Many of these difficulties, however, remain in dispute (e.g., Desch and Cuzzi 1999). Better understanding of terrestrial lightning and of the nebula's electrical properties will be needed to judge finally whether nebular lightning could have formed chondrules.

2. Shock Wave Model. The idea that chondrule and CAI materials were processed within the protoplanetary nebula by a shock wave has been discussed for more than 30 years (Wood 1962; Stolper and Paque 1986; Hood and Horanyi 1991, 1993; Boss 1996; Cassen 1996; Hood and Kring

1996; Connolly and Love 1998; Hood 1998; Weidenschilling et al. 1998). Shock waves have recently gained favor with the meteorite community as a potential chondrule/CAI-forming mechanism (Boss 1996). As discussed below, the predictions of the model agree well with observations of chondrules and their host meteorites.

One of the biggest problems for the shock wave model is determining how the hypothetical shocks were produced. There is no current observational evidence for producing powerful, reliable, repeatable, and astrophysically realistic shocks. Several recent suggestions, largely theoretical, include clumpy accretion to the nebula (Boss and Graham 1993), outbursts from the protosun analogous to FU Orionis events (Boss 1996), spiral arm instabilities in the disk (Morfill et al. 1993; Wood 1996), and eccentric planetesimals moving at hypersonic speeds through the protoplanetary disk (Hood 1998; Weidenschilling et al. 1998). Although the lateral size and distance traveled by shock waves may have almost any values, their spatial scales must be limited to events that can produce the localized processes indicated by the observational constraints for chondrule formation.

A shock wave is a sharp discontinuity between hot, compressed, high-speed gas (moving faster than the local speed of sound) and cooler, less dense, slower-moving gas. The nebular shock wave model envisions nebula gas that is overrun by a shock wave and becomes abruptly heated, compressed, and accelerated. The kind of shock is assumed to be a thin, flat surface (a plane) that moved through an initially cool, quiet (turbulent velocities of $\sim$50 m s^{-1} or less, Cuzzi et al. 1996) region of the nebula composed dominantly of H_2 gas and silicate dust. For simplicity the shock is assumed to be normal (i.e., traveling in a direction perpendicular to its front surface).

Given the Mach number of the shock and the ideal gas equation of state, analytical relations govern the postshock density, pressure, velocity, and temperature in the gas, as illustrated in Fig. 6 (Hood and Horanyi 1993; Cassen 1996; Hood and Kring 1996; Scott et al. 1996; Connolly and Love 1998). Temperature and density increase moderately behind the shock wave, while pressure increases significantly. For example in a Mach 5 shock, pressure increases by a factor of 29, density by a factor of 5, and temperature by a factor of 5.8 (Fig. 6).

The major hurdle to overcome with the shock wave model is showing rigorously that such a mechanism could have melted chondrule and CAI precursors. As was determined by Hood and Horanyi (1993) and discussed by Cassen (1996) and Connolly and Love (1998), a solid particle that is overrun by a shock suddenly finds itself in a blast of wind moving at several km s^{-1}. At this time the particle begins to heat due to friction with the postshock gas, which is forcing the particle to match speeds with the gas. Heating by thermal radiation from hot neighboring particles also occurs. In addition, particles can be heated radiatively and conductively

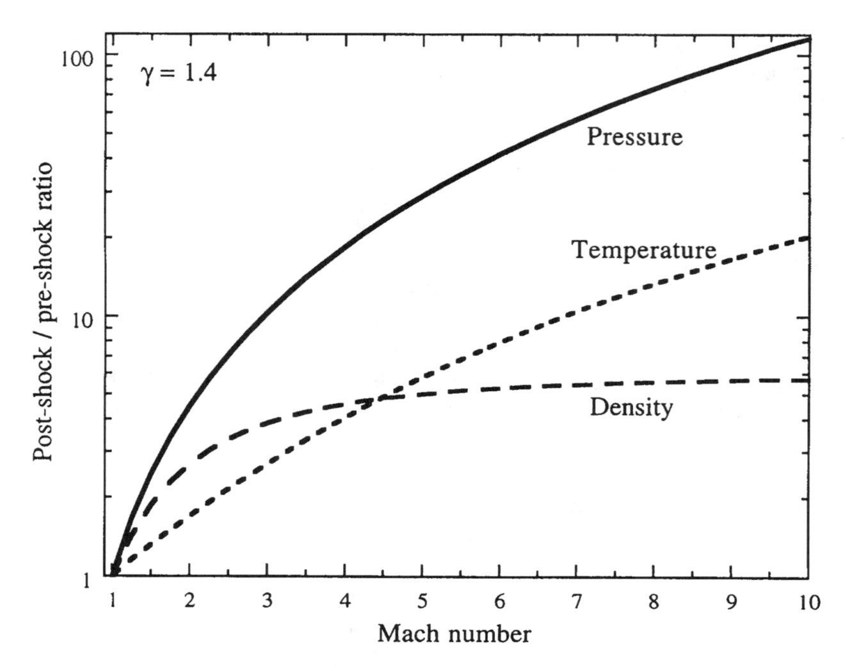

Figure 6. Post- to pre-shock pressure, temperature, and gas density ratios as a function of shock strength expressed in Mach number M. Shocks of M 3 to 8 are thought to be capable of forming chondrules. An M 5 shock increases the density by a factor of 5, the temperature by a factor of 5.8, and the pressure by a factor of 29.

by the hot postshock gas until cooling begins or a postshock rarefaction wave follows.

During the heating process, the flux of kinetic energy absorbed by a particle (drag) must at least balance the radiative loss of heat from that particle if melting is to be achieved. At the melting point, this relation can be expressed by the equation $\frac{1}{2}\rho_s V^3 \pi a^2 > 4\pi a^2 \sigma T_{\mathrm{melt}}^4$, where V is the velocity of the particle relative to the postshock gas, a is the particle radius, T_{melt} is the particle temperature, and ρ_s is the density of shocked gas (Cassen 1996). Initially the heating is mainly due to drag or friction, which will begin to approximate zero once the particle has achieved the speed of the shock. In theory, frictional heating occurs only for a finite distance and time, which is related to the stopping distance of a particle, expressed as $l_{\mathrm{stop}} = d_p \rho_p / 2\rho_s$, where d_p is the particle diameter, ρ_p is the particle density, and ρ_s is the postshock gas density. Inserting typical values into the above relations yields drag heat pulse durations of a few tens of seconds. Chondrule formation via shocks in an optically thin medium, however, cannot prevent the solidification of the molten chondrules within seconds (Cassen 1996; Hood and Horanyi 1993). For drag heating alone to produce cooling rates comparable to those of chondrules, shock waves with unreasonable speeds (~Mach 60) are required (F. Shu, personal communication, 1998).

The petrologic and geochemical constraints on chondrule formation require, however, that chondrules formed within the presence of other chondrules and fine-grained silicate dust. The mass density of chondrules and other dust particles may have exceeded that of the gas before chondrules were heated and may have provided an opaque "blanket" that slowed radiative heat loss rates to values comparable to those experimentally determined for chondrules (e.g., Hood and Kring 1996). In this optically thick scenario, shock waves of Mach 3 to 8 (a physically realistic range) produce peak temperatures and cooling rates that match those of chondrules (Cassen 1996; Hood and Horanyi 1993; Hood and Kring 1996). It should be pointed out, however, that a full analysis of cooling within an optically thick medium composed of different-sized particles has yet to be performed.

Because of space limitation, it is impossible for us to describe in detail the comparison between the predictions of the shock wave model and the meteorite evidence. Connolly and Love (1998) and Hood and Kring (1996) showed that the predictions of the model agree well with the petrologic and geochemical constraints on chondrule formation. Although the predictions of the shock wave model for chondrule formation also agree with many observations and constraints on the formation of igneous CAIs, the longer cooling times required for CAIs are difficult to reconcile with the model. Below we highlight some of the more important issues on forming chondrules by the nebular shock wave model.

The first constraint we examine is the intensity and duration of particle heating. Nebular shocks of Mach 3 to 8 can heat initially cold chondrules (as mandated by the petrologic observations discussed above) to melting temperatures for times consistent with those determined experimentally. Shock wave heating is also virtually instantaneous; particles are heated to their centers on timescales limited by their size and thermal diffusivities, ~1 s for 1-mm silicate spheres (Love and Brownlee 1991), satisfying the need to melt chondrules quickly to prevent the loss of volatile phases or elements. In addition to melting, postshock cooling rates similar to those experienced by chondrules have been determined (Hood and Kring 1996). A potential weakness of the shock wave model is that if shock waves stronger than Mach 8 occurred, they might have heated chondrules to temperatures above their experimentally derived formation temperatures. This weakness may be offset by hydrogen dissociation (see section II), which absorbs orders of magnitude more energy than simple caloric heating (Wasson 1996), and may be effective in limiting chondrule peak temperatures to the observed range even in shock waves substantially stronger than Mach 8.

An important petrological test of the shock wave model is the fact that chondrules have been recycled. Because shock waves need not be singular or identical events, they can provide the multiple heating episodes of varying intensity indicated by the texture and chemistry of some chondrules. Collisions within shock waves would also have broken up chondrules, assisting in the recycling process and potentially producing their observed fine-grained accretion rims.

3. X-wind Model. The suggestion that chondrules and CAIs may have an astrophysical origin has stimulated the study of solid material in relation to the observed phenomena of young sunlike stars. Well-documented physical processes in young, low-mass stars may contribute to the formation of chondrules and CAIs. Sorby (1877) first proposed that melting of chondrule precursors occurred near the surface of the Sun and those molten spheroids were later flown outwards in solar flares. Herbig (1977) proposed that the enhanced stellar radiation in an FU Orionis outburst melts chondrule or CAI precursors, and the ensuing mass loss drags the grains to interplanetary space. Energetic events occurring in the vicinity of young stars, such as jets and outflows, have motivated more modern investigations (e.g. Skinner 1990*a*,*b*; Liffman 1992; Liffman and Brown 1996). The x-wind model (see the chapter in this volume by Shu et al.) has the potential to meet the astrophysical criteria to be a jet-bearing wind that drives the molecular outflow, and is therefore worthy of more detailed discussions for its meteoritic connections.

The x-wind arises as a result of the interaction between a disk and a strongly magnetized central star in the early evolution of a sunlike star. The stellar magnetosphere truncates the surrounding accretion disk at an inner edge of radius R_x as the gas presses onto the magnetic fields. A fraction of the accreting mass blows open the magnetic field lines to form the x-wind, and the rest funnels onto the star on the closed field lines that connect the star and the disk. Making a star of 1 $M_\odot$ requires a total mass of between 0.33 and 0.5 $M_\odot$ to be lost in the x-wind. A detailed treatment and a schematic drawing of the x-wind environment are presented in the chapter by Shu et al., this volume.

In the steady state, the stellar dipole rotates at the same Keplerian angular speed as the inner edge of the disk at R_x. On the other hand, in the time-dependent x-wind environment the edge of the disk fluctuates about the equilibrium position as the magnetosphere undergoes cycles, analogous to the solar example (Shu et al. 1997). In the high states of the magnetic cycles, the magnetosphere is relatively stronger than the accretion gas flow, and the inner edge of the disk is pushed outward until a balance is found, at which point the star rotates faster than the inner edge of the disk. In the low states the disk can intrude further inward, and the inner edge rotates faster than the star. Violent magnetic reconnections can occur, releasing the wrapped-up magnetic fields accumulated by inequalities in rotation rates, on sites of magnetic reversals (Linker and Mikic 1995). In addition to the soft X-rays produced steadily from the coronal fields, flares contribute many more hard X-rays, UV radiations, and MeV-energy cosmic ray particles powered by the magnetic energy release (see the chapters by Shu et al. and Glassgold et al., this volume).

At the truncation point R_x of the disk, only gas will be stopped by the magnetosphere, while rock/dust continues spiraling in until the evaporation point of the most refractory material. Millimeter- to centimeter-sized molten droplets may have dropped out from the funnel gas on the way to the star. These rocks make a ring of particles under the funnel flows, known as the reconnection ring. The extension of the dust/rock disk inward of the gas disk is supported by the observations that some classical T Tauri stars require the inner edges of optically thick dust disks to be as close as $0.5R_x$ (Meyer et al. 1997). If left alone, the ring of particles would eventually spiral inward through the equatorial plane and accrete onto the Sun as vapor.

We now outline the theoretical framework of the x-wind model for forming chondritic meteorites (Shu et al. 1996, 1997). A continuous flow of gas and dust accretes through the disk. The disk is self-shielded from the direct rays of the protosun and is relatively cool. The dust component becomes the precursors of chondrules and CAIs. When the accreting material meets the magnetosphere at the x-point, the gas is launched from the inner edge, R_x, as the x-wind. Precursors are also lifted with the gas into the wind by strong gas-grain coupling in the x-region. Out of the shade of the disk, they are immediately exposed to full sunlight. In the protosolar radiation field, if heated to temperatures above ~1700 K lasting for several hours to days, molten CAIs form and cool in the wind, in radiative equilibrium with the sunlight as the spheroids move further from the protosun. However, the prolonged high temperatures of the mean radiation field would have evaporated all chondrule components. In the steady state, the production of chondrules may need subsolar heating on the rotating dustballs to reach local high melting temperatures when the mean radiation field becomes weaker. In the time-dependent x-wind model, chondrules may be irradiated by flares during the high states of the magnetic cycles when the x-region is further from the Sun. They are flash-heated by enhanced luminosity in all wavelengths, on a timescale of minutes to hours, to temperatures approaching 2200 K. The cooled and solidified particles are thrown into space by the wind like stones from a sling. Large particles fall back into the disk not far from the launch point, whereas small particles remain well coupled to the gas and are carried to interstellar space. Intermediate-sized particles are well enough coupled to the x-wind that they fall back to the disk at interplanetary distances, where they then accumulate along with the ambient dust grains in the nebular disk to form the parent bodies of chondritic meteorites.

To quantify the scenario outlined, we refer to two sets of fiducial numerical examples that correspond to typical evolutionary stages of low-mass star formation. The first set represents the embedded stage, when the young star is still deeply embedded inside its infalling envelope of gas and dust; the second one represents the revealed stage, when the outflowing wind has reversed the infall over almost all solid angles and revealed the star and the disk as optical and infrared objects (Shu et al. 1987).

As the system evolves, as a result of stronger dynamo action, reduced accretion rates, or both, the magnetic fields become relatively stronger, and R_X gradually recedes. The parameters are listed in Tables 1 and 2 of Shu et al. (1997). The embedded stage (mass of protostar, $M_\star = 1 \times 10^{33}$ g; stellar radius, $R_\star = 2.1 \times 10^{11}$ cm; $R_X = 4R_\star$; mass loss rate at x-wind outflow, $M_w = 3 \times 10^{19}$ g s^{-1}) lasts for about 2×10^5 yr, and the revealed stage ($M_\star = 1.6 \times 10^{33}$ g; $R_\star = 2.1 \times 10^{11}$ cm; $R_X = 5.3R_\star$; $M_w = 2 \times 10^{18}$ g s^{-1}) for about 3×10^6 yr, in rough agreement with the observational lifetimes of the young stellar objects.

CAIs can form in the steady-state x-wind environment. The surface-averaged temperature of a millimeter-sized sphere can be estimated from the general radiation equilibrium condition with the contributions from the direct and the diffuse radiation. The diffuse radiation field can be estimated as half of a blackbody from the underlying disk characterized by the disk temperature T_X at the x-point. When R_X is very small, no solid material can survive. As R_X increases, the temperature around the launch point falls and the most refractory material starts to survive. Peak temperatures reached in the fiducial embedded and revealed cases are 1800 K and 1300 K for T_X equal to 1200 K and 800 K, respectively. The first value is high enough to melt CAIs and some chondrules. Lifted molten droplets will stay near liquidus temperatures for an interval of a few days, when the radial distances of the CAIs increase from about 1 R_X to 1.5 R_X. The inferred cooling rates along the trajectories in a direct radiation field are >3 K hr^{-1} in the embedded phase and >2 K hr^{-1} in the revealed phase.

For chondrules, we need a fluctuating system (Shu et al. 1997). In varying magnetic cycles, the high states and the low states of the star-disk system can create significantly different thermal environments corresponding to the same average states discussed in the previous paragraph. In the embedded phase the disk temperature has a range of 750 K $< T_X <$ 1600 K, and in the revealed phase the range is 500 K $< T_X <$ 1100 K. The peak temperature oscillates between 1300 K and 2200 K in the embedded phase and between 950 K and 1600 K in the revealed phase. In the low magnetic and average states, T_X is high enough to drive off moderately volatile elements from chondrule precursors. When launched, the rocks remain near peak temperatures for days. On the other hand, in the high states of the revealed phase, T_X (500 K) is low enough to retain volatiles in the chondrule precursors. During flight, the peak temperature reached is insufficient to melt the magnesium-iron silicates completely. The production of chondrules may peak slightly towards the high states, although it would be mixed with the production of CAIs in the low states.

Observations of young stellar objects show that typical X-ray flares rise in a short time and decay on a timescale of up to few hours (chapter by Glassgold et al., this volume). Flare heating can result in doubling or tripling of the base temperatures because of a favorable geometry that does not involve very oblique rays (from the helmet streamers) and is close to chondrule precursors (in the reconnection ring). In the embedded state, tripling of T_x results in a temperature rise to 2250 K, and cooling to 1300 K in an hour equals a cooling rate of 950 k hr^{-1}. In the revealed state, the peak temperature is 1500 K, and the cooling rate is 550 k hr^{-1}. Near the x-region both in and out of the disk, the particle density is high, and collisions of (partially) molten chondrule droplets may occur fairly frequently. After launch in the wind, chondrules would acquire magnetization, recording magnetic fields with intensities of 0.09–0.3 mT. This scenario is consistent with several of the observational constraints for chondrules.

For prolonged flares or an extended heating environment in the reconnection ring, the protochondrules may vaporize, and the constituent atoms will condense as rims onto preexisting CAIs. These CAIs would have already undergone at least one launch cycle, when they were formed, and subsequently dropped back down into the midplane. This process may occur many times before the particles are relaunched by an encroaching x-wind, resulting in accumulation of the thick layered mantles seen in many CAIs. Many flares in the right range of strengths may also reheat chondrule material, accounting for recycling observations.

The aerodynamic drag provides a size sorting mechanism for both chondrules and CAIs. The x-wind sprays particles according to the inverse product of densities and radii. The sizes of the solidified spheroids that return to the disk in the region of the asteroid belt can be calculated. Particle diameters are 3.0 and 0.14 mm from the high states and 4.9 and 0.22 mm from the low states, for the embedded phase and revealed phase, respectively. Within the same average evolutionary stage, subsequent (and rapid) inclusion of CAIs and chondrules in larger (sub)planetary bodies will preserve the narrow size distribution. The x-wind theory specifically predicts that small chondrules and CAIs should exist in comets.

The x-wind environment has the potential of reviving the explanation of producing meteoritic-level ^{26}Al by young solar cosmic ray bombardment (Lee et al. 1998; but see also the chapter by Goswami and Vanhala, this volume). The proton and α particle fluence sufficient to synthesize ^{41}Ca and ^{53}Mn at levels corresponding to those observed in chondrites fails, by one to two orders of magnitude, to produce enough ^{26}Al. However, impulsive flares arising from the reconnection ring accelerate numerous ^{3}He nuclei to MeV energy per nucleon, and this process may yield an enhanced abundance of ^{26}Al. Producing ^{26}Al within the solar system would remove the difficulties of having to form CAIs a few million years earlier than chondrules.

The general framework of the x-wind model addresses several of the observational constraints for CAIs and chondrules. However, some important and essential constraints still remain to be examined in the context of the model. These include the O isotopic compositions of chondrules and CAIs as well as the proportion of material in a chondrite that would be derived from dust that had not been processed through the x-wind.

IV. DISCRIMINATORY TESTS

We now turn our attention to suggesting some discriminatory tests that would be helpful in evaluating the viable models. One such test requires understanding the relationship between chondrule size and the degree of heating experienced. Although chondrule recycling might complicate this matter somewhat, if every heating event produces the same size dependence, the sum of many such events should also. In shock wave models, larger chondrules are predicted to be heated more strongly than smaller ones; in lightning and x-wind models, smaller ones are heated more; and in most other models there is no predicted correlation between size and heating intensity. Lightning and x-winds heat material with radiation, light, or ion bombardment. The heat input is proportional to the cross-sectional area, but the energy required to heat and melt is proportional to the particle's volume. In contrast, in a shock wave, larger particles take longer to come up to speed with the postshock gas than smaller ones, thus experiencing a longer drag-heat pulse duration and a higher integrated T/t.

A second discriminatory test that would be useful in evaluating formation models is the relationship between the abundances of opaque minerals (metal and sulfides) in chondrules and the degree of heating experienced. In a model that heats chondrules with light (lightning, x-wind), heating effects should be strongest in objects rich in opaque minerals because these would absorb more light than transparent materials (silicates). In models such as the shock wave model, there would be no expected correlation between opaque abundance and degree of heating, although particles with high opaque abundance would be denser and have longer stopping distances (and thus heating durations) than particles of similar diameter that were poor in opaques.

Unfortunately, these two potentially valuable constraints are difficult to evaluate from the chondrule record. It is hard to create a semiquantitative scale for the degree of heating experienced by an individual chondrule, because the unique texture and composition presently observed is a function of the grain size of precursor material, bulk composition, peak T relative to the liquidus, the influence of fine-grained dust as nucleation seeds, and many other variables. One type of chondrule that may provide an opportunity to examine these questions is the so-called agglomeratic olivine (AO) or dark-zoned (DZ) chondrules (Weisberg and Prinz 1996; Zanda et al. 1996). These chondrules are very fine-grained and have not been melted extensively, but their bulk compositions span a wide range from FeO-poor to FeO-rich. The problem of determining initial opaque abundances is complicated by the possibilities of volatilization and centrifugal loss of metal beads during formation. Nevertheless, an attempt to address these parameters would be a profitable exercise for future research efforts.

V. SUMMARY

Despite several significant advances in our understanding of chondrules and CAIs and in our theoretical models, the puzzle of providing a viable and acceptable model for formation of these objects is still not resolved. The authors of this paper do not all agree on some of the points presented, and it is clear that the debate about heating mechanisms will continue for some time. Lightning matches observational constraints for chondrule formation well, but there are still some concerns about the feasibility of lightning in the nebula. Shock wave models are well developed and are consistent with many of the observational constraints, but the source(s) of the shock waves needs to be defined. The x-wind model has the advantage of being an observed phenomenon in disks, and it addresses many of the constraints for CAI formation. However, it needs to be considerably better developed (by addressing more of the observational constraints) if it is to be considered a viable model for chondrule formation. Other models fail to satisfy observational constraints in one or more critical aspects and must therefore be considered to be flawed in their current formulations.

Acknowledgments We are most grateful to Jeff Cuzzi and John Wasson for thoughtful and constructive reviews of this manuscript. This work was partially funded (R. H. J.) by the Institute of Meteoritics, University of New Mexico, and NASA grant MRA-97-282 (J. J. Papike, PI). H. C. C.'s contribution was supported by NASA grant NAG5-4319 (D. S. Burnett, PI), and H. S.'s contribution by NASA grant NAG5-4851.

REFERENCES

Allègre, C. J., Manhès, G., and Göpel, C. 1995. The age of the earth. *Geochim. Cosmochim. Acta* 59:1445–1456.

Beckett, J. R., Live, D., Tsay, F.-D., Grossman, L., and Stolper, E. 1988. Ti^{+3} in meteoritic and synthetic hibonite. *Geochim. Cosmochim. Acta* 52:1479–1495.

Bell, K. R., Lin, D. N. C., Hartmann, L. W., and Kenyon, S. J. 1995. The FU Orionis outburst as a thermal accretion event: Observational constraints for protostellar disk models. *Astrophys. J.* 444:376–395.

Bischoff, A., and Keil, K. 1984. Al-rich objects in ordinary chondrites: Related origin of carbonaceous and ordinary chondrites and their constituents. *Geochim. Cosmochim. Acta* 48:693–709.

Boss, A. P. 1996. A concise guide to chondrule formation models. In *Chondrules and the Protoplanetary Disk,* ed. R. H. Hewins, R. H. Jones, and E. R. D. Scott (Cambridge: Cambridge University Press), pp. 257–263.

Boss, A. P., and Graham, J. A. 1993. Clumpy disk accretion and chondrule formation. *Icarus* 106:168–178.

Brearley, A. J., and Jones, R. H. 1998. Chondritic meteorites. In *Planetary Materials,* Reviews in Mineralogy Vol. 36, ed. J. J. Papike (Washington D.C.: Mineralogical Society of America), pp. 3-1–3-398.

Brownlee, D. E., Bates, B., and Beauchamp, R. H. 1983. Meteor ablation spheres as chondrule analogs. In *Chondrules and their Origins,* ed. E. A. King (Houston: Lunar and Planetary Institute), pp. 10–25.

Cameron, A. G. W. 1966. The accumulation of chondritic material. *Earth Planet. Sci. Lett.* 1:93–96.

Cassen, P. 1996. Overview of models of the solar nebula: Potential chondrule-forming environments. In *Chondrules and the Protoplanetary Disk,* ed. R. H. Hewins, R. H. Jones, and E. R. D. Scott (Cambridge: Cambridge University Press), pp. 21–28.

Chen, J. H., and Wasserburg, G. J. 1981. The isotopic composition of uranium and lead in Allende inclusions and meteoritic phosphates. *Earth Planet. Sci. Lett.* 52:1–15.

Clarke, R. S., Jarosewich, E., Nelen, J., Gomez, M., and Hyde, J. R. 1970. The Allende, Mexico, meteorite shower. *Smithsonian Contrib. Earth Sci.* 5:53pp.

Clayton, R. N. 1993. Oxygen isotopes in meteorites. *Ann. Rev. Earth Planet. Sci.* 21:115–149.

Clayton, R. N., Grossman, L., and Mayeda, T. K. 1973. A component of primitive nuclear composition in carbonaceous chondrites. *Science* 182:485–488.

Clayton, R. N., Onuma, N., Ikeda, Y., Mayeda, T. K., Hutcheon, I., Olsen, E. J., and Molini-Velsko, C. 1983. Oxygen isotopic compositions of chondrules in Allende and ordinary chondrites. In *Chondrules and Their Origins,* ed. E. A. King (Houston: Lunar and Planetary Institute), pp. 37–43.

Connolly, H. C., Jr., and Burnett, D. S. 1998. Minor element distributions in and among spinels from type B CAIs. *Lunar Planet. Sci.* 29:Abstract #1487.

Connolly, H. C., Jr., and Hewins, R. H. 1992. Chondrule modification as a possible indicator of rim-forming mechanisms. *Lunar Planet. Sci.* 23:239–240.

Connolly, H. C., Jr., and Hewins, R. H. 1995. Chondrules as products of dust collisions with totally molten droplets within a dust-rich nebular environment: An experimental investigation. *Geochim. Cosmochim. Acta* 59:3231–3246.

Connolly, H. C., Jr., and Love, S. G. 1998. The formation of chondrules: Petrologic tests of the shock wave model. *Science* 280:62–67.

Connolly, H. C., Jr., Hewins, R. H., Ash, R. D., Zanda, B., Lofgren, G. E., and Bourot-Denise, M. 1994. Carbon and the formation of reduced chondrules. *Nature* 371:136–139.

Connolly, H. C., Jr., Jones, B. D., and Hewins, R. H. 1998. The flash melting of chondrules: An experimental investigation into the melting history and physical nature of chondrule precursors. *Geochim. Cosmochim. Acta,* in press.

Cuzzi, J. N., Dobrovolskis, A. R., and Hogan, R. C. 1996. Turbulence, chondrules, and planetesimals. In *Chondrules and the Protoplanetary Disk,* ed. R. H. Hewins, R. H. Jones, and E. R. D. Scott (Cambridge: Cambridge University Press), pp. 35–43.

Davis, A. M., and MacPherson, G. J. 1996. Thermal processing in the solar nebular: Constraints from refractory inclusions. In *Chondrules and the Protoplanetary Disk,* ed. R. H. Hewins, R. H. Jones, and E. R. D. Scott (Cambridge: Cambridge University Press), pp. 71–76.

DeHart, J. M., and Lofgren, G. E. 1996. Experimental studies of group A1 chondrules. *Geochim. Cosmochim. Acta* 60:2233–2242.

Desch, S. J., and Cuzzi, J. N. 1999. The generation of lightning in the solar nebula. *Icarus,* in revision.

Ebel, D. S., and Grossman, L. 1997. Direct condensation of ferromagnesian liquids from cosmic gases. *Lunar Planet. Sci.* 28:317–318.

Ebel, D. S., and Grossman, L. 1998. Effect of dust enrichment on solid and liquid compositions in equilibrium with cosmic gases. In *Lunar and Planetary Science Conference XXIX,* Abstract #1421 (Houston: Lunar and Planetary Institute), CD-ROM.

Gibbard, S. G., Levy, E. H., and Morfill, G. E. 1997. On the possibility of lightning in the protosolar nebula. *Icarus* 130:517–533.

Göpel, C., Manhès, G., and Allègre, C. J. 1994. U-Pb systematics of phosphates from equilibrated ordinary chondrites. *Earth Planet. Sci. Lett.* 121:153–171.

Greenwood, J. P., and Hess, P. C. 1996. Congruent melting kinetics: Constraints on chondrule formation. In *Chondrules and the Protoplanetary Disk,* ed. R. H. Hewins, R. H. Jones, and E. R. D. Scott (Cambridge: Cambridge University Press), pp. 205–211.

Grossman, J. N. 1988. Formation of chondrules. In *Meteorites and the Early Solar System,* ed. J. F. Kerridge and M. S. Matthews (Tucson: University of Arizona Press), pp. 680–696.

Grossman, J. N. 1996. Chemical fractionations of chondrites: Signatures of events before chondrule formation. In *Chondrules and the Protoplanetary Disk,* ed. R. H. Hewins, R. H. Jones, and E. R. D. Scott (Cambridge: Cambridge University Press), pp. 243–253.

Grossman, J. N., Rubin, A. E., Nagahara, H., and King, E. A. 1988. Properties of chondrules. In *Meteorites and the Early Solar System,* ed. J. F. Kerridge and M. S. Matthews (Tucson: University of Arizona Press), pp. 619–659.

Herbig, G. 1977. Eruptive phenomena in early stellar evolution. *Astrophys. J.* 217:693–715.

Hewins, R. H. 1989. The evolution of chondrules. *Proc. Nat. Inst. Polar Res. Symp. Antarctic Meteorites* 2:200–220.

Hewins, R. H. 1997. Chondrules. *Ann. Rev. Earth Planet. Sci.* 25:61–83.

Hewins, R. H., and Connolly, H. C., Jr. 1996. Peak temperatures of flash-melted chondrules. In *Chondrules and the Protoplanetary Disk,* ed. R. H. Hewins, R. H. Jones, and E. R. D. Scott (Cambridge: Cambridge University Press), pp. 197–204.

Hewins, R. H., and Radomsky, P. M. 1990. Temperature conditions for chondrule formation. *Meteoritics* 25:309–318.

Hewins, R. H., Jones, R. H., and Scott, E. R. D. 1996. *Chondrules and the Protoplanetary Disk* (Cambridge: Cambridge University Press).

Hewins, R. H., Yu, Y., Zanda, B., and Bourot-Denise, M. 1997. Do nebular fractionations, evaporative losses, or both, influence chondrule compositions? *Antarctic Meteorite Res.* 10:275–298.

Hood, L. L. 1998. Thermal processing of chondrule precursors in planetesimal bow shocks. *Meteorit. Planet. Sci.* 33:97–107.

Hood, L. L., and Horanyi, M. 1991. Gas dynamic heating of chondrule precursor grains in the solar nebula. *Icarus* 93:259–269.

Hood, L. L., and Horanyi, M. 1993. The nebular shock wave model for chondrule formation: One-dimensional calculations. *Icarus* 106:179–189.

Hood, L. L., and Kring, D. A. 1996. Models for multiple heating mechanisms. In *Chondrules and the Protoplanetary Disk,* ed. R. H. Hewins, R. H. Jones, and E. R. D. Scott (Cambridge: Cambridge University Press), pp. 265–276.

Horanyi, M., and Robertson, S. 1996. Chondrule formation in lightning discharges: Status of theory and experiments. In *Chondrules and the Protoplanetary Disk,* ed. R. H. Hewins, R. H. Jones, and E. R. D. Scott (Cambridge: Cambridge University Press), pp. 303–310.

Horanyi, M., Morfill, G., Goertz, C. K., and Levy, E. H. 1995. Chondrule formation in lightning discharges. *Icarus* 81:174–185.

Huss, G. R. 1988. The role of interstellar dust in the formation of the solar system. *Earth Moon Planets* 40:165–211.
Jones, R. H. 1996. FeO-rich, porphyritic olivine chondrules in unequilibrated ordinary chondrites. *Geochim. Cosmochim. Acta* 60:3115–3138.
Jones, R. H., and Danielson, L. R. 1997. A chondrule origin for dusty relict olivine in unequilibrated chondrites. *Meteorit. Planet. Sci.* 32:753–760.
Jones, R. H., and Lofgren, G. E. 1993. A comparison of FeO-rich porphyritic olivine chondrules in unequilibrated chondrites and experimental analogues. *Meteoritics* 28:213–221.
Jones, R. H., Saxton, J. M., Lyon, I. C., and Turner, G. 1998. Oxygen isotope analyses of chondrule and isolated olivine grains in the CO3 chondrite, ALHA77307. In *Lunar and Planetary Science Conference XXIX,* Abstract #1795 (Houston: Lunar and Planetary Institute), CD-ROM.
Keil, K., Stöffler, D., Love, S. G., and Scott, E. R. D. 1997. Constraints on the role of impact heating and melting in asteroids. *Meteorit. Planet. Sci.* 32:349–363.
Keith, W. D., and Saunders, C. P. R. 1989. Charge transfer during multiple large ice crystal interactions with a riming target. *J. Geophys. Res.* 94:13013–13106.
Krot, A. N., and Rubin, A. E. 1996. Microchondrule-bearing chondrule rims: Constraints on chondrule formation. In *Chondrules and the Protoplanetary Disk,* ed. R. H. Hewins, R. H. Jones, and E. R. D. Scott (Cambridge: Cambridge University Press), pp. 181–184.
Krot, A. N., Rubin, A. E., Keil, K., and Wasson, J. T. 1997. Microchondrules in ordinary chondrites: Implications for chondrule formation. *Geochim. Cosmochim. Acta* 61:463–473.
Lee, T., Shu, F. H., Shang, H., Glassgold, A. E., and Rehm, K. E. 1998. Protostellar cosmic rays and extinct radioactivities in meteorites. *Astrophys. J.* 506:898–912.
Liepmann, H. W., and Roshko, A. 1957. *Elements of Gasdynamics* (New York: Wiley).
Liffman, K. 1992. The formation of chondrules by ablation. *Icarus* 100:608–619.
Liffman, K., and Brown, M. J. I. 1996. The protostellar jet model of chondrule formation. In *Chondrules and the Protoplanetary Disk,* ed. R. H. Hewins, R. H. Jones, and E. R. D. Scott (Cambridge: Cambridge University Press), pp. 285–302.
Linker, J. A., and Mikic, Z. 1995. Disruption of a helmet streamer by photospheric shear. *Astrophys. J. Lett.* 438:L45–L48.
Lofgren, G. E. 1996. A dynamic crystallization model for chondrule melts. In *Chondrules and the Protoplanetary Disk,* ed. R. H. Hewins, R. H. Jones, and E. R. D. Scott (Cambridge: Cambridge University Press), pp. 187–196.
Lofgren, G. E., and Lanier, A. 1990. Dynamic crystallization study of barred olivine chondrules. *Geochim. Cosmochim. Acta* 54:3537–3551.
Lofgren, G. E., and Russell, W. J. 1986. Dynamic crystallization of chondrule melts of porphyritic and radial pyroxene composition. *Geochim. Cosmochim. Acta* 50:1715–1726.
Love, S. G., and Brownlee, D. E. 1991. Heating and thermal transformation of micrometeorites entering the Earth's atmosphere. *Icarus* 89:26–43.
Love, S. G., Keil, K., and Scott, E. R. D. 1995. Electrical discharge heating of chondrules in the solar nebula. *Icarus* 115:97–108.
MacPherson, G. J., and Russell, S. S. 1997. Origin of aluminum-rich chondrules: Constraints from major element chemistry. *Meteorit. Planet. Sci. Suppl.* 32:A83.

MacPherson, G. J., Wark, D. A., and Armstrong, J. T. 1988. Primitive material surviving in chondrites: Refractory inclusions. In *Meteorites and the Early Solar System,* ed. J. F. Kerridge and M. S. Matthews (Tucson: University of Arizona Press), pp. 746–807.

MacPherson, G. J., Davis, A. M., and Zinner, E. K. 1995. The distribution of aluminum-26 in the early solar system—A reappraisal. *Meteoritics* 30:365–386.

McKeegan, K. D., Leshin, L. A., Russell, S. S., and MacPherson, G. J. 1998. Oxygen isotopic abundances in calcium-aluminum-rich inclusions from ordinary chondrites: Implications for nebular heterogeneity. *Science* 280:414–418.

Meeker, G. P. 1995. Constraints on formation processes of two coarse-grained calcium-aluminum-rich inclusions: A study of mantles, islands and cores. *Meteoritics* 30:71–84.

Metzler, K., Bischoff, A., and Stöffler, D. 1992. Accretionary dust mantles in CM chondrites: Evidence for solar nebula processes. *Geochim. Cosmochim. Acta* 56:2873–2897.

Meyer, M. R., Calvet, N., and Hillenbrand, L. A. 1997. Intrinsic near-infrared excesses of T Tauri stars: Understanding the classical T Tauri star locus. *Astron. J.* 114:288–300.

Morden, S. J., and Collinson, D. W. 1992. The implications of the magnetism of ordinary chondrite meteorites. *Earth Planet. Sci. Lett.* 109:185–204.

Morfill, G., Spruit, H., and Levy, E. H. 1993. Physical processes and conditions associated with the formation of protoplanetary disks. In *Protostars and Planets III,* ed. E. H. Levy and J. I. Lunine (Tucson: The University of Arizona Press), pp. 939–978.

Morfill, G., Durisen, R. H., and Turner, G. W. 1998. An accretion rim constraint on chondrule formation theories. *Icarus* 134:180–184.

Nagata, T., and Funaki, M. 1981. The composition of natural remanent magnetization of an Antarctic chondrite, ALHA76009 (L6). *Lunar Planet. Sci.* 12:747–748.

Paque, J. M. 1990. Relict grains in a Ca-Al-rich inclusion from Allende. *Lunar Planet. Sci.* 21:932–933.

Paque, J. M. 1995. Effect of residence time at maximum temperature on the texture and phase compositions of a type B Ca-Al-rich inclusion analog. *Lunar Planet. Sci.* 26:1099–1100.

Paque, J. M., and Cuzzi, J. N. 1997. Physical characteristics of chondrules and rims, and aerodynamic sorting in the solar nebula. *Lunar Planet. Sci.* 28:1071–1072.

Paque, J. M., and Lofgren, G. E. 1993. Comparison of experimental studies on chondrule and Ca-Al-rich inclusions. *Lunar Planet. Sci.* 22:1025–1026.

Paque, J. M., Le, L., and Lofgren, G. E. 1998. Experimentally produced spinel rims on Ca-Al-rich inclusion bulk compositions. In *Lunar and Planetary Science Conference XXIX,* Abstract #1221 (Houston: Lunar and Planetary Institute), CD-ROM.

Pilipp, W., Hartquist, T. W., Morfill, G. E., and Levy, E. H. 1998. Chondrule formation by lightning in the protosolar nebula? *Astron. Astrophys.* 331:121–146.

Podosek, F. A., and Cassen, P. 1994. Theoretical, observational, and isotopic estimates of the lifetime of the solar nebula. *Meteoritics* 29:6–25.

Prinz, M., Weisberg, M. K., and Nehru, C. E. 1988. Gunlock, a new type 3 ordinary chondrite with a golfball-sized chondrule. *Meteoritics* 23:297.

Radomsky, P. M., and Hewins, R. H. 1990. Formation conditions of pyroxene-olivine and magnesian olivine chondrules. *Geochim. Cosmochim. Acta* 54:3475–3490.

Rubin, A. E., and Krot, A. N. 1996. Multiple heating of chondrules. In *Chondrules and the Protoplanetary Disk,* ed. R. H. Hewins, R. H. Jones and E. R. D. Scott (Cambridge: Cambridge University Press), pp. 173–180.

Rubin, A. E., Sailer, A. L., and Wasson. 1999. Troilite in the chondrules of unequilibrated ordinary chondrites: Implications for chondrule formation. *Geochim. Cosmochim. Acta,* in review.

Ruzmaikina, T. V., and Ip, W. H. 1996. Chondrule formation in the accretional shock. In *Chondrules and the Protoplanetary Disk,* ed. R. H. Hewins, R. H. Jones, and E. R. D. Scott (Cambridge: Cambridge University Press), pp. 277–284.

Sahagian, D. L., and Hewins, R. H. 1992. The size of chondrule-forming events. *Lunar Planet. Sci.* 23:1197–1198.

Sanders, I. S. 1996. A chondrule-forming scenario involving molten planetesimals. In *Chondrules and the Protoplanetary Disk,* ed. R. H. Hewins, R. H. Jones, and E. R. D. Scott (Cambridge: Cambridge University Press), pp. 327–334.

Scott, E. R. D., Love, S. G., and Krot, A. N. 1996. Formation of chondrules and chondrites in the protoplanetary nebula. In *Chondrules and the Protoplanetary Disk,* ed. R. H. Hewins, R. H. Jones, and E. R. D. Scott (Cambridge: Cambridge University Press), pp. 87–96.

Sears, D. W. G., Huang, S., and Benoit, P. H. 1996. Open-system behavior during chondrule formation. In *Chondrules and the Protoplanetary Disk,* ed. R. H. Hewins, R. H. Jones, and E. R. D. Scott (Cambridge: Cambridge University Press), pp. 221–231.

Sentman, D. D., Wescott, E. M., Osborne, D. L., Hampton, D. L., and Heavner, M. J. 1995. Preliminary results from the Sprites94 aircraft campaign: 1. Red sprites. *Geophys. Res. Lett.* 22:2105–2108.

Shu, F. H., Adams, F. C., and Lizano, S. 1987. Star formation in molecular clouds: Observation and theory. *Ann. Rev. Astron. Astrophys.* 25:23–81.

Shu, F. H., Shang, H., and Lee, T. 1996. Toward an astrophysical theory of chondrites. *Science* 271:1545–1552.

Shu, F. H., Shang, H., Glassgold, A. E., and Lee, T. 1997. X-rays and fluctuating X-winds from protostars. *Science* 277:1475–1479.

Skinner, W. R. 1990*a*. Bipolar outflows and a new model of the early solar system. Part I: Overview and implications of the model. *Lunar Planet. Sci.* 21:1166–1167.

Skinner, W. R. 1990*b*. Bipolar outflows and a new model of the early solar system. Part II: The origins of chondrules, isotopic anomalies, and chemical fractionations. *Lunar Planet. Sci.* 21:1168–1169.

Skinner, W. R., and Leenhouts, J. M. 1993. Size distributions and aerodynamic equivalence of metal chondrules and silicate chondrules in Acfer 059. *Lunar Planet. Sci.* 24:1315–1316.

Sorby, H. C. 1877. On the structure and origin of meteorites. *Nature* 15:495–498.

Stolper, E. 1982. Crystallization sequences of Ca-Al-rich inclusions from Allende: An experimental study. *Geochim. Cosmochim. Acta* 46:2159–2180.

Stolper, E., and Paque, J. M. 1986. Crystallization sequences of Ca-Al-rich inclusions from Allende: The effects of cooling rate and maximum temperature. *Geochim. Cosmochim. Acta* 50:1785–1806.

Sugiura, N., and Strangway, D. W. 1982. Magnetic properties of low-petrologic grade non-carbonaceous chondrites. *Proceedings of the Seventh Symposium on Antarctic Meteorites. Mem. Natl. Inst. Polar Res.* 25 (special issue):260–280.

Sugiura, N., Lanoix, M., and Strangway, D. W. 1979. Magnetic fields of the solar nebula as recorded in chondrules from the Allende meteorite. *Phys. Earth Planet. Interiors* 20:342–349.

Swindle, T. D., Davis, A. M., Hohenberg, C. M., MacPherson, G. J., and Nyquist, L. E. 1996. Formation times of chondrules and Ca-Al-rich inclusions: Constraints from short-lived radionuclides. In *Chondrules and the Protoplanetary Disk,* ed. R. H. Hewins, R. H. Jones, and E. R. D. Scott (Cambridge: Cambridge University Press), pp. 77–86.

Symes, S. J., Sears, D. W. G., Akridge, D. G., Huang, S., and Benoit, P. H. 1998. The crystalline lunar spheres: Their formation and implication for the origin of meteoritic chondrules. *Meteorit. Planet. Sci.* 33:13–30.

Taylor, G. J., Scott, E. R. D., and Keil, K. 1983. Cosmic setting for chondrule formation. In *Chondrules and Their Origins,* ed. E. A. King (Houston: Lunar and Planetary Institute), pp. 262–278.

Thiemens, M. H. 1996. Mass-independent isotopic effects in chondrites: The role of chemical processes. In *Chondrules and the Protoplanetary Disk,* ed. R. H. Hewins, R. H. Jones, and E. R. D. Scott (Cambridge: Cambridge University Press), pp. 107–118.

Wasson, J. T. 1985. *Meteorites: Classification and Properties* (New York: W.H. Freeman).

Wasson, J. T. 1993. Constraints on chondrule origins. *Meteoritics* 28:14–28.

Wasson, J. T. 1996. Chondrule formation: Energetics and length scales. In *Chondrules and the Protoplanetary Disk,* ed. R. H. Hewins, R. H. Jones, and E. R. D. Scott (Cambridge: Cambridge University Press), pp. 45–54.

Wasson, J. T., and Kallemeyn, G. W. 1990. Allan Hills 85085: A subchondritic meteorite of mixed nebula and regolithic heritage. *Earth Planet. Sci. Lett.* 101:148–161.

Wasson, J. T., Krot, A. N., Lee, M. S., and Rubin, A. E. 1995. Compound chondrules. *Geochim. Cosmochim. Acta* 59:1847–1869.

Weidenschilling, S. J. 1996. Production of chondrules by lightning in the solar nebula. *Lunar Planet. Sci.* 27:1397–1398.

Weidenschilling, S. J. 1997. Production of chondrules by lightning in the solar nebula? Not so easy! *Lunar Planet. Sci.* 27:1515–1516.

Weidenschilling, S. J., Marzari, F., and Hood, L. L. 1998. The origin of chondrules at jovian resonances. *Science* 279:681–684.

Weinbruch, S., and Müller, F. W. 1995. Constraints on the cooling rates of chondrules from the microstructure of clinopyroxene and plagioclase. *Geochim. Cosmochim. Acta* 59:3221–3230.

Weisberg, M. K., and Prinz, M. 1996. Agglomeratic chondrules, chondrule precursors and incomplete melting. In *Chondrules and the Protoplanetary Disk,* ed. R. H. Hewins, R. H. Jones, and E. R. D. Scott (Cambridge: Cambridge University Press), pp. 119–127.

Weisberg, M. K., Prinz, M., and Nehru, C. E. 1988. Macrochondrules in ordinary chondrites: Constraints on chondrule forming processes. *Meteoritics* 23:309–310.

Whipple, F. L. 1966. Chondrules: Suggestion concerning the origin. *Science* 153:54–56.

Wood, J. A. 1962. Chondrules and the origins of the terrestrial planets. *Nature* 194:127–130.

Wood, J. A. 1984. On the formation of meteoritic chondrules by aerodynamic drag heating in the solar nebula. *Earth Planet. Sci. Lett.* 70:11–26.

Wood, J. A. 1996. Processing of chondritic and planetary material in spiral density waves in the nebula. *Meteorit. Planet. Sci.* 31:641–646.

Wood, J. A., and Hashimoto, A. 1993. Mineral equilibrium in fractionated nebular systems. *Geochim. Cosmochim. Acta* 57:2377–2388.

Yu, Y., and Hewins, R. H. 1998. Transient heating and chondrule formation: Evidence from sodium loss in flash heating simulation experiments. *Geochim. Cosmochim. Acta* 62:159–172.

Yu, Y., Hewins, R. H., and Zanda, B. 1996. Sodium and sulfur in chondrules: Heating time and cooling curves. In *Chondrules and the Protoplanetary Disk,* ed. R. H. Hewins, R. H. Jones, and E. R. D. Scott (Cambridge: Cambridge University Press), pp. 213–220.

Zanda, B., Yu, Y., Bourot-Denise, M., and Hewins, R. H. 1996. The history of metal and sulfide in chondrites. In *Workshop on Parent-Body and Nebular Modification of Chondritic Materials,* LPI Technical Report No 97–02, Part 1 (Houston: Lunar and Planetary Institute), pp. 68–70.

EXTINCT RADIONUCLIDES AND THE ORIGIN OF THE SOLAR SYSTEM

JITENDRA N. GOSWAMI
Physical Research Laboratory, Ahmedabad, India

and

HARRI A. T. VANHALA
Carnegie Institution of Washington

Isotopic studies of meteorites provide evidence for the presence of several short-lived, now extinct nuclides in the early solar system. These nuclides could represent freshly synthesized stellar products injected into the protosolar cloud just prior to its collapse. Alternatively, some of them could be products of energetic particle interactions with gas and dust present in the protosolar cloud or in the solar nebula. Recent experimental data on some of these nuclides (^{41}Ca, ^{26}Al, and ^{60}Fe) appear to support the first alternative, with a supernova, a thermally pulsing asymptotic giant branch star, or a Wolf-Rayet star being a plausible source. If true, this will constrain to less than a million years the timescale for the collapse of the protosolar cloud to form the Sun and some of the first solar system solids. This short timescale argues for a triggered collapse of the protosolar cloud, initiated by the same stellar event that produced and injected some of the radionuclides into the cloud. Recent numerical simulations studying the interaction of shock waves with molecular cloud cores show that collapse can be initiated and sufficient mixing can take place, thus confirming the viability of the scenario.

I. INTRODUCTION

Isotopic abundances in solar system objects provide information on the possible makeup of the protosolar cloud material from which the Sun and the solar system objects have formed. The abundances of stable nuclides as well as some of the long-lived radionuclides (e.g., isotopes of U and Th) can be used as tracers to identify the stellar sources as well as galactic cosmochemical processes that contributed to the mix represented by the material in the protosolar cloud (see, e.g., Symbalisty and Schramm 1981; Clayton 1988; Timmes et al. 1995).

Isotopic studies of meteorites have also provided evidence for the one-time presence of several short-lived nuclides in the material from which the solar system objects have formed. The half-life of these nuclides varies from $\sim 10^5$ yr to $\sim 10^8$ yr, and they provide very important constraints on the timescale of processes taking place during the early evolutionary history of the solar system (see the chapter by Wadhwa and Russell, this volume). For all practical purposes, these short-lived nuclides may be considered as "now extinct" nuclides, and their one-time presence is inferred from the observed excess in their decay products, correlated with the abundance of the parent element, in early solar system objects. These objects must have incorporated the short-lived nuclides "live" at the time of their formation, and *in situ* decay of these nuclides results in the observed excess. Although a "fossil" origin for this excess has also been proposed (Clayton 1982), the meteorite data are not consistent with this suggestion (see section III).

^{129}I (half-life $\sim 1.6 \times 10^7$ years) was the first extinct nuclide whose one-time presence in the early solar system was inferred from the observed excess of its decay product ^{129}Xe and the correlation of the excess with the element iodine in the Richardton meteorite (Reynolds 1960; Jeffery and Reynolds 1961; see also Hohenberg et al. 1998). Isotopic studies of primitive meteorites have now provided evidence for the presence of several other now extinct short-lived nuclides in the early solar system; these are listed in Table I. Since the last *Protostars and Planets* conference, evidence has been found for three new short-lived nuclides (Shukolyukov and Lugmair 1993*a*,*b*; Srinivasan et al. 1994, 1996; Lee and Halliday 1995, 1996, 1997; Harper and Jacobsen 1996), and there are strong hints for the presence of a few additional nuclides, as well.

Several sources for these nuclides have been proposed. Some nuclei may represent the ambient inventory of the local interstellar medium. However, those with mean life $<10^7$ years and present in substantial abundance could represent freshly synthesized nuclides from a stellar source that were injected into the protosolar cloud just prior to its collapse (Cameron 1993; Wasserburg et al. 1994, 1995; Cameron et al. 1995; Arnould et al. 1997*a*,*b*). The possibility that these nuclides are produced by energetic particle interactions with gas and dust present in the protosolar molecular cloud has also been proposed (Clayton 1994; Clayton and Jin 1995*a*; see also Ramaty et al. 1996); alternatively, they could have been produced by irradiation of gas and dust in the solar nebula by energetic particles from an active early (T Tauri) Sun (Clayton et al. 1977; Heymann et al. 1978; Lee 1978; Wasserburg and Arnould 1987; Clayton and Jin 1995*b*; Goswami et al. 1997; see also Shu et al. 1996, 1997; Lee et al. 1998). It is important to identify the most plausible source of these nuclides and, in particular, whether they were locally produced in the solar nebula or whether they were already present in the protosolar cloud that collapsed to form the Sun and the solar system objects. In the latter case, these nuclides provide a very strong constraint on the timescale for the protosolar cloud collapse; the strongest constraint comes, obviously, from the nuclide with the shortest half-life. On the other hand, if these nuclides were locally produced in the solar nebula, they may be used only as a measure of early solar system nuclear processes and not as an indicator of

TABLE I
Short-lived Nuclides in the Early Solar System

Nuclide	Half-life (Myr)	Daughter Nuclide	Initial Abundance	Ref.[a]	Stellar Production Site[b]
^{41}Ca	0.1	^{41}K	$\sim1.5\times10^{-8}$ (^{41}Ca/^{40}Ca)	[1]	SN, AGB, WR
^{26}Al	0.7	^{26}Mg	$\sim5\times10^{-5}$ (^{26}Al/^{27}Al)	[2]	SN, N, AGB, WR
^{60}Fe	1.5	^{60}Ni	$\sim4\times10^{-9}$ (^{60}Fe/^{56}Fe)	[3]	SN, AGB
			$\sim2\times10^{-8}$ [c]	[5]	
^{53}Mn	3.7	^{53}Cr	$\sim4.4\times10^{-5}$ (^{53}Mn/^{55}Mn)	[4]	SN
			$\sim4.7\times10^{-6}$, $\sim10^{-5}$ [c]	[5]	
^{107}Pd	6.5	^{107}Ag	$\sim2\times10^{-5}$ (^{107}Pd/^{108}Pd)	[6]	SN, AGB, WR
			$\sim4.5\times10^{-5}$ [c]		
^{182}Hf	9	^{182}W	$\sim2\times10^{-4}$ (^{182}Hf/^{180}Hf)	[7]	SN
^{129}I	15.7	^{129}Xe	$\sim10^{-4}$ (^{129}I/^{127}I)	[8]	SN
^{244}Pu	82	α, SF[d]	$\sim7\times10^{-3}$ (^{244}Pu/^{238}U)	[9]	SN
^{99}Tc[e]	0.21	^{99}Ru	$\sim10^{-4}$ (^{99}Tc/^{99}Ru)	[10]	AGB, WR
^{36}Cl[e]	0.3	^{36}Ar	$\sim1.4\times10^{-6}$ (^{36}Cl/^{35}Cl)	[11]	SN, AGB, WR
^{205}Pb[e]	15	^{205}Tl	$\sim3\times10^{-4}$ (^{205}Pb/^{204}Pb)	[12]	AGB, WR
^{92}Nb[e]	35	^{92}Zr	$\sim2\times10^{-5}$ (^{92}Nb/^{93}Nb)	[13]	SN

[a] References: [1] Srinivasan et al.(1994, 1996);[2] Lee et al. (1977); see also MacPherson et al. (1995); [3] Shukolyukov and Lugmair (1993*a,b*) and Lugmair et al. (1995); [4] Birck and Allegre (1985); [5] Lugmair et al. (1995) and Lugmair and Shukolyukov (1998); [6] Kelly and Wasserburg (1978); see also Chen and Wasserburg (1996); [7] Lee and Halliday (1995, 1996) and Harper and Jacobsen (1996); [8] Jeffery and Reynolds (1961); see also Hohenberg et al. (1998); [9] Rowe and Kuroda (1965) and Hudson et al. (1988); [10] Yin et al. (1992); [11] Murty et al. (1997); [12] Chen and Wasserburg (1987); [13] Harper (1996).
[b] SN = supernova; N = nova; AGB = asymptotic giant branch star; WR = Wolf-Rayet star.
[c] Extrapolated value at the time of formation of CAI.
[d] SF = spontaneous fission products.
[e] Suggestive evidence is present; needs confirmation.

presolar processes. Further, if these nuclides represent freshly synthesized products from a specific stellar source, one must consider the possibility that the very process of injection of these nuclides into the protosolar cloud could also have triggered its collapse (Cameron and Truran 1977). This idea is especially important, because the timescale for standard star formation through ambipolar diffusion and unassisted collapse, ~5–10 Myr (Mouschovias 1989; Shu 1995), is too long for the shortest-lived of the radioactivities to have survived in the measured amounts (see discussion by Cameron 1995; Cameron et al. 1997).

In recent years, the formulation of two- and three-dimensional numerical codes studying the induced collapse of molecular cloud cores has enabled the investigation of the viability of the scenario of the triggered origin of the solar system (Boss 1995; Foster and Boss 1996, 1997; Boss and Foster 1998; Vanhala and Cameron 1998). Some of these aspects have been discussed in the earlier *Protostars and Planets* volumes (Wasserburg 1985; Cameron 1993; Swindle 1993) and also in a recent conference proceedings (Podosek and Nichols 1997; Cameron et al. 1997; Boss and Foster 1997). In this chapter we will briefly review our current understanding in this field, emphasizing the progress made since the last *Protostars and Planets* conference. We will try to identify areas that need further attention for a better resolution of the existing problems.

II. SHORT-LIVED NUCLIDES IN THE EARLY SOLAR SYSTEM

The one-time presence of the short-lived nuclides in the early solar system is based on the observed excess in their decay products in samples from various meteorites. The case for their presence is greatly strengthened if one can show that the observed excess in the abundance of their decay product (e.g., ^{129}Xe in the case of ^{129}I) is also correlated with the abundance of the parent element (iodine), and this was indeed found to be true for ^{129}I (Jeffery and Reynolds 1961; see also Swindle and Podosek 1988; Hohenberg et al. 1998). Evidence for the one-time presence of the relatively longer-lived ^{129}I (half-life = 15.7 Myr) was found in bulk samples of many meteorites. However, similar attempts to detect the presence of the shorter-lived nuclide ^{26}Al (half-life = 0.7 Myr), which was proposed to be a potential heat source for the thermal differentiation and melting of planetesimals, were not successful (see, e.g., Schramm et al. 1970). The real breakthrough in the field came following the identification of Ca-Al-rich refractory inclusions (CAIs) in the Allende meteorite, which fell in Mexico in 1969. The mineralogical and chemical compositions of the CAIs closely resemble those expected for some of the first solid phases to condense from a high-temperature gaseous reservoir of solar composition (Grossman 1980). Unambiguous evidence for the one-time presence of the short-lived nuclide ^{26}Al in the Allende CAIs was reported by Lee et al. (1976, 1977), following strong hints for its presence found in some earlier attempts (Gray and Compston 1974; Lee and Papanastassiou 1974). As in the case of ^{129}I, the excess in ^{26}Mg (due to the decay of ^{26}Al) in Allende CAIs was shown to be correlated with the abundance of the parent element, aluminum (Fig. 1). This correlation can be expressed as a linear relation between the measured parameters as

$$\left[\frac{^{26}\mathrm{Mg}}{^{24}\mathrm{Mg}}\right]_{\mathrm{meas}} = \left[\frac{^{26}\mathrm{Mg}}{^{24}\mathrm{Mg}}\right]_{\mathrm{init}} + \left[\frac{^{26}\mathrm{Al}}{^{24}\mathrm{Mg}}\right]$$
$$= \left[\frac{^{26}\mathrm{Mg}}{^{24}\mathrm{Mg}}\right]_{\mathrm{init}} + \left[\frac{^{26}\mathrm{Al}}{^{27}\mathrm{Al}}\right]_{\mathrm{init}} \times \left[\frac{^{27}\mathrm{Al}}{^{24}\mathrm{Mg}}\right]_{\mathrm{meas}}$$

Thus, in Fig. 1, the slope of the correlation line gives the initial value of ^{26}Al/^{27}Al, and the intercept gives the initial ^{26}Mg/^{24}Mg at the time of formation of the analyzed CAI.

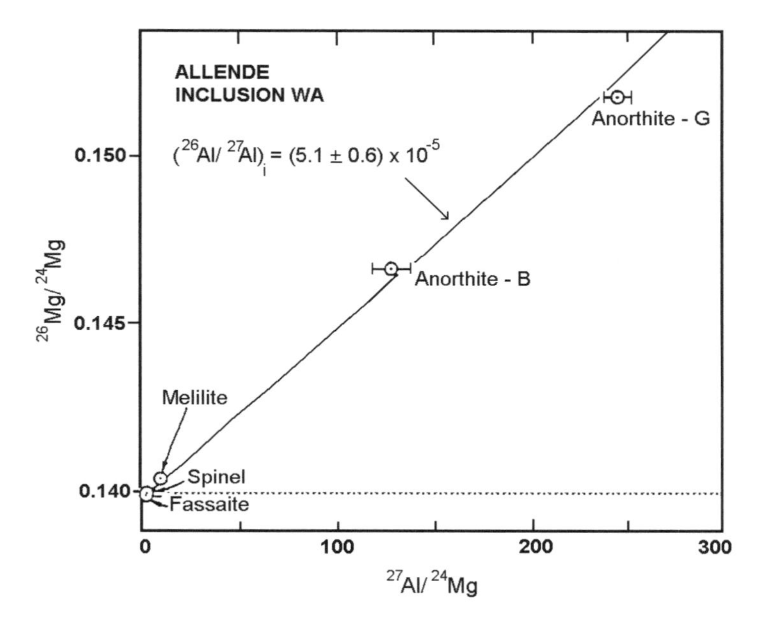

Figure 1. Magnesium isotopic ratio, ^{26}Mg/^{24}Mg, measured in different mineral phases in the Allende CAI WA, plotted as a function of their ^{27}Al/^{24}Mg ratios. The dotted line represents the reference (normal) ^{26}Mg/^{24}Mg ratio. The linear correlation of excess ^{26}Mg with the Al/Mg ratio indicates that the excess ^{26}Mg is related to Al and resulted from *in situ* decay of "live" ^{26}Al (mean life 1 Myr) present in these phases at the time of their formation in the early solar system. (Data from Lee et al. 1977.)

The discovery of ^{26}Al was followed by that of ^{107}Pd (half-life = 7 Myr) (Kelly and Wasserburg 1978; see also Chen and Wasserburg 1996) and ^{53}Mn (half-life = 3.7 Myr) (Birck and Allegre 1985; see also Lugmair and Shukolyukov 1998). Since the last *Protostars and Planets* conference, evidence for the presence of three more short-lived nuclides (^{60}Fe, ^{41}Ca, and ^{182}Hf) in the early solar system has been found (Table I).

A. ^{60}Fe

The one-time presence of ^{60}Fe (half-life = 1.5 Myr) in the early solar system was inferred from the observed excess in the abundance of ^{60}Ni in samples of several meteorites belonging to the eucrite group (Shukolyukov and Lugmair 1993*a,b*). The measured excess in ^{60}Ni/^{58}Ni ratio (above its normal value of 0.382192) in bulk samples of two eucrites, Chervony Kut and Juvinas, show a correlation with their ^{56}Fe/^{58}Ni ratios (see Fig. 2a), indicating that the observed excess in ^{60}Ni could be due to the decay of ^{60}Fe. However, the isotopic data for mineral separates (Fig. 2a) suggest redistribution of Ni relative to Fe following the decay of ^{60}Fe. From the presently available data, Lugmair et al. (1995) have deduced an initial

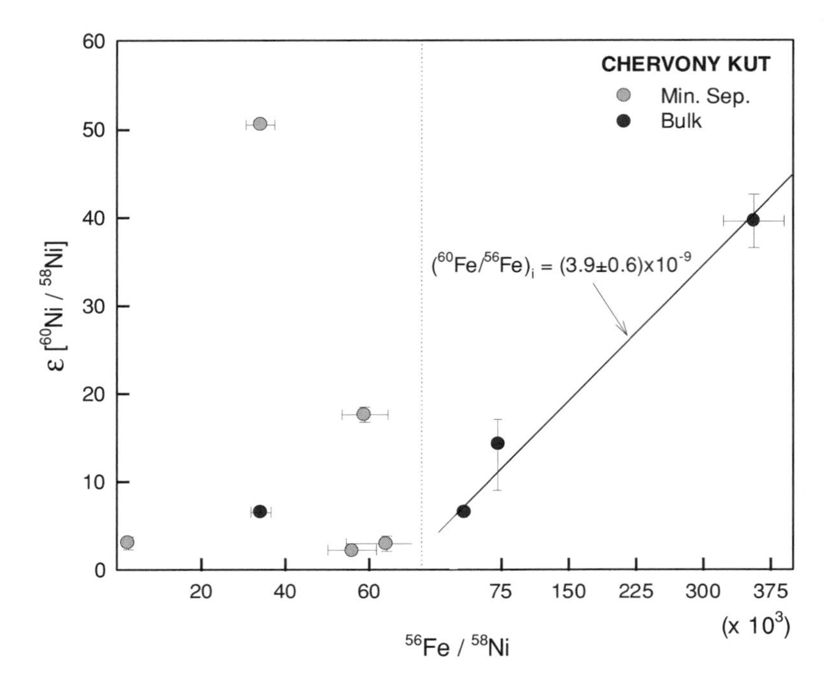

Figure 2. (a) Deviation in the measured nickel isotopic ratio [$^{60}Ni/^{58}Ni$] from normal value (expressed in ϵ units; parts per ten thousand) in mineral separates ("Min. Sep.") and bulk samples ("Bulk") of the Chervony Kut meteorite, plotted as a function of their Fe/Ni ratios.

$^{60}Fe/^{56}Fe$ ratio of $\sim 4 \times 10^{-9}$ at the time of eucrite formation. It will be important, however, to have additional experimental data for a more precise estimate of this value. The eucrites are products of melting and recrystallization taking place in large-sized (10–100 km) planetesimals, and the one-time presence of ^{60}Fe in these objects suggests that the formation of planetesimals and their thermal differentiation took place within about ten million years (Lugmair et al. 1995; Lugmair and Shukolyukov 1998; see also the chapter by Wadhwa and Russell, this volume).

B. ^{41}Ca

^{41}Ca (half-life = 0.1 Myr) has the shortest half-life among the extinct nuclides whose presence in the early solar system has been established (Srinivasan et al. 1994, 1996). Although a hint for the possible presence of ^{41}Ca in the Allende meteorite was reported earlier (Hutcheon et al. 1984), it was Srinivasan et al. (1994) who found a clear and correlated excess of ^{41}K in CAIs from the carbonaceous chondrite Efremovka. The presence of ^{41}Ca in the early solar system is now confirmed from additional studies of CAIs and refractory phases from the Allende and the Murchison meteorites (Srinivasan et al. 1996; Sahijpal 1997; Sahijpal et al. 1998). The

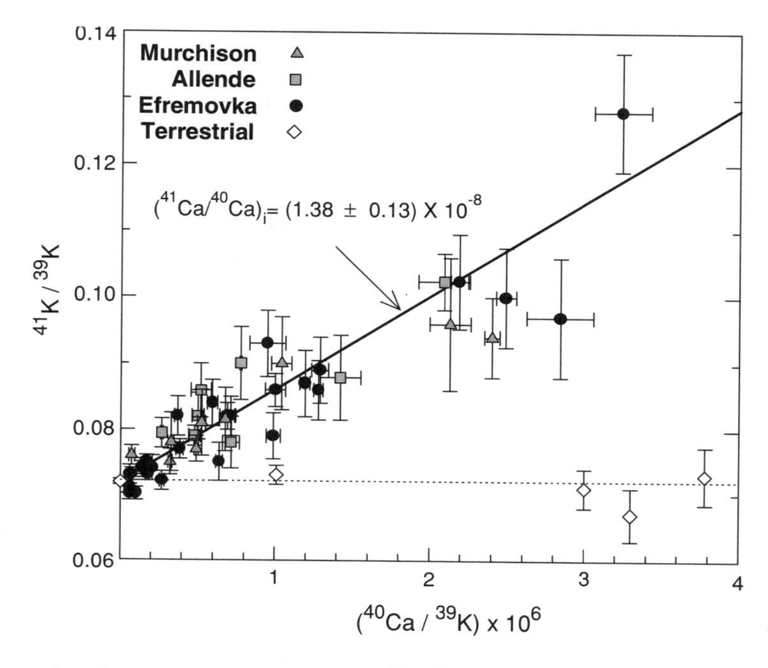

Figure 2. (b) Potassium isotopic ratio, $^{41}K/^{39}K$, in refractory oxide and silicate minerals from several primitive meteorites plotted as a function of their Ca/K ratios. The dotted horizontal line indicates normal $^{41}K/^{39}K$ ratio of 0.072, and the measured values in terrestrial samples lie along this line. The excesses in ^{60}Ni in bulk samples (a; right panel) and in ^{41}K in refractory phases (b) show a linear correlation with Fe/Ni and Ca/K ratios, respectively, indicating that the excesses are due to *in situ* decay of "live" ^{60}Fe and ^{41}Ca present in the analyzed samples at the time of their formation (from Shukolyukov and Lugmair 1993*a*; Srinivasan et al. 1996; Sahijpal 1997).

data shown in Fig. 2b demonstrate a clear excess in the measured $^{41}K/^{39}K$ ratio over the normal value of 0.072. This excess also correlates well with the $^{40}Ca/^{39}K$ ratios in the analyzed phases, indicating that the excess ^{41}K in them is due to the decay of ^{41}Ca initially present in these objects at the time of their formation. The short half-life of ^{41}Ca constrains to less than a million years the timescale for the collapse of the protosolar cloud to form the Sun and some of the first solar system solids, provided one can show that ^{41}Ca was present in the protosolar cloud and was not produced later by energetic particle interactions with gas and dust in the solar nebula.

C. ^{182}Hf

The one-time presence of the short-lived nuclide ^{182}Hf (half-life = 9 Myr) in the early solar system has been inferred from isotopic studies of both meteoritic and planetary objects (Lee and Halliday 1995, 1996, 1997; Harper and Jacobsen 1996). ^{182}Hf decays to ^{182}W. Because tungsten (W) has a strong affinity for metals whereas hafnium (Hf) prefers to go with silicates, the ^{182}Hf-^{182}W system is an extremely important tracer for delineating the timescale for the silicate/metal fractionation that took place in planetesimals as well as in larger objects like the Earth, Moon, and Mars. The ^{182}W/^{180}W ratio in metal and silicate phases in these objects will be very different depending on whether the silicate/metal fractionation took place prior to or after significant decay of ^{182}Hf. Although a direct correlation between the excess in the daughter nuclide (^{182}W) with the abundance of the parent element (Hf) has not been established, the experimental data show a widespread deficiency of ^{182}W in metallic phases of different meteorites as well as a large excess in silicate phases that conclusively suggest the presence of ^{182}Hf in the early solar system. These aspects are discussed in the accompanying chapter by Wadhwa and Russell, this volume.

A complete list of the short-lived nuclides whose one-time presence in the early solar system has been inferred from the isotopic studies of meteorites is given in Table I. The inferred initial abundances at the time of formation of the analyzed objects are also included. When the observation is made in late-forming objects, such as differentiated meteorites (e.g., in iron meteorites for ^{107}Pd or in eucrites for ^{60}Fe and ^{53}Mn), we also tabulate the inferred abundance at the time of the formation of CAIs, which are considered to be some of the first solid objects to have formed in the solar system. The extrapolated values, particularly for ^{53}Mn and ^{60}Fe, should be treated with some caution. There is a distinct mismatch between the value obtained from the eucrite data for the initial ^{53}Mn/^{55}Mn ratio at the time of CAI formation, and that obtained directly from the analysis of CAIs. As already emphasized, additional data are needed for a more precise determination of the initial ^{60}Fe/^{56}Fe ratio. We have also included in this table several other short-lived nuclides whose presence in the early solar system has been suggested from results obtained from meteorite studies; at present these results can be considered only as weak to strong hints that need further confirmation.

III. SOURCES OF THE SHORT-LIVED NUCLIDES IN THE EARLY SOLAR SYSTEM

Although the "live" presence of several short-lived nuclides in the early solar system at the time of formation of the CAIs is now well accepted, there has been a suggestion that the observed excess in their decay products seen in primitive meteorites may simply represent a "fossil" record. In this scenario (Clayton 1975, 1982, 1986), stellar condensates will incorporate freshly synthesized short-lived nuclides (e.g., ^{26}Al, ^{41}Ca, etc.) at the time of their condensation in specific stellar environments, and *in situ* decay of these nuclides will lead to enhanced abundances of their respective decay products (^{26}Mg, ^{41}K, etc.) within the condensates. These will constitute an important component in the initial mix of material from which the solar system has formed. Early solar system objects such as the

CAIs may inherit "fossil" excess of ^{26}Mg or ^{41}K from these pre-existing stellar condensates from a much earlier epoch. In this proposal, the observed correlation between the excess in the daughter nuclide and the abundance of the parent element (e.g., Figs. 1 and 2) can be explained as a two-component mixture: one representing the normal solar system ratio and the other an enhanced ratio present in the presolar grains. Although stellar condensates (circumstellar grains) with ^{26}Mg and ^{41}K excess have indeed been found in meteorites (Hoppe et al. 1994; Amari et al. 1996), the "fossil" hypothesis appears to be untenable in the context of CAIs, for two important reasons. First, the end member composition for the presolar grains needs to be adjusted arbitrarily to match the meteorite observation, both in terms of their isotopic composition and ratios. Second, and more importantly, the analyzed CAIs, which are often products of crystallization from a refractory melt, would have undergone complete isotopic homogenization, leading to a uniform isotopic composition of Mg and K, rather than the well-preserved correlation seen in Figs. 1 and 2. Podosek and Swindle (1988) provide a detailed discussion of this aspect, and we shall not further address the question of the possible "fossil" nature of these nuclides. We believe the meteorite data provide definitive evidence for the "live" presence of the short-lived nuclides in the early solar system and shall make an attempt to trace their possible source and origin.

The most probable sources of the short-lived nuclides present in the early solar system are (i) freshly synthesized nuclides from specific presolar stellar source(s), (ii) products of energetic particle interactions with gas and dust in the protosolar cloud, and (iii) products of energetic particle interactions with gas and dust in the solar nebula.

Energetic particle irradiation could be a plausible source only for a subset of the short-lived nuclides listed in Table I (e.g., ^{41}Ca, ^{36}Cl, ^{26}Al, and ^{53}Mn). In contrast, all the short-lived nuclides could be stellar nucleosynthesis products; some of them, like ^{129}I, ^{182}Hf, and ^{244}Pu, are distinct products of long-term *r*-process nucleosynthesis, whereas the low-mass nuclides could be generated by nucleosynthesis processes operating in different stellar sources. In Table I we have shown the stellar sites that could be responsible for synthesizing the short-lived nuclides.

All the meteorites are exposed to galactic cosmic rays (GCR) while in interplanetary space as small (meter-sized) objects following their ejection from their parent bodies and prior to their eventual capture by Earth. This exposure duration could be millions of years, and it is necessary to consider the possibility that some of the short-lived nuclides may be produced by interactions of energetic GCR particles with meteorites. Although production by primary GCR particles (chiefly protons) can be ruled out on account of their low intensity ($\sim$2 protons cm^{-2} s^{-1} at 1 AU), production by secondary neutrons (generated within the meteorite by interacting primary protons) could be particularly important for the production of nuclides such as ^{41}Ca and ^{36}Cl, which have high neutron production (n,γ)

cross sections. This question has been examined in detail, and it has been shown that the secondary neutron fluence experienced by meteorites such as Allende and Efremovka are more than an order of magnitude lower than those necessary for producing the observed excesses of ^{41}K and ^{36}Ar found in these meteorites (Göbel et al. 1982; Srinivasan et al. 1996; Murty et al. 1997).

It has also been suggested that several of the short-lived nuclides, particularly ^{41}Ca, ^{36}Cl, and ^{107}Pd, could have been produced by secondary low-energy neutrons interacting with nebular gas and dust in the early solar system (Marti and Lingenfelter 1995). However, detailed calculations have not been attempted yet, and it is not feasible to produce ^{26}Al, ^{53}Mn, and ^{60}Fe in this scenario.

A. Production by Energetic Particles

The production of any short-lived nuclide by energetic particles can be expressed as

$$P[i](t) = \sum_j \int_0^t [F(E,t) \times \sigma(j \rightarrow i, E) \times N(j)] \qquad (1)$$

where i is the nuclide of interest, $F(E,t)$ is the time variation in the flux of the interacting energetic particles, the j are the target nuclides, $\sigma(j \rightarrow i, E)$ is the production cross section as a function of energy, and $N(j)$ is the target abundance. If there are more than one type of interacting particles, one has to consider contribution from each of the interacting species (e.g., contribution from both proton- and alpha-particle-induced reactions in the case of solar energetic particles). Production by secondary particles may be ignored if the energies of the interacting particles are low (a few to a hundred MeV/n), as is the case with solar energetic particles and also with the low-energy heavy ions postulated to be present in a star-forming molecular cloud complex (see section III.C). In such a case, one needs to consider only ionization loss processes suffered by the interacting particles within the target and in traversing the intervening medium between the source and the target.

B. Production by Solar Energetic Particles

Soon after the discovery of the presence of ^{26}Al in the Allende CAIs, an effort was made to see whether this nuclide could have been produced in the early solar system by an enhanced flux of energetic particles from an active early Sun interacting with gas and dust in the solar nebula (Heymann and Dziczkaniec 1976). The basic approach used in this and other later efforts was to consider appropriate energy spectra for the solar energetic particles (SEP), take gas and/or dust as target with "solar" composition, invoke or ignore self-shielding of the energetic particles by gas and dust, and finally estimate the various parameters, particularly for the energy spectra

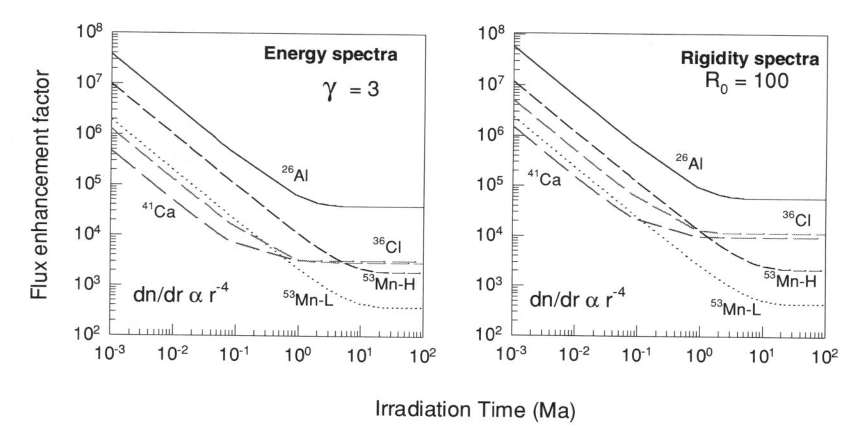

Figure 3. Enhancement factor in solar energetic particle flux, relative to its long-term average value of $N(>10\text{MeV}) = 100$ protons cm^{-2} sec^{-1}, required to explain the observed initial abundances of the short-lived nuclides, ^{26}Al, ^{41}Ca, ^{53}Mn, and ^{36}Cl, plotted as a function of irradiation duration. Two values for initial abundance of ^{53}Mn are used (see Table I). Results are presented for SEP flux representation both in kinetic energy ($dN/dE = KE^{-\gamma}$) and in rigidity ($dN/dR = C\exp(-R/R_0)$)

and particle fluence that may explain the meteorite data on the initial abundances of the short-lived nuclides (Table I). The initial attempts to explain the data for ^{26}Al led to suggestions for very special spectral shapes for the SEP (to avoid abundance anomalies in other nuclides) and extremely high fluences that would require the SEP flux to be more than 10^5 times the contemporary value (Heymann and Dziczkaniec 1976; Clayton et al. 1977; Heymann et al. 1978; Lee 1978). Self-shielding of the SEP by gas and dust in the solar nebula, leading to selective irradiation of nebular material, was also proposed, and candidate nuclides that may be produced by SEP in the early solar system have been identified (Clayton et al. 1977; Heymann et al. 1978; Lee 1978).

Unfortunately, none of these initial proposals can be considered satisfactory. Their inadequacy was made evident by the attempt to match the data for the initial abundance of ^{53}Mn, following the discovery of its presence in the early solar system. Detailed calculations on the relative production of ^{26}Al and ^{53}Mn by SEP (Wasserburg and Arnould 1987) clearly showed that it is not possible to produce both nuclides in the required amounts with the same set of parameters for particle flux and irradiation duration. In Fig. 3 we show the results obtained in a recent study on the production of the four short-lived nuclides, ^{41}Ca, ^{36}Cl, ^{26}Al, and ^{53}Mn, by SEP interacting with an ensemble of micron- to millimeter-sized objects in the early solar system (Goswami et al. 1997, 1999). Production by both protons and alpha particles was considered, and self-shielding of the energetic particles by nebular material was ignored to maximize production. Enhancement factors for the flux of energetic particles from the early Sun compared to the contemporary long-term (million-year) averaged flux (based on lunar sample data) necessary to match the initial abundances of the short-lived nuclides are shown in this figure as a function of the irradiation duration. It is obvious that no combination of enhancement factor and irradiation duration can coproduce ^{26}Al with any of the other three nuclides. If one tries to match the ^{26}Al data, the other nuclides will be overproduced by a factor of 10 or more, and *vice versa.* It may, however, be noted that if the enhancement factor is $\sim 10^4$, irradiation for a million-year timescale could in principle lead to the coproduction of ^{41}Ca, ^{36}Cl, and ^{53}Mn to match the meteorite data. Although the possibility of such an enhancement in the flux of particles from an active early (T Tauri) Sun cannot be ruled out (see, e.g., Caffee et al. 1991; and the chapter by Glassgold et al., this volume), new experimental data that show that ^{41}Ca and ^{26}Al are cogenetic (Sahijpal et al. 1998; see also section III.D) rule out the possibility of SEP production of the short-lived nuclides in the early solar system. Further, SEP production of ^{60}Fe is insignificant, and an external source for this nuclide appears to be essential.

Certain variations of the SEP irradiation scenario have been proposed in recent years. Clayton and Jin (1995*b*) invoke shielding of energetic particles by material in the protostellar disk, thus restricting the production of radionuclides close to the boundary of the disk (a variation of the proposal made earlier by Lee 1978). They also pointed out the possibility of a highly enhanced flux of anomalous cosmic rays (particularly ^{4}He) interacting with material in the protostellar disk and producing ^{26}Al and ^{41}Ca. In another scenario, Shu et al. (1996, 1997) proposed the x-wind irradiation model, in which interaction of the accreting disk material with the magnetic field of the proto-Sun will lead to the formation of refractory CAIs very close to the Sun. Some of these CAIs will be subjected to an intense irradiation by a very high flux of SEP for short durations (a few years to tens of years), leading to the production of the short-lived nuclides before the CAIs are ejected further out into the protostellar disk. However, the problem of coproduction of ^{41}Ca and ^{26}Al in the required amounts, as well as that of extreme underproduction of ^{60}Fe, remains. An effort made by Lee et al. (1998) to remove these problems by invoking production by an enhanced flux of ^{3}He particles, emitted during impulsive flares, did not meet with success. Thus, even though the solar energetic particles are capable of coproducing several of the short-lived nuclides that were present in the early solar system, they fail to produce them in amounts that can match their initial abundances as inferred from the meteorite data.

C. Irradiation in a Molecular Cloud Environment

The protosolar cloud fragment that collapsed to form the Sun and the solar system was part of a large molecular cloud complex. Emission of gamma-ray lines from such clouds, which may be attributed to nuclear deexcitation, provides evidence for the presence of a steady-state population of low-energy particles within such a cloud complex. A major discovery in this field took place in 1994, when enhanced emission of 4.44-MeV and 6.13-MeV gamma rays was detected from a star-forming region in the Orion Molecular Cloud (Bloemen et al. 1994). The measured flux of these gamma rays, identified as deexcitation of ^{12}C and ^{16}O, is 100-fold higher than previously predicted (Ramaty et al. 1979) and demands the presence of an enhanced flux of low-energy carbon, oxygen, and other heavy ions within the molecular cloud complex (Bloemen et al. 1994). The sources of the energetic particles inside the molecular cloud are not unambiguously known, but they could be T Tauri stars and mass ejecta from stars or exploding supernovae. If the protosolar cloud evolved in a similar environment about 4.6 Gyr ago, it is obvious that it would have been subjected to irradiation by these energetic particles and that production of nuclides, including some of the short-lived nuclides, would have taken place. If this irradiation continued until the time of cloud collapse (with the possibility of continuing throughout the collapse), the early solar system would have been endowed with these short-lived nuclides. The possibility that such an irradiation may account for the short-lived nuclides present in the early solar system was first proposed by Clayton (1994, see also Clayton and Jin 1995*a*). A very detailed analysis of the problem was carried out by Ramaty et al. (1996) by assuming different compositions for both the irradiating and irradiated particles and various plausible energy spectra, as well. However, as in the case of SEP irradiation, the problem of matching the initial abundances of the short-lived nuclides ^{41}Ca, ^{26}Al, and ^{53}Mn remains. In fact, the problem of this mismatch in all the irradiation scenarios results primarily from the combined effect of the target composition, which has to be close to "solar," and the reaction cross sections of interest, and it is difficult to overcome this problem by changing parameters such as the spectral shapes and the particle flux. Irrespective of the problems associated with the irradiation scenarios, it is not possible to rule out any contribution from such a source toward the inventory of the short-lived nuclides present in the early solar system. A plausible way to check the magnitude of this contribution will be to look for excess abundance of nuclides that are produced primarily by energetic particle interaction. ^{138}La and ^{50}V appear to be two such possibilities (Lee et al. 1998).

D. Stellar Source(s) for the Short-Lived Nuclides

A stellar origin for the short-lived nuclides and their injection into the protosolar cloud prior to its collapse has remained a widely favored proposal from the very beginning, because all of the radionuclides present in the early solar system could be produced in distinct stellar nucleosynthesis sites (see Tables I and II). In the case of the relatively longer-lived nuclides such as ^{244}Pu, ^{182}Hf, and ^{129}I, it is important to consider contributions from continuous galactic nucleosynthesis toward the inventory of

TABLE II

Plausible Stellar Sources for the Short-lived Nuclides

Source [Ref.]	Nuclide	Processes	Remarks
TP-AGB (Thermally pulsing asymptotic giant branch star) [1,2]	^{26}Al	H-burning	* Self-consistent dilution factor with $\Delta = 0.6$ Myr
	^{41}Ca, ^{60}Fe ^{107}Pd, ^{36}Cl	He-burning, He-S process	* Cannot account for ^{53}Mn * Predict ^{205}Pb and ^{135}Cs at detectable level * Precise estimate of neutron source function difficult
Massive supernova ($M > 25$ M$_\odot$) [3]	^{26}Al, ^{41}Ca, ^{53}Mn, ^{60}Fe, ^{36}Cl	Shell-burning, NSE, and explosive synthesis	*^{26}Al underproduced (contribution from WR stage could help) * Yield of ^{107}Pd uncertain * Within 2–20 pc of the protostellar cloud * Dynamics of explosion and material ejection poorly understood
Nonexploding Wolf-Rayet star ($M \sim 60$ M$_\odot$) [4]	^{26}Al	Core H-burning	* Self-consistent dilution factor with $\Delta = 0.2$ Myr
	^{41}Ca, ^{36}Cl ^{107}Pd	Core He-burning and He-S process	* Cannot account for ^{53}Mn * Extreme underproduction of ^{60}Fe * Predict very high ^{205}Pb abundance; not seen yet * Dynamics of material ejection poorly understood

[1] Cameron (1993); [2] Wasserburg et al. (1994 1995); [3] Cameron et al. (1995); [4] Arnould et al. (1997*a*, *b*)

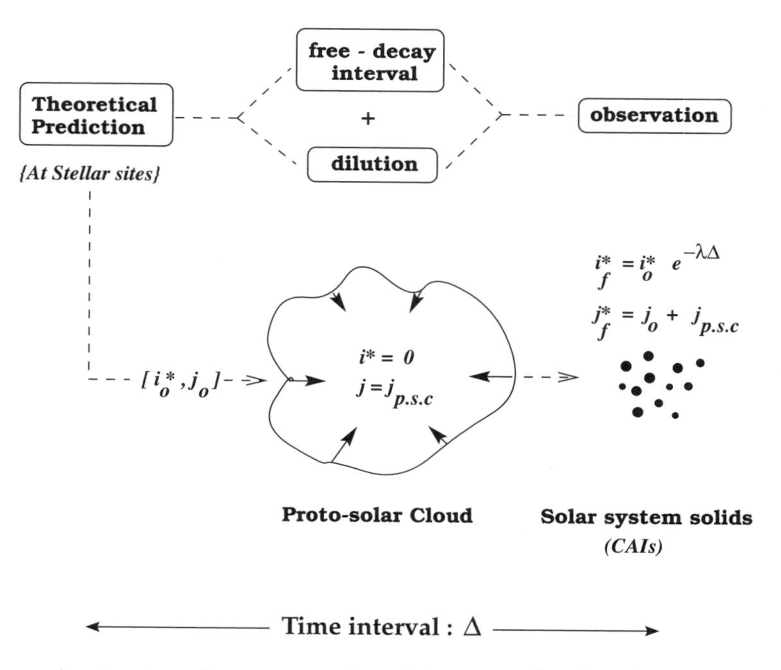

Figure 4. A schematic representation of the approach adopted to explain the abundances of the short-lived nuclides in the early solar system, assuming their injection into the protosolar cloud from a single stellar source.

these nuclides in the protosolar cloud. However, for most of the shorter-lived nuclides, the emphasis is to look for a self-consistent solution considering a single stellar object, such as a supernova, a thermally pulsing asymptotic giant branch (TP-AGB) star, or a Wolf-Rayet (W-R) star, that could have injected these nuclides into the protosolar cloud just prior to its collapse. The basic approach followed in these calculations is shown schematically in Fig. 4. The production ratio of the radioactive nuclides to their stable counterparts, $[P_i/P_j]$ (e.g., $P[^{26}Al/^{27}Al]$), in the stellar site is estimated from stellar nucleosynthesis calculations. The dilution factor includes possible dilution of the freshly synthesized material by material in the intervening medium between the source and the protosolar cloud and in the protosolar cloud that are devoid of the radioactive species. The symbol "Δ" represents the time interval between production of the nuclides in the stellar source and their incorporation by the early solar system solids. If one assumes the same dilution factor for all the nuclides, it is possible to do a self-consistency check by suitably adjusting the value of "Δ" that will match the initial abundances of the short-lived nuclides at the time of formation of specific early solar system objects like the CAIs.

Several attempts have been made in recent years to obtain such self-consistent solutions by choosing specific stellar sources for the short-lived nuclides (with mean life $<10^7$ years) present in the early solar system. The essential features of these studies are summarized in Table II. A TP-AGB star could be a plausible source for ^{41}Ca, ^{26}Al, ^{60}Fe, ^{107}Pd, and ^{36}Cl, while a W-R star yields a self-consistent solution only for ^{41}Ca, ^{107}Pd, and ^{36}Cl (Wasserburg et al. 1994, 1995; Arnould et al. 1997*a,b*). Neither of these sources can account for ^{53}Mn. A nonexploding W-R star underproduces ^{26}Al and ^{60}Fe and predicts very high initial abundance of the short-lived nuclide ^{205}Pb (Arnould et al. 1997*a,b*), which has not been observed yet. On the other hand, the prediction for the initial abundances for the nuclides ^{205}Pb and ^{135}Cs (Wasserburg et al. 1994) in the TP-AGB model are not inconsistent with the inferred upper limits of their abundances in the early solar system. A Type II supernova can in principle account for the four short-lived nuclides (^{41}Ca, ^{26}Al, ^{60}Fe, and ^{53}Mn) present in the early solar system. However, it underproduces ^{26}Al, and it is not possible to obtain a self-consistent dilution factor for all the four nuclides (Cameron et al. 1995; Wasserburg et al. 1998). A very massive supernova, where the initial star has gone through the Wolf-Rayet state before evolving to a supernova, has been proposed to overcome this problem (Cameron et al. 1995). The yield of ^{107}Pd during such a supernova event is also uncertain. If one considers the possibility that continuous galactic nucleosynthesis might have contributed to the inventory of ^{53}Mn in the protosolar cloud (Wasserburg et al. 1996) or that it was a product of energetic particle interactions in the early solar system (Goswami et al. 1997, 1999), an AGB star appears to be a plausible source, with its capability to account for all the other short-lived nuclides up to ^{107}Pd. Studies of the interaction of stellar mass outflows with their surroundings might also provide clues to the most likely stellar source, but at present such studies are not at the level at which they could differentiate between the possible sources.

The question of the close association of an evolved star with the nascent protosolar cloud also requires scrutiny. A survey of AGB stars during the present epoch suggests a very small probability for the association of such a star with a molecular cloud (Kastner and Myers 1994). The association of a massive supernova with the protosolar cloud, even though it could be at a distance of a few to 20 parsecs, is yet to be studied quantitatively. It is generally thought that most core collapse supernovae occur in the vicinity of their natal molecular clouds, because of the short lifetimes of their progenitor high-mass stars. Indeed, there is some evidence for this behavior in the Galaxy (e.g., Wootten 1977; Wootten 1981; Dubner and Arnal 1988; Tatematsu et al. 1990*a,b*; Feldt and Green 1993; Reynolds and Moffett 1993; Wilner et al. 1998) and especially in the Large Magellanic Cloud, where most of the observed supernova remnants appear to be associated with molecular clouds (Cohen et al. 1988; Banas et al. 1997). It is important to determine, however, whether a supernova event will lead to the disruption of its natal molecular cloud or whether it can also induce collapse within the cloud. Studies of the interaction between supernova remnants and molecular clouds, especially in the context of "mixed-morphology" supernova remnants (Rho and Petre 1998; Reach and Rho 1999), may be able to provide answers to this question.

The case for a stellar source for the short-lived nuclides received a boost from a recent experiment, which showed that the initial abundances of the two short-lived nuclides ^{41}Ca and ^{26}Al in the early solar system are very strongly correlated (see Fig. 5), indicating them to be cogenetic (Sahijpal et al. 1998). This rules out the energetic-particle irradiation scenarios, because they cannot coproduce these nuclides in amounts necessary to match their initial abundances in the early solar system. As already discussed, matching the initial abundance of ^{26}Al leads to more than an order-of-magnitude overabundance of both ^{41}Ca and ^{53}Mn. It may be noted here that along with the short-lived nuclides, the stellar source will also inject some freshly synthesized stable nuclides in "non-solar" proportions into the protosolar cloud, and this process could lead to abundance anomalies in particular stable isotopes as well. The possible magnitude of such anomalies has been evaluated for several low-mass nuclides only in the case of a TP-AGB star, and the effect was found to be very small [$<\epsilon$ (parts per ten thousand) level; R. Gallino, personal communication] compared to the parts-per-thousand to percent level anomalies observed in the case of the decay products of the short-lived nuclides.

If we consider a stellar source for the short-lived nuclides, the time interval "Δ" between their synthesis at a stellar site and the formation of some of the first solar system solids (CAIs) has to be less than a million years. Model calculations suggest this timescale to be 0.6 Myr for a TP-AGB star and 0.2 Myr for a WR star (Wasserburg et al. 1995; Arnould et al. 1997*a*,*b*), indicating a high density for the protosolar cloud ($>10^4$ H_2 cm^{-3}) at the time of its collapse. This time scale is much shorter than the nominal duration of 5–10 Myr for unassisted collapse in the standard scenario (Mouschovias 1989; Shu 1995) and is suggestive for an assisted or triggered collapse of the protosolar cloud. It therefore appears highly probable that the very process that injected radionuclides from a stellar source could have triggered the collapse of the protosolar cloud.

IV. THE VIABILITY OF THE TRIGGERED COLLAPSE SCENARIO

The idea of the triggered origin of the solar system, where the collapse of the protosolar cloud was initiated by the impact of an interstellar shock wave propagating from a stellar source (Cameron and Truran 1977; Cameron 1993; Wasserburg et al. 1994, 1995; Boss 1995; Cameron et al. 1995, 1997; Boss and Foster 1997), is one aspect of the more general

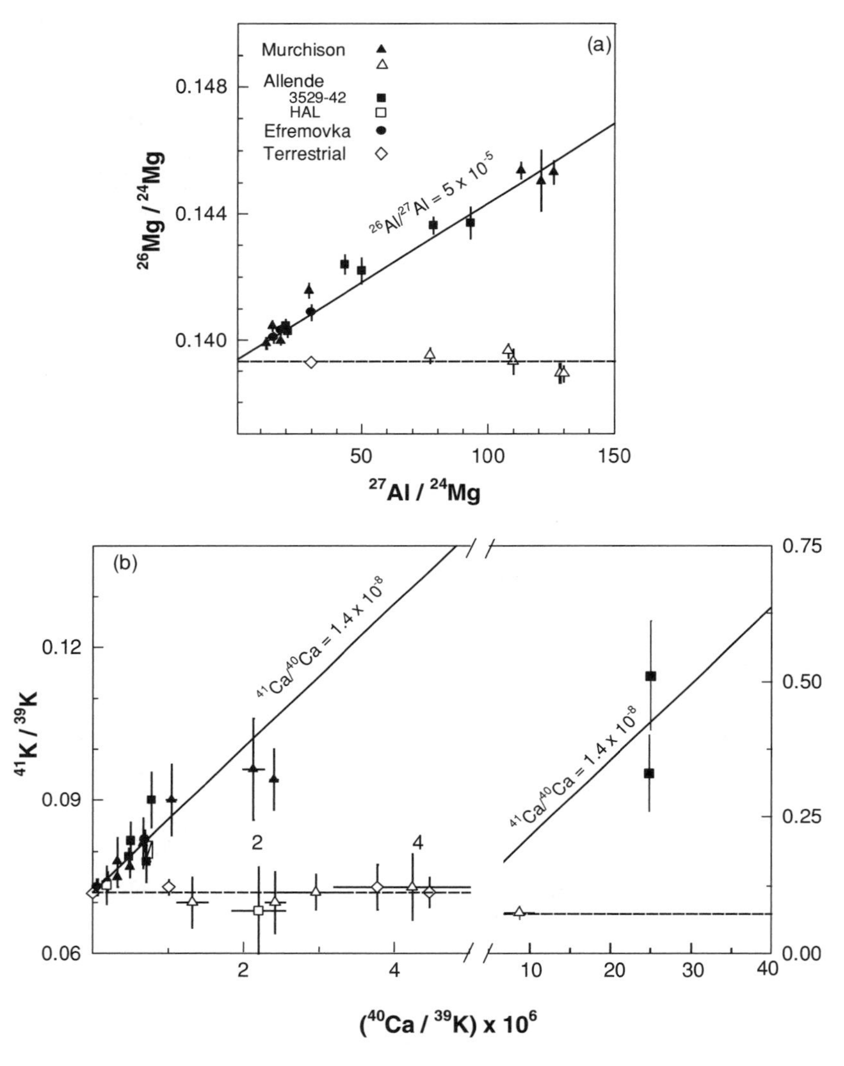

Figure 5. Measured Mg and K isotopic ratios in refractory oxide (hibonite) from several primitive meteorites plotted as a function of their Al/Mg and Ca/K ratios. The same symbols are retained for representing data for individual hibonites in both panels. The dotted horizontal lines represent normal $^{41}K/^{39}K$ ratio of 0.072, and the measured values in terrestrial analog samples are consistent with this value. Correlated presence of excess ^{26}Mg and ^{41}K is clearly evident, indicating a common source for their parent nuclides ^{26}Al and ^{41}Ca (data from Sahijpal et al. 1998).

idea of assisted star formation (Cameron et al. 1997; Elmegreen 1998). Numerical simulations can be used to study the viability of the scenario by addressing the following basic questions:

1. Can the collapse of a molecular cloud core be induced by the impact of an interstellar shock wave?
2. Can radioactivities carried by the shock wave be injected into the collapsing system?
3. Is the timescale sufficiently short for the survival of the radioactivities?

A. Triggered Collapse of Molecular Cloud Cores

The early studies of the interaction between shocks and interstellar clouds were principally concerned with high-velocity shock waves impacting molecular clouds or traveling through ambient interstellar material (e.g., Nittman et al. 1982; Heathcote and Brand 1983; Krebs and Hillebrandt 1983; Różycka and Tenorio-Tagle 1987; Bedogni and Woodward 1990; Stone and Norman 1992; Klein et al. 1994; Mac Low et al. 1994; Xu and Stone 1995). The main result of these studies is that the clouds are usually destroyed by the Rayleigh-Taylor (RT) and Kelvin-Helmholtz (KH) instabilities created at the contact surface between the shock wave and the molecular cloud. However, these calculations usually did not include self-gravity, so the results are not directly applicable to the question of triggered collapse. Instead, they are important for describing cloud shredding at high velocities, such as in observed supernova remnants, and for examining in detail the instabilities created during the interaction of the shock wave with molecular cloud material.

The problem of triggered collapse has been addressed directly by recent three-dimensional hydrocode calculations (Boss 1995), two-dimensional piecewise-parabolic method (PPM) simulations (Foster and Boss 1996, 1997; Boss and Foster 1997, 1998), and three-dimensional smoothed particle hydrodynamics (SPH) calculations (Cameron et al. 1997; Vanhala and Cameron 1998). These studies concentrated on moderately slow (10–50 km s^{-1}) shock waves impacting centrally condensed molecular cloud cores with densities $\sim 10^{-19}$–10^{-16} g cm^{-3} and masses $\sim$1–10 $M_{\odot}$. Apart from the simulation methods themselves, the calculations also differed in their choice of thermodynamics. The hydrocode and PPM calculations used an isothermal (adiabatic exponent $\gamma = 1$) or adiabatic (with $\gamma = \frac{5}{3}$) equation of state, whereas the SPH calculations employed an equation-of-state solver, which yielded pressure and specific internal energy at the desired density and temperature. Subsequently, the adiabatic exponent γ ($[d(\log P)/d(\log \rho)]_{ad}$) did not have a single value but varied as a function of density and temperature. In addition, the SPH code included cooling due to molecules, atoms, and dust as well as effects from magnetic pseudofluid.

TABLE III
Results from Simulations of Triggered Collapse

Adiabatic exponent γ	Shock strength	Result
1	Momentum $\leq$ 0.1 $M_\odot$ km s^{-1} [a]	No collapse
	Momentum $\geq$ 0.1 $M_\odot$ km s^{-1} [a]	Collapses
$\frac{5}{3}$	All velocities	No collapse; cloud shredded apart
Variable	velocity < 20 km s^{-1}	May collapse[b]
	velocity = 20–45 km s^{-1}	Collapses[c]
	velocity $\geq$ 45 km s^{-1}	Cloud shredded apart[d]

[a] For 10-K, 1-$M_\odot$ cloud.
[b] Collapses only if very dense/massive core.
[c] If core temperature rises above ~27 K.
[d] Very dense/massive cores may survive and collapse.

The basic results of these simulations are summarized in Table III. Molecular clouds can be triggered into collapse if the momentum of the shock wave is sufficient to compress the core to the point of collapse but not so high that the core will be torn apart. Foster and Boss (1996) formulated the condition for successful triggering in terms of the critical momentum, which divides cases that induce collapse and those that do not. For the isothermal top-hat model of the shock wave used in their calculations, the critical value was found to be 0.1 $M_\odot$ km s^{-1} for a 10-K, 1-$M_\odot$ cloud and to scale as the mass of the cloud times its sound speed.

If the shock wave properties are derived from the Rankine-Hugoniot jump conditions, the SPH calculations using variable-γ thermodynamics (Vanhala and Cameron 1998) suggest that there are three realms of results:

1. Low-velocity shocks ($v_s < 20$ km s^{-1}): The core is compressed by a factor of ~10, but it usually bounces back and is torn apart.
2. Intermediate-velocity shocks (20 km $s^{-1} < v_s <$ 45 km s^{-1}): The core is compressed and stretched and may collapse (see discussion below).
3. High-velocity shocks ($v_s > 45$ km s^{-1}): The core is destroyed.

The lower velocity limit for successful triggering in the variable-γ calculations arises from the postshock temperature becoming sufficiently high to destroy the principal coolants: molecular hydrogen and carbon monoxide (Cameron et al. 1997; Vanhala and Cameron 1998). At velocities higher than 20 km s^{-1}, the gas remains hot, and the pressure exerted on the core is sufficient to compress it to the point of collapse. At lower velocities ($v_s < 20$ km s^{-1}) the coolants survive, the postshock flow cools efficiently, and the pressure is usually insufficient to compress the core to the point of collapse. However, the results are strongly dependent on

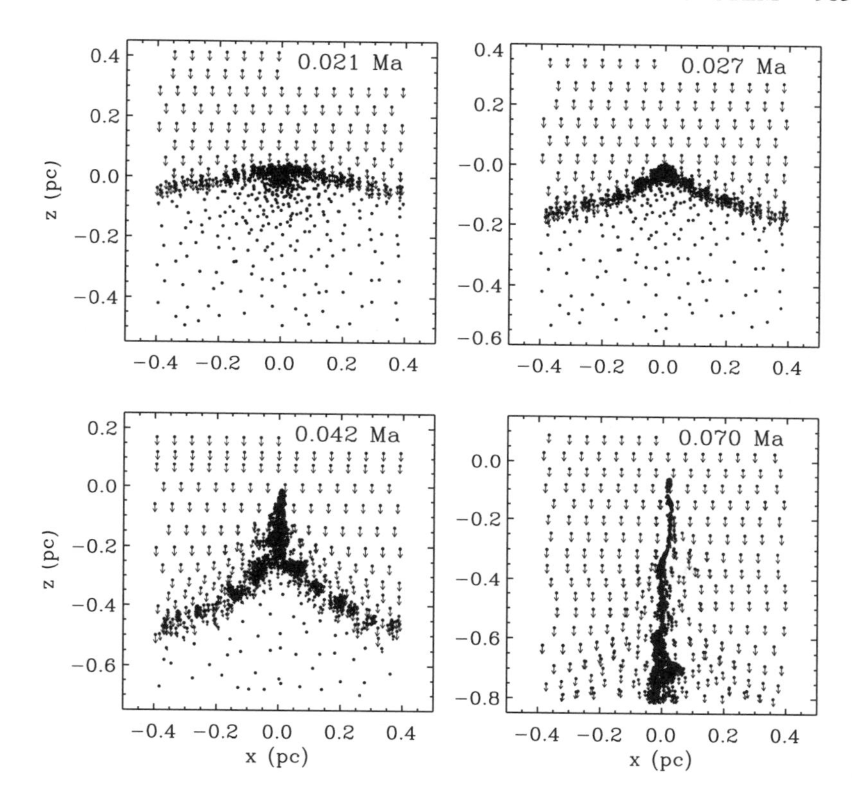

Figure 6. Interaction between the shock wave and the molecular cloud core in a 3D SPH simulation run with 4496 particles in the core and 7772 in the shock wave. The initial peak density in the core is 7.35×10^{-17} g cm^{-3}. The shock approaches from the $+z$ direction with a velocity 25 km s^{-1}. The system is shown in the xz plane with the units in parsecs. The times marked in the panels indicate the time from the initial approach of the shock wave. The head of the filament in the last frame is collapsing, and the peak density is 1.72×10^{-11} g cm^{-3}, 2.4×10^{6} times the original core density.

the initial properties of the core; the velocity limits quoted above apply to typical low-mass cores considered in the calculations. The triggered collapse of higher-mass cores requires further study. It also must be noted that these limits are derived from calculations using the variable-γ scheme; if an isothermal equation of state is used, the only requirement for successful triggering is the momentum requirement, and shock waves with velocities lower than 20 km s^{-1} can also trigger collapse.

Figure 6 shows a sequence of images from a typical SPH simulation run leading to triggered collapse (Vanhala and Cameron 1998). After initial planar flattening of the core (Fig. 6a), the flow settles to a standing bow shock (Fig. 6b). With continuing arrival of shocked material, the facing side of the core is flattened further while its sides suffer erosion. As a result, the core is stretched to a thin filament (Fig. 6c), the head of which collapses, while its tail merges with the postshock flow (Fig. 6d).

The thermodynamics employed in the calculations appears to be crucial in determining the outcome of the simulations. Calculations using identical initial conditions and differing only in the value of the adiabatic exponent γ found that a shock with $\gamma = \frac{5}{3}$ results in the destruction of the cloud, whereas $\gamma = 1$ pushes the core into collapse (Foster and Boss 1996; Vanhala and Cameron 1998). Calculations performed with a variable γ further illustrate the important role of thermodynamics: Sustained collapse is found to take place only when the temperature in the compressed region rises above ~27 K (Cameron et al. 1997; Vanhala and Cameron 1998), which corresponds to the point at which the adiabatic exponent first falls below the stability value of $\gamma = \frac{4}{3}$ (e.g., Low and Lynden-Bell 1976; Tohline 1981; Whitworth 1981). However, the behavior of γ, including the temperature at which the value of γ falls below $\frac{4}{3}$, depends on the ortho-to-para ratio of hydrogen. The SPH calculations assume an equilibrium mix between the two states of hydrogen, but the sensitivity of the results on thermodynamics stresses the need for a more thorough investigation of this aspect of the problem.

The final outcome of the interaction also depends on the evolutionary state of the preimpact core. If the core is assumed to have evolved under the control of ambipolar diffusion, cores at later evolutionary stages have higher central densities and smaller radii at the time of impact (e.g., Mouschovias 1991). The SPH simulations indicate that well-evolved cores can be triggered at lower velocities (at least down to 10 km s^{-1}) than less evolved cores, and they usually collapse to a single mass concentration. Less evolved cores may fragment during the interaction and form a multiple system. The details of this behavior have not been fully examined, and more calculations are needed to study the effect of different initial densities and core masses.

B. Injection of Radioactivities into the Collapsing Core

An important aspect of the numerical studies is to determine how the radioactivities carried by the shock wave are injected into the collapsing molecular cloud core. The 2D PPM simulations of Foster and Boss (1997) and Boss and Foster (1998) reveal that shock wave material can be injected into the core through Rayleigh-Taylor fingers (Fig. 7), when dense shocked material gathers on the surface of the cloud and penetrates into the core in fingerlike structures. Injection appears to occur at a roughly constant rate and usually takes place after the central regions of the original core have gone into collapse. The injection efficiency, defined as the ratio of the shock wave material contained in the central regions of the collapsing system to that originally incident on the core, is typically 10–20%. The efficiency does not decrease significantly even if the radioactivities are lagging far behind the immediate shock front (Boss and Foster 1998).

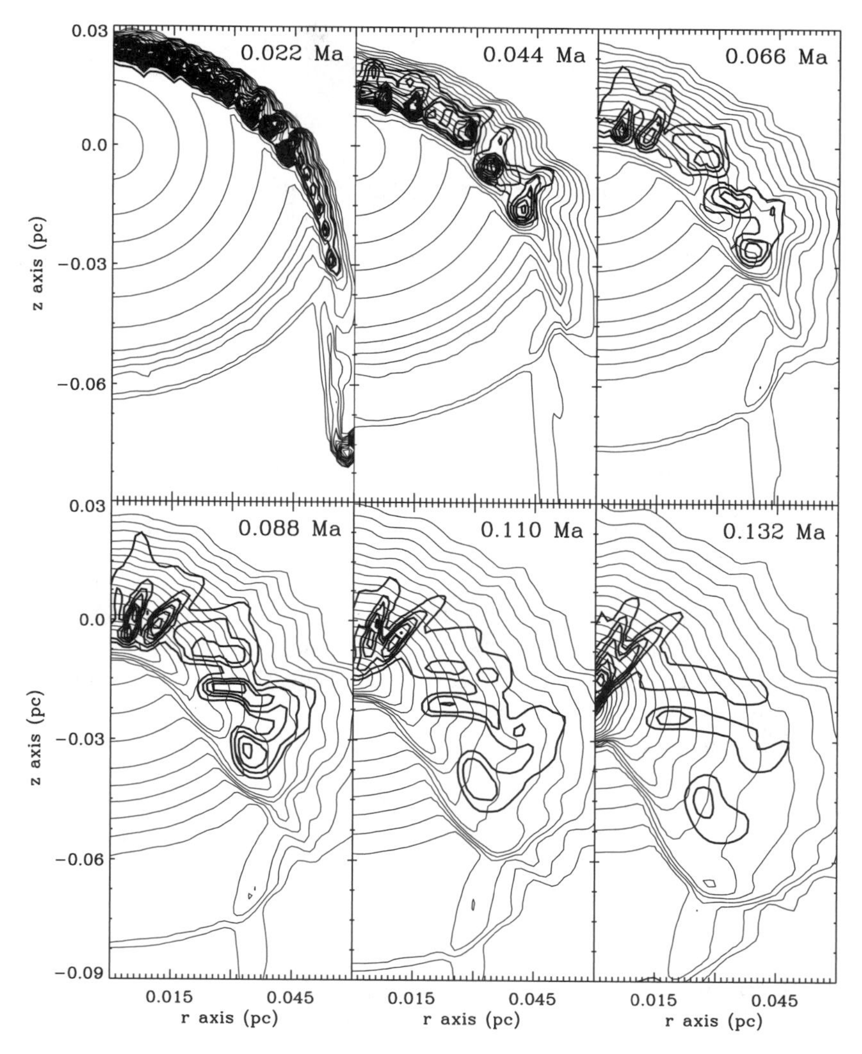

Figure 7. Injection of shock wave material in the 2D PPM calculations. The thin contours represent density of the molecular cloud core during the impact of the shock wave, which travels at the velocity of 20 km s^{-1}, and they range from 1.7×10^3 cm^{-3} to 2×10^8 cm^{-3}. The thick contours trace the changes in the color field, which is used to follow the behavior of the shock flow material. The times marked in the panels indicate the time from the initial approach of the shock wave. The panels show the development of fingerlike structures, which inject the shock wave material into the collapsing system.

If an isothermal equation of state is used, the SPH calculations verify the 2D PPM results (Vanhala and Cameron 1998). However, the details of injection (the timescale, the depth of injection, and the degree of mixing within the collapsing system) require further study. It is especially important to study the details of injection with high spatial resolution so that the results are directly applicable for interpreting meteorite observations.

The steady injection rate found by the PPM calculations suggests that temporal inhomogeneities are unlikely to occur within the mixing layer between the shock wave and the collapsing core. The calculations also revealed that most of the shock flow material deposited into the collapsing system goes to the outer half (by mass) of the cloud. Spatial gradients of the isotopes might therefore occur in the solar nebula, because the last arriving material, the outer parts of the cloud, would carry most of the radioactivities. However, further work is required to determine whether the possible heterogeneities would persist on the scale of the solar nebula to the level where they can be measured in the primitive meteoritic material.

Simulations employing the variable-γ scheme (Vanhala and Cameron 1998) have further complicated the question of mixing. According to these calculations, the hot shocked gas is unable to penetrate into the cold core because of buoyancy and entropy effects. Instead, injection is expected to occur through small-scale instabilities at the contact surface between the shock wave and the core, but the resolution of the 3D SPH simulations is inadequate to observe this effect.

These results suggest that mixing of the shock wave material into the core is most efficient when the shocked gas cools efficiently, giving rise to approximately isothermal shocks. Subsequently, the details of mixing are sensitive to the structure of the shock wave, such as shock thickness and the postshock temperature and cooling, as well as the thermodynamics employed. These details need to be examined more carefully before a comprehensive answer to the injection problem can be provided.

C. Timescale of the Triggered Collapse

The timescale for the triggered collapse of the molecular cloud core from the first approach of the shock wave to the collapse of the core is $\sim 10^4$–10^5 years (Foster and Boss 1996; Vanhala and Cameron 1998). Travel time from the most distant object considered as a candidate for the stellar source of the radioactivities, a supernova explosion occurring at the distance of a few parsecs, is a few times 10^4 years (Cameron et al. 1997). The transportation time is even shorter if the object is nearby, as is the case for the TP-AGB or a W-R star near the cloud, the other suggested stellar sources. The total time from the production of the radioactivities to the end of core collapse is therefore typically less than 1.5×10^5 years, the mean life of ^{41}Ca. Even though the effective time scale of the injection process is not currently well constrained and may take place some 2–4 $\times$ 10^5 years after the initial impact of the shock front (Foster and Boss 1997), the total time span is still consistent with the short time constraints set by the presence of short-lived radionuclides in the early solar system.

V. SUMMARY AND FUTURE DIRECTIONS

Isotopic studies of primitive meteorites have provided evidence for the presence of several now extinct nuclides in the early solar system. The relatively long-lived and high-mass nuclides such as ^{129}I, ^{182}Hf, and ^{244}Pu are products of r-process nucleosynthesis, and their initial abundances in the early solar system can be explained in terms of long-term galactic nucleosynthesis. On the other hand, the relatively short-lived nuclides, with mean lives $<$10 Myr, could represent freshly synthesized material from a specific stellar source, or they could have been produced by energetic particle irradiation of gas and dust in the protosolar cloud or in the solar nebula. The presently available experimental data suggest a single stellar object, such as a supernova, a TP-AGB star, or a W-R star, as the most plausible source of the short-lived nuclides with mean-life $<$10 Myr. Although possible contribution from energetic particle irradiation cannot be ruled out completely, it appears to be at a much lower level compared to the stellar contribution.

A stellar origin for the short-lived nuclides in the early solar system constrains the timescale for the collapse of the protosolar cloud to form the Sun and some of the first solar system objects to less than a million years. This short timescale argues for a triggered collapse of the protosolar cloud, with the stellar source responsible for synthesizing and injecting the radionuclides also acting as the trigger. Numerical simulations have shown that molecular clouds can be triggered into collapse by the impact of interstellar shock waves traveling at a velocity of up to a few tens of km s^{-1} and that the timescale for the process is sufficient for the shortest-lived detected radioactivities to have survived. Preliminary calculations of the mixing of shock wave material into the collapsing system indicate that radioactivities can be injected through instabilities at the contact surface between the core and the shocked material.

Although a self-consistent scenario for a stellar source for the short-lived nuclides and a triggered collapse of the protosolar cloud appears viable, there are several aspects that need further scrutiny before a firm conclusion can be reached. For example, it is important to quantify the possible contribution from energetic particle irradiation toward the inventory of the intial abundances of the short-lived nuclides in the early solar system. Experimental determination of possible excess in abundances of nuclides that may be attributed primarily to production by energetic particle interactions (e.g., ^{138}La, ^{50}V) will be very important in this regard. A precise determination of the initial abundance of ^{60}Fe in samples of CAIs will be extremely useful. Production of this nuclide by energetic particles is extremely inefficient, and its presence in CAIs will necessitate a stellar source for this nuclide. The choice among the plausible stellar sources can be narrowed down by further studies of ^{53}Mn, whose initial abundance in the solar system is not well constrained by the presently available experimental data. If the lower initial abundance of ^{53}Mn inferred from the data for eucrites (Lugmair et al. 1995; Lugmair and Shukolyukov 1998) is indeed correct, there is a distinct possibility that it can be produced by solar energetic particles in the early solar system without concurrently producing significant amounts of ^{26}Al and ^{41}Ca. The question of a possible contribution to the inventory of ^{53}Mn by galactic nucleosynthesis also needs attention (see, e.g., Wasserburg et al. 1996). If both these sources can be ruled out, a supernova will be the most likely source of the short-lived nuclides present in the early solar system. On the other hand, if we consider a TP-AGB or W-R star as a viable source, a determination of the initial abundances of the short-lived nuclide ^{205}Pb will be crucial to choosing between these two alternatives. Evidence for the presence of other short-lived nuclides such as ^{99}Tc and ^{135}Cs and confirmation of the hints for the presence of nuclides such as ^{36}Cl, ^{92}Nb, ^{205}Pb, and for excess abundance of ^{138}La will also be crucial for a final resolution of the problem of the source and origin of the short-lived nuclides in the early solar system.

Apart from the need for additional efforts in the experimental front, it is important to have better stellar evolution and nucleosynthesis models to remove some of the shortcomings listed in Table II. An improved understanding of the dynamics of ejection and mixing of stellar material with its surrounding environment is also essential. Astronomical observations that may provide evidence for triggered star formation will bolster the proposal for a similar mode of formation for the solar system.

The numerical studies of triggered collapse also require further work. Processes that may influence the results but have been ignored in the calculations described above include the full treatment of magnetic effects, the alignment of the rotation axis of the core with respect to the approaching shock front, and the detailed structure of the shock wave. As discussed in section IV.B, the last of these effects may be especially important for the injection of shock wave material into the collapsing core. The calculations described in this chapter provide preliminary answers, but more work is required for a more comprehensive answer to the problem posed by the short-lived radioactivities detected in primitive meteoritic material.

Acknowledgments We thank A. Boss, A. G. W. Cameron, D. Clayton, P. Foster, G. Lugmair, F. Shu, R. Ramaty, and G. J. Wasserburg for helpful comments. We are grateful to G. J. Wasserburg for a comprehensive review of the manuscript. This work was supported by the Dept. of Space, Govt. of India (J. N. G.) and NASA Origins of Solar Systems Program (H. A. T. V.).

REFERENCES

Amari, S., Zinner, E., and Lewis, R. S. 1996. ^{41}Ca in presolar graphite of supernova origin. *Astrophys. J. Lett.* 470:L101–L104.

Arnould, M., Meynet, G., and Paulus, G. 1997*a*. Wolf-Rayet stars and their nucleosynthetic signatures in meteorites. In *Astrophysical Implications of the Laboratory Study of Presolar Materials*, ed. T. J. Bernatowicz and E. Zinner (Woodbury: AIP), pp. 179–202.

Arnould, M., Paulus, G., and Meynet, G. 1997*b*. Short-lived radionuclide production by non-exploding Wolf-Rayet stars. *Astron. Astrophys.* 321:452–464.

Banas, K. R., Hughes, J. P., Bronfman, L., and Nyman, L.-Å. 1997. Supernova remnants associated with molecular clouds in the Large Magellanic Cloud. *Astrophys. J.* 480:607–617.

Bedogni, R., and Woodward, P. R. 1990. Shock wave interactions with interstellar clouds. *Astron. Astrophys.* 231:481–498.

Birck, J. L., and Allegre, C. J. 1985. Evidence for the presence of ^{53}Mn in the early solar system. *Geophys. Res. Lett.* 12:745–748.

Bloemen, H., Wijnands, R., Bennett, K., Diehl, R., Hermsen, W., Lichti, G., Morris, D., Ryan, J., Schönfelder, V., Strong, A. W., Swanenburg, B. N., de Vries, C., and Winkler, C. 1994. COMPTEL observations of the Orion complex: Evidence for cosmic-ray induced gamma-ray lines. *Astron. Astrophys.* 281:L5–L8.

Boss, A. P. 1995. Collapse and fragmentation of molecular cloud cores. II. Collapse induced by stellar shock waves. *Astrophys. J.* 439:224–236.

Boss, A. P., and Foster, P. N. 1997. Triggering presolar cloud collapse and injecting material into the presolar nebula. In *Astrophysical Implications of the Laboratory Study of Presolar Materials*, ed. T. J. Bernatowicz and E. Zinner (Woodbury: AIP), pp. 649–664.

Boss, A. P., and Foster, P. N. 1998. Injection of short-lived isotopes into the presolar cloud. *Astrophys. J. Lett.* 494:L103-L106.

Caffee, M. W., Hohenberg, C. M., Nichols, R. H., Jr., Olinger, C. T., Wieler, R., Pedroni, A., Signer, P., Swindle, T. D., and Goswami, J. N. 1991. Do meteorites contain irradiation records from exposure to an enhanced-activity sun? In *The Sun in Time*, ed. C. P. Sonett, M. S. Giampapa, and M. S. Matthews (Tucson: University of Arizona Press), pp. 413–425.

Cameron, A. G. W. 1993. Nucleosynthesis and star formation. In *Protostars and Planets III*, ed. E. H. Levy and J. I. Lunine (Tucson: University of Arizona Press), pp. 47–73.

Cameron, A. G. W. 1995. The first ten million years in the solar nebula. *Meteoritics* 30:133–161.

Cameron, A. G. W., and Truran, J. W. 1977. The supernova trigger for formation of the solar system. *Icarus* 30:447–461.

Cameron, A. G. W., Höflich, P., Myers, P. C., and Clayton, D. D. 1995. Massive supernovae, Orion gamma rays, and the formation of the solar system. *Astrophys. J. Lett.* 447:L53–L57.

Cameron, A. G. W., Vanhala, H., and Höflich, P. 1997. Some aspects of triggered star formation. In *Astrophysical Implications of the Laboratory Study of Presolar Materials*, ed. T. J. Bernatowicz and E. Zinner (Woodbury: AIP), pp. 665–693.

Chen, J. H., and Wasserburg, G. J. 1987. A search for evidence of extinct lead 205 in iron meteorites. *Lunar Planet. Sci.* 18:165–166.

Chen, J. H., and Wasserburg, G. J. 1996. Live ^{107}Pd in the early solar system and implications for planetary evolution. In *Earth Processes: Reading the Isotopic Code* (Washington, DC: American Geophysical Union) Geophys. Mon. Ser. 95, ed. A. Basu and S. Hart, pp. 1–20.
Clayton, D. D. 1975. Extinct radioactivities: Trapped residuals of presolar grains. *Astrophys. J.* 199:765–769.
Clayton, D. D. 1982. Cosmic chemical memory: A new astronomy. *Q. J. Roy. Astron. Soc.* 23:174–212.
Clayton, D. D. 1986. Interstellar fossil ^{26}Mg and its possible relationship to excess meteoritic ^{26}Mg. *Astrophys. J.* 310:490–498.
Clayton, D. D. 1988. Stellar nucleosynthesis and chemical evolution of the solar neighbourhood. In *Meteorites and the Early Solar System*, ed. J. F. Kerridge and M. S. Matthews (Tucson: University of Arizona Press), pp. 1021–1062.
Clayton, D. D. 1994. Production of ^{26}Al and other extinct radionuclides by low-energy heavy cosmic rays in molecular clouds. *Nature* 368:222–224.
Clayton, D. D., and Jin, L. 1995*a*. Gamma rays, cosmic rays, and extinct radioactivity in molecular cloud. *Astrophys. J.* 451:681–699.
Clayton, D. D., and Jin, L. 1995*b*. A new interpretation of ^{26}Al in meteoritic inclusions. *Astrophys. J. Lett.* 451:L87–L91.
Clayton, D. D., Dwek, E., and Woosley, S. E. 1977. Isotopic anomalies and proton irradiation in the early solar system. *Astrophys. J.* 214:300–315.
Cohen, R. S., Dame, T. M., Garay, G., Montani, J., Rubio, M., and Thaddeus, P. 1988. A complete CO survey of the Large Magellanic Cloud. *Astrophys. J. Lett.* 331:L95–L99.
Dubner, G. M., and Arnal, E. M. 1988. Neutral hydrogen and carbon monoxide observations towards the SNR Puppis A. *Astron. Astrophys. Suppl.* 75: 363–369.
Elmegreen, B. G. 1998. Observations and theory of dynamical triggers for star formation. In *Origins*, ed. C. E. Woodward, J. M. Shull, and H. A. Thronson, Jr. (San Francisco: Astronomical Society of the Pacific), pp. 150–183.
Feldt, C., and Green, D. A. 1993. CO and H I associated with the supernova remnant G84.2-0.8? *Astron. Astrophys.* 274:421–426.
Foster, P. N., and Boss, A. P. 1996. Triggering star formation with stellar ejecta. *Astrophys. J.* 468:784–796.
Foster, P. N., and Boss, A. P. 1997. Injection of radioactive nuclides from the stellar source that triggered the collapse of the presolar nebula. *Astrophys. J.* 489:346–357.
Göbel, R., Begemann, F., and Ott, U. 1982. On neutron-induced and other noble gases in Allende inclusions. *Geochim. Cosmochim. Acta* 46:1777–1792.
Goswami, J. N., Marhas, K. K., and Sahijpal, S. 1997. Production of short-lived nuclides by solar energetic particles in the early solar system (abstract). *Lunar Planet. Sci.* 28:439–440.
Goswami, J. N., Marhas, K. K., and Sahijpal, S. 1999. Did solar energetic particles produce the short-lived nuclide in the early solar system. *Astrophys. J.*, in preparation.
Gray, C. M., and Compston, W. 1974. Excess ^{26}Mg in the Allende meteorite. *Nature* 251:495–497.
Grossman, L. 1980. Refractory inclusions in the Allende meteorite. *Ann. Rev. Earth Planet. Sci.* 8:559–608.
Harper, C. L., Jr. 1996. Evidence for ^{92g}Nb in the early solar system and evaluation of a new *p*-process cosmochronometer from ^{92g}Nb/^{92}Mo. *Astrophys. J.* 466:437–456.

Harper, C. L., Jr., and Jacobsen, S. B. 1996. Evidence for ^{182}Hf in the early solar system and constraints on the timescale of terrestrial accretion and core formation. *Geochim. Cosmochim. Acta* 60:1131–1153.

Heathcote, S. R., and Brand, P. W. J. L. 1983. The state of clouds in a violent interstellar medium. *Mon. Not. Roy. Astron. Soc.* 203:67–86.

Heymann, D., and Dziczkaniec, M. 1976. Early irradiation of matter in the solar system: Magnesium (proton, neutron) scheme. *Science* 191:79–81.

Heymann, D., Dziczkaniec, M., Walker, A., Huss, G., and Morgan, J. A. 1978. Effects of proton irradiation on a gas phase in which condensation takes place. I. Negative ^{26}Mg anomalies and ^{26}Al. *Astrophys. J.* 225:1030–1044.

Hohenberg, C. M., Brazzle, R. H., Pravdivtseva, O. V., and Meshik, A. P. 1998. The I-Xe chronometer. *Proc. Indian Acad. Sci.* 107:413–423.

Hoppe, P., Amari, S., Zinner, E., Ireland, T., and Lewis, R. S. 1994. Carbon, nitrogen, magnesium, silicon, and titanium isotopic compositions of single interstellar silicon carbide grains from the Murchison carbonaceous chondrite. *Astrophys. J.* 430:870–890.

Hudson, G. B., Kennedy, B. M., Podosek, F. A., and Hohenberg, C. M. 1988. The early solar system abundance of ^{244}Pu as inferred from the St. Severin chondrite. *Proc. LPSC* 19:547–557.

Hutcheon, I. D., Armstrong, J. T., and Wasserburg, G. J. 1984. Excess ^{41}K in Allende CAI: A hint reexamined. *Meteoritics* 19:243–244.

Jeffery, P. M., and Reynolds, J. H. 1961. Origin of excess Xe^{129} in stone meteorites. *J. Geophys. Res.* 66:3582–3583.

Kastner, J. H., and Myers, P. C. 1994. An observational estimate of the possibility of encounters between mass-losing evolved stars and molecular clouds. *Astrophys. J.* 421:605–614.

Kelly, W. R., and Wasserburg, G. J. 1978. Evidence for the existence of ^{107}Pd in the early solar system. *Geophys. Res. Lett.* 5:1079–1082.

Klein, R. I., McKee, C. F., and Colella, P. 1994. On the hydrodynamic interaction of shock waves with interstellar clouds. I. Nonradiative shocks in small clouds. *Astrophys. J.* 420:213–236.

Krebs, J., and Hillebrandt, W. 1983. The interaction of supernova shockfronts and nearby interstellar clouds. *Astron. Astrophys.* 128:411–419.

Lee, D.-C., and Halliday, A. N. 1995. Hafnium-tungsten chronometry and the timing of terrestrial core formation. *Nature* 378:771–774.

Lee, D.-C., and Halliday, A. N. 1996. Hf-W isotopic evidence for rapid accretion and differentiation in the early solar system. *Science* 274:1876–1879.

Lee, D.-C., and Halliday, A. N. 1997. Core formation on Mars and differentiated asteroids. *Nature* 388:854–857.

Lee, T. 1978. A local proton irradiation model for isotopic anomalies in the solar system. *Astrophys. J.* 224:217–226.

Lee, T., and Papanastassiou, D. A. 1974. Mg isotopic anomalies in the Allende meteorite and correlation with O and Sr effects. *Geophys. Res. Lett.* 1: 225–228.

Lee, T., Papanastassiou, D. A., and Wasserburg, G. J. 1976. Demonstration of ^{26}Mg excess in Allende and evidence for ^{26}Al. *Geophys. Res. Lett.* 3: 109–112.

Lee, T., Papanastassiou, D. A., and Wasserburg, G. J. 1977. Aluminum-26 in the early solar system: Fossil or fuel? *Astrophys. J. Lett.* 211:L107–L110.

Lee, T., Shu, F. H., Shang, H., Glassgold, A. E., and Rehm. K. E. 1998. Protostellar cosmic rays and extinct radioactivities in meteorites. *Astrophys. J.* 506: 898–912.

Low, C., and Lynden-Bell, D. 1976. The minimum Jeans mass or when fragmentation must stop. *Mon. Not. Roy. Astron. Soc.* 176:367–390.

Lugmair, G. W., and Shukolyukov, A. 1998. Early solar system timescales according to ^{53}Mn-^{53}Cr systematics. *Geochim. Cosmochim. Acta.* 62:2863–2886.

Lugmair, G. W., Shukolyukov, A., and MacIsaac, Ch. 1995. The abundance of ^{60}Fe in the early solar system. In *Nuclei in the Cosmos III*, ed. M. Busso, R. Gallino, and C. M. Raiteri (New York: AIP), pp. 591–594.

Mac Low, M.-M., McKee, C. F., Klein, R. I., Stone, J. M., and Norman, M. L. 1994. Shock interactions with magnetized interstellar clouds. I. Steady shocks hitting nonradiative clouds. *Astrophys. J.* 433:757–777.

MacPherson, G. J., Davis, A. M., and Zinner, E. K. 1995. The distribution of aluminum-26 in the early solar system: A reappraisal. *Meteoritics* 30: 365–386.

Marti, K., and Lingenfelter R. E. 1995. The Orion phenomenon: Particle fluences in the solar nebula. In *Nuclei in the Cosmos III*, ed. M. Busso, R. Gallino and C. M. Raiteri (New York: AIP), pp. 549–552.

Mouschovias, T. Ch. 1989. Magnetic fields in molecular clouds: Regulators of star formation. In *The Physics and Chemistry of Interstellar Molecular Clouds*, ed. G. Winnewasser and J. T. Armstrong (Berlin: Springer-Verlag), pp. 297–312.

Mouschovias, T. Ch. 1991. Magnetic braking, ambipolar diffusion, cloud cores, and star formation: Natural length scales and protostellar masses. *Astrophys. J.* 373:169–186.

Murty, S. V. S., Goswami, J. N., and Shukolyukov, Yu. A. 1997. Excess ^{36}Ar in the Efremovka meteorite: A strong hint for the presence of ^{36}Cl in the early solar system. *Astrophys. J. Lett.* 475:L65–L68.

Nittman, J., Falle, S. A. E. G., and Gaskell, P. H. 1982. The dynamical destruction of shocked gas clouds. *Mon. Not. Roy. Astron. Soc.* 201:833–847.

Podosek, F. A., and Nichols, R. H., Jr. 1997. Short-lived radionuclides in the solar nebula. In *Astrophysical Implications of the Laboratory Study of Presolar Materials,* ed. T. J. Bernatowicz and E. Zinner (Woodbury: AIP), pp. 617–647.

Podosek, F. A., and Swindle, T. D. 1988. Extinct radionuclides. In *Meteorites and the Early Solar System,* ed. J. F. Kerridge and M. S. Matthews (Tucson: University of Arizona Press), pp. 1093–1113.

Ramaty, R., Kozlovsky, B., and Lingenfelter, R. E. 1979. Nuclear gamma-rays from energetic particle interactions. *Astrophys. J. Suppl.* 40:487–526.

Ramaty, R., Kozlovsky, B., and Lingenfelter, R. E. 1996. Light isotopes, extinct radioisotopes and gamma ray lines from low-energy cosmic-ray interactions. *Astrophys. J.* 456:525–540.

Reach, W. T., and Rho, J. 1999. Excitation and disruption of a giant molecular cloud by the supernova remnant 3C 391. *Astrophys. J.* 511:836–846.

Reynolds, J. H. 1960. Determination of the age of the elements. *Phys. Rev. Lett.* 4:8–10.

Reynolds, S. P., and Moffett, D. A. 1993. High-resolution radio observations of the supernova remnant 3C 391: Possible breakout morphology. *Astron. J.* 105:2226–2230.

Rho, J., and Petre, R. 1998. Mixed-morphology supernova remnants. *Astrophys. J. Lett.* 503:L167–L170.

Rowe, M. W., and Kuroda, P. K. 1965. Fissiogenic xenon from the Pasamonte meteorite. *J. Geophys. Res.* 70:709–714.

Różyczka, M., and Tenorio-Tagle, G. 1987. The hydrodynamics of clouds overtaken by supernova remnants. II. Attrition shocks, condensation and ejection of clouds. *Astron. Astrophys.* 176:329–337.

Sahijpal, S. 1997. Isotopic studies of early solar system objects in meteorites by an ion microprobe. Ph.D. Thesis, Gujarat University, India.

Sahijpal, S., Goswami, J. N., Davis, A. M., Grossman, L., and Lewis, R. S. 1998. A stellar origin for the short-lived nuclides in the early solar system. *Nature* 391:559–561.

Schramm, D. N., Tera, F., and Wasserburg, G. J. 1970. The isotopic abundance of ^{26}Mg and limits on ^{26}Al in the early solar system. *Earth Planet. Sci. Lett.* 10:44–59.

Shu, F. H. 1995. The birth of sunlike stars. In *Molecular Clouds and Star Formation*, ed. Chi Yuan and Junhan You (Singapore: World Scientific), pp. 97–148.

Shu, F. H., Shang, H., and Lee, T. 1996. Towards an astrophysical theory of chondrites. *Science* 271:1545–1552.

Shu, F. H., Shang, H., Glassgold, A. E., and Lee, T. 1997. X-rays and fluctuating x-winds from protostars. *Science* 277:1475–1479.

Shukolyukov, A., and Lugmair, G. W. 1993*a*. Live iron-60 in the early solar system. *Science* 259:1138–1142.

Shukolyukov, A., and Lugmair, G. W. 1993*b*. ^{60}Fe in eucrites. *Earth Planet. Sci. Lett.* 119:159–166.

Srinivasan, G., Ulyanov, A. A., and Goswami, J. N. 1994. ^{41}Ca in the early solar system. *Astrophys. J. Lett.* 431:L67–L70.

Srinivasan, G., Sahijpal, S., Ulyanov, A. A., and Goswami, J. N. 1996. Ion microprobe studies of Efremovka CAIs: II. Potassium isotope composition and ^{41}Ca in the early solar system. *Geochim. Cosmochim. Acta.* 60:1823–1835.

Stone, J. M., and Norman, M. L. 1992. The three-dimensional interaction of a supernova remnant with an interstellar cloud. *Astrophys. J. Lett.* 390: L17–L19.

Swindle, T. D. 1993. Extinct radionuclides and evolutionary time scales. In *Protostars and Planets III*, ed. E. H. Levy and J. I. Lunine (Tucson: University of Arizona Press), pp. 867–881.

Swindle, T. D., and Podosek, F. A. 1988. Iodine-xenon dating. In *Meteorites and the Early Solar System,* ed. J. F. Kerridge and M. S. Matthews (Tucson: University of Arizona Press), pp. 1127–1146.

Symbalisty, E. M. D., and Schramm, D. N. 1981. Nucleocosmochronology. *Rep. Prog. Phys.* 44:293–328.

Tatematsu, K., Fukui, Y., Iwata, T., Seward, F. D., and Nakano, M. 1990*a*. A further study of the molecular cloud associated with the supernova remnant G109.1-1.0. *Astrophys. J.* 351:157–164.

Tatematsu, K., Fukui, Y., Landecker, T. L., and Roger, R. S. 1990*b*. The interaction of the supernova remnant HB 21 with the interstellar medium: CO, H I, and radio continuum observations. *Astron. Astrophys.* 237:189–200.

Timmes, F. X., Woosley, S. E., and Weaver, T. A. 1995. Galactic chemical evolution: Hydrogen through zinc. *Astrophys. J. Suppl.* 98:617–658.

Tohline, J. E. 1981. The collapse to equilibrium of rotating, adiabatic spheroids. I. Protostars. *Astrophys. J.* 248:717–726.

Vanhala, H. A. T., and Cameron, A. G. W. 1998. Numerical simulations of triggered star formation. I. Collapse of dense molecular cloud cores. *Astrophys. J.* 508:291–307.

Wasserburg, G. J. 1985. Short-lived nuclei in the early solar system. In *Protostars and Planets II*, ed. D. C. Black and M. S. Matthews (Tucson: University of Arizona Press), pp. 703–737.

Wasserburg, G. J., and Arnould, M. 1987. A possible relationship between extinct ^{26}Al and ^{53}Mn in meteorites and early solar system. In *Lecture Notes in Physics 287, 4th Workshop on Nuclear Astrophysics*, ed. W. Hillebrandt,

R. Kuhfuß, E. Müller, and J. W. Truran (Heidelberg: Springer-Verlag), pp. 262–276.
Wasserburg, G. J., Busso, M., Gallino, R., and Raiteri, C. M. 1994. Asymptotic giant branch stars as a source of short-lived radioactive nuclei in the solar nebula. *Astrophys. J.* 424:412–428.
Wasserburg, G. J., Gallino, R., Busso, M., Goswami, J. N., and Raiteri, C. M. 1995. Injection of freshly synthesized ^{41}Ca in the early solar nebula by an asymptotic giant branch star. *Astrophys. J. Lett.* 440:L101–L104.
Wasserburg, G. J., Busso, M., and Gallino, R. 1996. Abundances of actinides and short-lived nonactinides in the interstellar medium: Diverse supernova sources for the *r*-processes. *Astrophys. J. Lett.* 466:L109–L113.
Wasserburg, G. J., Gallino, R., and Busso, M. 1998. A test of the supernova trigger hypothesis with ^{60}Fe and ^{26}Al. *Astrophys. J. Lett.* 500:L189–L193.
Whitworth, A. 1981. Global gravitational stability for one-dimensional polytrope. *Mon. Not. Roy. Astron. Soc.* 195:967–977.
Wilner, D. J., Reynolds, S. P., and Moffett, D. A. 1998. CO observations toward the supernova remnant 3C 391. *Astron. J.* 115:247–251.
Wootten, A. 1981. A dense molecular cloud impacted by the W28 supernova remnant. *Astrophys. J.* 245:105–114.
Wootten, H. A. 1977. The molecular cloud associated with the supernova remnant W44. *Astrophys. J.* 216:440–445.
Xu, J., and Stone, J. M. 1995. The hydrodynamics of shock-cloud interactions in three dimensions. *Astrophys. J.* 454:172–181.
Yin, Q., Jagoutz, E., and Wanke, H. 1992. Re-search for extinct ^{99}Tc and ^{98}Tc in the early solar system. *Meteoritics* 27:310.

TIMESCALES OF ACCRETION AND DIFFERENTIATION IN THE EARLY SOLAR SYSTEM: THE METEORITIC EVIDENCE

MEENAKSHI WADHWA
Field Museum, Chicago

and

SARA S. RUSSELL
Natural History Museum, London

We review the ages of a variety of meteoritic materials that formed at different times during the early history of the solar system. The oldest known solids that formed during the nebular disk phase and are preserved in meteorites are the calcium-aluminum-rich inclusions (CAIs). Chondrules, also thought by many researchers to have formed in the disk, yield ages up to a few million years younger than CAIs, suggesting that the disk may have lasted for at least this long. However, these data are open to alternative interpretations, including disturbance of isotopic systematics in chondrules and formation of chondrules in an environment that postdated the nebula. Information about the timescales involved in planetesimal aggregation and differentiation can be obtained from meteorites from the HED (Howardite-Eucrite-Diogenite) parent body, which underwent extensive melting. Most of these samples formed within 10 Myr of CAIs, although some ages are much younger (up to 100 Myr after CAIs). Iron meteorites, which are also products of planetesimal differentiation in the early solar system, mostly formed within a 12-Myr time interval subsequent to CAI formation. The total accretion interval for the Earth may have extended over ~100 Myr. Assuming a simple two-stage model, core formation for the Earth must have occurred late (>50 Myr after differentiation of iron meteorite parent bodies); however, if accretion and core formation were continuous, the latter process may have had a mean time of only ~10–20 Myr. The accretion and differentiation of Mars apparently took place within the first 30 Myr of solar system history. The timescales over which inner solar system objects accreted and differentiated are, in many cases, poorly defined, because isotopic signatures are often disturbed and initial isotope distributions are not well constrained. Data from different chronometers and a wider variety of samples are required to obtain a more complete picture.

I. INTRODUCTION

Meteorites and their components provide the only record of events that occurred early in the history of the solar system. In particular, absolute and relative radiometric dating of meteoritic materials provides a quantitative indicator of nebular and planetary differentiation timescales. Different meteorite types formed at various stages in solar system history. Some chondritic meteorites have suffered little geological processing since the earliest epoch of the solar system and contain components formed during its nebular disk phase. Analysis of these nebular components yields information about the chemical and thermal environments in the protosolar disk and the timescales involved in its evolution. Achondritic, stony-iron, and iron meteorites, on the other hand, have experienced melting processes and allow investigation into planetary differentiation and evolution processes. These studies of meteoritic materials are complementary to theoretical determinations of nebular lifetimes and processes and astronomical observations of young, T Tauri stars and their associated disks.

In this chapter we will discuss and reevaluate early solar system timescales of (1) condensation and formation of chondritic components in the solar nebula and (2) subsequent accretion and differentiation of planetesimals and protoplanets. Our discussions are based mainly on recent studies of the systematics of various extinct radionuclides, which can provide a high-resolution chronology of events in the early solar system; some of these radionuclides, such as ^{53}Mn and ^{182}Hf, have found extensive application only within the last decade. Note that the chapter by Goswami and Vanhala (this volume) provides an updated review of all extinct radionuclides that have so far been detected, including discussion of possible production mechanisms, detection, and implications of each (which will not be repeated here).

It is noted that extinct radionuclides alone can provide only *relative* age constraints. Therefore, to obtain high-resolution absolute age constraints, it is necessary to anchor the relative ages derived from short-lived chronometers to the absolute timescale with an appropriate long-lived chronometer. The only chronometer that has the required precision to provide such a time anchor is the Pb-Pb chronometer (e.g., Lugmair and Galer 1992; Göpel et al. 1994). However, the ability to anchor the relative ages to an absolute timescale is restricted, because few meteorites have been precisely dated with the Pb-Pb chronometer, and only a small subset of these objects were analyzed to determine initial extinct radionuclide abundances. Additionally, the validity of highly precise ages (often quoted with errors of only ± 1–2 Myr) obtained by the Pb-Pb chronometer has recently been questioned (Tera and Carlson 1999). Nevertheless, in many instances, the *relative* intervals between discrete events in early solar system history are of interest. For detailed accounts of how radioactive isotopes are used in dating meteorites, see Wasserburg (1985), Tilton (1988), and Swindle and Podosek (1988).

The "age" of a meteoritic component, whether recorded by a long- or short-lived chronometer, represents the time of last chemical equilibration, when the temperature of the object passed below that at which the elements involved easily diffuse (i.e., the "closure temperature"). Also, certain assumptions apply to all radiometric dating techniques. The first is that, at a given time, the ratio of the initial abundance of the radioactive isotope to that of the reference isotope was the same in all objects being measured (i.e., that the source region was isotopically homogeneous). Another is that the elements of interest were last redistributed completely during the event that is being dated and that there has been no subsequent disturbance of the isotopic system under consideration. Finally, all of the radiogenic material must have accumulated in the object after it formed (i.e., there is no "chemical memory" effect). The validity of each of these assumptions has to be evaluated on a case-by-case basis (i.e., for the particular chronometer being utilized and the sample to which it is applied).

II. FORMATION TIMESCALES OF NEBULAR PRODUCTS

Timescales for nebular processes and the age of the protosolar disk can be estimated by dating various components formed in the solar nebula, particularly the calcium-aluminum-rich inclusions (CAIs) found in different types of chondrites. These inclusions, which are composed of refractory minerals and can be up to several cm across, are thought to have formed by various processes that are still debated. Some CAIs, called Type Bs, have textures indicative of igneous processes, suggesting that they crystallized from a melt. Others may have formed by evaporation or condensation (see MacPherson et al. 1988 for a review of formation models of CAIs). All CAIs are believed to be relicts from the solar nebula.

Chondrules are silicate spherules, ranging in size from ~50 to ~5000 μm across. Their igneous textures suggest that their formation mechanism involved rapid heating followed by fast cooling (100–2000°C/hr; Hewins, 1988). Most researchers have interpreted them as nebular products, although the heating process involved in their formation is not known. Formation models for chondrules are numerous. Suggested environments for their production include shock events (Weidenschilling et al. 1998; Connolly and Love 1998), stellar jets (Shu et al. 1996), magnetic flares, and nebular lightning (for a review, see the chapter by Jones et al., this volume, and chapters by various authors in Hewins et al. 1996). Several researchers have also advocated that chondrules formed in early planetary bodies (e.g., Sanders 1996; Hutchison 1996).

In chondrites, between the chondrules and CAIs, there is a fine-grained material called matrix, which includes silicates, oxides, organic material, metal, and presolar dust grains.

A. Absolute Age of CAI Formation

The most precise absolute Pb-Pb age of CAIs can be used as a "time anchor" for the relative ages obtained by short-lived isotope dating techniques. Pb-Pb systematics for bulk CAIs have yielded ages of 4.559 ± 0.004 Gyr (Chen and Wasserburg 1981), 4.565 ± 0.004 Gyr (Chen and Tilton 1976), and 4.553 ± 0.004 Gyr (Tatsumoto et al. 1976). More recently, a precise Pb-Pb age of 4.566 ± 0.002 Gyr has been estimated for

Allende CAIs (Manhès et al. 1988; Göpel et al. 1991, 1994; Allègre et al., 1995). Ion microprobe Pb isotopic measurements of individual perovskite grains in Murchison and Allende CAIs give ages of 4.565 $\pm$ 0.034 Gyr (Allende grain) and 4.569 $\pm$ 0.026 Gyr (Murchison grains) (Ireland et al. 1990). The similarity of these ages to those obtained for bulk inclusions indicates that there is no evidence for any CAI material predating the solar system. It is further noted that all these ages (from bulk CAIs and from *in situ* analyses of individual grains) are largely consistent with the most precise Pb-Pb age estimated so far for these refractory inclusions (4.566 $\pm$ 0.002 Gyr; see above). Finally, these ages are the oldest recorded for any solar system material, indicating that CAIs are the earliest solids sampled from the solar nebula.

B. Relative Dating of CAIs

Recently, compelling evidence has emerged for the presence of live ^{41}Ca (which decays to ^{41}K with a half-life of merely 0.103 Myr) at the time of formation of CAIs from the Efremovka carbonaceous chondrite (Srinivasan et al. 1994, 1996*a*). Furthermore, correlation between excesses in ^{41}K and ^{26}Mg in these objects indicates that the short-lived radionuclides ^{41}Ca and ^{26}Al were synthesized in a single stellar source and were subsequently injected into the solar nebula (Sahijpal et al. 1998). This suggests that the time interval between the injection of freshly synthesized material into the nebular disk and the formation of CAIs was well within a million years (see chapter by Goswami and Vanhala, this volume).

The results of the extensive Al-Mg studies of chondritic components, undertaken over the last two decades, were recently reviewed by MacPherson et al. (1995). It is evident that such studies have focused on objects with specific compositions. High Al/Mg ratios are essential to obtain data of sufficient quality, so the Type B CAIs from carbonaceous CV meteorites are the favored candidates for Mg isotopic measurements, because these objects contain abundant anorthite, which has Al/Mg ratios of up to several hundred. The initial ^{26}Al/^{27}Al ratio in refractory inclusions has an upper limit of 5×10^{-5} (e.g., Gray and Compston 1974; Lee et al. 1976; Hutcheon 1982; Podosek et al. 1991; MacPherson and Davis 1993; Caillet et al. 1993; Russell et al. 1996) (Fig. 1); most CAIs yield values between 4×10^{-5} and 5×10^{-5}, indicating that they formed within a time interval of less than ~0.3 Myr in an environment in which ^{26}Al was homogeneously distributed. However, there are other CAIs that yield even lower values, or do not show evidence for having contained any radiogenic ^{26}Mg. MacPherson et al. (1995) suggested that such CAIs typically fall into one of two categories: Either they were altered (during an event that may have reset their Mg isotope composition), or else they contain isotope anomalies in other elements (e.g., Ti and Ca). The latter type, called FUN (Fractionation and Unidentified Nuclear isotopic anomalies) inclusions, is rare in CV meteorites but more common in CM chondrites. The wide range of stable isotope anomalies in FUN CAIs points to their formation in the

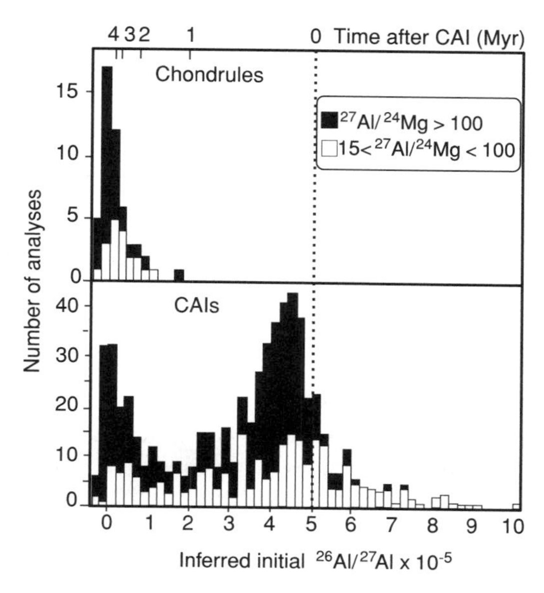

Figure 1. Histogram of inferred initial $^{26}Al/^{27}Al$ ratios for individual spots in chondrules and CAIs. For CAIs, most objects had initial $^{26}Al/^{27}Al$ values of 3.5 to 5×10^{-5}, with a subset of samples containing no measurable initial ^{26}Al. Chondrules yield lower initial $^{26}Al/^{27}Al$ ratios, suggesting that they mostly formed more than ~3 Myr after the oldest CAIs. Note that the distribution of inferred initial $^{26}Al/^{27}Al$ ratios in objects with high $^{27}Al/^{24}Mg$ ratios (i.e., >100) is similar to that in objects with relatively low Al/Mg ratios ($15<{}^{27}Al/^{24}Mg<100$). The complete data also include 27 points from CAIs and 7 points from chondrules that have been excluded from this figure because they have negative $\delta^{26}Mg$ values which cannot be interpreted in terms of live ^{26}Al; however, a few apparently "negative" values for the $^{26}Al/^{27}Al$ ratios are included, because each of these measurements are within error of 0. Based on Fig. 2 from Swindle et al. (1996); updated using data from Russell et al. (1996, 1997) and Huss et al. (1998).

very earliest stages of the solar system, before extensive isotope homogenization occurred. Thus, MacPherson et al. (1995) suggested that the lack of radiogenic ^{26}Mg in FUN inclusions did not indicate a late formation age but, rather, formation in an isotopically anomalous environment that was devoid of ^{26}Al. More recently, Sahijpal and Goswami (1998) have also suggested that these "anomalous" inclusions were the first solids to have formed in the solar nebula, prior to the injection of freshly synthesized ^{26}Al (and ^{41}Ca). Therefore, it may be inferred that ^{26}Al was extensively homogenized subsequent to its injection into the nebular disk and that its use as a chronometer is valid for most samples. Support for a model of largely homogeneous distribution of ^{26}Al in the early solar system is additionally provided by the work of Podosek et al. (1991), who found that CAIs with the "canonical" initial $^{26}Al/^{27}Al$ ratio (i.e., $\sim 5 \times 10^{-5}$) also typically contain the lowest (primordial) initial $^{87}Sr/^{86}Sr$ ratios.

Rigorous evaluation of the homogeneity of ^{26}Al in the early solar system requires measurement of objects formed at various heliocentric distances. While there is an extensive amount of data from the carbonaceous chondrite group, little has been acquired from other meteorite groups. Al-Mg data have been reported for only a few CAIs from ordinary chondrites, and none are reported from enstatite chondrites; CAIs from ordinary chondrites show similar Al-Mg characteristics to those from carbonaceous chondrites (Russell et al. 1996, 1997), suggesting either that the chondrite-forming region contained widespread ^{26}Al or that CAIs were formed in a restricted area and then were dispersed throughout the "meteorite-forming" region.

C. CAI Alteration Timescales

Many CAIs have not remained pristine since initial formation but have experienced multiple heating episodes that altered their primary mineralogy. Whether these episodes of alteration occurred in the nebula (in which case their timing would have implications for constraining nebular timescales) or in the parent body is not always well established. In all probability, both nebular and parent body conditions have affected CAIs, and distinguishing between the effects of these two is key to the interpretation of radiometric data.

In general, the "canonical" initial ^{26}Al/^{27}Al ratio of $\sim 5\times10^{-5}$ in most CAIs suggests formation over a short period of time, but the secondary minerals in CAIs have a range of ages. Secondary phases in CAIs contain lower ^{26}Mg excesses, indicating formation times from ~2 to >5 Myr after the formation of CAIs (Fig. 2). Traditionally, the later formation of several of the secondary phases in CAIs has been interpreted as requiring

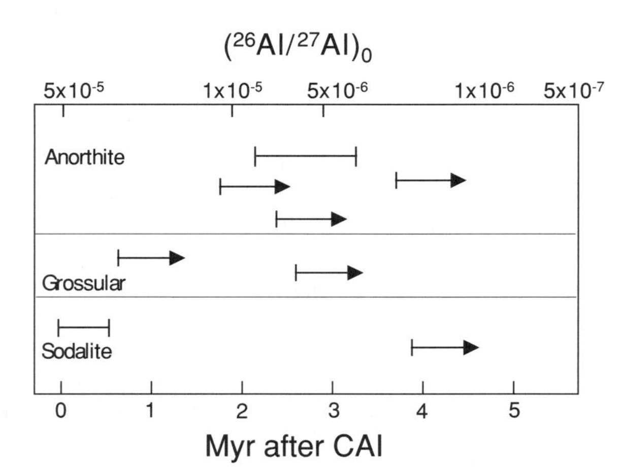

Figure 2. Relative ages of secondary minerals from CAIs. Arrows indicate upper limits. The range is from less than 1 Myr after initial CAI formation to more than ~4 Myr after CAIs. Data from Hutcheon and Newton (1981); Caillet et al. (1993); Davis et al. (1994); and Russell et al. (1997).

a long nebular history for these objects, because it was thought unlikely that parent body processes could raise temperatures high enough to produce the secondary minerals and textures observed in CAIs. For example, MacPherson and Davis (1993) showed that a CAI from Vigarano experienced a melting event several million years after initial formation and, on a textural basis, interpreted this event to have occurred in the nebula. Although this remains a hotly debated issue (e.g., Krot et al. 1997; MacPherson and Davis 1997), recent work suggests that some secondary phases in CAIs may have formed in a parent body (Krot et al. 1995), and it is thus conceivable that the age of some secondary alteration in CAIs may reflect postaccretionary rather than nebular processes.

I-Xe ages of secondary phases in several CAIs show a wide range, up to ~50 Myr after CAI formation (Swindle et al. 1988). These "ages" are interpreted to reflect the mobility of I in the parent body, because such long nebular timescales are at odds with theoretical determinations. More recently, however, Hohenberg et al. (1998) have shown that I-Xe systematics of sodalite from Allende CAIs indicate formation of this secondary phase 4 ± 2 Myr after CAI formation; this time interval for the formation of secondary minerals is consistent with that obtained from Al-Mg systematics (Fig. 2).

D. Chondrule Ages

Chondrule ages have been estimated using the short-lived isotopes ^{53}Mn, ^{26}Al, and ^{129}I. Chondrules from Chainpur and Bishunpur contain ^{53}Cr excesses corresponding to an initial ^{53}Mn/^{55}Mn ratio of $\sim 9.4 \times 10^{-6}$ (Nyquist et al. 1999). Birck and Allègre (1985) had reported an initial ^{53}Mn/^{55}Mn ratio of $(6.7 \pm 2.2) \times 10^{-5}$ for bulk CAIs from Allende; mineral separates from one Type B CAI (BR1), however, gave an initial ^{53}Mn/^{55}Mn ratio of $(3.7 \pm 1.2) \times 10^{-5}$. Therefore, these authors suggested that the average of these two values (i.e., $\sim 4.4 \times 10^{-5}$) be considered as the initial value of the ^{53}Mn/^{55}Mn ratio at the time of CAI production. The difference in the ^{53}Mn/^{55}Mn ratios between Chainpur and Bishunpur chondrules and Allende CAIs could indicate a chondrule formation time of ~8 Myr after CAIs, if (i) ^{53}Mn was homogeneously distributed in the region of the solar nebula where CAIs and chondrules were formed, (ii) the anomalous Cr isotopic composition in Allende CAIs does not contain contributions from a nucleosynthetic or a "fossil" component, and (iii) Mn-Cr systematics in the analyzed chondrules and CAIs are undisturbed. However, it has been argued that the ^{53}Cr excesses in Allende CAIs may represent a complex superposition of nucleosynthetic anomalies and ^{53}Mn decay (Lugmair and Shukolyukov 1998; Papanastassiou 1986). In fact, Birck et al. (1998) have recently reevaluated their previously reported data and have suggested that the Mn-Cr systematics in Allende CAIs may be the result either of an earlier stage of evolution in a low-Mn/Cr environment or of mixing of freshly nucleosynthesized ^{53}Mn with the "average" solar nebula.

It should be noted that if the ^{53}Mn/^{55}Mn ratio estimated by Lugmair and Shukolyukov (1998) for the time of CAI formation (i.e., $\sim 9 \times 10^{-6}$) is used for comparison with the chondrule data, it indicates essentially contemporaneous formation of Chainpur and Bishunpur chondrules and Allende CAIs (even though a time difference of 1–2 Myr between formation of these objects is feasible within the uncertainties). However, as noted by Nyquist et al. (1997), if the Mn-Cr systematics in chondrules were established during condensation of their solid precursors in the solar nebula rather than during the later melting event that produced the chondrule textures, the Mn-Cr data for chondrules would provide only upper limits on their formation ages.

Typical ferromagnesian chondrules have not been dated using the Al-Mg chronometer (with the exception of one Type II chondrule from Semarkona; Kita et al. 1998), because they usually do not contain phases with Al/Mg ratios high enough that excesses in ^{26}Mg can be detected by techniques currently available. However, numerous Al-Mg ion microprobe analyses have been made on a rare subset of chondrules that contain abundant primary, Al-rich minerals and therefore can be dated using this system. Radiogenic ^{26}Mg in these chondrules either is present at a level much lower than in CAIs or is undetectable (Hutcheon and Jones 1995; Srinivasan et al. 1996*b*; Russell et al. 1996). With the exception of one Semarkona chondrule that yielded an initial ^{26}Al/^{27}Al ratio of 2×10^{-5} (Russell et al. 1997), the initial ^{26}Al/^{27}Al ratio in Al-rich chondrules does not exceed $\sim 1 \times 10^{-5}$ (Figs. 1 and 3). Additionally, recent analyses of

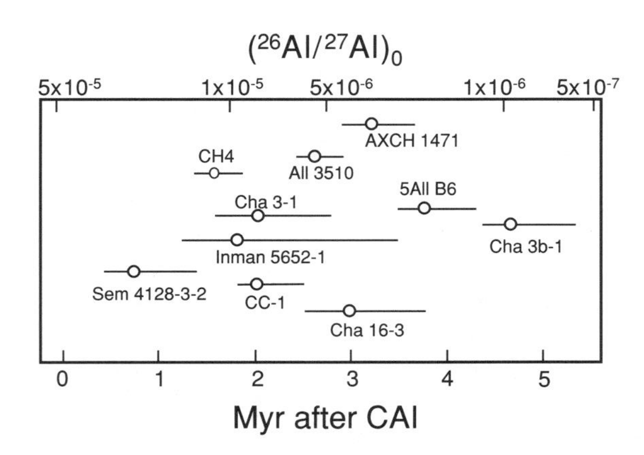

Figure 3. Relative ages of chondrules using the ^{26}Al chronometer. Objects for which only upper limits on the ^{26}Al/^{27}Al ratios could be constrained are not shown here. The data suggest that chondrule production proceeded for several millions of years after initial CAI formation. Data from Hutcheon and Hutchison (1989) (CC1); Sheng et al. (1991) (All 3510; 5 All B6); Russell et al. [1996 (Cha 3-1; Inman 5652-1), 1997 (Sem 4128-3-2; Cha 16-3; Cha 3b-1)]; Kita et al. (1998) (CH4); Srinivasan et al. (1996*c*) (AXCH 1471).

mesostasis areas (having relatively high Al/Mg ratios of ~200) in a Type II Semarkona chondrule provide evidence for a $^{26}Al/^{27}Al$ ratio of $\sim 1.2 \times 10^{-5}$ (Kita et al. 1998), indicating formation ~1.5 Myr after CAIs. These data suggest that most chondrules were likely formed >3 Myr after CAIs (Fig. 3). The lowest upper limit recorded, measured on the Chainpur chondrule 1251-14-2, indicates a formation time of >5 Myr after CAIs, suggesting that the protosolar disk (or nebula) stage of the solar system may have lasted longer than 5 Myr (Russell et al. 1996). Alternatively, the chondrule formation event(s) may have postdated the nebula by a few million years, Al-Mg systematics in these chondrules may have undergone reequilibration subsequent to their formation, or ^{26}Al distribution in the CAI- and chondrule-forming regions was heterogeneous.

Swindle et al. (1991) measured the I-Xe characteristics of 18 chondrules from the ordinary chondrite Chainpur (L3.4). They found that the initial $^{129}I/^{127}I$ ratio varied by a factor of 10 between chondrules, suggesting a range in age of ~50 Myr. Swindle et al. (1991) interpreted this time span as too long to represent timing of nebular processes, so they concluded that parent body effects had compromised the data. In contrast, Caffee et al. (1982) showed that large chondrules from Bjurbole exhibit a spread in apparent age of only 1.6 Myr; smaller chondrules from the same meteorite yield values in approximate agreement with this (Gilmour et al. 1995). I-Xe dates for chondrules from CV meteorites also tend to give younger ages than for CAIs (Swindle et al. 1998), although the range in ages is large.

E. Implications for the Lifetime of the Protosolar Disk

Although no isotope chronometer provides an ideal dating scheme, the data from each system investigated point to relative ages of CAIs and chondrules that seem to be, for the most part, internally consistent. CAIs formed ~4.566 Gyr ago over a short time interval (of the order of a million years or less) and appear to be older than chondrules by at least a few million years. Additionally, chondrule "ages" span several million years, possibly indicating formation over an extended period of time. Therefore, data from short-lived radionuclides indicate that nebular products (i.e., CAIs and possibly chondrules, as well as some alteration products in CAIs) may have formed over several million years. The implication is that the nebular disk may have existed over this period and that high-temperature events occurred within or around the disk during this period. If it is assumed that CAIs and chondrules are nebular in origin, then their "age" recorded by radiometric techniques probably represents the time at which primordial gaseous material in the protosolar disk condensed and accreted into larger particles. However, a problem with this interpretation could arise if the assumption of a nebular origin for chondrules is not strictly correct. As mentioned earlier, other potential sources of error affecting these conclusions are parent body alteration, which may have reset the chronometer after accretion, and possible heterogeneity of the short-lived isotopes in the solar system. For example, the apparent CAI-chondrule age gap may be caused by the formation of CAIs in an "anomalous" nebular region enriched in newly nucleosynthesized material containing short-lived isotopes.

Can the formation of solids in the nebula over several million years be accommodated into theoretical constraints of solar system formation? Cameron (1995) has attempted to incorporate the CAI-chondrule time gap into a model of stellar evolution. He suggested that CAIs were possibly formed in outflow jets or gas streams while the Sun was in its accumulation phase. Chondrules may have formed during the final T Tauri phase of the Sun, several million years later. However, it is recognized that long nebular timescales for CAIs may be problematic, because the drift time (i.e., the time taken to drift into the Sun) of such small objects in the nebular disk is estimated to be short (Weidenschilling 1977); assuming no turbulence in the inner solar system, a 1-cm CAI would drift into the Sun within ~30,000 years. Therefore, a long time gap in the formation of nebular products would require that CAIs be stored, perhaps in the outer solar system, until final parent body accretion in the meteorite-forming region.

Observational studies indicate that the nebular timescales inferred from dating of chondritic components may be reasonable. The survival time of protostellar disks as optically thick structures can be estimated by counting the number of optically visible and invisible young stellar objects or by the location of the star on the HR diagram. Strom (1994) found that the infall stage of young stars in the mass range 0.1–1.5 $M_\odot$ lasts 0.1 to 0.2 Myr. The average lifetime of the subsequent protostellar disk phase is determined by identifying the presence or absence of disks as a function of stellar age. In the mass range 0.2–1.5 $M_\odot$, and for ages <3 Myr, ~50% of observed stars have disks; only 10% of stars in the same mass range and with ages of 3–10 Myr have disks (Strom 1994). These results have been confirmed by more recent observations that suggest that disks around young intermediate-mass stars (Herbig Ae type) have lifetimes of the order of $\sim 10^7$ years (see the chapter by Natta et al., this volume). Therefore, it seems reasonable that a solar-type star could have a disk for as long as ~10 Myr, a timescale compatible with the data provided by meteoritic nebular components.

III. TIMESCALES OF ACCRETION AND DIFFERENTIATION

As noted earlier, some meteorite types (i.e., achondrites, stony-iron, and iron meteorites) are clearly products of melting and differentiation of their parent bodies. Precise radiometric dating of these types of meteorites and comparison with the formation age of CAIs (which represent the earliest solids to form in the solar nebula; see above) can provide constraints on the timescales over which planetesimal and protoplanet accretion and subsequent differentiation took place. The particular event being dated, however, depends on (i) the type of objects being analyzed (e.g., ages for noncumulate basaltic eucrites provide constraints on the timescales over which their parent body was volcanically active, whereas ages for "magmatic" iron meteorites may indicate timescales for crystallization of the Fe-Ni cores in asteroidal bodies); (ii) the postformation history of these objects (e.g., a highly equilibrated sample will provide constraints on the timing of its thermal metamorphism rather than its formation time); and (iii) the radiometric technique being applied (e.g., Mn-Cr systematics are less likely to be disturbed by thermal metamorphism than I-Xe systematics). Therefore, ages obtained by different radiometric techniques for various differentiated classes of meteorites can provide constraints on the time required for parent body accretion as well as insight into the subsequent differentiation and evolution of planetesimals and protoplanets in the early solar system.

The isotopic systems that are currently being investigated to address questions specific to the precise timing of differentiation and core formation, and the duration of igneous activity on asteroidal bodies, are the ^{53}Mn-^{53}Cr, ^{60}Fe-^{60}Ni, ^{107}Pd-^{107}Ag, ^{146}Sm-^{142}Nd, and ^{182}Hf-^{182}W short-lived chronometers. The ^{53}Mn-^{53}Cr, ^{107}Pd-^{107}Ag, and ^{182}Hf-^{182}W systems, in particular, have good potential for addressing the above issues because (i) the radionuclides ^{53}Mn, ^{107}Pd, and ^{182}Hf appear to have been widespread in meteorite-forming regions in the inner solar system, and evidence for their presence has been found in a wide variety of meteorites (e.g., Lugmair and Shukolyukov 1998; Lee and Halliday 1995, 1996, 1997; Chen and Wasserburg 1996), and (ii) they have half-lives ($\sim$3.7 Myr for ^{53}Mn, $\sim$6.5 Myr for ^{107}Pd, and $\sim$9 Myr for ^{182}Hf) that are applicable to objects that formed within tens of millions of years of solar system formation, allowing a maximum resolution of $\leq$1 Myr. However, before these chronometers can be confidently applied, it has to be demonstrated that the initial distribution of these extinct radionuclides was homogeneous, at least for the objects being analyzed. The recent work of Lugmair and Shukolyukov (1998) has shown that although there may have been a dependence in the distribution of ^{53}Mn with heliocentric distance, any difference in the original ^{53}Mn distribution among objects thought to have originated in the region of the asteroid belt is minimal. Therefore, Mn-Cr chronometry appears to be feasible for these objects, which include most differentiated meteorite types that will be considered here (with the exception of the rare ones thought to originate on the Moon and Mars). Homogeneity in the distributions of ^{182}Hf and ^{107}Pd has so far not been explicitly demonstrated, although it is an assumption that is made when applying the chronometers based on these radionuclides.

A. Timing of Accretion, Differentiation, and Duration of Igneous Activity on Planetesimals

Relative age differences between the CAIs and achondritic meteorites can provide constraints on when the earliest melting of planetesimals took place and how long igneous activity persisted on the parent bodies of these achondrites. Such constraints also have implications for potential heat sources for melting and differentiation. Moreover, age constraints on the stony-iron pallasites (which may represent the core-mantle boundary of their asteroidal parent bodies) and magmatic iron meteorites (i.e., those that formed by fractional crystallization of Fe-Ni melts, most likely in the cores of asteroids) can provide information on when core formation took place on their respective parent asteroids. Constraints on the timing of differentiation and core formation in turn provide limits on the timescales required for accretion of planetesimals in the early solar system.

The highly differentiated HED (Howardite-Eucrite-Diogenite) meteorites are mafic to ultramafic samples that are believed to have originated as crustal rocks on a common parent body (which has been suggested to be the asteroid 4 Vesta; Binzel and Xu 1993), based on characteristics such as oxygen isotope systematics (Clayton and Mayeda 1996) and similar FeO/MnO ratios in their pyroxenes. Attempts to find evidence for live ^{26}Al in HEDs and other differentiated meteorites (such as angrites and silicate clasts in mesosiderites) have, by and large, been unsuccessful (e.g., Schramm et al. 1970; Lugmair and Galer 1992; Hsu and Crozaz 1996), although useful upper limits have been established in several cases. However, preliminary results of a recent Al-Mg study of the Piplia Kalan basaltic eucrite suggest an ^{26}Al/^{27}Al ratio in this meteorite at the level of $\sim 7 \times 10^{-7}$, indicating formation ~5 Myr after CAIs (Srinivasan et al. 1999). This recent study suggests that igneous differentiation occurred very early on the HED parent body and seems to confirm previous suggestions that ^{26}Al may indeed have been an important heat source for melting and differentiation on asteroids.

Early igneous differentiation on the HED parent body is also supported by Mn-Cr systematics. Lugmair and Shukolyukov (1998) have recently reported the results of Cr isotopic analyses on several noncumulate and cumulate eucrites (which are brecciated as well as unbrecciated basaltic rocks), as well as diogenites (orthopyroxene-rich brecciated rocks). Mn-Cr systematics in the bulk rocks of these selected samples are shown in Fig. 4. It is evident from these data that the bulk rock Mn-Cr systematics of these samples define a good correlation line, indicating that the source regions for these HED meteorites were formed contemporaneously. This *bulk HED isochron* has a slope corresponding to a ^{53}Mn/^{55}Mn ratio of $(4.7 \pm 0.5) \times 10^{-6}$, with an initial ^{53}Cr/^{53}Cr ratio of ~0.25 ϵ-units (1 ϵ-unit = 1 part in 10^4). When the ^{53}Mn/^{55}Mn ratio for the bulk HED isochron is compared to the ^{53}Mn/^{55}Mn ratio in angrites of $(1.25 \pm 0.07) \times 10^{-6}$ (Lugmair and Shukolyukov 1998), whose absolute Pb-Pb age is established as 4557.8 ± 0.5 Myr (Lugmair and Galer 1992), a Mn-Cr model age of 4564.8 ± 0.9 Myr is obtained. This dates the time of last Mn/Cr fractionation in the mantle of the HED parent body and most likely corresponds to the timing of core formation. Additionally, this age is rather close to the Pb-Pb age of Allende CAIs (4566 ± 2 Myr; Manhès

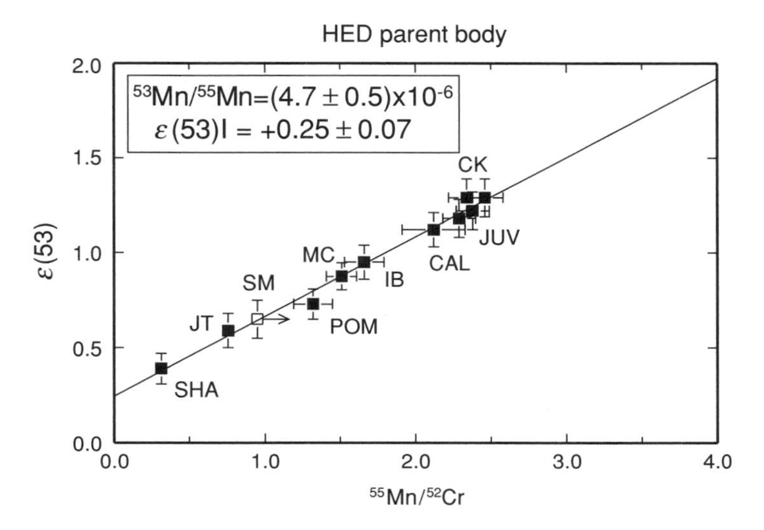

Figure 4. ^{53}Mn-^{53}Cr in the HED parent body. The ^{53}Cr/^{52}Cr ratios (expressed in ϵ units, defined as deviations in the ^{53}Cr/^{52}Cr ratios in the samples relative to this ratio in a terrestrial standard, in parts per 10^4) in the bulk meteorites are plotted versus their respective ^{55}Mn/^{52}Cr ratios. Data shown here are for the eucrites Chervony Kut (CK), Juvinas (JUV), Caldera (CAL), Ibitira (IB), Moore County (MC), Pomozdino (POM), and Serra de Magé (SM), and for the diogenites Johnstown (JT) and Shalka (SHA). Figure from Lugmair and Shukolyukov (1998): copyright 1998 Elsevier Science Ltd. (reprinted with permission).

et al. 1988; Göpel et al. 1991, 1994; Allègre et al. 1995). Thus, the significance of the age defined by the bulk HED isochron lies in the implication that the HED parent body was accreted and underwent complete differentiation within a maximum of ~4 Myr from the time of formation of the first known solids (i.e., CAIs) in the solar nebula.

Note that the bulk HED isochron obtained from Mn-Cr systematics does not provide any information on the ages (or rather the time of last equilibration of the Mn-Cr system) of the individual meteorites that define it. This information is given by the individual internal isochrons for each sample, which provide a range of Mn-Cr model ages for these samples (Lugmair and Shukolyukov 1998). For example, the ^{53}Mn/^{55}Mn ratio for the noncumulate eucrite Chervony Kut [i.e., $(3.7 \pm 0.4) \times 10^{-6}$; Lugmair and Shukolyukov 1998] corresponds to an age of 4563.6 ± 0.9 Myr, implying crystallization almost contemporaneous with the HED parent body mantle fractionation. In contrast, another noncumulate eucrite, Caldera, does not show any evidence for live ^{53}Mn; in fact, Wadhwa and Lugmair (1996) estimated, based upon a combination of Mn-Cr and ^{147,146}Sm-^{143,142}Nd systematics, that Caldera formed 17–41 Myr after CAIs. Taken at face value, this could indicate that the igneous activity resulting in the emplacement of basalts on the HED parent body lasted for tens of millions of years, which would have implications for the heat source for formation of such differentiated achondrites. It could suggest that the young age for Caldera is a cooling age and that even though essentially no ^{26}Al was present ~6 Myr after formation of the first solids, the HED parent body may have been large enough to retain heat for tens of millions of years thereafter. Alternatively, the parent melt for Caldera may have formed by another (external) heat source such as a large impact. Although the former possibility cannot be ruled out, the latter alternative may be more likely, because it does not require storage of the enormous amount of heat that would be required for prolonged igneous activity and basalt formation on the HED parent body. Additionally, based on Mn-Cr systematics in numerous eucrites, it appears likely that melting of basalt source regions and emplacement of volcanic flows on the HED parent body may have occurred very early, i.e., within a few million years of accretion and differentiation of the HED parent body (Lugmair and Shukolyukov 1998). If this is indeed the case, the apparently young ages of some noncumulate eucrites (e.g., Lugmair and Shukolyukov 1998; Tera et al. 1997; Galer and Lugmair 1996; Prinzhofer et al. 1992) and cumulate eucrites (e.g., Tera et al. 1997; Jacobsen and Wasserburg 1984) may reflect either secondary events, such as metamorphic reequilibration and impact melting or a protracted cooling history. Note, however, that, based on the cooling history of cumulate eucrites determined by Miyamoto and Takeda (1994*a*,*b*), Tera et al. (1997) have argued that the young (4.40–4.48 Gyr) Pb-Pb ages of three cumulate eucrites analyzed by them are, in fact, their true crystallization ages.

Short timescales of accretion and subsequent core formation on asteroidal bodies are additionally supported by Hf-W systematics (summarized in Fig. 5; Lee and Halliday 1995, 1996, 1997). The Hf-W system is particularly amenable toward the dating of metal segregation (silicate-metal fractionation) events in the early solar system, because Hf is strongly lithophile while W is moderately siderophile (even though both are highly refractory and are expected to be initially present in chondritic abundances). Tungsten isotopic data for various classes of iron meteorites, which indicate a deficit in ^{182}W relative to carbonaceous chondrites (Fig. 5), imply metal segregation before ^{182}Hf had decayed completely to ^{182}W and indicate that the parent bodies were accreted and core formation occurred within only a few million years. Moreover, Hf-W systematics in eucrites (which indicate large excesses in ^{182}W relative to chondrites; Fig. 5) point towards rapid accretion, differentiation, and core formation on the HED parent body; that is, within ~5–15 Myr of solar system history (Lee and Halliday 1996).

Finally, the ^{60}Fe-^{60}Ni and the ^{107}Pd-^{107}Ag systems also show evidence for early differentiation. ^{60}Fe, with a half-life of ~1.5 Myr, has been shown to have been extant at the time of differentiation and basalt formation on the HED parent body (Shukolyukov and Lugmair 1993*a*,*b*). Addi-

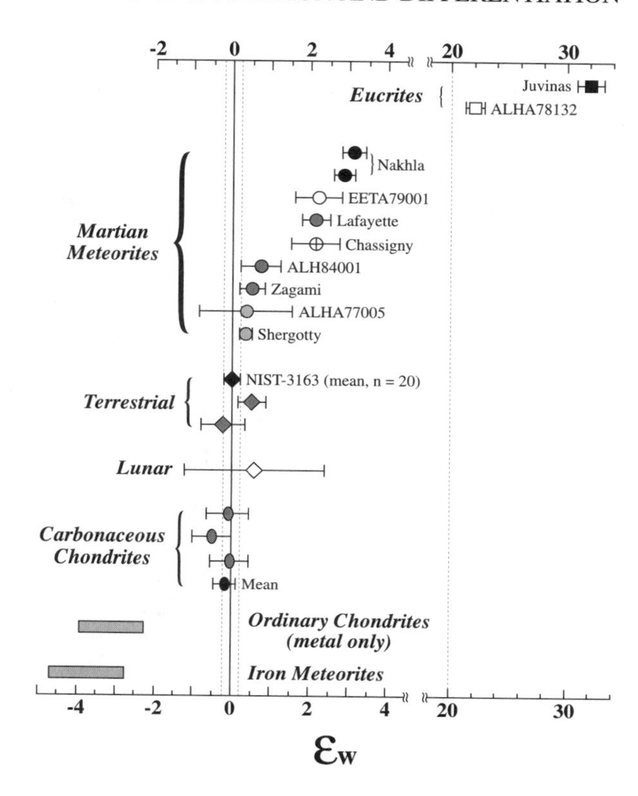

Figure 5. ϵ_W values (defined as deviations of measured $^{182}W/^{184}W$ ratios in the samples relative to this ratio in a NIST W standard, in parts per 10^4) of eucrites, martian meteorites, carbonaceous chondrites, metal in ordinary chondrites, iron meteorites, and lunar and terrestrial samples. Figure from Lee and Halliday (1997): copyright 1997 *Nature* (reprinted with permission).

tionally, a wide variety of iron and stony-iron meteorites show widespread excesses of ^{107}Ag, which are attributable to the decay of live ^{107}Pd. If the variations in the inferred initial $^{107}Pd/^{108}Pd$ ratios among various meteorite classes are due to time differences (ΔT) in condensation and metal segregation alone, it has been suggested that in most cases they imply a maximum ΔT of only $\sim$12 Myr (Chen and Wasserburg 1996).

B. Timescales of Accretion and Core Formation for Planetary Bodies

From the above discussion it seems evident that accretion, differentiation, and core formation occurred rather rapidly for asteroidal bodies (most likely within the first few million years of solar system history). However, what are the timescales for accretion and core formation of planetary bodies such as the Earth and Mars? Meteoritic evidence may be used to constrain the timing of these events as well.

It is generally thought that the accretion of the Earth to a mass approaching its present value took ~50–100 Myr, although a large fraction of this mass may well have accreted within ~10 Myr (Wetherill 1986). The issue of timing of core formation on the Earth is more controversial, part of the problem being that it is difficult to define the nature of this core formation event; for example, was it a single global melting event that took place in the accreting planet, or were the accreting planetesimals already differentiated so that core formation involved continual aggregation of the metal from the cores of these planetesimals over an extended time period? Recent work that compares the Hf-W systematics of chondrites and iron meteorites with those in samples representative of the silicate portion of the Earth (which are essentially identical to chondritic values; Fig. 5) indicates that core formation in the Earth, assuming a simple two-stage model in which this event occurs at a well-defined point in time, took place >50 Myr after differentiation of the planetesimals in which iron meteorites formed (Lee and Halliday 1995; Halliday et al. 1996). According to Halliday et al. (1996), this late core formation age implies that, unless accretion was also late, metal segregation did not occur during the major portion of the accretionary history for the Earth; this interpretation may be broadly consistent with suggestions of relatively young U-Pb model ages for the Earth (Allègre et al. 1995). However, Jacobsen and Harper (1996) have argued that core formation in the Earth is a process that is primarily rate-limited by accretion, which may have extended for a total time interval of over ~100 Myr but had a significantly shorter *mean* time (by a factor of ~8). Assuming continuous accretion and core formation, Hf-W systematics in the bulk silicate Earth (BSE) may be explained if the mean time of core formation was 10–20 Myr; in this case, the late tail of accretion (i.e., ~10% occurring after ^{182}Hf was completely decayed) would be sufficient to effectively erase any difference in the ϵ_W values between the BSE and chondrites (Jacobsen and Harper 1996).

The timing of differentiation and core formation on Mars may be estimated from isotopic systematics in SNCs, the meteorites widely believed to represent samples of the martian crust (McSween 1994 and references therein). Based on Hf-W systematics of these meteorites (a few of which have resolvable excesses in radiogenic ^{182}W; Fig. 5), Lee and Halliday (1997) concluded that core formation on Mars took place within the first ~30 Myr of solar system history. Lee and Halliday (1997) further showed that ^{182}W excesses in the SNC meteorites appear to be roughly correlated with ^{142}Nd excesses in these samples (Harper et al. 1995) (Fig. 6). This not only is indicative of rapid accretion and differentiation of Mars but may also suggest that silicate melting and core formation were "coeval and cogenetic" (Lee and Halliday, 1997), such that Hf/W and Sm/Nd essentially fractionated together at an early stage, most likely during metal segregation from a shallow magma ocean.

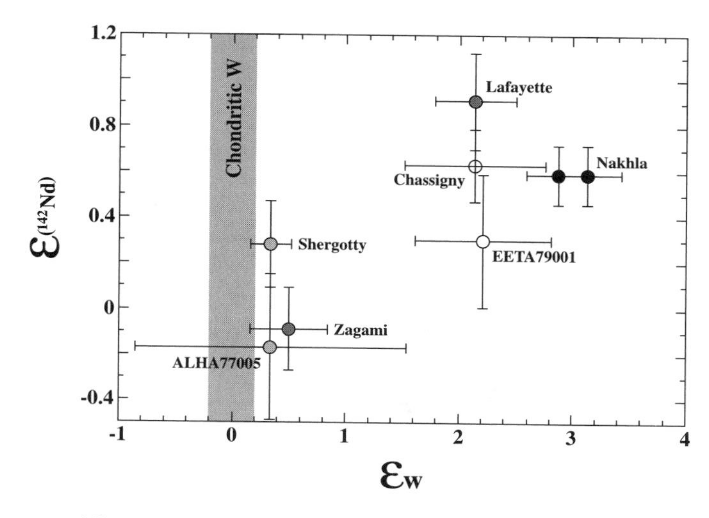

Figure 6. ϵ^{142}Nd v. ϵ_W for martian meteorites. ϵ^{142}Nd is the deviation in parts per 10^4 of the ^{142}Nd/^{144}Nd ratio from the terrestrial value. The vertical shaded bar indicates the uncertainty for the W isotopic composition of carbonaceous chondrites. Figure from Lee and Halliday (1997): copyright 1997 *Nature* (reprinted with permission).

IV. CURRENT CONUNDRUMS AND FUTURE CHALLENGES

In summary, Fig. 7 presents a schematic representation of timescales of events in the early solar system, as discussed in detail in the above sections. As is evident in section II.D, the ages obtained for some chondrules are >5 Myr younger than the CAI formation event. By this time, as is apparent from the discussion in section III.A, the HED parent body (Vesta?) had already accreted and differentiated (Lugmair and Shukolyukov 1998). If chondrules are assumed to have formed in a nebular environment, these observations are apparently contradictory, because dating of chondritic components indicates a prolonged nebular lifetime (i.e., several million years), whereas data from some achondrites suggest a relatively short-lived nebular phase and early accretion of solids into planetesimals and protoplanets. There are several ways in which this apparent contradiction may be explained, including disturbance of isotopic systematics in chondrules by reequilibration; chondrule formation in planetary environments; and a chaotic nebula with an extended parent body formation period.

Metamorphic reequilibration of Mg isotopes after parent body accretion would result in erroneously young ages for chondrules. LaTourrette and Wasserburg (1998) measured the rate of diffusion of Mg in plagioclase. They found that the minimum temperature required for Mg isotopic homogenization over the age of the solar system is 450°C. It is unlikely that the lowest petrologic type 3 chondrites (3.0 to 3.3) experienced higher temperatures; therefore, metamorphic equilibration probably cannot

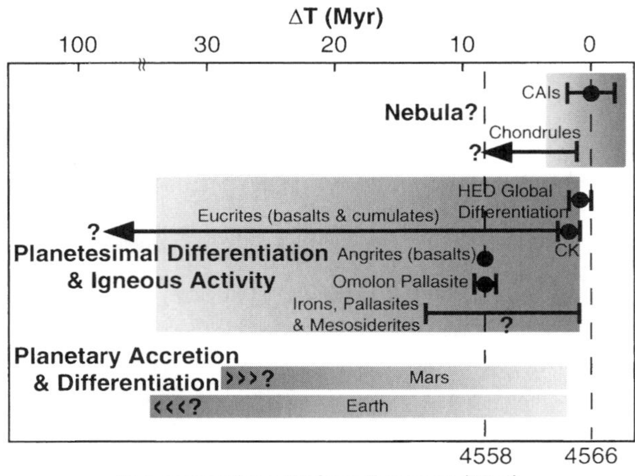

Figure 7. Schematic diagram illustrating the timescales of events that occurred within the first ~100 Myr of solar system history. Relative ages (ΔT) are anchored to the absolute Pb-Pb ages for Allende CAIs and angrites, indicated by the vertical dashed lines (references provided in the text). CAIs are thought to have formed during the protosolar disk (or nebular) phase. If chondrules are also nebular products, the time difference between CAIs and chondrules, indicated by Al-Mg systematics, could be suggestive of nebular lifetimes of several million years; however, there may be alternative explanations (see text). Timing of global differentiation of the HED parent body (4564.8 ± 0.9 Myr) and formation of the Chervony Kut (CK) basaltic eucrite (4563.6 ± 0.9 Myr) and the Omolon pallasite (4558.0 ± 1.0 Myr) are given by their Mn-Cr model ages (Lugmair and Shukolyukov 1998). Although most basaltic eucrites were likely formed by igneous activity on their parent body within ~10 Myr of CAIs, others may have crystallized later from impact melts; cumulate eucrites may have formed as much as ~100 Myr after the oldest basaltic eucrites (Tera et al. 1997). A variety of iron-rich meteorites have Pd-Ag relative ages (which likely represent the timing of condensation and segregation of the Fe-Ni metal phase) that span ~12 Myr (Chen and Wasserburg 1996); this time span may be mapped onto the absolute scale, assuming that ^{107}Pd and ^{53}Mn were homogeneously distributed in the meteorite-forming region and that a ^{107}Pd/^{108}Pd ratio of $\sim 1 \times 10^{-5}$ corresponds to a ^{53}Mn/^{55}Mn ratio of $\sim 1 \times 10^{-6}$ (Lugmair and Shukolyukov 1998). Finally, accretion and core formation for Mars appear to have occurred within the first ~30 Myr of solar system history (Lee and Halliday 1997), whereas these processes may have extended >50 Myr for Earth (although their mean time may have been significantly shorter; Jacobsen and Harper 1996).

account for all the data. However, peak metamorphic temperatures reached by ordinary chondrites are poorly constrained, so this possibility cannot be definitely excluded.

Alternatively, if chondrules formed in a planetary rather than a nebular environment, their age would reflect postaccretional formation. A parent body environment might produce chondrule-like textures and could account for their typically unfractionated chemistry (e.g., Hutchison 1996). However, this model cannot easily account for the diversity of chemical and isotopic compositions observed in chondrules (see the chapter by Jones et al., this volume), and therefore it may be an unlikely explanation for the apparently late (relative to CAIs) formation ages of chondrules.

Finally, the asteroid belt may have experienced a complex early history that involved prolonged accretion and disruption of planetesimals. One model that describes such a scenario was recently proposed by Weidenschilling et al. (1998). These authors suggested that although planetesimals formed early, resonances from the newly formed Jupiter increased their eccentricity, causing collisional disruption and dust melting by bow shocks in this region. In this way, at a relatively late stage in the nebula (several Myr after CAI formation), chondrule formation and planetesimal reaggregation may have occurred.

Currently, it cannot be ascertained which of the three possibilities discussed above is the most likely one. A better understanding of the mechanism and environment of chondrule formation is clearly needed to resolve this issue. Moreover, it will be necessary to obtain more rigorous constraints by precisely dating additional chondrules that appear to have remained isotopically "pristine" since their formation.

More broadly, to advance our understanding of the high-resolution chronology of events in the early solar system further, there clearly is a need for additional studies involving short-lived isotope measurements. In particular, studies of different short-lived chronometers in the same samples will allow the evaluation of the applicability of each chronometer. Additionally, analyses of a wider variety of samples will also expand our understanding of the initial distribution of isotopes and of the formation timescales of different solar system objects.

Acknowledgments We thank G. W. Lugmair and A. Shukolyukov for providing Fig. 4, and A. Halliday and D.-C. Lee for Figs. 5 and 6. Helpful comments and suggestions by A. P. Boss, G. Crozaz, J. N. Goswami, and A. Shukolyukov, as well as thorough reviews by A. Davis and G. W. Lugmair, are much appreciated and resulted in significant improvement. This work was partly supported by a NASA grant to M. W.

REFERENCES

Allègre, C. J., Manhès, G., and Göpel, C. 1995. The age of the Earth. *Geochim. Cosmochim. Acta* 59:1445–1456.

Binzel, R. P., and Xu, S. 1993. Chips off of asteroid 4 Vesta: Evidence for the parent body of basaltic achondrite meteorites. *Science* 260:186–191.

Birck, J.-L., and Allègre, C. J. 1985. Evidence for the presence of ^{53}Mn in the early solar system. *Geophys. Res. Lett.* 12:745–748.

Birck, J.-L., Rotaru, M., and Allègre, C. J. 1998. The evolution of ^{53}Mn-^{53}Cr in planet formation. *Paul Pellas Symposium Abstracts* (Fayetteville, AR: The Meteoritical Society), pp. 79–80.

Caffee, M. W., Hohenberg, C. M., and Swindle, T. D. 1982. I-Xe ages of individual Bjurbole chondrules. Proc. Lunar Planet. Sci. Conf. 13, *J. Geophys. Res.* 87:A303–A317.

Caillet, C., MacPherson, G. J., and Zinner, E. K. 1993. Petrologic and Al-Mg isotopic clues to the accretion of two refractory inclusions to the Leoville parent body: One was hot, the other wasn't. *Geochim. Cosmochim. Acta* 57:4725–4743.

Cameron, A. G. W. 1995. The first ten million years in the solar nebula. *Meteoritics* 30:133–161.

Chen, J. H., and Tilton, G. R. 1976. Isotopic lead investigations on the Allende carbonaceous chondrite. *Geochim. Cosmochim. Acta* 40:635–643.

Chen, J. H., and Wasserburg, G. J. 1981. The isotopic composition of uranium and lead in Allende inclusions and meteoritic phosphates. *Earth Planet. Sci. Lett.* 52:1–15.

Chen, J. H., and Wasserburg, G. J. 1996. Live ^{107}Pd in the early solar system and implications for planetary evolution. In *Earth Processes: Reading the Isotopic Code*, AGU Geophys. Mon. Ser. 95, ed. A. Basu and S. Hart (Washington, DC: American Geophysical Union), pp. 1–20.

Clayton, R. N. and Mayeda, T. K. 1996. Oxygen isotope studies of achondrites. *Geochim. Cosmochim. Acta* 60:1999–2017.

Connolly, H. C., and Love, S. G. 1998. The formation of chondrules: Petrologic tests of the shock wave model. *Science* 280:62–67.

Davis, A. M., Simon, S. B., and Grossman, L. 1994. Alteration of Allende Type B CAIs: When, where and how? *Lunar Planet. Sci.* 25:315–316.

Galer, S. J. G. and Lugmair, G. W. 1996. Lead isotope systematics of non-cumulate eucrites. *Meteoritics* 31:A47–A48.

Gilmour, J. D., Ash, R. D., Hutchison, R., Bridges, J. C., Lyon, I. C., and Turner, G. 1995. Iodine-xenon studies of Bjurbole and Parnallee using RELAX. *Meteoritics* 30:405–411.

Göpel, C., Manhès, G., and Allègre, C. J. 1991. Constraints on the time of accretion and thermal evolution of chondrite parent bodies by precise U-Pb dating of phosphates. *Meteoritics* 26:338.

Göpel, C., Manhès, G., and Allègre, C. J. 1994. U-Pb systematics of phosphates from equilibrated ordinary chondrites. *Earth Planet. Sci. Lett.* 121:153–171.

Gray, C. M., and Compston, W. 1974. Excess ^{26}Mg in Allende meteorite. *Nature* 251:495–497.

Halliday, A. N., Rehkämper, M., Lee, D.-C., and Wen, Y. 1996. Early evolution of the Earth and Moon: New constraints for Hf-W isotope geochemistry. *Earth Planet. Sci. Lett.* 142:75–89.

Harper, C. L., Nyquist, L. A., Bansal, B., Weismann, H., and Shih, C.-Y. 1995. Rapid accretion and early differentiation of Mars indicated by ^{142}Nd/^{144}Nd in SNC meteorites. *Nature* 267:213–217.

Hewins, R. H. 1988. Experimental studies of chondrules. In *Meteorites and the Early Solar System*, ed. J. F. Kerridge and M. S. Matthews (Tucson: University of Arizona Press), pp. 660–679.

Hewins, R. H., Jones, R. H. and Scott, E. R. D. 1996. *Chondrules and the Protoplanetary Disk* (Cambridge: Cambridge University Press).

Hohenberg, C. M., Brazzle, R. H., Pravdivtseva, O. V., and Meshik, A. P. 1998. Iodine-xenon chronometry: The verdict. *Meteorit. Planet. Sci.* 33 Supp.:A69–A70.

Hsu, W., and Crozaz, G. 1996. Mineral chemistry and the petrogenesis of eucrites: I. Noncumulate eucrites. *Geochim. Cosmochim. Acta* 60:4571–4569.

Huss, G. R., MacPherson, G. J., Russell, S. S., Srinivasan, G., and Wasserburg, G. J. 1998. ^{26}Al-containing Al-rich chondrules from unequilibrated ordinary chondrites. *Geochim. Cosmochim. Acta.,* in preparation.

Hutcheon, I. D., 1982. Ion probe magnesium isotopic measurements of Allende inclusions. *Amer. Chem. Soc. Symp. Ser.* 176:95–128.

Hutcheon, I. D., and Hutchison, R. H. 1989. Evidence from the Semarkona ordinary chondrite for the ^{26}Al heating of small planets. *Nature* 337:238–241.

Hutcheon, I. D., and Jones, R. H. 1995. The ^{26}Al-^{26}Mg record of chondrules: Clues to nebular chronology. *Lunar Planet. Sci.* 26:647–648.

Hutcheon, I. D., and Newton, R. C. 1981. Mg isotopes, mineralogy and mode of formation of secondary phases in C3 refractory inclusions. *Lunar Planet. Sci.* 12:491–493.

Hutchison, R. 1996. Chondrules and their associates in ordinary chondrites: A planetary connection? In *Chondrules and the Protoplanetary Disk,* ed. R. H. Hewins, R. H. Jones, and E. R. D. Scott (Cambridge: Cambridge University Press), pp. 311–318.

Ireland, T. R., Compston, W., Williams, I. S., and Wendt, I. 1990. U-Th-Pb systematics of individual perovskite grains from the Allende and Murchison carbonaceous chondrites. *Earth Planet. Sci. Lett.* 101:379–387.

Jacobsen, S. B., and Harper, C. L., Jr. 1996. Accretion and early differentiation history of the Earth based on extinct radionuclides. In *Earth Processes: Reading the Isotopic Code,* AGU Geophys. Monog. 95, ed. A. Basu and S. R. Hart (Washington, DC: American Geophysical Union), pp. 47–74.

Jacobsen, S. B., and Wasserburg, G. J. 1984. Sm-Nd isotopic evolution of chondrites and achondrites. *Earth Planet. Sci. Lett.* 67:137–150.

Kita, N. T., Nagahara, H., Togashi, S., and Morishita, Y. 1998. New evidence of aluminum-26 from a ferrous-oxide-rich chondrule in Semarkona (LL 3.0). *Meteorit. Planet. Sci.* 33:A83–A84.

Krot, A. N., Scott, E. R. D., and Zolensky, M. E. 1995. Mineralogical and chemical modification of components in CV3 chondrites: Nebular or asteroidal processing? *Meteoritics* 30:748–776.

Krot, A. N., Scott, E. R. D., and Zolensky, M. E. 1997. Mineralogical and chemical modification of CV3 chondrites during fluid-assisted metamorphism in the CV3 asteroid. In *Workshop on Parent-body and Nebular Modification of Chondritic Materials,* LPI Tech. Rep. 97-02, Part 1 (Houston: Lunar and Planetary Institute), pp. 34–36.

LaTourrette, T., and Wasserburg, G. J. 1998. Mg diffusion in anorthite: Implications for the formation of early solar system planetesimals. *Earth Planet Sci. Lett.* 158:91–108.

Lee, D.-C., and Halliday, A. N. 1995. Hafnium-tungsten chronometry and the timing of terrestrial core formation. *Nature* 378:771–774.

Lee, D.-C., and Halliday, A. N. 1996. Hf-W isotopic evidence for rapid accretion and differentiation in the early solar system. *Science* 274:1876–1879.

Lee, D.-C., and Halliday A. N. 1997. Core formation on Mars and differentiated asteroids. *Nature* 388:854–857.

Lee, T., Papanastassiou, D. A., and Wasserburg, G. J. 1976. Demonstration of ^{26}Mg excess in Allende and evidence for ^{26}Al. *Geophys. Res. Lett.* 3:109–112.

Lugmair, G. W., and Galer, S. J. G. 1992. Age and isotopic relationships among the angrites Lewis Cliff 86010 and Angra dos Reis. *Geochim. Cosmochim. Acta* 56:1673–1694.

Lugmair, G. W., and Shukolyukov, A. 1998. Early solar system timescales according to ^{53}Mn-^{53}Cr systematics. *Geochim. Cosmochim. Acta* 62:2863–2886.

MacPherson, G. J. and Davis, A. M. 1993. A petrologic and ion microprobe study of a Vigarano Type B refractory inclusion: Evolution by multiple stages of alteration and melting. *Geochim. Cosmochim. Acta* 57:231–243.

MacPherson, G. J. and Davis, A. M. 1997. Parent-body metamorphism of CV3 chondrites: Counterarguments based on accretionary rims and calcium-aluminum-rich inclusions. In *Workshop on Parent-body and Nebular Modification of Chondritic Materials*, LPI Tech. Rep. 97-02, Part 1 (Houston: Lunar and Planetary Institute), pp. 42–43.

MacPherson, G. J., Wark, D. A., and Armstrong, J. T. 1988. Primitive material surviving in chondrites: Refractory inclusions. In *Meteorites and the Early Solar System,* ed. J. F. Kerridge and M. S. Matthews (Tucson: University of Arizona Press), pp. 746–807.

MacPherson, G. J., Davis, A. M., and Zinner, E. K. 1995. The distribution of aluminum-26 in the early solar system: A reappraisal. *Meteoritics* 30:365–386.

Manhès, G., Göpel, C., and Allègre, C. J. 1988. Systématique U-Pb dans les inclusions réfractaires d'Allende: le plus vieux matériau solaire. In *Comptes Rendus de l'ATP Planétologie,* pp. 323–327.

McSween, H. Y., Jr. 1994. What we have learned about Mars from SNC meteorites. *Meteoritics* 29:757–779.

Miyamoto, M., and Takeda, H. 1994*a*. Evidence of excavation of deep crustal material of Vesta-like body from Ca compositional gradients in pyroxene. *Earth Planet. Sci. Lett.* 122:343–349.

Miyamoto, M., and Takeda, H. 1994*b*. Cooling rates of several cumulate eucrites. *Meteoritics* 29:505–506.

Nyquist, L., Lindstrom, D., Shih, C.-Y., Wiesmann, H., Mittlefehldt, D., Wentworth, S., and Martinez, R. 1997. Mn-Cr systematics of chondrules from the Bishunpur and Chainput meteorites. *Lunar Planet. Sci.* 28:1033–1304.

Nyquist, L., Shih, C.-Y., Wiesmann, H., Reese, Y., Ulyanov, A. A., and Takeda, H. 1999. Towards a Mn-Cr timescale for the early solar system. *Lunar Planet Sci.* 30: no. 1604.

Papanastassiou, D. A. 1986. Cr isotopic anomalies in the Allende meteorite. *Astrophys. J. Lett.* 308:L27–L30.

Podosek, F. A., Zinner, E. K., MacPherson, G. J., Lundberg, L. L., Brannon, J. C., and Fahey, A. J. 1991. Correlated study of initial $^{87}Sr/^{86}Sr$ and Al-Mg isotopic systematics and petrologic properties in a suite of refractory inclusions from the Allende meteorite. *Geochim. Cosmochim. Acta* 55:1083–1110.

Prinzhofer, A., Papanastassiou, D. A., and Wasserburg, G. J. 1992. Samarium-neodymium evolution of meteorites. *Geochim. Cosmochim. Acta* 56:797–815.

Russell, S. S., Srinivasan, G., Huss, G. R., Wasserburg, G. J., and MacPherson, G. J. 1996. Evidence for widespread ^{26}Al in the solar nebula and constraints for nebula time scales. *Science* 273:757–762.

Russell, S. S., Huss, G. R., MacPherson, G. J., and Wasserburg, G. J. 1997. Early and late chondrule formation: New constraints for solar nebula chronology from $^{26}Al/^{27}Al$ in unequilibrated ordinary chondrites. *Lunar Planet. Sci.* 28:1209–1210.

Sahijpal, S., and Goswami, J. N. 1998. Refractory phases in primitive meteorites devoid of ^{26}Al and ^{41}Ca: Representative samples of first solar system solids? *Astrophys. J. Lett.* 509:L137–L140.

Sahijpal, S., Goswami, J. N., Davis, A. M., Grossman, L., and Lewis, R. S. 1998. A stellar origin for the short-lived nuclides in the early solar system. *Nature* 391:559–561.

Sanders, I. S. 1996. A chondrule-forming scenario involving molten planetesimals. In *Chondrules and the Protoplanetary Disk,* ed. R. H. Hewins, R. H. Jones, and E. R. D. Scott (Cambridge University Press), pp. 327–334.

Schramm, D. N., Tera, F., and Wasserburg, G. J. 1970. The isotopic abundance of ^{26}Mg and limits on ^{26}Al in the early solar system. *Earth Planet. Sci. Lett.* 10:44–59.

Sheng, Y. J., Hutcheon, I. D., and Wasserburg, G. J. 1991. Origin of plagioclase olivine inclusions in carbonaceous chondrites. *Geochim. Cosmochim. Acta* 55:581–599.

Shu, F. H., Shang, H., and Lee, T. 1996. Toward an astrophysical theory of chondrites. *Science* 271:1545-1552.

Shukolyukov, A., and Lugmair, G. W. 1993*a*. Live iron-60 in the early solar system. *Science* 259:1138–1142.

Shukolyukov, A., and Lugmair, G. W. 1993*b*. ^{60}Fe in eucrites. *Earth Planet. Sci. Lett.* 119:159–166.

Srinivasan, G., Ulyanov, A. A., and Goswami, J. N. 1994. ^{41}Ca in the early solar system. *Astrophys. J. Lett.* 431:L67–L70.

Srinivasan, G., Sahijpal, S., Ulyanov, A. A., and Goswami, J. N. 1996*a*. Ion microprobe studies of Efremovka CAIs: II. Potassium isotope composition and ^{41}Ca in the early solar system. *Geochim. Cosmochim. Acta* 60:1823–1835.

Srinivasan, G., Russell, S. S., MacPherson, G. J., Huss, G. R., and Wasserburg, G. J. 1996*b*. New evidence for ^{26}Al in CAI and chondrules from type 3 ordinary chondrites. *Lunar Planet. Sci.* 27:1257–1258.

Srinivasan, G., Huss, G. R., and Wasserburg, G. J. 1996*c*. Aluminum-26 timescales and processes connecting plagioclase-rich chondrules and CAIs. *Meteorit. Planet Sci.* 31:A133.

Srinivasan, G., Goswami, J. N., and Bhandari, N. 1999. ^{26}Al in eucrite Piplia Kalan: Plausible heat source and formation chronology. *Science* 284:1348–1350.

Strom, S. E. 1994. The early evolution of stars. *Rev. Mex. Astron. Astrofis.* 29:23–29.

Swindle, T. D. and Podosek, F. A. 1988. Extinct radionuclides. In *Meteorites and the Early Solar System,* ed. J. F. Kerridge and M. S. Matthews (Tucson: University of Arizona Press), pp. 1093–1113.

Swindle, T. D., Caffee, M.W., and Hohenberg, C. M. 1988. Iodine-xenon studies of Allende inclusions: Eggs and the Pink Angel. *Geochim. Cosmochim. Acta* 52:2215–2227.

Swindle, T. D., Caffee, M.W., Hohenberg, C. M., Lindstrom, M. M., and Taylor, G. J. 1991. Iodine-xenon studies of petrographically and chemically characterized Chainpur chondrules. *Geochim. Cosmochim. Acta* 55:861–880.

Swindle, T. D., Davis, A. M., Hohenberg, C. M., MacPherson, G. J., and Nyquist, L. E. 1996. Formation times of chondrules and Ca, Al-rich inclusions: Constraints from short-lived nuclides. In *Chondrules and the Protoplanetary Disk,* ed. R. Hewins, R. Jones, and E. Scott (Cambridge: Cambridge University Press), pp. 77–86.

Swindle, T. D., Cohen, B., Li, B., Olsen, E., and Krot, A. N. 1998. Iodine-xenon studies of separated components of the Efremovka (CV3) meteorite. *Lunar Planet. Sci.* 29:1005–1006.

Tatsumoto, M., Unruh, D. M., and Desborough, G. A. 1976. U-Th-Pb and Rb-Sr systematics of Allende and U-Th-Pb systematics of Orgueil. *Geochim. Cosmochim. Acta* 40:617–634.

Tera, F., and Carlson, R. W. 1999. Assessment of the Pb-Pb and U-Pb chronometry of the early solar system. *Geochim. Cosmochim. Acta* 63:1877–1889.

Tera, F., Carlson, R. W., and Boctor, N. Z. 1997. Radiometric ages of basaltic achondrites and their relation to the early history of the solar system. *Geochim. Cosmochim. Acta* 61:1713–1731.

Tilton, G. R. 1988. Principles of radiometric dating. In *Meteorites and the Early Solar System,* ed. J. F. Kerridge and M. S. Matthews (Tucson: University of Arizona Press), pp. 249–258.

Wadhwa, M., and Lugmair, G. W. 1996. The age of the eucrite Caldera from convergence of long- and short-lived chronometers. *Geochim. Cosmochim. Acta* 60:4889–4893.

Wasserburg, G. J. 1985. Short-lived nuclei in the early solar system. In *Protostars and Planets II,* ed. D. C. Black and M. S. Matthews (Tucson: University of Arizona Press), pp. 703–754.

Weidenschilling, S. J. 1977. Aerodynamics of solid bodies in the solar nebula. *Mon. Not. Roy. Astron. Soc.* 180:57–70.

Weidenschilling, S. J., Marzari, F., and Hood, L. L. 1998. The origin of chondrules at jovian resonances. *Science* 279:681–684.

Wetherill, G. W. 1986. Accumulation of the terrestrial planets and implications concerning lunar origin. In *Origin of the Moon,* ed. W. K. Hartmann, R. J. Phillips, and G. J. Taylor (Houston: Lunar and Planetary Science Institute), pp. 519–550.

METEORITICAL AND ASTROPHYSICAL CONSTRAINTS ON THE OXIDATION STATE OF THE SOLAR NEBULA

ALEXANDER N. KROT
University of Hawai'i at Manoa

BRUCE FEGLEY, JR., AND KATHARINA LODDERS
Washington University, St. Louis, Missouri

and

HERBERT PALME
Universität zu Köln

We review efforts to constrain the oxidation state of the solar nebula by using the chemistry and mineralogy of primitive chondritic meteorites (i.e., type 3 carbonaceous, ordinary, and enstatite chondrites). Our review shows that the nebular redox state varied from several orders of magnitude more oxidizing than a solar gas to several orders of magnitude more reducing than a solar gas. Mechanisms to explain such large variations are reviewed, and their implications for planetary chemistry are also discussed. An important part of our effort is the attempt to disentangle the signatures of parent body and nebular processes. Parent body processes, such as thermal metamorphism and fluid/rock interactions, have altered, sometimes significantly, the original nebular signatures preserved in primitive (e.g., type 3 carbonaceous) chondrites and make interpretation of the nebular record difficult.

I. INTRODUCTION

In 1971 Ed Anders wrote, "Chondrites contain a unique archeological record of physical and chemical conditions in the solar nebula. It is the job of meteoriticists to decipher this record—to translate structure, composition, and mineralogy into temperature, pressure, time, and chemical environment" (Anders 1971). Anders goes on to emphasize the "fortunate coincidence" that theoretical models have, at the same time, evolved to a point where they allow detailed predictions of pressure, temperature and composition at various times in the evolution of the solar nebula. In the period since 1971, meteoriticists have learned that both tasks are more difficult than initially thought. Meteorites are far more complex, and the theoretical treatment of the protosolar nebula and young stellar objects with circumstellar disks (e.g., Hartmann 1996) is much more complicated, than was anticipated in the early 1970s.

The problems, however, remain. Under what conditions did the components of meteorites form? What do these conditions tell us about the solar nebula on a local or on a global scale? What are the effects of asteroidal processing (fluid-rock interaction and thermal metamorphism) on nebular records in meteorite components? How do we distinguish between nebular and asteroidal records in meteorites?

The most primitive (i.e., the least altered and thermally metamorphosed) meteorites are carbonaceous (CO, CV, CK, CH), ordinary (H, L, LL), and enstatite (EH, EL) chondrites of petrologic type 3, which largely consist of various amounts of chondrules, Ca-Al-rich inclusions (CAIs), and fine-grained matrices. Detailed mineralogical, experimental, and isotopic studies have shown that these chondritic components have experienced a complex formation history in the solar nebular and asteroidal environments (e.g., Kerridge and Matthews 1988; Hewins et al. 1996; Zolensky et al. 1997).

CAIs are products of high-temperature processes in the early solar nebula. By "high-temperature" we mean temperatures so high that materials with approximately solar compositions for nonvolatile elements would not survive very long (minutes to hours) if exposed to them. Some CAIs are condensates and residues from these events; others crystallized from refractory melts, which may themselves represent condensates (e.g., MacPherson et al. 1988; Palme and Boynton 1993). Both types may have experienced subsequent alteration at lower temperatures by gas-solid reactions in the solar nebula, by fluid-rock interaction in an asteroidal environment, or both (MacPherson et al. 1988; Russell et al. 1998). Chondrules probably formed by multiple, localized, brief episodes of melting of preexisting solids; some suffered subsequent low-temperature nebular or asteroidal alteration similar to that observed in CAIs (Kimura and Ikeda 1995). The origins of matrix material remain obscure; a minor component of matrix is presolar (Huss and Lewis 1995), and the remainder is a complex mixture of material resulting from both nebular and asteroidal processes (Brearley 1996).

The solar nebula did not produce chemically and isotopically uniform objects. The most primitive type 3 carbonaceous, ordinary, and enstatite chondrites contain components that formed under different temperatures and redox conditions. Preservation of this apparent disequilibrium in these meteorites for billions of years indicates that their parent asteroids were not heated to above 600–700 K at any length of time and that the extent of reaction with liquid water was moderate, although there is some diversity in opinion about the exact degree of water/rock interaction (see section III.G). Meteorites of higher petrologic types, from 4 to 6 (CK, H, L, LL, R, EH, EL), reflect increasingly more intense parent body heating, whereas meteorites of lower petrologic types, 2 and 1 (CM, CR, CI), indicate increasing degrees of reaction with water, erasing the high-temperature record of various minerals.

An important parameter for the mineralogy of meteorites is the oxygen fugacity (fO_2, which is identical to the oxygen partial pressure for an ideal gas) under which the minerals formed or equilibrated. The solar nebula gas is very reducing (see section II), producing on condensation only metallic iron and almost no FeO in silicate or oxide condensates. Parent body processes, if water is present, may imprint a high fO_2 on the minerals that were produced or that have equilibrated with the nebular gas. Some of the oxidized meteorite components may have been produced in the solar nebula environment.

In this chapter we will review the evidence constraining the fO_2 under which the components of meteorites were formed. We will also attempt to disentangle nebular and parent body effects to constrain the variations in fO_2 of the solar nebula in the formation location of the type 3 carbonaceous, ordinary, and enstatite chondrites. We also discuss mechanisms responsible for possible variations in redox conditions in the solar nebula.

II. OXYGEN FUGACITY OF THE SOLAR NEBULA

The elemental abundances of the ten most abundant elements in a solar gas (gas of a solar composition) are listed in Table I, which shows that H, O, and C are the three elements we expect (by virtue of their large abundances and their chemistry) to control the fO_2 of solar gas. To a very good first approximation, the fO_2 of solar gas is regulated by the reaction

$$H_2 + 0.5O_2 = H_2O \qquad (1)$$

TABLE I
Elemental Abundances and Major Species in a Solar-Composition Gas[a]

Element	Atomic abundances	Major gas(es)
H	2.82×10^{10}	H_2, H
He	2.82×10^{9}	He
O	2.09×10^{7}	CO, H_2O
C	1.00×10^{7}	CO, CH_4
Ne	3.31×10^{6}	Ne
N	2.63×10^{6}	N_2, NH_3
Mg	1.023×10^{6}	Mg
Si	1.00×10^{6}	SiO, SiS
Fe	8.91×10^{5}	Fe
S	4.47×10^{5}	H_2S, HS
All other elements combined	4.16×10^{5}	Various

[a] from Lodders and Fegley (1998).

The temperature-dependent equilibrium constant (K_1) for reaction (1) is given by the equation

$$\log_{10} K_1 = -2.836 + 12{,}832/T \quad (2)$$

with temperature (T) in kelvins. The fO_2 (bar) for reaction (1)

$$\log_{10} fO_2 = 2\log_{10}(H_2O/H_2) + 5.67 - 25{,}664/T \quad (3)$$

is valid from 298 to 2500 K, where the (H_2O/H_2) number ratio (i.e., the molar ratio) is used in equation (3). The amount of water vapor in solar gas depends on the partitioning of carbon between CO and CH_4 via the net thermochemical reaction

$$CO + 3\,H_2 = CH_4 + H_2O \quad (4)$$

and on the amount of oxygen consumed by rock-forming elements. As we shall discuss below, essentially all carbon is bound in CO in the solar nebula, and mass balance dictates that ~15% of total oxygen is bound in rock (i.e., the amount of O bound in $MgO + SiO_2 + CaO + Al_2O_3 + Na_2O$). Thus, under conditions where negligible dissociation of water vapor and hydrogen occurs, the H_2O/H_2 ratio in solar nebula gas is $\sim 5\times10^{-4}$, and equation (3) can be rewritten as

$$\log_{10} fO_2 = -0.85 - 25{,}664/T \quad (5)$$

Equation (5) gives $fO_2 \sim 10^{-17.9}$ bar at 1500 K, vs. $10^{-17.8}$ bar at the same temperature from a chemical equilibrium calculation using the CONDOR code (Fegley and Lodders 1994; Lodders and Fegley 1995) considering several thousand species in a solar gas. Comparisons of CONDOR code results and equation (5) indicate that $\log_{10} fO_2$ from equation (5) reproduces the code results within ±0.2 log bar units below 2000 K and within ±0.3 log bar units from 2000 to 2500 K.

Equation (5) shows that the fO_2 of a solar gas is extremely reducing; as a consequence, iron alloy is more stable than iron oxides and FeO in minerals in the solar nebula. For example, at 1600 K, solar gas is 6 log bar units more reducing than the iron-wüstite (IW) buffer (i.e., firmly inside the iron metal stability field). However, the difference between the fO_2 of solar gas and the fO_2 needed to stabilize Fe oxides decreases with decreasing temperature until FeO-bearing minerals become stable at 500 K and below. At these low temperatures, however, solid-state diffusion is so slow that Fe cannot be incorporated into the Mg silicates within the nebular lifetime. Thus, the dominance of FeO-bearing minerals in chondritic meteorites indicates either that they were not formed by equilibrium condensation from a solar gas or that they were later modified on a parent body.

A. Carbon Chemistry in a Solar Gas

As can be seen from Table I, the C/O atomic ratio in solar gas is ~0.5; as a consequence, the oxygen not bound in minerals is almost evenly divided between CO and H_2O. Carbon monoxide is the dominant carbon gas at high temperatures and low pressures in solar-composition material (cf. Lewis et al. 1979). However, at complete chemical equilibrium the CO/CH_4 ratio decreases dramatically with decreasing temperature, because reaction (4) proceeds toward the right. Eventually, $CO/CH_4 \ll 1$ and CH_4 becomes the dominant carbon gas. The oxygen released from CO increases the H_2O abundance almost twofold. However, as first demonstrated by Lewis and Prinn (1980), the kinetics of the $CO \rightarrow CH_4$ conversion are so slow that reaction (4) is quenched at very high temperatures. As a consequence, CO remains the dominant carbon gas throughout the solar nebula, and the H_2O/H_2 ratio remains constant at $\sim 5 \times 10^{-4}$ until water ice condenses. The water ice condensation point is pressure dependent and varies from $\sim$150 K at 10^{-7} bar total pressure to $\sim$190 K at 10^{-3} bar total pressure. As discussed later in this chapter, the nebular snowline was at $\sim$5.2 AU throughout most of the time that nebular gas was around, because water ice condensation led to runaway accretion of Jupiter (Lissauer 1987; Stevenson and Lunine 1988). As the nebula cooled, the snowline moved inward (e.g., Ruden and Lin 1986), and some ice condensed in the asteroid belt, providing a source of water for subsequent hydration reactions on asteroids.

III. OXYGEN BAROMETERS IN CHONDRITES

Chondritic meteorites (chondrites) largely consist of chondrules, CAIs, and fine-grained matrices. There is general agreement that these chondritic components originated in the solar nebula; hence, they potentially give us insights into the processes that were operative in the early solar system. In this section we review indicators of fO_2 of the solar nebula gas that have been proposed for CAIs, chondrules, and matrices in carbonaceous chondrites (Table II). Because all chondrites were altered to some degree after accretion into planetesimals and asteroids by fluid-rock interaction and thermal metamorphism, these processes may have modified the nebular features. As a result, some of the mineral and chemical indicators of fO_2 discussed in this section may reflect conditions during asteroidal alteration. There are two fundamentally different types of oxygen fugacity indicators: (a) mineral indicators and (b) chemical indicators. In mineral indicators the valence state of an element reflects the fO_2 of the ambient gas or fluid from which the mineral formed; the valence states either are determined directly or are reflected in physical properties of minerals, such as color. Chemical indicators are based on elemental abundances (in most cases normalized to solar abundances) in minerals; for example, the mineral sequence condensing from a gas and the trace element patterns in the condensed minerals are strongly dependent on the oxygen fugacity of the gas. Also, element patterns produced during evaporation or in igneous processes (e.g., crystallization) will be influenced by fO_2. In general, mineral indicators are much more susceptible to later alterations, either in the solar nebula or on a parent body, than trace element patterns.

TABLE II
Proposed Oxygen Fugacity Indicators in Chondrites

Proposed indicator	Reversible	References
Oxidized solar nebula gas:		
Color of hibonite	Yes	1–3
(Ti^{3+}/Ti^{4+}) ratios in hibonite	Yes	4
(Ti^{3+}/Ti^{4+}) ratios in fassaite	Yes	5,6
Rhönite	No	6
Sr, Ba, U, V depletions in CAIs	No	7
Ce depletions in CAIs	No	7–10
Mo and W depletions in CAIs	No	11
Fe, Ni, Co abundances in fine-grained CAIs	No	12
Secondary Ca-rich minerals in CAIs (e.g., andradite, wollastonite, grossular, kirschsteinite)	No	13–17
Fremdlinge minerals (e.g., powellite, molybdenite)	No	18,19
Fayalite	No	20–22
Fayalitic olivine	No	23–25
Reduced solar nebula gas:		
Mineralogy of enstatite chondrites e.g., oldhamite, niningerite, alabandite, Ti-rich troilite	No	26, 27

References 1: Stolper and Ihinger (1983); 2: Ihinger and Stolper (1986); 3: Live et al. (1986); 4: Beckett et al. (1988); 5: Stolper et al. (1982); 6: Beckett and Grossman (1986); 7: Boynton and Cunningham (1981); 8: Davis et al. (1982); 9: Boynton (1978); 10: Fegley (1986); 11: Fegley and Palme (1985); 12: Kornacki and Wood (1985); 13: Allen et al. (1978); 14: MacPherson et al. (1981); 15: Hashimoto and Grossman (1987); 16: McGuire and Hashimoto (1989); 17: Kimura and Ikeda (1995); 18: Armstrong et al. (1987); 19: Bischoff and Palme (1987); 20: Brigham et al. (1986); 21: Wood and Holmberg (1994); 22: Hua and Buseck (1995); 23: Palme and Fegley (1990); 24: Nagahara et al. (1988); 25: Dohmen et al. (1998); 26: Keil (1968); 27: Larimer (1968).

A. Color and Ti^{3+}/Ti^{4+} Ratios in Hibonite, Fassaite, and Rhönite

1. Hibonite [$Ca_2(Al,Ti)_{24}O_{38}$]. Hibonite is a Ca-Al-Ti oxide occurring in several types of CAIs. Some hibonites condensed from solar gas, whereas others are evaporative residues and crystallization products from a melt at high temperatures (Kornacki and Fegley 1984; Ireland et al. 1992). In meteorites hibonite occurs in three colors: blue, orange, and colorless. Ihinger and Stolper (1986) showed that the color of hibonite heated at 1430°C (close to its condensation temperature) depends on fO_2. An fO_2 increase changes the color of hibonite from blue to orange and to colorless; these changes are reversible and extremely rapid even under quenching. These authors proposed that the color of hibonite can be used as an fO_2 indicator of the environment where hibonite equilibrated last. Beckett et al. (1988) found that the color of hibonite is a function of the Ti^{3+}/Ti^{4+} ratio, which also can be used as an fO_2 indicator. Titanium is present in terrestrial and extraterrestrial samples as Ti^{4+}. Under the extremely reducing conditions of the solar nebula, Ti^{4+} is partly converted to Ti^{3+}. The Ti^{3+}/Ti^{4+} ratio is thus an indicator of the fO_2 that prevailed during condensation and/or crystallization of Ti-bearing minerals.

Beckett et al. (1988) measured Ti^{3+} contents of a blue hibonite in a CAI from the CM2 chondrite Murchison and concluded that this hibonite equilibrated with a solar gas. Ihinger and Stolper (1986) estimated the fO_2 recorded by blue hibonites in the Blue Angel CAI from Murchison and concluded that the hibonite equilibrated in a gas that was four or five orders of magnitude more oxidizing than a gas of solar composition. However, this CAI experienced asteroidal alteration under oxidizing conditions (Armstrong et al. 1982), so it seems possible that the color and Ti^{3+}/Ti^{4+} ratio in its hibonite reflect reequilibration during alteration (Beckett et al. 1988).

The coexistence of colorless or orange hibonite, reflecting oxidizing conditions, with fassaite with high Ti^{3+}/Ti^{4+} ratios, indicative of reducing nebular fO_2 (see subsection 2 below), was also found in other CAIs and is interpreted to indicate later oxidizing conditions that affected the color of hibonite but not the Ti^{3+}/Ti^{4+} ratios in fassaite (Ihinger and Stolper 1986).

2. *Fassaite [$Ca(Mg,Ti,Al)(Al,Si)_2O_6$] and Rhönite [$Ca_2(Mg,Al,Ti)_6(Si,Al)_6O_{20}$].* Fassaite is a Ti-rich Al-diopside occurring in several types of CAIs. Fassaites in Type B CAIs crystallized from a melt at ~1200°C and contain variable amounts of Ti^{3+} and Ti^{4+}. Beckett and Grossman (1986) obtained a relationship between the Ti^{3+}/Ti^{4+} ratio in fassaite and fO_2 in the coexisting gas and concluded that crystallization of fassaite in Type B CAIs from Allende took place in solar gas.

The mineral rhönite, although less commonly observed in CAIs, can be used in a similar fashion. Beckett and Grossman (1986) calculated fO_2 for compact Type A CAIs from Allende and concluded that crystallization of rhönite took place in solar gas.

To summarize, the observed Ti^{3+}/Ti^{4+} ratios in fassaite and rhönite crystallizing from CAI melts are consistent with their origin in a canonical solar gas. The observed colors and Ti^{3+}/Ti^{4+} ratios of hibonites in Allende may not reflect the redox conditions during their crystallization but instead may reflect a later equilibration with oxidized gases, either in nebular or asteroidal environments.

B. Chemical Indicators: Trace Element Concentrations in CAIs

A variety of factors, including fO_2, host phase, equilibration temperature, gas-solid fractionation, and low-temperature alteration may affect the abundances of trace elements in CAIs (Boynton 1975; Palme and Wlotzka 1976; Davis and Grossman 1979; Boynton and Cunningham 1981; Ekambaram et al. 1984; Fegley and Palme 1985). If the effects of fO_2 can be decoupled from those due to other factors, then trace elements can be used as oxygen barometers. Below we review several trace element oxygen barometers proposed for CAIs.

1. Ce Depletion. At fO_2 three to five orders of magnitude higher than a solar gas, Ce becomes highly volatile (the major gas species of Ce is CeO_2) relative to other rare earth elements (REEs), which occur as monoxides in the gas (e.g., LaO). If REEs condensed in a mineral from such a gas, large negative Ce anomalies are expected (Boynton 1975; Boynton and Cunningham 1981; Davis et al. 1982; Fegley 1986). It was found that five hibonite-rich CAIs and C1 (a type B CAI), all of which are FUN (Fractionation and Unidentified Nuclear isotopic anomalies) inclusions, exhibit strong negative anomalies in Ce (Fig. 9 in Floss et al. 1996). Thermodynamic calculations for the CAI C1 (Fegley 1986) show that its Ce depletion could have been produced under oxidizing conditions, similar to those proposed to explain Mo and W depletions (see subsection III.B.3 below). Because one of the best-studied CAIs exhibiting Ce depletion, the HAL inclusion, was originally interpreted as an aggregate of nebular condensates (Allen et al. 1980), this depletion was proposed as an indicator of oxidizing conditions in the solar nebula (Davis et al 1982; Rubin et al. 1988; Palme and Boynton 1993). However, several isotopic studies and vacuum evaporation experiments (Ireland et al. 1992; Davis et al. 1995; Floss et al. 1996) showed that the FUN inclusions formed by kinetically controlled evaporation (Fig. 1 from Davis et al. 1995). Because the evaporating CAI material was not in equilibrium with a solar gas, the Ce depletions in the FUN inclusions cannot be used as indicators of the redox conditions in the solar nebula. Apparently, conditions during evaporation are generally much more oxidizing than reflected in the ambient gas phase into which evaporation occurs.

With a few rare exceptions (e.g., a small Ce anomaly in a fine-grained CAI from Efremovka; Boynton et al. 1986), there is no evidence for Ce anomalies in most CAIs (e.g., Wänke et al. 1974; Grossman and Ganapathy 1976*a,b*; Davis et al. 1978; Mason and Taylor 1982; Ekambaram et al. 1984). This suggests either that condensation of the REEs in most CAIs took place in solar gas or that CAIs equilibrated with the nebular gas at such a low temperature that complete condensation of Ce occurred.

2. Yb Depletion in the HAL-type CAIs. At solar gas fO_2, most REEs are stable as monoxide gas species except for Yb, whose predominant gas species is metallic Yb. This leads to a depletion of Yb relative to the other REEs in condensed oxide or silicate phases. Under more oxidizing conditions, however, Yb approaches the behavior of the other REEs, and the anomaly disappears. Ytterbium depletions were found in five HAL-type hibonites that also showed strong Ce depletions (Fig. 9 from Floss et al. 1996). The presence of both Ce and Yb anomalies in the same mineral grain is incompatible with equilibrium between gas and solid at any fO_2. Ireland et al. (1992) suggested that to produce anomalies in both, there must be two evaporation episodes, one under oxidizing conditions, the other reducing. Floss et al. (1996) inferred instead that Yb anomalies were inherited from the precursor materials.

3. Mo and W Depletions. Alloys of metals with low vapor pressures (W, Os, Re, Ir, Mo, Ru, Pt, Rh) are expected as the first metal condensates from a cooling gas of solar composition. CAIs commonly have high concentrations of these elements, which now reside in mineralogically complex inclusions that probably formed from initially homogeneous alloys by exsolution and reaction with ambient gases. However, in one Allende CAI, submicrometer homogeneous alloys of all refractory metals, including W and Mo, with solar relative abundances, were found. These are presently the best candidates for primary unaltered condensates (Eisenhour and Buseck 1992). Tungsten and molybdenum form volatile oxides under moderately oxidizing conditions. Their presence in the refractory metal alloys reflects the reducing conditions of formation that are created by the solar gas. Larger CAIs often show depletions of W and Mo (Fegley and Palme 1985). The observed depletions have a very characteristic pattern, with Mo always being more depleted than W, and have been interpreted as indicating condensation of refractory metals under more oxidizing conditions than those of a solar gas. Fegley and Palme (1985) found that the same Mo and W depletion pattern occurs in refractory metal alloy condensation calculations done under increasingly oxidizing conditions (Fig. 7.7.3 from Rubin et al. 1988). They also showed that observed patterns could be matched by refractory metal alloy condensation calculations done at oxygen fugacities 10^3 to 10^4 times greater than in a solar gas at the same temperature and pressure (Fig. 1).

Subsequent work showed that Mo and W depletions were found in ~80% of the 30 samples from the CV3 Allende, Grosnaja, Leoville, and CO_3 Ornans carbonaceous chondrites analyzed (Fegley and Kong 1989;

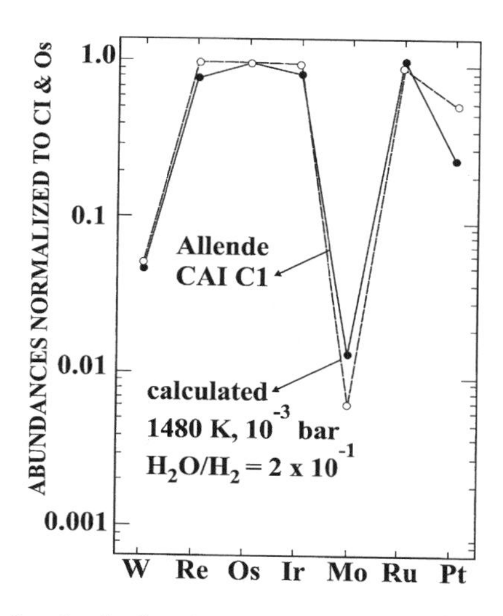

Figure 1. Observed and calculated refractory metal abundances in Allende FUN inclusion C1 (after Rubin et al. 1988).

Fegley and Prinn 1989). Although this database is still limited in size, several important conclusions can be drawn: (1) Mo is always more depleted than W. (2) The depletions are found in both normal and isotopically anomalous (i.e., FUN) CAIs. (3) The depletions are not simply a volatility effect, because W is the most refractory of these seven metals, and Mo has a vapor pressure very similar to that of Ir. (4) The Mo and W depletions do not simply reflect redistribution of these two metals inside CAIs; they are bulk depletions of Mo and W from CAIs.

Fegley and Palme (1985) considered several potential mechanisms, including fractional condensation, reactions with sulfur compounds, and metamorphic alteration, to form the Mo and W depletions. However, none of these processes reproduced the observed depletions. They concluded that the Mo and W depletions were produced by high-temperature oxidation during condensation or vaporization in the solar nebula. However, Fegley and Palme (1985) also noted that the rarity of Ce anomalies in CAIs seemed to be inconsistent with oxidizing conditions in the solar nebula, because Ce depletions should also form under the oxidizing conditions that produce Mo and W depletions. Fegley and Palme (1985) suggested that the REEs and refractory metals last equilibrated under different conditions at a different time or place. According to this scenario, CAIs accumulated as mixtures of Mo- and W-depleted refractory metals that condensed under oxidized conditions and silicate-lithophile components that formed under reduced conditions. This explanation is consistent with the fact that refractory metals and REE were carried into CAIs in different phases with different chemical and physical properties.

Beckett et al. (1988) also noticed the differences between the higher fO_2 indicated by Mo and W depletions in CAIs and the lower fO_2 indicated by Ti^{3+}/Ti^{4+} ratios in oxide and silicate minerals in CAIs. They speculated that the Mo and W depletions formed at low temperatures during oxidation in the solar nebula or during asteroidal metamorphism. The latter alternative appears to be contradicted by the characteristic pattern of the Mo and W depletions, because Fegley and Palme (1985) calculated that metamorphic alteration gave W depletions equal to or larger than Mo depletions.

In this regard we note that Palme et al. (1994) studied one CAI (Egg-6 from Allende) that has variable W and Mo depletions unlike the characteristic patterns produced by high-temperature oxidation (Fig. 7.7.3 from Rubin et al. 1988 and Fig. 2a, this chapter). However, the W in this inclusion has also been redistributed from refractory metal alloys into the surrounding silicate, and Mo is more depleted than W. Palme et al. (1994) concluded that Egg-6 was affected by later oxidation at moderately high temperatures. This conclusion is supported by vaporization experiments by Köhler et al. (1988) and Wulf (1990), which showed that W is redistributed into the surrounding silicates, while Mo is lost to the ambient gas, when chondritic material is heated under oxidizing conditions (Fig. 2b).

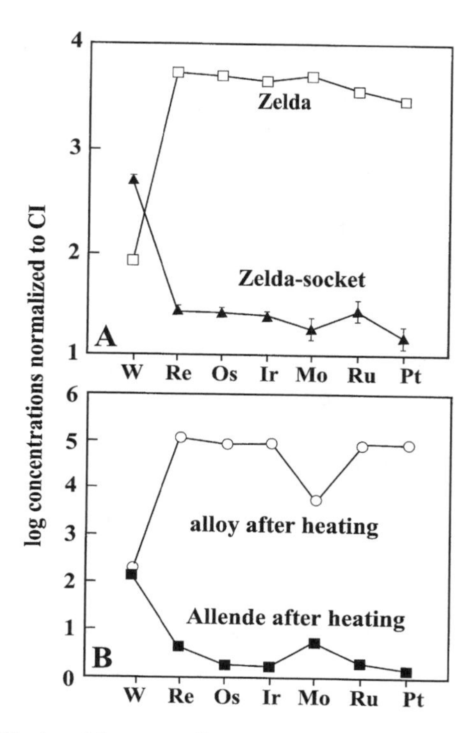

Figure 2. (a) CI chondrite-normalized refractory siderophile element abundances in Fremdling (see section III.E) Zelda in the Egg-G CAI from Allende and silicate portion of the host CAI (socket). The complementary patterns of the Fremdling and its socket indicate *in situ* mobilization of W from the Fremdling into silicates. (b) CI chondrite-normalized refractory siderophile element abundances of the heating experiments (at 1150°C for one week at Ni-NiO) on a refractory metal alloy embedded in Allende reproduces the pattern in Fig. 2a (from Palme et al. 1994).

The overall picture that seems to emerge is that Mo and W depletions in CAIs resulted from high-temperature oxidation in the solar nebula, which is the only process shown to produce the characteristic patterns in refractory metal alloys, with Mo always more depleted than W.

C. Fe, Ni, and Co Abundances in Fine-grained CAIs in Allende

Kornacki and Wood (1985) showed that spinel grains in the fine-grained CAIs in Allende are enriched in FeO, whereas bulk compositions of these CAIs have Fe/Ni and Fe/Co ratios significantly higher than chondritic. These results suggest that the major enrichment of spinels in FeO did not result from oxidation of Fe, Ni-metal grains inside the CAIs; an external source of Fe is required. In a gas of solar composition, the oxidation of Fe metal takes place at low temperatures, where Fe transport in a gas form is insignificant. Kornacki and Wood (1985) inferred that if FeO concentrations in the CAIs were largely established in the hot nebula, the nebular fO_2 locally may have approached the iron-wüstite buffer, which is ~5–6 orders of magnitude more oxidizing than the equilibrium fO_2 of a gas of solar composition at high temperatures (>1000 K).

High-temperature experiments (1300–1400°C for 10–20 hr) in a Knudsen cell with metal and spinel by Dohmen et al. (1998*b*) show that Fe, Ni, and Co could have been transported into spinel by the gas phase. The reaction forming Fe-rich spinel is

$$MgAl_2O_4(s) + Fe(g) = FeAl_2O_4(s) + Mg(g) \tag{6}$$

Dohmen et al. (1998*b*) concluded that the enrichment of the Allende spinels in FeO took place by reaction (6) at high temperatures under oxidizing conditions.

Hashimoto and Grossman (1987) found that the FeO content of spinel in the Allende CAIs increases with the degree of low-temperature alteration of associated phases (e.g., melilite or anorthite) to nepheline, sodalite, and phyllosilicates. Because it is commonly accepted that phyllosilicates formed during aqueous activity in an asteroidal environment, it seems plausible that oxidation of Fe and its diffusion into spinel occurred during planetary metamorphism (McSween 1977). It is, therefore, unclear to what extent spinels in Allende inclusions have acquired FeO during high-temperature oxidation and to what extent during low-temperature alteration.

D. Loss of Ca from CAIs and Chondrules and Formation of Secondary Ca-Fe-bearing Silicates

It has been shown that CAIs and chondrules in Allende experienced alteration resulting in formation of secondary Ca-, Na- and Fe-bearing minerals, such as wollastonite ($CaSiO_3$), andradite ($Ca_3Fe_2Si_3O_{12}$), diopside-hedenbergite pyroxenes ($CaMgSi_2O_6$-$CaFeSi_2O_6$), grossular ($Ca_3Al_2Si_3O_{12}$), anorthite ($CaAl_2Si_2O_8$), nepheline ($NaAlSiO_4$), sodalite ($Na_4(AlSiO_4)_3Cl$), and phyllosilicates (MacPherson et al. 1988; Kimura and Ikeda 1995; Krot et al. 1998*d*). Largely based on textural observations, Allen et al. (1978) and MacPherson et al. (1985) inferred that these minerals formed by rapid growth from a gaseous phase. Because several of these minerals cannot condense from a canonical solar nebula gas (Grossman 1972), it has been postulated that they condensed from a gas of nonsolar composition ($H_2O/H_2 \sim 5 \times 10^{-1}$–$5 \times 10^{-2}$). However, subsequent thermodynamic calculations of condensation from nonsolar gas under various oxidizing conditions failed to reproduce these secondary Ca-rich minerals (Wood and Hashimoto 1993; Petaev and Wood 1998).

During the alteration, Ca and Al were lost, whereas Si, Mg, Fe, Na, Cl, and H_2O were introduced into CAIs (e.g., MacPherson et al. 1981; Hashimoto and Grossman 1987; McGuire and Hashimoto 1989). Hashimoto (1992) showed that in oxidizing gases Ca, Al, Si, Mg, and

Fe can be volatilized as hydroxides [$Ca(OH)_2$, $Al(OH)_3$, etc.]. Assuming vapor-phase transport, Hashimoto (1992) estimated a lower limit for the Allende CAI alteration time. His calculations, done for a highly oxidized gas ($H_2O/H_2 \sim 500 \times$ solar) at 10^{-3} bar total pressure, gave alteration times of 2 yr (1500 K) to 500 yr (1200 K). Lower temperature (< 950 K; Hutcheon and Newton 1981), lower total pressure ($\sim 10^{-5}$–10^{-6} atm; Wood and Morfill 1988) and lower fO_2 translate into much longer reaction times (10^5–10^6 yr). Hashimoto (1992) concluded that this timescale is implausibly long, because a zone of locally oxidized gas could be expected to dissipate within hours or days by convection in the nebula.

An alternative model of Ca mobilization from CAIs and chondrules involves low-temperature reactions in an asteroidal environment in the presence of aqueous solution (Armstrong et al. 1982; Meeker et al. 1983; Greenwood et al. 1994; Krot et al. 1998*d*). The minor phyllosilicates in Allende CAIs and chondrules (Hashimoto and Grossman 1987; Brearley 1997) and the high mobility of Ca, Si, Mg, Fe, and Al in aqueous solutions (Brearley and Duke 1998) are consistent with this model. Krot et al. (1998*d*) described wollastonite, diopside-hedenbergite pyroxenes, and andradite in veins in and rims around chondritic clasts (dark inclusions) in Allende (Fig. 3). Because veins in and rims around chondritic clasts can be formed only after aggregation and lithification of these rocks, Krot et al. (1998*d*) concluded that wollastonite, diopside-hedenbergite pyroxenes, and andradite formed in the presence of aqueous solution below 300°C in an asteroidal environment.

To summarize, the observed loss of Ca from CAIs and chondrules, and formation of the secondary Ca- and Fe-rich minerals in Allende, require the presence of an oxidizing fluid. Arguments favoring nebular and asteroidal environments have been presented, but no agreement has been reached yet.

E. Metal-Oxide-Sulfide Assemblages in CAIs (Fremdlinge)

Metal-oxide-sulfide assemblages in CAIs, largely studied in Allende, were originally inferred to have an exotic origin unrelated to the host CAIs and were called *Fremdlinge* (German for "strangers"; El Goresy et al. 1978). Fremdlinge commonly contain minerals requiring oxidizing conditions much higher than in a solar gas during formation, such as magnetite (Fe_3O_4), awaruite (Ni_3Fe), wairuite (CoFe), powellite ($CaWO_4$), scheelite ($CaMoO_4$), and phosphates. Fremdlinge were once believed to have formed as pristine nebular condensates of Pt-group element-rich metallic phases, Ni-Fe alloys, sulfides, and oxides that were later mixed and incorporated into CAIs (Armstrong et al. 1985, 1987; Bischoff and Palme 1987). As a result, Fremdling mineralogy has been considered as indicator of oxidizing conditions in the solar nebula prior to CAI formation (Rubin et al. 1988). Although it was subsequently shown (Blum et al. 1989; Palme et al. 1994) that Fremdlinge formed from homogeneous metallic

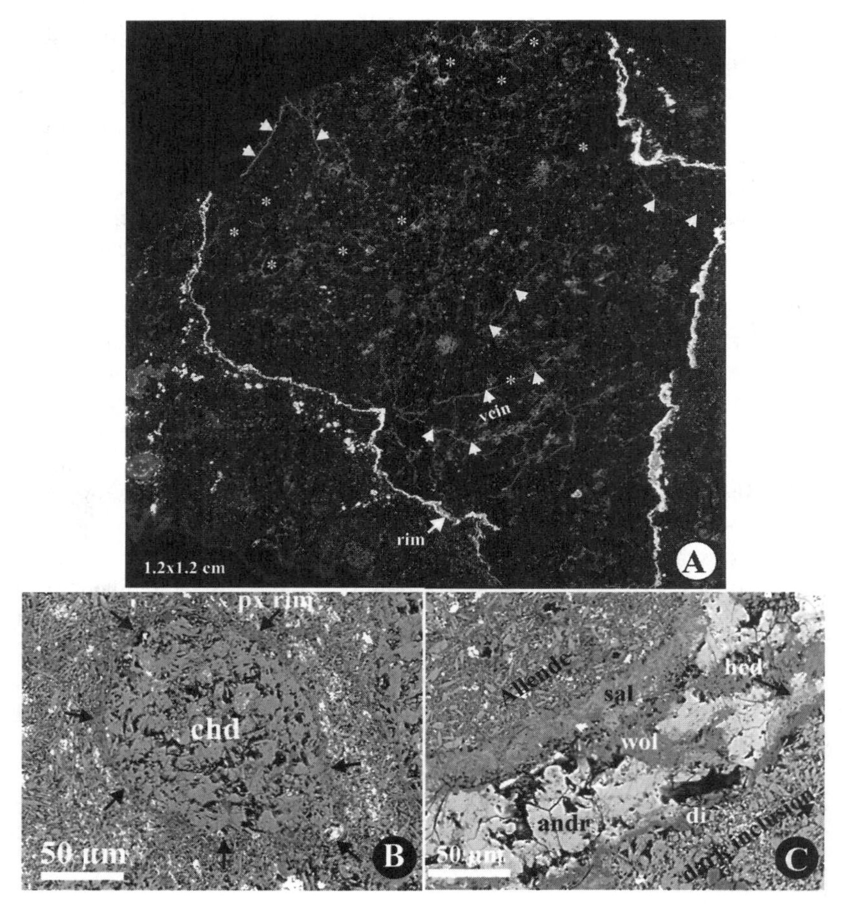

Figure 3. X-ray elemental map in Ca Kα (a) and backscattered electron images (b, c) of the lithified chondritic clast in Allende. (a) The clast consists of chondrules (indicated by stars) surrounded by Ca-rich rims; it is crosscut by Ca-rich pyroxene veins (indicated by arrows) and surrounded by a continuous Ca-rich rim. (b) Chondrule (chd) replaced entirely by fayalitic olivine and surrounded by a Ca-rich pyroxene rim (indicated by black arrows). (c) Ca-rich rim around the clast consists of andradite (andr), wollastonite (wol), hedenbergitic (hed), salitic (sal) and diopsidic (di) pyroxenes. Krot et al. (1998*b*) inferred that the clast experienced two stages of low-temperature alteration after lithification. An early alteration resulted in replacement of chondrules by fayalitic olivine. Calcium lost from the chondrules was redeposited as rims around chondrules and veins. Then clast was excavated from its original location (probably CV3 parent body) and introduced into Allende. *In situ* alteration resulted in dissolution of Ca-bearing phases inside the clast and their precipitation along the boundary with the host meteorite.

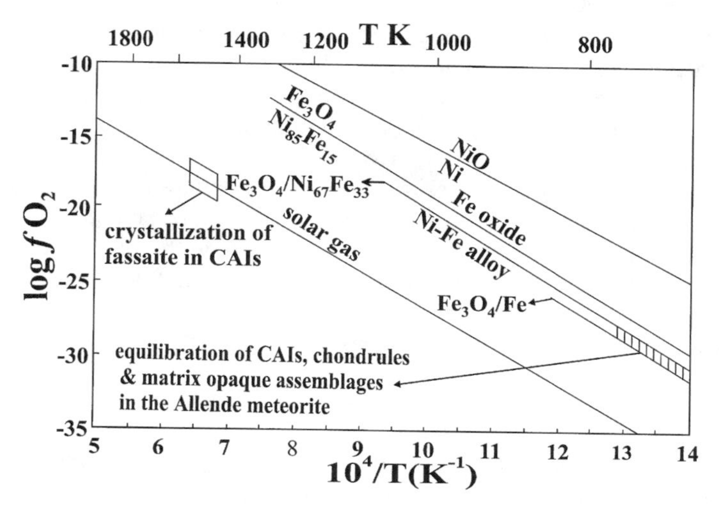

Figure 4. Equilibrium conditions (fO_2 *vs.* temperature) for opaque assemblages in CAIs, chondrules, and matrix in Allende. Opaque assemblages in all three components reflect low-temperature conditions, which are consistent with an asteroidal environment. The $Ni_{67}Fe_{33}$ alloy and coexisting magnetite observed in most Allende CAIs define an fO_2 1.5 log units above the FeO-Fe_3O_4 buffer and 6 log units above the fO_2 defined by the solar gas at temperatures below 800 K (from Blum et al. 1989).

nodules as a result of exsolution, oxidation, and sulfidization reactions *in situ*, there is no agreement on the place and temperature of this oxidation. Based on the similar ranges of equilibration temperatures and fO_2 of the opaque assemblages in CAIs, chondrules, and matrix in Allende, which are significantly higher than those in solar gas (Fig. 4), Blum et al. (1989) concluded that oxidation of Fremdlinge occurred in the CV3 asteroid. On the other hand, Palme et al. (1994) argued that conditions of oxidation of the Allende CAIs were quite variable and inferred that oxidation occurred before final accretion of the meteorite.

To summarize, Fremdling mineralogy reflects late-stage oxidation of metal assemblages either in the solar nebula or in an asteroidal environment; no consensus has been reached yet.

F. Fayalite

In a gas of solar composition, FeO-poor olivine (forsterite, Mg_2SiO_4) is the first major mineral to condense. At about the same temperature, Fe will condense as metal (FeNi alloy). With decreasing temperatures, most of the forsterite is converted to enstatite [$Mg_2(SiO_3)_2$]. At temperatures of 500 K, enstatite is calculated to react with FeNi and H_2O to produce FeO-rich olivine. At this temperature, however, reaction of enstatite with metal, which is diffusion controlled, is so slow that formation of FeO-rich olivine grains larger than 1 μm appears impossible (Palme and Fegley 1990).

There are two possible ways to circumvent this problem. Formation of FeO-rich olivine at higher temperatures by condensation from a gas more oxidized than the solar gas or by reaction of fluids on the parent body at elevated temperatures. These two endmember models will be used below in discussing the various occurrences of FeO-rich olivine.

1. Fayalite in Carbonaceous Chondrites. Hua and Buseck (1995) described large grains (up to 100 μm) of nearly pure fayalite (Fe_2SiO_4) containing up to 1 wt% MnO and associated with magnetite in the Kaba and Mokoia CV3 chondrites. They inferred that these fayalite grains formed by replacement of magnetite in an oxidized nebular gas ($H_2O/H_2 \sim 2 \times 10^3$–$10^4$ times solar) at 800–1200 K:

$$2\,Fe_3O_4(s) + 3\,SiO(g) + H_2O(g) = 3\,Fe_2SiO_4(s) + H_2(g) \qquad (7)$$

Based on the textural observations, thermodynamic calculations, and O-isotopic compositions of the coexisting fayalite and magnetite, which plot close to the terrestrial fractionation line with $\Delta^{18}O_{fayalite-magnetite}$ fractionation ~20%, Krot et al. (1997, 1998*c*) concluded that fayalite formed at relatively low temperatures ($< 300°C$) during fluid-rock interaction in an asteroidal environment:

$$2\,Fe_3O_4(s) + 3\,SiO_2(aq) + 2\,H_2(g) = 3\,Fe_2SiO_4(s) + 2\,H_2O(g) \quad (8)$$

Recently, Hutcheon et al. (1998) studied the $^{53}Mn/^{53}Cr$ isotopic systematics of the Mokoia fayalites. They discovered large excesses of ^{53}Cr, which provide evidence for *in situ* decay of ^{53}Mn and show that these fayalites formed ~7–10 Myr after crystallization of Allende CAIs. Hutcheon et al. (1998) concluded that the Mokoia fayalites must have formed in an asteroidal environment and, hence, cannot be used as an oxygen barometer for nebular gas.

2. Fayalite in Silica-bearing Chondrules in Ordinary Chondrites. Brigham et al. (1986) and Wood and Holmberg (1994) described fayalite (Fa_{70-100}) replacing silica (SiO_2) in ordinary chondrite chondrules and suggested that fayalite formed by reaction between silica and a hot (>1300°C), oxidized nebular gas by the reaction

$$SiO_2(s) + Fe(s) + H_2O(g) = Fe_2SiO_4(s) + H_2(g) \qquad (9)$$

Based on the presence of troilite in the fayalite-bearing chondrules, Wasson and Krot (1994) and Krot and Wasson (1994) argued against the high-temperature formation of fayalite, because troilite must have evaporated under such conditions. These authors inferred instead that fayalite formed by reaction (9) at low temperatures in an asteroidal environment. We conclude that fayalite in ordinary chondrite chondrules cannot be used as a strong indicator of oxidizing conditions in the solar nebula.

G. Fayalitic Olivine

1. Mineralogical Observations Nebular and Asteroidal Models. Fayalitic olivine [$(Fe,Mg)_2SiO_4$] is a ubiquitous constituent of carbonaceous and ordinary chondrites. The reactions controlling the fayalite content of olivine in the presence or absence of enstatite are

$$2\ Fe(s) + Mg_2Si_2O_6(s) + H_2O(g) = 2\ FeMgSi_2O_4(s) + 2H_2(g) \quad (10)$$

and

$$Mg_2SiO_4(s) + x\ Fe(g) = Mg_{2-x}Fe_x SiO_4(s) + x\ Mg(s) \quad (11)$$

respectively. Reaction (11) has been used by Larimer (1967, 1968), who tabulated fayalitic contents (X_{Fa}) in olivine, assuming ideal solid solution of fayalite and forsterite in olivine, as a function of temperature and of the H_2O/H_2 ratio in a gas of solar composition:

$$\log_{10} X_{Fa} = \log_{10}(H_2O/H_2) - 0.74 + 1690/T \quad (12)$$

This equation shows that fayalitic content in olivine increases with decreasing temperature and with increasing oxygen fugacity of a gas in equilibrium with the olivine. In a gas of solar composition, the calculated fayalite content in olivine is ~0.4 mol% at 1000 K and ~22 mol% at 500 K. Palme and Fegley (1990) concluded that the low-temperature formation of fayalitic olivine in a solar gas is kinetically inhibited and suggested that fayalitic olivine formed at high temperature under highly oxidized conditions (see section III.G.2). High-temperature formation of fayalitic olivine has been reproduced in laboratory experiments by Nagahara et al. (1988, 1994) and Dohmen et al. (1998*a*). These experiments are discussed below (see sections III.G.2 and 4 below).

High-temperature mechanisms of fayalitic olivine formation are largely used to explain the postulated (and controversial) nebular origin of fayalitic olivine in Allende (e.g., Peck and Wood 1987; Hua et al. 1988; Weinbruch et al. 1990; Krot et al. 1995, 1997, 1998*c, d*; Brearley and Prinz 1996; Brearley 1997). Fayalitic olivine in Allende occurs as lath-shaped grains in the matrix; it also rims and veins forsterite, enstatite, and magnetite and forms halos around Ni-rich metal inclusions in forsterite (Fig. 5). Both nebular and asteroidal models have been proposed to explain these textural types of fayalitic olivine. According to the nebular models, fayalitic olivine rims, veins, and halos formed by high-temperature gas-solid reactions in the oxidized solar nebula; matrix olivine condensed from the gas (Peck and Wood 1987; Hua et al. 1988; Weinbruch et al. 1990). According to the asteroidal models, fayalitic olivine in Allende resulted from low-temperature alteration in the presence of aqueous solutions and hence does not reflect redox conditions in the solar nebula (Krot et al. 1995, 1997, 1998*a,c,d*; Brearley and Prinz 1996; Brearley 1997).

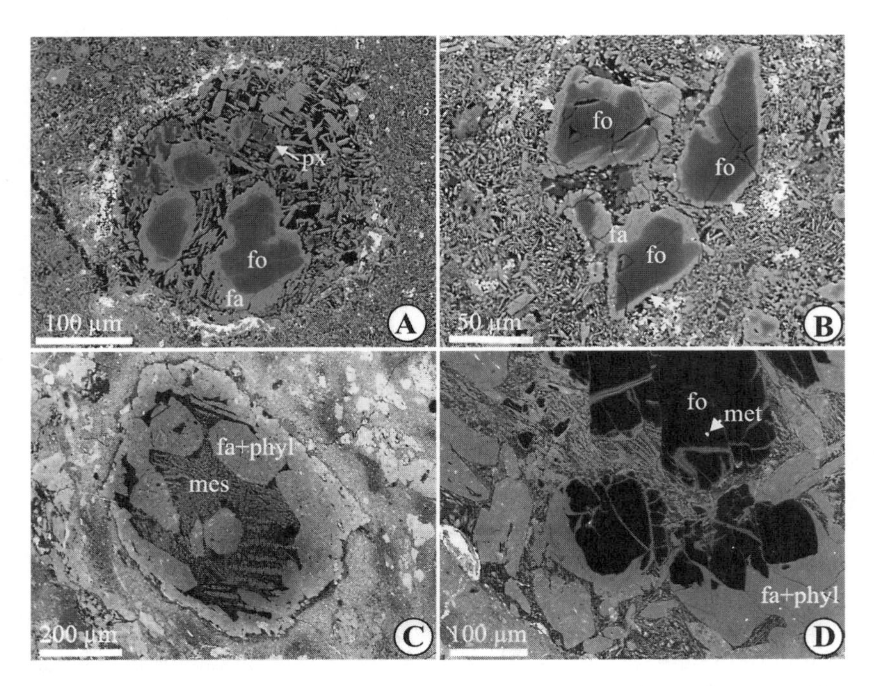

Figure 5. Backscattered electron images of fayalitic olivine in the CV3 chondrite ALHA81258 (a) and lithified chondritic clasts in the CV3 chondrites Allende (b) and Efremovka (c, d). (a) Magnesian chondrule with low-Ca pyroxene (px) and forsteritic olivine phenocrysts (fo) replaced by fayalitic olivine (fa). Fayalitic olivines around the chondrule's periphery contain inclusions of Fe,Ni sulfides. Since sulfides are not stable above ~700 K, high-temperature formation of fayalitic olivine is unlikely. (b) Fayalitic olivine rims surround three parts of a single forsterite grain, indicating that these rims formed by *in situ* replacement of the forsterite. (c, d) Chondrules replaced to various degrees by a fine-grained mixture of fayalitic olivine and phyllosilicates (phyl), suggesting that fayalitic olivine formed by low-temperature alteration in the presence of aqueous solution (modified from Krot et al. 1998*c*).

2. Condensation Calculations from an Oxidized Gas. Thermodynamic calculations by Palme and Fegley (1990) show that the first condensing olivine is always forsteritic ($Fa_{<1}$), independent of the H_2O/H_2 ratio (Fig. 6). At lower temperatures, when the major fraction of Mg and Si is condensed and at appropriately high fO_2, the Fa content of the condensed olivine will gradually increase until a temperature is reached where iron metal, wüstite (FeO), or magnetite would be stable as a separate phase. According to these calculations, the condensation of fayalitic olivine must have occurred before condensation of other Fe-bearing phases. Palme and Fegley (1990) applied these calculations to the origin of all textural types of fayalitic olivine in Allende.

Brearley and Prinz (1996) and Krot et al. (1997) found that fayalitic olivine in the matrix of Allende commonly contains inclusions of

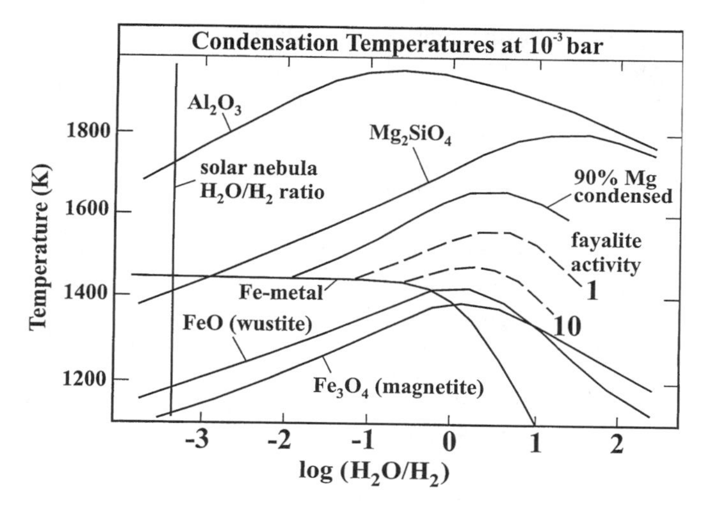

Figure 6. Calculated condensation temperatures of corundum, forsterite, fayalite, Fe-metal, wüstite, magnetite as a function of log(H_2O/H_2) ratios. Temperatures for fayalite activities of 1 mol% and 10 mol% are indicated (from Weinbruch et al. 1990).

pentlandite [$(Fe,Ni)_9S_8$] and poorly graphitized carbon, suggesting that both phases predated formation of fayalitic olivine. However, neither pentlandite nor poorly graphitized carbon are predicted to be high-temperature condensates, even at more oxidizing conditions than those in Palme and Fegley's (1990) calculations (Wood and Hashimoto 1993; Petaev and Wood 1998). Brearley (1997) described talc and biopyriboles in the Allende chondrules and inferred that these hydrous minerals predated formation of fayalitic olivine replacing forsterite and enstatite in these chondrules. Since hydrous minerals are most likely to be asteroidal in origin (Prinn and Fegley 1989) and cannot survive high-temperature processing, Brearley (1997) concluded that high-temperature condensation origin of fayalitic olivine in Allende is unlikely.

3. Vaporization-Recondensation of Forsteritic Olivine. Nagahara et al. (1988) heated forsteritic olivine (Fo_{89-94}) 1450–1600°C in a Knudsen cell kept in a vacuum chamber at $P_T = 3\times10^{-9}$ bar for 6–72 hr. The pressure and fO_2 in the Knudsen cell were not controlled. The gas flowed from the cell to lower pressure and temperature, where the gas recondensed at 1200–500°C. The residue after the experiments was stoichiometric olivine (Fo_{92-98}). The condensing phases were forsterite at ~1200–1150°C, enstatite at ~1100°C, silica at ~1000°C, mixture of Fe-bearing pyroxene and silica at ~800°C, and fayalitic olivine with lesser amounts of silica and metallic Fe at ~500°C. Nagahara et al. (1988) concluded that the low-temperature condensates are quench products from the gas vaporized from olivine of the starting composition, not equilibrium condensates; the latter should consist exclusively of fayalitic olivine (Nagahara et al. 1994).

Mineralogical observations (e.g., Nagahara 1984; Keller 1997) indicate that matrices in type 3 ordinary and carbonaceous chondrites consist of disequilibrium mineral assemblages similar to those produced by Nagahara et al. (1988). If fayalitic olivine in these matrices formed as a quench product (not as equilibrium condensate) of vaporization-recondensation of forsteritic olivine, it cannot be used as indicator of oxidizing conditions in the solar nebula.

In contrast to the matrices of most type 3 chondrites, the Allende matrix consists largely of fayalitic olivine; enstatite, forsterite, and metal are absent (Krot et al. 1995). Although matrix olivine in Allende may have formed as equilibrium condensate during vaporization-recondensation of forsteritic olivine (Nagahara et al. 1994), mineralogical observations indicate that forsteritic olivine in Allende was enriched in Fe, not depleted (e.g., Krot et al. 1997). Additional arguments against high-temperature formation of fayalitic olivine in Allende are discussed in subsection 2 above.

4. Gas-Solid Exchange Reactions. Dohmen et al. (1998*a*) heated forsterite and Fe metal at 1400°C in a vacuum for 2 hr and showed that forsterite incorporates gaseous Fe produced by evaporation of Fe metal, while Mg is released to the gas phase. The formation of the Fe-rich olivine is explained by the following two steps. First, forsterite evaporates and dissociates into gaseous Mg, SiO, and O_2:

$$MgSi_{0.5}O_2(s) = Mg(g) + 0.5\ SiO(g) + \tfrac{3}{4}\ O_2(g) \quad (13)$$

Second, mixing of evaporation products with Fe vapor produced by evaporation of Fe metal results in condensation of fayalitic olivine:

$$Mg(g) + Fe(g) + 0.5\ SiO(g) + \tfrac{3}{4}\ O_2(g) = (Mg, Fe)Si_{0.5}O_2(s) \quad (14)$$

Summing the two reactions, the net reaction may be written as

$$MgSi_{0.5}O_2(s) + Fe(g) = FeSi_{0.5}O_2(s) + Mg(g) \quad (15)$$

The composition of the olivine produced is proportional to the ratio of partial pressures of Fe and Mg: $X_{Fa}/X_{Fo} = K \times (P_{Fe}/P_{Mg})$. Providing that P_{Fe} is buffered by Fe metal, it follows that the fayalitic content of the olivine is inversely proportional to P_{Mg} in the gas phase. Although P_{Mg} is controlled by variations in fO_2 (the higher the fO_2 the lower P_{Mg} and the higher X_{Fa}), it was found (Dohmen et al. 1998*a*) that X_{Fa} in olivine is largely controlled by variable Fe/Mg ratios in the gas phase, which result from the difference in kinetics of evaporation of Fe metal (fast) and forsterite (slow). Since the kinetics of the forsterite evaporation plays a key role in these experiments, it is difficult to derive from the fayalite content of the olivine the fO_2 at which these olivines formed.

Dohmen et al. (1988*a*) inferred that their experiments simulated growth of fayalitic olivine rims around forsterite grains in Allende, which are illustrated in Fig. 5. Arguments against high-temperature formation of fayalitic olivine rims in Allende are discussed in section III.G.2 above.

H. Variations in fO_2 during Chondrule Formation

There are two major groups of chondrules: FeO-poor and FeO-rich. It has been suggested that FeO-poor chondrules formed in solar gas, whereas FeO-rich chondrules formed under oxidizing conditions ($H_2O/H_2 \sim 10^4$–$10^6 \times$ solar) (e.g., Huang et al. 1996; Hewins et al. 1997). Mineralogical observations indicate that chondrules formed by multiple episodes of incomplete melting of mineralogically and chemically heterogeneous precursor materials (e.g., Grossman 1988, 1996). Although brief, localized variations in fO_2 may have accompanied chondrule formation (e.g., Connolly and Love 1998), there is no evidence for an equilibrium between nebular gas and chondrule melts (e.g., Jones 1990, 1996; Zanda et al. 1994). Some FeO-poor chondrules show evidence for internal reduction of FeO-bearing silicates by carbon-rich solid precursors (e.g., Connolly et al. 1994; Hanon et al. 1996). As a result, the inferred fO_2 recorded by chondrules could be an artifact of their precursor composition; that is, chondrules may have behaved as internally buffered systems. We infer that chondrules cannot be used as reliable indicators of redox conditions in the solar nebula.

I. Mineral Indicators of the Reduced Solar Nebula

The chalcophile behavior of typically lithophile elements such as Mg, Ca, Ti, Mn, Cr, K, and Na, which form various sulfides in enstatite chondrites [e.g., oldhamite, CaS; niningerite, MgS; alabandite, MnS; Ti-bearing troilite, FeS; heideite, $Fe_{1+x}(Ti,Fe)_2S_4$; caswellsilverite, $NaCrS_2$; djerfisherite [$K_3(Cu,Na)(Fe,Ni)_{12}S_{14}$]; daubreelite, ($FeCr_2S_4$)]; osbornite (TiN); and high contents of Si in Fe, Ni metal provide clear evidence for reducing conditions in the region of enstatite chondrite formation (e.g., Keil 1968). However, model-independent estimates of fO_2 of a gas equilibrated with the enstatite chondrite minerals in the solar nebula are absent.

IV. MECHANISMS PROPOSED TO ACCOUNT FOR LARGE fO_2 VARIATIONS IN THE SOLAR NEBULA

The foregoing discussion shows that a wide spectrum of oxidation states, ranging from several orders of magnitude more oxidizing than a solar gas to several orders of magnitude more reducing than a solar gas is recorded in chondritic meteorites. We now consider two key questions: (1) How were variations in nebular fO_2 produced? and (2) What were the spatial and temporal extents of these more oxidizing and reducing regions of the solar nebula?

A. Oxidized Nebular Gas

1. Enhancement of the Dust/Gas and Ice/Gas Ratios. One possible mechanism for production of oxidizing conditions in the solar nebula is enhancement of the dust/gas ratio, the ice/gas ratios, or both. In a solar gas, oxygen is partitioned between CO (48 at%), H_2O (37 at%), and rock-forming minerals (15 at%). The oxygen in CO is essentially locked up and unavailable for reactions because of the very strong triple bond in CO, whereas the oxygen in water vapor and rock-forming minerals can potentially take part in other chemistry. We know that the dust/gas and ice/gas ratios were enriched over the solar values during accretion of planets, satellites, asteroids, and comets. It is also possible that dust- and/or ice-enriched parcels of presolar material fell into the nebula during earlier stages of nebular evolution. Furthermore, laboratory studies of CAI and chondrule cooling rates indicate that some of these objects cooled more slowly than if they had radiated into free space. Dust-enriched regions of the nebula are needed to provide the slow cooling rates.

Table 5 of Fegley and Palme (1985) illustrates how enrichments of ice and dust lead to higher fO_2 in the solar nebula. The H_2O/H_2 ratios are tabulated there because this ratio is a temperature-independent indicator of how oxidizing (or reducing) the nebular gas was. The respective fO_2 at a given temperature can be calculated from the H_2O/H_2 ratio using equation (3). The H_2O/H_2 ratios inferred from the Mo and W depletions in CAIs correspond to dust (or ice enrichments) in the range of 50–500 times the solar dust/gas or ice/gas ratios. Similar enrichments are needed to explain the high-temperature formation of fayalitic olivines (if high-temperature model of fayalitic olivine formation is correct; for discussion see section III.G).

Dust and ice enrichments have other consequences in addition to increasing the nebular oxygen fugacity. These include increases in the opacity, viscosity, and coagulation rates in the dust- or ice-enriched regions. Furthermore, dust and ice enrichments also lead to changes in the condensation chemistry of the major rock-forming elements, although condensation calculations by different groups give different predictions (Wood and Hashimoto 1993; Yoneda and Grossman 1995). In this respect the difference between dust and ice enrichments is that dust enrichments also lead to enrichments in Mg, Si, Fe, and so on, whereas ice enrichments lead only to increases in fO_2.

2. *Nebular Photooxidation.* Alternatively, the Mo and W depletions found in CAIs may be the result of photooxidation of nebular water vapor in x-winds (Shu et al. 1996) or in low opacity regions of the inner solar nebula, such as the nebular photosphere, where backscattered solar H I Lyα (121.6 nm), He I (58.4 nm), and He II (30.4 nm) UV radiation may drive nebular photochemistry (Gladstone 1993; Gladstone and Fegley 1997). Photolysis of water vapor and hydroxyl radicals via the reactions

$$H_2O + h\nu \rightarrow OH + H \quad (16)$$

$$OH + h\nu \rightarrow O + H \quad (17)$$

may produce OH radical and O atom concentrations significantly higher than the thermochemical equilibrium values in a solar gas, while the H atoms recombine with each other via

$$H + H + M \rightarrow H_2 + M \quad (18)$$

where M is any third body (statistically H_2, He, or H). The enhanced OH and O concentrations could then drive reactions exemplified by

$$Mo(g) + O \rightarrow MoO(g) \quad (19)$$

$$W(g) + O \rightarrow WO(g) \quad (20)$$

which increase the Mo and W oxide vapor concentrations above the thermochemical equilibrium values in a solar gas. Shu et al. (1996) note that this general concept is supported by observations of atomic O in the outflows from young T Tauri stars. However at present, no quantitative modeling has been done to work out the details of a photochemical oxidation scheme for Mo and W, and this must be done to test this model.

B. Reduced Nebular Gas

There are two principal mechanisms to reduce the solar nebula gas: addition of C-rich dust or removal of water from the inner solar system (Larimer 1968, 1975; Larimer and Bartholomay 1979). A measure of the redox state is the C/O ratio, which is ~0.5 for a solar gas (Grevesse and Noels 1993). At high temperatures and low pressures in a solar gas, all carbon is in CO while oxygen is about evenly divided between CO and H_2O, and oxide condensation occurs via net reactions such as

$$2\,Al(g) + 3\,H_2O = Al_2O_3(s) + 3\,H_2 \quad (21)$$

An increase in C/O ratio leads to the formation of more CO, so less H_2O is available for oxide forming reactions. Oxides then form at lower temperatures when reaction (4) proceeds to the right and supplies H_2O. Therefore, equilibrium condensation temperatures of oxides decrease with increasing C/O ratio. At some critical C/O ratio, the suite of initial condensates for the major elements switches from oxides and silicates to carbides, nitrides, and sulfides.

We used the CONDOR code (Lodders and Fegley 1993) to study the effect of increasing the C/O ratio (by either pathway) on types of condensates and condensation temperatures and computed the condensation temperatures for major elements at a total pressure of 10^{-3} bar.

Major element condensations as a function of C/O ratio as the O abundance is systematically decreased from the solar value are illustrated in Fig. 7. Initial condensates from a solar gas (C/O ~ 0.5) are corundum; hibonite; grossite, $CaAl_4O_7$; perovskite, $CaTiO_3$; melilite,

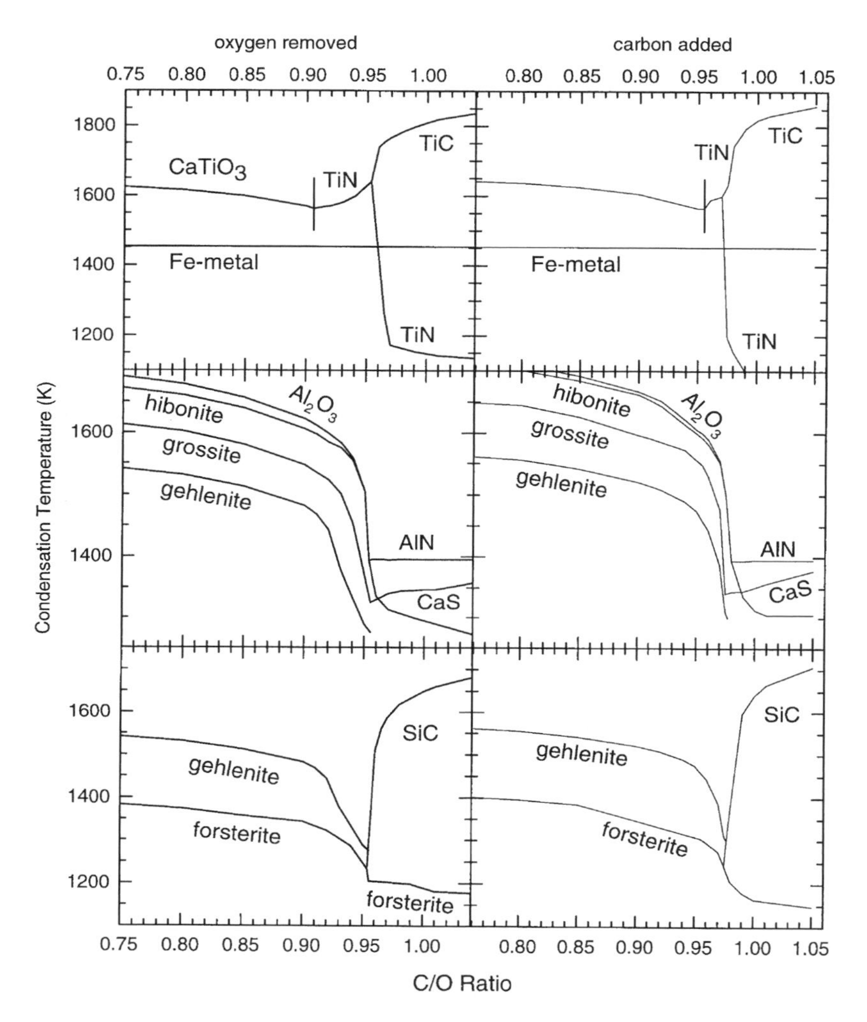

Figure 7. Condensation temperatures for initial major element condensates as a function of C/O ratio at a total pressure of 10^{-3} bar. The left panel shows the condensation temperatures as the C/O ratio is increased by removing oxygen from a solar-composition gas, and the right panel shows the respective condensation temperatures as the C/O ratio is altered by adding carbon (from Lodders and Fegley 1997).

$(Ca,Na)_2(Al,Mg)(Si,Al)_2O_7$ [shown by gehlenite, $Ca_2Al(Si,Al)_2O_7$]; spinel ($MgAl_2O_4$); forsterite; and enstatite. Iron metal condensation is independent of C/O. However, as the C/O ratio increases, the condensation temperatures of the oxides and silicates decrease. At C/O $\sim$ 0.91 to $\sim$0.95, osbornite forms instead of perovskite. Oxides of Al, Ca, and Si remain the initial condensates of these elements, because enough H_2O is still available for reactions such as equation (21). At C/O > 0.95, TiC replaces TiN as an initial condensate, AlN and CaS replace Ca-Al oxides, and SiC replaces gehlenite. Graphite and cohenite [$(Fe,Ni)_3C$] condensation requires $C/O > 1$.

The condensation temperatures for major element-forming minerals as a function of C/O ratio for the case in which carbon is increased above solar values are shown in the right panel in Fig. 7. In general, the condensation temperatures are somewhat higher, and the new suite of reduced condensates appears at higher C/O ratios than in the case of oxygen removal. Osbornite appears at $C/O \sim 0.95$, and TiC, AlN, and CaS start forming at $C/O \sim 0.98$.

The C/O range over which enstatite chondrites (ECs) may have formed is constrained by their mineralogy. Graphite and cohenite are not very abundant in these meteorites, indicating that the upper bound in C/O was $\sim$1, because graphite and cohenite form at $C/O > 1$. The occurrence of CaS puts the lower bound at $C/O \sim 0.95$ if the C/O ratio was altered by oxygen removal from the solar gas. On the other hand, the lower bound of the C/O ratio necessary for the suite of reduced condensates is 0.98 if carbon was added to a solar gas; this is very close to the assumed upper bound of $C/O \sim 1$.

How were reducing conditions obtained in the solar nebula? Removal of oxygen by oxide condensation increases C/O in the gas from 0.48 to $\sim$0.57, but this is too small an increase to produce reduced condensates. Another problem with removal of oxygen into silicates is that the major elements are then also condensed and are no longer available to form the reduced minerals by direct condensation. It is necessary to reduce the nebular gas before the onset of condensation to obtain enstatite chondrite minerals (e.g., CaS) as primary condensates. Plausible mechanisms to increase the C/O ratio are injection of carbonaceous dust or removal of water (Larimer 1968, 1975; Larimer and Bartholomay 1979; Baedecker and Wasson 1975). As seen above, both cases (i.e., O removal from and C addition to a solar gas) lead to reduced condensate formation, and thermodynamics alone does not unambiguously constrain which alternative is responsible for the reduction of the solar gas.

1. Adding Carbonaceous Dust to Solar Gas. Isotopic variations in chondrites indicate that the solar nebula was not homogeneous, and it is not inconceivable that there may have been "pockets" enriched in carbonaceous interstellar dust. Heating and evaporation of such carbon-rich pockets during their infall to the nebular midplane could have locally produced more reducing conditions than those obtained from evaporation of the standard presolar cloud material. Condensation from a solar gas reduced by carbonaceous dust would then produce the reduced minerals observed in ECs. The larger abundances of presolar SiC in the enstatite chondrite Qingzhen (EH3) than in the carbonaceous chondrite Orgueil (CI) (Huss and Lewis 1995) may indicate that larger amounts of carbonaceous dust, which could raise the C/O ratio, were present in the EC-forming region.

An indication that carbonaceous dust (including SiC dust) may have been responsible for the reducing conditions in the EC-forming region is the Si/Mg mass ratio of ~1.5 in enstatite chondrites, which is higher than in CI chondrites (Si/Mg ~ 1.1). The increased Si/Mg ratio in ECs could thus reflect the addition of Si from SiC. However, enstatite chondrites have normal Si isotopic compositions (Molini-Velsko et al. 1986), which argues against large amounts of SiC and, by implication, also against larger amounts of other carbonaceous dust in the formation region of enstatite chondrites.

Larimer (1988) has pointed out that the higher Si/Mg ratio in ECs points toward Mg fractionation rather than enhancement of Si, because the Si/Fe ratio in EH and CI chondrites is very similar. If so, the argument that Si was added from carbonaceous dust to ECs is weakened. The amount of carbonaceous dust required to reduce the nebular gas is very high. For example, mass returned to the interstellar medium (ISM) in our galaxy by evolved intermediate-mass stars is estimated to be 0.35 solar masses per year and 10–50% of that mass is reduced dust from carbon stars (Thronson et al. 1987). Although we do not have any observations of the ratio of carbon dust and SiC dust in the ISM, the major sources of the reduced dust are cool carbon stars, which preferentially show SiC dust emission, and there is evidence in only a few cases for carbon dust in their stellar outflows to the ISM (Lodders and Fegley 1997). The reduced dust may undergo processing in the ISM, but the survival of presolar SiC and carbon dust in meteorites suggests that SiC is not "lost" and is expected to be present in any carbonaceous dust pocket in the nebula. We may reasonably expect Si isotopic anomalies in ECs if the nebula gas were reduced by carbonaceous dust in the EC formation location.

Thus, although the presence of reduced presolar dust is ubiquitous in all chondrites, there is no clear evidence that enhanced carbonaceous dust addition was responsible for the increased C/O ratio in the solar nebula that is required for the condensation of EC minerals. Another drawback of dust addition to solar nebula gas is that the C/O range over which EC minerals form is very narrow (~0.98 to ~1; see above), and it may appear as mere coincidence that enstatite chondrites ever formed. In the case of oxygen removal, the range in C/O ratios over which reduced EC minerals can form (C/O $\geq$ 0.95) is somewhat wider and possibly more favorable than in the case of carbonaceous dust addition (C/O $\geq$ 0.98).

2. Removing Oxygen from Solar Gas. Oxygen removal from the EC formation region is an alternative pathway to increase the C/O ratio (Baedecker and Wasson 1975). About half of the oxygen had to be lost to increase the C/O ratio from solar to about 1, possibly by H_2O ice condensation. Solar nebula models place water ice condensation at 5.2 AU and beyond (Cameron 1995). Early accretion of ice and rock is believed to have caused a runaway accretion of Jupiter (Lissauer 1987), so that the jovian planets served as cold traps for water ice (Stevenson and Lunine 1988).

In the case of water removal from the inner to the outer solar system, the jovian planets should be relatively enriched in water. One indication of water enrichment comes from the observed CO concentrations in the atmospheres of the jovian planets. For example, about 0.6 parts per million by volume (ppmv) CO is observed on Neptune. About 440 times the solar abundance of oxygen as water is required to explain this CO, which is formed by the reaction of water and methane in Neptune's deep atmosphere and subsequent upwelling (see Lodders and Fegley 1994). Similarly, the upper limit of ~0.01 ppmv CO on Uranus implies an overabundance of oxygen as water of up to 260 times solar. However, other heavy elements such as carbon (as CH_4) are only 41 and 32 times solar abundance in Uranus and Neptune, respectively. Thus, Neptune and (possibly) Uranus have an overabundance of water as compared to carbon, and their amount of water is also larger than that available from a solar-composition gas. Interior structure models and atmospheric chemistry models indicate that water abundances in Jupiter, Saturn, Uranus, and Neptune increase as a function of heliocentric distance. This trend may reflect more efficient cold trapping in the coolest regions of the nebula, which also implies that water was redistributed within the solar system.

Water removal from the inner to the outer solar nebula constrains the formation location of the ECs to be closer to the sun. In contrast, C dust addition could have happened anywhere in the solar system. The large enrichments of water on the outer planets also implies that the removal of water from the inner solar system was quite efficient. Early accretion of reduced matter in the inner solar nebula is also consistent with the two-component accretion models for the terrestrial planets, which postulate that terrestrial planet accretion started from material as reduced as enstatite chondrites (e.g., Wänke 1981; see also Lodders 1995). Water removal from the inner solar system could have been an efficient mechanism to produce this reduced component. The terrestrial planets now clearly do not appear to be as reduced as enstatite chondrites (with the possible exception of Mercury); this fact indicates that after accretion of the reduced components, more oxidized matter was accreted and current oxidation states of the terrestrial planets evolved.

In fact, the production of highly oxidizing and highly reducing regions in the solar nebula may be linked. During the infall of presolar gas and grains into the solar nebula, icy grains and grain mantles evaporated, first producing local enrichments of water vapor. Mixing of this water- enriched gas into the hotter inner nebula then produced highly oxidized regions. As nebular evolution and heating continued, all (or at least most) of the icy grains evaporated, and the H_2O/H_2 ratio decreased back to the solar value.

Later cooling of the nebular cloud eventually led to recondensation of icy material in outer regions of the nebula (the snowline). The position of the nebular snowline must have varied with nebular cooling, but it is generally believed that formation of ice at about 5.2 AU was responsible for the runaway accretion of Jupiter. During the cooling of the nebula, water-bearing gas from the inner nebula was desiccated as it was transported outward and ice condensed at the snowline (e.g., Stevenson and Lunine 1988). As a consequence, water depletion in the inner solar nebula led to more reducing conditions and to the formation of the highly reduced enstatite chondrites (e.g., Larimer and Bartholomay 1979) and to reduced planetesimals later accreted by the proto-Earth (e.g., Lodders 1995). The formation of highly reduced material was plausibly limited by the smaller extent of hotter regions in the inner nebula during this stage of nebular evolution. However, more theoretical modeling is needed to quantitatively address the timing and dynamics of this model for formation of highly oxidized and highly reduced regions.

V. CONCLUSIONS AND SUMMARY

Chondritic meteorites show a wide spectrum of oxidation states, ranging from highly oxidizing relative to solar gas (e.g., Mo and W depletions in CAIs and fayalitic rims in carbonaceous chondrites) to highly reducing relative to solar gas (the mineralogy of enstatite chondrites). At present it is possible to recognize (at least partially) the signatures of nebular and planetary processing in chondritic meteorites. In the carbonaceous chondrites, chemical indicators generally preserve the nebular record with greater fidelity than do mineralogical indicators (such as valence state ratios and colors). Several mineralogical indicators in carbonaceous chondrites indicate that thermal metamorphism and fluid/rock interactions have altered the original nebular signatures and left an overprint of planetary processing. However, in the enstatite chondrites, the existence of highly reduced minerals, which would be destroyed by fluid/rock and thermal processing on parent bodies, provide the best evidence of nebular processes and of the nebular oxidation state in the enstatite chondrite formation region.

Although the origin of fO_2 variations in the nebula is unknown, several plausible mechanisms have been proposed. It may be that the production of highly oxidizing regions (relative to a solar gas) is linked to the initial evaporation of icy presolar grains and that the production of highly reducing regions (relative to a solar gas) is linked to the later recondensation of icy grains and to the runaway accretion of Jupiter. However, the relative timing of fO_2 variations in the solar nebula, as well as the spatial extents of highly oxidized and highly reduced regions, remain matters of conjecture and cannot be quantified. The issue of timing is difficult to resolve because of inherent limitations in the time intervals that can be distinguished using different short-lived radionuclides (about 50% of a half-life), which are generally long relative to the estimates of 0.1–10 million years for the nebular lifetime (Podosek and Cassen 1994).

Further studies along the following lines may help to understand better the origin of oxidized components in carbonaceous chondrites and allow better estimates of fO_2 variations in the solar nebula:

1. Study of the time required at various temperatures (1000–300°C) for the blue-to-orange color transition in hibonite, to constrain alteration conditions experienced by hibonite-bearing CAIs.
2. Laboratory studies of iron transport and incorporation into olivine and enstatite at low temperatures in the presence of aqueous solutions and in dry systems.
3. Good quantitative models of oxidizing regions and reducing regions in the solar nebula.
4. More detailed theoretical models of nebular radial mixing and thermal evolution.
5. Absolute dating of secondary and primary mineralization in chondrules and CAIs using short-lived nuclides (^{53}Mn, ^{26}Al, ^{129}I), to constrain the time interval between primary and secondary mineralization and the duration of alteration.

Acknowledgments This work was supported by NASA grants NAG 5-4212 (K. Keil, P.I.) and NAG 5-4323 (B. Fegley, P.I.). We thank H. Y. McSween, Jr., and H. C. Connolly, Jr., for constructive and helpful reviews. This is HIGP publication number 1070 and SOEST publication number 4855.

REFERENCES

Allen, J. M., Grossman, L., Davis, A. M., and Hutcheon, I. D. 1978. Mineralogy, textures and mode of formation of a hibonite-bearing Allende inclusion. *Proc. Lunar Planet. Sci. Conf.* 9:1209–1233.

Allen, J. M., Grossman, L., Lee, T., and Wasserburg, G. J. 1980. Mineralogy and petrography of HAL, an isotopically-unusual Allende inclusion. *Geochim. Cosmochim. Acta* 44:685–699.

Anders, E. 1971. Meteorites and the early solar system. *Ann. Rev. Astron. Astrophys.* 9:1–34.

Armstrong, J. T., Meeker, G. P., Huneke, J. C., and Wasserburg, G. J. 1982. The Blue Angel: I. The mineralogy and petrogenesis of a hibonite inclusion from the Murchison meteorite. *Geochim. Cosmochim. Acta* 46:575–595.

Armstrong, J. T., El Goresy, A., and Wasserburg, J. T. 1985. Willy: A prize noble UR-Fremdlinge: Its history and implications for the formation of Fremdlinge and CAI. *Geochim. Cosmochim. Acta* 49:1001–1022.

Armstrong, J. T., Hutcheon, I. D., and Wasserburg, G. J. 1987. Zelda and company: Petrogenesis of sulfide-rich Fremdlinge and constraints on solar nebula processes. *Geochim. Cosmochim. Acta* 51:3155–3173.

Baedecker, P. A., and Wasson, J. T. 1975. Elemental fractionations among enstatite chondrites. *Geochim. Cosmochim. Acta* 39:735–765.

Beckett, J. R., and Grossman, L. 1986. Oxygen fugacities in the solar nebula during crystallization of fassaite in Allende inclusions (abstract). *Lunar Planet. Sci.* 17:36–37.

Beckett, J. R., Live, D., Tsay, F.-D., Grossman, L., and Stolper, E. 1988. Ti^{3+} in meteoritic and synthetic hibonite. *Geochim. Cosmochim. Acta* 52:1479–1495.

Bischoff, A., and Palme, H. 1987. Compositions and mineralogy of refractory-metal-rich assemblages from a Ca-, Al-rich inclusion in the Allende meteorite. *Geochim. Cosmochim. Acta* 51:2733–2748.

Blum, J. D., Wasserburg, G. J., Hutcheon, I. D., Beckett, J. R. and Stolper, E. M. 1989. Origin of opaque assemblages in C3V meteorites: Implications for nebular and planetary processes. *Geochim. Cosmochim. Acta* 53:483–489.

Boynton, W. V. 1975. Fractionation in the solar nebula: Condensation of yttrium and the rare earth elements. *Geochim. Cosmochim. Acta* 39:569–584.

Boynton, W. V. 1978. Rare-earth elements as indicators of supernova condensation (abstract). *Lunar Planet. Sci.* 9:120–122.

Boynton, W. V., and Cunningham, C. C. 1981. Condensation of refractory lithophile trace elements in the solar nebula and in supernovae (abstract). *Lunar Planet. Sci.* 12:106–108.

Boynton, W. V., Wark, D. A., and Ulyanov, A. A. 1986. Trace elements in Efremovka fine-grained inclusion E14: Evidence for high temperature, oxidizing fractionations in the solar nebula (abstract). *Lunar Planet. Sci.* 17:78–79.

Brearley, A. J. 1996. Nature of matrix in unequilibrated chondrites and its possible relationship to chondrules. In *Chondrules and the Protoplanetary Disk*, ed. R. H. Hewins, R. H. Jones, and E. R. D. Scott (Cambridge: Cambridge University Press), pp. 137–153.

Brearley, A. J. 1997. Disordered biopyriboles and talc in chondrules in the Allende meteorite: Possible origins and formation conditions (abstract). *Lunar Planet. Sci.* 28:153–154.

Brearley, A. J., and Duke, C. L. 1998. Aqueous alteration of chondritic meteorites: Insights from experimental low temperature hydrothermal alteration of Allende (abstract). *Lunar Planet. Sci.* 28:CD-ROM 1274.

Brearley, A. J., and Prinz, M. 1996. Dark inclusions in the Allende meteorite: New insights from transmission electron microscopy (abstract). *Lunar Planet. Sci.* 28:153–154.

Brigham, C. A., Yabuki, H., Ouyang, Z., Murrell, M. T., El Goresy, A., and Burnett, D. S. 1986. Silica-bearing chondrules and clasts in ordinary chondrites. *Geochim. Cosmochim. Acta* 50:1655–1666.

Cameron, A. G. W. 1995. The first ten million years in the solar nebula. *Meteoritics* 30:133–161.

Connolly, H. C., Jr., and Love, S. G. 1998. The formation of chondrules: Petrologic tests of the shock wave model. *Science* 280:62–67.

Connolly, H. C., Jr., Hewins, R., Ash, R. D., Zanda, B., Lofgren, G. E., and Bourot-Denise, M. 1994. Carbon and the formation of reduced chondrules. *Nature* 371:136.

Davis, A. M., and Grossman, L. 1979. Condensation and fractionation of rare earths in the solar nebula. *Geochim. Cosmochim. Acta* 43:1611–1632.

Davis, A. M., Grossman, L., and Allen, J. M. 1978. Major and trace element chemistry of separated fragments from a hibonite-bearing Allende inclusion. *Proc. Lunar Planet. Sci. Conf.* 9:1235–1247.

Davis, A. M., Tanaka, T., Grossman L., Lee, T., and Wasserburg, G. J. 1982. Chemical composition of HAL, an isotopically-unusual Allende inclusion. *Geochim. Cosmochim. Acta* 46:1627–1651.

Davis, A. M., Hashimoto, A., Clayton, R. N., and Mayeda, T. 1995. Isotopic and chemical fractionation during evaporation of $CaTiO_3$ (abstract). *Lunar Planet. Sci.* 26:317–318.

Dohmen, R., Chakraborty, S., Palme, H., and Rammensee, W. 1998*a*. Solid-solid reactions mediated by a gas phase: An experimental study of reaction progress and the role of surfaces in the system olivine + Fe-metal. *Am. Mineral.* 83:970–984.

Dohmen, R., Palme, H., Chakraborty, S., and Rammensee, W. 1998*b*. Experimental evidence for gas phase transport of siderophile elements in the solar nebula (abstract). *Lunar Planet. Sci.* 29:CD-ROM 1588.

Eisenhour, D. D., and Buseck, P. R. 1992. Transmission electron microscopy of RMNs: Implications for single-phase condensation of the refractory siderophile elements (abstract). *Meteoritics* 27:217–218.

Ekambaram, V., Kawabe, I., Tanaka, T., Davis, A. M., and Grossman, L. 1984. Chemical compositions of refractory inclusions in the Murchison C2 chondrite. *Geochim. Cosmochim. Acta* 48:2089–2105.

El Goresy, A., Nagel, K., and Ramdohr, P. 1978. Fremdlinge and their noble relatives. *Proc. Lunar Planet. Sci. Conf.* 9:1279–1303.

Fegley, B., Jr. 1986. A comparison of REE and refractory metal oxidation state indicators for the solar nebula (abstract). *Lunar Planet. Sci.* 25:220–221.

Fegley, B., Jr., and Kong, D. 1989. Mo and W depletions in CAIs in carbonaceous chondrites: A theoretical study of the effects of nebula total pressure (abstract). *Lunar Planet. Sci.* 20:279–280.

Fegley, B., Jr., and Lodders, K. 1994. Chemical models of the deep atmospheres of Jupiter and Saturn. *Icarus* 110:117–154.

Fegley, B., Jr., and Palme, H. 1985. Evidence for oxidizing conditions in the solar nebula from Mo and W depletions in refractory inclusions in carbonaceous chondrites. *Earth Planet. Sci. Lett.* 72:311–326.

Fegley, B., Jr., and Prinn, R. G. 1989. Meteoritic evidence for oxidizing conditions in the solar nebula. Presented at the Intl. Geologic Congress, Washington, D.C.

Floss, C., El Goresy, A., Zinner, E., Kransel, G., Rammensee, W., and Palme, H. 1996. Elemental and isotopic fractionations produced through evaporation of the Allende CV chondrite: Implications for the origin of HAL-type hibonite inclusions. *Geochim. Cosmochim. Acta* 60:1975–1997.

Gladstone, G. R. 1993. Photochemistry in the primitive solar nebula. *Science* 261:1058.

Gladstone, G. R., and Fegley, B., Jr. 1997. Lyα photochemistry in the primitive solar nebula. *EOS Trans. AGU* 78:F538–F539.

Greenwood, R. C., Lee, M. R., Hutchison, R., and Barber, D. J. 1994. Formation and alteration of CAIs in Cold Bokkeveld (CM2). *Geochim. Cosmochim. Acta* 58:1913–1935.

Grevesse, N., and Noels, A. 1993. Cosmic abundances of the elements. In *Origin and Evolution of the Elements*, ed. N. Prantzos, E. Vangioni-Flam, and M. Cassé (Cambridge: Cambridge University Press), pp. 14–25.

Grossman, J. N. 1988. Formation of chondrules. In *Meteorites and the Early Solar System*, ed. J. F. Kerridge and M. S. Matthews (Tucson: University of Arizona Press), pp. 680–696.

Grossman, J. N. 1996. Chemical fractionations of chondrites: Signatures of events before chondrule formation. In *Chondrules and the Protoplanetary Disk*, ed. R. H. Hewins, R. Jones, and E. R. D. Scott (Cambridge: Cambridge University Press), pp. 243–256.

Grossman, L. 1972. Condensation in the primitive solar nebula. *Geochim. Cosmochim. Acta* 36:597–619.

Grossman, L. and Ganapathy, R. 1976*a*. Trace elements in the Allende meteorite. I. Coarse-grained, Ca-rich inclusions. *Geochim. Cosmochim. Acta* 40:331–344.

Grossman, L. and Ganapathy, R. 1976*b*. Trace elements in Allende meteorite. II. Fine-grained, Ca-rich inclusions. *Geochim. Cosmochim. Acta* 40:967–977.

Hanon P., Chaussidon, M., and Robert, F. 1996. The redox state of the solar nebula: C and H concentrations in chondrules (abstract). *Meteorit. Planet. Sci.* 31:A57.

Hartmann, L. 1996. Astronomical observations of phenomena in protostellar disks. In *Chondrules and the Protoplanetary Disk*, ed. R. H. Hewins, R. H. Jones, and E. R. D. Scott (Cambridge: Cambridge University Press), pp. 13–21.

Hashimoto, A. 1992. The effect of H_2O gas on volatiles of planet-forming major elements: I. Experimental determination of thermodynamic properties of Ca-, Al-, and Si-hydroxide gas molecules and its application to the solar nebula. *Geochim. Cosmochim. Acta* 56:511–532.

Hashimoto, A. and Grossman, L. 1987. Alteration of Al-rich inclusions inside amoeboid olivine aggregates in the Allende meteorite. *Geochim. Cosmochim. Acta* 51:1685–1704.

Hewins, R.H., Jones, R. H., and Scott, E. R. D., eds. 1996. *Chondrules and the Protoplanetary Disk* (Cambridge: Cambridge University Press).

Hewins, R. H., Yu, Y., Zanda, B., and Bourot-Denise, M. 1997. Do nebular fractionations, evaporative losses or both, influence chondrule compositions? *Antarct. Meteorite Res.* 10:275–298.

Hua, X. and Buseck, P. R. 1995. Fayalite in the Kaba and Mokoia carbonaceous chondrites. *Geochim. Cosmochim. Acta* 59:563–578.

Hua X., Adam, J., Palme, H., and El Goresy, A. 1988. Fayalite-rich rims, veins, and halos around and in forsteritic olivines in CAIs and chondrules in carbonaceous chondrites: Types, compositional profiles and constraints of their formation. *Geochim. Cosmochim. Acta* 52:1389–1408.

Huang, S., Jie L., Prinz, M., Weisberg, M. K., Benoit, P. H., and Sears, D. W. G. 1996. Chondrules: Their diversity and the role of open-system processes during their formation. *Icarus* 122:316–346.

Huss, G. R., and Lewis, R. S. 1995. Presolar diamond, SiC, and graphite in primitive chondrites: Abundances as a function of meteorite class and petrologic type. *Geochim. Cosmochim. Acta* 59:115–160.

Hutcheon, I. D., and Newton, R. C. 1981. Mg isotopes, mineralogy, and mode of formation of secondary phases in C3 refractory inclusions (abstract). *Lunar Planet. Sci.* 12:492–493.

Hutcheon, I. D., Krot, A. N., Keil, K., Phinney, D. L., and Scott, E. R. D. 1998. ^{53}Mn-^{53}Cr dating of fayalite formation in the CV3 chondrite Mokoia: Evidence for asteroidal alteration. *Science* 282:1865–1867.

Ihinger, P. and Stolper, E. 1986. The color of meteoritic hibonite: An indicator of oxygen fugacity. *Earth Planet. Sci. Lett.* 78:67–79.

Ireland, T. R., Zinner, E. K., Fahey, A. J., and Esat, T. M. 1992. Evidence for distillation in the formation of HAL and related hibonite inclusions. *Geochim. Cosmochim. Acta* 56:2503–2520.

Jones, R. H. 1990. Petrology and mineralogy of type II, FeO-rich chondrules in Semarkona (LL3.0): Origin by closed-system fractional crystallization, with evidence for supercooling. *Geochim. Cosmochim. Acta* 54:1785–1802.

Jones, R. H. 1996. FeO-rich, porphyritic pyroxene chondrules in unequilibrated ordinary chondrites. *Geochim. Cosmochim. Acta* 60:3115–3138.

Keil, K. 1968. Mineralogical and chemical relationships among enstatite chondrites. *J. Geophys. Res.* 73:6945–6976.

Keller, L. P. 1997. A transmission electron microscope study of the matrix mineralogy of the Leoville CV3 (reduced-group) carbonaceous chondrite: Nebular and parent-body features (abstract). In *LPI Tech. Rep. No. 97-02*, part 1 (Houston: Lunar and Planetary Institute), pp. 31–33.

Kerridge, J. F., and Matthews, M. S., eds. 1988. *Meteorites and the Early Solar System* (Tucson: University of Arizona Press).

Kimura, M., and Ikeda, Y. 1995. Anhydrous alteration of Allende chondrules in the solar nebula. II: Alkali-Ca exchange reactions and formation of nepheline, sodalite and Ca-rich phases in chondrules. *Proc. NIPR Symp. Antarct. Meteorites* 8:123–138.

Köhler, A. V., Palme, H., Spettel, B., and Fegley, B., Jr. 1988. Fractionation of refractory metals by high temperature oxidation. *Lunar Planet. Sci.* 19:627–628.

Kornacki, A. S., and Fegley, B., Jr. 1984. Origin of spinel-rich chondrules and inclusions in carbonaceous and ordinary chondrites. Proc. 14th Lunar Planet. Sci. Conf. *J. Geophys. Res.* 89:B588–B596.

Kornacki, A. S., and Wood, J. A. 1985. Mineral chemistry and origin of spinel-rich inclusions in the Allende CV3 chondrite. *Geochim. Cosmochim. Acta* 49:1219–1237.

Krot, A. N., and Wasson, J. T. 1994. Silica- and merrihueite/roedderite-bearing chondrules and clasts in ordinary chondrites: New occurrences and possible origin. *Meteoritics* 29:707–717.

Krot, A. N., Scott, E. R. D., and Zolensky, M. E. 1995. Mineralogical and chemical modification of components in CV3 chondrites: Nebular or asteroidal processing? *Meteoritics* 30:748–776.

Krot, A. N., Scott, E. R. D., and Zolensky, M. E. 1997. Origin of fayalitic olivine rims and lath- shaped matrix olivine in the CV3 chondrite Allende and its dark inclusions. *Meteorit. Planet. Sci.* 32:31–49.

Krot, A. N., Brearley, A. J., Ulyanov, A. A., Biryukov, V. V., Swindle, T. D., Keil, K., Mittlefehldt, D. W., Scott, E. R. D., Clayton, R. N., and Mayeda, T. K. 1998*a*. Mineralogy, petrography and bulk chemical, I-Xe, and oxygen isotopic compositions of dark inclusions in the reduced CV3 chondrite Efremovka. *Meteorit. Planet. Sci.*, in press.

Krot, A. N., Petaev, M. I., and Meibom, A. 1998*b*. Mineralogy and petrography of Ca-Fe-rich silicate rims around heavily altered Allende dark inclusions. *Lunar Planet. Sci.* 29:CD-ROM 1555.

Krot, A. N., Petaev, M. I., Scott, E. R. D., Choi, B.-G., Zolensky, M. E., and Keil, K. 1998*c*. Progressive alteration in CV3 chondrites: More evidence for asteroidal alteration. *Meteorit. Planet. Sci.* 33:1065–1085.

Krot, A. N., Petaev, M. I., Zolensky, M. E., Keil, K., Scott, E. R. D., and Nakamura, K. 1998*d*. Secondary Ca-Fe-rich minerals in the Bali-like and Allende-like oxidized CV3 chondrites and Allende dark inclusions. *Meteorit. Planet. Sci.* 33:623–645.

Larimer, J. W. 1967. Chemical fractionation of meteorites. I. Condensation of the elements. *Geochim. Cosmochim. Acta* 31:1215–1238.

Larimer, J. W. 1968. An experimental investigation of oldhamite, CaS, and the petrologic significance of oldhamite in meteorites. *Geochim. Cosmochim. Acta* 32:965–982.

Larimer, J. W. 1975. The effect of C/O ratio on the condensation of planetary material. *Geochim. Cosmochim. Acta* 39:389–392.

Larimer, J. W. 1988. The cosmochemical classification of the elements. In *Meteorites and the Early Solar System*, ed. J. F. Kerridge and M. S. Matthews (Tucson: University of Arizona Press), pp. 375–390.

Larimer, J. W., and Bartholomay, M. 1979. The role of carbon and oxygen in cosmic gases: Some applications to the chemistry and mineralogy of enstatite chondrites. *Geochim. Cosmochim. Acta* 43:1453–1466.

Lewis, J. S., and Prinn, R. G. 1980. Kinetic inhibition of CO and N_2 reduction in the solar nebula. *Astrophys. J.* 238:357–364.

Lewis, J. S., Barshay, S. S., and Noyes, B. 1979. Primordial retention of carbon by the terrestrial planets. *Icarus* 37:190–206.
Lissauer, J. J. 1987. Timescales for planetary accretion and the structure of the protoplanetary disk. *Icarus* 69:249–265.
Live, D., Beckett, J. R., Tsay, F. D., Grossman, L., and Stolper, E. 1986. Ti^{3+} in meteoritic and synthetic hibonite: A new oxygen barometer (abstract). *Lunar Planet Sci.* 17:488–489.
Lodders, K. 1995. Alkali elements in the Earth's core: Evidence form enstatite meteorites. *Meteoritics* 30:93–101.
Lodders, K., and Fegley, B., Jr. 1993. Lanthanide and actinide chemistry at high C/O ratios in the solar nebula. *Earth Planet. Sci. Lett.* 117:125–145.
Lodders, K., and Fegley, B., Jr. 1994. The origin of carbon monoxide in Neptune's atmosphere. *Icarus* 112:368–375.
Lodders, K., and Fegley, B., Jr. 1995. The origin of circumstellar silicon carbide grains found in meteorites. *Meteoritics* 30:661–678.
Lodders, K., and Fegley, B. Jr. 1997. Condensation chemistry of carbon stars. In *The Astrophysical Implications of the Laboratory Study of Presolar Materials*, AIP Conf. Proc. Vol. 402, ed. E. Zinner and T. Bernatowicz (Woodbury, NY: The American Institute of Physics), pp. 391–423.
Lodders, K., and Fegley, B., Jr. 1998. *The Planetary Scientist's Companion*. (New York: Oxford University Press).
MacPherson, G. J., Grossman, L., Allen, J. M., and Beckett, J. R. 1981. Origin of rims on coarse-grained inclusions in the Allende meteorite. *Proc. Lunar Planet. Sci. Conf.* 12B:1079–1091.
MacPherson, G. J., Hashimoto, A., and Grossman, L. 1985. Accretionary rims on inclusions in the Allende meteorite. *Geochim. Cosmochim. Acta* 49:2267–2279.
MacPherson, G. J., Wark, D. A., and Armstrong, J. T. 1988. Primitive material surviving in chondrites: Refractory inclusions. In *Meteorites and the Early Solar System*, ed. J. F. Kerridge and M. S. Matthews (Tucson: University of Arizona Press), pp. 746–807.
Mason, B., and Taylor, S. R. 1982. Inclusions in the Allende meteorite. *Smithsonian Contrib. Earth Sci.* 25:30.
McGuire, A. V. and Hashimoto, A. 1989. Origin of zoned fine-grained inclusions in the Allende meteorite. *Geochim. Cosmochim. Acta* 53:1123–1133.
McSween, H. Y., Jr. 1977. Petrographic variations among carbonaceous chondrites of the Vigarano type. *Geochim. Cosmochim. Acta* 41:1777–1790.
Meeker, G. P., Wasserburg, G. J., and Armstrong, J. T. 1983. Replacement textures in CAI and implications regarding planetary metamorphism. *Geochim. Cosmochim. Acta* 47:707–721.
Molini-Velsko, C., Mayeda, T. K., and Clayton, R. N. 1986. Isotopic composition of silicon in meteorites. *Geochim. Cosmochim. Acta* 50:2719–2726.
Nagahara, H. 1984. Matrices of type 3 ordinary chondrites; primitive nebular records. *Geochim. Cosmochim. Acta* 48:2581–2595.
Nagahara, H., Kushiro, I., Mysen, B. O., and Mori, H. 1988. Experimental vaporization and condensation of olivine solid solution. *Nature* 331:516–518.
Nagahara, H., Kushiro, I., and Mysen, B. O. 1994. Evaporation of olivine: Low pressure relations of the olivine system and its implication for the origin of chondritic components in the solar nebula. *Geochim. Cosmochim. Acta* 58:1951–1963.
Palme, H., and Boynton, W. V. 1993. Meteoritic constraints on conditions in the solar nebula. In *Protostars and Planets III*, ed. E. Levy and J. I. Lunine (Tucson: University of Arizona Press), pp. 979–1004.

Palme, H., and Fegley, B., Jr. 1990. High-temperature condensation of iron-rich olivine in the solar nebula. *Earth Planet. Sci. Lett.* 101:180–195.

Palme, H., and Wlotzka, F. 1976. A metal particle from a Ca-Al-rich inclusion from the meteorite Allende, and condensation of refractory siderophile elements. *Earth Planet. Sci. Lett.* 33:45–60.

Palme, H., Hutcheon, I., and Spettel, B. 1994. Composition and origin of refractory-metal-rich assemblages in a Ca, Al-rich Allende inclusion. *Geochim. Cosmochim. Acta* 58:495–513.

Peck, J. A., and Wood, J. A. 1987. The origin of ferrous zoning in Allende chondrule olivine. *Geochim. Cosmochim. Acta* 51:1503–1510.

Petaev, M. I., and Wood, J. A. 1998. The condensation with partial isolation (CWPI) model of condensation in the solar nebula. *Meteorit. Planet. Sci.*, in press.

Podosek, F., and Cassen, P. 1994. Theoretical, observational, and isotopic estimates of the lifetime of the solar nebula. *Meteoritics* 29:6–25.

Prinn, R. G., and Fegley, B., Jr. 1989. Solar nebula chemistry; origin of planetary, satellite, and cometary volatiles. In *Origin and Evolution of Planetary and Satellite Atmospheres*, ed. S. K. Atreya, J. B. Pollack, and M. S. Matthews (Tucson: University of Arizona Press), pp. 78–136.

Rubin, A. E., Fegley, B., Jr., and Brett, R. 1988. Oxidation state in chondrites. In *Meteorites and the Early Solar System*, ed. J. F. Kerridge and M. S. Matthews (Tucson: University of Arizona Press), pp. 488–511.

Ruden, S. P., and Lin, D. N. C. 1986. The global evolution of the primordial solar nebula. *Astrophys. J.* 308:883–901.

Russell, S. S., Huss, G. R., Fahey, A. J., Greenwood, R. C., Hutchison, R., and Wasserburg, G. J. 1998. An isotopic and petrologic study of calcium-aluminum-rich inclusions in CO3 meteorites. *Geochim. Cosmochim. Acta* 62:689–714.

Shu, F. H., Shang, H., and Lee, T. 1996. Toward an astrophysical theory of chondrites. *Science* 271:1545–1552.

Stevenson, D. J., and Lunine, J. I. 1988. Rapid formation of Jupiter by diffusive redistribution of water vapor in the solar nebula. *Icarus* 75:146–155.

Stolper, E., and Ihinger, P. 1983. The color of meteoritic hibonite: An indicator of oxygen fugacity (abstract). *Lunar Planet. Sci.* 14:749–750.

Stolper, E., Paque, L., and Grossman, G. R. 1982. The influence of oxygen fugacity and cooling rate on the crystallization of Ca-Al-rich inclusions from Allende (abstract). *Lunar Planet. Sci.* 13:773–774.

Thronson, H. A., Latter, W. B., Black, J. H., Bally, J., and Hacking, P. 1987. Properties of evolved mass-losing stars in the Milky Way and variations in the interstellar dust composition. *Astrophys. J.* 322:770–786.

Wänke, H. 1981. Constitution of terrestrial planets. *Philosoph. Trans. Roy. Soc. Lond.* A303:287–302.

Wänke, H., Baddenhausen, H., Palme, H., and Spettel, B. 1974. On the chemistry of the Allende inclusions and their origin as high-temperature condensates. *Earth Planet. Sci. Lett.* 23:1–7.

Wasson, J. T., and Krot, A. N. 1994. Fayalite-silica association in unequilibrated ordinary chondrites—evidence for aqueous alteration on a parent body. *Earth Planet. Sci. Lett.* 122:403–416.

Weinbruch, S., Palme, H., Müller, W. F., and El Gorsey, A. 1990. FeO-rich rims and veins in Allende forsterite: Evidence for high temperature condensation at oxidizing conditions. *Meteoritics* 25:115–125.

Wood, J. A., and Hashimoto, A. 1993. Mineral equilibrium in fractionated nebular systems. *Geochim. Cosmochim. Acta* 57:2377–2388.

Wood, J. A., and Holmberg, B. B. 1994. Constraints placed on the chondrule-forming process by merrihueite in the Mezö-Madaras chondrite. *Icarus* 108:309–324.

Wood, J. A., and Morfill, G. E. 1988. A review of solar nebular models. In *Meteorites and the Early Solar System*, ed. J. F. Kerridge and M. S. Matthews (Tucson: University of Arizona Press), pp. 329-348.

Wulf, A. V. 1990. Experimentelle Untersuchungen zum Flüchtigkeitsverhalten von Spurenelementen in primitiven Meteoriten. Ph.D. Thesis, Universität Mainz.

Yoneda, S., and Grossman, L. 1995. Condensation of CaO-MgO-Al_2O_3-SiO_2 liquids from cosmic gases. *Geochim. Cosmochim. Acta* 57:3413–3444.

Zanda, B., Bourot-Denise, M., Perron, C., and Hewins, R. H. 1994. Origin and metamorphic redistribution of silicon, chromium and phosphorous in the metal of chondrites. *Science* 265:1846–1849.

Zolensky, M. E., Krot, A. N., and Scott, E. R. D., eds. 1997. *Workshop on Nebular and Asteroidal Modification of Chondritic Components*, LPI Technical Report Number 97-02, Part I (Houston, TX: Lunar and Planetary Institute).

THE OUTER SOLAR SYSTEM: CHEMICAL CONSTRAINTS AT LOW TEMPERATURES ON PLANET FORMATION

JONATHAN I. LUNINE
CNR-Istituto di Astrofisica Spaziale and University of Arizona

TOBIAS C. OWEN
University of Hawaii

and

ROBERT H. BROWN
University of Arizona

The past decade has seen powerful ground- and space-based techniques applied to the investigation of the composition of outer solar system bodies, including the spectacular measurements of He/H, isotope ratios, and noble gas abundances in the jovian atmosphere by the *Galileo* probe; improved deuterium and trace species measurements in the atmospheres of the giant planets and Titan; and first detection of absorption features on the surface of a Kuiper Belt object. In this chapter we review the new observations and discuss how they constrain the composition of primordial reservoirs supplying material to the giant planets, their moons, and other outer solar system bodies. Planned or anticipated planetary missions in the next decade, along with improved ground-based capability on giant telescopes, will enable models of the early evolution of the outer solar system to be tested and improved, ultimately to provide a picture of how the four giant planets and their moons came to be.

I. INTRODUCTION

The outer solar system is the province of four giant planets, their retinue of more than 60 moons, and an assortment of small bodies that represent leftover planet-forming material scattered by the giant planets. The condensation front of water ice in the later stages of evolution of the protosolar disk (also called the solar nebula) provides a convenient inner boundary for the region. This front existed roughly in the region 4–5 AU from the Sun. There is a corresponding compositional threshold in present-day objects. Inward of the front lie bodies whose composition is largely rock and metal, with water present primarily as water of hydration. The ice that sits on the surfaces of Earth, Mars, and, apparently, the Moon can be considered a volatile veneer rather than a principal planet-building material.

From 5 AU outward, however, the situation is different. Most of the solid bodies beyond 4 AU, with the principal exceptions of rocky inner satellites and rings, have bulk densities or other indicators pointing to water ice as an important or primary constituent. (We do not yet know how much, if any, water ice is contained within the Trojan asteroids or Jupiter's irregular satellites.) The giant planets themselves probably accreted spectacular amounts of water, many tens of Earth masses, though this is an inference from gravitational field data because water is elusive in the jovian atmosphere and buried too deep for quantitative detection in the other giants. Beyond Neptune lie at least two dynamically distinct "junkyards" containing remnants of the rocky and icy planetesimals that fed the voracious appetites of the growing giants; today these regions supply comets to the inner solar system.

The grand scale of the outer solar system, spanning 10 times the linear extent of the inner solar system, and the presence of water as a condensable may well be interrelated. Although still poorly understood, the formation of giant planets probably was aided, if not triggered by, the large amount of condensed material afforded by the presence of water ice (Stevenson and Lunine 1988; Boss 1995). Although giant planets exist at very small orbital distances from other solar-type stars, the consensus is that these bodies formed further out and spiraled inward to their current orbits (Lin et al. 1996; Trilling et al. 1998). As these bodies grew to their present masses, they opened gaps in the surrounding protoplanetary disk, scattered planetesimals gravitationally, and in general made large annular swaths of the nebula unstable for smaller solid bodies. Thus, intrinsically the building of planets tens or hundreds of times more massive than the terrestrial planets involved great distances.

These great distances, in turn, make detailed study of the outer solar system difficult either from Earth or by spacecraft. Therefore, the outer solar system is more poorly understood than the inner, despite the important clues it must hold to planet formation. Understanding the assembly of giant planets has become more imperative as the list of extrasolar giant planets has grown. The discovery of the Kuiper Belt has opened to direct study an important reservoir of planetesimals not ejected or destroyed by the giant planets. The molecular and isotopic compositions of icy phases in the outer solar system provide bridges both to the physical properties in the nascent molecular clouds and to volatile materials that seeded the Earth with water and organics. For these reasons and others, the outer solar system is worthy of increased efforts toward exploration and characterization.

The purpose of this chapter is to summarize some of the highlights of the compositional information derived from observational activities since the publication of *Protostars and Planets III*, and to interpret them in terms of models of solar nebula processes and giant planet formation. Limitations of space demand selectivity in the choice of data to emphasize and models to present. However, the references provided at the end are adequate to open the door to more detailed research by interested readers.

II. DATA

A. *Galileo* Probe and Other Jupiter Measurements

The analysis of the rich harvest of results from the 1995 descent of the *Galileo* Probe through Jupiter's atmosphere is far from complete, and remote sensing data on the jovian atmosphere continue to be returned by the Orbiter. Some basic measurements have been established, however, that tend to support the current paradigm for giant planet formation: Contributions from solid planetesimals are mixed with captured solar nebula gases, and the mixture is subsequently modified by chemistry occurring within the planet itself. The solar nebula should have contributed the overwhelming proportions of hydrogen, helium, and neon, as these gases are not easily retained in planetesimals. Just what the planetesimals carried and in what proportions can in principle be deduced from the composition of the jovian atmosphere today. One can then compare this derived planetesimal composition with that of extant remnants from the early solar system such as comets and meteorites whose volatiles are reasonably well characterized.

1. Hydrogen and Helium. The ratio of helium to hydrogen in Jupiter's atmosphere has been established by both the *Galileo* Probe Mass Spectrometer (GPMS) (Niemann et al. 1996) and the helium abundance detector (HAD) (Von Zahn and Hunten 1996) as $^4He/H_2 = 0.157$. The mass fraction is then $Y/(X+Y) = 0.238 \pm 0.007$, where X and Y are the mass fractions of $H + H_2$ and He, respectively. This is significantly smaller than the protosolar value of $Y/(X+Y) = 0.280$ that would have existed in the solar nebula (Proffitt 1994). The difference may be a result of the formation of helium "raindrops" in the planet's deep interior at a pressure of one to a few megabars. These fall to deeper levels before dissolving, thereby depleting Jupiter's outermost envelope in He (Stevenson and Salpeter 1977). The process is far more evident in Saturn than in Jupiter, because the former's luminosity is much greater than expected after 4.5 Gyr of cooling, but it is at least plausible that the process started (albeit later) in Jupiter as well. Roulsten and Stevenson (1995) predicted that neon would dissolve in the helium drops and would therefore also be deficient. Detecting this gas for the first time, the GPMS indeed found the neon abundance to be only $\frac{1}{10}$ the solar value.

The isotope ratios in these abundant gases should correspond to values that existed in the interstellar cloud from which the solar system formed. Jupiter is too cool for nuclear reactions to occur in its interior, and the helium rainout is not predicted to fractionate the isotopes. The value of D/H measured by the GPMS in jovian hydrogen is D/H $= 2.6 \pm 0.7 \times 10^{-5}$ (Mahaffy et al. 1998). Over the intervening 4.5 Gyr since solar system formation, D/H in interstellar hydrogen should have steadily decreased as a result of the conversion of D to ^{3}He in stars. Its present value in the local interstellar medium (ISM) is D/H $= 1.6 \pm 0.2 \times 10^{-5}$ (Linsky 1996), confirming this expectation.

The jovian value of ^{3}He/^{4}He $= 1.66 \pm 0.05 \times 10^{-4}$ (Mahaffy et al. 1998) agrees closely with the value for primitive helium determined in meteorites: ^{3}He/^{4}He $= 1.5 \pm 0.3 \times 10^{-4}$ (Geiss 1993). This indicates that possible fractionation processes during the trapping of helium in the meteorites or during the helium "rainout" in Jupiter's interior have indeed been negligible. There is not yet a well-determined value for ^{20}Ne/^{22}Ne from the *Galileo* data.

If there are no additional nuclear reactions supplying ^{3}He or depleting D, and no mixing with other parts of the galaxy, we expect (^{3}He + D)/H in the interstellar gas to be time invariant at our distance from the galactic center. Measurements by the *Ulysses* spacecraft yield ^{3}He/H $= 2.48^{+0.85}_{-0.79} \times 10^{-5}$ in the local ISM (Geiss and Gloeckler 1998). Combining this with Linsky's (1996) measurement of D/H, we have the current interstellar ratio (^{3}He + D)/H $= (2.5 + 1.6) \times 10^{-5} = 4.1 \pm 1 \times 10^{-5}$. The jovian values give (^{3}He + D)/H $= 3.9 \pm 0.7 \times 10^{-5}$. Within the uncertainties they indeed agree with each other over 4.5 Gyr, about one-third of the age of the Galaxy.

2. Carbon and Sulfur. The abundances of the heavier elements in Jupiter's atmosphere are still under review. It is clear that carbon is enriched by a factor of 2.9 ± 0.2 times the solar value (Niemann et al. 1996). This determination was made in methane, which does not condense in Jupiter's atmosphere. The next best-determined abundance is that of sulfur, in the form of H_2S. This gas will combine with NH_3 to form NH_4SH (Wildt 1937), which in turn can condense to make clouds (Lewis 1969). These clouds should occur at 2.2 bar for a solar value of S/H and 2.7 bar for 3 times the solar sulfur abundance (Weidenschilling and Lewis 1973; Atreya and Romani 1995). Below the clouds, one expects H_2S to reach the globally mixed value for the planet, yet the GPMS did not even detect H_2S at altitudes above the 4-bar pressure level, setting an upper limit of 6×10^{-7} (Niemann et al. 1998). This was consistent with the nephelometer showing only traces of clouds, not the expected thick cloud bank (Ragent et al. 1996), and the depletion of NH_3 and H_2O as well (Niemann et al. 1996, 1998; Sromovsky et al. 1996). Evidently the *Galileo* Probe's descent through a 5-micron "hot spot" revealed a local region on Jupiter in which all condensable species were strongly depleted. This would be consistent with a model for these hot spots as massive cells of descending air that have been thoroughly desiccated by ascent to high altitudes (Owen et al. 1997; Atreya et al. 1997). However, the details of this process for the spatial scales and abundances found by *Galileo* remain to be worked out (Showman and Ingersoll 1996).

Further support for the downdraft idea comes from the H_2S abundance found by the GPMS at lower levels: It rises from a value of 0.23 relative to solar at 8.7 bar to about 2.5 times solar at 16 bar. It then remains constant until termination of probe communications at 22 bar (Niemann et al. 1998). The constancy at the end indicates that the probe was then sampling the globally mixed value of H_2S, presumably as a result of entrainment of "normal" air into the downdraft (Owen et al. 1997; Atreya et al. 1997). Unfortunately, it was not possible to measure the ammonia abundance with the GPMS. Also, the water vapor value, while rising by a factor of 10 between 10 and 20 bar pressure, never reached a constant mixing ratio with increasing depth and remained well below solar even at the last received data point. Thus, the GPMS cannot give us the mixing ratios of N and O at this time.

3. Nitrogen and Oxygen. The Net Flux Radiometer derived a value of N from NH_3 of solar to twice solar at 3–6 bar and 20 times less than this at 1 bar (Sromovsky et al. 1996, 1998). At 3–6 bar, the measurement is roughly consistent with the global abundance of $\approx$1.3 times solar derived from Earth-based observations of Jupiter's thermal spectrum at radio wavelengths (0.6 mm–12 cm) (de Pater and Massie 1985). In striking contrast, Folkner et al. (1998) have reported a mixing ratio for NH_3 of 3.6 times solar at 9 bar from an analysis of the attenuation of the Probe's radio transmission. Indeed, even three times the solar abundance of NH_3 will not fit the radio data, so either there is some unexplained error in the probe transmission data, there is an additional source of opacity that becomes important at pressures greater than 5 bar, or ammonia is significantly depleted globally at pressure levels sampled by the ground-based measurements. Steffes (private communication) has just discovered that phosphine has a high opacity at the wavelengths corresponding to the probe's transmission. It thus appears highly probable that this will contribute to the Folkner et al. (1998) result, which must then be seen as an upper limit on the global ammonia abundance. Clearly more work is needed on this important problem.

The NIMS instrument on the *Galileo* Orbiter has demonstrated that the H_2O abundance increases dramatically outside the 5-micron hot spots, but it is not possible to derive quantitative global mixing ratios for H_2O from the results (Roos-Serote et al. 1998). The microwave observations from Earth do not probe sufficiently deeply to contribute an answer to this question. Thus we do not yet have a definitive value for the global oxygen abundance on Jupiter.

B. Recent Bright Comets

Comets provide an opportunity to examine outer solar system bodies at relatively close range to the Earth. However, the points of origin of comets are difficult to ascertain, so the significance of their molecular, elemental, and isotopic compositions remains in dispute. So-called long-period comets appear to originate from a region of space known as the Oort Cloud, ranging in distance from the Sun from 10^4 to 10^5 AU. The consensus view, though not necessarily a unanimous one, is that this is the region to which comets were ejected by gravitational interactions with the giant planets; their birthplace is thought to be within the zone of the giant planets (Weissman 1991). Short-period comets may arrive from two or more reservoirs. Some may be Oort Cloud objects, and others may originate in a so-called inner Oort Cloud ranging from 10^3 to 10^4 AU from the Sun. However, the majority appear to be derived from a region stretching roughly from 30 AU out to 100 AU or so, predicted by Edgeworth and more quantitatively by Kuiper (Jewitt and Luu 1998), and referred to in the current literature as the Kuiper Belt. It is understood now to consist of remnant planetesimals left over from solar system formation, residing (with some important caveats) largely in place since the beginning (Malhotra 1996). In particular, the inner part of the Belt has evolved dynamically under the influence of the giant planets, and a broader swath via collisions among objects, but fresh objects sampled from this zone are certainly less evolved dynamically than Oort Cloud comets (Duncan et al. 1995). Further discussion of the Kuiper Belt in the context of spectroscopic data is provided in the next subsection; a more comprehensive review of the current state of knowledge regarding the region is provided in the chapter by Jewitt and Luu in this volume.

The recent apparitions of Comets C/1996 B2 (Hyakutake) and C/1995 O1 (Hale-Bopp) have permitted detection of many new species and a better determination of abundances of species previously seen in other comets. Both comets are on eccentric, long-period orbits typical of Oort Cloud comets. Of particular interest from the viewpoint of the origin of outer solar system material (and terrestrial water) are the deuterium enrichments measured in H_2O and HCN. The ratio D/H in HCN in Hale-Bopp, using the James Clerk Maxwell Telescope (JCMT) in Hawaii at 362 GHz, is $2.3 \pm 0.4 \times 10^{-3}$ (Meier et al. 1998*a*). The D/H ratio in cometary H_2O is essentially identical in three Oort Cloud comets: In Halley D/H$= 3.08^{+0.38}_{-0.53} \times 10^{-4}$ from the analysis of Balsiger et al. (1995) and $3.16 \pm 0.34 \times 10^{-4}$ from Eberhardt et al. (1995). In Hyakutake D/H $= 2.9 \pm 1.0 \times 10^{-4}$ (Bockelée-Morvan et al. 1998). In Hale-Bopp D/H $= 3.3 \pm 0.8 \times 10^{-4}$ (Meier et al. 1998*b*). The Halley measurements were from the ion neutral mass spectrometer on the *Giotto* spacecraft, the Hyakutake observations were made with the Caltech Submillimeter Observatory (CSO) on Mauna Kea, and those for Hale-Bopp were made with the JCMT at the same site. The close correspondence among the values of D/H for water, from in *situ* and remote measurements on three different comets, suggests that this is a typical value for Oort Cloud comets. A detailed compilation of isotopic abundances in comets is given in the chapter by Irvine et al., this volume.

The D/H values are significant in two respects. First, D/H in HCN is significantly higher than in the H_2O, and the values for both exceed the protosolar value by more than an order of magnitude. The difference in value between HCN and H_2O indicates the absence of significant exchange of hydrogen with a warm solar nebula or giant planet subnebula (Meier et al. 1998*a*). Indeed, the deuterium enrichment is perhaps the most reliable indicator that processing of grains in the outermost solar nebula was limited. The enrichment could be preserved under sublimation and recondensation of molecular cloud grains at low temperatures (Chick and Cassen 1997; Lunine et al. 1991) but not in the face of extensive radial mixing of the solar nebula (Prinn 1993).

On the other hand, the enrichment in HCN is actually less than that generally found for this molecule in molecular clouds (van Dishoeck et al. 1993). Enrichment is most plausibly obtained through ion-molecule reactions in the diffuse cloud regions, where the ion population is significant, and these reactions are strongly temperature dependent. The HCN deuterium abundance in Hale-Bopp implies processing of the material at temperatures of 30 ± 10 K (Meier et al. 1998*a*), warmer than the canonical 10 K in the coldest, most diffuse parts of the cloud but colder than the Neptune-forming zone of the solar nebula itself, which might have been at 50 K or so (Lunine et al. 1991). Dilution of the diffuse cloud value through ion-molecule reactions in the solar nebula itself is unlikely because of the high densities and low ion fractions. Perhaps in the grains of Hale-Bopp we are seeing the results of ion-molecule reactions in somewhat denser, warmer clumps of gas evolving toward collapse and disk formation. Support for this possibility is provided by the recent observations of Hatchell et al. (1998), who found DCN/HCN values in hot molecular cores comparable to that measured in Hale-Bopp.

We do not use this correspondence to argue that cometary species were processed in a hot core region; indeed, such regions, associated as they are with massive stars, are not likely analogs for the site of solar system formation. However, they are an example of the strong environmental variations (temporal and spatial) found in molecular clouds, and one cannot rule out the possibility that molecular cloud environments modestly warmer than the canonical 10 K set the D/H abundance in HCN-bearing grains. Alternatively, the cometary HCN deuterium abundance may reflect the release of that molecule from different populations of grains in Hale-Bopp, some of which formed at 10 K and others formed or modified at 50 K or more. Indeed, the presence of two very different D/H components in the Semarkona and Bishunpur meteorites (Bockelée-Morvan et al. 1998) is a warning that grains from quite distinct source environments or histories may be contained within the same body (see in this regard the chapter by Irvine et al., this volume). Direct sampling of individual volatile-bearing grains on a cometary nucleus, and further measurements of D/H variations within molecular clouds, will eventually resolve these ambiguities. Such sampling of isotopic ratios in species in individual grains could also provide a powerful technique for mapping the history of outer solar system temperatures complementary to meteoritical investigations pertinent to the inner solar system (see the chapters by Wadhwa and Russell and by Bell et al., this volume).

The deuterium enrichment in water is significant as well, beyond the observation that it is identical in three Oort Cloud comets. The cometary value is approximately twice that of Earth's ocean, the so-called standard mean ocean water value (SMOW) of 1.6×10^{-4}. While this difference might not be considered significant if it were measured in just one comet, we have confidence that it is a genuine distinction between SMOW and Oort Cloud comets based on the several objects and different techniques. The higher value in comets rules out the idea that Earth's ocean was supplied exclusively or even principally by comets in unaltered form (Owen and Bar-Nun 1995). One possibility is that dilution of the cometary deuterium abundance occurred after terrestrial impact through exchange of the water with a massive, hot, nebular hydrogen reservoir in contact with the primordial atmosphere. However, such a process would result in loss of water to space and, indeed, might even enhance the terrestrial deuterium abundance through selective retention of deuterium during escape. Thus a more plausible model is one in which Earth's ocean received part of its inventory (tens of percent) from Oort Cloud comets and the rest from a warmer, more processed source of water.

Other sources of water for the Earth remain problematic. We do not know D/H for water ice in Kuiper Belt comets; although we expect this to be similar to or higher than D/H in the Oort Cloud comets (because they formed from grains farther out in the nebula), we cannot rule out a lower value. Meteorite parent bodies containing water of hydration or small amounts of water ice, gravitationally scattered from more distant orbits, are a possibility. Owen (1998) pointed out that new rate coefficients for deuterium exchange between nebular hydrogen and water (Lecluse and Robert 1994) permit a low D/H ratio in water of 8×10^{-5} to be achieved in less than a million years at temperatures above 170 K. The deuterium-poor water, adsorbed on silicate grains, might then be accreted into the Earth and outgassed to dilute the deuterium enrichment supplied by comets. The high value of $^{20}Ne/^{22}Ne$ found in mantle rocks suggests that solar-composition gases in the Earth survived differentiation and formation of the Moon. This, in turn, lends support to the idea that H_2O vapor from the inner nebula was trapped in rocks that formed the Earth. Yet another possibility is water ice condensed in the 3–5 AU region, with a D/H ratio consistent with processing in the inner nebula (Delsemme 1999). We argue later that water vapor transport in the inner nebula might enhance the abundance of such planetesimals.

C. Pluto/Triton/Kuiper Belt Spectra

Optical and near infrared reflectance spectra of Pluto and Triton have been obtained since the late 1970s, but only over the past decade has sufficient spectral resolution been available to make assessment of the inventory of surface ices practical. Additional data come from multiple instruments operating during the 1989 *Voyager* flyby of Triton, ground-based observations of stellar occultations by both Pluto and Triton, mutual events observations of Pluto and its moon Charon, and *Hubble Space Telescope* measurements of the barycentric wobble of the Pluto-Charon system (see the review by Brown and Cruikshank 1997). The first ground-based spectrum of a Kuiper Belt object (KBO) was obtained in 1997 at the Keck I telescope and shows features interpreted to be hydrocarbons (Brown et al. 1997).

Pluto's place in the outer solar system makes a compelling case for its membership as a KBO, one whose size suggests accretion from smaller objects. Many of the directly detected KBOs are in the same 3:2 resonance with Neptune as is Pluto, suggesting a similar early dynamical evolution, and all observed KBOs are large enough to be products of accretion in an early, more populous, primordial Kuiper Belt (Jewitt and Luu, this volume). We present further discussion of such a primordial belt later. Triton may have been a Kuiper Belt body captured by Neptune early in the solar system's history. Despite their evidently very different dynamical histories, Pluto and Triton show surface abundances of key nitrogen- and carbon-bearing species that are remarkably similar to each other (Cruikshank et al. 1993; Owen et al. 1993). These abundances, in turn, appear to differ from the molecular abundances seen in comets to date; in particular, the abundance of N_2 relative to the carbon-bearing volatiles is much larger than that inferred for comets. Perhaps Pluto and Triton reflect a class of objects formed at very low nebular temperatures (30 K) and directly captured N_2. Alternatively, the ratio of nitrogen to methane and carbon monoxide may reflect more efficient photochemical and cosmic ray conversion of these last two species into refractory compounds that are difficult to detect spectroscopically. It is also possible that selective outgassing and escape early in Triton's history led to the current composition of volatile ices. A much more detailed inventory for Pluto and another KBO will be available from the *Pluto/Kuiper Express* mission, but not until 2013 (results to be reviewed in *Protostars and Planets VI*).

It is also instructive to compare the surface compositions of the Centaurs and KBOs. The orbits of the Centaurs (of which seven are known) are dynamically unstable on timescales of about 10^6–10^7 years (Gladman and Duncan 1990; Holman and Wisdom 1993; Dones et al. 1996). This implies a source of bodies to replenish those lost from the Centaur region by collisions or catastrophic gravitational encounters with the jovian planets. The presumed source of the Centaurs is the Kuiper Belt (Gladman and Duncan 1990).

Spectra of three Centaurs have been published: 5145 Pholus (Luu et al. 1994; Cruikshank et al. 1998), 2060 Chiron (Luu et al. 1994), and 1997 CU26 (Brown et al. 1998). Pholus shows a complex spectrum that has been modeled with a mixture of refractory organics, olivine, H_2O ice, CH_3OH ice, and amorphous carbon (Cruikshank et al. 1998). Chiron's spectrum is virtually devoid of color and detectable absorption features in existing data (Luu et al. 1994). CU26 shows evidence in its spectrum of water ice.

From the standpoint of dynamics it is quite compelling that the Kuiper Belt is the source of the Centaurs (Holman and Wisdom 1993); thus, it might be expected that these two groups of objects would show some common compositional patterns. The KBO 1996 TL_{66} (Luu and Jewitt 1998; R.H. Brown, D. P. Cruikshank, Y. J. Pendleton, and G. J. Veeder, unpublished data) shows a relatively flat spectrum that is almost featureless and resembles the spectrum of the Centaur Chiron. The KBO 1993 SC (Brown et al. 1997) shows a number of near infrared spectral features and has no direct spectral analog among the Centaurs, except that it may have light hydrocarbons on its surface, like Pholus. At present, there is no known spectral analog in the Kuiper Belt for the Centaur 1997 CU26.

An explanation for the differences may be that when objects leave the Kuiper Belt and more closely approach the Sun, their surfaces are altered, both chemically and physically. For example, if ices such as methane or methanol are irradiated with photons or charged particles, chemical bonds are broken, and the material proceeds to molecules with increasingly higher carbon abundances (Sagan et al. 1984; Andronico et al. 1987; Thompson et al. 1987; Strazzulla 1997). If the photolytic or radiolytic dose is large enough, in time the color of a high-albedo surface initially composed of light hydrocarbon and water ices can proceed from bluish to very red and eventually back to neutral with a very low albedo 3–5%. Thus the combined photolysis, radiolysis, and thermal escape of volatile material could lead to substantial devolatilization and carbonization of an originally volatile-rich surface (where water ice at the temperatures of interest is considered nonvolatile). In some cases enough dark material may be left behind that absorption bands due to residual water ice in the surface layers would be completely masked.

Centaur spectra might be a result of faster and more complete photochemical alteration and devolatilization of the Centaurs' near-surface layers (because they are closer to the Sun), so that a possible end state is a flat, featureless spectrum similar to that of amorphous carbon (see also Cruikshank et al. 1998). Crucial parameters are the initial inventory of primordial dark material and light hydrocarbons. If those inventories are too low, the final albedo of the surface would be expected to be relatively high (where 3–5% would be considered low), and the water ice absorption bands in the 1–2.5 μm spectral region would never be completely masked. However, we do not fully understand the long-term effects of photolysis, radiolysis, and heating of the surfaces of objects in the outer solar system, nor do we have a complete understanding of the expected initial inventories of amorphous carbon, organics, and light hydrocarbons. Significantly, the Trojan asteroids are very red and yet have spent gigayears well inward of the Centaurs, and the jovian irregular satellites vary in color from red to neutral. Hence, this picture must be considered speculative at present.

D. Titan

Saturn's giant moon Titan possesses a thick atmosphere of molecular nitrogen with secondary amounts of methane, higher hydrocarbons, and hydrogen. The chemical state of the atmosphere and surface have evolved through time as solar ultraviolet radiation photolyzes methane, producing less volatile, heavier compounds and allowing hydrogen to escape (Lunine et al. 1989). Though chemically evolved, the atmosphere retains clues to the environment within which Titan formed. A key question is whether the nitrogen composing Titan's atmosphere today originated as N_2 or in significantly more reduced form, for example as NH_3. If it was the latter, then photochemistry or other chemical processes converted this molecule to molecular nitrogen (Atreya et al. 1978). When the question was originally posed, it was thought that this might distinguish between a cometary origin for Titan's volatiles and a more local, saturnian-processed formation. However, the apparent dominance of ammonia over nitrogen in several comets (Wyckoff et al. 1991*a,b*) suggests that we might expect ammonia to have been the source in either case.

More than fifteen years ago Owen (1982) proposed a test for the origin of the nitrogen. Because of the similar volatilities of N_2 and Ar, it is expected they would be trapped in the ice to a similar extent, but proportional also to their relative abundances in the gas phase; hence in the atmosphere today Ar/N_2 would be of order 10%. A refinement of this ratio based on trapping of gases in clathrate leads to $Ar/N_2 \approx 1$–10%, although this number needs rework if the ices were amorphous in the protosaturnian nebula (possible) or in the solar nebula (highly probable). Because ammonia, on the other hand, hydrogen-bonds with water, it will be trapped in the water ice to a much greater extent than either N_2 or Ar; hence, if NH_3 is the source of Titan's nitrogen, we expect $Ar/NH_3 \ll 1\%$ at present. The current upper limit to argon is 7% from careful analysis of *Voyager* data (Strobel et al. 1993); it will require the measurements of *Cassini/Huygens* to refine this number and decide between the two cases. In any event we can rule out capture of Titan's atmosphere directly from the nebular gas, because the upper limit on neon is 0.2% (at the surface), far less than solar elemental abundance would yield (Owen 1982). Hidayat and Marten (1998) have detected $HC^{15}N$ in Titan's atmosphere and found $HC^{15}N/HC^{14}N$ to be strongly enhanced relative to the solar value. One way to obtain this enrichment, they argue, is through escape of nitrogen equivalent to many tens of times the current atmosphere. Such an escape process, early in Titan's history, might have involved other species as well and suggests the possibility of a massive early atmosphere on this object (Lunine 1985).

The origin of the methane in Titan's atmosphere is a more difficult issue. The rate of photolysis is such as to deplete the current atmospheric inventory in a time of the order of 1% the age of the solar system. Because of condensation, the atmosphere is in fact not capable of holding more than several times the current inventory without a substantial increase in temperature; hence the idea of "stoking up" the methane early on and storing it in the atmosphere for gigayears is not tenable. The suspicion since the *Voyager 1* encounter is that a large reservoir of methane might exist at Titan's surface, mixed with products of photolysis in a liquid hydrocarbon ocean (Lunine et al. 1983), or just beneath the surface (Stevenson 1992). Remote sensing data do not show evidence for widespread oceans on Titan (for example, Griffith 1993; Smith et al. 1996). Thus, outstanding correlative questions are the following: How much methane has been processed on Titan since its formation, and where is the remainder?

The deuterium abundance on Titan hints at a much larger methane reservoir in the past. The most recently determined ratio of deuterated methane (CH_3D) to normal methane (CH_4) is equivalent to a deuterium enrichment approximately 2–5 times the generally accepted primordial solar value, $D/H = 2.6 \pm 0.7 \times 10^{-5}$ (Gautier 1999). However, it is much lower than that seen in comets (Meier et al. 1998*a*) and suggests that comets could not have been the principle source of the volatile reservoir from which the methane was derived. Much of the deuterated methane enhancement could be the result of photochemical breaking of C–H bonds in preference to C–D bonds, progressively concentrating deuterium in the volatile atmospheric methane molecule while the hydrogen escapes (Pinto et al. 1986). Thus, provided the initial ratio of deuterated to normal methane can be estimated, the photochemical enrichment of D/H provides a means of estimating just how much methane has been photolyzed in the surface-atmosphere system of Titan over geologic time. Other sources of enrichment of CH_3D, such as differential vapor pressure effects of the deuterated species over a surface ocean or near-surface crustal reservoir, have been shown to be very small, changing enhancements by 10–20% at most (Pinto et al. 1986). Using the current D/H value in methane on Titan and assuming less than a factor of 2 enhancement in the original protosaturnian nebula, an original methane reservoir some one to two orders of magnitude larger than the current atmospheric abundance is obtained from a model of photochemical enrichment (Lunine et al. 1999). Because some methane may yet be stored in the crust or on the surface of Titan, the calculation yields a lower limit to the total original methane inventory prior to the start of photolysis.

The conclusion from the foregoing analysis is supported by a simple calculation based on the absolute photolysis rate as inferred from multiple types of *Voyager* data (Yung et al. 1984). The total amount of methane in the atmosphere at present will be irreversibly converted to higher hydrocarbons (with upward loss of hydrogen) in a time approximately 1% the age of the solar system. Thus, if Titan's current state with respect to methane and photochemistry is typical of that over its history, the original reservoir of methane must have been of order 100 times the present atmospheric abundance. It is important to recognize that the ways by which we have derived this number are in large part independent: One depends on knowing the absolute photolysis rate of methane, the other on the ratio of deuterated to normal methane and the relative photolysis rates of the two. The agreement between the two suggests that Titan did not begin with a large enhancement of deuterium in methane and, hence, was composed of planetesimals that were significantly processed relative to the nascent molecular cloud. This is consistent with Titan having been formed within a well-mixed protosaturnian nebula.

III. PROCESSES

In what follows we attempt a selective discussion of nebular processes that may have played a role in altering outer solar system material from molecular cloud composition, in delivering processed water ice to the terrestrial planets, and in seeding giant planet atmospheres. The intention is not to be comprehensive but to be detailed enough to provide a flavor for the kind of modeling required to connect the data described above with the formation of the giant planets and dispersal of planetesimal material throughout the solar system.

A. Grain and Gas Dynamical Heating during Infall

Infall of material from the molecular cloud to the solar nebula, associated with the collapse of the cloud clump from which the solar system was derived, leads to heating and chemical modification of molecular cloud material. This has been modeled extensively beginning with Ziglina and Ruzmaikina (1991) and Hood and Horanyi (1991). Gas dynamical heating is a consequence of the tendency of grains to decouple from the gas as they grow in size or encounter a shock front at the surface of the nebular disk. As the grains partly decouple from the gas and encounter a steepening density gradient associated with passage into the nebula, frictional heating leads to a temperature rise in the grains and possible sublimation of volatile species. The temperature rise and amount of sublimation are rather sensitively dependent on the grain's infrared emissivity or, equivalently, its ability to reradiate the frictional energy. This in turn is a function of the grain size and, secondarily, its composition and fluffiness.

Because typical interstellar grains are so small that they would remain embedded in the gas even through a shock front, growth of grains is essential for gas dynamical heating to be a significant process. Work by Weidenschilling and Ruzmaikina (1994), building on earlier results (Volk and Morfill 1991), show rather convincingly that the collapse of dense cloud clumps leads to significant grain growth during infall to the central protostar or protoplanetary disk. Grains of hundreds of microns or even millimeters in size grow during infall, and these will partly decouple from the gas and more fully decouple at the nebular surface shock front.

Lunine et al. (1991) quantified the heating of crystalline water ice grains and showed that infall to the Jupiter/Saturn region, and possibly even the Uranus/Neptune radial zone, engendered sufficient dynamical heating to permit sublimation of a large fraction of the water ice mass of the grains. Like miniature comets, the sublimating water ice would release volatile gases and entrain silicate dust particles. Lunine et al. (1995) quantified the gas dynamical process for a wider range of particle properties and composition, including amorphous ice, ice composed of species more volatile than water ice, and very fluffy grains. Also potentially important is the effect of solar heating, which is relevant for grains at large distances from the center of the nebula and still high above the midplane. The flared nature of the nebula ensures that both direct and scattered sunlight heat the grains (Simonelli et al. 1997). The most comprehensive study of grain heating and evolution within, and during entry into, the solar nebula is that of Chick and Cassen (1997). All models produce results that are broadly consistent, considering the various assumptions about the nebular environment and grain properties.

The sublimation of much of the grain's icy component suggests that the composition and physical trapping of the volatiles inherited from the molecular cloud is lost for grains that fall inward of 10–20 AU from the center of the nebula. Once in the nebula, cooling of the grain is rapid, and its surface becomes a cold finger on which water ice and other volatiles can condense or be trapped by various means. This natural cold trapping experiment is reproduced by the laboratory work of Bar-Nun and colleagues (Owen and Bar-Nun 1995), which suggests a sensitive temperature dependence to the final compositional state of the ice. Thus, at least in the realm of Jupiter and Saturn, and maybe beyond, much of the nature of solar nebula grains may be determined not by equilibrium condensation, nor by properties derived completely unaltered from the nascent molecular cloud, but by a dynamical two-step process of sublimation during infall followed by recondensation and trapping on the rapidly cooling remnant of the grain.

B. Water Transport in the Nebula: From Vapor to Cometary Delivery

Building on initial work by Morfill and Volk (1984) as well as Stevenson and Lunine (1988), several recent modeling efforts have focused on the effects of the transport of water throughout the disk on nebular physical and chemical state, grain evolution, and planet formation. As the principal condensable, water vapor is moved to cold-trapping regions by nebular diffusion and advection, condensed out, and recycled back to the inner nebula through ice grain growth and radial drift of grains (Stepinski and Valageas 1997; Cuzzi et al. 1993). The different dependencies of these processes on various timescales lead to a potentially complex spatial and temporal history of water vapor in the nebula, one that is also sensitive to specific nebular processes. In particular, the existence of the enstatite chondrites, as well as possible water ice in P and D asteroids are consistent with the water diffusion model of Cyr et al. (1998), in which the inner nebula is partly depleted of water vapor and the 3–5 AU region enriched in water ice. The model also predicts a higher abundance of reduced carbon species, such as methane, produced in the inner, chemically active part of the nebula than do standard solar nebula models.

The D/H ratio in water vapor processed in the warm inner nebula and transported outward toward the 3–5 AU region is perhaps half that of the terrestrial ocean value (Lecluse and Robert 1994). This water could serve to dilute the deuterium-rich water coming to early Earth from outer solar system comets. As noted above, such deuterium-poor vapor might be carried to the forming Earth adsorbed on silicate grains. Alternatively, since much of this water vapor ends up condensed as ice in the 3–5 AU region of the nebula (Cyr et al. 1998), gravitational stirring by Jupiter of the orbits of the resulting planetesimals might have delivered some of the water ice to Earth (Delsemme 1999). Indeed, Owen and Bar-Nun (1995) invoke comets formed at the orbit of Jupiter as a way of bringing water to the Earth without altering terrestrial noble gas abundances. Lecluse and Robert (1994) show that the vapor-ice fractionation effect during condensation might lead to a D/H ratio in these planetesimals close to the terrestrial ocean value. Thus they would seem not to be a very effective diluting agent for the high-D/H water brought in by outer solar system comets. Additionally, such Jupiter-region planetesimals might still trap carbon- and nitrogen-bearing species in their grains during formation at 160 K, potentially causing a budget problem in those elements at the Earth. Nonetheless, since large amounts of icy planetesimals are likely to have existed in the 3–5 AU region, it is important to ask how much of this material could have been scattered back to the inner solar system as opposed to being ejected by or accreted into Jupiter. Detailed simulations of the gravitational scattering of such planetesimals during and after the formation of Jupiter must be undertaken to evaluate to what extent these planetesimals contributed to Earth's ocean.

C. Origin of Giant Planet Atmospheres

The currently favored model of the formation of the giant planets invokes nucleation and collapse (Mizuno 1980; Pollack and Bodenheimer 1989; chapter by Wuchterl et al., this volume). According to this model, a core consisting of rock and ice formed first by accretion of planetesimals. When the mass of the core grew sufficiently large, it began to attract gas, eventually leading to the collapse of the surrounding solar nebula. Bombardment by planetesimals would continue even beyond this phase. The atmosphere therefore resulted from three components: volatiles released from the core during accretional heating, those released by infalling planetesimals dissolving in the gaseous envelope, and those contributed directly by the solar nebula. The condensed-phase contribution to the giant planets is largely based on interior models and consideration of nebular dynamics (Hubbard 1989; Pollack and Bodenheimer 1989; Stevenson 1983). This picture implies that those elements whose volatile compounds were initially trapped in the planetesimals that produced the core and dissolved in the envelope should be enhanced in the atmosphere.

To determine expected enhancements requires a model for the composition of the icy planetesimals in the outer solar system. In the interstellar medium, the reservoir of carbon is mostly in the form of grains and refractory organics, rather than volatiles such as CO and CH_4 (see the chapter by Irvine et al., this volume). On the other hand, nitrogen is mostly in the volatile form, that is, N_2 (Irvine and Knacke 1989; van Dishoeck et al. 1993). Most of the comets in the Oort Cloud are thought to have formed at 55 $\pm$15 K in the vicinity of Uranus and Neptune, though some may have been at warmer temperatures in the Jupiter-Saturn part of the nebula.

Temperatures of 30 $\pm$ 10 K are required for ion-molecule reactions to produce the observed ratios of D/H in H_2O and HCN in comet Hale-Bopp (Meier et al. 1998*a,b*). However, most of the grains that went into Jupiter and Saturn were likely exposed to (indeed, may have retrapped volatiles at) higher temperatures, above 40 K. Laboratory experiments have demonstrated that ice at temperatures near 55 K does not trap N_2, CO, or CH_4 well but easily traps refractory organics and NH_3 as well as compounds of other heavy elements (Bar-Nun and Kleinfeld 1989). Unless the temperature is below 25 K, ice does not trap Ne; temperatures below 10 K are required to trap He and significant amounts of H_2, although modest amounts of the latter are trapped at higher temperature (Dissly et al. 1994). The light gases are not readily absorbed in rock, either, so in Jupiter's atmosphere we expect H, He, and Ne to be contributed essentially solely by solar nebula gas and hence to exhibit relative abundances that are solar. Contribution of nitrogen to the atmosphere by the core or by infalling planetesimals is primarily by ammonia, which is present in amounts of order 10^{-3} relative to water in some comets (Wyckoff et al. 1991*a,b*). In this picture, considerable enrichment in the abundances of carbon, sulfur, and oxygen, with much less (or no) enhancement in the nitrogen elemental ratio, should be present in the jovian atmosphere.

These predictions rest on the assumption that most of the icy planetesimals contributing to Jupiter's growth and development were subjected to temperatures above 40 K. This is a reasonable assumption, based on the nebular physics discussed above. However, we cannot rule out *a priori* a population of Kuiper Belt comets that contain ices unaltered from lower molecular cloud temperatures and hence would carry more nitrogen and heavy noble gases, which in turn would lead to some enrichment of even these species over their solar abundances.

The *Galileo* Probe mass spectrometer measured C/H = 2.9 times solar in Jupiter, in accord with ground-based and *Voyager* results. It also measured sulfur, in the form of H_2S, for the first time and found that S/H at the deepest levels probed (20 bar) is 2.5 times solar. The nearly equal enrichment of carbon and sulfur is consistent with the predictions of a model in which icy planetesimals provide the major contribution of heavy elements (Owen and Bar-Nun 1995). Even the carbonaceous chondrites, the most carbon-rich of the meteorites, have C/S = 0.1 times solar. Unless there are rocky bodies unsampled by our collection of meteorites, it appears that only icy planetesimals have the correct composition to enrich these elements equally (Owen et al. 1997). The icy planetesimal model also predicts that oxygen (as H_2O) should be similarly enhanced, perhaps even somewhat more than carbon. Thus we expect water in Jupiter to be significantly greater than solar in abundance, an assertion supported by very recent evolution models of Jupiter (T. Guillot, unpublished manuscript). However, as discussed above, *Galileo* has not provided a quantitative number for the deep water value.

The nitrogen abundance in Jupiter's atmosphere challenges this particular picture of icy planetesimal composition. If the ammonia mixing ratio determined by Folkner et al. (1998) is correct, it suggests that nitrogen is at least as enriched as carbon and sulfur in Jupiter's atmosphere. Although cometary ammonia is a possible source, it is not abundant enough to provide nitrogen abundances comparable to the sulfur and carbon values (Wyckoff et al.1991*a*). Molecular nitrogen is effectively incorporated in ices only at temperatures below about 35 K (Owen and Bar-Nun 1995). Thus, a value of N on Jupiter of approximately 3 times solar would require building up the core and/or dominating the infalling ices with solid material coming directly from the cold regions of the molecular cloud that formed the solar system, without modification by a wealth of possible physical processes (Chick and Cassen 1997).

As we have seen, however (see section II.A.3), there is a fundamental contradiction between the Folkner et al. (1998) mixing ratio and the radio spectrum of the planet; at least a partial solution appears to lie in the existence of an additional absorber. Depending on how strong this additional absorption is, nitrogen may indeed be as deficient as the icy planetesimal model predicts. Another constraint on the problem is provided by the abundances of the heavy noble gases. If N_2 was captured by the icy planetesimals that enriched carbon and sulfur on Jupiter in sufficient amounts to provide a 3-times-solar enhancement of N, the same enrichment should have occurred for argon, krypton, and xenon. Otherwise we would expect these gases to be present in roughly solar proportions, enhanced to a limited extent by any colder material from the outermost disk or molecular cloud. The argon mixing ratio of 1.7 ± 0.6 times solar measured by the *Galileo* mass spectrometer is marginally consistent with this picture, but it must be remembered that some argon may have dissolved along with neon in helium raindrops sinking into the Jovian deep interior.

We must await the determination of mixing ratios for both Kr and Xe to test this point further. However, the present upper limit on Xe from the GPMS yields Xe/Ar $\leq$ 5 times solar in Jupiter, which appears to rule out the carrier phases in comets as being clathrate hydrate, because for clathrate phases Xe/Ar exceeds 9 times solar for C/H greater than 3 times solar in Jupiter (Lunine and Stevenson 1985). The xenon upper limit is thus consistent with the theoretical and laboratory arguments for amorphous water ice as the relevant carrier phase of adsorbed volatiles at the temperatures appropriate to the outer solar nebula. For icy planetesimals composed of amorphous water ice we expect mixing ratios of Xe and Kr in Jupiter's atmosphere to be between solar and twice solar. Again, however, a significant uncertainty in the use of jovian atmospheric noble gas abundances is the poorly determined relative solubilities of these species in the helium raindrops that may be selectively removing elements from the observable atmosphere (Roulston and Stevenson 1995). Indeed, we cannot even be certain that the atmospheric composition is a true reflection of that of the bulk interior, as we do not yet know quantitatively the extent of elemental partitioning at the molecular-metallic phase boundary within Jupiter and Saturn. This uncertainty will fade with time only if experimental and theoretical work enable progress in understanding the phase transition in detail.

IV. SOME CONCLUSIONS

1. *The amount and variation of processing of outer solar system grains remain uncertain.* The protoplanetary disk from which the solar system formed was not isolated; it was embedded within a nascent molecular cloud, and there was a continuum of processing of material between the two. Even grains destined for the outermost reaches of the planetary formation zone underwent significant growth in size, as well as heating by direct exposure to the Sun and by frictional effects, during their infall. Some grains may have completely sublimated and recondensed as far out as the Uranus-Neptune zone of the nebula. Thus, the relative and absolute abundances of the adsorbed phases trapped in the water ice may have been altered from molecular cloud values. Because the efficiency of volatile trapping varies from one molecular species to another, the relative abundances of carbon-, sulfur-, oxygen- and nitrogen-bearing compounds trapped in the ice could have changed. A counterindication lies in the jovian nitrogen abundance. Should it prove to be three times solar, then the source of such large amounts of nitrogen must lie in very cold ($\leq$35 K) icy grains preserved from colder regions of the molecular cloud. In any event, the isotopic composition of molecular species was probably not changed by processing of icy grains; this should remain preserved in comets from the

presolar values, unless there was a significant (and, we believe unlikely) exchange of material between the outer disk and the warm, inner nebula.

2. *The water abundance in the nebula varied significantly in time and space.* Advective and diffusive transport of water vapor from the warm inner disk to the water condensation boundary, accompanied by radial drift of icy planetesimals inward through the boundary, created a complex profile of water vapor and ice that changed substantially on timescales short compared to planet formation. In the inner disk the water profile overall was depleted relative to the bulk starting value (usually assumed to be solar), but in selected locations it might have had a substantial peak. There were always water ice planetesimals venturing significantly inward of the condensation boundary because of gas drag drift and gravitational perturbations. Outward of the condensation front, water was mostly in the form of ice because of the very steep falloff of ice vapor pressure with temperature at low temperature. The ice was a time-varying mixture of native condensed water vapor from the inner disk (significant only near Jupiter), ice moved inward from more distant parts of the nebula by drift and gravitational perturbations, and ice falling directly from the surrounding molecular cloud.

3. *Earth's ocean was not principally derived from Oort Cloud comets.* The D/H ratios determined for the three Oort Cloud comets Hyakutake, Hale-Bopp and Halley are inconsistent with the significantly lower value for the Earth's ocean. Thus, unless Earth's water budget exchanged deuterium with a large hydrogen reservoir after delivery (possible but difficult to accomplish without loss of the water to the nebula), the source had to be less deuterium rich than these three comets. Under the assumption that these comets are typical of the Oort Cloud reservoir, we must look elsewhere for the source of terrestrial ocean water. Determination of deuterium abundance in KBOs or Jupiter-family short-period comets (thought to be derived from the Kuiper Belt) is required to assess whether these could be sources. We expect, however, such objects to have high D/H values as well. Eliminating Oort Cloud and Kuiper Belt sources would rule out much or all of the outer solar system as candidate sources, because the Oort Cloud comets formed in the giant planet zone. Potential alternatives include water of hydration or adsorbed nebular water in silicates, and inner solar system water vapor condensed at the water condensation front and brought back inward by drag-induced radial drift and gravitational scattering.

4. *Surface volatile budgets of Triton and Pluto remain unexplained.* Whether by origin, chemical processing, or escape processes, the abundances of volatiles on the surfaces of Triton and Pluto differ significantly from those measured to date in comets. Models must account for the remarkable similarity in volatile composition of the two bodies, despite their different early dynamical histories. Also, the atmospheric compositions of Triton and Pluto are not all that different from what one would achieve

by lowering the temperature on Titan to Triton or Pluto values. Whether this resemblance is carried through to the bulk volatile budget must await *Cassini/Huygens* measurements of Titan and future, more detailed observations of Triton and Pluto.

5. *There were multiple sources of organics in the outer solar nebula.* Organic molecules in molecular clouds are highly enriched in deuterium. Data from Hale-Bopp indicate that this comet, at least, preserved that enrichment in the one organic molecule measured, HCN. The D/H value in methane in Titan's atmosphere, on the other hand, is only modestly enhanced relative to solar. Much of this may be the result of photochemical processing, over gigayears, of methane originally formed in a compact subnebula around Saturn. However, the fact that Titan's deuterium enrichment is essentially the same as that measured on Uranus and Neptune suggests that a solar nebular origin might be considered as well: for example, exchange of deuterium between hydrogen and water, followed by formation of methane during impacts of accreting planetesimals. Deciding between these two possibilities is important because it lies at the heart of the issue of how Titan acquired its prodigious volatile budget. It will require a determination of the deuterium abundance in multiple hydrogen-bearing species, noble gas abundances, and more precise (or de novo) determinations of isotopic ratios in carbon- and nitrogen-bearing species, all in Titan's atmosphere. The *Cassini/Huygens* mission represents a unique opportunity to make these measurements *in situ* and with high precision.

V. FUTURE PROSPECTS

In the coming two decades, seven missions are expected or proposed to return data from outer solar system bodies. These include *Cassini/Huygens* to orbit the Saturn system; *Pluto Kuiper Express* to fly past Pluto and a KBO; *Europa Orbiter Mission*; and *Rosetta, Stardust, Contour,* and *Deep Impact* to explore several comets. *Stardust* will collect coma dust samples, and *Deep Impact* will expose subsurface layers of the nucleus for spectroscopic analysis. Each of these missions has a well-designed Web site, to which the reader is referred in place of descriptive material here, for which space does not allow.

Continued progress in ground-based studies of the outer solar system depends in large part on access to large telescopes and increases in detector performance. The challenge for the Kuiper Belt is to probe deeper in terms of larger distances and smaller sizes. Earth-based spectroscopy will continue to be a principal tool for determining the composition of small bodies. Increasing sensitivity and aperture are essential for determining isotopic ratios and abundances of atoms and molecules that are relatively inactive spectroscopically. The future looks bright for availability of large-aperture systems. Government-funded projects such as Gemini in the United States; the 8-meter Japanese National Telescope, Subaru; and the European Very Large Telescope promise broader access to systems comparable to Keck in capability. The technology of large mirrors and structures has progressed to the point where smaller consortia of universities and research organizations can afford 6-meter apertures and larger. At least four such systems in the 6-to-10-meter class are scheduled to come on line between now and the time of the next *Protostars and Planets* book.

Acknowledgments The authors thank Drs. T. Guillot and S. Russell for helpful reviews. J. L. is most grateful to Dr. A. Coradini and the CNR/Istituto di Astrofisica Spaziale in Rome, Italy, for hosting him during the writing of this chapter. He is also grateful to Dr. G. Valsecchi for unselfishly troubleshooting TEX. Support for the preparation of this chapter came from NASA and CNR.

REFERENCES

Andronico, G., Baratta, G. A., Spinella, F., and Strazzulla, F. 1987. Optical evolution of laboratory-produced organics: Applications to Phoebe, Iapetus, outer belt asteroids and cometary nuclei. *Astron. Astrophys.* 184:333–336.

Atreya, S. K., and Romani, P. N., 1995. Photochemistry and clouds of Jupiter, Saturn, and Uranus. In *Recent Advances in Planetary Meteorology*, ed. G. E. Hunt (Cambridge: Cambridge University Press), pp. 17–68.

Atreya, S. K., Donahue, T. M., and Kuhn, W. R. 1978. Evolution of a nitrogen atmosphere on Titan. *Science* 201:611–613.

Atreya, S. K., Wong, M., Owen, T. C., Niemann, H. B., and Mahaffy, P. 1997. Chemistry and clouds of the atmosphere of Jupiter: A Galileo perspective. In *Three Galileos: The Man, the Spacecraft, the Telescope*, ed. C. Barbieri, J. Rahe, T. Johnson, and A. Sohus (Dordrecht: Kluwer), pp. 249–260.

Balsiger, H., Altwegg, K., and Geiss, J. 1995. D/H and $^{18}O/^{16}O$ ratio in the hydronium ion and in neutral water from *in situ* ion measurements in comet P/Halley. *J. Geophys. Res.* 100:5827–5834.

Bar-Nun, A., and Kleinfeld, I. 1989. On the temperature and gas composition in the region of comet formation. *Icarus* 80:243–253.

Bockelée-Morvan, D., Gautier, D., Lis, D. C., Young, K., Keene, J., Phillips, T. G., Owen, T., Crovisier, J., Goldsmith, P. F., Bergin, E. A., Despois, D., and Wooten, A. 1998. Deuterated water in comet C/1996 B2 (Hyakutake) and its implications for the origin of comets. *Icarus* 133:147–162.

Boss, A. P. 1995. Proximity of Jupiter-like planets to low-mass stars. *Science* 267:360–362.

Brown, R. H., and Cruikshank, D. P. 1997. Determination of the composition and state of icy surfaces in the outer solar system. *Ann. Rev. Earth Planet. Sci.* 25:243–277.

Brown, R. H., Cruikshank, D. P., Pendleton, Y., and Veeder, G. J. 1997. Surface composition of Kuiper Belt object 1993SC. *Science* 276:937–939.

Brown, R. H., Cruikshank, D. P., Pendleton, Y., and Veeder, G. J. 1998. Identification of water ice on the Centaur 1997 CU26. *Science* 280:1430–1432.

Chick, K. M., and Cassen, P. 1997. Thermal processing of interstellar dust grains in the primitive solar environment. *Astrophys. J.* 477:398–409.

Cruikshank, D. P., Roush, T. L., Owen, T. C., Geballe, T. R., de Bergh, C., Schmitt, B., Brown, R. H., and Bartholomew, M. J. 1993. Ices on the surface of Triton. *Science* 261:742–745.

Cruikshank, D. P., Roush, T. L., Bartholomew, M. J., Geballe, T. R., Pendleton, Y. J., White, S. M., Bell, J. F. I., Daviews, J. K., Owen, T. C., de Bergh, C., Tholen, D. J., Bernstein, M. P., Brown, R. H., Tryka, K. A., and Dalle Ore, C. M. 1998. The composition of Centaur 5145 Pholus. *Icarus* 135:389–407.

Cuzzi, J. N., Dobrovolskis, A. R., and Champney, J. M. 1993. Particle-gas dynamics in the midplane of a protoplanetary nebula. *Icarus* 106:102–134.

Cyr, K. E., Sears, W. D., and Lunine, J. I. 1998. Distribution and evolution of water ice in the solar nebula: Implications for solar system body formation. *Icarus* 135:537–548.

de Pater, I., and Massie, S. T. 1985. Models of the millimeter-centimeter spectra of the giant planets. *Icarus* 62:143–171.

Delsemme, A. H. 1999. The deuterium enrichment observed in recent comets is consistent with the cometary origin of seawater. *Planet. Space Sci.* 47:125–131.

Dissly, R. W., Allen, M., and Anicich, V. G. 1994. H_2-rich interstellar grain mantles: An equilibrium description. *Astrophys. J.* 435:685–692.

Dones, L., Levison, H. F., and Duncan, M. 1996. On the dynamical lifetimes of planet-crossing objects. In *Completing the Inventory of the Solar System*, ed. T. W. Rettig and J. M. Hahn (San Francisco: Astronomical Society of the Pacific), pp. 233–244.

Duncan, M. J., Levison, H. F., and Budd, S. M. 1995. The dynamical structure of the Kuiper Belt. *Astron. J.* 110:3073–3081.

Eberhardt, P., Reber, M., Krankowsky, D., and Hodges, R. R. 1995. The D/H and $^{18}O/^{16}O$ ratios in water from comet P/Halley. *Astron. Astrophys.* 302:301–316.

Folkner, W. M., Woo, R., and Nandi, S. 1998. Ammonia abundance at the Galileo Probe site derived from absorption of its radio signal. *J. Geophys. Res.* 103:22847–22855.

Gautier, D. 1999. Deuterium in the solar system and cosmogonical implications. In *Planetary Systems: The Long View*, ed. L. Celnikier (Gif-sur-Yvette: Editions Frontières), in press.

Geiss, J. 1993. Primordial abundance of hydrogen and helium isotopes. In *Origin and Evolution of the Elements*, ed. N. Prantzos, L. Vangioni-Flam, and M. Casse (New York: Cambridge University Press), pp. 89–106.

Geiss, J., and Gloeckler, G. 1998. Measurement of the abundance of helium-3 in the local interstellar cloud and the Sun. *Space Sci. Rev.* 84:239–250.

Gladman, B., and Duncan, M. 1990. On the fates of minor bodies in the outer solar system. *Astron. J.* 100:1680–1683.

Griffith, C. A. 1993. Evidence for surface heterogeneity on Titan. *Nature* 364:511–514.

Hatchell, J., Millar, T. J., and Rodgers, S. D. 1998. The DCN/HCN abundance ratio in hot molecular cores. *Astron. Astrophys.* 332:695–702.

Hidayat, T., and Martin, A. 1998. Evidence for a strong $^{15}N/^{14}N$ enrichment in Titan's atmosphere from millimetre observations. *Ann. Geophys.* 16 (suppl. 3) C998.

Holman, M. J., and Wisdom, J. 1993. Dynamical stability in the outer solar system and the delivery of short period comets. *Astron. J.* 105:1987–1999.

Hood, L. L., and Horanyi, M. 1991. Gas dynamic heating of chondrule precursor grains in the solar nebula. *Icarus* 93:259–269.
Hubbard, W. B. 1989. Structure and composition of giant planet interiors. In *Origin and Evolution of Planetary and Satellite Atmospheres*, ed. S. K. Atreya, J. B. Pollack, and M. S. Matthews (Tucson: University of Arizona Press), pp. 539–563.
Irvine, W. M., and Knacke, R. F. 1989. The chemistry of interstellar gas and grains. In *Origin and Evolution of Planetary and Satellite Atmospheres*, ed. S. K. Atreya, J. B. Pollack, and M. S. Matthews (Tucson: University of Arizona Press), pp. 3–34.
Lecluse, C., and Robert, F. 1994. Hydrogen isotopic exchange reaction rates: Origin of water in the inner solar system. *Geochim. Cosmochim. Acta* 58:2927–2939.
Lewis, J. S., 1969. Observability of spectroscopically active compounds in the atmosphere of Jupiter. *Icarus* 10:393–409.
Lin, D. N. C., Bodenheimer, P., and Richardson, D. C. 1996. Orbital migration of the planetary companion of 51 Pegasi to its present location. *Nature* 380:606–607.
Linsky, J. L. 1996. GHRS observations of the LISM. *Space Sci. Rev.* 78:157–164.
Lunine, J. I. 1985. Volatiles in the outer solar system. Ph.D. Dissertation, California Institute of Technology.
Lunine, J. I., Yung, Y. K., and Lorenz, R. D. 1999. On the volatile inventory of Titan from isotopic abundances in nitrogen and methane. *Planet. Space Sci.* in press.
Lunine, J. I., and Stevenson, D. J. 1985. Evolution of Titan's coupled ocean-atmosphere system and interaction of ocean with bedrock. In *Proc. of the NATO Workshop on Ices in the Solar System,* ed. J. Klinger, D. Benest, A. Dollfus, and R. Smoluchowski (Dordrecht: D. Reidel), pp. 741–757.
Lunine, J. I., Stevenson, D. J., and Yung, Y. L. 1983. Ethane ocean on Titan. *Science* 222:1229–1230.
Lunine, J. I., Atreya, S. K., and Pollack, J. B. 1989. Evolution of the atmospheres of Titan, Triton and Pluto. In *Origin and Evolution of Planetary and Satellite Atmospheres*, ed. S. K. Atreya, J. B. Pollack, and M. S. Matthews (Tucson: University of Arizona Press), pp. 605–665.
Lunine, J. I., Engel, S., Rizk, B., and Horanyi, M. 1991. Sublimation and reformation of icy grains in the primitive solar nebula. *Icarus* 94:333–343.
Lunine, J. I., Dai, W., and Ebrahim, F. 1995. Solar system formation and the distribution of volatile species. In *Deep Earth and Planetary Volatiles*, ed. K. Farley (New York: AIP Press), pp. 117–122.
Luu, J. X., and Jewitt, D. C. 1998. Optical and infrared reflectance spectrum of Kuiper Belt object 1996TL66. *Astrophys. J. Lett.* 494:L117.
Luu, J. X., Jewitt, D., and Cloutis, E. 1994. Near-infrared spectroscopy of primitive solar system objects. *Icarus* 109:133–144.
Mahaffy, P. R., Donahue, T. M., Atreya, S. K., Owen, T. C., and Niemann, H. B. 1998. Galileo Probe measurements of D/H and ^{3}He/ ^{4}He in Jupiter's atmosphere. *Space Sci. Rev.* 84:251–263.
Malhotra, R. 1996. The phase space structure near Neptune resonances in the Kuiper Belt. *Astron. J.* 111:504–516.
Meier, R., Owen, T. C., Jewitt, D. C., Matthews, H. E., Senay, M., Biver, N., Bockelée-Morvan, D., Crovisier, J., and Gautier, D. 1998*a*. Deuterium in comet C/1995 O1 (Hale-Bopp): Detection of DCN. *Science* 279:1707–1710.

Meier, R., Owen, T. C., Matthews, H. E., Jewitt, D. C., Bockelée-Morvan, D., Biver, N., Crovisier, J., and Gautier, D. 1998*b*. A determination of the HDO/H_2O ratio in comet C/1995 O1 (Hale-Bopp). *Science* 279:842–844.

Mizuno, H. 1980. Formation of the giant planets. *Prog. Theor. Phys.* 64:544–557.

Morfill, G. E., and Volk, H. J. 1984. Transport of dust and vapor and chemical fractionation in the early protosolar cloud. *Astrophys. J.* 287:371–395.

Niemann, H. B., Atreya, S. K., Carignan, G. R., Donahue, T. M., Haberman, J. A., Harpold, D. N., Hartle, R. E., Hunten, D. M., Kasprzak, W. T., Mahaffy, P. R., Owen, T. C., Spencer, N. W., and Way, S. H. 1996. The Galileo Probe mass spectrometer: Composition of Jupiter's atmosphere. *Science* 272:846–849.

Niemann, H. B., Atreya, S. K., Carignan, G. R., Donahue, T. M., Haberman, J. A., Harpold, D. N., Hartle, R. E., Hunten, D. M., Kasprzak, W. T., Mahaffy, P. R., Owen, T. C. and Way, S. H. 1998. The composition of the Jovian atmosphere as determined by the Galileo Probe mass spectrometer. *J. Geophys. Res.* 103:22831–22845.

Owen, T. 1998. The Origin of the atmosphere. In *The Molecular Origins of Life*, ed. A. Brack (Cambridge: Cambridge University Press), pp. 13–34.

Owen, T., and Bar-Nun, A. 1995. Comets, impacts and atmospheres. *Icarus* 116:215–226.

Owen, T., and Bar-Nun, A. 1996. Comets, meteorites and atmospheres. *Earth Moon Planets* 72:425–432.

Owen, T., Bar-Nun, A., and Kleinfeld, I. 1991. Noble gases in terrestrial planets: Evidence for cometary impacts. In *Comets in the Post-Halley Era*, Vol. 1, ed. R. L. Newburn, Jr., M. Neugebauer, J. Rahe (Dordrecht: Kluwer), pp. 429–437.

Owen, T. C. 1982. The composition and origin of Titan's atmosphere. *Planet. Space Sci.* 30:833–838.

Owen, T. C., Roush, T. L., Cruikshank, D. P., Elliot, J. L., Young, L. A., de Bergh, C., Schmitt, B., Geballe, T. R., Brown, R. H., and Bartholomew, M. J. 1993. Surface ices and the atmospheric composition of Pluto. *Science* 261:745–748.

Owen, T. C., Atreya, S. K., Mahaffy, P., Niemann, H. B., and Wong, M. H. 1997. On the origin of Jupiter's atmosphere and the volatiles on the Medicean stars. In *Three Galileos: The Man, the Spacecraft, the Telescope*, ed. C. Barbieri, J. Rahe, T. V. Johnson, and A. Sohus (Dordrecht: Kluwer), pp. 289–298.

Pinto, J. P., Lunine, J. I., Kim, S.-J., and Yung, Y. L. 1986. D to H ratio and the origin and evolution of Titan's atmosphere. *Nature* 319:388–390.

Pollack, J. B., and Bodenheimer, P. 1989. Theories of the origin and evolution of the giant planets. In *Origin and Evolution of Planetary and Satellite Atmospheres*, ed. S. K. Atreya, J. B. Pollack, and M. S. Matthews (Tucson: University of Arizona Press), pp. 564–604.

Prinn, R. G. 1993. Chemistry of the protosolar nebula. In *Protostars and Planets III*, ed. E. H. Levy and J. I. Lunine (Tucson: University of Arizona Press), pp. 1005–1028.

Proffitt, C. R. 1994. Effects of heavy-element settling on solar neutrino fluxes and interior structure. *Astrophys. J.* 425:849–855.

Ragent, B., Colburn, D. S., Avrin, P., and Rages, K. A. 1996. Results of the Galileo Probe nephelometer experiment. *Science* 272:854–855.

Roos-Serote, M., Drossart, P., Encrenaz, T., Lellouch, E., Carlson, R. W., Baines, K. H., Kamp, L., Mehlman, R., Orton, G. S., Calcutt, S., Irwin, P., Taylor, F., and Wier, A. 1998. Analysis of Jupiter north equatorial belt hot spots in the 4–5 micron range from Galileo/near-infrared mapping spectrometer observations: Measurements of cloud opacity, water and ammonia. *J. Geophys. Res.* 103:23023–23041.

Roulston, M. S., and Stevenson, D. J. 1995. Prediction of neon depletion in Jupiter's atmosphere. *EOS* 76:343.
Sagan, C., Khare, B. N., and Lewis, J. S. 1984. Organic matter in the Saturn system. In *Saturn*, ed. T. Gehrels and M. S. Matthews. (Tucson: University of Arizona Press), pp. 788–807.
Showman, A. P., and Ingersoll, A. P. 1996. Deep dry downdraft as an explanation for low Jovian water observations. *Bull. Am. Astron. Soc.* 28:1141.
Simonelli, D. P., Pollack, J. B., and McKay, C. P. 1997. Radiative heating of interstellar grains falling toward the solar nebula: 1-D diffusion calculations. *Icarus* 125:261–280.
Smith, P. H., Lemmon, M. T., Lorenz, R. D., Sromovsky, L. A., Caldwell, J. J., and Allison, M. D. 1996. Titan's surface, revealed by HST imaging. *Icarus* 119:336–339.
Sromovsky, L. A., Best, F. A., Collard, A. D., Fry, P. M., Revercomb, H. E., Freedman, R. S., Orton, G. S., Hayden, J. L., Tomasko, M. G., and Lemmon, M. T. 1996. Solar and thermal radiation in Jupiter's atmosphere: Initial results of the Galileo Probe net flux radiometer. *Science* 272:851–853.
Sromovsky, L. A., Collard, A. D., Fry, P. M., Orton, G. S., Lemmon, M. T., Tomasko, M. G., and Freedman, R. S. 1998. Galileo Probe measurements of thermal and solar radiation fluxes in the Jovian atmosphere. *J. Geophys. Res.* 103:22929–22977.
Stepinski, T. F., and Valageas, P. 1997. Global evolution of solid matter in turbulent protoplanetary disks: II. Development of icy planetesimals. *Astron. Astrophys.* 319:1007–1019.
Stevenson, D. J., 1983. Structure of the giant planets: Evidence for nucleated instabilities and post-formational accretion. *Lunar Planet. Sci.* 14:770–771.
Stevenson, D. J. 1992. The surface of Titan. In *Proceedings of the Symposium on Titan*, ESA SP-338 (Noordwijk, The Netherlands: ESA), pp. 29–33.
Stevenson, D. J., and Lunine, J. I. 1988. Rapid formation of Jupiter by diffusive redistribution of water vapor in the solar nebula. *Icarus* 75:146–155.
Stevenson, D. J., and Salpeter, E. E. 1977. The dynamics and helium distribution in hydrogen-helium planets. *Astrophys. J. Suppl.* 35:239–261.
Strazzulla, G. 1997. Ion bombardment of comets. In *From Stardust to Planetesimals*, ed. Y. J. Pendleton and A. G. G. M. Tielens (San Francisco: Astronomical Society of the Pacific), pp. 423–433.
Stroble, D. F., Hall, D. T., Zhu, X., and Summers, M. E. 1993. Upper limit on Titan's atmospheric argon abundance. *Icarus* 103:333–336.
Thompson, W. R., Murray, B. G. J. P. T., Khare, B. N., and Sagan, C. 1987. Coloration and darkening of methane clathrate and other ices by charged particle irradiation: Applications to the outer solar system. *J. Geophys. Res.* 92:14933–14947.
Trilling, D. E., Benz, W., Guillot, T., Lunine, J. I., Hubbard, W. B., and Burrows, A. 1998. Orbital evolution and migration of giant planets: Modeling extrasolar planets. *Astrophys. J.* 500:428–439.
van Dishoeck, E. F., Blake, G. A., Draine, B. T., and Lunine, J. I. 1993. Chemical evolution of protostellar and protoplanetary matter. In *Protostars and Planets III*, ed. E. H. Levy and J. I. Lunine (Tucson: University of Arizona Press), pp. 163–241.
Volk, H. J., and Morfill, G. E. 1991. Physical processes in the protoplanetary disk. *Space Sci. Rev.* 56:65–73.
Von Zahn, U., and Hunten, D. M. 1996. The helium mass fraction in Jupiter's atmosphere. *Science* 272:849–850.
Weidenschilling, S. J., and Lewis, J. S. 1973. Atmospheric and cloud structures of the Jovian planets. *Icarus* 20:465–476.

Weidenschilling, S. J., and Ruzmaikina, T. V. 1994. Coagulation of grains in static and collapsing protostellar clouds. *Astrophys. J.* 430:713–726.
Weissman, P. R. 1991. Dynamical history of the Oort cloud. In *Comets in the Post-Halley Era*, Vol. 1. ed. R. L. Newburn Jr., M. Neugebauer, and J. Rahe (Dordrecht: Kluwer), pp. 463–486.
Wildt, R. 1937. Photochemistry of planetary atmospheres. *Astrophys. J.* 86:321–336.
Wyckoff, S., Tegler, S. C., and Engel, L. 1991*a*. Ammonia abundances in four comets. *Astrophys. J.* 368:279–286.
Wyckoff, S., Tegler, S. C., and Engel, L. 1991*b*. Nitrogen abundance in comet Halley. *Astrophys. J.* 367:641–648.
Yung, Y. L., Allen, M. A., and Pinto, J. P. 1984. Photochemistry of the atmosphere of Titan: Comparison between model and observations. *Astrophys. J. Suppl.* 55:465–506.
Ziglina, I. N., and Ruzmaikina, T. V. 1991. The influx of interstellar matter onto a protoplanetary disk. *Astron. Vest.* 25:53–60 (in Russian).

GIANT PLANET FORMATION

GÜNTHER WUCHTERL
Institut für Astronomie der Universität Wien

TRISTAN GUILLOT
Observatoire de la Côte d'Azur

and

JACK J. LISSAUER
NASA-Ames Research Center

Giant planet formation is closely interrelated with star formation, protoplanetary disks, the growth of dust and solid planets in those nebula disks, and finally nebula dispersal. Models of the interiors and evolution of giant planets in our solar system point to a bulk enrichment of heavy elements more than a factor of 2 above solar composition and imply heavy element cores up to 17 times the mass of Earth. Detailed models of giant planet formation explain the diversity of solar system and extrasolar giant planets by variations in the core growth rates caused by a coupling of the dynamics of planetesimals and the contraction of the massive envelopes into which they dive, as well as by changes in the hydrodynamical accretion behavior of the envelopes resulting from differences in nebula density, temperature, and orbital distance.

I. INTRODUCTION

Our four giant planets contain 99.5% of the angular momentum of the solar system but only 0.13% of its mass. On the other hand, more than 99.5% of the mass of the planetary system is in those four largest bodies. The angular momentum distribution can be understood on the basis of the "nebula hypothesis" (Kant 1755), which assumes concurrent formation of a planetary system and a star from a centrifugally supported flattened disk of gas and dust with a pressure-supported central condensation (Laplace 1796; Safronov 1969; Lissauer 1993). Theoretical models of the collapse of slowly rotating molecular cloud cores have demonstrated that such preplanetary nebulae are the consequence of the observed cloud core conditions and the hydrodynamics of radiating flows, provided there is a macroscopic angular momentum transfer process (chapter by Stone et al., this volume; Cassen and Moosman 1981; Morfill et al. 1985; Laughlin and Bodenheimer 1994; Podosek and Cassen 1994). Assuming turbulent viscosity to be that process, dynamical models have shown how mass and angular momentum separate by accretion through a viscous disk onto a growing central protostar (Tscharnuter 1987; Tscharnuter and Boss 1993; see chapter by Stone et al., this volume, for viscosity mechanisms). Those calculations, however, do not yet reach to the evolutionary state of the nebula where planet formation is expected. Observationally inferred disk sizes and masses are overlapping theoretical expectations and fortify the nebula hypothesis. High-resolution observations at millimeter wavelengths are now sensitive to disk conditions at orbital distances $>$50 AU (see, e.g., chapter by Wilner and Lay, this volume; Dutrey et al. 1998; Guilloteau and Dutrey 1998). However, observations thus far provide little information about the physical conditions in the respective nebulae on scales of 1 to 40 AU, where planet formation is expected to occur.

Planet formation studies therefore obtain plausible values for disk conditions from nebulae that are reconstructed from the present planetary system and disk physics. The so obtained "minimum reconstituted nebula masses," defined as *the total mass of solar-composition material needed to provide the observed planetary/satellite masses and compositions by condensation and accumulation*, are a few percent of the central body, both for the solar nebula and for the circumplanetary protosatellite nebulae (Kusaka et al. 1970; Hayashi 1980; Stevenson 1982*a*). The total angular momenta of the satellite systems, however, are only about 1% of the spin angular momenta of the respective giant planets (Podolak et al. 1993), in strong contrast with the planetary system/Sun ratio. Assembling planets from a nebula disk and advecting the angular momentum due to Keplerian shear until the present giant planet masses are reached results in total angular momenta overestimating the present spin angular momenta of the giant planets only by small factors (Götz 1993). Even if giant planets had kept this angular momentum, they still would not rotate critically! Giant planets, unlike stars, therefore do not have an angular momentum problem. This may justify why most studies of proto-giant planets neglect rotation or treat it as a small perturbation.

We discuss new results on interior models in section II. We review recent work on planetesimal formation and growth of solid planets in section III. The "nucleated instability hypothesis" is the only model for the formation of Uranus and Neptune at the moment, whereas other models also exist for Jupiter and Saturn; in section IV, we review these various models. We put emphasis on envelope evolution and gas accumulation using the "nucleated instability" model in section V. We apply the formation theories to extrasolar planets in section VI.

II. INTERIORS OF THE GIANT PLANETS

Our knowledge of the mechanisms that led to the formation of the giant planets is essentially based on numerical models and on the constraints provided by studies of the internal structure and composition of Jupiter,

Saturn, Uranus, and Neptune. This involves the calculation of interior models matching the observed gravitational fields. Each of the four giant planets of our solar system are thus believed to consist of a central, dense core and a surrounding envelope composed of hydrogen, helium, and small amounts of heavy elements. The cores of Jupiter and Saturn are very small compared to the total masses of the planets, whereas Uranus and Neptune are mostly core and possess small (i.e., low-mass) envelopes.

The giant planets, with the exception of Uranus, emit significantly more energy than received from the Sun, a consequence of their progressive cooling and contraction. Two important consequences can be drawn from this:

1. They have inner temperatures of a few thousand kelvins or more; therefore, their hydrogen-helium envelopes are *fluid*.
2. They are mostly convective (see Hubbard 1968; Stevenson and Salpeter 1977). The convective hypothesis has been challenged (Guillot et al. 1994*b*), but the regions where convection could be suppressed due to radiative transport are limited to a small fraction of the envelope, at temperatures between 1500 and 2000 K, or in low-temperature regions where the abundance of water is small.

These two conclusions are also expected to hold for Uranus for essentially two reasons: First, it is highly unlikely that its interior has cooled much more than that of Neptune (thus, one can expect that its intrinsic heat flux is small but larger than zero; see Marley and McKay 1999); second, it possesses a magnetic field of similar strength to that of the other giant planets, a sign of convective activity in its interior.

It seems, therefore, logical to assume that the envelopes of all four giant planets are homogeneously mixed. Some caveats are necessary, however:

1. Condensation and chemical reactions alter chemical composition (these should be confined to the external regions).
2. A first-order phase transition (such as the one between molecular and metallic hydrogen) imposes an abundance discontinuity across itself.
3. Hydrogen-helium phase separation might occur and lead to a variation of the abundance of helium in the planet.
4. The envelopes of Uranus and Neptune are small and enriched in heavy elements; it is thus conceivable that molecular weight gradients inhibit convection and yield nonhomogeneous envelopes.

On this basis, a three-layer structure is generally adopted for the four giant planets (Fig. 1). In the case of the less massive Uranus and Neptune, in which hydrogen is believed to remain in molecular phase, the planets are divided into a "rock" core (a mixture of the most refractory elements including silicates and iron), an "ice" layer (consisting of H_2O, CH_4, and NH_3) and a hydrogen-helium envelope. The latter is substantially enriched in heavier elements, as demonstrated by the $\sim$30 times

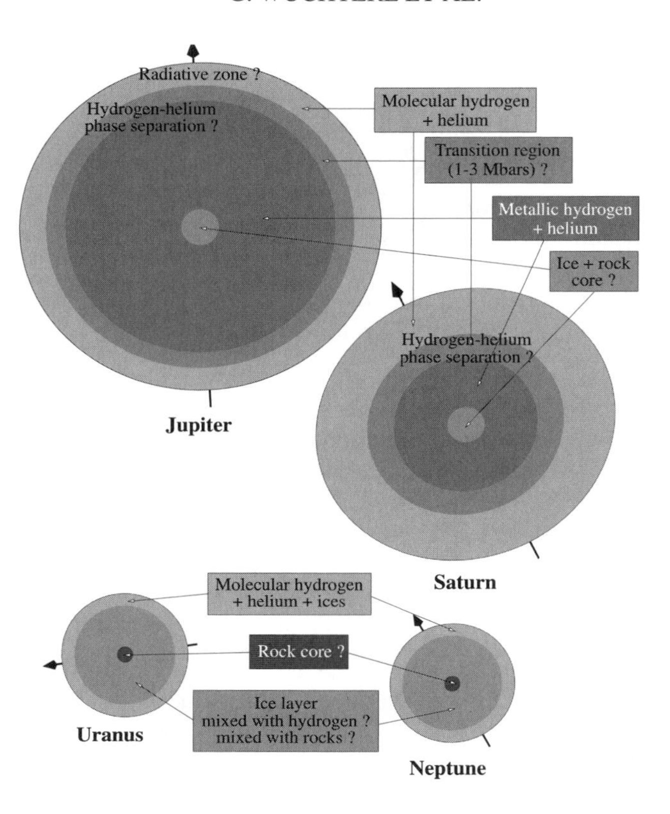

Figure 1. The interiors of Jupiter, Saturn, Uranus, and Neptune, according to the conventional wisdom. The sizes and oblatenesses of the planets are represented to scale. Inside Jupiter and Saturn, hydrogen, which is in molecular form (H_2) at low pressures, is thought to become metallic in the 1- to 3-Mbar region. This transition could be abrupt or gradual. The equation of state is very uncertain for a substantial portion of the interiors of both planets. Uranus and Neptune, contain, in a relative sense, more heavy elements. There are indications that their interiors are partially mixed (see text).

solar enrichment in carbon, in the form of CH_4, spectroscopically measured in the tropospheres of Uranus and Neptune (e.g., Fegley et al. 1991; Gautier et al. 1995). Other elements, in particular oxygen (mostly in the form of H_2O), are also believed to be substantially enriched compared to a solar-composition mixture but are hidden deep in the atmosphere because of condensation.

Although the two planets share many similarities (mass, magnetic field, atmospheric structure), several factors point toward some differences in their internal structure. Uranus emits scarcely more energy than received from the Sun, whereas Neptune possesses a very significant intrinsic heat flux, and Uranus's gravitational field indicates that it is more centrally condensed (see Hubbard et al. 1995). Furthermore, three-layer models of these planets that assume homogeneity of each layer and adiabatic temperature profiles succeed in reproducing Neptune's gravitational field but not that of Uranus (Podolak et al. 1995). The difficulty is circumvented by using slightly reduced (by $\sim$10%) densities in the ice layer, which is interpreted either as hydrogen mixed to the ice or as higher temperatures (superadiabatic temperature gradients). Both explanations imply that substantial parts of the planetary interior are not homogeneously mixed. The existence of such compositional gradients could also explain the fact that Uranus's heat flux is so small: part of its internal heat would not be allowed to escape to space by convection, but had to escape through a much slower diffusive process in the regions of high molecular-weight gradient (Podolak et al. 1991). Such regions would also be present in Neptune, but deeper, thus allowing more heat to be transported outward. This could also explain the fact that the magnetic fields of these two planets possess a very significant quadrupolar component, by allowing a hydromagnetic dynamo to form only in a relatively thin shell rather than in a sphere (Ruzmaikin and Starchenko 1991; Hubbard et al. 1995).

The existence of these nonhomogeneous regions is further confirmed by the fact that if hydrogen is supposed to be confined solely to the hydrogen-helium envelope, models predict ice/rock ratios of the order of 10 or more, much larger than the protosolar value of $\sim$2.5. On the other hand, if we impose the constraint that the ice/rock ratio is protosolar, the overall composition of both Uranus and Neptune is, by mass, about 25% rock, 60–70% ices, and 5–15% hydrogen and helium (Hubbard and Marley 1989; Podolak et al. 1991, 1995; Hubbard et al. 1995). The formation of these nonhomogeneous regions is certainly contemporaneous with the accretion of the planets (Hubbard et al. 1995). The importance of stochastic processes during that epoch is shown by the 98° obliquity of Uranus, a strong sign that giant impacts shaped the actual structure of these ice giants (Lissauer and Safronov 1991; cf., however, Tremaine 1991 for an alternative explanation of giant planet obliquities).

The structure of the much more massive Jupiter and Saturn, which are mostly formed from hydrogen and helium, is comparatively simpler. Most interior models (Hubbard and Marley 1989; Chabrier et al. 1992; Guillot et al. 1994*a*) of these planets assume a three-layer structure: a core, an inner envelope where hydrogen is in metallic phase, and an outer one where hydrogen is mostly in the form of H_2. More complex models (e.g., Zharkov and Gudkova 1991) can be calculated, but these further divisions into multiple layers do not qualitatively affect the main results.

Each layer is assumed to be globally homogeneous (i.e., neglecting condensation and chemical reactions), a consequence of efficient mixing by convection. Because less helium is observed in the external layers of Jupiter and Saturn than was present in the protosolar nebula (von Zahn et al. 1998; Gautier and Owen 1989), it is believed that the metallic regions of these planets contain more helium than the molecular ones. The difference is thought to be due to a first-order molecular-metallic phase transition of hydrogen, a hydrogen-helium phase separation, or both (e.g., Stevenson and Salpeter 1977; Hubbard and Marley 1989). Both phenomena are expected to occur in similar regions (e.g., Stevenson 1982*b*), and therefore we do not differentiate one from the other. We emphasize, however, that a first-order transition such as that suggested by Saumon and Chabrier (1989), would lead to a discontinuity of abundance of all chemical elements, whereas a phase *separation* would principally affect helium and other minor species, such as neon, that tend to be dissolved into helium droplets (unless, e.g., water is present in large enough abundances and can also separate from hydrogen). The lack of neon measured by the *Galileo* Probe (Niemann et al. 1998) suggests that, in Jupiter, helium phase separation has begun (Roulston and Stevenson 1995), and that, consequently, it also occurs in Saturn, which is colder.

With these hypotheses, Guillot et al. (1997) and Guillot (1999) calculate the ensemble of interior models of Jupiter and Saturn that match the gravitational moments within the error bars of the measurements. Using the inferred mass mixing ratio of helium in the protosolar nebula and the presently observed one in the atmospheres of Jupiter and Saturn, they retrieve the possible abundances of heavy elements in the metallic and molecular regions. Their calculations include uncertainties in the hydrogen-helium and heavy elements equations of state, in the inner temperature profile (convective/radiative), and regarding the internal rotation (solid/differential).

The resulting constraints on the core mass and total mass of heavy elements in Jupiter, Saturn, Uranus and Neptune are summarized in Fig. 2. (The cases of Uranus and Neptune are relatively trivial; these planets contain little hydrogen and helium.) A first result is that the gravitational fields of Jupiter and Saturn do not necessarily imply that these planets have ice/rock cores. In the case of Jupiter, models without a core are obtained only in the case of the less favored interpolated equation of state of hydrogen, whose calculation is not completely thermodynamically consistent (see Saumon et al. 1995). In the case of Saturn, it is difficult to distinguish between heavy elements in the core and those in the metallic region, hence yielding an even larger uncertainty in the core mass. As a result, Jupiter has a core whose mass lies between 0 and 14 $M_\oplus$, and Saturn's core is between 0 and 22 $M_\oplus$. We stress, however, that larger core masses are possible if gravitational layering occurs and the cores, possibly eroded by convective mixing, extend into the metallic envelope.

A second result concerns the total mass of heavy elements. The models of Saturn show that the planet is significantly enriched in heavy elements, by a factor of 10 to 15 compared to the solar value (corresponding to 20–30 $M_\oplus$, including the core) and by at least a factor of 5 when considering only the envelope. The constraints are much weaker in the case of Jupiter because of the larger metallic region, where the equation of state is considerably more uncertain. Figure 2 shows that Jupiter contains 10 to 42 $M_\oplus$

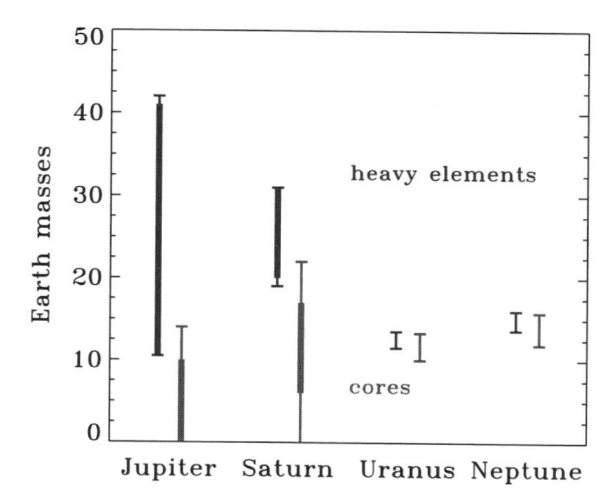

Figure 2. Limits on the abundances of heavy elements in the four jovian planets in our solar system. For each planet, the point on the left represents the total amount of high-Z material, whereas the (lower) point on the right shows the amount of heavy elements segregated into the planet's core. For Jupiter and Saturn, the thick lines represent solutions with additional constraints obtained from evolution models. Note the high level of uncertainty, especially regarding the core masses of Jupiter and Saturn. Models of Jupiter with small cores (i.e., less than 2 $M_{\oplus}$) require significant enrichments in heavy elements (i.e., more than 20 $M_{\oplus}$).

of heavy elements, implying that it is moderately to significantly enriched in heavy elements compared to the protosolar nebula.

Recent interior models calculated with different assumptions (Hubbard and Marley 1989; Zharkov and Gudkova 1991; Chabrier et al. 1992) generally predict core masses and heavy-elements abundances that fall within the ranges given in Fig. 2. Larger core masses (10–30 $M_{\oplus}$) were found in previous calculations (see Stevenson 1982*b* for a review), but the largest core masses also yielded helium mass fractions well below the protosolar value and therefore are unrealistic. The main reason for the discrepancy with today's values is, however, that the calculation of core masses, especially in the case of Jupiter, is very sensitive to changes in the equation of state. At present, we can only hope that advances in our understanding of the behavior of hydrogen and helium at high pressures have led us in the right direction. Progress in compression experiments on liquid deuterium (Weir et al. 1996; Collins et al. 1998) should allow us to check that assertion in the near future.

In the case of Jupiter and Saturn, further constraints on today's internal structure can be sought from evolution models that account for the progressive sedimentation of helium (Hubbard et al. 1999; Guillot 1999). Models with small cores tend to require a more pronounced helium differentiation and therefore yield longer cooling times. The time to cool to the present temperature is constrained by the age of the solar system

(4.56 Gyr). Some static solutions are thus ruled out. Figure 2 shows that an upper limit of 10 $M_\oplus$ is obtained for Jupiter's core, and that Saturn's core mass lies between 6 and 17 $M_\oplus$.

Finally, observations of the atmospheric composition give two important clues: First, the C/H ratio steadily increases from Jupiter to Neptune, a fact that has to be explained by formation models (we refer the reader to the review by Podolak et al. 1993 for further details). Second, the D/H isotopic ratios recently measured in Jupiter by the *Galileo* Probe (Mahaffy et al. 1998), and in Saturn by the *Infrared Space Observatory* (ISO) (Griffin et al. 1996) are, within the error bars, consistent with the protosolar value derived from ^{3}He/^{4}He in the solar wind, namely $2.1 \pm 0.5 \times 10^{-5}$ (Geiss and Gloecker 1998), whereas they are about three times larger in Uranus and Neptune (Feuchtgruber et al. 1999). This has to be compared to the D/H values measured in comets Halley, Hyakutake, and Hale-Bopp, which are all about 10 times larger than the protosolar value (Bockelée-Morvan et al. 1998). If Uranus and Neptune contained mixtures of cometlike ices and protosolar H_2 that were isotopically homogenized within these planets, their large ice fractions would have produced a more deuterium-rich atmospheric composition than that observed. Thus, either a significant isotopic exchange between vaporized ices and hydrogen took place in an early hot turbulent solar nebula (Drouart et al. 1999), or Uranus and Neptune formed from high-D/H, cometlike ices that had never been fully mixed with hydrogen in their interiors. The precise determination of D/H in the giant planets is thus an important tool for constraining their formation. We leave a more thorough discussion of this problem to the chapter by Lunine et al., this volume.

III. FORMATION OF PLANETESIMALS AND GROWTH OF SOLID PLANETS

A. Formation of Planetesimals

Even a very slowly rotating molecular cloud core has far too much rotational angular momentum to collapse down to an object of stellar dimensions, so a significant fraction of the material in a collapsing core falls onto a rotationally supported disk in orbit about the pressure-supported star. Such a disk has the same elemental composition as the growing star; that is, primarily H and He, with ~1–2% heavier elements. Sufficiently far from the central star, it is cool enough for some of this material to be in solid form, either remnant interstellar grains or condensates formed within the disk. Dust agglomerates via inelastic collisions and gradually settles towards the disk midplane as particles grow large enough to be able to drift relative to the surrounding gas (Weidenschilling and Cuzzi 1993).

Although to a first approximation the gas in the disk is centrifugally supported in balance with the star's gravity, negative radial pressure gradients provide a small, outwardly directed force that acts to reduce the effective gravity, so the gas rotates at slightly less than the Keplerian velocity. Small solid bodies (dust grains) rotate with the gas. Large solid bodies orbit at the Keplerian velocity, and medium-sized particles move at a rate intermediate between the gas velocity and the Keplerian velocity; thus, macroscopic solid bodies are subjected to a headwind from the gas (Adachi et al. 1976). This headwind removes angular momentum from the particles, causing them to spiral inward towards the central star. This inward drift can be very rapid, especially for particles whose coupling time to the gas is similar to their orbital period. Smaller particles drift less rapidly because the headwind they face is not as strong, whereas large particles drift less because they have a greater mass-to-surface-area ratio. Orbital decay times for meter-sized particles at 1 AU from the Sun have been estimated to be only ~100 years (Weidenschilling 1977). The large radial velocities of bodies in this size range, relative to both larger and smaller particles, implies frequent collisions, so it is possible that most solid bodies grow through the critical size range quickly without substantial radial drift. However, it is also possible that a large amount of solid planetary material is lost from the disk in this manner.

Solid bodies larger than ~1 km in size face a headwind only slightly faster than meter-sized objects (for parameters thought to be representative of the planetary region of the solar nebula), and because of their much greater mass-to-surface-area ratio they suffer far less orbital decay from interactions with the gas in their path. The growth of solid bodies from the meter-sized "danger zone" to the kilometer-sized "safe zone" could occur by collective gravitational instabilities in a thin subdisk of solids (Safronov 1960; Goldreich and Ward 1973) in regions of protoplanetary disks that are not too turbulent, or (more likely) via continued binary accretion (Weidenschilling and Cuzzi 1993). Kilometer-sized planetesimals appear to be reasonably safe from loss (unless they are ground down to smaller sizes via disruptive collisions) until some of these planetesimals grow into planetary-sized bodies.

B. Growth of Solid Planets

The primary perturbations on the Keplerian orbits of kilometer-sized and larger bodies in protoplanetary disks are mutual gravitational interactions and physical collisions (Safronov 1969). These interactions lead to accretion (and in some cases erosion and fragmentation) of planetesimals. Gravitational encounters are able to stir planetesimal random velocities up to the escape speed from the largest common planetesimals in the swarm (Safronov 1969). The most massive planets have the largest gravitationally enhanced collision cross sections and accrete almost everything with which they collide. If the random velocities of most planetesimals remain much smaller than the escape speed from the largest bodies, then these large "planetary embryos" grow extremely rapidly (Safronov 1969; cf. Greenzweig and Lissauer 1990, 1992 for three-body growth rates). The size distribution of solid bodies becomes quite skewed, with a few large bodies growing much faster than the rest of the swarm in a process known as *runaway accretion* (Greenberg et al. 1978; Wetherill and Stewart 1989; Kokubo and Ida 1996). Eventually, planetary embryos accrete most of the (slowly moving) solids within their gravitational reach, and the runaway growth phase ends. Planetary embryos can continue to accumulate solids rapidly beyond this limit if they migrate radially relative to planetesimals as a result of interactions with the gaseous component of the disk (Tanaka and Ida 1999).

The eccentricities of planetary embryos in the inner solar system were subsequently pumped up by long-range mutual gravitational perturbations; collisions between these embryos eventually formed the terrestrial planets (Wetherill 1990; Chambers and Wetherill 1998). However, timescales for this type of growth in the outer solar system (at least 10^8 years; Safronov 1969) are longer than the lifetime of the gaseous disk (cf. Lissauer et al. 1995). Moreover, unless the eccentricities of the growing embryos are damped substantially, embryos will eject one another from the star's orbit (Levison et al. 1998). Thus, runaway growth, possibly aided by migration (Tanaka and Ida 1999), appears to be the way by which solid planets can become sufficiently massive to accumulate substantial amounts of gas while the gaseous component of the protoplanetary disk is still present (Lissauer 1987).

Most models of the accumulation of giant planet atmospheres have assumed a constant accretion rate for planetesimals. The models of Pollack et al. (1996) calculate the planetesimal accretion rate together with that of gas; however, these models neglect growth of competing planetary cores as well as radial migration. Models of giant planet growth will improve once atmospheric accumulation models are coupled to sophisticated models of solid planet growth, such as the multizoned numerical accretion code of Weidenschilling et al. (1997), and when radial migration of planetesimals and planets is better understood and included in the models.

IV. GAS ACCUMULATION THEORIES

The key problem in giant planet formation is that preplanetary disks are only weakly self-gravitating equilibrium structures, supported by centrifugal forces augmented by gas pressure (see chapters by Stone et al., Hollenbach et al., Calvet et al., and Beckwith et al., this volume). Any isolated, orbiting object below the Roche density is pulled apart by the stellar tides. Typical nebula densities are more than two orders of magnitude below the Roche density, so compression is needed to confine a condensation of mass M inside its tidal or Hill radius R_T at orbital distance a

$$R_T = a\left(\frac{M}{3\ \mathrm{M}_\odot}\right)^{1/3} \qquad (1)$$

A local enhancement of self-gravity is needed to overcome the counteracting gas pressure. Giant planet formation theories may be classified by how they provide this enhancement.

1. The *nucleated instability* model relies on the extra gravity field of a sufficiently large solid core (condensed material represents a gain of ten orders of magnitude in density, and therefore self-gravity, compared to the nebula gas).
2. A *disk instability* may operate on lengthscales between short-scale pressure support and long-scale tidal support.
3. An *external perturber* could compress an otherwise stable disk on its local dynamical timescales, e.g., by accretion of a clump onto the disk or rendezvous with a stellar companion.

If the gravity enhancement is provided by a dynamical process as in the latter two cases, the resulting nebula perturbation (say, of a Jupiter mass, M_{Jup}, of material) is compressionally heated, because it is optically thick under nebula conditions. Giant planet formation would then involve a transient phase of tenuous giant gaseous protoplanets, which would be essentially fully convective and would contract on a timescale of $\sim 10^6$ yr (see Bodenheimer 1985).

Another mechanism of forming stellar companions, *fragmentation during collapse*, is plausible for binary stars and possibly brown dwarfs, but it is unlikely to form objects of planetary masses, because opacity limits the process to masses above $\sim 10\ M_{Jup}$ (cf. the chapter by Bodenheimer et al. in this volume; and Bodenheimer et al. 1993).

A. Nebula Stability

Preplanetary nebulae with minimum reconstituted mass are stable. Substantially more massive disks resulting from the collapse of cloud cores are self-stabilizing by transfer of disk mass to the stabilizing central protostar (Bodenheimer et al. 1993). Nevertheless, a moderate-mass nebula disk might be found that can develop a disk instability leading to a strong density perturbation, especially when forced with a finite external perturbation. Giant gaseous protoplanets (GGPPs) might form when the instability has developed into a clump (DeCampli and Cameron 1979; Bodenheimer 1985). Boss (1997, 1998) has constructed such an unstable disk with $0.13\ M_\odot$ within 10 AU and obtained maximum density enhancements (by a factor ~ 20) with $\sim 10\ M_{Jup}$ above the background for a few orbital periods. (The density enhancement at the surface of a $1\text{-}M_\oplus$ core is between 10^5 and 10^7, for comparison.) These clumps, provided they are stable on a few cooling times, are candidates to become proto-giant planets via an intermediate state as tenuous GGPPs.

A key issue, as in any theory involving an instability of the disk gas, is then the *a posteriori* formation of a core. Only metals that are present initially would rain out to form a core, whereas material added later by impacts of small bodies after the GGPP had formed would be soluble in the envelope (Stevenson 1982*a*). Boss (1998) outlines how a core corresponding to the solar-composition high-Z material (6 $M_\oplus$ and 2 $M_\oplus$ for Jupiter and Saturn mass, respectively) might form if the density enhancements are long-lived, need no more pressure confinement, and evolve into GGPPs that are *nonturbulent*. It should be noted here (see section II) that although interior models of Saturn do not rule out the possibility that the planet has no core (or, equivalently a 2-$M_\oplus$ core), this is not the favored solution. Also, GGPP models would probably predict that Jupiter should have a bigger core than Saturn, which is only marginally consistent with present interior models. Finally, Jupiter and certainly Saturn contain a lot of heavy elements (see Fig. 2). To account for these bulk heavy element compositions, planetesimal accretion must occur anyway after the GGPPs have formed their cores.

If GGPPs need pressure confinement, they also require the presence of an (undepleted) nebula and pose a lifetime constraint for the nebula, namely that nebula dispersal can begin only after a cooling time, that is $\sim 10^6$ yr (Bodenheimer 1985). To determine whether GGPPs are convectively stable, so that the nonturbulent core growth scenario can be applied, a detailed calculation of their thermal structure during contraction is necessary. DeCampli and Cameron (1979) found largely convective GGPPs.

One of us (G. W.) checked convective stability of GGPPs by a radiation hydrodynamical calculation. Alexander and Ferguson (1994) opacities and time-dependent MLT convection were used in the description of energy transfer. The initial condition was a Jeans-critical nebula condensation of M_{Jup} and a temperature of 10 K. Initially the GGPP had similar properties as Boss's (1998) banana-shaped density enhancements (mean density 8×10^{-10} g cm^{-3}, central density 3.3×10^{-8} g cm^{-3}). According to the new calculation, the GGPP needs 1.8×10^4 yr to contract into the tidal radius and is essentially fully convective from <100 yr to 2×10^5 yr, when a radiative zone spreads out from the planet's center.

B. Nucleated Instability

Planetesimals in the solar nebula are small bodies surrounded by gas. A rarefied equilibrium atmosphere forms around such objects. Early work in the nucleated instability hypothesis, which assumes that such solid "cores" trigger giant planet formation, was motivated by the idea that at a certain critical core mass the atmosphere could not be sustained, and isothermal, shock-free accretion (Bondi and Hoyle 1944; Bondi 1952) would set in. Determinations of this critical mass were made for increasingly detailed description of the envelopes: adiabatic (Perri and Cameron 1974), isothermal (Sasaki 1989), isothermal-adiabatic (Harris 1978; Mizuno et al. 1978), and with radiative and convective energy transfer (Mizuno 1980). By then, modeling the formation and evolution of a proto-giant planet had become essentially a miniature stellar structure calculation, with energy dissipation of impacting planetesimals replacing the nuclear reactions as the energy source. Present results on the critical mass are reviewed in the next section. Already, Safronov and Ruskol (1982) pointed out that, even after the instability at the critical mass, *"the rate of gas accretion is determined not by the rate of delivery of mass to the planet [as in Bondi accretion] but by the energy losses from the contracting envelope."* The planet's accretion rate is limited by the delivery of mass only when $M_P \gtrsim 100\ \mathrm{M}_\oplus$. Consequently, the energy budget of the envelope has been modeled more carefully, taking into account the heat generated by gravitational contraction (quasihydrostatic models by Bodenheimer and Pollack 1986).

Major progress since the *Protostars and Planets III* conference has been made by a detailed treatment of planetesimal accretion to calculate the core growth rate and the capture, dissolution, and sinking that determines how much and where in the envelope the planetesimal kinetic energy is liberated (Pollack et al. 1996). That made possible the first study of the coupling between gas accretion and solid accretion. Additionally, the description of the mechanics of contraction has been improved by hydrodynamic studies that determine the flow velocity of the gas by solving an equation of motion for the envelope gas in the framework of convective radiation-fluid dynamics (e.g., Wuchterl 1993, 1995*a*, 1999). That allows the study of collapse of the envelope, accretion with finite Mach number, and an access to the study of linear adiabatic (Tajima and Nakagawa 1997) and nonlinear, nonadiabatic pulsational stability and pulsations of the envelope. Furthermore, the treatment of convective energy transfer has been improved by calculations using a time-dependent mixing length theory of convection (Wuchterl 1995*b*, 1996, 1997) in hydrodynamics. The first hydrodynamic calculations with rotation in the quasispherical approximation have been undertaken by Götz 1993.

Most aspects of early envelope growth, up to $\sim 10\ \mathrm{M}_\oplus$, can be understood on the basis of a simplified analytical model given by Stevenson (1982*a*) for a protoplanet with constant opacity κ_0, core mass accretion rate $\dot{M}_{\mathrm{core}}$, and core density ρ_{core}, inside the tidal radius R_T. The key properties of Stevenson's model come from the "radiative zero solution" for spherical protoplanets with static, fully radiative envelopes, in hydrostatic and thermal equilibrium. We present here the solution relevant to the structure of an envelope in the gravitational potential of a constant mass, for zero external temperature and pressure and using a generalized opacity law of the form $\kappa = \kappa_0 P^a T^b$. The critical mass, defined as the largest mass to which a core can grow while forced to retain a static envelope, is then given by

$$M_{\mathrm{core}}^{\mathrm{crit}} = \left[\frac{3^3}{4^4} \left(\frac{\mathcal{R}}{\mu} \right)^4 \frac{1}{4\pi G} \frac{4-b}{1+a} \frac{3\kappa_0}{\pi\sigma} \left(\frac{4\pi}{3} \rho_{\mathrm{core}} \right)^{1/3} \frac{\dot{M}_{\mathrm{core}}}{\ln\left(R_T / r_{\mathrm{core}}\right)} \right]^{3/7} \qquad (2)$$

where $M_{\mathrm{core}}^{\mathrm{crit}} / M_{\mathrm{tot}}^{\mathrm{crit}} = \frac{3}{4}$; and $\mathcal{R}$, G, and σ denote the gas constant, the gravitational constant, and the Stefan-Boltzmann constant, respectively. The critical mass depends on neither the midplane density ρ_{Neb}, nor on the temperature T_{Neb} of the nebula in which the core is embedded. The outer radius, R_T, enters only logarithmically. The strong dependence of the analytic solution on molecular weight, μ, led Stevenson (1984) to propose "superganymedean puffballs" with atmospheres assumed to be enriched in heavy elements. Such objects would have low critical masses, providing a way to form giant planets rapidly (see also Lissauer et al. 1995). Equation (2) permits a glimpse of the effect of the run of opacity via the power law exponents a and b. Except for the weak dependences discussed above, a proto-giant planet essentially has the same global properties for a given core wherever it is embedded in a nebula. Even the dependence on $\dot{M}_{\mathrm{core}}$ is relatively weak; detailed radiative/convective envelope models show that a variation of a factor of 100 in $\dot{M}_{\mathrm{core}}$ leads only to a 2.6 variation in the critical core mass.

This similarity in the static structure of proto-giant envelopes yields similar dynamical behaviors characterized by pulsation-driven mass loss for solar-composition nebula opacities (see section V.B). However, other static solutions are found for protoplanets with *convective* outer envelopes, which occur for somewhat larger midplane densities than in minimum mass nebulae (Wuchterl 1993). These largely convective proto-giant planets have larger envelopes for a given core and a reduced critical core mass. Their properties can be illustrated by a simplified analytical solution for fully convective, adiabatic envelopes with constant first adiabatic exponent, Γ_1:

$$M_{\mathrm{core}}^{\mathrm{crit}} = \frac{1}{\sqrt{4\pi}} \frac{\sqrt{\Gamma_1 - \frac{4}{3}}}{(\Gamma_1 - 1)^2} \left(\frac{\Gamma_1}{G} \frac{\mathcal{R}}{\mu} \right)^{3/2} T_{\mathrm{Neb}}^{3/2} \rho_{\mathrm{Neb}}^{-1/2} \tag{3}$$

and $M_{\mathrm{core}}^{\mathrm{crit}}/M_{\mathrm{tot}}^{\mathrm{crit}} = \frac{2}{3}$. In this case, the critical mass depends on the nebula gas properties and therefore the location in the nebula, but it is independent of the core accretion rate. Of course, both the radiative zero and fully convective solutions are approximate, because they only roughly estimate envelope gravity, and all detailed calculations show radiative *and* convective regions in proto-giant planets. In Fig. 3 the transition from "radiative" to "convective" protoplanets is shown by results from detailed static radiative/convective calculations for $\dot{M}_{\mathrm{core}} = 10^{-6}\ \mathrm{M}_{\oplus}\ \mathrm{yr}^{-1}$ (Wuchterl 1993). Nebula conditions are varied from low densities, resulting in radiative outer envelopes, to enhanced densities that result in largely convective proto-giant planets. The critical mass can be as low as 1 $\mathrm{M}_{\oplus}$, and subcritical static envelopes can grow to 48 $\mathrm{M}_{\oplus}$. Calculations with updated opacity and improved, mixing-length convection (Wuchterl 1999) and the inclusion of rotational effects in the quasispherical approximation (Götz 1993) show a reduction of the critical core mass from 13 to 7 $\mathrm{M}_{\oplus}$ for the "radiative" proto-giant planets at the low nebula densities. The new, lower values are in better agreement with the new interior models.

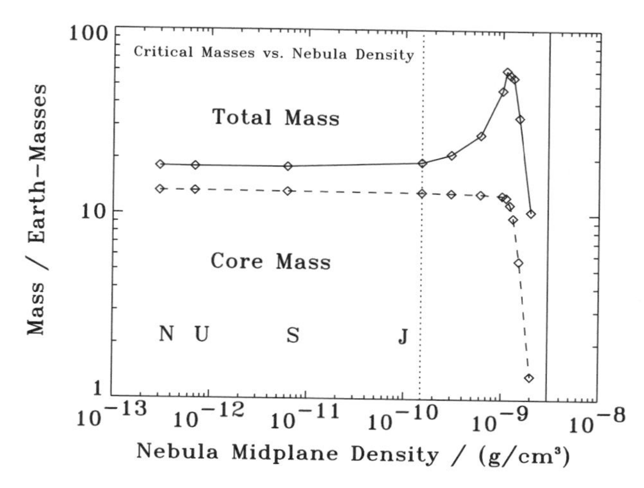

Figure 3. Critical masses of static protoplanets as a function of nebula midplane density. Critical total mass and core mass values are connected by a solid and a dashed curve, respectively. Observe the increased envelope masses and decreased core masses for the convective outer envelopes occurring at larger nebula densities. The conditions in the nebula correspond to Mizuno's minimum-mass nebula (Mizuno 1980); densities at the Neptune, Uranus, Saturn, and Jupiter positions are labeled by N, U, S, and J, respectively. They illustrate the constancy of the critical mass in the case of radiative outer envelopes. Densities to the right of the dotted vertical line are arbitrarily enhanced relative to the minimum-mass values, so that the outer envelopes become convective (see text). The solid vertical line gives an estimate for the critical density of a Jupiter-mass nebula fragment at Jupiter's position. The value plotted is the mean density of a condensation that is Jeans critical *and* fits into its Hill sphere.

The early phases of giant planet formation discussed above are dominated by the growth of the core. The envelopes adjust rapidly to the changing size and gravity of the core. As a result, the envelopes of proto-giant planets remain very close to static and in equilibrium below the critical mass (Mizuno 1980; Wuchterl 1993). This must change when the envelopes become more massive and cannot reequilibrate as rapidly as the cores grow. The nucleated instability was assumed to set in at the critical mass, originally as a hydrodynamic instability analogous to the Jeans instability. With the recognition that energy losses from the proto-giant planet envelopes control the further accretion of gas, it followed that quasihydrostatic contraction of the envelopes would play a key role.

V. DETAILED NUCLEATED INSTABILITY MODELS FOR THE GIANT PLANETS IN THE SOLAR SYSTEM

Major progress has been made since *Protostars and Planets III* by calculating the growth of the cores from planetesimal dynamics and the growth of the envelopes using hydrodynamics. We review these results below.

A. Quasihydrostatic Models with Detailed Core Accretion

Pollack et al. (1996) constructed models in which they simulated the concurrent accretion rates of both the gaseous and solid components of giant planets. Pollack et al. (1996) used an evolutionary model having three major components: a calculation of the three-body accretion rate of a single dominant-mass protoplanet surrounded by a large number of planetesimals, a calculation of the interaction of accreted planetesimals with the gaseous envelope of the growing giant protoplanet, and a calculation of the gas accretion rate using a sequence of quasihydrostatic models having a core/envelope structure. These three components of the calculation were updated every time step in a self-consistent fashion in which relevant information from one component was used in the other components.

The model of Pollack et al. (1996) is very detailed in many respects (core accretion rate, planetesimal dissolution in the envelope, treatment of energy loss via radiation and convection, equation of state), but it includes the following simplifying assumptions:

1. The planet is assumed to be spherically symmetric.
2. Hydrodynamic effects are not considered in the evolution of the envelope.
3. The opacity in the outer envelope is determined by a solar mixture of small grains in most of the simulations. Solar abundances are also used to calculate the opacity in deeper regions of the envelope, where molecular opacities dominate.
4. The equation of state for the envelope is that for a solar mixture of elements.
5. During the entire period of growth of a giant planet, it is assumed to be the sole dominant mass in the region of its feeding zone, i.e., there are no competing embryos, and planetesimal sizes and random velocities remain small. A corollary of this assumption is that accretion can be described as a quasicontinuous process, as opposed to a discontinuous one involving the occasional accretion of a massive planetesimal.
6. Planetesimals are assumed to be well-mixed within the planet's feeding zone, which grows as the planet's mass increases, but planetesimals are not allowed to migrate into or out of the planet's feeding zone as a consequence of their own motion. Tidal interaction between the protoplanet and the disk, or migration of the protoplanet (see the chapters by Lubow and Artymowicz, Ward and Hahn, and Lin et al. in this volume), are not considered.

It is not at all obvious that these various assumptions are valid, but no well-defined, quantitatively justifiable alternative assumptions are available.

The parameters in the calculations of Pollack et al. (1996) were adjusted to fit the properties of giant planets in the solar system and observations of disks around young stars. They judged the applicability of a given simulation to planets in our solar system using two basic criteria. One criterion is provided by the time required to reach the runaway gas accretion phase. This time interval should be less than the lifetime of the *gas* component of the solar nebula, t_{sn}, for successful models of Jupiter and Saturn and greater than t_{sn} for successful models of Uranus and Neptune. Limited observations of accretion disks around young stars suggest that $t_{sn} \lesssim 10^7$ yr, based on observations of the dust component (see the chapters by Calvet et al., Natta et al., and Lagrange et al., this volume). The lifetime of the gas component is less well constrained observationally (Strom et al. 1993). See the chapter by Wadhwa and Russell, this volume, and see also Podosek and Cassen (1994) for a review of nebula-lifetime estimates.

A second criterion is provided by the amount of high-Z mass accreted, M_Z. In the case of Jupiter and Saturn, M_Z at the end of a successful simulation should be comparable to, but somewhat smaller than, the current high-Z masses of these planets, because additional accretion of planetesimals occurred between the time they started runaway gas accretion and the time they contracted to their current dimensions and were able to scatter planetesimals gravitationally out of the solar system. Updated values of the constraints on high-Z material in the jovian planets are discussed in section II of this chapter.

In the models of Pollack et al. (1996), there are three main phases to the accretion of Jupiter and Saturn. Phase 1 is characterized by rapidly varying rates of planetesimal and gas accretion. Throughout phase 1, dM_Z/dt exceeds the rate of gas accumulation, dM_{XY}/dt. Initially, there is a very large difference (many orders of magnitude) between these two rates. However, they become progressively more comparable as time advances. Over much of phase 1, dM_Z/dt increases steeply. After a maximum at 5×10^5 years, it declines sharply. Meanwhile, dM_{XY}/dt grows steadily from its extremely low initial value. Phase 2 of accretion is characterized by relatively time-invariant values of dM_Z/dt and dM_{XY}/dt, with $dM_{XY}/dt > dM_Z/dt$. Finally, phase 3 is defined by rapidly increasing rates of gas and planetesimal accretion, with dM_{XY}/dt exceeding dM_Z/dt by steadily increasing amounts. The accretion of Uranus and Neptune was terminated during phase 2, presumably as a result of the dissipation or dispersal of the gas in the protoplanetary disk.

The models of Pollack et al. (1996) imply that the crossover mass, at which the solid and gas components of the planet are equal in mass, depends almost exclusively on the surface mass density of solids and the distance from the Sun. The crossover time is a rapidly decreasing function of the initial surface mass density of solids. A surface mass density of ≈ 10 g cm^{-2} at Jupiter yields both a small enough condensables mass and rapid enough gas accretion to be consistent with observations for nominal values of other parameters. Good fits for Saturn and Uranus are obtained if surface density of solids drops off with distance from the Sun as r^{-2}. Constraints on the surface density are quite restrictive in the "baseline" case, but a lower value is allowed if the opacity of the outer envelope is low because grains sink, if planetesimal heating is reduced because accreted planetesimals dissolve well above the core and their residue does not sink to the core, or if planetesimal accretion stops during phase 2, e.g., as a result of accretion by neighboring embryos (Pollack et al. 1996). Increasing the mean molecular weight of the envelope also increases the gas accretion rate [cf. equation (4)], but this parameter variation was not modeled by Pollack et al. (1996). The model results are relatively insensitive to moderately large changes in the gas density and temperature. Planetesimal size (which affects the velocity dispersion) is important in determining the duration of phase 1; for nominal parameters this has a small effect on the overall growth time for Jupiter, but the accretion time of Uranus is more profoundly affected by changes in planetesimal size.

B. Hydrodynamic Accretion beyond the Critical Mass

The static and quasihydrostatic models discussed so far rely on the assumptions that gas accretion from the nebula onto the core is very subsonic and that the inertia of the gas and dynamical effects such as dissipation of kinetic energy do not play a role. To check whether hydrostatic equilibrium is achieved and whether it holds, especially beyond the critical mass, hydrodynamical investigations are necessary. Two types of hydrodynamical investigations have been undertaken since *Protostars and Planets III*: (1) linear adiabatic dynamical stability analysis of envelopes evolving quasihydrostatically (Tajima and Nakagawa 1997) and (2) nonlinear, convective radiation hydrodynamical calculations of core-envelope proto-giant planets (Wuchterl 1993, 1995*a*, 1996, 1997, 1999) that follow the evolution of a proto-giant planet without *a priori* assuming hydrostatic equilibrium and which *determine* whether envelopes are hydrostatic, pulsate, or collapse and at what rates mass flows onto the planet. Wuchterl's models solve the flow equations for the envelope gas, essentially assuming only that spherical symmetry holds. They determine the net gain and loss of mass from the equations of motion for the gas in spherical symmetry, whereas quasihydrostatic calculations add mass according to some prescription and then calculate the structure for the updated mass, yielding a new equilibrium. Although the other assumptions made in the hydrodynamic calculations agree with those listed in the previous section for the quasihydrostatic models, there is a second important difference: The core accretion rate is, for simplicity, assumed to be either constant or calculated according to the particle-in-a-box approximation (see, e.g., Lissauer 1993).

The first hydrodynamical calculation of the nucleated instability (Wuchterl 1989, 1991*a,b*) started at the static critical mass and brought a surprise: Instead of collapsing, the proto-giant planet envelope begins to pulsate after a very short contraction phase (see Wuchterl 1990 for a simple discussion of the driving κ mechanism). The pulsations of the inner protoplanetary envelope expanded the outer envelope, and the outward-traveling waves caused by the pulsations resulted in a mass loss from the envelope into the nebula. The process can be described as a pulsation-driven wind. After a large fraction of the envelope mass has been pushed back into the nebula, the dynamical activity fades, and a new quasiequilibrium state is found that resembles Uranus and Neptune in core and envelope mass (Wuchterl 1991*a,b*). The mass loss process occurs in a very similar way for nebula conditions at Jupiter to Neptune positions and for core mass accretion rates from 10^{-7} to 10^{-5} $M_\oplus$ yr^{-1}. Starting the hydrodynamics at low core mass rather than at the critical mass does not change the eventual mass loss (Wuchterl 1995*a*).

Pulsations and mass loss do not occur when "no dust," zero-metallicity opacities are used; the lack of dust makes conditions most favorable for energy loss from the envelope and therefore for accretion. It is interesting to note that even for zero-metallicity opacities the static critical core mass is between 1.5 and 3 $M_\oplus$ for $\dot{M}_{core} = 10^{-8}$ to 10^{-6} $M_\oplus$ yr^{-1}, respectively. Envelope accretion becomes independent of core accretion at about 15 $M_\oplus$; the quasihydrostatic assumptions hold until inflow velocities reach a Mach number of 0.01 at about 50 $M_\oplus$. At a total mass of about 100 $M_\oplus$ the nebula gas influx approaches the Bondi accretion rate, and at 300 $M_\oplus$ the envelope collapses overall (cf. Wuchterl 1995*a*). This result shows that there must be an opacity-dependent transition from pulsation-driven winds to efficient gas accretion at the critical mass.

The main question concerning the hydrodynamics was then to ask for conditions that allow gas accretion (i.e., damp envelope pulsations) for "realistic" solar-composition opacities that include dust. Wuchterl (1993) derived conditions for the breakdown of the radiative zero solution by determining nebula conditions that would make the outer envelope of a "radiative" critical mass proto-giant planet convectively unstable. The resulting criterion gives a minimum nebula density that is necessary for a convective outer envelope. Protoplanets that grow under nebula conditions above that density have larger envelopes for a given core and a reduced critical mass as described in section IV.B. Convection is of great importance in damping stellar pulsations of RR Lyrae and δ-Cepheïd stars at the cool, so-called "red" end of the stellar instability strip; similar behavior may be expected in proto-giant planet envelopes. Wuchterl (1995*a*) calculated the growth of giant planets from low core masses hydrodynamically for a set of nebula conditions ranging from below the critical density to somewhat above. As the density was increased, the envelopes became increasingly more convective at the critical mass but still showed the mass loss. At a nebula density of 10^{-9} g cm^{-3} (i.e., greater by a factor of 6.7 than Mizuno's (1980) minimum reconstituted mass nebula value), the dynamical behavior was different: The pulsations were damped, and rapid accretion of gas set in and proceeded to 300 $M_\oplus$. Apparently the spreading of convection in the outer envelope had damped the pulsations, thereby inhibiting the onset of a wind and leading to accretion. The critical core masses required for the formation of this class of proto-giant planets are significantly smaller than for the Uranus/Neptune-type (see Wuchterl 1993, 1995*a*).

Improved Convective Energy Transfer and Opacities. Most giant planet formation studies use zero-entropy-gradient convection; that is, they set the temperature gradient to the adiabatic value in convectively unstable layers of the envelope. That is done for simplicity but can be inaccurate, especially when the evolution is rapid and hydrodynamical waves are present (see Wuchterl 1991*b*). It was, therefore, important to develop a time-dependent theory of convection that can be solved together with the equations of radiation hydrodynamics. Such a time-dependent convection model (Kuhfuß 1987) has been reformulated for self-adaptive grid radiation hydrodynamics (Wuchterl 1995*b*) and applied to giant planet formation (Götz 1993; Wuchterl 1996, 1997). In a reformulation by Wuchterl and Feuchtinger (1998), it closely approximates standard mixing length theory in a static local limit and accurately describes the solar convection zone and RR Lyrae light curves. In addition, updated molecular opacities (Alexander and Ferguson 1994) are used in a compilation by Götz (1993) to improve the accuracy of radiative transfer in the proto-giant planet envelopes. The effect of these improvements in energy transfer is that the core mass needed to initiate gas accretion to a few hundred Earth masses at various orbital radii is reduced to 8.30, 9.48, and 9.56 $M_\oplus$ at 0.052, 5.2, and 17.2 AU, respectively (see Fig. 4), even in a minimum-mass nebula.

VI. FORMATION OF EXTRASOLAR PLANETS

More than a dozen planets have thus far been discovered to orbit main-sequence stars other than the Sun; all of these objects are more massive than Saturn, and most are more massive than Jupiter (Mayor and Queloz 1995; chapter by Marcy et al., this volume, and references therein). The extrasolar planets currently known all orbit nearer to their stars than Jupiter does to the Sun (this is primarily an observational selection effect; high-precision radial velocity surveys have not been in operation long enough to have observed a full orbit of more distant planets). Some of these planets orbit on highly eccentric paths, suggesting that after they formed they were subjected to close encounters with other giant planets (Weidenschilling and Marzari 1996; Lin and Ida 1997; Levison et al. 1998) or, in the case of the companion to 16 Cygni B, secular perturbations from the star 16 Cyg A (Holman et al. 1997). Some of the extrasolar planets are separated from their stars by less than 1% of the Jupiter-Sun distance. Guillot et al. (1996) showed that giant planets are stable over the main-sequence lifetime of a 1-$M_\odot$ star even if they are as close as 0.05 AU. Models involving migration caused by disk-planet interactions are favored by many researchers for the formation of these objects (e.g., Lin et al. 1996; Trilling et al. 1998; see also the chapters by Ward and Hahn and by Lin et al., this volume). However,

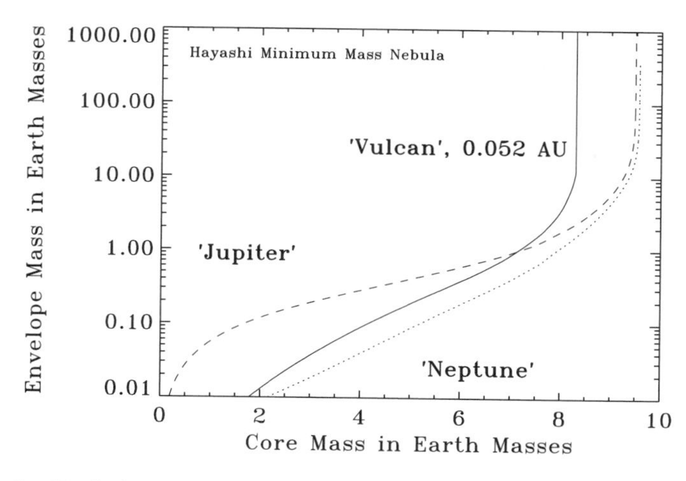

Figure 4. Evolution of proto-giant planets with mixing length convection. Envelope masses as obtained from hydrodynamic accretion calculations are plotted as functions of core mass for locations at 0.05 AU in the Hayashi et al. (1985) nebula (full) and for Mizuno's (1980) "Jupiter" (dashed) and "Neptune" (dotted) cases. Nebula temperatures and densities for the three cases are "Vulcan": 1252 K, 5.3×10^{-6} g/cm^3; "Jupiter": 97 K, 1.5×10^{-10} g/cm^3; and "Neptune": 45 K, 3.0×10^{-13} g/cm^3. The core accretion rate is 10^{-6} $M_\oplus$/yr.

simulations also show that it may be possible to form giant planets very close to stars, and we review these models in this section.

A. Hydrostatic Models for *In Situ* Formation

Bodenheimer et al. (2000) have modeled the formation and evolution of the planets recently discovered in orbit about the stars 51 Pegasi, ρ Coronae Borealis, and 47 Ursae Majoris, assuming that these planets formed in or near their current orbits. They used updated versions of the quasihydrostatic codes developed by Bodenheimer and Pollack (1986) and Pollack et al. (1996). The isolated protoplanet/no migration model of Pollack et al. (1996) requires high surface mass density of solids for giant planets to form close to stars within the observed lifetimes of protoplanetary disks. The primary cause of this restriction is that the larger Kepler shear near the star decreases the solid core's isolation mass unless the amount of solids is large; the increase in temperature closer to the star has only a very small effect (Mizuno 1980; Bodenheimer and Pollack 1986), and the higher density of gas acts in the opposite sense (Wuchterl 1996). The planet orbiting 2.1 AU from 47 UMa can form in ~2 Myr for a surface density of condensed material $\sigma = 90$ g cm^{-2} but requires ~18 Myr for $\sigma = 50$ g cm^{-2} (Bodenheimer et al. 2000). A value of $\sigma = 90$ g cm^{-2} at 2.1 AU is well above that used by Pollack et al. (1996), but still well below that required for local axisymmetric gravitational instabilities (Toomre 1964), assuming a solar-composition mix.

The surface mass density of solids required to form giant planets at 0.23 AU (ρ CrB) and 0.05 AU (51 Peg) is prohibitively large unless orbital decay of planetesimals is incorporated into the models. On the other hand, Ruzmaikina (1998) has given a model to provide the required amounts of both gas and solids. As no well-constrained method to quantify core growth close to stars is available, Bodenheimer et al. (2000) made the *ad hoc* assumption of a constant rate of solid-body accretion for these inner planets. Model results for 51 Peg indicate that if the growth rate of the core is 1×10^{-5} $M_\oplus$ yr^{-1}, then the planet takes $\sim 4 \times 10^6$ years to form and has a final high-Z mass of ~ 40 $M_\oplus$. Using the same definition for the planetary radius and the same planetesimal accretion rate as used by Bodenheimer et al. (2000), Wuchterl obtained, in a comparison calculation undertaken for this work, a critical core mass of about 25 $M_\oplus$. The two groups are currently attempting to resolve this discrepancy, an effort that will include calculations with identical opacities.

B. Hydrodynamical Models of Giant Planet Formation Near Stars

A major result of the hydrodynamical studies is that proto-giant planets may pulsate and develop pulsation-driven mass loss. Only if the pulsations are damped can gas accretion produce Jupiter-mass envelopes. Since all extrasolar planets discovered so far have minimum masses $\gtrsim 0.5$ M_{Jup}, they probably require efficient gas accretion and should satisfy the convective outer envelope criterion (Wuchterl 1993). A glance at Wuchterl's (1993) Fig. 2 shows that proto-giant planets located somewhat inside of Mercury's orbit in the Hayashi et al. (1985) minimum-mass nebula fulfill this condition. Convective radiation hydrodynamical calculations of core-envelope growth at 0.05 AU, for particle-in-a-box core mass accretion at nebula temperatures of 1250 and 600 K, show gas accretion beyond 300 $M_\oplus$ at core masses of 13.5 $M_\oplus$ and 7.5 $M_\oplus$, respectively (Wuchterl 1996, 1997).

It is interesting to apply the arguments based on the convection-controlled bifurcation in hydrodynamic accretional behavior to an ensemble of preplanetary nebula models, to simulate a variety of initial conditions for planet formation that might have been present around other stars. Wuchterl (1993) has shown that almost all nebula conditions, from a literature collection of nebula models, result in radiative outer envelopes at the critical mass. Nonlinear radiation hydrodynamical calculations with zero-entropy gradient convection show that Uranus/Neptune-type giant planets are produced under such circumstances (section V.B). Jupiter-mass planets should then be the exception. The first calculations with time-dependent mixing length convection, discussed in section V.B, show gas accretion to beyond a Jupiter mass for a much wider range of nebula conditions. Apparently the improved description of convection (and the updated opacities) have shifted the instability strip for pulsations and mass loss at the critical mass. Further calculations and a reanalysis of the conditions for efficient gas accretion for mixing length convection have to be undertaken before an updated expectation concerning the mass distribution of extrasolar planets can be given. An important requirement for that is an extensive theoretical and observational study of plausible preplanetary nebulae.

VII. CONCLUSIONS

Jupiter and Saturn are composed primarily of hydrogen and helium, yet the heavy elements that they contain may hold the key to the problem of their formation. The density profiles of these planets derived from interior models, as well as the composition of their atmospheres, clearly indicate a significantly larger fraction of heavy elements than was present in the protosolar gas. Were the heavy elements the first to accrete, or did the enrichment occur at later stages? Depending on the scenario, Jupiter and Saturn might have received very different amounts of planetesimals, thereby providing a way to differentiate a very rapid formation (such as in the nebula instability mechanism) from a slower one (such as in the nucleated instability).

Interior and evolution models for Jupiter and Saturn tend to favor core masses that lie within the range of acceptable critical core masses predicted within the nucleated instability hypothesis. The models based on this hypothesis also explain why Uranus and Neptune are mostly core: either because (i) gas accretion is limited to $\sim 1\ M_{\oplus}$ by a hydrodynamic instability that operates under certain nebula conditions, low gas density being the dominant factor (Wuchterl 1993, 1995*a*); (ii) their cores grew more slowly than those of Jupiter and Saturn because orbital timescales are longer farther from the Sun, and thus they did not achieve sufficient mass to accrete large quantities of gas before the solar nebula gas was dispersed (Pollack et al. 1996); or (iii) the gas in the Uranus/Neptune region of the nebula was dispersed rapidly via photoevaporation, whereas gas remained in the Jupiter/Saturn region for a much longer period of time (Shu et al. 1993).

The nucleated instability hypothesis thus provides a viable model for the formation of the giant planets observed in our solar system and beyond. Presently known extrasolar planets may have accreted *in situ* if their preplanetary nebulae provided sufficient amounts of gas and solids. Alternatively, according to studies of disk-induced migration (chapters by Ward and Hahn and by Lin et al., this volume) and gravitational encounters with other planets, they could have formed elsewhere and moved into the present positions. In that case the orbits of most if not all planets known to be bound to main-sequence stars *other than the Sun* suffered substantial orbital evolution.

The next few years will be dedicated to the development of a synoptic understanding of giant planet formation processes for a variety of preplanetary nebulae to work out predictive elements of the formation theories. These theories will be confronted by a representative observational census of giant planets orbiting neighboring stars.

Acknowledgments T. G.'s work is supported by CNRS (UMR 6529) and the Programme National de Planétologie. T. G. thanks Daniel Gautier for stimulating discussions and comments. G. W.'s work on this article has been supported by the Österreichischer Fonds zur Förderung der wissenschaftlichen Forschung (FWF) under project numbers S-7305-AST, S-7307-AST. We thank W. B. Hubbard for a constructive review of this manuscript.

REFERENCES

Adachi, I., Hayashi, C., and Nakazawa, K. 1976. The gas drag effect on the elliptic motion of a solid body in the primordial solar nebula. *Prog. Theor. Phys.* 56:1756–1771.

Alexander, D. R., and Ferguson, J. W. 1994. Low-temperature Rosseland opacities. *Astrophys. J.* 437:879–891

Bockelée-Morvan, D., Gautier, D., Lis, D. C., Young, K., Keene, J., Phillips, T., Owen, T., Crovisier, J., Goldsmith, P. F., Bergin, E. A., Despois, D., and Wootten, A. 1998. Deuterated water in comet C/1996 B2 (Hyakutake) and its implications for the origin of comets. *Icarus* 133:147–162.

Bodenheimer, P. 1985. Evolution of the giant planets. In *Protostars and Planets II*, ed. D. C. Black and M. S. Matthews (Tucson: University of Arizona Press), pp. 493–533.

Bodenheimer, P., and Pollack, J. B. 1986. Calculations of the accretion and evolution of giant planets: The effects of solid cores. *Icarus* 67:391–408.

Bodenheimer, P., Ruzmaikina, T., and Mathieu, R. D. 1993. Stellar multiple systems: Constraints on the mechanism of origin. In *Protostars and Planets III*, ed. E. H. Levy and J. I. Lunine (Tucson: University of Arizona Press), pp. 367–404.

Bodenheimer, P., Hubickyj, O., and Lissauer, J. J. 2000. Models of the *in situ* formation of detected extrasolar giant planets. *Icarus*, in press.

Bondi, H. 1952. On spherically symmetric accretion. *Mon. Not. Roy. Astron. Soc.* 112:195–204.

Bondi, H., and Hoyle, F. 1944. On the mechanism of accretion by stars. *Mon. Not. Roy. Astron. Soc.* 104:273–282.

Boss, A. P. 1997. Formation of giant gaseous protoplanets by gravitational instability. *Science* 276:1836–1839.

Boss, A. P. 1998. Evolution of the solar nebula IV—giant gaseous protoplanet formation. *Astrophys. J.* 503:923–937.

Cassen, P. M., and Moosman, A. 1981. On the formation of protostellar disks. *Icarus* 48:353–376.

Chabrier, G., Saumon, D., Hubbard, W. B., and Lunine, J. I. 1992. The molecular-metallic transition of hydrogen and the structure of Jupiter and Saturn. *Astrophys. J.* 391:817–826.

Chambers, J., and Wetherill, G. 1998. Making the terrestrial planets: N-body integrations of planetary embryos in three dimensions. *Icarus* 136:304–327.

Collins, G. W., Da Silva, L. B., Celliers, P., Gold, D. M., Foord, M. E., Wallace, R. J., Ng, A., Weber, S. V., Budil, K. S., and Cauble, R. 1998. Measurements of the equation of state of deuterium at the fluid insulator-metal transition. *Science* 282:89–92.
DeCampli, W. M., and Cameron, A. G. W. 1979. Structure and evolution of isolated giant gaseous protoplanets. *Icarus* 38:367–391.
Drouart, A., Dubrulle, B., Gautier, D., and Robert, F. 1999. Structure and transport in the solar nebula from constraints on deuterium enrichment and giant planets formation. *Icarus*, submitted.
Dutrey, A., Guilloteau, S., Prato, L., Simon, M., Duvert, G., Schuster, K., and Ménard, F. 1998. CO study of the GM Aurigae Keplerian disk. *Astron. Astrophys.* 339:467–476.
Fegley, B., Jr., Gautier, D., Owen, T., and Prinn, R. G. 1991. Spectroscopy and chemistry of the atmosphere of Uranus. In *Uranus*, ed. J. T. Bergstralh, E. D. Miner, and M. S. Matthews (Tucson: University of Arizona Press), pp. 147–203.
Feuchtgruber, H., Lellouch, E., Bézard, B., Encrenaz, T., De Graauw, T., and Davis, G. R. 1999. Detection of HD in the atmospheres of Uranus and Neptune: A new determination of the D/H ratio. *Astron. Astrophys.* 341:L17–L21.
Gautier, D., and Owen, T. 1989. Composition of outer planet atmospheres. In *Origin and Evolution of Planetary and Satellite Atmospheres*, ed. S. K. Atreya, J. B. Pollack, and M. S. Matthews (Tucson: University of Arizona Press), pp. 487–512.
Gautier, D., Conrath, B. J., Owen, T., De Pater, I., and Atreya, S. K. 1995. The troposphere of Neptune. In *Neptune and Triton*, ed. D. P. Cruikshank (Tucson: University of Arizona Press), pp. 547–611.
Geiss, J., and Gloecker, G. 1998. Abundances of deuterium and helium-3 in the protosolar cloud. *Space Sci. Rev.* 84:239-250.
Goldreich, P., and Ward, W. R. 1973. The formation of planetesimals. *Astrophys. J.* 183:1051–1061.
Götz, M. 1993. Die Entwicklung von Proto-Gasplaneten mit Drehimpuls—Strahlungshydrodynamische Rechnungen [The evolution of proto-giant planets with angular momentum—Radiation-hydrodynamical calculations], Dissertation, University of Heidelberg.
Greenberg, R., Hartmann, W. K., Chapman, C. R., and Wacker, J. F. 1978. Planetesimals to planets—numerical simulation of collisional evolution. *Icarus* 35:1–26.
Greenzweig, Y., and Lissauer, J. J. 1990. Accretion rates of planets. *Icarus* 87:40–77.
Greenzweig, Y., and Lissauer, J. J. 1992. Accretion rates of protoplanets. II. Gaussian distribution of planetesimal velocities. *Icarus* 100:440–463.
Griffin, M. J., Naylor, D. A., Davis, G. R., Ade, P. A. R., Oldham, P. G., Swinyard, B. M., Gautier, D., Lellouch, E., Orton, G. S., Encrenaz, T., De Graauw, T., Furniss, H., Smith, I., Armand, C., Burgdorf, M., Di Giorgio, A., Ewart, D., Gry, C., King, K. J., Lim, T., Molinari, S., Price, M., Sidher, S., Smith, A., Texier, D., Trams, N., Unger, S. J., and Salama, A. 1996. First detection of the 56-μm rotational line of HD in Saturn's atmosphere. *Astron. Astrophys. Lett.* 315:L389–L392.
Guillot, T. 1999. A comparison of the interiors of Jupiter and Saturn. *Planet. Space Sci.*, in press.
Guillot, T., Chabrier, G., Morel, P., and Gautier, D. 1994*a*. Non-adiabatic models of Jupiter and Saturn. *Icarus* 112:354–367.
Guillot, T., Gautier, D., Chabrier, G., and Mosser, B. 1994*b*. Are the giant planets fully convective? *Icarus* 112:337–353.

Guillot, T., Burrows, A., Hubbard, W. B., Lunine, J. I., and Saumon, D. 1996. Giant planets at small orbital distances *Astrophys. J. Lett.* 459:L35–L38.

Guillot, T., Gautier, D., and Hubbard, W. B. 1997. New constraints on the composition of Jupiter from Galileo measurements and interior models. *Icarus* 130:534–539.

Guilloteau, S., and Dutrey, A. 1998. Physical parameters of the Keplerian protoplanetary disk of DM Tau. *Astron. Astrophys.* 339:467–476.

Harris, A. W. 1978. The formation of the outer planets (abstract). *Lunar Planet. Sci.* 9:459–461.

Hayashi, C. 1980. Structure of the solar nebula, growth and decay of magnetic fields and effects of turbulent and magnetic viscosity on the nebula. *Prog. Theor. Phys. Suppl.* 70:35–53.

Hayashi, C., Nakazawa, K., and Nakagawa, Y. 1985. Formation of the solar system. In *Protostars and Planets II*, ed. D. C. Black and M. S. Matthews (Tucson: University of Arizona Press), pp. 1100–1153.

Holman, M., Touma, J., and Tremaine, S. 1997. Chaotic variations in the eccentricity of the planet orbiting 16 Cyg B. *Nature* 386:254–256.

Hubbard, W. B. 1968. Thermal structure of Jupiter. *Astrophys. J.* 152:745–753.

Hubbard, W. B., and Marley, M. S. 1989. Optimized Jupiter, Saturn and Uranus interior models. *Icarus* 78:102–118.

Hubbard, W. B., Podolak, M., and Stevenson, D. J. 1995. The interior of Neptune. In *Neptune and Triton*, ed. D. P. Cruikshank (Tucson: University of Arizona Press), pp. 109–138.

Hubbard, W. B., Guillot, T., Marley, M. S., Burrows, A., Lunine, J. I., and Saumon, D. 1999. Comparative evolution of Jupiter and Saturn. *Planet Space Sci.*, in press.

Kant, I. 1755. *Allgemeine Naturgeschichte und Theorie des Himmels* (Königsberg und Leipzig: Johann Friederich Petersen). English trans. W. Hastie, *Universal Natural History and Theories of the Heavens*. In *Kant's Cosmology*, (New York: Greenwood Publishing), 1968.

Kokubo, E., and Ida, S. 1996. On runaway growth of planetesimals. *Icarus* 123:80–191.

Kuhfuß, R. 1987. Ein Modell für zeitabhängige, nichtlokale Konvektion in Sternen [A model for time-dependent, non-local convection in stars], Ph.D. Thesis, Technical University of Munich.

Kusaka, T., Nakano, T., and Hayashi, C. 1970. Growth of solid particles in the primordial solar nebula. *Prog. Theor. Phys.* 44:1580–1596.

Laplaçe, P. S. 1796. *Exposition du système du monde* (Paris: Circle-Sociale), English trans. J. Pond, *The System of the World* (London: Richard Phillips), 1809.

Laughlin, G., and Bodenheimer, P. 1994. Nonaxisymmetric evolution in protostellar disks. *Astrophys. J.* 436:335–354.

Levison, H. F., Lissauer, J. J., and Duncan, M. J. 1998. Modeling the diversity of outer planetary systems. *Astron. J.* 116:1998–2014.

Lin, D. N. C., and Ida, S. 1997. On the origin of massive eccentric planets. *Astrophys. J.* 477:781–791.

Lin, D. N. C., Bodenheimer, P., and Richardson, D. 1996. Orbital migration of the planetary companion of 51 Pegasi to its present location. *Nature* 380:606–607.

Lissauer, J. J. 1987. Timescales for planetary accretion and the structure of the protoplanetary disk. *Icarus* 69:249–265.

Lissauer, J. J. 1993. Planet formation. *Ann. Rev. Astron. Astrophys.* 31:129–174.

Lissauer, J. J., and Safronov, V. S. 1991. The random component of planetary rotation. *Icarus* 93:288–297.

Lissauer, J. J., Pollack, J. B., Wetherill, G. W., and Stevenson, D. J. 1995. Formation of the Neptune system. In *Neptune and Triton*, ed. D. P. Cruikshank (Tucson: University of Arizona Press), pp. 37–108.

Mahaffy, P. R., Donahue, T. M., Atreya, S. K., Owen, T. C., and Niemann, H. B., 1998. Galileo probe measurements of D/H and ^{3}He/^{4}He in Jupiter's atmosphere. *Space Sci. Rev.* 84:251–263.

Marley, M. S., and McKay, C. P. 1999. Thermal structure of Uranus' atmosphere. *Icarus* 138:268–286.

Mayor, M., and Queloz, D. 1995. A Jupiter-mass companion to a solar-type star. *Nature* 378:355–359.

Mizuno, H. 1980. Formation of the giant planets. *Prog. Theor. Phys.* 64:544–557.

Mizuno, H., Nakazawa, K., and Hayashi, C. 1978. Instability of a gaseous envelope surrounding a planetary core and formation of giant planets. *Prog. Theor. Phys.* 60:699–710.

Morfill, G., Tscharnuter, W. M., and Völk, H. 1985. Dynamical and chemical evolution of the protoplanetary nebula. In *Protostars and Planets II*, ed. D. C. Black and M. S. Matthews (Tucson: University of Arizona Press), pp. 493–533.

Niemann, H. B., Atreya, S. K., Carignan, G. R., Donahue, T. M., Haberman, J. A., Harpold, D. N., Hartle, R. E., Hunten, D. M., Kasprzak, W. T., Mahaffy, P. R., Owen, T. C., and Way, S. H. 1998. The composition of the jovian atmosphere as determined by the *Galileo* probe mass spectrometer. *J. Geophys. Res.* 103:22831–22846.

Perri, F., and Cameron, A. G. W. 1974. Hydrodynamic instability of the solar nebula in the presence of a planetary core. *Icarus* 22:416–425

Podolak, M., Hubbard, W. B., and Stevenson, D. J. 1991. Models of Uranus' interior and magnetic field. In *Uranus*, ed. J. T. Bergstralh, E. D. Miner, and M. S. Matthews (Tucson: University of Arizona Press), pp. 29–61.

Podolak, M., Hubbard, W. B., and Pollack, J. B. 1993. Gaseous accretion and the formation of giant planets. In *Protostars and Planets III*, ed. E. H. Levy and J. I. Lunine (Tucson: University of Arizona Press), pp. 1109–1147.

Podolak, M., Weizman, A., and Marley, M. S. 1995. Comparative models of Uranus and Neptune. *Planet. Space Sci.* 43:1517–1522.

Podosek, F. A., and Cassen, P. 1994. Theoretical, observational, and isotopic estimates of the lifetime of the solar nebula. *Meteoritics* 29:6–25.

Pollack, J. B., Hubickyj, O., Bodenheimer, P., Lissauer, J. J., Podolak, M., and Greenzweig, Y. 1996. Formation of the giant planets by concurrent accretion of solids and gases. *Icarus* 124:62–85.

Roulston, M. S., and Stevenson, D. J. 1995. Prediction of neon depletion in Jupiter's atmosphere. *EOS* 76:343.

Ruzmaikin, A. A., and Starchenko, S. V. 1991. On the origin of the Uranus and Neptune magnetic fields. *Icarus* 93:82–87.

Ruzmaikina, T. V. 1998. Formation of 51 Peg type systems (abstract). *Lunar Planet. Sci. Abs.* 29:1873.

Safronov, V. S. 1960. On the gravitational instability in flattened systems with axial symmetry and non-uniform rotation. *Ann. d'Astrophys.* 23:901–904.

Safronov, V. S. 1969. *Evolution of the Protoplanetary Cloud and Formation of the Earth and Planets* (Moscow: Nauka Press) (Also NASA–TT–F–677, 1972.)

Safronov, V. S., and Ruskol, E. L. 1982. On the origin and initial temperature of Jupiter and Saturn. *Icarus* 49:284-296.

Sasaki, S. 1989. Minimum planetary size for forming outer Jovian-type planets: Stability of an isothermal atmosphere surrounding a protoplanet. *Astron. Astrophys.* 215:177–180.

Saumon, D., and Chabrier, G. 1989. Pressure ionization of hydrogen: the plasma phase transition. *Phys. Rev. Lett.* 62:2397–2400.

Saumon, D., Chabrier, G., and Van Horn, H. M. 1995. An equation of state for low-mass stars and giant planets. *Astrophys. J. Suppl.* 99:713–741.

Shu, F. H., Johnstone, D., and Hollenbach, D. 1993. Photoevaporation of the solar nebula and the formation of the giant planets. *Icarus* 106:92–101.

Stevenson, D. J. 1982*a*. Formation of the giant planets. *Planet. Space Sci.* 30:755–764.

Stevenson, D. J. 1982*b*. Interiors of the giant planets. *Ann. Rev. Earth Planet. Sci.* 10:257–295.

Stevenson, D. J. 1984. On forming the giant planets quickly ("superganymedean puffballs") (abstract) *Lunar Planet. Sci.* 15:822–823.

Stevenson, D. J., and Salpeter, E. E. 1977. The dynamics and helium distribution in hydrogen-helium fluid planets. *Astrophys. J. Suppl.* 35:239–261.

Strom, S. E., Edwards, S., and Skrutskie, M. F. 1993. Evolutionary timescales for circumstellar disks associated with intermediate and solar-type stars. In *Protostars and Planets III*, ed. E. H. Levy and J. I. Lunine (Tucson: University of Arizona Press), pp. 837–866.

Tajima, N., and Nakagawa, Y. 1995. Evolution and dynamical stability of the proto-giant-planet envelope. *Icarus* 126:282–292.

Tanaka, H., and Ida, S. 1999. Growth of a migrating protoplanet. *Icarus* 139:350–366.

Toomre, A. 1964. On the gravitational stability of a disk of stars. *Astrophys. J.* 139:1217–1238.

Tremaine, S. 1991. On the origin of the obliquities of the outer planets. *Icarus* 89:85–92.

Trilling, D., Benz, W., Guillot, T., Lunine, J. I., Hubbard, W. B., and Burrows, A. 1998. Orbital evolution and migration of giant planets: Modeling extrasolar planets. *Astrophys. J.* 500:428–439.

Tscharnuter, W. M. 1987. A collapse model of the turbulent presolar nebula. *Astron. Astrophys.* 188:55–73.

Tscharnuter, W. M., and Boss, A. P. 1993. Formation of the protosolar nebula. In *Protostars and Planets III*, ed. E. H. Levy and J. I. Lunine (Tucson: University of Arizona Press), pp. 921–938.

von Zahn, U., Hunten, D. M., and Lehmacher, G. 1998. Helium in Jupiter's atmosphere: Results from the *Galileo* Probe helium interferometer experiment. *J. Geophys. Res.* 103:22815–22830.

Weidenschilling, S. J. 1977. Aerodynamics of solid bodies in the solar nebula. *Mon. Not. Roy. Astron. Soc.* 180:57–70.

Weidenschilling, S. J., and Cuzzi, J. N. 1993. Formation of planetesimals in the solar nebula. In *Protostars and Planets III*, ed. E. H. Levy and J. I. Lunine, (Tucson: University of Arizona Press), pp. 1031–1060.

Weidenschilling, S. J., and Marzari, F. 1996. Gravitational scattering as a possible origin for giant planets at small stellar distances. *Nature* 384:619–621.

Weidenschilling, S. J., Spaute, D., Davis, D. R., Marzari, F., and Ohtsuki, K. 1997. Accretional evolution of a planetesimal swarm. *Icarus* 128:429–455.

Weir, S. T., Mitchell, A. C., and Nellis, W. J. 1996. Metallization of fluid molecular-hydrogen at 140 GPa (1.4 Mbar). *Phys. Rev. Lett.* 76:1860–1863.

Wetherill, G. W. 1990. Formation of the Earth. *Ann. Rev. Earth Planet. Sci.* 18:205–256.

Wetherill, G. W., and Stewart, G. R. 1989. Accumulation of a swarm of small planetesimals. *Icarus* 77:330–357.

Wuchterl, G. 1989. Zur Entstehung der Gasplaneten. Kugelsymmetrische Gasströmungen auf Protoplaneten [On the formation of giant planets. Spherically symmetric gas flows onto protoplanets], Dissertation, Universität Wien.

Wuchterl, G. 1990. Hydrodynamics of giant planet formation. I: Overviewing the κ-mechanism. *Astron. Astrophys.* 238:83–94.
Wuchterl, G. 1991*a*. Hydrodynamics of giant planet formation. II: Model equations and critical mass. *Icarus* 91:39–52.
Wuchterl, G. 1991*b*. Hydrodynamics of giant planet formation. III: Jupiter's nucleated instability. *Icarus* 91:53–64.
Wuchterl, G. 1993. The critical mass for protoplanets revisited: Massive envelopes through convection. *Icarus* 106:323–334.
Wuchterl, G. 1995*a*. Giant planet formation: A comparative view of gas accretion. *Earth Moon Planets* 67:51–65.
Wuchterl, G. 1995*b*. Time-dependent convection on self-adaptive grids. *Comp. Phys. Commun.* 89:119–126.
Wuchterl, G. 1996. Formation of giant planets close to stars. *Bull. Am. Astron. Soc.* 28:1108.
Wuchterl, G. 1997. Giant planet formation and the masses of extrasolar planets. In *Science with the VLT Interferometer,* ed. F. Paresce (Berlin: Springer) pp. 64–71.
Wuchterl, G. 1999. Convection and giant planet formation. In *Euroconference on Extrasolar Planets* (Dordrecht: Kluwer), in press. (Special issue of *Earth, Moon and Planets.*)
Wuchterl, G., and Feuchtinger, M. U. 1998. A simple convection model for self-gravitating fluid dynamics—time-dependent convective energy transfer in protostars and nonlinear stellar pulsations. *Astron. Astrophys.* 340:419–430.
Zharkov, V. N., and Gudkova, T. V. 1991. Models of giant planets with a variable ratio of ice to rock. *Ann. Geophys.* 9:357–366.

ORBITAL EVOLUTION AND PLANET-STAR TIDAL INTERACTION

D. N. C. LIN
Lick Observatory

J. C. B. PAPALOIZOU
Queen Mary & Westfield College, London

C. TERQUEM
Lick Observatory

G. BRYDEN
Lick Observatory

and

S. IDA
Tokyo Institute of Technology

We consider several processes operating during the late stages of planet formation that can affect observed orbital elements. Disk-planet interactions, including recent work on the flow induced by embedded protoplanets and gap formation, are reviewed. Recent results on phenomena caused by the tidal interaction of an orbiting companion with a central star, such as orbital circularization and spin synchronization, are reviewed and applied to extrasolar planets. Dynamical processes that could produce short-period planets or planets in highly eccentric orbits, such as long-term orbital instability and the Kozai mechanism, are discussed.

I. THE FORMATION OF PROTOPLANETS

It is generally accepted that the planets in our solar system were formed in a flattened gaseous nebula centered around the Sun. In typical star-forming molecular clouds, dense cores are observed to have specific angular momentum greater than 6×10^{20} cm^2 s^{-1} (Goodman et al. 1993) such that their collapse leads to rotationally supported disks analogous to the primordial solar nebula (Terebey et al. 1984). Between 25 and 75% of the young stellar objects (YSOs) in the Orion Nebula appear to have disks (Prosser et al. 1994; McCaughrean and Stauffer 1994), with typical mass $M_D \sim 10^{-2\pm1}$ M$_\odot$, temperature $\sim 10^{2\pm1}$ K, and size $\sim 40 \pm 20$ AU

(Beckwith and Sargent 1996; see also the chapters by McCaughrean et al., Calvet et al., and Wilner and Lay, this volume). The common existence of protostellar disks around YSOs, with properties similar to those expected for the solar nebula, suggests that the conditions for planetary formation may be generally satisfied.

Recent observational breakthroughs have led to the discovery of Jupiter-mass (M_{Jup}) planets around at least a few percent of nearby solar-type stars (see chapter by Marcy et al., this volume). With the present data, we can assert that planetary formation is robust.

In conventional planetary formation models, the first stage of protoplanetary formation is the rapid buildup of solid cores through the coagulation of planetesimals (Safronov 1969; Hayashi et al. 1985; Lissauer and Stewart 1993). When the core mass increases above a critical value ($\sim$15 $M_\oplus$), quasistatic evolution is no longer possible, and a rapid accretion phase begins (Mizuno 1980; Bodenheimer and Pollack 1986), leading to the formation of gaseous giant planets (Pollack et al. 1996; also see the chapter by Wuchterl et al., this volume).

The details of this model are not yet fully worked out, but if we suppose that the protoplanet can accrete gas as efficiently as possible, it will first take in gas in the neighborhood of its orbit, assumed circular at radius r_p, until it fills its Roche radius $R_R = (q/3)^{1/3} r_p$, where $q \equiv M_p/M_\star$ is the protoplanet central star mass ratio, while orbiting in an empty annulus. The mass of the protoplanet is then given by $M_p = 3[4M_D(r)/3M_\star]^{3/2} M_\star$, where $M_D(r) = \pi\Sigma r^2$ is the characteristic disk mass within radius r, with Σ being the surface density. At 5.2 AU and $\Sigma = 200$ g cm^2, this gives a mass $M_p \sim 0.4$ M_{Jup}.

Further mass growth now depends on whether the disk has a kinematic viscosity ν capable of producing a mass accretion rate $\dot{M}_p$ onto the protoplanet. If ν is finite, then all of the mass flow through the outer disk should go to the protoplanet. To model the disk viscosity, we adopt the prescription of Shakura and Sunyaev (1973), in which $\nu = \alpha H^2 \Omega$, where α is a dimensionless constant, H is the disk semithickness, and Ω is the disk angular velocity. The most likely mechanism for providing an effective viscosity in stellar accretion disks is MHD turbulence (Balbus and Hawley 1991), which produces $\alpha \gtrsim 10^{-2}$ in a fully ionized disk. Note, however, that α may vary throughout the disk and be much smaller in its intermediate parts (see the chapter by Stone et al., this volume).

Observationally inferred values of the disk accretion rate $\dot{M}_D \sim 10^{-8\pm1}$ $M_\odot$ yr^{-1} (Hartmann et al. 1998) are model dependent and highly uncertain. Nonetheless, for such a fiducial $\dot{M}_D$, a protoplanet may attain a mass $M_p > 10$ M_{Jup} within 10^6 yr. The mass of extrasolar planets is $\sim$1 M_{Jup} (see chapter by Marcy et al., this volume). Unless these planets are preferentially formed in low-mass disks, their growth needs to be terminated or inhibited such that they are unable to accept all the mass that flows through the disk. In this paper we focus on disk-protoplanet tidal interactions as a mechanism for accomplishing this.

A. Protoplanet-Disk Tidal Interactions

A protoplanet exerts tidal perturbations that, for $M_p \sim 1\ \mathrm{M_{Jup}}$, may be adequate to induce the formation of a gap in the disk near its orbit and thereby start to limit accretion flow onto it (Lin and Papaloizou 1993). In general, the gravitational potential due to the protoplanet may be Fourier decomposed in the form

$$\psi = \sum_{l} \sum_{m=0}^{\infty} \psi_{l,m} \cos[m(\phi - \omega_{l,m} t)] \tag{1}$$

where ϕ is the azimuthal angle, l and the azimuthal mode number m are integers, and $\omega_{l,m} = \omega + (l - m)\kappa_p/m$ is the pattern speed, with ω and κ_p being the angular and epicyclic frequency of the protoplanet, respectively. If the orbit of the planet has a small eccentricity e, then $\psi_{l,m} \propto e^{|l-m|}$ (Goldreich and Tremaine 1980; Shu 1984). For circular orbits, only $l = m$ need be considered.

Both outgoing and ingoing density waves are excited in the disk at the Lindblad resonances, located at $r = r_L$ and where, for $l = m$, $\Omega = \omega \pm \kappa/m$ (Goldreich and Tremaine 1978). Here κ is the epicyclic frequency of the gas and Ω is the disk angular velocity. In a Keplerian disk, $\kappa = \Omega$. The ingoing (outgoing) waves carry a negative (positive) angular momentum flux measured in their direction of propagation as they move away from the protoplanet into the disk interior (exterior). The waves thus carry a positive, outward-propagating, conserved angular momentum flux or wave action F_H. In most cases it is reasonable to assume that the waves are dissipated at some location in the disk, where their angular momentum density is deposited (see Lin and Papaloizou 1993 and references therein). In the limit of a cold two-dimensional disk, Goldreich and Tremaine (1978) found for a particular m that

$$F_H = \left[\frac{m\pi^2 \Sigma r^2}{3\Omega^2(m-1)} \left(\frac{d\psi_{m,m}}{dr} + \frac{2\Omega\, \psi_{m,m}}{r\left(\Omega - \omega_{m,m}\right)} \right)^2 \right]_{r=r_L} \tag{2}$$

The back reaction torque exerted on the disk interior (exterior) to the planet is $-F_H$ (F_H). Thus, the inner disk loses angular momentum while the outer disk gains it; hence, the tendency is to form a gap. Evaluation of the total torque acting on each side of the disk requires the summation of contributions from all the resonances, which, for circular orbits, amounts to summing over m. As $m \to \infty$, the location of the resonance approaches the orbit.

However, for a non-self-gravitating disk and large m, the waves are sonic in character and thus can exist only farther than a distance $\Delta \sim 2H/3$ from the orbit, beyond which the relative disk flow is supersonic. This results in a torque cutoff (Goldreich and Tremaine 1980; Artymowicz 1993*b*; Korycansky and Pollack 1993; Ward 1997) for $m > r_p/H$. The dynamics in the coorbital region $\Delta \lesssim 2H/3$ are not wavelike and are considered below. Summing over all resonances, the total angular momentum flux carried by the waves is essentially the same as that given by Papaloizou and Lin (1984), namely

$$\dot{H}_T = 0.23q^2\Sigma r_p^4\omega^2(r_p/H)^3 \tag{3}$$

which was obtained by direct calculation of the torques for a model disk in which the protoplanet orbited in a gap of radial width $\sim H$.

Using equation (3) to estimate tidal torques and then considering the competition between viscous torques, which tend to fill a gap, and tidal torques, which tend to empty it, Lin and Papaloizou (1979*b*, 1980, 1986*a*, 1993) found the following viscous condition for gap opening: $q > 40\nu/(\Omega r_p^2)$. They proposed that gap formation would lead to the limitation of accretion, but they were able to consider only empty gaps. With the introduction of powerful numerical finite-difference techniques and computers, it has recently become possible to study gap formation numerically more fully than previously (Artymowicz et al. 1998; Bryden et al. 1998; Kley 1998; and see section I.B). The first results of this work indicate that there is a transitional regime in viscosity for which a gap exists but some accretion still occurs, which is then essentially switched off for small enough viscosity.

In addition, Lin and Papaloizou (1993) pointed out that in a disk with very small viscosity, to form a gap it is necessary that $H < R_R = (q/3)^{1/3}r_p$. This thermal condition can be viewed in several ways. It means that the protoplanet's gravity is more important than pressure at a distance R_R from the protoplanet; generation of large enough hydrostatic pressure forces would require gradients that would cause a violation of Rayleigh's criterion. From Korycansky and Papaloizou (1996), it is apparent that the thermal condition is required so that the flow in the neighborhood of the protoplanet is nonlinear enough that shocks are produced, which can provide the dissipation associated with gap opening (see below). This condition was found for a simple 2D model disk with a barotropic equation of state.

With more complicated physics and the introduction of three-dimensional effects (Lin et al. 1990*a,b*), it may be possible to alter wave dissipation patterns and, thus, angular momentum deposition. However, because of the difficulty of clearing material near the protoplanet (it is difficult to see how linear waves could be dissipated closer than several vertical scale heights), it is likely that gap formation will occur much less readily if the thermal condition is not satisfied.

B. Embedded Protoplanets and Gaps

The flow around a partially embedded protoplanet has been simulated in the low-M_p limit in a shearing-sheet approximation (Korycansky and Papaloizou 1996; Miyoshi et al. 1999). These simulations provide useful clues on the flow pattern near the protoplanet. A global illustrative model is shown in Color Plate 16, in which we set $q = 10^{-4}$, $H/r = 0.07$, and $\alpha = 10^{-3}$. The flow pattern can be divided into three regions:

1. A circumplanetary disk is formed within the Roche radius of the protoplanet, with semithickness $\sim|r - r_p|$. In such a geometrically thick disk, the tidal perturbations of the host star induce a two-arm spiral shock wave with an open pitch angle. Angular momentum is efficiently transferred from the disk to the star's orbit around the planet (equivalent to the planet's orbit around the star) so that gas is accreted onto the core of the protoplanet within a few orbital periods $P_o = 2\pi/\omega$.
2. An extended arc is formed in the coorbital region near r_p, with gas streaming in horseshoe orbits around the L_4 and L_5 points. Because these are local potential maxima, viscous dissipation leads to the depletion of this region (Lin et al. 1987).
3. In those regions of the disk that have $|r - r_p| > R_R$, the protoplanet's tidal perturbation results in the convergence of streamlines to form pronounced trailing wakes, both inside and outside r_p. These high-density ridges have been identified by some as "stream accretion," through which material flows from the disk to the protoplanet (Artymowicz and Lubow 1996). However, the velocity field viewed in a frame rotating with the companion (Color Plate 16) clearly indicates that gas in the postshock region along the ridge line is moving away from the protoplanet.

Simulations of gap formation need to consider model evolution over many orbital periods and have to deal with a large density contrast between the gap region and other parts of the disk. Special care is needed to minimize the tendency of numerical viscosity to produce spurious diffusion into the gap region and thus significantly affect the results. In Color Plate 17, we illustrate the excitation and propagation of waves and the existence of a clear gap for a planet (with $q = 10^{-3}$) interacting with a disk (with $H/r = 0.04$ and $\alpha = 10^{-3}$).

In Figure 1, we plot the growth timescale $M_p/\dot{M}_p$ as a function of α and H/r for protoplanets of mass 1 and 10 M_{Jup}. We comment that it is not necessary for accretion to stop entirely for gap formation to affect the final mass of a protoplanet. All one needs is that the mass doubling time should become longer than either the disk lifetime or the time for the interior disk to accrete. In the latter case, the planet would approach the central star before it could increase its mass significantly (Ivanov et al. 1998). From Fig. 1 we see that, for $\alpha \sim 10^{-3}$ and $H/r = 0.04$, the mass doubling time for 1 M_{Jup} is 10^6 yr. For 10 M_{Jup} the same results apply for $H/r = 0.07$. The results in Fig. 1 thus suggest that the tidal truncation process operates during planetary formation and is important in determining the final mass of the planet M_p.

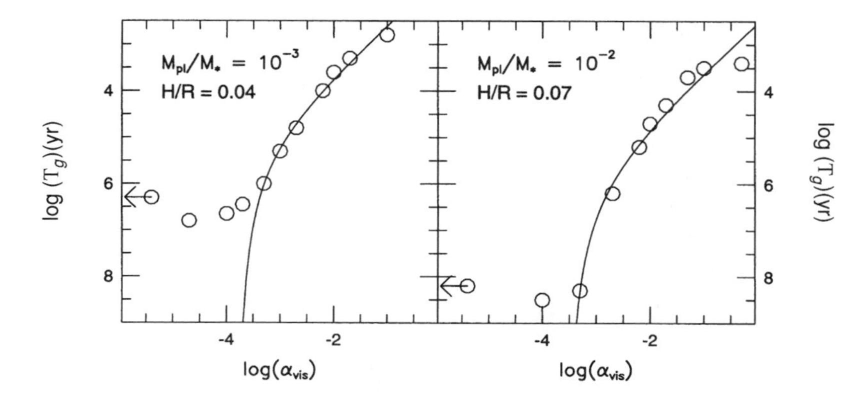

Figure 1. The mass doubling timescale τ_g as a function of viscosity parameter α for one Jupiter mass with $H/r = 0.04$ (left) and 10 Jupiter masses with $H/r = 0.07$ (right). As the gas viscosity is decreased, the accretion rate onto the protoplanet drops off sharply until reaching a numerical limit for $\alpha < 10^{-4}$.

II. PROTOPLANETARY ORBITAL EVOLUTION

A. Eccentricity Evolution

Resonant interaction between the disk and the planet can lead to changes in the orbital eccentricity e (Lin and Papaloizou 1993; Artymowicz 1993*a*). The way in which Lindblad torques cause a small eccentricity to grow has been reviewed by Lin and Papaloizou (1993). If only the Lindblad torques are considered, the disk matter is able to pump eccentricity tidally in a similar way to a rapidly rotating star (see below). However, corotation resonances in the disk (at which the pattern speed of a Fourier component of the tidal perturbation corotates with the disk) also need to be considered (see section I.A). When these resonances exist, a torque is applied to the disk at the corotation radius r_C, where $\Omega = \omega_{l,m}$ at a rate given by

$$T^C_{l,m} = -\frac{m\pi^2}{2}\left[\Sigma\left(\frac{d\,\Omega}{dr}\right)^{-1}\frac{d\,\varsigma^{-1}}{dr}(\psi_{l,m})^2\right]_{r=r_C} \tag{4}$$

where $\varsigma = \kappa^2/(2\Omega\Sigma)$ is the vortensity. The role of corotation resonances depends strongly on the ς distribution. When Σ vanishes at some disk edge, or when there is a gap near a low-mass perturber, Goldreich and Tremaine (1980) showed that the action of torques due to corotation resonances damps e more effectively than effects due to Lindblad resonances excite it. However, this result does not apply to the situation where a protoplanet is embedded in a disk with no edges or gap and where tidal interaction is linear.

Assuming $d \ln \varsigma / d \ln r \sim 1$ in the unperturbed disk, Artymowicz (1993*a*) suggested that the eccentricity of low-mass embedded protoplanets would be damped by tidal interaction. However, in the minimum-mass solar nebula model where $\Sigma \propto r^{-3/2}$, vortensity is independent of r, and then all corotation torques vanish (Ward 1993), although the torque associated with coorbital Lindblad resonances may persist. For a planet with a sufficiently large mass to open a wide gap, both corotation and coorbital Lindblad resonances would be severely weakened and e could then increase on a timescale $e^2/(de^2/dt) \sim q^{-1}(M_\star/M_d)P_o$ (Lin and Papaloizou 1993) which, for massive planets, can be shorter than the inferred disk evolution timescale.

If the growth of protoplanets is limited by the formation of a gap of width Δ_g, tides may cause e to increase until $e \sim \Delta_g/r_p$ when a potential expansion valid for small e assumed in the above analysis breaks down. In a cold disk, the protoplanet's angular velocity at peri-aphelion would then be greater/less than $\Omega(r_p \mp \Delta_g)$, resulting in a torque reversal. Such a process could provide a limit to the growth of e. However, a protoplanet's increased radial excursion may also cause the gap to widen, in which case e might increase to yet larger values. Analytic treatments and self-consistent numerical simulations in the large-e limit have not been carried out so far.

The characterization of extrasolar planets' orbits from forthcoming observations will provide some useful constraints on the origin of eccentricity. The coexistence of several planets with similar e but very different M_p (or similar M_p but very different e) is not a natural outcome of the excitation of eccentricity by disk tidal interaction, because this would tend to produce an (e, M_p) relation (Artymowicz 1992).

B. Orbital Migration

Planetary orbital migration is induced by the difference between the inner and outer disk torques, ΔT, which react back on the planet, assumed to remain in a circular orbit, such that

$$\frac{dr_p}{dt} = \frac{2\Delta T}{r_p \Omega M_p} \tag{5}$$

Based on recent work indicating that ΔT is almost always negative, Ward (1997) suggested that embedded planets migrate inward on a timescale $\sim 10^5 (M_p/\mathrm{M}_\oplus)^{-1}$ yr. If this process occurs in protostellar disks, protoplanets would migrate towards the stellar surface at $r = R_\star$ once $M_p > 1\ \mathrm{M}_\oplus$, because their growth timescale (Lissauer and Stewart 1993) would then become longer than this migration timescale. Resolution of this rapid migration dilemma may require the complete and nonlinear analysis of the disk response to the protoplanet in the corotation regions which may be quite complex (see Color Plates 16 and 17). However, supposing that migration occurs and can be stopped near the star (see below), a scenario leading to the formation of short-period planets has been proposed (see chapter by Ward and Hahn, this volume).

After gap formation, there may still be some residual accretion depending on the value of α (see section I.B). Torques are still exerted between disk and protoplanet, and the imbalance due to differing properties of the disk on either side of the protoplanet causes orbital migration (Goldreich and Tremaine 1980; Lin and Papaloizou 1986*b*; Takeuchi et al. 1996). The estimated effects of advection of angular momentum through the accretion flow on this process are usually found to be small. For small M_p, the protoplanet behaves like a disk particle and so migrates towards the star. For larger masses, the evolution is slower. Nonetheless, the protoplanet is always expected to reach the star before it has time to double its mass (Ivanov et al. 1998).

The above discussion naturally leads to the suggestion that short-period planets were formed at several AU away from their host stars and subsequently migrated to their present location (Lin et al. 1996). However, *in situ* formation cannot be completely ruled out (Bodenheimer 1997).

A protoplanet's migration may be terminated near $R_\star$ if (1) the disk does not extend down to $R_\star$ or (2) the host star induces angular momentum transfer to the protoplanet's orbit via tidal effects (see section III). For the first possibility, interaction between the disk and an intense (>1 kG) stellar magnetic field has been suggested (Konigl 1991) as the cause for the modest observed rotation period ($P_\star \sim 8$ days) of classical T Tauri stars (Bouvier et al. 1993; Choi and Herbst 1996). Such a magnetic field strength may be consistent with recent measurements (Guenther 1997). In this scenario, the stellar field is assumed to induce a cavity out to the magnetospheric radius ($r_m \sim$ a few 10^{11} cm) where $P_\star$ is equal to the local Keplerian period of the disk. When a protoplanet migrates interior to the 2:1 resonance of the gas at r_m, the protoplanet is decoupled from the disk, and its migration is stalled. The main uncertainties here are in the observed distribution of $P_\star$ (Stassun et al. 1999).

III. PLANET-STAR TIDAL INTERACTION

As the planet approaches $R_\star$, the tides raised on the star or planet by its companion become strong enough that their dissipation leads to orbital evolution. If the distribution of mass of the perturbed object has the same symmetry as the perturbing potential, no tidal torque results from the interaction. However, in general, dissipative processes (e.g., radiative damping or turbulent viscosity) acting on the tides produce a lag between the response of the star or planet and the perturbing potential, enabling mechanical energy to be lost and angular momentum to be exchanged between the rotation of the perturbed object and the orbital motion. Only if the system is circular, synchronous, and coplanar (i.e., the spin of the binary components and that of the orbit are parallel) does the tidal torque vanish (Hut 1980). However, such an equilibrium state is not necessarily stable (Counselman 1973; Hut 1980, 1981).

The response of a star or planet to a tidal perturbation is the sum of two terms: an equilibrium tide and a dynamical tide. The equilibrium tide

(Darwin 1879) is the shape that the perturbed object would have if it could adjust instantly to the tidal potential; it is obtained by balancing the pressure and gravity forces. The dynamical tide contains the oscillatory response of the star or planet. It takes into account the fact that gravity or *g* modes can be excited in the convectively stable layers of the star or planet and that resonances between the tidal disturbance and the normal modes of the star or planet can occur (Cowling 1941).

In massive close binaries, which have a convective core and a radiative envelope, the dynamical tide cannot be neglected, because tidal friction is caused predominantly by the radiative damping of the tidally excited modes (Zahn 1975, 1977; Savonije and Papaloizou 1983, 1984, 1997; Papaloizou and Savonije 1985, 1997; Goldreich and Nicholson 1989; Savonije et al. 1995; Kumar et al. 1995). In that case, *g* modes, which are excited mainly near the convective core boundary, propagate out through the envelope to the atmosphere, carrying energy and angular momentum. They are damped close to the surface, where the radiative diffusion time becomes comparable to the forcing period, thus enabling a net tidal torque to be exerted on the star or planet.

In the case of solar-type binaries, which have a radiative core and a convective envelope, it was thought until recently that tidal friction could be well described by the theory of the equilibrium tide, in which only turbulent dissipation of the equilibrium tide in the convective envelope is taken into account (Zahn 1977). However, recent studies have indicated that this is not the case. Even if radiative damping can be ignored, the torque derived using only the equilibrium tide is 4–6 times larger than that taking into account the dynamical tide for binary periods of several days (Terquem et al. 1998*b*). The reason is that the theory of the equilibrium tide can in principle be applied only when the characteristic timescales of the perturbed object are small compared to the forcing period. In a solar-type star, however, the convective timescale in the interior region of the convective envelope is as large as a month. Furthermore, because of the uncertainty over the magnitude of the turbulent viscosity associated with convection, it is not clear that the torque due to turbulent dissipation acting on the full tide in the convective envelope is more important than that due to radiative damping acting on the *g* modes that propagate inwards in the radiative core (Terquem et al. 1998*b*; Goodman and Dickson 1998). Here, we focus on the case of a planet orbiting around a solar-type star.

Tides raised on the star by the planet can be analyzed in the limit that the rotational frequency of the star is small compared to the orbital frequency. Then, as a result of tidal friction, the star spins up, the orbit decays (the planet spirals in), and the orbit's eccentricity, if any, decreases. To calculate the timescales on which this evolution occurs, we need to quantify the dissipative mechanisms that act on the tides. Recent studies (Claret and Cunha 1997; Goodman and Oh 1997; Terquem et al. 1998*b*) have found that the turbulent viscosity in the convective envelope that is required to provide the observed circularization rates of main-sequence solar-type binaries (Mathieu 1994) is at least 50 times greater than that simply estimated from mixing length theory for nonrotating stars. This indicates either (1) that the observations are questionable, (2) that solar-type binaries are not circularized through turbulent viscosity acting on tidal perturbations (see Tassoul 1988 and Kumar and Goodman 1996 for other suggested tidal mechanisms), or (3) that dissipation in the convective envelope of solar-type stars is significantly more efficient than is currently estimated (see Terquem et al. 1998*b* for a more detailed discussion of the uncertainties involved). Here we assume that circularization of solar-type binaries does occur through the action of turbulent viscosity on the tides, and we then calibrate its magnitude so as to account for the observed timescales. Under these circumstances, when the response of the star is not in resonance with one of its global normal modes, the tides are dissipated more efficiently by turbulent viscosity than by radiative damping. In a resonance, radiative damping dominates and limits the response of the star at its surface. Although the planetary companion may go through a succession of resonances as it spirals in under the action of the tides, for a fixed spectrum of stellar normal modes its migration is controlled essentially by the nonresonant interaction. For a nonrotating star, and with the calibration mentioned above, the orbital decay timescale, spinup timescale of the star, and circularization timescale, in Gyr, are (Terquem et al. 1998*b*):

$$t^*_{\rm orb}\left(\mathrm{Gyr}\right) = 2.763 \times 10^{-4} \frac{\left(M_p/M_\star + 1\right)^{5/3}}{M_p/M_\star} \left(\frac{P_o}{1\ \mathrm{day}}\right)^{13/3} \tag{6}$$

$$t^*_{\rm sp}\left(\mathrm{Gyr}\right) = 1.725 \times 10^{-6} \left(\frac{M_p + M_\star}{M_p}\right)^2 \left(\frac{P_o}{1\ \mathrm{day}}\right)^3 \tag{7}$$

$$t^*_{\rm circ}\left(\mathrm{Gyr}\right) = 4.605 \times 10^{-5} \frac{\left(M_p/M_\star + 1\right)^{2/3}}{M_p/M_\star} \left(\frac{P_o}{1\ \mathrm{day}}\right)^{13/3} \tag{8}$$

where P_o is the orbital period. This circularization timescale is valid only if the initial eccentricity is not too large. If only the convective envelope of the star, where tidal dissipation occurs, is spun up during tidal evolution, then the spinup timescale has to be multiplied by I_c/I, where I_c and I are the moments of inertia of the convective envelope and the entire star, respectively. For the Sun, $I_c/I \simeq 0.14$.

Tides are also raised on the planet by the star. In contrast to the giant planets of our solar system, Jupiterlike planets on a close orbit are expected to have an isothermal, and thus radiative (convectively stable) envelope (Guillot et al. 1996; Saumon et al. 1996). For these planets, both turbulent dissipation in the convective core and radiative damping in the envelope act on the tides. So far it is not clear which mechanism is more important.

We express the timescales associated with turbulent dissipation of the tides in terms of the parameter Q, the inverse of which is the effective tidal dissipation function (MacDonald 1964). Damping of the tides in the convective core of the planet leads to the synchronization of the planet rotation with the orbital revolution on the following characteristic timescale, in Gyr (Goldreich and Soter 1966):

$$t_{\rm sp}^{p,1}\left(\mathrm{Gyr}\right) \sim 4.4 \times 10^{-13} Q \left| \frac{1\ \mathrm{day}}{P_p} - \frac{1\ \mathrm{day}}{P_o} \right| \frac{M_p}{M_\star} \left(\frac{P_o}{1\ \mathrm{day}} \right)^2 \left(\frac{a}{R_p} \right)^3 \tag{9}$$

where R_p is the planet radius, a is the semimajor axis of the orbit, and P_p is the initial value of the planet rotational period (when it first undergoes tidal interaction with the star). Because the rotational angular momentum of the planet is in general small compared to its orbital angular momentum, synchronization of the planet occurs before any significant orbital evolution can take place. Once synchronization is achieved, damping of the tides raised in the planet always leads to the decay of the orbital eccentricity on a characteristic timescale, which is, in Gyr (Goldreich and Soter 1966),

$$t_{\rm circ}^{p,1}\left(\mathrm{Gyr}\right) \sim 2.8 \times 10^{-14} Q \frac{M_p}{M_\star} \frac{P_o}{1\ \mathrm{day}} \left(\frac{a}{R_p} \right)^5 \tag{10}$$

The tides raised on the planet do not lead to the decay of the orbit once the planet is synchronized. We note that Q may depend on the tidal frequency as seen by the planet, and therefore on the rotational frequency of the planet. Since $t_{\rm circ}^{p,1}$ is calculated assuming synchronization, the value of Q in the above equation may be different from that used for calculating $t_{\rm sp}^{p,1}$. Orbital evolution and tidal heating of Jupiter's satellites leads to an estimate of $Q \sim 10^5$–10^6 for this planet (Goldreich and Soter 1966; Lin and Papaloizou 1979*a*; Gailitis 1982). However, it is not yet understood where this value of Q comes from, because turbulent viscosity arising from convection would produce $Q \simeq 5 \times 10^{13}$ (Goldreich and Nicholson 1977; see Stevenson 1983 for an alternative). There is, of course, no reason to assume that the Q values of the extrasolar planets are similar to that of Jupiter. First, some of these planets orbit very close to their parent stars and therefore are much hotter than Jupiter. Also, Q may depend on the magnitude and the frequency of the tidal oscillation, in which case it would be different for Jupiter if this planet were synchronously rotating on a closer orbit. Therefore we can only speculate when applying the above formulae to extrasolar planets.

Radiative damping of the tides in the envelope, as described above for massive binaries, gives rise to the synchronization and circularization timescales $t_{\rm sp}^{p,2}$ and $t_{\rm circ}^{p,2}$, respectively. These timescales have been evaluated (in particular for 51 Pegasi) by Lubow et al. (1997). However, as they pointed out, the asymptotic analysis they use is valid only if the initial spin rate of the planet is less than half that of Jupiter. Besides, this analysis neglects the effect of rotation on the tides, which is important for near-synchronous planets.

In the context of extrasolar planets we first comment on the magnitude of the perturbed velocity induced by the tides at the stellar surface. Terquem et al. (1998*a,b*) have found that, in the case of 51 Peg, this velocity is too small to be observed. This result is insensitive to the magnitude of the stellar turbulent viscosity and is not affected by the possibility of resonance. It also holds for the other extrasolar planets that have been detected so far.

As indicated above, the rotation of planets on a close orbit is almost certainly synchronous with the orbital revolution. For the star to be synchronized in less than 5 Gyr (i.e., $t_{sp}^{*} < 5$ Gyr), a 1-(3)-Jupiter-mass planet would have to be on an orbit with a period less than 1.4 (3) day(s). We note that in the case of τ Bootis, the rotation of the star may then have been synchronized as a result of tidal effects. If only the convective envelope of the star is spun up, these periods have to be multiplied by about a factor of 2. Therefore, the observation of anomalous rapid rotators could give further evidence of the presence of close planets and provide some indication on how the dissipation of tides affects the rotation of the star (Marcy et al. 1997).

The circularization timescale for the orbit is $t_{\rm circ}$ such that $1/t_{\rm circ} = 1/t_{\rm circ}^{*} + 1/t_{\rm circ}^{p,1} + 1/t_{\rm circ}^{p,2}$. We note that, according to the expression of the timescales we have given above, $t_{\rm orb}^{*} \simeq 6t_{\rm circ}^{*}$ and $t_{\rm orb}^{p,1} > 10t_{\rm circ}^{p,1}$, so that circularization can be achieved without a significant orbital decay taking place (in contrast to the statement by Rasio et al. 1996). So far it is not clear which of the timescales $t_{\rm circ}^{*}$, $t_{\rm circ}^{p,1}$, and $t_{\rm circ}^{p,2}$ is the shortest. For 51 Peg, $t_{\rm circ}^{p,1} < t_{\rm circ}^{*}$ requires $Q < 2 \times 10^{7}$ (see also Rasio et al. 1996 and Marcy et al. 1997 for estimates of $t_{\rm circ}^{p,1}$). We point out that for a Jupiter-mass planet, $t_{\rm circ}^{*} < 5$ Gyr for $P_o < 3$ days. Therefore, all the Jupiter-mass planets orbiting with a period less than about 3 days should be on a circular orbit. This cutoff period is a lower limit because $t_{\rm circ}^{*}$ is an upper limit of the circularization timescale.

Planets found on eccentric orbits at smaller periods are indicative that the stellar rotation frequency (assumed zero up to now) is large compared to the orbital frequency. Indeed, in that case, tidal friction in the star pumps up the orbital eccentricity, opposing the effect of tidal friction in the planet. However, the above calculations indicate that the planet may have to be on a very close orbit for the tidal friction in the star to be able to increase the orbital eccentricity significantly. Whether the planet can get to such a close orbit is questionable if the disk in which the planet formed did not extend down to $R_\star$.

The timescale on which the orbit decays is $t_{\rm orb}^{*}$. The condition $t_{\rm orb}^{*} < 5$ Gyr requires $P_o < 2$ days for a Jupiter-mass planet. If a planet has not been able to get to such a close orbit (e.g., because the inner parts of the disk were truncated), orbital decay may not occur as a result of tidal friction, and the planet may not plunge into its parent star. Note too that a giant planet that spirals in towards the star may lose mass through Roche lobe overflow (Trilling et al. 1998). Apart from reducing the planetary mass, conservation of angular momentum results in an outward torque on the orbit, which slows the inward migration. A low-mass planet in a very small orbit may result in this way.

However, several planets may attain short periods. After the firstborn protoplanet emerges and migrates to a few $R_\star$, subsequently formed protoplanets (beginning farther out because the formation timescale increases with r) may migrate inward, pushing protoplanetary cores ahead of their inward path, until they become trapped at Lindblad resonances. Such resonant trapping is found for the Galilean satellites (e.g., Goldreich and Peale 1966; Lin and Papaloizou 1979*a*) and it occurs if the migration rate is sufficiently slow. It raises the intriguing possibility that planets temporarily parked close to the star because of absence of disk material could be forced to plunge into the host star.

On the other hand, if the star is a rapid rotator (with period shorter than that of the orbit) and has a weak magnetic field, tidal effects transfer angular momentum from the star to the orbit, with increasing efficiency as the orbit is pushed out, until a tidal barrier is produced such that the migration time and inverse orbital decay time balance. Tentatively, using estimates derived from equation (7), an inward migration time of 10^6 yr could be balanced by tidal effects acting on a 1-M_{Jup} planet only if the period were ~0.3 days, which increases to ~0.7 days for a 3-M_{Jup} planet and a migration time of 10^7 yr. If planets can survive at such short periods (more likely for massive planets and slow migration rates) and if eccentricity growth is suppressed by tides within the planet, tidal effects might temporarily cause the halting of migration by transferring angular momentum from the star to the inner orbit and from there to any residual disk material via Lindblad torques and so on to outer planets (cf. the situation for Saturn's rings). Such processes have not yet been worked out in detail. However, a solar-type star is likely to have a small amount of angular momentum in comparison to the disk/planet system in total, and once the star slows down sufficiently, the inward migration must continue.

As a planet eventually becomes engulfed by the host star, it is disrupted by tidal breakup, heating, and ram pressure stripping (Sandquist et al. 1998). Supporting evidence for such phenomena may be found in the supersolar metal abundance of 51 Pegasi (G2 V), 55 ρ Cancri (G8 V), and τ Bootis (F7 V) (Butler et al. 1997; Gonzalez 1997, 1998). The convective envelope for each of these stars contains a few 10^{-2} $M_\odot$. The mixing of 10–40 $M_\oplus$ of planetary heavy element material within the stellar envelope (Zharkov and Gudkova 1991) would lead to a significant metal enrichment there. Because the depth of the convection zone of main-sequence stars decreases significantly with increasing stellar mass, the routine engulfing of planets by their hosts might lead to a tendency for hotter planetary hosts to show a general overall metallicity enhancement with respect to cooler ones.

IV. LONG-TERM STABILITY OF PLANETARY SYSTEMS

In a relatively massive disk, several giant planets may be formed with $M_p \sim 1$–3 M_{Jup} and $a > 1$ AU (Lin and Ida 1997). Their long-term orbital stability determines the dynamical evolution of the system. After the depletion of the disk gas, mutual gravitational perturbation between the planets may gradually increase their eccentricities until their orbits cross each other on a timescale τ_x. Extrapolations of existing numerical results (e.g., Franklin et al. 1990; Chambers et al. 1996) give $\tau_x \sim 10^{12-18}$ yr for the solar system. However, τ_x would reduce to $\sim 10^{8-12}$ or 10^{5-8} yr if all the gas giants had $M_p = 1$ or 2 M_{Jup}, respectively. More recent simulations and crossing times for subsets of planets, configured as in the solar system but with a reduced solar mass, are given in Duncan and Lissauer (1998). These authors find that the orbits of the giant planets remain stable under the expected solar mass loss for up to 10^{10} yr or more. The more the central mass is reduced, the shorter is τ_x.

Since systems of massive planets formed with similar values of a and small e eventually suffer orbit crossing (Lin and Ida 1997), massive eccentric planets may have acquired their orbital properties as a consequence of orbit crossing (Rasio and Ford 1996; Weidenschilling and Marzari 1996; Lin and Ida 1997; Levison et al. 1998). Once orbit crossing occurs, close encounters eventually take place. When these involve equal-mass planets, they produce a velocity perturbation V with magnitude limited by the surface escape velocity $V_{esc} \simeq 60(M_p/M_{Jup})^{1/3}(\rho_p/1 \text{ g cm}^{-3})^{1/6}$ km s^{-1}, where ρ_p is the planet's internal density (Safronov 1969). This results in $e \lesssim V_{esc}/(a\Omega) \sim 2(M_p/M_{Jup})^{1/3}(a/1 \text{ AU})^{1/2}$. Thus, observed eccentricities can be accounted for.

Supposing that giant planets form at $\sim$1 AU, massive planets ($M_p \gtrsim 5$ M_{Jup}) with moderately high $e \sim 0.3$ and moderately small $a \sim 0.5$ AU can be accounted for by orbit crossing and merging. Eccentric planets with $a > 1$ AU and $M_p \sim M_{Jup}$ can be produced by orbit crossing followed by ejection. Merger, resulting in $e \lesssim 0.5$, is favored at small r, whereas ejection occurs preferentially at large r, because the cross section for direct collisions is independent of r whereas that for close encounters is $\propto V^{-4} \propto (r\Omega)^{-4} \propto r^2$ (Lin and Ida 1997). Lin and Ida (1997) find that the orbital properties of a merged body are consistent with those of the planets in eccentric orbits around 70 Virginis and HD 114762.

Numerical simulations of systems with many planets indicate that although some may be ejected, a residual population of eccentric planets ($e \gtrsim 0.5$) may remain bound to the central star at large distances ($a > 100$ AU). Close encounters also excite the relative inclination up to $\sim e/2$

radians. Detection of additional companions around stars with an eccentric planet would provide tests of these scenarios.

Rasio and Ford (1996) suggested that the short-period planets were dynamically scattered into a region close to the central star. Although the short-period planets may be circularized (see section III), numerical simulations indicate that the scattering scenario would lead to a population of planets with high eccentricity at large distances (10–100 AU) from their central stars. A comprehensive search for these planets would provide a useful test of this scenario as against that of disk-planet interaction, which produces orbits with smaller eccentricity.

V. EFFECTS OF SECULAR PERTURBATIONS DUE TO A DISTANT COMPANION

A. Kozai Effect

Some extrasolar planets are found in binary systems. For example, a planet is found around 16 Cyg B that orbits at a distance of 1.7 AU from the central star, which is known to have a binary companion, 16 Cyg A. It has been suggested that the high eccentricity ($e = 0.67$) of 16 Cyg B is excited by the gyroscopic perturbation due to 16 Cyg A (Holman et al. 1997). A similar effect may be caused by the gravitational perturbations from other planetary bodies. Here we discuss the effect of secular perturbations on a long timescale such that a time average may be performed. The orbiting bodies may then be considered as having their mass distributed continuously around their orbits as in the classical theories of Laplace and Lagrange (see Hagihara 1972 and references therein). Semimajor axes do not change under secular perturbation. However, changes to the eccentricity and inclination may be produced.

If orbits are coplanar, in general only modest changes to eccentricities are produced if the perturbing bodies are widely separated. An exception to this might occur if secular resonances sweep through the system because of a changing gravitational potential due to disk dispersal or changing stellar oblateness (e.g., Ward 1981). However, if orbits are allowed to have high inclination, large eccentricity changes can occur to the orbit of an inner planet perturbed by a distant body or binary companion.

To consider this gyroscopic perturbation effect, let us consider the simplest case of the motion of an inner planet with mass M_p perturbed by an outer companion with mass M' assumed to be in a circular orbit with semimajor axis a'. The outer companion is assumed to contain sufficient angular momentum that its orbit remains fixed, defining a reference plane to which the inner planet orbit has inclination i. We note ω_a, the longitude of the apsidal line measured in the orbital plane from the line of nodes.

If the distance of M_p from $M_\star$ is r, its speed is v, the angle between its position vector and that of M' is Θ, and only the dominant quadrupole term in the interaction potential multipole expansion is retained, its motion is governed by the Hamiltonian (Kozai 1962; Hagihara 1972; Innanen et al. 1997):

$$H = \frac{1}{2}M_\star v^2 - \frac{GM_\star M_p}{r} - \frac{GM' M_p r^2}{2a'^3}\left(3\cos^2\Theta - 1\right) \tag{11}$$

After performing a time average, appropriate for secular perturbations,

$$H = -\frac{GM_\star M_p}{2a} - \frac{GM' M_p a^2}{16a'^3}\left[\left(3\cos^2 i - 1\right)(2 + 3e^2) + 15e^2 \sin^2 i \cos 2\omega_a\right] \tag{12}$$

The fact that both H and the component of angular momentum parallel to the outer orbital axis, $L\cos i = \sqrt{GM_\star a(1 - e^2)}\cos i$, are constant enables elimination of i and ω_a. A complete solution can then be found from the equation for the rate of change of e, which can be derived from the canonical equation $dL/dt = -\partial H / \partial\omega_a$ in the form

$$\frac{de}{dt} = \frac{15nM' a^3 e \sin^2 i \sin 2\omega_a}{8M_\star a'^3} \tag{13}$$

with $2\pi/n$ being the orbital period of the planet M_p.

The above equation indicates that e can oscillate between extremes occurring when $\sin 2\omega_a = 0$. There has been particular interest in finding conditions under which e can start from very small values and then increase unstably to values close to 1. To examine conditions for this to occur, we use $H = $ const and $L\cos i = $ const to find ω_a and i in terms of e. Assuming initial values $i = i_0$ and $e = e_0$, when $\omega_a = \omega_0$, for infinitesimally small but nonzero values of e_0 that can be neglected, we find

$$\cos 2\omega_a = \frac{1 - 5\cos^2 i_0 - e^2}{5\left(1 - e^2 - \cos^2 i_0\right)} \tag{14}$$

From equation (14), we see that when $e \to 0$,

$$\cos 2\omega_a = \frac{\left(1 - 5\cos^2 i_0\right)}{5\left(1 - \cos^2 i_0\right)} \tag{15}$$

Clearly, for solutions of the type we seek, we require $\cos 2\omega_a > -1$, which requires that the initial inclination exceed a critical value (see Innanen et al. 1997) such that

$$\cos^2 i_0 < \tfrac{3}{5} \tag{16}$$

If this inequality is satisfied (if it is not, then e_0 cannot be neglected), equation (13) indicates that e grows from a very small value up until the value obtained by setting $\omega_a = \pi/2$, namely

$$e^2 = 1 - 5\cos^2 \frac{i_0}{3} \tag{17}$$

attained when $\cos^2 i = \frac{3}{5}$. A range of eccentricities may be generated in this way. However, values close to unity may be attained for initial inclinations close to 90°. There are then solutions that have large-amplitude oscillations in eccentricity (Kozai effect), and adding the nonsecular terms back in can lead to chaotic behavior. The large eccentricity changes occur independently of the size of the perturbation. However, the characteristic timescale for the eccentricity changes to occur is given from equation (13) as $\sim (M_\star a'^3)/(M'a^3)$ orbital periods of the inner planet, which is long for small perturbations.

Note, however, that the effect requires the motion of the apsidal angle ω_a be governed only by the perturbation considered. Other effects may disrupt this, such as general-relativistic corrections (e.g., Holman et al. 1997) and the oblateness of the central star. Both could be considered in the following discussion, but here we limit ourselves to considering the effects of relativistic apsidal precession, which may lead to constraints on the orbital elements of planets with short periods.

B. Relativistic Effects

To incorporate relativistic apsidal precession to lowest order, we modify the Newtonian potential such that Keplerian elliptic orbits are induced to precess at the correct rate. We modify the potential due to the central mass such that

$$\frac{-GM_\star}{r} \rightarrow \frac{-GM_\star}{r}\left(1 + \frac{3GM_\star}{rc^2}\right) \tag{18}$$

with c being the speed of light. The additional term can be added into the Hamiltonian H and averaged so that equation (12) becomes

$$H \rightarrow H - \frac{3(GM_\star)^2 M_p}{2a^2c^2\sqrt{(1-e^2)}} \tag{19}$$

The same procedure as described above can be used to find ω_a and i in terms of e and hence conditions that must be satisfied in order that large values of e might be attained. The condition analogous to (16) for the Kozai effect to work is

$$\cos^2 i_0 < \frac{3-2f}{5} \tag{20}$$

Here the parameter

$$f = \left(\frac{GM_\star}{ac^2}\right)\left(\frac{M_\star a'^3}{M' a^3}\right) \tag{21}$$

measures the importance of relativistic precession relative to the perturbation due to the outer companion. From this, if $f > \frac{3}{2}$, the effect is completely suppressed. When the Kozai effect occurs, the maximum possible eccentricity that can be generated is obtained by setting $\omega_a = \pi/2$ and $i_0 = 90°$. This gives

$$e^2 = \frac{1}{2} + \sqrt{\frac{4f}{3} + \frac{1}{4}} - \frac{4f}{3} \tag{22}$$

which indicates that, for $f = 0.1$, e as high as 0.99 can be generated for $i_0 = 90°$. The parameter f may be conveniently expressed as

$$f = 5.0519 \times 10^{-7} \frac{1}{q'} \left(\frac{P'_o}{P_o}\right)^2 \left(\frac{1\ \mathrm{day}}{P_o}\right)^{2/3} \tag{23}$$

Here, P_o and P'_o denote the orbital periods of M_p and M', respectively, and $q' = M'/M_\star$.

Planets with periods less than 3–4 days are likely to have their orbits circularized as a result of tidal effects (see above). From the requirement $f > \frac{3}{2}$, we see that for inner planets with somewhat larger periods (but still on the order of days), binary companions with mass ratio of order unity need orbital periods shorter than about 40 yr in order to pump significant eccentricity. If a massive planet with $q' \sim 10^{-2}$ is considered, it must orbit with a period ten times shorter, ~4 yr. Such an object should be readily detectable.

We comment that for a given distant binary companion with inclination close to 90° to the planet orbit, and no other perturbers, there is a planet with a period long enough such that the Kozai effect could produce eccentricities close enough to unity that the planet has arbitrarily close approaches to the central star. Then tidal effects may act towards circularization, so decreasing the distance to apoapse until a circular orbit is produced. The final period would have to be a few days. Alternatively, should the closest approach to the star be beyond the largest radius at which orbits can be circularized, the orbit would retain a Kozai cycle with high maximum eccentricity.

VI. SUMMARY

Here we summarize the various processes discussed in this chapter, their influence on the evolution of planetary systems, and their observational implications.

1. *Planet-disk tidal interaction* excites potentially observable spiral density waves and can create clean gaps in the disk, limits mass growth, and drives orbital migration. Growth limitation by gap formation, together with inward migration, may naturally account for the upper limit in the observed mass ratio distribution of extrasolar planets, provided they were formed in disks with $H \lesssim 0.1 r_p$ and $\alpha \lesssim 10^{-3}$. Because the critical mass for gap formation is an order of magnitude larger than for dynamical gas accretion, a bimodal mass distribution with a depression between these masses may be anticipated. After gap formation, massive protoplanets are expected to migrate with the viscous evolution of the disk until they encounter a stellar magnetospheric cavity or tidal barrier. The existence of short-period planets around classical T Tauri stars would imply that the local disk viscous evolution timescale is shorter than the typical age ($\sim 10^6$ yr). Orbital migration may lead to stellar consumption of protoplanets with surviving planets locked into commensurable orbits. Contamination of host stars (during the latter's main-sequence evolution) by plunging planets may have led to a relatively high metallicity. Planet-disk interaction is likely to excite only small $e \lesssim 0.3$ and lead to a relation between M_p and e that could be observed. A planet with larger eccentricity would overrun the gap and induce nonlinear dissipation.

2. *Stellar tides raised on short-period planets* should rapidly synchronize their spin with that of the orbit. Short-period planets with $M_p \sim M_{Jup}$ and $P \lesssim$ 3–4 days (e.g., τ Boo) should have their orbits circularized by *tides raised on a slowly rotating star* within a few Gyr. In the case of massive planets, the stellar rotation may also be synchronized with the orbit. Only very short-period planets ($P \lesssim 2$ days), however, are expected to undergo significant orbital decay.

3. *Long-term gravitational interaction between planets* with initially small eccentricity leads to orbit crossing on a timescale that is sensitively determined by their masses and initial separations. Subsequent close encounters between comparable masses can lead to high orbital eccentricity and planets scattered into extended orbits. This mechanism may provide a supply of planets with a range of short periods, some of which may undergo tidal circularization or plunge into the star, together with outlying accomplices. The latter could be imaged by the next generation of IR interferometers.

4. *Secular perturbations by distant binary*, or massive outlying planetary, companions in relatively highly inclined orbits can also excite high eccentricity through the Kozai mechanism. Potentially this process might produce a supply of short-period planets ($P \lesssim$ 2–3 days) that have undergone tidal circularization. Because of the very high relative inclinations required, this mechanism is likely to work only in a small number of cases.

Acknowledgments This work is supported by NSF and NASA through grants AST-9618548, NAG5-4277, and NAG5-7515.

REFERENCES

Artymowicz, P. 1992. Dynamics of binary and planetary-system interaction with disks—eccentricity changes. *Pub. Astron. Soc. Pacific* 104:769–774.

Artymowicz, P. 1993*a*. Disk-satellite interaction via density waves and the eccentricity evolution of bodies embedded in disks. *Astrophys. J.* 419:166–180.

Artymowicz, P. 1993*b*. On the wave excitation and a generalized torque formula for Lindblad resonances excited by external potential. *Astrophys. J.* 419:155–165.

Artymowicz, P., and Lubow, S. 1996. Mass flow through gaps in circumbinary disks. *Astrophys. J. Lett.* 467:L77–L80.

Artymowicz, P., Lubow, S., and Kley, W. 1998. Planetary systems and their changing theories. In *Planetary Systems: The Long View*, ed. L. Celnikier (Gif-sur-Yvette: Editions Frontières), in press.

Balbus, S. A., and Hawley, J. F. 1991. A powerful local shear instability in weakly magnetized disks. I. Linear analysis. *Astrophys. J.* 376:214–233.

Beckwith, S. V. W., and Sargent, A. I. 1996. Circumstellar disks and the search for neighboring planetary systems. *Nature* 383:139–144.

Bodenheimer, P. 1998. Formation of substellar objects orbiting stars. In *Brown Dwarfs and Extrasolar Planets,* ASP Conf. Ser. 134, ed. R. Rebolo, E. L. Martin, and M. R. Zapatero Osorio (San Francisco: Astronomical Society of the Pacific), p. 115.

Bodenheimer, P., and Pollack, J. B. 1986. Calculations of the accretion and evolution of giant planets: The effects of solid cores. *Icarus* 67:391–408.

Bouvier, J., Cabrit, S., Fernandez, M., Martin, E. L., and Matthews, J. M. 1993. COYOTES-I: The photometric variability and rotational evolution of T-Tauri stars. *Astron. Astrophys.* 272:176–206.

Bryden, G., Chen, X., Lin, D. N. C., Nelson, R., and Papaloizou, J. C. B. 1998. Tidally induced gap formation in protostellar disks: Gap clearing and suppression of protoplanetary growth. *Astrophys. J.*, submitted.

Butler, R. P., Marcy, G. W., Williams, E., Hauser, H., and Shirts, P. 1997. Three new "51 Pegasi-type" planets. *Astrophys. J. Lett.* 474:L115–L118.

Chambers, J. E., Wetherill, G. W., and Boss, A. P. 1996. The stability of multi-planet systems. *Icarus* 119:261–268.

Choi, P. I., and Herbst, W. 1996. Rotation periods of stars in the Orion nebula cluster: The bimodal distribution. *Astron. J.* 111:283–298.

Claret, A., and Cunha, N. C. S. 1997. Circularization and synchronization times in main-sequence of detached eclipsing binaries. II. Using the formalisms by Zahn. *Astron. Astrophys.* 318:187–197.

Counselman, C. C., III. 1973. Outcomes of tidal evolution. *Astrophys. J.* 180:307–316.

Cowling, T. G. 1941. The non-radial oscillations of polytropic stars. *Mon. Not. Roy. Astron. Soc.* 101:367–375.

Darwin, G. H. 1879. Bodily tides of viscous and semi-elastic spheroids. *Phil. Trans. Roy. Soc. London* 170:1.

Duncan, M. J., and Lissauer, J. J. 1998. The effects of post-main-sequence solar mass loss on the stability of our planetary system. *Icarus* 134:303–310.

Franklin, F., Lecar, M., and Quinn, T. 1990. On the stability of orbits in the solar system: A comparison of a mapping with a numerical integration. *Icarus* 88:97–103.

Gailitis, A. 1982. Tidal heating of Io and orbital evolution of the Jovian satellites. *Mon. Not. Roy. Astron. Soc.* 201:415–420.

Goldreich, P., and Nicholson, P. D. 1977. Turbulent viscosity and Jupiter's tidal *Q*. *Icarus* 30:301–304.

Goldreich, P., and Nicholson, P. D. 1989. Tidal friction in early-type stars. *Astrophys. J.* 342:1079–1084.
Goldreich, P., and Peale, S. 1966. Spin-orbit coupling in the solar system. *Astron. J.* 71:425–438.
Goldreich, P., and Soter, S. 1966. *Q* in the solar system. *Icarus* 5:375–389.
Goldreich, P., and Tremaine, S. 1978. The velocity dispersion in Saturn's rings. *Icarus* 34:227–239.
Goldreich, P., and Tremaine, S. 1980. Disk-satellite interactions. *Astrophys. J.* 241:425–441.
Gonzalez, G. 1997. The stellar metallicity–giant planet connection. *Mon. Not. Roy. Astron. Soc.* 285:403–412.
Gonzalez, G. 1998. Spectroscopic analyses of the parent stars of extrasolar planetary system candidates. *Astron. Astrophys.* 334:221–238.
Goodman, A. A., Benson, P. J., Fuller, G. A., and Myers, P. C. 1993. Dense cores in dark clouds. VIII. Velocity gradients. *Astrophys. J.* 406:528–547.
Goodman, J., and Dickson, E. S. 1998. Dynamical tide in solar-type binaries. *Astrophys. J.* 507:938–944.
Goodman, J., and Oh, S. P. 1997. Fast tides in slow stars: The efficiency of eddy viscosity. *Astrophys. J.* 486:403–412.
Guenther, E. W. 1997. Magnetic fields of T Tauri stars. In *Herbig-Haro Flows and the Birth of Low-mass Stars*, IAU Symp. 182, ed. B. Reipurth and C. Bertout (Dordrecht: Kluwer), pp. 465-474.
Guillot, T., Burrows, A., Hubbard, W. B., Lunine, J. I., and Saumon, D. 1996. Giant planets at small orbital distances. *Astrophys. J. Lett.* 459:L35–L38.
Hagihara, Y. 1972. *Celestial Mechanics, Vol. II, Part 1: Perturbation Theory* (Cambridge: MIT Press), p. 447.
Hartmann, L., Calvet, N., Gullbring, E., and D'Alessio, P. 1998. Accretion and the evolution of T Tauri disks. *Astrophys. J.* 495:385–400.
Hayashi, C., Nakazawa, K., and Nakagawa, Y. 1985. Formation of the solar system. In *Protostars and Planets II,* ed. D. C. Black and M. S. Matthew (Tucson: University of Arizona Press), pp. 1100–1153.
Holman, M., Touma, J., and Tremaine, S. 1997. Chaotic variations in the eccentricity of the planet orbiting 16 Cyg B. *Nature* 386:254.
Hut, P. 1980. Stability of tidal equilibrium. *Astron. Astrophys.* 92:167–170.
Hut, P. 1981. Tidal evolution in close binary systems. *Astron. Astrophys.* 99:126–140.
Innanen, K. A., Zheng, J. Q., Mikkola, S., and Valtonen, M. J. 1997. The Kozai mechanism and the stability of planetary orbits in binary star systems. *Astron. J.* 113:1915–1919.
Ivanov, P. B., Papaloizou, J. C. B., and Polnarev, A. G. 1998. The evolution of supermassive binary caused by an accretion disc. *Mon. Not. Roy. Astron. Soc.*, submitted.
Kley, W. 1998. Mass flow and accretion through gaps in accretion disks. *Mon. Not. Roy. Astron. Soc.*, submitted.
Konigl, A. 1991. Disk accretion onto magnetic T Tauri stars. *Astrophys. J. Lett.* 370:L39–L43.
Korycansky, D. G., and Papaloizou, J. C. B. 1996. A method for calculations of nonlinear shear flow: Application to formation of giant planets in the solar nebula. *Astrophys. J. Suppl.* 105:181–190.
Korycansky, D. G., and Pollack, J. B. 1993. Numerical calculations of the linear response of a gaseous disk to a protoplanet. *Icarus* 102:150–165.
Kozai, Y. 1962. Secular perturbations of asteroids with high inclination and eccentricity. *Astron. J.* 67:591–598.
Kumar, P., and Goodman, J. 1996. Nonlinear damping of oscillations in tidal-capture binaries. *Astrophys. J.* 466:946–956.

Kumar, P., Ao, C. O., and Quataert, E. J. 1995. Tidal excitation of modes in binary systems with applications to binary pulsars. *Astrophys. J.* 449:294–309.

Levison, H. F., Lissauer, J. J., Duncan, M. J. 1998. Modeling the diversity of outer planetary systems. *Astron. J.* 116:1998–2014.

Lin, D. N. C., and Ida, S. 1997. On the origin of massive eccentric planets. *Astrophys. J.* 477:781–791.

Lin, D. N. C., and Papaloizou, J. C. B. 1979*a*. On the structure of circumbinary accretion disks and the tidal evolution of commensurable satellites. *Mon. Not. Roy. Astron. Soc.* 188:191–201.

Lin, D. N. C., and Papaloizou, J. C. B. 1979*b*. Tidal torques on accretion discs in binary systems with extreme mass ratios. *Mon. Not. Roy. Astron. Soc.* 186:799–830.

Lin, D. N. C., and Papaloizou, J. C. B. 1980. On the structure and evolution of the primordial solar nebula. *Mon. Not. Roy. Astron. Soc.* 191:37–48.

Lin, D. N. C., and Papaloizou, J. C. B. 1986*a*. On the tidal interaction between protoplanets and the primordial solar nebula. II. Self-consistent nonlinear interaction. *Astrophys. J.* 307:395–409.

Lin, D. N. C., and Papaloizou, J. C. B. 1986*b*. On the tidal interaction between protoplanets and the primordial solar nebula. III. Orbital migration of protoplanets. *Astrophys. J.* 309:846–857.

Lin, D. N. C., and Papaloizou, J. C. B. 1993. On the tidal interaction between protostellar disks and companions. In *Protostars and Planets III*, ed. E. H. Levy and J. I. Lunine (Tucson: University of Arizona Press), pp. 749–835.

Lin, D. N. C., Papaloizou, J. C. B., and Ruden, S. P. 1987. On the confinement of planetary arcs. *Mon. Not. Roy. Astron. Soc.* 227:75–95.

Lin, D. N. C., Papaloizou, J. C. B., and Savonije, G. J. 1990*a*. Propagation of tidal disturbance in gaseous accretion disks. *Astrophys. J.* 365:748–756.

Lin, D. N. C., Papaloizou, J. C. B., and Savonije, G. J. 1990*b*. Wave propagation in gaseous accretion disks. *Astrophys. J.* 364:326–334.

Lin, D. N. C., Bodenheimer, P., and Richardson, D. C. 1996. Orbital migration of the planetary companion of 51 Pegasi to its present location. *Nature* 380:606–607.

Lissauer, J. J., and Stewart, G. R. 1993. Growth of planets from planetesimals. In *Protostars and Planets III*, ed. E. H. Levy and J. I. Lunine (Tucson: University of Arizona Press), pp. 1061–1088.

Lubow, S. H., Tout, C. A., and Livio, M. 1997. Resonant tides in close orbiting planets. *Astrophys. J.* 484:866–870.

MacDonald, G. J. F. 1964. Tidal friction. *Rev. Geophys.* 2:467–541.

Marcy, G. W., Butler, R. P., Williams, E., Bildsten, L., Graham, J. R., Ghez, A. M., and Jernigan, J. G. 1997. The planet around 51 Pegasi. *Astrophys. J.* 481:926–935.

Mathieu, R. D. 1994. Pre-main-sequence binary stars. *Ann. Rev. Astron. Astrophys.* 32:465–530.

McCaughrean, M. J., and Stauffer, J. R. 1994. High resolution near-infrared imaging of the trapezium: A stellar census. *Astron. J.* 108:1382–1397.

Miyoshi, K., Takeuchi, T., Tanaka, H., and Ida, S. 1999. Gravitational interaction between a protoplanet and a protoplanetary disk. I. Local three-dimensional simulations. *Astrophys. J.* 516:451–464.

Mizuno, H. 1980. Formation of the giant planets. *Prog. Theor. Phys.* 64:544–557.

Papaloizou, J. C. B., and Lin, D. N. C. 1984. On the tidal interaction between protoplanets and the primordial solar nebula. I. Linear calculation of the role of angular momentum exchange. *Astrophys. J.* 285:818–834.

Papaloizou, J. C. B., and Savonije, G. J. 1985. On the tidal evolution of massive X-ray binaries—the tidal evolution timescales for very long orbital periods. *Mon. Not. Roy. Astron. Soc.* 213:85–96.

Papaloizou, J. C. B., and Savonije, G. J. 1997. Non-adiabatic tidal forcing of a massive, uniformly rotating star. III. Asymptotic treatment for low frequencies in the inertial regime. *Mon. Not. Roy. Astron. Soc.* 291:651–657.

Pollack, J. B., Hubickyj, O., Bodenheimer, P., Lissauer, J. J., Podolak, M., and Greenzweig, Y. 1996. Formation of the giant planets by concurrent accretion of solids and gas. *Icarus* 124:62–85.

Prosser, C. F., Stauffer, J. R., Hartmann, L., Soderblom, D. R., Jones, B. F., Werner, M. W., and McCaughrean, M. J. 1994. HST photometry of the Trapezium cluster. *Astrophys. J.* 421:517–541.

Rasio, F. A., and Ford, E. B. 1996. Dynamical instabilities and the formation of extrasolar planetary systems. *Science* 274:954–956.

Rasio, F. A., Tout, C. A., Lubow, S. H., and Livio, M. 1996. Tidal decay of close planetary orbits. *Astrophys. J.* 470:1187–1191.

Safronov, V. S. 1969. *Evoliutsiia Doplanetnogo Oblaka* (Moscow: Izdatel'stvo Nauka).

Sandquist, E., Taam, R. E., Lin, D. N. C., and Burkert, A. 1998. Planet stripping and stellar metallicity enhancements. *Astrophys. J.*, in press.

Saumon, D., Hubbard, W. B., Burrows, A., Guillot, T., Lunine, J. I., and Chabrier, G. 1996. A theory of extrasolar giant planets. *Astrophys. J.* 460:993–1018.

Savonije, G. J., and Papaloizou, J. C. B. 1983. On the tidal spin-up and orbital circularization rate for the massive X-ray binary systems. *Mon. Not. Roy. Astron. Soc.* 203:581–593.

Savonije, G. J., and Papaloizou, J. C. B. 1984. On the tidal evolution of massive X-ray binaries: The spin-up and circularization rates for systems with evolved stars and the effects of resonances. *Mon. Not. Roy. Astron. Soc.* 207:685–704.

Savonije, G. J., and Papaloizou, J. C. B. 1997. Non-adiabatic tidal forcing of a massive, uniformly rotating star. II. The low-frequency, inertial regime. *Mon. Not. Roy. Astron. Soc.* 291:633–650.

Savonije, G. J., Papaloizou, J. C. B., and Alberts, F. 1995. Nonadiabatic tidal forcing of a massive uniformly rotating star. *Mon. Not. Roy. Astron. Soc.* 277:471–496.

Shakura, N. I., and Sunyaev, R. A. 1973. Black holes in binary systems: Observational appearance. *Astron. Astrophys.* 24:337–355.

Shu, F. H. 1984. Waves in planetary rings. In *Planetary Rings*, ed. R. Greenberg and A. Brahic (Tucson: University of Arizona Press), pp. 513–561.

Stassun, K. G., Mathieu, R. D., Mazeh, T., and Vrba, F. J. 1999. The rotation period distribution of pre-main-sequence stars in and around the Orion Nebula. *Astron. J.* 117:2941–2979.

Stevenson, D. J. 1983. Anomalous bulk viscosity of two-phase fluids and implications for planetary interiors. *J. Geophys. Res.* 88:2445–2455.

Takeuchi, T., Miyama, S. M., and Lin, D. N. C. 1996. Gap formation in protoplanetary disks. *Astrophys. J.* 460:832–847.

Tassoul, J.-L. 1988. On orbital circularization in detached close binaries. *Astrophys. J. Lett.* 324:L71–L73.

Terebey, S., Shu, F. H., and Cassen, P. 1984. The collapse of the cores of slowly rotating isothermal clouds. *Astrophys. J.* 286:529–551.

Terquem, C., Papaloizou, J. C. B., Nelson, R. P., and Lin, D. N. C. 1998*a*. Oscillations in solar-type stars tidally induced by orbiting planets. In *Planetary Systems: The Long View*, ed. L. Celnikier (Gif-sur-Yvette: Editions Frontières), in press.

Terquem, C., Papaloizou, J. C. B., Nelson, R. P., and Lin, D. N. C. 1998*b*. On the tidal interaction of a solar-type star with an orbiting companion: Excitation of g-mode oscillation and orbital evolution. *Astrophys. J.* 502:788–801.

Trilling, D. E., Benz, W., Guilot, T., Lunine, J. I., Hubbard, W. B., and Burrows, A. 1998. Orbital evolution and migration of giant planets: Modelling extrasolar planets. *Astrophys. J.* 500:428–439.
Ward, W. R. 1981. Solar nebula dispersal and the stability of the planetary system. I. Scanning secular resonance theory. *Icarus* 47:234–264.
Ward, W. R. 1993. Disk-planet interactions: Torques from the coorbital zone. *Ann. N.Y. Acad. Sci.* 675:314–323.
Ward, W. R. 1997. Protoplanet migration by nebula tides. *Icarus* 126:261–281.
Weidenschilling, S. J., and Marzari, F. 1996. Gravitational scattering as a possible origin for giant planets at small stellar distances. *Nature* 384:619–621.
Zahn, J. P. 1975. The dynamical tide in close binaries. *Astron. Astrophys.* 41:329–344.
Zahn, J. P. 1977. Tidal friction in close binary stars. *Astron. Astrophys.* 57:383–394.
Zharkov, V. N., and Gudkova, T. V. 1991. Models of giant planets with a variable ratio of ice to rock. *Ann. Geophys.* 9:357–366.

DISK-PLANET INTERACTIONS AND THE FORMATION OF PLANETARY SYSTEMS

WILLIAM R. WARD
Southwest Research Institute

and

JOSEPH M. HAHN
Lunar and Planetary Institute

Several aspects of the interaction physics between an emerging planetary system and its precursor nebula are reviewed. The principal interaction mechanism is through density waves and their associated torques. Two types of orbital migration of protoplanets are distinguished, and the conditions favoring each are delineated. The effects of migration on the style and timescales of planetary formation and on protoplanet survival are discussed.

I. INTRODUCTION

Aside from the primary, the most massive component of a forming planetary system is the precursor circumstellar disk. And yet, the gravitational influence of a disk on an emerging planetary system has been largely ignored throughout most of the cosmogonical literature. The disk's role as a source of planetary material is, of course, obvious at the outset, and its influence on the orbits of small objects through aerodynamic drag has been recognized since the work of Whipple (1972) and Weidenschilling (1977), but direct gravitational effects were not considered until almost a decade later. Even then, an appreciation of the many aspects of disk-planet gravitational coupling has only slowly worked its way into cosmogonical modeling, over the last fifteen years or so. Lately, however, this process has accelerated because of the discovery of extrasolar planets, particularly those in tight orbits around their primaries. Given the apparent difficulty of forming planets of a jovian mass or more in close stellar proximity, their existence seems to constitute strong circumstantial evidence of large-scale migration of planetary objects, a suggestion that seemed almost heretical a few years ago. The angular momentum change associated with large-scale migration is considerable. Except for the possibility of mutual scattering between giant planets, only the disk provides a large enough angular momentum reservoir to effect such modifications.

The first study of collective gravitational interaction with a disk in a planetary context was the proposal by Goldreich and Tremaine (1978) that the Cassini division in Saturn's ring system could have been opened and be maintained by Mimas through its 2:1 mean-motion resonance. This is a Lindblad resonance, in which the forcing frequency, seen from a reference frame moving with disk material, matches the disk's natural oscillation frequency, which is (nearly) the local epicycle frequency κ (e.g., Lynden-Bell and Kalnajs 1972). The calculations employed a fluid-mechanical description of a self-gravitating ring that demonstrated the importance of wave action in redistributing the (negative) angular momentum deposited at resonance to nearby portions of the ring, thereby accounting for the large width of the Cassini division compared to the nominal width of the resonance. In a closely related problem, Lin and Papaloizou (1979) utilized an impulse model of disk-planet interaction to show that a large enough secondary could truncate a circumstellar disk at a low-order Lindblad site.

Modification of the perturber's orbit by disk torques was first addressed fully by Goldreich and Tremaine (1980). The resonant forcing of the disk results in the launching of a spiral density wave train (e.g., Goldreich and Tremaine 1979; Papaloizou and Lin 1984; Ward 1986; Lin and Papaloizou 1993; Artymowicz, 1996*b*). The attraction of the perturber for this nonaxisymmetric surface density, σ, results in the torque, sometimes collectively referred to as disk *tidal* torques in analogy to solid-body tides between a planet and its satellite. In general, the interactions at various mean-motion resonances result in *work* being done on the planet as well as angular momentum exchange. From the Jacobi constant, the work done on the planet is $\dot{E} = \Omega_{\rm ps}\dot{L}$, where $\Omega_{\rm ps}$ represents the pattern speed of the resonant term, and where

$$E = \frac{-GM_pM_\star}{2a_p} \qquad L = M_p\sqrt{GM_\star a_p(1-e_p^2)} \tag{1}$$

are the energy and angular momentum of the planet, respectively (Goldreich and Tremaine 1980). This leads to equations for the rates of change of the planet's a_p and e_p due to resonant torque $T_{l,m}$;

$$\frac{\dot{a}_p}{a_p} = \frac{2T_{l,m}}{M_pa_p^2\Omega_p}\frac{\Omega_{\rm ps}}{\Omega_p} \qquad \frac{\dot{e}_p}{e_p} = \frac{-T_{l,m}(1-e_p^2)^{1/2}}{e_p^2M_pa_p^2\Omega_p}\left(1-\frac{\Omega_{\rm ps}}{\Omega_p}\sqrt{1-e_p^2}\right) \tag{2}$$

with Ω_p denoting the planet's mean motion.

II. DISK TORQUES

Each $\{m, l\}$ term in the planet's Fourier-expanded disturbing potential

$$\begin{aligned}\phi &= -\frac{GM_p}{|\mathbf{r}-\mathbf{r}_p|} + \frac{M_p}{M_\star}\Omega^2(\mathbf{r}\cdot\mathbf{r}_p) \\ &= \sum_{l=-\infty}^{\infty}\sum_{m=0}^{\infty}\phi_{l,m}\cos[m(\theta-\Omega_pt)-(l-m)\kappa_pt]\end{aligned} \tag{3}$$

has a pattern speed $\Omega_{\text{ps}} \equiv \Omega_p + (l - m)\kappa_p/m$ and gives rise to a corotation resonance (CR, $\Omega = \Omega_{\text{ps}}$) plus an inner [ILR, $m(\Omega - \Omega_{\text{ps}}) = \kappa$] and an outer [OLR, $m(\Omega - \Omega_{\text{ps}}) = -\kappa$] Lindblad resonance. In most applications, the disk's epicycle frequency is nearly its angular velocity $\Omega(r)$. The wave train launched from each Lindblad resonance carries an angular momentum flux (e.g., Goldreich and Tremaine 1979)

$$\mathcal{F} = -\text{sgn}(k)\frac{mr\Phi^2}{4G}\left[1 - \frac{c^2|k|}{\pi G\sigma}\right] \tag{4}$$

where Φ is the portion of the disk's gravitational potential associated with the wave, c is the gas sound speed, and k is the wavenumber that satisfies the dispersion relation

$$m^2(\Omega - \Omega_{\text{ps}})^2 = \kappa^2 - 2\pi G\sigma|k| + c^2k^2 \tag{5}$$

We have used the sign convention of Goldreich and Tremaine (1979), where the waves are trailing (leading) for $k > 0$ ($k < 0$), and the group velocity of the waves is given by

$$c_g = \text{sgn}(k)\left(\frac{\pi G\sigma - c^2|k|}{m(\Omega - \Omega_{\text{ps}})}\right) \tag{6}$$

Therefore, the angular momentum density of the waves as defined by $2\pi rc_g\mathcal{H} = \mathcal{F}$, is (e.g., Goldreich and Tremaine 1978)

$$\mathcal{H} = -\left(\frac{\Omega - \Omega_{\text{ps}}}{2\sigma}\right)\left(\frac{m\Phi}{2\pi G}\right)^2 \tag{7}$$

and has the same sign as the torque on the disk. OLRs exert a positive torque, whereas ILRs exert a negative torque. The corresponding reaction torques $T_{l,m}$ on the planet are of opposite sign.

The torque exerted on an object by the disk due to interaction at a Lindblad resonance is (Goldreich and Tremaine 1979),

$$T^L_{l,m} = f_L(\xi)\text{sgn}(\Omega - \Omega_{\text{ps}})\frac{\pi^2\sigma\Psi^2_{l,m}}{|rdD/dr|} \tag{8}$$

where $\Psi_{l,m} \equiv rd\Phi_{l,m}/dr + 2\Omega\Phi_{l,m}/(\Omega - \Omega_{\text{ps}})$ is the forcing function, and the so-called cutoff function $f_L(\xi)$ has been introduced, with $\xi \equiv mc/r\Omega$. The cutoff function approaches unity in a cold disk ($c = 0$) but kills the torque at high ξ. The high-ξ behavior is due to a pressure-induced shift of the resonance sites away from the perturber when the azimuthal wavelength $\sim r/m$ of the forcing term is less than the scale height $\sim c/\Omega$ of the disk (Ward 1988, 1997*a*; Artymowicz 1993*b*). The quantity D that appears in the denominator is the frequency distance,

$$D \equiv \kappa^2 - m^2(\Omega - \Omega_{\text{ps}})^2 \tag{9}$$

which vanishes at the Lindblad resonance. From the dispersion relationship [equation (5)], *long* waves for which $\pi G\sigma > c^2|k|$ must propagate where $D > 0$. Causality requires that the group velocity be directed away from the resonance, or $\text{sgn}(c_g) = \text{sgn}(dD/dr)$ when evaluated at the resonance. For $m \geq 2$, $\text{sgn}(dD/dr) = \text{sgn}\left(\Omega - \Omega_{\text{ps}}\right)$, and comparing this with equation (6) reveals that the waves are trailing, $k > 0$. The $m = 1$ resonances are a special case; the OLR waves are trailing, but for the ILR the waves are launched from a secular resonance where the apsidal precession rate $d\tilde{\varpi}/dt = g(r)$ of the disk matches that of the planet; i.e., $g(r) = g_p$ (Ward and Hahn 1998*a,b*). For this case, $\text{sgn}(k) = \text{sgn}(dg/dr)$, and waves can be either leading or trailing depending on the gradient of $g(r)$.

Finally, the torque on the perturber due to a corotation resonance is (Goldreich and Tremaine 1979, 1980)

$$T^C_{l,m} = f_C(\xi)\frac{m\pi^2\Phi_{l,m}}{-d\Omega/dr}\frac{d}{dr}\left(\frac{\sigma}{B}\right) \tag{10}$$

where $B \equiv (2r)^{-1}d\left(r^2\Omega\right)/dr$ is the vorticity parameter and f_c is the cutoff function for this type of resonance (e.g., Ward 1989). For a review of many of these particulars the reader is directed to the chapters on this subject in the earlier volumes of the *Protostars and Planets* series, notably Lin and Papaloizou (1985, 1993).

III. PROTOPLANET MIGRATION

A. Type I: Dynamical Friction

The most important resonances for changing the semimajor axes are those associated with terms in the Fourier-expanded disturbing function that are zero-order in eccentricity. From equations (2) and (8) we see that the inner resonances expand the orbit, whereas the outer ones contract it. Unless these various torques all exactly cancel, a net torque ΔT on the planet is expected and, with it, possible orbital evolution of the perturbing body. Goldreich and Tremaine (1980) pointed out that a mismatch in the cumulative inner and outer tidal torques was likely and reasoned that the fractional torque difference should be of order h/r, where $h \sim c/\Omega$ is the vertical scale height of a gas disk. They refrained from predicting the sign, cautioning that this was hard to ascertain without detailed knowledge of the disk. Ward (1986) argued that this is overly pessimistic and showed that the near-Keplerian rotation results in a negative torque bias ΔT for a wide range of disk models, a result since confirmed by the numerical integrations of Korycansky and Pollack (1993). The resulting orbital decay rate is

$$\dot{r} = \frac{2\Delta T}{M_p r\Omega} \approx c_{\text{I}}\mu\left(\frac{\sigma r^2}{M_\star}\right)\left(\frac{r\Omega}{c}\right)^3 r\Omega \tag{11}$$

where $\mu \equiv M_p/M_\star$, and $c_{\rm I} < 0$ is an asymmetry constant proportional to $c/r\Omega$. In earlier discussions of this problem (Ward 1997*a*, 1998*b*), $c_{\rm I}$ of order unity was assumed, which is appropriate for a nebula scale height about $h \sim 10^{-1}r$. In order to reveal the functionality better, here we use the relatively constant parameter $C_a \equiv |c_{\rm I}|(r\Omega/c)$ introduced by Ward (1986). Note that this implies $\dot{r}$ itself scales as $(r\Omega/c)^2$, as described by Goldreich and Tremaine (1980). We designate this sort of behavior as *Type I*. Both analytic and numerical calculations indicate that $C_a \approx O(1\text{–}10)$; e.g., Ward (1986, 1997*a*); Korycansky and Pollack (1993), Takeuchi and Miyama (1998). The orbital drift of a planet in a minimum-mass solar nebula has a characteristic decay time $\tau_{\rm I} \sim |r/\dot{r}|$ of

$$\tau_{\rm I} = \frac{\Omega^{-1}}{C_a}\left(\frac{M_\star}{M_p}\right)\left(\frac{M_\star}{\sigma r^2}\right)\left(\frac{c}{r\Omega}\right)^2 \tag{12}$$

B. Type II: Coevolution

Papaloizou and Lin (1984) and Lin and Papaloizou (1985, 1993) concentrated on the reaction of the disk, including the case where a planet truncates the disk both inside and outside its orbit, thereby creating a gap in which it then resides. The ability of a planet to open and maintain a gap depends in part on the viscosity ν of the disk (Hourigan and Ward 1984; Papaloizou and Lin 1984, 1993; Ward and Hourigan 1989) as well as the planet's mass and the damping characteristics of the waves (Takeuchi et al. 1996). When conditions are met, the gap presents a barrier to any radial flow of disk material that may occur due to global viscous diffusion. The planet in effect locks itself into the angular momentum transport process of the disk by acting as a sort of "relay station" that transmits angular momentum across the gap via tidal torques. As the disk evolves, the planet will maintain its *relative* position to the disk material. For this to happen, the disk must adjust its local configuration so that the torque differential ΔT is just that required for the planet to drift at the same rate, $\sim O(\nu/r)$, as the gas. For a Sakura-Sunyaev viscosity prescription, the planet's migration rate is

$$\dot{r} \approx c_{\rm II}\alpha\left(\frac{c}{r\Omega}\right)^2 r\Omega \tag{13}$$

where $c_{\rm II}$ is also a constant of order unity. This is the migration type proposed as a delivery mechanism for emplacement of close stellar companions such as 51 Pegasi b (Lin et al. 1996). Trilling et al. (1998) have presented simulations of this behavior. We designate this sort of migration as *Type II*.

Although both types of orbital drift are described in the literature, they have not always been clearly distinguished from one another. In fact, these cases represent the weak and strong coupling limits of a range of disk-planet interactions (Ward 1997*a*,*b*), and type-casting these distinctly

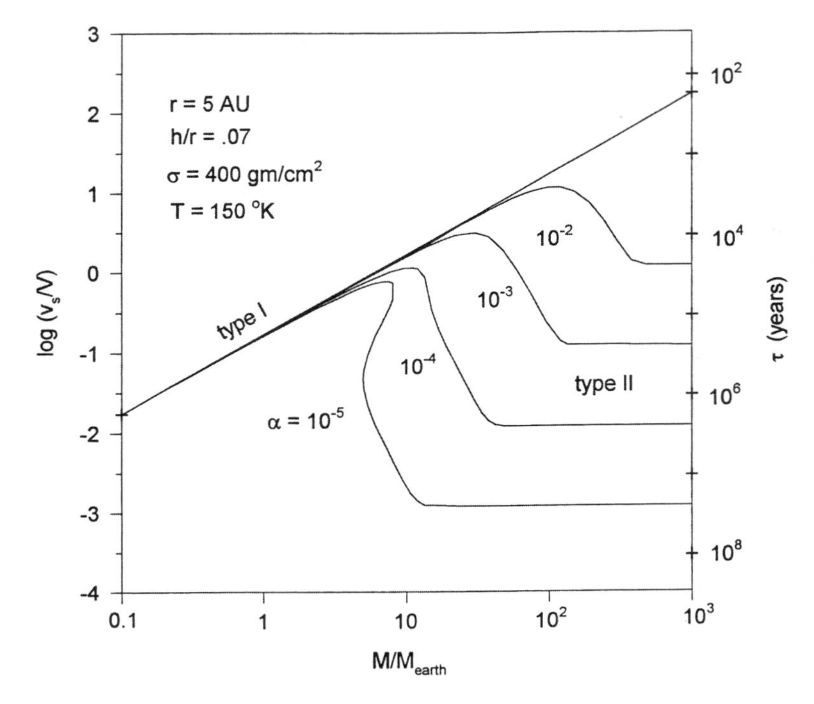

Figure 1. Characteristic decay times $\tau \equiv r/\dot{r}$ (right-hand scale) as a function of protoplanet mass, measured in Earth masses. Curves are labeled by the viscosity parameter $\alpha \equiv \nu/c^2\Omega^{-1}$ and constructed for a minimum-mass solar nebula at a distance of 5 AU. The timescale decreases inversely with mass (Type I) until it reaches a threshold size M_S (Shiva limit), past which the time scale increases (Type II) as a gap progressively opens. When a gap is fully established, the planet coevolves with the disk with a timescale inversely proportional to α. The right-hand scale gives the migration rate normalized to $V \equiv 2\mu_\oplus\mu_d(r/h)^3 r\Omega$, which is comparable to the Type I drift velocity for an Earth-sized object. Taken from Ward (1997*a*).

different behaviors proves useful in clarifying discussions. For a given disk, there is a critical mass where a transition from Type I to Type II occurs (Ward 1997*a*). Figure 1 shows the radial drift velocity as a function of protoplanet mass evaluated at ~5 AU in a minimum-mass solar nebula. The different curves are labeled by the assumed value of α. The velocity drops by as much as two orders of magnitude at transition.

C. Gap Formation

The type of behavior executed by a protoplanet depends on whether it can open and maintain a gap. A gap can be maintained against viscous diffusion if the cumulative torque $T_o \sim O[\mu^2(\sigma r^2)(r\Omega)^2(r/h)^3]$ that the planet exerts on one side of the disk exceeds the viscous couple $g \approx 3\pi\nu\sigma r^2\Omega$

(e.g., Lin and Papaloizou 1985; Ward 1986). For this to be satisfied, the protoplanet mass must exceed

$$\mu_\nu \sim c_\nu \left(\frac{\nu}{r^2\Omega}\right)^{1/2} \left(\frac{c}{r\Omega}\right)^{3/2} \sim c_\nu \sqrt{\alpha}\left(\frac{h}{r}\right)^{5/2} \tag{14}$$

where $c_\nu \sim O(1 - 10)$. Adopting a scale height $h \sim c/\Omega \sim 0.07r$, this mass limit becomes $\mu_\nu \sim c_\nu \sqrt{\alpha} \times 10^{-3}$. Lin and Papaloizou (1985) suggest that gap development terminates a giant planet's gas accretion phase and sets an upper mass limit for protoplanets. If Jupiter ($\mu_{\rm Jup} = 10^{-3}$) represents such a limit, a turbulence parameter of $\alpha \sim$ few $\times 10^{-2}$ is inferred. However, a much smaller value, $\alpha \sim 10^{-4}$–10^{-3}, is usually quoted from attempts to estimate nebula turbulence (e.g., Cabot et al. 1987; Dubrulle 1993). Note also that the Type II decay time for the higher value of α is much shorter than the probable lifetime of disks inferred from observation. How do we reconcile this with the existence of Jupiter and the numerous extrasolar planets of comparable size or larger that have been recently discovered?

Lin and Papaloizou (1993) assert that in addition to a sufficiently strong torque, it is necessary that the protoplanet's Hill radius, $R_H = r(\mu/3)^{1/3}$ exceed the scale height of the disk. This would set a lower limit for gap formation of $\mu_h \sim 3(h/r)^3 \sim O(10^{-3})$ even if α is small. The justification given for this condition is that a gap narrower than $O(h)$ would be Rayleigh unstable, which seems to implicitly assume that all objects would tend to open gaps of width comparable to their respective Hill radii. In fact, the scale height sets the minimum gap width for objects smaller than μ_h because pressure effects prevent high-order Lindblad resonances from lying closer than that to the protoplanet (Artymowicz 1993*a*; Ward 1988, 1997*a*). Accordingly, the validity of this so-called "necessary" condition is suspect.

On the other hand, the ability of a planet to open a gap also depends on the propagation distance of the waves, because the angular momentum density of the waves is not permanently transferred to the disk until the waves damp; before that happens, the disk cannot undergo secular adjustment in structure. If one assumes that the damping length $\ell \gtrsim h$ of the waves better represents the minimum gap width that can be opened, then (provided the gap is not so large as to exclude all resonances) the torque exerted on a disk with an edge distance ℓ is down by a factor $\sim(\ell/h)^{-3}$ from T_o, and the planet mass needed to maintain the gap increases to $\mu_{\nu,\ell} \sim c_\nu \sqrt{\alpha}(h/r)^{5/2}(\ell/h)^{3/2}$. This still leaves a quandary: If gap formation is delayed, what keeps the planet from spiraling into the primary via Type I migration? Perhaps the solution is the onset of gas accretion. If accretion can deplete the local surface density of the disk sufficiently to override the intrinsic torque asymmetry, but not so much as to shut off accretion, a survival strategy might be found. Models of gas accretion by a migrating planet have not yet been devised but should be a high-priority problem.

D. Wave Damping

Several processes can damp density waves:

(i) *Viscosity*. Goldreich and Tremaine (1980) and Ward (1986) both considered damping of waves in a high-Q disk by turbulent viscosity. In a more thorough treatment, Takeuchi et al. (1996) find that viscosity can dissipate the waves over a distance $\ell_\nu \sim \alpha^{-2/5}h$. In this case the critical mass is $\mu_{\nu,\ell} = 10^{-3}c_\nu\alpha^{-1/10}$, which, considering our order of magnitude approach, is still comparable to the masses of extrasolar planets for $\alpha < 10^{-2}$. However, other processes may supersede viscous damping for small α.

(ii) *Nonlinearity*. If $\mu \gtrsim (h/r)^3$, the perturbed surface density achieves nonlinear status locally and may shock-dissipate relatively close ($\sim h$) to the planet (e.g., Ward 1986). Indeed, this seems a more likely cause of gap formation when an object's Hill radius exceeds h, rather than the above-mentioned Rayleigh argument.

Other promising damping mechanisms include (iii) *radiation damping* (Cassen and Woolum 1996), in which the temperature rise during wave compression increases the radiation rate from the surface of the disk, and (iv) *channeling* (Lubow and Ogilvy 1998), a process where the wave energy becomes concentrated near the surface of the disk in a layer of thickness $\sim 1/k$, with k being the wavenumber. This occurs when there is a vertical temperature gradient. As the wave propagates, the wavenumber increases, confining the energy to a progressively thinner layer until nonlinearities set in.

Both of these latter mechanisms would appear to work even if α is very low. Indeed, recent work by Stone and coworkers (1996) claims that vertical convection in the disk (the principally invoked cause of turbulence) does not couple well to the horizontal motions, as has usually been assumed, and will not significantly contribute to the viscous evolution of the disk. This has led Gammie (1996) to suggest a layered model of the solar nebula, where turbulence is generated by magnetohydrodynamic (MHD) mechanisms but is confined to the surface, where cosmic rays can maintain a requisite degree of ionization. Coupled with channeling, this suggestion indicates that it may be possible to damp waves fairly locally in a disk with a very low effective α.

Even in an inviscid disk, however, there is a lower limit to the mass of an object that can open a gap, because of inertial effects. If the planet drifts across the gap in less time than it can excavate it, the gap will fail to open (Hourigan and Ward 1984; Ward and Hourigan 1989; Takeuchi et al. 1996). This so-called "inertial" limit is the inviscid limit of the Shiva mass (see below) and is typically of order

$$\mu_i = \frac{C_a^2}{4\pi}\left(\frac{\pi\sigma r^2}{M_\star}\right)\left(\frac{c}{r\Omega}\right)^3 \approx 8\mu_d\left(\frac{h}{r}\right)^3 \tag{15}$$

where $\mu_d \equiv \pi\sigma r^2/M_\star$ is a normalized measure of the disk mass. For $\mu_d \sim$ few $\times\, 10^{-3}$, $h/r = 0.07$, the inertial mass is a few Earth masses.

E. Torque at Corotation

So far we have paid little attention to the $m = l$ corotation resonances that fall at the perturber's semimajor axis. The disk's response to this forcing is to execute horseshoe-type behavior (see chapter by Lin et al., this volume). Horseshoe streamlines in an annulus of half-width w result in a torque on the perturber, of the form (Ward 1991)

$$T_{hs} = 4|A|B^2 w^4 r \frac{d}{dr}\left(\frac{\sigma}{B}\right) \tag{16}$$

where $A \equiv (r/2)d\Omega/dr$; $B \equiv (2r)^{-1}d\left(r^2\Omega\right)/dr$ are the Oort parameters of the disk. As one should expect, this is the same functional dependence on the vortensity as equation (10). For a Keplerian flow, $B = \Omega/4$, implying that T vanishes if $\sigma \propto r^{-3/2}$. Korycansky and Pollack (1993) have numerically integrated the linear response of a disk to an mth-order term and determined which portion of the torque is due to corotation by subtracting the downstream angular momentum flux of the waves from the whole torque. This exploits the fact that only angular momentum deposited at the inner and outer Lindblad resonances will be carried away by wave action. The residual corotation torques for each m are then summed to find the total at corotation, obtaining the approximate result

$$T_C \approx \frac{4}{3}\mu^2\left(r\Omega\right)^2\left(\frac{r\Omega}{c}\right)^2\left[r^3 B\frac{d}{dr}\left(\frac{\sigma}{B}\right)\right] \tag{17}$$

This result can also be obtained from equation (16) if one assumes that the particles have an eccentricity $e \sim h/a$ and that the horseshoe streamline coming closest to the perturber is turned around at a distance $\sim 2ea$ equal to half the axis of the epicycle in the azimuthal direction. The ratio $T_C/T_o \sim 4h/3r \ll c_{\rm I}$ implies that it is unlikely that the corotation torque could balance out the differential Lindblad torque. Furthermore, the corotation torque is subject to saturation on the horseshoe libration timescale $t_{\rm sat} \sim (r/w)\Omega^{-1}$ and so could, at best, reduce the drift rate to $\dot{r} \sim O[\mu(r/h)r\Omega]$, unless turbulence prevents saturation. This could be a limiting factor only if $\mu_d > O(10^{-2})$.

The above discussion notwithstanding, the corotation zone is among the least well-understood resonant regions. Lin et al. (this volume) point out that there are complex nonlinear effects, especially on the scale of the protoplanet's Hill radius, $R_H = r(\mu/3)^{1/3}$, that manifest themselves in recent numerical simulations of the flow. They suggest that these features may be important contributors to the protoplanet's orbital evolution, perhaps countering the torques inferred from linear analyses. We await such a possible revelation with interest, because current estimates of dynamical friction seem almost too powerful, and as yet no device other than gap formation or the onset of gas accretion seems capable of short-circuiting it.

IV. ACCRETION

Ward and coworkers have (e.g., Hourigan and Ward 1984; Ward 1989*b*, 1993*a*) stressed that a source of orbit drift for large objects would constitute an as yet unexploited degree of radial mobility for accretion models. Here we discuss a number of ways in which disk tides could modify our understanding of the planet building process.

A. Runaway Limit

One of the tightest constraints on models of solar system formation is the suspected 10^{6-7} year lifetime of the nebula, inferred from observation of T Tauri stars (e.g., Walter 1986; Walter et al. 1988). Obviously, the existence of gas giants like Jupiter and Saturn establish that the planet-building process for these objects was essentially completed before nebula dispersal. These planets are believed to have acquired their H/He component by gas accretion onto preexisting solid cores with estimated masses of 10–20 $M_\oplus$ (Mizuno et al. 1978; Bodenheimer and Pollack 1986; Podolak et al. 1993). If this model is correct, we must account for the accretion of 10^{29}-g cores within the lifetime of the gas disk.

Current models of solid body accretion indicate that large embryos can form in a relatively short timescale as a result of the onset of accretion runaway (Greenberg et al. 1978; Wetherill and Stewart 1989, 1993; Ida and Makino 1993). This runaway is due to a strong feedback loop in the growth rate, $\dot{M} \sim \sigma_d \Omega \pi R^2 F_g$, through the gravitational enhancement factor F_g where σ_d is the surface density of accretable material in the disk and M and R are the embryo's mass and radius, respectively (e.g., Greenzweig and Lissauer 1993; Lissauer and Stewart 1993). The implied characteristic growth time $\tau_{\rm growth} \equiv M/\dot{M}$ is

$$\tau_{\rm growth} \approx \Omega^{-1} \left(\frac{\rho_p R}{\sigma_d} \right) F_g^{-1} \tag{18}$$

where ρ_p is the body density of the embryo. The enhancement factor is the ratio of the effective collision cross section to the geometrical cross section. If the relative velocities v are dominated by velocity dispersion instead of disk shear, the enhancement factor becomes $F_g = 1 + (v_e/v)^2$ where $v_e \equiv (2GM/R)^{1/2}$ is the embryo's escape velocity. When the smaller field particles establish their velocity dispersion via equilibrium between mutual gravitational scattering and inelastic collisions, v will be comparable to the escape velocity, v_e', of a typical field particle, R'. Since the escape velocity of any object is proportional to its radius, $F_g \sim (v_e/v_e')^2 \sim (R/R')^2$ for $R \gg R'$, and the characteristic growth time for the runaway is inversely proportional to R. There is a limit, however, to the size of F_g. At some point the embryo begins to stir the surrounding swarm of particles and to contribute to their dispersion velocities (Lissauer 1987). In this case the enhancement factor approaches a limiting value of order

$F_g \rightarrow O(R_H/R)$, where R_H is the protoplanet's Hill radius. This ratio is dependent only on heliocentric distance and protoplanet density and, for a solar-mass star, is $\sim 133(\rho_p/1\text{g/cm}^{-3})^{1/3}(r/\text{AU})$. At 5 AU, a 3-g/cm^3 protoplanet would have an enhancement factor of order $\sim 10^3$. Once this limit is reached, the growth timescale becomes proportional to R [see, e.g., Lissauer and Stewart (1993) and Ward (1996) for recent readable reviews of solid body accretion].

This runaway phase has been generally thought to stall down by isolation of the embryo; that is, when it has cannibalized the disk locally, it simply runs out of material to sustain its growth (e.g., Wetherill 1990; Lissauer and Stewart 1993). Further accretion would then seem to depend on long-range gravitational interactions with other embryos, which will eventually generate crossing orbits. In this case, the enhancement factor reverts to a value of order unity, and the growth timescale lengthens accordingly. The so-called runaway mass limit is

$$M_{\text{run}} \equiv 3^{1/4}\left(\frac{8\pi\sigma_d r^2}{M_\star}\right)^{3/2} M_\star \tag{19}$$

which, for a minimum model of the solar nebula, is well short of a giant planet core. Binary accretion with $F_g \sim O(1)$ is too slow to form cores within the probable $\sim 10^{6-7}$ year duration of the nebula. However, it now seems that this concern was probably unfounded, because disk tides will prevent isolation by causing the embryo to migrate into undepleted regions of the disk (Ward and Hahn 1995; Tanaka and Ida 1998). The migrating embryo sweeps past fresh material equal to its own mass in a time $\tau_{\text{swp}} \equiv (M_p/2\pi\sigma_d r^2)\tau_{\text{I}}$. If this is less than τ_{growth}, accretion continues unabated.

B. Dispersion Velocities

For the eccentricity of a perturber, the most important potential terms are those that are first-order in e, i.e., those with $l = m \pm 1$ (Goldreich and Tremaine 1980). From equation (1), the sign of the eccentricity change is equal to $\text{sgn}(T_{l,m}) \times \text{sgn}(\Omega_{\text{ps}}\sqrt{1-e^2} - \Omega_p)$. For Lindblad resonances, $\text{sgn}(T_{l,m}) = \text{sgn}(\Omega - \Omega_{\text{ps}})$, so that inner (outer) resonances exert positive (negative) torques on the perturber. With the faster pattern speed, $\Omega_{\text{ps}} = \Omega_p + \kappa_p/m$, negative torques from outer Lindblad resonances damp the eccentricity while positive torques from inner resonances excite it. Just the reverse is true for the slower pattern speed, $\Omega_{\text{ps}} = \Omega_p - \kappa_p/m$. Consequently, the eccentricity is excited by those resonances that fall well inside and outside of the orbit (external) but damped by those that fall in the vicinity of the perturber's orbit (coorbiting). These behaviors are summarized in Table I.

For corotation resonances, $\text{sgn}(T_{l,m}) = \text{sgn}(d(\sigma/B)/dr)$. If there is no gap and the gradient in the vortensity, σ/B, does not switch sign across the orbit, the fast and slow corotation resonances oppose each other and

TABLE I
Zero- and First-Order Resonances: Locations and Signs

	Radius				
Ω_{ps}	$a\left(1-\frac{4}{3m}\right)$	$a\left(1-\frac{2}{3m}\right)$	a	$a\left(1+\frac{2}{3m}\right)$	$a\left(1+\frac{4}{3m}\right)$
$\Omega_p + \frac{\kappa_p}{m}$	ILR	CR	OLR		
$\phi \propto e^1$	$\dot{e} > 0$	$\dot{e} = \pm$	$\dot{e} < 0$		
Ω_p		ILR	CR	OLR	
$\phi \propto e^0$		$\dot{a} > 0$	$\dot{a} = \pm$	$\dot{a} < 0$	
$\Omega_p - \frac{\kappa_p}{m}$			ILR	CR	OLR
$\phi \propto e^1$			$\dot{e} < 0$	$\dot{e} = \mp$	$\dot{e} > 0$

Zero-order terms are the most important for semimajor axes, while eccentricities are most influenced by terms proportional to e. The former have pattern speeds Ω_{ps} equal to the planet's mean motion Ω_p; the latter have pattern speeds either slightly faster or slower than Ω_p by an increment κ_p/m. Each Fourier term has a corotation and two Lindblad resonances. The resonances due to the fast and slow terms have their sites shifted inward and outward respectively from the zero-order terms. The signs of da/dt and de/dt due to Lindblad resonances depend on the frequencies involved, as explained in the text. For corotation torques, the sign depends on the gradient of σ/B. Adapted from a figure by Goldreich and Tremaine (1980).

largely cancel out (Ward 1988). However, if the protoplanet occupies a gap, $d(\sigma/B)/dr$ will be negative (positive) inside (outside) the orbit, so *both* corotation resonances damp the eccentricity (Goldreich and Tremaine 1980). Also, in this case, the coorbiting Lindblad resonances shut off, because they fall within the gap. Goldreich and Tremaine (1980) compared the excitation rate of external Lindblad resonances to the damping rate of corotation resonances caused by a ring of material and concluded that the latter dominated slightly; that is, $\left|\dot{e}/e\right|_L / \left|\dot{e}/e\right|_C = 0.95$. However, this result assumes that there is no saturation among the resonances. Indeed, the stability of the orbit of a gap-confined protoplanet against eccentricity growth is still not well determined.

Recently, Ward and Hahn (1998*a,b*) have shown that strong eccentricity damping accompanies the $m = 1$, ILR of the slow-pattern-speed term. The pattern speed is $\Omega_{ps} = \Omega_p - \kappa_p \equiv d\,\tilde{\omega}_p/dt$, which is the precession rate of the planet's longitude of perihelion, $\tilde{\omega}_p$. The pattern's slow rotation rate results in a very long wavelength for the one-arm spiral wave (called an apsidal wave), which, in turn, allows for inordinately strong coupling to the planet's gravitational potential. The damping rate for this resonance is given by

$$\frac{1}{e^2}\frac{de^2}{dt} = -\frac{\pi}{4}\beta^{3/2}\left[b_{3/2}^{(2)}(\beta)\right]^2 \mu_p \mu_d \left(\frac{\Omega}{\left| r\, dg(r)/dr \right|}\right)\Omega_p \qquad (20)$$

where $\beta \equiv r_{res}/a_p$. This is a factor of order $\sim \Omega_p/\left|g_p\right| \gg 1$ greater than other low-order Lindblad resonances for $m \gtrsim 2$. Furthermore, since the location of a secular resonance is sensitive to the mass of the protoplanet and the disk, it may fall within the disk even if the planet occupies a gap. This situation may have application to the orbital circularization of extrasolar planets (Ward 1998*a*). Tremaine (1998) has since derived equation (20) in his discussion of "resonant friction" in a planetesimal disk. We prefer the name "secular resonant damping" (SRD), because the term "friction" usually implies energy dissipation, whereas very little energy [specifically, $\dot{E} = (d\tilde{\omega}_p/dt)L$] is carried by apsidal waves launched from a secular resonance.

Returning to the case where no gap exists, the situation is more transparent. The strength of the coorbiting Lindblad resonances exceeds that of external resonances by a comfortable margin (Ward 1988) as shown in Fig. 2. Consequently, the eccentricity decays with a characteristic timescale of

$$\tau_e = \frac{\Omega^{-1}}{C_e}\left(\frac{M_\star}{M_p}\right)\left(\frac{M_\star}{\sigma r^2}\right)\left(\frac{c}{r\Omega}\right)^4 \tag{21}$$

The constant C_e depends on the disk model but is of order unity (Artymowicz 1993*a*). Note that the stronger dependence on $c/r\Omega$ makes this timescale much shorter than the migration time of equation (12). For large objects, this timescale can be short compared to gas drag or collisional damping (Ward 1993*a*). Figure 3 shows the equilibrium velocities found by equating various damping rates to the gravitational relaxation rate of

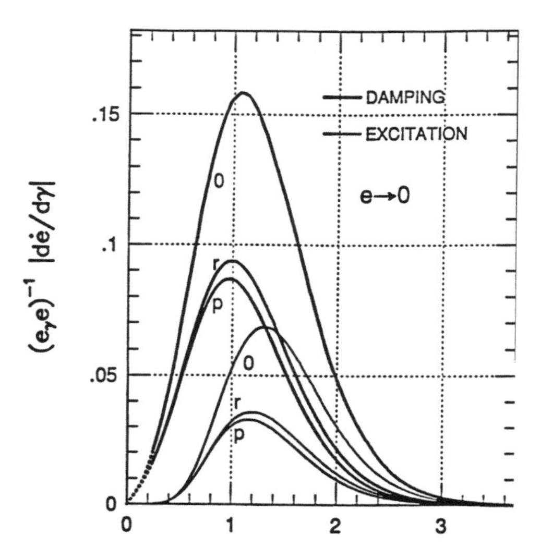

Figure 2. Comparison of the damping rate from coorbiting Lindblad resonances to the excitation rate due to external Lindblad resonances. The parameter, $\gamma \equiv 2(mc/r\Omega)^{2/3}/3^{1/3}$. The three pairs of curves are labeled by different treatments of the disk thickness. "o" is a 2D disk, "r" is a vertically averaged response, and "p" is a vertically averaged potential. In each case, the damping strength comfortably exceeds excitation. Taken from Artymowicz (1996*b*).

a disk composed of equal-mass objects (Ward 1993*a*). The masses are normalized to a so-called gravitational mass, M_G, which is the mass that has an escape velocity equal to the gas sound speed. The calculations are performed in a 150-K minimum-mass nebula at 5 AU, for which M_G = 1.5×10^{25} g, or about a fifth of a lunar mass. The disk tidal damping begins to dominate over a widening range of eccentricities as M rises above M_G and tends to limit the dispersion velocities to the sound speed c. This keeps the planetesimal disk flatter, which, in turn, shortens the collision timescale. At first blush, this would seem to eliminate the accretion timescale problem (that it takes too long to form the giant planet cores by the conventional accretion model). A little reflection, however, reveals that things are not so simple. When objects achieve Earth size, the eccentricities are so strongly damped that orbits become noncrossing and collisions cease. Close encounters no longer produce strong scattering events, which explains why the equilibrium velocity drops precipitously in Fig. 3. In this situation, further accretion would have to await disk-induced changes in the planetesimals' semimajor axes.

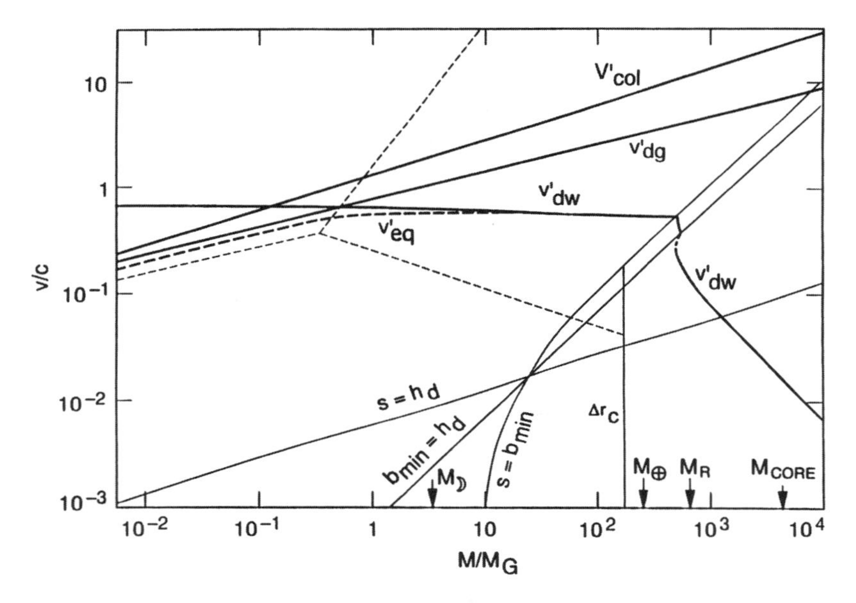

Figure 3. Equilibrium velocities in a unimodal disk as a function of mass for three different sources of damping. The mass is normalized to the value M_G for which the escape velocity equals the sound speed; velocities are normalized to the sound speed c. The short dashed lines partition the diagram into regions dominated by each damping mechanism. They are clockwise from the lower left: collisions, gas drag, and disk torques. Disk torques keep velocities mildly subsonic until object spacing renders the disk collisionless (indicated by Δr_c); then velocities drop quickly with mass. Also shown are boundaries where various combinations of disk thickness h_d, accretion radius S, and minimum impact parameter $b_{\min}$, are equal. Taken from Ward (1993*a*).

C. Nonisotropic Dispersion

In a real disk, where there is a range of object sizes, there will be differential decay in semimajor axes that can force further interactions. Recall that the time it takes a protoplanet to sweep past an equal mass of smaller, nearly stationary solid debris of surface density σ_d is

$$\tau_{\text{swp}} = \frac{\Omega^{-1}}{2\pi f_d C_a}\left(\frac{M_\star}{\sigma r^2}\right)^2\left(\frac{c}{r\Omega}\right)^2 \tag{22}$$

which is independent of protoplanet mass (Ward 1986). Here, $f_d = \sigma_d/\sigma$ represents the solid-to-gas ratio of the nebula. The ratio $\tau_{\text{swp}}/\tau_{\text{growth}} \equiv \mathscr{E}$ represents an accretion efficiency (Tanaka and Ida 1998). In terms of the "optical depth," $\tau_p = \sigma_d/\rho_p R_p$, of embryos, $\mathscr{E} = (\tau_p F_g/2\pi f_d C_a) \times (M_\star/\sigma r^2)^2(c/r\Omega)^2$ unless this expression exceeds unity, in which case $\mathscr{E} = 1$.

In the most general terms, the accretion rate can be written as $\dot{M} \sim A\times(\rho v)$, where A is an effective collision cross section and ρv represents a mass flux. The spatial density of accretable material is of order $\rho \sim \sigma_d/h_d$, where h_d is the scale height of the particle disk. The effective collision radius due to gravitational focusing is $S \sim R\sqrt{1+v_{\text{esc}}^2/v^2}$. We can define an accretable column density as $\tilde{\sigma} \equiv \rho S \sim \sigma_d\left(S/h_d\right)$ unless $S > h_d$, in which case $\tilde{\sigma} \to \sigma_d$. Considering the situation where the particle dispersion velocity v is much less than the protoplanet's escape velocity, $S \sim R_p v_{\text{esc}}/v$, and the accretion rate can be approximated by $\dot{M} \sim \pi R_p v_{\text{esc}}\tilde{\sigma}$. The scale height is roughly $h_d \sim v_z/\Omega$, where $v_z \sim Ir\Omega$ is the *vertical* component of the dispersion velocity, with I being the characteristic orbital inclination of the disk particles. Accordingly, $S/h_d \sim R_p v_{\text{esc}}\Omega/vv_z$. For a particle disk in collisional equilibrium, $I \sim e/2$, and v_z is comparable to v (e.g., Lissauer and Stewart 1993). The accretion then reduces to the usual expression, $\dot{M} \sim \pi R_p^2\sigma\Omega(v_{\text{esc}}/v)^2$, with enhancement factor $F_g \sim (v_{\text{esc}}/v)^2$. However, if the particle disk is *not* in equilibrium with $I \ll e$, then $v_z \ll v$, and the enhancement has an additional factor of $e/I \gg 1$. We will return to this possibility in the next section.

V. PROTOPLANET SURVIVAL

An irony that comes with an appreciation of protoplanet mobility is the worry that they could be too mobile for their own good. Comparison of the two rate expressions reveals that Type I exceeds Type II for masses greater than

$$\mu_o \sim \alpha\left(\frac{c_{\text{II}}}{c_{\text{I}}}\right)\left(\frac{M_\star}{\sigma r^2}\right)\left(\frac{c}{r\Omega}\right)^5 = \alpha\left(\frac{c_{\text{II}}}{C_a}\right)\left(\frac{M_\star}{\sigma r^2}\right)\left(\frac{c}{r\Omega}\right)^4 \tag{23}$$

This implies that such objects decay relative to disk material and could eventually be lost to the star. For example, if common values $\alpha \sim 10^{-3}$, $\sigma r^2/M_\star \sim 10^{-2}$, $c/r\Omega \sim 10^{-1}$, $c_{\text{II}}/c_{\text{I}} \sim 1$, are used, the nondimensional mass is $\mu_o \sim 10^{-6}$, which for a one-solar-mass primary corresponds to a fraction of an Earth mass ($\mu_\oplus = 3 \times 10^{-6}$). A longer lifetime can be enjoyed through Type II behavior, since the orbital decay is linked to the disk's evolution timescale.

A. The Threshold (Shiva) Mass

We expect Type I motions to be exhibited by small protoplanets and Type II motions exhibited by large protoplanets. A key question is, at what mass does a transition take place? Is it near $M_o = \mu_o M_\star$, where the rates are comparable, or does the transition threshold "overshoot" this value, so that there is a range of masses with orbital lifetimes *less* than the evolution timescale of the disk? Figure 1 shows a model calculation for protoplanet behavior at 5 AU in a minimum-mass solar nebula (Ward 1997*a*). The characteristic decay times are displayed as a function of mass measured in Earth masses. The curves are labeled by the strength of the turbulent viscosity. The transition from I to II occurs at a mass μ_S that depends on the specifics of the disk. In the limit of small viscosity, it approaches the inertial limit, μ_i [equation (15)]; in the high-viscosity limit, it approaches μ_ν [equation (14)]; but at intermediate values it is given approximately by (Ward 1997*b*)

$$\mu_S \approx c_S \, \alpha^{2/3} \left(\frac{M_\star}{\sigma r^2}\right)^{1/3} \left(\frac{c}{r\Omega}\right)^3 \tag{24}$$

where $c_S \sim O(1)$. This lies in the regime where Type I decay is up to two orders of magnitude faster than Type II. Thus, disk tides render the mass range ($\mu_o M_\star < M < \mu_S M_\star$) an especially precarious stage in the growth of a planet. [For this reason, the upper limit of this range has been named the "Shiva mass" after the Hindu god of destruction (Ward 1997*b*)]. For the range of α shown, planetary embryos between $\sim$0.1 and 10 Earth masses are in danger of decaying out of the disk.

How does a planetary system survive this process? The characteristic growth time of an embryo that has run away in size from neighboring planetesimals is given by equation (15). Equating this to the Type I timescale gives us the protoplanet size that will decay out of the disk before significantly more growth can occur (Ward 1997*b*, 1998*b*):

$$\mu_{\rm crit} \approx \left(\frac{f_d F_g}{C_a}\right)^{3/4} \left(\frac{M_\star}{\rho_p r^3}\right)^{1/2} \left(\frac{c}{r\Omega}\right)^{3/2} \tag{25}$$

If $\mu_S < \mu_{\rm crit}$, the growing embryo can transition to Type II behavior before being lost from the system. The enhancement factor from stirring (Lissauer 1987; Ida and Makino 1993) is roughly, $F_g \sim F_{\rm stir} \approx 10^3(r/5\ \mathrm{AU})$. With this, equation (25) can be recast as

$$\frac{M_{\rm crit}}{\mathrm{M}_\oplus} \approx \left[\left(\frac{T}{150\ \mathrm{K}}\right)\left(\frac{1}{C_a}\right)\left(\frac{f_d}{0.01}\right)\left(\frac{F_g}{F_{\rm stir}}\right)\right]^{3/4} \tag{26}$$

and is independent of r except through the temperature gradient.

It is instructive to apply these concepts to our own solar system. If C_a is not too large, $M_{\rm crit}$ is comparable to an Earth mass, so that the terrestrial planets may have outlasted the nebula. For the outermost planets, it is even more likely that accretion was too slow for the critical mass to have been achieved during the disk's lifetime. However, neither of these alternatives are available for the giant planets. This seems to be a conundrum. They must form in the presence of the gas, and yet, the predicted decay time of their $\sim$10-$M_\oplus$ cores is much less than the generally assumed lifetime of the disk.

B. Nonequilibrium Accretion

In section IV.C, we discussed the accretion rate increase due to a decrease of planetesimal inclinations compared to eccentricities. We now want to consider how such a state could be produced. As a protoplanet migrates through the disk, it may encounter planetesimals with equilibrium (i.e., $I \sim e/2$) dispersion velocities lower than would be induced by stirring. The protoplanet will quickly force the eccentricities up, but, because its perturbations are primarily horizontal, the inclinations will increase more slowly from planetesimal interactions as the swarm seeks equipartition. If the protoplanet migrates rapidly enough compared to the relaxation time of the disk, the inclinations may be close to the pre-encounter values by the time they enter the protoplanet's feeding zone. Recent numerical experiments by Tanaka and Ida (1998) have confirmed this trend, as illustrated in Fig. 4. Their model calculations do reveal an enhanced accretion rate over the usual runaway prediction. Furthermore, the characteristic growth time is rather insensitive to protoplanet mass, in a manner similar to $\tau_{\rm swp}$. This implies that the ratio I/e must deviate more and more ($\propto 1/R$) from the equilibrium value as the protoplanet gets larger and quickens its migration. However, the efficiencies, $\mathscr{E} = \tau_{\rm swp}/\tau_{\rm growth}$ found by Tanaka and Ida (1998) tend to be low, i.e., $\lesssim$10%, so core formation appears possible only if the embryo migrates a distance considerably larger than its final orbit radius, and the nebula is at least a factor of 5 more massive than the minimum model. The possibility of planet loss weakens the rationale for the minimum-mass nebula model, so such conditions cannot be ruled out. On the other hand, these results are somewhat sensitive to the assumed starting inclinations and masses of the planetesimal swarm, and more study of this important issue is warranted.

VI. CONCLUSION

The discovery of extrasolar planets (see Marcy and Butler 1998) together with advances in our understanding of disk-planet interactions are leading to important revisions in the planetary formation paradigm. Before resonant interactions with the disk were considered, accretion models usually assumed that protoplanets spent most of their formative life in the general vicinity of their final orbits. The existence of close stellar companions

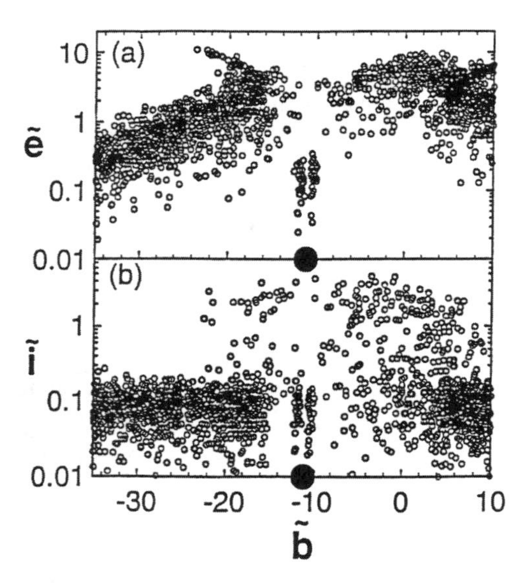

Figure 4. (a) Eccentricities $\tilde{e}$ and (b) inclinations $\tilde{i}$ of planetesimals normalized to $(\mu/3)^{1/3}$ produced by the stirring action of a migrating protoplanet. As the protoplanet approaches from the right, it pumps up their eccentricities while their inclinations are hardly enhanced from their initial values before they enter the feeding zone. Taken from Tanaka and Ida (1998).

provides persuasive evidence that this is not necessarily so. We have stressed that there are two distinct types of migrational behavior: Type I, wherein orbits decay relative to disk material due to an intrinsic imbalance of inner and outer disk torques, and Type II, wherein the protoplanet opens a gap and subsequently coevolves with the disk as it responds to its own angular momentum transport processes, such as viscous stresses.

Type II migration has been invoked by Lin et al. (1996) as the agent responsible for close stellar companions. Parking a planet just short of plunging it into the primary requires either a countertorque, such as stellar tides; a disk torque shutoff mechanism, such as a magnetospheric cavity around the star; or disk removal. Trilling et al. (1998) have presented detailed numerical models of Type II orbit decay, exploring the eventual outcomes for various star, planet, and disk combinations. Roche lobe overflow is a common event for gaseous planets not stabilized by stellar tides, during which the planet temporarily drifts away from the star. Although a providential dispersal of the nebula may save a very small fraction of these objects, most do not survive this process. It should be pointed out that the stabilizing mechanisms suggested by Lin et al. (1996) would also abort Type I orbit decay. Since these objects tend to be planets or cores that did not initiate gas accretion, Type I migration could result in a concentration of solids near the star. This may, in turn, be reflected in a high percentage of rock and CNO in the composition of close stellar companions (Ward 1997*b*).

Orbit migration due to disk torques may prevent embryos from becoming isolated, and thereby speed up their formation process, but it may also be an agent of destruction, driving newly formed protoplanets into their primaries. Thus, migration removes one motivation to assume a greater-than-minimum-mass nebula, but it supplies another by revealing a possible protoplanet loss mechanism that could decrease the overall accretion efficiency. Both migration types would eventually drive any planet into the primary if allowed to operate without restriction. Our own solar system is obvious evidence that planets can survive, but those recently discovered extrasolar planets occupying tight orbits suggest that it is sometimes a close call. Indeed, we cannot rule out a substantial mortality rate for newly formed planets. Understanding how these migration mechanisms are terminated is a crucial element for our comprehension of the planet-building process. At the very least, the mobility of large planetary objects introduces another degree of freedom in accretion modeling. In addition, disk torques provide a further damping mechanism for eccentricities and inclinations. This may affect dispersion velocities and alter the timescale and style of the accretion process. Although we do not yet understand all aspects of the interaction physics, even less is understood about their complex ramifications for cosmogonical models.

Acknowledgments The work was supported in part by funds from NASA's Origins of Solar Systems and Planetary Geology and Geophysics Programs. This paper represents contribution 967 from the Lunar and Planetary Institute. J. M. H. thanks the Southwest Research Institute for their hospitality during a portion of this project.

REFERENCES

Artymowicz, P. 1993*a*. On the wave excitation and a generalized torque formula for Lindblad resonances excited by evolution of bodies embedded in disks. *Astrophys. J.* 419:155–165.

Artymowicz, P. 1993*b*. Disk-satellite interactions via density waves and the eccentricity evolution of bodies embedded in disks. *Astrophys. J.* 419:166–180.

Bodenheimer, P., and Pollack, J. B. 1986. Calculations of the accretion and evolution of giant planets: The effects of solid cores. *Icarus* 67:1101–1113.

Cabot, W., Canuto, V. M., Hubickyj, O., and Pollack, J. B. 1987. The role of turbulent convection in the primitive solar nebula. I. Theory. *Icarus* 69:387–422.

Cassen, P., and Woolum, D. S. 1996. Radiatively damped density waves in optically thick protostellar disks. *Astrophys. J.* 472:789.

Dubrulle, B. 1993. Differential rotation as a source of angular momentum transfer in the solar nebula. *Icarus* 106:59–76.

Gammie, C. F. 1996. Layered accretion in T Tauri disks. *Astrophys. J.* 457:355.

Goldreich, P., and Tremaine, S. 1978. The formation of the Cassini division in Saturn's rings. *Icarus* 34:240–253.

Goldreich, P., and Tremaine, S. 1979. The excitation of density waves at the Lindblad and corotation resonances by an external potential. *Astrophys. J.* 233:857–871.

Goldreich, P., and Tremaine, S. 1980. Disk-satellite interactions. *Astrophys. J.* 241:425–441.

Greenberg, R., Wacker, J. F., Hartmann, W. L., and Chapman, C. R. 1978. Planetesimals to planets: Numerical simulations of collisional evolution. *Icarus* 35:1–26.

Greenzweig, Y., and Lissauer, J. J. 1993. Accretion rates of protoplanets. *Icarus* 87:40–77.

Hourigan, K., and Ward, W. R. 1984. Radial migration of preplanetary material: Implications for the accretion timescale problem. *Icarus* 60:29–39.

Ida, S., and Makino, J. 1993. Scattering of planetesimals by a protoplanet: Slowing down of runaway growth. *Icarus* 106:210–227.

Korycansky, D. G., and Pollack, J. B. 1993. Numerical calculations of the linear response of a gaseous disk to a protoplanet. *Icarus* 102:150–165.

Lin, D. N. C., and Papaloizou, J. 1979. Tidal torques on accretion discs in binary systems with extreme mass ratios. *Mon. Not. Roy. Astron. Soc.* 186:799–812.

Lin, D. N. C., and Papaloizou, J. 1985. On the dynamical origin of the solar system. In *Protostars and Planets II*, ed. D. C. Black and M. S. Matthews (Tucson: University of Arizona Press), pp. 981–1072.

Lin, D. N. C., and Papaloizou, J. 1993. On the tidal interaction between protoplanetary disks and companions. In *Protostars and Planets III*, ed. E. H. Levy and J. I. Lunine (Tucson: University of Arizona Press), pp. 749–836.

Lin, D. N. C., Bodenheimer, P., and Richardson, D. C. 1996. Orbital migration of the planetary companion of 51 Pegasi to its present location. *Nature* 380:606–607.

Lissauer, J. J. 1987. Timescales for planetary accretion and the structure of the protoplanetary disk. *Icarus* 69:249–265.

Lissauer, J. J., and Stewart, G. R. 1993. Growth of planets from planetesimals. In *Protostars and Planets III*, ed. E. H. Levy and J. I. Lunine (Tucson: University of Arizona Press), pp. 1061–1088.

Lubow, S. H., and Ogilvy, G.I. 1998. Three dimensional waves generated at Lindblad resonances in thermally stratified disks. *Astrophys. J.* 504:983.

Lynden-Bell, D., and Kalnajs, A. J. 1972. On the generating mechanism of spiral structure. *Mon. Not. Roy. Astron. Soc.* 157:1–30.

Marcy, G. W., and Butler, R. P. 1998. Detection of extrasolar giant planets. *Ann. Rev. Astron. Astrophys.* 36:57–98.

Mizuno, H., Nakazawa, K. and Hayashi, C. 1978. Instability of gaseous envelope surrounding a planetary core and formation of giant planets. *Prog. Theor. Phys.* 60:699–710.

Papaloizou, J. and Lin, D. N. C. 1984. On the tidal interactions between protoplanets and the primordial solar nebula. I. Linear calculation of the role of angular momentum exchange. *Astrophys. J.* 285:818–834.

Podolak, M., Hubbard, W. B., and Pollack, J. B. 1993. Gaseous accretion and the formation of giant planets. In *Protostars and Planets III*, ed. E. H. Levy and J. I. Lunine (Tucson: University of Arizona Press), pp. 1109–1147.

Stone, J. M., and Balbus, S. A. 1996. Angular momentum transport in accretion disks via convection. *Astrophys. J.* 464:364.

Takeuchi, T., and Miyama, S. M. 1998. Wave excitation in isothermal disks by external gravity. *Pub. Astron. Soc. Japan* 50:141–148.

Takeuchi, T., Miyama, S. M., and Lin, D. N. C. 1996. Gap formation in protoplanetary disks. *Astrophys. J.* 460:832.

Tanaka, H., and Ida, S. 1999. Growth of a migrating planet. *Icarus*, 139:350–366.
Tremaine, S. 1998. Resonant relaxation in protoplanetary disks. *Astron. J.* 116: 2015–2022.
Trilling, D. E., Benz, W., Guillot, T., Lunine, J. I., Hubbard, W. B., and Burrows, A. 1998. Orbital evolution and migration of giant planets: Modeling extrasolar planets. *Astron. J.* 500:428–439.
Walter, F. M. 1986. X-ray sources in regions of star formation. I. The naked T Tauri stars. *Astrophys. J.* 306:573–586.
Walter, F. M., Brown, A., Mathieu, R. D., Meyers, P. C., and Vrba, F. J. 1988. X-ray source in regions of star formation. III. Naked T Tauri stars associated with the Taurus-Auriga complex. *Astron. J.* 96:297–325.
Ward, W. R. 1986. Density waves in the solar nebula: Differential Lindblad torque. *Icarus* 67:164–180.
Ward, W. R. 1988. On disk-planet interactions and orbital eccentricities. *Icarus* 73:330–348.
Ward, W. R. 1989*a*. Corotation torques in the solar nebula: The cut-off function. *Astrophys. J.* 336:526–538.
Ward, W. R. 1989*b*. On the rapid formation of giant planet cores. *Astrophys. J. Lett.* 345:L99–L102.
Ward, W. R. 1991. Horseshoe orbit drag. *Lunar Planet Sci.* 22:1463–1464.
Ward, W. R. 1993*a*. Density waves in the solar system: Planetesimal velocities. *Icarus* 106:274–287.
Ward, W. R. 1993*b*. Disk-protoplanet interactions: Torques from the coorbital zone. In *Astrophysical Disks*, Vol. 675, ed. S. F. Dermott, J. H. Hunter, and R. E. Wilson (New York: New York Academy of Sciences), pp. 314–323.
Ward, W. R. 1996. Planetary accretion. In *Completing the Inventory of the Solar System*, ASP Conf. Ser. 107, ed. T.W. Rettig and J.M. Hahn (San Francisco: Astronomical Society of the Pacific), pp. 337–361.
Ward, W. R. 1997*a*. Protoplanet migration by nebula tides. *Icarus* 126:261.
Ward, W. R. 1997*b*. Survival of planetary systems. *Astrophys. J. Lett.* 482:L211–L214.
Ward, W. R. 1998*a*. Eccentricity damping of extra-solar planets. *Bull. Am. Astron. Soc. Abstr.* 30:1057.
Ward, W. R. 1998*b*. On planet formation and migration. In *Origins*, ASP Conf. Ser. 148, ed. C. E. Woodward, J. M. Shull, and H. A. Thronson (San Francisco: Astronomical Society of the Pacific), pp. 338–346.
Ward, W. R., and Hahn, J. M. 1995. Disk tides and accretion runaway. *Astrophys. J. Lett.* 440:L25.
Ward, W. R., and Hahn, J. M. 1998*a*. Dynamics of the trans-Neptune region: Apsidal waves in the Kuiper belt. *Astron. J.* 116:489–498.
Ward, W. R., and Hahn, J. M. 1998*b*. Neptune's eccentricity and the nature of the Kuiper belt. *Science* 280:2104–2106.
Ward, W. R., and Hourigan, K. 1989. Orbital migration of protoplanets: The inertial limit. *Astrophys. J.* 347:490–495.
Weidenschilling, S. J. 1977. Aerodynamics of solid bodies in the solar system. *Mon. Not. Roy. Astron. Soc.* 180:57–70.
Wetherill, G. W. 1990. Formation of the Earth. *Ann. Rev. Earth Planet. Sci.* 18:205–256.
Wetherill, G. W., and Stewart, G. R. 1989. Accumulation of a swarm of small planetesimals. *Icarus* 77:330–357.
Wetherill, G. W., and Stewart, G. R. 1993. Formation of planetary embryos: Effects of fragmentation, low relative velocity, independent variation of eccentricity and inclination. *Icarus* 106:190–209.
Whipple, F. L. 1972. In *From Plasma to Planet*, ed. A. Elvius (London: Wiley), p. 211.

PART VI
Comets and the Kuiper Belt

COMETS: A LINK BETWEEN INTERSTELLAR AND NEBULAR CHEMISTRY

WILLIAM M. IRVINE, F. PETER SCHLOERB
University of Massachusetts at Amherst

JACQUES CROVISIER
Observatoire de Paris-Meudon

BRUCE FEGLEY, JR.
Washington University, St. Louis, Missouri

and

MICHAEL J. MUMMA
NASA Goddard Space Flight Center

The chemical and isotopic composition of comets is reviewed, with emphasis on results obtained since the *Protostars and Planets III* conference. Observations from the apparitions of comets Hyakutake (C/1996 B2) and Hale-Bopp (C/1995 O1) have almost tripled the number of known cometary parent molecules, including nonpolar hydrocarbons observed in the infrared. Large D/H fractionation has been observed for water and HCN, whereas the carbon, nitrogen, and sulfur isotopic ratios appear to be solar. Cometary dust has been shown to include both crystalline and amorphous silicates, including both olivines and pyroxenes. The dust thus appears similar to that observed in some interplanetary dust particles (IDPs) and to some circumstellar grains. The chemistry of potential cometary volatiles in the solar nebula is reviewed and is contrasted with that expected for preserved interstellar matter. We conclude that cometary nuclei likely contain interstellar material that has been partially processed in the nebula and partially diluted with nebular condensates. The chapter concludes with recommendations for future observational, laboratory, and theoretical work that could clarify the origin of cometary matter.

I. INTRODUCTION

The structure and chemical composition of comets clearly provide key data on the processes and conditions in the outer solar system at the time of its formation and potentially in the interstellar molecular cloud from which the solar nebula condensed (the "natal cloud"). Cometary nuclei are the most volatile-rich, and hence the least thermally processed, material that enters the inner solar system. This does not mean that comets are pristine samples of the ancient solar system. Numerous processes over the last 4.5 Gyr can affect their nature: cosmic rays; the interstellar radiation field; heating pulses from passing stars and from supernovae while the nuclei are in the Oort Cloud or Edgeworth-Kuiper Belt; and, of course, solar irradiation as the comets approach the Sun. Nonetheless, the record that comets preserve should be substantially easier to read than that in the planets, satellites, asteroids, and meteorites, which have been chemically differentiated or at least volatile depleted relative to the nebula as a whole.

Of crucial importance to many issues concerning the formation of solar-type stars and planetary systems is the question of whether or not comets contain relatively unprocessed interstellar molecular material. In the classical view of planetesimal formation, the solar nebula began in a sufficiently hot state that any preexisting matter was vaporized to its atomic constituents and was well mixed. Subsequently, as the nebula cooled, solids condensed in accordance with thermochemical equilibrium, resulting in a gradient in composition with respect to distance from the Sun that reflected the corresponding temperature and density gradients (Lewis 1972). In the outer portions of the nebula the cold temperature led to efficient condensation of volatiles, producing the nuclei of the comets (and other icy objects). However, as noted long ago by Urey (1953), "at lower temperatures [in the solar nebula] thermodynamic equilibrium may not be reached even in periods of time that are long compared to the age of the universe, and at these temperatures the kinetics of thermal reactions or of photochemical reactions become important." Lewis and Prinn (1980) subsequently showed that the conversion of the dominant high-temperature equilibrium species, CO and N_2, to methane and ammonia is so slow that only minor amounts of NH_3 and CH_4 should be present in the solar nebula itself. However, in the higher-density subnebulae around the forming jovian planets, these conversions can proceed rapidly, so CH_4 and NH_3 are expected to be the dominant carbon- and nitrogen-bearing gases in those environments (Prinn and Fegley 1981, 1989).

In contrast, Greenberg (1982, 1998) has long argued that comets are fundamentally aggregates of ice-mantled interstellar grains that survived the formation of the nebula. If this view is correct, it has profound implications for the character of the processes by which matter accreted into the outer nebula, requiring that passage through any accretion shock be a rather gentle process. Moreover, it would mean that comets were direct probes of the nature of the natal cloud and of the interstellar medium (ISM) in a more general sense. It is now generally accepted that meteorites contain refractory presolar grains such as SiC, diamond, graphite, and Si_3N_4 (Zinner 1997), and isotopic data also strongly suggest that organic material in carbonaceous chondrites contains interstellar molecules or their immediate derivatives (Cronin and Chang 1993; Cronin et al. 1995; Messenger and Walker 1997). If the latter view is indeed correct, then it would seem hard to avoid the conclusion that even more interstellar material is preserved in comets. The striking similarities in the composition of cometary and interstellar ices point in this direction, as will be discussed in this review, but the alternative possibility (that such similarities simply reflect general properties of low-temperature, disequilibrium chemistry in solar-composition material) must be borne in mind.

Since there have been as yet no *in situ* analyses of cometary nuclei, and only one set of close spacecraft encounters with a comet (Halley), deducing the nature of comets remains a matter of interpreting the complex and interlinked physics and chemistry of the processes that produce the coma and tail as the nucleus approaches the Sun. Much progress has been made in recent years as advances in instrumentation have allowed the detection of "parent" molecules, which presumably have sublimated from the nucleus, in addition to their photodissociation and photoionization products ("daughter" molecules) traditionally observed at optical and ultraviolet wavelengths. However, many questions remain concerning chemistry in the coma, fractionation during the complex sublimation process, release of molecules from solid grains, and other processes that influence the interpretation of coma observations. Some of these will be discussed below. It remains clear, however, that space missions that actually sample cometary nuclear material are extremely important.

The period since *Protostars and Planets III* has produced extremely important results for our understanding of comets. In part this has been due to the fortuitous apparitions of two very bright comets: C/1996 B2 (Hyakutake), a moderately active comet that passed very close to the Earth, and C/1995 O1 (Hale-Bopp), a very active comet. The impact of D/1993 F2 (Shoemaker-Levy 9) with Jupiter provided a unique opportunity to study the internal strength of a comet as well as other phenomena associated with this encounter (Noll et al. 1996). These and other observations were facilitated by new astronomical instrumentation, including spacecraft, and have resulted in major discoveries in the composition and nature of comets. Among these results has been the discovery of a whole new set of solar system objects in the Kuiper Belt that are certainly related to comets (see chapters by Jewitt and Luu, Farinella et al., and Malhotra et al., this volume), as well as the unexpected detection of X-ray emission from comets, which, however, is probably more related to the solar wind or to the solar UV spectrum than to the nature of comets themselves (e.g., Lisse et al. 1996, 1998; Mumma et al. 1997*b*; Dennerl et al. 1997; Cravens 1997; Krasnopolsky 1997; Krasnopolsky et al. 1997).

Although no single review can cover all aspects of comets, that at *Protostars and Planets III* by Mumma et al. (1993) was very comprehensive. Consequently, the present paper will focus on recent results rather than seeking great breadth. We do point out the following other recent reviews concerning comets: A'Hearn (1998), A'Hearn et al. (1998), Eberhardt (1999), Campins (1998), Sekanina et al. (1998), and the references that begin section III.A.

II. INTERSTELLAR GRAINS

In considering the question of whether comets consist of or contain interstellar molecular material, it is obviously crucial to investigate the nature of interstellar solids. The latter include quite refractory material such as silicates and silicon carbide, both refractory and more volatile organic matter, and ices. There has been significant progress toward understanding the nature of interstellar grains in the last several years, in part because of the success of the *Infrared Space Observatory* (ISO), but very fundamental questions remain.

Direct investigation of the interstellar grains can be carried out by infrared spectroscopy as well as by the traditional studies of the interstellar extinction and polarization in the ultraviolet, visible, and near infrared. In addition, it remains possible that some of the Diffuse Interstellar Bands (DIBs) may originate in the grains, and DIBs as well as other phenomena strongly suggest the presence of interstellar polycyclic aromatic hydrocarbon molecules (PAHs), which may simply be the small end of a continuous spectrum of grain sizes (e.g., Salama 1996; Snow 1997). Moreover, because there is certainly interchange of material between the gas and solid phases in dense interstellar clouds, the composition of the gas provides clues to that of the grains in various environments; this is very important, because the gas can be observed at high spectral resolution and sensitivity at millimeter and submillimeter wavelengths.

The precise nature of the bulk of the organic matter in interstellar grains remains unknown, however (cf. Tielens et al. 1996). This is perhaps not surprising, given the difficulty in characterizing the organic matter in the comet Halley CHON particles (e.g., Fomenkova 1997), in the matrix of carbonaceous chondrites (Cronin and Chang 1993), and in terrestrial kerogens, in spite of the analytical tools that can be brought to bear on these samples. The apparent preservation of interstellar organic molecules in carbonaceous chondrites and the mass spectroscopy of the CHON particles suggest, however, that comets contain complex, heteropolymeric organic matter, some of which may include the corresponding portion of interstellar grains.

Knowledge of interstellar and circumstellar silicates has grown significantly over the last few years with the ability of ISO to obtain complete infrared spectra over the range $2.4 \leq \lambda \leq 45$ μm (and, at lower resolution, $43 \leq \lambda \leq 196$ μm). Circumstellar silicates are seen to include both amorphous and crystalline material, including both olivines [$(Mg,Fe)_2SiO_4$] and pyroxenes [$(Mg,Fe,Ca)SiO_3$] (Waelkens et al. 1998;

Malfait et al. 1998). Curiously, in the interstellar environment (away from circumstellar envelopes and disks), crystalline silicates have not been identified. The greater similarity of the Hale-Bopp silicates to those in circumstellar as opposed to interstellar environments is puzzling (see section III.B).

The volatile content of comets provides the most stringent constraints on their origin, so the comparison with interstellar ices is particularly interesting. Here two sources of new data are especially important: (i) ISO and ground-based spectra of icy grain mantles in molecular clouds and (ii) millimeter/submillimeter observations of gas-phase molecules in "hot cores": regions of high-mass star formation in molecular clouds where the grain mantles have been sublimated and/or sputtered from grains. These topics are reviewed in van Dishoeck and Blake (1998) and in the chapter by Langer et al., this volume. A definite similarity exists between interstellar ices and comets in both chemical composition and deuterium/hydrogen fractionation. However, several limitations in such comparisons should be borne in mind.

1. The sensitivity of the infrared spectra limits the identifiable ice constituents to those with abundances $\geq 0.5\%$ relative to the principal component, H_2O.
2. The gas in hot cores includes the products of prestellar, cold, predominantly ion-molecule chemistry; molecules synthesized on the grains and subsequently released to the gas phase; and products of further chemical processing of the first two components in the gas heated by the young stars. Distinguishing among these components is not easy (e.g., it is unclear whether ethanol is synthesized on the grains or by high-temperature gas-phase reactions; Millar and Hatchell 1998), so the deduced composition of the icy grain mantles is model dependent.
3. Moreover, the hot core environment itself, representing a region of high-mass star formation, differs in unclear ways from that in regions of low-mass star formation, which may be more similar to that in which the solar nebula formed.

Isotopic abundances provide a particularly useful probe of chemical processes. The large deviations from terrestrial isotopic ratios for several elements in portions of meteoritic material provide the principal argument for the preservation of interstellar matter in these objects (e.g., Cronin and Chang 1993; Cronin et al. 1995). In comparing differences in isotopic ratios between the interstellar medium and comets, it is important to differentiate between overall nuclear isotope abundances (e.g., $[^{12}C]/[^{13}C]$), which are the result of ongoing nucleosynthesis in stars and typically show gradients with respect to both galactocentric distance and time, and chemical fractionation, which concentrates a particular isotope in specific molecular species under certain physical conditions. The Sun formed about 4.5 Gyr ago and, it is now believed, about 2 kiloparsecs closer to the center of the Galaxy than its current location (Wielen et al. 1996), so the most relevant comparisons for solar system objects are with conditions at that position at that time, not with the local ISM today. Although there is considerable scatter in the data, perhaps as a result of local star formation history, the trends vs. galactocentric distance agree with current models for the relative abundances of the isotopes of C, N, and O (cf. Wielen and Wilson 1997, and references therein); the observed gradients for sulfur isotopes may require modifications to current theories (Chin et al. 1996). Estimates of the temporal evolution, in contrast, are based primarily on comparisons with current solar system values for all these elements.

Superimposed on these general trends of galactic isotopic evolution are chemical fractionation among and within clouds, due to variations in factors such as the UV field, electron density, kinetic temperature, overall density, and stochastic fluctuations such as might result from recent nearby supernovae. The very large D/H fractionation observed for many interstellar molecular species (up to 4 orders of magnitude) is reasonably well understood to be the result of both gas-phase and grain surface processes at low temperatures (e.g., Rodgers and Millar 1996; Irvine and Knacke 1989; Tielens 1983). The effects are much smaller for heavier atoms, but carbon isotope fractionation (up to a factor of a few) in particular sources (e.g., Langer et al. 1984; Taylor and Dickman 1989), and even differential $^{13}C/^{12}C$ fractionation among the carbons at different positions in HC_3N in a given cloud (Takano et al. 1998), are observed. Chemical fractionation of oxygen is not thought to be significant under most conditions in the ISM (Langer et al. 1984); in addition, it would be difficult to separate from carbon isotopic fractionation, because the principal observed gas-phase reservoir for both C and O is CO. Although some fractionation might in principle occur at very low temperatures for nitrogen, the large $^{14}N/^{15}N$ ratio would make it very difficult to measure (Guélin and Lequeux 1980). Fractionation is also expected to be small for sulfur, although there are some theoretical and observational suggestions that cold, dark clouds may be enhanced in ^{34}S relative to ^{32}S (Chin et al. 1996; Pratap et al. 1997) compared to warmer sources, perhaps by of order 40%. Comparisons to isotope ratios observed in comets are described below. It is unfortunate that the difficulty in observing through the terrestrial atmosphere has thus far precluded measurement of the HDO/H_2O fractionation in regions of low-mass star formation.

Likewise, the interesting results on the ortho-to-para abundance ratio for water in several recent comets (see below) cannot yet be compared to interstellar values, although such ratios have been measured for other molecular species in cold interstellar clouds (e.g., H_2CO, H_2CCO, H_2CS, C_3H_2; Irvine 1992; Minh et al. 1995). Interestingly, a difference in the ratio for formaldehyde may exist between cold cloud cores with and without embedded young low-mass stars (Dickens 1998).

III. THE COMPOSITION OF COMETS

A. The Composition of Cometary Volatiles

The composition of comets, as it was known at the time of the previous *Protostars and Planets* conference, was reviewed by Mumma et al. (1993), Festou et al. (1993), and Crovisier (1994). More recent results, including those from comets Hyakutake and Hale-Bopp, are described by Bockelée-Morvan (1997), Mumma (1997), Eberhardt (1998), Rauer (1999), and Bockelée-Morvan and Crovisier (1998); see also Crovisier (1998*a,b*; 1999) and Despois (1998). The improvement coming from the observations of these last two comets is really significant, because it increased the number of known *parent* molecules from about 8 to about 22. Our present knowledge of the composition of cometary volatiles is summarized in Table I. A sample spectrum is shown in Fig. 1.

Most of these new molecules have been detected through their rotational lines at radio wavelengths. This technique is very sensitive even to minor species, provided they are polar molecules. In addition to the previously known HCN, important new cyanide-containing molecules have been identified: HNC, an isomeric form of HCN (Irvine et al. 1996; see discussion in section V.B); and CH_3CN and HC_3N, pointing to the possible existence of carbon chain molecules such as those found in interstellar clouds (Bockelée-Morvan et al. 1998*b*). Ammonia has been firmly identified for the first time through its centimetric lines (Bird et al. 1997; Palmer et al. 1996).

Several new sulfur-bearing molecules have been identified to complement the already known H_2S and CS_2 (the probable parent of CS). Surprisingly, SO and SO_2 were observed with a significant abundance, much higher than the upper limits previously inferred from UV data (Kim and A'Hearn 1991), which indicates that the UV excitation of these molecules has perhaps not been understood (Bockelée-Morvan et al. 1998*b*; Lis et al. 1998). OCS and H_2CS were observed for the first time as minor constituents (Woodney et al. 1997, 1998; Dello Russo et al. 1998*a*). The mysterious S_2, previously observed only in C/1983 H1 (IRAS-Araki-Alcock), was observed again in the near UV in comet Hyakutake (Weaver et al. 1996), the detection of this short-lived molecule benefiting in these two cases from the close approach of the comet to the Earth. Interferometric observations of Hale-Bopp suggest that SO may be in part a daughter radical produced from SO_2 photolysis in the coma (Wink et al. 1998).

Several CHON (organic) species have been detected with minor abundances relative to the previously known methanol and formaldehyde: formic acid (HCOOH), isocyanic acid (HNCO), methyl formate ($HCOOCH_3$), and formamide (NH_2CHO); see Bockelée-Morvan et al. (1998*b*) and Lis et al. (1997, 1998). In addition, significant upper

TABLE I
Abundances of Cometary Volatiles

Molecule	Abundance[a]		Method[b]	N[c]	Comments
	Hale-Bopp	Other Comets			
H_2O	= 100	= 100	IR	6	Also indirect (from OH, O, H)
CO	20	1–20	UV, radio, IR	>5	Extended source?
CO_2	20[d]	3–10	IR	3	
H_2CO	0.1–1	0.1–1	Radio, IR	>5	Extended source
CH_3OH	2	1–7	Radio, IR	>5	
HCOOH	~0.05		Radio	1	
HNCO	0.1		Radio	2	
NH_2CHO	~0.01		Radio	1	
$HCOOCH_3$	~0.05		Radio	1	
CH_4	~0.6		IR	2	
C_2H_2	~0.1		IR	2	
C_2H_6	~0.3		IR	2	
NH_3	0.6		Radio, IR	3	
HCN	0.2	0.05–0.2	Radio, IR	>5	
HNC	0.04		Radio	2	Extended source?
CH_3CN	0.02		Radio	2	
HC_3N	0.03		Radio	1	
H_2S	1.5	0.2–1.5	Radio	>5	
H_2CS	~0.02		Radio	1	
CS	0.2	0.2	UV, radio	>5	From CS_2?
OCS	0.5		Radio, IR	2	Extended source?
SO	~0.5		Radio	1	From SO_2?
SO_2	~0.1		Radio	1	
S_2		0.005	UV	2	

[a] Abundance relative to water. See text for references. All abundances were measured at $r_h \sim 1$ AU except for CO_2 in Hale-Bopp. Listed abundances may be uncertain by a factor of 2 or more for some species and may not pertain to nucleus production for "extended sources."
[b] Method of observation.
[c] Number of comets in which this species was reliably and directly observed.
[d] Measured at $r_h = 2.9$ AU.

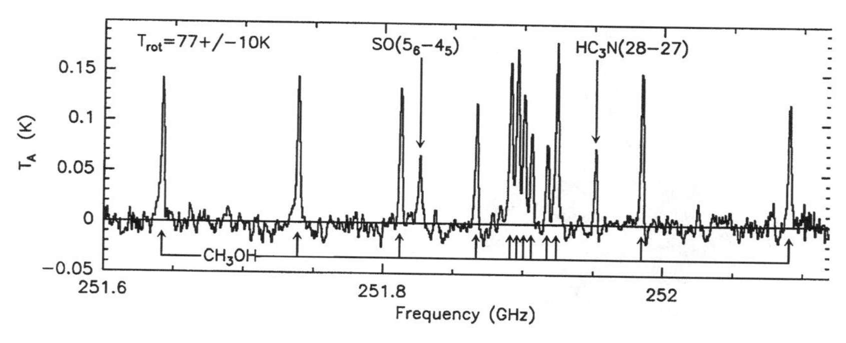

Figure 1. A spectrum of comet Hale-Bopp observed at the Caltech Submillimeter Observatory (CSO) on February 21, 1997, showing a series of lines of methanol and the corresponding rotational temperature, and the first observations of SO and HC_3N in a comet (Lis et al. 1998). Such data illustrate the power of millimeter and submillimeter-wavelength observations for cometary studies.

limits have been obtained for molecules such as methylenimine (CH_2NH), ethanol (C_2H_5OH), ketene (CH_2CO), and many others.

The infrared is a unique domain for observing nonpolar species that do not have allowed rotational transitions as well as for obtaining new information on molecules also observed at radio wavelengths. Identifications of hydrocarbons in comets Hyakutake and Hale-Bopp from new ground-based observations (Mumma et al. 1996; Brooke et al. 1996; Weaver et al. 1998) include C_2H_2, C_2H_6, and CH_4 (Fig. 2). Numerous other parent volatiles were detected at infrared wavelengths (e.g., CO, CH_3OH, HCN, H_2CO, OCS, NH_3), and sensitive searches were performed for other species (e.g., C_2H_4, C_3H_6, C_3H_8, O_3), although actual upper limits are not yet available.

Water has been detected from ground-based observatories only at infrared wavelengths. The approach of hot-band fluorescence developed for earlier comets was applied to comet Hyakutake (Mumma et al. 1996) and for Hale-Bopp was used to obtain production rates for various heliocentric distances (Dello Russo et al. 1999; Weaver et al. 1998). Observations of water in Hale-Bopp and 103P/Hartley 2 from space (with ISO) provided production rates and the ortho-para ratio, which agree well with those obtained for 1P/Halley (see section III.D).

The high spatial resolution afforded by long-slit spectrometers permits detailed study of the distribution of molecular emissions about the nucleus. These can discriminate nuclear from extended sources of "parent" volatiles. CO was studied in this way; similarly to comet 1P/Halley, when comet Hale-Bopp was at a heliocentric distance of approximately 1 AU, up to 50% of the CO was found to be produced from an extend ed source (DiSanti et al. 1999; Weaver et al. 1998). Water and HCN,

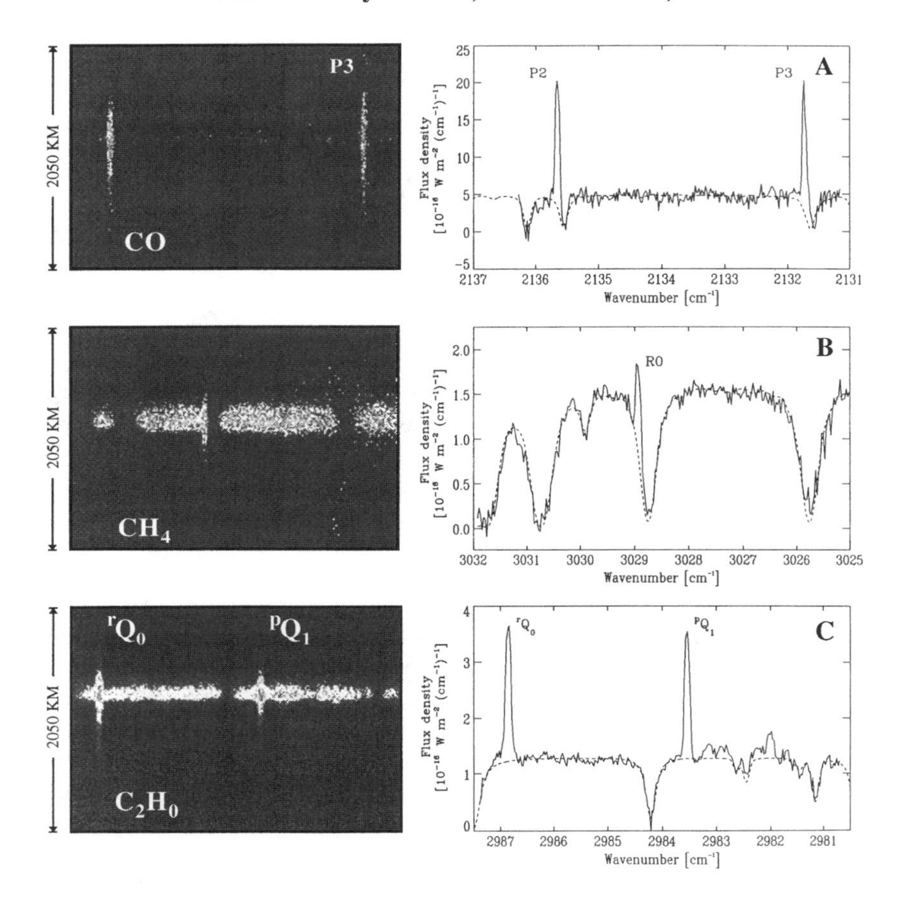

Figure 2. High-dispersion (R = 20,000), long-slit spectra (0.2 arcsec per pixel) of comet Hyakutake, showing emission from parent volatiles and dust. Full spectral-spatial frames are shown at left, and corresponding spectra extracted for seven rows centered on the nucleus are shown at right. Telluric lines are seen in absorption against the cometary continuum, and their cometary counterparts are Doppler-shifted to higher frequencies by the comet's motion. (*top*) The P2 and P3 lines of the carbon monoxide 1–0 vibrational band near 4.69 μm; (*middle*) the R0 line of the methane ν_3 band near 3.30μm: (*bottom*) The $^rQ_0$ and $^PQ_1$ branches of the ethane ν_7 band near 3.351 μm. These are the discovery spectra for saturated hydrocarbons in comets (Mumma et al. 1996).

however, appear to be released directly from the nucleus (Dello Russo et al. 1999, Magee-Sauer et al. 1998; Weaver et al. 1998). A comparison of the spatial distribution for OCS relative to those of water and dust in Hale-Bopp provided evidence that most OCS was produced from an extended source (Dello Russo et al. 1998*a*). The rotational temperature measured for CO and HCN in Hale-Bopp revealed changes in the coma temperature with distance from the nucleus, similar to those predicted by models of photolytic heating (DiSanti et al. 1999; Magee-Sauer et al. 1998; Bockelée-Morvan and Crovisier 1987). The ability to distinguish direct and extended sources for "parent" volatiles is key to understanding the composition of cometary ices.

The ISO provided the second direct observation of CO_2 in a comet (Crovisier et al. 1996, 1997*b*) through its ν_3 band at 4.25 μm. CO_2 can also be traced following its photodissociation from the CO Cameron system in the UV (e.g., Weaver et al. 1997). The direct detection is important; CO_2 production rates cannot be reliably derived from the UV observations because the rates of the relevant mechanisms are still poorly known.

Some words of caution are necessary concerning the relationship between reported "production rates" of cometary volatiles observed in the coma (as listed in Table I) and the actual molecular abundances in cometary ices:

1. *Sublimation fractionation* affects all production rates, depending on the heliocentric distance. This is now clearly shown by the evolution of production rates observed over the range of distance 1–7 AU in comet Hale-Bopp, which show significant variations relative to water for several species (Biver et al. 1997, 1998).
2. *Chemical reactions* in the coma may be able to synthesize some species of minor abundance in highly productive comets where the coma density could be high. This was evident from the [HNC]/[HCN] ratio varying with the heliocentric distance in Hale-Bopp (see the discussion in section V.B). Thus, one might question the significance of the abundances of some of the minor species detected in comet Hale-Bopp near its perihelion.
3. The existence of *extended sources* within the coma has been demonstrated for molecules such as CO, H_2CO, OCS, and perhaps others. The sources of such molecules are poorly known. They might be grains with CHON mantles, but the desorption mechanism of molecules is still uncertain (see, e.g., Crovisier 1998*a*; Greenberg and Li 1998). For these molecules the production rates may not pertain to sublimation from the nucleus and thus may not be relevant to nuclear ice abundances.

B. The Composition of Cometary Dust

In addition to the *in situ* analysis of P/Halley's dust particles by mass spectroscopy (recently reanalyzed by Schulze et al. 1997 and Eberhardt 1999), which revealed the *elemental* rather than the *chemical* composition, the composition of cometary dust can be investigated by spectroscopy in the thermal infrared. Until recently, most of the clues came from the analysis of the intricate band around 10 μm. It revealed the presence of silicates, part of them being crystalline in some comets (e.g., Hanner et al. 1994).

ISO gave us, for the first time, access to the full infrared spectrum of a comet, from 2.4 to 196 μm, covering thermal emission of cometary dust. In addition to the already well-known silicate bands around 10 and 20 μm, the observations of comet Hale-Bopp revealed peaks at 16, 23.5, 27.5, and 33.5 μm (Crovisier et al. 1997*a,b*). These were identified with Mg-rich olivine (forsterite, Mg_2SiO_4). The Hale-Bopp spectrum appears strikingly similar to those of circumstellar dust around Vega-type stars (e.g., that of HD 100546, a Herbig Ae star; Waelkens et al. 1996; Malfait et al. 1998). Quantitative fits of the ISO spectra (with still preliminary calibration) with laboratory spectra of terrestrial and IDP (interplanetary dust particles) analogs reveal that, in addition to forsterite, amorphous silicates as well as a featureless component (such as amorphous carbon) are necessary to reproduce the spectra (Hanner et al. 1998; Brucato et al. 1999; Wooden et al. 1999).

ISO could observe comet Hale-Bopp only at heliocentric distances greater than 2.9 AU. In ground-based spectra in the 8–13 μm window, features characteristic of pyroxene (at 9.3 and 10.0 μm) appeared at smaller heliocentric distances, implying the existence of two crystalline grain components with different temperatures: the hotter including olivines, the cooler, pyroxenes (Wooden et al. 1999). The dust of comet Hale-Bopp, with Mg-rich pyroxenes and olivines, thus seems to have a composition comparable with that of IDPs (Bradley et al. 1988, 1997). This composition is also compatible with that predicted for the first Mg silicates to condense from a solar-composition gas (Grossman and Larimer 1974).

Do Jupiter-family comets, presumed to have formed in the Edgeworth-Kuiper Belt, have the same composition as Oort Cloud comets? The former are fainter and more difficult to observe. Until recently, silicate bands were undetected or gave inconclusive results in the few Jupiter-family comets that were observed (e.g., Hanner et al. 1996). However, observations of 103/P Hartley 2 with ISO have now revealed the 11.3 μm feature characteristic of crystalline olivine (Crovisier et al. 1999).

C. Isotopic Abundances

Isotopic abundances (and especially the [D]/[H] ratio in water) in comet Halley from the *Giotto* mass spectrometers have now been refined (Balsiger et al. 1995; Eberhardt et al. 1995). In addition, new results were obtained from radio spectroscopic observations of comets Hyakutake and Hale-Bopp, resulting in determinations of the [D]/[H] ratios in water (Bockelée-Morvan et al. 1998*a*; Meier et al. 1998*b*) and HCN (Lis et al. 1998; Meier et al. 1998*a*), and of C, N, and S isotopic ratios in HCN and CS (Jewitt et al. 1997). All these results are summarized in Table II. A

TABLE II
Isotopic Ratios in Comets

Isotopes	Molecule	Comet	Method	Cosmic Value	Comet Value	Reference
[D]/[H]	H_3O^+	Halley	Mass spect.	1.5×10^{-5}	$3.08 \pm 0.53 \times 10^{-4}$	Balsiger et al. (1995)
					$3.02 \pm 0.22 \times 10^{-4}$	Eberhardt et al. (1995)
	H_2O	Hyakutake	Radio		$2.9 \pm 1.0 \times 10^{-4}$	Bockelée-Morvan et al. 1998*a*
		Hale-Bopp	Radio		$3.3 \pm 0.8 \times 10^{-4}$	Meier et al. (1998*b*)
	HCN	Hale-Bopp	Radio		$2.3 \pm 0.4 \times 10^{-3}$	Meier et al. (1998*a*)
	CH_3OH[a]	Halley	Mass spect.		$< 1 \times 10^{-2}$	Eberhardt et al. (1994)
$[^{18}O]/[^{16}O]$	H_3O^+	Halley	Mass spect.	2.0×10^{-3}	$1.93 \pm 0.12 \times 10^{-3}$	Balsiger et al. (1995)
					$2.13 \pm 0.18 \times 10^{-3}$	Eberhardt et al. (1995)
$[^{13}C]/[^{12}C]$	CN	Halley	Visible	1.1×10^{-2}	$1.05 \pm 0.13 \times 10^{-2}$	Kleine et al. (1995)
	HCN	Hyakutake	Radio		$2.9 \pm 1.0 \times 10^{-2}$	Lis et al. (1997)
		Hale-Bopp	Radio		$1.11 \pm 0.18 \times 10^{-2}$	Lis et al. (1998)
					$0.90 \pm 0.09 \times 10^{-2}$	Jewitt et al (1997)
$[^{15}N]/[^{14}N]$	CN	Halley	Visible	3.6×10^{-3}	$< 3.6 \times 10^{-3}$	Kleine et al. (1995)
	HCN	Hale-Bopp	Radio		$3.1 \pm 0.4 \times 10^{-3}$	Jewitt et al. (1997)
$[^{34}S]/[^{32}S]$	Atomic S	Halley	Mass spect.	4.2×10^{-2}	$4.5 \pm 1.0 \times 10^{-2}$	Krankowsky et al. (1986)
	CS	Hale-Bopp	Radio		$3.7 \pm 0.4 \times 10^{-2}$	Jewitt et al. (1997)

[a] $[CH_3OD + CDH_2OH]/[CH_3OH]$.

detailed discussion of the significance of the [D]/[H] ratio in comets is given in section IV.A below; cf. also Mumma (1997) and Bockelée-Morvan et al. (1998a).

The mean C, N, and S isotopic ratios are not significantly different from the solar values. This confirms, with an improved accuracy, previous determinations for volatiles from optical spectra and mass spectroscopy. Some individual grains measured *in situ* by the *Giotto* and *Vega* spacecraft did, however, show anomalous [^{12}C]/[^{13}C] ratios. Whereas the low ("heavy") values might represent interference from $^{12}CH^+$, the high ratios (up to 5000) are similar to some values seen for micron-sized SiC grains in carbonaceous meteorites and are taken as evidence for the preservation of essentially unaltered presolar organic material (Eberhardt 1998, although there is no evidence for SiC in Halley from the high-fidelity mass spectra obtained with the PUMA-1 instrument; Schulze et al. 1997).

The three comets in which the [D]/[H] ratio has been determined are Oort Cloud comets. One could wonder what this ratio might be in Jupiter-family comets, perhaps formed at larger heliocentric distances; the answer might come from future cometary missions toward short-period comets.

D. The Spin Temperature of Cometary Water

As reviewed by Mumma et al. (1993), the ortho/para ratio (OPR) of cometary water is characterized by the spin temperature of this species. This parameter is believed to be of primordial character, because conversions between the ortho (hydrogen nuclear spins parallel) and para (spins opposite) states through collisions or radiative transitions are strictly forbidden. The real meaning of the spin temperature is not understood, however; it could be the temperature of water at the moment of its chemical formation, or it may reflect re-equilibration with the internal temperature of the nucleus.

The initial observations gave OPR $= 2.5 \pm 0.1$ in 1P/Halley, corresponding to $T_{spin} \approx 29$ K, and OPR $= 3.2 \pm 0.2$ in C/1986 P1 (Wilson), consistent with $T_{spin} > 50$ K (Mumma et al. 1993). However, these determinations are hampered by the difficulty in modeling opacity effects and because only part of the ν_3 vibrational band of water was observed.

ISO observed the ν_3 band of water in comets C/1995 O1 (Hale-Bopp) and 103P/Hartley 2 (Crovisier et al., 1997*b*, 1999). The full band was observed with a spectral resolution allowing resolution of the rotational structure, and the opacity was found to be moderate; see Fig. 3. Values of OPR $= 2.45 \pm 0.10$ and 2.70 ± 0.10, corresponding to $T_{spin} \approx 25$ and 35 K, were determined for Hale-Bopp and Hartley 2, respectively. Thus, the existence of OPR values significantly lower than the high-temperature limit of 3 is confirmed in both an Oort Cloud (Hale-Bopp) and a Jupiter-family comet. The issue is discussed further near the end of section IV.A.

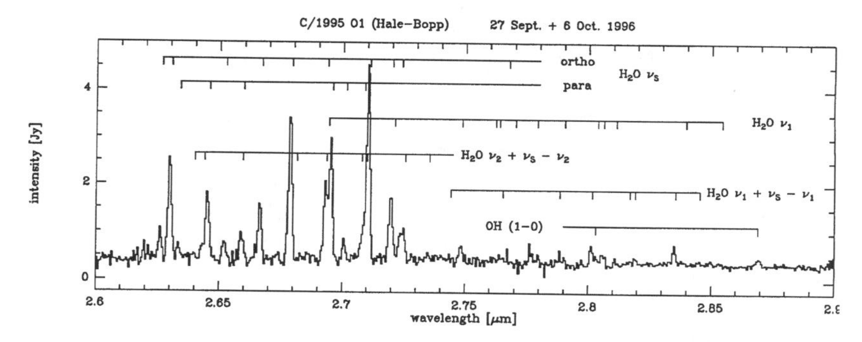

Figure 3. The region of the ν_3 band of water observed in the comet Hale-Bopp with the short-wavelength spectrometer of ISO (average of observations on September 26 and October 6, 1996, when the comet was at 2.9 AU from the Sun). The resolution of the rotation structure permits an evaluation of the rotational temperature (28 K) and of the ortho-to-para ratio (2.45 ± 0.10, corresponding to a spin temperature ~25 K; Crovisier et al. 1997*a,b*).

IV. NEBULAR CHEMISTRY OF COMETARY VOLATILES

A. Water

Because water is the dominant volatile in those comets for which its abundance has been determined, its origin is of prime importance for assessing how much comets represent pristine presolar material and how much presolar material reprocessed to a greater or lesser extent in the solar nebula. (We use the term "presolar" rather than "interstellar" to emphasize that some of this material may have been formed in circumstellar shells, while other portions may have been formed or modified in interstellar space.) As noted by Fegley (1993), water ice in comets could be pristine presolar ice, presolar water that was vaporized and recondensed in the solar nebula, or water synthesized in jovian protoplanetary subnebulae via thermochemical reactions such as

$$CO + 3H_2 = CH_4 + H_2O \tag{1}$$

Even if all water in comets resulted from equation (1), which is not being advocated, this does not imply that CH_4 and H_2O would be retained in a 1:1 ratio in comets, because of the greater volatility of CH_4; the consequences of equation (1) for the [D]/[H] ratio of cometary water, and the production of intermediates such as H_2CO and CH_3OH, will be discussed below. In fact, the water in comets could be a mixture, with different comets having different proportions of pristine presolar water and reprocessed nebular water (cf. Fegley 1993; Mumma 1997).

Until *unaltered* samples of comet nuclei are accessible to chemical and physical measurements by spacecraft experiments or by laboratory analyses of returned samples, the source of cometary water has to be inferred primarily from its hydrogen and oxygen isotopic composition. The available data are given in Tables II and III. The measured [D]/[H] ratios in Halley, Hyakutake, and Hale-Bopp cluster around 30×10^{-5}. This [D]/[H] ratio is about 20 times larger than the (present-day) [D]/[H] ratio in the local interstellar medium (Pisunkov et al. 1997) and about twice as large as in terrestrial SMOW (Standard Mean Ocean Water), thus challenging the hypothesis that terrestrial water is derived primarily from cometary bombardments. What do the [D]/[H] ratios in Halley, Hyakutake, and Hale-Bopp tell us about the origin of water in these three comets?

One possible interpretation is that the water in these three comets is pristine interstellar water that was incorporated into the comets. The [D]/[H] ratios in cometary water are similar to the [D]/[H] ratios reported for water in the hot cores of molecular clouds, where massive stars are forming. For example, radio observations of HDO and $H_2^{18}O$ by Jacq et al. (1990), Gensheimer et al. (1996), and Helmich et al. (1996) gave [D]/[H] ratios of order $(20–60) \times 10^{-5}$ for hot core regions. This water vapor is thought to preserve the fractionation in the ice mantles of grains present in the cores before they were heated by the embedded young stars (Rodgers and Millar 1996). The accretion of unaltered water-bearing interstellar grains into comets may also explain the observed ortho/para ratios of cometary water vapor (discussed later in this section).

An important implication of the unmodified-interstellar-grain scenario is that other interstellar molecules will also be accreted unaltered into comets. For example, hydrogen cyanide with a [D]/[H] ratio of $(230 \pm 40) \times 10^{-5}$ was observed in Hale-Bopp by Meier et al. (1998*a*), which is in the range reported in hot cores by Hatchell et al. (1998). Meier et al. (1998*a*) note that the [DCN]/[HCN] ratio is consistent with the [HDO]/[H_2O] ratio within the context of interstellar ion-molecule chemistry, provided that the chemistry has reached steady state and the temperature is of order 30–40 K (Millar et al. 1989). It should be noted, however, both that the interstellar hot cores are quite young, so steady-state models may not be applicable, and that the observed HCN ratio in the gas may have been diluted from an originally higher value on the presumed icy grain mantles (Hatchell et al. 1998).

Methanol (CH_3OH) is also observed in comets (including Hale-Bopp, Hyakutake, and Halley), and it is present in hot cores, with [CH_3OD]/[CH_3OH] ratios of 0.01–0.06 (Mauersberger et al. 1988), while the [CH_2DOH]/[CH_3OD] ratio is of order unity (Jacq et al. 1993; Charnley et al. 1997). However, Eberhardt et al. (1994) reported a [deuterated CH_3OH]/[CH_3OH] ratio < 0.01 for Halley, where deuterated CH_3OH includes both CH_3OD and CDH_2OH, which could not be distinguished by the mass spectrometry. The cometary results thus seem to disagree with the deuterium fractionation for the hot cores. There are no reports of

TABLE III

[D]/[H] Ratios in Comets, the Solar System, and Interstellar Space

Object or Species	D/H Ratio ($\times 10^5$)	δD (%)[a]	Enrichment[b]	Reference
Present-day local ISM	1.6 ± 0.12	−900	1	Mahaffy et al. (1998)
Solar wind	2.1 ± 0.5	−865	1.3	Geiss and Gloeckler (1998)
Orion hot core: HDO/H_2O	100–400	$(5.4–24.7)\times 10^3$	60–250	van Dishoeck et al. (1993)
Several hot cores: HDO/H_2O	30	925	19	Gensheimer et al. (1996)
Several hot cores: DCN/HCN	90–400	$(5–25)\times 10^3$	56–250	Hatchell et al. (1998)
TMC-1: C_3HD/C_3H_2	$(8–16)\times 10^3$	$(5–10)\times 10^6$	$(5–10)\times 10^3$	van Dishoeck et al. (1993)
TMC-1: CH_2DC_2H/CH_3C_2H	6000	385,000	3750	van Dishoeck et al. (1993)
TMC-1: DCN/HCN	2300	146,600	1440	van Dishoeck et al. (1993)
TMC-1: DC_3N/HC_3N	1500	95,280	940	van Dishoeck et al. (1993)
Halley: HDO/H_2O	31.6 ± 3.4	1030	20	Eberhardt et al. (1995)
Hyakutake: HDO/H_2O	29 ± 10	860	18	Bockelée-Morvan et al. (1998)
Hale-Bopp: HDO/H_2O	33 ± 8	1120	21	Meier et al. (1998*a*)
Hale-Bopp: DCN/HCN	230 ± 40	13,760	144	Meier et al. (1998*b*)
IDPs (maximum value)[c]	400–800	24,800–50,000	250–500	Messenger and Walker (1997)
Venus: atm. H_2O	2200 ± 300	140,200	1400	Donahue (1995)
Mars: atm. H_2O	81 ± 3	4200	51	Donahue (1995)
Earth: SMOW	15.58	0	9.7	Lodders and Fegley (1998)
Jupiter: CH_3D/CH_4	2.6 ± 1.0	−830	1.6	Fegley and Prinn (1988)
Jupiter: HD/H_2	2.6 ± 0.7	−830	1.6	Mahaffy et al. (1998)
Saturn: HD/H_2	2.5 ± 1.0	−840	0.9–2.2	Lodders and Fegley (1998)[d]
Titan: CH_3D/CH_4	~7.5	~ −519	~5	Coustenis et al. (1998)
Neptune: HD/H_2	6.5 (−1.5, +2.5)	~ − 580	~4	Feuchtgruber et al. (1999)
Uranus: HD/H_2	5.5 (−1.5, +3.5)	~ −650	~3.4	Feuchtgruber et al. (1999)
Chondrites: Semarkona (LL3) phyllosilicates	75 ± 12	3815 ± 170		Deloule and Robert (1995)
Renazzo (CR) phyllosilicates	>31	>990		Deloule and Robert (1995)
Chondrules (LL3, CR)	12–23	−230 to 480		Deloule and Robert (1995)
Orgueil (CI) kerogen	36.6	1360		Halbout et al. (1990)

[a] The delta notation is defined as $\delta D (\%) = [(D/H)_{sample}/(D/H)_{SMOW} - 1] \times 1000$ where SMOW is Standard Mean Ocean Water.

[b] The enrichment factor is defined as $(D/H)_{sample}/(D/H)_{ISM}$, taking $(D/H)_{ISM} = (1.6 \pm 0.12) \times 10^{-5}$ as the average present day local ISM value. Measurements of the local ISM D/H ratio fall in the range of $(1.4–2.2)\times 10^{-5}$ (Pisunkov et al. 1997).

[c] Lower value for a particular "cluster" interplanetary dust particle, higher value for a "hotspot" on that particle.

[d] Griffin et al. (1996) give 2.3 (−0.8, +1.2).

deuterated CH_3OH in Hale-Bopp or Hyakutake. Finally, for formaldehyde both HDCO and D_2CO have been observed in hot cores (Sutton et al. 1995; Turner 1990; see also Rodgers and Millar 1996), whereas [D]/[H] ratios for H_2CO in comets have not been reported. Because the unmodified-interstellar-grain model implies high [D]/[H] ratios for formaldehyde in comets, it is important to see whether an upper limit for [HDCO]/[H_2CO] can be derived from the mass spectrometer data for Halley or from remote sensing observations of other H_2CO-bearing comets.

Another implication of the unmodified-interstellar-grain model is that infalling interstellar grains do not lose their volatiles during incorporation into the solar nebula. In other words, ices on infalling grains either are not vaporized or, if vaporized, do not undergo significant chemical or isotopic alteration before recondensing onto grain surfaces. In contrast, recent calculations (Chick and Cassen 1997; Cassen and Chick 1997; Engel et al. 1990; Lunine et al. 1991) indicate that a significant fraction of infalling icy grains may have been vaporized during incorporation into the solar nebula. For example, Chick and Cassen (1997) predict that the water ice vaporization distance lies between 2 and 30 AU, and Lunine et al. (1991) predict that $\sim 90\%$ of water ice grains are vaporized at 30 AU and $< 10\%$ of water ice grains are vaporized beyond 100 AU. While the computational results are somewhat dependent on how the solar nebula is modeled, including such poorly constrained parameters as the area/mass ratio for the infalling grains, it would seem that the vaporization of infalling icy grains may pose a hurdle for the unmodified-interstellar-grain model of cometary volatiles. These theoretical considerations and the possible lack of consistent D enrichments in cometary species such as methanol lead us to consider the alternative: that the D/H ratio of cometary water was produced either by fractionation within the solar nebula itself or by partial reprocessing of presolar water in the solar nebula.

In principle, deuterium isotopic exchange in the solar nebula between hydrogen and water (or other hydrides) via net reactions such as

$$\mathrm{HD\ (g) + H_2O\ (g) = H_2\ (g) + HDO\ (g)} \tag{2}$$

leads to more HDO (or more CH_3D, NH_2D, HDS) with decreasing temperature. Values of the D/H fractionation as a function of temperature are given for water in Fig. 11 of Prinn and Fegley (1989); comparison with the observed values in comets Hyakutake, Halley, and Hale-Bopp (Table III) indicates temperatures of order 140 K, close to the calculated water ice condensation temperatures in the solar nebula (where pressures are much higher than in the ISM). Although the correspondence with the nebular snowline is suggestive of the cometary water having been reprocessed in the solar nebula, isotope exchange kinetics at low temperatures are probably too slow for reaction (2) to have enriched water vapor in D within the lifetime of the solar nebula (Grinspoon and Lewis 1987; Fegley and Prinn 1989; Lecluse and Robert 1994).

As noted by Fegley (1993, 1997), it may be more realistic to consider the deuterium exchange process as a back reaction in which D-rich presolar water is losing deuterium to the surrounding H_2-rich nebular gas. The derived D/H exchange temperatures would then be the maximum temperatures at which the water in Hyakutake, Halley, and Hale-Bopp last exchanged deuterium with nebular H_2. This could occur, for example, via reactions such as

$$H_2O + h\nu \rightarrow OH + H \qquad (\lambda < 200\ \text{nm}) \tag{3}$$

$$HDO + H \rightarrow HD + OH \tag{4}$$

$$H_2O + H \rightarrow H_2 + OH \tag{5}$$

$$OH + H_2 \rightarrow H_2O + H \tag{6}$$

$$H + H + M \rightarrow H_2 + M \tag{7}$$

driven by scattered solar UV and the interstellar UV radiation field in optically thin regions (e.g., the outer skin) of the solar nebula. This is essentially the reverse of the scheme proposed by Yung et al. (1988) and is also supported by recent calculations (Gladstone and Fegley 1997) that indicate that photochemistry was an important disequilibrating process in the outer solar nebula. Ion-molecule reactions driven by galactic cosmic rays in the outer skin of the solar nebula (e.g., see the suggestions of Yung et al. 1988; Deloule and Robert 1995; Aikawa et al. 1998) might also drive the back reactions, but ion-molecule chemistry may only be significant in very low-density regions of the nebula. Grain-catalyzed thermochemistry, e.g., in the subnebulae surrounding the gas giant planets during their formation, probably also played a role both by facilitating the back reaction and also by producing D-poor water [via reaction (1)] which then diluted D-rich presolar water evaporated from presolar ice grains. As discussed later, the giant protoplanetary subnebulae may also have been the sources for the CH_4 and NH_3 in comets.

In principle, the oxygen isotopic composition of cometary water can also be used to constrain its origin (Table II). The $[^{18}O]/[^{16}O]$ ratio for water in Halley was measured by the *Giotto* NMS and is the same within error as the $[^{18}O]/[^{16}O]$ ratio for terrestrial SMOW. More precise measurements, however, would be needed to show that water in Halley and other comets has the same oxygen isotopic composition as water on Earth. Extensive measurements of the 16–17–18 oxygen isotope ratios in meteorites by Clayton and colleagues (Clayton 1993) show that different types of meteorites have oxygen isotopic compositions that differ from each other and from the oxygen isotopic composition of the Earth-Moon system. The SNC meteorites, which are widely believed to come from Mars, and the eucrite meteorites, which are widely believed to come from 4 Vesta, also have different oxygen isotopic compositions than the Earth and Moon or than other types of meteorites. The observed differences are not due to mass-dependent isotopic fractionations and can be very small, on the order of parts per thousand or less. The origin of the oxygen isotope differences is controversial.

It is very important to measure the bulk 16–17–18 oxygen isotopic composition of comets to see where comets plot on the oxygen isotope composition diagram. However, the required measurements are extremely difficult and may be possible only with *in situ* spacecraft experiments or by laboratory analyses of returned samples from comet nuclei. Measurements on individual cometary grains or on mineralogical entities within grains would be particularly revealing, as would isotopic ratios for individual volatile species.

Finally, we consider some implications of the measured ortho/para ratios in cometary water. As discussed earlier, spin temperatures derived from the OPR are 25–35 K for Halley, Hale-Bopp, and Hartley 2, and >50 K for Wilson. Mumma et al. (1993) argued that the OPR reflected the formation (or condensation) temperature of cometary water. Alternatively, if some process is able to equilibrate the OPR for water inside the cometary nucleus, the observed ratio may refer to this low temperature (Mumma et al. 1993). However, the orbital periods for Hale-Bopp, Halley, and Hartley 2 span the range from 4000 to 6 years, so it seems very unlikely that all three would have internal temperatures near 30 K; for Wilson, the only dynamically new comet in this sample, it was argued that the OPR in its outermost layer was reset to the high temperature limit of 3 by radiation damage while the comet was in the Oort Cloud. In either case, the OPR values for Halley and Hale-Bopp give temperatures that are significantly lower than those derived from the [D]/[H] ratios assuming reprocessing of interstellar water in the solar nebula. How can the apparent discrepancy be explained within the scenario that some reprocessing occurs in the nebula?

One possibility is that the OPR of cometary water was retained during back reaction of the D-rich cometary water with D-poor H_2 in the solar nebula. This scenario is supported by studies of ortho/para exchange and D/H exchange in hydrogen (e.g., Farkas 1935), but the large difference in energy between the lowest para and the lowest ortho state for H_2 compared with the corresponding difference for H_2O would suggest that preserving the OPR in water would be much more difficult. Further experimental studies of the ortho/para conversion kinetics for water vapor and ice are needed to test this suggestion. Another possibility is that in the solar nebula the cometary water was physically diluted by nebular water having a lower [D]/[H] ratio but a similar OPR. This scenario requires temperatures <50 K in the comet-formation zone of the outer solar nebula to preserve an OPR <3 and is consistent with the low temperatures needed to condense CO and N_2 on outer solar system bodies such as Triton and Pluto. Alternatively, the possibility that the cometary OPR might be reset in the coma, where near-nucleus temperatures drop to $\sim$30 K and proton transfer can occur with H^+ and H_3O^+, needs to be investigated.

To summarize, the observed [D]/[H] and ortho/para ratios in cometary water may be consistent either with unmodified interstellar water or with some physical mixing and limited reprocessing of interstellar water in the solar nebula. The upper limit on [D]/[H] in methanol from comet Halley may also be consistent with some physical mixing and limited reprocessing of interstellar methanol. However, the preservation of unmodified interstellar ices may be difficult, because water ice grains are calculated to have been partially or totally vaporized during infall into the solar nebula (Lunine et al. 1991; Cassen and Chick 1997), although the efficiency of this process might well be different for, and hence distinguish between, Oort Cloud and Jupiter-family (Edgeworth-Kuiper Belt) comets. Keeping interstellar water ice separate from nebular water ice also seems difficult. It is clear to us that laboratory studies of both [D]/[H] and ortho/para exchange under conditions relevant to comets are essential to the correct interpretation of the observational results.

At present the $[^{18}O]/[^{16}O]$ ratio of cometary water, which agrees with terrestrial SMOW within large errors, does not provide strong constraints on its origin. High-resolution measurements of the $^{16-17-18}O$ isotopic ratios in comets (and their constituent volatile and involatile components) will be very useful for constraining their origin and to see if comets are isotopically linked to the volatile-rich CI and CM2 chondrites.

B. Carbon Compounds

As discussed earlier in this chapter, a diverse suite of carbon compounds has been observed in comets. Observations from the *Giotto* spacecraft indicate that P/Halley has about the solar [C]/[O] ratio (e.g., Schulze et al. 1997; Anders and Grevesse 1989) and that $\sim$75% of all carbon is sequestered in the CHON grains, with the other $\sim$25% present as volatile species (Delsemme 1988). However, these conclusions and the published bulk compositions for P/Halley depend upon several assumptions about the analysis of the PUMA and PIA data and the applicability of observed gas/dust ratios and dust chemistry to the larger particles not measured by the *Giotto* instruments (e.g., see McDonnell et al. 1997; Schulze et al. 1997; and references therein).

The volatile carbon compounds in comets are generally dominated by oxidized carbon species such as CO, CO_2, H_2CO, CH_3OH, with smaller amounts of CH_4 and other hydrocarbons. If extended sources are considered, CO is generally the most abundant carbon gas, and the $[CO]/[H_2O]$ ratio reaches several tens of percent. However, present observations show $[CO]/[H_2O]$ ratios $<7\%$ for nuclear sources of CO. The $[H_2CO]/[H_2O]$ ratio reached 7% in Halley when the extended source was considered but is typically much less when only nuclear production is considered (Eberhardt 1998). CH_3OH seems to be more important than H_2CO as a volatile present in cometary ices (Table I). Based on the data for nuclear sources of cometary volatiles compiled by Eberhardt (1998), ratios of the oxidized carbon gases are ~0.3 to >3 for [CO_2]/[CO], ~0.004 to 0.08 for [H_2CO]/[CO], and ~0.1 to 2.4 for [CH_3OH]/[CO].

Reduced carbon species are generally less abundant than oxidized carbon species in comets, but they are potentially very important for constraining the origin of cometary carbon volatiles. The CH_4 abundance in Halley is somewhat controversial, but a recent reanalysis of the *Giotto* ion mass spectrometer data by Altwegg et al. (1994) reports an upper limit of 0.01 for the [CH_4]/[H_2O] ratio. The [CO]/[H_2O] ratio of 0.035 for Halley (Eberhardt 1998) then gives an upper limit of <0.3 for the [CH_4]/[CO] ratio in Halley, consistent with the [CH_4]/[CO] ratio of about 0.1 in Hyakutake, the first comet in which CH_4 was definitely observed (Mumma et al. 1996). Acetylene and C_2H_6 were also observed in Hyakutake, and their ratios relative to CO are ~0.05–0.15 and ~0.07, respectively (Brooke et al. 1996; Mumma et al. 1996). These three species were also detected in Hale-Bopp, with similar relative abundances (Mumma et al. 1997*a*; Weaver et al. 1998); the reductions are still preliminary, but only about 50% of the total CO comes from the nuclear source (DiSanti et al. 1999).

The approximate 3:1 ratio of refractory to volatile carbon species in Halley (Delsemme 1988) is the same as the 3:1 ratio for graphite to carbon gases (CO ~ CO_2 ~ CH_4) inside the graphite stability field at $T < 470$ K and $P < 10^{-7.6}$ bar (Lewis et al. 1979). This similarity was noted by Fegley (1993), who suggested that Fe-grain-catalyzed chemistry, such as Fischer-Tropsch type (FTT) reactions, may be responsible for the carbon inventory in Halley, because CH_4, CO_2, other oxidized carbon compounds, and volatile and involatile hydrocarbons can all be produced from mixtures of CO + H_2 by FTT reactions. It is probably difficult to precipitate graphite from a solar gas at the low temperatures and pressures where Lewis et al. (1979) calculated that graphite is stable, and organic matter, such as the CHON particles in Halley, may be formed instead.

Recent experimental work by Fegley and Hong (1998) demonstrates that CO reduction to CH_4 in H_2-rich gas is indeed catalyzed by Fe metal and alloys. They studied reduction of CO in an approximately solar-composition gas and found that high-purity Fe metal, Fe meteorite metal, and ordinary chondrite meteorite powder all catalyzed the production of large amounts of CH_4, with efficiencies up to 40% of the equilibrium values. The experiments with the ordinary chondrite meteorite powder were important because chondritic meteorites contain sulfides, which are potential catalyst poisons. Although the catalytic efficiency was less than that for high-purity Fe metal or iron meteorite metal, the powdered chondritic meteorite was still catalytically active and reduced CO to methane. The experimental work was done at 500–1000 K and ambient atmospheric pressure; it is relevant to giant protoplanetary subnebulae where CH_4 production is predicted in the outer solar nebula (Prinn and Fegley 1989; Fegley and Prinn 1989; Fegley 1993).

A separate set of experiments by Llorca and Casanova (1998) is also relevant. They heated a CO–H_2 gas mixture containing 4000 ppmv CO with a catalyst (Fe-Ni particles supported on silica) in a sealed vessel for 1000 hr (~42 d) at 473 K and 5×10^{-4} bars total pressure. They found that carbonaceous material and $(Fe,Ni)_2C$ were formed on the catalyst and that CH_4 (90%), C_2H_6 (~5%), C_2H_4 (~2%), C_3H_8 (~2%), C_3H_6 (0.5%), and C_4H_{10} (0.4%) were formed in the gas phase.

These two studies show that Fe-grain-catalyzed chemistry is capable of producing CH_4, higher hydrocarbons, and carbonaceous material under conditions relevant to the solar nebula and giant protoplanetary subnebulae (i.e., low [CO]/[H_2] ratios, moderate T, low P to ~1 bar). Further experimental work needs to be done to see whether the 3:1 ratio of refractory to volatile carbon and the relative abundances of oxidized and reduced carbon gases observed in comets can be produced by Fe-grain-catalyzed chemistry. Further theoretical work is also needed to quantify the efficiency of mixing between the giant protoplanetary subnebulae and the surrounding solar nebula and the efficiency of radial mixing from the inner to outer solar nebula. It is also important to extend studies of grain-catalyzed chemistry to the even lower temperatures characterizing the outer nebula and to include determinations of possible deuterium fractionation.

Alternatively, the carbon chemistry of comets may reflect the carbon chemistry of unmodified interstellar grains (e.g., Greenberg 1993). All the molecular species found in comets (Table I) except the nonpolar ethane are detected in interstellar clouds, and observable ethane derivatives such as ethyl cyanide are widespread (e.g., Irvine 1998). Furthermore, the relative abundances for the organic molecules found in both comets and interstellar icy grain mantles are quite similar (within an order of magnitude if cometary HNCO, CH_3CN, and HCN are combined for comparison with interstellar "XCN"; see the chapter by Langer et al., this volume); an exception is HCOOH, but this assignment in the case of the grain mantles may be tentative.

However, CO and CH_4 are highly volatile and may be vaporized from infalling interstellar grains during their accretion by the solar nebula, even at radial distances where water ice can be preserved. Moreover, although the production of CH_4 and other hydrides by laboratory UV irradiation of simulated interstellar ices (e.g., Greenberg 1993) has been used to argue for an interstellar origin of the carbon gases in Halley, the recent calculations by Gladstone and Fegley (1997) show that UV photochemistry is also important in the outer solar nebula. The relatively low D/H ratio in methane on Titan (see Table III) suggests that it may be possible to use D/H ratios to distinguish nebular from interstellar hydrides in comets.

To summarize, the carbon chemistry of Halley and other comets is dominated by refractory carbon species with smaller amounts of volatile carbon compounds. The volatiles are mainly oxidized carbon compounds (CO, CH_3OH, H_2CO) with smaller amounts of reduced carbon gases

(CH_4, C_2H_2, C_2H_6). Production of refractory and volatile carbon species by Fe-grain-catalyzed FTT reactions is supported by experimental studies done at low [CO]/[H_2] ratios, moderate temperatures, and low P to ~1 bar. However, the Fe-grain-catalyzed reactions have not reproduced all features of cometary carbon chemistry, and the efficiency of mass transport between possible FTT reaction sites (giant protoplanetary subnebulae, the inner solar nebula) and cometary formation sites (presumably in the Uranus-Neptune region of the outer solar nebula) is unknown. Conversely, the carbon chemistry of cometary ices is rather similar to the chemistry of interstellar ices. However, calculations of presolar grain heating during solar nebula accretion suggest that the preservation of (unmodified) volatile carbon compounds in interstellar grains during their infall into the solar nebula may be difficult.

C. Nitrogen Compounds

The recent observations of Hale-Bopp and Hyakutake have greatly expanded the number of nitrogen compounds observed in comets. Several of these (HCN, CH_3CN, HC_3N, HNCO, NH_2CHO) must be products of disequilibrium chemistry, such as UV-driven photochemistry and ion-molecule chemistry, although whether this processing occurred in interstellar space or in optically thin and low-density regions of the outer solar nebula is still unknown (Gladstone and Fegley 1997; Aikawa et al. 1998). Other compounds (N_2, NH_3) may be nebular products. At least part of the observed HNC is probably due to chemical processing in the cometary coma (Rodgers and Charnley 1998; Irvine et al. 1998*a,b*), raising the possibility that other trace constituents may also be influenced by coma chemistry.

Considering NH_3 and N_2 first, several lines of evidence summarized by Eberhardt (1998) show that the [NH_3]/[H_2O] ratio in Halley is ~1%, whereas the radio detections of NH_3 in Hyakutake and Hale-Bopp yield about 0.6% (Table I). The [N_2]/[H_2O] ratio in Halley is less well constrained, but observations of N_2^+ by Wyckoff et al. (1991) give [N_2]/[H_2O] ~ 0.02%. The derived [NH_3]/[N_2] ratio of ~30–50 is intermediate between the predicted [NH_3]/[N_2] ratios of ~200 in giant protoplanetary subnebulae and ~0.01 in the solar nebula. Prinn and Fegley (1989) argued that the intermediate ratio indicated a mixture of solar nebula and giant protoplanetary subnebula condensates in Halley, whereas Lunine (1989) and Engel et al. (1990) suggested an interstellar source for the NH_3 in Halley. The two alternatives (nebular versus interstellar) can probably be tested by measuring the D/H ratio in cometary NH_3, because the interstellar source should give a significantly higher D/H ratio than the nebular sources. Experimental tests of NH_3 production by Fe-grain-catalyzed chemistry may also be useful in distinguishing between the two sources.

Nebular and interstellar sources have also been proposed for the origin of cometary HCN. Meier et al. (1998*a*) measured the [DCN]/[HCN] ratio in Hale-Bopp and argued that the D/H ratio of $\sim(230\pm40)\times10^{-5}$ in the HCN is consistent with a low-temperature interstellar source and the [HDO]/[H_2O] ratio in the same comet. Moreover, the high [DCN]/[HCN] ratio is inconsistent with the lightning shock chemistry source proposed by Prinn and Fegley (1989) and Fegley and Prinn (1989). It is very important to measure (or set upper limits on) the D/H ratio in HCN in other comets to see whether interstellar or nebular sources are implied. Methyl cyanide, but not HC_3N, HNCO, or NH_2CHO, were included in prior nebular chemistry modeling. However, the CH_3CN abundances from disequilibrium chemistry in the nebula (quenched thermochemistry or lightning chemistry) are very much less than the [CH_3CN]/[HCN] ratio of $\sim$0.1 in Hale-Bopp. Thus, other disequilibrium processes (UV photochemistry or ion-molecule chemistry) appear necessary to produce these species, either in the interstellar medium or in the nebula. All of these nitrogen-containing species observed in comet Hale-Bopp are also found in the dense cores of interstellar dark clouds, except NH_2CHO, which is found in hot cores (Irvine 1999; Ohishi et al. 1992).

D. Sulfur Compounds

Here we focus on the sulfur compounds H_2S, H_2CS, CS_2, S_2, OCS, and SO_2 that are presumably parent (or primary) species emitted from comet nuclei. Sulfur monoxide (SO) is also observed in comets, but, as noted in section III.A, its source (nuclear emission vs. coma production) is not yet certain.

The [H_2S]/[H_2O] ratio in several comets ranges from $\sim$0.2 to 1.6% (this chapter, Eberhardt 1998). This is smaller than the solar S/O ratio, which corresponds to [H_2S]/[H_2O] $\sim$ 2.5%, but the total S-gas/[H_2O] ratio in Hale-Bopp is $\sim$2.8%. The other major sulfur species are OCS and CS_2, with SO_2, H_2CS, and S_2 being less abundant. However, this listing reflects the relative abundances of these compounds in only a few comets and may not be representative. Taken at face value, it appears that reduced sulfur compounds are more abundant than oxidized sulfur compounds, which is also true for nitrogen compounds but the opposite of the situation for volatile carbon compounds.

Fegley (1993) proposed that H_2S and NH_3 in Halley formed by the decomposition of solid NH_4SH. Ammonium hydrosulfide is a volatile solid that can be formed in the solar nebula at 134 K if rapid (i.e., heterogeneous) accretion isolates Fe metal from H_2S gas (Fegley and Lewis 1980). The similar NH_3 and H_2S abundances (Table I) are generally consistent with this idea. Ultraviolet photolysis of NH_4SH also leads to disequilibrium sulfur chemistry (Lebofsky and Fegley 1976), but further work is needed to see how much S_2 can be produced. Two tests which may distinguish between interstellar H_2S ice or nebular NH_4SH as sources of cometary H_2S are (1) the D/H ratio in H_2S and (2) the presence of an extended source (NH_4SH grains) vs. a point source (H_2S ice).

With the exception of H_2S, the major sulfur gas over a wide *P-T* range in a solar-composition gas, the sulfur species in comets (H_2CS, CS_2, S_2, OCS, and SO_2) appear to be products of disequilibrium chemistry. The sources of this disequilibrium are less certain. The relative abundances in Hale-Bopp of the polar sulfur-containing species (H_2S, H_2CS, OCS, SO_2, SO), which can be observed in interstellar clouds, are quite similar to those measured in the region of low-mass star formation IRAS 16293−2422 (Blake et al. 1994), with one exception: The relative H_2S abundance is an order of magnitude larger in the comet. In warmer cores the H_2S abundance increases, and the relative composition is also quite similar to that in Hale-Bopp, except that the hot cores tend to have higher (and quite variable) SO_2 and SO abundances (van Dishoeck and Blake 1998; Helmich and van Dishoeck 1997). It should be noted, however, that the H_2S abundance in these regions is typically not well constrained, because of the paucity of observable lines (cf. Minh et al. 1990). Likewise, the fraction of cometary SO corresponding to parent molecules in the nucleus, rather than daughter molecules produced in the coma, is unclear.

Equilibrium chemistry in the solar nebula is unable to produce H_2CS, CS_2, S_2, OCS, and SO_2 in the amounts observed in Hale-Bopp. However, Zahnle et al. (1995) showed that high-temperature shock chemistry during the impacts of fragments of Shoemaker-Levy 9 on Jupiter produced CS_2, CS, and S_2 from H_2O-poor jovian gas above the cloud tops. Oxidized sulfur gases, including SO and SO_2, were calculated to form from impact-induced shock chemistry of H_2O-bearing jovian gas. These results raise the possibility of producing some of the disequilibrium sulfur gases via lightning-induced shock chemistry in the outer solar nebula, but further work is needed to quantify this suggestion.

V. FROM NUCLEUS TO COMA: MODELS AND RELATIONS

To infer the composition of the cometary nucleus from observations of the coma, the nature of the nucleus, the complexities of sublimation, and the degree of chemical processing in the coma must all be understood. Some recent research on these topics is described in this section (see also Rauer 1999).

A. Nuclear Models, Fragmentation, and Activity

Because no macroscopic cometary samples have been studied in the laboratory, the nature of the nucleus must still be inferred from remote observations. Many very fundamental questions remain in this regard, including the density, structure, and degree of heterogeneity of the nucleus and the intrinsic similarities or differences among comets.

Potentially significant new information on the nature of at least one cometary nucleus has been provided by comet Shoemaker-Levy 9 (D/1993 F2), both in its close approach to Jupiter and consequent fragmentation in 1992 and in its subsequent impact on that planet in 1994 (Noll et al. 1996). Models of the tidal breakup (Asphaug and Benz 1996) reproduced the observed fragmentation by starting with a nucleus that has zero strength: that is, a nucleus consisting of hundreds of pieces with dimensions of order 100 m each, that are bound together only gravitationally. Olson and Mumma (1994) obtained similar results for slightly smaller pieces. Interestingly enough, this is the size range found for the typical structural elements of cometary nuclei in a numerical simulation of cometesimal growth by Weidenschilling (1997). Such a model also seems to fit the prominent crater chains observed on Callisto and Ganymede, if the chains are the results of impacts from tidally disrupted comets (Schenk et al. 1996). There is not universal agreement on such a strengthless, "rubble pile" model of the nucleus, however. Sekanina (1996), for example, favors a cohesive nucleus that can fragment along weak points or cracks. A mechanism for forming deep cracks at low nuclear temperatures has been proposed by Jenniskens and Blake (1996), based on laboratory demonstration of a phase change in ice between two amorphous forms. The competing models for the structure of Shoemaker-Levy 9 are discussed in a recent review by A'Hearn (1998). Note that although detailed spectroscopic studies of the impacts with Jupiter were carried out, they do not unambiguously provide information on the chemical composition of Shoemaker-Levy 9 because of the interaction with jovian material (see Crovisier 1996 and Lellouch 1996).

References for various earlier models of the structure of cometary nuclei are given by Mumma et al. (1993). Since that time, in addition to the work on Shoemaker-Levy 9 described above, the structure and composition of the nucleus has been considered theoretically, by Bockelée-Morvan and Rickman (1998), Prialnik (1997), Enzian et al. (1998), Møhlmann (1996), Hughes (1996), and Greenberg (1998), among others. Laboratory work on the evolution of structure due to radiative heating in near surface layers has demonstrated the formation of compositional and structural stratification in comet analogs (e.g., Kossacki et al. 1997).

B. Chemistry in the Coma

Chemistry in cometary comae has traditionally been assumed to involve (a) the sublimation of volatiles ("parent molecules") as the nuclear ices are heated by solar radiation and phase transitions and (b) the subsequent photolysis of these species to produce the radicals and ions ("daughter molecules") observed in visible wavelength spectra. It has been recognized that gas-phase reactions (ion-neutral and neutral-neutral) would occur in the coma of an active comet (e.g., Schmidt et al. 1988), but there had been little direct evidence for such processes, apart from the detection of mass spectrometry peaks in the spacecraft flyby of comet Halley that probably correspond to such protonated species as H_3O^+ and HCO_2^+ (Huebner et al. 1991).

This situation has changed significantly with the apparition of comet Hale-Bopp. The first unambiguous identification of H_3O^+ was made from submillimeter observations (Lis et al. 1998). The first cometary detection of the HCO^+ ion, a classic indicator of ion-molecule chemistry (Herbst and Klemperer 1973), was followed by maps that showed an emission pattern and intensity that could be understood in terms of gas-phase chemistry in the coma and interaction with the solar wind (Lovell et al. 1998*a,b*; Wright et al. 1998).

The origin of cometary HNC, first detected in comet Hyakutake (Irvine et al. 1996), raised the question: Was this neutral but "unstable" molecule produced by chemistry in the coma, or was it present in the nuclear ices? Observations of comet Hale-Bopp showed that the [HNC]/[HCN] ratio in the coma varied with heliocentric distance in a manner that could be explained if HNC were produced by the same gas-phase chemistry model that matched the HCO^+ observations, but not if the HNC were intrinsic to the nucleus or if it were a simple photodissociation product of a larger parent molecule (Irvine et al. 1998*a,b*). Curiously, though, the coma chemistry models have been unable to reproduce the HNC abundance measured in comet Hyakutake, for which the gas production and hence density in the coma were very much less than in comet Hale-Bopp, with a correspondingly drastically reduced efficiency for gas-phase chemistry (cf. Rodgers and Charnley 1998).

C. Nuclear Ices: Phase and Sublimation

Because the principal volatile constituent of comets is water, the phase or phases of H_2O ice present in the nucleus, its rate of sublimation, and the degree to which the sublimation of other volatiles is controlled by that of water are all key questions that must be solved to infer the nature of the nucleus correctly from observations of the coma. These issues have often been oversimplified in the past, such as by using equilibrium condensation theory for pure substances. In reality, all these problems involve the complex microphysics of inhomogeneous, poorly characterized solid mixtures.

For example, it is often assumed that the phase of H_2O ice formed by deposition of water vapor onto substrates at low temperature is determined solely by the temperature, with amorphous ice forming below 130 K, crystalline cubic ice for $130 < T < 170$ K, and hexagonal ice for $T > 170$ K. However, both theoretical and laboratory results indicate that the kinetics of the deposition and crystallization may control the crystallinity of ices in astrophysical situations. Thus, Kouchi et al. (1994) conclude that if interstellar icy grain mantles are preserved during the formation of the solar nebula, comets formed from these grains would contain amorphous water ice. In contrast, if such mantles were sublimed and then recondensed as the solar nebula cooled, the H_2O ice would be crystalline. The situation is even more complex, with the discoveries that more than one amorphous phase of water ice exist (Jenniskens et al. 1995) and that the phase of deposited ice can be affected by the morphology (microstructure) of the underlying substrate (Trakhtenberg et al. 1997).

The sublimation of the volatiles in a cometary nucleus as it approaches the Sun is likewise a very complex process. Laboratory experiments for a water matrix deposited at low temperatures have demonstrated, for example, that the percentage of trace constituents in the sublimate depends on ice temperature, ice phase, and whether the release is primarily from the surface rather than the interior of a porous layer (Notesco and Bar-Nun 1997). The radical change in the composition of the subliming gas as comet Hale-Bopp approached the Sun was demonstrated by the [CO]/[H_2O] ratio, which varied from >5 to $<\frac{1}{5}$ for heliocentric distances >4.5 and <2.8 AU, respectively (Biver et al. 1997, 1998). An interesting result reported by Jenniskens et al. (1997) is the presence of a viscous phase of liquid water that coexists with cubic crystalline ice for $120 < T < 210$ K when a thin layer of amorphous ice is warmed above 120 K. The presence of liquid water in cometary nuclei might have interesting and significant chemical effects.

In comparing abundances in interstellar clouds to those in comets, it must be remembered that chemical fractionation occurs not only during sublimation, but also during the initial trapping of gases as ice particles form. For example, laboratory experiments show that condensation can enhance the abundance of CO relative to N_2 and of C_2H_6 relative to CH_4, compared to the initial gas phase (Notesco and Bar-Nun 1996; Notesco et al. 1997). The measured fractionation at a temperature of 50–60 K seems to match abundance ratios observed in comets Halley and Hyakutake, although this match involves assumptions about the composition of the initial gas. This temperature is significantly higher than the water spin temperature (OPR) reported for comet Halley.

VI. SUMMARY

Comets are clearly chemically disequilibrium assemblages, which cannot be understood in terms of equilibrium models. On the one hand, the very few comets that have been observed in detail resemble interstellar material in several ways, both in their composition and their isotopic fractionation. On the other hand, accretion models suggest that the preservation of pristine interstellar material in the nebula may be difficult. Moreover, there must have been a variety of disequilibrium processes at work in the outer regions of the solar nebula where comets formed.

There has been considerable discussion in the literature concerning the ortho/para ratio measured for water in several comets, as discussed above in sections III.D and IV.A. The low spin temperatures that characterize this ratio in several comets, 25–35 K, are usually considered to represent the temperature of the cometary ices at the time of their formation, which might provide constraints on the formation site. As noted by Mumma et al. (1993), however, there exists relatively little data on ortho/para conversion for water, so that it is unclear how the OPR will be affected by the condensation of water vapor onto grains, by storage in the Oort Cloud or Edgeworth-Kuiper Belt for 4.5 Gyr, by the warming of the ices prior to and during sublimation, and by possible processing in the coma. We do note that fractionation of the water OPR during condensation on various matrixes has been reported by Konyukhov et al. (1988).

We have discussed the compositional similarity between comets and interstellar volatiles at several points in this chapter. This applies to both the composition of icy grain mantles and to the gas-phase molecules that may have sublimated from the mantles in hot cores (compare Table I with the interstellar abundances in the chapter by Langer et al., this volume, and in van Dishoeck and Blake 1998). There are several caveats in such comparisons, however. As noted in section III.A, the observed composition of the coma differs in subtle ways from the composition of the nucleus. One might wonder, for example, whether some of the minor species detected in Hale-Bopp (perhaps NH_2CHO, $HCOOCH_3$) might not be produced by chemical processes in the coma. Likewise, the composition of hot cores certainly includes products of gas-phase chemistry in a warm environment, as well as the former constituents of grain mantles.

Isotopic fractionations are generally regarded as good tracers of the origin and evolution of cometary matter. Even here, though, questions arise. What is the significance of the agreement of the [HDO]/[H_2O] and [DCN]/[HCN] ratios between comets and hot cores if, as has been suggested, the interstellar [DCN]/[HCN] ratio no longer reflects the probable solid-phase value (Hatchell et al. 1998)? In other words, if comets preserve interstellar HCN, why is the D/H fractionation so *low* compared with that in cold clouds such as TMC-1? Likewise, is the failure to date to detect cometary HDCO and CH_3OD significant? Could the cometary isotopic ratios reflect nonequilibrium processes carried out in the cold, low-density regions of the nebula rather than in the interstellar environment?

We believe that the answers to these questions are not yet clear but that developments over the next several years should help to remedy this situation. In Table IV we list some recommendations for future observations, laboratory studies, and theoretical modeling that will significantly clarify the nature of comets. Thus, measurement of the D/H fractionation in additional species (including H_2CO, CH_3OH, NH_3, and H_2S) and in Edgeworth-Kuiper Belt as well as in Oort Cloud comets may distinguish among possible sources for cometary volatiles. Both ground-based and spacecraft studies can help to discriminate between nuclear and extended sources of molecules such as CO, OCS, H_2S, and H_2CO, and returned samples should eventually provide an unbiased inventory of cometary composition and allow for the comparison of accurate isotopic ratios with

TABLE IV
Recommendations

Measurement	Goal
Earth-based remote sensing	
D/H ratios in H_2O from more comets	Constrain source of cometary water
O-P ratios in H_2O from more comets	Constrain source of cometary water
Spacecraft or Earth-based experiments	
D/H ratios in CH_3OH, H_2CO for comets	Constrain source of cometary volatiles
D/H ratios in NH_3 and H_2S	Constrain nitrogen and sulfur chemistry
Mapping of volatile emission	Resolve nuclear vs. extended sources (CO, H_2CO, etc.)
In situ analyses of volatiles and grains	Constrain origin of cometary material
Return cold (< 20 K) nuclei samples	Composition and structure of nucleus Unbiased inventory of cometary volatiles
Laboratory studies of returned comet samples	
Study presolar grains in rock	Nature and abundance of presolar grains
$^{16-17-18}O$ isotopic ratios of rock	Plot comets on Clayton's 3 O isotope diagram Pristine presolar rock or nebular condensate?
$^{16-17-18}O$ isotopic ratios of water	Determine if comets provided Earth's oceans
Ti isotopic analyses of rock	See if comets are related to C chondrites; pristine presolar rock or nebular condensate?
^{26}Al and other extinct radionuclides in rock	High presolar values or nebulalike values?
Radiometric dating of rock	Look for ages > age of solar system
Experimental studies for data analysis	
Conversion kinetics of ortho/para water	Interpret observed *o-p* ratios of cometary H_2O
D/H exchange kinetics (H_2O, CH_4, NH_3)	Interpret observed D/H ratios of cometary volatiles
Grain-catalyzed CO $\rightarrow$ organics reaction	Nebular vs. interstellar origin of organic molecules
Grain-catalyzed $N_2 \rightarrow NH_3$ reaction	Nebular vs. interstellar origin of ammonia
Theoretical modeling for data analysis	
UV, X-ray photochemistry of gas and grains in outer solar nebula at and beyond snowline	Steady-state abundances of volatile carbon species (CO_2, CH_3OH, H_2CO, HCOOH, etc.)
Ion-molecule chemistry in outer solar nebula	Compare composition and isotopic ratios with interstellar values

terrestrial values (e.g., for $^{16-17-18}O$). Laboratory study is urgently needed of D/H exchange kinetics for cometary volatiles, of ortho/para conversion for water, and of low-temperature, grain-catalyzed reactions for the reduction of CO and N_2 to organics and NH_3. Finally, realistic theoretical modeling of nonequilibrium chemistry in the outer solar nebula is needed to evaluate the extent to which such processes could mimic those that occur in interstellar clouds, accompanied by correspondingly detailed analyses of coma chemistry. Fortunately, such work is beginning (e.g., Aikawa et al. 1998; Willacy et al. 1998).

Acknowledgments This work was supported in part by grants from the Planetary Atmospheres, Exobiology, and Planetary Astronomy Programs of the National Aeronautics and Space Administration. We are grateful for helpful discussions with K. Lodders, D. Whittet, E. Herbst, E. Bergin, and an anonymous referee.

REFERENCES

A'Hearn, M. F. 1998. Recent results in the study of comets. In *Asteroids, Comets, Meteors 1996*, ed. A. C. Levasseur-Regourd and M. Fulchignoni, COSPAR Colloq. 10 (Oxford: Pergamon-Elsevier), in press.

A'Hearn, M. F., Boehnhardt, H., Kidger, M., and West, R. M., eds. 1998. Proceedings of the First International Conference on Comet Hale-Bopp. *Earth Moon Planets*, in press.

Aikawa, Y., Umebayashi, T., Nakano, T., and Miyama, S. 1998. Molecular evolution in planet forming circumstellar disks. *Faraday Discuss.* 109:281–301.

Altwegg, K., Balsiger, H., and Geiss, J. 1994. Abundance and origin of the CH_n^+ ions in the coma of comet P/Halley. *Astron. Astrophys.* 290:318–323.

Anders, E., and Grevesse, N. 1989. Abundances of the elements: Meteoritic and solar. *Geochim. Cosmochim. Acta* 53:197–214.

Asphaug, E., and Benz, W. 1996. Size, density and structure of comet Shoemaker-Levy 9 inferred from the physics of tidal breakup. *Icarus* 121:225–248.

Balsiger, H., Altwegg, K., and Geiss, J. 1995. D/H and $^{18}O/^{16}O$ ratio in the hydronium ion and in neutral water from *in situ* ion measurements in comet Halley. *J. Geophys. Res.* 100:5827–5834.

Bird, M. K., Huchtmeier, W. K., Gensheimer, P., Wilson, T. L., Janardhan, P., and Lemme, C. 1997. Radio detection of ammonia in comet Hale-Bopp. *Astron. Astrophys. Lett.* 325:L5–L8.

Biver, N., Bockelée-Morvan, D., Colom, P., Crovisier, J., Davies, J. K., Dent, W. R. F., Despois, D., Gerard, E., Lellouch, E., Rauer, H., Moreno, R., and Paubert, G. 1997. Evolution of the outgassing of comet Hale-Bopp (C/1995 O1) from radio observations. *Science* 275:1915–1918.

Biver, N., Bockelée-Morvan, D., Colom, P., Crovisier, J., Germain, B., Lellouch, E., Davies, J. K., Dent, W. R. F., Moreno, R., Paubert, G., Wink, J., Despois, D., Lis, D. C., Mehringer, D., Benford, D., Gardner, M., Phillips, T. G., Gunnarson, M., Rickman, H., Winnberg, A., Bergman, P., Johansson, L. E. B., and Rauer, H. 1998. Long-term evolution of the outgassing of comet Hale-Bopp from radio observations. *Earth Moon Planets,* in press.

Blake, G. A., van Dishoeck, E. F., Jansen, D. J., Groesbeck, T. D., and Mundy, L. G. 1994. Molecular abundances and low-mass star formation. I. Si- and S-bearing species toward IRAS 16293–2422. *Astrophys. J.* 428:680–692.

Bockelée-Morvan, D. 1997. Cometary volatiles: The status after C/1996 B2 Hyakutake. In *Molecules in Astrophysics: Probes and Processes,* ed. E. F. van Dishoeck (Dordrecht; Kluwer), pp. 219–236.

Bockelée-Morvan, D., and Crovisier, J. 1987. The role of water in the thermal balance of the coma. In *Proceedings of the International Symposium on the Diversity and Similarity of Comets,* ed. E. J. Rolfe and B. Battrick (Paris: European Space Agency), pp. 235–240.

Bockelée-Morvan, D., and Crovisier, J. 1998. New results on the composition of comets. In *Planetary Systems: The Long View,* ed. L. H. Celnikier and J. Tran Than Van, IXth Rencontres de Blois, in press.

Bockelée-Morvan, D., Gautier, D., Lis, D. C., Young, K., Keene, J., Phillips, T. G., Owen, T., Crovisier, J., Goldsmith, P. F., Bergin, E. A., Despois, D., and Wootten, A. 1998*a*. Deuterated water in comet C/1996 B2 (Hyakutake) and its implications for the origin of comets. *Icarus* 133:147–162.

Bockelée-Morvan, D., and Rickman, H. 1998. C/1996 O1 (Hale-Bopp): Gas production curves and their interpretation. *Earth Moon Planets,* in press.

Bockelée-Morvan, D., Wink, J., Despois, D., Colom, P., Biver, N., Crovisier, J., Gautier, D., Gerard, E., Lellouch, E., Moreno, R., Paubert, G., Rauer, H., Davies, J. K., and Dent, W. R. F. 1998*b*. A molecular survey of comet C/1995 O1 (Hale- Bopp) at the IRAM telescopes (abstract). *Earth Moon Planets,* in press.

Bradley, J. P., Sandford, S. A., and Walker, R. M. 1988. Interplanetary dust particles. In *Meteorites and the Early Solar System,* ed. J. F. Kerridge and M. S. Matthews (Tucson: University of Arizona Press), pp. 861–895.

Bradley, J. P., Brownlee, D. E., and Snow, T. P. 1997. GEMS and other pre-accretionally irradiated grains in interplanetary dust particles. In *From Stardust to Planetesimals,* ASP Conf. Ser. 122, ed. Y. J. Pendleton and A. G. G. M. Tielens, (San Francisco: Astronomical Society of the Pacific), pp.369–396.

Brooke, T. Y., Tokunaga, A. T., Weaver, H. A., Crovisier, J., Bockelée-Morvan, D., and Crisp, D. 1996. Detection of acetylene in the infrared spectrum of comet Hyakutake. *Nature* 383:606–608.

Brucato, R., Colangeli, L., Mennella, V., Palumbo, P., and Bussoletti, E. 1999. Silicates in Hale-Bopp: Hints from laboratory studies. *Planet. Space Sci.,* 47:773–779.

Campins, H. 1998. Comets Hale-Bopp (1995 O1) and Hyakutake (1996 B2): Records of presolar chemistry and more. In *Solar System Formation and Evolution,* ASP Conf. Ser. 149, ed. D. Lazzaro et al. (San Francisco: Astronomical Society of the Pacific), pp. 117–133.

Cassen, P., and Chick, K. M. 1997. The survival of presolar grains during the formation of the solar system. In *Astrophysical Implications of the Laboratory Study of Presolar Materials,* ed. T. Bernatowicz and E. Zinner (New York: AIP), pp. 697–719.

Charnley, S. B., Tielens, A. G. G. M., and Rodgers, S. D. 1997. Deuterated methanol in the Orion compact ridge. *Astrophys. J. Lett.* 482:L203–L206.

Chick, K. M., and Cassen, P. 1997. Thermal processing of interstellar dust grains in the primitive solar environment. *Astrophys. J.* 477:398–409.

Chin, Y.-N., Henkel, C., Whiteoak, J. B., Langer, N., and Churchwell, E. B. 1996. Interstellar sulfur isotopes and stellar oxygen burning. *Astron. Astrophys.* 305:960–969.

Clayton, R. N. 1993. Oxygen isotopes in meteorites. *Ann. Rev. Earth Planet. Sci.* 21:115–149.

Coustenis, A., Salama, A., Lellouch, E., Encrenaz, T., de Graauw, T., Bjoraker, G. L., Samuelson, R. E., Gautier, D., Feuchtgruber, H., Kessler, M. F., and Orten, G. S. 1998. Titan's atmosphere from ISO observations: Temperature, composition and detection of water vapor. *Bull. Am. Astron. Soc.* 30:1060 (abstract).

Cravens, T. E. 1997. Comet Hyakutake X-ray source: Charge transfer of solar wind heavy ions. *Geophys. Res. Lett.* 24:105–108.

Cronin, J. R., and Chang, S. 1993. Organic matter in meteorites: Molecular and isotopic analyses of the Murchison meteorite. In *The Chemistry of Life's Origins,* ed. J. M. Greenberg, C. X. Mendoza-Gómez, and V. Pirronello (Dordrecht: Kluwer), pp. 209–258.

Cronin, J. R., Cooper, G. W., and Pizzarello, S. 1995. Characteristics and formation of amino acids and hydroxy acids of the Murchison meteorite. *Adv. Space Res.* 15(3):91–97.

Crovisier, J. 1994. Molecular abundances in comets. In *Asteroids, Comets, Meteors 1993,* ed. A Milani et al. (Dordrecht: Kluwer), pp. 313–326.

Crovisier, J. 1996. Observational constraints on the composition and nature of comet Shoemaker-Levy 9. In *The Collision of Comet Shoemaker-Levy 9 and Jupiter,* IAU colloquium 156, ed. K. S. Noll, H. A. Weaver, and P. D. Feldman (Cambridge: Cambridge University Press), pp. 31–54.

Crovisier, J. 1998*a*. Infrared observations of volatile molecules in comet Hale-Bopp. *Earth Moon Planets,* in press.

Crovisier, J. 1998*b*. Physics and chemistry of comets: Recent results from comets Hyakutake and Hale-Bopp; answers to old questions and new enigmas. *Faraday Discuss.* 109:437–452.

Crovisier, J. 1999. Solids and volatiles in comets. In *Formation and Evolution of Solids in Space,* ed. J. M. Greenberg and A. Li (Dordrecht: Kluwer), pp. 389–426.

Crovisier, J., Brooke, T. Y., Hanner, M. S., Keller, H. U., Lamy, P. L., Altieri, B., Bockelée-Morvan, D., Jorda, L., Leech, K., and Lellouch, E. 1996. The infrared spectrum of comet C/1995 O1 (Hale-Bopp) at 4.6 AU from the Sun. *Astron. Astrophys. Lett.* 315:L385–L388.

Crovisier, J., Leech, K., Bockelée-Morvan, D., Brooke, T. Y., Hanner, M. S., Altieri, B., Keller, U., and Lellouch, E. 1997*a*. The infrared spectrum of comet Hale- Bopp. In *First ISO Workshop on Analytical Spectroscopy,* ESA SP-419 (Paris: European Space Agency), pp. 137–140.

Crovisier, J., Leech, K., Bockelée-Morvan, D., Brooke, T.Y., Hanner, M.S., Altieri, B., Keller, U., and Lellouch, E. 1997*b*. The spectrum of comet Hale-Bopp (C/1995 O1) observed with the Infrared Space Observatory at 2.9 AU from the Sun. *Science* 275:1904–1907.

Crovisier, J., Encrenaz, T., Lellouch, E., Bockelée-Morvan, D., Altieri, B., Leech, K., Salama, A., Griffin, M., de Graauw, T., van Dishoeck, E., Knacke, R., and Brooke, T. Y. 1999. ISO spectroscopic observations of short-period comets. In *The Universe as Seen by ISO,* ESA SP-427 (Paris: European Space Agency), pp. 161–164.

Dello Russo, N., DiSanti, M. A., Mumma, M. J., Magee-Sauer, K., and Rettig, T. W. 1998*a*. Carbonyl sulfide in comets C/1996 B2 (Hyakutake) and C/1995 O1 (Hale-Bopp): Evidence for an extended source in Hale-Bopp. *Icarus* 135:377–388.

Dello Russo, N., Mumma, M. J., DiSanti, M. A., Magee-Sauer, K., Novak, R., and Rettig, T. W. 1999. Water production and release in comet C/1995 O1 Hale-Bopp, *Icarus,* submitted.

Deloule, E., and Robert, F. 1995. Interstellar water in meteorites? *Geochim. Cosmochim. Acta* 59:4695–4706.

Delsemme, A. H. 1988. The chemistry of comets. *Phil. Trans. Roy. Soc. London* A325:509–523.

Dennerl, K., Englhauser, J., and Truemper, J. 1997. X-ray emissions from comets detected in the Roentgen X-ray Satellite All-Sky Survey. *Science* 277:1625–1630.

Despois, D. 1998. Radio line observations of molecular and isotopic species in comet C/1995 O1 (Hale-Bopp): Implications on the interstellar origin of cometary ices. *Earth Moon Planets,* in press.

Dickens, J. E. 1998. Chemical studies of molecular clouds. Ph.D. Dissertation, University of Massachusetts, Amherst.

DiSanti, M. A., Mumma, M. J., Dello Russo, N., Magee-Sauer, K., Novak, R., and Rettig, T. W. 1999. Identification of two sources of carbon monoxide in comet Hale-Bopp. *Nature* 339:662–665.

Donahue, T. M. 1995. Water on Mars and Venus. In *Volatiles in the Earth and Solar System,* ed. K. A. Farley (New York: AIP), pp. 154–166.

Eberhardt, P. 1998. Volatiles, isotopes and origin of comets. In *Asteroids, Comets, Meteors 1996,* ed. A. C. Levasseur-Regourd and M. Fulchignoni, COSPAR Colloq. 10 (Oxford: Pergamon-Elsevier), in press.

Eberhardt, P. 1999. Composition of comets: The *in situ* view. In *Cometary Nuclei in Space and Time,* IAU Colloq. 168, ed. M. A'Hearn (San Francisco: Astronomical Society of the Pacific), in press.

Eberhardt, P., Meier, R., Krankowsky, D., and Hodges, R. R. 1994. Methanol and hydrogen sulfide in comet P/Halley. *Astron. Astrophys.* 288:315–329.

Eberhardt, P., Reber, M., Krankowsky, D., and Hodges R. R. 1995. The D/H and $^{18}O/^{16}O$ ratios in water from comet P/Halley. *Astron. Astrophys.* 302:301–316.

Engel, S., Lunine, J. I., and Lewis, J. S. 1990. Solar nebula origin for volatile gases in Halley's Comet. *Icarus* 85:380–393.

Enzian, A., Cabot, H., and Klinger, J. 1998. Simulation of the water and carbon monoxide of comet Hale-Bopp using a quasi-3D nucleus model. *Planet. Space Sci.* 46:851–858.

Farkas, A. 1935. *Orthohydrogen, Parahydrogen and Heavy Hydrogen* (New York: Cambridge University Press).

Fegley, B., Jr. 1993. Chemistry of the solar nebula. In *The Chemistry of Life's Origins,* ed. J. M. Greenburg, C. X. Mendoza-Gómez, and V. Pirronello (Dordrecht: Kluwer), pp. 75–147.

Fegley, B., Jr. 1997. Disequilibrium chemistry in the solar nebula and early solar system: Implications for the chemistry of comets. In *Analysis of Returned Comet Nucleus Samples,* NASA CP 10152, ed. S. Chang, pp. 73–92.

Fegley B., Jr., and Hong, Y. 1998. Experimental studies of grain catalyzed reduction of CO to methane in the solar nebula. *EOS Trans AGU* 79:S361–362.

Fegley, B., Jr., and Lewis, J. S. 1980. Volatile element chemistry in the solar nebula: Na, K, F, Cl, Br, and P. *Icarus* 41:439–455.

Fegley, B., Jr., and Prinn, R. G. 1988. The predicted abundances of deuterium-bearing gases in the atmospheres of Jupiter and Saturn. *Astrophys. J.* 326:490–508.

Fegley, B., Jr., and Prinn, R. G. 1989. Solar nebula chemistry: Implications for volatiles in the solar system. In *The Formation and Evolution of Planetary Systems,* ed. H. Weaver and L. Danley (Cambridge: Cambridge University Press), pp. 171–211.

Festou, M. C., Rickman, H., and West, R. M. 1993. Comets. *Astron. Astrophys. Rev.* 5:37–163.

Feuchtgruber, H., Lellouch, E., Bezard, B., Encrenaz, T., de Graauw, T., and Davis, G. R. 1999. Detection of HD in the atmospheres of Uranus and

Neptune: A new determination of the D/H ratio. *Astron. Astrophys. Lett.* 341:L17–L21.
Fomenkova, M. 1997. Organic components of cometary dust. In *From Stardust to Planetesimals,* ASP Conf. Ser. 122, ed. Y. J. Pendleton and A. G. G. M. Tielens, (San Francisco: Astronomical Society of the Pacific), pp. 415–422.
Geiss, J., and Gloeckler, G. 1998. Abundances of deuterium and ^{3}He in the protosolar cloud. *Space Sci. Rev.* 84:239–250.
Gensheimer, P. D., Mauersberger, R., and Wilson, T. L. 1996. Water in galactic hot cores. *Astron. Astrophys.* 314:281–294.
Gladstone, G. R., and Fegley, B., Jr. 1997. Lyα photochemistry in the primitive solar nebula. *EOS Trans. AGU* 78:F538–F539.
Greenberg, J. M. 1982. What are comets made of? A model based on interstellar dust. In *Comets,* ed. L. L. Wilkening (Tucson: University of Arizona Press), pp. 131–163.
Greenberg, J. M. 1993. Physical and chemical composition of comets—from interstellar space to the Earth. In *The Chemistry of Life's Origins,* ed. J. M. Greenberg, C. X. Mendoza-Gómez, and V. Pirronello (Dordrecht: Kluwer), pp. 195–207.
Greenberg, J. M. 1998. Making a comet nucleus. *Astron. Astrophys.* 330:375–380.
Greenberg, J. M., and Li, A. 1998. From interstellar dust to comets: The extended CO source in comet Halley. *Astron. Astrophys.* 332:374–384.
Griffin, M. J., Naylor, D. A., Davis, G. R., Ade, P. A. R., Oldham, P. G., Swinyard, B. M., Gautier, D., Lellouch, E., Orton, G. S., Encrenaz, T., de Graauw, T., Furniss, I., Smith, H., Armand, C., Burgdorf, M., Di Giorgio, A., Ewart, D., Gry, C., King, K. J., Lim, T., Molinari, S., Price, M., Sidher, S., Smith, A., Texier, D., Trams, N., Unger, S. J., and Salama, A. 1996. First detection of the 56-μm rotational line of HD in Saturn's atmosphere. *Astron. Astrophys. Lett.* 315:L389–L392.
Grinspoon, D. H., and Lewis, J. S., 1987. Deuterium fractionation in the presolar nebula: Kinetic limitations on surface catalysis. *Icarus* 72:430–436.
Grossman, L., and Larimer, J. W. 1974. Early chemical history of the solar system. *Rev. Geophys. Space Phys.* 12:71–101.
Guélin, M., and Lequeux, J. 1980. Interpretation of isotopic abundances in interstellar clouds. In *Interstellar Molecules,* ed. B. H. Andrew (Dordrecht: D. Reidel), pp. 427–438.
Halbout, J., Robert, F., and Javoy, M. 1990. Hydrogen and oxygen isotopic compositions in kerogen from the Orgueil meteorite: Clues to a solar origin. *Geochim. Cosmochim. Acta* 54:1453–1462.
Hanner, M. S., Lynch, D. K., and Russell, R. W. 1994. The 8–13 micron spectra of comets and the composition of silicate grains. *Astrophys. J.* 425:274–285.
Hanner, M. S., Lynch, D. K., Russell, R. W., Hackwell, J. A., Kellogg, R., and Blaney, D. 1996. Mid-infrared spectra of comets P/Borrelly, P/Faye, and P/Schaumasse. *Icarus* 124:344–351.
Hanner, M. S., Gehrz, R. D., Harker, D. E., Hayward, T. L., Lynch, D. K., Mason, C. G., Russell, R. W., Williams, D. M., Wooden, D. H., and Woodward, C. E. 1998. Thermal emission and dust properties of comet Hale-Bopp. *Earth Moon Planets,* in press.
Hatchell, J., Millar, T. J., and Rodgers, S. D. 1998. The DCN/HCN abundance ratio in hot molecular cores. *Astron. Astrophys.* 332:695–702.
Helmich, F. P., and van Dishoeck, E. F. 1997. Physical and chemical variations within the W3 star-forming region. *Astron. Astrophys. Suppl.* 124:205–253.
Helmich, F. P., van Dishoeck, E. F., and Jansen, D. J. 1996. The excitation and abundance of HDO toward W3(OH)/(H_2O). *Astron. Astrophys.* 313:657–663.

Herbst, E., and Klemperer, W. 1973. The formation and depletion of molecules in dense interstellar clouds. *Astrophys. J.* 185:505–533.

Huebner, W. F., Boice, D. C., Schmidt, H. U., and Wegmann, R. 1991. Structure of the coma: Chemistry and solar wind interaction. In *Comets in the Post-Halley Era,* Vol. 2, ed. R. L. Newburn et al. (Dordrecht: Kluwer), pp. 907–936.

Hughes, D. W. 1996. The interior of a cometary nucleus. *Planet. Space Sci.* 44:705–710.

Irvine, W. M. 1992. Cold, dark interstellar clouds: Can gas phase reactions explain the observations? In *Chemistry and Spectroscopy of Interstellar Molecules,* ed. N. Kaifu (Tokyo: University of Tokyo Press), pp. 47–55.

Irvine, W. M. 1998. Extraterrestrial organic matter: A review. *Origins Life Evol. Biosphere* 28:365–383.

Irvine, W. M. 1999. The composition of interstellar molecular clouds. In *The Origin and Composition of Cometary Material,* ed. K. Altwegg and J. Geiss (Dordrecht: Kluwer), in press, and *Space Sci. Rev.,* in press.

Irvine, W. M., and Knacke, R. F. 1989. The chemistry of interstellar gas and grains. In *Origin and Evolution of Planetary and Satellite Atmospheres,* ed. S. K. Atreya, J. B. Pollack, and M. Matthews (Tucson: University of Arizona Press), pp. 3–34.

Irvine, W. M., Bockelée-Morvan, D., Lis, D., Matthews, H. E., Biver, N., Crovisier, J., Davies, J. K., Dent, W. R. F., Gautier, D., Godfrey, P. D., Keene, J., Lovell, A. J., Owen, T. C., Phillips, T. G., Rauer, H., Schloerb, F. P., Senay, M., and Young, K. 1996. Spectroscopic evidence for interstellar ices in comet Hyakutake. *Nature* 383:418–420.

Irvine, W. M., Bergin, E. A., Dickens, J. E., Jewitt, D., Lovell, A. J., Matthews, H. E., Schloerb, F. P., and Senay, M. 1998*a*. Chemical processing in the coma as the source of cometary HNC. *Nature* 393:547–550.

Irvine, W. M., Dickens, J. E., Lovell, A. J., Schloerb, F. P., Senay, M., Bergin, E. A., Jewitt, D., and Matthews, H. E., 1998*b*. Chemistry in cometary comae. *Faraday Discuss.* 109:475–492, 510–512.

Jacq, T., Walmsley, C. M., Henkel, C., Baudry, A., Mauersberger, R., and Jewell, P. R. 1990. Deuterated water and ammonia in hot cores. *Astron. Astrophys.* 228:447–470.

Jacq, T., Walmsley, C. M., Mauersberger, R., Anderson, T., Herbst, E., and De Lucia, F. C. 1993. Detection of interstellar CH_2DOH. *Astron. Astrophys.* 271:276–281.

Jenniskens, P., and Blake, D.F. 1996. A mechanism for forming deep cracks in comets. *Planet. Space Sci.* 44:711–713.

Jenniskens, P., Blake, D. F., Wilson, M. A., and Pohorille, A. 1995. High-density amorphous ice, the frost on interstellar grains. *Astrophys. J.* 455:389–401.

Jenniskens, P., Banham, S. F., Blake, D. F., and McCoustra, M. R. S. 1997. Liquid water in the domain of cubic crystalline ice I_c. *J. Chem. Phys.* 107:1232–1241.

Jewitt, D. C., Matthews, H. E., Owen, T. C., and Meier, R. 1997. Measurements of $^{12}C/^{13}C$, $^{14}N/^{15}N$, and $^{32}S/^{34}S$ ratios in comet Hale-Bopp (C/1995 O1). *Science* 278:90–93.

Kim, S. J., and A'Hearn, M. F. 1991. Upper limits of SO and SO_2 in comets. *Icarus* 90:79–95.

Kleine, M., Wyckoff, S., Wehinger, P. A., and Peterson, B. A. 1995. The carbon isotope abundance ratio in comet Halley. *Astrophys. J.* 439:1021–1033.

Konyukhov, V. K., Prokhorov, A. M., Tikhonov, V. I., and Faizulaev, V. N. 1988. Rotationally selective condensation and spin-modification separation for isotopic water species. *Izv. Akad. Nauk SSSR Ser. Fiz.* 52:1059–1065.

Kossacki, K. J., Kømle, N.I., Leliwa-Kopystynski, J., and Kargl, G. 1997. Laboratory investigation of the evolution of cometary analogs: Results and interpretation. *Icarus* 128:127–144.
Kouchi, A., Yamamoto, T., Kozasa, T., Kuroda, T., and Greenberg, J. M. 1994. Conditions for condensation and preservation of amorphous ice and crystallinity of astrophysical ices. *Astron. Astrophys.* 290:1009–1018.
Krankowsky, D., Eberhardt, P., Dolder, U., Schulte, W., Laemmerzahl, P., Hoffman, J. H., Hodges, R. R., Berthelier, J. J., and Illiano, J. M. 1986. In-situ gas and ion measurements at Comet Halley. *Nature* 321:326–329.
Krasnopolsky, V. 1997. On the nature of soft X-ray radiation in comets. *Icarus* 128:368–385.
Krasnopolsky, V. A., Mumma, M. J., Abbott, M., Flynn, B. C., Yeomans, D. K., Feldman, P. D., and Cosmovici, C. B. 1997. Detection of soft X-rays and a sensitive search for noble gases in comet Hale-Bopp (C/1995 O1). *Science* 277:1488–1491.
Langer, W. D., Graedel, T. E., Frerking, M. A., and Armentrout, P. B. 1984. Carbon and oxygen isotopic fractionation in dense interstellar clouds. *Astrophys. J.* 277:581–604.
Lebofsky, L. A., and Fegley, M. B., Jr. 1976. Laboratory reflection spectra for the determination of chemical composition of icy bodies. *Icarus* 28:379–387.
Lecluse, C., and Robert, F. 1994. Hydrogen isotope exchange reaction rates: Origin of water in the inner solar system. *Geochim. Cosmochim. Acta* 58:2927–2939.
Lellouch, E. 1996. Chemistry induced by the impacts: Observations. In *The Collision of Comet Shoemaker-Levy 9 and Jupiter,* IAU Colloq. 156, ed. K. S. Noll, H. A. Weaver, and P. D. Feldman (Cambridge: Cambridge University Press), pp. 31–54.
Lewis, J. S. 1972. Low temperature condensation from the solar nebula. *Icarus* 16:241–252.
Lewis, J. S., and Prinn, R. G. 1980. Kinetic inhibition of CO and N_2 reduction in the solar nebula. *Astrophys. J.* 238:357–364.
Lewis, J. S., Barshay, S. S., and Noyes, B. 1979. Primordial retention of carbon by the terrestrial planets. *Icarus* 37:190–206.
Lis, D. C., Keene, J., Young, K., Phillips, T. G., Bockelée-Morvan, D., Crovisier, J., Schilke, P., Goldsmith, P. F., and Bergin, E. 1997. CSO observations of comet C/1996 B2 (Hyakutake). *Icarus* 130:355–372.
Lis, D. C., Mehringer, D. M., Benford, D., Gardner, M., Phillips, T. G., Bockelée-Morvan, D., Biver, N., Colom, P., Crovisier, J., Despois, D., and Rauer, H. 1998. New molecular species in comet C/1995 (Hale-Bopp) observed with the Caltech Submillimeter Observatory. *Earth Moon Planets,* in press.
Lisse, C. M., Dennerl, K., Englehauser, J., Harden, M., Marshall, F. E., Mumma, M. J., Petre, R., Pye, J. P., Ricketts, M. J., Schmit, J., Truemper, J., and West, R. G. 1996. Discovery of X-ray and extreme ultraviolet emission from comet Hyakutake C/1996 B2. *Science* 274:205–209.
Lisse, C. M., Dennerl, K., Englhauser, J., Trumper, J., Marshall, F. E., Petre, R., and Valina, A. 1998. X-ray emission from comet Hale-Bopp. *Earth Moon Planets,* in press.
Llorca, J., and Casanova, I. 1998. Formation of carbides and hydrocarbons in chondritic interplanetary dust particles: A laboratory study. *Meteorit. Planet. Sci.* 33:243–251.
Lodders, K., and Fegley, B., Jr. 1998. *The Planetary Scientist's Companion* (New York: Oxford University Press).
Lovell, A. J., Schloerb, F. P., Bergin, E. A., Dickens, J. E., DeVries, C. H., Senay, M., and Irvine, W. M. 1998*a*. HCO^+ ion-molecule chemistry in comet C/1995 O1 Hale-Bopp. *Earth Moon Planets,* in press.

Lovell, A. J., Schloerb, F. P., Dickens, J. E., DeVries, C. H., Senay, M. C., and Irvine, W. M. 1998*b*. HCO^+ imaging of comet Hale-Bopp (C/1995 O1). *Astrophys. J. Lett.* 497:L117–L121.

Lunine, J. I. 1989. Primitive bodies: Molecular abundances in comet Halley as probes of cometary formation environments. In *The Formation and Evolution of Planetary Systems,* ed. H. Weaver and L. Danly (Cambridge: Cambridge Univ. Press), pp. 213–242.

Lunine, J. I., Engel, S., Rizk, R., and Horanyi, M. 1991. Sublimation and reformation of icy grains in the primitive solar nebula. *Icarus* 94:333–344.

Magee-Sauer, K., Mumma, M. J., DiSanti, M. A., Dello Russo, N., and Rettig, T. W. 1998. Infrared spectroscopy of the ν_3 band of hydrogen cyanide in C/1995 O1 Hale-Bopp. *Icarus,* submitted.

Mahaffy, P. R., Donahue, T. M., Atreya, S. K., Owen, T. C., and Niemann, H. B. 1998. Galileo probe measurements of D/H and $^3He/^4He$ in Jupiter's atmosphere. *Space Sci. Rev.* 84:251–263.

Malfait, K., Waelkens, C., Waters, L. B. F. M., Vandenbussche, B., Huygen, E., and de Graauw, M. S. 1998. The spectrum of the young star HD 100546 observed with the Infrared Space Observatory. *Astron. Astrophys.* 332:L25–L28.

Mauersberger, R., Henkel, C., Jacq, T., and Walmsley, C. M. 1988. Deuterated methanol in Orion. *Astron. Astrophys. Lett.* 194:L1–L4.

McDonnell, J. A. M., Pankiewicz, G. S., Birchley, P. N. W., Green, S. F., and Perry, C. H. 1997. The *in situ* particulate size distribution measured for one comet: P/Halley. In *Analysis of Returned Comet Nucleus Samples,* NASA CP 10152, ed. S. Chang, pp. 205-216.

Meier, R., Owen, T. C., Jewitt, D. C., Matthews, H. E., Senay, M., Biver, N., Bockelée-Morvan, D., Crovisier, J., and Gautier, D. 1998*a*. Deuterium in comet C/1995 O1 (Hale-Bopp): Detection of DCN. *Science* 279:1707–1710.

Meier, R., Owen, T. C., Matthews, H. E., Jewitt, D. C., Bockelée-Morvan, D., Biver, N., Crovisier, J., and Gautier, D. 1998*b*. A determination of HDO/H_2O in comet C/1995 O1 (Hale-Bopp). *Science* 279:842–844.

Messenger, S., and Walker, R. M. 1997. Evidence for molecular cloud material in meteorites and interplanetary dust. In *Astrophysical Implications of the Laboratory Study of Presolar Materials,* ed. T. Bernatowicz and E. Zinner (New York: Academic Press), pp. 545–564.

Millar, T. J., and Hatchell, J. 1998. Chemical models of hot molecular cores. *Faraday Discuss.* 109:15–30.

Millar, T. J., Bennett, A., and Herbst, E. 1989. Deuterium fractionation in dense interstellar clouds. *Astrophys. J.* 340:906–920.

Minh, Y. C., Ziurys, L. M., Irvine, W. M., and McGonagle, D. 1990. Observations of H_2S toward OMC-1. *Astrophys. J.* 360:136–141.

Minh, Y. C., Dickens, J. E., Irvine, W. M., and McGonagle, D., 1995. Measurements of the $H_2{}^{13}CO$ ortho/para ratio in cold, dark molecular clouds. *Astron. Astrophys.* 298:213–218.

Møhlmann, D. 1996. Cometary activity and nucleus modeling: A new approach. *Planet. Space Sci.* 44:541–546.

Mumma, M. J. 1997. Organic volatiles in comets: Their relation to interstellar ices and solar nebula material. In *From Stardust to Planetesimals,* ASP Conf Ser. 122, ed. Y. J. Pendleton and A. G. G. M. Tielens (San Francisco: Astronomical Society of the Pacific), pp. 369–396.

Mumma, M. J., Weissman, P. R., and Stern, S. A. 1993. Comets and the origin of the solar system: Reading the Rosetta stone. In *Protostars and Planets III,* ed. E. H. Levy and J. I. Lunine (Tucson: University of Arizona Press), pp. 1177–1252.

Mumma, M. J., DiSanti, M. A., Dello Russo, N., Fomenkova, M., Magee-Sauer, K., Kaminski, C. D., and Xie, D. X. 1996. Detection of abundant ethane and methane, along with carbon monoxide and water, in comet C/1996 B2 (Hyakutake): Evidence for interstellar origin. *Science* 272:1310–1314.

Mumma, M. J., DiSanti, M. A., Dello Russo, N., Magee-Sauer, K., and Fomenkova, M. 1997*a*. Detection of parent volatiles in comet C/1995 O1 (Hale-Bopp). *Intl. Astron. Union Circ.* 6573 and 6568.

Mumma, M. J., Krasnopolsky, V. A., and Abbott, M. J. 1997*b*. Soft X-rays from four comets observed with EUVE. *Astrophys. J. Lett.* 49:L125–L128.

Noll, K. S., Weaver, H. A., and Feldman, P. D., ed. 1996. *The Collision of Comet Shoemaker-Levy 9 and Jupiter,* IAU Colloq. 156 (Cambridge: Cambridge University Press).

Notesco, G., and Bar-Nun, A. 1996. Enrichment of CO over N_2 by their trapping in amorphous ice and implications to comet P/Halley. *Icarus* 122:118–121.

Notesco, G., and Bar-Nun, A. 1997. Trapping of methanol, hydrogen cyanide, and *n*-hexane in water ice, above its transformation temperature to the crystalline form. *Icarus* 126:336–341.

Notesco, G., Laufer, D., and Bar-Nun, A. 1997. The source of the high C_2H_6/CH_4 ratio in comet Hyakutake. *Icarus* 125:471–473.

Ohishi, M., Irvine, W. M., and Kaifu, N. 1992. Molecular abundance variations among and within cold, dark molecular clouds. In *Astrochemistry of Cosmic Phenomena,* ed. P. D. Singh (Dordrecht: Kluwer), pp. 171–177.

Olson, K. M., and Mumma, M. J. 1994. Simulations of the breakup and dynamical evolution of comet Shoemaker-Levy 9 employing a swarm model. *Bull. Am. Astron. Soc.* 26:1574–1575; *Icarus,* submitted.

Palmer, P., Wootten, A., Butler, B., Bockelée-Morvan, D., Crovisier, J., Despois, D., and Yeomans, D. K. 1996. Comet Hyakutake: First secure detection of ammonia in a comet. *Bull. Am. Astron. Soc.* 28:927–928.

Pisunkov, N., Wood, B. E., Linsky, J. L., Dempsey, R. C., and Ayres, T. R. 1997. Local interstellar properties and deuterium abundances for the lines of sight toward HR 1099, 31 Comae, β Ceti, and β Cassiopeiae. *Astrophys. J.* 474:315–328.

Pratap, P., Dickens, J. E., Snell, R. L., Miralles, M. P., Bergin, E. A., Irvine, W. M., and Schloerb, F. P. 1997. A study of the physics and chemistry of TMC-1. *Astrophys. J.* 486:862–885.

Prialnik, D. 1997. A model for the distant activity of comet Hale-Bopp. *Astrophys. J. Lett.* 478:L107–L110.

Prinn, R. G., and Fegley, B., Jr. 1981. Kinetic inhibition of CO and N_2 reduction in circumplanetary nebulae: Implications for satellite composition. *Astrophys. J.* 249:308–317.

Prinn, R. G., and Fegley, B., Jr. 1989. Solar nebula chemistry: Origin of planetary, satellite, and cometary volatiles. In *Origin and Evolution of Planetary and Satellite Atmospheres,* ed. S. K. Atreya, J. B. Pollack, and M. Matthews (Tucson: University of Arizona Press), pp. 78–136.

Rauer, H. 1999. Remote observations of cometary volatiles and implications on comet nuclei. In *Cometary Nuclei in Space and Time,* IAU Colloq. 168, ed. M. A'Hearn (San Francisco: Astronomical Society of the Pacific), in press.

Rodgers, S. D., and Charnley, S. B. 1998. HNC and HCN in comets. *Astrophys. J. Lett.* 501:L227–L230.

Rodgers, S. D., and Millar, T. J.. 1996. The chemistry of deuterium in hot molecular cores. *Mon. Not. Roy. Astron. Soc.* 280:1046–1054.

Salama, F. 1996. Low temperature spectroscopy: From ground to space. In *Low Temperature Molecular Spectroscopy,* ed. R. Fausto (Dordrecht: Kluwer), pp. 169–191.

Schenk, P. M., Asphaug, E., McKinnon, W. B., Melosh, H. J., and Weissman, P. R. 1996. Cometary nuclei and tidal disruption: The geologic record of crater chains on Callisto and Ganymede. *Icarus* 121:249–274.

Schmidt, H. U., Wegmann, R., Huebner, W. F., and Boice, D. C. 1988. Cometary gas and plasma flow with detailed chemistry. *Comput. Phys. Comm.* 49:17–59.

Schulze, H., Kissel, J., and Jessberger, E. K. 1997. Chemistry and mineralogy of comet Halley's dust. In *From Stardust to Planetesimals,* ed. Y. J. Pendleton and A. G. G. M. Tielens, ASP Conf. Ser. 122, (San Francisco: Astronomical Society of the Pacific), pp. 397-414.

Sekanina, Z. 1996. Tidal breakup of the nucleus of comet Shoemaker-Levy 9. In *The Collision of Comet Shoemaker-Levy 9 and Jupiter,* ed. K. S. Noll, H. A. Weaver, and P. D. Feldman (Cambridge: Cambridge University Press), pp. 55–80.

Sekanina, Z., Hanner, M. S., Jessberger, E. K., and Fomenkova, M. N. 1998. Cometary dust. In *Interplanetary Dust,* ed. S. F. Dermott et al. (Tucson: University of Arizona Press), in press.

Snow, T. P. 1997. The diffuse interstellar bands and interstellar molecules. In *From Stardust to Planetesimals,* ed. Y. J. Pendleton and A. G. G. M. Tielens, ASP Conf. Ser. 122, (San Francisco: Society of the Pacific), pp. 147–154.

Sutton, E. C., Peng, R., Danchi, W. C., Jaminet, P. A., Sandell, G., and Russell, A. P. G. 1995. The distribution of molecules in the core of OMC-1. *Astrophys. J. Suppl.* 97:455–496.

Takano, S., Masuda, A., Hirahara, Y., Suzuki, H., Ohishi, M., Ishikawa, S., Kaifu, N., Kasai, Y., Kawaguchi, K., and Wilson, T. L. 1998. Observations of ^{13}C isotopomers of HC_3N and HC_5N in TMC-1: Evidence for isotopic fractionation. *Astron. Astrophys.* 329:1156–1169.

Taylor, D. K., and Dickman, R. L. 1989. A reassessment of the double isotope ratio $[^{13}C]/[C^{18}O]$ in molecular clouds. *Astrophys. J.* 341:293–298.

Tielens, A. G. G. M. 1983. Surface chemistry of deuterated molecules. *Astron. Astrophys.* 119:177–184.

Tielens, A. G. G. M., Wooden, D. H., Allamandola, L. J., Bregman, J., and Witteborn, F. C. 1996. The infrared spectrum of the galactic center and the composition of interstellar dust. *Astrophys. J.* 461:210–222.

Trakhtenberg, S., Naaman, R., Cohen, S. R., and Benjamin, I. 1997. Effect of the substrate morphology on the structure of adsorbed ice. *J. Phys. Chem. B* 101:5172–5176.

Turner, B. E. 1990. Detection of doubly deuterated interstellar formaldehyde (D_2CO)—An indicator of active grain surface chemistry. *Astrophys. J. Lett.* 362:L29–L33.

Urey, H. C. 1953. Chemical evidence regarding the Earth's origin. In *XIIIth International Congress on Pure and Applied Chemistry and Plenary Lecture* (Stockholm: Almquist and Wiksells), pp. 188–217.

van Dishoeck, E. F., and Blake, G. A. 1998. Chemical evolution of star-forming regions. *Ann. Rev. Astron. Astrophys.* 36:317–368.

van Dishoeck, E. F., Blake, G. A., Draine, B. T., and Lunine, J. I. 1993. The chemical evolution of protostellar and protoplanetary matter. In *Protostars and Planets III,* ed. E. H. Levy and J. I. Lunine (Tucson: University of Arizona Press), pp. 163–241.

Waelkens, C., Waters, L. B. F. M., de Graauw, M. S., Huygens, E., Malfait, K., Plets, H., Vandenbussche, B., Beintema, D. A., Boxhoorn, D. R., Habing, H. J., Heras, A. M., Kester, D. J. M., Lahuis, F., Morris, P. W., Roelfsema, P. R., Salama, A., Siebenmorgen, R., Trams, N. R., van der Bliek, N. R., Valentijn, E. A., and Wesselius, P. R. 1996. SWS observations of young main-sequence stars with dusty circumstellar disks. *Astron. Astrophys. Lett.* 315: L245–L248.

Waelkens, C., Malfait, K., and Waters, L. B. F. M. 1998. Comet Hale-Bopp, circumstellar dust, and the interstellar medium. *Earth Moon Planets,* in press.

Weaver, H. A., Feldman, P. D., McPhate, J. B., A'Hearn, M. F., Arpigny, C., Brandt, J. C., and Randall, C. E. 1996. Ultraviolet spectroscopy and optical imaging of comet Hyakutake (1996 B2) with HST (abstract). In *ACM 96,* COSPAR Colloq. 10, p. 37.

Weaver, H. A., Feldman, P. D., A'Hearn, M. F., Arpigny, C., and Brandt, G. P. 1997. The activity and size of the nucleus of comet Hale-Bopp (C/1995 O1). *Science* 275:1900–1904.

Weaver, H. A., Brooke, T. Y., Chin, G., Kim, S. J., Bockelée-Morvan, D., and Davies, D. 1998. Infrared spectroscopy of comet Hale-Bopp. *Earth Moon Planets,* in press.

Weidenschilling, S. J. 1997. The origin of comets in the solar nebula: A unified model. *Icarus* 127:290–306.

Wielen, R., and Wilson, T. L. 1997. The evolution of the C, N, and O isotope ratios from an improved comparison of the interstellar medium with the Sun. *Astron. Astrophys.* 326:139–142.

Wielen, R., Fuchs, B., and Dettbarn, C, 1996. On the birthplace of the Sun and the places of formation of other nearby stars. *Astron. Astrophys.* 314:438–447.

Willacy, K., Klahr, H. H., Millar, T. J., and Henning, T. 1998. Gas and grain chemistry in a protoplanetary disk. *Astron. Astrophys.,* 338:995–1005.

Wink, J., Bockelée-Morvan, D., Despois, D., Colom, P., Biver, N., Crovisier, J., Gerard, E., Lellouch, E., Davies, J. K., Dent, W. R. F., and Jorda, L. 1998. Evidence for extended sources and temporal modulations in molecular observations of C/1995 O1 (Hale-Bopp) at the IRAM interferometer (abstract). *Earth Moon Planets,* in press.

Wooden, D. H., Harker, D. E., Woodward, C. E., Butner, H. M., Koike, C., Witteborn, F. C., and McMurtry, C. M. 1999. Silicate mineralogy of the dust in the inner coma of comet C/1995 O1 (Hale-Bopp) pre- and post-perihelion. *Astrophys. J.* 517:1034–1058.

Woodney, L. M., McMullin, J., and A'Hearn, M. F. 1997. Detection of OCS in comet Hyakutake (C/1996 B2). *Planet. Space Sci.* 45:717–719.

Woodney, L. M., A'Hearn, M. F., McMullin, J., and Samarasinha, N. 1998. Sulfur chemistry at millimeter wavelength in C/Hale-Bopp (abstract). *Earth Moon Planets,* in press.

Wright, M. C. H., de Pater, I., Foster, J. R., Palmer, P., Snyder, L. E., Veal, J. M., A'Hearn, M. F., Woodney, L. M., Jackson, W. M., Kuan, Y.-J., and Lovell, A. J. 1998. Mosaiced images and spectra of $J = 1$–0 HCN and HCO^+ emission from comet Hale-Bopp (1995 O1). *Astron. J.* 116:3018–3028.

Wyckoff, S., Tegler, S., and Engel, L. 1991. Nitrogen abundance in comet Halley. *Astrophys. J.* 367:641–648.

Yung, Y. L., Friedl, R. R., Pinto, J. P., Bayes, K. D., and Wen, J. S. 1988. Kinetic isotopic fractionation and the origin of HDO and CH_3D in the solar system. *Icarus* 74:121–132.

Zahnle, K., MacLow, M. M., Lodders, K., and Fegley, B., Jr. 1995. Sulfur chemistry in the wake of comet Shoemaker-Levy 9. *Geophys. Res. Lett.* 22:1593–1596.

Zinner, E. 1997. Presolar material in meteorites: An overview. In *Astrophysical Implications of the Laboratory Study of Presolar Materials,* ed. T. Bernatowicz and E. Zinner (New York: AIP), pp. 3–26.

PHYSICAL NATURE OF THE KUIPER BELT

DAVID C. JEWITT
University of Hawai'i at Manoa

and

JANE X. LUU
Sterrewacht, Leiden University

Recent ground-based observations have unveiled a large number of bodies in orbit beyond Neptune, in a region now widely known as the Kuiper (or, less commonly, Edgeworth-Kuiper) Belt. About 10^5 Kuiper Belt objects (KBOs) with diameters larger than 100 km exist in the 30-AU to 50-AU trans-Neptunian region. Their combined mass is about 10% of that of Earth. The orbits of KBOs fall into at least three distinct dynamical classes: the "Classical" objects, the Plutinos, and the "Scattered" objects. Each throws light on physical processes operating in the solar system prior to and during the formation of the planets. The Kuiper Belt is significant both as the likely source of the short-period comets (and the dynamically intermediate Centaurs) and as a repository of the solar system's most primitive (least thermally processed) material. KBOs show an unexpected and presently unexplained diversity of surface colors, possibly reflecting intrinsic compositional variations and transient resurfacing by impacts. The present-day Kuiper Belt is probably the surviving remnant of a once much more massive (10 $M_\oplus$?) preplanetary disk. It is very likely that collisions and disk-planet interactions played a major role in shaping this early precursor. Although the collisional production of dust is presently modest ($\sim 10^3$ kg s^{-1}) and the optical depth small ($\sim 10^{-7}$), the early Belt was probably very dusty and may have sustained a disk analogous to those reported around some nearby main-sequence stars.

I. HISTORY

The Kuiper Belt has a long and mostly speculative history, dating back at least to papers by Edgeworth (1943, 1949) and Kuiper (1951). These authors noticed that prevailing theories of planetary accretion provided no reason to expect that the planetary system should end at the orbit of Neptune (or of Pluto, then believed to be a massive and bona fide planet in its own right; Tombaugh 1961). Edgeworth believed that the comets might be derived from the trans-Plutonian region (MacFarland 1996), an idea that is now widely accepted. Kuiper argued that kilometer-sized comets would grow in the trans-Neptunian region on timescales of 10^9 yr and that Pluto was responsible for scattering these objects to the Oort Cloud. Subsequently, these comets would be sent back to the planetary region by stellar perturbations to appear as long-period comets (those having periods >200 yr), according to Oort's (1950) prescription. These ideas were further developed by other researchers, including Whipple (1964), who used anomalies in the motion of Neptune to infer the mass of trans-Neptunian objects (the anomalies are now known to be due to systematic astrometric errors). Later workers realized that the observed flux of short-period comets was too large to be derived from planetary perturbations of long-period comets falling from the Oort Cloud (Joss 1973), raising the possibility of separate sources for the long-period and short-period comets. The Kuiper Belt was explicitly advanced as the source of short-period comets by Fernandez (1980). Numerical work by Duncan et al. (1988) confirmed that the short-period comets could not be captured from the spherical Oort Cloud, providing indirect support for the existence of a beltlike cometary source.

Observationally, the search for objects in the outer solar system was impeded by their expected faintness. The flux of sunlight scattered from a solid body and received at Earth varies approximately as $p_R r^2 R^{-4}$, where p_R is the geometric albedo, r is the radius, and R is the heliocentric distance. Objects of a given size and albedo are 10^4 times fainter when beyond Neptune (R = 30 AU) than when in the main asteroid belt (R = 3 AU). This simple fact provided a daunting and, for many years, insurmountable obstacle to the direct observation of small bodies in the outer solar system. Kowal (1989), for example, used the Palomar Schmidt to conduct a 6000-square-degree photographic survey of the ecliptic to red magnitude 19.5 but found nothing beyond the planetary region. The turning point came with the application of large-format, high-quantum-efficiency charge-coupled device (CCD) detectors in the late 1980s. Although the first CCD survey was also negative (Luu and Jewitt 1988), the new technology spurred increasing observational effort (Levison and Duncan 1990; Cochran et al. 1991; Tyson et al. 1992), culminating with the detection of the first Kuiper Belt object (KBO), 1992 QB1, on August 30, 1992, using a 2048 $\times$ 2048 pixel CCD on the University of Hawaii 2.2-m telescope (Jewitt and Luu 1993). Intensive surveys since 1992 have increased the observational sample to about 160 objects at the time of writing (Jewitt and Luu 1995; Irwin et al. 1995; Williams et al. 1995; Jewitt et al. 1996, 1998; Luu and Jewitt 1998*a*, Gladman et al. 1998; Chiang and Brown 1999). Our knowledge of the Kuiper Belt is based primarily on these 160 objects, most particularly on the 54 whose orbital elements are known with confidence from astrometry at more than one opposition. An updated list of these orbits is maintained by the Minor Planet Center at the web site http://cfa-www.harvard.edu/cfa/ps/lists/TNOs.html; furthermore, updated information on Kuiper Belt research is maintained at http://www.ifa.hawaii.edu/faculty/jewitt/kb.html.

This chapter aims to present the observational status of the Kuiper Belt and is organized as follows. In section II we discuss the optical surveys; in section III we consider dust production in the Kuiper Belt;

and in section IV we remark on the dynamically related Centaur class. Physical properties are the subject of section V. We end with a list of topical questions in section VI.

II. RESULTS FROM THE OPTICAL SURVEYS

Figure 1 shows a plan view of the Kuiper Belt based on survey observations up to mid-1999. Evidence for dynamical substructure in the Kuiper Belt is provided in Fig. 2, which shows orbital semimajor axis a vs. eccentricity e for the known KBOs. The KBOs display a highly nonuniform distribution of orbital elements in a-e space, leading to the definition of three distinct dynamical groups within the Kuiper Belt.

A. Classical Objects

About $\frac{2}{3}$ of the well-observed KBOs have semimajor axes $a \geq 42$ AU and perihelia $q > 35$ AU. Together these characteristics define the "classical" Kuiper Belt. Numerical simulations suggest that the orbits are stable over periods comparable to the age of the planetary system (Holman and Wisdom 1993; Duncan et al. 1995; Morbidelli et al. 1995), essentially because the perihelia are well separated from Neptune. The classical KBOs possess modest eccentricities but can have inclinations as large as 32° (Table I). Both the inclination and eccentricity distributions are wider than expected

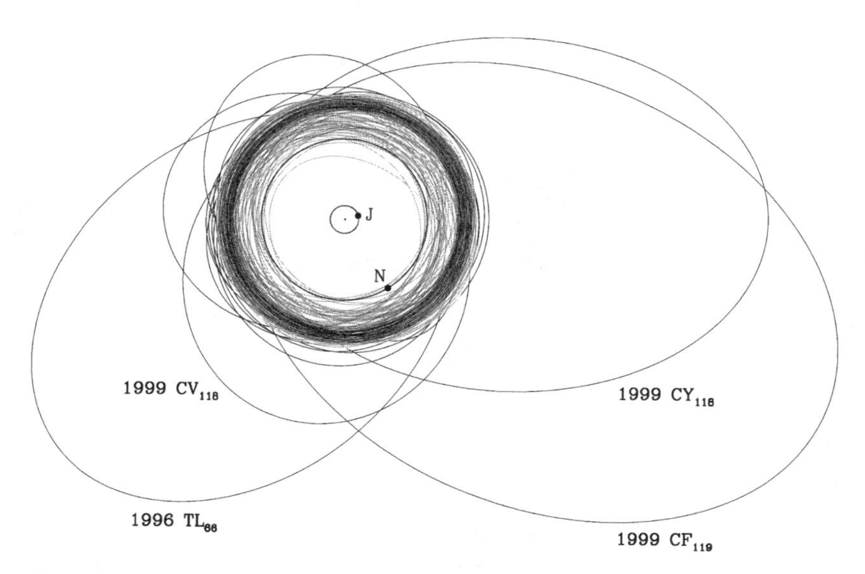

Figure 1. Plan view of the Kuiper Belt showing the orbits of the Kuiper Belt objects known as of June 1999. Orbits of Jupiter (J) and Neptune (N) are marked, as are the first four scattered KBOs. The majority of the KBOs follow orbits in the 30- to 50-AU radius band. The region shown is 260 AU × 340 AU. Figure prepared by Chad Trujillo, University of Hawaii.

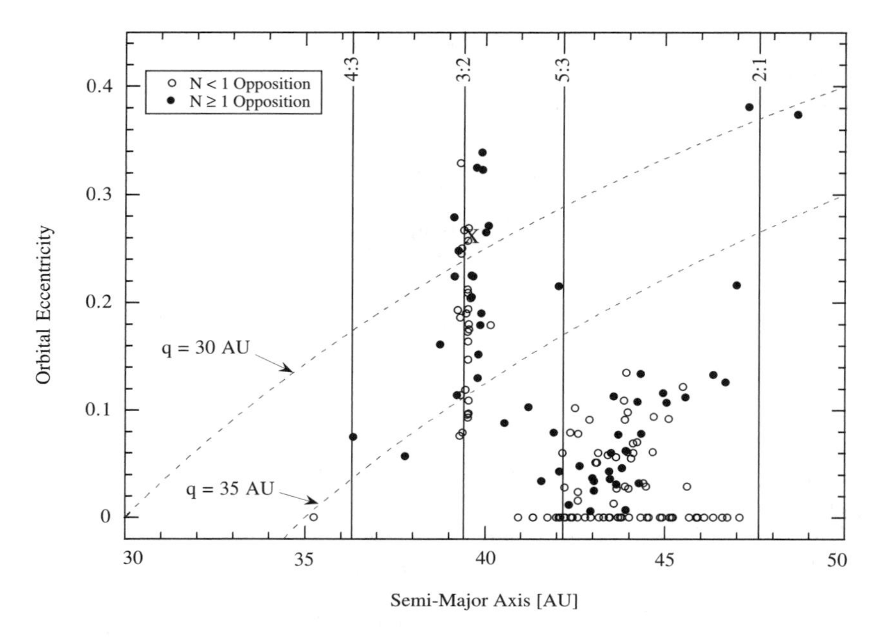

Figure 2. Semimajor axis eccentricity for Kuiper Belt objects. Filled circles denote multiopposition orbits. Open circles denote orbits computed from astrometry taken within a single opposition. Pluto is marked by X. Neptune is at $(a,e) = (30,0)$. The dashed lines mark $q = 30$ AU and $q = 35$ AU. Objects above the upper line are Neptune-crossers. Scattered Kuiper Belt object 1996 TL_{66} $(a,e = 85$ AU, 0.59) is off scale and not plotted. Vertical lines mark the locations of mean-motion resonances.

TABLE I
Mean Orbital Elements ($N_{opp} > 1$ orbit)

Group	N	Semimajor axis		Eccentricity		Inclination	
		Mean	Median	Mean	Median	Mean	Median
3:2 Resonance[a]	15	39.56 ± 0.13	39.66	0.220 ± 0.018	0.226	8.0 ± 1.5	5.7
Classical[b]	34	43.24 ± 0.33	43.49	0.086 ± 0.012	0.070	8.9 ± 1.6	4.4

[a] Includes Pluto.
[b] Omits all resonant plus scattered objects.

on the basis of long-term simulations that account only for planetary perturbations (e.g., Figs. 3 and 4 of Holman and Wisdom 1993 show typical eccentricities and inclinations of only 0.02 and 1 degree, respectively), implying excitation by another mechanism.

B. Resonant Objects (Plutinos)

Approximately $\frac{1}{3}$ of KBOs appear clustered near the 3:2 mean-motion resonance with Neptune at $a = 39.4$ AU (Fig. 2). These objects have large eccentricities ($0.1 \le e \le 0.34$) and inclinations up to ~20° (Table I). Many of them cross the orbit of Neptune (the critical eccentricity for crossing is $e = 0.24$) and would be subject to strong Neptune perturbations were they not dynamically protected by the resonance. In this regard, the 3:2 objects resemble Pluto, which is located within the resonance (plotted in Fig. 2 with an X) and which is also Neptune-crossing. Like Pluto, the "Plutinos" (a name chosen to symbolize this remarkable dynamical similarity) are aphelion librators, which reach perihelion when near ±90° from Neptune. Pluto itself occupies several other resonances, notably the "argument of perihelion libration," which maintains its perihelion at high ecliptic latitude and therefore further minimizes Neptune perturbations (e.g., Malhotra and Williams 1998). It is not yet known what fraction of the Plutinos might occupy these other resonances, because the sky-plane coverage of published surveys remains very small. The upper limit to the eccentricity of the Plutinos (the most eccentric is 1996 TP_{66}, with $e = 0.34$, $q = 26.4$ AU) is presumably set by destabilizing perturbations from Uranus. There is a deficiency of Plutinos with $e \le 0.1$.

The apparent fraction of Plutinos is overestimated as a result of observational bias (the Plutinos have smaller orbits than classical KBOs, so they appear brighter and are easier to detect at a given size). Bias-corrected statistics suggest that the "true Plutino fraction" is closer to 10–15% (Jewitt et al. 1998). Thus, the Plutinos appear to be a significant but distinct minority component of the Kuiper Belt. Other mean-motion resonances are also populated (Fig. 2), notably the 2:1 resonance at 47.6 AU (1996 TR_{66} and 1997 SZ10), the 4:3 at 36.4 AU (1995 DA_2), and the second-order 5:3 resonance at 42.3 AU (1994 JS).

The mechanism by which the resonances were populated has not yet been firmly established. The leading hypothesis invokes radial migration of the planets due to the exchange of angular momentum with planetesimals scattered from the protoplanetary disk. Simulations suggest that planetesimals scattered among the planets lead to an asymmetry that drives Saturn, Uranus, and Neptune away from the sun. Jupiter provides the ultimate source of the angular momentum and, accordingly, its orbit contracts (Fernandez and Ip 1984, Malhotra 1995, Hahn and Malhotra 1999). As Neptune migrated outward, its mean-motion resonances would have swept through the Kuiper Belt, collecting objects. Malhotra (1995) finds especially efficient trapping of KBOs in the 3:2 and 2:1 resonances. The trapping efficiency is high (of order unity) for initial orbits of low eccentricity and inclination and for slow and smooth planetary migration. The relative populations of the resonances depend partly, in this resonance sweeping model, on the rate of migration. This raises the intriguing possibility that measurements of the population ratios may allow an estimate of the rate at which Neptune's orbit expanded. One problem for the resonance sweeping hypothesis, at least in its simplest form, is that most known KBOs are not trapped in mean-motion resonances. For example, the classical KBOs lie in a region that should have been swept clean by the outwardly migrating

2:1 resonance. It is likely that effects such as the stochastic migration of Neptune in response to the ejection of massive planetesimals would decrease the trapping efficiency, possibly to levels consistent with the data.

C. Scattered Objects

Four KBOs in the known sample possess large, eccentric, inclined orbits with perihelia near 35 AU (Fig 3). The prototype "scattered Kuiper Belt object" (SKBO) is 1996 TL_{66} (Luu et al. 1997), with $a = 85$ AU, $e = 0.59$, $i = 24°$. These objects may represent a chaotic swarm of bodies scattered outward by Neptune in the early stages of the solar system (Torbett 1989; Ip and Fernandez 1991; Duncan and Levison 1997). With perihelia near 35 AU, Neptune is able to exert weak dynamical control. The number of scattered KBOs is highly uncertain. A population of order 6×10^3 (500-km diameter or larger) is suggested by the discovery of 1996 TL_{66} (Luu et al. 1997), but this estimate is good at best to order of magnitude. The SKBOs appear rare in flux-limited surveys, because their large, eccentric orbits render them invisible except when near perihelion. In absolute numbers the SKBOs may dominate the trans-Neptunian region.

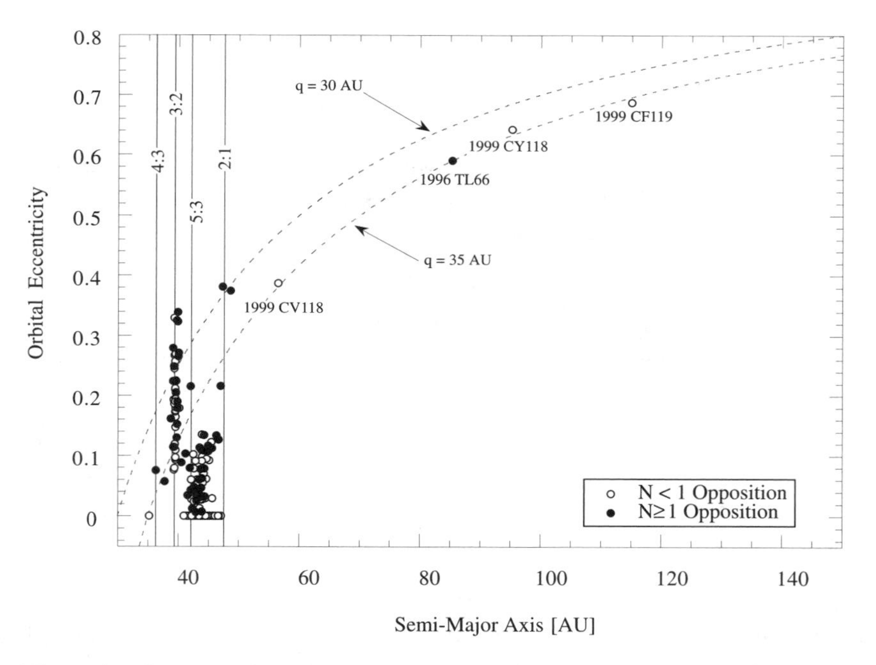

Figure 3. Same as Figure 2 but on a larger scale to show the scattered Kuiper Belt objects. All four scattered KBOs fall along the perihelion distance $q =$ 35 AU line.

D. Cumulative Luminosity Function

The cumulative luminosity function (CLF) of the Kuiper Belt is shown in Fig. 4. In the figure we have omitted measurements based on photographic plates, principally the survey by Kowal (1989) and the photographic portion of Luu and Jewitt (1988). Photographic surveys are difficult to calibrate and require long integrations that permit excessive trailing loss even on slow-moving objects. The *Hubble Space Telescope* (HST) measurement (Cochran et al. 1995) at red magnitude 28.1 is controversial and deserves special mention. The central issue is that the HST images are undersampled, so astronomical objects and cosmic rays cannot be readily distinguished by their morphology alone. Consequently, the KBOs must be sought against a large background due to cosmic rays. The measurement has been criticized by M. Brown et al. (1997) on the basis that noise in the data would prevent the detection of objects at the claimed limiting magnitude (but see Cochran et al. 1998). Furthermore, the authors searched the HST data only for Plutino-type orbits. Inclusion of other dynamical classes would, by analogy with the ground-based surveys, increase the derived surface density by a factor of order 3. For these reasons the usefulness of the HST measurement is presently not clear.

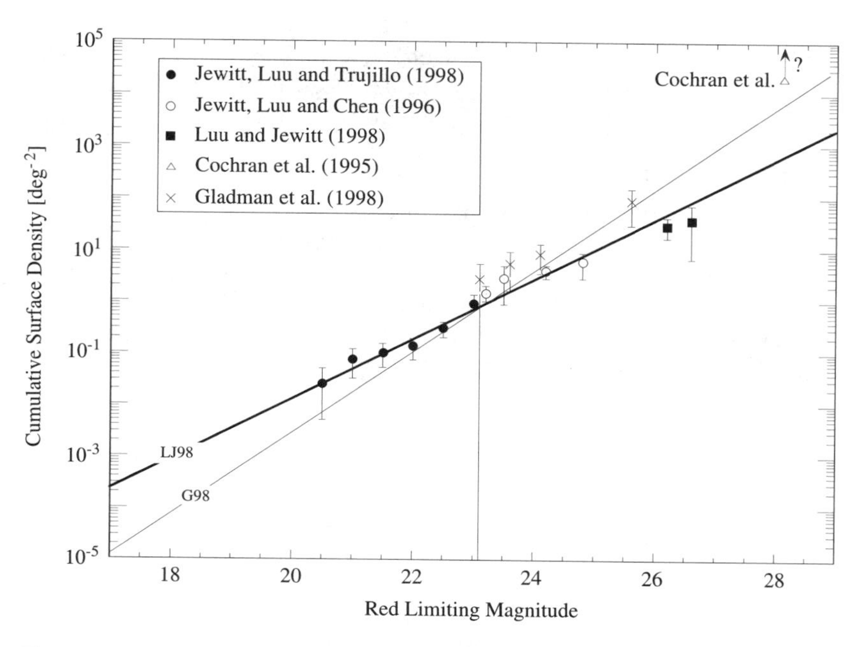

Figure 4. Luminosity function of the Kuiper Belt excluding photographic surveys. The lines show fits to the data from Luu and Jewitt (1998*a*) and Gladman et al. (1998), discussed in the text. Chiang and Brown (1999) report a CLF close to that of Luu and Jewitt.

The CLF is well described by a power law relation of the form

$$\log[\Sigma(m_R)] = \alpha(m_R - m_0) \tag{1}$$

where $\Sigma(m_R)$ is the cumulative surface density measured to limiting magnitude m_R, and α and m_0 are constants. Fits to the CLF give $\alpha = 0.58 \pm 0.05$, $m_0 = 23.27 \pm 0.11$ ($20 \leq m_R \leq 25$; Jewitt et al. 1998), $\alpha = 0.54 \pm 0.04$, $m_0 = 23.20 \pm 0.10$ ($20 \leq m_R \leq 26.6$; Luu and Jewitt 1998*a*), $\alpha = 0.76^{+0.10}_{-0.11}$, $m_0 = 23.40^{+0.20}_{-0.18}$ ($20 \leq m_R \leq 26$; Gladman et al. 1998), and $\alpha = 0.52 \pm 0.02$, $m_0 = 23.50 \pm 0.06$ ($20 \leq m_R \leq 27$; Chiang and Brown 1999). Gladman et al. (1998) observed that least-squares fits are not appropriate when the data are cumulative and the distribution of the uncertainties is Poisson, rather than Gaussian. The slope they derived using a maximum-likelihood method differs from those found by other investigators using least-squares, but Chiang and Brown (1999) showed that this difference is due to their selective use of only a subset of the available data, not to the fitting method. In any case, differences among the fitted parameters are at the 2σ level for α and m_0 and are therefore statistically insignificant. Furthermore, the assumption that the uncertainties are Poisson in distribution is probably not correct. We already know, for example, that the azimuthal distribution of Plutinos is nonuniform (Malhotra 1995, Jewitt et al. 1998), meaning that systematic sky-plane surface density variations are folded into the survey data but have not been taken into account. More survey observations, especially at $m_R \leq 20$ and $m_R \geq 26$, will soon resolve these ambiguities.

E. Inclination Distribution

The apparent full width at half maximum (FWHM) of the belt is $\sim 10°$ (Jewitt et al. 1996, cf. Fig. 5), the most extreme inclination being 39.1° for 1999 CZ_{118} (not plotted in Fig. 5 because it is still a single-opposition object). However, this must be considered as a lower limit to the width of the intrinsic inclination distribution, because of the effects of observational bias. Objects of high orbital inclination spend a larger fraction of each orbit far from the ecliptic than do objects of small inclination. Existing surveys, which have mainly been focused on the ecliptic, are therefore biased against finding objects of high inclination.

Models that attempt to correct for the inclination bias so far do not yield unique values of the intrinsic inclination distribution because of the limited sample size. The most secure result from these models is that the intrinsic inclination distribution must be very broad, perhaps 2 to 3 times the apparent FWHM (Jewitt et al. 1996), to give rise to the apparent distribution. An immediate implication of the broad inclination distribution is that the velocity dispersion among KBOs is $\Delta V \approx$ 1–1.5 km s^{-1}. Collisions occur with specific energies that are high enough to guarantee erosion of all but the largest objects (Stern 1995). Indeed, it has been suggested

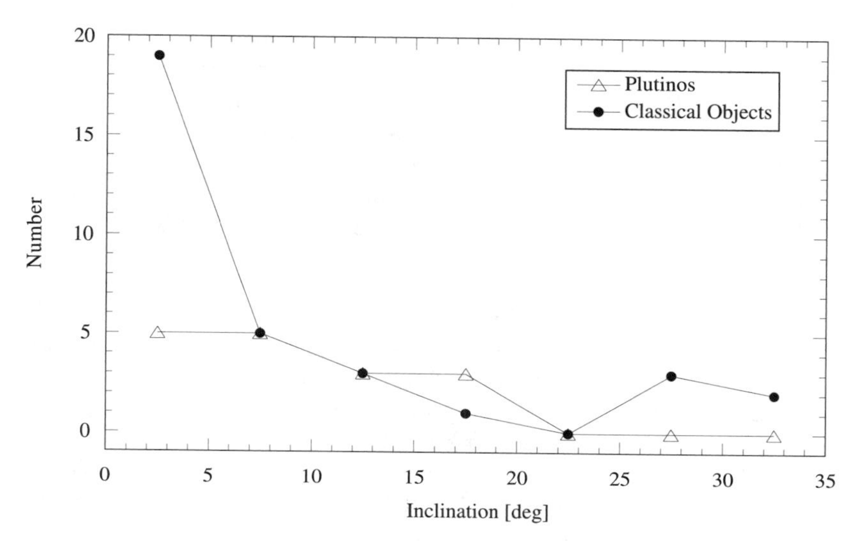

Figure 5. Histogram of inclinations of Kuiper Belt objects, plotted with a 5-degree bin size. The Plutinos and classical KBOs are shown separately. Only multiopposition orbits have been used in the histogram. The two populations have similar mean inclinations. However, the Plutinos are deficient in both high- ($i > 20°$) and low- ($i < 5°$) inclination orbits relative to the classical objects.

that the nuclei of short-period comets are collisionally produced fragments of large KBOs (Farinella and Davis 1996).

Numerical simulations show that the inclinations of resonant objects can be pumped to maximal values near 20° (Malhotra 1996). This is in good agreement with the maximum inclination among the known Plutinos, namely 19.5° (1995 QZ9). What is more surprising is that the classical KBOs have a mean inclination formally consistent with the Plutinos and that the most highly inclined KBOs are found outside of mean-motion resonances (Fig. 5). The origins of the high inclinations of classical KBOs are not understood. One possibility is that the classical Belt might have been stirred by a small number of bodies of Earth- and sub-Earth mass, scattered outward by Neptune and now lost from the planetary system (Morbidelli and Valsecchi 1997; Petit et al. 1998). Although this suggestion is interesting in its own right (and possibly even correct!), it has not been shown how the massive scatterers could excite the classical Belt without simultaneously depopulating the mean-motion resonances. Some KBOs could be scattered into resonances (the fractional volume of phase space occupied by resonances is $\approx 7\%$), albeit with low probability of long-term capture (Levison and Stern 1995).

F. Size Distribution and Statistics

The optical surveys are flux limited rather than volume limited, leading to a bias toward large (bright) objects near the inner edge of the belt and against small (faint) objects far away. As a result, the size distribution of KBOs cannot be measured directly but must be inferred from the CLF using a model of the spatial distribution. We consider the size distribution of KBOs as represented by a differential power law, $n(r)dr = \Gamma r^{-q} dr$, where r (km) is radius and Γ and q are constants. Assuming a fixed albedo 0.04 for all objects and an inverse square radial distance distribution, Monte Carlo simulations give best-fit values $\Gamma = 3.8 \times 10^{10}$ and $q \sim 4$ (Jewitt et al. 1998; Luu and Jewitt 1998*a*). In a $q = 4$ distribution, the cumulative number of objects with radius $>r$ is $N(r) = \Gamma/(3r^3)$. The objects discovered in ground-based surveys number $N(50 \text{ km}) \sim 10^5$. The distribution allows $N(1000 \text{ km}) \approx 10$; that is, ten Pluto-scale objects, all but one as yet undiscovered (cf. Stern 1991). At the other end of the size distribution, the effective radius of the well-studied short-period comets is in the range 1 km to 5 km. The fitted distribution allows $N(5 \text{ km}) = 1 \times 10^8$, while $N(1 \text{ km}) \sim 10^{10}$. Clearly, the Kuiper Belt is likely to hold large numbers of undetected small bodies. Models of the growth of KBOs (Kenyon and Luu 1999) give $q = 3.5$ ($r < 0.3$–3 km) and $q = 4$ ($r > 1$–3 km), in excellent agreement with inferences from the gradient of the CLF.

The total mass of a $q = 4$ distribution is

$$M = \frac{4 \times 10^9 \pi \rho \Gamma}{3} \left(\frac{0.04}{p_R}\right)^{3/2} \ln\left(\frac{r_{\text{max}}}{r_{\text{min}}}\right) \tag{2}$$

where ρ (kg m^{-3}) is the bulk density and p_R is the red geometric albedo. The bulk density is uncertain to within a factor of a few. Cometary nuclei may provide useful analogs of small KBOs; they have bulk density $\rho \approx$ 500 to 1000 kg m^{-3}. Conversely, Pluto and Charon are examples of large KBOs; they have $\rho \sim 2000$ kg m^{-3} (Foust et al. 1997). Taking $\rho = 1000$ kg m^{-3} as a plausible average effective density, and with $p_R = 0.04$, we find that the observable KBOs, with $r_{\text{min}} = 50$ km and $r_{\text{max}} = 1150$ km (the radius of Pluto), have mass $M \sim 5 \times 10^{23}$ kg (0.1 $M_\oplus$). A comparable mass is contained in bodies smaller than 50-km radius in this distribution; however, the size distribution of the smaller bodies is observationally unconstrained, and their combined mass could potentially be larger if the distribution is steeper than $q = 4$. Higher geometric albedos would decrease the mass estimate. With $p_R = 0.4$, for example, the mass given by equation (2) is reduced by a factor of 30. The optical survey data thus suggest a mass within the 30–50-AU zone that is a few tenths of an Earth mass, but which is quite uncertain. The optical mass is smaller than the dynamical limit ($\sim$1 $M_\oplus$) set by Hamid et al. (1968) and far smaller than the mass expected by Kuiper (1951) based on an extrapolation of the surface mass density of the solar system. Significantly, the modern Kuiper Belt contains too little mass for KBOs to have grown in the $\sim 10^8$ yr available prior to the disruptive emergence of Neptune (Stern and Colwell 1997; Kenyon and Luu 1998, 1999). If the observed KBOs formed *in situ*, then the original mass must have been 100 times the present mass. Scattering by Neptune (Holman and Wisdom 1993) appears incapable of clearing the Kuiper Belt except in the region close to the planet ($R < 42$ AU). Self-destruction by collisional grinding has been suggested (Stern 1996*b*) and partially modeled (Kenyon and Luu 1999). KBOs larger than ~100 km are largely immune to collisional disruption (Farinella and Davis 1996). The ground-based KBOs are thus likely to be true survivors from the earliest days of the solar system.

G. Radial Extent of the Belt

Depletion of the Kuiper Belt by Neptune perturbations is largely restricted to the $30 \leq R \leq 42$ AU range (Holman and Wisdom 1993; Duncan et al. 1995), suggesting that the more distant parts of the Belt might be more densely populated than the region so far observed. Using an R^{-2} extrapolation of the surface mass density of the solar nebula, Stern (1996*b*) has estimated that the density of the Kuiper Belt beyond 50 AU might be 100 times higher than in the $30 \leq R \leq 50$ AU annulus. On the other hand, it is not obvious that the Kuiper Belt necessarily extends to very large radii. An R^{-2} disk with a finite mass must be truncated at some definite outer radius, and several physical processes are known that could provide an edge. Empirically, two of six protoplanetary disks seen in silhouette against the Orion Nebula have diameters ≤ 100 AU (McCaughrean and O'Dell 1996) and are thus comparable in scale to the known Kuiper Belt. These disks are thought to have been tidally truncated during close stellar encounters in an early, dense stellar environment.

The semimajor axes of the orbits of known KBOs span the range 35 AU to about 115 AU, whereas the heliocentric distances at which KBOs have been discovered are restricted to the more limited range 26 AU to about 50 AU. Clearly, observational selection discriminates against the discovery of distant objects because they are faint. However, detailed Monte Carlo simulations (Fig. 6) show that, with the parameters of our own Mauna Kea CCD surveys, we should expect to have discovered substantial numbers of classical KBOs beyond 50 AU, whereas none has yet been found (Dones 1997; Jewitt et al. 1998). This result may be explained in several ways. First, the maximum object size (r_{max}) might be a decreasing function of semimajor axis. This would seem physically plausible as a consequence of lower surface density at larger distances, resulting in longer timescales for growth in the protoplanetary disk. However, the surface density in an R^{-2} disk at 45 AU (where we see many objects) is only 50% larger than at 55 AU (where we see none), and it is hard to imagine that r_{max} could be so sensitive to the local surface density. Second, the size distribution might be steeper than $q = 4$, so reducing the number of large objects visible beyond 50 AU (Gladman et al. 1998). Third, the Kuiper Belt surface density might decline faster than R^{-2} beyond the observed region. In fact, our data are consistent with a discrete outer edge to the

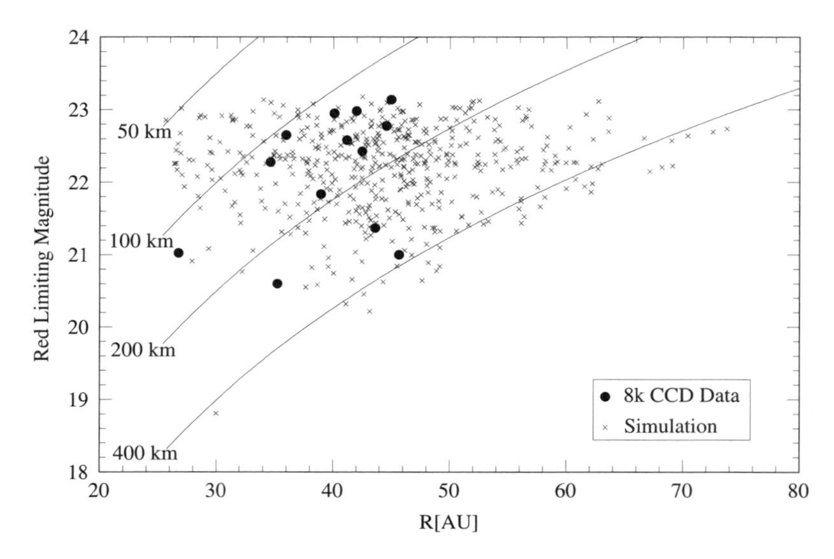

Figure 6. Heliocentric distance at discovery vs. apparent red magnitude. Solid circles mark objects found in the 8k survey (Jewitt et al. 1998, from which this figure is taken). Crosses denote simulated objects that passed the survey detection criteria. The classical Kuiper Belt is taken to extend to maximum semimajor axis $a_{\max} = 200$ AU. The KBOs obey a power law size distribution with index $q = 4.0$, and all have albedo $p_R = 0.04$. Plutinos have been included in proportions needed to yield an apparent Plutino fraction $P_a \approx 0.35$. The radial density index in the classical Kuiper Belt is taken to be $p = 2$ in both models; in the particular simulation shown here we have chosen $r_{\max} = 500$ km. Diagonal lines show the apparent magnitude as a function of heliocentric distance for KBO radii 50, 100, 200, and 400 km.

classical belt at $a \sim 50$ AU, although we remain uncomfortable with this interpretation. Only the scattered KBOs occupy a volume that is clearly more extended than the known classical belt.

III. DUST AND THE RELATION TO CIRCUMSTELLAR DISKS

Collisions among KBOs generate dust. Unfortunately, in a $q = 4$ distribution the collisional cross section is dominated by the smallest (optically invisible) KBOs, and the dust production rate is therefore difficult to assess with confidence. A crude upper limit, $\dot{M}_+ \sim 10^7$ kg s^{-1}, may be obtained by dividing the current mass of the Belt ($\sim$0.1 $M_\oplus$) by the 4.5-Gyr age of the solar system. A model-dependent lower limit is set by erosion of KBOs by interstellar dust grains, with potential production rates estimated at $\dot{M}_- \sim 10^2$ to 10^4 kg s^{-1} (Yamamoto and Mukai 1998).

Surprisingly, Kuiper Belt dust may already have been detected. The *Voyager 1* and *2* spacecraft recorded significant dust impact rates when beyond Neptune's orbit (Gurnett et al. 1997). The *Voyager* experimenters did not themselves cite the Kuiper Belt as the source of dust. However, in the absence of other identifiable sources in the trans-Neptunian region, this seems to us a plausible hypothesis.

The reported average number density of micron-sized grains is $N_1 = 2 \times 10^{-8}$ m^{-3} (Gurnett et al. 1997). The normal optical depth due to these grains is of order $\tau \sim N_1 Q_s \pi a^2 H$, where a is the grain radius, Q_s is the dimensionless scattering efficiency, and H is the vertical thickness of the Kuiper Belt. We take $Q_s \sim 1$, $a \sim 1$ μm, and $H \sim 10$ AU to find $\tau \sim 10^{-7}$. This is much smaller than the optical depths of dust disks around other nearby main-sequence stars. For example, the β Pictoris and α Lyrae dust disks have normal optical depths $\tau \sim 10^{-3}$ and $\tau \sim 10^{-5}$, respectively (Backman and Paresce 1993).

We take the average mass of a grain as $m_d \sim 2 \times 10^{-14}$ kg (cf. Gurnett et al. 1997), and the volume of the Kuiper Belt (represented as an annular slab with inner and outer radii 30 AU and 50 AU, respectively, and a thickness of 10 AU) as $V \sim 2 \times 10^{38}$ m^3. The total mass of dust in micron-sized particles is then $M_d \sim m_d N_1 V \sim 8 \times 10^{16}$ kg (10^{-8} $M_\oplus$). With a $\tau_c \sim 10^6$ yr lifetime against collisional destruction (Jewitt and Luu 1997), the implied production rate in micron-sized particles is of order $M_d/\tau_c \sim 3 \times 10^3$ kg s^{-1}. This constitutes a realistic estimate provided the dust size distribution extends to particles no larger than those measured by *Voyager*'s plasma wave analyzers, as would be the case for dust grains produced by the small interstellar dust impacts. It is intriguing to observe that the empirical dust production rate is within the range of estimates for interstellar erosion (Yamamoto and Mukai 1998).

On the other hand, if the dust is produced by collisional grinding among KBOs, the size distribution must extend to sizes much larger than those measured by *Voyager*, and the mass production rate should be corrected for the presence of larger but less abundant particles. The plasma wave analyzer is sensitive to only a narrow range of particle sizes; impacts below the threshold mass 1.2×10^{-14} kg (corresponding to particle radius 1.4 μm at unit density) do not excite measurable plasma waves (Gurnett et al. 1997), whereas substantially larger particles are too rare to be counted. We assume that the *Voyager* spacecraft detected particles in the radius range $1.4 \leq a \leq 10$ μm and adopt a differential power law size distribution with index -3.5 (Dohnanyi 1969) in order to estimate the mass in larger particles. The mass and mass production rates must then be augmented by a factor

$$f = \frac{\int_{1.4}^{a_{\max}} a^{-1/2} da}{\int_{1.4}^{10} a^{-1/2} da} \approx \frac{\sqrt{a}_{\max}}{2} \tag{3}$$

where $a_{\max} \gg 1.4\mu$m is the maximum particle size expressed in μm. From equation (3) we find $f \sim 16$ for $a_{\max} = 1$ mm, $f \sim 500$ for $a_{\max} = 1$ m, and $f \sim 1.6 \times 10^4$ for $a_{\max} = 1$ km. The debris production rates therefore approach 5×10^4 kg s^{-1} for millimeter-sized particles, 1.5×10^5 kg s^{-1} for meter-sized particles, and 5×10^7 kg s^{-1} for kilometer-sized particles. Clearly, these values involve huge extrapolations from the *Voyager* dust detections and are highly uncertain. Nevertheless, they seem broadly compatible with our crude upper limit to the dust production rate, and with at least the lower estimates (10^6 kg s^{-1}) of Stern (1996*b*).

Perhaps the principal unexplained feature of the *Voyager* data is that no impacts were detected beyond 51 AU for *Voyager 1* or beyond 33 AU for *Voyager 2*. If future detailed models of dust dynamics cannot explain this cutoff, then it is possible that the detected dust grains have another source. In this case, the above mass and mass production rates must be regarded as upper limits to the collisionally produced Kuiper Belt dust content.

The survival of a 42-cm-diameter fuel tank on the spacecraft *Pioneer 10* during its 10-year flight through the Kuiper Belt (at relative velocity 12 km s^{-1}) sets an independent but less tightly constraining limit to the dust content of the Kuiper Belt (Anderson et al. 1998).

Three other opportunities exist for the detection of Kuiper Belt dust.

First, one might search for diffuse thermal emission from Kuiper Belt dust. From an observational point of view, the detection of diffuse Kuiper Belt dust is complicated by the foreground Zodiacal Cloud as well as by the background due to galactic dust. Backman et al. (1995) used *Cosmic Background Explorer* (COBE) observations at 140 μm and 240 μm wavelength to set an upper limit to the mass of Kuiper Belt dust $M_d < 10^{-5}$ $M_\oplus$. This limit, which is itself quite model dependent, is compatible with the 10^{-8} $M_\oplus$ estimated above from the *Voyager* dust impacts.

Second, recent impacts in the Kuiper Belt should produce observable optical (Alcock and Hut 1993) and infrared (Stern 1996*a*) dust signatures. Consider a body of initial radius a_0 that is pulverized by impact into particles of mean radius a. The resulting dust cloud will contain $(a_0/a)^3$ particles of combined geometric cross section $C \sim a^2(a_0/a)^3$. The cloud will expand and brighten in reflected light until it reaches the critical radius $a_C = a\,(a_0/a)^{3/2}$, at which the optical depth is unity. This expansion takes a time $t = a_C/v$, where v is the mean speed of ejection of the dust. For example, with $a_0 = 1$ km, $a = 1$ μm, and $v \sim 1$ km s^{-1} we find $C = 10^{15}$ m^2, $a_C = 3 \times 10^7$ m (corresponding to angular radius 1″ at 40 AU distance) and $t \sim 1$ day. Such a debris cloud would reach peak apparent red magnitude ~7 on a timescale of 1 day and thereafter expand into a diffuse cloud of steadily decreasing surface brightness. After 100 days, the cloud would have angular diameter ~100″, a mean surface brightness of order 17 mag arcsec^{-2}, and would still be an easy observational target, with 50 to 100 times the surface brightness of the moonless night sky. The timescale for radiation pressure to deflect particles in the expanding cloud is $t_{rp} \sim v/(\beta g)$, where $\beta \sim 1$ for micron-sized grains and $g = 4 \times 10^{-6}$ m s^{-2} is the solar gravity at 40 AU. We find $t_{rp} \sim 10$ yr, meaning that no perceptible radiation pressure-swept tail would grow. The number of such spherical, expanding debris clouds visible at one time is highly uncertain. Estimates range from a few to many thousands (Alcock and Hut 1993; Stern 1996*a*). No convincing examples have yet been reported.

Third, the Poynting-Robertson and plasma drag forces may carry Kuiper Belt dust particles into the inner solar system, where direct samples might be found in existing interplanetary dust particle collections (Flynn 1996). The dust lifetime against collisional destruction by interstellar impacts ($\sim 10^6$ yr; Jewitt and Luu 1997) is short compared to the $\sim 10^7$-yr transport time (Liou et al. 1996, 1999) for particle sizes from $0.1 \leq a \leq 100$ μm. Therefore, most Kuiper Belt particles will be collisionally destroyed while en route to the Earth. On the other hand, the rate of production of Zodiacal dust is only $\sim 10^4$ kg s^{-1} (Leinert et al. 1983), so even small amounts of surviving Kuiper Belt debris might contribute measurably to the Zodiacal dust complex (cf. Table II).

Some nearby stars possess disks in which the dust lifetimes (to Poynting-Robertson drag and collisional shattering) are short compared to the stellar age (Smith and Terrile 1984; Jayawardhana et al. 1998; Koerner et al. 1998; Trilling and Brown 1998; Schneider et al. 1999). In these disks, dust might be generated by collisional grinding of larger parent bodies, just as in the Kuiper Belt. Images of HR 4796A show a 70-AU radius, ringlike morphology that is highly suggestive of our own Kuiper Belt. The famous disk of β Pictoris extends to $\sim$1000 AU. As mentioned above, the quantities of dust (estimated from the scattering optical depths) are three to four orders of magnitude larger than the dust content of the Kuiper Belt. It is tempting but conjectural to associate this difference with a collisionally more active phase in the lives of circumstellar debris rings. Future observations with sensitive ground- and space-based coronagraphs will presumably show that Kuiper Belts are common if not ubiquitous around other stars.

TABLE II
Dust Production Rates

Estimate	Production Rate (kg s^{-1})	Reference
Kuiper Belt:		
M_+	10^7	Text
M_-	10^2 to 10^4	Yamamoto and Mukai 1998
Voyager ($a_{\max} = 10$ μm)	3×10^3	Text
Voyager ($a_{\max} = 1$ mm)	5×10^4	Text
Voyager ($a_{\max} = 1$ km)	5×10^7	Text
Zodiacal Cloud	10^4	Leinert et al. 1983

IV. RELATION TO THE SHORT-PERIOD COMETS AND CENTAURS

Objects that are dislodged from the Kuiper Belt as a result of intrinsic dynamical instabilities (Duncan et al. 1995; Morbidelli 1997) or mutual gravitational scattering (Ip and Fernandez 1997) eventually fall under the gravitational control of the gas giant planets. The prime candidates for these ex-KBOs are the "Centaurs," whose orbits intersect those of the gas giant planets and are consequently highly chaotic and short-lived. Their eventual fate is (1) to be scattered close to the sun, where they sublimate and are recorded as short-period comets, (2) to be scattered out of the solar system, or (3) to impact a planet or the sun. A useful practical definition is that the Centaurs are objects having both perihelion distance and semimajor axis between the orbits of Jupiter (5 AU) and Neptune (30 AU). By this definition, there are 11 known Centaurs (Table III). Three (P/Schwassmann-Wachmann 1, P/Oterma, and 2060 Chiron) display comae at least part of the time, indicating their volatile-rich cometary nature (P/Schwassmann-Wachmann 1 is a prodigious source of carbon monoxide, indicating formation at temperature $<$50 K; Senay and Jewitt 1994). The remaining six have been stellar in appearance in all observations to date but presumably contain buried ice. The Centaurs are important because they provide the dynamical link between the KBOs and the short-period comets; on the more practical side, they are much closer (brighter) than the typical KBOs and therefore make much easier targets for physical study.

TABLE III
Properties of the Centaurs

Name	a(AU) (1)	e (2)	i(°) (3)	q(AU) (4)	Q(AU) (5)	Class (6)	r(km) (7)	p_V(%) (8)	Reference (9)
P/SW1	6.00	0.05	9	5.7	6.3	C	20?	?	a
P/Oterma	7.28	0.25	2	5.5	9.1	C	?	?	
1998 SG_{35}	8.37	0.30	15.7	5.85	10.88	A	?	?	
2060 Chiron	13.65	0.38	7	8.5	18.8	C	90 ± 7	$14^{+6}/_{-3}$	b,c,d
1997 CU_{26}	15.71	0.17	23	13.1	18.4	A	151 ± 15	4.5 ± 1.0	e
1994 TA	16.84	0.30	5	11.7	22.0	A	11	?	f
1995 GO	18.07	0.62	18	6.9	29.3	A	37	?	f
1998 QM_{107}	20.13	0.14	9.4	17.3	22.9	A	?	?	
5145 Pholus	20.22	0.57	25	8.7	31.8	A	95 ± 13	4.4 ± 1.3	g
7066 Nessus	24.59	0.52	16	11.8	37.4	A	39	?	f
1995 DW_2	24.92	0.24	4	18.9	31.0	A	35	?	f

Column list: (1) semimajor axis; (2) eccentricity; (3) inclination; (4) perihelion distance; (5) aphelion distance; (6) morphological class (C = comet, A = asteroid); (7) radius; (8) geometric albedo; and (9) references to the measurements.

References: a: Meech et al. 1993; b: Campins et al. 1994; c: Altenhoff and Stumpff 1995; d: Bus et al. 1996; e: Jewitt and Kalas 1998; f: Radius calculated from optical photometry alone assuming geometric albedo 0.04; g: Davies et al. 1993.

The populations of Kuiper Belt Objects, Centaurs, and short-period comets should be in the approximate ratio $n_{\mathrm{KBO}} : n_{\mathrm{C}} : n_{\mathrm{SPC}} \sim t_{\mathrm{KBO}} : t_{\mathrm{C}} : t_{\mathrm{SPC}}$. Here, $t_{\mathrm{KBO}} \sim 10^{10}$ yr, $t_{\mathrm{C}} \sim 10^6$–10^7 yr (Asher and Steel 1993; Hahn and Bailey 1990; Dones et al. 1997; Levison and Duncan 1997), and $t_{\mathrm{SPC}} \sim 10^5$ yr (Levison and Duncan 1994) are the respective dynamical lifetimes. With 10^5 KBOs larger than 100 km in diameter, we would expect to find $n_{\mathrm{KBO}}{:}n_{\mathrm{C}}{:}n_{\mathrm{SPC}} \sim 10^5{:}(10^2{-}10^1) : 1$. Currently, only three Centaurs with diameters >100 km are known (1997 CU_{26}, Chiron, and Pholus; Table III) compared with 10 to 100 expected. The majority of such objects should be located just interior to Neptune (Levison and Duncan 1997), and sky surveys sensitive to these objects are far from complete (the most telling evidence of incompleteness is that the largest Centaur, 1997 CU_{26}, is also one of the most recently discovered). There is presently no short-period comet with diameter near 100 km.

V. PHYSICAL PROPERTIES

A. Albedos

The apparent optical brightness of a body viewed in reflected sunlight is proportional to the product of the geometric albedo with the cross section and is given by

$$p_R r^2 \phi(\alpha) = 2.25 \times 10^{22} R^2 \Delta^2 10^{0.4(m_\odot - m_R)} \tag{4}$$

Here, p_R is the geometric albedo, α is the phase (Earth-object-Sun) angle, the heliocentric and geocentric distances R and Δ are expressed in AU, and $m_\odot$ and m_R are the apparent magnitudes of the sun and Kuiper Belt object, respectively. In equation (4) $\phi(\alpha)$ is a phase function that describes the angular dependence of the scattered light, reasonably approximated by $\phi(\alpha) = 10^{-0.04\alpha}$. At the small phase angles attained by KBOs ($\alpha < 2°$), $\phi(\alpha) = 1$ provides a good first approximation.

The thermal flux density is given by

$$S_\nu = \frac{\epsilon_\nu B_\nu(T) \pi r^2}{\Delta^2} \tag{5}$$

where ϵ_ν is the emissivity, and $B_\nu(T)$ (W m^{-2} sr^{-1} Hz^{-1}) is the Planck function. The effective surface temperature T(K) is a complicated function of the emissivity, Bond albedo, and thermal diffusivity of the body, as well as of the heliocentric distance and rotation vector (Spencer et al. 1989). Equations (4) and (5) contain too many unknowns to allow a unique solution for radius or albedo. Nevertheless, measurements of main-belt asteroids have shown that useful estimates for radius and albedo may be obtained from equations (4) and (5) using judiciously assumed values of the unknown parameters.

In the case of the KBOs, thermal emission has yet to be detected, and the albedos remain unknown. At $T \sim 50$ K, the Planck maximum falls near 60 μm, a wavelength inaccessible from the ground. Marginal ($\leq 3\sigma$) detections of 1993 SC and 1996 TL_{66} obtained using the *Infrared Space Observatory* (ISO) satellite imply low (few percent) albedos (N. Thomas, private communication). Future orbiting infrared satellites (e.g., the *Space Infrared Telescope Facility*) may be able to detect thermal radiation from KBOs with greater significance. Rayleigh-Jeans emission (800-μm wavelength) may also be detected using submillimeter detectors on ground based interferometers (such as the Submillimeter Array, nearing completion on Mauna Kea).

In principle, occultations of background stars might be used to determine the sizes of KBOs (cf. Bailey 1976). Optical observations would then give the albedos directly, through equation (4). Astrometry of KBOs is presently good at the $\pm 0.3''$ level, while the largest angular diameters subtended by KBOs are of order $0.02''$ (for 1996 TO_{66}). Consequently, meaningful predictions of occultations by KBOs cannot yet be made, and to date no occultation observation has been attempted. The Taiwanese-American Occultation Survey (TAOS) experiment should detect occultations by previously unseen KBOs within the next five years. However, most of these objects will be so small that optical detections will be impossible, and the albedos will remain unmeasured.

All KBO diameters reported in the literature have been derived under the assumption of a red geometric albedo $p_R = 0.04$. This is not unreasonable, given that both short-period comets and Centaurs probably originated in the Kuiper Belt, and these objects have red geometric albedos mostly in the range $0.02 \leq p_R \leq 0.05$. The existence of a wide diversity of colors (Luu and Jewitt 1996; Jewitt and Luu 1998; Tegler and Romanishin 1998) suggests that different KBOs very possibly could have different albedos. On the other hand, the extreme color differences between Centaurs 2060 Chiron, 5145 Pholus, and 1997 CU_{26} are not matched by a wide range in the measured albedos ($0.14 +0.06/-0.03$, 0.044 ± 0.013, and 0.045 ± 0.010, respectively; Campins et al. 1994; Jewitt and Kalas 1998).

B. Colors

Most KBOs are too faint for spectroscopy, and observers have resorted to broadband colors as low-resolution (but higher in signal-to-noise ratio) substitutes. Optical colors for $\sim$20 KBOs are now available (Luu and Jewitt 1996; Green et al. 1997; Tegler and Romanishin 1998), as well as near-infrared colors for a handful of KBOs (Jewitt and Luu 1998).

Surprisingly, the KBOs (Luu and Jewitt 1996; Green et al. 1997; Jewitt and Luu 1998) and Centaurs (Romanishin et al. 1997; Davies et al. 1998) exhibit a wide spread of colors, ranging from nearly neutral to very red (Fig. 7a). Tegler and Romanishin (1998) confirm this diversity of KBO colors but further report that the $B - V$ and $V - R$ colors of KBOs and Centaurs are bimodally distributed (Fig. 7b), suggesting the existence of two distinct surface types. The optical-infrared $V - J$ index varies widely among KBOs but is not obviously bimodal (Jewitt and Luu 1998). A statistically significant (3σ) but unexpected and unexplained correlation between the absolute red magnitude and $V - J$ has also been reported (Jewitt and Luu 1998).

Figure 7 shows that the KBOs and Centaurs are indistinguishable in broadband color-color plots, consistent with a common origin. However,

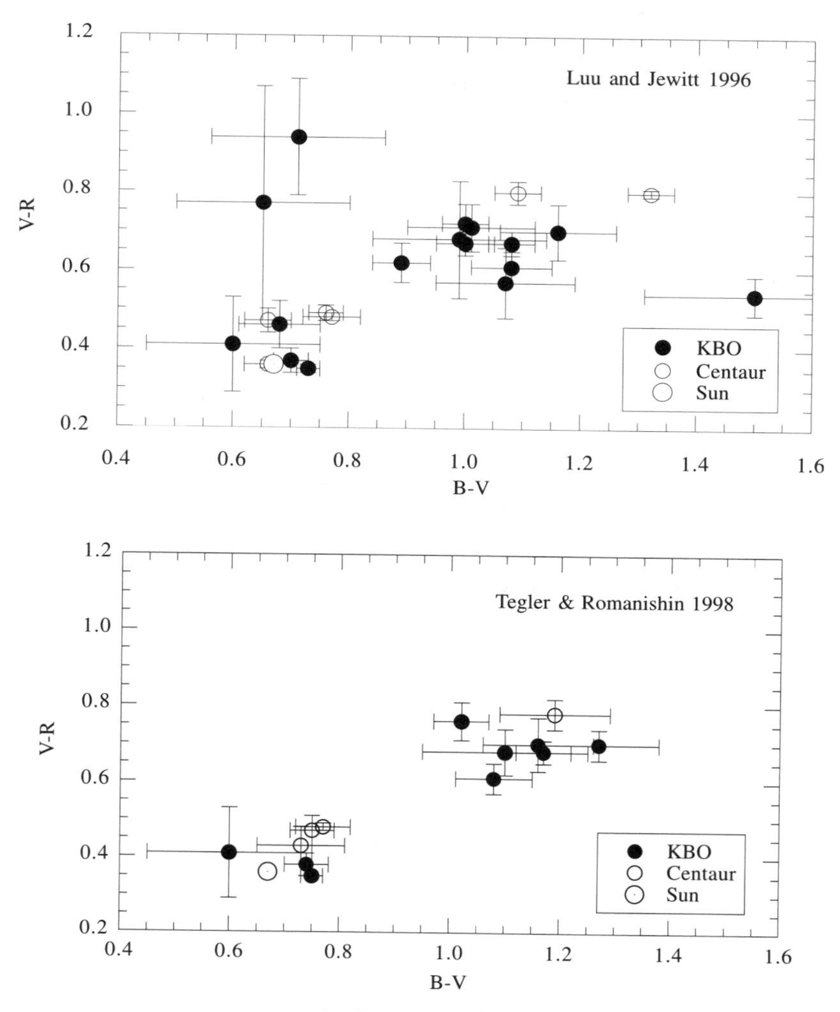

Figure 7. (a) $B - V$ color plotted against $V - R$ color for KBOs (filled circles) and Centaurs (empty circles). The color of the sun is marked. Error bars are 1σ. Data from Luu and Jewitt (1996). (b) Same as (a) but showing data from Tegler and Romanishin (1998). Note that many objects are common to both panels.

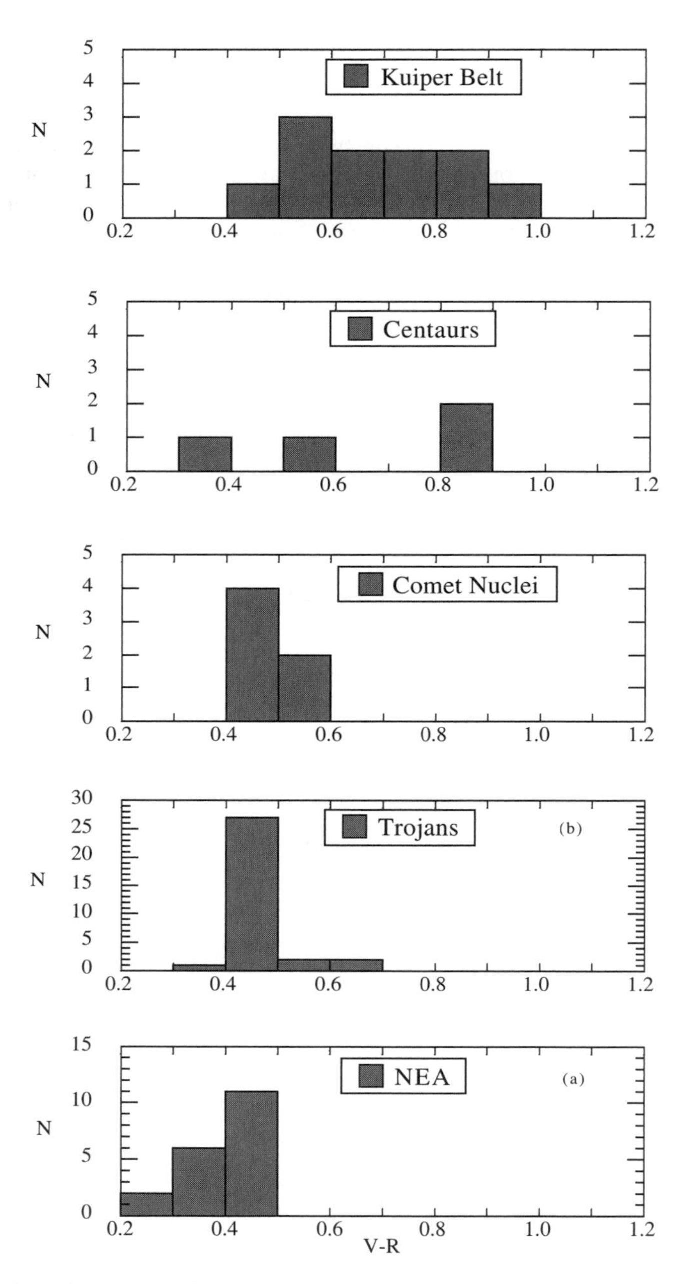

Figure 8. Histograms of the $V - R$ color index computed for five different dynamical classes of object. KBOs and Centaurs share a common, large color range. The nuclei of comets and the jovian Trojans display a smaller color dispersion and seem to lack the extremely red material present on some KBOs and Centaurs. The near-earth (NEA) objects, included here only for reference, are comparatively blue and quite distinct from the other classes. Figure modified from Luu and Jewitt (1996).

their color distributions are unmatched by those of comet nuclei and asteroids (Fig. 8), thanks to some extremely red KBO and Centaur surface materials ($V - R > 0.6$) that seem to be rare or absent among comets and asteroids. The absence of extremely red material on the cometary nuclei is consistent with progressive destruction, or burial, of irradiation mantle formed in the Kuiper Belt (cf. Cruikshank et al. 1998), but it might also be an observational artifact, given that few nuclei have yet been measured.

Laboratory experiments suggest that KBOs might be covered by "irradiation mantles," produced by high-energy particle irradiation of surface ices. When carbon-containing ices (e.g., methane or CO) are irradiated by high-energy particles (cosmic rays and UV photons), this irradiation leads to the selective loss of hydrogen while encouraging the formation of complex carbon compounds (e.g., Moroz et al. 1998 and references therein). The result is a solid residue (having a column density of $\sim 10^3$ kg m^{-2}, corresponding to a thickness ~ 1 m in solid ice) that is dark because of its complex carbon compounds. Exactly how the color and albedo change with time depends on the initial composition and the irradiation fluence (Moroz et al. 1998). The composition of the mantle is quite different from the interior, which retains pristine ice. The mantle thickness and strength are uncertain but may be sufficient to survive the object's first entry into the inner solar system.

C. Spectra

Wilson et al. (1994) were able to fit the optical spectrum of Centaur 5145 Pholus with "Titan tholins," chemically complex hydrocarbon mixtures produced by UV irradiation of simple ices. Cruikshank et al. (1998) obtained a fit for both the optical and near-infrared spectra of Pholus with a model consisting of carbon black (61.5%) and an intimate mixture (38.5%) of olivine, Titan tholin, H_2O ice, and CH_3OH ice (Fig. 9). Water ice has been reported on Centaur 1997 CU_{26} (Brown et al. 1998; Brown and Koresco 1998), although its surface is very dark ($p_R = 0.045 \pm 0.010$; Jewitt and Kalas 1998). On the other hand, 2060 Chiron is spectrally bland (Luu et al. 1994). Spectral fitting models are nonunique, but there is consensus that organics of some type are present on the surface of Pholus, and the identification of water on CU_{26} also appears secure. Color diversity among the KBOs is thus matched by spectral and compositional diversity among the Centaurs.

Spectra of only three KBOs have been reported. The smoothed reflectance spectrum of 1993 SC (R. Brown et al. 1997) shows prominent absorptions that have been qualitatively interpreted as hydrocarbon features (Fig. 10). In contrast, the spectrum of 1996 TL_{66} is neutral, with no apparent absorption features (Luu and Jewitt 1998*b*). The spectrum of 1996 TO66 shows absorptions at 1.5 and 2.0 μm, indicative of water ice (R. Brown et al. 1999). The differences among the KBOs are intriguing,

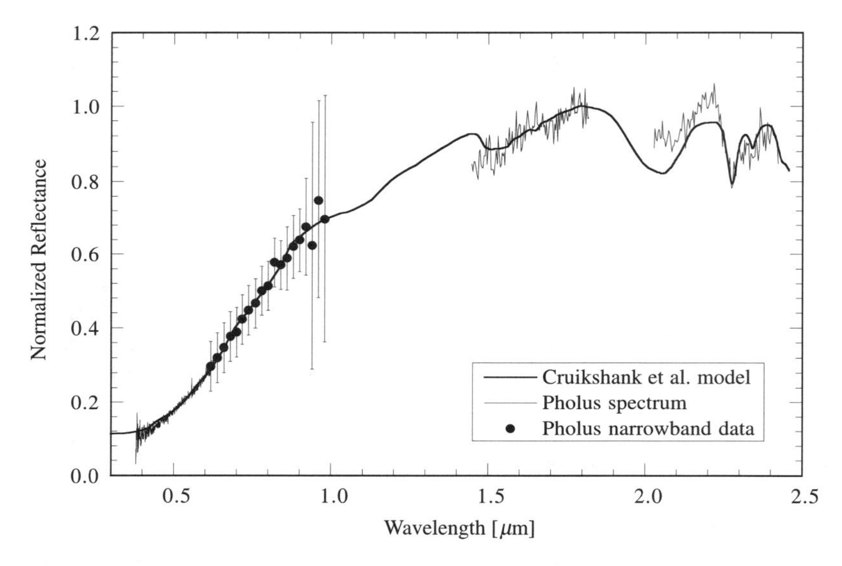

Figure 9. Composite optical and near-infrared spectrum of Centaur 5145 Pholus. The model is an intimate mixture of olivine (to provide absorption near 1.5 μm), Titan tholin (the red optical slope), water ice (the 2 μm absorption), methanol ice (the 2.27 μm absorption), and carbon black (Cruikshank et al. 1998).

but it would be premature to attempt an interpretation. Spectra of higher quality are urgently needed in order to make progress in this field.

D. Origin of Color Diversity

The diversity of the colors is a surprise in the sense that all KBOs should be exposed to cosmic rays and thus should have irradiation mantles with similar colors. Three simple explanations have been suggested for the origin of the color diversity (Luu and Jewitt 1996). First, it is possible that the KBOs possess intrinsically different compositions and that the different colors are tracers of compositional variations. For example, asteroids in the Mars-Jupiter belt show different compositions that seem to be related to their sites and temperatures of formation. However, the KBOs probably formed *in situ*, in the presence of a very slight radial temperature gradient ($T \propto R^{-1/2}$). Temperature differences across the 35- to 50-AU zone likely amounted to only ~10 K. It is hard to see how strong compositional differences might arise from such small differences in the formation temperature. A second explanation is that occasional collisions in the Kuiper Belt puncture the irradiation mantles on some objects, excavating craters and showering nearby regions with impact debris. The freshly excavated material would be unirradiated and thus should be of a different composition (more ice-rich, less red?) than the irradiated mantle. This mechanism requires that the timescales for resurfacing and for radiation damage of the surface layers are comparable, to avoid saturation of the color at one or the

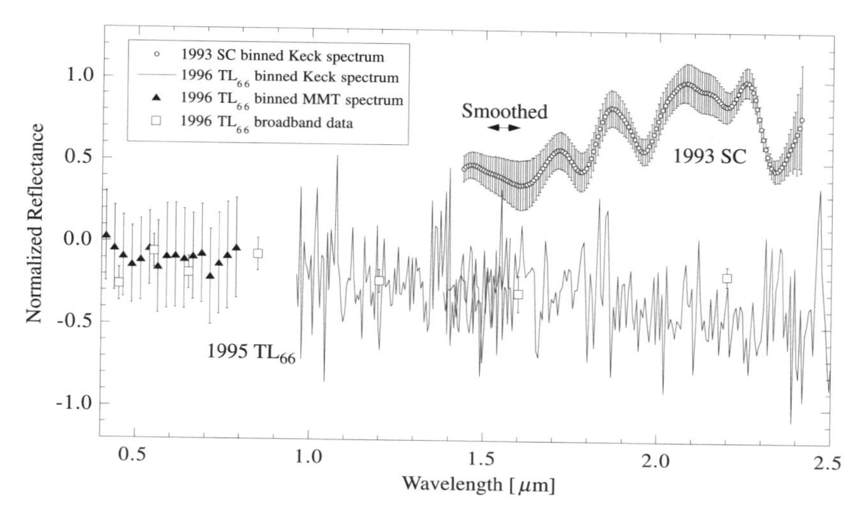

Figure 10. Optical-infrared reflectance spectra of 1993 SC and 1996 TL_{66}. The 1993 SC spectrum has been smoothed to a resolution of 0.08 μm (by R. Brown et al. 1997). The 1996 TL_{66} spectrum is shown unsmoothed (Luu and Jewitt 1998*b*) and is featureless, ruling out a surface of pure water ice. The spectra have been vertically displaced for clarity.

other extreme. One testable prediction of the resurfacing hypothesis is that KBOs should show rotational modulation of albedo and color, corresponding to local blanketing by excavated debris. Another is that albedo and color should be closely related. The bimodal color distribution of KBOs (Tegler and Romanishin 1998), if independently confirmed, would seem to rule out impact resurfacing as an explanation for the color diversity. Third, Moroz et al. (1998) have observed that the color of light reflected from laboratory bitumen samples is a function of the effective particle size in the scattering surface.

E. Structure

The observed few-hundred-km diameter KBOs have thermal diffusion times [$\tau \sim r^2/\kappa$, where r (m) is radius and κ (m^2 s^{-1}) is thermal diffusivity] comparable to or longer than the age of the solar system. Energy liberated in these bodies by the decay of radioactive elements is trapped and must contribute to an increase in the central temperature of magnitude

$$\Delta T \sim Hr^2/(\kappa c_p) \qquad (6)$$

where H [W kg^{-1}] is the power production per unit mass from radioactivity and c_p (J kg^{-1} K^{-1}) is the specific heat capacity of the bulk material. With $H \sim 10^{-12}$ W kg^{-1} (the heating rate measured for carbonaceous chondrite material, diluted by a factor of 4 to account for the presence of substantial ice), $\kappa \sim$(1 to 10)$\times 10^{-7}$m^{2}s^{-1}, and $c_p \sim 10^3$J kg^{-1} K^{-1}, we estimate ΔT(K)$\sim$(10 to 100)$(r/100\text{km})^2$. Objects of a few $\times$ 100 km radius may experience internal heating sufficient to mobilize interior volatiles, giving rise to a compositionally layered structure in the KBOs. This, in turn, would complicate our understanding of the compositional significance of short-period comets, if these are indeed collisional fragments from the upper layers of 100-km-scale KBO parents. One might even expect that the larger KBOs should be predisposed to outgas their supervolatiles (CO, N_2), giving rise to surface frosts and systematic variations in albedo and surface color with radius. The limiting case, Pluto, has long been known to sport a high-albedo, patchily frosted surface that is probably not typical of the smaller KBOs. Outgassing due to mobilization of internal volatiles might be responsible for some of the color and spectral variation observed on Pluto and amongst the other KBOs.

VI. MAJOR QUESTIONS

We end with ten leading questions about the Kuiper Belt.

1. What is the size distribution of KBOs at the smallest (<50-km) and largest (>500-km) scales? Are the size distributions of the classical, resonant, and scattered object subclasses equal? Is there observational evidence for a cutoff in the size distribution at large radii? Is Pluto unique?
2. What are the albedos of KBOs? Is there an albedo-color relation as expected from collisional resurfacing?
3. What is the origin of the spectral diversity exhibited by the KBOs? Are the optical colors bimodal? Are the colors compositionally diagnostic or indicative of other effects (e.g., particle size variations)?
4. Can compelling evidence be found for compositional differences between the Kuiper Belt (short-period) and Oort Cloud (long-period) comets? Different temperatures in the source zones of the long-period comets (5–30 AU) and the short-period comets (30–40 AU) should, for example, lead to major variations in the abundances of supervolatiles (e.g., N_2, CO) between the two groups, yet no such differences have been reported.
5. Are small (1–10 km) KBOs present in numbers sufficient to supply short-period comets to the inner solar system for 4.5 Gyr? If the comets originate in the Kuiper Belt, precisely where (in chaotic zones associated with resonances, among the scattered objects, or elsewhere)?
6. Was the ancient Kuiper Belt much more massive than the present-day Kuiper Belt, and if so, how was the original mass depleted? What, if anything, can the cratering rates on the outer planet satellites tell us about the time dependence of the Kuiper Belt mass? Was the early Kuiper Belt massive enough to produce dust collisionally

in quantities sufficient to rival the dust rings circling some nearby, main-sequence stars (e.g., β Pic)?

7. How common is binarity in the Kuiper Belt? Presently, only Pluto is known to hold a satellite. If Charon's presence indicates a formerly more active collisional regime, then it is reasonable to suppose that other KBO binaries await discovery.
8. What are the relative populations in the mean-motion resonances, and what do these tell us about the population mechanism (cf. Malhotra 1995, 1996)?
9. What processes were responsible for the excitation of the velocity dispersion in the classical Kuiper Belt?
10. What role has been played by collisions in shaping the present-day Kuiper Belt? Are the nuclei of short-period comets the collision fragments of larger bodies (cf. Farinella and Davis 1996)?

Acknowledgments This work was supported by grants from NASA's Origins Program.

REFERENCES

Alcock, C. and Hut, P. 1993. Preprint.

Altenhoff, W. J., and Stumpff, P. 1995. Size estimate of "asteroid" 2060 Chiron from 250GHz measurements. *Astron. Astrophys.* 293:L41–42.

Anderson, J., Lau, E., Scherer, K., Rosenbaum, D., and Teplitz, V. 1998. Kuiper belt constraint from Pioneer 10. *Icarus* 131:167–170.

Asher, D. J., and Steel, D. I. (1993). Orbital evolution of the large outer solar system object 5145 Pholus. *Mon. Not. Roy. Astron. Soc.* 263:179–190.

Backman, D. E., and Paresce, F. 1993. Main-sequence stars with circumstellar solid material: The Vega phenomenon. *Protostars and Planets III*, ed. E. H. Levy and J. I. Lunine (Tucson: University of Arizona Press), pp. 1253–1304.

Backman, D. E., Dasgupta, A., and Stencel, R. E. 1995. Model of a Kuiper Belt small grain population and resulting far-IR emission. *Astrophys. J. Lett.* 450:L35–L38.

Bailey, M. 1976. Can 'invisible' bodies be observed in the solar system? *Nature* 259:290–291.

Brown, M. E., and Koresko, C. D. 1998. Detection of water ice on the Centaur 1997 CU_{26}. *Astrophys. J. Lett.* 505:65–67.

Brown, M., Kulkarni, S. and Liggett, T. 1997. An analysis of the statistics of the Hubble Space Telescope Kuiper Belt object search. *Astrophys. J. Lett.* 490:L119.

Brown, R. H., Cruikshank, D. P., Pendleton, Y., and Veeder, G. J. 1997. Surface composition of Kuiper Belt object 1993SC. *Nature* 276:937–939.

Brown, R. H., Cruikshank, D. P., and Pendleton, Y. 1999. Water ice on Kuiper Belt Object 1996 TO_{66}. *Astrophys J. Lett.* 519:L101–L104.

Brown, R. H., Cruikshank, D. P., Pendleton, Y., and Veeder, G. J. 1998. Identification of water ice on the Centaur 1997 CU_{26}. *Science* 280:1430–1432.

Bus, S. J., Buie, M. W., Schleicher, D. G., Hubbard, W. B., Marcialis, R. L., Hill, R., Wasserman, L. H., Spencer, J. R., Millis, R. L., Franz, O. G., Bosh, A. S., Dunham, E. W., Ford, C. H., Young, J. W., Elliott, J. L., Meserole, R., Olkin, C. B., Mcdonald, S. W., Foust, J. A., Sopata, L. M., and Bandyopadhyay, R. M. 1996. Stellar Occultation by 2060 Chiron. *Icarus* 123:478–490.

Campins, H., Telesco, C., Osip, D., Rieke, G., Rieke, M., and Schulz, B. 1994. The color temperature of (2060) Chiron: A warm and small nucleus. *Astron. J.* 108:2318–2322.

Chiang, E., and Brown, M. 1999. Keck pencil-beam survey for faint Kuiper Belt Objects. *Astron. J.* 118:1411–1422.

Cochran, A., Levison, H., Stern, S., and Duncan, M. 1995. The discovery of Halley-sized Kuiper Belt objects. *Astrophys. J. Lett.* 455:L342.

Cochran, A. L., Levison, H. F., Tamblyn, P., Stern, S. A., and Duncan, M. J. 1998. Calibration of the HST Kuiper Belt object search: Setting the record straight. *Astrophys. J.* 503:L89.

Cochran, J. A., Cochran, J. W., and Torbett, J. M. 1991. A deep imaging search for the Kuiper disk of comets. *Bull. Amer. Astron. Soc.* 23:131 (abstract).

Cruikshank, D. P., Roush, T. L., Bartholomew, M. J., Moroz, L. V., Geballe, T. R., Pendleton, Y. J., White, S. M., Bell, J. F., III, Davies, J. K., Owen, T. C., de Bergh, C., Tholen, D. J., Bernstein, M. P., Brown, R. H., Tryka, K. A., and Dalle Ore, C. M. 1998. The composition of Centaur 5145 Pholus. *Icarus* 135:389–407.

Davies, J., Spencer, J., Sykes, M., Tholen, D., and Green, S. 1993. IAU Circular 5698 (January 23).

Davies, J. K., McBride, N., Ellison, S., Green, S. F., and Ballantyne, D. R. 1998. Visible and infrared photometry of six Centaurs. *Icarus* 134:213–227.

Dohnanyi, J. 1969. Collisional models of asteroids and their debris. *J. Geophys. Res.* 74:2531–2554.

Dones, L. 1997. Origin and evolution of the Kuiper Belt. In *From Stardust to Planetesimals*, ASP Conf. Ser. 122, ed. Y. J. Pendleton and A. Tielens (San Francisco: Astronomical Society of the Pacific), pp. 347–365.

Duncan, M., and Levison, H. 1997. A disk of scattered icy objects and the origin of Jupiter-family comets. *Science* 276:1670.

Duncan, M., Quinn, T., and Tremaine, S. 1988. The origin of the short period comets. *Astrophys. J.* 328:L69–L73.

Duncan, M., Levison, H. F., and Budd, S. M. 1995. The dynamical structure of the Kuiper Belt. *Astron. J.* 110:3073–3081.

Edgeworth, K. E. 1943. The evolution of our planetary system. *J. Brit. Astron. Soc.* 53:181–88.

Edgeworth, K. E. 1949. The origin and evolution of the solar system. *Mon. Not. Roy. Astron. Soc.* 109:600–609.

Farinella, P., and Davis, D. 1996. Short-period comets: Primordial bodies or collisional fragments? *Science* 273:938.

Fernandez, J. 1980. On the existence of a comet belt beyond Neptune. *Mon. Not. Roy. Astron. Soc.* 192:481–491.

Fernandez, J., and Ip, W. 1984. Some dynamical aspects of the accretion of Uranus and Neptune. *Icarus* 58:109–120.

Flynn, G. J. 1996. Collisions in the Kuiper Belt and the production of interplanetary dust particles. *Meteorit. Plan. Sci.* 31:A45.

Foust, J., Elliot, J., Olkin, C., McDonald, S., Dunham, E., Stone, R., McDonald, J., and Stone, R. 1997. Determination of the Charon/Pluto mass ratio from center of light astrometry. *Icarus* 126:362–372.

Gladman, B., Kavelaars, J., Nicholson, P., Loredo, T., and Burns, J. 1998. Pencil-beam surveys for faint trans-Neptunian objects. *Astron. J.* 116:2042–2054.

Green, S. F., McBride, N., O'Ceallaigh, D. P., Fitzsimmons, A., Williams, I. P., and Irwin, M. J. 1997. Surface reflectance properties of distant solar system bodies. *Mon. Not. Roy. Astron. Soc.* 290:186–192.

Gurnett, D. A., Ansher, J. A., Kurth, W. S., and Granroth, L. J. 1997. Micron-sized dust particles detected in the outer solar system by the Voyager 1 and 2 plasma wave instruments. *Geophys. Res. Lett.* 24:3125–3128.

Hahn, G., and Bailey, M. E. 1990. Rapid dynamical evolution of giant comet Chiron. *Nature* 348:132–136.

Hahn, J. M., and Malhotra, R. 1999. Orbital evolution of planets embedded in a planetesimal disk. *Astron. J.* 117:3041–3053.

Hamid, S. E., Marsden, B., and Whipple, F. 1968. Influence of a comet belt beyond Neptune on the motions of periodic comets. *Astron. J.* 73:727–729.

Holman, M., and Wisdom, J. 1993. Dynamical stability in the outer solar system and the delivery of short-period comets. *Astron. J.* 105:1987–1999.

Ip, W.-H., and Fernandez, J. A. 1991. Steady-state injection of short-period comets from the trans-Neptunian cometary belt. *Icarus* 92:185–193.

Ip, W.-H., and Fernandez, J. 1997. On dynamical scattering of Kuiper Belt objects in 2:3 resonance with Neptune into short-period comets. *Astron. Astrophys.* 324:778–84.

Irwin, M., Tremaine, S., and Zytkow, A. N. 1995. A search for slow-moving objects and the luminosity function of the Kuiper Belt. *Astron. J.* 110:3082–3092.

Jayawardhana, R., Fisher, S., Hartmann, L., Telesco, C., Pina, R., and Fazio, G. 1998. A dust disk surrounding the young A star HR 4796A. *Astrophys. J.* 503:L79.

Jewitt, D., and Kalas, P. 1998. Thermal infrared observations of Centaur 1997 CU_{26}. *Astrophys. J.* 499:L103–106.

Jewitt, D. C., and Luu, J. X. 1993. Discovery of the candidate Kuiper Belt object 1992 QB_1. *Nature* 362:730–732.

Jewitt, D. C., and Luu, J. X. 1995. The solar system beyond Neptune. *Astron. J.* 109:1867–1876.

Jewitt, D. C., and Luu, J. X. 1997. The Kuiper Belt. In *From Stardust to Planetesimals*, ASP Conf. Ser. 122, ed. Y. J. Pendleton and A. Tielens (San Francisco: Astronomical Society of the Pacific), pp. 335–345.

Jewitt, D., and Luu, J. 1998. Optical-infrared spectral diversity in the Kuiper Belt. *Astron. J.* 115:1667–1670.

Jewitt, D. C., Luu, J. X., and Chen, J. 1996. The Mauna Kea–Cerro Tololo (MKCT) Kuiper Belt and Centaur survey. *Astron. J.* 112:1225–1238.

Jewitt, D. C., Luu, J. X. and Trujillo, C. 1998. Large Kuiper Belt objects: The Mauna Kea 8k CCD survey. *Astron. J.* 115:2125–2135.

Joss, P. 1973. On the origin of short-period comets. *Astron. Astrophys.* 25:271–273.

Kenyon, S. J., and Luu, J. X. 1998. Accretion in the early Kuiper Belt I. Coagulation and velocity evolution. *Astron. J.* 115:2136–2160.

Kenyon, S. J., and Luu, J. X. 1999. Accretion in the early Kuiper Belt II. Fragmentation. *Astron. J.* 118:1101–1119.

Koerner, D., Ressler, M., Werner, M., and Backman, D. 1998. Mid-infrared imaging of a circumstellar disk around HR 4796: Mapping the debris of planetary formation. *Astrophys. J.* 803:L83.

Kowal, C. 1989. A solar system survey. *Icarus* 77:118–123.

Kuiper, G. 1951. On the origin of the solar system. In *Astrophysics*, ed. J. A. Hynek (New York: McGraw-Hill), pp. 357–424.

Leinert, C., Roser, S., and Buitrago, J. 1983. How to maintain the spatial distribution of interplanetary dust. *Astron. Astrophys.* 118:345–357.

Levison, H. F., and Duncan M. J. 1990. A search for proto-comets in the outer regions of the solar system. *Astron. J.* 100:1669–675.
Levison, H. F., and Duncan, M. J. 1994. The long-term dynamical behavior of short-period comets. *Icarus* 108:18–36.
Levison, H., and Duncan, M. 1997. From the Kuiper Belt to Jupiter family comets. *Icarus* 127:13–32.
Levison, H. F., and Stern, S. A. 1995. Possible origin and early dynamical evolution of the Pluto-Charon binary. *Icarus* 116:315–339.
Liou, J.-C., Zook, H. A., and Dermott, S. F. 1996. Kuiper Belt dust grains as a source of interplanetary dust particles. *Icarus* 124:429–440.
Liou, J.-C., and Zook, H. A. 1999. Signatures of the gas giant planets imprinted on the Edgeworth-Kuiper Belt dust disk. *Astron. J.* 118:580–590.
Luu, J. X., and Jewitt, D. 1988. A two-part search for slow moving objects. *Astron. J.* 95:1256–1262.
Luu, J. X., Jewitt, D. C., and Cloutis, E. 1994. Near-infrared spectroscopy of primitive solar system objects. *Icarus* 109:133–144.
Luu, J. X., and Jewitt, D. C. 1996. Color diversity among the Centaurs and Kuiper Belt objects. *Astron. J.* 112:2310–2318.
Luu, J. X., Marsden, B., Jewitt, D., Trujillo, C., Hergenrother, C., Chen, J., and Offutt, W. 1997. A new dynamical class of object in the outer solar system. *Nature* 387:573–575.
Luu, J. X., and Jewitt, D. C. 1998*a*. Deep imaging of the Kuiper Belt with the Keck 10m telescope. *Astrophys. J. Lett.*, L91–L94.
Luu, J. X., and Jewitt, D. C. 1998*b*. Optical and infrared reflectance spectrum of Kuiper Belt object 1996TL_{66}. *Astrophys. J. Lett.* 494:L117–L120.
MacFarland, J. 1996. Kenneth Essex Edgeworth—Victorian polymath and founder of the Kuiper Belt? *Vistas Astron.* 40:343–354.
Malhotra, R. 1995. The origin of Pluto's orbit: Implications for the solar system beyond Neptune. *Astron. J.* 110:420–429.
Malhotra, R. 1996. The phase space structure near Neptune resonances in the Kuiper Belt. *Astron. J.* 111:504–516.
Malhotra, R., and Williams, J. G. 1998. Pluto's heliocentric orbit. In *Pluto and Charon*, ed. S. A. Stern and D. Tholen (Tucson: University of Arizona Press).
McCaughrean, M. J., and O'Dell, C. R. 1996. Direct imaging of circumstellar disks in the Orion Nebula. *Astron. J.* 111:1977–1986.
Meech, K., Belton, M., Mueller, B., Dicksion, M., and Li, H. 1993. Nucleus properties of P/Schwassmann-Wachmann 1. *Astron. J.* 106:1222–1236.
Morbidelli, A., and Valsecchi, G. B. 1997. Neptune-scattered planetesimals could have sculpted the primordial Edgeworth-Kuiper Belt. *Icarus* 128:464–468.
Morbidelli, A., Thomas, F., and Moons, M. 1995. The resonant structure of the Kuiper belt and the dynamics of the first five trans-Neptunian objects. *Icarus* 118:322–340.
Morbidelli, A. 1997. Chaotic diffusion, and the origin of comets from the 3/2 Resonance in the Kuiper Belt. *Icarus* 127:1–12.
Moroz, L., Arnold, G., Korochantsev, A., and Wasch, R. 1998. Natural solid bitumens as possible analogs for cometary and asteroid organics. *Icarus* 134:253–268.
Oort, J. H. 1950. The structure of the cloud of comets surrounding the solar system and a hypothesis concerning its origin. *Bull. Astron. Inst. Netherl.* 11:91–110.
Petit, J.-M., Morbidelli, A., and Valsecchi, G. 1998. Large scattered planetesimals and the excitation of the small body belts. *Icarus*, in press.
Romanishin, W., Tegler, S., Levine, J., and Butler, N. 1997. *BVR* photometry of Centaur objects 1995 GO, 1993 HA_2, and 5145 Pholus. *Astron. J.* 113:1893–1898.

Schneider, G., Smith, B. A., Becklin, E. E., Koerner, D. W., Meier, R., Hines, D. C., Lowrance P. J., Terrile, R. J., Thompson, R. I., and Rieke, M. 1999. NICMOS imaging of the HR 4796A circumstellar disk. *Astrophys. J. Lett.* 513:L127-L130.

Senay, M., and Jewitt, D. 1994. Activity in a distant comet: First detection of carbon monoxide. *Nature* 371:229-231.

Smith, B. A., and Terrile, R. J. 1984. A circumstellar disk around β Pictoris. *Science* 226:1421-1424.

Spencer, J. R., Lebofsky, L. A., and Sykes, M. V. 1989. Systematic biases in radiometric diameter determinations. *Icarus* 78:337–354.

Stern, S. A. 1991. On the number of planets in the solar system: Evidence for a substantial population of 1000 km bodies. *Icarus* 90:271–281.

Stern, S. A. 1995. Collisional time scales in the Kuiper Disk and their implications. *Astron. J.* 110:856–868.

Stern, S. A. 1996*a*. Signatures of collisions in the Kuiper Belt. *Astron. Astrophys.* 310:999–1010.

Stern, S. A. 1996*b*. On the collisional environment of the massive, primordial Kuiper Belt. *Astron. J.* 112:1203–1211.

Stern, S. A., and Colwell, J. 1997. Accretion in the Edgeworth-Kuiper Belt: Forming 100–1000 km radius bodies at 30 AU and beyond. *Astron. J.* 114:841–849.

Tegler, S. C., and Romanishin, W. 1998. Two distinct populations of Kuiper Belt objects. *Nature* 392:49–51.

Tombaugh, C. W. 1961. The trans-Neptunian planet search. In *Planets and Satellites*, ed. G. P. Kuiper and B. M. Middlehurst (Chicago: University of Chicago Press), pp. 12–30.

Torbett, M. 1989. Chaotic motion in a comet disk beyond Neptune: The delivery of short-period comets. *Astron. J.* 98:1477–1481.

Trilling, D., and Brown, R. 1998. Imaging a circumstellar disk around a star with a radial-velocity planetary companion. *Nature* 395:775-777.

Tyson, J., Guhathakurta, P., Bernstein, G., and Hut, P. 1992. Limits on the surface density of faint Kuiper Belt objects. *Bull. Amer. Astron. Soc.* 24:1127 (abstract).

Whipple, F. 1964. Evidence for a comet belt beyond Neptune. *Proc. Nat. Acad. Sci.* 51:711–718.

Wilson, P. D., Sagan, C., and Thompson, W. R. 1994. The organic surface of 5145 Pholus: Constraints set by scattering theory. *Icarus* 107:288–303.

Williams, I. P., O'Ceallaigh, D. P., Fitzsimmons, A., and Marsden, B. G. 1995. The slow-moving objects 1993 SB and 1993 SC. *Icarus* 116:180–185.

Yamamoto, S., and Mukai, T. 1998. Dust production by impacts of interstellar dust on Edgeworth-Kuiper objects. *Astron. Astrophys.* 329:785–791.

DYNAMICS OF THE KUIPER BELT

RENU MALHOTRA
Lunar and Planetary Institute

MARTIN J. DUNCAN
Queen's University

and

HAROLD F. LEVISON
Southwest Research Institute

Our current knowledge of the dynamical structure of the Kuiper Belt is reviewed here. Numerical results on long-term orbital evolution and dynamical mechanisms underlying the transport of objects out of the Kuiper Belt are discussed. Scenarios about the origin of the highly nonuniform orbital distribution of Kuiper Belt objects are described, as well as the constraints these provide on the formation and long-term dynamical evolution of the outer solar system. Possible mechanisms include an early history of orbital migration of the outer planets, a mass loss phase in the outer solar system, and scattering by large planetesimals. The origin and dynamics of the scattered component of the Kuiper Belt are discussed. Inferences about the primordial mass distribution in the trans-Neptune region are reviewed. Outstanding questions about Kuiper Belt dynamics are listed.

I. INTRODUCTION

In the middle of this century, Edgeworth (1943) and Kuiper (1951) independently suggested that our planetary system is surrounded by a disk of material left over from the formation of planets. Both authors considered it unlikely that the protoplanetary disk was abruptly truncated at the orbit of Neptune. Each also suggested that the density in the solar nebula was too small beyond Neptune for a major planet to have accreted but that this region may be inhabited by a population of planetesimals. Edgeworth (1943) even suggested that bodies from this region might occasionally migrate inward and become visible as short-period comets. These ideas lay largely dormant until the 1980s, when dynamical simulations (Fernández 1980; Duncan et al. 1988; Quinn et al. 1990) suggested that a disk of transneptunian objects, now known as the Kuiper Belt, was a much more likely source of the Jupiter-family short-period comets than was the distant and isotropic Oort comet cloud.

With the discovery of its first member in 1992 by Luu and Jewitt (1993), the Kuiper Belt was transformed from a theoretical construct to a bona fide component of the solar system. By now, on the order of 100 Kuiper Belt objects (KBOs) have been discovered, a sufficiently large number to permit first-order estimates about the mass and spatial distribution in the transneptunian region. A comparison of the observed orbital properties of these objects with theoretical studies provides tantalizing clues to the formation and evolution of the outer solar system. Other chapters in this volume describe the observed physical and orbital properties of KBOs (Jewitt and Luu) and their collisional evolution (Farinella et al.); here we focus on the dynamical structure of the transneptunian region and the dynamical evolution of bodies in it. The outline of this chapter is as follows: In section II we review numerical results on long-term dynamical stability of small bodies in the outer solar system; in section III we review our current understanding of the Kuiper Belt phase space structure and dynamics of orbital resonances and chaotic transport of KBOs; section IV provides a discussion of resonance sweeping and other mechanisms for the origin and properties of KBOs at Neptune's mean motion resonances; the origin and dynamics of the "scattered disk" component of the Kuiper Belt is discussed in section V; current ideas about the primordial Kuiper Belt are described in section VI; we conclude in section VII with a summary of outstanding questions in Kuiper Belt dynamics.

II. LONG-TERM ORBITAL STABILITY IN THE OUTER SOLAR SYSTEM

A. Trans-Neptune Region

We first consider the orbital stability of test particles in the transneptunian region and the implications for the resulting structure of the Kuiper Belt several Gyr after its presumed formation. Torbett (1989) performed direct numerical integration of test particles in this region, including the perturbative effects of the four giant planets, although the latter were taken to be on fixed Keplerian orbits. He found evidence for chaotic motion with an inverse Lyapunov exponent (i.e., timescale for divergence of initially adjacent orbits) on the order of Myr for moderately eccentric, moderately inclined orbits with perihelia between 30 and 45 AU (a "scattered disk"). Torbett and Smoluchowski (1990) extended this work and suggested that even particles with initial eccentricities as low as 0.02 are typically on chaotic trajectories if their semimajor axes are less than 45 AU. Except in a few cases, however, the authors were unable to follow the orbits long enough to establish whether or not most chaotic trajectories in this group led to encounters with Neptune. Holman and Wisdom (1993) and Levison and Duncan (1993) showed that indeed some objects in the belt were dynamically unstable on timescales of Myr–Gyr, evolving onto Neptune-encountering orbits, thereby potentially providing a source of Jupiter-family comets at the present epoch.

Duncan et al. (1995) performed integrations of thousands of particles for up to 4 Gyr in order to complete a dynamical survey of the transneptunian region. The main results can be seen in Color Plate 18 and include the following features:

1. For nearly circular, very low-inclination particles there is a relatively stable band between 36 and 40 AU, with essentially complete stability beyond 42 AU. The lack of observed KBOs in the region between 36 and 42 AU suggests that some mechanism besides the dynamical effects of the planets in their current configuration must be responsible for the orbital element distribution in the Kuiper Belt. This mechanism is almost certainly linked to the formation of the outer planets. Several possible mechanisms are described in later sections.

2. For higher eccentricities (but still very low inclinations) the region interior to 42 AU is largely unstable except for stable bands near mean-motion resonances with Neptune (e.g., the well-known 2:3 near 39.5 AU, within which lies Pluto). The boundaries of the stable regions for each resonance have been computed independently by Morbidelli et al. (1995) and Malhotra (1996) and are in good agreement with the results shown in Color Plate 18. The dynamics within the mean-motion resonances are discussed in detail in section III.

3. The dark vertical bands between 35–36 AU and 40–42 AU are particularly unstable regions in which the particles' eccentricities are driven to sufficiently high values that they encounter Neptune. These regions match very well the locations of the ν_7 (see section IV.B) and ν_8 secular resonances as computed analytically by Knezevic et al. (1991): the test particles precess in these regions with frequencies very close to two of the characteristic secular frequencies of the planetary system.

4. A comparison of the observed orbits of KBOs with the phase space structure (Color Plate 18, Fig. 1) shows that virtually all of the bodies interior to 42 AU are on moderately eccentric orbits and located in mean-motion resonances, whereas most of those beyond 42 AU appear to be in nonresonant orbits of somewhat lower eccentricities. In addition, there is one observed object, 1996 TL_{66}, with semimajor axis $\approx$80 AU, beyond the range of Color Plate 18 and Fig. 1. It is thought to be a member of a third class of KBOs representing a "scattered disk" (see section V). Attempts to understand the origin of these three broad classes of KBOs will occupy much of what follows in this chapter.

B. Test Particle Stability between Uranus and Neptune

Gladman and Duncan (1990) and Holman and Wisdom (1993) performed long-term integrations, up to several hundred million years' duration, of the evolution of test particles on initially circular orbits in between the giant planets' orbits. The majority of the test particles were perturbed

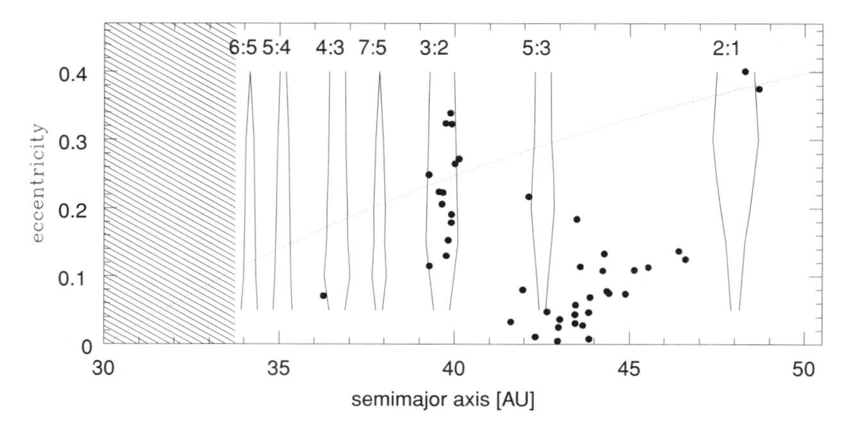

Figure 1. The locations and widths in the (a, e) plane of Neptune's low-order mean-motion resonances in the Kuiper Belt. Orbits above the dotted line are Neptune-crossing; the hatched zone on the left indicates the chaotic zone of first-order resonance overlap. The dots indicate the orbits of KBOs with reasonably well-determined orbits in January 1999. (Adapted from Malhotra 1996.)

into a close approach to a planet on timescales of 0.01–100 Myr, suggesting that these regions should largely be clear of residual planetesimals. However, Holman (1997) has shown that there is a narrow region, 24–26 AU, lying between the orbits of Uranus and Neptune, in which roughly 1% of minor bodies could survive on very low-eccentricity and low-inclination orbits for the age of the solar system. He estimated that a belt of mass totaling roughly 10^{-3} $M_\oplus$ cannot be ruled out by current observational surveys. This niche is, however, extremely fragile. Brunini and Melita (1998) have shown that any one of several likely perturbations (e.g., mutual scattering, planetary migration, and Pluto-sized perturbers) would have largely eliminated such a primordial population. We note also that there are similarly stable (possibly even less "fragile"), but apparently unpopulated dynamical niches in the outer asteroid belt (Duncan 1994) and the inner Kuiper Belt (Duncan et al. 1995).

C. Neptune Trojans

The only observational survey of which we are aware specifically designed to search for Trojans of planets other than Jupiter covered 20 square degrees of sky to limiting magnitude $V = 22.7$ (Chen et al. 1997). Although 93 jovian Trojans were found, no Trojans of Saturn, Uranus, or Neptune were discovered. Although this survey represents the state of the art, it lacks the sensitivity and areal coverage to reject the possibility that Neptune holds Trojan swarms similar in magnitude to those of Jupiter (D. Jewitt, personal communication, 1998). Further searches clearly need to be done.

Several numerical studies of orbital stability in Neptune's Trojan regions have been published. Mikkola and Innanen (1992) studied the behavior of 11 test particles initially near the Neptune Trojan points for 2×10^6 years. Holman and Wisdom (1993) performed a 2×10^7-year integration of test particles initially in near-circular orbits near the Lagrange points of all the outer planets. Most recently, Weissman and Levison (1997) integrated the orbits of 70 test particles in the L4 Neptune Trojan zone for 4 Gyr. In all these studies, some Neptune Trojan orbits were found to be stable. Weissman and Levison (1997) found that stable Neptune Trojans must have libration amplitudes $D \lesssim 60°$ and proper eccentricities $e_p \lesssim 0.05$. It is interesting to note that this range for D is similar to that Levison et al. (1997) found for the Jupiter Trojans, but the maximum stable e_p for the Neptune regions is a factor of 3 smaller than that of the Jupiter regions. Holman and Wisdom (1993) reported a curious asymmetric displacement of the L4 and L5 Trojan libration centers of Neptune whose cause remains unknown.

III. RESONANCE DYNAMICS AND CHAOTIC DIFFUSION

It is evident from the numerical analysis of test particle stability in the transneptunian region that the timescales for orbital instability span several orders of magnitude and are very sensitive to orbital parameters. For example, the map of stability time (i.e., time to first encounter within a Hill sphere radius of Neptune) for initially circular orbits is very patchy, with short dynamical lifetimes interspersed among very stable regions (see Color Plate 18). Most particles that have a close approach to Neptune are removed from the Kuiper Belt shortly thereafter by means of a quick succession of close approaches to the giant planets. However, a small fraction evolve into anomalously long-lived chaotic orbits beyond Neptune that do not have a second close approach to the planet on timescales comparable to the age of the solar system (see section V). The nature and origin of the long-timescale instabilities (which are most relevant for understanding the origin of short-period comets from the Kuiper Belt) is not well understood at present.

In general, we understand that the mean-motion resonances of Neptune form a "skeleton" of the phase space (Fig. 1), with the perturbations of the other giant planets, including secular resonances, forming a web of superstructure on that skeleton. The phase space in the neighborhood of Neptune's 3:2 mean-motion resonance is the best studied, following the discovery of Pluto and its myriad of resonances (cf. Malhotra and Williams 1997). Color Plate 19*a* shows the dynamical features that have been identified at the 3:2 resonance, in the (a, e) plane. The boundary of this resonance, determined by means of a seminumerical analysis of the averaged perturbation potential of Neptune, is shown (dark solid lines) as well as the loci of the apsidal ν_8 and nodal ν_{18} secular resonances (see section IV.B) and the Kozai resonance in this neighborhood (Morbidelli 1997). The stable libration zone, determined from an inspection of many surfaces of section of the planar restricted three-body model of the Sun-Neptune-Plutino (Malhotra 1996), is indicated by the blue shaded region. The stable resonance libration boundary is significantly different from the formal perturbation theory result, because averaging is not a good approximation in the vicinity of the resonance separatrix: the separatrix broadens into a chaotic zone because of the interaction with neighboring mean-motion resonances. The width of the chaotic separatrix increases with eccentricity, eventually merging with the chaotic separatrices of neighboring mean-motion resonances. The ν_8 secular resonance is mostly embedded in the chaotic zone, while the ν_{18} occurs at large libration amplitudes close to the chaotic separatrix; the Kozai resonance occurs at large libration amplitude for low-eccentricity orbits, and at smaller libration amplitude for eccentricities near 0.2–0.25.

Color Plate 19*b* shows a "map" of the diffusion speed of test particles determined by numerical integrations of up to 4 Gyr by Morbidelli (1997). It is clear that instability timescales in the vicinity of the 3:2 resonance range from less than a million years to longer than the age of the solar system. In general, within the resonance, higher-eccentricity orbits are less stable than those of lower eccentricities; the stability times are longest deep in the resonance and shorter near the boundaries. We note that the uncolored regions exterior and adjacent to the colored zones at eccentricities below ~ 0.15 are stable for the age of the solar system, but those above ~ 0.15 are actually chaotic on timescales shorter than the shortest indicated in the colored zones.

The short stability timescales in the most unstable zones are due to dynamical chaos generated by the interaction with neighboring mean-motion resonances; this can be directly visualized in surfaces of section of the planar restricted three-body model (Malhotra 1996). Test particles in these zones suffer large chaotic changes in semimajor axis and eccentricity on short timescales, $\mathcal{O}(10^{5-7})$ yr, and are not protected from close encounters with Neptune. In a small zone near semimajor axis 39.5 AU, initially circular low-inclination orbits are excited to high eccentricity *and* high inclination on timescales of $\mathcal{O}(10^7)$ yr (Holman and Wisdom 1993; Levison and Stern 1995); this instability is possibly due to overlapping secondary resonances (Morbidelli 1997). The finite diffusion timescale, comparable to but shorter than the age of the solar system, in a large area (green zone) inside the resonance is not understood at all; possibly higher-order secondary resonances are the underlying cause. The numerical evolution of test particles in this region follows initially a slow diffusion in semimajor axis with nearly constant mean eccentricity and inclination, until the orbit eventually reaches a strongly chaotic zone (Morbidelli 1997). The diffusion timescales in this region are comparable to or only slightly less than the age of the solar system, so it is likely to be an active source region for short-period comets via purely dynamical instabilities.

The dynamical structure in the vicinity of other mean-motion resonances has not been obtained in as much detail as that of the 3:2. We expect differences in details (different profile of the libration zone, differing secular resonance effects, etc.) but generally similar qualitative features. We also note that because the orbital evolution obtained in nondissipative models of Kuiper Belt dynamics is time reversible, the transport of particles from strongly chaotic regions to weakly chaotic regions is also allowed. In the most general terms, this is the likely explanation for the putative scattered disk (section V).

IV. RESONANT KUIPER BELT OBJECTS

The origin of the great abundance of resonant KBOs and, equally importantly, their high orbital eccentricities is an interesting question whose understanding may lead to significant advances in our understanding of the early history of the solar system. In this section, we discuss current ideas pertinent to this class of KBOs.

A. Planet Migration and Resonance Sweeping

An outward orbital migration of Neptune in early solar system history provides an efficient mechanism for sweeping up large numbers of transneptunian objects into Neptune's mean-motion resonances. We describe this scenario in some detail here; its importance stems from the linkage it provides between the detailed orbital distribution in the Kuiper Belt and the early orbital migration history of the outer planets. This theory, which was originally proposed for the origin of Pluto's orbit (Malhotra 1993), supposes that Pluto formed in a common low-eccentricity, low-inclination orbit beyond the (initial) orbit of Neptune. It was captured into the 3:2 resonance with Neptune and had its eccentricity excited to its current Neptune-crossing value as Neptune's orbit expanded outward as a result of angular momentum exchange with residual planetesimals. The theory predicts that resonance capture and eccentricity excitation would be a common fate of a large fraction of transneptunian objects (Malhotra 1995).

The process of orbital migration of Neptune (and, by self-consistency, migration of the other giant planets) invoked in this theory is as follows. Consider the late stages of planet formation, when the outer solar system had reached a configuration close to its present state, namely four giant planets in well-separated, near-circular, coplanar orbits; the nebular gas had already dispersed; the planets had accreted most of their mass; but there remained a residual population of icy planetesimals and possibly larger planetoids. The subsequent evolution consisted of the gravitational scattering and accretion of these small bodies. Circumstantial evidence for this exists in the obliquities of the planets (Lissauer and Safronov 1991; Dones and Tremaine 1993). Much of the Oort Cloud, the putative source of long-period comets, would have formed during this stage by the scattering of planetesimals to wide orbits by the giant planets and the subsequent action of galactic tidal perturbations and perturbations from passing stars and giant molecular clouds (e.g., Fernández 1985; Duncan et al. 1987). During this era, the back reaction of planetesimal scattering on the planets could have caused significant changes in their orbital energy and angular momentum. Overall, there was a net loss of energy and angular momentum from the planetary orbits, but the loss was not extracted uniformly from the four giant planets. Jupiter, by far the most massive of the planets, likely provided all of the lost energy and angular momentum, and more; Saturn, Uranus, and Neptune actually gained orbital energy and angular momentum and their orbits expanded, while Jupiter's orbit decayed sufficiently to balance the books. This was first pointed out by Fernández and Ip (1984), who noticed it in numerical simulations of the late stages of accretion of planetesimals by the proto-giant planets.

1. Migration of the Jovian Planets. The reasons for this rather nonintuitive result can be understood from the following heuristic picture of the clearing of a planetesimal swarm from the vicinity of Neptune. Suppose that the mean specific angular momentum of the swarm is initially equal to that of Neptune. A small fraction of planetesimals is accreted as a result of physical collisions. Of the remaining, there are approximately equal numbers of inward and outward scatterings. To first order, these cause no net change in Neptune's orbit. However, the subsequent fate of the inward- and outward-scattered planetesimals is not symmetrical. Most of the inward-scattered objects enter the zones of influence of Uranus, Saturn, and Jupiter. Of those objects scattered outward, some are eventually lifted into wide, Oort Cloud orbits, while most return to be rescattered; a fraction of the latter is again rescattered inward, where the inner jovian planets, particularly Jupiter, control the dynamics. The massive Jupiter is very effective in causing a systematic loss of planetesimal mass by ejection into solar system escape orbits. As Jupiter preferentially removes the inward-scattered planetesimals from Neptune's influence, the planetesimal population encountering Neptune at later times is increasingly biased toward objects with specific angular momentum and energy greater than Neptune's. Encounters with this planetesimal population produce effectively a negative drag on Neptune, which increases its orbital radius. Uranus and Saturn, also being much less massive than and exterior to Jupiter, experience a similar orbital migration, but smaller in magnitude than that of Neptune. Thus, Jupiter is in effect the source of the angular momentum and energy needed for the orbital migration of the outer giant planets as well as for the planetesimal mass loss. However, owing to its large mass, its orbital radius decreases by only a small amount.

The magnitude and timescale of the radial migration of the jovian planets due to their interactions with residual planetesimals is difficult to determine without a full-scale N-body model. The work of Fernández and Ip (1984) is suggestive but not conclusive, on account of several limitations of their numerical model that produced highly stochastic results:

they modeled a small number of planetesimals, ~ 2000, and the masses of individual planetesimals were in the rather exaggerated range of 0.02–0.3 $M_\oplus$; and, perhaps most significantly, they neglected long-range gravitational forces. Current studies attempt to overcome these limitations by using the faster computers now available and more sophisticated integration algorithms (Hahn and Malhotra 1999). Still, fully self-consistent high-fidelity models remain a distant goal at this time. Remarkably, an estimate for the magnitude and timescale of Neptune's outward migration is possible from an analysis of the orbital evolution of KBOs captured in Neptune's mean-motion resonances.

2. Resonance Sweeping. Capture into resonance is a delicate process, difficult to analyze mathematically. Under certain simplifying assumptions, Malhotra (1993, 1995) showed that resonance capture is very efficient for adiabatic orbital evolution of KBOs whose initial orbital eccentricities were smaller than ~ 0.05. Resonance capture leads to an excitation of orbital eccentricity whose magnitude is related to the magnitude of orbital migration:

$$\Delta e^2 \simeq \frac{k}{j+k} \ln \frac{a_f}{a_i} = \frac{k}{j+k} \ln \frac{a_N}{a_{N,i}} \tag{1}$$

where a_i and a_f are the initial and current semimajor axes of a KBO, a_N is Neptune's current semimajor axis, and $a_{N,i}$ is its value in the past at the time of resonance capture; j and k are positive integers defining a $j+k:j$ mean-motion resonance. From this equation and the observed eccentricities of Pluto and the Plutinos (see the chapter by Jewitt and Luu, this volume), it follows that Neptune's orbit has expanded by ~ 9 AU.

Numerical simulations of resonance sweeping of the Kuiper Belt have been carried out assuming adiabatic giant-planet migration of specified magnitude and timescale. The orbital distribution of Kuiper Belt objects obtained in one such simulation is shown in Fig. 2. The main conclusions from such simulations are that (i) few KBOs remain in circular orbits of semimajor axis $a \lesssim 39$ AU, which marks the location of the 3:2 Neptune resonance; (ii) most KBOs in the region up to $a = 50$ AU are locked in mean-motion resonances; the 3:2 and 2:1 are the dominant resonances, but the 4:3 and 5:3 also capture noticeable numbers of KBOs; (iii) there is a significant paucity of low-eccentricity orbits in the 3:2 resonance; (iv) the maximum eccentricities in the resonances are in excess of Neptune-crossing values; (v) inclination excitation is not as efficient as eccentricity excitation; only a small fraction of resonant KBOs acquire inclinations in excess of 10°. Not shown in Fig. 2 are other dynamical features, such as the resonance libration amplitude and the argument-of-perihelion behavior (libration, as for Pluto, or circulation), which are also reflective of the planet migration/resonance sweeping process.

More detailed discussion of numerical results on resonance sweeping is given in Malhotra (1995, 1998*a*, 1999) and Holman (1995). Two

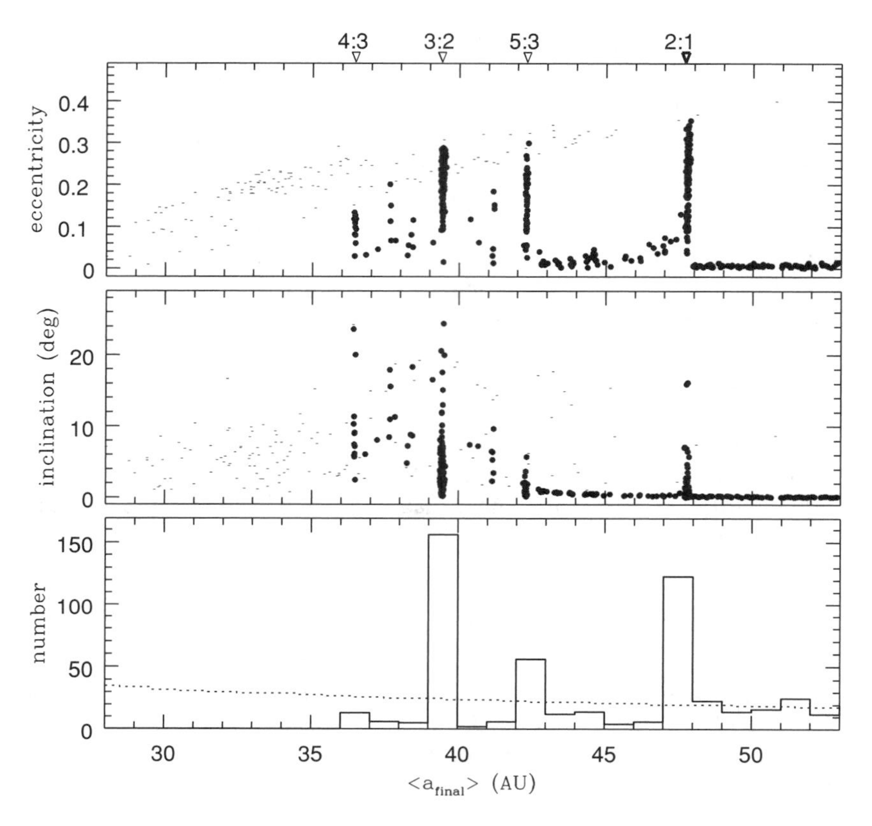

Figure 2. The distribution of orbital elements of transneptunian test particles after resonance sweeping, as obtained from a 200-Myr numerical simulation in which the jovian planets' semimajor axes evolve according to $a(t) = a_f - \Delta a \exp(-t/\tau)$, with timescale $\tau = 4$ Myr and $\Delta a = \{-0.2, 0.8, 3.0, 7.0\}$ AU for Jupiter, Saturn, Uranus, and Neptune, respectively (Malhotra, 1999). The test particles were initially distributed smoothly in circular, zero-inclination orbits between 28 AU and 63 AU, as indicated by the dotted line in the lower panel. In the upper two panels, the larger filled circles represent particles that remain on stable orbits at the end of the simulation, and the smaller open circles represent the elements of those particles that had a close encounter with a planet and subsequently move on scattered chaotic orbits ("removed" from the simulation at the instant of encounter). Neptune's mean-motion resonances are indicated at the top of the figure.

additional points are worthy of note here. One is that, owing to the longer dynamical timescales associated with vertical resonances, the magnitude of inclination excitation of KBOs is sensitive to both the magnitude and the timescale of planetary migration; it is estimated that a timescale on the order of 1–3 $\times 10^7$ yr would account for Pluto's inclination (Malhotra 1998*a*). The second is that the total mass of residual planetesimals required for a ~9-AU migration of Neptune is estimated to be $\sim 50\ M_\oplus$ (Malhotra

1999); this estimate is supported by recent numerical simulations of planet migration (Hahn and Malhotra 1999).

An outstanding issue is the apparent paucity of KBOs at Neptune's 2:1 mean-motion resonance in the observed sample (Fig. 1),[a] whereas the planet migration/resonance sweeping theory predicts comparable populations in the 3:2 and 2:1 resonances (Fig. 2). Possible explanations are (i) observational incompleteness (cf. Gladman et al. 1998), (ii) significant leakage out of the 2:1 resonance on Gyr timescales by means of weak instabilities or perturbations by larger members of the Kuiper Belt, or (iii) that the planet migration/resonance sweeping did not occur as postulated. However, the success of the resonance sweeping mechanism in explaining the orbital eccentricity distribution of Plutinos, and the difficulty of explaining the distribution by other means, argues strongly in favor of this model. Further work is needed to refine the relationship between the orbital element distributions and the detailed characteristics of the planet migration process, including the overall efficiency of resonance capture and retention.

If such planet migration and resonance sweeping occurred, then it follows that the KBOs presently resident in Neptune's mean-motion resonances formed closer to the sun than their current semimajor axes would suggest. If there were a significant compositional gradient in the primordial transneptunian planetesimal disk, it may be preserved in a rather subtle manner in the present orbital distribution. Because each resonant KBO retains memory of its initial orbital radius in its final orbital eccentricity [equation (1)], there would exist a compositional gradient with orbital eccentricity within each resonance; nonresonant KBOs in near-circular low-inclination orbits between 30 and 50 AU most likely formed at their present locations and would reflect the primordial conditions at those locations. However, if there were significant orbital mixing by processes other than resonance sweeping, these systematics would be diluted or erased.

B. Secular Resonance Sweeping

The combined, orbit-averaged perturbations of the planets on each other cause a slow precession of the direction of perihelion and of the pole of the orbit plane of each planet. Of particular importance to the long-term dynamics of the Kuiper Belt are the perihelion and orbit pole precession of Neptune, both of which have periods of about 2 Myr in the present planetary configuration. The perihelion direction and orbit pole orientation of the orbits of Kuiper Belt objects also precess slowly at rates that depend upon their orbital parameters. For certain ranges of orbital parameters, the perihelion precession rate matches that of Neptune; this condition is termed

---

[a] In December 1998, while this article was in the review process, the Minor Planet Center reported new observations yielding revised orbits for 1997 SZ10 and 1996 TR66, identifying these as the first two KBOs in the 2:1 resonance with Neptune (Marsden 1998).

the ν_8 *secular resonance*. Similarly, the 1:1 commensurability of the rate of precession of the orbit pole with that of Neptune's orbit pole is called ν_{18} *secular resonance*. (See Malhotra 1998*b* for an analytical model of secular resonance.) The ν_8 and ν_{18} secular resonances occur at several regions in the Kuiper Belt, where they cause strong perturbations of the orbital eccentricity and orbital inclination, respectively (cf. Color Plates 18 and 19*a*).

The secular effects are sensitive to the mass distribution in the planetary system (see Ward 1981 and references therein). Levison et al. (1999) have noted that a primordial massive transneptunian disk would have significantly altered the locations of the ν_8 and ν_{18} secular resonances. From a suite of numerical simulations, they estimate that a $\sim$10 $M_\oplus$ primordial disk between 30 and 50 AU would have the ν_8 secular resonance near $a \lesssim 36$ AU, and that it would have moved outward to its current location near 42 AU as the disk mass declined. Such sweeping by the ν_8 secular resonance excites the orbital eccentricities of KBOs sufficiently to cause them to encounter Neptune and be removed from the Kuiper Belt. Only objects fortuitously trapped in Neptune's mean-motion resonances remain stable. The simulations also find that the ν_{18} secular resonance sweeps *inward* from well beyond its current location as the Kuiper Belt mass declines, thereby moderately increasing the inclinations (up to $\sim$15 degrees) of KBOs beyond 42 AU. However, this mechanism does not produce orbital eccentricities in Neptune's 2:3 mean-motion resonance as large as those observed. Furthermore, damping of the eccentricity and inclination by density waves is to be expected in a massive primordial disk (Ward and Hahn 1998) but has not yet been included in the simulations. We conclude that the sweeping of secular resonances has probably played some role in the excitation of the Kuiper Belt, but its quantitative effects remain to be determined.

C. Stirring by Large Neptune-scattered Planetesimals

A third mechanism to explain the observed orbital properties of KBOs is to invoke the orbital excitation produced by close encounters with large Neptune-scattered planetesimals on their way out of the solar system or to the Oort Cloud. The observed excitation in the Kuiper Belt requires the prior existence of planetesimals with masses $\sim$1 Earth mass, according to Morbidelli and Valsecchi (1997). There is circumstantial evidence (e.g., the obliquity of Uranus's spin axis) that a population of objects this massive might have formed in the region between Uranus and Neptune and are now gone (Safronov 1966; Stern 1991). Many of these objects must have spent some time orbiting through some parts of the Kuiper Belt. A much more massive initial belt might then have been sculpted to its presently observed structure because of the injection of most of the small bodies into dynamically unstable regions in the inner belt and the enhanced role of mutually catastrophic collisions among small planetesimals in the outer belt.

In this picture, then, the observed KBOs interior to ~ 42 AU are the lucky ones ending up in the small fraction of phase space (a few percent; see Fig. 1) protected from close encounters with Neptune by mean-motion resonances. This mechanism is similar to one involving large Jupiter-scattered planetesimals proposed earlier to explain the excitation of the asteroid belt (Ip 1987; Wetherill 1989).

Petit et al. (1998) have combined three-body integrations and semi-analytic estimates of scattering to model the effects of large planetesimals in the Kuiper Belt. They argue that the best reconstruction of the observed dynamical excitation of the Kuiper Belt requires the earlier existence of two large bodies. The first is a body of half an Earth mass on an orbit of large eccentricity with a dynamical lifetime of several times 10^8 yr. The second is a body of one Earth mass that evolves for ~ 25 Myr on a low-eccentricity orbit spanning the 30–40 AU range.

An attractive feature of this mechanism is that it yields an overall mass depletion in the inner Kuiper Belt and accounts for the fact that the outer Kuiper Belt ($a > 42$ AU) is moderately excited. It does not, however, appear to explain the lack of low-eccentricity objects in Neptune's 2:3. In addition, the models performed to date require a specific set of very large objects, for which there is no direct evidence, to be at the right place for the right length of time. The presumed eventual removal of these objects by means of a final close encounter with Neptune would perturb Neptune's orbit significantly and also jeopardize the stability of resonant KBOs; this is in conflict with the observed evidence.

V. THE SCATTERED DISK

As noted previously, the current renaissance in Kuiper Belt research was prompted by the suggestion that the Jupiter-family comets (JFCs) originated there (Fernández 1980; Duncan et al. 1988). Thus, as part of the research intended to understand the origin of these comets, a significant amount of effort has gone into understanding the dynamical behavior of objects that are on orbits that can encounter Neptune (Duncan et al. 1988; Quinn et al. 1990; Levison and Duncan 1997). It is somewhat ironic, therefore, that these studies have led to the realization that a structure known as the *scattered comet disk,* rather than the Kuiper Belt, could be the dominant source of the Jupiter-family comets.

For our purposes, scattered-disk objects are distinct from Kuiper Belt objects in that they evolved out of their primordial orbits beyond Uranus early in the history of the solar system. These objects were then dynamically scattered by Neptune into orbits with perihelion distances near Neptune but semimajor axes in the Kuiper Belt (Duncan and Levison 1997). Finally, some process, usually interactions with Neptune's mean-motion resonances, raised their perihelion distances, thereby effectively storing the objects for the age of the solar system. Scattered-disk objects (hereafter

SDOs) occupy the same physical space as KBOs but can be distinguished from KBOs by their orbital elements. In particular, as we describe in more detail below, SDOs tend to be on much more eccentric orbits than KBOs.

Of the 60 or so Kuiper Belt objects thus far cataloged, only one, 1996 TL_{66}, is an obvious SDO. 1996 TL_{66}, discovered in October 1996 by Jane Luu and colleagues (Luu et al. 1997), is estimated to have a semimajor axis of 85 AU, a perihelion of 35 AU, an eccentricity of 0.59, and an inclination of 24°. Such a high-eccentricity orbit most likely resulted from gravitational scattering by a giant planet, in this case Neptune.

The idea of the existence of a scattered comet disk dates to Fernández and Ip (1989). Their numerical simulations indicated that some objects on eccentric orbits with perihelia inside the orbit of Neptune could remain on such orbits on gigayear timescales and hence might be present today. However, their simulations were based on an algorithm that incorporated only the effects of close gravitational encounters (Öpik 1951) and hence severely overestimates the dynamical lifetimes of bodies such as those in their putative disk (Dones et al. 1998). As a result, the dynamics of their scattered disk bears little resemblance to the structure found in more recent direct integrations to be discussed below.

The only investigation of the scattered disk that uses modern direct numerical integrations is the one by Duncan and Levison (1997). This investigation was a followup to Levison and Duncan (1997), which was an investigation of the behavior of 2200 small, massless objects that initially were encountering Neptune. The latter's focus was to model the evolution of these objects down to Jupiter-family comets and followed the system for only 1 Gyr. Duncan and Levison (1997) extended these integrations to 4 Gyr. Most objects that encounter Neptune have short dynamical lifetimes. Usually, they either (1) are ejected from the solar system, (2) hit the sun or a planet, or (3) are placed in the Oort Cloud, in less than $\sim 5 \times 10^7$ years. It was found, however, that 1% of the particles remained in orbits beyond Neptune after 4 Gyr. So, if at early times there was a significant amount of material from the region between Uranus and Neptune or the inner Kuiper Belt that evolved onto Neptune-crossing orbits, there could be a significant amount of this material remaining today. What is meant by "significant" is the main question when it comes to the current importance of the scattered comet disk.

Duncan and Levison (1997) found that some of the long-lived objects were scattered to very long-period orbits, where encounters with Neptune became infrequent. However, at any given time, the majority of them were interior to 100 AU. Their longevity is due in large part to their being temporarily trapped in or near mean-motion resonances with Neptune. The "stickiness" of the mean-motion resonances, which was mentioned by Holman and Wisdom (1993), leads to an overall distribution of semimajor axes for the particles that is peaked near the locations of many of the mean-motion resonances with Neptune. Occasionally, the longevity is enhanced by the presence of the Kozai resonance. In all long-lived cases, particles had their perihelion distances increased so that close encounters with Neptune no longer occurred. Frequently, these increases in perihelion distance were associated with trapping in a mean-motion resonance, although in many cases it has not yet been possible to identify the exact process that was involved. On occasion, the perihelion distance can become large, but 81% of scattered-disk objects have perihelia between 32 and 36 AU.

Figure 3 shows the dynamical behavior of a typical particle. This object initially underwent a random walk in semimajor axis because of encounters with Neptune. At about 7×10^7 yr it was temporarily trapped in Neptune's 3:13 mean-motion resonance for about 5×10^7 yr. It then performed a random walk in semimajor axis until about 3×10^8 years, when it was trapped in the 4:7 mean-motion resonance, where it remained for 3.4×10^9 years. Notice the increase in the perihelion distance near the time of capture. While trapped in this resonance, the particle's eccentricity became as small as 0.04. After leaving the 4:7, it was trapped temporarily in Neptune's 3:5 mean-motion resonance for $\sim 5 \times 10^8$ yr and then went through a random walk in semimajor axis for the remainder of the simulation.

Duncan and Levison (1997) estimated an upper limit on the number of possible SDOs by assuming that they are the sole source of the Jupiter-family comets. Duncan and Levison computed the simulated distribution

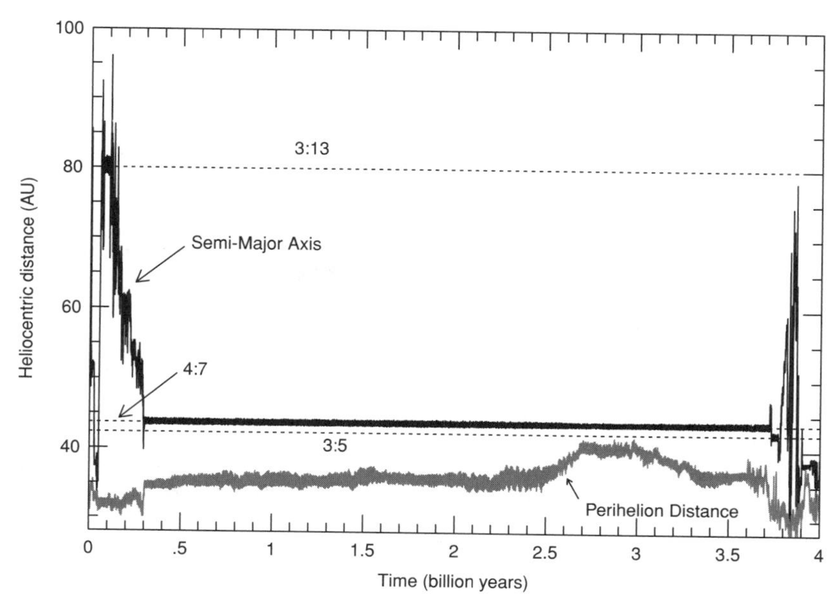

Figure 3. The temporal behavior of a long-lived member of the scattered disk. The black curve shows the behavior of the comet's semimajor axis. The gray curve shows the perihelion distance. The three dotted curves show the location of the 3:13, 4:7, and 3:5 mean-motion resonances with Neptune.

of comets throughout the solar system at the current epoch (averaged over the last Gyr for better statistical accuracy). They found that the ratio of scattered-disk objects to visible Jupiter-family comets[b] is 1.3×10^6. There are currently estimated to be 500 visible JFCs (Levison and Duncan 1997), so there are $\sim 6 \times 10^8$ comets in the scattered disk if it is the sole source of the JFCs. It is quite possible that the scattered disk could contain this much material. Figure 4 shows the spatial distribution for this model.

To review the above findings, $\sim 1\%$ of the objects in the scattered disk remain after 4 Gyr, and 6×10^8 comets are currently required to supply all of the Jupiter-family comets. Thus, a scattered comet disk requires an initial population of only 6×10^{10} comets (or $\sim 0.4\ M_\oplus$, Weissman 1990) on Neptune-encountering orbits. Because planet formation is unlikely to have been 100% efficient, the original disk could have resulted from the scattering of even a small fraction of the tens of Earth masses of cometary

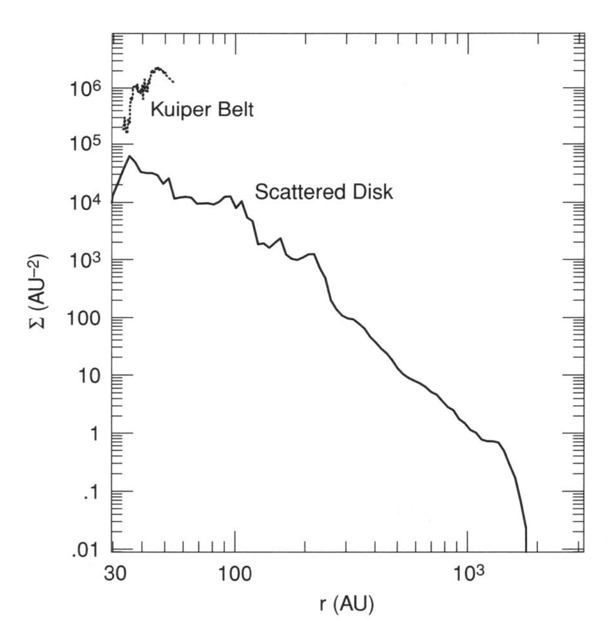

Figure 4. The surface density of comets beyond Neptune for two different models of the source of Jupiter-family comets. The dotted curve is a model assuming that the Kuiper Belt is the current source (Levison and Duncan 1997). There are 7×10^9 comets in this distribution between 30 and 50 AU. This curve ends at 50 AU because the models are unconstrained beyond this point, not because it is believed that there are no comets there. The solid curve is the model of Duncan and Levison (1997), assuming that the scattered disk is the sole source of the Jupiter-family comets. There are 6×10^8 comets currently in this distribution.

[b] We define a "visible" Jupiter-family comet as one with a perihelion distance less than 2.5 AU.

material that must have populated the outer solar system in order to have formed Uranus and Neptune.

VI. THE PRIMORDIAL KUIPER BELT

As with many scientific endeavors, the discovery of new information tends to raise more questions than it answers. Such is the case with the Kuiper Belt. Even the original argument that suggested the Kuiper Belt is in doubt. Edgeworth's (1949) and Kuiper's (1951) speculations were based on the idea that it seemed unlikely that the disk of planetesimals that formed the planets would have abruptly ended at the current location of the outermost known planet, Neptune. An extrapolation into the Kuiper Belt of the current surface density of nonvolatile material in the outer planetary region implies that there should be about 30 $M_{\oplus}$ of material there. However, the recent KBO observations indicate only a few $\times$ 0.1 $M_{\oplus}$ between 30 and 50 AU. Were Kuiper and Edgeworth wrong? Is there a sharp outer edge to the planetary system? One line of theoretical arguments suggests that the answer may be no to both questions; we discuss these below, but we note at the end some contrary arguments as well.

Over the last few years, several points have been made supporting the idea of a massive primordial Kuiper Belt; see the chapter by Farinella et al., this volume. Stern (1995) and Davis and Farinella (1997) have argued that the inner part of the Kuiper Belt ($\lesssim$50 AU) is currently eroding away due to collisions and therefore must have been more massive in the past. Stern (1995) also argued that the current surface density in the Kuiper Belt is too low to grow bodies larger than about 30 km in radius by means of two-body collisional accretion over the age of the solar system. The observed $\sim$100-km-size KBOs could have grown in a more massive Kuiper Belt, of at least several Earth masses, if the mean eccentricities of the accreting objects were small, on the order of a few times 10^{-3} or perhaps as large as 10^{-2} (Stern 1996; Stern and Colwell 1997*a*; Kenyon and Luu 1998). Models of the collisional evolution of a massive Kuiper Belt suggest that 90% of the mass inside of $\sim$50 AU could have been been lost through collisions over the age of the solar system (Stern and Colwell 1997*b*; Davis and Farinella 1998). Thus, it is possible for a massive primordial Kuiper Belt to grind itself down to the levels that we see today.

With these arguments, it is possible to build a "straw man" model of the Kuiper Belt, which is depicted in Fig. 5. Following Stern (1996), there are three distinct zones in the Kuiper Belt. Region A is a zone where the gravitational perturbations of the outer planets have played an important role, tending to pump up the eccentricities of objects. About half of the objects in this region could have been dynamically removed from the Kuiper Belt (Duncan et al. 1995). The remaining objects have eccentricities that are large enough that accretion has ceased and erosion is dominant. Thus, we expect that a significant fraction of the mass in region A has been removed by collisions. The dotted curve shows an estimate of the initial

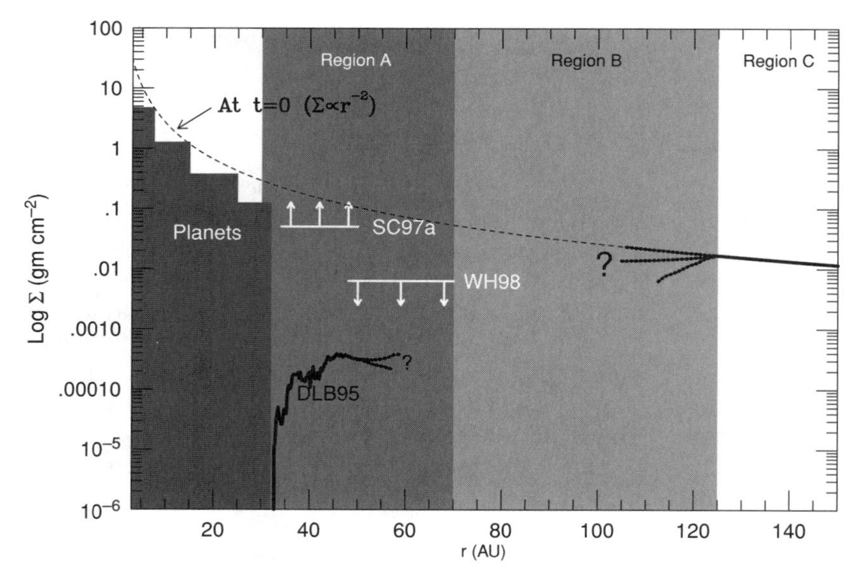

Figure 5. A "straw man" model of the Kuiper Belt. The dark area at left marked "Planets" shows the distribution of solid material in the outer planets region; it follows a power law with a slope of about -2 in heliocentric distance. The dashed curve is an extension into the Kuiper Belt of the power law found for the outer planets and illustrates the likely initial surface density distribution of solid material in the Kuiper Belt. The dashed curve has been scaled so that it is twice the surface density of the outer planets, under the assumption that planet formation was 50% efficient. The solid curve shows a model of the mass distribution by Duncan et al. (1995, see Fig. 4), scaled to an estimated population of 5×10^9 Kuiper Belt objects between 30 and 50 AU. The Kuiper Belt is divided into three regions; see text for a description. The dotted curves illustrate the unknown shape of the surface density distribution in region B.

surface density of solids, extrapolated from that of the outer planets, and the solid curve (marked DLB95) shows an estimate of the current Kuiper Belt surface density from Duncan et al. (1995, reproduced from their Fig. 7). Region B in Fig. 5 is a zone where collisions are important but the perturbations of the planets are not. The orbital eccentricities of objects in this region will most likely remain small enough that collisions lead to accretion of bodies, rather than erosion. Large objects could have formed here (see Stern and Colwell 1997*a*), but the surface density may not have changed very much in this region over the age of the solar system. Observational constraints based on *Cosmic Background Explorer* (COBE), Diffuse Infrared Background Experiment (DIRBE), data put the transition between regions A and B beyond 70 AU (Teplitz et al. 1999). The outer boundary of region B is very uncertain but will likely be beyond 100 AU (S. A. Stern, personal communication). Region C is a zone where collision rates are low enough that the surface density of the Kuiper Belt and the size distribution of objects in it have remained virtually unchanged over the history of the solar system.

If the above model is correct, then the Kuiper Belt that we have so far observed may be a low-density region that lies inside a much more massive outer disk. In other words, we may now be seeing a "Kuiper Gap" (Stern and Colwell 1997*b*).

The idea of an increase in the surface density of the Kuiper Belt beyond 50 AU may explain a puzzling feature of the dynamics of the planets: Neptune's small eccentricity. Ward and Hahn (1998) have shown that if the Kuiper Belt smoothly extends to a couple of hundred AU and if the eccentricities in the disk are smaller than ~ 0.05, then Neptune can excite apsidal waves in the disk that will carry away angular momentum from Neptune's orbit and damp that planet's eccentricity. They estimate an upper limit of $\sim 2\ M_{\oplus}$ on the mass of material between 50 and 75 AU. This upper limit is marked WH98 in Fig. 5.

We note that Ward and Hahn (1998) estimate more mass than a simple extension of the Kuiper Belt's *observed* surface density would imply (i.e., we are seeing a Kuiper Gap). However, their upper limit on the surface density is much less than that required by the accretion models to grow the observed KBOs. There are three possible resolutions to this problem:

1. It could be a matter of timing. As Stern and Colwell (1997*a*) describe, the KBOs must have been formed before Neptune in order for their eccentricities to be small enough for accretion. Conversely, the Ward and Hahn (1998) upper limit applies only after Uranus and Neptune have cleared away any local objects that could excite their eccentricities. Perhaps the surface density of the Kuiper Belt decreased significantly between these two times. For example, it could have decreased because of Neptune and Uranus injecting massive objects into the 50–100-AU region, stirring up the disk and greatly increasing the collisional erosion (Morbidelli and Valsecchi 1997).
2. The uncertainties in the models could be larger than the discrepancy between them. Stern and Colwell (1997*a*) make assumptions about the velocity evolution of their objects and the physics of the collisions. Ward and Hahn (1998) make assumptions about the shape of the surface density distribution and eccentricities. Perhaps when more detailed models are constructed, the discrepancy will vanish.
3. There really was an edge to the disk from which the planets formed, at about 50 AU. This would explain why no classical KBOs have been discovered beyond this point. Although the lack of discoveries may be due to observational incompleteness (Gladman et al. 1998), Monte Carlo simulations of the detection statistics of observational surveys are consistent with a Kuiper Belt edge at 50 AU (Jewitt et al. 1998). Clearly, more work needs to be done on this topic.

VII. CONCLUDING REMARKS

Studies of Kuiper Belt dynamics offer the exciting prospect of deriving constraints on dynamical processes in the late stages of planet formation that have hitherto been considered beyond observational constraint. Many questions have been raised by the intercomparisons between observations of the Kuiper Belt and theoretical studies of its dynamics; these outstanding issues are listed below.

1. It is likely that the Kuiper Belt defines an outer boundary condition for the primordial planetesimal disk and, by extension, for the primordial solar nebula. What new constraints does it provide on the solar nebula, its spatial extent and surface density, and on the timing and manner of formation of the outer planets Uranus and Neptune? How does the Kuiper Belt fit in with observed dusty disks around other sunlike stars, such as β Pictoris?

2. What is the spatial extent of the Kuiper Belt: its radial and inclination distribution? What are the relative populations in the scattered disk and the classical Kuiper Belt? What mechanisms have given rise to the large eccentricities and inclinations in the transneptunian region?

3. What are the relative proportions of the resonant and nonresonant KBOs and their eccentricity, inclination, and libration amplitude distributions? These provide constraints on the orbital migration history of the outer planets.

4. The phase space structure in the vicinity of Neptune's mean motion resonances is reasonably well understood only for the 3:2 resonance. Similar studies of the other resonances are warranted.

5. Given the apparent highly nonuniform orbital distribution of KBOs, precisely what is the source region of short-period comets and Centaurs? What is the nature of the long-term instabilities that provide dynamical transport routes from the Kuiper Belt to the short-period comet and Centaur population?

6. Was the primordial Kuiper Belt much more massive than at present? What, if any, were the mass loss mechanisms (collisional grinding, dynamical stirring by large KBOs or lost planets)?

7. Is there a significant population of Neptune Trojans or a belt between Uranus and Neptune? What can we learn from its presence/absence about the dynamical history of Neptune?

8. What is the distribution of spins of KBOs? It may help constrain their collisional evolution.

9. What is the frequency of binaries in the Kuiper Belt? (How unique is the Pluto-Charon binary?)

10. What is the relationship between the Kuiper Belt and the Oort Cloud? How does the mass distribution in the Kuiper Belt relate to the formation of the Oort Cloud? Is there a continuum of small bodies spanning the two regions?

Acknowledgments The authors acknowledge support from NASA's Planetary Geosciences and Origins of Solar Systems Research Programs. M. J. D. is also grateful for the continuing financial support of the Natural Science and Engineering Research Council of Canada. H. L. thanks P. Weissman and S. A. Stern for discussions. Part of the research reported here was done while R. M. was a Staff Scientist at the Lunar and Planetary Institute, which is operated by the Universities Space Research Association under contract no. NASW-4574 with the National Aeronautics and Space Administration. This paper is Lunar and Planetary Institute Contribution no. 959.

REFERENCES

Brunini, A., and Melita, M. D. 1998. On the existence of a primordial cometary belt between Uranus and Neptune. *Icarus*, in press.

Chen, J., Jewitt, D., Trujillo, C., and Luu, J. 1997. Mauna Kea Trojan survey and statistical studies of L4 Trojans. American Astronomical Society, DPS meeting #29, #25.08.

Davis, D. R., and Farinella, P. 1997. Collisional evolution of Edgeworth-Kuiper Belt objects. *Icarus* 125:50–60.

Davis, D. R., and Farinella, P. 1998. Collisional erosion of a massive Edgeworth-Kuiper Belt: Constraints on the initial population. 29th Annual Lunar and Planetary Science Conference, paper no. 1437.

Dones, L., and Tremaine, S. 1993. On the origin of planetary spins. *Icarus* 103(1):67–92.

Dones, L., Gladman, B., Melosh, H. J., Tonks, W. B., Levison, H., and Duncan, M. 1998. Dynamical lifetimes and final fates of small bodies: Orbit integrations vs. Öpik calculations. *Icarus*, submitted.

Duncan, M. J. 1994. Orbital stability and the structure of the solar system. In *Circumstellar Dust Disks and Planet Formation,* ed. R. Ferlet and A. Vidal-Madjar (Paris: Editions Frontieres), pp. 245–256.

Duncan, M., and Levison, H. 1997. A scattered disk of icy objects and the origin of Jupiter-family comets. *Science* 276:1670–1672.

Duncan, M., Quinn, T., and Tremaine, S. 1987. The formation and extent of the solar system comet cloud. *Astron. J.* 94:1330–1338.

Duncan, M., Quinn, T., and Tremaine, S. 1988. The origin of short-period comets. *Astrophys. J. Lett.* 328:L69–L73.

Duncan, M. J., Levison, H. F., and Budd, S. M. 1995. The dynamical structure of the Kuiper Belt. *Astron. J.* 110:3073–3081.

Edgeworth, K. E. 1943. The evolution of our planetary system. *J. Brit. Astron. Assoc.* 20:181–188.

Edgeworth, K. E. 1949. The origin and evolution of the solar system. *Mon. Not. Roy. Astron. Soc.* 109:600–609.

Fernández, J. A. 1980. On the existence of a comet belt beyond Neptune. *Mon. Not. Roy. Astron. Soc.* 192:481–491.

Fernández, J. A. 1985. Dynamical capture and physical decay of short period comets. *Icarus* 64(2):308–319.

Fernández, J. A., and Ip, W. H. 1984. Some dynamical aspects of the accretion of Uranus and Neptune: The exchange of orbital angular momentum with planetesimals. *Icarus* 58:109–120.

Fernández, J. A., and Ip, W.-H. 1989. Dynamical processes of macro-accretion of Uranus and Neptune: A first look. *Icarus* 80:167–178.

Gladman, B., and Duncan, M. J. 1990. On the fates of minor bodies in the outer solar system. *Astron. J.* 100:1669–1675.

Gladman, B., Kavelaars, J. J., Nicholson, P. D., Loredo, T. J., and Burns, J. A. 1998. Pencil-beam surveys for faint trans-Neptunian objects. *Astron. J.* 116:2042–2054.

Hahn, J. M., and Malhotra, R. 1999. Orbital evolution of planets embedded in a massive planetesimal disk. *Astron. J.* 117:3041–3053.

Holman, M. J. 1995. The distribution of mass in the Kuiper Belt. *Proceedings of the 27th Symposium on Celestial Mechanics*, ed. H. Kinoshita and H. Nakai (Mitaka: National Astronomical Observatory of Japan).

Holman, M. J. 1997. A possible long-lived belt of objects between Uranus and Neptune. *Nature* 387:785–788.

Holman, M. J., and Wisdom, J. 1993. Dynamical stability of the outer solar system and the delivery of short-period comets. *Astron. J.* 105:1987–1999.

Ip, W. H. 1987. Gravitational stirring of the asteroid belt by Jupiter zone bodies. *Gerl. Beitr. Geoph.* 96:44–51.

Jewitt, D., Luu, J., and Trujillo, C., 1998. Large Kuiper Belt objects: The Mauna Kea 8K CCD survey. *Astron. J.* 115:2125–2135.

Kenyon, S. J., and Luu, J. X. 1998. Accretion in the early Kuiper Belt. I. Coagulation and velocity evolution. *Astron. J.* 115:2136–2160.

Knezevic, Z., Milani, A., Farinella, P., Froeschle, Ch., and Froeschle, Cl. 1991. Secular resonances from 2 to 50 AU. *Icarus* 93:316–330.

Kuiper, G. 1951. On the origin of the solar system. In *Astrophysics: A Topical Symposium*, ed. J. A. Hynek (New York: McGraw Hill), pp. 357–414.

Levison, H., and Duncan, M. J. 1993. The gravitational sculpting of the Kuiper Belt. *Astrophys. J. Lett.* 406:L35–38.

Levison, H., and Duncan, M. 1997. From the Kuiper Belt to Jupiter-family comets: The spatial distribution of ecliptic comets. *Icarus* 127:13–32.

Levison, H. F., and Stern, S. A. 1995. Possible origin and early dynamical evolution of the Pluto-Charon binary. *Icarus* 116:315–339.

Levison, H., Shoemaker, E. M., and Shoemaker, C. S. 1997. The dispersal of the Trojan asteroid swarm. *Nature* 385:42–44.

Levison, H. F., Stern, A., and Duncan, M. J. 1999. The role of a massive primordial Kuiper Belt on its current dynamical structure. Preprint.

Lissauer, J. J., and Safronov, V. S. 1991. The random component of planetary rotation. *Icarus* 93(2):288–297.

Luu, J., and Jewitt, D. 1993. Discovery of the candidate Kuiper Belt object 1992 QB1. *Nature* 362:730–732.

Luu, J., Jewitt, D., Trujillo, C. A., Hergenrother, C. W., Chen, J., and Offutt, W. B. 1997. A new dynamical class in the trans-Neptunian solar system. *Nature* 287:573–575.

Malhotra, R. 1993. The origin of Pluto's peculiar orbit. *Nature* 365:819–821.

Malhotra, R. 1995. The origin of Pluto's orbit: Implications for the solar system beyond Neptune. *Astron. J.* 110:420–429.

Malhotra, R. 1996. The phase space structure near Neptune resonances in the Kuiper Belt. *Astron. J.* 111:504–516.

Malhotra, R., 1998*a*. Pluto's inclination excitation by resonance sweeping. *29th Annual Lunar and Planetary Science Conference*, paper no. 1476.

Malhotra, R., 1998*b*. Orbital resonances and chaos in the solar system. In *Solar System Formation and Evolution*, ASP Conf. Ser. 149, ed. D. Lazzaro, R. Vieira Martins, S. Ferraz-Mello, J. Fernández, and C. Beaugé. (San Francisco: Astronomical Society of the Pacific). pp. 57–63.

Malhotra, R. 1999. Implications of the Kuiper Belt structure for the solar system. *Planet. Space Sci.*, in press.

Malhotra, R. and J. G. Williams, 1997. Pluto's heliocentric orbit. In *Pluto and Charon*, ed. D. J. Tholen and S. A. Stern. (Tucson: University of Arizona Press), pp 127–157.

Marsden, B. 1998. 1997 SZ10 and 1996 TR66. IAU Circular no. 7073.

Mikkola, S., and Innanen, K. 1992. A numerical exploration of the evolution of Trojan type asteroidal orbits. *Astron. J.* 104:1641–1649.

Morbidelli, A. 1997. Chaotic diffusion and the origin of comets from the 2/3 resonance in the Kuiper Belt. *Icarus* 127:1–12.

Morbidelli, A., Thomas, F., and Moons, M. 1995. The resonant structure of the Kuiper Belt and the dynamics of the first five trans-neptunian objects. *Icarus* 118:322–340.

Morbidelli, A., and Valsecchi, M. 1997. Neptune-scattered planetesimals could have sculpted the primordial Edgeworth-Kuiper Belt. *Icarus* 128:464–468.

Öpik, E. J. 1951. Collision probabilities with the planets and the distribution of interplanetary matter. *Proc. Roy. Irish Acad.* 54A:165–199.

Petit, J. M., Morbidelli, A., and Valsecchi, G. 1998. Large scattered planetesimals and the excitation of the small body belts. *Icarus*, in press.

Quinn, T., Tremaine, S., and Duncan, M. J. 1990. Planetary perturbations and the origin of short-period comets. *Astrophys. J.* 355:667–679.

Safronov, V. S. 1966. Sizes of the largest bodies falling onto the planets during their formation. *Sov. Astron.* 9:987–991.

Stern, S. A. 1991. On the number of planets in the outer solar system: Evidence of a substantial population of 1000-km bodies. *Icarus* 90:271–281.

Stern, S. A. 1995. Collisional timescales in the Kuiper Disk and their implications. *Astron. J.* 110:856–868.

Stern, S. A. 1996. On the collisional environment, accretion time scales, and architecture of the massive, primordial Kuiper Belt. *Astron. J.* 112:1203–1211.

Stern, S. A., and Colwell, J. E. 1997*a*. Accretion in the Edgeworth-Kuiper Belt: Forming 100–1000 km radius bodies at 30 AU and beyond. *Astron. J.* 114:841–849.

Stern, S. A., and Colwell, J. E. 1997*b*. Collisional erosion in the primordial Edgeworth-Kuiper Belt and the generation of the 30–50 AU Kuiper Gap. *Astrophys. J.* 490:879–882.

Teplitz, V., Stern, S.A., Anderson, J. D., Rosenbaum, D.C., Scalise, R., and Wentzler, P. 1999. IR Kuiper Belt constraints. *Astrophys. J.*, in press.

Torbett, M. 1989. Chaotic motion in a comet disk beyond Neptune: The delivery of short-period comets. *Astron. J.* 98:1477–1481.

Torbett, M., and Smoluchowski, R. 1990. Chaotic motion in a primordial comet disk beyond Neptune and comet influx to the solar system. *Nature* 345:49–51.

Ward, W. 1981. Solar nebula dispersal and the stability of the planetary system. I. Scanning secular resonance theory. *Icarus* 47:234–264.

Ward, W. R., and Hahn, J. M. 1998. Neptune's eccentricity and the nature of the Kuiper Belt. *Science* 280:2104–2106.

Weissman, P. R. 1990. The cometary impactor flux at the Earth. In *Global Catastrophes in Earth History*, ed. V. L. Sharpton and P. D. Ward (Boulder, CO: Geological Society of America), Special Paper 247, pp.171–180.

Weissman, P., and Levison, H. 1997. The population of the trans-neptunian region: The Pluto-Charon environment. In *Pluto and Charon*, ed. S. A. Stern and D. J. Tholen (Tucson: University of Arizona Press), pp. 559–604.
Wetherill, G. W. 1989. Origin of the asteroid belt. In *Asteroids II*, ed. R. P. Binzel, T. Gehrels, and M. S. Matthews (Tucson: University of Arizona Press), pp. 661–680.

FORMATION AND COLLISIONAL EVOLUTION OF THE EDGEWORTH-KUIPER BELT

PAOLO FARINELLA
University of Trieste

DONALD R. DAVIS
Planetary Science Institute

and

S. ALAN STERN
Southwest Research Institute

We provide a summary of current research concerning the formation and the collisional history of the Edgeworth-Kuiper Belt. Collisions appear to have first built up sizable (up to Pluto-sized) bodies in a primordial, massive planetesimal population. Then, following the formation of Neptune, collisional grinding has been eroding the population at diameters smaller than about 100 km, at a variable extent depending on heliocentric distance. In both phases, collisional evolution has interacted in a complex way with a variety of subtle dynamical processes, and this interplay has been responsible for stopping accretion and for ejecting bodies (including the currently observed Jupiter-family comets) from the stable regions of orbital element space. We compare the properties and history of the transneptunian belt to those of main-belt and Trojan asteroids, and we discuss the recent evidence for similar disks of planetesimals and debris around both newly formed and main-sequence stars.

I. INTRODUCTION

The discovery of the transneptunian Edgeworth-Kuiper Belt (hereinafter EKB) has expanded in both space and time our knowledge of the planetary system. Pluto is not seen any longer only as the outermost planet but also as the largest member of an immense population of small icy bodies forming a vast disk-shaped structure that surrounds a central "hole" containing the planets. This disk is the remnant from the planetesimal accretion process, which occurred in the first 10^8–10^9 yr of solar system history, a remnant subjected to varying degrees of collisional and dynamical reprocessing at different heliocentric distances.

The EKB provides a fascinating empirical bridge between planetary science and stellar astrophysics, because many similar structures have been discovered around solar-type stars, both in their formative phase and in their main-sequence maturity. Thus, comparative studies based on similar modeling concepts and techniques have now become possible.

What is known about the detailed architecture of the EKB is reviewed in the chapter by Jewitt and Luu, this volume (see also references therein). Here we summarize the main data that provide the key observational constraints for models of the formation and evolution of the belt. Between 30 and 50 AU from the Sun, besides Pluto and Charon (about 2300 and 1200 km in diameter, respectively), there are about 70–140 $\times 10^3$ objects in the 100- to 500-km diameter range (only 0.1% discovered so far), probably $\approx 10^8$ "comets" 10–100 km in diameter, and $\approx 10^{10}$ objects 1–10 km in size. The total mass is $\approx$0.1–0.2 $M_\oplus$. About 40% of the observed bodies (but only 10 to 20% of the real population, due to observational selection) have orbits with relatively large (>0.1) eccentricities and inclinations, locked like Pluto in the 2:3 mean-motion resonance with Neptune, at a semimajor axis of about 39.5 AU. The remaining bodies have typical eccentricities of 0.02–0.08 and typical inclinations of 2–8 degrees. The abundance of bodies in and beyond the 1:2 neptunian mean-motion resonance at 47.6 AU is still uncertain. Ongoing and future surveys should ascertain whether the belt extends beyond its currently observed outer limits near 50 AU or, alternatively, the outer belt population is deficient (or reaches smaller maximum sizes) compared to that present in the inner region. As we will see, this is a critical issue for reconstructing the formation and history of the EKB population.

In the remainder of this chapter we will describe some recent models for the current and past erosive collisional evolution of the EKB (section II) and for the accretion process, both in a primordial, presumably more massive, disk inside 50 AU (section III) and for the putative population beyond 50 AU (section IV). In section V we will outline some intriguing analogies between the EKB and the other populations of small bodies located in dynamically stable regions of the solar system (main-belt asteroids and Trojans), whereas in section VI we will draw some comparisons with the disk-shaped structures observed around other stars. In the final section we will briefly describe some outstanding open problems that still need to be addressed at the theoretical as well as the observational level.

II. CURRENT AND PAST COLLISIONAL EVOLUTION

The collision rate within a population of orbiting objects depends on two factors: (i) the orbital distribution of the bodies and (ii) their number as a function of size. After the discovery of 1992 QB_1 (about 200 km in diameter) and its siblings in the early 1990s, supplemented by the detection of 10-km-sized bodies by Cochran et al. (1995), and the requirement for a $\approx 10^{10}$ reservoir of Jupiter-family comets[a] (Holman and Wisdom 1993; Levison and Duncan 1993; Duncan et al. 1995), it became clear that collisions are an important process, at least in the inner portion of the EKB. This was first pointed out by Stern (1995), who carried out simple particle-in-a-box calculations and showed that collisions occur frequently on the astronomical timescale in the EKB and that, at least for comet-sized bodies, impact velocities are high enough to promote comminution rather than binary accretion in the present conditions.

Based on these results, Stern (1996*a*) also suggested that such collisions could generate both discrete dust clouds and a smooth dust background. Backman et al. (1995), Stern (1996*a*), and Teplitz et al. (1999) have evaluated the importance of this dust (as well as larger debris) for creating detectable IR signatures, finding several potential diagnostic signatures that both the *Infrared Space Observatory* (ISO) and the *Space Infrared Telescope Facility* (SIRTF) are in principle capable of detecting. In addition, Liou et al. (1996), Flynn (1996), and Yamamoto and Mukai (1998) have evaluated the possibility that EKB-derived dust, produced either by collisions in the EKB or by interstellar grain impacts into EKB objects (EKOs), will contribute significantly to the interplanetary dust complex. Liou et al.'s numerical integrations of dust dynamics after leaving the EKB revealed that ~80% of the test particles they tracked are lost to the interstellar medium (ISM), but ~20% enter the planetary region as a result of Poynting-Robertson (P-R) drag. All three papers agree that EKB-derived dust should form a significant source, contributing perhaps 10–30%, of the 10-μm class dust particles at 1 AU. Flynn (1996) even tentatively identified a particular interplanetary dust particle (IDP) with an anomalously long exposure time and low atmospheric entry speed as potentially being derived from the EKB. Further evidence for EKB-derived dust comes from the *Pioneer 10/11* and *Voyager 1/2* spacecraft. Humes (1980) published results of *Pioneer 10/11* dust detector impact rates out to 18 AU (where the detectors ceased operating) and found that the rate of dust impacts detected by the *Pioneers* did not diminish significantly beyond 4 AU, indicating that an additional, external source is present beyond 18 AU. Gurnett et al. (1997) reported results from the *Voyager 1* and *2* plasma wave detectors, which are capable of detecting certain types of dust impacts on the spacecraft. Their results, which tracked dust impacts as the *Voyager* spacecraft traveled from 6 to 60 AU, detected a small but persistent level of impacts, which Gurnett et al. (1997) modeled to reveal

[a] This number assumes that most of the Jupiter-family comets come directly from the EKB. Another source that may supply some of these comets is the so-called "scattered disk," a long-lived population of bodies that has been scattered onto very long-period orbits following encounters with the outer planets (Duncan and Levison 1997). The relative contributions of these two sources are not known at present.

a mean dust particle density for μm-sized impactors of 2×10^{-14} cm^{-3}, which they attribute to a solar system source. Interestingly, no dust impacts were recorded by either spacecraft between 50 and 60 AU.

The next generation of collisional models was time dependent. Similar codes were developed independently by Farinella and Davis (1996; see also Davis and Farinella 1997) and by Stern and Colwell (1997*a,b*). These codes compute the statistical results of collisions among a population of objects at a given heliocentric distance, taking into account the energetics of the collisions, so that both erosion and accretion may be represented (in the current regime of impact speeds $\approx$0.5–1 km s^{-1}, however, accretion is negligible). Collision rates were computed by Farinella and Davis (1996) by applying a collision probability theory developed by Wetherill (1967) and widely applied to the asteroid case (Farinella and Davis 1992), and by Stern and Colwell (1997*b*) by using a locally averaged, particle-in-a-box formalism, with collision cross sections appropriately scaled for gravitational focusing but limited by Keplerian shear. Farinella and Davis (1996) tracked the evolution of the EKB population by distributing it over a set of discrete size bins and calculating the collisional interactions of each bin with every other one during a sequence of small time steps (see, e.g., Davis et al. 1989). Stern and Colwell (1997*b*) applied instead a moving-bin ("batch") technique like the one developed by Wetherill (1990). In these models, dynamical communication is not allowed between masses at different heliocentric distances, but results for different heliocentric distances can be computed in separate runs. Collisional outcomes are computed from the energy of the impact and the size and velocity distribution of the resulting debris. The outcome of any given collision can range from complete accretion (no debris achieves escape velocity from the colliding pair) to cratering (local target damage with escape of some ejecta) to complete disruption (in which the object is destroyed because greater than half the target mass has escape velocity). Model runs have been performed with differing assumptions about the initial distribution and the collisional response parameters of EKOs.

The results derived from the two codes have been in broad agreement. First we shall summarize the results reported by Davis and Farinella (1997), who studied in more detail the zone within 50 AU and assumed a baseline scenario tailored on the observed orbital distribution and a moderate starting mass of 3 $M_\oplus$, an order of magnitude larger than for the current population. Then we shall report on the results of Stern and Colwell (1997*b*), who explored in more detail a "massive" starting population scenario (up to several tens of $M_\oplus$) and the radial dependence of the collisional process.

The main conclusions of Davis and Farinella (1997) confirmed Stern's (1995) estimates and provided some further insights. Here is a summary of them:

1. *The population of EKOs larger than about 100 km diameter has not been significantly altered by collisions over the age of the solar system* (see Fig. 1). The size distribution in this range must be the original (accretional) population. Many of these bodies, however, have probably been converted into "rubble piles," because there is a significant energy gap between the projectile energy needed to shatter the target and that required to disrupt it, i.e., to disperse most of the target mass to infinity (Davis et al. 1989). Using the baseline initial population and an icelike impact strength $S = 3 \times 10^6$ erg cm^{-3}, it can be shown that each 200 km diameter EKO near 40 AU is likely to have undergone several shattering (albeit not disrupting) collisions over the solar system's lifetime. The growing importance of self-gravitational effects for larger sizes may explain the varying slope of the observed size distribution.
2. *As shown in Fig. 1, smaller EKOs are mostly fragments undergoing a collisional cascade, with a (differential) size distribution exponent*

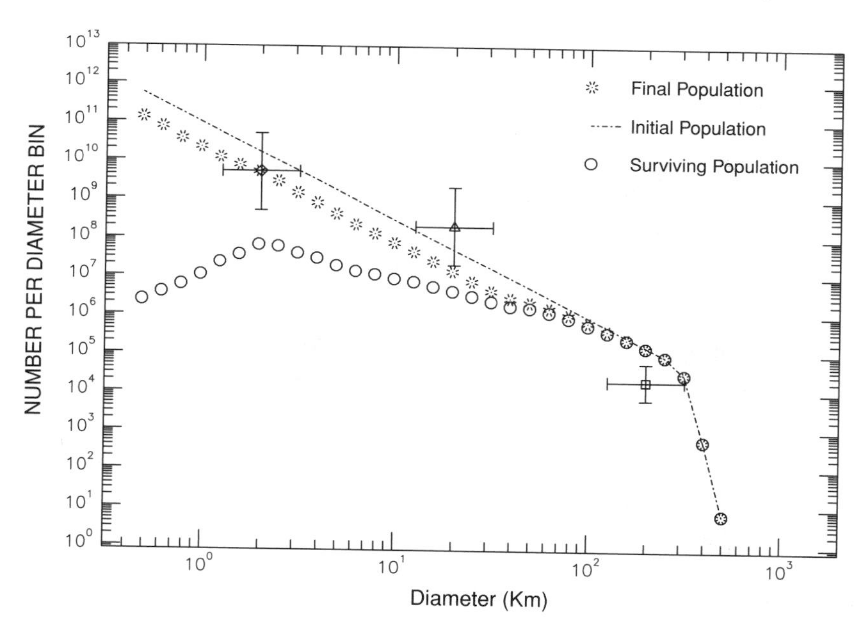

Figure 1. The collisionally evolved population of EKOs after 4.5 Gyr, starting from a hypothetical initial population having a mass of about 3 $M_{\oplus}$ and a power law size distribution for diameters <300 km (dashed-dotted line). The number of "survivors" from the original population is also shown as a function of diameter. The bars correspond to the observational constraints discussed in the text and span a factor of 100 in the number of bodies and a factor of 4 in their sizes. The observational point for the large EKOs has a smaller uncertainty in the population, reflecting a more reliable estimate of their numbers. (Adapted from Davis and Farinella 1997.)

close to the −3.5 equilibrium value (Dohnanyi 1969). However, their current abundance implies that there was a substantial original (accretionary) population down to a few tens of kilometers in size; otherwise, there would have been a shortage of projectiles, and the current EKB reservoir of comets would be too small. On the other hand, for initial values of the size distribution index ≤ -2 at diameters <300 km, the final population is quite insensitive to initial conditions, reflecting collisional relaxation to the −3.5 equilibrium value. If observational surveys show that the current index has a significantly different value in the 1–50 km size range, that result will probably require an interpretation in terms of size-dependent collisional response parameters.

3. *As a by-product of the collisional process, about 10 fragments per year 1 to 10 km in size are currently produced in the EKB at 40 AU.* This estimate refers to the baseline case, with an uncertainty factor of $\simeq$4 depending on the assumed collisional response parameters. With ejection speeds of 10–100 m s^{-1}, similar to those inferred for asteroids, these fragments have semimajor axes that differ by about 0.1–1.0 AU from those of their parent bodies. This is sufficient to cause at least a few percent of these fragments (say, 0.2 yr^{-1}) to fall into the resonant "escape hatches" from the EKB (Duncan et al. 1995; Morbidelli et al. 1995) and to evolve chaotically into the planetary region of the solar system. This is roughly in agreement with the required flux to replenish the short period comets, which have an estimated population (including extinct/dormant nuclei) of 2×10^4 bodies and a dynamical lifetime of about 3×10^5 years (Levison and Duncan 1994).
4. *However, disruptive collisions cannot explain how large bodies, such as Chiron and other sizable members of the so-called Centaur population, originate, because collisional fragments cannot be as large as 100 km or more.* Some purely dynamical delivery mechanism is probably at work, such as slow diffusion from the vicinity of the resonance zones (e.g., Morbidelli 1997). An additional mechanism for the insertion of big Centaurs into the delivery routes is through nondisruptive collisions: a 10^{-2} projectile-to-target mass ratio is not enough to disrupt a 100-km-sized EKO but can change its orbital velocity by several m s^{-1}; hence, its semimajor axis by $\approx$0.2% $\approx$ 0.1 AU. Such events may be frequent enough to provide a significant influx of Chiron-sized bodies. As for the short-period comets, it remains to be seen whether the dynamical or the collisional delivery mechanism is the dominant one.
5. *Relatively large changes in semimajor axis may result from the (rare) breakup of a sizable EKO of the order of 100 km in diameter, because a small relative speed goes a long way in the outer solar system in terms of changing the orbits* (see Fig. 2). Therefore, it will be very

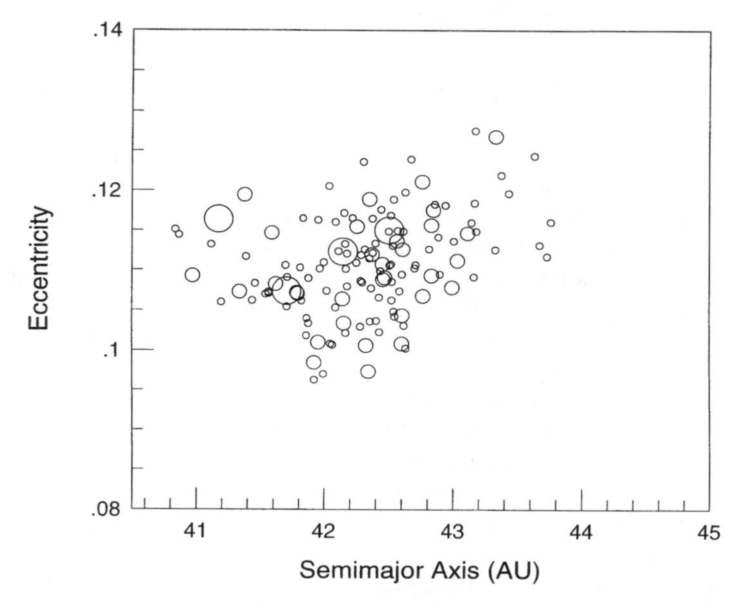

Figure 2. The distribution in the semimajor axis-eccentricity plane of a simulated EKB "family," the fragments resulting from a collisional breakup event involving two bodies of 100 and 55 km in diameter, initially orbiting at $a = 42.5$ AU and $e = 0.1$. We assumed that a fraction 0.2 of the initial kinetic energy in the center-of-mass reference frame is partitioned into kinetic energy of the ejected fragments. The larger open circles represent fragments 15–40 km in diameter, and the intermediate and smaller ones fragments 8–15 and 3–8 km in diameter, respectively. Note the relatively large spread in semimajor axis, which may result in a significant fraction of the fragments "falling" into chaotic zones associated with resonances. (Adapted from Farinella and Davis 1996.)

difficult to distinguish EKB dynamical families from background objects, even when a large catalog of well-determined orbits becomes available (but see Stern et al. 1999 with regard to the speculative possibility of a Pluto family in the 2:3 mean-motion resonance, resulting from the Pluto-Charon binary fission). Another consequence of this fact is that collisions occurring far away from the resonant escape hatches can inject many fragments into them, so if the collisional injection mechanism is important, the source regions of the short-period comets should sample a fairly large portion of orbital phase space in the EKB.

Although we have clear evidence of both widespread accretion and mass loss having occurred in the now familiar EKB 30–50 AU zone, there is little information about the degree to which these processes operated farther out. This can be traced in large measure to the fact that, as of mid-1998, no objects have been detected on near-circular orbits beyond 48 AU.[b]

Despite this lack of observational data, some modeling results concerning this region have been reported, leading to predictions about what the disklike region beyond 50 AU may contain. Given the apparent need, discussed below, for a much larger mass (say, 15–50 $M_\oplus$, see section III) in the ancient 30–50 AU zone, it would seem plausible that the protoplanetary disk was not truncated at about 50 AU but probably extended much farther out. After all, simply displacing the "edge" of the planetary region and the accompanying region of high surface mass density from the 30-AU limit of Kuiper's day to the present-day observational limit at 50 AU seems an *ad hoc* solution.

Presently, there are not good constraints on the amount of material that may lie beyond 50 AU. Hamid et al.'s (1968) study of perturbations on long-period comet orbits constrains the mass inside 50 AU to be $<1.3\ M_\oplus$; Yeomans (1986) obtained a similar constraint. Given the recent advances in observational knowledge about the transneptunian region, and the fact that a simple extrapolation of the surface mass density profile of solids alone in the giant planets would yield 50–100 $M_\oplus$ of material out to 100 AU, this is not a very significant constraint. On the other hand, dynamical studies by Knežević et al. (1991) and Duncan et al. (1995) have provided strong evidence that the role of the giant planets is negligible in exciting eccentricities beyond ~50 AU. This in turn suggests that erosion-driven mass loss may not have been very important beyond ~50 AU (or perhaps somewhat farther, depending on the degree to which internal velocity evolution excited eccentricities) and that any primordial material beyond this distance could have survived to the present.

Stern and Colwell (1997*b*) investigated this possibility by running their time-dependent model of collisional evolution of the EKB between 30 and 70 AU. Their primary motivation was to determine whether collisional erosion due to the perturbing effects of Neptune and coupled mass-velocity evolution internal to the EKB could sufficiently deplete the massive primordial EKB, believed necessary to build the observed population, down to its present low-mass state. A secondary objective was to determine how far collision-generated erosion may have occurred. The main assumptions and techniques of this collisional model have been summarized earlier, and some more details will be given in section III.

Color Plate 20 shows the results of model simulations by Stern and Colwell (1997*b*) for the time-dependent evolution of EKB mass, assuming an initial mass of 35 $M_\oplus$ and different values of the mean random

[b] There are objects with semimajor axes $a > 48$ AU, such as 1996 TL_{66} (see Luu et al. 1997), but their high eccentricities indicate that they were most likely formed interior to this region and scattered outward by Neptune in the distant past (Duncan and Levison 1997).

eccentricity. In the model, EKOs were broken up and eroded by collisions and lost from the system as a result of P-R drag and radiation pressure blowout of small particles resulting from two-body collisions. The size distribution had a higher cutoff at a radius of 200 km, similar to the current-day EKB population. The key result is the evolution of a deep gap in the mass distribution outward from 30 AU to about 50–60 AU[c], the limit of Neptune's significant gravitational influence. This evolution produced characteristic mass loss rates of 10^{20}–10^{21} g yr^{-1} during the collisional erosion phase, which was found to be most severe for 30–100 Myr after perturbed eccentricities rose high enough to induce significant mass loss. Further, Stern and Colwell (1997*b*) found that during this period, radial optical depths of order 10^{-3} to 10^{-4} prevailed in the 30–50-AU zone of the EKB. They concluded that collisions alone could have caused the 30–50-AU zone to have been depleted from a few tens of $M_\oplus$ to a value near its present-day state, creating a severe surface density decrease in this region, similar to what is observed. The main requirement for this is a relatively high mean orbital eccentricity (larger than about 0.1) over a period of at least 100 Myr. If this were not the case, other mechanisms (dynamical losses or interactions with Neptune–scattered bodies) had to play a significant role in depleting the EKB.

Although many details of the picture are still missing, one fundamental implication of this work is that the mass-depleted EKB region is an *evolved* rather than primordial structure. This result in turn implies that unless the primordial solar nebula coincidentally truncated near 50 AU, then the 30–50-AU region is plausibly actually a *local* depletion in what could be a far more extensive Edgeworth-Kuiper "disk." Figure 3 shows a schematic view of the implied structure of the entire 30–100-AU region, based on these findings and assuming that the primordial nebula indeed extended outward smoothly beyond 50 AU.

If confirmed, this result has other implications. Davis and Farinella (1998) have also explored the massive-belt scenario using their code described earlier. They found that in the 30–50-AU region a massive primordial EKB resulting from accretion can be collisionally eroded to $\approx 1\%$ of its initial mass only if the initial population had a very steep mass distribution. In other words, the accretion process had to be stopped at an early phase, when no more than $\approx 10^5$ bodies (roughly, the current number) had grown to >100 km diameter and most ($>90\%$) of the mass was still residing in the original 1–10-km planetesimals and their fragments. In the subsequent, disruptive collisional process, almost the entire mass loss must have been accounted for by bodies smaller than a few tens of kilometers in diameter. Thus, the large EKOs have never been much more

[c] Erosion may have occurred somewhat farther out if other effects, not modeled in Stern and Colwell (1997*b*), such as the role of a scattered disk, were important.

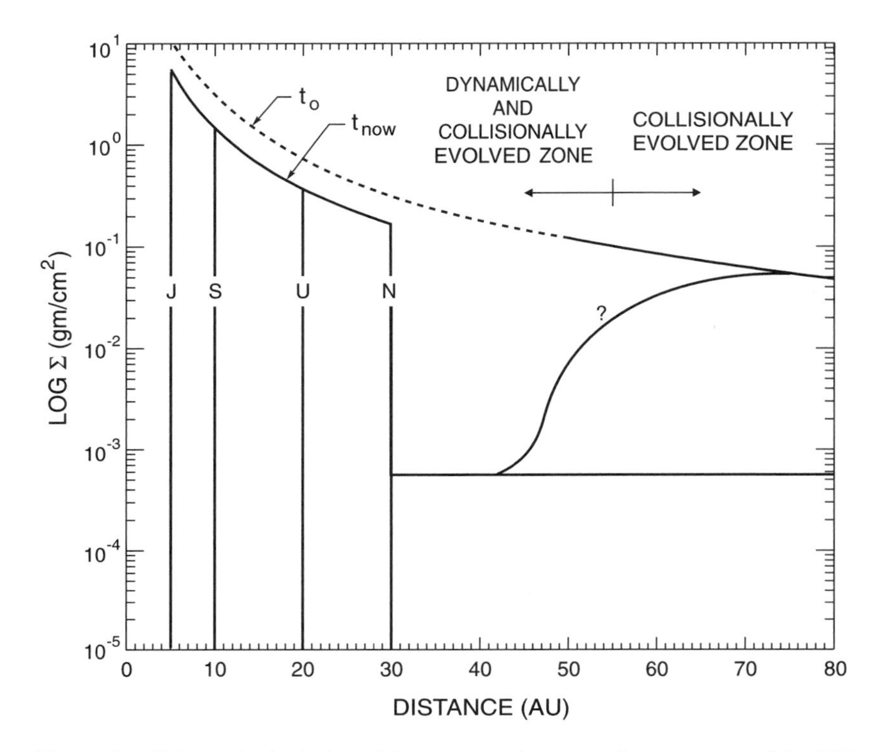

Figure 3. Schematic depiction of the suggested present-day structure of the EK "disk." The 30–50-AU zone is shown as both collisionally and dynamically evolved, because dynamics acted to destabilize many orbits with $a < 42$ AU. Beyond some distance, perhaps as great as 80 AU, where dynamical and collisional erosion had little effect, we expect there to be a zone in which accretion has occurred but perturbations by the giant planets (and internal velocity evolution) were too ineffective to have initiated erosion, and beyond that, a primordial zone in which the accretion rates hardly modified the initial population of objects. The dashed line indicates the suggested primordial structure. The question-marked line indicates a possible present-day structure. No attempt is made here to depict the high-frequency structure of mass density in the 30–42 AU zone (cf. Duncan et al. 1995). The current surface mass density "edge" at 30 AU first noticed by Edgeworth and Kuiper may in fact represent only the inner boundary of a trough resulting from a combination of dynamical and collisional evolution. (Adapted from Stern 1996*a*.)

abundant than they are now (although losses by dynamical diffusion and nondisruptive collisions may have somewhat decreased their numbers, as discussed earlier). Of course, it remains to be seen under which conditions the accretionary process can meet these constraints and whether a plausible mechanism related to the growth of Neptune could have interrupted accretion at the required early stage.

III. THE ACCRETIONARY PHASE OF THE EKB

In recent decades, the accretion hypothesis has emerged as the reigning paradigm for the fundamental process by which the solar system formed, although unanswered questions still remain. Excellent recent reviews of this topic can be found in Lissauer et al. (1995) and Ward (1996). Here, we will restrict our discussion to the formation of the outer planetary system, specifically Neptune, Pluto, and the EKOs.

Understanding the formation of EKOs is not a simple matter, a characteristic shared with other small-body populations (see section V). As noted by Stern (1995), a straightforward calculation of the timescale for the formation of the currently observed bodies from kilometer-sized planetesimals, assuming that the mass available for accretion in the EKB was the same as it is today, shows that it takes much longer than the age of the solar system to grow even a 100-km-sized body. Clearly, conditions were much different early on from what they are today.

Stern (1995) suggested that the easiest way to solve the timescale problem is to assume that there was significantly more mass in the EKB region during the accretionary phase than there is today. Under this assumption, Stern and Colwell (1997*a*), using a single-zone accretion model (hereafter called SC97*a*) including fragmentation, and running both fixed-impact-speed cases and an empirical velocity evolution case, showed that QB_1 and even Pluto-sized bodies can be formed within ≈1 Gyr provided that:

i. An initial mass of the order of 15–50 $M_\oplus$ (about a factor of 100 in excess of the current value) resided in the region 30–50 AU.
ii. Initial planetesimals were strong, i.e., quite resistant to collisional disruption.
iii. Mean random orbital eccentricities were less than ≈0.0025 during accretion.

Stern and Colwell (1997*a*) assumed two scenarios for how EKOs respond to collisions: a "strong-body" model, which assumes that EKOs are like rocky bodies, and a "weak-body" model, appropriate for weakened or porous ice bodies. According to their results, when bodies are assumed to be "weak," growth becomes much slower, with the largest body that formed after 2 Gyr reaching only about 200 km in diameter. Accretion is also hindered if the initial planetesimal disk is dynamically excited to eccentricities >0.0025.

Kenyon and Luu (1998) also recently calculated the timescale for growth of bodies up to Pluto size in the region 32–38 AU. Their model (hereafter called KL98) also used the "batch" method (Wetherill 1990) to approximate the continuous particle mass distribution, and they numerically solved the coagulation and energy equations to evolve the mass and velocity distributions in time. They found a range of timescales for growth, reflecting different starting sizes; interestingly, the smaller the initial planetesimals, the shorter the accretion time. One point to note in comparing results between different workers is that one needs to compensate for differing assumptions. KL98 used a heliocentric zone smaller by about a factor of 3 than did SC97*a*; hence, the initial mass must be reduced accordingly to compare timescales. Thus, the 35-$M_\oplus$ initial mass of SC97*a* should be compared to a 10-$M_\oplus$ starting mass under KL98. However, the growth times are dramatically shorter in the KL98 case, by nearly an order of magnitude. The largest body grows to 500-km diameter in 25–90 Myr and to 2000-km diameter in 32–150 Myr. This difference is possibly because the KL98 model neglected fragmentation events, which hinder accretion.

Davis and Farinella (1999) used a multizone accretion code, which simultaneously evolves the mass and velocity distribution and includes fragmentation effects to study the growth of bodies in the EKB in the presence of a growing Neptune. They found growth times of several 10^8 yr for both the "weak" and "strong" cases of the strength of particles: that is, intermediate to the previous two studies. Clearly, further work remains to be done to understand the timescales for the formation of bodies in the outer solar system and how sensitively they depend on the input parameters and assumptions of the models.

Although assuming a massive initial EKB can solve the growth time problem for the observed EKOs larger than 100 km, there is then the question as to what happened to the extra mass that was added to solve the accretion timescale problem. Perhaps dynamical mechanisms can account for a significant (size-independent) mass loss, but certainly not for a $\approx$99% depletion. Also, what was the mechanism that transformed the low-eccentricity and -inclination orbits that characterized the accretionary period into the much more dynamically excited orbits that we see today and produce the collisionally destructive environment?

As discussed in section II above, Stern and Colwell (1997*b*) argue that collisions would grind down the massive initial population to dust, which would then be removed by radiation forces. This model solves the timescale problem and results in sufficient erosion to transform the massive primordial EKB into the present-day one, provided some significant constraints on the size distribution of the EKB population at the end of the accretion phase are met. These constraints have been analyzed by Davis and Farinella (1998). They used the massive (35 $M_\oplus$) initial population preferred by Stern and Colwell (1997*a,b*) as the initial conditions for carrying out collisional evolution studies of the EKB population, to determine the starting size distribution that collisionally evolves to the present EKB size distribution. A robust conclusion appears to be that the starting population must have had most of its mass in small ($<$10 km) bodies, and the population of QB_1-sized EKOs could only have been moderately larger than it is today (see Fig. 4).

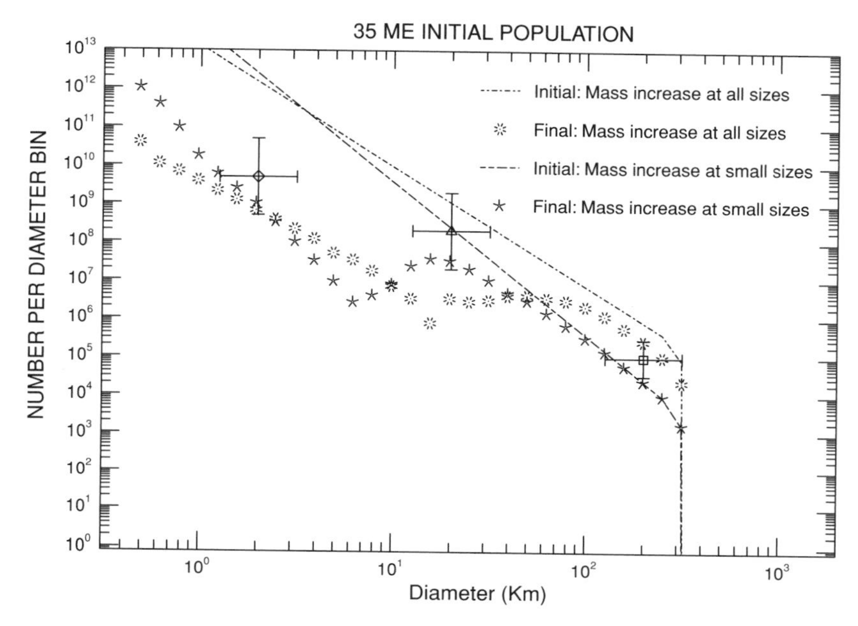

Figure 4. The same as Fig. 1 but for two "massive" (about 35 $M_\oplus$) initial populations. In one case the current population was just scaled up by a factor 100, with all the parameters of the models chosen in order to maximize collisional erosion. The final size distribution does not match the current one, because there are too many QB_1-sized bodies and too few 10–20-km sized objects. A better match is found in the other case, where the lower initial population at large sizes is compensated by a much steeper power law slope at small sizes, such that most of the mass is accounted for by 1–10-km sized bodies. (Adapted from Davis and Farinella 1998.)

Given these constraints on the total mass as well as the distribution of mass with size, we next ask whether it is plausible to have such a size distribution as a result of accretion and what the mechanism is for terminating growth at this particular stage. The most frequently invoked mechanism for terminating growth in the EKB is the formation of Neptune (Stern 1996*a*). Stern and Colwell (1997*a*) show that gravitational perturbations by Neptune are sufficient to "heat up" the orbits of EKOs enough to terminate the accretion era and initiate the collisional grinding period that continues to the present day. They further suggest that the size distribution of EKOs must bear the signature of accretion, although somewhat modified by subsequent collisional evolution.

An alternative scenario has been proposed by Morbidelli and Valsecchi (1997) and Petit et al. (1999). These authors argue that encounters of EKOs with large (0.1–5 $M_\oplus$) Neptune-scattered planetesimals (LNSPs) may have terminated growth in the EKB. In this scenario, Earth-mass planetesimals in Neptune's zone were gravitationally scattered by the growing proto-Neptune into the EKB. These planetesimals were numerous and large enough to stir up in turn the orbits of the growing EKOs to the point where accretion was halted and disruption began. According to this scenario, after a time of the order of 100 Myr, all of the LNSPs were in turn removed by ejection from the solar system, scattered onto very long-period orbits, or impacted onto Neptune.

Other dynamical mechanisms have also been proposed for sculpting the EKB and accounting for different features of its distribution in mass and orbital elements. Malhotra (1995) argued that a migrating Neptune was responsible for the abundance of objects (including Pluto) in the 2:3 neptunian mean-motion resonance. Ward and Hahn (1998) have suggested that spiral density waves may have damped growth in the EKB. According to Levison et al. (1999), sweeping secular resonances with Neptune may have depleted the inner part of the belt. For more details on these models, we refer the reader to the chapter by Malhotra et al., this volume.

All of the above scenarios, however, depend upon Neptune having already grown to a significant fraction of its present mass, although the timescale for doing this is not well constrained as it is for the gas giant planets, Jupiter and Saturn. Uranus and Neptune contain a relatively small amount (5–20%) of hydrogen and helium. Most models predict that too much gas would be accreted if the nebula existed for times much longer than $\approx$10 Myr after the formation of bodies about 10 $M_\oplus$ in mass (Lissauer et al. 1995). However, an alternative is that a number of growing embryos 0.1–1 $M_\oplus$ could trap the nebular gas quasistatically, and that Neptune's gas reservoir arrived via accretion of such bodies (Stevenson 1984). In that case, the timescale for the final formation of Neptune could have been much longer than the nebular lifetime.

A possible constraint on the timescale of Neptune formation might be inferred from the lunar cratering record. This record shows a significant decline in the cratering projectile flux, starting about 3.8 Gyr ago and declining to nearly the present flux by 3.2 Gyr ago. The final growth of Neptune would almost certainly have been accompanied by the gravitational scattering of residual planetesimals in the Neptune zone; this scattering would have greatly contributed to the lunar cratering flux (Lissauer et al. 1995). If the decline in the lunar cratering flux is related to the decline in the Neptune-scattered flux, then Neptune must have formed over a time scale of $\approx$1 Gyr after the solar system's origin. However, it must be said that the origin of the late heavy bombardment is still poorly understood, and other sources of impactors cannot be ruled out.

In the broad view, our theoretical understanding of the formation of Neptune has come full circle. Pioneering work by Safronov (1969) on the formation of Neptune based on the planetesimal accretion hypothesis yielded an embarrassing result: It took longer than the age of the solar system to form a Neptune! Subsequent work by Greenberg et al. (1984) and by Wetherill and Stewart (1989) pointed out that the "runaway growth" mechanism, whereby a single body would grow rapidly in a zone, could partially alleviate this dilemma. However, once the dominant planetesimal (planetary embryo) in a region has depleted the mass within its gravitational reach, its growth rate slows dramatically. Similar growth processes are occurring throughout the planet formation region, so many embryos are forming; the net result of this stage of planet formation is to transform the circumstellar mass from a distribution in which the mass is in bodies smaller or $\approx$10–20 km in size (10^{18}–10^{19} g) to one in which most of the mass is in much larger bodies, 10^{26}–10^{27} g in mass, which are moving on low-eccentricity, low-inclination orbits. The final stage of planet formation (the "end game") involves the accumulation of these large bodies into a few planet-sized objects, on a timescale that is typically longer than it takes the embryos to grow in the previous phase.

Current work in progress indicates that major difficulties still remain in understanding formation timescales in the outer solar system. One paper that claimed to solve the growth time problem of Neptune is that of Ip (1989), who used a Monte Carlo code to model the final stage of Neptune's growth and claimed to form a fully grown Neptune within $\simeq 10^8$ yr. A similar result was found, with a larger swarm of smaller (Mars-sized) planetesimals starting on near-circular and coplanar orbits, by Fernández and Ip (1996). However, this result may be an artifact of the simplified dynamical model that was adopted. Recently, Stewart and Levison (1998) attempted to reproduce the Monte Carlo results by a full N-body simulation of the "end game" of accretion in the outer solar system, including Jupiter and Saturn, but were unable to do so. In their work, the region interior to 20 AU was rapidly cleared out by Jupiter's and Saturn's long-range effects, and very little accumulation occurred outside of 20 AU within 10^8 yr. Thus, the timescale for Neptune's formation is a problem that continues to vex the theorists, and the role of Neptune in terminating growth in the EKB clearly depends on its solution. Perhaps the argument should be reversed, and the existence and properties of the EKB will be used in the future to constrain models of planetary formation in the outer solar system.

IV. ACCRETION IN THE REGION BEYOND 50 AU

In the distant region of the EKB where severe mass loss never occurred because Neptune and internal velocity evolution were not effective in stirring up eccentricities, growth may have proceeded unabated to the present time. To investigate accretion in this region, Stern and Colwell (1997*a*) applied their time-dependent collisional evolution model. Their model runs explored a wide range of initial EKB masses, putting 1 to 50 $M_\oplus$ in the 30–50-AU zone and extrapolating outward as far as 120 AU under the assumption that the surface mass density is proportional to the $-\frac{3}{2}$ or -2 power of heliocentric distance. Once an initial EKB mass was specified,

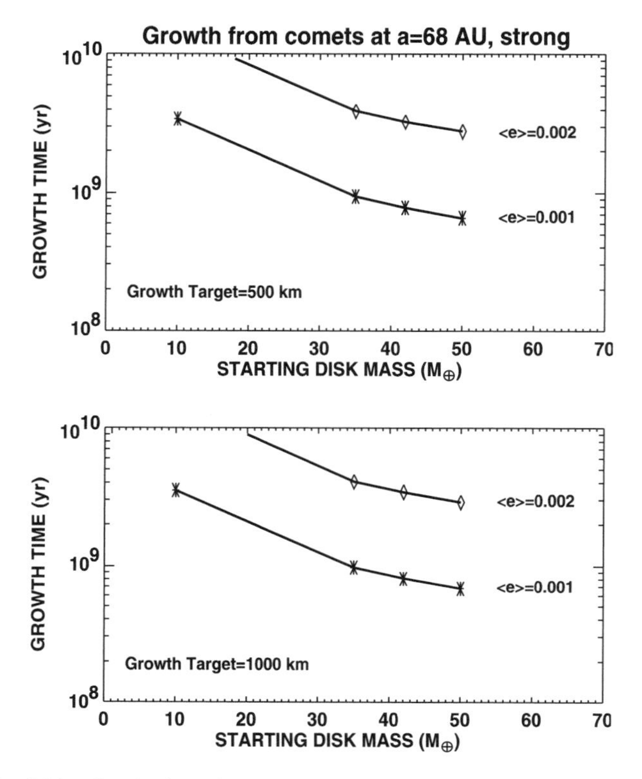

Figure 5. This plot depicts the timescale required to grow 500- and 1000-km radius objects at 68 AU, as a function of (i) the starting disk mass in the 30–50-AU zone and (ii) the assumed mean random eccentricity of the swarm. These results are based on a starting condition of 1–10-km-radius building blocks and a mechanically strong case. Data are shown only for cases in which the growth target was reached in <5 Gyr. (Adapted from Stern and Colwell 1997*a*.)

the mass at each heliocentric distance was partitioned among a suite of logarithmic size bins, separated by factors of 2 in mass. They began the runs with an initial population consisting only of building blocks 1–10 km in radius distributed with equal mass per size bin.

Relevant results concerning accretional growth are shown in Color Plate 21 and Fig. 5, which refer to a 68-AU distance. Although Stern and Colwell (1997*a*) found that growth times are about 4 to 5 times longer than at 40 AU, QB_1-scale and even Pluto-class bodies can be grown well beyond 50 AU, purely from 1–10 km building blocks, in a small fraction of the age of the solar system if the average eccentricity $\langle e \rangle \approx 10^{-3}$ and if the primordial EKB mass interior to 50 AU was $>15\ M_\oplus$, extended outward as described above. Once again, however, we stress that the validity of this conclusion depends on the details of the actual internal velocity evolution that took place in the region beyond 50 AU. The SC97*a* model did not include a sophisticated treatment of these effects, so these conclusions certainly bear testing by a more realistic simulation. Also, in the Neptune-scattered-planetesimals scenario of Morbidelli and Valsecchi (1997) and Petit et al. (1999), massive bodies from the Neptune zone temporarily crossing the EKB may have stirred up eccentricities to a few percent up to $\approx$70 AU; according to SC97*a* this may be enough to inhibit growth.

Despite this caveat, it does seem that if the massive primordial EKB extended beyond 50 AU, then the growth of QB_1-scale and larger bodies could plausibly have occurred in the region between 50 and 100 AU. As a result, Pluto-scale and possibly larger objects may reside there today. Because Neptune itself never induced significant eccentricities on most orbits in this region, the dynamical conditions necessary for growth may have persisted for the whole age of the solar system.

So what does the observational database say regarding this? Consider first IR surveys as a guide to dust production, as an indirect proxy for the infrared surveys by *Infrared Astronomy Satellite* (IRAS), *Cosmic Background Explorer* (COBE), and even ISO results. As we already mentioned in section II, modeling work (Backman et al. 1995; Stern 1996*b*, 1998) indicates that such observations are in principle capable of detecting thermal emission from both macroscopic objects and cold dust at large heliocentric distance. However, the actual data reported to date are not yet constraining (Backman et al. 1995; Anderson et al. 1998).

Next, what about large objects that may have accreted beyond 50 AU? The sky coverage of recent EKB surveys (i.e., a few square degrees covered to red magnitude $R \sim 24$) is far too small to expect the discoveries of large, Pluto-scale objects. Although such objects would be relatively bright (visual magnitude $V \sim 18$–22), their density on the sky is expected to be small (e.g., about 10^{-2} deg^{-2} for a population of 100 bodies). Concerning historical surveys, Tombaugh's (1961) extensive study did not reach sufficient limiting magnitude; and despite a reported search depth of $V \approx 19$, Kowal's (1989) extensive (6400 deg^2) 1970s–1980s ecliptic survey probably did not in fact reach that deep.

Finally, consider the existing limits on smaller, presumably more numerous objects. Detection of comet-scale building blocks at 40 AU ($V \sim 28.5$) is at the limit of present-day capabilities (Cochran et al. 1995, 1998), and their detection farther out is not yet feasible. QB_1-scale objects, with $R \approx 23$ at 40 AU and $R \approx 25$ at 60 AU, are reachable by the surveys performed to date. However, the most recent and extensive deep surveys by Gladman et al. (1998) and Fletcher et al. (1999) have obtained somewhat conflicting conclusions on the likelihood that these bodies do exist. According to Fletcher et al. (1999), the observed brightness distribution indicates that there may be a decline in the density of 100-km radius EKOs in the 50–60 AU zone compared to the 30–50 AU zone and that any rise in the surface mass density of solids may lie beyond 60 AU. A deficiency of sizable EKOs beyond 50 AU, relative to Monte Carlo simulations of power law disks, has also been suggested by Jewitt et al. (1998), based on their large-area ecliptic survey. On the other hand, Gladman et al. (1998) claim that the lack of detection of distant EKOs is not surprising, given the currently limited available data set and the uncertainty on the steepness of the luminosity function. In any case, current observations are not significantly constraining farther out, where, for example, a QB_1 at 90 AU would display $R \approx 27$.

Still, despite the fact that past optical and IR surveys are not strongly constraining, the confluence of large-format charge-coupled device (CCD) technology, computer-aided moving-object detection, and the recent addition of so many 4-m and larger telescopes to the world astronomical inventory opens up new possibilities for meaningful search work. Since objects of the QB_1-class at 70 AU would be only $\approx$2.5 magnitudes fainter than the R = 23–24 magnitude bodies presently being discovered inside 45 AU by 2-m class telescopes, there is the prospect that future searches using larger (e.g., 8–10-m class) telescopes could detect these more distant bodies. Their number density should provide a powerful constraint on the total mass of the disk beyond 50 AU, in turn providing a strong test of whether the 35–50 AU region has indeed evolved to its present state through mass loss. Any detection of individual objects or the IR signature of dust beyond 50 AU would provide valuable constraints on the initial surface mass density distribution of the outer solar system.

V. COMPARISON WITH OTHER SMALL-BODY POPULATIONS

Three populations of small bodies are now known in dynamically stable regions of the solar system: the EKOs, the main-belt asteroids (MBAs), and the Trojans. These populations share several properties: (1) They formed in the early solar system environment and contain the signature of primordial processes; (2) they are sources of other transient populations that collide with the planets and the Sun, namely, Jupiter-family comets from the EKB (and possibly also from Trojans) and near-Earth asteroids and meteoroids from the MBAs; and (3) these bodies are the remnants of materials that failed to build large planets in their respective zones. These three populations allow comparative studies to discern similarities and differences and explore reasons for these. Here follows a list of remarkable similarities and some differences:

- There are vast numbers of bodies in all of these populations. As shown in Table I, the number of bodies larger than 1 km in diameter is quite uncertain but is of the order of millions in the MBA and Trojan cases and of billions in the EKB case. The current size distributions are also similar, with the number of bodies growing rapidly for decreasing sizes. Roughly, this behavior can be matched by power laws, with superimposed "bumps" or "dips" in specific size ranges (in particular,

TABLE I
Comparison of the Major Collisional Parameters for MBAs, Trojan Asteroids, and EKOs

Population	$\langle P_i \rangle$[a] (km^{-2} yr^{-1})	N[b] (10^6)	Collision rate[c] (Myr^{-1})	$\langle V \rangle$[d] (km/s)	Loss rate[e] (Myr^{-1})
MBAs[f]	2.8×10^{-18}	3 ± 2	1/47	5.8 ± 1.9	2,000
Trojans[g]	5.8×10^{-18}	1.5 ± 1	1/66	4.9 ± 2.6	1,000
EKOs[h]	1.3×10^{-21}	$10{,}000 \pm 5{,}000$	1/30	0.5 ± 0.3	200,000

[a] The intrinsic mean collisional probability $\langle P_i \rangle$ follows Wetherill (1967) and gives the mean impact rate per unit cross section for a typical member of the population (times a factor π).
[b] The numbers N refer to objects of diameter larger than 1 km and are estimates from the available observational surveys or theoretical arguments.
[c] The collision rate is the actual rate at which a 100-km diameter target body is hit by a projectile 1 km or larger in diameter.
[d] $\langle V \rangle$ is the mean impact velocity (with the corresponding r.m.s. deviation) for each population.
[e] The loss rate is the estimated rate at which fragments larger than 1 km in diameter are leaking out of the population after having been injected into resonances or chaotic regions of orbital element space.
[f] Farinella and Davis (1992, 1994) and Menichella et al. (1996).
[g] Marzari et al. (1995, 1996, 1997).
[h] Davis and Farinella (1997).

"bumps" tend to appear at sizes for which self-gravitational effects become important).

- However, the total mass in each swarm is relatively small, substantially less than 1 $M_\oplus$, much less than what would be expected by assuming a smoothly varying mass distribution in the primordial solar nebula (a possible exception is the most distant part of the EKB, as discussed in section IV). The mass deficit corresponds to a factor 10^2 to 10^3 for MBAs and EKOs inside 50 AU, whose current total masses are of about 10^{-3} and 10^{-1} $M_\oplus$, respectively; it is more difficult to estimate it for Trojans, but still their total mass (probably some 10^{-5} $M_\oplus$) is extremely low compared to that which was accreted in the jovian core (≈ 10 $M_\oplus$).
- As we have seen in section III for the EKB, if one assumes that the largest of these bodies had to form by binary accretion from smaller planetesimals, then the low mass density that is found today implies that the timescale needed to grow bodies as large as Ceres or Pluto is long compared to the age of the solar system. There are only two ways to escape this paradox: Either there was much more mass in each of these zones when bodies were accreting there, or the standard accretion scenario is incorrect and some peculiar or "exotic" mechanism must be invoked instead.
- A large planet formed just adjacent to these regions: Jupiter in the case of MBAs/Trojans and Neptune for the EKOs. It is widely (but

not universally) believed that the formation of these planets may have interrupted the accretion process before a full-sized planet could form, pumping up orbital eccentricities and inclinations and eventually leaving behind the debris that we observe today as small-body populations. As for the mechanism by which this may have happened, however, we have several competing hypotheses: growing long-range perturbations by the major planets, sweeping resonances triggered by migrating planets or nebular gas dissipation, or encounters with massive scattered planetesimals.

- Currently all the three populations are undergoing a disruptive collisional process that grinds down macroscopic bodies into small particles (later removed by nongravitational draglike effects). This process also forms dynamical families and injects some of the fragments into chaotic zones of orbital element space, often associated with resonances, whence they can be transported into short-lived planet-crossing orbits (typical of near-Earth asteroids, comets, and Centaurs).
- Subtle dynamical mechanisms, including both mean-motion and secular resonances, are at work in all the three regions. In various circumstances these can be associated either with long-term stability, with rapid and strong instability (effective on timescales, say, of 10^5–10^6 yr), or with slow diffusionlike chaotic processes (which may take 10^8 yr or more to yield substantial orbital changes).
- Both collisions and dynamical diffusion are effective in removing bodies from these three populations. These two mechanisms work with about the same efficiency in the EKB. Collisions were thought to dominate in the depletion of the asteroid belt, but recent work by Migliorini et al. (1998) has shown that weak, high-order resonances cause a "leaking" of bodies onto Mars-crossing orbits, an escape hatch that could explain the observed population of sizable (multi-km) planet-crossing asteroids in the inner solar system. In the asteroid case, the nongravitational Yarkovsky force also plays a role in feeding the resonances (Farinella and Vokrouhlický 1999). Further work is needed to quantify the relative roles of dynamics and collisions in removing bodies from the asteroid belt and from the Trojan populations. Note, however, that collisional injection should be increasingly efficient when moving outward in the solar system, because orbital speeds are lower but fragment ejection velocities are approximately constant.
- One significant difference is that collisional grinding has been proposed as the mechanism to remove a significant amount of mass from the EKB (see section II), whereas the amount of collisional grinding that has occurred in the asteroid belt over most of solar system history must have been rather modest. This result stems from the relatively well-preserved basaltic crust of Vesta, which would have

been completely stripped away if collisions had substantially ground down a massive initial asteroid belt (Davis et al. 1985, 1994).

Future observations will reveal whether or not other symmetries between the asteroids and the EKB exist. In the inner solar system, there are the terrestrial planets orbiting interior to the asteroids. Whether or not there are substantial bodies (comparable to the largest EKOs or perhaps even to Pluto) orbiting exterior to the immediate transneptunian zone remains to be determined.

Table I provides a quantitative comparison among the major collisional parameters for MBAs, Trojan asteroids, and EKOs. From this table, it is apparent that the numbers of bodies, intrinsic collision probabilities, and mean impact speeds are very different in the three cases. The difference in the intrinsic collision probabilities is mainly due to the different volumes of space through which the three populations move, whereas impact velocities scale roughly as the corresponding Keplerian orbital speeds (because average eccentricities and inclinations are not very different). However, intrinsic collision probabilities and projectile abundances balance each other almost perfectly, in such a way that absolute collision rates are the same within a factor of 2. Therefore, we may expect that bodies of similar size have undergone similar collisional fluxes during their history. As a consequence, in all the three cases we probably have a transition from a small-size population dominated by fragments from catastrophic disruptions (at diameters $<$50 km), to a large-size population (at diameters $>$100 km) in which most bodies are survivors from primordial accretionary processes, possibly converted into "rubble piles" by the most energetic collisions.

VI. RELATIONSHIP TO DISKS AROUND MAIN-SEQUENCE STARS

Studies of nearby star-forming regions such as ρ Ophiuchi and Taurus have now well established that pre-main-sequence and young stars commonly produce a nebular disk consisting of gas and dust, and that these disks typically dissipate (largely owing to stellar winds) within 1–10 Myr after the star reaches the main sequence. Millimeter-wave observations of CO at high enough frequency resolution to resolve velocity structures (e.g., Weintraub et al. 1989) have revealed that the gas in these disks (and by extension, the dust) follows Keplerian orbits. It is also widely recognized that dusty disklike assemblages of macroscopic objects are common around main-sequence stars (e.g., Backman and Paresce 1993; see also the chapters by Natta et al., Grady et al., and Lagrange et al., this volume).

Thus, the study of debris disks around stars and the study of the EKB have proven intellectually symbiotic. Indeed, this relationship extends back even to the way in which the discovery of IR excess signals of both

T Tauri (cf., Chandler's 1998 review) and nearby main-sequence stars by IRAS (Aumann 1985), and the coronagraphic imaging of the dusty disk surrounding β Pictoris by Smith and Terrile (1984), motivated observers searching for the EKB in the late 1980s and early 1990s (Aumann and Good 1990).

Some of the main attributes of main-sequence stellar disks can be summarized as follows:

- *Frequency of main-sequence stellar disks:* IRAS observations revealed that 10–20% of the nearby main-sequence stars have detectable far IR excess, indicative of a substantial orbiting assemblage of cool dust. The frequencies of such IR excesses appear (e.g., Mannings and Barlow 1998) to be much higher around A-type and later stars than for O and B stars.
- *Nature of main-sequence disks:* The key clue to the nature of main-sequence disks is the fact that the dynamical (and in some cases physical) lifetime of the dust that makes these disks observable is short. Dynamical removal processes include radiation pressure and P-R drag (Aumann et al. 1984), erosion and sweeping by ISM clouds (Stern 1990), and gravitational removal by planets or planet-induced resonances. The obvious implication of the dust's short lifetime is that the dust must be continuously replenished, presumably by an underlying population of unseen, macroscopic objects which generate dust primarily through collisions.
- *Evidence for comets in main-sequence disks:* Owing to the need for macroscopic bodies in the cool distant regions around the central stars in which disks are observed, many workers have suspected colliding comets as a natural source of the observed dust. In at least one case (β Pic), however, there is strong additional evidence for comets. This comes in the form of the frequent, transient appearance of a wide variety of metal-line ion spectral line absorption complexes in the stellar spectra, whose velocities and equivalent widths are commonly interpreted (e.g., Beust et al. 1990; Beust and Morbidelli 1996) as cometary close approaches to or impacts onto, β Pic itself.
- *Mass of main-sequence stellar disks:* Using present-day estimates for dust opacities, millimeter-wave work indicates that the median pre-main-sequence stellar disk dust mass is of scale 0.003–0.03 $M_\odot$, quite comparable to minimum-mass solar nebula models. Similar studies of the dust in disks around main-sequence stars are far less massive, 10^{-5} $M_\odot$ to 10^{-6} $M_\odot$; and Zuckerman et al. (1995) have shown that less than a Jupiter mass of gas is likely to be contained in the main-sequence disks studied to date. As noted above, because the dust in such disks has a short dynamical (and in many cases, collisional) lifetime, it must be continuously created. As such, an underlying, more massive population of small bodies with masses of several to several

hundred $M_\oplus$ is implied. The natural implication is that a major transition occurs in many extrasolar disks, causing their masses to decline severely as they age.

- *Radial extent of extrasolar main-sequence stellar disks:* The extent of a disk can be detected either (i) through direct imaging of scattered light from the disk, typically involving a coronagraph to shield away light from the central star, or (ii) by inverting the spectral energy distribution (SED) of the IR excess flux above the stellar blackbody signal to derive the amount of radiating material as a function of astrocentric distance. Optical and mid-IR images of β Pictoris, the best-known and best-studied main-sequence disk, have revealed a massive disk extending more than 1500 AU in radius. As noted above, IRAS and millimeter-wave IR data have by now resolved disk structures extending several hundred AU around the IR excess stars α Piscis Austrini (Fomalhaut) and α Lyrae (Vega); the detected diameters of these dusty disks are 200 to 2000 AU.
- *Structure within main-sequence stellar disks:* The three brightest and nearest IRAS main-sequence IR excess sources (β Pictoris, Fomalhaut, and Vega) all now have been shown to reveal evidence for a disk structure, and it is widely anticipated that such geometries are prevalent among other, unresolved IRAS main-sequence IR excess sources. Further, inversion of the SEDs of main-sequence sources reveals a prominent characteristic for them to have a central hole, as indicated by a lack of hot material. Such a central hole, often inferred from the SED to have a radius of 20–100 AU (i.e., comparable to expectations for a planetary system), is normally interpreted as evidence for clearing, presumably by planets that have accreted in these systems. Additionally, in the case of β Pictoris, the disk has been observed to be warped, flared, and asymmetric at large distance (e.g., Artymowicz et al. 1989), though a widely accepted cause for these structures has not been well established.

The attributes of the detected population of disks around main sequence stars is reminiscent in many ways of our present-day view of the EKB. Relatively secure similarities include (i) the ancient origin of the structure; (ii) its disklike geometry; (iii) its steep early mass loss; (iv) its continuing mass loss; (v) the underlying population of comets and other small bodies losing mass via collisions; (vi) an inner "hole" carved out by planets; and (vii) the infeeding of objects from the debris system to the inner, "planetary" region, where they become visible, primarily as comets.

At the same time, the EKB is already teaching us some things that may bear strongly on our expectations for, and even detailed understanding of, extrasolar main-sequence disks. These include (i) the role of planetary migration and planetary resonances in shaping the fine structure of the embedded population orbit distribution (see the chapter by Malhotra et al., this volume); (ii) the significant degree of accretion into 100–1000-km objects that has taken place there over time; and (iii) the role of an interior giant planet (Neptune) in generating both dynamical and collisional mass loss.

Less well understood is the issue of the radial extent and total mass of the EKB. At the crux of this is the issue of whether the EKB extends significantly beyond 50 AU or not (see section IV). If it does, the EKB (or more properly, the EK *disk*) will be seen as a true analog to the disks being detected and studied around main-sequence stars. If the EKB is truly limited to only 50 AU in extent, then it may represent a class of main-sequence debris structure that is not as yet widely recognized. In either case, the result would be interesting in its own right. At the same time, future studies concentrating on the physical nature of the objects in the EKB, the present-day rate of mass loss from the belt, and the way in which it internally evolved at late times (i.e., 1–4.5 Gyr) should prove of substantial value to studies of its extrasolar analogs.

VII. OPEN PROBLEMS

Eight years after the discovery of 1992 QB_1, many crucial pieces of observational evidence on the structure of the EKB and the properties of EKOs are still missing. For instance, the results of collisional evolution studies should be compared to the observed size distribution in the range 1 to 100 km, possibly as a function of heliocentric distance. The radial structure of the disk should be better determined and disentangled from observational selection effects. In particular, the existence of bodies beyond 50 AU and their size distribution would provide a critical constraint to models of the primordial disk surrounding the Sun and its relationships to those observed around other stars. Since the physical properties (albedos, compositions, shapes, rotations, binarity) of EKOs have been probably affected by collisional processing as well as by the accretion environment, they may provide important insights both on the properties of the original planetesimals and on their subsequent history. Laboratory work on the collisional response parameters of icy bodies at impact velocities up to ≈ 1 km s^{-1} is also needed to make the collisional models more realistic and reliable (for some results of such experiments, see Ryan et al. 1999).

From the theoretical point of view, a crucial question concerns the mechanism that terminated accretion and started the current erosive process (at least within 50 AU). Most probably, this was related to the formation of Neptune, but whether primordial EKO orbits were stirred up by long-range perturbations, sweeping resonances, or scattered massive planetesimals is still uncertain. Neither do we know how much mass loss occurred in the belt during this primordial phase, and how much in the subsequent collisional regime, which is continuing today. Other puzzles concern the Pluto-Charon pair: Are there other bodies of similar (or perhaps larger) size in the belt? Where and how were these bodies formed? Was the formation of a binary an anomalous fluke or the predictable result of the primordial environment? Clearly, answering these questions would be much easier if we had a quantitative understanding (when, where, how fast?) of Neptune's formation process, which in turn may be related to those of the other giant planets. On the other hand, the vast population of small transneptunian objects can provide us with useful constraints on the process that formed the giant planets and shaped the overall architecture of the outer solar system.

REFERENCES

Anderson, J. D., Scalise, R., Teplitz, V., Rosenbaum, D., and Stern, S. A. 1998. COBE DIRBE Kuiper Belt constraints. *Icarus* 131:167–176.

Artymowicz, P., Burrows, A., and Paresce, F. 1989. The structure of the β Pic circumstellar disk from combined IRAS and coronagraphic observations. *Astrophys. J.* 347:494–513.

Aumann, H. H., Gillett, F. C., Beichmann, G. A., de Jong, T., Houck, J., Low, F. J., Neugebauer, G., Walker, R., and Wesselius, P. R. 1984. Discovery of a shell around Alpha Lyrae. *Astrophys. J. Lett.* 278:L23–L27.

Aumann, H. H. 1985. IRAS observations of matter around nearby stars. *Pub. Astron. Soc. Pacific* 97:885–991.

Aumann, H. H., and Good, J. C. 1990. IRAS constraints on a cold cloud around the solar system. *Astrophys. J.* 350:408–412.

Backman, D. E., and Paresce, F. 1993. Main sequence stars with circumstellar material. In *Protostars and Planets III*, ed. E. H. Levy and J. I. Lunine (Tucson: University of Arizona Press), pp. 1253–1304.

Backman, D. E., Dasgupta, A., and Stencel, R. E. 1995. A model of Kuiper Belt small grain emission and resulting far-IR emission. *Astrophys. J. Lett.* 450:L35–L39.

Beust, H., and Morbidelli, A. 1996. Mean-motion resonances as a source for infalling comets toward β Pictoris. *Icarus* 120:358–370.

Beust, H., Vidal-Madjar, A., Ferlet, R., and Lagrange–Henri, A. H. 1990. The Beta Pic circumstellar disk. X. Numerical simulations of infalling evaporating bodies. *Astron. Astrophys.* 236:202–216.

Chandler, C. J. 1998. Observations of circumstellar disks. In *Cosmic Origins: Galaxies, Stars, Planets, and Life*, ASP Conf. Ser. 148, ed. J. M. Shull, C. E. Woodward, and H. A. Thronson (San Francisco: Astronomical Society of the Pacific), pp. 237–260.

Cochran, A. L., Levison, H. F., Stern, S. A., and Duncan, M. J. 1995. The discovery of Halley-sized Kuiper Belt objects using the Hubble Space Telescope. *Astrophys. J.* 455:342–346.

Cochran, A. L., Levison, H. F., Tamblyn, P. M., Stern, S. A., and Duncan, M. 1998. The calibration of the HST Kuiper Belt object search: Setting the record straight. *Astrophys. J. Lett.* 503:L89–L94.

Davis, D. R., and Farinella, P. 1997. Collisional evolution of Edgeworth–Kuiper Belt objects. *Icarus* 125:50–60.

Davis, D. R., and Farinella, P. 1998. Collisional erosion of a massive Edgeworth–Kuiper Belt: Constraints on the initial population. *Lunar Planet. Sci.* 29:1437.

Davis, D. R., and Farinella, P. 1999. Accretion and collisional erosion of a massive Edgeworth-Kuiper Belt: Constraints on the initial population. In preparation.

Davis, D. R., Chapman, C. R., Weidenschilling, S. J., and Greenberg, R. 1985. Collisional history of asteroids: Evidence from Vesta and the Hirayama families. *Icarus* 62:30–53.

Davis, D. R., Farinella, P., Paolicchi, P., Weidenschilling, S. J., and Binzel, R. P. 1989. Asteroid collisional history: Effects on sizes and spins. In *Asteroids II*, ed. R. P. Binzel, T. Gehrels, and M. S. Matthews (Tucson: University of Arizona Press), pp. 805–826.

Davis, D. R., Ryan, E. V., and Farinella, P. 1994. Asteroid collisional evolution: Results from current scaling algorithms. *Planet. Space Sci.* 42:599–610.

Dohnanyi, J. W. 1969. Collisional models of asteroids and their debris. *J. Geophys. Res.* 74:2531–2554.

Duncan, M., and Levison, H. F. 1997. A scattered comet disk and the origin of Jupiter-family comets. *Science* 276:1670–1672.

Duncan, M. J., Levison, H. F., and Budd, S. M. 1995. The dynamical structure of the Kuiper Belt. *Astron. J.* 110:3073–3081.

Farinella, P., and Davis, D. R. 1992. Collision rates and impact velocities in the main asteroid belt. *Icarus* 97:111–123.

Farinella, P., and Davis, D. R. 1994. Will the real asteroid size distribution please step forward? *Lunar Planet. Sci.* 25:365–366.

Farinella, P., and Davis, D. R. 1996. Short-period comets: Primordial bodies or collisional fragments? *Science* 273:938–941.

Farinella, P., and Vokrouhlický, D. 1999. Semimajor axis mobility of asteroidal fragments. *Science* 283:1507–1510.

Fernández, J. A., and Ip, W.–H. 1996. Orbital expansion and resonant trapping during the late accretion stages of the outer planets. *Planet. Space Sci.* 44:431–439.

Fletcher, E., Fitzsimmons, A., and Irwin, M. 1999. A deep survey of distant Kuiper Belt material. In preparation.

Flynn, G. J. 1996. Sources of 10 micron interplanetary dust: The contribution from the Kuiper Belt. In *Physics, Chemistry, and Dynamics of Interplanetary Dust*, ASP Conf. Ser. 104, ed. B. A. S. Gustafson and M. S. Hanner (San Francisco: Astronomical Society of the Pacific), pp. 171–175.

Gladman, B., Kavelaars, J. J., Nicholson, P., Loredo, T., and Burns, J. 1998. Pencil-beam surveys for faint transneptunian comets. *Astron. J.* 116:2042–2054.

Greenberg, R., Weidenschilling, S. J., Chapman, C. R., and Davis D. R. 1984. From icy planetesimals to outer planets and comets. *Icarus* 59:87–113.

Gurnett, D. A., Ansher, J. A., Kurth, W. S., and Granroth, L. J. 1997. Micron-sized dust particles detected in the outer solar system by Voyager 1 and 2 plasma wave instruments. *Geophys. Res. Lett.* 24:3125–3128.

Hamid, S. E., Marsden, B. G., and Whipple, F. L. 1968. Influence of a comet belt beyond Neptune on the motions of periodic comets. *Astron. J.* 73:727–729.

Holman, M. J., and Wisdom, J. 1993. Dynamical stability in the outer solar system and the delivery of short-period comets. *Astron. J.* 105:1987–1999.

Humes, D. H. 1980. Results of Pioneer 10 and 11 meteoroid experiments—Interplanetary and near-Saturn. *J. Geophys. Res.* 85:5841–5852.

Ip, W.–H. 1989. Dynamical processes of macro-accretion of Uranus and Neptune: A first look. *Icarus* 80:167–178.

Jewitt, D., Luu, J., and Trujillo, C. 1998. Large Kuiper Belt objects: The Mauna Kea 8K CCD survey. *Astron. J.* 115:2125–2135.

Kenyon, S. J., and Luu, J. X. 1998. Accretion in the early Kuiper Belt. I. Coagulation and velocity evolution. *Astron. J.* 115:2136–2160.

Knežević, Z., Milani, A., Farinella, P., Froeschlé, Ch., and Froeschlé, Cl. 1991. Secular resonances from 2 to 50 AU. *Icarus* 93:316-330.

Kowal, C. 1989. A solar system survey. *Icarus* 77:118–123.

Levison, H. F., and Duncan, M. J. 1993. The gravitational sculpting of the Kuiper Belt. *Astrophys. J. Lett.* 406:L35–L38.

Levison, H. F., and Duncan, M. J. 1994. The long-term dynamical behavior of short-period comets. *Icarus* 108:18–36.

Levison, H. F., Stern, S. A., and Duncan, M. J. 1999. The role of a massive primordial Kuiper Belt on its current dynamical structure. *Icarus*, submitted.

Liou, J. C., Zook, H. A., and Dermott, S. F. 1996. Kuiper Belt dust grains as a source of interplanetary dust particles. *Icarus* 124:429–440.

Lissauer, J. J., Pollack, J. B., Wetherill, G. W., and Stevenson, D. J. 1995. Formation of the Neptune system. In *Neptune and Triton*, ed. D. P. Cruikshank and M. S. Matthews (Tucson: University of Arizona Press), pp. 37–108.

Luu, J., Marsden, B. G., Jewitt, D., Trujillo, C. A., Hergenrother, C. W., Chen, J., and Offutt, W. B. 1997. A new dynamical class of objects in the outer solar system. *Nature* 387:573–575.

Malhotra, R. 1995. The origin of Pluto's orbit: Implications for the solar system beyond Neptune. *Astron. J.* 110:420–429.

Mannings, V., and Barlow, M. J. 1998. Candidate main sequence stars with debris disks: A new sample of Vega-like sources. *Astrophys. J.* 497:330–341.

Marzari, F., Farinella, P., and Vanzani, V. 1995. Are Trojan collisional families a source for short-period comets? *Astron. Astrophys.* 299:267–276.

Marzari, F., Scholl, H., and Farinella, P. 1996. Collision rates and impact velocities in the Trojan asteroid swarms. *Icarus* 119:192–201.

Marzari, F., Farinella, P., Davis, D. R., Scholl, H., and Campo Bagatin, A. 1997. Collisional evolution of Trojan asteroids. *Icarus* 125:39–49.

Menichella, M., Paolicchi, P., and Farinella, P. 1996. The main belt as a source of near-Earth asteroids. *Earth Moon Planets* 72:133–149.

Migliorini, F., Michel, P., Morbidelli, A., Nesvorný, D., and Zappalà, V. 1998. Origin of multikilometer Earth- and Mars-crossing asteroids: A quantitative simulation. *Science* 281:2022–2024.

Morbidelli, A. 1997. Chaotic diffusion and the origin of comets from the 2/3 resonance in the Kuiper Belt. *Icarus* 127:1–12.

Morbidelli, A., and Valsecchi, G. B. 1997. Neptune scattered planetesimals could have sculpted the primordial Edgeworth-Kuiper Belt. *Icarus* 128:464–468.

Morbidelli, A., Thomas, F., and Moons, M. 1995. The resonant structure of the Kuiper Belt and the dynamics of the first five transneptunian objects. *Icarus* 118:322–340.

Petit, J. -M., Morbidelli, A., and Valsecchi, G. B. 1999. Large scattered planetesimals and the excitation of the small body belts. *Icarus*, submitted.

Ryan, E. V., Davis, D. R., and Giblin, I. 1999. A laboratory impact study of simulated Edgeworth-Kuiper Belt objects. *Icarus*, submitted.

Safronov, V. S. 1969. *Evolution of the Protoplanetary Cloud and Formation of the Earth and the Planets* (Moscow: Nauka Press); English transl. (1972), NASA TT F-677.

Smith, B. A., and Terrile, R. J. 1984. A circumstellar disk around β Pic. *Science* 226:1421–1424.

Stern, S. A. 1990. ISM–induced erosion and gas dynamical drag in the Oort cloud. *Icarus* 84:447–466.

Stern, S. A. 1995. Collisional time scales in the Kuiper disk and their implications. *Astron. J.* 110:856–868.

Stern, S. A. 1996*a*. Signatures of collisions in the Kuiper disk. *Astron. Astrophys.* 310:999–1005.

Stern, S. A. 1996*b*. The development and status of the Kuiper disk. In *Completing the Inventory of the Solar System*, ASP Conf. Ser. 107, ed. T. Rettig and J. M. Hahn (San Francisco: Astronomical Society of the Pacific), 209–232.

Stern, S. A., 1998. The Sun's Kuiper Belt and its surrounding disk. In *Cosmic Origins: Galaxies, Stars, Planets, and Life*, ASP Conf. Ser., ed. J. M. Shull, C. E. Woodward, and H. A. Thronson (San Francisco: Astronomical Society of the Pacific), 516:425–435.

Stern, S. A., and Colwell, J. E. 1997*a*. Accretion in the Edgeworth–Kuiper Belt: Forming 100–1000 km radius bodies at 30 AU and beyond. *Astron. J.* 114:841–849.

Stern, S. A., and Colwell, J. E. 1997*b*. Collisional erosion in the primordial Edgeworth-Kuiper Belt and the generation of the 30–50 AU Kuiper gap. *Astrophys. J.* 490:879–882.

Stern, S. A., Canup, R., and Durda, D. D. 1999. Pluto's family: Debris from the binary-forming collision in the 2:3 resonance? *Lunar Planet. Sci.* 30:1213.

Stevenson, D. J. 1984. On forming giant planets quickly (superganymedean puffballs). *Lunar Planet. Sci.* 15:822–823.

Stewart, G. R., and Levison, H. F. 1998. On the formation of Uranus and Neptune. *Lunar Planet. Sci.* 24:abstract 1960.

Teplitz, V., Stern, S. A., Anderson, J. D., Rosenbaum, D. C., Scalise, R., and Wentzler, P. 1999. IR Kuiper Belt constraints. *Astrophys. J.*, 516:425–435.

Tombaugh, C. W. 1961. The transneptunian planet search. In *Planets and Satellites,* ed. G. P. Kuiper and B. M. Middlehurst (University of Chicago Press).

Ward, W. R. 1996. Planetary accretion. In *Completing the Inventory of the Solar System*, ASP Conf. Ser. 107, ed. T. Rettig and J. M. Hahn (San Francisco: Astronomical Society of the Pacific), pp. 337–361.

Ward, W. R., and Hahn, J. M. 1998. Dynamics of the trans-Neptune region: Apsidal waves in the Kuiper Belt. *Astron. J.* 116:489–498.

Weintraub, D. A., Masson, C. R., and Zuckerman, B. 1989. Measurements of the Keplerian rotation of the gas in the circumbinary disk around T Tauri. *Astrophys. J.* 344:915–924.

Wetherill, G. W. 1967. Collisions in the asteroid belt. *J. Geophys. Res.* 72:2429–2444.

Wetherill, G. W. 1990. Comparison of analytical and physical modeling of planetesimal accumulation. *Icarus* 88:336–354.

Wetherill, G. W., and Stewart, G. R. 1989. Accumulation of a swarm of small planetesimals. *Icarus* 77:330–357.

Yamamoto, S., and Mukai, T. 1998. Dust production by impacts of interstellar dust in the Edgeworth-Kuiper disk. *Astron. Astrophys.* 329:785–791.

Yeomans, D. K. 1986. The intermediate comets and nongravitational effects. *Astron. J.* 91:971–973.

Zuckerman, B., Foreville, T., and Kastner, J. H. 1995. Inhibition of giant planet formation by rapid gas depletion around young stars. *Nature* 373:494–504.

PART VII
Extrasolar Planets and Brown Dwarfs

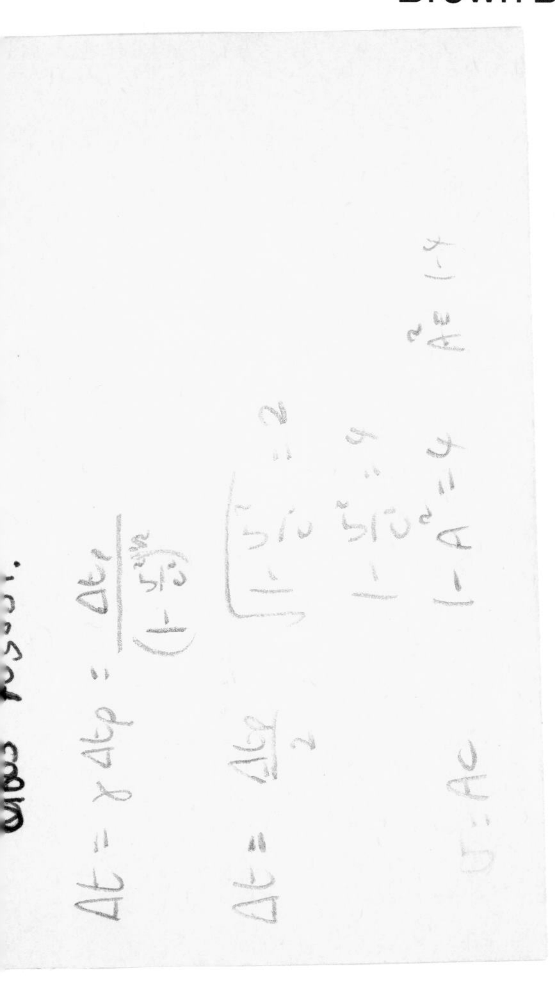

EXTRASOLAR PLANETS AROUND MAIN-SEQUENCE STARS

GEOFFREY W. MARCY
San Francisco State University and University of California at Berkeley

WILLIAM D. COCHRAN
University of Texas

and

MICHEL MAYOR
Geneva Observatory

The mass distribution of substellar companions exhibits a steep rise for masses below 5 M_{Jup}. Thus, the 14 companions having $M \sin i = 0.5$–$5\ M_{Jup}$ are considered the best candidate planets around main-sequence stars. The occurrence of such planets within 3 AU is $\sim$4%, but two-thirds of them orbit within just 0.3 AU. This "pileup" of planets near stars suggests that inward orbital migration occurred after formation. The planet candidates orbiting within 0.1 AU have nearly circular orbits, but all nine of those orbiting outside 0.2 AU have $e > 0.1$, i.e., greater than that of Jupiter, $e_{Jup} = 0.048$. Thus, eccentric orbits predominate for Jupiter-mass companions from 0.2 to 2.5 AU and may arise from gravitational interactions with other planets, stars, or the protoplanetary disk. The planet-bearing stars are systematically metal-rich, as is the Sun, compared to the solar neighborhood. The occurrence of "brown dwarf" companions in the next higher mass decade, 5–50 M_{Jup}, is at most 1% within 3 AU, based on surveys of $\sim$600 stars. Thus, brown dwarfs represent a minimum in the mass function of companions, between the masses of stars and of planets.

I. INTRODUCTION

Our Sun is a very common and ordinary star. There is really nothing to distinguish it from a myriad of other similar stars in this region of the Galaxy. Yet, the Sun possesses a marvelous system of nine diverse planets. This has led us to believe that the formation of planetary systems should be a natural, common result of the process of star formation. We expect that a significant fraction of solar-type stars should have some type of planetary system in orbit around them. The discovery by the Hubble Space Telescope (HST) of disks of dust around many stars in the Orion nebulae certainly reinforces that feeling. The quest to discover and explore these extrasolar planetary systems has proven to be a difficult and elusive task that has occupied astronomers for decades. The early phases of the search for substellar companions are chronicled by van de Kamp (1977, 1982, 1986).

However, much to the surprise of everybody, the first confirmed detection of an extrasolar planetary system was not found around a "normal" main-sequence star like the Sun but rather around a millisecond pulsar, PSR 1257+12 (Wolszczan and Frail 1992; Wolszczan 1994). It was not until three years later, with the discovery of the companion to 51 Pegasi in the astonishing 4.23-day orbit (Mayor and Queloz 1995), that the first planetary-mass companion was found in orbit around a solar-type star. Further, the detection and spectrum of the brown dwarf, Gliese 229b, portends the wealth of information obtainable by future direct images of extrasolar planets, which we hope will be made possible with space-borne interferometers (Oppenheimer et al. 1998; Beichman 1998).

In this chapter, we will discuss the quest for planets around stars like the Sun, for it is these systems that shed light on the physics of star formation and allow us to test the elaborate paradigm that has developed for the process of planetary system formation.

II. TECHNIQUES

A. Radial Velocities

In a typical high-dispersion optical spectrometer, a Doppler shift of one pixel corresponds to 2000 m s^{-1}, which is much larger than the wobble velocity of the Sun, 13 m s^{-1}, due to Jupiter. A secure detection of a Jupiter analog requires that Doppler shifts be measured with a precision of $\sim \frac{1}{1000}$ pixel or, correspondingly, that wavelengths be measured with a precision of $1{:}10^8$.

Radial velocity precision of $\sim$10 m s^{-1} is accomplished by three basic approaches. Several groups use a glass container filled with iodine gas, which is placed at the entrance of the spectrometer. The resulting iodine absorption lines are superimposed on the stellar spectrum to provide the wavelength calibration (Marcy and Butler 1992; Cochran and Hatzes 1994; Noyes et al. 1997). The iodine lines also reveal the asymmetries in the instrumental profile that induce spurious Doppler shifts. The iodine permits removal of such instrumental effects. The iodine cells were developed as a safer and more versatile alternative to the HF gas absorption cell used by Campbell et al. (1988).

In the second approach, two fiber optic cables feed the stellar light and the calibration emission source (typically thorium) into the spectrometer (Brown et al. 1994; Mayor and Queloz 1995). The fiber approach carries two advantages. The entire spectrum can be used (not just the region containing iodine lines), and the stellar spectrum is not contaminated by iodine lines. One disadvantage is that the optical path of the calibration light is not identical to that of the starlight, thus rendering the approach vulnerable to zonal optical aberrations.

A third method is to use a tunable Fabry-Perot etalon as the velocity metric (McMillan et al. 1990, 1994). A cross-dispersed echelle spectrograph spatially separates the Fabry-Perot interference maxima on a charge-coupled device (CCD) detector. Small Doppler shifts of a star will then become evident through changes in the intensities of those Fabry-Perot orders falling on the steep slopes of stellar absorption lines.

B. Astrometry

The angular wobble of a star in response to its companion is proportional to both planet mass and orbital radius and inversely proportional to the distance to the star. Reviews of the astrometric approach to planet detection are provided by Gatewood (1987) and Colavita and Shao (1994). As a benchmark, a Jupiter analog orbiting 5 AU from a solar-type star that is located 10 pc away would produce an astrometric amplitude of 0.5 milliarcsec (mas).

The astrometric technique offers two great prospects: (1) determination of an unambiguous mass and orbital inclination of a planet and (2) detection of sub-Jupiter-mass planets for future astrometric precision below 0.1 mas. An astrometric planet detection provides a secure mass, which is not offered by direct detection.

Gatewood (1987) has demonstrated an annually averaged astrometric precision of 1 mas, which he expects to improve by using the Keck II telescope. Pravdo and Shaklan (1996) have demonstrated a precision of 0.1 mas with direct, short CCD exposures from the Palomar 5-m and Keck telescopes. With the Palomar testbed interferometer, Colavita and Shao (1994) currently achieve precision of 60–70 μas/hr$^{-1/2}$, portending a bright future for next-generation interferometric astrometry. The two Keck telescopes and the European Very Large Telescope Interferometer should yield astrometric precision of 20 μas (Colavita and Shao 1994). A planned NASA space-borne astrometric interferometer called the *Space Interferometry Mission* (SIM) has a goal of 4 μas for global astrometry (Unwin et al 1999; Boden et al. 1996), and perhaps better for planet searches. Due for launch in 2006, SIM should achieve many-σ detections of planets having Neptune-like masses ($\frac{1}{20}$ M_{Jup}) at 5 AU for stars within 10 pc. A mission lifetime of $\sim$12 years will be required. Interferometric astrometry offers a clear path to statistically valuable ensembles of gas giants through the Neptune-mass regime, to constrain the planetary mass function.

To date, no definitive detection of a planetary companion has been accomplished by the astrometric method. Gatewood (1996) has noted strongly suggestive accelerations in Lalande 21185 that may be due to two low-mass orbiting companions, but further data are required to confirm the orbits and masses (G. D. Gatewood, personal communication, 1997).

C. Transits and Microlensing

Planets may be detected as they transit in front of the disk of a star. The fractional reduction in the light from the star is simply the ratio of the areas of the planet and the star. For a benchmark Jupiter having radius 1 R_{Jup}, a solar-type star will dim by 1% with a duration of order hours, depending on the orbital radius (Hale and Doyle 1994). Such photometry is possible from the ground, made efficient with automated wide-field telescopes that acquire CCD images of thousands of stars simultaneously (Borucki and Summers 1984). The 1% dimming is easily distinguished from other effects such as starspots, flares, and fluctuations in photospheric granulation, which would rarely exceed 1%. Actual transits would cause no change in color, would exhibit a flat-bottomed minimum (with limb-darkened ingress and egress), and would repeat like clockwork.

A transit requires a special geometry, such that

$$\tan i > \frac{a}{R_\star}$$

where i is the orbital inclination, a is the planet's semimajor axis, and $R_\star$ is the star's radius.

For randomly oriented orbital planes, the probability P that i will reside between 90° (edge-on) and i' is simply: $P(i' \text{ to } 90°) = \cos i'$. For a population of Jupiters at 0.1 AU (51 Peg-like), 4.7% of them will transit, and ~2% of solar-type stars have such close Jupiters (Marcy and Butler 1998). This implies that one in a thousand solar-type stars should exhibit transits from close Jupiters. To enrich the stellar sample with edge-on planetary orbits, eclipsing binary stars may be selected. For CM Draconis, companions larger than 2.5 Earth radii can be ruled out for periods less than 60 days (Deeg et al. 1998).

A transit, followed by Doppler measurements of the star, would yield the planet mass directly, because $\sin i \approx 1.0$. Transits also yield the radius of the planet and hence its density, which distinguishes gas giants from solid planets. The photometric transit approach should be pursued vigorously.

Transits by Earth-sized planets would dim the star 0.01%. The requisite photometric precision requires a space-borne platform, wide-field camera, and a detector capable of photometric precision of 3:100,000. Such a mission should reveal transits in ~1% of solar-type stars if terrestrial planets at ~1 AU are common (Borucki et al. 1996).

The French space mission COROT, to be launched in 2001, is planned to monitor about 25,000 stars photometrically and permit the detection of a few dozen extrasolar planets (Schneider et al. 1998).

Gravitational microlensing of stars in the galactic bulge may also reveal the presence of planets in orbit around the intervening lensing objects. Intensive follow-up photometry of microlensing events by a global telescope network can reveal the short-term perturbations on the standard microlensing light curve caused by an attendant planet (Peale 1997; Griest and Safizadeh 1998). Microlensing is most sensitive to planets at a projected distance from the lensing star of about an Einstein radius, which corresponds to 3–6 AU for a typical galactic bulge microlensing event. The duration of the planetary perturbation on the light curve is proportional to $\sqrt{M_P}$. Microlensing is unique in its ability to detect Earth-mass planets in orbits with semimajor axes of several AU around main-sequence stars from the ground (hence inexpensively) and would yield statistics on the occurrence of such planets. Follow-up study of these planets would be difficult owing to their large distance ($\sim$5 kpc) and ambiguities in identifying the lensing object.

III. PLANETARY RESULTS TO DATE

Several previous Doppler surveys for planets and brown dwarfs done from 1980 to 1990 foreshadowed and constrained the results to come later. Doppler surveys of $\sim$200 K- and M-type dwarfs at a precision of $\sim$250 m s^{-1} (Marcy and Benitz 1989; Tokovinin 1992) would have revealed 5–80-M_J companions within 3 AU. No such substellar companions were found. Another Doppler survey of 570 G and K dwarfs, done with a precision of 300 m s^{-1}, revealed 10 brown dwarf candidates (Duquennoy and Mayor 1991; Mayor et al. 1997). With similar precision, Latham et al. (1989) identified another one having $M \sin i = 11\ M_{\rm Jup}$, though Cochran et al. (1991) have suggested that sin i may be small. Studies of the distribution of sin i show that some of these companions are likely to be simply H-burning stars, not brown dwarfs (Mazeh et al. 1992; Marcy and Butler 1995). Mayor et al. (1998*a*) have indeed shown that almost all have $M > 0.075\ M_\odot$, based on Hipparcos satellite astrometry that constrains sin i (see section IV.C). Murdoch et al. (1993) pushed "classical" radial velocity techniques to a very impressive precision limit of 55 m s^{-1} in their survey of 29 solar-type stars for low-mass companions. In a survey that pioneered the field of high-precision radial velocity work, Walker et al. (1995) monitored 21 late F, G, and K dwarfs for 12 yr at a precision of 13 m s^{-1}. No companions were found above a threshold of 2$M_{\rm Jup}$ within 5 AU.

These Doppler surveys impose a firm upper limit on the occurrence of brown dwarf and planetary companions. At most 1% of 0.3–1.2 $M_\odot$ stars harbor a brown dwarf (10–80 $M_{\rm Jup}$) within 3 AU. Similarly, at most 5% have giant planets above 2 $M_{\rm Jup}$ within 3 AU. By 1994, it had become clear that companions having more than twice Jupiter's mass are rare within 5 AU.

Several teams responded to this absence of detections by monitoring Doppler shifts of larger samples of F, G, K, and M dwarfs at higher precision ($\sim$10 m s^{-1}). Leading surveys were carried out by McMillan et al. (1993), Cochran et al. (1997), Mayor and Queloz (1995), Noyes et al. (1997), and Marcy and Butler (1998). Mayor and Queloz (1995) have now monitored 140 main-sequence stars for 4 yr, and Cochran and Hatzes (1994) have monitored 33 stars for 11 yr. Marcy and Butler (1997) have monitored 107 F, G, K, and M dwarfs for 11 yr, and Noyes et al. (1997) have searched ~100 additional solar-type dwarfs and subgiants for ~3 yr. In all cases, the stellar selection avoided binaries with separations less than 2 arcsec (typically ~20 AU), to avoid double spectra and to reject stars for which detectable planets could not persist dynamically. A total of ~300 individual stars were surveyed.

Starting with the detection of the companion to 51 Peg (Mayor and Queloz 1995; Marcy et al. 1997), the Doppler surveys above have now revealed 17 companions that have $M \sin i < 11\ M_{Jup}$ (see Table I). Figures 1a and 1b show the Doppler measurements for the eight planet candidates known just prior to the *Protostars and Planets IV* meeting in June 1998, reviewed by Marcy and Butler (1998). Figures 2 and 3 show the Doppler measurements for two planets announced at that meeting, namely around Gliese 876 (M4 V; Marcy et al. 1998, Delfosse et al. 1998) and 14 Herculis (K0 V; Mayor et al. 1998*d*). Gliese 876 is an M4 dwarf, suggesting that planets are common around stars from 1.3 to 0.3 $M_{\odot}$. The planet around 14 Her has a period of at least 4.5 yr, implying that $a > 2.5$ AU, the largest orbit yet found for a planet candidate. Figure 4 shows the Doppler results for HD 187123, which has a planet with the shortest period to date, $P = 3.09$ d and $M \sin i = 0.59\ M_{Jup}$ (Butler et al. 1998). A summary of all planet candidates known as of January 1999 is provided by Marcy et al. (1999). Table II shows the stellar characteristics of those planet-bearing stars for which careful spectroscopic synthesis has been carried out (Gonzalez 1998*a,b*).

The 5-M_{Jup} benchmark may represent a physically meaningful upper limit to planetary masses, because the companion mass distribution rises discontinuously at that mass (Fig. 6). Thus, companions with $M \sin i < 5\ M_{Jup}$ represent the best extrasolar planet candidates. Table I shows the orbital characteristics, P, a, K, e, and minimum masses of these planet candidates, along with the few that have $M \sin i = 5$–$11\ M_{Jup}$. Here, P is orbital period, a is semimajor axis, K is the velocity semi-amplitude, and e is orbital eccentricity. We shall refer to these 17 companions as the extrasolar planet candidates. Planets may also be distinguished from brown dwarfs by their inability to ignite deuterium as brown dwarfs do (Burrows et al. 1997).

The orbital periods of the 17 extrasolar planets range from 3.1 d to 4.5 yr, corresponding to semimajor axes of 0.04 to 2.5 AU. However, 11 of the 17 planets have $a < 0.3$ AU. This "piling-up" of planets near their host stars appears to be a real effect, although enhanced by the selection effect that favors detection of small orbits (see section IV.C).

Upsilon (υ) Andromedae, after subtraction of the clear wobble that is caused by the planet with $P = 4.6$ d, exhibits velocity residuals with two

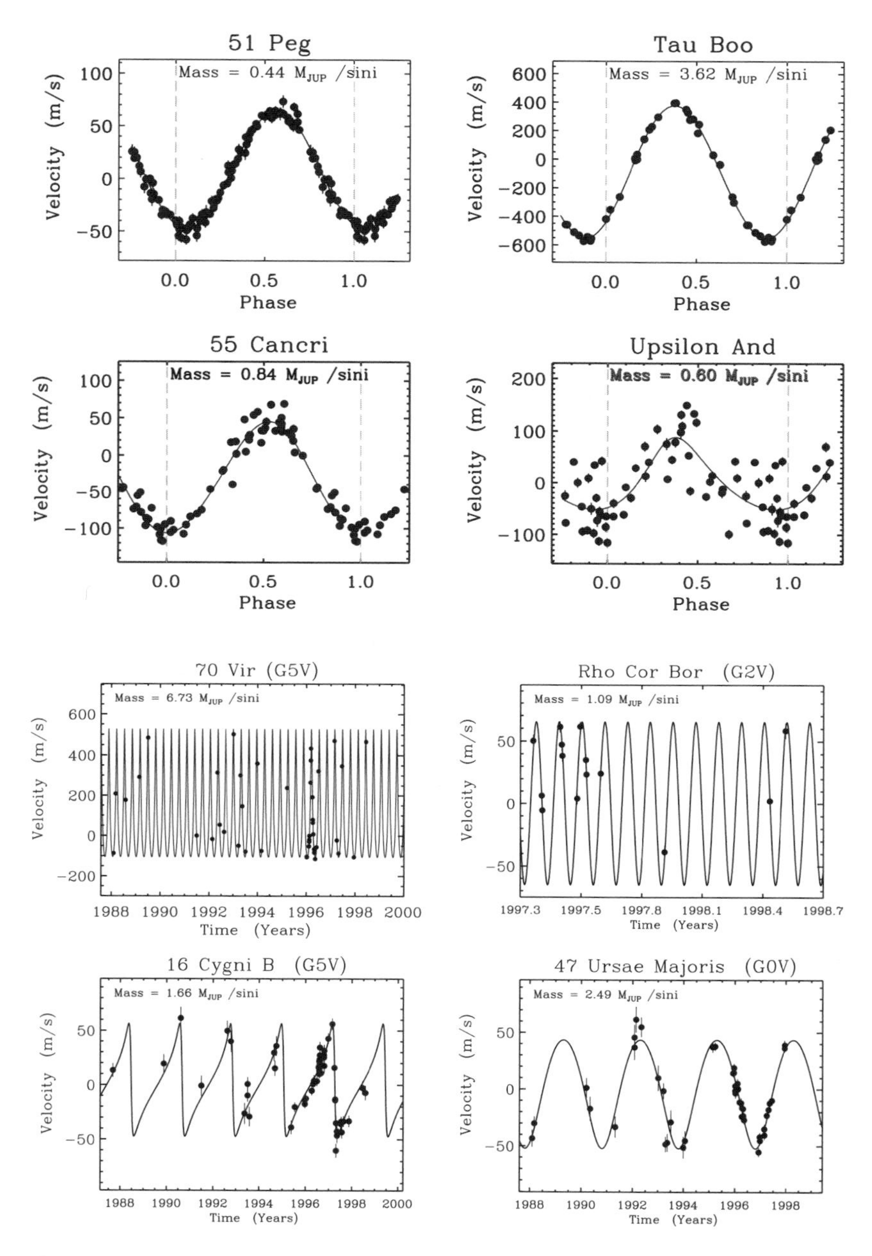

Figure 1. Doppler data for the first eight planet candidates that have $M \sin i < 5\ M_{\rm Jup}$, including the marginal 70 Virginis, with $M \sin i = 6.8\ M_{\rm Jup}$. (a) The short–period planets, having $P < 15$ d. All have nearly circular orbits, possibly induced by tidal effects. b) The long–period planets, having $P = 39$ d–3 yr. Two have eccentric orbits and two have $e < 0.15$. Tidal effects should play no role for these.

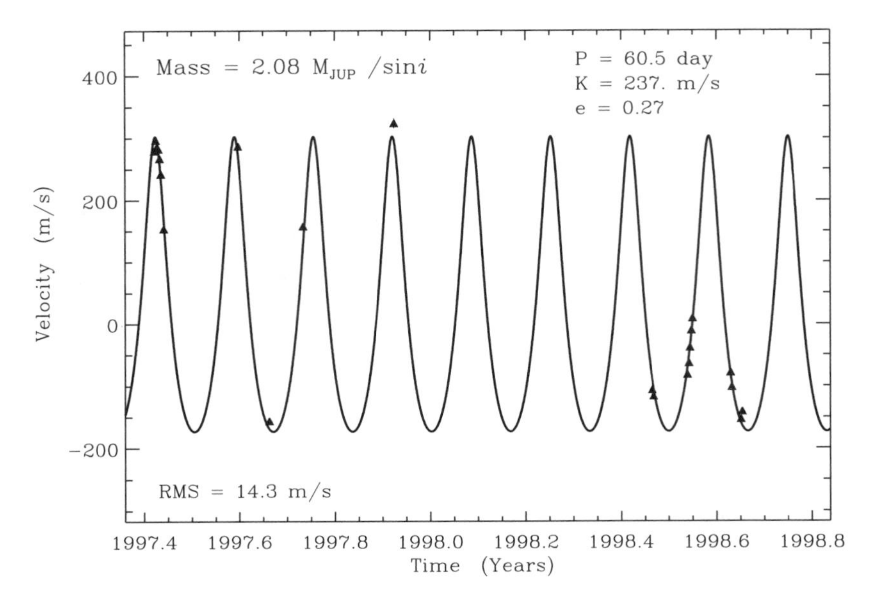

Figure 2. The Doppler data for Gliese 876 from the Keck telescope, displayed at the *Protostars and Planets* IV meeting. The implied planet has $P = 61$ d, $a = 0.20$ AU, and $e = 0.27$. Gliese 876 is spectral type M4 V, with mass = 0.32 $M_\odot$, suggesting that planets may be common around low-mass stars.

additional periods of 240 ± 10 and 1300 ± 30 d. The raw velocities are well fitted by three companions orbiting at 0.05, 0.83, and 2.5 AU, with minimum ($M \sin i$) masses of 0.75, 1.9, and 4.4 Jupiter masses, respectively (Butler et al. 1999). This multiple-planet system appears to be the first ever found around a main-sequence star outside the solar system.

IV. INTERPRETATION OF VELOCITY RESULTS

A. Orbital Inclinations and sin *i*

The actual masses of the planetary candidates remain unknown pending determination of the orbital inclination i. Upper limits to i come from Hipparcos astrometry in a few cases (Perryman et al. 1996). For example, the lack of astrometric wobble of 47 UMa imposes an upper mass limit of 7 M_{Jup} for that planet (Table I).

Nonetheless, a hypothetical population of 10–80-M_{Jup} brown dwarfs could serve, in principle, as the reservoir from which nearly face-on orbits would masquerade as planets. In an ensemble of randomly oriented orbital planes, the fraction whose normal vector points toward us within angle i (double cone) is $P(i) = 1 - \cos i$. For example, $\sin i$ lies between 0 and 0.1 in only 1:200 random orbits. In comparison, the ~300 stars surveyed probably contain only ~3 brown dwarf companions, too small a brown dwarf reservoir to expect any having $\sin i < 0.1$. Thus, it is un-

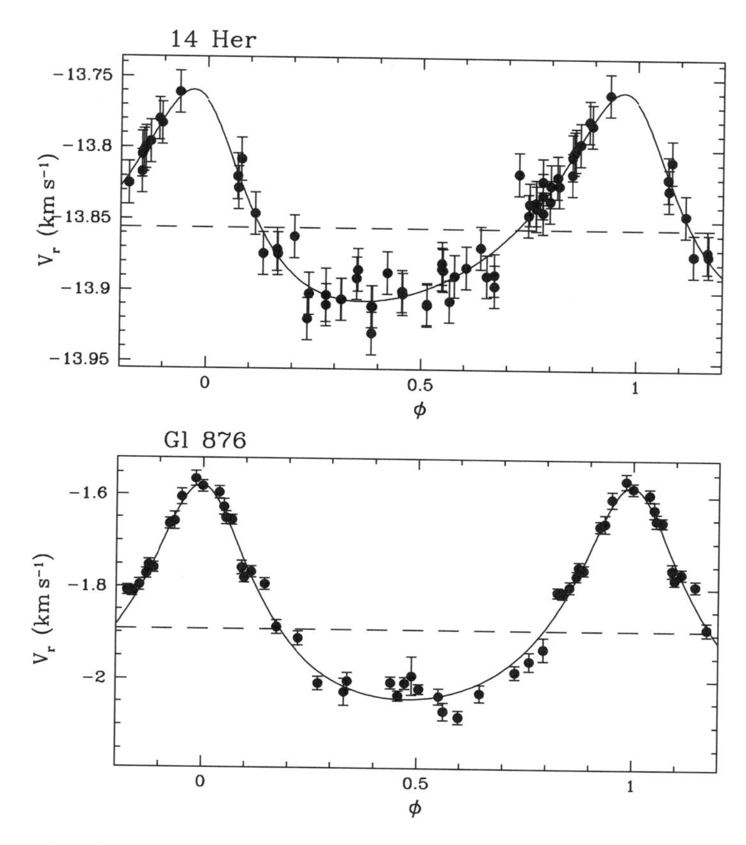

Figure 3. The two new planets announced at the IAU 170 and *Protostars and Planets IV* meetings, based on Haute-Provence Observatory measurements. (a) Phased Doppler data for 14 Her (K0 V). The planet has P >1630 d, a >2.5 AU, and $e \approx 0.34$. This is the longest-period extrasolar planet yet found, and more measurements are needed to determine the period more precisely. (b) Phased Doppler data for Gl 876.

likely that any of the candidate planets have masses 10 times greater than their $M \sin i$.

It is tempting to adopt an "expectation" value of $\sin i$, given by the mathematical mean of its distribution: $\langle \sin i \rangle = \pi/4$. However, selection effects render this approach dangerous, because the most easily detectable companions reside in nearly "edge-on" orbits ($i \approx 90°$), which would produce the highest velocity amplitudes. Thus, edge-on orbital inclinations are favored, implying that the actual planet mass resides close to $M \sin i$. Put differently, the bias toward detecting edge-on orbits implies that the largest values of the $M \sin i$ distribution (near 5 M_{Jup}) represent the maximum planetary masses themselves.

For the shortest-period planets at least, we could expect collinear spin axes for the stellar rotation and planet orbital motion. From a precise

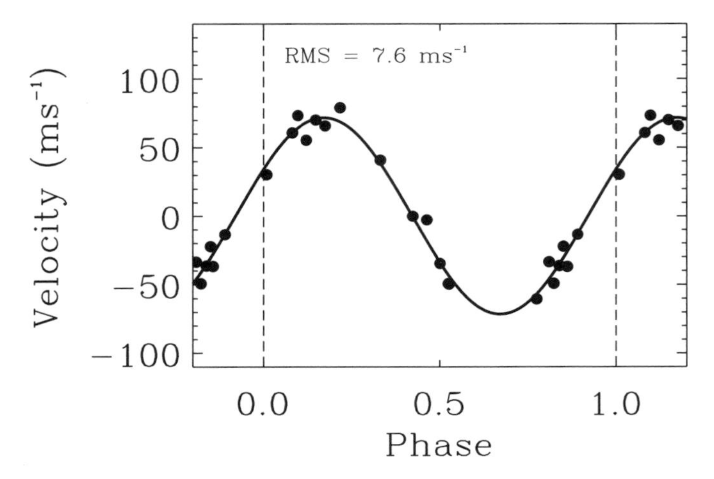

Figure 4. The phased Doppler data for HD 187123 (G3 V) from the Keck 1 Observatory. The planet has $M \sin i = 0.57\ M_{Jup}$, $P = 3.09$ d, $a = 0.42$ AU, and $e \approx 0.00$. This is the smallest planetary orbit known. Tidal effects presumably circularized the orbit, if not primordial.

TABLE I
Orbits of Extrasolar Planet Candidates[a]

Star	P (d)	a (AU)	K ($\mathrm{m\,s^{-1}}$)	e	$m \sin i$ (M_{Jup})	Companion Mass[b] (M_{Jup})
HD 187123	3.097	0.042	83.	0.03	0.57	
τBoo	3.3125	0.047	468.	0.00	3.66	$5.9^{+43.9}_{-1.8}$
51 Peg	4.231	0.051	56.0	0.01	0.44	0.49 ± 0.03
υ And	4.62	0.054	71.9	0.15	0.61	$0.76^{+0.19}_{-0.09}$
HD 217107	7.11	0.072	141	0.14	1.28	
55 Cnc	14.65	0.11	75.9	0.04	0.85	>0.66
Gliese 86	15.84	0.11	379.	0.04	3.6	
HD 195019	18.3	0.14	269	0.05	3.43	
Gliese 876	61	0.21	239	0.27	2.1	
ρ CrB	39.6	0.23	67.	0.11	1.1	$2.9^{+13.6}_{-1.3}$
HD 168443	57.9	0.28	330	0.54	5.04	
HD 114762	84.0	0.35	618	0.33	11.0	
70 Vir	116.5	0.47	316	0.40	7.4	>9.4
HD 210277	437	1.10	41.5	0.45	1.28	
16 Cyg B	799	1.6	50.3	0.687	1.67	$2.0^{+1.1}_{-0.3}$
47 UMa	1092	2.1	47.3	0.10	2.45	$3.4^{+3.1}_{-1.1}$
14 Her	1620	2.5	75.	0.36	3.3	

[a] All planet candidates as of January 1, 1999.
[b] Mass estimates from Gonzalez (1998*a,b*).

TABLE II
Parent Stars of Planet Candidates[a]

Star	T_{eff} (K)	log g (cgs)	ξ_t (km s^{-1})	[Fe/H]	M_V
υ And	6250 ± 100	4.30 ± 0.10	1.40 ± 0.10	0.17 ± 0.08	3.45 ± 0.03
τ Boo	6600 ± 100	4.50 ± 0.15	1.60 ± 0.10	0.34 ± 0.09	3.53 ± 0.03
55ρ^1 Cnc	5250 ± 70	4.40 ± 0.17	0.80 ± 0.09	0.45 ± 0.03	5.47 ± 0.03
ρ CrB	5750 ± 75	4.10 ± 0.05	1.20 ± 0.10	−0.29 ± 0.06	4.18 ± 0.03
16 Cyg B	5700 ± 75	4.35 ± 0.05	1.00 ± 0.10	0.06 ± 0.06	4.60 ± 0.04
51 Peg	5750 ± 75	4.40 ± 0.10	1.00 ± 0.10	0.21 ± 0.06	4.52 ± 0.04
47 UMa	5800 ± 75	4.25 ± 0.05	1.00 ± 0.10	0.01 ± 0.06	4.29 ± 0.03
70 Vir	5500 ± 75	3.90 ± 0.05	1.00 ± 0.01	−0.03 ± 0.06	3.68 ± 0.04
14 Her	5300 ± 90	4.27 ± 0.16	0.80 ± 0.12	0.49 ± 0.05	5.31 ± 0.03
Gliese 876	–	–	–	–	11.80 ± 0.03
HD 187123	5830 ± 40	4.40 ± 0.07	1.00 ± 0.08	0.16 ± 0.03	4.43 ± 0.03
HD 114762	5950 ± 75	4.45 ± 0.05	1.00 ± 0.10	−0.60 ± 0.06	4.23 ± 0.13

[a] Planet-bearing stars for which detailed atmospheric analysis has been done (Gonzalez 1998*a,b*)

measurement of stellar radius and $V \sin i$, we can derive an orbital inclination estimate and then estimate the mass of the planet. Different values have been proposed by Gonzalez (1998*a,b*) and Fuhrmann et al. (1998), and the Gonzalez mass estimates are listed in Table I.

B. Detection Efficiency and Selection Effects

The Doppler approach is most sensitive to planets that impart the largest reflex velocity to the host star. The velocity semiamplitude, K, is given by

$$K = \left(\frac{2\pi G}{P}\right)^{1/3} \frac{m_p \sin i}{(M_\star + m_p)^{2/3}} \frac{1}{\sqrt{1-e^2}}$$

where m_p is the mass of the unseen companion and $M_\star$ is the stellar mass. The period can be replaced by the semimajor axis with Kepler's third law.

In practice the detectability of reflex velocities depends on both the Doppler errors and the density of sampling in orbital phase. As a rule of thumb, with about 20–30 data points well spread in phase, one can reliably detect a velocity wobble if $K > 4\sigma$, where σ represents the velocity uncertainty in a given measurement.

This detectability criterion is apparent empirically in Fig. 5, which shows the detectability in the two-parameter space of semimajor axis and companion mass ($M \sin i$). Two curves show the locus of points of constant velocity amplitude for 10 and 40 m s^{-1}, assuming circular orbits. For reference, a Jupiter mass orbiting at 1 AU induces a reflex velocity of 28 m s^{-1} in a 1-$M_\odot$ star. Figure 5 also shows many of the detected planet candidates to date, as dots. These dots clearly show that amplitudes over 40 m s^{-1} have been required for detection, owing to the Doppler precision of 10 m s^{-1} that characterized the Doppler surveys during the past 10 yr.

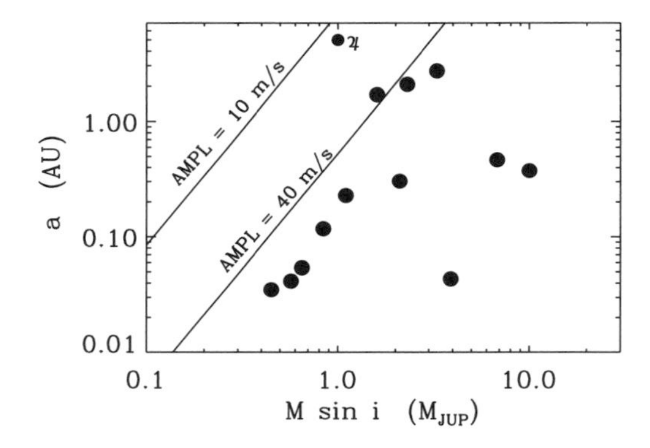

Figure 5. Detectability of companions by the induced reflex velocity, as a function of the orbital radius and companion mass (for circular orbits). The lines represent constant reflex velocities of 10 and 40 m s^{-1}. Planetary companions are shown as dots, including Jupiter. Past precision of 10 m s^{-1} enabled detection of amplitudes above 40 m s^{-1}, suggesting that 3 m s^{-1} (annual average) is required to detect a true Jupiter at 5 AU.

Figure 5 suggests that the detection of true Jupiter analogs at 5 AU will require a precision of under 10 m s^{-1}.

C. Planetary Masses, Semimajor Axes, and Eccentricities

Three specific major surveys can be examined to elucidate the mass distribution of substellar companions from 0 to 70 M_{Jup} within 5 AU. These three surveys include the modest-precision (300 m s^{-1}) survey of ~600 G and K dwarfs by Mayor, Duquennoy, and Udry (Duquennoy and Mayor 1991; Halbwachs et al. 1998). The other two are high-precision Doppler surveys of Mayor and Queloz (1995) and Marcy and Butler (1998), which surveyed 140 and 107 G and K stars respectively. The first survey of modest precision revealed 10 "brown dwarf" candidates, characterized by $M \sin i = 15$–60 M_{Jup} (Mayor et al. 1997). However, Hipparcos now enables the detection of wobbles for those cases in which the companion is massive enough and the semimajor axis is large enough that the wobble would be over a few milliarcsec. For every one of the 10 brown dwarf candidates for which the semimajor axis was large enough that a stellar companion could have been detected by Hipparcos, an astrometric wobble indicative of a pole-on orbital orientation was indeed seen. Thus, 7 of the 10 brown dwarf candidates have extreme values of $\sin i$, rendering the companion more massive than 0.075 $M_{\odot}$.

The M $\sin i$ values for the surviving substellar candidates are shown in Fig. 6, drawn from all three surveys mentioned above. This histogram shows eight companions in the mass bin from 0–5 M_{Jup} (i.e., those planet candidates from Table I that came from only these three surveys). From 5 to 10 M_{Jup}, there is only one companion (70 Vir b). The remaining mass range from 10 to 70 M_{Jup} contains only four companions. Thus, there is

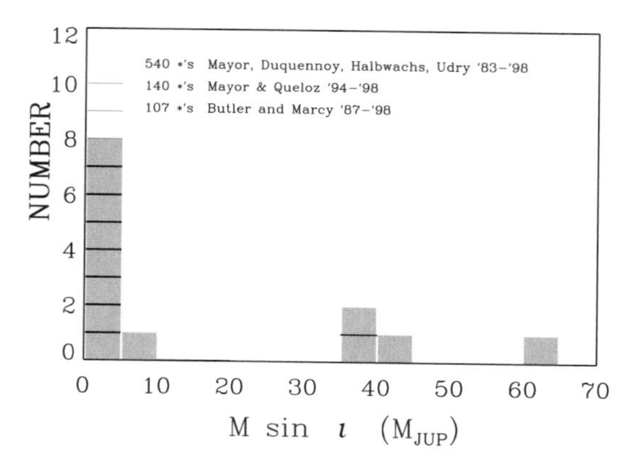

Figure 6. Histogram of M sin i for all companions known around solar-type stars, in the domain 0–70 M_{Jup}, within 5 AU, drawn from two large surveys of Mayor et al. (1997) and Marcy and Butler (1998). Companions for which Hipparcos astrometry has shown that $M > 70$ M_{Jup} have been removed. The tallest peak resides at the lowest masses (0–5 M_{Jup}), and signals a jump in the mass function. Selection effects favor detection of the more massive companions; conversely, companions having $M < 5$ M_{Jup} suffer incompleteness, suggesting that the current peak underestimates their relative occurrence. The brown dwarf regime is sparsely populated, indicating that at most 1% of G and K stars have brown dwarfs within 5 AU. In logarithmic mass intervals, the planetary regime is still more populated than the brown dwarf regime.

a remarkable spike in the mass function for masses from 0 to 5 M_{Jup}, in contrast to the "desert" of brown dwarfs of higher mass. Indeed, those four brown dwarf candidates may also be H-burning stars, but their periods are too short to permit astrometric detection of the wobble even by Hipparcos.

Thus, nature manufactures companions having masses below 5 M_{Jup} more prodigiously than it does companions from 5 to 70 M_{Jup} within 3 AU. Note that all selection effects favor detection of the most massive companions. Therefore, the spike at low masses suffers incompleteness and is likely to grow even taller as Doppler precision improves with time. This discontinuity in the mass function at 5 M_{Jup} permits a taxonomic segregation in the populations on either side. A common nomenclature is found in the terms "planet" and "brown dwarf" for the two species. Clearly, the formation processes for both types await further work, notably to determine kinship with the planets in our solar system.

Eleven of the extrasolar planets detected to date reside in close orbits, with $a < 0.3$ AU. The full histogram of semimajor axes of the extrasolar planets is shown in Fig. 7. Such small orbits were not predicted by theory (Lissauer 1995; Boss 1995). The surprisingly small orbits stand in apparent contrast to the prediction that the giant planets formed first from ice grains, which exist only beyond ~3 AU. Such grain growth provides the supposed requisite solid core around which gas could accrete. In section V.A we describe current migration models to explain these close orbits.

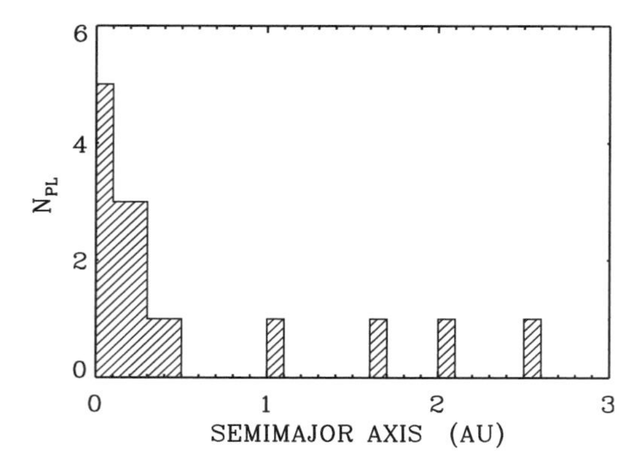

Figure 7. Histogram of semimajor axes of extrasolar planets to date. The preponderance of planets within 0.3 AU appears to be a real effect, because detectability of Jupiters is complete within 1.5 AU, as evidenced by the detections of three companions beyond 1.5 AU. Orbital migration inward may explain this piling-up of planets.

Nonetheless, it appears that giant planets in general reside systematically close to the host star, despite the obvious selection effect that enhances detectability for the closest planets. For example, giant planets that reside 0.5–1.5 AU from their stars would impart a reflex velocity of 25–40 m s^{-1}, which is clearly detectable with current precision. Indeed, Marcy and Butler (1998) have achieved precision better than 5 m s^{-1} for 60 chromospherically quiet stars from Lick Observatory during the past 3 yr. None exhibit planetary-mass companions with a semimajor axis between 0.5 and 1.5 AU. Thus, the distribution of planetary semimajor axes contains an apparent maximum within 0.1 AU, in terms of dN/da.

The eccentricities of the extrasolar planet candidates range from 0.00 to 0.68, and nine out of 17 have an eccentricity above that of Jupiter, $e = 0.05$ (see Table I). Remarkably, all nine planet candidates that orbit farther than 0.2 AU from their host stars (and hence are immune to tidal circularization) reside in noncircular orbits. The large orbital eccentricities have been interpreted by some to imply that they are simply "brown dwarfs" (Mazeh et al. 1996; Black 1997), defined as objects that form "as stars do." If so, brown dwarfs extend from 75 M_{Jup} toward arbitrarily lower masses, with only theoretical ideas (such as opacity-limited cloud fragmentation, e.g., Boss 1988) remaining to constrain the lowest mass at which "star formation" can occur. Instead, it is more likely that these Jupiter-mass companions form in protostellar disks and acquire eccentricities by gravitational perturbations.

It is interesting to compare the distributions of orbital eccentricities for companions to solar-type stars when the mass rises from the domain of giant planets to that of stellar secondaries. Figure 8 illustrates the distri-

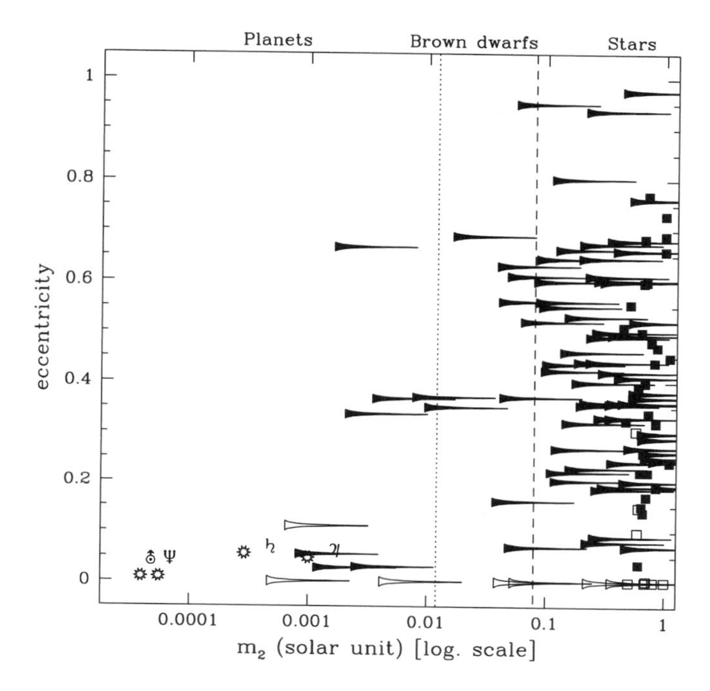

Figure 8. The eccentricity-mass ($M_\odot$) distribution of companions to solar-type stars. Starlike symbols are used to identify the four giant planets of our solar system. Binaries among G and K dwarfs of the solar neighborhood are plotted with squares if their actual masses are known and with wedges otherwise; the width of a wedge at a given mass is proportional to the probability of having such a mass due to the $\sin i$ distribution. Empty symbols indicate probable circularization by tidal interaction with the central star.

butions (e, $\log m_2$) (updated version of Fig. 9 of Mayor et al. 1997). The physical interpretation of the diagram is discussed in section V.B.

Although brown dwarfs presumably form at masses approaching 10 M_{Jup} and perhaps lower, the observed mass function (dN/dM) exhibits a dramatic rise at 5 M_{Jup} (Fig. 6). The peak of the mass function distribution is still more impressive if we consider an enlarged domain of mass (up to 0.6 $M_\odot$) for the companions to solar-type stars (Fig. 9). One may adopt the view that a continuum of formation processes, from 75 to 1 M_{Jup}, somehow yields a sharp increase in the mass distribution near 5 M_{Jup}. Alternatively, one may interpret the discontinuous increase in the mass function for $M < 5$ M_{Jup} (Fig. 6) as an indication that new formation processes are operating. We adopt this latter view. Thus while the orbital eccentricities of the sub-5 M_{Jup} objects seem similar to those of the more massive stellar companions, the significance of the jump in the mass function at 5 M_{Jup} cannot be ignored. Whether identifications of the members residing on either side of 5 M_{Jup} as "planets" and "brown dwarfs" proves valuable will require additional detections and more sophisticated theory.

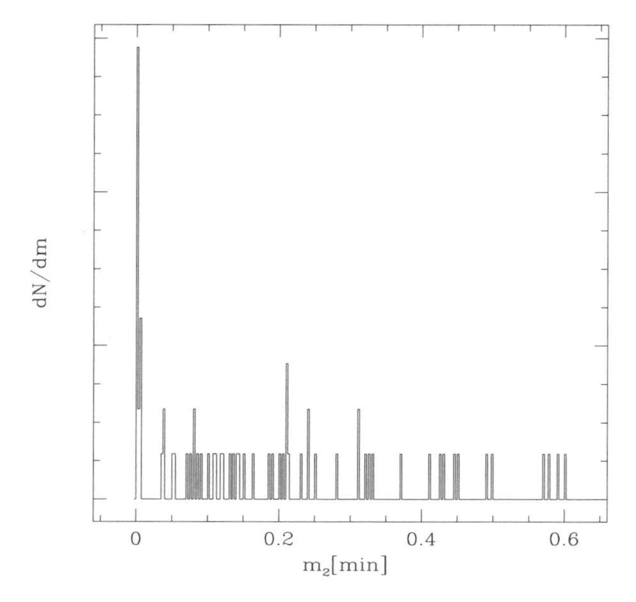

Figure 9. The distribution of companion masses ($M_{\odot}$) to solar-type stars within the solar vicinity (G–K V). The number of Jupiter-mass detections has been weighted according to the relative sizes of the observed samples.

D. Metallicity

The metallicities of the planet-bearing stars are shown in Fig. 10 and come from detailed local thermodynamic equilibrium (LTE) spectroscopic synthesis of high-resolution spectra (Gonzalez 1997, 1998*a,b,* Gonzalez and Vanture 1998). The histogram of metallicities [Fe/H] = log([Fe]/[H])/([Fe]/[H])$_{\odot}$ of both field and planet-bearing stars are shown (the latter as tick marks at top), except for that of Gliese 876, for which a spectral synthesis is yet to be done. The field star measurements come from Table 12 of Gonzalez (1998*b*). Compared to stars in the surrounding galactic field, the planet-bearing stars appear to be metal rich. Figure 10 also includes a tick mark at top for the star HD 114762 with its companion of $M \sin i = 11\ M_{\rm Jup}$, a brown dwarf candidate. That star indeed has low metallicity, [Fe/H] = −0.6, in contrast to the metal-rich stars that harbor the best planet candidates, which have $M \sin i < 5\ M_{\rm Jup}$.

The correlation between the presence of planets and high stellar metallicity represents a remarkable (and the only) astrophysical tie between their existence and independent stellar properties. The correlation thus lends support to the existence of the planets but alternatively could signify some stellar pathology that produces spurious Keplerian Doppler shifts, a prospect that seems unlikely.

The most remarkable cases are 55 Cnc and 14 Her, which have metallicities of [Fe/H] = 0.45 ± 0.03 and [Fe/H] = 0.50 ± 0.05, respectively

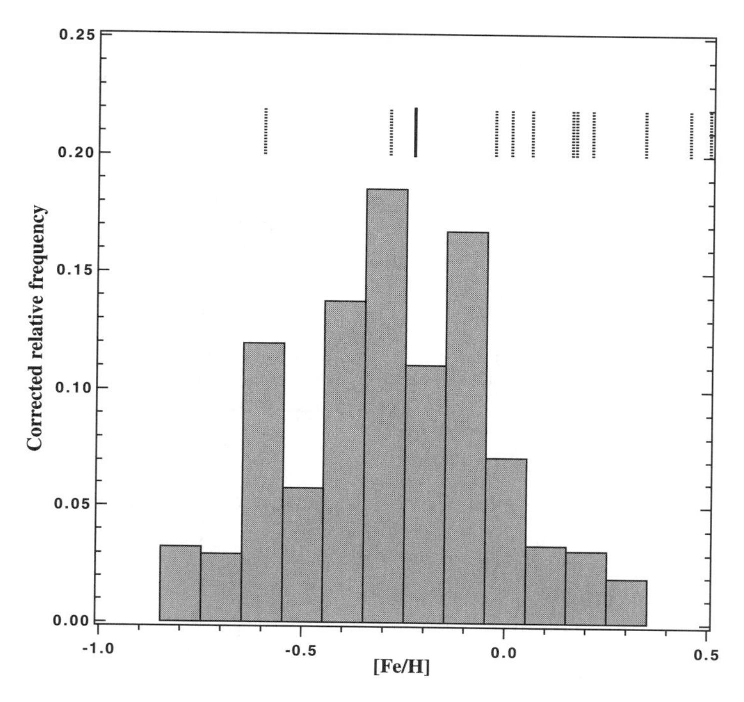

Figure 10. The histogram of [Fe/H] for field stars (shaded) and the planet-bearing stars (upper tick marks) presented by Gonzalez (1998*a,b*). The planet-bearing stars are systematically metal rich. The [Fe/H] for HD 114762 is represented by the tick mark at far upper left, indicating that star to be significantly metal poor relative to the planet-bearing stars. Indeed, its companion has $M \sin i = 10\ M_{Jup}$, suggesting brown dwarf status. The heavy tick mark at top shows the mean [Fe/H] of the nearby star distribution. The solar metallicity, [Fe/H] = 0, is clearly metal rich compared to the field.

(Gonzalez 1998*a,b*). Taylor (1996) finds [Fe/H] $= 0.38 \pm 0.07$ for 14 Her. These two stars had been identified as remarkably metal rich prior to the planet detection (Taylor 1996), and they are two of the seven "Super Metal Rich" stars known in the solar vicinity (Taylor 1996). The metallicity of HD 187123 is also above solar, [Fe/H] $= 0.16$. From Fig. 10, the Sun itself is metal rich compared to the field stars nearby.

It is not clear what the causal relationship is between planets and high metallicity, if indeed the correlation is real. High metallicity may enhance growth rates of rocky cores. Alternatively, giant planets may pollute the host star with metals (Laughlin and Adams 1998).

E. Fraction of Stars with Planets

The detections of planetary companions by different surveys provides estimates of the fraction of stars that harbor them. The Lick Observatory of 107 F, G, K, and M stars revealed 6 companions having $M \sin i < 7\ M_{Jup}$ (Marcy and Butler 1998). Among these, 70 Vir with its large $M \sin i =$

6.8 M_{Jup} may be considered a planet or brown dwarf. The Lick survey has lasted for 11 yr, providing sensitivity to companions having semimajor axes as large as 5 AU. The historical precision of $\sim$10 m s^{-1} implies that companions having masses greater than 2 M_{Jup} would have been detected at 5 AU, but companions having masses of $\sim$1 M_{Jup} are detectable only within 1 AU. Roughly, the Lick survey implies that within 5 AU, $\sim$5% of solar-type stars harbor companions having masses 0.5–5 M_{Jup}.

The Haute Provence Observatory project (Mayor and Queloz 1995; Mayor et al. 1998*b*) has surveyed 140 F, G, and K dwarfs for 4 yr. It has revealed two planetary companions (51 Peg and 14 Her) and was sensitive to companions having $M \sin i \sim 1$ M_{Jup} at 1 AU. Ironically, although carried out with shorter duration than the Lick survey, this project has discovered the longest period yet found, 14 Her, with $a \approx 2.5$ AU. Thus from this survey, the occurrence of Jupiters within 2.5 AU appears to be $\sim$2%.

The surveys by Walker et al. (1995) and by Cochran and Hatzes (1994) add constraints on the planetary occurrence rate that are consistent with those from Lick and Haute Provence. An approximate average of these surveys, with no weighting, yields an occurrence of Jupiter-mass companions within 3 AU of 4%.

V. ASTROPHYSICAL IMPLICATIONS

A. Orbital Migration

Migration of protoplanets in disks had been predicted prior to the detection of the extrasolar planets by Goldreich and Tremaine (1980), Ward (1981), Lin and Papaloizou (1986), Ward and Hourigan (1989), and Artymowicz (1993); see the chapters by Lin et al., and by Ward and Hahn, this volume. There appear to be two possible modes of angular momentum loss:

1. Small giant planet cores of a few Earth masses, which cannot clear a gap in the protostellar disk, lose angular momentum via interactions with the Lindblad resonances induced in the disk (Artymowicz 1993; Ward 1997). The timescale for orbital decay in a nominal disk may be as short as $\sim 3\times10^5$ yr.
2. The second migration mode occurs for protoplanets with mass $M > \sim 0.1$ M_{Jup}, which clear a gap in the disk. The clearing of the gap may curtail the growth of the planet, thereby setting the upper limit to the planet mass function (Artymowicz and Lubow 1996; Lin et al., this volume). Gravitational tidal torques between the planet and disk material at the inner and outer gap edges transfer angular momentum outward and thus force the planet to flow inward along with the viscous disk material toward the star (Ward 1997; Lin and Papaloizou 1995; Trilling et al. 1998). The time for material at 5 AU to drain into a T Tauri star appears to be $\sim 1\times10^6$ yr, from measured values of the mass accretion rate and the disk mass (Valenti et al. 1993; Hartmann et al. 1997; Beckwith and Sargent 1996).

Migration of Jupiters implies that they may form and spiral into the star within 1×10^6 yr. Successive Jupiters may form, each meeting the same fate (Lin 1986). Finally, the viscous timescale is matched by the remaining lifetime of the disk, at which point the last Jupiter survives, left at some intermediate point in its travel. Inward migration implies that giant planets will be found at a range of orbital distances, 0–20 AU from the host star. Their final distribution may not be uniform, because the migration speed, *dr/dt,* depends on the surface density profile of the disk, for a given *dM/dt.*

The deposited Jupiters might be statistically located with a probability distribution that is proportional to the timescale for inward migration, which in turn is proportional to the local surface mass density, $\sigma(r)$ for a fixed mass accretion rate. If $\sigma(r)$ increases inward for typical T Tauri disks, then Jupiters may be found preferentially close to the star.

Lin et al. (1996) suggested that the 51 Peg-type planets (now including υ And, τ Boo, and HD 187123) might have originally formed at ~5 AU according to the conventional planet formation paradigm (Boss 1995; Lissauer 1995). These planets then may have suffered inward orbital migration as described above. The inward migration may be stopped at about 0.05 AU, either by tidal interactions with the spin of the star or by the clearing of the inner disk by the stellar magnetosphere (Lin et al. 1996). Trilling et al. (1998) consider the possibility that Roche lobe overflow by the planet can halt the migration, though the mechanism is difficult to implement.

B. Origin of Orbital Eccentricities

Four mechanisms appear capable of generating eccentricities as added processes to the standard paradigm of planet formation. Protoplanets excite spiral density waves in the protoplanetary disk, which can gravitationally perturb the planet, pumping the eccentricity to modest values (Artymowicz 1998; Lubow and Artymowicz, this volume). It remains to be seen how high the eccentricities can be augmented by this process.

Multiple giant planets can mutually perturb each other, causing orbit crossings and large gravitational scatterings (Weidenschilling and Marzari 1996; Rasio and Ford 1996; Lin and Ida 1997). Arbitrarily high eccentricities can result (see the chapter by Lin et al., this volume). These scattering models predict that Doppler measurements should eventually reveal additional eccentric companions. None has yet been found. Further, planetesimals and other planets may gently perturb the orbits of giant planets over timescales of 10 Myr during runaway accretion (Lissauer et al. 1995) or over timescales of the "heavy bombardment" period (~100 Myr) as computed by Levison et al. (1998).

Eccentric giant planets could be formed in open star clusters as a result of purely dynamical processes. Up to 50% of giant planets around G stars in open clusters can develop high eccentricities due to multibody gravitational interactions, as shown by de la Fuente Marcos and de la Fuente Marcos (1997) and by Laughlin and Adams (1999).

Finally, a stellar companion (such as 16 Cyg A) can gravitationally pump the eccentricity of a planet, even for ratios of orbital separation (star-star vs. star-planet) as great as 1000 (Holman et al. 1997; Mazeh et al. 1997; Cochran et al. 1997). Indeed, the eccentric orbit of 16 Cyg A and B brings them within 1000 AU and possibly within 200 AU, making stellar perturbations plausible (Hauser and Marcy 1998; Romanenko 1994).

C. Formation in Clusters

Our standard paradigm for star and planet formation assumes that the system forms in isolation; once the collapse of the protostar begins, it does not interact with its surroundings in any way. However, it is quite likely that most (if not all) stars form in clusters, where the stellar density is much higher than in the field where we find the older stars (Lada et al. 1991). A simple comparison of the sizes of protostellar disks (100–1000 AU) with the mean stellar separation in protoclusters (5000–10,000 AU) suggests that star-disk and disk-disk encounters must be important. Kroupa (1995*a,b,c)* explored the dynamical evolution of star-forming clusters and concluded that the observed frequency and properties of field binary stars is consistent with all stars having formed in clusters of binary systems. Gravitational encounters between a disk and a nearby star (Boffin et al. 1998) or between two disks (Watkins et al. 1998*a,b)* can drastically alter the evolution of the disk from the theoretically idealized case of a completely isolated disk. At the densities of most current star-forming regions, most protostars will undergo at least one significant interaction with a neighboring star before its protostellar disk is dissipated. Star-disk encounters are unlikely to lead to capture, but coplanar disk-disk encounters can result in the formation of multiple-star systems. More important for the problem of planet formation, however, is what happens to the disk material during such encounters. Star-disk encounters can truncate the disk, strip newly formed planets from the system, or introduce tilts and twists of up to 30° in the disk (Boffin et al. 1998). In disk-disk encounters, disk material can be swept into a shock layer, which can fragment to produce new condensations (Watkins et al. 1998*a,b).* Clearly, the types of planetary systems formed under these conditions may differ drastically from our current idealized models.

D. Miscellaneous

The Doppler detection of planets is subject to the alternative interpretation that the stars are pulsating rather than undergoing reflex motion (Gray 1997; Gray and Hatzes 1997). Under the assumption of radial pulsation, the time integral of the velocity during $\frac{1}{4}$ period indicates the net physical displacement of the photosphere. For most stars listed in Table I, this hypothetical displacement due to pulsation is more than 1% $R_\odot$. Such large changes in stellar radius (at optical depth unity) should cause photometric variations over 1% unless nature conspires to hide the oscillations with temperature changes that cancel the radius changes. Barring that, the millimagnitude-level photometric constancy of the stars in Table I (Baliunas et al. 1997; Henry et al. 1997) renders radial oscillations unlikely. Nonradial sectoral-mode oscillations can produce effective shifts of spectral lines while causing small photometric variability (Gray and Hatzes 1997). In addition, the observed stellar line profile may be slightly distorted by starlight that is reflected at particular orbital phases by a close-in planetary companion (Charbonneau et al. 1998). The line profiles of τ Boo and of 51 Peg exhibit constant shape when observed at high spectral resolution and high signal-to-noise ratios, which also argues against oscillations (Brown et al. 1998*a,b;* Gray 1998; Hatzes et al. 1998*a,b;* Hatzes and Cochran 1998). However, one must beware of the possibility of oscillations that might maintain line shape while displacing the entire line. Further, extraordinarily new physics is always possible, so pulsations can never be ruled out. However, more work is required to securely understand stellar pulsation and surface velocity fields in general.

VI. CONCLUSIONS

Jovian-mass companions have been found around other solar-type stars. Radial velocity surveys have now reached the level of precision at which such planet candidates can be found routinely. Other detection techniques should reach this point in the near future. The mass distribution of substellar companion objects reveals the tail of the stellar mass function in the 10–70-M_{Jup} regime, and a significant excess of objects at masses less than 5 M_{Jup}. This indicates that two different physical processes operate. The star-formation process forms the more massive group of substellar companions that are true brown dwarfs. The objects with masses of a few Jupiter masses that orbit within a few AU are most probably formed in the protoplanetary disk and thus are planets according to most standard definitions. It is not yet clear whether there is some overlap in the 10–70-M_{Jup} mass ranges of brown dwarfs and of planets; that answer will require better statistics on the mass ranges of both types of objects.

Remarkably, all nine planet candidates that orbit beyond 0.2 AU reside in eccentric orbits, unlike the circular orbits of the solar system gas giants. Such eccentric orbits predominate among Jupiter-mass companions from 0.2–2.5 AU and may stem from perturbations imposed by the other planets, passing stars, or by the protoplanetary disk.

None of the systems found so far resembles ours, with a massive planet at a distance from the central star at which ice would have condensed in the protostellar disk. While this is partly a result of detection sensitivity (close-in planets give a larger Doppler signal), there is a significant tendency for planets to be found within 0.3 AU of the star. Formation of objects in these close-in orbits is extremely difficult but not necessarily impossible.

Early dynamical evolution of the system provides an explanation for the observed distribution of planetary semimajor axes. Tidal torques between the newly formed planet and the remnant disk can cause an inward migration of the gas giant planet from the 3+ AU region, where it can be formed easily, to the inner regions, where such planets reside today. The piling-up of planets near 0.05 AU is quite remarkable and suggests that there is some efficient mechanism to stop this inward migration of planets just before they plunge into the star.

Stars that have detected planets in orbit around them appear to be slightly metal-rich compared to the general population of disk solar-type stars. This can be understood within our current understanding of the planet formation process. The core of a jovian planet is thought to grow by collisional growth and accretion of solid planetesimals in the disk. The speed of this process should scale roughly as the square of the metallicity, because collisional rates proceed as density squared. Gas giant planets should form more rapidly around metal-rich stars. If the planet forms while there is still a viscous disk, it will migrate inward. It is quite possible that some planets may not magically stop their inward migration at 0.05 AU and may spiral all of the way into the star. This would only slightly pollute the stellar convective zone with metal-rich material.

The true test of our paradigm of solar system formation requires the characterization of many systems of multiple planets. The three planets around υ Andromedae offer the first indication of possibly extensive diversity among such systems.

Acknowledgments We give special thanks to R. Paul Butler, without whom most of the detections of planet candidates would not have been made. We thank D. Queloz, A. Hatzes, S. Udry, S. Vogt, G. Gonzalez, T. Brown, S. Horner, R. Noyes, S. Baliunas, G. Gatewood, D. Black, P. Bodenheimer, A. Cumming, D. Lin, J. Lissauer, and L. Hartmann for valuable discussions. Funding for this work came from NSF (AST 9520443 to G. W. M.; AST 9808980 to W. D. C.) and NASA (NAGW 5125 to G. W. M.; NAG5 4384 to W. D. C.) and from the Swiss National Foundation for Scientific Research (FNRS).

REFERENCES

Artymowicz, P. 1993. Disk-satellite interaction via density waves and the eccentricity evolution of bodies embedded in disks. *Astrophys. J.* 419:166–180.

Artymowicz, P. 1998. On the formation of eccentric superplanets. In *Brown Dwarfs and Extrasolar Planets*, ASP Conf. Ser. 134, ed. R. Rebolo, E. L. Martín, and M. R. Zapatero-Osorio (San Francisco: Astronomical Society of the Pacific), pp. 152–161.

Artymowicz, P., and Lubow, S. H. 1996. Mass flow through gaps in circumbinary disks. *Astrophys. J. Lett.* 467:L77–L80.
Baliunas, S. L., Henry, G. W., Donahue, R. A., Fekel, F. C., and Soon, W. H. 1997. Properties of sun-like stars with planets: ρ^1 Cancri, τ Bootis, and υ Andromedae. *Astrophys. J. Lett.* 474:L119–L122.
Beckwith, S. V. W., and Sargent, A. 1996. Circumstellar disks and the search for neighbouring planetary systems. *Nature* 383:139–144.
Beichman, C. A. 1998. Terrestrial planet finder: The search for life-bearing planets around other stars. *Proc. SPIE* 3350:719.
Black, D. C. 1997. Possible observational criteria for distinguishing brown dwarfs from planets. *Astrophys. J. Lett.* 490:L171–L174.
Boden, A., Milman, M., Unwin, S., Yu, J., and Shao, M. 1996. Astrometry with the Space Interferometry Mission. *Bull. Amer. Astron. Soc.* 189:1909 (abstract).
Boffin, H. M. J., Watkins, S. J., Bhattal, A. S., Francis, N., and Whitworth, A. P. 1998. Numerical simulations of protostellar encounters. I. Star-disc encounters. *Mon. Not. Roy. Astron. Soc.,* in press.
Borucki, W. J., and Summers, A. L. 1984. The photometric method of detecting other planetary systems. *Icarus* 58:121–134.
Borucki, W. J., Cullers, D. K., Dunham, E. W., Koch, D. G., Cochran, W. D., Rose, J. A., Granados, A., and Jenkins, J. M. 1996. FRESIP: A mission to determine the character and frequency of extra-solar planets around solar-like stars. *Astrophys. Space Sci.* 241:111–134.
Boss, A. P. 1988. Protostellar formation in rotating interstellar clouds. VII. Opacity and fragmentation. *Astrophys. J.* 331:370–376.
Boss, A. P. 1995. Proximity of Jupiter-like planets to low-mass stars. *Science* 267:360–362.
Brown, T. M., Noyes, R. W., Nisenson, P., Korzennik, S. G., and Horner, S. 1994. AFOE: A spectrograph for precise Doppler studies. *Pub. Astron. Soc. Pacific* 86:1285–1297.
Brown, T. M., Kotak, R., Horner, S. D., Kennelly, E. J., Korzennik, S., Nisenson, P., and Noyes, R. W. 1998*a*. Exoplanets or dynamic atmospheres? The radial velocity and line shape variations of 51 Pegasi and τ Bootis. *Astrophys. J. Suppl.* 117:563–585.
Brown, T. M., Kotak, R., Horner, S. D., Kennelly, E. J., Korzennik, S., Nisenson, P., and Noyes, R. W. 1998*b*. A search for line shape and depth variations in 51 Pegasi and τ Bootis. *Astrophys. J. Lett.* 497:L85–L88.
Burrows, A., Marley, M., Hubbard, W. B., Lunine, J. I., Guillot, T., Saumon, D., Freedman, R., Sudarsky, D., and Sharp, C. 1997. A nongray theory of extrasolar giant planets and brown dwarfs. *Astrophys. J.* 491:854–875.
Butler, R. P., Marcy, G. W., Vogt S. S., and Apps, K. 1998. A planet with a 3.1 day period around a solar twin. *Pub. Astron. Soc. Pacific* 110:1389-1395.
Butler, R. P., Marcy, G. W., Fischer, D. A., Brown, T. W., Contos, A. R., Korzennik, S. G., Nisenson, P., Noyes, R. W. 1999. Evidence for multiple companions to Upsilon Andromedae. *Astrophys. J.* to appear Dec. 1, 1999, vol. 526.
Campbell, B., Walker, G. A. H., Yang, S. 1988. A search for substellar companions to solar-type stars. *Astrophys. J.* 331:902–921.
Charbonneau, D., Jha, S., and Noyes, R. W. 1998. Spectral line distortions in the presence of a close-in planet. *Astrophys. J. Lett.* 507:153–156.
Cochran, W. D., and Hatzes, A. P. 1994. A high precision radial-velocity survey for other planetary systems. *Astrophys. Space Sci.* 212:281–291.
Cochran, W. D., Hatzes, A. P., and Hancock, T. J. 1991. Constraints on the companion object to HD114762. *Astrophys. J. Lett.* 380:L35–L38.
Cochran, W. D., Hatzes, A. P., Butler R. P., and Marcy, G. W. 1997. The discovery of a planetary companion to 16 Cygni B. *Astrophys. J.* 483:457–463.

Colavita, M. M., and Shao, M. 1994. Indirect planet detection with ground-based long-baseline interferometry. *Astrophys. Space Sci.* 212:385–390.
Deeg, H. J., Doyle, L. R., Kozhevnikov, V. P., Martín, E. L. Oetiker, B., Palaiologou, E., Schneider, J., Afonso, C., Dunham, E. W., Jenkins, J. M., Ninkov, Z., Stone, R. P. S., and Zakharova, P. E. 1998. Near-term detectability of terrestrial extrasolar planets: TEP network observations of CM Draconis. *Astron. Astrophys.* 338:479–490.
de la Fuente Marcos, C., and de la Fuente Marcos, R. 1997. Eccentric giant planets in open star clusters. *Astron. Astrophys. Lett.* 326:L21–L24.
Delfosse, X., Forveille, T., Mayor, M., Perrier, C., Naef, D., and Queloz, D. 1998. The closest extrasolar planet. A giant planet around the M4 dwarf Gl 876. *Astron. Astrophys. Lett.* 338:L67–L70.
Duquennoy, A., and Mayor, M. 1991. Multiplicity among solar type stars in the solar neighbourhood. II. Distribution of the orbital elements in an unbiased sample. *Astron. Astrophys.* 248:485–524.
Fuhrmann, K., Pfeiffer, M., and Bernkopf, J. 1998. F- and G-type stars with planetary companions: υ Andromedae, ρ (1) Cancri, τ Bootis, 16 Cygni and ρ Coronae Borealis. *Astron. Astrophys.* 336:942–952.
Gatewood, G. D. 1987. The multichannel astrometric photometer and atmospheric limitations in the measurement of relative positions. *Astron. J.* 94:213–214.
Gatewood, G. D. 1996. Lalande 21185. *Bull. Am. Astron. Soc.* 28:885 (abstract).
Goldreich, P., and Tremaine, S. 1980. Disk-satellite interactions. *Astrophys. J.* 241:425–441.
Gonzalez, G. 1997. The stellar metallicity-giant planet connection. *Mon. Not. Roy. Astron. Soc.* 285:403–412.
Gonzalez, G., 1998*a*. Parent stars of extrasolar planets IV: 14Let Herculis, HD187123, and HD210277. Submitted. *Astrophys. J. Lett.*
Gonzalez, G. 1998*b*. Spectroscopic analyses of the parent stars of extrasolar planetary system candidates. *Astron. Astrophys.* 334:221–238.
Gonzalez, G., and Vanture, A. D. 1998, *Astron Astrophys.*, submitted.
Gray, D. F. 1997. Absence of a planetary signature in the spectra of the star 51 Pegasi. *Nature* 385:795–796.
Gray, D. F. 1998. A planetary companion for 51 Pegasi implied by absence of pulsations in the stellar spectra. *Nature* 391:153–154.
Gray, D. F, and Hatzes, A. P. 1997. Non-radial oscillation in the solar-temperature star 51 Pegasi. *Astrophys. J.* 490: 412–424.
Griest, K., and Safizadeh, N. 1998. The use of high-magnification microlensing events in discovering extrasolar planets. *Astrophys. J.* 500:37–50.
Halbwachs, J.-L., Mayor, M., and Udry, S. 1998. On the distribution of mass ratios of late-type main sequence spectroscopic binaries. In *Brown Dwarfs and Extrasolar Planets*, ASP Conf. Ser. 134, ed. R. Rebolo, E. L. Martín, and M. R. Zapatero-Osorio (San Francisco: Astronomical Society of the Pacific), pp. 308–311.
Hale, A., Doyle, L. R. 1994. The photometric method of extrasolar planet detection revisited. *Astrophys. Space Sci.* 212:335–348.
Hartmann, L., Calvet, N., Gullbring, E., and D'Alessio, P. 1997. Accretion and the evolution of T Tauri disks. *Astrophys. J.* 495:385–400.
Hatzes, A. P., and Cochran, W. D. 1998. A search for variability in the spectral line shapes of τ Bootis: Does this star really have a planet? *Astrophys. J.* 502:944–950.
Hatzes, A. P., Cochran, W. D., and Bakker, E. J. 1998*a*. Further evidence for the planet around 51 Pegasi. *Nature* 391:154–155.
Hatzes, A. P., Cochran, W. D., and Bakker, E. J. 1998*b*. The lack of spectral line variability in 51 Pegasi: Confirmation of the planet hypothesis. *Astrophys. J.*, in press.

Hauser, H., and Marcy, G. W. 1998. The orbit of 16 Cygni AB. *Pub. Astron. Soc. Pacific.*, submitted.

Henry, G. W., Baliunas, S. L., Donahue, R. A., Soon, W. H., and Saar, S. H. 1997. Properties of Sun-like stars with planets: 51 Pegasi, 47 Ursae Majoris, 70 Virginis, and HD 114762. *Astrophys. J.* 474:503–510.

Holman, M., Touma, J., and Tremaine, S. 1997. Chaotic variations in the eccentricity of the planet orbiting 16 Cygni B. *Nature* 386:254–356.

Kroupa, P. 1995*a*. The dynamical properties of stellar systems in the galactic disc. *Mon. Not. Roy. Astron. Soc.* 277:1507–1521.

Kroupa, P. 1995*b*. Inverse dynamical population synthesis and star formation. *Mon. Not. Roy. Astron. Soc.* 277:1491–1506.

Kroupa, P. 1995*c*. Star cluster evolution, dynamical age estimation and the kinematical signature of star formation. *Mon. Not. Roy. Astron. Soc.* 277:1522–1540.

Lada, E. A., Evan, N. J., II, DePoy, D. L., and Gatley, I. 1991. A 2.2 micron survey in the L1630 molecular cloud. *Astrophys. J.* 371:171–182.

Latham, D. W., Mazeh, T., Stefanik, R. P., Mayor, M., and Burki, G. 1989. The unseen companion of HD114762: A probable brown dwarf. *Nature* 339: 38–40.

Laughlin G., and Adams F. C. 1998. Possible stellar metallicity enhancements from the accretion of planets. *Astrophys. J. Lett.* 491:L51–L54.

Laughlin, G., and Adams, F. C. 1999. The modification of orbits in dense stellar clusters. *Astrophys. J. Lett.* submitted.

Levison, H. F., Lissauer, J. J., and Duncan, M. J. 1998. Modeling the diversity of outer planetary systems. *Astron. J.*, in press.

Lin, D. N. C. 1986. The nebular origin of the solar system. In *The Solar System: Observations and Interpretation*, ed. M. G. Kivelson (Englewood Cliffs: Prentice-Hall), pp. 68–69.

Lin, D. N. C., and Ida, S. 1997. On the origin of massive eccentric planets. *Astrophys. J.* 477:781–791.

Lin, D. N. C., and Papaloizou, J. C. B. 1986. On the tidal interaction between protoplanets and the protoplanetary disk. III. Orbital migration of protoplanets. *Astrophys. J.* 309, 846–857.

Lin, D. N. C., and Papaloizou, J. C. B. 1995. Theory of accretion disks. *Ann. Rev. Astron. Astrophys.* 34:703–747.

Lin, D. N. C., Bodenheimer, P., and Richardson, D. C. 1996. Orbital migration of the planetary companion of 51 Pegasi to its present location. *Nature* 380:606–607.

Lissauer, J. J. 1995. Urey Prize lecture: On the diversity of plausible planetary systems. *Icarus* 114:217–236.

Lissauer, J. J., Pollack, J. B., Wetherill, G. W., and Stevenson, D. J. 1995. Formation of the Neptune system. In *Neptune and Triton*, ed. D. P. Cruikshank (Tucson: University of Arizona Press), pp.42—59.

Marcy, G. W., and Benitz, K. J. 1989. A search for substellar companions to low-mass stars. *Astrophys. J.* 344:441–453.

Marcy, G. W., and Butler, R. P. 1992. Precision radial velocities with an iodine absorption cell. *Pub. Astron. Soc. Pacific.* 104:270–277.

Marcy, G. W., and Butler, R. P. 1995. Brown dwarfs and planets around solar-type stars: Searches by precise velocities. In *The Bottom of the Main Sequence—and Beyond*, ed. C. G. Tinney (Berlin: Springer), p. 98.

Marcy, G. W., and Butler, R. P. 1998. Detection of extrasolar giant planets. *Ann. Rev. Astron. Astrophys.* 36:57.

Marcy, G. W., Butler, R. P., Williams, E., Bildsten, L. Graham, J. R., Ghez, A., and Jernigan, G. 1997. The planet around 51 Pegasi. *Astrophys. J.* 481:926–935.

Marcy, G. W., Butler, R. P., Vogt, S. S., Fischer, D., and Lissauer, J. J. 1998. A planetary companion to a nearby M4 dwarf, Gliese 876. *Astrophys. J. Lett.* 505:L147–L149.

Marcy, G. W., Butler, R. P., Vogt, S. S., Fischer, D. A., and Liu, M. C. 1999. Two new candidate planets in eccentric orbits. *Astrophys. J.*, in press.

Mayor, M., and Queloz, D. 1995. A Jupiter-mass companion to a solar-type star. *Nature* 378:355–359.

Mayor, M., Queloz, D., Udry, S., and Halbwachs, J.-L. 1997. From brown dwarfs to planets. In *Astronomical and Biochemical Origins and the Search for Life in the Universe*, IAU Colloq. 161, ed. C. Cosmovici, S. Bowyer, and D. Werthimer. (Bologna: Editrice Compositori) pp. 313–330.

Mayor, M., Arenou, F., Halbwachs, J.-L., Udry, S., and Queloz, D. 1998*a*. In *Extrasolar Planets: Formation, Detection and Modelling*, proceedings of the conference held in Lisbon, Portugal, in press.

Mayor, M., Beuzit, J.-L., Mariotti, J.-M., Naef, D., Perrier, C., Queloz, D., and Sivan, J.-P. 1998*b*. Searching for giant planets at the Haute-Provence Observatory. In *Precise Stellar Radial Velocities*, IAU Colloquium 170, in press.

Mayor, M., Udry, S., Queloz, D. 1998*c*. In *Cools Stars, Stellar Systems and the Sun,* Proceedings of the 10th Cambridge Workshop, held in Boston, ed. R. Donahue and A. Dupree, in press.

Mayor, M., Beuzit, J.-L., Mariotti, J.-M., Naef, D., Perrier, C., Queloz, D., and Sivan, J.-P. 1998*d*. The planet around the very metal rich star 14 Herculis. *Protostars and Planets conf. IV*, oral announcement.

Mazeh, T., Goldberg, D., Duquennoy, A., and Mayor, M. 1992. On the mass-ratio distribution of spectroscopic binaries with solar-type primaries. *Astrophys. J.* 401:265–268.

Mazeh, T., Latham, D. W., and Stefanik, R. P. 1996. Spectroscopic orbits for three binaries with low-mass companions and the distribution of secondary masses near the substellar limit. *Astrophys. J.* 466:415–426.

Mazeh, T., Krymolowsky, Y., and Rosenfeld, G. 1997. The high eccentricity of the planet around 16 Cyg B. *Astrophys J. Lett.* 477:L103–L106.

McMillan, R. S., Smith, P. H., Perry, M. L., Moore, T. S., and Merline, W. J. 1990. Long-term stability of a Fabry-Perot interferometer used for measurement of stellar Doppler shift. *Proc. Society of Photo-optical Instrumentation Engineers* 1235:601.

McMillan, R. S., Moore, T. S., Perry, M. L., and Smith, P. H. 1993. Radial velocity observations of the sun at night. *Astrophys. J.* 403:801–809.

McMillan, R. S., Moore, T. L., Perry, M. L., and Smith, P. H. 1994. Long, accurate time series measurements of radial velocities of solar-type stars. *Astrophys. Space Sci.* 212:271–280.

Murdoch, K. A., Hearnshaw, J. B., and Clark, M. 1993. A search for substellar companions to southern solar-type stars. *Astrophys. J.* 413:349–363.

Noyes, R. W., Jha, S., Korzennik, S. G., Krockenberger, M., Nisenson, P., Brown, T. M., Kennelly, E. J., and Horner, S. D. 1997. A planet orbiting the star ρ Coronae Borealis. *Astrophys. J. Lett.* 483:L111–L114.

Oppenheimer, B., Kulkarni, S. R., Matthews, K.; and Van Kerkwijk, M. H. 1998. The spectrum of the brown dwarf Gliese 229B. *Astrophys. J.* 502:932.

Peale, S. J. 1997. Expectations from a microlensing search for planets. *Icarus* 127:269–289.

Perryman, M. A. C., Lindegren, L., Arenou, F., Bastian, U., Bernstein, H. H., Van Leeuwen, F., Schrijver, H., Bernacca, P. L., Evans, D. W., Falin, J. L., Froeschle, M., Grenon, M., Hering, R., Hog, E., Kovalevsky, J., Mignard, F., Murray, C. A., Penston, M. J., Petersen, C. S., Le Poole, R. S., Soderhjelm, S., and Turon, C. 1996. Hipparcos distances and mass limits for the planetary candidates: 47 UMa, 70 Vir, and 51 Peg. *Astron. Astrophys.* 310:L21–L24.

Pravdo, S. H., and Shaklan, S. B. 1996. Astrometric detection of extrasolar planets: Results of a feasibility study with the Palomar 5 meter telescope. *Astrophys. J.* 465:264–277.

Rasio, F. A., and Ford, E. B. 1996. Dynamical instabilities and the formation of extrasolar planetary systems. *Science* 274:954–956.

Romanenko, L. G. 1994. Determination of the orbital elements of the wide double stars ADS 10759 (Psi Dra) and ADS 12815 (16 Cyg) by the method of apparent-motion parameters. *Astron. Rep.* 38:779–785.

Schneider, J., Auvergne, M., Baglin, A., et al. 1998. The COROT mission: From structure of stars to origin of planetary systems. In *Origins*, ASP Conf. Ser. 148, ed. C. Woodward, J. M. Shull, and H. A. Thronson (San Francisco: Astronomical Society of the Pacific), pp. 298–303.

Taylor, B. 1996. Supermetallicity at the quarter-century mark: A conservative statistician's review of the evidence. *Astrophys. J. Suppl.* 102:105–128.

Tokovinin, A. A. 1992. The frequency of low-mass companions to K and M stars in the solar neighborhood. *Astron. Astrophys.* 256:121–132.

Trilling, D., Benz, W., Guillot, T., Lunine, J. I., Hubbard, W. B., and Burrows, A. 1998. Orbital evolution and migration of giant planets: modeling extrasolar planets. *Astrophys. J.* 500:428–439.

Unwin, S., Pitesky, J. and Shao, M. 1999. Science with the Space Interferometry Mission. In *Harmonizing Cosmic Distance Scales in a Post-HIPPARCOS Era,* ASP Conf. Ser. 167, ed. D. Egret and Andre Heck (San Francisco: Astronomical Society of the Pacific), p.38.

Valenti, J. A., Basri, G., and Johns, C. M. 1993. T Tauri stars in blue. *Astron. J.* 106:2024–2050.

van de Kamp, P. 1977. Barnard's star 1916–1976: A sexagintennial report. *Vistas Astron.* 20:501–521.

van de Kamp, P. 1982. The planetary system of Barnard's star. *Vistas Astron.* 26:141–157.

van de Kamp, P. 1986. Dark companions of stars. *Space Sci. Rev.* 43:211–327.

Walker, G. A. H., Walker, A. R., Irwin, A. W., Larson, A. M., Yang, S. L. S., and Richardson, D. C. 1995. A search for Jupiter-mass companions to near by stars. *Icarus* 116:359–375.

Ward, W. R. 1981. Solar nebula dispersal and the stability of the planetary system: Scanning secular resonance theory. *Icarus* 47:234–264.

Ward, W. R. 1997. Survival of planetary systems. *Astrophys. J. Lett.*,482:L211–L214.

Ward, W. R., and Hourigan, K. 1989. Orbital migration of protoplanets—the inertial limit. *Astrophys. J.* 347:490–495.

Watkins, S. J., Bhattal, A. S., Boffin, H. M. J., Francis, N., and Whitworth, A. P. 1998*a*. Numerical simulations of protostellar encounters. II. Coplanar disc-disc encounters. *Mon. Not. Roy. Astron. Soc.,* in press.

Watkins, S. J., Bhattal, A. S., Boffin, H. M. J., Francis, N., and Whitworth, A. P. 1998*b*. Numerical simulations of protostellar encounters. III. Non-coplanar disc-disc encounters. *Mon. Not. Roy. Astron. Soc.*, in press.

Weidenschilling, S. J., and Marzari, F. 1996. Gravitational scattering as a possible origin for giant planets at small stellar distances. *Nature* 384:619–621.

Wolszczan, A. 1994. Confirmation of Earth-mass planets orbiting the millisecond pulsar PSR B1257+12. *Science* 264:538–542.

Wolszczan, A., and Frail, D. A. 1992. A planetary system around the millisecond pulsar PSR1257+12. *Nature* 355:145–147.

BROWN DWARFS

B. R. OPPENHEIMER, S. R. KULKARNI
Palomar Observatory

and

J. R. STAUFFER
Harvard-Smithsonian Center for Astrophysics

After a discussion of the physical processes in brown dwarfs, we present a complete, precise definition of brown dwarfs and of planets inspired by the internal physics of objects between 0.1 and 0.001 $M_\odot$. We discuss observational techniques for characterizing low-luminosity objects as brown dwarfs, including the use of the lithium test and cooling curves. A brief history of the search for brown dwarfs leads to a detailed review of known isolated brown dwarfs, with emphasis on those in the Pleiades star cluster. We also discuss brown dwarf companions to nearby stars, paying particular attention to Gliese 229B, the only known cool brown dwarf.

I. WHAT IS A BROWN DWARF?

A main-sequence star is to a candle as a brown dwarf is to a hot poker recently removed from the fire. Stars and brown dwarfs, although they form in the same manner, out of the fragmentation and gravitational collapse of an interstellar gas cloud, are fundamentally different because a star has a long-lived internal source of energy: the fusion of hydrogen into helium. Thus, like a candle, a star will burn constantly until its fuel source is exhausted. A brown dwarf's core temperature is insufficient to sustain the fusion reactions common to all main-sequence stars. Thus brown dwarfs cool as they age. Cooling is perhaps the single most important salient feature of brown dwarfs, but an understanding of their definition and their observational properties requires a review of basic stellar physics.

A. Internal Physics

In stellar cores, nuclear fusion acts as a strict thermostat maintaining temperatures very close to the nuclear fusion temperature, $T_{\rm nucl} = 3 \times 10^6$ K, the temperature above which hydrogen fusion becomes possible. In the stellar core, the velocities of the protons obey a Maxwell-Boltzmann distribution. However, the average energy of one of these protons is only $kT_{\rm nucl} = 8.6 \times 10^{-8} T$ keV ~ 0.1 keV, where k is the Boltzmann constant. In contrast, the Coulomb repulsion between these protons is on the order of MeV. Despite this enormous difference in energies, fusion is possible because of quantum mechanical tunneling. The nuclear reaction rate is governed by the proton pair energy E, at the high-energy tail of the Maxwell-Boltzmann distribution, which scales as $\exp(-E/kT)$, and the nuclear cross section due to quantum mechanical tunneling through the Coulomb repulsion, which scales as $\exp(-E^{-1/2})$. The product of these two factors defines a sharp peak, the Gamow peak, at a critical energy $E_{\rm crit}$. $E_{\rm crit}$ is approximately 10 keV for the reactions in the proton-proton (pp) chain, the most basic form of hydrogen fusion. Because $E_{\rm crit} \gg kT$, the nuclear reactions involve the tiny minority of protons in the high-velocity tail of the Maxwell-Boltzmann distribution. In terms of temperature, this reaction rate is proportional to $(T/T_{\rm nucl})^n$, where $n \approx 10$ for temperatures near $T_{\rm nucl}$ and reactions in the pp chain. The large value of n ensures that the core temperature is close to $T_{\rm nucl}$.

For the low-mass main-sequence stars in which the above discussion holds, the mass is roughly proportional to the radius. This can be shown with a simplified argument by appealing to the virial theorem. In equilibrium, the thermal energy and the gravitational potential energy are in balance: $GM^2/R \sim (M/m_{\rm p})kT_{\rm nucl}$, where $m_{\rm p}$ is the mass of the proton, M is the mass of the star, R is its radius, and G is the gravitational constant. Therefore, $R \propto M$.

If radius is proportional to mass, then the density ρ increases with decreasing mass: $\rho \propto MR^{-3} \propto M^{-2}$. At a high enough density a new source of pressure becomes important. Electrons, because they have half-integer spins, must obey the Pauli exclusion principle and are accordingly forbidden from occupying identical quantum energy states. This requires that electrons successively fill up the lowest available energy states. The electrons in the higher energy levels contribute to degeneracy pressure, because they cannot be forced into the filled lower energy states. The degeneracy pressure, which scales as $\rho^{5/3}$, becomes important when it approximately equals the ideal gas pressure: $\rho T \propto \rho^{5/3}$. Explicit calculation of this relation shows that degeneracy pressure dominates when $\rho >$ 200 g cm^{-3} and $T < T_{\rm nucl}$. For the Sun, $\rho \approx 1$ g cm^{-3}. Using the scaling relations above, one finds that degeneracy pressure becomes important for stars with $M < 0.1$ $M_\odot$. Objects supported *primarily* by some sort of degeneracy pressure are called "compact."

An examination of Fig. 1 (Burrows and Liebert 1993) demonstrates the key elements described above. This plot of mass vs. radius shows the main sequence (labeled "M Dwarfs"), where $R \propto M$, and the line that white dwarfs must obey because they are completely supported by electron degeneracy pressure. For such objects the energy density of the degenerate electrons ($\propto \rho^{5/3}$) must match the gravitational potential energy density, $GM^2/R/R^3$. In that case, $R \propto M^{-1/3}$. Note that the white dwarf sequence meets the main sequence at about 0.1 $M_\odot$. At this point the main-sequence curve turns and remains at a roughly constant radius for all the masses

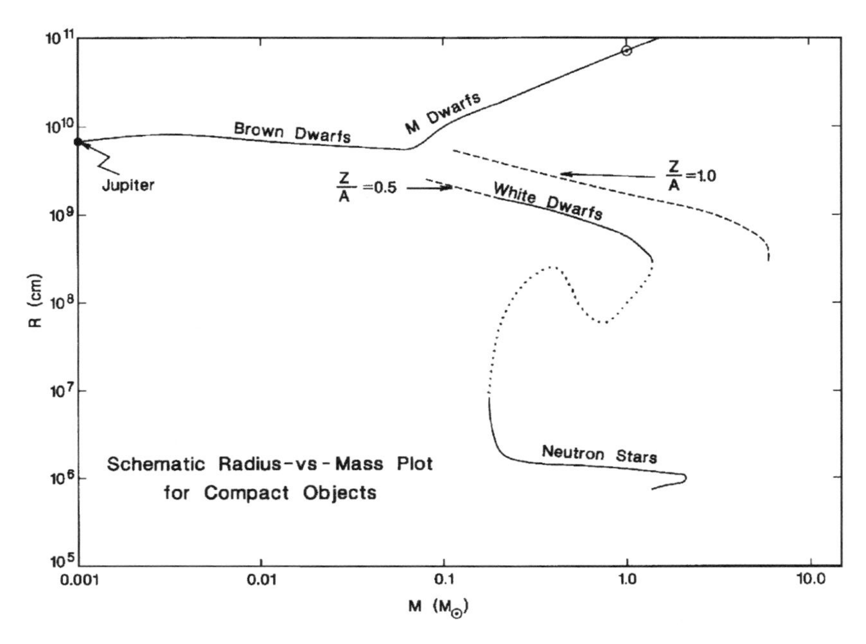

Figure 1. Mass vs. radius. This plot shows that brown dwarfs have a roughly constant radius as a function of mass. This is important because it makes $T_{\rm eff}$ a function only of luminosity. If $T_{\rm eff}$ is observable then an object can be classified as a brown dwarf or star based on its $T_{\rm eff}$ or luminosity. (See Color Plate 22.) (Courtesy A. Burrows. From Burrows and Liebert 1993.)

down to the mass of Jupiter. This mass range, from about 0.1 $M_\odot$ to 0.001 $M_\odot$, has an essentially constant radius, because at the high-mass end the degeneracy pressure leads to the slow function $R \propto M^{-1/3}$, whereas at the low-mass end the Coulomb pressure, which is characterized by constant density ($\rho \propto M/R^3$, which implies $R \propto M^{+1/3}$), begins to dominate over degeneracy; the net result is that approximately $R \propto M^0$ (Burrows and Liebert 1993).

Kumar (1963) calculated the mass at which "starlike" objects could be stable against gravitational collapse through electron degeneracy pressure instead of ideal gas pressure maintained by energy input from fusion reactions. This mass, the lowest mass at which a star can fuse hydrogen, is now called the hydrogen-burning mass limit (HBML). Modern calculations of the HBML place it, for objects of solar metallicity, between 0.080 $M_\odot$ and 0.070 $M_\odot$ (84 to 73 $M_{\rm Jup}$, where $M_{\rm Jup}$ is the mass of Jupiter; Burrows and Liebert 1993; Baraffe et al. 1995).

B. Definition of "Brown Dwarf" and of "Planet"

The canonical definition of a brown dwarf is a compact object that has a core temperature insufficient to support sustained nuclear fusion reactions.

As shown above, this temperature requirement translates directly into a mass requirement: a brown dwarf's mass must be below the HBML.

This definition does not distinguish planets from brown dwarfs, however. Consensus in the literature on this issue suggests that planets and brown dwarfs can be distinguished by their formation processes. Planets form in circumstellar disks, while brown dwarfs form out of interstellar gas cloud collapse. This distinction is problematic because there is no simple observable of the birth process. In the case of brown dwarf companions of stars, one might expect the orbit to be rather eccentric about the central star, whereas the orbit of a planet might be roughly circular (Black 1997). However, as reviewed by Lissauer et al. (this volume), planets formed in circumstellar disks can be in highly eccentric orbits.

We propose a new definition of planets as *objects for which no nuclear fusion of any kind takes place during the entire history of the object.* As far as we know, the only other attempt to define the term "planet" in the literature is that of Basri and Marcy (1997), which suggested a scheme similar to the one presented here. Though Burrows et al. (1997) use this same classification scheme, they present it as a purely semantic definition only, for the purposes of their paper; they do not advocate its replacement of the standard formation-motivated definition. We do.

According to Burrows et al. (1997), objects (at solar metallicity) with masses between 0.08 $M_\odot$ and 0.013 $M_\odot$ fuse deuterium when they are young. Deuterium undergoes fusion reactions at lower temperatures than hydrogen, primarily because the reaction $D(p, \gamma)^3He$ is extremely rapid, being driven by the electromagnetic force. In contrast, the pp chain, driven by the weak nuclear force, is much slower and therefore less efficient, requiring higher temperatures. A plot of the luminosity evolution of objects between 0.2 $M_\odot$ and 0.0003 $M_\odot$ (Color Plate 22) illuminates this issue. The very highest-mass objects, stars, start out bright but eventually reach an equilibrium luminosity at the right side of the plot. The lower curves, for brown dwarfs and planets, continue to drop in luminosity past 10^{10} yr. The first bump in the upper curves of Color Plate 22 indicates the age at which deuterium fusion ends, having completely depleted the deuterium fuel. However, for masses below 0.013 $M_\odot$, the curves are devoid of this bump because the objects are not even capable of fusing deuterium. These objects are planets. With these precise definitions, planets, brown dwarfs, and stars occupy a hierarchy based on their internal physics. Stars fuse hydrogen in equilibrium; brown dwarfs do not fuse hydrogen in equilibrium but do fuse deuterium for some portion of their evolution; and planets never fuse anything. Table I provides a summary of this hierarchy.

Another justification of these definitions comes from the untested theoretical notion that deuterium fusion and convection, which may lead to magnetic fields and thus mass outflows, might halt the accretion process as these low-mass objects form (Shu et al. 1987). This process would then place the lower limit to the mass of an object formed in isolation (not in a circumstellar disk) at 0.013 $M_\odot$, because a lower-mass object could con-

TABLE I
Summary of Definitions of Star, Brown Dwarf, and Planet

Object Type	Mass[a] ($M_\odot$)	H Fusion	D Fusion	Contains Li	Contains D
Star	0.1–0.075	Sustained	Evanescent	No	No
Brown dwarf	0.075–0.065	Evanescent	Evanescent	Yes[b]	No
Brown dwarf	0.065–0.013	Never	Evanescent	Yes	No
Planet	< 0.013	Never	Never	Yes	Yes

[a] Masses given here assume that the objects have solar metallicity.
[b] Brown dwarfs in this mass range have lithium abundances that are age dependent.

tinue to accrete mass until it exceeded this limit, when deuterium burning, and consequently mass outflow, would ensue. Thus, planets could not form the way stars and brown dwarfs do. However, if confirmed, this would be an outcome of the definition, not the overriding principle.

C. Observational Identification of Brown Dwarfs

There are three principal methods for confirming that a candidate brown dwarf is, in fact, substellar.[a]

1. $L \propto T_{\rm eff}^4$. The definitions presented above have direct consequences for the observations of brown dwarfs, stars, and planets. The most obvious distinction between stars and brown dwarfs is illustrated in the cooling curves mentioned above (Color Plate 22). An object of solar metallicity below 10^{-4} $L_\odot$ cannot be a star, regardless of its age. Intrinsic luminosity, L, is an observable only for objects with known distances. A good, but less sensitive and yet more practical, surrogate for luminosity is effective temperature, $T_{\rm eff} = (L/4\pi R^2\sigma)^{1/4}$, where σ is the Stefan-Boltzmann constant, because, as we demonstrated above, R is essentially constant for objects below the HBML (except for very low-mass planets, where only the Coulomb force is important; see Fig. 1). Spectral synthesis models are complete enough at this point that comparison of spectra with the models constrains $T_{\rm eff}$ to better than 10% in most cases. A luminosity of 10^{-4} $L_\odot$ corresponds to $T_{\rm eff} = 1800$ K. The cooling curves show that a 0.013-$M_\odot$ brown dwarf reaches this temperature at an age of approximately 100 Myr. Therefore, this technique for identifying brown dwarfs only works for relatively old and cool objects.

2. Lithium. Distinguishing young, hot brown dwarfs and planets from stars is easiest with the "lithium test."

The lithium test as proposed by Rebolo et al. (1992) relies on the fact that objects without hydrogen fusion retain their initial lithium

[a] In our discussion of the observations of brown dwarfs we discuss only directly detected objects. For this reason we do not include substellar objects discovered in radial velocity studies. (See the chapter by Marcy et al., this volume.)

abundances forever. This is a direct result of one of the nuclear fusion reactions: $Li^7(p,\alpha)He^4$. This reaction effects the complete destruction of lithium in the cores of very low-mass stars in 50 Myr and in brown dwarfs with masses between 0.08 and 0.065 $M_\odot$, which have short-lived hydrogen fusion reactions, in 50 to 250 Myr (D'Antona and Mazzitelli 1994; Bildsten et al. 1997). Below 0.065 $M_\odot$, brown dwarfs retain their initial lithium abundances forever, because they never host any hydrogen fusion reactions.

Theoretical models show that brown dwarfs and very low-mass stars are fully convective. Thus, the elemental abundances in the core, where the putative fusion reactions happen, are reflected on the convective timescale in their observable atmospheres. The convective timescale for a brown dwarf is on the order of decades but scales proportional to $L^{1/3}$. In contrast, the evolutionary timescale is 6 to 8 orders of magnitude larger (Burrows and Liebert 1993; Bildsten et al. 1997), so core abundances can be assumed identical to atmospheric abundances.

Figure 2 (from Rebolo et al. 1996) shows lithium abundance measurements as a function of $T_{\rm eff}$ for objects in the Pleiades. G and K stars have

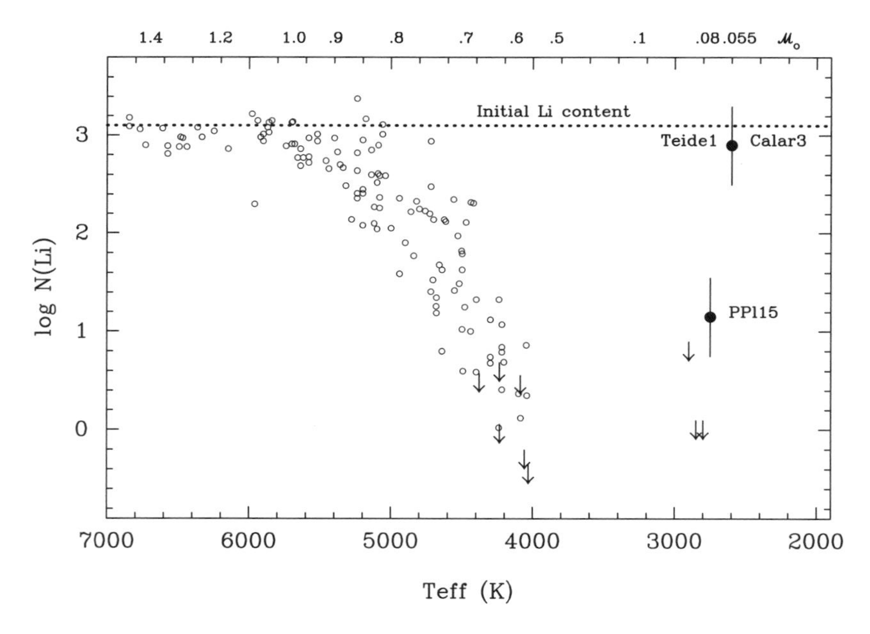

Figure 2. Lithium abundance vs. $T_{\rm eff}$ for stars and brown dwarfs in the Pleiades. This plot shows how the presence of lithium in a low-temperature object can be used to establish its classification as a brown dwarf. The G and K dwarfs have lithium, but M dwarfs do not, because they are fully convective and the hydrogen fusion reactions destroy lithium. Brown dwarfs contain lithium because they do not support fusion reactions. PPL 15, Teide 1, and Calar 3, members of the Pleiades star cluster, are the first brown dwarfs confirmed in this manner. (Courtesy of M. Zapatero-Osorio. From Rebolo et al. 1996.)

cosmic lithium abundances, but once $T_{\rm eff}$ reaches the M dwarf regime, the lithium abundance plummets for the reasons explained above. Below 3000 K young brown dwarfs, such as PPL 15, Teide 1, and Calar 3 (described in section I.D), have measurable lithium abundances.

An interesting outcome of the lithium test is that one can accurately determine the age of an open cluster by finding the "lithium depletion boundary," which is an imaginary line that separates faint objects without lithium from slightly fainter objects with lithium. This is demonstrated in Fig. 3 for the Pleiades (Stauffer et al. 1998). After 250 Myr, this boundary remains indefinitely at 0.065 $M_\odot$. In older clusters, the brightest cool objects with lithium will have a mass of 0.065 $M_\odot$, which can be used with the measured luminosity to place the object in a well-constrained part of Color Plate 22. Objects above the HBML deplete their lithium within 100 Myr, so as long as the cluster being studied is older than 100 Myr, all of the cool objects in the cluster that show lithium absorption must be brown dwarfs.

Figure 3 demonstrates the application of this technique to determine the age of the Pleiades. The lithium depletion boundary is indicated by a line perpendicular to the zero-age main sequence (ZAMS) and is defined

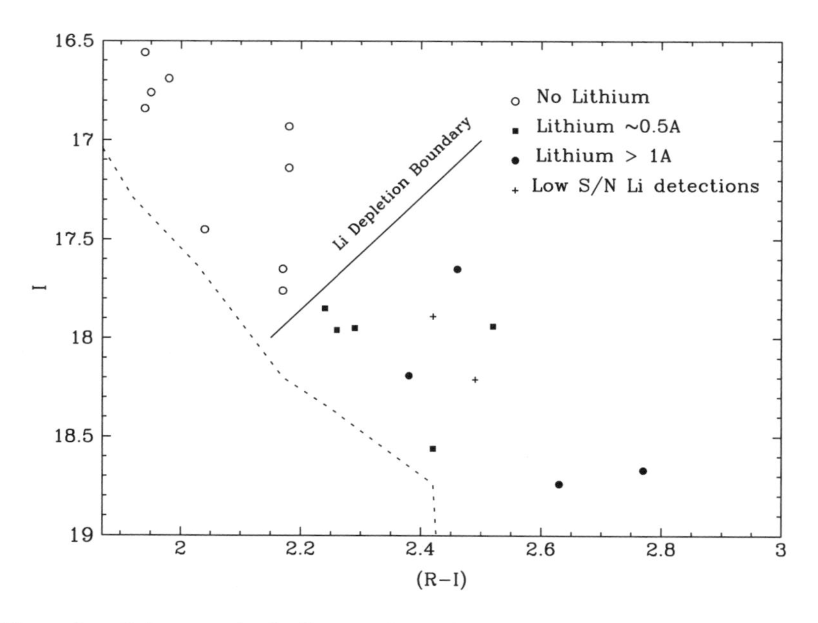

Figure 3. Color-magnitude diagram for Pleiades very low-mass star and brown dwarf members with available spectra capable of detecting the lithium 6708 Å doublet at equivalent widths greater than 0.5 Å. The dotted line is an empirical main sequence at Pleiades distance. The location of the "lithium depletion boundary" is indicated by the solid line and is used to determine a precise age for the cluster of 125 ± 8 Myr (Stauffer et al. 1998).

by extensive spectroscopic observations of all the Pleiads near the line. (See section II.A.1 for more discussion.)

3. Molecules. Once a brown dwarf cools to $T_{\rm eff} = 1500$ K, lithium begins to form molecules, and the Li I spectral signature weakens (Pavlenko 1998; Burrows and Sharp 1998). Fortunately, a new diagnostic becomes available. Below 1500 K, chemical equilibrium between CO and CH_4 strongly favors CH_4 (Tsuji 1964; Fegley and Lodders 1994; Burrows and Sharp 1998). CH_4 has a number of extremely strong absorption features in the range of 1 to 5 μm. As a result, the spectroscopic detection of methane means that the effective temperature of the object must be below 1500 K, requiring that it be less massive than the HBML (section I.C.1). Ammonia forms at temperatures slightly lower than 1000 K, and a host of more exotic species appear at even cooler temperatures. See the chapter by Burrows et al., this volume, for a more complete discussion of this progression. These spectral signatures allow observers to classify brown dwarfs by $T_{\rm eff}$. In sections II and III we deal with this subject in greater depth.

4. Deuterium. Distinguishing brown dwarfs from planets, as defined here, involves a search for deuterium. Several of the planets in the solar system have measured deuterium abundances. (See, for example, Krasnopolsky et al. 1997.) In brown dwarfs, deuterium should be depleted to unmeasurable quantities, even in their atmospheres, because convection causes a complete reflection of the core abundances in the atmosphere, as we reasoned in section I.C.2. Spectroscopic signatures of deuterium in the 1–8-μm region include absorption lines of HDO with numerous features between 1.2 and 2.1 μm (Toth 1997) and possibly CH_3D with strong features at 3.7 and 4.4 μm (Noll 1993; Krasnopolsky et al. 1997). This proposed classification scheme is hampered only by the current state of technology, in that spectra of the known extrasolar "planets" indirectly detected through radial velocity studies (Marcy et al., this volume) cannot be obtained yet.

5. Caveat: Dust. As brown dwarfs cool, theory predicts that dust will form in the atmosphere. Even some cool main-sequence stars seem to contain dust (Jones and Tsuji 1997). Dust formation occurs at the ridge in Color Plate 22 at approximately 10^{-4} $L_\odot$ (1800 K). As the dust forms, the luminosity drops more precipitously. Below this ridge, a progression of species with important features in the near-IR appears. These species will have two effects: (1) weakening of the molecular absorption features and (2) reorganization of the broad-band spectral energy distribution toward a blackbody spectrum.

D. Observational History

Ever since Kumar's pioneering work, astronomers have searched for brown dwarfs primarily because they were regarded as "terra incognita."

Some of the initial discussions of brown dwarfs suggested that they could be the "missing" matter implied by the dynamics of the galaxy. For example, simple extension of the Salpeter initial mass function (IMF), in which $dN/dM \propto M^{-2.35}$, to brown dwarf masses suggests that brown dwarfs ought to outnumber stars by two or three orders of magnitude. Whether an appreciable percentage of the dark matter is brown dwarfs or planets is still the subject of some debate. Most researchers agree that based on the microlensing experiments of the MACHO (Massive Compact Halo Object) Collaboration (Alcock et al. 1998), most of the dark matter is not made of brown dwarfs. However, by adopting unusual parameters, it is still possible to construct galactic models consistent with the MACHO results and with more than 50% of the dynamical mass in brown dwarfs (Kerins and Wyn Evans 1998).

From 1984 through 1994, approximately 170 refereed papers on brown dwarfs were written. By the end of 1996 that number doubled, and based on current publication rates, 1998 alone will see approximately 170 more papers submitted. This sudden explosion in observational and theoretical results was due to the discovery in late 1995 of Gliese 229B, the first cool brown dwarf detected (described in section III) and the confirmation of lithium in the brown dwarf candidate PPL 15 in early 1996 (and in Teide 1 and Calar 3 slightly later). The sustained publication rate is largely due to new large-scale surveys, which are now turning up brown dwarfs by the dozen. These dramatic successes, however, were preceded by several decades of unsuccessful searches and two conferences whose proceedings are punctuated with the wreckage of disproven brown dwarf candidates and steady improvements in theoretical work.

In 1985, when the first conference solely devoted to brown dwarfs was held at George Mason University (Kafatos et al. 1986), a new breed of infrared and optical detectors had enabled the first searches designed to detect brown dwarfs directly. No brown dwarfs were found, however, and in retrospect this was because of a lack in sensitivity. The basic strategies behind these early searches are imitated to this day. One can look for brown dwarfs in isolation or as companions of nearby stars. Isolated brown dwarfs can be found in all-sky surveys or smaller surveys of star clusters. Companion searches employ techniques to prevent the bright nearby star from washing out the faint companion.

In 1987 the results of the *Infrared Astronomical Satellite* (IRAS), launched in 1984 to survey the entire sky at far-infrared wavelengths, were presented. One of the principal goals of this satellite mission was to find brown dwarfs. None were detected (Beichman 1987).

The first direct searches for brown dwarf companions of nearby stars were coincident with the development of sensitive electronic infrared photodetectors. Probst (1983) used a single-pixel device on NASA's Infrared Telescope Facility (IRTF) to search for infrared excess around nearby white dwarf stars. The search was sensitive to companions within 15 arcsec of the stars. Probst targeted white dwarf stars because they are intrinsically fainter than main-sequence stars, so that any infrared excess would be easier to detect. No brown dwarfs were detected.

Using the same single-pixel detector, McCarthy et al. (1985) applied the technique of speckle interferometry. This technique uses rapid exposures to compensate for the blurring effects that the turbulent atmosphere has on astronomical images. In principle this permits the detection of a companion fainter and closer to the primary star than is possible in a standard direct image. McCarthy et al. (1985) reported a faint companion of the red dwarf vB 8. Their results suggested that this object had a luminosity of about 10^{-5} $L_{\odot}$. However, the putative companion was never detected again and remains an irreproducible result.

Forrest et al. (1988), using a 32×32 pixel InSb array on the IRTF, found several stellar companions of red dwarfs in the solar neighborhood but still uncovered nothing faint enough to be considered a brown dwarf.

In 1988, Becklin and Zuckerman, who extended the work of Probst (1983) to survey nearby white dwarfs for faint companions, found the object known as GD 165B. GD 165B has a temperature of 1800 K (Jones et al. 1994) and until 1995 was the best candidate brown dwarf known. Subsequent spectroscopy by Kirkpatrick et al. (1998) has demonstrated that GD 165B has no lithium and is therefore either a star right at the HBML or a high-mass brown dwarf a few Gyr old.

Henry and McCarthy (1990) conducted a search for infrared companions of the stars within 8 pc of the Sun using the speckle interferometry technique and an array of pixels. Although they found no brown dwarfs, their sensitivity was somewhat limited, reaching a maximum of 7.5 magnitudes of difference between the central star and the faintest object detectable (Henry et al. 1992). For reference, Gliese 229AB (a red dwarf/cool brown dwarf system) has a contrast of over 10 magnitudes in the near-infrared (Matthews et al. 1996; see section III).

Substantial gains in detecting stars of lower and lower mass had been made by the next conference on brown dwarfs, held in Garching, Germany (see Tinney 1995 for proceedings). However, still no definitive brown dwarfs were presented. Determinations of the mass function presented at this meeting suggested that brown dwarfs are extremely rare, but, in retrospect, the surveys used were still not sensitive enough (D'Antona 1995).

During the course of a coronagraphic survey of nearby stars, Nakajima et al. (1995) reported the discovery of an object with the same proper motion as Gliese 229. At a distance of 5.7 pc from the Sun, the inferred intrinsic luminosity of the companion, Gliese 229B, is 6.4×10^{-6} $L_{\odot}$ (Matthews et al. 1996). Spectroscopy of this object revealed deep methane features and implied a temperature below 1200 K (Oppenheimer et al. 1995, 1998). Although many astronomers had become inured to brown dwarf announcements that were retracted months later, the spectrum of Gliese 229B is sufficiently distinctive that, when shown briefly at the 9th "Cool Stars, Stellar Systems and the Sun" conference in Florence, Italy (1995), it was unanimously taken as proof that the object was indeed a brown dwarf.

Parallel to the searches for companion brown dwarfs, several searches for isolated brown dwarfs were conducted between 1989 and 1995. The first deep imaging charge-coupled device (CCD) surveys of the Pleiades believed to have reached below the HBML were those of Jameson and Skillen (1989, hereafter JS89) and Stauffer et al. (1989, hereafter S89). These initial surveys covered only very small portions of the cluster. JS89 imaged just 225 square arcmin and identified seven objects as likely Pleiades brown dwarfs; S89 surveyed 1000 square arcmin and identified just four brown dwarf candidates. Subsequent analysis indicated that due to an error in the photometric calibration, only one of the JS89 objects was faint enough to be a possible Pleiades brown dwarf (Stringfellow 1991; Stauffer et al. 1994).

After the lithium test was proposed in 1992, several attempts to apply it to brown dwarf candidates in the Pleiades revealed no lithium (Martín et al. 1994; Marcy et al. 1994). This was largely because the Pleiades is older than these studies presumed, so they selected candidates that were not faint enough. By obtaining a spectrum of a fainter object, PPL 15 (a Pleiades brown dwarf candidate identified by Stauffer et al. 1994), Basri et al. (1996) made the first detection of lithium in a brown dwarf candidate. Rebolo et al. (1996) soon after detected lithium in two other Pleiades brown dwarf candidates, Teide 1 and Calar 3.

By the beginning of 1996 the first brown dwarfs had been found, and in March 1997 a conference was held in Tenerife, Spain (Rebolo et al. 1998), where the wealth of positive observational results was a direct testament to the sudden change in the field.

We separate our discussion of successful observations of brown dwarfs into two sections: one on isolated brown dwarfs (section II) and the other on brown dwarf companions of nearby stars (section III). We also recommend the useful reviews by Kulkarni (1997), Allard et al. (1997), and Hodgkin and Jameson (1997).

II. ISOLATED BROWN DWARFS

The principal reason for studying isolated brown dwarfs is to acquire a complete census of objects with masses below the HBML (i.e., to measure the mass function). The relative number of brown dwarfs of a given mass compared to the number of higher-mass objects has important implications for star formation theories. Indeed, the eventual mass of an object formed out of an interstellar cloud fragment would seem to be entirely independent of the HBML, so that objects with masses well below the HBML ought to form out of interstellar cloud fragmentation (Burkert and Bodenheimer 1993; Shu et al. 1987). Observations of star formation regions in which very low-mass clumps of gas exist certainly suggest that brown dwarfs can form out of this process (Pound and Blitz 1995). Measuring the mass function would determine whether there is a lower limit to the mass of an object formed like a star and not in a circumstellar disk.

Because brown dwarfs cool, and a given brown dwarf can have a huge range of luminosities over its lifetime, making a complete census of them is greatly simplified by examining a sample of the same age. In such a sample, mass will be solely a function of luminosity, which is directly observable, as discussed in section I.C.1. By far the best means to find a population of brown dwarfs with the same age is to identify low-luminosity members of well-studied open clusters, where, in principle, the age, distance, and metallicity should be known accurately.

Surveys for field brown dwarfs, outside stellar associations, require immense sky coverage because of the intrinsic faintness of the brown dwarfs, which effectively limits the volume of space that the surveys probe. For example, a 1000-K brown dwarf is detectable by the new near-infrared all-sky surveys (2MASS and DENIS; see section II.B) out to a distance of only about 6 pc.

Critical to both types of surveys is the certification that a given object is a brown dwarf. The principal method for this is the lithium test. However, as the surveys probe fainter and fainter limits, certification through the molecular features described in section I.C.3 will become equally important.

A. Brown Dwarfs in Open Clusters and Star-forming Regions

Brown dwarf candidates are identified in photometric surveys based on their lying above the ZAMS (Fig. 3). To confirm that they are in fact cluster members (and not reddened background stars) requires accurate proper motion measurements or accurate radial velocity measurements. However, to date at most one of the likely cluster brown dwarfs has a sufficiently accurate proper motion (Rebolo et al. 1995). The lithium test has therefore provided the primary means to confirm the substellar nature of the young brown dwarf candidates.

1. The Substellar Mass Population of the Pleiades. The Pleiades is the richest nearby open cluster (see Table II). For a nominal age of

TABLE II
Nearby Clusters ($d < 200$ pc)

Cluster Name	Distance (pc)	Age (Myr)	No. of Known Members	Area on Sky Sq. Deg.
Ursa Major	25	300	25	20
Hyades	46	600	550	100
Coma	80	500	50	20
Pleiades	130	125[a]	800	25
IC 2602	155	30	125	10
IC 2391	160	30	100	8
Praesepe	170	600	800	25
Alpha Persei	175	75	350	30

[a] The Pleiades age is based on the lithium depletion boundary, which is 25 to 60% higher than the age determined from the upper main-sequence turnoff but is probably more accurate. The other clusters' ages are based on the upper main sequence turnoff. However, Alpha Persei's age agrees with its lithium depletion age.

100 Myr (Meynet et al. 1993), objects at the HBML are predicted to have effective temperatures of about 2500 K, corresponding to spectral class M6 V on the main sequence. Because of the cluster's proximity to the Sun (Table II), these brown dwarfs should be detectable with modern optical CCDs. The cluster half-mass radius is ~2 pc, and the tidal radius is about 16 pc (Raboud and Mermilliod 1998; Pinsonneault et al. 1998). The areas on the sky corresponding to circles with these radii are 2.5 and 150 square degrees, respectively. This is important because it indicates that it is necessary to search a large area to sample a significant portion of the cluster. Because the Hyades is three times closer, it is spread over a much larger area on the sky than the Pleiades. For these reasons, the Pleiades has been the principal hunting ground for isolated brown dwarfs.

Since 1989, at least 10 deep imaging surveys of the Pleiades other than those described in section I.D have been conducted; a summary is provided in Bouvier et al. (1998). By using redder filters and more sensitive, larger-format CCDs, these surveys have been able to reach lower inferred mass limits and cover larger portions of the cluster. A conservative assessment of the current surveys suggests that at least 40 substellar members of the Pleiades have now been identified (Fig. 3).

Using the "lithium depletion boundary" described above, Basri et al. (1996) and Rebolo et al. (1996) estimated the age of the Pleiades at about 120 Myr, with PPL 15 and HHJ 3 (Hambly et al. 1993), the faintest Pleiad without lithium, defining the boundary. However, Basri and Martín (1998) subsequently discovered that PPL 15's luminosity was overestimated because it is an approximately equal-mass, short-period binary, with each component being about 0.7 mag fainter than the composite and having a mass of approximately 0.06 $M_\odot$. PPL 15 is the first brown dwarf binary system found.

Spectra of 10 additional Pleiades brown dwarf candidates have recently allowed Stauffer et al. (1998) to define the lithium depletion boundary in the Pleiades to ± 0.1 mag and thus to derive an age for the cluster of $\tau \sim 125 \pm 8$ Myr (Fig. 3). By coincidence, at this age the lithium depletion boundary corresponds to 0.075 ± 0.005 $M_\odot$, and therefore all Pleiades members fainter than the lithium depletion boundary are brown dwarfs (Ventura et al. 1998; Chabrier and Baraffe 1997). The faintest Pleiades candidates identified to date have masses on the order of 0.035 $M_\odot$ (Martín et al. 1998).

Zapatero-Osorio et al. (1997) and Bouvier et al. (1998) have used their surveys to estimate the Pleiades mass function in the substellar regime. Both groups obtain slightly rising mass functions for the range $0.045 \leq M \leq 0.2$ $M_\odot$, with $dN/dM \propto M^{-1.0}$ and $M^{-0.7}$, respectively. However, the relatively small fraction of the cluster that has been surveyed to date make these estimates fairly uncertain.

2. Brown Dwarfs in Other Open Clusters. Basri and Martín (1999) have reported a lithium detection for the faintest known member of the α Persei open cluster (and nondetection of lithium in one or two brighter, probable members), thus allowing them to place the age of the cluster between 60 and 85 Myr (Table II).

Two deep imaging surveys of Praesepe (see Table II) have been conducted (Pinfield et al. 1997; Magazzù et al. 1998). Magazzù et al. report one object in their survey with $I \sim 21$ and a spectral type of about M8.5, which would indicate a mass near the substellar limit if the object is indeed a Praesepe member and if the cluster age is as expected ($\sim$600 Myr).

The deepest survey of the Hyades to date is that provided by Leggett and Hawkins (1988) and Leggett et al. (1994). The faintest objects in this survey may also be approximately at the substellar mass boundary; however, no spectra for the faintest candidates have yet been reported.

3. Brown Dwarfs in Star-Forming Regions. Brown dwarfs in star-forming regions (age $<$ 1 Myr) will be much more intrinsically luminous than those in the open clusters discussed above and should be easier to discover. However, it is in fact more difficult to "certify" that any given object is substellar in a star-forming region than in an open cluster. First, the "lithium test" is of limited value at this age, because all low-mass stars should still have their original lithium abundances. However, as Basri (1998) points out, if a candidate object lacks lithium, it can be discarded as a member of the star-forming region. Second, the theoretical isochrones for young, low-mass objects are quite uncertain. Thus, determining the ages of these objects is difficult. Even if the age can be determined, the intrinsic luminosity of a candidate object is difficult to measure because of uncertainties in extinction parameters for these star-forming regions.

However, based on the existing models and using the Pleiades as a reference, Basri (1998) has argued that any object with spectral type M7 or later must be a brown dwarf if it contains lithium. This is because stars more massive than the HBML deplete lithium before they can cool to the M7 effective temperature (i.e., before they reach the ZAMS). Brown dwarfs, on the other hand, can cool to the M7 spectral type when they are much younger than the stars and still retain their lithium.

Luhman et al. (1997) have identified an apparent brown dwarf member of the ρ Ophiuchi star-forming region based on its spectral type of M8.5. This object was originally thought to be a foreground star (Rieke and Rieke 1990); however, the new data indicate that it is much more likely to be a member of the cluster (in particular, it has very strong Hα emission and relatively low surface gravity). Comeron et al. (1998) have identified three other members of this region with spectral classes later than M7, based on new data with the ISO satellite and spectroscopy by Wilking et al. (1998).

Luhman et al. (1998) have also obtained spectra for a large number of faint candidate members of the star-forming region IC 348 and have identified three good brown dwarf candidates: two with spectral type M7.5 and one with spectral type M8.

B. Brown Dwarf Members of the Field Population

The search for isolated brown dwarfs in the field has also seen dramatic progress in the past two years. The primary sources of the newly discovered field brown dwarfs are the wide-field near-IR imaging surveys DENIS and 2MASS; however, several objects have also been identified using other techniques.

1. Brown Dwarfs from DENIS and 2MASS. DENIS (Deep Near Infrared Survey) obtains simultaneous images at *I, J,* and *K* of the southern sky to limiting magnitudes of 18.5, 16, and 14.0 (3σ), respectively. Based on the photometry of previously identified very low-mass stars, the DENIS project uses a color criterion of $I - J > 2.5$ to select "interesting" objects, with the most interesting objects being those with colors like that of GD165B. The DENIS search contains no color criteria to distinguish analogs of Gliese 229B.

The DENIS team has reported about five objects with GD165B colors after analyzing only 500 square degrees. Optical spectra have been obtained for three of those objects, with one of them showing a strong lithium absorption feature (Delfosse et al. 1997; Tinney et al. 1997; Martín et al. 1997). All of these objects have spectra in the 0.8-μm region similar to that of GD165B. The lithium feature in combination with the very late spectral type for DENIS-P J1228.2−1547 indicate that this object is undoubtedly a brown dwarf. No lithium has been detected in two of the objects with very late spectral types. They could be old substellar objects in the mass range 0.065 to 0.075 $M_{\odot}$.

2MASS (Two-Micron All Sky Survey) obtains simultaneous images at *J, H,* and *K* of the entire sky to limiting magnitudes of 17, 16.5, and 15.5 (3σ), respectively. Digitized scans of the E or N plates from the Palomar Sky Survey are used to derive *R* or *I* magnitudes for objects detected in the infrared. 2MASS uses color criteria of $J - K > 1.3$ and $R - K > 6$, or $J - K < 0.4$, to select brown dwarf candidates; the latter criterion is designed to find analogs of Gliese 229B (see section III).

More than a dozen candidates from 420 square degrees have had spectroscopic follow-up observations (Kirkpatrick et al. 1998). Six of these objects have extremely late spectral type and show lithium absorption and thus are substellar; an approximately equal number of objects are similarly late but do not show lithium. None of the objects observed spectroscopically show methane in their spectra, and none have been found with colors similar to those of Gliese 229B.

Considering the small fraction of the sky analyzed so far, it appears likely that 2MASS and DENIS will eventually provide a list of hundreds of field brown dwarfs. Due to the correlation of age, mass, effective temperature, and luminosity, it is inevitable that this sample of brown dwarfs will favor relatively young objects with masses not far below the HBML.

The spectra of the coolest DENIS and 2MASS objects are sufficiently different from previously known objects that a new spectral class must be defined. Kirkpatrick (1997) and Martín et al. (1997) have suggested use of the letter L for this class. Kirkpatrick et al. (1998) have begun to define this class through the weakening of TiO bands to later types; the presence of resonance lines of alkali metals (in particular, potassium, rubidium, and cesium, with these lines becoming extremely strong at later types); the presence of other molecular species such as CrH, FeH, and VO; and the absence of methane. One brown dwarf, Gliese 229B (section III) does not fit in the L class because it is several hundred degrees cooler than the coolest L dwarf. Another spectral class must be created once Gliese 229B analogs are discovered.

2. Other Field Brown Dwarfs. Two field brown dwarfs have been identified from proper motion surveys. The first of these, Kelu-1, was identified as part of a survey of 400 square degrees of the southern sky using deep Schmidt plates (Ruiz et al. 1997). Kelu-1 has a spectrum similar to GD165B's and has lithium in absorption and Hα in emission. At $K = 11.8$, Kelu-1 is comparatively very bright (and hence, presumably quite nearby) and so is a good target for detailed study.

The other field brown dwarf identified via proper motion is LP 944−20, originally cataloged as a high-proper-motion object by Luyten and Kowal (1975). It was rediscovered in a search for very red objects by Irwin et al. (1991) and identified as a very late-type M dwarf by Kirkpatrick et al. (1997). Tinney (1998) subsequently showed that lithium was present with an equivalent width of about 0.5 Å. LP 944−20 has a measured parallax from which the intrinsic luminosity can be derived. Tinney (1998) used the inferred luminosity of 1.4×10^{-4} $L_{\odot}$ combined with an estimate of the lithium abundance to derive a mass estimate of 0.06 ± 0.01 $M_{\odot}$ and an age of about 500 Myr.

One other possible field brown dwarf has been identified from a deep photographic *RI* imaging survey by Thackrah et al. (1997). The object, 296A, was selected because it was quite bright ($I \sim 14.5$) and reasonably red ($R - I \sim 2.5$). Spectroscopy revealed a spectral type of M6 and a lithium absorption equivalent width of about 0.5 Å. These features suggest that it could be a Pleiades-age star at the HBML.

III. COMPANION BROWN DWARFS

Searching for brown dwarf companions of nearby stars is attractive mainly because it is the most effective way to identify the coolest (lowest-luminosity; see section I.C.1) objects. The principal difficulty in this is the scattered light of the primary star. There are several techniques to circumvent this difficulty. First, one can search at the longer wavelengths, where the contrast between the brown dwarf and the star is lowest. Second, one can search for companions of white dwarf stars, where the contrast is very small because of the intrinsic faintness of the star. Third, the use of a coronagraph artificially suppresses the starlight with a series of optical stops (Nakajima et al. 1994).

A. Gliese 229B

A large survey of nearby stars using a coronagraph (with a tip-tilt image motion compensator) was carried out at the Palomar 60-inch telescope. This proved successful when Nakajima et al. (1995) showed that the star Gliese 229 has a companion with a luminosity of less than 10^{-5} $L_\odot$. The discovery image of Gliese 229B is shown in Fig. 4 with a subsequent image taken by the *Hubble Space Telescope* (HST) (Golimowski et al. 1998). The impact of this discovery was far reaching. Not only did it validate the immense effort of the astronomers who persisted in working on brown dwarfs despite all the nondetections, but it also excited considerable interest among planetary scientists, partly because Gliese 229B is orbiting a nearby star but also because its spectrum (Oppenheimer et al. 1995, 1998) looks remarkably like Jupiter's, with major features due to water and methane. Methane dissociates at temperatures between 1200 and 1500 K, a fact that further implies the extremely low luminosity of Gliese 229B.

Matthews et al. (1996) made photometric measurements of Gliese 229B from r band at 0.7 μm through N band at 12 μm. These data account for between 75 and 80% of the bolometric luminosity of the brown dwarf (depending on the model used for the unmeasured flux). The observed luminosity is $(4.9 \pm 0.6) \times 10^{-6}$ $L_\odot$, implying a bolometric luminosity of 6.4×10^{-6} $L_\odot$ and an effective temperature of 900 K, assuming the brown dwarf radius is 0.1 $R_\odot$, as was argued in section I.A.

Allard et al. (1996) and Marley et al. (1996), using the photometry of Matthews et al. (1996) and the spectrum from Oppenheimer et al. (1995),

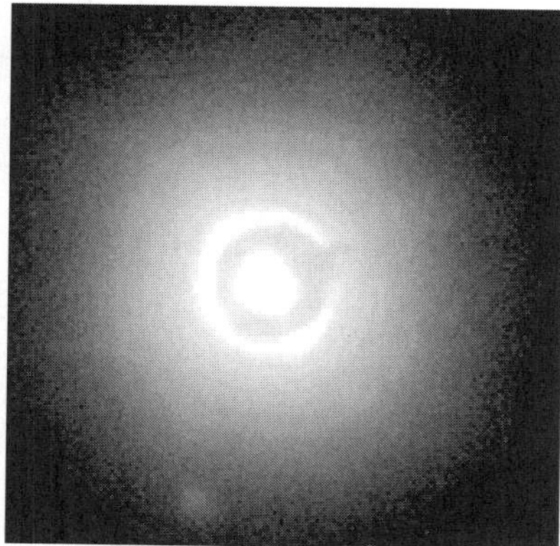

Figure 4. Two images of the Gliese 229 system. The left panel shows a direct image from the *Hubble Space Telescope* (HST)'s Wide-Field Planetary Camera (WFPC2). The brown dwarf is at the bottom left of the image. The right panel shows the discovery image from the Palomar 60-inch telescope fitted with a coronagraph. The coronagraphic stop is visible, obscuring most of the light from the primary star. The stop is 4 arcsec in diameter and is semitransparent. The brown dwarf is visible in the bottom left of the image. Both images are oriented with N up and E to the left and are approximately 17 arcsec on a side.

constrained the mass of Gliese 229B between 0.02 and 0.055 $M_\odot$ by fitting nongray atmosphere models. The mass is so uncertain because (i) the age of Gliese 229B is unknown and could be from 0.5 to 5 Gyr based on the spectrum of Gliese 229A and (ii) the gravity has not been accurately measured.

Geballe et al. (1996) obtained a high-resolution spectrum of Gliese 229B in the 1–2.5-μm region and showed that there are hundreds of very fine spectral features due to water molecules. These may be the most gravity-sensitive features in the spectrum (Burrows et al. 1997), and with higher-resolution spectra the gravity of Gliese 229B may be constrained to within 10%.

Another important conclusion of the Matthews et al. (1996) paper is that the photometry indicates a complete lack of silicate dust in the atmosphere of the brown dwarf. Figure 5 (from Matthews et al. 1996) is a plot of the photometric measurements along with a model spectrum from Tsuji et al. (1996) and three blackbody curves, assuming the brown dwarf has a radius of 0.1 $R_\odot$ and is at 5.7 pc. The model spectrum has no dust included in the calculation. By adding even a minute quantity of silicate dust, Tsuji et al. (1996) find that the spectrum no longer fits the photometric data. In

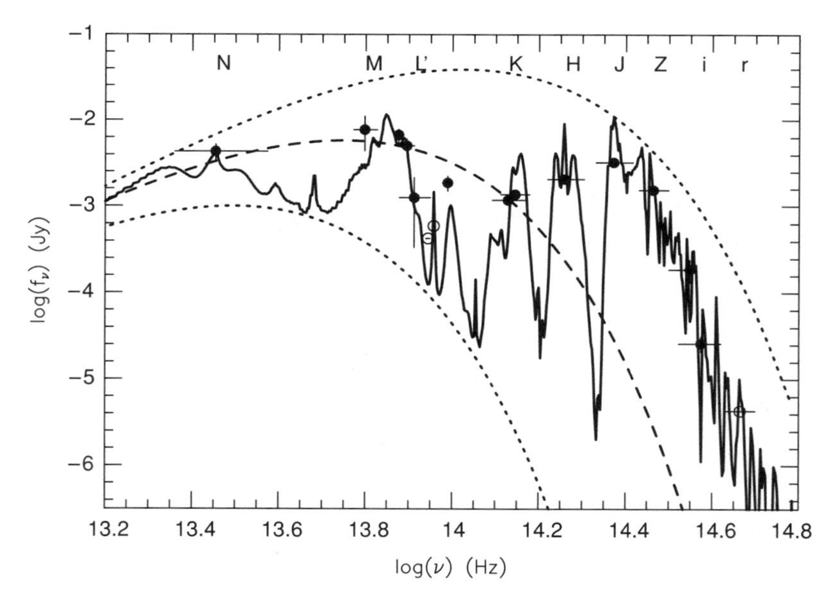

Figure 5. Photometric measurements of Gliese 229B (from Matthews et al. 1996). The photometry shown here represents 75 to 80% of the bolometric luminosity (6.4×10^{-6} $L_\odot$) of the brown dwarf and that the atmosphere is devoid of dust grains that affect the infrared spectrum. The solid line is a dust-free model spectrum from Tsuji et al. (1996). The long-dashed line indicates a blackbody spectrum for $T_{\rm eff}$ = 900 K, the estimated value for Gliese 229B. The dotted curves are blackbody spectra for $T_{\rm eff}$ = 500 K (bottom) and 1700 K (top).

contrast, the L dwarfs of Kirkpatrick et al. (1998) and the spectrum of GD 165B (Jones et al. 1994; Jones and Tsuji 1997) appear to be considerably affected by the presence of dust.

Oppenheimer et al. (1998) confirmed these conclusions with a high-signal-to-noise spectrum of Gliese 229B (shown in Fig. 6) in the near infrared but also reported a smooth, almost featureless spectrum in the optical (0.85–1.0 μm) region that could not be fitted by any of the models. An optical spectrum was also obtained from the HST by Schultz et al. (1998). There is excellent overall agreement between the Keck and the HST spectra. However, the HST spectrum lacks the resolution and sensitivity to reveal the fine features seen in the Keck spectrum (Fig. 6) and discussed here.

The fact that the optical spectrum is smooth (Fig. 6) and that the water band at 0.92 μm is shallower than expected (Allard et al. 1997) indicates that there may be an additional source of continuum opacity not previously discussed by theorists. Indeed, Griffith et al. (1998) show that a haze of photochemical aerosols might be responsible for the relative smoothness of this part of the spectrum. These aerosols may be activated by ultraviolet

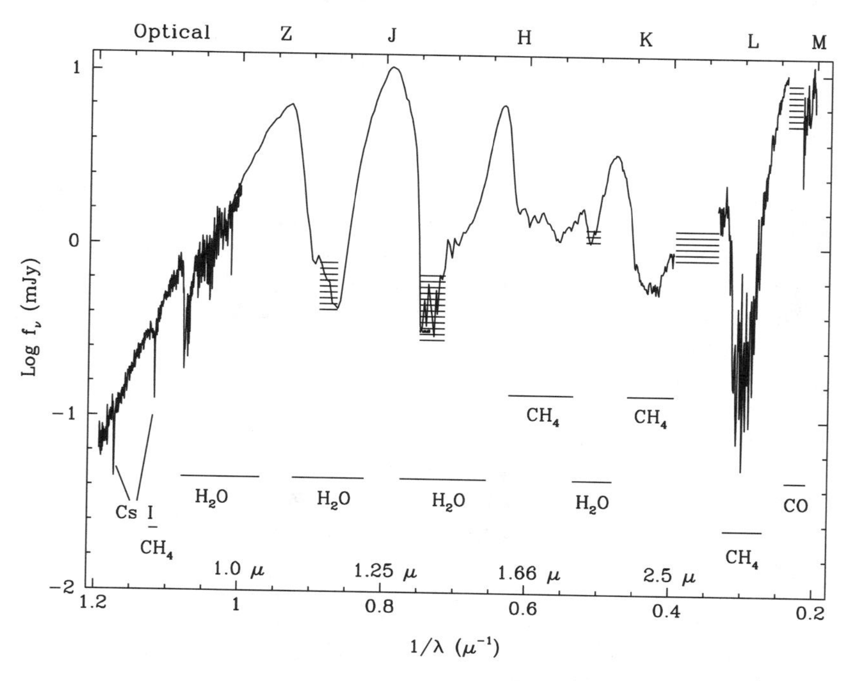

Figure 6. The spectrum of Gliese 229B from 0.8 μm to 5.0 μm. The principal absorption features are, as indicated, from water and methane, which is not present in stellar atmospheres. Additional features due to cesium and carbon monoxide are also marked. The relative smoothness of the 0.8–1.0-μm region indicates the presence of an unpredicted continuum opacity source. (From Oppenheimer et al. 1998.)

radiation from Gliese 229A (known to flare from time to time). A possible test of this conjecture is that field brown dwarfs with effective temperatures near 900 K should not have these aerosols in their atmosphere because they have no nearby source of incident ultraviolet radiation. (See Griffith et al. 1998.)

This problem of whether silicate dusts, sulfides, hazes, or even polyacetylenes appear in the atmospheres of objects below the HBML remains a subject of debate. It is, however, extremely important, because dust can have a substantial effect on the emergent spectrum. The issue of whether a parent star can cause photochemical reactions that greatly affect the spectra of its companions has implications for planet and brown dwarf searches. Indeed, the design of new searches and instruments will need to take heed of this work. In the case of Gliese 229B, a spectrum in the 5–12-μm region might yield some answers, because incident ultraviolet radiation can also produce certain organic molecules, such as C_4H_2, with spectral features in the mid-infrared, as have been observed in spectra of Titan (Griffith et al. 1998; Khlifi et al. 1997; Raulin and Bruston 1996).

Noll et al. (1997) reported the detection of a feature due to carbon monoxide in the 4–5-μm spectrum of Gliese 229B (Fig. 6). Confirmed by Oppenheimer et al. (1998), this feature shows that carbon monoxide exists in nonequilibrium abundances. At 900 K, the balance between CO and CH_4 strongly favors CH_4. Convection must, therefore, dredge up appreciable quantities of CO from deeper, hotter parts of the atmosphere.

However, Oppenheimer et al. (1998) suggested that the convection cannot be efficient even deeper, where species such as VO and TiO ought to be present, because there are no spectral signatures of these constituents. VO and TiO are observed in all of the late M and L-type dwarfs. This may indicate that cool, old brown dwarfs are not fully convective but rather have an inner radiative zone below the outer convective layer, as the models of Burrows et al. (1997) show.

In Fig. 6 several absorption features due to neutral cesium are indicated. Oppenheimer et al. (1998) argued that this should be the last of the neutral atomic species present in atmospheres as one proceeds toward lower and lower temperature. Of all the known elements, cesium has the lowest ionization potential, so at temperatures well above Gliese 229B's it is ionized and its signature is hidden in the extremely faint ultraviolet regions of the spectrum. In addition, it, along with the other alkali metals, is less refractory than the more familiar stellar atomic species (Al, Mg, Fe), so it survives in atomic form to lower temperatures. Indeed, its presence also in some of the L dwarf spectra (Kirkpatrick et al. 1997) shows that it exists for effective temperatures from 1800 K to below 900 K. Oppenheimer et al. (1998) suggest that neutral cesium, along with the other alkali metals, be used as an indicator of effective temperature for brown dwarfs.

Other than Gliese 229B, no other brown dwarfs have been confirmed as companions of stars. Oppenheimer et al. (2000) have surveyed all of the northern stars within 8 pc and found only one substellar companion. That survey was sensitive to objects up to 4 magnitudes fainter than Gliese 229B, with separations from the star between 3 and 30 arcsec. The survey implies a star/brown dwarf binary frequency of less than 1%, although more than one specimen is needed to make this statement significantly meaningful. Several other searches are under way, however, including that of Krist et al. (1998) with the HST. It seems clear from the lack of numerous brown dwarf companions that hundreds or thousands of stars must be surveyed before a substantial population of these elusive objects can be studied in detail.

Acknowledgments S. R. K. would like to particularly thank T. Nakajima for getting him interested in brown dwarfs during his postdoctoral stay at Caltech. We would also like to thank M. Zapatero-Osorio, E. Martín, and R. Rebolo for being so helpful with figures; A. Burrows for ever useful and engaging discussions and the use of several of his figures; and G. Basri, E. Martín, B. Brandl, G. Vasisht, and K. Adelberger for thorough comments on the draft. We also thank the NSF and NASA for support of our brown dwarf research.

REFERENCES

Alcock, C., Allsman, R. A., Alves, D., Ansari, R., Aubourg, E., Axelrod, T. S., Bareyre, P., Beaulieu, J.-Ph., Becker, A. C., Bennett, D. P., Brehin, S., Cavalier, F., Char, S., Cook, K. H., Ferlet, R., Fernandez, J., Freeman, K. C., Griest, K., Grison, Ph., Gros, M., Gry, C., Guibert, J., Lachieze-Rey, M., Laurent, B., Lehner, M. J., Lesquoy, E., Magneville, C., Marshall, S. L., Maurice, E., Milsztajn, A., Minniti, D., Moniez, M., Moreau, O., Moscoso, L., Palanque-Delabrouille, N., Peterson, B. A., Pratt, M. R., Prevot, L., Queinnec, F., Quinn, P. J., Renault, C., Rich, J., Spiro, M., Stubbs, C. W., Sutherland, W., Tomaney, A., Vandehei, T., Vidal-Madjar, A., Vigroux, L., and Zylberajch, S. 1998. EROS and MACHO combined limits on planetary-mass dark matter in the galactic halo. *Astrophys. J. Lett.* 499:L9–L12.

Allard, F., Hauschildt, P. H., Baraffe, I., and Chabrier, G. 1996. Synthetic spectra and mass determination of the brown dwarf Gliese 229B. *Astrophys. J. Lett.* 465:L123–127.

Allard, F., Hauschildt, P. H., Alexander, D. R., and Starrfield, S. 1997. Model atmospheres of very low-mass stars and brown dwarfs. *Ann. Rev. Astron. Astrophys.* 35:137–177.

Baraffe, I., Chabrier, G., Allard, F., and Hauschildt, P. H. 1995. New evolutionary tracks for very low-mass stars. *Astrophys. J. Lett.* 446:L35–L38.

Basri, G. 1998. The lithium test for young brown dwarfs. In *Brown Dwarfs and Extrasolar Planets*, ASP Conf. Ser. 134, ed. R. Rebolo, E. L. Martín, and M. Zapatero-Osorio (San Francisco: Astronomical Society of the Pacific) pp. 394–404.

Basri, G., and Marcy, G. 1997. Early hints on the substellar mass function. In *Star Formation Near and Far*, ed. S. S. Holt and L. G. Mundy (New York: AIP Press) pp. 228–240.

Basri, G., and Martín, E. L. 1998. PPL 15: The first binary brown dwarf system? In *Brown Dwarfs and Extrasolar Planets*, ASP Conf. Ser. 134, ed. R. Rebolo, E. L. Martín, and M. Zapatero-Osorio (San Francisco: Astronomical Society of the Pacific) pp. 284–287.

Basri, G., and Martín, E. L. 1999. The mass and age of very low-mass members of the open cluster α Persei. *Astrophys. J.,* in press.

Basri, G., Marcy, G., and Graham, J. 1996. Lithium in brown dwarf candidates: The mass and age of the faintest Pleiades stars. *Astrophys. J.* 458:600–609.

Becklin, E. E., and Zuckerman, B. 1988. A low-temperature companion to a white dwarf star. *Nature* 336:656–658.

Beichman, C. A. 1987. The IRAS view of the galaxy and the solar system. *Ann. Rev. Astron. Astrophys.* 25:521–563.

Bildsten, L., Brown, E. F., Matzner, C. D., and Ushomirsky, G. 1997. Lithium depletion in fully convective pre-main-sequence stars. *Astrophys. J.* 482:442–447.

Black, D. C. 1997. Possible observational criteria for distinguishing brown dwarfs from planets. *Astrophys. J. Lett.* 490:L171–L174.

Bouvier, J., Stauffer, J., Martín, E., Barrado y Navascués, D., Wallace, B., and Bejar, V. 1998. Brown dwarfs and very low-mass stars in the Pleiades cluster: A deep wide-field imaging survey. *Astron. Astrophys.* 335:183–198.

Burkert, A., and Bodenheimer, P. 1993. Multiple fragmentation in collapsing protostars. *Mon. Not. Roy. Astron. Soc. 264:798–806.*

Burrows, A., and Liebert, J. 1993. The science of brown dwarfs. *Rev. Mod. Phys.* 65:301–336.

Burrows, A., and Sharp, C. M. 1998. Chemical equilibrium abundances in brown dwarf and extrasolar giant planet atmospheres. *Astrophys. J.*, in press.

Burrows, A., Marley, M., Hubbard, W. B., Lunine, J. I., Guillot, T., Saumon, D., Freedman, R., Sudarsky, D., and Sharp, C. 1997. A nongray theory of extrasolar giant planets and brown dwarfs. *Astrophys. J.* 491:856–875.

Chabrier, G., and Baraffe, I. 1997. Structure and evolution of low-mass stars. *Astron. Astrophys.* 327:1039–1053.

Comeron, F., Rieke, G., Claes, P., and Torra, J. 1998. ISO observations of candidate young brown dwarfs. *Astron. Astrophys.* 335:522–532.

D'Antona, F. 1995. Luminosity functions and the mass function. In *The Bottom of the Main Sequence--and Beyond*, ed. C. G. Tinney (Berlin: Springer), pp. 13–23.

D'Antona, F., and Mazzitelli, I. 1994. New pre-main-sequence tracks for $M < 2.5\ M_{\odot}$ as tests of opacities and convection model. *Astrophys. J. Suppl.* 90:467–500.

Delfosse, X., Tinney, C., Forveille, T., Epchtein, N., Bertin, E., Borsenberger, J., Copet, E., de Batz, B., Fouque, P., Kimeswenger, S., LeBertre, T., Lacombe, F., Rouan, D., and Tiphene, D. 1997. Field brown dwarfs found by DENIS. *Astron. Astrophys. Lett.* 327:L25–L28.

Fegley, B., Jr., and Lodders, K. 1994. Chemical models of the deep atmospheres of Jupiter and Saturn. *Icarus* 110:117–154.

Forrest, W. J., Skrutskie, M. F., and Shure, M. 1988. A possible brown dwarf companion to Gliese 569. *Astrophys. J. Lett.* 330:L119–L123.

Geballe, T. R., Kulkarni, S. R., Woodward, C. E., and Sloan, G. C. 1996. The near-infrared spectrum of the brown dwarf Gliese 229B. *Astrophys. J. Lett.* 467:L101–L104.

Golimowski, D. A., Burrows, C. J., Kulkarni, S. R., Oppenheimer, B. R., and Brukardt, R. A. 1998. Wide Field Planetary Camera 2 observations of the brown dwarf Gliese 229B: Optical colors and orbital motion. *Astron. J.* 115:2579–2586.

Griffith, C. A., Yelle, R. V., and Marley, M. 1998. The dusty atmosphere of Gliese 229B. *Science*, in press.
Hambly, N., Hawkins, M. R. S., and Jameson, R. F. 1993. Very low mass members of the Pleiades. *Astron. Astrophys.* 100:607–640.
Henry, T. J., and McCarthy, D. W., Jr. 1990. A systematic search for brown dwarfs orbiting nearby stars. *Astrophys. J.* 350:334–347.
Henry, T. J., and McCarthy, D. W., Jr. 1993. The mass-luminosity relation for stars of mass 1.0 to 0.08 $M_\odot$. *Astron. J.* 106:773–789.
Henry, T. J., McCarthy, D. W., Jr., Freeman, J., and Christou, J. C. 1992. Nearby solar-type star with a low-mass companion—New sensitivity limits reached using speckle imaging. *Astron. J.* 103:1369–1373.
Hodgkin, S., and Jameson, R. F. 1997. Brown dwarfs. *Cont. Phys.* 38:395–407.
Irwin, M., McMahon, R., and Hazard, C. 1991. APM optical surveys for high redshift quasars. In *The Space Distribution of Quasars*, ASP Conf. Ser. 21, ed. D. Crampton (San Francisco: Astronomical Society of the Pacific) pp. 117–126.
Jameson, R. F., and Skillen, I. 1989. A search for low-mass stars and brown dwarfs in the Pleiades. *Mon. Not. Roy. Astron. Soc.* 239:247–253.
Jones, H. R. A., and Tsuji, T. 1997. Spectral evidence for dust in late-type M dwarfs. *Astrophys. J. Lett.* 48:L39–L42.
Jones, H. R. A., Longmore, A. J., Jameson, R. F., and Mountain, M. 1994. An infrared spectral sequence for M dwarfs. *Mon. Not. Roy. Astron. Soc.* 267:413–423.
Kafatos, M. C., Harrington, R. S., and Maran, S. P., ed. 1986. *Astrophysics of Brown Dwarfs* (New York: Cambridge University Press).
Kerins, E., and Wyn Evans, N. 1998. Microlensing halo models with abundant brown dwarfs. *Astrophys. J. Lett.* 503:L75–L78.
Khlifi, M., Paillous, P., Bruston, P., Guillemin, J. C., Benilan, Y., Daoudi, A., and Raulin, F. 1997. Gas infrared spectra, assignments, and absolute IR band intensities of C_4N_2 in the 250–3500 cm(−1) region: Implications for Titan's stratosphere. *Spectrochim. Acta A* 53:707–712.
Kirkpatrick, J. D. 1997. Spectroscopic properties of ultra-cool dwarfs and brown dwarfs. In *Brown Dwarfs and Extrasolar Planets*, ASP Conf. Ser. 134, ed. R. Rebolo, E. Martín, and M. Zapatero-Osorio (San Francisco: Astronomical Society of the Pacific), pp. 405–415.
Kirkpatrick, J. D., Henry, T., and Irwin, M. 1997. Ultra-cool M dwarfs discovered by QSO surveys. I: The APM objects. *Astron. J.* 113:1421–1428.
Kirkpatrick, J. D., Allard, F., Bida, T., Zuckerman, B., Becklin, E. E., Chabrier, G., and Baraffe, I. 1998. An improved optical spectrum and new model fits of the likely brown dwarf GD 165B. *Astrophys. J.*, submitted.
Krasnopolsky, V. A., Bjoraker, G. L., Mumma, M. J., and Jennings, D. E. 1997. High-resolution spectroscopy of Mars at 3.7 and 8 μm: A sensitive search for H_2O_2, H_2CO, HCl, and CH_4, and detection of HDO. *J. Geophys. Res.* 102:6525–6534.
Krist, J. E., Golimowski, D. A., Schroeder, D. J., and Henry, T. J. 1998. Characterization and subtraction of well-exposed HST/NICMOS Camera 2 point spread functions for a survey of very low-mass companions to nearby stars. *Pub. Astron. Soc. Pacific,* in press.
Kulkarni, S. R. 1998. Brown dwarfs: A possible missing link between stars and planets. *Science* 276:1350–1354.
Kumar, S. S. 1963. The structure of stars of very low mass. *Astrophys. J.* 137:1121-1126.
Leggett, S., and Hawkins, M. R. 1988. The infrared luminosity function for low-mass stars. *Mon. Not. Roy. Astron. Soc.* 234:1065–1090.

Leggett, S., Harris, H., and Dahn, C. 1994. Low-mass stars in the central region of the Hyades cluster. *Astron. J.* 108:944–963.

Luhman, K. L., Liebert, J., and Rieke, G. H. 1997. Spectroscopy of a young brown dwarf in the Rho Ophiuchi cluster. *Astrophys. J. Lett.* 489:L165–L168.

Luhman, K. L., Rieke, G. H., Lada, C., and Lada, E. 1998. Spectroscopy of low-mass members of IC348. *Astrophys. J.,* submitted.

Luyten, W., and Kowal, C. 1975. *Proper Motion Survey with the 48 Inch Schmidt Telescope* (Minneapolis: Observatory of the University of Minnesota).

Magazzù, A., Rebolo, R., Zapatero-Osorio, M., Martín, E., and Hodgkin, S. 1998. A brown dwarf candidate in the Praesepe open cluster. *Astrophys. J. Lett.* 497:L47–L51.

Marcy, G., Basri, G., and Graham, J. 1994. A search for lithium in Pleiades brown dwarf candidates using the Keck HIRES Echelle. *Astrophys. J. Lett.* 428:L57–L60.

Marley, M. S., Saumon, D., Guillot, T., Freedman, R. S., Hubbard, W. B., Burrows, A., and Lunine, J. I. 1996. Atmospheric, evolutionary, and spectral models of the brown dwarf Gliese 229B. *Science* 272:1919–1921.

Martín, E. L., Rebolo, R., and Magazzù, A. 1994. Constraints to the masses of brown dwarf candidates from the lithium test. *Astrophys. J.* 436:262–269.

Martín, E., Basri, G., Delfosse, X., and Forveille, T. 1997. Keck HIRES spectra of the brown dwarf DENIS-PJ1228.2-1547. *Astron. Astrophys. Lett.* 327:L29–L32.

Martín, E. L., Basri, G., Gallegos, J. E., Rebolo, R., Zapatero-Osorio, M. R., and Bejar, V. J. S. 1998. A new Pleiades member at the lithium substellar boundary. *Astrophys. J. Lett.* 499:L61–L64.

Matthews, K., Nakajima, T., Kulkarni, S. R., and Oppenheimer, B. R. 1996. Spectral energy distribution and bolometric luminosity of the cool brown dwarf Gliese 229B. *Astron. J.* 112:1678–1682.

McCarthy, D. W., Jr., Probst, R. G., and Low, F. J. 1985. Infrared detection of a close cool companion to van Biesbroeck 8. *Astrophys. J. Lett.* 290:L9–L13.

Meynet, G., Mermilliod, J.-C., and Maeder, A. 1993. New dating of galactic open clusters. *Astron. Astrophys. Suppl.* 98:477–504.

Nakajima, T., Durrance, S. T., Golimowski, D. A., and Kulkarni, S. R. 1994. A coronagraphic search for brown dwarfs around nearby stars. *Astrophys. J.* 428:797–804.

Nakajima, T., Oppenheimer, B. R., Kulkarni, S. R., Golimowski, D. A., Matthews, K., and Durrance, S. T. 1995. Discovery of a cool brown dwarf. *Nature* 378:463–465.

Noll, K. S. 1993. Spectroscopy of outer solar system atmospheres from 2.5 to 7.0 microns. In *Astronomical Infrared Spectroscopy: Future Observational Directions*, ASP Conf. Ser. 41, ed. S. Kwok (San Francisco: Astronomical Society of the Pacific), pp. 29–40.

Noll, K. S., Geballe, T. R, and Marley, M. S. 1997. Detection of abundant carbon monoxide in the brown dwarf Gliese 229B. *Astrophys. J. Lett.* 489:L87–L90.

Oppenheimer, B. R., Kulkarni, S. R., Matthews, K., and Nakajima, T. 1995. Infrared spectrum of the cool brown dwarf GL 229B. *Science* 270:1478–1479.

Oppenheimer, B. R., Golimowski, D. A., Kulkarni, S. R., and Matthews, K. 2000. An optical coronagraphic and near IR imaging survey of stars for faint companions. In preparation.

Oppenheimer, B. R., Kulkarni, S. R., Matthews, K., and van Kerkwijk, M. H. 1998. The spectrum of the brown dwarf Gliese 229B. *Astrophys. J.* 502:932–943.

Pavlenko, Y. V. 1998. Lithium lines in the spectra of late M dwarfs. *Astron. Rep.* 42:501–507.

Pinfield, D., Hodgkin, S., Jameson, R., Cossburn, M., and von Hippel, T. 1997. Brown dwarf candidates in Praesepe. *Mon. Not. Roy. Astron. Soc.* 287:180–188.

Pinsonneault, M., Stauffer, J., Soderblom, D., King, J., and Hanson, R. 1998. The problem of Hipparcos distances to open clusters: I. Constraints from multicolor main sequence fitting. *Astrophys. J.,* in press.

Pound, M. W., and Blitz, L. 1995. Proto-brown dwarfs 2: Results in the Ophiuchus and Taurus molecular clouds. *Astrophys. J.* 444:270–287.

Probst, R. G. 1983. An infrared search for very low mass stars: The luminosity function. *Astrophys. J.* 274:237–244.

Raboud, D., and Mermilliod, J.-C. 1998. Investigation of the Pleiades cluster. IV. The radial structure. *Astron. Astrophys.* 329:101–114.

Raulin, F., and Bruston, P. 1996. Photochemical growing of complex organics in planetary atmospheres. *Life Sci.* 18:41–49.

Rebolo, R., Martín, E., and Magazzù, A. 1992. Spectroscopy of a brown dwarf candidate in the α Persei open cluster. *Astrophys. J. Lett.* 389:L83–L87.

Rebolo, R., Zapatero-Osorio, M., and Martín, E. 1995. Discovery of a brown dwarf in the Pleiades star cluster. *Nature* 377:129–131.

Rebolo, R., Martín, E. L., Basri, G., Marcy, G., and Zapatero-Osorio, M. R. 1996. Brown dwarfs in the Pleiades cluster confirmed by the lithium test. *Astrophys. J. Lett.* 469:L53–L57.

Rebolo, R., Martín, E. L., and Zapatero-Osorio, M. R., eds. 1998. *Brown Dwarfs and Extrasolar Planets,* ASP Conf. Ser. 134 (San Francisco: Astronomical Society of the Pacific).

Rieke, G. H., and Rieke, M. J. 1990. Possible substellar objects in the Rho Ophiuchi cloud. *Astrophys. J. Lett.* 362:L21–L24.

Ruiz, M. T., Leggett, S. K., and Allard, F. 1997. Kelu-1: A free-floating brown dwarf in the solar neighborhood. *Astrophys. J. Lett.* 491:L107–L111.

Schultz, A. B., Allard, F., Clampin, M., McGrath, M., Bruhweiler, F. C., Valenti, J. A., Plait, P., Hulbert, S., Baum, S., Woodgate, B. E., Bowers, C. W., Kimble, R. A., Mara, S. P., Moos, H. W., and Roesler, F. 1998. First results from the Space Telescope Imaging Spectrograph: Optical spectra of Gliese 229B. *Astrophys. J. Lett.* 492:L181–L182.

Shu, F. H., Adams, F. C., and Lizano, S. 1987. Star formation in molecular clouds: Observation and theory. *Ann. Rev. Astron. Astrophys.* 25:23–81.

Simons, D. A., Henry, T. J., and Kirkpatrick, J. D. 1996. The solar neighborhood. III. A near-infrared search for widely separated low-mass binaries. *Astron. J.* 112:2238–2247.

Stauffer, J., Hamilton, D., Probst, R., Rieke, G., and Mateo, M. 1989. Possible Pleiades members with $M \sim 0.07$ solar mass: Identification of brown dwarfs of known age, distance and metallicity. *Astrophys. J. Lett.* 344:L21–L25.

Stauffer, J., Liebert, J., Giampapa, M., MacIntosh, B., Reid, I. N., and Hamilton, D. 1994. Radial velocities of very low mass stars and candidate brown dwarf members of the Hyades and Pleiades. *Astron. J.* 108:160–174.

Stauffer, J., Schultz, G., and Kirkpatrick, J. D. 1998. Keck spectra of Pleiades brown dwarf candidates and a precise determination of the lithium depletion edge in the Pleiades. *Astrophys. J. Lett.* 499:L199–L203.

Stringfellow, G. 1991. Brown dwarfs in young stellar clusters. *Astrophys. J. Lett.* 375:L21–L25.

Thackrah, A., Jones, H., and Hawkins, M. 1997. Lithium detection in a field brown dwarf candidate. *Mon. Not. Roy. Astron. Soc.* 284:507–512.

Tinney, C. G., ed. 1995. *The Bottom of the Main Sequence--and Beyond* (Berlin: Springer).

Tinney, C. G. 1998. The intermediate age brown dwarf LP 944-20. *Mon. Not. Roy. Astron. Soc.* 296:L42–L44.
Tinney, C. G., Delfosse, X., and Forveille, T. 1997. DENIS-P J1228.2-1547—A new bench mark brown dwarf. *Astrophys. J. Lett.* 490:L95–L98.
Toth, R. A. 1997. Measurements of HDO between 4719 and 5843 cm(−1). *J. Mol. Spect.* 186:276–292.
Tsuji, T. 1964. *Ann. Tokyo Astron. Obs. Ser. II* 9:1–26.
Tsuji, T., Ohnaka, K., Aoki, W., and Nakajima, T. 1996. Evolution of dusty photospheres through red to brown dwarfs: How dust forms in very low-mass objects. *Astron. Astrophys. Lett.* 308:L29–L32.
Ventura, P., Zeppieri, A., Mazzitelli, I., and D'Antona, F. 1998. Pre-main-sequence lithium burning: The quest for a new structural parameter. *Astron. Astrophys.* 331:1011–1021.
Wilking, B. A., Green, T. P., and Meyer, M. R. 1998. Spectroscopy of brown dwarf candidates in the ρ Oph molecular core. *Astron. J.,* submitted.
Zapatero-Osorio, M. R., Rebolo, R., Martin, E. L., Basri, G., Magazzu, A., Hodgkin, S. T., Jameson, R. F., and Cossburn, M. R. 1997. New brown dwarfs in the Pleiades cluster. *Astrophys. J. Lett.* 491:L81–L84.
Zuckerman, B., and Becklin, E. E. 1992. Companions to white dwarfs—very low-mass stars and the brown dwarf candidate GD 165B. *Astrophys. J.* 386:260–264.

NEW IDEAS IN THE THEORY OF EXTRASOLAR GIANT PLANETS AND BROWN DWARFS

ADAM BURROWS, WILLIAM B. HUBBARD,
JONATHAN I. LUNINE
University of Arizona

MARK S. MARLEY
New Mexico State University

and

DIDIER SAUMON
Vanderbilt University

I. INTRODUCTION

The study of extrasolar giant planets (EGPs; the terms "exoplanet" or "super-Jupiter" are equally appropriate) and brown dwarfs via reflex stellar motion, broad-band photometry, and spectroscopy has finally come into its own. Doppler spectroscopy alone has revealed about 15 objects in the giant planet/brown dwarf regime, including companions to τ Bootis, 51 Pegasi, υ Andromedae, 55 Cancri, ρ Coronae Borealis, 70 Virginis, 16 Cygni B, and 47 Ursae Majoris (Latham et al. 1989; Mayor and Queloz 1995; Marcy and Butler 1996; Butler and Marcy 1996; Noyes et al. 1997; Butler et al. 1997; Cochran et al. 1997; and the chapter by Marcy et al., this volume).

The direct detection of Gliese 229 B (Oppenheimer et al. 1995; Nakajima et al. 1995; Matthews et al. 1996; Geballe et al. 1996; Marley et al. 1996; Allard et al. 1996; Tsuji et al. 1996; and the chapter by Oppenheimer et al., this volume) was a milestone, because Gl 229 B displays methane spectral features and low surface fluxes that are unique to objects with effective temperatures (in this case, $T_{\rm eff} \sim 950$ K) below those at the solar-metallicity main-sequence edge ($\sim$1750 K, Burrows et al. 1993). In addition, the almost complete absence of spectral signatures of metal oxides and hydrides (such as TiO, VO, FeH, and CaH) is in keeping with theoretical predictions that these species are depleted in the atmospheres of all but the youngest (hence, hottest) substellar objects and are sequestered in condensed form below the photosphere (Lunine et al. 1989; Marley et al. 1996). The wide range in mass and period, as well as the proximity of many of the planets/brown dwarfs to their primaries, was not anticipated by most planetary scientists. Though the technique of Doppler spectroscopy used to find these companions selects for massive, nearby objects, their variety and existence is a challenge to conventional theory. Because direct detection is now feasible and has been demonstrated by the recent acquisition of Gl 229 B, it is crucial for the future of extrasolar giant planet searches that the spectra, colors, evolution, and physical structure of objects from Saturn's mass (0.3 M_{Jup}) to 70 M_{Jup} be theoretically investigated.

II. EARLY CALCULATIONS OF THE EVOLUTION AND STRUCTURE OF EXTRASOLAR GIANT PLANETS

A. Gray Models

EGPs radiate in the optical by reflection and in the infrared by the thermal emission of both absorbed stellar light and the planet's own internal energy. In Burrows et al. (1995) and Saumon et al. (1996), the EGPs were assumed to be fully convective at all times. We included the effects of "insolation" by a central star of mass M_* and considered semi-major axes (a) between 2.5 and 20 AU. Giant planets may form preferentially near 5 AU (Boss 1995), but, as the new data dramatically affirm, a broad range of semimajor axes cannot be excluded. In these calculations, we assumed that the Bond albedo of an EGP is that of Jupiter ($\sim$0.35). For the Burrows et al. (1995) study, we evolved EGPs with masses from 0.3 M_{Jup} (the mass of Saturn) through 15 M_{Jup}. Whether a 15-M_{Jup} object is a planet or a brown dwarf is largely a semantic issue, though one might distinguish gas giants and brown dwarfs by their mode of formation (e.g., in a disk or "directly").

If 51 Peg b is a gas giant, its radius is only 1.2 R_{Jup} and its luminosity is about 3.5×10^{-5} $L_{\odot}$. This bolometric luminosity is more than 1.5×10^4 times the present luminosity of Jupiter and only a factor of two below that at the edge of the main sequence. The radiative region encompasses the outer 0.03% in mass, and 3.5% in radius. The study by Guillot et al. (1996) demonstrated that 51 Peg b is well within its Roche lobe and is not experiencing significant photoevaporation. Its deep potential well ensures that, even so close to its parent, 51 Peg b is stable. If 51 Peg b were formed beyond 1 AU and moved inward over a timescale greater than $\sim 10^8$ years, it would closely follow a $R_p \sim R_{Jup}$ trajectory to its equilibrium position.

B. Nongray Models

However, to credibly estimate the infrared band fluxes and improve upon the blackbody assumption made in Burrows et al. (1995) and Saumon et al. (1996), we have recently performed nongray simulations at solar metallicity of the evolution, spectra, and colors of isolated EGPs/brown dwarfs down to T_{eff} values of 100 K (Burrows et al. 1997). Color Plate 22 portrays the luminosity vs. time for objects from Saturn's mass (0.3 M_{Jup}) to 0.2 $M_{\odot}$ for this model suite. The early plateaus between 10^6 years and

10^8 years are due to deuterium burning, where the initial deuterium mass fraction was taken to be 2×10^{-5}. Deuterium burning occurs earlier, is quicker, and is at higher luminosity for the more massive models but can take as long as 10^8 years for a 15-M_{Jup} object. The mass below which less than 50% of the "primordial" deuterium is burnt is ~13 M_{Jup} (Burrows et al. 1995). On this figure, we have arbitrarily classed as "planets" those objects that do not burn deuterium and as "brown dwarfs" those that do burn deuterium, but not light hydrogen. While this distinction is physically motivated, we do not advocate abandoning the definition based on origin. Nevertheless, the separation into M dwarfs, "brown dwarfs," and giant "planets" is useful for parsing by eye the information in the figure.

In Color Plate 22, the bumps between 10^{-4} $L_\odot$ and 10^{-3} $L_\odot$ and between 10^8 and 10^9 years, seen on the cooling curves of objects from 0.03 $M_\odot$ to 0.08 $M_\odot$, are due to silicate and iron grain formation. These effects, first pointed out by Lunine et al. (1989), occur for T_{eff} values between 2500 and 1300 K. The presence of grains affects the precise mass and luminosity at the edge of the main sequence. Because grain and cloud models are problematic, there still remains much to learn concerning their role and how to model them (Lunine et al. 1989; Allard et al. 1997; Tsuji et al. 1996).

III. NEW INSIGHTS

The studies of Burrows et al. (1997) and Marley et al. (1996) reveal major new aspects of EGP/brown dwarf atmospheres that bear listing and that uniquely characterize them. Below T_{eff} of about 1300 K (subject to pressure), the dominant equilibrium carbon molecule is CH_4, not CO, and below about 600 K the dominant nitrogen molecule is NH_3, not N_2 (Fegley and Lodders 1996). In objects with $T_{eff} \leq 1300$ K, the major opacity sources are H_2, H_2O, CH_4, and NH_3. For T_{eff} below ~400 K, water clouds form at or above the photosphere (defined where $T = T_{eff}$), and for T_{eff} below 200 K, ammonia clouds form (viz., Jupiter). Collision-induced absorption of H_2 partially suppresses emissions longward of ~10 μm. The holes in the opacity spectrum of H_2O that define the classic telluric IR bands also regulate much of the emission from EGP/brown dwarfs in the near-infrared. Importantly, the windows in H_2O and the suppression by H_2 conspire to force flux to the blue for a given T_{eff}. The upshot is an exotic spectrum enhanced relative to the blackbody value in the J and H bands (~1.2 μm and ~1.6 μm, respectively) by as much as *two* to *ten* orders of magnitude, depending upon T_{eff}. The enhancement at 5 μm for a 1-Gyr-old, 1-M_{Jup} extrasolar planet is by four orders of magnitude. As T_{eff} decreases below ~1000 K, the flux in the M band (~5 μm) is progressively enhanced relative to the blackbody value. Whereas at 1000 K there is no enhancement, at 200 K it is near 10^5. The J, H, and M bands are the premier bands in which to search for cold substellar objects. The Z band (~1.05 μm) is also in excess of the blackbody value over this T_{eff}

range. Even though K band ($\sim$2.2 μm) fluxes are generally higher than blackbody values, H_2 and CH_4 absorption features in the K band decrease its importance *relative* to J and H. As a consequence of the increase of atmospheric pressure with decreasing T_{eff}, the anomalously blue $J - K$ and $H - K$ colors get bluer, not redder. The K and J vs. $J - K$ infrared H-R diagrams loop back to the blue below the edge of the main sequence and are not continuations of the M dwarf sequence into the red. The difference between the blackbody curves and the model curves is between 3 and 10 magnitudes for J vs. $J - K$, more for K vs. $J - K$. Gl 229 B fits nicely among these theoretical isochrones. The suppression of K by H_2 and CH_4 features is largely responsible for this anomalous blueward trend with decreasing mass and T_{eff}.

IV. ATMOSPHERIC COMPOSITIONS OF EGPs, CONDENSATION, AND CLOUDS

The molecular compositions of the exotic, low-ionization atmospheres of EGPs and brown dwarfs can serve as diagnostics of temperature, mass, and elemental abundance and can help define a spectral sequence, just as the presence or absence of spectral features associated with various ionization states of dominant, or spectroscopically active, atoms and simple molecules does for M through O stars. However, the multiplicity of molecules that appear in their atmospheres lends an additional complexity to the study of substellar-mass objects that is both helpfully diagnostic and confusing. Nowhere is the latter more apparent than in the appearance at low temperatures of refractory grains and clouds. These condensed species can contribute significant opacity and can alter an atmosphere's temperature/pressure profile and its albedo. Grain and cloud droplet opacities depend on the particle size and shape distribution, and these are intertwined with the meteorology (convection) in complex ways. Furthermore, condensed species can rain out and deplete the upper atmosphere of heavy elements, thereby changing the composition and the observed spectrum. In brown dwarf and EGP atmospheres, abundance and temperature/pressure profiles, particle properties, spectra, and meteorology are inextricably linked.

The formation of refractory silicate grains below 2500 K was already shown by Lunine et al. (1989) and Burrows et al. (1989) to influence the evolution of late M dwarfs and young brown dwarfs through their "Mie" opacity. The blanketing effect they provide lowers the effective temperature (T_{eff}) and luminosity (L) of the main-sequence edge mass from about 2000 K and 10^{-4} $L_\odot$ to about 1750 K and 6×10^{-5} $L_\odot$, an effect recently verified by Chabrier et al. (1998). In addition, grain opacity slightly delays the cooling of older brown dwarfs, imprinting a slight bump on their luminosity/age trajectories (see Color Plate 22). The presence of grains in late M dwarf spectra was invoked to explain the weakening of the TiO bands and the shallowing of their H_2O troughs in the near-infrared (Tsuji et al. 1996; Jones and Tsuji 1997). Tsuji and collaborators concluded that titanium was being depleted into refractories, a conclusion with which we agree.

With a $T_{\rm eff}$ of ~950 K, a luminosity below 10^{-5} $L_\odot$, and spectra or photometry from the *R* band through 5 μm, Gl 229 B hints at or exemplifies all of the unique characteristics of the family: metal (Fe, Ti, V, Ca, Mg, Al, Si) depletions, the dominance of H_2O vapor, the appearance of CH_4 and alkali metals, and the signatures of clouds. Clouds of low-temperature condensable species above the photosphere are the most natural explanation for the steep drop below 1 μm in the Keck spectra between 0.83 μm and 1 μm (Oppenheimer et al. 1998). These clouds might not be made up of the classic silicate refractories formed at much higher temperatures, because these species may have rained out. From simple Mie theory, their mean particle size must be small (~0.1 μm) in order to influence the "optical" without much perturbing the near-infrared. In addition, such a population of small droplets can help explain why Gl 229 B's near-infrared troughs at 1.8 μm and 3.0 μm are not as deep as theory would otherwise have predicted. Just as Tsuji and collaborators (1996) have shown that silicate grains at higher temperatures can shallow out the H_2O troughs, so too can species that condense at lower temperatures ($\leq$1000 K?) explain the shallower-than-predicted Gl 229 B H_2O troughs. What those species might be can be illuminated by chemical abundance studies (Fegley and Lodders 1996; Burrows and Sharp 1999). Note that a cloud surface density in these small-radius, low-temperature refractories of only ~10^{-5} g $\rm cm^{-2}$ would be adequate to explain the anomalies.

A. Abundance and Condensate Results

At temperatures of ~1500–2000 K, the refractories, such as the silicates, spinel, and iron, condense out into grain clouds, which by their large opacity lower the $T_{\rm eff}$ and luminosity of the main-sequence edge (Burrows et al. 1993; Fegley and Lodders 1996; Burrows and Sharp 1999) and alter in detectable ways the spectra of objects around the transition mass (Jones and Tsuji 1997). As $T_{\rm eff}$ decreases below that at the stellar edge, the high-temperature refractories are buried progressively deeper below the photosphere and less refractory condensates and gas-phase molecules come to dominate (Marley et al. 1996; Burrows et al. 1997).

Below temperatures of ~1500 K, the calculations of Burrows and Sharp (1999) indicate that the alkali metals, which are not as refractory as Fe, Al, Ca, Ti, V, and Mg, emerge as important atmospheric and spectral constituents. At still lower temperatures, chlorides and sulfides appear, some of which will condense in the cooler upper atmosphere and form clouds that will affect emergent spectra and albedos. Cloud decks of many different compositions at many different temperature levels are expected, depending upon $T_{\rm eff}$ (and weakly upon gravity). Clouds of chlorides and sulfides at temperature levels below ~1000 K may be responsible for the steeper slope observed in the spectrum of Gl 229 B at the shorter wavelengths (Oppenheimer et al. 1998). At slightly higher temperatures, MnS, ZnS, $NaAlSi_3O_8$, $KAlSi_3O_8$, V_2O_3, and $MgTi_2O_5$ may play a role, but only if their constituents are not scavenged into more refractory compounds and rained out deeper down.

As $T_{\rm eff}$ decreases (either as a given mass cools or, for a given age, as we study objects with lower masses), the major atmospheric constituents of brown dwarfs and EGPs change. This change affects which spectral features are most prominent as well as the albedos of substellar objects near their primaries. Equilibrium chemical sequences are predominantly a function of temperature and can help to define a spectral sequence for substellar objects from the main-sequence edge near 2000 K to EGPs with $T_{\rm eff}$s of a few hundred kelvins. The appearance and disappearance of various molecules and refractories delineates an effective temperature sequence, and the new proposed "L" dwarf spectral classification, suggested by Kirkpatrick et al. (1999), may correspond to a subset of such a compositional sequence. Very crudely, the L spectral type would correspond to $T_{\rm eff}$ between about ~1500 K and ~2200 K. All but the youngest and most massive brown dwarfs and only the very youngest EGPs could have this proposed spectral designation. Most brown dwarfs and EGPs will be of an even later spectral type yet to be coined, a spectral type that would include Gl 229 B.

B. Clouds

Cloud formation consists of the production of solid or liquid particles in the atmosphere of a brown dwarf or planet. Cloud formation in the Earth's troposphere is a complex phenomenon involving a single species (water), two phases, multiple particle size distributions, incompletely known particle properties, and myriad cloud morphologies. The range of radiative properties of terrestrial clouds is commensurately broad, and different types of clouds can serve to either warm or cool the surface of the Earth. Much of the complexity lies in the heterogeneous distribution of such clouds in vertical and horizontal directions; the problem of treatment of the radiative transfer of broken clouds remains a difficult one (Goody and Yung 1989). Furthermore, cloud formation itself contributes latent heat to an atmosphere and hence can destabilize a radiative region, requiring treatment of the heat transfer by moist convection (Emanuel 1994). Cloud-forming species are known or suspected to be present in eight of the nine planets of the solar system as well as two of the moons of the outer solar system (Titan and Triton).

The giant planets of our solar system remain the best guide to the study of clouds in brown dwarfs and extrasolar giant planets, because the background gas is hydrogen-helium in all cases, and the atmospheres are not thin layers atop a solid surface. Principal cloud-forming species in the upper jovian and saturnian atmospheres are ammonia, water, and possibly ammonium hydrosulfide; in Uranus and Neptune they are methane and either ammonium hydrosulfide or hydrogen sulfide (Baines et al. 1995).

However, a variety of minor species may participate in cloud formation, and stratospheric photochemical hazes are generated by the action of solar ultraviolet radiation. *In situ* observations of Jupiter's cloud structure by the *Galileo* entry probe revealed the extent of heterogeneity across the surface of a gas giant planet; the probe failed to detect anything but very a tenuous and narrow cloud layer (Ragent et al. 1996). Because ample visual and spectroscopic evidence exists for widespread clouds on Jupiter (Carlson et al. 1995), the *Galileo* probe must have found a region of unusual dryness. This notion is supported by the observation that the probe fell into a "hot spot," a region of unusually high 5-micron emission (Orton et al. 1996). Furthermore, the results from the mass spectrometer on board the entry probe demonstrate just how difficult it is to theoretically fit and understand the composition of the atmospheres of the giant planets (Niemann et al. 1998). This should serve as a caution to EGP theorists as well as to those who would extrapolate from Jupiter to EGPs.

Modeling a potentially broad range of compositions and properties from the small number of detailed observations of clouds available to date is daunting (Lewis 1969; Weidenschilling and Lewis 1973). One strategy is to consider end-member cloud models, characterized by simple condensation in a radiative atmosphere on the one hand and by large-scale vertical motions driven by thermal convection or other processes on the other (Lunine et al. 1986, 1989). Such models can be reduced to a few parameters, which can then be constrained, albeit weakly, with spectral observations of the atmosphere. In the radiative case the cloud mass density vs. altitude is determined by the saturation vapor pressure and an assumed value for the meteorological supersaturation, which determines the condensable available to form clouds, but is determined for the Earth empirically (Emanuel 1994). This factor is difficult to determine *a priori* for an object in which the clouds cannot be directly studied, but it can be crudely estimated (Rossow 1978). A simple prescription for the cloud number density for a convectively unstable brown dwarf atmosphere is given by Lunine et al. (1989). In each case the mean particle size can be determined semi-analytically (Rossow 1978) or by use of a numerical particle growth model (Yair et al. 1995). The problem with the former approach is that the mean particle size can be significantly underestimated, based on terrestrial experience, and a size distribution is not obtained. The numerical approach suffers from the problem of poor specificity of input parameters when applied to extrasolar planets and brown dwarfs.

Spectroscopy of Gl 229 B suggests that grain formation may be occurring in its atmosphere. Grains both alter the spectral contrast of molecular lines and, through condensation, remove gas-phase species that may be directly responsible for various absorption features. Because of condensation, silicate or iron features in the spectrum may disappear at around or above 1000 K, with gaseous water bands doing the same around 400 K. The effects of clouds on the albedo or reflectivity of a brown dwarf or extrasolar planet are also large. Modeling by Marley et al. (1998)

suggests large variations (factors of 2 or more) in the albedo of a brown dwarf or extrasolar planet, depending on the type of cloud (large-particle convective vs. small-particle laminar clouds). The four giant planets all have surface reflectivities influenced by a combination of cloud and gas opacity sources, to varying extents.

Further progress in quantifying the extent of the effects of clouds on brown dwarf atmospheres will require more realistic models of cloud formation, transport and growth of particles, data on indexes of refraction of cloud particles, and incorporation of moist convective instability into radiative-convective models. These models will require additional computational resources beyond the already substantial speed and memory requirements of fully nongray radiative-convective codes. Perhaps most challenging from a mathematical point of view is to characterize and incorporate the size-frequency spectrum of broken clouds. The problem is somewhat akin to the classical one of computing Rosseland mean absorption coefficients in gray atmosphere models, where the greatest flux contribution comes from the frequency windows of smallest opacity. Quantifying a broken cloud atmosphere through a single number characterizing percentage of area covered by clouds is likely to be highly inaccurate, because the amount of thermal radiation escaping will be highly sensitive to the gaps between the clouds. In the near term, the most promising observations for constraining such a complex situation will be high-spectral-resolution data obtained from large telescopes with sensitive electronic detectors. However, the theoretical challenge of accurately modeling broken clouds with existing computational resources remains unsolved.

V. MODELS OF THE BROWN DWARF GLIESE 229 B

The wide gap between stars and brown dwarfs near the edge of the main sequence on the one hand, and Jupiter on the other hand, is fortuitously occupied by the cool brown dwarf Gl 229 B. It is sufficiently bright to allow high-resolution and high-signal-to-noise spectroscopy. This fascinating object displays phenomena common to giant planets and to very low-mass stars and represents a benchmark for modeling atmospheres of EGPs. Model spectra for Gl 229 B (Marley et al. 1996; Allard et al. 1996; Tsuji et al. 1996) reproduce the overall energy distribution fairly well, and all agree that (i) $T_{\rm eff} \sim 950$ K, (ii) the silicate opacity is small compared to the gaseous molecular opacity and can be ignored in a first approximation, and (iii) the gravity of Gl 229 B is poorly constrained at present. The models, however, fail to reproduce the visible flux, the observed depth of the strongest molecular absorption bands, or the detailed structure of the observed spectrum. The calculation of better models for Gl 229 B is currently limited by the inadequate knowledge of CH_4 opacities and a very limited understanding of the role of grain opacity in such a cool object.

Aside from possible grain absorption and scattering, the spectrum of Gl 229 B is shaped entirely by absorption by H_2, H_2O, CH_4, and, in the mid-infrared, NH_3. While the opacities of H_2 and H_2O are now well understood over the entire temperature range of interest for Gl 229 B ($500 \lesssim T \lesssim 2000$ K), the current databases for CH_4 and NH_3 are limited to $T \leq 300$ K (HITRAN database, Rothman et al. 1992) and have an incomplete wavelength coverage. In the case of CH_4, this incompleteness is partly remedied by complementing the line list with frequency-averaged opacity (Karkoschka 1994; Strong et al. 1993). Because CH_4 plays an important role in shaping the 1–4-μm spectrum of Gl 229 B, the near-infrared fluxes of the present models are of limited accuracy, precisely where Gl 229 B is brighter and most easily observable (Z, J, H, and K bands).

The models of Allard et al. (1996) and Tsuji et al. (1996) include opacities of TiO and VO (as well as metal hydrides), which have very strong bands in the red part of the spectrum. As a consequence, their synthetic spectra reproduce the rapid decrease of the flux shortward of 1 μm, as evidenced by R and I photometry (Matthews et al. 1996), quite nicely. However, the visible spectra of Schultz et al. (1998) and Oppenheimer et al. (1998) show a strong H_2O band but *not* the bands of oxides and hydrides predicted by those models. As anticipated by Marley et al. (1996), refractory elements condense in the atmosphere of Gl 229 B and are removed from the gas phase (Burrows and Sharp 1998; Fegley and Lodders 1996), and a significant fraction of the condensed particles may settle to unobservable depths in the atmosphere. Fluxes from models that do not include grain opacity in the visible region predict visible fluxes that are grossly in error.

The lack of any other important molecular spectral features in the visible (Griffith et al. 1998; Saumon et al. 1998) suggests that a source of continuum opacity is required. An opacity source that fits these requirements is submicron grains. Mie scattering theory predicts that 0.1-micron radius grains can provide substantial opacity below about 1 micron yet still be transparent at longer wavelengths (Fig. 1). In fact, such behavior is commonly observed in the solar system and has been the solution for such diverse puzzles as the radar scattering behavior of Saturn's rings particles and the spectrum of Venus's atmosphere.

Figure 1 demonstrates the dependence of the opacity of small particles on their size. Submicron particles interact strongly with optical radiation, while affecting longer wavelengths only slightly. Griffith et al. (1998) investigated a range of possible particle optical properties and sizes and found that very red, absorbing, submicron particles can indeed lower the optical flux of Gliese 229 B while only slightly affecting the near infrared spectrum. Griffith et al. (1998) speculate that since Gl 229 B receives twice the UV flux of Titan, an object whose atmosphere is dominated by a photochemical smog of condensates, the grains in the brown dwarf's atmosphere could also consist of photochemically derived nonequilibrium species. Such carbonaceous material is seen throughout the solar system and is generally recognizable by its very red color.

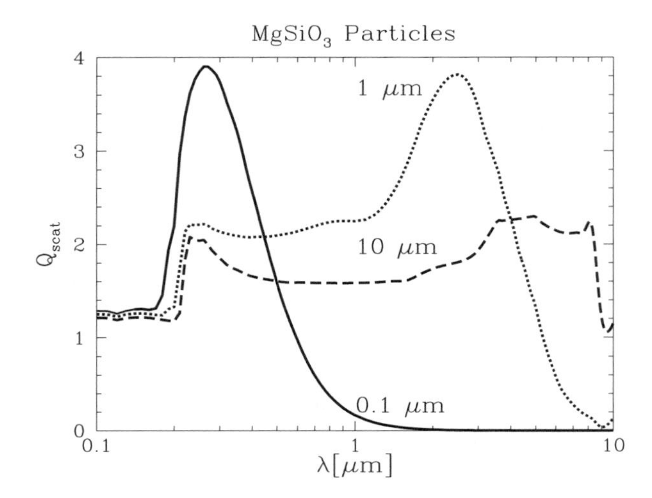

Figure 1. Approximate Mie scattering efficiency of various-sized enstatite dust grains. Submicron grains can produce large scattering and extinction opacity at optical wavelengths and yet remain essentially invisible at longer wavelengths. However, because they scatter almost conservatively in the optical, silicates are poor candidates for the grains responsible for the lower optical flux of Gl 229 B.

A. Gravity

It is highly desirable to constrain the value of the surface gravity of Gl 229 B to an astrophysically useful range. As reported by Allard et al. (1996), the spectral energy distribution of Gl 229 B models is fairly sensitive to the gravity. The most gravity-sensitive colors are $H - K$ and $J - H$ with $\Delta(H - K)/\Delta(\log g) = -0.39$ and $\Delta(J - K)/\Delta(\log g) = -0.26$, respectively. However, the uncertainties in the photometry of Matthews et al. (1996) in these color indexes are comparable to this gravity sensitivity. The present uncertainties in CH_4 opacities and in the role of grains also limit the ability of the models to predict reliable near-infrared colors.

Nevertheless, it is possible to constrain the gravity by analyzing the spectrum in the 1.9–2.1-μm region (Saumon et al. 1998). In this unique part of the spectrum of Gl 229 B, the absorption is completely dominated by two well-understood opacity sources: H_2O and the collision-induced absorption by H_2. Methane absorption is more than two orders of magnitude weaker and unlikely to become significant even when high-temperature CH_4 opacities become available. The importance of H_2O opacity in this region is confirmed by the remarkable correspondence of the features of the observed spectrum and of the opacity of water (Geballe et al. 1996). The detailed features of the synthetic spectrum are therefore far more reliable in the 1.9–2.1-μm region than in any other part of the spectrum. Figure 2 shows how the gravity affects the modeled features of water in this narrow spectral range. The models shown all have the bolo-

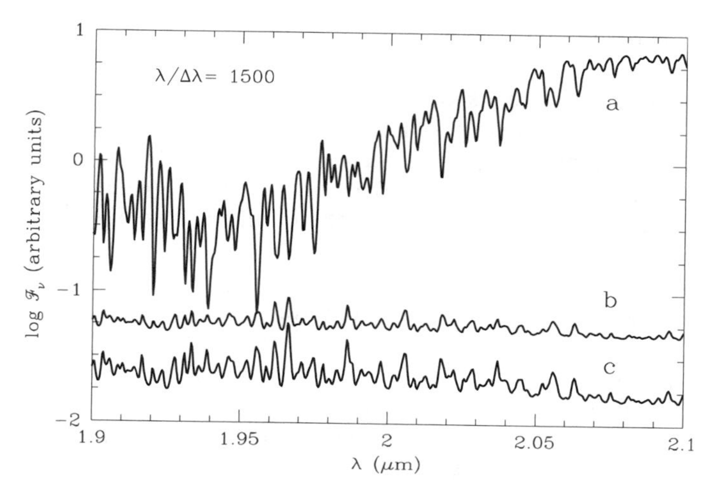

Figure 2. Gravity dependence of the H_2O features in the 1.9–2.1-micron region. The three synthetic spectra correspond to models that all have the bolometric luminosity of Gl 229 B. Curve "a" shows the $T_{\rm eff} = 870$ K, $\log g = 4.5$ spectrum, used as a reference spectrum in this figure. The two lower curves are ratios of synthetic spectra to that of curve "a": Curve "b" shows the $T_{\rm eff} = 940$ K, $\log g = 5$ spectrum divided by the reference spectrum "a," and the ratio of the $T_{\rm eff} = 1030$ K, $\log g = 5.5$ spectrum to the reference spectrum is shown by "c." Note that the figure shows the logarithm of the flux. Departures from a straight line in curves "b" and "c" reveal the features that are the most sensitive to a change in gravity. All spectra have been shifted vertically for clarity. The synthetic spectra are shown at a spectral resolution of 1500.

metric luminosity of Gl 229 B given by Matthews et al. (1996). The top curve shows the $T_{\rm eff} = 870$ K, $\log g = 4.5$ model. To emphasize the effects of changing the gravity, the flux *ratios* between this model and the $T_{\rm eff} = 940$ K, $\log g = 5$ model (middle curve) and between the model and the $T_{\rm eff} = 1030$ K, $\log g = 5.5$ model (lower curve) are shown. If there were no gravity sensitivity, the two lower curves would be flat. This region contains thousands of H_2O features, and their number and depth increases with the spectral resolution. Correspondingly, the gravity sensitivity washes out at the lower resolutions presently attained ($R \gtrsim 700$; Geballe et al. 1996). High-signal-to-noise spectroscopy at a resolution of $R \gtrsim 1500$ should show this effect very well and has the potential of reducing the uncertainty in the surface gravity of Gl 229 B to ± 0.25 decades.

B. Metallicity

The metallicity of late M dwarfs is notoriously difficult to determine, and modern efforts are still limited to a small number of stars (see, for example, Leggett et al. 1996; Schweitzer et al. 1996; Viti et al. 1997). The metallicity of the primary star, Gl 229 A, appears to be approximately solar but is rather uncertain. Published values give [Fe/H] = −0.2 (Schiavon, et al. 1997) and [M/H] = 0.2 (Mould 1978). The case of Gl 229 B is even more problematic. The relative success of the current models indicates that Gl 229 B is also approximately solar in composition. However, if the brown dwarf formed from a dissipative accretion disk, it may be analogous to the giant planets of the solar system and be enriched in heavy elements. On the other hand, phase separations in the interior may deplete the atmosphere of its heaviest elements.

Grain formation complicates the definition of the metallicity of an atmosphere. Elements are selectively removed from the gas phase and introduced in condensed phases whose composition is far more difficult to establish by spectroscopy than that of the gas. Since the gas-phase abundance of refractory elements is quite sensitive to the physics of condensation, it is best to focus on the abundant metals C, N, and O, which are not significantly depleted by condensed phases in the atmosphere of Gl 229 B.

The limitations of the CH_4 opacity database currently prevents a reliable determination of [C/H] in Gl 229 B. While the opacity of H_2O, the main oxygen-bearing molecule, is now well understood, the metallicity dependence of the synthetic spectra is muddled by the presence of dust affecting the near infrared. The need to untangle the veiling due to dust and the effects of metallicity on the H_2O absorption bands make it very difficult to determine the [O/H] ratio in Gl 229 B, although Griffith et al. (1998) find that a best fit to the optical spectrum of Gl 229 B is achieved with a subsolar oxygen abundance ([O/H] = −0.2).

Our model spectra predict a strong feature of NH_3 near 10.5 μm, which has not yet been sought by observers (Marley et al. 1996; Saumon et al. 1998). The identification of NH_3 in the spectrum of Gl 229 B would represent the first detection of this molecule in a compact object outside of the solar system. It also offers a good possibility of measuring the [N/H] ratio because dust opacity is negligible in Gl 229 B at this long wavelength.

The model atmospheres of Burrows et al. (1997) all assumed a solar abundance of the elements. In fact, the metallicity of the Sun is somewhat higher than that of the average star in the solar neighborhood, and it is appropriate to consider a greater variety of atmospheric metallicities. We have begun this process by computing atmosphere models for brown dwarfs in the mass and temperature range of Gliese 229 B. These exploratory models were constructed by artificially varying the mixing ratio of all molecules uniformly away from that predicted for a solar mixture of the elements in thermochemical equilibrium. In actuality, the relative mixing ratios of the molecules will not change uniformly as the overall metallicity changes. However, such departures will be slight compared to the large range in overall metallicity that we have considered.

Low-resolution spectra for three models, each with $T_{\rm eff}$ = 1000 K and g = 1000 m sec^{-2}, but with varying metallicities, are presented in Fig. 3.

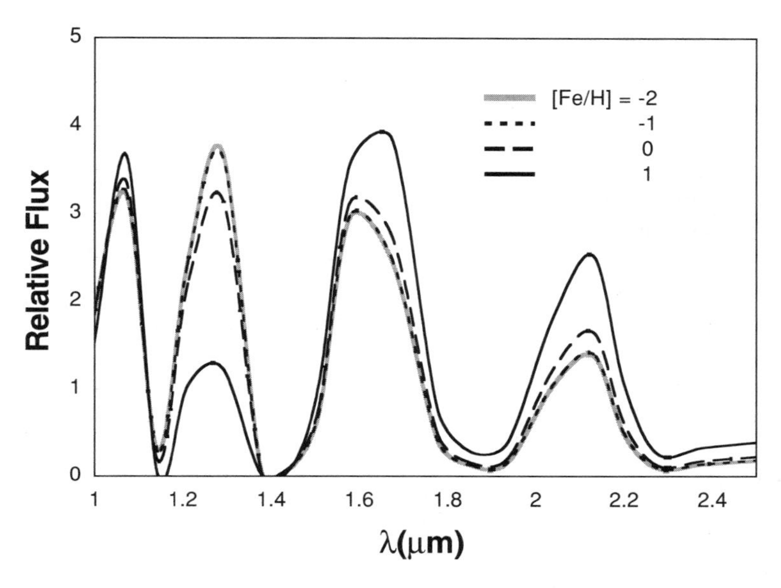

Figure 3. Model thermal emission spectra for model atmospheres ($g = 1000$ m sec^{-2}, $T_{\text{eff}} = 1000$ K) of varying metallicity. These low-resolution spectra demonstrate that broadband near-IR flux measurements will not be able to distinguish differences in metallicity below about about 1/10 solar. For larger metallicities, the J band is most sensitive to increasing metallicity.

These models were computed following the procedures in Burrows et al. (1997), although the treatment of Rayleigh scattering has been improved. In the figure, the metallicities are varied over the exceptionally large range of [Fe/H] $= -2$ to 1 to demonstrate the overall trends. Generally, as the metallicity decreases, the temperature profile adjusts. More flux emerges in the 3–5-μm band as the continuum molecular opacity falls. Surprisingly, this redistribution of flux results in a decrease in the flux emerging from the depths of the near infrared water bands. Instead, the lower molecular opacity in the window regions allows more flux to emerge from deeper, hotter layers of the atmosphere, resulting in a larger flux in the windows, a cooler upper atmosphere, and less flux in the depths of the near-infrared water and methane bands. Beyond about 2 μm, the overall flux rises as the metallicity increases. The larger metallicity closes off the near-infrared windows, raising the upper atmosphere temperature and increasing the continuum flux.

The changes in metallicity considered above produce relatively few changes in the color differences of Gl 229 B-like models, particularly for solar and subsolar abundances. Significant changes are found only for increases in metallicity above solar. $J - H$ and $J - K$ are most sensitive to such metallicity variations, because both colors become redder as the metallicity increases above solar. $H - K$ and $K - L'$ are relatively insensitive.

C. Convection

The molecules found in the atmosphere of a brown dwarf or EGP constrain atmospheric structure, dynamics, and chemistry. By identifying the atmospheric composition, spectroscopy provides information on the physical processes that govern the atmosphere.

Departures of atmospheric composition from equilibrium are especially interesting. CO, PH_3, GeH_4, and AsH_3 have all been detected in Jupiter's atmosphere at abundances many orders of magnitude higher than expected from equilibrium chemistry (see review by Fegley and Lodders 1994). The presence of these nonequilibrium molecules is taken to be evidence of convection. Since convective timescales are shorter than chemical equilibrium timescales, these molecules can be dredged up from deeper in Jupiter's interior and transported to the visible atmosphere.

In Gl 229 B, the detection of CO in abundances in excess of that predicted for chemical equilibrium (Noll et al. 1997) implies that the visible atmosphere (near 800 to 1400 K) must also be convective. Yet many atmosphere models find that the radiative-convective boundary lies far deeper, below 1700 K. However, the models of Marley et al. (1996) and Burrows et al. (1997) predict an additional, detached, upper convection zone. Such a zone would transport CO to the visible atmosphere from depths where it is more abundant. Other molecules may also trace convection, including PH_3. The chemical equilibrium profile of Cs is very similar to that of CO (Burrows and Sharp 1998). Thus, the same convection that dredges CO must also be dredging Cs. However, the lack of TiO and other refractory diatomics in the spectrum suggests that the atmosphere is not fully convective to the depth (below 2000 K) where these molecules condense. Taken together, these results may support the presence of a detached convection zone. Thus, CO and Cs may be tracing the vertical convective structure of the brown dwarf. A similar radiative zone, lying below Jupiter's visible turbulent atmosphere, has been predicted by Guillot et al. (1994). A confirmation of such a zone at Gl 229 B would strengthen the argument for such a zone in Jupiter. PH_3 is also potentially detectable in Gl 229 B by space-based platforms and will also act as a tracer of convection (Noll et al. 1997). Measurements of the abundances of this suite of molecules in a variety of objects may map out the atmospheric dynamics of substellar objects.

Photochemistry driven by incident radiation can also produce important nonequilibrium species. Thus, many hydrocarbons are found in the atmospheres of the solar jovian planets, including C_2H_2 and C_2H_6, that would not otherwise be expected. A rich variety of photochemical products will likely be found in the atmospheres of the extrasolar planets, particularly those with warm atmospheres and large incident fluxes. Hazes produced by the condensation of some species can produce signatures in the spectra of these objects far in excess of what might be expected given their small mixing ratios.

VI. ALBEDOS AND THE REFLECTIVITY OF EGPs

Both scattered light from the primary star and thermal emission by a planet contribute to a planet's spectrum. These two components can be crudely modeled as the sum of the reflection of a high-temperature Planck function, characterizing light from the primary, plus a second, lower-temperature Planck function, representing the thermal emission from the planet itself (Saumon et al. 1996). For the planets of our solar system, the two Planck functions are well separated in wavelength. This can be seen from the Wien displacement law, $\lambda T = 0.29$ cm K. For a 6000-K primary and a planet radiating at 200 K, the Planck functions of the primary and planet peak at 0.48 μm and 14.5 μm, respectively. This separation in the bulk of the radiation field from the planets and the Sun has led to a specialized nomenclature in which "solar" and "planetary" radiation are often treated separately.

However, for an arbitrary planet orbiting at an arbitrary distance from its primary, there can be substantial overlap of the two Planck functions. A general theory of extrasolar atmospheres must consistently compute the absorbed and scattered incident light. Marley (1998) has generated exploratory EGP atmosphere models that include deposition of incident radiation. He finds that in typical EGP atmospheres absorption in the strong near-IR water and methane bands produces temperature inversions above the tropopause, similar to Jupiter's stratosphere. Generation of a comprehensive suite of model reflected and emitted spectra will require that a large range of primary stellar types, orbital distances, and planetary masses and ages be investigated.

Planets are brightest near the peak in the solar Planck function, and the reflected flux falls off at shorter and longer wavelengths. To remove the effects of the solar spectrum and understand more clearly the processes acting in the planet's atmosphere, the reflected spectra of planets are commonly presented as geometric albedo spectra. The geometric albedo is essentially the planetary spectrum divided by the solar spectrum. Formally, it is the ratio of the flux received at Earth at opposition to the flux that would be received by a Lambert disk of the same size as the planet at its distance from the sun.

Reflected planetary spectra can also be approximated by computing wavelength-dependent geometric albedos of EGP atmospheres. The resultant planetary spectrum is then the sum of the emitted flux plus the product of the incident radiation times the geometric albedo with a phase correction. Figure 4 compares the geometric albedo spectra of Jupiter and Uranus. A purely Rayleigh scattering planet would have a geometric albedo of 0.75 at all wavelengths.

As Fig. 4 demonstrates, planets are not gray reflectors. They reflect best near 0.5 μm, where Rayleigh scattering dominates the reflected flux. At shorter wavelengths, Raman scattering, which shifts some UV photons to longer wavelengths, and absorption by the ubiquitous high-altitude

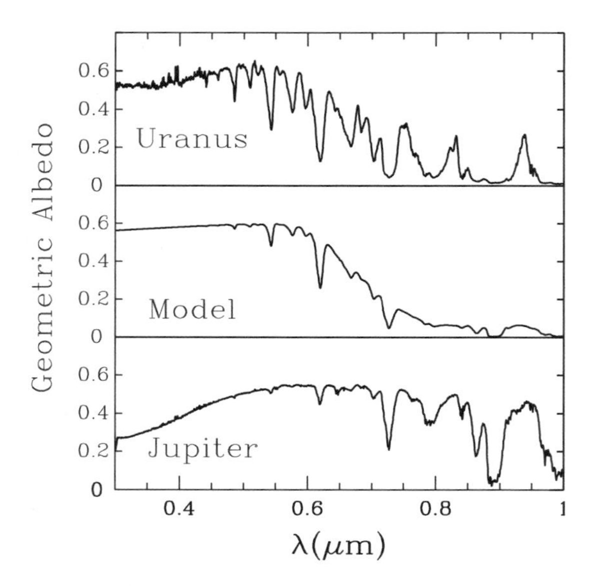

Figure 4. Geometric albedo spectra for Uranus, Jupiter, and a model 7-Jupiter-mass planet with $T_{\rm eff}$ = 400 K. The model includes no clouds and is thus darker than either planet, longward of about 0.6 μm. Absorption by methane and water removes incident photons before they can be Rayleigh scattered in the model. In contrast, Jupiter's and Uranus's cloud decks reflect brightly in between strong methane absorption bands. Jupiter is much darker than either the model or Uranus at blue and UV wavelengths. Dark photochemical hazes are predominantly responsible for lowering the reflectivity of the planet in this spectral range.

haze found throughout the outer solar system, lower the geometric albedo. Longward of 0.6 μm the strength of molecular rotational-vibrational bands increases, and molecular absorption, rather than scattering, begins to dominate the spectrum. Methane and the pressure-induced bands of hydrogen are the most important absorbers. In between the molecular bands, solar photons reach bright cloud decks and are scattered, so the planets remain bright in some bandpasses.

Marley et al. (1998) computed geometric albedo spectra for a large variety of EGPs, ranging from planets of less than 1 Jupiter mass to the most massive brown dwarfs. They considered objects with effective temperatures between 100 and 1200 K and found that the UV and optical spectra of extrasolar giant planets are generally similar to those of the solar giant planets. At longer wavelengths, however, the reflected flux depends critically on the presence or absence of atmospheric condensates. When condensates are present, photons have the opportunity to scatter before they are absorbed.

The most important condensate in EGP atmospheres is water. Water clouds appear as EGPs cool through effective temperatures of about 400 K.

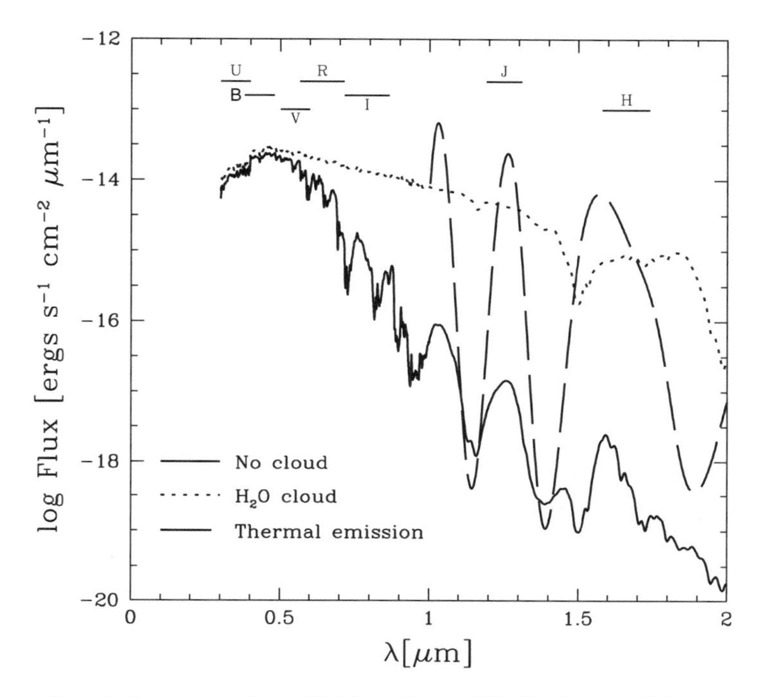

Figure 5. Model spectra for a 12-M_{Jup}, T_{eff} = 300 K planet orbiting at 1 AU from a G2 V star. The solid line shows reflected flux received at 10 pc if the planet's atmosphere is cloud-free. The dotted line demonstrates enhancement in flux when water clouds are added to the model. Both models are from Marley et al. (1998). The long dashed line gives thermal emission in the case of no clouds (Burrows et al. 1997). For this object, reflected near-IR flux begins to dominate thermal emission when clouds form as the object cools through about 400 K.

The sudden appearance of water clouds brightens the planets in reflected red and IR light, as shown in Figure 5. This figure presents a computed spectrum for an extrasolar planet orbiting at 1 AU from a G2 V primary. Also shown is the emitted flux for an object with T_{eff} = 400 K. It is apparent from this plot that the thermal emission dominates in the infrared for cloud-free objects. However, once clouds form, they will both attenuate the emitted flux and reflect a far larger proportion of the incident radiation. Of course, there will be a continuum, such that for objects with the same effective temperature but that are closer to their primary, the reflected flux will surpass the emitted flux. The emitted flux will continue to be important at lower temperatures for objects further from their primaries. Again, individual models are required for each specific case.

The ratio of the total reflected light to the total light incident on a planetary atmosphere (Bond albedo) depends sensitively on the spectral distribution of the incident radiation. For example, most of the flux of an A star emerges in the UV and blue. When incident on a giant planet, most of this light may be Rayleigh scattered, resulting in a large Bond albedo. In contrast, the predominantly red and infrared photons from an M star are far more likely to be absorbed. Thus, the Bond albedo of the same planet, when illuminated by two different stars, could vary by up to an order of magnitude as the stellar type of the primary varies (Marley et al. 1998). Note that nonequilibrium photochemical hazes can darken the planets in the UV, further complicating the reflected spectra and energy budgets.

VII. THE RELATIONSHIP BETWEEN GIANT PLANETS OF THE SOLAR SYSTEM AND EGPs

The advent of the science of EGPs and brown dwarfs, with its many new discoveries outside the solar system, should not cause us to lose sight of the central role of our local giant planets as exemplars that can be studied in unrivaled detail. Many questions still remain, concerning Jupiter's and Saturn's formation and cooling histories, that we must answer before we can confidently tackle their analogs beyond the solar system. One issue that still surrounds the study of our gas giants is the character, radial distribution, and origin of their compositions. Do they have a common formation mechanism with the newly discovered substellar objects, or is there a distinct mass-related boundary between the formation of such giant planets ($\sim$0.001 $M_\odot$) and brown dwarfs ($\sim$0.01 $M_\odot$) (see the chapter by Marcy et al., this volume) that is reflected in their compositions? Jupiter and Saturn provide unique laboratories by which we can address these and related questions of fundamental interest to planetary science and astronomy.

A. Nonsolar Metallicities and Implications for Formation

The solar system exhibits a progressive change in the bulk composition of its four giant planets as a function of heliocentric distance. According to interior models constrained by improvements in the hydrogen-helium equation of state and by the *Galileo* entry probe results for the abundance of helium in the jovian atmosphere, Jupiter is close to bulk solar composition, with a probable enhancement of C, N, and O by about a factor of $\sim$5, distributed uniformly in its envelope, and with limits on a dense central core amounting to $\geq$0.02 by mass fraction, or about 6 Earth masses maximum (Guillot et al. 1997). Models of Saturn, on the other hand, are distinctly nonsolar, with envelope enhancements of C, N, and O by about twice the jovian factor and a similar dense core of $\sim$6 to 8 Earth masses (Guillot et al. 1994). Uranus and Neptune models seem to resemble a larger version of the Saturn or Jupiter core with a thin ($\sim$0.1 by mass) envelope of hydrogen and helium (Hubbard et al. 1995).

The traditional interpretation of this sequence, which is now open to revision in the light of recent detections of extrasolar giant planets and recent downward revisions of the Jupiter and Saturn core masses, is that the giant planets Jupiter and Saturn formed by the capture of nebular hydrogen and helium onto dense Uranus- or Neptune-like cores (Mizuno et al. 1978), with subsequent significant accretion of icy cometesimals being responsible for enrichment of C-, N-, and O-bearing molecules in their envelopes. Presumably, the capture of nebular hydrogen onto proto-Uranus and proto-Neptune was less efficient owing to lower nebular density at their orbital radii and slower accretion of their dense icy cores.

The detection of EGPs at very small orbital radii has strongly suggested the possibility of significant radial migration of giant planets during the nebular phase (Trilling et al. 1998). Recent observations (Terebey et al. 1998) may support the possibility of direct formation of giant planets without the necessity of an initial dense core to trigger hydrodynamic accretion of hydrogen-helium gas.

B. Constraints on Cooling Mechanisms

The luminosity of a giant planet is determined by the heat radiated into space by means of depletion of its interior entropy as determined by the atmospheric boundary condition. The latter is strongly affected by metallicity. Moreover, the heat evolved can be strongly increased by a redistribution of hydrogen (with high specific entropy) toward shallower, lower-temperature regions of the planet, while denser components (with lower specific entropy) sink to deeper, hotter regions. Thus, atmospheric depletion of metals and helium may be accompanied by higher luminosity. It is this effect that is believed to be responsible for the anomalously high luminosity of Saturn (Stevenson and Salpeter 1977).

In principle, the detection of EGPs, whose values of luminosity, age, mass, and atmospheric composition can be determined, will provide a test of this concept along with constraints on the hydrogen phase diagram. The latter predicts that helium separation may be important for objects with $M < 0.001\ M_\odot$, whereas possible segregation of other species via a first-order phase transition in hydrogen may occur in objects with masses $M < 0.01\ M_\odot$.

VIII. CONCLUSION

With the recent discoveries of numerous L dwarfs in the solar neighborhood, EGPs around nearby stars, many brown dwarfs in young star clusters, and Gl 229 B, the study of substellar-mass objects is in full bloom. Searches for extrasolar planets and brown dwarfs from the ground and from space are entering a new phase as a large fraction of the astronomical community redirects its efforts to study this fascinating class, so long the subject of uninformed speculation and skepticism. The calculations and studies we have presented here are in aid of these searches. They are meant to illuminate some fruitful research paths and the theoretical character of the objects that are the missing link between the solar planets and the stars. Given the exotic nature of substellar atmospheres, the future cannot but hold many surprises.

Acknowledgments We thank David Sudarsky, Christopher Sharp, Richard Freedman, Shri Kulkarni, Jim Liebert, Davy Kirkpatrick, France Allard, Gilles Chabrier, Ben Oppenheimer, Chris Gelino, and Tristan Guillot for a variety of useful contributions. This work was supported under NASA grants NAG5-7499, NAG5-7073, NAG5-4988, and NAG5-2817 and under NSF grant AST93-18970.

REFERENCES

Allard, F., Hauschildt, P. H., Baraffe, I., and Chabrier, G. 1996. Synthetic spectra and mass determination of the brown dwarf Gl 229 B. *Astrophys. J. Lett.* 465:L123–L127.

Allard, F., Hauschildt, P. H., Alexander, D. R., and Starrfield, S. 1997. Model atmospheres of very low mass stars and brown dwarfs. *Ann. Rev. Astron. Astrophys.* 35:137–177.

Baines, K. H., Hammel, H. B., Rages, K. A., Romani, P. N., and Samuelson, R. E. 1995. Clouds and hazes in the atmosphere of Neptune. In *Neptune and Triton*, ed. D. P. Cruikshank (Tucson: University of Arizona Press), pp. 489–546.

Boss, A. 1995. Proximity of Jupiter-like planets to low-mass stars. *Science* 267:360–362.

Burrows, A., and Sharp, C. M. 1999. Chemical equilibrium abundances in brown dwarf and extrasolar giant planet atmospheres. *Astrophys. J.* 512:843–863.

Burrows, A., Hubbard, W. B., and Lunine, J. I. 1989. Theoretical models of low mass stars and brown dwarfs. *Astrophys. J.* 345:939–958.

Burrows, A., Hubbard, W. B., Saumon, D., and Lunine, J. I. 1993. An expanded set of brown dwarf and very low mass star models. *Astrophys. J.* 406:158–171.

Burrows, A., Saumon, D., Guillot, T., Hubbard, W. B., and Lunine, J. I. 1995. Prospects for detection of extrasolar giant planets by next generation telescopes. *Nature* 375:299–301.

Burrows, A., Marley, M. S., Hubbard, W. B., Lunine, J. I., Guillot, T., Saumon, D., Freedman, R. S., Sudarsky, D., and Sharp, C. 1997. A nongray theory of extrasolar giant planets and brown dwarfs. *Astrophys. J.* 491:856–875.

Butler, R. P., and Marcy, G. W. 1996. A planet orbiting 47 Ursae Majoris. *Astrophys. J. Lett.* 464:L153–L156.

Butler, R. P., Marcy, G. W., Williams, E., Hauser, H., and Shirts, P. 1997. Three new "51 Pegasi-type" planets. *Astrophys. J. Lett.* 474:L115–L118.

Carlson, R. W., Baines, K. H., Orton, G. S., Encrenaz, T., Drossart, P., Roos-Serote, M., Taylor, F. W., Irwin, P., Wier, A., Smith, S., and Calcutt, S. 1995. Near-infrared spectroscopy of the atmosphere of Jupiter. *EOS* 78 (suppl.):F413.

Chabrier, G. 1999. Is the galactic dark matter white? *Astrophys. J. Lett.* 513: L103–L106.

Cochran, W. D., Hatzes, A. P., Butler, R. P., and Marcy, G. 1997. The discovery of a planetary companion to 16 Cygni B. *Astrophys. J.* 483:457–463.

Emanuel, K. A. 1994. *Atmospheric Convection* (New York: Oxford University Press).

Fegley, B., Jr., and Lodders, K. 1994. Chemical models of the deep atmospheres of Jupiter and Saturn. *Icarus* 110:117–154.

Fegley, B., Jr., and Lodders, K. 1996. Atmospheric chemistry of the brown dwarf Gliese 229B: Thermochemical equilibrium predictions. *Astrophys. J. Lett.* 472:L37–L39.

Geballe, T. R., Kulkarni, S. R., Woodward, C. E., and Sloan, G. C. 1996. The near-infrared spectrum of the brown dwarf Gliese 229B. *Astrophys. J. Lett.* 467:L101–L104.

Goody, R. M., and Yung, Y. L. 1989. *Atmospheric Radiation: Theoretical Basis* (New York: Oxford University Press).

Griffith, C. A., Marley, M., and Yelle, R. 1998. The dusty atmosphere of the brown dwarf Gliese 229B. *Science* 282:2063–2067.

Guillot, T., Chabrier, G., Morel, P., and Gautier, D. 1994. Nonadiabatic models of Jupiter and Saturn. *Icarus* 112:354–367.

Guillot, T., Burrows, A., Hubbard, W. B., Lunine, J. I., and Saumon, D. 1996. Giant planets at small orbital distances. *Astrophys. J. Lett.* 459:L35–L38.

Guillot, T., Gautier, D., and Hubbard, W. B. 1997. Note: New constraints on the composition of Jupiter from Galileo measurements and interior models. *Icarus* 130:534–539.

Hubbard, W. B., Podolak, M., and Stevenson, D. J. 1995. The interior of Neptune. In *Neptune and Triton*, ed. D. P. Cruikshank (Tucson: University of Arizona Press), pp. 109–138.

Jones, H. R. A., and Tsuji, T. 1997. Spectral evidence for dust in late-type M dwarfs. *Astrophys. J. Lett.* 480:L39–L41.

Karkoschka, E. 1994. Spectrophotometry of the jovian planets and Titan at 300 to 1000 nm wavelength: The methane spectrum. *Icarus* 111:174–192.

Kirkpatrick, J. D., Reid, I. N., Liebert, J., Cutri, R. M., Nelson, B., Beichman, C. A., Dahn, C. C., Monet, D. G., Skrutskie, M. F., and Gizis, J. 1999. Dwarfs cooler than "M": The definition of spectral type "L" using discoveries from the 2-micron All-Sky Survey (2 Mass), *Astrophys. J.* 519:802–833.

Latham, D. W., Mazeh, T., Stefanik, R. P., Mayor, M., and Burki, G. 1989. The unseen companion of HD114762—a probable brown dwarf. *Nature* 339:38–40.

Leggett, S. K., Allard, F., Berriman, G., Dahn, C. C., and Hauschildt, P. H. 1996. Infrared spectra of low-mass stars: Towards a temperature scale for red dwarfs. *Astrophys. J. Suppl.* 104:117–143.

Lewis, J. S. 1969. The clouds of Jupiter and the NH_3-H_2O and NH_3-H_2S systems. *Icarus* 10:365–378.

Lunine, J. I., Hubbard, W. B., and Marley, M. 1986. Evolution and infrared spectra of brown dwarfs. *Astrophys. J.* 310:238–260.

Lunine, J. I., Hubbard, W. B., Burrows, A. S., Wang, Y.-P., and Garlow, K. 1989. The effect of gas and grain opacity on the cooling of brown dwarfs. *Astrophys. J.* 338:314-337.

Marcy, G. W., and Butler, R. P. 1996. A planetary companion to 70 Virginis. *Astrophys. J. Lett.* 464:L147–L151.

Marley, M. 1998. Atmospheres of giant planets from Neptune to Gliese 229B. In *Brown Dwarfs and Extrasolar Planets*, ASP Conf. Ser. 134, ed. R. Rebolo, E. L. Martín, and M. R. Zapatero-Osorio (San Francisco: Astronomical Society of the Pacific), pp. 383–393.

Marley, M. S., Saumon, D., Guillot, T., Freedman, R. S., Hubbard, W. B., Burrows, A., and Lunine, J. I. 1996. Atmospheric, evolutionary, and spectral models of the brown dwarf Gliese 229B. *Science* 272:1919–1921.

Marley, M. S., Gelino, C., Stephens, D., Lunine, J. I., and Freedman, R. 1998. Reflected spectra and albedos of extrasolar giant planets. I: Clear and cloudy atmospheres. *Astrophys. J.*, in press.

Matthews, K., Nakajima, T., Kulkarni, S. R., and Oppenheimer, B. R. 1996. Spectral energy distribution and bolometric luminosity of the cool brown dwarf Gliese 229B. *Astron. J.* 112:1678–1682.

Mayor, M., and Queloz, D. 1995. A Jupiter-mass companion to a solar-type star. *Nature* 378:355–357.

Mizuno, H., Nakazawa, K., and Hayashi, C. 1978. Instability of a gaseous envelope surrounding a planetary core and formation of giant planets. *Prog. Theor. Phys* 60:699-710.
Mould, J. R. 1978. Infrared spectroscopy of M dwarfs. *Astrophys. J.* 226:923–930.
Nakajima, T., Oppenheimer, B. R., Kulkarni, S. R., Golimowski, D. A., Matthews, K., and Durrance, S. T. 1995. Discovery of a cool brown dwarf. *Nature* 378:463–465.
Niemann, H. B., Atreya, S. K., Carignan, G. R., Donahue, T. M., Haberman, J., Harpold, D., Hartle, R. E., Hunten, D. M., Kasprzak, W., Mahaffy, P., Owen, T. C., and Way, S. 1998. The composition of the jovian atmosphere as determined by the Galileo Probe mass spectrometer. *J. Geophys. Res.* 103:22831–22845.
Noll, K. S., Geballe, T. R., and Marley, M. S. 1997. Detection of abundant carbon monoxide in the brown dwarf Gl 229 B. *Astrophys. J. Lett.* 489:L87–L90.
Noyes, R. W., et al. 1997. A planet orbiting the star Rho Coronae Borealis. *Astrophys. J. Lett.* 483:L111–L114.
Oppenheimer, B. R., Kulkarni, S. R., Matthews, K., and Nakajima, T. 1995. Infrared spectrum of the cool brown dwarf Gl 229 B. *Science* 270:1478–1479.
Oppenheimer, B. R., Kulkarni, S. R., Matthews, K., and van Kerkwijk, M. H. 1998. The spectrum of the brown dwarf Gliese 229 B. *Astrophys. J.* 502:932–943.
Orton, G., Ortiz, J. L., Baines, K., Bjoraker, G., Carsenty, U., Colas, F., Dayal, A., Deming, D., Drossart, P., Frappa, E., Froedson, J., Goluen, J., Golisch, W., Griep, D., Hernandez, C., Hoffman, W., Jennings, D., Kaminski, C., Kohn, J., Laques, P., Limaye, S., Lin, H., Lecacheux, J., Martin, T., McCabe, G., Momary, T., Parker, D., Peutter, R., Ressler, M., Reyes, G., Sada, P., Spencer, J., Spitale, J., Stewart, S., Varsik, J., Warell, J., Wild, W., Yanamandra-Fisher, P., Fazio, G., Hora, J., and Deutch, L. 1996. Earth-based observations of the Galileo Probe entry site. *Science* 272:839–840.
Ragent, B., Colburn, D. S., Avrin, P., and Rages, K. A. 1996. Results of the Galileo Probe nephelometer experiment. *Science* 272:854–856.
Rossow, W. B. 1978. Cloud microphysics: Analysis of the clouds of Earth, Venus, Mars and Jupiter. *Icarus* 36:1–50.
Rothman, L. S., Gamache, R., Tipping, R. H., Rinsland, C. P., Smith, M. A. H., Benner, P. C., Devi, U. M., Flaud, J.-M., Camy-Peyret, C., and Perrin, A. 1992. The HITRAN molecular database—editions of 1991 and 1992. *J. Quant. Spectrosc. Rad. Transf.* 48:469–507.
Saumon, D., Hubbard, W. B., Burrows, A., Guillot, T., Lunine, J. I., and Chabrier, G. 1996. A theory of extrasolar giant planets. *Astrophys. J.* 460:993–1018.
Saumon, D., Marley, M. S., Guillot, T., and Freedman, R. S. 1998. Spectral diagnostics for the brown dwarf Gliese 229 B. *Astrophys. J.*, submitted.
Schiavon, R. P., Barbuy, B., and Singh, P. D. 1997. The FeH Wing-Ford band in spectra of M stars. *Astrophys. J.* 484:499–510.
Schultz, A. B., Allard, F., Clampin, M., McGrath, M., Bruhweiler, F. C., Valenti, J. A., Plait, P., Hulbert, S., Baum, S., Woodgate, B. E., Bowers, C. W., Kimble, R. A., Maran, S. P., Moos, W. H., and Roesler, F. 1998. First results from the space telescope imaging spectrograph: Optical spectra of Gl 229B. *Astrophys. J. Lett.* 492:L181–L184.
Schweitzer, A., Hauschildt, P. H., Allard, F., and Basri, G. 1996. Analysis of Keck high resolution spectra of VB10. *Mon. Not. Roy. Astron. Soc.* 283:821–829.
Stevenson, D. J., and Salpeter, E. E. 1977. The dynamics and helium distribution in hydrogen-helium fluid planets. *Astrophys. J. Suppl.* 35:239–261.
Strong, K., Taylor, F. W., Calcutt, S. B., Remedios, J. J., and Ballard, J. 1993. Spectral parameters of self and hydrogen broadened methane from 2000 to 9500 cm^{-1} for remote sounding of the atmosphere of Jupiter. *J. Quant. Spectrosc. Rad. Transf.* 50:363–429.

Terebey, S., Van Buren, D., Padgett, D. L., Hancock, T., and Brundage, M. 1998. A candidate protoplanet in the Taurus star-forming region. *Astrophys. J. Lett.* 507:L71–L74.
Trilling, D. E., Benz, W., Guillot, T., Lunine, J. I., Hubbard, W. B., and Burrows, A. 1998. Orbital evolution and migration of giant planets: Modeling extrasolar planets. *Astrophys. J.* 500:428–439.
Tsuji, T., Ohnaka, K., Aoki, W., and Nakajima, T. 1996. Evolution of dusty photospheres through red to brown dwarfs: How dust forms in very low mass objects. *Astron. Astrophys. Lett.* 308:L29–L32.
Viti, S., Jones, H. R. A., Schweitzer, A., Allard, F., Hauschildt, P. H., Tennyson, J., Miller, S., and Longmore, A. J. 1997. The effective temperature and metallicity of CM Draconis. *Mon. Not. Roy. Astron. Soc.* 291:780–796.
Weidenschilling, S. J., and Lewis, S. J. 1973. Atmospheric and cloud structures of the jovian planets. *Icarus* 20:465-476.
Yair, Y., Levin, Z., and Tzivion, S. 1995. Microphysical processes and dynamics of a jovian thundercloud. *Icarus* 114:278–299.

PART VIII
Initial Conditions for Astrobiology

PLANETARY HABITABILITY AND THE ORIGINS OF LIFE

CHRISTOPHER F. CHYBA
SETI Institute

DANIEL P. WHITMIRE
University of Louisiana at Lafayette

and

RAY REYNOLDS
NASA-Ames Research Center

A general definition of life that is useful for remote exploration has proven elusive. As a practical matter, the search for life becomes the search for life as we know it, based on organic molecules in liquid water and a source of free energy. Organics are common in much of the solar system, so liquid water has become the focus of exobiological searches. Recent years have seen the discovery of the deep subsurface biosphere on Earth, some components of which appear to be entirely independent of surface conditions. It remains an open question whether life may in fact have originated in the subsurface. That possibility, however, suggests that deep liquid water environments, such as we now suspect exist on Mars and Europa, are plausible locales for extraterrestrial life. The most conservative requirement to set for extraterrestrial habitable environments is to require liquid water not at depth but, as with Earth, at a world's surface. The "circumstellar habitable zone" is defined as the volume of space around a star or star system within which an Earth-like planet could support surface liquid water. Although these zones change with evolving stellar luminosity, there is a significant habitable zone lasting for gigayears for stars with masses between 0.1 and 1.5 times the mass of the Sun. Therefore, even a conservative definition of habitability suggests many locales for life around other stars. Possible subsurface environments for life only expand the possibilities for biology even further.

I. HABITABILITY AND LIFE

A. Habitable for Whom?

The notion of planetary "habitability" was originally used to refer to conditions suitable for human life (Dole 1964). The word has since come to imply less stringent conditions, roughly those necessary for stability of liquid water at a world's surface (see section III below for a precise formulation). Prior to the elucidation of Earth's subsurface biosphere (discussed in section II), this could be viewed as the key necessary condition for simple microscopic life. Clearly, whether we regard a world as habitable depends in part on who or what will be doing the inhabiting. If we were interested in conditions specifically suitable for, say, forests (Heath 1996), still different requirements might be applicable.

On Earth, life had existed for about 3 Gyr before metazoa (multicellular animals) appeared (Runnegar 1992). Animals, plants, and fungi are multicellular, eukaryotic organisms. Eukaryotes are cells containing well-organized nuclei and other characteristic internal structures. Most multicellular life on Earth is eukaryotic, though there are also many eukaryotes that are single-celled. In addition to the eukaryotes, there are two other known broad domains of terrestrial life, the bacteria and the archaea, which are prokaryotes: cells lacking nuclei. Most prokaryotes are single-celled, but multicellular prokaryotes also exist, such as the filamentous bacteria and fruiting myxobacteria; some of the latter can form differentiated structures of $\sim 10^9$ cells approaching a millimeter in height (Madigan et al. 1997). Some of the oldest known microfossils (dating from 3.4 to 3.5 Gyr ago) resemble filamentous bacteria (Schopf and Walter 1983). Prokaryotes exhibit a wide variety of metabolic styles, deriving their energy from diverse electron donors and acceptors (fuels and oxidants), but eukaryotes require free molecular oxygen as their terminal electron acceptor (Nealson 1997*a,b*). This gives eukaryotes a substantial advantage in energy produced per carbohydrate molecule metabolized over those many prokaryotic organisms that cannot use O_2 (Day 1984), but at the price of requiring an oxic environment. In this sense, Earth may not have been habitable for eukaryotes until the rise of free oxygen in the atmosphere, beginning about 2 Gyr ago (Des Marais 1994, 1996).

A terrestrial atmosphere containing abundant free oxygen seems to have required aeons of oxygen production by photosynthesizing bacteria, as well as the oxidation of reductant sinks, and ultimately the burial of large quantities of organic carbon (Des Marais 1994, 1996). One is free to imagine worlds in which these processes required far less time than they did on Earth (McKay 1996), with eukaryotic analogs therefore able to arise much more quickly. Even so, it seems likely that at a given moment many more worlds will be home to single-celled organisms than, say, to metazoa. The most broadly useful definition of "habitability" will therefore be one appropriate to the simplest organisms that we would consider to be alive. But how do we distinguish life from nonlife?

B. Definitions of Life

A satisfactory general definition of life has proven elusive (Chyba and McDonald 1995), and one philosopher of science has even argued that a complete definition is impossible (Küppers 1990). Surely one reason for this pessimism is the historical fact that most attempts at a definition have failed. Sagan (1970) reviewed physiological, metabolic, biochemical, genetic, and thermodynamic definitions of life, but, as he pointed out, each of these definitions faces problems, often in the form of counterexamples: either entities that fit the definition but that we would be reluctant to call alive, or organisms that are clearly alive but that the definition excludes. Nor is it sufficient to claim that a particular counterexample is "unimportant"; such a claim merely implicitly invokes other criteria in addition to the ostensible definition that has been proposed.

Consider, for example, a metabolic definition of life. A metabolic definition "describes a living system as an object with a definite boundary, continually exchanging some of its materials with its surroundings, but without altering its general properties, at least over some period of time" (Sagan 1970). But fire seems to fulfill these properties, and indeed, the chemical reaction by which flames maintain themselves, combining organics with molecular oxygen, is similar to the metabolic reaction that fuels eukaryotic organisms.

1. The Darwinian Definition. There is a working definition of life, which we call the "Darwinian definition," that is becoming accepted within the origins-of-life community (Chyba and McDonald 1995). A careful formulation (Joyce 1994) is: "Life is a self-sustained chemical system capable of undergoing Darwinian evolution." The heart of this definition is its insistence on the importance of Darwinian evolution: self-reproduction, genetic variation, and natural selection. Its strength is its distillation of the objectives of ongoing laboratory experiments that seek to devise, through directed molecular evolution, a system of RNA or other molecules that are capable of self-replication, and evolve according to natural selection (von Kiedrowski 1986; Joyce 1993; Breaker and Joyce 1994). The view that "the origin of life is the same as the origin of Darwinian evolution" is becoming commonplace. Once Darwinian evolution is established, other diverse attributes of life may arise through natural selection.

Whatever its utility for laboratory experiments, the Darwinian definition faces serious drawbacks in the context of a remote search for life (Fleischaker 1990; Chyba and McDonald 1995). How long would we wait for a system to demonstrate whether it is "capable" of undergoing Darwinian evolution? The Darwinian definition, while useful in a laboratory setting, must give way in spacecraft exploration to a much less precise, but operationally more useful definition.

2. Lessons from the Viking Biology Package. Controversies over the definition of life have implications for laboratory and spacecraft experiments. Consider the three experiments in the *Viking* biology package (Klein et al. 1976; Klein 1977, 1978; Horowitz 1986). The experiments established broad criteria for the detection of life, based on the assumption that martian life could be recognized through its metabolism. For example, the labeled release (LR) experiment looked for organisms that would ingest any of a brew of organics provided by the experiment and metabolize one or more of them into a carbon-containing gas (Levin and Straat 1981*b*).

The results of the biology package, and especially those of the LR experiment, gave tantalizing results that in some respects mimicked life. Indeed, the head of the *Viking* biology team has noted that "if information from other experiments on board the two *Viking* landers had not been available, this set of data would almost certainly have been interpreted as presumptive evidence for biology" (Klein 1978). But a biological interpretation was undercut by an instrument not formally part of the biology package. The *Viking* gas chromatograph/mass spectrometer (GCMS) searched for organic molecules but found none at the ppm (for one- and two-carbon compounds) to the ppb (for more than two carbons) level, not even those that had been expected from estimates of micrometeorite infall (Biemann et al. 1977). Levin and Straat (1981*a*) have argued that some Antarctic soil samples contain no organics detectable at the level of the *Viking* GCMS but nevertheless yield a biological response from the LR experiment, suggesting a low level of microbial life that might be analogous to the situation on Mars. Nevertheless, the absence of organics, and the results of the *Viking* biology experiments, are now widely viewed as due to the action of organic-destroying oxidants produced in the martian atmosphere and surface (Hunten 1979).

The use of the GCMS as a life detection experiment approached the search for life on Mars from a perspective different from that of the biology package experiments. Whereas the latter relied upon a metabolic definition of life, the GCMS allowed a search for life based on a biochemical definition. Viewed as a life detection experiment, the GCMS simply assumed that life elsewhere would, like life on Earth, be based on organic molecules: No organics, no life. [The converse is certainly not true: The mere presence of organics does not imply the presence of life. To the contrary, it is now clear that organics are common molecules throughout the solar system and indeed the interstellar medium (Cruikshank 1997; Pendleton and Chiar 1997).]

One lesson from the *Viking* experience is the value of searching for life from the perspectives of different definitions. A second lesson from the *Viking* missions is the importance of the chemical and geological context for the interpretation of biological experiments. A third is the value of designing life detection experiments to provide useful information even in the case of a negative result.

Extrapolating from the *Viking* experience, the biochemical definition of life seems likely to trump all others in a remote sensing context. In the absence of compelling organic biomarkers, other biologically suggestive experimental results are likely to be distrusted or dismissed. One possible exception to this general conclusion might come from a microscopic exploration (Lederberg 1960, 1965) that imaged entities doing something unambiguously alive, such as propelling themselves or reproducing. In any case, but especially in the absence of detailed biochemical information, the importance of excluding the possibility of forward contamination is abundantly clear.

II. LIFE AS WE KNOW IT

Various techniques for the remote detection of life have been discussed since the dawn of the space age (see, e.g., Lederberg 1960, 1965). Even at the close of the 20th century, however, it may still be wisest to emphasize the central requirements for life as we know it: a source of free energy, a source of carbon, and liquid water. There were abundant terrestrial and solar sources of energy available for the origin of life on Earth (Miller and Urey 1959; Chyba and Sagan 1992, 1997), and analogs of many of these could be available and significant on other worlds within or beyond our solar system. Carbon-based molecules are also common, though not ubiquitous, throughout the solar system. But the *sine qua non* of life as we know it, and the requirement that appears the hardest to find, is liquid water.

A. Why Water?

Even given an appropriate carbon source and copious sunlight or other energy, terrestrial life apparently absolutely requires liquid water (Mazur 1980; Kushner 1981; Horowitz 1986). This also seems to be the lesson of the dry valleys of Antarctica, which are among the harshest deserts on Earth and include areas where no microorganisms seem to exist outside of specific, protected habitats (McKay 1986; Campbell and Claridge 1987), although these claims largely rest on traditional microbial culture methods rather than on modern DNA amplification techniques (Vishniac 1993). From an operational point of view, then, the search for life beyond the Earth begins with the search for liquid water. Where there is liquid water, there is at least the possibility that life as we know it could exist. (Of course, life elsewhere might not prove to be at all "as we know it," but in that event it is hard to provide useful criteria to select among possible places to search.) As we will see in section III below, the requirement for liquid water is the key to the usual definition of habitability.

Why is life so dependent on liquid water? Cellular life requires an internal medium within which molecules may dissolve and chemical reactions may occur. Water has been called the "universal solvent," because of its ability to form hydrogen bonds with polar solutes (Blum 1962). A few other comparably good polar solvents exist, such as liquid ammonia (Barrow and Tipler 1986), and on much colder worlds, where water would be frozen but ammonia a liquid, ammonia might play the role for life that water plays on Earth. However, a biochemistry employing ammonia as a solvent would likely proceed far more slowly than its terrestrial counterpart, because of the typical Arrhenius exponential dependence of chemical reaction rates with temperature. One is also free to imagine an exotic biology (perhaps on Saturn's moon Titan?) whose solvent was liquid hydrocarbons. But hydrocarbons are nonpolar molecules, so any such biochemistry would be so different from that of terrestrial life that such speculations are currently scarcely constrained.

Water is also fundamental to terrestrial life because of its effect on the three-dimensional shapes of enzymes. Enzymes are proteins that catalyze chemical reactions, and like all proteins, they are made by linking together amino acids. Amino acids have side chains of atoms that may themselves be polar or nonpolar. In solution, different amino acids in the sequence of amino acids constituting the enzyme will be influenced by the surrounding water molecules according to the polarity of their side chains. Water is therefore crucial to enzymes' assumption of specific three- dimensional shapes, the shapes that confer to the enzymes their catalytic function.

Finally, liquid water has many important global effects. For example, water has an extremely high specific heat, so Earth's oceans provide an important moderating influence on the climate. Ice floats, so lakes and oceans freeze from the top down and are therefore more likely to thaw annually. [Solid ammonia, incidentally, does not float on its liquid phase (Wald 1964; Barrow and Tipler 1986).] These global effects no doubt seem pleasant to creatures such as ourselves that have evolved on a water-covered planet, but it is hard to see how any of them represents a requirement for the origin of life. This contrasts with the "microscopic" attributes of liquid water previously discussed, which do suggest that life based on liquid water may be the cosmochemically most likely version. Of course, this conclusion, based as it is on extrapolations from a single example, could ultimately prove to be mistaken (see, e.g., Feinberg and Shapiro 1980), but it gives us a place to start, and a conservative one: Were it possible to base life on liquid ammonia or hydrocarbons, for example, a liquid-water criterion for life would at worst have underestimated the scope for biology.

B. Liquid Water and Habitability in the Solar System

Earth provides the example of the kind of world on which traditional definitions of habitability (see section III below) rest: Liquid water on Earth is stable at the surface, and has probably been stable (with the possible exception of brief intermittent freezes; Hoffman et al. 1998) since early in Earth's history. Mars, on the other hand, presents a world that may once have allowed liquid water at its surface but has since apparently become a dry and frozen desert (Carr 1996; McKay and Stoker 1989; Goldspiel and Squyres 1991). If life exists on Mars today, it must have retreated to special surface or subsurface niches where liquid water remains possible (Boston et al. 1992).

Jupiter's moon Europa may well harbor a volume of liquid water equal to that in Earth's oceans, hidden beneath kilometers of surface ice (Carr et al. 1998). The geological evidence for a subsurface ocean on Europa is suggestive, but not decisive (Pappalardo et al. 1999). The hypothesis is, however, consistent with data from the *Galileo* spacecraft's magnetometer experiment (Khurana et al. 1998). Results from the latter experiment also imply a subsurface salty ocean within Jupiter's moon Callisto (Khurana et al. 1998), suggesting that subsurface liquid water may be a common feature of large icy moons. The planned 2003 Europa orbiter mission (Johnson et al. 1999) and possible subsequent lander missions (Chyba et al. 1999) should determine whether an ocean really exists on Europa, and begin the search for signs of life.

Could there be sites of hydrothermal activity on Saturn's moon Titan, a world the size of Mercury? Titan is rich in organic molecules, produced in its nitrogen-methane atmosphere by ultraviolet and charged-particle processing (Sagan et al. 1992). If hydrothermal sites were also present, Titan would become an extremely important candidate not only for prebiotic chemistry, but for extraterrestrial biology as well. The upcoming *Cassini/Huygens* mission should begin to tell us what lies beneath Titan's pervasive organic haze layer (Matson 1997; Lebreton 1997).

Other liquid water environments may exist, or may once have existed, in the solar system. In particular, large asteroids and comets may have harbored liquid water environments in their deep interiors for up to the first $\sim 10^8$ yr of solar system history (Zolensky and McSween 1988, Podolak and Prialnik 1997) . This timescale is of interest because it is comparable to that available for the origin of life on Earth (Sleep et al. 1989). However, it is unlikely that any putative asteroidal or cometary life that may have originated during this early liquid water phase could still be viable, because of the subsequent 4.5 Gyr of radioactive decay in the minerals composing the object and the resulting accumulated radiation damage in a frozen organism unable to repair itself (Clark et al. 1999).

C. Life at Depth

If life could exist only on worlds where liquid water were stable at the surface, then the only environment for life in our solar system would be Earth. However, the last two decades have revealed the existence of a terrestrial "deep, hot biosphere" (Gold 1992) in addition to the more evident one at Earth's surface. Estimates of the total mass of the subsurface biosphere (Whitman et al. 1998) suggest that it may be comparable to the surface biomass ($\sim 8 \times 10^{17}$ g, of which $\sim 10^{15}$ g resides in the oceans).

The mass of the surface biosphere is dominated by multicellular eukaryotic organisms (mostly trees), dependent on free oxygen (Hayes et al. 1983). As one moves down into the subsurface, the eukaryotes quickly drop away as the oxygen level declines, and one enters the realm of the prokaryotes (Nealson 1997*a*,*b*). The extensive deep biosphere implies that eukaryotes may not dominate the terrestrial biomass. Primary production of organic matter remains likely to be dominated by the surface biosphere, which has access to orders of magnitude more free energy (as sunlight) than does the subsurface biosphere.

The critical question for the exobiological implications of the deep biosphere is the extent to which subsurface ecosystems are independent of surface life and conditions. In 1995, Stevens and McKinley (1995) reported the discovery of methanogenic microbes in samples drawn from 1.5 km beneath the surface of the Columbia River basin in Washington State.

These organisms seemed to obtain their energy from the oxidation by carbon dioxide of hydrogen derived from the weathering of basalt and ground water. If correct, this would provide evidence for subsurface microbial life that could exist entirely independently of surface photosynthesis. No free oxygen, ultimately derived from photosynthesis, appeared to be needed. Neither were any previously synthesized organics (also likely to be ultimately dependent on surface photosynthesis) needed to serve as electron donors; hydrogen produced by the interaction of anoxic water and rock seemed to fulfill this role. Carbon dioxide served as the sole carbon source. The Sun could disappear, and with it Earth's surface biosphere, and at least these microbes would continue to thrive, provided only that the terrestrial geothermal gradient maintained a subsurface region of liquid water.

However, subsequent experimental work found no significant H_2 production from the interaction of water with basalts from Snake River, Idaho (Anderson et al. 1998), contradicting conclusions from the initial experiments by Stevens and McKinley (1995). If correct, this suggests that the Stevens and McKinley methanogens may in fact be fueled by some other source of H_2, which may or may not be independent of the surface biosphere. The extent to which the subsurface biosphere is independent of surface photosynthesis remains a critical question. To date, we have only begun to sample the deep biosphere (Nealson 1997*a*,*b*; Whitman et al. 1998), and probably have nothing approaching a comprehensive understanding of its organisms or its metabolic diversity.

If some terrestrial life exists or can exist independently of surface photosynthesis, then the possibilities for deep biospheres on worlds such as Mars greatly expand. If life originated on Mars during its earlier, apparently more clement period (McKay and Stoker 1989), it would be possible that its progeny remain in subsurface niches associated with ongoing hydrothermal activity.

D. Can Life Originate at Depth?

In the preceeding section we asked whether life, once having originated at the surface of a world, could expand into subsurface niches and evolve at least some ecosystems that became entirely independent of surface conditions. A more fundamental question is whether life can *originate* at depth, independently from the Sun. If not (if, for example, the origin of life were to require the abundant energy available from sunlight), then only worlds that have clement surfaces (such as Earth) or that once did (such as Mars) could host extant ecosystems. But if the origin of life could occur at depth, then any world with subsurface liquid water would become a candidate for Earthlike (at least prokaryotic) life. In particular, worlds such as Europa would become prime candidates for contemporary biospheres.

A small amount of sunlight, filtering through the ice at young cracks, may be available to drive biology at Europa (Reynolds et al. 1983, 1987; Oró et al. 1992). Because the energy flux from sunlight at these cracks could be several orders of magnitude greater than the average energy flux from Europa's core, even if life first originated at depth on Europa, there would have been a powerful energy incentive for it to evolve to areas of recent cracks or other areas of thin ice.

1. Hydrothermal Vents as Sites for Origins. Soon after the discovery of hydrothermal ecosystems at submarine thermal springs (Corliss et al. 1979), hydrothermal vents were proposed as a possible site for the origin of life on early Earth (Corliss et al. 1981). Suboceanic vents are attractive sites for the origin of life because of the associated presence of chemical reducing power in the form of minerals (Wächtershäuser 1988*a*,*b*; Holm 1992), and the protection they afford from all but the largest giant impacts (Maher and Stevenson 1988; Oberbeck and Fogleman 1989; Sleep et al. 1989). Controversy persists over whether the high temperatures characteristic of vents allow the synthesis (Shock 1990, 1992; Hennet et al. 1992) or require the thermal decomposition (Miller and Bada 1988) of prebiotic organic polymers. But substantial experimental progress is beginning to be made using the sort of iron and nickel sulfide minerals that currently collect near the vents as catalytic agents for prebiotically interesting reactions.

Wächtershäuser (1988*a*,*b*, 1990) proposed that synthesis of prebiotic organics on early Earth could have occurred through the reduction of aqueous dissolved CO_2 to organic compounds on iron and nickel sulfide mineral surfaces; the energy for these reductions would, for example, be provided by the exergonic formation of pyrite (FeS_2) from FeS and hydrogen sulfide (H_2S). Negatively charged organic acids would be electrostatically held to the surface of the positively charged pyrite, and a kind of "surface metabolism" would develop. This hypothesis faces a number of objections (de Duve and Miller 1991; Chyba and McDonald 1995), but it has inspired a list of successful experiments. For example, metallic iron can reduce N_2 to ammonia (NH_3) via the formation of iron oxides in the presence of liquid water at temperatures between 300 and 800°C (Brandes et al. 1998). Activated acetic acid can be synthesized by mixing carbon monoxide (CO) and H_2S with a slurry of nickel and iron sulfide particles at 100°C (Huber and Wächtershäuser 1997); that is, this environment can catalyze the fixation of inorganic carbon to organic carbon. While amino acid synthesis has not been demonstrated, it has been shown (Huber and Wächtershäuser 1998) that if amino acids were present under these conditions, they could be linked together to form short peptides. The problem of peptide formation has been a fundamental problem in the origin of life that had largely resisted solution under prebiotically plausible conditions. While these experiments remain far from demonstrating the origin of life at depth, they do represent an ongoing research program, motivated by this hypothesis, that is showing significant successes.

2. Meteoritic Evidence. Certain meteorites provide us with the only extraterrestrial examples, apart from ancient Mars, where liquid water was once available. There is substantial evidence for preterrestrial aqueous alteration in carbonaceous chondrite meteorites, both from mineralogy (Grimm and McSween 1989) and from their amino and hydroxy acid profiles, which suggest that they formed by Strecker synthesis, requiring liquid water (Peltzer et al. 1984; Cronin 1989). While models of the thermal history of asteroids suggest that large ($\sim$100 km) objects could have maintained liquid water interiors for $\sim 10^8$ yr, it seems likely that the Murchison meteorite, whose organic chemistry is the best studied of all the meteorites, experienced liquid water for much less time than this, perhaps $\sim 10^4$ yr (Peltzer et al. 1984; Lerner 1995).

Carbonaceous chondrites are of potential interest because they represent actual examples of prebiotic organic synthesis in subsurface hydrothermal environments. Early in their history, carbonaceous chondrites hosted liquid water in the presence of the organic monomers (such as amino acids) needed for biomolecular synthesis, as well as potentially catalytic clay mineral surfaces. Yet abiotically produced peptides or oligonucleotides have not been reported. Cronin (1976) searched for small peptides in Murchison and concluded that no more than 9 mole percent of the total acid-labile amino acid precursors in the meteorite were peptides. Does Murchison therefore mean that liquid water, prebiotic organic monomers, and mineral catalysts together for $\sim 10^4$ yr are insufficient for further progress toward the origin of life? If so, does this argue against the possibility of a deep subsurface origin, or does it point instead to some critical ingredient that Murchison is missing? Are there other carbonaceous chondrites in our collections that experienced liquid water for timescales orders of magnitude longer than did Murchison, and what do they show? These are important questions at the interface of planetary science and prebiotic chemistry (Chyba and McDonald 1995).

3. The RNA World. The prebiotic formation of peptides from individual amino acids is far from the only dilemma facing hypotheses concerning the origin of terrestrial life. As a second example, consider the "RNA world" hypothesis (Gilbert 1986). The discovery that certain sequences of ribonucleic acid (RNA) could exhibit catalytic activity (Zaug and Cech 1986) resulted in a proposed solution to an outstanding "chicken-or-egg" problem in origin-of-life research (Dyson 1985). The problem had been that in modern organisms deoxyribonucleic acid (DNA) carries the genetic information that codes for the sequence of amino acids that make up biological proteins, whereas proteins are needed for the replication of DNA. Which could possibly have come first? The RNA world hypothesis suggested that our current DNA-protein world was preceded on early Earth by one in which both the functions of genetic information storage and catalytic activity were carried out by RNA alone.

Although it is an important step forward (Joyce 1991; Joyce and Orgel 1993), the RNA world hypotheses faces numerous difficulties (Joyce 1989). In sum, under prebiotically plausible circumstances, it is hard to synthesize the right components for RNA; it is hard to put these components together to make RNA; and it is hard for RNA to resist thermal decomposition for very long once it has formed (Chyba and McDonald 1995).

The existence of so many uncertainties in the RNA world and other hypotheses for the origin of life renders it difficult to use these hypotheses to reach compelling conclusions about the kinds of worlds on which Earth-like life could or could not originate. In the face of these uncertainties, it is probably best to return to simple fundamental criteria, such as the presence of liquid water, in order to select places to pursue the search for extraterrestrial life.

E. An Expansive View of Habitability

Given our current, still quite tentative knowledge about the prebiotic syntheses needed for the origin of life, we cannot say with confidence whether life would in general originate at depth, at the surface of a world, in either location, or whether both are needed together (Chyba 1998). The most conservative course remains to search for worlds like the Earth where surface liquid water exists. The remainder of this chapter will be devoted to estimating the extent and frequency of such environments. If even this conservative definition of habitability suggests many locales for life around other stars, then whatever we have ignored will only broaden an already substantial potential biological arena (Sagan 1996). However, we should bear in mind that this view of habitability may in fact prove far too conservative. As a practical matter, within our own solar system our gaze should linger upon any world that harbors liquid water at any depth. Beyond our solar system, those worlds that we are most likely remotely to recognize as habitable will probably be worlds that satisfy the more conservative definition.

III. CIRCUMSTELLAR HABITABLE ZONES

A. A Conservative Definition of Habitable Zone

We define the circumstellar habitable zone (HZ) as the volume of space around a single or multiple-star system within which an Earth-like planet could support surface liquid water. An Earth-like planet is one similar in mass and composition to Earth and having comparable surface inventories of CO_2, H_2O, and N_2 (Kasting et al. 1993; Whitmire and Reynolds 1996). The upper planetary mass limit is not well constrained, but the lower mass limit must be large enough to maintain sufficient geological activity to power the climate-stabilizing carbonate-silicate negative feedback cycle (Walker et al. 1981; Kasting 1988). If we require stability for ≥ 1 Gyr, then a mass greater than $\sim 0.1\ M_{\oplus}$ is necessary, by analogy to the geological history of Mars (see III.B). The HZ evolves in time as the star's luminosity evolves (Kasting et al. 1993; Whitmire and Reynolds 1996).

If there were a real Earth-like planet in the HZ, then, by definition, it would have the compounds necessary for life. However, just the existence of a ~1-$M_\oplus$ terrestrial planet in the HZ would not imply that it is inhabited or even habitable. For example, the role of impacts in planetary habitability is only just becoming understood (Chyba 1996*a*,*b*). Other considerations (see section V) are the possible extreme obliquity variations in the absence of a large stabilizing moon (Laskar et al. 1993; Williams and Kasting 1997) and synchronous rotation of planets around M stars (Joshi et al. 1997).

The continuously habitable zone (CHZ) is defined as that volume of space around a single or multiple-star system within which an Earth-like planet could support surface liquid water for a specified period of time. The CHZ is smaller than the HZ and it also evolves in time (Kasting et al. 1993; Whitmire and Reynolds 1996).

B. Evolution of the Sun's Habitable Zone

In the standard solar model, the Sun began its life on the zero-age main-sequence (ZAMS) with a luminosity of 0.7 $L_\odot$ and a photospheric temperature a few hundred degrees less than the present value. Compared to today, the early Sun probably rotated much faster, had a stronger and more active magnetic field and solar wind, a greater emission of X-ray and UV radiation, and a larger fraction of its surface covered with sunspots. The Sun's main-sequence lifetime is 11 Gyr (Sackmann et al. 1993), by the end of which its luminosity will have increased to 2.2 $L_\odot$. Upon core hydrogen depletion, the Sun will ascend the first red giant branch. Our interest here is primarily the evolution of the Sun's HZ during the main-sequence phase. The evolution of the HZ during the post-main-sequence phases was discussed and illustrated in Kasting et al. (1993) and Whitmire and Reynolds (1996).

Given the luminosity and temperature as a function of time, the evolution of the HZ can be calculated for specific values of critical fluxes defining the inner and outer boundaries from $S_1(T) = L(t)/r_1^2$ and $S_2(T) = L(t)/r_2^2$, where $r_1(r_2)$ is the inner (outer) HZ boundary radius, $L(t)$ is the solar luminosity as a function of time, obtained from the stellar evolutionary model, and $S_1(S_2)$ is the appropriate critical solar flux, which in general is a function of stellar temperature. The HZ radii will be given in units of AU when the luminosity is in solar units and the flux is in units normalized to the present solar constant at Earth's orbit, taken to be 1360 W m^{-2}.

Kasting et al. (1993) identified three critical fluxes for the inner radius of the HZ. In flux units normalized to the present solar constant these limits are 1.1, 1.4, and 1.76. The first two result from the loss of planetary water by the moist greenhouse and runaway greenhouse, respectively (Kasting 1988). These are theoretical values obtained from a one-dimensional radiative-convective climate model in which the radiative effect of clouds is parameterized by use of a high surface albedo. This assumption is expected to result in conservative limits; a real planet can probably maintain liquid water at a higher solar flux than calculated. The third limit is empirical and corresponds to the flux at Venus 1 Gyr ago. Spacecraft observations show no evidence that liquid water has flowed on the surface of Venus for the last 1 Gyr (Solomon and Head 1991). This flux limit is an upper bound, because Venus could not have had significant water since then. The inner HZ radius is therefore reasonably well bracketed. The evolution of these inner HZ boundary radii for the Sun are shown in the bottom three curves of Figs. 1a and 1b.

Three critical fluxes corresponding to the outer radius of the HZ were also identified by Kasting et al. (1993). In normalized units these flux limits are 0.32, 0.36, and 0.53. The smallest and least conservative limit corresponds to the flux at early Mars 3.8 Gyr ago in the standard solar model. This limit is based on arguments (e.g., Pollack et al. 1987) that early Mars had a warm and wet climate and therefore must have been in the Sun's HZ. (Because of its low mass of 0.1 $M_{\oplus}$, Mars is not an Earth-like planet as we define it. An Earth-mass planet at the distance of Mars from the Sun might have maintained a stable climate for a much longer period.) The intermediate theoretical flux limit of 0.36 corresponds to the maximum possible CO_2 greenhouse heating, which occurs for a CO_2 partial pressure of ~8 bar (Kasting et al. 1993; Kasting 1991; Whitmire and Reynolds 1996). At higher CO_2 partial pressures, the increase in planetary albedo outweighs the increase in greenhouse heating. Since the actual solar flux at Mars 3.8 Gyr ago was 0.32 in the standard model, there is a theoretical discrepancy between these two limits. This discrepancy is currently unresolved but might be explained by the presence of other greenhouse gases in Mars' early atmosphere (Kasting 1991; Sagan and Chyba 1997), by a scattering variant of the greenhouse effect (Forget and Pierrehumbert 1997), or by a nonstandard solar model in which the Sun's ZAMS mass was 4–7% greater than the present value (Graedel et al. 1991; Whitmire et al. 1995). Other possible astronomical explanations for an increase in the solar flux at early Mars have been discussed by Whitmire et. al. (1995).

The most conservative flux limit for the outer boundary is the "first condensation" limit of 0.53. This is the flux at which CO_2 first begins to condense and thus to increase the planetary albedo. This limit is probably too conservative but is difficult to improve on without a climate model that can treat CO_2 clouds. Kasting et al. (1993) used these three estimates of the critical minimum flux necessary to sustain a mean surface temperature of 273 K, along with a standard solar evolution model, to generate the evolution of the outer HZ boundary radii (three upper curves) in Figs. 1a and 1b.

These figures show the evolution of the Sun's HZ, i.e., the inner and outer radii versus time, for sets of the paired critical fluxes. The three sets are paired in the following way. Case 1 corresponds to the two most conservative flux limits, the moist greenhouse (1.1) and the first condensation point (0.53). Case 2 is the intermediate case and corresponds to limits determined by the runaway greenhouse (1.4) and the maximum greenhouse (0.36). Case 3 is our least conservative case and corresponds to limits inferred from the lack of water on Venus as of 1 Gyr ago (1.76) and the evidence for the presence of liquid water on early Mars (0.32).

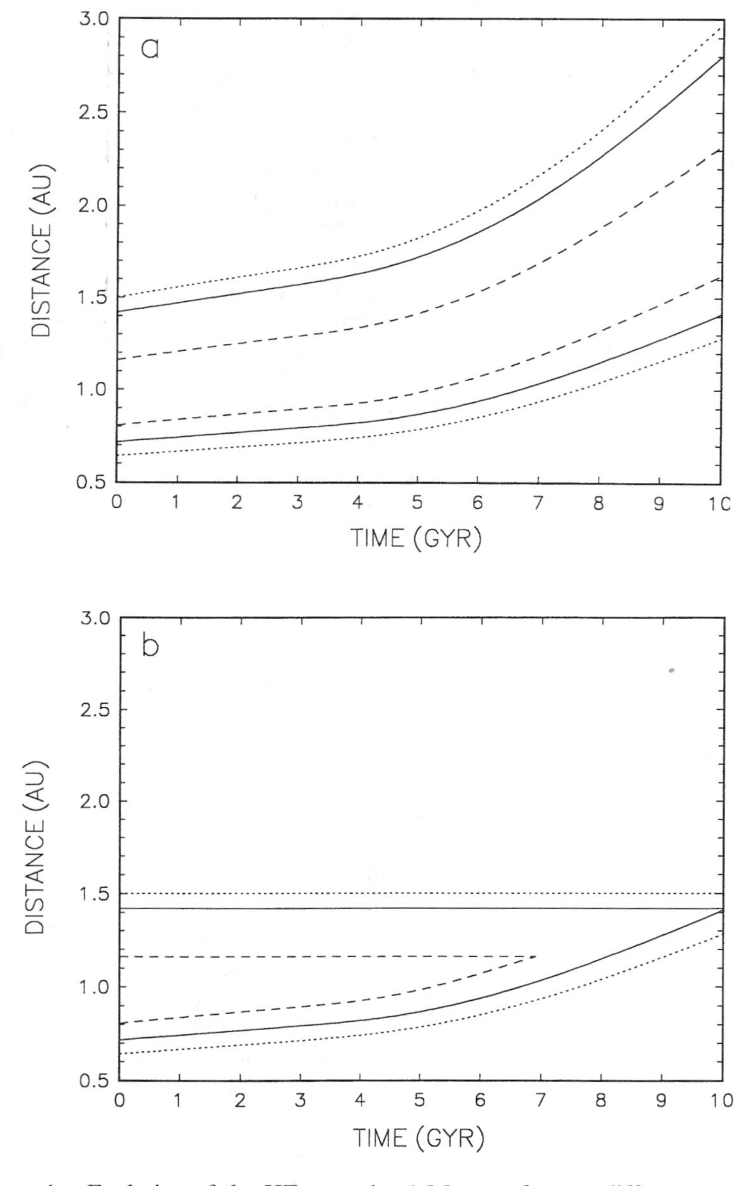

Figure 1. Evolution of the HZ around a 1-$M_\odot$ star for two different assumptions concerning the possibility of planetary "cold starts": (a) cold starts permitted; (b) no cold starts allowed. The term "cold start" refers to whether or not a planet that is initially beyond the outer edge of the HZ will deglaciate once the stellar luminosity increases to the appropriate critical value. The three pairs of curves correspond to the different habitability estimates discussed in the text: long dashes, "water loss" and "first CO_2 condensation" limits (most conservative); solid curves, "runaway" and "maximum greenhouse" limits; dotted curves, "recent Venus" and "early Mars" limits (most optimistic). (Figure from Kasting et al. 1993.)

The time evolution of the three HZ cases shown in Fig. 1a is based on the assumption that a planet initially beyond the outer boundary of the HZ and therefore frozen, will deglaciate once the solar flux reaches a value equal to the calculated climatic limits. On the other hand, Caldeira and Kasting (1992) have argued that a planet that formed beyond the outer boundary of the HZ would develop a reflective blanket of ice, reducing the absorption of solar radiation. Such a planet might also develop thick, highly reflective, CO_2 clouds (Kasting 1991). A planet that undergoes global glaciation soon after accretion might thereby remain in that state even when the solar flux increases beyond the appropriate (lower) critical value discussed above. Thus, it may be impossible to "cold start" an initially frozen planet before the solar flux increases by a factor of $\sim$2 and exceeds the critical greenhouse value (Caldeira and Kasting 1992).

However, recent evidence that the Earth may have completely glaciated during the Neoproterozic and subsequently deglaciated supports the view that "cold starts" are in fact possible (Hoffman et al. 1998). Moreover, Forget and Pierrehumbert (1997) have shown that the high-albedo properties of CO_2 clouds may be countered by an increased greenhouse effect due to the clouds' scattering of infrared radiation. It might also be possible that a major climatic perturbation, such as a massive impact or extensive volcanism, could "jump start" a frozen planet. Assuming that cold starts are possible, it can be seen from Fig. 1a that the HZ moves outward and becomes broader in time.

Figure 1b shows the same evolution as Fig. 1a but with the more conservative assumption that cold starts are not possible during the main-sequence phase. In this case the outer boundary of the HZ is constant and equal to its ZAMS value. The inner boundary radius moves outward in time as before, so the HZ now becomes narrower in time.

Continuously habitable zones for any assumed time period can be obtained from Figs. 1a and 1b for the three limiting cases. For our most conservative Case 1 fluxes without cold starts, the 4.5-Gyr CHZ extends from 0.95 to 1.15 AU, which is a factor of 5 larger than that found in earlier work (Hart 1979). For our least conservative Case 3, the 4.5-Gyr CHZ extends from 0.75 to 1.9 AU. For simplicity of discussion here and in the next section we shall focus on the intermediate Case 2. For this case the 4.5-Gyr CHZ extends from 0.84 to 1.77 AU if cold starts are possible, and from 0.84 to 1.43 AU if they are not.

C. Implications for SETI

Figure 1 has interesting implications for the Search for Extraterrestrial Intelligence (SETI). One long-standing objection to the prospects for intelligent life elsewhere (e.g., Mayr 1995) is the claim that the evolution of intelligence on Earth was an extraordinarily unlikely event, contingent on very special circumstances. If we could rewind the tape of life and start again from the beginning, this argument runs, nothing like human intelligence would again evolve. Although the Earth has seen the multiple independent evolution of such highly selected capabilities as flight and sight, technical intelligence is argued not to be analogous to these. Metazoa arose on Earth over 550 Myr ago (Runnegar 1992), but technical intelligence has only just arisen for the first time, in only one of the billions to tens of billions of species that have existed on Earth.

One objection to this argument is that it seems unsurprising that the first intelligent species on a planet would look back over the history of life and ask why it, among all others, was first. But the objection (Chyba 1999) can be expressed more precisely with the help of Fig. 1. The total expanse of time over which metazoa are likely to exist on Earth is much greater than the ~600 Myr during which they have so far existed. Figure 1 shows that Earth will likely remain habitable for another ~2 Gyr, after which it will become uninhabitable due to the increasing luminosity of the Sun. Until that time it seems probable (not certain) that metazoa are here to stay: It would be a remarkable extinction indeed for every worm, every clam, every cockroach, and every rat to disappear (although the elimination of photosynthesis might accomplish this). Metazoa therefore seem likely to endure *in toto* for over 2.5 Gyr of Earth history. From this perspective, technical intelligence arose on Earth *early* in the history of metazoa, and it is less clear that we should be impressed by the observation that, from our point of view, technical intelligence is a latecomer on the scene. Such an argument appears strongly rooted in the contemporary human perspective. Nevertheless, none of these arguments seems likely to be compelling in the absence of additional data, arguing that SETI remains fundamentally a problem for empirical investigation.

D. Evolution of the Habitable Zone around Other Stars

The same type of HZ calculations can be performed for stars with masses different from the Sun's. Main-sequence stars much more massive than the Sun have large HZs on an absolute scale, but their lifetimes are arguably too short to be of interest as sites for the evolution of complex organisms (Huang 1960; but see McKay 1996). For the massive O-type stars it is doubtful whether there is even sufficient time for planets to form during their $\sim 10^6$-yr stellar lifetimes (Bodenheimer 1989). The main-sequence lifetime of a star, τ_{ms}, is proportional to M/L. Since $L \propto M^{4.75}$ over the mass range of most interest (Iben 1967), $\tau_{ms} \propto M^{-3.75}$. Here we restrict ourselves to stellar lifetimes greater than 2 Gyr, which correspond to masses less than 1.5 $M_\odot$. At the other extreme, the low-mass main-sequence M stars with masses ≤ 0.5 $M_\odot$ have lifetimes longer than the age of the universe. Their evolution in 10 Gyr is negligible, so their 4.5-Gyr CHZ is identical to their ZAMS HZ.

Figures 2a and 2b show the evolution of the HZ (for Case 2 fluxes) around stars of selected masses between 0.5 and 1.5 $M_\odot$, with and without cold starts, respectively. The effect of stellar temperature on the blackbody spectrum of light and, hence, on planetary albedo has been taken into account. The fluxes were corrected for stellar temperature. Details of this modeling may be found in Kasting et al. (1993). For the stars in Fig. 2, the widths of various CHZs can be obtained for any assumed relevant timescale. Assuming no cold starts, the CHZ for a given τ is equal to the HZ at time $t = \tau$.

IV. ACCRETION OF PLANETS IN THE HABITABLE ZONE

A. Accretion of Planets in the Habitable Zone of Single Stars

Wetherill (1996) numerically investigated the problem of accretion of terrestrial planets around stars of masses between 0.5 and 1.5 $M_\odot$ and found that, for circumstellar disk parameters that are not too different from those of the solar system, the number and radial distribution of final terrestrial planets are insensitive to stellar mass and that these planets concentrate in the vicinity of 1 AU. However, the location of the HZ is quite sensitive to stellar mass; it lies inside the 1-AU planet formation region for low-mass stars and beyond the terrestrial planet formation region for high-mass stars. That is, terrestrial planets would be too cold around low-mass stars and too hot around high-mass stars. In the case of 1-$M_\odot$ stars there was nearly always a planet of mass $\geq \frac{1}{3}$ $M_\oplus$ formed in the HZ. In addition to the assumption of circumstellar disk parameters similar to those of the solar system, these simulations also assumed the existence of Jupiter-like planets. In the absence of jovian planets, the median terrestrial planetary mass was 2 $M_\oplus$ for the same initial surface density used in the models characteristic of the solar system.

The fundamental assumption of that study, namely, an initial circumstellar accretion disk similar to those used in modeling the solar system, may not be generally valid. For example, the disk mass distribution may scale with stellar mass (Kasting et al. 1993; Whitmire and Reynolds 1996) in such a manner that terrestrial planets form closer than 1 AU around low-mass stars and farther than 1 AU around high-mass stars, making it more likely that planets would form in the HZ. Kasting et al. (1993) gave arguments suggesting that planet formation distances should scale with stellar mass. If it is further assumed that planet spacing is generally logarithmic (i.e., equal numbers of planets in each decade of semimajor axis), as is true in the solar system and the jovian system and as found in various numerical studies (Isaacman and Sagan 1977; Lissauer 1995; Wetherill 1996), Whitmire and Reynolds (1996) showed that the number of planets in the HZ would be independent of stellar mass. If this conclusion is correct, most habitable planets in the galaxy would reside around K and high-mass M stars.

However, one important uncertainty in these results is our need to understand the origin of "hot Jupiters" (Marcy and Butler 1998), Jupiter-mass planets well within 1 AU of their stars. If these objects arise as a result of one or another type of planetary migration (Weidenschilling and Marzari 1996; Murray et al. 1998; Trilling et al. 1998), the implications of

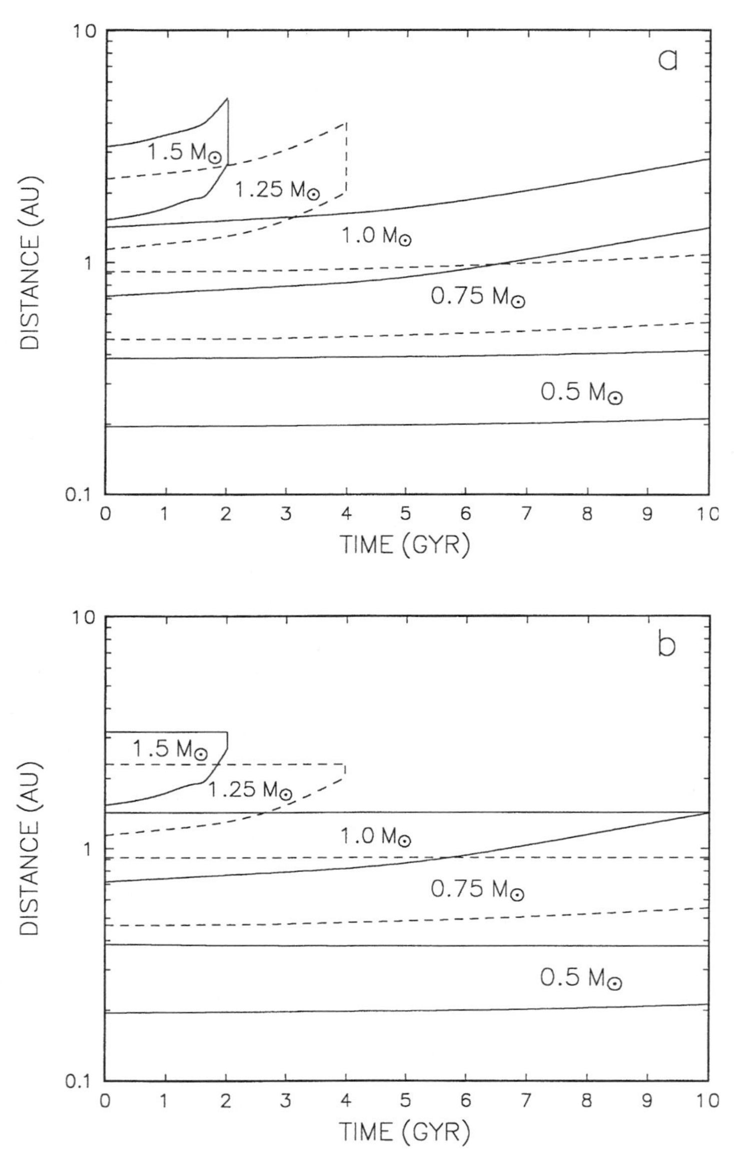

Figure 2. Evolution of the HZ around stars of different masses for two assumptions about the possibility of planetary cold starts: (a) cold starts permitted; (b) no cold starts allowed. The intermediate (runaway and maximum greenhouse) estimates for the width of the HZ were used, and the evolution was truncated at the end of the main-sequence phase. (Figure from Kasting et al. 1993.)

these migrations for the formation of terrestrial planets in the HZ are an important concern that needs to be understood.

B. Accretion of Planets in the Habitable Zone of Binary Star Systems

Whitmire et al. (1998) numerically studied terrestrial planet formation in the HZs of binary star systems. Assuming current models of terrestrial planet formation in the solar system, they investigated the conditions under which the secondary star would inhibit planet growth in the HZ. Runaway accretion was assumed to be precluded (i) if the secondary caused the planetesimal orbits to cross within the runaway accretion timescale and (ii) if, during crossing, the relative velocities of the planetesimals were accelerated beyond a certain critical value that results in disruptive collisions rather than accretion. For a two-solar-mass binary with planetesimals in circular orbits about one star at 1 AU, and a typical wide binary eccentricity of 0.5, the minimum binary semimajor axis a_c that would not inhibit planet formation is 32 AU. Similar results were presented for a range of binary eccentricities e_B, secondary masses m, critical disruption velocities U_c, and mean planetesimal locations $\bar{a}$. A semiempirical relation between these parameters and the critical semimajor axis was found to be

$$a_c = 16(1 - e_B)^{-1}(m/1\ \mathrm{M_\odot})^{0.31}(\bar{a}/1\ \mathrm{AU})^{0.80}(100\ \mathrm{m\ s^{-1}}/U_c)^{0.30}\ \mathrm{AU}$$

For fixed m the e_B dependency corresponds to a constant periastron distance. For two 1-$\mathrm{M_\odot}$ stars this constant is 16 AU. Based on the distributions of orbital elements of a bias-corrected sample of nearby G dwarfs (Duquennoy and Mayor 1991), it was found that at least 60% of solar-type binaries cannot be excluded, *solely* on the basis of the perturbative effect of the secondary star, from having a habitable planet.

Whitmire et al. (1998) found that 12 of 18 systems containing newly discovered giant extrasolar planets or brown dwarfs can be excluded from having a habitable planet at 1 AU, *assuming* the secondaries formed *in situ* in their present orbits. If they formed further than 1 AU from the star and then migrated or were scattered inward, a greater fraction of these systems might be excluded. We note however that only $\sim$5% of observed stars have giant planets or brown dwarfs within detectable (less than $\sim$3 AU) limits (Marcy and Butler 1998).

V. HABITABILITY OF WORLDS WITHIN THE HABITABLE ZONE

A. Planetary Obliquities

The Earth's obliquity would vary chaotically from 0 to 85 degrees were it not for the presence of the Moon (Laskar and Robutel 1993; Laskar et al.

1993). The Moon is believed to have formed as the result of a stochastic glancing collision with a Mars-sized embryo. This type of collision may be uncommon in the planet formation process. The fact that Earth is the only terrestrial planet with a major moon lends support to this idea. Noting the likely extreme climatic fluctuations that would occur on planets with high obliquities, Laskar et al. (1993) suggested that habitable planets may be rare. We note that the large chaotic variation of the Earth's spin axis is due to the presence of the other planets, primarily Jupiter, as well as Earth's planetary quadrupole moment.

To test the assertion that extreme obliquities would lead to climatic variations that would make a planet uninhabitable, Williams and Kasting (1997) used an energy balance climate model to simulate the Earth's climate at obliquities up to 90 degrees. They found that Earth's climate would become regionally severe, with large seasonal cycles and accompanying temperature extremes on middle- and high-latitude continents, which might be detrimental to some forms of life. Whether such variations would pose a problem for the origin of life or for microbial life seems less likely (Chyba 1996*b*). The response of other, hypothetical, Earth-like planets to large obliquity fluctuations depends on their land-sea distribution and on their position within the HZ. Planets with several modest-sized continents or equatorial supercontinents are more climatically stable than those with polar supercontinents. Planets farther out in the HZ are less affected by high obliquities because their atmospheres should accumulate CO_2 in response to the carbonate-silicate cycle. Dense CO_2 atmospheres transport heat very effectively and therefore limit the magnitude of both seasonal cycles and latitudinal temperature gradients. Therefore, a significant fraction of extrasolar Earth-like planets may still be habitable, even if they were subject to large obliquity fluctuations.

B. Synchronously Rotating Planets

Dole (1964) noted that habitable planets around M dwarfs are likely to be synchronously rotating, with one hemisphere always illuminated by the parent star and the other hemisphere in perpetual darkness. Planets in the HZ of late K dwarfs can also be synchronously rotating (Kasting et al. 1993). The M dwarfs are believed to be the most numerous ($\sim$75%) of stars, having masses in the range 0.1–0.5 $M_\odot$ (Rodonò 1986). [However, recent observations indicate that M stars in the mass range 0.1–0.3 $M_\odot$ are more than an order of magnitude less abundant than previously believed (Gould et al. 1997).] If the atmosphere of a synchronously rotating planet is in radiative-convective equilibrium, the surface temperature on the dayside will be very high, while the nightside will be so cold that the major atmospheric constituent will condense out on the surface.

Joshi et al. (1997) used a three-dimensional climate model to investigate the conditions under which atmospheric collapse would occur on synchronously rotating planets in the HZ. They find that if CO_2 partial pressure is controlled by the carbon-silicate cycle, these planets would need a minimum surface pressure of 1–1.5 bar of CO_2 in order to support liquid water on the dark hemisphere. The minimum pressure necessary to prevent atmosphere collapse is ~30 mbar. Thus, they conclude that synchronously rotating planets in the HZ of M stars could support atmospheres over a large range of conditions, and even surface liquid water under certain conditions.

C. Habitable Moons around Extrasolar Giant Planets

The jovian-mass planet around the solar-type star 16 Cygni B has an elliptical orbit that lies in the HZ part of the time. The star 47 Ursae Majoris has a jovian-mass planet in a circular orbit that lies near the least conservative outer limit of the HZ. It seems likely that these jovian planets are themselves not habitable (Chyba 1997); however, they may have large habitable moons. Williams et al. (1997) have investigated constraints on the habitability of such moons and find that a rocky moon orbiting jovian planets or brown dwarfs lying in the stellar HZ could be habitable provided that (1) the moon's mass is $\geq 0.12\ M_\oplus$, in order to retain a substantial and long-lived atmosphere, and (2) the moon has a Ganymede-like magnetic field sufficient to prevent the atmosphere from being sputtered away by the bombardment of energetic ions from the planet's magnetosphere. Long-term habitability requires the operation of the carbonate-silicate cycle and therefore some form of geological activity. If driven by radiogenic heating, a minimum mass of $0.23\ M_\oplus$ is required. If the moon is in an Io-like orbit in resonance with another more distant moon, then tidal heating could supply the internal heat source necessary to drive plate tectonics or other geological activity and thus maintain surface liquid water for Gyr timescales.

VI. CONCLUSIONS: AN EXPANDING ARENA FOR LIFE

As a practical matter, at least for the present, the search for extraterrestrial life is the search for life as we know it, life made of organic molecules in liquid water. Since organics are common in much of the solar system (Cruikshank 1997) and in interstellar space (Pendleton and Chiar 1997), liquid water has become the focus of exobiological searches.

It has become clear in recent years that there is a deep subsurface biosphere on Earth (Gold 1992; Whitman et al. 1998). It is still an open question whether life may in fact have originated in the subsurface. Therefore, deep liquid water environments, such as we now suspect may exist on Mars and Europa, could be plausible locales for extraterrestrial life in our own solar system or elsewhere.

The most conservative and observationally relevant requirement to set for extraterrestrial habitable environments is not to require liquid water at depth, but at a world's surface. Assuming this, the "circumstellar habitable zone" is defined as the volume of space around a star (or star system) within which an Earth-like planet could support surface liquid water. Although these zones change with evolving stellar luminosity, there are significant habitable zones lasting for billions of years around stars with masses between 0.1 and 1.5 times the mass of the Sun. Even a conservative definition of habitability suggests abundant locales for life around other stars.

The detection of oxygen and water in the spectrum of a terrestrial planet lying in a star's habitable zone could provide important remote evidence of life (Woolf and Angel 1998). The presence of methane, simultaneously with oxygen and water, would be even more suggestive. However, it should be cautioned that both Ganymede and Europa show spectroscopic evidence for molecular oxygen (Spencer et al. 1995; Hall et al. 1995). Indeed, Europa has a tenuous O_2 atmosphere (Hall et al. 1995), thought to be produced by the sputtering of ice by energetic particles from the jovian magnetosphere (Sieger et al. 1998). One lesson of the *Viking* missions to Mars is that anticipated unambiguous signatures of biology may prove to be less certain in the face of unanticipated nonbiological processes.

Nonetheless, it is exciting that the detection of spectroscopic signatures highly suggestive of life on nearby extrasolar planets is within the capabilities of existing or foreseeable technology (Woolf and Angel 1998). Such measurements have been proposed using a system of orbiting IR telescopes operating as an interferometer. Ongoing SETI searches offer an important complementary approach in the search for extraterrestrial life (Tarter and Michaud 1990; Dick 1996; Tarter 1997). While our knowledge remains too limited to assess quantitatively the likelihood of finding extraterrestrial biology, the combination of new discoveries about Earth's deep biosphere, the apparent subsurface liquid water environments elsewhere in our solar system, the theoretical recognition that circumstellar habitable zones are more expansive than once thought, and the discovery of extrasolar planets all combine to make the next decades the most exobiologically promising in the history of astronomy.

Acknowledgments D. W. thanks John Matese for suggestions. C. F. C. acknowledges support from NASA Exobiology and a Presidential Early Career Award for Scientists and Engineers.

REFERENCES

Anderson, R. T., Chapelle, F. H., and Lovley, D. R. 1998. Evidence against hydrogen-based microbial ecosystems in basalt aquifers. *Science* 281:976–977.

Barrow, J. D., and Tipler, F. J. 1986. *The Anthropic Cosmological Principle* (New York: Oxford University Press).

Biemann, K., Oró, J., Toulmin, P., Orgel, L. E., Nier, A. O., Anderson, D. M., Simmonds, P. G., Flory, D., Diaz, A. V., Rushneck, D. R., Biller, J. E., and Lafleur, A. L. 1977. The search for organic substances and inorganic volatile compounds in the surface of Mars. *J. Geophys. Res.* 82:4641–4658.

Blum, H. F. 1962. *Time's Arrow and Evolution* (New York: Harper and Brothers).

Bodenheimer, P. 1989. The impact of stellar evolution on planetary system development. In *The Formation and Evolution of Planetary Systems*, ed. H. A. Weaver and L. Danly (Cambridge: Cambridge University Press), pp. 243–273.

Boston, P. J., Ivanov, M. V., and McKay, C. P. 1992. On the possibility of chemosynthetic ecosystems in subsurface habitats on Mars. *Icarus* 95:300–308.

Brandes, J. A., Boctor, N. Z., Cody, G. D., Cooper, B. A., Hazen, R. M., and Yoder, H. S. 1998. Abiotic nitrogen reduction on the early Earth. *Nature* 395:365–367.

Breaker, R. R., and Joyce, G. F. 1994. Emergence of a replicating species from an *in vitro* RNA evolution reaction. *Proc. Natl. Acad. Sci. USA* 91:6093–6097.

Caldeira, K., and Kasting, J. F. 1992. Susceptibility of the early Earth to glaciation caused by carbon dioxide clouds. *Nature* 359:226–228.

Campbell, I. B., and Claridge, G. G. C. 1987. *Antarctica: Soils, Weathering Processes and Environment* (Amsterdam: Elsevier).

Carr, M. H. 1996. *Water on Mars* (New York: Oxford University Press).

Carr, M. H., Belton, M. J. S., Chapman, C. R., Davies, M. E., Geissler, P., Greenberg, R., McEwen, A. S., Tufts, B. R., Greeley, R., Sullivan, R., Head, J. W., Pappalardo, R. T., Klaasen, K. P., Johnson, T. V., Kaufman, J., Senske, D., Moore, J., Neukum, G., Schubert, G., Burns, J. A., Thomas, P., and Veverka, J. 1998. Evidence for a subsurface ocean on Europa. *Nature* 391:363–365.

Chyba, C. F. 1996*a*. Are comets necessary for a habitable zone? In *Circumstellar Habitable Zones: Proceedings of the First International Conference*, ed. L. R. Doyle (Menlo Park, CA: Travis House Publications), pp. 277–281.

Chyba, C. F. 1996*b*. Catastrophic impacts and the Drake equation. In *Astronomical and Biochemical Origins and the Search for Life in the Universe*, ed. C. B. Cosmovici, S. Bowyer, and D. Werthimer (Bologna: Editrice Compositori), pp. 157–164.

Chyba, C. F. 1997. Life on other moons. *Nature* 385:201.

Chyba, C. F. 1998. Buried beginnings. *Nature*, 395:329–330.

Chyba, C. F. 1999. Some thoughts on SETI. *The Planetary Report* 19(3):6.

Chyba, C. F., and McDonald, G. D. 1995. The origin of life in the solar system: Current issues. *Ann. Rev. Earth Planet. Sci.* 23:215–249.

Chyba, C., and Sagan, C. 1992. Endogenous production, exogenous delivery and impact-shock synthesis of organic molecules: An inventory for the origins of life. *Nature* 355:125–132.

Chyba, C. F., and Sagan, C. 1997. Comets as a source of prebiotic organic molecules for the early Earth. In *Comets and the Origin and Evolution of Life*, ed. P. J. Thomas, C. F. Chyba, and C. P. McKay (New York: Springer), pp. 147–173.

Chyba, C. F., McKinnon, W. B., Coustenis, A., Johnson, R. E., Kovach, R. L., Khurana, K., Lorenz, R., McCord, T. B., McDonald, G. D., Pappalardo, R. T., Race, M., and Thomson, R. 1999. Europa and Titan: Preliminary recommendations of the Campaign Strategy Working Group on Prebiotic Chemistry in the Outer Solar System. Proceeding of the 30th Annual Lunar and Planetary Science Conference 1537.

Clark, B. C., Baker, A. L., Cheng, A. F., Clemett, S. J., McKay, D., McSween, H. Y., Pieters, C., Thomas, P., and Zolensky, M. 1999. Survival of life on asteroids, comets and other small bodies. *Orig. Life Evol. Biosph.*, in press.

Corliss, J. B., Dymond, J., Gordon, L. I., Edmond, J. M., von Herzen, R. P., Ballard, R. D., Green, K., Williams, D., Bainbridge, A., Crane, K., and van Andel, T. H. 1979. Submarine thermal springs on the Galapagos rift. *Science* 203:1073–1083.

Corliss, J. B., Baross, J. A. and Hoffman, S. E. 1981. An hypothesis concerning the relationship between submarine hot springs and the origin of life. *Oceanol. Acta* SP:59–69.

Cronin, J. R. 1976. Acid-labile amino acid precursors in the Murchison meteorite II: A search for peptides and amino acyl amides. *Orig. Life Evol. Biosph.* 7:343–348.

Cronin, J. R. 1989. Origin of organic compounds in carbonaceous chondrites. *Adv. Space Res.* 9(2):59–64.

Cruikshank, D. P. 1997. Organic matter in the outer solar system: From the meteorites to the Kuiper Belt. In *From Stardust to Planetesimals*, ed. Y. J. Pendleton and A. G. G. M. Tielens (San Francisco: Astronomical Society of the Pacific), pp. 315–333.

Day, W. 1984. *Genesis on Planet Earth*, 2nd ed. (New Haven: Yale University Press).

de Duve, C. and Miller, S. L. 1991. Two-dimensional life? *Proc. Natl. Acad. Sci. USA* 88:10014–10017.

Des Marais, D. J. 1994. Tectonic control of the crustal organic carbon reservoir during the Precambrian. *Chem. Geol.* 114:303–314.

Des Marais, D. J. 1996. Has life made its global environment more habitable? In *Circumstellar Habitable Zones; Proceedings of the First International Conference*, ed. L. R. Doyle (Menlo Park, CA: Travis House Publications), pp. 371–377.

Dick, S. J. 1996. *The Biological Universe* (New York: Cambridge University Press).

Dole, S. H. 1964. *Habitable Planets for Man* (New York: Blaisdell Publishing Co.).

Duquennoy, A., and Mayor, M. 1991. Multiplicity among solar-type stars in the solar neighborhood. II. Distribution of the orbital elements in an unbiased sample. *Astron. Astrophys.* 248:485–521.

Dyson, F. 1985. *Origins of Life* (Cambridge: Cambridge University Press).

Feinberg, G., and Shapiro, R. 1980. *Life beyond Earth* (New York: William Morrow and Co.).

Fleischaker, G. R. 1990. Origins of life: An operational definition. *Orig. Life Evol. Biosph.* 20:127–137.

Forget, F., and Pierrehumbert, R. T. 1997. Warming early Mars with carbon dioxide clouds that scatter infrared radiation. *Science* 278:1273–1276.

Gilbert, W. 1986. The RNA world. *Nature* 319:618.

Gold, T. 1992. The deep, hot biosphere. *Proc. Natl. Acad. Sci. USA* 89:6045–6049.

Goldspiel, J. M., and Squyres, S. W. 1991. Ancient aqueous sedimentation on Mars. *Icarus* 89:392–410.

Gould, A., Bahcall, J. N., and Flynn, C. 1997. M dwarfs from the Hubble Space Telescope star counts: The Groth strip. *Astrophys. J.* 482:913–918.

Graedel, T., Sackmann, I., and Boothroyd, A. 1991. Early solar mass loss: A potential solution to the weak Sun paradox. *Geophys. Res. Lett.* 18:1881–1884.

Grimm, R. E., and McSween, H. Y. 1989. Water and thermal evolution of carbonaceous chondrite bodies. *Icarus* 82:244–280.

Hall, D. T., Strobel, D. F., Feldman, P. D., McGrath, M. A., and Weaver, H. A. 1995. Detection of an oxygen atmosphere on Jupiter's moon Europa. *Nature* 373:677–679.

Hart, M. H. 1979. Habitable zones around main-sequence stars. *Icarus* 37:351–357.

Hayes, J. M., Kaplan, I. R., and Wedeking, K. W. 1983. Precambrian organic geochemistry: Preservation of the record. In *Earth's Earliest Biosphere,* ed. J. W. Schopf (Princeton: Princeton University Press), pp. 93–134.

Heath, M. 1996. The forest-habitability of Earthlike planets. In *Circumstellar Habitable Zones: Proceedings of the First International Conference*, ed. L. R. Doyle (Menlo Park, CA: Travis House Publications), pp. 445–457.

Hennet, R. J. C., Holm, N. G., and Engel, M. H. 1992. Abiotic synthesis of amino acids under hydrothermal conditions and the origin of life: A perpetual phenomenon? *Naturwissenschaften* 79:361–365.

Hoffman, P. F., Kaufman, A. J., Halverson, G. P., and Schrag, D. P. 1998. A neoproterozoic snowball Earth. *Science* 281:1342–1346.

Holm, N. G. 1992. Why are hydrothermal systems proposed as plausible environments for the origin of life? *Orig. Life Eval. Biosph.* 22:5–14.

Horowitz, N. H. 1986. *To Utopia and Back: The Search for Life in the Solar System* (New York: W. H. Freeman and Co.).

Huang, S.-S. 1960. Life outside the solar system. *Sci. Amer.* 202:55–63.

Huber, C., and Wächtershäuser, G. 1997. Activated acetic acid by carbon fixation on (Fe,Ni)S under primordial conditions. *Science* 276:245–247.

Huber, C., and Wächtershäuser, G. 1998. Peptides by activation of amino acids with CO on (Ni,Fe)S surfaces: Implications for the origin of life. *Science* 281:670–672.

Hunten, D. M. 1979. Possible oxidant sources in the atmosphere and surface of Mars. *J. Mol. Evol.* 14:71–78.

Iben, I. 1967. Stellar evolution within and off the main sequence. *Ann. Rev. Astron. Astrophys.* 5:571–626.

Isaacman, R., and Sagan, C. 1977. Computer simulation of planetary accretion dynamics: Sensitivity to initial conditions. *Icarus* 31:510–533.

Johnson, T. V., Chyba, C., Klaasen, K. P., and Terrile, R. J. 1999. Outer planets/solar probe project: Europa orbiter. Proceedings of the 30th Annual Lunar Planetary Science Conference 30:1423.

Joshi, M. M., Haberle, R. M., and Reynolds, R. T. 1997. Simulations of the atmospheres of synchronously rotating terrestrial planets orbiting M dwarfs: Conditions for atmospheric collapse and the implications for habitability. *Icarus* 129:450–465.

Joyce, G. F. 1989. RNA evolution and the origins of life. *Nature* 338:217–224.

Joyce, G. F. 1991. The rise and fall of the RNA world. *New Biol.* 3:399–407.

Joyce, G. F. 1993. Climbing Darwin's ladder. *Curr. Biol.* 3:703–704.

Joyce, G. F. 1994. The RNA world: Life before DNA and protein. In *Extraterrestrials—Where Are They?* ed. B. Zuckerman and M. Hart (Cambridge: Cambridge University Press), pp. 139–151.

Joyce, G. F., and Orgel, L. E. 1993. Prospects for understanding the RNA world. In *The RNA World*, ed. R. F. Gesteland and J. F. Atkins (Cold Spring Harbor, NY: Cold Spring Harbor Laboratory Press), pp. 1–25.

Kasting, J. 1988. Runaway and moist greenhouse atmospheres and the evolution of Earth and Venus. *Icarus* 74:472–494.

Kasting, J. 1991. CO_2 condensation and the climate of early Mars. *Icarus* 94: 1–13.

Kasting, J. F., Whitmire, D. P., and Reynolds, R. T. 1993. Habitable zones around main sequence stars. *Icarus* 101:108–128.

Khurana, K. K., Kivelson, M. G., Stevenson, D. J., Schubert, G., Russell, C. T., Walker, R. J., and Polanskey, C. 1998. Induced magnetic fields as evidence for subsurface oceans in Europa and Callisto. *Nature* 395:777–780.

Klein, H. P. 1977. The Viking biological investigation: General aspects. *J. Geophys. Res* 82:4677–4680.

Klein, H. P. 1978. The Viking biological experiments on Mars. *Icarus* 34:666–674.
Klein, H. P., Horowitz, N. H., Oyama, V. I., Lederberg, J., Rich, A., Hubbard, J. S., Hobby, G. L., Straat, P. A., Berdahl, B. J., Carle, G. C., Brown, F. S., and Johnson, R. D. 1976. The Viking biological investigation: Preliminary results. *Science* 194:99–105.
Küppers, B. 1990. *Information and the Origin of Life* (Cambridge, MA: MIT Press).
Kushner, D. 1981. Extreme environments: Are there any limits to life? In *Comets and the Origin of Life*, ed. C. Ponnamperuma (Hingham, MA: Reidel), pp. 241–248.
Laskar, J., and Robutel, P. 1993. The chaotic obliquity of the planets. *Nature* 361:608–614.
Laskar, J., Joutel, F., and Robutel, P. 1993. Stabilization of The Earth's obliquity by the Moon. *Nature* 361:615–617.
Lebreton, J. 1997. Huygens Titan atmospheric probe. In *Encyclopedia of Planetary Sciences*, ed. J. H. Shirley and R. W. Fairbridge (New York: Chapman and Hall), pp. 311–314.
Lederberg, J. 1960. Exobiology: Approaches to life beyond the Earth. *Science* 132:393–398.
Lederberg, J. 1965. Signs of life: Criterion-system of exobiology. *Nature* 207:9–13.
Lerner, N. R. 1995. Influence of Murchison or Allende minerals on hydrogen-deuterium exchange of amino acids. *Geochim. Cosmochim. Acta* 59:1623–1631.
Levin, G. V. and Straat, P. A. 1981*a*. Antarctic soil No. 726 and implications for the Viking labeled release experiment. *J. Theor. Biol.* 91:41–45.
Levin, G. V., and Straat, P. A. 1981*b*. A search for a non-biological explanation of the *Viking* labeled release life detection experiment. *Icarus* 45:494–516.
Lissauer, J. J. 1995. Urey Prize Lecture: On the diversity of plausible planetary systems. *Icarus* 114:217–236.
Madigan, M. T., Martinko, J. M., and Parker, J. 1997. *Brock Biology of Microorganisms*, 8th ed. (Upper Saddle River, NJ: Prentice Hall).
Maher, K. A., and Stevenson, D. J. 1988. Impact frustration and the origin of life. *Nature* 331:612–614.
Marcy, G. W., and Butler, R. P. 1998. Detection of extrasolar giant planets. *Ann. Rev. Astron. Astrophys.* 36:57–97.
Matson, D. L. 1997. Cassini mission. In *Encyclopedia of Planetary Sciences*, ed. J. H. Shirley and R. W. Fairbridge (New York: Chapman and Hall), pp. 84–87.
Mayr, E. 1995. The Search for extraterrestrial intelligence. In *Extraterrestrials–Where Are They?*, 2nd ed., ed. B. Zuckerman and M. H. Hart (Cambridge: Cambridge University Press), pp. 152–156.
Mazur, P. 1980. Limits to life at low temperatures and at reduced water contents and water activities. *Orig. Life Evol. Biosph.* 10: 137–159.
McKay, C. P. 1986. Exobiology and future Mars missions: The search for Mars' earliest biosphere. *Adv. Space Res.* 6:269–285.
McKay, C. P. 1996. Time for intelligence on other planets. In *Circumstellar Habitable Zones: Proceedings of the First International Conference*, ed. L. R. Doyle (Menlo Park, CA: Travis House Publications), pp. 405–419.
McKay, C. P., and Stoker, C. R. 1989. The early environment and its evolution on Mars: Implications for life. *Rev. Geophys.* 27:189–214.
Miller, S. L., and Bada, J. L. 1988. Submarine hot springs and the origin of life. *Nature* 334:609–611.
Miller, S. L., and Urey, H. C. 1959. Organic compound synthesis on the primitive Earth. *Science* 130:245–251.

Murray, N., Hansen, B., Holman, M., and Tremaine, S. 1998. Migrating planets. *Science* 279:69–72.

Nealson, K. H. 1997*a*. The limits of life on Earth and searching for life on Mars. *J. Geophys. Res.* 102:23675–23686.

Nealson, K. H. 1997*b*. Sediment bacteria: Who's there, what are they doing, and what's new? *Ann. Rev. Earth Planet. Sci.* 25:403–434.

Oberbeck, V. R., and Fogleman, G. 1989. Impacts and the origin of life. *Nature* 339:434.

Oró, J., Squyres, S. W., Reynolds, R. T., and Mills, T. M. 1992. Europa: Prospects for an ocean and exobiological implications. In *Exobiology in Solar System Exploration*, ed. G. C. Carle, D. E. Schwartz, and J. L. Huntington (NASA SP-512), (Washington, D.C.: U. S. Government Printing Office) pp. 103–125.

Pappalardo, R. T., Belton, M. J. S., Breneman, H. H., Carr, M. H., Chapman, C. R., Collins, G. C., Denk, T., Fagents, S., Geissler, P. E., Giese, B., Greeley, R., Greenberg, R., Head, J. W., Heifenstein, P., Hoppa, G., Kadel, S. D., Klaasen, K. P., Klemaszewski, J. E., Magee, K., McEwen, A. S., Moore, J. M., Moore, W. B., Neukum, G., Philips, C. B., Prockter, L. M., Shubert, G., Senske, D. A., Sullivan, R. J., Tufts, B. R., Turtle, E. P., Wagner, R., and Williams, K. K. 1999. Does Europa have a subsurface ocean? Evaluation of the geological evidence. *J. Geophys. Res.*, in press.

Peltzer, E. T., Bada, J. L., Schlesinger, G., and Miller, S. L. 1984. The chemical conditions on the parent body of the Murchison meteorite: Some conclusions based on amino, hydroxy and dicarboxylic acids. *Adv. Space Res.* 4:69–74.

Pendleton, Y. P., and Chiar, J. E. 1997. The nature and evolution of interstellar organics. In *From Stardust to Planetesimals*, ed. Y. J. Pendleton and A. G. G. M. Tielens (San Francisco: Astronomical Society of the Pacific), pp. 179–200.

Podolak, M., and Prialnik, D. 1997. ^{26}Al and liquid water environments in comets. In *Comets and the Origin and Evolution of Life,* eds. P. J. Thomas, C. F. Chyba, and C. P. McKay (New York: Springer-Verlag), pp. 259–272.

Pollack, J. B., Kasting, J. F., Richardson, S. M., and Poliakoff, K. 1987. The case for a wet warm climate on early Mars. *Icarus* 71:203–224.

Reynolds, R. T., Squyres, S. W., Colburn, D. S., and McKay, C. P. 1983. On the habitability of Europa. *Icarus* 56:246–254.

Reynolds, R. T., McKay, C. P., and Kasting, J. F. 1987. Europa, tidally heated oceans, and habitable zones around giant planets. *Adv. Space Res.* 7(5):125–132.

Rodonò, M. 1986. Rotational modulation and flares on RS CVn and BY Dra-type stars. *Astron. Astrophys.* 165:135–156.

Runnegar, B. 1992. Evolution of the earliest animals. In *Major Events in the The History of Life*, ed. J. W. Schopf (Boston: Jones and Barlett), pp. 65–93.

Sackmann, I.-J., Boothroyd, A. I., and Kraemer, K. E. 1993. Our Sun. III. Present and future. *Astrophys. J.* 418:457–468.

Sagan, C. 1970. Life. In *Encyclopaedia Britannica.*

Sagan, C. 1996. Circumstellar habitable zones: An introduction. In *Circumstellar Habitable Zones: Proceedings of the First International Conference*, ed. L. R. Doyle (Menlo Park, CA: Travis House Publications), pp. 3–14.

Sagan, C., and Chyba, C. 1997. The early faint Sun paradox: Organic shielding of ultraviolet-labile greenhouse gases. *Science* 276:1217–1221.

Sagan, C., Thompson, W. R., and Khare, B. N. 1992. Titan: A laboratory for prebiological organic chemistry. *Acc. Chem. Res.* 25:286–292.

Schopf, J. W., and Walter, M. R. 1983. Archean microfossils: New evidence of ancient microbes. In *Earth's Earliest Biosphere, Its Origin and Evolution*, ed. J. W. Schopf (Princeton: Princeton University Press), pp. 214–239.

Shock, E. L. 1990. Do amino acids equilibrate in hydrothermal fluids? *Geochim. Cosmochim. Acta* 54:1185–1189.
Shock, E. L. 1992. Stability of peptides in high-temperature aqueous solutions. *Geochim. Cosmochim. Acta* 56:3481–3491.
Sieger, M. T., Simpson, W. C., and Orlando, T. M. 1998. Production of O_2 on icy satellites by electronic excitation of low-temperature water ice. *Nature* 394:554–556.
Sleep, N. H., Zahnle, K. J., Kasting, J. F., and Morowitz, H. J. 1989. Annihilation of ecosystems by large asteroid impacts on the early Earth. *Nature* 342:139–142.
Solomon, S. C., and Head, J. W. 1991. Fundamental issues in the geology and geophysics of Venus. *Science* 252:252–260.
Spencer, J. R., Calvin, W. M., and Person, M. J. 1995. Charge-coupled device spectra of the Galilean satellites: Molecular oxygen on Ganymede. *J. Geophys. Res.* 100:19049–19056.
Stevens, T. O., and McKinley, J. P. 1995. Lithotrophic microbial ecosystems in deep basalt aquifers. *Science* 270:450–453.
Tarter, J. C. 1997. Results from Project Phoenix: Looking up from down under. In *Astronomical and Biochemical Origins and the Search for Life in the Universe*, ed. C. B. Cosmovici, S. Bowyer, and D. Werthimer (Bologna: Editrice Compositori), pp. 633–643.
Tarter, J. C., and Michaud, M. A., eds. 1990. *Acta Astronaut.* 21(2):69–154.
Trilling, D. E., Benz, W., Guillot, T., Lunine, J. I., Hubbard, W. B., and Burrows, A. 1998. Orbital evolution and migration of giant planets: Modeling extrasolar planets. *Astrophys. J.* 500:428–439.
Vishniac, H. S. 1993. The microbiology of antarctic soils. In *Antarctic Microbiology*, ed. E. I. Friedmann (New York: Wiley-Liss), pp. 297–341.
von Kiedrowski, G. 1986. A self-replicating hexadeoxynucleotide. *Angew. Chem. Int. Ed. Engl.* 25:932–935.
Wächtershäuser, G. 1988*a*. Before enzymes and templates: Theory of surface metabolism. *Microbiol. Rev.* 52:452–484.
Wächtershäuser, G. 1988*b*. Pyrite formation, the first energy source for life: A hypothesis. *Syst. Appl. Microbiol.* 10:207–210.
Wächtershäuser, G. 1990. The case for the chemoautotrophic origin of life in an iron-sulfur world. *Orig. Life Evol. Biosph.* 20:173–176.
Wald, G. 1964. The origins of life. *Proc. Natl. Acad. Sci. USA* 52:595.
Walker, J. C. G., Hays, P. B., and Kasting, J. F. 1981. A negative feedback mechanism for the long-term stabilization of Earth's surface temperature. *J. Geophys. Res.* 86:9776–9782.
Weidenschilling, S. J., and Marzari, F. 1996. Gravitational scattering as a possible origin for giant planets at small stellar distances. *Nature* 384:619–621.
Wetherill, G. W. 1996. The formation and habitability of extra-solar planets. *Icarus* 119:219–238.
Whitman, W. B., Coleman, D. C., and Wiebe, W. J. 1998. Prokaryotes: The unseen majority. *Proc. Natl. Acad. Sci. USA* 95:6578–6585.
Whitmire, D. P., and Reynolds, R. T. 1996. Circumstellar habitable zones: Astronomical considerations. In *Circumstellar Habitable Zones: Proceedings of the First International Conference*, ed. L. R. Doyle (Menlo Park, CA: Travis House Publications), pp. 117–142.
Whitmire, D. P., Doyle, L. R., Reynolds, R. T., and Matese, J. J. 1995. A slightly more massive young Sun as an explanation for warm temperatures on early Mars. *J. Geophys. Res.* 100:5457–5464.
Whitmire, D. P., Matese, J. J., Criswell, L., and Mikkola, S. 1998. Habitable planet formation in binary star systems. *Icarus* 132:196–203.

Williams, D. W., and Kasting, J. F. 1997. Habitable planets with high obliquities. *Icarus* 129:254–267.

Williams, D. W., Kasting, J. F., and Wade, R. A. 1997. Habitable moons around extrasolar giant planets. *Nature* 385:234–236.

Woolf, N., and Angel, R. 1998. Astronomical searches for Earth-like planets and signs of life. *Ann. Rev. Astron. Astrophys.* 36:507–537.

Zaug, A. J., and Cech, T. R. 1986. The intervening sequence RNA of *Tetrahymena* is an enzyme. *Science* 231:470–475.

Zolensky, M., and McSween, H. Y. 1988. Aqueous alteration. In *Meteorites and the Early Solar System,* eds. J. F. Kerridge and M. S. Matthews (Tucson: University of Arizona Press), pp. 114–143.

GLOSSARY

GLOSSARY*

AGB star	asymptotic giant branch star; a large red giant, 100–1000 solar radii, with a hot carbon-oxygen white dwarf as its core. Such stars are precursors of planetary nebulae and get their name from the path they take on an HR diagram (plot of luminosity vs. effective temperature).
Alfvén surface	geometric surface in a magnetized wind where the poloidal components of the flow speed and of the local Alfvén speed are equal.
Alfvén velocity (v_A)	the speed at which hydromagnetic waves are propagated along a magnetic field: $\nu_A = B/(4\pi\rho)^{1/2}$, where B is the magnetic field strength and ρ is the density.
AU	astronomical unit; the mean distance between Earth and the Sun (1.496×10^{13} cm $\simeq$ 500 light-seconds).
BN-KL	the Becklin-Neugebauer Object/Kleinman-Low Nebula region, a small dense region in the Orion Molecular Cloud that contains luminous infrared sources, interstellar and circumstellar masers, hot gas, and young stars.
bolometric	referring to electromagnetic radiation received from an object at all wavelengths of the spectrum.
brown dwarf	a starlike object that is not massive enough to sustain hydrogen fusion as its main source of luminosity, but

* We have used some definitions from *Glossary of Astronomy and Astrophysics* by J. Hopkins (by permission of the University of Chicago Press, copyright 1980 by the University of Chicago), from *Astrophysical Quantities* by C. W. Allen (London: Athlone Press, 1973), and from *The Planetary System* by David Morrison and Tobias Owen (Reading, Mass.: Addison-Wesley Publishing Co., 1988).

	instead radiates by virtue of the energy of collapse of the original gas and dust from which it formed and, for a short time, by deuterium fusion.
CAI	calcium and aluminum-rich inclusion; a millimeter- to centimeter-sized, white-colored component of a chondritic meteorite, composed of high-temperature minerals. CAIs are thought to be among the oldest sampled solids from the early solar system. Compositions are greatly depleted from cosmic abundances in the relatively volatile elements and hence are enriched in Ca, Al, and other involatile elements.
carbonaceous chondrites types 1,2,3	a rare type of chondrite (about 30 are known) characterized by the presence of carbon compounds. Carbonaceous chondrites are the most primitive samples of matter known in the solar system. They resemble the solid material expected when a gas cloud of solar composition cools to a temperature of about 300 K at low pressure. Wiik (1956) classification: Type C1 are the least dense and are strongly magnetic. They are the most primitive and show only minimal chemical alteration. Type C2 are weakly magnetic or non-magnetic. Type C3 are the densest, are water-poor, and usually are largely olivine.
chondrite	a type of meteorite composed mainly of silicates that aggregated in the solar nebula. Chondrites have not undergone extensive melting, so they preserve some of their initial components and textures.
chondrules	almost-spherical silicate objects up to $\sim$1 cm in size that solidified from the quick cooling of melt droplets. They formed in the early solar system and are the dominant component of most chondrites.
Class 0 source	a protostar near the beginning of its main accretion phase; characterized observationally by strong, compact continuum emission and by little or no emission at wavelengths shorter than about 10 μm.
Class I and II sources	a Class I source is a deeply embedded, usually invisible, young stellar object with a spectral energy

distribution that rises with increasing wavelength longward of 2.2 microns. A Class II source is a young stellar object whose spectral energy distribution decreases with increasing wavelength longward of 2.2 microns. Classical T Tauri stars are typically Class II sources. The energy distribution of a Class II source is often well modeled by emission from a star and a circumstellar disk.

Coulomb barrier — the potential energy "hill" due to the repulsive electrostatic force between like-charged particles. For nuclear reactions, the Coulomb barrier acts to bar low-energy (energies less than the height of the barrier) positively charged particles from approaching the positively charged atomic nucleus to the point where a nuclear reaction can proceed.

FWHM — full width at half maximum.

Hayashi phase — the period of pre-main-sequence stellar evolution during which a star has negligible nuclear energy production and has sufficiently low internal temperatures that convective energy transport dominates through most or all of the interior. The star evolves in the HR diagram with decreasing luminosity and nearly constant effective temperature.

Herbig Ae/Be stars (HAeBe stars) — pre-main-sequence stars of spectral types A and B. Named after George Herbig, who was first to identify these high-mass counterparts to the approximately solar-mass T Tauri stars. Like T Tauri stars, Herbig Ae/Be stars are often characterized by strong infrared-millimeter continuum excesses, which in some cases arise as thermal emission in circumstellar disks.

High-*Z* material — in stellar and planetary models, that component composed of atoms with an atomic number higher than 2.

HR diagram — Hertzsprung-Russell diagram; a plot of stars' luminosity vs. surface temperature (which can be indicated as spectral type or color). In present usage, a plot of absolute bolometric magnitude against effective temperature for a population of stars.

HST	*Hubble Space Telescope;* an orbiting observatory that provides high-spatial-resolution measurements in the ultraviolet, visible, and near infrared.
Initial mass function (IMF)	the distribution of stellar masses at birth; relative number of stars formed per unit mass.
IRAS	*Infrared Astronomical Satellite;* a joint UK-USA-Dutch satellite that provided the first surveys of the sky at thermal infrared wavelengths with moderate spatial and spectral resolution.
ISO	*Infrared Space Observatory:* a European infrared space telescope launched in 1995 and operated until May 1998.
jansky (Jy)	unit of flux density adopted by the International Astronomical Union (IAU) in 1973. 1 Jy = 10^{-26} W m^{-2} Hz^{-1}; named after Karl Jansky, who discovered galactic radio waves in 1931.
Jeans instability	a gravitational instability in an idealized, infinite, homogeneous medium.
Jeans length	the critical wavelength at which the oscillations in an infinite, homogeneous medium become gravitationally unstable. Any disturbance whose characteristic dimension is greater than the Jeans length will decouple by self-gravitation from the rest of the medium to become a stable, bound system.
Jeans mass	the mass necessary for protostellar collapse, $M_j \geq 1.795(RT/G)^{3/2}\rho^{-1/2}$, where R is the gas constant, T the cloud temperature, G the gravitational constant, and ρ the cloud density.
Kelvin-Helmholtz instability	the tendency of waves to grow on a shear boundary between two regions in relative motion parallel to the boundary.
Kelvin-Helmholtz time	the time over which a self-gravitating star radiates away enough thermal energy to contract by a significant fraction of its initial radius. It is assumed that the star is not being supported by internal nuclear fusion.

Kuiper Belt	population of cometary objects extending beyond the planetary region of our solar system. Sometimes known as the Edgeworth-Kuiper Belt.
Lagrangian points	the five equilibrium points in the restricted three-body problem.
Lindblad resonance	a resonance of first order in the eccentricity of a perturbed particle. This terminology comes from galactic dynamics; in the 1920s P. Lindblad invoked such resonances to explain spiral arms.
MHD	magnetohydrodynamics, sometimes called hydromagnetics; the study of the collective motions of charged particles in a magnetic field.
Mie scattering theory	the scattering of light by particles with size comparable to a wavelength of the light. Exact albedo and phase functions are highly oscillatory with direction and wavelength; however, slight irregularities and distributions of size smooth out the features, leaving predominantly forward-scattering behavior.
minimum-mass solar nebula	the minimum mass for the solar nebula (~ 0.01 solar masses) is computed from the total mass of the planets, adjusting for the difference in chemical composition between the planets and the interstellar medium.
MRN grain size distribution	distribution of grain sizes introduced by J. S. Mathis, W. Rumpl, and K. H. Nordsieck (1977) to account for observations of interstellar extinction from 0.11 μm to 1.0 μm. The distribution has the form $N(a)da \propto a^{-3.5}da$, where a is the grain radius. It extends from 5 nm to about 1 μm for graphite and over a narrower range for other materials.
OMC-1	Orion Molecular Cloud 1; a very dense north-south ridge about 1 pc in length near the Orion nebula. It contains the active BN-KL region.
Orion-A molecular ridge	a giant molecular cloud near the Orion Nebula within which the Orion Molecular Cloud 1 (OMC-1) and the BN-KL region are located.

PAH	polycyclic aromatic hydrocarbon; a class of organic compounds thought to be important in interstellar and possibly protoplanetary chemistry.
parsec (pc)	the distance where 1 AU subtends 1 arcsec = 206,265 AU = 3.26 lightyear = 3.086×10^{18} cm.
P Cygni line profile	spectral line profile characterized by asymmetric emission and absorption components indicative of massive winds around a star. Named for a particular B-type supergiant star with a massive wind.
Planck blackbody function	the dependence of emitted radiance on wavelength for an object that behaves as a perfect blackbody, i.e., is capable of absorbing all radiation incident upon it.
position angle	angle in the sky (for example, of the direction of linear polarization). It is normally denoted by θ and is measured with $\theta = 0°$ toward the north and θ increasing towards the east.
proplyds	*pro*to*pl*anetar*y* *d*isk*s*, specifically compact nebulae surrounding low-mass stars in the Orion Nebula that are externally ionized in the radiation fields of nearby OB stars.
Rayleigh-Jeans side	the low-frequency portion of the Planck blackbody emission law, for which $h\nu/kT \ll 1$.
Reynolds number	a dimensionless number (Re $= Lv/\nu$, where L is a typical dimension of the system, v is a measure of the velocities that prevail, and ν is the kinematic viscosity) that governs the conditions for the occurrence of turbulence in fluids.
SIRTF	*Space Infrared Telescope Facility;* a cryogenically cooled telescope with sensitive, large-format array detectors for infrared imaging, photometry, and spectroscopy. Due to be launched in late 2001.
SOFIA	*Stratospheric Observatory for Infrared Astronomy;* a 2.5-m diameter telescope that will fly in an aircraft above most of the water vapor in the Earth's atmosphere.

T Tauri stars	pre-main-sequence stars of approximately solar mass. T Tauri stars are often characterized by strong emission-line spectra, infrared-millimeter continuum excesses (from circumstellar disks), and photometric and spectroscopic variability.
virial theorem	the principle that for a bound gravitational system the long-term average of the kinetic energy is one-half of the potential energy.
Wien side	the high-frequency portion of the Planck blackbody emission law, for which $h\nu/kT \gg 1$.
YSO	young stellar object; general term for stars that are in the several phases of evolution before reaching the main sequence.
ZAMS	zero-age main sequence; the curve on the HR diagram occupied by stars that have just begun stable fusion of hydrogen. A star's position on the ZAMS depends on its mass.

COLOR SECTION

t=0.2 t_g β=0.01, n_J=3, $E_{w,\ init}/c_s^2$=100

t=0.1 t_g β=0.1, n_J=3, $E_{w,\ init}/c_s^2$=100

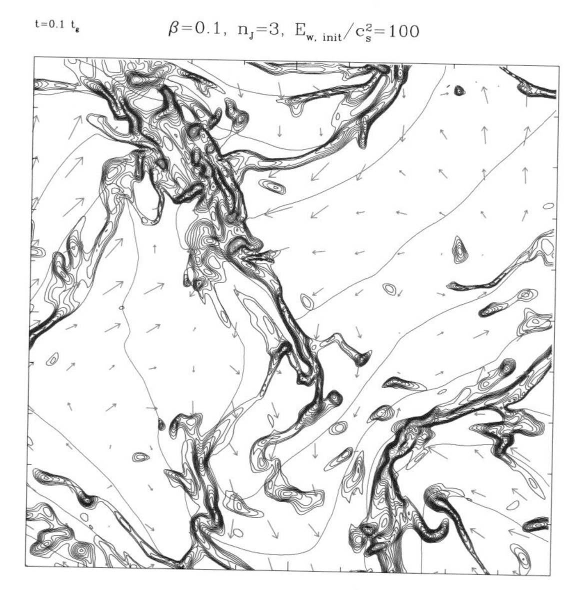

Color Plate 1. Sample density, velocity, and magnetic field structure in two turbulent cloud simulations. Density contours are logarithmic, with dark blue running from the mean density to ten times the mean, and light blue 10–100 times the mean; magnetic field lines are shown in green, and the velocity field is shown in red. For both models, the Mach number is ten; for the upper (lower) figure, β = 0.01 (0.1), corresponding to stronger (weaker) mean magnetic fields. See Ostriker et al. (1998) for more details. (See the chapter by Vázquez-Semadeni et al., this volume.)

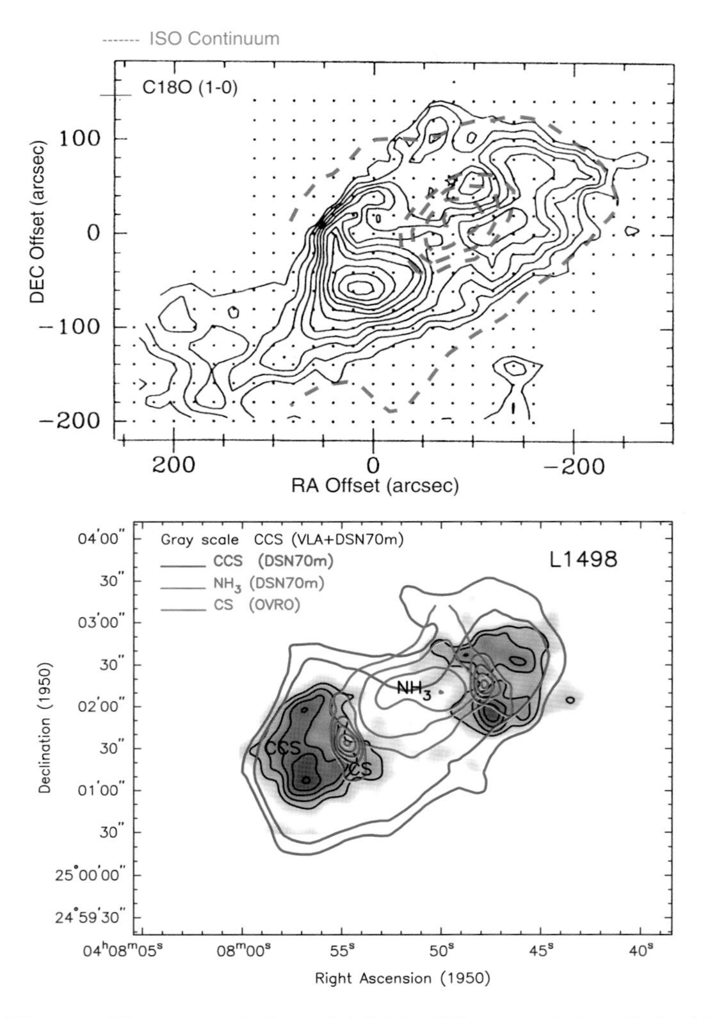

Color Plate 2. The top panel shows L1498 in 200-μm emission (dashed lines) measured by the *Infrared Space Observatory* (ISO), which traces out the dense core (Willacy et al. 1998a). The solid lines are a $C^{18}O(1\text{–}0)$ map showing a minimum at this position. This indicates that even CO is strongly depleted in the center of cold dense cores. The bottom panel traces the structure of the L1498 core in CS, CCS, and NH_3 emission (Kuiper et al. 1996). The molecules and telescopes used are indicated. These interferometer data highlight limb-brightened emission, indicating that these molecules trace shells around the center. The ammonia data also peak at the continuum position. (See the chapter by Langer et al., this volume.)

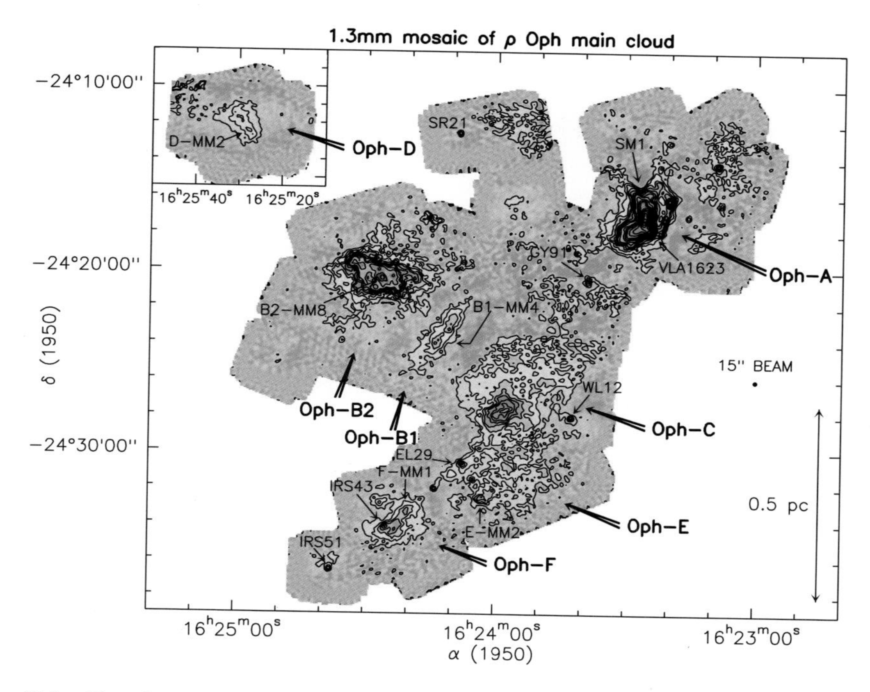

Color Plate 3. Dust continuum mosaic of the ρ Oph cloud taken at 1.3 mm with the IRAM 30-m telescope and the MPIfR bolometer array (Motte et al. 1998). (See the chapter by André et al., this volume.)

Color Plate 4. The Mon R2 cluster imaged in the near-infrared (from Carpenter et al. 1997). The field of view is ~9×6 arcmin, corresponding to ~2.2×1.4 pc for a distance of 830 pc. The bright nebulous stars near the image periphery are part of a larger chain of reflection nebulae. The central cluster is completely embedded and contains >300 stars within a 0.4-pc diameter. (See the chapter by Clarke et al., this volume.)

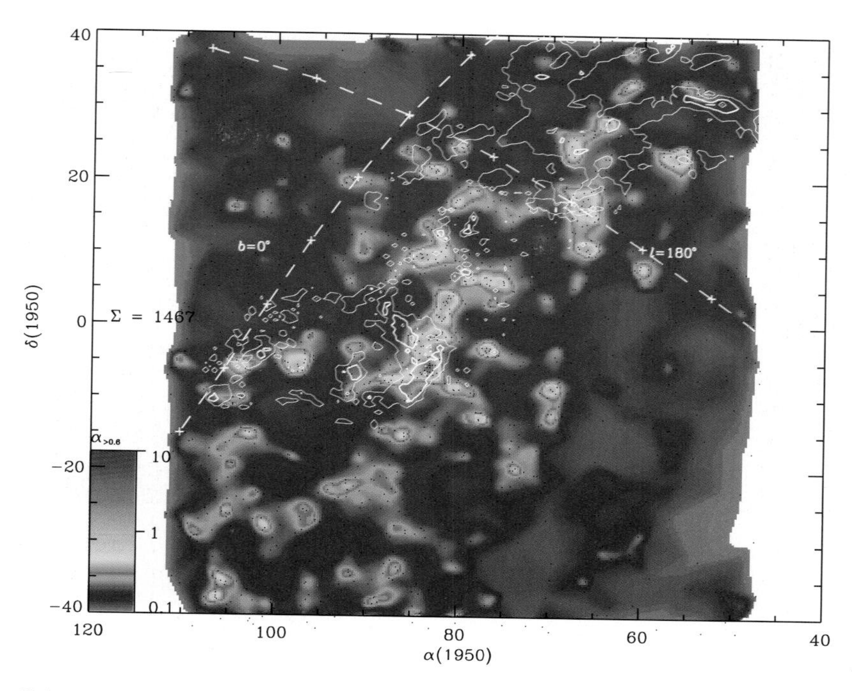

Color Plate 5. Large-scale spatial distribution of candidate young stars in Orion and Taurus-Auriga. The X-ray sources (dots) have been selected from the ROSAT All-Sky Survey. Their space density is color coded, in units of X-ray sources per square degree. Contours of molecular clouds (Maddalena et al. 1986) are overlain. The dashed lines show the galactic plane and galactic longitude = 180°. The regions of enhanced young star candidate density south of the galactic plane coincide with Gould's Belt. The Orion Nebula Cluster (ONC) is the strongest enhancement at the center of the image. The enhancements immediately to the north of the ONC are Ori OB1b and Ori OB1a. NGC 1788 is west of Ori OB1b. The Pleiades cluster is visible at the northwestern edge of the field. (See chapter by Walter et al., this volume.)

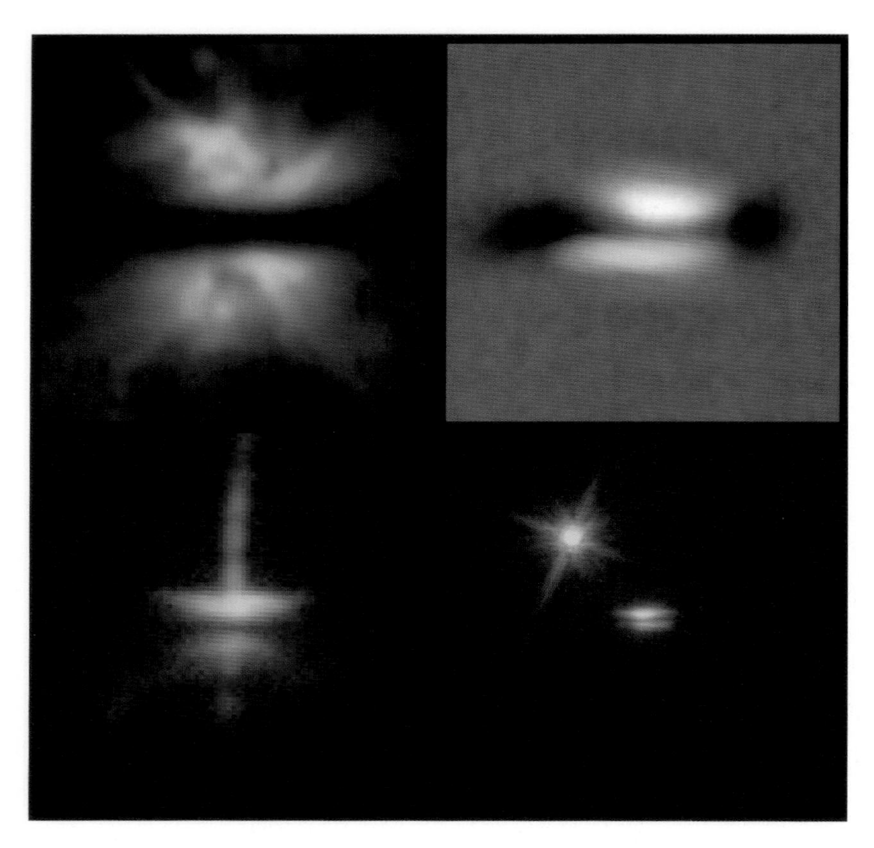

Color Plate 6. Continuum HST images of four edge-on disks, all plotted at the same equivalent linear scale and orientation for direct comparison. IRAS 04302+2247 (top left; Padgett et al. 1999*a*) and Orion 114-426 (top right; McCaughrean et al., in preparation) are shown as near-infrared true-color *JHK* composites. Both HH 30 (bottom left; Burrows et al. 1996) and HK Tau/c (bottom right; Stapelfeldt et al. 1998) are shown as optical pseudocolor *RI* composites. Each panel is 1200 AU square, and all intensities are logarithmically scaled. (See the chapter by McCaughrean et al., this volume.)

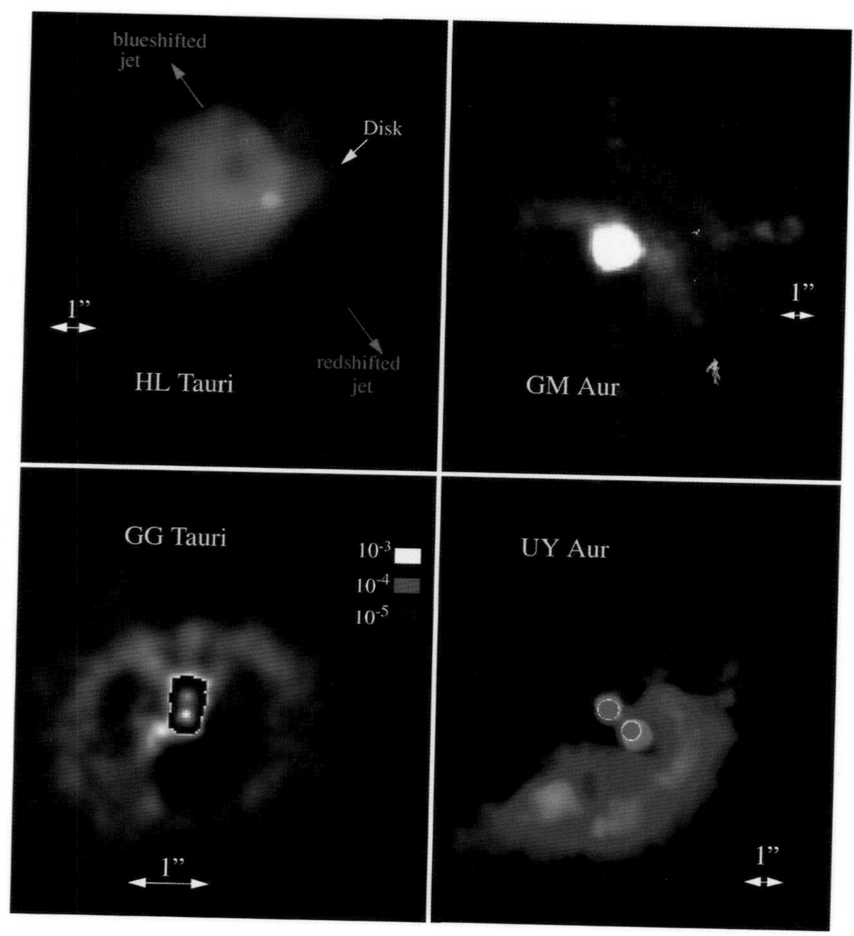

Color Plate 7. Four circumstellar disks detected in scattered light. The HL Tau image (top left) is a composite of an HST *R*-band image as blue (from Stapelfeldt et al. 1995) and AO images at *J* and *H* as green and red respectively (from Close et al. 1997*a*). The green extension is likely to be infrared light scattered off the front surface of the disk. Both GM Aur (top right) and GG Tau (bottom left) are represented with *J*-band AO images; note that the GG Tau logarithmic intensity scale is wrapped in order to show both the faint nebulosity and the fully resolved central binary (from Roddier et al. 1996). UY Aur (lower right) is shown as a composite *JHK* (as blue, green, red) AO image (from Close et al. 1998a). This circumbinary disk is somewhat closer to edge-on than that seen in GG Tau, and thus emission from the nearside of the disk is strongly scattered into the line of sight. (See the chapter by McCaughrean et al., this volume.)

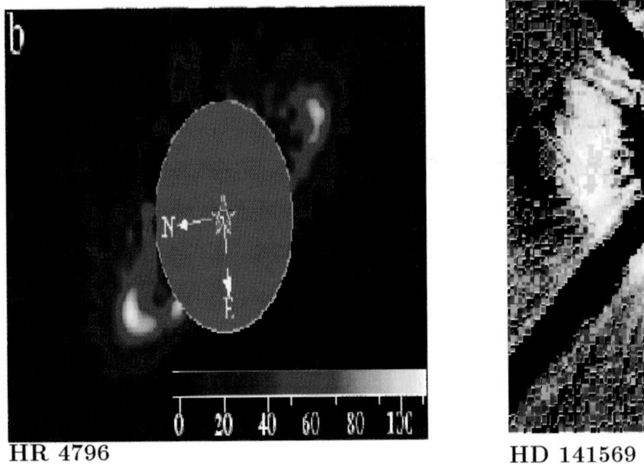

HR 4796
AOV, $\tau = 5\,10^{-3}$, 8–10 Myr
(Schneider et al. 1999).

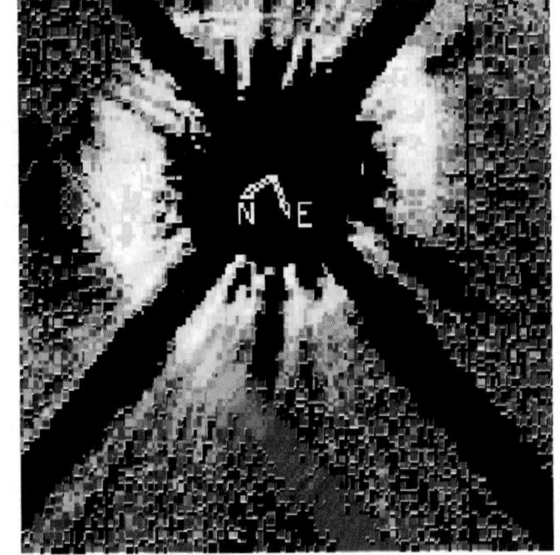

HD 141569
AOV, $\tau = 8.4\,10^{-3}$, $\geq$ 10 Myr
(Augereau et al., in prep).

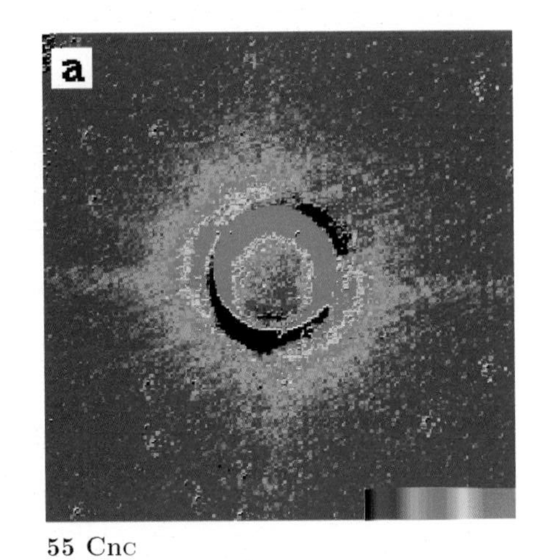

55 Cnc
G8V, $\tau = 5\,10^{-5}$, $\simeq$ 5 Gyr
(Trilling and Brown, 1998).

Color Plate 8. Images of recently detected disks around stars of various ages and fractional disk luminosities. (a) HR 4796; (b) HD 141569; (c) 55 Cnc. (See the chapter by Lagrange et al., this volume.)

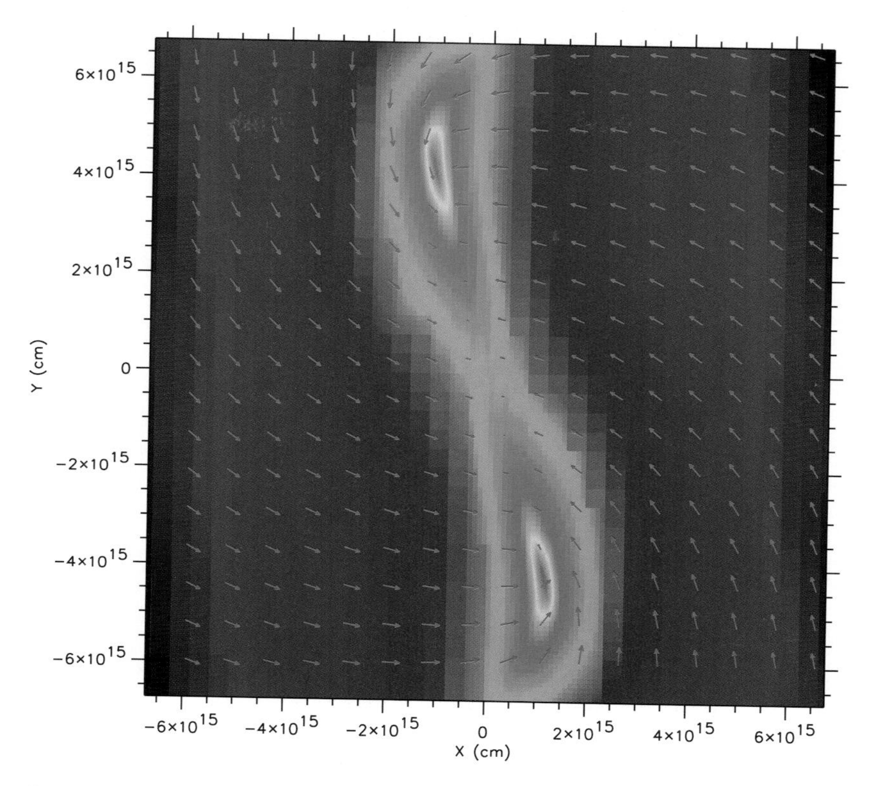

Color Plate 9. Equatorial slice of log density (in g cm^{-3}) at a time $t = 1.415 \times 10^{12}$ s after the onset of collapse of a uniform isothermal cloud subject to a 10% $m \geq 2$ perturbation (AMR code). The fragments have begun to evolve into filamentary structures. Density range: log $\rho_{\rm max} = -12.41$ (red); log $\rho_{\rm min} = -15.57$ (violet); resolution 2.44×10^{13} cm. Velocity vectors are shown with length proportional to speed; maximum velocity 0.824 km s^{-1}. (See the chapter by Bodenheimer et al., this volume.)

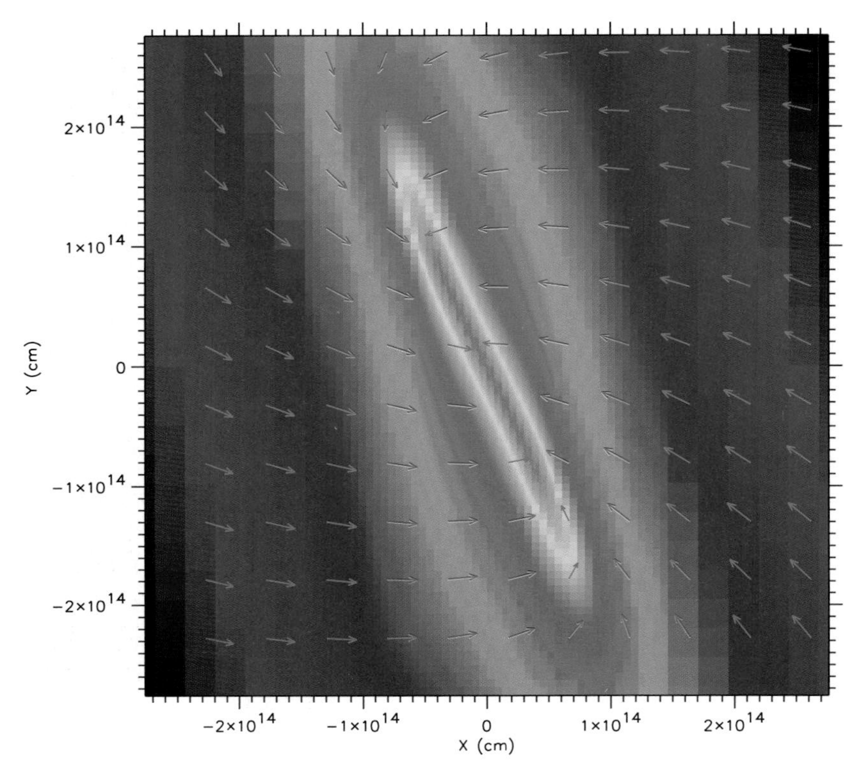

Color Plate 10. Equatorial slice of log density (in g cm^{-3}) at a time $t = 7.23 \times 10^{11}$ s after the onset of collapse of an isothermal cloud, initially with a Gaussian density profile subject to a 10% $m = 2$ perturbation (AMR code). Density range: $\log \rho_{\rm max} = -10.33$ (red); $\log \rho_{\rm min} = -13.11$ (violet); resolution 1.526×10^{12} cm; maximum velocity 0.742 km s^{-1}. The single-filament morphology was reproduced in a run at eight times the spatial resolution (twice in linear dimension), verifying the convergence of the filamentary structure. (See the chapter by Bodenheimer et al., this volume.)

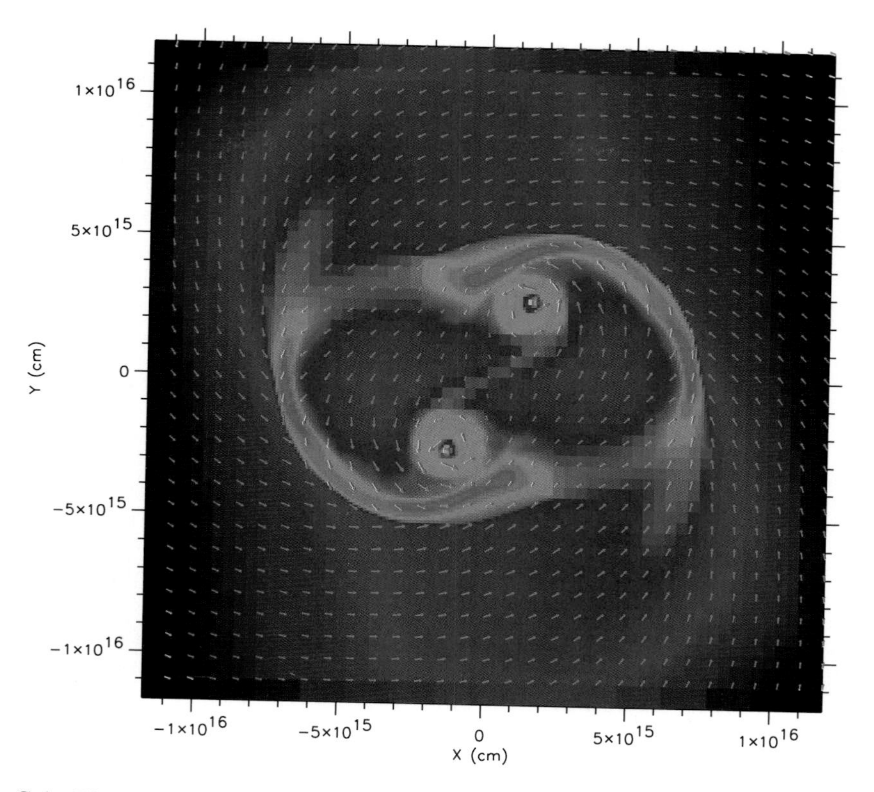

Color Plate 11. Equatorial slice in log density (in g cm^{-3}) at $t = 1.60 \times 10^{12}$ s for a nonisothermal cloud initially having uniform density and an $m = 2$ perturbation of 10% amplitude. Density range: log $\rho_{\rm max} = -10.28$ (red); log $\rho_{\rm min} = -16.48$ (violet); resolution 9.76×10^{13} cm; maximum velocity 2.75 km s^{-1}. The protostellar disks have interacted with the surrounding spiral arms. Fifty percent of the initial mass of the cloud has accreted into the binary cores, protostellar disks, and spiral arms at this time. (See the chapter by Bodenheimer et al., this volume.)

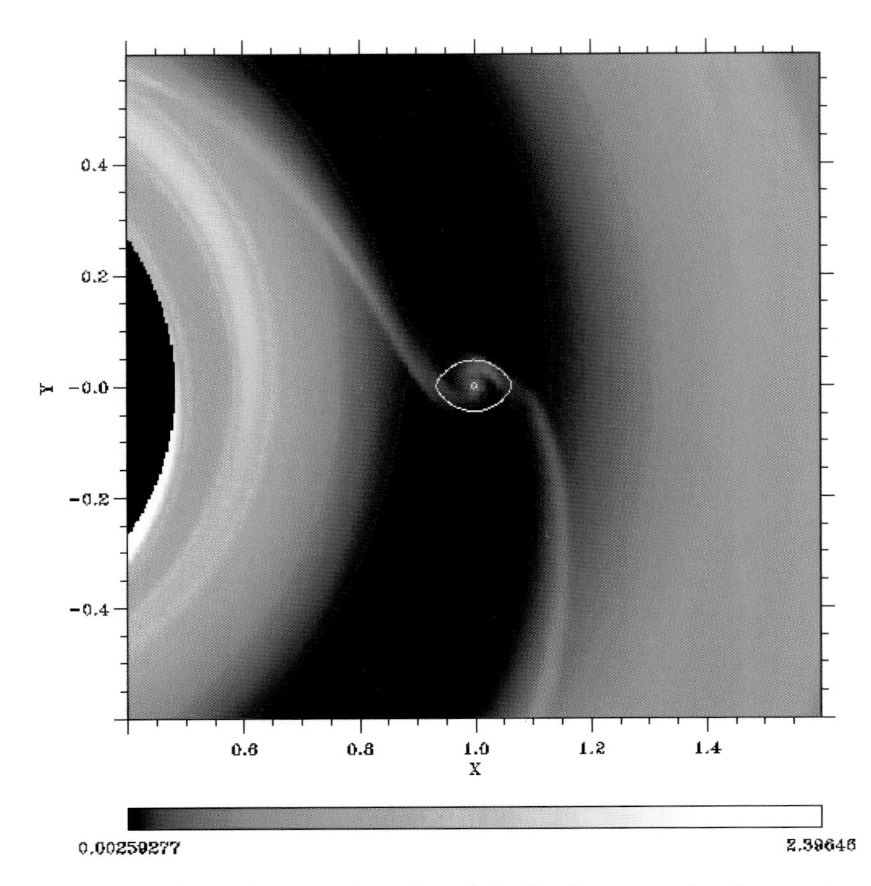

Color Plate 12. Jupiter grows in a solar nebula, despite gap opening, by accreting gas from each side of the disk. In this simulation (see text for details), Jupiter would double the mass in less than 1 Myr (Artymowicz et al. 1999). (See the chapter by Lubow and Artymowicz, this volume.)

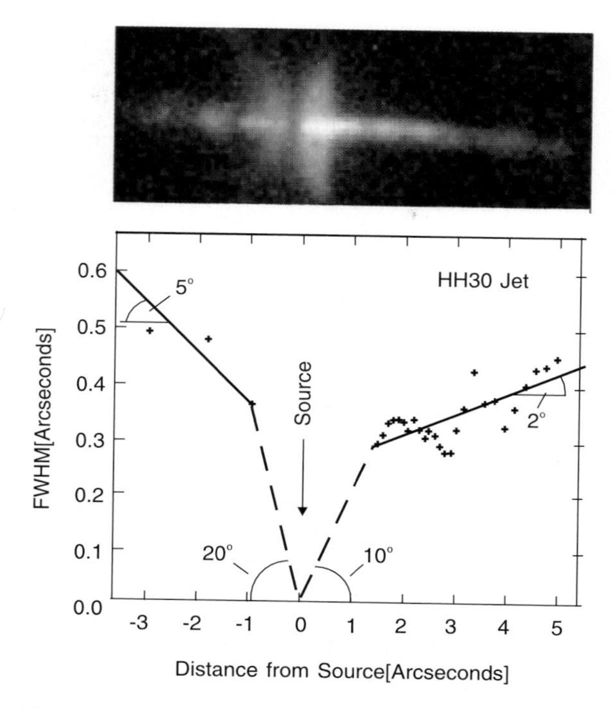

Color Plate 13. Top: Composite HST image of the HH 30 jet based on data from Ray et al. 1996. [S II]$\lambda\lambda$ 6716, 6731 emission is in red, Hα in green, and continuum light (F547M filter). Of the faint counterjet, only the two brightest knots are recognizable. Bottom: Measured width (FWHM) of the jet and counterjet as a function of distance from the source. Note the relatively large inferred angles (about 10°/20°) between the source and the innermost parts of the jet and counterjet as compared to the relatively small opening angles of the jet and counterjet (2° and 5°, respectively) farther out. (See the chapter by Eislöffel et al., this volume.)

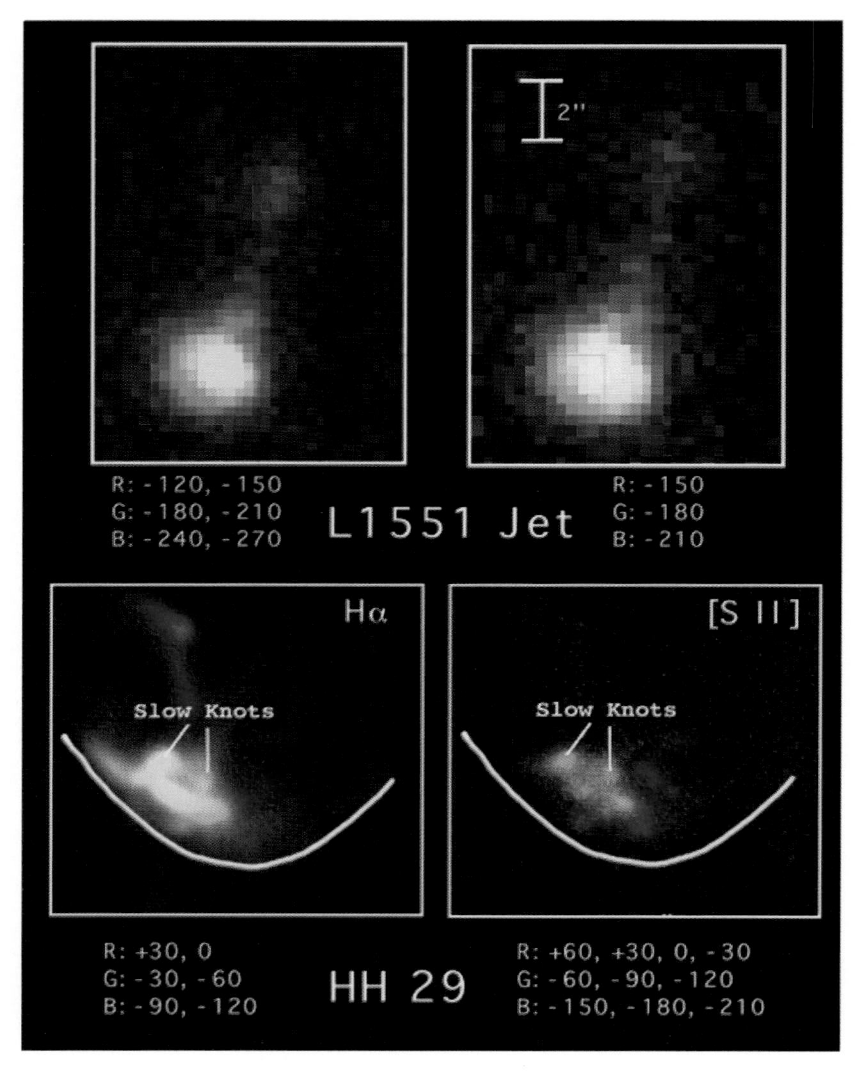

Color Plate 14. The velocity field of the L1551 jet and HH 29 obtained from Fabry-Perot data. The R, G, and B labels refer to the velocities of the emission, in km s^{-1} with respect to the molecular cloud, that are loaded into the red, green, and blue channels, respectively, for display. Hence, areas in white have emission in each of the channels and denote areas of large linewidths. Top: The jet emerges from an infrared source located above the top of the frame in these images of the L1551 jet. The highest blueshifted radial velocities occur in the jet. The bright knot at the end of the jet has a large linewidth (color white) and is surrounded by a shell of lower-velocity material, as expected for a bow shock. Bottom: The bow shock visible in the Hα image of HH 29 (left) appears to have wrapped around two of the slowest knots, as indicated by the red and green colors in the [S II] image (right). (See the chapter by Hartigan et al., this volume.)

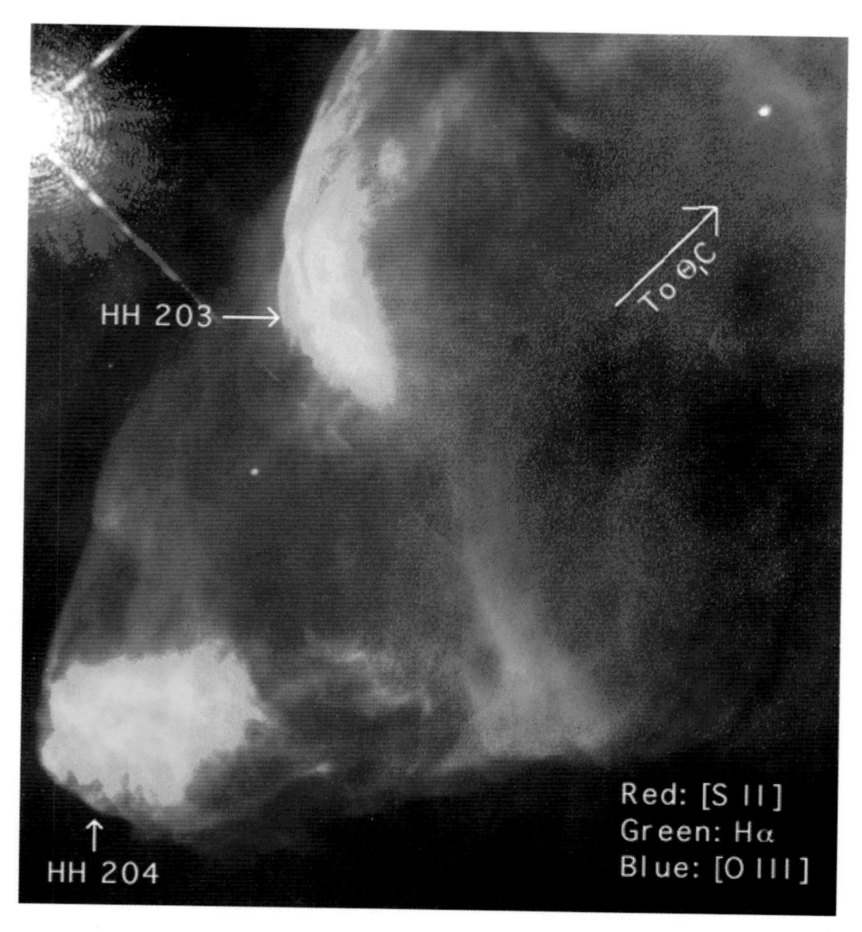

Color Plate 15. HST images of the HH 203/204 complex from O'Dell et al. (1997*a*). The HH objects seem to mark the apexes of bow shocks that move to the lower left in the figure. Unlike normal HH objects, where the highest ionization lines occur near the apexes of bow shocks, in this case the high-ionization line [O III] λ 5007 emits at the right and top of the figure, in the direction of the ionizing source θ^1C. This region shows an example of a mixture of photoionization and shock excitation. (See the chapter by Hartigan et al., this volume.)

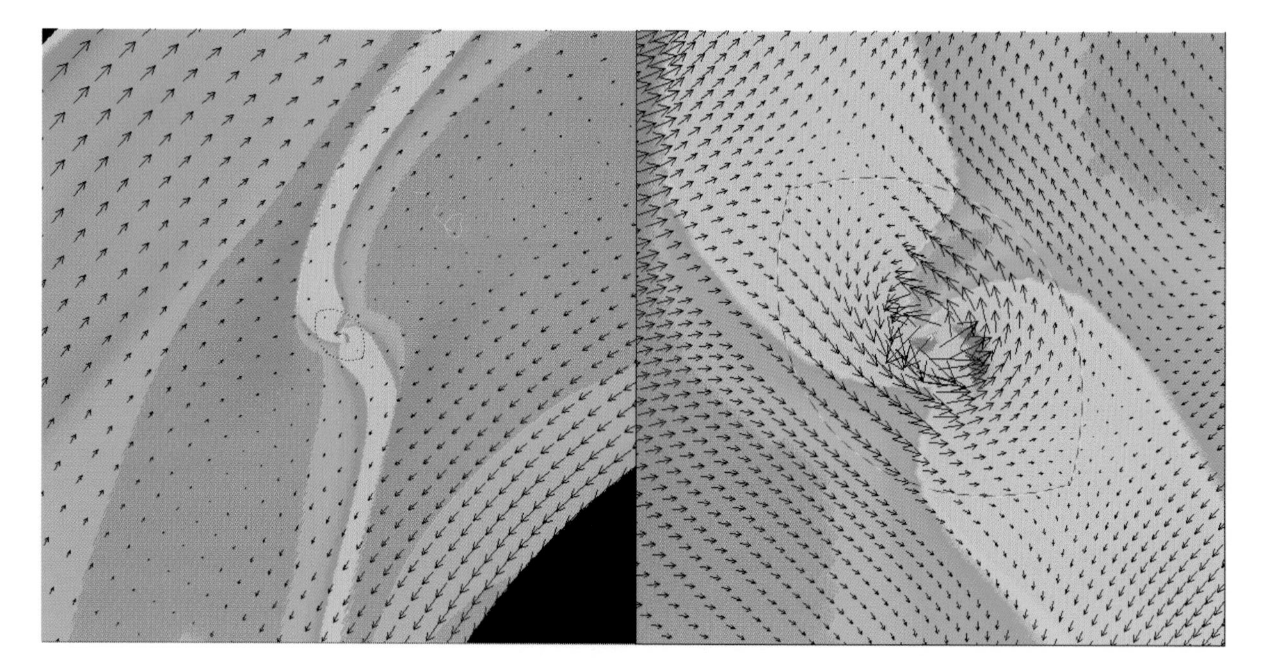

Color Plate 16. The flow pattern around an embedded protoplanet. On the left is the global trailing wave (the star is located toward the lower right). On the right is a close-up view of the circulatory region immediately surrounding the protoplanet. Color scale indicates surface density, and the protoplanet's Roche lobe is shown with a dotted line. Model parameters are $q = 10^{-4}$, $H/r = 0.07$ and $\alpha = 10^{-3}$. (See the chapter by Lin et al., this volume.)

Color Plate 17. Surface density distribution for a gap-opening protoplanet. The surface density near the planet (indicated by a white sphere) has been reduced by four orders of magnitude. Waves are clearly seen propagating both inward and outward away from the protoplanet. Model parameters are $q = 10^{-3}$, $H/r = 0.04$, and $\alpha = 10^{-3}$. (See the chapter by Lin et al., this volume.)

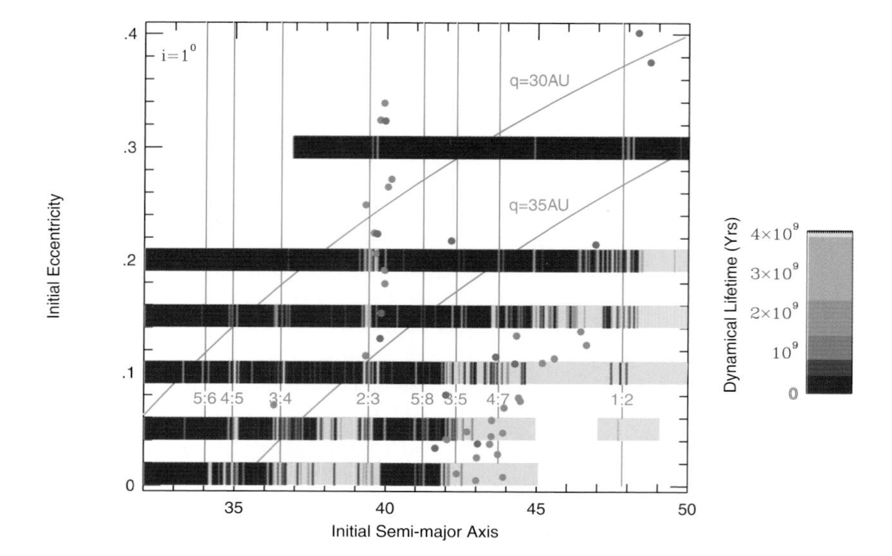

Color Plate 18. Dynamical lifetime before first close encounter with Neptune for test particles with a range of semimajor axes and eccentricities, and with initial inclinations of one degree (based on Duncan et al. 1995). Each particle is shown as a narrow vertical strip, centered on the particle's initial eccentricity and semimajor axis. The lightest-colored (yellow) strips represent particles that survived the length of the integration (4 Gyr). Dark regions are particularly unstable. The dots indicate the orbits of Kuiper Belt objects with reasonably well-determined orbits in January 1999. (Orbits with inclinations less than 10 degrees are shown in red; those with higher inclinations are displayed in green.) The locations of low-order mean-motion resonances with Neptune and two curves of constant perihelion distance q are shown. (See the chapter by Malhotra et al., this volume.)

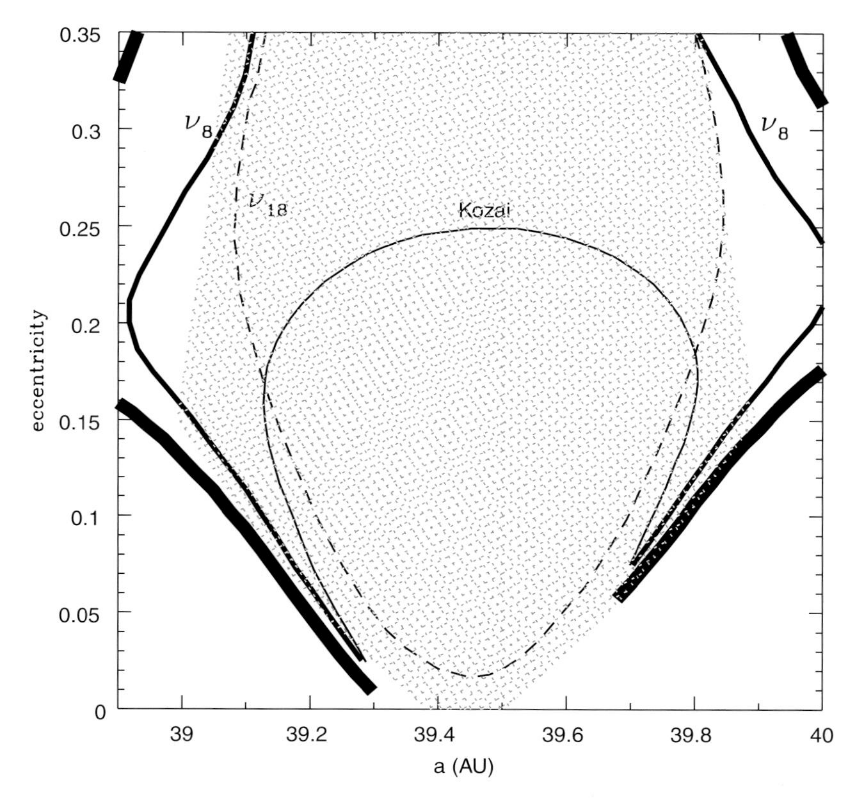

Color Plate 19. (a) The major dynamical features in the vicinity of Neptune's 3:2 mean-motion resonance. The locations of the resonance separatrix (dark solid lines), two secular resonances, apsidal ν_8 and nodal ν_{18}, and the Kozai resonance were obtained by a seminumerical analysis of the averaged perturbation potential of Neptune and the other giant planets. The blue-shaded region is the stable resonance libration zone in the unaveraged potential of Neptune. (Adapted from Morbidelli 1997 and Malhotra 1995.) (See the chapter by Malhotra et al., this volume.)

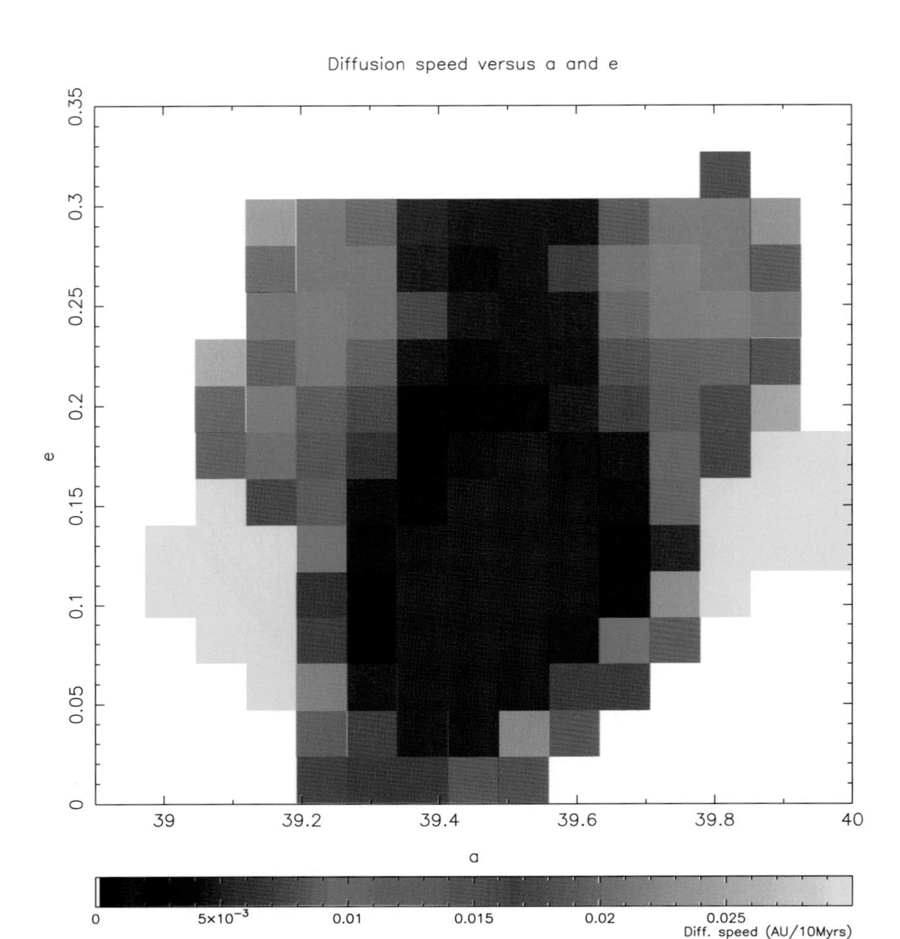

Color Plate 19. (b) Map of the dynamical diffusion speed of the semimajor axis of test particles in Neptune's 3:2 mean-motion resonance. The test particles in this numerical study had initial inclinations less than 5 degrees. The color scale is indicated on the bottom. (From Morbidelli 1997.) (See the chapter by Malhotra et al., this volume.)

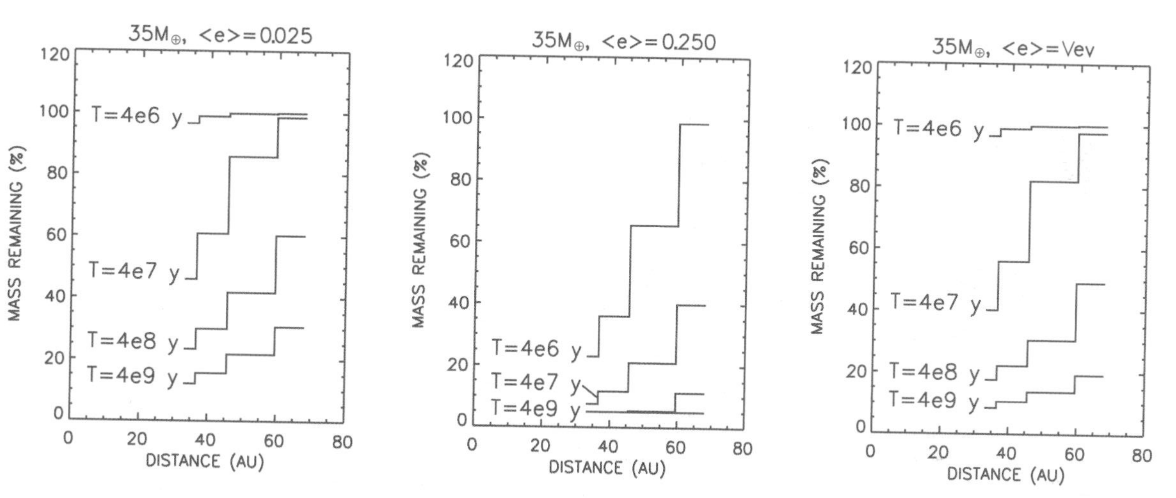

Color Plate 20. The collisional time evolution of surface mass density in the EKB as a function of heliocentric distance, time, and the mean random orbital eccentricity. These results assume an initial mass of 35 $M_{\oplus}$ and strong, icelike target bodies; weaker targets produce even more pronounced erosion and therefore deeper gaps in the inner part of the EKB between 30 and 50–60 AU. In the "Vev" case a crude model for the time evolution of the mean eccentricity was adopted. (Adapted from Stern and Colwell 1997*b*.) (See the chapter by Farinella et al., this volume.)

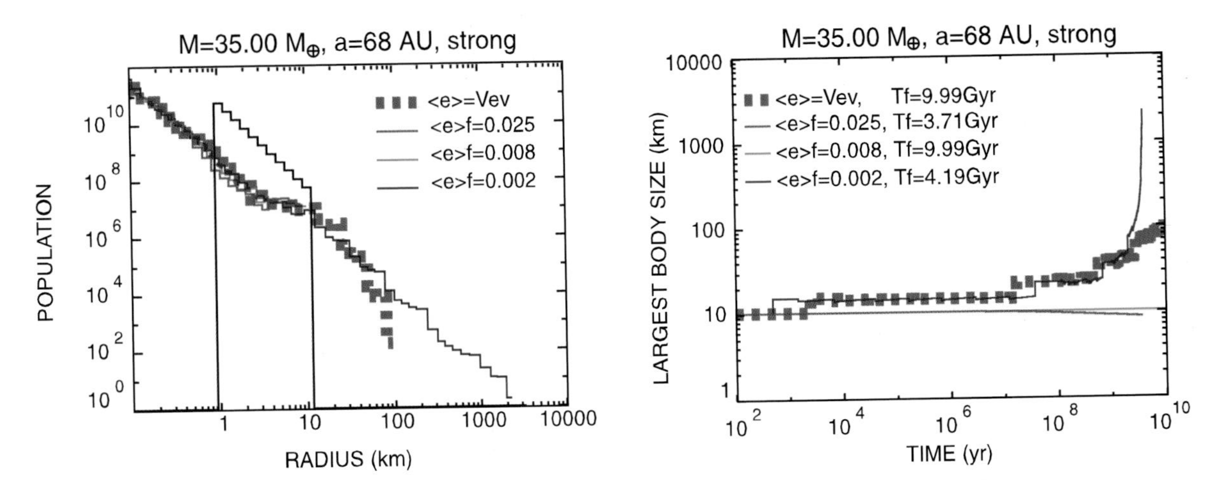

Color Plate 21. Results for model runs at 68 AU, assuming mechanically strong seed objects and a disk mass of 35 $M_\oplus$ in the 30–50-AU zone. For the mean eccentricity $\langle e \rangle$ constant values of 0.002, 0.008, and 0.025 were assumed, or alternatively a crude velocity evolution model ("Vev"). The lefthand panel shows the starting population of 1–10-km building blocks distributed in a constant mass-per-bin power law (bold line), and the final populations for each eccentricity case run for a 1-AU-wide slice of the belt centered at 68 AU. The righthand panel shows the size of the largest object accreted in a 1-AU-wide bin at 68 AU, for each run case, as a function of run time. (Adapted from Stern and Colwell 1997*a*.) (See the chapter by Farinella et al., this volume.)